I0066181

A BIO-BIBLIOGRAPHY FOR THE HISTORY OF THE BIOCHEMICAL SCIENCES SINCE 1800

A BIO-BIBLIOGRAPHY FOR THE HISTORY OF THE BIOCHEMICAL SCIENCES SINCE 1800

SECOND EDITION

JOSEPH S. FRUTON

American Philosophical Society
Independence Square Philadelphia

CONTENTS

PREFACE

Since the publication of the previous version of this bio-bibliography (1982) and a supplement (1985), many books and journal articles have provided valuable information about persons who have played a role in the development of the biochemical sciences during the nineteenth and twentieth centuries. New collective biographical works (or new installments of established ones) have also appeared. Moreover, recent discoveries have called attention to individuals whose research forms part of the historical background of these advances in biochemical knowledge. For these reasons, this volume includes additional citations (as well as correction of errors) for the many of the men and women listed in the previous edition, and the number of separate entries has been greatly increased. There are more entries for present-day investigators in the biochemical sciences, and also for individuals who are not usually listed among biologists or chemists. As in the previous edition, among the latter are people from disciplines ranging from physics to pharmacy, especially inventors of new instruments and physicians who studied particular human diseases.

With a few exceptions, each of the entries includes references to biographical or bibliographical reference works (abbreviations in capital letters) or citations of books and of articles in serial publications, or both. The abbreviations of the titles of reference works and of serial publications are defined in a section at the front of the volume.

New Haven, September 1992 Joseph S. Fruton

ABBREVIATIONS

Reference Works

AD [Das Akademische Deutschland (G. Zieler and T. Scheffer, Eds.) Leipzig, 1905-1906]
ADB [Allgemeine Deutsche Biographie, Leipzig 1875-1912]
AMS [American Men of Science, New York 1906- (after Ed.12, Men and Women)]
AO [The Annual Obituary, New York 1980-]
AARGAU [O. Mittler and G. Boner, Biographisches Lexikon des Aargaus 1803-1957, Aarau 1858]
ABBOTT [D. Abbott, The Biographical Dictionary of Scientists: Biologists, London 1983]
ABBOTT-C [D. Abbott, The Biographical Dictionary of Scientists: Chemists, London 1984]
AGRIFOGLIO [L. Agrifoglio, Igienisti Italiani degli Ultimi Cento Anni, Milan 1952]
ALBRECHT [H. Albrecht, Catalogus Professorum der Technischen Universität Carolo-Wilhelmina zu Braunschweig, Braunschweig 1986]
APPEL [W. Appel, Personalbibliographien von Professoren und Dozenten der Psychiatrie und Neurologie an der Medizinischen Fakultät der Universität München...1870-1945, Erlangen-Nürnberg 1970]
ARBUZOV [A.E. Arbuzov, Izbranniye Raboti po Istorii Khimii, Moscow 1975]
ARCIERI [G.P. Arcieri, Figure della Medicina Contemporanea Italiana, Milan 1952]
ARNSBERG [P. Arnsberg, Die Geschichte der Frankfurter Juden seit der Zeit der Französischen Revolution, vol.3, Biographisches Lexikon, Darmstadt 1983]
ASEN [J. Asen, Gesamtverzeichnis des Lehrkörpers der Universität Berlin, Leipzig 1955]
ATUYANA [E. Atuyana, Personalbibliographien der Professoren und Dozenten der Kinderklinik and der Medizinischen Fakultät der Universität Wien von 1850-1920, Erlangen-Nürnberg 1972]
AUERBACH [I. Auerbach, Catalogus Professorum Academiae Marburgensis, vol.2 (1911-1971), Marburg 1979]
BAD [Badische Biographien, Karlsruhe-Heidelberg, 1875-1935; Neue Folge, Stuttgart, 1982-]
BAF [Belgian and American C.R.B. Fellows 1920-1950, New York 1950]
BHDE [W. Roder and H.A. Strauss, Biographisches Handbuch der Deutsch sprachigen Emigration nach 1933, Munich 1980,1983]
BJN [Biographisches Jahrbuch und Deutscher Nekrolog, Berlin 1897-1917]
BK [Biologen-Kalender (B. Schmid and C. Thesing, Eds.), Leipzig 1914]
BME [Bolshaya Meditsinskaya Entsiklopedia, Moscow 1956-1971]
BNB [Biographie Nationale de Belgique, Brussels 1866-]
BSE [Bolshaya Sovietskaya Entsiklopedia, Moscow 2nd Ed. 1950-1958; 3rd Ed. 1970-1978]
BWN [Biografisch Woordenboek van Nederland, Amsterdam 1978-]
BACHMANN [G. Bachmann, Personalbibliographien der Professoren und Dozenten der Inneren Medizin an der Medizinischen Fakultät der Ludwig- Maximilians Universitität zu München...1870-1920, Erlangen- Nürnberg 1971]
BALLAND [A. Balland, Travaux Scientifiques des Pharmaciens Militaires Français, Paris 1882]
BALLAND(2) [A. Balland, Les Pharmaciens Militaires Français, Paris 1913]
BARNHART [J.H. Barnhart, Biographical Notes upon Botanists, Boston 1965]
BARRY [J. Barry, Notable Contributions to Medical Research by Public Health Service Scientists, Washington 1960]
BELFAST [T.W. Moody and J.C. Beckett, Queens Belfast 1845-1949. London 1959]
BERG [C.P. Berg, The University of Iowa and Biochemistry, Iowa City 1980]
BERLIN [Verzeichnis der Berliner Universitätsschriften 1810-1885, Berlin 1899]
BERWIND [M. Berwind, Personalbibliographien von Professoren und Dozenten der Anatomie, Pathologie, Pharmakologie, Physiologie, Physiologischen Chemie, Hygiene und Bakteriologie an der Medizinischen Fakultät der Universität Erlangen-Nürnberg...1900-1965, Erlangen-Nürnberg 1968]
BERGER-LEVRAULT [O. Berger-Levrault, Annales des Professeurs des Académies et Universités Alsaciennes 1523-1871, Nancy 1892]
BETTANY [G.T. Bettany, Eminent Doctors: their Life and Work, London 1885]
BIBEL [D.J. Bibel, Milestones in Immunology, Madison, Wis. 1988]
BINDSEIL [G. Bindseil, Personalbibliographien von Professoren und Dozenten der Gerichtsmedizin an der Medizinischen Fakultät Wien...1804 bis 1900, Erlangen-Nürnberg 1970]
BIRK [A. Birk, Die Deutsche Technische Hochschule in Prag 1806-1931, Prague 1931]
BLOKH [M.A. Blokh, Biograficheski Spravochnik, Leningrad 1929,1931]
BLUM [M. Blum, Biographie Luxembourgeoise, Luxembourg 1902-1938]
BOCK [R. Bock, Personalbibliographien von Professoren und Dozenten der Hygiene und Pharmakologie an der Medizinischen Fakultät der Universität Würzburg...1900-1945, Erlangen-Nürnberg 1971]
BOEDEKER [E. Boedeker and M. Meyer-Plath, 50 Jahre Habilitation von Frauen in Deutschland, Göttingen 1974
BOEGERSHAUSEN [H.M. Boegershausen, Personalbibliographien von Professoren und Dozenten der Inneren Medizin an der Medizinischen Fakultät der Universität zu Wien...1790-1850, Erlangen-Nürnberg 1972]
BOHM [W.E. Böhm, Forscher und Gelehrte, Stuttgart 1966]

BONN [F. Milkau, Verzeichniss der Bonner Universitätsschriften 1818-1885, Bonn 1897]
BOSSART [C. Bossart, Schweizer Aerzte als Naturforscher im 19. Jahrhundert, Zurich 1979]
BOURQUELOT [E. Bourquelot, Le Centenaire du Journal de Pharmacie et Chimie, Paris 1910]
BRAZIER [M.A. Brazier, A History of Neurophysiology in the 19th Century, New York 1988]
BRIQUET [J. Briquet, Biographies des Botanistes a Genève de 1500 a 1931 (F. Cavillier, Ed.), Geneva 1940]
BROBECK [J.R. Brobeck et al., History of the American Physiological Society 1887-1987, Bethesda 1987]
BRUNKHORST [N. Brunkhorst, Personalbibliographien von ordentlischen und ausserordentlichen Professoren der Chirurgie und Orthopädie an der Medizinischen Fakultät der Universität Würzburg...1900-1945, Erlangen-Nürnberg 1969]
BUGGE [G. Bugge, Ed., Das Buch der Grossen Chemiker, Berlin 1929,1930]
BUICAN [D. Buican, Histoire de la Génétique et d'Evolutionisme en France, Paris 1984
BULLOCH [W. Bulloch, The History of Bacteriology, London 1938]
BURSEY [M.M. Bursey, Carolina Chemists, Chapel Hill 1982]
BUSCHHUTER [H. Buschhüter, Personalbibliographien von Professoren und Dozenten der Anatomie und Pathologie an der Medizinischen Fakultät der Universität München...1945-1968, Erlangen- Nürnberg 1971]
CB [Current Biography, New York 1940-]
CE [The Canadian Encyclopedia, Edmonton 1985-]
CALLISEN [A.C.P. Callisen, Medicinisches Schriftsteller-Lexikon, Copenhagen 1830-1845]
CAMPBELL [W.A. Campbell and N.N. Greenwood, Contemporary British Chemists, London 1961]
CASARETTO [D. Casaretto, Personalbibliographien von Professoren und Dozenten der II. Universitäts-Hautklinik in Wien...1849-1969, Erlangen- Nürnberg 1972]
CHARLE [C. Charle and E. Telkes, Les Professeurs du Collège de France, Paris 1988]
CHARLE(2) [C. Charle and E. Telkes, Les Professeurs de la Faculté des Sciences de Paris, Paris 1989]
CHEN [K.K. Chen, Ed., The American Society for Pharmacology and Experimental Therapeutics: The First Sixty Years 1908-1969, Washington 1969]
CHITTENDEN [R.H. Chittenden, The Development of Physiological Chemistry in the United States, New York 1930]
COHEN [H. Cohen and I.J. Carmin, Jews in the World of Science, New York 1956]
COLE [W.A. Cole, Chemical Literature 1700-1860, London 1988]
CULE [J. Cule, Wales and Medicine, Cardiff 1980]
CURINIER [C.E. Curinier, Dictionnaire National des Contemporains, Paris 1899-1906,1914-1918]
DAB [Dictionary of American Biography, New York 1928-]
DAMB [M. Kaufman et al., Dictionary of American Medical Biography Westport, Conn. 1984]
DBA [Nouveau Dictionnaire de Biographie Alsacienne, Strasbourg 1982-]
DBF [Dictionnaire de Biographie Française, Paris 1933-]
DBI [Dizionario Biografico degli Italiani, Rome 1960-]
DBJ [Deutsches Biographisches Jahrbuch, Stuttgart 1925-1932]
DBL [Dansk Biografisk Leksikon, Copenhagen 1933-1944; 3nd Ed. 1979-1984]
DFI [Dizionario Storico Biografico dei Farmacisti Italiani, Turin 1984]
DGB [Deutsches Geschlechterbuch, Berlin etc. 1889-]
DHB [Dictionnaire Historique et Biographique de Suisse, Neuchatel 1921-1934]
DNB [Dictionary of National Biography, London 1885-]
DNBI [Dictionary of National Biography (India), Calcutta 1972-]
DSB [Dictionary of Scientific Biography, New York 1970-1990]
DZ [Deutsches Zeitgenossenlexikon, Leipzig 1905]
DANIS [C. Danis, Personalbibliographien von Professoren und Dozenten der Pathologie an der Medizinischen Fakultät der Universität Würzburg... 1900 bis 1968, Erlangen-Nürnberg 1971]
DAVIS [B.H. Davis and W.T. Hettinger, Jr., Heterogeneous Catalysis, Washington 1983]
DECHAMBRE [A. Dechambre et al., Dictionnaire Encyclopédique des Sciences Médicales, Paris 1864-1889]
DECKER [H. Decker, Personalbibliographien von Professoren und Dozenten des Pharmakologischen Institutes der Universität Wien...1850-1970, Erlangen-Nürnberg 1975]
DENOTH [R. Denoth, Kurzbiographien 1938 Verstorbener Aerzte des Franzosischen Sprachraums, Zurich 1979]
DESMOND [R. Desmond, Dictionary of British and Irish Botanists and Horticulturists, London 1977]
DIEPGEN [P. Diepgen, Unvollendete vom Leben und Wirken: Frühverstorbener Forscher und Aerzte aus anderthalb Jahrhunderten, Stuttgart 1960]
DIERGART [P. Diergart, Ed., Beiträge aus der Geschichte der Chemie, Leipzig 1909]
DOETSCH [R.N. Doetsch, Microbiology, Historical Contributions from 1776 to 1908, New Brunswick 1960]
DORVEAUX [P. Dorveaux, Catalogue des Thèses de Pharmacie Soutenues en Province, Paris 1894]
DRULL [D. Drüll, Heidelberger Gelehrtenlexikon 1803-1932, Berlin 1986]
DRUM [G. Drum, Geschichte der Deutschen Pharmazeutischen Gesellschaft (1890-1986), Stuttgart 1990]
DUELUND [H. Duelund, Medicina Monacensis, Professoren im 19. Jahrhundert Erlangen-Nürnberg 1976]
DUNSCHELE [H.B. Dünschele, Tropenmedizinische Forschung bei Bayer, Düsseldorf 1970]
EI [Enciclopedia Italiana, Milan-Rome 1929-1939]
EBEL [W. Ebel, Catalogus Professorum Gottingensium 1734-1962, Göttingen 1962]
EBERT [C. Ebert, Personalbibliographien der Ordinarien und Extraordinarien der Anatomie mit Histologie und Embryologie, der Physiologie und der Physiologischen Chemie an der Medizinischen Fakultät der Julius Maximilian-Universität Würzburg...1900-1945, Erlangen-Nürnberg 1971]

ECKERT [D. Eckert, Personalbibliographien der Professoren und Dozenten der Pathologie an der Medizinischen Fakultät an der Ludwig-Maximilian- Universität in München...1870-1945, Erlangen-Nürnberg 1971]
EGERER [G. Egerer, Personalbibliographien von Professoren und Dozenten der Anatomie and der Medizinischen Fakultät der Universität München... 1870-1945, Erlangen-Nürnberg 1970]
ELLIOTT [C.A. Elliott, Biographical Dictionary of American Science. The Seventeenth through the Nineteenth Centuries, Westport, Conn. 1979]
ELLIOTT(2) [Biographical Index to American Science, New York 1990]
ENKVIST [T. Enkvist, The History of Chemistry in Finland 1828-1918, Helsinki 1972]
ETTRE [L.S. Ettré and A. Zlatkis, 75 Years of Chromatography, Amsterdam 1979]
EUI [Enciclopedia Universal Illustrada Europeo-Americana, Madrid 1907-]
FARBER [E. Farber, Great Chemists, New York 1961]
FELLER [F.X. Feller, Biographie Universelle, Lyon-Paris 1860]
FERCHL [F. Ferchl, Chemisch-Pharmazeutisches Bio- und Bibliographicon, Mittenwald 1937-1938]
FINDLAY [A. Findlay and W.H. Mills, British Chemists, London 1947]
FISCHER [I. Fischer, Biographisches Lexikon der Hervorragenden Aerzteletzten Fünfzig Jahren, Berlin 1932-1933]
FLUELER [U. Fleuler-Ambühl, Kurzbiographien im Jahre 1939 Verstorbene Aerzte des Englischen Sprachraums, Zürich 1978]
FORBES [E.W. Forbes and J.H. Finley, Eds., The Saturday Club: A Century Completed 1920-1956, Boston 1958]
FRANKEN [F.H. Franken, Die Leber und ihre Krankheiten, Zwei Hundert Jahre Hepatologie, Stuttgart 1968]
FRANKENTHAL [K. Frankenthal, Der dreifache Fluch: Jüdin, Intellektuelle, Sozialistin, Frankfurt a.M. 1981]
FREUND [H. Freund and A. Berg, Geschichte der Mikroskopie, Frankfurt a.M. 1963-1966]
FRITZ [H. Fritz, Personalbibliographien von Professoren und Dozenten der I. und II. Universitäts-Frauenklinik und der III. Geburtshilflichen Klinik der Universität Wien...1875-1905, Erlangen-Nürnberg 1971]
FROHNE [S. Frohne, Personalbibliographien der Professoren und Dozenten der Inneren Medizin an der Medizinischen Fakultät der Universität Wien...1850-1875, Erlangen-Nürnberg 1972]
FUIERER [J. Fuierer, Personalbibliographien von Professoren und Dozenten der Hals-, Nasen- und Ohrenheilkunde an der Medizinischen Fakultät der Universität Wien...1861-1910, Erlangen-Nürnberg 1972]
GE [Grande Encyclopédie, Paris 1886-1902]
GEPB [Grande Enciclopedia Portugesa e Brasiliera, Lisbon 1936-1960]
GGTA [Gothaisches Genealogisches Taschenbuch der Adeliger Häuser, Gotha 1907-]
GGTF [Gothaisches Genealogisches Taschenbuch der Freiherrlichen Häuser, Gotha 1848-]
GAUSS [C.J. Gauss and B. Wilde, Die Deutsche Geburtshelferschulen, Munich 1956]
GEDENKBUCH [Gedenkbuch: Opfer der Verfolgung der Juden unter der National sozialistischer Gewaltschaft in Deutschland, Koblenz 1986]
GEISSLER [K.H. Geissler, Personalbibliographien von ordentlichen und ausserordentlichen Professoren der Medizinischen Poliklinik an der Universität Würzburg...1900-1945, Erlangen-Nürnberg 1971]
GENEVA [Catalogue des Ouvrages, Articles et Mémoires Publies par les Professeurs et Privat-Docents de l'Université de Genève, Geneva 1883-]
GENTY [M. Genty, Les Biographies Médicales, Paris 1927-1930]
GERNETH [G.M. Gerneth, Personalbibliographien von Professoren und Dozentender Neurologie und Psychiatrie, der Arbeitsmedizin und der Physiologischen Chemie der Universität Erlangen-Nürnberg 1900-1968,Erlangen-Nürnberg 1970]
GERRITS [G.C. Gerrits, Grote Nederlanders bij de Opbouw de Naturwetenschapen Leiden 1948]
GHENT [Liber Memorialis Rijksuniversiteit te Gent 1913-1960, Ghent 1960]
GIESE [E. Giese and B. von Hagen, Geschichte der Medizinischen Fakultät Universität Jena, Jena 1958]
GLOSAUER [R. Glosauer, Personalbibliographien der Mitglieder des zu München von 1826 bis 1850 mit biographischen Angaben gesichtet im Hinblick auf die Beziehung zur Lehre und Forschung in der Medizinischen Fakultät, Erlangen-Nürnberg 1971]
GOCZOL [M. Goczol, Personalbibliographien von Mitgliedern der Kaiserlichen Akademie der Wissenschaften Sankt Petersburg unter besonderer Berück-sichtigung der Anatomie, Physiologie, Zoologie und Botanik.. 1725-1875, Erlangen-Nürnberg 1971]
GORIS [A. Goris, Centenaire de l'Internat en Pharmacie des Hôpitaux et Hospices de Paris, Paris 1929]
GOULD [R.F. Gould, Eminent Chemists of Maryland, Maryland Historical Magazine 80 (1985), 19-47]
GRAETZER [J. Graetzer, Lebensbilder Hervorragender Schlesischer Aerzte, Breslau 1889]
GRAML [W. Graml, Personalbibliographien von Professoren und Dozenten der Hals, Nasen und Ohrenkunde und Neurologie und Psychiatrie an der Medizinischen Fakultät der Julius-Maximilian-Universität Würzburg...1900-1945, Erlangen-Nürnberg 1970]
GREIFSWALD [Festschrift zur 500-Jahrfeier der Universität Greifswald, Greifswald 1956]
GRMEK [M.D. Grmek, Hrvtska Medicinska Bibliografija 1876-1918, Zagreb 1970]
GROTE [L.R. Grote, Ed., Die Medizin der Gegenwart in Selbstdarstellungen Leipzig 1923-1929]
GRUETTER [R. Gruetter, Kurzbiographien von Aerzten des Deutschen Sprachraums die 1934 Verstorben sind, Zürich 1978]
GUIART [J. Guiart, L'École Médicale Lyonnaise, Paris 1941]
GUNDEL [H.G. Gundel et al., Eds., Giessener Gelehrte in der ersten Hälfte des 20. Jahrhunderts, Marburg 1982]
GUNDLACH [F. Gundlach, Catalogus Professorum Academiae Marburgensis 1527-1910, Marburg 1927]
HB [Hessische Biographien, Darmstadt 1918-1934]
HD [Hommes et Destins, Paris 1975-1988]
HABERLING [W. Haberling, Die Geschichte der Düsseldorfer Aerzte und Krankenhäuser bis zum Jahre 1907, Düsseldorfer Jahrbuch 38 (1936), 1-441]
HAGEL [K.H. Hagel, Personalbibliographien von Professoren und Dozenten der Medizinischen Klinik und Poliklinik der Universität Erlangen-Nürnberg ...1900-1965, Erlangen-Nürnberg 1968]
HAKUSHI [K.R. Iseki, Who's Who in "Hakushi" in Great Japan, Tokyo 1921-1930]
HALLE [W. Suchier, Bibliographie der Universitätsschriften von Halle- Wittenberg 1817-1885, Berlin 1953]

HAMON [P. Hamon, Nouvelle Biographie du Dauphiné, Marseilles 1980-]
HANNOVER [Der Lehrkörper der Technischen Hochschule Hannover 1831-1956, Hannover 1956]
HARTMANN [W. Hartmann, Personalbibliographien der Professoren und Dozenten der Geburtshilfe und Gynäkologie und der Pharmakologie und Pharma kognosie an der Medizinischen Fakultät der Deutschen Karl Ferdinands-Universität in Prag...1900-1945, Erlangen-Nürnberg 1972]
HARTWIG [P. Hartwig, Personalbibliographien von Professoren und Dozenten der Gynäkologie der Julius-Maximilian-Universität Würzburg...1900-1945, Erlangen-Nürnberg 1970]
HASENEDER [M. Haseneder, Personalbibliographien von Professoren und Dozenten der Inneren Medizin an der Medizinischen Fakultät der Universität Wien...1890-1940, Erlangen-Nürnberg 1971]
HAYMAKER [W. Haymaker and W. Schiller, The Founders of Neurology, 2nd Ed., Springfield, Ill. 1970]
HEID [D. Heid, Personalbibliographien der Professoren und Dozenten der Augenheilkunde an der Medizinischen Fakultät der Universität Wien... 1812-1884, Erlangen-Nürnberg 1972]
HEIM [R. Heim et al., A la Mémoire de quinze Savants Francais Lauréats de l'Institut Assassinés par les Allemands, Paris 1959]
HEIN [W.H. Hein and H.D. Schwarz, Deutsche Apotheker-Biographie, 1975,1978]
HEIN-E [W.H. Hein and H.D. Schwartz, Deutsche Apotheker-Biographie Ergänzungsband, Stuttgart 1986]
HEINDEL [W. Heindel, Personalbibliographien von Professoren und Dozenten des Histologisch-Embryologischen Institutes der Universität Wien... 1848-1968, Erlangen-Nürnberg 1971]
HES [H.S. Hes, Jewish Physicians in the Netherlands 1600-1940, Assen 1980]
HIENSTOFER [I. Hienstofer, Personalbibliographien von Professoren der Philosophischen Fakultät zu Prag...1800-1860, Erlangen-Nürnberg 1970]
HINK [R. Hink, Ed., Archiv der Geschichte der Naturwissenschaften, Vienna 1981-]
HIRSCH [A. Hirsch et al., Biographisches Lexikon der Hervorragenden Aerzte aller Zeiten und Völker, 2nd Ed. by W. Haberling et al., Berlin 1929-1935]
HIS [E. His, Basler Gelehrte des 19. Jahrhunderts, Basle 1941]
HOEFER [F. Hoefer, Nouvelle Biographie Générale, Paris 1852-1866]
HUBER [S. Huber, Kurzbiographien 1940 Verstorbener Aerzte des Englischen Sprachraums, Zurich 1976]
HUHLE-KREUTZER [G. Huhle-Kreutzer, Die Entwicklung arzneilichen Produktions-stätten aus Apothekenlaboratorien, Stuttgart 1989]
HUMBERT [G. Humbert, Contribution à l'Histoire de la Pharmacie Strasbourgeoise, Mulhouse 1938]
HUMPHREY [H.B. Humphrey, Makers of North American Botany, New York 1961]
HUPKA [H. Hupka, Ed., Grosse Deutsche aus Schlesien, Munich 1969]
HUTER [F. Huter, Hundert Jahre Medizinische Fakultät Innsbruck 1869 bis 1969, Innsbruck 1969]
IB [G.C. Hirsch, Ed., Index Biologorum, Berlin 1928]
ICC [A.M. Harvey, The Interurban Clinical Club, Philadelphia 1978]
IWW [International Who's Who, London 1935-]
IKONNIKOV [V.S. Ikonnikov, Biograficheski Slovar Professorov i Prepodava telei Imperatorskavo Universitita Vladimira, Kiev 1884]
ILEA [T. Ilea et al., Invatamintel Medical si Farmacutic din Bucuresti, Bucharest 1963]
JBE [Japan Biographic Encyclopedia 3rd Ed., Tokyo 1964-1965]
JV [Jahres-Verzeichnis der an den Deutschen Universitäten Erschienen Schriften, Berlin 1887-]
JOHN [W. John, Personalbibliographien von Professoren und Dozenten der Physiologie, Psychiatrie und Ohrenheilkunde an der Medizinischen Fakultät der Universität Wien...1790-1878, Erlangen-Nürnberg 1971]
KBB [Kraks Blaa Bog, Copenhagen 1910-]
KALIN [P. Kälin, Kurzbiographien von Aerzten des Deutschen Sprachraums die 1930 Verstorben sind, Zurich 1977]
KAGAN [S.R. Kagan, Modern Medical World, Boston 1945]
KAGAN(JA) [S.R. Kagan, Jewish Contributors to Medicine in America, 2nd Ed., Boston 1939]
KALLMORGEN [W. Kallmorgen, Siebenhundert Jahre Heilkunde in Frankfurt am Main, Frankfurt a.M. 1936]
KAMP [F.W. von Kamp, Personalbibliographien von Professoren und Dozenten der Haut- und Geschlechtskrankheiten an der Medizinischen Fakultät der Universität München...1870-1945, Erlangen-Nürnberg 1971]
KAZAN [Biograficheski Slovar Professorov i Prepodavatelei Imperatorskavo Kazanskavo Universiteta 1804-1904, Kazan 1904]
KELLY [H.A. Kelly and W.L. Burrage, Dictionary of American Medical Biography, New York 1928]
KERNBAUER [A. Kernbauer, Das Fach Chemie an der Philosophischen Fakultät Universität Graz, Graz 1985]
KERVILLER [R. Kerviller, Répertoire Général de Bio-bibliographie Bretonne, Rennes 1886-1907]
KHARKOV [I.L. Skvortsov and D.I. Bagalia, Eds., Meditsinski Fakultet Kharkovskovo Universiteta 1805-1905, Part 3, Biograficheski Slovar, Kharkov 1905-1906]
KHARKOV-F [I.P. Osipov and D.I. Bagalia, Eds., Fiziko-Matematikecheski Fakultet Kharkovskovo Universiteta 1805-1905, Part 3, Biograficheski Slovar, Kharkov 1908]
KIEL [F. Volbehr and R. Weyl, Professoren und Dozenten der Christian Albrechts-Universität in Kiel, 4th Ed., Kiel 1956]
KIHL [M. Kihl, Oberoesterreichisches Landesmuseum Biographisches Lexikon, Linz 1955-]
KILLIAN [H. Killian, Meister der Chirurgie, 2nd Ed., Stuttgart 1980]
KIRCHENBERGER [S. Kirchenberger, Lebensbilder Hervorragender Oesterreich ungarischer Militär- und Marinärzte, Vienna 1913]
KLEIN [E. Klein, Die Akademische Lehrer der Universität Hohenheim, Stuttgart 1968]
KNOLL [F. Knoll, Oesterreichische Naturforscher, Aerzte und Techniker, Vienna 1957]
KOBRO [I. Kobro, Norges Laeger 1800-1908, Kristiania 1915]
KOERTING [W. Koerting, Die Deutsche Universität in Prag. Die letzten Hundert Jahre ihrer Medizinischen Fakultät, Munich 1968]
KOHUT [A. Kohut, Berühmte Israelitische Männer und Frauen, Leipzig 1901]
KOLLE [K. Kolle, Grosse Nervenärzte, Stuttgart 1956-1963]

KONOPKA [S. Konopka, Polska Bibliografia Lekarska 19. Wieku, Warsaw 1974-1984]
KOREN [N. Koren, Jewish Physicians. A Biographical Index, Jerusalem 1973]
KOSMINSKI [A. Kosminski, Slownik Lekarzow Polskich, Warsaw 1883-1888]
KOVACSICS [H. Kovacsics, Personalbibliographien der Lehrer der Heilkunde der Universität Erlangen 1792-1900, Erlangen-Nürnberg 1967]
KOZLOV [V.V. Kozlov, Ocherki Istorii Khimicheskikh Obshchestv SSSR, Moscow 1958]
KROLLMANN [C. Krollmann, Altpreussische Biographie, Königsberg, 1941,1943, Marburg 1975; continued by K. Forstreuter and F. Gaule, 1984-]
KURSCHNER [Kürschners Gelehrten-Kalender, Berlin 1925-]
KUKULA [R.Kukula, Allgemeiner Deutscher Hochschul-Almanach, Vienna 1888]
KUZNETSOV [I.V. Kuznetsov, Ed., Lyudi Russkoi Nauki, Moscow 1961-1963]
LDGS [List of Displaced German Scholars, London 1936]
LF [Lebensläufe aus Franken, Munich 1919-1960]
LKW [Lebensbilder aus Kurhessen und Waldeck 1830-1930, Giessen 1939-1958]
LABARRE [D. Labarre de Raillecourt, Nouveau Dictionnaire des Biographies Françaises et Étrangères, Paris 1961-]
LACROIX [A. Lacroix, Figures des Savants, Paris 1932]
LADIS [C. Ladis, Personalbibliographien von Professoren und Dozenten der Pharmazie an der Universität Erlangen...1851-1900. Erlangen-Nürnberg 1974]
LANGHANS [P.M. Langhans, Personalbibliographien von Professoren der Philosophischen Fakultät zu Würzburg von 1803-1852, mit biographischen Angaben, gesichtet im Hinblick auf die Beziehungen zu Lehre und Forschung in der Medizinischen Fakultät, Erlangen-Nürnberg 1971]
LAROUSSE [A. Larousse, Grand Dictionnaire Universel du XIXe Siècle, Paris 1866-1890]
LEHMANN [E. Lehmann, Schwäbische Apotheker und Apothekergeschlechter in ihrer Beziehung zur Botanik, Stuttgart 1951]
LEIPZIG [Bedeutende Gelehrte in Leipzig, Leipzig 1965]
LESKY [E. Lesky, Die Wiener Medizinische Schule im 19. Jahrhundert, Graz 1965]
LEVITSKI [G.V. Levitski, Biograficheski Slovar Professorov i Prepodavatelei Imperatorskavo Yurievskavo Universiteta, Yuriev (Dorpat) 1902-1903]
LIMLEY [C. Limley, Personalbibliographien von Professoren und Dozenten der Gerichtsmedizin an der Medizinischen Fakultät der Universität Wien ...1900-1968, Erlangen-Nürnberg 1971]
LINDEBOOM [G.A. Lindeboom, Dutch Medical Biography, Amsterdam 1984]
LIPSHITS [S.I. Lipshits, Russkie Botaniki, Moscow 1947-1952]
LOMMATZSCH [H. Lommatzsch, Personalbibliographien der Professoren und Dozenten der Anatomie, Histologie und Pathologie, Pharmakologie und Physiologie an der Medizinischen Fakultät der Deutschen Karl- Ferdinands-Universität in Prag...1880-1900, Erlangen-Nürnberg 1969]
LORENZSONN [B. Lorenzsonn, Personalbibliographien von Professoren und Dozenten der I. und II. Universitäts-Frauenklinik und der III. Geburthilflichen Klinik in Wien...1905-1930, Erlangen-Nürnberg 1973]
LUDY [W. Ludy, Personalbibliographien von Professoren der Philosophie, Zoologie und Botanik an der Philosophischen Fakultät der Karl-Ferdinands-Universität in Prag...1860-1918, Erlangen-Nürnberg 1970]
LYONS [J.B. Lyons, Brief Lives of Irish Doctors 1600-1965, Dublin 1978]
MEL [Magyar Eletrajzi Lexikon, Budapest 1967-1981]
MH [McGraw-Hill Modern Men of Science, New York 1966,1968]
MR [W. Munk et al., The Roll of the Royal College of Physicians of London, 2nd Ed., London 1878; later entitled Lives of the Fellows of the Royal College of Physicians of London]
MSE [McGraw-Hill Modern Scientists and Engineers, New York 1980]
MAASS [K. Maass, Personalbibliographien der Professoren und Dozenten der Anatomie, Histologie und Physiologie an der Medizinischen Fakultät der Deutschen Karl-Ferdinands-Universität in Prag...1900-1945, Erlangen-Nürnberg 1971]
MACHEK [G. Machek, in Die Fächer Mathematik, Physik und Chemie an der Philosophischen Fakultät zu Innsbruck bis 1945, Innsbruck 1971]
McLACHLAN [D. McLachlan, Jr. and J.P. Glusker, Eds., Crystallography in North America, New York 1983]
MATSCHOSS [C. Matschoss, Männer der Technik, Berlin 1925]
MAYERHOFER [J. Mayerhöfer, Lexikon der Geschichte der Naturwissenschaften, Vienna 1959-1975]
MEDVEI [V.C. Medvei, A History of Endocrinology, Lancaster 1982]
MEINEL [C. Meinel, Die Chemie an der Universität Marburg seit Beginn des 19. Jahrhunderts, Marburg 1978]
MEISEN [V. Meisen, Ed., Prominent Danish Scientists, Copenhagen 1932]
MEISTER [M. Meister, Personalbibliographien der Professoren und Dozenten des Physiologisch-Chemischen Institutes an der Medizinischen Fakultät der Universität Würzburg...1900-1970, Erlangen-Nürnberg 1974]
MEITES [J. Meites et al., Pioneers in Neuroendocrinology, New York 1975]
MERKLE [W. Merkle, Personalbibliographien von Professoren und Dozenten der Frauenklinik und Poliklinik Würzburg 1945-1970, Erlangen-Nürnberg 1975]
MILES [W.D. Miles, Ed., American Chemists and Chemical Engineers, Washington 1976]
MOTSCH [W. Mötsch, Personabibliographien der Professoren und Dozenten der Gerichtsmedizin, Geburtshilfe und Frauenheilkunde, Hygiene und Kinderheilkunde an der Medizinischen Fakultät der Karl-Ferdinands- Universität in Prag...1880-1910, Erlangen-Nürnberg 1972]
NAW [Notable American Women, Cambridge, Mass. 1971,1980]
NB [O. Renkhoff, Nassauische Biographie, Wiesbaden 1985]
NBL [Norsk Biografisk Leksikon, Oslo 1923-1976]
NBW [National Biografisch Woordenboek , Brussels 1964-]

NCAB [National Cyclopedia of American Biography, New York 1892-]
NDB [Neue Deutsche Biographie, Munich 1953-]
NDBA [V.O. Cutelo, Nuevo Diccionario Biografico Argentino 1750-1930, Buenos Aires 1968-]
NL [R. Vaupel, Ed., Nassauische Lebensbilder, Wiesbaden 1940-1961]
NNBW [Niew Nederlandisch Biografisch Woordenboek, Leiden 1911-1937]
NOB [Neue Oesterreichische Biographie, Vienna 1923-]
NSL [Niedersächsische Lebensbilder, Hildesheim 1939-]
NUC [National Union Catalog Pre-1956 Imprints, London 1968-1981]
NAVRATIL [M. Navratil, Almanach Českych Lekaru, Prague 1931]
NUESCH [H. Nuesch, Kurzbiographien 1936 Verstorbener Aerzte des Englischen Sprachraums, Zurich 1976]
OBL [Oesterreichisches Biographisches Lexikon 1815-1950, Graz 1957-]
O'CONNOR [W.J. O'Connor, Founders of British Physiology; A Biographical Dictionary 1820-1885, Manchester 1988]
O'CONNOR(2) [W.J. O'Connor, British Physiologists 1885-1914, Manchester 1991]
OEHRI [A. Oehri, Kurzbiographien Amerikanischer Aerzte und Naturwissenschaftler die zwischen 1930 und 1940 Verstorben sind, Zurich 1987]
OGILVIE [M.B. Ogilvie, Women in Science, Cambridge, Mass. 1986]
OLPP [G. Olpp, Hervorragende Tropenärzte, Munich 1932]
POGG [J.C. Poggendorff, Biographisch-literarisches Handwörterbuch zur Geschichte der exacten Naturwissenschaften, Leipzig 1863-]
PSB [Polski Slownik Biograficzny, Warsaw 1935-]
PAGEL [J. Pagel, Biographisches Lexikon Hervorragenden Aerzte des neunzehnten Jahrhunderts, Berlin 1901]
PARASCANDOLA [J. Parascandola and E. Keeney, Sources in the History of American Pharmacology, Madison, Wis. 1983]
PELZNER [K. Pelzner, Personalbibliographien von Professoren und Dozenten der Inneren Medizin und der Kinderheilkunde der Deutschen Karl-Ferdinands-Universität in Prag...1900-1945, Erlangen-Nürnberg, 1972]
PETRY [H. Petry, Personalbibliographien von Professoren und Dozenten der Inneren Medizin an der Medizinischen Fakultät zu Wien...1850-1925, Erlangen-Nürnberg 1972]
PFISTER [W. Pfister, Die Einbürgerung der Ausländer in der Stadt Basel im 19. Jahrhundert, Basle 1976]
PHILIPPE [A. Philippe, Geschichte der Apotheker, transl. by H. Ludwig, 2nd Ed., Wiesbaden 1859]
PICCO [C. Picco, Das Biochemische Institut der Universität Zürich 1931-1981, Aarau 1981]
PICOT [C. Picot and E. Thomas, Centenaire de la Société Médicale de Genève 1823-1923, Geneva 1923]
PITTROFF [R. Pittroff, Die Lehrer der Heilkunde der Universität Erlangen 1843-1943 und ihr Werdegang, Erlangen-Nürnberg 1964]
PLANER [F. Planer, Das Jahrbuch der Wiener Gesellschaft, Vienna 1929]
POHL [D. Pohl, Zur Geschichte der Pharmazeutischen Privat Institute in Deutschland von 1779 bis 1873, Marburg 1972]
POINTNER [G. Pointner, Personalbibliographien von Professoren und Dozenten der Psychiatrie und Neurologie an der Wiener Medizinischen Fakultät...1880-1920, Erlangen-Nürnberg 1972]
POTSCH [W.R. Pötsch et al., Lexikon bedeutender Chemiker, Leipzig 1988]
PRAG [F. Stark et al., Eds., Die Deutsche Technische Hochschule in Prag 1806-1906, Prague 1906]
PRANDTL [W. Prandtl, Deutsche Chemiker in der ersten Hälfte des Neunzehnten Jahrhundert, Weinheim 1956]
PROVENZAL [G. Provenzal, Profili Bio-bibliografici di Chimici Italiani Sec. XV-Sec.XIX, Rome 1938]
PUDOR [F. Pudor, Nekrologe aus dem Rheinisch-Westfälischen Industriegebiet, Düsseldorf 1955]
PUSCHEL [E. Püschel, 75 Jahre Kinderheilkunde in Rheinland und Westfalen, Düsseldorf 1975]
PUTZ [H. Putz, Personalbibliographien der Professoren und Dozenten der Chirurgie, Orthopädie und Dermatologie an der Medizinischen Fakultät der Deutschen Karl-Ferdinands-Universität in Prag...1900-1945, Erlangen-Nürnberg 1969]
QQ [Qui et Qui en France, Paris 1953]
RBS [Russki Biographicheski Slovar, St. Petersburg 1896-1918]
RSC [Royal Society Catalogue of Scientific Papers 1800-1900, London 1867-1925]
RAUSCH [U. Rausch, Das Medicinal- und Apothekerwesen des Landsgrafschaft Hessen-Darmstadt etc., Darmstadt 1978]
RAZINGER [Kurzbiographien anno 1940 Verstorbener Aerzte des Deutschen Sprachraums, Zürich 1977]
RIGA [Album Academicum des Polytechnicums zu Riga 1862-1912, Riga 1912]
ROEDER [C. Roeder in Scudari, Personalbibliographien der Lehrer der Heilkunde an der Medizinischen Fakultät der Julius-Maximilians-Universität zu Würzburg...1850-1900, Teil I, Theoretische Fächer, Erlangen-Nürnberg 1973]
ROTH [R.Roth, Personalbibliographien von Professoren und Dozenten der Augen-heilkunde, Dermatologie, Gerichtsmedizin, Gynäkologie, Inneren Medizin, Pädiatrie, Pharmakologie, Psychiatrie und Zahnheilkunde an der Medizinischen Fakultät der Karl-Ferdinands-Universität in Prag...1848-1880, Erlangen-Nürnberg 1972]
RUSTLER [P. Rustler, Personalbibliographien von Professoren und Dozenten der Augenheilkunde, Dermatologie, Oto- und Rhinologie, Psychiatrie und Zahnheilkunde an der Medizinischen Fakultät der Deutschen Karl-Ferdinands-Universität in Prag...1880-1900, Erlangen-Nürnberg 1971]
SBA [W. Keller, Schweizer Biographisches Archiv, Zürich 1952-1958]
SBL [Svenskt Biografiskt Lexikon, Stockholm 1918-1956]
SG [Index-Catalogue of the Library of the Surgeon-General's Office, United States Army, Washington 1880-1961]
SHBL [Schleswig-Holsteinisches Biographisches Lexikon, Neumünster 1970-]
SL [Schwäbische Lebensbilder, after vol.6 Lebensbilder aus Schwaben und Franken, Stuttgart 1940-]
SMK [Svenska Män och Kvinnor, Stockholm 1942-1955]
SRS [Record of the Science Research Scholars of the Royal Commission for the Exhibition of 1851, London 1961]
STC [Scienzati e Technologi Contemporanei, Milan 1974]
SACKMANN [W. Sackmann, Biographische und Bibliographische Materialen zur Geschichte der Mikrobiologie und zur Bakteriologischen Nomenklatur, Frankfurt a.M. 1985]

SANDRITTER [W. Sandritter und Kasten, 100 Years of Histochemistry in Germany, Stuttgart 1964]
SCHAEDLER [C. Schaedler, Biographisch-literärisches Handworterbuch der Wissenschaftlich Bedeutenden Chemiker, Berlin 1881]
SCHALI [J. Schäli, Kurzbiographien von Aerzten des Franzosischen Sprachraums die 1939 Verstorben sind, Zürich 1977]
SCHELENZ [H. Schelenz, Geschichte der Pharmazie, Berlin 1904]
SCHIEBER [A. Schieber, Personalbibliographien der Professoren und Dozenten der Chirurgie und Inneren Medizin an der Medizinischen Fakultät der Karl-Ferdinands-Universität in Prag...1880-1900, Erlangen-Nürnberg 1969]
SCHMITZ [R. Schmitz, Die Naturwissenschaften an der Philipps-Universität Marburg 1527-1977, Marburg 1978]
SCHNACK [I. Schnack, Marburger Gelehrte in der ersten Hälfte des 20. Jahrhunderts, Marburg 1977]
SCHOLZ [H. Scholz and P. Schroeder, Aerzte in Ost- und Westpreussen, Würzburg 1970]
SCHULTE [W. Schulte, Westfälische Kopfe, Münster 1963]
SCHULZ-SCHAEFFER [J. Schulz-Schaeffer, A Short History of Cytogenetics, Biologisches Zentralblatt 95 (1976), 193-221]
SCHWARTZ [C. Schwartz, Personalbibliographien der Lehrstuhlinhaber der Fächer Anatomie, Physiologie, Pathologie und Pathologische Anatomie, Pharmakologie, Innere Medizin, Chirurgie, Frauenheil kunde, Psychiatrie, Zahnheilkunde an der Medizinischen Fakultät der Universität Erlangen...1850-1900, Erlangen-Nürnberg 1969]
SCHWARZ [K.H. Schwarz, Personabibliographien von Professoren und Dozenten des Pathologisch-Anatomischen Institutes der Universität Wien... 1875-1936, Erlangen-Nürnberg 1973]
SCHWEINITZ [A. Schweinitz, Personalbibliographien der Professoren und Dozenten der II. Chirurgischen Klinik der Universität Wien... 1880-1930, Erlangen-Nürnberg 1974]
SCHWERTE [H. Schwerte and W. Spengler, Eds., Forscher und Wissenschaftler im Heutigen Europa, Oldenburg 1955]
SIEGL [N. Siegl, Personalbibliographien von Professoren und Dozenten der I. Wiener Hautklinik...1845-1969, Erlangen-Nürnberg 1971]
SIGRIST [J. Sigrist, Kurzbiographien 1937 Verstorbener Aerzte des Französischen Sprachraums, Zürich 1977]
SILLIMAN [B. Silliman, American Contributions to Chemistry, Philadelphia 1874]
SITZMANN [E. Sitzmann, Dictionnaire de Biographie des Hommes Célèbres de l'Alsace, Rixheim 1909; repr. Paris 1973]
SKOROKHODOV [L.Y. Skorokhodov, Materiali po Istorii Meditsinskoi Mikrobiologii v Dorevolutionni Rossii, Moscow 1948]
SKULINA [C. Skulina, Personalbibliographien der Aerztlichen Mitglieder der Prager Privatgesellschaft der Wissenschaften und der Königlichen Böhmischen Gesellschaft der Wissenschaften mit Biographischen Angeben...1772-1884, Erlangen- Nürnberg 1976]
SKVARC [M. Skvarc, Kurzbiographien 1938 Verstorbener Aerzte des Deutschen Sprachraums, Zürich 1976]
STALDER [A. Stalder, Kurzbiographien von Aerzten des Franzosischen Sprachraums die 1933 Verstorben sind, Zürich 1976]
STANGL [T. Stangl, Personalbibliographien von Professoren und Dozenten der Inneren Medizin an der Medizinischen Fakultät der Universität Wien ...1890-1950, Erlangen-Nürnberg 1972]
STELLING-MICHAUD [S. Stelling-Michaud, Le Livre du Recteur de l'Académie de Genève 1559-1878, Geneva 1959-1980]
STOBER [M. Stober, Personalbibliographien der Professoren und Dozenten der Anatomie an der Medizinischen Fakultät der Universität Wien... 1845-1969, Erlangen-Nürnberg 1971]
STRAHLMANN [Mitteilungen aus dem Gebiete der Lebensmitteluntersuchung und Hygiene 53 (1962), 459-482]
STUBLER [E. Stübler, Geschichte der Medizinischen Fakultät der Universität Heidelberg, Heidelberg 1926]
STUMPF [J. Stumpf, Personalbibliographien von Professoren und Dozenten der Inneren Medizin an der Medizinischen Fakultät der Universität Wien ...1885-1935, Erlangen-Nürnberg 1972]
STUPP-KUGA [I. Stupp-Kuga, Personalbibliographien von Professoren und Dozenten der Chemie an der Universität Erlangen...1851-1900, Erlangen-Nürnberg 1971]
STURM [H. Sturm et al., Biographisches Lexikon zur Geschichte der Böhmischen Länder, Munich-Vienna 1975-]
SURREY [A.R. Surrey, Name Reactions in Organic Chemistry, New York 1954]
TLK [Technischer Literatur Kalender, Munich 1918,1920,1929]
TETZLAFF [W. Teztlaff, 2000 Kurzbiographien beudetender deutscher Juden des 20. Jahrhunders, Lindhorst 1982]
TSCHIRCH [A. Tschirch, Handbuch der Pharmakognosie, 2nd Ed., vol.1, Biographicon, by J.A. Häfliger, pp.1009-1151, Leipzig 1932]
ULLMANN [V.J. Ullmann, Kurzbiographien von anno 1934 Verstorbenen Aerzte des Englischen Sprachraumes, Zürich 1977]
UKRAINE [Istoria Akademii Nauk Ukrainskoi RSR, Kiev 1967]
UTRECHT [G. ten Doesschate, De Utrechtse Universiteit en de Geneeskunde 1636-1900, Niewkoop 1963]
VAD [Vem är Dat, Stockholm 1912-]
VACCARO [G. Vaccaro, Panorama Biografico degli Italiani d'Oggi, Rome 1956]
VAPEREAU [C. Vapereau, Dictionnaire Universel des Contemporains, Paris 1858-1895]
VEIBEL [S. Veibel, Kemien i Danmark, Copenhagen 1939,1943,1968]
VENGEROV [S.A. Vengerov, Kritiko-biograficheski Slovar Russkikh Pisatelei i Uchennikh, St. Petersburg 1886-1904]
VERSO [M.L. Verso, The Evolution of Hemoglobinometry, Melbourne 1981]
VOGEL [H. Vogel, Personalbibliographien der Professoren und Dozenten der Botanik, Chemie, Diätetik, Pharmazeutische Warenkunde, Physiologie, Physiologie, Tierheilkunde und Zoologie an der Medizinischen Fakultät der Karl-Ferdinands-Universität in Prag...1800-1850, Erlangen-Nürnberg 1972]
VOIGT [B.F. Voigt, Neuer Nekrolog der Deutschen, Weimar 1824-1856]
VORONTSOV [D.S. Vorontsov et al., Narisi z Istorii Fiziologii na Ukranii, Kiev 1950]
W [T.I. Williams, A Biographical Dictionary of Scientists, 2nd Ed., London-New York 1974]
WE [Wielka Encyklopedia Powszechna PWN, Warsaw 1962-1970]
WL [Westfälische Lebensbilder, Münster 1930-1962]
WW [Who's Who, London 1849-]
WWA [Who's Who in America, Chicago 1899-]
WWAH [Who Was Who in American History, Science and Technology, Chicago 1976]

WWR [Who Was Who in the USSR, Metuchen, N.J. 1972]
WWWS [World Who's Who in Science, Chicago 1968]
WAGENITZ [G. Wagenitz, Göttinger Biologen 1737-1945, Göttingen 1988]
WARSAW [T. Manteuffel, Universytet Warszawski w Latach 1915/1916-1934/1935, Warsaw 1936]
WEBER [M. Weber, Kurzbiographien 1935 Verstorbener Aerzte des Deutschen Sprachraums, Zürich 1976]
WEINMANN [J. Weinmann, Egerländer Biographisches Lexikon, Bayreuth 1985-]
WELDING [O. Welding et al., Deutschbaltisches Biographisches Lexikon 1710-1960, Vienna 1970]
WENIG [O. Wenig, Verzeichnis der Professoren und Dozenten der Rheinischen Friedrich-Wilhelm-Universität zu Bonn 1818-1968, Bonn 1968]
WERSTLER [F. Werstler, Personalbibliographien von Professoren und Dozenten der Medizinischen Fakultät zu Prag...1853-1880, Erlangen-Nürnberg 1972]
WIDMANN [H. Widmann, Exil und Bildungshilfe: Die Deutschsprachige Akademische Emigration in die Türkei nach 1933, Berne 1973]
WIDSTRAND [A. Widstrand, Sveriges Läkarehistorie, Stockholm 1930-1935]
WIEN [Die Feierliche Inauguration des Rektors der Wiener Universität]
WILKS [S. Wilks and G.T. Bettany, A Biographical History of Guy's Hospital, London 1892]
WININGER [S. Wininger, Grosse Jüdische National Biographie, Czernowitz 1925- 1936]
WOLF [W. Wolf, Personalbibliographien von Professoren und Dozenten der Röntgenologie und Radiologie an der Medizinischen Fakultät der Universität Wien...1900-1945, Erlangen-Nürnberg 1972]
WOLF(2) [C. Wolf, Verzeichnis der Hochschullehrer der Technischen Hochschule Darmstadt, Part I, Kurzbiographien 1836-1945, Darmstadt 1977]
WRANY [A. Wraný, Geschichte der Chemie und der auf chemischer Grundlage beruhende Betriebe in Böhmen bis zur Mitte des 19. Jahrhunderts, Prague 1902]
WREDE [R. Wrede and H. von Reinfels, Das Geistige Berlin, vol.3, Berlin 1898]
WURZBACH [C. von Wurzbach, Biographisches Lexikon des KaiserthumsOesterreich, Vienna 1856-1891]
YOUNG [E.G. Young, The Development of Biochemistry in Canada, Toronto 1976]
ZMEEV [L.F. Zmeev, Russkie Vrachi Pisateli, St. Petersburg 1886-1892]
ZURICH [75 Jahre Chemische Forschung an der Universität Zürich, Zürich 1909]
ZURICH-D [Verzeichnis Zürichischer Universitätsschriften 1833-1892, Zürich 1904]
ZWEIFEL [M. Zweifel, Kurzbiographien von Aerzten des Deutschen Sprachraums die 1937 Verstorben sind, Zürich 1976]

Serial Publications

Abh. Braunschw. Wiss. Ges. [Abhandlungen der Braunschweigerischen Wissen schaftlichen Gesellschaft]
Abh. Nat. Ges. Nürnberg [Abhandlungen der Naturhistorischen Gesellschaft zu Nürnberg]
Abh. Nat. Ver. Bremen [Abhandlungen des Naturwissenschaftlichen VereinsBremen]
Acta Anat. [Acta Anatomica]
Acta Biochim. Polon. [Acta Biochimica Polonica]
Acta Biol. Med. Germ. [Acta Biologica et Medica Germanica)
Acta Bot. Neer. [Acta Botanica Neerlandica]
Acta Chem. Scand. [Acta Chemica Scandinavica]
Acta Chim. Acad. Sci. Hung. [Acta Chemica Academiae Scientarum Hungaricae]
Acta Cryst. [Acta Crystallographica]
Acta Embryol. Exp. [Acta Embryologiae Experimentalis]
Acta Gast. Belg. [Acta Gastroenterologica Belgica]
Acta Gen. Stat. Med. [Acta Genetica et Statistica Medica]
Acta Hist. Leop. [Acta Historica Leopoldina]
Acta Hist. Rerum Nat. [Acta Historiae Rerum Naturalium necron Technicarum]
Acta Med. Hist. Pat. [Acta Medicae Historiae Patavina]
Acta Med. Scand. [Acta Medica Scandinavica]
Acta Microbiol. Pol. [Acta Microbiologica Polonica]
Acta Orthop. Scand. [Acta Orthopaedica Scandinavica]
Acta Path. Scand. [Acta Pathologica et Microbiologica Scandinavica]
Acta Pharm. Tox. [Acta Pharmacologica et Toxicologica]
Acta Physiol. Neer. [Acta Physiologica et Pharmacologica Neerlandica]
Acta Physiol. Pol. [Acta Physiologica Polonica]
Acta Physiol. Scand. [Acta Physiologica Scandinavica]
Acta Pont. Acad. Sci. [Acta Pontificiae Academiae Scientarum]
Acta Soc. Bot. Pol. [Acta Societatis Botanicorum Polonicae]
Actes Acad. Bord. [Actes de l'Académie Nationale des Sciences,

Belles-Lettres et Arts de Bordeaux]
Adv. Carbohydrate Chem. [Advances in Carbohydrate Chemistry]
Adv. Catalysis [Advances in Catalysis]
Adv. Chem. [Advances in Chemistry Series]
Adv. Coll. Sci. [Advances in Colloid Science]
Adv. Genetics [Advances in Genetics]
Adv. Met. Dis. [Advances in Metabolic Disorders]
Adv. Protein Chem. [Advances in Protein Chemistry]
Aerzt. Int. Bl. [Aerztliches Intelligenz-Blatt]
Ala. J. Med. Sci. [Alabama Journal of Medical Sciences]
Allgem. prakt. Chem. [Allgemeine und praktische Chemie]
Alm. Akad. Wiss. Wien [Almanach der Akademie der Wissenschaften in Wien]
Alm. Bayer. Akad. Wiss. [Almanach der Bayerischen Akademie der Wissen schaften]
Am. Chem. J. [American Chemical Journal]
Am. J. Anat. [American Journal of Anatomy]
Am. J. Bot. [American Journal of Botany]
Am. J. Cancer [American Journal of Cancer]
Am. J. Card. [American Journal of Cardiology]
Am. J. Clin. Nutr. [American Journal of Clinical Nutrition]
Am. J. Clin. Path. [American Journal of Clinical Pathology]
Am. J. Human Gen. [American Journal of Human Genetics]
Am. J. Ophthalm. [American Journal of Ophthalmology]
Am. J. Path. [American Journal of Pathology]
Am. J. Pharm. [American Journal of Pharmacy]
Am. J. Pharm. Ed. [American Journal of Pharmaceutical Education]
Am. J. Phys. [American Journal of Physics]
Am. J. Physiol. [American Journal of Physiology]
Am. J. Psychol. [American Journal of Psychology]
Am. J. Pub. Hth. [American Journal of Public Health]
Am. J. Roentgenol. [American Journal of Roentgenology]
Am. J. Sci. [American Journal of Science]
Am. J. Surg. [American Journal of Surgery]
Am. J. Trop. Med. [American Journal of Tropical Medicine]
Am. Med. Mon. [American Medical Monthly]
Am. Min. [American Mineralogist]
Am. Nat. [American Naturalist]
Am. Phil. Soc. Year Book [American Philosophical Society Year Book]
Am. Pract. [American Practitioner and Digest of Treatment]
Am. Rev. Tub. [American Review of Tuberculosis]
Am. Scient. [American Scientist]
Am. Soc. Microbiol. News [American Society of Microbiologists News]
Anal. Biochem. [Analytical Biochemistry]
Anal. Proc. [Analytical Proceedings]
Anat. Anz. [Anatomischer Anzeiger]
Anat. Nachr. [Anatomische Nachrichten]
Anat. Rec. [Anatomical Record]
Ang. Chem. Int. Ed. [Angewandte Chemie, International Edition]
Ann. Acad. Roy. Belg. [Annuaire de l'Académie des Sciences, des Lettres et des Beaux-Arts de Belgique]
Ann. Accad. Ital. [Annuario della Accademia d'Italia]
Ann. Appl. Biol. [Annals of Applied Biology]
Ann. Biol. [Année Biologique]
Ann. Biol. Clin. [Annales de Biologie Clinique]
Ann. Chem. [Annalen der Chemie]
Ann. Chim. [Annales de Chimie]
Ann. Chim. Anal. [Annales de Chimie Analytique]
Ann. Embr. Morph. [Annales d'Embryologie et de Morphogénèse]
Ann. Endocrin. [Annales d'Endocrinologie]
Ann. Ent. Soc. Am. [Annals of the Entomological Society of America]
Ann. Falsif. [Annales des Falsifications]
Ann. Ferment. [Annales des Fermentations]
Ann. Histochim. [Annales d'Histochimie]
Ann. Hyg. Publ. [Annales d'Hygiène Publique]
Ann. Inst. Actin. [Annales de l'Institut d'Actinologie]
Ann. Inst. Agron. [Annales de l'Institut Agronomque]

Ann. Inst. Pasteur [Annales de l'Institut Pasteur]
Ann. Int. Med. [Annals of Internal Medicine]
Ann. Mag. Nat. Hist. [Annals and Magazine of Natural History]
Ann. Med. Hist. [Annals of Medical History]
Ann. Med. Nancy [Annales Médicales de Nancy]
Ann. Med. Psych. [Annales Médico-Psychologiques]
Ann. N.Y. Acad. Sci. [Annals of the New York Academy of Sciences]
Ann. Ocul. [Annales d'Oculistique]
Ann. Parasit. [Annales de Parasitologie]
Ann. Pharm. Fran. [Annales Pharmaceutiques Françaises]
Ann. Phys. [Annalen der Physik]
Ann. Physiol. [Annales de Physiologie]
Ann. Rep. Ferm. Proc. [Annual Reports on Fermentation Processes]
Ann. Rev. Biochem. [Annual Review of Biochemistry]
Ann. Rev. Biophys. [Annual Review of Biophysics and Bioengineering]
Ann. Rev. Gen. [Annual Review of Genetics]
Ann. Rev. Immunol. [Annual Review of Immunology]
Ann. Rev. Microbiol. [Annual Review of Microbiology]
Ann. Rev. Pharm. [Annual Review of Pharmacology, later and Toxicology]
Ann. Rev. Phys. Chem. [Annual Review of Physical Chemistry]
Ann. Rev. Physiol. [Annual Review of Physiology]
Ann. Rev. Plant Physiol. [Annual Review of Plant Physiology]
Ann. Sci. [Annals of Science]
Ann. Sci. Agron. [Annales de la Science Agronomique]
Ann. Sci. Nat. [Annales des Sciences, Botanique]
Ann. Scient. [L'Année Scientifique et Industrielle]
Ann. Soc. Agr. Lyon [Annales de la Société d'Agriculture, Sciences et Industrie de Lyon]
Ann. Soc. Belg. Micr. [Annales de la Société Belge de Microscopie]
Ann. Soc. Belge Med. Trop. [Annales de la Société Belge de Medecine Tropicale]
Ann. Soc. Entomol. [Annales de la Société Entomologique de France]
Ann. Soc. Zool. Belg. [Annales de la Société Royale Zoologique de Belgique]
Ann. Univ. Lyon [Annales de l'Université de Lyon]
Ann. Univ. Paris [Annales de l'Université de Paris]
Apoth. Z. [Apotheker Zeitung]
Arb. Ehrlich Inst. [Arbeiten aus dem Paul Ehrlich Institut]
Arbeitsphys. [Arbeitsphysiologie]
Arch. Anat. [Archives d'Anatomie, d'Histologie et d'Embryologie]
Arch. Anat. Entw. [Archiv für Anatomie und Entwicklungsgeschichte]
Arch. Anat. Micr. [Archives d'Anatomie Microscopique]
Arch. Anthr. [Archiv für Anthropologie]
Arch. Antr. Etnol. [Archivio per l'Antropologia e la Etnologia]
Arch. Bioch. Biophys. [Archives of Biochemistry and Biophysics]
Arch. Biog. Cont. [Archives Biographiques Contemporains]
Arch. Biol. [Archives de Biologie]
Arch. Chem. Mikr. [Archiv für Chemie und Mikroskopie]
Arch. Dermatol. [Archives of Dermatology and Syphilology]
Arch. Entwickl. [Archiv für Entwicklungsmechanik der Organismen]
Arch. Env. Hth. [Archives of Environmental Health]
Arch. exp. Path. Pharm. [Archiv für experimentelle Pathologie und Pharmakologie]
Arch. exp. Zellforsch. [Archiv für experimentelle Zellforschung]
Arch. Fisiol. [Archivio di Fisiologia]
Arch. Fr. Nat. Meckln. [Archiv des Vereins der Freunde der Naturgeschichte in Mecklenburg]
Arch. Gen. Med. [Archives Générales de Médecine]
Arch. ges. Virusforsch. [Archiv für gesamte Virusforschung]
Arch. Gesch. Naturwiss. [Archiv für die Geschichte der Naturwissenschaften und der Technik]
Arch. Hist. Exact Sci. [Archives of the History of the Exact Sciences]
Arch. Hist. Fil. Med. [Archiwum Historii i Filozofii Medycyny]
Arch. Hist. Med. [Archiwum Historii Medycyny]
Arch. Hyg. [Archiv für Hygiene und Bakteriologie]
Arch. Inst. Pasteur Alg. [Archives de l'Institut Pasteur d'Algérie]
Arch. Int. Cl. Bern. [Archives Internationales Claude Bernard]
Arch. Int. Med. [Archives of Internal Medicine]
Arch. Inter. Hist. Sci. [Archives Internationales d'Histoire des Sciences]
Arch. Inter. Pharmacodyn. [Archives Internationales de Pharmacodynamie et de Thérapie]

Arch. Inter. Physiol. [Archives Internationales de Physiologie et de Biochimie]
Arch. Ital. Biol. [Archives Italiennes de Biologie]
Arch. Ital. Sci. Farm. [Archivio Italiano de Scienze Farmacologiche]
Arch. Ital. Sci. Med. Trop. [Archivio Italiano di Scienze Mediche Tropicale]
Arch. Kind. [Archiv für Kinderheilkunde]
Arch. klin. Chir. [Archiv für klinische Chirurgie]
Arch. klin. Med. [Archiv für klinische Medizin, same as Deutsches Archiv für klinische Medizin]
Arch. Med. Navale [Archives de Médecine Navale]
Arch. mikr. Anat. [Archiv für mikroskopische Anatomie]
Arch. Neer. Physiol. [Archives Néerlandaises de Physiologie]
Arch. Neurol. [Archives of Neurology and Psychiatry]
Arch. Ophthalm. [Archiv für Ophthalmologie]
Arch. Otolaryng. [Archives of Otolaryngology]
Arch. Parasit. [Archives de Parasitologie]
Arch. Pat. Clin. Med. [Archivio di Patologia e Clinica Medica]
Arch. Path. [Archives of Pathology]
Arch. path. Anat. [Archiv für pathologische Anatomie, same as Virchow's Archiv]
Arch. Ped. [Archives of Pediatrics]
Arch. Pharm. [Archiv der Pharmazie]
Arch. Pharm. Chem. [Archiv for Pharmaci og Chemi]
Arch. Physiol. [Archives de Physiologie]
Arch. Protist. [Archiv für Protistenkunde]
Arch. Psychiat. [Archiv für Psychiatrie und Nervenkrankheiten]
Arch. Sci. Biol. [Archives des Sciences Biologiques]
Arch. Sci. Med. Torino [Archivio per le Scienze Mediche Torino]
Arch. Sci. Phys. Nat. [Archives des Sciences Physiques et Naturelles]
Arch. Vecchi Anat. Pat. [Archivio de Vecchi per Anatomia Patalogica]
Arch. Verdauubgskr. [Archiv für Verdauungskrankheiten]
Arch. Zellforsch. [Archiv für Zellforschung]
Arch. Zool. Exp. [Archives de Zoologie Expérimentale et Générale]
Archivio Sci. Biol. [Archivio di Scienze Biologiche]
Arkh. Anat. Gist. Emb. [Arkhiv Anatomii, Gistologii i Embriologii]
Arkh. Biol. Nauk [Arkhiv Biologicheskikh Nauk]
Arkh. Klin. Eksp. Med. [Arkhiv Klinicheskoi i Ekspermentalnoi Meditsini]
Arkh. Pat. [Arkhiv Patologii]
Arzneimitt. [Arzneimittelforschung]
Atti Accad. Ferrara [Atti della Accademia delle Scienze Mediche e Naturali di Ferrara]
Atti Accad. Gioenia [Atti della Accademia Gioenia di Scienze Naturali]
Atti Accad. Lincei [Atti della Accademia Nazionale dei Lincei]
Atti Accad. Pontiniana [Atti della Accademia Pontiniana]
Atti Accad. Torino [Atti della Accademia delle Scienze di Torino]
Atti Accad. Sci. Med. Palermo [Atti della Accademia delle Scienze Mediche di Palermo]
Atti Ist. Bot. Pavia [Atti del Istituto Botanico della Universita di Pavia]
Atii Ist. Veneto [Atti del Reale Istituto Veneto di Scienze Lettere ed Arti]
Atti Soc. Ital. Sci. Nat. [Atti della Società Italiana di Scienze Naturali]
Atti Soc. Lomb. Sci. Med. [Atti della Società Lombarda di Scienze Mediche e Biologiche]
Atti Soc. Sci. Genova [Atti della Società Ligustica di Scienze Naturali e Geografiche di Genova]
Austral. J. Exp. Biol. [Australian Journal of Experimental Biology]
Bact. Revs. [Bacteriological Reviews]
Bayer. Akad. Wiss. Jahrbuch [Bayerische Akademie der Wissenschaften Jahrbuch]
Behring Inst. Mitt. [Behring Institut Mitteilungen]
Beitr. ger. Med. [Beiträge zur gerichtlichen Medizin]
Beitr. path. Anat. [Beiträge zur pathologischen Anatomie und allgemeinen Pathologie]
Beitr. physiol. path. Chem. [Beiträge zur physiologischen und pathologischen Chemie]
Beitr. Wurtt. Apothekgesch. [Beiträge zur Württembergischen Apotheken Geschichte]
Ber. Bayer. bot. Ges. [Berichte der Bayerischen botanischen Gesellschaft]
Ber. bot. Ges. [Berichte der deutschen botanischen Gesellschaft]
Ber. Bunsen Ges. [Berichte der Bunsen Gesellschaft]
Ber. chem. Ges. [Berichte der deutschen chemischen Gesellschaft]
Ber. Komm. Meeresforsch. [Berichte der deutschen Kommission für Meeresforschung]
Ber. Nat. Ges. Freiburg [Berichte der Naturforschenden Gesellschaft zu Freiburg i.B.]
Ber. Nat. Med. Innsb. [Berichte des Naturwissenschaftlich-Medizinischen Vereins in Innsbruck]
Ber. pharm. Ges. [Berichte der deutschen pharmazeutischen Gesellschaft]
Ber. phys. Ges. [Berichte der deutschen physikalischen Gesellschaft]

Ber. Senck. Nat. Ges. [Berichte der Senckenbergischen Naturforschenden Gesellschaft]
Berl. klin. Wchschr. [Berliner klinische Wochenschrift]
Berl. Med. [Berliner Medizin]
Berl. med. Z. [Berliner medizinische Zeitschrift]
Bibl. Acad. Louvain [Bibliographie Académique de l'Université de Louvain]
Bibl. Paed. [Bibliotheca Paediatrica]
Bioch. Biophys. Acta [Biochimica et Biophysica Acta]
Bioch. Bull. [Biochemical Bulletin]
Bioch. Ed. [Biochemical Education]
Bioch. Soc. Symp. [Biochemical Society Symposia]
Bioch. Soc. Trans. [Biochemical Society Transactions]
Bioch. Ter. Sper. [Biochimica e Terapia Sperimentale]
Biochem. J. [Biochemical Journal]
Biochem. Pharm. [Biochemical Pharmacology]
Biochem. Z. [Biochemische Zeitschrift]
Biog. Mem. Fell. Nat. Inst. India [Biographical Memoirs of Fellows of the National Institute of India]
Biog. Mem. Fell. Roy. Soc. [Biographical Memoirs of Fellows of the Royal Society of London]
Biog. Mem. Nat. Acad. Sci. [Biographical Memoirs, National Academy of Sciences of the United States]
Biol. Gen. [Biologia Generalis]
Biol. Listy [Biologicke Listy]
Biol. Med. [Biologie Médicale]
Biol. Revs. [Biological Reviews]
Biol. Unter. [Biologische Untersuchungen]
Biol. Z. [Biologisches Zentralblatt]
Biol. Zhur. [Biologicheski Zhurnal]
Boll. Accad. Med. Genova [Bollettino della Accademia Medica di Genova]
Boll. Sci. Med. Bologna [Bollettino delle Scienze Mediche Bologna]
Boll. Soc. Med. Chir. Pavia [Bollettino della Società Medico-Chirurgica di Pavia]
Boll. Soc. Nat. Napoli [Bollettino delle Società dei Naturalisti in Napoli]
Bot. Arch. [Botanisches Archiv]
Bot. Jahrb. [Botanische Jahrbücher]
Bot. Not. [Botaniska Notiser]
Bot. Rev. [Botanical Review]
Bot. Tid. [Botanisk Tidsskrift]
Bot. Zb. [Botanisches Zentralblatt]
Bot. Zhur. [Botanicheski Zhurnal]
Bot. Zt. [Botanische Zeitung]
Bristol Med. Chir. J. [Bristol Medico-Chirurgical Journal]
Brit. J. Dermatol. [British Journal of Dermatology]
Brit. J. Exp. Path. [British Journal of Experimental Pathology]
Brit. J. Hist. Sci. [British Journal for the History of Science]
Brit. J. Ind. Med. [British Journal of Industrial Medicine]
Brit. J. Nutrition [British Journal of Nutrition]
Brit. J. Pharm. [British Journal of Pharmacology]
Brit. J. Phil. Sci. [British Journal for the Philosophy of Science]
Brit. J. Phys. Med. [British Journal of Physical Medicine]
Brit. J. Surg. [British Journal of Surgery]
Brit. Med. Bull. [British Medical Bulletin]
Brit. Med. J. [British Medical Journal]
Bull. Acad. Med. [Bulletin de l'Académie Nationale de Médecine]
Bull. Acad. Med. Belg. [Bulletin de l'Académie Royale de Médecine de Belgique]
Bull. Acad. Roy. Belg. [Bulletin de l'Académie Royale de Belgique, Classe des Sciences]
Bull. Acad. Sci. Nancy [Bulletin de l'Académie et Société Lorraines des Sciences Nancy]
Bull. Acad. Sci URSS [Bulletin de l'Académie des Sciences URSS, Division des Sciences Chimiques]
Bull. Biol. [Bulletin Biologique de la France et de la Belgique]
Bull. Can. Bioch. Soc. [Bulletin of the Canadian Biochemical Society]
Bull. Ecol. Soc. Am. [Bulletin of the Ecological Society of America]
Bull. Hist. Chem. [Bulletin for the History of Chemistry]
Bull. Hist. Med. [Bulletin of the History of Medicine]
Bull. Indian Inst. Hist. Med. [Bulletin of the Indian Institute of the History of Medicine]
Bull. Inst. Hist. Med. [Bulletin of the Institute of the History of Medicine]
Bull. Inst. Pasteur [Bulletin de l'Institut Pasteur]
Bull. Johns Hopkins Hosp. [Bulletin of the Johns Hopkins Hospital]
Bull. Med. Nord [Bulletin Médical du Nord]

Bull. Mus. Hist. Mulhouse [Bulletin du Musée Historique des Sciences Humaines de Mulhouse]
Bull. N.Y. Acad. Med. [Bulletin of the New York Academy of Medicine]
Bull. Res. Coun. Israel [Bulletin of the Research Council of Israel]
Bull. Schw. Akad. Med. Wiss. [Bulletin der Schweizerischen Akademie der Medizinischen Wissenschaften]
Bull. Sci. Fr. Belg. [Bulletin Scientifique de la France et de la Belgique]
Bull. Sci. Hist. Auvergne [Bulletin Scientifique et Historique de l'Auvergne]
Bull. Sci. Pharm. [Bulletin des Sciences Pharmaceutiques]
Bull. Soc. Bot. [Bulletin de la Société Botanique de France]
Bull. Soc. Bot. Belg. [Bulletin de la Société Botanique de Belgique]
Bull. Soc. Bot. Gen. [Bulletin de la Société Botanique de Geneve]
Bull. Soc. Chim. [Bulletin de la Société Chimique de France]
Bull. Soc. Chim. Belg. [Bulletin de la Société Chimique de Belgique]
Bull. Soc. Chim. Biol. [Bulletin de la Société Chimie Biologique]
Bull. Soc. Chir. [Bulletin et Mémoires de la Société de Chirurgie de Paris]
Bull. Soc. Fran. Min. [Bulletin de la Société Francaise de Minéralogie]
Bull. Soc. Frib. Sci. Nat. [Bulletin de la Société Fribourgeoise des Sciences Naturelles]
Bull. Soc. Hist. Med. [Bulletin de la Société Française de l'Histoire de la Médecine]
Bull. Soc. Hist. Nat. Afr. [Bulletin de la Société d'Histoire Naturelle d'Afrique du Nord]
Bull. Soc. Ind. Mulhouse [Bulletin de la Société Industrielle de Mulhouse]
Bull. Soc. Langres [Bulletin de la Société Historique et Archeologique de Langres]
Bull. Soc. Med. Paris [Bulletin et Mémoires de la Société Médicale des Hôpitaux de Paris]
Bull. Soc. Med. Suisse Rom. [Bulletin de la Société Médicale de la Suisse Romande]
Bull. Soc. Med. Yonne [Bulletin de la Société Médicale de l'Yonne]
Bull. Soc. Nat. Moscou [Bulletin de la Société de Naturalistes de Moscou]
Bull. Soc. Mycol. [Bulletin de la Société Mycologique de France]
Bull. Soc. Path. Exot. [Bulletin de la Société de Pathologie Exotique]
Bull. Soc. Pharm. Stras. [Bulletin de la Société de Pharmacie de Strasbourg]
Bull. Soc. Sci. Nancy [Bulletin de la Société Lorraines des Sciences Nancy]
Bull. Soc. Sci. Tarn [Bulletin de la Société des Sciences, Arts et Belles-Lettres du Tarn]
Bull. Soc. Vaud. Sci. Nat. [Bulletin de la Société Vaudoise des Sciences Naturelles]
Bull. Soc. Vet. [Bulletin de la Société Veterinaire de France]
Bull. Soc. Zool. [Bulletin de la Société Zoologique de France]
Bull. Torrey Bot. Club [Bulletin of the Torrey Botanical Club]
C.R. Acad. Agr. [Comptes Rendus de l'Académie d'Agriculture de France]
C.R. Acad. Sci. [Comptes Rendus Hebdomadaires de l'Académie des Sciences Paris]
C.R. Lab. Carlsberg [Comptes Rendus des Travaux du Laboratoire Carlsberg]
C.R. Soc. Biol. [Comptes Rendus Hebdomadaires des Séances et Mémoires de la Société de Biologie]
Can. J. Biochem. [Canadian Journal of Biochemistry]
Can. Med. Assoc. J. [Canadian Medical Association Journal]
Cas. Lek. Cesk. [Casopis Lekaru Českych]
Cent. allgem. Path. [Centralblatt (Zentralblatt) für allgemeine Pathologie und pathologische Anatomie]
Cent. prakt. Augenheilk. [Centralblatt (Zentralblatt) für praktische Augenheilkunde]
Cesk. Fysiol. [Československa Fysiologie]
Chem. Ber. [Chemische Berichte]
Chem. Brit. [Chemistry in Britain]
Chem. Eng. News [Chemical and Engineering News]
Chem. Ind. [Chemistry and Industry]
Chem. Listy [Chemicke Listy]
Chem. News [Chemical News]
Chem. Pharm. Bull. [Chemical and Pharmaceutial Bulletin]
Chem. Phys. Lipids [Chemistry and Physics of Lipids]
Chem. Soc. Revs. [Chemical Society Reviews]
Chem. Wkbl. [Chemisch Weekblad]
Chem. Z. [Chemiker Zeitung]
Chim. Ind. [Chimie et Industrie]
Chim. Ind. Agr. Biol. [Chimica nell l'Industria, nell l'Agricoltora, nella Biologia]
Chron. Med. [Chronique Médicale Paris]
Ciba Z. [Ciba Zeitschrift]
Clin. Chem. [Clinical Chemistry]
Coll. Czech. Chem. Comm. [Collection of Czechoslovak Chemical Communications]
Comp. Bioch. Physiol. [Comparative Biochemistry and Physiology]
Conn. Agr. Sta. Bull. [Connecticut Agricultural Station Bulletin]
Conn. Tiss. Res. [Connective Tissue Research]
Contr. Boyce Thomp. Inst. [Contributions from the Boyce Thompson Institute]

Coord. Chem. Revs. [Coordination Chemistry Reviews]
Corr. Nat. Ver. Bonn [Correspondenzblatt des Naturhistorischen Vereins Bonn]
Corr. Schw. Aerzte [Correspondenzblatt für Schweizer Aerzte]
Croat. Chim. Acta [Croatica Chimica Acta]
Dan. Med. Bull. [Danish Medical Bulletin]
Derm. Wchschr. [Dermatologische Wochenschrift]
Deutsche Apoth. Z. [Deutsche Apotheker Zeitung]
Deutsche Leb. Rund. [Deutsche Lebensmittel Rundschau]
Deutsche med. Wchschr. [Deutsche medizinische Wochenschrift]
Deutsche Z. Chir. [Deutsche Zeitschrift für Chirurgie]
Deutsche Z. ger. Med. [Deutsche Zeitschrift für gesamte gerichtliche Medizin]
Deutsche Z. Thiermed. [Deutsche Zeitschrift für Thiermedizin]
Deutsche Z. Verdauung. [Deutsche Zeitschrift für Verdauung- und Stoffwechselkrankheiten]
Deutsches Arch. klin. Med. [Deutsches Archiv für klinische Medizin]
Deutsches Arch. prakt. Med. [Deutsches Archiv für praktische Medizin]
Deutsches Med. J. [Deutsches Medizinisches Journal]
Dev. Biol. [Developmental Biology]
Dublin J. Med. Sci. [Dublin Journal of Medical Science]
Echo Med. Nord [Echo Médical du Nord]
Edinburgh Med. J. [Edinburgh Medical Journal]
Entom. Bl. [Entomologische Blätter]
Entom. News [Entomological News]
Entomol. Z. [Entomologische Zeitschrift]
Erg. Hyg. [Ergebnisse der Hygiene, Bakteriologie, Immunitätsforschung und experimentellen Therapie]
Erg. Physiol. [Ergebnisse der Physiologie; after 1974, Reviews of Physiology, Biochemistry and Pharmacology]
Eur. J. Med. Chem. [European Journal of Medical Chemistry]
Eur. J. Nucl. Med. [European Journal of Nuclear Medicine]
Eur. J. Ped. [European Journal of Pediatrics]
Exp. Eye Res. [Experimental Eye Research]
Exp. Med. Surg. [Experimental Medicine and Surgery]
Exp. Neurology [Experimental Neurology]
Farm. Toks. [Farmakologia i Toksikologia]
Fed. Proc. [Federation Proceedings; after 1986, The FASEB Journal]
Fiziol. Zhur. [Fiziologicheski Zhurnal]
Fol. Biol. [Folia Biologica, Cracow]
Fol. Haemat. [Folia Haematologica]
Fol. Psych. Jap. [Folia Psychiatrica et Neurologica Japonica]
Forsch. Fort. [Forschung und Fortschritte]
Fort. Geb. Rontgenstr. [Fortschritte auf dem Gebiet der Röntgenstrahlung]
Fort. Med. [Fortschritte der Medizin]
Fort. Neurol. [Fortschritte der Neurologie, Psychiatrie und ihrer Grenz gebiete]
Freib. Univ. Bl. [Freiburger Universitätsblätter]
Gastroent. [Gastroenterologia]
Gaz. Hop. Paris [Gazette des Hôpitaux de Paris]
Gaz. Med. Fran. [Gazette Médicale de France]
Gaz. Med. Paris [Gazette Médicale de Paris]
Gaz. Sci. Med. Bord. [Gazette Hebdomadaire des Sciences Médicales Bordeaux]
Gazz. Chim. Ital. [Gazzetta Chimica Italiana]
Gazz. Med. Ital. [Gazzetta Medica Italiana]
Gior. Batt. Immun. [Giornale di Batterologia e Immunologia]
Glasgow Med. J. [Glasgow Medical Journal]
Guy's Hosp. Gaz. [Guy's Hospital Gazette]
Guy's Hosp. Rep. [Guy's Hospital Reports]
Hannah Inst. Hist. Sci. [Hannah Institute for the History of Science]
Harvard Med. Alumni Bull. [Harvard Medical Alumni Bulletin]
Heid. Akad. Jahrb. [Heidelberger Akademie der Wissenschaften Jahrbuch]
Helv. Chim. Acta [Helvetica Chimica Acta]
Helv. Paed. Acta [Helvetica Paeditrica Acta]
Helv. Phys. Acta [Helvetica Physica Acta]
Helv. Physiol. Pharm. Acta [Helvetica Physiologica et Pharmacologica Acta]
Hist. Med. [Histoire de la Médecine]
Hist. Phil. Life Sci. [History and Philosophy of the Life Sciences]
Hist. Sci. [History of Science]
Hist. Sci. Med. [Histoire des Sciences Médicales]

Hist. Stud. Phys. Sci. [Historical Studies in the Physical Sciences]
Horm. Prot. Pep. [Hormonal Proteins and Peptides]
Hosp. Tid. [Hospitalstidende]
Ill. Land. Z. [Illustrierte Landwirtschaftliche Zeitung]
Ind. Eng. Chem. [Industrial and Engineering Chemistry]
Ind. J. Hist. Sci. [Indian Journal of the History of Science]
Inf. Psychiat. [Information Psychiatrique]
Int. Arch. Allergy [International Archives of Allergy and Applied Immunology]
Int. J. Biochem. [International Journal of Biochemistry]
Int. J. Neuropharm. [International Journal of Neuropharmacology]
Int. J. Pep. Res. [International Journal of Peptide and Protein Research]
Int. J. Psychoanal. [International Journal of Psychoanalysis]
Int. Z. Vitaminforsch. [Internationale Zeitschrift für Vitaminforschung]
Inv. Ophthalm. [Investigative Ophthalmology]
Inv. Urol. [Investigative Urology]
Irish J. Med. Sci. [Irish Journal of Medical Science]
Israel J. Chem. [Israel Journal of Chemistry]
Israel J. Med. Sci. [Israel Journal of Medical Sciences]
Ist. Est. Tekhn. Pribalt. [Iz Istorii Estestvoznania i Tekhnika Pribaltiki]
Ital. J. Biochem. [Italian Journal of Biochemistry]
J. Agr. Food Chem. [Journal of Agricultural and Food Chemistry]
J. Am. Acad. Dermatol. [Journal of the American Academy of Dermatology]
J. Am. Chem. Soc. [Journal of the American Chemical Society]
J. Am. Diet. Assn. [Journal of the American Dietetic Association]
J. Am. Leather Assn. [Journal of the American Leather Chemists Association]
J. Am. Med. Assn. [Journal of the American Medical Association]
J. Am. Oil Chem. Soc. [Journal of the American Oil Chemists' Society]
J. Am. Pharm. Assn. [Journal of the American Pharmaceutical Association]
J. Am. Speech Assn. [Journal of the American Speech and Hearing Association]
J. Anat. Physiol. [Journal de l'Anatomie et de la Physiologie]
J. Antibiot. [Journal of Antibiotics]
J. Assn. Agr. Chem. [Journal of the Association of Official Agricultural Chemists]
J. Atheroscl. Res. [Journal of Atherosclerosis Research]
J. Bact. [Journal of Bacteriology]
J. Biochem. [Journal of Biochemistry]
J. Biol. Chem. [Journal of Biological Chemistry]
J. Biophys. Bioch. Cytol. [Journal of Biophysical and Biochemical Cytology]
J. Cancer Res. [Journal of Cancer Research]
J. Cell Biol. [Journal of Cell Biology]
J. Chem. Ed. [Journal of Chemical Education]
J. Chem. Phys. [Journal of Chemical Physics]
J. Chem. Soc. [Journal of the Chemical Society, London]
J. Chem. Thermodyn. [Journal of Chemical Thermodynamics]
J. Chim. Phys. [Journal de Chimie Physique]
J. Chromat. [Journal of Chromatography]
J. Clin. Chem. [Journal of Clinical Chemistry and Clinical Biochemistry]
J. Clin. Endocrin. [Journal of Clinical Endocrinology and Metabolism]
J. Clin. Path. [Journal of Clinical Pathology]
J. Coll. Sci. [Journal of Colloid and Interface Science]
J. Comp. Neurol. [Journal of Comparative Neurology]
J. Conn. Med. Prat. [Journal des Connaissances Médicales Pratiques]
J. Dairy Res. [Journal of Dairy Research]
J. Emb. Exp. Morph. [Journal of Embryology and Experimental Morphology]
J. Endocrin. [Journal of Endocrinology]
J. Exp. Biol. [Journal of Experimental Biology]
J. Exp. Med. [Journal of Experimental Medicine]
J. Franklin Inst. [Journal of the Franklin Institute]
J. Gen. Microbiol. [Journal of General Microbiology]
J. Gen. Physiol. [Journal of General Physiology]
J. Gen. Psychol. [Journal of General Psychology]
J. Heredity [Journal of Heredity]
J. Hist. Biol. [Journal of the History of Biology]
J. Hist. Ideas [Journal of the History of Ideas]
J. Hist. Med. [Journal of the History of Medicine and Allied Sciences]

J. Histochem. [Journal of Histochemistry and Cytochemistry]
J. Hygiene [Journal of Hygiene]
J. Immunol. [Journal of Immunology]
J. Ind. Chem. Soc. [Journal of the Indian Chemical Society]
J. Ind. Eng. Chem. [Journal of Industrial and Engineering Chemistry]
J. Inorg. Chem. [Journal of Inorganic and Nuclear Chemistry]
J. Inst. Brew. [Journal of the Institute of Brewing]
J. Inst. Fuel [Journal of the Institute of Fuel]
J. Int. Soc. Leather Chem. [Journal of the International Society of Leather Trades Chemists]
J. Inv. Derm. [Journal of Investigative Dermatology]
J. Lab. Clin. Med. [Journal of Laboratory and Clinical Medicine]
J. Med. Bordeaux [Journal Médical de Bordeaux]
J. Med. Ed. [Journal of Medical Education]
J. Med. Gen. [Journal of Medical Genetics]
J. Med. Microbiol. [Journal of Medical Microbiology]
J. Med. Vet. [Journal de Médecine Vétérinaire]
J. Mich. Med. Soc. [Journal of the Michigan State Medical Society]
J. Microscopie [Journal de Microscopie]
J. Microscopy [Journal of Microscopy]
J. Morphol. [Journal of Morphology]
J. Mt. Sinai Hosp. [Journal of the Mount Sinai Hospital, New York]
J. Nat. Cancer Inst. [Journal of the National Cancer Institute]
J. Nat. Med. Assn. [Journal of the National Medical Association]
J. Nerv. Ment. Dis. [Journal of Nervous and Mental Diseases]
J. Neurochem. [Journal of Neurochemistry]
J. Neurophysiol. [Journal of Neurophysiology]
J. Nutrition [Journal of Nutrition]
J. Opt. Soc. Am. [Journal of the Optical Society of America]
J. Parasit. [Journal of Parasitology]
J. Path. Bact. [Journal of Pathology and Bacteriology]
J. Pediat. [Journal of Pediatrics]
J. Pharm. Alsace [Journal de Pharmacie d'Alsace et de Lorraine]
J. Pharm. Chim. [Journal de Pharmacie et de Chimie]
J. Pharm. Exp. Ther. [Journal of Pharmacology and Experimental Therapeutics]
J. Phys. Chem. [Journal of Physical Chemistry]
J. Physiol. [Journal of Physiology]
J. Pol. Sci. [Journal of Polymer Science]
J. prakt. Chem. [Journal für praktische Chemie]
J. Protozool. [Journal of Protozoology]
J. Radiol. Elect. [Journal de Radiologie et d'Électrologie]
J. Repr. Fert. [Journal of Reproduction and Fertility]
J. Roy. Coll. Phys. [Journal of the Royal College of Physicians]
J. Roy. Inst. Chem. [Journal of the Royal Institute of Chemistry]
J. Roy. Micr. Soc. [Journal of the Royal Microscopical Society]
J. Roy. Soc. Arts [Journal of the Royal Society of Arts]
J. Sci. Res. Inst. [Journal of the Scientific Research Institute Tokyo]
J. Soc. Chem. Ind. [Journal of the Society of Chemical Industry]
J. South Afr. Vet. Assn. [Journal of the South African Veterinary Medical Association]
J. Thor. Surg. [Journal of Thoracic Surgery]
J. Trop. Med. Hyg. [Journal of Tropical Medicine and Hygiene]
J. Wash. Acad. Sci. [Journal of the Washington Academy of Sciences]
Jaarb. Akad. Wet. [Jaarboek Akademie van Wetenschappen Amsterdam]
Jaarb. Vlaam. Akad. [Jaarboek van de Vlaamische Akademie voor Weten schappen van Belgie]
Jahrb. Akad. Wiss. Berlin [Jahrbuch der deutschen Akademie der Wissen schaften in Berlin]
Jahrb. Akad. Wiss. Gott. [Jahrbuch der Akademie der Wissenschaften in Göttingen]
Jahrb. Kinderheil. [Jahrbuch für Kinderheilkunde]
Jahrb. Morph. [Jahrbuch der Morphologie und mikroskopische Anatomie]
Jahrb. Nass. Ver. Nat. [Jahrbuch des Nassauischen Vereins für Naturkunde]
Jahrb. Psychiat. [Jahrbuch für Psychiatrie und Neurologie]
Jahrb. Sachs. Akad. [Jahrbuch der Sächsischen Akademie der Wissenschaften zu Leipzig]
Jahrb. Ver. Meckl. Gesch. [Jahrbuch des Vereins für Mecklenburgische Geschichte]
Jahresb. Aerzt. Ver. [Jahresbericht des Aerztlichen Vereins zu Frankfurt am Main]
Jahresb. Fort. Tierchem. [Jahresberichte über die Fortschritte der Tierchemie]
Jahresb. Nat. Ges. Graub. [Jahresbericht der Naturforschenden Gesellschaft Graubündens]

Jahresb. Phys. Ver. Frank. [Jahresbericht des Physikalischen Vereins zu Frankfurt]
Jahresb. Schles. Ges. [Jahresbericht der Schlesischer Gesellschaft für Vaterlandische Kultur]
Jahresb. Ver. Nat. Braun. [Jahresbericht des Vereins für Naturkunde in Braunschweig]
Jahresb. Ver. Nat. Mann. [Jahresbericht der Vereins für Naturkunde inMannheim]
Jahreshefte Luneberg [Jahreshefte des Naturwissenschaflichen Vereins für das Fürstentum Lüneburg]
Jahreshefte Wurtt. [Jahreshefte des Vereins für Vaterlandische Naturkunde in Württemberg]
Jap. Stud. Hist. Sci. [Japanese Studies in the History of Science]
Jena Z. Naturw. [Jenaische Zeitschrift für Naturwissenschaften]
Johns Hopkins Med. Bull. [Johns Hopkins Medical Bulletin]
Johns Hopkins Med. J. [Johns Hopkins Medical Journal]
Journ. Physiol. [Journal de Physiologie]
Kazan. Med. Zhur. [Kazanskii Meditsinski Zhurnal]
Kharkov. Med. Zhur. [Kharkovskii Meditsinski Zhurnal]
Klin. Med. [Klinicheskaia Meditsina]
Klin. Mon. Augenheilk. [Klinische Monatsblätter für Augenheilkunde]
Klin. Wchschr. [Klinische Wochenschrift]
Koll. Z. [Kolloid Zeitschrift]
Kryt. Lek. [Krytyka Lekarska]
Kwart. Hist. Nauki [Kwartalnik Historii Nauki i Techniki]
Land. Jahrb. [Landwirtschaftilche Jahrbücher]
Land. Vers. [Landwirtschaftlichen Versuch-Stationen]
Lep. News [Lepidopterists' News]
Lev. Sven. Vet. [Levadsteckningar over Svenska Vetenskapsakademiens Ledamöter]
Ljet. Jugoslav. Akad. [Ljetopis Jugoslavenske Akademije]
Max-Planck-Ges. Ber. [Max-Planck-Gesellschaft Berichte]
Mayo Clin. Proc. [Mayo Clinic Proceedings]
Med. Chir. Trans. [Medico-Chirurgical Transactions]
Med. Conv. Hannover [Medicinisches Conversations- und Correspondenzblatt Hannover]
Med. Forum [Medicinsk Forum]
Med. Hist. [Medical History]
Med. Hist. J. [Medizin-Historisches Journal]
Med. Ital. [Medicina Italiana]
Med. J. Austral. [Medical Journal of Australia]
Med. J. Rec. [Medical Journal and Record]
Med. Klin. [Medizinische Klinik]
Med. Lab. Sci. [Medical Laboratory Science]
Med. Mon. [Medizinische Monatsschrift]
Med. Parazit. [Meditsinskaia Parazitologia]
Med. Welt [Medizinische Welt]
Mem. Acad. Lyon [Mémoires de l'Académie des Sciences, Belles-Lettres et Arts de Lyon]
Mem. Acad. Med. [Mémoires de l'Académie de Médecine Paris]
Mem. Acad. Med. Belg. [Mémoires de l'Académie Royale de Médecine de Belgique]
Mem. Acad. Sci. [Mémoires de l'Académie des Sciences Paris]
Mem. Acad. Stanislas [Mémoires de l'Académie de Stanislas]
Mem. Acad. Toulouse [Mémoires de l'Académie des Sciences, Inscriptions et Belles Lettres de Toulouse]
Mem. Lit. Phil. Soc. [Memoirs and Proceedings of the Manchester Literary and Philosophical Society]
Mem. Nat. Acad. Sci. [Memoirs of the National Academy of Sciences of the United States]
Mem. Soc. Med. Strasbourg [Mémoires de la Société de Médecine de Strasbourg]
Mem. Soc. Nat. Odessa [Mémoires de la Société des Naturalistes de la Nouvelle-Russie, Odessa]
Mem. Soc. Phys. Hist. Nat. Gen. [Mémoires de la Société de Physique et d'Histoire Naturelle de Genève]
Mem. Soc. Sci. Bordeaux [Mémoires de la Société des Sciences Physiques et Naturelles de Bordeaux]
Microchem. J. [Microchemical Journal]
Mikr. Zhur. [Mikrobiologichni Zhurnal]
Min. Mag. [Mineralogical Magazine]
Minerva Farm. [Minerva Farmaceutica]
Minerva Med. [Minerva Medica]
Mitt. Dendrol. Ges. [Mitteilungen der deutschen Dendrologischen Gesellschaft]
Mitt. Gesch. Med. [Mitteilungen zur Geschichte der Medizin und der Naturwissenschaften]
Mitt. Lebensmitt. [Mitteilungen aus dem Gebiet der Lebensmitteluntersuchungen und Hygiene]
Mitt. Nat. Ges. Bern [Mitteilungen der Naturforschenden Gesellschaft Bern]
Mitt. Nat. Ver. Neu-Vorpomm. [Mitteilungen des Naturwissenschaftlichen Vereins Neu-Vorpommern]
Mitt. Nat. Ver. Steier. [Mitteilungen des Naturwissenschaftlichen Vereins für Steiermark]
Mitt. Pharm. Ges. [Mitteilungen der deutschen Pharmazeutischen Gesellschaft]
Mitt. Schw. Ent. Ges. [Mitteilungen der Schweizerischen Entomologischen Gesellschaft]

Mitt. Ver. Aerzte Steier. [Mitteilungen des Vereins der Aerzte in Steiermark]
Mitt. Vers. Gär. Wien [Mitteilungen der Versuchstation für das Gärung gewerbe in Wien]
Mol. Cell. Biochem. [Molecular and Cellular Biochemistry]
Mon. Geburt. [Monatsschrift für Geburtshilfe]
Mon. Kind. [Monatsschrift für Kinderheilkunde]
Mon. Ohren. [Monatsschrift für Ohrenheilkunde]
Mon. Sci. [Moniteur Scientifique]
Mon. Vet. Med. [Monatshefte für Veterinär Medizin]
Mon. Zool. Ital. [Monitore Zoologico Italiano]
Mons. Hipp. [Monspelliensis Hippocrates]
Montpell. Med. [Montpellier Médical]
Mt. Sinai J. Med. [Mount Sinai Journal of Medicine]
Münch. med. Wchschr. [Münchener medizinische Wochenschrift]
NTM [Schriftenreihe für Geschichte der Naturwissenschaften, Technik und Medizin]
N.Y. J. Med. [New York State Journal of Medicine]
N.Y. Med. J. [New York Medical Journal]
N.Y. Med. Phys. J. [New York Medical and Physical Journal]
Nachr. Chem. Techn. [Nachrichten aus Chemie und Technik]
Nachr. Giessener Ges. [Nachrichten der Giessener Hochschulgesellschaft]
Naturw. Rund. [Naturwissenschaftliche Rundschau]
Naturwiss. [Naturwissenschaften]
Ned. Tijd. Gen. [Nederlands Tijschrift voor Geneeskunde]
Neues Rep. Pharm. [Neues Repertorium für die Pharmazie]
Neuropat. Pol. [Neuropatologia Polska]
New Eng. J. Med. [New England Journal of Medicine]
Nord. Med. [Nordisk Medicin]
Nord. Med. Hist. Arsbok [Nordisk Medicinhistorisk Årsbok]
Norske Vid. Akad. Arbok [Norske Videnskaps-Akademi i Oslo Årbok]
Not. Acad. Sci. [Notices et Discours, Académie des Sciences Paris]
Notes Roy. Soc. [Notes and Records of the Royal Society of London]
Nouv. Arch. Hist. Nat. [Nouvelles Archives du Muséum d'Histoire Naturelle]
Nouv. Pr. Med. [Nouvelle Presse Médicale]
Nouv. Rev. Hem. [Nouvelle Revue Française d'Hematologie]
Nova Acta Leop. [Nova Acta Leopoldina]
Nutr. Revs. [Nutrition Reviews]
Obit. Fell. Roy. Soc. [Obituary Notices of Fellows of the Royal Society of London]
Obs. Gyn. [Obstetrics and Gynecology]
Orv. Het. [Orvosi Hetilep]
Orv. Kozl. [Orvostorteneti Kozlemenyek]
Ost. Apoth. Z. [Oesterreichische Apotheker Zeitung]
Ost. bot. Z. [Oesterreichische botanische Zeitung]
Ost. Chem. Z. [Oesterreichische Chemiker Zeitung]
Pag. Stor. Med. [Pagine di Storia della Medicina]
Pat. Fiz, Eksp. Ter. [Patologicheskaiia Fiziologia i Eksperimentalnaia Terapia]
Pavlovian J. Biol. Sci. [Pavlovian Journal of Biological Science]
Penn. Med. J. [Pennsylvania Medical Jornal]
Persp. Biol. Med. [Perspectives in Biology and Medicine]
Pflügers Arch. [Pflügers Archiv für die gesamte Physiologie des Menschen und der Tiere]
Pharm. Hist. [Pharmacy in History]
Pharm. Ind. [Pharmazeutische Industrie]
Pharm. J. [Pharmaceutical Journal]
Pharm. Weekbl. [Pharmaceutisch Weekblad]
Pharm. Z. [Pharmazeutische Zeitung]
Pharm. Zent. [Pharmazeutische Zentralhalle]
Phil. J. [Philosophical Journal]
Phil. Mag. [Philosophical Magazine]
Phys. Bl. [Physikalische Blätter]
Phys. Z. [Physikalische Zeitschrift]
Physiol. Bohem. [Physiologia Bohemoslovenica]
Pol. Med. Sci. [Polish Medical Sciences Historical Bulletin]
Pol. Tyg. Lek. [Polski Tygodnik Lekarski]
Pop. Sci. Mon. [Popular Science Monthly]
Post. Bioch. [Postępy Biochimii]
Prag. med. Wchschr. [Prager medizinische Wochenschrift]

Précis Anal. Acad. Rouen [Précis Analytique de l'Académie des Sciences, Belles-Lettres et Arts de Rouen]
Presse Med. [Presse Médicale]
Proc. Acad. Nat. Sci. Phila. [Proceedings of the Academy of Natural Sciences of Philadephia]
Proc. Am. Acad. Arts Sci. [Proceedings of the American Academy of Arts and Sciences]
Proc. Am. Phil. Soc. [Proceedings of the American Philosophical Society]
Proc. Chem. Soc. [Proceedings of the Chemical Society London]
Proc. Geol. Soc. [Proceedings of the Geological Society]
Proc. Geol. Soc. Am. [Proceedings of the Geological Society of America]
Proc. Indian Acad. Sci. [Proceedings of the Indian Academy of Science]
Proc. Indiana Acad. Sci. [Proceedings of the Indiana Academy of Science]
Proc. Inst. Med. Chicago [Proceedings of the Institute of Medicine, Chicago]
Proc. Iowa Acad. Sci. [Proceedings of the Iowa Academy of Science]
Proc. Linn. Soc. [Proceedings of the Linnean Society]
Proc. Phys. Soc. [Proceedings of the Physical Society]
Proc. Roy. Inst. [Proceedings of the Royal Institution of Great Britain]
Proc. Roy. Irish Acad. [Proceedings of the Royal Irish Academy]
Proc. Roy. Micr. Soc. [Proceedings of the Royal Microscopical Society]
Proc. Roy. Soc. [Proceedings of the Royal Society of London]
Proc. Roy. Soc. Canada [Proceedings of the Royal Society of Canada]
Proc. Roy. Soc. Edin. [Proceedings of the Royal Society of Edinburgh]
Proc. Roy. Soc. Med. [Proceedings of the Royal Society of Medicine, Section on History of Medicine]
Proc. Roy. Soc. N.Z. [Proceedings of the Royal Society of New Zealand]
Proc. Soc. Anal. Chem. [Proceedings of the Society for Analytical Chemistry]
Proc. Soc. Exp. Biol. Med. [Proceedings of the Society of Experimental Biology and Medicine]
Proc. Verb. Aveyron [Procés-Verbaux de la Société de Lettres, Sciences et Arts de l'Aveyron]
Proc. Verb. Linn. Bordeaux [Procés-Verbaux de la Société Linnéenne de Bordeaux]
Proc. Virchow Med. Soc. [Proceedings of the Virchow Medical Society]
Prog. Med. [Progrès Médical]
Prog. Rei Bot. [Progressus Rei Botanicae]
Psychiat. Neurol. [Psychiatria et Neurologia]
Quart. J. Exp. Biol. [Quarterly Journal of Experimental Biology]
Quart. Rev. Biol. [Quarterly Review of Biology]
Rass. Clin. Terap. [Rassegna di Clinica Terapia e Scienze Affini]
Rec. Austral. Acad. Sci. [Records of the Australian Academy of Science]
Rec. Chem. Prog. [Record of Chemical Progress]
Rec. Med. Vet. [Recueil de Médecine Vétérinaire]
Rec. Prog. Hormone Res. [Recent Progress in Hormone Research]
Rec. Trav. Chim. Pays-Bas [Recueil des Travaux Chimiques des Pays-Bas]
Reg. Biochim. [Regard sur la Biochimie]
Rend. Accad. Bologna [Rendiconti dell'Accademia delle Scienze Bologna]
Rend. Accad. Lincei [Rendiconti dell'Accademia dei Lincei]
Rend. Ist. Lombardo [Rendiconti dell'Istituto Lombardo di Scienze e Lettere]
Rend. Soc. Chim. [Rendiconti della Società Chimica di Roma]
Rep. Pharm. [Repertorium der Pharmazie]
Resp. Physiol. [Respiration Physiology]
Rev. Can. Biol. [Revue Canadienne de Biologie]
Rev. Cytol. [Revue de Cytologie]
Rev. Endocrin. [Revue Française d'Endocrinologie]
Rev. Ferment. [Revue des Fermentations et des Industries Alimentaires]
Rev. Gen. Bot. [Revue Générale de Botanique]
Rev. Gen. Chim. [Revue Générale de Chimie Pure et Appliquee]
Rev. Gen. Sci. [Revue Générale des Sciences]
Rev. Hemat. [Revue d'Hématologie]
Rev. Hist. Med. Heb. [Revue d'Histoire de la Médecine Hébraique]
Rev. Hist. Pharm. [Revue d'Histoire de la Pharmacie]
Rev. Hist. Sci. [Revue d'Histoire des Sciences]
Rev. Int. Electr. [Revue Internationale d'Électrotherapie]
Rev. Med. [Revue de Médecine]
Rev. Med. Brux. [Revue Médicale de Bruxelles]
Rev. Med. Est [Revue Médicale de l'Est]
Rev. Med. Liege [Revue Médicale de Liège]
Rev. Med. Louvain [Revue Médicale de Louvain]
Rev. Med. Nancy [Revue Médicale de Nancy]
Rev. Med. Suisse Rom. [Revue Médicale de la Suisse Romande]

Rev. Med. Toulouse [Revue Médicale de Toulouse]
Rev. Mod. Phys. [Reviews of Modern Physics]
Rev. Pol. Acad. Sci. [Review of the Polish Academy of Science]
Rev. Quest. Sci. [Revue des Questions Scientifiques]
Rev. Roum. Physiol. [Revue Roumaine de Physiologie]
Rev. Russe Ent. [Revue Russe d'Entomologie]
Rev. Sci. [Revue Scientifique]
Rhein. Westf. Wirt. Biog. [Rheinisch-Westfälisch Wirtschaftsbiographien]
Rhode Isl. Med. J. [Rhode Island Medical Journal]
Ric. Sci. [La Ricerca Scientifica]
Riv. Biol. [Rivista di Biologia]
Riv. Storia Med. [Rivista di Storia della Medicina]
Riv. Storia Sci. Med. [Rivista di Storia Critica delle Scienze Mediche e Naturali]
Rock. Univ. Notes [Rockefeller University Notes and News]
Rock. Univ. Rev. [Rockefeller University Review]
Roy. Soc. Edin. Year Book [Royal Society of Edinburgh Year Book]
Rubber Chem. Techn. [Rubber Chemistry and Technology]
Samml. anat. Vortr. [Sammlung anatomischer und physiologischer Vorträge]
Samml. chem. Vortr. [Sammlung chemischer und chemisch-technischer Vorträge]
Scand. J. Clin. Inv. [Scandinavian Journal of Clinical and Laboratory Investigation]
Scand. J. Resp. Dis. [Scandinavian Journal of Respiratory Diseases]
Schr. Nat. Ges. Danzig [Schriften der Naturforschenden Gesellschaft in Danzig]
Schr. Phys. Ges. Konigsberg [Schriften der Physikalischen-Oekonomischen Gesellschaft zu Königsberg]
Schw. Apoth. Z. [Schweizerische Apotheker Zeitung]
Schw. Arch. Neurol. [Schweizer Archiv für Neurologie und Psychiatrie]
Schw. Chem. Z. [Schweizerische Chemiker Zeitung]
Schw. med. Jahrb. [Schweizerisches medizinisches Jahrbuch]
Schw. med. Wchschr. [Schweizerische medizinische Wochenschrift]
Schw. Milchz. [Schweizerische Milchzeitung]
Schwab. Akad. Wiss. Jahrbuch [Schwäbische Akademie der Wissenschaften Jahrbuch]
Sci. Med. Ital. [Scientia Medica Italiana]
Sci. Mon. [Scientific Monthly]
Sci. Pharm. [Scientia Pharmaceutica]
Sitz. Akad. Wiss. Leipzig [Sitzungsberichte der Sächsischen Akademie der Wissenschaften zu Leipzig, Math.-Naturwiss. Kl.]
Sitz. Bayer. Akad. [Sitzungsberichte der Bayerischen Akademie der Wissen schaften zu München, Math.-Physik. Kl.]
Sitz. Ges. Nat. Berlin [Sitzungsberichte der Gesellschaft der Naturforschenden Freunde zu Berlin]
Sitz. Ges. Nat. Marburg [Sitzungsberichte der Gesellschaft zur Beförderung der Naturwissenschaften zu Marburg]
Sitz. Nat. Ges. Dorpat [Sitzungsberichte der Naturforschenden Gesellschaft Dorpat]
Sitz. Nat. Ges. Isis [Sitzungsberichte der Naturwissenschaftlichen Gesell schaft Isis, Dresden]
Sitz. Nat. Ver. Rhein.-Westf. [Sitzungsberichte des Naturhistorischen Vereins Rheinland-Westfalen, Bonn]
Sitz. Phys. Med. Erlangen [Sitzungsberichte der Physikalisch-Medizinischen Sozietät in Erlangen]
Sitz. Phys. Med. Würzburg [Sitzungsberichte der Physikalisch-Medizinischen Gesellschaft zu Würzburg]
Sitz. Preuss. Akad. Wiss. [Sitzungsberichte der Preussischen Akademie der Wissenschaften zu Berlin]
Skand. Arch. Physiol. [Skandinavisches Archiv für Physiologie]
Soc. Stud. Sci. [Social Studies of Science]
South. Med. J. [Southern Medical Journal]
Sov. Bot. [Sovietskaia Botanika]
Sov. Med. [Sovietskaia Meditsina]
Sov. Zdrav. [Sovietskoe Zdravokhranienie]
St. Barts. Hosp. J. [St. Bartholomew's Hospital Journal]
St. Barts. Hosp. Rep. [St. Bartholomew's Hospital Reports]
Strassburg Med. Z. [Strassburger Medizinische Zeitschrift]
Stud. Biophys. [Studia Biophysica]
Stud. Hist. Biol. [Studies in the History of Biology]
Stud. Hist. Phil. Sci. [Studies in the History and Philosophy of Science]
Stud. Nauk. Pol. [Studia i Materialy z Dziejow Nauki Polskiej]
Sudhoffs Arch. [Sudhoffs Archiv für Geschichte der Medizin und der Naturwissenschaften]
Süddeutsche Apoth. Z. [Süddeutsche Apotheker Zeitung]
Sven. Farm. Tid. [Svensk Farmaceutisk Tidskrift]
Sven. Kem. Tid. [Svensk Kemisk Tidskrift]
Sven. Vet. Akad. Arsbok [Svenska Vetenskaps Akademiens Årsbok]
Symp. Soc. Dev. Biol. [Symposia of the Society of Developmental Biology]
TIBS [Trends in Biochemical Sciences]
Tech. Cult. [Technology and Culture]

Ter. Arkh. [Terapevtichski Arkhiv]
Thromb. Diath. Haemorrh. [Thrombosis et Diathesis Haemorrhagica]
Thromb. Res. [Thrombosis Research]
Tijd. Diergen. [Tijdschrift voor Diergeneeskunde]
Tijd. Gesch. Gennesk. [Tijdschrift voor de Geschiedenes der Geneeskunde, Naturwetenschappen, Wiskunde und Techniek]
Tohoku J. Exp. Med. [Tohoku Journal of Experimental Medicine]
Tow. Nauk. Lwow. [Towarzystwo Naukowe we Lwowie Sprawozdania]
Trans. Am. Clin. Assn. [Transactions of the American Clinical and Climatological Association]
Trans. Am. Micr. Soc. [Transactions of the American Microscopical Society]
Trans. Am. Neurol. Assn. [Transactions of the American Neurological Association]
Trans. Am. Phil. Soc. [Transactions of the American Philosophical Society]
Trans. Assoc. Am. Phys. [Transactions of the Association of AmericanPhysicians]
Trans. Bot. Soc. Edin. [Transactions and Proceedings of the Botanical Society of Edinburgh]
Trans. Coll. Phys. Phila. [Transaction and Studies of the College of Physicians, Philadelphia]
Trans. Conn. Acad. Arts Sci. [Transactions of the Connecticut Academy of Arts and Sciences]
Trans. Faraday Soc. [Transactions of the Faraday Society]
Trans. Kansas Acad. Sci. [Transactions of the Kansas Academy of Science]
Trans. Med. Soc. N.Y. [Transactions of the Medical Society of the State of New York]
Trans. N.Y. Acad. Sci. [Transactions of the New York Academy of Sciences]
Trans. Soc. Trop. Med. [Transactions of the Royal Society of Tropical Medicine and Hygiene]
Trans. Roy. Soc. Canada [Transactions of the Royal Society of Canada]
Trans. Wis. Acad. [Transactions of the Wisconsin Academy of Sciences Arts and Letters]
Trudy Inst. Fiziol. Pavlov [Trudy Instituta Fiziologii Pavlov]
Trudy inst. Ist. Est. [Trudy Instituta Istorii Estestvoznania i Tekhniki]
Ukr. Khem. Zhur. [Ukrainski Khemichni Zhurnal]
Ung. med. Presse [Ungarische medizinische Presse]
Univ. Manitoba Med. J. [University of Manitoba Medical Journal]
Univ. Mich. Med. Bull. [University of Michigan Medical Bulletin]
Usp. Khim. [Uspekhi Khimii]
Usp. Sov. Biol. [Uspekhi Sovremennoi Biologii]
Verhandl. Akad. Weten. Amst. [Verhandlingen der Akademie van Wetenschappen Amsterdam]
Verhandl. Berl. med. Ges. [Verhandlungen der Berliner mediizinischen Gesellschaft]
Verhandl. Int. Ver. Limn. [Verhandlungen der Internationalen Vereinigung für Limnologie]
Verhandl. Nat. Ges. Basel [Verhandlungen der Naturforschenden Gesellschaft in Basel]
Verhandl. Nat. Med. Heidel. [Verhandlungen des Naturhistorisch-Medizinischen Vereins zu Heidelberg]
Verhandl. path. Ges. [Verhandlungen der deutschen pathologischen Gesellschaft]
Verhandl. phys. Ges. [Verhandlungen der deutschen physikalischen Gesell schaft]
Verhandl. Phys. Med. Würz. [Verhandlungen der Physikalisch-MedizinischenGesellschaft zu Würzburg]
Verhandl. Schw. Nat. Ges. [Verhandlungen der Schweizerischen Natur forschenden Gesellschaft]
Verhandl. zool. bot. Ges. Wien [Verhandlungen der zoologisch-botanischen Gesellschaft in Wien]
Vest. Akad. Nauk [Vestnik Akademii Nauk SSSR]
Vest. Akad. Med. Nauk [Vestnik Akademii Meditsinskikh Nauk SSSR]
Vest. Cesk. Akad. Zem. [Vestnik Československa Akademie Zemedelske]
Vest. Khir. [Vestnik Khirurgii]
Vest. Leningrad Univ. [Vestnik Leningradskavo Universiteta]
Vest. Mikr. Ep. [Vestnik Mikrobiologii i Epidemiologii]
Vest. Mosk. Univ. [Vestnik Moskovskavo Universiteta, Khimia]
Vest. Oftalm. [Vestnik Oftalmologicheski]
Vid. Med. Nat. For. [Videnskabelige Meddelelser fra Dansk Naturhistorisk Forening i Kjøbenhavn]
Viert. Nat. Ges. Zürich [Vierteljahrschrift des Naturforschenden Gesell schaft in Zürich]
Virchows Arch. [Archiv für pathologische Anatomie und Physiologie und fürklinische Medizin]
Virchows Jahresber. [Jahresbericht über die Leistungen und Fortschritte der gesamten Medicin]
Vop. Ist. Est. Tekhn. [Voprosi Istorii Estestvoznania i Tekhniki]
Vop. Med. Khim. [Voprosi Meditsinskoi Khimii]
Vop. Onk. [Voprosi Onkologii]
Warsz. Czas. Lek. [Warszawskie Czasopismo Lekarskie]
Wiad. Chem. [Wiadomosci Chemiczne]
Wiad. Lek. [Wiadomosci Lekarskie]
Wiener Chem. Z. [Wiener Chemiker Zeitung]
Wiener klin. Wchschr. [Wiener klinische Wochenschrift]
Wiener med. Bl. [Wiener medizinische Blätter]
Wiener med. Wchschr. [Wiener medizinische Wochenschrift]
Wiss. Fort. [Wissenschaft und Fortschritt]
Wiss. Z. Berlin [Wissenschaftliche Zeitschrift der Humboldt Universität

Berlin, Math.-Naturw. Reihe]
Wiss. Z. Greifswald [Wissenschaftliche Zeitschrift der Universität Greifswald]
Wiss. Z. Halle [Wissenschaftliche Zeitschrift der Martin-Luther Universität Halle-Wittenberg, Math.-Naturw. Reihe]
Wiss. Z. Jena [Wissenschaftliche Zeitschrift der Friedrich-Wilhelm Universität Jena, Math.-Naturw. Abteil.]
Wiss. Z. Leipzig [Wissenschaftliche Zeitschrift der Karl-Marx Universität Leipzig]
Würz. Nat. Z. [Würzburger Naturwissenschaftliche Zeitung]
Yale J. Biol. Med. [Yale Journal of Biology and Medicine]
Z. Ackerbau [Zeitschrift für Acker- und Pflanzenbau]
Z. ärzt. Fortbild. [Zeitschrift für ärztlicher Fortbildung]
Z. Agrikulturchem. [Zeitschrift für Agrikulturchemie]
Z. allgem. Path. [Zentralblatt für allgemeine Pathologie]
Z. allgem. Physiol. [Zeitschrift für allgemeine Physiologie]
Z. anal. Chem. [Zeitschrift für analytische Chemie]
Z. Anat. Ent. [Zeitschrift für Anatomie und Entwicklungsgeschichte]
Z. angew. Chem. [Zeitschrift für angewandte Chemie; after 1947, Angewandte Chemie]
Z. angew. Ent. [Zeitschrift für angewandte Entomologie]
Z. anorg. Chem. [Zeitschrift für anorganische und allgemeine Chemie]
Z. Bakt. [Zentralblatt für Bakteriologie]
Z. Bayer. Landesgesch. [Zeitschrift für Bayerische Landesgeschichte]
Z. Biol. [Zeitschrift für Biologie]
Z. Bot. [Zeitschrift für Botanik]
Z. Chem. [Zeitschrift für Chemie]
Z. Chir. [Zentralblatt für Chirurgie]
Z. Elektrochem. [Zeitschrift für Elektrochemie]
Z. Entomol. [Zeitschrift für Entomologie]
Z. Fisch. [Zeitschrift für Fischerei]
Z. Gärungsphysiol. [Zeitschrift für Gärungsphysiologie]
Z. Geb. [Zeitschrift für Geburtshilfe und Gynäkologie]
Z. geol. Ges. [Zeitschrift der deutschen geologischen Gesellschaft]
Z. ges. inn. Med. [Zeitschrift für die gesamte innere Medizin]
Z. ges. Neurol. [Zeitschrift für die gesamte Neurologie und Psychiatrie]
Z. Gyn. [Zentralblatt für Gynäkologie]
Z. Haut Geschl. [Zeitschrift für Haut- und Geschlechtskrankheiten]
Z. Heilkunde [Zeitschrift für Heilkunde]
Z. Hyg. [Zeitschrift für Hygiene und Infektionskrankheiten]
Z. Immunitätsforsch. [Zeitschrift für Immunitätsforschung und Experimentelle Therapie]
Z. Instrum. [Zeitschrift für Instrumentenkunde]
Z. Kinderheil. [Zeitschrift für Kinderheilkunde]
Z. klin. Chemie [Zeitschrift für klinische Chemie]
Z. klin. Med. [Zeitschrift für klinische Medizin]
Z. Krebsforsch. [Zeitschrift für Krebsforschung]
Z. Kreis. [Zeitschrift für Kreislaufforschung]
Z. Kristall. [Zeitschrift für Kristallographie]
Z. Land. Ver. Ost. [Zeitschrift für das Landwirtschaftliche Versuchwesen in Oesterreich]
Z. Laryngol. [Zeitschrift für Laryngologie]
Z. Leb. Unt. [Zeitschrift für Lebensmitteluntersuchung]
Z. med. Chem. [Zeitschrift für medizinische Chemie]
Z. mikr. anat. Forsch. [Zeitschrift für mikroskopisch-anatomische Forschung]
Z. Nahrung [Zeitschrift für Nahrungs- und Genussmittel]
Z. Naturforsch. [Zeitschrift für Naturforschung]
Z. Pflanzenern. [Zeitschrift für Pflanzenernährung]
Z. Pflanzenzücht. [Zeitschrift für Pflanzenzüchtung]
Z. physik. Chem. [Zeitschrift für physikalische Chemie]
Z. Physiologie [Zentralblatt für Physiologie]
Z. physiol. Chem. [(Hoppe-Seylers) Zeitschrift für physiologische Chemie]
Z. Psych. Physiol. Sinn. [Zeitschrift für Psychologie und Physiologie de Sinnesorgane]
Z. Sinnesphysiol. [Zeitschrift für Sinnesphysiologie]
Z. tech. Phys. [Zeitschrift für technische Physik]
Z. Tierernährung [Zeitschrift für Tierernährung]
Z. Tierphysiol. [Zeitschrift für Tierphysiologie]
Z. Tierpsychol. [Zeitschrift für Tierpsychologie]
Z. Tropenmed. [Zeitschrift für Tropenmedizin]
Z. Tub. [Zeitschrift für Tuberkulose]
Z. Urol. [Zeitschrift für Urologie]

Z. Ver. Hess. Gesch. [Zeitschrift des Vereins für Hessische Geschichte und Landeskunde]
Z. Ver. Zuckerind. [Zeitschrift des Vereins der deutschen Zuckerindustrie]
Z. Vit. Forsch. [Zeitschrift für Vitamin-, Hormon- und Fermentforschung]
Z. wiss. Mikr. [Zeitschrift für wissenschaftliche Mikroskopie]
Z. wiss. Phot. [Zeitschrift für wissenschaftliche Photographie]
Z. wiss. Zool. [Zeitschrift für wissenschaftliche Zoologie]
Z. Zellforsch. [Zeitschrift für Zellforschung und mikroskopische Anatomie]
Z. Zuckerind. [Zeitschrift der Zuckerindustrie]
Zbl. Zuckerind. [Zentralblatt für die Zuckerindustrie]
Zdrav. Prac. [Zdravotnicka Pracovnice]
Zhur. Fiz. Khim. [Zhurnal Russkavo Fiziko-Khimicheskavo Obshchestva, Khimicheski Otdel; later, Zhurnal Fizicheskoi Khimii]
Zhur. Mikr. Ep. Imm. [Zhurnal Mikrobiologii, Epidemiologii i Immunobiologii]
Zhur. Nevropat. Psikh. [Zhurnal Nevropatologii i Psikhiatrii]
Zhur. Ob. Biol. [Zhurnal Obshchei Biologii]
Zhur. Ob. Khim. [Zhurnal Obshchei Khimii]
Zhur. Prikl. Khim. [Zhurnal Prikladnoi Khimii]
Zool. Anz. [Zoologischer Anzeiger]
Zool. Jahrb. [Zoologische Jahrbücher]

al-ABBASSY, AHMAD [1910-]
Who's Who in the Arab World 1974-1975, p.1203

ABBE, ERNST [1840-1905]
DSB 1,6-9; NDB 1,2-4; BJN 10,3-16; FREUND 1,45-63; MAYERHOFER 1,115-116;ALLMORGEN, 203; W, 1; WWWS, 2; MATSCHOSS, 1; POGG 3,2-3; 4,1-2; 5,1; 6,1; 7aSuppl.,7-14
C. von Voit, Sitz. Bayer. Akad. 35 (1905), 346-355
S. Czapski, Ber. phys. Ges. 3 (1905), 89-121
F. Auerbach, Ernst Abbe. Leipzig,1918
F. Löwe, Z. angew. Chem. 33 (1920), 17-18
H. Hartinger, Naturwiss. 18 (1930), 49-63
M. von Rohr, Abbe's Apochromats. Jena 1936; Ernst Abbe. Jena 1940
N. Gunther, Ernst Abbe (2nd Ed.). Stuttgart 1951
F. Stier, Ernst Abbe's akademische Tätigkeit an der Universität Jena. Jena 1955
P.G. Esche, Ernst Abbe. Leipzig 1963
E. Brüche, Phys. Bl. 21 (1965), 261-269
H. Volkmann, Applied Optics 5 (1966), 1720-1731

ABBENE, ANGELO [1799-1865]
DBI 1,37; DFI, 7; TSCHIRCH, 1010; FERCHL, 1; POGG 3,3
G. Ostino, Minerva Farm. 6 (1957), 43-45

ABBOTT, ALEXANDER CREVER [1860-1935]
AMS 5,1; IB,1
Anon., Science 82 (1935), 270

ABDERHALDEN, EMIL [1877-1950]
NDB 1,5-6; MAYERHOFER 1,117; FISCHER 1,2-3; PICCO, 24-30; POTSCH, 7; WWWS, 3; IB, 1; POGG 5,2-3; 6,2-11; 7a(1),1-5
R.E. Oesper, J. Chem. Ed. 14 (1937), 201,237
D. Ackermann, Münch. med. Wchschr. 92 (1950), 1389-1392
K. Wezler, Jahrbuch der Akademie der Wissenschaften und Literatur Mainz 1950, pp.148-152; 1951, pp.163-204
F. Leuthardt, Viert. Nat. Ges. Zürich 95 (1950), 282-283
K. Heyns, Pflügers Arch. 253 (1951), 229-237
Anon., Z. Vit. Forsch. 4 (1951), 1-18
O. Schiller et al., Nova Acta Leop. NS14 (1952), 147-189
J. Methfessel, Wiss. Z. Halle 14 (1965), 49-78
H. Hanson, Nova Acta Leop. NS36 (1970), 257-317
W. Kaiser and W. Piechocki, Z. ges. inn. Med. 32 (1977), 445-453
W. Kaiser et al., Wiss. Z. Halle 26 (1977), 37-74
G. Döring, Münch. med. Wchschr. 120 (1978), 478-479
V. Schmidt, Dansk Medicinhistorisk Arbog 1979, pp.116-129
W. Heese, Leopoldina [3]24 (1981), 133-176
W. Sackmann, Gesnerus 38 (1981), 215-224; Schweizer Archiv für Tierheilkunde 124 (1982), 1-10

ABDERHALDEN, RUDOLF [1910-1965]
KURSCHNER 10,1; SBA 5,9; POGG 7a(1),5-7

ABEGG, RICHARD [1869-1910]
NDB 1,7; BJN 15,5; KROLLMANN, 1; BLOKH 1,1-3; W, 1; POTSCH, 7; WWWS, 3; POGG 4,2-3; 5,3-4; 7aSuppl.,14
W. Herz, Chem. Z. 34 (1910), 369
W.R., J. Chem. Soc. 99 (1911), 599-602
W. Nernst, Ber. chem. Ges. 46 (1913), 619-628

ABEL, EMIL [1875-1958]
BHDE 2,3; TETZLAFF, 1; POGG 4,3; 5,4; 6,11-12; 7a(1),7-9
F. Halla, Ost. Chem. Z. 38(1935),90-93; 48(1947),54-55; 51(1950),91-92
G.M. Schwab, Z. Elektrochem. 59 (1955), 591-592
O. Redlich, Ost. Chem. Z. 59 (1958), 149-151
P. Gross, Z. Elektrochem. 62 (1958), 831-833
H. Tompa, Nature 181 (1958), 1765-1766
O. Kratky, Alm. Akad. Wiss. Wien 110 (1960), 417-437

ABEL, GERHARD [1903-]
LDGS, 16; NUC 1,560

ABEL, JOHN JACOB [1857-1938]
DSB 1,9-12; DAB 22,4-5; PARASCANDOLA, 26-28; CHEN, 17-20; MILES, 1-2; FISCHER 1,3; CHITTENDEN, 392-394,396-397,403-405;411-416; DAMB, 4-5; NCAB 28,23-24; MEDVEI, 514-516; POTSCH, 7-8; WWAH, 2; WWWS, 4; IB, 1; POGG 6,12-13; 7b,11
E.K. Marshall, Ind. Eng. Chem. 18 (1926), 984; Science 87 (1938),566-569; Trans. Assoc. Am. Phys. 54 (1939), 7-8
E.M.K. Geiling and E.A. Evans, Arch. Inter. Pharmacodyn. 60(1938),241-250
W.H. Howell, Am. Phil. Soc. Year Book 1938, pp.355-358
H.H. Dale, Obit. Not. Fell. Roy. Soc. 2 (1939), 577-585
C. Voegtlin, J. Pharm. Exp. Ther. 67 (1939), 373-406
P.D. Lamson, Bull. Johns Hopkins Hosp. 68 (1941), 119-157
W. MacNider, Biog. Mem. Nat. Acad. Sci. 24 (1946), 231-257
T. Sollmann et al., Bull. Johns Hopkins Hosp. 101 (1957), 297-338
H.H. Swain et al., Univ. Mich. Med. Bull. 29 (1963), 1-14
J. Murnaghan and P. Talalay, Persp. Biol. Med. 10 (1967), 334-380
A.M. Harvey, Johns Hopkins Med. J. 135 (1974), 245-258
J. Parascandola, Bull. Hist. Med. 56 (1982), 512-527
R.A. Becker, Pharm. Hist. 24 (1982), 115-116
H.W. Davenport, Physiologist 25 (1982), 76-82

ABEL, MARY HINMAN [1850-1938]
E. Todhunter, J. Am. Diet. Assn. 32 (1956), 706

ABEL, RUDOLF [1868-1942]
KURSCHNER 5,2; FISCHER 1,3; TLK, 3; IB, 1
Anon., Nature 151 (1943), 217-218

ABEL-MUSGRAVE, CURT [1860-1938]
BHDE 2,841; JV 23,306
New York Times, 5 November 1938

ABELES, MARCUS [1837-1894]
PAGEL, 3; HIRSCH Suppl., 5; KOREN, 145; WININGER 1,10
E. Ludwig, Wiener klin. Wchschr. 8 (1895), 35-36

ABELES, ROBERT HEINZ [1926-]
AMS 15(1),6; BHDE 2,3

ABELIN, ISAAK [1883-1965]
KURSCHNER 10,3; FISCHER 1,4; KOREN, 145; IB, 1; COHEN, 3; TLK 3,1; SBA 5,9; POGG 4,13; 7a(1),9-10
H. Aebi, Mitt. Nat. Ges. Bern NF23 (1966), 211-212

ABELJANZ, HARATHIUN [1849-1921]
DHB 1,33-34; ZURICH,47-49; BLOKH 1,3; SCHAEDLER, 1; ZURICH-D, 118; POGG 5,4
P. Karrer, Viert. Nat. Ges. Zürich 66 (1921), 353-356

ABELL, ROBERT DUNCOMBE [1874-1957]
SRS, 25; NUC 1,598
Anon., J. Roy. Inst. Chem. 1957, p.597

ABELOUS, JACQUES ÉMILE [1864-1940]
J. Gautrelet, Bull. Acad. Med. 124 (1941), 34-36; Presse Med. 1941, pp.325-326

ABELSON, PHILIP HAUGE [1913-]
AMS 15(1),7; IWW 1982-1983, p.4; WWWS, 4

ABENIUS, PER WILHELM [1864-1956]
SBL 1,10-12; SMK 1,2; POGG 4,3-4

ABERCROMBIE, JOHN [1780-1844]
DNB 1,37-38; HIRSCH 1,6-8; WWWS, 4-5; DECHAMBRE 1,200-201; CALLISEN 1,4-7; 26,3-4
J.R. Watson, Ann. Med. Hist. 4 (1942), 468-472

ABERCROMBIE, MICHAEL [1912-1979]
WW 1979, p.4; WWWS, 5
Anon., J. Emb. Exp. Morph. 54 (1979), 1-3
P. Medawar, Biog. Mem. Fell. Roy. Soc. 26 (1980), 1-15

ABERNETHY, JOHN [1764-1831]
DNB 1,49-52; BETTANY, 226-241; HIRSCH 1,9-11; FERCHL, 1; WWWS, 5; CALLISEN 1,8-17; 26,4-6; POGG 1,4

G. MacIlwain, Memoirs of John Abernethy (3rd Ed.). London 1856
J.L. Thornton, John Abernethy. London 1953

ABERNETHY, JOHN LEO [1915-1987]
AMS 15(1),8
Anon., J. Chem. Ed. 64 (1987), 410

ABERSON, JOHANNES HENDRIKUS [1857-1935]
NUC 1,680; RSC 13,12-13; POGG 4,4; 5,4; 6,13; 7b,11
S.C.J. Olivier, Chem. Wkbl. 25 (1928), 582-585
J. Hudig, Chem. Wkbl. 32 (1935), 738-739

ABNEY, WILLIAM de WIVELESLIE [1843-1920]
DSB 1,21-23; DNB[1912-1921], 1-2; POGG 3,6-7; 4,4-5; 5,5
E.H. Grove-Hills, Proc. Roy. Soc. A99 (1921), i-v

ABRAHAM, ARTHUR [1894-]
JV 38,268; NUC 2,88

ABRAHAM, EDWARD PENLEY [1913-]
WW 1981, p.6; IWW 1982-1983, p.5; MSE 1,2-3; WWWS, 5

ABRAMOV, SERGEI SEMENOVICH [1875-]
IB,2
Almanach na Sofiiski Universitet (2nd Ed.), pp.1-2. Sofia 1940
Kratka Bulgarska Entsikdopedia 1,3

ABRAMS, ALBERT [1863-1924]
AMS 2,2; WININGER 1,43-45

ABRAMSON, HAROLD ALEXANDER [1899-1980]
AMS 11,11; NCAB E,190; AO 1980, pp.585-586; WWA 1978-1979, p.8; WWWS, 6; KOREN, 145; KAGAN(JA), 435,750; POGG 6,15; 7b,15-17

ACCUM, FRIEDRICH CHRISTIAN [1769-1838]
DSB 1,43-44; DNB 1,57; HEIN, 1; BLOKH 1,3; FERCHL, 1-2; COLE, 1-7; POTSCH, 8; WWWS, 7; CALLISEN 1,22-25; 26,7-8; POGG 1,6-7; 7aSuppl.,15
C.A. Browne, J. Chem. Ed. 2 (1925), 828-851,1008-1034,1140-1149; Chymia 1 (1948), 1-9
P.J. Cole, Ann. Sci. 7 (1951), 128-143
F.C. Bing, J. Nutrition 89 (1956), 3-8

ACH, FRIEDRICH [1864-1902]
JV 11,286
E. Fischer, Ber. chem. Ges. 35 (1902), 3849

ACH, LORENZ [1868-1948]
JV 8,272; NUC 2,620; RSC 14,1009

ACHALME, PIERRE JEAN [1866-1936]
CURINIER 4,299-300; POGG 6,15
Anon., Presse Med. 44 (1936), 400

ACHARD, (ÉMILE) CHARLES [1860-1944]
FISCHER 1,6
G. Roussy, Presse Med. 52 (1944), 257-258
H. Benard, Paris Médical 34 (1944), 125-127
M. Loeper, Bull. Acad. Med. 128 (1944), 504-515; 136 (1952), 613-619

ACHARD, FRANZ KARL [1753-1821]
DSB 1,44-45; NDB 1,27-28; ADB 1,27-28; MAYERHOFER 1,128-129; FERCHL,2;POTSCH,8;BLOKH1,3-4;BRIQUET,1-2; SCHAEDLER, 1; MATSCHOSS, 1; WWWS,7; POGG 1,7; 7aSuppl., 15-19
K. Groba, Schlesische Lebensbilder 4 (1931), 200-218
K. Ulrich, Z. Zuckerind. 3 (1953), 145-150
J. Baxa and G. Bruhns, Zucker im Leben der Völker, pp.99-119,128-139. Berlin 1967

ACHELIS, JOHANN DANIEL [1898-1963]
KURSCHNER 4,6; IB, 2; POGG 7a(1),10-11
Anon., Nachr. Chem. Techn. 11 (1963), 423; Arzneimitt. 13 (1963), 1018
K. Kramer, Arzneimitt. 23 (1973), 1355-1358

H. Schäfer, Heid. Akad. Jahrb. 1985, pp.51-53

ACKERKNECHT, ERWIN HEINZ [1906-1988]
KURSCHNER 13,9; BHDE 1,3-4
Anon., Gesnerus 23 (1966), 6-12
E.H. Thomson, J. Hist. Med. 27 (1972), 3-4
H.H. Walser, Gesnerus 33 (1976), 3-7; 45 (1988), 309-312
E. Lesky, Gesnerus 43 (1986), 3-10

ACKERMANN, DANKWART [1878-1965]
MEISTER, 5-29; EBERT, 163-181; GUNDLACH, 249; BK, 157-158; TLK 3,3; IB, 2; WWWS, 8; POGG 6,16; 7a(1),12-13
Anon., Chem. Z. 82 (1958), 776
H. Reinwein, Leopoldina [3]1 (1965), 243-253
E. Klenk, Deutsche med. Wchschr. 91 (1966), 280-281
J. Kutsche and J.V. Parkin, Agents and Actions 18 (1986), 19-22

ACKERMANN, EDWIN [1863-1926]
STRAHLMANN, 468, 470; RSC 13,21

ACKERMANN, ERNST [1874-1945]
JV 23,327; NUC 3,20
Anon., Z. angew. Chem. 59 (1947), 96

ACKERMANN, FRITZ [1883-1939]
JV 23,307

ACKERMANN, THEODOR [1825-1896]
BJN 1,149-150; 3,123*; HIRSCH 1,18-19; Suppl.,7; KUKULA, 1-2; FRANKEN, 173
Anon., Leopoldina 32 (1896), 189-190
F. Marckwald, Münch. med. Wchschr. 43 (1896), 1234-1235
A. Wilhelmi, Mecklenburgische Aerzte, p.139. Schwerin 1901

ACKROYD, HAROLD [1877-1917]
F.G. Hopkins, Biochem. J. 12 (1918), 1-3
J.A. Venn, Alumni Cantabrigensis II,1,5. Cambridge 1940

ACLAND, HENRY WENTWORTH [1815-1900]
DNB [Suppl.1],10-12; HIRSCH 1,21; Suppl.,8; WWWS, 9
Anon., Lancet 1900(II), pp.1158-1160; Proc. Roy. Soc. B75(1905),169-174
J.B. Atlay, Sir Henry Wentworth Acland. London 1903
H.C. Harley, Oxford Medical School Gazette 18 (1966), 9-22

ACREE, SALOMON FARLEY [1875-1957]
AMS 8,7; WWAH, 3; WWWS, 9; POGG 5,6-7; 6,16-17; 7b,17-18
Anon., Chem. Eng. News 35(45) (1957), 127; Science 126 (1957), 1010

ADAIR, GILBERT SMITHSON [1896-1979]
DNB [1971-1980], 4-5; WW 1978, p.10; POGG 6,17-18; 7b,20-21
H. Gutfreund, Nature 284 (1980), 198
P. Johnson and M.F. Perutz, Biog. Mem. Fell. Roy. Soc. 27 (1981), 1-27
A. Schejter and E. Margoliash, TIBS 10 (1985), 490-492

ADAM, ALFRED [1888-1956]
H. Kleinschmidt, Mon. Kind. 104 (1956), 466-467
W. Tharau, Kinderärztliche Praxis 25 (1957), 102-103

ADAM, FRANÇOIS ARMAND [1820-1886]
DBF 1,448-449; GORIS, 209

ADAM, NEIL KENSINGTON [1891-1973]
WW 1965, p.13; WWWS, 10; POGG 6,18; 7b,21-22
G. Hills and K. Webb, Chem. Brit. 9 (1973), 571
Anon., Nature 245 (1973), 344
A. Carrington et al., Biog. Mem. Fell. Roy. Soc. 20 (1974), 1-26
J.F. Danielli and M.C. Phillips, Adv. Chem. 144 (1975), 1-13

ADAM, PAUL GABRIEL [1856-1916]
POGG 4,7; 5,7; 6,18; 7b,22
C. Poulenc, Bull. Soc. Chim. [4]19 (1916), 346

ADAMCZEWSKI, BOLESLAW [1885-1962]
JV 26,392

ADAMETZ, LEOPOLD [1861-1941]
NDB 1,54-55; OBL 1,5; WWWS, 10
A. Himmelbauer, Alm. Akad. Wiss. Wien 91 (1941), 203-210

ADAMI, JOHN GEORGE [1862-1926]
DNB [1922-1930], 6-7; FISCHER 1,7-8; O'CONNOR(2), 63-65; WWWS, 10
E. Glynn, Proc. Roy. Soc. Edin. 48 (1926), 349-351
Anon., Lancet 1926(II), pp.522-524; Brit. Med. J. 1926(II), pp.507-510
C.F.M., J. Path. Bact. 30 (1927), 151-167
A.E.B., Proc. Roy. Soc. B101 (1927), xv-xvi
M. Adami, George Adami. London 1930

ADAMKIEWICZ, ALBERT [1850-1921]
OBL 1,5; DBJ 3,289; PSB 1,25-26; HIRSCH 1,24-25; Suppl.,9; WWWS, 10; KOSMINSKI, 2-3,586; KONOPKA 1,10-23; WININGER 1,61-63
Anon., Chem. Z. 45 (1921), 1171
E. Herman, Neuropat. Pol. 6 (1968), 1-10

ADAMS, ELLIOT QUINCY [1888-1971]
AMS 8,9; POGG 6,19; 7b,23
D.S. Tarbell, J. Chem. Ed. 67 (1990), 7-8

ADAMS, ELLIOTT TORREY [1899-]
AMS 10,14

ADAMS, MARK HOPKINS [1912-1956]
AMS 9(I),8
E. Racker, Science 125 (1957), 434-435

ADAMS, MILDRED [1899-1982]
AMS 11,22; WWWS, 12

ADAMS, ROGER [1889-1971]
DSB 15,1-3; CB 1947, pp.1-2; MILES, 4-5; NCAB G,336-337; AMS 11,23; MH 1,4-5;MSE 1,5-6; STC 1,13-14; POTSCH, 9; WWAH, 5; POGG 6,20-22; 7b,25-31
C.S. Marvel, Am. Phil. Soc. Year Book 1974, pp.111-114
E.J. Corey, Welch Foundation Conferences on Chemical Research 20 (1977), 204-228
D.S. Tarbell, J. Chem. Ed. 56 (1979), 163-165; Isis 71 (1980), 620-626
D.S. Tarbell and A.T. Tarbell, Roger Adams, Scientist and Statesman. Washington 1981; Biog. Mem. Nat. Acad. Sci. 53 (1982), 3-47

ADDIS, THOMAS [1881-1949]
AMS 8,13; FISCHER 1,8; IB, 2; WWAH, 6
Anon., Ann. Int. Med. 32 (1950), 822-824
A.L. Bloomfield, Trans. Assoc. Am. Phys. 63 (1950), 7-8
S.J. Peitzman, Kidney International 37 (1990), 833-840

ADDISON, THOMAS [1793-1860]
DSB 1,59-61; DNB 1,133-134; MAYERHOFER 1,134-135; BETTANY 2,2-14; W, 3-4; WILKS, 221-234; MR 3,205-211; O'CONNOR, 20-22; HIRSCH 1,30; PAGEL, 809; WWWS, 13; ABBOTT, 7
W. Hale-White, Guy's Hosp. Rep. [4]6 (1926), 253-279
G.A.R. Winston and W. Hill, Guy's Hosp. Rep. 109 (1960), 280-283
H. Dale, Brit. Med. J. 1949(II), pp.347-352
K.D. Keele, Med. Hist. 13 (1969), 195-202

ADDISON, WILLIAM [1802-1881]
HIRSCH Suppl.,10; WWWS, 13
H.A. McCallum, Lancet 1907(I), pp.182-183
L.J. Rather, Addison and the White Corpuscles. Berkeley 1972

ADEANE, CHARLES ROBERT WHORWOOD [1863-1943]
WW 1943, p.20ADELBERG, EDWARD ALLEN [1920-]
AMS 15(1), 31

ADELMANN, HOWARD BERNHARDT [1898-1988]
AMS 13,22; WWA 1980-1981, p.21; IB, 2; WWWS, 13

ADENEY, WALTER ERNEST [1857-1935]
POGG 4,8; 5,9; 6,24; 7b,33
A.G.G. Leonard, J. Chem. Soc. 1935, pp.1891-1892

ADERHOLD, RUDOLF [1865-1907]
NDB 1,65-66; BJN 12,143-145,6*; WWWS, 14
J. Behrens, Ber. bot. Ges. 25 (1907), (47)-(56)
O. Appel, Jahresb. Schles. Ges. 85 (1907), 10-18

ADICKES, FRANZ [1897-1973]
TLK 3,4; POGG 6,24; 7a(1),14
Anon., Chem. Z. 96 (1972), 635; 97 (1973), 627

ADKINS, HOMER BURTON [1892-1949]
DAB [Suppl.4],5-7; MILES, 5-7; AMS 8,14; WWAH, 6; WWWS, 14; POGG 6,24-25; 7b,34-36
F. Daniels, Biog. Mem. Nat. Acad. Sci. 27 (1952), 293-317

ADLER, ERICH [1889-]
BHDE 1,8; JV 30,319; NUC 4,234; SG [4]1,148

ADLER, ERICH [1905-1985]
VAD 1981, p.4; BHDE 2,8-9; LDGS, 16; WWWS, 14

ADLER, JULIUS [1920-]
AMS 15(1),35; WWA 1980-1981, p.23; BHDE 2,10

ADLER, LEO [1886-1925]
FISCHER 1,9

ADLER, LUDWIG [1879-1958]
FISCHER 1,9-10; LORENZSONN, 58-66; GAUSS, 249; KOREN, 146

ADLER, OSKAR [1879-1932]
FISCHER 1,10; PELZNER, 70-75; KURSCHNER 4,12; POGG 6,25; 7a(1),14

ADLUNG, ALFRED [1875-1937]
HEIN, 2-3; TSCHIRCH, 1010
Anon., Deutsche Apoth. Z. 52 (1937), 1280-1282; 53 (1938), 8-9

ADOLPH, EDWARD FREDERICK [1895-1986]
AMS 11,29; BROBECK, 162-163; IB, 3; WWWS, 15
E.F. Adolph, Physiologist 22(5) (1979), 11-15
M.J. Fregly and M.S. Fregly, Physiologist 25 (1982), 1; 30 (1987), 43

ADOLPH, WILLIAM HENRY [1890-1958]
AMS 9(II),6; NCAB 44,290
Anon., Science 128 (1958), 831
E.N. Todhunter, Nutr. Revs. 32 (1974), 318

ADOLPHI, HERMANN [1863-1919]
FISCHER 1,10; LEVITSKI 2,40-41
J. Brennsohn, Die Aerzte Estlands, p.415. Riga 1922

ADOLPHI, WILHELM [1867-1945]
HEIN-E, 2; NUC 4,315; RSC 13,32
Anon., Chem. Z. 56 (1932), 298
G. Adelheim, Baltische Totenschau 1939-1947, p.7. Göttingen 1947

ADOR, ÉMILE [1845-1920]
DHB 1,82; ZURICH-D, 119; POGG 3,11-12; 4,9; 6,26
A.P., Arch. Sci. Phys. Nat. [5]2 (1920), 449-451

ADRIAN, CARL [-1937]
JV 8,237; RSC 13,33; NUC 4,350; SG [3]1,176
Anon., Paris Médical 106 (1937), Suppl.51,1

ADRIAN, EDGAR DOUGLAS [1889-1977]
DNB [1971-1980],7-9; FISCHER 1,10; STC 1,15-17; W(3rd Ed.), 577-578; WWWS, 15; ABBOTT, 8; IB, 2
A.L. Hodgkin, Nature 269 (1977), 543-544; Biog. Mem. Fell. Roy. Soc. 25 (1979), 1-73
R. Keynes, TIBS 3 (1978), N16
G. Moruzzi, Erg. Physiol. 87 (1980), 1-24; Trends in Neurosciences 5 (1982), 262-265

ADRIANI, JOHANNES HERMANUS [1874-1948]
NUC 4,358; RSC 13,33

AEBI, HUGO [1921-1983]
KURSCHNER 14,18
J.P. von Wartburg, Chimia 35 (1981), 161-162; 37 (1983), 360

AEBY, CHRISTOPH [1835-1885]
NDB 1,87-88; DHB 1,87; STURM 1,7; LOMMATZSCH, 9-11; PAGEL, 12-14; WWWS,15
L. Hirzel, Verhandl. Schw. Nat. Ges. 68 (1885), 111-127
W. His, Corr. Schw. Aerzte 15 (1885), 521-523
Anon., Leopoldina 21 (1885), 211-212
R.H. Laeng, Mitt. Nat. Ges. Bern NF30 (1973), 8-10

AFANASIEV, MIKHAIL IVANOVICH [1850-1910]
BME 2,1196-1197; SKOROKHODOV, 186-192; HIRSCH 1,40; Suppl.,13; ZMEEV 4,17; VENGEROV 1,877
V.N. Nikitin, Prakticheski Vrach 9 (1910), 266
A.A. Efremenko and K.Z. Levtova, Zhur. Mikr. Ep. Imm. 32 (1961),145-147

AFANASIEV, NIKOLAI SERGEEVICH [1842-1878]
IKONNIKOV, 29-31; ZMEEV 4,17; VENGEROV 1,988; KHARKOV, 62; RSC 7,11

AFANASIEV, VASILI IVANOVICH [1849-1903]
BME 2,1194-1195; FISCHER 1,11; ZMEEV 4,15-16; VENGEROV 1,874-875
W. Pagel, Virchows Jahresber. 1903(I), p.409
Anon, Leopoldina 40 (1904), 33

AFANASIEV, VYACHESLAV ALEKSEEVICH [1859-1942]
BME 2,1195-1196; WWR, 5; LEVITSKI 2,107-111; ZMEEV 4,16; IB, 3
A.A. Lager, Arkh. Pat. 41(9) (1979), 57-58

AGAR, WILFRED EADE [1882-1951]
WW 1949, p.20; IB, 3; WWWS, 16
O.W. Teigs, Obit. Not. Fell. Roy. Soc. 8 (1952), 3-11

AGNEW, CORNELIUS REA [1830-1888]
DAB 1,123-124; NCAB 8,205-206; WWAH, 7; WWWS, 17
I.K. Rhodes, Bull. Hist. Med. 32 (1958), 438-445

ÅGREN, KARL GUNNAR [1907-1982]
VAD 1983, pp.1146-1147

AGUIAR, ANTONIO AUGUSTO [1838-1887]
GEPB 1,652-653; BLOKH 1,4; DIERGART, 472-473
Anon., Leopoldina 23 (1887), 162

AHLGREN, GUNNAR [1898-1962]
VAD 1963, p.12; SMK 1,38; IB, 3
K.O. Møller, Acta Pharm. Tox. 19 (1962), 392-394

AHRENS, EDWARD HAMBLIN, Jr. [1915-]
AMS 15(1),46; WWA 1980-1981, p.28

AHRENS, FELIX BENJAMIN [1863-1910]
NDB 1,113; BJN 15,7; BLOKH 1,4-5; KROLLMANN, 5; WWWS, 19; POGG 4,10-11; 5,11
W. Herz, Chem. Z. 34 (1910), 1245
Anon., Leopoldina 47 (1911), 37

AICKELIN, HANS [1885-1944]
JV 23,537; NUC 5,552
New York Times, 17 July 1944, p.15

AIRILA, YRJÖ ELIAS [1887-1949]
FISCHER 1,13; IB, 3; NUC 6,61
B. von Bonsdorff, The History of Medicine in Finland 1828-1918, pp.43-44. Helsinki 1975

AITKEN, JOHN MORRISON [1880-1958]
JV 26,412
Anon., Proc. Chem. Soc. 1959, p.64AITKEN, WILLIAM [1825-1892]
DNB 22,26-27; HIRSCH 1,54; Suppl.,16; PAGEL, 16-17; WWWS, 20
Anon., Brit. Med. J. 1892(II), p.54; Leopoldina 28 (1892), 157-158
J.F., Proc. Roy. Soc. 55 (1894), xiv-xvi

AKABORI, SHIRO [1900-1992]
POTSCH, 10; IB, 16
Anon., Leopoldina [3]12 (1966), 9

ALBERS, HENRY [1904-1987]
KURSCHNER 13,24; HANNOVER, 41; POGG 7a(1),19
Anon., Chem. Z. 111 (1987), 284

ALBERS, JOHANN FRIEDRICH HERMANN [1805-1867]
NDB 1,126; ADB 1,180; HIRSCH 1,63; Suppl.,17; TSCHIRCH, 1011; WWWS, 23
R. Rassmann, Münsterländische Schriftsteller 1866, p.2; 1888, p.1
G. Rath, Sudhoffs Arch. 37 (1953), 122-130

ALBERT, ADRIEN [1907-1989]
WWWS, 23
E.E. Campaigne, J. Chem. Ed. 63 (1986), 860-863
D.J. Brown, Historical Studies of Australian Science 8 (1990), 63-75; Chem. Brit. 27 (1991), 1038

ALBERT, AUGUST [1882-1951]
TLK 3,5; JV 24,22; POGG 6,32; 7a(1),20
Anon., Z. angew. Chem. 63 (1951), 584

ALBERT, HEINRICH [1835-1908]
NDB 1,138; NB, 4-5; NL 1,220-231; HUHLE-KREUTZER, 39-41; POGG 7a Suppl., 28-29
H. Albert, Mein Leben. Wiesbaden 1952

ALBERT, KURT [1881-1945]
NB, 5

ALBERT, ROBERT [1869-1952]
TLK 3,6; POGG 7a(1),21-22

ALBERT, TALBOT JONES [1888-]
AMS 5,10; JV 32,325; NUC 7,174

ALBERTONI, PIETRO [1849-1933]
DBI 1,762; HIRSCH 1,67; Suppl.,17; FISCHER 1,16; GRUETTER, 14; IB, 3-4; PAGEL, 20-21; TSCHIRCH, 1011; REBER, 171-173
G. Pugliese, Bioch. Ter. Sper. 20 (1933), 584-589
L.M. Patrizi, Erg. Physiol. 36 (1934), 11-14

ALBERTY, ROBERT ARNOLD [1921-]
AMS 15(1),57; IWW 1982-1983, p.18; WWA 1980-1981, p.34

ALBITSKI, ALEKSEI ANDREEVICH [1860-1920]
ARBUZOV, 178-179; BLOKH 1,5-6; KAZAN 1,255-257; POGG 4,13; 5,13

ALBITSKI, PETR MIKHAILOVICH [1853-1922]
BME 1,829-830; WWR, 15-16; ZMEEV 4,4
S.N. Sorinson, Arkh. Pat. 1955(3), pp.68-71

ALBRECHT, EUGEN [1872-1908]
BJN 13,6*; FISCHER 1,16-17; KALLMORGEN, 204
J. Oberndorfer, Münch. med. Wchschr. 55 (1908), 1539-1542
G. Herxheimer, Z. allgem. Path. 19 (1908), 657-662
B. Fischer, Verhandl. path. Ges. 13 (1909), 416-422

ALBRECHT, HEINRICH [1866-1922]
NDB 1,181; OBL 1,13; FISCHER 1,17; LESKY, 573,576; WWWS, 24
R. Paltauf, Wiener klin. Wchschr. 35 (1922), 692-694
K., Wiener med. Wchschr. 72 (1922), 1189
L. Arzt, Verhandl. path. Ges. 29 (1937), 349-351

ALBRIGHT, FULLER [1900-1969]
DSB 17,8-11; AMS 9(II),9-10; MEDVEI, 524-529; ICC, 247-248; DAMB, 8-9; WWWS, 24-25
A.P. Forbes, Metabolism 11 (1962), 3-5; Harvard Med. Alumni Bull. 44(4) (1970), 36-37
L. Axelrod, New Eng. J. Med. 283 (1970), 964-970
J.E. Howard, Trans. Assoc. Am. Phys. 83 (1970), 10-11; Persp. Biol. Med. 24 (1981), 374-381
F.C. Bartter, Endocrinology 87 (1970), 1109-1112
V.A. McKusick, Birth Defects 7(6) (1971), 1-4
A. Leaf, Biog. Mem. Nat. Acad. Sci. 48 (1976), 3-22

ALBU, ALBERT [1867-1921]
DBJ 3,289; FISCHER 1,18; WREDE, 1-2; WININGER 6,406-407
C. Neuberg, Riv. Biol. 3 (1921), 851-852
I. Boas, Berl. klin. Wchschr. 58 (1921), 141
A. Alexander, Arch. Verdauungskr. 27 (1921), 224
Anon., Chem. Z. 45 (1921), 78,99

ALCOCK, NATHANIEL HENRY [1871-1913]
O'CONNOR(2), 212-213; FISCHER 1,18; WWWS, 25
Anon., Brit. Med. J. 1913(I), p.1353; Lancet 1913(I), p.1835
D. Zuck, Anaesthesia 43 (1988), 972-980

ALCOCK, ROBERT SAXELBY [1907-]
SRS, 89

ALDER, ALBERT [1888-1980]
KURSCHNER 11,22
H.U. Späth, Der Hämatologe Albert Alder. Zürich 1983

ALDER, KURT [1902-1958]
DSB 1,105-106; MH 1,6-7; MSE 1,8; STC 1,17-18; W, 8; WWWS, 25-26; POTSCH, 11; ABBOTT-C, 7-8; KIEL, 196; POGG 6,34; 7a(1),25-26;(4),135*
M. Günzl-Schumacher, Chem. Z. 82 (1958), 489-490
S. Goldschmidt, Bayer. Akad. Wiss. Jahrbuch 1958, pp.197-198

ALDRICH, THOMAS BELL [1861-1938]
AMS 4,11; FISCHER 1,18; OEHRI,13; JV 9,132

ALEKSANDROV, NIKOLAI ALEKSANDROVICH [1858-1935]
RSC 18,386

ALEKSANDROV, PETR ALEKSANDROVICH [1816-1867]
BLOKH 1,6

ALEKSEEV, ALEKSEI GEORGIEVICH [1882-1950]
BME 36,69; WWR,12
D.N. Zasukhin and G.A. Alekseev, Parazitologia 16(3) (1982), 262-264

ALEKSEEV, PETR PETROVICH [1840-1891]
BSE 2,90; VENGEROV 1,402-406; IKONNIKOV, 14-17; ARBUZOV, 179-180; KOZLOV, 174-176; BLOKH 1,8-10; POGG 3,19
Anon., Chem. Z. 15 (1891), 1898

ALEKSEEV, VLADIMIR FEDOROVICH [1852-1919]
BLOKH 1,6-8; KOZLOV, 61-62
N.S. Kurnakov, Zhur. Fiz. Khim. 54 (1923), 1-4

ALEXANDER, ALBERT ERNEST [1914-1970]
D.H. Naffer, J. Coll. Sci. 34 (1970), 33-332
R.J. Hunter, Chem. Brit. 7 (1971), 343
R.J.W. LeFevre, Rec. Austral. Acad. Sci. 2 (1972), 61-81

ALEXANDER, BENJAMIN [1909-1978]
AMS 13,44; NCAB 61,223-225; ICC, 345-346
S. Niewarowski and R.W. Colman, Thromb. Res. 13 (1978), 1-4
A.S. Friedberg, Trans. Assoc. Am. Phys. 92 (1979), 18-21
K.M. Brinkhous, Thromb. Diath. Haemorrh. 40 (1979), 271-272

ALEXANDER, ERNST [1902-]
BHDE 2,16-17; LDGS, 94; NUC 8,234; POGG 6,34; 7a(1),27
Who's Who in Israel 1969-1970, p.21

ALEXANDER, FRANZ [1872-]
KALLMORGEN, 205

ALEXANDER, HATTIE ELIZABETH [1901-1968]
NAW 4,10-11
New York Times, 25 June 1968
R. McIntosh, Pediatrics 42 (1968), 544

ALEXANDER, JEROME [1876-1959]
AMS 9(I),18; NCAB 52,479-480; WWAH, 10; WWWS,28; POGG 6,35; 7b,55-56
F.M. Turner, Chemist 28 (1951), 12-16
Anon., Chem. Eng. News 37(5) (1959), 94

ALEXANDER, PETER [1922-]
BHDE 2,18; WWWS, 28

ALEXANDER, WALTER [1871-]
JV 9,15; NUC 8,308; RSC 13,60

ALFREY, TURNER, Jr. [1918-1981]
AMS 14,56
R.F. Boyer (Ed.), Turner Alfrey, Jr. Midland,Mich. 1983

ALIBERT, JOSEPH [1814-1866]
HAMON 1,75

ALLAN, JAMES [1825-1866]
W.A. Miller, J. Chem. Soc. 20 (1867), 386-387

ALLARD, HARRY ARDELL [1880-1963]
AMS 10,45
Anon., Science 139 (1963), 1191
E.W. Brandes, Phytopathology 54 (1964), 125-126

ALLARD, ROBERT WAYNE [1919-]
AMS 15(1),72; WWA 1980-1981, p.46; WWWS, 32

ALLEE, WARDER CLYDE [1885-1955]
DSB 17,16-18; AMS 8,29; NCAB 42,159; IB, 4; WWAH, 11; WWWS, 32
A.E. Emerson and T. Park, Science 121 (1955), 686-687
K.P. Schmidt, Biog. Mem. Nat. Acad. Sci. 30 (1957), 3-40

ALLEN, ALFRED HENRY [1847-1904]
BLOKH 1,10; RSC 13,64-65; POGG 3,22-23; 4,15

ALLEN, CHARLES ELMER [1872-1954]
AMS 8,30; NCAB 42,124; WWWS, 32-33; IB, 4
R.E. Cleland, Am. Phil. Soc. Year Book 1954, pp.392-393
G.M. Smith, Biog. Mem. Nat. Acad. Sci. 29 (1956), 3-15

ALLEN, CHESTER HARMON [1889-1971]
AMS 11,59

ALLEN, EDGAR [1892-1943]
DSB 1,123-124; DAB [Suppl.3],6-7; FISCHER 1,22; DAMB, 12-13; STC 1,26; WWAH, 12; WWWS, 33
G.W. Corner et al., Anat. Rec. 86 (1943), 595-598
W.U. Gardner, Yale J. Biol. Med. 15 (1943), 641-644; Science 97 (1943), 368-369
Anon., Yale J. Biol. Med. 17 (1944-45), 2-12
A.W. Diddle, Obs. Gyn. 38 (1971), 631-637

ALLEN, FRANK WORTHINGTON [1903-1964]
AMS 10,48

ALLEN, FREDERICK MADISON [1879-1964]
AMS 10,48; NCAB 50,519-520
Anon., Metabolism 13 (1964), 383-385
G.G. Duncan, Diabetes 13 (1964), 318-319
A.R. Henderson, Academy of Medicine of New Jersey Bulletin 16 (1970), 40-49

ALLEN, RICHARD SWEETNAM [1896-1977]
AMS 11,65; WWA 1978-1979, p.51; WWWS, 34

ALLEN, WILLARD MYRON [1904-]
AMS 14,66; WWWS, 35
W.M. Allen, Gynecologic Investigation 5 (1974), 142-182

ALLEN, WILLIAM [1770-1843]
DNB 1,322-323; FERCHL, 7; WWWS, 35; CALLISEN 1,113-113; POGG 1,31
Anon., Proc. Roy. Soc. 5 (1843-50), 532-533
M.E. Weeks, J. Chem. Ed. 35 (1958), 70-73

ALLERS, RUDOLF [1883-1964]
KURSCHNER 11,25; FISCHER 1,23; KOREN, 146

ALLES, GORDON ALBERT [1901-1963]
AMS 10,54
Anon., Science 139 (1963), 1191
R.M. Featherstone, Pharmacologist 5 (1963), 120-122

ALLFREY, VINCENT GEORGE [1921-]
AMS 12,83; WWA 1980-1981, p.54; WWWS, 35

ALLIHN, FELIX RICHARD [1854-1915]
RSC 9,33; 13,71
Anon., Chem. Z. 39 (1915), 754; Z. angew. Chem. 28(III) (1915), 560

ALLISON, FRANKLIN ELMER [1892-1972]
AMS 12,84; IB, 5; WWWS, 36
Anon., Soil Science 114 (1972), 413

ALLISON, JAMES BOYD [1901-1964]
AMS 10,56
Anon., Chem. Eng. News 42(49) (1964), 102
J.W. Green, J. Nutrition 107 (1977), 703-707

ALLMAND, ARTHUR JOHN [1885-1951]
WW 1949, p.44; SRS, 43; WWWS, 36; POGG 5,16; 6,41; 7b,62-63
G. Temple, Nature 168 (1951), 408-409
F.A. Freeth, J. Chem. Soc, 1954, pp.4706-4709; Obit. Not. Fell. Roy. Soc. 9 (1954), 3-13

ALLOTT, ERIC NEWMARCH [1899-1980]
WW 1981, pp.42-43
Anon., Chem. Brit. 17 (1981), 362

ALLOWAY, JAMES LIONEL [1900-1954]
NCAB 48,180
Anon., J. Am. Med. Assn. 154 (1954), 1365

ALMAGIA, MARCO [1876-]
VACCARO, 28
Chi É 1948, pp.16-17

ALMASY, FELIX [1901-1984]
KURSCHNER 13,33; SBA 3,12; WWWS, 37; POGG 7a(1),27-28

ALMÀN, AUGUST THEODOR [1833-1903]
SBL 1,418-423; HIRSCH 1,98; Suppl.,24; POGG 2,24; 4,15
K. Linroth, Hygiea 66 (1904), 1-17

ALMQUIST, HERMAN JAMES [1903-]
AMS 11,73; WWWS, 37
H.J. Almquist, Fed. Proc. 38 (1979), 2687-2689
T. Jukes, Journal of Applied Biochemistry 7 (1985), 1-2; J. Nutrition 117 (1987), 409-415

ALMS, AUGUST [1803-1847]
HEIN, 5
C. Lüdtke, Deutsche Apoth. Z. 11 (1959), 14-15

ALOSHIN, BORIS VLADIMIROVICH [1901-]
VORONTSOV, 84-86

ALOY, FRANÇOIS JULES [1866-1928]
POGG 5,16-17; 6,42
A. Tiffeneau, Bull. Acad. Med. 99 (1928), 459-460

ALSBERG, CARL LUCAS [1877-1940]
AMS 6,23; NCAB 30,31-32; MILES, 7-8; WWAH, 15; WWWS, 38; IB, 5; KAGAN(JA), 338,736; POGG 6,42; 7b,65
H.M.A., Bioch. Bull. 2 (1913), 211-216
E.C. Voorhees, Science 93 (1941), 32-33
Anon., Chronica Botanica 6 (1941), 329-330
M.K. Bennett, Cereal Chemistry 19 (1942), 721-723
J.S. Davis (Ed.), Carl Alsberg, Scientist at Large. New York 1949

ALT, FERDINAND [1867-1923]
STURM 1,13; FISCHER 1,25; FUIERER, 107-116; WININGER 7,512
I. Mayer, Mon. Ohren. 57 (1923), 240

ALT, HOWARD LANG [1900-1972]
AMS 12,91; WWWS, 39
D.P. Earle, Trans. Assoc. Am. Phys. 85 (1972), 8

ALTENBURG, EDGAR [1888-1967]
AMS 11,76
L.S. Browning, Genetics 60 (1968), s20-s21

ALTER, DAVID [1807-1881]
DAB 1,230; ELLIOTT, 16; W, 9; WWAH, 15; WWWS, 39
W.A. Hamor, Isis 22 (1935), 507-510
E.C. Watson, Am. J. Phys. 25 (1957), 630-632

ALTHAUSSE, MAX [1861-]
JV 4,222; RSC 13,275

ALTMAN, SIDNEY [1939-]
AMS 17(1),101

ALTMANN, KARL [1880-1968]
KURSCHNER 12,30; FISCHER 1,25; KALLMORGEN, 205; KOREN, 147; LDGS, 61
Deutscher Dermatologen Kalender 1929, pp.5-6

ALTMANN, MANFRED [1900-1954]
NUC 10,570; SG [4]1,276
Anon., Brit. Med. J. 1954(II), p.1172

ALTMANN, RICHARD [1852-1900]
FISCHER 1,26; KROLLMANN, 11; WWWS, 39; POGG 4,16
W. Pagel, Virchows Jahresber. 1900(I), p.328
Anon., Anat. Anz. 18 (1900), 589-590

ALTSCHUL, EMIL (ELIAS) [1812-1865]
OBL 1,17; STURM 1,15; WERSTLER, 230-234

ALZONA, FEDERICO [1888-1958]
L. Bacialli, Boll. Sci. Med. Bologna 130 (1958), 287-298

AMADORI, MARIO [1886-1941]
DBI 2,612; POTSCH, 11; POGG 5,17; 6,44-45; 7b,67
C. Sandonnini, Chimica e Industria 24 (1942), 32

AMBARD, LÉON [1876-1962]
DBA 1,35; FISCHER 1,28
Anon., C. R. Soc. Biol. 156 (1962), 1530-1532
A. Monnier, Bull. Acad. Med. 147 (1963), 315-318
A. Plichet, Presse Med. 71 (1963), 2542

AMBERSON, WILLIAM RUTHRAUFF [1894-1976]
AMS 12,81; WWWS, 41

AMBRONN, HERMANN [1856-1927]
NDB 1,242; FREUND 2,1-12; TSCHIRCH, 1012; POGG 4,18; 5,19; 6,46; 7a,33
A. Frey, Ber. bot. Ges. 45 (1927), (60)-(71); Z. mikr. anat. Forsch. 44 (1927), 129-133; Koll. Z. 44 (1928), 6-8

AMBROS, OTTO [1901-]
JV 42,504; NUC 11,348
Anon., Nachr. Chem. Techn. 9 (1961), 162; Chem. Z. 105 (1981), 154

AMBROSIONI, FELICE [1790-1843]
DFI, 9; CALLISEN 1,129; 26,42
V. Bianchi, Minerva Med. 40I (1949), 590-592

AMBÜHL, GOTTWALT [1850-1923]
DHB 1,296
H. Rehsteiner, Verhandl. Schw. Nat. Ges. 105 (1924), 10-12
B. Strahlmann, Mitt. Lebensmitt. 53 (1962), 470

AMEND, BERNARD GOTTWALD [1820-1911]
New York Times, 7 April 1911
O.P. Amend, The Chemist 8(1) (1930), 5-6

AMES, BRUCE NATHAN [1928-]
AMS 15(1),102; WWA 1980-1981, p.63; IWW 1982-1983, pp.28-29; WWWS, 42; MSE 1,14-15; POTSCH, 12

AMICI, GIOVANNI BATTISTA [1784-1863]
DSB 1,135-137; DBI 2,780-781; FREUND 3,1-14; MAYERHOFER 1,210; W, 9-10; WWWS, 42-43; POGG 1,37; 3,26
V. Ronchi, Physis 11 (1969), 520-523

AMIS, EDWARD STEPHEN [1905-]
AMS 14,84; WWA 1980-1981, p.64; WWWS, 43

AMLONG, HANS ULRICH [1909-1943]
LDGS, 11
H. Borris, Ber. bot. Ges. 68a (1955), 200-202

AMMANN, KARL [1883-1957]
JV 26,393

AMMELBURG, ALFRED [1864-1939]
NDB 1,252; NB, 9; KALLMORGEN, 206-207; POGG 6,47-48; 7a(1),34
Anon., Chem. Z. 68 (1939), 463

AMMON, FRIEDRICH von [1799-1861]
NDB 1,254; ADB 1,406; HIRSCH 1,117-121; Suppl.,28; WWWS, 1728; RSC 1,57-58; CALLISEN 1,139-144; 26,46-48

AMMON, FRIEDRICH CARL von [1916-1985]
KURSCHNER 14,48
Anon., Nachr. Chem. Techn. 33 (1985), 741; Chem. Z. 109 (1985), 315

AMMON, ROBERT [1902-]
KURSCHNER 13,43; POGG 6,48; 7a(1),34-35
W. Dirscherl, Arzneimitt. 17 (1967), 1090
G. Werth, Arzneimitt. 22 (1972), 1560-1561

AMOROSO, EMMANUEL CIPRIAN [1901-1982]
WW 1982, p.41
R.V. Short, Biog. Mem. Fell. Roy. Soc. 31 (1985), 3-30

AMOSS, HAROLD LINDSAY [1886-1956]
AMS 9(II),20; NCAB 43,252; F,118-119; ICC, 175-176; WWWS, 43-44; IB, 6
J.B. Paul, Trans. Am. Clin. Assn. 69 (1957), xli-xlii
H.J. Morgan, Trans. Assoc. Am. Phys. 70 (1957), 9-10

AMPÀRE, ANDRÉ MARIE [1775-1836]
DSB 1,139-147; DBF 2,713-720; GE, 814-817; MATSCHOSS, 4; W, 10-11; LAROUSSE, 1,126; MAYERHOFER 1,212-213; FERCHL, 8; BLOKH 1,10-12; POTSCH, 12; SCHAEDLER, 2; POGG 1,39-40,1528; 6,49
L. de Launoy, Le Grand Ampère. Paris,1925
M. Lewandowski, André-Marie Ampère, la Science et la Foi. Paris,1936
P. Costabel et al. Rev. Hist. Sci. 30 (1977), 113-167
P. Laszlo, Nouveau Journal de Chemie 4 (1980), 699-701

AMSLER, CAESAR HANS [1881-1965]
DECKER, 101-105; SBA 2,12-13; IB, 6; POGG 7a(1),37

AMTHOR, CARL [1853-1939]
HEIN-E, 4; NUC 15,24; POGG 4,21; 7a(1),38
Anon., Z. angew. Chem. 46 (1933), 284-285; 52 (1939), 412

ANCEL, PAUL [1873-1961]
DSB 1,152-153; DBA 1,38; FISCHER 1,28; WWWS, 45
Anon., Bull. Acad. Med. Belg. [7]1 (1961), 293-300
A. Giroud, Bull. Acad. Med. 145 (1961), 335-337
E. Wolff, Arch. Anat. 44 Suppl. (1961), 5-27
A. Beau, Rev. Med. Nancy 86 (1961), 407-412

ANDERLINI, FRANCESCO [1844-1933]
RSC 13,92; POGG 4,21; 5,21; 6,50
A. Cappodoro, Chimica e Industria 15 (1933), 309; Studi Trentini di Scienze Naturali 14 (1933), 141

ANDERSAG, HANS [1902-1955]
DUNSCHELE, 134; POTSCH, 12

ANDERSEN, ALFRED CHRISTIAN [1882-1970]
KBB 1969, p.28; VEIBEL 1,236-237; 2,23-27

ANDERSON, ALAN BRUCE [1902-1985]
Who's Who in British Science 1953, p.7
Anon., Chem. Brit. 21 (1985), 721

ANDERSON, ALEXANDER PIERCE [1862-1943]
NCAB F,303; AMS 6,27; WWAH, 16

ANDERSON, EDGAR [1897-1969]
DSB 17,25-27; AMS 11,95
G.L. Stebbins, Biog. Mem. Nat. Acad. Sci. 49 (1978), 3-23

ANDERSON, GEORGE WASHINGTON [1913-]
AMS 14,83

ANDERSON, HUGH KERR [1865-1928]
DNB [1922-1930],19-20; O'CONNOR(2), 15-17; WWWS, 47
Anon., Lancet 1928(II), p.1000; Brit. Med. J. 1928(II), p.871
C.S.S., Proc. Roy. Soc. B104 (1929), xx-xxv

ANDERSON, JOHN FLEETEZELLE [1873-1958]
AMS 9(II),23; BARRY, 1-4; WWWS, 47; IB, 6
F.C. Wood, Trans. Assoc. Am. Phys. 72 (1959), 13-14

ANDERSON, RICHARD JOHN [1848-1914]
NUC 15,610; RSC 13,95
Anon., Brit. Med. J. 1914(II), p.418

ANDERSON, RUBERT SIGFRED [1898-1974]
AMS 12,128; WWWS, 48

ANDERSON, RUDOLPH JOHN [1879-1961]
AMS 10,82; CHITTENDEN, 256-266; WWAH, 17; POGG 6,51-52; 7b,77-78
H.B. Vickery, Biog. Mem. Nat. Acad. Sci. 36 (1962), 19-50

ANDERSON, THOMAS [1819-1874]
DSB 1,155-156; DNB 1,392; W, 12; SCHAEDLER, 2-3; TSCHIRCH, 1012; BLOKH 1,12; FERCHL, 9; POGG 1,43; 3,27-28
E.J. Mills, J. Chem. Soc. 28 (1875), 1309-1313
Anon., Chem. Brit. 28 (1992), 442-444

ANDERSON, THOMAS FOXEN [1911-1991]
AMS 15(1),130; IWW 1982-1983, p.33; WWA 1980-1981, p.78; WWWS,

48-49
T.F. Anderson, Ann. Rev. Microbiol. 29 (1975), 1-18
New York Times, 13 August 1991

ANDERSSON, BENGT [1912-]
VAD 1981, p.31

ANDERVONT, HOWARD BANCROFT [1898-1981]
AO 1981, pp.171-172; IB, 6
Anon., J. Nat. Cancer Inst. 67 (1981), 1

ANDOUARD, AMBROISE [1839-1914]
DBF 2,868-869; GORIS, 211-212; TSCHIRCH, 1012; POGG 4,22; 5,22
Anon., J. Pharm. Chim. [7]9 (1914), xxviii; Chem. Z. 38 (1914), 560

ANDRAL, GABRIEL [1797-1876]
DBF 2,871-874; GENTY 4,193-208; HIRSCH 1,131-133; PAGEL, 36-37; VERSO, 93-94; WWWS, 49; FRANKEN, 173-174; GE 2,1032-1033
L. Figuier, Ann. Scient. 20 (1876), 542-544
Anon., Proc. Am. Acad. Arts Sci. 11 (1876), 363-365
C. Dreyfus, Nouv. Rev. Hem. 3 (1963), 261-276
J. Perpignan, La Vie et l'Oeuvre de Gabriel Andral. Paris 1964
M.L. Verso, Med. Hist. 15 (1971), 55-56
P. Huard and M.J. Imbault-Huard, Rev. Hist. Sci. 35 (1982), 131-153

ANDRÉ, ÉMILE [1877-1969]
POGG 6,53-54; 7b,82-84

ANDRÉ, GUSTAVE [1856-1927]
DBF 2,916-917; WWWS, 49; POGG 4,24; 5,23-24; 6,54
E. Demoussy, Bull. Soc. Chim. Biol. 9 (1927), 857-860; Ann. Inst. Agron. 21 (1928), 38-45
C. Barrois, C. R. Acad. Sci. 184 (1927), 1141-1144
P. Boischot, Rev. Sci. 1927, pp.439-440

ANDREAE, EDWARD PHILIP [1879-1975]
JV 20,20; NUC 16,184

ANDREAE, HANS CARL [1854-1920]
HANNOVER, 43

ANDREASCH, RUDOLF [1857-1928]
TLK 3,10; POGG 4,25; 5,24; 6,54
Anon., Ost. Chem. Z. 31 (1928), 218; Chem. Z. 53 (1929), 12

ANDREEV, FEDOR ANDREEVICH [1879-1952]
BME 2,6-7; WWR, 19
V.A. Negovsky, Pat. Fiz. Eksp. Ter. 1980(2), pp.91-92
E.N. Kuzmina, Sov. Med. 25(5) (1962), 149-150

ANDREEV, IVAN IVANOVICH [1880-1919]
BSE 2,430; BLOKH 1,12-13
M.A. Blokh, Zhur. Fiz. Khim. 60 Suppl. (1928), 132-133
V.A. Volkov, Vop. Ist. Est. Tekh. 1980(3), pp.120-122

ANDREEV, LEONID ALEKSANDROVICH [1891-1941]
BSE 2,430; BME 2,5-6; WWR, 19
K.M. Bykov, in Materiali po Fiziologii Retseptorov (K.M. Bykov and I.P. Razenkov, Eds.), pp.5-6. Moscow 1948

ANDREEV, NIKOLAI NIKOLAEVICH [1852-]
VENGEROV 1,69; ZMEEV, Suppl.,1; SG [2]1,434ANDREEV, PAVEL NIKOLAEVICH [1872-1949] WWR, 20
V.I. Kalugin, Veterinaria 48(12) (1972), 104-107

ANDREOCCI, AMERICO [1863-1899]
BLOKH 1,13; POGG 4,26
Anon., Ber. chem. Ges. 32 (1899), 2545
G. Grassi, Bolletino della Accademia Gioenia di Scienze Naturale NS62 (1900), 27-33
J. Grassi, Annales de la Société d'Émulation du Departement des Vosges 78 (1902), 568-577

ANDREWES, CHRISTOPHER HOWARD [1896-1988]
WW 1981, p.56; MH 2,9-10; MSE 1,21-22; STC 1,37-38; WWWS, 50-51
C.H. Andrewes, Ann. Rev. Microbiol. 27 (1973), 1-11
E.D. Kilbourne, Am. Phil. Soc. Year Book 1989, pp.147-152

ANDREWES, FREDERICK WILLIAM [1859-1932]
DNB [1931-1940],14-15; BULLOCH, 349; O'CONNOR(2), 269-270; IB,7; WWWS, 51
E.B. Poulton, Obit. Not. Fell. Roy. Soc. 1 (1932), 37-44
H.T., St. Barts. Hosp. Rep, 65 (1932), 1-11

ANDREWS, DONALD HATCH [1898-1973]
AMS 12,135; WWWS, 51; POGG 6,56-57; 7b,88

ANDREWS, JAMES CLARENCE [1892-1975]
AMS 11,114; IB. 7; POGG 6,57; 7b,88-89
Anon., Chem. Eng. News 53(48) (1975), 31

ANDREWS, LAUNCELOT WINCHESTER [1856-1938]
AMS 5,26; NCAB 46,398-399; MILES, 8-9; RSC 13,110; NUC 16,428-429; POGG 4,26-27; 5,24-25; 6,57; 7b,89
New York Times, 15 April 1938
E. Bartow, J. Chem. Soc. 1938, p.1735

ANDREWS, THOMAS [1813-1885]
DSB 1,160-161; DNB 22,49-51; W, 12-13; BLOKH 1,14; POTSCH, 13; WWWS, 52; FERCHL, 9; ABBOTT-C, 8-9; BELFAST, 582-583; POGG 1,46; 3,30-31
Anon., Nature 33 (1885), 157-159; 39 (1889), 554-556
H. Müller, J. Chem. Soc. 49 (1886), 342-344
A.C.B., Proc. Roy. Soc. 41 (1886), xi-xv
H. Mackle and C.L. Wilson, Endeavour 30 (1971), 8-10
W.J. Davis, Proc. Roy. Irish Acad. 77B (1977), 309-315

ANDRUS, EDWIN COWLES [1896-1978]
AMS 11,115; ICC, 208-209; WWWS, 52
B.M. Baker et al., Trans. Assoc. Am. Phys. 91 (1978), 10-12

ANFINSEN, CHRISTIAN BOEHMER [1916-]
AMS 15(1),139-140; MH 2,10-11; MSE 1,23; STC 1,38-40; POTSCH, 13; WWWS,52; IWW 1982-1983, p.37; WWA 1980-1981, p.85

ANGELESCU, EUGEN [1896-1968]
POTSCH, 13; POGG 6,58-59; 7b,93-96

ANGELI, ANGELO [1864-1931]
PROVENZAL, 291-301; TSCHIRCH, 1013; POTSCH, 14; POGG 4,27-28; 5,25; 6,59-60; 7b,97
R. Poggi and Z. Jolles, Ber. chem. Ges. 64A (1931), 129-130
R. Willstätter, Bayer. Akad. Wiss. Jahrbuch 1931-1932, pp.64-65
L. Cambi, Gazz. Chim. Ital. 63 (1933), 527-560
G. Provenzal, Rass. Clin. Terap. 37 (1938), 364-371

ANGERER, ERNST [1881-1951]
NDB 1,292; POGG 6,61; 7a(1),43

ANGERER, KARL von [1883-1945]
BERWIND, 56-60; KURSCHNER 5,27; PITTROFF, 109-110; IB, 7
H.Eyer and B. Freytag, Münch. med. Wchschr. 97 (1955), 1598-1599

ANICHKOV, NIKOLAI NIKOLAEVICH [1885-1964]
BSE 2,453; FISCHER 1,31; WWR, 21; WWWS, 54
E.N. Pavlovski, Priroda 35(4) (1946), 80-82
W. Raab, Cardiologia 47 (1965), 207-208
T.N. Khavkin et al., J. Atheroscl. Res. 6 (1966), 199-200
I.R. Petrov, Pat. Fiz. Eksp. Ter. 12(1) (1968), 92-95

ANICHKOV, SERGEI VIKTOROVICH [1892-]
BSE (3rd Ed.) 2,35; BME 2,192-193; IWW 1982-1983, p.37
V.M. Karasik, Farm. Toks. 25 (1962), 379-383
S.N. Golichov, Farm. Toks. 35 (1972), 517-529
S.V. Anickov, Ann. Rev. Pharm. 15 (1975), 1-10

ANKER, HERBERT [1912-1976]
AMS 13,101; BHDE 2,28

ANNAHEIM, JOSEPH [1843-1914]
ZURICH, 49; ZURICH-D, 121; POGG 3,33
Anon., Ber. chem. Ges. 47 (1914), 1326

ANREP, GLEB [1891-1955]
WW 1949, p.65; IB, 7; WWWS, 1728
J.H. Gaddum, Biog. Mem. Fell. Roy. Soc. 2 (1956), 19-34

ANREP, VASILI KONSTANINOVICH [1852-1918?]
VENGEROV 1,619-621,967-969; SG [3]1,559

d'ANS, JEAN [1881-1969]
POTSCH, 107; WOLF, 14; POGG 5,260; 6,62; 7a(1),43-45
H. Autenrieth, Z. angew. Chem. 63 (1951), 365-367
Anon., Chem. Z. 75 (1951), 417; 80 (1956), 528; 93 (1969), 151
J. Schormüller, Z. Elektrochem. 60 (1956), 641-642

ANSBACHER, STEFAN [1905-]
AMS 14,116; WWWS, 54

ANSCHÜTZ, LUDWIG [1889-1954]
AUERBACH, 764-765; MEINEL, 318-319,496; POGG 6,62; 7a(1),45
F. Krollpfeiffer, Chem. Ber. 90 (1957), xv-xviii

ANSCHÜTZ, OTTOMAR [1846-1907]
NDB 1, 308; BJN 12,6*; BLOKH 1,14; MATSCHOSS, 5

ANSCHÜTZ, RICHARD [1852-1937]
DSB 1,168-169; NDB 1,308; DZ, 19-20; POTSCH, 14; SCHAEDLER, 3; TLK 3,11; POGG 3,33-34; 4,30-32; 5,27-28; 6,62-63; 7a(1),45
H. Meerwein, Ber. chem. Ges. 74A (1941), 29-74
E.H. Huntress, Proc. Am. Acad. Arts Sci. 81 (1952), 37-39
G.B. Kauffman, J. Chem. Ed. 59 (1982), 627-629,745-751

ANSELMINO, KARL JULIUS [1900-1978]
KURSCHNER 12,45; POGG 7a(1),46-47
L. Beck, Z. Geb. 182(2) (1978), 97

ANSELMINO, OTTO HERMANN [1873-1955]
HEIN-E, 5-6; KURSCHNER 3,33; TLK 3,11; POGG 6,63; 7a(1),47; (4),135*
H. Valentin, Pharmazie 11 (1956), 160
Anon., Chem. Z. 80 (1956), 289; Deutsche Apoth. Z. 113 (1973),1109-1110
C. Friedrich and H.J. Seidlein, Pharmazie 36 (1981), 846-852

ANSLOW, GLADYS AMELIA [1892-1969]
AMS 11,120; WWAH, 20; WWWS, 55

ANSLOW, WILLIAM PARKER, Jr. [1912-1966]
AMS 11,120

ANSON, MORTIMER LOUIS [1901-1968]
AMS 11,120
C.O. Chichester et al., Advances in Food Research 17 (1969), 1-2
J.T. Edsall, Adv. Protein Chem. 24 (1970), vii-x

ANTON, RICHARD [1887-]
JV 32,297

ANTONI, WILHELM [1879-]
JV 18,14; NUC 18,167

ANTONINI, ERALDO [1931-1983]
DBI 34,154-155
A. Rossi Fanelli, Riv. Biol. 76 (1983), 375-378
F. Bruneri et al., TIBS 9 (1984), 12-13ANTRICK, OTTO [1862-1942]
POGG 7a(1),48
Anon., Chem. Z. 66 (1942), 107
P. Korn, Ber. chem. Ges. 75A (1942), 137-139

ANTROPOFF, ANDREAS von [1878-1956]
WELDING, 15; WENIG, 5; RIGA, 397; POGG 5,28-29; 6,65; 7a(1),48; (4),135*
W. Fahrbach, Album der Landsleute der Fraternitas Baltica (3rd Ed.), pp. 165-166. Aschaffenburg,1961

ANTUSCH, AUGUSTIJN CONSTANTIJN [1862-]
NUC 18,248

AOKI, KAORU [1877-1938]
FISCHER 1,33; HAKUSHI 2,248-249; IB, 8
Y. Satake, Tohoku J. Exp. Med. 34 (1938), 605

APÁTHY, ISTVÁN [1863-1922]
DSB 1,176-177; FISCHER 1,34; FREUND 1,65-75; PAGEL, 1920-1921
B. Farkas, Riv. Biol. 5 (1923), 565-570
A. Bethe, Klin. Wchschr. 2 (1923), 811
A. Abraham, Acta Biologica Szeged 20 (1974), 27-35

APELT, ERNST FRIEDRICH [1812-1859]
NDB 1,323-324

APITZ, KURT [1906-1945]
KURSCHNER 6,31
R. Rössle, Verhandl. path. Ges. 32 (1950), 419-420

APITZSCH, HERMANN [1868-1937]
POGG 7a(1),49
M. Busch, Ber. chem. Ges. 70A (1937), 124-125

APJOHN, RICHARD [1846-1877]
POGG 3,36-37
Anon., J. Chem. Soc. 33 (1878), 227
J.A. Venn, Alumni Cantabrigensis II 1, p.63. Cambridge,1940

APOLANT, HUGO [1866-1915]
ARNSBERG, 20-22; KALLMORGEN, 209; BK, 159-160; WININGER 6,422
H. Sachs, Jahresb. Aerzt. Ver. 1915, pp.150-152
Anon., Chem. Z. 39 (1915); Z. angew. Chem. 28(III) (1915), 123

APPEL, HERBERT [1907-]
LDGS, 16
International Chemistry Directory 1969-1970, p.178

APPERT, FRANÇOIS NICOLAS [1750-1841]
DBF 3,139-145; GE 3,421; LAROUSSE 1,512; W, 14; WWWS, 56; CALLISEN 1,205-206; 26,73-74
A.W. Bitting, Appertizing or the Art of Canning. San Francisco 1937
J.C. Graham, J. Roy. Soc. Med. 74 (1981), 374-381

APPLEMAN, CHARLES ORVILLE [1878-1964]
AMS 10,97; WWAH, 20; IB, 8

APRISON, MORRIS HERMAN [1923-]
AMS 15(1),152; WWA 1981-1982, pp.93-94; WWWS, 57

ARAKI, TORASABURO [1866-1942]
FISCHER 1,35; JBE, 58-59; IB, 8; JV 6,255; RSC 13,138
Y. Kotake, Kagaku 16 (1961), 335-337
Anon., Japanese Physiology Past and Present, p.62. Tokyo,1965

ARATA, PEDRO [1849-1922]
NDBA 1,197-198; TSCHIRCH, 1013; POGG 4,35; 6,68
Anon., Semana Medica (Buenos Aires) 29 (1922), 941-944
E. Herrero, Revista de la Facultad de Ciencias Quimicas del Universidad

Nacional de la Plata 1 (1923), 265-283
E. Pennini de la Vega, Rev. Hist. Sci. 22 (1975), 501-506

ARBER, AGNES [1879-1960]
DSB 1,205-206; DNB [1951-1960], 28-30; WW 1959, p.78; DESMOND,16; WWWS,58
W.T. Stearn, Nature 186 (1960), 847-848; Taxon 9(9) (1960), 261-263

H.H. Thomas, Biog. Mem. Fell. Roy. Soc. 6 (1960), 1-11

ARBER, WERNER [1929-]
MSE 1,25-26; IWW 1981-1982, p.41; POTSCH, 14

ARBUCKLE, HOWARD BELL [1870-1945]
MILES, 11; AMS 7,46; WWAH, 21; WWWS, 58

ARBUZOV, ALEKSANDR ERMININGELDOVICH [1877-1968]
BSE (3rd Ed.) 2,165; ARBUZOV, 128-130; STC 1, 42-43; POTSCH, 14-15; WWWS, 58; POGG 6,68-69; 7b,106-110
V.W. Zoroastrova and F.G. Valitova, Zhur. Ob. Khim. 38 (1968),1205-1207
N.P. Grechi and V.I. Kuznetsov, A.E. Arbuzov. Moscow 1977

ARBUZOV, BORIS ALEKSANDROVICH [1903-]
STC 1,43-44; IWW 1982-1983, p.43; WWWS, 58-59; POGG 6,69; 7b,110-117
G.C. Kamay and V.A. Kurtin, Zhur. Ob. Khim. 33 (1963), 3455-3460

ARCHIBALD, REGINALD MacGREGOR [1910-]
AMS 14,126; WWA 1976-1977, p.91; WWWS, 59

ARCHINARD, JOHANN ISAAC FRANZ [1819-1890]
A. Wankmüller, Der Schweizer Familienforscher 33 (1966), 46-50

ARDENNE, MANFRED von [1907-]
KURSCHNER 11,49-50; TLK 3,12; POGG 6,72; 7a(1),50-52
M. von Ardenne, Ein glückliches Leben fUr Technik und Forschung. Munich 1972
S. Reball, Phys. Bl. 38 (1982), 19-20

ARENDS, GEORG [1862-1946]
NDB 1,344; HEIN, 11-12; KURSCHNER 4,48-49; TSCHIRCH, 1013; TLK 3,13; SCHELENZ, 722; WWWS, 60
Anon., Pharm. Z. 83 (1947), 9-11

ARENDT, RUDOLF [1828-1902]
NDB 1,345-346; BJN 7,7'; 8,379-381; BLOKH 1,15; POGG 3,39; 4,36; 5,32; 7a(1),53
F. Etzold, Ber. chem. Ges. 35 (1903), 4542-4549

ARENSON (PEISAKHOV-ABRAMOV), YAKOV ABRAMOVICH [1862-]
SG [3]1,724

ARFVEDSON, JOHANN AUGUST [1792-1841]
DSB 15,7-8; SBL 2,165-166; BLOKH 1,15; FERCHL, 13; SCHAEDLER, 3; POTSCH, 15; WWWS, 61; POGG 1,59-60

ARGUTINSKY-DOLGORUKOV, PETR MIKHAILOVICH [1850-1911]
KAZAN, 2,109-111; ZMEEV 4,11; Suppl.,1; HALLE, 294; RSC 13,146

ARHEIDT, RICHARD [-1924]
RSC 13,146
Anon., Ber. chem. Ges. 58A (1925), 28

ARIGONI, DUILIO [1928-]
KURSCHNER 13,64; POTSCH, 15

ARINKIN, MIKHAIL INNOKENTIEVICH [1876-1948]
BME 2,581-582
G.N. Chekulaev, Klin. Med. 51(10) (1973), 142-143
V.A. Beier, Klin. Med. 54(10) (1976), 137-140

ARKEL, ANTON EDUARD van [1893-1976]
BWN 2,13-14; POTSCH, 15-16; WWWS, 1706; POGG 6,74-75; 7b,123-124
E.W. Gorter and F.C. Romeyn, Chem. Wkbl. 60 (1964), 298-308
E.J.W. Verwey, Jaarb. Akad. Wet. 1976, pp.184-192

ARKHANGELSKY, KONSTANTIN FEDOROVICH [1870-1905]
KAZAN 2,124-126

ARKWRIGHT, JOSEPH ARTHUR [1864-1944]
DNB [1941-1950], 14-15; WW 1944, p.74; WWWS, 62
C.J. Martin, Obit. Not. Fell. Roy. Soc. 5 (1945), 127-140
S.P. Bedson, J. Path. Bact. 58 (1946), 134-147

ARLOING, SATURNIN [1846-1911]
DBF 3,648-649; FISCHER 1,37; GUIART, 127-128; WWWS, 62
J.F. Heymans, Arch. Inter. Pharmacodyn. 22 (1911), 1-25
O.M. Lannelongue, Bull. Acad. Med. 65 (1911), 418-420
Anon., Lyon Médical 1 (1911), 624-629,729

ARLT, FERDINAND CARL [1812-1887]
NDB 1,352-353; ADB 46,38; OBL 1,28; STURM 1,25-26; ROTH, 1-19; GE 3,981 HIRSCH 1,197-198; Suppl.,40; PAGEL, 42-43; HEID, 34-35
F. Arlt, Meine Erlebnisse. Wiesbaden 1887
E. Meyer, Ann. Ocul. 50 (1887), 1-24
F. Dimmer, Wiener klin. Wchschr. 25 (1912), 607-608
E. Fuchs, Wiener med. Wchschr. 75 (1925), 2477-2478
A. Elschnig, Sudetendeutsche Lebensbilder 3 (1934), 205-207

ARMES, HENRY PERCY [1884-1951]
AMS 7,48; SRS, 36; NUC 21,345

ARMSBY, HENRY PRENTISS [1853-1921]
DAB 1,349-350; CHITTENDEN, 60-61; WWWS, 63
F.G. Benedict, Biog. Mem. Nat. Acad. Sci. 19 (1938), 271-284
R.W. Swift, J. Nutrition 54 (1954), 3-16

ARMSTRONG, EDWARD FRANKLAND [1878-1945]
DSB 1,286-287; WW 1945, pp.77-78; POGG 6,76; 7b,126-127
C.S. Gibson and T.P. Hilditch, Obit. Not. Fell. Roy. Soc. 5 (1948), 619-631

ARMSTRONG, HENRY EDWARD [1848-1937]
DSB 1,288-289; DNB [1931-1940], 16-17; FINDLAY, 58-95; POTSCH, 16; W, 18-19; FARBER, 877-906; POGG 4,37; 5,33; 6,127-128; 7b,127-128
C.A. Browne, Ind. Eng. Chem. News Ed. 15 (1937), 344
Anon., Chem. Ind. 1937, pp.668-670; Pharm. J. 139 (1937), 63
E.H. Rodd, J. Chem. Soc. 1940, pp.1418-1439
E.F. Armstrong, Chem. Ind. 1941, pp.80-88
K.W. Keeble, Obit. Not. Fell. Roy. Soc. 3 (1941), 229-245
H. Hartley, Chem. Ind. 1945, pp.398-402,406-410
J.V. Eyre, Henry Edward Armstrong. London 1958
W.H. Brock, H.E. Armstrong and the Teaching of Science. London 1973

ARMSTRONG, PHILIP BROWNELL [1898-1980]
AMS 12,167

ARMSTRONG, WALLACE DAVID [1905-1984]
AMS 15(1),165

ARNAUD, ALBERT [1853-1915]
DBF 3,783-784; POGG 4,37; 6,33
L. Maquenne, L'Oeuvre Scientifique de M. Albert Arnaud. Paris 1915

ARNDT, FRITZ [1885-1969]
BHDE 2,31-32; WIDMANN, 254; POTSCH, 16-17; LDGS, 16; POGG 5,33-34; 6,76-77; 7a(1),55-56
B. Eistert, Chem. Z. 79 (1955), 442-444
E. Campaigne, J. Chem. Ed. 36 (1959), 336-339
W. Walter and B. Eistert, Chem. Ber. 108 (1975), i-xliv

ARNDT, KURT [1873-1946]
POTSCH, 17; TLK 3,14; POGG 5,34; 6,77; 7a(1),56
Anon., Z. angew. Chem. 56 (1943), 223; 59 (1947), 64
A. Pollack, Chem. Z. 67 (1943), 326-327

ARNETH, JOSEPH [1873-1955]
KURSCHNER 8,41; FISCHER 1,39
A. Piney, Medical Bookman 1(10) (1947), 19-21

ARNHOLD, EDUARD [1849-1925]
TETZLAFF, 10; NUC 22,23

[J. Arnhold], Ein Gedenkbuch: Eduard Arnhold. Berlin 1928
W. Treue, Tradition 5 (1960), 65-85,97-115

ARNOLD, FRIEDRICH [1803-1890]
NDB 1,383-384; DRULL, 4-5; BAD 1,8-10; HIRSCH 1,210-211; Suppl.,43
K. von Bardeleben, Anat. Anz. 5 (1890), 397-405
Anon., Leopoldina 26 (1890), 165-167
M. Fürbringer, in Heidelberger Professoren des 19. Jahrhunderts, Vol.2, pp.51-67. Heidelberg,1903
R. Hildebrand, Würzburger medizinhistorische Mitteilungen 6 (1988), 323-326

ARNOLD, HERBERT [1909-1973]
KURSCHNER 11,56-57; POGG 7a(1),57-58
Anon., Chem. Z. 97 (1973), 274

ARNOLD, JULIUS [1835-1915]
DBJ 1,321; DRULL, 5; HIRSCH 1,211-212; Suppl.,43; BK, 161-162; DZ, 27
E. Schwalbe, Beitr. path. Anat. Suppl.7 (1905), 777-795
P. Ernst, Münch. med. Wchschr. 62 (1915), 370-372; Fol. Haemat. 19 (1915), 220-225
E. von Gierke, Verhandl. path. Ges. 28 (1935), 337-340

ARNOLD, KARL [1853-1929]
DBJ 11,343; HEIN, 13-14; TSCHIRCH, 1014; TLK 2,17; RSC 12,23; 13,159; POGG 4,38; 5,35; 6,78
P.W. Danckwortt, Ber. chem. Ges. 62A (1929), 90-91

ARNOLD, VINCENZ [1864-1942]
STURM 1,27; FISCHER 1,39; RSC 13,161
W. Mozolowski, Post. Bioch. 12 (1966), 587-596

ARNOLD, WILLIAM ARCHIBALD [1904-]
AMS 14,137; WWWS, 65

ARNON, DANIEL ISRAEL [1910-]
AMS 14,138; MH 2,11-13; MSE 1,29-30; STC 1,50-51; CB 1955, pp.28-30
D.I. Arnon, TIBS 9 (1984), 258-262

ARNSTEIN, KARL AUGUSTOVICH [1840-1919]
BSE 3,111; KAZAN 2,115-124; VENGEROV 1,751-752,978-980

ARNY, HENRY VINECOME [1868-1943]
AMS 7,52; NCAB 33,88; E,128; WWAH, 22; WWWS, 66; TSCHIRCH, 1014
Anon., Science 98 (1943), 423; Am. J. Pharm. 115 (1943), 446

ARON, HANS [1881-1958]
FISCHER 1,40; KOREN, 148; LDGS, 76; POGG 7a(1),58; (4),135*
Anon., J. Am. Med. Assn. 169 (1959), 967

ARON, MAX [1892-1974]
DBA 1,64; IB, 9
E. Legait, Ann. Med. Nancy 14 (1975), 549-550

ARONHEIM, BERTHOLD [1850-1881]
NUC 22,383; RSC 7,48-49; 9,72; 13,163; POGG 3,42
A.W. Hofmann, Ber. chem. Ges. 14 (1881), 863

ARONHEIM, FELIX [1843-1913]
NUC 22,383; RSC 7,49; SG [1]1,562

ARONSON, HANS [1865-1919]
FISCHER 1,40; PAGEL, 1922-1923; KOREN, 148; RSC 13,163
Anon., Chem. Z. 43 (1919), 922

ARONSTEIN, LUDWIG [1841-1913]
BLOKH 1,16; POGG 3,43; 4,39; 5,35

ARPPE, ADOLF EDUARD [1818-1894]
ENKVIST, 53-62; BLOKH 1,16; POTSCH, 17; WWWS, 66; FERCHL, 14; POGG 1,66; 3,43; 4,39

ARRHENIUS, SVANTE AUGUST [1859-1927]
DSB 1,296-302; BUGGE 2,443-462; MAYERHOFER 1,282-283; POTSCH. 17-18; W, 19-20; WWWS, 66; ABBOTT-C, 9-10; POGG 4,39-40; 5,35-36; 6,79
W. Ostwald, Z. phys. Chem. 69 (1909), v-xxvii
R. Lorenz, Z. angew. Chem. 40 (1927), 1461-1465
J.W., Proc. Roy. Soc. A119 (1928), ix-xix
J. Walker, J. Chem. Soc. 1928, pp.1380-1401
E.H. Riesenfeld, Ber. chem. Ges. 63A (1930), 1-40; Svante Arrhenius. Leipzig 1931
H. von Euler, Chem. Ind. 1959, pp.245-249
S. Rochietta, Minerva Med. 50 (1959), 2050-2053
L.P. Rubin, J. Hist. Med. 35 (1980), 397-425d'

ARSONVAL, JACQUES ARSÈNE [1851-1940]
DSB 1,302-305; FISCHER 1,40; WWWS, 410; POGG 4,40-41; 5,261; 6,79; 7b,130
L. Chauvois, d'Arsonval. Paris,1941
A. Strohl, Bull. Acad. Med. 124 (1941), 213-222 (1941); Presse Med. 49 (1941), 383-396
P. Chevenard et al., Not. Acad. Sci. 3 (1957), 286-307

ARTELT, WALTER [1906-1976]
KURSCHNER 11,59; POGG 7a(1),58-59
H.H. Eulner et al., Medizingeschichte in unserer Zeit, pp.466-477. Stuttgart 1971
Anon., J. Hist. Med. 31 (1976), 467-468

ARTH, GEORGES MARIE FLORENT [1853-1909]
DBF 3,1163; BLOKH 1,16; POGG 4,41; 5,36
A. Haller, Bull. Soc. Chim. [4]7 (1910), i-x

ARTHUS, NICOLAS MAURICE [1862-1945]
DSB 17,33-35; FISCHER 1,40-41; IB, 9; WWWS, 67
M. Arthus, Bull. Hist. Med. 14 (1943), 366-390
H. Roger, Presse Med. 53 (1945), 261-262
L.R. Binet, Bull. Acad. Med. 129 (1945), 374-376

ARTOM, CAMILLO [1893-1970]
AMS 11,150; NCAB 55,477-478; COHEN, 10; IB, 9; WWAH, 22; WWWS, 68
Anon., J. Am. Med. Assn. 212 (1970), 1388

ARTUS, WILLIBALD [1809-1880]
HEIN, 14-15; FERCHL, 14; TSCHIRCH, 1014; SCHELENZ, 720; POGG 1,67; 3,45

ARZT, LEOPOLD [1883-1955]
FISCHER 1,41; SIEGL, 67-97; WIEN 1956, pp.47-49

ASAHINA, YASUHIKO [1881-1975]
JBE, 69; IB, 9; POGG 6,81-82; 7b,134-137
S. Shibata, Chem. Pharm. Bull. 19 (1971), v-vii
S. Kurokawa, Lichenologist 8 (1976), 93-94

ASAI, TOSHINOBU [1902-]
JBE, 71; WWWS, 69

ASANO, MITIZO [1894-1948]
POGG 6,82; 7b,137-138

ASAYAMA, CHUAI [1879-]
HAKUSHI 2,234

ASBRAND, KARL [1897-1925]
BLOKH 1,11
W. Biltz, Ber. chem. Ges. 59A (1926), 1

ASCHAFFENBURG, RUDOLF [1902-]
LDGS, 16; NUC 23,246

ASCHAN, OSSIAN [1860-1939]
ENKVIST, 84; POSCH, 18; WWWS,69; POGG 4,42-43; 5,37; 6,82
W. Hückel, Ber. chem. Ges. 74A (1941), 189-220

ASCHER, MAX [-1908]
RSC 7,51; NUC 23,272
Anon., Chem. Z. 32 (1908), 1271

ASCHERSON, PAUL [1834-1913]
NDB 1,412-413; BJN 18,75*-76*
E. Roth, Leopoldina 49 (1913), 35-37
L. Wittmack, Ber. bot. Ges. 31 (1913), (102)-(110)

ASCHHEIM, SELMAR [1878-1965]
KURSCHNER 10,48; FISCHER 1,43; BHDE 2,37; LDGS, 63; WININGER 7,514; TETZLAFF, 11; KOREN, 148; WWWS, 69
W. Hohlweg, Z. Gyn. 87 (1965), 1025-1026
A. Ravina, Presse Med. 73 (1965), 1375-1376

ASCHNER, BERNHARD [1883-1960]
FISCHER 1,43; MEDVEI, 703-704; LORENZSONN, 162-179; WININGER 7,514; TETZLAFF, 12
A.W. Bauer, Brit. Med. J. 1960(II), p.73
K.B. Absolon, Surgery 48 (1960), 979-983

ASCHOFF, ERNST FRIEDRICH [1792-1863]
HEIN, 16-17; HIRSCH 1,222; FERCHL, 15; TSCHIRCH, 1016; WAGENITZ, 18
Anon., Arch. Pharm. 179 (1863), 1-5

ASCHOFF, JÜRGEN [1913-]
KURSCHNER 13,75; POGG 7a(1),60
Anon., Leopoldina [3]24 (1981), 45

ASCHOFF, LUDWIG [1866-1942]
NDB 1,413; FREUND 2,13-21; BAD NF2,7-9; MAYERHOFER 1,289; FISCHER 1,44; PAGEL, 53-54; SCHWERTE 2,60-67; FRANKEN, 174; W, 20; DZ, 29; IB,9-10
W. Koch, Deutsche med. Wchschr. 68 (1942), 785-786
J.W. McNee et al., J. Path. Bact. 55 (1943), 229-236
L. Aschoff, Ein Gelehrtenleben in Briefen an die Familie. Freiburg 1966
F. Büchner, Verhandl. path. Ges. 50 (1966), 475-489

ASCHOFF, LUDWIG ADOLPH [1807-1861]
HEIN, 16-17; FERCHL, 15; POGG 3,46
Anon., Arch. Pharm. 168 (1861), 209-211; 179 (1863), 507

ASCOLI, ALBERTO [1877-1957]
DBI 34,187-190; FISCHER 1,44; GRMEK, 27; KOREN,148; COHEN, 10-11
Anon., Presse Med. 65 (1957), 1830; Ann. Inst. Pasteur 93(1957),681-682; Science 126 (1957), 1010; Bull. Acad. Med. Belg. [6]22(1957),490-491

ASCOLI, MAURIZIO [1876-1958]
FISCHER 1,44; KOREN, 148; IB, 10
Chi É? 1948, p.43
G. Frada, Riforma Medica 72 (1958), 1259-1262
C. Bevilacqua, Lanternino 11(6) (1988), 6-7

ASENJO, CONRADO FEDERICO [1908-1989]
AMS 14,145; WWWS, 70

ASHBY, JOHN EYRE [1820-1863]
FERCHL, 15; POGG 3,46

ASHBY, WINIFRED MAYER [1879-1975]
AMS 10,118; WWWS, 70
V.F. Fairbanks, The Mayo Alumnus 11(2) (1975), 28-33; Blood 46 (1975), 977-978

ASHDOWN, HERBERT HARDING [1859-1893]
O'CONNOR(2), 381-382
Anon., Brit. Med. J. 1893(II), pp.875-976

ASHER, LEON [1865-1943]
FISCHER 1,45-46; KOREN, 149; IB, 10; POGG 7a(1),60-61
A. von Muralt, Mitt. Nat. Ges. Bern NF1 (1944), 168-170; Erg. Physiol. 46 (1950), 1-5

ASHFORD, BAILEY KELLY [1873-1934]
DAB [Suppl.3],32-33; FISCHER 1,46; NCAB A,29-30; DAMB, 19-20; IB, 10; ULLMANN, 15-16; WWAH, 22-23; WWWS, 70
G.W. Bachmann, Science 80 (1934), 471
R. Ciferri and P. Redaelli, Riv. Biol. 21 (1936), 481-490

ASHFORD, CHARLES AMOS [1907-]
Who's Who of British Scientists 1971-1972, p.26

ASHWELL, GEORGE GILBERT [1916-]
AMS 15(1),183; WWA 1980-1981, p.116

ASKANASY, SELLY [1866-]
KURSCHNER 4,63; WININGER 6,427; KOREN, 149

ASKANAZY, MAX [1865-1940]
KURSCHNER 4,62-63; FISCHER 1,47; GENEVA 5,153-156; 6,184-186; KOREN, 149; WININGER 1,172; SCHOLZ, 263-264; KROLLMANN, 856; RAZINGER, 10-14; IB, 10; WWWS, 71
K. Staehelin, Gastroent. 66 (1940), 305-306
E. Rütishauser, Rev. Med. Suisse Rom. 61 (1941), 245-251; Schw. med. Jahrb. 1941, pp.xix-xxv
P. Huebschmann, Verhandl. path. Ges. 40 (1956), 359-377

ASKENASY, EUGEN [1845-1903]
NDB 1,417; BJN 8,8*; BAD 6,70-74; DRULL, 7-8; WININGER 1,173; WWWS, 71
M. Möbius, Ber. bot. Ges. 21 (1903), (47)-(66)

ASKONAS, BRIGITTE ALICE [1925-]
WW 1987, p.56; IWW 1987-1988, p.60
B. Askonas, Ann. Rev. Immunol. 8 (1990), 1-21

ASO, KEIJIRO [1875-1953]
JBE, 87,2311

ASTASHEVSKI, PAVEL PETROVICH [1845-]
KAZAN 2,126; VENGEROV 1,844; ZMEEV 4,15; RSC 13,188

ASTAUROV, BORIS LVOVICH [1904-1974]
DSB 17,35-39; BSE (3rd Ed.) 2,333; IWW 1974-1975, p.69; STC 1,55-57; WWWS, 1851
A.E. Gaissinovich, Folia Mendeliana 10 (1975), 247-252
R.L. Berg, Quart. Rev. Biol. 54 (1979), 397-416

ASTBURY, WILLIAM THOMAS [1898-1961]
DSB 1,319-320; DNB [1961-1970],41-42; W, 20-21; POSCH, 19; WWWS, 72; POGG 6,84; 7b,145-148
K. Lonsdale, Chem. Ind. 1961, pp.1174-1175
I. McArthur, Nature 191 (1961), 331-332
K. Bailey, Adv. Protein Chem. 17 (1962), x-xiv
J.D. Bernal, Biog. Mem. Fell. Roy. Soc. 9 (1963), 1-35
R. Olby, The Path to the Double Helix, pp.417-470. London 1974
J.A. Witkowski, Notes Roy. Soc. 35 (1980), 195-219
M. Davies, Ann. Sci. 47 (1990), 607-618

ASTIÉ, HERMANN [1860-1903]
JV 4,223; RSC 13,188

ASTIER, CHARLES BERNARD [1771-1837]
DBF 3,1343-1344; LAROUSSE 1,826; BALLAND, 3-4; BALLAND(2), 251-252; HAMON 1,132-133; CALLISEN 1,267; 26,95; POGG 1,70-71,1530
A. Rochas, Biographie du Dauphiné 1,43-44. Paris 1856

ASTON, EMILY ALICIA [1866-]
SRS, 18; RSC 13,188
M.R.S. Creese, Brit. J. Hist. Sci. 24 (1991), 288

ASTON, FRANCIS WILLIAM [1877-1945]
DSB 1,320-322; DNB [1941-1950],24-26; STC 1,57-58; POSCH, 19; W, 21-22; WWWS, 73; ABBOTT-C, 10-12; POGG 5,39; 6,84; 7b,148-149
G.P. Thomson, Nature 157 (1946), 290-292
G. Hevesy, Obit. Not. Fell. Roy. Soc. 5 (1948), 635-650;

J. Chem. Soc. 1948, pp.1468-1475

ASTRE, CHARLES [1851-1929]
DORVEAUX, 55,72,94; TSCHIRCH, 1014; POGG 4,45; 5,39; 6,84-85
A. Astruc, J. Pharm. Chim. [8]9 (1929), 592-594

ASTRUC, ALBERT [1875-1956]
POGG 6,85; 7b,149-150
J. Giroux, Ann. Pharm. Fran. 14 (1956), 728-737
M.M. Janot, Bull. Acad. Med. 140 (1956), 536-540

ASTRUP, POUL [1915-]
KBB 1981, pp.52-53
R. Dybkaer, J. Clin. Chem. 19 (1981), 903-910

ASTRUP, TAGE [1908-]
KBB 1981, P.53; WWWS, 73

ASTWOOD, EDWIN BENNETT [1909-1976]
AMS 13,134; MH 2,13-14; MSE 1,30-31; STC 1,58-59; WWWS, 73
C.E. Cassidy, Endocrinology 99 (1976), 1155-1160
R.O. Greep, Rec. Prog. Hormone Res. 33 (1977), xiii-xix
R.O. Greep and M.A. Greer, Biog. Mem. Nat. Acad. Sci. 55 (1985), 3-42

ATKINS, WILLIAM RINGROSE GELSTON [1884-1959]
WW 1959, p.108; DESMOND, 22; IB, 10; WWWS, 74; POGG 6,85-86; 7b,151-153
F.S. Russell, Nature 183 (1959), 1228
H. Poole, Biog. Mem. Fell. Roy. Soc. 5 (1960), 1-22

ATKINSON, DANIEL EDWARD [1921-]
AMS 15(1),190; WWA 1980-1981, P.121; WWWS, 74

ATLEE, WALTER FRANKLIN [1828-1910]
R. Jansen, Penn. Med. J. 64 (1961), 386-387

ATTERBERG, ALBERT [1846-1916]
SBL 2,411-416; SMK 1,152-153
Anon., Sven. Kem. Tid. 28 (1916), 188-189

ATTFIELD, JOHN [1835-1911]
REBER, 367-368; TSCHIRCH, 1014; WWWS, 74; POGG 3,48-49; 5,40
A.C., J. Chem. Soc. 101 (1912), 688-691;
Proc. Roy. Soc. A86 (1912), xliv-xlvi
M.B. Mrtek, Pharm. Hist. 29 (1987), 55-59

ATWATER, WILBUR OLIN [1844-1907]
DSB 1,325-326; DAB 1,417-418; CHITTENDEN, 54-60; DAMB, 21-22; WWAH, 24; WWWS, 74
F.G. B[enedict], Science 26 (1907), 523-524
L.A. Maynard, J. Nutrition 78 (1962), 3-9
P.W. Wilson, Bact. Revs. 27 (1963), 405-416
A.J. Ihde and J.F. Janssen, Mol. Cell. Biochem. 5 (1974), 11-16
W.J. Darby, Nutr. Revs. 14 (1976), 1-14
N. Aronson, Social Problems 29 (1982), 474-487

ATZLER, EDGAR [1887-1938]
FISCHER 1,49; SKVARC, 14-16; TLK 3,15; IB, 10; POGG 7a(1),63-64
Anon., Chem. Z. 62 (1938), 774
E. von Skramlik, Deutsche med. Wchschr. 64 (1938), 1660-1661
G. Lehmann, Med. Welt 12 (1938), 1549-1550;
Erg. Physiol. 41 (1939), v-vi; Arbeitsphys. 10 (1939), 351-352

AUB, JOSEPH CHARLES [1890-1973]
AMS 12,197; IB, 10; KOREN, 149
P.C. Zamecnik, Trans. Assoc. Am. Phys. 87 (1974), 12-14
A.M. Harvey, American Journal of Medicine 71 (1981), 13-15
N.L.R. Bucher, Harvard Med. Alumni Bull. 61(3) (1987), 46-51

AUBEL, EUGÈNE [1884-1975]
IB, 10; POGG 6,87; 7b,155-156
M. Grunberg-Manago and R. Wurmser, Reg. Biochim. 1977, pp.2-5

AUBERGIER, PIERRE HECTOR [1809-1884]
DBF 3,1498-1499; RSC 1,112; 7,57; NUC 25,370

AUBERT, HERMANN [1826-1892]
ADB 46,81; HIRSCH 1,238; Suppl.,48; PAGEL, 58-59; POGG 3,49
Anon., Leopoldina 28 (1892), 58
A. Wilhelmi, Mecklenburgische Aerzte, pp.137-138. Schwerin,1901

AUBRY, ANDRÉ [1889-]
GORIS, 214
Who's Who in Europe 1972, p.1155

AUBRY, LUDWIG [1844-1901]
BJN 6,8*; BLOKH 1,18
Anon., Ber. chem. Ges. 34 (1901), 4061; Chem. Z. 25 (1901), 1081

AUDOUIN, JEAN VICTOR [1797-1841]
DSB 1,328-329; DBF 4,432-433; FERCHL, 16; WWWS, 75; POGG 1,73-74
P.M.T. de Serre et al., Ann. Sci. Nat. (Zool.) [2]16 (1841), 356-378
P.A.J. Duponchel, Ann. Soc. Entomol. 11 (1842), 95-164
Anon., Abeille 26 (1889), 232-239
W. Schoenfeld, Deutsche med. Wchschr. 82 (1957), 1235-1237

AUDRIETH, LUDWIG FREDERICK [1901-1967]
AMS 11,165; POSCH, 19-20; WWAH, 24; WWWS, 75
J. Kleinberg, J. Inorg. Chem. 35 (1973), 1757-1768

AUER, ALOYS [1888-1948]
KALLMORGEN, 211; NUC 25,611; SG [4]1,863

AUER, JOHN [1875-1948]
DAB [Suppl.4],34; AMS 7,58; NCAB 37,302-303; PARASCANDOLA, 28; CHEN, 29; DAMB, 22-23; WWAH, 24; WWWS, 75; IB, 10
R. Kinsella, Trans. Assoc. Am. Phys. 61 (1948), 6-7
G.B. Roth, J. Pharm. Exp. Ther. 95 (1949), 285-286

AUERBACH, ALEXANDER [1852-1904]
WREDE, 6; RSC 9,83; 13,198
Anon., Leopoldina 40 (1904), 94

AUERBACH, CHARLOTTE [1899-]
WW 1981, pp.95-96; IWW 1982-1983, pp.57-58; MSE 1,32-33; BHDE 2,39
F.H. Sobels, Mutation Research 29 (1975), 171-180
C. Auerbach, Heredity 40 (1978), 177-187;
Persp. Biol Med. 21 (1978), 319-334

AUERBACH, FRIEDRICH [1870-1925]
NDB 1,433-434; BLOKH 1,18; BHDE 2,39; WININGER 1,187; WWWS, 76; POGG 5,40-41; 6,88-89
H. Pick, Z. Elektrochem. 31 (1925), 569-571
M. Mugdan, Chem. Z. 49 (1925), 689; Ber. chem. Ges. 60A (1927), 141-152

AUERBACH, LEOPOLD [1828-1897]
NDB 1.434; ADB 46,85-87; BJN 2,34-35; 4,54*-55*; WININGER 1,189-190; HIRSCH 1,243-244; Suppl.,49; KOREN, 149; WWWS, 76
G. Born, Anat. Anz. 14 (1898), 257-267
B. Kisch, Trans. Am. Phil. Soc, 44 (1954), 297-313

AUERBACH, LEOPOLD [1857-1936]
FISCHER 1,50; ARNSBERG, 32; KALLMORGEN, 211; WININGER 7,515; KOREN, 149

AUFHÄUSER, DAVID [1878-1949]
LDGS, 16; NUC 25,678; POGG 7a(1),65

AUGER, VICTOR [1864-1949]
CHARLE, 28-30; POGG 6,28-30; 7b,161-162
L. Hackspill, Ann. Univ. Paris 20 (1950), 65-68

AUGUSTIN, BÉLA [1877-1954]
MEL 1,62; IB, 11; TSCHIRCH, 1015
Das Geistige Ungarn 1 (1918), 34

P. Gulyas, Magyar Irok 1 (1939), 910-912

AUHAGEN, ERNST [1904-]
KURSCHNER 13,84
Anon., Chem. Z. 93 (1969), 875; 98 (1974), 565
E. Auhagen, TIBS 7 (1982), 225-226

AURBACH, GERALD DONALD [1927-1991]
AMS 18(1),227

AUSTIN, JAMES HAROLD [1883-1952]
AMS 8,83; WWAH, 25; IB, 11
W.C. Stadie, Trans. Assoc. Am. Phys. 65 (1952), 7-8
O. Pepper and H. Perry, Trans. Coll. Phys. Phila. [4]21 (1953), 25-26

AUSTRIAN, CHARLES ROBERT [1885-1956]
AMS 8,84
A.M. Chesney, Trans. Assoc. Am. Phys. 70 (1957), 11-12

AUTENRIETH, JOHANN HEINRICH FERDINAND von [1772-1835]
NDB 1,460-461; ADB 1,695-696; SL 5,149-160; HIRSCH 1,249-252; POTSCH, 20;KILLIAN, 205; CALLISEN 1,301-306; 26,105; POGG 1,76; 7aSuppl.,42-43
E. Stübler, J.H.F. von Autenrieth. Stuttgart 1948
P. Bihl, Der Gerichtsmediziner F.J.H. Autenrieth. Tübingen 1974
J. Neumann, Med. Hist. J. 20 (1985), 66-82

AUTENRIETH, WILHELM LUDWIG [1863-1926]
NDB 1,461; HEIN, 20-21; BLOKH 1,18-19; TSCHIRCH, 1015; POGG 4,47; 5,43; 6,93; 7a(1),67
H. Kiliani, Ber. chem. Ges. 59A (1926), 33-34

AUTERHOFF, HARRY [1915-1983]
HEIN-E, 12-13; DRUM, 304-306; POTSCH, 20-21; POGG 7a(1),67

AUWERS, KARL FRIEDRICH von [1863-1939]
DSB 1,340-341; NDB 1,463; SCHMITZ, 294-302; MEINEL, 270-304,497; POTSCH, 21; AUERBACH, 766-767; DRULL, 8; TLK 3,16; WWWS, 1728; DZ, 33; POGG 4,49-50; 5,44-46; 6,93-95; 7a(1),68-69
H. Meerwein, Ber. chem. Ges. 72A (1939), 111-121
J. Thorpe, Chem. Ind. 58 (1939), 838-840

AUZIAS-TURENNE, ALEXANDRE [1812-1870]
DBF 4,801; HIRSCH 1,253-254

AVAEV, PETR MIKHAILOVICH [1867-1925]
BLOKH 1,19-20

AVERBECK, HEINRICH [1884-]
JV 25,385; NUC 27,473

AVERY, OSWALD THEODORE [1877-1955]
DSB 1,342-343; DAB [Suppl.5], 25-26; DAMB, 23-24; MH 2,15-17; MSE 1,34-35; STC 1,60-61; NCAB 44,491-492; ABBOTT, 10-11; WWWS, 78
R.J. Dubos, Biog. Mem. Fell. Roy. Soc. 2 (1956), 35-48;
The Professor, the Institute, and DNA. New York 1976
C. MacLeod, J. Gen. Microbiol. 17 (1957), 539-549
A.R. Dochez, Biog. Mem. Nat. Acad. Sci. 32 (1958), 32-49
R.D. Hotchkiss, Genetics 51 (1965), 1-10
A.F. Coburn, Persp. Biol. Med. 12 (1969), 623-630
W.M. Stanley, Arch. Env. Hlth. 21 (1970), 256-262
A.W. Downie, J. Gen. Microbiol. 73 (1972), 1-11
H.V. Wyatt et al., Nature 235 (1972), 86-89; 239 (1972), 234,295-296
N.W. Pirie, Nature 240 (1972), 572
W.F. Goebel, Persp. Biol. Med. 18 (1975), 419-426M. McCarty, Ann. Rev. Gen. 14 (1980), 1-15; Proc. Am. Phil. Soc. 128 (1984), 20-26; The Transforming Principle. New York 1985
A.M. Diamond, Persp. Biol. Med. 26 (1982), 132-136
N. Russell, Brit. J. Hist. Sci. 21 (1988), 393-400

AVERY, ROY CROWDY [1885-1971]
AMS 11,174

AVOGADRO, AMEDEO [1776-1856]
DSB 1,343-350; DBI 4,689-707; PROVENZAL, 113-120; MAYERHOFER 1,350-351; W, 23; SCHAEDLER, 4; FERCHL, 17; BLOKH 1,20-22; ABBOTT-C, 12-13; POTSCH, 21; WWWS, 78; POGG 3,52-53
I. Guareschi, Amedeo Avogadro. Basle 1903
A. Meldrum, Avogadro and Dalton. Aberdeen 1904
C.N. Hinshelwood and L. Pauling, Science 124 (1956), 708-713
N.G. Coley, Ann. Sci. 20 (1964), 195-210
S.H. Mauskopf, Isis 60 (1969), 61-74
J.H. Brooke, Hist. Sci. 19 (1981), 235-273
N. Fisher, Hist. Sci. 20 (1982), 77-102,212-231
M. Morselli, Amedeo Avogadro. Dordrecht 1984

AWE, WALTHER [1900-1968]
HEIN-E, 13-14; KURSCHNER 10,56; POGG 7a(1),69-70
W. Schneider, Arzneimitt. 10 (1960), 695-696
Anon., Chem. Z. 92 (1968), 366
W. Schulze, Arzneimitt. 18 (1968), 629-630

AWENG, EUGÈNE [1859-1929]
TSCHIRCH, 1015
G. Humbert, J. Pharm. Alsace 56 (1929), 188-192

AXELROD, JULIUS [1912-]
AMS 15(1),207; MSE 1,35-36; STC 1,61-62; IWW 1982-1983, p.61; POTSCH, 21; WWWS, 78; WWA 1980-1981, p.131
J. Axelrod, Ann. Rev. Pharm. 28 (1988), 1-23

AXHAUSEN, WALTER [1882-]
JV 20,20; NUC 27,680

AYCOCK, WILLIAM LLOYD [1889-1951]
AMS 8,86
Anon., New Eng. J. Med. 246 (1952), 158-159

AYKROYD, WALLACE RUDDELL [1899-1979]
WW 1979, p.97
R. Passmore, Brit. J. Nutrition 43 (1980), 245-250

AYRES, PHILIP BURNARD [1813-1863]
DESMOND, 24-25; RSC 1,129

AYRES, WILLIAM C [1853-1896]
NUC 28,135
Anon., N.Y. Med. J. 64 (1896), 491

AYRTON, WILLIAM EDWARD [1847-1908]
DNB [Suppl.2] 1,72-75; W, 23-24; WWWS, 80; POGG 3,53-54; 4,50; 5,46
J. Perry, Proc. Roy. Soc. A85 (1911), i-viii

B

BAAS, KARL [1866-1944]
FISCHER 1,53; KURSCHNER 4,71-72

BAAS-BECKING, LOURENS GERHARD MARINUS [1895-1963]
BWN 3,20-22; IB, 11
F.W. Went, Nature 198 (1963), 134

BABAK, EDWARD [1873-1926]
OBL 1,40; FISCHER 1,53; NAVRATIL, 7-9
F. Studnicka et al., Biol. Listy 12 (1926), 161-205,307-320
V. Suk, Anthropologie 5 (1927), 1-3
V. Kruta, Scripta Medica 39 (1966), 285-288
J. Krecek, Cesk. Fisiol. 22 (1973), 505-520

BABCOCK, ERNEST BROWN [1877-1954]
DSB 17,42-43; AMS 8,88; IB, 11; NCAB D,356; WWAH, 26; WWWS, 81
G.L. Stebbins, Biog. Mem. Nat. Acad. Sci. 32 (1958), 50-66

BABCOCK, STEPHEN MOULTON [1843-1931]
DSB 1,356-357; DAB 21,37; MILES, 14-15; WWAH, 26; WWWS, 81; W, 24-25; FARBER, 808-813
H.L. Russell, Science 74 (1931), 86-88
Anon., Ind. Eng. Chem. News Ed. 9 (1931), 213
E.B. Hart, J. Nutrition 37 (1949), 3-7
A.J. Ihde, in Perspectives in the History of Science and Technology (D.H.D. Roller, Ed.), pp.271-282. Norman, Okla. 1971

BABÉS, VICTOR [1854-1926]
HIRSCH 1,265-266; Suppl.,52; OLPP, 12-13; WWWS, 81
P. Bar, Bull. Acad. Med. 96 (1926), 178
B. Möllers, Klin. Wchschr. 6 (1927), 527
N. Cajal and V.T. Babés, Virologie 27 (1976), [2],79-81
J. Spielman and G.L. Georgescu, Noesis 5 (1979), 95-103

BABINSKI, JÓZEF [1873-1921]
PSB 1,195; RIGA, 325

BABKIN, BORIS PETROVICH [1877-1950]
AMS 8,89; FISCHER 1,54; IB, 12
Anon., Rev. Can. Biol. 10 (1951), 3-7
I. deB. Daly et al., Obit. Not. Fell. Roy. Soc. 8 (1952), 13-23

BABO, LAMBERT von [1818-1899]
NDB 1,480-481; ADB 46,151-154; BJN 4,126*; BAD 5,6-11; POTSCH, 21-22; W,25; WWWS,25; BLOKH 1,22-23; FERCHL, 28; SCHAEDLER, 4; POGG 1,83; 3,55; 4,51
H. Landolt, Ber. chem. Ges. 32 (1899), 1163-1164
Anon., Chem. Z. 23 (1899), 3552

BABUKHIN, ALEKSANDR IVANOVICH [1827-1891]
BSE 4,16; BME 3,9-11; HIRSCH 1,268; ZMEEV 1,10; 3,3; 4,18
A. Voltov, Meditsina 3 (1891), 357
A.N. Metelkin et al., A.I. Babukhin. Moscow,1955

BACH, ALEKSEI IVANOVICH [see BAKH]

BACH, STEPHEN JOSEPH [1897-1973]
BHDE 2,44; LDGS, 59
Who's Who of British Scientists 1971-1972, p.34
Anon., Chem. Z. 92 (1968), 13; Chem. Brit. 10 (1974), 189

BACHARACH, ALFRED LOUIS [1891-1966]
DNB [1961-1970], 59060; WW 1966, P.121
H. Jephcott, Chem. Ind. 1966, pp.1651-1653
S.K. Kon, Brit. J. Nutrition 21 (1967), 235-236
W.F.J. Cuthbertson, Proc. Soc. Anal. Chem. 4 (1967), 67-69

BACHE, FRANKLIN [1792-1864]
DAB 1,463-464; MILES, 15-16; NCAB 5,346; ELLIOTT, 20; COLE, 13-14; WWAH, 27; FERCHL, 18; POGG 3,56
E.F. Smith, J. Chem. Ed. 20 (1943), 367-368

BACHELARD, GASTON [1884-1962]
DSB 1,365-366; STC 1,65-67; WWWS, 82
G. Canghuilhem, Ann. Univ. Paris 33 (1963), 24-39
F. Dagognet, Gaston Bachelard. Paris 1985
D. Lecourt, L'Épistémologie Historique de Gaston Bachelard. Paris 1969

BACHEM, CARL [1880-1935]
KURSCHNER 4,75; FISCHER 1,55; WENIG, 8-9; TSCHIRCH, 1015; IB,12; TLK 3,18; GRUETTER, 17

BACHMANN, RUDOLF [1910-]
KURSCHNER 13,96; BUSCHHUTER, 66-73

BACHMANN, WERNER EMMANUEL [1901-1951]
DAB [Suppl.5], 28-29; MILES, 16-17; POTSCH, 22; WWWS, 83; POGG 6,100; 7b,172-175
R.C. Elderfield, Biog. Mem. Nat. Acad. Sci. 34 (1960), 1-30

BACKER, HILMAR JOHANNES [1882-1959]
BWN, 2,19; WWWS, 83-84; POGG 5,48; 6,101; 7b,176-179
F. Challenger, Nature 184 (1959), 1187; Proc. Chem. Soc. 1960,pp.126-128
G. Bertrand, C. R. Acad. Sci. 248 (1959), 3503-3506
H.J. Scheltema, Jaarboek der Rijksuniv. te Groningen 1959, pp.33-53
J.H. de Boer, Jaarb. Akad. Wet. 1958-1959, pp.297-301

BACKMAN, EUGENE LOUIS [1883-1965]
SBL 2, 541-544; SMK 1, 166; VAD 1965, pp.65-66; FISCHER 1,56

BACMEISTER, ADOLF [1882-1945]
FISCHER 1,56-57; KURSCHNER 6(I),49

BACON, JOHN STANLEY DURRANT [1917-]
Who's Who of British Scientists 1980-1981, p.17

BACOT, ARTHUR WILLIAM [1866-1922]
WW 1922, p.98; FISCHER 1,57; OLPP, 14-16
Anon., Nature 109 (1922), 618-620; Brit. J. Exp. Path. 3 (1922),117-125
A.H. Smith, Sci. Mon. 15 (1923), 359-363
M. Greenwood, J. Hygiene 22 (1924), 265-305

BACQ, ZENON MARCEL [1903-1983]
BAF, 2-3
R.A. Peters, Biochem. Pharm. 23 (1974), 1031
J. Lecomte, Rev. Med. Liège 38 (1983), 627-630; Ann. Acad. Roy. Belg. 151 (1985), 53-99

BADDILEY, JAMES [1918-]
WW 1981, pp.104-105; IWW 1982-1983, p.67; CAMPBELL, 11-12; WWWS, 84

BADGER, RICHARD McLEAN [1896-1974]
DSB 17,43-44; AMS 12,224; MH 2,19-20; MSE 1,40; WWWS, 85; POGG 6,102; 7b,182-183
O.R. Wolf, Biog. Mem. Nat. Acad. Sci. 56 (1987), 3-20

BĄDZYNSKI (BONDZYNSKI), STANISLAW [1862-1929]
PSB 1,379-381; IB, 12; KONOPKA 1,185-187; WARSAW, 138; RSC 13,670
S. Dąbrowski, Kosmos 49 (1924), 969-994

BAEKELAND, LEO HENDRIK [1863-1944]
DSB 1,384-385; DAB [Suppl.3], 25-27; MILES, 18-19; W, 26-27; WWWS, 85-86; WWAH, 27; FARBER, 1183-1190; ABBOTT-C, 13; POTSCH, 24
W.P. Cohoe, Am. Phil. Soc. Year Book 1944, pp.339-344;
Chem. Eng. News 23 (1945), [3],228-232,276
J. Gillis, Leo Hendrik Baekeland. Brussels,1965

BAER, ERICH [1901-1975]
AMS 11,189; BHDE 2,46; WWWS, 86
M. Kates, Chem. Phys. Lipids 17 (1976), 89-107
C.C. Lucas, Proc. Roy. Soc. Canada [4] 4 (1976), 41-47

BAER, FRIEDRICH [1908-]
KURSCHNER 13,104; POGG 7a(1), 76BAER, JULIUS [1876-1961]
ARNSBERG, 34; KURSCHNER 4,93; KALLMORGEN, 213-214

BAER, KARL ERNST von [1792-1876]
DSB 1,385-389; NDB 1,524; ADB 46,207-212; FREUND 1,77-91; W, 27; WWWS, 86; WELDING, 22-23; GOCZOL, 83-110; KROLLMANN, 26; PAGEL, 72-73; KUZNETSOV (1963), 56-72; MAYERHOFER 1,363-364; ABBOTT, 130-131; POGG 1,86; 3,59-60
B.I. Raikov, Karl Ernst von Baer. Leipzig 1968
F. Kuhn-Schnyder, Naturw. Rund. 30 (1977), 432-436

BAERENSPRUNG, FELIX von [1822-1864]
NDB 1,526; ADB 2,59-60; PAGEL, 75-76; HIRSCH 1,284-285; Suppl.,56; DIEPGEN, 127-141
O. Veit, Charité Annalen 12 (1865), 74-85
D.F. King, American Journal of Dermopathology 4 (1982), 39-48
A. Scholz, Z. ärzt. Fortbild. 83 (1989), 1137-1140

BAERWIND, HEINRICH [1892-1968]
JV 37,1097

Anon., Nachr. Chem. Techn. 5 (1957), 276

BÄUMER, JOHANNA [1887-]
JV 30,715; NUC 30,211

BÄUML, FRITZ [1887-]
JV 28,68; NUC 30,214

BÄUMLER, CHRISTIAN [1836-1933]
HIRSCH 1, 286; PAGEL, 76-77; KOVACSICS, 7-10; NUC 30,217-218

BAEYER, ADOLF von [1835-1917]
DSB 1,389-391; NDB 1,534-536; DBJ 2,215-218,646; DZ, 66; W, 27-28; MAYERHOFER 1,365-366; BUGGE 2,321-325; MATSCHOSS, 11-12; POTSCH, 24; BLOKH 1,23-27; ABBOTT-C, 13-14; SCHAEDLER, 5; BERLIN, 660; POGG 3,60-61; 4,54; 5,50; 6,105; 7a Suppl.,45-48
A. Baeyer, Gesammelte Werke. Braunschweig 1905
H. Wieland, Chem. Z. 39 (1915), 829-832
R. Willstätter et al., Naturwiss. 3 (1915), 559-589
R. Willstätter, Z. angew. Chem. 30 (1917) [I], 229-231; Bayer. Akad. Wiss. Jahrbuch 1918, pp.33-59
W.H. Perkin, J. Chem. Soc. 123 (1923), 1520-1546
K. Henrich, J. Chem. Ed. 7 (1930), 1231-1248
H. Rupe, Adolf von Baeyer als Lehrer und Forscher. Stuttgart 1932
F. Richter, Ber. chem. Ges. 68A (1935), 175-180
J.R. Partington, Nature 136 (1935), 669-670
K. Schmorl, Adolf von Baeyer. Stuttgart 1952
R. Huisgen, Z. angew. Chem. (Int. Ed.) 25 (1986), 297-311
J.S. Fruton, Contrasts in Scientific Style, pp.118-162. Philadelphia 1990

BAEYER, ERICH OTTO von [1909-]
GGTA 23 (1931), 20

BAEYER, HANS von [1875-1941]
DRULL, 9-10; FISCHER 1,58; BRUNKHORST, 147-151; JV 16,233; NUC 30,221

BAEYER, OTTO von [1877-1946]
NDB 1,537; POGG 5,50; 6,105; 7a(1),80
O. Hahn and G. Hettner, Naturwiss. 34 (1947), 193-194

BAGH, ALEXANDER von [1882-]
JV 24,568; NUC 30,294

BAGINSKY, ADOLF [1843-1918]
DBJ 2,680; HIRSCH 1,287-288; Suppl.,56; PAGEL,77-78; KOHUT 2,272; WREDE,7
A. Schlossmann, Münch. med. Wchschr. 60 (1913), 1095-1096
O. Winkelmann, Berl. Med. 16 (1965), 739-744

BAGLIONI, SILVESTRO [1876-1957]
DBI 5,247-249
A. Ciminata, Minerva Med. 48 (1957), 1660-1662

BAIERLACHER, EDUARD [1825-1889]
HIRSCH 1,289; Suppl.,57
O. Fuchs, Münch. med. Wchschr. 36 (1889), 863-865

BAIL, OSKAR [1869-1927]
FISCHER 1,58-59; MOTSCH, 143-157; KOREN, 146; KOERTING, 127
H. Braun, Klin. Wchschr. 7 (1928), 719

BAILAR, JOHN CHRISTIAN [1904-1991]
AMS 14,186; MH 1,18-19; MSE 1,42; CB 1959, pp.20-21; WWWS, 88; WWA 1980-1981, p.144; POGG 6,107; 7b,188-190
Anon., Chem. Eng. News 37 (1959) [1],103; 39 (1961) [11],94; 69 (1991) [43],46-47

BAILEY, CLYDE HAROLD [1887-1968]
AMS 11,194; IB, 12
Anon., Cereal Chemistry 45 (1968), [2],iii-iv

BAILEY, JACOB WHITMAN [1811-1857]
DAB 1,498; ELLIOTT, 21; HUMPHREY, 9-10; WWWS, 88; POGG 3,62

BAILEY, JAMES ROBINSON [1868-1941]
AMS 6,56; JV 13,222; NUC 30,557; RSC 13,239; 19,73
W.A. Felsing, Ind. Eng. Chem. News Ed. 17 (1939), 154

BAILEY, KENNETH [1909-1963]
WW 1959, p.128; POTSCH, 24-25
R.R. Porter, Nature 200 (1963), 520-521
A.C. Chibnall, Biog. Mem. Fell. Roy. Soc. 10 (1964), 1-13
S.V. Perry, Adv. Protein Chem. 20 (1965), 11-18

BAILEY, KENNETH CLAUDE [1896-1951]
WW 1949, p.118; WWWS, 89; POGG 6,107-108; 7b,191
J.H.J. Poole, Nature 168 (1951), 679-680

BAILEY, LIBERTY HYDE [1858-1954]
DSB 1,395-397; DAB [Suppl.5], 30-32; AMS 8,98; HUMPHREY, 10-15; NCAB 43,514-515; WWWS, 89; WWAH, 28
A.D. Rodgers, Liberty Hyde Bailey. Princeton,1949
W.J. Robbins, Am. Phil. Soc. Year Book 1955, pp.429-431
G.H.M. Lawrence, Nature 175 (1955), 451-452; Baileya 3 (1955), 27-40
P.A. Munz, Baileya 6 (1958), 85-89
C. Zirkle, J. Hist. Biol. 1 (1968), 205-218

BAILLY, OCTAVE [1886-1962]
GORIS, 215; POGG 6,109-110; 7b,197

BAIN, WILLIAM [1855-1936]
O'CONNOR(2), 186-187; NUC 31,84
Anon., Lancet 1936(II), pp.47-48

BAINBRIDGE, FRANCIS ARTHUR [1874-1921]
O'CONNOR(2), 262-264; FISCHER 1,59; WWWS, 90-91
E.H. Starling, Brit. Med. J. 1921 (II), pp.770-771
Anon., Lancet 1921 (II), p.980

BAIRD, SPENCER FULLERTON [1823-1887]
DSB 1,404-406; DAB 1,513-515; NCAB 3,405-406; ELLIOTT, 21; W, 29; WWWS, 91; WWAH, 29
Anon., Am. J. Sci. [3]34 (1887), 319-322
J.S. Billings, Biog. Mem. Nat. Acad. Sci. 3 (1895), 141-160
W.H. Dall, Spencer Fullerton Baird. Philadelphia 1915
H.M. Smith, Science 42 (1915), 425-428
E. Linton, Science 48 (1918), 25-34

BAISCH, KARL [1869-1943]
NDB 1,546; FISCHER 1,59-60; WWWS, 91

BAIST, LUDWIG [1825-1899]
HB 1,379-381
H.W. Flemming, Ludwig Baist, der Gründer der chemischen Fabrik Griesheim. Munich 1965 (Tradition 4. Beiheft)

BAITSELL, GEORGE ALFRED [1885-1971]
AMS 11,202; IB, 13

BAKER, HERBERT BRERETON [1862-1935]
DNB [1931-1940], 32-33; WWWS, 93; POGG 4,57; 5,51; 6,111; 7b,200
J.F. Thorpe, Obit. Not. Fell. Roy. Soc. 1 (1935), 523-526
J.C. Philip, J. Chem. Soc. 1935, pp.1893-1896
W.V. Farrar, Proc. Chem. Soc. 1963, pp.125-128

BAKER, JOHN RANDAL [1900-1984]
WW 1981, p.115; IB, 13; WWWS, 94
S.M. McGee-Russell and K.F.A. Ross (Eds.), Cell Structure and its Interpretation, pp.407-417. London,1968
The Times, 16 June 1984, p.12
E.N. Willmer and P.C.J. Brunet, Biog. Mem. Fell. Roy. Soc. 31 (1985), 33-63

BAKER, JOHN WILLIAM [1898-1967]
POGG 6,112; 7b,202-204
E. Rothstein, Chem. Brit. 4 (1968), 74

BAKER, JULIAN LEVETT [1873-1958]
R.H. Hopkins, Proc. Chem. Soc. 1958, pp.266-267

BAKER, LILLIAN ELOISE [1890-]
AMS 10,161

BAKER, PETER FREDERICK [1939-1987]
WW 1987, p.77
D.E. Knight and A.L. Hodgkin, Biog. Mem. Fell. Roy. Soc. 35 (1990), 3-35.

BAKER, WILLIAM OLIVER [1915-]
AMS 15(1),250-251; MSE 1,45-46; IWW 1982-1983, pp.71-72; WWA 1980-1981, p.156

BAKER, WILSON [1900-]
WW 1981, P.117; CAMPBELL, 15-16; WWWS, 95; POGG 6,113; 7b,204-207

BAKH, ALEKSEI NIKOLAEVICH [1857-1946]
DSB 1,360-363; BSE 4,322-323; BME 3,500-502; KUZNETSOV (1963), 288-301; POTSCH, 22; WWR, 43; BRIQUET, 11-14; LIPSHITS 1,135-141; POGG 5,46-47; 6,98-99; 7b,169-171
A.I. Oparin, Usp. Khim. 6 (1937), 315-327
D. Mikhlin, Bull. Acad. Sci. URSS 1937, pp.517-528
N.M. Sisakyan, Usp. Khim. 10 (1941), 648-653
A.I. Oparin and A.N. Frumkin, Aleksei Nikolaevich Bakh. Moscow 1950
A.N. Frumkin, Bull. Acad. Sci. URSS 1957, pp.795-799
W.L. Kretovich, Selected Topics in Biochemistry (G. Semenza, Ed.), pp.353-364. Amsterdam 1983

BALANDIN, ALEKSEI ALEKSANDROVICH [1898-1967]
DSB 1,414-415; STC 1,72-73; WWR, 46; WWWS, 96; POTSCH, 25; POGG 6,114; 7b,208-211
B.M.W. Trapnell, Adv. Catalysis 3 (1951), 1-25
N.B. Polyakova and E.I. Klabunovski, A. A. Balandin. Moscow 1984

BALARD, ANTOINE JÉROME [1802-1876]
DSB 1,416-417; DBF 4,1400-1402; GE 5,84-85; LAROUSSE 2,92; BLOKH 1,27; MAYERHOFER 1,373; W, 30; WWWS, 96; SCHAEDLER, 5-6; FERCHL, 20-21; POTSCH, 25; DORVEAUX, 34; POGG 1,91-92; 3,64
L. Figuier, Ann. Scient. 20 (1876), 530-536
C.A. Wurtz, J. Pharm. Chim. [4]23 (1976), 375-379
Anon., J. Chem. Soc. 31 (1877), 512-514
E.H. Huntress, Proc. Am. Acad. Arts Sci. 81 (1952), 39-40
J.F. Jeanjean, Mons. Hipp. 43 (1969), 21-28
H. Bonnemain et al., Rev. Hist. Pharm. 24 (1977), 5-96,137-189,203-217

BALBIANI, ÉDOUARD GÉRARD [1823-1899]
DSB 1,417-418; DBF 4,1414-1416; WWWS, 96
F.L. Henneguy, Arch. Anat. Micr. 3 (1900), i-xxxvi

BALBIANO, LUIGI [1852-1917]
DBI 5,387-388; TSCHIRCH, 1016; POGG 3,64; 4,58; 5,54

BALCOM, REUBEN WILFRED [1877-1929]
AMS 4,44-45; JV 21,272

BALDAMUS, ALFRED FERDINAND [1820-1886]
M. Schwarz, Biographisches Handbuch der Reichstage, p.259. Hannover 1965

BALDES, KARL [1877-1938]
KALLMORGEN, 215

BALDRIDGE, ROBERT CRARY [1921-]
AMS 15(1),255

BALDWIN, EDWARD ROBINSON [1864-1947]
DAB [Suppl.4], 48-49; AMS 6,62
J.A. Miller, Am. Rev. Tub. 56 (1947), 261-265
J.J. Waring, Trans. Assoc. Am. Phys. 61 (1948), 7-8
E.R. Long, Am. Rev. Tub. 62 (1950), 3-12

BALDWIN, ERNEST [1909-1969]
WW 1959, p.138; WWWS, 97
Anon., Nature 225 (1970), 569-570

BALDWIN, IRA LAWRENCE [1895-]
AMS 11,216; WWA 1980-1981, p.159; WWWS, 98

BALDWIN, ROBERT LESH [1927-]
AMS 15(1),257; WWA 1980-1981, p.160; WWWS, 98

BALE, WILLIAM FREER [1911-]
AMS 15(1),257; WWWS, 98

BALFOUR, FRANCIS MAITLAND [1851-1882]
DSB 1,420-422; DNB 1,970-972; O'CONNOR, 247-248; MAYERHOFER 1,374; W, 30-31; WWWS, 99
W. Waldeyer, Arch. mikr. Anat. 21 (1882), 828-835
H.F. Osborn, Science 2 (1883), 299-301
M.F[oster], Proc. Roy. Soc. 35 (1883), xx-xxvii

BALKE, PAUL [1871-]
JV 8,174; NUC 32,295; RSC 13,263

BALL, BENJAMIN [1833-1893]
DBF 4,1450; HIRSCH 1,305; Suppl.,61; PAGEL, 81
Anon., Lancet 1893 (I), p.562; Brit. Med. J. 1893 (I), p.613

BALL, ERIC GLENDENNING [1904-1979]
AMS 13,185; MH 2,22-24; MSE 1,46-47; WWA 1976-1978, pp.150-151; WWWS, 100
J.M. Buchanan et al., Harvard Gazette 75 [39] (1980), 6-7; Biog. Mem. Nat. Acad. Sci. 58 (1989), 49-73

BALLANCE, CHARLES ALFRED [1856-1936]
DNB [1931-1940], 37-38; O'CONNOR(2), 494-496; NUC 32,398-399
Who Was Who 1929-1940, p.59
Anon., Lancet 1936(I), pp.396,450-452; Brit. Med. J. 1936(I), pp.339-341

BALLAND, ANTOINE [1845-1927]
DBF 4,1458-1459; TSCHIRCH, 1016; WWWS, 101
C. Barrois, C. R. Acad. Sci. 184 (1927), 49-50

BALLING, KARL [1805-1868]
NDB 1,562-563; ADB 2,23; STURM 1,44-45; STARK, 342-343; BLOKH 1,27; SCHAEDLER, 6; FERCHL, 20-21; PRAG, 342-343; WWWS, 101; POTSCH, 25; POGG 1,95; 3,66
J.V. Divis, Beiträge zur Geschichte der Zuckerindustrie in Boehmen 1830-1860, pp.125-128. Kolin 1891

BALLING, KARL ALBERT [1835-1896]
STURM 1,44; POGG 3,66; 4,59; 5,54-55
Anon., Chem. Z. 20 (1896), 337; Leopoldina 32 (1896), 103

BALLOU, CLINTON EDWARD [1923-]
AMS 15(1),263-264

BALLOWITZ, EMIL [1859-1936]
FISCHER 1,63; PAGEL, 82-83; IB, 14; DZ, 41; KURSCHNER 4,86
H. Becher, Anat. Anz. 91 (1941), 33-69

BALLS, ARNOLD KENT [1891-1966]
AMS 11,224; WWWS,102; POGG 6,117; 7b,225-227
W.Z. Hassid, Biog. Mem. Nat. Acad. Sci. 41 (1970), 1-22

BALOGH, KALMAN [1835-1888]
MEL 1,98-99
Anon., Pharm. Z. 33 (1888), 465
D. Karasszon, Orv. Het. 129 (1988), 1163-1165

BALTIMORE, DAVID [1938-]
AMS 15(1),265; MSE 1,47-48; IWW 1982-1983, p.75;
WWA 1980-1981, p.165

BALTZER, FRITZ [1884-1974]
KURSCHNER 11,92; IB, 14; WWWS, 102-103
M. Fischberg, C. R. Soc. Phys. Hist. Nat. Gen. 10 (1975), 7-9

BALTZER, OTTO [1861-]
JV 6,67; NUC 33,87; RSC 13,271; 17,756

BALY, EDWARD CHARLES CYRIL [1871-1948]
WW 1945, p.129; SRS, 15; POGG 5,55,1415; 6,117-118; 7b,227-228
F.G. Donnan, Obit. Not. Fell. Roy. Soc. 6 (1948), 7-21

BAMANN, EUGEN [1900-1981]
HEIN-E, 15-17; KURSCHNER 13,121; DRUM, 300-301; WWWS, 103;
POGG 6,118; 7a(1),84-86
H. Thies and G. Kallinich, Pharmazie 15 (1960), 1-3
B. Reichert, Arzneimitt. 10 (1960), 63; 15 (1965), 96
G. Kallinich, Pharm. Z. 110 (1965), 35-36
Anon., Leopoldina [3]11 (1965), 49-50; Deutsche Apoth. Z. 110 (1970),
7-11; Chem. Z. 105 (1981), 95

BAMBACH, ADOLF [1880-]
JV 20,257

BAMBERGER, EUGEN [1857-1932]
DSB 1,426-427; NDB 1,572; WININGER 6,436; WWWS, 103; POTSCH,
26; POGG 5,55-56,1415; 6,118
L. Blangey, Helv. Chim. Acta 16 (1933), 644-685

BAMBERGER, HEINRICH von [1822-1888]
NDB 1,573; ADB 46,192-193; OBL 1,47; ROEDER, 135-146;
KOERTING, 165-166; STURM 1,46; PETRY, 70-79; HIRSCH 1,314;
Suppl.,61; PAGEL, 84-85; FRANKEN, 174-175; WININGER 1,233;
KOREN, 150
Anon., Wiener klin. Wchschr. 38 (1888), 1537-1538,1619-1622,1719-1721,
1745-1763; Leopoldina 24 (1888), 226
E. Neusser, Wiener klin. Wchschr. 12 (1899), 1084-1086
M. Becker, Das Leben und Wirken Heinrich von Bambergers. Düsseldorf
1937

BAMBERGER, MAX [1861-1927]
NDB 1,574; OBL 1,47-48; TSCHIRCH, 1016; TLK 2,24; WWWS, 103;
POGG 4,62; 5,56; 6,118
W. Schlenk, Ber. chem. Ges. 60A (1927), 178

BAMMANN, JOHANNES (HANS) [1866-1910]
JV 4,223; RSC 13,275
Anon., Chem. Z. 34 (1910), 1353

BANCROFT, FRANK WATTS [1871-1923]
AMS 3,33BANCROFT, WILDER DWIGHT [1867-1953]
DSB 1,430-431; DAB [Suppl.5],35-37; MILES, 22-23; POTSCH, 26-27;
AMS 8,116; WWAH, 33; WWWS, 103-104; NCAB 42,234-235; POGG
4,62-63; 5,57; 6,119-120; 7b,230-231
A. Findlay, J. Chem. Soc. 1953, pp.2506-2514
C.W. Mason, J. Am. Chem. Soc. 76 (1954), 2601-2602
J.W. Servos, Isis 73 (1982), 207-232

BANDROWSKI, ERNEST [1853-1920]
PSB 1,257-258; WE 1,586; KONOPKA 1,148-150; WWWS, 1729;
POGG 4,63; 5,57; 6,120
Anon., Roczniki Chemii 1 (1921), 351-359

BANG, BERNHARD [1848-1932]
DBL 2,69-73; MEISEN, 165-168; HIRSCH 1,317; BULLOCH, 350;
WWWS, 104
Anon., Brit. Med. J. 1931 [II], p.303
T.F. Williams and V.A. McKusick, Bull. Hist. Med. 28 (1954), 60-72
F. Bang, Historia Medicinae Veterinariae 4 (1979), 25-30
A. Jepsen, Dansk Veterinaer-Historisk Aarbog 31 (1985), 1-101

BANG, FREDERICK BARRY [1916-1981]
AMS 14,218; AO 1981, pp.624-625; ICC, 346-347; WWWS, 104
Anon., Rock. Univ. Notes 13 (1982) [2],2

BANG, IVAR CHRISTIAN [1869-1918]
SMK 1,179; FISCHER 1,65; BLOKH 1,28
O. Hammarsten, Erg. Physiol. 18 (1920), xi-xiii
D.D. Van Slyke, Scand. J. Clin. Inv. 10 (1958), Suppl.31, 18-26
V. Schmidt, Clin. Chem. 32 (1986), 213-215

BANGA, ILONA (Mrs. JOSEPH BALO) [1906-]
WWWS, 104
Anon., Leopoldina [3]8/9 (1962-1963), 23-24
Who's Who in the Socialist Countries of Europe 1989, vol.1, p.54

BANNOW, ADOLPH [1844-1919]
BLOKH 1,28-29
S. Gabriel, Ber. chem. Ges. 52A (1919), 65-67

BANTI, GUIDO [1852-1925]
DSB 1,438-440; DBI 5,759-764; EI 6,94-95; ARCIERI, 29-36; FISCHER
1,66; FRANKEN, 175; WWWS, 106; RSC 13,283-284
G. Barrier, Bull. Acad. Med. 93 (1925), 113
A. Lustig, Sperimentale 79 (1925), i-xxxi
E. Marchiafava, Atti Accad. Lincei [6]323 (1926), 103-112
P. Franceschini, Riv. Storia Sci. Med. 43 (1952), 157-167
F.W. Grannis, Mayo Clin. Proc. 50 (1975), 41-48

BANTING, FREDERICK GRANT [1891-1941]
DSB 1,440-443; DNB [1941-1950], 53-55; YOUNG, 78-81; FISCHER
1,66; W, 32-33; ABBOTT, 12-13; STC 1,76-78; POGG 6,121; 7b,234-235
C.A. Best, Science 93 (1941), 248-249; Obit. Not. Fell. Roy. Soc.
4 (1942), 21-26
W.E.B. Hall, Arch. Path. 31 (1941), 657-662
L. Stevenson, Sir Frederick Banting (2nd Ed.). Toronto 1947
J.H. Pratt, J. Hist. Med. 9 (1954), 281-289
W.R. Feasby, J. Hist. Med. 13 (1958), 68-84
P. Stein, Gesnerus 31 (1974), 107-172
M. Bliss, Banting: A Biography. Toronto 1984

BANUS, MARIO GARCIA [1893-1953]
AMS 8,118

BARANETSKI, OSIP VASILIEVICH [1834-1905]
BSE 4,220; LIPSHITS 1,113-115; VENGEROV 2,101-102; IKONNIKOV,
34-35

BARANOWSKI, THADDEUS [1910-]
Who's Who in the Socialist Countries 1978, p.45

BARANY, MICHAEL [1921-]
AMS 15(1),176; WWWS, 107

BARANY, ROBERT [1876-1936]
DSB 1,446-447; NDB 1,581; SBL 2,716-724; SMK 1,180-181; WWWS,
107; FISCHER 1,67-68; WININGER 1,241; 6,437-438; TETZLAFF, 18
H. Wyklicky, Wiener klin. Wchschr. 98 (1986), 622-625

BARBACKI, STEFAN [1903-1979]
WE 1,600; WWWS, 107
G. Kurhanska, Theoretical and Applied Genetics 55 (1979), 241-242

BARBAGLIA, GIOVANNI ANGELO [1838-1891]
POGG 3,68; 4,63
A.W. Hofmann, Ber. chem. Ges. 25 (1892), 137-139
Anon., Annuario Universita di Pisa 1892-1893, pp.219-222

BARBER, MARSHALL ALBERT [1868-1953]
AMS 7,119
D.A. Terreros and J.J. Grantham, Am. J. Physiol. 242 (1982), F293-F296

BARBIER, PHILIPPE ANTOINE [1848-1922]
POTSCH, 27; WWWS, 108; POGG 3,68-69; 4,64; 5,57; 6,123
H. Rheinboldt, J. Chem. Ed. 27 (1950), 476-488

BARBIERI, JOHANN [1852-1926]
ZURICH-D, 131
Eidgenössische Technische Hochschule 1855-1955, p.242. Zurich 1955

BARBOUR, HENRY GRAY [1886-1943]
DSB 1,449-450; DAB [Suppl.3], 32-33; PARASCANDOLA, 29; IB, 15; WWWS, 108; WWAH, 34
W.T. Salter, Science 98 (1943), 442-443

BARBOUR, THOMAS [1884-1946]
DAB [Suppl.4], 51-53; IB, 15; WWWS, 108; WWAH, 34-35
Anon., Am. Nat. 80 (1946), 214-216
J.L. Peters, Auk 65 (1948), 432-438
H.B. Bigelow, Biog. Mem. Nat. Acad. Sci. 27 (1952), 13-45
A.S. Romer, Systematic Zoology 13 (1964), 227-234

BARCLAY, JOHN [1758-1826]
DSB 1,452; DNB 1,1086-1087; HIRSCH 1,326; Suppl.,64; WWWS, 109

BARCROFT, HENRY [1904-]
WW 1982, P.113; IWW 1982-1983, p.78; BELFAST, 580-581

BARCROFT, JOSEPH [1872-1947]
DSB 1,452-455; DNB [1941-1950], 57-58; WW 1945, pp.138-139; WWWS, 109; O'CONNOR(2), 46-50; FISCHER 1,69
D.H. Barron, Science 106 (1947), 160-161
F. Verzar, Experientia 3 (1947), 298-300
F.J.W. Roughton, Obit. Not. Fell. Roy. Soc. 6 (1949), 315-345
K.J. Franklin, Joseph Barcroft 1872-1947. Oxford 1953
F.L. Holmes, J. Hist. Biol. 2 (1969), 89-122

BARD, PHILIP [1898-1977]
AMS 13,200; NCAB M,443-445; MEITES, 16-33; BROBECK, 151-152; WWWS, 109
P. Bard, Ann. Rev. Physiol. 35 (1973), 1-16
V.B. Mountcastle, Physiologist 20 (1977), [3],1-2
J.E. Howard, Trans. Assoc. Am. Phys. 91 (1978), 13-14

BARDAKH, YAKOV YULIEVICH [1857-1929]
BME 3,423-424; WWR, 49; WININGER 7,518; IB, 15
A.A. Efremenko, Zhur. Mikr. Ep. Imm. 36 (1959), [8],119-124
V.P. Tulchinskaya, Mikrobiologia 28 (1959), [1],140-143
K.G. Vassiliev and A.I. Zaitseva, Mikr. Zhur. 35 (1973), 670-673

BARDELEBEN, ADOLF von [1819-1895]
NDB 1,583; ADB 46,214-215; HB 1,299-302; HIRSCH 1,328-329; Suppl.,64; PAGEL, 86-87; KILLIAN, 334-335; WWWS, 1729
W. Waldeyer, Anat. Anz. 11 (1895), 303-305
P. Kohler, Gedächtnissrede auf Adolf von Bardeleben. Berlin 1895

BARDELEBEN, KARL HEINRICH [1849-1918]
DBJ 2,680-681; FISCHER 1,70; PAGEL,88-90; GIESE, 483-485
W. Waldeyer-Hartz, Anat. Anz. 51 (1918), vii-xii
W. Linss, Anat. Anz. 150 (1981), 5-10

BARDET, ÉDOUARD GODEFROY [1852-1923]
DBF 5,399-400; LABARRE 2,336; TSCHIRCH, 1016

BARDIER, ÉMILE [1870-1949]
IB, 15; NUC 35,248; RSC 13,296; SG [3]2,347
H. Gaussen, Mem. Acad. Toulouse [3]8 (1956), 16-17

BARENDRECHT, HENDRIK PIETER [1871-1942]
RSC 13,297
Anon., Chem. Wkbl. 39 (1942), 191

BARFOED, CHRISTIAN THOMSEN [1815-1889]
DBL 2,162-164; VEIBEL 1,193-194; 2,40-43; 3,94-114; BLOKH 1, 30; TSCHIRCH, 1016; SCHAEDLER, 6; FERCHL, 22; POGG 3,69-70; 4,64-65
Anon., Leopoldina 25 (1889), 113

BARFURTH, DIETRICH [1849-1927]
NDB 1,588; GROTE 2,1-22; FISCHER 1,71; PAGEL, 92; LEVITSKI 2,45-47; WAGENITZ, 20; DZ, 48; IB, 15; BONN, 352; WWWS, 110
O. Dragendorff, Anat. Anz. 69 (1930), 47-49

BARGELLINI, GUIDO [1879-1927]
DBI 6,347-349; POGG 5,58; 6,124; 7b,240
C.B. Marini-Bettolo, Chimica e Industria 45 (1963), 1558-1559

BARGER, ABRAHAM CLIFFORD [1917-]
AMS 17(1),303; BROBECK, 205-209; WWWS, 110

BARGER, GEORGE [1878-1939]
DSB 15,10-11; DNB [1931-1940], 39-40; FINDLAY, 419-431; POTSCH, 28; FLUELER, 14-16; STC 1,80-82; W, 34-35; WWWS, 110; POGG 6,124-125; 7b,241-242
H.H. Dale, Nature 143 (1939), 107-108; Obit. Not. Fell. Roy. Soc. 3 (1940), 63-85
C.R. Harington, J. Chem. Soc. 1939, pp.715-721; Biochem. J. 33 (1939), 859-864

BARGMANN, WOLFGANG [1906-1978]
KURSCHNER 12,105; MEITES, 36-43; KIEL, 90; WWWS, 110; POGG 7a(1),88-89
H. Leonhardt, Christiana Albertina NF9 (1978), 171-176
K. Fleischhauer, Anat. Anz. 146 (1979), 209-234
B. Scharrer, Anat. Rec. 195 (1979), 145-147

BARKAN, GEORG [1889-1945]
WININGER 7,518-519; IB, 15; POGG 6,125; 7b,243-244
Anon., Chem. Eng. News 23 (1945), 586; Science 101 (1945), 265

BARKER, GEORGE FREDERICK [1835-1910]
MILES, 23-24; NUC 35,481-484
E.F. Smith, Am. J. Sci. 180 (1910), 225-232
E. Thomson, Proc. Am. Phil. Soc. 50 (1911), xiii-xxix

BARKER, HORACE ALBERT [1907-]
AMS 14,232; MH 2,24-25; MSE 1,49-50; WWA 1978-1979, p.171; IWW 1982-1983, p.80
H.A. Barker, Reflections on Biochemistry (A. Kornberg et al., Eds.), pp.95-104. Oxford 1976; Ann. Rev. Biochem. 47 (1978), 1-33

BARKER, LEWELLYS FRANKLIN [1867-1943]
AMS 6,71; FISCHER 1,72-73; NCAB 32,308-309; A,265-266; ICC, 52-53; DAMB, 36; IB, 15
L.F. Barker, Time and the Physician. New York 1942
C.R. Austrian, Bull. Johns Hopkins Hosp. 73 (1943), 401-404
W.T. Longcope, Science 98 (1943), 316-318
J.D. Rolleston, Nature 152 (1943), 407

BARKER, NELSON WAITE [1899-1968]
AMS 11,242; WWWS, 111; WWAH, 35

BARKER, SIDNEY ALAN [1926-]
Who's Who of British Scientists 1971-1972, p.47

BARKLA, CHARLES GLOVER [1877-1944]
DSB 1,456-459; DND [1941-1950], 60-62; STC 1,82-83; W, 35; SRS, 25; POGG 5,59-60; 6,126; 7b,245
H.S. Allen, Obit. Not. Fell. Roy. Soc. 5 (1947), 341-366
R.J. Stephenson, Am. J. Phys. 35 (1967), 140-152
B. Wynne, Soc. Stud. Sci. 6 (1976), 307-347

BARLOW, THOMAS [1845-1945]
DNB [1941-1950], 62-63; FISCHER 1,73; PAGEL, 93-94; WWWS, 112
T.R. Elliott, Obit. Not. Fell. Roy. Soc. 5 (1947), 159-167

BARLOW, WILLIAM [1845-1934]
DSB 1,460-463; POGG 4,65; 5,60; 6,127; 7b,246
W.J. Pope, Obit. Not. Fell. Roy. Soc. 1 (1935), 367-370
G.T. Moody and W. Mills, J. Chem. Soc. 1940, pp.697-715

BARLOW, WILLY [1869-]
JV 11,231; NUC 35,622; RSC 19,74

BARNARD, HAROLD LESLIE [1868-1908]
FISCHER 1,73; O'CONNOR(2), 280; NUC 35,663
Anon., Brit. Med. J. 1908(II), pp.538-539

BARNARD, JOSEPH EDWIN [1868-1949]
WW 1949, p.149; W, 35-36; WWWS, 113; POGG 6,127; 7b,246
J.A. Murray, Obit. Not. Fell. Roy. Soc. 7 (1950), 3-8
W.J.P., J. Roy. Micr. Soc. 71 (1951), 104-113

BARNES, JAMES HECTOR [1879-1917]
BLOKH 1,30
A.S., J. Chem. Soc. 115 (1919), 409-412

BARNES, RICHARD HENRY [1911-1979]
AMS 14,240; WWA 1976-1977, P.170; WWWS, 113
D.A. Roe, J. Nutrition 109 (1979), 1509-1514

BARNHART, MARION ISABEL [1921-1985]
AMS 14,243
W.H. Seegers, Thromb. Res. 42 (1986), 427-432; Ann. N.Y. Acad. Sci. 509 (1987), xi-xii

BARON, DENIS NEVILLE [1924-]
Who's Who of British Scientists 1980-1981, p.27

BÁRON, JULIUS (GYULA) [1891-1973]
P. Gulyas, Magyar Irok 2 (1940), 491
Anon., N.Y. J. Med. 74 (1974), 576

BARR, DAVID PRESWICK [1889-1977]
AMS 11,255
D.E. Rogers, Trans. Assoc. Am. Phys. 91 (1978), 15-20
G.D. Whedon, J. Clin. Endocrin. 48 (1979), 509-510

BARR, MURRAY LLEWELLYN [1908-]
WW 1984, p.126; AMS 15(1),303; ABBOTT, 13-15; WWWS, 115

BARRAL, ÉTIENNE VICTOR [1860-1938]
DENOTH, 9; POGG 4,66; 5,66; 6,129; 7b,250
M. Delépine, Bull. Acad. Med. 119 (1938), 404-405

BARRAL, JEAN AUGUSTIN [1819-1884]
DBF 5,543-544; GE 5,471; LAROUSSE 2,254; VAPEREAU 5,127; FERCHL, 23; POGG 1,106; 3,72; 4,65-66

BARRENSCHEEN, HERMANN KARL [1887-1958]
IB, 16; POGG 6,130; 7a(1),89-90; (4),136*

BARRESWIL, CHARLES LOUIS [1817-1870]
DSB 1,471; DBF 5,593-594; GE 5,487; LAROUSSE 2,261; BLOKH 1,30; FERCHL, 23; BOURQUELOT, 68; SCHAEDLER, 6; POGG 1,105; 3,72-73
L. Figuier, Ann. Scient. 15 (1870-1871), 535-536
J.J. Peumery, Hist. Sci. Med. 20 (1986), 243-248

BARRINGTON, ERNEST WILLIAM [1909-1985]
WW 1985, pp.107-108
A.J. Matty, Biog. Mem. Fell. Roy. Soc. 35 (1990), 39-54

BARRNETT, RUSSELL JOFFREE [1920-1989]
AMS 15(1),308; WWWS, 117

BARRON, DONALD HENRY [1905-1993]
AMS 14,251; WWA 1980-1981, P.190; WWWS, 117
L.H. Nahum, Yale J. Biol. Med. 42 (1969), 130-138
L.D. Longo, Fetal and Newborn Cardiovascular Physiology (L.D. Longo and D.D. Reneau, Eds.), pp.xxvii-xxxiv. New York 1978

BARRON, ELEAZAR SEBASTIAN GUZMAN [1898-1957]
AMS 9(I),95; WWWS, 117; WWAH, 37
A.B. Hastings, Science 126 (1957), 964
Anon., Anales de la Facultad de Medicina Lima 49 (1957), 798-813
F.C. McLean, Trans. Assoc. Am. Phys. 71 (1958), 11-12
H.B. Steinbach, Biol. Bull. 115 (1958), 9-10

BARRON, MOSES [1883-1974]
AMS 11,261; DAMB, 38-39; KOREN, 151; WWWS, 117
M. Barron, Minnesota Medicine 49 (1966), 689-690,861-862

BARROW, FRED [1882-1964]
SRS, 36; NUC 37,106
D.J.G. Ives, Proc. Chem. Soc. 1964, pp.379-380

BARRUEL, FRANÇOIS CLAUDE [1783-]
FERCHL, 23; POGG 1,107

BARRY, MARTIN [1802-1855]
DSB 1,476-478; DNB 1,1246-1247; WWWS, 118
Anon., Proc. Roy Soc. 7 (1855), 577-582
J.B., Edinburgh Med. J. 1 (1856), 81-91

BARSCHALL, HERMANN [1879-1966]
JV 17,10; BHDE 2,54

BARTELL, FLOYD EARL [1883-1961]
AMS 10,205; NCAB 49,602; POGG 6,130-131; 7b,251-252
Anon., Chem. Eng. News 37 (1959), [16],55; Science 133 (1961), 747

BARTELS, CARL [1822-1878]
HIRSCH 1,349-350; Suppl.,67-68; SHBL 1,64-65; PAGEL, 95-96; KIEL, 79
J. Cohnheim, Deutsches Arch. prakt. Med. 1878, p.334

BARTH, MAX [1855-1899]
BJN 4,126*
Anon., Chem. Z. 23 (1899), 1085; Leopoldina 35 (1899), 157-158

BARTH von BARTHENAU, LUDWIG [1839-1890]
OBL 1,51; MACHEK, 179-184; BLOKH 1, 30-31; POTSCH, 28-29; SCHAEDLER, 6; POGG 3,74; 4,67
C. Senhofer and G. Goldschmidt, Ber. chem. Ges. 24 (1891), 1089-1115

BARTHE, LÉONCE [1857-1941]
DBF 5,650; BALLAND, 255,306; TSCHIRCH, 1016; POGG 4,67-68; 5,67; 6,132; 7b,255
Anon., Bull. Sci. Pharm. 48 (1941), 260-269

BARTHELÉMY, AIMÉ [1831-1885]
DBF 5,658; POGG 3,74-75; 4,68
L. Figuier, Ann. Scient. 29 (1885), 540-541

BARTHEZ, ERNEST [1811-1891]
DBF 5,688-689; HIRSCH 1,355-356; PAGEL, 99; VAPEREAU 5,136
P. Fischer, F. Rilliet und E. Barthez und ihre Traité. Zurich,1966

BARTLETT, PAUL DEVERE [1911-1981]
AMS 14,257

BARTLETT, PAUL DOUGHTY [1907-]
AMS 14,257; MH 1,22-24; MSE 1,56-58; STC 1,89-91; WWA 1978-1979, p.189; WWWS, 120; POTSCH, 29
Anon., Nachr. Chem. Techn. 2 (1954), 114; Leopoldina [3]15 (1969),39-41
J.M. McBride et al., P. D. and the Bartlett Group at Harvard 1934-1974. Fort Worth, Tex. 1975

BARTON, DEREK HAROLD RICHARD [1918-]
WW 1981, p.153; IWW 1982-1983, pp.85-86; MH 1,24-25; MSE 1,58-59; STC 1,91-92; CAMPBELL, 23-24; WWWS, 121; POTSCH, 29-30
P. Farago, J. Chem. Ed. 50 (1973), 234-237
T. Shiori, Heterocycles 28 (1989), 1-14
D.H.R. Barton, Some Recollections of Gap Jumping. Washington, D.C. 1991

BARTON-WRIGHT, EUSTACE CECIL [1902-1976]
Who's Who of British Scientists 1969-1970, pp.52-53
Anon., Chem. Brit. 12 (1976), 330

BARY, ANTON de [1831-1888]
DSB 3,611-614; NDB 1,616; ADB 46,225-228; DBA 2,120;

MAYERHOFER 1,386; BULLOCH, 350-351; TSCHIRCH, 1017; KALLMORGEN, 216; WWWS, 425; LEHMANN, 180-181
M. Rees, Ber. bot. Ges. 6 (1888), (viii)-(xxvi)
F. Cohn, Deutsche med. Wchschr. 14 (1888), 98-99,118-119
Anon., Nature 37 (1888), 297-299
C. von Voit, Sitz. Bayer. Akad. 18 (1888), 187-192
H. Solms-Laubach, Bot. Zt. 47 (1889), 33-49
L. Jost, Z. Bot. 24 (1930), 1-74
E. Fischer, Naturwiss. 19 (1931), 97-102
V.A. Parnes, Anton de Bary. Moscow 1972
K. Sparrow, Mycologia 70 (1978), 222-252
K. Mägdefrau, Naturw. Rund. 34 (1981), 413-415
J.G. Horsfall and S. Wilhelm, Annual Review of Phytopathology 20 (1982), 27-32

BARY, JOHANN JAKOB de [1840-1915]
KALLMORGEN, 216-217; RSC 7,95
R. Fridberg, Jahresb. Aerzt. Ver. 1915, pp.163-164

BARYKIN, VLADIMIR ALEKSANDROVICH [1879-1942]
BME 3,465-466; WWR, 51; FISCHER 1,76; IB, 15
S.S. Dischenko, Mikr. Zhur. 42 (1980), 260-263

BASCH, KARL [1859-1913]
FISCHER 1,76
Anon., Jahrb. Kinderheil. 78 (1913), 123-124
R. Raudnitz, Prag. med. Wchschr. 38 (1913), 305

BASCH, SAMUEL SIEGFRIED KARL von [1837-1905]
NDB 1,617; BJN 10,222-224,142*; STURM 1,53-54; HIRSCH 1,367; Suppl.,72; PAGEL, 99-100; WININGER 1,257; 6,411; KOREN, 151; WWWS, 1729
A. Fröhlich, Deutsche med. Wchschr. 31 (1905), 878
A. Biedl, Wiener klin. Wchschr. 18 (1905), 498

BASEDOW, KARL ADOLPH von [1799-1854]
NDB 1,620; ADB 46,230; HIRSCH 1,367-368; PAGEL, 100; HALLE, 100; WWWS, 1728; CALLISEN 1,474-478; 26,165
K. Sudhoff, Münch. med. Wchschr. 57 (1910), 749-750
K. Wolff, Med. Welt 9 (1935), 34-36

BASHFORD, ERNEST FRANCIS [1873-1923]
WWWS, 123
Anon., Lancet 1923 (II), p.536; Brit. Med. J. 1923 (II), p.440

BASKERVILLE, CHARLES [1870-1922]
DAB 2,34; MILES, 25-26; WWWS, 123; WWAH, 40; POGG 5,69
F.P. Venable, Ind. Eng. Chem. 14 (1922), 247
W.A. Hamor, Science 65 (1922), 693-694
Anon., J. Franklin Inst. 193 (1922), 566-567

BASLER, ADOLF [1878-1945]
KURSCHNER 4,103; FISCHER 1,77; IB, 16; NUC 38,379

BASOLO, FRED [1920-]
AMS 15(1),324; MSE 1,59-60; WWA 1980-1981, p.200; WWWS, 124

BASOV, VASILI ALEKSANDROVICH [1812-1879]
BSE 4,287-288; BME 3,479-480; HIRSCH 1,372; ZMEEV 1, 17-18; WWWS, 124; VENGEROV 2,218-219
E.I. Zakharov, V. A. Basov. Moscow 1953

BASS, LAWRENCE WADE [1898-1982]
AMS 12,325; WWA 1978-1979, pp.193-194; WWWS, 124

BASSALIK, KAZIMIERZ [1879-1960]
WE 1,633; IB, 16; WARSAW, 229
P. Strebeyko, Acta Soc. Bot. Pol. 30 (1961), 5-13
A. Drozdowicz, Stud. Nauk. Pol. B13 (1967), 87-101

BASSERMANN, HEINRICH [1886-1965]
JV 28,365; NUC 38,487
Alte Mannheimer Familien 3/4 (1923), 123-124

BASSETT, HENRY [1881-1965]
WW 1965, p.179; SRS, 30; POGG 6,136-137; 7b,258-259
Anon., Nature 157 (1946)
G.H. Cheesman, Chem. Brit. 2 (1966), 351-352

BASSETT-SMITH, PERCY WILLIAM [1861-1927]
WW 1927, pp.171-172; FISCHER 1,77; OLPP, 28-29
Anon., Lancet 1928 (I), pp.56-57; Brit. Med. J. 1928 (I), p.35

BASSI, AGOSTINO [1773-1856]
DSB 1,492-494; DBI 7,121-122; BULLOCH, 351; W, 37; WWWS, 125
S. Calandruccio, Arch. Parasit. 6 (1902), 97-107
U. Faucci, Riv. Storia Sci. Med. 27 (1936), 1-26,59-102,153-206,286-326, 371-423; 28 (1937), 24-37,283-305; 30 (1939), 85-98,209-228; 32 (1941), 1-32
R.H. Major, Bull. Hist. Med. 16 (1944), 97-107
G.P. Arcieri, Agostino Bassi in the History of Medical Thought. Florence 1956
J.R. Porter, Bact. Revs. 37 (1973), 284-288

BASTIAN, HENRY CHARLTON [1837-1915]
DSB 1,495-498; BULLOCH, 351; DESMOND, 45; HAYMAKER, 405-407
Anon., Brit. Med. J. 1915(II),pp.795-796; Lancet 1915(II),pp.1220-1224
F.W. M[ott], Proc. Roy. Soc. B89 (1917), xxi-xxiv
E.G. Jones, J. Hist. Med. 27 (1972), 298-311

BASTICK, WILLIAM [1818-1903]
NUC 38,563; RSC 1,204
Anon., Pharm. J. 71 (1903), 904

BASWITZ, MAX [1854-1886]
RSC 9,138; 13,338
Anon., Chem. Z. 10 (1886), 1618

BATAILLON, JEAN EUGÈNE [1864-1953]
DSB 1,498-499; IB, 17; WWWS, 126
R. Courrier, Not. Acad. Sci. 3 (1954), 1-26
J.L. Fischer and J. Smith, Hist. Phil. Life Sci. 6 (1984), 23-29

BATALIN, ALEKSANDR FEDOROVICH [1847-1896]
LIPSHITS 1,129-132; VENGEROV 2,219-220
G. Winkler, Ber. bot. Ges. 15 (1897), (43)-(46)

BATE-SMITH, EDGAR CHARLES [1900-1989]
WW 1982, p.134

BATES, FREDERICK JOHN [1877-1958]
WWWS, 127; WWAH, 41; POGG 6,138; 7b,263
Anon., J. Opt. Soc. Am. 49 (1959), 311-312

BATES, ROBERT WESLEY [1904-]
AMS 14,269; WWWS, 127

BATESON, WILLIAM [1861-1926]
DSB 1,505-506; DNB [1922-1930], 66-68; MAYERHOFER 1,391-392; DESMOND, 46; W, 37-38; ABBOTT, 15-16; WWWS, 127
Anon., J. Heredity 17 (1926), 433-449
T.H. Morgan, Science 63(1926),531-535; Proc. Linn. Soc. 138(1926),66-74
B. Bateson, William Bateson FRS, Naturalist. Cambridge,1928
W. Coleman, Centaurus 15 (1970), 228-314
A.G. Cock, J. Hist. Biol. 6 (1973), 1-36; Ann. Sci. 40 (1983), 19-59
L. Darden, J. Hist. Biol. 10 (1977), 87-106
R.D. Harvey, Mendel Newsletter 25 (1985), 1-11
R. Olby, Brit. J. Hist. Sci. 20 (1987), 399-420

BATTELLI, FEDERICO [1867-1941]
IB, 17; GENEVA 5,355-361; 6,179-184; 7,263-265; 8,258
M.Monnier, Riv. Biol. 33 (1942),267-280; Erg. Physiol. 45 (1944),12-15
G. de Morsier and M. Monnier, La Vie et l'Oeuvre de Frédéric Battelli. Basle 1977

BATTERSBY, ALAN RUSHTON [1925-]
WW 1981, p.162; IWW 1982-1983, p.89; MSE 1,67-68; CAMPBELL, 25-

26; POTSCH, 30-31

BAUBIGNY, HENRI [1842-1912]
POGG 4,74-75; 5,72
R. Lespieau, Bull. Soc. Chim. [4]13 (1913), i-xviii

BAUDISCH, OSCAR [1881-1950]
AMS 8,145; STURM 1,55-56; POGG 5,73; 6,141; 7b,269-271
Anon., Chem. Eng. News 28 (1950), 1413; Science 111 (1950), 395
F. Feigl, Microchimica Acta 36 (1951), 33-37

BAUDOUIN, ALPHONSE [1876-1957]
FISCHER 1,79
H. Bénard, Bull. Acad. Med. 141 (1957), 178-186

BAUDRIMONT, ALEXANDRE ÉDOUARD [1806-1880]
DSB 1,517-519; GE 5,881; BLOKH 1,31; POTSCH, 31; FERCHL, 27; TSCHIRCH, 1017; COLE, 19-20; WWWS, 128; POGG 3,79-80; 4,75
L. Mice, Actes Acad. Bord. [3]42 (1880), 729-760; 44 (1882), 587-624
S.C. Kapoor, Actes du XIIe Congrès Internationale de l'Histoire des Sciences 6 (1968), 37-42
G. Dillemann, Produits et Problemes Pharmaceutiques 28 (1973), 294-295

BAUDRIMONT, MARIE VICTOR ERNEST [1821-1885]
GORIS, 220-222; TSCHIRCH, 1017; WWWS, 128
C. Méhu, Bull. Acad. Med. 14 (1885), 1252-1254;
J. Pharm. Chim. [5]12 (1885), 331-334

BAUER, ALEXANDER [1836-1921]
NDB 1,636; OBL 1,54; DBJ 3,290; POTSCH, 31; POGG 3,80-81; 4,75; 5,73; 6,141; 7a(1),97 F. Strunz, Mitt. Gesch. Med. 9 (1910), 1-11; Arch. Gesch. Naturwiss. 11 (1929), 276-292
F. Diergart, Z. angew. Chem. 34 (1921), 402-403
F. Bock, Ost. Chem. Z. 24 (1921), 99-104

BAUER, ÉDOUARD [1879-1915]
RSC 13,350
A. Haller, Bull. Soc. Chim. [4]19 (1916), 70-71
Anon., Z. angew. Chem. 29[III] (1916), 324

BAUER, ERWIN [1895-1942]
JV 43,409
W. Saute, Die Lichte, 11 August 1964, pp.18-19

BAUER, FERDINAND LUCAS [1760-1826]
DSB 1,520; NDB 1,634; ADB 2,140-141; DESMOND, 47; WWWS, 129
W.T. Stearn, Endeavour 19 (1960), 27-35

BAUER, FRIEDRICH ALFRED [1883-1957]
SBA 5,13; JV 23,596; NUC 39,516
J.F., Davoser Revue 32[4/5] (1957), 88

BAUER, HANS [1904-1988]
KURSCHNER 13,149; BOHM, 187-188

BAUER, HUGO [1883-1968]
AMS 10,222; KALLMORGEN, 217; LDGS, 17; POGG 6,142; 7a(1),99
Anon., Chem. Eng. News 46 (1968), [21],84

BAUER, JOSEPH von [1845-1912]
NDB 1,641-642; BJN 18,8*; HIRSCH 1,381; Suppl., 75; BACHMANN, 75-82; WWWS, 1729
F. Müller, Münch. med. Wchschr. 59 (1912), 1329-1332

BAUER, JULIUS [1879-]
FISCHER 1,80; KALLMORGEN, 217; HABERLING, 58-59; LDGS, 76; KOREN, 151

BAUER, JULIUS [1887-1979]
AMS 11,641-642; FISCHER 1,80; MEDVEI, 707-708; BHDE 2,58-59; WININGER 7,520; KOREN, 151-152
J. Bauer, Medizingeschichte des 20. Jahrhunderts im Rahmen einer Autobiographie. Vienna 1964
H. Wyklicky, Osterreichische Aerztezeitung 32 (1977), 870-873

BAUER, KARL FRIEDRICH [1904-1985]
KURSCHNER 14,166
H. Haug, Anat. Anz. 162 (1986), 313-315

BAUER, KARL HEINRICH [1890-1978]
KURSCHNER 11,113; FISCHER 1,81; IB, 17
D. Schmähl, Arzneimitt. 15 (1965), 1096-1097E. Holder, Med. Welt 38 (1965), 2188-2189
G.H. Ott, Münch. med. Wchschr. 120 (1978), 1236-1237
F. Linder, Heid. Akad. Jahrb. 1979, pp.63-65

BAUER, KARL HUGO [1874-1944]
HEIN, 27-28; FISCHER 1,81; TSCHIRCH, 1017; TLK 3,26; POGG 5,73-74; 6,142; 7a(1),99-100
T. Boehm, Pharm. Zent. 85 (1944), 97-98

BAUER, MAX HERMANN [1844-1917]
NDB 1,692-693; DBJ 2,646; AUERBACH, 770-771; NUC 39,577-578; POGG 3,81; 4,75-76; 6,143
G.F. Kunz, Science 41 (1915), 392-395
R. Brauhauser, Zeitschrift für Mineralogie 1918, pp.73-84

BAUER, OTTO [1909-1979]
S. Kozlowski et al., Acta Physiol. Polon. 35 (1984), Suppl.27, 23-35

BAUER, VICTOR [1881-1927]
IB, 17
E. Wagler, Verhandl. Int. Ver. Limn. 5 (1931), 701-704

BAUM, FRITZ [1872-1939]
POGG 4,77; 5,74; 6,144; 7a(1),103

BAUM, WILHELM [1799-1883]
NSL 5,13-27; KROLLMANN, 33
Anon., Leopoldina 19 (1883), 168

BAUMAN, LOUIS [1880-1954]
AMS 9(II),106; IB, 17
Anon., J. Am. Med. Assn. 156 (1954), 1620

BAUMANN, ARNO [1877-]
JV 19,258; NUC 40,2

BAUMANN, ARTUR [1885-1945]
JV 27,641; NUC 40,2

BAUMANN, CARL AUGUST [1906-]
AMS 14,279; WWA 1978-1979, p.203

BAUMANN, EMIL JACOB [1890-1962]
AMS 10,225; IB, 17
Anon., Chem. Eng. News 40 (1962), [40],96

BAUMANN, EUGEN [1846-1896]
NDB 1,651; BJN 1,93-94; 3,89*; BAD 5,34-36; HEIN, 29-30; FISCHER 1,81; BLOKH 1,31-34; SCHELENZ, 701; SCHAEDLER,6; TSCHIRCH, 1017; POTSCH, 31; WWWS, 130; PAGEL, 103-104; POGG 4,77-78; 7a Suppl.,51
F. Tiemann, Ber. chem. Ges. 29 (1896), 2575-2580
N. Zuntz, Deutsche med. Wchschr. 22 (1896), 748-749
C.A. Ewald, Berl. klin. Wchschr. 33 (1896), 1015-1016
L.B. Mendel, Science 5 (1897), 51-53
A. Kossel, Z. physiol. Chem. 23 (1897), 1-22
J.M. Orten, J. Nutrition 58 (1956), 3-10
M. Spaude, Eugen Albert Baumann (1846-1896). Zurich 1973
A.B. Roy, TIBS 1 (1976), N233-N234

BAUMANN, OTTO [1891-1964]
JV 37,1281

BAUMBERGER, JAMES PERCY [1892-1973]
AMS 11,290

BAUMEISTER, EDUARD [1865-]
JV 13,222; RSC 14,804

BAUMERT, FRIEDRICH MORITZ [1818-1865]
ADB 2,157; HIRSCH 1,386; BLOKH 1,34; FERCHL,28; SCHAEDLER, 7; POGG 1,117; 3,83

BAUMERT, GEORG [1852-1927]
KURSCHNER 2,75; TLK 2,33; HALLE, 534,647; SCHAEDLER, 7; POGG 4,78; 5,75; 6,144
Anon., Chem. Z. 40 (1927), 1333

BAUMGÄRTEL, CONRAD [1868-]
JV 12,242; NUC 40,74; RSC 13,234

BAUMGARTEN, PAUL von [1848-1928]
FISCHER 1,82; PAGEL, 104-106; DZ, 62-63
M. Askanazy, Münch. med. Wchschr. 65 (1918), 973-974
A. Wolff-Eisner, Münch. med. Wchschr. 75 (1928), 1507-1508
A. Dietrich, Verhandl. path. Ges. 24 (1928), 361-368

BAUMGARTEN, PAUL [1896-1943]
POGG 6,144-145; 7a(1),105
E. Tiede, Ber. chem. Ges. 76A (1943), 117-118
Anon., Z. angew. Chem. 56 (1943), 348

BAUMHAUER, EDOUARD HENRI [1820-1885]
DSB 1,527-528; NNBW 1,253-254; BLOKH 1,35; SCHAEDLER, 7; FERCHL, 28; TSCHIRCH, 1018; POGG 1,118; 3,83-84
J.W. Gunning, Jaarb. Akad. Wet. 1887, pp.1-57

BAUMHAUER, HEINRICH ADOLF [1848-1926]
DSB 1,528-529; NDB 1,667-668; WWWS, 131; POGG 3,84; 4,78-79; 5,75; 6,145
L. Weber, Verhandl. Schw. Nat. Ges. 107 (1926), 3-15

BAUMSTARK, FERDINAND [1839-1889]
BLOKH 1,35; SCHAEDLER, 7; POGG 3,85; 4,79
Anon., Leopoldina 26 (1890), 54

BAUP, SAMUEL [1791-1862]
BLOKH 1,35; FERCHL, 28; TSCHIRCH, 1018; POGG 1,118-119; 3,85; CALLISEN 1,499-501; 26,183
F. Roux, Verhandl. Schw. Nat. Ges. 46 (1862), 233-250
W. Robert, Bull. Soc. Vaud. Sci. Nat. 29 (1893), 185-210

BAUR, EMIL [1873-1944]
TLK 3,27-28; POGG 5,75-76; 6,145; 7a(1),106-107
W.D. Treadwell, Helv. Chim. Acta 27 (1944), 1302-1313

BAUR, ERWIN [1875-1933]
DSB 17,53-58; NDB 1,669-670; KURSCHNER 4,124; FISCHER 1,82; STC 1,95-96; WWWS, 131
E. Schiemann, Ber. bot. Ges. 52 (1934), (51)-(114)
R.R. Gates, Nature 133 (1934), 239-240
E. Tschermak-Seysenegg, Alm. Bayer. Akad. Wiss. 84 (1934), 243-246
M. Hartmann et al., Naturwiss. 22 (1934), 258-288
R. Hagemann, Leopoldina [3]21 (1978), 179-187

BAVINK, BERNHARD [1879-1947]
NDB 1,676; NSL 5,28-39; TLK 3,28; POGG 6,146; 7a(1),108-109
A. Hellbrück, Kosmos 43 (1947), 241-242
Anon., Naturwiss. 33 (1947), 375; Naturw. Rund. 1 (1947), 82

BAWDEN, FREDERICK CHARLES [1908-1972]
DSB 17,58-61; DNB [1971-1980], 41-42; WW 1965, p.187; DESMOND, 48; WWWS, 132
D.J. Watson, Ann. Appl. Biol. 70 (1972), 106-108
N.W. Pirie, J. Gen. Microbiol. 72 (1972), 1-7; Biog. Mem. Fell. Roy. Soc. 19 (1973), 19-63
Anon., Nature 236 (1972), 128-129

BAXTER, GREGORY PAUL [1876-1953]
AMS 8,149; NCAB 52,598; WWWS, 132; WWAH, 43; POGG 5, 76-77; 6,146-147; 7b,271-272
G.S. Forbes, J. Chem. Ed. 11 (1934), 444-447
Anon., Chem. Eng. News 31 (1953), 1228
W.M. McNevin, J. Chem. Ed. 31 (1954), 303-305

BAYER, FRIEDRICH [1851-1920]
DBJ 4,313-314; POGG 6,147; 7a Suppl.,52
C. Duisberg, Ber. chem. Ges. 53A (1920), 145-146
B. Heymann, Z. angew. Chem. 33 (1920), 169

BAYER, GUSTAV [1879-1938]
HUTER, 160-163; FISCHER 1,83; POGG 7a(1),109-110
T. Wense, Wiener klin. Wchschr. 59 (1947), 735-736

BAYER, HEINRICH [1853-1926]
FISCHER 1,83; AD 3,160-161; GAUSS, 145
R. Freund, Z. Gyn. 50 (1926), 1431-1435

BAYER, JOSEPH (FRANZ LUDWIG) [1889-]
JV 30,727; NUC 40,519

BAYER, KARL [1854-1930]
STURM 1,62; FISCHER 1,83-84; PUTZ, 2-11

BAYER, KURT [1888-]
JV 31,221

BAYER, OTTO [1902-1982]
KURSCHNER 14,185; POTSCH, 32-33; POGG 7a(1),110
Anon., Deutsche Apoth. Z. 102 (1962), 1457; Chem. Z. 106 (1982), 348
H. Holtschmidt, Arzneimitt. 22 (1972), 2012
K.H. Buechel et al., Chem. Ber. 120 (1987), xxi-xxxv

BAYLISS, LEONARD ERNEST [1900-1964]
DSB 1,533-535; IB, 18
F.R. Winton and A.V. Hill, Nature 204 (1964), 327-328

BAYLISS, WILLIAM MADDOCK [1860-1924]
DSB 1,535-538; DNB [1921-1930], 69-70; FISCHER 1,84; O'CONNOR(2),135-139; W, 38-39; WWWS, 133; MEDVEI, 708-709; POGG 5,77-78; 6,147-148
W.D. Halliburton, Biochem. J. 18 (1924), 1185-1186
J.B., Proc. Roy. Soc. B99 (1926), xxvii-xxxii
E.H. Starling, Erg. Physiol. 25 (1926), xx-xxiv
F.A.E. Crew, Biol. Gen. 1 (1926), 163-166
L.E. Bayliss, Persp. Biol. Med. 4 (1961), 460-479
A.V. Hill, J. Physiol. 204 (1969), 1-13
H.H. Simmer, Med. Welt 29 (1978), 1991-1996

BAYNE-JONES, STANHOPE [1888-1970]
WWA 1968-1969, P.165; NCAB 55,173-174; DAMB, 46-47; IB, 18; WWWS, 43
C.F. Keefer, Trans. Assoc. Am. Phys. 83 (1970), 14-15
J.R. Paul, Yale J. Biol. Med. 45 (1972), 22-32
M.C. Leikind, Bull. N. Y. Acad. Med. 48 (1972), 584-595
G.W. Corner, Am. Phil. Soc. Year Book 1975, pp.118-123
A.M. Harvey, Johns Hopkins Med. J. 149 (1981), 150-166

BAYON, HENRY PETER GEORGE [1876-1952]
Anon., Brit. Med. J. 1952(II), 1260-1261,1424

BAZETT, HENRY CUTHBERT [1885-1950]
DAB [Suppl.4],58-59; AMS 8,151; BROBECK, 157; IB, 18; WWWS, 133
Anon., Lancet 1950(II), pp.309-310; Brit. Med. J. 1950(II), pp.220-221

BAZHENOV, EVGENI IVANOVICH [1856-]
ZMEEV 4,19; SG [2]2,169

BAZHENOV, NIKOLAI NIKOLAEVICH [1857-1923]
BME 3,15-16; WWR, 54-55
M.V. Korkina, Zhur. Nevropat. Psikh. 57 (1957), 1033-1034

BEACH, ELIOT FREDERICK [1911-1988]
AMS 15,352

BEADLE, GEORGE WELLS [1903-1989]
WWA 1978-1979, p.208; IWW 1982-1983, p.92; AMS 14,287; NCAB I,372-373; MH 1,28-30; MSE 1,68-70; STC 1,96-98; CB 1956, pp.37-39; WWWS, 134; SCHULZ-SCHAEFFER, 211; POTSCH, 33
G.W. Beadle, Ann. Rev. Biochem. 43 (1974), 1-13
L.E. Kay, J. Hist. Biol. 22 (1989), 73-101
N.H. Horowitz, Am. Phil. Soc. Year Book 1989, pp.161-171; Genetics 124 (1990), 1-6; Biog. Mem. Nat. Acad. Sci. 59 (1990), 27-52

BEALE, LIONEL SMITH [1828-1906]
DSB 1,539-541; DNB [1901-1911] 1,118-120; HIRSCH 1,399-400; Suppl.,78; O'CONNOR, 42-44; PAGEL, 107-108; BULLOCH, 351; DESMOND, 49
Anon., Lancet 1906(I), pp.1004-1007; Brit. Med. J. 1906(I), pp.836-837
P.H. P.-S., Proc. Roy. Soc. B79 (1907), lvii-lxiii
W.D. Foster, Med. Hist. 2 (1958), 269-273

BEAMS, JESSE WAKEFIELD [1898-1977]
WWA 1976-1977, pp.203-204; AMS 13,255; MH 2,31-33; MSE 1,70-72; NCAB I,346; WWWS, 135; POGG 6,148-149; 7b,275-277
D.W. Kupke, TIBS 2 (1977), N284
E.J. McShane, Am. Phil. Soc. Year Book 1978, pp.45-50
W. Gordy, Biog. Mem. Nat. Acad. Sci. 54 (1983), 3-49

BEAN, JOHN WILLIAM [1901-1987]
WWA 1976-1977, p.204; AMS 14,290; WWWS, 135

BEARD, HOWARD HORACE [1894-1984]
AMS 9(I),112-113

BEARD, JOSEPH WILLIS [1901-1983]
AMS 14,292; WWWS, 136
R.M. Bridges, North Carolina Medical Journal 46 (1985), 303-309

BEARD, RICHARD OLDING [1856-1936]
AMS 5,74
E.P. Lyon, Minnesota Medicine 19 (1936), 683
J.A. Myers, Journal-Lancet 85 (1965), 302-308

BEATTY, WALLACE APPLETON [1877-1938]
New York Times, 2 December 1938, p.23

BEAUCLAIR, LOUIS THEODOR [1813-1846]
R. Bonnet, Genealogisches Lexikon Nassauischen Beamtenfamilien des 19. Jahrhundert, p.72. Frankfurt/M 1930

BEAUMONT, WILLIAM [1785-1853]
DSB 1,542-545; DAB 2,104-110; KELLY, 82-85; W, 39-40; WWWS, 137; WWAH,45; DAMB, 50; ELLIOTT, 27-28; HIRSCH 1,404; CALLISEN 2,8-9; 26,189-190)J.S. Myer, Life and Letters of William Beaumont (2nd Ed.). St.Louis 1939
G. Rosen, The Reception of William Beaumont's Discovery in Europe. New York 1942
J.J. Bylebyl, J. Hist. Med. 25 (1970), 3-21

BÉCHAMP, ANTOINE [1816-1908]
DSB 15,11-12; DBF 5,1236-1237; DBA, 143; HIRSCH 1,407; Suppl.,80; PAGEL, 1926-1927; GE 5,1107; TSCHIRCH, 1018; HUMBERT, 202; POTSCH, 33; DORVEAUX, 78; WWWS, 137; POGG 3,88-89; 4,82-83
Anon., Mon. Sci. 22 (1908), 790-800; Nature 78 (1908), 13-14
E.D. Hume, Béchamp or Pasteur. Chicago 1923
F. Guermonprez, Béchamp, Etudes et Souvenirs. Paris 1927
A. Valerie, Béchamp et l'Evolution Européenne. Paris 1927; Béchamp et le Bon Sens. Paris 1958; De Béchamp a Lazzaro Spallanzani. Paris 1963
P. Pages, Mons. Hipp. 2 (1959), 13-59
P. Decourt, Arch. Int. Cl. Bern. 1 (1972), [2],23-118; Hist. Sci. Med. 18 (1984), 147-151
M. Villemin, Bull. Acad. Sci. Nancy 11 (1972), 276-284
M. Nonclerq, Rev. Hist. Pharm. 25 (1978), 257-262; Antoine Béchamp 1816-1908. Paris 1982

BÉCHAMP, MARIE JOSEPH [1847-1893]
DBF 5,1237; DORVEAUX, 83

BECHER, ERNST SIEGFRIED [1884-1926]
NDB 1,689; FREUND 1,93-103; WWWS, 137-138
E. Merker, Zool. Jahrb. 43 (1927), 431-538

BECHER, FRIEDRICH ERWIN [1890-1944]
FISCHER 1,84-85; KALLMORGEN, 218; POGG 6,151; 7a(1),111-112
K.H. Hildebrand, Deutsche med. Wchschr. 71 (1946), 73

BECHHOLD, HEINRICH JAKOB [1866-1937]
NDB 1,691-692; FISCHER 1,85; ARNSBERG, 43-44; KALLMORGEN, 218; POTSCH, 34; TLK 3,29; WININGER 7,522; TETZLAFF, 20; ZWEIFEL, 13-14; IB, 18; WWWS, 138; POGG 4,83; 5,78; 6,151-152; 7a(1),113-114
H. Bechhold, Z. angew. Chem. 39 (1926), 440,770

BECK, ADOLF [1863-1942]
FISCHER 1,86; BRAZIER, 194-204,209-211
B. Zernicki, Acta Physiol. Pol. 38 (1987), 114-122

BECK, JOHN CHRISTIAN [1924-]
AMS 15(1),368; WWA 1980-1981, p.97; WWWS, 138

BECK, LOUIS CALEB [1798-1853]
DAB 2,116; MILES, 26-27; ELLIOTT, 28; WWWS, 138; WWAH, 45; FERCHL, 32; SILLIMAN, 55; POGG 1,125-126
L.F. Kebler, Ind. Eng. Chem. 16 (1924), 968-970

BECK, LUDWIG [1880-]
JV 23,538; NUC 42,263BECK, LYLE VIBERT [1906-]
AMS 14,299

BECK, WILHELM [1887-1964]
JV 27,422; NUC 42,287

BECKER, FERDINAND WILHELM [1805-1834]
HIRSCH 1,414

BECKER, GÜNTHER [1912-1980]
KURSCHNER 13,182

BECKER, PAUL [-1925]
JV 15,75; NUC 42,421; RSC 13,391

BECKMANN, ERNST OTTO [1853-1923]
DSB 1,553-554; NDB 1,725-726; DBJ 5,422; HEIN, 35-36; POTSCH, 34-35; BLOKH 1,36; DZ, 71; LADIS, 30-60; W, 40-41; WWWS, 141; POGG 4,86-87; 5,81-82; 6,156-157; 7aSuppl.,55-56
C.R. Platzmann, Chem. Z. 47 (1923), 629
G. Lockemann, Z. angew. Chem. 36 (1923); Ber. chem. Ges. 61A (1928), 87-130; Chem. Z. 77 (1953), 433-434

BECKMANN, OTTO [1832-1860]
HIRSCH 1,416; DIEPGEN, 66-68; PAGEL, 115-116
R. Virchow, Arch. path. Anat. 19 (1860), 557-562

BECKURTS, HEINRICH AUGUST [1855-1929]
HEIN, 36-37; REBER, 281-285; TSCHIRCH, 1018; SCHAEDLER, 8; TLK 3,32; DZ, 72; POGG 3,91; 4,87-88; 5,82; 6,157
C.A. Rojahn, Arch. Pharm. 267 (1929), 509-514
W. Schneider, Arch. Pharm. 287 (1954), 357-360
C. Friedrich and K. Koehle, Pharmazie 44 (1989), 287-291

BÉCLARD, JULES AUGUSTE [1817-1887]
DBF 5,1244-1245; HIRSCH 1,417; PAGEL, 116-117; WWWS, 141
C. Sappey, Bull. Acad. Med. [2]17 (1887), 199-203

BÉCLÈRE, ANTOINE [1856-1939]
DBF 5,1246; FISCHER 1,88; SCHALL, 12-16
M. Kahn et al., J. Radiol. Elect. 20 (1936), 533-588
G. Forssall, Acta Radiologica 20 (1939), 521-536
E. Rist, Presse Med. 47 (1939), 801-804

A.M. Baudouin, Bull. Acad. Med. 121 (1939), 351-356; 132 (1948), 684-689
A. Béclère, Antoine Béclère. Paris 1973

BECQUEREL, ALEXANDRE EDMOND [1820-1891]
DSB 1,555-556; DBF 5,1249-1251; BLOKH 1,37; WWWS, 141; POGG 1,129-130; 3,93; 4,88
Anon., Nouv. Arch. Hist. Nat. [3]3 (1891), i-xviii
W.C., Proc. Roy. Soc. 51 (1892), xxi-xxiv

BECQUEREL, ALFRED [1814-1862]
DBF 5,1248; HIRSCH 1,417-418; Suppl.,81; PAGEL, 117-118; FRANKEN, 175; DECHAMBRE 8,690; WWWS, 141

BECQUEREL, ANTOINE CÉSAR [1788-1878]
DSB 1,557-558; DBF 5,1248-1249; BLOKH 1, 36-37; SCHAEDLER, 8; FERCHL, 73; WWWS, 141; POGG 1,128-129; 3,91-93
A. Fizeau and G. Dubrée, C. R. Acad. Sci. 86 (1878), 125-131
Anon., Nature 17 (1878), 244-245; Am. J. Sci. [3]15 (1878), 239-240

BECQUEREL, HENRI [1852-1908]
DSB 1,558-561; DBF 5,1251-1252; BLOKH 1,37-38; POTSCH, 35; GE 5,1123; W, 41; WWWS, 141; POGG 3,93; 4,88; 5,82-83
G. Darboux et al., C. R. Acad. Sci. 147 (1908), 443-451
W.C., Proc. Roy. Soc. A83 (1909-1910), xx-xxiii
O. Lodge, J. Chem. Soc. 101 (1912), 2005-2042
E.H. Huntress, Proc. Am. Acad. Arts Sci 81 (1952), 40-42
A.B. Garrett, J. Chem. Ed. 39 (1962), 533
L. Badash, Arch. Inter. Hist. Sci. 18 (1965), 55-66

BECQUEREL, PAUL [1879-1955]
DSB 1,561-562; WWWS, 141
R. Combes, C. R. Acad. Sci. 241 (1955), 137-140
E. Boureau, Rev. Gen. Bot. 64 (1957), 133-139

BEDALL, KARL [1827-1895]
HEIN, 37-38; TSCHIRCH, 1019
Anon., Pharm. Z. 38 (1893), 189-190; 40 (1895), 639-640

BEDALL, KARL [1858-1930]
HEIN, 38; TSCHIRCH, 1019; NUC 42,602; RSC 9,167
F. Ferchl, Pharm. Z. 73(1928),785-787; Chem. Z. 52(1928),539;54(1930),385
Anon., Süddeutsche Apoth. Z. 70 (1930), 262

BEDSON, SAMUEL PHILLIPS [1886-1969]
DNB [1961-1970], 86-87; WW 1965, p.211
A.W. Downie, Biog. Mem. Fell. Roy. Soc. 16 (1970), 15-35

BEEBE, SILAS PALMER [1876-1930]
Obituary Record of Yale Graduates No.90, pp.261-262. New Haven,1931

BEENSCH, LEO [1869-]
NUC 43,138; RSC 14,1009

BEER, AUGUST [1825-1863]
NDB 1,734; ADB 2,245-246; POTSCH, 36; POGG 1,132-133,1534; 3,94-95
H.G. Pfeiffer and H.A. Liebhafsky, J. Chem. Ed. 28 (1951), 123-125

BEER, GEORG JOSEF [1763-1821]
NDB 1,735; ADB 2,248-249; OBL 1,63-64; WURZBACH 1,222; HEID, 13-18; HIRSCH 1,422-423; Suppl.,81-82; DECHAMBRE 8,693-694; WWWS, 143 H.M. Koelbing, Clio Medica 5 (1970), 225-248
D.M. Albert and F.C. Blodi, Documenta Ophthalmologica 68 (1988), 79-103

BEESON, PAUL BRUCE [1908-]
AMS 14,310; WWA 1976-1977, p.218; ICC, 347-349

BEEVERS, HARRY [1924-]
AMS 15(1),382; WWA 1980-1981, p.234; IWW 1982-1983, p.97; WWWS, 143

BÉGUIN, CHARLES [1874-1941]
P. de Chastenay, Schw. Apoth. Z. 79 (1941), 357-359

BÉGUIN, CHARLES [1900-1974]
POGG 7a(1),126-127
A. Bédal, Schw. Apoth. Z. 112 (1974), 168

BÉHAL, AUGUSTE [1859-1941]
DBF 5,1285; GORIS, 224-227; TSCHIRCH, 1019; WWWS, 144; POGG 4,90; 5,84; 6,160; 7b,286-287
E. Choay, J. Pharm. Chim. [9]1 (1941), 512-518
M. Tiffeneau, Bull. Acad. Med. 105 (1941), 342-353
M. Sommelet, Presse Med. 49 (1941), 349-350
M. Delépine, Rev. Hist. Pharm. 14 (1960), 249-254,301-308
G. Dillemann, Produits et Problèmes Pharmaceutiques 27 (1972), 918-920

BEHN, ULRICH [1868-1908]
BJN 13,11*; POGG 4,90; 5,84
Anon., Chem. Z. 34 (1908), 478

BEHR, ARNO [1846-1921]
AMS 3,50
Anon., Z. angew. Chem. 34 (1921), 596; Chem. Z. 45 (1921), 1142

BEHREND, PAUL [1853-1905]
KLEIN, 47-48; BLOKH 1,38-39; RSC 9,172; 12,63; 13,410
E. Glinn, Z. angew. Chem. 18 (1905), 849-852
Anon., Chem. Z. 29 (1905), 379

BEHREND, ROBERT [1856-1926]
DSB 15,12-13; NDB 2,11; DZ, 75-76; HANNOVER, 30; POTSCH, 37; WWWS, 144; POGG 4,90-91; 5,84-85; 6,162; 7a(1),130
A. Skita, Ber. chem. Ges. 59A (1926), 158-164
J.T. Stock, J. Chem. Ed. 69 (1992), 197-199

BEHRENDT, HANS JOSEPH [1894-1974]
KALLMORGEN, 219
Anon., N. Y. J. Med. 74 (1974), 1679
M. Green, Proc. Virchow Soc. 30 (1976), 11-12

BEHRENS, BEHREND [1895-1969]
KURSCHNER 11,150; DRULL, 16; POGG 7a(1),130-131
G. Zetler, Arzneimitt. 10 (1960), 416; 20 (1970), 1823-1824
O. Orzechowski, Arzneimitt. 15 (1965), 582-583

BEHRENS, MARTIN GERHARD [1899-1971]
KURSCHNER 13,200; WWWS, 144; POGG 7a(1),131-132

BEHRENS, OTTO KARL [1911-1989]
AMS 15(1),385

BEHRENS, THEODOR HEINRICH [1842-1905]
NNBW 4,102-103; BJN 10,144*; KIEL,210; POGG 3,96; 4,91; 5,85
S. Hoogewerff, Rec. Trav. Chim. Pays-Bas 24 (1905), 147-164
P. Kley, Chem. Wkbl. 2 (1905), 131-134

BEHRENS, WILHELM JULIUS [1854-1903]
NDB 2,12-13; BJN 8,11*; FREUND 1,105-109; WAGENITZ, 24
E. Oppermann, Z. wiss. Mikr. 20 (1903), 273-278
E. Küster, Ber. bot. Ges. 22 (1904), (39)-(44)

BEHRING, EMIL von [1854-1917]
DSB 1,574-578; NDB 2,14-15; DBJ 2,21-26,647; LKW 1,10-14; BULLOCH, 352; GUNDLACH, 271; DZ, 77; FISCHER 1,90-91; AUERBACH, 193-194; W, 42-43; WWWS, 1729; OLPP, 30-32; PAGEL, 125-126; POGG 7aSuppl., 59-65
H. Zeiss and R. Bieling, Behring, Gestalt und Werk. Berlin 1940
Anon., Mitt. Gesch. Med. 39 (1940), 260-265
H. Schadewaldt, Deutsche med. Wchschr. 100 (1975), 2172-2178; Med. Welt 30 (1979), 1795-1801

BEHRINGER, HANS [1911-]
KURSCHNER 15,238; POGG 7a(1),132

BEIJERINCK, MARTINUS WILLEM [1851-1931]
DSB 15,13-15; BWN 2,32-34; LINDEBOOM, 130-131; BULLOCH, 352; ABBOTT, 18-19; SACKMANN, 27-30; DOETSCH, 179-181; POGG 6,162
J. Smith, Chem. Wkbl. 28 (1931), 94-97
A. Mayer, Naturwiss. 19 (1931), 302-305
S.A. Waksman, Sci. Mon. 33 (1931), 285-288
A. Trotter, Marcellia 27 (1932), 120-124
W.B., Proc. Roy. Soc. B109 (1932), i-iii
G. van Iterson, Ber. bot. Ges. 52 (1934), (115)-(158)
G. van Iterson et al., Martinus Willem Beijerinck, his Life and Work. The Hague 1940 (reprinted Madison, Wis. 1983)

BEILSTEIN, FRIEDRICH KONRAD [1838-1906]
DSB 1,578-579; NDB 2,20; ARBUZOV, 102; BLOKH 1,39-40; POTSCH, 37; SCHAEDLER, 9; WWWS, 145; POGG 3,96-97; 4,91; 5,85; 7aSuppl.,65-66
O. Lutz, Z. angew. Chem. 19 (1906), 2058-2060
P. Jacobson, Chem. Z. 30 (1906), 1057-1058
E. Hjelt, Ber. chem. Ges. 40 (1907), 5041-5078
O.N. Witt, J. Chem. Soc. 99 (1911), 1646-1649
F. Richter, Ber. chem. Ges. 71A (1938), 35-55; Z. angew. Chem. 51 (1938), 101-107

BEIN, SIGISMUND [1859-1933]
TLK 2,38; RSC 13,415; POGG 4,91-92
Anon., Öst. Chem. Z. 36 (1933), 144-145; Ber. chem. Ges. 66A(1933),74; Z. angew. Chem. 46 (1933), 570

BEIN, WILLY [1869-1943]
KURSCHNER 4,149-150; TLK 3,34; POGG 4,92; 5,85-86; 6,162; 7a(1),133-134

BEINERT, HELMUT [1913-]
AMS 15(1),387
R. Camack, Bioch. Soc. Trans. 13 (1985), 541

BEISSENHIRTZ, FRIEDRICH WILHELM [1779-1831]
HEIN,42-43; TSCHIRCH, 1019; FERCHL, 34
J. Faber, Arch. Pharm. 99 (1847), 105-109

BEITLER, KARL IVANOVICH [1872-]
SG [3]2,426

BEKETOV, ANDREI NIKOLAEVICH [1825-1902]
BSE 4,401; HIRSCH 1,435; Suppl.,83; KUZNETSOV(1963), 116-125; LIPSHITS 1,153-157
A.A. Shcherbakova, Trudy Inst. Ist. Est. 5 (1953), 211-214;
A.N. Beketov. Moscow 1958
A.G. Voronov, Bull. Soc. Nat. Moscou NS80 (1975), [5],136-140

BEKETOV, NIKOLAI NIKOLAEVICH [1827-1911]
DSB 1,579; BSE 4,401-402; BME 3,601-602; ARBUZOV, 175-176; BLOKH 1,40-42; KHARKOV-F, 103-111; KOZLOV, 141-142; VENGEROV 2,366-374; POTSCH, 37; POGG 4,92-93; 5,86
I.P. Osipov et al., Zhur. Fiz. Khim. 45 (1913), 383-433
Y.I. Turchenko, N.N. Beketov. Moscow 1954

BEKHTEREV, VLADIMIR MIKHAILOVICH [1857-1927]
DSB 1,579-581; BSE 5,124-126; BME 3,880-881; KAZAN 2,130-137; WWR, 56-57; FISCHER 1,85-86; KUZNETSOV(1963), 592-604; HAYMAKER, 167-171; BRAZIER, 227-233,246-247; WWWS, 145

BĚLAŘ, KARL [1895-1931]
DSB 17,64-66; IB, 19-20
C. Stern, Naturwiss. 19 (1931), 921-923

BELCHER, RONALD [1909-1982]
CAMPBELL, 29-30
W.I. Stephen, Anal. Proc. 19(1982),455-460; Chem. Brit. 19(1983),325-326

BELEHRADEK, JAN [1896-1980]
IB, 20; AO 1980, pp.289-291BELFANTI, SERAFINO [1860-1939]
DBI 7,563-565; AGRIFOGLIO, 59-64; FISCHER 1,92; IB, 20
Anon., Biochem. J. 33 (1939), 864-865
A. Coppadoro, Chimica e Industria 21 (1939), 171
D. Carbone, Riv. Biol. 30 (1940), 411-434

BELITSER, VLADIMIR ALEKSANDROVICH [1906-1988]
BSE(3rd Ed.) 3,110; UKRAINE 2,187-188; WWWS, 146
I. Talberg, Who's Who in Soviet Science and Technology, p.22
T.V. Varetskaya and A.P. Demchenko, TIBS 13 (1988), 248; Thromb. Res. 55 (1989), 1-4

BELL, CHARLES [1774-1842]
DSB 1,583-584; DNB 2,154-157; HIRSCH 1,438-441; Suppl.,85; W, 45; BETTANY 1,242-263; O'CONNOR, 3-7; BRAZIER, 47-50,61-62; WWWS, 147; CALLISEN 2,73-85; 26,218-223
Anon., Proc. Roy. Soc. 4 (1837-1843), 402-406
A. Pinchot, The Life and Labours of Sir Charles Bell. London 1860
G. Gordon-Taylor and E.W. Wells, Sir Charles Bell. London 1958

BELL, CHARLES JOHN [1854-1903]
NUC 44,576; RSC 9,176
Anon., Science 17 (1903), 119

BELL, DAVID JAMES [1905-1972]
SRS, 90
D.J. Manners, Chem. Brit. 9 (1973), 278
J.S.D. Bacon and D.J. Manners, Adv. Carbohydrate Chem. 39 (1974), 1-8

BELL, FREDERICK KELLER [1895-1974]
AMS 11,334

BELL, GEORGE HOWARD [1905-1987]
WW 1987, p.123; WWWS, 147
Who's Who of British Scientists 1971-1972, p.65

BELL, JAMES [1824-1908]
WWWS, 147
T.E.T., Nature 77 (1908), 539-540

BELL, PAUL HADLEY [1914-]
AMS 15(1),395

BELL, RICHARD DANA [1887-1925]
O.F. Rogers, Harvard Class of 1908, 25th Anniversary Report, pp.42-43. Cambridge, Mass. 1933

BELL, RONALD PERCY [1907-]
WW 1981, p.188; IWW 1982-1983, p.100; CAMPBELL, 31-32; WWWS, 147

BELL, WILLIAM BLAIR [1871-1936]
FISCHER 1,93
Anon., Am. J. Cancer 26 (1936), 787

BELLAMY, FÉLIX [1828-]
KERVILLER 2,359-361; NUC 697,73; RSC 13,423

BELLANI, ANGELO [1776-1852]
DSB 1,585-587; EI 6,547; WWWS, 148; POGG 1,139-140,1535
G. Cantoni, Rend. Ist. Lombardo 10 (1877), 17-23; 11 (1878), 873-880

BELLENOT, GUSTAVE [1858-1935]
NUC 45,178; RSC 17,793
P. Vouga, Bulletin de la Société Neuchateloise des Sciences Naturelles 60 (1935), 229-230

BELLING, JOHN [1866-1933]
AMS 5,81; SCHULZ-SCHAEFFER, 209; DESMOND, 55-56
C.B. Davenport, Science 77 (1933), 250-251
E.B. Babcock, J. Heredity 24 (1933), 296-300; Genetics 21 (1936), i-iii
A.F. Blakeslee, in H.J. Conn, The History of Staining, pp.113-115. Geneva, N.Y. 1933

BĚLOHOUBEK, ANTONIN [1845-1910]
A. Nyrdle, Chem. Listy 5 (1911), 42-43

Anon., Chem. Z. 35 (1911), 34

BĚLOHOUBEK, AUGUST [1847-1908]
TSCHIRCH, 1019; POGG 3,103; 4,94
Anon., Chem. Z. 32 (1908), 509

BELOUSOV, NIKOLAI FEDOROVICH [1863-1942]
VORONTSOV, 54-55; VENGEROV 1,100

BELOZERSKI, ANDREI NIKOLAEVICH [1905-1972]
DSB 17, 68-69; BSE (3rd Ed.) 3,122; MH 2,33-34; MSE 1,73-75; STC 1,105-106
Anon., Leopoldina [3]17 (1971), 33-34; Usp. Sov. Biol. 80 (1975), 3-4
A.S. Spirin and A.N. Shamin, Priroda 1976, [10],pp.72-82

BEMMELEN, JAKOB MAARTEN van [1830-1911]
BWN 1,37; BLOKH 1,43-44; WWWS, 1706; POTSCH, 38; POGG 3,105-106; 4,95; 5,1288; 6,169
C. Liebermann, Ber. chem. Ges. 44 (1911), 811-812
W.P. Jorissen, Chem. Z. 35 (1911), 321
W.B.H., Nature 86 (1911), 116-117
H.A.M. Snelders, Tijd. Gesch. Geneesk. 9 (1986), 10-24

BENACERRAF, BARUJ [1920-]
AMS 15(1),402; WWA 1980-1981, p.245; IWW 1982-1983, p.103; MSE 1,75-76
B. Benacerraf, Immunological Reviews 84 (1985), 7-27;
Ann. Rev. Immunol. 9 (1991), 1-26

BENARIO, JACOB [1868-1916]
KALLMORGEN, 220; ARNSBERG, 45
H. Sachs, Jahresb. Aerzt. Ver. 1916, pp.111-113
Anon., Chem. Z. 40 (1916), 1044; Z. angew. Chem. 29 (1916), [III],674

BENARY, ERICH [1881-1941]
GEDENKBUCH 1,92; JV 21,404; POGG 5,90; 6,171-172

BENCKISER, ANNELIESE [1896-]
JV 39,516

BENCKISER, EDMUND [1818-1836]
DGB 81 (1934), 27

BENCKISER, OSKAR [1828-1912]
STELLING-MICHAUD 2,168

BENCOWITZ, ISAAC [1896-1972]
AMS 12,404

BENDA, CARL [1857-1932]
FISCHER 1,94-95; PAGEL, 129-130; WININGER 7,525; TETZLAFF, 22; KOREN, 153; IB, 20; BERLIN, 575
G. Levi, Arch. exp. Zellforsch. 13 (1932), i-iv
R. Jaffe, Anat. Anz. 76 (1933), 277-285

BENDA, LOUIS LUDWIG [1873-1945]
KALLMORGEN, 220; BHDE 2,76-77; LDGS, 16; POGG 5,90; 6,172; 7a(1),135
Anon., Z. angew. Chem. 37 (1924), 608
H. Ritter, Chem. Ber. 90 (1957), i-xiii

BENDER, FRITZ [1859-1908]
BLOKH 1,44-45
G. Kraemer, Ber, chem. Ges. 41 (1908), 1103

BENDER, GEORG [1838-1918]
RSC 13,435
Anon., Chem. Z. 42 (1918), 630

BENDER, MYRON LEE [1924-1988]
AMS 15(1),404; WWA 1980-1981, p.246; IWW 1982-1983, pp.103-104; WWWS, 150
J.B. Lambert, Chem. Brit. 25 (1989), 70

BENDICH, AARON [1917-1979]
AMS 14,329M.E. Balis et al., Cancer Research 40 (1980), 493

BÉNECH, ÉLOPHE [1865-1944]
J. Guérin and B. Guérin, Des Hommes et des Activités autour d'un Demisiècle, p.65. Bordeaux 1957

BENECKE, BERTHOLD HEINRICH [1843-1886]
KROLLMANN, 46-47
Anon., Leopoldina 22 (1886), 59; Schr. Phys. Ges. Königsberg 27 (1887), 17-18

BENECKE, FRANZ [1857-1903]
A. Wieler, Ber. bot. Ges. 21 (1903), (23)-(31)

BENECKE, WILHELM [1868-1946]
NDB 2,41*; KURSCHNER 5,71; WENIG, 17-18; SACKMANN, 31-32; IB, 21;KIEL, 183

BENEDEN, EDOUARD van [1846-1910]
DSB 1,600-602; BNB 26,174-184; HIRSCH 1,453-454; Suppl.,85; PAGEL, 130; MAYERHOFER 1,411-412; SCHULZ-SCHAEFFER, 200; W, 46-47; WWWS, 1706
A.S., Nature 83 (1910), 344-345
A. Brachet, Anat. Anz. 36 (1910), 598-607; Arch. Biol. 33 (1923), 1-59
K. Rabl, Arch. mikr. Anat. 88 (1915), 1-470
G. Hamoir, Rev. Med. Liège 41 (1986), 779-785

BENEDETTI-PICHLER, ANTON ALEXANDER [1894-1964]
AMS 10,268; MILES, 29-30; POGG 6,172; 7b,307-309
Anon., Chem. Eng. News 42[52] (1964), 53
H.K. Alber and W.R. Alber, Nature 205 (1965), 1158-1159
M.K. Zacherl, Mikrochimica Acta 1965, pp.204-206

BENEDICT, FRANCIS GANO [1870-1957]
DSB 1,609-611; AMS 8,175; MILES, 30-31; CHITTENDEN, 61-65; DAMB, 87-88; IB, 20; WWWS, 150; POGG 5,92; 6,173-174; 7b,311
O. Riddle, Am. Phil. Soc. Year Book 1957, pp.109-113
E.F. DuBois and O. Riddle, Biog. Mem. Nat. Acad. Sci. 32 (1958), 66-98
L.A. Maynard, J. Nutrition 98 (1969), 1-8

BENEDICT, STANLEY ROSSITER [1884-1936]
DAB 22,35-36; MILES, 31-32; CHITTENDEN, 138-143; OEHRI, 20-21; WWWS,151; POGG 6,174; 7b,311-312
H.D. Dakin, Science 85 (1937), 65-66
Anon., Ind. Eng. Chem. News Ed. 15 (1937), 38
E.V. McCollum, Biog. Mem. Nat. Acad. Sci. 27 (1952), 155-172
H. Pollack, Diabetes 2, (1953), 420-421

BENEDIKT, MORIZ [1835-1920]
DBJ 2,739; JOHN, 93-128; HIRSCH 1,455; Suppl.,86; PAGEL, 130-132; HAYMAKER, 408-410; KOREN, 153; TLK 2,39-40
M. Benedikt, Aus meinem Leben. Vienna 1906
J.P. Karplus, Wiener klin. Wchschr. 33 (1920), 387
P. Gulyas, Magyar Irok 2 (1940), 1094-1095
H.F. Ellenberger, Rev. Hist. Med. Heb. 27 (1974), 133-142

BENEDIKT, RUDOLF [1852-1896]
BJN 1,322-324; 3,89*; OBL 1,69; BLOKH 1,45; WININGER 1,305; POGG 3,107; 4,97
Anon., Leopoldina 32 (1896), 59; Chem. Z. 20 (1896), 1036
C. Liebermann, Ber. chem. Ges. 29 (1896), 407-409

BENEKE, FRIEDRICH WILHELM [1824-1882]
ADB 46,355; LKW 1,15-20; HIRSCH 1,456-457; PAGEL, 132-133
C.A. Ewald, Berl. klin. Wchschr. 20 (1883), 15-16

BENEKE, RUDOLF [1861-1946]
KURSCHNER 6,93; FISCHER 1,95-96; IB, 21
M.B. Schmidt, Verhandl. path. Ges. 33 (1950), 403-406
W. Kaiser and W. Piechowski, Zeitschrift für die gesamte Hygiene 18 (1972), 438-442

BENESCH, REINHOLD [1919-1986]
AMS 15(1),408; WWA 1980-1981, p.248
New York Times, 1 January 1987

BENETATO, GRIGORE ALEXANDRU [1905-1972]
WWWS, 151
Anon., Rev. Roum. physiol. 9 (1972), 261-263

BENJAMIN, RICHARD [1871-1943]
GEDENKBUCH 1,97; NUC 46,543; RSC 13,442; SG [2]2,229

BENNET-CLARK, THOMAS ARCHIBALD [1903-1975]
DNB [1971-1980], pp.46-47; WW 1976, p.177; DESMOND, 683; IB, 21
R. Brown, Biog. Mem. Fell. Roy. Soc. 23 (1977), 1-18

BENNETT, ALEXANDER HUGHES [1848-1901]
FISCHER 1,97
Anon., Brit. Med. J. 1901(II), p.1444

BENNETT, GEORGE MacDONALD [1892-1959]
WW 1958, p.228; WWWS, 152; POGG 6,177; 7b,315-316
R.D. Haworth, Biog. Mem. Fell. Roy. Soc. 5 (1959), 23-26
E.E. Turner, Proc. Chem. Soc. 1959, pp.197-200
A.D. Mitchell, Nature 183 (1959), 717-718

BENNETT, HENRY STANLEY [1910-1992]
AMS 17(1),440; WWWS,152-153
New York Times, 22 August 1992

BENNETT, IVAN LOVERIDGE, Jr. [1922-1990]
AMS 17(1),440; WWA 1988-1989, p.224; WWWS, 153

BENNETT, JOHN HUGHES [1812-1875]
DNB 2,244-246; HIRSCH 1,461-462; Suppl.,87; PAGEL, 135-136; WWWS, 153; BETTANY 2,209-216; O'CONNOR, 96-98
J.G. McKendrick, Brit. Med. J. 1875(II), pp.473-478;
Edinburgh Med. J. 21 (1875), 466-474
D.H. Scott, Proc. Roy. Micr. Soc. 2 (1900), 145-148
J.H. Warner, Med. Hist. 24 (1980), 241-248

BENRATH, ALFRED [1878-1969]
KURSCHNER 11,162; TLK 3,35; POGG 5,94; 6,178; 7a(1),137-138
W. Klemm et al., Z. anorg. Chem. 295 (1958), 153-155
E. Baum, Chem. Z. 93 (1969), 152

BENSCH, AUGUST [1817-]
FERCHL, 35; RSC 1,276; POGG 1,144

BENSLEY, ROBERT RUSSELL [1867-1956]
DSB 1,613-614; DAB [Suppl.6], 53-54; AMS 9(II),81; WWAH, 49
H.T. Ricketts, Diabetes 4 (1955), 334-335
E.V. Cowdry, Science 124 (1956), 972-973
N.L. Hoerr et al., Anat. Rec. 128 (1957), 1-18
F. Wassermann, Anat. Anz. 104 (1957), 443-450

BENTHEIM, ALFONS von [1862-]
JV 23,538; NUC 47,336

BENTHIN, WALTHER [1882-1950]
FISCHER 1,97; GAUSS, 145-146

BENZ, MAX [1879-1939]
JV 22,515; NUC 47,468

BENZER, SEYMOUR [1921-]
AMS 14,346; WWA 1980-1981, p.260; IWW 1982-1983, pp.108-109; COHEN, 20; MH 2,34-37; MSE 1,77-80; STC 1,106-108; WWWS, 155-156

BÉRARD, JACQUES ÉTIENNE [1789-1869]
DSB 1,616-617; DBF 5,1464-1465; GE 6,255; WWWS, 156; CALLISEN 2,122-123; POGG 1,146; 3,108

BÉRARD, PIERRE HONORÉ [1797-1858]
DBF 5,1466; DBA 1,171

BERBERICH, JOSEPH [1897-1969]
KURSCHNER 11,165; FISCHER 1,98; ARNSBERG, 45; KALLMORGEN, 221; IB, 21; KOREN, 153; BHDE 2,81; LDGS, 74

BERBLINGER, WALTHER [1882-1966]
KURSCHNER 10,137-138; FISCHER 1,98; KIEL, 99; BHDE 2,81; IB, 21; WWWS, 156
I. Kracht, Endokrinologie 50 (1966), 207-208

BERCZELLER, LASZLO [1885-1955]
MEL 3,74; IB, 21; POGG 6,179
F. Arnould, Rev. Hist. Med. Heb. 13 (1960), 153-168
H. Baruk, Hist. Sci. Med. 8 (1974), 235-239

BERDEZ, JULES [1858-1925]
DHB 2,63
Schweizerisches Zeitgenossenlexikon 1921, p.46
Anon., Schw. med. Wchschr. 55 (1925), 748

BERENBLUM, ISAAC [1903-]
STC 1,108-109; IWW 1981-1982, p.107
I. Berenblum, Cancer Research 37 (1977), 1-7

BEREND, MAX [-1888]
RSC 1,287
Anon., Ber. chem. Ges. 21 (1888), 2189

BERENDES, JULIUS [1837-1914]
NDB 2,59; HEIN, 44-45; TSCHIRCH, 1020; SCHELENZ, 697
H. Salzmann, Apoth. Z. 22 (1907), 225
H. Peters, Chem. Z. 38 (1914), 977-978

BERENDES, JULIUS [1907-]
KURSCHNER 13,221-222; WWWS, 156

BEREZKIN, PETR KUZMICH [1867-]
SG [2]2,241

BERG, ARMAND [1860-1934]
POGG 4,100; 5,94; 6,180; 7b,317
A. Tian, Bull. Soc. Chim. [5] 1 (1934), 1451-1452

BERG, CARL FRIEDRICH [1774-1835]
HEIN, 45; FERCHL, 36; POGG 1,147

BERG, CLARENCE PETER [1900-]
AMS 13,308; BERG, 59-60; WWWS, 157

BERG, OTTO CARL [1815-1866]
NDB 2,76-77; HEIN, 45-46; FERCHL, 36; TSCHIRCH, 1020

BERG, PAUL [1926-]
AMS 14,350; WWA 1980-1981, p.263; IWW 1982-1983, p.110; MH 2,37-38; POTSCH, 38; MSE 1,82-83; ABBOTT, 21; WWWS, 157-158

BERG, RAGNAR [1873-1956]
KURSCHNER 6,1046; FISCHER 1,99; POTSCH, 38-39; POGG 6,181; 7a(1),144-145; (4),136*
W. Zabel, Hippokrates 27 (1956), 331BERG, RICHARD [1889-1944]
POGG 6,181; 7a(1),145
Anon., Z. angew. Chem. 57 (1944), 108

BERG, WALTHER [1878-]
FISCHER 1,100; KURSCHNER 4,162

BERGEAT, ALFRED [1866-1924]
NDB 2,77; JV 7,213; POGG 4,100; 6,182; 7a(1),145

BERGEIM, OLAF [1888-1980]
AMS 11,367; IB, 22

BERGEL, FRANZ [1900-1987]
WW 1981, p.201; CAMPBELL, 33-34; BHDE 2,84; LDGS, 16; POGG 6,182; 7a(1),145-146
A.R. Todd, Biog. Mem. Fell. Roy. Soc. 34 (1988), 3-19

BERGEL, SALO [1868-1937]
WININGER 6,458; KOREN, 154

BERGELL, PETER [1875-]
KURSCHNER 4,162; TLK 3,37; AD 3,41; ASEN, 13

BERGEMANN, CARL WILHELM [1804-1884]
WENIG, 19; FERCHL, 36; POGG 1,147; 3,110; 4,100

BERGER, ARIEH [1920-1972]
BHDE 2,85
E. Katzir-Katchalski, Israel J. Chem. 12 (1974), 1-14

BERGER, ERWIN [1898-1975]
KURSCHNER 12,186; POGG 7a(1),147-148

BERGER, HANS [1873-1941]
DSB 2,1-2; KURSCHNER 6(I),102; FISCHER 1,101; HAYMAKER, 171-175; W, 48; WWWS, 158
Anon., Lancet 1941(II), p.706
A.E. Kornmüller, Münch. med. Wchschr. 88 (1941), 889-890
R. Ginzburg, J. Hist. Med. 4 (1949), 361-371
J. Sayk, Leopoldina [3]19 (1975), 140-147
R. Jung and W. Berger, Arch. Psychiat. 227 (1979), 279-300

BERGER, HERBERT [1888-1973]
JV 28,193; NUC 48,254

BERGER, JEAN FRANÇOIS [1779-1833]
C.W.P. MacArthur, Hist. Sci. Med. 20 (1986), 49-55

BERGH, ALBERT ABRAHAM HIJMANS van den [1869-1943]
BWN 3,283-285; LINDEBOOM, 868-870; HES, 77-79; FRANKEN, 175
K.F. Winkelblech, Ned. Tijd. Gen. 81 (1937), 239
E. Gorter, Jaarb. Akad. Wet. 1943-1944, pp.194-207

BERGHAUS, WILHELM [1873-]
KALLMORGEN, 221-222

BERGHAUSEN, OSCAR [1879-1959]
NCAB 50,233; NUC 48,254

BERGIUS, FRIEDRICH [1884-1949]
DSB 2,3-4; NDB 2,84; W, 48-49; WWWS, 159; HUPKA, 290-296; POTSCH, 39; TLK 3,37; HANNOVER, 44-45; POGG 6,183-184; 7a(1),149-150
P.A. Thiessen, Z. Elektrochem. 50 (1944), 241-242
K. Freudenberg, Sitzungsberichte der Heidelberger Akademie der Wissenschaften 1943-1955, pp.77-79
K. Schmorl, Naturw. Rund. 2 (1949), 271-272
A.N. Stranges, Isis 75 (1984), 643-667
R. Haul, Chemie in Unserer Zeit 19 (1985), 59-67

BERGLUND, EMIL [1846-1887]
SBL 3,585-586; POGG 3,111-112; 4,101; 5,95

BERGLUND, HILDING [1887-1962]
IB, 22
L. Werko, Nord. Med. 68 (1962), 1423-1426
H. Reimann, Trans. Assoc. Am. Phys. 76 (1963), 9-11

BERGMANN, ERNST von [1836-1907]
NDB 2,88-89; BJN 12,11*; NB, 25; HIRSCH 1,480-481; Suppl.,91-92; WELDING, 55-56; KILLIAN, 265-266,349-351; DZ, 88-89; PAGEL, 141-144; WREDE, 13-14; LEVITSKI 2,276-283
J. Brennsohn, Die Aerzte Livlands, pp.98-99. Mitau 1905.
H. Fischer, Deutsche med. Wchschr. 32 (1906), 2039-2041
M. Borchardt, Berl. klin. Wchschr. 44 (1907), 737-742
O. von Angerer, Münch. med. Wchschr. 54 (1907), 837-839
A. Buchholtz, Ernst von Bergmann, 4th Ed. Leipzig 1925

BERGMANN, ERNST DAVID [1903-1975]
BHDE 1,55-56; LDGS, 16; COHEN, 22; TETZLAFF, 26; WWWS, 159; POGG 6,184-185; 7b,328-335
Anon., Nature 256 (1975), 606
K. Ascher and E. Shaya, Phytoparasitica 3 (1975), 145-147

BERGMANN, FELIX [1908-]
BHDE 2,89-90

BERGMANN, GUSTAV von [1878-1955]
FISCHER 1,101-102; KURSCHNER 8,131; KALLMORGEN, 222; FRANKEN, 175-176; AUERBACH, 197G.
Katsch, Münch. med. Wchschr. 97 (1955), 1398-1400
H. Schwiegk, Klin. Wchschr. 33 (1955), 1063-1064
T. von Uexküll, Nervenarzt 27 (1956), 1-2

BERGMANN, MAX [1886-1944]
DSB 15,15-16; DAB [Suppl.3], 60-61; MILES, 32-33; POTSCH, 40; BHDE 2,90; LDGS, 16; TETZLAFF, 26; TLK 3,38; POGG 6,185-187; 7b,335-337
H.T. Clarke, Science 102 (1945), 168-170
C.R. Harington, J. Chem. Soc. 1945, pp.716-718
A. Neuberger, Nature 155 (1945), 419-420
B. Helferich, Chem. Ber. 102 (1969), i-xxvi
G. Reich, Leder Schuhe Lederwesen 21 (1986), 210-213

BERGMANN, WERNER [1904-1959]
AMS 10,287
Anon., Chem. Eng. News 37(47) (1959), 110;
Nachr. Chem. Techn. 7 (1959); Naturw. Rund. 13 (1960), 120

BERGSMA, CORNELIS ADRIAAN [1798-1859]
NNBW 4,121-122; LINDEBOOM, 117-118; UTRECHT, 78; FERCHL, 38; POGG 3,112

BERGSTRÖM, SUNE [1916-]
VAD 1981, p.97; POTSCH, 40

BERKEFELD, WILHELM [1836-1897]
NDB 2,93; FISCHER 1,103; OLPP, 32-33

BERL, ERNST [1877-1946]
NDB 2,93-94; BHDE 2,93; LDGS, 16; TETZLAFF, 26; POTSCH, 40-41; WOLF, 22; WWWS, 160; WWAH, 51; POGG 5,97; 6,188-190; 7a(1),152-154
M. Isler, Helv. Chim. Acta 29 (1946), 957-973
Anon., Chem. Eng. News 24 (1946), 511

BERLÉ, BERNHARD [1866-]
JV 6,237; RSC 13,275

BERLIN, NILS JOHANN [1812-1891]
SBL 3,759-764; SMK 1,280; HIRSCH 1,484; PAGEL, 147; TSCHIRCH, 1020; FERCHL, 38; CALLISEN 26,255; POGG 1,153; 3,113; 4,102
Anon., Hygiea 54 (1892), 97

BERLIN, WILLEM [1825-1902]
NNBW 4,141; LINDEBOOM, 118-119; RSC 1,298

BERLINER, MAX [1888-1961]
KURSCHNER 4,172; ASEN, 13
Anon., N.Y. J. Med. 61 (1961), 2143

BERLINER, ROBERT WILLIAM [1915-]
AMS 17(1),471; BROBECK, 193-197

BERNAL, JOHN DESMOND [1901-1971]
DSB 15,16-20; DNB [1971-1980], 53-54; MH 2,39-40; STC 1,115-117; POTSCH, 41; MSE 1,84-85; WWWS, 161-162; POGG 6,191; 7b,341-343
D. Hodgkin, Birkbeck, Science and History. London 1969; Biog. Mem. Fell. Roy. Soc. 26 (1980), 17-84; Proc. Roy. Irish. Acad. 81B (1981), 11-24
H.F.W. Taylor, Acta Cryst. 28 (1972), 359-360

J.G. Crowther, Brit. J. Hist. Sci. 6 (1972), 104-105
C.P. Snow, in M. Goldsmith and A. Mackay, Society and Science. pp.19-29. New York 1964
M. Goldsmith, Sage: A Life of J.D. Bernal. London 1980.
A. Synge, Notes Roy. Soc. 46 (1992), 267-278

BERNARD, CLAUDE [1813-1878]
DSB 2,24-34; DBF 6,52-54; GENTY 3,129-160; BRAZIER, 50-56,62-63; W,49-50; HIRSCH 1,486-488; Suppl.,93; MAYERHOFER 1,423-424; PAGEL, 148-150; GE 6,362-365; BOURQUELOT, 95-96; HAYMAKER, 175-178; FRANKEN, 176; WWWS, 162; GUIART, 220-221; FERCHL, 39; POTSCH, 41-42
L. Figuier, Ann. Scient. 22 (1878), 487-497
A. Vulpian, C.R. Acad. Sci. 86 (1978), 407-415
P. Bert, Rev. Sci. [2]16 (1879), 741-755
G. Hahn, Rev. Quest. Sci. 7 (1880), 71-112
M. Duval, L'Oeuvre de Claude Bernard. Paris 1881
J. Béclard, Bull. Acad. Med. [2]14 (1885), 714-739
P. van Tieghem, Rev. Sci. 49 (1911), 513-521
A. Dastre, C.R. Soc. Biol. 76 (1914), 2-10
E. Gley, Rev. Sci. 53 (1915), 257-264
H.M.D. Olmsted, Claude Bernard Physiologist. New York 1938
J. Godart, Les Reliques de Claude Bernard. Villefranche 1939
J.M.D. Olmsted and E.H. Olmsted, Claude Bernard. New York 1952
R. Virtanen, Claude Bernard and his Place in the History of Ideas. Lincoln, Neb. 1960; J. Hist. Ideas 47 (1986), 275-286
N. Mani, Z. klin. Chem. 2 (1964), 97-104
B. Halpern, Rev. Hist. Sci. 19 (1966), 97-114
R. Heim et al., Les Concepts de Claude Bernard sur le Milieu Intérieur. Paris 1967
E. Wolff et al., Philosophie et Méthodologie Scientifique de Claude Bernard. Paris 1967
M.D. Grmek, Catalogue des Manuscrits de Claude Bernard. Paris 1967; J. Hist. Biol. 1 (1968), 141-154; Raisonnement Expérimental et Recherches Toxicologiques chez Claude Bernard. Paris 1973; Scientia 111 (1976), 135-155
J. Schiller, Claude Bernard et les Problèmes Scientifiques de son Temps. Paris 1967
F.L. Holmes, Claude Bernard and Animal Chemistry. Cambridge, Mass. 1974; Hist. Phil. Life Sci. 8 (1986), 3-25
N. Roll-Hanson, J. Hist. Biol. 9 (1976), 59-91
L.J. Jordanova, Hist. Sci. 16 (1978), 214-221
C. Debru, Rev. Hist. Sci. (1979), 143-162
E.D. Robin (ed.), Claude Bernard and the Internal Environment. New York 1979
J. Montaux, Cahiers Philosophiques 18 (1984), 53-81
P. Huard, Histoire et Nature 26/27 (1985), 49-61
W. Coleman, Isis 76 (1985), 49-70
L. Leahu, Physiologie 25 (1988), 151-157

BERNARD, LUCIEN [1882-1937]
SIGRIST, 98; NUC 49,455
Anon., Ann. Univ. Paris 13 (1938), 518

BERNARDI, ALESSANDRO [1886-1953]
POGG 6,191; 7b,343-344
G. Rossi, Chimica e Industria 35 (1953), 355

BERNATZIK, WENZEL [1821-1902]
OBL 1,76; BJN 7,13*; HIRSCH 1,490; Suppl.,94; PAGEL, 150-151;
A. Vogel, Wiener klin. Wchschr. 16 (1903), 283

BERNAYS, ALBERT JAMES [1823-1892]
NUC 49,571
Anon., J. Chem. Soc. 61 (1892), 488; Chem. Z. 16 (1892), 1973; Leopoldina 28 (1892), 54

BERNE, ROBERT MATTHEW [1918-]
AMS 17(1),477; BROBECK, 213-217; WWWS, 163

BERNER, ENDRE QVIE [1893-1983]
IWW 1982-1983, p.114; WWWS, 163; POGG 6,192-193; 7b,344-345
Hver er Hvem? 1979, pp.47-48

BERNERT, RICHARD [-1910]
NUC 49,627; RSC 13,482; SG [3]2, 477
W.Pagel, Virchows Jahresb. 1910, p.422

BERNFELD, PETER [1912-]
AMS 15(1),450; BHDE 2,105; POGG 7a(1),156

BERNHARD, KARL [1904-]
KURSCHNER 13,240; POTSCH, 42; POGG 7a(1),156-158
Anon., Chimia 18 (1964), 60-61; 28 (1974), 94

BERNHARD, SIDNEY [1927-1988]
M.F. Dunn and G.L. Rossi, TIBS 15 (1990), 84-85

BERNHART, CARL [1851-1920]
RSC 17,793
Anon., Z. angew. Chem. 33(II) (1920), 123

BERNHAUER, KONRAD [1900-1975]
KURSCHNER 11, 185; POGG 6,193; 7a(1),158-159
Anon., Chem. Z. 99 (1975), 335
R. Brunner, Mitt. Vers. Gär. Wien 30 (1976), 22-28

BERNHEIM, FREDERICK [1905-]
AMS 14,366; WWA 1978-1979, p.266
F. Bernheim, North Carolina Medical Journal 46 (1985), 147-148

BERNHEIM, HIPPOLYTE [1840-1919]
DBF 6,107; RSC 13,483

BERNOUILLI, FRIEDRICH [1824-1913]
Schweizerisches Geschlechterbuch 1 (1905), 42
A. Wankmüller, Der Schweizer Familienforscher 33 (1966), 46-50

BERNSTEIN, FELIX [1878-1956]
BHDE 2,98; POGG 5,99-100; 6,195; 7a(1),160
Anon., Gerontologia 1 (1957), 123-124
C. Gini, Genetics 60 (1968), s22-s23

BERNSTEIN, JULIUS [1839-1917]
DSB 15,20-22; DBJ 2,647; FISCHER 1,104-105; BRAZIER, 88-91,104-105; PAGEL, 153-154; DRULL, 18-19; BK, 166-167; DZ, 96; KOHUT 2,235-236; WININGER 1,353; 7,528; WWWS, 164; POGG 3,114; 4,103; 6,195
E. Abderhalden, Med. Klin. 13 (1917), 260
A. von Tschermak, Julius Bernsteins Lebensarbeit. Berlin 1919; Pflügers Arch. 174 (1919), 1-89
H. Grundfest, Arch. Ital. Biol. 103 (1965), 483-490

BERNSTEIN, RICHARD BARRY [1923-1990]
AMS 15(1),455; WWWS, 164

BERNTHSEN, AUGUST [1855-1931]
DSB 2,59; NDB 2,142-143; DRULL, 19; DZ, 98; TLK 3,41; POTSCH, 42; WWWS, 164-165; POGG 3,114-115; 4,103; 5,100; 6,196-197
P. Julius, Z. angew. Chem. 38 (1925), 737-739
M. Bodenstein, Ber. chem. Ges. 65A (1932), 21-22
K. Holdermann, Z. angew. Chem. 45 (1932), 141-143

BERRY, ARTHUR JOHN [1886-1967]
POGG 6,197; 7b,349-350
Anon., The Times, 20 January 1967; Chem. Brit. 4 (1968), 225

BERSIN, THEODOR [1902-1967]
AUERBACH, 774-775; MEINEL, 408-409,498; POGG 7a(1),162-163
P. Marquardt, Arzneimitt. 17 (1967), 790
Anon., Nachr. Chem. Techn. 15 (1967), 229; Chem.Z. 91 (1967), 547; Chem. Eng. News 45(41) (1967), 112

BERSON, JEROME ABRAHAM [1924-]
MSE 1,87-88; AMS 15(1),460; IWW 1982-1983, p.115; WWA 1980-1981, p.281

BERSON, SALOMON AARON [1918-1972]
AMS 11,390; BIBEL, 294-297; WWWS, 165; ICC, 511-513
D.T. Krieger et al., Mt. Sinai J. Med. 40 (1972), 260-297
W.H. Daughaday, Trans. Assoc. Am. Phys. 85 (1972), 9-10
J.E. Rall, Biog. Mem. Nat. Acad. Sci. 59 (1990), 55-70

BERT, PAUL [1833-1886]
DSB 2,59-63; DBF 6,171-172; GE 6,429-432; HIRSCH 1,498-500; W, 52-53; HD 6,29-32; PAGEL, 154-155; WWWS, 165
N. Gréhant, C.R. Soc. Biol. 39 (1887), 17-24
L. Dubreuil, Paul Bert. Paris 1935
N. Mani, Gesnerus 23 (1966), 109-166
P. Mercier, Bull. Acad. Med. 162 (1978), 886-889
P. Huard, Hist. Sci. Med. 13 (1979), 159-169
C. Jacquemin, Arch. Inter. Physiol. 96 (1988), A34-A42

BERTAGNINI, CESARE [1827-1857]
DBI 9,442-444; PROVENZAL, 203-205; BLOKH 1,47; POGG 3,115
G. Provenzal, Vita e Opere di Cesare Bertagnini. Rome 1928;
Rass. Clin. Terap. 36 (1937), 245-246

BERTALANFFY, LUDWIG von [1901-1972]
KURSCHNER 11,188-189; WWWS, 165
F. Gessner, Naturw. Rund. 4 (1951), 456-457
Anon., Leopoldina [3]14 (1968), 10-11
G. Nierhaus, Sudhoffs Arch. 65 (1981), 144-172
M. Davidson, Uncommon Sense. Los Angeles 1983

BERTHEIM, ALFRED [1879-1914]
DBJ 1,274; KALLMORGEN, 223; BLOKH 1,47; WININGER 6,469; POGG 5,101; 6,198
Anon., Deutsche med. Wchschr. 40 (1914), 1606; Chem. Z. 38 (1914), 1041
E. Beckmann, Ber. chem. Ges. 47 (1914), 2666-2667

BERTHELOT, ALBERT [1881-1947]
IB, 23
J.Magrou, Ann. Inst. Pasteur 74 (1948), 81-83

BERTHELOT, DANIEL [1865-1927]
DBF 6,193-194; DORVEAUX, 92,94; TSCHIRCH, 1020; POGG 4,108-109; 5,102-103; 6,199
C. Barrois, C.R. Acad. Sci. 184 (1927), 637-641
H.F.L. Coutiére, Bull. Acad. Med. 97 (1927), 338-343
E. Tassily, Bull. Sci. Pharm. 34 (1927), 372-390
R. Fabre, Bull. Soc. Chim. Biol. 9 (1927), 629-632

BERTHELOT, MARCELIN [1827-1907]
DSB 2,63-72; DBF 6,197-199; GE 6,441-444; LAROUSSE 2,615-616; W, 53-54; BUGGE 2,190-199; LACROIX 2,331-334; MAYERHOFER 1,432-433; POTSCH, 42-43; ABBOTT-C, 15-16; BLOKH 1,47-50; SCHAEDLER, 9-10; POGG 1,165; 3,115-119; 4,105-108; 5,101-102; 6,199
P. Walden, Chem. Z. 31 (1907), 367-373
G. Bredig, Z. angew. Chem. 20 (1907), 689-694
W.R., Proc. Roy. Soc. A80 (1908), iii-x
C. Graebe, Ber. chem. Ges. 41 (1908), 4805-4872
E. Jungfleisch, Bull. Soc. Chim. [4]13 (1913), i-cclx
E.J. Holmyard, Isis 6 (1924), 479-499
P. Sabatier, J. Chem. Ed. 3 (1926), 1099-1102
C. Moureau, Revue des Deux Mondes [13]41 (1927), 912-934
H.E. Armstrong, Nature 120 (1927), 659-663
G. André, Chim. Ind. 17 (1927), 673-680
A.A. Ashdown, J. Chem. Ed. 4 (1927), 1217-1232
E. Gley, Bull. Acad. Med. 98 (1927), 260-265
A.M.A. Boutaric, Marcelin Berthelot. Paris 1927
A. Ranc, La Pensée de Marcelin Berthelot. Paris 1948
L. Velluz, Vie de Berthelot. Paris 1964
G. Dillemann, Produits et Problèmes Pharmaceutiques 27 (1972), 561-563
M.J. Nye, Ann. Sci. 38 (1981), 585-590
G. Ciancia, Arch. Inter. Hist. Sci. 36 (1986), 54-83
J. Jacques, Berthelot. Paris 1987
A. Rosu, Sudhoffs Arch. 74 (1990), 186-209

BERTHO, ALFRED [1899-1977]
KURSCHNER 12,204; POGG 6,199; 7a(1),163-164
Anon., Chem. Z. 101 (1977), 269

BERTHOLD, ARNOLD ADOLPH [1803-1861]
DSB 2,72-73; NDB 2,166; ADB 2,512; HIRSCH 1,501-502; PAGEL, 155-157;NL 4,23-30; POGG 1,166,1538; CALLISEN 2,188-190; 26,269-271
H.P. Rush, Ann. Med. Hist. 1 (1929), 208-214
T.R. Forbes, Bull. Hist. Med. 23 (1949), 263-267
G.B. Gruber, Münch. med. Wchschr. 92 (1950), 137-140
H. Simmer and I. Simmer, Deutsche med. Wchschr. 86 (1961), 2189-2192

BERTHOLD, GOTTFRIED [1854-1937]
NDB 2,167; WAGENITZ, 26; WWWS, 166; POGG 6,199-200
E. Küster, Ber. bot. Ges. 54 (1936), (100)-(121)

BERTHOLLET, CLAUDE LOUIS [1748-1822]
DSB 2,73-82; DBF 6,224-225; GE 6,449; LAROUSSE 2,617-618; BLOKH 1, 50-52; W, 54-55; WWWS, 166; PHILIPPE, 707-709; FERCHL,40-42; SCHAEDLER, 10; BUGGE 1,342-349; MAYERHOFER 1, 434-435; ABBOTT-C, 16-17; POTSCH, 43-44; COLE, 48-54; POGG 1,166
P. Lemay and R.E. Oesper, J. Chem. Ed. 23 (1946), 158-165,230-236
S.C. Kapoor, Chymia 10 (1965), 51-110
T.S. Patterson, Chem. Ind. 1944, pp.99-102
M. Sadoun-Goupil, Le Chimiste Claude Louis Berthollet. Paris 1977
B.W. Keyser, Ann. Sci. 47 (1990), 213-260

BERTHOUD, ALFRED [1874-1939]
POGG 6,200; 7a(1),164-165
E. Briner, Helv. Chim. Acta 22 (1939), 1227-1228

BERTIN-SANS, HENRI [1862-1932]
RSC 13,503
P. Betouillières, Mons. Hipp. 4(13) (1961), 23-28

BERTRAM, JULIUS [1851-1925]
HEIN-E, 28-29; POGG 4,110; 6,201

BERTRAM, PAUL [1867-]
JV 8,148; NUC 50,556

BERTRAND, GABRIEL [1867-1962]
DSB 2,86-87; CHARLE(2), 34-37; DORVEAUX, 94; POTSCH, 44; WWWS, 167; IB, 23; POGG 4,111-112; 5,103-104; 6,201-202; 7b,351-358
M. Delépine, C.R. Acad. Sci. 255 (1962), 217-222;
Ann. Pharm. Fran. 20 (1962), 930-934
Y. Raoul, Bull. Soc. Chim. Biol. 44 (1962), 1051-1056
P. Fleury, Bull. Acad. Med. 146 (1962), 536-545
L. de Saint-Rat, Ann. Falsif. 56 (1963), 65-77;
Bull. Soc. Chim. 1969, pp.4197-4219
M. Zamansky, Ann. Univ. Paris 33 (1963), 90-92
R. Courrier, Not. Acad. Sci. 5 (1972), 111-171
J.M. Lehn et al., TIBS 11 (1986), 228-230

BERTSCHINGER, ALFRED [1846-1920]
STRAHLMANN, 471,481; RSC 13,511

BERZELIUS, JÖNS JACOB [1779-1848]
DSB 2,90-97; BUGGE 1,428-449; FARBER, 387-402; PHILIPPE, 719-728; BLOKH 1,52-55; ABBOTT-C, 17-19; FERCHL, 42-43; SCHAEDLER, 10; POTSCH, 45; W, 55-56; WWWS, 167; MATSCHOSS, 19-20; POGG 1,171-175; 6,202-203; COLE, 55-63; CALLISEN 2,199-212; 26,274-279
H. Rose, Am. J. Sci. [2]16 (1853), 1-15,173-186,305-313;
17 (1854), 103-113
F. Wöhler, Ber. chem. Ges. 8 (1875), 838-852
H.G. Söderbaum, Berzelius' Werden und Wachsen. Leipzig 1899
P. Walden, Z. angew. Chem. 43 (1930), 325-329,351-354,366-370
J.J. Berzelius, Autobiographical Notes. Baltimore 1934
B.S. Jørgensen, Centaurus (1965), 258-281; J. Chem. Ed.
42 (1965), 394-397
J.E. Jorpes, Jac. Berzelius, his Life and Work. Stockholm 1966
P.A. Chalmers and F. Szabadvary, Talanta 27 (1980), 1029-1035

E.M. Melhado, Jacob Berzelius. Stockholm 1981
L. Dünsch, Jöns Jacob Berzelius. Leipzig 1986
E.M. Melhado and T. Frängsmyr (eds.), Enlightenment Science in the Romantic Era. Cambridge 1992

BESEMFELDER, EDUARD [1863-1929]
TLK 2,47; JV 6,237; RSC 13,513
Anon., Chemische Fabrik 2 (1929), 414; Chem. Z. 53 (1929), 703

BESREDKA, ALEXANDRE (ALEKSANDR MIKHAILOVICH) [1870-1940]
FISCHER 1,107-108; WININGER 6,469-470; WWWS, 168
D. Nai, Bioch. Ter. Sper. 9 (1922), 344-351
Anon., Ann. Inst. Pasteur 64 (1940), 269-274
E. Wollman, Presse Med. 48 (1940), 428
E. Lagrange, Rev. Hist. Med. Heb. 10 (1957), 101-110
A. Delaunay, Bull. Inst. Pasteur 69 (1971), 73-78
P. Nicolle, Bull. Inst. Pasteur 69 (1971), 189-192
M. Wainwright, Med. Hist. 34 (1990), 79-85

BESSAU, GEORG [1884-1949]
FISCHER 1,108

BESSEY, CHARLES EDWIN [1845-1915]
DSB 2,102-104; DAB 2,229; NCAB 8,361-362; HUMPHREY, 27-20; WWWS, 168
R.J. Pool, Am. J. Bot. 2 (1915), 505-518

BESSMAN, SAMUEL PAUL [1921-]
AMS 15(1),466; WWA 1980-1981, p.283; WWWS, 168

BEST, CHARLES HERBERT [1899-1978]
DSB 17,80-81; MH 1,36-37; STC 1,121-123; MSE 1,89-90; AMS 13,334-335; YOUNG, 81-82; WW 1978, p.199; CB 1957, pp.52-54
C.H. Best, Adv. Met. Dis. 7 (1974), 141-154; TIBS 3 (1978), N155-N156
G.W. Corner, Am. Phil. Soc. Year Book 1978, pp.50-54
T.H. Jukes, Nature 274 (1978), 99-100; J. Nutrition 110 (1980), 18-21; Can. J. Biochem. 57 (1979), 455-459
R.E. Haist, Proc. Roy. Soc. Canada [4]16 (1979), 43-47
B. Leibel, Trans. Assoc. Am. Phys. 93 (1980), 15-17
F. Young and C.N. Hales, Biog. Mem. Fell. Roy. Soc. 28 (1982), 1-25

BESTHORN, EMIL [1858-1921]
NDB 2,184; BLOKH 1,56; WWWS,168; POGG 4,114; 5,106; 6,203
Anon., Chem. Z. 45 (1921), 1127
R. Pummerer, Ber. chem. Ges. 55A (1922), 20-22

BETHE, ALBRECHT [1872-1954]
FISCHER 1,109; KOLLE 3,47-58; DBA 1,202; KALLMORGEN, 223; KIEL, 83
R. Thauer, Münch. med. Wchschr. 97 (1955), 707-708; Pflügers Arch. 261 (1955), i-xiv

BETHMANN, FRITZ [1874-]
JV 14,12; NUC 51,328; RSC 13,519

BETHMANN, HEINRICH GEORG [1868-]
JV 5,186; NUC 51,328; RSC 13,519

BETTELHEIM, KARL [1840-1895]
ADB 46,493; HIRSCH 1,515; PAGEL, 161-162; WININGER 1,366; KOREN, 155
Anon., Leopoldina 31 (1895), 169-170; Münch. med. Wchschr. 45 (1895),1415

BETTI, MARIO [1875-1942]
DBI 9,719-721; STC 1,125; POGG 4,114-115; 5,106; 6,204-205; 7b,365-366
A. Coppadoro, Chimica e Industria 24 (1942), 187
S. Berlingozzi, Gazz. Chim. Ital. 83 (1953), 693-719

BEUDANT, FRANÇOIS SULPICE [1787-1850]
DSB 2,106; DBF 5,358-359; FERCHL, 44; POTSCH, 46; WWWS, 169; POGG 1,179-180
A. Lacroix, Mem. Acad. Sci. 60 (1931), i-xlviii,lxxxvii-lxxxix

BEUMER, HANS [1884-1945]
KURSCHNER 6(I),117; FISCHER 1,110
Anon., Deutsche med. Wchschr. 71 (1946), 40

BEUTLER, HANS [1896-1942]
KURSCHNER 4,195; BHDE 2,101; LDGS, 17; POGG 6,206; 7a(1),171

BEUTLER, RUTH [1897-1959]
KURSCHNER 8,147; BOEDEKER, 25

BEUTNER, REINHARD [1885-1964]
IB, 23; POGG 6,206-208; 7b,366-368
Anon., Naturw. Rund. 17 (1964), 248
J.S. Hepburn, Pharmacologist 7 (1965), 9-10

BEVAN, EDWARD JOHN [1856-1921]
W, 7; WWWS, 170; POTSCH, 46-47
C.F. Cross, J. Chem. Soc. 119 (1921), 2121-2123

BEVERIDGE, WILLIAM IAN BEARDMORE [1908-]
WW 1981, p.213; WWWS, 170

BEYER, CARL [1859-1891]
BLOKH 1,56-57; RSC 9,232; 13,518
L. Claisen, Ber. chem. Ges. 24(3) (1891), 1117-1121

BEYER, HANS [1905-1971]
KURSCHNER 12,216; POGG 7a(1),172
Anon., Chem. Z. 95 (1971), 296

BEYER, HERBERT [1910-1988]
Anon., Chem. Z. 94 (1970), 303

BEYER, ROBERT EDWARD [1928-]
AMS 15(1),473; WWWS, 170

BEZNÁK, ALADÁR [1901-1959]
IB, 23-24

BEZOLD, ALBERT von [1836-1868]
DSB 2,110-111; NDB 2,210; ADB 2,607; HIRSCH 1,518; DIEPGEN, 34-37; PAGEL, 164-165; ROEDER, 152-156; GIESE, 491; WWWS, 1729; POGG 3,125
F. von Recklinghausen, Sitz. Phys. Med. Würzburg 1868, pp.xli-xlviii
R. Herrlinger and I. Krupp, Albert von Bezold. Stuttgart 1964

BEZSSONOFF, NIKOLAI ALEKSEEVICH [1885-1951]
LIPSHITS 1,150-151; IB, 24; POGG 6,209; 7b,368-370
P. Grabar, Bull. Soc. Chim. Biol. 34 (1952), 842-843

BHATNAGAR, SHANTI SWARUPA [1895-1955]
WWWS, 171; POGG 6,210-211; 7b,373-375
Anon., Chem. Eng. News 33 (1955), 252; Chem. Ind. 1955, p.42; Nachr. Chem. Techn. 3 (1955), 101
T.R. Seshadri, Biog. Mem. Fell. Roy. Soc. 8 (1962), 1-17

BIACH, OTTO [1881-1923]
NUC 52,204
Anon., Z. angew. Chem. 36 (1923), 388; Chem. Z. 47 (1923), 565

BIAL, MANFRED [1870-1908]
FISCHER 1,112
W. Pagel, Virchows Jahresber. 1908(I), p.404

BIALASZEWICZ, KAZIMIERZ [1882-1943]
WE 1,751; IB, 24
S. Niemerko, Acta Physiol. Pol. 38 (1987), 177-185

BIBERFELD, JOHANNES DAVID [1872-1922]
JV 15,32; NUC 52,337

BIBERGEIL, ARTUR [1880-1922]
BLOKH 1,57-58
A. Erlenbach, Ber. chem. Ges. 55A (1922), 81-82

BIBERSTEIN, HANS [1889-1965]
AMS 11,402; KURSCHNER 10,168; KOREN, 156; LDGS, 61
F. Herrmann, Derm. Wchschr. 150 (1964), 617-618

BIBRA, ERNST von [1806-1878]
NDB 2,216; ADB 47,758-759; HIRSCH 1,521; BLOKH 1,58; SCHAEDLER, 11-12; FERCHL,45; WWWS, 1729; POGG 1,186-187; 3,127; 7a Suppl.,88
F. von Kobell, Sitz. Bayer. Akad. 9 (1879), 129-131
Anon., Aus dem Leben Schweinfurter Männer und Frauen, pp.56-58. Schweinfurt 1968

BICHAT, XAVIER [1771-1802]
DSB 2,122-123; DBF 6,396-397; GE 6,690-692; LAROUSSE 2,703-704; W, 58; HIRSCH 1,521-523; Suppl.,98-99; DIEPGEN, 22-24; FRANKEN, 176-177
E. Gley, C.R. Soc. Biol. 54 (1902), 17-22
A. Cartaz, Les Médecins Bressans, pp.34-36. Paris 1902
M. Genty, Biographies Médicales et Scientifiques, pp.181-318. Paris 1972
W.R. Albury, Stud. Hist. Biol. 1 (1977), 47-131
J.V. Pickstone, Hist. Sci. 19 (1981), 115-142
G. Sutton, Bull. Hist. Med. 58 (1984), 53-71

BICKEL, ADOLF [1875-1946]
NB, 28; FISCHER 1,112-113; KURSCHNER 6(I),122-123; ASEN, 15; EBEL, 94

BIDDER, FRIEDRICH HEINRICH [1810-1894]
DSB 2,123-125; NDB 2,217-218; ADB 46,538-540; WELDING, 63; FRANKEN, 177; HIRSCH 1,524-525; Suppl.,99; PAGEL, 165-167; LEVITSKI 2,13-15; WWWS, 173
T. Acard, Der Physiologe Friedrich Bidder. Zurich 1969
F.C. Bing, J. Nutrition 103 (1973), 637-648
A.N. Khazanov, Ist. Est. Tekhn. Pribalt. 5 (1976), 68-91

BIECHELE, MAX [1839-1922]
HEIN, 54; TSCHIRCH, 1021; SCHELENZ, 699
Anon., Pharm. Z. 67 (1922), 613

BIEDERMANN, RICHARD [1843-1880]
NDB 2,224; POGG 3,127
L.K. Biedermann, Z. Agrikulturchem. 9 (1880), 393

BIEDERMANN, RUDOLF [1845-1929]
DBJ 11,345; TLK 3,46; SCHAEDLER, 12; WININGER 1,373; POGG 3,127-128; 4,119; 6,214
H. Freundlich, Ber. chem. Ges. 62A (1929), 82
Anon., Ost. Chem. Z. 32 (1929), 85

BIEDERMANN, WILHELM [1852-1929]
DBJ 11,345; STURM 1,92; FISCHER 1,113; LOMMATZSCH, 110-115; IB, 24; GIESE, 494-496; WININGER 7,531; POGG 6,214-215
F.N. Schulz, Erg. Physiol. 30 (1930), xi-xxviii
W. Winterstein, Ergebnisse der Biologie 6 (1930), 1-3

BIEDERT, PHILIPP [1847-1916]
DBA 1,220; DBJ 1,348; FISCHER 1,113-114; PAGEL, 167-169; DZ, 109-110
A. Holtzmann, Strasbourg Médical 13 (1916), 121-123
M. von Pfaundler, Z. Kinderheil. 15 (1917), 95-99

BIEDL, ARTUR [1869-1933]
FISCHER 1,114; STURM 1,92; MEDVEI, 711-712; KOERTING, 154-155; KOREN, 156 TETZLAFF, 31
L. Asher, Endokrinologie 13 (1933), 153-155
Anon., Wiener med. Wchschr. 83 (1933), 1078-1079
R. Abderhalden, Ciba Z. 5 (1951), 1602-1616
L. Feher and W. Kaiser, NTM 7 (1970), 99-108
J. Kenez, Med. Mon. 24 (1970), 259-262

BIEHRINGER, JOACHIM [1858-1920]
BLOKH 1,58-59; POGG 4,120; 5,109-110; 6,215
R. Meyer, Ber. chem. Ges. 53A (1920), 101-105

BIELIG, HANS JOACHIM [1912-1987]
KURSCHNER 13,268; POGG 7a(1),179-180
Anon., Chem. Z. 96 (1972), 41; 109 (1985), 315; 111 (1987), 383; Nachr. Chem. Techn. 20 (1972); Naturw. Rund. 38 (1985), 352

BIELING, RICHARD [1888-1967]
FISCHER 1,115; KURSCHNER 10,172-173; KALLMORGEN, 223; AUERBACH,199-200; WWWS, 174; IB, 24; POGG 6,215-216; 7a(1),180-181
F. Brücke and J. Bock, Wiener klin. Wchschr. 70 (1958), 633-634
Anon., Z. Bakt. 172 (1958), 353-362
H. Flamm, Z. Bakt. 204 (1967), 449-451

BIELSCHOWSKY, ALFRED [1871-1940]
NDB 2,227; FISCHER 1,115; BHDE 2,105; RAZINGER, 17-18; AUERBACH, 200; TETZLAFF, 31; WWWS, 174; KOREN, 156
Anon., Archives of Ophthalmology 23 (1940), 1354-1365
R. Herzberg, Australian Journal of Ophthalmology 10 (1982), 282-284

BIELSCHOWSKY, FRANZ [1902-1965]
BHDE 2,105; LDGS, 64; KOREN, 156
C.M. Goodall et al., Cancer Research 26 (1966), 347-348
G.M. Bonser and R.A. Willis, J. Path. Bact. 93 (1966), 357-364
N.L.E., Proc. Roy. Soc. N.Z. 94 (1966), 121-124

BIELSCHOWSKY, MAX [1869-1940]
NDB 2,227-228; FISCHER 1,115-116; HAYMAKER, 319-322; RAZINGER, 19-22; BHDE 2,105-106; KOREN, 156
M. Ostertag, Deutsche med. Wchschr. 84 (1959), 765-766
J. Hallervorden, Nervenarzt 30 (1959), 325-327

BIEN, ZOLTAN [1891-]
IB, 24

BIENSTOCK, BARTHOLD [1861-1940]
RSC 13,543
C. Durand, Paris Médical 116 (1940), 271

BIER, AUGUST [1861-1949]
NDB 2,230-231; KURSCHNER 6(I),126-127; FISCHER 1,116; PAGEL, 169-170; KILLIAN, 351-353; WININGER 7,531; WWWS, 174
K. Vogeler, August Bier. 2nd Ed. Munich 1942
W. Stoecke et al., Z. Chir. 74 (1949), 1122-1131
W. Baetzner, Münch. med. Wchschr. 103 (1961), 2343-2347
W. Block, Landarzt 37 (1961), 1249-1252
G. Waas, Z. ärzt. Fortbild. 74 (1980), 982-987; 76 (1982), 72-74
H. Wolff, Charité Annalen NF6 (1986), 331-335

BIERBAUM, KURT [1881-1945]
FISCHER 1,116; KALLMORGEN, 224; ASEN, 16; IB, 24

BIERMER, ANTON [1827-1892]
NDB 2,232-233; ADB 46,546; HIRSCH 1,530-531; Suppl.,100; PAGEL, 170-171; WWWS, 174
Anon., Jahresb. Schles. Ges. 70 (1892), 1-3
H. Ziemssen, Münch. med. Wchschr. 39 (1892), 521-522
A. Adler, Berl. klin. Wchschr. 29 (1892), 738-739
G. Rosenfeld, Med. Klin. 23 (1927), 1916-1918

BIERNACKI, EDMUND [1866-1911]
PSB 2,78-79; FISCHER 1,116-117; KONOPKA 1,275-283
S. Ebel, Deutsche med. Wchschr. 38 (1912), 175; Münch. med. Wchschr. 59 (1915), 313-314
M. Paciorkiewicz, Pol. Tyg. Lek. 34 (1979), 1365-1367
W. Lisowski, Lekarz Wojsowy 56 (1980), 401-406
J.W. Guzek, Acta Physiol. Polon. 38 (1987), 170-176
E.J. Kucharz, J. Lab. Clin. Med. 112 (1988), 279-280

BIERRY, HENRI GEORGES [1876-1948]
IB, 24; POGG 6,216-218; 7b,379-380
A.M.B. Lacassagne, Bull. Acad. Med. 132 (1948), 589-590

BIGELOW, HENRY BRYANT [1879-1967]
AMS 11,409; WWWS, 175; IB, 24
A.C. Redfield, Biog. Mem. Nat. Acad. Sci. 48 (1976), 51-80

BIGELOW, WILLARD DELL [1866-1939]
NCAB 38,631-632; AMS 6,115; WWWS, 175; WWAH, 54

BIGGS, HERMAN MICHAEL [1859-1923]
DAB 2,262-263; NCAB 19,219-221; FISCHER 1,117; DAMB, 65-66; WWWS, 176
Anon., Science 58 (1923), 413-415
C.E.A. Winslow, The Life of Herman M. Biggs. Philadelphia 1929
M. Terris, Bull. N.Y. Acad. Med. 51 (1975), 242-257

BIGWOOD, EDOUARD JEAN [1891-1975]
BAF, 5; POGG 6,221; 7b,384-387

BIHAN, RICHARD [1867-]
JV 10,211; RSC 19,74

BIILMANN, EINAR CHRISTIAN SAXTORPH [1873-1946]
VEIBEL 1,231-232; 2,49-55; 3,69-94; POGG 5,112,1416; 6,221-222; 7b,387
S. Veibel and J.S. Simonsen, J. Chem. Soc. 1949, pp.534-535
J.T. Stock, J. Chem. Ed. 66 (1989), 910-912

BIJL, HENRIK CONRAAD [1877-1951]
Anon., Chem. Wkbl. 47 (1951), 245

BIJLERT, ALBERTUS van [1864-1925]
RSC 13,555
Anon., Chem. Wkbl. 15 (1918), 381-382; 22 (1925), 106

BIJVANCK, HENDRIK [1875-]
JV 14,223; RSC 13,518

BIJVOET, JOHANNES MARTIN [1892-1980]
BWN 2,89-90; POGG 6,222; 7b,388-389
C.H. MacGillavry, Acta Cryst. A36 (1980), 837-838
M.P. Groenwege and A.F. Peerdemann, Biog. Mem. Fell. Roy. Soc. 29 (1983), 27-41

BILETSKI, MIKOLA FEDOROVICH [1851-1882]
VORONTSOV, 23-33

BILHARZ, THEODOR [1825-1862]
DSB 2,127-128; NDB 2,237-238; ADB 2,636-637; HIRSCH 1,538; Suppl.,101; PAGEL, 173; DIEPGEN, 74-81; OLPP, 37-42; WWWS, 176
E. Ebstein, Münch. med. Wchschr. 72 (1925), 480-481
E. Senn, Theodor Bilharz. Stuttgart 1931
H. Schadewaldt, Deutsche med. Wchschr. 80 (1955), 1053-1055; Lebensbilder aus Schwaben und Franken 7 (1960), 337-345; Münch. med. Wchschr. 104 (1962), 1730-1734; Berliner Medizin 14 (1963), 244-250

BILLETER, OTTO [1851-1927]
ZURICH-D, 123; POGG 4,123; 5,112; 6,223
H. Rivier, Helv. Chim. Acta 11 (1928), 700-710
H. Rivier and O. Billeter, Verhandl. Schw. Nat. Ges. 110 (1929), 3-9
E.M. Huntress, Proc. Am. Acad. Arts Sci. 79 (1951), 7-8

BILLING, ARCHIBALD [1791-1881]
HIRSCH 1,540
Anon., Brit. Med. J. 1881(I), p.466

BILLINGS, JOHN SHAW [1838-1913]
DAB 2,266-269; NCAB 4,78; HIRSCH 1,540-541; Suppl.,102; DAMB, 67-68; WWWS, 177; WWAH, 54
F.H. Garrison, John Shaw Billings. New York 1915
S.W. Mitchell, Biog. Mem. Nat. Acad. Sci. 8 (1919), 375-416
S.V. Larkey, Bull. Inst. Hist. Med. 6 (1938), 360-376
D.F. Cowan, McGill Medical Journal 28 (1959), 87-96
A.M. Harvey, Persp. Biol. Med. 21 (1977), 35-57

BILLS, CHARLES EVERETT [1900-1972]
AMS 12,490; WWWS, 177; POGG 6,223-224; 7b,394
Anon., Chem. Eng. News 50(31) (1972), 24

BILTZ, ERNST CHRISTIAN [1822-1903]
HEIN, 55-56; BLOKH 1,59; FERCHL, 46; TSCHIRCH, 1021; SCHELENZ, 674
Anon., Deutsche Apoth. Z. 18 (1903), 38-40

BILTZ, FRIEDRICH HEINRICH [1790-1835]
HEIN, 56; TSCHIRCH, 1021
H.R. Abe, Pharmazie 30 (1975), 191-193

BILTZ, HEINRICH JOHANN [1865-1943]
NDB 2,241-242; KIEL, 182; TLK 3,48; POTSCH, 47; WWWS, 177; POGG 4,123-124; 5,113-115; 6,224-225; 7a(1),184-185
E. Wilke-Dörfurt, Z. angew. Chem. 38 (1925), 457-458
W.A. Roth, Jahrb. Akad. Wiss. Gott. 1943-1944, pp.93-96
W. Hückel, Chem. Ber. 82 (1949), lxvii-lxxxviii
G. Schiemann, Chem. Z. 75 (1951), 83-84

BILTZ, WILHELM [1877-1943]
NDB 2,242; TLK 3,48; HANNOVER, 31-32; POTSCH, 47-48; WWWS, 177; POGG 4,124; 5,115-116; 6,225-226; 7a(1),185-187
R. Juza, Z. Elektrochem. 50 (1944), 1-2
W. Fischer, Chem. Ber. 82 (1949), lxxxix-cxiii

BINEAU d'ALIGNY, ARMAND [1812-1861]
DBF 6,492; GE 6,885; POGG 1,193; 3,132

BINDEWALD, HUGO [1820-]
H. Bindewald, Das Geschlecht Bindewald aus Nieder-Ohmen, p.137. Berlin-Zehlendorf 1935

BING, FRANKLIN CHURCH [1902-1988]
AMS 12,482

BINGEL, ADOLF [1879-1963]
SG [3]2,540
T.H. Newton, Radiology of Skull and Brain 1(1),5. St. Louis 1971

BINGOLD, KONRAD [1886-1955]
FISCHER 1,120; KURSCHNER 8,161

BINOT, JEAN [1867-1909]
Anon., Bull. Inst. Pasteur 7 (1909), p.1049

BINZ, ARTHUR HEINRICH [1868-1943]
NDB 2,250; TLK 3,48-49; WWWS, 178; POTSCH, 48; POGG 4,124; 5,117; 6,227; 7a(1),187-188
O. von Schickh, Z.angew. Chem. 51 (1938), 779-782
P. Duden, Ber. chem. Ges. 76A (1943), 63-70

BINZ, CARL [1832-1913]
NDB 2,250; BJN 18,79*; HIRSCH 1,545; REBER, 29-33,354-355; WWWS, 178; TSCHIRCH, 1021; PAGEL, 179-181; DZ, 115; POGG 3,132;, 4,124; 7a(1),188
Anon., Arch. Internat. Pharmacodyn. 15 (1905), 5-20
R.M. Berfling, Der Pharmacologe Carl Binz. Bonn 1969

BIONDI, CESARE [1867-1936]
DBI 10,524; FISCHER 1,121
G. Bianchini, Studi Senesi 50 (1936), 223-262

BIOT, JEAN BAPTISTE [1774-1862]
DSB 2,133-140; DBF 6,506; GE 6,899-901; LAROUSSE 2,762; BLOKH 1,59-60; POTSCH, 48; W, 60; WWWS, 178; MAYERHOFER 1,473-474; FERCHL, 47; POGG 1,195-199; 3,136; CALLISEN 2,261-263; 26,301
Anon., Proc. Roy. Soc. 12 (1863), xxxv-xlii
T.M. Lowry, Nature 117 (1926), 271-275
E. Picard, La Vie et l'Oeuvre de J.B. Biot. Paris 1927
E. Frankel, Brit. J. Hist. Sci. 11 (1978), 36-48

BIRCH, ARTHUR JOHN [1915-]
MSE 1,95-96; WW 1981, p.222; SRS, 70; POTSCH, 48
A.J. Birch, To See the Obvious. Washington, D.C. 1991

BIRCH, THOMAS W. [1874-1939]
J. Needham and E. Baldwin, Hopkins and Biochemistry, p.336. Cambridge 1949

BIRCH-HIRSCHFELD, FELIX VICTOR [1842-1899]
NDB 2,252; ADB 46,555-559; BJN 4,229-231; PAGEL, 181-183; KOREN, 156; WININGER 1,378
M. Seiffert, Berl. klin. Wchschr. 36 (1899), 1135-1136

BIRCHARD, FREDERICK JAMES [1875-1940]
AMS 5,95; YOUNG, 70

BIRCHER, EUGEN [1882-1956]
AARGAU, 67-70; SBA 1,25; FISCHER 1,121; KURSCHNER 8,161-162
O. Häuptli, Verhandl. Schw. Nat. Ges. 137 (1957), 301-305
Anon., Leopoldina [3]4/5 (1958-1959), 85-86

BIRD, GOLDING [1814-1854]
DNB 2,536-537; HIRSCH 1,548; WILKS, 245-250; DESMOND, 65; WWWS, 179; POGG 1,200
W. Hale-White, Guy's Hosp. Rep. [4]6 (1926), 1-20
N.G. Coley, Med. Hist. 13 (1969), 363-376

BIRKENBACH, LOTHAR [1876-1962]
TLK 3,49; NUC 58,428
Anon., Nachr. Chem. Techn. 9 (1961), 132; 10 (1962), 320

BIRKOFER, LEONHARD [1911-]
KURSCHNER 13,285; POGG 7a(1),189

BIRNBAUM, KARL [1839-1887]
BLOKH 1.60; SCHAEDLER, 12; POGG 3,134
Anon., Leopoldina 23 (1887), 57; Ber. chem. Ges. 20 (1887), 473-478

BIRON, EVGENI VLADISLAVOVICH [1874-1919]
BSE 5,250-251; BLOKH 1,60-62
B.N. Menshutkin, Zhur. Fiz. Khim. 62 (1930), 1749-1776

BISCHKOPFF, EDUARD [1875-1932]
JV 13,128; NUC 58,592
Anon., Z. angew. Chem. 45 (1932); Chem. Z. 56 (1932), 665

BISCHLER, AUGUST [1865-]
ZURICH, 85-86; ZURICH-D, 158; SURREY, 11; RSC 18,569; NUC 58,592

BISCHOF, CARL GUSTAV CHRISTOPH [1792-1870]
DSB 2,158-159; NDB 2,261-262; ADB 2,665-669; BLOKH 1,62; FERCHL, 48-49; POTSCH, 49; SCHAEDLER, 12-13; WWWS, 181; POGG 1,202-203; 3,134-136; 7aSuppl., 90-91
P. Diergart, Studien zur Geschichte der Chemie, pp.195-203. Berlin 1927
M. Nierenstein, Isis 21 (1934), 123-130
G. Rath, Sudhoffs Arch. 37 (1953), 122-130

BISCHOFF, CARL [1851-1912]
BLOKH 1,63-64
C. Liebermann, Ber. chem. Ges. 45 (1912), 853

BISCHOFF, CARL ADAM [1855-1908]
NDB 2,264; DBJ 13,13*; BLOKH 1,63; ARBUZOV, 197-198; RIGA, 694; POGG 3,135-136; 4,126-127; 5,119
P. Walden, Chem. Z. 32 (1908), 1053
I.S. Telesov, Zhur. Fiz. Khim. 42 (1910), 1501-1516
G.V. Bykov, Ist. Est. Tekhn. Pribalt. 7 (1984), 55-62

BISCHOFF, CHRISTIAN HEINRICH ERNST [1781-1861]
ADB 2,672; HIRSCH 1,550; TSCHIRCH, 1022
G. Rath, Sudhoffs Arch. 37 (1953), 122-130

BISCHOFF, HENRI [1813-1899]
DHB 2,193; STRAHLMANN, 468,470

BISCHOFF, THEODOR LUDWIG WILHELM [1807-1882]
DSB 2,160-162; NDB 2,264-266; ADB 46,570; HB 3,1-11; FREUND 2,23-29; HIRSCH 1,550-551; PAGEL, 185-187; EGERER, 9-10; DRULL, 21-22; WWWS, 181
K. von Kupffer, Gedächtnisrede auf T.L.W. Bischoff. Munich 1884

BISHOP, ARTHUR WRIGHT [1867-1920]
JV 5,233; RSC 13,572
Anon., Chem. Z. 44 (1920), 992

BISHOP, GEORGE HOLMAN [1889-1973]
AMS 12,503; DAMB, 69
J.L. O'Leary, Trans. Am. Neurol. Assn. 99 (1974), 278-279
W.M. Landau, Biog. Mem. Nat. Acad. Sci. 55 (1985), 45-66

BISHOP, KATHERINE SCOTT [1889-1976]
AMS 8,213
Anon., J. Am. Med. Assn. 235 (1976), 2438

BISTRZYCKI, AUGUSTIN [1862-1936]
TLK 3,50-51; POGG 4,128; 5,119; 6,232; 7a(1),191

C. Gyr, Helv. Chim. Acta 20 (1937), 477-489

BITTER, HEINRICH [1863-1918]
FISCHER 1,124; RSC 13,574; NUC 59,164

BITTING, ARVILL WAYNE [1870-1946]
AMS 6,121-122

BITTNER, JOHN JOSEPH [1904-1961]
AMS 10,331; WWWS,182
L.C. Strong, Cancer Research 22 (1962), 392

BITTORF, ALEXANDER [1876-1949]
FISCHER 1,124; KURSCHNER 6(I),138
H. Franke, Z. ärzt. Fortbild. 43 (1949), 227

BIZIO, BARTOLOMEO [1791-1862]
DBI 10,734-736; PROVENZAL, 123-132; DFI, 21; CALLISEN 2,287-288; POGG 1,204; 3,136-137
G. Provenzal, Rass. Clin. Terap. 36 (1937), 371-377

BIZZOZERO, GIULIO [1846-1901]
DSB 2,164-166; DBI 10, 747-751; FISCHER 1,124-125; PAGEL, 187-188
R. Fusari, Anat. Anz. 19 (1901), 313-319
C. Golgi, Arch. Sci. Med. Torino 25 (1901), 205-234
P. Foà, Arch. Ital. Biol. 35 (1901), 303-312; Atti Accad. Torino 36 (1901), 452-480; Beitr. path. Anat. 30 (1901), 175-178
A. Baserga, Sci. Med. Ital. 7 (1958), 44-63
G. DiMacco, Arch. Sci. Med. Torino 112 (1961), 390-406
P. Franceschini, Physis 4 (1962), 227-267

BJERRUM, JANNIK [1909-1992]
KBB 1980, p.99; VEIBEL 2,60-61; WWWS, 183
G.B. Kauffman, J. Chem. Ed. 62 (1985), 1002-1005

BJERRUM, NIELS [1879-1958]
DSB 2,160-171; DBL 3,183-185; VEIBEL 2,61-67; 3,95-114; WWWS, 183; POTSCH, 49-50; POGG 6,233-234; 7b,406-407
E.A. Guggenheim, Proc. Chem. Soc. 1960, pp.104-114
H. Nowotny, Alm. Akad. Wiss. Wien 110 (1961), 438-446
G.B. Kauffman, J. Chem. Ed. 57 (1980), 779-782,863-867

BJØRN-ANDERSEN, HAKON ARTHUR WOLFGANG [1874-1932]
VEIBEL 2,68-69

BLACHER, KARL von [1867-1939]
KURSCHNER 4,222-223; WELDING, 71-72; RIGA, 694; POGG 4,130; 6,234

BLACHSTEIN, ARTHUR [1863-1940]
WWWS, 183; SG [2]2,361; RSC 13,580; NUC 59,380

BLACK, OTIS FISHER [1867-1933]
AMS 5,99; IB, 25
Anon., J. Wash. Acad. Sci. 23 (1933), 580

BLACKETT, PATRICK [1897-1974]
DNB [1971-1980], pp.60-62; MH 1,43-44; STC 1,135-137; MSE 1,100-101; WW 1965, pp.274-275
B. Lovell, Biog. Mem. Fell. Roy. Soc. 21 (1975), 1-115

BLACKLEY, CHARLES HARRISON [1820-1900]
E.B. Leech, Brit. Med. J. 1929(II), pp.1171-1172
A.E. Lowndes, Isis 37 (1947), 21-24
G. Taylor and J. Walker, Clinical Allergy 3 (1973), 103-108

BLACKMAN, FREDERICK FROST [1866-1947]
DSB 2,183-185; DNB [1941-1950], pp.82-83; DESMOND, 67; WW 1947, p.251
F.C. Steward, Plant Physiology 23(3) (1947), iii-viii
G.E. Briggs, Obit. Not. Fell. Roy. Soc. 5 (1948), 651-658

BLACKMAN, VERNON HERBERT [1872-1967]
DNB [1961-1970], pp.113-114; DESMOND, 67-68; WW 1965, p.276; WWWS, 184
H.K. Porter, Biog. Mem. Fell. Roy. Soc. 14 (1968), 37-60
W. Brown, Annals of Botany 32 (1968), 233-235

BLADIN, JOHAN ADOLF [1856-1902]
SBL 4,746-748; SMK 1,355-356; POGG 4,130,1702; 5,121

BLAGOVESHCHENSKI, ANDREI VASILIEVICH [1889-]
LIPSHITS 1,193-197; IB, 25; POGG 6,235; 7b,410-412
Z.V. Vasilieva, Bull. Soc. Nat. Moscou NS74(5) (1969), 5-7

BLAINVILLE, HENRI MARIE DUCROTAY de [1777-1850]
DSB 2,186-188; DBF 6,563-564; GE 6,989-990; LAROUSSE 2,785; WWWS, 426; MAYERHOFER 1,478-479; POGG 1,207,1539
P. Nicard, Etude sur la Vie et les Travaux de M. Ducrotay de Blainville. Paris 1890
E. Schuster-Aziza, Rev. Hist. Sci. 25 (1972), 191-200
B. Ballan et al., Rev. Hist. Sci. 32 (1979), 5-96
T.A. Appel, J. Hist. Biol. 13 (1980), 291-319

BLAISE, ÉMILE EDMOND [1872-1939]
CHARLE(2), 41-42; GORIS, 232-234; SCHALI, 17-18; POGG 4,131; 5,121-122; 6,235; 7b,412
G. Gault, Bull. Soc. Chim. [5]8 (1941), 269-346

BLAKE, FRANCIS GILMAN [1887-1952]
DAB [Suppl.5], 63-64; DAMB, 73-74; ICC, 126-128; WWWS, 186
J.R. Paul, Yale J. Biol. Med. 24 (1952); Trans. Assoc. Am. Phys. 65 (1952), 9-13; Biog. Mem. Nat. Acad. Sci. 28 (1954), 1-29
E. Goodpasture, Am. Phil. Soc. Year Book 1952, pp.302-307

BLAKE, JAMES [1815-1893]
MILES, 36; HIRSCH 1,561
C.D. Leake, Gesnerus 8 (1951), 114-123
M.A. Devereux et al., J. Chem. Ed. 33 (1956), 340-343
W.F. Bynum, Bull. Hist. Med. 44 (1970), 518-538

BLACHSTEIN, ARTHUR [1863-1940]
WWWS, 183; SG [2]2,361; RSC 13,580; NUC 59,380

BLACK, OTIS FISHER [1867-1933]
AMS 5,99; IB, 25
Anon., J. Wash. Acad. Sci. 23 (1933), 580

BLACKETT, PATRICK [1897-1974]
DNB [1971-1980], pp.60-62; MH 1,43-44; STC 1,135-137; MSE 1,100-101; WW 1965, pp.274-275
B. Lovell, Biog. Mem. Fell. Roy. Soc. 21 (1975), 1-115

BLACKLEY, CHARLES HARRISON [1820-1900]
E.B. Leech, Brit. Med. J. 1929(II), pp.1171-1172
A.E. Lowndes, Isis 37 (1947), 21-24
G. Taylor and J. Walker, Clinical Allergy 3 (1973), 103-108

BLACKMAN, FREDERICK FROST [1866-1947]
DSB 2,183-185; DNB [1941-1950], pp.82-83; DESMOND, 67; WW 1947, p.251
F.C. Steward, Plant Physiology 23(3) (1947), iii-viii
G.E. Briggs, Obit. Not. Fell. Roy. Soc. 5 (1948), 651-658

BLACKMAN, VERNON HERBERT [1872-1967]
DNB [1961-1970], pp.113-114; DESMOND, 67-68; WW 1965, p.276; WWWS, 184
H.K. Porter, Biog. Mem. Fell. Roy. Soc. 14 (1968), 37-60
W. Brown, Annals of Botany 32 (1968), 233-235

BLADIN, JOHAN ADOLF [1856-1902]
SBL 4,746-748; SMK 1,355-356; POGG 4,130,1702; 5,121

BLAGOVESHCHENSKI, ANDREI VASILIEVICH [1889-]
LIPSHITS 1,193-197; IB, 25; POGG 6,235; 7b,410-412
Z.V. Vasilieva, Bull. Soc. Nat. Moscou NS74(5) (1969), 5-7

BLAINVILLE, HENRI MARIE DUCROTAY de [1777-1850]
DSB 2,186-188; DBF 6,563-564; GE 6,989-990; LAROUSSE 2,785; WWWS, 426; MAYERHOFER 1,478-479; POGG 1,207,1539
P. Nicard, Etude sur la Vie et les Travaux de M. Ducrotay de Blainville. Paris 1890
E. Schuster-Aziza, Rev. Hist. Sci. 25 (1972), 191-200
B. Ballan et al., Rev. Hist. Sci. 32 (1979), 5-96
T.A. Appel, J. Hist. Biol. 13 (1980), 291-319

BLAISE, ÉMILE EDMOND [1872-1939]
CHARLE(2), 41-42; GORIS, 232-234; SCHALI, 17-18; POGG 4,131; 5,121-122; 6,235; 7b,412
G. Gault, Bull. Soc. Chim. [5]8 (1941), 269-346

BLAKE, FRANCIS GILMAN [1887-1952]
DAB [Suppl.5], 63-64; DAMB, 73-74; ICC, 126-128; WWWS, 186
J.R. Paul, Yale J. Biol. Med. 24 (1952); Trans. Assoc. Am. Phys. 65 (1952), 9-13; Biog. Mem. Nat. Acad. Sci. 28 (1954), 1-29
E. Goodpasture, Am. Phil. Soc. Year Book 1952, pp.302-307

BLAKE, JAMES [1815-1893]
MILES, 36; HIRSCH 1,561
C.D. Leake, Gesnerus 8 (1951), 114-123
M.A. Devereux et al., J. Chem. Ed. 33 (1956), 340-343
W.F. Bynum, Bull. Hist. Med. 44 (1970), 518-538

BLAKE, WILLIAM PHIPPS [1826-1910]
NCAB 25,202-203; GOULD, 20; ELLIOTT, 32; WWWS, 186
Anon., Obituary Records of Yale Graduates 1900-1910, pp.1303-1305. New Haven 1910

BLAKESLEE, ALBERT FRANCIS [1874-1954]
NCAB 42,662-663; AMS 8,222; CB 1941, pp.86-87; W, 62-63; WWWS, 186; HUMPHREY, 31-35; SCHULZ-SCHAEFFER, 206-207; WWAH, 58
E.W. Sinnott, Am. Phil. Soc. Year Book 1954, pp.394-398; Biog. Mem. Nat. Acad. Sci. 33 (1959), 1-38
R. Summers, Nature 174 (1954), 1037
M. Demerec, Genetics 44 (1959), 1-4

BLANC, GUSTAVE LOUIS [1872-1927]
POTSCH, 50; POGG 4,131; 5,123; 6,236
R. Delange, Bull. Soc. Chim. [4]43 (1928), 32-34

BLANCHARD, ARTHUR ALPHONZO [1876-1956]
AMS 9(I),166; WWWS, 187; POGG 5,123; 6,236; 7b,413

BLANCHARD, KENNETH CLARK [1900-1982]
AMS 11,443

BLANCHET, RODOLPHE [1807-1864]
STELLING-MICHAUD 2,225; 6,328; RSC 1,415-416; 7,189; POGG 1,208-209
Anon., Bull. Soc. Vaud. Sci. Nat. 10 (1868-1870), 756-757

BLANCHETIÈRE, ALEXANDRE [1875-1934]
DBF 6,629-630; IB, 26; POGG 6,237; 7b,413
E. Darmois, Bull. Soc. Chim. [5]1 (1934), 1614-1616
L. Cornil, Presse Med. 42 (1934), 1587

BLANCK, FREDERICK CONRAD [1881-1965]
AMS 11,443; NCAB 51,669-670

BLANDIN, PHILIPPE FRÉDÉRIC [1798-1849]
DBF 6,633; HIRSCH 1,563-564; WWWS, 188
J.B.I. Bourdon, Bull. Acad. Med. 14 (1849), 733-737

BLANGEY, JEAN LOUIS [1879-1960]
POGG 7a(1),195; (4),137*
Anon., Schweizer Biographisches Archiv 1 (1952), 26

BLANK, ALBERT [1865-1956]
JV 2,60
Anon., Nachr. Chem. Techn. 5 (1957), 10

BLANK, HUGO [1824-1898]
W. Blank, Über die Familie Blank. Elberfeld 1910

BLANK, PAUL [1882-]
JV 21,16; NUC 60,555

BLANKENHORN, ERNST [1853-1917]
DBJ 2,648; RSC 9,261

BLANKSMA, JAN JOHANNES [1875-1950]
POGG 5,123-124; 6,238-239; 7b,413-414
L. Seekles, Chem. Wkbl. 42 (1946), 14-18
J. van Alphen, Chem. Wkbl. 46 (1950), 793-795

BLARINGHEM, LOUIS [1878-1957]
CHARLE(2), 42-44; BUICAN, 247-261; IB, 26
R. Combes, C.R. Acad. Sci. 246 (1958), 22-27

BLAS, CHARLES [1839-1919]
TSCHIRCH, 1022; POGG 3,141; 4,133
R.V.M., Bull. Soc. Chim. Belg. 30 (1921), 88-89
Anon., Bibl. Acad. Louvain 6 (1937), 216

BLASCHKO, HERMANN [1900-]
STC 1,138-139; WW 1981, p.242; BHDE 2,114; LDGS, 79; WWWS, 188
H.K.F. Blaschko, Ann. Rev. Pharm. 20 (1980), 1-14; Selected Topics in the History of Biochemistry (G. Semenza, ed.), pp.189-231. Amsterdam 1983

BLASDALE, WALTER CHARLES [1871-1960]
AMS 8,224; POGG 6,224; 7b,414

BLATHERWICK, NORMAN ROBERT [1888-1961]
AMS 10,345; NCAB 46,70
Anon., Science 133 (1961), 268

BLATT, ALBERT HAROLD [1903-1986]
AMS 14,431; WWWS, 189
N. Rabjohn, Organic Syntheses 65 (1987), xi-xii

BLAU, FRITZ [1865-1929]
NDB 2,293; DBJ 11,345; TETZLAFF, 33; WWWS, 189; POGG 4,133-134; 6,240
M. Pirani, Naturwiss. 18 (1930), 97-101

BLEIBTREU, HERMANN [1821-1881]
NDB 2,297-298

BLEIBTREU, HERMANN [1855-]
BONN, 330

BLEIBTREU, LEOPOLD [1862-1932]
RSC 13,603-604
Anon., Deutsche med. Wchschr. 58 (1932), 1144

BLEIBTREU, MAX [1861-1939]
FISCHER 1,129
Anon., Klin. Wchschr. 18 (1939), 628
J. Ziewitz, Das Leben und Wirken von Max Bleibtreu. Greifswald 1969

BLEICHER, PAUL ALFRED [1892-]
JV 38,437

BLENDERMANN, HERMANN LUDWIG [1858-1884]
BERLIN, 589; RSC 12,88; NUC 61,158; SG [2]2,415
Anon., Deutsche med. Wchschr. 10 (1884), 624; Pharm. Z. 29 (1884), 529

BLEY, FRANZ LUDWIG [1801-1868]
HEIN, 60-61; FERCHL, 51; TSCHIRCH, 1022; CALLISEN 2,329-331; 26,324-325; POGG 1,211-212,1539-1540; 3,143
T. Geiseler, Arch. Pharm. 186 (1868), 1-15
C. Schümann et al., Pharmazie 46 (1991), 663-666

BLEYER, BENNO [1885-1945]
NDB 2,301; HEIN, 61-62; TSCHIRCH, 1022; TLK 3,53; IB, 26; WWWS, 190; POGG 5,125; 6,241; 7a(1),199-200
Anon., Z. angew. Chem. 59 (1947), 184
S.W. Souci, Z. Leb. Unt. 88 (1948), 122-124
W. Diemair, Deutsche Leb. Rund. 52 (1956), 32-47

BLICKE, FREDERICK FRANKLIN [1891-1968]
AMS 11,451; POGG 6,241-242; 7b,415-417
Anon., Chem. Eng. News 46(35) (1968), 58

BLINKS, LAWRENCE ROGERS [1900-1989]
MH 2,42-44; MSE 1,103-104; AMS 14,437; WWA 1978-1979, p.313

BLISH, MORRIS JOSLIN [1889-1975]
AMS 10.349; WWWS, 191

BLISS, SIDNEY WILLIAM [1892-1960]
AMS 10,350; YOUNG, 10

BLIX, GUNNAR [1894-1981]
SMK 1,360; VAD 1981, p.123; FISCHER 1,130; POGG 6,242; 7b,418-419

BLIX, MAGNUS [1849-1904]
SBL 4,779-783; SMK 1,360; FISCHER 1,130
R. Tigerstedt, Skand. Arch. Physiol. 16 (1904), 334-347
J.E. Johansson, Hygiea 66 (1904), 191-196

BLIX, MARTIN [1877-1933]
JV 17,11; NUC 61,314
New York Times, 19 January 1933

BLOCH, BRUNO [1878-1933]
NDB 2,306; AARGAU, 80-82; FISCHER 1,130-131; KURSCHNER 4,299; IB, 26; WWWS, 191; WININGER 7,632-533; TETZLAFF, 34; KOREN, 157
L.G. Miescher, Schw. med. Wchschr. 14 (1933), 576-578
F. Guggenheim, Bruno Bloch. Zurich 1969
M.B. Sulzberger, American Journal of Dermopathology 2 (1980), 321-325

BLOCH, EUGÈNE [1878-1944]
DBF 6,677-678; WWWS, 191; POGG 5,126; 6,243; 7b,240-241
J. Cabannes, Ann. Univ. Paris 22 (1952), 374-378; Cahiers de Physique 47 (1954), 1-4

BLOCH, HUBERT [1913-1974]
AMS 12,537; WWWS, 191-192
Anon., Chimia 28 (1974), 315

BLOCH, IGNAZ [1878-1942]
STURM 1,109; GEDENKBUCH 1,133; WEINMANN 1,77; NUC 61,344; POGG 6,243
Anon., Chem. Z. 52 (1928), 462

BLOCH, KONRAD EMIL [1912-]
MH 1,46-47; STC 1,141; MSE 1,106-107; AMS 15(1),540; POTSCH, 51;BHDE 2,120; IWW 1982-1983, p.132; WWA 1980-1981, p.327; WWWS, 192
K. Bloch, Reflections on Biochemistry (A. Kornberg et al., eds.), pp.143-150. Oxford 1976; Ann. Rev. Biochem. 56 (1987), 1-19

BLOCH, ROBERT [1898-]
AMS 10,351; LDGS, 11; IB, 26

BLOCH, SIEGFRIED [1882-]
JV 20,464; NUC 61,373

BLOCHMANN, REINHARD [1848-1920]
NDB 2,308; DBJ 2,740; BLOKH 1,64; SCHAEDLER, 13; POGG 4,135-136; 6,245
R. Saar, Chem. Z. 44 (1920), 613

BLOCHMANN, RICHARD HERMANN [1877-]
JV 18,357; NUC 61,383; POGG 4,135

BLOCK, RICHARD JOSEPH [1906-1962]
AMS 10,352
G.L. McNew, Trans. N.Y. Acad. Sci. 24 (1962), 670-674
J.A. Stekol and H.W. Howard, Science 136 (1962), 1039-1040
G.L. McNew and H.W. Howard, Contr. Boyce Thomp. Inst. 21 (1962), 247-250
J.A. Stekol, Advances in Chemistry Series 44 (1964), xiii-xxiii

BLODGETT, KATHERINE BURR [1898-1979]
AMS 10,352; WWWS, 192
V.J. Schaefer and G.L. Gaines, J. Coll. Sci. 76 (1980), 269-271
K. Davis, J. Chem. Ed., 61 (1984), 437-439

BLOEM, FRIEDRICH [1860-1915]
E. Gersten, Z. angew. Chem. 28(III) (1915), 400
Anon., Chem. Z. 39 (1915), 541

BLÖMER, ALFRED [1888-]
JV 30,727; NUC 61,439
Anon., Nachr. Chem. Techn. 11 (1963), 10

BLOKH, MAKS ABRAMOVICH [1882-1941]
BSE 5,312; WWR, 72; POGG 6,244-245; 7b,422-423
A.E. Fersman, Priroda 30(5) (1941), 117-119
V.G. Khlopin, Usp. Khim. 10 (1941), 501-504
H.S. Klickstein and H.M. Leicester, J. Chem. Ed. 23 (1946), 451-453

BLOM, JAKOB [1898-1965]
KBB 1965, p.141; VEIBEL 2,69-70; IB, 27; WWWS, 193
Anon., Chem. Eng. News 44(12) (1966), 88

BLOMSTRAND, CHRISTIAN WILHELM [1826-1897]
DSB 2,199-200; SBL 5,64-73; SMK 1,370-371; BLOKH 1,64-66; POTSCH, 51
P.Klason, Ber. chem. Ges. 30 (1897), 3227-3241; Lev. Sven. Vet. 4 (1909), 67-157
F. Elander et al., Sven. Kem. Tid. 38 (1926), 233-314
O. Krätz, Physis 15 (1973), 157-177
G.B. Kauffman, Ann. Sci. 32 (1975), 13-37

BLONDEAU, CHARLES [1810-1878]
RSC 2,433; 7,199; 9,267-268
M. Constans, Proc. Verb. Aveyron 26 (1917), 84-96
L. Balsan, Revue de Rouerge 32 (1978), 5-10; Proc. Verb. Aveyron 42 (1978), 456-461

BLONDLOT, NICOLAS [1808-1877]
DBF 6,713; GE 6,1173; HIRSCH 1,574; FERCHL, 51; POGG 3,145
L. Poincaré, Mem. Acad. Stanslas [4]9 (1877), lxxxix-xciii

BLOOM, WILLIAM [1899-1972]
AMS 12,542; NCAB 57, 673-674; IB, 27; WWWS, 194

BLOOMFIELD, ARTHUR LEONARD [1888-1962]
AMS 10,355; NCAB 50,499-500
A.J. Cox et al., Stanford Medical Bulletin 20 (1962), 137-139
L.A. Rantz, Trans. Assoc. Am. Phys. 76 (1963), 12-13

BLOOR, WALTER RAY [1877-1966]
AMS 11,459-460; CHITTENDEN, 201-206; IB, 27; WWAH, 59; POGG 6,247; 7b,424-425
Anon., Chem. Eng. News 44(12) (1966), 88

BLOUT, ELKAN ROGERS [1919-]
AMS 15(1),548; WWA 1980-1981, p.332; IWW 1982-1983, p.133

BLOW, DAVID MERVIN [1931-]
WW 1981, p.247

BLOXAM, CHARLES LOUDON [1831-1887]
Anon., J. Chem. Soc. 53 (1888), 508-509
D.I. Davies et al., Ambix 33 (1986), 11-32

BLÜCHER, HANS HELMUTH [1805-1862]
FERCHL, 51-52; POGG 3,145

BLUM, FERDINAND [1865-1959]
KURSCHNER 8,175; KALLMORGEN, 225; BHDE 2,122; ARNSBERG, 47; TETZLAFF, 35; POGG 7a(1),202-204
Anon., Chem. Z. 83 (1959), 793; Nachr. Chem. Techn. 7 (1959), 413
H. Göring, Arzneimitt. 10 (1960), 61-62

BLUM, JACOB JOSEPH [1926-]
AMS 15(1),550; WWA 1980-1981, p.333

BLUM, LÉON [1878-1930]
DBA 1,262; FISCHER 1,133; KOREN, 157
K. Spiro, Schw. med. Wchschr. 60 (1930), 379-380
E. Vaucher, Presse Med. 38 (1930), 531-532
R. Kohn, Rev. Hist. Med. Heb. 1952, pp.200-201

BLUM, LOUIS [1858-1920]
M. Blum, Bibliographie Luxembourgoise 1,84-85

BLUMANN, ARNOLD [1885-1969]
POGG 6,248; 7a(1),204
Anon., Chem. Brit. 6 (1970), 183

BLUMBERG, BARUCH SAMUEL [1925-]
MSE 1,108-110; AMS 15(1),551; WWA 1980-1981, p.334

BLUMENTHAL, FERDINAND [1870-1941]
FISCHER 1,134; KURSCHNER 5,109; FRANKENTHAL, 276-277; BHDE 2,124; IB, 27; LDGS,64; PAGEL, 1931; TETZLAFF, 35; POGG 4,138; 7a(1),204
P. Kraus and E. Pütter, Z. Krebsforsch. 32 (1930), 3-9

BLUMENTHAL, FRANZ [1878-1971]
FISCHER 1,134; ASEN, 17; KOREN, 157-158; LDGS, 61; KAGAN(JA), 649-650
C.A. Hoffmann, Hautarzt 4 (1953), 394-395
E. Hoffmann, Derm. Wchschr. 138 (1958), 786-787
I.I. Edgar, A History of Early Jewish Physicians in the State of Michigan, pp.69-71. New York 1982

BLUMENTHAL, GEORG [1888-1964]
KURSCHNER 9,161; POGG 7a(1),204-205
Anon., Chem. Z. 88 (1964), 364

BLUMENTHAL, HERBERT [1885-1952]
JV 23,17
Anon., Z. angew. Chem. 64 (1952), 660

BLUMER, ARTHUR [1868-1924]
JV 8,162; NUC 62,593

BLUMGART, HERRMAN LUDWIG [1895-1977]
AMS 13,396; ICC, 203-205
Anon., New Eng. J. Med. 296 (1977), 1117-1118
A.S. Freedberg, Journal of Nuclear Medicine 19 (1978), 569

BLUNDELL, JAMES [1790-1877]
DNB 2,728; HIRSCH 1,580; CALLISEN 2,359-360; 26,342
Anon., Lancet 1878(I), p.255
H.W. Jones and G. Mackmull, Ann. Med. Hist. 10 (1928), 242-248
J.H. Young, Med. Hist. 8 (1964), 159-169

BLYTH, ALEXANDER WYNTER [1844-1921]
FISCHER 1,135
J.K.C., J. Chem. Soc. 119 (1921), 545-546
Anon., Analyst 46 (1921), 177-178; J. Roy. Inst. Chem. 1921, pp.222-223

BLYTH, JOHN [1814-1871]
RSC 1,443; POGG 1,215; 3,146
Anon., J. Chem. Soc. 25 (1872), 343-344

BOAS, FRIEDRICH [1886-1960]
KURSCHNER 9,162; IB, 27

BOAS, ISMAR (ISIDORE) [1858-1938]
NDB 2,338-339; GROTE 7,51-98; FISCHER 1,135-136; PAGEL, 196-197; KURSCHNER 4,298; SKVARC, 27-28; OLPP, 45-47; WININGER 1,412-413;TETZLAFF, 36; KOREN, 158; WREDE, 225
R. Ehrmann, Deutsche med. Wchschr. 54 (1928), 492-493
R. Schorlemmer, Münch. med. Wchschr. 75 (1928), 525
M. Einhorn, Arch. Verdaungskr. 63 (1930), 102
A. Eichmann, Ismar Boas. Zurich 1970
L.J. Hoenig, Journal of Clinical Gastroenterology 11 (1989), 586-587

BOCHALLI, RICHARD [1878-1966]
POGG 7a(1),206
P. Diepgen, Deutsche med. Wchschr. 78 (1953), 930
F. Oldenburg, Z. Tub. 120 (1963), 203-204
K. Meister, Aerztlicher Dienst 28 (1967), 64-67

BOCK, CARL [1842-1873]
RSC 7,203; NUC 62,559

BOCK, CARL ERNST [1809-1874]
NDB 2,343-344; ADB 2,767-768; NB, 32; HIRSCH 1,496-497; PAGEL, 198-199

BOCK, JOHANNES CARL [1867-1953]
DBL(3rd Ed.) 2,285-286; KBB 1952, p.157; VEIBEL 2,71

BOCK, ROBERT MANLEY [1923-1991]
AMS 15(1),559; WWA 1980-1981, p.338
New York Times, 4 July 1991

BOCKMÜHL, AUGUST [1881-]
JV 24,336; NUC 62,593

BOCKMÜHL, MAX [1882-1949]
JV 25,603; POGG 6,251; 7a(1),208-209
Anon., Chem. Z. 59 (1935), 666; 66 (1942), 393; Z. angew. chem. 61 (1949), 80

BODANSKY, AARON [1887-1960]
AMS 10,364; KAGAN(JA), 738
Anon., Chem. Eng. News 38(16) (1960), 160

BODANSKY, MEYER [1896-1941]
AMS 6,134; CB 1941, p.89; KAGAN(JA), 350,745; WWWS, 197; WWAH, 60; IB, 27; POGG 6,252; 7b,434-435
I. Davidson, Am. J. Clin. Path. 11 (1941), 789-794
Anon., Science 93 (1941), 613

BODANSKY, OSCAR [1901-1977]
AMS 13,401; WWA 1976-1977, pp.314-315; WWWS, 197
Anon., Science 198 91977), 478
M.K. Schwartz, Advances in Clinical Chemistry 20 (1978), xv-xix; Pharmacologist 21 (1979), 102-103

BODANSZKY, MIKLOS [1915-]
AMS 15(1),560; WWA 1980-1981, p.338; WWWS, 197-198

BODASZEWSKI, LUKASZ JAN [1849-1908]
WE Suppl.,61

BODE, OTTO [1913-1981]
WAGENITZ, 31; NUC 62,642

BODEA, CORNEL [1903-1985]
POTSCH, 51; WWWS, 198

BODENDORF, KURT [1898-1976]
KURSCHNER 12,268; POGG 6,252; 7a(1),211
W. Schneider, Arzneimitt. 8 (1958), 781-782
H. Vogt, Arzneimitt. 13 (1963), 1118-1119; 18 (1968), 1620
R. Neidlein, Arzneimitt. 26 (1976), 1727-1728; Deutsche Apoth. Z. 116 (1976), 1140-1141
G. Schwenker, Pharmazie Unserer Zeit 8(2) (1979), 35-45

BODENHEIMER, FRITZ SIMON [1897-1959]
DSB 2,221-222; WININGER 6,481
F.S. Bodenheimer, Studies in Biology and its History, pp.119-144. Jerusalem 1957; A Biologist in Israel. Jerusalem 1959
B.P. Uvarov, Nature 184 (1959), 937-938
R. Zaunick, Physis 2 (1960), 270-273
A. Virieux-Reymond, Revue de Synthèse 81 (1960), 439-441

BODENHEIMER, WOLF [1905-1975]
BHDE 1,75-76; LDGS, 17
S. Yariv, Israel J. Chem. 8(3) (1970), i-iv

BODENSTEIN, DIETRICH [1908-1984]
AMS 14,455; WWA 1976-1977, p.315; IWW 1983-1984, pp.136-137; BHDE 2,127; LDGS, 13

BODENSTEIN, MAX [1871-1942]
DSB 15,36-38; NDB 2,357-358; TLK 3,56; HANNOVER, 31; POTSCH, 51-52; WWWS, 198; POGG 4,141; 5,129-130; 6,252-253; 7a(1),211
H.J. Schumacher, Z. angew. Chem. 54 (1941), 329-333
K.F. Bonhoeffer, Naturwiss. 30 (1942), 737-739
P. Günther, Z. Elektrochem. 48 (1942), 585-587
G.M. Schwab, Wiener Chem. Z. 46 (1943), 1-11; Rete 1 (1971), 125-134
K. Clusius, Bayer. Akad. Wiss. Jahrbuch 1944-1948, pp.248-249
E. Cremer, Chem. Ber. 100 (1967), xcv-cxxvi
Z.G. Szabo, Naturw. Rund. 22 (1969), 69-72

BODFORSS, SVEN [1890-1978]
POGG 6,253-254; 7b,435

BODIAN, DAVID [1910-1992]
AMS 15(1),561; IWW 1982-1983, p.137; WWWS, 198

BODLÄNDER, GUIDO [1855-1904]
BJN 10,14*; BLOKH 1,67-68; WWWS, 198; POGG 4,141-142
A. Coehn, Ber. chem. Ges. 38 (1905), 4263-4290
J. Bieringer, Z. angew. Chem. 18 (1905), 561-569
W. Nernst, Z. Elektrochem. 11 (1905), 157-161

BODNAR, JANOS [1888-1953]
MEL 1,227; IB, 27; POGG 6,254; 7b,435-437

BODROUX, FERNAND [1873-1968]
POGG 5,180-181; 6,254
P. Boissonade et al., Histoire de l'Universite de Poitiers, pp.456-457. Poitiers 1932

BÖCK, HERMANN von [1834-1885]
HIRSCH 1,589-590; DUELUND, 26-28
C. Voit, Aertzliches Intelligenzblatt 32 (1895), 739-740

BOECKEL, THÉODORE [1802-1869]
DBF 6,766; DBA 1,272; SITZMANN, 179

BOECKLER, AUGUST [1871-]
JV 10,264; NUC 63,104

BOECKMANN, EMIL [1811-]
FERCHL, 52; RSC 1,449; POGG 1,221

BOECKMANN, OTTO [1860-]
RSC 13,275; NUC 63,326

BÖDECKER, FRIEDRICH [1883-]
IB, 27; POGG 6,255; 7a(1),212
Anon., Nachr. Chem. Techn. 1 (1953), 47; 6 (1858), 110; 11 (1963),105

BÖDEKER, CARL [1815-1895]
HEIN, 65-66; TSCHIRCH, 1022; FERCHL, 52-53; POGG 1,221-222; 3,148
E. Prael, Pharm. Z. 40 (1895), 186-187
Anon., Leopoldina 31 (1895), 59

BOEHM, JOSEPH ANTON [1831-1893]
NDB 2,383-384; ADB 47,75-77; WWWS, 199
K. Wilhelm, Ber. bot. Ges.12 (1894), (14)-(28)

BÖHM, GUSTAV [1827-1900]
HB 1,64-66

BÖHM, RUDOLF [1844-1926]
NDB 2,379; FISCHER 1,138; PAGEL, 203-204; ROEDER, 160-163; DZ, 137; TSCHIRCH, 1022; REBER, 242-244; IB, 27; LEVITSKI 2, 191-193; POGG 3,148; 4,142-143; 5,134; 6,258; 7a(1),216
E. Brennsohn, Die Aerzte Estlands, p.432. Riga 1922
K. Thomas, Sitz. Akad. Wiss. Leipzig 78 (1926), 348-357
A. Schüller, Münch. med. Wchschr. 73 (1926), 2170
W. Wiezorek and M. Müller, NTM 12 (1975), 97-107

BOEHM, THEODOR [1892-1969]
HEIN-E, 37-38; KURSCHNER 12,278; DRUM, 331-332; POGG 6,258; 7a(1),216-217
Anon., Nachr. Chem. Techn. 17 (1969), 79

BÖHME, ARTHUR [1878-1962]
FISCHER 1,138; KURSCHNER 9,170; KIEL, 118
H. Laut, Deutsche med. Wchschr. 87 (1962), 120-121

BÖHME, HORST [1908-]
AUERBACH, 776-777; KURSCHNER 13,337; DRUM, 306-308; POTSCH, 53; POGG 7a(1),217-218
B. Unterhalt, Deutsche Apoth. Z. 128 (1988), 1081-1082

BOEHNCKE, KARL ERNST [1874-1929]
JV 27,468
Anon., Schw. med. Wchschr. 59 (1929), 1072; Münch. med. Wchschr. 76 (1929), 1664

BÖHNER, GEORG [-1915]
Anon., Ber. chem. Ges. 48 (1915), 730

BOEHNER, REGINALD [1880-1945]
Anon., Chem. Eng. News 23 (1945), 1554

BOEHRINGER, ERNST [1896-]
NUC 63,300

BOEKER, GILBERT F. [1905-]
AMS 11,474

BOELL, EDGAR JOHN [1906-]
AMS 14,458; WWWS, 199

BÖRNSTEIN, ERNST [1854-1932]
NDB 2,406; TLK 3,62; WWWS, 213; POGG 3,149-150; 4,145; 5,134; 6,261-262; 7a(1),220
F. Frank, Ber. chem. Ges. 65A (1932), 67-68

BOESEKEN, JACOB [1868-1949]
BWN 2,41-42; POGG 4,146; 5,131-133; 6,255-256; 7b,437-438
J.J. Benedictus and C. van Loon, Chem. Wkbl. 47 (1951), 805-806

BÖSLER, MAGNUS [1852-1924]
RSC 9,300; NUC 63,487
Anon., Chem. Z. 48 (1924), 512; Z. angew. Chem. 37 (1924), 584

BÖTTCHER, ARTHUR [1831-1889]
HIRSCH 1,605; Suppl.,113; WELDING, 85-86; LEVITSKI 2,95-102
W. Stricker, Virchows Arch. 119 (1890), 337
J. Brennsohn, ie Aerzte Estlands, p.423. Riga 1922

BÖTTGER, RUDOLPH CHRISTIAN [1806-1881]
DSB 2,340; NDB 2,410; ADB 47,143-144; KALLMORGEN, 228; POTSCH, 58; BLOKH 1,77-78; FERCHL, 54-55; SCHAEDLER, 14-15; WWWS, 1730; CALLISEN 26,358; POGG 1,224-225; 3,150-151; 7a Suppl.,96-97
T. Peterson, Ber. chem. Ges. 14 (1881), 2913-2919; Leopoldina 17 (1881), 146-148,166-168,182-188
E. Fischer, Chem. Z. 88 (1964), 630-631

BÖTTGER, WILHELM [1871-1949]
NDB 2,411' HEIN-E, 38-39; TLK 3,65; POTSCH, 58; TSCHIRCH, 1024; WWWS, 216; POGG 4,147; 5,136; 6,262; 7a(1),222
E. Brennecke, Z. angew. Chem. 62 (1950), 279-280

BÖTTINGER, CARL [1851-1901]
RSC 13,641-642; POGG 4,147-148
Anon., Ber. chem. Ges. 34 (1901), 3210; Pharm. Z. 46 (1901), 793

BÖTTINGER, HEINRICH [1820-1874]
GGTA 3 (1909), 65
Anon., Ber. chem. Ges. 7 (1874), 1805
A. Wankmüller, Beitr. Wurtt. Apothekgesch. 7 (1966), 83

BÖTTINGER, HENRY THEODORE von [1848-1920]
DBJ 2,500-501; BLOKH 1,78-79
B. Lepsius, Ber. chem. Ges. 53A (1920), 111-116
Genealogisches Handbuch des Adels 83 (1984), 58

BOETZELEN, ERNST [1869-]
JV 14,140; NUC 63,571

BOGENDÖRFER, LUDWIG [1892-1963]
FISCHER 1,140; GEISSLER, 71-76; KURSCHNER 5,114

BOGERT, MARSTON TAYLOR [1868-1954]
AMS 8,237-238; MILES, 38-39; NCAB 42,624-625; WWWS, 200; WWAH, 61; POGG 5,135; 6,264-265; 7b,443-446
L.P. Hammett, J. Chem. Soc. 1954, pp.4709-4710; Biog. Mem. Nat. Acad. Sci. 45 (1974), 97-126
G.B. Pegram and J.M. Nelson, Am. Phil. Soc. Year Book 1954, pp.399-402
J.M. Nelson, J. Am. Chem. Soc. 76 (1954), 5009-5011
D. Price, J. Chem. Ed. 32 (1955), 506-509

BOGOMOLETS, ALEKSANDR ALEKSANDROVICH [1881-1946]
BSE 5,353-354; BME 4,10-11; WWR, 77-78; KUZNETSOV(1963), pp.415-422; VORONTSOV, 123-126; UKRAINE 2,195-197; FISCHER 1,140; IB, 28
I.M. Neiman, Arkh. Pat. 18 (1956), 3-9
N.N. Sirotinin, A.A. Bogomolets. Moscow 1967
N.N. Zaiko et al., Fiziol. Zhur. (Kiev) 27(3) (1981), 291-443

BOGOMOLOV, TIMOFEI IVANOVICH [1843-1897]
VENGEROV 1,293; SG [2]1,517; RSC 12,97; 13,646

BOGUE, ROBERT [1889-]
AMS 12,566-567; POGG 6,266; 7b,448-449

BOHLAND, KARL [1861-1945]
FISCHER 1,141; PAGEL, 206; WENIG, 29; AD 3,95
Anon., Deutsche med. Wchschr. 71 (1946), 40

BOHN, GEORGES [1868-1948]
IB, 28; WWWS, 202
M. Caullery, Bull. Biol. 82 (1948), i-iii

BOHN, RENÉ [1862-1922]
NDB 2,421; DBJ 4,350; DBA 1,288; BLOKH 1,70-72; POTSCH, 53; WWWS, 202; POGG 5,137
M.A. Kunz, Chem. Z. 46 (1922), 297-298
E. Noelting, Helv. Chim. Acta 5 (1922), 566-570
P.Julius and M.A. Kunz, Ber. chem. Ges. 56A (1923), 13-30

BOHR, CHRISTIAN [1855-1911]
DSB 17,92-93; MEISEN, 173-176; FISCHER 1,141; VEIBEL 2,72-74; WWWS, 202; POGG 4,149-150; 5,138; 7b,449
. Tigerstedt, Skand. Arch. Physiol. 25 (1911), v-xvii
N. Zuntz, Med. Klin. 7 (1911), 434-435

BOHR, NIELS [1885-1962]
DSB 2,239-254; MAYERHOFER 1,512-513; SCHWERTE 1,71-78; MSE 1,113-115; W, 64-66; WWWS, 202; POGG 5,138-139; 6,268-269; 7b,451-455
J. Cockcroft, Biog. Mem. Fell. Roy. Soc. 9 (1963), 37-53
R. Oppenheimer, Am. Phil. Soc. Year Book 1963, pp.107-117
W. Heisenberg, Bayer. Akad. Wiss. Jahrbuch 1963, pp.204-207
S. Rozenthal (Ed.), Niels Bohr, his Life and Work as Seen by his Friends and Colleagues. New York 1967
C.F. von Weizsäcker, Phys. Bl. 41 (1985), 308-315

BOHRISCH, PAUL [1871-1952]
HEIN-E, 40-41; TSCHIRCH, 1023; TLK 3,59
Anon., Chem. Z. 47 (1923), 115; 48 (1924), 172; Apoth. Z. 44 (1929), 1630-1631

BOISSONAS, ROGER [1921-]
POGG 7a(1),225

BOIVIN, ANDRÉ [1895-1949]
DBA 1,290; WWWS,203; POGG 6,269; 7b,455-459
J. Roche, Bull. Soc. Chim. Biol. 31 (1949), 1564-1567
M. Polonovski, Bull. Acad. Med. 133 (1949), 608-613
M. Javillier, C.R. Acad. Sci. 229 (1949), 153-157
R. Vendrely, Paris Médical 39 (1949), 447-448

BOJANOWSKI, KAROL [1835-1898]
PSB 2,237; KOSMINSKI, 39,590; SG [1]2,218; NUC 700,173

BOKÁY, ARPÁD [1856-1919]
MEL 1,236; FISCHER 1,142; PAGEL, 1932-1933
P. Gulyas, Magyar Irok 3 (1941), 784-786
A. Szallasi, Orvosi Hetilap 117 (1976), 2191-2193

BOKORNY, THOMAS [1856-1929]
KURSCHNER 3,203; TSCHIRCH, 1023; POGG 4,151; 6,269
Anon., Pharm. Z. 32 (1887), 592; Chem. Z. 33 (1909), 616; 53 (1929),170

BOKS, ALBERT JOHAN [1871-]
NUC 64,410

BOLAM, HERBERT WILLIAM [1871-1949]
SRS, 14; NUC 64,418
Anon., J. Roy. Inst. Chem. 1949, p.446

BOLD, HAROLD CHARLES [1909-1987]
AMS 14,465; WWWS,203

BOLDINGH, GERRIT HONDIUS [1865-1936]
NUC 64,441
Wie is Dat? 1935, p.164
C.G. van Arkel, Pharm. Weekbl. 73 (1936), 1197-1198

BOLDYREV, VASILI NIKOLAEVICH [1872-1946]
IB, 29
A.C. Ivy, Gastroent. 6 (1946), 613-614
D.G. Kavsov and A.K. Federova-Grot, Fiziol. Zhur. 46 (1960), 126-132

BOLL, FRANZ [1849-1879]
HIRSCH 1,614; PAGEL, 211-212; POGG 4,151
Anon., Nature 21 (1880), 214
G. Haltenhoff, Ann. Ocul. 83 (1880), 90-102
C. Baumann, Vision Research 17 (1977), 1267-1268
L. Belloni, Med. Hist. J. 17 (1982), 129-137

BOLLENBACH, HERMANN FRIEDRICH [1877-1940]
JV 18,174; NUC 64,643

BOLLEY, ALEXANDER POMPEIUS [1812-1870]
ADB 3,109-110; DHB 2,228; AARGAU, 90-91; BLOKH 1,72; SCHAEDLER, 13-14; TSCHIRCH, 1025; FERCHL, 55; PHILIPPE, 775; POGG 1,228-229; 3,154-155
E. Kopp and J. Scherr, Ber. chem. Ges. 3 (1870), 813-820
Anon., Verhandl. Schw. Nat. Ges. 54 (1871), 265-268
H.R. Feldmann, Pharm. Z. 125 (1980), 72-73

BOLLINGER, OTTO von [1843-1909]
NDB 2,432-433; BJN 14,13*; FISCHER 1,143; PAGEL, 212; ECKERT, 8-20;DZ, 142; WWWS, 204
H. Dürck, Münch. med. Wchschr. 56 (1909), 2058-2064
R. Rössle, Z. allgem. Path. 20 (1909), 961-966; Verhandl. path. Ges. 14 (1910), 368-370
O. Everbusch, Berl. klin. Wchschr. 46 (1909), 1996-2000

BOLLMAN, JESSE LOUIS [1896-1979]
AMS 14,468; WWWS, 204

BOLTON, BENJAMIN MEADE [1857-1929]
AMS 4,99; SG [2]2,530; RSC 13,662; NUC 65 67; WWWS, 205
Anon., J. Am. Med. Assn. 93 (1929), 1006

BOLTON, ELMER KEISER [1886-1968]
AMS 11,485; MH 1,52; MSE 1,115-116; MILES, 39-40; WWWS, 205; WWAH, 62
R.M. Joyce, Biog. Mem. Nat. Acad. Sci. 54 (1983), 51-72

BOLTON, HENRY CARRINGTON [1843-1903]
DAB 2,422-423; MILES, 40-41; ELLIOTT, 34; BLOKH 1,72-73; WWAH, 62; SILLIMAN, 154-155; POGG 3,155; 4,152; 5,139
D.S. Martin, Ann. N.Y. Acad. Sci. 16 (1905), 75-81
C.A. Browne, J. Chem. Ed. 17 (1940), 457-461

BOLTON, WERNER von [1868-1912]
NDB 2,435; BJN 18,11*; BLOKH 1,73; POTSCH, 54; POGG 4,152-153; 5,139; 7a(1),229
Anon., Chem. Z. 36 (1912), 1344

BOLTWOOD, BERTRAM BORDEN [1870-1927]
DSB 2,257-260; DAB 2,424-425; NCAB 15,138-139; POTSCH, 54; W, 66-67; WWWS, 205; POGG 4,153; 5,139; 6,272
A.F. Kovarik, Am. J. Sci. [5]15 (1928), 188-198; Biog. Mem. Nat. Acad. Sci. 14 (1929), 69-96
E. Rutherford, Nature 121 (1928), 64-65
L. Badash, Proc. Am. Phil. Soc. 112 (1968), 157-168

BOLTZMANN, LUDWIG [1844-1906]
DSB 2,260-268; NDB 2,436-437; NOB 2,117-137; BJN 11,96-104,11*; WWWS,205;BLOKH 1,73-75; W, 67; MAYERHOFER 1,515-517; POGG 3,155-156; 4,153-154; 5,140; 6,272; 7aSuppl.,102-104
T. Des Coudres, Sitz. Akad. Wiss. Leipzig 58 (1906), 615-628
H.A. Lorenz, Ber. phys. Ges. 5 (1907), 206-238
C. Voit, Sitz. Bayer. Akad. 37 (1907), 262-267
E. Brode, Ludwig Boltzmann. Berlin 1957
P. Urban, Phys. Bl. 38 (1982), 259-264

BOMMER, MAX [1892-1964]
NUC 65,242
Anon., Verhandl. Nat. Ges. Basel 76 (1964), 378

BONANNI, ATTILIO [1869-1938]
N. Spano, L'Università di Roma, p.278. Rome 1935
P. di Mattei, Arch. Inter. Pharmacodyn. 60 (1938), 115-117

BONDI, AMEDEO [1912-]
AMS 15(1),583; WWWS, 107

BONDI, SAMUEL [1878-1959]
FISCHER 1,144-145; KURSCHNER 4,257; KOREN, 158

BONE, WILLIAM ARTHUR [1871-1938]
DNB [1931-1940], 85-86; WWWS, 207; POGG 4,156; 5,141-142; 6,274; 7b,466-467
G.I. Finch and A.G. Egerton, Obit. Not. Fell. Roy. Soc. 2 (1939), 587-611

BONHOEFFER, KARL FRIEDRICH [1899-1957]
KALLMORGEN, 229; WWWS, 208; TLK 3,60; POTSCH, 54; POGG 6,275; 7a(1),231-232; (4),138*
J. Eggert, Bayer. Akad. Wiss. Jahrbuch 1958, pp.189-196
H. Staude, Schwab. Akad. Wiss. Jahrbuch 1957-1959, pp.338-345
G.M. Schwab, Z. Elektrochem. 62 (1958), 221-224

BONHOFF, HEINRICH [1864-1940]
FISCHER 1,145; KURSCHNER 3,208; GUNDLACH, 272; AUERBACH, 206; ASEN, 20
R. Wigand, Münch. med. Wchschr. 87 (1940), 456-457

BONINO, GIOVANNI BATISTA [1899-]
STC 1,151-152; POGG 6,275-276; 7b,469-471
Lui Chi É (2nd Ed.), 352-353

BONNE, GEORG [1859-1945]
R. Wald, Fort. Med. 50 (1932), 167-174; Münch. med. Wchschr. 81 (1934), 1233-1234; Z. ärzt. Fortbild. 35 (1938), 596-598

BONNER, DAVID MAHLON [1916-1964]
AMS 10,381; WWAH, 63
Anon., Chem. Eng. News 42(23) (1964), 89
J.A. deMoss et al., Genetics 52 (1965), s15-s16

BONNER, JAMES FREDERICK [1910-]
MH 1,53-54; STC 1,152-153; MSE 1,117-118; IWW 1982-1983, p.144; WWA 1978-1979, p.341; AMS 14,476; WWWS, 208

BONNET, ROBERT [1851-1921]
DBJ 3,41-43,291; EGERER, 26-34
J. Sobotta, Anat. Anz. 56 (1922), 145-158

BONNICHSEN, ROGER [1913-1986]
VAD 1986, p.157
R. Andreasson and A.W. Jones, American Journal of Forensic Medicine and Pathology 10 (1989), 353-359

BONNIER, GASTON [1853-1922]
DSB 2,290-291; DBF 6,1039-1040; CHARLE(2), 46-48; WWWS, 209
M.H. Jumelle, Rev. Gen. Bot. 36 (1924), 289-307
A. Davy de Virville, in Essays in Biochemistry (P. Smit and R.J.C.V. ter Laage, Eds.), pp.1-13. Utrecht 1970

BOOTH, JAMES CURTIS [1810-1888]
MILES, 42-43; SILLIMAN, 85-86; COLE, 73
P. duBois, Proc. Am. Phil. Soc. 25 (1888), 204-211
E.F. Smith, J. Chem. Ed. 20 (1943), 315-318,357
W.D. Miles, Chymia 11 (1966), 139-149

BOOTH, VERNON HOLLIS [1903-]
WWWS, 210
Who's Who of British Scientists 1971-1972, p.91

BOOTHBY, WALTER MEREDITH [1880-1953]
AMS 8,247; NCAB 45,58; CHITTENDEN, 273-275; IB, 30; WWAH, 63
W.W. Spink, Ann. Int. Med. 39 (1953), 673
S.F. Haines, Trans. Assoc. Am. Phys. 67 (1954), 10-11
D.B. Dill, Science 120 (1954), 688
L.D. Vandam, New Eng. J. Med. 276 (1967), 558-563

BOPP, FRIEDRICH [1824-1849]
HB 3,261-264; RSC 1,492; POGG 1,237

BOQUET, ALFRED [1879-1947]
J.L.A. Verge, Bull. Acad. Med. 131 (1947), 474-476
Anon., Ann. Inst. Pasteur 73 (1947), 617-621

BORCHARDT, HELENE [1904-]
LDGS, 59; NUC 66,627

BORCHARDT, LEO [1879-]
FISCHER 1,148; KURSCHNER 4,263-264

BORDET, JULES [1870-1961]
DSB 2,300-301; BNB 38,26-36; FISCHER 1,149; SACKMANN, 33-36; W, 69-70;WWWS, 211; POGG 6,278; 7b,473-474
E. Renaux et al., Ann. Inst. Pasteur 73 (1950), 479-520
P. Vallery-Radot, Presse Med. 69 (1961), 1369-1370
A.R. Dujarric de la Rivière, Bull. Acad. Med. 145 (1961), 402-410
J. Trefouël, C.R. Acad. Sci. 252 (1961), 2969-2972
M. Welsch, Rev. Med. Liège 16 (1961), 269-272
C.L. Oakley, Biog. Mem. Fell. Roy. Soc. 8 (1962), 19-25
J. Beumer, J. Gen. Microbiol. 29 (1962), 1-13
A. Delaunay, Hist. Med. 21 (1971), 2-45
A. Nisonoff, Mem. Acad. Med. Belg. 7 (1972), 369-376
A.B. Laurell, Scandinavian Journal of Immunology 35 (1990), 429-432

BORDT, FRIEDRICH [1851-]
JV 4,223; RSC 13,275,690

BOREK, ERNEST [1911-1986]
AMS 14,484
New York Times, 21 February 1986

BORESCH, KARL [1886-1947]
STURM 1,125; KURSCHNER 5,92; TLK 3,61; IB, 30

BORGMANN, EUGEN [1843-1895]
RSC 7,218; 9,295; NUC 67,204
C. Liebermann, Ber. chem. Ges. 28 (1895), 721-722

BORISOV, PETR YAKOVLEVICH [1864-1916]
RSC 13,694
L.I. Grabovskaya, Farm. Toks. 27 (1964), 745-746

BORN, GUSTAV [1851-1900]
BJN 5,221-222,83*; PAGEL, 215; RSC 13,695; WININGER 1,425-428; KOREN, 158
W. Gebhardt, Anat. Anz. 18 (1900), 139-143
W. Roux, Arch. Entwickl. 10 (1900), 256-262BORNET, ÉDOUARD [1828-1911] DBF 6,1111-1112; WWWS, 213
L. Guignard, Bull. Soc. Bot. 59 (1912), 257-301

BORNHARDT, CARL [1886-]
JV 29,722; NUC 67,322

BORNHEIM, WILHELM [1891-]
JV 37,1281

BORNSTEIN, ARTHUR [1881-1932]
FISCHER 1,150; TLK 3,62; IB, 30

BORNTRÄGER, AUGUST FRIEDRICH [1820-1905]
HEIN, 72; BJN 10,149*; DRULL, 26; BLOKH 1,75; SCHAEDLER, 15

BORODIN, ALEKSANDR PORFIREVICH [1833-1887]
DSB 2,316-317; BSE 5,593-595; ARBUZOV, 105-106; BLOKH 1,75-76; POTSCH, 57; ZMEEV 1,28; POGG 3,164-165; 4,160
A.P. Dianin, Zhur. Fiz. Khim. 20 (1888), 367-379
G. Sarton, Osiris 7 (1939), 225-260
H.B. Friedman, J. Chem. Ed. 18 (1941), 521-525
N.A. Figurovsky and Y.I. Soloviev, A.P. Borodin. Moscow 1950 (English translation 1987)
M.M. Eisenberg, J. Hist. Med. 15 (1960), 78-81
W.B. Ober, N.Y. J. Med. 68 (1967), 836-845
W. Kwasnik, Chem. Z. 91 (1967), 312-313
G.P. Shultsev, Klin. Med. 62(7) (1984), 144-147
C.B. Hunt, Chem. Brit. 23 (1987), 547-550
I.D. Rae, Ambix 36 (1989), 121-137

BORODIN, IVAN PARFENIEVICH [1847-1930]
BSE 5,595-596; WWR, 85; LIPSHITS 1,248-254; BK, 169-170;

VENGEROV 5,272-278; UKRAINE 2, 198-199
V.N. Lyubimenko, Priroda 19 (1930), 1055-1067
T.S. Paskina, Biokhimia 16 (1951), 374-381

BORREL, AMÉDÉE [1867-1936]
DBF 6,1115; DBA 1,305-306; SACKMANN, 37-39
J. Magrou, Presse Med. 44 (1936), 1697-1698
Anon., Ann. Inst. Pasteur 57 (1936), 337-342
L. Nègre, Biologie Médicale 46 (1957), 1-39
P. Pagès, Mons. Hipp. 8(27) (1965), 23-31; 8(28) (1965), 19-27
R. LeGuyon, Bull. Acad. Med. 151 (1967), 585-593

BORRIES, BODO von [1905-1956]
DSB 2,318; POGG 7a(1),237-239; (4),138*
B. von Borries, Phys. Z. 45 (1944), 314-326
E. Ruska, Z. wiss. Mikr. 63 (1956), 129
Anon., Chem. Z. 80 (1956), 489

BORSCHE, WALTHER [1877-1950]
NDB 2,476; KALLMORGEN, 230; TLK 3,63; POGG 4,160; 5,145-146; 6,282-283; 7a(1),239
W. Ried, Chem. Ber. 85 (1952), xxxi-liv

BORSOOK, HENRY [1897-1984]
AMS 14,488; COHEN, 32

BORST, MAXIMILIAN [1869-1946]
NDB 2,477-478; FISCHER 1,151; ECKERT, 26-41; IB, 31; WWWS, 214
H. Groll, Deutsche med. Wchschr. 65 (1939), 1694-1695
A. Schminke, Z. allgem. Path. 84 (1948), 1-3
B. Romeis, Bayer. Akad. Wiss. Jahrbuch 1949, pp.126-133
W. Hueck, Verhandl. path. Ges. 32 (1950), 422-431

BORTELS, HERMANN [1902-1979]
KURSCHNER 10,237; WAGENITZ, 35; WWWS, 214

BORUTTAU, HEINRICH JOHANNES [1869-1923]
FISCHER 1,152; BK, 172-173; TLK 2,71
M. Gildemeister, Deutsche med. Wchschr. 49 (1923), 993-994

BOSART, LOUIS WILLIAM [1874-1947]
AMS 7,179; JV 13,18; NUC 67,559

BOSCH, CARL [1874-1940]
DSB 2,323-324; NDB 2,478-479; TLK 3,63; W, 71; WWWS, 214; POGG 6,286; 7a(1),239-240
C. Krauch, Z. angew. Chem. 53 (1940), 285-288
K. Holdermann, Chem. Ber. 90(11) (1957), xix-xxxix
A. von Nagel, Ludwigshafener Chemiker, pp.109-136. Ludwigshafen 1958

BOSE, JAGADIS CHUNDER [1858-1937]
DSB 2,325; 15,46-47; POGG 4,161; 6,286; 7b,485
P. Gedder, Life and Work of Jagadis Chunder Bose. London 1920
Anon., Nature 140 (1937), 1041-1043
M.N. Saha, Obit. Not. Fell. Roy. Soc. 3 (1940), 2-12
S. Basu, Jagadis Chunder Bose. New Delhi 1970

BOSE, PRAFULLA KUMAR [-1960]
POGG 6,286-287; 7b,485-486

BOSSHARD, EMIL [1860-1937]
TLK 3,64; ZURICH-D, 141; POGG 6,289; 7a(1),240
H.E. Fierz-David, Helv. Chim. Acta 20 (1937), 1335-1344
G. Müller-Schollhorn, Viert. Nat. Ges. Zurich 82 (1937), 482-486

BOSTOCK, JOHN [1773-1846]
DSB 2,335-336; DNB 2,884-885; DESMOND, 77; FERCHL, 59; WWWS, 216; COLE, 74-75; CALLISEN 2,472-478; POGG 1,249-250
Anon., Proc. Roy. Soc. 5 (1846), 636-638

BOSTROEM, EUGEN [1850-1928]
KURSCHNER, 2,174; FISCHER 1,154; WELDING, 92-93; GUNDEL, 99-104
W. Krumholz, Eugen Woldemar Bostroem. Giessen 1983

BOSWORTH, ALFRED WILLSON [1879-1963]
AMS 10,395

BOTKIN, SERGEI PETROVICH [1832-1889]
BSE 5,642-644; BME 4,230-239; HIRSCH 1,644; KUZNETSOV(1963), 518-526; ZMEEV 1,28-29; 3,6-7
V.B. Farber, Sergei Petrovich Botkin. Leningrad 1948
D.Y. Shurygin, Klin. Med. 60(4) (1982), 112-115

BOTTAZZI, FILIPPO [1867-1941]
DSB 2,339-340; DBI 13,420; FISCHER 1,154-155; IB, 31; WWWS, 216; POGG 5,149; 6,290; 7b,489-490
G. Quagliariello, Chimica e Indistria 23 (1941), 465; Erg. Physiol. 45 (1944), 16-33
P. Rondoni. Ann. Accad. Ital. 14 (1942). 156-169

BOTTERI, ALBERT [1879-1955]
FISCHER 1,155
S. Hondl et al., Ljet. Jugosl. Akad. 43 (1931), 108-111
L. Jese, Zdravstveni Vestnik 24 (1955), 164-165
Z. Pavisic, Lijecnicki Vestnik 77 (1955), 361-363

BOTTLER, HANS [1897-]
JV 41,427; NUC 68,664

BOTTOMLEY, WILLIAM BEECHCROFT [1863-1922]
DESMOND, 78
R.R.G., Nature 109 (1922), 524-525

BOUCHARD, CHARLES [1837-1915]
DBF 6,1183; HIRSCH 1,646; Suppl.,122; GUIART, 225; WWWS, 217
E. Perrier, C.R. Acad. Sci. 161 (1915), 509-513
A. Desgrez, Rev. Sci. 58 (1920), 577-586
A.M. Baudouin, Bull. Acad. Med. 134 (1950), 732-739

BOUCHARDAT, APOLLINAIRE [1806-1886]
DBF 6,1187-1188; GE 7,524-525; LAROUSSE 2,1044; GORIS, 239-245; PAGEL, 218-219; BLOKH 1,79-80; TSCHIRCH, 1024; SCHAEDLER, 15; COLE, 75-76; FERCHL, 60; POGG 3,167-168; 4,163
L. Figuier, Ann. Scient. 30 (1886), 568-571
J. Cheymol, Bull. Acad. Med. 159 (1976), 760-769; Comptes Rendus... de l'Académie de Pharmacie 1977, pp.48-69

BOUCHARDAT, GUSTAVE [1842-1918]
DBF 6,1188; FISCHER 1,155; TSCHIRCH, 1024; WWWS, 217; POGG 3,168; 4,163; 5,292M. Delépine, Bull. Soc. Chim. [4]25 (1919), 521-541; Bull. Sci. Pharm.26 (1919), 227-249
G. Dillemann, Produits et Problèmes Pharmaceutiques 27 (1972), 208

BOUCHUT, EUGÈNE [1818-1891]
DBF 6,1242; HIRSCH 1,647-648; PAGEL, 219-221; VAPEREAU 4,273
Anon., Leopoldina 28 (1892), 27

BOUDET, FÉLIX HENRI [1806-1878]
DBF 6,1251-1252; LAROUSSE 2,1068; VAPEREAU, 4,274; TSCHIRCH, 1024; SCHAEDLER, 15; FERCHL, 60; POGG 1,252-253; 3,168
A. Riche, Bull. Acad. Med. 7 (1878), 360-365

BOUDET, JEAN PIERRE [1778-1849]
GE 7,610; BOURQUELOT, 37; DECHAMBRE 10,274; TSCHIRCH, 1024; FERCHL, 60; WWWS, 217; POGG 1,252

BOUGAULT, JOSEPH [1870-1955]
GORIS, 248-250; POGG 5,149-150; 6,293; 7b,493-494
P. Cordier, Bull. Soc. Chim. 1956, pp.1088-1097; Ann. Pharm. Fran. 14 (1956), 133-153

BOUILLON-LAGRANGE, EDMÉ JEAN BAPTISTE [1764-1844]
DBF 6,1331-1332; GE 7,651-652; LAROUSSE 2,1085; BOURQUELOT, 45-46; HIRSCH 1,652; PHILIPPE, 756-757; TSCHIRCH, 1024,1077; FERCHL, 60-61; COLE, 76-78; CALLISEN 2,492-499; 26,392-393; POGG 1,255-256
H. Buignet, J. Pharm. Chim. [3]6 (1844), 447-463
C. Clerk, Ann. Chim. Anal. 27 (1945), 137-138

BOUIN, ANDRÉ POL [1870-1962]
DSB 2,344-346; DBA 1,315; FISCHER 1,156; IB, 31; WWWS, 217
J. Benoit, Bull. Acad. Med. 146 (1962), 204-210
Anon., Bull. Acad. Med. Belg. 2 (1962), 199-209
E. Fauré-Fremiet, C.R. Acad. Sci. 254 (1962), 1361-1364
R. Courrier, Not. Acad. Sci, 4 (1964), 590-624

BOUIS, JULES [1822-1886]
DBF 6,1335; BLOKH 1,80; SCHAEDLER, 15; WWWS, 217-218; POGG 3,170; 4,164
P. Schützenberger, Bull. Acad. Med. 15 (1886), 340-343
Anon., Bull. Soc. Chim. [2]47 (1887), xiii-xiv

BOUISSON, ÉTIENNE FRÉDÉRIC [1813-1884]
DBF 6,1336; HIRSCH 1,652-653; Suppl.,123; FRANKEN, 177; WWWS, 218
J. Bouillat, Bouisson, sa Vie, son Oeuvre. Paris 1890

BOULANGER, PAUL [1905-]
QQ 1978-1979, P.209

BOULLAY, PIERRE FRANÇOIS GUILLAUME [1777-1869]
DBF 6,1363; LAROUSSE 2,1098-1099; TSCHIRCH, 1024; FERCHL, 61; WWWS, 218; BOURQUELOT, 35-36; CALLISEN 2,502-507; 26,394; POGG 1,257
H. Buignet, Bull. Acad. Med. 34 (1869), 1066-1071
M. Wruble, Am. J. Pharm. 105 (1933), 244-247,289-295,340-352

BOULLAY, POLYDORE [1806-1835]
GE 7,690; BOURQUELOT, 58; TSCHIRCH, 1025; BLOKH 1,80; FERCHL,61; WWWS, 218; COLE, 78-79; CALLISEN 26,394-395; POGG 1,257
Anon., J. Pharm. Chim. [2]21 (1835), 334

BOUQUET, JEAN PIERRE [1818-]
POGG 1,259; 3,171-172

BOURCART, EDMUND [1872-1952]
JV 15,129; NUC 69,413

BOURDILLON, ROBERT BENEDICT [1889-1972]
Who's Who in British Science 1953, p.35
Anon., Chem. Brit. 8 (1972), 268

BOURDON, JEAN BAPTISTE [1795-1861]
DBF 6,1453-1454; HIRSCH 1,605; DECHAMBRE 10,351; CALLISEN 3,2-4; 26,396
F. Boudet, Bull. Acad. Med. 27 (1861), 160-162
J. Roger, Les Médecins Normands, 2,244-245. Paris 1895

BOURGET, LOUIS [1856-1913]
DHB 2,270; FISCHER 1,157; NUC 69,514-515; SG [3]3,274
S. Rabow, Münch. med. Wchschr. 60 (1913), 2180-2182
Anon., Brit. Med. J. 1913 (II), p.1413; Rev. Med. Suisse Rom. 33 (1913), 657-659

BOURGOIN, EDMÉ ALFRED [1836-1897]
DBF 6,1497-1498; GORIS, 251-254; VAPEREAU 6,212; TSCHIRCH, 1025; POGG 3,174; 4,165-166
L. Prunier and M. François, J. Pharm. Chim. [6]5 (1897), 265-270
L. Guignard, Bull. Acad. Med. 37 (1897), 167-171

BOURGUEL, MAURICE [1893-1932]
POGG 6,296; 7b,499-500
R. Lespieau, Bull. Soc. Chim. [4]53 (1933), 1145-1153

BOURNE, EDWARD JOHN [1922-1974]
CAMPBELL, 35-36; WW 1974, p.342
L. Butler, Chem. Brit. 11 (1975), 297
M. Weigel, Adv. Carbohydrate Chem. 34 (1977), 1-22

BOURNE, GEOFFREY HOWARD [1909-1988]
WW 1981, p.274; IWW 1982-1983, P.154; WWWS, 219

BOURQUELOT, ÉMILE [1851-1921]
DBF 7,6; GORIS, 254-264; TSCHIRCH, 1025; WWWS, 220; POGG 4,166-167; 5,151,1417
J. Bougault and H. Hérissey, Notice sur la Vie et les Travaux de Emile Bourquelot. Paris 1921
Anon., Nature 106 (1921), 836-837
J.E. Courtois, Adv. Carbohydrate Chem. 18 (1963), 1-8
G. Dillemann, Produits et Problèmes Pharmaceutiques 28 (1973), 121-122

BOURQUIN, ALCIDE [1858-1934]
RSC 13,737; NUC 69,658
Anon., Schw. Apoth. Z. 73 (1934), 24

BOURSNELL, JOHN COLIN [1913-1977]
Anon., Chem. Brit. 15 (1979), 325

BOUSFIELD, WILLIAM ROBERT [1854-1943]
POGG 5,151-152; 6,297; 7b,500
H. Dale, Proc. Roy. Soc. A182 (1944), 223-224
W.C.D. Dampier, Obit. Not. Fell. Roy. Soc. 4 (1944), 571-576

BOUSSINGAULT, JEAN BAPTISTE [1802-1887]
DSB 2,356-357; DBF 7,32-33; DBA 1,325-326; LACROIX 2,115-117; BLOKH 1,81; FERCHL,62; TSCHIRCH, 1025; PHILIPPE, 784-785; SCHAEDLER, 15-16; POTSCH, 59-60; W, 72-73; WWWS, 220; POGG 1,262-264; 3,177; 4,168
N. Pringsheim, Ber. bot. Ges. 5 (1887), (ix)-(xxxiii)
Anon., J. Chem. Soc. 53 (1888), 509-513; Pop. Sci. Mon. 33(1888), 836- 841
A. Lacroix, Notice Historique sur Jean Baptiste Boussingault. Paris 1926
E.H. Huntress, Proc. Am. Acad. Arts Sci. 81 (1952), 43-44
G.R. Cowgill, J. Nutrition 84 (1964), 3-9
R.P. Aulie, Proc. Am. Phil. Soc. 114 (1970), 435-479
F.W.J. McCosh, Jean Baptiste Boussingault, his Life and Work. London 1974; Ann. Sci. 32 (1975), 475-490; Boussingault, Chemist and Agriculturist. Dordrecht 1984
J. Lavollaye et al., C.R. Acad. Agr. 73(6) (1987), 3-37
E. Kahane, Boussingault entre Lavoisier et Pasteur. Elbeuf 1989

BOUSSINGAULT, JOSEPH [1842-1908]
DBF 7,32-33; DBA 1,326

BOUTARIC, AUGUSTIN [1885-1949]
POGG 5,154; 6,297-298; 7b,500-508
R. Amiot, Rev. Sci. 87 (1949), 127-128

BOUTRON-CHARLARD, ANTOINE FRANÇOIS [1796-1878]
LAROUSSE 2,1162; BOURQUELOT, 63-64; FERCHL, 63; TSCHIRCH, 1025;CALLISEN 3,21-22; 26,401; POGG 1,265; 3,178
F. Planchon, Bull. Acad. Med. 8 (1879), 1146-1151

BOUTROUX, LÉON [1851-1921]
POGG 3,178; 4,169

BOUVEAULT, LOUIS [1864-1909]
DBF 7,71; CHARLE(2),54-56; BLOKH 1,81-82; WWWS,221; POTSCH, 60; POGG 4,170; 5,155-156
E. Bodtker, Chem. Z. 33 (1909), 1273-1274
O.N. Witt, Ber. chem. Ges. 42 (1909), 3561-3562
A. Béhal, Bull. Soc. Chim. [4]7 (1910), i-xxi

BOUVET, MAURICE [1885-1964]
C. Bedel, Ann. Pharm. Fran. 22 (1964), 49-50
J. Nauroy, Rev. Hist. Pharm. 17 (1964-1965), 61-71,319-324
C. Warolin et al., Rev. Hist. Pharm. 32 (1985), 282-361

BOUWDIJK BASTIAANSE, FRANS SUSAN van [1885-1953]
LINDEBOOM, 80
E. Hoelen, Ned. Tijd. Gen. 97 (1953), 3022-3023

BOVARNICK, MAX [1909-1974]
NUC 70,178
Anon., Harvard Med. Alumni Bull. 49(3) (1975), 63

BOVERI, THEODOR [1862-1915]
DSB 2,361-365; NDB 2,493-494; DBJ 1,323; LF 2,40-56; FREUND 1,121-132; W, 73; WWWS, 221; BK, 173; MAYERHOFER 1,540-541; SCHWERTE 2,183-192; DZ, 157; SCHULZ-SCHAEFFER, 201-202
R. Hertwig, Bayer. Akad. Wiss. Jahrbuch 1916, pp.118-135
R. Goldschmidt, Science 43 (1916), 263-270
F. Baltzer, Theodor Boveri. Stuttgart 1962 (English translation, Berkeley 1967); Science 144 (1964), 809-815

BOVET, DANIEL [1907-1992]
MH 1,57-58; STC 1,161-162; MSE 1,121-122; IWW 1982-1983, p.155; POTSCH, 60; WWWS,221
New York Times, 11 April 1992

BOVET, VICTOR [1853-1922]
Schweizerisches Zeitgenossenlexikon 1921, p.88
P.B., Annales Valaisannes 4 (1922), 111

BOWDITCH, HENRY PICKERING [1840-1911]
DSB 2,365-368; DAB 2,494-496; HIRSCH 1,662; DAMB, 84-85; WWWS, 221-222
C.S. Minot et al., Science 33 (1911), 598-601,651-652
W.B. Cannon, Proc. Am. Arts Sci. 46 (1911), 739-747; Biog. Mem. Nat. Acad. Sci. 17 (1922), 183-196; Science 87 (1938), 471-474
F.W. Ellis, New Eng. J. Med. 219 (1938), 819-828
W.B. Fye, Bull. Hist. Med. 56 (1982), 19-29; Harvard Med. Alumni Bull. 57(2) (1983), 46-49; The Development of American Physiology, pp.92-129. Baltimore 1987
D.L. Crandall, Physiologist 32 (1989), 88-96

BOWEN, EDMUND JOHN [1898-1980]
DNB [1971-1980], 72-73; WW 1980, p.274; AO 1980, pp.703-704; WWWS, 222; MSE 1,122-123; CAMPBELL, 37-38
R.P. Bell, Biog. Mem. Fell. Roy. Soc. 27 (1981), 83-101

BOWER, FREDERICK ORPEN [1855-1948]
DSB 2,370-372; DNB [1941-1950], 94-95; DESMOND, 80; W, 73; WWWS, 222
W.H. Lang, Obit. Not. Fell. Roy. Soc. 6 (1949), 347-374
S. Williams, Ber. bot. Ges. 68a (1955), 217-220

BOWERS, ROY ANDERSON [1913-]
AMS 15(1),621; WWA 1980-1981, p.375; WWWS, 223

BOWLES, PERCY EWART [1882-1960]
JV 24,336; NUC 70,431
Anon., J. Roy. Inst. Chem. 85 (1961), 121

BOWMAN, WILLIAM [1816-1892]
DSB 2,375-377; DNB 22,242-243; HIRSCH 1,663-664; BETTANY 2,261-267; O'CONNOR, 37-40; W, 73-74; WWWS, 224
Anon., Brit. Med. J. 1892(I), pp.742-745
J.P., Proc. Roy. Soc. 52 (1893), i-vii
B. Chance, Ann. Med. Hist. 10 (1924), 143-158
K.B. Thomas, Med. Hist. 10 (1966), 245-256
F. Grondona, Clio Medica 6 (1971), 195-204

BOXER, GEORGE ERNST [1915-1968]
AMS 11,527; BHDE 2,138; WWWS, 224
Anon., Chem. Eng. News 46(21) (1968), 84

BOYCOTT, ARTHUR EDWIN [1877-1938]
DNB [1931-1940], 95-96; O'CONNOR(2), 114-117
H.R.D. et al., J. Path. Bact. 47 (1938), 161-194
C.J. Martin, Obit. Not. Fell. Roy. Soc. 2 (1939), 561-571

BOYD, JOHN SMITH KNOX [1891-1981]
WW 1981, p.285; SACKMANN, 40-41; WWWS, 225
L.G. Goodwin, Biog. Mem. Fell. Roy. Soc. 28 (1982), 27-57

BOYD, WILLIAM CLOUSER [1903-1983]
MH 1,60-61; STC 1,167-168; MSE 1,125-126; WWA 1978-1979, p.368; AMS 14,512; WWWS, 225
New York Times, 28 February 1983

BOYD-BARRETT, HERBERT STANISLAUS [1904-]
Who's Who in British Science 1953, p.36

BOYER, HERBERT WAYNE [1936-]
AMS 15(1),632; WWA 1980-1981, p.383

BOYER, PAUL DELOS [1918-]
AMS 15(1),632-633; WWA 1980-1981, p.383; IWW 1981-1982, p.154
P.D. Boyer, in Of Oxygen, Fuels and Living Matter Part 1 (G. Semenza, ed.), pp.229-264. Chichester 1981

BOYLAND, ERIC [1905-]
WW, 1981, p.287; IWW 1982-1983, p.158; WWWS, 226
M.B. Shimkin, Cancer Research 40(6) (1980), iv
D.V. Parke, Xenobiotica 16 (1986), 887-898

BOYSEN-JENSEN, PETER [1883-1959]
KBB 1959, p.728; VEIBEL 2,77-78; BK, 173; IB, 32; WWAH, 68
L. Brauner, Bayer. Akad. Wiss. Jahrbuch 1960, pp.172-176
D. Müller, Ber. bot. Ges. 74 (1961), (76)-(79)

BRACHET, ALBERT [1869-1930]
DSB 2,383-385; BNB 30,196-209; FISCHER 1,160; IB, 32
J. Boeke, Anat. Anz. 72 (1931), 307-316
H. de Winiwarter, Arch. Biol. 42 (1931), 1-40; Ann. Acad. Roy. Belg. 99 (1933), 143-192
J.P. Hill, Obit. Not. Fell. Roy. Soc. 1 (1932), 64-70
G. Leboucq, Bull. Acad. Med. Belg. [5]13 (1933), 568-578

BRACHET, JEAN LOUIS [1789-1858]
DBF 7,129; HIRSCH 1,671; DECHAMBRE 10,428; CALLISEN 3,44; 26,406
G. Legee, Histoire et Nature 26/27 (1985), 27-45

BRACHET, JEAN LOUIS [1909-1988]
MH 2,46-47; STC 1,169-171; MSE 1,126-127; IWW 1977-1978, p.207; BAF, 11; WWWS, 227
J. Brachet, Comp. Bioch. Physiol. 67B (1980), 367-372; TIBS 12 (1987), 244-246
A. Burny, Bull. Acad. Med. Belg. 144 (1989), 237-241
N.W. Pirie, Biog. Mem. Fell. Roy. Soc. 36 (1990), 85-99
H. Chantrenne, in Selected Topics in the History of Biochemistry (G. Semenza and R. Jaenicke, eds.), pp.201-213. Amsterdam 1990
R. Thomas, Genetics 131 (1992), 515-518

BRACONNOT, HENRI [1780-1855]
DSB 2,385-386; DBF 7,132-133; GE 7,948; PHILIPPE, 785-786; FERCHL, 65; POTSCH, 61; TSCHIRCH, 1025; BLOKH 1,82-83; CALLISEN 3,48-53; 26,408-409; POGG 1,270-271; 3,180
J. Nicklès, Mem. Acad. Stanislas [3]22 (1856), xxiii-cxlix; Am. J. Sci. [2]21 (1856), 118-119
J. Aubry, Mem. Acad. Stanislas [7]7 (1978-1979), 87-98

BRADFORD, JOHN ROSE [1863-1935]
DNB [1931-1940], 96-98; O'CONNOR(2), 124-127; WWWS, 228
T.R. Elliott, Obit. Not. Fell. Roy. Soc. 1 (1935), 527-535
Anon., Lancet 1935(I), p.906; Brit. Med. J. 1935(I), p.805

BRADLEY, HAROLD CORNELIUS [1878-1976]
AMS 10,419-420; CHITTENDEN, 206-211; IB, 32; WWWS, 228
Anon., Science 193 (1976), 469; Chem. Eng. News 54(24) (1976), 69

BRADY, OSCAR LISLE [1890-1968]
POGG 6,307; 7b,534
C. Ingold, Chem. Brit. 4 (1968), 554

BRADY, ROSCOE OWEN [1923-]
AMS 15(1),632; WWA 1980-1981, p.392; WWWS, 229

BRAGG, WILLIAM HENRY [1862-1942]
DSB 2,397-400; DNB [1941-1950], 99-101; W, 77-78; WWWS, 229; POTSCH, 61-62; POGG 5,157-158,1417; 6,307-308; 7b,534-536
P.P. Ewald et al., Naturwiss. 20 (1932), 527-544
E.N. daC. Andrade et al., Nature 149 (1942), 346-351
E.N. daC. Andrade, Obit. Not. Fell. Roy. Soc. 4 (1943), 277-300
L. Bragg and G.M. Caroe, Notes Roy. Soc. 16 (1961), 169-182
P.P. Ewald, Nature 195 (1962), 320-325
R.H. Stuewer, Brit. J. Hist. Sci. 5 (1971), 258-281
G.M. Caroe, William Henry Bragg. Cambridge 1978
R.W. Home, Historical Records of Australian Science 6 (1984), 19-30

BRAGG, WILLIAM LAWRENCE [1890-1971]
DSB 15,61-64; DNB [1971-1980], 77-79; MH 1,61-62; MSE 1,127-128; POTSCH, 62; STC 1,171-173; W, 578-579; POGG 5,158; 6,308-309; 7b,536-538
W.L. Bragg, Proc. Roy. Inst. 41 (1967), 92-100
D. McLachlan, Acta Cryst. 26A (1970), 189-193
L. Bragg, The Development of X-ray Analysis. London 1975
D. Phillips, Biog. Mem. Fell. Roy. Soc. 25 (1979), 75-143
M.F. Perutz, Proc. Roy. Inst. 62 (1990), 183-198

BRAINARD, DANIEL [1812-1866]
DAB 2,589-590; HIRSCH 1,673; KELLY, 140-141; WWWS, 230
E.F. Ingals, Bull. Alumni Assoc. Rush Med. Coll. 8 (1912), 1-13

BRAKELEY, ELIZABETH [1894-1973]
Anon., J. Am. Med. Assn. 225 (1973), 646

BRAMBELL, FRANCIS WILLIAM ROGERS [1901-1970]
DNB [1961-1970], 129-130; WW 1970, pp.347-348; MSE 1,128-129; WWWS, 230
Anon., Nature 228 (1970), 694
J.S. Perry, J. Repr. Fert. 27 (1971), 1-3

BRAMWELL, BYROM [1847-1931]
DNB [1931-1940], 100-101; FISCHER 1,162; MEDVEI, 714-715; WWWS, 231
Anon., Edinburgh Med. J. NS38 (1931), 444-447
D. Drummond and R. Hutchison, Brit. Med. J. 1931(I), pp.823-826
E.B., Proc. Roy. Soc. Edin. 51 (1932), 224-231
B. Ashworth, Scottish Medical Journal 26 (1981), 364-370

BRAN, FRIEDRICH [1871-]
JV 15,142; NUC 72,278

BRAND, ERWIN [1881-]
JV 21,4; SG [3]3,444; NUC 72,331

BRAND, ERWIN [1891-1953]
AMS 8,274
O.E. Reynolds and J. Greenstein, Science 119 (1954), 144-145

BRAND, KURT [1877-1952]
HEIN-E, 45-46; AUERBACH, 778-779; TLK 3,66; POGG 5,158-159; 6,310;7a(1),243-244

BRAND, THEODOR (Freiherr von) [1899-]
BHDE 2,140; LDGS, 13; IB, 33; NUC 642,258-259

BRANDE, WILLIAM THOMAS [1788-1866]
DSB 2,420; DNB 2,1124; TSCHIRCH, 1025; FERCHL, 66; WWWS, 231; COLE, 80-82; CALLISEN 3,62-66; 26,412; POGG 1,276; 3,181
Anon., J. Chem. Soc. 19 (1866), 509-511; Chem. News 13 (1866), 107- 108
A.S.T., Proc. Roy. Soc. 16 (1868), ii-vi
E. Ironmonger, Proc. Roy. Inst. 38 (1961), 450-461
C.H. Spiers, Ann. Sci. 25 (1969), 179-201
A. Tulley and E. Ironmonger, Proc. Roy. Inst. 44 (1971), 259-273

BRANDES, RUDOLPH [1795-1842]
NDB 2,522; ADB 3,244-245; HEIN, 75-76; HIRSCH 1,676; POTSCH, 62-63; FERCHL, 66-67; TSCHIRCH, 1026; PHILIPPE, 804; WWWS, 231; CALLISEN 3,70-86; 26,414-416; POGG 1,279; 7aSuppl.,107-108
J.A. Buchner, Rep. Pharm. 79 (1843), 85-95
L.F. Bley, Leben und Wirken des Dr. Rudolph Brandes. Hannover 1844;Arch. Pharm. 87 (1844), 165-226
H. Grunewald, Arch. Pharm. 280 (1942), 421-424
H. Zimmermann, Simon Rudolph Brandes. Stuttgart 1985

BRANDIS, ERNST [1861-1921]
DGB 171 (1975), 108; JV 4,223; RSC 13,765
Anon., Chem. Z. 45 (1921), 980; Z. angew. Chem. 34 (1921), 524

BRANDL, JOSEF [1858-1925]
RSC 9,330; 13,765-766
Anon., Chem. Z. 47 (1923), 547; 49 (1925), 259,678; Z. angew. Chem. 31(III) (1918), 256; 38 (1925), 734

BRANDT, ALEKSANDR FEDOROVICH [1844-1894]
VORONTSOV, 54; RSC 7,242; 13,766

BRANTL, JOSEF [1855-]
JV 13,223

BRASCH, WALTER [1878-1918]
FISCHER 1,163; BACHMANN, 110-112
A. Eskuchen, Münch. med. Wchschr. 65 (1918), 1356

BRASS, ARNOLD [1854-]
BK, 174; RSC 9,332; 13,770; SG [2]2,747; NUC 72,694-696
L. Szyfman, Bull. Biol. 113 (1979), 375-406

BRASS, KURT [1880-1964]
STURM 1,136; BIRK, 118-119; TLK 3,67; POGG 6,312; 7a(1),247-248
R. Pummerer, Z. angew. Chem. 62 (1950), 470

BRASS, WILHELM [1861-]
BONN, 372

BRAUELL, FRIEDRICH [1807-1882]
HIRSCH 1,681; Suppl.,128; BULLOCH, 354; SKORODOKHOV, 84-88; WWWS, 233
Anon., Leopoldina 18 (1882), 210
H. Vierordt, Medizingeschichtliches Hilfsbuch, pp.40-41. Tübingen 1916.

BRAUER, KURT [1888-1950]
TLK 3,67; POGG 6,613; 7a(1),250
W. Roth, Chem. Z. 75 (1951), 95

BRAUN, ALEXANDER [1805-1877]
DSB 2,425-427; NDB 2,548; ADB 47,186-193; WWWS, 233
C. Mettenius, Leopoldina 13 (1877), 50-60,66-72; Alexander Brauns Leben. Berlin 1882
R. Caspary, Flora 60 (1877), 433-442,449-457,465-471,497-507,513-519
W.M. Montgomery, J. Hist. Biol. 3 (1970), 299-323
B. Hoppe, in Medizingeschichte in unserer Zeit (H.H. Eulner et al., Eds.), pp.393-421. Stuttgart 1971
M. Guédès, Episteme 7 (1973), 32-51

BRAUN, ARMIN CHARLES JOHN [1911-1986]
AMS 15(1),661; WWA 1980-1981, p.401; WWWS, 233

BRAUN, FERDINAND [1850-1918]
DSB 2,427-428; NDB 2,554-555; DBJ 2,682; LKW 2, 51-63; BLOKH 1,83-84; WWWS, 233; POGG 3,184; 4,175; 5,159; 6,313; 7aSuppl., 108-109
H. Rohmann, Phys. Z. 19 (1918), 537-539
L. Mandelstam and N. Papalix, Naturwiss. 16 (1928), 621-626
E.H. Huntress, Proc. Am. Acad. Arts Sci. 78 (1950), 22-23
F. Kurylo, Ferdinand Braun. Munich 1965
G. Schiers, Scientific American 230(4) (1974), 92-101
F. Kurylo and C. Susskind, Ferdinand Braun. Cambridge, Mass. 1981

BRAUN, HUGO [1881-1963]
STURM 1,137; KURSCHNER 9,206; KALLMORGEN, 231; BHDE 2,144; LDGS, 56; WININGER 7,536; WIDMANN, 81-82; POGG 7a(1),252-253
H. Schlossberger, Arzneimitt. 1 (1951), 147-148
C. Hiller, Med. Welt 21 (1961), 1172-1173
G. Lebek, Deutsche med. Wchschr. 86 (1961), 748-749
G. Hohmann, Münch. med. Wchschr. 106 (1964), 130-131

BRAUN, JULIUS von [1875-1939]
KALLMORGEN, 231; TLK 3,68; POTSCH, 63; POGG 5,159-161; 6,313-315; 7a(1),253
Anon., Z. angew. Chem. 52 (1939), 399-400
P. Kurtz, Chem. Ber. 99(6) (1966), xxxv-lxxxvi

BRAUN, WERNER [1914-1972]
AMS 11,559; BHDE 2,145; LDGS, 56; WAGENITZ, 36-37

BRAUNE, CHRISTIAN WILHELM [1831-1892]
ADB 47,206-208; HIRSCH 1,685; Suppl.,129; PAGEL, 234; AD 3,220
W. His, Arch. Anat. Entw. 1892, pp.231-256; Anat. Anz. 13 (1897), 331-333
K. von Bardeleben, Anat. Anz. 7 (1892), 440-445,476

BRAUNE, HERMANN [1886-1977]
HANNOVER, 33; JV 26,393; POGG 6,315-316; 7a(1),255
Wer ist Wer 14 (1962), 161

BRAUNER, ALEKSANDR ALEKSANDROVICH [1857-1941]
BSE 6,60; WWR, 87; IB, 33
I.I. Puzanov, Trudy Inst. Ist. Est. 32 (1960), 309-371

BRAUNER, BOHUSLAV [1855-1935]
DSB 2,428-430; WWWS, 234; POTSCH, 63-64; POGG 3,184; 4,175-176; 5,161; 6,316; 7b,542
Anon., Coll. Czech. Chem. Comm. 2 (1930), 211-218
J. Heyrovsky, Coll. Czech. Chem. Comm. 7 (1935), 51-56
S.I. Levy, J. Chem. Soc. 1935, pp.1876-1890
G. Druce, Nature 150 (1943), 623-624; Two Czech Chemists. London 1944

BRAUNER, LEO [1898-1974]
KURSCHNER 11,320; BHDE 2,146-147; WIDMANN, 257-258; IB, 33; POGG 7a(1),255-256

H. Ziegler, Bayer. Akad. Wiss. Jahrbuch 1974, pp.235-239
H.D. Zinsmeister, Ber. bot. Ges. 93 (1980), (459)-(466)

BRAUNITZER, GERHARD [1921-1989]
KURSCHNER 14,416; POTSCH, 64
Who's Who in Science in Europe 1978, p.382
Anon., Chem. Z. 113 (1989), 272
J. Godovaz-Zimmermann, TIBS 15 (1990), 3-4

BRAUNS, FRIEDRICH EMIL [1890-1982]
AMS 9(I),213; JV 32,238
Anon., Chem. Eng. News 60(41) (1982), 40

BRAUNSTEIN, ALEKSANDR EVSEEVICH [1902-1986]
BSE(3rd Ed.) 4,12; BME 4,322-323; MH 2,47-48; STC 1,178-179; MSE 1,133-134; IWW 1982-1983, p.163; POTSCH, 64
A.E. Braunstein, Leopoldina [3]4/5(1958-1959),12-13; in Of Oxygen, Fuels and Living Matter (G. Semenza, Ed.), pp.251-313. Chichester 1982;
Na Styke Khimii i Biologii. Moscow 1987
Anon., Vop. Med. Khim. 8 (1962), 325-328; Biokhimia 37 (1972), 668-669
A.J.L. Cooper and A. Meister, Biochimie 71 (1989), 387-404

BRAUS, HERMANN [1868-1924]
NDB 2,562-563; FISCHER 1,167; GIESE, 475-478; EBERT, 72-81; DRULL, 29-30
H. Spemann, Naturwiss. 13 (1925), 253-261; Arch. Entwickl. 106 (1925), i-xxv
F. König et al., Verhandl. Phys. Med. Würzburg NF50 (1925), 100-123
H. von Eggeling, Anat. Anz. 62 (1927), 255-291

BRAUTLECHT, CHARLES ANDREW [1881-1964]
AMS 11,561

BRAUTLECHT, JULIUS [1837-1883]
W. Blasius, Jahresb. Ver. Nat. Braun. 4 (1883), 207-212; Lebensbeschreibungen Braunschweiger Naturforscher und Naturfreunde, pp.20-21. Braunschweig 1887

BRAVAIS, AUGUSTE [1811-1863]
DSB 2,430-432; DBF 7,170-171; GE 7,1034-1035; LAROUSSE 2,1212-1213; MAYERHOFER 1,549; WWWS, 234; POGG 1,283-284; 3,184-186
J. Nicklès, Am. J. Sci. [2]36 (1863), 401-402
J. Messié, Revue du Vivarois 67 (1963), 7-11

BRAWERMAN, GEORGE [1927-]
AMS 18(1),736

BRAY, WILLIAM CROWELL [1879-1946]
NCAB 35,122; AMS 7,201; SRS, 30; WWWS, 234; WWAH, 70; POGG 5,162; 6,317; 7b,543
W.M. Latimer, Science 104 (1946), 500
J.H. Hildebrand, Biog. Mem. Nat. Acad. Sci. 26 (1951), 13-24

BRDICKA, RUDOLF [1906-1970]
POTSCH, 64; POGG 6,318; 7b,545-548
Anon., Leopoldina [3]4/5 (1958-1959), 37-41
J. Dvorak and V. Hanus, Chem. Listy 64 (1970), 1081-1087

BRECHT, KARL [1912-1982]
KURSCHNER 13,418; POGG 7a(1),257
Anon., Naturw. Rund. 35 (1982), 343

BRECHT, MAX [1859-]
BERLIN, 601; SG [2]2,780; NUC 73,664

BREDERECK, HELMUTH [1904-1981]
KURSCHNER 13,420; WWWS, 235; POTSCH, 64-65; POGG 7a(1),260-261
Anon., Nachr. Chem. Techn. 14 (1966), 412; Leopoldina [3]15 (1969), 44-45; Naturw. Rund. 34 (1981), 316; Chem. Z. 105 (1981), 235

BREDIG, GEORG [1868-1944]
DRULL, 30; TLK 3,69; BHDE 2,149-150; WININGER 7,536; TETZLAFF, 41; POTSCH, 65; ABBOTT-C, 21-22; POGG 4,178; 5,162-163; 6,318-319; 7a(1),261
M. Trautz, Z. Elektrochem. 36 (1930), 348-350
W. Kuhn, Chem. Ber. 95 (1962), xlii-lxii

BREDT, JULIUS [1855-1937]
NDB 2,568; DZ, 175; TLK 3,69; WWWS, 235; POTSCH, 65; POGG 4,178-179; 5,163-164; 6,319; 7a(1),262
R. Lipp, Ber. chem. Ges. 70A (1937), 150-151
G.B. Kauffman, J. Chem. Ed. 60 (1983), 341-342

BREED, DANIEL [1825?-]
ELLIOTT(2), 28

BREEST, FRITZ [1879-1923]
JV 21,480; NUC 74,68
Anon., Chem. Z. 47 (1923), 231; Z. angew. Chem. 36 (1923), 140

BREFELD, OSCAR [1839-1925]
DSB 2,436-438; HEIN, 82; FISCHER 1,168; KURSCHNER 1,102; TSCHIRCH, 1026; DZ, 175-176
R. Falck, Bot. Arch. 11 (1925), 1-25

BREHME, PAUL [1883-]
JV 24,336

BREIDENBACH zu BREIDENSTEIN, EBERHARD von [1803-1872]
GGTF 8 (1858), 73
F.W. Weitershaus, Mitteilungen des Oberhessisches Geschichtsvereins NF62 (1977), 185

BREINL, FRIEDRICH [1888-1936]
STURM 1, 141; FISCHER 1,168; KURSCHNER 4,304; OLPP, 51-52; IB, 34; KOERTING, 129-130
Anon., Science 84 (1936), 128; Deutsche med. Wchschr. 62 (1936), 1134

BREITENBACH, WOLFGANG [1908-1978]
KURSCHNER 12,351; WWWS, 236; POGG 7a(1),263
O. Kratky, Alm. Akad. Wiss. Wien 128 (1979), 311-330

BREKKE, BÅRD JOHANNES [1916-]
Hver er Hvem? 1979, p.77

BREMER, GUSTAV JACOB WILHELM
[1847-1909] NNBW 3,169
Anon., Chem. Z. 33 (1909), 1349
S. Birnie, Chem. Wkbl. 7 (1910), 28-32

BRENNER, MAX [1915-]
KURSCHNER 13,427; POGG 7a(1),265
Anon., Chimia 29 (1974), 758-759

BRENNER, SYDNEY [1927-]
WW 1981, p.305; IWW 1982-1983, p.165; ABBOTT, 23-24

BRESCHET, GILBERT [1784-1845]
DSB 2,442-443; DBF 7,218; WWWS, 238; CALLISEN 3,136; 26,435-436; POGG 1,293,1543

BRESLOW, RONALD CHARLES [1931-]
AMS 15(1),675; MSE 1,136-137; IWW 1982-1983, p.165; WWA 1980-1981, p.410; WWWS, 238

BRESSLAU, ERNST LUDWIG [1877-1935]
KALLMORGEN, 232; BHDE 2,152
C.B. Aust, Staden Jahrbuch 11/12 (1963-1964), 197-211

BRETONNEAU, PIERRE [1778-1862]
DSB 2,444-445; DBF 7,249-250; GENTY 6,209-234; HIRSCH 1,695; Suppl.,131; GE 7,1169; LAROUSSE 2,1243; DECHAMBRE 10,627-

629; PAGEL, 239-240; BULLOCH, 354-355; WWWS, 238;
CALLISEN 3,152-153; 26,440
J.D. Rolleston, Proc. Roy. Soc. Med. 18 (1924), 1-12
U. Mutzner-Scharplatz, Pierre Betonneau. Zurich 1965
E. Aron, Bretonneau, le Médecin de Tours. Paris 1979

BRETSCHNEIDER, HERMANN [1905-1985]
KURSCHNER 13,429; POGG 7a(1),267-268
Anon., Nachr. Chem. Techn. 7 (1959), 236; Chem. Z. 104 (1980), 75; Naturw. Rund. 39 (1986), 188

BRETTAUER, ERWIN [1884-]
JV 24,570; NUC 74,584

BREUER, HEINZ [1926-1982]
KURSCHNER 14,465
J. Büttner, J. Clin. Chem. 21 (1983), 65-68

BREUER, JOSEF [1842-1925]
DSB 2,445-450; NDB 2,606-607; OBL 1,113-114; NOB 5,30-47; 15,126-130; FISCHER 1,170; WNININGER 1,457-458; TETZLAFF, 42; WWWS. 238
G. Pilleri and J.J. Schnyder, Josef Breuer. Berne 1983

BREUSCH, FRIEDRICH LUDWIG [1903-1983]
KURSCHNER 13,432; BHDE 2,154; WIDMANN,258; POGG 7a(1),268- 269

BREWER, WILLIAM HENRY [1828-1910]
DAB 3,25-26; NCAB 13,561; WWWS, 239
E.H. Jenkins, Am. J. Sci. [4]31 (1911), 71-74
R.H. Chittenden, Biog. Mem. Nat. Acad. Sci. 12 (1929), 289-323

BRIDEL, MARC [1883-1931]
DBF 7,291; GORIS, 267-270; POGG 6,325-326; 7b,559
M. Nicloux, Bull. Soc. Chim. Biol. 14 (1932), 197-206
G.M., J. Pharm. Chim. [8]15 (1932), 386-390

BRIDGES, CALVIN BLACKMAN [1889-1938]
DSB 2,455-457; DAB 22,60-62; AMS 6,163; NCAB 30,374-375; STC 1,180-181; SCHULZ-SCHAEFFER, 207; IB, 35; WWWS, 240; WWAH, 72
T.H. Morgan, J. Heredity 30 (1939), 355-359; Science 89 (1939), 118-119;
Genetics 25 (1940), i-v; Biog. Mem. Nat. Acad. Sci. 22 (1941), 31-48

BRIDRÉ, JULES [1869-1950]
Anon., Ann. Inst. Pasteur 80 (1951), 4-6

BRIEGER, HEINRICH [1895-1972]
BHDE 2,155-156; LDGS, 56; NUC 75,379

BRIEGER, FRIEDRICH GUSTAV [1900-1985]
LDGS, 11
H.P. Linskens, Ber. bot. Ges. 99 (1986), 137-143

BRIEGER, LUDWIG [1849-1919]
NDB 2,612; DBJ 2,713; FISCHER 1,171; PAGEL, 240-243; BLOKH 1,85; SCHAEDLER, 17; OLPP, 52; WININGER 1,458; KOHUT 2,272; WWWS, 240
Anon., Ber. chem. Ges. 52A (1919), 166
A. Laqueur, Med. Klin. 15 (1919), 757
F. Blumenthal, Deutsche med. Wchschr. 45 (1919), 1334
C. Neuberg, Chem. Z. 43 (1919), 769

BRIEGER, WALTER [1891-]
TLK 2,81; JV 30,502; NUC 75,384

BRIEGLEB, GUNTHER [1905-1991]
KURSCHNER 13,433; POGG 7a(1),269-270
Anon., Chem. Z. 115 (1991), 259

BRIEN, PAUL [1894-1975]
BUICAN, 187-192; IB, 35
J. Brachet, C.R. Acad. Sci. 280 (1975), 96
M. Poll and H. Herlant-Meewis, Ann. Acad. Roy. Belg. 145 (1979), 39-141

BRIGGS, ALFRED POYNEER [1888-1969]
AMS 11,579

BRIGGS, DAVID REUBEN [1899-1988]
AMS 12,683; POGG 6,327-328; 7b,563-564

BRIGGS, GEORGE EDWARD [1893-1985] WW 1981, p.311; IB, 35; WWWS, 240
R. Robertson, Biog. Mem. Fell. Roy. Soc. 32 (1986), 37-64

BRIGGS, ROBERT WILLIAM [1911-1983]
AMS 15(1),685; IWW 1982-1983, p.168; WWWS, 241
S. Hennen and G. Malacinski, Dev. Biol. 98 (1983), 255-256

BRIGHT, RICHARD [1789-1858]
DSB 2,463-465; DNB 2,1242-1245; HIRSCH 1,790; Suppl.,132; WILKS, 212-221; BETTANY 2,14-23; O'CONNOR, 19-20; DECHAMBRE 10,634-635; W, 81-82; ABBOTT, 24; CALLISEN 3,171-172; 26,447-448
Anon., Proc. Roy. Soc. 10 (1860), i-iv
W.S. Thayer, Guy's Hosp. Rep. 77 (1927), 253-301
J.S. Cameron et al., Guy's Hosp. Rep. 107 (1958), 263-352 (1958); 113 (1964), 159-171
R.M. Kirk and D.T. Moore, Archives of Natural History 10 (1981), 119-151
S.J. Peltzmann, Bull. Hist. Med. 55 (1981), 307-321
P. Bright, Dr. Richard Bright. London 1983

BRIGL, PERCY [1885-1945]
KURSCHNER 6(I),198; TLK 3,72; KLEIN, 54; IB, 35; POGG 5,166; 6,328-329; 7a(1),271
Anon., Z. angew. Chem. 59 (1947), 96

BRILL, HARVEY CLAYTON [1881-1972]
AMS 11,583; NCAB 57,577-578; WWAH, 72

BRILL, LUDWIG [1814-1876]
DGB 96 (1937), 22; RAUSCH, 424

BRILL, RUDOLF [1899-1989]
KURSCHNER 13,434; WOLF, 32; POGG 6,329-330; 7a(1),272
Anon., Nachr. Chem. Techn. 8 (1960), 51; 17 (1969), 309
H. Mark, Ber. Bunsen Ges. 68 (1964), 613-614

BRILLOUIN, LÉON NICOLAS [1889-1969]
DSB 17,104-109; CHARLE, 38-41; POGG 6,330; 7b,565-568
L.H. Thomas, Biog. Mem. Nat. Acad. Sci. 55 (1985), 69-92

BRINER, ÉMILE [1879-1965]
POTSCH, 66; POGG 5,167-168; 6,332; 7a(1),273-275
H. Paillard and B. Susz, Chimia 3 (1949), 29-32
B. Susz, Helv. Chim. Acta 49 (1966), 1041-1048

BRINK, FRANK, Jr. [1910-]
AMS 14,559; WWA 1976-1977, p.387; IWW 1982-1983, pp.166-167; WWWS, 242

BRINK, ROYAL ALEXANDER [1897-1984]
AMS 14,560; WWA 1978-1979, p.404; IWW 1982-1983, p.167; MH 2,50-51;MSE 1,139-140; IB, 35; WWWS, 243
R.D. Owen and O.E. Nelson, Genetics 112 (1986), 1-10

BRINKHOUS, KENNETH MERLE [1908-]
AMS 14,560; WWA 1976-1977, p.388; IWW 1982-1983, p.167
S. Sherry, Thrombosis and Haemostasis 40 (1979), 3-23
D.M. Surgenor, Progress in Clinical and Biological Research 72 (1981), 249-258

BRINTZINGER, HERBERT [1898-1969]
KURSCHNER 11,337; POGG 6,353-354; 7a(1),276-278

BRION, ALBERT AUGUSTE [1874-1936]
DBA 1,364; FISCHER 1,172

BRISSAUD, ÉDOUARD [1852-1909]
DBF 7,357; FISCHER 1,172-173; WWWS, 244; SG [3]3,511
L. Labbe, Bull. Acad. Med. 62 (1909), 488-489

BROBECK, JOHN RAYMOND [1914-]
AMS 17(1),731; BROBECK, 209-212; WWWS, 245

BROCA, PIERRE PAUL [1824-1880]
DSB 2,477-478; DBF 7,383-384; GENTY 9,209-224; HAYMAKER, 259- 263; HIRSCH 1,705-708; Suppl.,133-134; W, 82-83; WWWS, 245
E.C. Achard, Bull. Acad. Med. 92 (1924), 1347-1366
P. Huard, Rev. Hist. Sci. 14 (1961), 47-86; Clio Medica 1 (1966), 289-301
F. Schiller, Paul Broca. Berkeley, Cal. 1979
A. Delmas et al., Cahiers d'Anthropologie 1980, pp.15-93,245-283

BROCKLESBY, HORACE NICHOLAS [1901-1963]
AMS 10,457; YOUNG, 62-63
N.M. Carter, Proc. Roy. Soc. Canada [4]2 (1964), 83-87

BROCKMANN, HANS [1903-1988]
KURSCHNER 13,441; POTSCH, 66-67; POGG 7a(1),279-281
Anon., Chem. Z. 112 (1988), 250

BRODE, JOHANNES [1876-1930]
POGG 6,336-337
A. Johannsen, Z. Elektrochem. 36 (1930), 283

BRODE, WALLACE REED [1900-1974]
DSB 17,109-111; AMS 12,697; MH 2,53-54; MSE 1,141-143; CB 1958, pp.60-61; POTSCH, 67; WWWS, 246; POGG 6,337; 7b,577-579
R. Adams, Science 125 (1957), 279-280

BRODIE, ARNOLD FRANK [1923-1981]
AMS 14,567; WWA 1980-1981, p.426; WWWS, 246

BRODIE, BENJAMIN COLLINS [1783-1862]
DSB 2,482-484; DNB 2,1286-1288; PAGEL, 251-252; BETTANY 1,286-303; W, 83; WWWS, 246
Anon., Proc. Roy. Soc. 12 (1863), xlii-lvi
H.W. Acland, Biographical Sketch of Sir Benjamin Brodie. London 1864
B.C. Brodie, Autobiography. London 1865
T. Holmes, Sir Benjamin Collins Brodie. London 1898
W.R. LeFanu, Notes Roy. Soc. 19 (1964), 42-52
K.B. Thomas, Med. Hist. 8 (1964), 286-291

BRODIE, BENJAMIN COLLINS, Jr. [1817-1880]
DSB 2,484-486; DNB 2,1288; BLOKH 1,85-86; POTSCH, 67; WWWS,246
H.E. Roscoe, Nature 23 (1880), 126-127; J. Chem. Soc. 39 (1881), 182- 184
W.V. Farrar, Chymia 9 (1964), 169-179
W.H. Brock and D.M. Knight, Isis 56 (1965), 5-25
W.H. Brock, Ed., The Atomic Debates. Leicester 1967

BRODIE, BERNARD BERYL [1909-1989]
AMS 14,567; WWA 1976-1977, p.393; IWW 1982-1983, p.169; WWWS, 246
New York Times, 2 March 1989
E. Costa et al., Ann. Rev. Pharm. 29 (1989), 1-21

BRODIE, THOMAS GREGOR [1866-1916]
FISCHER 1,174; O'CONNOR(2), 243-245
Anon., Nature 98 (1916), 9-10; Science 44 (1916), 349
W.D.H., Proc. Roy. Soc. B91 (1920), xxviii-xxx
T.L. Sourkes, J. Chem. Ed. 29 (1952), 383-384

BRODY, SAMUEL [1890-1956]
AMS 9(II),133; IB, 36; WWWS, 247
A.F. Morgan, J. Nutrition 70 (1960), 3-9

BROEK, JOHANNES HUBERTUS van den [1815-1896]
LINDEBOOM, 259-260; RSC 1,642-643; NUC 77,374

BRÖMEL, HEINZ [1914-1942]

BRÖMME, EDUARD [1860-1893]
JV 4,223; RSC 13,832; NUC 77,387
Anon., Ber. chem. Ges. 26 (1893), 310

BRÖMME, EDUARD CHRISTIAN [1859-1931]
JV 3,197
Anon., Chem. Z. 55 (1931), 785; Chemische Fabrik 4 (1931), 411

BROEMSER, PHILIPP [1886-1940]
NDB 2,630; NB, 43; FISCHER 1,175; RAZINGER, 30-32; DRULL, 31; IB, 36; WWWS, 247; POGG 6,338-339; 7a(1),282-283
O.F.Ranke, Erg.Physiol.44 (1941), 1-17
K. Wezler, Münch. med. Wchschr. 88 (1941), 102-105
A. Hahn, Bayer. Akad. Wiss. Jahrbuch 1944-1948, pp.186-187

BRØNSTED, JOHANNES NICOLAUS [1879-1947]
DSB 2,498-499; VEIBEL 1,235; 2,80-88; 3,69-94; W, 84; ABBOTT-C, 22-23; POTSCH, 67-68; WWWS, 249; POGG 5,170; 6,342; 7b,585-586
J.A. Christiansen et al., Acta Chem. Scand. 3 (1949), 1187-1276
R.P. Bell, J. Chem. Soc. 1950, pp.409-419

BROILI, FERDINAND [1874-1946]
DSB 2,489-490; IB,36; POGG 7a(1),283
B. Peyer, Verhandl. Schw. Nat. Ges. 126 (1946), 358-360
D.M.S. Watson, Nature 158 (1946), 16-17

BROMAN, IVAR [1868-1946]
SBL 6,354-358; SMK 1,477; FISCHER 1,175-176
C. Häggquist, Anat. Nachr. 1 (1950), 260-269

BROMBERG, OTTO [1870-]
JV 10,19; RSC 13,833; NUC 77,477

BROMEIS, JOHANN CONRAD [1820-1862]
ADB 3,351; MEINEL, 117-119,498-499; GUNDLACH, 466; SCHMITZ, 266-267; TSCHIRCH, 1027; FERCHL, 70; POGG 1,305,1544

BRONFENBRENNER, JACQUES [1883-1953]
AMS 8,297; NCAB 42,244-245; KAGAN(JA), 302-303; IB, 36

BRONGNIART, ADOLPHE THÉODORE [1801-1876]
DSB 2,491-493; DBF 7,418-419; GE 8,129-130; LAROUSSE 2,1307; WWWS, 248; MAYERHOFER 1,554-555
M. Cornu, Rev. Sci. 10 (1876), 564-574
Anon., Proc. Roy. Soc. 28 (1879), iv-vii
L. de Launay, Une Grande Famille des Savants. Les Brongniarts. Paris 1940

BRONK, DETLEV WOLF [1897-1975]
DSB 17,111-113; MH 2,56-57; MSE 1,144-145; AMS 12,704-705; DAMB, 97-98; WWWS, 248
F. Seitz, Nature 259 (1976), 516
E.D. Adrian, Biog. Mem. Fell. Roy. Soc. 22 (1976), 1-9
B. Chance, Am. Phil. Soc. Year Book 1978, pp.54-66
F. Brink, Biog. Mem. Nat. Acad. Sci. 50 (1979), 3-87
D.Y. Cooper, Trans. Coll. Phys. Phila. [5]6 (1984), 113-124

BRONK, JOHN RAMSEY [1929-]
AMS 15(1),705; WWWS, 248

BROOK, FRANCIS WILLIAM [1877-]
RSC 15,934

BROOKENS, NORRIS L. [1911-1969]
AMS 11,599

BROOKER, LESLIE GEORGE SCOTT [1902-1971]
MILES, 49; AMS 12,706

BROOKS, CHANDLER McCUSKEY [1905-1989]
AMS 17(1),744; WWWS, 249

BROOKS, MATILDA MOLDENHAUER [1900-]
AMS 14,576; IB, 36; POGG 6,1763; 7b,3361-3362

BROOKS, WILLIAM KEITH [1848-1908]
DSB 2,501-502; DAB 3,90-91; WWWS, 250; WWAH, 74
E.A. Andrews and T.B. Comstock, Science 28 (1908), 777-786; 29 (1909), 614-616
E.G. Conklin, Biog. Mem. Nat. Acad. Sci. 7 (1913), 23-88
R.P. Bigelow, Proc. Am. Acad. Arts Sci. 71 (1937), 489-492
D.M. McCullough, J. Hist. Biol. 2 (1969), 411-438
K.R. Benson, William Keith Brooks. Corvallis, Ore. 1979; J. Hist. Med.
18 (1985), 163-205; Mendel Newsletter 23 (1983), 7-11

BROOM, JOHN CONSTABLE [1902-1960]
Anon., Lancet 1960(I), p.710

BROQUIST, HARRY PEARSON [1919-]
MSE 1,145-146; AMS 15(1),711; WWA 1980-1981, p.435

BROUSSAIS, FRANÇOIS [1772-1838]
DSB 2,507-509; DBF 7,449-450; HIRSCH 1,716-717; GENTY 1,53-80
J.D. Rolleston, Proc. Roy. Soc. Med. 32 (1939), 405-413
E.H. Ackerknecht, Bull. Hist. Med. 27 (1953), 320-343
F. Guido, Pag. Stor. Med. 12 (1968), 42-50
J.F. Braunstein, Broussais et le Materialisme. Paris 1986
M. Valentin, François Broussais. Dinard 1988

BROWN, ADRIAN JOHN [1852-1919]
WW 1919, p.311; DESMOND, 93
H.E. A[rmstrong], Nature 103 (1919), 369; J. Inst. Brew. 27 (1922), 197-260; Proc. Roy. Soc. B93 (1922), iii-ix; J. Chem. Soc. 121 (1922), 2899-2907
A. Harden, Biochem. J. 14 (1920), 1-3

BROWN, ALEXANDER CRUM [1838-1922]
DSB 2,514-516; BLOKH 1,86; W, 85; POTSCH, 68; POGG 4,189; 5,172
J. Walker, J. Chem. Soc. 123 (1923), 3422-3431; Proc. Roy. Soc. A105 (1924), i-v
J.E. Mackenzie, Chem. Ind. 1949, pp.461-463
D.F. Larder, Ambix 14 (1967), 112-132

BROWN, AMOS PEASLEE [1864-1917]
AMS 2,59
W. Stone, Proc. Am. Phil. Soc. 57 (1918), iii-xv

BROWN, DUGALD EDMUND SMITH [1901-1980]
AMS 11,612

BROWN, GENE MONTE [1926-]
AMS 14,586; WWA 1980-1981, p.443

BROWN, GEORGE BOSWORTH [1914-1985]
AMS 15(1),721; WWA 1980-1981, P.443; WWWS, 253
New York Times, 28 April 1985

BROWN, GEORGE LINDOR [1903-1971]
DNB [1971-1980], 94-96; WW 1971, p.401
F.C. MacIntosh and W.D.M. Paton, Biog. Mem. Fell. Roy. Soc. 20 (1974),41-73

BROWN, HERBERT CHARLES [1912-]
MSE 1,150-151; AMS 15(1),724; IWW 1982-1983, p.173; WWA 1980-1981, p.445; POTSCH, 68-69; WWWS, 254
H.C. Brown, Israel J. Chem. 25 (1985), 84-94

BROWN, HORACE TABBERER [1848-1925]
WW 1925, p.368; DESMOND, 94; POGG 3,203-204; 4,189-190; 5,173; 6,345
J.L. Baker, Biochem. J. 19 (1925), 165-167
J.B.F., Proc. Roy. Soc. A109 (1925), xxiv-xxvii
H.E. Armstrong, J. Chem. Soc. 131 (1928), 1061-1066

BROWN, JAMES CAMPBELL [1843-1910]
H.B.D. and J.N.C., J. Chem. Soc. 100 (1911), 1457-1460

BROWN, JAMES HOWARD [1884-1956]
AMS 9(II),139
R.D. Reid et al., J. Bact. 72 (1956), 281-282

BROWN, PERCY EDGAR [1885-1937]
AMS 5,143; IB, 37
B.J. Firkins, Proc. Iowa Acad. Sci. 10 (1938), 32-35

BROWN, RACHEL FULLER [1898-1980]
AMS 11,623; AO 1980, PP.33-35
R.S. Baldwin, The Fungus Fighters: Two Women Scientists and their Discovery. Ithaca, N.Y. 1981

BROWN, ROBERT [1773-1858]
DSB 2,516-522; DNB 3,25-27; DESMOND, 96; MAYERHOFER 1,560-561; WWWS, 256; POTSCH, 69; ABBOTT, 24-25; TSCHIRCH, 1027; PHILIPPE, 842
Anon., Proc. Roy. Soc. 9 (1858), 527-532
C.F.P. Martius, Flora 42 (1859), 10-15,25-31
J. Ramsbottom et al., Proc. Linn. Soc. 144 (1931), 17-54
D.C. Goodman, Episteme 6 (1972), 12-29

BROWN, SPENCER WHARTON [1918-1977]
AMS 13,525; WWA 1976-1977, p.477
E. Dempster et al., Genetics 88 (1978), s137-s138

BROWN, WADE HAMPTON [1878-1942]
FISCHER 1,178; AMS 6,179; WWWS, 256; WWAH, 77
H.S.N. Greene, Science 96 (1942), 221-222

BROWN, WILLIAM MICHAEL COURT [1918-1968]
DNB [1961-1970], 152-153
Anon., Lancet 1969(I), pp.57-58; Brit. Med. J. 1969(I), p.59

BROWN-SÉQUARD, CHARLES ÉDOUARD [1817-1894]
DSB 2,524-526; DBF 7,459-460; GE 8,186-187; HIRSCH 1,725-726; Suppl.,138; PAGEL, 256-258; HAYMAKER, 181-186; W, 85; WWWS, 256-257
E. Gley, Arch. Physiol. [5]6 (1894), 501-516
E. Dupuy, C.R. Soc. Biol. [10]1 (1894), 759-770
M. Berthelot, Rev. Sci. [4]10 (1898), 801-812
H.P. Bowditch, Biog. Mem. Nat. Acad. Sci. 4 (1902), 93-97
F. Rouget, Brown-Séquard et son Oeuvre. Port-Louis 1930
L. Delhoume, De Claude Bernard a d'Arsonval. Paris 1939
J.M.D. Olmsted, Charles Édouard Brown-Séquard. Baltimore 1946
A.T. Kenyon, Endocrinology 58 (1956), 284-291
H.B. Hoag, Virginia Medical Monthly 86 (1959), 311-313
W. Gooddy, Proc. Roy. Soc. Med. 57 (1964), 189-192
M. Borell, Bull. Hist. Med. 50 (1976), 309-320

BROWNE, CHARLES ALBERT [1870-1947]
DAB [Suppl.4],113-115; AMS 7,228; MILES, 52-54; NCAB E,250; 35,56-57; WWAH, 77; POGG 6,346-347; 7b,602-604
W.D. Bigelow, Ind. Eng. Chem. News Ed. 14 (1936), 431-432
A.K. Balls, J. Assn. Agr. Chem. 30 (1947), vi-x
C.K. Deischer, Chymia 1 (1948), 11-22
H.S. Klickstein and H.M. Leicester, J. Chem. Ed. 25 (1948), 315-317,343

BROWNE, JOHN SYMONDS LYON [1904-1984]
AMS 13,528; YOUNG, 14
P.T. Macklen, Trans. Assoc. Am. Phys. 98 (1985), c-ci
A. Carbelleira, Endocrinology 122 (1988), 376-378

BROWNING, CARL HAMILTON [1881-1972]
WW 1972, pp.414-415; FISCHER 1,179; POGG 6,348; 7b,604
C.L. Oakley, Biog. Mem. Fell. Roy. Soc. 19 (1973), 173-215

BROWNSDON, HENRY WINDER [1876-1950]
JV 18,190
Anon., Chem. Ind. 1951, p.565; J. Roy. Inst. Chem. 1951, p.61

BROZEK, ARTHUR [1882-1934]
IB, 37; WWWS, 257
Anon., Science 81 (1935), 65

BRUCE, DAVID [1855-1931]
DSB 2,527-530; DNB [1931-1940], 109-110; FISCHER 1,180; SACKMANN, 44-48; OLPP, 52-57; W, 86; WWWS, 258
J.R. Bradford, Obit. Not. Fell. Roy. Soc. 1 (1932), 79-85
J. Boyd, Notes Roy Soc. 28 (1973), 93-110

BRUCE, JAMES [1871-1952]
JV 24,570; NUC 80,131
Anon., J. Chem. Soc. 1952, p.64

BRUCHHAUSEN, FRIEDRICH von [1886-1966]
HEIN-E, 52-53; KURSCHNER 10,277; AUERBACH, 780; TLK 3,74; POGG 6,349; 7a(1),284
C. Rohmann, Deutsche Apoth. Z. 91 (1951), 707-709
G. Zinner, Pharmazie Unserer Zeit 16(3) (1987), 65-68

BRUCK, CARL [1879-1944]
KURSCHNER 4,327; FISCHER 1,181

BRUCKNER, VICTOR [1900-1980]
K. Medzihradszky, Acta Chim. Acad. Sci. Hung. 107 (1981), 287-314

BRÜCKE, ERNST THEODOR von [1880-1941]
NDB 2,654-655; FISCHER 1,182; JOHN, 19-37; BHDE 2,161; IB, 37; WWWS, 1730; POGG 7a(1),186-187
A. Jarisch, Erg. Physiol. 45 (1944), 1-11
F. Scheminzky, Alm. Akad. Wiss. Wien 95 (1945), 393-398

BRÜCKE, ERNST WILHELM von [1819-1892]
DSB 2,530-532; NDB 2,655; ADB 47,273-275; NOB 5,66-73; LESKY, 258-268; HIRSCH 1,728-730; PAGEL, 258-262; MAYERHOFER 1,563-564; WWWS, 1730; BRAZIER, 80-83,87; BLOKH 1,86; SCHAEDLER, 17-18; FERCHL, 71; POGG 1,312; 3,204; 4,191; 7a Suppl.,110-111
S. Exner, Wiener klin. Wchschr. 3 (1890), 807-812
C. von Voit, Sitz. Bayer. Akad. 22 (1892), 203-207
E.T. Brücke, Ernst Brücke. Vienna 1928

BRÜCKE, FRANZ THEODOR [1908-1970]
DECKER, 120-135; WWWS, 258; POGG 7a(1),287-288
O. Kraupp, Wiener med. Wchschr. 120 (1970), 311-313; Arzneimitt. 20 (1970), 728-730
A. Lindner, Wiener klin. Wchschr. 82 (1970), 301-303
H. Konzett, Alm. Akad. Wiss. Wien 121 (1972), 296-312

BRÜHL, JULIUS WILHELM [1850-1911]
NDB 2,663; BJN 16,133-136,12*; DRULL, 32; BLOKH 1,87-88; POTSCH, 69-70; WWWS, 259; POGG 3,204-205; 4,192; 5,175; 7a Suppl.,111
Anon., Chem. Z. 35 (1911), 167
K. Auwers, Ber. chem. Ges. 44 (1912), 3757-3794
O. Bütschli and E. Ebler, Verhandl. Nat. Med. Heidel. NF11 (1912), 330-358
E.H. Huntress, Proc. Am. Acad. Arts Sci. 78 (1950), 11

BRÜNER, HERMANN [1907-]
KURSCHNER 13,458; POGG 7a(1),290

BRÜNING, ADOLF von
[1837-1884]
ADB 47,770-772; NDB 2,665; NB, 44; NL 6,255-257
Anon., Chem. Z. 8 (1884), 595; Pharm. Z. 29 (1884), 277

BRÜNING, GUSTAV von [1864-1913]
NB, 44; BJN 18,82*; RSC 13,863; POGG 6,350-351
H. Will, Ber. chem. Ges. 46 (1913), 390
Anon., Chem. Z. 37 (1913), 199

BRÜNING, GUSTAV von [1898-1938]
JV 39,516

BRUES, AUSTIN MOORE [1906-1991]
AMS 14,606-607; WWA 1976-1977, p.424; WWWS, 259
New York Times, 6 March 1991

BRUGNATELLI, GASPARE [1795-1852]
DBI 14,496; COLE, 86-87; POGG 3,205; CALLISEN 3,245; 26,469

BRUGNATELLI, LUIGI VALENTINO [1761-1818]
DBI 14,494-496; DFI, 26-27; PROVENZAL, 91-95; HIRSCH 1,734; Suppl.,139; POTSCH, 69; FERCHL, 72; PHILIPPE, 772; POGG 1,316-317,1544
Anon., Phil. Mag. 53 (1819), 321-326
G. Provenzal, Rass. Clin. Terap. 36 (1937), 68-72

BRUGSCH, THEODOR [1878-1963]
FISCHER 1,185; KURSCHNER 9,231; POGG 6,351-352; 7a(1),292-293
T. Brugsch, Arzt seit fünf Jahrzehnten. Berlin 1959
W. Kaiser and W. Piechocki, Z. ges. inn. Med. 25 (1970), 1028-1034
W. Kaiser, Z. ges. inn. Med. 33 (1978), 641-650
D. Schmidt and R. Lange-Pfautsch, Charité-Annalen NF5 (1985), 304-309
J. Konert, Theodor Brugsch: Internist und Politiker. Leipzig 1988

BRUHNS, GUSTAV [1864-1945]
POGG 5,175-176; 6,352-353; 7a(1),293

BRUICE, THOMAS CHARLES [1925-]
AMS 15(1),747; WWA 1980-1981, p.460

BRUMPT, ÉMILE [1877-1951]
DSB 2,533-534; DBF 7,506; HD 1,99-102; FISCHER 1,186; IB, 37; WWWS, 260
A. Giroud, Bull. Acad. Med. 135 (1951), 517-519
H. Gaillard, Ann. Parasit. 27 (1952), 5-46
L.W. Hackett, J. Parasit. 38 (1952), 271-273
J. Théodoridès, Clio Medica 12 (1977), 269-278

BRUNCK, HEINRICH von [1847-1911]
NDB 2,677; BJN 16,12*; BUGGE 2,360-373; POTSCH, 70; BLOKH 1,88; POGG 5,176
P. Julius, Z. angew. Chem. 24 (1911), 2417-2420
A. Bernthsen, Chem. Z. 35 (1911), 1385-1386
C. Glaser, Ber. chem. Ges. 46 (1913), 353-389

BRUNCK, RUDOLPH [1867-1942]
RSC 13,868; NUC 80,525
Anon., Z. angew. Chem. 56 (1943), 20

BRUNI, GUISEPPE [1873-1946]
DBI 14,616-618; POGG 5,177-178; 6,355; 7b,609-610
A. Quilico, Chimica e Industria 28 (1946), 1-2
M.A. Rollier, J. Am. Chem. Soc. 71 (1949), 381-383

BRUNN, ALBERT von [1849-1895]
FISCHER 1,186; PAGEL, 262-263
W. Waldeyer, Anat. Anz. 11 (1895), 481-485

BRUNNENGRABER, CHRISTIAN [1832-1893]
HEIN, 86-87; HUHLE-KREUTZER, 207-217; SCHELENZ, 776-777
Anon., Ber. chem. Ges. 26 (1893), 395
H.J. Böttger, Pharm. Z. 38 (1893), 187-189,195-197,203-204,225-226

BRUNNER, ARNOLD [1880-1940]
JV 20,22
Anon., Z. angew. Chem. 53 (1940), 220; Chem. Z. 64 (1940), 189

BRUNNER, HEINRICH [1847-1910]
DHB 2,319-320; BLOKH 1,89; ZURICH-D,117; TSCHIRCH,1027; POGG 4,196; 5,179
E. Chuard, Verhandl. Schw. Nat. Ges. 93 (1910), 17-26

BRUNNER, KARL [1855-1935]
MACHEK, 190-194; HUTER, 237,242; POTSCH, 70-71; TLK 3,75; POGG 4,196; 5,179; 6,356; 7a(1),296
E. Philippi, Ber. chem. Ges. 68A (1935), 181; Ost. Chem. Z. 38 (1935),190

BRUNNER, KARL EMMANUEL [1796-1867]
TSCHIRCH, 1027; FERCHL, 72-73; POTSCH, 70; POGG 1,321; 3,207
B. Strahlmann, Chimia 21 (1967), 566-572

BRUNNER, OTTO [1899-1977]
KURSCHNER 12,383; POGG 7a(1),296-297

BRUNS, PAUL von [1846-1916]
NDB 2,686-687; DBJ 1,349; FISCHER 1,188; PAGEL, 164-165; KILLIAN, 208;
RSC 7,287; 9,381
F. Hofmeister, Z. Chir. 43 (1916), 969-973
M. von Brunn, Münch. med. Wchschr. 63 (1916), 1155
H. Küttner, Med. Klin. 12 (1916), 764

BRUNS, WILHELM [1864-1945]
JV 5,71; RSC 13,876; NUC 81,76
Anon., Chem. Z. 38 (1914), 49; Z. angew. Chem. 59 (1947), 35

BRUNTON, THOMAS LAUDER [1844-1916]
DSB 2,547-548; DNB [1912-1921], 75-76; HIRSCH 1,740; Suppl.,141; O'CONNOR, 218-220; PAGEL, 1936-1937; WWWS, 262
J.A. MacDougall, Edinburgh Med. J. NS17 (1916), 345-349
Anon., Lancet 1916(II), pp.572-575; Brit. Med. J. 1916(II), pp.440-442
C.A., Proc. Roy. Soc. B89 (1917), xliv-xlviii
J.L. Thornton et al., St. Barts. Med. J. 71 (1967), 289-299
W.F. Bynum, Bull. Hist. Med. 44 (1970), 518-538
W.B. Fye, Circulation 74 (1986), 222-229

BRUSCHETTINI, ALESSANDRO [1868-1932]
DBI 14,699-700; IB, 38
Anon., Med. Ital., 6 (1925), 245-248
M. Mariotti, Minerva Med. 48 (1957), 3953-3954

BRUSH, GEORGE JARVIS [1831-1912]
DAB 3,187-188; NCAB 28,197-198; WWWS, 262; WWAH, 78; POGG 3,208; 4,197; 5,179
E.S.Dana, Am. J. Sci. [4]33 (1912), 389-396; Biog. Mem. Nat. Acad. Sci. 17 (1924), 107-112

BRUTON, OGDEN CARR [1907-]
BIBEL, 60-63

BRUYLANTS, GUSTAVE [1850-1925]
BNB 39,173-176; RSC 13,879
F. Daels, Bull. Acad. Med. Belg. [5]8 (1928), 101-107

BRYK, ANTONI [1820-1881]
PSB 3,27; HIRSCH 1,743; Suppl.,142; KOSMINSKI, 54-55; KONOPKA 1,421-423
Anon., Leopoldina 17 (1881), 159; Przeglad Lekarski 20 (1881), 413
J. Gawlik, Arch. Hist. Fil. Med. 9 (1929), 218-224

BUBNOV, NIKOLAI ALEKSANDROVICH [1851-1884]
KUZNETSOV(1963), 572-577; ZMEEV 4,42; RSC 7,289; 12,129
K. Gambaroglu, Klin. Med. 42 (1962), 152-153

BUBNOV, SERGEI FEDOROVICH [1851-1909]
BME 4,700-701; LEVITSKI 2,223-232; VENGEROV 1,359

BUCHANAN, ANDREW [1798-1882]
O'CONNOR, 102-103; HIRSCH 1,746
Anon., Glasgow Med. J. 18 (1882), 134-140

BUCHANAN, JOHN MACHLIN [1917-]
AMS 15(1),761; WWA 1980-1981, p.469; IWW 1982-1983,p.179; WWWS, 264
J.M. Buchanan, in Selected Topics in the History of Biochemistry (G. Semenza, Ed.), pp.1-69. Amsterdam 1986

BUCHANAN, JOHN YOUNG [1844-1925]
DSB 2,557-558
A.E.S., Proc. Roy. Soc. A110 (1926), xii-xiii

BUCHER, KARL [1912-]
KURSCHNER 13,470; POGG 7a(1),299-300

BUCHERER, HANS THEODOR [1869-1949]
NDB 2,700-701; TLK 3,77; WWWS, 264; POTSCH, 71; POGG 4,199; 5,181-162; 6,361; 7a(1),300

BUCHET, CHARLES [1848-1933]
TSCHIRCH, 1027; STALDER, 11
E.H.-G., Rev. Hist. Pharm. 21 (1933), 1-3

BUCHHEIM, RUDOLF [1820-1879]
NDB 2,701; HB 1,33-37; HIRSCH 1,747-748; Suppl.,143; PAGEL, 268-269; WELDING, 116; BLOKH 1,89; FERCHL, 74; LEVITSKI 2, 187-189
O. Schmiedeberg, Arch. exp. Path. Pharm. 67 (1912), 1-54
T.C. Butler, Bull. Hist. Med. 44 (1970), 168-172
M. Bruppacher-Cellier, Rudolf Buchheim [1820-1879] und die Entwicklung einer Experimentellen Pharmakologie. Zurich 1971
E.R. Habermann, Ann. Rev. Pharm. 14 (1974), 1-8
J. Benedum, Med. Hist. J. 15 (1980), 103-119

BUCHKA, FRANZ ANTON [1828-1896]
Deutsche Stammtafeln in Listenform 1 (1927), 30

BUCHKA, KARL von [1856-1917]
BLOKH 1,88; POGG 3,210; 4,199; 5,182; 6,361-362
Anon., Z. angew. Chem. 30(III) (1917), 116
H. Boruttau, Arch. Gesch. Naturwiss. 8 (1918), 124-128

BUCHNER, EDUARD [1860-1917]
DSB 2,560-563; NDB 2,705; DBJ 2,650; MAYERHOFER 1,574-575; KIEL, 213; BLOKH 1,89-91; SACKMANN, 51-54; W, 87-88; WWWS, 264-265; POTSCH, 71-72; ABBOTT-C, 23; POGG 4,200; 5,182,1417; 6,362
C. Harries, Ber. chem. Ges. 50 (1917), 1843-1876; Chem. Z. 41 (1917),753
H. Biltz, Jahresb. Schles. Ges. 95 (1917), 2-4
N. Gordon, J. Chem. Ed. 6 (1929), 1849-1850
D. Ackermann, Sitz. Phys. Med. Würzburg NF70 (1960-1961), 87-93
R. Buchner, Z. Bayer. Landesgesch. 26 (1963), 631-645
R. Kohler, J. Hist. Biol. 4 (1971), 35-61; 5 (1972), 327-353
G. Westphal and B. Lipke, Lebensmittelindustrie 28 (1981), 401-405
A. Neubauer, Wiss. Z. Berlin 39 (1990), 247-250

BUCHNER, HANS [1850-1902]
NDB 2,705; BJN 7,316-320,19*; HIRSCH 1,750; Suppl.,143; PAGEL, 270-271; BULLOCH, 355-356; WWWS, 265
F. Hueppe, Münch. med. Wchschr. 49 (1902), 844-847

BUCHNER, JOHANN ANDREAS [1783-1852]
NDB 2,706; ADB 3,487-488; HEIN, 88-89; POHL, 121-131; BLOKH

1,89-91; POTSCH, 72; PHILIPPE, 804-805; TSCHIRCH, 1027-1028; WWWS, 265; POGG 1,327,1545; RSC 1,695-696; CALLISEN 3,274-287; 26,478-480
M. von Pettenkofer, Neues Rep. Pharm. 1 (1852), 342-347
L.A. Buchner, Arch. Pharm. 225 (1887), 889-890
B. Beyerlein, Pharmazie als Hochschuldisziplin. Stuttgart 1991

BUCHNER, KARL HERMANN GEORG [1886-]
JV 24,570; NUC 82,90

BUCHNER, LUDWIG ANDREAS [1813-1897]
NDB 2,706; ADB 47,329; BJN 2,49-50; 4,78*; HEIN, 89-90; TSCHIRCH, 1028; REBER, 85-86; BLOKH 1,91; FERCHL, 74; POGG 1,328; 3,210; 4,199-200
C. Bedall, Pharm. Z. 37 (1892), 157-158
Anon., Leopoldina 33 (1897), 160
C. von Voit, Sitz. Bayer. Akad. 28 (1898), 431-440

BUCHNER, MAX [1833-1899]
KERNBAUER, 30-32; POGG 3,210; 4,200

BUCHNER, MAX [1866-1934]
NDB 2,708; WWWS, 265; POTSCH, 72
B. Rassow, Z. angew. Chem. 39 (1926), 813-814
H. Bretschneider, Ber. chem. Ges. 67A (1934), 80-81

BUCHNER, PAUL ERNST CHRISTOF [1886-1978]
FISCHER 1,190; KURSCHNER 12,391; IB, 38; WWWS, 265

BUCHOLZ, CHRISTIAN FRIEDRICH [1770-1818]
DSB 2,564-565; ADB 3,491-492; HEIN, 90-91; BLOKH 1,91; TSCHIRCH, 1028; POTSCH, 72-73; COLE, 90; FERCHL, 74; SCHAEDLER, 18; POGG 1,329-330; 7a Suppl.,116

BUCHTHAL, FRITZ [1907-]
KBB 1981, pp.153-154; IWW 1982-1983, p.180; BHDE 2,166; LDGS, 79;KOREN, 160; WWWS, 265
Anon., Muscle and Nerve 1 (1978), 442-449

BUCK, JOHANNES SYBRANDT [1895-1956]
AMS 9(I),250; SRS 84; NUC 82,167; POGG 6,362-363; 7b,619-620
R. Baltzly, Proc. Chem. Soc. 1957, p.183

BUCK, JOHN BONNER [1912-]
AMS 15(1),765; WWA 1980-1981, p.471; WWWS, 265

BUCKLAND, FRANCIS TREVELYAN [1826-1880]
DNB 3,204-205
Anon., Nature 23 (1880), 175; Pop. Sci. Mon. 18 (1881), 812-820
G. Bompas, Life of Frank Buckland. London 1885

BUCKMAN, THOMAS ELLWOOD [1891-1945]
Anon., Harvard Class of 1912, 25th Anniversary Report, pp.106-109. Cambridge, Mass. 1937
New York Times, 27 March 1945

BUCKMASTER, GEORGE ALFRED [1859-1937]
WW 1938, p.456; FISCHER 1,191
J.A.G., Nature 141 (1938), 190-191

BUCKTON, GEORGE BOWDLER [1818-1905]
DNB [Suppl.2], 248; BLOKH 1,91; FERCHL, 75; WWWS, 269; POGG 3,211; 5,183
W.F. Kirby, Nature 72(1905),587-588; Proc. Roy. Soc. B79(1907),xlv-xlviii
J. Spiller, J. Chem. Soc. 91 (1907), 663-665

BUCURA, CONSTANTIN [1874-1936]
FISCHER 1,191; KURSCHNER 5,169-170; LORENZSONN, 37-46; PLANER, 43; WEBER, 13-14
H. Kahr, Wiener klin. Wchschr. 48 (1935), 1529
R. Hofstätter, Z. Gyn. 60 (1936), 130-134

BUDD, GEORGE [1808-1882]
DNB 3,219; HIRSCH 1,754-755; Suppl.,145; BETTANY 2,125-130; FRANKEN, 178
J. Paget, Proc. Roy. Soc. 34 (1883), i-iii
R.E. Hughes, Med. Hist. 17 (1973), 127-135

BUDD, WILLIAM [1811-1880]
DSB 2,574-576; DNB 3,220-221; HIRSCH 1,754; Suppl.,145
W.M. Clarke, Brit. Med. J. 1880(I), pp.163-166
W.C. Rucker, Bull. Johns Hopkins Hosp. 28 (1916), 208-215
E.W. Goodall, William Budd. Bristol 1936
R.C. Wolfinden, Bristol Med. Chir. J. 82 (1967), 63-67

BUDDE, VILHELM CHRISTIAN [1844-1893]
DBL(3rd Ed.) 3,56; VEIBEL 2,90-91

BUDDENBROCK, WOLFGANG von [1884-1964]
KURSCHNER 10,297
F. Scaller and D. Buckmann, Med. Hist. J. 20 (1985), 109-134

BUDGE, ALBRECHT [1846-1885]
PAGEL, 273; KOREN, 160
Anon., Leopoldina 21 (1885), 161

BUDGE, JULIUS LUDWIG [1811-1888]
ADB 47,337-339; NB, 46; PAGEL, 272-273; WININGER 1,476; KOREN, 160; POGG 3,212
Anon., Anat. Anz. 3 (1888), 651-652; Deutsche med. Wchschr. 14 (1888),607
A. Cantani, Leopoldina 24 (1888), 167-168
C. Lindenmeyer, Ludwig Julius Budge. Zurich 1966

BUDGETT, SIDNEY PAYNE [1862-]
AMS 2,65; NUC 82,486; SG [3]3,585-586
Who Was Who in America 4,133

BÜCHER, THEODOR [1914-]
KURSCHNER 13,480; AUERBACH, 211; POGG 7a(1),304BÜCHI, GEORGE HERMANN [1921-] AMS 15(1),763; MSE 1,156-157; IWW 1982-1983, p.180; WWWS, 264

BÜCHNER, JOHANN AUGUST WILHELM [1790-1849]
HEIN, 92-93; HB 2,348-349; TSCHIRCH, 1028; FERCHL, 76; POGG 1,334; CALLISEN 3,292-294; 26,482

BÜCHNER, (LOUIS) WILHELM [1816-1892]
K. Hechler, Pfungstätter Zeitung, 16 January, 26 March 1976

BÜCHNER, LUDWIG [1824-1899]
DSB 2,563-564; NDB 2,722; ADB 55,459-461; BJN 4,191*; HB 1,49-56; PAGEL, 273-274; WWWS, 265; POGG 3,212; 4,201; 7aSuppl.,118-119
F. Gregory, Scientific Materialism in Nineteenth-Century Germany. pp.100-121. Dordrecht 1977

BÜCHNER, PHILIPP THEODOR [1821-1890]
HEIN, 93-94; HB 3,311-313; WOLF, 33; RSC 1,697; 7,296; POGG 5,184
Anon., Chem. Z. 14 (1890), 1782

BUEDING, ERNST [1910-1986]
AMS 14,627; WWA 1976-1977, p.437; BHDE 2, 168; WWWS, 267
New York Times, 20 April 1986

BÜLBRING, EDITH [1903-1990]
WW 1984, p.314; BHDE 2,168; LDGS, 78

BÜLOW, KARL [1857-1933]
TLK 3,90; POGG 4,202; 5,184-185; 6,365
F. Seidel, Ber. chem. Ges. 66A (1933), 61-62

BÜLOW, MARGARETE [1902-]
JV 45,433; 51,49; NUC 82,664

BÜNGELER, WALTER HERMANN [1900-1987]
KURSCHNER 13,487; IWW 1979-1980, p.176; BUSCHHUTER, 178-196; KIEL, 90
Anon., Naturw. Rund. 40 (1987), 80

BÜNNING, ERWIN [1906-1990]
KURSCHNER 13,487-488; IWW 1982-1983, p.183; WWWS, 271

BUERGER, MARTIN JULIAN [1903-1986]
AMS 14,628; WWA 1976-1977, p.437; IWW 1982-1983, p.181; McLACHLAN, 48-50; WWWS, 268
New York Times, 11 March 1986
L.V. Azaroff, Journal of Applied Crystallography 19 (1986), 205-207

BÜRGER, MAX [1885-1966]
FISCHER 1,194; KURSCHNER 4,366-367; KIEL, 99
K. Seidel, Deutsche Z. Verdauung 25 (1965), 257-259
R. Emmrich, Jahrb. Sachs. Akad. 1966-1968, pp.337-338
K. Seege, Z. ges. inn. Med. 41 (1986), 568-571BÜRGI, EMIL [1872-1947] NDB 2,747; FISCHER 1,194; TLK 3,81; TSCHIRCH, 1029; IB, 39; WWWS, 274
J. Klaesi, Mitt. Nat. Ges. Bern NF5 (1948), 61-63

BÜRKER, KARL [1872-1957]
FISCHER 1,194; KURSCHNER 4,3427; GUNDEL, 131-140; IB, 39; POGG 7a(1),311-312; (4),139*
E. Koch, Münch. med. Wchschr. 89 (1942), 703
V. Horn, Nachr. Giessener Ges. 26 (1957), 5-6

BÜSCHELBERGER, CARL [1879-1914]
JV 18,191; NUC 83,171
Anon., Chem. Z. 38 (1914), 1085; Z. angew. Chem. 30(III) (1917), 95

BÜTSCHLI, OTTO [1848-1920]
DSB 2,625-628; NDB 3,6; DBJ 2,743; MAYERHOFER 1,577; DRULL, 33-34; FISCHER 1,195; DZ, 216; POGG 4,202-203; 5,185; 6,366
R. Goldschmidt et al., Naturwiss. 8 (1920), 543-570; in Science, Medicine and History (E.A. Underwood, Ed.), vol.2, pp.223-232. London 1953
J. Spek, Protoplasma 39 (1949-1950), 99-102

BUFALINI, MAURIZIO [1787-1875]
DBI 14,799-802; HIRSCH 1,761-762
G. Garin, Sperimentale 78 (1924), 377-406
C. Giachetti, Riv. Storia Sci. Med. 15 (1924), 257-299
A. Spallicci et al., Riv. Storia Sci. Med. 42 (1951), 140-215
A. Simili, Arch. Pat. Clin. Med. 45 (1968), 114-129
P. Berri, Minerva Med. 63 (1972), 2536-2541

BUFF, FRIEDRICH [1826-1871]
F. Buff, Die Buff, Table 5. Darmstadt 1934

BUFF, HEINRICH [1805-1878]
NDB 3,8-9; ADB 47,774-779; HB 1,438-446; BLOKH 1,92; TSCHIRCH, 1028 POTSCH, 73; FERCHL, 76-77; SCHAEDLER, 18-19; CALLISEN 3,313; 26,487-488; COLE, 92-93; POGG 1,337-338; 3,213
F. von Kobell, Sitz. Bayer. Akad. 9 (1879), 132-133
H. Kopp and C. Bohn, Ber. chem. Ges. 14 (1881), 2867-2886

BUFF, HEINRICH [1845-1902]
NDB 3,8*; BLOKH 1,93
G. Kraemer, Ber. chem. Ges. 35 (1902), 2763-2764

BUFF, HEINRICH LUDWIG [1828-1872]
ADB 3,503; HEIN-E, 57-58; NB, 48; STARK, 344-345; TSCHIRCH, 1028; BLOKH 1,93; SCHAEDLER, 19; POGG 3,213-214
K. Kraut, Ber. chem. Ges. 6 (1873), 688-693

BUFLEB, HERMANN [1875-]
JV 17,169; NUC 83,322

BUGARSZKY, STEFAN [1868-1941]
POGG 4,203; 5,185-186; 6,367; 7b,643
F. Szabadvary, Waage 13 (1974), 154-156

BUGGE, GÜNTHER [1885-1944]
JV 23,539; TLK 3,79; POGG 6,367; 7a(1),315
Anon., Chem. Z. 69 (1945); Z. angew. Chem. 62 (1950), 200

BUHL, LUDWIG von [1816-1880]
NDB 3,11; HIRSCH 1,763; Suppl.,146; PAGEL, 277-278; WWWS, 1731
M. von Pettenkofer and C. Voit, Z. Biol. 16 (1880), 411-412
F. von Kobell, Sitz. Bayer. Akad. 11 (1881), 372

BUHLMANN, OTTO LUDWIG [1877-]
JV 18,320; NUC 83,391

BUIGNET, HENRI [1815-1876]
DBF 7,641; GE 8,387; BOURQUELOT, 69-70; GORIS, 272-274; TSCHIRCH, 1028
A. Riche et al., J. Pharm. Chim. [4]24 (1876), 69-78
L. Figuier, Ann. Scient. 20 (1876), 547-550

BUJARD, BENJAMIN LOUIS [1824-1862]
STELLING-MICHAUD 2,371

BUJWID, ODON [1857-1942]
WE 2,208; KOSMINSKI, 56,591; KONOPKA 1,439-453; RSC 13,908-909
J. Chomiczewski, Acta Microbiol. Pol. 9 (1960), 9-32
J. Mostowski et al., Gruzlica 32 (1964), 387-398

BULANKIN, IVAN NIKOLAEVICH [1901-1960]
UKRAINE 2,204-205; WWR, 95
V.N. Nikitin and A.M. Utevski, Biokhimia 25 (1961), 1122-1123

BULIGINSKY, ALEKSANDR DMITRIEVICH [1838-1907]
BME 4,734; RSC 7,301

BULL, BENJAMIN SAMUEL [1866-1910]
JV 12,242; RSC 13,910; 14,804; NUC 83,591
Anon., J. Roy. Inst. Chem. 1910(4), p.25

BULL, CARROLL GIDEON [1884-1931]
AMS 4,138; NCAB 31,115; FISCHER 1,197; IB, 39; WWAH, 81; WWWS, 269
I.W. Pritchett, J. Immunol. 22 (1932), 245-249

BULL, HENRY BOLIVAR [1905-1982]
AMS 14,632; WWA 1978-1979, p.461; BERG, 72-73; WWWS, 269

BULLER, ARTHUR HENRY REGINALD [1874-1944]
DSB 2,582-583; DNB [1941-1950], 116-117; HUMPHREY, 42-45; DESMOND, 103; SRS, 23; IB, 39; WWWS, 270
G.R. Bisby, Nature 154 (1944), 173
W.F. Hanna et al., Phytopathology 35 (1945), 577-584
F.T. Brooks, Obit. Not. Fell. Roy. Soc. 5 (1945), 51-59

BULLOCH, WILLIAM [1868-1941]
DSB 2,583-585; DNB [1941-1950], 117-118; O'CONNOR(2), 294-296; FISCHER 1,198-199; WWWS, 270
P. Fildes, J. Path. Bact. 53 (1941), 297-308
J.C.G. Ledingham, Obit. Not. Fell. Roy. Soc. 3 (1941), 819-852
C.E. Dolman, Clio Medica 3 (1968), 65-84

BULLOCK, JOHN LLOYD [1812-1905]
Anon., J. Soc. Chem. Ind. 24 (1905), 575; Pharm. J. 74 (1905), 843; Chemist and Druggist 67 (1905), 339

BUMM, ERNST von [1858-1925]
NDB 3,16; FISCHER 1,200; PAGEL, 279; WWWS, 270-271
W. Stoeckel, Z. Gyn. 49 (1925), 177-183
F. von Müller, Deutsche med. Wchschr. 51 (1925), 280

W. Leibbrand, Berl. Med. 16 (1965), 51-57
P. Schneck, Z. Gyn. 105 (1983), 662-665

BUMM, ERWIN [1901-1957]
POGG 7a(1),315-316; (4),139*
Anon, Chem. Z. 81 (1957), 370; Nachr. Chem. Techn. (1957), 167

BUNGE, ALEKSANDR ANDREEVICH [1803-1890]
BSE 6,285-286; WELDING, 131; LIPSHITS 1,301-304; LEVITSKI 1,450-354; TSCHIRCH, 1028; WWWS, 1731
Anon., Leopoldina 26 (1890), 213; Sitz. Nat. Ges. Dorpat 9 (1892), 359-373

BUNGE, GUSTAV von [1844-1920]
DSB 2,585-586; DBJ 2,742; FISCHER 1,200-201; HIS, 283-284; WELDING, 133; PAGEL, 279-280; LEVITSKI 2,306-308; TSCHIRCH, 1029; POGG 3,214; 4,204-205; 5,187; 6,369
A. Kossel, Chem. Z. 44 (1920), 889
C.M. McCay, J. Nutrition 49 (1953), 3-19
M.L. Portmann, Gesnerus 31 (1974), 39-46
G. Schmidt, Das Geistige Vermächtnis von Gustav von Bunge. Zurich 1974

BUNGE, NIKOLAI ANDREEVICH [1842-1914]
BSE 6,286; BLOKH 1,94-96; ARBUZOV, 180-182; IKONNIKOV, 71-74; KOZLOV, 247; WWWS, 271; POGG 3,214; 4,205; 5,187
S. Reformatski, Zhur. Fiz. Khim. 48 (1916), 373-411

BUNGENBERG de JONG, HENDRIK GERARD [1893-1977]
BWN 3,304-306; IB, 39; POGG 6,399; 7b,645-650
P.A. de Wilde, Ned. Tijd. Gen. 79 (1935), 937
J.T.G. Overbeck, Jaarb. Akad. Wet. 1977, pp.158-163

BUNN, CHARLES WILLIAM [1905-1990]
Who's Who of British Scientists 1980-1981, p.64

BUNSEN, ROBERT WILHELM [1811-1899]
DSB 2,586-590; NDB 3,18-20; ADB 47,369-376; BJN 4,192-198; BUGGE 2,78-91; BAD 5,860-862; MAYERHOFER 1,579-580; GUNDLACH, 463-494; DRULL, 35; MEINEL, 26-42,499; SCHMITZ, 232-240; BLOKH 1,96-100; ABBOTT-C,23-24; W, 88-89; WWWS, 271-272; SCHAEDLER, 19-20; FERCHL,77; POTSCH 73-74; POGG 1,340-341; 3,214-215; 4,205; 6,369; 7a Suppl.,121-126
H.E. Roscoe, J. Chem. Soc. 77 (1900), 513-554
T. Curtius, J. prakt. Chem. NF61 (1900), 381-407; Robert Bunsen als Lehrer und Forscher. Heidelberg 1906; Ber. chem. Ges. 41 (1908), 4875-4910
W. Ostwald, Z. Elektrochem. 7 (1900), 608-618
C. von Voit, Sitz. Bayer. Akad. 30 (1900), 359-369
H. Debus, Erinnerungen an Robert Wilhelm Bunsen. Kassel 1901
R.E. Oesper, J. Chem. Ed. 4 (1927), 431-439; 18 (1941), 253-260
M. Bodenstein, Naturwiss. 24 (1936), 193-196
G. Lockemann, Robert Wilhelm Bunsen. Stuttgart 1949
H. Rheinboldt, Chymia 3 (1950), 223-241
K. Freudenberg, Z. Elektrochem. 64 (1960), 777-784; Heidelberger Jahrbücher 6 (1962), 111-184
K. Danzer, Robert W. Bunsen und Gustav R. Kirchhoff: Die Begründer der Spektralanalyse. Leipzig 1972
R. Schmitz, Ber. Bunsen Ges. 85 (1981), 932-937
S. Lotze, Z. Ver. Hess. Gesch. 91 (1986), 105-131

BUNTE, HANS [1848-1925]
NDB 3,20; BLOKH 1,100-101; SCHAEDLER, 20; TLK 2,91; DZ, 203; POTSCH, 74; POGG 3,215; 4,205-206; 6,369-370; 7a Suppl.,126
A. Sander, Chem. Z. 42 (1918), 621-622
B. Lepsius, Ber. chem. Ges. 58A (1925), 39-42
G. Keppeler, Z. angew. Chem. 38 (1925), 977-980

BUNTE, KARL [1878-1944]
KURSCHNER 4,361; TLK 3,80; POGG 6,370; 7a(1),316-317

BURCH, GEORGE JAMES [1852-1914]
O'CONNOR(2), 79-80; WWWS, 272; POGG 4,207; 5,188
H.M.V., Nature 93 (1914), 114-115

BURCKER, ÉMILE EUGÈNE [1846-1908]
DBF 7,679; DBA 1,420; SITZMANN 2,1076-1077; HEIN, 94; BALLAND(2), 81-82; TSCHIRCH, 1029; POGG 4,207
Anon., J. Pharm. Chim. [6]28 (1908), 143-144

BURCKHARDT, RUDOLF [1866-1908]
G. Imhof, Verhandl. Nat. Ges. Basel 1910, pp.1-32; Zoologische Annalen
3 (1910), 156-176
P. Smit, Gesnerus 42 (1985), 67-83

BURDACH, KARL FRIEDRICH [1776-1847]
DSB 2,594-597; ADB 3,578-580; HIRSCH 1,771-772; MAYERHOFER 1,580-581; KROLLMANN, 95; LEVITSKI 2,3-8; WWWS, 273; CALLISEN 3,328-336;26,494-496
T.H. Bast, Ann. Med. Hist. 10 (1928), 34-46

BURDICK, CHARLES LALOR [1892-]
AMS 14,640; WWA 1978-1979, p.468; McLACHLAN, 33-34; NUC 84,579; WWWS, 273

BURDON-SANDERSON, JOHN SCOTT [1828-1905]
DSB 2,598-599; DNB [1901-1911], 267-269; PAGEL, 1474-1475; BULLOCH, 395; O'CONNOR, 141-146; BRAZIER, 161-165,180-181; DESMOND, 107
Anon., Brit. Med. J. 1905(II), pp.1481-1492
C.L. Taylor, Berl. klin. Wchschr. 43 (1906), 149-151
P. Daser, Münch. med. Wchschr. 53 (1906), 174-175
F. G[otch], Proc. Roy. Soc. B79 (1907), iii-xviii
G. Burdon-Sanderson et al., Sir John Burdon-Sanderson. Oxford 1911

BURG, OTTO [-1884]
RSC 9,400
Anon., Ber. chem. Ges. 17 (1884), 2701,3048; Chem. Z. 9 (1885), 21

BURGDORF, CHRISTIAN [1868-]
JV 6,182; RSC 13,275

BURGER, KARL [1893-1962]
MERKLE, 5-20; GAUSS, 216
L. Neuhaus, Z. Gyn. 84 (1962), 1506-1508

BURGER, OSKAR [1883-]
JV 22,515; NUC 85,51

BURIAN, RICHARD [1871-1954]
FISCHER 1,203; IB, 40; KOREN, 161
Who's Who in Central and East Europe 1935, p.144
Srpska Akademiya Nauka [Year Book 1938] 48 (1939), 211-219
Anon., Science 119 (1954), 600

BURK, DEAN [1904-1988]
AMS 14,644-645; WWA 1978-1979, p.470; WWWS, 275; POGG 6,373; 7b,659-663
D. Burk, TIBS 9 (1984), 202-204
New York Times, 10 October 1988

BURKE, OLIVER WALLIS [1910-]
NUC 85,341

BURKHARDT, ADOLF [1870-]
JV 12,182; NUC 85,384

BURKHARDT, LUDWIG [1903-]
KURSCHNER 13,507; BUSCHHUTER, 197-209; POGG 7a(1),320-321BURKHOLDER, PAUL RUFUS [1903-1972] AMS 12,800;

WWAH, 83; WWWS. 275-276
J.G. Horsfall, Biog. Mem. Nat. Acad. Sci. 47 (1975), 3-25

BURN, JOSHUA HAROLD [1892-1981]
WW 1981, pp.364-365; MH 2,61-62; MSE 1,165-166; AO 1981, pp.446-447; WWWS, 276
J.H. Burn, Ann. Rev. Pharm. 9 (1969), 1-20
Anon., Brit. Med. J. 283 (1981), 444; Lancet 1981(II), p.212
J.R. Vane, Brit. J. Pharm. 75 (1982), 3-7
M. Vogt, Erg. Physiol. 94 (1982), 1-10
E. Bülbring and J.M. Walker, Biog. Mem. Fell. Roy. Soc. 30 (1984), 45-89

BURNE, BENJAMIN [1875-]
JV 22,298
J.A. Venn, Alumni Cantabrigensis II 2,457. Cambridge 1940

BURNET, ÉTIENNE [1873-1960]
G. Ramon, Bull. Acad. Med. 145 (1961), 189-191
J.C. Levaditi, Ann. Inst. Pasteur 100 (1961), 401-405

BURNET, FRANK MACFARLANE [1899-1985]
MH 1,81-82; STC 1,198-200; MSE 1,166; CB 1954, pp.134-135; ABBOTT, 26-27; WW 1981, PP.370-371; IWW 1982-1983, p.187; SACKMANN,60-62; WWWS, 276
F.M. Burnet, Changing Patterns. Melbourne 1968
G.J.V. Nossal, Nature 317 (1985), 108
Anon., Lancet 1985(II), p.620; Brit. Med. J. 291 (1985), 747,980
F. Fenner, Am. Phil. Soc. Year Book 1986, pp.91-95; Historical Records of Autralian Science 7 (1987), 39-77; Biog. Mem. Fell. Roy. Soc. 33 (1987), 101-162

BUROWOY, ABRAHAM [1905-1959]
LDGS, 17
G. Baddiley and J.E. Bloor, Proc. Chem. Soc. 1960, p.158

BURR, GEORGE OSWALD [1896-1993]
AMS 14,655; WWWS, 278
G.O. Burr, Progress in Lipid Research 20 (1981), xxvii-xxix

BURRIS, ROBERT HARZA [1914-]
AMS 15(1),806; IWW 1982-1983, pp.188-189; WWA 1980-1981, p.498; WWWS, 279

BURROWS, HAROLD [1875-1955]
WW 1949, p.400; NUC 86,391-392
E.E. Horning, Nature 176 (1955), 859-860
Anon., Lancet 1955(II), p.780; Brit. Med. J. 1955(II), p.914
R.H.O.B. Robinson and W.R. LeFanu, Lives of the Fellows of the Royal College of Surgeons of England 1952-1964, pp.58-59. London 1970

BURROWS, MONTROSE THOMAS [1884-1947]
AMS 7,253; NCAB G,174; IB, 40; WWAH, 85; WWWS, 279
Anon., J. Am. Med. Assn. 135 (1947), 656

BURTON, BEVERLY SCOTT [1846-1904]
ELLIOTT(2), 28; RSC 13,934; NUC 86,467
Anon., Chem. Z. 28 (1904), 1276; Science 19 (1904), 117
Obituary Record of Graduates of Yale University 1900-1910, pp.497-498; New Haven 1910

BURTON, ELI FRANKLIN [1879-1948]
SRS, 32; WWWS, 280; POGG 6,377; 7b,670-671
J. Satterly, Nature 162 (1948), 880-881
Anon., Science 108 (1948), 56

BURTON, HAROLD [1901-1966]
WW 1965, pp.438-439; POGG 6,377; 7b,671-672
R.S. Cahn, Chem. Ind. 1967, pp.100-101

BURTON-OPITZ, RUSSELL [1875-1954]
AMS 8,350; WWAH, 86; WWWS, 280
Anon., Science 120 (1954), 970

BURY, CHARLES RUGELEY [1890-1968]
DSB 17,132-134; POGG 6,377-378; 7b,672
M. Davies, J. Chem. Ed. 63 (1986), 741-743; Arch. Hist. Exact Sci. 36 (1986), 75-90; Chem. Brit. 23 (1987), 118

BUSCH, ALBERT [1867-1902]
JV 7,213; RSC 13,936
Anon., Ber. chem. Ges. 35 (1902), 4481

BUSCH, AUGUST [1876-]
JV 20,258

BUSCH, HARRIS [1923-]
AMS 18(1),894

BUSCH, MAX [1865-1941]
NDB 3,64; STUPP-KUGA, 214-231; TLK 3,82; WWWS, 281; POGG 4,209; 5,193; 6,378-379; 7a(1),324
G. Scheibe, Z. angew. Chem. 38 (1925), 710
Anon., Chem. Z. 65 (1941), 366
R. Pummerer, Sitz. Phys. Med. Erlangen 72 (1942), xxxiii-li
R. Dietzel, Arch. Pharm. 280 (1942), 16-20

BUSCH, WILHELM [1826-1881]
ADB 47,406-407; HIRSCH 1,783-784; PAGEL, 288-289
Anon., Brit. Med. J. 1881(II), pp.1001-1002
O.W. Madelung, Arch. klin. Chir. 27 (1881-1882), 490-528

BUSCHKE, ABRAHAM [1868-1943]
NDB 3,68; GEDENKBUCH 1,188; FISCHER 1,205-206; BHDE 2,175; LDGS, 62; TETZLAFF, 45; WININGER 6,501BUSCHKE, FRANZ [1902-] AMS 11,697; BHDE 2,175; LDGS, 81

BUSS, CARL EMIL [1849-1878]
DHB 2,370
H. Buess and H. Balmer, Gesnerus 19 (1962), 130-154

BUSSE, OTTO [1867-1922]
NDB 3,76; DBJ 4,351; FISCHER 1,206; WWWS, 282
B. Bloch, Viert. Nat. Ges. Zurich 67 (1922), 401-407
H. von Meyenburg, Verhandl. path. Ges. 31 (1939), 528-531

BUSSY, ANTOINE ALEXANDRE [1794-1882]
DBF 7,720; GE 8,505-506; BOURQUELOT, 53-54; BLOKH 1,100-101; POTSCH, 75; FERCHL, 78; TSCHIRCH, 1029; SCHAEDLER, 20; POGG 1,351-352; 3,220
L. Figuier, Ann. Scient. 26 (1883), 526-530
M. Picon, Figures Pharmaceutiques Francaises, pp.65-70. Paris 1953

BUTENANDT, ADOLF [1903-]
MH 2,67-69; STC 1,204-205; MSE 1,173-174; KURSCHNER 13,517; POTSCH, 75; WWWS, 283; IWW 1982-1983, p.191; BOHM, 175; POGG 6, 379; 7a(1),325-328
J. Schmidt-Thomé, Chem. Z. 87 (1963), 211-212
A. Butenandt, TIBS 4 (1979), 215-216; Das Werk eines Lebens. Göttingen 1981
K. Mothes, Naturw. Rund. 36 (1983), 93-94
P. Karlson, Adolf Butenandt. Stuttgart 1990

BUTKEVICH, VLADIMIR STEPANOVICH [1872-1942]
BSE 6,377-378; WWR, 98; POGG 6,380; 7b,675-676
V. Kretovich, Biokhimia 7 (1942), 283-284

BUTLER, ALLAN MACY [1894-1986]
AMS 12,825; WWA 1974-1975, p.456; ICC, 236-238; WWWS, 283
New York Times, 10 October 1986
J.D. Crawford, Harvard Med. Alumni Bull. 60(4) (1986), 61-63

BUTLER, ELMER GRIMSHAW [1900-1972]
AMS 11,703; IB, 41; WWAH, 86-87; WWWS, 283
Anon., Am. Phil. Soc. Year Book 1972, pp.129-131

BUTLER, GORDON CECIL [1913-]
AMS 15(1),821; YOUNG, 60; SRS, 70; WWWS, 283

BUTLER, JOHN ALFRED VALENTINE [1899-1977]
WW 1976, p.350; CAMPBELL, 45-46; POTSCH, 75-76; POGG 6,381; 7b,677-681
E.W. Johns, Nature 269 (1977), 735-736
W.V. Mayneord, Biog. Mem. Fell. Roy. Soc. 25 (1979), 145-178

BUTLER, THOMAS HOWARD [1887-1959]
JV 24,358
Anon., Chem. Ind. 1959, p.1458; J. Roy. Inst. Chem. 1960, p.117

BUTLEROV, ALEKSANDR MIKHAILOVICH [1828-1886]
DSB 2,620-625; BSE 6,378-383; KUZNETSOV(1961), 448-455; MAYERHOFER 1,583;KAZAN 1,278-286; ARBUZOV, 115-124; FARBER, 689-696; BLOKH 1,102-106; POTSCH, 76; W, 90; WWWS, 220,284; POGG 1,352; 3,221-222; 4,210
Anon., J. Chem. Soc. 51 (1887), 472-473; Usp. Khim. 22 (1953),1033-1063
W.N. Dawydoff, Uber die Entstehung der Chemischen Strukturlehre. Berlin 1957
H.M. Leicester, J. Chem. Ed. 17 (1940), 203-209; 36 (1959), 328-329
G.V. Bykov, Arch. Inter. Hist. Sci. 14 (1961); J. Chem. Ed. 39 (1962), 220-224
D.F. Larder, Ambix 18 (1971), 26-48
N.P. Grechkin, Vop. Ist. Est. Tekh. 1980(3), pp.115-120
A.J. Rocke, Brit. J. Hist. Sci. 14 (1981), 27-57

BUTTERFIELD, HERBERT [1900-1979]
WW 1979, p.374
J.H. Elliott and H.G. Koenigsberger, The Diversity of History, pp.315-325. London 1970

BUU-HOI, NG PH [1915-1972]
HD 6,63
Anon., Nature 237 (1972), 470-471; Chem. Eng. News 50(13) (1972), 54
G. Lambelin, Arzneimitt. 22 (1972), 950-951

BUXTON, BERTRAM HARRINGTON [1883-1947]
NUC 87,662
New York Times, 11 February 1947

BUXTON, BERTRAM HENRY [1852-1934]
NUC 87,662

BUXTON, PATRICK ALFRED [1892-1955]
DNB [1951-1960], 169-170; IB, 41; WWWS, 285
V.B. Wigglesworth, Biog. Mem. Fell. Roy. Soc. 2 (1956), 69-84

BUYTENDIJK, FREDERIK [1887-1974]
BWN 3,87-89; LINDEBOOM, 312-314; IB, 41; POGG 6,381-382; 7b,682
J.J. Prick, Jaarb. Akad. Wet. 1974, pp.207-229

BYK, ALFRED [1878-1942?]
GEDENKBUCH 1,190; LDGS, 95; POGG 5,194; 6,383; 7a(1),329

BYKOV, KONSTANTIN MIKHAILOVICH [1886-1959]
BSE 6,424-425; BME 4,815-817; STC 1,205-206; IB, 41
S. Kozlowski, Acta Physiol. Polon. 10 (1959), 759-761
G. Misgeld, Z. arzt. Fortbild. 53 (1959), 1207-1208
E. Matrosova et al., Trudy Inst. Fiziol. Pavlov 9 (1960), 24-31

BYWATERS, HUBERT WILLIAM [1881-1966]
O'CONNOR(2), 332; NUC 88,308
Who Was Who 1961-1970, p.168

C

CAAN, ALBERT [1882-]
KALLMORGEN, 235; LDGS, 80

CABOT, RICHARD CLARKE [1868-1939]
DAB [Suppl.2], 83-85; FISCHER 1,210; NCAB A,223; ICC, 11-15; WWWS, 287; DAMB, 112-113; FLUELER, 24-26
T.F. Franklin, Bull. Hist. Med. 24 (1950), 462-481
C.R. Burns, Bull. Hist. Med. 51 (1977), 353-368
L. O'Brien, New England Quarterly 58 (1985), 533-553

CADET de GASSICOURT, CHARLES LOUIS [1769-1821]
DSB 3,6; DBF 7,794; LAROUSSE 3,45; HIRSCH 1,793-794; BOURQUELOT, 31-32; GE 8,694; DECHAMBRE 11,450-451; TSCHIRCH, 1029,1051; FERCHL, 79-80; COLE, 93-94; CALLISEN 27,1
J.J. Virey, J. Pharm. Chim. 8 (1822), 1-15
E. Salverte, Notice sur la Vie et les Ouvrages de Ch. L. Cadet Gassicourt. Paris 1822
L.G. Toraude, Bull. Sci. Pharm. 6 (1902), 50-94,217-232,248-277
A. Berman, Bull. Hist. Med. 40 (1966), 101-111
S. Flahaut, Rev. Hist. Pharm. 27 (1980), 53-61

CADET de GASSICOURT, CHARLES LOUIS FÉLIX [1789-1861]
DBF 7,794-795; TSCHIRCH, 1029-1030; FERCHL, 80; POGG 3,224

CADET de VAUX, ANTOINE ALEXIS [1743-1828]
DSB 3,6-8; DBF 7,796; HOEFER 8,68-70; HIRSCH 1,794; DECHAMBRE 11,451-453; TSCHIRCH, 1030; FERCHL, 80
A. Vaquier, Paris et Ile-de-France, pp.367-467. Paris 1958
A. Berman, Bull. Hist. Med. 40 (1966), 101-111

CADY, HAMILTON PERKINS [1874-1943]
MILES, 60-61; POGG 6,385; 7b,691
R. Taft, J. Chem. Ed. 10 (1933), 34-39
A.W. Davidson, Ind. Eng. Chem. News Ed. 17 (1939), 660-661
R.Q. Brewster, Science 98 (1943), 190-191

CAESAR, KARL [-1891]
Anon., Pharm. Z. 36 (1891), 597

CAESAR, WILHELM [1878-1973]
JV 17,155; NUC 89,66

CAGNIARD-LATOUR, CHARLES [1777-1859]
DSB 3,8-10; DBF 7,823-824; GE 8,759; MEYERHOFER 1,591; FERCHL, 81; W, 311; WWWS, 287-288; CALLISEN 4,246; POGG 1,358-359; 3,224
J. Nicklès, Am. J. Sci. [2]28 (1859), 424-425; 29 (1850), 266-268

CAHILL, GEORGE FRANCIS [1890-1959]
AMS 9(II),161; NCAB 44,218-219; WWAH, 88; WWWS, 288
Anon., Science 130 (1959), 381; Brit. Med. J. 1959(I), p.700

CAHN, ARNOLD [1858-1927]
DBA 1,441; PAGEL, 194; KOREN, 161; WININGER 1,487; 6,502
P. Humbert, Strasbourg Médical 85 (1927), 180

CAHN, HEINZ [1892-]
JV 35,375; NUC 89,165

CAHN, ROBERT SIDNEY [1899-1981]
F.A. Robinson, Chem. Brit. 18 (1982), 359-361

CAHNMANN, HANS JULIUS [1906-]
AMS 15(2),6; LDGS, 59

CAHOURS, AUGUSTE [1813-1891]
DSB 3,10-11; DBF 7,833; GE 8,770-771; LAROUSSE 3,76; POTSCH, 76-77; BLOKH 1,107; TSCHIRCH, 1030; FERCHL, 81; POGG 1,360-361; 3,224
A. Étard, Bull. Soc. Chim. [3]7 (1892), i-xii

CAILLAU, JEAN MARIE [1765-1820]
DBF 7,846; GENTY 4,353-368; HIRSCH 1,799-800; DECHAMBRE 11,558-560

CAILLIOT, AMÉDÉE [1805-1884]
DBF 7, 867; DBA 1,443; RSC 1,761; 9,422; CALLISEN 3,401-402; 27,4
A. Gautier, Bull. Soc. Chem. [2]42 (1884), 610-612

CAIN, JOHN CANNELL [1871-1921]
WW 1919, p.381; SRS, 15; POGG 4,212; 5,195
J.F.T., J. Chem. Soc. 119 (1921), 533-537

CAIRNS, HUGH JOHN FORSTER [1922-]
WW 1983, p.345; ABBOTT, 27

CAJORI, FLORIAN ANTON [1892-1978]
AMS 11,720

CALDERON Y ARANA, LAUREANO [1847-1894]
EUI 10,664; RSC 9,423

CALDWELL, CHARLES [1772-1853]
DAB 3,406; HIRSCH 1,803; KELLY, 193; DAMB, 113-114; WWWS, 289
C. Caldwell, Autobiography of Charles Caldwell M.D. (H.W. Warner, Ed.).
Philadelphia 1855 (reprinted, New York 1968)
W.S. Middleton, Ann. Med. Hist. 3 (1921), 156-178
A.H. Bartley, Ann. Med. Hist. NS7 (1935), 141-146
H.S. Klickstein, Chymia 4 (1953), 129-157

CALDWELL, GEORGE CHAPMAN [1834-1907]
MILES, 61-62; NCAB 26,142-143; RSC 1,765; 9,424; 12,138; 14,13-14; NUC 89,569-570
Who Was Who in America 1,183
C.A. Browne, J. Am. Chem. Soc. 48(8A) (1926), 184

CALDWELL, MARY LETITIA [1890-1972]
MILES, 62-63; AMS 11,723; IB, 42
Anon., Chem. Eng. News 50(30) (1972), 64

CALDWELL, PETER CHRISTOPHER [1927-1979]
WW 1979, pp.385-386
E.J. Denton, Biog. Mem. Fell. Roy. Soc. 27 (1981), 153-172

CALDWELL, WILLIAM [1875-1939]
SRS, 25; POGG 6,388-389

CALKINS, GARY NATHAN [1869-1943]
DSB 3,16-17; DAB [Suppl.3], 1126-127; NCAB 33,50-51; IB, 42; WWAH, 89; WWWS, 289

CALLAN, THOMAS [1885-1946]
JV 26,393; NUC 91,38
Anon., J. Roy. Inst. Chem. 1946, p.1514

CALLOUD, FABIEN [1791-1855]
DBF 7,910; RSC 1,769
Anon., Mémoires de l'Académie Imperiale de Savoie [2]4 (1855), xxiii-xxvii

CALLOW, ANNIE BARBARA CLARK [1894-1948]
SG [5]1,109

CALLOW, ERNEST HAROLD [1895-1951]
Anon., Chem. Wkbl. 47 (1951), 245

CALLOW, ROBERT KENNETH [1901-1983]
WW 1981, p.401; POGG 6,389; 7b,699-701
Anon., Chem. Brit. 19 (1983), 558; Lancet 1983(I), p.1000
A. Neuberger, Biog. Mem. Fell. Roy. Soc. 30 (1984), 93-116

CALLSEN, JÜRGEN [1873-]
NUC 91,137

CALM, ARTHUR [1859-1885]
BLOKH 1,108
V. Meyer, Ber. chem. Ges. 18(3) (1885), 835-837

CALMAN, ALBERT [1862-]
RSC 13,275; 17,793; NUC 91,141

CALMETTE, ALBERT [1863-1933]
DSB 3,22-23; DBF 7,915-916; FISCHER 1,212; BULLOCH, 356-357; OLPP, 58-59; HD 2,148-159; IB, 42; STALDER, 12-16; WWWS, 290
K.E. Birkhaug, Ann. Med. Hist. 6 (1934), 291-300
C.J. Martin, Obit. Not. Fell. Roy. Soc. 1 (1934), 315-325
P.N. Bernard and L. Nègre, Albert Calmette, sa Vie, son Oeuvre Scientifique. Paris 1939
N. Bernard, La Vie et l'Oeuvre d'Albert Calmette. Paris 1961
R. Kervran, Albert Calmette et le B.C.G. Paris 1962
R. Dujarric de la Rivière, Bull. Acad. Med. 148 (1964), 648-655

CALVERT, FEREDERICK CRACE [1819-1873]
DNB 3,721; BLOKH 1,108-109; SCHAEDLER, 21; FERCHL, 107; POGG 3,307-308
E. Sell, Ber. chem. Ges. 6 (1873), 1587
Anon., J. Chem. Soc. 27 (1874), 1198-1199; Chem. News 31 (1875), 56-57
J.K. Crellin, Veröffentlichungen der Internationalen Gesellschaft für Geschichte der Pharmazie 28 (1966), 61-67

CALVERY, HERBERT ORION [1897-1945]
AMS 7,269; WWAH, 90; WWWS, 290; POGG 6,389; 7b,701-702
Anon., Chem. Eng. News 23 (1945), 2121

CALVIN, MELVIN [1911-]
MH 1,85-86; STC 1,208-209; MSE 1,180-181; AMS 15(2),20; NCAB I,191-192; IWW 1982-1983, pp.199-200; WWA 1981-1982, p.526; CB 1962, pp.68-70; ABBOTT-C, 24-25; POTSCH, 77
D. Ridgway, J. Chem. Ed. 50 (1973), 811-817
M. Calvin, Following the Trail of Light: A Scientific Odyssey. Washington, D.C. 1991

CALZADO, HECTOR [1910-]
JV 50,220; NUC 91,335

CAMERER, WILHELM [1842-1910]
NDB 3,107; BJN 15,11-16,18*; SL 3,45-60; FISCHER 1,213; PAGEL, 300-301; WWWS, 291
P. Grützner, Jahreshefte Wurtt. 66 (1910), xxxvi-xli
O. Heubner, Jahrb. Kinderheil. 71 (1910), 651-654
M. Pfaundler, Münch. med. Wchschr. 57 (1910), 967-968
A. Keller, Mon. Kinderheil. 9 (1910), 1-7
O. Zehender, Johann Friedrich Wilhelm Camerer. Zurich 1969

CAMERON, ALEXANDER THOMAS [1882-1947]
YOUNG, 8-9; SRS, 36; IB, 42; POGG 5,197-198; 6,390; 7b,704
G.B. Reed, Rev. Can. Biol. 7 (1948), 626-628
F.D. White and J.B. Collip, Biochem. J. 43 (1948), 1-2
J.B. Collip, Proc. Roy. Soc. Canada [3]42 (1948), 83-85

CAMERON, CHARLES ALEXANDER [1830-1921]
C.A. Cameron, Autobiography. Dublin 1920
B. Dyer, Analyst 46 (1921), 175-176
T. Burns, Anal. Proc. 14 (1977), 173

CAMERON, GORDON ROY [1899-1966]
DNB [1961-1970], 167-169; MH 1,86-87; MSE 1.181-182; WW 1965, p.469; IB, 42; WWWS, 292
C.L. Oakley, Biog. Mem. Fell. Roy. Soc. 14 (1968), 83-116

CAMERON, WILLIAM [1822-1855]
Medical Officers in British Army 1660-1910, vol.1, p.337. London 1968

CAMMARATA, PETER S. [1920-]
AMS 18(2),25

CAMMIDGE, PERCY JOHN [1872-1961]
WW 1959, p.467; FISCHER 1,214

CAMPBELL, ALAN NEWTON [1899-1987]
AMS 14,698; POGG 6,392; 7b,707-708
Anon., Chem. Brit. 24 (1988), 213

CAMPBELL, ALFRED WALTER [1868-1937]
L.R. Parker et al., Med. J. Austral. 1938(I), pp.181-185

CAMPBELL, DAN HAMPTON [1907-1974]
AMS 12,863; WWWS, 293
Anon., Science 186 (1974), 1058
I.L. Trapani et al., Immunochemistry 12 (1975), 441-447

CAMPBELL, DOUGLAS HOUGHTON [1859-1953]
DSB 3,29-31; DAB [Suppl.5], 97-98; NCAB A,284; WWAH, 91; WWWS, 293
G.M. Smith, Biog. Mem. Nat. Acad. Sci. 29 (1956), 45-63

CAMPBELL, DUGALD [1818-1882]
FERCHL, 83; POGG 3,229-230
J.H. Gilbert, J. Chem. Soc. 43 (1883), 252-253

CAMPBELL, GEORGE FLAVIUS [1870-1902]
RSC 17,640
Obituary Records of Graduates of Yale University 1900-1910, p.282. New Haven 1910

CAMPS, RUDOLF [1860-]
SURREY, 25; JV 7,224; RSC 14,39; NUC 92,503

CAMUS, JEAN [1872-1924]
FISCHER 1,215
H.A. Bénard, Bull. Acad. Med. 144 (1960), 768-776

CANDOLLE, ALPHONSE de [1806-1893]
DSB 3,42-43; DHB 2,398; BRIQUET, 130-147; GENEVA 4,6-21; TSCHIRCH, 1037; WWWS, 295; POGG 1,352; 3,338; 4,304
G.L. Goodale, Am. J. Sci. [3]46 (1893), 236-239
W.G. Farlow, Proc. Am. Acad. Arts Sci. 28 (1893), 406-411
A. Engler, Ber. bot. Ges. 11 (1893), (46)-(61)
G. Bonnier, Verhandl. Schw. Nat. Ges. 76 (1893), 203-211
J.D.H., Proc. Roy. Soc. 57 (1895), xiv-xx
S. Mikulinsky et al., Alphonse de Candolle. Jena 1980

CANDOLLE, AUGUSTIN PYRAMUS de [1778-1841]
DSB 3,43-45; DHB 2,398; BRIQUET, 114-130; TSCHIRCH, 1037; FERCHL, 116; MAYERHOFER 1,598-599; PHILIPPE, 831-832; W, 91-92; WWWS, 295; POGG 1,532; CALLISEN 3,439-443; 27,15-18
G.B. Emerson, Am. J. Sci 42 (1841-1842), 217-227
C.F.P. Martius, Am. J. Sci. 44 (1842-1843), 217-239
A.A. de la Rive, Notice sur la Vie et les Ecrits de A.P. de Candolle. Geneva 1851
M. Guésdès, Histoire et Nature NS4(2) (1974), 35-46
F. Merke, Gesnerus 32 (1975), 215-222
L. Dulieu, Hist. Sci. Med. 13 (1979), 461-462
P.F. Stevens, J. Hist. Biol. 17 (1984), 49-82

CANELLAKIS, ZOE NAKOS [1927-]
AMS 18(2),36

CANN, JOHN RUSWEILER [1920-]
AMS 15(2),36; WWWS, 295

CANNAN, ROBERT KEITH [1894-1971]
AMS 11.744; NCAB 56,106-108; IB, 43
Anon., Chem. Eng. News 49(26) (1971), 35
J.T. Edsall, Biog. Mem. Nat. Acad. Sci. 55 (1985), 107-133

CANNIZZARO, STANISLAO [1826-1910]
DSB 3,45-49; DBI 18,1131-141; PROVENZAL, 195-200; MAYERHOFER 1,599; BUGGE 2,173-189; FARBER, 663-674; W,92; WWWS, 295; POTSCH, 77-78; BLOKH 1,109-110; ABBOTT-C, 25-26; POGG 4,216-217; 5,200-201; 6,395
A. Gautier, Bull. Soc. Chim. [4]7 (1910), i-xiii
A. Miolati, Chem. Z. (1910), 593-594
A. von Baeyer, Sitz. Bayer. Akad. 1911, pp.33-36
R. Nasini, Rend. Soc. Chim. 3 (1911), 181-199
W.A. Tilden, J. Chem. Soc. 101 (1912), 1677-1693
L.C. Newell, J. Chem. Ed. 3 (1926), 1361-1367
N. Parravano, J. Chem. Ed. 4 (1927), 835-844
S. Baglioni, Arch. Stor. Sci. 7 (1926), 67-79
G. Provenzal, Rass. Clin. Terap. 36 (1937), 378-382
D. Marotta, Gazz. Chim. Ital. 69 (1939), 689-717
C. de Milt, Chymia 1 (1948), 153-169
H. Hartley, Notes Roy. Soc. 20 (1966), 56-63
J.W. van Spronsen, Chymia 11 (1966), 125-137
L. Cerutti, Chimica e Industria 64 (1982), 667-673; 65 (1983), 645-650

CANNON, PAUL ROBERTS [1892-1986]
AMS 14,709
E.R. Long, Arch. Path. 74 (1962), 263-266

CANNON, WALTER BRADFORD [1871-1945]
DSB 15,71-77; DAB [Suppl.3], 133-137; FISCHER 1,215; HAYMAKER, 279-281; BROBECK, 135-137; MEDVEI, 507-509; DAMB, 119-120; IB, 43; WWAH, 92-93; WWWS, 296
W.B. Cannon, The Way of an Investigator. New York 1945
C.K. Drinker, Science 102 (1945), 470-472
A. Forbes, Am. Phil. Soc. Year Book 1945, pp.349-354
H. Dale, Obit. Not. Fell. Roy. Soc. 5 (1947), 407-423
Anon., Leopoldina [3]1 (1955), 14-15
J. Mayer, J. Nutrition 87 (1966), 3-8
H.W. Davenport, Gastroent. 63 (1972), 878-892
C.M. Brooks et al., The Life and Contributions of Walter Bradford Cannon. New York 1975
H.W. Davenport and A.C. Barger, Physiologist 24(5) (1981), 1-14
D. Fleming, Social Research 51 (1984), 609-640
S. Benison et al., Walter B. Cannon: Life and Times of a Young Scientist. Cambridge, Mass. 1987; Med. Hist. 35 (1991), 217-249.
S. Benison, Bull. Hist. Med. 65 (1991), 234-251

CANTACUZÈNE (CANTACUZINO), JEAN [1863-1934]
FISCHER 1,215-216; GRUETTER, 26; IB, 43; WWWS, 296
A. Tiffeneau, Bull. Acad. Med. 111 (1934), 884-904
P. Desfosses, Presse Med. 42 (1934), 427

CANTANI, ARNALDO [1837-1893]
DBI 18,237-239; EI 8,772; HIRSCH 1,820; PAGEL, 304-305; NAVRATIL, 29
Anon., Leopoldina 29 (1893), 110
L. Kleinwächter, Janus 9 (1904), 325-330
E. Greco, Arch. Pat. Clin. Med. 38 (1961), 192-194
C. Maggiore et al., Lancet 1982(I), p.1133

CANTONI, GAETANO [1815-1887]
DBI 18,319-323

CANTONI, GIULIO LEONARD [1915-]
AMS 11,746; WWWS, 296

CAP, PAUL ANTOINE [1788-1877]
DBF 7,1056-1057; BOURQUELOT, 61-62; TSCHIRCH, 1030; FERCHL, 83

CAPLOW, MICHAEL [1935-]
AMS 18(2),45

CAPPARELLI, ANDREA [1855-1921]
RSC 12,141; 14,56-57
Anon., Atti Accad. Gioneia 98-99 (1921-1922), i-ix

CAPRANICA, STEFANO [1850-1899]
RSC 9,443; 12,141; 14,58
R. Oddi, Boll. Accad. Med. Genova 14 (1899), 140-152

F. Faggioli, Atti Soc. Sci. Genova 10 (1899), 139-159

CAPUTTO, RANWEL [1914-]
WWWS, 197

CARIUS, GEORG LUDWIG [1829-1875]
ADB 3,781-782; HEIN, 100-101; GUNDLACH, 464-465; MEINEL, 140- 148,499; DRULL, 36; SCHMITZ, 267-271; BLOKH 1,110-111; SCHAEDLER, 21; POTSCH, 78; POGG 3,325-326
H. Kämmerer, J. prakt. Chem. 120 (1875), 455-458
A. Ladenburg, Ber. chem. Ges. 9 (1876), 1992-1997

CARL, HANS (JOHANNES) [1880-1966]
JV 22,18; NUC 95,366

CARL, RICHARD WALDEMAR [1868-1933]
JV 7,213; RSC 16,377
Anon., Z. angew. Chem. 46 (1933), 750; Chem. Z. 57 (1933), 952

CARLISLE, ANTHONY [1768-1840]
DSB 3,67-68; DNB 3,1012-1013; HIRSCH 1,832-833; FERCHL, 85; POTSCH, 78; WWWS, 299; POGG 1,380-381,1547
Anon., Proc. Roy. Soc. 4 (1837-1843), 260-261
R.J. Cole, Ann. Sci. 8 (1952), 255-270
B. Hill, Practitioner 201 (1968), 950-955

CARLSON, ANTON JULIUS [1875-1956]
DSB 3,68-70; DAB [Suppl.6], 99-100; AMS 9(II),170; NCAB 45,486-487; BROBECK, 141-142; CB 1948, pp.93-95; FISCHER 1,219; IB, 44; WWAH, 94; WWWS, 99-100
V. Johnson, Science 124 (1956), 713-714
L.R. Dragstedt, Am. Phil. Soc. Year Book 1956, pp.106-109; Biog. Mem. Nat. Acad. Sci. 35 (1961), 1-12; Geriatrics 16 (1961), 541-550; Persp. Biol. Med. 7 (1964), 145-148
D.J. Ingle, Persp. Biol. Med. 22 (1979), S114-S136

CARLSON, LOREN DANIEL [1915-1972]
AMS 11,757; BROBECK, 197-201; WWWS, 299

CARLSSON, ARVID [1923-]
VAD 1991,195
A. Carlsson, Annual Review of Neuroscience 10 (1987), 19-40

CARMICHAEL, HENRY [1846-1924]
AMS 3,113; RSC 7,335; 9,446; 14,68

CARMINATI, BASSANIO [1750-1830]
HIRSCH 1,834-835; HOEFER 8,770; WURZBACH 2,287; POGG 1,381,1547

CARNELLEY, THOMAS [1852-1890]
BLOKH 1,111; SCHAEDLER, 21-22; POGG 3,237; 4,222
H.E.R. and P.P.B., Nature 42 (1890), 522-523
Anon., J. Chem. Soc. 59 (1891), 455-460

CARNELUTTI, GIOVANNI [1850-1901]
BLOKH 1,111; RSC 14,70
Anon., Ber. chem. Ges. 34 (1901), 3210; Chem. Z. 25 (1901), 1168

CARNOT, SADI [1796-1832]
DSB 3,79-84; DBF 7,1185-1186; GE 9,480; LAROUSSE 3,429; POTSCH, 78;MAYERHOFER 1,603-504; BLOKH 1,111-112; W, 84-95; POGG 1, 382
E. Aries, L'Oeuvre Scientifique de Sadi Carnot. Paris 1921
E. Mendoza, Arch. Inter. Hist. Sci. 12 (1959), 377-396; Brit. J. Hist. 14 (1981), 75-78
V.V. Raman, J. Chem. Ed. 47 (1970), 331-337
M.J. Klein, Physics Today 27 (1974), 23-28

CARNOY, JEAN BAPTISTE [1836-1899]
BNB 29,421-426; WWWS, 301
E.D., Ann. Soc. Belge Micr. 26 (1899-1900), 165-168
G. Gilson, La Cellule 17 (1900), i-xxiv

CARO, HEINRICH [1834-1910]
DSB 3,84-85; NDB 3,152-153; BJN 15,18*; BAD 6,692-699; BUGGE 2,298-309; POTSCH, 78-79; BLOKH 1,112-114; MATSCHOSS, 37; POGG 5,205; 6,401; 7aSuppl.,130-131
A. Bernthsen, Ber. chem. Ges. 45 (1912), 1987-2042
C. Duisberg et al., Z. angew. Chem. 24 (1911), 1059-1073
C. Schuster, Tradition 5 (1960), 49-64
R.S. Travis, Ambix 38 (1991), 113-134

CARO, NIKODEMUS [1871-1935]
NDB 3,153; WININGER 1,501; 6,506; 7,542; POTSCH, 79; TETZLAFF, 48; TLK 3,85
Anon., Chem. Z. 55 (1931), 393

CAROTHERS, ESTRELLA ELEANOR [1883-1957]
AMS 8,387; OGILVIE, 52-53; NUC 96,216-217

CAROTHERS, WALLACE HUME [1896-1937]
DSB 3,85-86; DAB 22,96-97; MILES, 65-66; STC 1,222-223; ABBOTT-C, 26-27; POTSCH, 79-80; W, 95; WWWS, 301; WWAH, 95; POGG 6,402; 7b,722-723
R. Adams, Biog. Mem. Nat. Acad. Sci. 20 (1939), 293-309
J.R. Johnson, J. Chem. Soc. 1940, pp.100-102
J.W. Hill, Welch Foundation Conferences on Chemical Research 20 (1977),232-250

CARPENTER, FREDERICK HILTMAN [1918-1982]
AMS 15(2),65; WWWS, 302

CARPENTER, THORNE MARTIN [1878-1971]
AMS 10,595; WWWS, 301; IB, 44; POGG 6,403; 7b,724-725

CARPENTER, WILLIAM BENJAMIN [1813-1885]
DSB 3,87-89; DNB 3,1075-1077; PAGEL, 307-308; DESMOND, 119; WWWS, 303; O'CONNOR, 107-109; POGG 3,239
E.R. Lankester, Nature 33 (1885), 83-85; Proc. Roy. Soc. 41 (1887), ii-ix
C. von Voit, Sitz. Bayer. Akad. 16 (1886), 45-49
V.M.D. Hall, Med. Hist. 23 (1979), 129-155

CARR, EMMA PERRY [1880-1972]
AMS 11,768; MILES, 66-67; NCAB F,364-365; CB 1959, pp.55-57; WWWS, 303
Anon., Chem. Eng. News 50(4) (1972), 20
B.H. Jennings, J. Chem. Ed. 63 (1986), 923-927

CARR, FRANCIS HOWARD [1874-1969]
WW 1965, p.498; WWWS, 303; POGG 6,403; 7b,725
F.J. Griffin, Chem. Ind. 1969, p.196

CARR, JOHN GARDNER [1913-]
Who's Who in British Science 1953, p.51
Directory of British Scientists 1966-1967, 1,276

CARRACIDO, JOSÉ RODRIGUEZ [1856-1928]
J. Giral, Bull. Soc. Chim. Biol. 10 (1928), 972-974
Anon., Chem. Z. 52 (1928), 39; Z. angew. Chem. 41 (1928), 37
A. Sanchez-Moscoso, Bolletin de la Sociedad de Historia de la Farmacia 21 (1970), 111-131,153-171; 22 (1971), 14-36,54-84

CARREL, ALEXIS [1873-1944]
DSB 3,90-92; DAB [Suppl.3], 139-142; FISCHER 1,222; FREUND 2,31-35; SCHWERTE 2,45-51; W, 95-96; WWWS, 304; DAMB, 121-122; IB, 44
S. Flexner, Am. Phil. Soc. Year Book 1944, pp.344-349
R. Soupault, Alexis Carrel. Paris 1952
W.S. Edwards and P.D. Edwards, Alexis Carrel, Visionary Surgeon, Springfield, Ill. 1974
J.A. Witkowski, Med. Hist. 23 (1979), 279-296; 24 (1980), 129-142

CARRIÈRÉ, GEORGES LEON [1872-1944]
FISCHER 1,222
C. Laubry, Bull. Acad. Med. 129 (1945), 439-442

CARROLL, DENIS CHARLES [1901-1956]
Anon., Lancet 1956(II), p.116

CARSTANJEN, ERNST [1836-1884]
BLOKH 1,114; SCHAEDLER, 22; POGG 3,241
H. Kolbe, J. prakt. Chem. 30 (1884), 96
Anon., Chem. Z. 8 (1884), 1037

CARSTENSEN, MAREN (INGEBORG) [1901-]
SG [4]3,244

CARTER, CYRIL WILLIAM [1898-1974]
NUC 97,143
Anon., Lancet 1974(I), p.1298; Brit. Med. J. 1974(II), pp.733-774

CARTER, HENRY ROSE [1852-1925]
DAB 3,535-536; NCAB 25,346-347; OLPP, 65-66; FISCHER 1,223; DAMB, 124; WWWS, 306
E.D. Richter, New Eng. J. Med. 277 (1967), 734-735
E.B. Carmichael, Ala. J. Med. Sci. 6 (1969), 348-353

CARTER, HERBERT EDMUND [1910-]
MH 1,88-89; MSE 1,184; WWA 1978-1979, p.545; IWW 1982-1983, p.210; AMS 14,653
H.E. Carter, Fed. Proc. 38 (1979), 2684-2866

CARTER, PETER GEORGE [1902-1948]
SRS, 61
W.A. Cowdrey, J. Chem. Soc. 1949, p.3082

CARTER, SYDNEY RAYMOND [1889-1966]
POGG 6,408-409; 7b,735
T.C. Tatlow, Chem. Brit. 2 (1966), 446
M. Stacey, Chem. Ind. 1966, pp.697-698

CARVER, GEORGE WASHINGTON [1864-1943]
DAB [Suppl.3], 145-147; MILES, 68-69; NCAB 33,316-317; E,170-171; POTSCH, 80; WWAH, 98; WWWS, 307

CASAMAJOR, PAUL [1831-1887]
MILES, 70-71; POGG 3,241-242
H. Endemann, J. Am. Chem. Soc. 9 (1887), 206-208

CASARES-GIL, JOSÉ [1866-1961]
EUI Suppl. 1961-1962, p.162; NUC 97,524

CASH, JOHN THEODORE [1854-1936]
DNB [1931-1940], 152-153; WW 1936, p.573; O'CONNOR(2), 420-421; NUESCH,42-44
C.R. Marshall, Nature 138 (1936), 1087; Obit. Not. Fell. Roy. Soc. 2 (1938), 295-300

CASPARI, ERNEST WOLFGANG [1909-]
AMS 14,751; BHDE 2,182; WWWS, 308
E.M. Eicher, Adv. Genetics 24 (1987), xv-xxxi

CASPARI, ERNST WILHELM [1872-1944]
FISCHER 1,225; KURSCHNER 5,191; KALLMORGEN, 236; BHDE 2,182; LDGS, 13 WARGENITZ, 41; WININGER 6,507
E. Schwartz and R. Chambers, Science 105 (1947), 613

CASPARI, WILLIAM AUGUSTUS [1877-1951]
SRS, 21; POGG 6,410

CASPARIS, PAUL [1889-1964]
KURSCHNER 9,267; TSCHIRCH, 1031; TLK 3,86; POGG 7a(1), 336-337
Anon., Chem. Z. 88 (1964), 609

CASPERSSON, TORBJÖRN OSKAR [1910-] VAD 1981, p.185; IWW 1982-1983, p.213; SCHULZ-SCHAEFFER, 210-211,216; WWWS, 308
B.H. Mayall et al., Cytometry 5 (1984), 314-318

CASELLA, LEOPOLD [1766-1847]
NDB 3,167

CASSELMANN, OSCAR [1820-1872]
ADB 4,61-62; NB, 53; FERCHL, 88; POGG 1,388; 3,244

CASSIDY, HAROLD GOMES [1906-]
AMS 15(2),93; WWA 1980-1981, pp.574-575; WWWS, 309

CASTALDI, LUIGI [1890-1945]
DSB 3,112-114; DBI 21,556-558; FISCHER 1,226; IB, 45
C. Maxia, Anat. Nachr. 1 (1951), 281-302

CASTELLANI, ALDO [1877?-1971]
DBI 21,605-609; FISCHER 1,226-227; OLPP, 67-69; WWAH, 99-100; WWWS, 309
A. Castellani, Microbes, Men and Monarchs. London 1968
Anon., Lancet 1971(II), p.883; Brit. Med. J. 1971(IV), p.175
J. Boyd, Notes Roy Soc. 28 (1973), 93-110
R. Vanbreugshem, Bull. Acad. Med. Belg. 128 (1973), 69-78
K. Ito, Bull. N.Y. Acad. Med. 60 (1984), 1011-1024

CASTELLI, GIORGIO [1884-1937]
Anon., Bioch. Ter. Sper. 24 (1937), 362; Scienza e Technica 1 (1937), 254-255

CASTIGLIONI, ARTURO [1874-1953]
DBI 22,117-122; FISCHER 1,228; WININGER 6,510; 7,543; WWWS, 310
J.F. Fulton, Yale J. Biol. Med. 17 (1944); J. Hist. Med. 8(1953),129-132
H.E. Sigerist, Bull. Hist. Med. 27 (1953), 387-389
A. Corsini et al., Riv. Stor. Sci. 45 (1954), 1-103

CASTLE, WILLIAM BOSWORTH [1897-1990]
AMS 14,755; WWA 1978-1979, p.557; IWW 1982-1983, p.215; ICC, 199-201
H.L. Blumgart, Ann. Int. Med. 101 (1958), 173-183
P. Karlson, TIBS 4 (1979), 286
D.G. Nathan, Proc. Am. Phil. Soc. 135 (1991), 593-597
J.H. Jandl, J. Lab. Clin. Med. 118 (1991), 614-616

CASTLE, WILLIAM ERNEST [1867-1962]
DSB 3,120-124; AMS 10,614-615; NCAB 16,297; IB, 45; WWAH, 100; WWWS, 310
E.S. Russell, J. Heredity 45 (1954), 210-213
L.C. Dunn, Am. Phil. Soc. Year Book 1962, pp.115-119; Biog. Mem. Nat. Acad. Sci. 38 (1965), 31-80
S. Wright, Genetics 48 (1963), 1-5
C.C. Li, Am. J. Hum. Gen. 19 (1967), 70-74

CASTOLDI, ARTURO [1864-1923]
DFI, 32-33

CASTORO, NICOLA [1869-1928]
POGG 5,208; 6,413; 7b,742

CATCHESIDE, DAVID GUTHRIE [1907-]
WW 1982, p.377; IWW 1982-1983, p.216

CATHCART, EDWARD PROVAN [1877-1954]
DNB [1951-1960], 195-196; WW 1949, p.471; O'CONNOR(2), 409-411; WWWS, 311
G.M. Wishart, Obit. Not. Fell. Roy. Soc. 9 (1954), 35-53
D. Smith and M. Nicolson, Soc. Stud. Sci. 19 (1989), 195-238

CATON, RICHARD [1842-1926]
O'CONNOR, 235-236; BRAZIER, 185-194,208-209; RSC 14,100-101; WWWS, 312
Anon., Lancet 1926(I), p.102; Brit. Med. J. 1926(I), pp.71-72
Cohen, Proc. Roy. Soc. Med. 52 (1959), 645-651
B.S. Schoenberg, Mayo Clin. Proc. 49 (1974), 474-481

CATTANEO, ANTONIO [1786-1845]
DFI, 33; FERCHL, 89; POGG 1,397-398

CATTANEO, GIACOMO [1857-1925]
DBI 22,464-465
G. Montalenti, Arch. Stor. Sci. 6 (1925), 270-271
A. Berzolari, Rend. Ist. Lombardo [2]58 (1925), 695-696
R. Issel, Riv. Biol. 8 (1926), 128-135

CATTANI, GUISEPPE [1859-1915]
RSC 14,102
R. Gurrieri, Boll. Sci. Med. Bologna [9]3 (1915), 123-128

CATTELAIN, EUGÈNE [1887-1955]
POGG 6,414; 7b,743-745
R. Weitz, Ann. Pharm. Fran. 13 (1955), 17-18

CATTELL, JAMES McKEEN [1860-1944]
DSB 3,130-131; DAB [Suppl.3], 148-151; NACB 34,337-338; D,94; IB, 45
F.R. Moulton, Sci. Mon. 58 (1944), 249-252
E.G. Conklin et al., Science 99 (1944), 151-165
W.B. Pillsbury, Biog. Mem. Nat. Acad. Sci. 25 (1947), 1-16
A.I. Poffenberger, James McKeen Cattell. Lancaster, Pa. 1947
M.M. Sokal, American Psychologist 26 (1971), 525-635
S.G. Kohlstedt et al., Science 209 (1980), 33-60

CATTELL, McKEEN [1891-1983]
AMS 14,759; CHEN, 59-61
New York Times, 11 February 1983

CAULLERY, MAURICE [1868-1958]
DSB 3,148-149; CHARLE(2), 67-71; BUICAN, 194-208; IB, 45; WWWS, 312
J. Théodoridès, Rev. Hist. Sci. 12 (1959), 60-62
G. Cousin, Bull. Biol. 93 (1959), 1-6
E. Fauré-Fremiet, Not. Acad. Sci. 4 (1964), 429-480

CAUSSE, HENRI EUGÈNE [1858-1907]
GORIS, 283-284; GUIART, 160; DORVEAUX, 93,94; POGG 4,229; 5,208

CAUVET, ÉMILIEN LUC DESIRÉ [1827-1890]
DBA 1,474-475; BALLAND, 18-20; HUMBERT, 202-203; BERGER-LEVRAULT, 37; GUIART, 112
A. Balland, J. Pharm. Chim. [5]21 (1890), 227-228

CAVALLO, WILHELM [1866-1950]
JV 5,293; RSC 15,993; 16,344; NUC 100,380

CAVAZZANI, EMILIO [1866-1922]
FISCHER 1,229
A. Stefani, Riv. Biol. 5 (1923), 148-155
V. Aducco, Arch. Ital. Biol. 72 (1923), 153

CAVENTOU, EUGÈNE [1824-1912]
DBF 7,1507; WWWS, 313
C. Gariel, Bull. Acad. Med. 67 (1912), 135-136
E. Gautier, Ann. Scient. 56 (1912), 338

CAVENTOU, JOSEPH BIENAIMÉ [1795-1877]
DSB 3,159-160; DBF 7,1507; GE 9,972; LAROUSSE 3,648; BALLAND(2), 226; MAYERHOFER 1,613; BLOKH 1,116; FERCHL, 89-90; TSCHIRCH, 1031; SCHAEDLER, 22-23; W, 100-101; GORIS, 285-286; POTSCH, 81-82; COLE, 98-100; CALLISEN 4,28-30; 27,52-53; POGG 1,407; 3,248
J. Bergeron, Mem. Acad. Sci. 38 (1899), 1-25
M. Delépine, J. Chem. Ed. 28 (1951), 454-461
M. Javillier et al., Not. Acad. Sci. 3 (1957), 185-201
C. Cabanis et al., Rev. Hist. Pharm. 19 (1969), 315-324
G. Dillemann, Produits et Problèmes Pharmaceutiques 28 (1973), 523-524
G. Valette, Bull. Acad. Med. 161 (1977), 333-338

CAYLEY, ARTHUR [1821-1895]
DSB 3,162-170; DNB 22,401-402; W, 101; WWWS, 313
A.R. Forsyth, Proc. Roy. Soc. 58 (1895), i-xliii
D.H. Rouvray, Endeavour 34 (1975), 28-33

CAZENEUVE, PAUL [1852-1934]
DBF 8,18; GORIS, 286-290; GUIART, 132; POGG 4,231-232; 5,209; 6,417; 7b,747-748
M. Chambon, Bull. Soc. Chim. [5]1 (1934), 911-912; Bull. Sci. Pharm. 14 (1934), 357-367

ČECH, CARL FRANZ OTTOKAR [1842-1895]
POGG 4,232-233
Anon., Leopoldina 31 (1895), 110-111

CECH, THOMAS ROBERT [1947-]
AMS 17(2),111

CENTANNI, EUGENIO [1863-1942]
DBI 23,583-584; EI 9,745; FISCHER 1,232-233; IB, 45
Anon., Pathologica 34 (1942), 345-347; Minerva Med. 56 Suppl. (1965), 1722-1724
G. Favilli, Arch. Vecchi Anat. Pat. 46 (1965), 819-836

CENTNERSZWER, MIECZYSLAW [1874-1944]
DSB 3,176-177; TLK 3,86-87; WARSAW, 230; RIGA, 696; POTSCH, 82; POGG 4,233; 5,209-210; 6,419; 7b,754-755
W. Lazniewski, Przemysl Chemiczny 37 (1958), 246-251
E.Y. Kaprovich, Ist. Est. Tekhn. Pribalt. 1 (1968), 169-177

CERECEDO, LEOPOLD RAYMOND [1898-]
AMS 11,801

CERNY, FRANTISEK [1864-1928]
J. Satava, Chem. Listy 23 (1929), 21-22
Anon., Chem. Z. 53 (1929), 91

CERNY, KAREL [1871-1922]
NAVRATIL, 37
B. Vecerek and J. Taizoch, Chem. Listy 70 (1976), 1009

CERTES, ADRIEN [1835-1903]
DBF 8,68-69
J. Guiart, Bull. Soc. Zool. 28 (1903), 176-180

CERVELLO, VINCENZO [1854-1919]
FISCHER 1,233
C. Lazzarro, Atti Accad. Sci. Med. Palermo 1920, pp.xiii-xxvi

CESARI, ÉMILE [1876-1956]
L.C., Ann. Inst. Pasteur 91 (1956), 928

CESARIS-DEMEL, ANTONIO [1866-1938]
DBI 24,214-217; FISCHER 1,233; IB, 46
A. Ascenci, Archivio Italiano di Anatomia e di Istologia Patologica 37 (1963), 485-499
F. Pascarella, Clinica Terapeutica 56 (1971), 369-387

CHABRIÉ, CAMILLE [1860-1928]
DBF 8,142; CHARLE, 72-73; POGG 4,236; 5,2110212; 6,421

CHABRY, LAURENT [1855-1894] DSB 3,184-185; DBF 8,158; WWWS, 316
M. Caullery, Rev. Sci. 78 (1940), 230-232
J.L. Fischer and J. Smith, Hist. Phil. Life Sci. 6 (1984), 23-29

CHAGAS, CARLOS [1879-1934]
DSB 3,185-186; OLPP, 73; GRUETTER, 27-28; WWWS, 316
E. Villela, Memorias do Instituto Oswaldo Cruz 29 (1934), i-xxiii
F.E.G. Marchoux, Bull. Acad. Med. 112 (1934), 786-792
R. Lewinsohn, J. Roy. Soc. Med. 74 (1974), 451-455
B.H. Kean, Am. J. Trop. Med. 26 (1977), 1084-1087

CHAGOVETS, VASILI YURIEVICH [1873-1941]
BSE 47,17; BME 34,571-572; WWR, 102; VORONTSOV,112-116; UKRAINE 2,455-456
A.V. Lebedinski and A.S. Mozhukhin, Fiziol. Zhur. 39 (1953), 250-256

CHAIKOFF, ISRAEL LYON [1902-1966]
AMS 11,804; KOREN, 162
T.H. Jukes, Nature 209 (1966), 963-964
L.L. Bennett, Persp. Biol. Med. 30 (1987), 362-383

CHAIN, ERNST BORIS [1906-1979]
DSB 17,148-150; DNB [1971-1980], 132-134; MH 1,93-94; STC 1,235-236; MSE 1,190-191; CB 1965, pp.77-79; WWWS, 316; WW 1978, p.426; W(3rd Ed), 585; SACKMANN, 63-65; ABBOTT, 28-29; BHDE 2,185; LDGS, 59; COHEN, 38; POTSCH, 82-83
K.R.L. Mansford, in Biologically Active Substances (D.A. Hems, Ed.), pp.xxi-xxvi. New York 1977; Nature 281 (1979), 715-717
A. Nuberger, TIBS 4 (1979), N247-N248
E.P. Abraham, J. Antibiot. 32 (1979), 1080-1081; Biog. Mem. Fell. Roy.Soc. 29 (1983), 43-91
Anon., Lancet 1979(II), p.505; Brit. Med. J. 1979(II), p.505;
Chemical Technology 10 (1980), 474-481
D.C. Hodgkin, Chem. Brit. 16 (1980), 267
B. Chain, Nature 353 (1991), 492-494

CHALLENGER, FREDERICK [1887-1983]
WW 1981, pp.452-453; SRS, 44; POGG 6,422-423; 7b,765-767

CHAMBERLAIN (MOORE), MARY MITCHELL [1892-1960]
AMS 7,1247
B.T. Scheer, Science 137 (1962), 411-412

CHAMBERLAND, CHARLES ÉDOUARD [1851-1908]
DSB 3,188-189; DBF 8,241; FISCHER 1,235; PAGEL, 1940-1941; BULLOCH, 357; WWWS, 317
E. Roux and G. Darboux, Ann. Inst. Pasteur 22 (1908), 370-380

CHAMBERS, ROBERT [1881-1957]
AMS 9(II),183; NCAB 46,14-15; JV 24,571; FREUND 1,133-147; WWAH, 102; IB, 46
J. Gray, Nature 180 (1957), 1167
W.R. Duryee, Science 126 (1957), 645; Acta Anat. 37 (1959), 1-20
B.W. Zweifel and M.J. Kopac, Protoplasma 51 (1960), 154-160
I.P. Goldring, American Zoologist 19 (1979), 1271-1273

CHAMBON, PIERRE [1931-]
DBA 1,484-485
Who's Who in Science in Europe 5,388

CHAMOT, ÉMILE MONNIN [1868-1950]
AMS 8,411; MILES, 74; IB, 46; WWWS, 230; POGG 5,213; 6,426; 7b,770
G.W. Cavanaugh, Ind. Eng. Chem. 25 (1933), 826-827
C.W. Mason, Ind. Eng. Chem. (Anal. Ed.) 11 (1938), 341-343
Anon., Chem. Eng. News 28 (1950), 4108

CHAMPY, CHRISTIAN [1885-1962]
FISCHER 1,235-236; IB, 47
H.A. Bénard, Bull. Acad. Med. 146 (1962), 552-559

CHANCE, BRITTON [1913-]
MH 1,95-97; MSE 1,191-193; AMS 15(2),126; WWA 1980-1981, pp.589-590; IWW 1982-1983, p.223; WWWS, 318; POTSCH, 83
S. Congdon, TIBS 3 (1978), N171-N172
B. Chance, Ann. Rev. Biophys. 20 (1991), 1-28

CHANCEL, GUSTAVE [1822-1890]
DSB 3,193-194; DBF 8,364-365; GE 10,471-472; BLOKH 1,117-118; POTSCH, 83;WWWS, 318; POGG 3,259; 4,238
R. de Fourcrand, Bull. Soc. Chim. [3]5 (1891), i-xx

CHANDELON, THÉODORE [1851-1921]
RSC 9,487; 14,148
E. Schoop, in L'Université de Liège (L. Halkin, Ed.), vol.3, pp.119-121. Liège 1936

CHANDLER, CHARLES FREDERICK [1836-1925]
DAB 3,611-613; MILES, 74-75; SILLIMAN, 138-141; DAMB, 128-129; POTSCH, 83-84
E. Hendrick, Ind. Eng. Chem. 17 (1925), 1090-1091
M.T. Bogert, Biog. Mem. Nat. Acad. Sci. 14 (1931), 127-181
A.W. Hixson, J. Chem. Ed. 32 (1955), 499-506

CHANDLER, JOSEPH PAXTON [1903-]
AMS 11,812

CHANIEWSKI, STANISLAW [1859-1920]
PSB 3,262-263; KONOPKA 2,29; POGG 3,262-263

CHANNON, HAROLD JOHN [1897-1979]
WW 1978, p.431; IB, 47
G.R. Tristram, Chem. Brit. 16 (1980), 326

CHANTEMESSE, ANDRÉ
[1851-1919]
DBF 8,393; FISCHER 1,236; WWWS, 319
E. Roux, Ann. Inst. Pasteur 33 (1919), 137
G. Thibierge, Bull. Acad. Med. 81 (1919), 231-324
Anon., Lancet 1919(I), p.433; Brit. Med. J. 1919(I), p.431

CHANUTIN, ALFRED [1897-1986]
AMS 12,964; WWWS, 319

CHAPLIN, EDWARD MITCHELL [1868-1948]
JV 8,273; RSC 14,154; 17,462; NUC 103,470
Anon., J. Roy. Inst. Chem. 1949, p.144

CHAPMAN, ALFRED CHASTON [1869-1932]
WWWS, 320; POGG 6,425-426; 7b,770
B. Dyer, J. Chem. Soc. 1932, pp.2980-2986
R. Robertson, Nature 130 (1932), 654-655
J.A. Voelker, Analyst 57 (1932), 745-749
B.D., Biochem. J. 26 (1932), 1715-1718

CHAPMAN, ARTHUR WILLIAM [1898-]
POGG 6,425; 7b,770

CHAPMAN, DAVID LEONARD [1869-1958]
DSB 3,197; DNB [1951-1960],207-208; WWWS, 320; POGG 5,213; 6,426; 7b,770
E.J. Bowen, Nature 181 (1958), 453; Biog. Mem. Fell. Roy. Soc. 4 (1958), 35-44
D.L. Hammick, Proc. Chem. Soc. 1959, pp.101-103

CHAPMAN, ERNEST THEOPHRON [1846-1872]
POGG 3,261
Anon., Ber. chem. Ges. 5 (1872), 1123; J. Chem. Soc. 26 (1873), 775-777

CHAPPUIS, JAMES [1854-1934]
WWWS, 321; POGG 4,239; 6,427; 7b,776
E. Bloch, Journal de Physique [7]5 (1934), 175

CHAPTAL, JEAN [1756-1832]
DSB 3,198-203; DBF 8,448-451; GE 10,577; LAROUSSE 3,969; GENTY 3,65-80; MAYERHOFER 1,618-619; BLOKH 1,118; TSCHIRCH, 1032; FERCHL, 91-92; POTSCH, 84; W, 104; WWWS, 321; COLE, 101-111;
POGG 1,420-421; CALLISEN, 4,73-78; 27,52-53
J. Pigeire, La Vie et l'Oeuvre de Chaptal. Paris 1931
E.V. Armstrong and H.S. Lukens, J. Chem. Ed. 13 (1936), 257-262
H.E. LeGrand, Brit. J. Hist. Sci. 17 (1984), 31-46

CHARAUX, CAMILLE [1861-1941]
POGG 6,127-128; 7b,776-777
V. Plouvier, Rev. Hist. Pharm. 1 (1991), 75-84

CHARDONNET, LOUIS MARIE HILAIRE de [1839-1924]
DSB 3,207; DBF 6,120; BLOKH 1,118; ABBOTT-C, 28; POTSCH, 84; WWWS,322; POGG 6,428
G. Bigourdan, C.R. Acad. Sci. 178 (1924), 977-978
J.L. Parsons, Ind. Eng. Chem. 17 (1925), 754-755

CHARGAFF, ERWIN [1905-]
MH 2,73-74; STC 1,240-241; MSE 1,196-197; AMS 14,793; BHDE 2,186; POTSCH, 84-85; LDGS,59; WWA 1978-1979, p.577; IWW 1982-1983, p.227; POGG 6,428; 7b,777-782
E. Chargaff, Ann. Rev. Biochem. 44 (1975), 1-18; TIBS 1 (1976), N171-N172; The Heraclitean Fire. New York 1978
P. Abir-Am, Hist. Phil. Life Sci. 2 (1980), 3-60

CHARLES, THOMAS WILLIAM CRANSTOUN [1849-1894]
O'CONNOR(2), 239; NUC 104,108
Anon., Brit. Med. J. 1894(I), p.275

CHARONNAT, RAYMOND [1894-1957]
R. Delaby and M. Lachaux, Bull. Soc. Chim. 1958, pp.1271-1278
R.C. Moreau, Ann. Pharm. Fran. 16 (1958), pp.261-284
G. Dillemann, Produits et Problèmes Pharmaceutiques 27 (1972), 1022-1023

CHARRIN, ALBERT [1857-1907]
DBF 8,661; FISCHER 1,238; GUIART, 222; WWWS, 323
E. Gley, C.R. Soc. Biol. 62 (1907), 926-928
H. Roger, Presse Med. 15 (1907), 329-331

CHASE, MERRILL WALLACE [1905-]
AMS 11,827; WWA 1978-1979, p.579
M.W. Chase, Ann. Rev. Immunol. 3 (1985), 1-29

CHASTAING, PAUL LOUIS [1847-1907]
GORIS, 292-294; TSCHIRCH, 1032; POGG 3,264; 4,242
L. Grimbert, J. Pharm. Chim. [6]26 (1907), 287-288

CHATIN, GASPARD ADOLPHE [1813-1901]
DBF 8,820; GORIS, 294-299; REBER, 384-385; VAPEREAU 6,324; TSCHIRCH,1032; WWWS, 324
L. Guignard, J. Pharm. Chim. [6]13 (1901), 151-160
G. Bonnier, C.R. Acad. Sci. 132 (1901), 105-110; Rev. Gen. Bot. 13 (1901), 97-108
J. Bornet, Bull. Soc. Bot. 48 (1901), 26-38
I. Greenwald, Science 111 (1950), 501-502

CHATIN, JOHANNÈS [1847-1912]
DBF 8,820-821; CHARLE, 73-76; WWWS, 324

CHATTAWAY, FREDERICK DANIEL [1860-1944]
POGG 4,242-243; 5,216-217; 6,429-430; 7b,785-786
G.D. Parkes, Chem. Ind. 1944, pp.113-114
G.R. Clemo, Obit. Not. Fell. Roy. Soc. 4 (1944), 713-716
J.J. Sudborough, J. Chem. Soc. 1944, pp.356-358

CHATTON, EDOUARD PIERRE LÉON [1883-1947]
CHARLE(2), 76-79; IB, 47; NUC 104,587-588
E. Chatton, Titres et Travaux Scientifiques. Sete 1938

CHAUFFARD, ANATOLE ÉMILE [1855-1932]
DBF 8,839-840; FISCHER 1,239; FRANKEN, 179; WWWS, 324
P. Legendre, Bull. Soc. Hist. Med. 26 (1932), 402-407
P. Ravaut, Bull. Acad. Med. 108 (1932), 1313-1321
G. Laroche, Bull. Acad. Med. 138 (1954), 572-581

CHAUTARD, JULES [1826-1901]
DBF 8,891; GORIS, 299-300; TSCHIRCH, 1032; POGG 3,264
R. Richebé, Jules Chautard. Paris 1903

CHAUVEAU, JEAN BAPTISTE AUGUSTE [1827-1917]
DSB 3,219-220; DBF 8,897-898; HIRSCH 1,897-898; Suppl.,171; BULLOCH, 358; SACKMANN, 66-69
Anon., Lancet 1917(I), pp.121-122
E. Gley, Journal de Physiologie 17 (1917), 1-11
H. Roger, Presse Med. 25Suppl. (1917), 9-11
G. Bouchardat, Bull. Acad. Med. 77 (1917), 56-61
G. Legée, Histoire et Nature 8 (1976), 53-73
H. Monod, Hist. Sci. Med. 20 (1986), 461-473

CHAUVOIS, LOUIS [1881-1972]
J. Théodoridès, Hist. Sci. Med. 6 (1972), 187-194
A.D. Wright, Med. Hist. 16 (1972), 407-408
P. Huard, Rev. Hist. Sci. 25 (1972), 278-282
T. Vetter, Clio Medica 8 (1973), 151-155

CHELDELIN, VERNON HENDRUM [1916-1966]
AMS 11,830; WWAH, 104
Anon., Chem. Eng. News 44(39) (1966), 100-101

CHELINTSEV, VLADIMIR VASILIEVICH [1877-1947]
BSE 47,91; WWR, 106; ARBUZOV, 169-170; POGG 5,1272; 6,2696
V.M. Rodionov and E.K. Nikitin, Zhur. Ob. Khim. 22 (1952), 1271-1278

CHEN, KO KUEI [1898-1988]
AMS 13,704; IB, 48; WWWS, 326; POGG 6,433; 7b,794-798
Anon., BioScience 16 (1966), 705,720
K.K. Chen, Ann. Rev. Pharm. 21 (1981), 1-6

CHENEVIX, RICHARD [1774-1830]
DSB 3,232-233; DNB 4,185-186; FERCHL, 93; COLE, 112-113; WWWS, 326; POGG 1,428-429
A.M. White and H.B. Friedman, J. Chem. Ed. 9 (1932), 236-245
D. Reilly, J. Chem. Ed. 32 (1955), 37-39

CHENOT, ADRIEN [1803-1855] DBF 8,992; FERCHL, 93; POGG 1,429,1549
J. Nicklès, Am. J. Sci. [2]21 (1856), 254-256

CHENOT, LUCIEN [1866-1951]
BUICAN, 81-135

CHEPURKOVSKI, IVAN VASILIEVICH [1861-]
SG [3]3,1037

CHERBULIEZ, ÉMILE [1891-1985]
SBA 4,35; GENEVA 7,32-34; 8,39-42; 9,26; 10,25-26; WWWS, 326-327; POGG 6,434; 7a(1),340-341
J. Rabinowitz and A.J.A. van der Wyk, Chimia 20 (1966), 1-2
J. Rabinowitz, Chimia 39 (1985), 367-368
E. Giovannini, Helv. Chim. Acta 69 (1986), 1-3

CHERINOV, MIKHAIL PETROVICH [1839-1905]
FISCHER 2,1587; ZMEEV 5,191

CHERNAYEV, ILYA ILYICH [1893-1966]
DSB 3,235-236; BSE(3rd Ed.) 29,111; WWR, 109; POTSCH, 82; POGG 6,2697
V.A. Golovna et al., Ambix 23 (1976), 187-198

CHESNEY, ALAN MASON [1888-1964]
AMS 10,650; FISCHER 1,240; DAMB, 132-133
T.B. Turner, Trans. Assoc. Am. Phys. 78 (1965), 17-20

CHETVERIKOV, SERGEI SERGEEVICH [1880-1959]
DSB 17,155-165

CHEVALIER, JOSEPH [1874-]
BK, 179-180
J. Chevalier, Titres et Travaux Scientifiques. Paris 1910

CHEVALLIER, ALPHONSE [1793-1879]
DSB 3,237-238; DBF 8,1076-1077; HIRSCH 2,9; GORIS, 300-317; BLOKH 1,120; TSCHIRCH, 1032; FERCHL, 93-94; CALLISEN 4,108-114; 27,81-83
A. Proust, Bull. Acad. Med. 8 (1879), 1219-1222
G. Sicard, Vie et Travaux de M. Chevallier. Paris 1880

G. Dillemann, Produits et Problèmes Pharmaceutiques 28 (1973), 39-40
A. Berman, Bull. Hist. Med. 52 (1978), 200-213

CHEVANDIER de VALDÔRME, EUGÈNE [1810-1878]
DBF 8,1081-1082; POGG 1,432; 3,367

CHEVREUL, MICHEL EUGÈNE [1786-1889]
DSB 3,240-244; DBF 8,1109-1110; GE 10,1168-1169; LAROUSSE 4,70; W, 106; FARBER,437-451; MAYERHOFER 1,631-632; BLOKH 1,120-122; FERCHL,93-94; POTSCH, 85; TSCHIRCH, 1032-1033; ABBOTT-C, 28-29; COLE, 113-114; POGG 1,432-434;3,267-268; 4,244; 6,346; CALLISEN 4,117-123; 27,84
G. Malloizel, Oeuvres Scientifiques de Michel Eugène Chevreul. Paris 1886
Celebration du Centenaire de M. Chevreul. Rouen 1886
A.W. Hofmann, Ber. chem. Ges. 22 (1889), 1163-1169
A. Bourgougnon, J. Am. Chem. Soc. 11 (1889), 71-79
C. Brongniart, Le Naturaliste [2]3 (1889), 89-91
C. von Voit, Sitz. Bayer Akad. 20 (1890), 418-427
M. Berthelot, Mem. Acad. Sci. 47 (1904), 388-413
H.E. Armstrong, Nature 116 (1925), 750-754
H. Metzger, Archeion 14 (1932), 6-11
G. Bouchard, Chevreul. Paris 1932
G. Sarton, Bull. Hist. Med. 8 (1940), 419-445
P. Lemay and R.E. Oesper, J. Chem. Ed. 25 (1948), 62-70
M. Javillier et al., Not. Acad. Sci. 3 (1957), 224-240
A.B. Costa, Michel Eugène Chevreul, Pioneer of Organic Chemistry. Madison, Wis. 1962
G. Kersaint, Bull. Soc. Chim. 1964, pp.1656-1665

CHEYMOL, JEAN [1896-1988]
J.E. Courtois, C.R. Soc. Biol. 182 (1988), 352-353
P. Lechat, Bull. Acad. Med. 173 (1989), 845-853

CHIARI, HANS [1851-1916]
OBL 1,142; DBJ 1,350; STURM 1,191; DBA 1,506; FISCHER 1,242; LOMMATZSCH, 59-79; KOERTING, 143-144
A. Weichselbaum, Alm. Akad. Wiss. Wien 66 (1916), 340-346

CHIARI, HERMANN [1897-1969]
KURSCHNER 10,329; WWWS, 329
M. Ratzenhofer, Verhandl. path. Ges. 54 (1970), 615-621
K. Meyer, Personalbibliographien von Professoren und Dozenten des Pathologisch-Anatomischen Institutes...der Universität Wien im ...1936-1969, pp.5-15. Erlangen-Nürnberg 1970
F.T. Brücke, Alm. Akad. Wiss. Wien 120 (1971), 315-330

CHIBNALL, ALBERT CHARLES [1894-1988]
WW 1981, p.472; IB, 48; POGG 6,436; 7b,805-806
A.C. Chibnall, Ann. Rev. Biochem. 35 (1966),1-22; TIBS 7(1982),191-192; My Early Days in Biochemistry. London 1987
S.V. Perry, The Biochemist 10(3) (1988), 4-8
E. Ashby, Cambridge Review 109 (1988), 90-93
R.L.M. Synge and E.F. Williams, Biog. Mem. Fell. Roy. Soc. 35 (1990), 57-96

CHICHIBABIN, ALEKSEI EVGENIEVICH [1871-1945]
DSB 3,246-247; BSE 47,416; WWR, 111; ARBUZOV, 156-157; POTSCH, 86-87; WWWS, 329; POGG 5,1273-1274; 6,2699-2700; 7b,825-826
M. Delépine and C. Barkovsky, Bull. Soc. Chim. 1946, pp.501-510
I. Marszak, J. Chem. Soc. 1946, pp.760-761
P.M. Evteeva, Trudy Inst. Ist. Est. 18 (1958), 296-356
E. Cerkovnikov, J. Chem. Ed. 38 (1961), 622-624

CHICK, HARRIETTE [1875-1977]
DSB 17,165-166; DNB [1971-1980], 142-143; WW 1976, p.433; O'CONNOR(2), 458-459; SRS, 35
Anon., Brit. Med. J. 1977(II), p.270
A.M. Copping, Brit. J. Nutrition 39 (1978), 3-4; Nutrition Historical Notes (Fall 1986), pp.1-6

CHILD, CHARLES MANNING [1869-1954]
DSB 3,247-248; DAB [Suppl.5], 109-111; AMS 8,425; IB, 48; ABBOTT, 30; WWAH, 106
L. Hyman, Science 121 (1955), 717-718; Biog. Mem. Nat. Acad. Sci. 30 (1957), 73-103

CHIOZZA, LUIGI [1828-1889]
DBI 25,39-41; PROVENZAL, 209-211; BLOKH 1,122; SCHAEDLER, 24
I. Guareschi, Atti Accad. Torino [2]58 (1905), 171-216

CHIRIEV, SERGEI IVANOVICH [1850-1915]
VORONTSOV, 108-111
D.G. Kvasov, Fiziol. Zhur. 52 (1966), 1379-1388

CHIRVINSKI, NIKOLAI PETROVICH [1848-1920]
BSE 47,383; KUZNETSOV(1963), 758-765
P. Tscherwinsky and W. Tscherwinsky, Zeitschrift für Tierzüchtung 6 (1926), 397-401
E.Y. Borisenko, Zhivotnovodstvo 1976(6), pp.88-91

CHISTOVICH, ALEKSEI NIKOLAEVICH [1905-1970]
WWWS, 331
N.I. Golshtein, Arkh. Pat. 17 (1955), 94-95

CHISTOVICH, FEDOR YAKOVLEVICH [1870-1942]
BSE 47,399
A.N. Metelkin, Zhur. Mikr. Ep. Imm. 47(8) (1972), 128-130

CHISTOVICH, NIKOLAI YAKOVLEVICH [1860-1926]
BSE(3rd Ed.) 29,218; BME 34,779-780
M.D. Tushinski, Klin. Med. 38(2) (1960), 135-137
V.A. Baier, Ter. Arkh. 33 (1961), 110-113
G.N. Chistovich, Zhur. Mikr. Ep. Imm. 32 (1961), 1357-1360

CHISTOVICH, YAKOV ALEKSEEVICH [1820-1885]
BME 34,781-782; ZMEEV 2,154-162; 3,74
Anon., Leopoldina 21 (1885), 214
P.D. Zikeev, Klin. Med. 49 (1971), 146-149

CHITTENDEN, RUSSELL HENRY [1856-1943]
DSB 3,256-258; DAB [Suppl.3], 162-164; FISCHER 1,244; POTSCH, 85; WWAH, 107; IB, 49; W, 106-107; DAMB, 134-135; WWWS, 331; POGG 3,268; 4,246
Anon., Pop. Sci. Mon. 53 (1898), 115-121; Nature 177 (1956), 311-312
Y. Henderson, Am. Phil. Soc. Year Book 1943, pp.373-379
H.B. Lewis, J. Biol. Chem. 153 (1944), 339-342
G.R. Cowgill, Science 99 (1944), 116-118; J. Nutrition 28 (1944), 2-6
H.B. Vickery, Biog. Mem. Nat. Acad. Sci. 24 (1945), 95-104
W.C. Rose, J. Chem. Ed. 46 (1969), 759-763

CHOAY, EUGÈNE [1861-1942]
GORIS, 317-318
G. Dupont, Bull. Soc. Chim. [5]9 (1942), 601
F. Debat, Figures Pharmaceutiques Françaises, pp.227-232. Paris 1953

CHODAT, ROBERT [1865-1934]
DSB 3,259; DHB 2,511; GENEVA 4,153-157; 5,62-70; 6,79-81; 7,35-42; 8,47-49;TSCHIRCH, 1033; IB, 49; WWWS, 331; POGG 6,439; 7b,813-814
E. Fischer, Bull. Soc. Bot. Gen. [2]25 (1932), 23-33
A. Lendner, Schw. Apoth. Z. 72 (1934); Verhandl. Schw. Nat. Ges. 115 (1934), 529-550
O. Jaag, Ber. bot. Ges. 52 (1934), (159)-(187)

CHODOUNSKY, FRANTISEK [1845-1924]
J. Satava, Chem. Listy 18 (1924), 314-315

CHODOUNSKY, KAREL [1843-1931]
OBL 1,146; FISCHER 1,245; PAGEL, 1941; NAVRATIL, 111-112
E. Babak, Cas. Lek. Cesk. 62 (1923), 529

CHOJNACKI, KASIMIR [-1904]
RSC 7,386
Anon., Chem. Z. 28 (1904), 1276

CHOLNOKY, LASZLO [1899-1967]
MEL 1,285; IB, 49
V. Bruckner, Acta Chim. Acad. Sci. Hung. 55 (1968), 129-136

CHOPRA, RAM NATH [1883-1973]
R.N. Chopra, Ann. Rev. Pharm. 5 (1965), 1-7
B. Mukerji, Pharmacologist 16 (1974), 33-36

CHOSSAT, CHARLES [1796-1875]
PICOT, 84-88; CALLISEN 4,144-145; POGG 1,411; 3,269
K. Kilgus, Charles Chossat. Zurich 1967
J.J. Dreifuss, Gesnerus 45 (1988), 239-262

CHOULANT, JOHANN LUDWIG [1791-1861]
ADB 4,139; HEIN, 104; HIRSCH 2,21-22; Suppl.,176; TSCHIRCH, 1033; FERCHL,95; WWWS, 332; POGG 1,441,1550; CALLISEN 4,146-152; 27,91-95
J. Grosse, Janus 6 (1901), 13-17,83-88
I.Propp, Nova Acta Leop. NF27 (1963), 327-338
R.J. ter Laage, in Essays in Biohistory (P. Smith and R.J. ter Laage, Eds.), pp.115-133. Utrecht 1970

CHRISTELLER, ERWIN [1889-1928]
FISCHER 1,246; FREUND 2,37-44; IB, 49
L. Pick, Verhandl. path. Ges. 24 (1929), 368-370
C. Kaiserling, Z. allgem. Path. 45 (1929), 65-66

CHRISTENSEN, ANDERS CHRISTIAN [1852-1923]
DBL 5,9; VEIBEL 2,93-97; TSCHIRCH, 1033; POGG 5,221; 6,441
Anon., Arch. Pharm. Chem. 31 (1924), 3-8

CHRISTENSEN, ERIK HOHWÜ [1904-]
VAD 1977, p,183; VEIBEL 2,98

CHRISTENSEN, HALVOR NIELS [1915-]
AMS 15(2),203; WWA 1980-1981, p.614; WWWS, 333

CHRISTENSEN, ODIN TIDEMAND [1851-1914]
DBL(3rd Ed.) 3,280-281; VEIBEL 1,220-221; 2,102-107; POGG 3,169; 4,247; 5,221

CHRISTIAN, HENRY ASBURY [1876-1951]
DAB (Suppl.5], 111-112; NACB 39,388-389; ICC, 79,82; IB, 49; WWAH, 107; WWWS, 333
J.V. Warren, J. Lab. Clin. Med. 112 (1988), 401-402

CHRISTIAN, WALTER [1907-1955]
Anon., Chem. Z. 79 (1955), 182; Nachr. Chem. Techn. 3 (1955), 43

CHRISTIANI, ARTHUR [1843-1887]
BLOKH 1,123; RSC 12,158; 14,216
Anon., Leopoldina 23 (1887), 218

CHRISTIANSEN, CHRISTIAN [1843-1917]
DSB 15,83-84; DBL 5,195-198; VEIBEL 2,108-109; WWWS, 334; POGG 3,269-270; 4,247; 5,221

CHRISTIANSEN, JENS ANTON [1888-1969]
KBB 1969, p.228; VEIBEL 2,109-112; WWWS, 334; POGG 5,222; 6,441; 7b,818-820
W. Jost, Ber. Bunsen Ges. 67 (1963), 619-620
R.P. Bell, Chem. Brit. 6 (1970), 491

CHRISTIANSEN, JOHANNE OSTENFELD [1882-1968]
KBB 1968, p.248; VEIBEL 2,113-114
P. Astrup, in Oxygen Affinity of Hemoglobin etc. (M. Rorth and P. Astrup, Eds.), pp.809-822. Copenhagen 1972

CHRISTIE, GEORGE HALLATT [1899-1965]
POGG 6,441-442
Who's Who in British Science 1953, p.56
G.E. Coates, Chem. Brit. 2 (1966), 24

CHRISTISON, ROBERT [1797-1882]
DNB 4,290-291; DESMOND, 129; BETTANY 2,286-291; TSCHIRCH, 1033; WWWS, 334
Anon., Nature 25 (1882), 339-340
J.H. Balfour et al., Trans. Bot. Soc. Edin. 14 (1883), 266-277
[R. Christison], Life of Sir Robert Christison (edited by his sons). Edinburgh 1885-1886

CHRISTMAN, CLARENCE CARL [1909-]
AMS 11,858

CHRISTOMANOS, ANASTASIOS [1841-1906]
BLOKH 1,123-124; POGG 4,248; 5,222
C. Zenghelis, Ber. chem. Ges. 39 (1906), 3788-3789

CHROBAK, RUDOLF [1843-1910]
OBL 1.147; STURM 1,200; FISCHER 1,247; FRITZ, 99-107; WWWS, 335
H. Peham, Wiener klin. Wchschr. 23 (1910), 1507-1510

CHROMETZKA, FRIEDRICH [1901-1943]
KURSCHNER 5,250-251; KIEL, 103

CHRZĄSZCZ, TADEUSZ [1877-1941]
POGG 6,442; 7b,820-821
Anon., Sprawozdanie Poznaniego Towarzystwa Przyjaciol Nauk 13 (1945-1946), 128-129

CHU, TSE-TSING TUNG-CHENG [1906-]
NUC 108,565

CHUARD, ERNEST LOUIS [1857-1942]
STRAHLMANN, 468,480; POGG 4,249; 6,442; 7a(1),343
Schweizetisches Zeitgenossen-Lexikon 2,185

CHUEVSKI, IVAN AFANASIEVICH [1858-1926]
VORONTSOV, 51-54

CHUGAEV, LEV ALEKSANDROVICH [1873-1922]
DSB 3,271-272; BSE 47,459-460; KUZNETSOV(1961), 556-563; ARBUZOV,163-166; POTSCH, 102-103; BLOKH 2,744-747; WWR, 114; WWWS, 335; POGG 4,1528-1529; 5,1274-1275; 6,2700-2701
M.A. Blokh, Zhur. Fiz. Khim. 60Suppl. (1928), 145-146
G.B. Kauffman, J. Chem. Ed. 40 (1963), 656-664

CHURCH, ARTHUR HERBERT [1834-1915]
DESMOND, 130; POGG 3,272; 4,249-250; 5,223
A.P. Laurie, J. Chem. Soc. 109 (1916), 374-379

CHURCHMAN, JOHN WOOLMAN [1877-1937]
AMS 5,200; IB, 50; WWAH, 109; WWWS, 336
Anon., Stain Technology 12 (1937), 175

CHURILOV, IVAN ALEKSANDROVICH [1864-]
RSC 14,433; SG [2]3,637

CHVOSTEK, FRANZ [1835-1884]
OBL 1,148; PETRY, 6-18; NAVRATIL, 114; KIRCHENBERGER, 24-25; WWWS, 336

CHVOSTEK, FRANZ [1864-1944]
NDB 3,253; OBL 1,148; FISCHER 1,248; PAGEL, 325; STUMPF, 168-173; KURSCHNER 4,397; WWWS, 336
L. Arzt, Wiener klin. Wchschr. 47 (1934), 1185
O. Gerke, Deutsche med. Wchschr. 70 (1944), 371-372
E. Risak, Wiener klin. Wchschr. 57 (1944), 235-237

CIAMICIAN, GIACOMO LUIGI [1857-1922]
DSB 3,279-280; DBI 25,118-122; PROVENZAL, 281-287; FARBER, 1087-1092; POTSCH, 86; BLOKH 1,124-125; POGG 4,250-252; 5,223-224,1418; 6,444-445

T.E. Thorpe, Nature 109 (1922), 245-246
W. McPherson, J. Am. Chem. Soc. 44 (1922), 101-106
G. Plancher, Gazz. Chim. Ital. 54 (1924), 1-22
C. Ravenna, Bioch. Ter. Sper. 11 (1924), 183-197
R. Nasini, J. Chem. Soc. 129 (1926), 996-1004
G. Provenzal, Rass. Clin. Terap. 37 (1938), 228-233
N.D. Heindel and M.A. Pfau, J. Chem. Ed. 42 (1965), 383-386

CINADER, BERNHARD [1919-]
AMS 15(2),222; WWA 1980-1981, p.622; BHDE 2,188; WWWS, 337

CIOCILTEU (CIOCALTEAU), VINTILA [1890-1947]
ILEA, 148-149

CISZKIEWICZOWA, TERESA [1848-1921]
PSB 4,86-87; WE 2,568; KOSMINSKI, 592-593
S. Bronowska, Gazeta Lekarska 55(5) (1921), 64

CIUSA, RICCARDO [1877-1965]
WWWS, 337; POGG 5,226; 6,447-448; 7b,834-835
L. Musajo, Chimica e Industria 47 (1965), 1249-1250

CIVIALE, JEAN [1792-1867]
DBF 8,1329-1330; HIRSCH 2,34-35; Suppl.,178; WWWS, 337
P. Huard, Prog. Med. 20 (1967), 1-8; Episteme 1 (1967), 239-254

CLAESSON, STIG MELKER [1917-1988]
VAD 1983, p.187; WWWS, 338

CLAISEN, LUDWIG [1851-1930]
DSB 3,286; NDB 3,257-258; KIEL, 153; W, 107; WWWS, 338; POTSCH, 87; TLK 3,88; DZ, 227; POGG 3,274-275; 4,253; 5,227; 6,448
R. Anschütz, Ber. chem. Ges. 69A (1936), 97-170
E.H. Huntress, Proc. Am. Arts Sci. 79 (1951), 10-11

CLAPARÈDE, RENÉ ÉDOUARD [1832-1871]
DHB 2,525; HIRSCH 2,36; WWWS, 338
A.G., Am. J. Sci. [3]2 (1871), 229-230
H. de Saussure, Arch. Sci. Phys. Nat. 42 (1871), 51-79; Mem. Soc. Phys. Hist. Nat. 22 (1873), i-xxviii
C. Vogt, Journal de Zoologie 2 (1873), 138-159

CLAPP, SAMUEL HOPKINS [1876-1952]
Yale University Doctors of Philosophy 1861-1960, p.37. New Haven 1961

CLAR, ERIC (ERICH JULIUS) [1902-1987]
POTSCH, 87-88; WWWS, 338-339; POGG 6,449; 7a(1),347
Anon., Chem. Z. 111 (1987), 251

CLARA, MAX [1899-1966]
KURSCHNER 10,333-334; BUSCHHUTER, 31-50; EGERER, 165-167; WWWS, 339
H. Ferner, Anat. Anz. 121 (1967), 220-230

CLARIDGE, PETER ROBERT PERCIVAL [1911-]
Who's Who of British Scientists 1971-1972, p.161

CLARK, ALFRED JOSEPH [1885-1941]
FISCHER 1,251; IB, 50; WWWS, 339
E.B. Verney and J. Barcroft, Obit. Not. Fell. Roy Soc. 3 (1941),969-984
Anon., Quarterly Journal of Experimental Pharmacology 31 (1941),99-101; Edinburgh Med. J. NS48 (1941), 699-702
J.H. Burn, J. Pharm. Exp. Ther. 75 (1942), 187-190
J. Parascandola, Trends in Pharmacological Sciences 3 (1982), 421-423

CLARK, EARL PERRY [1892-1943]
AMS 6,254
Anon., Chem. Eng. News 21 (1943), 2056
C.A. Browne, J. Assn. Agr. Chem. 27(2) (1944), x-xii

CLARK, GUY WENDELL [1877-]
AMS 10,679; IB, 51

CLARK, JANET HOWELL [1889-1969]
AMS 11,878
E. Fee and A.C. Rodman, Physiologist 28 (1985), 397-400

CLARK, LELAND CHARLES, Jr. [1918-]
AMS 15(2),246; WWWS, 340
J.W. Severinghaus and P.B. Astrup, Journal of Clinical Monitoring 2 (1986), 125-139

CLARK, ROBERT HARVEY [1880-1961]
AMS 10,683; SRS, 37; NUC 111,104
J.A.F. Gardner, Proc. Roy. Soc. Canada [3]56 (1962), 173-174

CLARK, THOMAS [1801-1867]
DSB 3,289-290; DNB 4,407-408; FERCHL, 97; POTSCH, 88; WWWS, 341; POGG 3,275
W. De la Rue, J. Chem. Soc. 21 (1868), viii-xvi
A. Bain, Transactions of the Aberdeen Philosophical Soc. 1 (1884),101-115
J.H.S. Green, Ann. Sci. 13 (1957), 164-179
R.A. Chalmers, Anal. Proc. 17 (1980), 463-466

CLARK, WILLIAM MANSFIELD [1884-1964]
DSB 3,290; MH 1,103-104; STC 1,244-245; MSE 1,205; MILES, 79-80; IB, 51; GOULD, 32-33; BARRY, 13-16; AMS 10,686; NCAB F,536-537; 52,425; WWAH, 111; WWWS, 341; POGG 6,451-452; 7b,842
W.M. Clark, Ann. Rev. Biochem. 31 (1962), 1-24
W.B. Wood, J. Bact. 87 (1964), 751-754
A.B. Hastings, Am. Phil. Soc. Year Book 1965, 116-120
H.B. Vickery, Biog. Mem. Nat. Acad. Sci. 39 (1967), 1-26
A.M. Harvey, Johns Hopkins Med. J. 139 (1976), 257-263

CLARK, WILLIAM SMITH [1826-1886]
MILES, 80-81
Anon., Proc. Am. Acad. Arts Sci. 21 (1886), 520-523
D.P. Penhallow, Science 27 (1908), 172-180

CLARK-LEWIS, JOHN WILLIAM [1920-]
Who's Who of British Scientists 1971-1972, p.165
Who's Who in Australia 1992, p.284

CLARKE, CHARLES HUGH [1888-1967]
SRS, 45
The Times, 17 November 1967, p.10h

CLARKE, DONALD ALSTON [1915-1987]
AMS 15(2),241; WWWS, 341

CLARKE, FRANK WIGGLESWORTH [1847-1931]
DSB 3,292-294; DAB [Suppl.1], 177-178; MILES, 82-83; POTSCH, 88; POGG 3,277-278; 4,254; 5,277; 6,452-453
L.M. Dennis and L.K. Williams, Biog. Mem. Nat. Acad. Sci. 15 (1934), 139-165
M. Gorman, J. Chem. Ed. 62 (1985), 233-235

CLARKE, HANS THACHER [1887-1972]
AMS 11,886; NCAB 57,53-54; SRS, 46; WWAH, 111; POGG 6,453; 7b,843-844
H.T. Clarke, The Chemist 31 (1954), 353-359; Ann. Rev. Biochem. 27 (1958), 1-14
D. Shemin, Am. Phil. Soc. Year Book 1974, pp.134-137
H.B. Vickery, Biog. Mem. Nat. Acad. Sci. 46 (1975), 3-20

CLASSEN, ALEXANDER [1843-1934]
DSB 15,84; NDB 3,265; SCHAEDLER, 34; DZ, 228; TLK 3,88; POTSCH, 88; WWWS, 342; POGG 3,278; 4,255; 5,227; 6,453
R. Fresenius, Chem. Z. 37 (1913), 449-450
H. Fischer, Z. angew. Chem. 47 (1934), 129-130

CLAUDE, ALBERT [1899-1983]
BAF, 18; AO 1983, pp.353-359; IWW 1982-1983, p.248; WWA 1978-1979, p.613; MSE 1,205-206
G.E. Palade and C. deDuve, J. Cell. Biol. 50 (1971), 5D-55D
M. Florkin, Arch. Inter. Physiol. 80 (1972), 632-647
C. de Duve and G. Palade, Nature 304 (1983), 588

C. de Duve, Cellule 74 (1987), 11-19
J. Brachet, Ann. Acad. Roy. Belg. 165 (1988), 93-135

CLAUDE, HENRI [1869-1945]
DBF 8,1377-1378; FISCHER 1,252-253
J. Lhermite, Bull. Acad. Med. 130 (1946), 63-66

CLAUS, ADOLF [1838-1900]
DSB 3,299-301; NDB 3,268; BJN 5,85*; 9,348-349; BAD 5,101-103; WWWS, 342; POTSCH, 89; BLOKH 1,125-126; POGG 3,279-280; 4,256-259
G.N. Vis, J. prakt. Chem. 62 (1900), 127-133

CLAUS, CARL ERNST (KARL KARLOVICH) [1796-1864]
DSB 3,301-302; NDB 3,269-270; ADB 4,284; BSE 21,394-395; WELDING,146-147; HEIN, 106-107; LEVITSKI 2,235-239; LIPSHITS 4,183-184; WWWS,342-343; POTSCH, 89; BLOKH 1,363-364; KAZAN 1,353-354; POGG 1,452;3,278-279
M.E. Weeks, J. Chem. Ed. 9 (1932), 1017-1034
M. Schofield, Pharm. J. 157 (1946), 135
S. Borström, Baltische Hefte 12 (1967), 298-303
N.N. Ushakova, Karl Karlovich Klaus. Moscow 1972

CLAUS, HANS [1873-1938]
FISCHER 1,253; SKVARC, 37-38
K. Brandenburg, Med. Klin. 34 (1938), 1019

CLAUS, RICHARD [1875-1933]
Anon., Chemische Fabrik 6 (1933), 458

CLAUSEN, JENS CHRISTIAN [1891-1969]
DSB 17,168-170; AMS 11,889; IB, 51

CLAUSEN, ROY ELWOOD [1891-1956]
AMS 9(II),201; NCAB 43, 352-353; WWAH, 112
J.A. Jenkins, Science 124 (1956), 1286; Biog. Mem. Nat. Acad. Sci. 39 (1967), 37-54

CLAUSIUS, RUDOLF [1822-1888]
DSB 3,303-311; NDB 3,276-278; ADB 55,720-729; MATSCHOSS, 42-43; POTSCH, 89; MAYERHOFER 1,664-665; BLOKH 1,126-127; W, 108-109; POGG 1,454-456; 3,281-282; 4,258; 6,454L 7aSuppl., 137-139
J.W. Gibbs, Proc. Am. Acad. Arts Sci. 16 (1889), 458-465
F. Folie, Rev. Quest. Sci. 27 (1890), 419-487
E.F.F.G., Proc. Roy. Soc. 48 (1890), i-viii
F. Krüger, Pommersche Lebensbilder 1 (1934), 208-211
C. Ronge, Gesnerus 12 (1955), 73-108
W. Ebeling and J. Orphal, Wiss. Z. Berlin 39 (1990), 210-223

CLAUTRIAU, GEORGES [1863-1900]
L. Errera, Ann. Soc. Belg. Micr. 27 (1901), 17-38

CLAY, REGINALD STANLEY [1868-1954]
POGG 6,455; 7b,849-850
W.L. Daughty, Nature 174 (1954), 16-17
L.C. Martin, Proc. Phys. Soc. A67 (1954), 1124-1125; B67 (1954),912-913
Who Was Who 1951-1960, pp.215-216

CLELAND, RALPH ERSKINE [1892-1971]
DSB 17,170-171; MH 1,106-108; MSE 1,208-209; AMS 11,894; IB, 51; WWAH, 113; WWWS, 334
T. Sonneborn, Am. Phil. Soc. Year Book 1971, pp.120-124
F.K. Daily, Proc. Indiana Acad. Sci. 81 (1971), 29-32
W. Stubbe, Ber. bot. Ges. 89 (1976), 91-96
E. Steiner, Biog. Mem. Nat. Acad. Sci. 53 (1982), 121-139

CLELAND, WILLIAM WALLACE [1930-]
AMS 15(2),251

CLÉMENT, NICOLAS [1779-1841]
DSB 3,315-317; DBF 8,1451; GE 11,640; LAROUSSE 4,426; MAYERHOFER 1,667;POTSCH, 90; BLOKH 1,127-128; CALLISEN 4,204-205; POGG 1,455-456
P. Lemay, Chymia 2 (1949), 45-49

CLEMENTI, ANTONINO [1888-1968]
DBI 26,376-378; POGG 6,456; 7b,850-852
G. Ricceri, Enzymologia 38 (1970), 1-2; Arch. Fisiol. 68 (1971), 113-130
E. Fadiga, Arch. Ital. Biol. 108 (1970), 207-212

CLEMENTS, FREDERIC EDWARD [1874-1945]
DSB 3,317-318; DAB [Suppl.3],168-170; NCAB 34,266-267; WWAH,113; WWWS,345
A.G. Tansley, Journal of Ecology 34 (1947), 194-196

CLEMM, AUGUST [1837-1910]
NDB 3,285-286; BJN 15,19*; WWWS, 345
Anon., Z. angew. Chem. 24(I) (1911), 78-80

CLEMM, CARL FRIEDRICH [1836-1899]
NDB 3,286-287; BJN 4,134*
Anon., Ber. chem. Ges. 32 (1899), 429; Chem. Z. 23 (1899), 159; Z. angew. Chem. 37 (1904), 1349

CLEMM, GUSTAV [1814-1866]
NDB 3,287

CLEMM, HANS [1872-1927]
NDB 3,286; WWWS, 345
V. Hottenroth, Chem. Z. 51 (1927), 912
H. Müller-Clemm, Tradition 6 (1961), 22-28

CLEMM-LENNIG, CARL [1817-1887]
NDB 3,287; HEIN-E, 67-68; RSC 1,952

CLEMMENSEN, ERIK CHRISTIAN [1876-1941]
DBL 16,273-274; VEIBEL 2,117-118; SURREY, 37
Anon., Ind. Eng. Chem. News Ed. 19 (1941), 694

CLEMO, GEORGE ROGER [1899-1983]
WW 1981, p.503; CAMPBELL, 59-60; POGG 6,456; 7b,852-853
G.A. Swan, Chem. Brit. 20 (1984), 918
B. Lythgoe and G.A. Swan, Biog. Mem. Fell. Roy. Soc. 31 (1985), 67-86

CLEVE, PER TEODOR [1840-1905]
DSB 3,321-322; SBL 8,636-649; SMK 2,111-112; BLOKH 1,129; WWWS,345-346; POTSCH, 90; POGG 3,283-284; 4,259
H. Euler and A. Euler, Ber. chem. Ges. 38 (1905), 4221-4238
T.E. Thorpe, J. Chem. Soc. 89 (1906), 1301-1317

CLEVER, AUGUST [1869-]
JV 11,232; RSC 14,270; NUC 112,522
Anon., Z. angew. Chem. 32 (1919), 87

CLIFTON, CHARLES EGOLF [1904-]
AMS 11,900; WWWS, 346

CLOETTA, ARNOLD [1828-1890]
HIRSCH 2,52; Suppl.,182; PAGEL, 332; ZURICH-D, 40; POGG 3,285
Anon., Corr. Schw. Aerzte 20 (1890), 297-299

CLOETTA, MAX [1868-1940]
NDB 3,293-294; FISCHER 1,255; RAZINGER, 35-38; ZURICH-D, 85; TLK 3,89; IB, 52; WWWS, 346
P. Wolfer, Viert. Nat. Ges. Zurich 85 (1940), 365-372
H. Fischer, Schw. med. Wchschr. 70 (1940), 749-752; Verhandl. Schw. Nat. Ges. 121 (1940), 423-432; Erg. Physiol. 44 (1941), 18-26; Schw. Med. Jahrb. 1941, pp.xxvii-xxxiii; Arch. exp. Path. Pharm. 200 (1942), 1-5

CLOËZ, FRANCOIS STANISLAS [1817-1883]
DBF 9,22; GORIS, 318-321; BLOKH 1,129-130; FERCHL,98-99; SCHAEDLER,24-25; POGG 3,285
L. Figuier, Ann. Scient. 27 (1883), 492
E. Grimaux. Bull. Soc. Chim. [2]41 (1884), 146-157

CLOWES, GEORGE HENRY ALEXANDER [1877-1958]
AMS 9(I),349; NCAB 46,12-13; G,94; WWAH, 114; WWWS, 347
Anon., Chem. Eng. News 36(36) (1958), 96
P.B. Armstrong, Biol. Bull. 119 (1960), 10-11

CLUSIUS, KLAUS [1903-1963]
POTSCH, 90-91; POGG 7(1),351-353
W. Jost, Z. Naturforsch. 18a (1963), 1240-1241
G.M. Schwab, Bayer. Akad. Wiss. Jahrbuch 1963, pp.225-229
K. Schleich, Helv. Chim. Acta 47 (1964), 234-246

CLUTTERBUCK, PERCIVAL WALTER [1897-1938]
H.R., Biochem. J. 32 (1938), 435-436

COAN, TITUS MUNSON [1801-1882]
DAB 4,236-237; NCAB 11,273; ELLIOTT, 57
L.B. Coan, Titus Coan. Chicago 1884

COATNEY, GEORGE ROBERT [1902-]
AMS 12,1076; WWA 1976-1977, p.600; WWWS, 348

COBLENTZ, VIRGIL [1862-1932]
AMS 4,188; WWAH, 115; WWWS, 348-349
Anon., Science 75 (1932), 630; Ind. Eng. Chem. News Ed. 10 (1932), 158

COBLINER, JESAIAH [1878-]
JV 18,321; NUC 113,393

COBLITZ, FRANZ PETER [1868-]
JV 11,232; RSC 14,277; NUC 113,393

COBURN, ALVIN FREDERICK [1899-1975]
AMS 12,1079
G.H. Stollerman, Journal of Infectious Diseases 133 (1976), 595

COCCIUS, ERNST ADOLF [1825-1890]
ADB 47,502; HIRSCH 2,60; Suppl.,184; PAGEL, 334
Anon., Leopoldina 26 (1890), 216; Ann. Ocul. 104 (1890), 273-274

COCHIN, DENYS [1851-1922]
DBF 9,80-81; GE 11,770
H.P. Thieme, Bibliographie de la Littérature Française 1 (1953), 459-460

COCHRAN, WILLIAM [1922-]
WW 1981, p.512; WWWS, 349

COCKER, WESLEY [1908-]
WWWS, 349
Who's Who of British Scientists 1971-1972, p.171

CODMAN, ERNEST AMORY [1869-1940]
AMS 6,267; NCAB 30,66-67; DAMB, 144; OEHRI, 30-31
F.D. Moore, Harvard Med. Alumni Bull. 49(3) (1975), 12-21
S. Reverby, Bull. Hist. Med. 55 (1981), 156-171

COE, WESLEY ROSWELL [1869-1960]
AMS 10,709-710; NCAB 47,648-649; WWAH, 116; IB, 52

COEHN, ALFRED [1863-1938] WININGER 7,544-545; TLK 3,90;
POGG 4,261; 5,230; 6,458-459; 7a(1),354
G. Jung, Z. Elektrochem. 35 (1929), 1-2

COGHILL, ROBERT DeWOLF [1901-]
AMS 12,1087; WWWS, 351

COHEN, AARON ARTHUR [1915-]
AMS 10,712

COHEN, BARNETT [1891-1952]
AMS 8,463; NCAB 42,290-291; BARRY, 17-18; KAGAN(JA), 348;
KOREN, 162;IB, 52; WWAH, 117; POGG 6,459-460; 7b,861
W.M. Clark, Bact. Revs. 16 (1952), 205-209

COHEN, ERNST JULIUS [1869-1944]
DSB 3,333-334; BWN 1,114-115; WININGER 6,520; WWWS,352;
POTSCH, 91; POGG 4,262-263; 5,230-232; 6,460-461; 7b,861-863
Anon., Chem. Wkbl. 36 (1939), 515-522
W.R. Kruyt, Chem. Wkbl. 41 (1945), 126-129
F.G. Donnan, J. Chem. Soc. 1947, pp.1700-1706; Obit. Not. Fell. Roy. Soc. 5 (1948), 667-687
C.A. Browne, J. Chem. Ed. 25 (1948), 302-307
A.L.T. Moesveld, J. Chem. Ed. 25 (1948), 308-314

COHEN, HERMANN [1874-]
JV 13,224

COHEN, JACOB ANTONIE [1915-1969]
LINDEBOOM, 356-357
E.M. Cohen, Chem. Wkbl. 65(30) (1969), 38
A. Querido, Jaarb. Akad Wet. 1969-1970, pp.261-264

COHEN, JAKOB [1832-]
RSC 2,8

COHEN, JULIUS BEREND [1859-1935]
WININGER 1,569; WWWS, 352; POGG 4,263; 5,232-233; 6,461;
7b,863
H.S. Raper, J. Chem. Soc. 1935, pp.1331-1337; Obit. Not. Fell. Roy. Soc. 1 (1935), 503-513

COHEN, LOUIS ARTHUR [1926-]
AMS 18(2),317

COHEN, MORRIS RAPHAEL [1880-1947]
DSB 3,335; DAB [Suppl.4], 168-170; NCAB 40,134-135
M.R. Cohen, A Dreamer's Journey. New York 1949
L.C. Rosenfeld, Portrait of a Philosopher. New York 1962

COHEN, PHILIP PACY [1908-1993]
AMS 14,901; WWA 1978-1979, p.639; IWW 1982-1983, p.254;
COHEN, 42; WWWS, 352

COHEN, SEYMOUR STANLEY [1917-]
MH 2,79-81; STC 1,251-253; MSE 1,214-215; AMS 15(2),287;
COHEN, 43; WWA 1980-1981, p.664; IWW 1982-1983, p.255;
WWWS, 353

COHEN, STANLEY [1922-]
AMS 15(2),288; WWA 1980-1981, p.665; POTSCH, 91

COHN, ALFRED [1890-]
LDGS, 62; NUC 114,305

COHN, ALFRED EINSTEIN [1879-1957]
DAB [Suppl.6], 117-118; FISCHER 1,258; DAMB, 148; IB, 52;
KOREN, 163; WWAH, 117
J.M. Steele, Trans. Assoc. Am. Phys. 71 (1958), 17-19

COHN, EDWIN JOSEPH [1892-1953]
DSB 3,335-336; DAB [Suppl.5], 121-123; AMS 8,465; KAGAN(JA),
350,739-740; POTSCH, 91; DAMB, 148-149; KOREN, 163; POGG
6,462; 7b,863-866
A.B. Hastings, Am. Phil. Soc. Year Book 1953, pp.336-340
J.T. Edsall, University Laboratory of Physical Chemistry Related to Medicine and Public Health Harvard University. Cambridge, MA 1950;
Erg. Physiol.
48(1955),23-48; Am. Scient. 38(1955),580-593; Biog. Mem. Nat. Acad. Sci. 35 (1961), 47-84; TIBS 6 (1981), 335-337
G. Scatchard, Vox Sanguinis 17 (1969), 37-44
L.K. Diamond, Vox Sanguinis 20 (1971), 433-440

COHN, FELIX [1869-1942?]
GEDENKBUCH 1,216; RSC 14,294; SG [2]3,727; NUC 114,323

COHN, FERDINAND [1828-1898]
DSB 3,336-341; NDB 3,313-314; ADB 47,503-505; BJN 3,284-296; PAGEL, 335; BERLIN, 647; SACKMANN, 73-76; BULLOCH,358-359; MAYERHOFER 1,670-671; W, 110-111; DOETSCH, 25-27; KOHUT 2,228; WWWS, 353
M. Neisser, Münch. med. Wchschr. 45 (1898), 1005-1007
C. Mez, Pharm. Z. 43 (1898), 456-457
P. Cohn, Ferdinand Cohn, Blätter der Erinnerung 2nd Ed. Breslau 1901
F. Rosen, Ber. bot. Ges. 17 (1899), (172)-(201); Schlesische Lebensbilder 1, 167-173
C.S. Dolley, Bull. Hist. Med. 7 (1939), 49-92
E.G. Pringsheim, Med. Mon. 25 (1971), 118-121
B. Hoppe, Sudhoffs Arch. 67 (1981), 158-189

COHN, FRANK [1898-]
SG [4]3,729

COHN, GEORG [1868-1942]
GEDENKBUCH 1,217; TLK 3,89-90; POGG 4,264-265; 5,234; 6,462

COHN, HERMANN [1838-1906]
WININGER 1,580; POGG 3,288-289; 4,263-264
E. Roth, Leopoldina 42 (1906), 136-144

COHN, LASSAR [1858-1922]
DSB 3,341-342; NDB 3,316-317; DBJ 4,351; WININGER 1,581; TETZLAFF, 52; TLK 2,411; WWWS, 354; POGG 4,842; 5,710-711; 6,462-463
Anon., Chem. Z. 46 (1922), 939

COHN, MARTHA [1887-]
JV 29,547; NUC 114,344

COHN, MELVIN [1922-]
AMS 15(2),288; WWA 1980-1981, p.665

COHN, MILDRED [1913-]
AMS 15(2),290; WWA 1980-1981, p.666; IWW 1982-1983, p.255

COHN, RUDOLF [1862-]
AD 3,61; WININGER 1,583-584; KOREN, 163; POGG 4,264; 6,463
Anon., Chem. Z. 56 (1932), 345

COHN, WALDO E. [1910-]
MH 2,81-82; MSE 1,217-218; AMS 14,904; WWWS, 354

COHNHEIM, JULIUS [1839-1884]
ADB 55,729-733; HIRSCH 2,66-67; Suppl.,185-186; DIEPGEN, 50-55; KIEL, 80; PAGEL, 338-340; SHBL 1,114-117; W, 111; WWWS, 354; BULLOCH, 359; WININGER 1,585-586; KOHUT 2,274; KOREN, 164
E. Klebs, Arch. exp. Path. Pharm. 18 (1884), i-x
F. Marchand, Deutsche med. Wchschr. 10 (1884), 577-579,596-599
C. Weigert, Berl. klin. Wchschr. 21 (1884), 564-565
R.C. Maulitz, Bull. Hist. Med. 52 (1978), 162-182
B. Wohlgemuth and G. Borte, Z. ärtz. Fortbild. 83 (1989), 743-745

COHNHEIM (later KESTNER), OTTO [1873-1953]
NDB 11,555-556; FISCHER 1,755; DRULL, 39; BHDE 2,618; LDGS, 79; IB, 150; BK, 182; TLK 3,325; WININGER 7,168; TETZLAFF, 174; KOREN, 201; POGG 6,1309-1310; 7a(2),733
D.M. Matthews, Brit. Med. J. 1978(II), pp.618-619

COINDET, JEAN CHARLES WALCKER [1796-1876]
DHB 2,535; GENEVA 4,50-52; CALLISEN 4,249; 27,120
Anon., Bull. Soc. Med. Suisse Rom. 11 (1877), 6-12

COINDET, JEAN FRANÇOIS [1774-1834]
DHB 2,534-535; HIRSCH 2,67-68; HOEFER 11,83; BOSSART, 113; WWWS, 354; DECHAMBRE 18,711; CALLISEN 4,251; 27,120
Anon., Am. J. Sci. 27 (1835), 404-405
B. Reber, Aesculape 3 (1913), 93-96
J.D. Wiener, Schw. med. Wchschr. 110 (1980), 1784-1787

COIRRE, JEAN [1884-1948]
NUC 114,393
Dictionnaire National des Contemporains 1936, p.172
Anon., Paris Médical 38 (1948), 342

COKER, ERNEST GEORGE [1869-1946]
DNB [1941-1950], 165-166; WWWS, 354
H.T. Jessup, Obit. Not. Fell. Roy. Soc. 8 (1952), 389-393

COLASANTI, GUISEPPE [1846-1903]
DBI 26,704-706; FISCHER 1,259
L. Luciani, Arch. Ital. Biol. 39 (1903), 493-500

COLDING, LUDWIG AUGUST [1815-1888]
DSB 15,84-87; MAYERHOFER 1,672; POGG 1,461-462; 3,289
P.F. Dahl, Centaurus 8 (1963), 174-188

COLE, KENNETH STEWART [1900-1984]
MH 2,83-85; STC 1,254-256; MSE 1,218-220; AMS 14,908; WWWS, 355;
AO 1984, pp.206-208; WWA 1978-1979, p.646; IWW 1982-1983, p.257
K.S. Cole, Ann. Rev. Physiol. 41 (1979), 1-24
New York Times, 20 April 1984

COLE, LEON JACOB [1877-1948]
DSB 17,340; AMS 7,340; WWWS, 355
G.E. Dickerson and A,B, Chapman, Journal of Animal Science 67 (1989),1653-1656

COLE, ROGER DAVID [1924-]
AMS 15(2),296-297; WWA 1980-1981, p.671; WWWS, 356

COLE, RUFUS [1872-1966]
AMS 11,932; FISCHER 1,259-260; DAMB, 151-152; ICC, 47-49
W.S. Tillett, Trans. Assoc. Am. Phys. 30 (1967), 9-10
C.P. Miller, Biog. Mem. Nat. Acad. Sci. 50 (1979), 119-139

COLE, SYDNEY WILLIAM [1877-1951]
O'CONNOR(2), 37; IB, 52
J.A. Venn, Alumni Cantabrigienses II 2,90. Cambridge 1944

COLEBROOK, LEONARD [1883-1967]
DNB [1961-1970], 230-232; WW 1965, p.618
C.L. Oakley, Biog. Mem. Fell. Roy. Soc. 17 (1971), 91-138
W.C. Noble, Coli, Great Healer of Men. The Biography of Dr. Leonard Colebrook. London 1974
E.J.L. Lowbury, Brit. Med. J. 287 (1983), 1981-1983

COLEMAN, JOSEPH EMORY [1930-]
AMS 18(2),335

COLEMAN, WARREN [1869-1948]
AMS 7,342; WWAH, 119

COLEMAN, WILLIAM [1934-1988]
O. Temkin, Am. Phil. Soc. Year Book 1989, pp.187-190

COLIN, HENRI ERNEST [1880-1943]
DBF 9,234; WWWS, 356-357; POGG 5,235; 6,465; 7b,867-869
H. Belval, Bull. Soc. Chim. Biol. 25 (1943), 431-433; Bull. Soc. Chim. [5]10 (1943), 493-502
A. Chevalier, Not. Acad. Sci. 2 (1949), 383-389

COLIN, JEAN JACQUES [1784-1865]
DBF 9,237-238; GE 11,918; LAROUSSE 4,588; TSCHIRCH, 1034; FERCHL, 100; WWWS, 357; COLE, 117-118; POGG 3, 289-290; CALLISEN 4,258-260

COLLADON, JEAN ANTOINE [1755-1830]
DHB 2,543; FERCHL, 100; TSCHIRCH, 1034
Anon., Verhandl. Schw. Nat. Ges. 1830, pp.108-109
R. Hahn, Rev. Hist. Sci. 12 (1959), 55-56
G. de Morsier and M. Cramer, Gesnerus 16 (1959), 113-123
R.M. Tecoz, Journal de Génétique Humaine 8 (1959), 208-217
G. de Morsier, Physis 7 (1965), 489-516
A.F. Corcos, J. Heredity 59 (1968), 373-374

COLLADON, THÉODORE [1792-1862]
FERCHL, 100; TSCHIRCH, 1034
Anon., Verhandl. Schw. Nat. Ges. 46 (1862), 278-279
J. Rostand, Rev. Hist. Sci. 8 (1955), 170-173; 10 (1957), 175-176

COLLANDER, PAUL RUNAR [1894-1973]
IB, 53; POGG 6,465-466; 7b,869-870
Vem Och Vad 1970, p.75
H. Borris, Wiss. Z. Greifswald 11 (1961), 1-5

COLLARD de MARTIGNY, C. P. [-1851]
RSC 2,18-19; CALLISEN 4,263-264
E. Simonin, Mem. Acad. Stanislas [3]19 (1853), x

COLLETT, MARY ELIZABETH [1888-1969]
AMS 10,729; IB, 53; NUC 115,444
Anon., Science 167 (1970), 1600

COLLIE, JOHN NORMAN [1859-1942]
DSB 3,347-348; DNB [1941-1950], 167-168; TSCHIRCH, 1034; POTSCH, 92; POGG 4,267-268,1703; 5,235; 6,467; 7b,871
E.C.C. Baly, Nature 150 (1942), 655-656; Obit. Not. Fell. Roy. Soc. 4 (1943), 329-356
R.W. Clark, Six Great Mountaneers, pp.87-116. London 1956
C. Mill, Norman Collie: A Life in Two Worlds. Aberdeen 1987

COLLIER, HERBERT BRUCE [1905-]
AMS 14,916; YOUNG, 34; WWWS, 357

COLLIER, LESLIE [1921-]
WW 1983, p.458

COLLIER, PETER [1835-1896]
DAB 4,304; MILES, 88; NCAB 8,356; ELLIOTT, 59; WWAH, 120
A.B. Prescott, Am. J. Sci. [4]2(1896),246; J. Am. Chem. Soc. 18(1896),748
Obituary Notices of Graduates of Yale University 1890-1900, pp.468-469. New Haven 1900

COLLIN, EUGÈNE [1845-1919]
GORIS, 322-235; TSCHIRCH, 1034; SCHELENZ, 716
E. Perrot, Bull. Sci. Pharm. 28 (1920), 98-104

COLLIP, JAMES BERTRAM [1892-1965]
DSB 3,351-354; STC 1,257-258; AMS 11,944; YOUNG, 16-18; IB, 53
D.L. Thomson, Can. J. Biochem. 35 (1957), 1-5
R.L. Noble, Can. Med. Assoc. J. 93 (1965), 1356-1364
R.J. Rossiter, Proc. Roy. Soc. Canada [4]4 (1966), 73-82
J.S. Browne et al., Endocrinology 79 (1966), 225-228
M.L. Barr and R.J. Rossiter, Biog. Mem. Fell. Roy. Soc. 19(1973), 235-267
M.L. Barr, Hannah Inst. Hist. Sci. 1 (1977), 6-15
M. Bliss, Bull. Hist. Med. 56 (1982), 554-568

COLMAN, HAROLD GOVETT [1866-1954]
RSC 14,315
H. Hollings, J. Chem. Soc. 1954, p.4059
T.F.E. Read, Analyst 79 (1954), 466

COLOSSER, RICHARD [-1905]
T. Curtius, J. prakt. Chem. 98 (1917), 340

COLOWICK, SIDNEY PAUL [1916-1985]
AMS 15(2),313-314; WWA 1980-1981, p.682

COMBES, ALPHONSE EDMOND [1858-1896]
BLOKH 1,130
C. Friedel, Bull. Soc. Chim. [3]17 (1897), i-xxii

COMBES, CHARLES [1801-1872]
DSB 3,358; DBF 9,363-364; WWWS, 360; POGG 1,468-469; 3,292
R. Locqueneux, Arch. Int. Hist. Sci. 40 (1990), 11-29

COMBES, RAOUL [1883-1964]
DSB 3,359; CHARLES(2), 81-83; SACKMANN, 77-79; IB, 53
R. Ulrich, Bull. Soc. Bot. 3 (1964), 190-192
L. Plantefol, C.R. Acad. Sci. 258 (1964), 2951-2956

COMESSATTI, GUISEPPE [1880-1964]

COMMAILLE, MARIE AUGUSTE ANTOINE [1826-1876]
DBF 9,388; BALLAND, 22-26

COMPTON, JACK [1909-]
AMS 13,807

COMROE, JULIUS HIRAM, Jr. [1911-1984]
AMS 15(2),321; AO 1984, pp.361-363; IWW 1982-1983, p.261; BROBECK, 171-172; WWWS, 361
R.J. Havel, Trans. Assoc. Am. Phys. 98 (1985), cii-civ

COMSTOCK, WILLIAM JAMES [1860-1922]
AMS 3,140; POGG 4,272; 5,239
Anon., Science 55 (1922), 289
Obituary Record of Yale Graduates 1922, pp.517-518. New Haven 1922

CONANT, JAMES BRYANT [1893-1978]
DSB 17,175-178; AMS 13,808; NCAB D,48-49; CB 1951, pp.129-132; WWWS, 362; POTSCH, 92; WWA 1976-1977, p.633; WW 1976, p.495; POGG 6,471-472; 7b,883-886
H.T. Clarke, The Chemist 11(5) (1934), 124-128
Anon., Nachr. Chem. Techn. 2 (1954), 152-153
J.B. Conant, My Several Lives. New York 1970
M. Saltzman, J. Chem. Ed. 49 (1972), 411-413

G.B. Kistiakowsky, Nature 273 (1978), 793-795
C.P. Haskins, Am. Phil. Soc. Year Book 1978, pp.66-73
P.D. Bartlett, Chem. Brit. 15 (1979), 251-252; Biog. Mem. Nat. Acad. Sci. 54 (1983), 91-124
G.B. Kistiakowsky and F. Westheimer, Biog. Mem. Fell. Roy. Soc. 25 (1979), 209-232

CONE, LEE HOLT [1880-1957]
AMS 4,199; 9(I),890-891

CONGDON, CHARLES C. [1920-]
AMS 15(2),324; WWWS, 362

CONKLIN, EDWIN GRANT [1863-1952]
DSB 3,389-391; DAB [Suppl.5], 127-128; AMS 8,842; NCAB 12,351; IB, 54; WWAH, 123; WWWS, 363
E.N. Harvey, Science 117 (1953), 703-705; Biog. Mem. Nat. Acad. Sci. 31 (1958), 54-91
H.H. Plough, Genetics 39 (1954), 1-3
E.G. Butler, Proc. Am. Phil. Soc. 108 (1964), 55-56
G.E. Allen and D.M. McCullough, J. Hist. Biol. 1 (1968), 325-331
A.C. Clement, American Zoologist 19 (1979), 1255-1259
J.T. Bonner, Proc. Am. Phil. Soc. 128 (1984), 79-84
J.W. Atkinson, J. Hist. Biol. 18 (1985), 31-50

CONN, HAROLD JOEL [1886-1975]
AMS 10,743; IB, 54
R.D. Lillie, Stain Technology 52 (1977), 3-4

CONN, HERBERT WILLIAM [1859-1917]
AMS 2,97; NCAB 20,409-410; WWAH, 123; WWWS, 363
C.E.N., Science 45 (1917), 451-452
H.J. Conn, Bact. Revs. 12 (1948), 275-296

CONN, JEROME W. [1907-]
AMS 14,933; WWA 1976-1977, pp.636-637; IWW 1982-1983, p.261; KOREN, 164; WWWS, 363

CONNE, FREDERIC [1862-1905]
STRAHLMANN, 473,481

CONNELL, ARTHUR [1794-1863]
FERCHL, 102; POGG 3,294
Anon., Proc. Roy. Soc. 13 (1864), i; Proc. Roy. Soc. Edin. 5 (1862-1866), 136-137

CONNEY, ALLAN HOWARD [1930-]
AMS 15(2),329; WWWS, 363

CONNOR, RALPH [1907-]
AMS 15(2),330; IWW 1982-1983, p.262; WWWS, 364

CONNSTEIN, WILHELM [1870-]
TLK 2,102
Anon., Z. angew. Chem. 35 (1922), 472

CONRAD, FRIEDRICH FERDINAND [1826-1857]
RAUSCH, 320

CONRAD, KARL FRIEDRICH [1892-]
AMS 9(I),371

CONRAD, LUDWIG [1878-]
JV 23,539; NUC 120,290

CONRAD, MAX [1848-1920]
BLOKH 1,130-131; DZ, 232-233; TLK 2,102; POTSCH, 93; POGG 3,294-295; 4,272-273; 5,239
B. Lepsius, Ber. chem. Ges. 54A (1921), 92-93

CONRADI, HEINRICH [1876-]
FISCHER 1,264; KURSCHNER 4,412; TLK 3,91; IB, 54

CONROY, JAMES TERENCE [1870-1944]
SRS, 13; NUC 120,331
Anon., J. Roy. Inst. Chem. 1944, p.73

CONSDEN, RAPHAEL [1911-]
Who's Who in British Science 1953, p.61

CONTARDI, ANGELO [1877-1951]
POGG 6,474; 7b,889
C. Ravazzini, Chimica e Industria 33 (1951), 163-164

CONWAY, EDWARD JOSEPH [1894-1968]
STC 1,262-263; WW 1965, p.642; WWWS, 365
M. Maizels, Biog. Mem. Fell. Roy. Soc. 15 (1969), 69-82

COOK, ARTHUR HERBERT [1911-]
WW 1981, p.542; CAMPBELL, 63-64; WWWS, 366

COOK, JAMES WILFRED [1900-1975]
DNB [1971-1980],174-175; WW 1974, p.691; CAMPBELL, 65-66; POTSCH, 93; POGG 6,475; 7b,890-893
K. Schofield, Chem. Brit. 12 (1976), 60-61
J.M. Robertson, Biog. Mem. Fell. Roy. Soc. 22 (1976), 71-103

COOK, ROBERT PERCIVAL [1906-]
Who's Who of British Scientists 1980-1981, p.99

COOKE, JOSIAH PARSONS [1827-1894]
DSB 3,397-399; DAB 4,387-388; MILES, 91-92; ELLIOTT, 61-62; BLOKH 1,131; POTSCH, 93; SILLIMAN, 131-134; WWWS, 367; POGG 3,296-297; 4,273
C.L. Jackson et al., Proc. Am. Acad. Arts Sci. 30 (1895), 514-517; Biog. Mem. Nat. Acad. Sci. 4 (1902), 175-183

COOLEY, THOMAS BENTON [1871-1945]
DAB [Suppl.3], 189-190; AMS 7,358; DAMB, 157; WWWS, 368
W.W. Zuelzer, J. Pediat. 49 (1956), 642-650

COOMBS, HERBERT ISAAC [1912-1973]
Anon., Brit. Med. J. 1973(I), p.425

COON, MINOR JESSER [1921-]
AMS 15(2),350; WWA 1980-1981, p.704; WWWS, 368

COONS, ALBERT HEWETT [1912-1978]
AMS 13,829; MH 1,113-114; MSE 1,225-226; BIBEL, 276-; WWWS, 368
A.H. Coons, J. Immunol. 87 (1961), 499-503
M.J. Karnovsky, J. Histochem. 27 (1979), 1117-1118
K.F. Austen, Harvard Med. Alumni Bull. 53(3) (1979), 35-36

COOPER, ASTLEY PASTON [1768-1841]
DNB 4,1062-1064; HIRSCH 2,97-98; Suppl.,192; WILKS, 317-329; BETTANY 1,202-206; WWWS, 368-369
Anon., Proc. Roy. Soc. 4 (1837-1843), 344-346
B.B. Cooper, The Life of Sir Astley Cooper. London 1843
S.J. Symonds, Guy's Hosp. Rep. 71 (1921), 1-18
G. Keynes, St. Barts Hosp. Rep. 55 (1922). 9-36
J. Needham et al., Guy's Hosp. Rep. 117 (1968), 139-255

COOPER, KENNETH ERNEST [1903-]
WWWS, 369
Who's Who of British Scientists 1971-1972, p.183

COOPER, THOMAS [1759-1839]
DSB 3,399-400; DAB 4,414-416; MILES, 92-93; ELLIOTT, 62; FERCHL, 103; POTSCH, 93-94; WWAH, 127; WWWS, 369-370; COLE, 119; POGG 1,475
E.F. Smith, Chemistry in America, pp.128-146. New York 1914; Chemistry in Old Philadelphia, pp.62-81. Philadelphia 1919
D. Malone, The Public Life of Thomas Cooper. New Haven 1926
E.V. Armstrong, J. Chem. Ed. 14 (1937), 153-158

W.J. Bell, J. Hist. Med. 8 (1953), 70-87
D.A. Davenport, J. Chem. Ed. 53 (1976), 419-422

COOPS, JAN [1894-1969]
WWWS, 370; POGG 6,476; 7b,895-896
K. von Nes, Chem. Wkbl. 50 (1954), 865-872
Anon., Chem. Wkbl. 65(30) (1969), 38; Chem. Brit. 6 (1970), 37

COPAUX, HIPPOLYTE [1872-1934]
DSB 3,400-401; WWWS, 370; POGG 5,240-241; 6,476; 7b,896-897
E. Darmois, Bull. Soc. Chem. [5]1 (1934), 1612-1614
H. Perperot, Bull. Soc. Chim. [5]2 (1935), 1777-1785

COPE, ARTHUR CLAY [1909-1966]
AMS 11,985; MILES, 93-94; NCAB 53,159-160
L.F. Fieser, Am. Phil. Soc. Year Book 1966, pp.113-115
J.D. Roberts and J.C. Sheehan, Biog. Mem. Nat. Acad. Sci. 60 (1991),17-30

COPE, EDWARD DRINKER [1840-1897]
DSB 15,91-93; DAB 4,420-421; ELLIOTT, 63; WWAH, 127; WWWS, 370
H.F. Osborn, Science 5 (1897), 705-717
T. Gill, Science 6 (1897), 225-243
H.F. Osborn et al., Biog. Mem. Nat. Acad. Sci. 13 (1930), 127-317
H.F. Osborn and H.W. Warren, Cope: Master Naturalist. Princeton 1931
P.J. Bowles, Isis 68 (1977), 249-265

COPE, FREEMAN WIDENER [1930-]
AMS 15(2),359; WWWS, 370

COPEMAN, SYDNEY ARTHUR MONCKTON [1862-1947]
WW 1944, pp.590-591; O'CONNOR(2), 241-242; WWWS, 370
A.S. MacNally and J. Craigee, Obit. Not. Fell. Roy. Soc. 6 (1948), 37-50

COPISAROW, MAURICE [1889-1959]
POGG 6,477; 7b,897-898
T.K. Walter, Nature 184 (1959), 315
Anon., Proc. Chem. Soc. 1960, pp.189-190

CORBELLINI, ARNALDO [1901-]
POGG 6,477; 7b,899

CORDIER-LOEWENHAUPT, VIKTOR [1874-1926]
TSCHIRCH, 1035; POGG 4,174; 5,242; 6,478
Anon., Ost. Chem. Z. 31 (1928), 178
F. Emich, Ber. chem. Ges. 61A (1928), 142-143

CORENWINDER, BENJAMIN [1820-1884]
DBF 9,650; FERCHL, 104; POGG 1,478-479; 3,300-301
A. Colas, Mémoires de la Société des Sciences, de l'Agriculture et des Arts de Lille [4]14 (1885), 381-382

COREY, ELIAS JAMES [1928-]
AMS 15(2),368; WWA 1980-1981, p.715; IWW 1982-1983, p.267; POTSCH, 94; WWWS, 372

COREY, ROBERT BRAINARD [1897-1971]
AMS 11,992; WWAH, 128

CORI, CARL FERDINAND [1896-1984]
MH 1,114-115; STC 1,266-267; MSE 1,231-232; AMS 14,966; NCAB H,312-313; WWA 1976-1977, p.658; IWW 1982-1983, p.269; AO 1984, pp.523-525; CB 1947, pp.135-137; IB, 55; WWWS, 372; POTSCH, 94-95
B.A. Houssay, Biochim. Biophys. Acta 20 (1956), 11-15
C.F. Cori, Ann. Rev. Biochem. 38 (1969), 1-20
H.M. Kalckar, in Selected Topics in Biochemistry (G. Semenza, Ed.), pp.1-24. Amsterdam 1983
S. Ochoa, TIBS 10 (1985), 147-150
J.T. Edsall, Am. Phil. Soc. Year Book 1985, pp.109-116
P. Randle, Biog. Mem. Fell. Roy. Soc. 32 (1986), 67-95
M. Cohn, Biog. Mem. Nat. Acad. Sci. 61 (1992), 79-109

CORI, CARL ISIDOR [1865-1954]
NDB 3,360; LUDY, 176-181; IB, 55; WWWS, 372
R. Rensch, Zool. Anz. Suppl.18 (1955), 464-466

CORI, GERTY THERESA [1896-1957]
DSB 3,415-416; DAB [Suppl.6], 126-127; MILES, 94-95; STURM 1,210; IB, 55; MH 1,115-116; STC 1,266-267; MSE 1,232; NCAB H,313; 48,327; POTSCH, 95; DAMB, 159-160; CB 1947, pp.135-137; NAW 4,165-167
S. Ochoa and H.M. Kalckar, Science 128 (1958), 16-17
E.A. Doisy, Am. Phil. Soc. Year Book 1958, pp.108-111
C.F. Cori, Ann. Rev. Biochem. 38 (1969), 1-20
J. Larner, Biog. Mem. Nat. Acad. Sci. 61 (1992), 111-135

CORI, OSVALDO [1921-1987]
IWW 1987-1988, pp.303-304
Anon., Archivos de Biologia y Medicina Exerimentales 21 (1988), 1-5

CORIN, GABRIEL [1864-1919]
E. van Ermengem, Bull. Acad. Med. Belg. [4]29 (1919), 388-392
J. Firket, in L'Université de Liège (L. Halkin, Ed.), vol.3, pp.197-205. Liège 1936

CORLEIS, EHRENFRIED [1855-1919]
RSC 14,354; NUC 122,671
Anon., Z. angew. Chem. 32(II) (1919), 232; Chem. Z. 43 (1919), 141

CORNELIUS, HANS [1863-1947]
NDB 3,362-363; NB, 58-59; JV 1,170; RSC 14,355; NUC 123,108
Anon., Chem. Z. 34 (1910), 850

CORNER, GEORGE WASHINGTON [1889-1981]
MH 1,116-117; STC 1,167-168; MSE 1,232-233; FISCHER 1,267; AMS 14,969; AO 1981, pp.614-616; WWA 1978-1979, p.690; IB, 55; WWWS, 373
C.G. Hartman and R.H. Shryock, Am. J. Anat. 98 (1956), 5-34
G.W. Corner, Anatomist at Large. New York 1958; Persp. Biol. Med. 21 (1978), 406-419; The Seven Ages of a Medical Scientist. Philadelphia 1981
L.G. Stevenson, Bull. Hist. Med. 43 (1969), 497-500
J.E. Rhoads, Trans. Coll. Phys. Phila. [5]4 (1982), 162-165
G.E. Erikson, Anat. Rec. 204 (1982), 46A-48A
J.M. Oppenheimer and W.J. Bell, Am. Phil. Soc. Year Book 1982, pp.460-468
S. Zuckerman, Biog. Mem. Fell. Roy. Soc. 29 (1983), 93-112

CORNET, GEORG [1858-1915]
NDB 3,366-367; DBJ 1,325; WWWS, 373
L.E. Reinert, Münch. Med. Wchschr. 62 (1915), 711-712

CORNFORTH, JOHN WARCUP [1917-]
MSE 1,233-234; CAMPBELL, 69-70; WW 1981, p.560; IWW 1082-1983, p.268; POTSCH, 95; SRS, 71; ABBOTT-C, 30-31; WWWS, 373

CORNIL, ANDRÉ VICTOR [1837-1908]
DBF 9,692-693; HIRSCH 2,112; Suppl.,195; VAPEREAU 6,578
L.C. Mallassez, Bull. Acad. Med. 59 (1908), 480-483
M. Letulle, Presse Med. 16 (1908), 273-275; Rev. Sci. 50 (1911), 40-44
K. Alterman, J. Hist. Med. 31 (1976), 431-447

CORNING, JAMES LEONARD [1855-1923]
NCAB 17,177-178; RSC 14,359; WWWS, 373
Anon., J. Am. Med. Assn. 81 (1923), 846

CORNWELL, RALPH [1898-]
AMS 10,776

CORONEDI, GUISTO [1863-1941]
RSC 14,362-363; SG [3]3,213; [4]3,925; NUC 128,311
P. Niccolini, Bioch. Ter. Sper. 22 (1935), 243-250
Anon., Chimica e Industria 24 (1942), 32

CORRAN, HENRY STANLEY [1915-]
Directory of British Scientists 1966-1967, vol.1, p.368

CORRENS, CARL [1864-1933]
DSB 3,421-423; NDB 3,368; FISCHER 1,268; SCHWERTE 2,193-202; LEHMANN,185; MAYERHOFER 1,668-669; IB, 55; WWWS, 374
E. Tschermak von Seysenegg, Alm. Akad. Wiss. Wien 83 (1933), 290-293
F. von Wettstein, Naturwiss. 22 (1934), 2-8; Ber. bot. Ges. 56 (1938), (140)-(160)G. Stein, Naturwiss. 37 (1950), 457-463
M.S. Saha, Mendel Newsletter 21 (1981), 1-6

CORRENS, ERICH [1896-1981]
AO 1981, p.323; KURSCHNER 12,461; POTSCH, 95-96; POGG 7a(1),359-360
Anon., Chem. Z. 105 (1981), 235; Naturw. Rund. 34 (1981), 439

CORTESE, FRANK [1898-]
AMS 8,508; NUC 123,657

CORTI, ALFONSO [1822-1876]
DSB 3,424-425; DBI 29,778-783; EI 11,551; HIRSCH 2,117; Suppl.,196; PAGEL, 352-353
J. Schaffer, Anat. Anz. 46 (1914), 368-382; Naturwiss. 10 (1922),537-542
B. Pincherle, La Vita e l'Opera di Alfonso Corti. Rome 1932
E. Hintsche, Alfonso Corti. Berne 1944
E.V. Ullman, Arch. Otolaryng. 54 (1951), 1-28
A. Corti, Riv. Storia Sci. Med. 46 (1955), 229-254
S. Iurato, in Per la Storia della Neurologia Italiana (L. Belloni, Ed.), pp.165-177. Milan 1963

CORTI, ARNOLD [1873-1932]
DHB, Suppl.,33; JV 14,224
E. Weilenmann, Viert. Nat. Ges. Zurich 77 (1932), 278-281

CORVISART, LUCIEN [1824-1882]
DBF 9,744-745; HIRSCH 2,118-119; PAGEL, 1943-1944; WWWS,375

CORWIN, ALSOPH HENRY [1908-]
AMS 13,846; WWWS, 375

CORYELL, CHARLES DUBOIS [1912-1971]
AMS 11,1001; NCAB 56,472-473; MILES, 95-96; WWAH, 129
Anon., Chem. Eng. News 49(4) (1971), 56
G.E. Gordon, J. Inorg. Chem. 34 (1972), 1-11

COSSA, ALFONSO [1833-1902]
DSB 17,182-185; BLOKH 1,132; POGG 3,303; 4,276; 5,243
A. Piccini, Atti Accad. Lincei [5]11 (1902), 235-237
G.B. Kauffman and E. Molayem, Ambix 37 (1990), 20-34

COSTE, VICTOR [1807-1873]
DBF 9,803-904; DBA 1,542; HIRSCH 2,122; HOEFER 12,80-81; DECHAMBRE 21,34
L. Figuier, Ann. Scient. 17 (1873), 526-529
M.E. Fauré-Fremiet, Not. Acad. Sci. 4 (1964), 343-389

COTTEREAU, E. [1823-1853]
FERCHL, 106; POGG 3,304

COTTEREAU, PIERRE LOUIS [1797-1847]
DBF 9,843; CALLISEN 4,360-361; 27,161-162
E. Rochambeau, Biographie Vendomoise, vol.1, p.321. Paris 1884

COTTON, AIMÉ AUGUSTE [1869-1951]
DSB 17,185-186; DBF 9,852-854; POGG 5,244-245; 6,482-483; 7b,909-910
P. Manigault, Revue d'Optique 30 (1951), 217-220
L. de Broglie, Not. Acad. Sci. 3 (1957), 448-477
A. Kastler, C.R. Acad. Sci. 269 (1969), 70-74
J. Rosmorduc, Pensée 1972, pp.112-126

COTTON, FRANK ALBERT [1930-]
AMS 15(2),383; MSE 1,234-235; WWA 1980-1981, p.275; IWW 1982-1983, p.271; WWWS, 376

COTTRELL, FREDERICK GARDNER [1877-1948]
DSB 3,436-437; DAB [Suppl.4], 183-185; NCAB 38,294-295; MILES, 96-97; POTSCH, 96-97; WWAH, 130; WWWS, 377; POGG 5,245; 6,483; 7b,911-912
F. Daniels, Science 110 (1949), 497-498; Am. Phil. Soc. Year Book 1950, pp.272-277
V. Bush, Biog. Mem. Nat. Acad. Sci. 27 (1952), 1-11

COTZIAS, GEORGE CONSTANTIN [1918-1977]
AMS 13,851
M.D. Yahr, Nature 268 (1977), 779-780
L. Thomas, Trans. Assoc. Am. Phys. 91 (1978), 23-24
L.C. Tang, Neurotoxicology 5 (1984), 5-12

COUCH, JAMES FITTON [1888-1951]
AMS 8,512; POGG 6,483; 7b,912-913
Anon., Chem. Eng. News 29 (1951), 3561
B.A. Brice, Science 115 (1952), 301-302

COUCH, JAMES RUSSELL [1909-1991]
AMS 14,978; WWA 1976-1977, p.668; WWWS, 377

COUERBE, JEAN PAUL (or PIERRE?) [1807-1867]
DBF 9,890-891; GE 13,38; RSC 2,71-72; CALLISEN 4,362; 27,163
C. Chatagnon and P.A. Chatagnon, Ann. Med. Psych. 112(II) (1954),364-390

COULSON, CHARLES ALFRED [1910-1974]
DNB [1971-1980], 182-183; WW 1970, p.683; MSE 1,237-238; CAMPBELL, 73-74; ABBOTT-C, 31-32; WWWS, 378; POTSCH, 97

COULTER, CALVIN BREWSTER [1888-1940]
AMS 6,297
Anon., Science 91 (1940), 469; J. Am. Med. Assn. 115 (1940), 146

COULTER, JOHN MERLE [1851-1928]
DAB 4,467-468; AMS 4,209; NCAB 11,68-69; HUMPHREY, 57-59; IB,56; WWAH,130
G.D. Fuller, Science 69 (1929), 177-180
O.W. Caldwell et al., Science 70 (1929), 299-301
W. Trelease, Biog. Mem. Nat. Acad. Sci. 14 (1930), 99-123

COUNCILMAN, WILLIAM THOMAS [1854-1933]
DSB 3,447-448; DAB [Suppl.1], 205-206; NCAB 5,550; FORBES, 61-72; FISCHER 1, 270; DAMB, 161; WWWS, 378
H. Cushing, Science 77 (1933), 613-618; Biog. Mem. Nat. Acad. Sci. 18 (1938), 157-174
S.B. Wolbach, Arch. Path. 16 (1933), 114-119
A.M. Harvey, Johns Hopkins Med. J. 146 (1980), 185-192,199-201

COUNCLER, KONSTANTIN [1851-1910]
POGG 3,305; 4,278-279; 5,245
Anon., Ber. chem. Ges. 43 (1910), 985

COUPER, ARCHIBALD SCOTT [1831-1892]
DSB 3,448-450; FARBER, 705-715; BLOKH 1,131-132; MAYERHOFER 1,691; W, 118; WWWS, 378; POTSCH, 97
R. Anschütz, Life and Work of Archibald Scott Couper. Edinburgh 1909; Proc. Roy. Soc. Edin. 29 (1909), 193-273
L. Dobbin, J. Chem. Ed. 11 (1934), 331-338
J.H.S. Green, J. Roy. Inst. Chem. 82 (1958), 518-525
D.G. Duff, Chem. Brit. 23 (1987), 350-354

COURRIER, ROBERT [1895-1986]
QQ 1979-1980, p.405; IWW 1982-1983, p.273; CHARLE, 52-54; MH 2,97-98; STC 1,273; IB, 56; WWWS, 379
A. Jost, Annuaire Collège de France 1985-1986, pp.87-90
P. Dustin, Bull. Acad. Med. Belg. 143 (1988), 276-281
J.R. Tata, Biog. Mem. Fell. Roy. Soc. 36 (1990), 103-123

COURTOIS, BERNARD [1777-1838]
DSB 3,455; DBF 9,1038; GE 13,188; LAROUSSE 5,397; BLOKH 1,133; WWWS, 379; POTSCH, 97-98; TSCHIRCH, 1035; POGG 1,489;

CALLISEN 4,369-370
L.G. Toraude, Bernard Courtois et la Découverte de l'Iode. Paris 1921; Mémoires de l'Académie des Sciences de Dijon [5]3 (1921), 193-347

COURTOIS, JEAN ÉMILE [1907-1989]
QQ 1978-1979, p.407; WWWS, 1852-1853
Anon., Ann. Univ. Paris 26 (1956), 69-72
Y. Raoul, Bull. Acad. Med. 174 (1990), 1123-1128
F. Percheron, Rev. Hist. Pharm. 37 (1990), 331-332; Biochimie 74 (1992), 1

COURVOISIER, LUDWIG [1843-1918]
DHB 2,599; FISCHER 1,270-271; FRANKEN, 179; BOSSART, 65-66; WWWS, 379
E. Veillon, Corr. Schw. Aerzte 39 (1918), 1314-1318; Basler Jahrbuch 1920, pp.1-13
A. Lotz, Verhandl. Nat. Ges. Basel 30 (1919), 29-35

COUSIN, HENRI GEORGES [1863-1936]
GORIS, 327-328; TSCHIRCH, 1035; POGG 5,246; 6,485; 7b,917
J. Bougault, J. Pharm. Chim. [8]23 (1936), 483-486

COUTANCEAU, GODEFROY [1775-1831]
HIRSCH 1,129-130; DECHAMBRE 22,84; CALLISEN 4,371-372; 27,168
L. Peisse, in Les Médecins Français Contemporains, p.69. Paris 1827

COWARD, KATHERINE HOPE [1885-]
IB, 56; POGG 6,486; 7b,917-919
Directory of British Scientists 1966-1967, vol.1, p.374

COWDRY, EDMUND VINCENT [1888-1975]
AMS 11,1014; NCAB 61,145-146; CB 1948, pp.118-120; FISCHER 1,271; IB, 56; SACKMANN, 81-83; WWWS, 380
W.M. Cowan, Anat. Rec. 186 (1976), 237-239
J.T. Freeman, Gerontologist 24 (1984), 641-645

COWELL, STUART JASPER [1891-1971]
WW 1970, pp.690-691
Anon., Lancet 1971(II), p.384

COWGILL, GEORGE RAYMOND [1893-1973]
AMS 11,1014-1015; IB, 56; WWWS, 380
J.M. Orten, J. Nutrition 106 (1976), 1227-1234

COX, ERNEST GORDON [1906-]
WW 1981, p.577; CAMPBELL, 75-76; WWWS, 381

COX, HERALD REA [1907-1986]
AMS 14,989; CB 1961, pp.118-120; WWWS, 381
Anon., BioScience 16 (1966), 679
H.R. Cox, in Rickettsiae and Rickettsial Diseases (W. Burgdorfer and R.L. Anacker, eds.), pp.11-15. New York 1981
V.A. Harden, Rocky Mountain Spotted Fever, pp.175-196. Baltimore 1990

COYNE, FREDERICK PHILIP [1903-]
Who's Who in British Science 1953, p.64
Directory of British Scientists 1966-1967, vol.1, p.380

COZE, JEAN BAPTISTE ROZIER [1795-1875]
DBF 9,1155; DBA 1,548-549; HIRSCH 2,134; BERGER-LEVRAULT, 46-47; DECHAMBRE 22,297-298
T., Rev. Med. Est 3 (1875), 333

COZE, LÉON [1819-1896]
DBF 9,1155; DBA 1,549-550; HIRSCH 2,134-135; TSCHIRCH, 1035; WWWS, 382
Anon., Rev. Med. Est 28 (1896), 639-644

COZZI, ANDREA [1795-1856]
TSCHIRCH, 1035; FERCHL, 107; POGG 1,491

CRABTREE, HERBERT GRACE [1892-1966]
Anon., Chem. Brit. 3 (1967), 314CRAFTS, JAMES MASON [1839-1917] DAB 4,492-493; MILES, 99-100; NCAB 13,474-475; SILLIMAN, 147-148; W, 118; POTSCH, 98; WWAH, 132; WWWS, 382; POGG 3,308; 4,279; 5,247
T.W. Richards, Proc. Am. Acad. Arts Sci. 53 (1917-1918), 801-804
C.R. Cross, Biog. Mem. Nat. Acad. Sci. 9 (1919), 159-177
A. Ashdown, J. Chem. Ed. 5 (1928), 911-921

CRAIG, CHARLES FRANKLIN [1872-1950]
AMS 8,521; FISCHER 1,273; IB, 56; WWAH, 132; WWWS, 382-383
E.C. Faust, Am. J. Trop. Med. 31 (1951), 267-269
C.C. Bass and J.S. Simmons, Am. J. Trop. Med. 32 (1952), 5-26

CRAIG, LYMAN CREIGHTON [1906-1974]
AMS 12,1215; MH 2,98-99; STC 1,273-275; MSE 1,243-244; CB 1964, pp.92-93;WWWS, 383; POTSCH, 98
S. Moore, in Peptides: Chemistry, Structure and Biology (R. Walter and J. Meienhofer, Eds.), pp.5-16. Ann Arbor, Mich. 1975; Biog. Mem. Nat. Acad. Sci. 49 (1978), 49-77

CRAIGEE, JAMES [1899-1978]
Who Was Who 1971-1980, p.180
C. Andrewes, Biog. Mem. Fell. Roy. Soc. 25 (1979), 233-240

CRAIGIE, JOHN HUBERT [1887-]
AMS 12,1217; CE 1,436
G.J. Green et al., Annual Review of Phytopathology 18 (1980), 19-25

CRAM, DONALD JAMES [1919-]
AMS 15(2),408; MSE 1,245-246; WWA 1980-1981, p.740; IWW 1982-1983, p.277; WWWS, 383; POTSCH, 98-99
D.J. Cram, From Design to Discovery. Washington, D.C. 1990

CRAMER, ANTONIE [1822-1855]
NNBW 4,466; LINDEBOOM, 383-384; HIRSCH 2,136

CRAMER, AUGUST [1860-1912]
NDB 3,391; BJN 18,14*-15*; FISCHER 1,274; PAGEL, 355; WWWS, 383-384
H. Vogt. Arch. Psychiat. 50 (1912), iii-xi

CRAMER, CARL (CHARLES) [1886-1952]
AARGAU, 142-144; NUC 126,286
Neue Schweizer Biographie 1938, pp.108-109

CRAMER, FRIEDRICH [1923-]
KURSCHNER 13,654; WWWS, 384; POGG 7a(1),361

CRAMER, MARC [1892-1976]
M. Archinard, C.R. Soc. Phys. Hist. Nat. Gen. 12 (1977), 10-13

CRAMER, MORITZ EDUARD 1863-1902]
DRULL, 40; BJN 7,22*; FISCHER 1,274; PAGEL, 355
W.Pagel, Virchows Jahresber. 1902(I), p.412
Anon., Pharm. Z. 47 (1902), 82

CRAMER, WILHELM (WILLIAM) [1878-1945]
FISCHER 1,274-275; O'CONNOR(2), 394; WWAH, 133
E.V. Cowdry, Arch. Path. 40 (1945), 283-285
E.H. Woglom, Cancer Research 6 (1946), 31-35

CRAMPTON, HENRY EDWARD [1875-1956]
AMS 9(II),234; NCAB 42,186-187; IB, 57; WWAH, 133; WWWS, 384

CRANDALL, DANA IRVING [1915-]
AMS 15(2),410

CRANE, FREDERICK LORING [1925-]
AMS 15(2),412; WWWS,384

CRANE, ROBERT KELLOGG [1919-]
AMS 15(2),413; WWA 1980-1981, p.743; WWWS, 384-385
R.K. Crane, in Selected Topics in the History of Biochemistry (G. Semenza, Ed.), pp.43-69. Amsterdam 1983

CRASSO, GUSTAV LUDWIG [1810-]
FERCHL, 108; RSC 2,85; POGG 1,494

CRAW, JOHN [1873-1909]
Anon., Brit. J. Med. 1909 (II), p.117
J.Pagel, Virchows Jahresber. 1910, p.417

CRAWFORD, ALBERT CORNELIUS [1869-1921]
AMS 3,152; PARASCANDOLA, 30; CHEN, 9

CRAWFORD, MALCOLM [1909-]
JV 55,558
Who's Who in British Science 1953, p.65

CREMER, MAX [1865-1935]
FISCHER 1,275; WEBER, 16; IB, 57; POGG 7a(1),366-367
W. Trendelenburg, Erg. Physiol. 37 (1935), 1-11

CREMER, WERNER [1906-]
JV 45,36; NUC 127,74

CRÉPIEUX, PIERRE [1865-]
GENEVA 5,275-276

CRETCHER, LEONARD HARRISON [1888-1967]
AMS 10,810-811; WWWS, 387CREW, FRANCIS ALBERT ELEY [1886-1973] WW 1973, p.738
L. Hogben, Biog. Mem. Fell. Roy. Soc. 20 (1974), 135-153

CRICHTON, JAMES [-1868]
RSC 2,93
Anon., Medical Directory 1869, p.1014

CRICK, FRANCIS HARRY COMPTON [1916-]
MH 1,119-121; STC 1,276-278; MSE 1,248-249; WW 1981, p.593; WWWS, 387; IWW 1982-1983, p.280; POTSCH, 99-100
R. Olby, Daedalus 99 (1970), 938-987
F.H.C. Crick, What Mad Pursuit. New York 1989

CRIEGEE, RUDOLF [1902-1975]
AUERBACH, 784-785; MEINEL, 410-412,500; POTSCH, 100; POGG 7a(1),378-379
Anon., Nachr. Chem. Techn. 8 (1960), 168; Leopoldina [3]4 (1968), 12-13
G. Wittig, Heid. Akad. Jahrb. 1976, pp.75-76
R. Huisgen, Bayer. Akad. Wiss. Jahrbuch 1976, pp.234-238; J. Chem. Ed. 56 (1979), 369-374
G. Maier, Chem. Ber. 110 (1977), xxvii-xlvi

CRILE, GEORGE WASHINGTON [1864-1943]
DAB [Suppl.3], 200-203; NCAB 31,19-21; C,72-73; DAMB, 166; IB, 57; WWAH, 136; WWWS, 387-388
D. Marine, Science 97 (1943), 277-278
E.A. Graham, Am. Phil. Soc. Year Book 1943, pp.380-383
P.C. English, Shock, Physiological Surgery and George Washington Crile. Westport, Conn. 1980

CRIPPA, GIUNIO BRUTO [1892-]
POGG 6,492; 7b,929-930

CROCKER, WILLIAM [1876-1950]
DAB [Suppl.4], 191-192; NCAB D,330; AMS 8,532; IB, 57; WWAH,136; WWWS,388
O. Kunkel, Am. Phil. Soc. Year Book 1950, pp.277-280
E.W. Sinnott, Contr. Boyce Thomp. Inst. 16 (1950), 1-3

CRONHEIM, WALTER [1868-1912]
BLOKH 1,134
W.Will, Ber. chem. Ges. 45 (1912), 3646-3647

CROOK, ERIC MITCHELL [1914-1993]
Who's Who of British Scientists 1971-1972, p.201
W. Templeton, Chem. Brit. 18 (1982), 267

CROOKES, WILLIAM [1832-1919]
DSB 3,474-482; DNB [1912-1921], 136-137; FINDLAY, 11-29; BUGGE 2,288-297; BLOKH 1,134-137; SCHAEDLER, 25; W, 120; ABBOTT-C, 32-33; POTSCH, 101; POGG 3,313-314; 4,282; 6,496
C. Baskerville, Science 31 (1910), 100-103 W.A.T., Proc. Roy. Soc. A96 (1919-1920), i-ix; J. Chem. Soc. 117 (1920), 444-454
F. Greenaway, Proc. Roy. Inst. 39 (1962), 172-198
R.K. DeKosky, Isis 67 (1976), 36-60
D. Batteridge, Anal. Proc. 14 (1977), 179-183
S.B. Sinclair, Ambix 32 (1985), 15-31

CROSS, CHARLES FREDERICK [1855-1935]
DNB [1931-1940], 204-205; W, 120-121; WWWS, 390; POTSCH, 101-102; POGG 4,283-284; 5,250; 6,496; 7b,935
Anon., Chem. Ind. 1935, pp.905-906
C. Dorée, J. Chem. Soc. 1935, pp.1337-1340
E.F. Armstrong, Obit. Not. Fell. Roy. Soc. 1 (1935), 459-464

CROSS, PAUL CLIFFORD [1907-1978]
AMS 12,1244-1245; WWA 1978-1979, p.728; WWWS, 390

CROSS, WILLIAM ERNEST [1887-]
SRS, 40; NUC 128,111-112

CROSSE, ANDREW [1784-1855]
POGG 1,500,1533
O.G.W. Stallybrass, Proc. Roy. Inst. 41 (1967), 597-617
J.E. Secord, Nature 345 (1990), 471-472

CROSSLEY, ARTHUR WILLIAM [1869-1927]
WW 1927, p.693; TSCHIRCH, 1036; POGG 4,284; 5,250-251; 6,496
Anon., Chemical Age 16 (1927), 250,255
W.P. Wynne, J. Chem. Soc. 1927, pp.3165-3173; Proc. Roy. Soc. A117 (1928), vi-x

CROTOGINO, FRIEDRICH [1878-]
JV 16,91; RSC 14,416; NUC 128,160

CROW, JAMES FRANKLIN [1916-]
AMS 15(2),438; WWA 1980-1981, p.749; IWW 1982-1983, p.282; WWWS, 391
J.F. Crow, Ann. Rev. Gen. 21 (1987), 1-22

CROZIER, WILLIAM JOHN [1892-1955]
AMS 9(II),240; IB, 57; WWAH, 138
H. Hoagland and R.T. Mitchell, Am. J. Psychol. 69 (1956), 135-138

CRUIKSHANK, ERNEST WILLIAM HENDERSON [1888-1964]
WW 1965, p.720; WWWS, 391
R.C. Garry, Roy. Soc. Edin. Year Book 1966, pp.12-13

CRUM, ALEXANDER [1828-1893]
RSC 2,101
Anon., J. Soc. Chem. Ind. 12 (1893), 650
F. Boase, Modern English Biography Suppl.1,818. Truro 1908

CRUM, WALTER [1796-1867]
DSB 3,488-489; BLOKH 1,137; FERCHL, 137; POTSCH, 102; WWWS, 392; POGG 3,316
W. De La Rue, J. Chem. Soc. 21 (1868), xvii-xviii
Anon., Proc. Roy. Soc. 16 (1868), viii-x

CRUTO, ALFONSO [1892-1935]
POGG 6,497-498; 7b,937
C. Serono, Rass. Clin. Terap. 34 (1935), 1-2
Anon., Chimica e Industria 17 (1935), 185

CRUVEILHIER, JEAN [1791-1874]
DSB 3,489-491; DBF 9,1335-1336; GENTY 3,293-308; DECHAMBRE 24, 10-13; HIRSCH 2,150-151; Suppl.,203-204; HAYMAKER, 322-325; WWWS, 392; FRANKEN,179-180; CALLISEN 4,428-431; 27,182-186
J. Béclard, Mem. Acad. Med. 31 (1875), xxi-xl
P. Menetrier, Aesculape 1927, pp.182-187,212-216

A. Hermann, Das Leben Jean Cruveilhiers etc. Düsseldorf 1931
L. Delhoume, L'École de de Dupuytren. Jean Cruveilhier. Paris 1937
J.I. Waring, J. Hist. Med. 23 (1968), 349-355
P. Huard, Episteme 8 (1974), 46-57

CSONKA, FRANK ANTON [1889-1958]
AMS 9(I),412; IB, 57

CUÉNOT, LUCIEN [1866-1951]
DSB 3,492-494; DBF 9,1348-1349; SACKMANN, 84-87; WWWS, 393
R. Goldschmidt, Science 113 (1951), 309-310
L. Bounoure, Rev. Sci. 90 (1953), 155-164
J. Rostand, Genetics 42 (1957), 1-6
R. Courrier, Not. Acad. Sci. 3 (1957), 332-389
A. Beau, Annales de l'Est [5]25 (1973), 358-366
C. Limoges, J. Hist. Med. 31 (1976), 176-214
D. Buican, Scientia 117 (1982), 105-136

CUFFER. PAUL [1849-1906]
DBF 9,1352; FISCHER 1,280
A. Siredey, Bull. Soc. Med. Paris [3]23 (1906), 129-132,1376-1381

CULLEN, GLENN ERNEST [1890-1940]
AMS 6,314-315; FISCHER 1,280; IB, 58; WWWS, 393
D.D. Van Slyke, J. Biol. Chem. 134 (1940), 463-465
H.W. Robinson, Science 91 (1940), 468-469

CULLIS, WINIFRED CLARA [1875-1956]
DNB [1951-1960], 276-278; O'CONNOR(2), 314-315; FISCHER 1,281; NUC 129,176-177
Who Was Who 1951-1960, p.266
K.L.W., Brit. Med. J. 1956(II), p.1242

CULMANN, JULIUS [-1930]
JV 5,293; RSC 14,425
Anon., The Percolator 11 (1930), 200; The Chemist 8 (1930), 122

CUMMING, JAMES [1777-1861]
DSB 3,497; DNB 5,296-297; WWWS, 394; COLE, 127; POGG 1,503-504; 3,316

CUMMING, WILLIAM [1812-1886]
WWWS, 394
Anon., Edinburgh Med. J. 31 (1886), 1183-1184

CUMMINGS, MARTIN MARC [1920-]
AMS 15(2),453; WWA 1980-1981, p.767; IWW 1981-1982, p.275; WWWS, 394

CURIE, MARIE [1867-1934]
DSB 3,497-503; DBF 9,1400-1401; SCHWERTE 1,139-148; MAYERHOFER 1,696 CHARLE(2), 86-90; OGILVIE, 64-72; POTSCH, 103-104; ABBOTT-C, 33-34; W,122-123; WWWS, 395; POGG 4,286,1074; 5,254; 6,499-500; 7b,941-944
A.S. Russell, J. Chem. Soc. 1935, pp.654-663
R. Reid, Marie Curie. London 1974
R.L. Walker, J. Chem. Ed. 65 (1988), 561-573
R. Pflaum, Grand Obsession. New York 1989

CURIE, PIERRE [1859-1906]
DSB 3,503-508; DBF 9,1401-1403; BLOKH 1,137-138; MAYERHOFER 1,696-597; CHARLE(2), 90-93; WWWS, 395; POGG 4,286,1704; 5,253-254; 6,500
A. Laborde, Pierre Curie dans son Laboratoire. Paris 1956

CURME, GEORGE OLIVER, Jr. [1888-1976]
AMS 13,903-904; NCAB D,54; 60,47; WWWS, 396
Anon., Science 194 (1976), 169; Chem. Eng. News 54(34) (1976), 31
A.B. Kimmel, Biog. Mem. Nat. Acad. Sci. 52 (1980), 121-137

CURSCHMANN, HEINRICH [1846-1910]
NDB 3,442; BJN 15,20*; FISCHER 1,283

CURTIS, JOHN GREEN [1844-1913]
DAB 4,616-617; WWAH, 141
Anon., Medical Record 84 (1913), 579

CURTISS, RICHARD SYDNEY [1863-1944]
AMS 7,399; RSC 14,437
Obituary Record of Yale Graduates No.104, p.404. New Haven 1946

CURTIUS, FRIEDRICH [1896-1975]
KURSCHNER 10,353; WENIG, 49; NUC 130,170

CURTIUS, HANS [1878-1959]
JV 19,178; NUC 130,174

CURTIUS, LILLY [1912-]
SG [5]1,148

CURTIUS, THEODOR [1857-1928]
DSB 3,510-512; NDB 3,445-446; DBJ 10,320; STUPP-KUGA, 160-188; KIEL, 151; POTSCH, 104; DRULL, 42-43; SCHAEDLER, 26; TLK 2,104; WWWS, 398;POGG 4,287; 5,255-256,1418; 6,503-504; 7aSuppl.,142-143
A. Darapsky, Z. angew. Chem. 40 (1927), 581-583; J. prakt. Chem. 125 (1930), 1-22
H. Wieland, Z. angew. Chem. 41 (1928), 193-194
C. Duisberg, Z. angew. Chem. 43 (1930), 723-725
K. Freudenberg, Chem. Ber. 96 (1963), i-xxv

CURTZE, PHILIPP HEINRICH [1809-]
I. Stotz, Zur Geschichte der Apotheken in den freien Reichstädten Speyer und Worms etc., pp.160-161. Marburg 1976

CUSHING, HARVEY [1869-1939]
DSB 3,516-519; DAB [Suppl.2], 137-140; NCAB 32,402-403; C,36; W, 124-125; FORBES, 171-187; MEDVEI, 509-511; FLUELER, 36-45; DAMB, 171-172; WWAH, 141; WWWS, 398; ABBOTT, 33
J. Homans, Am. Phil. Soc. Year Book 1939, pp.436-440
W.G. MacCallum, Biog. Mem. Nat. Acad. Sci. 22 (1943), 49-70
J.F. Fulton, Harvey Cushing. Springfield, Ill. 1946
A.M. Harvey, Johns Hopkins Med. J. 138 (1976), 196-216

CUSHMAN, JOSEPH AUGUSTINE [1881-1949]
DSB 3,519-520; DAB [Suppl.4], 205-207; FREUND 3,99-107; NCAB 37,385-386; WWWS, 398

CUSHMAN, MARGARET [1905-]
AMS 11,1079

CUSHNY, ARTHUR ROBERTSON [1866-1926]
DSB 15,99-104; DNB [1922-1930], 234-235; O'CONNOR(2), 166-168; FISCHER 1,284; DAMB, 172-173; WWWS, 398
J.J. Abel, Science 63 (1926), 507-515; J. Pharm. Exp. Ther. 27 (1926), 265-286
B.G., Proc. Roy. Soc. Edin. 46 (1926), 354-356
H.H. D[ale], Proc. Roy. Soc. B100 (1926), xix-xxvii
H. MacGillivray, Ann. Rev. Pharm. 8 (1968), 1-24
J. Parascandola, J. Hist. Biol. 8 (1975), 145-165

CUSMANO, GUIDO [1882-1956]
POGG 5,256-257; 6,505; 7b,948

CUTBUSH, JAMES [1788-1823]
DAB 5,10; MILES, 105-106; ELLIOTT, 68; FERCHL, 111; WWWS, 398
E.F. Smith, James Cutbush. Philadelphia 1919
H.G. Wolfe, Am. J. Pharm. Ed. 12 (1948), 89-125

CUVIER, GEORGES [1769-1832]
DSB 3,521-528; DBF 9,1438-1442; GE 13,661-663; MAYERHOFER 1,697-699; W, 125; WWWS, 399; CALLISEN 4,462-480; 27,195-202; POGG 1,507 P. Flourens, Cuvier: Histoire de ses Travaux. Paris 1845
J. Vienot, Georges Cuvier. Paris 1932
W. Coleman, Georges Cuvier, Zoologist. Cambridge, Mass. 1964
C. Canguilhem et al., Rev. Hist. Sci. 23 (1970), 7-92

D. Outram, Hist. Sci. 14 (1976), 101-137
T.A. Appel, The Cuvier-Geoffroy Debate. New York 1987

CYBULSKI, NAPOLEON [1854-1919]
PSB 4,116-118; WE 2,644
F. Czubalski, Acta Physiol. Polon. 5 (1954), 3-14
J. Kaulbersz et al., Acta Physiol Polon. 6 (1955), 125-161; 38 (1987), 74-90

CYON, ÉLIE de [1843-1912]
BME 34,451-452; HIRSCH 2,161; Suppl.,208; PAGEL, 363-364; WWWS, 429; BRAZIER, 218-220,243-244; WININGER 7,547
L.Delhoume, Concours Médical 79 (1957), 1439-1442,1551-1553
J.R. Kagan, Rev. Hist. Med. Heb. 18 (1965), 149-154
G.F. Kennan, American Scholar 55 (1986), 449-475

CZAJA, ALPHONS THEODOR [1894-1984]
KURSCHNER 14,620; ASEN, 32

CZAPEK, FRIEDRICH [1838-1891]
KIRCHENBERGER, 26

CZAPEK, FRIEDRICH [1868-1921]
NDB 3,456; OBL 1,160; DBJ 3,61-64,293; STURM 1,217; KOERTING, 133-134; LUDY, 128-138; STARK, 349; TSCHIRCH, 1036; POGG 6,506
K. Boresch, Ber. bot. Ges. 39 (1921), (97)-(114); Lotos 69 (1921), 3-14

CZERMAK, JOHANN NEPOMUK [1828-1873]
DSB 3,530-531; NDB 3,456; ADB 4,672-673; OBL 1,161; STURM 1,222; HIRSCH 2,164;Suppl.,208; PAGEL, 366; NAVRATIL,34-35; GIESE, 491-493; KOERTING, 117; DECHAMBRE 24,279-280; VAPEREAU 5,497; WWWS, 399
M. Foster, Nature 9 (1873), 63-64
A. Schrötter, Alm. Akad. Wiss. Wien 24 (1874), 179-186
R. von Jaksch, Prag. med. Wchschr. 31 (1906), 571-573
A. Durig, Wiener med. Wchschr. 78 (1928), 791-792

CZERNY, ADALBERT [1863-1941]
NDB 3,460; STURM 1,224; DBA 1,172-173; FISCHER 1,286; MOTSCH, 208-218; PUSCHEL, 133-134; KOERTING, 183; WEINMANN, 103
A. Czerny, Pädiatrie meiner Zeit. Berlin 1939
A. Reuss, Deutsche med. Wchschr. 67 (1941), 1268-1269
E. Glanzmann, Schw. med. Wchschr. 23 (1942), 200-201
E. Schiff, J. Pediat. 48 (1956), 391-399
H. Opitz, Deutsche med. Wchschr. 88 (1963), 723-725
H. Kleinschmidt, Med. Welt 11 (1963), 598-600
P. Wunderlich, Arch. Hist. Med. 42 (1979), 337-343

CZERNY, VINCENZ [1842-1916]
NDB 3,461; OBL 1,163; STURM 1,225-226; FISCHER 1,286; PAGEL, 367-368; DRULL, 43; KILLIAN, 91-92; DZ, 243; WWWS,400
Anon., Brit. Med. J. 1916(II), p.606
R. Werner, Münch. med. Wchschr. 63 (1916), 1619
R. Gersuny, Wiener klin. Wchschr. 29 (1916), 1375

CZUBALSKI, FRANCISZEK [1885-1965]
WE 2,767
A. Trzebski, Acta Physiol. Polon. 38 (1987), 123-131

CZYHLARZ, ERNST von [1873-1950]
STURM 1,229; FISCHER 1,286; STANGL, 9-17
A. von Decastello, Wiener klin. Wchschr. 62 (1950), 88-89
Oesterreicher der Gegenwart 1951, p.370

D

DĄBROWSKI, WACLAW [1876-1962]
WE 2,839; NUC 131,109
E. Pijanowski, Acta Microbiol. Pol. 12 (1963), 83-90

DADIEU, ARMIN [1901-1878]
KERNBAUER, 361-366,422-423; POGG 7a(1),371

DAEHNHARDT, CHRISTIAN JOHANN [1844-1892]
KIEL, 112; SG [1]13,580

DAFERT, FRANZ WILHELM [1863-1933]
NDB 3,473; OBL 1,166; KURSCHNER 4,428; TLK 3,93-94; IB, 59; POGG 4,291;5,258; 6,509; 7a(1),374
Anon., Ost. Chem. Z. 36 (1933), 183-186
R. Wegscheider, Alm. Akad. Wiss. Wien 84 (1935), 229-236

DAGLEY, STANLEY [1916-1987]
AMS 14,1049
S. Dagley, Ann. Rev. Microbiol. 41 (1987), 1-23
P.J. Chapman, TIBS 13 (1988), 247-248

DAHL, FRIEDRICH [1856-1929]
JV 3,161; RSC 17,451-452,756; NUC 131,226-227; IB, 59

DAHLBERG, GUNNAR [1893-1956]
SMK 2,166-167
G. Dahlberg, Acta Gen. Stat. Med. 4 (1953), 101-131
L.C. Dunn, Science 124 (1956), 1195-1196

DAHLGREN, ULRIC [1870-1946]
AMS 7,403; IB, 59; WWAH, 143; WWWS, 401

DAHR, PETER [1906-1984] KURSCHNER 13,584; POGG 7a(1),376-378

DAKIN, HENRY DRYSDALE [1880-1952]
DSB 17,193-194; DAB [Suppl.5], 149-150; AMS 8,556; MILES, 107-108; O'CONNOR(2), 462-463; FISCHER 1,289; SRS, 29; CHITTENDEN, 182-187; WWAH, 143; POTSCH, 105; POGG 6,510; 7b,950-951
P. Hartley, Nature 169 (1952), 481-482; Obit. Not. Fell. Roy. Soc. 8 (1952), 129-148
H.T. Clarke, J. Chem. Soc. 1952, pp.3319-3324
L. Baguenier-Desormaux, Rev. Hist. Pharm. 28 (1981), 79-88
R.M. Hawthorne, Persp. Biol. Med. 26 (1983), 533-566

DAKIN, WILLIAM JOHN [1883-1950]
WW 1949, pp.669-670; SRS, 38; IB, 59; WWWS, 402

DALCQ, ALBERT [1893-1973]
BNB 44,351-359; FISCHER 1,289
Anon., Ann. Soc. Zool. Belg. 104 (1974), 3-6
J. Pasteels, Ann. Acad. Roy. Belg. 141 (1975), 3-70

DALE, HENRY HALLETT [1875-1968]
DSB 15,104-107; DNB [1961-1970], 262-265; WW 1965, p.743; MEDVEI,722-723; O'CONNOR(2), 468-476; MH 2,103-104; MSE 1,252-254; FISCHER 1,290; HAYMAKER, 282-285; W, 579-581; ABBOTT, 33-34; IB, 59; POTSCH, 105
E.D. Adrian et al., Brit. Med. J. 1955(I), pp.1355-1361
H.H. Dale, Persp. Biol. Med. 1 (1958), 125-137; Ann. Rev. Pharm. 3 (1963), 1-8; Adventures in Physiology. London 1965
G.E.W. Wolstenholme, Am. Phil. Soc. Year Book 1968, pp.118-123
J.H. Burn, Pharmacologist 11 (1969), 29-32
W. Feldberg, Biog. Mem. Fell. Roy. Soc. 16 (1970), 77-174
H.O. Schild, Erg. Physiol. 63 (1971),1-19; Brit. J. Pharm. 56 (1976),3-7
M.C. Fishman, Yale J. Biol. Med. 45 (1972), 104-118
W.D.M. Paton, Notes Roy. Soc. 30 (1976), 231-248
A. Burgen, Trends in Neurosciences 2 (1979), xii-xiii
G.R. Fick, J. Hist. Med. 42 (1987), 467-485

DALLINGER, WILLIAM HENRY [1843-1909]
DNB [1901-1911], 462-463; WWWS, 403
R.G.H., J. Roy. Micr. Soc. 1909, pp.699-702
Anon., Nature 82 (1909), 71-72
A.E.S., Proc. Roy. Soc. B82 (1910), iv-vi

DALTON, JOHN [1766-1844]
DSB 3,537-547; DNB 5,428-434; BUGGE 1,378-385; MAYERHOFER 1,700-701; DESMOND, 170-171; BLOKH 1,139-141; ABBOTT-C, 34-35; POTSCH, 106; SCHAEDLER, 26; W, 127-129; WWWS, 404;

COLE, 129-130; POGG 1.512-514
F. Greenaway, Mem. Lit. Phil. Soc. 100 (1958-1959), 1-98; John Dalton and the Atom. London 1966; Proc. Roy. Inst. 41 (1966), 162-177
A.L. Smyth, John Dalton. Manchester 1966
D.S.L. Cardwell (Ed.), John Dalton and the Progress of Science. Manchester 1968
J.T. March, Mem. Lit. Phil. Soc. 111 (1968-1969), 27-47
E. Patterson, John Dalton and the Atomic Theory. Garden City, N.Y. 1970
A. Thackray, John Dalton. Cambridge, Mass. 1972
T. Cole, Jr., Ambix 25 (1978), 117-130
H.T. Pratt, Ambix 39 (1992), 17-20

DALTON, JOHN CALL [1825-1889]
DSB 15,107-110; DAB 5,40; HIRSCH 2,173-174; PAGEL, 372; ELLIOTT, 70-71; KELLY, 187-288; DAMB, 176-177; WWAH, 143; WWWS, 404
S.W. Mitchell, Biog. Mem. Nat. Acad. Sci. 3 (1895), 179-185
W.B. Fye, The Development of American Physiology, pp.15-53. Baltimore 1987

DALY, IVAN de BURGH [1893-1974]
WW 1972, pp.777-778; WWWS, 404
H. Barcroft, Biog. Mem. Fell. Roy. Soc. 21 (1975), 197-226

DAM, HENRIK [1895-1976]
DSB 17,196-200; MH 1,125-126; STC 1,288-290; MSE 1,258-259; WWWS, 404; CB 1949, pp.135-136; KBB 1974, p.204; POTSCH, 106-107; VEIBEL 2,119-122; W(3rd Ed.), 588-589
R.A. Morton, Nature 261 (1976), 621

DAMBERGIS, ANASTASIOS [1857-1920]
TSCHIRCH, 1037; POGG 3,322; 4,293; 6,512
T. Komnenos, Chem. Z. 44 (1920), 585

DAMESHEK, WILLIAM [1900-1969]
AMS 11,1103; DAMB, 177-178; KOREN, 165; WWAH, 144
Anon., Lancet 1969(II), pp.913-914
F.W. Gunz, Blood 35 (1970), 577-582

DAMM, GUSTAV [1858-]
RSC 9,629; NUC 132,99

DAMODARAN, MANAYATH [1903-1957]
K. Ramamurti, Enzymologia 19 (1958), 329-334

DAMPIER-WHETHAM, WILLIAM CECIL [1867-1952]
DNB [1951-1960], 282-283; WW 1949, pp.676-677; WWWS, 405; POGG 4,1625; 5,1360; 6,513; 7b,954
G. Taylor, Obit. Not. Fell. Roy. Soc. 9 (1954), 55-63

DANA, JAMES FREEMAN [1793-1827]
DAB 5,56; DECHAMBRE 28,384-385; SILLIMAN, 40-41; TSCHIRCH, 1037; FERCHL, 114; COLE, 131; POGG 1,516,1553
Anon., N.Y. Med. Phys. J. 6 (1827), 314-318

DANA, SAMUEL LUTHER [1795-1868]
DAB 5,61; ELLIOTT, 72; NCAB 8,167; WWWS, 405DANCKWORTT, PETER [1876-1962] HEINE, 71-72; KURSCHNER 9,297; HANNOVER, 36-37; TLK 3,95; TSCHIRCH, 1037; POGG 6,514; 7a(1),378-380

DANE, ELISABETH [1903-1984]
KURSCHNER 13,589; BOEDEKER, 48; POGG 7a(1),380
Anon., Chem. Z. 108 (1984), 301

DANFORTH, CHARLES HASKELL [1883-1969]
DSB 3,555-556; FISCHER 1,291; IB, 60; WWWS, 406
W.W. Greulich, Am. Phil. Soc. Year Book 1969, pp.112-113
B.H. Willier, Biog. Mem. Nat. Acad. Sci. 44 (1974), 1-56

DANGEARD, PIERRE AUGUSTIN [1862-1947]
DBF 10,95-96; CHARLE(2), 95-97; IB, 60; WWWS, 406
R. Combes, Not. Acad. Sci. 3 (1957), 1-18

R. Cambar, Actes de l'Académie Nationale des Sciences, Arts et Belles-Lettres de Bordeaux [4]28 (1973), 80-87

DANGSCHAT, PAUL [1893-]
TLK 3,95
Anon., Nachr. Chem. Techn. 11 (1963), 146

DANIELL, JOHN FREDERIC [1790-1845]
DSB 3,556-558; DNB 5,483; MAYERHOFER 1,705; FERCHL, 114; W, 129; POTSCH, 107; WWWS, 407; COLE, 132-133; POGG 1,518-519
Anon., Proc. Roy. Soc. 5 (1845), 577-580
D.H. Hey, J. Roy. Inst. Chem. 79 (1955), 305-308
D.I. Davies, Chem. Brit. 26 (1990), 946-949

DANIELLI, JAMES FREDERIC [1911-1984]
AMS 15(2),495; WWA 1980-1981, p.792; WWWS, 407
W.D. Stein, Biog. Mem. Fell. Roy. Soc. 32 (1986), 117-135

DANIELS, FARRINGTON [1889-1972]
DSB 17,200-202; MH 1,126-127; MSE 1,260-261; AMS 11,1108; MILES,110-112; NCAB H,240-241; WWAH, 145; WWWS, 407; POGG 6,515; 7b,955-958
J.E. Willard, Am. Phil. Soc. Year Book 1972, pp.149-152
O.B. Daniels, Farrington Daniels. Madison, Wis. 1978
A.N. Stranges, Soc. Stud. Sci. 22 (1992), 317-337

DANILEVSKI, ALEKSANDR YAKOVLEVICH [1838-1923]
BSE 13,335; BME 8,659-661; WWR, 117; KHARKOV, 55; KAZAN 2,178-179; ZMEEV 1,84-85; 3,13
I.N. Bulankin, Biokhimia 15 (1950), 97-104
G.E. Vladimirov, Fiziol. Zhur. 39 (1953), 509-515
A.V. Valdman, Farm. Toks. 18(3) (1955), 56-60

DANILEVSKI, VASILI YAKOVLEVICH [1852-1939]
BSE 13,335-336; BME 8,661-662; WWR, 117-118; UKRAINE 2,252-253; IB, 60; VORONTSOV, 33-51; KHARKOV, 41-51; PAGEL, 375-376; ZMEEV 4,87-88; BRAZIER, 220-223,244-245
E.A. Finkelshtein, Vasili Yakovlevich Danilevski. Moscow 1955

DANILOV, STEPAN NIKOLAEVICH [1889-1978]
WWWS, 407; POGG 6,515-516; 7b,958-962
A.N. Anikeeva et al., Zhur. Ob. Khim. 48 (1978), 947-949

DANJOU, ÉMILE [1879-1954]
GORIS, 334
Anon., Bulletin de la Société Linnéenne de Normandie [9]8 (1957), 143

DANN, WILLIAM JOHN [1904-1948]
AMS 7,408
P. Handler, Science 110 (1949), 51

DANNENBERG, HEINZ [1912-1975]
KURSCHNER 11,451; KROLLMANN, 1092; POGG 7a(1),382-383

DANYSZ, JAN [1860-1928]
PSB 4,431-432; DBF 10,149; WWWS, 408
Anon., Bull. Inst. Pasteur 26 (1928), 97-98
A. Szwejcerowa, Arch. Hist. Med. 28 (1965), 371-380

DARAPSKY, AUGUST [1874-1942]
DRULL, 44; TLK 3,96; POGG 5,260; 6,518; 7a(1),384

DARBY, STEPHEN [1825-1911]
RSC 2,148; 7,485; NUC 193,135
Anon., J. Chem. Soc. 101 (1912), 640

DARBY, WILLIAM JEFFERSON [1913-]
AMS 15(2),502; IWW 1982-1983, p.295
W.J. Darby, Annual Review of Nutrition 5 (1985), 1-24

DARCET (or d'ARCET), FÉLIX [1814-1846]
DBF 3,334-336; FERCHL,114; POGG 1,522

DARCET (or d'ARCET), JEAN PIERRE JOSEPH [1777-1844]
DBF 3,339-340; HIRSCH 1,185-186; FERCHL, 12,114; WWWS, 409; CALLISEN 1,208-212; 26,76; POGG 1,521-522,1554

DARLINGTON, CYRIL DEAN [1903-1981]
DSB 17,203-209; WW 1981, p.637; MH 1,127-128; STC 1,291-293; SCHULZ-SCHAEFFER, 209; AO 1981, pp.210-211; WWWS, 409
D. Lewis, Biog. Mem. Fell. Roy. Soc. 29 (1983), 113-157

DARMOIS, EUGÈNE ÉMILE [1884-1958]
CHARLE(2), 99-101; WWWS, 409; POGG 6,519; 7b,967-970
J. Duclaux, J. Chim. Phys. 56 (1959), 517-525
G. Valansi, Bull. Soc. Chim. 1959, pp.673-677

DARMOIS, GEORGES [1888-1960]
DBF 10,200-201; POGG 6,519-520; 7b,970-971
A. Danjon, C.R. Acad. Sci. 250 (1960), 241-245
L. de Broglie, Not. Acad. Sci. 4 (1964), 390-412

DARMSTÄDTER, ERNST [1877-1938]
TLK 3,96; LDGS, 18; POTSCH, 108; POGG 5,261; 6,520; 7a(1),384-385
G. Sarton, Isis 30 (1939), 511-514

DARMSTÄDTER, LUDWIG [1846-1927]
NDB 3,516-517; ARNSBERG, 89-90; WININGER 2,7; TETZLAFF, 55; POTSCH, 108; TLK 2,108-109; POGG 4,296-297; 5,261; 6,250
W. Schlenk, Ber. chem. Ges. 60A (1927), 177
J. Ruska, Z. angew. Chem. 40 (1927), 1387

DARNELL, JAMES EDWIN [1930-]
AMS 15(2),504; MSE 1,264-265; WWA 1980-1981, p.799; WWWS, 409

DARROW, DANIEL CADY [1895-1965]
AMS 10,862; WWA 1964-1965, p.476; NCAB 52,162

DARWIN, CHARLES [1809-1882]
DSB 3,565-577; DNB 4,522-534; MAYERHOFER 1,718-726; DESMOND, 173; W, 130-132; ABBOTT, 34-35; WWWS, 410; POGG 3,327-328
F. Darwin (Ed.), The Autobiography of Charles Darwin and Selected Letters, 2nd Ed. New York 1965
G.L. Geison, J. Hist. Med. 24 (1969), 375-411
R.C. Stauffer, Charles Darwin's Natural Selection. London 1973
R.K. Freeman, The Works of Charles Darwin. An Annotated Bibliographical Handlist. London 1976
J.P. Regelmann, Med. Hist. J. 20 (1985), 185-194
A. Desmond and J. Moore, Darwin. London 1991

DARWIN, CHARLES GALTON [1887-1962]
DSB 3,563-565; DNB [1961-1970], 272-274; POGG 5,262; 6,521; 7b,972-974
G.P. Thomson, Biog. Mem. Fell. Roy. Soc. 9 (1963), 69-85

DARWIN, FRANCIS [1848-1925]
DSB 3,581-582; DNB [1922-1930], 237-238; WWWS, 410-411; DESMOND, 173-174; POGG 6,521
Anon., Nature 116 (1925), 583-584
A.C. Seward and F.F. Blackman, Proc. Roy. Soc. B110 (1932), i-xxi

DARZENS, GEORGES [1867-1954]
POTSCH, 108; POGG 5,262; 6,521-522; 7b,974-975
H. Moureu, Bull. Soc. Chim. 1955, p.169
J. Fourié, Bull. Soc. Sci. Aude 76 (1976), 263-270

DASTRE, ALBERT JULES [1844-1917]
DBF 10,236-237; FISCHER 1,294; CHARLE(2), 101-103; WWWS, 411
L. Lapique, Presse Med. 25 (1917),647-651; Rev. Sci. 57 (1919),641-651

DAUBENY, CHARLES [1795-1867]
DSB 3,585-586; DNB 5,544-545; BRIQUET, 202-204; DESMOND, 174; WWWS, 411-412; COLE, 134-135; POGG 1,525; 3,329
W. De La Rue, J. Chem. Soc. 21 (1868), xviii-xxi
J. Phillips, Proc. Roy. Soc. 17 (1868-1869), lxxiv-lxxx
R.T. Gunther, The Daubeny Laboratory Register. London 1904-1924
D.R. Oldroyd and D.W. Hutchings, Notes Roy. Soc. 33 (1979), 217-259

DAUGE, PAUL [1886-]
RIGA, 697-698

DAUSSET, JEAN [1916-]
QQ 1981-1982, p.414; MSE 1,264-266; IWW 1982-1983, p.297; BIBEL, 257-251

DAUTWITZ, FRITZ [1877-1932]
LESKY, 330

DAUWE, FERDINAND [1881-1948]
GHENT 2,94-98; NUC 134,44
E. Derom, Acta Gast. Belg. 11 (1948), 501-505

DAVAINE, CASIMIR JOSEPH [1812-1882]
DSB 3,587-589; DBF 10,322-323; GE 13,994; HIRSCH 2,188; Suppl.,214; DECHAMBRE 31, 400-406; PAGEL, 378; BULLOCH, 361; OLPP, 89-91; W, 132-133; WWWS, 413
E.C. Achard, Bull. Acad. Med. 100 (1928), 1343-1356
J. Théodoridès, Un Grand Medecin et Biologiste. Oxford 1968; Hist. Sci. Med. 8 (1974), 241-287; 16 (1983), 263-266
D. Wrotnowska, Hist. Sci. Med. 9 (1975-1976), 213-230

DAVENPORT, CHARLES BENEDICT [1866-1944]
DSB 3,589-591; DAB [Suppl.3], 214-216; NCAB 15,397-398; IB, 61; WWWS, 413
G.H. Parker, Am. Phil. Soc. Year Book 1944, pp.358-362
M. Steggerda, Eugenical News 29 (1944), 3-19
O. Riddle, Biog. Mem. Nat. Acad. Sci. 25 (1948), 75-110
E.C. MacDowell, Bios 17 (1946), 2-50
C.E. Rosenberg, Bull. Hist. Med. 35 (1961), 266-276

DAVENPORT, EUGENE [1856-1941]
DAB [Suppl.3], 216-217; IB, 61; WWAH, 146; WWWS, 413
D. Kinley, Science 94 (1941), 105-107

DAVENPORT, HAROLD ALVIN [1895-]
AMS 11,1122DAVENPORT, HORACE WILLARD [1912-]
AMS 15(2),512; WWA 1980-1981, p.802; BROBECK, 172-173; WWWS, 413
H.W. Davenport, Physiologist 23(2) (1980), 11-15; Ann. Rev. Physiol. 47 (1985), 1-14

DAVIDIS, ERNST [1873-]
JV 11,166; NUC 134,223

DAVIDOV, VLADIMIR [1887-]
SG [3]4,345; NUC 134,228

DAVIDSON, DAVID [1900-1959]
AMS 9(I),433; IB, 61

DAVIDSON, HEINRICH [1884-]
JV 24,8; NUC 134,241

DAVIDSON, JAMES NORMAN [1911-1972]
DNB [1971-1980], 215; WW 1965, p.764; WWWS, 415; POTSCH, 108-109
Anon., Nature 240 (1972), 59-60; Lancet 1972(II), p.664
A. Neuberger, Biog. Mem. Fell. Roy. Soc. 19 (1973), 281-303

DAVIDSON, NORMAN RALPH [1916-]
AMS 15(2),518; IWW 1982-1983, p.298

DAVIDSON, WILLIAM BROWN [1869-1930]
SRS,16; NUC 134,324
Anon., J. Roy. Inst. Chem. 1930, pp.188-189

DAVIE, EARL W. [1927-]
AMS 15(2),519

DAVIES, ROBERT ERNEST [1919-1993]
AMS 15(2),521; WWA 1980-1981, p.808; WWWS, 416

DAVIS, BERNARD DAVID [1916-1994]
AMS 14,1087; WWA 1980-1981, p.809; IWW 1982-1983, p.300; WWWS, 416-417

DAVIS, GEORGE KELSO [1910-]
AMS 14,1090-1091; WWA 1976-1977, p.747; WWWS, 417-418

DAVIS, HALLOWELL [1896-1992]
AMS 15(2),528; IB, 61
New York Times, 10 September 1992

DAVIS, JAY CONGER [1899-1968]
AMS 11,1136

DAVIS, TENNEY LOMBARD [1890-1949]
AMS 7,421; MILES, 112-113; WWAH, 149; POGG 6,531-532; 7b,991-992
H.M. Leicester and H.S. Klickstein, J. Chem. Ed. 27 (1950), 222-224; Chymia 3 (1950), 1-16

DAVY, EDMUND [1785-1857]
DNB 5,635; BLOKH 1,142-143; FERCHL, 115-116; WWWS, 421; CALLISEN 5,25; POGG 1,530,1555
Anon., J. Chem. Soc. 11 (1859); Nature 180 (1957), 890-891
J. Russell, J. Chem. Ed. 30 (1953), 302-303
M. MacSweeney and J. Reilly, J. Chem. Ed. 32 (1955), 348-352

DAVY, EDMUND WILLIAM [1826-1899]
POGG 3,332; 4,302
Anon., Proc. Roy. Irish Acad. [3]5 (1898-1900), 314-315
C.A. Cameron, History of the Royal College of Surgeons in Ireland, 2nd Ed., pp.590-591. Dublin 1916

DAVY, HUMPHRY [1778-1829]
DSB 3,598-604; DNB 5,637-643; BUGGE 1,405-416; FARBER, 371-384; MAYERHOFER 1,735-737; BLOKH 1,143-144; SCHAEDLER, 26-27; TSCHIRCH, 1037; FERCHL, 116; ABBOTT-C, 35-36; POTSCH, 109-110; W, 133-134; WWWS, 421; COLE, 136-143; POGG 1,528-530; 6,533
J.A. Paris, Life of Sir Humphry Davy. London 1831
J.Z. Fullmer, Science 155 (1955); Sir Humphry Davy's Published Works. Cambridge, Mass. 1969
R. Siegfried, Chymia 5 (1959); Proc. Roy. Inst. 43 (1970), 1-21
A. Treneer, The Mercurial Chemist. London 1963
H. Hartley, Sir Humphry Davy. London 1966

DAVY, JOHN [1790-1868]
DSB 3,604-605; DNB 5,645-646; FERCHL, 116; WWWS, 421; POGG 1,530; 3,333-334; CALLISEN 5,25-31; 27,226
Anon., Proc. Roy. Soc. 16 (1868), lxxix-lxxxi

DAWES, EDWIN ALFRED [1925-]
Who's Who of British Scientists 1980-1981, p.123

DAWSON, CHARLES REGINALD [1911-]
AMS 14,1103; WWA 1980-1981, p.822; WWWS, 422

DAWSON, HARRY MEDFORTH [1875-1939]
SRS, 19; WWWS, 422; POGG 4,303; 5,264; 6,533-534; 7b,995-996
R.W. Gray and G.F. Smith, Obit. Not. Fell. Roy. Soc. 3 (1940), 139-154

DAWSON, MARTIN HENRY [1896-1945]
AMS 7,424
Anon., J. Am. Med. Assn. 128 (1945), 222; Science 101 (1945), 478
R. West, Trans. Assoc. Am. Phys. 59 (1946), 7-8

DAWSON, RAY FIELDS [1911-]
AMS 14,1105; WWA 1980-1981, p.822; WWWS, 422

DAY, DAVID TALBOT [1859-1925]
DSB 3,609-610; DAB 5,156-157; MILES, 113-114; POTSCH, 110; W, 134-135; WWAH, 150
N.H. Darton, Proc. Geol. Soc. Am. 1933, pp.185-191
V. Heines, CHEM TECH 1971, pp.280-285

DAY, PAUL LOUIS [1899-1980]
AMS 14,1107; WWWS, 423
J.R. Trotter and W.J. Darby, J. Nutrition 114 (1984), 241-246

DEAN, GEORGE [1863-1914]
O'CONNOR(2), 422-423; WWWS, 424
J.C.G. Ledingham, J. Path. Bact. 19 (1914-1915), 114-119

DEAN, HENRY ROY [1879-1961]
O'CONNOR(2), 62-63
Anon., Lancet 1961(I), pp.457-458; Brit. Med. J. 1961(I), pp.595-596; J. Path. Bact. 83 (1962), 587-597

DEAN, JOHN [1831-1888]
RSC 7,500; NUC 135,615
Anon., Proc. Am. Acad. Arts Sci. 23 (1888), 319-320

DEBAINS, EDMOND [1864-1949]
DBF 10,413
C. Levaditi, Presse Med. 57 (1949), 314

DE BEER, GAVIN RYLANDS [1899-1972]
DSB 17,213-214; DNB [1971-1980], 227; WW 1972, p.824; STC 1,298-300; MSE 1,274-275; ABBOTT, 35-36; IB, 19
Anon., Nature 239 (1972), 179-180
E.J.W. Barrington, Biog. Mem. Fell. Roy. Soc. 19 (1973), 65-93

DEBRAY, HENRI [1827-1888]
DSB 3,617; DBF 10,446; SCHAEDLER,27; WWWS,427; POGG 3,337-338; 4,303-304
L. Figuier, Ann. Scient. 32 (1888), 590-593

DEBRÉ, ROBERT [1882-1978]
QQ 1975-1976, p.529; WININGER 6,536; IB, 62
Nouveau Dictionnaire National des Contemporains (3rd Ed.), p.264. Paris 1964

DEBUCH, HILDEGARD [1919-]
KURSCHNER 15,704; BOEDEKER, 115

DEBUS, HEINRICH [1824-1915]
WW 1916, p.580; GUNDLACH, 469; MEINEL, 501; BLOKH 1,144-145; POTSCH, 110-111; POGG 1,532; 3,338; 4,304; 5,266; 7aSuppl.,416
H. Schelenz, Mitt. Gesch. Med. 15 (1916), 80-81
G.C.F., J. Chem. Soc. 111 (1917), 325-331

DEBYE, PETER [1884-1966]
DSB 3,616-621; BWN 1,129-131; MH 1,132-133; STC 1,300-301; MSE 1,276-277; W, 135-136; AMS 11,1163; CB 1963, pp.102-104; MAYERHOFER 1,739;ABBOTT-C, 36-37; MILES, 114-115; BHDE 2,207; TLK 3,97; POTSCH, 111; POGG 5,266-267; 6,535-536; 7a(1),386-387
M. von Laue, Z. Elektrochem. 58 (1954), 151-153
D.R. Corson et al., Science 145 (1964), 554-559
F. Hund, Jahrb. Akad. Wiss. Gott. 1966, pp.59-64
F.A. Long, Science 155 (1967), 979-980
M. Davies, J. Chem. Ed. 45 (1968), 467-473; Biog. Mem. Fell. Roy. Soc. 16 (1970), 175-232; J. Phys. Chem. 88 (1984), 6461-6468
K.K. Darrow, Am. Phil. Soc. Year Book 1968, pp.123-130
J.W. Williams, Biog. Mem. Nat. Acad. Sci. 46 (1975), 23-68
W.O. Baker, Welch Foundation Conferences on Chemical Research 20 (1977), 154-199
H.F. Eicke, Phys. Bl. 40 (1984), 106-108; Chimia 38 (1984), 347-353
G. Busch, Viert. Nat. Ges. Zurich 130 (1985), 19-33

DECASTELLO, ALFRED von [1872-1960]
KURSCHNER 5,716; FISCHER 1,298; STANGL, 18-26

DECKER, HERMANN [1869-1939]
HANNOVER, 44; POGG 5,267-268; 6,536-536; 7a(1),387

DECKER, PETER [1916-1983]
KURSCHNER 14,647; POGG 7a(1),387-388
Anon., Naturw. Rund. 36 (1983); Chem. Z. 107 (1983), 312

DÉCLAT, GILBERT [1827-1896]
DBF 10,488; WWWS, 428
Anon., Chron. Med. 2 (1896), 754-758
D. Wrotnowska, Hist. Sci. Med. 16 (1982), 115-123

DECOURT, PHILIPPE [1902-1990]
J. Postel and E. Farjon, Hist. Sci. Med. 25 (1991), 97-100

DEDICHEN, GEORG MARIA [1870-1942]
NBL 3,290-291; JV 9,145; NUC 136,403
Hver er Hvem? 1938, p.122

DE DUVE, CHRISTIAN RENÉ [1917-]
AMS 15(2),563; MSE 1,277-278; WWA 1980-1981, p.835; IWW 1982-1983, p.305; ABBOTT, 36; WWWS, 430

DeEDS, FLOYD [1894-1985]
AMS 11,1167

DEEN, IZAAK van [1805-1869]
NNBW 3,279-281; LINDEBOOM, 416-417; HES, 37-38; VEIBEL, 122-123
M.A. Herwerden, Janus 20 (1915), 174-201
E. Sluiter, Arch. Neerl. Physiol. 28 (1948), 563-571
P.J. Koehler and L.J. Endtz, Neurology 39 (1989), 446-448

DEETJEN, HERMANN [1867-1915]
FISCHER 1,299; KIEL, 114
C. Czerny, Münch. med. Wchschr. 62 (1915), 702
O. Teutschlaender, Verhandl. path. Ges. 28 (1935), 345-347

DeFRATES, JOSEPH SCARBOROUGH [1901-]
AMS 11,1170

DEGEN, JOSEF [1861-1942]
HEIN, 113-114; JV 2,244; RSC 14,525; NUC 137,25
J. Geuenich, Dürener Geschichtsblätter 37 (1965), 859

DEGENER, PAUL [1851-1901]
HEIN-E, 75-76; BJN 6,22*; POGG 3,341; 4,407; 5,539
Anon., Chem. Z. 25 (1901), 1169; Zbl. Zuckerind. 9 (1901), 945-946

DEHÉRAIN, PIERRE PAUL [1830-1902]
DBF 10,565-566; WWWS, 455; POGG 3,342; 4,308
F.D., Nature 67 (1902), 179

DEHIO, KARL KONSTANTINOVICH [1851-1927]
FISCHER 1,299; PAGEL, 382; WELDING, 159-160; LEVITSKI 2,178-185
J. Brennsohn, Die Aerzte Livlands, pp.134-135. Mitau 1905

DEHN, FRANK BERNARD [1886-1964]
NUC 137,126
Who's Who in British Science 1953, p.73
Anon., J. Roy. Inst. Chem. 1964, p.445

DEIBEL, WILHELM [1877-1936]
JV 20,258

DEITERS, OTTO [1834-1863]
HIRSCH 2,206; Suppl.,220; DIEPGEN, 24-28; PAGEL, 383
E. Deiters and R.W. Guillery, Exp. Neurol. 9(1) (1964), iii-vi
E. Nieschlag, Med. Welt 4 (1965), 222-226
H. Schierhorn, Z. mikr. Anat. 100 (1986), 308-336

DEJONG, RUSSELL NELSON [1907-]
AMS 14,1127; WWA 1976-1977, p.773; WWWS, 433

DEKKER, CHARLES ABRAM [1920-]
AMS 18(2),636

DE KRUIF, PAUL HENRY [1890-1971]
CB 1942, pp.186-188; 1963, pp.104-106; DAMB, 190; WWWS, 434
P. De Kruif, The Sweeping Wind. New York 1962
E. Chernin, Reviews of Infectious Diseases 10 (1988), 661-667

DELABY, RAYMOND [1891-1958]
POGG 6,543; 7b,1008-1011
M. Delépine, Bull. Acad. Med. 142 (1958), 700-706
Anon., Ann. Pharm. Fran. 18 (1960), 145-159
G. Champetier et al., Bull. Soc. Chim. 1961, pp.2041-2093
G. Dillemann, Produits et Problèmes Pharmaceutiques 28 (1973), 452-454

DELACRE, MAURICE [1862-1938]
BNB 41,169-182; GHENT 2,49-52; POGG 5,273-274; 6,543; 7b,1011
A. Bruylants, Ann. Acad. Roy. Belg. 140 (1974), 3-86

DELAFIELD, FRANCIS [1841-1915]
DAB 5,208; NCAB 10,278-279; FISCHER 1,301; DAMB, 191-192; WWWS, 435
T.C. Janeway, Medical Record (New York) 88 (1915), 929

DELAFOSSE, GABRIEL [1796-1878]
DSB 15,114-115; DBF 10,632-633; FERCHL, 118; POGG 1,538-539; 3,344
C. Friedel, Rev. Sci. [2]8 (1878), 481-484
L. Figuier, Ann. Scient. 22 (1878), 494
H.W. Schuett, Sudhoffs Arch. 69 (1985), 1-7

DELAGE, YVES [1854-1920]
DSB 4,11-13; DBF 10,639; CHARLE(2), 104-107; WWWS, 435
L. Boutan, Proc. Verb. Soc. Linn. Bordeaux 72 (1919), 129-160
M. Goldsmith, Ann. Biol. [2]1 (1920-19210, v-xix
L. Joubin, Not. Acad. Sci. 1 (1937), 1-22
J.L. Fischer, Revue de Synthèse [3]100 (1979), 443-461

DE LA HABA, GABRIEL LUIS [1926-]
AMS 18(2),637

DE LA RUE, WARREN [1815-1889]
DSB 4,18-19; DNB 17,387-389; BLOKH 2,767; SCHAEDLER, 146; W, 138-139; WWWS, 436; POGG 3,344-345; 4,309
A.W. Hofmann, Ber. chem. Ges. 22 (1889), 1169-1170
Anon., Nature 40 (1889), 26-28; J. Chem. Soc. 57 (1890), 441-445

DELAUNAY, ALBERT [1910-]
QQ 1978-1979, p.459; WWWS, 436

DELAUNAY, HENRI [1881-1937]
IB, 63; SIGRIST, 26-27
L. Launoy and P. Mauriac, Biol. Med. 27 (1937), 420-426
P. Mauriac, J. Med. Bordeaux 114 (1937), 637-640; Presse Med. 45 (1937), 812
L. Genevois, Enzymologia 2 (1937-1938), 192

DELAUNAY, PAUL [1878-1958]
DBF 10,743-744; FISCHER 1,301-302
P. Huard and J. Théodoridès, Rev. Hist. Sci. 12 (1959), 263-266

DELBRÜCK, KONRAD [1884-1915]
BLOKH 1,146; JV 23,20
H. Wichelhaus, Ber. chem. Ges. 48 (1915), 1741-1742
Anon., Z. angew. Chem. 28(3) (1915), 604; Chem. Z. 39 (1915), 823,850

DELBRÜCK, MAX [1850-1919]
NDB 3,580; DBJ 2,355-360,715; BLOKH 1,145-146; SACKMANN, 88-91; WWWS, 437; POGG 6,544
F. Hayduck, Ber. chem. Ges. 52A (1919), 101-103; 53A (1920), 47-62

B. Rassow, Z. angew. Chem. 32 (1919), 313-315
W. Hückel, Pommersche Lebensbilder 2 (1936), 362-369

DELBRÜCK, MAX [1906-1981]
MH 1,133-134; STC 1,301-303; MSE 1,280; AMS 14,1131; AO 1981, pp.167-169; WWA 1978-1979, p.810; BHDE 2,208; WWWS, 437; POTSCH, 111-112
G.S. Stent, TIBS 6(5) (1981), iii-iv; Genetics 101 (1982), 1-16
W. Hayes, Biog. Mem. Fell. Roy. Soc. 28 (1982), 59-90; Social Research 51 (1984), 641-673; Journal of Genetics 64 (1985), 69-84
M. Eigen, Phys. Bl. 38 (1982), 106-107
S.W. Golomb, American Scholar 51 (1982), 351-367
R.S. Edgar, Ann. Rev. Gen. 16 (1982), 501-505
L.E. Kay, J. Hist. Biol. 18 (1985), 207-246
P. Fischer, Licht und Leben. Ein Bericht über Max Delbrück etc. Konstanz 1985
E.P. Fischer and C. Lipson, Thinking about Science. New York 1989

DELÉPINE, MARCEL [1871-1965]
DSB 4,20-21; GORIS, 336-340; CHARLE, 55-58; POTSCH, 112; WWWS, 437-438; POGG 4,311-312; 5,275,1419; 6,545-546; 7b,1013-1016
E.H. Guitard, Rev. Hist. Pharm. 17 (1965), 435-440
C. Dufraisse, C.R. Acad. Sci. 261 (1965), 4931-4935
A. Horeau, Ann. Chim. [14]1 (1966), 5-6
A. Chrétien, Revue de Chimie Minérale 3 (1966), 187-200
G. Dillemann, Produits et Problèmes Pharmaceutiques 28 (1972), 208
G.B. Kauffman, Coord. Chem. Revs. 21 (1976), 181-219

DELEZENNE, CAMILLE [1868-1932]
DBF 10,819; WWWS, 438
L. Hallion, Ann. Physiol. 8 (1932), 785-805; Bull. Acad. Med. 108 (1932), 943-948
G. Schaeffer, Bull. Soc. Chim. Biol. 14 (1932), 1242-1246
H. Pieron, C.R. Soc. Biol. 110 (1932), 867-869

DELFFS, FRIEDRICH WILHELM HERMANN [1812-1894]
STUBLER, 329; DRULL, 45; BAD 5,907; SCHAEDLER, 27-28; FERCHL, 118; POGG 1,541-542; 3,348
Anon., Chem. Z. 18 (1894), 2061; Leopoldina 30 (1894), 107

DELIUS, LUDWIG [1908-1980]
KURSCHNER 13,608
Anon., Deutsche med. Wchschr. 105 (1980), 420

DELLSCHAFT, (LUDWIG JOHANN FRIEDRICH) HERMANN [1877-]
JV 15,130; NUC 138,244

DE LUCA, HECTOR FLOYD [1930-]
AMS 15(2),581; WWWS, 440

DEMARÇAY, EUGENE [1852-1903]
DBF 10,966; WWWS, 440; POGG 3,350; 4,314; 5,277
A. Étard, Bull. Soc. Chim. [3]31 (1904), i-viii
E.H. Huntress, Proc. Am. Acad. Arts Sci. 81 (1952), 44-45

DEMARÇAY, HORACE [1813-1866]
DBF 10,966; VAPEREAU 4,513; POTSCH, 113; RSC 2,234

DEMBOWSKI, JAN [1889-1963]
WE 2,873; IB, 63
Anon., Kosmos (Warsaw) A13 (1964), 3-19

DEMEREC, MILISLAV [1895-1966]
DSB 17,217-219; MH 1,135-136; STC 1,303-304; MSE 1,283-284; AMS 11,1184; IB, 63; WWAH, 155; WWWS, 441
T. Dobzhansky, Am. Phil. Soc. Year Book 1966, pp.115-121
B. Glass, Biog. Mem. Nat. Acad. Sci. 42 (1971), 1-27
T. Dobzhansky et al., Adv. Genetics 16 (1971), xv-xl,23-26,349-361
B. Wallace, Genetics 67 (1971), 1-3
M.J. Miller, Mendel Newsletter 16 (1978), 1-4
P.E. Hartman, Genetics 120 (1988), 615-620

DEMOLE, VICTOR [1887-1974]
POGG 7a(1),393
T. Hürny, Bull. Schw. Akad. Med. Wiss. 30 (1974), 251-252

DEMOOR, JEAN [1867-1941]
BNB 40,175-193; FISCHER 1,305
R. Bruynoghe, Bull. Acad. Med. Belg. [6]6 (1941), 268-271

DEMYANOV, GRIGORI STEPANOVICH [1885-1958]
WWR, 124

DEMYANOV, NIKOLAI YAKOVLEVICH [1861-1938]
BSE (3rd Ed.) 8,86; WWR,124; ARBUZOV, 152-156; WWWS, 441; POTSCH, 113; POGG 4,316; 5,278; 6,548-549; 7b,1023-1024
V.V. Feofilatov, Zhur. Ob. Khim. 8 (1938), 1492-1502
A.S. Onishchenko, Usp. Khim. 17 (1948), 608-623

DENHAM, HENRY GEORGE [1880-1943]
POGG 5,279; 6,550; 7b,1028-1029
F.G. Donnan, Nature 152 (1943), 529-530
F.G. Soper, J. Chem. Soc. 1944, pp.41-42

DENHAM, WILLIAM SMITH [1878-1964]
POGG 6,550; 7b,1029
Anon., J. Roy. Inst. Chem. 1964, p.443

DENIGÈS, GEORGES [1859-1951]
DBF 10,1025; FISCHER 1,307; DORVEAUX, 92; TSCHIRCH, 1038; IB, 64; POGG 4,317; 5,279; 6,550-551; 7b,1029-1032
M. Macheboeuf and F. Tayeau, Bull. Soc. Chim. Biol. 33 (1951), 1636-1637
M. Javillier, C.R. Acad. Sci. 232 (1951), 773-775
P. Mesnard, Figures Pharmaceutiques Francaises, pp.215-220. Paris 1953
M. Casaux-Bussière, Georges Denigès. Bordeaux 1981

DENIS, PROSPER SYLVAIN [1799-1863]
DBF 10,1048; HIRSCH 2,225; Suppl.,224; FERCHL,121; CALLISEN 5,100-101; 27,231-232
C. Minel, Union Médicale [2]19 (1863), 127-128
P. Caffe, J. Conn. Med. Prat. 30 (1863), 335-336
H. Carnoy, Dictionnaire Biographique de l'Est, p.125. Paris 1895

DENIS, WILLEY GROVER [1879-1929]
AMS 4,246; FISCHER 1,307; CHITTENDEN, 290-293; IB, 64
S. Meites, Clin. Chem. 31 (1985), 774-778

DENISON, ROBERT BECKETT [1879-1951]
WW 1949, p.728; SRS, 28; NUC 139,131; POGG 5,279-280; 6,551; 7b,1033
Anon., Nature 168 (1951), 856

DENNIS, LOUIS MONROE [1863-1936]
AMS 5,278; NUC 139,391-392; WWWS, 444; POTSCH, 114

DENNSTEDT, INGEFROH [1895-]
JV 43,470; NUC 139,412

DENNSTEDT, MAXIMILIAN [1852-1931]
BERLIN, 698; POGG 4,318; 5,282; 6,552
M. Bodenstein, Ber. chem. Ges. 64A (1931), 163-164
Anon., Chem. Z. 55 (1931), 811

DENSTEDT, ORVILLE FREDERICK [1899-1975]
AMS 12,1409; YOUNG, 11-12
F.C. Macintosh, Proc. Roy. Soc. Cnada [4]13 (1975), 49-52

DENT, CHARLES ENRIQUE [1911-1976]
DNB [1971-1980], 234-235; WW 1976, p.628; MH 2,115-116; MSE 1,285-286; WWWS, 444
O.W., Lancet 1976(II), pp.813-814
F.V. Flynn, Nature 265 (1977), 571
A. Neuberger, Biog. Mem. Fell. Roy. Soc. 24 (1978), 15-31

DENT, FRANKLAND [1869-1929]
SRS, 16; JV 13,223; RSC 18,820; 19,74
Anon., J. Roy. Inst. Chem. 1929, p.119DENYS, JOSEPH [1857-1932]
Bibl. Acad. Louvain 4,139-146; 5,39; 6,335
Anon., Rev. Med. Louvain 1932, pp.113-118; Scalpel 85 (1932), 220-222

DE RENZO, EDWARD CLARENCE [1925-1988]
AMS 15(2),598; WWWS, 446

DERICK, CLIFFORD LAMBLE [1894-1972]
AMS 11,1196; IB, 64
E.C. Eppinger, Trans. Assoc. Am. Phys. 86 (1973), 10

DERLON, HANS [1900-]
JV 40,464

DERNBY, KARL GUSTAV [1893-1929]
VAD 1929, P.167; IB, 64
Anon., Chem. Z. 53 (1929), 989

DEROSNE, CHARLES LOUIS [1780-1846]
DSB 4,41-42; DBF 10,1143; HOEFER 13, 718; GE 14,197; LAROUSSE 6,513; POTSCH, 114; TSCHIRCH, 1038; FERCHL, 121; POGG 1,553,1556; CALLISEN 5,109

DERRIEN, EUGÈNE [1879-1931]
DBF 10,1149; FISCHER 1,309; IB, 64
G. Fontes, Bull. Soc. Chim. Biol. 13 (1931), 555-570
M. Tiffeneau, Bull. Acad. Med. 105 (1931), 695-703

DERX, HENRI GEORGES [1894-1953]
SACKMANN, 91-93
L.G.M. Baas-Becking, Annales Bogorienses 1 (1954), 159-164

DERYAGIN, BORIS VLADIMIROVICH [1902-]
BSE (3rd Ed.) 8,131; WWWS, 447; POTSCH, 114

DESCHAMPS, JEAN BAPTISTE [1804-1866]
DECHAMBRE 28,287
J.E. Courtois, Rev. Hist. Pharm. 33 (1986), 159-167

DESCHAUER, JOSEF [1879-1949]
JV 20,258

DESGREZ, ALEXANDRE [1863-1940]
DBF 10,1372-1373; FISCHER 1,310; GORIS, 345-347; IB, 64; WWWS, 449; POGG 6,554; 7b,1040
M. Tiffeneau, Bull. Acad. Med. 123 (1940), 164-176
M. Polonovski, Bull. Soc. Chim. [5]8 (1941), 1-27

DESMAZIÈRES, JEAN BAPTISTE [1786-1862]
DBF 10,1456; LAROUSSE 6,569; WWWS, 450; CALLISEN 27,267DESNUELLE, PIERRE [1911-1986]
QQ 1981-1982, p.452; IWW 1982-1983, p.319
P. Desnuelle, in Selected Topics in the History of Biochemistry (G. Semenza, Ed.), pp.283-331. Amsterdam 1983

DESORMES, CHARLES BERNARD [1777-1862]
DSB 4,74; DBF 10,1501; GE 14,263; LAROUSSE 6,576; FERCHL, 122; W, 143; POTSCH, 115; WWWS, 450; POGG 1,562,1557; CALLISEN 5,139-140
P. Lemay, Chymia 2 (1949), 45-49

DESPRETZ, CÉSAR [1792-1863]
DBF 11,29; GE 14,272; LAROUSSE 6,584-585; HOEFER 13,397-398; POTSCH, 115; BLOKH 1,148; SCHAEDLER, 28; WWWS, 450; POGG 1,562-563; 3,356
K.F.P. Martius, Sitz. Bayer. Akad. 1863(II), pp.385-388
J. Nicklès, Am. J. Sci. [2]36 (1983), 398-401
Anon., Proc. Roy. Soc. 13 (1864), viii-ix

DESSAIGNES, VICTOR [1800-1885]
DSB 4,75-76; DBF 11,70; GE 14,274-275; BLOKH 1,148-149; TSCHIRCH, 1038; POTSCH, 115; FERCHL, 122-123; WWWS, 450; POGG 1,563-564; 3,356
W.H. Perkin, J. Chem. Soc. 47 (1885), 309-310
L. Figuier, Ann. Scient. 29 (1885), 533-534

DESSAUER, ERWIN von [1861-1938]
NUC 140,618
Wer ist Wer? 10,285

DESSAUER, FRIEDRICH [1881-1963]
KURSCHNER 9,309-310; KALLMORGEN, 245; BHDE 1,126; WIDMANN, 82-83,259; WININGER 2,36; LDGS, 80; TLK 3,101; POGG 5,285; 6,555; 7a(1),395-396
H. Hartmann, Schöpfer des neuen Weltbilds, pp.291-315. Berlin 1952
B. Rajewsky, Strahlentherapie 121 (1963), 1-4
C. Kleinholz-Boerner, Friedrich Dessauer. Berlin 1968

DESSAUER, HANS [1869-1926]
RSC 13,890; NUC 140,621
Anon., Z. angew. Chem. 39 (1926), 1364; Chem. Z. 50 (1926), 852

DESTOUCHES, PIERRE [1779-1859]
BOURQUELOT, 38; CALLISEN 5,149-150
P.F.G. Boullay, J. Pharm. Chim. [3]35 (1859), 376-377

DETMER, WILHELM [1850-1930]
KURSCHNER 3,365; BK, 185; IB, 64
A. Heilbronn, Ber. bot. Ges. 49 (1931), (126)-(138)

DETOROS, GEORG JEAN [1880-]
JV 22,298; NUC 141,22

DETRE, LÁSZLO [1876-1939]
MEL 1,373; NUC 141,23
D. Karasszon, Orv. Het. 131 (1990), 1089-1090

DETWEILER, SAMUEL RANDALL [1890-1957]
AMS 9(II),273; IB, 65; WWAH, 157; WWWS, 452
L.S. Stone, Science 127 (1958), 227-228
R.L. Carpenter, Anat. Rec. 131 (1958), 5-18
J.S. Nicholas, Am. Phil. Soc. Year Book 1959, pp.116-122; Biog. Mem. Nat. Acad. Sci. 35 (1961), 85-111

DEUEL, HANS [1916-1962]
KURSCHNER 9,311; BHDE 2,211; POGG 7a(1),398-399
H. Neukomm, Chimia 16 (1962), 47

DEUEL, HARRY JAMES [1897-1956]
AMS 9(I),462; IB, 65; WWAH, 157
A.C. Frazer, Nature 177 (1956), 872
J.M. Luck, Science 124 (1956), 209
J.H. Roe, Ann. Rev. Biochem. 46 (1957), vii

DEULOFEU, VENANCIO [1902-]
POGG 6,555-556; 7b,1041-1042

DEUSSEN, ERNST [1868-1944]
HEIN-E, 78-79; TSCHIRCH,1039; TLK 3,103; POGG 5,285-286; 6,556; 7a(1),400

DEUTICKE, HANS JOACHIM [1898-1976]
KURSCHNER 11,473; POGG 7a(1),400-401
Anon., Chem. Z. 101 (1977), 160
G.F. Domagk and H.J. Bretschneider, Jahrb. Akad. Wiss. Gott. 1980,pp. 45-47

DEUTSCH, ADAM [1907-]
LDGS, 59

DEUTSCH, HAROLD FRANCIS [1918-]
AMS 15(2),610

DEVAUX, HENRI [1862-1956]
DSB 4,76-77; DBF 11,180-181; WWWS, 452
J.G. Kaplan, Science 124 (1956), 1017-1018
L. Genevois, Rev. Gen. Bot. 63 (1956), 341-346; Proc. Verb. Linn. Bordeaux 96 (1956), 79-83

DEVENTER, CHARLES MARIUS van [1860-1931]
BWN 3,141-142; POGG 4,1549-1550; 5,1293; 6,557
W. Kloos et al., Chem. Wkbl. 6 (1909), 1005-1014
J.J. van Laar, Chem. Wkbl. 28 (1931), 547-550

DEVERGIE, ALPHONSE [1798-1879]
DBF 11,189-190; HIRSCH 2,253; Suppl.,230; DECHAMBRE 28,497-499; WWWS,452; CALLISEN 5,161-163; 27,276-277
G. Lagneau, Bull. Acad. Med. 8 (1879), 1016-1019

DEVILLE, HENRI ÉTIENNE SAINTE-CLAIRE [1818-1881]
DSB 4,77-78; HD 1,548-549; LAROUSSE 14,61; LACROIX 1,45-57; WWWS, 1467; BLOKH 1,149-150; TSCHIRCH, 1039; FERCHL, 464-465; SCHAEDLER, 110; POTSCH, 375-376; ABBOTT-C, 125-126; POGG 2,737-738; 3,1162-1163
L. Pasteur, C.R. Acad. Sci. 93 (1881), 6-9
T.H.N., Nature 24 (1881), 219-221
R.E. Oesper and P. Lemay, Chymia 3 (1950), 205-221
M. Goupil-Sadoun, Actualités Chimiques 1983(4), pp.9-14

DE VOTO, LUIGI [1864-1936]
FISCHER 1,312
E. Schwarz, Medicina Italiana 17 (1936), 484

DEWAN, JOHN GEORGE [1907-]
Anon., Postgraduate Medicine (Minneapolis) 22 (1957), 1,3

DEWAR, JAMES [1842-1923]
DSB 4,78-81; DNB [1922-1930], 255-257; FINDLAY, 30-57; BLOKH 1,151-152; POTSCH, 115-116; MAYERHOFER 1,800-801; W,144; POGG 3,357-358; 4,325-326; 5,287; 6,558
H.E. Armstrong, Proc. Roy. Soc. A111 (1926), xiii-xxiii; J. Chem. Soc. 131 (1928), 1066-1076
K. Mendelssohn, Proc. Roy. Inst. 41 (1966), 212-233

DEWAR, MICHAEL JAMES STEUART [1918-]
WW 1981, p.692; IWW 1982-1983, p.322; STC 1,305-306; WWWS, 454
M.J.S. Dewar, A Semiempirical Life. Washington, D.C. 1992

DEY, BIMAN BIHARI [1889-1959]
POGG 6,558; 7b,1045-1048
P. Ray, J. Ind. Chem. Soc. 36 (1959), 294-298

DEYEUX, NICOLAS [1745-1837]
DBF 11,230; LAROUSSE 6,685; TSCHIRCH, 1039; FERCHL, 123; WWWS, 455; POGG 1,566; CALLISEN 5,178-184; 27,282
C. Clerk, Ann. Chim. Anal. 27 (1945), 118-120

DEZANI, SERAFINO [1884-1953]
TSCHIRCH, 1039; IB, 65; POGG 6,558-559; 7b,1048

DHAR, NIL RATAN [1892-]
WWWS, 455; POGG 5,287-288; 6,559; 7b,1048-1053

DHÉRÉ, CHARLES [1876-1955]
DHB 2,666; IB, 65; POGG 6,559-560; 7a(1),402-403
J. Roche, Bull. Acad. Med. 139 (1955), 108-109
L. Laszt, Bull. Soc. Frib. Sci. Nat. 44 (1955), 304-313
L.S. Ettre, J. Chromat. 600 (1992), 3-15

DIAKONOV, KOSTANTIN SERGEEVICH [1839-1868]
KAZAN 2,191-192; ZMEEV 4,107-108; RSC 7,530; 12,212
V.E. Anisimov, Kazan. Med. Zhur. 1961(1), pp.90-92

DIANIN, ALEKSANDR PAVLOVICH [1851-1918]
BSE (3rd Ed.)8, 237-238; BLOKH 1,152; WWWS, 456
A.D. Petrov, Materiali po Istorii Otechestvennoi Khimii, pp.97-104. Moscow 1953

DIATROPOV, PETR NIKOLAEVICH [1859-1934]
BSE 14,311; BME 9,248-249; WWR, 128
T.G. Kirilenko, Gigiena i Sanitaria 1977(10), pp.46-50

DICE, LEE RAYMOND [1887-1977]
AMS 11,1216; IB, 65; WWWS, 456

DICK, GEORGE FREDERICK [1881-1967]
AMS 10,937; NCAB 54,240; FISCHER 1,314; DAMB, 200-201; WWWS, 456
Anon., Proc. Inst. Med. Chicago 26 (1967), 325
L.O. Jacobson, Trans. Assoc. Am. Phys. 82 (1969), 32

DICKENS, FRANK [1899-1986]
WW 1981, p.697; IWW 1982-1983, p.326; POGG 6,560; 7b,1053-1055
R.H. Thompson and P.N. Campbell, Biog. Mem. Fell. Roy. Soc. 33 (1987), 189-210

DICKINSON, ROSCOE GILKEY [1894-1945]
DSB 4,82; AMS 7,448; WWAH, 158; POGG 6,562; 7b,1055-1056
L.Pauling, Science 102 (1945), 216
Anon., Chem. Eng. News 23 (1945), 1453

DICKINSON, WILLIAM LEE [1863-1904]
O'CONNOR(2), 305
Anon., Lancet 1904(II), pp.926,929

DIECKMANN, WALTER [1869-1925]
POTSCH, 116-117; BLOKH 1,152-153; POGG 4,328; 5,290; 6,563
R. Willstätter, Ber. chem. Ges. 58A (1925), 7-8
Anon., Z. angew. Chem. 38 (1925), 104

DIEFFENBACH, ERNST JOHANN [1811-1855]
ADB 5,120; HB 2,146-150; POGG 1,568
G.E. Bell, Ernst Dieffenbach - Rebel and Humanist. Palmerston North, N.Z. 1976

DIEFFENBACH, JOHANN FRIEDRICH [1792-1847]
NDB 3,641-643; ADB 5,120-126; HIRSCH 2,262-264; DECHAMBRE 29,279-281;KROLLMANN, 131; KILLIAN,346-347; CALLISEN 5,196-200; 27,287-290;
H. Fischer, Deutsche med. Wchschr. 38 (1912), 2179
E. Melchior, Deutsche med. Wchschr. 39 (1913), 373-374
R. Lampe, Dieffenbach. Leipzig 1934
B. Valentin, Dieffenbach an Stromeyer, Briefe. Leipzig 1934

DIEFFENBACH, OTTO [1827-1900]
NDB 3,640DIEHL, CLAUS [1879-]
JV 22,516; NUC 143,373

DIEHL, THEODOR [1865-1921]
BLOKH 1,153
H. Alexander, Chem. Z. 34 (1921), 421-422

DIELS, LUDWIG [1874-1945]
NDB 3,645-646; TLK 3,104; IB, 65; WWWS, 458
L.J. Mildbraed, Bot. Jahrb. 74 (1949), 173-198
H. Melchior, Ber. bot. Ges. 68a (1955), 281-287

DIELS, OTTO [1876-1954]
DSB 4,90-92; NDB 3,647; SHBL 1,126-127; MH 1,136-137; MSE 1,289-290; MAYERHOFER 1,817-818; TLK 3,104-105; W, 145-146; ABBOTT-C, 37-38; POTSCH, 117; KIEL, 160; POGG 5,290-291; 6,564; 7a(1),405-406
H. Wieland, Bayer. Akad Wiss. Jahrbuch 1954, pp.200-202
S. Olsen, Chem. Ber. 95 (1962), v-xlvi
J.T. Wilcox, J. Chem. Ed. 53 (1976), 7-9

DIENER, THEODOR OTTO [1921-]
AMS 15(2),636; WWA 1980-1981, p.873

DIÉNERT, FRÉDÉRIC VINCENT [1874-1948]
DBF 11,306; POGG 6,564-566; 7b,1059-1060
Anon., Arch. Biog. Cont. 5 (1911), 99-100; Chem. Ind. 1948, p.398

DIENES, LOUIS LADISLAUS [1885-1974]
AMS 11,1225

DIENSTBACH, OSKAR [1883-1914]
JV 23,540; NUC 143,434
Anon., Chem. Z. 38 (1914), 1299

DIEPGEN, PAUL [1878-1966]
KURSCHNER 10,384-384; FISCHER 1,315; POGG 7a(1),409-411
W. Artelt et al., Sudhoffs Arch. 37 (1953), 193-194,438-445
E. Heischkel and W. Artelt, Clio Medica 1 (1966), 357-359
G. Mann, Arch. Inter. Hist. Sci. 19 (1966), 365-367
H. Schadewaldt, Med. Welt 17 (1966), 313-315
A. Mayer, Z. Gyn. 88 (1966), 554-556

DIERGART, PAUL [1875-1943]
TLK 3,105-106; POGG 6,566-567; 7a(1),412
Anon., Proteus 2 (1937), 327-354

DIESBACH, HENRI de [1880-1970]
POGG 6,567; 7a(1),413
L. Chardonnens et al., Chimia 19 (1965), 191-200
L. Chardonnens, Helv. Chim. Acta 53 (1970); Bull. Soc. Frib. Sci. Nat. 59 (1970), 104-106

DIETERICH, EUGEN [1840-1904]
NDB 3,670; HEIN, 119-120; TSCHIRCH, 1039; POTSCH, 117; POGG 7aSuppl.,153
G.E. Dann, Eugen und Karl Dieterich. Karlsruhe 1969

DIETERICH, HERMANN [1887-1972]
JV 27,394

DIETERICH, KARL [1869-1920]
DBJ 2,744; HEIN, 120-121; BLOKH 1,153; TSCHIRCH, 1039; POGG 5,292; 7aSuppl.,153-154
Anon., Chem. Ind. 44 (1920), 237; Ber. chem. Ges. 53A (1920), 79-80
H. Thoms, Ber. pharm. Ges. 31 (1921), 113-115
G.E. Dann, Eugen und Karl Dieterich. Karlsruhe 1969

DIETERLE, HUGO [1881-1952]
HEIN-E, 82-83; LEHMANN, 146-147; DRUM, 289-290; AUERBACH, 788; TLK 3,106; POGG 6,567; 7a(1),413-414
L., Deutsche Apoth. Z. 56 (1941), 156-157
Anon., Arzneimitt. 1 (1951), 196-197
A. Wankmüller, Beitr. Wurtt. Apothekgesch. 11 (1977), 142-149

DIETL, JOSEF [1804-1878]
PSB 5,158-166; HIRSCH 2,267-268; Suppl.,233; KONOPKA 2,391-401; KOSMINSKI, 86-88,594; BOEGERSHAUSEN, 115-128; WWWS, 459
A. Grimm, Prag. med. Wchschr. 25 (1900), 445-448
J. Latkowski, Gazeta Lekarska 7 (1928), 813-815
A. Wrzosek and J. Gawlik, Arch. Hist. Fil. Med. 9 (1929), 81-90,218-224
T. Tempka, in Szescetlecie Medycyny Krakowskiej, pp.79-112. Cracow 1963
J.A. Mezyk, Pol. Med. Sci. 7 (1964), 99-102
E. Kucharz, Clio Medica 16 (1981), 25-35

DIETL, MICHAEL [1847-1887]
STURM 1,250; RSC 7,534; 9,699; 12,198
Anon., Leopoldina 23 (1887), 162

DIETZ, EMMA MARGARET [1905-]
AMS 7,451

DIETZEL, RICHARD [1891-1962]
KURSCHNER 9,322; TLK 3,107; POGG 6,567-568; 7a(1),416-417
Anon., Pharm. Zent. 95 (1956), 268-269; Nachr. Chem. Techn. 10 (1962), 299

DIETZSCH, OSKAR [1825-1890]
STRAHLMANN, 469-470,481

DIJKEN, BONNO van [1866-]
RSC 14,611; NUC 143,693

DIJKSTERHUIS, EDUARD JAN [1892-1965]
BWN 1,159-161; POGG 6,569; 7b,1063-1065
R. Hooykaas, Isis 58 (1967), 223-225

DILL, DAVID BRUCE [1891-1986]
AMS 14,1184; WWA 1966-1967, p.545; BROBECK, 158-159; WWWS, 460
S.M. Horvath and E.C. Horvath, Physiologist 30 (1987), 84-85

DILLING, WALTER JAMES [1886-1950]
WW 1949, p.754; IB, 66; WWWS, 460
R.W. Brookfield, Nature 166 (1950), 587

DILLON, ROBERT TROUTMAN [1904-1986]
AMS 12,1458

DILTHEY, ALFRED [1877-1915]
JV 16,308
Anon., Chem. Z. 40 (1916), 649; Z. angew. Chem. 29(III) (1916), 426

DILTHEY, WALTHER [1877-1955]
TLK 3,107; ZURICH, 96-97; POTSCH, 117-118; POGG 5,293; 6,569-570; 7a(1),418
Anon., Chem. Z. 79 (1955), 527

DIMOND, ALBERT EUGENE [1914-1972]
AMS 11,1233
J.G. Horsfall, Annual Review of Phytopathology 29 (1991), 29-33

DIMROTH, KARL [1910-]
KURSCHNER 13,645; AUERBACH, 789; MEINEL, 501; POGG 7a(1),418-419
Anon., Nachr. Chem. Techn. 23 (1975), 362

DIMROTH, OTTO [1872-1940]
NDB 3,726; TLK 3,107; WWWS, 461; POTSCH, 118; POGG 5,293,1420; 6,570; 7a(1),419
Anon., Chem. Z. 64 (1940), 208; Z. angew. Chem. 53 (1940), 263
F. Harms et al., Ber. chem. Ges. 74A (1941), 1-23

DINGEMANSE, ELIZABETH [1886-1952]
LINDEBOOM, 447-448
J.W. Everse, J. Clin. Endocrin. 12 (1952), 981

DINGLE, JOHN HOLMES [1908-1973]
AMS 12,1462; WWWS, 461
C.H. Rammelkamp, Trans. Assoc. Am. Phys. 87 (1974), 17-18
W.S. Jordan, Jr., Biog. Mem. Nat. Acad. Sci. 61 (1992), 137-163

DINGLER, EMIL MAXIMILIAN [1806-1874]
NDB 3,730*; ADB 5,239; BLOKH 1,154; MATSCHOSS, 57; POGG 3,363

DINGLER, GOTTFRIED [1778-1855]
NDB 3,730; ADB 5,239-240; HEIN, 121-122; BLOKH 1,153-154; MATSCHOSS, 57; FERCHL, 125; SCHAEDLER, 28; TSCHIRCH, 1040; POGG 1,573-574

DINGLER, HUGO [1881-1954]
DSB 4,100-102; NDB 3,729-730; TLK 3,107; WWWS, 461-462; POGG 5,294; 6,571-572; 7a(1),420-421
W. Leibbrand, Arch. Inter. Hist. Sci. 7 (1954), 345-346

DINGLER, MAX [1883-1961]
JV 25,734; IB, 66

DIPPEL, LEOPOLD [1827-1914]
NDB 3,738-739; DBJ 1,279; WOLF, 39; FREUND 1,161-174; TSCHIRCH, 104C; WWWS, 463
L.H. Schenk, Mitt. Dendrol. Ges. 23 (1914), 305-310

DIPPY, JOHN FREDERICK JAMES [1906-]
Who's Who of British Scientists 1971-1972, pp.234-235

DIRR, KARL [1894-1970]
KURSCHNER 13,648; POGG 7a(1),421-422
Anon., Chem. Z. 93 (1969), 906; Nachr. Chem. Techn. 18 (1970), 423

DIRSCHERL, WILHELM [1899-1982]
KURSCHNER 13,648-649; POGG 7a(1),422-423
R. Ammon, Arzneimitt. 9 (1959), 722; 14 (1964), 1269
Anon., Chem. Z. 106 (1982), 382

DISCHE, ZACHARIAS [1895-1988]
AMS 11,1237; BHDE 2,218-219; KOREN, 166; IB, 66; POGG 6,573; 7b,1069-1072
Z. Dische, Inv. Ophthalm. 4 (1965), 749-758; Exp. Eye Res. 4 (1965), 265-282; TIBS 1 (1976), N269-N-270
New York Times, 23 January 1988

DISCHENDORFER, OTTO [1890-1967]
KURSCHNER 9,325; HEIN-E, 87; POGG 6,573; 7a(1),424
W. Limontschew, Ost. Chem. Z. 51 (1950), 143

DiSOMMA, AUGUST ADRIAN [1905-1964]
AMS 10,953
Anon., Chem. Eng. News 42(51) (1964), 59

DISQUÉ, LUDWIG [1854-1928]
NUC 144,555; SG [1]3,854; [3]4,690
L. Disqué and R.M. Snethlage, Stammliste der Familie Disqué aus Siebeldingen, 2nd Ed., p.21. Aachen 1963

DITMAR, RUDOLF [1878-1939]
KURSCHNER 4,485-486; KERNBAUER, 385-388; TLK 3,108; POGG 5,295; 6,574; 7a(1),425
Anon., Chem. Z. 63 (1939), 123

DITT, FRIEDRICH WILHELM [1904-]
NUC 144,673DITTLER, RUDOLF [1881-1959]
FISCHER 1,318; NB, 75; AUERBACH, 219; IB, 66; POGG 6,574-575; 7a(1),426
C. Cüppers, Klin. Mon. Augenheilk. 130 (1957), 123-124

DITTMAR, CARL [1844-1920]
FISCHER 1,318-319; PAGEL, 400

DITTMAR, WILLIAM [1833-1892]
DSB 4,127-128; POGG 3,365-367; 4,333
A.C.B., Nature 45 (1892), 493-494

DITTRICH, MAX [1864-1913]
BJN 18,86; DRULL, 49-50; BLOKH 1,154; POGG 4,333-334; 5,296-297
Anon., Ber. chem. Ges. 46 (1913), 1892
E. Ebler, Chem. Z. 37 (1913), 745

DITTRICH, PAUL [1859-1936]
KURSCHNER 4,487; FISCHER 1,319; MOTSCH, 17-18; BINDSEIL, 67-77
R. Fischl, Med. Klin. 25 (1929), 1529-1530
Anon., Deutsche Z. ger. Med. 14 (1929), 199-204
A.M. Marx, Deutsche Z. ger. Med. 26 (1936), i-iii

DIVERS, EDWARD [1837-1912]
WWWS, 464; POGG 3,366; 4,334; 5,297
J.M., Proc. Roy. Soc. A88 (1913), viii-x
A.S., J. Chem. Soc. 103 (1913), 746-749
D. Reilly, J. Chem. Ed. 30 (1953), 234-237

DIXEY, FREDERICK AUGUSTUS [1855-1935]
WW 1935, pp.908-909; O'CONNOR(2), 80-81; IB, 66; WWWS, 464
G.D. Hale-Carpenter, Nature 135 (1935), 213

DIXON, HAROLD BAILY [1852-1930]
DSB 4,130; FINDLAY, 126-145; MAYERHOFER 1,846-847; WWWS, 464; POGG 4,335-336; 5,298; 6,576-577
H.B. Baker and W.A. Bone, J. Chem. Soc. 1931, pp.3349-3368; Proc. Roy. Soc. A134 (1932), i-xvii
E.H. Huntress, Proc. Am. Acad. Arts Sci. 81 (1952), 46-47

DIXON, HENRY HORATIO [1869-1953]
DSB 4,130-131; DNB [1951-1960], 302-303; DESMOND, 188; IB, 66; WWWS, 465; WW 1953, p.801; POGG 6,577; 7b,1076-1077
W.R.G. Atkins, Plant Physiology 14 (1939), 615-619; Obit. Not. Fell. Roy. Soc. 9 (1954), 79-97
N.G. Ball, Proc. Linn. Soc. 165 (1955), 213-216

DIXON, KENDAL CARTWRIGHT [1911-1990]
WW 1981, p.709DIXON, MALCOLM [1899-1985]
WW 1981, p.709; SRS, 85; POGG 6,577; 7b,1077-1078
C.J.R. Thorpe, Bioch. Soc. Trans. 8 (1980), 241-242
H.B.F. Dixon, TIBS 11 (1986), 266-268
R.N. Perham, Biog. Mem. Fell. Roy. Soc. 34 (1988), 99-131

DIXON, SAMUEL GIBSON [1851-1918]
AMS 2,123; NCAB 35,134-135; WWAH, 160; WWWS, 465
E.G. Conklin et al., Proc. Acad. Nat. Sci. Phila. 70 (1918), 115-126

DIXON, WALTER ERNEST [1870-1931]
DNB [1931-1940], 231; FISCHER 1,319-320; O'CONNOR(2), 52-54; WWWS, 465
Anon., Lancet 1931(II), p.429; Brit. Med. J. 1931(II), pp.361-363
F.B.P., Nature 128 (1931), 401-402
H.H. Dale, Brit. Med. J. 1931(II), p.405
J.A. Gunn, J. Pharm. Exp. Ther. 44 (1932), 1-21

DIZÉ, MICHEL JEAN JERÔME [1764-1852]
DBF 11,404-405; BALLAND, 32-33; TSCHIRCH, 1040; FERCHL, 126; WWWS, 465;POGG 1,580-581
A. Pallas and A. Balland, Le Chimiste Dizé. Paris 1906
M. Roquette, Rev. Hist. Pharm. 17 (1965), 411-418

DJERASSI, CARL [1923-]
AMS 15(2),660; MH 2,121-123; STC 1,312-313; MSE 1,294-296; BHDE 2,129; IWW 1982-1983, p.131; WWWS, 465; POTSCH, 119
Anon., Leopoldina [3]14 (1968), 16
E. Garfield, CHEM TECH 13 (1983), 534-538
C. Djerassi, Steroids 43 (1984), 351-361; Steroids Made it Possible. Washington, D.C. 1990; The Pill, Pygmy Chimps and Degas' Horse. New York 1992

DMOCHOWSKI, ANTONI [1896-1983]
WE 3,67; WARSAW, 230; IB, 66
Czy Wiesz Kto to Jest 1938, p.143
Anon., Acta Biochim. Polon. 31 (1984), 205; 32 (1985), 83-85; Postepy Biochemii 30 (1984), 219-223

DOAN, CHARLES AUSTIN [1896-1990]
AMS 11,1242; IWW 1982-1983, p. 332; IB, 66; WWWS, 465

DOBBIE, JAMES JOHNSTON [1852-1924]
WWWS, 465; POGG 4,336; 5,298; 6,578
W.N. H[aworth] and A. L[auder], J. Chem. Soc. 125 (1924), 2681-2690
E.H. Huntress, Proc. Am. Acad. Arts Sci. 81 (1952), 47-49

DOBELL, CECIL CLIFFORD [1886-1949]
DSB 4,132-133; LINDEBOOM, 451-452; IB, 66; WWWS, 466
C.A. Hoare and D.L. Mackinnon, Obit. Not. Fell. Roy. Soc. 7 (1950),35-61

DOBRINER, KONRAD [1902-1952]
AMS 8,628; WWAH, 160; WWWS, 466

Anon., J. Clin. Endocrin. 12 (1952), 1256-1258

DOBZHANSKY, THEODOSIUS [1900-1975]
DSB 17,232-242; AMS 12,1475; MH 1,139-140; MSE 1,296-297; CB 1962, pp.105-107; IB, 67; W (3rd Ed), 589; ABBOTT, 37-38
B. Glass, Am. Phil. Soc. Year Book 1976, pp.49-53
L. Ehrman and B. Wallace, Nature 260 (1976), 179
F.J. Ayala, Ann. Rev. Gen. 10 (1976), 1-6; J. Heredity 68 (1977), 3-10; Biog. Mem. Nat. Acad. Sci. 55 (1985), 163-213
E.B. Ford, Biog. Mem. Fell. Roy. Soc. 23 (1977), 59-89
M.B. Fuller, Mendel Newsletter 18 (1980), 1-8
J.R. Powell, Genetics 117 (1987), 363-366
R.E. Kohler, Hist. Sci. 29 (1991), 335-375

DOCHEZ, ALPHONSE RAYMOND [1882-1964]
AMS 10,958-959; FISCHER 1,320-321; NCAB E,325-326; DAMB, 205-206
Y. Kneeland, Trans. Assoc. Am. Phys. 78 (1965), 21-25
M. Heidelberger et al., Biog. Mem. Nat. Acad. Sci. 42 (1971), 29-46
A.M. Harvey, Johns Hopkins Med. J. 147 (1980), 59-63

DOCTERS van LEEUWEN, JAN HENDRIK KAREL [1863-1935]
RSC 16,674
Anon., Chem. Wkbl. 32 (1935), 751

DODDS, EDWARD CHARLES [1899-1973]
WW 1965, pp.844-845; DNB [1971-1980], pp.244-245; MEDVEI, 725-727; WWWS, 467
Anon., Lancet 1973(II), pp.1506-1507; Nature 249 (1974), 95-96
A.E. Kellie, Chem. Brit. 10 (1974), 304-305
F. Dickens, Biog. Mem. Fell. Roy. Soc. 21 (1975), 227-267

DODGE, BERNARD OGILVIE [1872-1960]
AMS 10,960; WWAH, 160
W.J. Robbins, Science 133 (1961), 741-742; Biog. Mem. Nat. Acad. Sci. 36 (1962), 85-124; Neurospora Newsletter 20 (1973), 10-14
W.J. Robbins et al., Bull. Torrey Bot. Club 88 (1961), 111-121

DÖBEREINER, JOHANN WOLFGANG [1780-1849]
DSB 4,133-135; NDB 4,11-12; ADB 5,268-270; HEIN, 123-126; PRANDTL, 37-77;MAYERHOFER 2,4-5; PHILIPPE, 757-758; TSCHIRCH, 1040; SCHAEDLER, 29; POTSCH, 119-120; BLOKH 1,155-156; FERCHL, 127-128; COLE, 153; CALLISEN 5,244-252; 27,317-319; POGG 1,582-584; 7a Suppl.,157-160
F. Henrich, Z. angew. Chem. 36 (1923), 482-484
E. Theis, Z. angew. Chem. 50 (1937), 46-50
E.H. Huntress, Proc. Am. Acad. Arts Sci. 77 (1949), 40-41
W. Prandtl, J. Chem. Ed. 27 (1950), 176-181
P. Collins, Ambix 23 (1976), 96-115
D. Linke, Z. Chem. 21 (1981), 309-319

DOEBNER, OSCAR [1850-1907]
NDB 4,13-14; BJN 12,21*; MAYERHOFER 2,5-6; BLOKH 1,156-157; POTSCH, 120; FERCHL, 132; TSCHIRCH, 1040; SCHAEDLER, 30; WWWS, 468; POGG 3,367; 4,337; 5,299
C. Schotten, Ber. chem. Ges. 40 (1907), 5131-5140
D. Vorländer, Z. angew. Chem. 20 (1907), 736
Anon., Chem. Z. 31 (1907), 361
E.H. Huntress, Proc. Am. Acad. Arts Sci. 78 (1950), 32-33

DÖLLINGER, IGNAZ [1770-1841]
DSB 4,146-147; NDB 4,20-21; ADB 5,315-318; LF 3,79-95; LANGHANS, 67-73; HIRSCH 2,283-285; Suppl.,237-238; MAYERHOFER 2,8-9; WWWS, 470; DECHAMBRE 30,361-362; POGG 1,585; CALLISEN 5,253-256; 27,319-320

DÖNITZ, WILHELM [1838-1912]
BJN 18,15*; FISCHER 1,321-322; KALLMORGEN, 247; OLPP, 97-98; WWWS, 472 BULLOCH, 361; PAGEL, 402-403
G. Gaffky, Deutsche med. Wchschr. 38 (1912), 718-719
G.H.F. Nuttall, Parasitology 5 (1913), 253-261

DÖPPING, OTTO [1814-1863]
FERCHL, 128; RSC 2,322; POGG 1,585

DOERING, WILLIAM von EGGERS [1917-]
AMS 15(2),670; WWA 1980-1981, p.891; IWW 1982-1983, p.334
Anon., Nachr. Chem. Techn. 4 (1956), 234

DOERMANN, AUGUST HENRY [1918-1991]
AMS 15(2),670

DÖRPINGHAUS, WILHELM THEODOR [1878-]
JV 18,16; NUC 146,32

DÖRR, GUSTAV [1874-1918]
JV 17,282; NUC 146,34

DÖERR, ROBERT [1871-1952]
NDB 4,36-37; KURSCHNER 5,247; FISCHER 1,322; IB, 67; WWWS, 469
F. Reuter, Wiener klin. Wchschr. 64 (1952), 129-130

DORR, WILHELM [1899-]
JV 41,428

DOFLEIN, FRANZ THEODOR [1873-1924]
NDB 4,40; FISCHER 1,322; WWWS, 469
A. Pratje, Münch. med. Wchschr. 71 (1924), 1543-1544
R. Hesse, Zool. Jahrb. (Abt. Anatomie) 47 (1925), 191-211

DOGEL, VALENTIN ALEKSANDROVICH [1882-1955]
DSB 4,142-143; BSE 14,621-622; KUZNETSOV(1963), 423-433; WWR, 136; IB, 67
Anon., Trudy Inst. Ist. Est. 27 (1955), 404-405
Y.I. Polyansky, V.A. Dogel. Leningrad 1969

DOGIEL (DOGEL), ALEKSANDR STANISLAVOVICH [1852-1922]
BSE 14,621; BME 9,666-667; PSB 5,279; FISCHER 1,322; KAZAN 2,183-185;KUZNETSOV(1963), 223-232; HAYMAKER, 108-111; KONOPKA 2,444-446;KOSMINSKI, 94,595; WWR, 136
D.I. Deineka, Arkh. Anat. Gist. Emb. 3(2) (1924), 117-124
N.G. Khlopin and V.P. Mikhailov, Usp. Sov. Biol. 36 (1953), 79-90

DOGIEL (DOGEL), JAN (IVAN MIKHAILOVICH) [1830-1916]
PSB 5,279-280; BME 9,688-689; HIRSCH 2,287-288; PAGEL, 403-405; KAZAN 2,185-188; KONOPKA 2,446-450; KOSMINSKI, 93-94,595
D. Kharkovich, Farm. Toks. 20(3) (1957), 89-91

DOHRN, ANTON [1840-1909]
DSB 15,122-125; NDB 4,54-56; BJN 14,19*-20*; FREUND 1,251-261
E.R. Lankester, Nature 81 (1909), 429-431
F. Raffaele, Arch. Ital. Biol. 52 (1909), 315-320
Anon., Leopoldina 45 (1909), 126; Pop. Sci. Mon. 76 (1910), 98-101
W. Waldeyer, Anat. Anz. 35 (1910), 596-603
C. Herbst, Pommersche Lebensbilder 1 (1934), 293-303
A. Kühn, Pubblicazione della Stazione Zoologica di Napoli 22 Suppl. (1950), 1-205
T. Heuss, Anton Dohrn. Berlin 1991

DOHRN, MAX [1874-1943]
NDB 4,56-57; IB, 67; WWWS, 469; POTSCH, 120; POGG 7a(1),431-432
Anon., Z. angew. Chem. 56 (1943), 44,184

DOIJER, JACOB WIJBRAND [1852-1917]
RSC 14,640
Anon., Chem. Wkbl. 14 (1917), 253

DOINIKOV, BORIS SEMENOVICH [1879-1948]
BME 9,700-701; WWR,141
Y.M. Zhabotinsky, Arkh. Anat. Gist. Emb. 36(4) (1959), 94-99
G.A. Akimov et al., Arkh. Pat. 42(6) (1980), 67-70; Vrachebnoe Delo 1980(4), pp.116-119

DOISY, EDWARD ADALBERT [1893-1986]
AMS 14,1208; MH 1,140; STC 1,314-315; MSE 1,298-299; CB 1949, pp.161-162; POTSCH, 120; IWW 1982-1983, p.334; IB,67; WWWS,469; POGG 6,584-585; 7b,1097-1100
E.A. Doisy, Ann. Rev. Biochem. 45 (1976), 1-9
R.W. McKee, Am. Phil. Soc. Year Book 1988, pp.153-157

DOLBEAR, AMOS EMERSON [1837-1910]
WWWS, 469-470; POGG 3,368-369; 4,339; 5,300
Anon., Pop. Sci. Mon. 76 (1910), 415-416

DOLD, HERMANN [1882-1962]
KURSCHNER 9,337; FISCHER 1,323-324; AUERBACH, 221-222; KIEL, 87; IB, 67; TLK 3,109; POGG 7a(1),432-434
Anon., Deutsche med. Wchschr. 87 (1962),2454; Naturw. Rund. 16 (1963),82

DOLE, MALCOLM [1903-1990]
AMS 14,1209; WWA 1978-1979, p.861
Anon., Chem. Eng. News 69(4) (1991), 57

DOLE, VINCENT PAUL [1913-]
AMS 15(2),672; WWA 1980-1981, p.894

DOLEZALEK, FRIEDRICH [1873-1920]
NDB 4,59; DBJ 2,744; MAYERHOFER 2,19-20; BLOKH 1,157-158; WWWS, 470; POTSCH, 120-121; POGG 5,301; 6,586-587
K.A. Hofmann, Ber. chem. Ges. 54A (1921), 21-25
K. Arndt, Chem. Z. 45 (1921), 85
H. Schulze, Z. Elektrochem. 27 (1921), 89-92

DOLINSKI, GUSTAW [1846-1906]
PSB 5,286; KOSMINSKI, 94,595; KONOPKA 2,451-453

DOLLFUS, CHARLES [1828-1907]
DBF 11,454; RSC 2,309

DOLLFUS, CHARLES ÉMILE [1805-1858]
DBF 11,454; WININGER 2,65

DOLLFUS-AUSSET, DANIEL [1797-1870]
DBF 11,458; WININGER 2,65

DOLLO, LOUIS [1857-1931]
DSB 4,147-148; BNB 34, 233-242
P. Brien, Ann. Acad. Roy. Belg. 117 (1951), 69-138
S.J. Gould, J. Hist. Biol. 3 (1970), 189-212

DOMAGK, GERHARD [1895-1964]
DSB 4,153-156; MH 2,123-125; MSE 1,299-300; SCHWERTE 2,143-150; W, 148; FREUND 2,45-58; MAYERHOFER 2,25-26; CB 1958, pp.124-126; WWWS, 471; POTSCH, 121; IB, 68; ABBOTT, 38-39; POGG 7a(1),435-436
J. Colebrook, Biog. Mem. Fell. Roy. Soc. 10 (1964), 39-50
H. Chiari, Alm. Akad. Wiss. Wien 114 (1964), 340-355
F. Grundmann, Verhandl. path. Ges. 49 (1965), 380-386
F.J. Beer, Hist. Sci. Med. 8 (1974), 435-444
H. Schreiber and H. Schadewaldt, Deutsche med. Wchschr. 110 (1985), 1138-1142,1179-1181
H. Otten, Journal of Antimicrobial Chemotherapy 17 (1985), 689-696
M.H. Bickel, Gesnerus 45 (1988), 67-86

DOMBASLE, MATTHIEU de [1777-1843]
DBF 11,467-468; POGG 1,589
R. Cercle, Matthieu de Dombasle. Paris 1946

DONALDSON, HENRY HERBERT [1857-1938]
DSB 4,160-161; DAB [Suppl.2], 156-157; NCAB 28,374; DAMB, 209; WWWS, 472
E.G. Conklin, Am. Phil. Soc. Year Book 1938, pp.364-370; Science 88 (1939), 72-74; Biog. Mem. Nat. Acad. Sci. 20 (1939), 229-243

DONATH, EDUARD [1848-1932]
OBL 1,195; STURM 1,271; KURSCHNER 4,500; WININGER 2,67; 6,545; TLK 3,112; POGG 3,370-371; 4,341; 5,301; 6,587-588
A. Lissner, Chem. Z. 52 (1928), 954-955
Anon., Chem. Z. 56 (1932), 455; Ost. Chem. Z. 35 (1932), 106-107

DONATH, JULIUS [1870-1950]
FISCHER 1,326; STUMPF, 30-36; LESKY, 324
P. Gulyas, Magyar Irok 6 (1944), 120-123
D. Goltz, Clio Medica 16 (1982), 193-217
A.M. Silverstein, Cellular Immunology 97 (1986), 173-188

DONATH, WILLEM FREDERIK [1889-1957]
LINDEBOOM, 463-464
J. Ruttink, Chem. Wkbl. 53 (1957), 449-454

DONAU, JULIUS [1877-1960]
KURSCHNER 9,339; TLK 3,112; POTSCH, 121; POGG 5,302; 6,588; 7a(1),437
G. Gorbach, Mikrochemie 33 (1948), 273-277
Anon., Microchem. J. 1 (1957), 175-176

DONCASTER, LEONARD [1877-1920]
WW 1920, p.724; WWWS, 472
Anon., Brit. Med. J. 1920(I), p.813
W.B., Proc. Roy. Soc. B92 (1921), xli-xlvi

DONDERS, FRANCISCUS CORNELIS [1818-1889]
DSB 4,162-164; NNBW 1,727-729; LINDEBOOM, 464-465; HIRSCH 2,291-294;UTRECHT, 111-116; WWWS, 472; POGG 3,371-372;4,341
C. von Voit, Sitz. Bayer. Akad. 19 (1889), 118-124
Anon., Proc. Am. Acad. Arts Sci. 24 (1889), 465-470
P.J. Nuel, Ann. Ocul. 52 (1889), 5-107
W. Bowman, Proc. Roy. Soc. 49 (1891), vii-xxiv
C.A. Pekelharing, Janus 24 (1919), 57-76
E.C. van Leesum, Het Levenswerk van Franciscus Cornelis Donders. Haarlem 1932

DONEGAN, JOSEPH FRANCIS [1893-1985]
SRS, 54; IB, 68
Anon., Lancet 1985(II), p.677

DONKER, HENDRICK JEAN LOUIS [1899-]
Wie is Dat 1956, p.146

DONNAN, FREDERICK GEORGE [1870-1956]
DSB 4,165; DNB [1951-1960], 305-306; WW 1949, pp.770-771; SRS, 15; POTSCH, 121-122; W, 149; MAYERHOFER 2,35-36; POGG 5,302; 6,590;7b,1104-1105
F.A. Freeth, Biog. Mem. Fell. Roy. Soc. 3 (1957), 23-39

DONNÉ, ALFRED [1801-1878]
DBF 11,530-531; DBA 1,684; HIRSCH 2,294-295; VERSO, 94-96; FERCHL, 129; BERGER-LEVRAULT, 56; WWWS, 472; POGG 3,372
L. Figuier, Ann. Scient. 22 (1878), 509-510
C. Dreyfus, Nouv. Rev. Hem. 2 (1962), 241-255
L. Alizard, Alfred Donné, un Precurseur d'Hématologie. Paris 1963
M.L. Verso, Med. Hist. 15 (1971), 56-58
A.F. La Berge, J. Hist. Med. 46 (1991), 20-43

DONNY, FRANÇOIS MARIE LOUIS [1822-1896]
NBW 12, 243-248; POGG 1,593; 3,372
G. van den Mensbrugge, Bull. Acad. Roy. Belg. [3]32 (1896), 496-498
D.H. Travenna, Ann. Sci. 37 (1980), 379-386

DONOHUE, JERRY [1920-]
AMS 15(2),684; WWA 1980-1981, p.901

DONOVAN, MICHAEL [1790-1876]
HIRSCH 2,295; COLE, 154; TSCHIRCH, 1041; FERCHL, 129; WWWS, 473; CALLISEN 27,327-328; POGG 1,590-591
Anon., Medical Press and Circular NS21 (1876), 391

DONSELT, WALTER [1873-]
JV 19,179; NUC 149,51

DONY-HÉNAULT, OCTAVE [1875-1952]
POGG 6,590-591; 7b,1105-1106
J. Timmermans, Ann. Acad. Roy. Belg. 130 (1964), 3-17

DOPTER, CHARLES [1873-1950]
DBF 11,550-551; FISCHER 1,326-327
Anon., Ann. Inst. Pasteur 79 (1950), 241-245
L. Launoy, Bull. Soc. Path. Exot. 44 (1951), 8-15

DORÉE, CHARLES [1875-1972]
POGG 6,592
Who's Who in British Science 1953, p.76
Directory of British Scientists 1966-1967, vol.1, p.478

DOREMUS, ROBERT OGDEN [1824-1906]
DAB 5,376-377; NCAB 28,275-276; WWAH, 162; WWWS, 474
C.F. Chandler, Science 23 (1906), 513-514
Anon., Medico-Legal Journal 24 (1906-1907), 81-100

DORFMAN, ALBERT [1916-1982]
AMS 15(2),688; WWA 1980-1981, p.904

DORFMAN, RALPH ISADORE [1911-1985]
AMS 15(2),689
New York Times, 9 December 1985DORFMÜLLER, THEODOR
[1899-] NUC 147,202

DORMANN, EDMUND [1884-1959]
JV 32,326; NUC 147,241

DORN, FRIEDRICH ERNST [1848-1916]
DBJ 1,351; WWWS, 474-475; POTSCH, 122; RSC 14, 658-659; POGG 3,372; 4,593; 6,593

DORNOW, ALFRED [1909-1966]
KURSCHNER 10,414; POGG 7a(1),439

DOROGI, STEFAN ANTON [1885-]
NUC 147,288

DORP, WILLEM ANNE van [1847-1914]
BLOKH 1,159-160; WWWS, 475; POGG 3,373-374; 4,344; 5,1294
S. Hoogewerf, Chem. Wkbl. 8 (1911), 461-468; Rec. Trav. Chim. Pays-Bas 34 (1915), 353-390
H.A. Lorentz, Chem. Wkbl. 11 (1914), 1014-1017
C. Liebermann, Ber. chem. Ges. 47 (1914), 2667-2670

DORRER, EUGEN [1902-]
JV 43,410; NUC 147,315

DORRER, OTTO [1907-]
JV 48,577; NUC 147,315-316

DORSET, MARION [1872-1935]
AMS 5,294; DAB 21, 258; NCAB 26,300; OEHRI, 43-44; WWAH, 162
Anon., J. Wash. Acad. Sci. 25 (1935), 428
J.R. Mohler, Science 82 (1935), 118

DORVAULT, FRANÇOIS LAURENT MARIE [1815-1879]
DBF 11,624; GORIS, 349; TSCHIRCH, 1041; FERCHL, 130; POGG 3,374
M. LePrince, Figures Pharmaceutiques Françaises, pp.119-124. Paris 1953
P. Julien, Bulletin de Pharmacie 29 (1976), 49-57
P. Boussel, Dorvault, sa Vie et son Oeuvre. Honfleur 1979
B. Mory et al., Rev. Hist. Pharm. 27 (1980), 79-122

DORVEAUX, PAUL [1851-1938]
DBF 11,625; FISCHER 1,328; REBER, 385; TSCHIRCH, 1041; POGG 4,344; 5,595; 7b,110
P. Delaunay, Bull. Soc. Hist. Med. 32 (1938), 20-25
E. Wickersheimer, Janus 42 (1938), 65-68
M. Speter, Isis 30 (1939), 46-51

DOTY, PAUL MEAD [1920-]
AMS 15(2),693-694; IWW 1982-1983, p.341; WWWS, 476

DOTZENRODT, HEINRICH [1910-]
JV 51,90

DOUDOROFF, MICHAEL [1911-1975]
AMS 12,1506; SACKMANN, 96-97
R.Y. Stanier, Am. Soc. Microbiol. News 41 (1975), 737-738

DOUGLAS, CLAUDE GORDON [1882-1963]
DNB [1961-1970], 303-304; WW 1959, pp.847-848; O'CONNOR(2),105-108; WWWS, 477
D.J. Cunningham, Biog. Mem. Fell. Roy. Soc. 10 (1964), 51-74

DOUGLAS, STEWART RANKEN [1871-1936]
WW 1936, pp.941-942; BULLOCH, 362; NUESCH, 56-59; WWWS, 477
A.F., J. Path. Bact. 42 (1936), 515-522
P. Laidlaw, Obit. Not. Fell. Roy. Soc. 2 (1936), 175-182
Anon., Lancet 1936(I), pp.229-231; Nature 137 (1936), 215;
Brit. Med. J. 1936(I), p.239

DOUNCE, ALEXANDER LATHAM [1909-]
AMS 14,1230-1231; WWWS, 478

DOWNES, ARTHUR HENRY [1851-1938]
WW 1938, p.945; WWWS, 479

DOWNES, HELEN RUPERT [1893-1992]
AMS 11,1279

DOWNEY, HAL [1877-1959]
AMS 9(II),290
O.P. Jones, J. Hist. Med. 27 (1972), 173-186

DOWNIE, ALLAN WATT [1901-1988]
WW 1981, p.729; IB, 69; WWWS, 479
D.A.J. Tyrrell and K. McCarthy, Biog. Mem. Fell. Roy. Soc. 35 (1990), 99-112

DOX, ARTHUR WAYLAND [1882-1954]
AMS 8,648; IB, 69; POGG 6,598; 7b,1114
Anon., Chemical Abstracts 48 (1954),13739; Chem. Eng. News. 33 (1955),158

DOYÈRE, LOUIS MICHEL FRANÇOIS [1811-1863]
DBF 11,719; HIRSCH 2,305; DECHAMBRE 30,550-551; WWWS, 480

DOYON, MAURICE [1863-1934]
FISCHER 1,329; GUIART, 138
P.J. Portier, Bull. Acad. Med. 112 (1934), 281-285
H. Roger, Presse Med. 42 (1934), 1381-1382

DRABKIN, DAVID LEON [1899-1980]
AMS 11,1284; KAGAN(JA), 742-743; COHEN, 51
J.B. Marsh, Trans. Coll. Phys. Phila. [5]3 (1981), 315-321

DRAGENDORFF, GEORG [1836-1898]
NDB 4,99; ADB 48,69-70; BJN 5,16*; HEIN, 129-130; LEVITSKI 2,239-246; WELDING, 174; BLOKH 1,160-161; REBER, 13-14; SCHAEDLER, 29; TSCHIRCH, 1041; WWWS, 480; POGG 3,377; 4,347; 7a Suppl.,163-164
Anon., Nature 57 (1898), 612-613; Leopoldina 34 (1898), 106-107;
Chem. Z. 22 (1899), 307
C. Franke-Schwerin, Arch. Fr. Nat. Meckl. 52 (1899), 42-45
K. Dragendorff, Pharmazie 7 (1952), 498-502
U. Kokoska, Johann Georg Noel Dragendorff. Berlin 1983

DRAGSTEDT, CARL ALBERT [1895-1983]
AMS 11,1284; WWWS, 480

C.A. Dragstedt, Persp. Biol. Med. 8 (1965), 218-229

DRAGSTEDT, LESTER REYNOLD [1893-1975]
AMS 12,1522; MH 2,126-128; STC 1,322-323; MSE 1,306-307; DAMB, 213-214
J.H. Landor, Surgery 81 (1977), 442-446; Gastroenterology 80 (1981), 846-853
O.H. and S.D. Wangensteen, Biog. Mem. Nat. Acad. Sci. 51 (1980), 63-95

DRAKE, NATHAN LINCOLN [1898-1959]
AMS 8,650; GOULD, 37
Anon., Chem. Eng. News 37(44) (1959), 121

DRALLE, EDUARD [1870-1940]
JV 9,207; RSC 19,74
P.T. Hoffmann, Neues Altona, vol.2, pp.286-288. Jena 1935

DRAPER, JOHN CHRISTOPHER [1835-1885]
MILES, 127; SILLIMAN, 129-130; POGG 3,378

DRAPER, JOHN WILLIAM [1811-1882]
DSB 4,181-183; DAB 5,438-441; KELLY, 346-347; MILES, 127-128; COLE, 157; ELLIOTT, 78; SILLIMAN, 78-82; MAYERHOFER 2,52-53; FERCHL, 131; POTSCH, 122; WWAH, 165; WWWS, 481; POGG 1,601-602; 3,377-378
Anon., Proc. Am. Acad. Arts Sci. 17 (1882), 424-429
G.F. Barker, Biog. Mem. Nat. Acad. Sci. 2 (1886), 351-388
D.H. Fleming, John William Draper and the Religion of Science. Philadelphia 1950
W.H. Waggoner, J. Chem. Ed. 60 (1983), 200-201

DRAWERT, HORST [1910-1976]
KURSCHNER 11,519; AUERBACH, 791-792; POGG 7a(1),442
E. Schnepf, Ber. bot. Ges. 92 (1979), 689-694

DRECHSEL, EDMUND [1843-1897]
NDB 4,104-105; ADB 48,77; BJN 4,55*; MAYERHOFER 2,53-54; BLOKH 1,161-162; POTSCH, 122; REBER, 315-316; PAGEL, 418; POGG 3,380; 4,347-348
M. Siegfried, Ber. chem. Ges. 30 (1897), 2169-2173
J.H. Graf, Verhandl. Schw. Nat. Ges. 80 (1897), 234-237
A. Tschirch, Naturw. Rund. 12 (1897), 632-635; Leopoldina 34 (1898), 43-46,61-68

DRESEL, KURT [1892-1951]
KURSCHNER 4,514; FISCHER 1,330; BHDE 2,226; LDGS, 65; KOREN, 267
O. Lowenstein, Proc. Virchow Med. Soc. 10 (1951), 101-104

DRESER, HEINRICH [1860-1924]
JV 1,102; POGG 7a(1),444-445
H. Meyer, Arch. exp. Path. Pharm. 106 (1925), i-vii

DRESSEL, OSKAR [1865-1941]
POGG 7a(1),446
Anon., Z. angew. Chem. 55 (1942), 20

DREW, CHARLES RICHARD [1904-1950]
AMS 8,653; DAB [Suppl.4], 242-243; CB 1944, pp.179-180; DAMB, 215-216
W.M. Cobb, J. Nat. Med. Assn. 42 (1950), 239-246
R. Hardwick, Charles Richard Drew, Pioneer in Blood Research. New York 1967
R. Lichello, Pioneer in Blood Plasma, Dr. Charles Richard Drew. New York 1968?
D. Parks, J. Nat. Med. Assn. 71 (1979), 893-895
C.E. Wynes, Charles Richard Drew: The Man and the Myth. Urbana, 1988
A. Malone-Lonesome, Charles Drew. New York 1990

DREWS, BRUNO [1898-1969]
KURSCHNER 10,420; POGG 7a(1),446-447
Anon., Chem. Z. 93 (1969), 437

DREWSEN, VIGGO BEUTNER [1858-1930]
POTSCH, 123; POGG 6,600
S. Schmidt-Nielsen, Paper Industry 13 (1931), 63-64

DREYER, GEORGE PETER [1866-1931]
AMS 4,264; WWAH, 166
H.A. McGuigen, Science 73 (1931), 355

DREYER, GEORGES [1873-1934]
DNB [1931-1940], 237-238; O'CONNOR(2), 92-93; ULLMANN, 46-48; WWWS, 483
Anon., J. Path. Bact. 39 (1934), 707-723
S.R. Douglas, Obit. Not. Fell. Roy. Soc. 1 (1934), 569-576
E.W.A. Walker, Brit. Med. J. 1934(II), p.946
M. Dreyer, Georges Dreyer. Oxford 1937

DREYFUS-BRISAC, LOUIS LUCIEN [1849-1903]
DBF 11,771
R. Kohn, Rev. Hist. Med. Heb. 1952, pp.45-46

DRIESCH, HANS [1867-1941]
DSB 4,186-189; NDB 4,125-126; DRULL, 51-52; MAYERHOFER 2,59-61; W, 151; SCHWERTE 2,218-227; LEIPZIG 1,183-189; ABBOTT, 39; IB, 70; WWWS, 483
E. Ungerer, Naturwiss. 29 (1941), 457-462
C. Herbst, Arch. Entwickl. 141 (1942), 111-153
H. Driesch, Lebenserinnerungen. Basle 1951
A. Wenzl (Ed.), Hans Driesch. Basle 1951
R. Mocek, Wilhelm Roux - Hans Driesch. Basle 1951
F.B. Churchill, J. Hist. Biol. 2 (1969), 165-185
J.M. Oppenheimer, Bull. Hist. Med. 44 (1970), 378-382
H.H. Freyhofer, The Vitalism of Hans Driesch. Frankfurt a.M. 1982

DRIGALSKI, KARL WILHELM von [1871-1950]
KURSCHNER 4,518-519; NB, 80-81; FISCHER 1,331-332; FRANKENTHAL, 281; WWWS, 1731-1732
B. Harms, Berl. med. Z. 1 (1950), 489
W. Schnell, Deutsche med. Wchschr. 75 (1950), 902-903

DRINKER, CECIL KENT [1887-1956]
DAB [Suppl.6], 174-175; FISCHER 1,332; WWAH, 166; WWWS, 483
J.H. Means, Trans. Assoc. Am. Phys. 69 (1956), 11-13
N.C. Staub, Lymphology 12(3) (1979), 115-117

DROBOTKO, VIKTOR GRIGORIEVICH [1885-1966]
WWR, 143; BME 9,791-792; UKRAINE 2,260-261
Anon., Mikr. Zhur. 27 (1965), 93-94

DROSSBACH, GEORG PAUL [1866-1903]
BJN 8,26*; POGG 4,348; 5,305
O. Brunck, Z. angew. Chem. 16 (1903), 855-856

DROZDOV, VIKTOR IVANOVICH [1846-1899]
VENGEROV 2,314; ZMEEV 4,104
W. Pagel, Virchows Jahresber. 1899(I), p.332

DRUCKER, CARL [1876-1959]
KURSCHNER 4,520; TLK 3,115; LDGS, 18; POGG 5,305-306; 6,602; 7a(1),449

DRUDE, PAUL [1863-1906]
DSB 4,189-193; NDB 4,138-139; BJN 11,17*; GUNDEL, 174-181; DZ, 283-284; MAYERHOFER 2,65-66; POGG 4,349-350; 5,306; 7a Suppl.,165-166
M. Planck, Ber. phys. Ges. 4 (1906), 599-630
Anon., Chem. Z. 30 (1906), 687

DRUMMOND, JACK CECIL [1891-1952]
DNB [1951-1960], 313-314; WW 1949, p.791; IB, 70; WWWS, 484; POGG 6,602-603; 7b,1119-1122
H.J. Channon, Chem. Ind. 1952, pp.905-906
F.G. Young, Obit. Not. Fell. Roy. Soc. 9 (1954), 99-129
D.F. Hollingsworth and N.C. Wright, Brit. J. Nutrition 8 (1954), 319-324

A.M. Copping, J. Nutrition 82 (1964), 3-9

DRURY, ALAN NIGEL [1889-1980]
DNB [1971-1980], 253-254; WW 1980, p.725
R.A. Kekwick, Biog. Mem. Fell. Roy. Soc. 27 (1981), 173-198

DRUSCHEL, WILLIAM ALLEN [1874-1931]
AMS 4,265; POGG 5,306
Anon., Science 73 (1931), 488

DRZNIEWICZ, KAZIMIERZ [1851-]
KONOPKA 2,503; SG [2]4,516

DUANE, WILLIAM [1872-1935]
DSB 4,194-197; DAB 21,266-267; NCAB 18,403-404; MAYERHOFER 2,75; WWWS, 484-485
H.C. Richards, Am. Phil. Soc. Year Book 1937, pp.349-351
P.W. Bridgman, Biog. Mem. Nat. Acad. Sci. 18 (1937), 23-41

DUBIN, HARRY ENNIS [1891-1981]
AMS 11,1298; IB, 70

DUBINI, ANGELO [1813-1902]
DSB 4, 197-198; HIRSCH 2,314; Suppl.,245; OLPP, 102-104; WWWS, 485
B. Galli-Valerio, Arch. Parasit. 7 (1903), 138-151
L. Castaldi, Riv. Stor. Sci. 28 (1937), 204-208
L. Belloni, Gesnerus 19 (1962), 101-118

DUBININ, MIKHAIL MIKHAILOVICH [1901-]
BSE (3rd Ed.)8,518; MH 2,130-132; MSE 1,311-312; IWW 1980-1981, p.333; WWWS, 485; POGG 6,604; 7b,1124-1129
B.P. Bering ans V.V. Serpinski, Zhur. Fiz. Khim. 35 (1961), 225-227

DUBININ, NIKOLAI PETROVICH [1907-]
BSE (3rd Ed.)8,518; IWW 1982-1983, p.347; WWWS, 485
N.P. Dubinin, Vechnoe Dvizhenie. Moscow 1973

DUBNOFF, JACOB WILLIAM [1909-1972]
AMS 12,1540

DU BOIS, DELAFIELD [1880-1965]
AMS 8,657
Anon., Science 148 (1965), 486

DUBOIS, EUGENE FLOYD [1882-1959]
AMS 9(II),294; NCAB 52,467-468; CHITTENDEN, 70-72; DAMB,216-217; IB, 28; WWAH, 167; WWWS, 485
D.B. Dill, Science 130 (1959), 1746-1747
J.H. Means, Trans. Assoc. Am. Phys. 72 (1959), 23-28
A. Forbes, Am. Phil. Soc. Year Book 1959, pp.122-127
H. Pollock, J. Nutrition 75 (1961), 306
J.C. Aub, Biog. Mem. Nat. Acad. Sci. 36 (1962), 125-145

DUBOIS, RAPHAEL HORACE [1849-1929]
DSB 17,244-245; DBF 11,970-971; GORIS, 351-359; GUIART, 174-175; WWWS, 486
H. Cardot, Rev. Sci. 66 (1928), 1-9
E.N. Harvey, A History of Luminescence, pp.242-243. Philadelphia 1957
G. Peres, Bulletin de l'Académie de Var (Toulon) 147 (1979), 273-282

DU BOIS-REYMOND, EMIL [1818-1896]
DSB 4,200-205; NDB 4,146-148; ADB 48,118-126; HAYMAKER, 178-181; W, 152 BJN 1,125-131; 3,90*-91*; HIRSCH 1,609-611; Suppl.,114; DRULL, 52-53; PAGEL, 207-210; BRAZIER, 71-80,85-86; POGG 1,228; 3,152-153; 4,150-151; 6,269; 7a Suppl.,98-102
C. von Voit, Sitz. Bayer. Akad. 27 (1897), 423-432
J. Bernstein, Naturw. Rund. 12 (1897), 87-92
I. Munk, Deutsche med. Wchschr. 23 (1897), 17-19
E. Metze, Emil du Bois-Reymond (3rd Ed.). Bielefeld 1918
H. Boruttau, Emil du Bois-Reymond. Munich 1922
E. Du Bois-Reymond and P. Diepgen, Zwei Grosse Naturforscher des 19. Jahrhunderts. Leipzig 1927 (English transl., Baltimore 1982)

K.E. Rothschuh and E. Tutte, Acta Hist. Leop. 9 (1975), 113-135
P.F. Cranefield, Gesnerus 45 (1988), 271-282

DU BOIS-REYMOND, RENÉ [1863-1938]
KURSCHNER 4,251; FISCHER 1,334; SKVARC, 44-45; IB, 28

DUBOS, RENÉ JULES [1901-1982]
AMS 14,1253; MH 1,144-146; STC 1,326-328; MSE 1,313-314; WWWS, 486; AO 1982, pp.81-85; CB 1952, pp.163-165; 1973, pp.105-109; NDAB I, 396-397
S. Benison, Bull. Hist. Med. 50 (1976), 459-476
J.G. Hirsch, Am. Phil. Soc. Year Book 1982, pp.473-481; Trans. Assoc. Am. Phys. 97 (1984), cvii-cix
J.G. Hirsch and C.L. Moberg, Biog. Mem. Nat. Acad. Sci. 58 (1989), 133-161
F.Lery, C.R. Acad. Agr. 75 (1989), 3-11
C.L. Moberg and Z.A. Cohn, Scientific American 264(1) (1991), 66-74

DUBOSCQ, JULES [1817-1886]
DBF 11,1015; FERCHL, 132; WWWS, 486; POGG 3,383; 4,351
L. Figuier, Ann. Scient. 31 (1886), 580
M.G. Ringler, Journal of Clinical Pathology 34 (1981), 287-291

DUBOSCQ, OCTAVE [1868-1943]
DSB 4,206-207; DBF 11,1015; CHARLE(2), 112-114; IB, 70; WWWS, 486
P.P. Grasse, Arch. Zool. Exp. 84 (1944), 1-45

DUBOUX, MARCEL [1883-1943]
POGG 7a(1),450-451
P. Dutoit, Helv. Chim. Acta 26 (1943), 2082-2089; Verhandl. Schw. Nat. Ges. 123 (1943), 305-308

DUBRUNFAUT, AUGUSTIN PIERRE [1797-1881]
DBF 11,1095-1096; LAROUSSE 6,1319; BLOKH 1,162-163; SCHAEDLER, 30;POTSCH, 124-125; FERCHL, 132; WWWS, 486; POGG 3,383
L. Figuier, Ann. Scient. 25 (1881), 532-534
W.A. Davis, Chem. Ind. 49 (1930), 641-644
H.G. Fletcher, J. Chem. Ed. 17 (1940), 153-156

DUBSKY, JAN VACLAV [1882-1946]
STURM 1,283; TLK 3,116; POGG 5,308; 6,607; 7b,1133-1136
G. Druce, Nature 157 (1946), 543
A. Okac, Chem. Listy 40 (1946), 198-207
F. Jursik, Chem. Listy 77 (1983), 625-633

DUBUISSON, MARCEL GEORGES [1903-1972]
BAF, 43; IWW 1974-1975, p.459
Z.M. Bacq, Ann. Acad. Roy. Belg. 146 (1980), 21-60

DUCCA, WILHELM [1880-]
JV 21,480; NUC 150,159

DUCCESCHI, VIRGILIO [1871-1952]
FISCHER 1,335
Anon., Arch. Fisiol. 53 (1953), i-iv
A. Roncato, Archivio Sci. Biol. 38 (1954), 542-548

DUCHACHEK, FRANTISEK [1875-1931]
IB, 71
Anon., Chem. Ind. 1931, p.289; Chem. Z. 55 (1931), 185; Chem. Listy 25 (1931), 97-101
O. Kopecky, Vest. Cesk. Akad. Zem. 7 (1931), 497-505

DUCHARTRE, PIERRE [1811-1894]
DBF 11,1174-1175; VAPEREAU 6,492; WWWS, 487
G. Bonnier, Rev. Gen. Bot. 6 (1894), 481-504
D. Clos, Bull. Soc. Bot. 42 (1895), 88-143

DUCHEK, ADALBERT [1824-1882]
OBL 1,201; STURM 1,284; HIRSCH 2,232; Suppl.,246; PAGEL, 419; DRULL, 53; FROHNE, 6-15; STUMPF, 6-15; NAVRATIL, 48

Anon., Wiener med. Wchschr. 32 (1882), 255-257

DUCHENNE, GUILLAUME [1806-1875]
DBF 11,1128; VAPEREAU 5,606; DECHAMBRE 30,626-628; WWWS, 487; HIRSCH 2,322; Suppl.,246
G. Guillain, Presse Med. 33 (1925), 1601-1606
V. Robinson, Medical Life 36 (1929), 287-306
P. Guilly, Duchenne de Boulogne. Paris 1936
J. Lhermite, Bull. Acad. Med. 130 (1946), 745-755
E. Joki, Episteme 1 (1967), 273-283
E.D. Campbell, Proc. Roy. Soc. Med. 66 (1973), 18-22

DUCLAUX, ÉMILE [1840-1904]
DSB 4,210-212; DBF 11,1262-1263; GE 14,1189; FISCHER 1,335; WWWS, 487;CHARLE(2), 114-116; BLOKH 1, 163-164; POGG 3,384; 4,351-352
C.J. Martin, Nature 70 (1904), 34-35
P.J. Tillaux, Bull. Acad. Med. 51 (1904), 399-401
E. Roux, Ann. Inst. Pasteur 18 (1904), 337-362
M.D. Duclaux, La Vie d'Émile Duclaux. Paris 1906

DUCLAUX, JACQUES [1877-1978]
QQ 1977-1978, p.607; CHARLE, 58-60; POGG 5,309,1421; 6,608; 7b,1135-1137
J. Roche, C.R. Soc. Biol. 172 (1979), 1053-1054

DUDEN, PAUL [1868-1954]
NB, 81; KALLMORGEN, 250; TLK 3,116; POTSCH, 125; POGG 4,352; 5,309; 6,608-609; 7a(1),451
R. Kuhn, Ber. chem. Ges. 71A (1938)
Anon., Chem. Z. 77 (1953), 728; 78 (1954), 157-158; Nachr. Chem. Techn. 2 (1954), 36

DUDGEON, LEONARD STANLEY [1876-1938]
Who Was Who 1929-1940, p.388
H.R. Dean, J. Path. Bact. 48 (1939), 231-235

DUDLEY, HAROLD WARD [1887-1935]
SRS, 43; WWWS, 488; POTSCH, 125-126; POGG 6,609; 7b,1137
Anon., Chem. Ind. 1935, p.905; Nature 136 (1935), 671-672
H.H. Dale, Obit. Not. Fell. Roy. Soc. 1 (1935), 595-606;
J. Chem. Soc. 1936, pp.541-546
A.C.C. and C.L.E., Biochem. J. 30 (1936), 1-4

DUDLEY, WILLIAM LOFLAND [1859-1914]
NCAB 8,227; WWAH, 167; POGG 4,352; 5,309
C. Baskerville, J. Ind. Eng. Chem. 6 (1914), 856-859

DÜNSCHMANN, MAX [1858-1923]
JV 3,59; RSC 14,707; NUC 150,381

DUERCK, HERMANN [1869-1941]
NDB 4,163; FISCHER 1,337; FREUND 2,59-71; ECKERT, 48-57; IB, 71; WWWS, 496
G. Gruber, Verhandl. path. Ges. 32 (1950), 431-433

DUFLOS, ADOLPH FERDINAND [1802-1889]
ADB 48,140-141; HEIN, 132-133; MAYERHOFER 2,99; BLOKH 1, 164; FERCHL,134; SCHAEDLER, 30; TSCHIRCH, 1042; POGG 1,612-613; 3,385
Anon., Apoth. Z. 4 (1889), 1091-1092; Chem. Z. 13 (1889), 1726

DUFOUR, ALEXANDRE EUGÈNE [1875-1942]
CHARLE, 116-117; POGG 5,310; 6,611; 7b,1144

DUFRAISSE, CHARLES [1885-1969]
IB, 71; POGG 6,611-612; 7b,1145-1149
M. Delépine, Chim. Ind. 61 (1949), 614-615
A. Etienne and R.E. Oesper, J. Chem. Ed. 29 (1952), 110-111
J. LeBras, Rubber Chem. Tech. 47 (1974), G16-G22

DUGGAR, BENJAMIN MINGE [1872-1956]
DSB 4,219-221; DAB [Suppl.6], 175-177; AMS 9(II),296; NCAB 46,236-237; HUMPHREY, 72-76; CB 1952, pp.166-169; IB, 71
G.C. Ainsworth, Nature 178 (1956), 834-835
E.C. Stakman, Science 126 (1957), 690-691
G.W. Keitt, Mycologia 49 (1957), 434-438
F. Daniels, Am. Phil. Soc. Year Book 1957, pp.117-121
J.C. Walker, Biog. Mem. Nat. Acad. Sci. 32 (1958), 113-131;
Annual Review of Phytopathology 20 (1982), 33-39

DUHEM, PIERRE [1861-1916]
DSB 4,225-233; DBF 12,28-30; MAYERHOFER 2, 108; WWWS, 490; POTSCH, 126; POGG 4,354-356; 5,310-311; 6,612-613
E. Jouget, Rev. Gen. Sci. 28 (1917), 40-49
E. Picard, Mem. Acad. Sci. [2]57 (1922), xcix-cxliii
O. Manville et al., L'Oeuvre Scientifique de Pierre Duhem. Bordeaux 1927
H. Duhem, Un Savant Français. Paris 1936
M. d'Ocagne et al., Archeion 19 (1937), 126-139
D.G. Miller, Physics Today 19(12) (1966), 47-53
H.W. Paul, J. Hist. Ideas 33 (1972), 497-512
R.N.D. Miller, Ann. Sci. 33 (1976), 119-129

DUISBERG, FRIEDRICH CARL [1861-1935]
NDB 4,181-182; KALLMORGEN, 251; DZ, 287-288; TLK 3,116; WWWS, 490; POTSCH, 126-127; POGG 3,386-387; 4,356; 6,613; 7a Suppl.,172-174
B. Heymann et al., Z. angew. Chem. 44 (1931), 797-813
H. Hummel, Chem. Z. 55 (1931), 741-742
K. Duisberg, Meine Lebenserinnerungen. Leipzig 1933
A. Stock, Ber. chem. Ges. 68A (1935), 111-148
Anon., Chem. Ind. 1935, pp.292-294
H.J. Flechtner, Carl Duisberg. Düsseldorf 1981

DUISBERG, HERWARTH [1901-]
JV 44,445; NUC 151,17

DUISBERG, WALTHER [1893-1964]
JV 38,777
Anon., Chem. Eng. News 42(27) (1964), 79

DUJARDIN, FÉLIX [1801-1860]
DSB 4,233-237; DBF 12,42; GE 15,27; LAROUSSE 6,1364-1365; W, 153; MAYERHOFER 2,108-109; TSCHIRCH, 1042; WWWS, 490
L. Joubin, Arch. Parasit. 4 (1901), 5-57

DUJARRIC de la RIVIÈRE, RENÉ [1885-1969]
J. Trefouël, Bull. Acad. Med. 154 (1970), 299-304

DUKE-ELDER, WILLIAM STEWART [1898-1978]
DNB [1971-1980], 255-256; WW 1978, pp.704-705; WWWS, 490
D. Vail and A.J. Goldsmith, Am. J. Ophthalm. 45(4) (1958), 5-14
T.K. Lyle et al., Biog. Mem. Fell. Roy. Soc. 26 (1980), 85-105

DULBECCO, RENATO [1914-]
AMS 15(2),741; STC 1,328-330; MSE 1,315; WWA 1980-1981, p.933; WWWS, 490;IWW 1982-1983, p.350
R. Dulbecco, Scienza, Vita e Aventura. Milan 1989

DULK, FRIEDRICH PHILIPP [1788-1852]
NDB 4,184; HEIN, 137-138; TSCHRICH, 1042; SCHELENZ, 618; POGG 1,619-620
Anon., Arch. Pharm. 129 (1854), 81-82

DULONG, PIERRE LOUIS [1785-1838]
DSB 4,238-242; DBF 12,83; GE 15,140; LAROUSSE 6,1448-1449; FERCHL, 135;BLOKH 1,164-166; SCHAEDLER, 30; TSCHIRCH, 1042; ABBOTT-C, 38-39; POTSCH, 127; MAYERHOFER 2,109-110; W, 153-154; WWWS, 490; POGG 1,620,1559; CALLISEN 5,379-382; 27,360
P. Lemay and R.E. Oesper, Chymia 1 (1948), 171-180
J. Lecomte, Précis Anal. Acad. Rouen 1964, pp.101-127
J.W. Spronson, Chymia 12 (1967), 157-159; Janus 58 (1971), 207-221
R. Fox, Brit. J. Hist. Sci. 4 (1968), 1-22
L. Medard, Rev. Hist. Sci. 35 (1982), 321-330

DUMANSKI, ANTON VLADIMIROVICH [1880-1967]
BSE (3rd Ed.)8,534; WWR, 146; LIPSHITS 3,217; UKRAINE 2,261-262; POGG 6,614-615; 7b,1151-1155
S.M. Lipatov, Usp. Khim. 19 (1960), 759-763

DUMAS, JEAN BAPTISTE [1800-1884]
DSB 4,242-248; DBF 12,129-133; BUGGE 2,53-68; MAYERHOFER 2,111-112; PHILIPPE, 788-791; BLOKH 1,164-166; FERCHL, 135; TSCHIRCH, 1042;POTSCH, 127-128; COLE, 159-162; ABBOTT-C, 39; W, 154-155; WWWS, 491; POGG 1,621-623; 3,387; CALLISEN 5,383-385; 27,361-362
A.W. Hofmann, Ber. chem. Ges. 17(3),629-760; Proc. Roy. Soc. 37 (1884), x-xxvii
F. LeBlanc, Bull. Soc. Chim. [2]42 (1884), 549-559
W. Perkin, J. Chem. Soc. 47 (1885), 310-323
C. von Voit, Sitz. Bayer. Akad. 15 (1885), 136-153
G. Urbain, Bull. Soc. Chim. [5]1 (1934), 1425-1447
M. Schofield, Chem. Ind. 1944, pp.333-335
E.H. Huntress, Proc. Am. Acad. Arts Sci. 78 (1950), 23-24
J. Cheymol and A. Soubiran, Presse Med. 76 (1968), 2366-2368
S. Kapoor, Ambix 16 (1969), 1-65
M. Chaigneau, J.B. Dumas, Chimiste et Homme Politique. Paris 1984
L.J. Klosterman, Ann. Sci. 42 (1985), 1-80
J. Roche, Bull. Acad. Med. 169 (1985), 50-53
R. Passmore, J. Nutrition 116 (1986), 491-498
B.B. Chastain, Bull. Hist. Chem. 8 (1990), 8-12

DU MÉNIL, AUGUST PETER JULIUS [1777-1852]
HEIN, 138; TSCHIRCH, 1087-1088; SCHELENZ, 653; BERENDES, 200; FERCHL,136; POGG 2,118-119; CALLISEN 5,386-393; 27,362-363
L.F. Bley, Arch. Pharm. 127 (1854), 83-109DU MEZ, ANDREW GROVER [1885-1948 AMS 7,480; NCAB 37,123; WWAH, 168; WWWS, 491
Anon., J. Am. Pharm. Assn. 9 (1948), 617

DUÑAITURRIA, SALUSTIANO [1896-]
JV 43,410; NUC 151,537

DUNBAR, WILLIAM PHILLIPS [1863-1922]
NDB 4,193-194; DBJ 4,353-354; FISCHER 1,340; BK, 190-191; WWWS, 492
O. Kammann, Gesundheitsingenieur 45 (1922), 253-255
M. Beninde, Med. Klin. 18 (1922), 615

DUNCAN, ANDREW [1773-1832]
DNB 6,163; DESMOND, 199; CALLISEN 5,407-411; 27,365-366

DUNGERN, EMIL von [1867-1961]
FISCHER 1,340-341; DRULL, 54; BK, 191-192; AD 3,71; IB, 71
Anon., Chem. Z. 30 (1906), 489; Gior. Batt. Immun. 14 (1937), 705-706

DUNHAM, EDWARD KELLOGG [1860-1922]
AMS 3,191; WWWS, 493
Anon., J. Am. Med. Assn. 78 (1922), 1332
S. Flexner, Science 57 (1923), 683-685

DUNHAM, FLORENCE MARGARET [1869-1949]
M.R.S. Creese, Brit. J. Hist. Sci. 24 (1991), 282

DUNHILL, THOMAS PEEL [1876-1957]
DNB [1951-1960], 324-325; WW 1949, p.810
I.D. Vellar, Med. Hist. 18 (1974), 22-56

DUNN, LESLIE CLARENCE [1893-1974]
DSB 17,248-250; AMS 12,1564; MH 2,133-134; STC 1,330-331; MSE 1,318;NCAB I,506-507
Anon., Nature 250 (1974), 451-452
T. Dobzhansky, Am. Phil. Soc. Year Book 1974, pp.150-156; Biog. Mem. Nat. Acad. Sci. 49 (1978), 79-104
G.E. Allen, Folia Mendeliana 10 (1975), 253-257
D. Bennett, Ann. Rev. Gen. 11 (1978), 1-12

DUNN, MAX SHAW [1895-1976]
AMS 11,1320; IB, 72
Anon., Chem. Eng. News 54(46) (1976), 42

DUNSTAN, ALBERT ERNEST [1878-1964]
WW 1959, p.889; POGG 5,313; 6,619; 7b,1161
G. Sell and W.H. Thomas, Proc. Chem. Soc. 1964, pp.270-271
H.M. Langton, Chem. Ind. 1964, pp.883-884

DUNSTAN, WYNDHAM ROWLAND [1861-1949]
DNB [1941-1950], 227-228; WWWS,495; POGG 4,356-357; 5,313; 6,619; 7b,1162
T.A. Henry, J. Chem. Soc. 1950, pp.1022-1026; Obit. Not. Fell. Roy. Soc. 7 (1950), 63-81

DUPARC, LOUIS [1866-1932]
DHB 2,724; FREUND 3,119-132; POGG 4,357-358; 5,313-314; 6,619
E. Joukowsky, Verhandl. Schw. Nat. Ges. 114 (1933), 488-489; Mem. Soc. Phys. Hist. Nat. Gen. 50 (1933), 9-13

DUPETIT, GABRIEL [1861-1886]
RSC 9,755

DUPONT, GEORGES [1884-1958]
DBF 12,440; POGG 6,619-620; 7b,1162-1165
R. Dulou et al., Bull. Soc. Chim. 1959, pp.1311-1324

DUPPA, BALDWIN FRANCIS [1828-1873]
BLOKH 1,170-171; WWWS, 495; POGG 3,390
Anon., Proc. Roy. Soc. 21 (1873), vi-ix
E. Sell, Ber. chem. Ges. 6 (1873), 1588-1590
W. Odling, J. Chem. Soc. 27 (1874), 1199-1200

DUPRÉ, AUGUST [1835-1907]
DNB [1901-1911], 535-536; WWWS, 495; POGG 3,391; 5,314
O.H., Analyst 32 (1907), 313-316
H.W. Hake, J. Chem. Soc. 93 (1908), 2269-2275; Proc. Roy. Soc. A80 (1908), xiv-xviii

DURAN-REYNALS, FRANCISCO [1899-1958]
AMS 9(II),300-301
A. Haddow, Nature 182 (1958), 1549
C.C. Little, Science 129 (1959), 881-882
M.P. Queralt del Herro, Historia y Vida 17 (1985), 28-33

DURDUFI, GEORGI NIKOLAEVICH [1860-1930]
RSC 14,744
U.A. Shinilis, Ter. Arkh. 33(8) (1961), 108-111

DURHAM, HERBERT EDWARD [1866-1945]
WW 1945, p.808; BIBEL, 263-265; O'CONNOR(2), 69-70; WWWS, 496
G. Le Page, Nature 156 (1945), 742
Anon., Lancet 1945(II), pp.654-655

DURIG, ARNOLD [1872-1961]
FISCHER 1,342; PLANER, 61; IB, 72; POGG 7a(1),455
W. Auerswald, Wiener klin. Wchschr. 73 (1961), 897-898
R. Wagner, Bayer. Akad. Wiss. Jahrbuch 1962, pp.198-200
F. Scheminzky, Alm. Akad Wiss. Wien 113 (1963), 495-500

DUSCH, THEODOR von [1824-1890]
NDB 4,205; BAD 4,91-93; HIRSCH 2,352; PAGEL, 430-431; STUBLER, 316; DRULL, 55; BULLOCH, 363; WWWS, 430-431
L.A. Hoche, Deutsche med. Wchschr. 16 (1890), 121-122
Anon., Leopoldina 26 (1890), 55-56

DUSPIVA, FRANZ [1907-]
KURSCHNER 13,706

DUSTIN, ALBERT [1884-1942]
BAF, 45-46; FISCHER 1,343; IB, 72
M.E. Varela, Revista de la Sociedad Argentina de Biologia 19 (1943), 1-2
G. Roussy, Presse Med. 51 (1943), 157

DUTCHER, RAYMOND ADAMS [1886-1962]
AMS 10,1023; IB, 72

DUTHIE, EDWARD STEPHENS [1907-1959]
C.L. Oakley, J. Path. Bact. 83 (1962), 300-306

DUTOIT, PAUL [1873-1944]
POGG 4,360; 5,315; 6,623; 7a(1),456-457
C. Haenny, Helv. Chim. Acta 27 (1944), 1414-1422

DUTROCHET, HENRI [1776-1847]
DSB 4,263-265; DBF 12,939-940; MAYERHOFER 2,130-135; FERCHL,127; WWWS,498 POGG 1,633-634; CALLISEN 5,460-464; 27,394-397
A.R. Rich, Bull. Johns Hopkins Hosp. 39 (1926), 330-365
J.W. Wilson, Isis 37 (1947), 14-21
J. Schiller, Episteme 5 (1971), 188-200; 6 (1972), 147
J. Schiller and T. Schiller, Henri Dutrochet: Le Matérialisme Mécaniste et la Physiologie Générale. Paris 1975
J.V. Pickstone, Bull. Hist. Med. 50 (1976), 191-212; Brit. J. Hist. Sci. 11 (1978), 49-64

DUVAL, MATHIAS [1844-1907]
DSB 4,266-267; DBF 12,982-983; HIRSCH 2,358; Suppl.,253; WWWS, 498
E. Retterer, J. Anat. Physiol. 43 (1907), 241-331
A. Gautier, Bull. Acad. Med. 57 (1907), 343-344

DU VIGNEAUD, VINCENT [1901-1978]
AMS 13,1121; MH 1,147-148; STC 1.331-332; MSE 1,318-319; NCAB I,285-286 POTSCH, 437-438; CB 1956, pp.160-162; WWA 1976-1977, p.87; POGG 6,2725; 7b,5742-5749
R. Plane, J. Chem. Ed. 53 (1976), 8-12
M. Bodanszky, Nature 279 (1979), 271-272
J.S. Fruton, Am. Phil. Soc. Year Book 1979, pp.54-58
K. Hofmann, in Peptides: Structure and Biological Function (E. Gross and J. Meienhofer, Eds.), pp.5-24. Rockford, Ill. 1979; Biog. Mem. Nat. Acad. Sci. 56 (1987), 543-595
F.C. Bing, J. Nutrition 112 (1982), 1465-1473

DYBKOVSKI, VLADIMIR IVANOVICH [1830-1870]
PSB 6,32; IKONNIKOV, 188-195; ZMEEV 1,99; RSC 7,587
M.L. Tarachowski, Farm. Toks. 16(4) (1953), 58-63

DYCKERHOFF, AUGUST [1868-1947]
NDB 4,212; JV 10,19; NUC 153,554

DYCKERHOFF, HANNS [1904-1965]
KURSCHNER 10,430; POGG 7a(1),457-458

DYE, CLAIR ROBERT [1869-1949]
AMS 8,676; NCAB 39,149-150; WWWS, 499

DYE, MARIE [1891-1974]
AMS 11,1334; WWWS, 499
Anon., Chem. Eng. News 52(24), 43

DYER, DON C. [1878-]
AMS 5,311; JV 21,273

DYER, ROLLA EUGEN [1886-1971]
AMS 11,1335; MH 1,148-149; MSE 1,319-320; WWWS, 500
Anon., Chem. Eng. News 49(28) (1971), 46
N. Topping, Trans. Assoc. Am. Phys. 86 (1973), 11-12

DYMOND, THOMAS SOUTHALL [1861-]
TSCHIRCH, 1043; NUC 153,661; POGG 4,362

DYSON, GEORGE MALCOLM [1902-1978]
POTSCH, 128; POGG 6,625; 7b,1176-1178
M.F. Lynch, Chem. Brit. 16 (1979), 396-397

DZIERZGOWSKI, SZYMON [1866-1928]
PSB 6,150; WARSAW, 140; ZURICH-D,163-164; KONOPKA 2,583-585
J. Zaleski, Roczniki Chemii 10 (1930), 401-410

DZIEWOŃSKI, KAROL [1876-1943]
PSB 6,173-174; POTSCH, 128; POGG 6,625-626; 7b,1178-1179
J. Moszew, Roczniki Chemii 21 (1947), 1-21

E

EADIE, GEORGE SHARP [1895-1976]
AMS 11,1338
F. Bernheim, Pharmacologist 19 (1977), 23

EAGLE, HARRY [1905-1992]
AMS 14,1291; MSE 1,323-324; WWA 1978-1979, p.918; IWW 1982-1983, p.358; KAGAN(JA), 725; COHEN, 53; WWWS, 501
A.M. Harvey, Journal of Chronic Diseases 33 (1980), 543-545
New York Times, 14 June 1992EAGLES, BLYTHE ALFRED [1902-]
AMS 12,1587; YOUNG, 37

EAKLE, ARTHUR STARR [1862-1931]
AMS 4,273; NUC 154,78; WWWS, 501-502; POGG 6,627

EASSON, LESLIE HILTON [1904-]
Who's Who in British Science 1953, p.81

EAST, EDWARD MURRAY [1879-1938]
DSB 4,270-272; DAB 22,167-168; IB, 73; WWAH, 173; WWWS, 502
W.E. Castle, Am. Phil. Soc. Year Book 1938, pp.370-373
R.A. Emerson, Science 89 (1939), 51
E. Anderson, Proc. Am. Acad. Arts Sci. 74(6) (1940), 117-118
D.F. Jones, Biog. Mem. Nat. Acad. Sci. 22 (1944), 217-242
M. Vicedo, Stud. Hist. Phil. Sci. 22 (1991), 201-221

EASTERFIELD, THOMAS HILL [1866-1949]
WW 1949, p.827; JV 10,264; POGG 4,363; 5,317; 6,628; 7b,1182-1183
D. Miller, Nature 163 (1949), 669
H.O. Askew, Proc. Roy. Soc. N.Z. 78 (1950), 381-384
E. Marsden, J. Chem. Soc. 1952, p.1557

EATON, JAMES HOWARD [1842-1877]
NUC 154,390

EATWELL, WILLIAM [1819-1899]
MR 4,212-213; NUC 154,426

EBASHI, SETSURO [1922-]
WWWS, 504
S. Ebashi, Ann. Rev. Physiol. 53 (1991), 1-16

EBBECKE, ULRICH [1883-1960]
FISCHER 1,344; IB, 73; POGG 7a(1),459
H. Schaefer, Deutsche med. Wchschr. 85 (1960), 2255-2257; Erg. Physiol. 51 (1961), 38-51

EBEL, JEAN PIERRE [1920-1992]
DBA 1,726

EBERLE, JOHANN NEPOMUK [1798-1834]
CALLISEN 27,410
E. Hickel, Naturw. Rund. 28 (1975), 14-18
H.W. Davenport, Gastrointerology International 4 (1991), 39-40

EBERMEYER, ERNST [1829-1908]
NDB 4,248; HEIN-E, 98-99; BJN 13,24*; BLOKH 1,171; DZ, 293-294; POGG 3,396-397; 4,364; 5,318EBERT, (JOHANN) GEORG von [1885-1956] KURSCHNER 6,343; JV 24,571; 29,222; NUC 154,569
Wer Ist's 10,335

EBERT, JAMES DAVID [1921-]
AMS 15(2),787; WWA 1980-1981, p.959; IWW 1982-1983, p.360; WWWS, 505

EBERT, LUDWIG [1894-1956]
NDB 4,258; WWWS, 505; POGG 6,630; 7a(1),462-463; (4),140*
O. Kratky, Ost. Chem. Z. 57 (1956), 329-334
G. Kortüm, Z. Elektrochem. 61 (1957), 457-459

EBERTH, CARL JOSEPH [1835-1926]
DSB 4,275-277; NDB 4,259; HIRSCH 2,273; Suppl.,256; MAYERHOFER 2,148-149; PAGEL, 437-438; BULLOCH, 363; DZ, 294-295; IB, 73; WWWS, 505
R. Beneke, Berl. klin. Wchschr. 52 (1915), 1010-1012; Münch. med. Wchschr. 82 (1935), 1536-1537
H. Schinz, Viert. Nat. Ges. Zurich 72 (1927), 416-421
D. Scheu, Carl Joseph Eberth. Zurich 1990

EBLE, MAX [1879-1939]
JV 20,574; NUC 154,589
Anon., Ber. chem. Ges. 73A (1940), 79

EBLER, ERICH [1880-1922]
DBJ 4,354; DRULL, 55; BLOKH 1,172; POGG 5,319; 6,630
F. Hahn, Ber. chem. Ges. 55A (1922), 43-44
F. Meyer, Chem. Z. 46 (1922), 133

EBNER, VICTOR von [1842-1925]
NDB 4,267-268; OBL 1,212-213; HIRSCH 2,373-374; HEINDEL, 19-30; WWWS,505; FREUND 1,175-188; LESKY, 514-519; PAGEL, 438-439; MAYERHOFER 2,149
J. Schaffer, Alm. Akad. Wiss. Wien 75 (1926), 184-194; Anat. Anz. 64 (1927), 1-50

EBSTEIN, ERICH [1880-1931]
KURSCHNER 4,544-545; FISCHER 1,345; TETZLAFF, 63

EBSTEIN, WILHELM [1836-1912]
NDB 4,270; BJN 17,57-60; 18,18*; FISCHER 1,345-346; PAGEL, 439-441; WININGER 2,93-94; 7,553; KOHUT 2,275; DZ, 295-296; KAGAN, 210; KOREN, 168; WWWS, 505
E. Ebstein, Deutsches Arch. klin. Med. 89 (1906), 367-368; Janus 17 (1912), iv-x [after p.484]
P. Fraenckel, Deutsche med. Wchschr. 38 (1912), 2421-2422
C. Hirsch, Deutsches Arch. klin. Med. 109 (1912), i-iii
M. Bartalos, Humangenetik 1 (1965), 396
R.J. Mann and J.T. Lie, Mayo Clin. Proc. 54 (1979), 197-204

ECCLES, JOHN CAREW [1903-]
WW 1981, p.772; MH 1,149-150; MSE 1,324-325; IWW 1982-1983, p.360; WWWS, 505
J.C. Eccles, Ann. Rev. Physiol. 39 (1977), 1-18

ECK (EKK), NIKOLAI VLADIMIROVICH [1849-1908]
BME 34,1259-1260
V.N. Melnikova, Vest. Khir. 77(10) (1956), 134-139
Anon., Lancet 1958(I), p.686
N.S. Epifanov, Klin. Med. 55(8) (1977), 145-147
J.M. Rocko and K.G. Swan, American Surgeon 51 (1985), 641-644

ECKER, ALEXANDER [1816-1887]
ADB 48,256-257; BAD 4,97-101; HIRSCH 2,375-376; Suppl.,257; PAGEL, 441; WWWS, 506
A. Ecker, 100 Jahre einer Freiburger Professor-Familie. Freiburg 1886
J. Ranke, Arch. Antr. 17 (1887), i-vi
Anon., Leopoldina 23 (1887), 113; Anat. Anz. 2 (1887), 436

ECKER, ENRIQUE EDUARDO [1887-1966]
AMS 11,1353; LINDEBOOM, 511; IB, 73; WWAH, 175

ECKERSDORFF, OTTO [1878-]
JV 19,321; NUC 155,102; SG [3]5,101

ECKERT, FRITZ [1888-]
KURSCHNER 4,547; POGG 6,632-633; 7a(1),465-466

ECKHARD, CONRAD [1822-1905]
NDB 4,293-204; BJN 10,161*; HIRSCH 2,376; Suppl.,258; WWWS, 506; POGG 3,398
F.A. Kehrer, Münch. med. Wchschr. 52 (1905), 1206
Anon., Brit. Med. J. 1905(I), p.1363

ECKHARDT, FRANZ [1861-]
TLK 3,121; JV 3,274; RSC 14,775; 16,802; NUC 155,125

ECKSTEIN, ALBERT [1891-1950]
KURSCHNER 5,264; FISCHER 1,347; BHDE 2,234; LDGS, 77; KOREN, 167; WIDMANN, 153-154,260-261
E. Freudenberg, Annales Paediatrici 175 (1950), 223
F. Bossert, Kinderärztliche Praxis 18 (1951), 587-588

ECKSTEIN, HENRY CHARLES [1890-1971]
AMS 11,1355; IB, 73
Anon., BioScience 22 (1972), 181

EDDY, NATHAN BROWNE [1890-1973]
AMS 12,1608; IB, 73; WWAH, 175; WWWS, 506-507
Anon., Chem. Eng. News 51(17) (1973), 44
E. May, Pharmacologist 16 (1974), 41-42

EDDY, WALTER HOLLIS [1877-1959]
AMS 9(I),521; WWWS, 507; POGG 6,634; 7b,1189-1190
Anon., Chem. Eng. News 37 (1959), 125

EDELMAN, GERALD MAURICE [1929-]
AMS 15(2),797; STC 1,338-340; MSE 1,327-328; WWA 1980-1981, p.963; IWW 1982-1983, pp.361-362; WWWS, 507; POTSCH, 129

EDELMANN, ADOLF [1885-1939]
FISCHER 1,348

EDENS, ERNST [1876-1944]
KURSCHNER 6,350; FISCHER 1,348; BACHMANN, 113-121
P. Martini, Deutsche med. Wchschr. 70 (1944), 343-344
K. Blumberger, Medizinische 14 (1956), 487-490
H. Zimmermann, Aerztliche Forschung 18 (1964), 169-171

EDER, HEINZ [1914-]
JV 54,426; NUC 155,427

EDER, JOSEF MARIA [1855-1944]
DSB 4,282-283; NDB 4,312-323; OBL 1,216; TLK 3,122-124; HINK 1,3-4; POTSCH, 129; PLANER, 62; WWWS, 507; POGG 3,399-400; 4,367-368; 5,321-322; 6,635-636; 7a(1),467-468
L. Ebert, Alm. Akad. Wiss. Wien 95 (1947), 332-336
F. Dworschak and O. Krumpel, Dr. Joseph Maria Eder. Vienna 1955

EDER, ROBERT EDMUND [1885-1944]
KURSCHNER 5,265; TLK 3,124; TSCHIRCH, 1043-1044; POGG 6,636; 7a(1),468
J. Büchi, Viert. Nat. Ges. Zurich 89 (1944), 225-228; Verhandl. Schw. Nat. Ges. 124 (1944), 329-339
H. Flück, Schw. Apoth. Z. 82 (1944), 325-333,362-367

EDGAR, EDWARD CHARLES [1881-1938]
WWWS, 507
F.P. Burt, J. Chem. Soc. 1939, pp.205-206

EDIE, EDWARD STAFFORD [1879-1927]
O'CONNOR(2), 516-517
H.E.R., Biochem. J. 22 (1928), 617-618

EDINGER, ALBERT [1865-1914]
DBJ 1,279; POGG 4,369; 5,322; 6,637

EDINGER, LUDWIG [1855-1918]
NDB 4,313; DBJ 2,685; FISCHER 1,349-350; PAGEL, 443-444; FREUND 2,73-78;ARNSBERG, 95-97; HAYMAKER, 111-116; KALLMORGEN, 252-253; KOREN, 168; WININGER 2,96

K. Goldstein, Z. ges. Neurol. 44 (1918), 114-169
E. Goppert, Anat. Anz. 52 (1919), 219-223
W. Krücke and H. Spatz, Ludwig Edinger. Wiesbaden 1959
W. Krücke, in 50 Jahre Neuropathologie in Deutschland (W. Scholz, Ed.), pp.21-33. Stuttgart 1961

EDKINS, JOHN SYDNEY [1863-1940]
WW 1940, p.952; FISCHER 1,350; O'CONNOR(2), 196-197; HUBER, 27; WWWS, 508
Anon., Brit. Med. J. 1940(II), pp.339-340
E.M. Lowicki, Surgery 58 (1965), 1044-1048

EDLBACHER, SIEGFRIED [1886-1946]
NDB 4,314; FISCHER 1,350; HINK 1,11; DRULL, 56; TLK 3,124; IB, 74; WWWS, 508; POGG 6,637-638; 7a(1),468-469
E. Rothlin, Helv. Chim. Acta 30 (1947), 1554-1561

EDMAN, PEHR [1916-1977]
VAD 1975, p.229; POTSCH, 130
S. Moore, TIBS 2 (1977), N160
H.D. Niall, Nature 268 (1977), 279-280
B. Blombäck, Thromb. Res. 11 (1977), 695-698
S.M. Partridge and B. Blombäck, Biog. Mem. Fell. Roy. Soc. 25 (1979), 241-265
J.S. Fruton, Int. J. Pep. Res. 39 (1992), 189-194

EDMUNDS, CHARLES WALLIS [1873-1941]
DAB [Suppl.3], 242-243; AMS 6,403; CHEN, 27
N.B. Eddy, Science 93 (1941), 366-367

EDRIDGE-GREEN, FREDERICK WILLIAM [1863-1953]
WW 1949, p.836; O'CONNOR(2), 153-154; WWWS, 509; POGG 6,638; 7b,1194
Anon., Brit. Med. J. 1953(I), p.998; Lancet 1953(I), pp.856-857

EDSALL, DAVID LINN [1869-1945]
DAB [Suppl.3], 243-244; WWA 1944-1945, p.605; ICC, 33-39; DAMB, 226-227;WWWS, 509
J.C. Aub and R.K. Hapgood, Pioneer in Modern Medicine: David Linn Edsall of Harvard. Boston 1970

EDSALL, JOHN TILESTON [1902-]
AMS 14,1813; MH 2,138-140; MSE 1,330-332; WWA 1978-1979, p.933; IWW 1982-1983, pp.362-363; POTSCH, 130
J.T. Edsall, J. Am. Med. Assn. 197 (1966), 799-802; Ann. Rev. Biochem. 40 (1971), 1-28; in Science and Scientists (M. Kageyama et al.,Eds.) pp.335-341. Tokyo 1981; TIBS 14 (1989), 310-312
K. Bloch, J. Biol. Chem. 243 (1968), 1333-1336

EDSON, NORMAN LOWTHER [1904-1970]
J.R.R., Lancet 1970(I), p.1296
R.D. Batt, Proc. Roy. Soc. N.Z. 99 (1971), 106-110

EDWARDS, WILLIAM FRÉDÉRIC [1776-1842]
DSB 4,285-286; DBF 12,1138; DESMOND, 206; FERCHL, 139; POGG 1,644; CALLISEN 5,529-530; 27,422EECKE, JOOST van [1860-1895]
LINDEBOOM, 513
W. Einthoven, Ned. Tijd. Gen. [2]31 (1895), 933

EFFRONT, JEAN [1856-1931]
POGG 6,639
Mémorial Jean Effront, Rev. Ferment. [2]1 (1933), 229-281

EGAMI, FUJIO [1910-1982]
STC 1,340-341; IWW 1981-1982, p.355

EGE, RICHARD [1891-1974]
KBB 1973, p.289; VEIBEL, 2,126-130; POGG 6,640; 7b,1197-1198

EGGELING, HEINRICH [1869-1954]
FISCHER 1,351; GIESE, 485-487
R. Herrlinger, Anat. Anz. 102 (1955-1956), 373-382

EGGLETON, MARION GRACE [1901-1970]
H.D., Brit. Med. J. 1970(III), p.592

EGGLETON, PHILIP [1903-1954]
G.F. Marrian, Nature 174 (1954), 952
W.O. Kermack, Roy. Soc. Edin. Year Book 1955, pp.17-19

EGLE, KARL [1912-1975]
KURSCHNER 12,600
G. Döhler, Ber. bot. Ges. 93 (1980), 467-476

EHLERS, ERNST HEINRICH [1835-1925]
NDB 4,346-347; WAGENITZ, 51; WWWS, 512
A. Kühn, Z. wiss. Zool. 128 (1926), i-iv
K. Grobben, Alm. Akad. Wiss. Wien 76 (1926), 198-200

EHLERS, HEINRICH [1875-1940]
IB, 74
K. Plenge, Verhandl. path. Ges. 32 (1950), 433-434

EHRENBERG, CHRISTIAN GOTTFRIED [1795-1876]
DSB 4,288-292; NDB 4,349-350; ADB 5,701-711; FREUND 3,141-148; W,159-160; HINK 1,25-27; HIRSCH 2,384; BULLOCH, 363; WWWS, 512; CALLISEN 4,537; 27,424-428; POGG 1,646-647; 3,401; 7aSuppl.,175-178
M. von Laue, Christian Gottfried Ehrenberg. Berlin 1895
R. Bolling, Das Leben und das Werk Christian Gottfried Ehrenbergs. Delitzsch 1976
W.G. Siesser, Centaurus 25 (1981), 166-188
A. Geus, Ber. bot. Ges. 100 (1987), 283-290; Med. Hist. J. 22 (1987), 228-245

EHRENBERG, PAUL [1875-1956]
NDB 4,351; TLK 3,127; IB, 74; POGG 5,323-324; 6,643; 7a(1),475

EHRENBERG, RUDOLF [1884-1969]
KURSCHER 10,484; FISCHER 1,352; IB, 74; POGG 6,643; 7a(1),746
Anon., Naturw. Rund. 22 (1969), 414

EHRENREICH, MOSES [1879-]
JV 19,419; NUC 156,687

EHRENSBERGER, EMIL [1858-1940]
PUDOR, 27-28; POGG 6,645; 7a(1),747
Anon., Chem. Z. 64 (1940), 274

EHRENSTEIN, MAXIMILIAN [1899-1968]
KURSCHNER 10,449; AMS 11,1373; BHDE 2,239-240; LDGS, 18; WWWS, 512;POGG 7a(1),477-478
W. Merkel, Helv. Chim. Acta 52 (1969), 2178-2182
Anon., Chem. Eng. News 47(20) (1969), 89

EHRENSTEIN, RICHARD [1873-1929]
JV 21,293; NUC 156,691
Anon., Ber. chem. Ges. 62A (1929), 80; Chem. Z. 53 (1929), 348
A. Jantzen, Z. angew. Chem. 43 (1929), 692

EHRENSVÄRD, GOSTA [1910-1980]
VAD 1979, p.237

EHRET, HERMANN [1869-1913]
JV 12,243; RSC 14,804
Anon., Z. angew. Chem. 27(III) (1914), 30

EHRHARDT, ERNEST FRANCIS [1866-1929]
JV 4,224; RSC 14,241; NUC 157,8

EHRHARDT, KARL [1895-]
KURSCHNER 12,607; KALLMORGEN, 253; WWWS, 512

EHRHARDT, (JOHANN) WILHELM (LUDWIG) [1825-]
RAUSCH, 238

EHRHART, GUSTAV [1894-1971]
KURSCHNER 11,560; KALLMORGEN,253; POGG 7a(1),478-479
Anon., Chem. Z. 96 (1972), 42

EHRHART, OSKAR [1888-]
JV 28,315; NUC 157,13

EHRICH, WILLIAM ERNST [1900-1967]
AMS 11,1373; WWWS, 512
F. Büchner, Verhandl. path. Ges. 54 (1970), 589-594

EHRLICH, FELIX [1877-1942]
NDB 4,362-363; HINK 1,29; LDGS, 59; WININGER 2,108; TLK 3,128; WWWS, 513; POGG 5,326; 6,645; 7a(1),479

EHRLICH, HANS (JOHANNES PAUL ALFRED) [1867-]
JV 5,6; SG [2]4,792; NUC 157,30

EHRLICH, PAUL [1854-1915]
DSB 4,295-305; NDB 4,364-365; DBJ 1,126-130,325-326; BUNGE 2,421-442; FREUND 2,79-89; SANDRITTER, 93-95; KALLMORGEN, 253-254; W, 160-161;HUPKA, 164-175; ARNSBERG, 98-99; OLPP, 106-109; BLOKH 1,173-175;SACKMANN, 107-110; FISCHER 1, 352-354; PAGEL, 446-447,1945-1948; BULLOCH, 363-364; DZ, 308-309; KOHUT 2,278; KOREN, 168; WWWS, 513; WININGER 2,109-110; TETZLAFF, 65; ABBOTT, 41-42; POTSCH, 1330-131; POGG 5,326-327; 6,645; 7aSuppl.,178-183
C. Oppenheimer et al., Naturwiss. 2 (1914), 243-283
F. Pinkus, Med. Klin. 11 (1915), 1116-1117,1143-1145
R. Muir, J. Path. Bact. 20 (1915), 350-360; Proc. Roy. Soc. B92 (1921), i-vii
H. Bechhold, Chem. Z. 39 (1915), 705-708
A. Neisser, Jahrb. Schles. Ges. 93 (1915), 13-16
L. Michaelis, Naturwiss. 7 (1919), 165-168
A. Lazarus, Paul Ehrlich. Vienna 1922
A. Weinberg, Ber. chem. Ges. 49 (1923), 1223-1248
H. Sachs, Schlesische Lebensbilder 2 (1926), 355-362
B. Heymann, Klin. Wchschr. 7 (1928), 1257-1260,1305-1309
M. Marquardt, Paul Ehrlich. london 1949
H. Loewe, Paul Ehrlich. Stuttgart 1950; Chem. Z. 78(1954),171-174,213-216
H. Bauer et al., Ann. N.Y. Acad. Sci. 59 (1954), 150-276
O. Temkin et al., Bull. N.Y. Acad. Med. 30 (1954), 953-987
P. Karrer, Gesnerus 12 (1955), 47-57
L. Vogel, Rev. Hist. Med. Hebr. 22 (1969), 75-85,107-117
L. Pelner, N.Y. J. Med. 72 (1972), 620-624
J. Parascandola, J. Hist. Med. 32 (1977), 151-171; 36 (1981), 19-43
L.P. Rubin, J. Hist. Med. 35 (1980), 397-425
J.G. Hirsch and B.I. Hirsch, in The Eosinophil in Health and Disease (A.A.F.Mahmoud and K.F. Austin, Eds.), pp.3-23. New York 1980
B. Witkop, Naturw. Rund. 34 (1981), 361-379
J. Kleeberg, Zeitschrift für Gatroenterologie 20 (1982), 424-428
G. Seifert and T. Stockel, Med. Hist. J. 18 (1983), 227-237
E. Bäumler, Paul Ehrlich. Frankfurt a.M. 1984
T. Lenoir, Minerva 26 (1988), 66-88
A.S. Travis, Science in Context 3 (1989), 383-408
J. Liebenau, Med. Hist. 34 (1990), 65-78

EHRMANN, FRANÇOIS CONRAD EUGÈNE [1804-1896]
DBA 1,775

EHRMANN, MARTIN [1800-1870]
OBL 1,231; WURZBACH 4,10-11; HEIN, 144-145; COLE, 164; TSCHIRCH, 1044
E. Lesky, Veröffentlichungen der Internationalen Gesellschaft für Geschichte der Pharmazie 18 (1961), 59-68

EHRMANN, RUDOLF [1879-1955]
FISCHER 1,354; WININGER 6,557; KOREN, 168EHRMANN, SALOMON [1854-1926] NDB 4,366; OBL 1,231; FISCHER 1,354; WININGER 2,111-113; TETZLAFF, 65
J. Fick, Wiener med. Wchschr. 74 (1924), 2763-2771

EHRSTRÖM, ROBERT [1874-1956]
FISCHER 1,354-355
H. Bergholm, Finlands Lakare, p.87. Tammerfors 1927
Kuka Kukin On 1950, p.93
A. Adlercreuz, Acta Med. Scand. 156 (1957), 333-335

EIBELER, HANNS [1903-]
JV 48,578
Anon., Nachr. Chem. Techn. 21 (1973), 35

EICHELBERGER, LILLIAN (Mrs. R.H. Cannon) [1897-]
AMS 12,1630; WWWS, 513

EICHENGRÜN, ERNST ARTHUR [1867-1949]
NDB 4,373-374; HINK 1,31; TLK 3,128; POTSCH, 131; POGG 7a(1),481
H.G. Bodenbender, Z. angew. Chem. 60 (1948), 111-112

EICHHOLTZ, ALFRED [1869-1933]
Anon., Lancet 1933(I), p.388

EICHHOLTZ, FRITZ [1889-1967]
KURSCHNER 12,610; TLK 3,128; POGG 7a(1),481-483
P. Marquardt, Arzneimitt. 9 (1959), 535-536
E. Baum, Chem. Z. 92 (1968), 14
A. Fleckenstein, Arch. Inter. Pharmacodyn. 173 (1968), 259-261
O. Eichler, Arzneimitt. 18 (1968), 259
W. Doerr, Heid. Akad. Jahrb. 1968, pp.56-60

EICHHORN, FRITZ [1901-]
JV 42,506

EICHHORST, HERMANN [1849-1921]
NDB 4,381; DBJ 3,295; FISCHER 1,355; KROLLMANN, 897; PAGEL, 447-448; DZ, 310-311; WWWS, 514
A. Huber, Schw. Med. Wchschr. 51 (1921), 881-883
E. Fasol, Der Internist Hermann Eichhorst. Zurich 1983

EICHLER, THEODOR [1885-1977]
JV 22,299

EICHLER, WALTER [1904-1942]
KURSCHNER 6(I),358
R. Jung, Z. Biol. 103 (1950), 123-126

EICHWALD, EDUARD [1839-1889]
WELDING, 184; HIRSCH 2,391-392; Suppl.,260; SKULINA, 105-106
Anon., Chem. Z. 13 (1889), 1727EICHWALD, EGON [1883-1943] TLK 3,129; POGG 6,646-647; 7a(1),485
H.H. Mendel, Chem. Wkbl. 52 (1956), 133-135

EICHWALD, KARL EDUARD IVANOVICH [1795-1876]
DSB 4,307-309; NDB 4,387-388; PSB 6,210-211; WELDING, 184; HINK 1, 89-90; LEVITSKI 1,270-275; KAZAN 1,544-547; POGG 1,650; 3,402

EICHWEDE, HEINRICH [1875-1956]
JV 15,212; RSC 19,74; NUC 157,146
Anon., Nachr. Chem. Techn. 4 (1956), 254

EIGEN, MANFRED [1927-]
KURSCHNER 13,756; MH 2,140-141; STC 1,341-342; MSE 1,333; ABBOTT-C,39-40; IWW 1982-1983, p.366; POTSCH, 131
R. Winkler-Oswaititsch, Biophysical Chemistry 26 (1987), 102-113

EIJK, CORNELIS van [187O-]
RSC 14,801

EIJKMAN, CHRISTIAAN [1858-1930]
DSB 4,310-312; LINDEBOOM, 520-523; FISCHER 1,356; HINK 1,39-40; W, 161;PAGEL, 451; OLPP, 110-114, KALIN, 42-44; WWWS, 515; POGG 6,647; 7b,1212-1213
B.C.P. Jansen, Het Levenswerk van Christiaan Eijkman. Haarlem 1959

EIMER, THEODOR [1843-1898]
DSB 17,261-264; NDB 4,393-394; ADB 48,300-301; PAGEL, 451-452; HINK 1,41-42; WWWS, 515
M. von Linden, Biol. Z. 18 (1898), 721-725
C.B. Klunzinger, Jahreshefte Wurtt. 55 (1898), 1-22
P.J. Bowler, J. Hist. Med. 34 (1979), 40-73

EINBECK, HANS [1873-]
JV 20,24; NUC 157,241

EINHOF, HEINRICH [1778-1808]
ADB 5,760; FERCHL, 140; WWWS, 515; POGG 1,651-652

EINHORN, ALFRED [1857-1917]
DBJ 2,652; WOLF, 47; BLOKH 1,175-177; WININGER 2,123; 7,555; POTSCH, 131-132; POGG 4,371-372; 5,647; 6,647
E. Uhlfeder, Ber, chem. Ges. 50 (1917), 668-670; Chem. Z. 41 (1917), 373-374
W. Schneider, Pharm. Ind. 18 (1956), 85-88

EINHORN, MAX [1862-1953]
NDB 4,397; FISCHER 1,357; GROTE 8,1-24; HINK 1,159; DAMB, 227-228; KAGAN(JA), 61-63,587-588; KOREN, 168; WWWS, 515
M.I. Goldstein, Journal of the International College of Surgeons 5 (1942), 343-346
C.S. Lilien and A.F.R. Andresen, Gastroenterology 26 (1954), 121-122

EINSTEIN, ALBERT [1879-1955]
DSB 4,312-333; NDB 4,404-408; DAB [Suppl.5], 202-204; SCHWERTE,1,47-55; STC 1,343-348; CB 1953, pp.178-181; BHDE 2,245-248; TLK 3,129; WININGER 2,123-126; TETZLAFF, 65-66; W, 161-163; POGG 5,328-329; 6,647-648; 7a(1),488-489
P. Frank, Einstein, his Life and Times. New York 1947
P.A. Schilpp (Ed.), Albert Einstein, Philosopher-Scientist. Evanston, Ill. 1949
J.A. Wheeler, Biog. Mem. Nat. Acad. Sci. 51 (1980), 97-117
A. Pais, Subtle is the Lord, The Science and Life of Albert Einstein. New York 1982

EINTHOVEN, WILLEM [1860-1927]
DSB 4,333-335; BWN 2,143-144; LINDEBOOM, 526-527; HINK 1, 197; W, 163-164; WWWS, 515; POGG 4,372-373; 5,329-330; 6,648
H. Winterberg, Wiener klin. Wchschr. 40 (1926), 1460-1461
A.V. Hill, Nature 120 (1927), 591-592
F. Wenckebach, Deutsche med. Wchschr. 51 (1927), 2176
T.L., Proc. Roy. Soc. B102 (1928), v-viii
A. de Waardt, Het Levenswerk van Willem Einthoven. Haarlem 1957
C.J. Wiggers, Circulation Research 9 (1961), 225-234
M. Fournier, Medica Mundi 21(2) (1976), 65-70
H.B. Burchell, British Heart Journal 57 (1987), 190-193
I. Erschler, Arch. Int. Med. 148 (1988), 453-455

EISECK, ERNST [1861-]
BERLIN, 614; SG [2]4,799; NUC 157,309

EISELSBERG, ANTON von [1860-1939]
NDB 4,410; NOB 9,107-112; OBL 1,236; KNOLL, 131-133; FISCHER 1,358;PAGEL, 452-453; LINDEBOOM, 527; UTRECHT, 158-159; KILLIAN, 72; PLANER, 67; WWWS, 1732
W. Denk, Wiener klin. Wchschr. 53 (1940), 55-60
A. Eiselsberg, Lebensweg eines Chirurgen, 4th Ed. Innsbruck 1949
H.H. Schmid, Wiener med. Wchschr. 110 (1960), 923-924
L. Schönbauer, Wiener klin. Wchschr. 72 (1960), 698-700
G. Rau, Personalbibliographien von Professoren und Dozenten der I. chirurgischen Klinik der Universität Wien...1901-1939, pp.6-29. Erlangen-Nürnberg 1972
U. Paul, Z. Chir. 107 (1982), 418-421

EISELT, JAN BOHUMIL [1831-1908]
OBL 1,236; STURM 1,304; NAVRATIL, 53-56; SCHIEBER, 37-43
J. Jeldicka, Cas. Lek. Cesk. 110 (1971), 881-886

EISEN, HERMAN NATHANIEL [1918-]
AMS 15(2),825; WWA 1980-1981, p.979; IWW 1982-1983, p.367; WWWS, 516

EISENBERG, HENRYK [1921-]
H. Eisenberg, in Selected Topics in the History of Biochemistry (G. Semenza and R. Jaenicke, eds.), pp.265-348. Amsterdam 1990

EISENBERG, PHILLIP [1876-1941]
A. Laskiewicz, Pol. Med. Sci. 13 (1970), 144-145

EISENBRAND, JOSEF [1901-1982]
KURSCHNER 13,762; HEIN-E, 103-104; POGG 7a(1),490-491
Anon., Chem. Z. 100 (1976), 242; 106 (1982), 103

EISENLOHR, FRITZ [1881-1957]
KROLLMANN, 897; TLK 2,143; POGG 5,331; 6,649; 7a(1),492-493
G. Geiseler, Z. angew. Chem. 63 (1951), 544

EISLER-TERRAMARE, MICHAEL [1877-1970]
FISCHER 1,359; POGG 6,649-650; 7a(1),493
J. Teichmann, Wiener klin. Wchschr. 69 (1957), 52-53;
Z. Immunitätsforsch. 124 (1962), 97-100
P. Speiser, Wiener klin. Wchschr. 82 (1970), 450-451

EISNER, GEORG [1885-]
JV 26,368; NUC 157,385

EISNER, HANS EDUARD [1892-]
AMS 11,1384; BHDE 1,152; LDGS, 18

EISTERT, BERND (BERNHARD) [1902-1978]
KURSCHNER 12,625; WOLF, 47; POTSCH, 132; POGG 7a(1),494-495
F. Arndt, Chem. Z. 86 (1962), 791-792
Anon., Chem. Z. 102 (1978), 240
M. Regitz et al., Chem. Ber. 113(3) (1980), xxix-lviii

EITEL, KARL HEINRICH [1856-1928]
NUC 157,403
Anon., Chem. Z. 52 (1928), 381

EKECRANTZ, THOR [1856-1936]
SBL 12,644-648; SMK 2,345; TSCHIRCH, 1044; POGG 5,331-332

EKENSTEIN, WILLEM ALBERDA van [1858-1937]
POTSCH, 132; POGG 6,651-652; 7b,1217

EKSTRAND, ÅKE GERHARD [1846-1933]
SBL 13,178-182; SMK 2,375; POGG 3,403-404; 4,374; 5,332-333; 6,652; 7b,1217
Anon., Sven Kem. Tid. 45 (1933), 299
A. Westgren, Sven. Vet. Akad. Årbok 1934, pp.259-264

EKWALL, PER [1895-1990]
POGG 6,652; 7b,1217-1220
K. Fontell, J. Coll. Sci. 143 (1991), 594-596ELBERS, ALFRED [1861-1936]
RSC 14,811; NUC 157,493
E. Winkhaus, Wir Stammen aus Bauern- und Schmiedegeschlecht, p.269. Gorlitz 1932

ELBERS, JOHANN CHRISTIAN [1824-1911]
RSC 2,475
W. Elbers, Hundert Jahre Baumwolltextilindustrie, pp.12-23. Braunschweig 1922

ELBS, CARL [1858-1933]
NDB 4,436; GUNDEL, 201-211; HINK 1,285; TLK 3,131; DZ, 314-315; POTSCH, 132; SCHAEDLER, 31-32; WWWS, 517; POGG 4,374-375; 5,333
F. Forster, Z. Elektrochem. 34 (1928), 420a-420b
E. Weitz, Ber. chem. Ges. 66A (1933), 74-75

ELDERFIELD, ROBERT COOLEY [1904-1979]
AMS 14,1338; WWWS, 517
Anon., Chem. Eng. News 58(2) (1980), 42

ELEMA, BENE [1901-]
NUC 157,631

ELENKIN, ALEKSANDR ALEKSANDROVICH [1873-1942]
BSE 15,498; WWR, 610; IB, 75; LIPSHITS 3,244-256
V.P. Savich, Sov. Bot. 1 (1944), 60-63
V.C. Asmous, Science 101 (1945), 166-167

ELFORD, WILLIAM JOSEPH [1900-1952]
WW 1952, pp.873-874; WWWS, 517
C.H. Andrewes, Nature 169 (1952); Obit. Not. Fell. Roy. Soc. 8 (1953), 149-158

ELIAS, HERBERT [1885-1975]
AMS 11,1389; FISCHER 1,359; KOREN, 169; COHEN, 55
B. Kisch and J. Gudemann, Exp. Med. Surg. 18 (1960), 4-16
D. Scherf, Proc. Virchow Med. Soc. 24 (1965), 44-46
H. Kaunitz, Pirquet Bull. Clin. Med. 24(6-8) (1975), 21-22

ELIAS, (MICHAEL) HANS [1907-1985]
AMS 14,1340; BHDE 2,256-257; TETZLAFF, 68; WWWS, 517
H. Haug, Anat. Anz. 161 (1986), 185-195

ELIAVA, GEORGI GRIGORIEVICH [1892-1937]
BME 35,306-307; FISCHER 1,359-360; WWR, 155

ELIEL, ERNEST LUDWIG [1921-]
AMS 15(2),838; BHDE 2,257-258; IWW 1982-1983, p.370
E.L. Eliel, From Cologne to Chapel Hill. Washington, D.C. 1990

ELION, GERTRUDE BELLE [1918-]
AMS 14,1341ELION, HARTOG [1853-1930]
POGG 4,375-376; 6,653-654
Anon., Chem. Wkbl. 27 (1930), 282-284; Chem. Z. 54 (1930), 338

ELLENBERGER, WILHELM [1848-1929]
NDB 4,453; DBJ 11,349; HINK 1,171; IB, 75; WWWS, 518
H. Baum, Anat. Anz. 67 (1929), 529-535
K. Thomas, Sitz. Akad. Wiss. Leipzig 81 (1929), 277-284
M. Stuerzbecher, Mon. Vet. Med. 14 (1959), 317-320

ELLER, WILHELM [1887-1943]
TLK 3,131; POGG 6,654; 7a(1),496
Anon., Chem. Z. 67 (1943); Z. angew. Chem. 56 (1943), 212,332
W. Voss, Ber. chem. Ges. 76A (1943), 112-113

ELLINGER, ALEXANDER [1870-1923]
NDB 4,457-458; DBJ 10,54-58,425; FISCHER 1,360-361; ARNSBERG, 107; BLOKH 1,177-178; KALLMORGEN, 257; KROLLMANN, 897; WWWS, 519; POTSCH, 133; WININGER 2,164-165; POGG 4,376; 6,655
G. Embden, Deutsche med. Wchschr. 49 (1923), 1450-1451
W. Marckwald, Ber. chem. Ges. 56A (1923), 84
O. Riesser, Klin. Wchschr. 2 (1923), 1869
P. Ellinger, Erg. Physiol. 23 (1924), 139-179

ELLINGER, FRIEDRICH [1900-1962]
AMS 10,1072; BHDE 2,259-260; LDGS, 80; POGG 6,655; 7b,1224-1226
E.B. Cook, Pharmacologist 5 (1963), 16-17

ELLINGER, PHILIPP [1887-1952]
FISCHER 1,361; DRULL, 58-59; BHDE 2,260; LDGS, 78; IB, 75-76; TLK 3,131 WININGER 7,555; POGG 6,655-656; 7a(1),496-497
W.T.J. Morgan, Nature 171 (1953), 502-503
Anon., J. Roy. Inst. Chem. 1953, p.105

ELLIOT, WALTER ELLIOT [1888-1958]
WW 1958, p.928
Boyd-Orr, Biog. Mem. Fell. Roy. Soc. 4 (1958), 73-80

ELLIOTT, KENNETH ALLAN CALDWELL [1903-1986]
AMS 13,1176; YOUNG, 12-13; WWWS, 519
K.A.C. Elliott, Bulletin of the Canadian Biochemical Society 17 (1980), 14-16
D.B. Tower and H.H. Jasper, Neurochemical Research 9 (1984), 285-289,449-460
Anon., Science 234 (1986), 754

ELLIOTT, THOMAS RENTON [1877-1961]
DNB [1961-1970], 329-330; WW 1959, p.932; FISCHER 1,361-362; O'CONNOR(2), 172-175; WWWS, 519
Anon., Lancet 1961(I), pp.567-568; Brit. Med. J. 1961(I), pp.752-754
H.P. Himsworth, Nature 190 (1961), 486-487
H.H. Dale, Biog. Mem. Fell. Roy. Soc. 7 (1961), 53-73
G. Pickering, Trans. Assoc. Am. Phys. 75 (1962), 21-23

ELLIS, EMORY LEON [1906-]
AMS 13,1178

ELLIS, STANLEY [1923-]
AMS 15(2),850; WWWS,520

ELLSWORTH, READ McLANE [1899-1936]
AMS 5,327

ELMAN, ROBERT [1897-1956]
AMS 9(II), 318; DAB [Suppl.6], 189-190; NCAB 42,447-448; WWAH, 180; WWWS,520-521

ELSBERG, LOUIS [1836-1885]
DAB 6,119; HIRSCH 2,404; Suppl.,263; KELLY, 377-378; PAGEL, 454; KAGAN(JA), 16-17; KOREN, 169; WWAH, 180; WWWS, 521
M.H. Henry, Trans. Med. Soc. N.Y. 1886, pp.601-608

ELSDEN, STANLEY REUBEN [1915-]
WW 1981, p.798

ELSINGHORST, GERHARD [1858-1929]
Anon., Z. angew. Chem. 28(III) (1915), 317; 42 (1929), 85; Chem. Z. 53 (1929), 81

ELSNER, FRANZ FRIEDRICH BERNHARD [1842-1921]
HEIN-E, 104-105; TSCHIRCH, 1044; SCHELENZ, 699; NUC 159,11; POGG 3,406-407; 4,378
Anon., Chem. Z. 45 (1921), 1263

ELSNER, HORST [1906-1972]
KURSCHNER 13,773; POGG 7a(1),499
Anon., Chem. Z. 96 (1972), 591

ELVEHJEM, CONRAD ARNOLD [1901-1962]
DSB 4,357-359; DAB [Suppl.7], 223-224; AMS 10,1082; MILES, 143-144; MH 1,152-153; MSE 1,335-336; DAMB, 230; CB 1948, pp.188-190; POTSCH, 133-134; POGG 6,658-659; 7b,1230-1244
F.O. Daniels, Am. Phil. Soc. Year Book 1962, pp.126-130
O.L. Kline and C.A. Baumann, J. Nutrition 101 (1971), 569-577
R.H. Burris et al., Biog. Mem. Nat. Acad. Sci. 59 (1990), 135-167
V.R. Potter, in One Hundred Years of Agricultural Chemistry and Biochemistry at Wisconsin (D.L. Nelson and B.C. Soltveldt, Eds.), pp.103-110. Madison, Wis. 1990

EMBDEN, GUSTAV [1874-1933]
DSB 4,359-360; NDB 4,473-474; KALLMORGEN, 257-258; ARNSBERG, 107-108; NB, 91; FISCHER 1,363; FRANKEN, 180; IB, 76; KOREN, 169; POTSCH, 134; WININGER 7,556; TETZLAFF, 70; POGG 6,660; 7a(1),500
A. Bethe, Klin. Wchschr. 12 (1933), 1471
H.J. Deuticke, Erg. Physiol. 35 (1933), 32-49
D. Schmitz, Deutsche med. Wchschr. 59 (1933), 1442-1443
G. Schmidt, Münch. med. Wchschr. 80 (1933), 1942-1944
E. Lehnartz, Arbeitsphys. 7 (1934), 475-483
K. Thomas, Z. physiol. Chem. 230 (1934), 3-11
F. Lipmann, Mol. Cell. Biochem. 6 (1975), 171-175
C.F. Cori, TIBS 8 (1983), 257-259

EMDE, HERMANN [1880-1935]
NDB 4,475; HEIN-E, 105-106; WWWS, 522; POTSCH, 134; POGG 5,335-336; 6,661; 7a(1),500

G. Wallrabe, Ber. chem. Ges. 68A (1935), 164-165
Anon., Z. angew. Chem. 48 (1935), 616

EMELÉUS, HARRY JULIUS [1903-1993]
WW 1983, p.697; MSE 1,336-337; CAMPBELL, 87-88; SRS, 28; ABBOTT-C, 40-41; POTSCH, 134-135; NUC 159,233-234; POGG 6,662; 7b,1246-1247
Anon., Nachr. Chem. Techn. 5 (1957), 3

EMERSON, ALFRED EDWARDS [1896-1976]
AMS 11,1410; WWA 1975-1976, p.923; WWWS, 522
E.O. Wilson and C.D. Michener, Biog. Mem. Nat. Acad. Sci. 53 (1982), 159-177

EMERSON, GLADYS ANDERSON [1903-1984]
AMS 14,1359; WWWS, 522
Anon., Chem. Eng. News 62(13) (1984), 61

EMERSON, OLIVER HUDLESTON [1900-1969]
AMS 11,1411
Anon., Chem. Eng. News 47(30) (1969), 52

EMERSON, ROBERT [1903-1959]
DSB 4,362-363; AMS 9(II),320; HINK 1,73; WWWS, 523; POTSCH, 135
C.S. French, Science 130 (1959), 437-438
E. Rabinowitch, Plant Physiology 34 (1959), 179-184; Biog. Mem. Nat. Acad. Sci. 35 (1961), 112-131

EMERSON, ROBERT LEONARD [1872-1951]
R.L. Emerson, in Harvard Class of 1894, pp.133-134. Norwood, Mass. 1919

EMERSON, ROLLINS ADAMS [1873-1947]
DAB [Suppl.4], 252-253; AMS 7,520; NCAB D,361; 39,297-298; IB, 76; SCHULZ-SCHAEFFER, 206; WWAH, 181; WWWS, 523
M.M. Rhoades, Biog. Mem. Nat. Acad. Sci. 25 (1949), 313-323

EMERSON, STERLING HOWARD [1900-1988]
AMS 14,1360; IB, 76

EMERY, THOMAS FRED [1931-1992]
AMS 18(2),951

EMICH, FRIEDRICH [1860-1940]
NDB 4,478; OBL 1,245; TLK 3,132; WWWS, 523; POTSCH, 135; POGG 4,380; 5,336-337; 6,663; 7a(1),501
G. Jantsch, Chem. Z. 54 (1930), 685-686
J.B. Niederl, Science 91 (1940), 376-377
H. Lieb, Ost. Chem. Z. 43 (1940), 43-47
A. Skrabal, Alm. Akad. Wiss. Wien 90 (1940), 195-199
G. Kainz, J. Chem. Ed. 35 (1958), 608-609
N.D. Cheronis, Microchem. J. 4 (1960), 423-444

EMMEL, VICTOR EMANUEL [1878-1928]
AMS 4,288; IB, 76
O.F. Kampmeier, Anat. Rec. 42 (1929), 75-90

EMMERICH, EMIL [1882-1937]
IB, 76-77
P. Rabl, Verhandl. path. Ges. 33 (1950), 407-409

EMMERICH, RUDOLF [1852-1914]
DBJ 1,280; FISCHER 1,364; PAGEL, 455-457; BULLOCH, 364; DZ, 317
H. Wehnert, Land. Vers. 64 (1906), 427-434; Z. angew. Chem. 19 (1906), 879
L., Münch. med. Wchschr. 61 (1914), 2342-2343

EMMERLING, ADOLF [1842-1906]
BJN 11,18*; BLOKH 1,178; KIEL, 210; POGG 4,380; 5,337
Anon., Chem. Z. 30 (1906), 265

EMMERLING, OSKAR [1853-1933]
POGG 3,408; 4,380-381; 5,337; 6,663; 7a(1),502

EMMETT, PAUL HUGH [1900-1985]
AMS 11,1413; MH 2,142-144; POTSCH, 135; WWWS, 523; POGG 6,663; 7b,1251-1254

EMMRICH, CURT [1897-1975]
Wer ist Wer? 1971-1972, p.221
P. Bamm, Eines Menschen Zeit. Munich 1972

EMMRICH, ROLF [1910-1974]
KURSCHNER 11,588-589; POGG 7a(1),503-504
H. Trenckmann, Jahrb. Sachs. Akad. 1973-1974, pp.227-242

ENDERLEN, EUGEN [1863-1940]
NDB 4,494-495; KURSCHNER 6,370; FISCHER 1,365; DRULL, 60; KILLIAN, 94-97; BRUNCKHORST, 5-22; RAZINGER, 46-48
M. Helferich, Münch. med. Wchschr. 80 (1933), 113-114
M. Borst, Verhandl. path. Ges. 32 (1950), 434-435
W. Wachsmuth, Münch. med. Wchschr. 104 (1962), 860-868

ENDERLIN, KARL FRIEDRICH [1819-1893]
PFISTER, 180; RSC 2,495

ENDERS, JOHN FRANKLIN [1897-1985]
AMS 14,1363; MH 1,154-155; STC 1,348-349; MSE 1,340-341; ABBOTT, 42-43; WWA 1978-1979, pp.965-966; IWW 1982-1983, p.376; WWWS, 524; CB 1955, pp.182-183
F.C. Robbins et al., Harvard Med. Alumni Bull. 59(4) (1985), 14-25
F.S. Rosen, Nature 317 (1985), 575
H. Amos et al., Harvard Gazette 82(20) (1987), 3,5
D.A. Tyrrell, Biog. Mem. Fell. Roy Soc. 33 (1987), 213-233
T.E. Weller and F.C. Robbins, Biog. Mem. Nat. Acad. Sci. 60 (1991), pp.47-65

ENDRES, GÜNTHER [1905-]
KURSCHNER 13,781; POGG 7a(1),507-508

ENGEL, HANS [1880-1959]
KURSCHNER 9,389; KALLMORGEN, 258

ENGEL, JOSEPH [1816-1899]
OBL 1,250; BJN 4,138*; HIRSCH 2,412-413; WERSTLER, 77-91
A. Weichselbaum, Wiener klin. Wchschr. 12 (1899), 468-470
Anon., Leopoldina 35 (1899), 135-136; Wiener med. Wchschr. 49 (1899), 732,877-881

ENGEL, LEWIS LIBMAN [1909-1978]
AMS 13,1198; WWA 1976-1977, p.927
K.J. Ryan, Endocrinology 104 (1979), 563-564
P.C. Zamecnik et al., Harvard Med. Alumni Bull. 53(6) (1979), 71-72
K. Savard, Rec. Prog. Hormone Res. 36 (1980), x-xviii

ENGEL, RODOLPHE [1850-1916]
DBF 12,1291; DBA 1,802; DORVEAUX, 82
C. Monod, Bull. Acad. Med. 75 (1916), 51-53
E. Persot, Bull. Sci. Pharm. 23 (1916), 50
A. Haller, Bull. Soc. Chim. [4]19 (1916), i-viii

ENGEL, STEFAN [1878-1968]
FISCHER 1,366; BHDE 2,264; LDGS, 77; KOREN, 170
K.Klinke, Kinderärztliche Praxis 36 (1968), 526

ENGELBACH, THEOPHIL [1823-1872]
ADB 6,119; HEIN, 148; BLOKH 1,178-179; RSC 2,496; 7,612; POGG 3,411

ENGELHARDT, ALEKSANDR NIKOLAEVICH [1832-1893]
BSE 49,42; KUZNETSOV(1963), 698-704; BLOKH 1,179-181; KOZLOV, 187-188;POGG 1,669; 3,411-412
Anon., Leopoldina 29 (1893), 58
N.A. Menshutkin and N.N. Beketov, Zhur. Fiz. Khim. 25 (1893), 43-46
N.S. Kozlov, Trudy Inst. Ist. Est. 30 (1960), 111-134

ENGELHARDT, VLADIMIR ALEKSANDROVICH [1894-1984]
BSE(3rd Ed.) 30,174; BME 35,386-388; MH 2,144-145; STC 1,349-351;

MSE 1,341-342; IWW 1982-1983, p.376; IB, 77; WWWS, 525; POTSCH, 136
S.A. Neifach, Fiziol. Zhur. 41 (1955), 3-8
A.E. Braunstein, Usp. Sov. Biol. 39 (1955), 18-24
N.S. Kozlov, Trudy Inst. Ist. Est. 30 (1960), 111-134
S. White, New Scientist 62 (1974), 327-329
E.C. Slater, TIBS 6 (1981), 226-227; 9 (1984), 504-505
V.A. Engelhardt, Ann. Rev. Biochem. 51 (1982), 1-19
A.A. Baev, Vospominanya o V.A. Engelgardta. Moscow 1989
L.V. Kisselev, in Selected Topics in the History of Biochemistry (G. Semenza and R. Jaenicke, eds.), pp.67-99. Amsterdam 1990

ENGELHART, JOHANN FRIEDRICH [1797-1837]
ADB 6,141; FERCHL, 143; WWWS, 525; POTSCH, 136-137; POGG 1,669; CALLISEN 6,69; 27,459

ENGELHORN, FRIEDRICH [1821-1902]
NDB 4,514-515; BAD 6,162-163

ENGELHORN, FRIEDRICH [1855-1911]
NDB 4,514*; BLOKH 1,181
C. Liebermann, Ber. chem. Ges. 44 (1911), 170

ENGELMANN, CHRISTIAN GOTTHOLD [1819-1884]
PFISTER, 181; RSC 2,498
A. Wankmüller, Beitr. Wurtt. Apothekgesch. 7 (1966), 84; 10 (1974), 120

ENGELMANN, THEODOR WILHELM [1843-1909]
DSB 4,371-373; NDB 4,517-518; BJN 14,213-219,21*-22*; FISCHER 1,367; PAGEL, 460-461; DZ,321-322; W, 164; WWWS, 525; POGG 3,412; 4,385-386; 6,666; 7aSuppl.,186-187
H. Piper, Münch. med. Wchschr. 56 (1909), 1797-1800; Naturw. Rund. 24 (1909), 437-439
K. Brandenburg, Med. Klin. 5 (1909), 829-830
H. Boruttau, Deutsche med. Wchschr. 35 (1909), 1110-1111
R. du Bois-Reymond, Berl. klin. Wchschr. 46 (1909), 1097-1099
M. Verworn, Z. allgem. Physiol. 10 (1910), i-vi; Jahrb. Akad. Wiss. Gott. 1910, pp.86-93
H. Kingreen, Theodor Wilhelm Engelmann. Münster 1972
M.D. Kamen, Proc. Am. Phil. Soc. 130 (1986), 232-246
H. Gest, Persp. Biol. Med. 34 (1991), 254-274

ENGER, RUDOLF [1897-]
KALLMORGEN, 258; IB, 77ENGLE, EARL THERON [1896-1957]
AMS 9(II),322; NCAB 48,252-253; WWAH, 182
Anon., Science 127 (1958), 78; J. Clin. Endocrin. 18 (1958), 670-672

ENGLER, ADOLF [1844-1930]
DSB 15,147-148; NDB 4,532; KIEL, 147; TSCHIRCH, 1045; W,164-165; WWWS,525
L. Diels, Ber. bot. Ges. 48 (1930), (146)-(163); Bot. Jahrb. 64 (1931),i-lvi

ENGLER, CARL [1842-1925]
NDB 4,533; BLOKH 1,181-184; TSCHIRCH, 1045; TLK 2,148-149; POTSCH, 137; DZ, 323; WWWS, 525; POGG 4,386-387; 5,339; 6,666-667; 7a(1),513
F. Haber, Chem. Z. 46 (1922), 2-3

ENKLAAR, JOHANNES ELIZA [1847-]
RSC 9,802; 14,853; NUC 160,426
W.P. Jorissen, Chem. Wkbl. 7 (1910), 399-404

ENKVIST, TERJE [1904-1975]
WWWS, 526
Vem och Vad 1975, p.113

ENNOR, ARNOLD HUGHES (HUGH) [1912-1977]
WW 1977, p.750
F.C. Courtice, Rec. Austral. Acad. Sci. 4(1) (1979), 105-130

ENRIQUES, PAOLO [1878-1932]
WININGER 6,569; 7,556; IB, 77

EÖTVÖS, ROLAND (LORAND) von [1848-1919]
DSB 4,377-381; MEL 1,432-433; W, 165; POGG 4,389; 5,340

EPHRAIM, FRITZ [1876-1935]
NDB 4,546; TLK 3,135-136; WININGER 7,556; TETZLAFF, 71; WWWS, 526; POGG 5,340-341,1421; 6,668-669; 7a(1),513-514
E. Michel, Helv. Chim. Acta 18 (1935), 1448-1464; Verhandl. Schw. Nat. Ges. 116 (1935), 440-445; Ber. chem. Ges. 68A (1935), 62-65

EPHRAIM, JULIUS [1867-1927]
TLK 3,136; POGG 4,389; 5,341; 6,669

EPHRUSSI, BORIS [1901-1979]
QQ 1977-1978, p.647; IWW 1974-1975, pp.499-500; BUICAN, 281-299; WWWS, 526
A. Lwoff, Somatic Cell Genetics 5 (1979), 677-679
H. Roman, Ann. Rev. Gen. 14 (1980), 447-450
J. Sapp, Beyond the Gene, pp.123-162. New York 1987

EPHRUSSI-TAYLOR, HARRIETT [1918-1968]
AMS 11,1426
A.W. Ravin, Genetics 60 (1968), s24EPPENS, AUGUST [1864-]
JV 8,215; RSC 16,377

EPPINGER, HANS [1846-1916]
NDB 4,551; DBJ 1,352; STURM 1,314; FISCHER 1,369; KOERTING, 143; LOMMATZSCH,48-53; FRANKEN, 180-181
W. Scholz, Mitt. Ver. Aerzte Steier. 53 (1916), 177-182
H. Bietzke, Verhandl. path. Ges. 29 (1937), 351-354

EPPINGER, HANS [1879-1946]
NDB 4,551-552; STURM 1,314; FISCHER 1,369; WWWS, 527
W. Pilgersdorfer and E. Rissel, Wiener med. Wchschr. 101 (1951),880-884
H. Beiglböck, Acta Hepatologica 5 (1957), 1-4
E. Deutsch, Wiener klin. Wchschr. 78 (1966), 674-675
L. Benda and E. Rissel, Wiener med. Wchschr. 116 (1966), 809-811
H.M. Spiro, Journal of Clinical Gastroenterology 6 (1984), 493-497

EPPINGER, PAUL [1879-]
JV 22,517; NUC 161,24

EPSTEIN, BERTHOLD [1890-1963]
KURSCHNER 5,287; KOERTING, 183; PELZNER, 214-220

EPSTEIN, EMIL [1875-1951]
KURSCHNER 4,602; IB, 77; WIEN 1952, p.43
Anon., Wiener klin. Wchschr. 63 (1951), 211

EPSTEIN, GERMAN VENIAMINOVICH [1889-1935]
BSE 49,138; BME 35,720-721; WWR, 156
Anon., Vest. Mikr. Ep. 15 (1936), 5-12
N. Koltsov, Biol. Zhur. 5 (1936), 179-182

ERBACHER, OTTO [1900-1950]
POGG 6,670; 7a(1),515-516
O.Hahn, Z. angew. Chem. 63 (1951), 83

ERBEN, FRANZ [1876-1937]
KURSCHNER 5,287; PELZNER, 37-43; PLANER, 69-70; POGG 6,680; 7a(1),516

ERBRING, HANS [1903-1982]
KURSCHNER 13,800; POGG 7a(1),516-517
P.W. Patt, Arzneimitt. 19 (1969), 246
Anon., Chem. Z. 106 (1982); Nachr. Chem. Techn. 30 (1982), 533

ERDHEIM, JAKOB [1874-1937]
OBL 1,261; FISCHER 1,371-372; MEDVEI, 730-731; ZWEIFEL, 31-32; KOREN, 170 WININGER 7,556-557
H. Chiari, Wiener klin. Wchschr. 50 (1937), 610-611
S.M. Rabson, Arch. Path. 68 (1959), 357-366
S. Roman, American Journal of Dermopathology 9 (1987), 447-450

ERDMANN, ERNST [1857-1925]
POGG 4,390; 5,142-143; 6,670-671
D. Vorländer, Ber. chem. Ges. 58A (1925), 38-39; Z. angew. Chem. 38 (1925), 980-981

ERDMANN, HUGO [1862-1910]
NDB 4,572; BJN 15,24*; BLOKH 1,185; WWWS, 527-528; POTSCH, 139; POGG 4,390-391; 5,343
H. Wichelhaus, Ber. chem. Ges. 43 (1910), 2073
Anon., Leopoldina 46 (1910), 79-89; Chem. Z. 34 (1910), 1391

ERDMANN, KARL GOTTLIEB HEINRICH [1798-1876]
HEIN-E, 106-107; FERCHL, 144; TSCHIRCH, 1045; NUC 161, 291-292; POGG 1,674; 3,414

ERDMANN, OTTO [1804-1869]
DSB 4,394-396; NDB 4,572-573; ADB 6,188-189; BLOKH 1,185-186; FERCHL,144; POTSCH, 139; TSCHIRCH, 1045; SCHAEDLER, 32; POGG 1,674-675; 3,414
H. Kolbe, J. prakt. Chem. 108 (1869), 449-458; Ber. chem. Ges. 3 (1870), 374-381

ERDMANN, RHODA [1870-1935]
NDB 4,573; JV 24,571; IB, 77; WEBER, 25-26; BOEDEKER, 20-21
P. Caffier, Arch. exp. Zellforsch. 18 (1936), 127-141

ERISMANN, FRIEDRICH HULDREICH (FEDOR FEDOROVICH) [1842-1915]
DBJ 1,326; FISCHER 1,372-373; PAGEL, 467-468; KUZNETSOV(1963), 542-549
H. Müller-Dietz, Clio Medica 4 (1969), 203-210
H. Wilk, Friedrich Huldreich Erismann. Zurich 1970

ERLANDSEN, ALFRED [1878-1918]
DBL 6,391-393; FISCHER 1,373; VEIBEL 2,132
J. Bock, Hosp. Tid. 61 (1918), 90-93

ERLANGER, BERNARD FERDINAND [1923-]
AMS 15(2),891; WWA 1980-1981, p.1014; WWWS, 529

ERLANGER, JOSEPH [1874-1965]
DSB 4,397-398; DAB [Suppl.7],225-227; MH 1,155-156; MSE 1,342; W,167-168; NCAB 51,547-548; HAYMAKER, 190-195; DAMB, 232; KAGAN(JA), 747
J. Erlanger, Ann. Rev. Physiol. 26 (1964), 1-14
E.A. Doisy, Am. Phil. Soc. Year Book 1966, pp.134-137
H. Davis, Biog. Mem. Nat. Acad. Sci. 41 (1970), 111-139
L.H. Marshall, Persp. Biol. Med. 26 (1983), 613-636

ERLENBACH, MICHAEL [1902-]
JV 43,411; NUC 161,448

ERLENMEYER, EMIL [1825-1909]
DSB 4,399-400; NDB 4,594-595; BJN 14,22*; HEIN-E, 107-108; DRULL, 64; NB, 95; BLOKH 1,186-188; SCHAEDLER, 33; POTSCH, 139-140; WWWS, 529; POGG 3,415-416; 4,391-392; 5,345
R. Meyer, Chem. Z. 33 (1909), 161-162
H. Kiliani, Z. angew. Chem. 22 (1909), 481-483
M. Conrad, Ber. chem. Ges. 43 (1910), 3645-3664
W.H. Perkin, J. Chem. Soc. 99 (1911), 1649-1651

ERLENMEYER, EMIL, Jr. [1864-1921]
DBJ 3,296; BLOKH 1,188-189; WWWS, 529; POGG 4,392; 5,345
P. Lepsius, Ber. chem. Ges. 54A (1921), 107-113

ERLENMEYER, HANS [1900-1967]
KURSCHNER 10,489; SBA 1,45; TLK 3,137; POGG 6,673; 7a(1),520-523
Anon., Chem. Eng. News 45(28) (1967), 72
H. Bloch, Basler Stadtbuch 1969, pp.37-40

ERLENVEIN, EVGENI ALEKSANDROVICH [1844-]
RSC 14,871; SG [2]5,104; NUC 161,451

ERMENGEM, ÉMILE van [1851-1932]
GHENT 2,26-30; PAGEL, 470; WWWS, 1709
Anon., Liber Memorialis Gand, vol.2, pp.558-563. Ghent 1913
A. Zimmern, Bull. Acad. Med. 108 (1932), 1465-1466

ERNST, HAROLD CLARENCE [1856-1922]
DAB 6,177-178; FISCHER 1,374; NCAB 20,89-90; WWWS, 529
Anon., J. Am. Med. Assn. 79 (1922), 1065

ERNST, PAUL [1859-1937]
KURSCHNER 4,612; FISCHER 1,374; DRULL, 64-65; ZWEIFEL, 33-35; IB, 78
A. Schminke, Verhandl. path. Ges. 31 (1939), 532-537
G.B. Gruber, Med. Klin. 18 (1959), 856-857

ERRERA, JACQUES [1896-1977]
POGG 6,674; 7b,1266-1267
P. Legrain, Le Dictionnaire des Belges, p.197. Brussels 1981

ERRERA, LÉO ABRAM [1858-1905]
DSB 4,401-402; BNB 32,177-185; WININGER 2,193
E. de Wildemann, Ber. bot. Ges. 23 (1905), (43)-(55); Bull. Soc. Bot. Belg. 44 (1907) 1-58; Ann. Soc. Belge Micr. 28 (1907), 65-114
L. Fredericq and J. Massart, Ann. Acad. Roy. Belg. 74 (1908), 131-277
J. Seide, Rev. Hist. Med. Heb. 34 (1956), 221-232
Univ. de Bruxelles, Commemoration Léo Errera. Brussels 1960

ERSPAMER, VITTORIO [1909-]
NUC 161,632ERWIG, EMIL [-1900]
RSC 14,876
Anon., Pharm. Z. 45 (1900), 183

ERXLEBEN, CHRISTIAN POLYKARP FRIEDRICH [1765-1831]
STURM 1,318; WRANY, 318-319

ESAU, KATHERINE [1898-]
AMS 14,1387; MH 1,157-159; STC 1,354-355; MSE 1,344-345; WWWS, 530; IWW 1982-1983, p.381
R.F. Evert, Plant Science Bulletin 31(5) (1985), 33-37

ESBACH, GEORGES HUBERT [1843-1890]
DBF 12,1407; FISCHER 1,375; WWWS, 530
Anon., Brit. Med. J. 1890(I), p.577
W.J. Hatcher and A.G.W. Webb, Med. Lab. Sci. 36 (1979), 185-190

ESCALES, RICHARD [1863-1924]
BLOKH 1,190; TLK 3,137-138; RSC 14,876
A. Stettenbacher, Chem. Z. 48 (1924), 825
Anon., Z. angew. Chem. 37 (1924), 835

ESCHENMEYER, ADOLPH KARL AUGUST [1768-1852]
NDB 4,644; ADB 6,349-350; HIRSCH 2,432-433; Suppl.,270; DECHAMBRE 38,655; FERCHL, 145; POGG 1,680-681; CALLISEN 6,111-114; 27,474-475
W. Wuttke, Sudhoffs Arch. 56 (1972), 253-296

ESCHENMOSER, ALBERT [1925-]
KURSCHNER 13,812; IWW 1982-1983, p.381; POTSCH, 140
V. Prelog, Aldrichemica Acta 23(3) (1990), 59-64

ESCHER, HEINRICH HERMANN [1884-1939]
NUC 162,64
A. Buchner, Neue Schweizer Biographie, p.137. Basle 1938
H. Steiner, Viert. Nat. Ges. Zurich 84 (1939), 379-381
Biographisches Lexikon Verstorbener Schweizer 2 (1948), 332

ESCHERICH, THEODOR [1857-1911]
DSB 4,403-406; NDB 4,649-650; BJN 16,45-56,20*-21*; FISCHER 1,375; PAGEL, 471-472; BULLOCH, 365; ATUYANA, 41-53; SACKMANN, 111-114
C. von Pirquet, Z. Kinderkrank. 1 (1911), 423-441
M. Pfaundler, Münch. med. Wchschr. 58 (1911), 521-523
J. Zappert, Wiener med. Wchschr. 61 (1911), 497-500
F. Hamburger, Wiener med. Wchschr. 82 (1932), 1216-1219

K. Kundratitz, Wiener klin. Wchschr. 73 (1961), 722-725

ESSON, WILLIAM [1838-1916]
DSB 4,411-412; WWWS, 531; POGG 4,394; 6,679
E.B.E., Proc. Roy. Soc. A93 (1917), liv-lvii

ESTABROOK, RONALD WINFIELD [1926-]
AMS 15(2),901; WWA 1980-1981, p.1018; WWWS, 531

ESTOR, ALFRED [1830-1886]
DBF 13,115-116; GE 16,429; HIRSCH 2,442
E. Masse, Gaz. Sci. Med. Bord. 7 (1886), 360

ESTREICHER von ROZBIERSKI, TADEUSZ [1871-1952]
WE 3,487; POTSCH, 140; POGG 4,394-395; 5,346; 6,680; 7b,1280-1281
J. Kamecki, Wiad. Chem. 6 (1952), 309-316
J. Read, Nature 170 (1952), 184-185

ÉTARD, ALEXANDRE LÉON [1852-1910]
DBF 13,176-177; W, 168; POGG 3,419; 4,395; 5,346
L. Olivier, Rev. Gen. Sci. 21 (1910), 581-606
P. Lebeau, Bull. Soc. Chim. [4]9 (1911), i-xix
E.H. Huntress, Proc. Am. Acad. Arts Sci. 81 (1952), 49-50

ETTI, KARL [1825-1890]
HEIN, 151; BLOKH 1,189-190; POGG 4,395
R. Wegscheider, Ber. chem. Ges. 23 (1890), 910-913

ETTISCH, GEORG [1890-]
LDGS, 59; POGG 6,681; 7a(1),532-533

ETTLING, FRIEDRICH [-1889]
Anon., Pharm. Z. 34 (1889), 751

ETTLING, KARL JACOB [1806-1856]
HB 3,76-78; FERCHL, 146; WWWS, 532; RSC 2,526; POGG 1,687

ETTLINGER, FRIEDRICH [1877-]
JV 17,282

EUCKEN, ARNOLD [1884-1950]
DSB 4,413-414; NDB 4,670; TLK 3,139; POTSCH, 140-141; POGG 5,346; 6,682-683; 7a(1),533-535
R. Suhrmann, Z. Elektrochem. 50 (1944), 169-170
E. Bartholomé, Naturwiss. 37 (1950), 481-483
G. Scheibe, Bayer. Akad. Wiss. Jahrbuch 1950, pp.200-203
R. Oesper, J. Chem. Ed. 27 (1950), 540-541
K. Schäfer, Göttinger Jahrbuch 1984, pp.263-266

EUGLING, MAX [1880-1950]
FISCHER 1,378
M. Kaiser, Wiener klin. Wchschr. 62 (1950), 728-729

EUGSTER, EDMUND [1859-1884]
RSC 9,815

EULENBURG, ALBERT [1840-1917]
NDB 4,683; DBJ 2,652-653; HIRSCH 2,446-447; Suppl.,272-273; DZ, 336; PAGEL, 477-478; WWWS, 533
I. Bloch, Med. Klin. 13 (1917), 774-776
H. Kron, Deutsche med. Wchschr. 43 (1917), 983

EULER (or EULER-CHELPIN), HANS von [1873-1964]
DSB 4,485-486; IB, 78; POTSCH, 141; POGG 4,396; 5,347-348; 6,683-687; 7a(1),536-543
W. Franke, Naturwiss. 40 (1953), 177-180; Brauwiss. 1953(3), pp.34-40
R. Lepsius, Chem. Z. 82 (1958), 101-102; 88 (1964), 933-936
H. Druckrey, Arzneimitt. 8 (1958), 162-163
F. Lynen, Bayer. Akad. Wiss. Jahrbuch 1965, pp.206-212
F. Wessely, Alm. Akad. Wiss. Wien 115 (1965), 408-415
K. Myrbäck, Enzymologia 29 (1965), 105-107

EULER, ULF SVANTE von [1905-1983]
VAD 1981, p.288; STC 1,359-361; MSE 1, 349-351; IWW 1982-1983, p.1371; AO 1983, p.106; WWWS, 1733-1734; POTSCH, 141
U.S. von Euler, Trends in Neurosciences 4(10) (1981), iv-ix
W. Paton, Nature 303 (1983), 662
J.P. Buckley, Pharmacologist 25 (1983), 329
D. Ottosen, Am. Phil. Soc. Year Book 1984, pp.108-110
H.K.F. Blaschko, Biog. Mem. Fell. Roy. Soc. 31 (1985), 145-170

EULNER, HANS HEINZ [1925-1980]
KURSCHNER 13,819
H. Winkelmann and S. Mildner, Med. Hist. J. 17 (1982), 148-155

EVANS, ALICE CATHERINE [1881-1975]
AMS 10,1113; FISCHER 1,378; CB 1943, pp.198-200; NAW 4,219-221; IB, 78-79
E.M. O'Hern, Am. Soc. Microbiol. News 39 (1973), 573-578
H. Machmann and W. Kohler, Z. ärzt. Fortbild. 82 (1988), 381-385

EVANS, CHARLES LOVATT [1884-1968]
DNB [1961-1970], 334-336; WW 1965, p.969; FISCHER 1,378-379; IB, 79; WWWS, 534
Anon., Nature 220 (1968), 1055-1056
I. deB. Daly and R.A. Gregory, Biog. Mem. Fell. Roy. Soc. 16 (1970), 233-252

EVANS, DAVID GWYNNE [1909-1984]
WW 1984, p.717
Anon., The Times, 21 June 1984; Brit. Med. J. 289 (1984), 258
A.W. Downie et al., Biog. Mem. Fell. Roy. Soc. 31 (1985), 173-196

EVANS, EARL ALISON [1910-]
AMS 14,1397; WWA 1978-1979, p.987; IWW 1978-1979, p.384; WWWS, 534

EVANS, HAROLD J. [1921-]
AMS 15(2),909; WWA 1980-1981, p.1024; IWW 1982-1983, p.384; WWWS,534-535

EVANS, HERBERT McLEAN [1882-1971]
AMS 11,1451; STC 1,361-362; MEDVEI, 520-524; FISCHER 1,379; WWWS, 535;CB 1959, pp.109-111; IB, 79; POGG 6,687-688; 7b,1284-1291
J.E. Zeitlin and V.F. Lenzen, Isis 62 (1971), 507-511
W.R. Lyons, Endocrinology 89 (1971), 947-950
E.C. Amoroso and G.W. Corner, Biog. Mem. Fell. Roy. Soc. 18 (1972), 83-186
G.W. Corner, Biog. Mem. Nat. Acad. Sci. 45 (1974), 153-192
L.L. Bennett, Horm. Prot. Pep. 3 (1975), 247-272; Persp. Biol. Med. 22 (1978), 90-103
I.D. Raacke, J. Hist. Biol. 9 (1976), 301-322; J. Nutrition 113 (1983), 927-943
M.M. Grumbach, J. Clin. Endocrin. 55 (1982), 1240-1247

EVANS, PERCY NORTON [1869-1925]
AMS 3,211; SRS, 13; NUC 164, 52

EVANS, ROBERT JOHN [1909-]
AMS 14,1400; WWA 1976-1977, p.949; WWWS, 535-536

EVANS, THOMAS BROWN [1863-1907]
JV 2,58; NUC 164, 76
Anon., Science 26 (1907), 29-30

EVANS, WILLIAM LLOYD [1870-1954]
MILES, 149-150; NCAB 40,354-355; POGG 6,690; 7b,1296-1297
Anon., Chem. Eng. News 32 (1954), 4408
M.L. Wolfrom and E. Mack, J. Am. Chem. Soc. 77 (1955), 4949-4955

EVE, ARTHUR STEWART [1862-1948]
POGG 5,348-349; 6,690-691; 7b,1297
Who Was Who 1941-1950, p.368
G.P. Thomson, Nature 161 (1948), 838
A.N. Shaw, Trans. Roy. Soc. Canada [3]42 (1948), 87-93
J.S. Foster, Obit. Not. Fell. Roy. Soc. 6 (1949), 397-407

EVEREST, ARTHUR ERNEST [1888-1983]
SRS, 46; NUC 164,194
Who's Who in British Science 1953, p.87
Anon., Chem. Brit. 19 (1983), 473

EVERETT, NARK REUBEN [1899-1983]
AMS 14,1404; IB, 79; WWWS, 536-537; POGG 6,691; 7b,1299

EWALD, AUGUST [1849-1924]
DRULL, 65-66; FISCHER 1,380; PAGEL, 481
Anon., Münch. med. Wchschr. 71 (1924), 222

EWALD, CARL ANTON [1845-1915]
NDB 4,695; DBJ 1,326; FISCHER 1,380-381; PAGEL, 479-481; FRANKEN, 181; WREDE, 44; DZ, 339; WWWS, 537
H. Strauss, Berl. klin. Wchschr. 52 (1915), 1054-1055
L. Kuttner, Deutsche med. Wchschr. 41 (1915), 1318
I. Güstemeyer, Carl Anton Ewald. Cologne 1969

EWALD, GOTTFRIED [1888-1963]
FISCHER 1,381; GERNETH, 13-20
G.E. Sterling, Psychiat. Neurol. 147 (1964), 65-67

EWALD, PAUL PETER [1888-1985]
DSB 17,272-275; AMS 12,1731; KURSCHNER 13,822; BHDE 2,275-276; IWW 1982-1983, p.386; POGG 6,692-693; 7a(1),544
P.P. Ewald, Physics Today 27(9) (1974), 42-49
G. Hildebrandt, Phys. Bl. 41 (1985), 412-413
M. Renninger, Z. Kristall. 173 (1985), 159-167
J.J. Dropkin and B. Post, Acta Cryst. A42 (1986), 1-5
H.A. Bethe and G. Hildebrandt, Biog. Mem. Fell. Roy. Soc. 34 (1988), 135-176

EWALD, (JULIUS) RICHARD [1855-1921]
NDB 4,695-696; DBJ 3,297; FISCHER 1,381; WWWS, 537
A. Bethe, Pflügers Arch. 193 (1922), 109-127

EWALD, WOLFGANG FELIX [1885-]
JV 25,605; SG [3]5,425; NUC 164,364

EWAN, THOMAS [1868-1955]
SRS, 12; POGG 4,397; 6,693; 7b,1301
L.J. Faulkner, Proc. Chem. Soc. 1957, p.236

EWART, ALFRED JAMES [1872-1937]
DNB [1931-1940], 263-264; IB, 79; WWWS, 537
T.G.B. Osborn, Proc. Linn. Soc. 150 (1938), 314-317;
Nature 141 (1938), 17
W. Stiles, Obit. Not. Fell. Roy. Soc. 2 (1939), 465-469

EWART, JAMES COSSAR [1851-1933]
O'CONNOR, 251-252; WWWS, 537
Anon., Nature 133 (1934), 165-166
F.H.A. Marshall, Obit. Not. Fell. Roy. Soc. 1 (1934), 189-195

EWING, JAMES [1866-1943]
DSB 4,498-500; DAB [Suppl.3], 257-258; FISCHER 1,382; WWAH, 186
F.E. Adair, Annals of Surgery 93 (1931), vii-xv
J.B. Murphy, Biog. Mem. Nat. Acad. Sci. 26 (1949), 45-60
J.A. del Regato, Journal of Radiology and Oncology 2 (1977), 185-198

EWINS, ARTHUR JAMES [1882-1957]
DNB [1951-1960], 344-345; WW 1949, p.869; W, 171
H.H. Dale, Biog. Mem. Fell. Roy. Soc. 4 (1958), 81-91

EXNER, SIGMUND [1846-1926]
NDB 4,701-702; OBL 1,277; HIRSCH 2,453; LESKY, 541-544; PAGEL, 422-423; JOHN, 48-59; WINIGER 2,208; WWWS, 538-539; POGG 3,422-423; 4,398
A. Durig, Wiener klin. Wchschr. 39 (1926), 221-224; Alm. Akad. Wiss. Wien 76 (1926), 184-190; Z. Sinnesphysiol. 57 (1926), 281-287

EXNER, WILHELM [1840-1931]
NDB 4,703; KURSCHNER 4,624-625; TLK 3,140-141; POGG 3,422; 4,398; 6,694
P. Ludwik, Alm. Akad. Wiss. Wien 81 (1931), 200-204

EYKMAN, JOHAN FREDERIK [1851-1915]
BWN 2,178-179; LINDEBOOM, 523-524; TSCHIRCH,1045; POGG 4,399; 5,351-352
A.F. Holleman, Rec. Trav. Chim. Pays-Bas 35 (1916), 365-420
E.H. Huntress, Proc. Am. Acad. Arts Sci. 79 (1951), 14-15

EYRING, HENRY [1901-1981]
DSB 17,279-284; AMS 14,1409; MH 1,161-163; STC 1,362-364; MSE 1,353-354; WWA 1978-1979, p.996; CB 1961, pp.151-152; POTSCH, 142; POGG 6,696; 7b,1304-1311
Anon., Nachr. Chem. Techn. 12 (1964), 4
R.C. Brasted, J. Chem. Ed. 53 (1976), 752-756
H. Eyring, Ann. Rev. Phys. Chem. 28 (1977), 1-13; Ber. Bunsen Ges. 86 (1982), 348-349
J.O. Hirschfelder, Am. Phil. Soc. Year Book 1982, pp.482-489
E.L. Kimball and S.H. Heath, Dialogue 15(3) (1982), 80-99
D. Henderson, J. Phys. Chem. 87 (1983), 2638-2640

F

FABBRONI, GIOVANNI [1752-1822]
DSB 4,503; HIRSCH 2,466-467; FERCHL, 147-148; COLE, 173-175; WWWS, 540; POGG 1,709-710
G. Cuvier, Eloges Historiques des Membres de l'Académie des Sciences, 2nd Ed., vol.3, pp.373-394. Paris 1861
H. Gliozzi, Archeion 18 (1936), 160-165

FABER, KARL [1822-1902]
RAUSCH, 319
Anon., Pharm. Z. 47 (1902), 921

FABER, KNUD [1862-1956]
DBL 6,506-509; GROTE 8,25-60; FISCHER 1,383-384; VEIBEL 2,136; WWWS, 540
O. Christensen, Acta Med. Scand. 78 (1932), 205-215
E. Meulengracht, Nord. Med. 55 (1956), 823-825
Anon., Leopoldina [3]4/5 (1958-1959), 89-90
J. Frandsen, Med. Forum 17(2) (1964), 33-42

FABERGÉ, ALEXANDER CYRIL [1912-]
AMS 15(2),924; WWWS, 540

FABINYI, RUDOLF [1849-1920]
POGG 3,424; 4,400; 5,352; 6,697
F. von Konek, Chem. Z. 44 (1920), 417FABRE, RENÉ [1889-1966]
DBF 13,414-415; POGG 6,697; 7b,1313-1319
J. Roche, C.R. Acad. Sci. 263 (1966), 94-97
J.E. Courtois et al., Bull. Soc. Chim. Biol. 49 (1967), 2-12

FAGRAEUS-WALLBOM, ASTRID ELSA [1913-]
VAD 1981, p.291

FAHLBERG, CONSTANTIN [1850-1910]
NDB 4,744; NB, 97-98; BJN 15,25*; MILES, 151; GOULD, 28-29; BLOKH 1,190; POTSCH, 142; WWWS, 541; POGG 3,425; 4,402; 5,353
W. Will, Ber. chem. Ges. 43 (1910), 2784
Anon., Chem. Z. 34 (1910), 888
G.B. Kauffman and P.M. Priebe, Ambix 25 (1978), 191-207

FÅHRAEUS, ROBIN SANNO [1888-1968]
SMK 2,649; FISCHER 1,385; WWWS, 541
E. Jorpes, Acta Med. Scand. 185 (1969), 23-26
H.L. Goldsmith et al., Am. J. Physiol. 257 (1989), H1005-H1015
A.L. Copley, Thromb. Res. 54 (1989), 521-559

FAHRIG, CARL [1882-1942]
ECKERT, 58-60
G. Gruber, Verhandl. path. Ges. 33 (1950), 409-413

FAILEY, CRAWFORD FAIRBANKS [1900-1981]
AMS 14,1415

Anon., Chem. Eng. News 59(43) (1981), 53

FAIRLEY, NEIL HAMILTON [1891-1966]
DNB [1961-1970], 344-346; WW 1965, pp.991-992; WWWS, 542
A.W. Woodruff, Nature 210 (1966), 1205
J. Boyd, Biog. Mem. Fell. Roy. Soc. 12 (1966), 123-145

FAIVRE, ERNEST [1827-1879]
DBF 13,491-492; NUC 166,73
G.A. Heinrich, Mem. Acad. Lyon 24 (1879), 117-132

FAJANS, KASIMIR [1887-1975]
DSB 17,284-286; AMS 11,1471; KURSCHNER 11,628; BHDE 2,278; LDGS, 18; ABBOTT-C, 41-42; WININGER 6,577; TETZLAFF, 74; COHEN, 59-60; POTSCH, 142; WWWS, 543; POGG 5,353-354; 6,701-702; 7a(2),4-5
T.M. Dunn, Nature 259 (1976), 611
G.M. Schwab, Bayer. Akad. Wiss. Jahrbuch 1976, pp.227-229
J. Hurwic, Kwart. Hist. Nauki 30 (1985), 215-245
R. Holmen, Bull. Hist. Chem. 4 (1989), 15-23; 6 (1990), 7-15

FALCK, CARL PHILIPP [1816-1880]
NDB 5,3; LKW 3,86-95; GUNDLACH, 250-253; HIRSCH 2,471-472; WWWS, 543
M.J. Rossbach, Berl. klin. Wchschr. 17 (1880), 590-591
E. Heischkel, Deutsche med. Wchschr. 69 (1943), 305-306
W. Schmid, Arzneimitt. 16 (1966), 577-578

FALCK, FERDINAND AUGUST [1848-1926]
FISCHER 1,386; REBER, 267-268; TSCHIRCH, 1045
E. Baur, Münch. med. Wchschr. 73 (1926), 2034

FALCK, RICHARD [1873-1955]
NDB 5,3-4; HEIN-E, 110-111; BHDE 2,279; LDGS, 12; TLK 3,143; WWWS, 543; WAGENITZ, 55; TETZLAFF, 74; POGG 6,702; 7a(2),5
L.J. Reichert, Palestine Journal of Botany 2 (1938-1939), 113-125
A. Hüttermann, Ber. bot. Ges. 100 (1987), 123-128,138-141

FALK, FRITZ [1878-1912]
JV 23,519; SG [3]5,630; NUC 166,200

FALK, ISIDORE SYDNEY [1899-1984]
AMS 14,1420; AO 1984, 528-530; IWW 1982-1983, pp.389-390; IB,80; WWWS,543
New York Times, 10 October 1984
M.I. Roemer, Am. J. Pub. Hlth. 75 (1985), 841-848

FALK, JOHN EDWIN ROGERS [1917-1970]
Anon., Nature 229 (1971), 143-144
M.R. Lemberg and O.H. Frankel, Rec. Austral. Acad. Sci. 2 (1972), 92-107

FALK, KAUFMAN GEORGE [1880-1953]
AMS 7,542; NCAB F,116; 42,39; WWAH, 188; WWWS, 544; POGG 5,355; 6,703-804; 7b,1324
Anon., Science 119 (1954), 35; Chem. Eng. News 32 (1954), 1048

FALKENHAUSEN, FRIEDRICH von [1908-]
JV 43,411; NUV 166,231

FALKENHEIM, KURT HERMANN [1893-1943]
KURSCHNER 4,632-633; BHDE 2,280; LDGS, 77

FALTA, WILHELM [1875-1950]
NDB 5,21-22; OBL 1,287; STURM 1,327; FISCHER 1,387; WEINMANN, 132; STANGL, 27-48; PLANER, 73; WWWS, 544
E. Pilgersdorfer, Wiener klin. Wchschr. 100 (1950), 547
Oesterreicher der Gegenwart 1951, p.371

FALTIS, FRANZ [1885-1963]
KURSCHNER 9,417; KERNBAUER, 270-280; WIEN 1964, pp.49-50; TLK 3,143; POGG 6,705; 7(2),8
R. Schwarz, Ost. Chem. Z. 56 (1955), 179
W.Winkler, Sci. Pharm. 31 (1963), 1-6

FAMINTZIN, ANDREI SERGEEVICH [1835-1918]
BSE 44,516; BLOKH 1,191-192; WWWS, 544
M.A. Blokh, Zhur. Fiz. Khim. 60Suppl. (1928), 144-145
V.M. Senchenkova, Bot. Zhur. 45 (1960), 309-317

FANCONI, GUIDO [1892-1979]
KURSCHNER 13,840; SBA 3,41
W.H. Hitzig, Münch. med. Wchschr. 104 (1962), 1189-1191
K.H. Schaefer, Deutsche med. Wchschr. 87 (1962), 45-47
G. Fanconi, Der Wandel der Medizin, wie ich ihn Erlebte. Berne 1970
H.R. Wiedemann, Eur. J. Ped. 132(3) (1979), 131-132
Anon., Schw. med. Wchschr. 109 (1979), 1720-1722
E. Rossi, Helv. Paed. Acta 34 (1979), 393-396
H. Zellweger, J. Pediat. 96 (1980), 674-675

FANKUCHEN, ISIDOR [1905-1964]
DSB 4,521-522; McLACHLAN, 54-58
J.D. Bernal, Nature 203 (1964), 916-917
P.P. Ewald, Acta Cryst. 17 (1964), 1091-1093
Anon., Chem. Eng. News 42(28) (1964), 113
J.D.H. Donnay, Am. Min. 50 (1965), 539-547

FANO, GIULIO [1856-1930]
EI 14,789; WININGER 2,231; 6,578; 7,558
E. Sereni, Riv. Biol. 12 (1930), 401-425
G. Rossi and I. Spadolini, Arch. Fisiol. 29 (1930), i-xv

FARADAY, MICHAEL [1791-1867]
DSB 4,527-540; DNB 6,1054-1066; BUGGE 1,417-427; FARBER, 467-480; COLE, 176-178; PHILIPPE, 779-782; BLOKH 1,192-193; FERCHL, 150; MATSCHOSS, 72; SCHAEDLER, 33-34; W, 174-176; ABBOTT-C, 42-43; POTSCH, 143; WWWS, 545; POGG 1,719-722; 3,427; 4,404-405; 6,706
H.B.J., Proc. Roy. Soc. 17 (1869), i-lxviii
L.P. Williams, Michael Faraday. London 1965
P. Dunsheath, Proc. Roy. Inst. 41 (1966), 76-91
H. Wolter, Chem. Z. 91 (1967), 677-682
R.H. Cragg, Chem. Brit. 3 (1967), 482-486
J. Agassi, Faraday as a Natural Philosopher. Chicago 1971
D. Gooding, Isis 73 (1982), 46-67
D. Gooding and F.A.J.L. James, Faraday Rediscovered. Basingstoke 1985
G. Cantor, Michael Faraday. Houndsmills 1991
J.M. Thomas, Analyst 116 (1991), 1205-1210
D.A. Davenport et al., Bull. Hist. Chem. 11 (1992), 3-104

FARBER, EDUARD [1892-1969]
AMS 11,1477-1478; MILES, 152; TLK 3,144; WININGER 7,558; POGG 6,698-699; 7a(2),9-10
W.D. Miles, Arch. Inter. Hist. Sci. 22 (1969), 63-65
R. Multhauf, Isis 62 (1971), 220-224

FARBER, SIDNEY [1903-1973]
AMS 12,1753
R. Toch, Harvard Med. Alumni Bull. 47(5) (1973), 56-57

FARKAS, ADALBERT [1906-]
AMS 11,1479; DAVIS, 89-117; LDGS, 18FARKAS, GÉZA [1872-1934]
MEL 1,468; 18, 80
E. Balogh, Orvosi Hetilap 78 (1934), 865

FARKAS, LASZLO (LADISLAUS) [1904-1948]
DSB 4,545; MEL 3,192; BHDE 2,282; LDGS, 18; POGG 6,706; 7b,1327-1329
E.K. Rideal, Nature 163 (1949), 313-314; in L. Farkas Memorial Volume (A. Farkas and E.P. Wigner, Eds.), pp.1-2,305-307. Jerusalem 1952

FARMER, CHESTER JEFFERSON [1886-1969]
AMS 11,1480; NCAB 54,470

FARMER, JOHN BRETLAND [1865-1944]
DSB 4,545-546; DNB [1941-1950], 245-246; O'CONNOR(2), 534-535; DESMOND, 217;IB, 80
V.H. Blackman, Nature 153 (1944); Obit. Not. Fell. Roy. Soc. 5 (1945), 17-31

R.J. Tabor, Proc. Linn. Soc. 156 (1944), 205-207

FARRINGTON, BENJAMIN [1891-1974]
WW 1974, p.1071
G.E.R. Lloyd, Arch. Inter. Hist. Sci. 26 (1976), 159-160

FARUP, PEDER [1875-1934]
POGG 5,365; 6,708; 7b,1333

FAURÉ-FREMIET, EMMANUEL [1883-1971]
DBF 13,780; CHARLE, 65-67; IB, 81
J. Tréfouël, C.R. Soc. Biol. 165 (1971), 2244-2246
J.O. Corliss, J. Protozool. 19 (1972), 389-400
E.N. Willmer, Biog. Mem. Fell. Roy. Soc. 18 (1972), 187-221
J. André and A. Dalcq, J. Microscopie 18 (1973), 1a-42a

FAURHOLT, CARL [1890-1972]
KBB 1971, p.378; VEIBEL 2,137-138; POGG 6,709-710; 7b,1334

FAUST, EDWIN STANTON [1870-1928]
KURSCHNER 3,504; FISCHER 1,389; BOCK, 87-90; TSCHIRCH, 1046; TLK 2,159; POGG 6,710; 7b,1334-1335
F. Flury, Münch. med. Wchschr. 75 (1928), 2053-2055

FAVORSKY, ALEKSEI EVGRAFOVICH [1860-1945]
DSB 4,553-554; BSE 44,487-488; WWR, 161; KUZNETSOV(1961), 516-529; POTSCH, 143-144; ARBUZOV, 135-138; POGG 4,407; 5,358; 6,712; 7b,1336-1339
S.N. Danilov, Usp. Khim. 14 (1945), 441-462
I.A. Pastac, Bull. Soc. Chim. 1947, pp.565-566
A.E. Arbuzov, Zhur. Ob. Khim. 25 (1955), 1441-1443
N.A. Domnin, Zhur. Ob. Khim. 30 (1960), 705-716

FAVRE, PIERRE ANTOINE [1813-1880]
DSB 4,554-555; DBF 13,873-874; GE 17,85-86; FERCHL, 151; WWWS, 550; POGG 1,726; 3,430
F. Le Blanc, Bull. Soc. Chim. [2]33 (1880), 390-400
H. Tachoire, Actualités Chimiques 1981(4), pp.33-36

FAWDINGTON, THOMAS [1795-1843]
HIRSCH 2,490; DECHAMBRE [4]1,316; CALLISEN 6,208; 28,18

FAY, IRVING WETHERBEE [1861-1936]
AMS 5,348; WWAH, 190; RSC 14,944
Anon., Ind. Eng. Chem. 14 (1936), 75

FAYOD, VICTOR [1860-1900]
Anon., Verhandl. Schw. Nat. Ges. 83 (1900), xxxii-xxxvi

FEARON, WILLIAM ROBERT [1892-1959]
WW 1959, p.992; WWWS, 550; POGG 6,714; 7b,1341-1342
W.J.E. Jessop, Nature 185 (1960), 283-284
Anon., Lancet 1960(I), p.128
V.M. Synge, Irish J. Med. Sci. 1960, p.45

FECHNER, GUSTAV THEODOR [1801-1887]
DSB 4,556-559; NDB 5,37-38; ADB 55,756-763; HIRSCH 2,493-494; Suppl.,280; FERCHL, 151; WWWS, 551; POGG 1,728-729; 3,433; 7aSuppl.,202-204
W. Wirth, Sächsische Lebensbilder 2 (1938), 95-113
J. Thiele, Centaurus 11 (1966), 222-235

FEDCHENKO, ALEKSEI PAVLOVICH [1844-1873]
BSE 44,583; BME 33,582-583; OLPP, 124
G. Lohde, Entomol. Z. 17 (1873), 236-238
E. Naust, Am. J. Trop. Med. 20 (1971), 511-523
N.I. Leonov, A.P. Fedchenko. Moscow 1972
O.V. Smirnov, Med. Parazit. 42 (1973), 735-736

FEDER, LUDWIG [-1882]
RSC 9,841; NUC 168,301
Anon., Ber. chem. Ges. 15 (1882), 3108

FEDERLEY, HARRY [1879-1951]
IB, 81
E. Suomalainen, Lep. News 6 (1952), 57-60
R.B. Goldschmidt, Science 115 (1952), 561-562

FEDEROV, EVGRAF STEPANOVICH [1853-1919]
DSB 5,210-214; BSE 44,572-574; KUZNETSOV(1962), 63-82; FREUND 3,157-162; BLOKH 1,194; POGG 4,409-410; 5,358-359; 6,715-716
A.E. Fersman, Priroda 8 (1919), 236-244
P. von Groth, Bayer. Akad. Wiss. Jahrbuch 1921, pp.35-38
I.I. Shafranovsky, Priroda 38(4) (1949), 61-65
J.J. Burkhardt, Arch. Hist. Exact Sci. 4 (1967), 235-246

FEDEROV, SERGEI PETROVICH [1869-1936]
BSE 44,575; BME 33,580-582; KUZNETSOV(1963), 641-651; WWR, 175
I.M. Kalman et al., Vest. Khir. 44 (1936), 4-13,171-178
V.A. Petrova et al., Khirurgia 45 (1969), 3-16,20-33
A.T. Ivanova, S.P. Fedorov. Moscow 1972
I. Kissin and A.J. Wright, Anesthesiology 69 (1988), 242-245

FEER, ADOLF [1862-1913]
JV 2,191; RSC 14,951
Anon., Z. angew. Chem. 26(III) (1913), 381; Chem. Z. 37 (1913), 643
E.A. Feer, Die Familie Feer, vol.2, pp.471-472. Aarau 1964

FEHLEISEN, FRIEDRICH [1854-1924]
HIRSCH 2,494; Suppl.,281; PAGEL, 490; WWWS, 552
C.D. Leake, Verhandlungen des 20. Internationalen Kongresses für Geschichte der Medizin, pp.123-124. Hildesheim 1968

FEHLING, HERMANN von [1811-1885]
NDB 5,47; ADB 48,508-510; HEIN, 155-156; SHBL 6,88-89; BLOKH 1,195-196; TSCHIRCH, 1046; FERCHL, 151-152; SCHAEDLER, 34; WWWS, 1733; POTSCH, 144; POGG 1,729-730; 2,434; 7a Suppl.,204
A.W. Hofmann, Ber. chem. Ges. 18 (1885), 1811-1818
C. von Voit, Sitz. Bayer. Akad. 16 (1886), 50-57
M.E. Cattelain, J. Pharm. Chim. [8]10 (1929), 405-413,449-458

FEIBELMANN, RICHARD [1883-1948]
BHDE 1,168; JV 23,541; POGG 6,716; 7a(2),16

FEIGEL, HEINRICH [1877-]
JV 21,480; NUC 168,622

FEIGL, FRIEDRICH [1891-1971]
HINK 1,797-803; BHDE 2,285; TLK 3,146; COHEN, 61; TETZLAFF, 76; POTSCH, 144; POGG 6,716-717; 7a(2),17; 7b,1344-1349
G. Kainz, J. Chem. Ed. 35 (1958), 609-611
A. Bondi, Israel J. Chem. 4 (1966), 167-182
F. Hecht. Alm. Akad. Wiss. Wien 121 (1971), 327-337
R. Belcher, Proc. Soc. Anal. Chem. 7 (1971), 172-173
J.A. Schufle and L.G. Ionescu, J. Chem. Ed. 53 (1976), 174
P. Hainberger, Quimica Nova 6(2) (1983), 55-60

FEILITZSCH, FABIAN CARL OTTOKAR [1817-1885]
NDB 5,57; POGG 1,730; 3,434

FEIST, FRANZ [1864-1941]
KURSCHNER 4,645-646; KIEL, 162; WININGER 6,582; TETZLAFF, 77; TLK 3,146; POGG 4,410; 5,359; 6,717-718; 7a(2),17

FEIST, KARL [1876-1952]
NDB 5,63-64; HEIN-E, 112-113; TSCHIRCH, 1046; WWWS, 533; POGG 6,718; 7a(2),17W. Awe, Pharm. Z. 87 (1951), 324-325
Anon., Chem. Z. 75 (1951), 274; 76 (1952), 226; Arzneimitt. 2 (1952),298

FELDBERG, WILHELM SIEGMUND [1900-1993]
WW 1981, p.847; KURSCHNER 13,855-856; IWW 1982-1983, p.396; COHEN, 62; BHDE 2,289-290; LDGS, 79; IB, 82; WWWS, 553
W. Feldberg, J. Physiol. 263 (1976), 89P-91P; Fifty Years On. Liverpool 1982

FELDHAUS, FRANZ MARIA [1874-1957]
NDB 5,68; TLK 3,146-147; POGG 6,719-720; 7a(2),19-20; (4),141*

FELIX, ARTHUR [1887-1956]
WW 1949, p.910; WININGER 6,582; WWWS, 554
J. Craigee, Biog. Mem. Fell. Roy. Soc. 3 (1957), 53-79
G.S. Wilson, J. Path. Bact. 73 (1957), 281-295

FELIX, KURT [1888-1960]
KURSCHNER 9,426; KALLMORGEN, 263; IB, 82; POGG 6,722; 7a(2),21-23; (4),141*
R. Duesberg, Deutsche med. Wchschr. 83 (1958), 1025-1026
W. Dirscherl, Münch. med. Wchschr. 103 (1961), 475-476

FELL, HONOR BRIDGET [1900-1986]
WW 1981, p.848; MH 2,149-150; MSE 1,360-361
Anon., Lancet 1986(I), p.1166
I. Lasnitzki, Nature 332 (1986), 214
J.A. Witkowski, TIBS 11 (1986), 486-488
J. Vaughan, Biog. Mem. Fell. Roy. Soc. 33 (1987), 237-259

FELLENBERG, THEODOR von [1881-1962]
TLK 3,147; POGG 6,722-723; 7a(2),23
B. Strahlmann, Alimenta 20 (1981), 159-164

FELLENBERG-RIVIER, LUDWIG RUDOLF von [1809-1878]
NDB 5,71-72; STELLING-MICHAUD 3,299; RSC 2,585-586; 7,650-651; POGG 1,732; 3,436

FELLMER, ERNST [1884-1958]
JV 27,395
Anon., Nachr. Chem. Techn. 7 (1959), 293

FELLNER, OTFRIED OTTO [1873-1942]
FISCHER 1,394; MEDVEI, 732; NUC 169,187
H.H. Simmer, Contraception 3 (1971), 1-20

FELSER, HEINRICH [1881-]
JV 23,22; NUC 169,236

FENN, WALLACE OSGOOD [1893-1971]
DSB 17,289-291; AMS 11,1511; WWA 1970-1971, p.705; MH 2,150-151; STC 1,369-370; IB, 82; MSE 1,362-363; WWAH, 191; WWWS, 555-556
W.O. Fenn, Ann. Rev. Physiol. 24 (1962), 1-10
F.D. Carlson et al., J. Gen. Physiol. 58Suppl. (1971), 1-3
G.W. Corner, Am. Phil. Soc. Year Book 1972, pp.152-156
H. Rahn, Physiologist 19 (1976), 1-10; Biog. Mem. Nat. Acad. Sci. 50 (1979), 141-173

FENNER, FRANK JOHN [1914-]
WW 1982, p.723

FENTON, HENRY JOHN HORSTMANN [1854-1929]
WW 1927, p.967; WWWS, 556; POGG 5,360-361; 6,723
W.H.M., J. Chem. Soc. 132 (1930), 889-894; Proc. Roy. Soc. A127 (1930),i-v

FERCHL, FRITZ [1892-1953]
NDB 5,81; HEIN-E, 114-116; TSCHIRCH, 1046; POGG 7a(2),26-28
Anon., Deutsche Apoth. Z. 92 (1952), 492-493; 102 (1962), 841
F. Schmidt, Pharmazie 8 (1953), 545-546
E.H. Guitard, Rev. Hist. Pharm. 21 (1972), 410-411

FERDMAN, DAVID LAZAREVICH [1903-1970]
BSE (3rd Ed.) 27,297; UKRAINE 2, 654-655; WWWS, 556

FERGUSON, JAMES KENNETH WALLACE [1907-]
AMS 14,1461; WWWS, 557

FERGUSON, JOHN [1837-1916]
J.M.T., J. Chem. Soc. 111 (1917), 333-342
J.W. Thomson, Proceedings of the Royal Philosophical Society of Glasgow 47 (1918), 103-114
E.H. Alexander, Records of the Glasgow Bibliographic Society 6 (1920), 39-63; 12 (1936), 82-127
G. Sarton, Isis 39 (1948), 60-61

FERGUSON, JOHN HOWARD [1902-1978]
AMS 14,1461; WWA 1976-1977, p.984; WWWS, 557
Who Was Who in America 7,191

FERGUSON, LEWIS KRAEER [1897-1968]
AMS 11,1516

FERMI, CLAUDIO [1862-1952]
FISCHER 1,396; AGRIFOGLIO, 89-91; IB, 82; ZURICH-D, 73

FERMI, ENRICO [1901-1954]
DSB 4,576-583; DAB [Suppl.5], 219-221; MH 1,168-169; STC 1,370-374; MSE 1,363-365; CB 1945, pp.179-181; W, 177-178; WWWS, 558
H.A. Bethe et al., Rev. Mod. Phys. 27 (1955), 249-275
E. Bretscher and J.D. Cockroft, Biog. Mem. Fell. Roy. Soc. 1 (1955), 69-78
S.K. Allison, Biog. Mem. Nat. Acad. Sci. 30 (1957), 125-155
E. Segré, Enrico Fermi Physicist. Chicago 1970

FERNBACH, AUGUSTE [1860-1939]
SCHALL, 38-39; IB. 82; POGG 4,413; 5,361; 6,726-727; 7b,1359-1360
G. Bertrand, Ann. Ferment. 5 (1939), 65-73; Ann. Univ. Paris 14 (1939), 206-208
Anon., Ann. Inst. Pasteur 62 (1939), 249-252

FERNER, HELMUT [1912-]
KURSCHNER 13,865; BUSCHHUTER, 7-22; WWWS, 558-559

FERNET, ÉMILE [1829-1905]
DBF 13,1050; POGG 3,437

FERNHOLZ, ERHARD [1909-1940]
G.A. Harrop and H.E. Stavely, Science 94 (1941), 130-131

FERNHOLZ, HANS [1915-]
KURSCHNER 13,865; POGG 7a(2),28

FERRÁN, JAIME [1852-1929]
FISCHER 1,396-397; WWWS, 559
E.W. Baader, Deutsche med. Wchschr. 56 (1930), 114-115
G.H. Bornside, Bull. Hist. Med. 55 (1981), 516-532; J. Hist. Med. 37 (1982), 399-422

FERRIER, DAVID [1843-1928]
DSB 4,593-595; DNB [1922-1930], 302-303; BRAZIER, 165-170,181-182; W, 179; WWWS, 559-560
C.S. S[herrington], Proc. Roy. Soc. B103 (1928), viii-xvi

FERRIÈRE, FRÉDÉRIC AUGUSTE [1848-1924]
DHB 3,95; STRAHLMANN, 470,481
G.W., Revue Internationale de la Croix-Rouge 6 (1924), 485-543

FERRY, EDNA LOUISE [1883-1919]
Obituary Record of Yale Graduates No.79, pp.1586-1587. New Haven 1921

FERRY, JOHN DOUGLASS [1912-]
AMS 15(2),995; WWA 1980-1981, pp.1069-1070; IWW 1982-1983, p.400; WWWS, 560
N.W. Tschoegl, Macromolecules 20 (1987), 909-910

FERRY, RONALD MANSFIELD [1891-1970]
AMS 11,1522; IB, 83
Anon., Chem. Eng. News 48(29) (1970), 49FERTIG, EDUARD [1872-1937] RSC 16,344
Anon., Chemische Fabrik 10 (1937), 370; Z. angew. Chem. 50 (1937), 754

FESTER, GUSTAV [1886-]
KURSCHNER 13,867; JV 26,636; TLK 3,148-149; POGG 6,731; 7a(2), 29-30

FETCHER, EDWIN STANTON [1909-]
AMS 15(2),996

FEULGEN, ROBERT [1884-1955]
DSB 4,603-604; NDB 5,115-116; FISCHER 1,401; SANDRITTER, 97-101; IB, 83; GUNDEL, 224-229; SCHULZ-SCHAEFFER, 207; TLK 3,149; WWWS,561; POTSCH, 144-145; POGG 6,732-733; 7a(2),31
Anon., Chem. Z. 79 (1955), 784; Nachr. Chem. Techn. 3 (1955), 213
K. Felix, Z. physiol. Chem. 307 (1957), 1-13
F.H. Kasten, Acta Histochimica 17 (1964), 88-99
H. Debuch, TIBS 3 (1978), 44-45

FEVOLD, HARRY LEONARD [1902-1981]
AMS 12,1807; WWWS, 561
Anon., Chem. Eng. News 59(28) (1981), 26

FEYRTER, FRIEDRICH [1895-1973]
KURSCHNER 11,651
M. Ratzenhofer, Verhandl. path. Ges. 58 (1974), 585-600; Wiener med. Wchschr. 126 (1976), 734-736

FIALA, SILVIO [1912-1983]
AMS 15(2),999

FIBIGER, JOHANNES [1867-1928]
DBL 7,17-20; FISCHER 1,401-402; W, 179-180; WWWS, 562
J.A. Murray, Nature 121 (1928), 250-251
Anon., Proc. Roy. Soc. B103 (1928), viii-xvi
C.O. Jensen, Hosp. Tid. 71 (1928), 125-136
K. Secher, The Cancer Researcher Johannes Fibiger. London 1947

FICHERA, GAETANO [1880-1935]
EI 15,218; FISCHER 1,402; SG [4]5,975; NUC 171,407-408
F. Fedeli, Boll. Soc. Med. Chir. Pavia 49 (1935), i-xxix
A. Pepere, Tumori 21 (1935), v-xi

FICHTER-BERNOUILLI, FRITZ [1869-1952]
SBA 1,49-50; TLK 3,149; WWWS, 562; POTSCH, 145; POGG 4,416-417; 5,362-363; 6,734-735; 7a(2),31-33
H. Erlenmeyer, Chimia 3 (1949), 157-158; Helv. Chim. Acta 36 (1953), 753-772; Verhandl. Schw. Nat. Ges. 133 (1954), 308-322
H. Mohler, Chimia 6 (1952), 170
A. Georg, Archives des Sciences (Geneva) 6 (1953), 40-43

FICK, ADOLF [1829-1901]
DSB 4,614-617; NDB 5,127-128; BJN 6,374-377,29*; LKW 4,82-90; WWWS, 562; LF 1,94-101; HIRSCH 2,515-519; Suppl.,285; PAGEL, 498-499; POTSCH, 145; ROEDER, 157-159; POGG 4,417
A.J. Kunkel, Münch. med. Wchschr. 48 (1901), 1705-1708
M. von Frey, Sitz. Phys. Med. Würzburg 1901, pp.65-82
F. Schenck, Pflügers Arch. 90 (1902), 313-361
C. von Voit, Sitz. Bayer. Akad. 32 (1902), 277-287
Anon., Nature 66 (1903), 180-182
E. Wöhlisch, Naturwiss. 29 (1938), 585-591
H.J.V. Tyrrell, J. Chem. Ed. 41 (1964), 397-400
R. Bezel, Der Physiologe Adolf Fick. Zurich 1979

FICK, RUDOLF [1866-1939]
NDB 5,129-130; FISCHER 1,402-403; HUTER, 205-206; STURM 1,341-342; IB,83; ROEDER, 63-65; MAASS, 8-24; KOERTING, 104-105; WWWS, 562
H. Stieve, Anat. Anz. 89 (1939), 98-127
A. Waldeyer, Deutsche med. Wchschr. 65 (1939), 1098-1099
Anon., Nature 144 (1939), 104

FICKER, MARTIN [1868-1950]
NDB 5,134-135; FISCHER 1,403; KURSCHNER 7,462; TLK 3,149; WWWS, 562
K.W. Jotten, Arch. Hyg. 134 (1951), 83-84

FIEDLER, ALBERT [1886-1961]
JV 26,34
Anon., Chem. Eng. News 40(4) (1962), 134

FIERZ-DAVID, HANS [1882-1953]
NDB 5,142-143; TLK 3,150; WWWS, 564; POGG 6,737; 7a(2),34-35
J. Read, Nature 172 (1953), 523
E.Z., Chimia 7 (1953), 230
A. Guyer and L. Blangey, Helv. Chim. Acta 27 (1954), 427-435

FIESER, LOUIS FREDERICK [1899-1977]
DSB 17,291-295; AMS 13,1287; MH 2,153-156; STC 1,378-379; MSE 1,372-374; POTSCH, 145-146; WWWS, 564; WWA 1976-1977, p.994; POGG 6,737;7b,1389-1395
Anon., Nachr. Chem. Techn. 2 (1954), 50
C.J. Brooks, Nature 270 (1977), 768-769
W.S. Johnson, Organic Syntheses 58 (1978), xiii-xvi
P.D. Bartlett, Am. Phil. Soc. Year Book 1980, pp.567-572

FIESER, MARY [1909-]
S. Pramer, J. Chem. Ed. 62 (1985), 186-191

FIESSELMANN, HANS [1909-1969]
KURSCHNER 11,529; POGG 7a(2),35
Anon., Chem. Z. 93 (1969), 831

FIESSINGER, NOËL [1881-1946]
DSB 4,617-618; DBF 13,1310; IB, 83
H.E. Gougerot, Bull. Acad. Med. 130 (1946), 128-133
C. Durand, Acta Med. Scand. 124 (1946), 417-420
G. Heuyer, Presse Med. 54 (1946), 155-156
J. Warter, Annuaire de la Société d'Histoire et d'Archéologie de Molsheim 1981, pp.103-114

FIGUIER, GUILLAUME LOUIS [1819-1894]
DBF 13,1325-1326; GE 17,434; LAROUSSE 8,355; VAPEREAU 6,583-584; FERCHL, 154; TSCHIRCH, 1046; WWWS, 564
M.B., Prog. Med. [2]20 (1894), 396-397
I.M. Tarbell, Pop. Sci. Mon. 51 (1897), 834-841
C. Clerk, Ann. Chim. Anal. 26 (1944), 97-99
S. Rochietta et al., Rev. Hist. Pharm. 20 (1971), 512-514
L.B. Hunt, Ambix 26 (1979), 221-223

FIGUROVSKY, NIKOLAI ALEKSANDROVICH [1901-1986]
G. Hartig and A. Mette, NTM 1(4) (1960), 116-117
Anon., Vop. Ist. Est. Tekhn. 1986(4), pp.163-164
Y.A. Pentin et al., Zhur. Fiz. Khim. 61 (1987), 1706-1707
T.A. Komarova et al., NTM 24 (1987), 139-141
A.T. Grigorian et al., Arch. Inter. Hist. Sci. 37 (1987), 151-152

FIKENTSCHER, WOLFGANG CASPAR [1770-1837]
NDB 5,145; HEIN, 162-163

FILATOV, DMITRI PETROVICH [1876-1943]
BSE 45,88
K. Kapronczay, Ontogenez 118 (1977), 1540-1641

FILATOV, NIL FEDOROVICH [1847-1902]
BSE 45,88-89; BME 33,774-775; FISCHER 1,404; KUZNETSOV(1963), 565-571;WWWS, 564
A. Hippius, Deutsche med. Wchschr. 28 (1902), 196

FILATOV, VLADIMIR PETROVICH [1875-1956]
BSE 45,87-88; BME 33,773-774; FISCHER 1,404-405; WWR, 166
M.A. Iasinovski and V.V. Skorodynskaya, Klin. Med. 33 (1955), 3-8
Anon., Science 124 (1956), 1021

FILDES, PAUL GORDON [1882-1971]
DSB 17,295-297; DNB [1971-1980], 315-316; WW 1965, p.1020
Anon., Lancet 1971(I), pp.401-402
G.P. Gladstone et al., Biog. Mem. Fell. Roy. Soc. 19 (1973), 317-347
R.E. Kohler, Bull. Hist. Med. 59 (1985), 54-74

FILEHNE, WILHELM [1844-1927]
NDB 5,146; SCHWARTZ, 49-57; PITTROFF, 191-192; WININGER 2,586;DZ, 355;WWWS, 564-565
J. Pohl, Med. Klin. 23 (1927), 781
K. Spiro, Deutsche med. Wchschr. 53 (1927), 1063
K. Brune, Agents and Actions 19Suppl. (1986), 19-29

FILETI, MICHELE [1851-1914]
POGG 3,441; 4,418-419; 5,364

FILHOL, ÉDOUARD [1814-1883]
DBF 13,1334; GORIS, 370-374; VAPEREAU 5,708; FERCHL, 155; POGG 3,441-442; 4,419
Anon., J. Pharm. Chim. [5]8 (1883), 165-169
V. Brustier, Figures Pharmaceutiques Françaises, pp.113-118. Paris 1953

FILIPCHENKO, YURI ALEKSANDROVICH [1882-1930]
DSB 17,297-303; BSE (3rd Ed.) 27,406

FILIPOVSKI, PETR IVANOVICH [1856-]
RSC 14,955; SG [2]5,785

FILIPOWICZ, BRONISLAW [1904-1988]
Anon., Acta Biochim. Polon. 35(3) (1988), i-ii; Post. Bioch. 35 (1989), 205-207

FILIPPOVICH, ARTEMI NIKITICH [1901-1961]
WWR, 167
Anon., Zdravokhranenie Belorussii 8 (1962), 76

FILOMAFITSKI, ALEKSEI MATVEEVICH [1807-1849]
BSE 45,121; BME 33,784-785; KUZNETSOV(1963), 78-88; ZMEEV 2,138-139
V.M. Merabishvili and S.N. Etin, Sov. Zdrav. 32 (1973), 69-72
A.A. Makarov, A.M. Filomafitski. Moscow 1986

FILOSOFOV, PETR IVANOVICH [1879-1935]
B.M. Prozorovski, Klin. Med. 13 (1935), 1749-1750

FINCKH, CARL [1878-1941]
JV 18,321; RSC 14,996; NUC 172,443
Anon., Chemische Fabrik 14 (1941), 110

FINDLAY, ALEXANDER [1874-1966]
WW 1965, pp.1021-1022; SRS, 23; POGG 4,419; 5,365; 6,740; 7b,1397
E.A. Moelwyn-Hughes, Chem. Ind. 1966, p.1880
R.B. Strathdee, Nature 212 (1966), 1302; Chem. Brit. 3 (1967), 24-25

FINDLAY, GEORGE MARSHALL [1893-1952]
WW 1949, p.925; WWWS, 565
R.R. Willcox, J. Roy. Micr. Soc. 72 (1952), 123-126

FINE, MORRIS S. [1886-1946]
AMS 7,563
Anon., Chem. Eng. News 24 (1946), 2384; Science 104 (1946), 207
Obituary Record of Yale Graduates No.106, p.148. New Haven 1948

FINGADO, RUDOLF [1901-1971]
JV 42,337; NUC 172,527

FINGER, ERNST [1856-1939]
OBL 1,316-317; STURM 1,349; KURSCHNER 4,666; FISCHER 1,405-406; CASARETTO, 55-72; LESKY, 342; PLANER, 77-78; WININGER 2,248; WWWS, 566
G. Schroeber, Wiener klin. Wchschr. 52 (1939), 538-539; Wiener med. Wchschr. 89 (1939), 543

FINGER, HERMANN [1864-1940]
WOLF, 49; TLK 3,150; POGG 5,365-366; 6,740; 7a(2),38
C. Schöpf, Ber. chem. Ges. 74A (1941), 107

FINGERLING, GUSTAV [1876-1944]
KLEIN, 143; TLK 3,150; IB, 84; NUC 172,542-543
W. Schneider, Z. Tierernährung 8 (1944), 97-106

FINK, HERMANN [1901-1962]
KURSCHNER 9,436; POGG 6,741-742; 7a(2),38-41
I. Schlie, Chem. Z. 85 (1961), 77-80; 86 (1962), 635

FINKELSTEIN, HEINRICH [1865-1942]
NDB 5,162-163; WWWS, 567
A. Ariztio, Rivista Chilena de Pediatrica 13 (1942), 485-491
I. Rosenstern, J. Pediatrics 49 (1956), 499-503
P. Wunderlich, Kinderärztliche Praxis 58 (1990), 587-592
P. Grossmann, Z. ärzt. Fortbild. 84 (1990), 733-735

FINKELSTEIN, JACOB [1910-]
AMS 14,1483; WWWS, 567

FINKENER, RUDOLF [1834-1902]
BJN 7,29*; BLOKH 1,196; POGG 3,443; 4,420
R. Toussaint, Ber. chem. Ges. 35 (1902), 4534-4535

FINKLE, PHILIP [1894-1942]
AMS 6,453; KAGAN(JA), 577

FINKLER, DITTMAR [1852-1912]
BJN 18,20*-21*; FISCHER 1,406-407
H. Selter, Deutsche med. Wchschr. 38 (1912), 662

FINSEN, NIELS RYBERG [1860-1904]
DSB 4,620-621; DBL 7,43-48; MEISEN, 181-185; DIEPGEN, 141-151; WWWS, 568; FISCHER 1,407-408; PAGEL, 508
H. Jansen, Münch. med. Wchschr. 51 (1904), 1879-1881
Anon., Lancet 1904(II), pp.1036-1039; Nature 70 (1904), 532-533
J. Rehns, Le Radium 1 (1904), 129-133
S.A. Heyerdahl, Pharmacia 1 (1904), 17-27
A. Reyn, Acta Radiologica 2 (1923), 207-209
H. Roesler, Ann. Med. Hist. 8 (1936), 353-356

FINZELBERG, HERMANN [1842-1922]
HEIN, 164; DRUM, 283-284

FIROR, WARFIELD MONROE [1896-]
AMS 11,1546
A.M. Harvey, Johns Hopkins Med. J. 146 (1980), 16-27

FISCHEL, ALFRED [1868-1938]
NDB 5,172; OBL 1,319; STURM 1,352; FISCHER 1,408-409; KOERTING, 103-104; MAASS, 43-51; HEINDEL, 176-182; SKVARC, 52-54; WWWS, 568
E.P. Pick, Wiener klin. Wchschr. 51 (1938), 121-122
G. Politzer, Anat. Anz. 100 (1953-1954), 290-300

FISCHEL, WILHELM [1852-1910]
FISCHER 1,409; PAGEL, 507-508; MOTSCH, 95-101; KOREN, 173

FISCHER, ALBERT [1891-1956]
KBB 1956, pp.414-415; VEIBEL 2,140-142; IB, 84
Anon., Leopoldina [3]4/5 (1958-1959), 90-91

FISCHER, ALFRED [1858-1913]
J. Behrens, Ber. bot. Ges. 31 (1913), 111-117

FISCHER, ANTON [1901-1978]
LDGS, 65,78; POGG 6,746; 7a(2),46-47

FISCHER, ARTHUR [1878-1922]
TLK 3,151; BLOKH 1,196; POGG 5,366; 6,746
A. Classen, Ber. chem. Ges. 56A (1923), 31

FISCHER, BERNHARD [1852-1915]
SACKMANN, 114-116; FISCHER 1,409-410

FISCHER, BERNHARD [1856-1905]
BJN 10,167*; HEIN, 164-165; FISCHER 1,410; REBER, 317-319; TSCHIRCH, 1046
Anon., Chem. Z. 29 (1905), 1326; Leopoldina 41 (1905), 102

FISCHER, CHARLES SUMNER [1866-1926]
JV 9,239; NUC 173,304
Anon., J. Am. Med. Assn. 86 (1926), 1785

FISCHER, EDMOND HENRI [1920-]
AMS 15(2),1023-1024; WWWS, 569

FISCHER, EMIL [1852-1919]
DSB 5,1-5; NDB 5,181-182; DBJ 2,378-379,716-717; SCHWERTE 1,158-166; BUGGE 2,408-420; STUPP-KUGA, 17-108; FARBER, 983-995; WREDE, 46-47; BLOKH 1,197-201; SCHAEDLER, 34; W, 180-181; DZ, 361; TSCHIRCH, 1047; POTSCH, 146; ABBOTT-C, 43-44; WWWS, 569; POGG 3,445; 4,422-424;5,366-368; 6,746-747; 7aSuppl.,204-207
C. Harries et al., Naturwiss. 7 (1919), 843-882
V. Kellogg, Science 50 (1919), 346-347
E. Abderhalden, Münch. med. Wchschr. 66 (1919), 938-940
P. Jacobson, Chem. Z. 43 (1919), 565-569
C. Wichelhaus et al., Ber. chem. Ges. 52A (1919), 129-164
R. Willstätter, Bayer. Akad. Wiss. Jahrbuch 1920, pp.17-24
M.O. Forster, J. Chem. Soc. 117 (1920), 1157-1201; Proc. Roy. Soc. A98 (1920-1921), I-viii
K. Hoesch, Emil Fischer. Berlin 1921
E. Fischer, Aus meinem Leben. Berlin 1922 (reprinted 1987)
C.S. Hudson, J. Chem. Ed. 18 (1941), 353-357; Adv. Carbohydrate Chem.3 (1948), 1-22
E.H. Huntress, Proc. Am. Acad. Arts Sci. 81 (1952), 50-55
K. Freudenberg, Adv. Carbohydrate Chem. 21 (1966), 1-38
F. Herneck, Z. Chem. 10 (1970), 41-48
G. Hilgetag, Z. Chem. 10 (1970), 281-289
G.D. Feldman, in Deutschland in der Weltpolitik des 19. und 20. Jahrhunderts (I. Geiss and B.J. Wendt, Eds), pp.341-362. Düsseldorf 1973
H. Scholz, NTM 21 (1984), 91-98; Wiss. Z. Jena 37(2) (1988), 231-238
H. Remane, Emil Fischer. Leipzig 1984
J.S. Fruton, Proc. Am. Phil. Soc. 129 (1985), 313-370
T.D. Moy, Ambix 36 (1989), 109-120
G.B. Kauffman and P.M. Priebe, J. Chem. Ed. 67 (1990), 93-101

FISCHER, ERNST [1896-1980]
AMS 12,1834; BHDE 2,298; LDGS, 79; KOREN, 173

FISCHER, ERNST OTTO [1918-]
KURSCHNER 13,888; STC 1,381-382; MSE 1,375; IWW 1982-1983, p.405; POTSCH, 147; ABBOTT-C, 44-45; WWWS, 569; POGG 7a(2),47
Anon., Nachr. Chem. Techn. 7 (1959), 305
W.A. Herrmann, Naturw. Rund. 41 (1988), 442-448

FISCHER, EUGEN [1854-1917]
NDB 5,187; NB, 101; DBJ 2,653; BLOKH 1,201-202; ZURICH-D, 131
J.O., Z. angew. Chem. 30(III) (1917), 476
H. Bucherer, Ber. chem. Ges. 51 (1918), 1-4

FISCHER, EUGEN [1874-1967]
FISCHER 1,410; EBERT, 37-60; IB, 84
Anon., Homo 18 (1967), 110-124

FISCHER, FERDINAND [1842-1916]
NDB 5,175; DBJ 1,353; BLOKH 1,202-203; SCHAEDLER, 34; POGG 4,421-422; 5,368
H. Precht, Z. angew. Chem. 29 (1916), 345-346

FISCHER, FRANZ [1877-1947]
NDB 5,184; PUDOR, 140-141; TLK 3,151; POTSCH, 147; POGG 5,369; 6,747-748;7a(2),47-48
F. Fischer, Leben und Forschung. Mülheim 1957
H. Pichler, Chem. Ber. 100(6) (1967), cxxvii-clviii

FISCHER, FRANZ GOTTWALD [1902-1960]
KURSCHNER 8,528; POGG 6,749; 7a(2),50; (4),141*
W. Grassmann, Bayer. Akad. Wiss. Jahrbuch 1964, pp.189-194

FISCHER, HANS [1878-]
JV 18,291; NUC 173,364

FISCHER, HANS [1881-1945]
DSB 15,157-158; NDB 5,187; NB, 102; STC 1,382-383; FRANKEN, 182; IB, 84; POTSCH, 147-148; TLK 3,152; ABBOTT-C, 45; WWWS, 569; POGG 5,369-370; 6,749-751; 7a(2),50-53
H. Wieland, Bayer. Akad. Wiss. Jahrbuch 1944-1945, pp.210-214; Z. angew. Chem. 62 (1950), 1-4
K. Zeile, Naturwiss. 33 (1946), 289-291
A. Treibs, Z. Naturforsch. 1 (1946), 476-479; Das Leben und Wirken von Hans Fischer. Munich 1971; TIBS 4 (1979), 71-72
S.F. MacDonald, Nature 160 (1947), 494-495
H. Kaemmerer, Münch. med. Wchschr. 103 (1961), 2164-2166
C.J. Watson, Persp. Biol. Med. 8 (1965), 419-435
Anon., Chemie in Unserer Zeit 1 (1967), 58-61
A.J. Stern, Ann. N.Y. Acad. Sci. 206 (1973), 754-761

FISCHER, HELMUTH [1902-1976]
KURSCHNER 12,726; POGG 6,751-752; 7a(2),53-55
Anon., Chem. Z. 100 (1976), 201

FISCHER, HERBERT [1919-1981]
KURSCHNER 13,893; POGG 7a(2),55
Anon., Chem. Z. 106 (1982), 103

FISCHER, HERMANN OTTO LAURENZ [1888-1960]
DSB 5,5-7; AMS 9(1),595; BHDE 2,300; ABBOTT-C, 45; WWAH, 196; POTSCH, 148; POGG 6,752; 7b,1407-1409
Anon., Nachr. Chem. Techn. 4 (1956), 2
H.O.L. Fischer, Ann. Rev. Biochem. 29 (1960), 1-14
J.C. Sowden, Adv. Carbohydrate Chem. 17 (1962), 1-14
W.M. Stanley and W.Z. Hassid, Biog. Mem. Nat. Acad. Sci. 40 (1969),91-112
F.W. Lichtenthaler, Carbohydrate Research 164 (1987), 1-22

FISCHER, HUGO [1865-1939]
IB, 84
B. Leisering, Ber. bot. Ges. 58 (1940), 55-69

FISCHER, ISIDOR [1868-1943]
WININGER 7,561
A. Castiglioni, Bull. Hist. Med. 14 (1943), 114-115

FISCHER, JOHANN HEINRICH [1842-1925]
AARGAU, 204-206

FISCHER, KARL [1865-1925]
AARGAU, 206FISCHER, LOUIS [1862-1921]
JV 6,238
Anon., Chem. Z. 45 (1921), 926

FISCHER, MARTIN HENRY [1879-1962]
AMS 10,1195; FISCHER 1,411-412; NCAB 58,535-536; WWAH, 196; IB, 84; POGG 5,370; 6,752-753; 7b,1409-1410
W. Ostwald, Koll. Z. 89 (1939), 1-12
H. Erbring and W.A. Mitchell, Koll. Z. 182 (1962), 4-7
H. Fabing, in Medical Portraits (C. Striker, Ed.), pp.87-94. Cincinnati 1963

FISCHER, MAX [1889-1967]
NUC 173,447

FISCHER, MAX HEINRICH [1872-1971]
KURSCHNER 11,670; FISCHER 1,412

FISCHER, NICOLAUS WOLFGANG [1782-1850]
DSB 5,7; STURM 1,358; FERCHL, 156; WWWS, 570; WININGER 2,258; POTSCH, 148; POGG 1,754-755; CALLISEN 6,297-299; 28,55-56
J. Schiff, Arch. Gesch. Naturwiss. 8 (1918), 225-231; 9 (1920), 29-38

FISCHER, OTTO [1852-1932]
DZ,364; STUPP-KUGA,132-159; WWWS,570; POTSCH, 148; POGG 3,445; 4,424-425; 5,371; 6,753
M. Bodenstein, Ber. chem. Ges. 65A (1932), 78
E.H. Huntress, Proc. Am. Acad. Arts Sci. 81 (1952), 55-56

FISCHER, PHILIPP WILHELM [1877-1946]
HEIN-E, 125; POGG 6,753-754; 7a(2),58
Anon., Aus dem Leben Schweinfurter Männer und Frauen, pp.163-164. Schweinfurt 1973

FISCHER, ROBERT [1903-]
KURSCHNER 13,898; POGG 6,754; 7a(2),58-60
H.F. Häusler, Arzneimitt. 13 (1963), 739-740

FISCHER, WALDEMAR [1881-1934]
RIGA, 699; POGG 5,371-372; 6,755
E. Eegriwe, Ber. chem. Ges. 67A (1934), 165
O. Lutz, Bull. Soc. Chim. [5]1 (1934), 1611-1612
M. Centnerszwer, Roczniki Chem. 15 (1935), 104-113
Y.P. Sradin, Ist. Est. Tekhn. Pribalt. 4 (1972), 143-159

FISCHER, WALTER WILLIAM [1842-1920]
BLOKH 1,203
B. Dyer, J. Chem. Soc. 117 (1920), 456-457

FISCHER, WILHELM [1868-]
JV 8,273; RSC 14,1019; NUC 173,489FISCHGOLD, HARALD [1903-] LDGS, 65; NUC 173, 512

FISCHL, RUDOLF [1862-1942]
KURSCHNER 4,686; FISCHER 1,413; WININGER 4,686; KOREN, 173

FISCHLER, FRANZ [1876-1957]
NDB 5,215-216; DRULL, 69; FISCHER 1,413; WWWS, 570; POGG 6,755-756;7a(2),63
E. Bamann and G. Kallinich, Deutsche Apoth. Z. 96 (1956), 225-226
G. Kallinich, Münch. med. Wchschr. 98 (1956), 506-507

FISHER, KENNETH TIMPERLEY [1882-1945]
JV 22,321; NUC 173,692
Anon., Chem. Ind. 1945, p.320

FISHER, REGINALD BRETTAUER [1907-1986]
WW 1981, p.869; WWWS, 572

FISHER, RONALD AYLMER [1890-1962]
DSB 5,7-11; DNB [1961-1970], 361-362; HINK 1,217-219; W, 181-182; ABBOTT, 45-46
F. Yates and K. Mather, Biog. Mem. Fell. Roy. Soc. 9 (1963), 92-129
J. Neyman, Science 156 (1967), 1456-1460
J.F. Box, R.A. Fisher: The Life of a Scientist. New York 1978
J.F. Crow, Genetics 124 (1990), 207-211
W.W. Piegorsch, Biometrics 46 (1990), 915-924

FISKE, CYRUS HARTWELL [1890-1978]
AMS 11,1561; WWWS, 573
E.G. Ball et al., Harvard Gazette, 12 October 1979, p.4

FITGER, PETER [1896-]
VAD 1981, p.300

FITTICA, FRIEDRICH BERNHARD [1850-1912]
BJN 18,21*; SCHMITZ, 279-282; GUNDLACH, 467; MEINEL, 213-217,501-502; AUERBACH, 800-801; BLOKH 1,203-204; TSCHIRCH, 1047; SCHAEDLER, 34-35; POGG 3,446;4,427; 5,372
Anon., Chem. Z. 36 (1912), 499

FITTIG, RUDOLF [1835-1910]
DSB 5,12-13; NDB 5,217; DBA 1,955; BJN 15,26*; BLOKH 1,204-205; W, 182;DZ, 366-367; SCHAEDLER, 35; WWWS, 574; POTSCH, 148; POGG 3,446-447;
4,427-428; 5,372
F. Fichter, Chem. Z. 34 (1910), 1277-1280; Ber. chem. Ges. 44 (1911), 1339-1401
R.M., J. Chem. Soc. 99 (1911), 1651-1653
I. Remsen, Am. Chem. J. 45 (1911), 210-215
A. von Baeyer, Sitz. Bayer. Akad. 1911, pp.36-39

FITZ, ALBERT [1842-1885]
BLOKH 1,205; POGG 4,428
A.W. Hofmann, Ber. chem. Ges. 18 (1885), 1505-1506

FITZ, REGINALD [1885-1953]
AMS 8,788; NCAB 39,87-88; ICC, 131-132
C.S. Burwell, Trans. Assoc. Am. Phys. 67 (1954), 12-13

FITZ, REGINALD HEBER [1843-1913]
DAB 6,433-434; NCAB 10,456; FISCHER 1,414; DAMB, 250; ICC, 82-83; WWWS, 574; SG [3]5,783
Anon., Johns Hopkins Hosp. Bull. 25 (1914), 87-89
S.D. Leach et al., Annals of Surgery 212 (1990), 109-113

FITZGERALD, MABEL PUREFOY [1872-1973]
O'CONNOR(2), 108-111; NUC 174,279

FLADE, LEONHARD FRIEDRICH [1880-1916]
MEINEL, 263-265,502; GUNDLACH, 472; AUERBACH, 801; POGG 5,372
K. von Auwers, Chem. Z. 41 (1917), 105-106

FLANDERS, FRED FORD [1876-1948]
AMS 7,573

FLASCHENTRÄGER, BONIFAZ [1894-1957]
PICCO, 15-22; IB, 85; POGG 6,760-761; 7a(2),67
Anon., Chem. Z. 81 (1957), 573; Nachr. Chem. Techn. 5 (1957), 221

FLATOW, LEOPOLD [1877-]
SG [3]5,786

FLAUM, MAXIMILIAN [1864-1933]
PSB 7,31-32; KONOPKA 3,301-303
Z. Srebrny, Warsz. Czas. Lek. 10 (1933), 689-690

FLAVITSKI, FLAVIAN MIKHAILOVICH [1848-1917]
BSE 45,225; ARBUZOV, 127; KAZAN 1,515-522; BLOKH 1,205-207; POGG 3,450; 4,429; 5,373
L.A. Chugaev, Zhur. Fiz. Khim. 49 (1917), 626-631
A. Dobroserdov, Bull. Soc. Chim. [4]23 (1918), 51-52
G.S. Vozdvyzhenski, Priroda 38 (1949), 90-92

FLECK, LUDWIK [1896-1961]
KURSCHNER 9,454
J. Gierasimiuk, Kwart. Hist. Nauki 26 (1981), 533-547
N. Tsoyopoulos, Med. Hist. J. 17 (1982), 20-36
T. Schnelle et al., Kwart. Hist. Nauki 28 (1983), 525-587
J. Neumann, Sudhoffs Arch. 73 (1989), 12-25
H. van den Belt and B. Gremmen, Stud. Hist. Phil. Sci. 21 (1990), 463-479

FLECK, WILHELM HUGO [1828-1896]
DBJ 1,422; 3,91*; HEIN, 167; RSC 2,635; 7,675-676; POGG 3,450; 4,429
Anon., Leopoldina 32 (1896), 101-102

FLECKENSTEIN, ALBRECHT [1917-1992]
KURSCHNER 13,910; WWWS, 577; POGG 7a(2),68-69
D.J. Triggle, Trends in Pharmacological Sciences 13 (1992), 265

FLEISCH, ALFRED [1892-1973]
KURSCHNER 11,681; SBA 4,48-49; IB, 85; WWWS, 577; POGG

7a(2),70-72
C. Perret, Verhandl. Schw. Nat. Ges. 153 (1973), 247-248
P.E. Pilet, Bull. Soc. Vaud. Sci. Nat. 72 (1974), 37-38

FLEISCHER, ANTON [1845-1877]
BLOKH 1,207; TSCHIRCH, 1047; POGG 3,451
A. Steiner, Ber. chem. Ges. 11 (1878), 2308-2310

FLEISCHER, RICHARD [1848-1909]
BJN 14,24*; FISCHER 1,416-417; PAGEL, 516-517; KOVACSICS, 15-21; PITTROFF, 191-192
F. Penzoldt, Münch. med. Wchschr. 56 (1909), 1285-1286

FLEISCHL von MARXOW, ERNST [1846-1891]
NDB 5,234; OBL 1,328; PAGEL, 517-518; LESKY, 538-540; JOHN, 81-92; BRAZIER, 204-208,211; KOREN, 174; POGG 4,429-430
E. Suess, Alm. Akad. Wiss. Wien 42 (1892), 192-193
S. Exner, Wiener klin. Wchschr. 11 (1898), 956

FLEISCHMANN, PAUL [1879-1957]
KURSCHNER 4,695; LDGS, 65; KOREN, 174
Anon., Brit. Med. J. 1957(I), pp.289-290

FLEISCHMANN, WALTER [1896-1979]
IB, 85
Anon., Naturw. Rund. 32 (1979), 262

FLEITMANN, THEODOR [1828-1904]
NDB 5,237; BJN 10,31*; SCHULTE, 78-79; BLOKH 1,207; RSC 2,636; 7,676; POGG 5,373-374
Anon., Chem. Z. 28 (1904), 1045
W. Schulte, Tradition 4 (1959), 205-217; Rhein. West. Wirt. Biog. 8 (1962), 56-58

FLEMING, ALEXANDER [1881-1955]
DSB 5,28-31; DNB [1951-1960], 361-364; FREUND 2,91-96; W, 184-185; SCHWERTE 2,256-262; CB 1944,pp.208-210; MH 1,172-174; MSE 1,380-381; STC 1,384-386; ABBOTT, 46-47; WWWS, 577
L. Colebrook, Biog. Mem. Fell. Roy. Soc. 2 (1956), 117-127
V.D. Allison, J. Gen. Microbiol. 14 (1956), 1-13
E. Chain, TIBS 4 (1979), 143-144E.P. Abraham, Review of Infectious Diseases 2 (1980), 129-140
R. Hare, Med. Hist. 26 (1982), 1-24; 27 (1983), 347-372
G. Macfarlane, Alexander Fleming. Cambridge, Mass. 1984

FLEMMING, WALTHER [1843-1905]
DSB 5,34-36; NDB 5,241-242; BJN 10,164-165,167*-168*; SHBL 4,72-73; FISCHER 1,417-418; PAGEL, 518-519; LOMMATZSCH, 27-30; KIEL, 80; SCHULZ-SCHAEFFER, 199; KOERTING, 100; W, 186; WWWS, 578
F. Meves, Münch. med. Wchschr. 52 (1905), 2232-2234
F. von Spee, Anat. Anz. 28 (1906), 41-59
C. von Voit, Sitz. Bayer. Akad. 36 (1906), 468-472
G. Peters, Walther Flemming: Sein Leben und Werk. Neumünster 1968

FLEROV, ALEKSANDR FEDOROVICH [1872-1960]
G.R. Matukhin and I.F. Lyashchenko, Bot. Zhur. 38 (1953),624-626; 46 (1961), 912-914

FLESCH, MAX [1852-1943]
ARNSBERG, 119-121; FISCHER 1,448; KALLMORGEN, 266; IB, 85
Anon., Fort. Med. 50 (1932), 84

FLETCHER, WALTER MORLEY [1873-1933]
DSB 5,36-38; DNB [1931-1940], 284-285; O'CONNOR(2), 35-37; WWWS, 579
T.R. Elliott, Obit. Not. Fell. Roy. Soc. 1 (1933), 153-163
J.C.G.L., Biochem. J. 27 (1933), 1333-1336
M. Fletcher, The Bright Countenance. London 1957

FLEURY, GUSTAVE CLÉMENT [1833-1910]
DBF 14,42; BALLAND, 190; HUMBERT, 204; BERGER-LEVRAULT, 74; DORVEAUX, 79
A. Barille, J. Pharm. Chim. [7]2 (1910), xxxii

FLEURY, PAUL [1885-1974]
DBF 14,55-56; GORIS, 375; IB, 85; POGG 6,764-765; 7b,1424-1428
J. Roche, C.R. Soc. Biol. 168 (1974), 1162-1163
J. Courtois, Bull. Hist. Med. 159 (1975), 531-538

FLEXNER, ABRAHAM [1866-1959]
DAB [Suppl.6], 207-208; NCAB D,61-62; 52,320-321; DAMB, 251-252; WININGER 6,591
A.N. Richards, Nature 186 (1960), 278-279
J.W. Gardner, Science 131 (1960), 594-595
R.P. Hudson, Bull. Hist. Med. 46 (1972), 545-561
C.B. Chapman, Daedalus 103 (1974), 105-117
D.M. Fox, Bull. Hist. Med. 54 (1980), 475-496

FLEXNER, LOUIS BARKHOUSE [1902-]
AMS 14,1516; WWA 1980-1981, p.1109; KOREN, 174; WWWS, 579

FLEXNER, SIMON [1863-1946]
DSB 5,39-41; DAB [Suppl.4], 286-289; FISCHER 1,419; SACKMANN, 117-119; OLPP, 134-135; DAMB, 252-253; KAGAN(JA), 294-297; WININGER 6,592 IB, 86; WWAH, 199; WWWS, 579
H.S. Gasser et al., Memorial Meeting for Simon Flexner. New York 1946
S. Bayne-Jones, Am. Phil. Soc. Year Book 1946, pp.289-297
R. Cole, Bull. N.Y. Acad. Med. 22 (1946), 546-552
E.L. Opie, Arch. Path. 42 (1946), 234-242
H.F. Swift, J. Immunol. 54 (1946), i-vii
P. Rous, Science 107 (1948), 611-613; Obit. Not. Fell. Roy. Soc. 6 (1948-1949), 409-445
S. Benison, Institute to University, pp.13-35. New York 1977
A.M. Harvey, Johns Hopkins Med. J. 142 (1978), 52-56
J.T. Flexner, An American Saga. Boston 1984

FLEXSER, LEO AARON [1910-]
AMS 17(2), 1109; NUC 175,383

FLINT, AUSTIN [1836-1915]
DAB 6,472-473; HIRSCH 2,546; PAGEL, 520-521; KELLY, 418; DAMB, 254-255; WWAH, 199-200; WWWS, 580
Anon., Science 42 (1915), 607-608
N. Shaftel, J. Med. Ed. 35 (1960), 1122-1135

FLÖSSNER, OTTO [1895-1948]
AUERBACH, 232-233; POGG 7a(2),74-75
Anon., Deutsche med. Wchschr. 73 (1948), 263

FLOREY, HOWARD WALTER [1898-1968]
DSB 5,41-44; DNB [1961-1970], 370-374; WW 1965, p.1047; MH 1,174-175; STC 1,388; MSE 1,381-382; CB 1944, pp.208-210; W, 581-582; IB, 86; ABBOTT, 47-48; WWWS, 581
N. Heatley, Am. Phil. Soc. Year Book 1968, pp.130-134
E.P. Abraham, Biog. Mem. Fell. Roy. Soc. 17 (1971), 255-302
L. Bickel, Rise up to Life. London 1972
G. MacFarlane, Howard Florey. Oxford 1979
T.I. Williams, Howard Florey: Penicillin and After. Oxford 1984

FLORKIN, MARCEL [1900-1979]
IWW 1974-1975, pp.545-546
E. Schoffeniels, Arch. Inter. Physiol. 78 (1970), i-v; Rev. Med. Liège 34 (1979), 482-487; Comp. Bioch. Physiol. 67B (1980), 359-365
C. Liebecq, TIBS 4 (1979), N239
E.H. Stotz, Nature 283 (1980), 704
J. Théodoridès, Hist. Sci. Med. 14 (1980), 145-150
J.T. Edsall, Isis 71 (1980), 286-288
Z.M. Bacq and J. Brachet, Ann. Acad. Roy. Belg. 147 (1981), 41-98

FLORY, PAUL JOHN [1910-1985]
AMS 14,1520; WWA 1978-1979, p.1071; IWW 1982-1983, p.410; MSE 1,382-383; CB 1975, pp.127-130; ABBOTT-C, 47; POTSCH, 149-150
R.C. Brasted and P. Farago, J. Chem. Ed. 54 (1977), 341-344
New York Times, 12 September 1985
H. Eisenberg, Nature 317 (1985), 474

W. Brostow, Chem. Brit. 22 (1986), 744
H. Taube, Am. Phil. Soc. Year Book 1986, pp.107-114

FLOSDORF, EARL WILLIAM [1904-1958]
AMS 9(I),607; WWWS, 581
Anon., Science 127 (1958), 1238; Chem. Eng. News. 36(20) (1958), 80

FLOURENS, MARIE JEAN PIERRE [1794-1867]
DSB 5,44-46; DBF 14,140-142; GE 17,659; LAROUSSE 8,508; HIRSCH 2,548-549; PAGEL, 521-522; HAYMAKER, 198-200; BRAZIER, 117-120,126; DBA 1,980; W, 186-187; BERGER-LEVRAULT,75; WWWS, 581; CALLISEN 6,343-345; 28,72
C.F.P. Martius, Sitz. Bayer. Akad. 1868(I), pp.458-465
J.M.D. Olmsted, in Science, Medicine and History (E.A. Underwood, Ed.),vol.2, pp.290-302. London 1953
G. Legée, Hist. Sci. Med. 7 (1973), 387-400; Pierre Flourens. Paris 1987

FLOYD, NORMAN FRANCIS [1911-1968]
AMS 10,1224

FLU, PAUL CHRISTIAN [1884-1945]
LINDEBOOM, 596-597; FISCHER 1,420-421; IB, 86; NUC 176,254

FLÜCKIGER, FRIEDRICH AUGUST [1828-1894]
NDB 5,258-259; DBA 1,981; HIRSCH 2,550; HEIN, 167-169; HUMBERT, 236-240; BLOKH 1,207-208; TSCHIRCH, 1047; REBER, 3-4,343-352; WWWS, 581; SCHELENZ, 713-714; STELLING-MICHAUD 3,331; POGG 3,454; 4,433
Anon., Chem. Z. 18 (1894), 1981-1982; Pharm. Z. 39 (1894), 869-870
A. Tschirch, Ber. pharm. Ges. 5 (1895), 3-46
P. Duquénois and F. Ludy-Tenger, Bull. Soc. Pharm. Stras. 16 (1973), 7-25
T. Haug, Friedrich August Flückiger. Stuttgart 1985

FLÜCKIGER, MAX [1863-1887]
JV 1,181; RSC 15,28; SG [2]5,870; NUC 176,263
Anon., Pharm. Z. 32 (1887), 557,562
T. Haug, Friedrich August Flückiger, pp.79-80. Stuttgart 1985

FLÜGGE, CARL [1847-1923]
NDB 5,261-262; DBJ 5,69-73,425; FISCHER 1,421; PAGEL, 522-523; BULLOCH, 366; DZ, 372-373; WWWS, 582
W. Kruse, Deutsche med. Wchschr. 43 (1917), 1542-1544
M. Hahn, Deutsche med. Wchschr. 50 (1924), 20
B. Heymann, Z. Tub. 39 (1924), 356-362

FLÜRSCHEIM, BERNARD [1874-1955]
POGG 6,766-767; 7a(2),78
C.K. Ingold, J. Chem. Soc. 1956, pp.1087-1089FLURY, FERDINAND [1877-1947] NDB 5,264-265; KURSCHNER 5,337; HEIN, 169-170; FISCHER 1,421; IB, 86;BOCK, 91-101; TSCHIRCH, 1047; WWWS, 582; POGG 6,767; 7a(2),78-79
F. Schwarz, Schw. med. Wchschr. 78 (1948), 187
H.W. Frickhinger, Naturw. Rund. 1 (1948), 83

FOÀ, CARLO [1880-1971]
EI 15,574; FISCHER 1,421; KOREN, 174
Lui Chi E? 2nd Ed. 1,1126. Turin n.d.
P.C. Federici, Minerva Med. 62Suppl. (1971), 45

FOÀ, PIO [1848-1923]
EI 15,574; FISCHER 1,422; WININGER 2,270
A.O. Zorini, Riv. Biol. 5 (1923), 801-808

FOCK, ANDREAS LUDWIG [1856-1928]
ASEN, 49; POGG 4,433; 5,375; 6,767; 7a(2),79

FOCKE, WILHELM OLBERS [1834-1922]
NDB 5,267
H. Klebahn, Ber. bot. Ges. 51 (1934), (128)-(156)

FODERÁ, MICHELE [1793-1848]
EI 15,579; HIRSCH 2,551; DECHAMBRE [4]2,469-470; WWWS, 582; CALLISEN 6,349-351; 28,74

FODOR, ANDOR [1884-1968]
FISCHER 2,1736; TLK 3,157; IB, 86; WININGER 2,271; COHEN, 68; POGG 6,768-769; 7b,1433
H. Meyer, Bull. Res. Coun. Israel 4 (1954), v-xv

FODOR, KALMAN von [1881-1931]
MEL 1,520-521
Anon., Chem. Z. 55 (1931), 542

FØLLING, IVAR ASBJØRN [1888-1973]
Hver er Hvem? 1973, p.168
W.R. Centerwell and S.A. Centerwell, J. Hist. Med. 16 (1961), 292-296
W.L. Nyhan, TIBS 9 (1984), 71-72
F. Güttler, Acta Pediatrica Scandinavica 73 (1984), 705-716

FOERSTER, AUGUST [1822-1865]
ADB 7,146-147; HIRSCH 2,554-555;Suppl.,291; PAGEL, 525-526; DIEPGEN,56-66
F. Boehmer, Würzb. Nat. Z. 6 (1866), xlv-lxxvii

FOERSTER, FRIEDRICH [1866-1931]
NDB 5,274; TLK 3,160; WWWS, 583; POTSCH, 150; POGG 4,434; 5,375-376; 6,770-771
G. Grube, Z. angew. Chem. 45 (1931), 57-59
H. Menzel, Chem. Z. 55 (1931), 769-770; Z. Elektrochem. 38 (1932), 1-7
E. Müller, Z. anorg. Chem. 204 (1932), 1-19
R. Luther, Sitz. Akad. Wiss. Leipzig 84 (1933), 325-330

FOERSTER, HANS [1877-1916]
JV 16,143; NUC 176,489
Anon., Z. angew. Chem. 29(III) (1916), 444

FÖRSTER, JOSEF [1844-1910]
BJN 15,26*; LINDEBOOM, 621-622; FISCHER 1,429
E. Levy, Deutsche med. Wchschr. 36 (1910), 2348
M. Rubner, Arch. Hyg. 73 (1910-1911), 245-246

FÖRSTER, THEODOR [1910-1974]
KURSCHNER 10,563-564; KLEIN, 63; POGG 7a(2),82-83
A. Weller, Ber. Bunsen Ges. 78 (1974), 969-971
G. Porter, Naturwiss. 63 (1976), 207-211

FÖRSTERLING, HANS [1872-1928]
JV 9,145; NUC 176,489
Anon., Chem. Z. 52 (1928), 731

FOGH, JOHAN BERTEL [1865-1925]
VEIBEL 2,143-144; RSC 15,37-38
Anon., Chem. Z. 49 (1925), 1033; Z. angew. Chem. 38 (1925), 1179

FOKIN, SERGEI ALEKSEEVICH [1865-1917]
BSE 45,277; RSC 15,38
M.A. Blokh, Zhur. Fiz. Khim. 60Suppl. (1928), 145

FOL, HERMANN [1845-1892]
DSB 5,51-53; DHB 3,132; DBF 14,266; PICOT, 93-97; BOSSART, 49-50; STELLING-MICHAUD 3,333-334; WWWS, 584; GENEVA 3,36-38; 4,169-177
M. Bedot, Arch. Sci. Phys. Nat. 31 (1894), 1-22

FOLCH-PI, JORDI [1911-1979]
AMS 13,1332
A. Pope et al., Neurochemical Research 6 (1981), 1039-1041

FOLIN, OTTO [1867-1934]
DSB 5,53; DAB 21,306-308; NCAB 25,197-198; FISCHER 1,423; DAMB, 257; POTSCH,150; CHITTENDEN, 79-82; OEHRI, 45-46; IB, 86; POGG 6,733; 7b,1442
W.R. Bloor, Ind. Eng. Chem. News Ed. 12 (1934), 454
P.A. Shaffer, Science 81 (1935), 35-38; Biog. Mem. Nat. Acad. Sci. 27 (1952); J. Nutrition 52 (1954), 1-11
S. Meites, Clin. Chem. 28 (1982), 2173-2177; 29 (1983), 1852-1853;

Otto Folin. Washington, D.C. 1989

FOLKERS, KARL AUGUST [1906-]
AMS 14,1527-1528; MH 1,175-176; STC 1,390-391; MSE 1,383; WWWS, 584; WWA 1978-1979, p.1076; IWW 1982-1983, p.412; CB 1962, pp.135-137
K. Folkers, J. Chem. Ed. 61 (1984), 747-756

FOLLEY, SYDNEY JOHN [1906-1970]
DNB [1961-1970], 375-376; WW 1965, p.1051
A.S. Parkes, Biog. Mem. Fell. Roy. Soc. 18 (1972), 241-265

FOMIN, ALEKSANDR VASILIEVICH [1869-1935]
BSE (3rd Ed)27,518-519; UKRAINE 2,448-449

FONTANA, FELICE [1730-1805]
DSB 5,55-57; EI 15,645; PROVENZAL, 55-62; HIRSCH 2,562; HAYMAKER,302-306; POTSCH, 150; COLE, 180-183; FERCHL,159; WWWS, 585; CALLISEN 28,81; POGG 1,767-768
A. Vedrani, Un Grande Naturalista Trentino: Felice Fontana. Lucca 1916
G. Bilancioni, Archeion 12 (1930), 296-362
G. Provenzal, Atti Ist. Veneto 90 (1930-1931), 89-110; Rass. Clin. Terap. 33 (1934), 205-210; 37 (1938), 228-233
F.H. Garrison, Bull. N.Y. Acad. Med. 11 (1935), 117-122
P.K. Knoefel, Physis 18 (1976), 185-197; Clio Medica 15 (1980), 35-65; Felice Fontana, an Annotated Bibliography. Trento 1980
G. Bani, Z. mikr. anat. Forsch. 100 (1986), 337-346

FORBES, ALEXANDER [1882-1965]
DSB 5,64-66; AMS 10,1234; NCAB 52,528-529; DAMB, 258-259; IB,87; WWAH,201
E.D. Adrian, Nature 206 (1965), 1095-1096
W.O. Fenn, Am. Phil. Soc. Year Book 1965, pp.140-145; Biog. Mem. Nat.Acad. Sci. 40 (1969), 113-141
J.C. Eccles, Persp. Biol. Med. 13 (1970), 388-404

FORBES, ROBERT JAMES [1900-1973}
L. White, Tech. Cult. 15 (1974), 438-439
D.A. Wittop Koning, Janus 62 (1975), 217-233
A.R. Hall, Arch. Inter. Hist. Sci. 26 (1976), 160-162

FORBES, STEPHEN ALFRED [1844-1930]
DSB 5,69-71; DAB 6,509-510; NCAB 22,291-292; WWAH, 202; WWWS, 587
H.B. Ward, Science 71 (1930), 378-381
L.O. Howard, Biog. Mem. Nat. Acad. Sci. 15 (1932), 3-54

FORBES, THOMAS ROGERS [1911-1988]
AMS 17(2),1132; WWWS, 587
R.J. Joy, Bull. Hist. Med. 64 (1990), 86-97

FORCRAND, ROBERT HIPPOLYTE de [1856-1933]
STALDER, 27; POTSCH, 150-151; POGG 4,439; 5,270-271; 6,776-777; 7b,1445
Anon., Montpell. Med. 76(III) (1933), 536-538
M. Godchot, Bull. Soc. Chim. [5]1 (1934), 1-30

FORD, WILLIAM WEBBER [1871-1941]
AMS 6,471; WWWS, 588
S.R. Damon, Science 93 (1941), 514-515

FORD-MOORE, ARTHUR HENRY [1896-1958]
E.A. Perren, Proc. Chem. Soc. 1958, pp.267-268

FORDOS, MATHURIN JOSEPH [1816-1878]
DSB 5,72-73; DBF 14,431; GORIS, 374-376; FERCHL, 159-160; WWWS,588; POGG 1,773; 3,460FORGEOT, PAUL EUGÈNE [1878-1957]
H.J., Ann. Inst. Pasteur 93 (1957), 405-406

FORGET, AMÉDÉE [1811-1869]
DBF 14,476; HIRSCH 2,570

FORGET, CHARLES POLYDORE [1800-1861]
DBF 14,476-477; BERGER-LEVRAULT, 76

FORKNER, CLAUDE ELLIS [1900-1992]
AMS 12,1896; WWA 1976-1977, p.1039; WWWS, 588
M.M. Wintrobe, Hematology, the Blossoming of a Science, pp.110-114. Philadelphia 1985

FORMANEK, EMANUEL [1869-1929]
STURM 1,369; FISCHER 1,427; NAVRATIL, 65; POGG 6,778; 7b,1449
E. Skarnitzel and O. Wagner, Cas. Lek. Cesk. 68 (1929), 539-543
J. Spinka, Chem. Listy 23 (1929), 208-212
Anon., Chem. Z. 53 (1929), 444

FORMANEK, JAROSLAV [1864-1936]
OBL 1,337; STURM 1,369-370; POGG 4,440-441; 5,379; 6,778-779; 7b,1449
J. Knop, Chem. Z. 48 (1924), 118; Chemicky Obzor 11 (1936), 201-208,241-244
Anon., Chem. Ind. 1936, p.695
A. Ernest, Chem. Listy 30 (1936), 225-226

FORNET, WALTER [1877-1970]
FISCHER 1,427; IB, 87

FORRER, CARL [1856-1921]
ZURICH-D, 140; RSC 9,902; 15,63; NUC 178,234
Anon., Chem. Z. 45 (1921), 810

FORSÉN, LENNART [1889-1943]
VAD 1941, p.256
F.W. Klingstedt, Finska Kemistsamfundets Medd. 53 (1944), 11-21
Anon., Chem. Z. 68 (1944), 15

FORSSMAN, JOHN [1868-1947]
SMK 2,574-575; VAD 1947, p.311; FISCHER 1,428; IB, 87
Anon., Acta Path. Scand. Suppl.16 (1933), ix-xv
A. Lindau, Acta Path. Scand. 25 (1948), 513-518

FORSSMANN, WERNER [1904-1979]
KURSCHNER 12,761; MH 1,178-179; STC 1,393-394; MSE 1,387; WWWS, 589
H. Schadewaldt, Deutsche med. Wchschr. 104 (1979), 1856-1857; Jahrbuch der Universität Düsseldorf 1979-1980, pp.35-38

FORST, AUGUST WILHELM [1890-1981]
KURSCHNER 13,931; POGG 7a(2),87
Anon., Chem. Z. 105 (1981), 271; Naturw. Rund. 34 (1981), 440

FORSTER, AQUILA [1889-1967]
SRS, 43; JV 28,93; NUC 178,320
Anon., Chem. Brit. 4 (1968), 225

FORSTER, MARTIN ONSLOW [1872-1945]
WW 1945, pp.946-947; WWWS, 590; POGG 5,380; 6,781; 7b,1453
E.F. Armstrong and J.L. Simonsen, Obit. Not. Fell. Roy. Soc. 5 (1945), 243-261; J. Chem. Soc. 1946, pp.550-557

FORSTER, ROBERT ELDER II [1919-]
AMS 17(2),1143; BROBECK, 189-193

FOSSE, RICHARD [1870-1949]
DBF 14,564; GORIS, 376-378; WWWS, 591; POGG 5,381-382,1422; 6,783-784; 7b,1456-1457
A. Mayer, Not. Acad. Sci. 3 (1957), 165-174
P. Chabert, Revue du Tarn [3]49 (1968), 15-23

FOSTER, GOODWIN LeBARON [1891]
AMS 10,1247

FOSTER, JOSEPH FRANKLIN [1918-1975]
AMS 12,1909; WWWS, 591
Anon., Chem. Eng. News 53(48) (1975), 31

J.T. Edsall, Nature 259 (1976), 433

FOSTER, MARY LOUISE [1865-1967]
AMS 9(II),620

FOSTER, MICHAEL [1836-1907]
DSB 5,79-84; DNB [1901-1911]2, 44-46; FISCHER 1,430-431; DESMOND, 233; O'CONNOR, 167-171
J.N. Langley, J. Physiol. 35 (1907), 233-246
Anon., Nature 75 (1907), 345-347
W.H. Gaskell, Proc. Roy. Soc. B80 (1908), lxxi-lxxxi
H. Dale, Notes Roy. Soc. 19 (1964), 10-32
G.L. Geison, Michael Foster and the Cambridge School of Physiology. Princeton 1978

FOURCAR, GEORG [1878-]
JV 14,224

FOURCROY, ANTOINE FRANÇOIS [1755-1809]
DSB 5,89-93; DBF 14,749-752; BUGGE 1,356-363; GE 17,905-906; W, 189-190; LAROUSSE 8,669-670; COLE, 184-199; BLOKH 1,209-211; FERCHL, 160-162; POTSCH, 151; TSCHIRCH, 1048; CALLISEN 28,91; POGG 1,782-783
J. Cheymol and A. Soubiran, Presse Med. 69 (1961), 2471-2472
W.A. Smeaton, Fourcroy, Chemist and Revolutionary. London 1962
G. Kersaint, Antoine François Fourcroy, sa Vie et son Oeuvre. Paris 1966
T.A. Kursanova, Istoria Biologicheskikh Issledovania 9 (1983), 176-186
T.L. Sourkes, J. Hist. Med. 47 (1992), 322-339

FOURNEAU, ERNEST [1872-1949]
DSB 5,99-100; DBF 14,792; FISCHER 1,431; GORIS,378-379; IB, 87; POTSCH, 151-152; WWWS, 592-593; POGG 6,786-788; 7b,1459-1461
R. Tiffeneau, Paris Médical 39 (1949), 470-471
M. Delépine, Bull. Soc. Chim. [5]17 (1950), 953-982
T.A. Henry, J. Chem. Soc. 1952, pp.261-266
J.P. Fourneau, Rev. Hist. Pharm. 34 (1987), 335-355
B. Drevon, Mem. Acad. Lyon [3]43 (1989), 41-49

FOWLER, GILBERT JOHN [1868-1953]
POGG 6,790-791; 7b,1462-1463
H.E. Watson, J. Chem. Soc. 1953, pp.4191-4192

FOWNES, GEORGE [1815-1849]
DSB 5,103-104; DNB 7,530-531; BLOKH 1,211; FERCHL, 162; SCHAEDLER, 35; WWWS, 594; POGG 1,785-786
Anon., J. Chem. Soc. 2 (1850), 184-187
J.S. Rowe, Ann. Sci. 6 (1950), 422-435

FOX, DENIS LLEWELLYN [1901-]
AMS 14,1554; WWWS, 594

FOX, EDWARD LAWRENCE [1859-1938]
Anon., Lancet 1938(II), pp.1493-1494; Brit. Med. J. 1938(II), p.1396

FOX, HAROLD MUNRO [1889-1967]
DNB [1961-1970], 385-386; WW 1966, p.1076
G.P. Wells, Nature 213 (1967), 974
J.E. Smith, Biog. Mem. Fell. Roy. Soc. 14 (1968), 207-222

FOX, SIDNEY WALTER [1912-]
AMS 15(2),1107; WWA 1980-1981, p.1137; STC 1,394-395; WWWS, 595
S.W. Fox, The Emergence of Life. New York 1988

FOX, WILSON [1831-1887]
DNB 7,580-581; HIRSCH 2,587-588; PAGEL, 533-534
Anon., Lancet 1887(I), pp.1011-1013; Brit. Med. J. 1887(I), p.1021; Med. Chir. Trans. 71 (1888), 253-296

FRÄNKEL, ALBERT [1848-1916]
NDB 5,311; DBJ 1,354; FISCHER 1,433-434; PAGEL, 534-535; BULLOCH, 306; KOREN, 175; WWWS, 596
Anon., Lancet 1916(II), p.344; Brit. Med. J. 1916(II), p.783
W. Huebner, A. Fränkel, Arzt und Forscher. Mannheim 1973

FRAENCKEL, PAUL [1874-]
FISCHER 1,433; ASEN, 50; LDGS, 63; KOREN, 175; NUC 180,59

FRAENKEL, ALBERT [1864-1938]
DRULL, 71; SKVARC, 59-61; LDGS, 65; TETZLAFF, 81; KOREN, 175
Anon., Arzneimitt. 14 (1964), 160
K. Blumberger, Hippokrates 35 (1964), 252-260
G. Weiss, Zeitschrift für Allgemeinmedizin 54 (1978), 826-833

FRAENKEL (later FRAENKEN), CARL [1861-1915]
NDB 5,310-311; DBJ 1,326-327; FISCHER 1,434; BULLOCH, 366-367; DZ, 383; WININGER 2,289; KOREN, 175; WWWS, 598-599
Anon., Chem. Z. 40 (1916), 41; J. Am. Med. Assn. 66 (1916), 587
C. Günther, Deutsche med. Wchschr. 42 (1916), 362,392

FRAENKEL, ERNST [1886-1948]
KURSCHNER 4,716; LDGS, 56; KOREN, 175
Anon., Brit. Med. J. 1948(I), p.958

FRAENKEL, EUGEN [1853-1925]
NDB 5,312; FISCHER 1,435; WININGER 2,287; WWWS, 596

FRAENKEL, GOTTFRIED SAMUEL [1901-1984]
AMS 11,1627; WWWS, 596; LDGS, 13
C.L. Prosser et al., Biog. Mem. Nat. Acad. Sci. 59 (1990), 169-195

FRAENKEL, HEINRICH WALTER [1879-1945]
ARNSBERG, 121; JV 22,299; POGG 6,793-794; 7a(2),89
Wer ist Wer 9 (1928), 421

FRAENKEL, LUDWIG [1870-1951]
KURSCHNER 5,345; FISCHER 1,435-436; BHDE 2,313-314; LDGS, 63; KOREN, 175; WININGER 2,291; MEDVEI, 732-733
H.H. Simmer, Sudhoffs Arch. 55 (1971), 392-417; 56 (1972), 76-99

FRÄNKEL, SIGMUND [1868-1939]
OBL 1,340; FISCHER 1,436; PAGEL, 539; TLK 3,164-165; TSCHIRCH, 1048; WININGER 2,292; KOREN, 175; POGG 5,383-384; 6,792; 7a(2),89

FRÄNKEL, WALTER [1879-1945]
LDGS, 18; POGG 5,384; 6,793-794; 7a(2),89

FRAENKEL-CONRAT, HEINZ [1910-]
AMS 14,1559; MH 1,180-181; STC 1,397-398; MSE 1,390-391; ABBOTT, 48; WWA 1978-1979, p.1097; TETZLAFF, 82; WWWS, 596

FRANCESCONI, LUIGI [1864-1939]
POGG 5,384-385; 6,794; 7b,1465
M. Garino, Chimica e Industria 22 (1940), 90

FRANCHIMONT, ANTOINE PAUL [1844-1919]
BLOKH 1,211-212; WWWS, 597; POTSCH, 152; POGG 3,468-469; 4,446-447
P. von Romberg, Chem. Wkbl. 8 (1911), 243-253
H.J. Backer, Chem. Wkbl. 11 (1914), 382-391; Chem. Z. 38 (1914), 613; 43 (1919), 645
W. Adriani, Chem. Wkbl. 16 (1919), 980-983
T.E. Thorpe, J. Chem. Soc. 117 (1920), 457-461

FRANCHINI, GUISEPPE [1879-1938]
FISCHER 1,436; IB, 88; SG [4]5,1164
G.T. Castellani, Arch. Ital. Sci. Med. Trop. 19 (1938), 385-387

FRANCIS, FRANCIS [1871-1941]
POGG 4,447; 5,385; 6,795; 7b,1467
W.E. Garner, Nature 147 (1941), 569
Who Was Who 1941-1950, p.405

FRANCIS, GORDON EDWARD CHARLES [1914-]
Who's Who of British Scientists 1971-1972, p.303
W. Templeton, Chem. Brit. 18 (1982), 267

FRANCIS, THOMAS, Jr. [1900-1969]
AMS 11,1630; NCAB I,432-433; DAMB, 263-264; WWAH, 205; WWWS, 597
W. McDermott, Trans. Assoc. Am. Phys. 83 (1970), 16-18
C.M. MacLeod et al., Arch. Env. Hlth. 21 (1970), 226-275
C.M. MacLeod, Am. Phil. Soc. Year Book 1970, pp.121-126
J.R. Paul, Biog. Mem. Nat. Acad. Sci. 44 (1974), 57-110

FRANCIS, WILLIAM [1817-1904]
RSC 2,696; NUC 182,296; POGG 1,788-789; 3,469

FRANCK, HANS HEINRICH [1888-1961]
POTSCH, 152; TLK 3,162; POGG 6,795-796; 7a(2),90-91

FRANCK, JAMES [1882-1964]
DSB 5,117-118; AMS 10,1263; MH 2,160-162; STC 1,398; MSE 1,392-393; BHDE 2,315; LDGS, 96; WININGER 2,278; 7,563; TETZLAFF, 82; W, 192; COHEN, 69; WWWS, 597; POGG 5,385-386;6,796; 7a(2),91-92
W. Kroebel, Naturwiss. 49 (1962), 361-363; 51 (1964), 421-423; Phys. Bl. 38 (1982), 269-270
H.G. Kuhn, Biog. Mem. Fell. Roy. Soc. 11 (1965), 53-74
G. Hertz, Ann. Phys. 15 (1965), 1-4
R. Courant, Am. Phil. Soc. Year Book 1966, pp.143-147

FRANK, ADOLF [1834-1916]
NDB 5,337-338; DBJ 1,204-208,354; BUGGE 2,310-320; MATSCHOSS, 78-79; POTSCH, 152-153; BLOKH 1,212-214; WININGER 2,281; WWWS, 598; POGG 6,797; 7aSuppl.,215-216
H. Grossmann, Z. angew. Chem. 29 (1916), 373-377
N. Caro, Chem. Z. 40 (1916), 569-571
H. Wichelhaus, Ber. chem. Ges. 49 (1916), 1533-1534
H. Andreae, Chem. Z. 90 (1966), 356-357

FRANK, ALBERT BERNHARD [1839-1900]
NDB 5,338-339; BJN 5,257-260,90*; SACKMANN, 121-122; WWWS, 598
Anon., Leopoldina 36 (1900), 170-171
C. Schröder, Z. Entomol. 5 (1900), 390-391
F. Krüger, Ber. bot. Ges. 19 (1901), (10)-(36)FRANK, ALBERT RUDOLF [1872-1965]
BHDE 1,186; TETZLAFF, 82; JV 14,13; NUC 182,570; POGG 6,797; 7a(2),92-93
K.Zicke, Chem. Ing. Techn. 24 (1952), 609-610

FRANK, ALFRED ERICH [1884-1957]
KURSCHNER 4,711; FISCHER 1,438; BHDE 2,315; LDGS, 65; WIDMANN, 87,261; WININGER 7,563; KOREN, 175
F. Reimann, Münch. med. Wchschr. 90 (1957), 893-894
K. Steinitz, Deutsche med. Wchschr. 82 (1957), 1138-1139

FRANK, ARMANDO [1885-1951]
KURSCHNER 4,710; JV 26,151; NUC 182,572
Mannheimer Morgen, 12 December 1951

FRANK, CHRISTIAN [1872-]
JV 13,224

FRANK, FREDERICK CHARLES [1911-]
WW 1981, p.908; IWW 1981-1982, pp.410-411; SRS, 92
R.G. Chambers et al., Sir Charles Frank. New York 1991

FRANK, FRITZ [1868-1949]
NDB 5,337; HEIN-E, 131-132; LDGS, 18; TLK 3,162-163; WININGER 7,564; LDGS, 18; POGG 6,797; 7a(2),93
A. Kind, J. Inst. Fuel 24 (1951), 31-32

FRANK, GLEB MIKHAILOVICH [1904-1976]
WWWS, 598
G. Wangermann, Stud. Biophys. 62 (1977), 81-82
W. Rich, Nature 267 (1977), 736

FRANK, JOHANN PETER [1745-1821]
NDB 5,341-342; ADB 7,254-257; OBL 1,344; WURZBACH 4,320-323; DBA 1,1008; NOB 16,25-33; HIRSCH 2,598-601; BOEGERSHAUSEN, 11-28; WWWS, 598; MEDVEI, 733-735
F. Koelsch, Münch. med. Wchschr. 107 (1965), 1958-1959
S. Rivoir, Johann Peter Frank und sein therapeutisches Werk. Zurich 1968
C. Probst, Sudhoffs Arch. 59 (1975), 20-53
R. Oberhauser, Pfälzer Lebensbilder 3 (1977), 145-168

FRANK, OTTO [1865-1944]
NDB 5,335-336; FISCHER 1,438-439; IB, 88; WWWS, 598; POGG 7a(2),94
A. Hahn, Bayer. Akad. Wiss. Jahrbuch 1944-1948, pp.202-205
K. Wezler, Z. Biol. 103 (1950), 91-122

FRANK, RICHARD [1908-]
JV 51,82; NUC 182,637

FRANK, WALTER [1901-]
JV 43,411; NUC 182,647

FRANKE, ADOLF [1874-1964]
PLANER, 83; POGG 4,448; 5,388; 6,798; 7a(2),94
F. Wessely, Ost. Chem. Z. 50 (1949), 43-44
F. Hecht, Ost. Chem. Z. 55 (1954), 37

FRANKE, EWALD [1879]
JV 21,163; SG [3]5,955; NUC 182,671

FRANKE, WILHELM [1903-1967]
KURSCHNER 10,574; WWWS, 598; POGG 7a(2),97-98
Anon., Chem. Z. 91 (1967), 940

FRANKEL, MAX [1900-1971]
WWAH, 206; WWWS, 599; COHEN, 69; POGG 6,798-799; 7b,1470-1471
E. Katchalski, Bull. Res. Coun. Israel 10 (1961), iii-x
A. Zilkha, Israel J. Chem. 10 (1972), 739-752

FRANKENBURGER, WALTER GUSTAV [1893-1957]
JV 40,730; POGG 6,799; 7a(2),98-99

FRANKENHEIM, MORITZ [1801-1869]
NDB 5,350; FERCHL, 163; POGG 1,792; 3,469-470

FRANKFURT, SALOMON [1866-1954]
ZURICH-D, 171; WININGER 7,564; RSC 18,611; NUC 183,29
New York Times, 20 November 1954

FRANKFURTER, FRITZ [1890-]
JV 30,730; NUC 183,68

FRANKL-HOCHWART, LOTHAR von [1862-1914]
NDB 5,361-352; OBL 1,347; DBJ 1,281; FISCHER 1,440
O. Marburg, Wiener med. Wchschr. 64 (1914), 2611-2612

FRANKLAND, EDWARD [1825-1899]
DSB 5,124-127; DNB 22,662-665; BLOKH 1,214-216; POTSCH, 153; W, 192-193; WWWS, 599; POGG 1,792-793; 3,470; 4,448-449
J. Wislicenus, Ber. chem. Ges. 33 (1900), 3847-3874
M.N. West and A.J. Colenso, Sketches from the Life of Edward Frankland. London 1902
H. McLeod, J. Chem. Soc. 87 (1905), 574-590
G. Porter, Proc. Roy. Inst. 40 (1965), 384-397
C.A. Russell, Chem. Brit. 13 (1977), 4-7,20; Ann. Sci. 35 (1978), 253-273; Lancastrian Chemist: The Early Years of Edward Frankland. Milton Keynes 1986
C. Hamlin, Bull. Hist. Med. 56 (1982), 56-76; A Science of Impurity. Bristol 1990

FRANKLAND, EDWARD PERCY [1884-1958]
JV 24,690; POGG 6,799; 7b,1472
Who Was Who 1951-1960, pp.393-394

FRANKLAND, PERCY FARADAY [1858-1946]
DSB 5,127-129; DNB [1941-1950], 270-271; WW 1945, p.960; WWWS, 599; POTSCH, 153-154; POGG 4,449-450; 5,388; 6,799-800; 7b,1472
W.E. Garner, Obit. Not. Fell. Roy. Soc. 5 (1947), 697-715
C.A. Russell, Chem. Brit. 13 (1977), 425-427

FRANKLIN, EDWARD CLAUS [1928-1982]
AMS 14,1567; AO 1982, pp.85-86; BHDE 2,322

FRANKLIN, EDWARD CURTIS [1862-1937]
DAB [Suppl.2],205-206; MILES, 158-159; NCAB A,411-412; 30,158-159; WWAH, 206; POGG 5,388-389; 6,800; 7b,1472-1473
C.A. Kraus, Science 85 (1937), 232-234
Anon., Ind. Eng. Chem. News Ed. 15 (1937), 90
A. Findlay, J. Chem. Soc. 1938, pp.583-595
H.M. Elsey, J. Am. Chem. Soc. 71 (1949), 1-5; Biog. Mem. Nat. Acad. Sci. 60 (1991), 67-79

FRANKLIN, KENNETH JAMES [1897-1966]
WW 1965, p.1082; IB, 88
F.J. Aumonier, Nature 211 (1966), 353
I. deB. Daly and R.G. McBeth, Biog. Mem. Fell. Roy. Soc. 14(1968),223-242

FRANKLIN, ROSALIND ELSIE [1920-1958]
DSB 5,139-142; POTSCH, 154
J.D. Bernal, Nature 182 (1958), 154
A. Klug, Nature 219 (1968), 808-810,843-844,879,1192; 248 (1974),787-788
A. Sayre, Rosalind Franklin and DNA. New York 1975
M.M. Julian, J. Chem. Ed. 60 (1983), 660-662

FRANQUÉ, OTTO von [1833-1879]
HIRSCH 2,606

FRANQUÉ, OTTO von [1867-1937]
NDB 5,352; STURM 1,378; FISCHER 1,440; KOERTING, 219-220; MOTSCH, 90-94; HARTWIG, 45-72; ZWEIFEL, 41-44
L. Martius, Z. Gyn. 61 (1937), 1265-1276

FRANTZ, THEODOR [1884-]
JV 25,363
Wer ist Wer 9 (1928), 428

FRANZEN, ANTON [1859-]
JV 2,244; NUC 183,331

FRANZEN, HANS [1884-]
JV 27,395; NUC 186,334

FRANZEN, HARTWIG [1878-1923]
DBJ 5,77-79,426; DRULL, 72; BLOKH 1,216; TLK 2,179; POGG 5,389-390; 6,801
H. Franzen, Ber. chem. Ges. 56A (1923), 57

FRAPOLLI, AGOSTINO [1824-1903]
BLOKH 1,216; RSC 2,702
Anon., Chem. Z. 27 (1903), 1265-1266

FRASER, HENRY [1873-1930]
Anon., Lancet 1930(II), p.220

FRASER, THOMAS RICHARD [1841-1920]
DNB [1912-1921], 199; HIRSCH 2,608; PAGEL, 541-542; REBER, 179-184,369; O'CONNOR(2), 383-385; TSCHIRCH, 1048; WWWS, 601
J.T.C., Proc. Roy. Soc. B92 (1921), xi-xvii

FRASER-HARRIS, DAVID FRASER [see HARRIS]

FRASEY, VICTOR [1869-1939]
SCHALI, 40
Anon., Ann. Inst. Pasteur 63 (1939), 313-314

FRAUDE, GEORG [1848-1899]
DGB 136 (1964), 65; RSC 9,920

FRAZER, ALISTAIR CAMPBELL [1909-1969]
DNB [1961-1970], 397-398; WW 1965, p.1088; WWWS, 601
Anon., Lancet 1969(I), pp.1323-1324

FRAZER, JOSEPH CHRISTIE WHITNEY [1875-1944]
AMS 7,598; NCAB 34,384; WWAH, 207; WWWS, 601; POGG 6,802-803; 7b,1477-1478
Anon., Ind. Eng. Chem. News Ed. 17 (1939), 115; Chem. Eng. News 22 (1944), 1412
W.A. Patrick, Science 102 (1945), 110-111

FRED, EDWIN BROUN [1887-1981]
AMS 14,1575; WWA 1978-1979, p.1109; NCAB F,244; CB 1950,pp.156-157; IB,88
D.O. Johnson, Edwin Broun Fred. Madison, Wis. 1971
H. Lardy, Am. Phil. Soc. Year Book 1985, pp.122-125
I.L. Baldwin, Biog. Mem. Nat. Acad. Sci. 55 (1985), 247-290

FREDENHAGEN, KARL [1877-1949]
NDB 5,387; KURSCHNER 6(I),454; TLK 3,166; WWWS, 602; POTSCH, 154-155; POGG 5,391-392; 6,805; 7a(2),104
K. Wiechert, Scientia Chimica 3 (1950), vii-xxxi

FREDERICQ, LÉON [1851-1935]
DSB 5,148-150; BNB 37,301-310; FISCHER 1,448; PAGEL, 542-544; WWWS, 602
M. Florkin, Liège Médical 28 (1935), 1373-1391; Florilège des Sciences en Belgique pendant le XIXe Siecle et le Debut du XXe (P. Brien, Ed.), pp.1015-1034. Brussels 1968; L'Ecole Liègeoise de Physiologie et son Maitre Léon Fredericq. Liège 1979
A. Mayer, Bull. Acad. Med. 114 (1935), 244-251
Z.M. Bacq and M. Florkin, Arch. Inter. Pharmacodyn. 52 (1936), 245-280
P. Nolf, Ann. Acad. Roy. Belg. 103 (1937), 47-100
J. Lecomte and E. Schoffeniels, Arch. Biol. 91 (1980), 267-271
J. Roskam et al., Rev. Med. Liège 41 (1986), 836-845

FREDGA, ARNE [1902-1992]
VAD 1981, p.314; SMK 2,589; IWW 1982-1983, p.424; WWWS, 603; POGG 6,805; 7b,1483-1485

FREED, SIMON [1899-]
AMS 14,1579; POGG 6,805; 7b,1485-1487

FREER, PAUL CASPAR [1862-1912]
MILES, 163-164; NCAB 19,423; WWAH, 208; JV 2,192; POGG 4,452-453; 5,392
J.T. White, Pop. Sci. Mon. 80 (1912), 521-529
M. Egan et al., Philippine Journal Of Science 7 (1912), v-xii
E.D. Campbell, History of the Chemical Laboratory of the University of Michigan, pp.58-64. Ann Arbor, Mich. 1916

FREESE, ERNST [1925-1990]
AMS 15(2),1140
J.W. Drake, Mutation Research 251 (1991), 165-169

FREISE, EDUARD [1882-1921]
FISCHER 1,443-444
A. Frank, Mon. Kind. 21 (1921), 195

FREMERY, WALTER [1898-]
JV 40,730

FRÉMY, EDMÉ FRANÇOIS [1774-1866]
TSCHIRSCH, 1048; FERCHL, 163; SCHELENZ, 621

FRÉMY, EDMOND [1814-1894]
DSB 5,157-158; DBF 14,1198; GE 18,140-141; BOURQUELOT, 65-66; WWWS, 605; FERCHL, 163-164; BLOKH 1,217-218; SCHAEDLER, 36; PHILIPPE, 791-792; POTSCH, 155; POGG 1,797-798; 3,473; 4,453
P.P. Dehérain, Nouv. Arch. Hist. Nat. [3]6 (1894), xvii-xxxii

C. Chatagnon and P.A. Chatagnon, Ann. Med. Psych. 112(II) (1954), 364-390
G. Kersaint, Bull. Soc. Chim. 1963, pp.639-641,3084; Rev. Hist. Pharm. 52 (1964), 165-172

FRENCH, CHARLES STACY [1907-]
AMS 14,1587; WWA 1978-1979, p.1118; IWW 1982-1983, p.425; MH 2,162-163; MSE 1,397-398; POTSCH, 155
C.S. French, Ann. Rev. Plant Physiol. 30 (1979), 1-26

FRENCH, DEXTER [1918-1981]
AMS 15(2),1144; WWA 1980-1981, p.1160; WWWS, 606
J.J. Marshall and W.J. Whelan, Carbohydrate Research 61 (1978), 1-4
Anon., Chem. Eng. News 60(8) (1982), 48
J.H. Pazur, Adv. Carbohydrate Chem. 42 (1984), 1-13

FRENKEL, ALBERT W. [1919-]
AMS 15(2),1146

FRENKEL, ISAAK [1877-]
JV 23,307

FRENZEL, KARL [1871-1945]
OBL 1,357; TLK 3,166-167; POGG 4,454; 5,392; 6,808; 7a(2),105

FRERICHS, FRIEDRICH THEODOR von [1819-1885]
NDB 5,404; ADB 21,782-790; NB, 106; HIRSCH 2,613-614; PAGEL, 543-545; FRANKEN, 182-183; FERCHL, 164; WWWS, 1733; POGG 1,7980799; 3,474
E. Leyden, Deutsche med. Wchschr. 11 (1885), 177-178
M. Litten, Wiener med. Wchschr. 35 (1885), 466-468,498-502,538-542
H. Kurz, Friedrich Theodor Frerichs, sein Leben und seine Werke. Düsseldorf 1938

FRERICHS, GEORG AUGUST [1873-1940]
KURSCHNER 4,729; TLK 3,167; TSCHIRCH, 1049; SCHELENZ, 707; POGG 5,392; 6,808; 7a(2),106

FRESENIUS, CARL REMIGIUS [1818-1897]
DSB 5,163-165; NDB 5,406-407; ADB 48,739-742; BJN 2,248-253; 4,56*; NL 1,191-204; NB, 107; HB 2,119-127; HEIN, 175-176; COLE, 201; BLOKH 1,218-220; TSCHIRCH, 1049; FERCHL, 164; POTSCH, 155-156; WWS, 606; POGG 1,799-800; 3,474-475; 4,454; 6,809; 7aSuppl.,222-224
E. Fischer, Ber. chem. Ges. 30 (1897), 1349-1355
R. Fresenius, Z. anal. Chem. 36 (1897), i-xviii
G. Schwedt, Z. anal. Chem. 315 (1983), 395-401

FRESENIUS, HEINRICH [1847-1920]
NDB 5,405-406; DBJ 2,746; NL 1,204-208; NB, 107; POTSCH, 156; POGG 4,454-455; 5,393; 7aSuppl.,224
W. Fresenius, Z. anal. Chem. 59 (1920), iii-ix; Ber. chem. Ges. 53A (1920), 75-77

FRESENIUS, LUDWIG [1886-1936]
NL 1,214-219; NB, 107; POGG 5,393; 6,808-809; 7a(2),107
R. Fresenius, Z. anal. Chem. 106 (1936), xi-xvii; Ber. chem. Ges. 69A (1936), 209-210

FRESENIUS, REMIGIUS [1878-1949]
NDB 5,406; NL 1,214; NB, 108

FRESENIUS, THEODOR WILHELM [1856-1936]
NL 1,209-213; NB, 108
R. Fresenius, Z. anal. Chem. 105 (1936), xi-xv; Ber. chem. Ges. 69A (1936), 91-93
Anon., Chem. Z. 60 (1936), 329

FRESSEL, HANS [1887-]
JV 27,645; NUC 185,56FRETWURST, FRITZ [1898-]
POGG 7a(2),108
Anon., Nachr. Chem. Techn. 21 (1973), 168

FREUDENBERG, ERNST [1884-1967]
KURSCHNER 10,586; FISCHER 1,445-446; SCHNACK, 64-74; AUERBACH, 235; BHDE 2,329; KOREN, 176
Anon., Bibl. Paed. 58 (1954), xx-xxvii

FREUDENBERG, KARL [1886-1983]
DSB 17,311-312; KURSCHNER 13,962; DRULL, 72; MH 2,163-165; STC 1,402-403; MSE 1,398-399; IWW 1982-1983, p.426; TLK 3,168; BOHM, 146; POTSCH, 156; WWWS, 607; POGG 5,394; 6,810-811; 7a(2),108-111
Anon., Chem. Z. 80 (1956), 43; 105 (1981), 12; 107 (1983), 139
T.S. Stevens, Biog. Mem. Fell. Roy. Soc. 30 (1984), 169-189
F. Cramer, Heidelberger Jahrbücher 28 (1984), 57-72
F.W. Lichtenthaler, Carbohydrate Research 164 (1987), 1-22

FREUDENREICH, EDOUARD von [1851-1906]
NDB 5,410-411; BJN 11,22*; RSC 15,121-122; NUC 185,100

FREUDWEILER, MAX [1871-1901]
BJN 6,30*
Anon., Leopoldina 37 (1901), 109

FREUND, AUGUST [1835-1892]
OBL 1,358; PSB 7,133; HEIN-E,133-134; SACKMANN, 125-126; TSCHIRCH, 1049; POTSCH, 156; WWWS, 607; POGG 4,456
Anon., Chem. Z. 16 (1892), 1975
I.Z. Siemion, Wiad. Chem. 37 (1983), 509-521

FREUND, CARL [1883-]
JV 26,394

FREUND, ERNST [1863-1946]
FISCHER 1,446; LESKY, 526-529; BHDE 2,332; WININGER 2,314-315; 6,601; TETZLAFF, 88
R. Willheim, Nature 158 (1946), 229-230
E.R. Zak, Wiener klin. Wchschr. 58 (1946), 721-722
C.R., Biochem. J. 41 (1947), 139

FREUND, HERMANN [1882-1944]
KURSCHNER 5,352; FISCHER 1,446; DRULL, 73
L. Lendle, Arch. exp. Path. Pharm. 218 (1953), i-viii

FREUND, IDA [1863-1914]
NUC 185,123-124
B. Stephen, Girton College 1869-1932, p.176. Cambridge 1933
M.R.S. Creese, Brit. J. Hist. Sci. 24 (1991), 287

FREUND, JULES [1890-1960]
AMS 9(II),373-374; BIBEL, 279-282; IB, 89
Anon., Lancet 1960(I), pp.1031-1032

FREUND, LEOPOLD [1868-1943]
NDB 5,413; OBL 1,359; FISCHER 1,447; WOLF, 21-37; BHDE 2,333; TLK 3,168; TETZLAFF, 88; WWWS, 607
K. Weiss, Wiener klin. Wchschr. 59 (1947), 189

FREUND, MARTIN [1863-1920]
DBJ 2,746; KALLMORGEN, 269; BLOKH 1,220-221; TLK 2,182; BERLIN, 718; POTSCH, 156-157; WININGER 2,317; TETZLAFF, 88; POGG 4,456-457; 5,394-395; 6,811
F. Mayer, Chem. Z. 44 (1920), 297-298
E. Speyer, Z. angew. Chem. 33 (1920), 121-122
L. Spiegel, Ber. chem. Ges. 54A (1921), 53-79

FREUND, RUDOLF [1896-]
ASEN, 52; LDGS, 65; KOREN, 177

FREUNDLICH, HERBERT [1880-1941]
DSB 15,159-160; NDB 5,413-414; STC 1,403-404; BHDE 2,334; LDGS, 19; WININGER 6,601-602; TETZLAFF, 89; W, 197; ABBOTT-C, 47-48; POTSCH, 157; WWWS, 607; POGG 5,396; 6,812-813; 7a(2),112-113
R.A. Gortner, Science 93 (1941), 414-416
F.G. Donnan, Obit. Not. Fell. Roy. Soc. 4 (1942), 27-50

J. Reitstötter, Koll. Z. 139 (1954), 1-11
H. Zocher, Z. Elektrochem. 59 (1955), 1-4
D.W. Gillings and R.H. Ottewill, Chem. Ind. 1980, pp.360-363,377-381

FREW, WILLIAM [1870-1910]
SRS,12; JV 9,207; RSC 13,275; 15,126

FREY, CHARLES N. [1885-1972]
AMS 11,1662; NCAB 56,491-493; IB, 89; WWWS, 607
Anon., Chem. Eng. News 50(46) (1972), 46

FREY, EMIL KARL [1888-1977]
KURSCHNER 12,792; FISCHER 1,448; KILLIAN, 379-380; NUC 185,168; WWWS, 608
K. Vosschulte, Deutsche med. Wchschr. 102 (1977), 1860-1861
R. Zenker, Münch. med. Wchschr. 119 (1977), 1237-1238
R. Frey, Anaesthetist 27 (1978), 506

FREY, ERNST [1878-1960]
KURSCHNER 8,581; TSCHIRCH, 1049; POGG 6,813-814; 7a(2),113; (4),142*
E. Ruickholdt, Arzneimitt. 10 (1960), 415-416
Anon., Arch. exp. Path. Pharm. 241 (1960), 3

FREY, HANS [1865-1939]
W. Hohn, Viert. Nat. Ges. Zurich 84 (1939), 362-367

FREY, HEINRICH [1822-1890]
NDB 5,417-418; ADB 48,742-743; HIRSCH 2,617; Suppl.,303-304; WWWS, 608
Anon., Lancet 1890(I), p.329; Leopoldina 26 (1890), 56; Chem. Z. 14 (1890), 1781
E. Handschin, Mitt. Schw. Ent. Ges. 31 (1958), 109-120
A. Jonecko, Wiad. Lek. 32 (1979), 1181-1184

FREY, KARL [1868-1947]
JV 8,215; RSC 14,804
G.A. Frey-Bally, Die Frey von Aarau, vol.2, p.21. Aarau 1980

FREY, MAX von [1852-1932]
DSB 5,184-185; NDB 5,419-420; FISCHER 1,448; PAGEL, 547-548; DZ, 389; EBERT, 107-121; BK, 201-202; IB, 89; WWWS, 1733; POGG 4,457; 5,396; 6,814; 7a(2),113
P. Hoffmann, Z. Biol. 92 (1932), i-v; Verhandl. Phys. Med. Würz. NF57 (1932), 56-66
H. Rein, Erg. Physiol. 35 (1933), 1-9

FREY-WYSSLING, ALBERT [1900-1988]
KURSCHNER 13,970; MH 2,165-166; STC 1,406-407; MSE 1,399-400; SBA 1,53; IWW 1982-1983, p.426; BOHM, 215-216; WWWS, 608; POGG 6,814 7a(2),114-115
A. Frey-Wyssling, J. Microscopy 100 (1974), 21-34; Lehre und Forschung. Stuttgart 1984
Anon., Naturw. Rund. 41 (1988), 424
P. Matile, Biog. Mem. Fell. Roy. Soc. 35 (1990), 115-126

FREYTAG, FRANZ [1865-1925]
Anon., Chem. Z. 49 (1925), 794

FREYTAG, PAUL [1876-]
JV 28,13; NUC 185,272

FRIDOVICH, IRWIN [1929-]
AMS 15(2),1153; WWA 1980-1981, p.1165
W.H. Bannister and J.V. Bannister, Free Radical Biology and Medicine 5 (1988), 371-376

FRIED, JOSEF [1914-]
AMS 15(2),1154; WWA 1980-1981, p.1165; IWW 1982-1983, p.428; BHDE 2,335

FRIEDBERG, FELIX [1921-]
AMS 15(2),1155; WWWS, 609

FRIEDBERG, HARALD [1907-]
LDGS, 19; NUC 185,411

FRIEDBERG, HERMANN [1817-1884]
HIRSCH 2,621; PAGEL, 548-549; KOREN, 177
Anon., Leopoldina 20 (1884), 60; Jahresb. Schles. Ges. 62 (1884), 391-392

FRIEDBERGER, ERNST [1875-1932]
KURSCHNER 4,745; FISCHER 1,449-450; GREIFSWALD 2,333-334; BULLOCH, 367; TLK 3,170; IB, 89; WININGER 2,326; 7,565; TETZLAFF, 90; KOREN, 177
F. Schiff, Deutsche med. Wchschr. 58 (1932), 265-266
Anon., Chem. Z. 56 (1932), 89; Chem. Ind. 1932, p.150

FRIEDEL, CHARLES [1832-1899]
DBF 14,1291-1292; DBA 1,1043-1044; BLOKH 1,221-225; ABBOTT-C, 47-48; POTSCH, 157; W, 197-198; WWWS, 609; POGG 3,475-476; 4,458-459
A. Ladenburg, Ber. chem. Ges. 32 (1899), 3721-3744
M. Hanriot, Bull. Soc. Chim. [3]23 (1900), i-lvi
J.M. Crafts, J. Chem. Soc. 77 (1900), 993-1019
A. Willemart, J. Chem. Ed. 26 (1949), 3-9
L. Hackspill and A. Willemart, Bull. Soc. Chim. 1964, pp.555-567

FRIEDEL, GEORGES [1865-1933]
DSB 5,185-187; DBF 14,1292-1293; WWWS, 609; POGG 6,816-818; 7b,1496
C. Richet, C.R. Acad. Sci. 197 (1933), 1545-1547
H.D.H. Donnay, Am. Min. 19 (1934), 329-335
F. Grandjean, Bull. Soc. Fran. Min. 57 (1934), 144-171
R. Weil, Z. Kristall. 89 (1934), 1-9

FRIEDEL, JEAN [1874-1941]
IB, 89; NUC 185,416
M. Humbert, Bull. Soc. Bot. 88 (1941), 457

FRIEDEMANN, ULRICH [1877-1949]
NDB 5,446-447; FISCHER 1,450; BHDE 2,336; LDGS, 56; KAGAN(JA), 733; WININGER 2,327; TETZLAFF, 90; WWWS, 609; POGG 7a(2),121-122
J. Berberich, Proc. Virchow Soc. 8 (1949), 146-148

FRIEDEN, CARL [1928-]
AMS 15(2),1155; WWA 1980-1981, p.1166; WWWS, 609

FRIEDENTHAL, HANS [1870-1942]
KURSCHNER 4,747; FISCHER 1,450; BHDE 2,336; TLK 3, 170-171; IB, 69; BK, 203; WININGER 2,328-329; TETZLAFF, 90-91; KOREN, 177; POGG 6,818; 7a(2),122

FRIEDHEIM, CARL [1858-1909]
BJN 14,26*; BLOKH 1,225; WWWS, 610; POGG 4,459; 5,397
Anon., Verhandl. Schw. Nat. Ges. 92 (1909), 124-129
F. Ephraim, Chem. Z. 33 (1909), 869
A. Rosenheim, Ber. chem. Ges. 44 (1911), 2787-2806

FRIEDHEIM, ERNST [1899-1989]
A. Buchner, Neue Schweizer Biographie, p.161. Basel 1938
Anon., Rock. Univ. Notes 9(8) (1978), 2
C. Chesterman, Acta Tropica 36 (1979), 201-202
New York Times, 31 May 1989

FRIEDLÄNDER, ALBERT [1869-1942?]
GEDENKBUCH 1,370; JV 8,215; RSC 14,804

FRIEDLÄNDER, CARL [1847-1887]
ADB 48,785; HIRSCH 2,623; FISCHER 1,451-452; PAGEL, 549-550; WININGER 2,336; KOREN, 178
C. Weigert, Fort. Med. 5 (1887), 321-329
Anon., Leopoldina 23 (1887), 112; Deutsche med. Wchschr. 13 (1887), 441
W. Koehler and H. Mochmann, Z. ärzt. Fortbild. 81 (1987), 615-618

FRIEDLÄNDER, PAUL [1857-1923]
DBJ 5,426; KROLLMANN, 910; BLOKH 1,225-226; WOLF, 53;

WININGER 2,340-341; POTSCH, 157-158; TETZLAFF, 92; WWWS, 610; POGG 4,459-460; 5,397; 6,818-819
F. Mayer, Chem. Z. 47 (1923), 765
A. Weinberg, Ber. chem. Ges. 57A (1924), 13-29

FRIEDLEBEN, ALEXANDER [1819-1878]
HIRSCH 2,623

FRIEDMANN, ERNST [1877-1956]
KURSCHNER 4,751; LDGS, 59
J.S. Mitchell, Nature 178 (1956), 397

FRIEDREICH, NIKOLAUS [1826-1882]
NDB 5,458-459; ADB 48,785-786; BAD 4,143-147; DRULL, 73-74; WWWS, 611; HIRSCH 2,624-626; Suppl.,303; HAYMAKER, 439-441; ROEDER, 132-135
A. Kussmaul, Deutsches Arch. klin. Med. 32 (1882), 181-208
R. Virchow, Virchows Arch. 90 (1882), 213-216
A. Weil, Leopoldina 19 (1883), 14-16,19-20
W. Erb, in Heidelberger Professoren aus dem 19. Jahrhundert (K. Friedrich Ed.), vol.2, pp.389-466. Heidelberg 1903

FRIEDRICH, WALTHER [1883-1968]
KURSCHNER 10,602; POGG 5,397-398; 6,820; 7a(2),124-125
F. Köhler, Ed., Walther Friedrich, Leben und Wirken. Berlin 1963
H. Kraatz et al., Stud. Biophys. 1969, pp.3-69
P. Forman, Arch. Hist. Exact Sci. 6 (1969), 38-71

FRIEDRICH-FREKSA, HANS [1906-1973]
KURSCHNER 10,603
F. Duspiva, Heid. Akad. Jahrb. 1974, pp.90-93

FRIES, HANS ERNST AUGUST [1881-]
JV 27,582; SG [3]5,976

FRIES, KARL [1875-1962]
KURSCHNER 9,495; MEINEL, 224-228; SCHMITZ, 279-282; AUERBACH, 804-805; POTSCH, 158; GUNDLACH, 472; TLK 3,171; POGG 5,398; 6,821-822; 7a(2),126
G. Wittig, Z. angew. Chem. 62 (1950), 152
H. Bestian, Chem. Ber. 117(11) (1984), xxiii-xli

FRISCH, KARL von [1886-1982]
DSB 17,312-320; KURSCHNER 13,990; MH 1,184-185; IWW 1982-1983, p.429; ABBOTT, 131-132; AO 1982, pp.267-270; IB, 90; WWWS, 1733
K. von Frisch, A Biologist Remembers (trans. by L. Gombrich). Oxford 1967
H. Autrum, Naturw. Rund. 35 (1982), 435-437; 40 (1987), 421-425
W.H. Thorpe, Biog. Mem. Fell. Roy. Soc. 29 (1983), 197-200

FRISELL, WILLIAM RICHARD [1920-]
AMS 15(2),1166; WWWS, 612-613

FRISONI, ERICH [-1905]
Anon., Ber. chem. Ges. 38 (1905), 4201

FRITSCH, FELIX EUGEN [1879-1954]
DNB [1951-1960], 378-380; DESMOND, 240; IB, 90; WWWS, 613
E.J. Salisbury, Obit. Not. Fell. Roy. Soc. 9 (1954), 131-140
T.T. Macan, Hydrobiologia 7 (1955), 1-7
M.O.P. Iyengar, Journal of the Indian Botanical Society 35 (1956), 522-532

FRITSCH, MARTIN [1868-]
JV 8,215; RSC 13,890

FRITSCH, PAUL [1859-1913]
NDB 5,629; MEINEL, 221-222,502-503; GUNDLACH, 471; AUERBACH, 805; POGG 4,462; 5,399

FRITZ, VICTOR [1872-1926]
JV 11,18; RSC 15,147
Anon., Chem. Z. 50 (1926), 87; Z. angew. Chem. 39 (1926), 124

FRITZSCHE, CARL (YULI FEDEROVICH) [1808-1871]
DSB 5,197-198; HEIN-E, 137-138; BLOKH 1,226-227; ARBUZOV, 101; WWWS, 613; POTSCH, 158-159; FERCHL, 165-166; POGG 1,808-809; 3,481
A. Butlerov, Ber. chem. Ges. 5 (1872), 132-136
F.E. Sheibley, J. Chem. Ed. 20 (1943), 115-117

FRITZSCHE, (FRIEDRICH) HERMANN [1884-1966]
MUC 186,336

FROBENIUS, AUGUST WILHELM [1866-1926]
RSC 17,756
Anon., Ber. chem. Ges. 60A (1927), 138

FRÖHLICH, ALFRED [1871-1953]
NDB 5,648; BHDE 2,334; TETZLAFF, 94; TLK 3,172; WWWS, 614
F. Brücke, Wiener klin. Wchschr. 65 (1953), 306-307
E.P. Pick et al., Science 118 (1953), 314

FRÖHLICH, EMIL [1876-]
RIGA, 700; JV 22,619; NUC 186,591FRÖHLICH, FRIEDRICH WILHELM [1879-1932] NDB 5,649-650; OBL 1,372; KURSCHNER 4,763-764; FISCHER 1,455; IB, 90; KOREN, 178; WWWS, 614
L.K. Vogelsang, Med. Klin. 28 (1932), 1765

FRÖLICH, THEODOR [1870-1947]
FISCHER 1,456
L. Salomonssen, Nord. Med. 36 (1947), 2271-2272
Anon., Lancet 1947(II), p.454
R. Nicolaysen, Nord. Med. 85 (1971), 69-74

FROESCHELS, EMIL [1886-1972]
KURSCHNER 13,999; FISCHER 1,456-457; PLANER, 90; KOREN, 178
K.S. Brodnitz, J. Am. Speech Assn. 14 (1972), 231

FROMAGEOT, CLAUDE [1899-1958]
POGG 6,826; 7b,1504-1508
P. Desnuelle, Bull. Soc. Chim. Biol. 40 (1958), 1688-1709
H. Blaschko, Nature 181 (1958), 454-455
A. Lacassagne, C.R. Soc. Biol. 152 (1958), 234-235

FROMHERZ, CARL [1794-1854]
DAB 8,138-139; BAD 1,268-269; FERCHL, 186; TSCHIRCH, 1049; POGG 1,810-811
K.H. Baumgärtner, Gedächtnisrede auf K. Fromherz. Freiburg i.B. 1855

FROMHERZ, HANS [1902-1972]
KURSCHNER 10,613; POGG 6,826; 7a(2),135
Anon., Chimia 26 (1972), 214

FROMHERZ, KONRAD [1883-1963]
KURSCHNER 9,502; IB, 90

FROMM, EMIL [1865-1928]
OBL 1,373-374; FISCHER 1,457; PLANER, 89; TLK 2,187; IB, 90; POGG 4,464-465; 5,399-400; 6,826-827
H. Barrenscheen, Wiener med. Wchschr, 78 (1928), 786-787
Anon., Chem. Z. 52 (1928), 462; Ost. Chem. Z. 31 (1928), 95

FROMM, FRITZ [1904-]
KURSCHNER 10,613; LDGS, 19

FROMMANN, CARL [1831-1892]
ADB 49,184; HIRSCH 2,633-634; PAGEL, 562; GIESE, 482-483; WWWS, 615
K. von Bardeleben, Anat. Anz. 7 (1892), 437-439

FRORIEP, AUGUST von [1849-1917]
NDB 5,663; DBJ 2,81-83,655; FISCHER 1,459; PAGEL, 563-564; WWWS, 1733
M. Heidenhain, Anat. Anat. 50 (1917), 410-422
Anon., Württemberger Nekrolog 1919, pp.131-140
J. Rückert, Bayer. Akad. Wiss. Jahrbuch 1919, pp.47-57

FROSCH, PAUL [1860-1928]
NDB 5,664; FISCHER 1,459; PAGEL, 564; WWWS, 615
K.K. Kleine, Deutsche med. Wchschr. 54 (1928), 1097
E. Uhlenhuth, Z. Immunitätsforsch. 58 (1928), i-iv

FROUIN, ALBERT [1870-1926]
H. Mouton, Bull. Soc. Chim. Biol. 8 (1926), 1221-1225
Anon., Bull. Inst. Pasteur 24 (1926), 569

FROWEIN, PIETER COENRAAD FREDERIK [1854-1917]
RSC 15,155; POGG 4,466
P.J. Meertens, Medelingsblad van de Sociaal-Historische Studiekring 23 (1963), 3-7

FRUMKIN, ALEKSANDR NAUMOVICH [1895-1976]
DSB 17,322-324; POTSCH, 159; POGG 6,827-828; 7b,1508-1516
Anon., Journal of Electroanalytical Chemistry 10 (1965), 349-359

FRUNDER, HORST [1919-]
KURSCHNER 12,823; POGG 7a(2),137

FRUTON, JOSEPH STEWART [1912-]
AMS 15(2),1175-1176; WWA 1980-1981, p.1176; IWW 1982-1983, p.431; MH 2,168-170; MSE 1,406-407; WWWS, 616
J.S. Fruton, in Of Oxygen, Fuels and Living Matter, Part 2 (G. Semenza, Ed.), pp.315-360. Chichester 1982

FRY, HARRY SHIPLEY [1878-1949]
AMS 8,842; NCAB 38,247; WWAH,211; WWWS, 616; POGG 6,828-829; 7b,1516-1517
Anon., School and Society 69 (1949), 383

FUBINI, SIMONE [1841-1898]
PAGEL, 564-565; WININGER 2,359-360; KOREN, 178
V. Aducco, Arch. Ital. Biol. 31 (1899), 479-484

FUCHS, BERTOLD [1890-]
JV 37,737

FUCHS, FRIEDRICH [-1926]
RSC 9,944; 15,160; NUC 187,182
Anon., Chem. Z. 50 (1926), 267

FUCHS, JOHANN NEPOMUK von [1774-1856]
DSB 5,202-203; NDB 5,680; ADB 8,165-168; BLOKH 1,227-228; POTSCH, 159-160 SCHAEDLER, 37; FERCHL, 167; WWWS, 1733; POGG 1,814
A. Schrötter, Alm. Akad. Wiss. Wien 7 (1857), 108-128
W. Prandtl, J. Chem. Ed. 28 (1951), 136-142

FUCHS, LEOPOLD [1899-1962]
KURSCHNER 9,507; POGG 7a(2),139-140
Anon., Naturw. Rund. 15 (1962), 491
R. Wasicky, Planta Medica 11 (1963), 1-7

FUCHS, RICHARD FRIEDRICH [1870-]
KURSCHNER 4,773; FISCHER 1,462; BERWIND, 181-184; PITTROFF, 208-209; LDGS, 79; IB, 91; KOREN, 178-179

FUCHS, SIEGMUND [1859-1903]
OBL 1,379; BJN 8,80,34*; FISCHER 1,462
Anon., Z. Land. Ver. Ost. 6 (1903), 783-796; Leopoldina 39 (1903), 101
S. Exner, Z. Physiol. 17 (1903), 250-251

FUCHS, WALTER [1891-1957]
NDB 5,683-684; BHDE 2,349; LDGS, 19; POGG 6,831; 7a(2),141-142
Anon., Chem. Ing. Techn. 29 (1957), 765-767

FUCHSIG, PAUL [1908-1977]
KURSCHNER 12,827; KILLIAN, 75-6; ZAHRAN, 96-104; WWWS, 617

FUDAKOWSKI, HERMAN BOLESLAW [1834-1878]
PSB 7,177; HIRSCH 2,642; BLOKH 1,228; KONOPKA 3,139-144; KOSMINSKI, 126-127
J.G. Boguski, Ber. chem. Ges. 12 (1879), 1038
E. Ostachowski, Stud. Nauk. Pol. C9 (1964), 73-78
E, Kucharz, Wiad. Lek. 33 (1980), 931-934
Z. Bednarski, Wiad. Lek. 38 (1985), 973-975
Z.J. Gielman, Kwart. Hist. Nauki 33 (1988), 145-167

FÜCHTBAUER, CHRISTIAN [1877-1959]
POGG 6,531-532; 7a(2),143

FÜHNER, HERMANN [1871-1944]
NDB 5,687; KURSCHNER 6,479; HEIN-E, 139-140; FISCHER 1,462; TLK 3,174; TSCHIRCH, 1050; IB, 91; WWWS, 618; POGG 6,832-833; 7a(2),143-144
L.E. Hoffmann, Münch. med. Wchschr. 91 (1944), 85-86

FÜRBRINGER, MAX [1846-1920]
NDB 5,690; DBJ 2,747; FISCHER 1,463; GIESE, 468-470; DRULL, 76-77; LINDEBOOM, 639-640; BERLIN, 684; WWWS, 621
H. Braus, Deutsche med. Wchschr. 46 (1920), 470
H. Blüntschli, Anat. Anz. 55 (1922), 244-255

FÜRBRINGER, PAUL WALTER [1849-1930]
NDB 5,690-691; KURSCHNER 3,625; FISCHER 1,463-464; GIESE, 567-568; PAGEL, 567-568; WREDE, 56-57; KALIN, 62-64; WWWS, 621
F. Kraus, Med. Klin. 26 (1930), 1173-1174
W. His, Deutsche med. Wchschr. 57 (1931), 591-593

FÜRST, VALENTIN [1870-1961]
Hvem er Hvem? 1959, p.199

FÜRTH, JULIUS [1859-1923]
A. Eiselsberg, Wiener klin. Wchschr. 36 (1923), 372
F. Hansy, Wiener med. Wchschr. 73 (1923), 945-946FÜRTH, OTTO von [1867-1938]
NDB 5,701; OBL 1,382; STURM 1,404; FISCHER 1,465; TLK 3,176; KOREN, 179; WININGER 6,609; TETZLAFF, 97; IB, 91; WWWS, 1733-1734; POGG 6,833-834; 7a(2),144-145
F. Lieben, Wiener klin. Wchschr. 60 (1948), 377-379

FUHRMANN, FRANZ [1877-1957]
KURSCHNER 4,775-776; TLK 3,175; IB, 91
Anon., Chem. Z. 81 (1957), 438; Naturw. Rund. 11 (1958), 36

FUJINAMI, AKIRA [1870-1934]
FISCHER 1,466
Anon., Gann 28 (1934), 547-548; Science 81 (1935), 331; J. Parasit. 22 (1936), 548

FUJITA, AKIJI [1894-1985]
HAKUSHI 4,1069

FUKUI, KENICHI [1918-]
IWW 1982-1983, p.434; POTSCH, 160

FULD, ERNST [1873-1955]
FISCHER 1,466-467; WININGER 2,363; KOREN, 179
Anon., J. Am. Med. Assn. 160 (1956), 224
A. Sonnenfeld, Proc. Virchow Med. Soc. 15 (1956), 117

FULMER, ELLIS INGHAM [1891-1953]
WWWS, 620; POGG 6,837-838; 7b,1520-1522
Anon., Chem. Eng. News 31 (1953), 1228; Science 117 (1953), 273

FULTON, JOHN FARQUHAR [1899-1960]
DSB 5,207-208; DAB [Suppl.6],222-224; AMS 9(II),381; DAMB, 268-269; IB, 91; WWAH, 211; WWWS, 620
H.E. Hoff, Yale J. Biol. Med. 28 (1955-1956), 165-190; J. Hist. Med. 17 (1962), 16-37
G.E. Hutchinson, Am. Phil. Soc. Year Book 1960, pp.140-142
Anon., Brit. Med. J. 1960(I), pp.1815-1816
K.J. Franklin, Nature 187 (1960), 110-111
W. LeFanu, J. Hist. Med. 17 (1962), 38-71
E.H. Thomson, J. Hist. Med. 36 (1981), 151-167

FUNK, CASIMIR [1884-1967]
DSB 5,208-209; MILES, 165-166; STC 1,418; AMS 11,1697; FISCHER 1,467-468; CHITTENDEN, 351-353; KAGAN(JA), 340-342,737-738; ABBOTT-C, 49; CB 1945, pp.210-212; KOREN, 179; IB, 91; WWWS, 621; POTSCH, 160-161; POGG 6,838-839; 7b,1522-1523
B. Harrow, Casimir Funk. New York 1955
Anon., Chem. Eng. News 46(3) (1968), 100
P. Griminger, J. Nutrition 102 (1972), 1107-1113
S. Wajs, Polish Medical Science and History Bulletin 15 (1974), 107-113
P.F. Ostrowski, Polish Review 31 (1986), 171-185

FUNKE, OTTO [1828-1879]
BAD 3,45-49; HIRSCH 2,647; Suppl.,307; PAGEL, 571-572; TSCHIRCH, 1050; WWWS, 621; POGG 1,819; 3,487

FUOSS, RAYMOND MATTHEW [1905-1987]
AMS 14,1625; WWA 1978-1979, p.1141; IWW 1982-1983, p.435; MH 2,170-171; MSE 1,409-411

FURBERG, SVEN VERNER [1920-]
AUERBACH, 940-941
Hver er Hvem? 1979, p.180

FURTH, JACOB [1896-1979]
AMS 13,1423; KAGAN(JA), 721
D.M. Angevine, Cancer Research 26 (1966), 351-356
D.W. King, Am. J. Path. 98 (1980), 293-294

FURUKAWA, SEIJI [1886-1955]
Y. Tsuzuki and A. Yamashita, Jap. Stud. Hist. Sci. 7 (1968), 24
Y. Tsuzuki, J. Chem. Ed. 47 (1970), 695-696

FUSON, REYNOLD CLAYTON [1895-1979]
AMS 14,1629; MH 2,172-174; MSE 1,412-413; WWWS, 622
Anon., Chem. Eng. News 57(39) (1979), 65
N. Rabjohn, Organic Syntheses 59 (1979), ix-x

FYFE, ANDREW [1792-1861]
DNB 7,780-781; HIRSCH 2,648; FERCHL, 168; CALLISEN 6,527; POGG 1,824-825

G

GABRIEL, SIEGMUND [1851-1924]
DSB 5,214-215; NDB 6,10-11; BLOKH 1,228-229; TLK 2,191; W, 200; WININGER 2,375-376; SCHAEDLER, 37; WWWS, 623; POTSCH, 161; POGG 3,488; 4,471-472; 5,406; 6,842
J. Colman and A. Albert, Ber. chem. Ges. 59A (1926), 7-26
E.H. Huntress, Proc. Am. Acad. Arts Sci. 79 (1951), 17-18

GAD, JOHANNES [1842-1926]
STURM 1,409-410; PAGEL, 573-575; LOMMATZSCH, 123-128; KOERTING, 119
R. DuBois-Reymond, Med. Klin. 22 (1926), 1020-1022

GADAMER, JOHANNES GOERG [1867-1928]
NDB 6,11-12; FISCHER 1,469; SCHNACK, 96-105; HEIN, 186-187; GUNDLACH,485; AUERBACH, 808; TSCHIRCH, 1050; TLK 2,191; WWWS, 623; POGG 4,472-473; 5,406-407; 6,842-843; 7a(2),153
E. Späth, Ost. Chem. Z. 31 (1928), 89-90
C. Mannich, Ber. chem. Ges. 61A (1928), 80-82
W. Schulemann, Z. angew. Chem. 41 (1928), 487-488)
G. Zimmer, Deutsche Apoth. Z. 98 (1958), 335-340
C. Friedrich and G. Rudolph, Pharmazie 43 (1988), 788-792
M. Kollmann-Hess, Die Erste Marburger Schule. Stuttgart 1988

GADDUM, JOHN HENRY [1900-1965]
DNB [1961-1970], 412-413; WW 1965, p.1106
J.H. Gaddum, Ann. Rev. Pharm. 2 (1962), 1-10
J.H. Burn, Pharmacologist 8 (1966), 14-18
W. Feldberg, Biog. Mem. Fell. Roy. Soc. 13 (1967), 57-77
M. Vogt, Erg. Physiol. 62 (1970), 1-5

GADEMANN, FERDINAND JENS [1880-1969]
JV 21,409; NUC 188,554
Anon., Aus dem Leben Schweinfurter Männer und Frauen, pp.183-190. Schweinfurt 1973

GAEBEL, GUSTAV OTTO [1877-1912]
HEIN-E, 144; TSCHIRCH, 1050
Anon., Chem. Z. 36 (1912), 1513

GAEBLER, OLIVER HENRY [1895-1985]
AMS 13,1428; WWWS, 624
Anon., Science 229 (1985), 573

GAEDCKE, FRIEDRICH [1828-1890]
HEIN, 187-188

GAEDECHENS, JULIUS HEINRICH [1820-1862]
DGB 21 (1912), 214

GÄHTGENS, KARL [1839-1915]
WELDING, 223; HIRSCH 2,653; PAGEL, 577; LEVITSKI 2,305-306; BK, 204
J. Brennsohn, Die Aerzte Estlands, p.440. Riga 1922
J. Benedum, Med. Hist. J. 15 (1980), 103-119

GAERTNER, AUGUST [1848-1934]
NDB 6,23-24; KURSCHNER 4,794-795; FISCHER 1,470; PAGEL, 575-576; IB, 92; TLK 3,179; WWWS, 634
R. Abel, Z. Bakt. 107 (1928), i-xvi
F. Konrich, Deutsche med. Wchschr. 61 (1935), 233

GAERTNER, FREDERICK [1861-1929]
JV 1,182; RSC 15,181; SG [2]6,6; NUC 188,610
New York Times, 8 February 1929, p.25

GAERTNER, GUSTAV [1855-1937]
NDB 6,25; OBL 1,389-390; FISCHER 1,470-471; PAGEL, 576-577; IB, 92; WININGER 2,376-377; WWWS, 634
J. Pal, Wiener med. Wchschr. 75 (1925), 2153-2155
J. Wagner-Jauregg, Wiener med. Wchschr. 85 (1935), 1077
E. Bozzi, Gazz. Med. Ital. 133 (1974), 636-639

GAERTNER, KARL FRIEDRICH von [1772-1850]
DSB 5,217-219; NDB 6,22-23; ADB 8,382-384; SL 3,190-198; HINK 1,379-381; LEHMANN, 24-25; TSCHIRCH, 1051; CALLISEN 7,6-7
Anon., Flora 34 (1851), 135-143

GÄUMANN, ERNST ALBERT [1893-1963]
KURSCHNER 9,517; SBA 1,58; IB, 92
E. Landolt, Verhandl. Schw. Nat. Ges. 143 (1963), 194-206
H. Kern, Ber. bot. Ges. 77 (1964), (238)-(248)
M.W. Gardner and H. Kern, Mycologia 57 (1964), 1-5
S. Blumer, Mitt. Nat. Ges. Bern 21 (1964), 245-249

GAFFKY, GEORG [1850-1918]
DSB 5,219-220; NDB 6,28; DBJ 2,687; FISCHER 1,471; PAGEL, 577-579; GUNDEL, 256-263; BULLOCH, 367-368; OLPP, 139-141; WWWS, 624
H. Kossel, Münch. med. Wchschr. 65 (1918), 1191-1192
E. Neufeld, Berl. klin. Wchschr. 55 (1918), 1062-1063

GAFFRON, HANS [1902-1982]
AMS 13,1429; BHDE 2,354; POGG 6,843; 7a(2),155-156
H. Gaffron, Ann. Rev. Plant Physiol. 20 (1969), 1-40
Anon., Plant Science Bulletin 29(1) (1983), 51

GAGE, SIMON HENRY [1851-1944]
AMS 7,618; NCAB E,149; 36,127-128; IB, 92; WWAH, 213; WWWS, 624
B.F. Kingsbury, Science 100 (1944), 420-421

GAGER, CHARLES STUART [1872-1943]
AMS 6,498; NACB 38,289-290; HUMPHREY, 89-90; IB, 92; WWAH, 213

W.J. Robbins, Science 98 (1943), 234-235
Anon., Brooklyn Botanic Garden Record 33 (1944), 69-168
G.M. Reed and A.H. Graves, Bull. Torrey Bot. Club 71 (1944), 193-198

GAGLIO, GAETANO [1858-1925]
FISCHER 1,471-472
A. Baldoni, Riv. Biol. 7 (1925), 601-607; Bioch. Ter. Sper. 12 (1925), 268-271
L. Sabbatani, Arch. Ital. Biol. 76 (1926), 76-87

GAIDUKOV, NIKOLAI MIKHAILOVICH [1874-1928]
BSE (3rd Ed.) 6,48; IB, 92
R. Kolkwitz, Ber. bot. Ges. 48 (1930), (197)-(200)

GAIER, JULIUS [1897-1978]
JV 40,464

GAIL, GEORG KARL [1819-1882]
Georg Philipp Gail, Giessen: Gedenkschrift zur hundert-jährigen Bestehen der Firma, p.43. Berlin 1912

GALE, ERNEST FREDERICK [1914-]
WW 1981, pp.935-936; IWW 1982-1983, pp.438-439; SRS, 92; WWWS, 625-626

GALEOTTI, GINO [1867-1921]
FISCHER 1,473-474
C. N[euberg], Biochem. Z. 117 (1921), 117-118
E. Centanni, Bioch. Ter. Sper. 8 (1921), 97
A. Lustig, Sperimentale 76 (1923), 87-95

GALINOVSKY, FRIEDRICH [1908-1957]
POGG 7a(2),157-158
F. Wessely, Ost. Chem. Z. 58 (1957), 320-321

GALIPPE, VICTOR [1848-1922]
DBF 15,176-177
P. Le Gendre, Bull. Acad. Med. 87 (1922), 210-214

GALL, JOSEPH GRAFTON [1926-]
AMS 15(3),16; WWA 1980-1981, p.1193; IWW 1982-1983, p.439; WWWS, 626

GALLAGHER, PATRICK HUGH [1896-]
SRS, 57

GALLAGHER, THOMAS FRANCIS [1905-1991]
AMS 13,1437; WWWS, 627

GALLINI, STEFANO [1756-1836]
EI 16,328; WURZBACH 5,72-73; HIRSCH 2,674; DECHAMBRE [4]6,533; CALLISEN 7,29-30; 28,148

GALLOIS, FRANÇOIS NARCISSE [1831-1896]
GORIS, 388
N. Gréhant, C.R. Soc. Biol. [10]4 (1897), xv-xix

GALLOWAY, ROBERT [1822-1896]
Anon., J. Chem. Soc. 69 (1896), 733-734

GALSTON, ARTHUR WILLIAM [1920-]
AMS 15(3),21; WWA 1980-1981, p.1196; WWWS, 628

GALTIER, PIERRE VICTOR [1846-1908]
DBF 15,278-279; SACKMANN, 128-131; GUIART, 170; RSC 9,957; 15,196-197
Y. Robin, Vie et l'Oeuvre de P.V. Galtier. Lyon 1957
P. Goret and P. Lepine, Bull. Acad. Med. 153 (1969), 75-81
J. Théodoridès, Hist. Sci. Med. 7 (1973), 336-343
A. Violet, Revue du Gevaudan NS21 (1975), 149-154

GALTON, FRANCIS [1822-1911]
DSB 5,265-267; DNB [1901-1911]2, 70-73; W, 203; ABBOTT, 50-51; WWWS, 628;POGG 3,490; 4,476; 5,410
G.H.D., Proc. Roy. Soc. B84 (1912), x-xvii
K. Pearson, Life, Letters and Labours of Francis Galton. Cambridge 1914-1924
B.S. Bramwell, Eugenics Review 39 (1948), 146-153
R.G. Swinburne, Ann. Sci. 21 (1965), 21 (1965), 15-31
P. Froggatt and N.C. Nevin, Hist. Sci. 10 (1965), 1-27
R.S. Cowan, J. Hist. Biol. 5 (1972), 389-412; Isis 63 (1972), 509-528
D.W. Forrest, Francis Galton. London 1974
R. DeMarrais, J. Hist. Biol. 7 (1974), 141-174

GAMALEYA, NIKOLAI FEDOROVICH [1859-1949]
DSB 5,269-271; BSE (3rd Ed.)6,82-83; WWR, 179; KUZNETSOV(1963), 605-616; SKORODOKHOV, 286-291; BME 6,326-328; WWWS, 628
Y.I. Milenushkin, Priroda 38(6) (1949), 81-86
M. Caullery, C.R. Soc. Biol. 144 (1950), 77
N.P. Gracheva, Mikrobiologia 28 (1959), 300-307
D. Bardell, J. Hist. Med. 37 (1982), 222-225

GAMBARYAN (GAMBAROFF), STEPAN PAVLOVICH [1879-1948]
JV 23,542; NUC 190,135
V,D. Azatyan, Izvestia Akademii Nauk (Armenia) 1962(2), pp.205-206

GAMBLE, JAMES LAWDER [1883-1959]
AMS 9(II),385; DAMB, 274; ICC, 182-184
A.M. Butler, Trans. Assoc. Am. Phys. 73 (1960), 13-16
C.A. Janeway, J. Pediat. 56 (1960), 701-708
R.F. Loeb, Biog. Mem. Nat. Acad. Sci. 36 (1962), 146-160
F.C. Bing, J. Nutrition 111 (1981), 203-207

GAMGEE, ARTHUR [1841-1909]
DNB [1901-1911]2, 73-74; HIRSCH 2,679; Suppl.,313; FISCHER 1,476; PAGEL, 581-582; O'CONNOR, 232-234; WWWS, 628
G.A.B., Nature 80 (1909), 194-196
Anon., Lancet 1909(I), pp.1141-1144
R. D'Arcy Thompson, The Remarkable Gamgees. Edinburgh 1974

GAMOW, GEORGE [1904-1968]
DSB 5,271-273; AMS 11,1722; MH 1,85-86; STC 1,422-423; MSE 1,419-420; CB 1951, pp.228-230; WWWS, 629; POGG 6,848; 7b,1540-1543
Anon., Nature 220 (1968), 723
G. Gamow, My World Line. New York 1970
G. Greenstein, American Scholar 59 (1990), 118-125

GANONG, WILLIAM FRANCIS [1924-]
AMS 17(3),28; BROBECK, 235-239; WWWS, 629

GARDNER, WILLIAM ULLMAN [1907-1988]
AMS 16(3),38; WWWS, 631

GANS, LEO [1843-1935]
NDB 6,64; ARNSBERG, 131-132; KALLMORGEN, 273
Anon., Chem. Z., 52 (1928), 609; Chemische Industrie 56 (1933), 538

GANS, OSCAR [1888-1983]
KURSCHNER 14,1124-1125; ARNSBERG, 132-133; FISCHER 1,477; KALLMORGEN,274; DRULL, 78-79; BHDE 2,357; KOREN, 179; LDGS, 62
A. Hollander, Journal of the American Academy of Dermatology, 11 (1984), 162-164; American Journal of Dermopathology 6 (1984), 87-88
O. Braun-Falco, Hautarzt 36 (1985), 647-649

GANS, RICHARD [1880-1954]
NDB 6,64-65; TETZLAFF, 98; WWWS, 629; POGG 5,410-411; 6,849-850; 7a(2),159-160
B. Mrowka, Phys. Bl. 10 (1954), 512-513

GANZENMÜLLER, KARL WILHELM [1882-1955]
NDB 6,68; POGG 7a(2),161

GARBEN, EDUARD [1873-1940]
JV 17,283; RSC 13,518

GARBSCH, PAUL [1901-]
JV 43,412; NUC 190,504

GARCIA-BANUS, MARIO (see BANUS)

GARCIA-GONZALEZ, FRANCISCO [1902-1983]
A. Gomez-Sanchez and J. Fernandez-Bolanos, Adv. Carbohydrate Chem. 45 (1987), 7-17

GARCIA-VALENZUELA, ADEODATO [1864-1936]
RSC 15,206; NUC 191,51
A. Calderon, Vida Medica 39(2) (1987), 72-77

GARD, SVEN [1905-]
VAD 1983, p.330

GARDINER, JOHN STANLEY [1872-1946]
WW 1945, p.996; IB, 93; WWWS, 631
C. Forster-Cooper, Obit. Not. Fell. Roy. Soc. 5 (1948), 541-543

GARDINER, WALTER [1859-1941]
O'CONNOR(2), 30; DESMOND, 243
A.W. Hill, Nature 148 (1941), 462-463; Obit. Not. Fell. Roy. Soc. 1 (1941), 985-1004

GARDNER, DANIEL PEREIRA [-1853]
ELLIOTT(2), 85; BARNHART 2,29; RSC 2,767-768; NUC 191,219-220

GARDNER, JOHN [1804-1880]
DNB 7,871-872; WWWS, 631

GARDNER, JOHN ADDYMAN [1867-1946]
O'CONNOR(2), 305-307; IB, 94
G.W. Ellis, Biochem. J. 41 (1947), 321-324; J. Chem. Soc. 1947, pp.865-866

GAREN, ALAN [1926-]
AMS 15(3),38

GARNER, WILLIAM EDWARD [1889-1960]
DNB [1951-1960], 301-302; SRS, 50; WWWS, 632; POGG 6,853; 7b,1553-1554
C.E.H. Bawn, Proc. Chem. Soc. 1960, pp.230-232; Nature 186 (1960), 760-761; Biog. Mem. Fell. Roy. Soc. 7 (1961), 85-94

GARNIER, CHARLES [1875-1958]
DBF 15,473-474
F. Haguenau, International Review of Cytology 7 (1958), 425-483
Anon., Rev. Med. Nancy 84 (1959), 881-883

GARNIER, LÉON [1855-1939]
DBF 15,512
Anon., Rev. Med. Nancy 67 (1939), 549

GARNIER, MARCEL [1870-1940]
DBF 15,513-514; IB, 94
H. Roger, Presse Med. 48 (1940), 115

GARNJOBST, LAURA FLORA [1895-1977]
AMS 11,1737; NCAB 61,27-28

GARRÉ, KARL [1857-1928]
FISCHER 1,479-480
T. Nägeli, Schw. med. Wchschr. 88 (1958), 622-624

GARREAU, LAZARE [1812-1892]
DSB 5,277-278; DBF 15,559-560; BALLAND, 42-43
A. Folet, Bull. Med. Nord 31 (1892), 545-550
Anon., J. Pharm. Chim. [5]27 (1893), 109

GARREY, WALTER EUGENE [1874-1951]
AMS 8,867; BROBECK, 149; IB, 94; WWAH, 216; WWWS, 633
Anon., Science 114 (1951), 26; J. Am. Med. Assn. 147 (1951), 334

GARRIGUES, SAMUEL [1828-1889]
NUC 191,668; RSC 2,773
C.T.B., Am. J. Pharm. 61 (1889), 319-320

GARROD, ALFRED BARING [1819-1907]
DNB [1901-1911], 84-85; HIRSCH 2,689-690; WWWS, 633
Anon., Lancet 1908(I), pp.65-67; Brit. Med. J. 1908(I), pp.58,121
J.P. zum Busch, Deutsche med. Wchschr. 34 (1908), 292
C. Rutz, The Garrods. Zurich 1970

GARROD, ARCHIBALD EDWARD [1857-1936]
DSB 17,333-336; DNB [1931-1940], 308-309; FISCHER 1,480-481; O'CONNOR(2), 267-269; NUESCH, 75-77; WWWS, 633
G.G., St. Barts. Hosp. Rep. 69 (1936), 12-26
F. Frazer, Lancet 1936(I), pp.807-809
F.G. Hopkins, Nature 137 (1936); Obit. Not. Fell. Roy. Soc. 2 (1938), 225-228
C.J.R. Hirt, St. Barts. Hosp. J. 53 (1949), 160-165,186-190
L.S. Penrose, Bioch. Soc. Symp. 4 (1950), 10-15
K. Vella, St. Barts. Hosp. J. 69 (1965), 138-145
W.E. Knox, Genetics 56 (1967), 1-6
B. Childs, New Eng. J. Med. 282 (1970), 71-77
H. Harris, Brit. Med. J. 1970(I), pp.321-327
A.G. Beam and E.D. Miller, Bull. Hist. Med. 53 (1979), 315-328

GARROD, LAWRENCE PAUL [1895-1979]
DNB [1971-1980], 333-334
Who Was Who 1971-1980, p.289
Anon., Lancet 1979(III), p.289; Brit. Med. J. 1979(II), p.740
L.P. Garrod, Journal of Antimicrobial Therapy 15Suppl.B (1985), 1-46

GARTEN, SIEGFRIED [1871-1923]
NDB 6,74-75; DBJ 5,100-103,426; FISCHER 1,481; WWWS, 633; POGG 5,413; 6,854; 7a(2),164-165
O. Frank, Z. Biol. 81 (1924), 1-4
K. Thomas, Sitz. Akad. Wiss. Leipzig 76 (1924), 159-160

GARTENMEISTER, RUDOLF [1857-1918]
RSC 15,219; POGG 4,479
Anon., Chem. Z. 42 (1918), 459; Z. angew. Chem. 31(III) (1918), 468

GARZAROLLI, KARL [1854-1906]
OBL 1,405; BJN 11,22*; KERNBAUER, 107-110; POGG 1,494; 4,480; 5,413

GASCARD, ALBERT [1861-1934]
DBF 15,588; GORIS, 389-390; TSCHIRCH, 1051; POGG 4,480; 5,413; 6,855; 7b,1556
J. Bougault, Bull. Acad. Med. 111 (1934), 793-796
Anon., Bull. Sci. Pharm. 41 (1934), 490-496

GASKELL, JOHN FOSTER [1878-1960]
O'CONNOR(2), 71-72; IB, 94; NUC 192,176
Anon., Lancet 1960(II), p.711; Munk's Roll 5,145-146
C.H. Whittle, J. Path. Bact. 82 (1961), 537-541

GASKELL, WALTER HOLBROOK [1847-1914]
DSB 5,279-284[DNB [1912-1921], 207-209; HAYMAKER, 210-213; W, 205-206 O'CONNOR, 174-176
F.H. Garrison and F.H. Pike, Science 40 (1914), 802-807
J.N. L[angley], Proc. Roy. Soc. B88 (1915), xxvii-xxxvi

GASKING, ELIZABETH BELMONT [1914-1973]
D. Dyason, Isis 66 (1975), 387-389GASPARD, BERNARD [1788-1871]
DBF 15,598; HIRSCH 2,692; DECHAMBRE 42,762-763; BULLOCH, 368; CALLISEN 7,63-65; 28,157

GASSER, HERBERT SPENCER [1888-1963]
DSB 5,290-291; AMS 10,1354; DAB [Suppl.7],279-281; HAYMAKER, 213-217; NCAB E,236; 61,243-244;MH 1,186-187; STC 1,425-426; MSE 1,420-421; CB 1945, pp.224-225; FISCHER 1,482; DAMB, 281-282; W, 206-207; IB, 94; WWAH, 216; WWWS, 634
J.C. Hinsey, Am. Phil. Soc. Year Book 1963, pp.144-149
Anon., Pharmacologist 5 (1963), 122-124

E.D. Adrian, Biog. Mem. Fell. Roy. Soc. 10 (1964), 75-82

GASSNER, GUSTAV [1881-1955]
NDB 6,83-84; KURSCHNER 4,798; BHDE 2,359-360; LDGS, 12; IB, 94
K. Hassebrauk, Ber. bot. Ges. 68a (1955), 189-192
H. Richter, Phytopathologische Zeitschrift 23 (1955), 221-232

GATENBY, JAMES BRONTÉ [1892-1960]
DNB [1951-1960], 397-398; WW 1959, p.1116; WWWS, 635
J.N. Grainger, Nature 187 (1960), 990
E.T. Freeman, Irish J. Med. Sci. 1960, p.437

GATES, FREDERICK LAMONT [1886-1933]
AMS 5,401; FISCHER 1,483; IB, 95
Anon., Science 78 (1933), 8

GATES, FREDERICK TAYLOR [1853-1929]
DAB 7,182-183; NCAB 23,250-251; DAMB, 282-283
F.T. Gates, Chapters in my Life. New York 1977

GATES, MARSHALL DeMOTTE, Jr. [1915-]
AMS 15(3),55; WWA 1980-1981, p.1214; IWW 1982-1983, p.447; WWWS, 365

GATES, REGINALD RUGGLES [1882-1962]
DSB 5,293-294; DNB [1961-1970], 424-425; IB, 95; DESMOND, 246; WWWS, 635
D.G. Catcheside, Nature 195 (1962), 1252-1253
J.A. Roberts, Biog. Mem. Fell. Roy. Soc. 10 (1964), 83-106

GATTERMANN, LUDWIG [1860-1920]
NDB 6,91; DBJ 2,747; DRULL, 79-89; BLOKH 1,231-232; DZ, 423; TLK 2,196; POTSCH, 162-163; W, 207; WWWS, 636; POGG 4,480-481; 5,413-414; 6,856
R. Schwarz, Chem. Z. 44 (1920), 513
P. Jacobson, Ber. chem. Ges. 54A (1921), 115-141

GAUCHER, ERNEST [1854-1918]
DBF 15,682-683; FISCHER 1,483; PAGEL, 584-585
A.B.J. Marfan, Bull. Acad. Med. 79 (1918), 78-82
L. Fiaux, Ernest Gaucher. Paris 1919
R.J. Desnick, Mt. Sinai J. Med. 49 (1982), 443-455

GAUDIN, MARC ANTOINE [1804-1880]
DSB 5,294-295; DBF 15,710-711; BLOKH 1,232; FERCHL, 173; WWWS, 636; POTSCH, 163; POGG 1,851-852; 3,496
M. Delépine, Bull. Soc. Chim. [5]2 (1935), 1-15
S.H. Mauskopf, Isis 60 (1969), 61-74
J.A. Miller, in van't Hoff-Le Bel Centennial (O.B. Ramsey, Ed.), pp.1-17. Washington, D.C. 1975
T.M. Cole, Jr., Isis 66 (1975), 334-360

GAUER, OTTO [1909-1979]
KURSCHNER 13,1951; POGG 7a(2),168-169

GAULE, JUSTUS [1849-1939]
FISCHER 1,484
W.R. Hess, Viert. Nat. Ges. Zürich 84 (1939), 375-377

GAULT, HENRY [1880-1967]
POGG 5,414; 6,858; 7b,1559-1562
G.K., Chim. Ind. 99 (1968), 1265

GAULTIER de CLAUBRY, HENRI FRANÇOIS [1792-1878]
DSB 5,297-298; DBF 15,775-776; HOEFER 19,679; LAROUSSE 8,1086; HIRSCH 2,698; FERCHL, 173; SCHAEDLER, 37; WWWS, 636; POTSCH, 163; POGG 1,852-853; 3,497-498
G. Dillemann, Produits et Problèmes Pharmaceutiques 28 (1973), 525-526

GAUNT, RUFUS [1881-1961]
SRS, 31; NUC 192,576
Who's Who in British Science 1953, p.101

GAUS, WILHELM [1876-1953]
NDB 6,101; WWWS, 636-637; POGG 4,481; 6,858; 7a(2),169
K. Schoemann, Chemie Ingenieur Technik 23 (1951), 489-490

GAUSE, GEORGI FRANTSOVICH [1910-1986]
WWWS, 637
N.N. Vorontsov and J. M. Gall, Nature 323 (1986), 113

GAUTIER, ARMAND [1837-1920]
DSB 5,315; DBF 15,821-822; FISCHER 1,485; BULLOCH, 368; BLOKH 1,233; POTSCH, 163; WWWS, 637; POGG 3,498-499; 4,482-483; 5,415; 6,859
E. Lebon, Armand Gautier. Paris 1912
T.E. Thorpe, Nature 106 (1920), 85-76; J. Chem. Soc. 119 (1921), 537-539
A. Desgrez, Bull. Acad. Med. 84 (1920), 74-77; Rev. Sci. 59 (1921), 639-641; Bull. Soc. Chim. [4]31 (1922), 193-219
M. Delépine, Bull. Soc. Chim. [5]5 (1938), 117-148

GAUTIER, HENRY [1862-1928]
DBF 15,829; TSCHIRCH, 1051; POGG 4,483
P. Lebeau, Bull. Soc. Pharm. 36 (1929), 148-154

GAUTRELET, JEAN [1878-1941] IB, 95
G. Bourguignon, Bull. Acad. Med. 125 (1941), 168-171

GAVARRET, JULES [1809-1890]
DBF 15,875-876; HIRSCH 2,700-701; Suppl.,317; FERCHL, 173; POGG 3,499-500
L.H. Petit, Union Médicale [3]1 (1890), 325-327
D. Laborde, Bulletin de la Société d'Anthropologie de Paris [4]1 (1890), 645-651

GAVRILOV, NIKOLAI GAVRILOVICH [1877-]
VENGEROV 1,159

GAWRON, OSCAR [1914-1990]
AMS 17(3),64

GAY, FRANÇOIS [1858-1898]
TSCHIRCH, 1051; POGG 4,483-484
F. Jadin, Bull. Soc. Bot. 45 (1898), 334-335

GAY, FREDERICK PARKER [1874-1939]
DSB 5,316-317; DAB [Suppl.2],224-226; NCAB B,268-269; FISCHER 1,486; DAMB, 283-284; FLEULER, 63-64; IB, 95; WWAH, 217; WWWS, 637
C.W. Jungeblut, Science 90 (1939), 290-291
A.R. Dochez, Biog. Mem. Nat. Acad. Sci. 38 (1954), 99-116

GAY, JULES [1838-]
DORVEAUX, 56; NUC 193,152

GAY-LUSSAC, JOSEPH LOUIS [1778-1850]
DSB 5,317-327; DBF 15,903-904; BUGGE 1,386-401; FARBER, 361-367; W, 209; BLOKH 1,233-235; FERCHL, 174-175; TSCHIRCH, 1051; ABBOTT-C, 49-50; POTSCH, 163-164; COLE, 205-207; POGG 1,860-864; CALLISEN 7,89-99; 28,163-164
Anon., Proc. Roy. Soc. 5 (1843-1850), 1013-1023
E. Blanc and L. Delhoume, La Vie Emouvante et Noble de Gay-Lussac. Paris 1950
E.H. Huntress, Proc. Am. Acad. Arts Sci. 78 (1950), 18-20
M.P. Crosland, Ann. Sci. 17 (1961), 1-26; Endeavour NS2 (1978), 52-56; Gay-Lussac, Scientist and Bourgeois. Cambridge 1978
H. Goldwhite, J. Chem. Ed. 55 (1978), 366-368

GAY-LUSSAC, JULES [1810-]
RSC 2,799; 7,748; 15,238
M.P. Crosland, Gay-Lussac, p.41. Cambridge 1978

GAYON, ULYSSE [1845-1929]
DBF 15,913; POGG 3,500; 4,484; 5,415; 6,860
L. Mangin, C.R. Acad. Sci. 188 (1929), 1017-1018
J. Dubaquie, Ann. Falsif. 22 (1929), 197-199
R. Marcard, Symposium International Oenologique, pp.5-43. Station

Agronomique et Oenologique Bordeaux 1963

GEELMUYDEN, HANS CHRISTIAN [1861-1945]
NBL 4,419-420; FRANKEN, 183; IB, 95; POGG 4,485; 6,861-862; 7b,1565
E. Langfeldt, Norske Vid. Akad Arbok 1946, pp.31-36

GEFTER, IULIA MARKOVNA [1888-1970]
Anon., Vop. Med. Khim. 5 (1959),73-74; 14 (1968),643-644; 16 (1970),650

GEGENBAUR, KARL [1826-1903]
DSB 15,165-171; NDB 6,130-131; BJN 8,324-329,36*; BAD 6,22-31; DRULL, 80; LF 2,144-157; HIRSCH 2,706; PAGEL, 587-588; GIESE, 459-461
C. Gegenbaur, Erlebtes und Erstrebtes. Leipzig 1901
M. Fürbringer, Anat. Anz. 23 (1903), 589-608; in Heidelberger Professoren aus dem 19. Jahrhundert (K. Friedrich, Ed.) 2,389-466. Heidelberg 1903
C. Voit, Sitz. Bayer. Akad. 34 (1904), 252-259
O. Hertwig, Deutsche med. Wchschr. 29 (1904), 525-526
A. Maurer, Jena Z. Naturw. 62 (1926), 501-518

GEHLEN, ADOLF FERDINAND [1775-1815]
DSB 15,171-173; NDB 6,132-133; ADB 8,497-498; HEIN, 194; SCHAEDLER, 39; TSCHIRCH, 1053; FERCHL, 176; DECHAMBRE 43,203-204; PHILIPPE,752-753; POTSCH, 165; WWWS, 639; POGG 1,865-866; CALLISEN 28,166
W. Prandtl, Die Geschichte des Chemischen Laboratoriums der Bayerischen Akademie der Wissenschaften in München, pp.9-16. Munich 1952

GEHRENBECK, CLEMENS [1859-1909]
JV 4,64; RSC 14,804; 15,246; NUC 193,516
Anon., Chem. Z. 33 (1909), 548

GEHRING, ALFRED [1892-1972]
KURSCHNER 11,802; WAGENITZ, 63; POGG 7a(2),174-175

GEHRKE, MAX [1893-]
Jahrbuch der Dissertationen der Philosophischen Fakultät...Berlin 1920-1921, p.374

GEIDUSCHEK, ERNEST PETER [1928-]
AMS 15(3),69; WWA 1980-1981, p.1222; BHDE 2,362; WWWS, 639-640

GEIGEL, RICHARD [1859-1930]
FISCHER 1,488-489; GEISSLER, 10-21
J. Müller, Münch. med. Wchschr. 78 (1931), 624-627

GEIGER, ERNEST [1896-1959]
AMS 9(II),394; NCAB 47,255-256; IB, 95

GEIGER, GUSTAV [1819-1900]
RSC 2,812
Anon., Pharm. Z. 45 (1900), 253
A. Wankmüller, Beitr. Württ. Apoth. Gesch. 7 (1966), 84; 13 (1981),43-45

GEIGER, HANS [1882-1945]
DSB 5,330-333; NDB 6,141-142; KIEL, 167; TLK 3,182; W, 211; WWWS, 640; POGG 5,417-418; 6,863-864; 7a(2),176-177
E. Stühlinger, Z. Naturforsch. 1 (1946), 50-52
M. von Laue, Jahrb. Akad. Wiss. Berlin 1946-1949, pp.150-159
O. Haxel, Phys. Bl. 38 (1982), 296-297
E. Schwinne, Hans Geiger: Spuren aus einem Leben für Physik. Leipzig 1988

GEIGER, PHILIPP LORENZ [1785-1836]
NDB 6,147-148; HEIN, 194-196; HIRSCH 2,709; BLOKH 1,236; DRULL, 80-81; FERCHL, 177; TSCHIRCH, 1052; PHILIPPE, 806; POTSCH, 165; WWWS, 640; POGG 1,867-868; CALLISEN 7,109-117; 28,166-171
J.H. Dierbach, Arch. Pharm. 57 (1836), 1-17
U. Thomas, Die Pharmazie im Spannungsfeld der Neuorientierung: Philipp Lorenz Geiger. Stuttgart 1985; Ambix 35 (1988), 77-90

GEIGER, ROLF [1923-1988]
KURSCHNER 15,1252
Anon., Chem. Z. 112 (1988), 389

GEIGER, WALTER [1866-]
JV 23,23; NUC 193,586

GEIGY, RUDOLF [1862-1943]
RSC 15,247; NUC 193,594
Schweizerisches Zeitgenossen Lexikon 1932, p.315
Anon., Z. angew. Chem. 56 (1943), 200

GEIGY, RUDOLF [1902-]
KURSCHNER 14,1149
Anon., Acta Tropica 29 (1972), 507-513
H. Schumacher, Zeitschrift für Tropenmedizin 23 (1972), 341

GEIKIE, ARCHIBALD [1835-1924]
DSB 5,333-338; DNB [1922-1930], 332-334; W, 211-212; WWWS, 640; POGG 3,501-502; 4,486-487; 5,864-865
A. Geikie, Scottish Reminiscences. Glasgow 1904
J.R.B. and J.H., Proc. Roy. Soc. B99 (1926), i-xvi
D.R. Oldroyd, Ann. Sci. 37 (1980), 441-462

GEILING, EUGENE MAXIMILIAN KARL [1891-1971]
AMS 11,1750; PARASCANDOLA, 31-32; CHEN, 49-50; IB, 95
F.O. Kelsey and H.C. Hodge, Pharmacologist 14 (1972), 24-26
I. Starr, Trans. Assoc. Am. Phys. 86 (1973), 17-18

GEIS, WILHELM THEODOR [1883-1952]
JV 24,337

GEISSLER, EWALD ALBERT [1848-1898]
BJN 5,22*; HEIN, 196-197; TSCHIRCH, 1052
A. Schneider, Pharm. Zent. 39 (1898), 770-772
H.D. Schwarz, Pharm. Z. 118 (1973), 602

GEISSLER, HEINRICH [1814-1879]
DSB 5,340-341; NDB 6,159; FERCHL, 177; SCHAEDLER, 40; WWWS, 640; POGG 3,503-504
A.W. Hofmann, Ber. chem. Ges. 12 (1879), 147-148
K. Eichhorn, Bulletin of the Scientific Instrument Society 27 (1990), 17-19

GEISSMAN, THEODORE ALBERT [1908-1978]
AMS 12,2085
E. Leete, Phytochemistry 18 (1979), 1259-1261

GEITEL, HANS [1855-1923]
DSB 5,341-342; NDB 6,164; DBJ 5,111-116,427; BLOKH 1,236-239; WWWS, 641; POGG 4,489; 5,419
R. Pohl, Naturwiss. 12 (1924), 685-688

GÉLIS, AMÉDÉE [1815-1882]
DBF 15,967; GE 18,697-698; GORIS,391-392; FERCHL, 177; POGG 1,870; 3,504

GELLERT, MARTIN FRANK [1929-]
AMS 15(3),74

GELLHORN, ERNST [1893-1973]
AMS 11,1762; FISCHER 1,489; IB, 96; WWAH, 218; WWWS, 641-642; POGG 6,687-688; 7b,1567-1572

GEMILIAN [see HEMILIAN]

GEMMILL, CHALMERS LAUGHLIN [1901-1983]
AMS 11,1753; WWWS, 642

GENEVOIS, LOUIS [1900-]
Who's Who in Europe 1972, p.1155

GENGOU, OCTAVE [1875-1957]
BNB 39,379-390

P. Giroud, Bull. Acad. Med. 141 (1957), 485-486
A.R.P., Ann. Inst. Pasteur 95 (1958), 732-733
M. Millet, Mem. Acad. Med. Belg. 7 (1969), 79-88

GENHART, HEINRICH [1853-1915]
O., Corr. Schw. Aerzte 14 (1915), 1588

GENSSLER, OTTO [1881-]
JV 19,338; NUC 194,638

GENTH, FREDERICK AUGUSTUS (FRIEDRICH AUGUST) [1820-1893]
DSB 5,349-350; DAB 7,209-210; MILES, 168-169; MEINEL, 503; GUNDLACH, 468 SILLIMAN, 98-100; ELLIOTT, 100; TSCHIRCH, 1052; FERCHL, 178; WWWS, 643; POGG 3,506-507; 4,490
Anon., Am. J. Sci. [3]45 (1893), 257-258
G.F. Barker, Proc. Am. Phil. Soc. 40 (1901), x-xvii; Biog. Mem. Nat. Acad. Sci. 4 (1902), 201-231
C.A. Browne, J. Chem. Ed. 9 (1932), 718-719

GENTH, FREDERICK AUGUSTUS, Jr. [1855-1910]
MILES, 168-169; WWAH, 218
Anon., J. Franklin Inst. 170 (1910), 320

GENTNER, WOLFGANG [1906-1980]
KURSCHNER 13,1073; WWWS, 643; POGG 7a(2),184-185
A. Citron, Phys. Bl. 36 (1980), 358-359
P. Brix, Physik Unserer Zeit 12 (1981), 1-2; Heid. Akad. Jahrb. 1981, pp.48-51
V. Weisskopf, Phys. Bl. 37 (1981), 281-282
Max-Planck Gesellschaft. Wolfgang Gentner. Stuttgart 1981
B. Karlik, Alm. Akad. Wiss. Wien 131 (1982), 313-317

GENTY, MAURICE [1886-1961]
DBF 15,1102-1103
P. Astruc, Prog. Med. 1961(15), pp.327-330

GEOFFROY SAINT-HILAIRE, ÉTIENNE [1772-1844]
DSB 5,355-358; DBF 15,1154-1156; HOEFER 20,54-55; HIRSCH 2,717; POGG 1,875; CALLISEN 7,138-150; 28,150-151
T. Cahn, La Vie et l'Oeuvre d'E. Geoffroy Saint-Hilaire. Paris 1962
T.A. Appel, The Cuvier-Geoffroy Debate. New York 1987

GEOGHEGAN, EDWARD GEORGE [1829-1881]
HIRSCH 2,718; CALLISEN 7,151-153; 28,153

GEORG, ALFRED [1897-1985]
POGG 6,870-871; 7a(2),185-186
Anon., Helv. Chim. Acta 68 (1985), 1079

GEORGE, PHILIP [1920-]
AMS 15(3),82; WWWS. 644

GEORGI, WALTER [1889-1920]
FISCHER 1,491; KALLMORGEN, 275; KOREN, 180

GEPPERT, JULIUS [1856-1937]
FISCHER 1,492; PAGEL, 592-593; GUNDEL, 264-266; DZ, 437; TLK 3,185; IB, 96; BERLIN, 580

GÉRARD, ERNEST [1863-1935]
DBF 15,1212-1213; GORIS, 392-393; DORVEAUX, 92; POGG 4,491-492
F. Morvillez, J. Pharm. Chim. [8]22 (1935), 480-486; Echo Med. Nord 5 (1936), 422-434
A. Goris, Bull. Acad. Med. 114 (1935), 334-335
R. Delaby, Bull. Soc. Chim. [5]3 (1936), 27-29GÉRARD, POL [1886-1961] IB, 96
R. Cordier and A. Dalcq, Arch. Biol. 67 (1956), 395-407

GERARD, RALPH WALDO [1900-1974]
AMS 11,1769; MH 2,180-181; MSE 1,427-428; NCAB 58,472-473; WWWS, 644-645
B. Agranoff, J. Neurochem. 23 (1974), 5
Anon., Physiologist 23 (1980), 3
R.W. Gerard, Persp. Biol. Med. 23 (1980), 527-540

S.S. Kety, Biog. Mem. Nat. Acad. Sci. 53 (1982), 179-210

GERBER, CHARLES [1865-1928]
GORIS, 394-399; TSCHIRCH, 1053
E. Maurin, Bull. Sci. Pharm. 36 (1929), 414-428

GERBER, NIKLAUS [1850-1914]
WWWS, 645; ZURICH-D, 122; RSC 9,988; 15,266
Anon., Chem. Z. 38 (1914); Z. angew. Chem. 27(III) (1914), 164
K. Schneider, Schw. Milchz. 91 (1965), 205-206

GERGELY, JOHN [1919-]
AMS 15(3),87

GERGENS, EMIL [1848-]
RSC 9,988; SG [1]5,382

GERHARDT, CARL [1833-1902]
NDB 6,284-285; BJN 7,87-88,35*; LF 1,116-119; HIRSCH 2,725; Suppl.,321; PAGEL, 593-594; OLPP, 143-144; WREDE, 59
F. Müller, Deutsches Arch. klin. Med. 74 (1902), iii-xxix
E. Grawitz, Berl. klin. Wchschr. 39 (1902), 721-723
F. Martius, Münch. med. Wchschr. 49 (1902), 1581-1583
E. von Leyden, Deutsche med. Wchschr. 28 (1902), 565-566
J. Gerlach, Würzburg Medizin-historische Mitteilungen 4 (1986), 105-134

GERHARDT, CHARLES FRÉDÉRIC [1816-1856]
DSB 5,369-375; DBF 15,1290-1292; DBA 1,1161-1163; BUGGE 2,92-98; BOURQUELOT, 91-92; BLOKH 1,239-240; HUMBERT, 204-205; POTSCH, 167; FERCHL, 181-182; TSCHIRCH, 1053; PHILIPPE, 792; SCHAEDLER, 40-41; BERGER-LEVRAULT, 86; DORVEAUX, 77; COLE, 209-210; POGG 1,881-882
E. Grimaux and C. Gerhardt, Charles Gerhardt. Paris 1900
M. Tiffeneau, Bull. Soc. Chim. 1916 Suppl., pp.13-103; J. Pharm. Chim. [7]14 (1916), 129-135,161-173,202-211,234-235
E.R. Riegel, J. Chem. Ed. 3 (1926), 1105-1109
J. Jacques, Bull. Soc. Chim. 1956, pp.1315-1324
H. Gault, Bull. Soc. Chim. 1956, pp.1533-1540
E. Kahane, Bull. Soc. Chim. 1968, pp.4733-4742
N.W. Fischer, Ambix 20 (1973), 209-233
J. Dickerson, J. Chem. Ed. 62 (1985), 323-325

GERHARDT, DIETRICH [1866-1921]
NDB 6,285; DBJ 3,299; FISCHER 1,493; PAGEL, 1955-1956; GIESE, 572-573; HAGEL, 137-142; GEISSLER, 35-47; PITTROFF, 154-155; WWWS, 645-646
E. Magnus-Alsleben, Münch. med. Wchschr. 68 (1921), 1160-1161
R. Massini, Schw. med. Wchschr. 51 (1921), 1073
R. Wessely et al., Sitz. Phys. Med. Würzburg NF46 (1921), 103-120
F. Jamin, Med. Klin. 17 (1921), 1261
F. Umber, Klin. Wchschr. 58 (1921), 1118
J. Gerlach, Würzburg Medizin-historische Mitteilungen 4 (1986), 105-134

GERISCHER, HEINZ [1919-]
KURSCHNER 13,1082; POGG 7a(2),190
W. Jaenicke, Ber. Bunsen Ges. 88 (1984), 323-325

GERKE, OTTO HEINRICH [1873-1946]
KURSCHNER 4,823; HEIN, 200-201; TSCHIRCH, 1053; TLK 3,196

GERLACH, FERDINAND [1885-]
JV 26,414; NUC 196,44

GERLACH, JOSEPH [1820-1896]
ADB 49,303-307; BJN 1,152; 3,124*; HIRSCH 2,726; FREUND 2,97-100; PAGEL, 594-595; SCHWARTZ, 15-20; PITTROFF, 7-8
C. von Voit, Sitz. Bayer. Akad. 27 (1897), 433-436
H. Adami, Anat. Anz. 135 (1974), 277-287

GERLACH, LEO [1851-1918]
DBJ 2,688; FISCHER 1,494; PAGEL, 595; SCHWARTZ, 21-25; DZ, 440-441
Anon., Münch. med. Wchschr. 65 (1918), 1256

GERLACH, MAX [1861-1940]
NDB 6,302; IB, 97; WWWS, 646

GERLAND, BALTHASAR WILLIAM [1831-1915]
POGG 3,508; 4,494
Anon., Chem. Z. 40 (1916), 31

GERMAN, LUDWIG [1855-]
RSC 9,881; NUC 196,104

GERNEZ, DÉSIRÉ JEAN BAPTISTE [1834-1910]
DBF 15,1344-1345; WWWS, 647; POGG 3,508-509;; 4,494-495; 5,421
C. Moureu, Bull. Soc. Chim. [4]7 (1910), 1066; 9 (1911), i-viii
Anon., Chem. Z. 34 (1910), 1205

GERNGROSS, OTTO [1882-1966]
KURSCHNER 10,663-664; BHDE 2,367; WIDMANN, 189-190,262,263; LDGS, 19; TLK 3,187; POGG 6,876-877; 7a(2),193-194
Anon., Chem. Z. 76 (1952), 118; 90 (1966), 192
B. Lepsius, Chem. Z. 86 (1962), 149

GERNSHEIM, ALFRED [1870- 1931]
JV 9,208; RSC 13,910; 14,804; NUC 197,234

GERÖ, ALEXANDER [1907-]
AMS 15(3),164; LDGS, 19

GERRARD, ALFRED WILLIAM [1844-1920]
REBER, 67-68; TSCHIRCH, 1053
J. Humphrey, Pharm. J. 123 (1929), 506-508; Chemist and Druggist 111 (1929), 648-649

GERSH, ISIDORE [1907-1980]
AMS 14,1700-1701; WWWS, 647; KOREN, 180
Anon., Science 212 (1980), 37

GERSHENFIELD, LOUIS [1895-1979]
AMS 12,2105; WININGER 7,219; KOREN, 180; IB, 97
Who Was Who in America 7,219

GERSHON, RICHARD K. [1932-1983]
AMS 14,1701; BIBEL, 138-143

GESCHWIND, IRVING [1923-1978]
AMS 13,1489
H. Papkoff and H.A. Bern, Endocrinology 104 (1979), 276-277

GESCHWIND, NORMAN [1926-1984]
AMS 15(3),95
A.M. Galaburda, Neuropsychologia 23 (1985), 297-304

GESELL, ROBERT [1886-1954]
AMS 8,885; IB, 97; WWAH, 219
Anon., Univ. Mich. Med. Bull. 20 (1954), 130-131

GESERICK, ARTHUR [1860-]
JV 23,23; NUC 197,566

GESSARD, CARLE [1850-1925]
DBF 15,1402-1403; FISCHER 1,496
A. Balland, J. Pharm. Chim. [7]2 (1925), 509
Anon., Bull. Inst. Pasteur 23 (1925), 969
S. Lamb, Figures Pharmaceutiques Françaises, pp.185-190. Paris 1953
E. Bossard and J. Nauroy, Hist. Sci. Med. 10 (1976), 72-79
J. Nauroy and E. Bossard, Rev. Hist. Pharm. 23 (1976), 175-180

GESSARD, LOUIS MARIE [1771-1855]
E. Bossard and J. Nauroy, Rev. Hist. Pharm. 26 (1979), 235-239

GESSNER, OTTO [1895-1968]
KURSCHNER 10,668; TLK 3,188; IB, 97; POGG 7a(2),196-197

GETMAN, FREDERICK HUTTON [1877-1941]
AMS 5,407-408; NCAB 30,339-340; WWAH, 220; WWWS, 648; POGG 5,422; 6,877-878; 7b,1589
Anon., Ind. Eng. Chem. News Ed. 19 (1941), 1474
E.C. Bingham, Science 95 (1942), 36-37

GEUNS, JOHN WATERLOO [1874-]
RSC 15,281; NUC 198,3

GEUTHER, ANTON [1833-1889]
DSB 5,380-381; NDB 6,353-354; BLOKH 1,242; SCHAEDLER, 41; WWWS, 649; POTSCH, 167-168; POGG 3,510-511; 4,495
R. Hübner, Pharm. Z. 33 (1888), 247-248
Anon., Chem. Z. 13 (1889), 1726
C. Duisberg and K. Hess, Ber. chem. Ges. 63A (1930), 145-157

GEUTHER, THEODOR [1865-]
JV 8,149; TLK 3,188-189; RSC 16,344; NUC 198,11

GEVEKOHT, HEINRICH [1858-]
RSC 9,996; NUC 198,18

GEY, GEORGE OTTO [1899-1970]
AMS 11,1781
S. Federoff, Anat. Rec. 7 (1971), 127-128
H.W. Jones et al., Obs. Gyn. 38 (1971), 945-949
A.M. Harvey, Adventures in Medical Research, pp.121-123. Baltimore 1976

GEYGER, ADOLF [1835-1887]
BLOKH 1,235-236; TSCHIRCH, 1052; SCHAEDLER, 39
A.W. Hofmann, Ber. chem. Ges. 20 (1887), 3025-3032
Anon., Leopoldina 23 (1887), 217

GHON, ANTON [1866-1936]
NDB 6,366-367; OBL 1,437; STURM 1,435; FISCHER 1,497; SACKMANN, 133-136; SCHWARZ, 20-38; KOERTING, 147-149; WWWS, 649
R. Wiesner, Wiener klin. Wchschr. 49 (1936), 604-606
R. Maresch, Wiener med. Wchschr. 86 (1936), 61
F. Hochstetter, Alm. Akad. Wiss. Wien 86 (1936), 257-259
H. Beitzke, Deutsche med. Wchschr. 62 (1936), 907-908
K. Terplan, Verhandl. path. Ges. 25 (1937), 355-367

GHOSH, BHUPENDRA NATH [1900-]
POGG 6,880; 7b,1597-1599

GHOSH, JNAN CHANDRA [1894-1959]
DNBI Suppl.2,55-56; POGG 6,880-881; 7b,1599-1603
N.R. Dhar, Nature 183 (1959), 645-646
P.C. Rakshit, J. Ind. Chem. Soc. 36 (1959), 289-293
J.N. Mukerjee, Biog. Mem. Fell. Nat. Inst. India 1 (1966), 32-43

GIACOSA, PIERO [1853-1928]
FISCHER 1,497; PAGEL, 597-598; REBER, 269-272; TSCHIRCH, 1053; IB, 97
G. Montalenti, Archeion 11 (1929), 227-228

GIANNUZZI, GIUSEPPE [1839-1876]
EI 16,969
V. Busacchi, Altamura 17-18 (1975-1976), 57-58

GIARD, ALFRED MATHIEU [1846-1908]
DSB 5,385-386; DBF 15,1437-1439; CHARLE(2), 136-137; WWWS, 650
F. LeDantec and M. Caullery, Bull. Sci. Fr. Belg. 42 (1909), i-lxxiii
G. Bohn, Alfred Giard et son Oeuvre. Paris 1910
G. Gohau, Revue de Synthèse [3]100 (1979), 393-406

GIAUQUE, WILLIAM FRANCIS [1895-1982]
DSB 17,337-344; AMS 14,1710-1711; WWA 1978-1979, p.1190; WWWS, 658; IWW 1982-1983, p.457; MH 1,191-193; STC 1,435-437; MSE 1,430-431; CB 1950, pp.170-171; AO 1982, pp.143-145; POTSCH, 168; POGG 6,883; 7b,1609-1613
Anon., Chem. Eng. News 60(24) (1982), 94
A.N. Stranges, J. Chem. Ed. 67 (1990), 187-193

GIBBS, FREDERICK WILLIAM [1913-1966]
R.E. Parker and D. McKie, Chem. Brit. 2 (1966), 285-286

GIBBS, JOSIAH WILLARD [1839-1903]
DSB 5,386-393; DAB 7,248-251; MILES, 170-171; ELLIOTT, 102; W, 214; POTSCH, 168; BLOKH 1,242-243; ABBOTT-C, 50-51; POGG 3,513-514; 4,496,1708; 5,424; 6,884
J. L[armor], Proc. Roy. Soc. 75 (1905), 280-296
P. Duhem, Rev. Quest. Sci. 63 (1908), 5-43
C.S. Hastings, Biog. Mem. Nat. Acad. Sci. 6 (1909), 375-393
F. Donnan, J. Franklin Inst. 199 (1925), 457-484
J. Johnston, J. Chem. Ed. 5 (1928), 507-514
E.B. Wilson, Sci. Mon. 32 (1931), 211-227
C.A. Kraus, Science 89 (1939), 275-282
L.P. Wheeler, Josiah Willard Gibbs (rev. ed.). New Haven 1952
M.J. Klein, Physics Today 43(9) (1990), 40-48

GIBBS, OLIVER WOLCOTT [1822-1908]
MILES, 171-172; SILLIMAN, 88-92; POTSCH, 168-169; POGG 1,892; 3,513; 4,496; 5,424
C.L. Jackson, Am. J. Sci. [4]27 (1909), 253-259
E.W. Morley, Proc. Am. Phil. Soc. 49 (1910), xix-xxxii
F.W. Clarke, Biog. Mem. Nat. Acad. Sci. 7 (1910), 3-22

GIBSON, ALEXANDER GEORGE [1875-1950]
WW 1949, p.1037; FISCHER, 498; WWWS, 651-652
Anon., Lancet 1950(I), pp.140-141; Brit. Med. J. 1950(I), p.193
M.A.G. Campbell, British Heart Journal 13 (1951), 255-257

GIBSON, CHARLES STANLEY [1884-1950]
WW 1950, p.1047; WWWS, 652; POGG 6,885; 7b,1614-1615
F.H. Brain, Chem. Ind. 1950, pp.635-636
J.L. Simonsen, Obit. Not. Fell. Roy. Soc. 7 (1951), 115-137; J. Chem. Soc. 1951, pp.628-637

GIBSON, DAVID TEMPLETON [1899-]
SRS,57; POGG 6,885; 7b,1615

GIBSON, JOHN [1855-1914]
POGG 5,424
A.P. Laurie, Proc. Roy. Soc. Edin. 34 (1914), 285-289

GIBSON, QUENTIN HOWIESON [1918-]
AMS 15(3),112; IWW 1982-1983, p.458

GIBSON, ROBERT BANKS [1882-1959]
AMS 9(II),401; NCAB 50,579-580

GICKLHORN, JOSEF [1891-1957]
STURM 1,436; WIEN 1959,pp.50-53; WEINMANN,165; POGG 6,887; 7a(2),198-199
H. Wyklicky, Wiener klin. Wchschr. 70 (1958), 355-356
H. Rohrich, Schriften des Vereins zur Verbreitung Naturwissenschaftlichen Kenntnisse in Wien 108 (1968), 1-24

GIEBE, GEORG [1874-1899]
JV 11,19; RSC 15,295
Anon., Ber. chem. Ges. 32 (1899), 1973

GIEMSA, GUSTAV [1867-1948]
NDB 6,371-372; FISCHER 1,499; HEIN, 202-203; OLPP, 144-145; IB, 98;
WWWS, 653; POGG 6,887; 7a(2),199-201
E.G. Nauck, Deutsche med. Wchschr. 73 (1948), 56
H. Kirchmair, Wiener klin. Wchschr. 60 (1948), 520-521
Anon., Z. angew. Chem. 60 (1948), 260

GIERKE, EDGAR von [1877-1945]
NDB 6,373; FISCHER 1,499; IB, 98; WWWS, 1734
R. Böhmig, Verhandl. path. Ges. 34Suppl. (1950), 17-19
Anon., Aus der Chronik der Aerzteschaft Karlsruhe 1715-1977, pp.123-125. Karlsruhe 1978

GIERTZ, KNUT HARALD [1876-1950]
VAD 1949, p.329; SMK 3,62; FISCHER 1,500

GIES, WILLIAM JOHN [1872-1956]
AMS 9(I),683; MILES, 173-174; CHITTENDEN, 132-133; WWWS, 653
P.E. Bomberger et al., Pennsylvania Dental Journal 18 (1951), 272-283
H.T. Clarke, Am. Phil. Soc. Year Book 1956, pp.111-115

GIESE, ARTHUR CHARLES [1904-1994]
AMS 14,1718; WWWS, 653GIESE, GERTRUD [1888-1973]
JV 37,737

GIESE, JOHANN EMMANUEL FERDINAND [1781-1821]
HEIN, 203-204; LEVITSKI 1,241-242; TSCHIRCH, 1053; FERCHL, 183-
184; COLE, 211; POGG 1,893-894

GIESE, MAX [1869-1915]
JV 12,169; NUC 199,172
Anon., Chem. Z. 39 (1915), 786

GIESE, OSKAR [1873-1917]
JV 19,596
Anon., Ber. chem. Ges. 50 (1917), 666

GIESEL, FRIEDRICH OSKAR [1852-1927]
DSB 5,394-395' NDB 6,387; WWWS, 653; POTSCH, 169; POGG 4,496-497; 5,425;6,887
O. Hahn, Phys. Z. 29 (1928), 353-357
S. Meyer, Naturwiss. 36 (1949), 129-132
H.W. Kirby, Isis 62 (1971), 290-308

GIESER, KARL [1881-]
JV 21,274

GIGLI, TORQUATO [1845-1936]
REBER, 296-297; TSCHIRCH, 1053; SCHELENZ, 701; POGG 4,497; 5,425; 6,888; 7b,1618

GIGON, ALFRED [1883-1975]
FISCHER 1,501; IB, 98
C. Hedinger et al., Schw. med. Wchschr. 93 (1963), 1345-1348
O. Gsell, Schw. med. Wchschr. 105 (1975), 1328-1329
R. Geigy, Acta Tropica 33 (1976), 1-2

GILBERT, AUGUSTIN NICOLAS [1858-1927]
FISCHER 1,501; FRANKEN, 183
P.A.A. Nobécourt, Bull. Acad. Med. 97 (1927), 332-338
E. Chabrol, Bull. Acad. Med. 140 (1956), 632-642

GILBERT, JOSEPH HENRY [1817-1901]
DSB 8,92-93; DNB [1901-1911]2, 106; DESMOND, 251; FERCHL, 184; WWWS, 654; POGG 3,515; 4,498
J.A.V., J. Chem. Soc. 81 (1902), 625-628
Anon., Nature 65 (1902), 205-206; Proc. Linn. Soc. 114 (1902), 34-35
R.W., Proc. Roy. Soc. 75 (1905), 236-245
J. Russell, Nature 111 (1923), 466-470
F.C. Bawden, J. Nutriiton 90 (1966), 3-12

GILBERT, RALPH DAVIS [1878-1919]
Obituary Record of Yale Graduates, No.78, pp.1261-1262. New Haven 1920

GILBERT, WALTER [1932-]
AMS 15(3),121; WWA 1980-1981, pp.1246-1247; IWW 1982-1983, p.460; ABBOTT, 51-52; POTSCH, 169

GILBODY, ALEXANDER WILLIAM [1870-]
JV 9,208; RSC 14,804; 15,304

GILDEMEISTER, EDUARD [1860-1938]
NDB 6,393; HEIN-E, 150; TSCHIRCH, 1053-1054; TLK 3,190; POTSCH, 169; POGG 6,889; 7a(2),204

H. Weinhaus, Z. angew. Chem. 43 (1930), 301-302
F. Heusler, Ber. chem. Ges. 71A (1938), 164-165

GILDEMEISTER, EUGEN [1878-1945]
FISCHER 1,502; IB, 98

GILDEMEISTER, MARTIN [1876-1943]
NDB 6,394-395; FISCHER 1,502; KROLLMANN, 922; IB, 98; WWWS, 655; POGG 6,889; 7a(2),204
A. Bethe, Pflügers Arch. 247 (1944), 618-622
K. Thomas, Sitz. Akad. Wiss. Leipzig 96 (1944), 61-66
M. Monje, Erg. Physiol. 46 (1950), 22-30

GILKINET, ALFRED [1845-1926]
NUC 199,692
F. Schoofs and S. Leclercq, Liber Memorialis Université de Liège 1867-1935, vol.3, pp.75-79. Liège 1936

GILLESPIE, LOUIS JOHN [1886-1941]
AMS 6,523; NCAB 30,286; CB 1941, p.32; POGG 6,690; 7b,1620
J.A. Beattie, Nucleus 18 (1941), 205
G. Scatchard, Proc. Am. Acad. Arts Sci. 75 (1944), 164-165

GILLIS, JAN BAPTISTA [1893-1978]
NBW 9,239-304; GHENT 2,170-177; WWWS, 656; POGG 6,891-892; 7b,1623-1627
R. Kuhn, Nachr. Chem. Techn. 13 (1965), 384-385

GILMAN, ALFRED [1908-1984]
AMS 14,1732; WWA 1980-1981, p.1253; IWW 1982-1983, p.461; CHEN, 84-85; WWWS, 657

GILMAN, HENRY [1893-1986]
AMS 14,1733; IWW 1982-1983, pp.461-462; ABBOTT-C, 51; WWWS, 657; POGG 6,892-894; 7b,1627-1638
C. Eaborn, Biog. Mem. Fell. Roy. Soc. 36 (1990), 33-71

GILSON, EUGÈNE [1852-1908]
JV 6,264
H. Leboucq, Liber Memorialis Université de Gand, vol.2, pp.575-577. Ghent 1913GILVARG, CHARLES [1925-]
AMS 15(3),136; WWWS, 657

GINDROZ, THÉOPHILE [1813-1872]
STELLING-MICHAUD 3,466

GINS, HEINRICH [1883-1968]
FISCHER 1,503-504
Anon., Gior. Batt. Immun. 14 (1937), 667-668

GINSBERG, WILHELM [1880-]
KALLMORGEN, 277

GINSBURG, DAVID [1920-1988]
M.B. Rabin, Israel J. Chem. 29:127-129

GINTL, WILHELM FRIEDRICH [1843-1908]
NDB 6,404-405; OBL 1,442-443; STURM 1,437-438; BJN 13,32*; HEIN, 205-206; BLOKH 1,245-246; PRAG, 379; BIRK, 120; SCHAEDLER, 41-42; WWWS, 658; MATSCHOSS, 90; POGG 3,518; 4,499; 5,425
H. von Jüptner, Chem. Z. 32 (1908), 253-254

GINZBERG, ALEKSANDR SEMENOVICH [1870-1937]
HEIN, 206; TSCHIRCH, 1054
A.M. Khaletsky, Zhur. Ob. Khim. 7 (1937), 2874-2878

GIRARD, AIMÉ [1830-1898]
DBF 16,136-137; WWWS, 659; POGG 3,520; 4,500
L. Lindet, Bull. Soc. Chim. [3]19 (1898), i-xxvi

GIRARD, CHARLES [1837-1918]
DBF 16,143-144; POTSCH, 169-170; POGG 3,520-521; 4,500; 5,426
C. Poulenc, Bull. Soc. Chim. [4]23 (1918), 214-215

GIRARD, PIERRE [1880-1952]
DBF 16,171; IB, 99; POGG 6,897-898; 7b,1643-1644
J. Duclaux, Hommage a Pierre Girard. Paris 1959

GIRARDIN, JEAN PIERRE LOUIS [1803-1884]
DBF 16,203-204; GORIS, 401-403; TSCHIRCH, 1054; FERCHL, 185; POGG 1,905; 3,521
Anon., J. Pharm. Chim. [5]10 (1884), 74-75
L. Figuier, Ann. Scient. 28 (1884), 549

GIRNDT, OTTO [1895-1948]
NB, 122; IB, 99; POGG 7a(2),207
O. Riesser, Arch. exp. Path. Pharm. 208 (1949), 56

GIROD, PAUL ÉMILE [1856-1911]
RSC 9,1015; 15,323-324; NUC 201,274-275

GITHENS, THOMAS STOTESBURY [1878-1966]
AMS 11,1815
Anon., J. Am. Med. Assn. 197 (1966), 832

GIULINI, LORENZ [1824-1898]
NDB 6,419-420

GIULINI, WILHELM [1886-1932]
NDB 6,418; JV 27,396; NUC 201,424
Anon., Chem. Z. 56 (1932), 327

GLADSTONE, JOHN HALL [1827-1902]
DSB 5,412-413; DNB [1901-1911]2, 116-117; BLOKH 1,247; WWWS, 660; POTSCH, 170; POGG 3,523-524; 4,501-502
W.A. Tilden, J. Chem. Soc. 87 (1905), 591-597
T.E.T., Proc. Roy. Soc. 75 (1905), 188-192

GLÄSEL, HANS [1883-]
JV 25,388; NUC 201,576

GLAESSNER, KARL (CHARLES) [1876-1944]
STURM 1,441; FISCHER 1,505; STANGL, 49-63; KOREN, 181
B. Kisch, Exp. Med. Surg. 2 (1944), 164-174; Proc. Virchow Med. Soc 3 (1944), 32-34

GLASER, FELIX [1874-1931]
KURSCHNER 4,846; FISCHER 1,181; WININGER 7,10-11

GLASER, FRITZ [1868-1927]
NB, 123

GLASER, KARL [1841-1935]
NDB 6,431; WWWS, 660; POTSCH, 170; POGG 3,526; 4,503; 6,901; 7a(2),210
R. Anschütz and C, Miller, Z. angew. Chem. 40 (1927), 273-281
M. Kunz, Ber. chem. Ges. 68A (1935), 166-168

GLASER, LUIS [1932-]
AMS 15(3),148; WWA 1980-1981, p.1262; BHDE 2,379-380

GLASER, OTTO CHARLES [1880-1951]
AMS 8,904; IB, 99; WWAH, 224; WWWS, 661
Anon., Science 113 (1951), 260

GLASER, RUDOLF [1872-]
JV 15,131; NUC 201,671

GLASER, RUDOLF WILHELM [1888-1947]
AMS 7,654
W. Trager, Science 107 (1948), 131-132
N.R. Stoll, Experimental Parasitology 33 (1973), 189-196

GLASS, HIRAM BENTLEY [1906-]
AMS 14,1747; WWA 1976-1977, p.1162; IWW 1982-1983, pp.465-466; WWWS, 661

GLÉNARD, ALEXANDRE [1818-1894]
DBF 16,360-361; GUIART, 110; RSC 2,913; 7,788; 10,9
Anon., Ann. Soc. Agr. Lyon 2 (1895), 475-478

GLENNY, ALEXANDER THOMAS [1882-1965]
DNB [1961-1970], 434-435; WW 1965, p.1171
C.L. Oakley, Nature 211 (1966), 1130; Biog. Mem. Fell. Roy. Soc. 12 (1966), 163-180

GLEY, EUGÈNE [1857-1930]
DSB 17,347-348; DBF 16,362-364; FISCHER 1,507; CHARLE, 88-89; WWWS, 663
Anon., Arch. Inter. Pharmacodyn. 38 (1930), vii-xxx
J. Jolly, Presse Med. 1930, pp.1565-1566
P.J. Portier, Bull. Acad. Med. 104 (1930), 392-398
M. Postel, Nouv. Pr. Med. 1 (1972), 1527-1528

GLEY, RICHARD [1875-1909]
BLOKH 1,248
Anon., Ber. chem. Ges. 42 (1909), 3559-3560

GLICK, DAVID [1908-]
AMS 14,1753; WWA 1976-1977, p.1166; VEIBEL 2,162; WWWS, 663
D. Glick, J. Histochem. 33 (1985), 720-728

GLIKIN, VLADIMIR [1869-1916]
NUC 202,324
Anon., Chem. Z. 40 (1916), 941; Z. angew. Chem. 29(III), 626

GLIMM, ENGELHARDT [1877-1960]
KURSCHNER 8,672; TLK 3,193; POGG 6,905-906; 7a(2),215; (4),143*
Anon., Nachr. Chem. Techn. 8 (1960), 205

GLINSKI, GRIGORI NIKOLAEVICH [1842-1884]
KAZAN 1,299-300

GLOGAU, HENRIK MORITZ [1821-1877]
ADB 49,397-399

GLOVER, JOHN [1817-1902]
POTSCH, 171
Anon., Pharm. Z. 47 (1902), 395
D.W. Hardie, Chemical Age 78 (1957), 816,825

GLÜCKSMANN, ALFRED [1904-1985]
LDGS, 54

GLUECKSOHN-WAELSCH, SALOME [1907-]
AMS 16(7),366; BHDE 2,383; WWWS, 665

GLUUD, WILHELM [1887-1936].
NDB 6,474-475; TLK 3,194; WWWS, 665; POGG 5,430; 6,908-909; 7a(2),218
G. Schneider, Ber. chem. Ges. 69A (1936), 212
H.R. Hoppe, Bremische Biographien 1912-1962, p.180. Bremen 1969

GMELIN, CHRISTIAN GOTTLOB [1792-1860]
NDB 6,476; ADB 9,266; HIRSCH 2,777; HEIN, 209; BLOKH 1,249; FERCHL, 189; POTSCH, 171-172; SCHAEDLER, 43; POGG 1,917,1759
F.A. Quenstedt, Jahreshefte Wurtt. 17 (1861), 24-40
H. Loewe, Chem. Z. 66 (1942), 476-477
A. Wankmüller, Pharm. Zent. 89 (1950), 8-12

GMELIN, ERWIN [1883-1961]
JV 25,606; NUC 202,625

GMELIN, LEOPOLD [1788-1853]
DSB 5,429-432; NDB 6,480-481; ADB 9,272-273; BAD 1,308-314; W, 217; BLOKH 1,249-250; FERCHL, 190-191; SCHAEDLER, 43; PHILIPPE, 759-760; STUBLER, 255-259; TSCHIRCH, 1054; FRANKEN, 183; DRULL, 85-86; POTSCH, 172; POGG 1,915-916; 7aSuppl.,238-240; CALLISEN 7,251-254; 28,220-221
P. Yorke, J. Chem. Soc. 7 (1855), 144-149
E. Pietsch and E. Beyer, Ber. chem. Ges. 72A (1939), 5-33
E. Pietsch, Chem. Z. 77 (1953), 237-240
P. Walden, J. Chem. Ed. 31 (1954), 534-541
N. Mani, Gesnerus 13 (1956), 190-214
A. Wankmüller, Deutsche Apoth. Z. 112 (1972), 2067-2069
B. Woebke, Chemie Unserer Zeit 22 (1988), 208-216
E. Fluck, Naturw. Rund. 42 (1989), 435-441

GNEHM, ROBERT [1852-1926]
DHB 3,471; TLK 3,194; ZURICH-D, 124-125; POGG 3,527-528; 4,506; 5,431; 6,909
E. Bosshard, Viert. Nat. Ges. Zurich 71 (1926), 305-311

GOBLEY, NICOLAS THÉODORE [1811-1876]
DSB 5,432-433; DBF 16,396-397; GE 18,1141; GORIS, 404-405; FERCHL, 191; POTSCH, 176; BOURQUELOT, 71-72; TSCHIRCH, 1055; POGG 3,528; 4,506
E. Delpech, J. Pharm. Chim. [4]24 (1876), 328-333; Bull. Acad. Med. 5 (1876), 892-896
G.A. Chatin, Bull. Acad. Med. 5 (1876), 947-951
C. Chatagnon and P. Chatagnon, Ann. Med. Psych. 115(II) (1957), 256-275

GOCKEL, HEINRICH [1876-1916]
JV 20,259; NUC 203,46
Anon., Z. angew. Chem. 29(III) (1916), 452

GODA, TOKUSUKE [1901-]
IB, 100GODCHOT, MARCEL [1879-1939]
DBF 16,421; POGG 6,910; 7b,1662-1663
M. Delépine, C.R. Acad. Sci. 208 (1939), 547-549
A. Meyer, Bull. Soc. Chim. [5]14 (1947), 933-944

GODDARD, DAVID ROCKWELL [1908-1985]
AMS 14,1760; WWA 1980-1981, p.1271; IWW 1982-1983, p.469; WWWS, 666
E. Stellar, Am. Phil. Soc. Year Book 1985, pp.126-131

GODET, CHARLES [1883-1951]
DHB 3,475
C. Godet, Mitt. Lebensmitt. 44 (1953), 128-129

GODLEE, RICKMAN JOHN [1849-1925]
DNB [1922-1930], 342-344; WWWS, 666-667
Anon., Brit. Med. J. 1925(I), pp.809-810

GODLEWSKI, EMIL [1875-1944]
PSB 8,173-174; IB, 100
S. Skowron, Pol. Tyg. Lek. 19 (1964), 884-885
S. Skowron and M. Singer, Fol. Biol. 23 (1975), 193-202
W. Lisowski, Lekarz Wojskowy 58 (1982), 431-436

GODNEV, TIKHON NIKOLAEVICH [1893-1982]
BSE (3rd Ed.)7,9; LIPSHITS 2,297-299; IB, 100; WWWS, 667; POGG 6,912; 7b,1674-1680

GODRON, DOMINIQUE ALEXANDRE [1807-1880]
DBF 16,479-480; WWWS, 667; RSC 2,927-928; 7,791-792; 10,13; 12,278
E. Bonnet, Le Naturaliste 1 (1879-1881), 310-311
Anon., Bull. Soc. Bot. 27 (1880), 92-93; Journal of Botany 19 (1881),127

GOEBEL, FRANZ [1864-1953]
JV 3,60; SG [2]6,276; NUC 203,326
Anon., Nachr. Chem. Techn. 1 (1953), 48

GOEBEL, HEINRICH [1818-1893]
NDB 6,503-504; BLOKH 1,250; MATSCHOSS, 91

GOEBEL, JOHANNES [1869-]
JV 13; NUC 203,332

GOEBEL, KARL von [1855-1932]
DSB 5,437-439; NDB 6,504-505; DZ, 455-456; IB, 100; WWWS, 1734
G. Karsten, Ber. bot. Ges. 50 (1932), (131)-(162)
F.O. Bower. Obit. Not. Fell. Roy. Soc. 1 (1933), 103-108
O. Renner, Flora 3 (1936), v-xi; Ber. bot. Ges. 68 (1955), 147-162
K. Napp-Zinn, Ber. bot. Ges. 100 (1987), 327-340

GOEBEL, KARL CHRISTOPH TRAUGOTT FRIEDEMANN [1794-1851]
ADB 9,299; HEIN, 212-213; WELDING, 250; LEVITSKI 1,244-246; FERCHL, 192; TSCHIRCH, 1055; POGG 1,920GOEBEL, WALTER FREDERICK [1899-] AMS 14,1762; WWA 1978-1979, p.1221; IWW 1982-1983, p.469; WWWS, 667

GÖDDERTZ, BERNHARD ALBERT [1888-1916]
JV 26,36
Anon., Chem. Z. 41 (1917); Z. angew. Chem. 30(III) (1917), 46,56

GOES, BRUNO [-1900]
RSC 10,16
Anon., Pharm. Z. 45 (1900), 865

GÖHRING, CARL FRIEDRICH [1857-]
RSC 15,349; NUC 203,383-384
K. Matton, Chem. Z. 51 (1927), 760

GOEPP, RUDOLPH MAXIMILIAN [1907-1946]
AMS 7,659
Anon., Chem. Eng. News 24 (1946), 2676
M.L. Wolfrom, Adv. Carbohydrate Chem. 3 (1948), xv-xxiii

GOEPPERT, HEINRICH ROBERT [1800-1884]
DSB 5,440-442; NDB 6,519; ADB 49,455-460; HEIN, 214-216; HIRSCH 2,784; GRAETZER, 107-113; TSCHIRCH, 1053; FERCHL,192; POGG 1,921-923; 3,529
J. Wortmann, Bot. Zt. 42 (1884), 480-481
F. Cohn, Leopoldina 20 (1884),196-199,211-214; 21 (1885),135-139,149-154
T. Poleck, Arch. Pharm. 223 (1885), 1-20
M. Heidendain and F. Cohn, Jahresb. Schles. Ges. 62 (1885), ii-xxvii
H. Conwentz, Schr. Nat. Ges. Danzig NF6 (1885), 253-285
C. von Voit, Sitz. Bayer. Akad. 15 (1885), 193-201

GOERDELER, JOACHIM [1912-]
KURSCHNER 13,1136; POGG 7a(2),223

GOESSMANN, CHARLES ANTHONY [1827-1910]
DAB 7,354-355; MILES, 175-176; NCAB 11,350; SILLIMAN, 122-125; WWWS, 668; POGG 3,529; 4,433

GOETSCH, EMIL [1883-1963]
AMS 10,1430; FISCHER 1,511; WWWS, 668
Anon., Journal of the International College of Surgeons 40(6) (1963), 42

GÖTTLER, MAXIMILIAN [1882-]
JV 24,572; NUC 204,70

GOHR, HANS [1897-1972]
KURSCHNER 13,1147; POGG 7a(2),232-233
Anon., Nachr. Chem. Techn. 20 (1972), 175

GOLD, HARRY [1899-1972]
AMS 11,1844; PARASCANDOLA, 33-34; WWAH, 226; WWWS, 669; KOREN, 181
D.E. Hutchinson, Journal of Clinical Pharmacology 12 (1972), 303-305
W.F. Riker, Pharmacologist 14 (1972), 104-105

GOLD, VICTOR [1922-1985]
WW 1985, p.733; WWWS, 669-670
W.J. Alberg, Biog. Mem. Fell. Roy. Soc. 33 (1987), 263-288

GOLDBERG, ALFRED [1880-]
JV 19,180; NUC 204,354

GOLDBERGER, JOSEPH [1874-1929]
DSB 5,451-453; DAB 7,363-364; NCAB 21,82-84; BARRY, 34-39; FISCHER 1,512 CHITTENDEN, 383-385; DAMB, 295-296; KAGAN(JA), 392-402; WWWS, 670; WININGER 7,14
W.H. Sebrell, J. Nutrition 55 (1955), 3-12
M.F. Goldberger, J. Am. Diet. Assn. 32 (1958), 724-727
W.B. Bean, Arch. Int. Med. 117 (1966), 319-322
L. Peiner and W.B. Ober, N.Y. J. Med. 69 (1969), 2936-2941,3050-3055

GOLDBLATT, HARRY [1891-1977]
AMS 13,1549; KAGAN(JA), 316-317; COHEN, 83; WWWS, 670
J. Laragh, Trans. Assoc. Am. Phys. 91 (1978), 34-37
E. Haas, Journal of Hypertension 4Suppl. (1986), S21-S25

GOLDENBERG, HERMANN [1849-1909]
BLOKH 1,250
Anon., Ber. chem. Ges. 42 (1909), 683-684

GOLDFINGER, PAUL [1905-1970]
BHDE 2,390; LDGS, 19
J. Drowart, Revue de Chimie Minérale 8 (1971), 385-389
G. Chiltz, Bull. Soc. Chim. Belg. 81 (1972), 3-6

GOLDMANN, EDWIN [1862-1913]
BJN 18,92*; FISCHER 1,513; PAGEL, 608-609; WININGER 2,449; KOREN, 182; WWWS, 672
E. Kreuter, Münch. med. Wchschr. 60 (1913), 2735-2736

GOLDMANN, MAX [1879-1919]
JV 17,13; NUC 204,511
Anon., Ber. Chem. Ges. 53A (1920), 70

GOLDMARK, JOSEF [1819-1881]
J.C. Goldmark, Pilgrims of 1848. New Haven 1930

GOLDNER, MARTIN GERHARD [1902-]
AMS 13,1555; BHDE 2,391; LDGS, 66

GOLDSCHEIDER, ALFRED [1858-1935]
NDB 6,608; KURSCHNER 4,864; FISCHER 1,513; PAGEL, 609-610; KOREN, 182; TETZLAFF, 104; WWWS, 672
F. Umber, Deutsche med. Wchschr. 54 (1928), 1279
G. von Bergmann, Deutsche med. Wchschr. 61 (1935), 1053-1054; Verhandl. path. Ges. 66 (1936), 194-197

GOLDSCHMID, EDGAR [1881-1957]
IB, 100; LDGS, 75; KOREN, 182; POGG 7a(2),234
E.A. Underwood, Brit. Med. J. 1957(I), p.1478
K. Reucker, Schw. med. Wchschr. 87 (1957), 1028

GOLDSCHMIDT, CARL [1867-]
JV 6,238; RSC 13,275; 15,359-360

GOLDSCHMIDT, EDGAR [1881-1957]
ARNSBERG, 158-159; KALLMORGEN, 280; FISCHER 1,513-514; KOREN, 182

GOLDSCHMIDT, FRANZ [1869-]
JV 14,247; NUC 204,583

GOLDSCHMIDT, FRANZ [1877-1926]
Anon., Z. angew. Chem. 39 (1926), 1236

GOLDSCHMIDT, HANS (JOHANNES) [1861-1923]
NDB 6,609; W, 217-218; BLOKH 1,250-251; POTSCH, 172-173; WWWS, 672; WININGER 2,454-455; TETZLAFF, 105; POGG 4,511; 6,917
F. Haber, Ber. chem. Ges. 56A (1923), 77-79
O. Neuss, Z. angew. Chem. 36 (1923), 365-366

GOLDSCHMIDT, HARALD [1857-1923]
DBL 8,200-201; WININGER 7,21

GOLDSCHMIDT, HEINRICH [1857-1937]
STURM 1,454; DRULL, 87; BHDE 2,394; POTSCH, 173; POGG 3,530; 4,510-511; 5,434-435; 6,917; 7a(2),235
M. Trautz, Z. Elektrochem. 36 (1930), 346-348
A. Stock, Ber. chem. Ges. 70A (1937), 149-150

GOLDSCHMIDT, KARL [1857-1926]
NDB 6,609-610; BLOKH 1,251-252
F. Feigl, Ber. chem. Ges. 59A (1926), 3-4

GOLDSCHMIDT, MARTIN [1834-1915]
BLOKH 1,252
Anon., Chem. Z. 39 (1915), 738
H. Wichelhaus, Ber. chem. Ges. 48 (1915), 1311-1312

GOLDSCHMIDT, MAX [1884-1972]
KURSCHNER 4,867; JV 26,710; LDGS, 73
Anon., N.Y. J. Med. 78 (1978), 1968

GOLDSCHMIDT, RICHARD BENEDICT [1878-1958]
DSB 5,453-455; NDB 6,611-612; AMS 9(II),414; STC 1,462-463; WWWS, 672; ARNSBERG, 161-162; BHDE 2,395; IB, 100-101; WININGER 2,462; 7,22; KOREN, 182-183; TETZLAFF, 107; LDGS, 13
R.B. Goldschmidt, Portraits from Memory. Seattle 1956; In and Out of the Ivory Tower. Seattle 1960.
C. Stern, Science 128 (1958), 1069-1070; Naturwiss. 45 (1958), 429-431;
Biog. Mem. Nat. Acad. Sci. 39 (1967), 141-192; Persp. Biol. Med. 12 (1969), 179-203
E. Witschi, Biol. Z. 78 (1959), 209-213
J. Seiler, Bayer. Akad. Wiss. Jahrbuch 1960, pp.153-157
G.E. Allen, J. Hist. Biol. 7 (1974), 49-92
L.K. Piternick (Ed.), Richard Goldschmidt. Basle 1980
E.W. Caspari et al., Experientia 35Suppl. (1980), 1-154

GOLDSCHMIDT, STEFAN [1889-1971]
KURSCHNER 10,697-698; BHDE 2,396; LDGS, 19; POTSCH, 173; JV 27,645; TLK 3,197; TETZLAFF, 107; COHEN, 85; POGG 6,918; 7a(2),235-236
Anon., Nachr. Chem. Techn. 7 (1959), 95
W. Hieber, Bayer. Akad. Wiss. Jahrbuch 1972, pp.294-299

GOLDSCHMIDT, THEODOR [1883-1965]
JV 23,508; NUC 204,611; POGG 7a(2),236
Anon., Chem. Z. 87 (1965), 352

GOLDSCHMIDT, VICTOR [1853-1933]
DSB 5,455-456; NDB 6,612; DRULL, 87-88; TLK 3,197; WININGER 2,463; 7,23; TETZLAFF, 107; WWWS, 672; POGG 3,530; 4,510; 5,435-436; 6,918; 7a(2),236
A.E.H. Tutton, Nature 131 (1933), 791-792
C. Palache, Am. Min. 19 (1934), 106-111

GOLDSCHMIDT, VIKTOR MORITZ [1888-1947]
DSB 5,456-458; NDB 6,618-619; BHDE 2,396; LDGS, 19; WININGER 7,23; POTSCH, 173; TETZLAFF, 107; ABBOTT-C, 52-53; WWWS, 672; POGG 5,435; 6,918-919; 7a(2),236-237
C.W. Correns, Naturwiss. 34 (1947), 129-131
W. Noll, Naturw. Rund. 1 (1948), 78-81
F. Machatschki, Alm. Akad. Wiss. Wien 97 (1948), 325-328
C.E. Tilley, Obit. Not. Fell. Roy. Soc. 6 (1948), 51-66
J.D. Bernal, J. Chem. Soc. 1949, pp.2108-2114

GOLDSCHMIEDT, GUIDO [1850-1915]
NDB 6,619-620; OBL 2,26; STURM 1,454; DBJ 1,328; BLOKH 1,252-253; WININGER 2,465; WWWS, 673; POGG 3,530-531; 4,511-512; 5,436
R. Wegscheider, Chem. Z. 39 (1915), 649-650
J. Herzig, Ber. chem. Ges. 49 (1916), 893-932
E.H. Huntress, Proc. Am. Acad. Arts Sci. 78 (1950), 21-22

GOLDSTEIN, HENRI VICTOR [1897-]
WWWS, 673; POGG 6,919-920; 7a(2),237-238

GOLDSTEIN, MENEK [1924-]
AMS 15(3),201; WWWS, 674

GOLDZIEHER, MAX ALEXANDER [1883-1969]
AMS 11,1862; NUC 205,113-114

GOLGI, CAMILLO [1843-1926]
DSB 5,459-461; EI 17,495; HIRSCH 2,791-792; Suppl.,330; ARCIERI, 165-192; KOLLE 2,3-12; FREUND 2,101-116; PAGEL, 613; BRAZIER, 145-146,150; OLPP, 145-146; W, 218-219
C. daFano, J. Path. Bact. 29 (1926), 500-514
F. Marcora, Riv. Biol. 18 (1927), 116-127
G. Légee, Clio Medica 17 (1982), 15-32

GOLLWITZER-MEIER, KLOTHILDE [1894-1954]
KURSCHNER 8,691; FISCHER 1,515; BOEDEKER, 96-97; IB, 101; POGG 7a(2),238-239
H. Schaefer, Deutsche med. Wchschr. 79 (1954), 1908-1909
Anon., Brit. Med. J. 1954(I), p.850

GOLTZ, FRIEDRICH LEOPOLD [1834-1902]
DSB 5,462-464; NDB 6,636-637; BJN 7,37'; HIRSCH 2,792-793; Suppl.,330; DBA 1,1246-1247; PAGEL,614-615; HAYMAKER, 217-221; BRAZIER, 120-127; WWWS, 675
H. Kraft, Münch. med. Wchschr. 49 (1902), 965-970
A. Bickel, Deutsche med. Wchschr. 28 (1902), 403
J.R. Ewald, Berl. klin. Wchschr. 39 (1902), 479-480; Pflügers Arch. 94 (1903), 1-64

GOMBERG, MOSES [1866-1947]
DSB 5,464-466; DAB [Suppl.4],335-337; MILES, 176-177; KAGAN(JA), 331; POTSCH, 174; POGG 4,512; 5,437; 6,921; 7b,1692
C.S. Schoepfle and W.E. Bachmann, J. Am. Chem. Soc. 69 (1947), 2921-2925
A.J. Ihde, Chem. Eng. News 44(41) (1966), 90-92; Pure and Applied Chemistry 15 (1967), 1-13
J. Bailar, Biog. Mem. Nat. Acad. Sci. 41 (1970), 141-173
J.M. McBride, Tetrahedron 30 (1974), 2009-2022
C. Walling, Welch Foundation Conferences on Chemical Research 20 (1977), 72-84

GOMOLKA, FRANZ [1884-]
JV 21,20; NUC 205,444

GOMORI, GEORGE [1904-1957]
AMS 9(II),415; NCAB 42,552-553; MEL 3,257
R.D. Lillie, Science 125 (1957), 728
R.E. Stowell, Am. J. Clin. Path. 28 (1957), 405-407
E.P. Benditt, Arch. Path. 65 (1958), 580-582
F. Wohlrab, Z. allgem. Path. 136 (1990), 211-217

GONDER, KARL LUDWIG [1880-]
JV 20,466; NUC 205,314

GONDER, RICHARD [1881-1917]
KALLMORGEN, 280
J. Gross, Arch. Protist. 38 (1917), 137-145
Anon., Chem. Z. 41 (1917), 134; Z. angew. Chem. 30(III) (1917), 90
H. Ritz, Ber. Senck. Nat. Ges. 49 (1919), 124-125

GOOCH, FRANK AUSTIN [1852-1929]
MILES, 178; POGG 3,531; 4,513-514; 5,437-438; 6,922
R.G. Van Name, Biog. Mem. Nat. Acad. Sci. 25 (1931), 105-135

GOOD, ROBERT ALAN [1922-]
AMS 16(3),216-217

GOODALL, ALEXANDER [1867-1941]
WW 1941, p.1206
Anon., Lancet 1941(II), P.504; Brit. Med. J. 1941(II), p.595

GOODALL, HARRY WINFRED [1878-1935]
Anon., J. Am. Med. Assn. 104 (1935), 2014

GOODEVE, CHARLES FREDERICK [1904-1980]
WW 1979, p.965; SRS, 62
F.D. Richardson, Biog. Mem. Fell. Roy. Soc. 27 (1981), 307-353

GOODMAN, DeWITT STETTEN [1930-1991]
AMS 17(3), 224; WWWS, 677

GOODMAN, EDWARD HARRIS [1880-1939]
AMS 6,536
Anon., J. Am. Med. Assn. 113 (1939), 76

GOODMAN, LEON [1920-]
AMS 15(3),216; WWA 1980-1981, p.1294; WWWS, 677-678

GOODMAN, LOUIS SANFORD [1906-]
AMS 14,1798; WWA 1980-1981, p.1294; CHEN, 83; KOREN, 183; WWWS, 678

GOODPASTURE, ERNEST WILLIAM [1886-1960]
AMS 10,1456-1457; MH 1,197-199; MSE 1,449-451; FISCHER 1,517-518; DAMB, 298-299; IB, 101; WWAH, 228; WWWS, 678
C.H. Andrewes, Nature 188 (1960), 623-624
J.R. Dawson, Arch. Path. 72 (1960), 126-128
J.B. Youmans, Trans. Assoc. Am. Phys. 74 (1961), 21-24
T. Francis, Am. Phil. Soc. Year Book 1964, pp.111-120
E.R. Long, Biog. Mem. Nat. Acad. Sci. 38 (1965), 111-144
M. Burnet, Persp. Biol. Med. 16 (1973), 333-347

GOODSIR, JOHN [1814-1867]
DSB 5,469-471; DNB 8,137-139; HIRSCH 2,800; Suppl.,331-332; BULLOCH, 367; DESMOND, 257; WWWS, 678
J.H. Balfour, Trans. Bot. Soc. Edin. 9 (1867), 118-127
R.H. Follis, Bull. Hist. Med. 18 (1945), 438-444
E.H. Ackerknecht, Gesnerus 25 (1968), 188-194
L.S. Jacyna, J. Hist. Biol. 16 (1983), 75-99

GOODWIN, TREVOR WALWORTH [1916-]
WW 1981, p.1007; WWWS, 679

GOODWIN, WILLIAM [1873-1953]
SRS, 30; WWWS, 679; NUC 206,574
R.L. Wain, Nature 173 (1954), 149

GOPPELSROEDER, CHRISTOPH FRIEDRICH [1837-1919]
NDB 6,645-646; KOVACSICS, 51-63; WWWS, 679; POTSCH, 174-175; POGG 3,532; 4,514-515; 5,439
W. Ostwald, Koll. Z. 10 (1912), 1-3
Anon., Chem. Z. 43 (1919), 732
F. Fichter, Verhandl. Nat. Ges. Basel 31 (1919-1920), 133-152; Verhandl. Schw. Nat. Ges. 101 (1920), 30-31
D. Kritchevsky, J. Chem. Ed. 36 (1959), 196
S.V. Heines, J. Chem. Ed. 46 (1969), 315-316

GORBACH, GEORG [1901-1970]
KURSCHNER 10,701; WWWS, 679; POGG 7a(2),241-242
Anon., Chem. Z. 94 (1970), 367

GORBOV, ALEKSANDR IVANOVICH [1859-1939]
BSE 12,70-71; ARBUZOV, 132-134; POGG 4,515
A.P. Okatov, Zhur. Prikl. Khim. 12 (1939), 795-801

GORDON, ALBERT SAUL [1910-1992]
AMS 18(3),248; WWWS, 680

GORDON, MERVYN HENRY [1872-1953]
DNB [1951-1960], 422-423; WW 1949, p.1080; IB, 101
L.P. Garrod, Obit. Not. Fell. Roy. Soc. 9 (1954), 153-163

GORDON, MILTON PAUL [1930-]
AMS 15(3),228; WWA 1980-1981, p.1302; WWWS, 680

GORDY, WALTER [1909-1985]
AMS 14,1809; WWWS, 681

GORER, PETER ALFRED ISAAC [1907-1961]
DNB [1961-1970], 443-445; BIBEL, 232-235
Anon., Lancet 1961(I), pp.1120-1121; Brit. Med. J. 1961(I), pp.1467-1468
P.B. Medawar, Biog. Mem. Fell. Roy. Soc. 7 (1961), 95-109
J. Klein et al., Immunogenetics 24 (1986), 331-351

GORGAS, WILLIAM CRAWFORD [1854-1920]
DAB 7,430-432; NCAB 32,4-6; FISCHER 1,518; OLPP, 147-156; W, 220; WWAH, 230; WWWS, 681

GORHAM, JOHN [1783-1829]
DAB 7,433; KELLY, 482-483; MILES, 182; ELLIOTT, 105-106; COLE, 220; SILLIMAN, 38-39; WWAH, 230 J.W., American Journal of Medical Science 4 (1829), 538-539

GORIANINOV, PAVEL FEDOROVICH
[1796-1865] BSE 12,274-275; HIRSCH 2,482-483; KUZNETSOV(1963), 73-77; ZMEEV 1,74-75
E.N. Pavlovsky, Trudy Inst. Ist. Est. 5 (1953), 363-378
Z.S. Katsnelson, Tsitologia 10 (1968), 275-277

GORIANSKI, GRIGORI IVANOVICH [1860-]
RSC 15,387; SG [2]6,373

GORINI, COSTANTINO [1865-1950]
WWWS, 682
R. Ciferri, Enzymologia 15 (1951), 49-56
C. Arnaudi, Annali di Microbiologia 4 (1951), 113-123

GORINI, LUIGI [1903-1976]
AMS 12,2241; NCAB 59,9-10
J. Beckwith and M. Duncan, Nature 265 (1977), 193-194
J. Beckwith and D. Fraenkel, Biog. Mem. Nat. Acad. Sci. 52 (1980),203-221
M. Yarmolinsky, TIBS 13 (1988), 77-78

GORIS, ALBERT [1874-1950]
DBF 16,621-622; GORIS,405-407; TSCHIRCH, 1055; POGG 6,926; 7b,1706-1707
J.M. Javillier, Bull. Acad. Med. 114 (1950), 542-545
M. Janot, Ann. Univ. Paris 21 (1951), 93-96
R. Fabre et al., Ann, Pharm. Fran. 9 (1951), 443-462
A. Goris, Bull. Soc. Bot. (Mem.) 100 (1953), 135-145
G. Dillemain, Produits et Problèmes Pharmaceutiques 28 (1973), 222-223

GORNALL, ALLAN GODFREY [1914-]
AMS 15(3),234; YOUNG, 45-46; WWA 1980-1981, p.1306; WWWS, 682

GORODISSKAYA, HENRIETTA YAKOVLEVNA [1898-1969]
IB, 102
Anon., Vopr. Med. Khim. 16 (1970), 444-445

GORTER, EVERT [1881-1954]
BWN 1,205-206; LINDEBOOM, 695-697; GHENT 2,304-306; POGG 6,926; 7b,1707-1709
W.A. Seeder, Koll. Z. 136 (1954), 99-102

GORTNER, ROSS AIKEN [1885-1942]
DAB [Suppl.3],314-315; MILES, 183-184; IB, 102; WWAH, 230; WWWS, 682; POGG 6,926-929; 7b,1709-1711
L.S. Palmer, Science 96 (1942), 395-397
C.A. Browne, Sci. Mon. 55 (1942), 570-573
S.C. Lind, Biog. Mem. Nat. Acad. Sci. 23 (1944), 149-180
F.D. Mann, Persp. Biol. Med. 20 (1976), 142-144

GORTNER, ROSS AIKEN, Jr. [1912-1988]
AMS 16(3),242; WWWS, 682-683
Anon., Chem. Eng. News 67(3) (1989), 53-54

GORTNER, WILLIS ALWAY [1913-]
AMS 15(3),235; WWA 1980-1981, p.1307; WWWS, 683

GORUP-BESANEZ, EUGEN FRANZ von [1817-1878]
NDB 6,648; ADB 49,465-469; HIRSCH 2,805; BLOKH 1,254-255; TSCHIRCH, 1055; FERCHL, 195-196; KOVACSICS, 51-63; PITTROFF, 180-181; WWWS, 1729; POGG 1,929; 3,535
F. Hoppe-Seyler, Z. physiol. Chem. 2 (1878), 363
A. Hilger, Ber. chem. Ges. 12 (1879), 1029-1038
F. von Kobell, Sitz. Bayer. Akad. 9 (1879), 134-135
H.H. Simmer, J. Clin. Chem. 19 (1981), 497-509

GOSIO, BARTOLOMEO [1863-1944]
OLPP, 156-157; AGRIFOGLIO, 100-105
C. Fermi, Rivista di Malarologia 23 (1944), 70-72
F. Jerace, Rivista di Malarologia 24 (1945), 221-222

GOSTING, LOUIS JOSEPH [1921-1971]
AMS 11,1890; WWWS, 683

GOTCH, FRANCIS [1853-1913]
O'CONNOR(2), 77-79; FISCHER 1,520; WWWS, 683
J.S. Macdonald, Nature 91 (1913), 534-535
Anon., Lancet 1913(II), pp.347-351; Brit. Med. J. 1913(II), pp.153,209

GOTO, MOTONOSUKE [1867-1946]
HAKUSHI 2,70-71; FISCHER 1,520; IB, 102; RSC 15,395

GOTSCH, KARL [1905-1972]
STURM 1,457
S. Sailer, Wiener med. Wchschr. 123 (1973), 178-179

GOTSCHLICH, EMIL [1870-1949]
NDB 6,657-658; FISCHER 1,520-521; WIDMANN, 156; WWWS, 684
P. Uhlenhuth, Z. Bakt. 116 (1930), i-iv
E. Rodenwaldt, Z. Bakt. 155 (1950), 83-84; Sitzungsberichte der Heidelberger Akademie der Wissenschaften 1943-1955, pp.83-85

GOTTLIEB, JOHANN [1815-1875]
OBL 2,37; STURM 1,458; HEIN,220-221; KERNBAUER,26-29,482-484; FERCHL,196; BLOKH 1,255; SCHAEDLER, 43-44; VOGEL, 55-60; POGG 1,931; 3,536
R. Maly, Ber. chem. Ges. 8 (1875), 448-451
J. Stefan, Alm. Akad. Wiss. Wien 25 (1875), 175-177,212-216
T. Morawski, J. prakt. Chem. 120 (1875), 436-449

GOTTLIEB, RUDOLF [1864-1924]
FISCHER 1,521; PAGEL, 617-618; BLOKH 1,255; DZ, 467; TLK 2,215; KOREN, 184; POGG 6,931
W. Straub, Münch. med. Wchschr. 71 (1924), 1757
H.H. Meyer, Arch. exp. Path. Pharm. 105 (1925), i-xv
S. Jansen, Arzneimitt. 14 (1964), 1067-1069

GOTTSCHALK, ALFRED [1894-1973]
KURSCHNER 11,868; BHDE 2,405-406; TETZLAFF, 110; COHEN,90; POGG 6,931-932; 7a(2),245-246
R.W. Zilliken and R. Cerutelli, Behring Inst. Mitt. 55 (1974), xv-xxiii
V.H. Trikojus, Rec. Austral. Acad. Sci. 3 (1975), 53-74
A. Neuberger, Adv. Carbohydrate Chem. 33 (1977), 1-9

GOUGEROT, HENRI EUGÈNE [1881-1955]
DBF 16,709-710; FISCHER 1,522; SACKMANN, 139-141
A. Touraine, Bull. Acad. Med. 129 (1955), 132-135

GOULD, ROBERT GORDON [1909-1978]
AMS 13,1593

GOUY, LOUIS GEORGES [1854-1926]
DSB 5,483-484; DBF 16,847-848; WWWS,686; POGG 4,520-521; 5,441-442; 6,933
E. Picard, C.R. Acad. Sci. 182 (1926), 293-295
H. Gauthier, Rev. Quest. Sci. 90 (1926), 143-148

GOWANS, JAMES LEARMONTH [1924-]
WW 1981, p.1020; MSE 1,453-454

GOWEN, JOHN WHITTEMORE [1893-1967]
AMS 11,1900; WWAH, 231; WWWS, 687
D. Graham, Genetics 60 (1968), s25-s26

GOWERS, WILLIAM RICHARD [1845-1915]
DNB [1912-1921], 221-222; HIRSCH 2,814; Suppl.,335; PAGEL, 619-620; VERSO, 100-101; WWWS, 687
Anon., Nature 95 (1915), 296-297
D.F., Proc. Roy. Soc. B89 (1917), i-iii
M. Critchley, Sir William Gowers. London 1949
J.B. Lyons, Med. Hist. 9 (1965), 260-267

GRAB, WERNER [1903-1965]
KURSCHNER 9,579-580; POGG 7a(2),249
Anon., Nachr. Chem. Techn. 13 (1965), 72
H.D. Cramer, Arzneimitt. 15 (1965), 302-303

GRABAR, PIERRE [1898-1986]
QQ 1981-1982, p.683; IWW 1982-1983, p.484; BIBEL, 290-293
Anon., Lancet 1986(II), p.234
J.E. Courtois, Bull. Acad. Med. 170 (1986), 635-639

GRABFIELD, GUSTAVE PHILIP [1892-1965]
AMS 10,1478; IB, 102
C.L. Derick, Trans. Assoc. Am. Phys. 79 (1966), 47-48

GRABFIELD, JOSEPH PHILIP [1861-1935]
JV 2,192; RSC 14,804; NUC 209,83
New York Times, 15 November 1935

GRABOWSKI, JULIAN [1848-1882]
PSB 8,498; KONOPKA 3,359-360; RSC 7,810; 10,40; 12,287

GRADSTEIN, MARCEL [1908-]
AMS 11,1902; LDGS, 20

GRAEBE, CARL [1841-1927]
DSB 5,488-489; NDB 6,705-706; NB,128; GENEVA 3,36-38; 4,113-117; 5,31-34; SCHAEDLER, 44; TSCHIRCH, 1056; TLK 2,216; W, 221; WWWS, 688; POTSCH, 176; POGG 3,539; 4,522-523; 5,442; 6,936
Anon., Z. angew. Chem. 40 (1927), 217-218
P.Duden and H. Decker, Ber. chem Ges. 61A (1928), 9-46
R. Willstätter, Bayer. Akad. Wiss. Jahrbuch 1929-1930, pp.24-26

GRÄFF, SIEGFRIED [1887-1966]
KURSCHNER 10,708-709; FISCHER 1,525; DRULL, 90; IB, 103
W. Selberg, Verhandl. path. Ges. 52 (1968), 583-591

GRÄNACHER, CHARLES [1895-1975]
TLK 3,203; POGG 6,936-937; 7a(2),252
Anon., Chimia 19 (1965), 169; 30 (1976), 116
H. Batzer, Chimia 29 (1975), 195

GRÄTER, ADOLF [1874-1946]
JV 14,224

GRAETZ, ERICH [1903-]
LDGS, 13; NUC 209,274

GRÄVENITZ, RICHARD von [1890-1918]
JV 30,507; NUC 209,291

GRAF, RODERICH [1906-]
POGG 7a(2),256

GRAFE, ERICH [1881-1958]
FISCHER 1,520; DRULL, 91; GEISSLER, 84-105; BK, 209; IB, 103
H. Reinwein, Münch. med. Wchschr. 100 (1958), 1941-1942; Diabetes 8 (1959), 393
K. Oberdisse, Deutsche med. Wchschr. 84 (1959), 523-524

GRAFE, VIKTOR [1878-1936]
OBL 2,45; STURM 1,462; IB, 103; POGG 6,939

Anon., Ost. Chem. Z. 39 (1936), 180

GRAHAM, CHARLES [1836-1909]
A.C. Chapman, J. Chem. Soc. 97 (1910), 677-680

GRAHAM, EVARTS AMBROSE [1883-1957]
DAB [Suppl.6],245-247; NCAB 42,644-645; DAMB, 300-301; WWAH,232; WWWS,689
P.A. Shaffer, Am. Phil. Soc. Year Book 1957, pp.121-126
R.C. Brock, Annals of Thoracic Surgery 9 (1970), 272-279
P.D. Olch, J. Hist. Med. 27 (1972), 247-261
L.R. Dragstedt, Biog. Mem. Nat. Acad. Sci. 48 (1976), 221-250

GRAHAM, GEORGE [1882-1971]
O'CONNOR(2), 271; NUC 209,442
Who Was Who 1971-1980, p.310
Anon., Brit. Med. J. 1971(IV), p.563

GRAHAM, THOMAS [1805-1869]
DSB 5,492-495; DNB 8,361-363; BUGGE 2,69-77; FARBER, 553-571; WWWS, 690; BLOKH 1,256-258; FERCHL, 197; ABBOTT-C, 53-54; SCHAEDLER, 44-45; W, 221-222; DESMOND, 262; TSCHIRCH, 1056; COLE, 220-222; POTSCH, 176-177; POGG 1,936-937; 3,540
A.W. Hofmann, Ber. chem. Ges. 2 (1869), 753-780; Proc. Roy. Soc. 18 (1870), xvii-xxvi
J.P. Cooke, Am. J. Sci. [3]1 (1871), 115-123
R.A. Smith, Life and Work of Thomas Graham. Glasgow 1884
R.A. Gortner, J. Chem. Ed. 11 (1934), 279-283
S.G. Mokrushin, Nature 195 (1962), 861
R.H. Cragg, Chem. Brit. 5 (1967), 567-569
E.A. Mason and B. Kronstadt, J. Chem. Ed. 44 (1967), 740-744
E.A. Mason, Phil. J. 7 (1970), 99-115
E. Frame, Phil. J. 7 (1970), 116-127
A.C. Munro, Phil. J. 9 (1972), 30-42
M. Stanley, Chem. Brit. 27 (1991), 239-242
R.J.H. Clark, Chemical Society Reviews 20 (1991), 405-424

GRAINGER, RICHARD DUGARD [1801-1865]
DNB 8,370-371; HIRSCH 2,827-828; DECHAMBRE [4]10,283-284; WWWS, 690; CALLISEN 7,358; 28,257
Anon., Lancet 1865(I), p.190; Brit. Med. J. 1865(I), p.176

GRALÉN, NILS [1912-]
VAD 1981, pp.352-353; NUC 209,607

GRAM, HANS CHRISTIAN JOACHIM [1853-1938]
DSB 5,495-496; DBL 8,251-252; FISCHER 1,527; PAGEL, 626; VEIBEL 2,166; BULLOCH, 369; W, 222-223; SKVARC, 75-77; IB, 103; WWWS, 690
C. Sonne, Acta Med. Scand. 98 (1939), 441-443
R. Scherrer, TIBS 9 (1984), 242-245

GRAM, JENS BILLE [1857-1934]
DBL (3rd Ed.)5,263-264; VEIBEL 2, 166; TSCHIRCH, 1056; SCHELENZ, 704
E.H. Madson, Arch. Pharm. Chem. 41 (1934), 739-742

GRAMENETSKI, MIKHAIL IVANOVICH [1882-]
NUC 209,625

GRAMMLING, FRANZ [1881-]
JV 25,606; NUC 209,647GRANDEL, GOTTFRIED [1877-1950]
JV 16,143; NUC 210,132

GRANGER, JAMES DARNELL [1872-1942]
SRS,15; JV 12,19; NUC 210,257
Anon., J. Roy. Inst. Chem. 1942, p.229

GRANICK, SAM [1909-1977]
AMS 13,1605
Anon., Rock. Univ. Notes 8(8) (1977), 2

GRANIT, RAGNAR [1900-1991]
VAD 1981, p.355; IWW 1982-1093, p.487; MH 2,193-194; STC 1,466-468; MSE 1,455-456; IB, 103; WWWS, 691

GRANSTRÖM, EDUARD ANDREEVICH [1879-]
SG [3]6,235

GRASSET, EDMOND [1895-1957]
A.R.P., Ann. Inst. Pasteur 94 (1958), 224-225

GRASSHEIM, KURT [1897-1948]
KURSCHNER 4,905; LDGS, 66
New York Times, 17 November 1948, p.27
Anon., Proc. Virchow Med. Soc. 7 (1948), 149

GRASSI, GIOVANNI BATTISTA [1854-1925]
DSB 5,502-504; FISCHER 1,530; ARCIERI, 195-206; OLPP, 158-161
C. Janicki, Naturwiss. 14 (1926), 225-231,261-269
Anon., Entomol. News 37 (1926), 127-128
E.S.G., Proc. Linn. Soc. 199 (1927), 85-86
A. Pazzini, Riv. Biol. 19 (1935), 1-46
G. Cotroni, Sci. Med. Ital. 3 (1955), 607-619

GRASSI, UGO [1879-1936]
POGG 5,445-446; 6,943; 7b,1730
L. Pucciarti, Nuovo Cimento NS14 (1937), 474-479

GRASSMANN, WOLFGANG [1898-1978]
KURSCHNER 12,961; WWWS, 692; POGG 6,944; 7a(2),266-268
Anon., Nachr. Chem. Techn. 6 (1958), 123
U. Hofmann and K. Kühn, Chem. Z. 87 (1963), 132-133
K. Kühn et al., Conn. Tiss. Res. 7 (1979), 57-59

GRATIA, ANDRÉ [1893-1950]
BNB 39,444-451; BAF, 59
M. Welsch, Rev. Med. Liège 5 (1950), 735-738; L'Université de Liège 1936-1966 (R. Dumoulin, Ed.), pp.613-651. Liège 1967
R. Lépine, Ann. Inst. Pasteur 80 (1951), 196-199

GRATIOLET, LOUIS PIERRE [1815-1865]
DSB 5,504-506; DBF 16,1093-1094; HIRSCH 2,835; DECHAMBRE [4]10,334-336; WWWS, 692
H. Milne-Edwards and M.E. Chevreul, Union Médicale [2]25 (1865), 353-362
P. Bert, Bull. Soc. Med. Yonne 1868, pp.17-37
J. Andrieu, Bibliographie Générale de l'Agenais 1 (1886), 338-341

GRAVELIUS, GEORG [1808-]
RAUSCH, 419

GRAVES, ROBERT JAMES [1796-1853]
DNB 8,436-437; HIRSCH 2,837-838; Suppl.,339; LYONS, 63-68; WWWS, 693; DECHAMBRE [4]10,345-346; BETTANY 2,202-209; CALLISEN 7,381-382; 28,267
J.F. Duncan, Dublin J. Med. Sci. 65 (1878), 1-12
C.E. Stellhorn, Am J. Surg. 28 (1935), 183-189
M.J. Whelton, Irish J. Med. Sci. 148(5/6) (1979), 161-167
S. Taylor, Robert Graves: The Golden Years of Irish Medicine. London 1989

GRAWITZ, PAUL ALBERT [1850-1832]
NDB 7,13-14; KURSCHNER 4,908-909; GROTE 2,23-75; FISCHER 1,531; DZ, 478; PAGEL, 628-629; WWWS, 693
A.E. Leopold, Münch. med. Wchschr. 79 (1932), 1404-1406
H. Loeschke, Verhandl. path. Ges. 27 (1934), 322-323
F.A. Lloyd, Inv. Urol. 4 (1967), 615-616

GRAY, ASA [1810-1888]
DSB 5,511-514; DAB 7,511-514; NCAB 3,407-408; ELLIOTT, 107-108; HUMPHREY, 96-99; ABBOTT, 53; W, 223-224; WWWS, 693-694
Anon., Proc. Am. Acad. Arts Sci. 23 (1888), 321-343; Annals of Botany 2 (1889), 400-414
W.G. Farlow, Biog. Mem. Nat. Acad. Sci. 3 (1895), 161-175
T. Gill, Proceedings of the American Association for the Advancement of Science 46 (1898), 1-30

A.H. Dupree, Asa Gray. Cambridge, Mass. 1959

GRAY, HARRY LeBRETON [1876-1936]
POGG 6,945
Anon., Ind. Eng. Chem. News Ed. 14 (1936), 305

GRAY, JAMES [1891-1975]
DNB [1971-1980], 356-357; WW 1975, p.1265; IB, 104
J.A. Ramsay, Nature 259 (1976), 433
H.W. Lissmann, Biog. Mem. Fell. Roy. Soc. 24 (1978), 55-70

GRAY, PHILIP PAUL [1896-1966]
AMS 11, 1924
Anon., Chem. Eng. News 44(30) (1966), 79

GRAY, THOMAS [1869-1932]
JV 16,157; NUC 211,363
F.J. Wilson, J. Chem. Soc. 1932, pp.2989-2992

GREAVES, RONALD IVAN NORREYS [1908-1990]
WW 1985, p.768
H.T. Meryman, Cryobiology 28 (1991), 306-313

GREEFF, RUDOLF [1877-]
JV 21,274

GREEN, ARDA ALDEN [1899-1958]
AMS 7,687
S.P. Colowick, Science 128 (1958), 519-521

GREEN, ARTHUR GEORGE [1864-1941]
FINDLAY, 247-269; WWWS, 696; POTSCH, 177; POGG 4,530-531; 5,446-448; 6,947; 7b,1734-1735
B. Brightman, Chem. Ind. 1941, pp.773-774
J. Baddiley, Obit. Not. Fell. Roy. Soc. 4 (1942-1944), 251-270; J. Chem. Soc. 1946, pp.842-852

GREEN, DAVID EZRA [1910-1983]
AMS 14,1849; WWA 1978-1979, p.1279; IWW 1982-1983, pp.490-491; MH 2,194-196; MSE 1,457-458; COHEN, 91; WWWS, 696
H. Beinert and P.K. Stumpf, TIBS 8 (1983), 434-436

GREEN, HARRY [1917-]
AMS 15(3),282; WWWS, 696

GREEN, JOSEPH REYNOLDS [1848-1914]
WW 1914, pp.855-856; DESMOND, 266; O'CONNOR(2), 29-30; WWWS, 697
S.H. Vines, Nature 93 (1914), 379-380; Proc. Roy. Soc. B88 (1915), xxxvi-xxxviii

GREEN, MAURICE [1926-]
AMS 15(3),285

GREENBAUM, LOWELL MARVIN [1928-]
AMS 15(3),288

GREENBERG, DAVID MORRIS [1895-1988]
AMS 14,1856; COHEN, 92; POGG 6,947-948; 7b,1735-1741

GREENE, CHARLES WILSON [1866-1947]
AMS 6,554; BROBECK, 146-147

GREENFIELD, WILLIAM SMITH [1846-1919]
HIRSCH 2,842; Suppl.,340
Anon., Brit. Med. J. 1919(II), pp.255-258
J.M.B., Lancet, 1919(II), p.351
H.R., Edinburgh Med. J. NS23 (1919), 258-262

GREENGARD, PAUL [1925-]
AMS 15(3),297

GREENSTEIN, JRSSE PHILIP [1902-1959]
AMS 9(I),737; COHEN, 93; WWAH, 236; WWWS, 699
A. Meister, Arch. Biochem. Biophys. 82 (1959), i-iv
J.T. Edsall, Science 130 (1959), 83-85
G.B. Mider, Cancer Research 19 (1959), 788-789
D. Hamer and A. Haddow, Nature 183 (1959), 1776-1777
J.T. Edsall and A. Meister, in Amino Acids, Proteins and Cancer Biochemistry (J.T. Edsall, Ed.), pp.1-8,213-227. New York 1960
M. Winitz et al., J. Nat. Cancer Inst. 24 (1960), vii-xxx

GREENWALD, ISIDOR [1887-1976]
AMS 11,1945; IB, 104; KOREN, 185; KAGAN(JA), 344; COHEN, 93; POGG 6,948-949; 7b,1741-1743
Anon., J. Hist. Med. 31 (1976), 375; Science 193 (1976), 988; Chem. Eng. News 54(11) (1976), 36

GREENWOOD, MAJOR [1853-1917]
Anon., Lancet 1917(I), p.853; Brit. Med. J. 1917(I), p.703

GREENWOOD, MAJOR [1880-1949]
FISCHER 1,533; O'CONNOR(2), 282-283
Who Was Who 1941-1950, pp.467-468
L. Hogben, Obit. Not. Fell. Roy. Soc. 7 (1950), 139-154

GREENWOOD, MARION [1862-1932]
W.H.B., Nature 130 (1932), 689-690

GREEP, ROY ORVAL [1905-]
AMS 17(3),315
R.O. Greep, Steroids 52 (1988), 447-514

GREGERSEN, JENS PETER [1881-1947]
KBB 1944-1946, p.408; VEIBEL 2,166-167; NUC 217,311

GREGG, ALAN [1890-1957]
DAB [Suppl.6],252-253; WWAH, 236
M.C. Winternitz, Science 126 (1957), 1279
W. Weaver, J. Med. Ed. 35 (1960), 313-318
W. Penfield, J. Med. Ed. 37 (1962), 90-96; The Difficult Art of Giving. Boston 1967

GRÉGOIRE, VICTOR [1870-1938]
P. Debassieux, Rev. Quest. Sci. 115 (1939), 349-369
P. Martens, Cellule 48 (1939), 7-46; Ann. Acad. Roy. Belg. 133 (1967), 145-189

GREGORY, FREDERICK GUGENHEIM [1893-1961]
DSB 5,523-524; DNB [1961-1970], 456-457; WW 1961, p.1241; DESMOND, 268
H.K. Porter, Nature 193 (1962), 118
F.C. Steward, Plant Physiology 37 (1962), 450
H.K. Porter and F.J. Richards, Biog. Mem. Fell. Roy. Soc. 9 (1963), 131-153

GREGORY, REGINALD PHILIP [1879-1918]
DESMOND, 268
A.C. Seward and W. Bateson, Nature (1918), 247-248,284

GREGORY, RICHARD ARMAN [1864-1952]
DNB [1951-1960], 433-435; WWWS, 701
H. Hartley, Nature 171 (1953), 1040-1046
F.J.M. Stratton, Obit. Not. Fell. Roy. Soc. 8 (1953), 411-417
W.H.G. Armytage, Sir Richard Gregory. London 1957

GREGORY, WILLIAM [1803-1858]
DSB 5,530-531; DNB 8,548; FERCHL, 199; TSCHIRCH, 1057; WWWS, 701;POGG 1,949-950,1570
Anon., J. Chem. Soc. 12 (1860), 172-175
G.C. Green, Nature 157 (1946), 465-469

GREGORY, WILLIAM KING [1876-1970]
DSB 17,368-369; AMS 11,1950; NCAB A,105; IB, 704; WWWS, 701-702
G.G. Simpson, Am. Phil. Soc. Year Book 1971, pp.124-127

E.H. Colbert, Biog. Mem. Nat. Acad. Sci. 46 (1975), 91-133
R. Rainger, J. Hist. Biol. 22 (1989), 103-139

GRÉHANT, LOUIS NESTOR [1838-1910]
DBF 16,1148; GE 19,373; WWWS, 702; RSC 3,10; 7,834; 10,56-57; 12,292; 15,373
A. Dastre, C.R. Soc. Biol. 68 (1910), 567-568
G. Dieulafoy, Bull. Acad. Med. 63 (1910), 335-336
Anon., Nouv. Arch. Hist. Nat. [5]2 (1910), ii-xix

GREIFF, PHILIPP [-1885]
RSC 10,57
A.W. Hofmann, Ber. chem. Ges. 18 (1885), 2753

GREMELS, HANS [1896-1949]
NDB 7,43-44; AUERBACH, 244-245; WWWS, 702; POGG 7a(2),270
F. Heim, Erg. Physiol. 48 (1955), 49-53

GRESHOFF, MAURITS [1862-1909]
NNBW 2,505-506; TSCHIRCH, 1058; SCHELENZ, 720
Anon., Kew Bulletin 1909, pp.424-425

GRESSLY, OSKAR [1854-1916]
F. Schubiger, Corr. Schw. Arzte 46 (1916), 1209

GREULICH, RICHARD [1881-1929]
JV 21,294; NUC 218,104
Anon., Chem. Z. 53 (1929), 404; Z. angew. Chem. 43 (1930), 535

GREVEN, KURT [1911-1988]
KURSCHNER 13,1190GREVILLE, GUY DRUMMOND [1907-1969]
Who's Who in British Science 1953, p.112
Anon., Chem. Brit. 6 (1970), 131

GREWE, RUDOLF [1910-1968]
KIEL, 174; POGG 7a(2),272
A. Mondon, Christiana Albertina 7 (1969), 94-95
H. Henecka and A. Mondon, Chem. Ber. 111(9) (1978), i-xxvii

GREY, EGERTON CHARLES [1887-1928]
SRS, 49
A. Harden, Nature 122 (1928), 486; Biochem. J. 23 (1929), 1-2

GRIEBEL, CONSTANT [1876-1965]
HEIN-E, 158-159; TSCHIRCH, 1057; TLK 3,207; POGG 6,950-951; 7a(2),272-273
W. Lindner, Chem. Z. 80 (1956), 78
Anon., Chem. Z. 85 (1961), 18; 90 (1966), 19

GRIEPENKERL, FRIEDRICH [1826-1900]
BJN 5,251-252,92*; RSC 3,14
Anon., Leopoldina 36 (1900), 171; Science 12 (1900), 533

GRIESBACH, WALTER EDWIN [1888-1968]
KURSCHNER 10,722; FISCHER 1,535-536; IB, 104-105; COHEN, 03-04; KOREN,185
Anon., New Zealand Medical Journal 68 (1968), 187-188

GRIESINGER, WILHELM [1817-1868]
ADB 9,669-670; NDB 7,64-65; HIRSCH 2,850-852; WWWS, 704
O.M. Narx, Bull. Hist. Med. 46 (1972), 519-544
B. Wahrig-Schmidt, Der Junge Griesinger im Spannungsfeld zwischen Philosophie und Physiologie. Tübingen 1985
H.H. Walser, Gesnerus 43 (1986), 197-204

GRIESS, PETER [1829-1888]
DSB 5,536-537; NDB 7,66-67; ADB 49,547-550; BUGGE 2,217-228; WWWS, 704POTSCH, 178; BLOKH 1,258-259; W, 226; SCHAEDLER, 45; POGG 3,548-549; 4,533
A.W. Hofmann et al., Ber. chem. Ges. 24(3) (1891), i-xxxviii
V. Heines, J. Chem. Ed. 35 (1958), 187-191
R. Wizinger-Aust, Z. angew. Chem. 70 (1958), 198-204
W.H. Cliffe, Chem. Ind. 1958, pp.616-621

GRIFFIN, AMOS CLARK [1918-1981]
AMS 14,1876; WWWS, 704

GRIFFIN, DONALD REDFIELD [1915-]
AMS 15(3),314; IWW 1982-1983, p.496; WWWS, 704

GRIFFIN, JOHN JOSEPH [1802-1877]
DNB 8,670-671; FERCHL, 200; POGG 1,953; 3,549
Anon., J. Chem. Soc. 3 (1851), 412-418
J.H. Gladstone, J. Chem. Soc. 33 (1878), 229
B. Gee and W.H. Brock, Ambix 38 (1991), 29-62

GRIFFITH, FREDERICK [1877-1941]
Anon., Lancet 1941(I), pp.588-589; Brit. Med. J. 1941(I), p.691
A.W. Downie, J. Gen. Microbiol. 73 (1972), 1-11
J. Tooze, TIBS 3 (1978), 261-262

GRIFFITH, WENDELL HORACE [1895-1968]
AMS 11,1961; IB, 105; WWAH, 238; WWWS, 704
Anon., Chem. Eng. News 46 (10) (1968), 68
R.E. Olsen, Fed. Proc. 30(1) (1971), 131-138; J. Nutrition 116 (1986), 2326-2338

GRIFFITHS, ARTHUR BOWER [1859-]
TSCHIRCH, 1057; RSC 10,61; 15,462-463; NUC 218,568-569; POGG 4,534; 5,499

GRIFFITHS, THOMAS [-1858]
NUC 218,600
Anon., Medical Directory 1859, p.972
W. Templeton, Chem. Brit. 18 (1982), 263

GRIGAUT, ADRIEN [1884-1960]
GORIS, 409-411
M. Delaville, Ann. Biol. Clin. 19 (1961)
P. Desnuelle, Bull. Soc. Chim. Biol. 43 (1961), 5-6

GRIGNARD, VICTOR [1871-1935]
DSB 5,540-541; DBF 16,1218-1219; POTSCH, 178-179; POGG 5,449-450; 6,953; 7b,1747-1748
C. Courtot, Bull. Soc. Chim. [5]3 (1936), 1433-1472
R. Locquin, Ber. chem. Ges. 69A (1936), 69-72
C.S. Gibson and W.J. Pope, J. Chem. Soc. 1937, pp.171-179
H. Rheinboldt, J. Chem. Ed. 27 (1950), 476-488
F. Runge, Sitz. Akad. Wiss. Leipzig 107(2) (1970), 1-17
J.A. Gautier, Rev. Hist. Pharm. 20 (1971), 521-530
D. Hudson, Chem. Brit. 23 (1987), 141-142

GRIGORIEV, PAVEL SEMENOVICH [1879-1940]
BSE 12,605-606; BME 8,270-271; WWR, 215; FISCHER 1,536; NUC 218,647
Anon., Sov. Med. 4(22) (1940), 48

GRIJNS, GERRIT [1865-1944]
DSB 5,541-542; LINDEBOOM, 724-726; FISCHER 1,536; IB, 105
M.C. Kik, Science 82 (1935), 408-409; J. Nutrition 62 (1957), 3-12

GRIMAUX, LOUIS ÉDOUARD [1835-1900]
DBF 16,1249-1250; GE 19,429; VAPEREAU 6,725; BLOKH 1,260; TSCHIRCH, 1057; WWWS, 705; POGG 3,550; 4,535-536
C. Lauth, Rev. Sci. 13 (1900), 577-580
P.Adam, Bull. Soc. Chim. [4]9 (1911), I-xxxvi

GRIMBERT, LÉON [1860-1931]
DBF 16,1250-1251; GORIS, 411-414; IB, 105; TSCHIRCH, 1057; POGG 4,536; 5,450; 6,954
H. Hérissey and P. Fleury, J. Pharm. Chim. [8]6 (1932), 273-320
J.A. Gautier, Bull. Sci. Pharm. 39 (1932), 87-98
P. Fleury, Figures Pharmaceutiques Françaises, pp.221-226. Paris 1953
R. Fabre, Bull. Acad. Med. 147 (1963), 693-702; Ann. Pharm. Fran. 22 (1964), 689-695

GRIMM, FERDINAND [1845-1919]
STELLING-MICHAUD 3,535; RSC 7,840; NUC 219,69

GRIMME, CLEMENS [1881-1975]
HEIN-E, 159-160; JV 23,330; TLK 3,207; NUC 219,149
Anon., Chem. Z. 46 (1922), 347,831

GRINDEL, DAVID HIERONYMUS [1776-1836]
HEIN, 226-227; LEVITSKI 1,238-240; WELDING, 260-261; BLOKH 1,260-261; FERCHL, 201; TSCHIRCH, 1057-1058; POGG 1,955
E. Brennsohn, Die Aerzte Livlands, pp.179-180. Mitau 1905
Y.P. Stradin, Ist. Est. Tekhn. Pribalt. 6 (1980), 85-103

GRINDLEY, HARRY SANDS [1864-1955]
AMS 7,699; WWAH, 238

GROAT, WILLIAM AVERY [1876-1945]
AMS 7,700; IB, 105; WWAH, 239; WWWS, 707-708
E.C. Reifenstein, Ann. Int. Med. 23 (1945), 1044

GROB, CYRIL [1917-]
KURSCHNER 13,1198-1199; SBA 3,55; WWWS, 708; POGG 7a(2),276-277
Anon., Chimia 31 (1977), 80

GROB, EUGENE CONSTANT [1914-1972]
Anon., Chimia 26 (1972), 395
S., Mitt. Nat. Ges. Bern NF30 (1973), 156

GROB, KURT [1920-1987]
B. Brechbühler and W. Giger, Chimia 41 (1987), 250-252
W. Blum, Viert. Nat. Ges. Zürich 132 (1987), 188-190

GROENEVELD, ANTON [1871-]
JV 15,213; RSC 14,60; NUC 219,473

GROH, GYULA (JULIUS) [1886-1952]
MEL 1,622-623; IB, 105; POGG 6,956; 7b,1748-1749
Anon., Ost. Chem. Z. 53 (1952), 93

GROLL, HERMANN [1888-1947]
ECKERT, 62-63; IB, 105
E. Müller, Z. allgem. Path. 84 (1948), 3-4; Verhandl. path. Ges. 33 (1950), 412-413

GROLLMAN, ARTHUR [1901-1980]
AMS 13,1653; KAGAN(JA),755; COHEN,94; WWWS,709; POGG 6,957; 7b,1749-1754

GROOM, THOMAS THEODORE [1863-1943]
RSC 15,477
H. Woods, Nature 152 (1943), 183
J.A. Venn, Alumni Cantabrigensis (2)3, p.162. Cambridge 1947

GROS, FRANÇOIS [1925-]
QQ 1981-1982, p.696

GROS, JAMES [1817-1893]
DBA 2,1293; FERCHL, 201; RSC 3,28; POGG 1,958

GROS, OSCAR [1877-1947]
NDB 7,136-137; KIEL, 86; TLK 3,209; WWWS, 709; POGG 4,538; 6,958-959; 7a(2),282

GROSCHUFF, ERICH [1874-1921]
BLOKH 1,261-262; POGG 5,452
Anon., Ber. chem. Ges. 55A (1922), 22-25

GROSS, ALFRED [1876-1904]
KIEL, 115

GROSS, EBERHARD [1888-1976]
KURSCHNER 12,987; DRULL, 92; WENIG, 97; IB, 106; NUC 220,45
H. Weichardt, Zentralblatt für Arbeitsmedizin 26(12) (1976), 280-281

GROSS, ERHARD [1928-1981]
AMS 15(3),333
Anon., The Peptides (E.Gross and J. Meienhofer, Eds.), vol.5, pp.v-x. New York 1983
J. Meienhofer, Peptides 1982 (K. Blaha and P. Malon, Eds.), pp.xlix-lv. Berlin 1983

GROSS, FABIUS [1906-1950]
LDGS, 13
J. Ritchie, Nature 166 (1950), 295

GROSS, LUDWIK [1904-]
AMS 15(3),335; WWA 1980-1981, p.1366; IWW 1982-1983, pp.499-500; WWWS,710
M. Bessis, Nouv. Rev. Hem. 16 (1976), 287-304

GROSS, OSCAR [1881-1967]
KURSCHNER 10,733; FISCHER 1,539; LDGS, 62

GROSS, PAUL MAGNUS [1895-1986]
AMS 13,1656; WWWS, 710
New York Times, 10 May 1986GROSS, PHILIPP [1899-1974]
BHDE 2,420-421; WIDMANN, 263-264; POGG 6,959; 7b,1758-1759
H.E. Suess, Z. Elektrochem. 68 (1964), 615

GROSS, RICHARD [1879-]
JV 21,274

GROSS, WALTER [1878-1933]
NDB 7,145-146; FISCHER 1,539; DRULL, 92-93; IB, 106; WWWS, 711
F. Klinge, Verhandl. path. Ges. 27 (1934), 349-350

GROSSE, ARISTID [1905-1985]
AMS 14,1893-1894; WWA 1976-1977, p.1257; WWWS, 711
New York Times, 23 July 1985

GROSSER, OTTO [1873-1951]
NDB 7,152; STURM 1,475; FISCHER 1,540; STOBER, 95-105; MAASS, 82-100; KOERTING, 105-107; IB, 106; WWWS, 711
M. Watzka, Anat. Anz. 98 (1951-1952), 208-216
F. Hochstetter, Alm. Akad. Wiss. Wien 101 (1952), 433-438

GROSSER, PAUL [1880-1934]
ARNSBERG, 166; BHDE 2,421; FISCHER 1,540; KALLMORGEN, 283; GRUETTER, 42; WININGER 7,34; TETZLAFF, 114; KOREN, 186
Anon., Deutsche med. Wchschr. 60 (1934), 300

GROSSMANN, HERMANN [1877-] (before 1902, ITZIG, HERMANN)
KURSCHNER 4,930; BHDE 2,422-423; LDGS, 20; WININGER 7,34-35; TLK 3,210-211; POGG 4,686; 5,453; 6,961-962; 7a(2),286

GROTH, PAUL HEINRICH von [1843-1927]
DSB 5,556-558; NDB 7,167-168; DBA 2,1300; DZ,488; TLK 2,226; WWWS, 1734; POTSCH, 179; POGG 3,553; 4,539; 5,454; 6,962; 7aSuppl.,244-245
K. Mieleitner, Z. Kristall. 58 (1923), 3-6
E. Kaiser, Bayer. Akad. Wiss. Jahrbuch 1927, pp.37-40
H. Steinmetz, Ber. chem. Ges. 61A (1928), 65-68
E.H. Kraus, Science 67 (1928), 150-152
H.A.M., Proc. Roy. Soc. A119 (1928), xx-xxii
Anon., Chem. Z. 67 (1943), 235

GROTTHUSS, THEODOR [1785-1822]
DSB 5,558-559; NDB 7,171-172; ADB 9,767; WELDING, 268; BLOKH 1,262-263; POTSCH, 179-180; COLE, 227; FERCHL, 202; WWWS, 712; POGG 1,959-960
O. Clemen, Arch. Gesch. Naturwiss. 7 (1916), 377-389

GROVE, WILLIAM ROBERT [1811-1896]
DSB 5,559-561; DNB 22,796-797; FERCHL, 202; WWWS, 713; POTSCH, 180; POGG 1,960-961; 3,554

A. Gray, Nature 54 (1896), 393-394
K.R. Webb, J. Roy. Inst. Chem. 85 (1961), 291-293
M.L. Cooper and V.D.M. Hall, Ann. Sci. 39 (1982), 229-254

GRUBE, ADOLF EDUARD [1812-1880]
NDB 7,174; ADB 49,575; LEVITSKI 1,270-275
W.C.M., Nature 22 (1880), 435-436
A. Schimmelpfennig, Jahresb. Schles. Ges. 58 (1881), 298-304

GRUBE, KARL ADOLPH [1866-1920]
O'CONNOR(2), 448-449; WENIG, 97; RSC 15,488

GRUBER, JOSEF [1827-1900]
OBL 2,82; STURM 1,477; BJN 5,92*; HIRSCH 2,871; Suppl.,346; WEINMANN, 179
Anon., Leopoldina 36 (1900), 131

GRUBER, MAX von [1853-1927]
DSB 5,563-565; NDB 7,177-178; OBL 2,83; FISCHER 1,542-543; LESKY,595-602; BULLOCH, 369; IB, 106; DZ, 490-491; WWWS, 1734
K.B. Lehmann, Münch. med. Wchschr. 70 (1923),879-881; 74 (1927),1838-1839
P. Uhlenhuth, Z. Immunitätsforsch, 54 (1927), 1-10
O. Frank, Max von Gruber. Munich 1928
F. Kudlein, Med. Hist. J. 17 (1982), 373-389
H. Machmann and W. Köhler, Z. ärzt. Fortbild. 83 (1989), 1029-1033

GRUBER, WENZEL (VENTSESLAV LEOPOLDOVICH) [1814-1890]
OBL 2,84; STURM 1, 477; BSE 13,24-25; BME 8,331-332; HIRSCH 2,871 PAGEL, 640-642; GOCZOL, 66-74; KOERTING, 98; WEINMANN, 179-180
Anon., Anat. Anz. 5 (1890), 587-588

GRUBER, WOLFGANG [1886-]
JV 30,738; NUC 220,498; POGG 7a(2),291-292
Anon., Nachr. Chem. Techn. 4 (1956), 174

GRUBY, DAVID [1810-1898]
DSB 5,565-566; DBF 16,1374-1375; HIRSCH 2,871-872; Suppl.,346; PAGEL,643; BULLOCH, 369-370; TSCHIRCH, 1058; KOHUT 2,277-278; WININGER 2,530-31
R. Blanchard, Arch. Parasit. 2 (1899), 43-74
L. Le Leu, Le Dr. Gruby, Notes et Souvenirs. Paris 1908
T. Rosenthal, Ann. Med. Hist. 4 (1932), 339-346
S.J. Zakon and T. Benedek, Bull. Hist. Med. 16 (1944), 155-168
B. Kisch, Trans. Am. Phil. Soc. 44 (1954), 193-226

GRÜBLER, GEORG [1850-1915]
HEIN, 230-231
D. Tutzke, Janus 59 (1972), 39-45

GRÜN, ADOLF [1877-1947]
BHDE 2,425; TLK 3,213; ZURICH, 98-99; POGG 5,455-456; 6,965; 7a(2),292
A. Chwala, Ost. Chem. Z. 48 (1947), 175-176
Anon., Z. angew. Chem. 59 (1947), 256

GRÜNBAUM (later LEYTON), ALBERT SIDNEY [1869-1921]
O'CONNOR(2), 363-364; FISCHER 2,909; WININGER 2,533; WWWS, 714,1039
Anon., Lancet 1921(II), p.825; Brit. Med. J. 1921(II), p.579
M.J. Stewart, J. Path. Bact. 25 (1922), 109-112

GRÜNEBERG, HANS [1907-1982]
BHDE 2,428; LDGS, 13; WWWS, 714
D. Lewis and D.M. Hunt, Biog. Mem. Fell. Roy. Soc. 30 (1984), 227-247

GRÜNHAGEN, ALFRED [1842-1912]
BJN 18,25*; FISCHER 1,544; PAGEL, 643-644

GRÜNHUT, LEO [1863-1921]
NDB 7,199-200; DBJ 3,300; TLK 2,229; WININGER 2,539; TETZLAFF, 115; WWWS, 715; POGG 4,541; 5,456-457
H. Beckurts, Z. Nahrung 42 (1921), 1-2
Anon., Chem. Z. 45 (1921), 55

GRÜNING, WERNER [1913-]
KURSCHNER 12,1001; POGG 7a(2),296-297

GRÜNSTEIN, NATHAN [1878-1940]
ARNSBERG, 168-169

GRÜSS, JOHANNES [1860-]
TLK 3,216; IB, 106; BERLIN, 724; NUC 220,609-610

GRÜTTNER, GERHARD [1889-1918]
POGG 5,457-458
K.A. Hofmann, Ber. chem. Ges. 51 (1918), 1205-1206

GRÜTZNER, PAUL [1847-1919]
NDB 7,207-208; DBJ 2,719; FISCHER 1,544-545; PAGEL, 644-645; DZ, 495-496; WWWS, 1734; POGG 4,542; 5,458
L. Asher, Erg. Physiol. 18 (1920), xxviii-xxix

GRUHL, WOLDEMAR [1880-]
JV 23,619; NUC 220,625

GRUITHUISEN, FRANZ von PAULA [1774-1852]
NDB 7,210-211; ADB 10,6; HIRSCH 2,873-874; GLOSAUER, 49-51; WWWS, 714; POGG 1,964-965; CALLISEN 7,470-474; 28,298-299

GRUNDFEST, HARRY [1904-1983]
AMS 14,1902; KAGAN(JA), 740-741
Anon., Rock. Univ. Notes 15(2) (1984), 4

GRUNDMANN, CHRISTOPH [1908-]
AMS 14,1902; KURSCHNER 13,1228-1229; POGG 7a(2),296-297

GRUNDZACH, IGNACY [1861-1940]
KONOPKA 3,411-413

GRUTTERINK, JAN ADOLF [1879-1949]
C.L. van Nes, Ingenieur 1949, pp.49-50GRUYTER, PAUL de [1866-] JV 9,208; RSC 13,275

GSCHEIDLEN, RICHARD [1842-1889]
HIRSCH 2,877; PAGEL, 646; SCHAEDLER, 45; BULLOCH, 370; WWWS, 716
Anon., Chem. Z. 13 (1889), 1724; Berl. klin. Wchschr. 26 (1889), 220

GUARESCHI, ICILIO [1847-1918]
DFI, 72-73; EI 18,24; BLOKH 1,264; REBER, 227-230; TSCHIRCH, 1058; POTSCH, 180; SCHELENZ, 701-702; POGG 3,359; 4,543-544; 5,459-460
A. Haller, Bull. Soc. Chim. [4]23 (1918), 313-314
R. Nasini, Arch. Stor. Sci. 1 (1919-1920), 101-112

GUARNIERI, GUISEPPE [1856-1918]
EI 18,19-20; OLPP, 166
A. Garressini, Riforma Medica 34 (1918), 764

GUBELMAN, IVAN [1886-]
AMS 7,705; NUC 221,399

GUBLER, ADOLPHE [1821-1879]
HIRSCH 2,879-880; Suppl.,348; FRANKEN, 183-184; WWWS, 716
C. Paul, Union Médicale [3]28 (1879), 13-20
A. Robin, C.R. Soc. Biol. [7]1 (1880), 179-187
E.J. Bergeron, Bull. Acad. Med. 38 (1899), 1-28

GUCKELBERGER, CARL GUSTAV [1820-1902]
HEIN, 232-233; RSC 3,55
A. Wankmüller, Beitr. Wurtt. Apothekgesch. 7 (1966), 82-83

GUDERNATSCH, JOST FRIEDRICH [1881-1962]
AMS 10,1548; NCAB 52,542
F. Gudernatsch, Anat. Anz. 108 (1960), 249-255

GUDZENT, FRIEDRICH [1878-1952]
KURSCHNER 7,655-656; FISCHER 1,547-548; TLK 3,216

GÜMBEL, THEODOR [1879-1938]
JV 18,369; SG [3]6,286; NUC 221,512

GÜNTHER, ALBERT KARL LUDWIG GOTTHELF [1830-1914]
WWWS, 720-721
W.C.M., Proc. Roy. Soc. B88 (1915), xi-xxvi
R.T. Gunther, Ann. Mag. Nat. Hist. [10]6 (1930), 233-236

GÜNTHER, FRITZ [1877-1957]
NDB 7,273; POTSCH, 181; POGG 7a(2),309-310; (4),143*
K. Saftien, Chem. Ber. 92 (1959), xxix-xxxv

GÜNTHER, HANS [1884-1956]
FISCHER 1,549
W. Stepp, Endokrinologie 33 (1956), 257-258

GÜNTHER, JOHANN JACOB [1771-1852]
HIRSCH 2,286-287; FERCHL, 204; POGG 1,971; CALLISEN 7,486-495; 28,305-307

GÜNTHER, OSCAR [1876-1917]
JV 17,283; NUC 221,296

GÜNZBURG, ALFRED [1861-1945]
FISCHER 1,549; KALLMORGEN, 285; KOREN, 186

GÜNZBURG, LUDWIG [1895-1977]
KALLMORGEN, 285

GUERARD, JACQUES ALPHONSE [1796-1874]
HIRSCH 2,892-893; DECHAMBRE [4]11,459-460
A. Devergie, Bull. Acad. Med. [2]3 (1874), 647-649
T. Gaillard, Ann. Hyg. Publ. [2]42 (1874), 458-478

GÜRBER, AUGUST [1864-1937]
EBERT, 158-162; GUNDLACH, 252; AUERBACH, 248; IB, 107; ZURICH-D, 160; ZWEIFEL, 166

GUERBET, MARCEL [1861-1938]
GORIS, 417-418; TSCHIRCH, 1058; POGG 5,462; 6,972; 7b,1769
R. Fabre, J. Pharm. Chim. [8]29 (1939), 280-288

GUÉRIN, CAMILLE [1872-1961]
DBF 16,1481-1482
R. Dujarric de la Rivière, Bull. Acad. Med. 145 (1961), 645-651
F. von Deinse, Presse Med. 69 (1961), 1769-1770
Anon., Lancet 1961(I), p.1357; Brit. Med. J. 1961(I), p.1834
I. Galloway, Nature 191 (1961), 851-852

GUÉRIN-VARRY, ROCH THÉOGÈNE [1800-]
RSC 3,69-70; CALLISEN 28,312
A. Maire, Catalogue des Thèses de Sciences Soutenues en France de 1810 à 1890 Inclusivement, p.6. Paris 1892

GUERMONPREZ, FRANÇOIS JULES OCTAVE [1849-1932]
DBF 16,1516

GUERRINI, GUIDO [1878-1970]
FISCHER 1,549-550; VACCARO, 790; MODENA, 259; IB, 107
C., Med. Ital. 11 (1930), 3690370
Chi E? 1957, p.280
G. Favillo, Annuario Università Bologna 1969-1970, pp.445-447

GUERTLER, WILHELM MINOT [1880-1959]
POGG 5,462-463; 6,973-974; 7a(2),314-315

GUEST, HERBERT HARTLEY [1884-1956]
AMS 8,979
Anon., Chem. Eng. News 34 (1956), 6309

GÜTERBOCK, LUDWIG [1814-1895]
ADB 49,646; HIRSCH 2,898
Anon., Leopoldina 31 (1895), 59

GUGGENHEIM, EDWARD ARMAND [1901-1970]
DNB [1961-1970], 462-463; WW 1970, p.1285; POGG 6,974; 7b,1770-1772
F.C. Tompkins and C. Goodeve, Biog. Mem. Fell. Roy. Soc. 17 (1971), 303-326

GUGGENHEIM, KARL YECHIEL [1906-]
BHDE 2,434
Who's Who in Israel 1972, p.145

GUGGENHEIM, MARKUS [1885-1970]
TLK 3,216
R. Silberschmidt, Verhandl. Schw. Nat. Ges. 150 (1970), 295-299
H. Fischer, Gesnerus 28 (1971), 83-89
W. Loffler, Bull. Schw. Akad. Med. Wiss. 27 (1971), 167-169
H. Balmer, Gesnerus 31 (1974), 237-266

GUGGENHEIMER, HANS [1886-]
KURSCHNER 4,952; FISCHER 1,550; JV 26,370; ASEN, 65

GUHA, BIRES CHANDRA [1904-1962]
DNBI Suppl.2,92-94
B.W. Ghosh, Enzymologia 24 (1962), 309; Science and Culture 28 (1962), 218-220

GUHA, PRAPHULLA CHANDRA [1894-1962]
POGG 6,975; 7b,1773-1774
Anon., Fellows of the Indian National Science Academy, p.187. New Delhi 1984

GUIART, JULES [1870-1965]
DBF 17,39-40; FISCHER 1,550
G. Lavier, Bull. Acad. Med. 150 (1966), 535-537

GUIBOURT, NICOLAS [1790-1867]
DBF 17,57; LAROUSSE 8,1611; GORIS, 420-424; BOURQUELOT, 49-50; TSCHIRCH, 1058; FERCHL, 205; DECHAMBRE 47,456-457; POGG 1,975; 3,562-563; CALLISEN 7,510-514; 28,316
L. Mialhe, Bull. Acad. Med. 32 (1867), 1012-1014
J.E. Planchon et al., J. Pharm. Chim. [4]6 (1867), 200-209
H. Buignet, J. Pharm. Chim. [4]15 (1872), 69-86

GUICHARD, MARCEL [1873-1960]
DBF 17,66; CHARLE, 141-143; POGG 5,465; 6,976; 7b,1777-1778
J. Pérès, Ann. Univ. Paris 30 (1960), 430-433
G. Chaudron, Bull. Soc. Chim. 1961, pp.867-869

GUIGNARD, LÉON [1852-1928]
DSB 5,581-582; DBF 17,83-84; GORIS, 425-428; TSCHIRCH, 1058; IB,
107; WWWS, 718 P. Guérin, Bull. Sci. Pharm. 35 (1928), 374-380; Bull.
Soc. Chim. Biol. 10 (1928), 1387-1389
J. Constantin, Ann. Sci. Nat. [10]16 (1934), xxviii-xxix

GUILLAIN, GEORGES CHARLES [1876-1961]
DBF 17,127-128; FISCHER 1,551
P. Molaret, Presse Med. 69 (1961), 1695-1696
A. Alajouanine, Bull. Acad. Med. 146 (1962), 18-26

GUILLEMIN, ROGER [1924-]
AMS 15(3),357; WWA 1980-1981, p.1376; IWW 1982-1983, p.505; WWWS, 718; ABBOTT, 54-55; POTSCH, 180-181
N. Wade, The Nobel Duel. Garden City, N.Y. 1981

GUILLIERMOND, ALEXANDRE [1876-1945]
DSB 5,584-585; SACKMANN, 146-148; CHARLE(2), 143-145; IB, 107; WWWS, 718
L. Emberger, Rev. Gen. Bot. 53 (1946), 337-361
Anon., Rev. Cytol. 9 (1946-1947), 1-47
O. Verona, Mycopathologia 42 (1948), 124-130
R. Heim, Not. Acad. Sci. 2 (1949), 639-662

GUINIER, ANDRÉ [1911-]
QQ 1983-1984, p.682

GUITARD, EUGÈNE HUMBERT [1884-1976]
TSCHIRCH, 1058
P. Julien et al., Rev. Hist. Pharm. 21 (1972), 169-172; 23 (1976), 215-224

GULDBERG, CATO MAXIMILIAN [1836-1902]
DSB 5,586-587; NBL 5,76-81; BLOKH 1,264-265; W, 229-230; WWWS, 719; POTSCH, 181; POGG 3,564-565; 4,551-552
H. Haraldsen, Ed., The Law of Mass Action: A Centenary Volume 1864-1964. Oslo 1964
E.W. Lund, J. Chem. Ed. 42 (1965), 548-550
P. Øhrstrøm, Centaurus 28 (1985), 277-287

GULEVICH, VLADIMIR SERGEEVICH [1867-1933]
BSE 13,191; BME 8,605-606; WWR, 223; KHARKOV, 57-59; POGG 6,980-981; 7b,1782-1783
J. Hefter, Arkh. Biol. Nauk 33 (1933), 615-625; Ber. chem. Ges. 67A (1934), 9-10; Usp. Sov. Biol. 4 (1935), 125-130
I.A. Smorodintsev, Zhur. Ob. Khim. [2]9 (1939), 471-478
V.S. Gulevich. Isbrannye Trudy. Moscow 1955
L.M. Broude et al., Biokhimia 33 (1968), 195-202
A. Bezkorovainy, J. Chem. Ed. 51 (1974), 652-654

GULICK, ADDISON [1882-1969]
AMS 11,1993; IB, 107
Anon., Chem. Eng. News 47(53) (1969), 79

GULL, WILLIAM WITHEY [1816-1890]
DNB 8,776-777; WILKS, 261-274; HIRSCH 2,911; PAGEL, 656; O'CONNOR, 59-60
P.H.P.S., Proc. Roy. Soc. 48 (1890), viii-xii

GULLAND, GEORGE LOVELL [1862-1941]
O'CONNOR(2), 400-401; NUC 223,213
Who Was Who 1941-1950, p.479
Anon., Brit. Med. J. 1941(I), pp.799-800

GULLAND, JOHN MASSON [1898-1947]
DSB 5,589-590; WW 1945, pp.1138-1139; POTSCH, 181; POGG 6,981; 7b,1783-1784
J.W. Cook, Nature 160 (1947), 702-703; Biochem. J. 43 (1948), 161-162
R.D. Haworth, Obit. Not. Fell. Roy. Soc. 6 (1948), 67-82; J. Chem. Soc. 1948, pp.1476-1482

GULLIVER, GEORGE [1804-1882]
DNB 8,777-778; HIRSCH 2,911-912; WWWS, 719
Anon., Lancet 1882(II), p.916; Brit. Med. J. 1882(II), p.124; Edinburgh Med. J. 28 (1882-1883), 668-672

GULLSTRAND, ALLVAR [1862-1930]
DSB 5,590-591; SMK 3,139-140; KALIN, 68-71; W, 230; POGG 5,468-469; 6,981-982
H. Boegehold, Naturwiss. 18 (1930), 822-823
M. Herzberger, Optica Acta 3 (1960), 237-241
C. Snyder, Archives of Ophthalmology 68 (1962), 139-141

GUMLICH, GEORG EWALD OTTO [1875-]
JV 19,180; RSC 15,534; NUC 223,249

GUMPRECHT, FERDINAND [1864-1941]
KURSCHNER 4,956; FISCHER 1,553

GUND, RUDOLF [1896-1978]
JV 39,517

GUNDELACH, CARL [1821-1878]
RSC 3,87; NUC 223,287
Anon., Ber. chem. Ges. 11 (1878), 157; 12 (1879), 2394

GUNDERMANN, KARL DIETRICH [1922-]
KURSCHNER 13,1249; POGG 7a(2),319

GUNN, JAMES ANDREW [1881-1958]
WW 1959, p.1258; FISCHER 1,553-554
H.R. Ing, Nature 182 (1958), 1411
Anon., Lancet 1958(II), pp.965-966; Brit. Med. J. 1958(II), pp.1107-1108

GUNNING, JAN WILLEM [1827-1900]
POGG 4,553
Anon., Chem. Z. 24 (1900), 1147; Leopoldina 36 (1900), 48

GUNSALUS, IRWIN CLYDE [1912-]
AMS 15(3),364; WWA 1980-1981, p.1380; IWW 1982-1983, p.508; WWWS, 720
I.C. Gunsalus, in Reflections on Biochemistry (A. Kornberg et al.,Eds.). pp.125-135. Oxford 1976; Ann. Rev. Microbiol. 35 (1984), xiii-xliv

GURD, FRANK ROSS NEWMAN [1924-]
AMS 15(3),368; WWA 1980-1981, p.1381; WWWS, 721

GURIN, SAMUEL [1905-]
AMS 14,1918; WWA 1978-1979, pp.1323-1324; WWWS, 722

GURVICH, ALEKSANDR GAVRILOVICH [1874-1954]
DSB 5,594-595; BSE 13,211; BME 8,620-521; WWR, 225; IB, 107; KOREN, 187; WININGER 7,42-43; WWWS, 722
L.B. Belousov et al., Aleksandr Gavrilovich Gurvich. Moscow 1970

GURVICH, LEV GAVRILOVICH [1871-1926]
BSE 13,211-212; WWR, 225; POGG 6,983
M.A. Blokh, Zhur. Fiz. Khim. 60Suppl. (1928), 135

GUSSEROW, ADOLF LUDWIG SIGISMUND [1836-1906]
DBA 2,1337; BJN 11,27*; HIRSCH 2,916; GAUSS, 192; WREDE, 66-67

GUSTAVSON, GABRIEL (GAVRIL GAVRILOVICH) [1842-1908]
ARBUZOV, 170-173; BLOKH 1,265-268; POTSCH, 181-182; POGG 4,554-555; 5,470
Anon., Chem. Z. 32 (1908), 465
N. Demyanov, Zhur. Fiz. Khim. 41 (1909), 549-569

GUTBIER, ALEXANDER [1876-1926]
DSB 5,595-596; NDB 7,337-338; POTSCH, 182; POGG 5,470-472; 6,983-984
L. Birkenbach, Ber. chem. Ges. 59A (1926), 115-117
G.F. Hüttig, Z. angew. Chem. 40 (1927), 41-42

GUTFREUND, HERBERT [1921-]
WW 1992, p.770

GUTHERZ, SIEGFRIED [1881-1927]
IB, 108

GUTHRIE, CHARLES CLAUDE [1880-1963]
AMS 10,1560; IB, 108
S.P. Harbison and P.L. McLain, Pharmacologist 6 (1964), 5
L.G. Walker, Surgery 76 (1974), 359-362

GUTHRIE, FREDERICK [1833-1886]
DNB 8,817-818; WWWS, 724; POGG 3,567-568; 4,555
G.C. Foster, Nature 35 (1886), 8-10; Proc. Phys. Soc. 8 (1887), 9-13

GUTHRIE, SAMUEL [1782-1848]
DAB 8,62; NCAB 11,406-407; ELLIOTT, 111-112; DAMB, 312-313; FERCHL, 206; POTSCH, 182; SILLIMAN, 48-50; WWWS, 724; POGG 1,980
T.L. Davis, Archeion 13 (1931), 11-23
F.H. Getman, J. Chem. Ed. 17 (1940), 253-259
J.R. Pawling, Dr. Samuel Guthrie. Watertown, NY 1947

GUTHZEIT, MAX [1847-1915]
POGG 4,555-556; 5,473-474
Anon., Chem. Z. 39 (1915), 765

GUTMAN, ALEXANDER BENJAMIN [1902-1973]
AMS 12,2382; NCAB I,190
M.E. Bader et al., Mt. Sinai J. Med. 40 (1973), 713-714
L.H. Smith, American Journal of Medicine 56 (1974), 693-694
L.R. Wasserman, Trans. Assoc. Am. Phys. 87 (1974), 26-28

GUTMANN, AUGUST [1868-1929]
HEIN-E, 163-164; TSCHIRCH, 1059; POGG 5,474; 6,985

GUTMANN, HELMUT RUDOLPH [1911-]
AMS 15(3),376; BHDE 2,440

GUTMANN, VIKTOR [1921-]
KURSCHNER 16,1171; POTSCH, 182; WWWS, 724; POGG 7a(2),323-324

GUTSCHE, JESSE [1881-]
JV 26,394; NUC 224,71

GUTTMANN, LEO F. [1879-]
JV 19,180; NUC 224,94
Anon., Chem. Z. 37 (1913), 668

GUTTMANN, PAUL [1834-1893]
ADB 49,652-653; HIRSCH 2,920; Suppl.,353; PAGEL, 662; WININGER 2,570;
Anon., Leopoldina 29 (1893), 111
A. Eulenburg, Paul Guttmann. Berlin 1893
C.A. Ewald, Berl. klin. Wchschr. 30 (1893), 535
A.F., Münch. med. Wchschr. 19 (1893), 539

GUTZEIT, HEINRICH WILHELM THEODOR [1845-1888]
HEIN, 236-237; BLOKH 1,268; TSCHIRCH, 1059; SCHAEDLER, 43-44
Anon., Chem. Z. (1888), 1553

GUYE, CHARLES EUGÈNE [1866-1942]
DSB 5,597-598; DHB 3,717; GENEVA 5,14-19; 6,18-21; 7,79-85; 8,69-71; POGG 4,557-558; 5,475; 6,986-987; 7a(2),324
J.W., Helv. Phys. Acta 9 (1936), 511-514
E. Briner, J. Chim. Phys. 40 (1943), 1-4
J. Weigle, Arch. Sci. Phys. Nat. [5]25 (1943), 57-79

GUYE, PHILIPPE AUGUSTE [1862-1922]
DHB 3,716-717; BLOKH 1,268-269; WWWS, 725; POTSCH, 182-183; GENEVA 4,117-120; 5,34-38; 6,54-59; 7,85-89; POGG 4,556-557; 5,475-477; 6,987
Anon., Helv. Chim. Acta 5 (1922), 411-431
R. Chodat, Verhandl. Schw. Nat. Ges. 103 (1922), 18-33
T.E. Thorpe, Nature 109 (1922), 523-524
Anon., Bull. Soc. Chim. [4]33 (1923), 661-672
E. Briner, J. Chim. Phys. 20 (1923), 1-17

GUYENOT, ÉMILE [1885-1963]
DBF 17,395; BUICAN, 262-281; IB, 108; WWWS, 725
A. Binet, C.R. Soc. Biol. 157 (1963), 1347-1348
K. Ponse, Verhandl. Schw. Nat. Ges. 143 (1963), 212-216; Archives des Sciences 17 (1964), 71-73
R. Matthey, Bull. Soc. Vaud. Sci. Nat. 68 (1963), 350-351
P. Dangeard, Not. Acad. Sci. 5 (1972), 199-210

GUYER, MICHAEL FREDERIC [1874-1959]
DSB 5,598-599; AMS 9(II),448; NCAB A,357-358; WWAH, 242

GWINNER, HANS von [1887-1959]
JV 27,41
Anon., Nachr. Chem. Techn. 7 (1959), 399

GYEMANT, ANDREAS [1895-]
NUC 224,317

GYÖRGY, PAUL [1893-1976]
AMS 11,2009; MSE 1,465; DRULL, 97; BHDE 2,443; LDGS, 66; WWWS, 726
P. Gyorgy, Vitamins and Hormones 22 (1964), 361-365; Am. J. Clin. Nutr. 24 (1971), 1250-1256; Nutrition Reviews 34 (1976), 141-144
C.S. Rose, Trans. Coll. Phys. Phila. 44 (1976), 95-96
L.A. Barnes and R.M. Tomarelli, J. Nutrition 109 (1979), 19-23

H

HAACK, ERICH [1904-1968]
POGG 7a(2),327
Anon., Nachr. Chem. Techn. 8 (1960), 184; Chem. Z. 92 (1968), 107

HAAF, CARL [1834-1906]
RSC 15,552
A. Farmer, Verhandl. Schw. Nat. Ges. 90 (1907), xxxiv-xxxix

HAAG, WALTHER [1900-]
JV 41,663

HAAGER, ERNST [1876-1969]
JV 15,131; NUC 224,437

HAAKE, BRUNO [1874-1942]
GEDENKBUCH 1,489; JV 20,361; SG [3]6,377; NUC 224,444

HAAGEN-SMIT, ARIE JAN [1900-1977]
AMS 13,1691; NCAB 59,285-286
J. Bonner, Biog. Mem. Nat. Acad. Sci. 58 (1989), 189-216

HAALAND, MAGNUS [1876-1935]
NBL 5,195-197; IB, 108
Anon., Lancet 1935(II), p.283
O. Berner, Norske Vid. Akad. Arbok 1935, pp.35-48

HAAR, ANNE WILLEM van der [1878-1931]
POGG 5,1291; 6,990
D.H. Wester, Chem. Wkbl. 28 (1931), 272-275

HAARMANN, WALTER [1901-]
POGG 7a(2),329

HAARMANN, WILHELM [1847-1931]
NDB 7,372; WWWS, 727; POTSCH, 183; POGG 5,990-991
Anon., Chem. Z. 51 (1927), 395; 55 (1931), 214
M. Bodenstein, Ber. chem. Ges. 64A (1931), 38-39

HAAS, EMMY [1893-]
JV 34,287; NUC 224,505

HAAS, ERWIN [1906-]
AMS 14,1929; KURSCHNER 13,1264; WWWS, 727

HAAS, FRIEDRICH WILHELM [1888-1960]
JV 31,328; NUC 224,509

HAAS, GEORG [1886-1971]
GUNDEL, 357-364; KURSCHNER 4,979; FISCHER 1,557
Anon., Deutsche med. Wchschr. 96 (1971), 2022
J. Benedum, Med. Hist. J. 14 (1979), 196-217

HAAS, HANS [1907-]
KURSCHNER 13,1264; POGG 7a(2),330
H. Kleinsorge, Arzneimitt. 22 (1972), 168-170
G. Kroneberg, Arzneimitt. 27 (1977), 166

HAAS, PAUL [1877-1960]
DESMOND, 277; IB, 109; POGG 6,991-992; 7b,1797-1798
E.J. Salisbury, Nature 186 (1960), 595
F. Challenger, Biochem. J. 80 (1961), 1-4

HAAS, RICHARD [1910-1988]
KURSCHNER 13,1266; AUERBACH, 249; POGG 7a(2),331-332

HABER, EDGAR [1932-]
AMS 15(3),386; WWA 1980-1981, p.1390; BHDE 2,446

HABER, FRITZ [1868-1934]
DSB 5,620-623; NDB 7,386-389; HUPKA, 228-234; BHDE 2,446-447; WWWS, 727; W, 231-232; GRUETTER,43-44; WININGER 2,574-575; 7,46; TETZLAFF,122; POTSCH, 183-184; TLK 3,224-225; ABBOTT-C, 55-56; POGG 4,561; 5,479-480; 6,993-994; 7aSuppl.,250-251
W. Schlenk, Ber. chem. Ges. 67A (1934), 20-23
R. Willstätter, Bayer. Akad. Wiss. Jahrbuch 1934-1935, pp.51-54
E. Berl, J. Chem. Ed. 14 (1937), 203-207
J.E. Coates, J. Chem. Soc. 1939, pp.1642-1672
M. von Laue et al., Z. Elektrochem. 57 (1953), 1-8
K. Lohs, NTM 1 (1963), 37-44
R.A. Stern, Leo Baeck Institute Year Book 8 (1963), 70-113
H. Sachsse, Chemie in Unserer Zeit 2 (1967), 144-148
M. Goran, The Story of Fritz Haber. Norman, Okla. 1967
C. Haber, Mein Leben mit Fritz Haber. Munich 1969
M.R. Feldman and M.L. Tarver, J. Chem. Ed. 60 (1983), 463-464

HABERLAND, HERMANN [1865-]
JV 4,283; NUC 224,615

HABERLANDT, GOTTLIEB [1854-1945]
DSB 5,623-624; NDB 7,394-395; OBL 2,124-125; FREUND 1,189-195; WWWS, 728; KNOLL, 111-113; BK, 213; DZ, 510; TSCHIRCH, 1059; IB, 109
G. Haberlandt, Erinnerungen, Bekenntnisse und Betrachtungen. Berlin 1933
A.C. Noe, Plant Physiology 9 (1934), 851-855
F. Weber, Alm. Akad. Wiss. Wien 95 (1945), 372-380
O. Renner, Bayer. Akad. Wiss. Jahrbuch 1944-1948, pp.258-261
H. von Guttenberg, Phyton 6 (1955), 1-14
A.D. Krikorian and D.L. Berquam, Bot. Rev. 35 (1969), 59-88

HABERLANDT, LUDWIG [1885-1932]
NDB 7,395; KURSCHNER 4,985-986; FISCHER 1,559; HUTER, 221; IB, 109; WWWS, 728; POGG 7a(2),337-338
E. Brücke, Forsch. Fort. 8 (1932), 327
H.H. Simmer, Contraception 1 (1970), 3-27

HABERMANN, JOSEF [1841-1914]
OBL 2,126; DBJ 1,285; POGG 4,562; 5,480; 6,994
C. Frenzel, Chem. Z. 38 (1914), 973

HABS, HORST [1902-1987]
KURSCHNER 14,1380; WENIG, 101; ASEN, 67; POGG 7a(2),338-339
Anon., Münch. med. Wchschr. 109 (1967), 2216; Deutsche med. Wchschr. 92 (1967), 1744; Naturw. Rund. 40 (1987), 248

HACK, WILHELM [1851-1887]
ADB 49,696; HIRSCH 3,5-6; DIEPGEN, 186-198; BAD 4,162-164
M. Bresgen, Deutsche med. Wchschr. 13 (1887), 394
F. Simon, Internat. Centralbl. Laryngol. 4 (1887), 1-4
F. Keimer, Mon. Ohrenheilk. 21 (1887), 172-176HACKSPILL, LOUIS [1880-1963] DBF 17,475; DBA 2,1361-1362; CHARLE(2), 146-148
A.P. Rollet, Bull. Soc. Chim. 1964, pp.1427-1437

HADDOW, ALEXANDER [1907-1976]
DNB [1971-1980], 372-373; WW 1976, p.993; MH 2,200-201; MSE 2,1-2; WWWS, 729
P. Alexander, Nature 260 (1976), 179-180
F. Bergel, Biog. Mem. Fell. Roy. Soc. 23 (1977), 133-191
P. Dustin, Bull. Acad. Med. Belg. 136 (1981), 217-225

HADEN, RUSSELL LANDRAM [1888-1952]
AMS 8,991; NCAB 45,473-474; WWAH, 243
J.M. Edmonson, J. Lab. Clin. Med. 115 (1990), 528-530

HADLEY, PHILIP BARDWELL [1881-1963]
AMS 9(II),451; IB, 109
O. Amsterdamska, J. Hist. Biol. 24 (1991), 203-211

HADORN, ERNST [1902-1976]
DSB 17,373-377; KURSCHNER 12,1045; SBA 1,66
P.S. Chen et al., Revue Suisse de Biologie 79Suppl. (1972), 5-28
P. Tardent, Verhandl. Schw. Nat. Ges. 156 (1976), 120-121
W.J. Gehring, Dev. Biol. 53 (1976), iv-vi
R. Nöthiger, Genetics 86 (1977), 1-4
H.K. Mitchell, Ann. Rev. Gen. 12 (1978), 1-3

HAECKEL, ERNST [1834-1919]
DSB 6,6-11; NDB 7,423-425; DBJ 2,397-412,719; FISCHER 1,561-562; W, 233; PAGEL, 673-675; ABBOTT, 55-56; DZ, 511; WWWS, 729
K. Heider et al., Naturwiss. 7 (1919), 945-971
R. Hertwig, Bayer. Akad. Wiss. Jahrbuch 1919, pp.61-67
M. Verworn, Z. allgem. Physiol. 19 (1921), i-xi
G. Uschmann, Ernst Haeckel, 3rd Ed. Leipzig 1961; Ernst Haeckel: Biographie in Briefen. Leipzig 1983
J. Hemleben, Ernst Haeckel. Hamburg 1964
L. Szyfman, Bull. Biol. 113 (1979), 375-406
D.S. Peters, Med. Hist. J. 15 (1980), 57-69
J.M. Oppenheimer, Proc. Am. Phil. Soc. 126 (1982), 347-355
K. Keitel-Herz, Natur und Museum 114 (1984), 57-68
E. Krausse, Ernst Haeckel. Leipzig 1984 (2nd ed., 1987)
R. Mann, Berichte zur Wissenschaftsgeschichte 13 (1990), 1-11

HAECKEL, SIEGFRIED [1877-1931]
JV 17,283; NUC 225,224
Anon., Chem. Z. 55 (1931), 877; Z. angew. Chem. 44 (1931), 916; Chemische Fabrik 4 (1931), 444

HAECKER, ROLAND [1900-]
NUC 225,227HÄFLIGER, JOSEF ANTON [1873-1954]
NDB 7,429-430; KURSCHNER 8,769; HEIN-E, 166-167; POGG 7a(2),343-344
K. Meyer, Schw. Apoth. Z. 92 (1954), 948-953
Anon., Deutsche Apoth. Z. 94 (1954), 1180-1181

HAEHN, HUGO [1880-1957]
NDB 7,431-432; IB, 109; WWWS, 730; POGG 6,999; 7a(2),344; (4),143*
Anon., Chem. Z. 74 (1950), 696
B. Drews, Brauerei 52 (1955), 589

HÄHNEL, OTTO [1884-]
KURSCHNER 4,1001-1002; JV 24,30; NUC 225,274; POGG 6,999-1000; 7a(2),344
Anon., Nachr. Chem. Techn. 3 (1955), 7; 8 (1960), 22

HÄMÄLÄINEN, YUHO HEIKKI [1883-1915]
ENKVIST, 107-108

HÄMMERLING, JOACHIM [1901-1980]
KURSCHNER 13,1284; WWWS, 745
H. Harris, Biog. Mem. Fell. Roy. Soc. 28 (1982), 111-124

HÄNDEL, LUDWIG [1869-1939]
KURSCHNER 4,1017; FISCHER 1,562; IB, 109
Anon., Z. Bakt. 114 (1929), 4-6

HAENISCH, VICTOR [1871-1940]
JV 9,17; NUC 225,470

HÄRLEIN, JULIUS [1835-]
RSC 3,111; 7,884
A. Wankmüller, Beitr. Württ. Apothekgesch. 6 (1964), 95

HAESER, HEINRICH [1811-1885]
NDB 7,453; ADB 50,53-54; HIRSCH 3,9-11; GIESE, 540-541; OLPP, 167-168; WWWS, 730
O. Temkin and C.L. Temkin, Bull. Hist. Med. 32 (1958), 97-104

HÄUSSERMANN, CARL [1853-1918]
NDB 7,459-460; DBJ 2,689; BLOKH 1,269-270; POGG 4,564-565; 5,481-482
P. Lepsius, Ber. chem. Ges. 51 (1918), 1683-1685
C. von Hell, Z. angew. Chem. 31 (1918), 413-414
Württemberger Nekrolog 1922, pp.47-51

HAFFKINE, WALDEMAR [1860-1930]
DSB 6,11-13; FISCHER 1,563; BULLOCH, 370-371; OLPP, 168-170; WWWS, 730; KAGAN, 157; WININGER 2,579-581; 7,47; KOREN, 187
S.A. Waksman, The Brilliant and Tragic Life of Waldemar Mordecai Wolff Haffkine. New Brunswick, N.J. 1964
E. Lutzker, Am. Phil. Soc. Year Book 1967, pp.577-580
H.I. Jkala, Ind. J. Hist. Sci. 2 (1967), 105-120
G.H. Bornside, J. Hist. Med. 37 (1982), 399-422
I. Löwy, J. Hist. Med. 47 (1992), 270-309

HAFFNER, FELIX [1886- 1953]
NDB 7,461-462; FISCHER 1,563; IB, 110; WWWS, 730; POGG 7a(2),347-348
W. Schmidt, Arzneimitt. 3 (1953), 316; Deutsche Apoth. Z. 93 (1953), 230

HAGA, TAMEMASA [1856-1914]
NUC 225,584; POGG 4,565; 5,482

HAGEDOORN, AREND LOURENS [1880-1953]
M. Pease, Nature 173 (1954), 60-61

HAGEDORN, HANS CHRISTIAN [1888-1971]
DBL 8,577-578; KBB 1971, pp.411-412; VEIBEL 2,171
P. Felig, J. Am. Med. Assn. 251 (1984), 389-396

HAGEMANN, JOHANNES [1884-1943]
JV 28; SG [3]6,394; NUC 225,637

HAGEN, ROBERT HERMANN [1815-1858]
FERCHL, 210; POGG 1,993,1571

HAGENBACH, RUDOLF [1875-1927]
NUC 225,695
H. Rupe, Verhandl. Schw. Nat. Ges. 108 (1927), 10-13

HAGER, HERMANN [1816-1897]
NDB 7,490-491; HEIN, 241-242; TSCHIRCH, 1059-1060; SCHAEDLER, 47; FERCHL, 210; WWWS, 732; POGG 4,567
Anon., Pharm. Z. 31 (1897), 353-354; Am. J. Pharm. 69 (1897), 182-189
W. Zimmermann, Süddeutsche Apoth. Z. 81 (1941), 39-41
F. Schmidt, Deutsche Apoth. Z. 105 (1956), 1829-1830
H. Löhr, Pharmazie 45 (1990), 130-133

HAGER, HERMANN [1887-]
BLOKH 1,270; JV 26,415; NUC 226,16
New York Times, 4 May 1946

HAGIWARA, SUSUMU [1922-1989]
AMS 17(3),418

HAHN, AMANDUS [1889-1952]
NDB 7,501-502; TLK 3,227; WWWS, 733; POGG 6,1002-1003; 7a(2),349-350
H. Niemer, Z. Biol. 105 (1952), 1-6
R. Wagner, Bayer. Akad. Wiss. Jahrbuch 1952, pp.204-208

HAHN, FRIEDRICH VINCENZ von [1897-]
IB, 110; POGG 6,1004; 7a(2),351

HAHN, FRITZ [1907-1982]
KURSCHNER 13,1303; WWWS, 733; POGG 7a(2),351-353
Anon., Naturw. Rund. 35 (1982), 343

HAHN, GEORG [1899-1979]
POGG 6,1004; 7a(2),353
Anon., Chem. Z. 98 (1974), 620; 104 (1980), 181

HAHN, HERMANN [1872-1912]
BJN 18,26*; FISCHER 1,565; EGERER, 66-68
J. Rückert, Anat. Anz. 41 (1912), 105-109
F. Wassermann, Münch. med. Wchschr. 59 (1912), 767

HAHN, MARTIN [1865-1934]
KURSCHNER 4,1001; FISCHER 1,565-566; TLK 3,228; IB, 110; GRUETTER, 45 KOREN, 188
M. Hahn, Münch. med. Wchschr. 55 (1908), 515-516
Anon., Ber. chem. Ges. 69A (1934), 173
E. Schütz, Deutsche med. Wchschr. 61 (1935), 69-70

HAHN, PAUL FRANCIS [1908-1967]
AMS 11,2033

HAHNENKAMM, WILHELM [1875-]
JV 19,338; NUC 226,268

HAIDLEN, PAUL JULIUS [1818-1883]
HEIN, 243; LEHMANN, 58-59; RSC 3,125; POGG 1,998
Anon., Leopoldina 19 (1883), 219
A. Wankmüller, Beitr. Württ. Apothekgesch. 7 (1966), 83

HAILER, EKKEHARD [1877-1939]
NDB 7,520-521; IB, 110; WWWS, 734; POGG 4,568; 5,485; 6,1006; 7a(2),359
H. Reiter, Reichsgesundheitsblatt 14 (1939), 421

HAIM, ARTHUR [1898-]
LDGS, 56

HAISER, FRANZ [1871-1945]
W. Ziegenfuss, Philosophen-Lexikon 1 (1949), 438

HAISS, AUGUST [1856-1905]
RSC 10,113; NUC 226,402

HAITINGER, MAX [1868-1946]
NDB 7,527-528; OBL 2,154-155; KNOLL, 58-60; FREUND 3,187-194; WWWS, 735; POGG 6,1007; 7a(2),360
L.F. Bräutigam, Mikroskopie 2 (1947), 84-89
H. Freund, Mikroskopie 25 (1969), 73-77

HALBAN, HANS von [1877-1947]
NDB 7,530; OBL 2,157-158; KURSCHNER 5,472; TLK 3,228-229; WWWS, 1735; POGG 5,486; 6,1007; 7a(2),361-362
L. Ebert, Ost. Chem. Z. 48 (1947), 210-211; Viert. Nat. Ges. Zürich 93 (1948), 144-149
M. Kofler, Helv. Chim. Acta 31 (1948), 120-128

HALBAN, JOSEF von [1870-1937]
NDB 7,530-531; OBL 2,158; KURSCHNER 5,472; FISCHER 1,567; PLANER, 116; MEDVEI, 741-742; ZWEIFEL, 53-55; WININGER 2,588; 7,48; WWWS, 1735;TETZLAFF, 124
R. Köhler, Z. Gyn. 61 (1937), 1457-1466
W. Latzko, Wiener med. Wchschr. 87 (1937), 626-628
H.H. Simmer, Wiener med. Wchschr. 121 (1971), 549-552

HALBERKANN, JOSEF [1880-1952]
HEIN-E, 169-170; POGG 6,1007-1008; 7a(2),362
Anon., Deutsche Apoth. Z. 92 (1952), 141-142

HALDANE, JOHN BURDON SANDERSON [1892-1964]
DSB 6,21-23; DNB [1961-1970], 473-475; MH 1,208-209; STC 1,477-479; MSE 2,9-10; WW 1959, p.1277; CB 1940, pp.357-359; ABBOTT, 56-57; W, 233-234; WWWS, 735; POTSCH, 186M.J.D. White, Genetics 52 (1965), 1-7
J.B.S. Haldane, Persp. Biol. Med. 9 (1966), 476-481; The Man with Two Memories. London 1976
N.W. Pirie, Biog. Mem. Fell. Roy. Soc. 12 (1966); TIBS 4 (1979), N273-N175
R.W. Clark, J.B.S.: The Life and Work of J.B.S. Haldane. London 1968
K.R. Dronamraju (Ed.), Haldane and Modern Biology. Baltimore 1968
P.G. Werskey, J. Hist. Biol. 4 (1971), 171-183
K.R. Dronamraju, Haldane: The Life and Work of J.B.S. Haldane with Special Reference to India. Aberdeen 1985; Notes Roy. Soc. 41 (1987), 211-237

HALDANE, JOHN SCOTT [1860-1936]
DSB 6,23-25; DNB [1931-1940], 389-391; O'CONNOR(2), 97-103; FISCHER 1,567-568; NUESCH, 88-92; W, 234-235; VERSO, 103-104; ABBOTT, 57; WWWS, 735-736; POGG 6,1008; 7b,1820-1821
J.G.P., Nature 137 (1936), 566-569
C.G. Douglas, Obit. Not. Fell. Roy. Soc. 2 (1936), 115-139
[J.F. Fulton], New Eng. J. Med. 214 (1936), 651-652
J.B.S. Haldane, Nature 187 (1960), 102-105
G.E. Allen, J. Hist. Med. 22 (1967), 392-412
N. Mitchison, Proc. Roy. Inst. 47 (1974), 1-22
S. Sturdy, Brit. J. Hist. Sci. 21 (1988), 315-340

HALDEN (before 1921, CSANYI), WILHELM [1892-1981]
KURSCHNER 13,1313; WWWS, 736; POGG 6,1008; 7a(2),362-363

HALL, ALFRED DANIEL [1864-1942]
DNB [1941-1950], 339-341; DESMOND, 279
J. Russell, Obit. Not. Fell. Roy. Soc. 4 (1944), 229-250

HALL, ARCHIBALD ALEXANDER [1879-1949]
JV 19; NUC 227,145
Anon., J. Roy. Inst. Chem. 1950, p.276

HALL, CECIL EDWIN [1912-]
AMS 11,2044

HALL, FRANK GREGORY [1896-1967]
AMS 11,2045; WWA 1966-1967, p.865; IB, 111
D.B. Dill, Case History of a Physiologist: F.G. Hall. Boulder City, Nev. 1971

HALL, HARLOW HOMER [1904-1967]
AMS 10,1590; WWWS, 738

HALL, LYMAN BEECHER [1852-1935]
AMS 5,458; MILES, 194-195; RSC 10,116-117; NUC 227,384

HALL, MARSHALL [1790-1857]
DSB 6,58-61; DNB 9,964-967; HIRSCH 3,28-30; Suppl.,356; HAYMAKER,221-225; BETTANY 1,264-285; O'CONNOR, 15-18; W,237-238; WWWS,738; CALLISEN 8,71-75; 28,359-362
R. Christison, Proc. Roy. Soc. Edin. 4 (1857-1862), 11-14
C. Hall, Memoirs of Marshall Hall. London 1861
J.H.S. Green, Med. Hist. 2 (1958), 120-133
R. Erez-Federbusch, Marshall Hall, Physiologe und Praktiker. Zurich 1963
D.E. Manuel, Notes Roy. Soc. 35 (1980), 135-166

HALL, PETER FRANCIS [1924-]
AMS 15(3),424-425; WWA 1980-1981, p.1408; WWWS, 738-739

HALL, ROSS HUME [1926-]
AMS 15(3),426; WWA 1980-1981, p.1408' WWWS, 739

HALL, VICTOR ERNEST [1901-1981]
AMS 11,2052-2053

HALLAUER, CURT [1900-]
KURSCHNER 13,1314; SBA 1,68
S. Gard et al., Arch. ges. Virusforsch. 31 (1970), i-iii
R.H. Regamey, Schw. med. Wchschr. 100 (1970), 1364-1365

HALLAWAY, ROBERT RAILTON [1874-1923]
SRS, 22; JV 16,143; NUC 227,531

HALLÉ, JEAN NOËL [1754-1822]
DBF 17,518-519; GENTY 1,341-351; GE 19,770; LAROUSSE 9,36; WWWS, 739; HIRSCH 3,32; DECHAMBRE 48,71-72; CALLISEN 28,363
G. Cuvier, Éloges Historiques des Membres de l'Académie des Sciences, 2nd Ed., vol.3, pp.8-18. Paris 1861
E.F. Dubois, Éloges...de l'Académie de Medecine 1845-1863, vol.1, pp.219-270. Paris 1864

HALLENSLEBEN, RICHARD [1877-]
JV 20,467; NUC 227,584

HALLER, ALBIN [1849-1925]
DBF 17,521-522; DBA 2,1387; HEIN, 243-244; BLOKH 1,271-272; WWWS, 740; CHARLE(2), 148-150; POTSCH, 187; POGG 4,569-571; 5,488-489; 6,1011-1012
J.B., J. Pharm. Chim. [7]1 (1925), 510-512
P. Ramart, Bull. Soc. Chim. [4]39 (1926), 1037-1092
J.F.T., Proc. Roy. Soc. A118 (1928), i-iii
C. Moureu and M. Molliard, Not. Acad. Sci. 1 (1937), 37-45
E.H. Huntress, Proc. Am. Acad. Arts Sci. 77 (1949), 38-39
J. Aubry, Mem. Acad. Stanislas [7]8 (1979-1980), 331-339
P. Labrude, Rev. Hist. Pharm. 29 (1982), 207-209

HALLER, HERBERT LUDWIG [1894-1972]
AMS 11,2054; WWAH, 246; POGG 6,1012; 7b,1825-1828
Anon., Chem. Eng. News 42(32) (1964), 66; 50(47) (1972), 26

HALLERVORDEN, EUGEN [1853-1914]
DBJ 1,283; RSC 10,119; 15,590
Anon., Münch. med. Wchschr. 61 (1914), 2023
A. Bader and M. Bader, Nervenarzt 56 (1985), 134-139

HALLERVORDEN, JULIUS [1882-1965]
KROLLMANN, 940
H. Spatz, Nervenarzt 23 (1952), 468
Anon., Leopoldina [3]6/7 (1960-1961), 33

HALLIBURTON, WILLIAM DOBINSON [1860-1931]
DSB 17,377-379; DNB [1931-1940], 391-392; WW 1931, p.1356; FISCHER 1,569; WWWS, 740
Anon., Lancet 1931(II), pp.1263-1265
J.A. Hewitt, Nature 127 (1931), 932,945
J.A.H., Biochem. J. 26 (1932), 269-271
N. Morgan, Notes Roy. Soc. 38 (1983), 129-145

HALLIER, ERNST [1831-1904]
DSB 6,72-73; NDB 7,563-564; BJN 10,42*; HIRSCH 3,37; BULLOCH, 371; TSCHIRCH, 1060; WWWS, 740; POGG 3,577; 4,571-572
Anon., Leopoldina 41 (1905), 38

HALLION, LOUIS [1862-1940]
DBF 17,527; FISCHER 1,569
L. Binet, Bull. Acad. Med. 123 (1940), 402-405

HALLWACHS, WILHELM [1834-1881]
POGG 3,577

HALPERN, BERNARD [1904-1978]
STC 1,480-481
J. Roche, C.R. Soc. Biol. 172 (1978), 607-608

HALSEY, JOHN TAYLOR [1870-1951]
AMS 8,1009; RSC 15,596
H. Cummins, Bulletin of the Tulane Medical Society 11 (1952), 183-184

HALSTED, WILLIAM STEWART [1852-1922]
DSB 6,77-78; DAB 8,164-165; FISCHER 1,570; DAMB, 320; WWWS, 741
H. C[ushing], Science 56 (1922), 461-462
W.G. MacCallum, William Stewart Halsted. Baltimore 1930; Biog. Mem. Nat. Acad. Sci. 17 (1936), 151-170
S.J. Crowe, Halsted of Johns Hopkins. Sprongfield, Ill. 1957
P.D. Olch, Bull. Hist. Med. 40 (1966), 495-510
W. Penfield, J. Am. Med. Assn. 210 (1969), 2114-2118

HALVORSON, HALVOR ORIN [1897-1975]
AMS 11,2060; WWWS,741
Anon., BioScience 26 (1976), 71

HALVORSON, HARLYN ODELL [1925-]
ANS 15(3),435; WWA 1980-1981, p.1414; WWWS, 741-742

HAM, THOMAS HALE [1905-1987]
AMS 14,1971; WWWS, 742
W.B. Castle and E.R. Jaffé, Seminars in Hematology 13(2) (1976), 87-101
V. Herbert, J. Am. Med. Assn. 251 (1984), 522-523

HAMANN, KARL [1906-]
KURSCHNER 13,1321; POGG 7a(2),365-366
Anon., Chem. Z. 77 (1953), 154

HAMBSCH, OTTO [1894-]
JV 37,1282

HAMBURGER, ALEXANDER [1880-1914]
JV 23,543; NUC 228,290

HAMBURGER, CARL [1870-1944]
NDB 7,581-582; KURSCHNER 4,1009
H. Friedenwald, Archives of Ophthalmology 31 (1944), 557

HAMBURGER, FRANZ ANTON [1874-1954]
FISCHER 1,570-571; ATUYANA, 119-142; IB, 111
I. Türk and E.H. Major, Wiener med. Wchschr. 104 (1954), 889-891

HAMBURGER, HARTOG JAKOB [1859-1924]
BWN 2,207-209; LINDEBOOM, 775-776; FISCHER 1,571; WININGER 2,601; 7,53; HES, 67-69; WWWS, 742; POGG 4,575-576; 5,490; 6,1014
E. Cohen, Biochem. Z. 11 (1908), i-xxxiii
E. Laqueur, Klin. Wchschr. 3 (1924), 383

HAMBURGER, VIKTOR [1900-]
AMS 14,1973; WWA 1978-1979, p.1358; IWW 1982-1983, p.521; MH 2,201-202; MSE 2,11; BHDE 2,456; LDGS, 14; IB, 111; WWWS, 742
J. Holtfreter, in The Emergence of Order in Developing Systems (M. Locke, Ed.), pp.ix-xx. New York 1968
V. Hamburger, Annual Review of Neuroscience 3 (1980), 269-278; 12 (1989), 1-12
R. Levi-Montalcini, in Studies on Developmental Neurobiology (M. Cowan, Ed.), pp.22-43. Oxford 1982

HAMER, WALTER JAY [1907-]
AMS 14,1974; WWA 1976-1977, p.1305; WWWS, 743

HAMERNIK, JOSEF [1810-1887]
OBL 2,164; WURZBACH 7,262; HIRSCH 3,41-42; PAGEL, 682-683; NAVRATIL, 77-79; WERSTLER, 113-121
Anon., Wiener med. Wchschr. 37 (1887), 740-741; Cas. Lek. Cesk. 26 (1887), 359

HAMILL, JOHN MOLYNEUX [1880-1960]
WW 1959, p.1291; O'CONNOR(2), 152; NUC 228,381
Who's Who in British Science 1953, p.117

HAMILL, PHILIP [1883-1959]
O'CONNOR(2), 266; NUC 228,383
Anon., Lancet 1959(I), pp.586-587,739

HAMILTON, CLIFF STRUTHERS [1889-1975]
AMS 11,2065; WWWS, 743; POGG 6,1015; 7b,1830-1832
W.E. Noland, Organic Syntheses 55 (1976), vii-viii

HAMILTON, PAUL BARNARD [1909-1989]
AMS 17(3),460

HAMILTON, WALTER CLARK [1931-1973]
AMS 12,2453; McLACHLAN, 121-122; WWWS, 744
Anon., Chem. Eng. News 51(8) (1973), 39

HAMMAR, AUGUST [1861-1946]
SMK 3,283; FISCHER 1,572; IB, 111; RSC 15,606
M. Wrete, Anat. Nachr. 1 (1950), 254-259

HAMMARSTEN, EINAR [1889-1968]
SBL 18,214-215; FISCHER 1,572-573; POGG 6,1015; 7b,1832-1833
E. Jorpes, Nord. Med. 80 (1969), 1701-1703

HAMMARSTEN, HARALD [1897-]
POGG 6,1015; 7b,1833

HAMMARSTEN, OLOF [1841-1932]
SBL 18,207-209; SMK 3,295-296; FISCHER 1,573; PAGEL, 683-684; IB,111-112; POGG 3,580; 4,576; 6,1015-1016; 7b,1833
H. von Euler, Naturwiss. 9 (1921), 639-643
G. Blix, Hygeia 94 (1932), 737-742
Anon., Sven. Kem. Tid. 44 (1932), 131 H. Wieland, Bayer. Akad. Wiss. Jahrbuch 1932-1933, pp.25-26
T. Thunberg, Erg. Physiol. 35 (1933), 13-31
E. Wohlisch, Schw. med. Wchschr. 84 (1954), 774-776

HAMMERSCHLAG, ALBERT [1863-1935]
KURSCHNER 4,1013-1014; FISCHER 1,573; WEBER, 38
J. Mannaberg, Wiener klin. Wchschr. 48 (1935), 1000

HAMMES, GORDON [1934-]
AMS 15(3),449; WWA 1980-1981, p.1422

HAMMETT, FREDERICK SIMONDS [1885-1935]
AMS 8,1015; NCAB 42,203; IB, 112
S.P. Reimann and P.R. White, Growth 17 (1953), 77-80

HAMMETT, LOUIS PLACK [1894-1987]
AMS 14,1982; IWW 1982-1983, p.523; MH 1,210-211; MSE 2,11-12; POTSCH, 187; POGG 6,1017; 7b,1833-1835
New York Times, 5 March 1987
F.A. Long, Chem. Brit. 24 (1988), 63
J. Shorter, Progress in Physical Organic Chemistry 17 (1990), 1-29

HAMMICK, DALZIEL LLEWELLYN [1887-1966]
WW 1965, p.1305; ABBOTT-C, 56-57; POGG 6,1017; POGG 7b,1835-1837
A.S. Russell, Nature 212 (1966), 674
B.R. Brown, Chem. Ind. 1967, pp.656-657
E.J. Bowen, Biog. Mem. Fell. Roy. Soc. 13 (1967), 107-124

HAMMOND, GEORGE SIMMS [1921-]
AMS 17(3),468; POTSCH, 187; WWWS, 746

HAMMOND, WILLIAM ALEXANDER [1828-1900]
DAB 8,210-211; NCAB 26,468-469; KELLY 1,380-382; HIRSCH 3,47; PAGEL, 684-685; TALBOTT, 808-810; DAMB,322-323
D. Roosa et al., Post-Graduate (New York) 15 (1900), 594-643
Anon., N.Y. Med. J. 71 (1900), 64

HAMPERL, HERWIG [1899-1976]
KURSCHNER 12,1087; AUERBACH, 253; POGG 7a(2),367-368
H. Hamperl, Werdegang und Lebensweg eines Pathologen. Stuttgart 1972
Anon., Naturw. Rund. 29 (1976), 290
P. Gedigk, Deutsche med. Wchschr. 102 (1977), 842-844

HAMPIL, BETTY LEE [1896-]
AMS 12,2461

HAMSIK, ANTONIN [1878-1963]
FISCHER 1,574; IB, 112; POGG 6,1018; 7b,1837-1838
A.F. Richter, Chem. Listy 47 (1953), 161-167

HANAHAN, DONALD JAMES [1919-]
AMS 15(3),456; WWA 1980-1981, p.1425; WWWS, 747

HANAU, ARTHUR [1858-1900]
BJN 5,93*; FISCHER 1,574-575; DIEPGEN, 55-56; SG [2]6,723
Anon., Leopoldina 36 (1900), 153

HANAUSEK, THOMAS FRANZ [1852-1918]
NDB 7,603-604; REBER, 49-56; TSCHIRCH, 1060; WWWS, 747
H. Pabisch, Ber. bot. Ges. 35 (1918), 108-118; Chem. Z. 42 (1918), 125
J. Wiese, Arch. Chem. Mikr. 11 (1918), 27-46
E. Gilg and H. Thoms, Ber. pharm. Ges. 28 (1918), 245-252

HANBURY, DANIEL [1825-1875]
DNB 8,1154-1155; DESMOND, 282; REBER, 1-3; TSCHIRCH, 1060; POGG 3,581
A.G., Am. J. Sci. [3]9 (1875), 475-476
Anon., Nature 11 (1875), 428-429; 14 (1876), 366-367; Proc. Roy. Soc. 24 (1876), ii-iii
D. Chapman-Huston and E.C. Cripps, Through a City Archway, pp.157-169. London 1954

HANDLER, PHILIP [1917-1981]
AMS 14,1990; WWA 1978-1979, p.1368; AO 1981, pp.773-775
E.L. Smith, Am. Phil. Soc. Year Book 1982, pp.490-496
T.H. Jukes, J. Nutrition 113 (1983), 1085-1094
E.L. Smith and R.L. Hill, Biog. Mem. Nat. Acad. Sci. 55 (1985), 305-353

HANDOVSKY, HANS [1888-1959]
NDB 7,610; FISCHER 1,575-576; BHDE 2,457; LDGS, 78; IB. 112; KOREN, 189; WWWS, 747; POGG 6,1019-1020; 7a(2),368-369; (4),144*
P. Marquardt, Arzneimitt. 10 (1960), 62

HANES, CHARLES SAMUEL [1903-1990]
WW 1981, p.1110; SRS, 61,87; YOUNG, 23; WWWS, 748
J. Wong, TIBS 4 (1979), N130-N131

HANGER, FRANKLIN McCUE [1894-1971]
AMS 11,2082; WWAH, 249; WWWS, 748
D.W. Atchley, Trans. Assoc. Am. Phys. 85 (1972), 15-16

HANKE, MARTIN [1898-1976]
AMS 11,2083; NUC 229,582
Anon., Chem. Eng. News 54(46) (1976), 42; Science 194 (1976), 972

HANKE, MILTON THEODORE [1893-1961]
AMS 10,1618

HANKES, LAWRENCE VALENTINE [1919-]
AMS 15(3),462; WWWS, 748

HANNOVER, ADOLPH [1814-1894]
DBL 9,61-64; HIRSCH 3,53-54; PAGEL, 685; MEISEN, 120-123; WWWS, 749; VEIBEL 2,175; WININGER 7,54; KOREN, 189; CALLISEN 28,376-377
Anon., Leopoldina 30 (1894), 169-160
J.W.S. Johnson, Mitt. Gesch. Med. 14 (1915), 109-111
J. Thoms, Med. Forum 1971, pp.91-98; Adolph Hannover. Copenhagen 1978

HANRIOT, MAURICE [1854-1933]
DBF 17,598; STALDER, 32-33; WWWS, 749; POGG 4,581-582; 5,494; 6,1021; 7b,1840
A. Kling, Presse Med. 41 (1933), 1723-1724; Bull. Soc. Chim. [5]2 (1935), 1753-1776
M. Delépine, Bull. Acad. Med. 110 (1933), 197-202

HANSEMANN, DAVID PAUL von [1858-1920]
NDB 7,629-630; DBJ 2,748; FISCHER 1,576; PAGEL,686; WREDE, 74; WININGER 2,608-609; 7,55; KOREN, 189; WWWS, 1735
C. Benda, Deutsche med. Wchschr. 46 (1920), 1088
B. Ostertag, Verhandl. path. Ges. 29 (1937), 370-378

HANSEN, ADOLPH [1851-1920]
DBJ 2,748; HEIN, 245-247; GUNDEL, 365-371; TSCHIRCH, 1060-1061
E. Küster, Ber. bot. Ges. 38 (1920), (66)-(77)

HANSEN, EMANUEL [1894-1964]
KBB 1964, pp.532-533

HANSEN, EMIL CHRISTIAN [1842-1909]
DSB 6,99-101; MEISEN, 161-154; BULLOCH, 371; TSCHIRCH, 1061; WWWS, 749
A. Klöcker, Ber. bot. Ges. 27 (1909), (73)-(84);
Chem. Z. 33 (1909), 957; C.R. Lab. Carlsberg 9 (1913), i-xxxvi
N.C. Ortved, J. Ind, Eng. Chem. 1 (1909), 733-734
E. Almquist, Hygiea 71 (1909), 1137-1153

HANSGIRG, ANTONIN [1854-1917]
OBL 2,183
B. Nemec, Almanach České Akademie 28 (1918), 85-92
F. Drouet, Annalen des Naturhistorischen Musuem Wien 61 (1957), 41-59

HANSON, ADOLPH MELANCHTON [1888-1959]
AMS 9(II),467; WWWS, 750

HANSON, EMMELINE JEAN [1919-1973]
DNB [1971-1980], 377-378; WW 1973, p.1387
J. Randall, Biog. Mem. Fell. Roy. Soc. 21 (1975), 313-344

HANSON, FRANK BLAIR [1886-1945]
AMS 7,739; IB, 112
F.O. Schmitt, Science 103 (1946), 143

HANSON, HORST [1911-1978]
KURSCHNER 11,996; WWWS, 750; POGG 7a(2), 374-375
Anon., Naturw. Rund. 31 (1978), 399

HANSSEN, OLAV MIKAL [1878-1965]
NBL 5,415-416
Hver er Hvem? 1964, p.230

HANSTEEN (CRANNER), BARTHOLD [1867-1925]
NBL 3,145-146

HANSTEIN, HEINRICH [1825-1871]
HEIN, 247-248; HB 2,481-483; WOLF, 73; NUC 230,295

HANSTEIN, JOHANNES von [1822-1880]
NDB 7,640-641; ADB 49,768-770; TSCHIRCH, 1061; WWWS, 1736
H. Vöchting, Bot. Z. 39 (1881), 233-242

HANTZSCH, ARTHUR [1857-1935]
DSB 6,107-109; NDB 7,641-642; FARBER, 1067-1083; TLK 3,235; WWWS, 751; POTSCH, 188; POGG 3,586; 4,582-584; 5,494-496; 6,1023-1024; 7a(2), 375-376
C. Paal, Z. angew Chem. 40 (1927), 301-303
B. Helferich, Sitz. Akad. Wiss. Leipzig 87 (1935), 213-222
T.S. Moore, J. Chem. Soc. 1936,pp.1051-1066; Proc. Chem. Soc. 1959 ,pp.1-4
F. Hein, Z. Elektrochem. 42 (1936), 1-4; Ber. chem. Ges. 74A (1941), 147-163

HANZLIK, PAUL JOHN [1885-1951]
AMS 8,1027; NCAB 39,366-367; PARASCANDOLA, 34-35; IB, 113; WWAH, 250

HAPPE, GUSTAV HEINRICH [1876-1967]
JV 19,339; NUC 230,347
Anon., Nachr. Chem. Techn. 14 (1966), 365; 15 (1967), 85

HAPPOLD, FRANK CHARLES [1902-1991]
WW 1981, p.1114; IB, 113; WWWS, 751

HARBURY, HENRY ALEXANDER [1927-]
AMS 17(3),500

HARCOURT, AUGUSTUS GEORGE VERNON [1834-1919]
DSB 6,109-110; DNB [1912-1921], 238-239; WWWS, 751-752; POTSCH, 436-437; POGG 4,1560; 5,1306; 6,2751
H.B. D[ixon], Proc. Roy. Soc. A97 (1920), vii-xi
C. King, New Scientist 82 (1979), 1110-1111
J. Shorter, J. Chem. Ed. 57 (1980), 411-416
M.C. King, Ambix 31 (1984), 16-31

HARDEGGER, EMIL [1913-1978]
KURSCHNER 12,1098; POGG 7a(2),377
Anon., Chem. Z. 102 (1978), 326

HARDEN, ARTHUR [1865-1940]
DSB 6,110-112; DNB [1931-1940], 395-397; WW 1940, pp.1375-1376; WWWS,752; O'CONNOR(2), 461-462; FINDLAY, 270-284; FISCHER 1,578-579; W, 242; ABBOTT-C, 57-58; POTSCH, 188-189; HUBER, 46-47; POGG 6,1024; 7b,1842
I. Smedley-MacLean, Biochem. J. 35 (1941), 1071-1081
F.G. Hopkins and C.J. Martin, Obit. Not. Fell. Roy. Soc. 4 (1942), 3-14; J. Chem. Soc. 1943, pp.334-340
R.E. Kohler, Bull. Hist. Med. 48 (1974), 22-40

HARDER, RICHARD [1888-1973]
KURSCHNER 10,811; WAGENITZ, 72-73
M. Steiner, Ber. bot. Ges. 93 (1980), 477-504

HARDESTY, IRVING [1866-1944]
AMS 7,741
Anon., Science 100 (1944), 444

HARDING, EVERHART PERCY [1870-1932]
AMS 3,289; JV 16,143; NUC 230,555

HARDING, VICTOR JOHN [1885-1934]
YOUNG, 10
Anon., Chem. Ind. 1934,p.662; Trans. Roy. Soc. Canada [3]29 (1935),iv-vi
E.J. K[ing], Biochem. J. 29 (1935), 1-4
W.N. Haworth, J. Chem. Soc. 1935, pp.1341-1343

HARDY, ALISTER CLAVERING [1896-1985]
WW 1981, p.1119; IWW 1982-1983, p.528; ABBOTT, 59-60; WWWS, 752
N.B. Marshall, Biog. Mem. Fell. Roy. Soc. 32 (1986), 223-273

HARDY, EDMUND [1816-1878]
NDB 7,670*; RSC 3,176; 7,907

HARDY, WILLIAM BATE [1864-1934]
DSB 15,201-202; DNB [1931-1940], 397-398; O'CONNOR(2), 23-25; WWWS, 753; POGG 4,585; 5,498; 6,1026; 7b,1845
F.G. H[opkins], Biochem. J. 28 (1934), 1149-1152
T.M., Chem. Ind. 1934, pp.133-134
F.G. H[opkins] and F.E. S[mith], Obit. Not. Fell. Roy. Soc. 1 (1934), 327-333; Nature 133 (1934), 281-283
E.K. Rideal, Trans. Faraday Soc. 60 (1964), 1681-1687; Proc. Roy. Inst. 40 (1964), 178-185
E.C. Bate-Smith, Sir W.B. Hardy, Biologist, Physicist and Food Chemist. Cambridge 1964

HARE (BERNHEIM), MARY LILIAS CHRISTIAN [1902-]
AMS 13,323

HARE, ROBERT [1781-1858]
DSB 6,114-115; DAB 8,263-264; MILES, 195-196; ELLIOTT, 116-117; WWAH,251; FARBER, 421-433; COLE, 238-242; W, 242-243; POGG 1,1018-1019
Anon., Am. J. Sci. [2]26 (1858), 100-105
E.F. Smith, The Life of Robert Hare. Philadelphia 1917; Archeion 8 (1927),330-335

HARE, RONALD [1899-1986]
WW 1983, p.973
Anon., Lancet 1986(II), p.352; Brit. Med. J. 293 (1986), 455
L. Young, St. Thomas's Hospital Gazette 84(3) (1986), 27-28

HÁRI, PAUL [1869-1933]
OBL 2,188; MEL 1,676; FISCHER 1,579; WININGER 7,56; IB, 113
J. Sós, Orv. Kozl. 51-53 (1969), 143-154
I. Szekacs, Orvosi Hetilap 124 (1983), 1889-1894

HARINGTON, CHARLES ROBERT [1897-1972]
DSB 17,380-381; DNB [1971-1980], 380-382; WW 1965, p.1329; POTSCH, 189; MEDVEI, 744-746; WWWS, 753-754; POGG 6,1026; 7b,1846
R. Pitt-Rivers, Mayo Clin. Proc. 39 (1964), 553-559
Anon., Nature 236 (1972), 186
H. Himsworth and R. Pitt-Rivers, Biog. Mem. Fell. Roy Soc. 18 (1972), 267-308
A. Neuberger, Biochem. J. 129 (1972), 801-804

HARKAVY, ALEXANDER [1863-1939]
WININGER 3,1-2; SG [1]5,851; NUC 231,191

HARKER, DAVID [1906-1991]
AMS 14,2106; IWW 1982-1983, p.529; McLACHLAN, 59-60
New York Times, 2 March 1991
W.L. Daux, Acta Cryst. A48 (1992), 1-3

HARKER, JOHN ALLEN [1870-1923]
POGG 4,585-586; 5,499; 6,1026
G.W.C. Kaye, Nature 112 (1923), 629
J.R. Partington, J. Chem. Soc. 125 (1924), 988-990
R.T.G., Proc. Roy. Soc. A105 (1924), xi-xiii

HARKINS, WILLIAM DRAPER [1873-1951]
DSB 6,117-119; DAB [Suppl.5],273-274; AMS 8,1033; MILES, 196-198; POTSCH, 189-190; NCAB 42,312-313; POGG 5,499; 6,1026-1027; 7b,1847-1851
R.S. Mulliken, Biog. Mem. Nat. Acad. Sci. 47 (1975), 49-81
G.B. Kauffman, J. Chem. Ed. 62 (1985), 758-761

HARKNESS, ROBERT ANGUS [1931-]
Who's Who of British Scientists 1980-1981, p.209

HARLAY, VICTOR ANDRÉ [1872-1922]
GORIS, 432-433
J. Bougault, Bull. Soc. Mycol. 38 (1923), 25-28

HARLESS, EMIL [1820-1862]
HIRSCH 3,60; DUELUND, 13-18

HARLEY, EDWARD VAUGHAN BERKELEY [1863-1923] WW 1923, p.1220; O'CONNOR(2), 160-161; RSC 15,645
Anon., Lancet 1923(I), p.1132; Brit. Med. J. 1923(I), p.956; Med. J. Rec. 129 (1929), 702-704

HARLEY, GEORGE [1829-1896]
DNB 22,817-818; HIRSCH 3,60-61; PAGEL, 688; O'CONNOR, 151-153; FRANKEN, 185
Anon., Lancet 1896(II),pp.1330-1333; Brit. Med. J. 1896(II), pp.1354-1355; Proc. Roy. Soc. 61 (1897), v-x
C. von Voit, Sitz. Bayer. Akad. 27 (1897), 421-423
A. Tweedie, Ed., George Harley. London 1899

HARMS, JÜRGEN WILHELM [1885-1956]
NDB 7,685-686; AUERBACH, 817-818; IB, 113
H. Friedrich-Freksa, Zool. Anz. 20Suppl. (1957), 485-489

HARNACK, ERICH [1852-1915]
DBJ 1,328; WELDING, 297; FISCHER 1,580; PAGEL, 688-689; DZ,

528-529; TSCHIRCH,1061; POGG 3,589; 4,586-587; 6,1028; 7aSuppl.,265
O. Schmiedeberg, Arch. exp. Path. Pharm. 79 (1915), i
M. Kochmann, Mitteldeutsche Lebensbilder 1 (1926), 427-432

HARNED, HERBERT SPENCER [1888-1969]
AMS 11,2115; WWWS, 755; POGG 6,1028; 7b,2050-2051
J.M. Sturtevant, Biog. Mem. Nat. Acad. Sci. 51 (1980), 215-244

HARNISCH, OTTO [1901-1961]
KURSCHNER 9,671-672; IB, 113; POGG 7a(2),379-380

HARPER, CHARLES ATHIEL [1868-]
JV 11,19; NUC 231,445

HARPER, ROBERT ALMER [1862-1946]
DSB 6,121-122; NCAB A,401; AMS 7,746-747; WWAH, 252; WWWS, 756
B.O. Dodge, Am. Phil. Soc. Year Book 1946, pp.304-313
C. Thom, Biog. Mem. Nat. Acad. Sci. 25 (1949), 227-240

HARRIES, CARL [1866-1923]
DBJ 5,148-151,429; KIEL, 155-156; BLOKH 1,272-275; MATSCHOSS, 105; POTSCH, 190; POGG 4,587-588; 5,500-502; 6,1029
W. Nagel et al., Z. angew. Chem. 37 (1924), 105-177
R. Willstätter, Ber. chem. Ges. 59A (1926), 123-157

HARRINGTON, WILLIAM FIELD [1920-1992]
AMS 15(3),504; WWA 1980-1981, p.1451
W.F. Harrington, in The Impact of Protein Chemistry on the Biochemical Sciences (A.N. Schechter et al., Eds.), pp.23-37. New York 1984

HARRIS (FRASER-HARRIS), DAVID FRASER [1867-1937]
O'CONNOR(2), 425-426; IB, 88; NUC 183,153-154
Anon., Nature 139 (1937), 184; Science 85 (1937), 68

HARRIS, ELIJAH PADDOCK [1832-1920]
AMS 2,199; NUC 231,675
New York Times, 11 December 1920

HARRIS, GEOFFREY WINGFIELD [1913-1971]
WW 1971, p.1369; MEDVEI, 746-748
M.L. Vogt, Biog. Mem. Fell. Roy. Soc. 18 (1972), 309-329
C. Fortier, Endocrinology 90 (1972), 851-854

HARRIS, HARRY [1919-]
WW 1981, p.1130; AMS 15(3),508

HARRIS, HENRY [1925-]
WW 1981, p.1130; IWW 1982-1983, p.532
H. Harris, The Balance of Improbabilities. New York 1987

HARRIS, ISAAC FAUST [1879-1953]
AMS 7,750; NCAB 42,171; BURSEY, 52
Encyclopedia of American Biography 14 (1942), 55-56

HARRIS, JOHN IEUAN [1924-1978]
F. Sanger, TIBS 3 (1978), N209
R.N. Perham, in Methods in Peptide and Protein Sequence Analysis, pp.1-10. Amsterdam 1980

HARRIS, LESLIE JULIUS [1898-1973]
COHEN, 101; POGG 6,1030; 7b,1856-1759
F.M. Cruickshank, Brit. J. Nutrition 39 (1978), 1

HARRIS, MILTON [1906-1991]
AMS 15(3),512; WWWS, 758
M.N. Breuer, Milton Harris. Washington, D.C. 1982
E. Pace, New York Times, 14 September 1991
R.N. Hader, Chem. Eng. News 69(44) (1991), 67

HARRIS, STANTON AVERY [1902-]
AMS 11,2131; WWWS, 758

HARRISON, DOUGLAS CREESE [1901-]
WW 1981, p.1135; BELFAST, 581; POGG 6,1030; 7b,1859

HARRISON, HAROLD EDWARD [1908-1989]
AMS 17(3),537; WWWS, 759

HARRISON, JOHN HOFFMANN [1808-1849]
HIRSCH 3,67

HARRISON, KENNETH PRITCHARD [1912-]
Who's Who In British Science 1953, p.122

HARRISON, LEONARD HUBERT [1885-]
JV 26,637; NUC 232,327

HARRISON, ROSS GRANVILLE [1870-1959]
DSB 6,131-135; DAB [Suppl.6],281-283; FREUND 2,117-126; HAYMAKER,123-128; AMS 9(II),477; NCAB 15,172; W, 245; DAMB, 331-332; WWWS, 759-760
H. Autrum, Bayer. Akad. Wiss. Jahrbuch 1960, pp.165-169
J.S. Nicholas, Biog. Mem. Nat. Acad. Sci. 35 (1961); Am. Phil. Soc. Year Book 1961, pp.114-120
M. Abercrombie, Biog. Mem. Fell. Roy. Soc. 7 (1961), 111-126
R.N. Wegner, Anat. Anz. 109 (1961), 458-465
J.M. Oppenheimer, Bull. Hist. Med. 40 (1966), 525-543
K.F. Russell, Clio Medica 4 (1969), 109-119
J.A. Maienschein, Ross Harrison's Crucial Experiment as a Foundation for Modern American Experimental Embryology. Bloomington, Ind. 1978
V. Hamburger, Persp. Biol. Med. 23 (1980), 600-616
J.A. Witkowski, Notes Roy. Soc. 35 (1980), 195-219

HARROP, GEORGE ARGALE [1890-1945]
AMS 7,754; WWAH, 255
G.W. Rake and J.F. Anderson, Science 102 (1945), 295-296
Anon., J. Am. Med. Assn. 129 (1945),146; Chem. Eng. News 23 (1945),1453
A.M. Chesney, Trans. Assoc. Am. Phys. 59 (1946), 21-22

HARROW, BENJAMIN [1888-1970]
AMS 11,2138; MILES, 200; KAGAN(JA), 345; COHEN, 101; WWAH, 255; WININGER 3,3-4; POGG 6,1031; 7b,1862-1863

HART, EDWIN BRET [1874-1953]
DSB 6,135-135; DAB [Suppl.5],275-276; AMS 8,1048; MILES, 202-203; NCAB 43,396-397; WWAH, 255; WWWS, 760
C.A. Elvehjem, Biog. Mem. Nat. Acad. Sci. 28 (1954), 117-161
H.T. Scott, Food Technology 9 (1955), 1-13
C.H. Trottman, Edwin Bret Hart. Madison, Wis. 1972

HARTE, ROBERT ADOLPH [1911-1977]
AMS 13,1794
Anon., Chem. Eng. News 55(44) (1977), 33
J.T. Edsall, J. Biol. Chem. 253 (1978), 3353-3354

HARTECK, PAUL [1902-1985]
AMS 15(3),527; WWWS, 761; POTSCH, 190; POGG 6,1032-1033; 7a(2),381-382
New York Times, 24 January 1985

HARTENECK, ANNA [1896-]
JV 42,507

HARTIG, THEODOR [1805-1880]
DSB 6,136-137; NDB 7,713; ADB 10,662; TSCHIRCH, 1061; WWWS, 761; CALLISEN 28,389
Anon., Leopoldina 16 (1880), 70-71
W. Blasius, Jahresb. Ver. Nat. Braun. 5 (1887), 132-145; Lebensbeschreibungen Braunschweiger Naturforscher und Naturfreunde, pp.34-47. Braunschweig 1887
A. Hüttermann, Ber. bot. Ges. 100 (1987), 116-120

HARTING, PIETER [1812-1885]
DSB 6,137-138; LINDEBOOM, 785-787; HIRSCH 3,30; FREUND 1,197-

205; UTRECHT, 96-97; TSCHIRCH, 1061; POGG 1,1021-1023; 3,591

HARTL, FERDINAND [1880-1907]
JV 22,518; NUC 233,55

HARTLEY, BRIAN SELBY [1926-]
WW 1981, p.1142

HARTLEY, GILBERT SPENCER [1906-]
Who's Who in Science in Europe 1967, p.690

HARTLEY, HAROLD BREWER [1878-1972]
DNB [1971-1980], 387-389; WW 1972, pp.1406-1407; POGG 6,1033-1034; 7b,1864-1866
A.G. Ogston, Biog. Mem. Fell. Roy. Soc. 3 (1973), 349-373
R.V. Jones, Notes Roy. Soc. 27 (1973), 181-184
E.J. Bowen, Brit. J. Hist. Sci. 6 (1973), 338-339

HARTLEY, PERCIVAL [1881-1957]
WW 1949, p.1230; O'CONNOR(2), 464-465; WWWS, 762
H.H. Dale, Biog. Mem. Fell. Roy. Soc. 3 (1957), 81-100
Anon., Brit. Med. J. 1957(I), pp.466-467
D.G. Evans, J. Path. Bact. 75 (1958), 487-495

HARTLEY, PERCIVAL HORTON-SMITH [1867-1952]
WW 1949, p.1230-1231
Anon., Brit. Med. J. 1952(II), pp.99-100

HARTLEY, WALTER NOEL [1846-1913]
BLOKH 1,275-276; WWWS, 762; POGG 4,589-591; 5,503
W.E. Adeney, Chem. Z. 37 (1913), 1377
J.Y.B., J. Chem. Soc. 105 (1914), 1207-1216
J.H.B., Proc. Roy. Soc. A90 (1914), vi-xiii

HARTLINE, HALDAN KEFFER [1903-1983]
AMS 13,1795; MAH 2,210-211; STC 1,483-484; MSE 2,24-25; WW 1978, p.1078; WWA 1978-1979, p.1407; IWW 1982-1983, p.534; WWWS, 762
F. Ratcliff, Am. Phil. Soc. Year Book 1984, pp.111-120; Biog. Mem. Nat. Acad. Sci. 59 (1990), 197-213
R. Granit and F. Ratcliff, Biog. Mem. Fell. Roy. Soc. 31 (1985). 263-292

HARTMAN, CARL GOTTFRIED [1879-1968]
AMS 11,2144; IB, 114; WWAH, 255-256; WWWS, 762
R.F. Vollman, Experientia 15 (1959), 407-408; Fifty Years of Research on Mammalian Reproduction. Washington, D.C. 1965
J.D. Biggers, Biology of Reproduction 2 (1970), 1-4

HARTMAN, FRANK ALEXANDER [1883-1971]
AMS 11,2145; IB, 114; WWAH, 256; WWWS, 762
F.A. Hartman, Persp. Biol. Med. 6 (1963), 280-290

HARTMANN, ALEXIS FRANK [1898-1964]
AMS 10,1666; DAMB 1,333-334; WWAH, 256
G.B. Forbes, J. Pediat. 64 (1964), 793-795
J.A. Lee, Anaesthesia 36 (1981), 1115-1121

HARTMANN, FRIEDRICH [1887-1918]
JV 29,630; NUC 233,189

HARTMANN, GERHARD [1868-1945]
JV 7,264; RSC 15,667; NUC 233,194

HARTMANN, HANS [1908-1937]
A. von Muralt, Erg. Physiol. 39 (1937), 408-412
H. Rein, Deutsche med. Wchschr. 63 (1937), 1380

HARTMANN, JULES ALBERT [1823-1905]
T. Schlumberger, Bull. Soc. Ind. Mulhouse 75 (1905), 175

HARTMANN, MAX [1876-1962]
NDB 8,1-2; KURSCHNER 8,818; FISCHER 1,583-584; IB, 114; WWWS, 763
J. Hämmerling, Naturwiss. 50 (1963), 365-366
H. Autrum, Bayer. Akad. Wiss. Jahrbuch 1963, pp.197-201

HARTMANN, MAX [1884-1952]
JV 24,573
K. Miescher, Verhandl. Schw. Nat. Ges. 133 (1954), 323-326

HARTMANN, NORBERT [1917-]
KURSCHNER 12,1110; POGG 7a(2),387

HARTMANN, WALTER [1883-1917]
JV 25,389
Anon., Z. angew. Chem. 30(III) (1917), 562

HARTREE, DOUGLAS RAYNER [1897-1958]
DSB 6,147-148; WW 1958, p.1342; WWWS, 763; POGG 6,1036; 7b,1868-1870
C. Darwin, Biog. Mem. Fell. Roy. Soc. 4 (1958), 103-116

HARTREE, EDWARD FRANCIS [1910-]
Who's Who of British Scientists 1971-1972, pp.380-381

HARTREE, WILLIAM [1870-1943]
IB, 114
J.A. Venn, Alumni Cantabrigensis 3,274. Cambridge 1947

HARTRIDGE, HAMILTON [1886-1976]
WW 1976, p.1049; FISCHER 1,584; IB, 114; POGG 6,1036-1037; 7b,1870-1871
R.A. Weale, Nature 259 (1976), 611
W.A.H. Rushton, Biog. Mem. Fell. Roy. Soc. 23 (1977), 193-211; Vision Research 17 (1977), 507-513

HARTUNG, WALTER HENRY [1895-1961]
AMS 10,1668; POGG 6,1037; 7b,1873-1875
W.E. Weaver, Am. J. Pharm. Ed. 26 (1962), 134-135

HARTWICH, CARL [1851-1917]
DHB 3,770; STRAHLMANN, 468,480
R. Eder, Verhandl. Schw. Nat. Ges. 99 (1917), 8-25
C. Schröter, Viert. Nat. Ges. Zürich 62 (1917), 702-708

HARVEY, ABNER McGEHEE [1911-]
AMS 15(3),536; ICC, 299-301

HARVEY, EDMUND NEWTON [1887-1959]
DSB 17,383-385; AMS 9(II),480; NCAB 45,478-479; CB 1952, pp.247-248; IB, 115; WWWS, 764; WWAH, 256-257; SACKMANN, 150-152; POGG 6,1038-1040; 7b,1875-1878
E.G. Butler, Am. Phil. Soc. Year Book 1959, pp.127-130
A.M. Chase, Biol. Bull. 119 (1960), 9-10
F.H. Johnson, Arch. Biochem. Biophys. 87 (1960), i-iii; Biog. Mem. Nat. Acad. Sci. 39 (1967), 193-266

HARWOOD, HENRY FRANCIS [1886-1974]
JV 26,394
Anon., Chem. Brit. 11 (1975), 118

HASEBROEK, KARL [1860-1941]
RSC 15,677
T. Albers, Entomol. Z. 56 (1942), 81-88,95-96

HASEGAWA, CHOHACHI [1893-1952]

HASEGAWA, SHUJI [1898-]
JBE, 246

HASENCLEVER, FRIEDRICH WILHELM [1809-1874]
NDB 8,25-26; HEIN, 251-252; FERCHL, 216; POGG 3,593
H. Landolt, Ber. chem. Ges. 8 (1875), 703-705

HASENCLEVER, ROBERT [1841-1902]
NDB 8,29; BJN 8,393-394; BLOKH 1,276; MATSCHOSS, 106-107

F. Quincke, Z. angew. Chem. 15 (1902), 797-801

HASHIMOTO, HAKARU [1881-1934]
HAKUSHI 3,281-282
D. Donaich and I.M. Roitt, Lancet 1962(I), p.1074

HASKINS, CARYL PARKER [1907-]
AMS 14,2957; WWA 1976-1977, pp.1362-1363; IWW 1982-1983, p.537; WWWS, 765

HASLAM, HENRY COBDEN [1870-1948]
O'CONNOR(2), 70
Who Was Who 1941-1950, p.512

HASS, HENRY BOHN [1902-1987]
AMS 16(3),553; WWWS, 765
New York Times, 14 February 1987

HASSALL, ARTHUR HILL [1817-1894]
HIRSCH 3,83; PAGEL, 693; DESMOND, 293; RSC 3,208-209; 7,918; 15,678
A.H. Hassall, An Autobiography. London 1893
Anon., Lancet 1894(I), p.977; Brit. Med. J. 1894(I), p.833; Journal of Botany 32 (1894), 190-191
E.G. Clayton, A.H. Hassall, his Work in Public Hygiene. London 1908
E. Gray, By Candlelight. London 1983

HASSALL, CEDRIC HERBERT [1919-]
SRS, 92; WWWS, 766

HASSE, GEORG [1900-]
JV 41,428; NUC 234,203

HASSE, KARL [1841-1922]
DBJ 4,356; FISCHER 1,585; PAGEL, 695-696
L. Gräper, Anat. Anz. 56 (1922), 209-221

HASSE, KURT [1911-1981]
KURSCHNER 13,1370; POGG 7a(2),391
Anon., Chem. Z. 105 (1981), 384; Naturw. Rund. 35 (1982), 45

HASSEL, ODD [1897-1981]
MSE 2,26-27; STC 1,484-485; WWWS, 766; POTSCH, 191; POGG 6,1044; 7b,1881-1884
E.W. Lund and C. Romming, in Selected Topics in Structural Chemistry (P. Anderson et al., Eds.), pp.13-23,277-289. Oslo 1967
Hver er Hvem? 1979, p.237
O. Bastiansen, Chem. Brit. 18 (1982), 442

HASSELBALCH, KARL ALBERT [1874-1962]
VEIBEL 2,186-187; IB, 115
E. Warburg, Ugeskrift for Laeger 124 (1962), 1550

HASSELMANN, MICHEL [1913-1986]
DBA 2,1429
J. Castang, Ann. Falsif. 79 (1986), 343-346

HASSENFRATZ, JEAN HENRI [1755-1827]
DSB 6,164-165; DBF 17,700-701; GE 19,906; LAROUSSE 9,100; TSCHIRCH, 1062; POTSCH, 191; WWWS, 766; CALLISEN 28,406; POGG 1,1029-1030

HASSID, WILLIAM ZEV [1899?-1974]
AMS 12,2553; COHEN, 102; WWWS, 766
C.E. Ballou and H.A. Barker, Adv. Carbohydrate Chem. 32 (1976), 1-14; Biog. Mem. Nat. Acad. Sci. 50 (1979), 197-230

HASTINGS, ALBERT BAIRD [1895-1987]
AMS 14,213-214; MH 1,213-214; MSE 2,27-28; IWW 1982-1983, pp.538-539; POGG 6,1045; 7b,1884-1889
H.N. Christensen et al., Fed. Proc. 25 (1966), 818-826
A.B. Hastings, Ann. Rev. Biochem. 39 (1970), 1-24
J.T. Edsall and J.M. Buchanan, Am. Phil. Soc. Year Book 1987, pp.166-172
R.W. McKee, FASEB Journal 2 (1988), 197-198
H.N. Christensen (Ed.), Crossing Boundaries. Grand Rapids, Mich. 1989

HASTINGS, JOHN WOODLAND [1927-]
AMS 15(3),547; WWA 1980-1981, p.1473; WWWS, 766-767

HATA, SAHACHIRO [1873-1938]
FISCHER 1,586; KALLMORGEN, 291; SKVARC, 91-92; WWWS, 767
Anon., Chem. Z. 62 (1938), 933
H. Schlossberger, Deutsche med. Wchschr. 65 (1939), 32-33
W. Koehler, Leopoldina [3]28 (1985), 99-122
K. Holubar, Wiener med. Wchschr. 100 (1988), 747-749

HATCHER, ROBERT ANTHONY [1868-1944]
DAB [Suppl.3],342-343; NCAB 33,289-290; PARASCANDOLA, 35-36; CHEN, 39; DAMB, 334; IB, 155; WWAH, 258
T. Koppanyi, Science 99 (1944), 420-421
H. Gold, Journal of Clinical Pharmacology 11 (1971), 245-248

HATCHETT, CHARLES [1765-1847]
DSB 6,166-167; DNB 9,153; COLE, 246; POTSCH, 191-192; CALLISEN 8,193-194; POGG 1,1031
E.M. Weeks, J. Chem. Ed. 15 (1938), 153-158

HATSCHEK, BERTHOLD [1854-1941]
DSB 6,167; NDB 8,56-57; OBL 2,207-208; STURM 1,548-549; TETZLAFF, 127; WININGER 7,57
O. Storch, Alm. Akad. Wiss. Wien 99 (1949), 284-296

HATSCHEK, EMIL [1868-1944]
POGG 5,506; 6,1046; 7b,1892
R. Lessing, Chem. Ind. 1944, p.287

HATT, JEAN DANIEL [1887-]
NUC 234,448

HATTORI, SHIZUO [1902-]
JBE, 262; POGG 6,1046; 7b,1892-1894

HAUBERISSER, GEORG [1869-1925]
JV 11,151; NUC 234,505
Anon., Chem. Z. 50 (1926), 22; Z. angew. Chem. 39 (1926), 60

HAUCK, FERDINAND [1845-1889]
ADB 50,61-62; OBL 2,210; STURM 1,550
S.Z., Ost. Bot. Z. 37 (1887), 1-6
Anon., Annals of Botany 3 (1889), 464

HAUENSTEIN, EMIL [1884-1949]
NUC 234,539

HAUFF, FRIEDRICH [1863-1935]
NDB 8,87
Z. angew. Chem. 46 (1933), 814; 48 (1935), 314; Chemische Industrie 1935, p.318

HAUFFE, KARL [1913-]
KURSCHNER 16,1281; POTSCH, 192; POGG 7a(2),394-395

HAUGAARD, GOTFRED HANS CHRISTIAN [1894-1969]
VEIBEL 2,189

HAUGAARD, NIELS [1920-]
AMS 15(3),552; WWA 1980-1981, p.1476; WWWS, 768

HAUPTMANN, ALFRED [1881-1948]
FISCHER 1,588; KOREN, 189; NUC 234,678-679
F.G. von Stockert, Arch. Psychiat. 180 (1948), 529-530

HAUPTMANN, HERBERT AARON [1917-]
AMS 16(3),563; POTSCH, 192
H.A. Hauptmann and R.H. Blessing, Naturw. Rund. 40 (1987), 463-470

HAUPTMANN, HEINRICH [1905-1960]
C.B. Aust, Staden Jahrbuch 11/12 (1963-1964), 197-211

HAUROWITZ, FELIX [1896-1987]
AMS 14,2066; WWA 1978-1979, p.1420; KURSCHNER 13,1382; MH 2,211-213; STC 1,485-487; MSE 2,28-29; BHDE 2,465; WIDMANN, 265; TLK 3,242; POTSCH, 192; IB, 116; WWWS, 768; POGG 6,1048; 7a(2),399-401
Anon., Nachr. Chem. Techn. 14 (1966), 93-94; New York Times, 7 December 1987; Chem. Eng. News 66(33) (1988), 33

HAUSCHILD, FRITZ [1908-]
KURSCHNER 12,1129; WWWS, 768; POGG 7a(2),401-402

HAUSER, ERNST [1896-1956]
MILES, 203-204; BHDE 2,466-467; WWAH, 259; POGG 6,1049-1050; 7b,1897-1900
Anon., Chem. Z. 80 (1956), 212; Chem. Eng. News 34 (1956), 1209

HAUSER, GUSTAV [1856-1935]
NDB 8,115-116; GROTE 6,141-204; FISCHER 1,588-589; BERWIND, 100-110; WEBER, 40-41; IB, 116; WWWS, 768
E. Kirch, Cent. allgem. Path. 63 (1935), 369-373; Verhandl. path. Ges. 29 (1937), 379-381

HAUSER, OTTO [1877-1915]
BLOKH 1,276-278; RSC 15,689; POGG 5,508
R.J. Meyer, Ber. chem. Ges. 48 (1915), 437-439; Chem. Z. 39 (1915), 273

HAUSMANN, WALTHER [1877-1938]
FISCHER 1,590; HASENEDER, 90-103; SKVARC, 93-94; TLK 3,243-244; KOREN, 189; POGG 6,1050-1051; 7a(2),404-405
Anon., Ann. Inst. Actin. 12 (1938), 97-98

HAUSSER, KARL [1887-1933]
NDB 8,128; POGG 6,1051; 7a(2),405
W. Kossel, Strahlentherapie 48 (1933), 205-222

HAUSSKNECHT, OTTO [1844-1914]
DGB 87 (1935), 143; RSC 7,925; 10,162; NUC 235,143

HAUTEFEUILLE, PAUL GABRIEL [1836-1902]
DSB 6,177-178; DBF 17,753; LACROIX 1,81-89; POGG 3,598-599; 4,599
D. Gernez, Bull. Soc. Chim.[3]29 (1903), i-xx
G. Lemoine, Rev. Quest. Sci. 55 (1904), 5-25

HAÜY, RENÉ JUST [1743-1822]
DSB 6,178-183; DBF 17,775-776; LACROIX 1,177-182; LAROUSSE 9,115; GE 19,939; BLOKH 1,278-279; COLE, 246-248; FERCHL, 218-219; POTSCH, 192-193; WWWS, 769; POGG 1,1038-1041
G.F. Kunz et al., Am. Min. 3 (1918), 61-138
F. Kaisin, Rev. Quest. Sci. 109 (1936), 408-420
A. Lacroix et al., Bull. Soc. Fran. Min. 67 (1944), 15-348
D.C. Goodman, Ambix 16 (1969), 152-166
S.H. Mauskopf, Ambix 17 (1970), 182-191; Arch. Inter. Hist. Sci. 23 (1970), 185-206
K.H. Wiederkehr, Centaurus 22 (1978), 131-156
P.F. Stevens, J. Hist. Biol. 17 (1984), 49-82

HAVARD, ROBERT EMLYN [1901-1985]
Anon., Brit. Med. J. 291 (1985), 609

HAVEMANN, ROBERT [1910-1982]
KURSCHNER 12,1136; AO 1982, pp.172-174; POGG 7a(2),406
R. Havemann, Fragen, Antworten, Fragen. Hamburg 1972
Anon., Chem. Z. 106 (1982), 273

HAVEZ, RAYMOND [1928-1973]
Anon., Ann. Biol. Clin. 31 (1973), 433-434

HAVINGA, EGBERTUS [1909-1988]
E. Havinga, Enjoying Organic Chemistry 1927-1987. Washington, D.C. 1991

HAWK, PHILIP BOVIER [1874-1966]
AMS 10,1687; NCAB C,216-217; CHITTENDEN, 170-171; IB, 116; WWAH, 259
Anon., Chem. Eng. News 44(45) (1966), 66
G.W. Corner, Am. Phil. Soc. Year Book 1966, pp.147-148

HAWORTH, EDWARD [1873-1940]
J.W. Moore and E.O. Glover, J. Chem. Soc. 1941, pp.241-242

HAWORTH, ROBERT DOWNS [1898-1990]
WW 1981, p.1158; CAMPBELL, 105-106; STC 1,487; WWWS, 770-771; POGG 6,1054; 7b,1908-1910

HAWORTH, WALTER NORMAN [1883-1950]
DSB 6,184-186; DNB [1941-1950], 368-369; W, 246-247; STC 1,487-488; POTSCH, 192-193; SRS, 42; ABBOTT-C, 58-59; WWWS, 770; POGG 6,1053-1054; 7b,1905-1908
E.G.V. Percival, Chem. Ind. 1950, p.351
E.L. Hirst, Obit. Not. Fell. Roy. Soc. 7 (1951), 373-404; J. Chem. Soc. 1951, pp.2790-2806; Adv. Carbohydrate Chem. 6 (1951), 1-9
H. Wieland, Bayer. Akad. Wiss. Jahrbuch 1952, pp.187-188
M. Stacey, Chem. Soc. Revs. 2 (1973), 145-161
H.S. Isbell, Chem. Soc. Revs. 3 (1974), 1-16
S.A. Barker and N. Baggett, TIBS 3 (1978), 140-141

HAWTHORNE, JOHN [1875-1958]
SRS, 28; JV 19,201
Anon., Analyst 83 (1958), 490; J. Roy. Inst. Chem. 1958, p.683

HAY, MATTHEW [1855-1932]
WW 1931, p.1429
Anon., Lancet 1932(II), pp.369-370; Brit. Med. J. 1932(II),pp.332-333,386

HAYAISHI, OSAMU [1920-]
IWW 1982-1983, p.554

HAYASHI, INOSUKE [1887-1945]
HAKUSHI 3,488-489; FISCHER 1,592

HAYCRAFT, JOHN BERRY [1857-1922]
O'CONNOR, 240-241; FISCHER 1,592
T.G.B., Nature 111 (1923), 124
Anon., Lancet 1923(I), pp.158-159; Brit. Med. J. 1923(I), p.86

HAYDON, DENIS ARTHUR [1930-1988]
Who Was Who 1981-1990, pp.337-338
R.D. Keynes, Biog. Mem. Fell. Roy. Soc. 36 (1990), 201-216

HAYDUCK, FRIEDRICH [1880-1961]
NDB 8,150; TLK 3,246; POGG 6,1055-1056; 7a(2),407; (4),144*
Anon., Chem. Z. 74 (1950), 737; 80 (1956), 15

HAYDUCK, MAX [1842-1899]
NDB 8,150-151; BJN 4,120; BLOKH 1,279; POGG 4,599
Anon., Leopoldina 35 (1899), 179-180
M. Delbrück, Ber. chem. Ges. 32 (1899), 2771-2775

HAYEM, GEORGES [1841-1933]
DBF 17,792-793; GENTY 5,49-60; FISCHER 1,592-593; VERSO, 96-97; STALDER, 34-36; KOREN, 190; WWWS, 771
L. Rivet, Sang 1 (1927), 59-68
F. Bezançon, Presse Med. 41 (1933), 1761-1765
M.R.M. Loeper, Bull. Acad. Med. 110 (1933), 145-155
C. Dreyfus, J. Lab. Clin. Med. 27 (1942), 855-865
R. Kohn, Rev. Hist. Med. Heb. 1952, pp.91-94
M.L. Verso, Med. Hist. 15 (1971), 58-59

HAYES, AUGUSTUS ALLEN [1806-1882]
DAB 8,443-444; ELLIOTT, 122; WWAH, 260; WWWS, 771; POGG 3,600
Anon., Proc. Am. Acad. Arts Sci. 18 (1883), 422-427

HAYES, WILLIAM [1912-]
WW 1981, p.1162

HAYMANN, CLÄRE [1897-]
KALLMORGEN, 292

HAZARD, RENÉ JULES PAUL [1896-1974]
DBF 17,797
J. Cheymol, Bull. Acad. Med. 158 (1974), 587-596

HAZURA, KARL [1859-1941]
OBL 2,229; STURM 1,563
Anon. and A.W. Unger, Ost. Chem. Z. 32 (1929), 180-182
A. Chwala, Ost. Chem. Z. 44 (1941), 100

HEAD, HENRY [1861-1940]
DNB [1931-1940], 410-411; FISCHER 1,694; O'CONNOR(2), 286-289; HAYMAKER, 449-452; HUBER, 50-52; WWWS, 773
Anon., Brain 63 (1941), 205-208; Brit. Med. J. 1940(II), pp.539-541
G.M. Holmes, Obit. Not. Fell. Roy. Soc. 3 (1941), 665-689
C.S. Breathnach, J. Roy. Soc. Med. 84 (1991), 107-109

HEALD, PETER JOSEPH [1925-]
Who's Who of British Scientists 1971-1972, p.389

HEARD, ROBERT DONALD HOSKIN [1908-1957]
AMS 9(II),489; YOUNG, 12; SRS, 65

HEATH, CHRISTOPHER [1835-1905]
DNB [Suppl.2]2,233-234; HIRSCH 3,101; WWWS, 774
Anon., Lancet 1905(II), p.490; Brit. Med. J. 1905(II), p.359

HEATH, EDWARD CHARLES [1926-1984]
AMS 15(3),584; WWA 1980-1981, p.1496
W.J. Lennarz, TIBS 10 (1985), 267-268

HEATLEY, NORMAN GEORGE [1911-]
Who's Who of British Scientists 1971-1972, pp.390-391
C.L. Moberg, Science 253 (1991), 734-735

HEBLER, FELIX [1891-]
JV 39,251; TLK 3,246; POGG 6,1058

HEBRA, FERDINAND von [1816-1880]
NDB 8,172-173; ADB 50,88-89; OBL 2,232; STURM 1,563-564; PAGEL, 698-700; HIRSCH 3,103-105; Suppl.,362; SIEGL, 4-10; WWWS, 1736
M. Kaposi, Wiener klin. Wchschr. 30 (1880), 905-907,927-931
Anon., Leopoldina 17 (1881), 90-93
J. Stefan, Alm. Akad. Wiss. Wien 31 (1881), 194-200
J. Tappeiner and K. Holuba, Wiener klin. Wchschr. 93 (1981), 503-506
V. Misgeld, 100 Jahre Dermatologie in Berlin (E. Klaschka and K. Rauhut, Eds.), pp.57-71. Berlin 1984

HEBRA, HANS von [1847-1902]
OBL 2,232; BJN 7,44*; FISCHER 1,594; SIEGL, 11-14
E. Lang, Wiener klin. Wchschr. 15 (1902), 451

HEBTING, JOSEF [1876-1932]
JV 16,92; NUC 237,525

HECHT, HANS [1890-]
JV 32,327; NUC 237,546
Anon., Nachr. Chem. Techn. 3 (1955), 192

HECHT, OTTO [1846-1914]
BLOKH 1,279-280; POGG 3,603; 4,602
Anon., Ber. chem. Ges. 47 (1914), 1950-1951

HECHT, OTTO [1900-1973]
KURSCHNER 11,1038; LDGS, 14; IB, 117; POGG 7a(2),408-409

HECHT, SELIG [1892-1947]
DAB [Suppl.4],366-367; AMS 7,777; NCAB 38,322-323; KAGAN(JA),349; KOREN, 190; WWAH, 263; POGG 6,1058-1059; 7b,1914-1916
B. O'Brien et al., Science 107 (1948), 105-106
G. Wald, J. Gen. Physiol. 32 (1948), 1-16; Biog. Mem. Nat. Acad. Sci. 60 (1991), 81-100
M.H. Pirenne, Nature 6 (1948), 673
C.H. Graham, Am. J. Psych. 61 (1948), 126-128
O. Glaser, Biol. Bull. 99 (1950), 14-18

HECHTENBERG, WILHELM [1890-]
JV 36,674; NUC 237,559

HECKER, RUDOLF [1868-1963]
KURSCHNER 4,1070; FISCHER 1,595; AD 3,133

HEDENIUS, ISRAEL [1868-1932] SMK 3,352; FISCHER 1,595-596; WIDSTRAND 2,307-310; RSC 15,715
A. Josefson, Hygeia 94 (1932), 289-296

HEDIN, SVEN GUSTAV [1859-1933]
SMK 3,363; FISCHER 1,596; IB, 117; POGG 4,603; 6,1061; 7b,1920
E. Jorpes, Hygeia 95 (1933), 689-692
H. Ostberg, Nord. Med. Hist. Arsbok 1977, pp.157-162

HEDLEY, EDGAR PERCY [1885-1960]
SRS, 37; JV 24,492; NUC 238,46

HÉDON, EMMANUEL [1863-1933]
FISCHER 1,597; STALDER, 37-38, IB, 117
M. Lisbonne, Presse Med. 41 (1933), 617-619
Anon., Montpell. Med. 76(IV,2) (1933), 95-164

HEEREN, FRIEDRICH [1803-1885]
BLOKH 1,280; HANNOVER, 28; SCHAEDLER, 48; FERCHL, 220; MATSCHOSS,109; POGG 1,1045; 3,604
Anon., Leopoldina 21 (1885), 114

HEERMANN, PAUL [1868-1945]
KURSCHNER 4,1073-1074; TLK 3,248; POGG 6,1062-1063; 7a(2),411

HEFFTER, ARTHUR [1859-1925]
NDB 8,201-202; FISCHER 1,597; PAGEL, 1962; GUNDLACH, 252; TSCHIRCH, 1062; DRUM, 321-322; TLK 2,268; POGG 4,604; 6,1063
W. Straub, Arch. exp. Path. Pharm. 105 (1925), i-iv
E. Rost, Deutsche med. Wchschr. 51 (1925), 617
G. Joachimoglu, Arzneimitt. 9 (1959), 391-392

HEFTER, JULIE [1884-]
JV 26,611; SG [3]5,516; NUC 238,189

HEFTMANN, ERICH [1918-]
AMS 15(3),596

HEGAR, ALFRED [1830-1914]
NDB 8,205-206; DBJ 1,286; HB 2,186-188; BAD NF1, 161-162; HIRSCH 3,118-119; PAGEL, 701-702; WWWS, 776
A. Mayer, Alfred Hegar. Freiburg i.B. 1961
E.F. Podach, Deutsches Aerzteblatt 62 (1964), 1665-1668
F. Rihner, Med. Welt 31 (1980), 1522-1523

HEGEL, SIGMUND [1863-]
JV 1,49; TLK 3,248; RSC 15,723

HEGEMANN, FRIEDRICH AUGUST [1813-1860]
Anon., Proceedings of the American Pharmaceutical Association 32 (1884), 311HEGER, HANS [1855-1940]
STURM 1,569; HEIN, 254-255; TSCHIRCH, 1062; NUC 238,233-234
Anon., Ost. Apoth. Z. 9 (1955), 536-541

HEGER, PAUL [1846-1925]
BNB 37,423-429; BAF, 63; FISCHER 1,598; WWWS, 776
E. van Ermengem, Bull. Acad. Med. Belg. [5]5 (1925), 585-592
L. Fredericq, Arch. Inter. Physiol. 25 (1925), 217-220

A. Slosse, Bull. Acad. Med. Belg. [5]7 (1927), 791-806

HEGSTED, DAVID MARK [1914-]
AMS 17(3),616
D.M. Hegsted, Journal of the American College of Nutrition 9 (1990), 280-287

HEHNER, OTTO [1853-1924]
BLOKH 1,280
C.A. Mitchell, Analyst 49 (1924), 501-505
B. Dyer, J. Chem. Soc. 125 (1924), 2690-2693

HEHRE, EDWARD JAMES [1912-]
AMS 15(3),597

HEIDE, CAREL CHRISTIAN van der [1872-1931]
LINDEBOOM, 808-809; IB, 117

HEIDELBERGER, CHARLES [1920-1983]
AMS 15(3),598; WWA 1980-1981, p.1502
P. Brookes, Nature 303 (1983), 22
V.R. Potter, Carcinogenesis 10 (1985), 1-13
J.A. Miller and E.C. Miller, Biog. Mem. Nat. Acad. Sci. 58 (1989), 259-302

HEIDELBERGER, MICHAEL [1888-1991]
AMS 14,2101; WWA 1978-1979, p.1445; IWW 1982-1983, p.549; MH 2,216-218; STC 1,492-493; MSE 2,38-39; KAGAN(JA), 313; COHEN, 102-103; POTSCH, 193-194; WININGER 3,21; POGG 6,1064; 7b,1930-1935
M. Heidelberger, Ann. Rev. Biochem. 36 (1967), 1-12; 48 (1979), 1-21; TIBS 2 (1977),116; Ann. Rev. Microbiol. 31 (1977), 1-12; Persp. Biol. Med. 24 (1981), 619-636; Immunological Reviews 82 (1984), 7-27
E.A. Kabat, Carbohydrate Research 40 (1975), 1-6; FASEB Journal 2 (1988), 2233-2234; J. Immunol. 148 (1992), 301-307
P. Gabar and E.A. Kabat, TIBS 3 (1978), N86-N87
M. Heidelberger and F.E. Kendall, TIBS 4 (1979), 168
J.M. Cruise, J. Immunol. 140 (1988), 2861-2863
R. Kessel, Immunology 74 (1991), 365-366

HEIDENHAIN, MARTIN [1864-1949]
DSB 6,223-224; NDB 8,247; FREUND 2,127-146; FISCHER 1,599; ROEDER, 72-74; IB, 117; WININGER 3,22; 7,60; KOHUT 2,278; KOREN, 190
W. Jacobj, Anat. Anz. 99 (1952), 80-94

HEIDENHAIN, RUDOLF [1834-1897]
DSB 6,224-227; NDB 8,247-248; ADB 50,122-127; BJN 2,75-76,56*; WWWS, 777; HIRSCH 3,122-123; PAGEL, 704-705; BLOKH 1,281; KROLLMANN, 258; BRAZIER, 99-100,109-110; KOHUT 2,236; WININGER 3,21-22; KOREN, 190; POGG 3,604-605; 4,605
E. Fischer, Ber. chem. Ges. 30 (1897), 2383-2385
P. Grützner, Pflügers Arch. 72 (1898), 221-265
W. Waldeyer, Anat. Anz. 14 (1898), 182-184
C. von Voit, Sitz. Bayer. Akad. 28 (1898), 460-470

HEIDENREICH, FRIEDRICH WILHELM [1798-1857]
HIRSCH 3,124-125; CALLISEN 8,261-262; 28,438-439
L. Feuerbach, Friedrich Wilhelm Heidenreich. Leipzig 1874

HEIDENREICH, KARL FRIEDRICH [1871-1924]
JV 10,151; RSC 14,437; 19,74; NUC 238,359
Anon., Chem. Z. 49 (1925), 58; Z. angew. Chem. 38 (1925), 40

HEIDEPRIEM, WILHELM [1876-1945]
JV 16,250; NUC 238,359

HEIDER, KARL [1856-1935]
NDB 8,252-253; IB, 118
A. Kühn, Naturwiss. 23 (1935), 791-796
R. Hertwig, Bayer Akad. Wiss. Jahrbuch 1935-1936, pp.52-53
K. Grobben, Alm. Akad. Wiss. Wien 86 (1936), 241-245
W. Ulrich, Sitz. Ges. Nat. Berlin NF9 (1969), 34-137

HEIDUSCHKA, ALFRED [1875-1957]
KURSCHNER 4,1080; HEIN-E, 177-178; TSCHIRCH, 1062-1063; TLK 3,249; POGG 5,513; 6,1065; 7a(2),415; (4),144*

HEIL, HEINRICH [1878-1961]
JV 27,396; NUC 238,427

HEILBRON, IAN MORRIS [1886-1959]
DSB 17,390-391; DNB [1951-1960], 469-470; WW 1959, pp.1379-1380; STC 1,493; W, 248-249; COHEN, 103; POGG 6,1066-1067; 7b,1937-1941
T.H. West, Chem. Ind. 1959, pp.1224-1225
A.H. Cook, Biog. Mem. Fell. Roy. Soc. 6 (1960), 65-86

HEILBRONN, ALFRED [1885-1961]
KURSCHNER 9,702; BHDE 2,475; WININGER 7,62; WIDMANN, 265-266; IB, 188

HEILBRUNN, LEWIS VICTOR [1892-1959]
AMS 9,492; IB, 118; KAGAN(JA),249; COHEN, 103; WWAH, 264
H.B. Steinbach, Science 131 (1960), 397-399; Biol. Bull. 121 (1961),10-11
A.K. Campbell, Cell Calcium 7 (1986), 287-296

HEILMEYER, LUDWIG [1899-1969]
KURSCHNER 11,1051; WWWS, 777; POGG 6,1067; 7a(2),416-419
R. Clotten and P. Marquardt, Arzneimitt. 9 (1959), 191-192
A.M. Walter, Arzneimitt. 14 (1964), 239-240
K.H. Bauer, Heid. Akad. Jahrb. 1970, pp.49-54
Anon., Ber. Nat. Ges. Freiburg 60 (1970), 97-98
L. Heilmeyer, Lebenserinnerungen. Stuttgart 1971
T.M. Fliedner, Ulmer Forum 41 (1977), 56-62

HEILNER, ERNST [1876-1939]
KURSCHNER 4,1084; FISCHER 1,600; GEDENKBUCH 1,528; LDGS, 79; KOREN, 190; WININGER 7,63

HEIM, FRITZ [1910-1979]
KURSCHNER 13,1416; AUERBACH,258; POGG 7a(2),419-420

HEIM, LUDWIG [1857-1939]
FISCHER 1,600; BERWIND, 49-55; PITTROFF, 107-109; TLK 3,251; IB, 118
M. Knorr, Z. Bakt. 101 (1927), 457-460
K. von Angerer, Deutsche med. Wchschr. 65 (1939), 1136

HEIM, ROGER [1900-1979]
DBF 17,841-842; STC 1,493-495; MSE 2,39-40; WWWS, 777
M. Chaudefaud, Revue de Mycologie 43 (1979), 323-329
J. Roche, C.R. Soc. Biol. 173 (1979), 685-686
L.R. Batra, Mycologia 72 (1980), 1063-1064
P. Joly, Bull. Soc. Bot. 127 (1980), 201-204

HEIM de BALSAC, FRÉDÉRIC [1869-1962]
DBF 17,842-843; CURINIER 4,343-344; RSC 15,730-731

HEIMROD, GEORGE WILLIAM [1876-1917]
NUC 238,557; POGG 4,606; 5,514

HEIMSTADT, OSKAR [1879-1944]
NDB 8,278; POGG 6,1068; 7a(2),421
L.F. Bräutigam, Mikroskopie 2 (1947), 59-63

HEIN, FRANZ [1892-1976]
KURSCHNER 11,1054; TLK 3,251; POTSCH, 194; POGG 6,1068; 7a(2),421-422
F. Wolf, Jahrb. Sachs. Akad. 1975-1976, pp.229-239

HEINE, LEOPOLD [1870-1940]
KURSCHNER 4,1091; FISCHER 1,601-602; RSC 15,732-733; NUC 238,641-642

HEINE, OTTO [1876-]
JV 19,339; NUC 238,643

HEINTZ, WILHELM [1817-1880]
HEIN-E,182; BLOKH 1,281-282; TSCHIRCH, 1063; FERCHL, 221-22; POTSCH, 194; SCHAEDLER, 48-49; POGG 1,1052-1053; 3,606-607
J. Wislicenus, Ber. chem. Ges. 16 (1883), 3121-3140

HEINZ, ROBERT [1865-1924]
FISCHER 1,602-603; BERWIND, 139-144; PITTROFF, 192-194; WWWS, 779; RSC 15,736; NUC 239,119
O. Schulz, Münch. med. Wchschr. 71 (1924), 406
F. Jamin, Deutsche med. Wchschr. 50 (1924), 476-477
F. Heim, Arzneimitt. 15 (1965), 582

HEINTZEL, KURT [1870-]
JV 15,59; NUC 239,111

HEISS, HUGO [1892-]
JV 37,1355

HEITLER, WALTER HEINRICH [1904-1981]
KURSCHNER 13,1438; TETZLAFF, 133; LDGS, 97; POGG 6,1071; 7a(2),430-431

HEITZ, EMIL [1892-1965]
KURSCHNER 10,871-872; SCHULZ-SCHAEFFER, 210; IB, 118; WWWS, 779

HEITZMANN, CARL [1836-1896]
NDB 8,459; ADB 50,158-159; OBL 2,254; LESKY, 564-566; HIRSCH 3,141-142; PAGEL, 709-710
C. Heitzmann, Wiener klin. Wchschr. 8 (1895), 561-564

HEKI, MUTSUO [1903-1957]
NUC 238,250

HEKMA, EBEL [1868-1940]
IB, 118

HEKTOEN, LUDVIG [1863-1951]
DSB 6,232-233; NCAB 18,146-147; FISCHER 1,604; DAMB, 339-340; IB, 118; WWWS, 779-780
M. Fishbein, Arch. Path. 26 (1938), 3-31
J.M. Dack, J. Bact. 62 (1951), 519-520
E.E. Irons, Trans. Assoc. Am. Phys. 80 (1952), 21-23
P.R. Cannon, Biog. Mem. Nat. Acad. Sci. 28 (1954), 163-197
W.K. Beatty, Proc. Inst. Med. Chicago 35 (1982), 7-9

HELBERGER, JOHANN HEINRICH [1905-1961]
POGG 7a(2),431; (4),144*
Anon., Nachr. Chem. Techn. 9 (1961), 59

HELD, ALBERT [1881-1949]
JV 29,630; NUC 239,267

HELD, CHARLES ALFRED [1858-1902]
TSCHIRCH, 1063; DORVEAUX, 82; POGG 4,607; 5,514
Anon., Bull. Sci. Pharm. 8 (1903), 277

HELD, HANS [1866-1942]
FISCHER 1,604-605; BK, 219
H. Stieve, Z. mikr. anat. Forsch. 40 (1936), 1-28
H. Voss, Anat. Nachr. 1 (1950), 106-108

HELDT, WILHELM [1823-1865]
TSCHIRCH, 1063; FERCHL, 222-223; RSC 3,264; 7,942; POGG 1,1054; 3,608

HELE, MARY PRISCILLA [1920-]
AMS 15(3),609

HELE, THOMAS SHIRLEY [1881-1953]
WW 1949, p.1266; O'CONNOR(2), 38-39; IB, 118
Anon., Lancet 1953(I), p.248; Brit. Med. J. 1953(I), p.277

HELFERICH, BURCKHARDT [1887-1982]
KURSCHNER 13,1442-1443; BOHM, 168; TLK 3,254; WWWS, 780; POTSCH, 194-195; POGG 6,1072-1073; 7a(2),431-432
R. Lepsius, Chem. Z. 81 (1952), 351-352
H. Bredereck, Z. angew. Chem. 69 (1957), 405-412
Anon., Chem. Z. 106 (1952), 350; Naturw. Rund. 35 (1982), 343
E. Fanghänel, Jahrb. Sachs. Akad. 1981-1982, pp.212-228
H. Stetter, Chem. Ber. 118 (1985), i-xix; Adv. Carbohydrate Chem. 45 (1987), 1-6
F.W. Lichtenthaler, Carbohydrate Research 164 (1987), 1-22

HELL, CARL MAGNUS von [1849-1926]
POGG 4,607-608; 5,514; 6,1073
A. Gutbier, Chem. Z. 43 (1919), 577
H. Kauffmann, Ber. chem. Ges. 60A (1927), 54-55
E.H. Huntress, Proc. Am. Acad. Arts Sci. 77 (1949), 48

HELL, GUSTAV [1843-1921]
OBL 2,258; STURM 1,587; HEIN, 257-258; NUC 239,366
H. Heger, Pharmazeutische Post 54 (1921), 409-413
K. Ganzinger, Ost. Apoth. Z. 22 (1968), 798-799

HELLAUER, HORST [1913-]
KURSCHNER 13,1443; POGG 7a(2),434

HELLER, ARNOLD [1840-1913]
HIRSCH 3,147; PAGEL, 712-713; VOLBEHR, 80; KOVACSICS, 66-69
E. Wilke, Münch. med. Wchschr. 60 (1913), 987-989

HELLER, FELIX [1883-]
JV 24,184; SG [3]6,631; NUC 239,418
Deutscher Dermatologen Kalender 1929, p.90

HELLER, FRIEDRICH [1883-1963]
JV 26,11; NUC 239,421

HELLER, GUSTAV [1886-1946]
TLK 3,254; POGG 4,609-610; 5,514-515; 6,1074; 7a(2),435

HELLER, HANS [1870-]
JV 9,17; RSC 15,741; NUC 239,422
Anon., Chem. Z. 45 (1921), 1254

HELLER, HANS [1905-1974]
KURSCHNER 12,1180; MEDVEI, 749-751; BHDE 2,484; WWWS, 780
Anon., J. Endocrin. 64 (1975), 399-402

HELLER, JOHANN FLORIAN [1813-1871]
OBL 2,259-260; STURM 1,589; HEIN, 258-259; HIRSCH 3,146; Suppl.,363-364; WURZBACH 8,271; LESKY, 252-254; FERCHL, 223; POGG 1,1056; 3,608
N. Mani, in Wien und die Weltmedizin (E. Lesky, Ed.), pp.170-182. Vienna 1974
S. Danielsson, Nord. Med. Hist. Arsbok 1978, pp.91-98
J. Schmalhofer, Das Werk von Johann Florian Heller. Bonn 1980

HELLER, JÓZEF [1896-1982]
WE 4,604; IB, 119
J.W. Szarkowski et al., Acta Biochem. Polon. 31 (1984), 5-16

HELLER, JULIUS [1864-1931]
KURSCHNER 4,1105; FISCHER 1,606; WININGER 3,45; 7,67

HELLER, KURT [1898-1947]
STURM 1,590; POGG 6,1074; 7a(2),435-436
G. Druce, Nature 160 (1947), 391

HELLERMAN, LESLIE [1896-1981]
AMS 11,2220
Anon., Chem. Eng. News 59(43) (1981), 53

HELLMAN, TORSTEN [1878-1944]
SMK 3,395; FISCHER 1,606-607

G. Glimstedt, Anat. Nachr. 1 (1950), 197-204

HELLMANN, HEINRICH [1913-]
KURSCHNER 13,1448; POGG 7a(2),436-437

HELLMANN, KARL [1892-1959]
GRAML, 162-165; BHDE 2,486; WIDMANN, 89,266; LDGS, 74

HELLRIEGEL, HERMANN [1831-1895]
DSB 6,237-238; NDB 8,488; ADB 50,169-171; HINK 1,717-719; BLOKH 1,282; WWWS, 752
Anon., Chem. Z. 19 (1895), 2318
A. Orth, Ber. bot. Ges. 14 (1896), (25)-(37)

HELLY, KONRAD [1875-1967]
KURSCHNER 9,721; FISCHER 1,607; DANIS, 18-27; KOERTING, 145
Anon., Schw. med. Wchschr. 97 (1967), 724

HELM, JACOB [1761-1831]
CALLISEN 8,439
B. Kisch, J. Hist. Med. 9 (1954), 311-328; 10 (1955), 230-232; 22 (1967), 54-81

HELMERT, ERICA [1899-]
POGG 7a(2),439

HELMHOLTZ, HERMANN von [1821-1894]
DSB 6,241-253; NDB 8,498-501; ADB 51,461-472; BAD 5,281-294; W, 249-250; HIRSCH 3,151-152; Suppl.,364; PAGEL, 713-715; KROLLMANN, 264; HAYMAKER, 225-229; BRAZIER, 65-71,83-85; DRULL, 108-109; POTSCH, 195; BLOKH 1,282-284; POGG 1,1059-1060,1574; 3,611-612;
4,612-613; 5,157; 6,1076; 7aSuppl.,267-277
E. Fischer, Ber. chem. Ges. 27 (1894), 2643-2652
C. von Voit, Sitz. Bayer. Akad. 25 (1895), 185-196
G.F. Fitzgerald, J. Chem. Soc. 69 (1896), 885-912
T.C. Mendenhall, Science 3 (1896), 189-195
L. Koenigsberger, Hermann von Helmholtz. Braunschweig 1902-1903
H. Ebert, Hermann von Helmholtz. Stuttgart 1949
Y. Elkana, Hist. Stud. Phys. Sci. 2 (1970), 263-298

HELMHOLZ, HENRY FREDERIC [1882-1958]
AMS 9(II),495; FISCHER 1,608
Anon., J. Pediat. 53 (1958), 634
E.H. Christopherson, Pediatrics 23 (1959), 172

HELMOLT, AUGUST von [1829-]
GGTA 2 (1908), 462

HELMOLT, OTTO von [1829-1901]
GGTA 4 (1910), 296

HELMREICH, ERNST [1922-]
AMS 15(3),616; KURSCHNER 13,1452; MEISTER, 62-76; WWWS, 782; POGG 7a(2),439-440

HELWIG, HERMANN [1864-1935]
JV 4,224; RSC 13,275

HEMILIAN (GEMILIAN), VALERIAN ALEKSANDROVICH [1851-1914]
RSC 15,749
V.I. Atroshchenko, Zhur. Prikl. Khim. 36 (1963), 933-934
I. Kolokotseva and A.A. Makarenya, Vop. Ist. Est. Tekh. 1969(3),pp.66-68

HEMMERICH, PETER [1929-1981]
KURSCHNER 13,1454
H. Beinert and V. Massey, TIBS 7 (1982), 43-44
P. Hemmerich, in Selected Topics in the History of Biochemistry (G. Semenza, Ed.), pp.399-436. Amsterdam 1986

HEMPEL, KARL WILHELM [1820-1898]
HEIN-E, 184; TSCHIRCH, 1063; FERCHL, 225; SCHELENZ, 690; RSC 2,272

HEMPEL, WALTER [1851-1916]
NDB 8,513-514; DBJ 1,358; DZ, 570; BLOKH 1,184-186; POTSCH, 196; POGG 3,612; 4,613-614; 5,517; 6,1076
F. Foerster, Sitz. Akad. Wiss. Leipzig 69 (1917), 553-580; Z. angew. Chem. 30 (1917), 1-5; Ber. chem. Ges. 53A (1920), 123-143
E. Graefe, Chem. Z. 41 (1917), 85-87
E.H. Huntress, Proc. Am. Acad. Arts Sci. 79 (1951), 18-19

HEMPTINNE, ALEXANDRE de [1866-1955]
POGG 4,614; 5,272; 6,1077; 7b,1946-1947
M. de Hemptinne, Ann. Acad. Roy. Belg. 126 (1960), 3-30

HEMPTINNE, AUGUST DONAT de [1781-1854]
BNB 9,27-35; TSCHIRCH, 1063; POGG 1,1063
J.S. Stas, Ann. Acad. Roy. Belg. 23 (1857), 91-136

HENCH, PHILIP SHOWALTER [1896-1965]
DAB [Suppl.7],340-341; AMS 10.1728-1729; MH 1,217-218; STC 1,501; MSE 2,42-43; DAMB, 341-342; WWAH, 266; WWWS, 783
E.H. Rynearson, Nature 206 (1965), 1195-1196
R.G. Sprague, Trans. Assoc. Am. Phys. 78 (1965), 26-29
C.H. Slocumb, Arthritis and Rheumatism 8 (1965), 573-576
F. Francon, Biol. Med. 56 (1967), 109-123

HENDERSON, DAVID WILLIS WILSON [1903-1968]
DNB [1961-1970], 504-506; WW 1965, p.1393
L.H. Kent and W.J.T. Morgan, Biog. Mem. Fell. Roy. Soc. 16 (1970), 331-341
L.H. Kent and H. Smith, J. Gen. Microbiol. 60 (1970), 145-149

HENDERSON, GEORGE GERALD [1862-1942]
DNB [1941-1950]. 375-376; POGG 4,615; 5,518; 6,1078; 7b,1949-1950
J.L. Simonsen, Obit. Not. Fell. Roy. Soc. 4 (1942-1944), 491-502
J.C. Irvine, J. Chem. Soc. 1944, pp.202-206

HENDERSON, JAMES [1870-1968]
SRS, 17; JV 12,244; RSC 15,753
Anon., Chem. Brit. 4 (1968), 413

HENDERSON, LAWRENCE JOSEPH [1878-1942]
DSB 6,260-262; DAB [Suppl.3],349-352; MILES, 208-209; FISCHER 1,608; AMS 6,631; CHITTENDEN, 103-105,153-158; FORBES, 207-214; POTSCH, 197; DAMB, 342-343; WWWS, 783; POGG 6,1078
R.M. Ferry, Science 95 (1942), 316-318
W.B. Cannon, Biog. Mem. Nat. Acad. Sci. 23 (1943), 31-58
D.W. Richards, Physiologist 1(2) (1958), 32-37
J.H. Talbott, J. Am. Med. Assn. 198 (1966), 1304-1306
J. Parascandola, J. Hist. Biol. 4 (1971), 63-113; Med. Hist. J. 6 (1971), 297-309; in Science at Harvard (C.A. Elliott and M.W. Rossiter, eds.), pp.167-190. Bethlehem, Pa. 1992
D.B. Dill, Physiologist 20(2) (1977), 1-15
J.T. Edsall, Isis 75 (1984), 11-13; Hist. Phil. Life Sci. 7 (1985), 105-120

HENDERSON, VELYIEN EWART [1877-1945]
WW 1945, p.1257; O'CONNOR(2), 513; CHEN, 40; IB, 119; WWWS, 783
A.D. Welch, Science 104 (1946), 285-286
J.K.W. Ferguson, Rev. Can. Biol. 5 (1946), 4-8
H. Wasteneys, Proc. Roy. Soc. Canada 40 (1946), 91-93

HENDERSON, YANDELL [1873-1944]
DSB 6,264-265; DAB [Suppl.3],352-354; AMS 6,632; CHITTENDEN, 163-167; NCAB D,86-87; 36,25-26; PARASCANDOLA, 36-37; DAMB, 342-343; IB, 119; WWAH, 266; WWWS, 784
Y. Henderson, Adventures in Respiration. Baltimore 1938
C.G. Douglas, Nature 153 (1944), 308-309
H. Haggard, Am. Phil. Soc. Year Book 1944, pp.369-374

HENDRICKS, STERLING BROWN [1902-1981]
AMS 14,2123; WWA 1978-1979, p.1461; AO 1981, pp.8-10; MH 1,218-219; MSE 2,43-44; WWWS, 784; POGG 6,1078-1079; 7b,1951-1955
S.B. Hendricks, Ann. Rev. Plant Physiol. 21 (1970), 1-10
Anon., Chem. Eng. News 59(11) (1981), 53

C.H. Wadleigh, Am. Phil. Soc. Year Book 1981, pp.458-462
L. Pauling, Am. Min. 67 (1982), 406-409
W.L. Butler and C.H. Wadleigh, Biog. Mem. Nat. Acad. Sci. 56 (1987), 181-212

HENGLEIN, FRIEDRICH AUGUST [1893-1968]
KURSCHNER 11,1087; TLK 3,256; POGG 6,1079; 7a(2),441-442
Anon., Chem. Z. 77 (1953), 220

HENKE, KARL [1895-1956]
NDB 8,526-527; KURSCHNER 8,874-875; WAGENITZ, 76-77; IB, 119
A. Kühn, Jahrb. Akad. Wiss. Gott. 1944-1950, pp.165-167
E. Caspari, Science 125 (1957), 1076

HENKEL, JOHANN BAPTIST [1825-1871]
HEIN, 261-262; LEHMANN, 187-188; TSCHIRCH, 1063
A. Wankmüller, Beitr. Wurtt. Apothekgesch. 7 (1967), 110-117; 8 (1968), 38-48

HENKEL, THEODOR [1855-1934]
NDB 8,528-529; KURSCHNER 4,1115; IB, 119

HENKING, HERMANN [1858-1942]
DSB 6,267-268; NDB 8,529-530; SCHULZ-SCHAEFFER,202; WAGENITZ, 77; IB, 119
O. Schubert, Z. Fisch. 26 (1928), 311-342
P.F. Meyer, Ber. Komm. Meeresforsch. 12 (1950), 115-128

HENLE, FRANZ WILHELM [1876-1944]
NDB 8,530-531; TLK 3,257; POGG 5,519; 6,1081; 7a(2),443

HENLE, JACOB [1809-1885]
DSB 6,268-270; NDB 8,531-532; ADB 50,190-191; FREUND 2,147-159; LF 1,183-188; HIRSCH 3,162-165; Suppl.,365; PAGEL, 716-719; FRANKEN, 185; DRULL, 109; STUBLER, 301-305; WAGENITZ, 77-78; WININGER 3,52; KOHUT 2,279; WWWS, 784; CALLISEN 28,473-475
W. Waldeyer, Arch. mikr. Anat. 26 (1885), i-xxxii
K. Bardeleben, Deutsche med. Wchschr. 11 (1885), 463-464,483-484
J.B.S., Proc. Roy. Soc. 39 (1886), iii-viii
C. von Voit, Sitz. Bayer. Akad. 16 (1886), 31-38
F. Merkel, Jacob Henle. Braunschweig 1891
H. Hoepke, Sudhoffs Arch. 53 (1969), 193-216
G.B. Gruber, Niedersächsische Lebensbilder 6 (1969), 224-239

HENLE, WERNER [1910-1987]
AMS 11,2236; BHDE 2,489; WWWS, 785

HENNEBERG, BRUNO [1867-1941]
KURSCHNER 4,1117; GUNDEL, 378-386; FISCHER 1,609

HENNEBERG, WILHELM [1825-1890]
NDB 8,540-541; ADB 50,193-195; SCHAEDLER, 49
C. von Voit, Sitz. Bayer. Akad. 21 (1891), 161-174

HENNEBERG, WILHELM [1871-1936]
NDB 8,541; KURSCHNER 4,1117; KIEL, 191; IB, 120
A. Schittenhelm, Münch. med. Wchschr. 83 (1936), 941

HENNEGUY, LOUIS FÉLIX [1850-1928]
DBF 17,904-905; IB, 120; WWWS, 785
E. Fauré-Fremiet, Arch. Anat. Micr. 25 (1929), 1-36
G. Levi, Rend. Accad. Lincei 7(App.) (1929), 40-55
M. Caullery, Not. Acad. Sci. 2 (1949), 360-377

HENNELL, HENRY [-1842]
FERCHL, 226; POGG 1,1066
Anon., Proc. Roy. Soc. 4 (1842-1843), 419

HENNINGER, ARTHUR [1850-1884]
BLOKH 1,286; POGG 3,614; 4,616
P. Schützenberger, Bull. Soc. Chim. [3]42 (1884), 547-549
Anon., Ber. chem. Ges. 17 (1884), 2812-2813

E.H. Huntress, Proc. Am. Acad. Arts Sci. 78 (1950), 26-27

HENOCH, EDUARD HEINRICH [1820-1910]
NDB 8,549; BJN 15,37*; HIRSCH 3,168-169; Suppl.,365-366; PAGEL, 719-720; WININGER 3,53; KOHUT 2,279; KOREN, 191
A. Baginsky, Deutsche med. Wchschr. 36 (1910), 1329-1331
A. Schlossmann, Münch. med. Wchschr. 57 (1910), 1504-1505
Anon., Leopoldina 46 (1910), 103
E. Schwechten, Jahrb. Kinderheil. 72 (1910), 519-521
P. Wunderlich, Med. Mon. 21 (1967), 454-459

HÉNOCQUE, ALBERT WILLIAM LÉON [1840-1903]
RSC 10,194; 15,761-763
J. Pagel, Virchows Jahresber. 1903, p.416
Anon., Science 17 (1903), 198

HENRI, VICTOR [1872-1940]
DSB 17,410-413; BNB 42,345-354; TLK 3,259; POGG 5,520; 6,1081-1082; 7b,1955-1956
H. von Halban, Viert. Nat. Ges. Zurich 86 (1941), 307-320
H.P. Stevens, India-Rubber Journal 113 (1947), 689-691
B. Rosen et al., J. Chim. Phys. 50 (1953), 601-616
J. Duchesne, Liber Memorialis Université de Liège 1936-1966, vol.2, pp.471-478. Liège 1967

HENRICH, FERDINAND [1871-1945]
KURSCHNER 6(I),695; KERNBAUER, 154-158; STUPP-KUGA, 232-249; TLK 3,259; POGG 4,617; 5,520-521; 6,1082; 7a(2),446
L. Birkhofer, Chem. Ber. 83 (1950), vii-xiii

HENRICI, ARTHUR TRAUTWEIN [1889-1943]
DAB [Suppl.3],354-355; AMS 6,634; FISCHER 1,610; SACKMANN, 153-154; WWAH, 267; WWWS, 785
R.L. Starkey and S.A. Waksman, J. Bact. 46 (1943), 489-490

HENRIOT, ÉMILE JEAN CHARLES [1885-1961]
DBF 17,969

HENRIQUES, OSCAR [1895-1953]
KBB 1953, p.581; VEIBEL 2,193-194

HENRIQUES, ROBERT [1857-1902]
RSC 9,932; 15,764; POGG 4,617
D. Holde, Ber. chem. Ges. 35 (1902), 4528-4533

HENRIQUES, VALDEMAR [1864-1936]
DBL 10,136-137; VEIBEL 2,194-197; FISCHER 1,610; KOREN, 191
E. Lundsgaard, Skand. Arch. Physiol. 76 (1937), 101-108

HENRY, EMMANUEL OSSIAN [1826-1867]
DBF 17,982-983; RSC 3,292

HENRY, ÉTIENNE OSSIAN [1798-1873]
DBF 17,983; GE 19,1119-1120; LAROUSSE 9,193; GORIS, 439-443; FERCHL, 226; TSCHIRCH, 1063; BOURQUELOT, 57; POGG 1,1070-1071; 3,615

HENRY, LOUIS [1834-1913]
BNB 41,414-427; BLUM 1,425-435; POGG 3,615-616; 4,617-618; 5,521-522
C. Aschman, Chem. Z. 37 (1913), 765
G. Lemoine, Bull. Soc. Chim. [4]15 (1914), i-xxxi
A. Bruylants, Bull. Acad. Roy. Belg. 70 (1984), 718-732

HENRY, NOËL ÉTIENNE [1769-1832]
DBF 17,994-995; GORIS, 437-439; BOURQUELOT, 49-50; DECHAMBRE 49,551; BLOKH 1,287; TSCHIRCH, 1063; FERCHL, 226; PHILIPPE, 812; POGG 1,1070; CALLISEN 8,364-369; 28,480
L.A. Planche, J. Pharm. Chim. [2]18 (1832), 520-522
G. Dillemann, Produits et Problèmes Pharmaceutiques 27 (1972), 415

HENRY, PAUL [1866-1917]
BNB 41,427-433; RSC 15,771; POGG 4,620

HENRY, THOMAS ANDERSON [1873-1958]
POGG 6,1084; 7b,1957-1958
T.M. Sharp, Nature 181 (1958), 1699
W. Solomon, Proc. Chem. Soc. 1959, pp.21-22

HENRY, WILLIAM [1774-1836]
DSB 6,284-286; DNB 9,250-251; W, 250-251; CULE, 67-69; BLOKH 1,287; COLE, 251-256; FERCHL, 226-227; ABBOTT-C, 59-60; WWWS, 786; POTSCH, 197; POGG 1,1069-1070
Anon., Proc. Roy. Soc. 3 (1837), 439-440
W.C. Henry, Mem. Lit. Phil. Soc. [2]6 (1842), 99-141
W.V. Farrar et al., Ambix 21 (1974), 179-228; 22 (1975), 186-204; 23 (1976), 27-52

HENRY, WILLIAM CHARLES [1804-1892]
RSC 3,295
Anon., Mem. Lit. Phil. Soc. [4]5 (1892), 178-179; Proc. Roy. Soc. 54 (1893), xix
W.V. Farrar et al., Ambix 24 (1977), 1-26

HENSELEIT, KURT [1907-1973]
JV 50,373

HENSEMANS, PIERRE JOSEPH [1802-1862]
TSCHIRCH, 1063; FERCHL, 227; POGG 1,1072; CALLISEN 8,378-379; 28,484
Liber Memorialis Université de Gand, vol.2, pp.462-465. Ghent 1913

HENSEN, VICTOR [1835-1924]
DSB 6,287-288; NDB 8,563-564; SHBL 4,97-99; HIRSCH 3,173; Suppl.,366; PAGEL, 721-722; KIEL, 80; DZ, 575-576; W, 251; WWWS, 786; POGG 4,621; 5,522-623; 6,1084
O. Damsch, Münch. med. Wchschr. 52 (1905), 912-913
G. Wolff, Münch. med. Wchschr. 102 (1960), 1203-1208
R. Porep, Der Physiologe und Planktonforscher Victor Hensen. Kiel 1969; Med. Mon. 25 (1971), 314-321
J. Lussenhop, J. Hist. Biol. 7 (1974), 319-337

HENTSCHEL, HERBERT [1896-1983]
NUC 241,303
Anon., Chem. Z. 90 (1966), 779; 100 (1976), 500; 107 (1983), 312

HENZE, MARTIN [1873-1956]
NDB 8,567-568; FISCHER 1,612; HUTER, 242-243; TLK 3,260; POGG 5,523; 6,1085; 7a(2),448-449
Anon., Science 124 (1956), 1143

HEPBURN, JOSEPH SAMUEL [1885-1977]
AMS 11,2244; WWA 1968-1969, p.1004; WWWS, 787
Who Was Who in America 7,269

HEPP, EDUARD [1851-1917]
NDB 8,569-570; BLOKH 1,287-288; POTSCH, 197; POGG 6,1086
Anon., Z. angew. Chem. 30(III) (1917), 384; Chem. Z. 42 (1918), 99
O. Fischer, Ber. chem. Ges. 51A (1918), 165-168
E.H. Huntress, Proc. Acad. Arts Sci. 79 (1951), 20-21

HEPPEL, LEON ALMA [1912-]
AMS 15(3),642; WWA 1980-1981, p.1526; IWW 1982-1983, p.557
L.A. Heppel, in Reflections on Biochemistry (A. Kornberg et al., Eds.), pp.377-383. Oxford 1976

HERAPATH, WILLIAM [1796-1868]
DNB 9,615; BLOKH 1,288; FERCHL, 227; POGG 1,1072-1073; 3,617
W.A. Campbell, Anal. Proc. 17 (1980), 346-348
R.M. Weller, Anaesthesia 38 (1983), 678-682

HERAPATH, WILLIAM BIRD [1820-1868]
DSB 6,293-294; DNB 9,615; FERCHL, 227; WWWS, 787; POTSCH, 198; POGG 1,1073; 3,617
W. De la Rue, J. Chem. Soc. 22 (1869), 6-7

HERB, JOSEPH [1861-]
JV 5,74; RSC 15,779; NUC 241,466

HERBERGER, JOHANN EDUARD [1809-1855]
HEIN, 264; TSCHIRCH, 1054; FERCHL, 227; SCHELENZ, 682; POGG 1,1074,1573; CALLISEN 8,384; 28,487-488

HERBERT, DENIS [1916-]
Who's Who in British Science 1953, p.128

HERBORN (AVENARIUS-HERBORN), HEINRICH [1873-1955]
JV 11,20; RSC 14,1010
Anon., Nachr. Chem. Techn. 1 (1953), 140; 3 (1955), 213

HERBST, CARL FRIEDRICH REINHOLD [1824-1868]
R. Schmitz, Ueber das Apothekenwesen der Stadt und des Kreises Wetzlar 1233-1900, pp.47-48. Wetzlar 1957

HERBST, CURT ALFRED [1866-1946]
DSB 17,413-415; NDB 8,593; DRULL, 109-110; BK, 220-221; IB, 120
Anon., Sitzungsberichte der Heidelberger Akademie der Wissenschaften 1943-1955, pp.41-42
J.M. Oppenheimer, Bull. Hist. Med. 44 (1970), 241-250

HERBST, ROBERT MAX [1904-1992]
AMS 14,2137

HERDMAN, WILLIAM ABBOTT [1858-1924]
DNB [1922-1930], 415-416; WWWS, 788
S.J.H., Proc. Roy. Soc. B98 (1925), x-xiv
A.E. Shipley, Proc. Linn. Soc. 137 (1925), 78-80

d'HÉRELLE, FÉLIX [1873-1949]
DSB 6,297-299; FISCHER 1,612; BULLOCH, 372
A. Compton, Nature 163 (1949), 984-985
P. Lepine, Ann. Inst. Pasteur 76 (1949), 457-460
P. Nicolle, Biol. Med. 38 (1949), 233-306
S.J. Peitzmann, Trans. Coll. Phys. Phia. [4]37 (1969), 115-123
D.H. Duckworth, Bact. Revs. 40 (1976), 793-802
T. van Helvoort, Tijd. Gesch. Geneesk. 8 (1985), 49-87

HEREMANS, JOSEPH [1927-1975]
J.P. Vaerman, European Journal of Immunology 6 (1976), 1-2

HERING, EDOUARD [1814-1893]
RSC 3,308-309
Anon., Pharm. Z. 38 (1893), 671

HERING, EWALD [1834-1918]
DSB 6,299-301; NDB 8,617-618; DBJ 2,258-263,689-690; STURM 1,603; HIRSCH 3,181-182; PAGEL, 723-724; KOERTING, 118-119; DZ, 577; LOMMATZSCH, 101-109; FRANKEN, 185; WWWS, 788; POGG 3,618; 4,622-623; 5,524; 6,1088
S. Garten, Pflügers Arch. 170 (1918), 501-522; Sitz. Akad. Wiss. Leipzig 70 (1918), 381-402
C. Hess, Naturwiss. 6 (1918), 305-308
F.B. Hofmann, Münch. med. Wchschr. 65 (1918), 539-542
A. von Tschermak, Münch. med. Wchschr. 81 (1934), 1230-1233

HERING, HEINRICH EWALD [1866-1948]
NDB 8,618; OBL 2,283; FISCHER 1,613; WWWS, 788; POGG 7a(2),451
H. Nies, Z. ges. inn. Med. 4 (1949), 319-320HÉRISSEY, HENRI [1873-1959] DBF 17,1070-1071; GORIS, 434-437; POGG 6,1088-1090; 7b,1959-1960
P. Fleury and J. Courtois, Bull. Soc. Chim. Biol. 41 (1959), 933-957
G. Dillemann, Ann. Univ. Paris 29 (1959), 450-452
P. Fleury, Bull. Acad. Med. 143 (1959), 229-233

HERKEN, HANS [1912-]
KURSCHNER 13,1482; POGG 7a(2),452-453
Anon., Leopoldina [3]23 (1980), 41
W. Kalow, Klin. Wchschr. 66 (1988), 229-235

HERLINGER, ERICH [1899-1950]
BHDE 1,285; JV 40,733; POGG 6,1089; 7b,1960

HERMANN, GÜNTHER [1924-1982]
KURSCHNER 13,1485
Anon., Naturw. Rund. 35 (1982), 507

HERMANN, LUDIMAR [1838-1914]
NDB 8,662-664; DBJ 1,287; HIRSCH 3,185; PAGEL, 724-726; KROLLMANN, 949; SCHOLZ, 13-17; BRAZIER, 91-97,105-108; DZ, 579-580; KOHUT 2,236; WININGER 3,59; POGG 3,619-620; 4,623; 5,524; 6,1090
F.B. Hoffmann, Samml. anat. Vortr. 27 (1914), 19-27
F. Rudie and C. Schroter, Viert. Nat. Ges. Zürich 59 (1914), 571-572
H. Boruttau, Deutsche med. Wchschr. 40 (1914), 1529
L. Hermann, Erinnerungen. Berlin 1915
O. Frank, Bayer. Akad. Wiss. Jahrbuch 1915, pp.105-114
P. Jensen, Jahrb. Akad. Wiss. Gott. 1915, pp.79-91
W.M.B., Proc. Roy. Soc. B91 (1920), xxxviii-xl
J.H. Schawalder, Der Physiologe Ludimar Hermann. Zurich 1990

HERMANN, LUDWIG [1882-1938]
NDB 8,664; POGG 6,1090; 7a(2),454
Anon., Ber. chem. Ges. 71A (1938), 166

HERMANN, OTTO JULIUS THEODOR [1806-1862]
C. Rammelsberg, Ber. chem. Ges. 3 (1870), 439-441

HERMANN, RUDOLPH [1805-1879]
HEIN-E, 187; FERCHL, 228-229; POGG 1,1080-1081; 3,619
Anon., Bull. Soc. Nat. Moscou 54 (1880), 152-182
I.I. Iskoldsky, Priroda 37 (10) (1948), 85-92

HERMANS, PETRUS HENDRIK [1898-1979]
POGG 6,1090-1091; 7b,1960-1964
G. Challa and Anon., Z. angew. Chem. 47 (1934), 208
G. Challa and D. Heikens, Chem. Wkbl, 59 (1963), 177-179
H. Mark, J. Pol. Sci. 1963(2), pp.1-3

HERMBSTÄDT, SIGISMUND FRIEDRICH [1760-1833]
DSB 15,205-207; NDB 8,666-667; ADB 12,190-192; HEIN, 266-267; BLOKH 1,289-290; TSCHIRCH, 1064; FERCHL, 229-230; SCHAEDLER, 50; PHILIPPE, 738-739; COLE, 257-258; MATSCHOSS, 115; POTSCH, 198 POGG 1,1082-1083; CALLISEN 8,420-421; 28,498-502
I. Mieck, Technik Geschichte 32 (1965), 325-382

HERON, JOHN [1850-1913]
RSC 15,792; BOURQUELOT, 57; POGG 1,1070-1071; 3,615
L.T. Thorne, Analyst 38 (1913), 185
Anon., Chem. Z. 37 (1913), 626

HERPIN, JEAN CHARLES [1798-1872]
DBF 17,115-116; FERCHL, 231; POGG 1,1086; 3,621
E. Grellois, Mémoires de l'Académie de Metz 54 (1872-1873), 102-114

HERRICK, CHARLES JUDSON [1868-1960]
DSB 6,320-322; NCAB 47,90-91; IB, 121; WWAH, 269; WWWS, 790
P. Bailey, Proc. Inst. Med. Chicago 23 (1961), 232-234
J.L. O'Leary et al., Persp. Biol. Med. 12 (1969) 492-513
G.W. Barthelmez, Biog. Mem. Nat. Acad. Sci. 43 (1973), 77-108

HERRICK, FRANCIS HOBART [1858-1940]
NCAB C,381; 31,276-277; IB, 272; WWAH, 269; WWWS, 790
W.G. Leutner, Science 92, (1940), 371-372

HERRICK, JAMES BRYAN [1861-1954]
AMS 8,1104; NCAB 42,595-596; DAMB, 345; WWWS 790
J.B. Herrick, Memories of Eighty Years. Chicago 1949
E.E. Irons, Trans. Assoc. Am. Phys. 67 (1954), 15-19
J.L. O'Leary et al., Persp. Biol. Med. 12 (1969), 492-513
J.H. Talbott, A Biographical History of Medicine, pp. 1158-1160. New York 1970.
P.S. Rhoads, Proc. Inst. Med. Chicago 35 (1982), 3-6
R.S. Ross, Circulation, 67 (1983), 955-959

HERRING, PERCY THEODORE [1872-1967]
O'CONNOR (2),427
Who Was Who 1961-1970, p.522
R. Walmsley, Roy. Soc. Edin. Year Book 1969, pp.37-39

HERRIOTT, ROGER MOSS [1908-1992]
AMS 14,2146; WWA 1976-1977, p.1422
New York Times, 9 March 1992

HERRMANN, ALBERT [1859-1921]
NUC 243,66
Anon., Z. angew. Chem. 34 (1921), 49

HERRMANN, AUGUST [1873-]
JV 11,105; RSC 15,796-797; NUC 243,123

HERRMANN, EDMUND [1875-1930] OBL 2,291; FISCHER 1,615; LORENZSONN, 120-126; KOREN, 191

HERRMANN, FELIX [1848-1913]
BLOKH 1,290-291; POGG 4,625; 5,524
W. Will, Ber. chem. Ges. 46 (1913), 1649-1651

HERRMANN, FRANZ [1898-1977]
AMS 11,2256; KURSCHNER 12,1217; KALLMORGEN, 296; BHDE 2,496; LDGS, 62
H. Tronnier, Dermatosen in Beruf und Umwelt 26 (1978), 37

HERRMANN, HEINZ [1911-]
AMS 15(3),654; BHDE 2,496

HERRMANN, KARL [1882-]
LDGS, 20; POGG 6,1092; 7a(2),457

HERRMANN, MAXIMILIAN [1834-]
RSC 3,322; 7,962-963; NUC 243,119

HERSCHBACH, DUDLEY ROBERT [1932-]
AMS 17(3),676; POTSCH, 199

HERSHEY, ALFRED DAY [1908-]
AMS 14,2148; WWA 1978-1979, p.1476; IWW 1982-1983, p.561; MH 1,220-221; STC 1,501-502; MSE 2,47; CB 1970, pp.175-177; WWWS, 791

HERTEL, EDUARD [1899-1954]
TLK 3,264; POGG 6,1093; 7a(2),458-459
Anon., Nachr. Chem. Techn. 2 (1954), 195

HERTER, CHRISTIAN ARCHIBALD [1865-1910]
DAB 8,597-598; MILES, 216-217; KELLY, 560-561; FISCHER 1,616; DAMB, 346; CHITTENDEN, 177-180; WWWS, 792
G. Lusk, Science 33 (1911), 846-847
O.T. Williams, Biochem. J. 5 (1911), xxi-xxxi
R.M. Hawthorne, Jr., Persp. Biol. Med. 18 (1974), 24-39

HERTER, ERWIN [1849-1908]
BJN 13,40*; RSC 10,209; 12,329; 15,800
R. Andreasch and K. Spiro, Jahresb. Fort. Tierchem. 37 (1908), i-iii

HERTWIG, CARL [1820-1896]
NDB 8,706*; RSC 3,330

HERTWIG, OSCAR [1849-1922]
DSB 6,337-340; NDB 8,706-707; DBJ 4,357; FREUND 1,207-215; DZ, 583-584; FISCHER 1,617-618; PAGEL, 626-728; GIESE, 464-466; WREDE, 79-80; BK, 221; WWWS, 792
F. Keibel, Anat. Anz. 56 (1923), 372-373
R. Weissenberg, Oskar Hertwig. Leipzig 1959
F.B. Churchill, Isis 61, (1970), 429-457

P.J. Weindling, Darwinism and Social Darwinism in Imperial Germany. Stuttgart 1991

HERTWIG, RICHARD [1850-1937]
DSB 6,336-337; NDB 8,707-708; FISCHER 1,618; KROLLMANN, 272; BK,221-222; SCHULZ-SCHAEFFER, 199; ZWEIFEL, 61-62; DZ, 584; IB, 121; WWWS, 1736
F. Doflein et al., Naturwiss. 8 (1920), 767-782
K. von Frisch, Münch. med. Wchschr. 84 (1937), 1785-1786
C. Dobell, Proc. Linn. Soc. 150 (1937-1938), 319-323

HERTZ, GUSTAV LUDWIG [1887-1975]
KURSCHNER 10,909; WWWS, 792; POGG 6,1094; 7a(2),461
V. Rich, Nature 258 (1975), 464

HERTZ, JOHANN NICOLAUS [1869-1908]
NDB 8,709; JV 7,264; RSC 15,812
Anon., Chem. Z. 32 (1908), 441

HERVÉ de la PROVOSTAYE, FRÉDÉRIC [1812-1863]
DBF 17,1146-1147; POGG 1,1093-1094; 3,624

HERXHEIMER, GOTTHOLD [1872-1936]
NDB 8,727; NB, 161; FISCHER 1,618-619; KALLMORGEN, 296-297; KOREN, 191
L. Aschoff, Verhandl. path. Ges. 29 (1937), 381-386

HERXHEIMER, HERBERT [1894-1985]
KURSCHNER 13,1498; ASEN, 78; BHDE 2,498; LDGS, 66; KOREN, 191

HERXHEIMER, KARL [1861-1942]
NB, 161; KURSCHNER 4,1145; ARNSBERG, 187-188; FISCHER 1,619; KOREN, 191; KALLMORGEN, 297; WININGER 7,73; TETZLAFF, 136

HERZ, ALBERT [1876-]
FISCHER 1,619; NUC 243,427

HERZ, RICHARD [1867-1936]
NDB 8,731; POTSCH, 200; POGG 6,1095; 7a(2),461-462
A. Lüttringhaus, Chem. Ber. 89 (1956), i-x

HERZ, WALTER [1875-1930]
KURSCHNER 3,916; TLK 3,265; POGG 4,628; 5,527-528; 6,1096-1097
H. Biltz, Ber. chem. Ges. 63A (1930), 162-164
J. Meyer, Chem. Z. 54 (1930), 773-774

HERZBERG, WILHELM [1861-1930]
POGG 6,1097
O. Schartenburg, Z. angew. Chem. 43 (1930), 839-840
M. Bodenstein, Ber. chem. Ges. 63A (1930), 158-159
P. Roders, Chem. Z. 55 (1931), 317

HERZFELD, ALEXANDER [1854-1928]
NDB 8,732-733; WININGER 3,82; 7,73; RSC 15,808-808; NUC 243,469; POGG 7a Suppl.,286-287
O. Spengler, Z. Zuckerind. 78 (1928), 255-260
H. Claasen, Zbl. Zuckerind. 36 (1928), 1019-1020
Anon., Chem. Z. 52 (1928), 709; Ost. Chem. Z. 31 1928, 161

HERZIG, JOSEF [1853-1924]
NDB 8,735; OBL 2,298; BLOKH 1,292; WININGER 3,85; POGG 4,629; 5,529; 6,1098-1099
J. Pollak, Ber. chem. Ges. 58A (1925), 55-75

HERZOG, ALOIS [1872-1956]
NDB 8,738-739; POGG 7a(2),464-465
P.A. Koch, Zeitschrift fürr die gesamte Textilindustrie 58 (1956), 678-682

HERZOG, GEORG [1884-1962]
GUNDEL, 392-399; FISCHER 1,620; NUC 243,501
Anon., Verhandl. path. Ges. 46 (1962), 399-402

HERZOG, REGINALD OLIVER [1878-1935]
NDB 8,740-741; STTURM 1,616; BHDE 2,502-503; WIDMANN, 266; TLK 3,267; WININGER 7,77; TETZLAFF, 137; POGG 5,529-530; 6,1099-1100
K.F. Herzfeld, Science 81 (1935), 607-608
H. F[reundlich], Nature 135 (1935), 534-535

HESS, ALFRED FABIAN [1875-1933]
DAB 21,397-398; FISCHER 1,621; CHITTENDEN, 369-371; DAMB, 347-348; OEHRI, 52-53; KAGAN(JA), 154-161,615-616; KOREN, 192; WWWS, 794;POGG 6,1101; 7b,1970-1971
Anon., Science 79 (1934), 70-72
S.R. Kagan, Medical Life 41 (1934), 612-619
M.H. Bass, in Pediatric Profiles (B.S. Veeder, Ed.), pp. 175-181. St. Louis 1957
W.J. Darby and C.W. Woodruff, J. Nutrition 71 (1960), 3-9

HESS, CARL von [1863-1923]
NDB 9,9-10; DBJ 5,171-175,430; FISCHER 1,621-622; PAGEL, 731
L. Asher, Erg. Physiol. 23(II) (1925), 277-283

HESS, FRITZ [1882-]
KALLMORGEN, 297-298

HESS, GEORGE PAUL [1926-]
AMS 15(3),664; BHDE 2,503

HESS, GERMAIN HENRI [1802-1850]
DSB 6,353-354; KUZNETSOV (1961), 428-433; ARBUZOV, 99; BLOKH 1,241-242; POTSCH, 200-201; SCHAEDLER, 50-51; WWWS, 794; POGG 1,1094-1095
H.M. Leicester, J. Chem. Ed. 28 (1951), 581-583
E.H. Huntress, Proc. Am. Acad. Arts Sci. 81 (1952), 62

HESS, HERMANN [1882-1915]
JV 25,606; NUC 243,303

HESS, KURT [1888-1961]
NDB 9,10; HANNOVER, 37; TLK 3,268; POGG 5,530-532; 6,1101-1103; 7a(2),467-470
O. Kratky, Koll. Z. 163 (1959), 97-98

HESS, LEO [1879-1963]
KURSCHNER 4,1152; FISCHER 1,622; BHDE 2,504; WININGER 7,77; KOREN, 192

HESS, LUDWIG CONRAD [1882-1956]
NDB 9,3*; TLK 3,268; JV 29,724; NUC 243,621
Wer Ist Wer 10,663

HESS, OTTO PAUL [1860-]
RSC 14,1010; NUC 243,626

HESS, WALTER RUDOLF [1881-1973]
SBA 5,60-61; FISCHER 1,622; STC 1,507-508; MSE 2,52; WWWS, 794
W.R. Hess, Persp. Biol. Med. 6 (1963), 400-425
W. Auerswald, Alm. Akad. Wiss. Wien 124 (1975), 415-433
R. Jung, Erg. Physiol. 88 (1981), 1-21
P.G. Weber and A. Huber, Gesnerus 39 (1982), 279-293

HESS, WILHELM [1862-1934]
JV 3,100; RSC 14,804; 15,813
Anon., Z. angew. Chem. 47 (1934), 208

HESS, WILHELM [1865-1917]
NUC 243,639
G. Scheibe, Z. angew. Chem. 30(III) (1917), 452

HESSE, ALBERT [1866-1924]
NDB 9,15-16; BLOKH 1,292-295; TLK 2,290; POTSCH, 201; POGG 4,631-632; 5,532; 6,1104
M. Pflücke, Ber. chem. Ges. 57A (1924), 49-52

P. Alexander, Z. angew. Chem. 37 (1924), 470-473

HESSE, ALBERT [1897-1963]
JV 40,733
Anon., Chem. Z. 87 (1963), 715

HESSE, AUGUST [1873-]
JV 12,182; NUC 243,648

HESSE, ERICH [1895-1971]
TLK 3,270; POGG 6,1104; 7a(2),473-474

HESSE, GERHARD [1908-]
KURSCHNER 13,1508; AUERBACH, 825-826; MEINEL, 409,504; POGG 7a(2),474-475

HESSE, JULIUS [1864-]
JV 2,8; 13,24; SG [2]7,95; RSC 15,813; NUC 243,687

HESSE, OSWALD [1835-1917]
NDB 9,20-21; DBJ 2,658; BLOKH 1,295; TSCHIRCH, 1064; POGG 3,625-626; 4,631; 5,532
A. Weller, Ber. chem. Ges. 50 (1917), 475-476; Chem. Z. 42 (1918), 29-31
A.H. Effler and B.H. Effler, Pharm. Z. 117 (1972), 1076-1079

HESSERT, JULIUS [1852-1898]
RSC 10,218
Anon., Ber. chem. Ges. 32 (1899), 3706

HESSLING, KARL THEODOR von [1816-1899]
BJN 4,147*; HIRSCH 3,203-204; PAGEL, 731-732; EGERER, 11-12
Anon., Leopoldina 35 (1899), 136-137

HESTRIN, SCHLOMO [1914-1962]
M. Schramm, Bull. Res. Coun. Israel 11 (1963), 243-248

HETTCHE, OTTO [1902-]
KURSCHNER 15,1785

HEUBNER, OTTO [1843-1926]
NDB 9,38-39; FISCHER 1,624-625; PAGEL, 732-733; GROTE 4,93-124; DZ, 592-593; WREDE, 80-81; WWWS, 793
J. Bokay, Wiener klin. Wchschr. 40 (1927), 1481-1483
O, Neustätter, Sächsische Lebensbilder 1 (1930), 151-159
A. Klingenberg-Straub, Otto Heubners Leben und Lehrbuch der Kinderheilkunde. Zürich 1968

HEUBNER, WOLFGANG [1877-1957]
NDB 9,39-40; FISCHER 1,625; DRULL, 112; IB, 122; POGG 6,1105-1106; 7a(2),477-479
D. Lendle, Deutsche med. Wchschr. 82 (1957), 667-668; Med. Hist. J. 4 (1969), 24-40
P. Marquardt, Arzneimitt. 7 (1957), 331-332
H. Herken, Göttinger Jahrbuch 1980, pp.193-197

HEUMANN, CARL [1850-1894]
NDB 9,44; WOLF, 83; BLOKH 1,295-296; POTSCH, 201; POGG 3,627; 4,632
Anon., Chem. Z. 18 (1894), 2062

HEUSER, EMIL [1882-1953]
WOLF, 83; TLK 3,271; LDGS, 20; WWAH, 271; POGG 6,1106-1107; 7a(2),480
G. Jayme, Papier 6 (1952), 347-355; 8 (1954), 61
L.E. Wise, Adv. Carbohydrate Chem. 15 (1960), 1-9

HEUSER, KARL [1869-]
JV 11,233; RSC 19,74; NUC 244,221HEUSINGER, KARL FRIEDRICH [1792-1883] NDB 9,46; ADB 50,293; LKW 1,144-157; HIRSCH 3,207-209; GUNDLACH, 215-216; KILLIAN, 225-226; ZADEMACH, 42-51; CALLISEN 8,459-467; 28,518-520
U. Malchau, Carl Friedrich Heusinger. Marburg 1973; Med. Hist. J. 9 (1974), 49-62
W. Speckner, Carl Friedrich Heusinger: Sein Leben und Wirken in Würzburg. Würzburg 1981

HEUSS, ERNST [1864-1912]
NDB 9,57; DHB 4,94

HEUSSER, HANS [1917-]
POGG 7a(2),481-482

HEVESY, GEORGE de [1885-1966]
DSB 6,365-367; MH 1,222-224; STC 1,508-509; MSE 2,52-54; ABBOTT-C, 61-62; VEIBEL 2,198-199; BHDE 2,505; CB 1959, pp.186-188; TLK 3,271; POTSCH, 201-202; W, 256; TETZLAFF, 138; POGG 5,534; 6,1108-1109; 7b,1972-1978
G. Hevesy, Persp. Biol. Med. 1 (1958), 345-365; Adventures in Radioisotope Research. Oxford 1962
Anon., Leopoldina [3]6/7 (1960-1961), 34-37
H. Levi, Nuclear Physics A98 (1967), 1-24; Eur. J. Nucl. Med. 1 (1976), 3-10; George de Hevesy. Bristol 1985
J. Cockcroft, Biog. Mem. Fell. Roy. Soc. 13 (1967), 125-166
R. Spence, Chem. Brit. 3 (1967), 527-532
G. Marx, Ed., George de Hevesy. Budapest 1988

HEWITT, JOHN THEODORE [1868-1954]
WWWS, 796; POTSCH, 202; POGG 4,633; 5,534; 6,1109; 7b,1978-1979
E.E. Turner, Biog. Mem. Fell. Roy. Soc. 1 (1955), 79-99;
J. Chem. Soc. 1955, pp.4493-4496

HEWITT, LESLIE FRANK [1901-1967]
POGG 6,1109; 7b,1979
F.T. Perkins, Lancet 1967(I),pp.684-685; J. Path. Bact. 95 (1968), 587-591

HEWLETT, ALBION WALTER [1874-1925]
AMS 3,312
A.M. Harvey, Johns Hopkins Med. J. 144 (1979), 202-214

HEWLETT, RICHARD TANNER [1865-1940]
O'CONNOR(2), 193
J.C.G. Ledingham, Nature 146 (1940), 552
Anon., Lancet 1940(II), p.407; Brit. Med. J. 1940(II), p.400

HEY, DONALD HOLROYDE [1904-1987]
WW 1981, p.1200; IWW 1982-1983, p.565; POGG 6,1109; 7b,1980-1984
J.I.G. Cadogan and D.I. Davies, Biog. Mem. Fell. Roy. Soc. 34 (1988), 295-320

HEYDE, HENRI CHRISTIAN van der [1898-1933]
IB, 122
A. Toman, Tijd. Diergen. 61 (1934), 101-102

HEYDE, WERNER [1902-1964]
KURSCHNER 5,544; GRAML, 94-96; KOLLE 1,276; JV 43,495; 48,661
R.J. Lifton, The Nazi Doctors, pp.117-119. New York 1986
Encyclopedia of the Third Reich 1,405-406. New York 1991

HEYDEN, FRIEDRICH [1838-1926]
NDB 9,68-69; POGG 7aSuppl.,290
O. Schlenk, Z. ärzt. Fortbild. 37 (1940), 215

HEYDENREICH, EDUARD [-1885]
Anon., Pharm. Z. 31 (1886), 37

HEYL, FREDERICK WILLIAM [1885-1968]
AMS 11,2273
Anon., Chem. Eng. News 46(10) (1968), 71

HEYL, FRIEDRICH KARL (FRITZ) [1877-1950]
JV 16,158; NUC 244,523

HEYL, GEORG PAUL [1866-1942]
KURSCHNER 4,1168; HEIN, 274-275; WOLF, 85; TSCHIRCH, 1064-1065; TLK 3,271; POGG 4,635; 6,1109-1110; 7a(2),483

HEYMANN, BERNHARD [1861-1933]
DUNSCHELE, 136-137; POGG 6,1110
Anon., Chem. Z. 53 (1929), 101; 55 (1931), 327; 57 (1933), 386
H. Lecher, Z. angew. Chem. 44 (1931), 355-356
W. Lommel, Ber. chem. Ges. 66A (1933), 65-67

HEYMANN, BRUNO [1871-1943]
KURSCHNER 4,1168; FISCHER 1,626-627; BHDE 1,294; WININGER 7,79; IB, 122

HEYMANN, ERICH [1901-1949]
LDGS, 21; POGG 6,1110-1111; 7a(2),483-484
E.J. Hartung, J. Chem. Soc. 1950, pp.2910-2912

HEYMANN, HANS [1915-1979]
AMS 14,2162; BHDE 2,508; WWWS, 797
Anon., Chem. Eng. News 57(38) (1979), 39

HEYMANN, KARL [1904-]
JV 45,437; NUC 244,523

HEYMANN, WALTER [1901-1985]
AMS 14,2162; WWWS, 797
S. Emancipator, J. Lab. Clin. Med. 111 (1988), 259-260

HEYMANS, CORNEILLE [1892-1968]
BNB 40,423-434; GHENT 2,144-158; STC 1,509-510; IB, 122; WWWS,797
J.J. Bouckaert, Rev. Quest. Sci. 117 (1940), 157-170
A.F. DeSchaepdryver et al., Arch. Inter. Pharmacodyn. 139 (1962), 1-16; 174 (1968), 251-276; Suppl.20 (1973), 3-293
Anon., Bull. Acad. Med. Belg. 8 (1968), 591-604
F. Brücke, Deutsche med. Wchschr. 93 (1968), 2452-2454
C.F. Schmidt, Pharmacologist 11 (1969), 34-36
C. Neil, Am. Phil. Soc. Year Book 1970, pp.130-137

HEYMANS, JEAN FRANÇOIS [1859-1932]
GHENT 2,37-42; FISCHER 1,627-628; PAGEL, 736; IB, 122
M. Tiffeneau, Arch. Inter. Pharmacodyn. 44 (1932), 1-30

HEYN, EMIL [1867-1922]
DSB 6,368-369; NDB 9,92-93; DBJ 4,104-108,357; POGG 6,1111-1112

HEYNEMANN, LUDWIG HANS [1876-]
JV 17,156; NUC 244,553

HEYNS, KURT [1908-]
KURSCHNER 13,1522; POGG 7a(2),484-485

HEYNSIUS, ADRIANUS (ADRIAAN) [1831-1885]
LINDEBOOM, 863-865; NNBW 1,1107-1108; HIRSCH 3,217-218
G.D.L., Ned. Tijd. Gen. 21 (1885), 857-859
Anon., Leopoldina 21 (1885), 213

HEYNSIUS, DANIEL [1874-1959]
JV 16,158; NUC 244,562

HEYROVSKY, HANS [1877-1945]
FISCHER 1,628; SCHWEINITZ, 66-70

HEYROVSKY, JAROSLAV [1890-1967]
DSB 6,370-376; STURM 1,619; MH 1,225-226; STC 1,510-512; MSE 2,56-57; POTSCH, 202; CB 1961, pp.202-204; W, 256-257; POGG 6,1112; 7b,1984-1988
P. Zuman and P.J. Elving, J. Chem. Ed. 37 (1960), 562-567
R. Brdicka and M. Heyrovska, Coll. Czech. Chem. Comm. 25 (1960), 2945-2957
J.A.V. Butler and P. Zuman, Biog. Mem. Fell. Roy. Soc. 13 (1967),167-182
J. Koruta, Chem. Listy 66 (1972), 113-127; J. Chem. Ed. 49 (1972),183-185
P. Zuman, American Chemical Society Symposium Series 300 (1989), 339-369
R. Sherman, Chem. Brit. 26 (1990), 1165-1168

HIBBERT, HAROLD [1877-1945]
AMS 7,804; NCAB 33,376-377; YOUNG, 16; WWWS, 798; POGG 6,1113-1114; 7b,1990-1993
C.G. Fink, Science 102 (1945), 268-269
R.H. Clark, Proc. Roy. Soc. Canada [3]40 (1946), 95-99
M.L. Wolfrom, Biog. Mem. Nat. Acad. Sci. 32 (1958), 146-180
R.S. Tipson, Adv. Carbohydrate Chem. 16 (1961), 1-11

HICKMAN, KENNETH CLAUDE DEVEREUX [1896-]
AMS 14,2166; WWWS, 798; POGG 6,1115; 7b,1995-1997
Anon., Chem. Eng. News 48 (1956), 497

HICKS, CEDRIC STANTON [1892-1976]
WW 1976, p.1103; IB, 122
B.S. Hanson, Med. J. Austral. 1977(I), p.936

HICKS, WILLIAM LONGTON [1884-1929]
JV 23,332
Anon., Chemical Age 20 (1929), 591

HIEBER, WALTER [1895-1976]
KURSCHNER 12,1247; DRULL, 113; POTSCH, 202; POGG 6,1116; 7a(2),486-487
R.E. Oesper, J. Chem. Ed. 31 (1954), 140-141
Anon., Nachr. Chem. Techn. 2 (1954), 104; Leopoldina [3]2 (1956), 2-3; Chem. Z. 100 (1976), 37; 101 (1977), 40

HIENDLMAIER, HEINRICH [1878-1967]
JV 22,519; NUC 245,181
Anon., Deutsche Apoth. Z. 98 (1958), 1214

HIGGINS, EDWIN STANLEY [1926-]
AMS 15(3),682; WWWS, 799

HILDEBRAND, JOEL HENRY [1881-1983]
AMS 14,2174-2175; WWA 1978-1979, p.1491; IWW 1982-1983, p.567; MH 1,226-227; MSE 2,57-59; AO 1983, pp.207-209; CB 1955, pp.280-282; POTSCH, 203; POGG 6,1118; 7b,1998-2002
J.H. Hildebrand, Ann. Rev. Phys. Chem. 14 (1963), 1-4; 32 (1981), 1-23; Persp. Biol. Med.16 (1972), 88-111
J.H. Hildebrand and D.W. Ridgway, J. Chem. Ed. 52 (1975), 46-50
K.S. Pitzer, J. Phys. Chem. 85(22) (1981), 7A-20A; Am. Phil. Soc. Year Book 1983, pp.408-416

HILDEBRANDT, FRITZ [1887-1961]
NDB 9,126-127; FISCHER 1,630; GUNDEL, 415-422; TLK 3,273; IB, 123
J. Dorner, Deutsche med. Wchschr. 86 (1961), 1973
M. Grab, Arzneimitt. 11 (1961), 508-509

HILDEBRANDT, HERMANN [1866-1912]
JV 5,54; RSC 15,839-840; NUC 245,514
Anon., Chem. Z. 36 (1912), 1281

HILDEBRANDT, PAUL [1877-]
JV 15,5; NUC 245,519

HILDESHEIMER, ARNOLD [1885-1955]
BHDE 1,295; JV 25,28; NUC 245,537

HILDITCH, THOMAS PERCY [1886-1965] DSB 6,398; DNB [1961-1970], 513-515; WW 1965, p.1426; SRS, 40; POTSCH, 204; POGG 6,1119; 7b,2002-2007
P.N. Williams, Nature 208 (1965), 730-731
W.D. Raymond, Chem. Ind. 1966, p.251
R.A. Morton, Biog. Mem. Fell. Roy. Soc. 12 (1966), 259-289

HILGARD, EUGENE WOLDEMAR [1833-1916]
NCAB 10,308; WWAH, 273; WWWS, 800
E.J. Wickson, Science 43 (1916), 447-453
F. Slate, Biog. Mem. Nat. Acad. Sci. 9 (1919), 95-115

HILGER, ALBERT [1839-1905]
NDB 9,141-142; BJN 10,184*-185*; HEIN, 275-276; REBER, 220-225,363; LADIS, 12-20; BLOKH 1,296; TSCHIRCH, 1065; SCHAEDLER, 51-52; POGG 3,630-631; 4,639; 5,537
Anon., Chem. Z. 29 (1905), 561-562,1325
V. Coblentz, Am. J. Pharm. 77 (1905), 569-571
C. Bedall, Apoth. Z. 20 (1905), 544-555
H. Thoms, Ber. pharm. Ges. 15 (1905), 163-166

HILL, ARCHIBALD VIVIAN [1886-1977]
DNB [1971-1980], 406-408; WW 1966, p.1426; STC 1,514-515; ABBOTT, 61-62 WWWS, 800; POGG 6,1120; 7b,2007-2010
A.V. Hill, Ann. Rev. Physiol. 21 (1959), 1-18; Persp. Biol. Med. 14 (1970), 27-42
D.K. Hill and R.C. Wooledge, J. Physiol. 263 (1976), 85P-86P
B. Katz, Nature 268 (1977), 777-778; Biog. Mem. Fell. Roy. Soc. 24 (1978), 71-149
J.H. Humphrey, TIBS 3 (1978), N15-N16

HILL, ARTHUR CROFT [1863-1947]
O'CONNOR(2), 32-33
J.A. Venn, Alumni Cantabrigienses II 3,367. Cambridge 1947

HILL, ARTHUR WILLIAM [1875-1941]
DNB [1941-1950], 390-391; DESMOND, 307-308; IB, 123; WWWS, 800
F.T. Brooks, Obit. Not. Fell. Roy. Soc. 4 (1942), 87-100

HILL, CHARLES ALEXANDER [1874-1948]
Who Was Who 1941-1950, p.541
F.H. Carr, Chem. Ind. 1932, pp.363-364; 1948, p.749

HILL, DOUGLAS WILLIAM [1904-1985]
WW 1984, pp.1063-1064; NUC 245,667-668
D.M. Jones, Chem. Brit. 22 (1986), 343

HILL, EDGAR SMITH [1907-1952]
AMS 8,1134

HILL, HENRY BARKER [1849-1903]
DAB 9,33-34; ELLIOTT, 125; BLOKH 1,297; WWAH, 274; POGG 3,632; 4,639-640; 5,538
C.L. Jackson, Am. Chem. J. 30 (1903), 80-86; Ber. chem. Ges. 36 (1903), 4573-4581; Biog. Mem. Nat. Acad. Sci. 5 (1905), 255-266
T.W.R., Science 17 (1903), 841-843
E.H. Huntress, Proc. Am. Acad. Arts Sci. 77 (1949), 42-43

HILL, JAMES PETER [1873-1954]
WW 1954, p.1375; WWWS, 801
Anon., Journal of Anatomy 88 (1954), 542

HILL, LEONARD ERSKINE [1866-1952]
WW 1949, p.1305; O'CONNOR(2), 276-279; IB, 123; WWWS, 801
L.E. Hill, Philosophy of a Biologist. London 1930
C.G. Douglas, Obit. Not. Fell. Roy. Soc. 8 (1952-1953), 431-443
A.B. Hill and B. Hill, Proc. Roy. Soc. Med. 61 (1968), 307-316

HILL, ROBERT [1899-1991]
WW 1981, p.1214; SRS, 86; ABBOTT, 62

HILL, ROBERT LEE [1928-]
AMS 15(3),698; WWA 1980-1981, p.1554; WWWS, 802

HILL, ROBERT McCLAUGHRY [1894-1988]
AMS 12,2711; WWWS, 802
Anon., Chem. Eng. News 67(3) (1989), 54

HILL, TERRELL LESLIE [1917-]
AMS 15(3),698; WWA 1980-1981, p.1554; IWW 1982-1983, p.568; WWWS, 802

HILL, THOMAS GEORGE [1876-1954]
WW 1949, p.1307; DESMOND, 309; IB, 123; WWWS, 802
D.J.B. White, Nature 174 (1954), 159-160

HILLE, HERMANN [1871-1962]
JV 15,131; NUC 246,226
New York Times, 1 May 1962

HILLER, ALMA ELIZABETH [1892-1958]
AMS 9(I),863
Anon., Science 128 (1958), 80

HILLER, ARNOLD [1847-]
PAGEL, 738-739; ASEN, 80; AD 3,67; RSC 10,233; NUC 246,252

HILLERS, DIETRICH [1885-]
JV 26,395

HILLMANN, GÜNTHER [1919-1976]
KURSCHNER 11,1153; POGG 7a(2),492

HILPERT, SIEGFRIED [1883-1951]
TLK 3,274; POGG 5,539-540; 6,1123; 7a(2),492-493
Anon., Z. angew. Chem. 63 (1951), 228

HIMLY, CARL [1811-1885]
BLOKH 1,297; KIEL, 140; FERCHL, 235; TSCHIRCH, 1065; SCHAEDLER, 52; POGG 1,1106; 3,633
Anon., Leopoldina 21 (1885), 58

HIMSWORTH, HAROLD PERCIVAL [1905-1993]
WW 1981, p.1219; IWW 1982-1983, p.570

HIMWICH, HAROLD EDWIN [1894-1975]
AMS 12,2720; IB, 124; KAGAN(JA), 747-748; KOREN, 192
E. Costa, J. Neurochem. 25 (1975), 735-736
A.J. Plummer, Pharmacologist 17 (1975), 106-107
J. Wortis, Biological Psychiatry 10 (1975), 681-683

HINDHEDE, MIKKEL [1862-1945]
DBL 10,237-240; FISCHER 1,631; VEIBEL 2,199; IB, 124
S. Postmus, Voeding 26 (1965), 129-137

HINES, MARION [1889-1982]
AMS 11,2305-2306; WWA 1964-1965, p.924; WWWS, 804

HINRICHS, GUSTAV DETHLEF [1836-1923]
WWWS, 805; POTSCH, 204; POGG 3,634-635; 4,643; 5,542; 6,1124
J.W. Spronsen, Janus 56 (1969), 46-52

HINRICHSEN, FRIEDRICH WILLY [1877-1914]
DBJ 1,289; BLOKH 1,297-298; POGG 4,643; 5,542-543
W. Esch, Koll. Z. 15 (1914), 209-210; Chem. Z. 38 (1914), 1273
Anon., Leopoldina 51 (1914), 39
W. Mecklenburg, Z. Elektrochem. 21 (1915), 35-36

HINSBERG, KARL [1894-1982]
KURSCHNER 13,1542; WWWS, 805; POGG 7a(2),495-496
Anon., Chem. Z. 106 (1982), 445; Naturw. Rund. 35 (1982), 507

HINSBERG, OSCAR [1857-1939]
GENEVA 4,122-124; POGG 4,643-644; 5,543-544; 6,1125; 7a(2),497
W.R. Bett, Pharm. J. [4]125 (1957), 322

HINSHELWOOD, CYRIL NORMAN [1897-1967]
DSB 6,404-405; DNB [1961-1970], 516-519; WW 1965, p.1442; W, 257-258; MH 1,229-231; MSE 2,63-64; CB 1957, pp.259-260; ABBOTT-C, 62-63; POTSCH, 204-205; WWWS, 805; POGG 6,1125; 7b,2017-2024
E.J. Bowen, Chem. Brit. 3 (1967), 534-536
H. Thompson, Biog. Mem. Fell. Roy. Soc. 19 (1973), 563-582

HINTERBERGER, FRIEDRICH [1826-1875]
OBL 2,322; POGG 1,1108; 3,635
A. Bauer, Ost. Chem. Z. 23 (1920), 24

HINTZ, ERNST [1854-1934]
POGG 3,635; 4,644; 5,544; 6,1125-1126

T.W. Fresenius, Ber. chem. Ges. 67A (1934), 164-165

HIPPEL, EUGEN von [1867-1939]
NDB 9,200-201; KURSCHNER 4,1187; FISCHER 1,632; DRULL, 114
M. Baurmann, Verhandl. path. Ges. 32 (1950), 437-441

HIPPMEIER, WILHELM [1866-1943]
JV 10,138; NUC 247,114
Anon., Chem. Z. 36 (1912), 1173; Z. angew. Chem. 56 (1943), 280

HIRAYAMA, KINZO [1876-1934]
HAKUSHI 2,178; JBE(2nd Ed.), 291

HIRD, FRANCIS JOHN RAYMOND [1920-]
Who's Who in Australia 1992, p.594

HIRN, GUSTAVE ADOLPHE [1815-1890]
DSB 6,431-432; DBF 17,1224-1225; WWWS, 806; POGG 1,1109-1110; 3,636-637; 4,645
Anon., Leopoldina 26 (1890), 56
C. Matschoss, Beiträge zur Geschichte der Technik und Industrie 3,20-60. Berlin 1911
A. Kastler, Rev. Gen. Sci. 65 (1958), 294-297

HIRS, CHRISTOPHE HENRI WERNER [1923-]
AMS 15(3),714

HIRSCH, BRUNO [1826-1902]
HEIN, 277-278; REBER, 305; SCHELENZ, 691-692
Anon., Pharm. Z. 31 (1886), 539-540,547-549; 47 (1902), 974

HIRSCH, CARL [1870-1930]
FISCHER 1,633; KOREN, 193; WININGER 3,110; 7,84

HIRSCH, JAMES GERALD [1922-1987]
AMS 15(3),715; WWWS, 807
New York Times, 26 May 1987

HIRSCH, JULIUS [1892-1962]
KURSCHNER 8,934-935; ARNSBERG, 192; BHDE 2,514-515; LDGS, 56; KOREN, 193; WIDMANN, 81,269
Anon., Deutsche med. Wchschr. 87 (1962), 1590

HIRSCH, PAUL [1885-1955]
KURSCHNER 7,820; NB, 172; JV 26,38; TLK 3,276-277; POGG 6,1126-1127; 7a(2),500HIRSCH, PAUL [1887-1942?]
GEDENKBUCH 1,588; JV 28,316; NUC 247,258

HIRSCH, RAHEL [1870-1953]
NDB 9,209-210; BHDE 2,516; WININGER 3,120; 7,82; TETZLAFF, 143; KOREN, 193
S. Muntner, Koroth 3 (1964), 340-341

HIRSCH, SAMSON RAPHAEL [1890-1960]
ARNSBERG, 202-203; KALLMORGEN, 301
G. Schultze-Werninghaus and J. Meyer-Sydow, Clinical Allergy 12 (1982), 211-215

HIRSCH-KAUFMANN, HERBERT [1894-1960]
LDGS, 77
Anon., Deutsche med. Wchschr. 85 (1960), 2267

HIRSCHBERG, JULIUS [1843-1925]
NDB 9,221; FISCHER 1,634; ASEN, 81; WININGER 3,125-126; 7,82

HIRSCHBERGER, JOSEF [1866-1954]
JV 5,293; RSC 14,1010; NUC 247,289
New York Times, 10 November 1954

HIRSCHFELD, FELIX [1863-]
KURSCHNER 4,1192; FISCHER 1,635; ASEN, 81

HIRSCHFELD, HANS [1873-1944]
FISCHER 1,635; LDGS, 66; IB, 124; WININGER 3,130-131; TETZLAFF, 144; KOREN, 193
V. Schilling, Fol. Haemat. 69 (1950), 3
P. Voswinckel, Fol. Haemat. 114 (1987), 707-736

HIRSCHFELDER, JOSEPH OAKLAND [1911-1990]
AMS 17(3),738; WWWS, 807
Anon., Chem. Eng. News 68(33) (1990), 34-35
R.B. Byron et al., Theoretica Chimica Acta 82 (1992), 3-6

HIRSCHLER, AGOSTON [1861-1911]
MEL 1,725; FISCHER 1,636; PAGEL, 1963; KOREN, 193; WININGER 3,138; RSC 15,866

HIRSCHMANN, HANS [1909-]
AMS 14,2198; WWA 1976-1977, p.1455; BHDE 2,520

HIRSCHMANN, RALPH [1922-]
AMS 15(3),717; BHDE 2,520-521

HIRST, EDMUND LANGLEY [1898-1975]
DNB [1971-1980], 411-412; WW 1975, p.1488; MH 1,232; MSE 2,67; WWWS, 807; POGG 6,1127-1128; 7b,2025-2029
M. Stacey, Chem. Brit. 12 (1976), 321-322
M. Stacey and E. Percival, Biog. Mem. Fell. Roy. Soc. 22 (1976), 137-168
M. Stacey and D.J. Manners, Adv. Carbohydrate Chem. 35 (1978), 1-29

HIRST, GEORGE KEBLE [1909-1994]
AMS 14,2199; WWWS, 808

HIRSZFELD, LUDWIK [1884-1954]
DSB 6,432-434; PSB 9,533-535; FISCHER 1,637; SACKMANN, 158-161; IB, 125; STC 1,521-522; KOREN, 194
L. Hirszfeld, Historia Jednogo Zycia. Warsaw 1945
R. Speiser, Wiener klin. Wchschr. 66 (1954), 394-395
P. Moulec, Rev. Hemat. 9 (1954), 759-760
A. Kelus, Schw. med. Wchschr. 84 (1954), 745
G. Blumenthal, Z. Bakt. 162 (1955), 1-3
H. Hirszfeldowa et al., Ludwik Hirszfeld. Wroclaw 1956
J.W. Gilsohn, Prof. Dr. Ludwik Hirszfeld. Munich 1965
M. Jaworski, Ludwik Hirszfeld. Leipzig 1980

HIRSZOWSKI, ALFRED [1882-1943]
JV 24,32
Anon., Roczniki Chemii 20 (1946), xxi

HIRZEL, CHRISTOPH HEINRICH [1828-1908]
NDB 9,244; BJN 13,41*; HEIN-E, 192-193; BLOKH 1,298; TSCHIRCH, 1065; SCHAEDLER, 52; POGG 1,1111; 3,638; 5,545
Anon., Leopoldina 44 (1908), 110

HIS, HANS [1866-1915]
JV 9,208; RSC 14,804
Anon., Jahresb. Nat. Ges. Graub. 56 (1916), xxiii-xxiv

HIS, WILHELM [1831-1904]
DSB 6,434-435; NDB 9,249; BJN 9,231-242; 10,49*; HIS, 218-226; W, 259; LEIPZIG 2,87-94; BOSSART, 42-43; HIRSCH 3,240; PAGEL, 747-748
W. Waldeyer, Deutsche med. Wchschr. 30 (1904), 1438-1441,1469-1471; 1509-1511
F. Marchand, Sitz. Akad. Wiss. Leipzig 56 (1904), 323-340
H. Held, Berl. klin. Wchschr. 41 (1904), 684-687
J. Kollmann, Verhandl. Schw. Nat. Ges. 87 (1904), xiii-xl; Verhandl. Nat. Ges. Basel 15 (1904), 434-464
R. Fick, nat. Anz. 25 (1904), 161-208
C. von Voit, Sitz. Bayer. Akad. 35 (1905), 328-337
W. His, Jr., Wilhelm His der Anatom. Berlin 1931
L. Picken, Nature 178 (1956), 1162-1165
E. Ludwig, Der Anatom Wilhelm His-Vischer. Olten and Lausanne 1959
W. His, Lebenserinnerungen (E. Ludwig, Ed.). Berne 1965

HIS, WILHELM [1863-1934]
FISCHER 1,637; GRUETTER, 50-51; WWWS, 808
Anon., Lancet 1934(II), p.1200
V. Schilling, Deutsches Arch. klin. Med. 177 (1935), i-viii
A. Lotz, Schw. med. Wchschr. 65 (1935), 267-268
T.H. Bost and W.D. Gardner, J. Hist. Med. 4 (1949), 170-187

HISAW, FREDREICK LEE [1891-1972]
AMS 11,2315; MH 2,228-229; MSE 2,68-69; IB, 125; WWWS, 808
R.O. Greep et al., Am. Phil. Soc. Year Book 1973, pp.122-125;
Anat. Rec. 177 (1973), 114-116; Horm. Prot. Pep. 8 (1980), 199-224
Anon., Endocrinology 93 (1973), 273-276

HISSINK, DAVID [1874-1956]
RSC 15,869; NUC 247,453; POGG 6,1128; 7b,2029-2031
E.J. Russell et al., Soil Research 4 (1934), 93-112
J. van der Spek, Chem. Wkbl. 36 (1939), 732-737
Wie is Dat? 1956, p.272
S. Tovberg-Jensen, Plant and Soil 8 (1956), 1-3

HITCHINGS, GEORGE HERBERT [1905-]
AMS 16(3),732; WWWS, 809

HITCHCOCK, ALBERT EDWIN [1898-1979]
AMS 12,2734
New York Times, 12 June 1979

HITCHCOCK, DAVID INGERSOLL [1893-1976]
AMS 11,2316; NCAB 59,395-396; IB, 125; POGG 6,1128-1129; 7b,2031-2032

HITTORF, WILHELM [1824-1914]
DSB 6,438-440; NDB 9,266-270; DBJ 1,41-44,289-290; WL 1,128-148; DZ, 615; POTSCH, 205; BLOKH 1,298-300; W, 259-260; SCHAEDLER, 51; POGG 1,1113; 3,639-640; 4,646-647; 5,545; 6,1129
H. Goldschmidt, Z. angew. Chem. 27 (1914), 657-658
G.C. Schmidt, Chem. Z. 38 (1914), 401; Wilhelm Hittorf. Münster 1924
Anon., J. Chem. Soc. 107 (1915), 582-586
U. Hoyer, in Die Universität Münster 1780-1980 (H. Dollinger, Ed.), pp.437-445. Münster 1980

HITZIG, EDUARD [1838-1907]
DSB 6,440-441; NDB 9,273-274; BJN 12,37*-38*; FISCHER 1,638-639; W, 260; PAGEL, 749-750; HAYMAKER, 229-233; BRAZIER, 157-158,177-178; KOREN, 194
Anon., Leopoldina 43 (1907), 87
L. Bruns, Münch. med. Wchschr. 54 (1907), 2144-2145
G. Fritsch, Berl. klin. Wchschr. 44 (1907), 1185
W. Weber, Deutsche med. Wchschr. 33 (1907), 1871-1872
R. Wollenber, Arch. Psychiat. 43 (1908), iii-xv
H.H. Eulner, Wiss. Z. Halle 6 (1957), 709-712
A. Stender, Deutsches Med. J. 19 (1968), 335-339

HIXON, RALPH MALCOLM [1895-1978]
AMS 13,1936; IB, 125

HJELT, EDVARD [1855-1921]
ENKVIST, 66-83; BLOKH 1,300-301; WWWS, 809; POGG 3,640; 4,647; 5,545
O. Aschan, Ber. chem. Ges. 55A (1922), 163-193

HJORT, AXEL MAGNUS [1889-1975]
AMS 10,1801; IB, 125; WWWS, 809

HJORT, JOHAN [1869-1948]
DSB 6,441-442; NBL 6,139-144; WWWS, 809
E.S. Russell and H.G. Maurice, Nature 162 (1948), 764-766
J.T. Ruud, Am. Phil. Soc. Year Book 1949, pp.311-316
K.A. Anderson, Hydrobiologia 2 (1949), 97-99
A.C. Hardy, Obit. Not. Fell. Roy. Soc. 7 (1950), 167-181

HLASIWETZ, HEINRICH [1825-1875]
NDB 9,277-278; ADB 12,513-516; STURM 1,636; HEIN, 278-279; POTSCH, 206; MACHEK, 174-179; BLOKH 1,301-302; TSCHIRCH, 1066; POGG 1,1115; 3,640
H. Kolbe, J. prakt. Chem. 120 (1875), 463-468
L. Barth, Ber. chem. Ges. 9 (1876), 1961-1992
J. Habermann, Alm. Akad. Wiss. Wien 26 (1876), 211-227
M. Kohn, J. Chem. Ed. 22 (1945), 55-56,73
V.F. McConnell, J. Chem. Ed. 30 (1953), 380-385

HOAGLAND, CHARLES LEE [1907-1946]
DAB [Suppl.4],380-381; ICC, 305-306
Anon., J. Am. Med. Assn. 131 (1946), 1379

HOAGLAND, DENNIS ROBERT [1884-1949]
DSB 6,442-444; DAB [Suppl.4],381-383; NCAB 47,598-5999; HUMPHREY, 111-113; IB, 125; WWAH, 277; WWWS, 810
D.I. Arnon., Plant Physiology 25 (1950), iv-xvi; Science 112 (1950), 739-742; Plant and Soil 2 (1950), 129-144
W.P. Kelley, Biog. Mem. Nat. Acad. Sci. 29 (1956), 123-143

HOAGLAND, HUDSON [1899-1982]
AMS 14,2204; WWA 1976-1977, p.1459; WWWS, 810

HOAGLAND, MAHLON BUSH [1921-]
AMS 15(3),726; WWA 1980-1981, pp.1568-1569; MSE 2,69-70; ABBOTT, 63-64; WWWS, 810
M.B. Hoagland, Toward the Habit of Truth. New York 1990

HOBEIN, MAX [1857-1940]
NUC 248,617
Anon., Chemische Technik 15 (1942), 40

HOCHEDER, FERDINAND [1881-1945]
JV 23,544; NUC 249,5

HOCHSCHWENDER, KARL [1883-]
JV 24,337; NUC 249,27

HOCHSTER, ROLF MARTIN [1922-1971]
AMS 11,2326; YOUNG, 72-73

HOCHSTETTER, CHRISTIAN FERDINAND [1787-1860]
NDB 9,291
M. Habacher, Esslinger Studien 16 (1970), 172-227

HOCHSTETTER, FERDINAND [1861-1954]
NDB 9,292; STURM 1,642; FISCHER 1,641; STOBER, 55-65; IB, 125; WIEN 1956, pp.41-42
O. Grosser, Wiener klin. Wchschr. 54 (1941), 108-111
E. Pernkopf, Alm. Akad. Wiss. Wien 104 (1954), 386-402
B. Romeis, Bayer. Akad. Wiss. Jahrbuch 1955, pp.184-187
C. Elze, Anat. Anz. 103 (1956), 305-317
H. Hofer, Gegenbaurs Morphologisches Jahrbuch 119 (1973), 346-357; Leopoldina [3]21 (1978), 273-286

HOCK, ANDREAS [1905-]
KURSCHNER 12,1275; POGG 7a(2),503-504

HOCK, HEINRICH [1887-1971]
POTSCH, 206; JV 27,647; POGG 6,1132; 7a(2),504-505
Wer ist Wer 15 (1967), 774
Anon., Chem. Z. 96 (1972), 116

HOCK, LOTHAR [1890-1978]
KURSCHNER 12,1276; MEINEL, 504; AUERBACH, 826; TLK 3,277; POGG 6,1133; 7a(2),505
F.H. Müller and H. Schmidt, Koll. Z. 119 (1950), 65-68
J. Eggert, Z. Elektrochem. 60 (1956), 1-2
Anon., Chem. Z. 102 (1978), 458

HOCKENHULL, DONALD JOHN DARLINGTON [1918-]
Who's Who of British Scientists 1980-1981, p.230

HODGE, HAROLD CARPENTER [1904-1990]
AMS 15(3),734

HODGES, JOHN FREDERICK [1815-1899]
BELFAST, 576-577; RSC 3,373; NUC 249,161
Anon., J. Chem. Soc. 77 (1900), 593-594; Centenary Volume, Belfast Natural History and Philosophical Society, pp.85-86. Belfast 1924

HODGKIN, ALAN LLOYD [1914-]
WW 1981, p.1231; IWW 1982-1983, p.575; MH 1,235-236; STC 1,524-525; MSE 2,72-73
A.L. Hodgkin, J. Physiol. 263 (1976), 1-21; Ann. Rev. Physiol. 45 (1983), 1-16; Chance and Design. Cambridge 1992

HODGKIN, DOROTHY MARY CROWFOOT [1910-]
WW 1981, p.1231; IWW 1982-1983, p.575; MH 1,236-237; STC 1,525-527; CAMPBELL, 109-111; ABBOTT-C, 63-65; WWWS, 812; POTSCH, 102
M. Perutz et al., in Structural Studies on Molecules of Biological Interest (G. Dodson et al., Eds.), pp.5-78. London 1981
J. Cornforth, Notes Roy. Soc. 37 (1982), 1-4

HODGKIN, THOMAS [1798-1866]
DNB 9,957-958; HIRSCH 3,246-247; WILKS, 380-386; DECHAMBRE [4]14,184; CALLISEN 9,8-9; 29,4
S. Wilks, Guy's Hosp. Rep. [3]23 (1878), 55-127
G.W. Jones, Ann. Med. Hist. 2 (1940), 471-481
G.E. Foxon et al., Guy's Hosp. Rep. 115 (1966), 243-303
J.D. Leibowitz, Clio Medica 2 (1967), 97-101
E.H. Kass, Bull. Hist. Med. 43 (1969), 138-175
B. Friedman, Ala. J. Med. Sci. 12 (1975), 250-251
M. Rose, Curator of the Dead: Thomas Hodgkin. London 1981
A.M. Kass and H. Kass, Perfecting the World: The Life and Times of Dr. Thomas Hodgkin. Boston 1988

HODGSON, HERBERT HENRY [1883-1967]
POGG 6,1134-1135; 7b,2040-2044
H.H. Hodgson, Chem. Ind. 1951, pp.826-830
W.E. Scott, Chem. Brit. 4 (1968), 167

HÖBER, RUDOLF [1873-1953]
DSB 17,423-425; NDB 9,301-302; FISCHER 1,641; BHDE 2,524; LDGS, 79; KIEL, 83-84; IB, 125; WININGER 7,83-84; TETZLAFF, 145; POGG 4,649-650; 5,547; 6,1135; 7a(2),505-506
W.R. Amberson, Science 120 (1954), 199-201
H. Netter, Pflügers Arch. 259 (1954), 4-13
W. Wilbrandt, Erg. Physiol. 49 (1957), 23-46

HÖBOLD, KURT (KARL ALBERT) [1886-]
JV 26,638; NUC 249,203
Anon., Nachr. Chem. Techn. 4 (1956), 355

HÖCHTLEN, FRIEDRICH [1878-1951]
JV 19,339; NUC 249,309
Anon., Z. angew. Chem. 63 (1951), 340

HOEFER, FERDINAND [1811-1878]
DBF 17,1247-1248; NDB 9,310; HOEFER 24, 845-854; HIRSCH 3,249; COLE, 264; FERCHL, 237; POGG 1,1119; 3,642
G. Sarton, Bull. Hist. Med. 8 (1940), 419-445
J. Meyer, Rete 1 (1971), 33-50

HOEFLE, MARK AUREL [-1855]
HIRSCH 3,249

HÖFLER, KARL [1893-1973]
KURSCHNER 11,1174; PLANER, 134; IB, 125; POGG 7a(2),507-508
A. Ziegler, Protoplasma 57 (1963), 817-827
B. Huber, Verhandl. zool. bot. Ges. Wien 103-104 (1964), 5-11
L. Hofmeister, Ber. bot. Ges. 88 (1975), 369-378
F. Knoll, Alm. Akad. Wiss. Wien 124 (1975), 387-408

HÖHNE, FRITZ [1878-1962]
KALLMORGEN, 302; SG [3]6,769; NUC 249,419
Deutscher Dermatologen Kalender 1929, p.95

HÖHNEL, FRANZ von [1852-1920]
NDB 9,320; OBL 2,357-358; DBJ 2,749; FREUND 1,227-234; REBER, 187-191; TSCHIRCH, 1067
J. Weese, Ber. bot. Ges. 38 (1920), 103-126
F. Becke, Alm. Akad. Wiss. Wien 71 (1921), 171-173

HOEK, HEINZ [1903-]
JV 43,414; NUC 249,436

HOELSTI, ÖSTEN [1887-1952]
FISCHER 1,656-657
H. Bergholm, Finlands Läkare, p.198. Tammerfors 1927
Kuka Kukin On 1950, p.204

HÖLZL, FRANZ [1892-1976]
KURSCHNER 11,1179; KERNBAUER, 416-422; TLK 3,284; POGG 6,1137; 7a(2), 510

HÖNIGSCHMID, OTTO [1878-1945]
DSB 6,480-481; NDB 9,343-345; OBL 2,363; STURM 1,653; WEINMANN, 222-223;POYSCH, 211; TLK 3,284-285; POGG 5,547-548; 6,1137-1138; 7a(2),511
E. Zintl, Z. anorg. Chem. 236 (1938), 1-11
J. Goubeau, Naturwiss. 33 (1946), 353-354
K. Clusius, Bayer. Akad. Wiss. Jahrbuch 1944-1948, pp.287-290
H. Wieland, Z. angew. Chem. 62 (1950), 3-4

HOEPPNER, MAX [1872-1922]
JV 14,224; RSC 16,377,509
Anon., Chem. Z. 46 (1922), 505; Z. angew. Chem. 35 (1922), 272

HÖRING, PAUL [1868-1919]
HEIN-E, 197; DBJ 2,722; BLOKH 1,303; POGG 5,548
H. Simonis, Ber. chem. Ges. 52A (1919), 53-55
A. Wankmüller, Beitr. Württ. Apothekgesch. 13 (1981), 71-73

HÖRLEIN, HEINRICH [1882-1954]
NDB 9,353-354; DUNSCHELE, 137-138; POGG 6,1138; 7a(2),514
Anon., Chem. Z. 66 (1942), 236; 78 (1954), 365; Arzneimitt. 4 (1954), 689-690; Naturw. Rund. 7 (1954), 309

HÖRLIN, JULIUS [1869-1955]
JV 9,208; RSC 16,377

HÖRMANN, AUGUST [1865-1950]
JV 7,66; RSC 17,666; NUC 249,568

HÖRNER, OTTO [1905-1941]
ECKERT, 67-68
M. Borst, Verhandl. path. Ges. 32 (1950), 441-442

HOERR, NORMAND LOUIS [1902-1958]
DAB [Suppl.6],295-296; AMS 9(II),521; DAMB, 355-356; WWAH, 279
G.H. Scott and A. Lazarow, Anat. Rec. 138 (1960), 7-9

HÖRSTADIUS, SVEN OTTO [1898-]
VAD 1981, p.493; IWW 1982-1983, p.588; MH 2,239-241; MSE 2,90-91; WWWS, 829

HOESCH, ALFRED [1878-1908]
DGB 123,195; JV 19,181; NUC 249,597

HOESCH, KURT [1882-1932]
POTSCH, 206-207; POGG 5,548; 6,1138
Anon., Chem. Z. 56 (1932), 1014
M. Bergmann, Ber. chem. Ges. 66A (1933), 16

HOESSLI, HANS [1883-1918]
DHB 4,132; NUC 249,609

HÖSSLIN, HEINRICH von [1852-1902]
GGTA 24 (1932), 263; RSC 10,248; 15,887-888

HOESSLIN, HEINRICH von [1878-1955]
NDB 9,367*; KURSCHNER 4,1256; FISCHER 1,643
A. Wunderwald, Münch. med. Wchschr. 98 (1956), 554-555

HOET, JOSEPH PIERRE [1899-1968]
BAF, 66; FISCHER 1,644
Bibl. Acad. Louvain 6,397-400; 8,61-63; 12,579-583
A.M. Dalcq, Bull. Acad. Med. Belg. [8]8 (1968); 125-132

HOF, ADOLF CARL [1873-]
JV 14,26; RSC 15,888; SG [2]7,198; NUC 249,650

HOFBAUER, LUDWIG [1873-1951]
KURSCHNER 4,1200-1201; FISCHER 1,645; STANGL, 71-87; KOREN, 194
A. Kaiser-Petersen, Z. Tub. 101 (1952), 200

HOFE, CHRISTIAN von [1871-1954]
TLK 3,277-278; POGG 6,1138-1139; 7a(2),515-516
J. Euler, Optik 8 (1951), 187-189; Phys. Bl. 10 (1954), 374-375

HOFE, KARL von [1898-1969]
KURSCHNER 11,1187
P. Carsten, Schriftum Koln 1938, pp.308-310

HOFER, BRUNO [1861-1916]
NDB 9,379-380; DBJ 1,359; RSC 15,888
Anon., Leopoldina 52 (1916), 90

HOFF, FERDINAND [1896-1988]
KURSCHNER 13,1584-1585; HAGEL, 94-98
W. Schrade, Deutsche med. Wchschr. 86 (1961), 761
F. Hoff, Erlebnis und Besinnung: Erinnerungen eines Arztes. Berlin 1971

HOFF, HERMINUS JOHANNES van't [1859-1939]
RSC 15,888-889
Anon., Chem. Wkbl. 36 (1939), 555-556

HOFF, JACOBUS HENRICUS van't [1852-1911]
DSB 13,575-581; NDB 9,384-386; BWN 1,246-248; BJN 16,185-194,33*; BUGGE 2,391-407; GERRITS, 354-380; WREDE, 84-85; DZ, 621-622; W, 526; WWWS, 1712; BLOKH 1,303-305; ABBOTT-C, 141-142; POTSCH, 434; POGG 3,644; 4,1558-1559; 5,1299; 7aSuppl.,291-295
G. Bredig, Z. angew. Chem. 24 (1911), 1074-1087
W. Ostwald, Abhandlungen und Vorträge, pp.404-417. Leipzig 1904; Ber. chem. Ges. 44 (1911), 2219-2252
F.G. Donnan, Nature 86 (1911), 84-86; Proc. Roy. Soc. A86 (1912), xxxix-xliii
E. Cohen, Jacobus Henricus van't Hoff. Leipzig 1912
W.P. Jorissen and L.T. Reicher, J.H. van't Hoffs Amsterdamer Periode 1877-1895. Helder 1912
J. Walker, J. Chem. Soc. 103 (1913), 1127-1143
H.J.C. Tenderloo et al., Chem. Wkbl. 48 (1952), 621-663
E.H. Huntress, Proc. Am. Acad. Arts Sci. 81 (1952), 62-67
H.S. Van Klooster, J. Chem. Ed. 29 (1952), 376-379
J. d'Ans, Chem. Z. 76 (1952), 545-548
R. Kuhn, Naturw. Rund. 15 (1962), 1-8
H.P. Engster, Science 173 (1971), 481-489
H.A.M. Snelders, J. Chem. Ed. 51 (1974), 207; Janus 60 (1974), 261-268; 71 (1984), 1-30; in van't Hoff-LeBel Centennial (O.B. Ramsay, Ed.), pp.55-73. Washington 1975; Tijd. Gesch. Geneesk. 8 (1985), 49-57; 9 (1986), 11-24; 10 (1987), 2-19
E. Fischmann, Janus 72 (1985), 131-156
R.B. Grossman, J. Chem. Ed. 66 (1989), 30-33

HOFF-JØRGENSEN, EGON [1909-1983]
KBB 1981, p.452

HOFFA, ERWIN FRIEDRICH [1875-1967]
JV 12,20; RSC 14,1010

HOFFER, MAX [1906-1983]
AMS 9(I),877; LDGS, 21
New York Times, 16 April 1983

HOFFMAN, FREDERICK [1832-1904]
HEIN, 281-282; BJN 10,50*; REBER, 73-79; SILLIMAN, 145-146; NUC 250,187
F.A. F[lückiger], Pharm. Z. 33 (1888), 461-464
Anon., Pharm. Z. 47 (1902), 476-477; 49 (1904), 1036-1037
G. Urdang, Die Schelenz Stiftung, pp.119-132. Eutin 1953
S.K. Schultze, Pharm. Hist. 33 (1991), 118-122

HOFFMANN, FELIX [1868-1946]
NDB 9,415; HEIN-E, 198-199; JV 8,215; RSC 13,275; 15,892

HOFFMANN, FRIEDRICH ALBIN [1843-1924]
FISCHER 1,647-648; PAGEL, 762-763; LEVITSKI 2,138-139; DZ, 624-625
E. Brennsohn, Die Aerzte Livlands, p.208. Mitau 1905
L. Krehl, Deutsches Arch. klin. Med. 146 (1925), i-iv
B. Naunyn, Klin. Wchschr. 4 (1925), 142

HOFFMANN, (HEINRICH KARL) HERMANN [1819-1891]
NDB 9,424-425; ADB 50,412-416; HB 1,16-25; FERCHL, 243; WWWS, 814; RSC 3,387-388; 7,997; 10,250-251; 15,892-893; NUC 250,224-226; POGG 3,644-645; 4,652-653
E. Ihne, Bot. Zb. 48 (1891), 159-160; Bericht der Oberhessischen Gesellschaft für Natur- und Heilkunde 29 (1893), 1-40
Anon., Leopoldina 28 (1892), 49
E. Ihne and J. Schroeter, Ber. bot. Ges. 10 (1892), (11)-(27)

HOFFMANN, HERMANN [1858-]
NUC 250,226

HOFFMANN, HERMANN [1860-]
JV 6,270; RSC 15,893; NUC 250,226

HOFFMANN, PAUL [1884-1962]
NDB 9,400-401; KURSCHNER 9,799; FISCHER 1,648; EBERT, 122-130; IB, 126; POGG 7a(2),517-519
R. Jung, Deutsche med. Wchschr. 79 (1954), 1098; Erg. Physiol. 61 (1969), 1-17
O.A.M. Wyss, Experientia 18 (1962), 478-480

HOFFMANN, REINHOLD [1831-1919]
NDB 9,396*; RSC 3,388-389; 10,251; POGG 4,653

HOFFMANN, RICHARD von [1797-1877]
HIRSCH 3,262-263; DECHAMBRE [4]14,188; CALLISEN 9,42-43; 29,20

HOFFMANN, ROALD [1937-]
AMS 17(3),772; POTSCH, 207

HOFFMANN, ROBERT [1835-1869]
OBL 2,379; STURM 1,659-660; PRAG, 343-344; POGG 3,654

HOFFMANN, WALTER [1907-1955]
POGG 7a(2),519
Anon., Nachr. Chem. Techn. 3 (1955), 242

HOFFMANN-OSTENHOF, OTTO [1914-]
KURSCHNER 13,1594; BHDE 2,529; POTSCH, 207; POGG 7a(2),519-521
P. Karlson, Naturwiss. 71 (1984), 491-492

HOFMAN, TAMME SABE [1864-]
JV 11,167; RSC 14,438

HOFMANN, ALBERT [1906-]
KURSCHNER 13,1595; WWWS, 814
A. Hofmann, Journal of Psychodelic Drugs 11 (1979), 53-60

HOFMANN, AUGUST WILHELM [1818-1892]
DSB 6,461-464; NDB 9,446-450; ADB 50,577-589; BUGGE 2,136-153; W, 262; BLOKH 1,306-311; FERCHL, 238-239; SCHAEDLER, 53-54; TSCHIRCH, 1066; SCHELENZ, 670-671; MATSCHOSS, 119-120; ABBOTT-C, 65-66; WWWS, 1736; POTSCH, 208-209; POGG 1,1128-1129; 3,645-647; 4,653-654
Anon., Chem. Z. 16 (1892), 1759-1762,1976
F.A. Abel et al., J. Chem. Soc. 69 (1896), 575-732
J. Volhard and E. Fischer, Ber. chem. Ges. 35Suppl. (1902), 1-284
B. Lepsius, Ber. chem. Ges. 51Suppl. (1918), 1-54
A. Friedrich, Chem. Z. 66 (1942), 204-207
J.J. Beer, J. Chem. Ed. 37 (1960), 248-251
R. Kuhn, Naturw. Rund. 18 (1965), 43-48
J. Bentley, Ambix 17 (1970), 153-181
W.H. Brock, Ed., Justus von Liebig und August Wilhelm Hofmann in ihren Briefen. Weinheim 1984
R. Oelsner, August Wilhelm von Hofmann. Berlin 1989
A.S. Travis, Brit. J. Hist. Sci. 25 (1992), 145-167

HOFMANN, EBERHARD [1930-]
KURSCHNER 16,1487-1488; POTSCH, 208

HOFMANN, EDUARD von [1837-1897]
NDB 9,450-451; ADB 50,434; BJN 2,81-82; 4,79*-80*; HUTER 2,267-269; OBL 2,380; HIRSCH 3,269-270; PAGEL, 763-764; BINDSEIL, 39-53
A. Haberda, Vierteljahrschr. gericht. Med. [3]8 Suppl.(1894), i-xiv
Anon., Wiener med. Wchschr. 47 (1897), 1680-1682
F. Reuter, Nova Acta Leop. NF9 (1940), 563-632

HOFMANN, EDUARD [1897-1980]
KURSCHNER 13,1596; POGG 7a(2),523-524
Anon., Nachr. Chem. Techn. 28 (1980), 340; Naturw. Rund. 33 (1980), 350

HOFMANN, FRANZ [1843-1920]
DBJ 2,749; FISCHER 1,649
C. Flügge, Deutsche med. Wchschr. 46 (1920), 1173
A. Foetter, Münch. med. Wchschr. 68 (1921), 114

HOFMANN, FRANZ BRUNO [1869-1926]
STURM 1,662; FISCHER 1,649; AUERBACH, 269; HUTER, 220; KOERTING, 120; MAASS, 25-42; DZ, 627
E.T. von Brücke, Klin. Wchschr. 5 (1926), 1398-1399
E. Mangold, Pflügers Arch. 216 (1927), 281-299
A. von Tschermak, Erg. Physiol. 26 (1928), 776-780

HOFMANN, FRITZ [1866-1956]
HEINE, 200-201; POTSCH, 208-209; POGG 6,1141; 7a(2),524

HOFMANN, HEINRICH [1909-1971]
KURSCHNER 11,1201; POGG 7a(2),524-525

HOFMANN, KARL ANDREAS [1870-1940]
NDB 9,457; TLK 3,280; POTSCH, 209; POGG 4,654-655; 5,548-550; 6,1141-1142; 7a(2),526
R. Schwarz, Z. angew. Chem. 53 (1940), 133-135
M. Bodenstein, Jahrb. Akad. Wiss. Berlin 42 (1941), 196-209
K. Schleede, Ber. chem. Ges. 74A (1941), 235-246
W. Hieber, Bayer. Akad. Wiss. Jahrbuch 1944-1948, pp.231-234

HOFMANN, KARL BERTHOLD [1842-1922]
OBL 2,382-383; TSCHIRCH, 1066; POGG 3,647-648; 4,654; 5,550; 6,1142

HOFMANN, KLAUS HEINRICH [1911-]
AMS 15(3),751; CB 1951, pp.204-206
J.B. Field, Metabolism 31 (1982), 635-637

HOFMANN, OTTO [1885-1918]
JV 25,364; NUC 250,438
Anon., Z. angew. Chem. 31(III) (1918), 608

HOFMANN-DEGEN, KONRAD [1883-1961]
JV 21,274

HOFMEISTER, FRANZ [1850-1922]
DSB 17,430-433; NDB 9,470; DBJ 4,122-126; STURM 1,664; FISCHER 1,650; LOMMATZSCH, 88-94; PAGEL, 767-768; EBERT, 181-187; BLOKH 1,311-312; KOERTING, 133; DZ, 631-632; TSCHIRCH, 1067; POGG 3,648; 4,655; 5,550; 6,1143
E. Abderhalden, Med. Klin. 18 (1922), 1167-1168
G. Embden, Klin. Wchschr. 1 (1922), 1974-1975
C. Neuberg, Biochem. Z. 134 (1922), 1-2
K. Spiro, Arch. exp. Path. Pharm. 95 (1922), i-vii
J. Pohl and K. Spiro, Erg. Physiol. 22 (1923), 1-50
E.H. Huntress, Proc. Am. Acad. Arts Sci. 78 (1950), 28-29
J.L. Abernethy, J. Chem. Ed. 44 (1967), 177-180
J.S. Fruton, Proc. Am. Phil. Soc. 129 (1985), 313-370

HOFMEISTER, LOTHAR [1910-1977]
KURSCHNER 12,1312
Anon., Naturw. Rund. 31 (1978), 306
A. Ziegler, Verhandl. zool. bot. Ges. Wien 118-119 (1980), 8-10

HOFMEISTER, VICTOR [1828-1894]
RSC 7,1003; 10,255; 12,341; 15,899
Anon., Sitz. Nat. Ges. Isis 1894, pp.15-16; Deutsche Z. Thiermed. 20 (1894), 215-218

HOFMEISTER, WILHELM [1824-1877]
DSB 6,464-468; NDB 9,468-469; ADB 12,644-648; DRULL, 117-118; W, 262-263; LEHMANN, 181-183; TSCHIRCH, 1067; WWWS, 815; POGG 1,1129; 3,648
G. Haberlandt, Ost. Bot. Z. 27 (1877), 113-117
K. von Goebel, Wilhelm Hofmeister. Leipzig 1924
W.O. Locy, Sci. Mon. 19 (1924), 380-389
A.H. Larsen, Plant Physiology 5 (1930), 613-616

HOFSTEE, BAREND HENDRIK JAN [1912-1980]
AMS 14,2225

HOGAN, ALBERT GARLAND [1884-1961]
AMS 10,1822; IB, 126; WWWS, 815
L.R. Richardson, J. Nutrition 97 (1969), 3-7

HOGBEN, LANCELOT THOMAS [1895-1975]
DNB [1971-1980], 417-418; WW 1975, p.1505; IB, 126; WWWS, 815
G.P. Wells, Biog. Mem. Fell. Roy. Soc. 24 (1978), 183-221

HOGEBOOM, GEORGE HALL [1913-1956]
AMS 9(II),523
W.C. Schneider, J. Nat. Cancer Inst. 17 (1956), iii-ix
Anon., J. Bioph. Bioch. Cytol. 2 (1956), ix-xvi

HOGGAN, GEORGE [1837-1891]
HIRSCH 3,271-272
Anon., Brit. Med. J. 1891(I), p.1411

HOGNESS, DAVID SWENSON [1925-]
AMS 15(3),755; WWA 1980-1981, p.1583

HOGNESS, THORFIN RUSTEN [1894-1976]
AMS 11,2348; POGG 6,1143; 7b,2047-2048
Anon., Science 193 (1976), 988

HOHLWEG, HERMANN [1879-1941]
JV 19,325; NUC 250,694
Anon., 50 Jahre Evangelisches Krankenhaus "Bethesda" in Duisburg 1904-1954. Duisburg 1954

HOKE, EDMUND [1874-1932]
KURSCHNER 4,1224; KOERTING, 173; PELZNER, 28-36

HOKIN, LOWELL EDWARD [1924-]
AMS 17(3),779

L.E. Hokin, Trends in Pharmacological Sciences 8 (1987), 53-56

HOLCH, LUDWIG [1887-1965]
JV 29,724; NUC 251,132
Anon., Nachr. Chem. Techn. 14 (1966), 10

HOLDE, DAVID [1864-1938]
KURSCHNER 4,1225; TLK 3,282; WININGER 7,88; POGG 5,282; 6,1144-1145; 7a(2),532

HOLDERMANN, EUGEN [1852-1906]
HEIN, 286-287; HUHLE-KREUTZER, 330
O. Anselmino, Ber. pharm. Ges. 16 (1906), 99-107

HOLDERMANN, KARL [1882-1969]
POGG 7a(2),532
Anon., Chem. Z. 93 (1969), 408

HOLIDAY, ENSOR ROSLYN [1903-]
Who's Who in British Science 1953, p.135

HOLL, MORITZ [1852-1920]
NDB 9,533-534; OBL 2,401; DBJ 2,749; FISCHER 1,652; STOBER, 66-75
Anon., Alm. Akad. Wiss. Wien 71 (1921), 173-174
H. Rabl, Anat. Anz. 55 (1922), 12-29

HOLLAENDER, ALEXANDER [1898-1986]
AMS 14,2233; WWA 1978-1979, p.1530; IWW 1982-1983, p.579; MH 2,229-231; STC 2,16-18; MSE 2,78-80; WWWS, 817
M.B. Fuller, Mendel Newsletter 24 (1984), 5-7
New York Times, 11 December 1986

HOLLANDER, CHARLES SAMUEL [1877-1962]
AMS 10,1831; JV 17,284
Anon., Science 139 (1963), 363

HOLLANDER, FRANKLIN [1899-1966]
AMS 11,2357; WWAH, 281; KOREN, 195
Anon., Chem. Eng. News 44(28) (1966), 79
H.D. Janowitz, J. Mt. Sinai Hosp. 33 (1966), 533-535; Gastroenterology 51 (1966), 273-274

HOLLANDT, FRIEDRICH [1868-]
JV 13,224; RSC 14,804
Anon., Süddeutsche Apoth. Z. 89 (1949), 236

HOLLBØLL, SVEND AAGE [1895-]
VEIBEL 2,201-203

HOLLBORN, KARL [1862-1942]
HEIN, 287-288
Anon., Chem. Z. 66 (1942), 236; Deutsche med. Wchschr. 68 (1942), 540
D. Tutzke, Janus 59 (1972), 39-45

HOLLEMAN, ARNOLD FREDERIK [1859-1953]
BWN 2,241; WWWS, 818; POGG 4,658-659; 5,551-552; 6,1146; 7b,2048-2049
S. Coffey, Nature 172 (1953), 706
J.P. Wibaut, Rec. Trav. Chim. Pays-Bas 74 (1955), 1371-1375

HOLLEY, ROBERT WILLIAM [1922-1993]
AMS 15(3),767; WWA 1980-1981, p.1590; IWW 1982-1983, p.580; MH 2,231-233; STC 2,18-19; MSE 2,80-81; POTSCH, 209-210

HOLLINGER, ADOLF [1878-1915]
JV 19,325; SG [3]7,786; NUC 251,537

HOLLMANN, SIEGFRIED [1914-]
KURSCHNER 13,1613; POGG 7a(2),535-536

HOLLUTA, JOSEF [1895-1973]
STURM 1,672; KURSCHNER 10,985; TLK 3,283; POGG 6,1146-1147; 7a(2),536

HOLLY, FREDERICK WILLIAM [1919-1980]
AMS 14,2241
Anon., Chem. Eng. News 59(1) (1981), 53

HOLMAN, RALPH THEODORE [1918-]
AMS 15(3),773; WWA 1980-1981, p.1593; WWWS, 819

HOLMBERG, BROR [1881-1966]
SBL 19,230-231; POGG 5,552-553; 6,1148; 7b,2050-2051
H. Erdtman, Sven. Kem. Tid. 78 (1966), 600-602

HOLMBERG, CARL GOTFRID [1902-1986]
VAD 1983, p.452

HOLMES, ERIC GORDON [1897-1972]
WW 1972, p.1534
The Times, 22,27 May 1972

HOLMES, FRANCIS OLIVER [1899-]
AMS 11,2366; WWWS, 819-820

HOLMES, HARRY NICHOLLS [1879-1958]
AMS 9(I),889; MILES, 222-223; WWWS, 820; POGG 6,1149; 7b,2054-2056
Anon., J. Am. Chem. Soc. 80 (1958), 5899-5900; Chem. Eng. News 36 (1958), 122-123

HOLMGREN, ALARIK FRITHIOF [1831-1897]
DSB 6,476-477; SMK 3,514-515; HIRSCH 3,280; PAGEL, 773-774; WWWS, 820; POGG 3,651; 4,659
R. Tigerstedt, Skand. Arch. Physiol. 7 (1897), v-x
J. Bernstein, Naturw. Rund. 12 (1897), 579-580
R. Granit, Lychnos 1944-1945, pp.132-138

HOLMYARD, ERIC JOHN [1891-1959]
DNB [1951-1960], 496-497; WW 1959, pp.1464-1465; POGG 6,1149; 7b,2054-2056
J.R. Partington, Nature 184 (1959), 1360
D. McKie, Ambix 8 (1960), 1-5

HOLSBAER, HENDRIK BERNARD [1875-]
RSC 15,921; NUC 252,296
Personlijkeden in het Koninkrijk der Nederlanden, pp.679-680. Amsterdam 1938

HOLST, AXEL [1860-1931]
NBL 6,277-280; FISCHER 1,655; PAGEL, 775-776; WWWS, 821
P. Schoorl, Voeding 23 (1962), 436-440
R. Nicolaysen, Nord. Med. 85 (1971), 69-74

HOLT, LUTHER EMMETT [1855-1924]
DAB 9,183-184; NCAB 20,46; FISCHER 1,657; DAMB, 359; WWWS,821
T.M. Prudden, Science 59 (1924), 452
R.L. Duffus and L.E. Holt, Jr., Luther Emmett Holt. New York 1940
E.A. Park and H.H. Mason, J. Pediat. 49 (1956), 342-349; in Pediatric Profiles (B.S. Veeder, Ed.), pp.164-174. St. Louis 1957

HOLT, LUTHER EMMETT, Jr. [1895-1974]
AMS 12,2797; ICC, 206-207; WWWS, 821
S. Krugman, Trans. Assoc. Am. Phys. 88 (1975), 22-25

HOLTER, HEINZ [1904-]
DBL (3rd Ed.) 6,549-550; KBB 1981, p.465; IWW 1982-1983, p.582; VEIBEL 2,205, WWWS, 821
E. Zeuthen, in The Carlsberg Laboratory (H. Holter and K.M. Moller, Eds), pp.239-258. Copenhagen 1976
C. deDuve, Carlsberg Laboratory Communications 49 (1984), 137-138

HOLTFRETER, JOHANNES [1901-1992]
AMS 14,2249; WWA 1978-1979, p.1541; IWW 1982-1983, p.582; BHDE 2,534; MH 2,234-235; MSE 2,83-85; WWWS, 821

HOLTZ, FRIEDRICH [1898-1967]
IB, 127; POGG 7a(2),537-539
P. Marquardt, Arzneimitt. 8 (1958), 676-677; 17 (1967), 919-920
A. Butenandt, Arzneimitt. 13 (1963), 930

HOLTZ, JULIUS FRIEDRICH [1836-1911]
NDB 9,555-556; BJN 16,34*; BLOKH 1,312-313
C. Liebermann and G. Kraemer, Ber. chem. Ges. 44 (1911), 1679,3395-3398

HOLTZ, PETER [1902-1970]
WWWS, 821; POGG 7a(2),539-540
K. Credner, Arzneimitt. 17 (1967), 251-252
G. Kroneberg, Arzneimitt. 21 (1971), 150-151
H.J. Schümann, Erg. Physiol. 66 (1972), 1-12

HOLZAPFEL, JULIUS [1883-1918]
JV 24,338; NUC 252,575
Anon., International Paints: Seventy-Five Years of Paintmaking. London 1957

HOLZAPFEL, LUISE [1900-1963]
KURSCHNER 9,816; POGG 7a(2),541-542

HOLZER, FRANZ JOSEF [1903-1974]
KURSCHNER 11,1222; HUTER, 273-275
L. Breitnecker, Beitr. ger. Med. 32 (1974), v-xi
H. Patscheider, Ber. Nat. Med. Innsb. 62 (1975), 155-157

HOLZER, HELMUT [1921-]
KURSCHNER 13,1623; POGG 7a(2),542-543

HOLZER, WOLFGANG [1906-1980]
KURSCHNER 13,1623; POGG 7a(2),543-544

HOLZINGER, OTTO [1869-]
JV 12,245; RSC 19,74
Anon., Deutsche Apoth. Z. 43 (1928), 414-415

HOLZKNECHT, GUIDO [1872-1931]
NDB 9,575-576; OBL 2,411; FISCHER 1,658-659; WOLF, 47-75; WWWS, 822
G. Forsell, Acta Radiologica 12 (1931), 516-521
R. Lenk, Strahlentherapie 43 (1932), 1-8
G. Schwarz, Wiener med. Wchschr. 82 (1932), 132-134
K.M. Walther, Münch. med. Wchschr. 115 (1973), 1564-1567;
Guido Holzknecht. Vienna 1975

HOME, EVERARD [1756-1832]
DSB 6,478-479; DNB 9,1121-1122; HIRSCH 3,288-289; WWWS, 822;
POGG 1,1136-1137; CALLISEN 9,79-101; 29,37-40

HOMOLKA, BENNO [1860-1925]
STURM 1,677; BLOKH 1,313; POGG 6,1153
M. Bodenstein, Ber. chem. Ges. 58A (1925), 20-21
A. Simon, Chem. Z. 49 (1925), 254
H.W. Flemming, Ed., Benno Homolka. Dokumente aus Hochster Archiven No.38. Frankfurt a.M. 1968

HOMOLLE, AUGUSTIN EUGÈNE [1808-1875]
HIRSCH 3,289-290; DECHAMBRE [4]14,255; SG [2]7,275

HONCAMP, FRANZ [1875-1934]
TLK 3,284; IB, 127; POGG 6,1153-1154
W. Wohlbier, Land. Ver. 120 (1934), 1-12
Anon., Z. angew. Chem. 48 (1935), 145-146

HONIGMANN, MORITZ [1844-1918]
NDB 9,600; DBJ 2,691; WININGER 7,90-91; POTSCH, 210-211

HOOBLER, ICIE MACY [1892-1984]
AMS 13,1986; NCAB H,334-335; AO 1984, p.26; IB, 186; WWWS, 1093
I. Hoobler, Boundless Horizons. Smithtown, N.Y. 1982
Anon., New York Times 13 January 1984; Chem. Eng. News 62(7) (1984), 31
H.H. Williams, J. Nutrition 114 (1984), 1351-1362

HOOGEWERFF, SEBASTIAN [1847-1933]
BWN 2,243-244; POGG 3,654; 4,662; 5,557
J. Boeseken, Rec. Trav. Chim. Pays-Bas 53 (1934), 433-442

HOOKER, JOSEPH DALTON [1817-1911]
DSB 6,488-492; DNB [Suppl.2]2, 294-299; DESMOND, 318; ABBOTT, 66-67; W, 265-266; WWWS, 824
Anon., Nature 88 (1911), 249-254; Proc. Roy. Soc. B85 (1912), i-xxxv
W.B. Turrill, Joseph Dalton Hooker. London 1964
M. Allen, The Hookers of Kew. London 1967

HOOPER, DAVID [1858-1947]
DESMOND, 319; TSCHIRCH, 1067
Who Was Who 1941-1950, pp.599-600
T.A. Henry, J. Chem. Soc. 1948, pp.253-255

HOOYKAAS, REIJER [1906-]
L. de Albuquerque, Janus 64 (1977), 1-13

HOPE, EDWARD [1886-1953]
POGG 6,1156; 7b,2068
S.G.P. Plant, Nature 171 (1953), 417-418; J. Chem. Soc. 1953, pp.3730-3732

HOPE, THOMAS CHARLES [1766-1844]
DSB 6,495-496; W, 266; POTSCH, 211
T.S. Traill, Trans. Roy. Soc. Edin. 16 (1848), 419-434
R.H. Cragg, Med. Hist. 11 (1967), 186-189
J.B. Morrell, Ambix 16 (1969), 66-80

HOPE, WALTER BAYARD [1877-1940]
RSC 15,934

HOPKINS, BARBARA [1899-1981]
Directory of British Scientists 1966-1967, vol.1, p.849

HOPKINS, CYRIL GEORGE [1866-1919]
DAB 9,207; MILES, 228-229
E. Davenport, Science 50 (1919), 387-3888

HOPKINS, FREDERICK GOWLAND [1861-1947]
DSB 6,498-502; DNB [1941-1950], 406-408; WW 1945 p.1334; FISCHER 1,161; POTSCH, 211-212; W, 266-267; ABBOTT, 67; POGG 6,1158-1159; 7b,2069-2071
H. Wieland, Nature 141 (1938), 989-993; Bayer. Akad. Wiss. Jahrbuch 1944-1948, pp.280-281
M. Dixon and C. Rimington, Nature 160 (1947), 44-48
H. Dale, Obit. Not. Fell. Roy. Soc. 6 (1948), 115-145
J. Needham and E. Baldwin, Hopkins and Biochemistry. Cambridge 1949
R.A. Peters, Biochem. J. 71 (1959), 1-9 T.H. Bishop, Pharm. J. 186 (1961), 527-528
J. Needham, Notes Roy. Soc. 17 (1962), 117-162; Persp. Biol. Med. 6 (1962), 2-46
N.W. Pirie, TIBS 4 (1979), N75-N77; in Selected Topics in the History of Biochemistry (G. Semenza, Ed.), pp.103-128. Amsterdam 1983

HOPPE, JOHANNES [1872-1949]
JV 19,340; NUC 254,401
Anon., Z. angew. Chem. 61 (1949), 464

HOPPE-SEYLER, FELIX [1825-1895]
DSB 6,504-506; NDB 9,615-616; ADB 50,464-465; HIRSCH 3,294-295; W, 267; PAGEL, 778-780; BLOKH 1,314-316; SCHAEDLER, 55; WWWS, 826; POTSCH, 212; POGG 1,1140; 3,656-657; 4,664
E. Baumann and A. Kossel, Z. physiol. Chem. 21 (1895), i-lxi
E. Fischer, Ber. chem. Ges. 23 (1895), 2333-2336
H. Thierfelder, Berl. klin. Wchschr. 32 (1895), 928-930; Felix Hoppe-Seyler. Stuttgart 1926
R. Virchow, Arch. path. Anat. [14]2 (1895), 386-388
I. Munk, Deutsche med. Wchschr. 21 (1895), 563

A. Gürber, Münch. med. Wchschr. 43 (1895), 274
A. Gamgee, Nature 52 (1895), 575-576,623-625
A.P. Mathews, Pop. Sci. Mon. 53 (1898), 542-552
J.S. Fruton, Contrasts in Scientific Style, pp.72-117. Philadelphia 1990

HOPPE-SEYLER, FELIX ADOLF [1898-1945]
POGG 6,1160-1161; 7a (2),549
W. Steinhausen, Z. physiol. Chem. 283 (1948), 101-105

HOPPE-SEYLER, GEORG [1860-1940]
FISCHER 1,661-662; PAGEL, 780; RAZINGER, 85-87; KIEL, 97; RSC 15,939
F. Hoff, Deutsche med. Wchschr. 66 (1940), 1307-1308

HOPWOOD, MORTIMER LLOYD [1908-]
AMS 14,2265-2266; WWWS, 826

HORBACZEWSKI, JAN [1854-1942]
DSB 6,506; OBL 2,418-419; STURM 1,681; UKRAINE 2,235-236; POTSCH, 212-213; NAVRATIL, 99-100; KONOPKA 3,648-649; POGG 3,657; 4,665
O. Wagner, Biol. Listy 19 (1934), 95-97

HORECKER, BERNARD LEONARD [1914-]
AMS 15(3),804; WWA 1980-1981, p.1608; IWW 1982-1983, p.587; COHEN, 108; MH 2,235-237; STC 2,22-24; MSE 2,86-88; WWWS, 826-827
B.L. Horecker, in Reflections on Biochemistry (A. Kornberg et al.,Eds.), pp.65-72. Oxford 1976; in Of Oxygen, Fuels and Living Matter Part 2 (G. Semenza, Ed.), pp.59-172. Chichester 1982

HORLACHER, THEODOR von [1873-]
RIGA, 704; JV 15,132; NUC 254,676

HORNER, LEOPOLD [1911-]
KURSCHNER 13,1638; WWWS, 828; POTSCH, 213; POGG 7a(2),552-553

HORNUS, GEORGES [1905-1940]
Anon., Ann. Inst. Pasteur 65 (1940), 279-281

HORODYNSKI, WITOLD [1865-1954]
KONOPKA 3,651
M. Demel, Arch. Hist. Med. 29 (1966), 443

HOROWITZ, NORMAN HAROLD [1915-]
AMS 15(3),811; WWA 1980-1981, p.1612; IWW 1982-1983, p.588; WWWS, 828

HORRMANN, PAUL [1878-1942]
HEIN-E, 203-204; KIEL, 168; DRUM, 287-288; TSCHIRCH, 1068; TLK 3,287; POGG 5,558; 6,1162; 7a(2),554

HORSFALL, FRANK LAPPIN [1906-1971]
AMS 11,2398; DAMB, 365-366; ICC, 308-309; WWAH, 286; WWWS, 829
C.M. MacLeod, Am. Phil. Soc. Year Book 1971, pp.127-132
G.K. Hirst, Biog. Mem. Nat. Acad. Sci. 50 (1979), 233-267

HORSFORD, EBEN NORTON [1818-1893]
DSB 6,517-518; DAB 9,236-237; MILES, 23-231; ELLIOTT, 129-130; WWAH, 286; SILLIMAN, 107-108; FERCHL, 250; WWWS, 829; POGG 1,1144; 3,658
C.L. Jackson, Proc. Am. Acad. Arts Sci. 28 (1893), 340-346
S. Resneck, Tech. Cult. 11 (1970), 366-398
M.W. Rossiter, The Emergence of Agricultural Science. New Haven 1975

HORSLEY, VICTOR [1857-1916]
DSB 6,518-519; DNB [1912-1921],270-271; FISCHER 1,663; HAYMAKER, 562-566; O'CONNOR(2), 485-490; KOLLE 2,189-197; BRAZIER, 170-174,182-183; WWWS, 829
F.W.M., Proc. Roy. Soc. B91 (1920), xliv-xlviii
S. Paget, Sir Victor Horsley. London 1949
G. Jefferson et al., Brit. Med. J. 1957(I), pp.903-917,1008,1065,1307
Anon., Brit. J. Surg. 51 (1964), 81-86
J.B. Lyons, The Citizen Surgeon: A Biography of Sir Victor Horsley. London 1966; Med. Hist. 11 (1967), 361-373

HORSTMANN, AUGUST [1842-1929]
DSB 6,519-520; NDB 9,644-645; DBJ 11,355; DRULL, 120-121; SCHAEDLER, 55; POTSCH, 213; BLOKH 1,316-317; DZ, 652; TLK 3,287; WWWS, 829; POGG 3,658-659; 4,666; 6,1163
M. Trautz, Ber. chem. Ges. 63A (1930), 61-86
A. Mayer, Naturwiss. 18 (1930), 261-264

HORTON, DEREK [1932-]
AMS 15(3),814

HORTON-SMITH-HARTLEY ee HARTLEYJHORWITT, MAX KENNETH [1908-]
AMS 14,2276; WWWS, 830

HOSHINO, TOSHIO [1894-1979]
JBE, 319

HOSKINS, ROY GRAHAM [1880-1964]
AMS 10,1869; IB, 128
W.B. Cannon, Endocrinology 30 (1942), 839-845
R.O. Greep, Endocrinology 76 (1965), 1007-1011

HOSKYNS-ABRAHALL, JOHN LEIGH [1865-1891]
JV 4,225; RSC 13,275; 15,950
P.T.H., J. Chem. Soc. 61 (1892), 486-488

HOTCHKISS, ROLLIN DOUGLAS [1911-]
AMS 15(3),821; WWA 1980-1981, p.1616; IWW 1982-1983, p.589; MH 2,241-243; STC 2,26-28; MSE 2,91-93; WWWS, 831

HOTTENROTH, VALENTIN [1878-1954]
TLK 3,288; NUC 255,695
Anon., Chem. Z. 77 (1953), 764; 78 (1954), 302

HOUBEN, JOSEF [1875-1940]
NDB 9,659-660; TLK 3,288-289; WWWS, 831; POTSCH, 213-214; POGG 4,666;5,559-560; 6,1165-1166; 7a(2),556
E. Pfankuch, Ber. chem. Ges. 73A (1940), 119-121
Anon., Z. angew. Chem. 54 (1941), 139

HOUGHTON, ELIJAH MARK [1867-1937]
AMS 6,681; IB, 129
Anon., Detroit Medical News 29(19) (1938), 7

HOUSSAY, BERNARDO ALBERTO [1887-1971]
DSB 15,228-229; CB 1948, pp.295-297; MH 1,242-243; STC 2,28-29; MSE 2,93-94; MEDVEI, 755-757; OLPP, 180-181; WWWS, 832
L.F. Leloir, Am. Phil. Soc. Year Book 1972, pp.197-202
F. Young and V.G. Foglia, Biog. Mem. Fell. Roy. Soc. 20 (1974), 247-270
V.G. Foglia, J. Hist. Med. 35 (1980), 380-396
F.G. Young, TIBS 8 (1983), 181-182

HOUSSAY, FRÉDÉRIC [1860-1920]
DBF 17,1354-1355; CHARLE(2), 155-156; RSC 15,955
L. Roule, Rev. Gen. Sci. 31 (1920), 773-774
L. Bounoure, Isis 7 (1925), 433-455

HOUTON de la BILLARDIÈRE, JACQUES [1755-1834]
FERCHL, 251; WWWS, 982; RSC 3,447; POGG 1,1148; CALLISEN 9,184; 29,63-64

HOVANITZ, WILLIAM [1915-]
AMS 10,1876; WWWS, 832HOWE, ALLEN BREWER [1854-1902]
RSC 15,963
Obituary Record of Yale Graduates No.62, pp.281-282. New Haven 1903

HOWE, PAUL EDWARD [1885-1974]
AMS 11,2420; IB, 129
F.C. Bing, J. Nutrition 115 (1985), 297-302

HOWE, PERCY ROGERS [1864-1950]
DAB [Suppl.4],401-403; AMS 8,1187; DAMB, 368-369; WWWS, 834
Anon., Science 111 (1950), 293

HOWELL, WILLIAM HENRY [1860-1945]
DSB 6,525-527; DAB [Suppl.3],369-371; AMS 7,856; NCAB F,478-479; IB, 129; BROBECK, 132-133; FISCHER 1,665-666; CHITTENDEN, 143-147; DAMB, 370-371; WWAH, 291; WWWS,834
J. Erlanger, Science 101 (1945), 575-576; Biog. Mem. Nat. Acad. Sci. 26 (1951), 153-180
A.M. Chesney et al., Bull. Johns Hopkins Hosp. 109 (1961), 1-19
H.W. Davenport, Physiologist 25(1)Suppl. (1982), 45-49
W.B. Fye, Circulation 69 (1984), 1198-1203
L.B. Jacques, Canadian Bulletin of Medical History 5 (1988), 143-165
J. Marcum, Persp. Biol. Med. 33 (1990), 214-230

HOWITZ, FRANTZ JOHANNES [1828-1912]
HIRSCH 3,315
L. Kraft, Hosp. Tid. [5]6 (1913), 20-24

HOWLAND, JOHN [1873-1926]
DAB 9,313-314; NCAB 21,392; FISCHER 1,666; CHITTENDEN, 368; ICC, 88-90; DAMB, 371-372; WWWS, 835
E.A. P[ark], Science 64 (1926), 80-83
W.C. Davidson, J. Hist. Med. 5 (1950), 197-205; in Pediatric Profiles (B.S. Veeder, Ed.), pp.161-174. St. Louis 1957
L.E. Holt, J. Pediat. 69 (1966), 865-875

HOYLE, JOHN CLIFFORD [1901-1976]
NUC 257,391
Who Was Who 1971-1980, p.388

HROMATKA, OTTO [1905-]
KURSCHNER 13,1648; POGG 7a(2),559-560

HRUBY, KAREL [1910-1962]
B. Nemec, Preslia 35 (1963), 246-254

HRUSCHAUER, FRANZ [1807-1858]
OBL 2,441; WURZBACH 9,362-363; KERNBAUER, 7-24; FERCHL, 251; POGG 1,1150-1151,1575
A. Schrötter, Alm. Akad. Wiss. Wien 9 (1859), 134-140

HRYNAKOWSKY, KONSTANTY [1878-1938]
PSB 10,55; POGG 6,1169-1170; 7b,2089-2091
F. Adamanis, Roczniki Chemii 19 (1939), 297-306

HUARD, PIERRE [1901-1983]
DBF 17,1380; QQ 1983-1984, p.724
J. Théodoridès, Rev. Hist. Sci. 36 (1983), 332-334
M.J. Imbault-Huard et al., Hist. Sci. Med. 17 (1983), 381-402
M.D. Grmek, Arch. Inter. Hist. Sci. 34 (1984), 220-228

HUBBARD, RUTH [1924-]
AMS 15(3),857; BHDE 2,544

HUBEL, DAVID HUNTER [1926-]
AMS 15(3),858; WWA 1980-1981, p.1631; IWW 1982-1983, p.597; WWWS, 837

HUBER, ADOLPHE [1865-]
RSC 15,973; SG [2]7,451; NUC 258,128

HUBER, FRANÇOIS [1750-1831]
DHB 4,173; HOEFER 25,347; STELLING-MICHAUD 4,93; FERCHL, 251; WWWS, 837
A.P. de Candolle, Am. J. Sci. 23 (1833), 117-129
S.B. Herrick, Pop. Sci. Mon. 6 (1875), 486-498
A. Maurizio, in Essays in Biohistory (P. Smit and R.J.C.V. ter Laage, Eds.), pp.145-148. Utrecht 1970

HUBER, GOTTHELF CARL [1865-1934]
DAB [Suppl.1],438-440; NCAB D,274; 42,693; OEHRI, 54-55; IB, 130
C.P. Huber et al., J. Comp. Neurol. 65 (1936), 1-41

HUBRECHT, AMBROSIUS ARNOLD WILHELM [1853-1915]
DSB 6,535-536; UTRECHT, 98; WWWS, 837
Anon., Nature 95 (1915), 121-122
R. Assheton, Proc. Linn. Soc. 127 (1915), 28-31

HUDSON, CLAUDE SILBERT [1881-1952]
DSB 6,538; DAB [Suppl.5],327-328; AMS 8,1197; MILES, 235-236; WWWS, 838;POTSCH, 215; BARRY, 40-42; NCAB 40,468-469; FARBER, 1537-1550; POGG 6,1171-1172; 7b,2096-2099
M.L. Wolfrom, Adv. Carbohydrate Chem. 9 (1954), xiii-xviii
L.F. Small and M.L. Wolfrom, Biog. Mem. Nat. Acad. Sci. 32 (1958),181-220

HUE, CLEMENT [1779-1861]
MR 3,65-66

HÜBENER, WILHELM ADRIAN [1867-]
JV 5,8; RSC 15,973; SG [2]7,451; NUC 258,473

HÜBL, ARTUR von [1853-1932]
NDB 9,714-715; OBL 2,447-448; KNOLL, 163-165; PLANER, 143; TLK 3,289-290; POGG 6,1172-1173
J. Lewkowitsch, Chem. Ind. 51 (1932), 634
J.M. Eder, Alm. Akad. Wiss. Wien 82 (1932), 301-308

HÜBNER, FRIEDRICH [1876-1953]
JV 12,20; RSC 15,650
Anon., Nachr. Chem. Techn. 1 (1953), 164

HÜBNER, HANS [1837-1884]
NDB 9,716-717; BLOKH 1,317-319; POGG 3,663; 4,669
F. Beilstein, Ber. chem. Ges. 17(3) (1884), 763-776

HUEBNER, ROBERT JOSEPH [1914-]
AMS 15(3),867; WWA 1980-1981, p.1635; IWW 1982-1983, pp.597-598; WWWS,838

HUECK, WERNER [1882-1962]
KURSCHNER 8,1005; FISCHER 1,669; ECKERT, 71-78; IB, 131
C. Krauspe, Deutsche med. Wchschr. 87 (1962), 1922-1925
H. Büngeler, Münch. med. Wchschr. 104 (1962), 1886-1888

HÜCKEL, ERICH [1896-1980]
KURSCHNER 13,1663; AUERBACH, 828; ABBOTT-C, 67-68; TLK 3,290; POTSCH, 214; POGG 6,1173; 7a(2),564
Anon., Nachr. Chem. Techn. 13 (1965), 382-383; 28 (1980), 186; Chem. Z. 104 (1980), 181
E. Hückel, Ein Gelehrtenleben. Weinheim 1975
H. Hartmann and H.C. Longuet-Higgins, Biog. Mem. Fell. Roy. Soc. 28 (1982), 153-162

HÜCKEL, WALTER [1895-1973]
KURSCHNER 10,1015; BOHM, 167; WWWS, 837; POTSCH, 214-215; POGG 6,1173-1174; 7a(2),564-566
Anon., Nachr. Chem. Techn. 8 (1960), 35-36
H. Auterhoff, Arzneimitt. 10 (1960), 193-194
R. Neidlein and M. Hanack, Chem. Ber. 113 (1980), i-xxviii

HÜFNER, GUSTAV [1840-1908]
NDB 9,729-730; BJN 13,43*; FISCHER 1,669; BLOKH 1,319; SCHAEDLER, 55; DZ, 658; WWWS, 1736; POGG 3,663-664; 4,670; 5,562
R. von Zeynek, Z. physiol. Chem. 58 (1908), 1-38
K. Bürker, Münch. med. Wchschr. 55 (1908), 916-919

HÜLLWECK, GUSTAV [1889-1954]
JV 30,660; NUC 258,569

HÜNEFELD, FRIEDRICH LUDWIG [1799-1882]
HIRSCH 3,325; PAGEL, 785; FERCHL, 251; CALLISEN 9,214-217; 29,73-74; POGG 1,1154

HÜNI, ERNST [1885-]
NUC 285,602HUEPPE, FERDINAND [1852-1938]

NDB 9,742-743; NB, 181; STURM 1,704; KURSCHNER 4,1277; FISCHER 1,669-670; PAGEL, 785-787; GROTE 2,77-138; MOTSCH, 125-142; SKVARC, 98-102; KOERTING, 125-127; BULLOCH, 374; TLK 3,292; IB, 131; WWWS, 839
J. Kamp, Münch. med. Wchschr. 69 (1922), 1547-1549
J.D.R., Nature 143 (1939), 108

HÜRTHLE, KARL [1860-1945]
KURSCHNER 3,1022; FISCHER 1,671; PAGEL, 787; DZ, 664; TLK 3,292; IB, 131
H. Winterstein, Deutsche med. Wchschr. 56 (1930), 449
G. Rosenfeld, Med. Klin. 26 (1930), 411-412

HÜTTEL, RUDOLF [1912-]
KURSCHNER 15,1973

HÜTTIG, GUSTAV FRANZ [1870-1957]
NDB 9,748-750; POTSCH, 217; POGG 6,1174-1175; 7a(2),570-573; (4),145*
E. Bischoff, Koll. Z. 117 (1950), 73-75
E. Hayek, Alm. Akad. Wiss. Wien 108 (1958), 399-424

HÜTZ, HUGO [1871-]
JV 10,21; RSC 14,1010; NUC 258,687

HÜTZ, RUDOLF [1877-]
JV 17,284

HUFELAND, WILHELM [1762-1836]
NDB 10,1-7; ADB 13,286-296; HIRSCH 3,329-332; FERCHL, 252; WWWS, 839
K. Pfeiffer, Christoph Wilhelm Hufeland. Halle 1968
W. Genschorek, Christoph Wilhelm Hufeland. Leipzig 1976

HUFFMAN, HUGH MARTIN [1898-1950]
AMS 8,1199
Anon., Chem. Eng. News 28 (1950), 346

HUG, ERNST [1884-1962]
NUC 259,32
Anon., Verhandl. Nat. Ges. Basel 73 (1962), 356

HUGGETT, ARTHUR [1897-1968]
Who Was Who 1961-1970, p.561
F.W.B. Brambell, Biog. Mem. Fell. Roy. Soc. 16 (1970), 343-364

HUGGINS, CHARLES BRENTON [1901-]
AMS 14,2321; WWA 1978-1979, p.1580; IWW 1982-1983, p.598; MH 1,244-245; STC 2,32-33; MSE 2,101; WWWS, 839

HUGGINS, CLYDE GRIFFIN [1922-]
AMS 15(3),873; WWWS, 839

HUGGINS, MAURICE LOYAL [1897-1981
AMS 14,2321; McLACHLAN, 307-313; WWWS, 839-840; POGG 6,1176; 7b,2101-2104
M.L. Huggins, Chemical Technology 10 (1980), 422-429
Anon., Chem. Eng. News 60(15) (1982), 82

HUGHES, ARTHUR FREDERICK [1908-1975]
MEDVEI, 757-758
Anon., Journal of Anatomy 12 (1976), 399

HUGHES, EDWARD DAVID [1906-1963]
DNB [1961-1970], 548; WW 1959, pp.1508-1509; POTSCH, 215
P.B.D. de la Mare, Proc. Chem. Soc. 1964, pp.97-100
C.K. Ingold, Chem. Ind. 1964, pp.96-98

HUGOUNENQ, LOUIS [1860-1942]
FISCHER 1,671; GUIART, 137-138; TSCHIRCH, 1068; IB, 131; POGG 4,672; 5,562-563; 6,1177; 7b,2105
M. Polonovski, Bull. Acad. Med. 127 (1943), 122-124

HUISGEN, ROLF [1920-]
KURSCHNER 13,1673; MH 2,248-250; MSE 2,101-103; WWWS, 841; POTSCH, 215-216; POGG 7a(2),574
R. Huisgen, Mechanisms, Novel Reactions, Synthetic Principles. Washington, D.C. 1991

HUIZINGA, DERK (DIRK) [1840-1903]
LINDEBOOM, 934-935; HIRSCH 3,335

HULDSCHINSKY, KURT [1883-1941]
FISCHER 1,672; KOREN, 195
Anon., Brit. J. Phys. Med. 7 (1932), 126

HULTGREN, ERNST OLOF [1866-1922]
SMK 3,551-552; RSC 15,991; NUC 260,3
I. Holmgren, Hygeia 84 (1922), 385-394

HULTIN, TORE [1919-]
VAD 1981, p.474

HUMBOLDT, ALEXANDER von [1769-1859]
DSB 6,549-555; NDB 10,33-34; ADB 13,358-383; COLE, 265-266; POTSCH, 216; WAGENITZ, 85-86; CALLISEN 9,291-304; 29,95-100; POGG 1,1157-1159; 7aSuppl.,295-301
F.A. Henglein, Chem. Z. 83 (1959), 290-298
K.R. Biermann, Alexander von Humboldt. Leipzig 1983

HUME, ELEANOR MARGARET [1887-1968]
Anon., Brit. J. Nutrition 24 (1970), 1

HUMMEL, JOHN JAMES [1850-1902]
BLOKH 1,319
R. Beaumont, Ber. chem. Ges. 35 (1902), 4559-4562
A.S., Nature 66 (1902), 520
E.H. Huntress, Proc. Am. Acad. Arts Sci. 78 (1950), 12-13

HUMMEL, JOHN PHILIP [1920-1967]
AMS 11,2460; BERG, 65-66

HUMMEL, KARL [1902-1987]
KURSCHNER 13,1675; POGG 7a(2),578
Anon., Deutsche Apoth. Z. 92 (1952), 300; Chem. Z. 112 (1988), 84; Naturw. Rund. 41 (1988), 87

HUMPHREY, JOHN HERBERT [1915-1987]
WW 1981, p.1294; IWW 1981-1982, p.585; WWWS, 843
J.H. Humphrey, Annual Review of Immunology 2 (1984), 1-21
B. Askonas, Biog. Mem. Fell. Roy. Soc. 36 (1990), 275-300

HUMPIDGE, THOMAS SAMUEL [1853-1887]
Anon., Chem. News 56 (1887), 148; Leopoldina 23 (1887), 218; Nature 37 (1888), 155-156; J. Chem. Soc. 53 (1888), 513-517

HUNDESHAGEN, FRANZ [1857-1940]
RSC 10,293; 15,994-995
Anon., Chem. Z. 64 (1940), 488; Z. angew. Chem. 54 (1941), 115-116

HUNDLEY, JAMES MANSON [1915-]
AMS 13,2052; WWWS, 844

HUNGER, FRIEDRICH WILHELM TOBIAS [1874-1952]
IB, 132
Wie is Dat? 1948, p.239
J.A. Vollgraff, Arch. Inter. Hist. Sci. 5 (1952), 361-362
T.J. Stamps, Ber. bot. Ges. 68a (1955), 33-36

HUNICKE, HENRY AUGUST [1861-1909]
O.N. Witt, Ber. chem. Ges. 42 (1909), 1637
Anon., Science 29 (1909), 696

HUNSALZ, PAUL [1871-]
JV 9,17; RSC 14,1010; NUC 260,600

HUNT, REID [1870-1948]
DSB 17,439-440; DAB [Suppl.4],410-412; AMS 7,873; FISCHER 1,673; PARASCANDOLA, 38-40; BARRY, 43-47; DAMB, 378-379; CHEN, 25; IB, 132
G.P. Grabfield, Ind. Eng. Chem. News Ed. 13 (1935), 170-171; Science 108 (1948), 127; J. Pharm. Exp. Ther. 93 (1966), 259-260
E.K. Marshall, Jr., Biog. Mem. Nat. Acad. Sci. 26 (1949), 25-44

HUNT, ROBERT [1807-1887]
DNB 10,277-278; FERCHL, 252-253; WWWS, 845; POGG 3,669-670
G.C. Boase and W.P. Courtney, Bibliotheca Cornubiensis 1 (1879), 259-260
A.G., Proc. Roy. Soc. 47 (1890), i-ii

HUNT, THOMAS STERRY [1826-1892]
DSB 6,564-566; DAB 9,393-394; MILES, 237-238; NCAB 3,254; ELLIOTT, 133; BLOKH 1,319-320; SILLIMAN, 103-107; WWWS, 845; POGG 3,670-672; 4,675
P. Frazer, American Geologist 11 (1893), 1-13
J. Douglas, Proc. Am. Phil. Soc. Memorial vol.1 (1900), 63-121
F.D. Adams, Biog. Mem. Nat. Acad. Sci. 15 (1934), 207-238
E.R. Atkinson, J. Chem. Ed. 20 (1943), 244-245

HUNTER, ANDREW [1876-1969]
AMS 11,2469; YOUNG, 20-21; O'CONNOR, 512-513; POGG 6,1182; 7b,2117-2118
Canadian Who's Who 11 (1969), 519
J.A. Dauphinee, Proc. Roy. Soc. Canada [4]8 (1970), 83-89

HUNTER, GEORGE [1894-]
AMS 8,1212; YOUNG, 18

HUNTER, GORDON DENIS [1927-]
Who's Who of British Scientists 1980-1981, p.243

HUNTER, ROBERT FERGUS [1904-1963]
POGG 6,1182-1183; 7b,2118-2120
N.J.L. Megson, Proc. Chem. Soc. 1964, pp.33-34

HUNTSMAN, ARCHIBALD GOWANLOCK [1883-1973]
AMS 12,2915; MH 1,247-248; MSE 2,106-107; WWWS, 847

HUPPERT, KARL HUGO [1832-1904]
NDB 10,76; OBL 3,13-14; BJN 9,314; 10,53*; STURM 1,708; SCHIEBER, 76-81; HIRSCH 3,348; PAGEL, 791; KOERTING, 122; SCHAEDLER, 56; POGG 4,675
R. von Zeynek, Prag. med. Wchschr. 29 (1904), 593-596

HURD, CHARLES DeWITT [1897-]
AMS 11,2475; WWWS, 847; POGG 6,1183-1184; 7b,2123-2126

HURST (HERTZ), ARTHUR FREDERICK [1879-1944]
DNB [1941-1950], 417-418; FISCHER 1,675; O'CONNOR(2), 228-229; WWWS, 792
J.A. Ryle, Guy's Hosp. Rep. 94 (1945), 1-11
A. Hurst, A Twentieth Century Physician. London 1949

HURTLEY, WILLIAM HOLDSWORTH [1865-1936]
O'CONNOR(2), 272
F.D.C., Biochem. J. 30 (1936), 1787-1788
Anon., Lancet 1936(I), p.1367; St. Barts. Hosp. J. 43 (1936), 203-205
W. Templeton, Chem. Brit. 18 (1982), 266-267

HURWITZ, GERARD [1928-]
AMS 15(3),906; IWW 1981-1982, p.588HUSCHKE, EMIL [1797-1858]
DSB 6,573-574; NDB 10,82; ADB 13,449-451; HIRSCH 3,349; GIESE, 457-458; DECHAMBRE [4]14,515; CALLISEN 9,324; 29,108

HUSEMANN, AUGUST [1833-1877]
NDB 10,82-83; ADB 13,452-453; HEIN, 296-297; TSCHIRCH, 1068; POGG 3,673
T. Husemann, Ber. chem. Ges. 10 (1877), 2297-2299

HUSEMANN, THEODOR [1833-1901]
BJN 6,50*; HIRSCH 3,349-350; PAGEL, 791-793; REBER, 61-66,364-365; BLOKH 1,320-321; TSCHIRCH, 1068; SCHAEDLER, 56; POGG 3,672; 4,677
Anon., Pharm. Z. 32 (1887), 443-445,451-453; 46 (1901), 147-148; Leopoldina 37 (1901), 35; Chem. Z. 25 (1901), 174

HUSFELDT, ERIK [1901-1984]
DBL (3rd Ed.)6,615-616; KBB 1983, p.486; VEIBEL 2,206

HUSLER, JOSEF [1885-1976]
FISCHER 1,675
A. Wiskott, Münch. med. Wchschr. 97 (1955), 572-573
H. Hilber, Münch. med. Wchschr. 118 (1976), 823

HUSMANN, AUGUST [1872-]
JV 13,29; RSC 16,377; NUC 261,583

HUSSONG, LUDWIG [1869-]
JV 19,181; NUC 261,635

HUSTIN, ALBERT [1882-1967]
BNB 38,323-330; FISCHER 1,675; WWWS, 848
M. van der Ghinst, Acta Chirurgica Belgica 66 (1967), 601-603

HUTCHINSON, ARTHUR [1866-1937]
DNB [1931-1940], 458-459; WWWS, 849; POGG 5,567; 6,1185; 7b,2128
W.C. Smith, Obit. Not. Fell. Roy. Soc. 2 (1939), 483-491
H.W. Mills, J. Chem. Soc. 1939, pp.210-213

HUTCHINSON, HENRY BROUGHAM [1880-]
WAGENITZ, 87; NUC 262,41-42

HUTCHINSON, JOHN [1811-1861]
DSB 6,575-576; HIRSCH 3,327-328
Anon., Lancet 1862(I), p.240; 1920(II), p.563
E.A. Spriggs, Med. Hist. 21 (1977), 357-364

HUTCHISON, ROBERT [1871-1960]
DNB [1951-1960], 526; FISCHER 1,675-676; O'CONNOR(2), 280-282; MR 5,208-209
Anon., Lancet 1960(I), pp.442-443; Brit. Med. J. 1960(I), pp.571-573,735-736
A. Moncrieff, J. Pediat. 58 (1961), 137-139
D. Hunter, Brit. Med. J. 1971(IV), pp.222-223; London Hospital Gazette 75(1) (1972), 7-13

HUXLEY, ANDREW FIELDING [1917-]
WW 1981, pp.1313-1314; IWW 1982-1983, p.605; MH 1,249-250; STC 2,34-35
A. Huxley, Notes Roy. Soc. 38 (1984), 146-151; Ann. Rev. Physiol. 50 (1988), 1-16

HUXLEY, HUGH ESMOR [1924-]
WW 1981, p.1314; IWW 1982-1983, pp.605-606; MH 2,254-256; MSE 2,111-112; ABBOTT, 68-70; WWWS, 850

HUXLEY, JULIAN SORELL [1887-1975]
DNB [1971-1980], 439-440; WW 1965, pp.1555-1556; CB 1942, pp.406-408; MH 2,256-258; STC 2,35-38; MSE 2,112-113; IB, 133; WWWS, 850
J. Huxley, Memories. London 1970, 1973
J. Needham et al., Nature 254 (1975), 2-5
J.R. Baker, Biog. Mem. Fell. Roy. Soc. 22 (1976), 207-238
N.L. Boothe, Mendel Newsletter 27 (1987), 1-10
J.C. Greene, J. Hist. Biol. 23 (1990), 39-55

HUXLEY, THOMAS HENRY [1825-1895]
DSB 6,589-597; DNB 22,894-903; PAGEL, 794-796; DESMOND, 334; W, 272-273; O'CONNOR, 110-113; ABBOTT, 70-71; WWWS, 850
M. Foster, Proc. Roy. Soc. 59 (1896), xlvi-lxvi
L. Huxley, Life and Letters of Thomas Henry Huxley. London 1900
C. Bibby, T.H. Huxley. London 1959
G.L. Geison, Isis 60 (1969), 273-292
A.F. Huxley, J. Physiol. 263 (1976), 41P-45P
J.F. McCartney, Southern Quarterly 14 (1976), 97-101
M.A. DiGregorio, T.H. Huxley's Place in Natural Science. New Haven

1984
J.V. Jensen, Thomas Henry Huxley: Communicating for Science. Newark 1991

HYATT, ALPHEUS [1838-1902]
DSB 6,613-614; DAB 9,446-447; NCAB 23,362-363; ELLIOTT, 134; WWWS, 851
W.K. Brooks, Biog. Mem. Nat. Acad. Sci. 6 (1909), 311-325
A.S. Packard, Proc. Am. Acad. Arts Sci. 38 (1903), 715-727

HYDE, IDA HENRIETTA [1857-1945]
NAW 2,247-249; OGILVIE, 103-104; NUC 262,449-450

HYMAN, LIBBIE HENRIETTA [1888-1969]
DSB 17,442-443; AMS 11,2493; IB, 133; WWAH, 300; WWWS, 851
H.W. Stunkard and W.K. Emerson, in Biology of the Turbellaria (N.W. Riser and M.P. Morse, Eds.), pp.ix-xxv. New York 1974
L.H. Hyman and G.E. Hutchinson, Biog. Mem. Mat. Acad. Sci. 60 (1991), 103-114

HYRTL, JOSEF [1810-1894]
DSB 6,618-619; NDB 10,109-110; OBL 3,23-24; STURM 1,715; HINK 1,311-313; HIRSCH 3,361-363; PAGEL, 796-798; LESKY, 240-251; STOBER, 1-17; WWWS, 852
K. von Bardeleben, Anat. Anz. 9 (1894), 773-776
N. Rüdinger, Münch. med. Wchschr. 41 (1894), 637-639
M. Holl, Wiener klin. Wchschr. 7 (1894), 549-551,557-559
E. Roth, Leopoldina 31 (1895), 190-192,214-217
J. Hann, Alm. Akad. Wiss. Wien 45 (1895), 265-272
J. Steudel, Med. Welt 18 (1944), 462-465
V. Patzelt, Anat. Anz. 103 (1956), 160-175
F. Wolf and G. Roth, Professor Josef Hyrtl. Vienna 1962
K.E. Rothschuh, Clio Medica 9 (1974), 81-92
G.E. Steyer, Anat. Anz. 148 (1980), 462-473

I

IBELE, JOSEF [1877-1956]
JV 21,484; NUC 263,118

ICKES, THEODOR [1907-]
SG [4]8,7; NUC 263,344

IDE, MANILLE [1866-1945]
BNB 44,645-656; IB, 133; NUC 263,425-426
Anon., Bibl. Acad. Louvain 6 (1937), 337-349
R. Bruynoghe, Bull. Acad. Med. Belg. [6]10 (1945), 176-180
P. van Gehuchten, Mem. Acad. Med. Belg. 4(5) (1964), 5-14

IERUSALEMSKI, NIKOLAI DMITRIEVICH [1901-1967]
WWR, 246; LIPSHITS 3,438-439; WWWS, 854
Anon., Mikrobiologia 40 (1971), 581-582

IGERSHEIMER, JOSEF [1879-1965]
KURSCHNER 9,851; FISCHER 1,768-769; BHDE 2,549; WIDMANN, 89,270; LDGS, 73; WININGER 7,105; KOREN, 196

IGLAUER, FRITZ [1876-]
JV 16,251; RSC 19,643; NUC 263,522

IGNATOVSKI, ALEKSANDR IOSIFOVICH [1875-1955]
FISCHER 1,679
Anon., Makedonski Meditsinski Preglad 10(9) (1955), 1

IKEDA, KIKUNAE [1864-1936]
JBE, 354-355
Y. Tsuzuki and A. Yamashita, Jap. Stud. Hist. Sci. 7 (1968), 1-26

ILIENKOV, PAVEL ANTONOVICH [1821-1877]
BSE 17,545-546; BLOKH 1,321-322; KOZLOV, 180; RSC 3,491; 8,2; POGG 3,676
K.A. Timiryazev, Zhur. Fiz. Khim. 10 (1878), 19-32

ILIIN, LEV FEDOROVICH [1871-1938]
WWR, 232; LIPSHITS 3,449
V.P. Kalashnikov, Farm. Toks. 2(2) (1939), 70-71

ILINSKI, MIKHAIL ALEKSANDROVICH [1856-1941]
BSE 17,547-548; WWR, 233; WWWS, 854
V.M. Zezyulinski, Usp. Khim. 18 (1949), 760-761

ILISCH, FRIEDRICH [1822-1867]
HEIN, 297; WELDING, 351; RSC 3,401; POGG 1,1168

ILOSVAY, LAJOS [1851-1936]
MEL 1,773; POTSCH, 218; POGG 4,681; 5,569; 6,1189; 7b,2137
A. Stock, Ber. chem. Ges. 70A (1937), 38-39
K. Kempler, Orv. Het. 127 (1986), 3202-3206

ILSE, DORA [1898-1979]
WAGENITZ, 98; LDGS, 14
K. Herter, Begegnungen mit Menschen und Tieren, pp.82-83. Berlin 1979

ILTIS, HUGO [1882-1952]
KURSCHNER 4,1328; BHDE 2,550-551

IMMERHEISER, CARL [1863-1946]
JV 6,298; RSC 16,12; 17,127; NUC 265,156
F.L. Arnold, Nordpfälzer Geschichtsverein 41 (1961), 557-559

IMRIE, CYRIL GRAY [1889-1961]
IB, 134
Anon., Lancet 1961(I), p.1120; Brit. Med. J. 1961(I), pp.1653-1654

INADA, RYOKICHI [1874-1950]
JBE, 373; FISCHER 1,682; OLPP, 189; WWWS, 855

INCE, WALTER HOLINSHED [1865-1907]
JV 4,284; RSC 16,13; NUC 265,465
Anon., J. Roy. Inst. Chem. 1907(4), p.17

ING, HARRY RAYMOND [1899-1974]
DNB [1971-1980], 441-442; WW 1973, p.1645
H.O. Schild and F.L. Rose, Biog. Mem. Fell. Roy. Soc. 22 (1976), 239-255

INGENCAMP, COSMAS [1850-]
BONN, 375

INGERSOLL, ARTHUR WILLIAM [1894-1969]
AMS 11,2503; WWWS, 856
Anon., Chem. Eng. News 47(20) (1969), 89

INGLE, DWIGHT JOYCE [1907-1978]
AMS 13,2085; WWA 1976-1977, p.1557; MH 2,260-262; STC 2,44-45; MSE 2,116-117; WWWS, 856-857
Anon., BioScience 16 (1966), 705; Persp. Biol. Med. 22 (1978), 1-2
M.B. Visscher, Biog. Mem. Nat. Acad. Sci. 61 (1992), 247-268

INGLE, HARRY [1869-1921]
SRS, 12; JV 8,215; POGG 4,682
Anon., J. Roy. Inst. Chem. 1921, p.356; Z. angew. Chem. 35 (1922), 96

INGOLD, CHRISTOPHER KELK [1893-1970]
DNB [1961-1970], 563-565; WW 1969, pp.1579-1580; MH 1,253-254; STC 2,44-46; MSE 2,117-118; CAMPBELL, 115-116; ABBOTT-C, 69-70; POTSCH, 219; POGG 5,569-570; 6,1192-1194; 7b,2147-2154
J.H. Ridd, Chem. Brit. 7 (1971), 163
C.W. Shoppee, Biog. Mem. Fell. Roy. Soc. 18 (1972), 349-411
M.D. Saltzman, J. Chem. Ed. 57 (1980), 484-488

INGRAM, VERNON MARTIN [1924-]
AMS 15(3),936; IWW 1982-1983, p.614; BHDE 2,551; WWWS, 857

INGVALDSEN, THORSTEN [1885-1930]
BERG, 111; NUC 267,466
A.T. Cameron, Can. Med. Assoc. J. 23 (1930), 116-117

INHOFFEN, HANS HERLOFF [1906-]
KURSCHNER 13,1699; AUERBACH, 830; POTSCH, 219; POGG 7a(2),591-593
Anon., Chem. Z. 105 (1981), 94

INMAN, ONDESS LAMAR [1890-1942]
AMS 6,712; IB, 134

INOKO, YOSHITO [1865?-1893]
Anon., Pharm. Z. 38 (1893), 601; Japanese Physiology Past and Present, p.60. Tokyo 1965

INOZEMTSOV, FEDOR IVANOVICH [1802-1869]
BSE 18,173; BME 11,507-508; HIRSCH 3,375; ZMEEV 1,126-128
D.M. Rossiski, Fram. Toks. 17(5) (1954), 56-59

IODIDI, SAMUEL LEO [1867-1944]
AMS 7,911; NCAB 33,536; IB, 141

IORDAN [see JORDAN, ARTUR PAVLOVICH]

IOSIFOV, GORDEI MAKSIMOVICH [1870-1933]
BSE 18,375; BME 11,921-922; WWR, 236-237
W.I. Oschkaderov, Anat. Anz. 83 (1936), 62-64
D.A. Zhdanov, Arkh. Anat. Gist. Emb. 59(12) (1970), 3-5

IPATIEFF, VLADIMIR NIKOLAEVICH [1867-1952]
DSB 7,21-22; MILES, 241-242; WWR, 237; FARBER, 1279-1298; DAVIS, 23-32; POTSCH, 219; W, 275-276; ABBOTT-C, 70; WWWS, 858; POGG 4,682-683; 5,570-571; 6,1196-1197; 7b,2172-2177
V.N. Ipatieff, The Life of a Chemist. Stanford 1946
L. Schmerling, Biog. Mem. Nat. Acad. Sci. 47 (1975), 83-140
H. Pines, CHEM TECH 11(2) (1981), 78-82

IRISAWA, TATSUKICHI [1867?-1935]
JBE, 2004; HAKUSHI 2,28; FISCHER 1,684; NUC 271,632

IRSCHICK, ALFRED [1885-]
JV 29,639; NUC 272,87

IRVIN, JOSEPH LOGAN [1913-1984]
AMS 15(3),943; WWA 1980-1981, p.672; WWWS, 858-859

IRVINE, JAMES COLQUHOUN [1877-1952]
DNB [1951-1960], 536-537; WW 1949, p.1427; SRS, 24; W, 276; WWWS, 859; POGG 5,571; 6,1198; 7b,2180-2181
J. Read, Obit. Not. Fell. Roy. Soc. 8 (1953), 459-489
E.L. Hirst, Adv. Carbohydrate Chem. 8 (1953), xi-xvii

IRVING, JAMES TUTIN [1902-]
AMS 14,2378; WW 1981, p.1329; WWWS, 859

IRVING, LAURENCE [1895-1979]
AMS 13,2092; WWWS,859
Who Was Who in America 7,295

IRWIN, MALCOLM ROBERT [1897-1987]
AMS 14,2379; BIBEL, 228-231; WWWS. 859

IRWIN, MARIAN [1889-1973]
AMS 7,889; IB, 134; POGG 6,1198-1199

ISAAC, SIMON [1881-1942]
KURSCHNER 4,1343; FISCHER 1,684-685; KALLMORGEN, 309; BHDE 2,552; WININGER 3,201; 7,106-107; TETZLAFF, 151; KOREN, 196

ISAACHSEN, HAAKON [1873-1936]
NBL 6,536-537; IB, 134-135
Hver er Hvem? 1934, pp.238-239

ISAACS, ALICK [1921-1967]
DNB [1961-1970], 569-570; STC 2,50-51; W, 276; ABBOTT, 72; WWWS, 859
C.H. Andrewes, Biog. Mem. Fell. Roy. Soc. 13 (1967), 205-222
C.H. Stuart-Harris, Nature 213 (1967), 555

ISAACS, CHARLES EDWARD [1811-1860]
DSB 7,23-24; HIRSCH 3,378; KELLY, 631-632
J.C. Hutchison, Am. Med. Mon. 18 (1862), 81-97
R.N. Beiter, Ann. Med. Hist. 1 (1929), 363-377

ISAACS, RAPHAEL [1891-1965]
AMS 10,1954; WININGER 7,108; IB, 135; WWAH, 303
C.A. Doan, Trans. Assoc. Am. Phys. 80 (1967), 11-12

ISACHENKO, BORIS LAVRENTIEVICH [1871-1948]
BSE 18,474-475; BME 11,965-966; WWR, 237-238; UKRAINE 2,276-277
V.P. Savich and A.E. Kriss, Priroda 38(11) (1949), 82-89
E.N. Mishustin, Bot. Zhur. 34 (1949), 547-551

ISAEV, VASILI ISAEVICH [1854-1911]
BSE 18,472; BME 11,961-962; SKORODOKHOV, 269-270; WWWS, 859-860
Anon., Morskoi Vrach 1912, pp.i-viii
A.A. Shmarov, Sov. Zdrav. 1979(7), pp.67-70

ISBELL, HORACE SMITH [1898-1992]
AMS 14,2381; WWWS, 860; POGG 6,1199; 7b,2181-2183

ISCOVESCO, HENRI [1859-1932]
FISCHER 1,685; NUC 272,451
Anon., Presse Med. 40 (1932), 980

ISHIHARA, SHINOBU [1879-]
JBE, 402,2381; HAKUSHI 2,244-245; FISCHER 1,686-687; WWWS, 860

ISHIKAWA, SEIICHI [1889-1973]
POGG 6,1200-1201; 7b,2185-2187

ISHIMASA, MOTARO [1899-1982]

ISHIMORI, KUNIOMI [1874-1955]
HAKUSHI 2,238-239
K. Kubota, Neuroscience Research 6 (1989), 497-518

ISHIZAKA, KAMISHIGE [1925-]
BIBEL, 112

ISLER, MAX [1888-1921]
NUC 272,583
Anon., Chem. Z. 45 (1921), 975; Z. angew. Chem. 34 (1921), 500,532

ISLER, OTTO [1910-]
KURSCHNER 14,1845-1846
P. Karrer, Int. Z. Vitaminforsch. 40 (1970), 236-248

ISSELBACHER, KURT JULIUS [1925-]
AMS 15(3),952; BHDE 2,554; WWWS, 861

ITANO, HARVEY AKIO [1920-]
AMS 15(3),953; WWA 1980-1981, p.1677

ITTNER, FRANZ von [1787-1823]
ADB 14,646-647; BAD 1,430; FERCHL, 255; POGG 1,1172-1173

IVANOV, GEORGI FEDOROVICH [1893-1956]
WWR, 241
Anon., Arkh. Anat. Gist. Emb, 33(2) (1956), 95-96; 51(11) (1966), 9-14

IVANOV, ILYA IVANOVICH [1870-1932]
DSB 7,31-33; BSE 17,277-278; KUZNETSOV(1963), 833-843; VORONTSOV, 95-96; WWR, 241
P.N. Shatkin, Trydy Inst. Ist. Est. 32(7) (1960), 268-308
G.V. Zvereva et al., Veterinaria 1970(7), pp.88-95

IVANOV, LEONID ALKSANDROVICH [1871-1962]
BSE 17,279; BME 10,1147-1148; WWR, 241-242; LIPSHITS 3,402-408;

IB, 136
I.L. Rabotnova, Mikrobiologia 22 (1953), 740-743

IVANOV, MIKHAIL FEDOROVICH [1871-1935]
BSE 17,280; KUZNETSOV(1963), 844-851; WWR, 242; IB, 136
N.K. Ivanova, M.F. Ivanov, 2nd Ed. Moscow 1953
Anon., Veterinaria 1971(11), pp.116-119

IVANOV, NIKOLAI NIKOLAEVICH [1884-1940]
BSE 17,280; WWR, 242; LIPSHITS 3,408-414; IB,136; POGG 6,1206; 7b,2209-2211
M.I. Knyaginichev, Bot. Zhur. 26 (1941), 241-248

IVANOV, PETR PAVLOVICH [1878-1942]
DSB 7,33-34; BSE 17,281; KUZNETSOV(1963), 381-390
N.G. Khlopin and A.G. Knorre, Usp. Sov. Biol. 36(3) (1953), 367-379
P.G. Svetlov, Trudy Inst. Ist. Est. 24(5) (1958), 151-176
A.G. Knorre, Arkh. Anat. Gist. Emb. 55(2) (1968), 81-92
E.V. Rozhdestvenski and E.S. Gerlovin, Arkh. Anat. Gist. Emb. 75(10) (1978), 107-114

IVANOV, SERGEI LEONIDOVICH [1880-1960]
BSE(3rd Ed.) 10,14; LIPSHITS 3,415-419; IB, 136
A.A. Prokofiev, Fiziologia Rastenii 7 (1960), 378-380
A.A. Pristupa, Bot. Zhur. 46 (1961), 744-745

IVANOV, VLADIMIR VLADIMIROVICH [1873-1931]
BSE 17,276; BME 10,1144-1145; WWR, 241; FISCHER 1,690
A. Jordan, Derm. Wchschr. 93 (1931), 1783
A.P. Dolgov and N. Smelov, Vestnik Dermatologii 31(1) (1957), 46-47; 47(8) (1973), 53-54

IVANOVSKI, DMITRI IOSIFOVICH [1864-1920]
DSB 7,34-36; BSE 17,292-293; BME 10,1149-1150; KUZNETSOV(1963), 319-329; LIPSHITS 3,427-430; WWR, 244-245
K.S. Koshtoyants, Mikrobiologia 11(4) (1942), 140-147
P.A. Genkel, Zhur. Ob. Biol. 1950(6), pp.405-415
O.I. Shchepkina, Mikrobiologia 20 (1951), 541-549
H. Lechevalier, Bact. Revs. 36 (1972), 135-145

IVY, ANDREW CONWAY [1893-1978]
AMS 12,1972; WWA 1976-1977, p.1568; NCAB E,89; BROBECK,152-153; WWWS, 863
M.I. Grossman, Physiologist 21 (1978), 11-12
D.B. Dill, Physiologist 22 (1979), 21
P.S. Ward, Bull. Hist. Med. 58 (1984), 28-52

IWASAKI, KEN [1891- 1978 JBE, 444

IWATSURU, RYUZO [1894-]
POGG 6,1207; 7b,2215-2216

IZMAILOV, NIKOLAI ARKADIEVICH [1907-1961]
WWR, 246
A.M. Shkodin et al., Ukr. Khim. Zhur. 28 (1962), 271-272

J

JABLONS, BENJAMIN [1887-1971]
KAGAN(IA), 372,572
Anon., N.Y. J. Med. 72 (1972), 174

JACCOUD, FRANÇOIS SIGISMOND [1830-1913]
DBF 18,249-250; HIRSCH 3,387; PAGEL,807; WWWS, 864
Anon., Brit. Med. J. 1913(I), p.1012
C. Perier, Bull. Acad. Med. 69 (1913), 345-348
E.C. Achard, Bull. Acad. Med. 104 (1930), 589-618

JACKSON, CHARLES LORING [1847-1935]
AMS 5,568; NCAB 11,416; WWAH, 304; WWWS, 864; POTSCH, 220-221; POGG 3,680; 4,687-688; 5,574; 6,1210; 7b,2227
G.P. Baxter et al., Science 83 (1936), 294-296
G.S. Forbes, Biog. Mem. Nat. Acad. Sci. 37 (1964), 97-128

JACKSON, CHARLES THOMAS [1805-1880]
DSB 7,44-46; DAB 9,536-538; MILES, 245-246; ELLIOTT, 136; DAMB, 383-384; POTSCH, 221; SILLIMAN, 86-88; WWWS, 864; POGG 1,1176,1575-1576; 3,680
Anon., Pop. Sci. Mon. 19 (1881), 404-407
J.B. Woodworth, American Geologist 20 (1897), 87-110
A.H. Miller, Ann. Med. Hist. 6 (1934), 110-123

JACKSON, DAVID HAMILTON [1869-1938]
SRS, 14; NUC 274,585

JACKSON, HENRY [1892-1968]
AMS 10,1965
W.B. Castle, Trans. Assoc. Am. Phys. 82 (1969), 37-39

JACKSON, HERBERT [1863-1936]
WW 1936, p.1736; WWWS, 865
R.B. Pilcher, Analyst 62 (1937), 83-86
H. Moore, Obit. Not. Fell. Roy. Soc. 2 (1938), 307-314

JACKSON, HOLMES CONDIT [1875-1927]
AMS 4,496; IB, 137
Obituary Record of Yale Graduates 1927-1928, pp.237-238. New Haven 1928

JACKSON, JOHN HUGHLINGS [1835-1911]
DSB 7,46-50; DNB [1901-1911]2,356-358; O'CONNOR, 60-62; KOLLE 1,135-144; HIRSCH 3,390; Suppl.,379; HAYMAKER, 308-311
W. Broadbent, Brain 26 (1903), 305-366
Anon., Brit. Med. J. 1911(II), pp.950-954,1551-1554

JACKSON, LOUIS LINCOLN [1861-1935]
RSC 13,209,234
Harvard Alumni Directory, p.580. Cambridge, Mass. 1937

JACKSON, OSCAR ROLAND [1855-1916]
RSC 9,97
Harvard College Class of 1876, 9th Report,pp.66-68. Cambridge, Mass. 1916

JACKSON, RICHARD FAY [1881-1943]
POGG 6,1210; 7b,2230-2231
C.A. Browne, J. Assn. Agr. Chem. 26(4) (1943), ii-iv
C.F. Snyder, J. Wash. Acad. Sci. 33 (1943), 287-288

JACKSON, RICHARD WILLET [1901-1990]
AMS 11,2530; WWWS, 865
Anon., Chem. Eng. News 69(4) (1991), 74

JACOB, FRANÇOIS [1920-]
QQ 1981-1982, p.624; WW 1981,p.1339; MH 1,254-255; STC 2,53-56; MSE 2,121-122; CB 1966, pp.191-193; ABBOTT,72-73
F. Jacob, La Statue Intérieure. Paris 1987

JACOBI, ABRAHAM [1830-1919]
DAB 9,563-564; NCAB 9,345-346; HIRSCH 3,393-394; PAGEL, 808-809; DAMB, 389-390; WININGER 3,228; 7,115
V. Robinson, Medical Life 35 (1928), 212-306

JACOBI, CONSTANTIN [1877-1959]
JV 17,171; NUC 275,229
Anon., Chem. Z. 61 (1937), 479; Nachr. Chem. Techn. 7 (1959), 269

JACOBI, FRIEDRICH WILHELM [1863-]
RSC 16,44; NUC 275,242

JACOBI, HERMANN [-1901]
JV 6,297; RSC 16,44; NUC 275,246
Anon., Z. angew. Chem. 14 (1901), 459-461

JACOBI, RICHARD [1888-]
JV 41,471; NUC 275,263

JACOBJ, CARL [1857-1944]
NDB 10,239-240; KURSCHNER 4,1289; FISCHER 1,693; JV 2,207; TLK 3,294; IB, 137

JACOBS, JOHN LESH [1904-]
AMS 11,2534

JACOBS, MERKEL HENRY [1884-1970]
AMS 11,2535; IB, 137; WWWS, 867
J.R. Brobeck, Am. Phil. Soc. Year Book 1970, pp.137-140
W.E. Love, Biol. Bull. 141 (1971), 15-17

JACOBS, WALTER ABRAHAM [1883-1967]
DSB 15,252-253; AMS 11,2536; MILES, 246-247; KAGAN(JA), 737; WWAH, 305; POTSCH, 221; IB, 137; POGG 6,1210-1212; 7b,2231-2233
L.C. Craig, Rock. Univ. Rev. Nov.-Dec.1967, pp.23-25; Pharmacologist 10 (1968), 20-21
R.C. Elderfield, Biog. Mem. Nat. Acad. Sci. 51 (1980), 247-278

JACOBSEN, HENDRIK CHRISTIAAN [1882-1952]
NUC 275,376
A. Knetman, Chem. Wkbl. 48 (1952), 333-334

JACOBSEN, JACOB CHRISTIAN [1811-1887]
C. Nyrop, J.C. Jacobsen. Copenhagen 1911
J. Pedersen, The Carlsberg Foundation. Copenhagen 1956
D. Wrotnowska, Hist. Sci. Med. 1970(3/4), pp.131-142

JACOBSEN, OSCAR GEORG [1840-1889]
HEIN-E, 214-215; BLOKH 1,323-324; KIEL, 210; TSCHIRCH,1069; SCHAEDLER,56; POTSCH, 221; POGG 3,681-682; 4,688-689
C. Liebermann, Ber. Chem. Ges. 22 (1889), 2387-2388

JACOBSOHN, KURT [1904-]
LDGS, 21; POGG 6,1212; 7b,2234-2236

JACOBSON, BERNARD MAX [1904-1989]
AMS 11,2537
M.N. Swartz, Harvard Med. Alumni Bull. 63(2) (1989), 62-63

JACOBSON, EDMUND [1888-1983]
AMS 14,2402; WWWS, 868

JACOBSON, JOHN (FRITZ EMIL) [1859-]
JV 6,9; RSC 16,47; SG [2]8,390; NUC 475,426

JACOBSON, PAUL [1859-1923]
NDB 10,247; DBJ 5,195-200,431-432; KROLLMANN, 964; TLK 2,317; BERLIN,705; DRULL, 122-123; BLOKH 1,324-325; WININGER 3,241; TETZLAFF, 154; POTSCH, 221-222; WWWS, 869; POGG 3,682; 4,689; 5,575-576; 6,1213
B. Rassow, Z. angew. Chem. 36 (1923), 100
B. Prager, Ber. chem. Ges. 57A (1924), 57-81

JACOBSON, WERNER [1906-]
LDGS, 54
Directory of British Scientists 1966-1967 (1), p.909

JACOBSTHAL, ERWIN [1879-]
KURSCHNER 4,1291; FISCHER 1,694; LDGS, 57; KOREN, 197; IB, 137

JACOBY, MARTIN [1872-1941]
KURSCHNER 4,1292-1293; FISCHER 1,694; FRANKENTHAL, 291-292; DRULL,123; WININGER 7,116; AD 3,60; TLK 3,294; IB,137; POGG 6,1214; 7a(2),597
Anon., Science 94 (1941), 184

JACQUELIN, VICTOR AUGUSTE [1804-1885]
POGG 1,1183; 3,682

JACQUEMIN, EUGÈNE THÉODORE [1828-1909]
DBA 3,1776-1777; HUMBERT,206-207; TSCHIRCH, 1069; BERGER-LEVRAULT,118; DORVEAUX, 78,79; POGG 3,682
T. Klobb, Bull. Sci. Pharm. 17 (1910),39-41

JACQUIN, JOSEPH FRANZ von [1766-1839]
NDB 10,257; ADB 13,631; WURZBACH 10,23-26; COLE, 272-274; POGG 1,1185-1186

JADASSOHN, JOSEF [1863-1936]
NDB 10,259-260; FISCHER 1,695; BHDE 2,561; KOREN, 197; TETZLAFF, 155; WININGER 3,246; 7,116
J. Darier, Presse Med. 44 (1936), 774
W. Lutz, Schw. med. Jahrb. 1937, pp.xvii-xliii
M.B. Sulzberger, American Journal of Dermopathology 7 (1985), 31-36

JÄDERHOLM, AXEL [1837-1885]
HIRSCH 3,403
E.R. Petersson, Hygiea 47 (1885), 612-615

JAEGER, CARL [1850-1928]
AARGAU, 396; ZURICH-D, 127; RSC 10,319
A. Hartmann, Mitteilungen der Aargauischen Naturforschenden Gesellschaft 18 (1928), xxix-xxxi

JAEGER, CARL [1872-]
JV 18,323; NUC 275,648

JAEGER, EMIL [1842-1922]
NUC 275,654
Anon., Z. angew. Chem. 35 (1922), 104

JAEGER, FRANS MAURITS [1877-1945]
DSB 7,59; BWN 2,256-257; POGG 5,577-579; 6,1215-1216; 7b,2238-2240
W.P. Jorissen et al., Chem. Wkbl. 31 (1934), 182-212; 43 (1947), 67-71
C.L. Hogardi, Tijd. Gesch. Geneesk. 7 (1984), 183-195

JAEGER, GUSTAV [1832-1916]
NDB 10,269; DBJ 2,659-660; KLEIN, 78-79; TSCHIRCH,1070; BK, 228; WWWS,870
M. Jaeger, G. Jaegers Lebens- und Entwicklungsgang. Stuttgart 1932
S. Schenkling, Entomol. Bl. 29 (1933), 28-30
E. Kaufmann, Gustav Jaeger, Arzt, Zoologe und Hygieniker. Zürich 1984

JAEGER, GUSTAV [1865-1938]
NDB 10,275-276; OBL 3,55-56; STURM 2,9; POGG 4,691-692; 5,579; 6,1216; 7a(2),600
S. Meyer, Alm. Akad. Wiss. Wien 88 (1938), 234-237
F.A.P., Nature 141 (1938), 402

JÄGER, ROLF [1905-1969]
NDB 10,279-280; KURSCHNER 10,1048-1049; WWWS, 870; POGG 7a(2),602-603

JAEGLÉ, GEORG [1872-1931]
JV 9,209; RSC 13,518

JÄNECKE, ERNST [1875-1957]
KURSCHNER 8,1041-1042; DRULL, 123-124; POGG 5,580-581; 6,1217; 7a(2),603-604

Anon., Nachr. Chem. Techn. 3 (1955), 50

JAENICKE, LOTHAR [1923-]
KURSCHNER 13,1721; AUERBACH, 831; POGG 7a(2),604

JAFFÉ, BENNO [1840-1923]
BLOKH 1,325-327; BERLIN, 675; RSC 16,59
W. Marckwald, Ber. chem. Ges. 56A (1923), 84-85
Anon., Chem. Z. 47 (1923), 791

JAFFÉ, MAX [1841-1911]
NDB 10,291-292; BJN 16,36*; SCHOLZ, 41-43; FISCHER 1,696-697; PAGEL, 814; KROLLMANN, 966; FRANKEN, 186; DZ, 668; WWWS, 870; KOREN, 197; WININGER 3,249; 7,117;POGG 4,692
B. Naunyn, Arch. exp. Path. Pharm. 66 (1911), 1-3
R. Cohn, Münch. med. Wchschr. 59 (1912), 92-93
O. Cohnheim, Z. physiol. Chem. 77 (1912), i-ii
A. Ellinger, Ber. chem. Ges. 46 (1913), 831-847
K. Jung, Z. Urol. 80 (1987), 65-68

JAFFÉ, RUDOLF [1885-1975]
KURSCHNER 12,1409; FISCHER 1,697; BHDE 2,562; LDGS, 75; IB, 138; WWWS,870
K. Brass, Verhandl. path. Ges. 59 (1975), 634-640

JAGENDORF, ANDRE TRIDON [1926-]
AMS 15(4),25; WWWS, 870

JAGER, LAMBERTUS de [1863-1944]
LINDEBOOM, 968-969; RSC 16,59

JAGIČ, NIKOLAUS [1875-1956]
FISCHER 1,697; WIEN 1958, p.57
R. Klima, Wiener med. Wchschr. 107 (1957), 89-90; Münch. Med. Wchschr. 99 (1957), 457-458
K. Fellinger, Wiener klin. Wchschr. 69 (1957), 145-147

JAHN, DETMAR DIETRICH WILHELM [1900-1969]
KURSCHNER 11,1299; PELZNER, 138-144; KOERTING, 172
E. Volhard, Med. Welt 14 (1970), 640

JAHN, HANS MAX [1853-1906]
DSB 7,65; BJN 11,34*; KERNBAUER, 117-122; BLOKH 1,327-328; WREDE, 88-89; POGG 4,693; 5,582
W. Nernst, Chem. Z. 30 (1906), 893
H. Landolt, Ber. chem. Ges. 39 (1906), 4453-4470; Z. Elektrochem. 13 (1907), 89-90
H.G. Bartels, Wiss. Z. Berlin 39 (1990), 230-237

JAHN, STEPHAN [1876-1911]
JV 17,284; NUC 276,172; POGG 5,582
Anon., Chem. Z. 35 (1911), 1107; Ber. chem. Ges. 45 (1912), 695

JAHN, THEODORE LOUIS [1905-1979]
AMS 13,2119
E.C. Bovee, J. Protozool. 26 (1979), 527-529

JAHNS, ERNST FRIEDRICH [1844-1897]
HEIN, 302-303; BLOKH 1,328; WAGENITZ, 89; POTSCH, 222; POGG 7aSuppl.,303
E. Fischer, Ber. chem. Ges. 30 (1897), 907-908
H. Dietmann, Süddeutsche Apoth. Z. 90 (1950), 557-564

JAHODA, RUDOLF [1862-]
TSCHIRCH, 1070; POGG 4,694

JAKOBY, WILLIAM BERNARD [1928-]
AMS 15(4),28; WWWS, 871

JAKOWSKI, MARIAN [1857-1921]
PSB 10,337-338; KONOPKA 4,47-52
A. Pulawski, Gazeta Lekarska 55 (1921), 37-39

JAKSCH, ANTON von [1810-1887]
NDB 10,324; ADB 50,627; OBL 3,65-66; STURM 2,17; HIRSCH 3,411; WERSTLER 1,104-107; KOERTING, 164-165
Anon., Prag. med. Wchschr. 12 (1887), 303

JAKSCH, RUDOLF von [1855-1947]
NDB 10,325-326; STURM 2,18; FISCHER 1,698-699; PAGEL, 815-817; KOERTING, 169-170; SCHIEBER, 92-115; WEINMANN, 243-244
J. Lowy, Wiener med. Wchschr. 75 (1925), 1629-1635
A. Pellegrini, Münch. med. Wchschr. 87 (1940), 829
Anon., Lancet 1947(I), p.274

JAMES, ANTHONY TRAFFORD [1922-]
ETTRE, 167-172

JAMES, MICHAEL NORMAN GEORGE [1940-]
WW 1992, p.967

JAMES, WILLIAM OWEN [1900-1978]
WW 1977, p.1250; POTSCH, 222-223
J.L. Harley, Biog. Mem. Fell. Roy. Soc. 25 (1979), 337-364

JANDER, GERHARDT [1892-1961]
NDB 10,331-332; TLK 3,297; POTSCH, 223; POGG 6,1222-1223; 7a(2),608-609
R.E. Oesper, J. Chem. Ed. 30 (1953), 348-349
H. Spandau, Z. anorg. Chem. 319 (1962), 113-119

JANDER, WILHELM [1898-1942]
TLK 3,297-298; POTSCH, 223; POGG 6,1223; 7a(2),610
W. Noddack, Z. angew. Chem. 56 (1943), 53
R. Fricke, Ber. chem. Ges. 77A (1944), 15-20

JANEWAY, CHARLES ALDERSON [1909-1981]
AMS 11,1982; WWA 1978-1979, p.1641
W. Berenberg et al., Harvard Gazette 4 March 1983, p.7
D.G. Nathan, Trans. Assoc. Am. Phys. 96 (1983), cii-ciii

JANEWAY, THEODORE CALDWELL [1872-1917]
DAB 9,608; NCAB 17,214-215; ICC, 30-32; WWAH, 307; WWWS, 873
L.F. Barker, Science 47 (1918), 273-278
F.J. Goodnow and B.P. Clark, Bull. Johns Hopkins Hosp. 29 (1918), 142-148
N. Flaxman, Bull. Hist. Med. 9 (1941), 505-516

JANI, CURT [-1882]
RSC 16,72

JANKE, ALEXANDER [1887-1974]
KURSCHNER 11,1307; IB, 138; POGG 7a(2),612-613
Anon., Naturw. Rund. 28 (1975), 70

JANNASCH, PAUL ERHARD [1841-1921]
DBJ 2,722; BLOKH 1,332; DRULL, 125; POTSCH, 223-224; POGG 3,684-685; 4,696-697; 5,584
W. Strecker, Ber. chem. Ges. 55A (1922), 194-210

JANOSI, FERENC [1819-1879]
MEL 1,800
Magyar Irok 5 (1897), 398-401

JANOVSKY, JAROSLAV [1850-1907]
OBL 3,75-76; POGG 4,698-699; 5,584
Anon., Chem. Z. 31 (1907), 681

JANSEN, BAREND COENRAAD PETRUS [1884-1962]
LINDEBOOM, 969-971; POTSCH, 224
H.G.K. Westenbrink et al., Chem. Wkbl. 34 (1937), 471-484
C. von Hartog, Chem. Wkbl. 58 (1962), 610-611
J. van Eys, J. Nutrition 100 (1970), 485-490

JANSEN, JOHANNES BAPTIST [1875-1954]
JV 15,132; NUC 277,368

JANSKY, JAN [1873-1921]
OBL 3,77; NAVRATIL, 119; WWWS, 873
M. Matousek, Cas. Lek. Cesk. 93 (1954), 806-807

JANSSEN, LOUIS WILLEM [1901-]
NUC 277,432

JANSSENS, FRANCISCUS (FRANS) ALPHONSIUS [1863-1924]
BNB 38,344-353; NBW 2,384-385; BK, 230; SCHULZ-SCHAEFFER, 205
Anon., Bibl. Acad. Louvain 6 (1937), 239-240

JAPP, FRANCIS ROBERT [1848-1925]
WWWS, 874; POGG 3,686-687; 4,700-701; 5,584-585; 6,1226
A. F[indlay], J. Chem. Soc. 129 (1926), 1008-1020
J.F.T[horpe], Proc. Roy. Soc. A118 (1928), iii-vi

JAQUET, ALFRED [1865-1937]
NDB 10,352-353; FISCHER 1,702-703; ZWEIFEL, 69-72; TLK 3,299; IB, 139
A. Gigon, Schw. med. Wchschr. 62 (1932), 513-518
J. Karcher, Schw. med. Wchschr. 67 (1937), 479-480
H. Staub, Schw. med. Jahrb. 1938, pp.xix-xxix
Anon., Biographisches Lexikon Verstorbener Schweizer 2 (1948), 102
K. Bucher, Arzneimitt. 15 (1965), 581

JAQUET, DANIEL [1889-1927]
NUC 278,115

JARISCH, ADOLF [1850-1902]
OBL 3,80-81; BJN 7,53*; FISCHER 1,703; HUTER, 371-372
H. Kobner, Münch. med. Wchschr. 49 (1902), 709
Anon., Leopoldina 38 (1902), 56-57
J. Rille, Deutsche med. Wchschr. 28 (1902), 267-268

JARISCH, ADOLF [1891-1965]
NDB 10,355-356; HUTER, 295-297
H. Konzett, Arzneimitt. 11 (1961), 133
F. Brücke, Deutsche med. Wchschr. 90 (1965); Wiener klin. Wchschr. 77 (1965), 939
C. Henze, Arneimitt, 15 (1965), 1477-1478
F. Scheminzky, Ber. Nat. Med. Innsb. 54 (1966), 173-176

JAVILLIER, JEAN MAURICE [1875-1955]
DBF 18, 552-553; GORIS, 448-449; CHARLE(2), 161-164; WWWS, 876; POGG 6,1229; 7b,2254-2257
M. Lemoigne, C.R. Acad. Sci. 240 (1955), 2461-2465
R. Delaby, Bull. Acad. Med. 139 (1955), 477-481
J. Polonovsky, Ann. Inst. Pasteur 91 (1956), 260-262
J. Lavollay et al., Maurice Javillier. Lons-le-Saunier 1956

JAWORSKI, WALERY [1849-1924]
PSB 11,113-115; FISCHER 1,705; KONOPKA 4,136-147; KOSMINSKI, 197,604-605
M. Paciorkiewicz, Arch. Hist. Med. 37 (1974), 201-217

JAY, RUDOLF [1865-1928]
JV 4,65; RSC 16,90; NUC 278,431
Anon., Z. angew. Chem. 32(III) (1919), 273; 41 (1928), 850; Chem. Z. 52 (1928), 593

JAYLE, MAX FERNAND [1913-1978]
DBF 18,563; QQ 1977-1978, p.909
J. Roche, C.R. Soc. Biol. 172 (1979), 1055-1056
M. Drosdowsky, Ann. Endocrin. 40 (1979), 368-370

JAYNE, HARRY WALKER [1857-1910]
NCAB 19,398
Anon., Science 31 (1910), 576

JEANES, ALLENE ROSALIND [1906-]
AMS 12,3018; WWWS, 876
P.A. Sandford, Carbohydrate Research 66 (1978), 3-5

JEANLOZ, ROGER WILLIAM [1917-]
AMS 15(4),48; WWWS, 876
M.C. Glick et al., Carbohydrate Research 151 (1978), 1-5

JEANMAIRE, PAUL [1851-1928]
A. Romann, Bull. Soc. Ind. Mulhouse 94 (1928), 381-384

JEANNEL, JULIEN FRANÇOIS [1814-1896]
DBF 18,618-619; BALLAND, 46-51; BALLAND(2), 72-73; TSCHIRCH, 1070; FERCHL,258; POGG 3,687
Anon., J. Pharm. Chim. [6]3 (1896), 431-432

JEGER, OSKAR [1917-]
KURSCHNER 13,1746; POGG 7a(2),628-629

JEHLE, HERBERT [1907-1983]
AMS 15(4),51; KURSCHNER 14,1887; BHDE 2,566-567; WWWS, 877; POGG 7a(2),629-630
W. Drechsler and H. Rechenberg, Phys. Bl. 39(3) (1983), 71

JELINEK, EDMUND [1852-1928]
OBL 3,98; FISCHER 1,708

JELLETT, JOHN HEWITT [1817-1888]
DNB 10,729; POGG 3,689; 4,702-703
Anon., Nature 37 (1888), 396-397

JELLIFFE, SMITH ELY [1866-1945]
DAB [Suppl.3],384-386; AMS 7,904; NCAB 33,360-361; DAMB, 394; WWAH, 308; WWWS, 878
A.A. Brill et al., J. Nerv. Ment. Dis. 106 (1947), 221-253

JELLINEK, KARL [1882-1971]
NDB 10,392-393; BHDE 2,567-568; TLK 3,300; WININGER 3,296; WWWS, 878; POGG 5,587; 6,1233-1234; 7a(2),630-631
W. Klemm, Z. Elektrochem. 66 (1962), 609-610
Anon., Chem. Z. 95 (1971), 708

JENCKS, WILLIAM PLATT [1927-]
AMS 15(4),53; WWA 1980-1981, p.1706; IWW 1982-1983, p.636; WWWS, 878;MSE 2,126-127

JENDRASSIK, LORAND [1896-1970]
MEL 3,344; IB, 140; POGG 6,1234; 7b,2268-2269

JENISCH, KARL ALBERT [1868-1936]
JV 7,214; RSC 17,756-757

JENKINS, GEORGE NEIL [1914-]
WWWS, 878-879
Who's Who of British Scientists 1971-1972, p.454

JENKINS, GLENN LLEWELLYN [1898-1979]
AMS 13,2138; WWA 1966-1967, p.1069; NCAB I,578-579; WWWS, 879
Who Was Who in America 7,300

JENNER, WILLIAM [1815-1898]
DNB 22,909-910; BETTANY 1,169-201; WWWS, 879
Anon., Proc. Roy. Soc. 75 (1905), 28-29

JENNESS, ROBERT [1917-]
AMS 15(4),57; WWA 1980-1981, p.1708; WWWS, 879

JENNINGS, HERBERT SPENCER [1868-1947]
DSB 7,98-100; DAB [Suppl.4],424-428; AMS 7,907; NCAB A,278-279; 47,92-93; B, 140; WWAH, 309; WWWS, 879
E.G. Conklin, Am. Phil. Soc. Year Book 1947, pp.262-264
T.M. Sonneborn, Genetics 33 (1948), 1-4; Biog. Mem. Nat. Acad. Sci. 47 (1975), 143-223
O.C. Glaser, Biol. Bull. 95 (1948), 24-27
P.J. Pauly, Journal of the History of the Behavioral Sciences17 (1981), 504-515

JENNINGS, WALTER LOUIS [1866-1944]
AMS 7,907; RSC 14,1010
Anon., Chem. Eng. News 22 (1944), 1737; Science 100 (1944), 242

JENNY, ALEXANDER [1871-1942]
JV 17,284
Anon., Chemische Technik 15 (1942), 272; Chem. Z. 66 (1942), 572

JENSEN, ELWOOD VERNON [1920-]
AMS 15(4),61; WWA 1980-1981, pp.1709-1710; WWWS, 880

JENSEN, HANS FRIEDRICH [1896-1959]
AMS 9(I),966; POGG 6,1235; 7b,2271-2273
Anon., Chem. Eng. News 37(42) (1959), 126
E.A. Evans, Jr., Science 3 (1960), 1084-1085
B.R. Johnston, Thromb. Diath. Haemorrh. 4 (1960), 289-291

JENSEN, PAUL [1868-1952]
KURSCHNER 7,921; FISCHER 1,709; IB, 140; POGG 7a(2),634

JERCHEL, DIETRICH [1913-1980]
KURSCHNER 13,1752; POGG 7a(2),636
Anon., Chem. Z. 104 (1980), 281

JEREMIAS, BRUNO [1894-]
JV 41,428; NUC 279,599

JERNE, NILS KAJ [1911-]
KBB 1981, p.542; BIBEL, 188-193
N.K. Jerne, Ann. Rev. Microbiol. 14 (1960), 341-358
J. Lindenmann, Chimia 34 (1980), 90-91

JEROME, WILLAIM JOHN SMITH [1839-1929]
O'CONNOR(2), 94
Who Was Who 1929-1940, p.712

JESIONEK, ALBERT [1870-1935]
NDB 10,420; KURSCHNER 4,1323; GUNDEL, 467-474; FISCHER 1,710-711; KAMP, 38-43

JESS, ADOLF [1883-1977]
KURSCHNER 12,1435; FISCHER 1,711; NUC 280,82
F. Wagner, Klin. Mon. Augenheil. 172 (1978), 402-404

JESSEL, HENRY ROSE [1867-1933]
AMS 5,579; JV 15,11; NUC 280,103

JESSNER, MAX [1887-]
NDB 10,427*; KURSCHNER 4,1325; FISCHER 1,711; LDGS, 62; KOREN, 198
W. Gottwalt, Med. Welt 1968, p.410

JIRGENSONS, BRUNO [1904-1985]
AMS 14,2443; WWWS, 882; POGG 6,1238-1239; 7a(2),638-639

JOACHIMOGLU, GEORG [1887-1979]
KURSCHNER 4,1331; FISCHER 1,712-713; TSCHIRCH, 1070; TLK 3,304; IB, 141
A.S. Dontas, Pharmacologist 22 (1980), 126

JOB, ANDRÉ [1870-1928]
CHARLE(2), 165-167; POTSCH, 224; POGG 4,705-706; 5,588; 6,1239
G. Urbain, Bull. Soc. Chim. [4]45 (1929), 185-194

JOB, PAUL [1886-1957]
CHARLE, 167-167; POGG 6,1239-1240; 7b,2282-2283
J. Pérès, Ann. Univ. Paris 28 (1958), 178-181
G. Champertier and J. Brigado, J. Chim. Phys. 56 (1959), 419-427

JOBLING, JAMES WESLEY [1876-1961]
AMS 10,2005; FISCHER 1,713; IB, 141; WWAH, 310
D.W. Richards, Trans. Assoc. Am. Phys. 75 (1962), 24

JOBST, CARL [1816-1896]
NDB 10,445-446; FERCHL, 259; RSC 3,551; POGG 1,1197
A. Wankmüller, Beitr. Wurtt. Apothekgesch. 7 (1966), 83

JOCHEM, EMIL [1873-1943]
JV 17,367; RSC 16,111; NUC 281,111

JOCHHEIM, (FRIEDRICH RICHARD HERMANN) ERNST [1870-1932]
JV 12,170; NUC 281,171
Anon., Chem. Z. 56 (1932), 317

JODLBAUER, ALBERT [1871-1945]
NDB 10,451; TLK 3,304; IB, 141; POGG 6,1240; 7a(2),640

JOEDICKE, FRIEDRICH [1864-]
JV 2,60; RSC 16,344; NUC 281,215

JØRGENSEN, ALFRED [1848-1925]
DBL 12,203-205; BK, 234
L.K. R[osenvinge], Bot. Tid. 39 (1926), 305-307

JØRGENSEN, SOPHUS MADS [1837-1914]
DSB 7,179-180; DBL 12,253-256; VEIBEL 1,210-216; 2,234-240; 3,46-68; POTSCH, 225; BLOKH 1,336; POGG 3,692; 4,706; 5,589
A. Werner, Chem. Z. 38 (1914), 557-559
G.B. Kauffman, J. Chem. Ed. 36 (1959), 521-527; Chymia 6 (1960), 180-204

JÖTTEN, KARL WILHELM [1886-1958]
FISCHER 1,714; POGG 7a(2),640-643
H. Reploh, Arch. Hyg. 140 (1956), 167-173; 142 (1958), 339-341

JOHANNESSOHN, FRITZ [1888-1948]
POGG 7a(2),643-644
W. Heubner, Arch. exp. Path. Pharm. 212 (1950), 3

JOHANNSEN, WILHELM LUDVIG [1857-1927]
DSB 7,113-115; MEISEN, 177-180; FISCHER 1,715; VEIBEL 2,224-225; IB, 141; SCHULZ-SCHAEFFER, 204-205; WWWS, 883
F. von Wettstein, Naturwiss. 16 (1928), 350-352
O. Winge, J. Heredity 49 (1958), 82-88
F.B. Churchill, J. Hist. Biol. 7 (1974), 5-30
J.H. Wanscher, Centaurus 19 (1975), 125-147
N. Roll-Hansen, Centaurus 22 (1978), 201-235
G.A.M. van Balen, Stud. Hist. Phil. Sci. 17 (1986), 175-204

JOHANNSON, JOHAN ERIK [1862-1938]
SMK 4,94-95; FISCHER 1,715; SKVARC, 109-110; IB, 141
C.G. Santesson, Hygeia 100 (1938), 225-235
R. Wagner, Bayer. Akad. Wiss. Jahrbuch 1949, pp.136-139

JOHLIN, JACOB MARTIN [1884-1954]
AMS 8,1265; JV 26,29; NUC 281,425; POGG 6,1241; 7b,2284-2285
Anon., Science 120 (1954), 338

JOHN, HANNS [1891-1942]
HEIN-E, 218-219; POGG 6,1241-1242; 7a(2),645
T. Sabalitschka, Ber. chem. Ges. 75A (1942), 143-144

JOHN, JOHANN FRIEDRICH [1782-1847]
ADB 14,489; PHILIPPE, 754-755; TSCHIRCH, 1070; FERCHL, 259-260; POGG 1,1198; CALLISEN 9,476-482; 29,164-165
[W.D. Koner], Verzeichniss im Jahre 1845 in Berlin Lebender Schriftsteller, pp.170-173. Berlin 1846

JOHN, WALTER [1910-1942]
POGG 7a(2),645
K. Dimroth, Ber. chem. Ges. 76A (1943), 21-27

JOHNS, CARL OSCAR [1870-1942]
AMS 5,581; NCAB 30,67-68; WWAH, 311

JOHNSON, ALAN WOODWORTH [1917-1982]
WW 1982, p.1166; CAMPBELL, 123-124
Anon., Chem. Brit. 1 (1965), 397-398; 19 (1983), 25
E.R.H. Jones, Chem. Brit. 19 (1983), 930,932
E.R.H. Jones and R. Bonnett, Biog. Mem. Fell. Roy. Soc. 30 (1984),319-384

JOHNSON, DUNCAN STARR [1867-1937]
AMS 5,582; NCAB A,237-238; 32,367; HUMPHREY, 129-131; IB, 142; WWAH, 312; WWWS, 885
W.C. Coker, Science 86 (1937), 510-512

JOHNSON, FRANK HARRIS [1908-1990]
AMS 14,2456; WWA 1976-1977, p.1608; WWWS, 885
New York Times, 26 September 1990

JOHNSON, GEORGE [1818-1896]
DNB 22,916-917; HIRSCH 3,444; WWWS, 886
W.D.H., Proc. Roy. Soc. 60 (1897), xvi-xx

JOHNSON, JAMES McINTOSH [1883-1953]
AMS 8,1272; WWWS, 886
Anon., Chem. Eng. News 31 (1953), 1120

JOHNSON, JOHN RAVEN [1900-1983]
AMS 14,2462; WWWS, 886; POGG 6,1243; 7b,2285-2287
Anon., Chem. Eng. News 61(31) (1983), 32

JOHNSON, MARVIN JOYCE [1906-1982]
AMS 13,2168
W.D. Maxon, Ann. Rep. Ferm. Proc. 3 (1979), xiii-xiv

JOHNSON, OTIS COE [1839-1912]
NCAB 19,94; WWAH, 313

JOHNSON, SAMUEL WILLIAM [1830-1909]
DAB 10,120-121; MILES, 249-250; POGG 3,693-694; 4,707; 5,590
H.L. Wells, Am. J. Sci. [4]28 (1909), 405-407
T.B. Osborne, Science 30 (1909), 385-389; Biog. Mem. Nat. Acad. Sci. 7 (1911), 205-222
H.P. Armsby, Science 39 (1914), 509-511
H.B. Vickery, Yale J. Biol. Med. 13 (1941), 563-569

JOHNSON, THEODORE [1884-]
JV 22,24; NUC 282,480

JOHNSON, TREAT BALDWIN [1875-1947]
DAB [Suppl.4],434-435; AMS 7,919; NCAB F,231-232; 35,478-479; MILES, 251-252; CHITTENDEN, 124-128; WWWS, 887-888; POGG 5,590-591; 6,478-479; 7b,2291-2292
H.B. Vickery, Biog. Mem. Nat. Acad. Sci. 27 (1952), 83-119

JOHNSON, WILLIAM ARTHUR [1913-1993]
Who's Who of British Scientists 1971-1972, p.461

JOHNSON, WILLIAM SUMMER [1913-]
AMS 17(4),111; WWA 1988-1989, p.1587; WWWS, 888
W.S. Johnson, A Fifty Year Love Affair with Organic Chemistry. Washington, D.C. 1991; Tetrahedron 47(41) (1991), xi-l

JOHNSTON, JAMES FINLAY WEIR [1796-1855]
DNB 10,953-954; DESMOND, 348; TSCHIRCH, 1070; FERCHL, 260; COLE, 275-276; POGG 1,1198-1199
Anon., J. Chem. Soc. 9 (1857), 157-159
H.A.M. Snelders, Ann. Sci. 38 (1981), 571-584

JOHNSTON, JOHN [1806-1879]
ELLIOTT, 140; RSC 3,564
Anon., Am. J. Sci. [3]17 (1879), 82
W.D. Williams, Bull. Hist. Chem. 6 (1990), 23-26

JOKLIK, WOLFGANG KARL [1926-]
AMS 15(4),114; WWA 1980-1981, p.1735; WWWS, 889

JOLIN, SEVERIN [1852-1919]
SMK 4,107; POGG 3,895; 4,708

JOLIOT, FRÉDÉRIC [1900-1958]
DSB 7,151-157; DBF 18,721-722; CB 1946, pp.294-296; STC 2,67-68; ABBOTT-C, 70-71; CHARLE, 105-109; POTSCH, 224; W, 279-280; WWWS, 889; POGG 6,1246; 7b,2294-2299
M. Hassinsky, J. Chim. Phys. 56 (1959), 617-621
P.M.S. Blackett, Biog. Mem. Fell. Roy. Soc. 6 (1960), 87-105
M. Goldsmith, Frédéric Joliot-Curie. London 1976
E.H.S. Burhop, Nature 262 (1976), 727-728
R. Pflaum, Grand Obsession. New York 1989

JOLIOT-CURIE, IRÈNE [1897-1956]
DSB 7,157-159; DBF 18,722-723; CB 1940, pp.435-436; STC 2,69-71; W, 280-281; WWWS, 889; POTSCH, 224-225; POGG 6,500-501; 7b,2299-2302
J. Teillac, Nuclear Physics 4 (1957), 497-502

JOLLES, ADOLF [1863-1942]
OBL 3,128; FISCHER 1,716-717; TLK 3,305; IB, 142; KOREN, 198; POGG 4,709-710; 5,592; 6,1246-1247; 7a(2),645

JOLLOS, VICTOR [1887-1941]
KURSCHNER 4,1335; FISCHER 1,717; LDGS, 14; IB, 142; WININGER 7,130

JOLLY, JUSTIN [1870-1953]
DBF 18,734; CHARLE, 109-111; IB, 142; WWWS, 890
R. Leriche, C.R. Soc. Biol. 147 (1953), 180-185
R. Courrier, Not. Acad. Sci. 5 (1972), 499-537

JOLLY, WILLIAM TASKER ADAM [1878-1939]
O'CONNOR(2), 516; FISCHER 1,718; FLUELER, 75; IB, 142; WWWS, 890
Anon., Nature 144 (1939), 143; Brit. Med. J. 1939(II), p.373
P.C. Belonje, South African Medical Journal 80 (1991), 156-158

JOLY, NICOLAS [1812-1885]
DBF 18,749-750; WWWS, 890
Anon., Rev. Med. Toulouse 19 (1885), 625-628
E. Alix, Mem. Acad. Toulouse [9]3 (1891), 491-524

JONAS, AUGUST [1866-1927]
JV 5,75; RSC 16,132; NUC 283,369
Anon., Chem. Z. 51 (1927), 712; Z. angew. Chem. 40 (1927),1022,1070

JONAS, AUGUST FREDERICK, Jr. [1909-1957]
Anon., J. Am. Med. Assn. 166 (1958), 171

JONAS, KARL [1886-]
KURSCHNER 4,1336; LDGS, 1336; NUC 283,380

JONAS, WILLI [1871-]
JV 11,109; NUC 283,387

JONAS (JONAS-BERLIN), WILLI [1884-]
JV 24,324; SG [3]7,88; NUC 283,387

JONES, CHESTER MORSE [1891-1972]
AMS 11,2624; WWAH, 315
W.L. Palmer, Trans. Assoc. Am. Phys. 86 (1973), 19-21

JONES, DAVID BREESE [1879-1954]
AMS 9(I),985; IB, 142; POGG 6,1248-1249; 7b,2304-2306
Anon., Chem. Eng. News 32 (1954), 3777
C.B. Jones, J. Nutrition 69 (1959), 11-17

JONES, DONALD FORSHA [1890-1963]
AMS 10,2040; NCAB 49,12-13; IB, 142; WWAH, 315
P.C. Mangelsdorf, Biog. Mem. Nat. Acad. Sci. 46 (1975), 135-156
S.L. Becker, Conn. Agr. Sta. Bull. 763 (1976), 1-9

JONES, EWART RAY HERBERT [1911-]
WW 1982, p.1179; IWW 1982-1983, p.646

JONES, GRINNELL [1884-1947]
AMS 7,926; NCAB 36,209-210; POGG 6,1249; 7b,2306-2307
M. Dole, Ann. N.Y. Acad. Sci. 51 (1948-1951), 719-726

JONES, HARRY CLARY [1865-1916]
DSB 7,161-162; DAB 10,173-174; BLOKH 1,335; GOULD, 29; WWAH, 316; WWWS, 892; POGG 4,711-712; 5,594-595
J.W., Nature 97 (1916), 283
P.G., J. Chim. Phys. 14 (1916), 488

JONES, HENRY BENCE [1813-1873]
DNB 10,998; HIRSCH 3,451-452; PAGEL, 827-828; BLOKH 1,334-335; POTSCH, 38; FERCHL, 35,260-261; POGG 1,1201; 3,696
E. du Bois-Reymond, Ber. chem. Ges. 6 (1873), 1585-1587
W. Odling, J. Chem. Soc. 27 (1874), 1201-1202
H.N. Segall, Can. Med. Assoc. J. 63 (1950), 605-606
N.G. Coley, Notes Roy. Soc. 28 (1973), 31-56
D.G. Schoenberg and B.S. Schoenberg, South. Med. J. 72 (1975), 605-606
L. Rosenfeld, Clin. Chem. 33 (1987), 1687-1692
W.B. Jensen, Bull. Hist. Chem. 7 (1990), 26-33

JONES, HUMPHREY OWEN [1878-1912]
POGG 5,595-596
K.J.P. Orton, J. Chem. Soc. 103 (1913), 755-759
J. Shorter, Notes Roy. Soc. 33 (1979), 261-277

JONES, JOHN KENYON NETHERTON [1912-1977]
WW 1977, p.1288; WWWS, 892; POTSCH, 225
W.A. Szarek, Proc. Roy. Soc. Canada [4]16 (1978), 81-84; Adv. Carbohydrate Chem. 41 (1983), 1-26
M. Stacey, Biog. Mem. Fell. Roy. Soc. 25 (1979), 365-389

JONES, MARY ELLEN [1922-]
AMS 15(4),129

JONES, THOMAS GILBERT HENRY [1895-1970]
Who's Who in Australia 1968, p.481
H. Gregory, Vivant Professores, pp.68-71. St. Lucia 1987

JONES, THOMAS WHARTON [1808-1891]
HIRSCH 3,450-451; PAGEL, 827; WWWS, 893; CALLISEN 29,171-172
Anon., Lancet 1891(II), pp.1256-1258; Brit. Med. J. 1891(II), pp.1175-1177

JONES, WALTER JENNINGS [1865-1935]
DSB 17,447-448; AMS 5,592; NCAB 43,110-111; CHITTENDEN, 111-116; IB, 143; WWAH, 317; WWWS, 893; POGG 6,1252; 7b,2313
W.M. Clark, Biog. Mem. Nat. Acad. Sci. 20 (1938), 79-139
A.M. Harvey, Johns Hopkins Med. J. 139 (1976), 257-263

JONG, ANNE WILLEM KAREL de [1871-1948]
POGG 5,273; 6,1252; 7b,2313-2314

JONG, SAMUEL ISRAELS de [1878-1928]
LINDEBOOM, 991; FISCHER 1,719-720; WININGER 7,130-131

JONGH, SAMUEL ELZEVIER de [1898-1976]
WWWS, 433
Who's Who in the Netherlands 1962-1963, p.370
P.J. Gaillard, Jaarb. Akad. Wet. 1976, pp.200-202

JORDAN, ARTHUR [1908-1975]
Anon., Lancet 1975(II), p.565; Brit. Med. J. 1975(III), p.1658

JORDAN (IORDAN), ARTUR PAVLOVICH [1866-1945]
WWR, 236; FISCHER 1,720
N. Torsuev, Hautarzt 18 (1967), 335-336

JORDAN, DAVID STARR [1851-1931]
DSB 7,169-170; DAB 10,211-214; NCAB 22,68-70; IB, 143; WWAH, 317; WWWS, 894
B.W. Everman, Science 74 (1931), 327-329
G.S. Myers, Stanford Ichthyological Bulletin 4 (1952), 1-6
C.L. Hubbs, Systematic Zoology 13 (1964), 195-200
K.R. Benson, Mendel Newsletter 26 (1986), 1-5

JORDAN, DENIS OSWALD [1914-1982]
M.I. Bruce, Chem. Brit. 18 (1982), 874
J.H. Coates, Historical Records of Australian Science 6 (1985), 237-246

JORDAN, EDWIN OAKES [1866-1936]
DSB 7,170-171; DAB [Suppl.2],352-354; NCAB 42,622-623; OEHRI, 56-57; DAMB, 401; IB, 143; WWWS, 894
L. Hektoen, Science 84 (1936), 411-413
W. Burrows, Biog. Mem. Nat. Acad. Sci. 20 (1939), 197-228

JORDAN, HEINRICH [1876-]
JV 14,142; NUC 284,613

JORDAN, HERMANN JACQUES [1877-1943]
BWN 2,267-268; KURSCHNER 4,1336-1337; FISCHER 1,721; BK, 233; IB, 143
J.J. Vonk, In Memoriam H.J. Jordan. Utrecht 1946
N. Postma and P.Smit (Eds.), Hermann Jacques Jordan. Nijmegen 1980

JORDAN, JOHANN WILHELM [1771-1853]
POGG 1,1202

JORDAN, PASCUAL [1902-1980]
DSB 17,448-454; KURSCHNER 13,1767; POGG 6,1254-1255; 7a(2),648-649
H. Hartmann, Schöpfer des neuen Weltbildes, pp.291-315. Bonn 1952
P. Jordan, Begegnungen. Oldenburg 1971

JORDAN-LLOYD [see LLOYD]

JORDIS, EDUARD [1868-1917]
DBJ 2,660; BLOKH 1,335-336; POGG 4,715; 5,597; 6,1255
H. Wichelhaus, Ber. chem. Ges. 50 (1917), 1773-1774
F. Henrich, Chem. Z. 41 (1917), 869-870
M.K. Hoffmann, Koll. Z. 23 (1918), 49-56

JORISSEN, WILLEM PAULINUS [1869-1959]
BWN 2,268-269; POGG 4,715; 5,597-598; 6,1255-1256; 7b,2316-2318
G.L. Voermann, Nature 145 (1940), 20
K. Posthumus, Chem. Wkbl. 55 (1959), 589

JORPES, ERIK [1894-1973]
SBL 20,408-410; VAD 1973, p.500; POGG 6,1256; 7b,2318-2323
Anon., J. Hist. Med. 29 (1974), 123
M. Blombäck, Thromb. Diath. Haemorrh. 33 (1975), 11-16
L. Roden and D.S. Feingold, TIBS 10 (1985), 407-409

JOSEPH, EUGEN [1879-1933]
NDB 10,625; FISCHER 1,722; WININGER 7,132; TETZLAFF, 158-159; KOREN, 198

JOSEPHSON, KARL [1900-1986]
VAD 1981, p.528; SMK 4,125; POGG 6,1256-1257; 7b,2323

JOSLIN, ELLIOTT PROCTOR [1869-1962]
AMS 10,2058; NCAB 46,98-99; FISCHER 1,723; WWAH, 318
G. Wolff, Münch. med. Wchschr. 102 (1960), 1203-1208
A. Marble, Trans. Assoc. Am. Phys. 75 (1962), 25-29
C.S. Keefer, Am. Phil. Soc. Year Book 1962, pp.143-145
E.N. Todhunter, J. Am. Diet. Assn. 46 (1965), 150
A.A. Horner, Harvard Med. Alumni Bull. 43 (1969), 42-43
A.C. Holt, Elliott Proctor Joslin. Worcester, Mass. 1969

JOST, HANS [1894-1977]
KURSCHNER 12,1446-1447; HUTER, 243-244; KALLMORGEN, 314; WWWS, 895; POGG 7a(2),650

JOST, LUDWIG [1865-1947]
NDB 10,630-631; DRULL, 129; IB, 144; POGG 7a(2),650-651
W. Ruhland, Bayer. Akad. Wiss. Jahrbuch 1944-1948, pp.276-277
E.G. Pringsheim and E.J. Maskell, Obit. Not. Fell. Roy. Soc. 6 (1949), 471-478
F. Overbeck, Ber. bot. Ges. 68a (1956), 157-163

JOST, (FRIEDRICH) WILHELM [1903-1988]
KURSCHNER 15,2087-2088; HANNOVER, 40; POTSCH, 225-226; POGG 6,1257-1258; 7a(2),651-652
Wer ist Wer? 1988-1989, p.641
Anon., Chem. Z. 112 (1988), 389
H.G. Wagner, Jahrb. Akad. Wiss. Gott. 1988, pp.79-85

JOUBERT, JULES FRANÇOIS [1834-1910]
GE 21,213; BULLOCH, 375; WWWS, 895; POGG 3,698; 4, 715-716
Anon., Ann. Inst. Pasteur 24 (1910), 241-242; Chem. Z. 34 (1910), 354

JOULE, JAMES PRESCOTT [1818-1889]
DSB 7,180-182; DNB 10,1096-1102; BLOKH 1,337-338; MATSCHOSS, 131-132; POTSCH, 226; W, 281-282; WWWS, 895-896; POGG 1,1203-1204; 3,699
O. Reynolds, Memoir of James Prescott Joule. London 1892
A. Wood, Joule and the Study of Energy. London 1925
G. Jones, Centaurus 13 (1968), 198-219
J.T. Lloyd, Notes Roy. Soc. 25 (1970), 211-225
D. Cardwell, Tech. Cult. 17 (1976), 674-687; James Joule. Manchester 1989
J. Merleau-Ponty, Rev. Hist. Sci. 32 (1979), 325-331

JOVIČIĆ, MILORAD [1868-1937]
POGG 4,716; 5,599; 6,1258-1259; 7b,2327

JOWETT, HOOPER ALBERT DICKINSON [1879-1936]
RSC 16,151-152; NUC 285,628
T.A. Henry, J. Chem. Soc. 1937, pp.1328-1329

JOY, CHARLES ARAD [1823-1891]
MILES, 256-257
Anon., Am. J. Sci. 142 (1891), 78
M. Benjamin, Pop. Sci. Mon. 43 (1893), 405-409

JOYET-LAVERGNE, PHILIPPE [1884-1967]
POGG 6,1259-1260; 7b,2329-2331

JUCHUM, DANIEL [1902-]
JV 45,438; NUC 286,94

JUCKENACK, ADOLF [1870-1939]
NDB 10,634-635; HEIN, 306-307; TSCHIRCH, 1070; POGG 6,1260; 7a(2),653-654
J. Grossfeld, Z. Leb. Unt. 79 (1940), 18-23

JÜDELL, GUSTAV [1847-1876]
KOVACSICS, 79-80; PITTROFF, 136-137; RSC 8,45

JUKES, THOMAS HUGHES [1906-]
AMS 14,2509; WWA 1976-1977, pp.1646-1647; WWWS, 897
T.H. Jukes, Journal of the American College of Nutrition 7 (1988), 93-99; Ann. Rev. Nutrition 10 (1990), 1-20

JULIAN, PERCY LAVON [1899-1975]
AMS 12,3126; NCAB 62,55-56; WWWS, 897
Anon., Nature 255 (1975), 662; Chem. Eng. News 53(19) (1975), 32

JULIUS, PAUL [1862-1931]
NDB 10,658; POTSCH, 226; POGG 6,1262
K.H. Meyer and L. Blangey, Ber. chem. Ges. 64A (1931), 49-57
A. Lüttringhaus, Z. angew. Chem. 44 (1931), 109-111

JUNG, ALBERT [1901-1972]
POGG 7a(2),655-656

JUNG, CARL GUSTAV [1794-1864]
HIRSCH 3,468; KILLIAN, 182-183
W. His, Gedenkschrift zur Eröffnung des Vesalianums, pp.40-48. Leipzig 1885

JUNG, FRIEDRICH [1915-]
KURSCHNER 12,1453; POTSCH, 226; POGG 7a(2),656-657

JUNGEBLUT, CLAUS W. [1897-1976]
AMS 12,3128; IB, 144

JUNGFLEISCH, ÉMILE [1839-1916]
CHARLE, 113-115; GORIS, 452-454; TSCHIRCH, 1071; POGG 3,703; 4,719-720; 5,602-603
H. Leroux, J. Pharm. Chim. [7]3 (1916), 305-313
C. Jordan, C.R. Acad. Sci. 162 (1916), 617-620
C.E. Monod, Bull. Acad. Med. 75 (1916), 452-455
J.B.C., Nature 97 (1916), 244-245
E. Léger, Bull. Soc. Chim. [4]21 (1917), i-xxxiv

JUNKERSDORF, PETER [1878-1934]
KURSCHNER 4,1354; WENIG, 137; IB, 144; POGG 7a(2),662

JUNKMANN, KARL [1897-1976]
KURSCHNER 12,1459; POGG 7a(2),663
Anon., Chem. Z. 100 (1976), 297

JUST, ERNEST EVERETT [1883-1941]
AMS 6,752; DAMB, 402-403
K.R. Manning, Black Apollo of Science. New York 1983
C.E. Wynes, Southern Studies 23 (1984), 60-70

JUST, FELIX [1912-1961]
POGG 7a(2),664-665
Anon., Nachr. Chem. Techn. 9 (1961), 79

JUST, GERHARD [1877-]
POGG 4,720; 5,603

JUST, GÜNTHER [1892-1950]
KURSCHNER 7,94-941; IB, 144
E. Kretschmer, Zeitschrift für menschliche Vererbung und Konstitutionslehre 30 (1952), 293-298

JUSTIN-MUELLER, EDOUARD [1867-1955]
POGG 6,1265-1266; 7b,2336-2337

K

KAAS, KARL [-1915]
Anon., Chem. Z. 39 (1915), 310; Z. angew. Chem. 28(III) (1915), 225

KABAT, ELVIN ABRAHAM [1914-]
AMS 15(3),159; WWA 1980-1981, p.1758; IWW 1982-1983, p.653; WWWS, 899;MSE 2,137-140
E.A. Kabat, Ann. Rev. Immunol. 1 (1983), 1-32
T. Faizi, Bioch. Soc. Trans. 20 (1992), 257-259

KABLUKOV, IVAN ALEKSEEVICH [1857-1942]
DSB 7,203-204; BSE 19,234-235; WWR, 247; KUZNETSOV(1961), 509-515; POTSCH, 227; POGG 4,711; 5,603; 6,1266; 7b,2337-2339
A.F. Kaputsinski, Bull. Acad. Sci. URSS 1942, pp.181-184
Y.I. Soloviev, I.A. Kablukov. Moscow 1957

KACHLER, JOSEF [1847-1890]
OBL 3,163-164; STURM 2,77; WEINMANN, 250
Anon., Chem. Z. 14 (1890), 1783; Leopoldina 26 (1890), 114

KADISADE, A. REFIK [1892-]
JV 34,236; NUC 287,421

KÄMMERER, HEINRICH [1881-]
JV 22,520; NUC 287,468

KÄMMERER, HERMANN [1840-1898]
BJN 5,32*; GUNDLACH, 469; MEINEL, 149,505; BLOKH 1,344; POGG 3,705; 4,720-721

KÄMMERER, HUGO [1878-1968]
KURSCHNER 10,1100; BACHMANN, 148-159
W.C. Meyer, Münch. med. Wchschr. 100 (1958), 1542-1543
Anon., Deutsche med. Wchschr. 93 (1968), 556

KAESBERG, PAUL JOSEPH [1923-]
AMS 15(4),162; WWWS, 900

KÄSWURM, AUGUST [1859-]
JV 1,50; RSC 16,170; NUC 287,528

KAFKA, VICTOR [1881-1955]
STURM 2,83; FISCHER 1,731-732; WWWS, 900; POGG 7a(2),673-676
H. Demme, Fort. Neurol. 24 (1956), 163-164

KAGAN, EZRA MOISEEVICH [1887-1948]
VORONTSOV, 79

KAHANE, ERNEST [1903-]
POGG 6,1268; 7b,2343-2346

KAHLBAUM, GEORG WILHELM AUGUST [1853-1905]
NDB 11,22-24; BJN 10,48-54,192*; WWWS, 900; POGG 3,706; 4,721-722;5,604; 6,1269
Anon., Chem. Z. 29 (1905), 915
F. Strunz, Ber. chem. Ges. 38 (1905), 4239-4248
F. Fichter, Verhandl. Schw. Nat. Ges. 88 (1905), xlvi-lxviii
E. Hagenbach-Bischoff, Verhandl. Nat. Ges. Basel 18 (1905), 379-402
P.W. Schmidt et al., in Beiträge zur Geschichte der Chemie (P. Diergart, Ed.), pp.3-35. Leipzig 1909

KAHLBAUM, WILHELM [1822-1884]
NDB 11,24; BLOKH 1,339
H. Landolt, Ber. chem. Ges. 17 (1884), 1582-1583

KAHLENBERG, LOUIS ALBRECHT [1870-1941]
DSB 7,208-209; MILES, 259; WWAH, 319; WWWS, 800-901; POGG 4,722; 5,604; 6,1269; 7b,2346
A.T. Lincoln, Ind. Eng. Chem. News Ed. 16 (1938), 336-337
N.F. Hall, Trans. Wis. Acad. 39 (1949), 83-96; 40 (1950), 173-183

KAHLER, OTTO [1849-1893]
NDB 11,26-27; ADB 50,747; OBL 3,174-175; STURM 2,85; FISCHER 1,732; PAGEL, 835-836; SCHIEBER, 82-91; KOERTING, 168-169
F. Kraus, Z. Heilkunde 14 (1893), 3-12

KAHLER, OTTO [1878-1946]
OBL 3,175; STURM 2,85
R. Brighetti, Policlinico 74 (1967), 702-708

KAHLSON, GEORG [1901-1982]
SMK 4,157; VAD 1981, pp.535-536; WWWS, 901

KAHN, ANSELM [1878-]
JV 19,340; NUC 287,633

KAHN, PAUL [1895-1942?]
JV 39,14

KAHN, REUBEN LEON [1887-1979]
AMS 12,3140; FISCHER 1,733; IB, 145; WININGER 7,144; COHEN, 118; WWWS, 901
W.M. Cobb, J. Nat. Med. Assn. 63 (1971), 388-394

KAHN, RICHARD [1876-1941]
STURM 2,85-86; FISCHER 1,733; MAASS, 101-116; KOERTING, 120; IB, 145; KOREN, 199

KAHN, ROBERT [1868-1944]
POGG 4,722; 5,605; 6,1269; 7a(2),676

KAHN, WALTER ERNST [1878-]
JV 20,468; NUC 287,678

KAI, SOTARO [1893-]
HAKUSHI 4,917-918

KAISER, ARMIN [1861-]
NUC 288,34

KAISER, ARMIN DALE [1927-]
AMS 15(4),34

KAISER, CAJETAN GEORG [1803-1871]
HEIN-E, 223; BLOKH 1,342; FERCHL, 265; SCHAEDLER, 56; POGG 1,1220

KAISER, EMIL THOMAS [1938-1988]
AMS 16(4),171
Anon., Chem. Eng. News 66(33) (1988), 33-34
W. DeGrado, Proteins 4 (1988), i-ii

KAISER, HANS [1890-1977]
HEIN-E, 223-224; KURSCHNER 12,1478; DRUM, 295-297; POGG 6,1270; 7a(2),680

KAISER, JOHANNES [1880-1917]
JV 23,308

KAISER, KARL [1861-1933]
DRULL, 130; STUBLER, 310

KAISERLING, CARL [1869-1942]
KURSCHNER 4,1369-1370; FISCHER 1,733-734; FREUND 2,161-167; IB, 145
C. Krauspe, Z. allgem. Path. 80 (1942), 49-52

KAKINUMA, KOSAKU [1892-1952]
JBE, 475; HAKUSHI 3,593

KAKIUCHI, SHIRO [1929-1984]
A.R. Means, Nature 311 (1984), 708
M. Appleman et al., Journal of Cyclic Nucleotide and Protein Phosphorylation Research 10 (1985), 417-421

KALANTHARIANZ, ANUSHAWAN [1867-]
JV 13,20; RSC 16,175; NUC 288,137

KALASHNIKOV, VIKTOR PETROVICH [1893-1959]
WWR, 250; LIPSHITS 4,31-32
Anon., Aptechnoi Delo 1957(4), pp.84-85

KALB, LUDWIG [1879-1958]
JV 21,484; TLK 3,312; POGG 5,606; 6,1271-1272; 7a(2),682; (4),146*
Anon., Nachr. Chem. Techn. 7 (1959), 48

KALBERLAH, FRITZ [1875-]
KALLMORGEN, 317; NUC 288,151

KALCKAR, HERMAN MORITZ [1908-1991]
AMS 14,2525; WWA 1978-1979, p.1707; IWW 1982-1983, p.656
H.M. Kalckar, BioEssays 3 (1985), 134-137; in Selected Topics in the History of Biochemistry (G. Semenza and R. Jaenicke, eds.), pp.101-176. Amsterdam 1991; Ann. Rev. Biochem. 60 (1991), 1-37
Anon., Chem. Eng. News 69(27) (1881), 43; New York Times, 22 May 1991

KALENICHENKO, IVAN IOSIFOVICH [1805-1876]
VORONTSOV, 10-14; RSC 3,601-602

KALISCHER, GEORG [1873-1938]
NDB 11,60; POTSCH, 227-228; POGG 6,1272; 7a(2),682
O. Beyer, Chem. Ber. 89 (1956), xliii-lviii

KALLE, FERDINAND [1870-1954]
NB, 194; POGG 7a(2),683
Anon., Chem. Z. 78 (1954), 662

KALLE, WILHELM [1838-1919]
NDB 11,65-68; POTSCH, 228; NB, 195; NL 2,274-279; BLOKH 1,343
K. Albrecht, Ber. chem. Ges. 52A (1919), 86-89

KALLINICH, GÜNTER [1913-]
KURSCHNER 13,1812; POGG 7a(2),683

KALLOS, PAUL [1902-1988]
VAD 1989, p.572; WWWS, 902
P. Dukor et al., Clinical Immunology and Immunopathology 49 (1990), 1-20

KALTENBRUNNER, GEORG [1803-1833]
HIRSCH 3,484-485; CALLISEN 10,96-97; 29,205

KAMEN, MARTIN DAVID [1913-]
AMS 15(4),179; WWA 1980-1981, p.1765; IWW 1982-1983, p.658; COHEN, 119; MSE 2,141-142; WWWS, 903
M.D. Kamen, Radiant Science, Dark Politics. Berkeley 1985; Ann. Rev. Biochem. 55 (1986), 1-34

KAMETAKA, TOKUHEI [1872-1935]
M. Onuma and T. Doke, Jap. Stud. Hist. Sci. 12 (1973), 5-14
S. Mitsui, Memoirs of Urawa Technical High School 1 (1970), 1-10

KAMIN, HENRY [1920-1988]
AMS 15(4),180; WWA 1980-1981, p.1766; WWWS, 903
B.S. Masters, TIBS 14 (1989), 134-135

KAMM, OLIVER [1888-1965]
AMS 10,2075; MILES, 261-262
Anon., Chem. Eng. News 44(3) (1966), 84

KAMMERER, PAUL [1880-1926]
OBL 3,209-210; WWWS, 904
Anon., Nature 118 (1926), 635-636
F. Lenz, Archiv für Rassen und Gesellschaftsbiologie 21 (1929), 311-318

KAMP, WALTER TE [1901-]
SG [4]9,14

KANDEL, ERIC RICHARD [1929-]
AMS 15(4),183; BHDE 2,590-591

KANE, ROBERT JOHN [1809-1890]
DSB 7,224; DNB 10,1126-1127; LYONS, 99-100; PAGEL, 838; BLOKH 1,344; POTSCH, 228; FERCHL, 265-266; TSCHIRCH,1072; COLE, 281; WWWS, 904; POGG 3,707-708
J.E.R., Proc. Roy. Soc. 47 (1890), xii-xviii
D. Reilly, J. Chem. Ed. 32 (1955), 404-406
W.I. Davis, Proc. Roy. Irish Acad. 77B (1977), 309-315
J. Grimshaw, Chem. Brit. 26 (1990), 764-766

KANITZ, ARISTIDES [1877-]
RSC 16,182; NUC 16,182

KANITZ, AUGUST (AGOSTON) [1843-1896]
OBL 3,215; MEL 1,850
Anon., Ost. Bot. Z. 24 (1874),1-16; Leopoldina 32 (1896), 137

KANONNIKOV, INNOKENTI IVANOVICH [1854-1902]
BSE 20,18-19; KAZAN 1,347-349; BLOKH 1,345-346; POGG 4,725

KANTHACK, ALFREDO ANTUNES [1863-1898]
O'CONNOR(2), 59-60; FISCHER 1,737; PAGEL, 838-839; RSC 16,183; WWWS, 905
C.R. Hewitt, St. Barts. Hosp. J. 6 (1898), 51-53
Anon., Lancet 1898(II), p.1817; Nature 59 (1899), 252-253
G.S. Woodhead, J. Path. Bact. 6 (1899), 89-91

KAPFHAMMER, JOSEPH [1888-1968]
NDB 11,132; KURSCHNER 10,1116; IB, 146; POGG 6,1278; 7a(2),688
H. Holzer, Freib. Univ. Bl. 20 (1968), 9-10

KAPLAN, ANN ESTHER [1926-]
AMS 15(4),191

KAPLAN, HENRY SEYMOUR [1918-1984]
AMS 15(4),192; WWA 1980-1981, p.1771; AO 1984, pp.88-91; WWWS, 905-906
Z. Furs and M. Feldman, Cancer Survey 4 (1988), 294-311

KAPLAN, NATHAN ORAM [1917-1986]
AMS 15(4),194; WWA 1980-1981,p.1772; IWW 1982-1983, p.662
N.O. Kaplan, in Selected Topics in the History of Biochemistry (G. Semenza, Ed.), pp.255-296. Amsterdam 1986
B. Allison et al., Anal. Biochem. 161 (1987), 229-230

KAPLANSKY, SAMUEL YAKOVLEVICH [1897-1965]
WWR, 255-256; WWWS, 906
Anon., Biokhimia 30 (1965),1292-1293; Vop. Med. Khim. 12 (1966), 120-122

KAPOSI (KOHN), MORITZ [1837-1902]
NDB 11,133-134; OBL 3,222; BJN 7,55*; HIRSCH 3,487; PAGEL, 839-840; SIEGL, 15-27; WININGER 3,399-400; WWWS, 906
Anon., Brit. Med. J. 1902(I), p.683
E. Finger, Münch. med. Wchschr. 49 (1902), 708-709
J. Frankl, Rev. Hist. Med. Heb. 28(5) (1975), 111-113

KAPPELMEIER, PAUL [1887-]
JV 26,638; NUC 289,477
New York Times, 15 January 1922, p.1

KAPPEN, HUBERT [1878-1949]
KURSCHNER 7,958; IB, 146; POGG 6,1279; 7a(2),688
Anon., Z. angew. Chem. 62 (1950), 84

KAPPUS, ADOLF [1900-]
AMS 11,2688; WWWS, 907

KARLE, ISABELLA LUGOWSKI [1921-]
AMS 14,2549; WWA 1980-1981, p.1775; MSE 2,147-148
I. Noble, Contemporary Women Scientists in America, pp.107-122. New York 1979

KARLE, JEROME [1918-]
AMS 15(4),202; WWA 1980-1981, p.1775; McLACHLAN, 277-283; WWWS, 908; POTSCH, 228-229

KARLSON, PETER [1918-]
KURSCHNER 13,1832; AUERBACH, 281; POTSCH, 229; POGG 7a(2),692
C.E. Sekeris, Z. physiol. Chem. 359 (1978), 1245-1246
A. Butenandt, TIBS 3 (1978), N219
P. Karlson, in Of Oxygen, Fuels and Living Matter Part 2 (G. Semenza, Ed.), pp.447-500. Chichester 1982

KARNOVSKY, MANFRED LESLIE [1918-]
AMS 15(4),204

KARPECHENKO, GEORGI DMITRIEVICH [1899-1942]
DSB 17,460-464; BSE (3rd Ed.) 11,455

KARPLUS, JOHANN PAUL [1866-1936]
OBL 3,249; STURM 2,111; FISCHER 1,739
J. Wagner-Jauregg, Wiener klin. Wchschr. 49 (1936), 282-283
O. Marburg, Jahrb. Psychiat. 53 (1936), 1-9

KARPLUS, MARTIN [1930-]
AMS 15(4),206; BHDE 2,597

KARPOV, GRIGORI [1867-]
SG [2]8,586; NUC 290,106

KARPOV, GRIGORI YAKOVLEVICH [1856-]
SG [2]8,586; RSC 16,192

KARPOV, LEV YAKOVLEVICH [1879-1921]
BSE 20,251; BLOKH 1,347

KARR, WALTER GERALD [1892-1946]
AMS 7,942; IB, 147

KARRER, PAUL [1889-1971]
DSB 15,257-258; NDB 11,297-298; KURSCHNER 11,1381-1382; STC 2,91-92; SBA 1,82; BOHM, 151-152; ABBOTT-C, 71-72; W, 584-585; WWWS, 909; POTSCH, 229; COHEN, 121; POGG 5,613; 6,1284-1285; 7a(2),694-700
R.E. Oesper, J. Chem. Ed. 23 (1946), 392
E. Jucker, Z. angew. Chem. 71 (1959), 253-259
C.H. Eugster, Viert. Nat. Ges. Zürich 116 (1971), 506-

511; Chimia 37 (1983), 213-216
A. Wettstein, Helv. Chim. Acta 55 (1972), 313-328
G. Hesse, Bayer. Akad. Wiss. Jahrbuch 1972, pp.282-286
Anon., Chimia 43 (1989), 153-160

KARSNER, HOWARD THOMAS [1879-1970]
AMS 11,2695; NCAB 55,207-208; IB, 147

KARSTEN, HERMANN [1817-1908]
NDB 11,305-306; BJN 13,46*-47*; TSCHIRCH, 1072; FERCHL, 267; POGG 1,1229-1230; 3,710
Anon., Leopoldina 44 (1908), 94-95

KARSTEN, KARL JOHANN BERNHARD [1782-1853]
DSB 7,254-255; NDB 11,306-307; ADB 15,427-430; COLE, 281-282; POTSCH, 230; POGG 1,1227; 7aSuppl.,313

KARSTRÖM, HENNING [1899-1989]
Kuka Kukin On 1966, p.380
Who's Who in Europe 1972, p.1612

KARUSH, FRED [1914-]
AMS 16(4),211

KARUZHAS, YURI VARFOLMOEVICH [1866-]
SG [2]8,587

KARUZIN, PETR IVANOVICH [1864-1939]
BSE 20,305; BME 12,415; WWR, 259; IB, 147
M. Spirov, Radianska Meditsina 4(8/9) (1939), 84
V. Ternovsky, Arkh. Anat. Gist. Emb. 24 (1940), 117

KASERER, HERMANN [1877-1955]
PLANER, 161; IB, 147; POGG 7a(2),701
S. Goldschmidt, Bayer. Akad. Wiss. Jahrbuch 1956, p.234

KASHIN, NIKOLAI IVANOVICH [1825-1872]
BME 12,545; HIRSCH 3,490; LIPSHITS 4,110-111
N.N. Malinovski, N.I. Kashin. Moscow 1957

KASSELL, BEATRICE [1912-1977]
AMS 13,2248
Anon., Chem. Eng. News 56(7) (1978), 54

KASSNER, GEORG [1858-1929]
KURSCHNER 3,1114; DBJ 11,356; HEIN, 312-313; TSCHIRCH, 1072; TLK 3,316; POTSCH, 230
H. Freundlich, Ber. chem. Ges. 62A (1929), 79
F. Sierp, Chem. Z. 53 (1929), 409-410

KASSOWITZ, MAX [1842-1913]
NDB 11,321-322; OBL 3,256-257; BJN 18,100*; FISCHER 1,741; PAGEL,844-845; ATUYANA, 20-28; BK, 359-361; WININGER 3,413; 7,158
K. Hochsinger, Wiener med. Wchschr. 63 (1913), 1657; 83 (1933), 941-945

KAST, ALFRED [1856-1903]
BJN 8,58*; FISCHER 1,741; PAGEL, 845; WWWS, 910
C. Weigert, Münch. med. Wchschr. 50 (1903), 383
W. Filehne, Jahresb. Schles. Ges. 81 (1904), 8-13

KASTLE, JOSEPH HOEING [1864-1916]
AMS 2,251; MILES, 262-263; CHITTENDEN, 293-297; BARRY, 48-49; WWAH, 320; POGG 4,730; 5,615
Anon., Science 44 (1916), 461

KASTNER, KARL WILHELM GOTTLOB [1783-1857]
NDB 11,324; ADB 15,439; HEIN, 313-314; DRULL, 132; FERCHL, 268; RSC 3,615-618; 8,54-55; POGG 1,1231

KASTNER, RICHARD [1876-1959]
JV 15,132; NUC 290,340

KATAGIRI, HIDEO [1897-]
JBE, 519,2327; IB, 147

KATAYAMA, KUNIYOSHI [1855-1931]
JBE, 524-525; HAKUSHI 2,12-13; FISCHER 1,741-742

KATCHALSKI (KATZIR), EPHRAIM [1916-]
IWW 1982-1983, p.669; MH 2,278-279; STC 2,93-94; MSE 2,152-153

KATCHALSKY, AHARON KATZIR [1914-1972]
STC 2,95-97; COHEN, 122
Anon., Nachr. Chem. Techn. 20 (1972), 247-248
F.O. Schmitt ans R.B. Livingston, Ann. Rev. Biophys. 4 (1973), 1-6

KATO, GEN-ICHI [1890-1979]
JBE, 527, 2247
G. Kato, Ann. Rev. Physiol. 32 (1970), 1-20

KATSCH, GERHARDT [1887-1961]
NDB 11,328-329; FISCHER 1,742; KALLMORGEN, 318; AUERBACH, 281
H. Barthelemer, Münch. med. Wchschr. 99 (1957), 955-958
W. Lueken, Hippokrates 32 (1961), 405-408
G. Mohnicke, Z. ärzt. Fortbild. 55 (1961), 670-671

KATZ, BERNARD [1911-]
WW 1981, p.1405; IWW 1982-1983, p.668; BHDE 2,599; TETZLAFF, 167; ABBOTT, 75; POTSCH, 230
B. Katz, J. Physiol. 370 (1986), 1-12

KATZ, JOHANN RUDOLF [1880-1938]
POGG 6,1289-1290; 7b,2398-2400
E. Heuser and B.W. Rowland, J. Chem. Ed. 16 (1939), 153-154
N.P. Badenhuizen and A. Weidinger, Chem. Wkbl. 36 (1939), 230-237

KATZMAN, PHILIP AARON [1906-]
AMS 14,2564; WWWS, 912

KAUFFMANN, FRITZ [1899-1978]
KURSCHNER 12,1509; KBB 1976, p.555; BHDE 2,604; LDGS. 57; WWWS, 912
F. Kauffmann, Erlebte Bakteriologie: Zur Geschichte der Salmonella- und Escherichia-Forschung. Copenhagen 1967; Erinnerungen eines Bakteriologen. Copenhagen 1969
R. Rohde et al., Z. Bakt. 243 (1979), 141-147

KAUFFMANN, HUGO [1890-1916]
JV 30,510; NUC 290,554
Anon., Chem. Z. 40 (1916), 1014; Z. angew. Chem. 30(III) (1917), 48

KAUFFMANN, HUGO JOSEF [1870-1957]
KURSCHNER 4,1394-1395; LDGS, 21; TLK 3,318; POGG 4,731-732; 5,616; 6,1290; 7a(2),704
Anon., Nachr. Chem. Techn. 5 (1957), 336; Chem. Eng. News 35(39) (1957), 99

KAUFLER, FELIX [1878-1957]
POGG 4,732; 5,616
Anon., Chem. Ind. 1958, p.74

KAUFMAN, SEYMOUR [1924-]
AMS 15(4),255; WWWS, 912

KAUFMANN, BERWIND PETERSON [1897-1975]
AMS 14,1569; WWA 1974-1975, p.1655; WWWS, 913

KAUFMANN, CONSTANTIN [1853-1934]
DHB 4,332; FISCHER 1,744
G. Kaemig, Schw. med. Wchschr. 64 (1934), 582

KAUFMANN, EDUARD [1860-1931]
NDB 11,345-346; KURSCHNER 4,1396; FISCHER 1,744; PAGEL, 847
R. Hückel, Z. allgem. Path. 54 (1932), 193-195
G.B. Gruber, Verhandl. path. Ges. 27 (1934), 309-314
K.W. Weihe, Eduard Kaufmann. Gottingen 1945

KAUFMANN, HANS PAUL [1889-1971]
KURSCHNER 11,1389; HEIN-E, 230-231; TSCHIRCH, 1072; TLK 3,318-319; WWWS, 913; POGG 5,617; 6, 1290-1291; 7a(2),704-708
G. Zeidler, Chem. Z. 83 (1959), 686
Anon., Chem. Z. 78 (1954), 694; 95 (1971), 966; Deutsche Farben Zeitschrift 18 (1964), 466; Arzneimitt. 19 (1969), 1767-1768

KAUFMANN, LUDWIG [1872-1942]
GEDENKBUCH 1,727; JV 12,298; RSC 15,633; 19,671; NUC 290,613

KAUFMANN, MAURICE [1856-1924]
G. Petit, Bull. Acad. Med. 91 (1924), 582-586; Rec. Med. Vet. 100 (1924), 260-268

KAULICH, ALOIS [1839-1901]
BJN 6,54*
P., Prag. med. Wchschr. 26 (1901), 539-540

KAULICH, JOSEF [1830-1886]
NDB 11,358-359; STURM 2,120-121; HIRSCH 3,491; MOTSCH, 164-167; KOERTING, 180
Anon., Leopoldina 22 (1886), 169
W. Hirth, Josef Kaulich. Berlin 1977

KAUSCH, OSCAR [1872-1945]
TLK 3,319; POGG 6,1292; 7a(2),708

KAUSCH, WALTHER [1867-1928]
FISCHER 1,745; JV 6,258; RSC 16,208
Anon., Med. Klin. 24 (1928), 643

KAUTSKY, HANS [1891-1966]
NDB 11,375; KURSCHNER 10,1128; AUERBACH, 837; POGG 6,1292; 7a(2),708-709
Anon., Chem. Z. 90 (1966), 646

KAUTZSCH, KARL [1879-1920]
JV 19,284
Anon., Z. angew. Chem. 34 (1921), 16

KAUZMANN, WALTER JOSEPH [1916-]
AMS 15(4),228; WWA 1980-1981, p.1786; IWW 1982-1983, p.670; WWWS, 913

KAWAI, SIN'ITI [1895-1969]
POGG 6,1293; 7b,2403-2404

KAWAMURA, RINYA [1879-1947]
JBE, 564; HAKUSHI 2,182-183; FISCHER 1,745; IB, 148; NUC 291,60

KAY, FRANCIS WILLIAM [1883-1967]
SRS, 26; POGG 5,618; 6,1293; 7b,2404
Anon., Chem. Ind. 1967, p.2053; Chem. Brit. 4 (1968), 81

KAY, HERBERT DAVENPORT [1893-1976]
WW 1976, p.1228; IB, 148; POGG 6,1293; 7b,2404-2408
M.E. Coates, Nature 265 (1977), 392
J.W. Porter, J. Dairy Res. 44 (1977), 191-194
K. Blaxter, Biog. Mem. Fell. Roy. Soc. 23 (1977), 283-310

KAYSER, EDUARD [1870-]
JV 13,154; NUC 291,131

KAYSER, GUSTAV ADOLF [1817-1878]
HEIN-E, 232

KAZANSKI, ALEKSANDR SERGEEVICH [1885-1937]
LIPSHITS 4,19

KAZNELSON, PAUL [1892-]
KURSCHNER 5,654; PELZNER, 119-122

KEBLER, LYMAN FREDERIC [1863-1955]
AMS 9(I),1015; MILES, 266-267
L.F. Kebler, J. Am. Pharm. Assn. 29 (1940), 379-383

KEDROVSKI, VASILI IVANOVICH [1865-1937]
BSE 20,486; BME 12,585-588; WWR, 264; FISCHER, 747; IB, 148
Anon., Zhur. Mikr. Ep. Imm. 20(2) (1938), 3; Med. Parazit. 7 (1938), 150
A.A. Iushenko and N.A. Ivanova, Arkh. Pat. 28(10) (1966), 65-68

KEDZIE, ROBERT CLARK [1823-1902]
DAB 10,277; NCAB 8,488; MILES, 267-268; HIRSCH 3,495; KELLY, 686-687
Anon., Am. J. Sci. [4]14 (1902),470; Medical Record (N.Y.) 62 (1902),776
C.A. Browne, J. Am. Chem. Soc. 48(8a) (1926), 177-201

KEEBLE, FREDERICK WILLIAM [1870-1952]
DNB [1951-1960], 564-565; DESMOND, 353; WWWS, 915
V.H. Blackman, Obit. Not. Fell. Roy. Soc. 8 (1953), 491-501

KEEFER, CHESTER SCOTT [1897-1972]
AMS 11,1720; ICC, 254-255; DAMB, 406; WWAH, 321; WWWS, 915
A.B. Hastings, Boston Medical Quarterly 14 (1963), 89-95
R.W. Wilkins, Trans. Assoc. Am. Phys. 85 (1972), 24-26
S. Warren, Am. Phil. Soc. Year Book 1972, pp.202-204

KEEN, WILLIAM WILLIAMS [1837-1932]
DAB [Suppl.1],459-460; NCAB 11,367-368; OEHRI, 59-60; WWAH, 321-322; WWWS, 915
W. Pickles, Rhode Isl. Med. J. 10 (1927), 1-10
D.C. Geist, Trans. Coll. Phys. Phila. 41 (1974), 304-314; 44 (1977), 182-193

KEESER, EDUARD [1892-1956]
FISCHER 1,747; IB, 148; POGG 6,1295; 7a(2),711
K. Soehring, Arzneimitt. 6 (1956), 155-156
Anon., Nachr. Chem. Techn. 4 (1956), 57

KEGELES, GERSON [1917-]
AMS 15(4),243; WWA 1980-1981, p.1795; WWWS, 916

KEHR, AUGUST ERNEST [1914-]
AMS 12,3214; WWWS, 916

KEHRER, EDUARD ALEXANDER [1849-1906]
BJN 11,36*; POGG 4,734; 5,620

KEHRER, ERWIN [1874-1959]
KURSCHNER 8,1116; FISCHER 1,748; GAUSS, 262; DRULL, 133
H. Naujoks, Deutsche med. Wchschr. 70 (1944), 252

KEHRMANN, FRIEDRICH [1864-1929]
TLK 3,320-321; IB, 148; GENEVA 4,124-128; 5,288-293; POGG 4,734-735; 5,620-621; 6,1297
H. Decker, Ber. chem. Ges. 62A (1929), 60
H. Goldstein, Helv. Chim. Acta 15 (1932), 315-349

KEIDEL, WOLF DIETER [1917-]
KURSCHNER 13,1855-1856; WWWS, 916; POGG 7a(2),714-715

KEIGHLEY, GEOFFREY LORRIMER [1901-1987]
AMS 14,2584

KEIL, WERNER [1902-1956]
POGG 6,1297; 7a(2),717
R. Muschawek, Arzneimitt. 6 (1956), 712-713

KEILIN, DAVID [1887-1963]
DSB 7,272-274; DNB [1961-1970], 604-605; WW 1959, p.1655; MH 2,282-283; STC 2,101-103; MSE 2,157-159; W, 284-285; WWWS, 916; POTSCH, 231;
E.F. Hartree, Biochem. J. 89 (1963), 1-5; in Of Oxygen, Fuels and Living Matter (G. Semenza, Ed.), pp.161-227. Chichester 1981
T. Mann, Biog. Mem. Fell. Roy. Soc. 10 (1964), 183-197
P. Tate, Parasitology 55 (1965), 1-28
D. Keilin, The History of Cell Respiration and Cytochrome. Cambridge 1966
E.C. Slater, TIBS 2 (1977), 138-139
M.F. Perutz, Cambridge Review 107 (1986), 152-156
T.E. King et al., in Oxidases and Related Redox Systems (T.E. King et al., Eds.), pp.3-90. New York 1988

KEILIN, JOAN [1920-]
Who's Who of British Scientists 1971-1972, p.477

KEISER, KARL [1871-1951]
HANNOVER, 43; POGG 7a(2),719

KEITH, ARTHUR [1866-1955]
DSB 7,278-279; DNB [1951-1960], 565-566; O'CONNOR(2), 285-286; IB, 148;FISCHER 1,749; WWWS, 917
W. Le Gros Clark, Biog. Mem. Fell. Roy. Soc. 1 (1955), 145-162
J.C. Brash and A,J,E, Cave, Journal of Anatomy 89 (1955), 403-418

KEITH, NORMAN MACDONNELL [1885-1976]
AMS 10,2113; IB, 148; WWWS, 917
R.D. Pruitt, Trans. Assoc. Am. Phys. 89 (1976), 23-25

KEKULÉ, AUGUST [1829-1896]
DSB 7,279-283; NDB 11,414-424; ADB 51,479-486; BUGGE 2,200-216; WWWS,917; BJN 1,412-444, 3,92*-93*; BLOKH 1,348-350; SCHAEDLER, 57; W,286-287; POTSCH, 231-232; ABBOTT-C, 72-73; TSCHIRCH, 1072; POGG 1,1237-1238;3,711-712; 4,737; 6,1298; 7aSuppl.,315-318
G. Schulz, Ber. chem. Ges. 23 (1890), 1265-1312
H. Landolt, Ber. chem. Ges. 29 (1896), 1971-1976
F.R. Japp, J. Chem. Soc. 73 (1898), 97-138
E. Rimbach, Ber. chem. Ges. 36 (1903), 4613-4640
R. Anschütz, August Kekulé. Berlin 1929
H.E. Armstrong, Chem. Ind. 48 (1929), 914-918
A. Bernthsen, Z. angew. Chem. 43 (1930), 719-722
O.J. Walker, Ann. Sci. 4 (1939), 34-46
H.E. Fierz-David, Gesnerus 1 (1944), 146-152
E.N. Hiebert, J. Chem. Ed. 36 (1959), 320-327
W. Ruske, Naturwiss. 52 (1965), 485-489
J. Gillis, Ann. Acad. Roy. Belg. 37(1) (1966), 5-40
N.W. Fisher, Ambix 21 (1974), 29-52
K. Hafner, Z. angew. Chem. 91 (1979), 685-696
A.J. Rocke, Brit.J. Hist. Sci. 14 (1981), 27-57; Ann. Sci. 42 (1985), 355-381
J.H. Wotiz and S. Rudofsky, Chem. Brit. 20 (1984), 720-723
O.B. Ramsey and A.J. Rocke, Chem. Brit. 20 (1984), 1093-1094
W. Goebel, August Kekulé. Leipzig 1984
S. Rudofsky and J.H. Wotiz, Ambix 35 (1988), 31-38

KEKWICK, RALPH AMBROSE [1908-]
WW 1981, p.1415

KELLAWAY, CHARLES HALLILEY [1889-1952]
DNB [1951-1960], 566-567; WW 1949, p.1513; IB, 148; WWWS, 918
H.H. Dale, Obit. Not. Fell. Roy. Soc. 8 (1953), 503-521

KELLER, BORIS ALEKSANDROVICH [1874-1945]
BSE 20,495-496; WWR, 265; KUZNETSOV(1963), 391-399; LIPSHITS 4,127-140; IB, 148
V.K. Sukachev, Agribiologia 1947(1), pp.114-120
B.M. Kozo-Polyansky, Bot. Zhur. 33 (1948), 545-558

KELLER, HUGO [1882-]
JV 26,638; NUC 292,169

KELLER, KARL [1880-1933]
JV 20,260
Anon., Chem. Z. 58 (1934), 169

KELLER, KONRAD THEODOR [1885-1966]
JV 24,338

KELLER, OSKAR [1877-1959]
NDB 11,463-464; KURSCHNER 8,1123; HEIN-E, 233-234; AUERBACH, 838-839; TLK 3,322; TSCHIRCH, 1072; POGG 6,1299; 7a(2),723; (4),147*
Anon., Pharmazie 12 (1957), 461-462; Chem. Z. 83 (1959), 301-302
P. Vesterling, Arzneimitt. 9 (1959), 392

KELLER, RUDOLF [1875-1964]
AMS 9(I),1020; STURM 2,128; POGG 6,1299; 7a(2),723-724
F. Fuchs and H. Waelsch, Exp. Med. Surg. 13 (1955), 2-8

KELLER, WILHELM [1818-]
FERCHL, 269; RSC 3,631; NUC 292,264; POGG 1,1239

KELLERMAN, WILLIAM ASHBROOK [1850-1908]
NCAB 26,159-160; NUC 292,276
Anon., Science 27 (1908), 118,479,858
Who Was Who in America 1,661

KELLNER, CARL [1851-1905]
NDB 11,476-477; POTSCH, 232
Anon., Chem. Z. 29 (1905), 677

KELLNER, OSKAR [1851-1911]
NDB 11,478-479; BJN 16,39*; BLOKH 1,353; POGG 3,712; 4,738; 5,622
J. König, Chem. Z. 35 (1911), 1157-1158
F. Honcamp, Land. Ver. 76 (1912), iii-xliv

KELLOGG, VERNON LYMAN [1867-1937]
DSB 7,285-286; NCAB A,203-204; 28,354-355; IB, 149; WWAH, 323; WWWS, 919

C.E. McClung, Science 87 (1938), 158-159; Biog. Mem. Nat. Acad. Sci .20 (1939), 243-257
R.W. Doane, Ann. Ent. Soc. Am. 33 (1940), 599-607

KELLY, AGNES [1875-]
JV 16,251; RSC 16,233

KELLY, FRANCIS CHARLES [1897-]
Who's Who in British Science 1953, p.155

KELNER, ALBERT [1912-]
AMS 15(4),262

KELSER, RAYMOND ALEXANDER [1892-1952]
DSB 7,286-287; DAB [Suppl.5],382-383; AMS 8,1326; NCAB 43,84-85; WWAH, 324; WWWS, 920
R.E. Shope, Biog. Mem. Nat. Acad. Sci. 28 (1954), 199-221

KELVIN, LORD (WILLIAM THOMSON) [1824-1907]
DSB 13,374-388; DNB [1901-1911]3, 508-517; BLOKH 1,350-351; W,511-513; POTSCH, 428; MATSCHOSS, 137; SCHAEDLER, 57; POGG 3,1341-1343; 4,1496-1498; 5,1254; 6,2655
A. Gray, Lord Kelvin. London 1908
J.L., Proc. Roy. Soc. A81 (1908), ii-lxxvi
S.P. Thompson, Life of William Thomson, Baron Kelvin of Largs. London 1910
A. Russell, Lord Kelvin. London 1938
C.W. Smith and M.N. Wise, Energy and Empire. Cambridge 1989

KEMPE, MARTIN [1884-]
JV 22,26; NUC 293,48

KEMPF, RICHARD [1879-1935]
TLK 3,323; JV 18,20; NUC 293,75; POGG 5,623; 6,1301; 7a(2),726-727

KEMPF, THEODOR [1838-1923]
BLOKH 1,354
Anon., Ber. chem. Ges. 56A (1923), 85

KEMPNER, WALTER [1903-]
AMS 14,2601; BHDE 1,360; 2,613; LDGS, 66; WWWS, 921

E.A. Stead, North Carolina Medical Journal 44 (1983), 237-240

KENDALL, EDWARD CALVIN [1886-1972]
DSB 15,258-259; AMS 11,2746; MH 1,262-263; STC 2,104-105; MSE 2,161-162; CB 1950, pp.292-294; CHITTENDEN, 275-284,399-402; MEDVEI, 516-518; POTSCH, 233; DAMB, 411; IB, 149; WWWS, 921-922; POGG 6,1302; 7b,2418-2421
E.C. Kendall, Chem. Eng. News 28 (1950), 2074-2077; Mayo Clin. Proc.39 (1964), 548-552; Cortisone. New York 1971
H. Taylor, Am. Phil. Soc. Year Book 1972, pp.216-220
D.J. Ingle, Biog. Mem. Nat. Acad. Sci. 47 (1975), 249-290

KENDALL, FORREST EVERETT [1898-1987]
AMS 12,3238

KENDALL, JAMES PICKERING [1889-1978]
WW 1978, p.1338; CAMPBELL, 127-128; SRS, 47; POGG 5,624; 6,1302;7b,2421-2422
N. Campbell, Nature 275 (1978), 79
N. Campbell and C. Kemball, Biog. Mem. Fell. Roy. Soc. 26 (1980), 255-273

KENDREW, JOHN COWDERY [1917-]
WW 1981, p.1423; IWW 1982-1983, p.677; MH 1,263-264; STC 2,105-106; MSE 2,162-163; CAMPBELL, 129-130; CB 1963, pp.215-217; WWWS, 922; ABBOTT-C, 73; POTSCH, 233-234
Anon., Leopoldina [3]11 (1965), 18-19

KENNARD, OLGA [1924-]
WW 1990, p.1000

KENNAWAY, ERNEST LAURENCE [1881-1958]
DNB [1951-1960], 571-572; WW 1958, p.1657; O'CONNOR(2), 231-232; WWWS, 922
J.W. Cook, Biog. Mem. Fell. Roy. Soc. 4 (1958), 139-154; Proc. Chem. Soc. 1959, pp.68-69
I. Hieger and G.M. Badger, J. Path. Bact. 78 (1958), 539-606
A. Lacassagne, Nature 191 (1961), 743-747
A. Haddow, Persp. Biol. Med. 17 (1974), 543-588

KENNEDY, EUGENE PATRICK [1919-]
AMS 15(4),271; WWA 1980-1981, p.1810; IWW 1982-1983, p.678; MH 2,286-287; MSE 2,163-164; WWWS, 992
E.P. Kennedy, Ann. Rev. Biochem. 61 (1992), 1-28

KENNEDY, JAMES ARTHUR [1894-]
AMS 11,2750; WWA 1964-1965, p.1080

KENELLY, MARY ANTONIUS [1901-]
AMS 7,958

KENNER, GEORGE WALLACE [1922-1978]
WW 1978, pp.1341-1342; CAMPBELL, 131-132; POTSCH, 234
C.W. Rees, Nature 275 (1978), 677-678; Chem. Brit. 14 (1978), 570-571
A.R. Todd, Biog. Mem. Fell. Roy. Soc. 25 (1979), 389-420

KENNER, JAMES [1885-1974]
WW 1965, p.1681; POGG 5,625; 6,1303-1304; 7b,2424-2425
A.R. Todd, Biog. Mem. Fell. Roy. Soc. 21 (1975), 389-405

KENNGOTT, ERWIN [1897-]
JV 39,517
Anon., Nachr. Chem. Techn. 10 (1962), 56; Chem. Z. 96 (1972), 175

KENRICK, FRANK BOTELER [1874-1951]
POGG 4,740; 5,625; 6,1304; 7b,2425
A.R. Gordon, Trans. Roy. Soc. Canada III(3) 46 (1952), 91-96

KENT, ALBERT FRANK STANLEY [1863-1958]
WW 1958, p.1662; O'CONNOR(2), 330-331; WWWS, 923
Anon., Brit. Med. J. 1958(I), pp.894-895
R.H. Anderson and A.E. Becker, Journal of Thoracic and Cardiovascular Surgery 81 (1981), 649-658

KENT, ANDREW [1898-1976]
J.C. Speakman, Chem. Brit. 13 (1977), 72

KENT, WALTER HENRY [1851-1907]
RSC 16,244; NUC 293,499
Who Was Who in America 1,669

KENYON, JOSEPH [1885-1961]
DNB [1961-1970], 608-610; WW 1959, p.1674; ABBOTT-C, 74-75; POGG 5,625; 6,1304; 7b,2426
J.W. Smith, Nature 193 (1962), 117-118

E.E. Turner, Biog. Mem. Fell. Roy. Soc. 8 (1962), 49-66; Proc. Chem. Soc. 1962, pp.193-195

KENYON, RICHARD LEE [1917-1976]
AMS 12,3250
Anon., Chem. Eng. News 54(17) (1976), 9-11

KERB, JOHANNES WOLFGANG [1884-1953]
IB, 149; JV 23,158; NUC 294,109

KERCKHOFF, PETRUS JOHANNES [1813-1876]
NNBW 4,830-831; UTRECHT, 62-63; FERCHL, 270; POGG 1,1246; 3,1714
J.M. van Bemmelen, Jaarb. Akad. Wet. 1879, pp.1-38
H.A.M. Snelders, Janus 69 (1982), 77-95

KERKOVIUS, BERTHOLD WOLDEMAR [1882-]
JV 27,648; NUC 294,169

KERMACK, WILLIAM OGILVY [1890-1970]
DNB [1961-1970], 610-611; WW 1965, p.1685; POGG 6,1307; 7b,2433-2435
J.N. Davidson et al., Biog. Mem. Fell. Roy. Soc. 17 (1971), 399-429
J.L. Simkin, Chem. Brit. 7 (1971), 343

KERN, ALFRED [1850-1893]
NDB 11,517-518; POTSCH, 234
Anon., Chem. Z. 17 (1893), 1928
R. Wizinger-Aust, Alfred Kern. Stuttgart 1970

KERN, WALTHER [1900-1965]
KURSCHNER 9,954; HEIN-E, 234-235; WWWS, 925; POGG 7a(2),729
Anon., Nachr. Chem. Techn. 13 (1965), 157

KERN, WERNER [1906-1985]
KURSCHNER 13,1877; WWWS, 925; POGG 7a(2),730-731
H. Kämmerer, Chem. Z. 105 (1981), 11-12; 109 (1985), 271-273
Anon., Naturw. Rund. 38 (1985), 255; Nachr. Chem. Techn. 33 (1985), 247

KERNDT, CARL HULDREICH THEODOR [1821-]
POGG 1,1247-1248

KERR, JOHN GRAHAM [1869-1957]
DSB 7,314; DNB [1951-1960],578-579; WW 1949, pp.152-153; IB, 149; WWWS, 925
E. Hindle, Biog. Mem. Fell. Roy. Soc. 4 (1958), 155-166

KERR, STANLEY ELPHINSTONE [1894-1976]
AMS 11,2763; WWWS, 925

KERRIDGE, PHYLLIS MARGARET TOOKEY [1902-1940]
NUC 294,386

KERRY, RICHARD [1862-1896]
RSC 16,251
Anon., Chem. Z. 20 (1896), 1038

KERSCHBAUM, MAX [1871-1962]
POGG 6,1308; 7a(2),731
Anon., Z. angew. Chem. 63 (1951), 320

KERSCHNER, LUDWIG [1859-1911]
OBL 3,308-309; STURM 2,132; BJN 16,39*; FISCHER 1,754-755; HUTER, 212-213
H. Rabl, Wiener klin. Wchschr. 24 (1911), 1691-1698

KERSTING, RICHARD GEORG [1821-1875]
WELDING, 370-371; RSC 3,644; 8,70
G. Schweder, Korrespondenzblatt des Naturforscher-Vereins zu Riga 22 (1877), 53-60

KERTESZ, ZOLTAN IMRE [1903-1968]
AMS 11,2764-2765; POGG 6,1308-1309; 7b,2437-2439

KESSLER, GEORG [1828-1873]
RSC 3,645
A. Wankmüller, Beitr. Wurtt. Apothekgesch. 7 (1966), 84

KESTNER, KARL [1803-1870]
BLOKH 1,354; FERCHL, 271; POGG 1,1251; 3,715

KESTNER, OTTO [see COHNHEIM, OTTO]

KESTON, ALBERT S. [1911-1992]
AMS 9(I), 1033; 17(4),303
New York Times, 4 March 1992

KETY, SEYMOUR SOLOMON [1915-]
AMS 15(4),294; IWW 1982-1983, p.682; ICC, 428-429; WWWS, 927-928
S. Kety, Annual Review of Neuroscience 2 (1979), 1-15

KEY, AXEL [1832-1901]
SMK 4,225-226; HIRSCH 3,516
C.G. Santesson, Münch. med. Wchschr. 49 (1902), 242-243

KEYNES, RICHARD DARWIN [1915-]
WW 1981, p.1436; IWW 1982-1983, p.683; WWWS, 928

KEYS, ANCEL BENJAMIN [1904-]
AMS 14,2625; WWA 1972-1973, p.1704; WWWS, 928

KHALATOV, SIMON SERGEEVICH [1884-1951]
BSE(3rd Ed.) 28,172; BME 33,1082-1083; WWR, 266; FISCHER 1,235; IB, 46
V.I. Sukharev and N.T. Shutova, Sov. Med. 18(6) (1954), 43-44

KHARASCH, MORRIS SELIG [1895-1957]
DSB 7,323; DAB [Suppl.6],333-335; AMS 9(I),1034; MILES,271-272; WWWS,929; POTSCH, 234; STC 2,106-107; KAGAN(JA), 741; POGG 5,627; 6,1311-1312; 7b,2449-2454
D.H. Hey, Proc. Chem. Soc. 1958, pp.361-362
F.H. Westheimer, Biog. Mem. Nat. Acad. Sci. 34 (1960), 123-152

KHLOPIN, GRIGORI VITALIEVICH [1863-1929]
BSE 46,213; BME 33,1215-1217; WWR, 269-270; LEVITSKI 2,223-232; POTSCH, 85-86
V.I. Vernadsky, Priroda 19(1) (1930), 94-95

KHLOPIN, NIKOLAI GRIGORIEVICH [1897-1961]
BSE 46,213; BME 33,1217-1218; WWR, 270; IB, 49; WWWS, 929
A.G. Knoppe, Tstitologia 1961(6), pp.629-643
M.G. Shcherbakova and I.G. Puchkov, Vop. Onk. 23(7) (1977), 99-1

KHLOPIN, VITALI GRIGORIEVICH [1890-1950]
BSE 46,212-213; WWR, 270; POGG 6,438-438; 7b,811-813
A.A. Grinberg, Usp. Khim. 19 (1950), 137-141
Anon., Zhur. Anal. Khim. 5 (1950), 259-261

KHODIN, ANDREI VASILIEVICH [1847-1905]
IKONNIKOV, 709-712; FISCHER 1,757
A. Shimanovski, Vest. Oftalm. 22 (1905), i-xi

KHODNEV, ALEKSEI IVANOVICH [1818-1883]
BSE 46,258; KHARKOV-F, 102-103; BLOKH 1,122-123; POGG 1,440; 3,269
L.M. Krasnyanski, Biokhimia 15 (1950), 191-196
N.A. Figurovski and Y.I. Soloviev, Trudy Inst. Ist. Est. 2 (1954), 19-45

KHODUKIN, NIKOLAI IVANOVICH [1896-1957]
BME 34,32-33; WWR, 271
K.M. Dzhalalova, Zhur. Mikr. Ep. Imm. 49(3) (1972), 149-151

KHOLODNY, NIKOLAI GRIGORIEVICH [1882-1953]
BSE 46,300; WWR, 272; UKRAINE 2,452-453; SACKMANN, 69-72
P.I. Belokon, Bot. Zhur. 38 (1953), 453-496
Y. Porutsky, Bull. Soc. Nat. Moscou NS61(2) (1956), 83-92
I.I. Mochalov, Vop. Ist. Est. Tekh. 1982(2), pp.51-65

KHORANA, HAR GOBIND [1922-]
AMS 15(4),302; WWA 1980-1981, p.1825; IWW 1981-1982, p.667; WWWS, 930; DNBI Suppl.2, 279-280; STC 2,107-108; MSE 2,167-168; CB 1970, p.222; POTSCH, 234-235

KHORVAT (HORVATH), ALEKSEI NIKOLAEVICH [1836-]
KAZAN 2,371-372; ZMEEV 5,185; RSC 7,1019; 12,348-349; 15,947; NUC 295,258

KHRUSHCHEV, GRIGORI KONSTANTINOVICH [1897-1962]
WWR, 274; IB, 50; WWWS, 930

KHUDYAKOV, NIKOLAI NIKOLAEVICH [1866-1927]
BSE 46,409; WWR, 274; IB, 50
M.V. Fedorov, Mikrobiologia 21 (1952), 608-610

KIDD, JOHN [1775-1851]
DSB 7,365-366; DNB 11,91-92; BLOKH 1,355; FERCHL, 272; RSC 3,648-649; WWWS, 930; CALLISEN 10,177-178; 29,240

KIDD, JOHN GRAYDON [1908-1991]
AMS 13,2311; WWA 1978-1979, p.1767; WWWS, 930
New York Times, 2 February 1991

KIDDER, GEORGE WALLACE [1902-]
AMS 14,2631; CB 1949, pp.322-324; WWWS, 930

KIELLEY, WILLIAM WAYNE [1916-1980]
AMS 13,2313; WWWS, 931

KIELMEYER, CARL FRIEDRICH [1765-1844]
DSB 7,366-369; NDB 11,581; ADB 15,721-723; SL 1,313-323; FERCHL, 272; HIRSCH 3,518-519; WWWS, 931; CALLISEN 10,181-182; 29,236-288; POGG 1,1253-1254; 7aSuppl.,242-243
C.F. von Martius, Akademische Denkreden, pp.181-209. Leipzig 1866
F. Buttersack et al., Sudhoffs Arch. 23 (1930), 236-288
M. Rauther, Sudhoffs Arch. 31 (1938), 345-350
I. Schumacher, Med. Hist. J. 14 (1979), 81-99

KIESE, MANFRED [1910-1983]
KURSCHNER 13,1894; AUERBACH, 285-286; KIEL, 105; POGG 7a(2),738-739
Anon., Naturw. Rund. 36 (1983), 243; Nachr. Chem. Techn. 32 (1984), 251

KIESEL, ALEKSANDR ROBERTOVICH [1882-1948]
BSE(3rd Ed.) 12,103; IB, 150; POGG 6,1313-1314; 7b,2488-2489
V.K. Kretovich, Aleksandr Robertovich Kiesel. Moscow 1962

KIESER, DIETRICH GEORG [1779-1862]
ADB 15,726-730; HIRSCH 3,519-521; WAGENITZ, 94; CALLISEN 10,183-192; 29,244-246
W. Brednow, Sudhoffs Arch, Beiheft 12 (1970), 1-176

KIESEWETTER, PAUL [1862-1903]
JV 4,65; RSC 16,266; NUC 295,575
Anon., Pharm. Z. 48 (1903), 65

KIESSELBACH, WILHELM [1839-1902]
BJN 7,127-128,57*; HIRSCH 3,521; PAGEL, 855-856; SCHWARTZ, 190-194
V. Urbantschitsch, Mon. Ohren. 37 (1903), 373-376

KIESSLING, WILHELM [1901-1958]

KIHARA, HITOSHI [1893-1986]
JBE, 580-581; IWW 1982-1983, p.689; MH 2,290-291; MSE 2,168-169; IB, 150; SCHULZ-SCHAEFFER, 208
H. Kihara, Nova Acta Leop. NF21 (1959), 270-271
K. Tsunewaki, Am. Phil. Soc. Year Book 1989, pp.211-223

KIJLSTRA, HENRIK JOHANNES [1862-]
JV 4,137; NUC 295,632

KIKKAWA, HIDEO [1908-]
JBE, 582
M. Ishidate, Kagakusi Kenkyu [2]19 (1980), 129-139

KIKUTH, WALTER [1896-1968]
NDB 11,602-603; WELDING, 381-382; DUNSCHELE, 138-143; WWWS, 932; POGG 7a(2),742-744
L. Grün, Med. Welt 1968, pp.1875-1876
H. Knothe and K. Bartmann, Arzneimitt. 18 (1968), 1354-1355

KILIANI, HEINRICH [1855-1945]
DZ, 728-729; TLK 3,327-328; POGG 3,717-718; 4,747; 5,628-629; 6,1315; 7a(2),744
H. Kiliani, J. Chem. Ed. 9 (1932), 1908-1914
W. Hückel, Chem. Ber. 82 (1949), i-ix

KIMBALL, GEORGE ELBERT [1906-1967]
AMS 11,2784; WWAH, 329; WWWS, 932
P.M. Morse, Biog. Mem. Nat. Acad. Sci. 43 (1973), 129-146

KIMICH, CARL [1852-1937]
ZURICH-D, 126; RSC 10,395-396; NUC 296,757

KIMMELSTIEL, PAUL [1900-1970]
AMS 11,2785; BHDE 2,619; LDGS, 76
P.M. LeCompte, Diabetes 20 (1971), 117-118
K.F. Wellmann, Am. J. Clin. Path. 56 (1971), 117-119; Verhandl. path. Ges. 55 (1971), 749-754

KINDLER, KARL [1891-1967]
NDB 11,619-620; KURSCHNER 10,1162; TLK 3,328; WWWS, 933; POGG 6,1316; 7a(2),746
Anon., Chem. Z. 80 (1956), 680; Pharm. Z. 112 (1967), 1474

KINDRED, JAMES ERNEST [1893-1974]
AMS 11,2787; IB, 150; WWWS, 933

KINDT, GEORG CHRISTIAN [1793-1869]
NDB 11, 621; ADB 15,769-770; HEIN, 318-319; FERCHL, 272
F. Buchenau, Abh. Nat. Ver. Bremen 2 (1871), 191-200;

3 (1873), 378-379
W.O. Focke, Bremische Biographien, pp.248-250. Bremen 1972
K. Liesche, Deutsche Apoth. Z. 98 (1958), 1011-1013

KINDT, HEINRICH HUGO [1775-1837]
HEIN-E, 236-237; FERCHL, 272; POGG 1,1255-1256

KING, CHARLES GLEN [1896-1988]
AMS 13,2324; WWA 1976-1977, p.1719; IWW 1982-1983, p.692; NCAB I,280-281; MH 2,291-293; MSE 2,169-170; CB 1967, pp.227-229; WWWS, 933-934
C.G. King, Fed. Proc. 38 (1979), 2681-2683
New York Times, 25 January 1988
F.J. Stare, Nature 332 (1988), 398; J. Nutrition 118 (1988), 1272-1277
T.H. Jukes, J. Nutrition 118 (1988), 1290-1293

KING, EARL JUDSON [1901-1962]
DNB [1961-1970], 612-613; YOUNG, 98-99; POGG 6,1317-1318; 7b,2464-2469
Anon., Lancet 1962(II), pp.998-999; Brit. Med. J. 1962(II), pp.1262-1263
A. Neuberger and W. Klyne, Biochem. J. 89 (1963), 401-404
A.G. Signy, J. Clin. Path. 16 (1963), 92-93
H. Gaertner, Deutsche med. Wchschr. 88 (1963), 246-247
C.V. Harrison, J. Path. Bact. 88 (1964), 601-614

KING, FRANKLIN HIRAM [1848-1911]
NCAB 19,292-293; WWWS, 934
A.W. Shorger, Passenger Pigeon 7(4) (1945), 117-121

KING, FREDERICK ERNEST [1905-]
WW 1992, p.1043

KING, HAROLD [1887-1956]
DSB 17,472-474; DNB [1951-1960], 585-587; WW 1949, p.1547; WWWS, 934; POTSCH, 235; POGG 5,630; 6,1318; 7b,2469-2470
C. Harington, Biog. Mem. Fell. Roy. Soc. 2 (1956), 157-172
T.S. Work, Nature 177 (1956), 604-605

KING, HELEN DEAN [1869-1955]
DSB 17,474-478; AMS 8,1352; OGLIVIE, 108-110; NUC 286, 377-379

KING, HUGH KIRKMAN [1917-]
Who's Who of British Scientists 1980-1981, p.278

KING, THOMAS WILKINSON [1809-1847]
SG [1]8,437
Anon., Brit. Med. J. 1936(I), p.1223

KING, VICTOR LOUIS [1886-1958]
AMS 8,1354; NUC 296,499-500

KINGSBURY, BENJAMIN FREEMAN [1872-1946]
AMS 7,972
Anon., Science 104 (1946), 110

KINGSBURY, FRANCIS BULLARD [1886-1964]
AMS 6,781; IB, 151

KINGZETT, CHARLES THOMAS [1852-1935]
WWWS, 936; POGG 3,719; 6,1319; 7b,2472
Anon., Analyst 60 (1935); J. Roy. Inst. Chem. 1935, p.328
G.T. Morgan, J. Chem. Soc. 1935, pp.1899-1902
E.H. Huntress, Proc. Am. Acad. Arts Sci. 81 (1952), 67-68

KINNICUTT, LEONARD PARKER [1854-1911]
DAB 10,418-419; NACB 25,160; MILES, 273-274; BLOKH 1,355-356; WWAH, 332; POGG 5,531
R. Anschütz, Ber. chem. Ges. 44 (1911), 3567-3570
W.L. Jennings, Science 33 (1911), 649-651; Proc. Am. Acad. Arts Sci. 53 (1918), 821-824

KIONKA, HEINRICH [1868-1941]
FISCHER 1,761; GIESE, 517-518; TLK 3,329; IB, 151; POGG 7a(2),748-749
U. Heintzelmann, Balneologe 8 (1941), 128

KIPPENBERGER, KARL [1868-1937]
KURSCHNER 5,670; TSCHIRCH, 1073; TLK 2,354-355; POGG 4,749-750; 5,631-632; 6,1320; 7a(2),749-750
A. Stock, Ber. chem. Ges. 70A (1937), 75-76

KIPPING, FREDERIC BARRY [1901-1965]
POGG 6,1320; 7b,2472
F.G. Mann, Chem. Brit. 1 (1965), 379-380
D.H. Peacock, Chem. Ind. 1965, pp.1019-1020

KIPPING, FREDERICK STANLEY [1863-1949]
DSB 7,372-373; DNB [1941-1950], 462-563; WW 1949, p.1554; JV 2,193; POTSCH, 236; W, 290; WWWS, 937; POGG 4,750; 5,632; 6,1320; 7b,2473
F. Challenger, Obit. Not. Fell. Roy. Soc. 7 (1950), 183-219; J. Chem. Soc. 1951, pp.849-862

KIRCH, EUGEN [1888-1973]
KURSCHNER 11,1438; NB, 203; BERWIND, 111-117
C. Dhorn, Verhandl. path. Ges. 57 (1973), 495-498

KIRCHHOF, CONSTANTIN [1764-1833]
DSB 7,378-379; HEIN-E, 237-238; ARBUZOV, 97; BLOKH 1,359-360;POTSCH, 236; SCHAEDLER, 57-58; POGG 1,1260; 7aSuppl.,326
W. Volksen, Die Stärke 1 (1949), 30-36

KIRCHHOFF, GUSTAV ROBERT [1824-1887]
DSB 7,379-383; NDB 11,649-650; ADB 51,165-167; BAD 4,218-222; W, 290-291; KROLLMANN, 335; DRULL,135; BLOKH 1,356-359; WWWS, 937; POTSCH, 236; POGG 1,1260-1261; 3,720-721; 4,750-751; 6,1321; 7aSuppl.,326-329
A.W. Hofmann, Ber. chem. Ges. 20 (1887), 2771-2777
C. von Voit, Sitz. Bayer. Akad. 18 (1888), 181-186
W. Voigt, Zum Gedächtnis von Gustav Kirchhoff. Gottingen 1888
Anon., Proc. Am. Acad. Arts Sci. 23 (1888), 370-375
E. Warburg, Naturwiss. 13 (1925), 205-212

KIRK, JOHN ESBEN [1905-1975]
AMS 12,3309-3310; VEIBEL 2,244; WWWS, 938

KIRK, PAUL LELAND [1902-1970]
AMS 11,2806; NCAB 56,490-491; WWWS, 938
Anon., Chem. Eng. News 48(46) (1970), 45

KIRK, ROBERT [1843-1907]
RSC 10,403; 16,289
Anon., Glasgow Med. J. 68 (1907), 33

KIRK, ROBERT [1905-1962]
A.J.S. McFadzean et al., J. Path. Bact. 88 (1964), 614-621

KIRKWOOD, JOHN GAMBLE [1907-1959]
DSB 7,387; DAB [Suppl.6],345-346; AMS 9(I),1049; MILES, 276-277; WWAH, 333; WWWS, 939
Anon., Chem. Eng. News 37(35) (1959), 106
G. Scatchard, J. Chem. Phys. 33 (1960), 1279-1281

S.N. Timasheff et al., Arch. Biochem. Biophys. 86 (1960), i-iii

KIRMREUTHER, HEINRICH [1884-1961]
JV 24,575; NUC 298,6
Anon., Nachr. Chem. Techn. 2 (1954), 204; 9 (1961), 58

KIRPAL, ALFRED [1867-1943]
TLK 3,330; POGG 4,752; 5,633; 6,1322; 7a(2),753-754
C.J. Cori, Ber. chem. Ges. 76A (1943), 38-39

KIRRMANN, ALBERT [1900-1974]
QQ 1973-1974, p.928; WWWS, 939; POGG 6,1322; 7b,2477-2479
Anon., Ann. Univ. Paris 25 (1955), 175-176

KIRSANOV, ALEKSANDR TROFIMOVICH [1866-1927]
WWR, 278; LIPSHITS 4,166-167
L.I. Prasolov, Priroda 32(6) (1943), 90-91

KIRSCHLEGER, FRÉDÉRIC [1804-1869]
HEIN, 321-322; RSC 3,664-665; 8,82
Anon., Bull. Soc. Bot. 40 (1869), 318-320
P. Bachoffner, Rev. Hist. Pharm. 19 (1969), 493-496
P. Jaeger, Saisons d'Alsace NS38 (1971), 189-221

KISCH, BRUNO [1890-1966]
NDB 11,680-682; STURM 2,150; AMS 11,1812; NCAB 55,33; WININGER 7,172-173; FISCHER 1,763-764; KOERTING, 155-156; BHDE 2,621-622; LDGS, 79; KOREN, 202; KAGAN, 194; COHEN, 126; TETZLAFF, 175; TLK 3,330; IB, 151; POGG 6,1323; 7a(2),757-758
B. Kisch, Wanderungen und Wandlungen. Cologne 1966
H. Jentgens, Deutsche med. Wchschr. 91 (1966), 1853-1854
L.G. Stevenson, J. Hist. Med. 22 (1967), 47-53

KISHNER (KIZHNER), NIKOLAI MATVEEVICH [1867-1935]
BSE 20,608; WWR, 280; ARBUZOV, 157-160; POTSCH, 237-238
S.S. Nametkin, Zhur. Ob. Khim. 6 (1936), 1379-1392
N.I. Lomov, Trudy Tomskovo Universiteta 126 (1954), 240-250

KISSER, JOSEF [1899-1984]
KURSCHNER 13,1914; TLK 3,330; IB, 152; WWWS, 939; POGG 6,1324-1325; 7a(2),758-759
F. Gabler, Mikroskopie 14 (1959), 129-130
G. Halbwachs et al., Mikroskopie 35 (1979), 181-182
H. Richter and G. Halbwachs, Verhandl. zool. bot. Ges. Wien 118-119 (1980), 12-14
Anon., Naturw. Rund. 37 (1984), 472

KISSKALT, KARL [1875-1962]
NDB 11,687; KURSCHNER 8,1154; FISCHER 1,764-765; KIEL, 84; TLK 3,331; IB, 152; POGG 7a(2),759-760
K.W. Jötten et al., Arch. Hyg. 115 (1936), 127-134
Anon., Chem. Z. 75 (1951), 132; 80 (1956), 44
E. Kanz, Med. Klin. 57 (1962), 2194-2195
M. Knorr, Arch. Hyg. 146 (1962), 161-170

KISTIAKOVSKY, VASILI FEDOROVICH [1841-]
IKONNIKOV, 260-261

KISTIAKOVSKY, VLADIMIR ALEKSANDROVICH [1865-1952]
BSE 21,165; WWR, 280; UKRAINE 2,288-289; POTSCH, 237; POGG 4,752; 5,633; 6,1325-1326; 7b,2485-2486
P.D. Dankov, Usp. Khim. 20 (1951), 265-169
K.M. Gorbunova, Zhur. Fiz. Khim. 26 (1952), 1717-1720

KISTIAKOWSKY, GEORGE BOGDAN [1900-1982]
AMS 14,2667; IWW 1982-1983, p.697; CB 1960, pp.219-220; AO 1982, pp.562-565; POGG 6,1325; 7b,2482-2485
Anon., Nature 183 (1959), 1639; 187 (1960), 112-113; Chem. Eng. News 59(5) (1981), 20-26
E.B. Wilson, J. Phys. Chem. 75(10) (1972), 5A-10A
W. Hylin, Chem. Eng. News 60(50) (1982), 5
J.K. Galbraith et al., Harvard Gazette, 21 December 1984, p.5
F.H. Westheimer, Am. Phil. Soc. Year Book 1984, pp.121-129
F. Dainton, Biog. Mem. Fell. Roy. Soc. 31 (1985), 377-408

KIT, SAUL [1920-]
AMS 15(4),350; WWA 1980-1981, p.1852; WWWS, 940

KITASATO, SHIBASABURO [1852-1931]
DSB 7,391-393; FISCHER 1,765; PAGEL, 860; FREUND 2,169-176; BULLOCH, 375; OLPP, 197-198; SACKMANN, 173-176; W, 291-292; ABBOTT, 96-97; POGG 6,1327; 7b,2487
W. Bulloch, Nature 128 (1931), 142-143; Proc. Roy. Soc. B109 (1931), xi-xvi; J. Path. Bact. 34 (1931), 597-602
M. Miyajima, Science 74 (1931), 124-125
H. Fox, Ann. Med. Hist. 6 (1934), 491-499
N. Howard-Jones, Persp. Biol. Med. 16 (1973), 292-307
W. Koehler, Leopoldina [3]28 (1985), 99-122

KITASATO, ZENJIRO [1897-]
JBE, 631; POGG 6,1327; 7b,2487-2488

KITSCHELT, MAX [1866-1939]
JV 5,234; RSC 13,275; 16,298
Anon., Z. angew. Chem. 52 (1939), 560

KIWISCH, FRANZ [1814-1851]
NDB 11,695-696; ADB 16,47-49; OBL 3,361-362; STURM 2,153-154; WWWS, 941; WURZBACH 11,343-346; HIRSCH 3,534-535; PAGEL, 860-861; DECHAMBRE [4]16, 754
B. Mueller, Franz Kiwisch. Würzburg 1980

KIZER, DONALD EARL [1921-]
AMS 15(4),353; WWWS, 941

KJELDAHL, JOHAN GUSTAV [1849-1900]
DSB 7,393-394; DBL 12,480-482; MEISEN, 169-172; BLOKH 1,361-362; VEIBEL 1,217-218; 2,245-246; 3,115-151; WWWS, 941; POTSCH, 238
W. Johannsen, Ber. chem. Ges. 33 (1900), 3881-3888; C.R. Lab. Carlsberg 5 (1900), i-viii
R.E. Oesper, J. Chem. Ed. 11 (1934), 457-462
H.B. Vickery, Yale J. Biol. Med. 18 (1946), 473-516
S. Veibel, J. Chem. Ed. 26 (1949), 459-461
E.H. Huntress, Proc. Am. Acad. Arts Sci. 77 (1949), 47-48
P. Morries, Journal of the Association of Public Analysts 21(2) (1983), 53-58
D.T. Burns, Anal. Proc. 21 (1984), 210-214
K.D. Wutzke and W. Heine, Zeitschrift für Medizinische Laboratorium-diagnostik 26 (1985), 383-388

KLAATSCH, HERMANN [1863-1916]
NDB 11,697-698; DBJ 1,360; DRULL, 136; WWWS, 941
R. Wegner, Anat. Anz. 48 (1916), 611-623

KLAGES, AUGUST [1871-1957]
DRULL, 136; POGG 7a(2),761

KLAGES, FRIEDRICH [1904-1988]

KURSCHNER 13,1917; WWWS, 941; POGG 7a(2),761-762

KLAGES, LUDWIG [1872-1956]
JV 16,251; NUC 298,408-412
W. Hager, Ludwig Klages. Munich 1957

KLAPROTH, MARTIN HEINRICH [1743-1817]
DSB 7,394-395; NDB 11,707-709; ADB 16,60-61; HEIN, 322-324; W, 292-293; BUGGE 1,334-341; BLOKH 1,362-363; COLE, 288-291; FERCHL, 274-275; SCHAEDLER, 58-59; PHILIPPE, 714; TSCHIRCH, 1073; ABBOTT-C, 95-96; POTSCH, 238-239; CALLISEN 29,260; POGG 1,1266-1268; 7aSuppl.,329-331
G.E. Dann, Martin Heinrich Klaproth. Berlin 1958

KLARER, JOSEF [1898-1953]
NDB 11,709-710; DUNSCHELE, 143; POTSCH, 239; POGG 7a(2),764

KLARMANN, EMIL [1900-1963]
AMS 10,2183; POGG 6,1328; 7b,2490-2491
Anon., Chem. Eng. News 31 (1953), 5288-5289; 41(41) (1963), 144

KLASON, JOHANN PETER [1848-1937]
SMK 4,265-266; FARBER, 867-874; WWWS, 941; POGG 3,274; 4,753-754;5,634-635; 6,1328; 7b,2492
C. Kullgren, Sven. Kem. Tid. 49 (1937), 323-325
Anon., Ind. Eng. Chem. News Ed. 15 (1937), 313

KLATTE, FRITZ [1880-1934]
HEIN-E, 239; POTSCH, 239

KLAVEHN, WILFRID [1898-]
JV 40,465
Anon., Nachr. Chem. Techn. 11 (1963), 204

KLEBS, EDWIN THEODOR [1834-1913]
NDB 11,719-720; BJN 18,101*; STURM 2,159; HIRSCH 3,539-540; Suppl.,386;PAGEL, 863-864; BULLOCH, 376; SACKMANN, 176-180; DAMB, 417-418; LOMMATZSCH, 31-47; KROLLMANN, 336; KOERTING, 141-143
B. Naunyn, Arch. exp. Path. Pharm. 74 (1913), i-iii
P. Ernst, Verhandl. path. Ges. 17 (1914), 588-597
F.H. Garrison, Science 38 (1914), 920-921
O.M. Rothlin, Edwin Klebs. Zürich 1962

KLEBS, GEORG [1857-1918]
DSB 7,395-396; NDB 11,720-721; DBJ 2,693; DRULL, 136-137; WWWS, 942
E. Küster, Ber. bot. Ges. 36 (1918), (90)-(116)
M. Bopp, Naturw. Rund. 22 (1969), 97-101

KLECZKOWSKI, ALFRED ALEXANDER PETER [1908-1970]
WW 1970, p.1744; WWWS, 942
F.C. Bawden, Biog. Mem. Fell. Roy. Soc. 17 (1971), 431-440

KLEIBER, MAX [1893-1976]
AMS 12,2822
A. Schürch, Z. Tierphysiol. 36 (1976), 291-293
Anon., Science 193 (1976), 988

KLEIN, ALEXANDER [1865-1946]
LINDEBOOM, 1047-1048; FISCHER 1,769; HES, 86-87

KLEIN, EDWARD EMANUEL [1844-1925]
FISCHER 1,770; BULLOCH, 376; O'CONNOR, 155-157; HEINDEL, 14-18; KOREN, 202; WWWS, 942
W. Bulloch, J. Path. Bact. 28 (1925), 684-697
Anon., Lancet 1925(I), pp.411-412; St. Barts. Hosp. Rep. 58 (1925), 1-7
F.W.A., Proc. Roy. Soc. B98 (1925), xxv-xxix

KLEIN, FRIEDRICH [1852-1922]
KIEL, 95-96; NUC 296,628

KLEIN, GEORGE [1925-]
VAD 1987, p.597; IWW 1987-1988, p.783
G. Klein and E, Klein, Ann. Rev. Immunol. 7 (1989), 1-33

KLEIN, GUSTAV [1892-1954]
PLANER, 168; DRULL, 137; TLK 3,333; IB, 152; POGG 6,1330; 7a(2),769-770
F.K.T. Schwarz, Planta Medica 2 (1954), 177-178
Anon., Chem. Z. 78 (1954), 542

KLEIN, MARC [1905-1975]
IB, 152
G. Canguilhem, Arch. Inter. Hist. Sci. 26 (1976), 163-164
J. des Cilleuls, Hist. Sci. Med. 10 (1976), 46-54

KLEIN, OTTO [1891-1968]
STURM 2,163; KURSCHNER 9,986; PELZNER, 123-137; KOERTING, 174; IB, 152

KLEIN, SALOMON [1845-1937]
FISCHER 1,770-771

KLEINER, ISRAEL SIMON [1885-1966]
AMS 11,2826-2827; IB, 152; WWAH, 334; WWWS, 943
Anon., Chem. Eng. News 44(30) (1966), 79
I. Neuwirth, Pharmacologist 9 (1967), 107-108

KLEINFELLER, HANS [1897-1973]
KURSCHNER 11,1466; KIEL, 196; POGG 7a(2),772
Anon., Chem. Z. 97 (1973), 571

KLEINMANN, HANS [1895-]
LDGS, 59; POGG 6,1330-1331; 7a(2),773

KLEINSCHMIDT, HANS [1885-1977]
NDB 12,6-7; KURSCHNER 12,1586; FISCHER 1,771; AUERBACH, 289; PUSCHEL, 134
H. Kleinschmidt, Hippokrates 39 (1968), 783-788
K.H. Schäfer et al., Mon. Kind. 125 (1977), 117-121
G. Ippich, Eur. J. Ped. 124 (1977), 165-166

KLEINZELLER, ARNOST [1914-]
AMS 14,2680; WWA 1980-1981, p.1859; WWWS, 943; POTSCH, 239-240
Anon., Leopoldina [3]12 (1966), 35-36

KLEMEN, RICHARD [1902-]
KURSCHNER 13,1938; POGG 7a(2),774

KLEMENC, ALFONS [1885-1960]
TLK 3,335; POGG 5,637; 6,1331-1332; 7a(2),774-775
E. Hayek, Ost. Chem. Z. 56 (1955), 249-253
V. Guttmann, Ost. Chem. Z. 61 (1960), 90-91

KLEMENSIEWICZ, ZYGMUNT [1886-1963]
PSB 12,598-600; POGG 5,637; 6,1332; 7b,2494-2495
M. Konopacki and J. Szpilecki, Wiad. Chem. 18 (1964), 137-145

KLEMM, WILHELM [1896-1985]
KURSCHNER 13,1940; TLK 3,335; HANNOVER, 40; KIEL, 174; WWWS, 944; POTSCH, 240; POGG 6,1332; 7a(2),779-781
R.E. Oesper, J. Chem. Ed. 29 (1952), 336-337

Anon., Z. anal. Chem. 149 (1956), 1; Naturw. Rund. 38 (1985),535; 39 (1986), 10

KLEMME, CARL JOSEPH [1898-1967]
AMS 11,2829-2830

KLEMPERER, FELIX [1866-1932]
NDB 12,34*; KURSCHNER 4,1464; FISCHER 1,773; WININGER 3,465; 7,177; TETZLAFF, 178

KLEMPERER, GEORG [1865-1946]
NDB 12,34; FISCHER 1,773; PAGEL, 867; DZ, 743; BHDE 2,630; KOREN, 202; WININGER 3,465-466; TETZLAFF, 178
K. Hausen, Deutsche med. Wchschr. 72 (1947), 362-363
R. Nissen, Proc. Virchow Soc. 6 (1947), 129-130

KLEMPERER, PAUL [1887-1964]
AMS 10,2191; KOREN, 202; KAGAN(JA), 312,726-737; WWWS, 944
E. Moschcowitz, J. Mt. Sinai Hosp. 24 (1957), 648-654
S. Jarcho, Bull. Hist. Med. 38 (1964), 278-280
A.D. Pollack, Arch. Path. 78 (1964), 306-312
E. Rubin, Laboratory Investigation 47 (1982), 110
G.E. Ehrlich, J. Am. Med. Assn. 251 (1984), 1593-1596

KLENK, ERNST [1896-1971]
DSB 17,484-485; NDB 12,42-43; KURSCHNER 10,1191; POGG 6,1333; 7a(2),781-782
E. Klenk, Chem. Phys. Lipids 5 (1970), 193-197
H. Debuch, J. Neurochem. 21 (1973), 725-727
A. Butenandt, Ernst Klenk (Kölner Universitätsreden 48). Krefeld 1973
T. Yamagawa, TIBS 13 (1988), 452-454

KLENK, LUDWIG [1903-]
JV 47,554

KLENSCH, HERBERT [1914-]
KURSCHNER 13,1941; POGG 7a(2),782-783

KLETT, MAX [1865-]
JV 8,68; RSC 13,355,397; NUC 299,220
A. Wankmüller, Beitr. Wurtt. Apothekgesch. 13 (1982), 146

KLETZINSKY, VINCENZ [1826-1882]
OBL 3,398; BLOKH 1,364-365; TSCHIRCH, 1073; POGG 3,725-726
A.E. Hoswell, Ber. chem. Ges. 15 (1882), 3310-3315
Anon., Wiener med. Wchschr. 32 (1882), 346-347

KLEYN, ADRIAAN de [1883-1949]
LINDEBOOM, 1049-1050; FISCHER 1,769
B. Brouwer, Ned. Tijd. Gen. 93(II) (1949), 1658-1661

KLIEGL, ALFRED [1877-1953]
LEHMANN, 191; TLK 3,336; POGG 5,637; 6,1333; 7a(2),783-784
W. Hückel, Chem. Ber. 92 (1959), xxi-xxviii

KLIENEBERGER, CARL [1876-1938]
SG [2]8,752; [3]7,261; [4]9,246
E. Klieneberger-Nobel, Memoirs. London 1980

KLIENEBERGER-NOBEL, EMMY [1892-1985]
KURSCHNER 4,1467; LDGS, 57; IB, 153; BOEDEKER, 95
G. Henneberg, Z. Bakt. 204 (1967), 305-308
H. Chick et al., War on Disease, pp.138-140. London 1971
E. Klieneberger-Nobel, Pioneerleistungen für die Medizinische Mikrobiologie. Stuttgart 1977; Memoirs. London 1980
Anon., Lancet 1985(II), pp.960-961; Brit. Med. J. 291 (1985), 1213

KLIMMER, MARTIN [1873-1943]
NDB 12,68; FISCHER 1,774; TLK 3,336; IB, 153; POGG 4,759; 7a(2),784

KLING, ANDRÉ [1872-1947]
POGG 5,638-639; 6,1334; 7b,2496-2497
A. Lassieur, Chimie Analytique 29 (1947), 195-196
E. Cherbuliez, Chimia 2 (1948), 40

KLINGER, HEINRICH [1853-1945]
KURSCHNER 4,1470; TLK 3,336-337; POGG 4,760-761; 5,639; 6,1334; 7a(2),786

KLINKE, KARL [1897-1972]
KURSCHNER 11,1476; LDGS, 77
H. Küster, Kinderaerztliche Praxis 40 (1972), 337

KLOB, JULIUS [1831-1879]
OBL 3,415; STURM 2,181; HIRSCH 3,547
Anon., Wiener med. Bl. 2 (1879), 727-729

KLOBB, CONSTANT TIMOTHÉE [1861-1912]
DORVEAUX, 82; TSCHIRCH, 1073; POGG 4,761; 5,639-640
Anon., Chem. Z. 36 (1912), 404

KLOPSTOCK, ALFRED [1896-1968]
KURSCHNER 4,1473; DRULL, 138-139; BHDE 2,632-633; LDGS, 57

KLOSA, JOSEF [1921-]
WWWS, 945; POGG 7a(2),789-790

KLOTZ, CARL [1862-1938]
RSC 16,327; NUC 299,525

KLOTZ, IRVING MYRON [1916-]
AMS 15(4),379; WWA 1980-1981, p.1864; IWW 1982-1983, p.702; COHEN, 128; WWWS, 946

KLÜPFEL, RICHARD [1848-1917]
RSC 8,89
T. Grünewald, Med. Korr. Wurtt. 88 (1918), 28-29

KLUG, AARON [1926-]
WW 1981, p.1661; POTSCH, 240
K.C. Holmes, TIBS 8 (1983), 3-5

KLUG, NÁNDOR (FERDINAND) [1845-1909]
MEL 1,938; FISCHER 1,776; PAGEL, 369-370; RSC 10,417; 12,390; 16,328
F. Tangl, Orv. Het. 53 (1909), 378-379

KLUYVER, ALBERT JAN [1888-1956]
DSB 7,405-407; BWN 2,305-307; MH 2,294-296; MSE 2,175; SACKMANN, 181-184; POTSCH, 240-241; IB, 153; POGG 6,1338; 7b,2501-2503
G. van Iterson et al., Chem. Wkbl. 36 (1939), 307-323
V.J. Koningsberger, Acta Bot. Neer. 5 (1956), 215-217
T.V. Kingma Boltjes et al., Chem. Wkbl. 52 (1956), 661-665; Enzymologia 18 (1957), 3-6
C.B. van Niel, J. Gen. Microbiol. 16 (1957), 499-521
D.D. Woods, Biog. Mem. Fell. Roy. Soc. 3 (1957), 109-128
A.F. Kamp et al., Albert Jan Kluyver: His Life and Work. Amsterdam 1959 (reprint, Madison, Wis. 1983)

C.J. Bulder et al., Antonie van Leeuwenhoek 56 (1989), 109-126

KLYNE, WILLIAM [1913-1977]
WW 1976, pp.1335-1336; CAMPBELL, 137-138
D.N. Kirk and P.M. Scopes, Nature 272 (1978), 567-568
O. Hoffmann-Ostenhof, TIBS 3 (1978), N132-N133

KNABE, JOACHIM [1921-]
KURSCHNER 16,1832; WWWS, 947; POTSCH, 241

KNAF, JOSEF [1801-1865]
STURM 2,187
L. Celakovsky, Lotos 16 (1866), 82-89

KNAFFL-LENZ, ERICH [1880-1962]
FISCHER 1,776-777; PLANER, 172; DECKER, 71-76; WIEN 1964, pp.41-42; TLK 3,338; POGG 6,1338-1339; 7a(2),799
F. Brücke, Wiener klin. Wchschr. 74 (1962), 928

KNAPP, EDGAR [1906-1978]
KURSCHNER 13,1965; WAGENITZ, 96
T. Butterfass, Ber. bot. Ges. 93 (1980), 505-515

KNAPP, FRIEDRICH LUDWIG [1814-1904]
NDB 12,151; BJN 10,58*-59*; HEIN, 329; BLOKH 1,365; FERCHL, 277; POTSCH, 241; ALBRECHT, 41; SCHAEDLER, 59; WWWS, 947; POGG 1,1278-1279; 3,728; 4,764;5,641
R. Meyer, Ber. chem. Ges. 37 (1904), 4777-4814

KNAPP, HERMANN [1832-1911]
NDB 12,155; DAB 10,449-450; NB, 207-208; HIRSCH 3,555-556; PAGEL, 870; FISCHER 1,777; WWWS, 947

KNAPP, LUDWIG [1868-1925]
OBL 3,432; STURM 2,188; FISCHER 1,777-778; MOTSCH, 102-112

KNAUER, EMIL [1867-1935]
NDB 12,159-160; OBL 3,432-433; FISCHER 1,778; GAUSS, 263; WEBER, 59
A. Schauenstein, Wiener med. Wchschr. 85 (1935), 701-703
H. Knaus, Archiv für Gynäkologie 1 (1935), 429-431

KNAUS, HERMANN HUBERT [1892-1970]
NDB 12,163-164; STURM 2,189; KURSCHNER 9,1008-1009; FISCHER 1,778; HARTMANN, 61-76; WWWS, 947
H. Husslein, Z. Gyn. 93 (1971), 177-179

KNECHT, EDMUND [1861-1925]
ZURICH-D, 138; POGG 5,641; 6,1340
A. Rée, J. Chem. Soc. 129 (1926), 1021-1024
W.M. Gardner, Nature 117 (1926), 164-165

KNEISEL, RUDOLF [1881-1962]
JV 22,323; NUC 300,104

KNELL, WILHELM [1876-1916]
JV 17,285

KNIEP, HANS [1881-1930]
NDB 12,182; TLK 3,339; IB, 158
R. Harder, Ber. bot. Ges. 48 (1930), (164)-(196)
R. Kolkwitz, Verhandl. Int. Ver. Limn. 5 (1931), 712-713

KNIERIEM, WOLDEMAR von [1849-1935]
KURSCHNER 4,1486; WELDING, 391; RIGA, 706; LEVITSKI 1,399-401

Anon., Chem. Z. 59 (1935), 190

KNIES, MAX [1851-1917]
NDB 12,182*; FISCHER 1,779

KNIETSCH, RUDOLF [1854-1906]
NDB 12,183-184; BLOKH 1,365-366; POGG 4,765; 5,642
H. von Brunck, Ber. chem. Ges. 39 (1906), 4479-4490
F. Raschig, Z. angew. Chem. 19 (1906), 1217-1221
H. Wolf, Chem. Z. 79 (1955), 229-243

KNIGHT, BERT CYRIL GABRIEL [1904-1981]
WW 1981, p.1462
L.J. Zatman, J. Gen. Microbiol. 129 (1983), 1261-1268

KNIPPING, PAUL [1883-1935]
WOLF, 106; TLK 3,340; POGG 5,643; 6,1341; 7a(2),805
Anon., Z. angew. Chem. 48 (1935), 712
H. Stintzing, Z. tech. Phys. 19 (1938), 104-105

KNOBLOCH, HEINRICH [1915-]
KURSCHNER 13,1976; POGG 7a(2),805-806

KNÖPFER, GUSTAV [1864-1937]
STURM 2,176; POGG 5,644; 6,1343

KNOEVENAGEL, HEINRICH EMIL ALBERT [1865-1921]
NDB 12,206-207; DRULL, 142; BLOKH 1,366-367; TLK 2, 368; POTSCH, 241-242; POGG 4,766-767; 5,644; 6,1343
P. Jacobson, Ber. chem. Ges. 54A (1921), 269-271
C. Wilke, Z. angew. Chem. 35 (1922), 29-30

KNOEVENAGEL, OSCAR [1862-1944]
JV 2,245; RSC 14,1010; NUC 300,451

KNOLL, MAX [1897-1969]
KURSCHNER 10,1216; POGG 6,1343-1344; 7a(2),808
M.M. Freundlich, Science 142 (1963), 185-188

KNOLL, PHILIPP [1841-1900]
OBL 3,447-448; STURM 2,197; FISCHER 1,780; LOMMATZSCH, 80-87
I. Herrnhauser, Prag. med. Wchschr. 33 (1898), 621-625
M. Lowit, Wiener klin. Wchschr. 13 (1900), 139
F. Hofmeister, Arch. exp. Path. Pharm. 44 (1900), i-v

KNOOP, FRANZ [1875-1946]
NDB 12,214; FISCHER 1,780; TLK 3,340-341; IB, 154; POTSCH, 242; POGG 6,1344; 7a(2),809
H. Wieland, Bayer. Akad. Wiss. Jahrbuch 1944-1948, pp.271-272
K. Thomas, Z. physiol. Chem. 283 (1948), 1-8
P. Ohlmeyer, Z. angew. Chem. 60 (1948), 29-33
C. Martius, Sitzungsberichte der Heidelberger Akademie der Wissen-schaften 1943-1955, pp.51-53

KNOP, CONRAD ALEXANDER [1828-1873]
A. Paalzow, Ber. chem. Ges. 6 (1873), 1581-1582

KNOP, JOSEF [1885-1964]
STURM 2,197; POGG 6,1344; 7a(2),810
Anon., Chem. Listy 59 (1965), 122-126

KNOP, WILHELM [1817-1891]
NDB 12,214-215; BLOKH 1,367-368; FERCHL, 277-278; SCHAEDLER, 59; POGG 1,1283-1284; 3,729-730; 4,767
W. Haan, Sächsischer Schriftsteller-Lexikon, pp.165-166. Leipzig 1875
Anon., Leopoldina 27 (1891), 57

KNOPF, SIGARD ADOLPHUS [1857-1940]
AMS 6,791; NCAB 29,54-55; WWAH, 335-336; WWWS, 949
Anon., Science 92 (1940), 74; J. Am. Med. Assn. 115 (1940), 548

KNORR, ANGELO [1864-1899]
BJN 4,153*-154*; SG [2]8,789
H. Buchner, Münch. med. Wchschr. 46 (1899), 523-525

KNORR, ANGELO [1882-1932]
JV 25,610; NUC 300,513
Anon., Ber. chem. Ges. 66A (1933), 60

KNORR, EDUARD [1867-1926]
JV 12,170; RSC 16,343; NUC 300,514
Anon., Z. angew. Chem. 39 (1926), 599

KNORR, LUDWIG [1859-1921]
DSB 17,488-489; NDB 12,218-220; STUPP-KUGA, 109-131; BLOKH 1,368-369; POTSCH, 242; DZ, 756; WWWS, 949; POGG 4,768-769; 5,645-646; 6,1345
W. Schneider, Chem. Z. 45 (1921), 609-610
R. Scholl, Sitz. Akad. Wiss. Leipzig 75 (1923), 155-165
P. Duden and H.P. Kaufmann, Ber. chem. Ges. 60A (1927), 1-34
H.W. Flemming (Ed.), Dokumente aus Hoechster Archiven No.8. Frankfurt a.M. 1965; No.31. Frankfurt a.M. 1967
K. Brune, Agents and Actions 19Suppl. (1986), 19-29

KNORR, MAXIMILIAN [1895-1985]
KURSCHNER 13,1983; BERWIND, 61-71; PITTROFF, 112; IB, 154
Anon., Naturw. Rund. 38 (1985), 215

KNORRE, GEORG von [1859-1910]
BJN 15,47*; WELDING, 392-393; POGG 3,731; 4,469-470; 5,646
K. Arndt, Chem. Z. 35 (1911), 41
C. Liebermann, Ber. chem. Ges. 44 (1911), 169-170

KNOWER, HENRY McELDERRY [1868-1940]
AMS 6,792; IB, 154; WWAH, 336; WWWS, 949
R.G. Harrison, Science 92 (1940), 419-421
Anon., Entomol. News 51(2) (1940), 51

KNOWLES, JEREMY RANDALL [1935-]
WW 1992, p.1057

KNOX, WALTER EUGENE [1918-1982]
AMS 15(4),395; WWA 1980-1981, p.1873

KNUDSON, ARTHUR [1889-1959]
AMS 9(I),1064; IB, 154
Anon., Chem. Eng. News 37(50) (1959), 112

KNY, LEOPOLD [1841-1916]
NDB 12,233; DBJ 1,361; WREDE, 98-99; DZ, 758; TSCHIRCH, 1073
W. Magnus, Ber. bot. Ges. 34 (1916), (58)-(71)

KOBEL, MARIA [1897-]
TLK 3,342; POGG 6,1346; 7a(2),812

KOBELL, FRANZ von [1803-1882]
NDB 12,238-240; ADB 16,787-797; GLOSAUER, 159-172; FERCHL, 278; POTSCH, 242; WWWS, 1737; POGG 1,1286-1287; 3,731-732
C. Voit, Sitz. Bayer. Akad. 13 (1883), 217-222
K. Haushofer, Franz von Kobell. Munich 1884
A. Dreyer, Oberbayerisches Archiv für Vaterländische Kultur 52 (1904), 124-127

KOBER, GEORGE MARTIN [1850-1931]
DAB 10,483-484; OEHRI, 61-62
R.T. Legge, Industrial Medicine 17 (1948), 301-302

KOBER, PHILIP ADOLPH [1884-1978]
AMS 11,2855

KOBERT, RUDOLF [1854-1918]
NDB 12,247; DBJ 2,694; FISCHER 1.782; LEVITSKI 2,196-201; PAGEL, 873-874; REBER, 131-134,362; TSCHIRCH, 1073-1074; SCHELENZ, 717; DZ, 759; POTSCH, 242; POGG 4,771-772; 5,647; 7a(2),814
E. Sieburg, Chem. Z. 43 (1919), 25-26; Ber. pharm. Ges. 29 (1919),295-299
I. Brennsohn, Die Aerzte Estlands, pp.459-460. Riga 1922
H. Buess, Schw. med. Wchschr. 84 (1954), 442-451

KOCH, ALFRED [1858-1922]
WAGENITZ, 96-97
A. Rippel, Ber. bot. Ges. 41 (1923), (67)-(74)

KOCH, EBERHARD [1892-1955]
FISCHER 1,783; IB, 154; POGG 7a(2),815-816
R. Thaurer, Z. Kreis. 44 (1955), 515-516

KOCH, ERICH [1882-1920]
F. Keim, Chem. Z. 44 (1920), 865

KOCH, FERENC [1853-]
RSC 16,358
Magyar Irok 3,645-647

KOCH, FRED CONRAD [1876-1948]
DAB [Suppl.4],459-461; AMS 7,988; NCAB 46,388; MILES, 278-279; IB, 154; WWWS, 951; POGG 6,1346-1347; 7b,2510-2511
T.L. McMeekin, Archives of Biochemistry 17 (1948), 207-209
M.E. Hanke, Science 107 (1948), 671-672
Anon., Chem. Eng. News 26 (1948), 402-403; J. Clin. Endocrin. 8 (1948), 207

KOCH, FRIEDRICH [1786-1865]
HEIN, 333

KOCH, FRIEDRICH [1901-1985]
KURSCHNER 11,1507; POGG 7a(2),817-818

KOCH, HENRI [1911-]
Anon., Bibl. Acad. Louvain 11 (1969), 143-150; 12 (1972), 746-747;14 (1980), 290-291

KOCH, HERMANN [1858-1939]
JV 6,22; RSC 16,359
Anon., Chem. Z. 63 (1939), 162; Z. angew. Chem. 52 (1939), 183,184

KOCH, KARL FRIEDRICH [1802-1871]
HIRSCH 3,565-566

KOCH, KARL JAKOB WILHELM [1827-1882]
ADB 16,398-399

KOCH, PAUL [1844-1911]
A. Wankmüller, Beitr. Wurttt. Apothekgesch. 8 (1969), 113

KOCH, REINHARD [1874-]
RIGA, 707; JV 15,86; NUC 301,308

KOCH, RICHARD [1900-1967]
KURSCHNER 12,1634; POGG 7a(2),821-822

Anon., Chem. Z. 91 (1967), 979

KOCH, RICHARD HERMANN [1882-1949]
NDB 12,274-275; KURSCHNER 4,1500; FISCHER 1,784; KALLMORGEN, 326; ARNSBERG, 250; LDGS, 64; KOREN, 203; WININGER 7,180-181
F.S. Bodenheimer, Arch. Inter. Hist. Sci. 4 (1951), 478-479
K.E. Rothschuh, Med. Hist. J. 15 (1980), 16-43,223-243
H. Schwann, NTM 46(1) (1982), 94-103
G. Preiser, Richard Koch und die ärztliche Diagnose. Hildesheim 1988

KOCH, ROBERT [1843-1910]
DSB 7,420-435; NDB 12,251-255; BJN 15,48*-49*; OLPP, 202-211; W, 294-295; FISCHER 1,784-786; PAGEL, 876-878; FREUND 2,177-200; ABBOTT, 77-78; BULLOCH, 376-377; WREDE, 99-100; WWWS, 951; POGG 7aSuppl.,333-341
C.J. M[artin], Proc. Roy. Soc. B83 (1911), xviii-xxiv
W.W. Ford, Bull. Johns Hopkins Hosp. 22 (1911), 415-425
G.B. Webb, Ann. Med. Hist. 4 (1932), 509-523
L. Brown, Ann. Med. Hist. 7 (1935), 99-112,292-304,385-401
L.I. Dublin, American Scholar 13 (1943-1944), 95-109
R. Bochalli, Robert Koch. Stuttgart 1954
G.B. Gruber, Deutsche med. Wchschr. 83 (1958), 1627-1631
W.D. Foster, Scientia 103 (1968), 53-71
W. Genschorek, Robert Koch. Leipzig 1975
M. Penn and M. Dworkin, Bact. Revs. 40 (1976), 276-283
H. Boisvert, Hist. Sci. Med. 16 (1982), 93-97
E. Shapiro, Pharos 46(4) (1983(, 19-22
U. Schneeweiss, Med. Hist. J. 18 (1983), 21-32
W. Kaiser and H. Hübner, Eds., Wissenschaftliche Beiträge der Martin-Luther Universität Halle-Wittenberg 5-R80 (1983), 3-321
T.D. Brock, Robert Koch, a Life in Medicine and Bacteriology. Madison, Wis. 1988
A. Grafe, Gesnerus 45 (1988), 411-418

KOCH, WALDEMAR [1875-1912]
AMS 2,264; CHITTENDEN, 216-220
A.P. Mathews, Bioch. Bull. 1 (1912), 372-376

KOCH, WALTER [1880-1962]
KURSCHNER 9,1023; IB, 154-155
W. Giese, Verhandl. path. Ges. 47 (1963), 423-426

KOCHAKIAN, CHARLES DANIEL [1908-]
AMS 15(4),401
C.D. Kochakian, How It Was. Birmingham, Ala. 1984

KOCHENDÖRFER, ERNST [1863-1918]
JV 5,75
Anon., Chem. Z. 42 (1918), 632; Z. angew. Chem. 31(III) (1918), 484

KOCHER, RUDOLF FRIEDRICH [1811-1875]
A. Wankmüller, Der Schweizer Familienforscher 33 (1966), 46-50

KOCHER, THEODOR [1841-1917]
NDB 12,282-283; DHB 4,379-380; DBJ 2,661; FISCHER 1,787; PAGEL, 878-879; KILLIAN, 399-402; W, 295-296; WWWS, 951-952
A. Kocher, Verhandl. Schw. Nat. Ges. 99 (1917), 70-85
P. Clairmont, Wiener klin. Wchschr. 30 (1917), 1050
F. Sauerbruch, Münch. med. Wchschr. 65 (1918), 78-80
E. Bonjour, Theodor Kocher. Berne 1950
P.S. McGreevy and F.A. Miller, Surgery 65 (1969), 990-999
U. Troehler, Der Nobelpreisträger Theodor Kocher. Basle 1984

KOCHMANN, MARTIN [1878-1936]
NDB 12,285; FISCHER 1,787; TLK 3,342; IB, 155; POGG 7a(2),824-825
R. Zaunick, Pharmazie 3 (1948), 335
H.H. Eulner, Arzneimitt. 5 (1955), 553-557

KOCHS, WILHELM [1852-1898]
BJN 5,35*; PAGEL, 879
Anon., Leopoldina 34 (1898), 171-172

KODAMA, KEIZO [1891-1972]
JBE, 654
N. Shimazono and H. Yoshikawa, J. Biochem. 49 (1961), i
Y. Miura, J. Biochem. 73Suppl. (1973), 1-2

KODAMA, SHINTARO [1885-1923]
Y. Tsuzuki and A. Yamashita, Jap. Stud. Hist. Sci. 7 (1968), 20

KODICEK, EGON HYNEK [1908-1982]
WW 1982, p.1230
D.R. Fraser, Brit. J. Nutrition 49 (1983), 167-170
D.R. Fraser and E.M. Widdowson, Biog. Mem. Fell. Roy. Soc. 29 (1983), 297-331

KODWEISS, FRIEDRICH [1803-1866]
NDB 12,287; OBL 4,26; STURM 2,205; HEIN-E, 243; RSC 3,708
J.V. Divis, Beiträge zur Geschichte der Zuckerindustrie in Bohmen, pp.21-31. Prague 1891
J. Baxa, Z. Zuckerind. 9 (1956), 625-627

KOEBNER, HEINRICH [1838-1904]
BJN 10,59*; HIRSCH 3,567-568; Suppl.,387; PAGEL, 881-883; KOREN, 203;WININGER 3,474-475
A. Wechselmann, Deutsche med. Wchschr. 30 (1904), 1391
J. Pagel, Virchows Jahresber. 1904(I), p.470

KOECHLIN, CAMILLE [1811-1890]
P. Schützenberger, Mon. Sci. [4]4 (1890), 773-776
Anon., Leopoldina 26 (1890), 114
A. Scheurer, Bull. Soc. Ind. Mulhouse 69 (1899), 99-155
P. Brandt, Bull. Soc. Ind. Mulhouse 100 (1934), 349-364; Bull. Mus. Hist. Mulhouse 87 (1980), 93-95

KOECHLIN, DANIEL [1785-1871]
ADB 51,297; FERCHL, 279
A. Penot, Bull. Soc. Ind. Mulhouse 41 (1871), 237-262
P. Brandt, Bull. Mus. Hist. Mulhouse 87 (1980), 87-100

KOECHLIN, HORACE [1839-1898]
Anon., Chem. Z. 22 (1898), 55; J. Soc. Chem. Ind. 17 (1898), 228

KOECHLIN, JEAN ALBERT [1818-1889]
RSC 3,707

KOECHLIN, NICOLAS [1781-1852]
ADB 51,297
A. Penot, Bull. Soc. Ind. Mulhouse 24 (1852), 193-217

KOECK, KARL [1875-]
JV 20,260

KOEFOED, HENNING EMIL [1858-1937]
DBL (3rd Ed.) 8,143-144; KBB 1937, p.625; VEIBEL 2,249-250; TSCHIRCH, 1074

Anon., Theriaca (Copenhagen) 17 (1974), 1-137

KÖGL, FRITZ [1897-1959]
KURSCHNER 8,1211; POTSCH, 243; POGG 6,1349; 7a(2),828-829; (4),147*
Anon., Chem. Wkbl. 34 (1937), 610
G.J.M. van der Kerk, Nature 184 (1959), 1609-1610
F. Lynen, Bayer. Akad. Wiss. Jahrbuch 1959, pp.187-188
W.A.J. Borg, Chem. Wkbl. 55 (1959), 493-494
P. Karlson, TIBS 7 (1982), 382-383

KÖHLER, ALFRED [1881-1917]
JV 21,296
Anon., Z. angew. Chem. 30 (III) (1917), 340

KÖHLER, AUGUST [1866-1948]
DSB 7,447-449; FREUND 1,235-243; TLK 3,343-344; POGG 5,659; 6,1350; 7a(2),830-831
K. Michel, Naturwiss. 24 (1936), 145-150
G.G. Reinert, Mikroskopie 4 (1949), 65-70

KÖHLER, ERICH [1889-1985]
KURSCHNER 14,2163-2164; IB, 155

KÖHLER, FRITZ [1881-]
JV 23,308

KÖHLER, HERMANN [1834-1879]
ADB 16,441-442; HIRSCH 3,571; PAGEL, 883; RSC 3,709; 8,101
R. Kobert, Z. ges. Naturwiss. [3]14 (1879), 148-153

KÖHLER, OTTO [1889-1974]
NDB 12,309-310; KURSCHNER 11,1518; KROLLMANN, 1118-1120; IB, 155; WWWS, 952
G. Osche, Journal für Ornithologie 115 (1974), 460-463
B. Hassenstein et al., Z. Tierpsychol. 35 (1974), 449-472

KOELKER, ARTHUR HEINRICH [1883-1911]
JV 23,32; NUC 301,642
Anon., Chem. Z. 33 (1909), 813; 35 (1911), 1447

KOELKER, WILHELM FRIEDRICH [1880-1911]
JV 20,31; NUC 301,642

KOELLE, GEORGE BRAMPTON [1918-]
AMS 15(4),405; WWA 1980-1981, p.1877; IWW 1982-1983, p.706; CHEN, 99-100; WWWS, 952

KÖLLIKER, ALBERT [1817-1905]
DSB 7,437-440; NDB 12,322-323; BJN 10,130-137,198*; FREUND 2,201-213; DHB 4,381-382; LF 1,247-266; W, 536; HIRSCH 3,571-572; Suppl.,387; PAGEL, 885-887; BOSSART, 38-40; ROEDER, 19-22; WWWS, 1737
A. Kölliker, Erinnerungen aus meinem Leben. Leipzig 1899
A. Lang, Viert. Nat. Ges. Zürich 50 (1905), 567-572
C. von Voit, Sitz. Bayer. Akad. 36 (1906), 444-456
E. Ehlers, Z. wiss. Zool. 84 (1906), i-xxvi
W. Waldeyer, Anat. Anz. 28 (1906), 539-552
G.R. Cameron, Ann. Sci. 11 (1955), 166-172
H. Zuppinger, Albert Kolliker und die Mikroskopische Anatomie. Zürich 1974
R. Hildebrand, Würzburg medizin-historische Mitteilungen 3 (1986), 127-151

KÖLLISCH, ANTON [1888-1916]
JV 26,41; NUC 301,655
Anon., Z. angew. Chem. 29(III) (1916), 608

KOELSCH, FRANZ [1876-1970]
KURSCHNER 11,521; GERNETH, 89-126; TLK 3,347; POGG 7a(2),833-834
M. Thür, Zentralblatt für Arbeitsmedizin 21 (1971), 1-2

KÖNIG, ADOLF [1881-1964]
STURM 2,211; TLK 3,349; POGG 5,650; 6,1351; 7a(2),834
Anon., Chem. Z. 89 (1965), 127

KÖNIG, ARTHUR [1856-1901]
DSB 7,457-458; BJN 6,57*; FISCHER 1,790; WREDE, 102-103; WWWS, 957; POGG 3,735; 4,777-778
Anon., Leopoldina 37 (1901), 109-110
H. Ebbinghaus, Z. Psych. Physiol. Sinn. 27 (1902), 145-147

KÖNIG, CARL [1838-1885]
NDB 12,345; BLOKH 1,377
H. Landolt, Ber. chem. Ges. 18 (1885), 773-777
Anon., Chem. Z. 10 (1886), 3

KÖNIG, ERNST [1869-1924]
NDB 12,335-336; BLOKH 1,377; POTSCH, 246; POGG 6,1352
A.H., Z. angew. Chem. 37 (1924), 1030-1032
J.M. Eder, Chem. Z. 48 (1924), 905

KÖNIG, FRANZ [1832-1910]
NDB 12,331; BJN 15,49*-50*; LKW 1,164-166; HIRSCH 3,575-576; PAGEL, 887-889
Anon., Lancet 1910(II), p.1863; Leopoldina 47 (1911), 45-46
W. Müller, Münch. med. Wchschr. 58 (1911), 203
O. Hildebrand, Franz Konig. Berlin 1911; Berl. med. Wchschr. 48 (1911), 24-25
F. König, Lebenserinnerungen. Berlin 1912

KÖNIG, FRANZ JOSEPH [1843-1930]
NDB 12,343-344; KALIN, 103-105; DZ, 775; TLK 3,350; SCHAEDLER, 61; POGG 3,734; 4,776; 6,1352
A. Behre, Chem. Z. 47 (1923), 837
A. Boemer, Ber. chem. Ges. 63A (1930), 137-139
W. Sutthoff, Z. angew. Chem. 43 (1930), 495-496

KOENIG, THEODOR [1862-]
JV 4,66; RSC 16,373; NUC 302,70
Anon., Chem. Z. 56 (1932), 118

KÖNIG, WALTER [1878-1964]
NDB 12,350-351; KURSCHNER 9,1033; TLK 3,351; POTSCH, 246; POGG 5,651-652; 6,1353; 7a(2),838
W. Langenbeck, Jahrb. Sachs. Akad. 1963-1965, pp.308-310
K. Schwabe, Jahrb. Akad. Wiss. Berlin 1964, pp.241-243

KOENIGS, ERNST [1878-1945]
KURSCHNER 4,1529; TLK 3,351; POGG 5,652; 6,1353-1354; 7a(2),839
W. Hückel, Chem. Ber. 83 (1950), xv-xviii

KOENIGS, WILHELM [1851-1906]
BJN 11,37*; BLOKH 1,378-379; POTSCH, 243; POGG 4,779-780; 5,653
C. von Voit, Sitz. Bayer. Akad. 37 (1907), 257-262
T. Curtius and J. Bredt, Ber. chem. Ges. 45 (1912), 3781-3830
E.H. Huntress, Proc. Am. Acad. Arts Sci. 79 (1951), 23-24

KÖNIGSBERGER, FRANZ [1899-]
JV 40,735

KOEPPE, HANS [1867-1939]
KURSCHNER 5,715; GUNDEL, 523-533
P. Frick, Arch. Kind. 117 (1939), 79-80

KÖRNER, WILHELM (GUGLIELMO) [1839-1925]
NDB 12,391; WWWS, 960; POTSCH, 243; POGG 5,657,1423; 6,1357
J.B. Cohen, J. Chem. Soc. 128 (1925), 2975-2982
R. Anschütz, Ber. chem. Ges. 59A (1926), 75-111
H.E. Armstrong, Chem. Ind. 48 (1929), 914-918
L. Dobbin, J. Chem. Ed. 11 (1934), 596-600
D.H. Rouvray, Endeavour 34 (1975), 28-33
H.W. Schuett, Physis 17 (1975), 113-125

KÖRÖSY, KORNEL von [1879-1948]
MEL 1,1006; IB, 155
M. Polanyi, Nature 162 (1948), 953-954

KÖSSLER, KARL CONRAD [1880-1928]
Anon., Proc. Inst. Med. Chicago 7 (1928), 81-84

KÖSTER, HANS [1896-1966]
POGG 7a(2),843
Anon., Chem. Z. 90 (1966), 224

KÖSTER, HUGO [1858-1939]
SBL 22,17-18; SMK 4,398; FISCHER 1,795

KÖTHNER, PAUL [1870-1932]
TLK 3,355; POGG 5,657; 6,1358
E. Tiede, Ber. chem. Ges. 66A (1933), 15

KÖTZ, ARTHUR [1871-1944]
POGG 4,783; 5,658; 6,1359; 7a(2),821-822

KÖTZLE, ARTHUR ALFRED [1865-1930]
JV 4,284

KOF, KARL [1876-]
JV 19,181; NUC 302,305

KOFFLER, HENRY [1922-]
AMS 15(4),410; BHDE 2,597

KOFLER, LUDWIG [1891-1951]
NDB 12,420; HEIN-E, 246-247; FREUND 3,241-252; HUTER, 298-299; TLK 3,343; TSCHIRCH, 1074; IB, 155; POGG 6,1359-1360; 7a(2),849-850
M. Brandstätter, Mikrochemie 38 (1951), 295-308
R. Opfer-Schaum, Z. angew. Chem. 63 (1951), 476
Anon., Chem. Z. 75 (1951), 482; Z. angew. Chem. 63 (1951), 448; Arzneimitt. 1 (1951), 425-426
F. Wessely, Alm. Akad. Wiss. Wien 102 (1953), 367-374

KOFOID, CHARLES ATWOOD [1865-1947]
DSB 7,447; DAB [Suppl.4],461-462; NCAB A,280-281; OLPP, 211-213; IB, 155; WWWS, 953
H. Kirby, Sci. Mon. 61 (1945), 415-418; Science 106 (1947), 462-463
R.B. Goldschmidt, Biog. Mem. Nat. Acad. Sci. 26 (1951), 121-151
G.H. Parker, Am. Phil. Soc. Year Book 1947, pp.262-264

KOHL, FRIEDRICH GEORG [1855-1910]
BJN 15,50*
Anon., Leopoldina 46 (1910), 52

KOHL, HANS [1902-1967]
KURSCHNER 10,1245; WWWS, 953

KOHLER, ELMER PETER [1865-1938]
DAB [Suppl.2],365-366; AMS 6,795; WWWS, 954; POGG 5,659; 6,1360-1361; 7b,2516
Anon., Ind. Eng. Chem. News Ed. 16 (1938), 338
A.B. Lamb et al., Science 89 (1939), 595-596
J.B. Conant, Biog. Mem. Nat. Acad. Sci. 27 (1952), 265-291

KOHLRAUSCH, ARNT [1884-1969]
KURSCHNER 12,1662; IB, 155; POGG 6,1361; 7a(2),854-855
Anon., Naturw. Rund. 22 (1969), 514

KOHLRAUSCH, FRIEDRICH [1840-1910]
DSB 7,449-450; NDB 12,430-431; BJN 15,50*-51*; BLOKH 1,371-373; W, 296; KALLMORGEN, 328; WOLF, 109; DZ, 767; WWWS, 954; POTSCH, 243-244; POGG 3,737-738; 4,784-785; 5,659; 6,1361
R. Abegg, Z. Elektrochem. 11 (1905), 193
W. Herz, Chem. Z. 34 (1910), 65
G.C.F[oster], Proc. Roy. Soc. A85 (1911), xi-xiii
D. Kahan, Osiris [2]5 (1989), 167-185

KOHLRAUSCH, KARL [1862-]
JV 4,284; RSC 16,391; NUC 302,433

KOHLRAUSCH, OTTO [1811-1854]
HIRSCH 3,580; CALLISEN 29,311-312
[C. Schneemann], Med. Conv. Hannover 5 (1854), 129-132

KOHLRAUSCH, RUDOLF [1809-1858]
DSB 7,450; ADB 16,452-453; BLOKH 1,371; FERCHL, 280; WWWS, 954; POGG 1,1299-1300

KOHLSCHÜTTER, HANS WOLFGANG [1902-1986]
KURSCHNER 14,2194; WOLF, 109; POGG 7a(2),857
Anon., Chem. Z. 106 (1982), 309

KOHLSCHÜTTER, VOLKMAR [1874-1938]
TLK 3,345; POTSCH, 244; POGG 4,785-786; 5,661-662; 6,1362-1363; 7a(2),857
W. Feitknecht, Koll. Z. 68 (1934), 129-134; Helv. Chim. Acta 22 (1939), 1059-1088
W.A. Endriss, Ind. Eng. Chem. News Ed. 16 (1938), 616

KOHN, ALFRED [1867-1959]
STURM 2,221; FISCHER 1,796; KOERTING, 109-113; MAASS, 52-59,154-155; KOREN, 203; WININGER 7,184; TETZLAFF, 181
M. Watzka, Anat. Anz. 106 (1959), 449-457

KOHN, EDMOND [1863-1929]
OBL 4,65; STURM 2,222

KOHN, HANS [1866-1935]
FISCHER 1,797; WININGER 3,490-491; 7,580

KOHN, MORITZ [1878-1955]
STURM 2,223-224; KURSCHNER 8,1229; TLK 3,345; POGG 5,662-663; 6,1363-1364; 7b,2516-2517

KOIKE, ITSUO [1883-1915]
JBE, 662

KOLB, ADALBERT [1863-1938]
HEIN-E, 249-250; WOLF, 110; TSCHIRCH, 1074; POGG 4,788; 5,664; 6,1365;7a(2),858
A. Stock, Ber. chem. Ges. 71A (1938), 115
Anon., Chem. Z. 62 (1938), 149

KOLBE, HERMANN [1818-1884]
DSB 7,450-453; NDB 12,446-451; ADB 51,321-329; LEIPZIG 2, 25-35; BUGGE 2,124-135; MEINEL, 48-139,506-507; GUNDLACH, 464; WWWS, 955; SCHMITZ, 241-264; BLOKH 1,373-374; FERCHL, 281-282; W, 296-297; POTSCH, 244; ABBOTT-C, 76-77; POGG 1,1301; 3,739-740; 4,788

H. Kolbe, J. prakt. Chem. NF23 (1881), 305-323,353-379,497-517; 24 (1881), 375-425
E. von Meyer, J. prakt. Chem. NF30 (1884), 417-466
A.W. Hofmann, Ber. chem. Ges. 17 (1884), 2809-2810
Anon., Chem. Z. 8 (1884), 1725-1726
C. von Voit, Sitz. Bayer. Akad. 15 (1885), 160-167
W.H. Perkin, J. Chem. Soc. 47 (1885), 323-327
W. Kerkovius, Chem. Z. 35 (1911), 1117-1119,1142-1145
J.P. Phillips, Chymia 11 (1966), 89-97
H. Remane et al., Z. Chem. 24 (1984), 393-403
A.J. Rocke, Ambix 34 (1987), 156-168

KOLLISCH, RUDOLF [1867-1922]
OBL 4,82; STURM 2,231; FISCHER 1,797; PAGEL, 893-894; KOREN, 204; WININGER 3,499; 7,187
A. von Decastello, Wiener klin. Wchschr. 35 (1922), 424

KOLL, WERNER [1902-1968]
NDB 12,463-464; KURSCHNER 10,1250; KIEL, 129; POGG 7a(2),860-861
L. Lendle, Arzneimitt. 18 (1968), 1618-1619
A.W. Forst, Münch. med. Wchschr. 111 (1969), 962-963
W. Vogt, Deutsche med. Wchschr. 94 (1969), 50-51

KOLLATH, WERNER [1892-1970]
KURSCHNER 11,1541; WWWS, 955; POGG 7a(2),862-865
E. Kollath, Werner Kollath. Munich 1972 (2nd Ed. 1978)

KOLLE, WILHELM [1868-1935]
NDB 12,464-465; FISCHER 1,798; KALLMORGEN, 328-329; OLPP, 214-215; WEBER, 63-65; TSCHIRCH, 1074; TLK 3,346-347; IB, 156; WWWS, 955
H. Hetsch, Deutsche med. Wchschr. 61 (1935), 849-850
K. Lautenheimer, Münch. med. Wchschr. 82 (1935), 919-920

KOLLER, CARL [1857-1944]
DAB [Suppl.3],430-431; NDB 12,465; OBL 4,88; STURM 2,232; DAMB, 424; FISCHER 1,798; W, 297; WWWS, 955; KOREN, 204; WININGER 3,500-501
C.D. Leake, Isis 23 (1935), 253-256
Anon., J. Mt. Sinai Hosp. 11 (1945), 308-310
H. Honegger and H. Hessler, Pharm. Z. 117 (1972), 1153-1159, 1188-1192,1195

KOLLI, ALEKSANDR ANDREEVICH [1840-1916]
BSE 21,626-627; BLOKH 1,374-375; RSC 7,417; 12,164
V.V. Shavrin, Zhur. Fiz. Khim. 49 (1917), 113-119
A.P. Terentyev and S.M. Gurvich, Usp. Khim. 9 (1950), 128-133

KOLLMANN, JULIUS [1834-1918]
DUELUND, 6-12; EGERER, 13-14; WWWS, 956
H.K. Corning, Anat. Anz. 52 (1919-1920), 65-80

KOLLS, ALFRED CONRAD [1891-]
NUC 303,25

KOLMER, WALTHER [1879-1931]
KURSCHNER 4,1521; FISCHER 1,799-800; HEINDEL, 56-70; IB, 156; KOREN, 204
A. Durig, Anat. Anz. 73 (1932), 278-287

KOLOSOV, ALEKSANDR ALEKSANDROVICH [1862-1937]
BSE 22,27; BME 13,656-657; WWR, 289; IB, 156
N.I. Zazybin, Arkh. Anat. Gist. Emb. 17(1) (1937), 3-4

KOLOTOV, SERGEI SILVESTROVICH [1858-1916]
BLOKH 1,376-377
V.E. Tishchenko, Zhur. Fiz. Khim. 48 (1916), 1032-1034

KOLTHOFF, IZAAC MAURITS [1894-1993]
AMS 14,2725; IWW 1982-1983, p.710; MH 1,268-269; MSE 2,181-182; WWWS,956;POTSCH, 245; COHEN, 130; POGG 5,665; 6,1370-1373; 7b,2529-2540
J.J. Lingane, Talanta 11 (1964), 67-73
H.A. Laitinen and E.J. Meehan, Analytical Chemistry 56 (1984), 248A-262A
M. Warner, Analytical Chemistry 61 (1989), 287A-291A
Y.A. Zolotov, Zhurnal Analiticheskoi Khimii 46 (1991), 1233-1238

KOLTSOV, NIKOLAI KONSTANINOVICH [1872-1940]
DSB 7,454-457; BSE(3rd Ed.) 12,484; BME 13,690-691; IB, 156
N.W. Timofeeff-Ressovsky, Naturwiss. 29 (1941), 121-124
D.N. Borodin, J. Heredity 32 (1941), 347-349
B.L. Astaurov, Bull. Soc. Nat. Moscou NS77(6) (1972), 127-137
B.L. Astaurov and P.F. Rokitski, N.K. Koltsov. Moscow 1975

KOMAROV, VALDIMIR LEONTOVICH [1869-1945]
BSE 22,108-110; WWR,291; KUZNETSOV(1963), 360-373; LIPSHITS 4,288-308; IB, 156
I.I. Meshchaninov, V.L. Komarov. Moscow 1946
H.H. Airy Shaw, Proc. Linn. Soc. 162 (1950), 114-116
B.K. Shishkin, Bot. Zhur. 35(2) (1950), 140-147
N.V. Pavlov, V.L. Komarov. Moscow 1951
A.A. Elenkin, Trudy Inst. Ist. Est. 16 (1957), 253-334

KOMATSU, SHIGERU [1883-1947]
JBE, 677; IB, 156; POGG 6,1373-1374; 7b,2540-2541

KOMM, ERNST [1899-1964]
TLK 3,348; IB, 157; POGG 6,1374-1375; 7a(2),871
Anon., Chem. Z. 89 (1965), 127; Nachr. Chem. Techn. 13 (1965), 53

KOMNENOS, TELEMACHOS [1862-]
BONN, 356

KOMPPA, GUSTAF [1867-1949]
ENKVIST, 121-127; POTSCH, 245; POGG 4.789; 5,666-667; 6.1375-1376; 7b,2542-2543
Anon., Chem. Eng. News 27 (1949), 696; Chem. Ind. 1949, p.95
W. Hückel, Chem. Ber. 85 (1952), i-xxx

KON, GEORGE ARMAND ROBERT [1892-1951]
WW 1949, p.1572; POGG 6,1376; 7b,2543-2545
R.P. Linstead, Obit. Not. Fell. Roy. Soc. 8 (1952), 171-192; J. Chem. Soc. 1952, pp.4550-4560

KON, STANISLAW KAZIMIERZ [1900-1986]
Who's Who in British Science 1953, p.159
Anon., Chem. Brit. 22 (1986), 1076

KONCHALOVSKI, MAKSIM PETROVICH [1875-1942]
BSE 22,506-507; BME 13,973-974; WWR, 292; FISCHER 2,802
G.P. Shultsev, Ter. Arkh. 45(12) (1973), 7-9
A.I. Nesterov, Klin. Med. 53 (10), (1975), 11-13
P.M. Alperin, Kardiologia 15(9) (1975), 150-153

KONDAKOV, IVAN LAVRENTIEVICH [1857-1931]
DSB 7,457; BSE 22,343-344; ARBUZOV, 196-197; LEVITSKI 2,250-254; POTSCH, 245-246; POGG 4,789-790; 5,667; 6, 1376; 7b,2545 A.M. Maksimenko et al., Ist. Est. Tekhn. Pribalt. 3 (1971), 109-119

KONDO, HEIZABURO [1877-1963]
JBE, 684,2333; HAKUSHI 1(Pharm.),24; IB, 157;

POGG 6,1376-1377; 7b,2545-2547
E. Ochiai, Chem. Pharm. Bull. 12 (1964), ii-iii

KONDO, KURA [1876-]
HAKUSHI 3,382-383

KONDRATIEV, VIKTOR NIKOLAEVICH [1902-1979]
BSE(3rd Ed.) 13,19; MH 2,300-301; STC 2,118-120; MSE 2,182-183; WWWS, 957; POGG 6, 1377; 7b,2547-2552
N.Y. Ruben et al., Usp. Khim. 21 (1952), 988-995
Anon., Z. angew. Chem. 79 (1967), 260
N.N. Semenov, International Journal of Chemical Kinetics 11 (1979), 933-934

KONEK-NORWALL, FRIGYES [1867-1945]
MEL 1,964; JV 8,216; POGG 5,667; 6,1377; 7b,2553-2554

KONINCK, LAURENT GUILLAUME de [1809-1887]
DSB 7,460-461; NBW 9, 431-435; BLOKH 1,379; FERCHL, 282; SCHAEDLER, 61; WWWS, 434; POGG 1,1302-1303; 3,741; 4,790
Anon., Leopoldina 26 (1890), 154-155
E. Dupont, Ann. Acad. Roy. Belg. 57 (1891), 437-483

KONING, CORNELIS JOHAN [1863-1932]
NUC 303,332
Anon., Chem. Wkbl. 11 (1914), 688-689; 29 (1932), 538
T. van der Wielen, Pharm. Weekbl. 69 (1933), 1061

KONOPACKI, MIECZYSLAW [1880-1939]
PSB 13,550-551; IB, 157; WWWS, 958
Anon., Folia Morphologica 1 (1950), 165-168

KONOPKA, STANISLAW JOZEF [1896-1982]
E. Kucharz, Clio Medica 18 (1983), 239-240
H. Bojczuk et al., Arch. Hist. Med. 46 (1983), 439-486
T. Kikta, Farmacja Polska 40 (1984), 750-754

KONOVALOV, DMITRI PETROVICH [1856-1929]
DSB 7,461-462; BSE 22,391-392; WWR, 294; KUZNETSOV(1961), 504-508; POTSCH, 246-247; POGG 4,791; 5,668; 6,1379
Y.I. Soloviev and A.Y. Kipnis, Dmitri Petrovich Konovalov. Moscow 1964

KONOVALOV, MIKHAIL IVANOVICH [1858-1906]
BSE 22,392; ARBUZOV, 150-152; BLOKH 1,379-381; POTSCH, 247
L. Jawain, Chem. Z. 31 (1907), 25
Anon., Leopoldina 43 (1907), 52
A.N. Reformatsky et al., Zhur. Fiz. Khim. 42 (1910), 5-63

KONOVALOV, NIKOLAI VASILIEVICH [1900-1966]
BSE 22,392; BME 13,845-846; WWR, 294; WWWS, 958
Anon., Klin. Med. 44(8) (1966), 164-165
I. Wald, Neurologia i Neurochirurgia Polska 17 (1967), 111-112

KONZETT, HERIBERT [1912-]
KURSCHNER 13,2044; DECKER, 136-150; POGG 7a(2),872

KOOIJ, DIEDERIK MARGUS [1867-1837]
RSC 16,415
Anon., Chem. Wkbl. 34 (1937), 502

KOPACZEWSKI, WLADISLAS [1886-1953]
WWWS, 958; POGG 6,1379-1380; 7b,2553-2556
G. Garnier, Wladislas Kopaczewski. Luneville 1959

KOPISCH, FRIEDRICH [1867-]
JV 9,18; NUC 303,540

KOPITZSCH, HANS [1879-]
JV 19,202; NUC 303,541

KOPP, ÉMILE [1817-1875]
HUMBERT, 208-209; BERGER-LEVRAULT, 178-179; BLOKH 1,381-383; FERCHL, 282; TSCHIRCH, 1075; SCHAEDLER, 61; POGG 3,741-742
R. Gnehm, Ber. chem. Ges. 9 (1876), 1950-1951

KOPP, HERMANN [1817-1892]
DSB 7,463-464; NDB 12,567-568; ADB 55,820-826; LKW 3,258-266; W, 297; DRULL, 145-146; COLE, 291-292; BLOKH 1,383-386; POTSCH, 247-248; FERCHL, 282-283; POGG 1,1304-1305; 3,742; 4,792; 7aSuppl.,341-342
A.W. Hofmann, Ber. chem. Ges. 25 (1892), 505-521
T.E. Thorpe, J. Chem. Soc. 63 (1893),776-815; Proc. Roy. Soc.60(1897),i-v
B. Bessmertny, Archeion 14 (1932), 62-68
J. Ruska, J. Chem. Ed. 14 (1937), 1937), 3-12
M. Speter, Osiris 5 (1938), 392-460

KOPPANYI, THEODORE [1901-1985]
AMS 11,2881; WWWS, 958
S. Krop and F.G. Standaert, Pharmacologist 27 (1985), 80-81

KOPPE, ROBERT KARLOVICH [1848-]
SG [2]8,825; [3]7,310; RSC 8,111; 10,443; NUC 303, 568

KOPPEL, MAX [1890-1916]
E. Meyer, Strassburg Med. Z. 13 (1916), 84-86
K. Spiro, Jahresb. Fort. Tierchem. 45 (1916), v-viii
A. Roos and W.F. Baron, Resp. Physiol. 40 (1980), 1-32

KOPROWSKI, HILARY [1916-]
AMS 15(4),430; WWWS, 959
D. Kritchevsky, J. Cell. Physiol. 2Suppl. (1982), vi-viii

KORÁNYI, ALEXANDER (SANDOR) [1866-1944]
MEL 1,968-969; FISCHER 2,803-804; GROTE 3,63-88; WININGER 3,508-509; 7,188; WWWS, 959
S. Rusznyak, Deutsche med. Wchschr. 62 (1936), 985
I. Magyar, Koranyi Sandor. Budapest 1970

KORCZYNSKI, ANTONI [1879-1929]
PSB 14,48-49; POGG 5,669; 6,1381
W. Sobiecki, Ber. chem. Ges. 62A (1929), 80
W. Bryd, Roczniki Chem. 10 (1930), 211-220

KORCZYNSKI, EDWARD [1844-1905]
PSB 14,49-51; FISCHER 2,804; KONOPKA 4,506-513; KOSMINSKI, 231-233
J. Aleksandrowicz, Pol. Tyg. Lek. 19 (1964), 310-311

KORENCHEVSKY, VLADIMIR [1880-1959]
IB, 158
Anon., Lancet 1959(II), p.83; Brit. Med. J. 1959(II), p.194
E.V. Cowdry, Science 130 (1959), 1391-1392
K. Hall, J. Path. Bact. 80 (1960), 451-461

KOREY, SAUL ROY [1918-1963]
AMS 10,2228
J.H. Nurnberger, Journal of Neuropathology and Experimental Neurology 24 (1965), 183-186

KORFF, JOSEPH von [1843-1876]
BERLIN, 676; RSC 8,112; NUC 304,33
Genealogisches Handbuch des Adels 37 (1966), 286

KORKES, SEYMOUR [1922-1955]
AMS 9(II), 629
S. Ochoa, Science 124 (1956), 617-618

KORN, ADOLF [1877-]
JV 18,384; NUC 304,59

KORN, EDGAR [1874-]
JV 16,195; NUC 304,64

KORN, EDWARD DAVID [1928-]
AMS 15(4),432; WWWS, 959

KORNBERG, ARTHUR [1918-]
AMS 15(4),433; WWA 1980-1981, p.1887; IWW 1982-1983, p.714; MH 1,269-270; STC 2,121-123; MSE 2,184-185; CB 1968, pp.210-212; ABBOTT-C, 77-78; WWWS, 959-960; POTSCH, 248
A. Kornberg, in Reflections on Biochemistry (A. Korneberg et al., Eds.), pp.243-251. Oxford 1976; For the Love of Enzymes. Cambridge, Mass. 1989; Ann. Rev. Biochem. 58 (1989), 1-30

KORNBERG, HANS LEO [1928-]
WW 1981, pp.1471-1472; IWW 1982-1983, p.714; BHDE 2,651; ABBOTT-C, 78; WWWS, 960

KORNDÖRFER, GEORG [1874-1951]
KALLMORGEN, 329
Anon., Chem. Z. 53 (1929); Z. angew. Chem. 63 (1951), 584

KORNFELD, GERTRUD [1891-1955]
NDB 12,590-591; STURM 2,253; BHDE 2,651; LDGS, 21; BOEDEKER, 45; POGG 6,1384; 7a(2),880
Anon., Chem. Eng. News 33 (1955), 3276

KORNICK, ERICH [1886-]
JV 27,649; NUC 304,104
Anon., Nachr. Chem. Techn. 4 (1956), 84

KOROCHUK, ABRAM YANKELEEVICH [1862-1898]
M.D. Tsetlin, Vrach (St. Petersburg) 19 (1898), 423

KOROTKOV, ALEKSEI ANDREEVICH [1910-1967]
BSE(3rd Ed.) 13,204; WWR, 299; WWWS, 960

KOROVIN, IVAN PAVLOVICH [1843-1908]
ZMEEV 4,179-180; RSC 12,404; VENGEROV 1, 400

KORR, IRVIN MORRIS [1909-]
AMS 17(4),451; WWWS, 960

KORSAKOV, SERGEI SERGEEVICH [1854-1900]
BSE 23,69-70; BME 13,1080-1082; KUZNETSOV(1963), 578-591; WWWS, 961; FISCHER 2,804-805
Anon., Leopoldina 36 (1900), 132
A. Marie, Revue de Psychiatrie NS3 (1900), 188-190
V. Rot, Zhur. Nevropat. Psikh. 1 (1901), 1-40
M. Victor and P.I. Yakovlev, Neurology 5 (1955), 394-406,509
C.A. Birch, Practitioner 210 (1973), 715-716
M.V. Ikonnikov, Sov. Med. 1979(7), pp.115-117

KORSCHELT, EUGEN [1858-1946]
KURSCHNER 3,1237-1238; FISCHER 2,805; SCHMITZ, 178-182; GUNDLACH, 475; AUERBACH, 847-848; DZ, 786; IB, 158; WWWS, 961
E. Korschelt, Das Haus an der Minne. Marburg 1939
K. von Frisch, Bayer. Akad. Wiss. Jahrbuch 1944-1948, pp.273-284;Am. Akad. Wiss. Wien 105 (1955), 409-412
G. Koller, Naturw. Rund. 1 (1948), 182-183
H. Querner, Sudhoffs Arch. 64 (1980), 313-329

KORSHUN, STEPAN VASILIEVICH [1868-1931]
BME 13,1127-1128; WWR, 300; FISCHER 2,805
P. Diatropov, Zhur. Mikr. Ep. Imm. 5(2) (1928), 81
H. Zeiss, Münch. med. Wchschr. 79 (1932), 838-839
S.S. Diachenko, Mikr. Zhur. 45(5) (1983), 101-105

KORTE, FRIEDHELM [1923-]
KURSCHNER 13,2053; POGG 7a(2),880-881

KORTE, REINHOLD [1876-]
JV 18,176; NUC 304,192

KORTÜM, GUSTAV [1904-1990]
KURSCHNER 13,2054; POTSCH, 248; POGG 6,1385; 7a(2),881-882
Anon., Nachr. Chem. Techn. 12 (1964), 246; Chem. Z. 115 (1991), 29-30
W.A.P. Luck, Chem. Z. 93 (1969), 435-436

KORZENOVSKY, MITCHELL [1914-]
AMS 10,2231

KOSCHARA, WALTER [1904-1945]
POGG 7a(2),882

KOSER, STEWART ARMENT [1894-1971]
AMS 11,2886; IB, 159; WWAH, 338; WWWS, 961

KOSHLAND, DANIEL EDWARD, Jr. [1920-]
AMS 15(4),438; WWA 1980-1981, p.1888; IWW 1982-1983, p.716; WWWS, 961

KOSHTOYANTS, KHACHATUR SEDRAKOVICH [1900-1961]
BSE 23,183; WWR, 301-302
Anon., Zhur. Ob. Biol. 22 (1961), 161-163

KOSMANN, CONSTANT PHILIPPE [1811-1881]
RSC 3,737; 8,114; 12,400
A. Maire, Catalogue des Thèses des Sciences Soutenues en France de 1810 à 1890 Inclusivement, p.148. Paris 1892

KOSSEL, ALBRECHT [1853-1927]
DSB 7,466-468; NDB 12,615-616; FISCHER 2,806-807; PAGEL, 903; DRULL, 147; GUNDLACH, 245-246; DZ, 788-789; TLK 2,384; IB, 159; W, 298; POTSCH, 248-249; WWWS, 962; POGG 4,795; 5,672; 6,1386-1387
O. Cohnheim, Münch. med. Wchschr. 57 (1910), 2644-2645
F. Müller, Naturwiss. 11 (1923), 785-787
C. Neuberg, Ber. chem. Ges. 60A (1927), 159-160
A.P. Mathews, Science 66 (1927), 293
S. Edlbacher, Z. physiol. Chem. 177 (1928), 1-14
E. Kennaway, Ann. Sci. 8 (1952), 393-397
M.E. Jones, Yale J. Biol. Med. 26 (1953), 80-97
K. Felix, Naturwiss. 42 (1955), 473-477
D. Doenecke and P. Karlson, TIBS 9 (1984), 404-405
G. Gerber and G. Sauer, Charité Annalen NF5 (1985), 355-365

KOSSEL, HERMANN [1864-1925]
DRULL, 147-148; FISCHER 2,807; RSC 16,428
K. Laubenheimer, Deutsche med. Wchschr. 51 (1925), 999
P. Uhlenhuth, Med. Klin. 21 (1925), 1105

KOSSEL, WALTHER [1888-1956]
DSB 7,468-470; NDB 12,616-617; KIEL, 161-162; TLK 3,354; WWWS, 962; POTSCH, 249; POGG 5,672-673; 6,1387; 7a(2),884-885

E.N.daC. Andrade, Nature 178 (1956), 568-569

KOSSOVICH, PETR SAMSONOVICH [1862-1915]
BSE 23,120; KUZNETSOV(1963), 781-787; LIPSHITS 4,387-391
D.N. Prianishnikov, Izbrannie Sochenenya 4,292-296. Moscow 1955

KOSSOWICZ, ALEXANDER [1874-1917]
OBL 4,151
Anon., Chem. Z. 41 (1917), 891
J. Weese, Z. Gärungsphysiol. 6 (1918), 161-165

KOSSWIG, KURT [1903-1982]
KURSCHNER 13,2060; BHDE 2,653; WIDMANN, 272-273

KOSTANECKI, STANISLAW [1860-1910]
DSB 7,470-471; PSB 14,334-335; BLOKH 1,387-388; KONOPKA 4,550-554; POTSCH, 249; POGG 4,795-796; 5,673
E. Noelting, Verhandl. Schw. Nat. Ges. 94 (1911), 74-128
J. Tambor, Ber. chem. Ges. 45 (1912), 1683-1707
W. Lampe, Stanislaw Kostanecki. Warsaw 1958

KOSTERLITZ, HANS WALTER [1903-]
WW 1981, p.1472; MEDVEI, 761-762; BHDE 2,653-654; LDGS, 66; WWWS, 962
H.W. Kosterlitz, Ann. Rev. Pharm. 19 (1979), 1-12

KOSTYCHEV, PAVEL ANDREEVICH [1845-1895]
BSE 23,135-136; KUZNETSOV(1963), 720-727
Anon., Nature 53 (1895), 160-161
P.A. Kostychev, Izbrannie Trudy. Leningrad 1951
V.V. Krasnikov, P.A. Kostychev. Moscow 1951
A.I. Metelkin, Zhur. Mikr. Ep. Imm. 46 (1969), 141-144

KOSTYCHEV, SERGEI PAVLOVICH [1877-1931]
BSE 23,136; WWR, 304; LIPSHITS 4,403-411; IB, 159; POGG 6,1388-1389
S.A. Waksman, Plant Physiology 7 (1932), 335-336
R.W. Kolbe, Ber. bot. Ges. 51 (1933), (208)-(219)
S.V. Soldatenkov, Biokhimia 16 (1951), 298-304
M.P. Korsakova, Mikrobiologia 20 (1951), 168-182

KOTAKE, MUNIO [1894-1976]
JBE, 705; POGG 6,1389; 7b,2570-2572
Anon., Chem. Eng. News 55(14) (1977), 85
F. Lynen, Bayer. Akad. Wiss. Jahrbuch 1978, pp.220-222

KOTAKE, YASHIRO [1879-1968]
JBE, 705; IB, 159; POGG 6,1389; 7b,2572-2573
Y. Matsamura, Wakayama Medical Reports 22 (1980), 109-114

KOTHE, RICHARD [1863-1925]
DUNSCHELE, 143-144

KOVACS, JOSEPH [1915-1985]
AMS 15(4),445
Anon., Chem. Eng. News 63(49) (1985), 51

KOVALENKO, YAKOV ROMANOVICH [1906-1980]
BSE (3rd Ed.) 12,358

KOVALEVSKAYA, EKATERINA FEDOROVNA [1874-1958]
N.N. Senyushkina and A.N. Shamin, Vop. Ist. Est. Tekh. 1975, pp.107-109

KOVALEVSKY, ALEKSANDR ONUFRIEVICH [1840-1901]
DSB 7,474-475; BSE 21,500-502; IKONNIKOV, 264-268; KAZAN 1,357-358; KUZNETSOV(1963), 157-172; GOCZOL, 134-136; W, 298-299; WWWS, 964
E.R. Lankester, Nature 66 (1902), 394-396
P. Buczinsky and Y. Bardakh, Mem. Soc. Nat. Odessa 24(2) (1902), 1-31
C. von Voit, Sitz. Bayer. Akad. 32 (1902), 288-291
N.G. Khlopin and A.G. Knorre, Usp. Sov. Biol. 32 (1951), 412-430
T.V. Makarova, Trudy Inst. Ist. Est. 24(5) (1958), 222-254
C. Davidov, Rev. Hist. Sci. 13 (1960); Trudy Inst. Ist. Est. 31(6) (1960), 326-363

KOVALEVSKY, NIKOLAI OSIPOVICH [1840-1891]
BSE 21,504; HIRSCH 3,595-596; PAGEL, 906; KAZAN 2,217-223; ZMEEV 3,26-27; 4,152
Anon., Leopoldina 27 (1891), 202
I.A. Grigorian, N.O. Kovalevsky. Moscow 1978

KOVARSKI, ARON OSIPOVICH [1872-]
SG [2]8,838; RSC 17,482

KOYRÉ, ALEXANDRE [1892-1964]
DSB 7,482-490; WININGER 7,191
P. Costabel and C.C. Gillispie, Arch. Inter. Hist. Sci. 17 (1964),149-156
S. Delorme et al., Rev. Hist. Sci. 18 (1965), 129-159
I.B. Cohen and M. Clagett, Isis 57 (1966), 157-166
R. Taton, Revue de Synthese 88 (1967), 7-20

KOZLOWSKI, ANTON [1889-]
IB, 159

KRÄMER, ADOLF [1883-1914]
JV 23,32
Anon., Chem. Z. 38 (1914), 1113; Z. angew. Chem. 28(III) (1915), 636

KRAEMER, ELMER OTTO [1898-1943]
AMS 6,800; NCAB 32,394-395; POTSCH, 249-250; POGG 6,1393; 7b,2582-2583
Anon., Chem. Eng. News 21 (1943), 1570; J. Franklin Inst. 236 (1943), 403-406
J.B. Nichols and E.B. Sanigar, Adv. Coll. Sci. 2 (1946), xiii-xxiv

KRAEMER, GEORG [1888-]
JV 33,277

KRAEMER, GUSTAV WILHELM [1842-1915]
DBJ 1,309; HEIN, 340-342; HEIN-E, 251-252; BLOKH 1, 389-390; POGG 4,798-799; 5,676
C. Göpner, Chem. Z. 39 (1915), 201-202
A. Bannow, Ber. chem. Ges. 49 (1916), 445-467

KRAEMER, HENRY [1868-1924]
DAB 10,499; NCAB 26,451-452; TSCHIRCH, 1075; WWAH, 338; WWWS, 965
Anon., J. Am. Pharm. Assn. 13 (1924), 980-982; Ind. Eng. Chem. News Ed. 2(21) (1924), 11

KRÄNZLEIN, GEORG [1881-1943]
NDB 12,638-639; NB, 216; POGG 6,1393; 7a(2),888
Anon., Z. angew. Chem. 56 (1943), 316
M. Corell, Ber. chem. Ges. 77A (1944), 45-55

KRAFFT, CARL FREDERICK [1892-]
NUC 305,111-112

KRAFFT, FRIEDRICH [1852-1923]
NDB 12,643-644; DBJ 5,234-237,433; DRULL, 149; TLK 2,385; SCHAEDLER, 62; POTSCH, 250; POGG 3,746; 4,799; 5,676; 6,1393-1394

KRAFT, FRITZ [1882-1955]
JV 24,338

KRAFT, GÖTZ [1909-]
SG [4]9,342

KRAFT, KURT [1907-]
KURSCHNER 13,2074; WWWS, 965; POGG 7a(2),890

KRAHL, MAURICE EDWARD [1908-]
AMS 14,2749-2750; WWA 1976-1977, p.1773; WWWS, 965-966

KRAMER, BENJAMIN [1888-1972]
AMS 11,1898; NCAB 57,173-174; WWWS, 966

KRAMER, KURT [1906-1985]
KURSCHNER 13,2078; POGG 7a(2),892-893
Anon., Naturw. Rund. 38 (1985), 215

KRAMER, PAUL JACKSON [1904-]
AMS 14,1783; WWA 1980-1981, p.1896; IWW 1982-1983, p.720; WWWS, 966; MH 1,273-274; MSE 2,188-189

KRAMER, VASILI VASILIEVICH [1876-1935]
BME 14,187-188; WWR, 310
Anon., Voprosy Neirokhirurgi 24(3) (1960), 1-2

KRAMPITZ, LESTER ORVILLE [1909-]
AMS 17(4),475

KRANEPUHL, ERICH [1889-1971]
Anon., Nachr. Chem. Techn. 12 (1964), 476; 19 (1971), 454

KRANNICH, WALTER [1890-]
JV 36,702; NUC 305,336

KRANTZ, JOHN CHRISTIAN [1899-1983]
AMS 11,2902; POGG 6,1396; 7b,2588-2593
C.J. Carr, Pharmacologist 26 (1984), 46-47

KRASHENINNIKOV, FEDOR NIKOLAEVICH [1869-1938]
BSE 23,296-297; WWR, 310; LIPSHITS 4,487-489; IB, 160

KRASNOSELSKAYA, TATIANA ABRAMOVA [1884-1950]
WWR, 311-312
M. Chaikakhan, Bot. Zhur. 36 (1951), 438-444

KRASNOVSKI, ALEKSANDR ABRAMOVICH [1913-]
BSE (3rd Ed.)13,332; WWWS, 966-967; POTSCH, 250

KRASSER, FRIDOLIN [1863-1922]
OBL 4,213-214; STURM 2,282; LUDY, 139-149
J. Greger, Ber. bot. Ges. 40 (1922), (112)-(121)

KRASUSKY, KONSTANTIN ADAMOVICH [1867-1937]
BSE 23,282; WWR, 312
M.M. Movsum-Zade, Zhur. Ob. Khim. 8 (1938), 381-388

KRATKY, OTTO [1902-]
KURSCHNER 13,2083; WWWS, 967; POTSCH, 250-251; POGG 6,1397; 7a(2),895-897
H. Nowotny, Ost. Chem. Z. 63 (1962), 91-93
Anon., Nachr. Chem. Techn. 20 (1972), 111-112; Chem. Z. 106 (1982), 147

KRAUCH, KARL [1853-1934]
NDB 12,679; HEIN, 342; POGG 7a(2),898
A. Wankmüller, Beitr. Wurtt. Apothekgesch. 10 (1973), 38-42

KRAUCH, KARL, Jr. [1887-1968]
NDB 12,679-681; POGG 7a(2),899
Anon., Chem. Z. 86 (1962), 232; 92 (1968), 127

KRAUS, CHARLES AUGUST [1875-1967]
DSB 7,497-498; MILES, 280-281; WWAH, 339; WWWS, 967; POTSCH, 251; POGG 5,677-678; 6,1397-1398; 7b,2594-2597
Anon., Chem. Eng. News 45 (30) (1967), 59
W.C. Johnson, Am. Phil. Soc. Year Book 1968, pp.145-151
R.M. Fuoss, Biog. Mem. Nat. Acad. Sci. 42 (1971), 119-159

KRAUS, ERIK JOHANNES [1887-1955]
STURM 2, 286-287; FISCHER 2,815; KOERTING, 146-147; BHDE 2,657; IB, 160

KRAUS, FRIEDRICH [1858-1936]
NDB 12,685; OBL 4,226-227; STURM 2,288; FISCHER 2,815-816; DZ, 797
A. Schittenhelm, Münch. med. Wchschr. 83 (1936), 529-530
G. von Bergmann, Deutsche med. Wchschr. 62 (1936), 482-484

KRAUS, GREGOR [1841-1915]
NDB 12,686-687; DBJ 1,332; TSCHIRCH, 1075
H. Kniep, Ber. bot. Ges. 33 (1915), (69)-(95)
Anon., Chem. Z. 39 (1915), 906

KRAUS, JOHANN [1869-]
JV 11,20; NUC 305,511

KRAUS, RUDOLF [1868-1932]
STURM 2,290; FISCHER 2,816; PAGEL, 910-911; PLANER, 185-186; BULLOCH, 379; BIBEL, 265-268; IB, 160; WININGER 7,192-193
M. Eisler, Wiener klin. Wchschr. 45 (1932), 1072-1073
O. Koref, Wiener med. Wchschr. 83 (1933), 60-61

KRAUS, WALTER MAX [1889-1944]
AMS 7,999; NCAB D,323-324; 34,28-29; WININGER 3,530

KRAUSE, ALFONS [1895-1972]
WWWS, 967; POGG 6,1399; 7b,2598-2610
Anon., Rev. Pol. Acad. Sci. 8 (1963), 78-79; 18(4) (1973), 204-205

KRAUSE, ALLEN KRAMER [1881-1941]
DAB [Suppl.3],431-432; AMS 6,802; ICC, 129-130; WWAH, 339
M. Pinner, Am. Rev. Tub. 44 (1941), 248-253
E.R. Long, Trans. Assoc. Am. Phys. 57 (1942), 22-23
H.S. Willis, Am. Rev. Tub. 45 (1942), 595-606

KRAUSE, CARL FRIEDRICH THEODOR [1797-1868]
ADB 17,79-81; HIRSCH 3,607-608; Suppl.,388; DECHAMBRE [4]16,770; CALLISEN 10,374-375; 29,338-339

KRAUSE, ERICH [1884-1925]
BLOKH 1,390-391; POGG 6,1399
H. Pringsheim, Ber. chem. Ges. 58A (1925), 51
G. Bugge, Chem. Z. 49 (1925), 965

KRAUSE, ERICH [1895-1932]
POTSCH, 251; POGG 5,678; 6,1399-1400; 7a(2),902
K.A. Hofmann and G. Renwanz, Ber. chem. Ges. 65A (1932), 29-30

KRAUSE, GEORG [1849-1927]
HEIN, 343-344; TSCHIRCH, 1075
H. Stadtlinger, Chem. Z. 80 (1956), 585-590

KRAUSE, PAUL [1871-1934]
FISCHER 2,817; GIESE, 573-574
K. Wohlenberg, Deutsche med. Wchschr. 60 (1934), 873
C. Kruchen, Fort. Geb. Röngtenstr. 50 (1934), 182-186

KRAUSE, ROBERT LOUIS [1885-1957]
JV 25,364
New York Times, 24 October 1957

KRAUT, HEINRICH [1893-]
KURSCHNER 13,2095-2096; TLK 3,361; WWWS, 968; POGG 6,1401-1402; 7a(2),906-908
H. Glatzel, Nutritio e Dieta 5(2) (1963), 83-86

KRAUT, JOSEPH [1926-]
AMS 15(4),464

KRAUT, KARL JOHANN [1829-1912]
HEIN-E, 254-255; BJN 18,36*; BLOKH 1,391; HANNOVER, 29; TSCHIRCH, 1075; POGG 3,747-748; 4,802; 5,678; 7a(2),908
K. Seubert, Chem. Z. 36 (1912), 157-158
H. Precht, Z. angew. Chem. 25 (1912), 175-176

KRAUTH, WILHELM [1863-1924]
JV 19,181; NUC 305,665

KRAVKOV, NIKOLAI PAVLOVICH [1865-1924]
BSE 23,190-191; BME 14,181-183; WWR, 313-314; BLOKH 1,389
O. Steppuhn, Münch. med. Wchschr. 72 (1925), 1117
A.I. Kuznetsov, N.P. Kravkov. Moscow 1948
S.Y. Arbuzov, Fiziol. Zhur. 40 (1954), 515-524
V.V. Zakusov, Int. J. Neuropharm. 4 (1965), 205-206; Farm. Toks.47(5) (1984), 5-10
A.K. Ovchinnikova, N.P. Kravkov. Moscow 1969
I. Kissin and A.J. Wright, Anesthesiology 69 (1988), 242-245

KRAVKOV, SERGEI PAVLOVICH [1873-1938]
WWR, 314; LIPSHITS 4,430-433
E.I. Shilova, Vest. Leningrad. Univ. 1949(1), pp.149-152

KRAVKOV, SERGEI VASILIEVICH [1893-1951]
BSE 23,191; BME 14,183; WWR, 314
Anon., Vest. Oftalm. 30(4) (1951), 45-46

KRAYBILL, HENRY REIST [1891-1956]
AMS 9(II),634-635; IB, 161; WWAH, 339
Anon., Science 124 (1956), 885-886; Chem. Eng. News 34 (1956), 5030
P.B. Curtis, J. Assn. Agr. Chem. 40(2) (1957), v-vi

KRAYER, OTTO HERMANN [1899-1982]
AMS 14,2760; WWA 1976-1977, p.1778; IWW 1982-1983, p.721; BHDE 2,660; LDGS, 78; CHEN, 71-73; WWWS, 968
Anon., Arch. exp. Path. Pharm. 248 (1964); Life Sciences 30(23) (1982), i-ii
U. Trendelenburg, Pharmacologist 25 (1983), 31-32
A. Goldstein, Biog. Mem. Nat. Acad. Sci. 57 (1987), 151-225

KREBS, EDWIN GERHARD [1918-]
AMS 15(4),466; WWA 1980-1981, p.1900; WWWS, 968

KREBS, HANS ADOLF [1900-1981]
DSB 17,496-506; WW 1981, p.1473; MH 1,274-276; STC 2,127-129;MSE 2,189-191; COHEN, 133; TETZLAFF, 186; KOREN, 205; BHDE 2,660-661; LDGS, 59; W(3rd Ed.), 597-598; POTSCH, 251-252; AO 1981, pp.703-705; CB 1954, pp.384-385; ABBOTT, 77-79; POGG 6,1403-1404; 7a(2),909-910
H.A. Krebs, Presp. Biol. Med. 14 (1970), 154-170; Bioch. Ed. 1 (1973), 19-23; in The Urea Cycle (S. Grisolia et al., Eds.), pp.1-12. New York 1976; Med. Hist. J. 15 (1980), 357-377; Reminiscences and Reflections. London 1981; TIBS 7 (1982), 76-78
F.L. Holmes, Fed. Proc. 39 (1980), 216-225; Am. Phil. Soc. Year Book 1982, pp.501-507; Isis 75 (1984), 131-142; Hans Adolf Krebs: The Life of a Scientist. New York 1991, 1993
D.H. Williamson, TIBS 5(8) (1980), vi-viii; Biochem. J. 204 (1982), 1-2
Anon., Lancet 1981(II), p.1299; Brit. Med. J. 284 (1982), 59,517
H. Blaschko, FEBS Letters 117Suppl. (1980), K11-K15; Erg. Physiol. 98 (1983), 1-9
H. Krebs et al., Naturw. Rund. 35 (1982), 225-235
P. Lund, Clinical Science 63 (1982), 225-230
J.R. Quayle, J. Gen. Microbiol. 128 (1982), 2215-2220
H. Kornberg and D.H. Williamson, Biog. Mem. Fell. Roy. Soc. 30 (1984), 351-385

KREHL, LUDOLF von [1861-1937]
NDB 12,733-734; FISCHER 2,818-819; PAGEL, 913; GIESE, 569-571; IB, 161; DRULL, 150; GUNDLACH, 218; ZWIFEL, 90-92
R. Siebert, Deutsches Arch. klin. Med. 181 (1928), i-vi; Deutsche med. Wchschr. 57 (1931), 2169-2172
H. Bohnenkamp, Klin. Wchschr. 16 (1937), 1262-1263; Med. Mon. 15 (1961), 803-807
E. Grafe, Deutsche med. Wchschr. 63 (1937), 980-981
H. Dennig et al., Münch med. Wchschr. 103 (1961), 2489-2502

KREIDER, LEONARD CALE [1910-]
AMS 14,2762

KREIDL, ALOIS [1864-1928]
OBL 4,244; STURM 2,297; FISCHER 2,819; IB, 161; KOREN, 205; WWWS, 969
A. Durig, Wiener med. Wchschr. 79 (1929), 68-70
G. Alexander, Mon. Ohren. 63 (1929), 727-730
G.H. Wang, Bull. Hist. Med. 39 (1965), 529-530

KREIENBERG, WALTER [1911-]
KURSCHNER 13,2102

KREIS, HANS [1861-1931]
TLK 3,361-362; POGG 4,803-804; 5,680; 6,1404
R. Viollier, Verhandl. Schw. Nat. Ges. 113 (1932), 499-504

KRELL, CARL [1875-]
JV 20,261

KREMANN, ROBERT [1879-1937]
STURM 2,302-303; KERNBAUER, 160,334-359; TLK 3,362; POGG 4,804; 5,680-681; 6,1405-1406; 7a(2),915-916
A. Skrabal, Ber. chem. Ges. 70A (1937), 152; Z. Elektrochem. 43 (1937), 851-852; Alm. Akad. Wiss. Wien 88 (1938), 283-287

KREMERS, EDWARD [1865-1941]
DAB [Suppl.3],432-433; AMS 5,636; NCAB 30,75-76; TSCHIRCH, 1075-1076; DAMB, 426-427
C.H. Rogers, Am. J. Pharm. Ed. 4 (1940), 539-546
G. Urdang, Science 94 (1941), 293-294; Trans. Wis.

Acad. Sci. 37 (1945), 111-135; Am. J. Pharm. Ed. 11 (1947), 631-658
E.W. Stieb, in Pharmaceutical Historiography (A. Berman, Ed.), pp.87-92. Madison, Wis. 1967
M.M. Weinstein and R.G. Mrtek, Pharm. Hist. 13 (1971), 77-88

KREMERS, PETER [1827-]
RSC 3,751; 8,125; NUC 306,161; POGG 1,1317-1318

KREMERS, ROLAND EDWARD [1894-]
AMS 11,2912

KRETSCHY, MICHAEL [1839-1884]
OBL 4,264; BLOKH 1,392
M. Gruber, Ber. chem. Ges. 17(3) (1884), 761-762

KRETZ, RICHARD [1865-1920]
OBL 4,264; FISCHER 2,820-821; DANIS, 8-17
R. Paltauf, Wiener med. Wchschr. 33 (1920), 526-527
H. Chiari, Verhandl. path. Ges. 30 (1937), 537-542

KREUSLER, ULRICH [1844-1921]
DBJ 3,306; DZ, 807; TLK 2,391; POGG 3,751; 4,805; 5,682
P. Koenig, Chem. Z. 45 (1921), 1145

KREY, HANS [1861-1931]
TLK 3,361-362

KREY, HERMANN [1851-1929]
NDB 13,33-34

KREY, KARL [1871-]
JV 14.70
E. Korn, Düsseldorfer Jahrbuch 54 (1972), 93-96

KREY, WALTHER [1864-]
JV 8,150; NUC 306,330

KRIEGER, HANS THEODOR [1874-]
JV 14,248; SG [2]8,858; NUC 306,396

KRIES, JOHANNES von [1853-1928]
NDB 13,46-48; DBJ 10,329; GROTE 4,125-187; FISCHER 2,821-822; IB, 161; KROLLMANN, 367-368; DZ, 808-809; WWWS, 1737; POGG 6,1408
O. Frank, Bayer. Akad. Wiss. Jahrbuch 1928-1929, pp.87-88

KRISCH, KLAUS [1928-1975]
KURSCHNER 11,1601
Who's Who in Science in Europe 1972, p.1294
H. Staudinger et al., Christiana Albertina NF4 (1976), 45-53

KRISTELLER, LEO [1879-1958]
JV 22,27
H. Lax, Proc. Virchow Med. Soc. 17 (1958), 147

KRITZMANN. MARIA GRIGORIEVNA [1904-1971]

KRÍZENECKÝ, JAROSLAV [1896-1964]
IB, 161
W. Nowak, Archiv für Hydrobiologie 62 (1966), 111-114
A. Matolova, Mendel Newsletter 17 (1979), 1-5

KROCKER, EUGEN OTTO FRANZ [1818-1891]
FERCHL, 285; RSC 3,755; POGG 3,751-752
Anon., Leopoldina 27 (1891), 106; Chem. Z. 15 (1891), 1898

KRÖHNKE, FRITZ [1903-1981]
KURSCHNER 13,2123; POGG 7a(2),923
Anon., Chem. Z. 105 (1981), 197; Naturw. Rund. 34 (1981), 270
H. Albrecht and R. Huisgen, Chem. Ber. 116(1) (1983), i-xxvii

KRÖHNKE, OTTO [1871-1940]
TLK 3,365-366
F. Krohnke, Ber. chem. Ges. 73A (1940), 158-159

KROGH, AUGUST [1874-1949]
DSB 7,501-504; DBL(3rd Ed.) 8,322-325; VEIBEL 2,255-257; W, 299-300; IB, 162; WWWS, 972; POGG 6,1411; 7b,2621-2624
C.K. Drinker, Science 112 (1950), 105-107
G. Liljestrand, Skand. Arch. Physiol. 20 (1950), 109-116
A.V. Hill, Obit. Not. Fell. Roy. Soc. 7 (1950), 221-237
P.B. Rehberg, Yale J. Biol. Med. 24 (1951), 83-102; Dansk Medicinsk Historisk Årbog 1974, pp.7-28
P. Dejours, Scand. J. Resp. Dis. 56 (1975), 337-346
C.B. Jørgensen, Biol. Revs. 51 (1976), 291-328
B. Schmidt-Nielsen, J. Appl. Physiol. 57 (1984), 293-303

KROLLPFEIFFER, FRIEDRICH [1892-1957]
NDB 13,73-74; AUERBACH, 850; MEINEL, 315-318,507; SCHMITZ, 305; TLK 3,366; POGG 6,1411-1412; 7a(2),925
F. Kröhnke, Chem. Ber. 92(10) (1959), ic-cxx

KROMPECHER, EDMUND (OEDON) [1870-1926]
MEL 1,1020; FISCHER 2,826; NUC 307,41
L.Aschoff, Beitr. path. Anat. 76 (1926), 2
B. Korompay, Orv. Kozl. 64/65 (1972), 87-102

KRONECKER, FRANZ [1856-1919]
RSC 10,467; 16,477
K., Deutsche medizinische Presse 23 (1919), 16

KRONECKER, HUGO [1839-1914]
DSB 7,504-505; NDB 13,81-82; DBJ 1,294; HIRSCH 3,616-617; PAGEL, 917-918; KOREN, 205; WININGER 3,542; WWWS, 972
S.J. Meltzer, Science 40 (1914), 441-444
P. Heger, Münch. med. Wchschr. 61 (1914), 1629-1631
H. Sahli, Verhandl. Schw. Nat. Ges. 1914, pp.54-81
E.A.S., Proc. Roy. Soc. B89 (1915-1917), xlix-l

KRONTOVSKY, ALEKSEI ANTONINOVICH [1885-1933]
BSE 23,482; BME 14,815-816; WWR, 318; FISCHER 2,827; VORONTSOV, 122-123
Anon., Arch. exp. Zellforsch. 14(4) (1933), i-ii
N.I. Velegzhanin, Pat. Fiz. Eksp. Ter. 1(5) (1957), 74-75
S.S. Dyachenko and E.P. Bernsovskaya, Mikr. Zhur. 21(4) (1959), 66-68

KROPP, WALTHER [1885-1939]
JV 23,32
Anon., Chem. Z. 63 (1939), 259; Z. angew. Chem. 52 (1939), 256

KRUBER, OTTO [1888-1958]
TLK 3,366; POGG 6,1412; 7a(2),918-919
A. Marx, Chem. Ber. 91 (1958), xv-xvii

KRÜCHE, ARNO [1854-1926]
FISCHER 2,828; PAGEL, 918-919
H.J. Schneider and D. Doberentz, Zur Geschichte der Jenaer Harnstein-forschung. Jena 1979

KRÜGER, ALBERT [1861-]
RSC 16,483

KRUEGER, ALBERT PAUL [1902-1982]
AMS 15(4),486; WWWS, 973
A.P. Weber, International Journal of Biometereology 29 (1985), 197-204

KRÜGER, FRIEDRICH von [1862-1938]
KURSCHNER 4,1599; WELDING, 419; FISCHER 2,828; LEVITSKI 2,308-310; SKVARC, 130; IB, 162; POGG 7a(2),929
E. Brennsohn, Die Aerzte Livlands, p.225. Mitau 1905
Anon., Chem. Z. 62 (1938), 142; Med. Welt 12 (1938), 260
J. Brockhausen, Album Fratrum Academicorum, p.57. Weidenau 1960

KRÜGER, FRIEDRICH [1902-1982]
KURSCHNER 13,2133-2134; WWWS, 973; POGG 7a(2),929-930
Anon., Naturw. Rund. 35 (1982), 431

KRÜGER, GERHARD [1878-1965]
JV 17,285
Anon., Nachr. Chem. Techn. 11 (1963), 69; 13 (1965), 490

KRÜGER, MARTIN [1866-1904]
BJN 10,63*; FISCHER 2,828; POGG 5,684
P. Schmidt, Ber. chem. Ges. 37 (1904), 4815-4826

KRÜGER, PAUL [1858-1917]
POGG 6,1413

KRÜGER, PAUL [1879-1929]
JV 19,202; NUC 307,234
Anon., Chem. Z. 53 (1929), 150

KRÜGER, PAUL [1886-1964]
KURSCHNER 10,1311; IB, 162
F. Duspiva, Zool. Anz. 29Suppl. (1966), 567-568

KRÜSS, GERHARD [1859-1895]
ADB 51,410-412; BLOKH 1,393-394; POGG 4,810
H. Moraht, Z. anorg. Chem. 8 (1895), 243-252; 19 (1899), 327
E. Fischer, Ber. chem. Ges. 28 (1895), 177-179

KRÜSS, HUGO [1853-1925]
TLK 3,369; RSC 16,486-487; NUC 307,255; POGG 5,684-685; 6,1413-1414

KRUIS, KAREL [1851-1917]
OBL 4,302; STURM 2,320
J. Hanus, Almanach Česke Akademie 28 (1918), 145-157
Anon., Chem. Z. 42 (1918), 105

KRUKENBERG, CARL FRIEDRICH WILHELM [1852-1889]
NDB 13,118-119; HEIN-E, 258-259; GIESE, 503-504; RSC 10,470; 12,414; 16,487-488
Anon., Leopoldina 25 (1889), 44,55

KRUKENBERG, RICHARD [1863-1924]
JV 3,136; SG [2]8,866; NUC 307,303
Anon., Münch. med. Wchschr. 71 (1924), 1490

KRUMBHAAR, EDWARD BELL [1882-1966]
AMS 11,2928; FISCHER 2,829; IB, 162; WWAH, 340
E.R. Long, Bull. Hist. Med. 31 (1957), 493-504; 41 (1967), 1-4; Trans. Coll. Phys. Phila. 34 (1967), 121-124

KRUMMACHER, OTTO [1864-1938]
KURSCHNER 4,1605; IB, 162; POGG 7a(2),934
Anon., Deutsche med. Wchschr. 65 (1939), 34

KRUSE, WALTHER [1864-1943]
KURSCHNER 6(I),1014; PAGEL, 919-920; OLPP, 215-218; IB, 163; WWWS, 974
H. Selter, Münch. med. Wchschr. 81 (1934), 1390
H. Bürgers, Deutsche med. Wchschr. 69 (1943), 759

KRUYT, HUGO RUDOLPH [1882-1959]
BWN 1,327-328; IB, 163; POGG 5,685-686; 6,1416; 7b,2628-2630
E. Cohen et al., Chem. Wkbl. 30 (1933), 414-451
H.R. Kruyt, Chem. Wkbl. 42 (1946), 264-270
J.T.G. Overbeek, Chem. Wkbl. 42 (1946), 246-254; Proc. Chem. Soc. 1960, pp.437-440
J.H. de Boer, Chem. Ind. 1959, p.1475

KUBELKA, VACLAV [1892-1977]
STURM 2,326; POGG 6,1418; 7b,2632-2633
A. Blazej, Chemicke Zvesti 31 (1977), 143-144

KUBIERSCHKY, KONRAD [1860-1949]
TLK 3,369-370; POGG 6,1419; 7a(2),939
P. Krische, Z. angew. Chem. 43 (1930), 1105

KUBLI, HEINRICH [1884-1954]
SBA 3,81; NUC 307,556
H. Spillmann, Schw. Apoth. Z. 92 (1954), 272

KUBOTA, SEIKO [1884-]
HAKUSHI 3,423-424; NUC 307,562

KUBY, STEPHEN A. [1925-1991]
AMS 17(4),514

KUCHEROV, MIKHAIL GRIGORIEVICH [1850-1911]
BSE 24,152; BLOKH 1,405; POTSCH, 252; RSC 16,492
Y.S. Musadekov, Zhur. Prikl. Khim. 24 (1951), 781-786
A.D. Petrov, Usp. Khim. 21 (1952), 250-259

KUCK, JULIUS ANTON [1907-]
AMS 15(4),496

KUCZYNSKI, MAX HANS [1890-]
KURSCHNER 4,1611; FISCHER 2,831; ASEN, 108; LDGS, 57; IB, 163; KOREN, 205; NUC 307,593

KUDICKE, ROBERT [1876-1961]
KALLMORGEN, 331-332
M. Wohlrab, Deutsche med. Wchschr. 86 (1961), 1882-1883

KUEHL, FREDERICK ALBERT [1916-1988]
AMS 15(4),497; WWWS, 975

KÜHLEWEIN, MALTE von [1887-]
GGTA 29 (1937), 344; JV 30,225; 31,245; SG [3]7,339; NUC 307,679

KÜHLING, OTTO [1862-1933]
TLK 3,371; RSC 16, 493; POGG 4,812-813; 5,688-689
Anon., Chem. Z. 34 (1910), 1213; Z. angew. Chem. 47 (1934), 389

KÜHN, ALFRED [1885-1968]
DSB 7,516-517; NDB 13,192-193; KURSCHNER 10,1318-1319; BOHM, 192; WAGENITZ, 102-103; IB, 163; WWWS, 976
K. Henke, Naturwiss. 42 (1955), 193-199
Anon., Nachr. Chem. Techn. 7 (1959), 5-6
A. Egelhaf, Naturwiss. 56 (1969), 229-232
H. Autrum, Bayer. Akad. Wiss. Jahrbuch 1969, pp.263-266

K. Herter, Sitz. Ges. Nat. Berlin NF9 (1969), 4-8
H. Rissler, Attempto 31/32 (1969), 88-91

KÜHN, JULIUS [1825-1910]
BJN 15,52*-53*; WWWS, 976
P. Holdefleiss, Mitteldeutsche Lebensbilder 2 (1927), 353-360
G. Konnecke, Kühn Archiv 74 (1960), 4-12

KÜHN, OTTO BERNHARD [1799-1863]
FERCHL, 286; POGG 1,1325-1326; 3,755

KÜHNAU, JOACHIM [1901-1983]
KURSCHNER 13,2155; POGG 7a(2),943-944
L. Ludwig, Arzneimitt. 16 (1966), 782
Anon., Chem. Z. 107 (1983), 348; Naturw. Rund. 36 (1983), 422

KÜHNE, WILLY [1837-1900]
DSB 7,519-521; NDB 13,202-203; BJN 5,102*; 8,418-423; BAD 5,446-451; HIRSCH 3,627; PAGEL, 922-923; LINDEBOOM, 1109-1110; BLOKH 1,395; DRULL, 151-152; W, 300-301; WININGER, 7,200; POGG 4,813
F. Hofmeister, Ber. chem. Ges. 33 (1900), 3875-3880
C. von Voit, Z. Biol. 40 (1900), i-viii; Sitz. Bayer. Akad. 32 (1902), 249-262
A. Kreidl, Wiener klin. Wchschr. 13 (1900), 648-650
P. Schultz, Berl. klin. Wchschr. 37 (1900), 606-608
T. Leber, in Heidelberger Professoren aus dem 19. Jahrhundert (K. Friedrich, Ed.), vol.2, pp.207-220. Heidelberg 1903
H. Kronecker, Deutsche Revue 32 (1907), 99-112
A. Schalck, Das Leben und Wirken des Heidelberg Professoren WillyKühne. Düsseldorf 1940
F. Crescitelli, Vision Research 17 (1977), 1317-1323
H. Neurath and R. Zwilling, in Semper Apertus (W. Doerr. Ed.), vol.2, pp.361-374. Berlin 1985
J.S. Fruton, Contrasts in Scientific Style, pp.72-117. Philadelphia 1990

KÜHNERT, ERNST [1818-]
RSC 3,769

KÜLZ, EDUARD [1845-1895]
NDB 13,210; LKW 1,170-176; FISCHER 2,831-832; PAGEL, 923-924; GUNDLACH, 245
M. Cremer, Münch. med. Wchschr. 42 (1895), 166-167
M. Rubner, Z. Biol. 32 (1895), 177-179
C. Haeberlin, Leopoldina 31 (1895), 156-160,178-184

KÜLZ, FRITZ [1887-1949]
FISCHER 2,832; KALLMORGEN, 332; KIEL, 87; POGG 7a(2),945-946
Anon., Z. angew. Chem. 62 (1950), 152

KÜLZ, RICHARD [1860-1889]
RSC 10,474; 12,417; 16,497-498; NUC 308,93
Anon., Pharm. Z. 34 (1889), 406,427

KÜNSTNER, GERHARD [1902-]
JV 46,483; NUC 308,135

KUENY, RENÉ [1883-1922]
HUMBERT, 247-249

KÜPPERS, GUSTAV [1875-1916]
JV 17,157; NUC 308,205
Anon., Z. angew. Chem. 29(III) (1916), 120

KÜSPERT, FRANZ [1875-1929]
JV 13,225; RSC 15,897
H. Hess, Abh. Nat. Ges. Nürnberg 23 (1930), 41-44

KÜSS, ÉMILE [1815-1871]
BERGER-LEVRAULT, 131; HIRSCH 3,628-629
F. Herrgott, Mem. Soc. Med. Strasbourg 9 (1872), 45-66
G. Hervé, Bull. Soc. Hist. Med. 25 (1932), 107-130

KÜSTER, EMIL [1877-1945]
KURSCHNER 6(I),1025-1026; NB, 222; FISCHER 2,833; KALLMORGEN, 332; IB,164
H. Heinecke, Zeitschrift für Versuchstierkunde 33 (1990), 19-22

KÜSTER, ERNST [1874-1953]
NDB 13,236-237; KURSCHNER 6(I),1026; GUNDEL, 576-589; KIEL, 184; IB, 164;TLK 3,374; WWWS, 981
R.E. Liesegang, Koll. Z. 107 (1944), 161-163
K. Hofler, Protoplasma 39 (1949), 23-36
F. Weber, Protoplasma 43 (1954), 1-2
Anon., Marcellia 30Suppl. (1954), 3-9
E. Ullrich et al., Nachr. Giessener Ges. 23 (1954), 5-59

KÜSTER, FRIEDRICH WILHELM [1861-1917]
DBJ 2,663; GUNDLACH, 470-471; MEINEL, 242-250,507-508; BLOKH 1,403-405; DZ, 825; POGG 4,815-816; 5,690; 6,1424
G. Dahmer, Chem. Z. 41 (1917), 805-806
K. Schaum, Ber. chem. Ges. 51 (1918), 1017-1024

KÜSTER, WILLIAM [1863-1929]
NDB 13,237-238; DZ, 825-826; TLK 3,375; IB, 164; TSCHIRCH, 1076; POTSCH, 254; POGG 4,816; 5,690-691; 6,1424-1426
P. Pfeiffer, Z. angew. Chem. 42 (1929), 785-787
K.H. Bauer, Chem. Z. 53 (1929), 353; Jahreshefte Wurtt. 85 (1929), xlv-xlvii
P. Brigl, Ber. chem. Ges. 64A (1931), 15-36

KÜTZING, FRIEDRICH TRAUGOTT [1807-1893]
DSB 7,533-534; ADB 51,460-461; HEIN, 350-351; TSCHIRCH, 1077; FERCHL,286; POGG 1,1326-1327; 3,755
Anon., Hedwigia 32 (1893), 329-333
W. Zopf, Leopoldina 30 (1894), 145-151
W. Schumann, Friedrich Traugott Kützing. Nordhausen 1907
F.T. Kützing, Aufzeichnungen und Erinnerungen (R.H.W. Müller and R. Zaunick, Eds.). Leipzig 1960

KUFFERATH, AUGUST [1875-]
JV 15,132; NUC 308,205

KUFFLER, STEPHEN WILLIAM [1913-1980]
AMS 14,2787; WWA 1976-1977, p.1791; AO 1980, pp.615-616; BHDE 2,671
T. Wiesel, Trends in Neurosciences 5 (1981), i,iii
H. Petsche, Alm. Akad. Wiss. Wien 131 (1982), 319-322
B. Katz, Biog. Mem. Fell. Roy. Soc. 28 (1982), 225-259
D.H. Hubel, Am. Phil. Soc. Year Book 1988, pp.185-188

KUGLER, LUDWIG [1827-1894]
RSC 3,764

KUHARA, MITSURU [1855-1919]
JBE, 720; RSC 16,503; NUC 308,232

KUHLEMANN, FRIEDRICH [1864-1926]
JV 9,209; RSC 13,276
Anon., Z. angew. Chem. 39 (1926), 1135

KUHLMANN, CHARLES FRÉDÉRIC (KARL FRIEDRICH) [1803-1881]
HEIN, 349-350; BLOKH 1,394; FERCHL, 286-287; SCHAEDLER, 63; WWWS, 976; POTSCH, 252; POGG 1,1327-1328; 3,756
L. Figuier, Ann. Scient. 25 (1881), 517-518
Anon., J. Pharm. Chim. [5]6 (1881), 379-384

KUHN, ALFRED [1895-1960]
NDB 13,258-259; POGG 6,1427-1428; 7a(2),954-955

KUHN, OTTO [1847-1919]
RSC 16,508
Anon., Chem. Z. 44 (1920), 529

KUHN, RICHARD [1900-1967]
DSB 7,517-518; NDB 13,266-268; HINK 1,473-497; STC 2,132-133; WWWS, 976; POTSCH, 253; DRULL, 153-154; ABBOTT-C, 78-79; POGG 6,1428-1429; 7a(2),958-962
Anon., Nachr. Chem. Techn. 13 (1965), 355; Chem. Z. 91 (1967), 613-614
O. Westphal, Z. angew. Chem. 80 (1968), 501-519; [Int.Ed.] 7 (1968),489-506
H.J. Bielig and H.J. Haas, Arzneimitt. 18 (1968), 502-504
W. Grassmann, Bayer. Akad. Wiss. Jahrbuch 1969, pp.231-253
H.H. Baer, Adv. Carbohydrate Chem. 24 (1969), 1-12
G. Quadbeck, in Semper Apertus (W. Doerr, Ed.), vol.3, pp.55-72. Berlin 1985

KUHN, WERNER [1899-1963]
DSB 7,518-519; NDB 13,268-269; SBA 1,89-90; KIEL, 171; POGG 6,1429-1430; 7a(2),962-964
H. Mark, J. Pol. Sci. 35 (1959), 1-2
W. Feitknecht, Verhandl. Schw. Nat. Ges. 143 (1963), 224-227
H.J. Kuhn, Verhandl. Nat. Ges. Basel 74 (1963), 239-258
M. Thürkauf, Chimia 17 (1963), 333-334
H. Kuhn, Helv. Chim. Acta 47 (1964), 690-695; Chimia 38 (1964),191-211; Chemie in Unserer Zeit 19 (1985), 86-94

KUKLA, ANTONIN [1858-1910]
RSC 16,506
Anon., Chem. Z. 34 (1910), 354

KULCHITSKY, NICHOLAS [1856-1925]
Anon., Brit. Med. J. 1925(I), p.340; Nature 177 (1964), 164
J.P.H., Journal of Anatomy 59 (1925), 336-339

KULLGREN, CARL FREDRIK [1873-1955]
SMK 4,371; POGG 4,817; 5,691; 6,1430; 7b.2460
E. Hagglund, Svensk Papperstidning 58 (1955), 600

KUMAGAWA, HACHIRO [1889-1968]
HAKUSHI 4,849-850; IB, 164

KUMAKAWA (KUMAGAWA), MUNEO [1858-1918]
JBE, 724; HAKUSHI 2,18; WWWS, 977

KUMPF, WALTHER [1899-1973]
KURSCHNER 11,1648

KUNA, MARTIN [1909-]
AMS 12,3471; WWWS, 978

KUNCKELL, FRANZ [1868-1915]
BLOKH 1,395-396; POGG 5,692
A. Michaelis, Ber. chem. Ges. 48 (1915), 1913-1914

KUNDE, FELIX TOBIAS [1827-]
BERLIN, 307; RSC 3,772; SG [1]7,576; NUC 308,523

KUNDRAT, HANS [1845-1893]
NDB 13,290-291; OBL 4,350; HIRSCH 3,633; PAGEL, 929-930; LESKY, 567-568; WWWS, 978
A. Albert, Wiener klin. Wchschr. 6 (1893), 323-325
R. Wittelshofer, Wiener med. Wchschr. 43 (1893), 810-811

KUNDT, AUGUST [1839-1894]
DSB 7,526; NDB 13,291; BLOKH 1,396; WWWS, 978; POGG 3,757-758; 4,818; 5,692
H. DuBois, Nature 50 (1894), 152-153

KUNITZ, MOSES [1887-1978]
DSB 17,515-518; AMS 13,2453; COHEN, 135; POGG 6,1430-1431; 7b,2641-2642
R.M. Herriott, Nature 275 (1978), 351-352; Biog. Mem. Nat. Acad. Sci. 58 (1989), 305-317

KUNKEL, ADAM JOSEF [1848-1905]
BJN 10,209-210,202*; FISCHER 2,838; PAGEL, 930; ROEDER, 210-218
R. Geigel, Münch. med. Wchschr. 52 (1905), 2130-2133
K.B. Lehmann, Verhandl. Phys. Med. Würz. NF38 (1906), 257-276

KUNKEL, HENRY GEORGE [1916-1983]
AMS 15(4),509; WWA 1982-1983, pp.1892-1893; AO 1983, pp.595-597; ICC, 494-495; BIBEL, 255-256
A.G. Bearn, Trans. Am. Assoc. Phys. 97 (1984), cxiv-cxvi
A.G. Bearn et al., J. Exp. Med. 161 (1985), 869-895

KUNKEL, LOUIS OTTO [1884-1960]
AMS 9(II),640
F.O. Holmes, Phytopathology 50 (1960), 777-778
W.M. Stanley, Am. Phil. Soc. Year Book 1962, pp.145-149; Biog. Mem. Nat. Acad. Sci. 38 (1965), 145-160

KUNZ, KARL JACOB [1899-1976]
KURSCHNER 9,1339; WOLF, 120; POGG 6,1432; 7a(2),969

KUNZ-KRAUSE, HERMANN [1861-1936]
HEIN, 355-356; TSCHIRCH, 1076; TLK 2,402-403; POGG 4,818-819; 5,692; 7a(2),969-970
C. Schnabel, Deutsche Apoth. Z. 51 (1936), 268-269
Anon., Chem. Z. 55 (1931), 793; 60 (1936), 179; Pharm. Z. 81 (1936), 209-210
G. Schramm, Beiträge zur Geschichte der Pharmazie 40 (1988), 37-38

KUPALOV, PETR STEPANOVICH [1888-1964]
BSE(3rd Ed.) 14,12; BME 14,1033-1034; WWR, 327; IB, 165; WWWS, 979
I.V. Danilov et al., Fiziol. Zhur. 44 (1958), 911-913
C. Giurgea, Pavlovian J. Biol. Sci. 9(4) (1974), 192-207

KUPCHAN, S. MORRIS [1922-1976]
AMS 12,3475; WWWS. 979
Anon., Chem. Eng. News 54(46) (1976), 42

KUPFFER, ADOLPH THEODOR de [1799-1865]
BSE 24,74; KAZAN 1,391-393; BLOKH 1,397-398; WWWS, 434; POGG 1,1332-1333; 3,759
Anon., Proc. Roy. Soc. 15 (1867), xlvi-xlvii

KUPFFER, KARL WILHELM von [1829-1902]
NDB 13,319-320; BJN 7,66*; WELDING, 431; HIRSCH 3,634-635; PAGEL,930-931; LEVITSKI 2,21-25; EGERER, 17-25; KIEL, 79-80
B. Dean, Science 11 (1900), 364-369

Anon., Leopoldina 39 (1903), 41-42
K. von Bardeleben, Arch. mikr. Anat. 62 (1903), 669-718;Deutsche med. Wchschr. 29 (1903), 58

KURAEV, DMITRI IVANOVICH [1869-1908]
KHARKOV, 59-61; FISCHER 2,839
A. Nyurenberg, Kharkov Med. Zhur. 6 (1908), 366-370
C. Neuberg, Biochem. Z. (1910), 1-2
C. von Voit, Sitz. Bayer. Akad. 33 (1903), 492-511

KURBATOV, APOLLON APOLLONOVICH [1851-1903]
BSE 24,81; BLOKH 1,398-399
C. Liebermann, Ber. chem. Ges. 36 (1903), 2371

KURCHINSKI, VASILI PALLADIEVICH [1855-]
LEVITSKI 2,311-312; VENGEROV 2,6

KURILOV, BENEDIKT VIKTOROVICH [1867-1921]
BSE 24,95; BLOKH 1,399-402; POGG 4,819-820; 5,693

KURLOV, MIKHAIL GEORGIEVICH [1859-1832]
WWR, 329; FISCHER 2,840; ZMEEV 4,182; RSC 16,518
D.D. Yablukov, Ter. Arkh. 31(6) (1959), 81-86
A.A. Demin and A.M. Ushakov, Klin. Med. 38(12) (1960), 131-134

KURNAKOV, NIKOLAI SEMENOVICH [1860-1941]
DSB 7,529-530; BSE 24,101-103; WWR, 329-330; KUZNETSOV(1961), 546-555; POTSCH, 254; UKRAINE 2,316-317; POGG 4,820; 5,693-694; 6,1434-1435; 7b,1652-2659H.V.A. Briscoe, Nature 148 (1941), 310
S.I. Volfkovich, Usp. Khim. 10 (1941), 757-762
G.G. Urazov et al., Usp. Khim. 21 (1952), 1019-1057,1068-1095
G.B. Kauffman and A. Beck, J. Chem. Ed. 39 (1962), 44-49
G.B. Kauffman, Polyhedron 2 (1983), 855-863

KURSANOV, ANDREI LVOVICH [1902-]
IWW 1982-1983, p.731; BSE(3rd Ed.) 14,40; MH 2,307-308; STC 2,136-137;MSE 2,194-195; LIPSHITS 4,627-631

KURSANOV, DMITRI NIKOLAEVICH [1899-1983]
BSE(3rd Ed.) 14,40; WWWS, 980
A.N. Nesmeyanov, Zhurnal Organicheskoi Khimii 5(5) (1969), i-vii

KURSANOV, LEV IVANOVICH [1877-1954]
BSE 24,112; SACKMANN, 186-188; IB, 165
K.I. Meier, Bull. Soc. Nat. Moscou NS57(1) (1952), 84-87
N.A. Komarnitski, Mikrobiologia 24 (1955), 121-122

KURSANOV, NIKOLAI IVANOVICH [1874-1921]
BSE 24,112; BLOKH 1,402-403; POGG 6,1435

KUSCHINSKY, GUSTAV [1904-]
KURSCHNER 13,2195; HARTMANN, 135-161; POGG 7a(2),972-974

KUSEL, HERMANN [1871-]
JV 18,21; NUC 309,275

KUSS, ERNST [1888-1956]
NDB 13,341-342; POGG 7a(2),974

KUSSMAUL, ADOLF [1822-1902]
NDB 13,344-345; BJN 7,66*; BAD 6,306-315; HIRSCH 3,636-637; Suppl.,389; PAGEL, 932-934; SCHWARTZ, 64-73; PITTROFF, 121-122; WWWS, 981
W. Fleiner, Deutsches Arch. klin. Med. 73 (1902), 1-89
C. Bäumler, Deutsche med. Wchschr. 28 (1902), 125-127
A. Cahn, Berl. klin. Wchschr. 39 (1902), 579,598-600,626-627
L. Edinger, Münch. med. Wchschr. 49 (1902), 281-286
A. Kussmaul, Aus meiner Dozentenzeit in Heidelberg. Stuttfart 1908; Jugenderinnerungen eines alten Arztes, 20th Ed. Munich 1953
T.H. Bast, Ann. Med. Hist. 8 (1926), 95-127; The Life and Times of Adolf Kussmaul. New York 1926
H.M. Koelbing, Praxis 62 (1973), 265-271

KUSUMOTO, CHOSABURO [1871-]
HAKUSHI 2,123
Anon., Ciba Z. (1935), 702

KUTSCHER, FRIEDRICH [1866-1942]
NDB 13,347-348; FISCHER 2,842; GUNDLACH, 247; AUERBACH, 299-300; SACKMANN, 190-192; POGG 6,1437; 7a(2),975
D. Ackermann, Klin. Wchschr. 15 (1936), 1151
Anon., Chem. Z. 66 (1943), 327

KUTSCHER, WALDEMAR [1898-1981]
KURSCHNER 13,2197; POGG 7a(2),975
Anon., Chem. Z. 105 (1981), 384; Naturw. Rund. 35 (1982), 45

KUZEL, HANS [1859-1921]
OBL 4,379
R. Krücke, Chem. Z. 45 (1921), 685

KUZNETSOV, PETR IZMAILOVICH [1868-]
KAZAN 1,386-387; VENGEROV 1,430

KUZNETSOV, SERGEI IVANOVICH [1900-]
BSE (3rd Ed.)13,563; SACKMANN, 192-193; LIPSHITS 4,581-583

KYES, PRESTON [1875-1949]
NCAB 39,103-104; FISCHER 2,844; IB, 165

KYLIN, JOHANN HARALD [1879-1949]
DSB 7,534-536; SMK 4,383; IB, 166; POGG 6,1437-1438; 7b,2662-2663
F.E. Fritsch, Nature 165 (1950), 588
S. Suneson and J. Tuneld, Bot. Not. 1950, pp.94-116

KYM, OTTO [1864-1919]
ZURICH, 58-59; ZURICH-D, 163; POGG 6,695

KYRIACOU, NIKOLAOS [1882-]
JV 24,339; NUC 309,542

L

LAAN, BERNHARD van der [1875-1935]
JV 17,157; NUC 309,620

LAAR, CONRAD [1853-1929]
HANNOVER, 43; TLK 3,376; POTSCH, 255; POGG 3,760; 4,822; 5,697; 6,1439; 7a(3),1
P. Pfeiffer, Z. angew. Chem. 42 (1929), 1117

LAAR, JOHANNES van der [1860-1938]
BWN 2,333-334; TLK 3,376; POTSCH, 255; POGG 4,1552; 5,1295-1297; 6,1439-1440; 7b,2665
W.P. Jorissen, Chem. Wkbl. 17 (1920), 362-367; 26 (1929), 548-550
F.E.C. Scheffer et al., Chem. Wkbl. 27 (1930), 418-427
H.R. Kruyt, Chem. Wkbl. 36 (1939), 19-20

Anon., Chem. Wkbl. 56 (1960), 509
H.S. van Klooster, J. Chem. Ed. 39 (1962), 74-76
H.A.M. Snelders, Centaurus 29 (1976), 53-71
E.P. van Emmerik, J.J. van Laar, a Mathematical Chemist. Delft 1991

LABARRAQUE, ANTOINE GERMAIN [1777-1850]
HOEFER 28,323-324; FERCHL, 290; TSCHIRCH, 1077; SCHAEDLER, 63; POTSCH, 255; WWWS, 982; POGG 1,1336

LABAT, JEAN ANDRÉ [1877-1954]
POGG 6,1440; 7b,2665-2666
R. Weitz, Ann. Pharm. Fran. 12 (1954), 121-123

LABBÉ, MARCEL [1870-1939]
FISCHER 2,845-846; SCHALI, 50-53
F. Bezancon, Bull. Acad. Med. 121 (1939), 804-811
M. Tiffeneau, Ann. Univ. Paris 14 (1939), 314-315

LABES, RICHARD [1889-1971]
KURSCHNER 11,1665; POGG 6,1440; 7a(3),2-3
Anon., Chem. Z. 83 (1959), 490

LABORDE, EUGÈNE [1863-1934]
HUMBERT, 253-254
J.E. Lobstein, Bull. Soc. Pharm. 41 (1934), 556-557

LABORDE, JEAN BAPTISTE VINCENT [1830-1903]
VAPEREAU 6,889; FISCHER 2,846; WWWS, 983
J. Deniker, Nature 68 (1903), 105-106
E. Lancereaux, Bull. Acad. Med. 49 (1903), 566-570

LACASSAGNE, ANTOINE [1884-1971]
STC 2,140-142; GUIART, 123-125; IB, 166; WWWS, 983
Anon., Nature 235 (1972), 291-292
J. Roche, Bull. Acad. Med. 156 (1972), 286-290
J. Bréhant, Nouv. Pr. Med. 1 (1972), 1616-1618
J. Tréfouël, C.R. Soc. Biol. 166 (1972), 6-8
R. Latarjet, Not. Acad. Sci. 6 (1974), 50-59
J.A. del Regato, International Journal of Radiology 12 (1986), 2165-2173

LACAZE-DUTHIERS, HENRI de [1821-1901]
DSB 7,545-546; VAPEREAU 6,892-893; WWWS, 983
R. von Hanstein, Naturw. Rund. 16 (1901), 541-543
Anon., Nature 64 (1901), 380-381
L. Boutan, Rev. Sci. [4]17 (1902), 33-40
G. Pruvot et al., Arch. Zool. Exp. [3]10 (1902), 1-78
S.F.H., Proc. Roy. Soc. 75 (1905), 146-150
D. Wrotnowska, Hist. Sci. Med. 1 (1967), 53-67
G. Petit and J. Théodoridès, Gesnerus 29 (1972), 19-32

LACHINOV, PAVEL ALEKSANDROVICH [1837-1891]
BSE 24,375; BLOKH 1,420-421; KOZLOV, 181; POGG 3,779; 4,842
M. Kucherov, Zhur. Fiz. Khim. 24 (1892), 567-614

LACHMAN, ARTHUR [1873-1957]
AMS 8,1417; JV 10,212; WWAH, 341; POGG 4,822; 5,697; 6,1442; 7b,2670

LACHOWICZ, BRONISLAW [1856-1903]
PSB 16,398-399; WE 6,349; KONOPKA 5,217-218
Anon., Chemik Polski 3 (1903), 298

LACHS, JAN HILARY [1881-1942]
PSB 16,401-402; POGG 5,697; 6,1443; 7b,2670
I. Stranski, Kwart. Hist. Nauki 4 (1959), 305-329

LACK, DAVID LAMBERT [1910-1973]
WW 1973, p.1836; WWWS, 983-984
W.H. Thorpe, Biog. Mem. Fell. Roy. Soc. 20 (1974), 271-293

LA COSTE, WILHELM [1854-1885]
BLOKH 1,132; POGG 4,823
A. Michaelis, Ber. chem. Ges. 19(3) (1886), 903-905

LA COUR, LEONARD FRANCIS [1907-1984]
WW 1983, p.1281
D. Lewis, Biog. Mem. Fell. Roy. Soc. 32 (1986), 357-375

LADE, FRIEDRICH GUSTAV [1821-1856]
DGB 49,219; RSC 3,793

LADENBURG, ALBERT [1842-1911]
DSB 7,551-552; NDB 13,390-391; BJN 16,171-178,45*; DRULL, 154-155; BLOKH 1,406-411; DZ, 829; KIEL, 145; WININGER 3,560; 7,207; POTSCH, 255-256; WWWS, 984-985; POGG 3,763; 4,825-826; 5,698
A. Ladenburg, Lebenserinnerungen. Breslau 1912
W. Herz, Chem. Z. 35 (1911), 933-934; Ber. chem. Ges. 45 (1919), 3597-3644
F.S. Kipping, J. Chem. Soc. 103 (1913), 1871-1895

LADENBURG, RUDOLF [1882-1952]
DSB 7,552-556; NDB 13,391-392; KURSCHNER 4,1646; BHDE 2,683; TETZLAFF, 191; POGG 5,699; 6,1444; 7a(3),4
H. Kupfermann, Naturwiss. 39 (1952), 289-290

LADISCH, KARL [1860-1925]
JV 14,225; RSC 14,804
Anon., Chem. Z. 49 (1925), 244; Z. angew. Chem. 38 (1925), 288

LAER, MARC HENRI van [1893-1951]
IB, 166; POGG 6,1445; 7b,2675-2676
R.H. Hopkins, Chem. Ind. 1952, pp.65-66

LAEVERENZ, PAUL [1905-]
JV 47,555

LaFORGE, FREDERICK BURR [1882-1958]
AMS 8,1419; MILES, 283-284; JV 23,33; POGG 6,1446; 7b,2681-2682
Anon., Science 128 (1958), 765; Chem. Eng. News 36(38) (1958), 120

LAGUESSE, GUSTAVE ÉDOUARD [1861-1927]
FISCHER 2,850-851; IB, 166
L. Lapicque, Bull. Acad. Med. [2]98 (1927), 448-450
P. Pagniez, C.R. Soc. Biol. 97 (1927), 1318-1319
A. Debeyre, Revue Française d'Endocrinologie 6 (1928), 1-9

LAHOUSSE, ÉMILE [1850-]
NUC 312,61-62
Liber Memorialis Université de Gand, vol.2, pp.569-570. Ghent 1913

LAIDLAW, PATRICK PLAYFAIR [1881-1940]
DNB [1931-1940], 520-521; WW 1939, p.1806; HUBER, 62-64; WWWS, 987 ;O'CONNOR(2), 476-478; SACKMANN, 194-196
H.H. Dale, Obit. Not. Fell. Roy. Soc. 3 (1939-1941), 427-447
C.H. Andrewes, J. Path. Bact. 51 (1940), 145-155
P.H., Biochem. J. 34 (1940), 781-782

LAIDLER, KEITH JAMES [1916-]
AMS 15(4),540; WWA 1980-1981, p.1928; WWWS, 987
B.E. Conway, International Journal of Chemical Kinetics 13 (1981), 787-788

LAIRE, GEORGES de [1836-1908]
BLOKH 1,411-412; POTSCH, 256
E. de Laire, Bull. Soc. Chim. [4]5 (1909), i-viii
O.N. Witt, Ber. chem. Ges. 42 (1909), 3-5

LAITINEN, HERBERT AUGUST [1915-1991]
AMS 15(4), 541; WWWS, 987-988
J. Winefordner, Talanta 38(10) (1991), v-vi

LAJTHA, ABEL [1922-]
AMS 15(4),541; WWWS, 988

LAKATOS, IMRE [1922-1974]
MEL 3,458
P. Feyerabend, Brit. J. Phil. Sci. 26 (1975), 1-18

LAKER, CARL [1859-]
RSC 16,563; SG [2]9,178; [3]7,432; NUC 312,274

LAKI, KOLOMAN [1909-1983]
AMS 14,2820-2821
Anon., Chem. Eng. News 61(18) (1983), 54

LALLEMAND, ALEXANDRE [1816-1886]
POGG 3,776-778
Anon., Leopoldina 22 (1886), 112

LALOU, SOCRATE [1875-]
IB, 166-167; NUC 312,401-402

LAMARCK, JEAN BAPTISTE [1744-1829]
DSB 7,584-594; GE 21,810-811; LAROUSSE 10,99-100; COLE, 295-296;W, 303-304; WWWS, 988; CALLISEN, 29,418-422; POGG 1,1353-1354
Y. Delange, Lamarck, sa Vie, son Oeuvre. Paris 1984
L.J. Jordanova, Lamarck. Oxford 1984

LAMB, ARTHUR BECKET [1880-1952]
DAB [Suppl.5],406-408; AMS 8,1422; NCAB 46,346-347; MILES, 284-285; WWAH, 343; WWWS, 989; POGG 5,702; 6,1449; 7b,2685-2686
G.B. Kistiakowsky, Am. Phil. Soc. Year Book 1952, pp.327-329
A.D. Bliss, J. Am. Chem. Soc. 77 (1955), 5773-5780
F.G. Keyes, Biog. Mem. Nat. Acad. Sci. 29 (1956), 202-234

LAMBOTTE, EMIL [1879-1955]
JV 19,182; NUC 313,145

LaMER, VICTOR KUHN [1895-1966]
AMS 11,2976-2977; NCAB 52,621; MILES, 285; IB, 196; WWAH, 343; WWWS, 990; POTSCH, 256; POGG 6,1451; 7b,2688-2693
E.K. Rideal, J. Coll. Sci. 21 (1966), 263-265
L.P. Hammett, Biog. Mem. Nat. Acad. Sci. 45 (1974), 193-214

LAMM, OLE ALBERT [1902-1964]
VAD 1963, p.605
Anon., Chem. Eng. News 44(49) (1954), 118

LaMOTTE, FRANK LINTON [1893-1977]
GOULD, 33
Baltimore Sun, 17 February 1977

LAMPADIUS, WILHELM AUGUST [1772-1842]
NDB 13,456-457; ADB 17,578-579; HEIN, 358-359; BLOKH 1,413-414; WWWS,990; POTSCH, 256-257; SCHAEDLER, 64; POGG 1,1361-1362; 7aSuppl.,350-351
A. Seifert, W.A. Lampadius. Berlin 1933
H.D. Schwarz, Pharm. Z. 117 (1972), 1223-1225

LAMPE, WIKTOR ALEXANDER [1875-1962]
PSB 16,431-432; POGG 6,1452; 7b,2694-2695

LAMPE, WILHELM [1867-1940]
JV 8,274; NUC 313,412

LAMPEN, JOSEPH OLIVER [1918-]
AMS 15(4),554

LAMPL, HANS [1889-1958]
M.M. Montessori, Int. J. Psychoanal. 41 (1960), 163-164

LAMY, CLAUDE AUGUSTE [1820-1878]
POTSCH, 257; FERCHL, 293; SCHAEDLER, 64; WWWS, 991; POGG 3,769

LANCEFIELD, DONALD ELWOOD [1893-1981]
AMS 11,2980

LANCEFIELD, REBECCA CRAIGHILL [1895-1981]
DSB 17,523-525; AMS 14,2831; AO 1981, pp.148-149
M. McCarty, Biog. Mem. Nat. Acad. Sci. 57 (1987), 227-246

LANCEREAUX, ÉTIENNE [1829-1910]
VAPEREAU 6,916; HIRSCH 3,658-659; Suppl.,390; PAGEL, 941-942; WWWS, 991; FRANKEN, 188
O. Lannelongue, Bull. Acad. Med. 64 (1910), 237-238
J.J. Poumery, Hist. Sci. Med. 23 (1989), 279-284

LANDA, STANISLAV [1898-1981]
POTSCH, 257; POGG 6,1453; 7b,2699-2704

LANDAU, MAX [1886-1915]
FISCHER 2,854; WININGER 7,212
L. Aschoff, Wiener klin. Wchschr. 28 (1915), 798

LANDAUER, PAUL [1881-]
JV 25,137; NUC 314,11

LANDAUER, WALTER [1896-1978]
AMS 13,2488; WWA 1964-1965, p.1150; IB, 167; WWWS, 992
J.M. Oppenheimer, Symp. Soc. Dev. Biol. 39 (1981), 1-13

LANDERGREN, ERNST [1867-1912]
SMK 4,453-454; FISCHER 2,855

LANDIS, EUGENE MARKLEY [1901-1987]
AMS 14,2835; MH 2,310-312; MSE 2,205-207; ICC, 205-206; WWWS, 993
J.R. Pappenheimer and A.C. Barger, Physiologist 30 (1987), 83

LANDMANN, GEORG RUDOLF [1888-]
JV 30,225; NUC 314,144

LANDOIS, LEONARD [1837-1902]
NDB 13,506-507; BJN 7,86-87,67*; WL 1,452-476; HIRSCH 3,660-661; PAGEL, 947-948; WWWS, 993
W. Peiper, Deutsche med. Wchschr. 28 (1902), 891-892

LANDOLT, ALEXIS [1853-1924]
AARGAU, 477-478

LANDOLT, HANS [1869-1951]
AARGAU, 478-479

LANDOLT, HANS HEINRICH [1831-1910]
DSB 7,619-620; NDB 13,508-509; BJN 15,54*; BLOKH 1,415-417; DZ, 834; SCHAEDLER, 64-65; W, 305; WWWS, 993; POTSCH, 257; POGG 1,1366; 3,771; 4,834-

835; 5,706; 7aSuppl.,351-352
W. Marckwald, Chem. Z. 34 (1910), 297-298
O. Schönrock, Z. Instrum. 30 (1910), 93-96
R. Pribram, Ber. chem. Ges. 44 (1911), 3337-3394
H.T.C., J. Chem. Soc. 99 (1911), 1653-1660

LANDOLT, HANS ROBERT GEORG [1865-1932]
NDB 13,508*; JV 9,142; RSC 16,583; SG [2]9,195; NUC 314,168

LANDSBERG, LUDWIG [1858-1923]
RSC 16,584
Anon., Chem. Z. 47 (1923), 255
F. Bergius, Z. angew. Chem. 36 (1923), 477

LANDSBERGER, RICHARD [1864-1939]
KOREN, 207; RSC 16,584

LANDSTEINER, KARL [1868-1943]
DSB 7,622-625; DAB [Suppl.3],440-442; NDB 13,521-523; FISCHER 2,856-857; AMS 6,818; NCAB D,403-40; KNOLL, 403-404; W, 302-303; WWWS, 993-994; ABBOTT, 80-81; DAMB, 432-433; KOREN, 207; WININGER 7,216-217; TETZLAFF, 195; POGG 6, 1455-1456; 7a(3),11
M. Heidelberger, Science 98 (1943); Biog. Mem. Nat. Acad. Sci. 40 (1959), 177-210
O.T. Avery, J. Path. Bact. 56 (1944), 592-603
W.C. Boyd, J. Immunol. 48 (1944), 1-16
P. Rous, Obit. Not. Fell. Roy. Soc. 5 (1947), 295-324
P. Speiser, Karl Landsteiner. Vienna 1961 (English ed., Vienna 1975); Wiener med. Wchschr. 80 (1968), 37-40
P. Levine, Transfusion 1 (1961), 45-52
G.R. Simms, The Scientific Work of Karl Landsteiner. Zurich 1963
P. Mazumdar, J. Hist. Biol. 8 (1975), 115-134; Karl Landsteiner and the Problem of Species. Baltimore 1976
D. Goltz, Clio Medica 16 (1982), 193-217

LANE, MALCOLM DANIEL [1920-]
AMS 15(4),564

LANG, ANTON [1913-]
AMS 15(4),566; IWW 1982-1983, pp.742-743; WWWS, 994
A. Lang, Ann. Rev. Plant Physiol. 31 (1980), 1-28

LANG, ARNOLD [1855-1914]
DSB 8,3-4; NDB 13,529-530; DBJ 1,295; AARGAU, 481; WWWS, 994-995
K. Hescheler, Verhandl. Schw. Nat. Ges. 97 (1915), 1-31
E. Kuhn-Schnyder, Aargovia 65 (1953), 391-397

LANG, GEORGI FEDOROVICH [1875-1948]
BSE 24,270; BME 15,216-218; WWR, 336
K.N. Zamyslova, Ter. Arkh. 30(6) (1958), 3-10; 45(12) (1973), 9-16
B.V. Ilinski, Klin. Med. 53(7) (1975), 5-15

LANG, KONRAD [1898-1985]
KURSCHNER 13,2221; POGG 7a(3),13
P. Marquardt, Arzneimitt. 13 (1963), 738-739
Anon., Naturw. Rund. 38 (1985), 535; Chem. Z. 110 (1986), 62

LANG, VICTOR von [1838-1921]
OBL 4,444-445; DBJ 3,172-178,307; TLK 2,408; POGG 3,772-773; 4,835; 5,706-707; 6,1457
E. Lecher, Alm. Akad. Wiss. Wien 72 (1923), 146-151

LANGE, BRUNO ALBERT [1903-1969]
NDB 13,552-553; POTSCH, 258; POGG 6,1457-1458; 7a(3),6

LANGE, CARL GEORG [1834-1900]
DSB 8,7-8; DBL(3rd Ed.) 8,478-479; HIRSCH 3,666; WWWS, 995
K. Faber, Hosp. Tid. [4]8 (1900), 573-578
Anon., Leopoldina 36 (1900), 132-133

LANGE, GÜNTHER [1922-1971]
Anon., Chem. Z. 95 (1971), 336

LANGE, HEINRICH [1873-1922]
JV 17,171; NUC 314,684

LANGE, HERMANN [1893-1929]
IB, 167
Anon., Schw. med. Wchschr. 59 (1929), 348

LANGE, OTTO [1875-1936]
OBL 4,448; STURM 2,377; TLK 3,379; POGG 6,1459; 7a(3),20
Anon., Chem. Z. 60 (1936), 706

LANGE, WILLY [1900-]
AMS 13,2496; KURSCHNER 4,1664; LDGS, 21; POGG 6,1459; 7a(3),20

LANGECKER, HEDWIG [1894-1989]
KURSCHNER 13,2229; HARTMANN, 119-134; BOEDEKER, 161-162; TLK 3,380; IB, 168; WWWS, 995; POGG 7a(2),20-21
Anon., Berliner Medizin 15 (1964), 79-84
M. Kramer, Arzneimitt. 14 (1964), 159-160

LANGENBECK, WOLFGANG [1899-1967]
NDB 13,583-584; KURSCHNER 11,1688; POTSCH, 258; POGG 6,1460; 7a(3),22-23
R.E. Oesper, J. Chem. Ed. 31 (1954), 94-95
Anon., Nachr. Chem. Techn. 7 (1959), 219; Chem. Z. 91 (1967), 547
W. Pritzkow, Z. Chem. 4 (1964), 202-203
W. Treibs, Jahrb. Sachs. Akad. 1966-1968, pp.348-349
A. Rieche, Forsch. Fort. 41 (1967), 285-286
M. Augustin, Wiss. Z. Halle 30(1) (1981), 123-126

LANGENDORFF, OSKAR [1853-1908]
BJN 13,53*; FISCHER 2,862; PAGEL, 958; DZ, 840
R. Tigerstedt, Erg. Physiol. 8 (1909), 797-812

LANGENSKIÖLD, FABIAN [1886-1957]
FISCHER 2,862
Kuka Kukin On 1950, pp.391-392
H. Bergholm, Finlands Lakare, p.286. Tammerfors 1927
K.E. Kallio, Acta Orthop. Scand. 27 (1957), 165-166

LANGENWALTER, JACOB [1868-1943]
JV 8,274; NUC 315,136

LANGER, CARL [1859-1935]
Anon., Chem. Ind. 1935, pp.273-274
C.M.W. Grieb, J. Chem. Soc. 1935, pp.1344-1345

LANGER, JOSEF [1866-1937]
OBL 5,5-6; STURM 2,381; KURSCHNER 4,1666; MOTSCH, 186-192; KOERTING, 181; ZWEIFEL, 101
B. Epstein, Jahrb. Kinderheil. 99 (1937), 265-266

LANGERHANS, PAUL [1847-1888]
DSB 8,8-9; NDB 13,593-594; ADB 51,588-589; FISCHER 2,862; PAGEL, 958-959; WWWS, 996
K. Bardeleben, Anat. Anz. 3 (1888), 850-851

H. Morrison, Bull. Inst. Hist. Med. 5 (1937), 259-297
H. Voss, Anat. Anz. 125 (1969), 333-335
V. Becher, Deutsche med. Wchschr. 95 (1970), 358-362
G. Wolff and H. Schadewaldt, Med. Welt 28 (1977), 1-7,91-96
H.M. Dittrich and H. Hahn von Borsche, Anat. Anz. 143 (1978), 231-241
F.J. Ebling, J. Inv. Derm. 75 (1980), 3-5
K. Holubar, Wiener klin. Wchschr. 100 (1988), 514-510
B.M. Hausen, Die Inseln von Paul Langerhans. Vienna 1988

LANGEVIN, PAUL [1872-1946]
DSB 8,9-14; CHARLE, 121-125; WWWS, 996; POGG 5,707; 6,1461; 7b,2714-2716
F. Joliot-Curie, Obit. Not. Fell. Roy. Soc. 7 (1950), 405-419
P. Biquard, Paul Langevin. Paris 1969
A. Langevin, Paul Langevin, mon Pere. Paris 1972

LANGHANS, THEODOR [1839-1915]
DBJ 1,333-334; FISCHER 2,863; PAGEL, 959; WWWS, 996
C. Wegelin, Corr. Schw. Aerzte 45 (1915), 1654-1659; Verhandl. path. Ges. 30 (1937), 542-545
Anon., Lancet 1916(I), p.161
F. Strauss, Mitt. Nat. Ges. Bern NF14 (1957), 33-48

LANGHELD, KURT [1880-1913]
BJN 18,104*; BLOKH 1,417; POGG 5,707
Anon., Chem. Z. 38 (1914), 10

LANGLET, NILS ABRAHAM [1868-1936]
SMK 4,462; POGG 4,837; 5,707-708; 5,1461

LANGLEY, JOHN NEWPORT [1852-1925]
DSB 8,14-19; DNB [1922-1930], 478-481; FISCHER 2,863-864; W, 306-307; O'CONNOR, 176-180; HAYMAKER, 289-292; WWWS, 996
W.M. Fletcher, J. Physiol. 61 (1926), 1-27; Proc. Roy. Soc. B101 (1927), xxxiii-xli
R. du Bois-Reymond, Erg. Physiol. 25 (1926), xv-xix
O. Frank, Bayer. Akad. Wiss. Jahrbuch 1927, pp.43-44

LANGLOIS, CHARLES [1800-1880]
BALLAND, 53-56; BERGER-LEVRAULT, 136; BLOKH 1,417; SCHAEDLER, 65; FERCHL, 295; POGG 1,1371; 3,774

LANGLOIS, JEAN PAUL [1862-1923]
FISCHER 2,864; WWWS, 996
E. Gley, Bull. Acad. Med. 89 (1923), 681-685; C.R. Soc. Biol. 89 (1923), 282-286

LANGMUIR, IRVING [1881-1957]
DSB 8,22-25; DAB [Suppl.6],363-365; CB 1940,pp.478-480; 1950,pp.320-322; NCAB C, 29-30; MILES, 288-289; FARBER, 1509-1523; DAVIS, 13-22; POTSCH, 258-259; STC 2,149-150; W, 307-308; POGG 5,708; 6,1461-1462; 7b,2716-2719
W.R. Whitney, Am. Phil. Soc. Year Book 1957, pp.129-133
V.J. Schaefer, Science 127 (1958), 1227-1229
H. Taylor, Biog. Mem. Fell. Roy. Soc. 4 (1958), 167-184
E. Rideal, Proc. Chem. Soc. 1959, pp.80-83
A. Rosenfeld et al., Langmuir the Man and the Scientist. Oxford 1962
C.G. Suits and M.J. Martin, Biog. Mem. Nat. Acad. Sci. 45 (1974), 215-247
R.E. Kohler, Hist. Stud. Phys. Sci. 6 (1975), 431-468
L.S. Reich, Tech. Cult. 24 (1983), 199-221

LANGSDORFF, WILHELM von [1827-1898]
NDB 13,610*; RSC 3,843; 16,599; NUC 315,346

LANGSDORFF, WILHELM von [1897-1960]
JV 43,415; NUC 315,346
Genealogisches Handbuch des Adels B17 (1986), 222

LANGSTEIN, LEO [1876-1933]
NDB 13,613-614; KURSCHNER 4,1677; FISCHER 2,864; WININGER 7,218-219; TETZLAFF, 195-196; KOREN, 207
H. Rietschel, Med. Klin. 29 (1933), 963-964
Anon., Chem. Z. 57 (1933), 467

LANGWORTHY, CHARLES FORD [1864-1932]
AMS 4,565-566; WWAH, 346
Anon., Science 75 (1932), 302; Ind. Eng. Chem. News Ed. 10 (1932), 106
P. Swanson, J. Nutrition 86 (1965), 3-16

LANHAM, SHIELA MARY [1929-]
Who's Who of British Scientists 1980-1981, pp.286-287

LANKESTER, EDWIN [1814-1874]
DNB 11,578-580; DESMOND, 371
Anon., Nature 11 (1875), 15-16; Trans. Bot. Soc. Edin. 12 (1876), 202-203; J. Roy. Micr. Soc. 1895, pp.16-17

LANKESTER, EDWIN RAY [1847-1929]
DSB 8,26-27; DNB [1922-1930], 481-483; BULLOCH, 379; O'CONNOR, 244-246; W, 308-309; WWWS,997
E.S. Goodrich et al., Nature 124 (1929), 309-314, 345-347
E.S. G[oodrich], Proc. Roy. Soc. B106 (1930), x-xv
S.F. Harmer, Proc. Linn. Soc. 142 (1931), 200-211
P. Austerfield, New Scientist 83 (1979), 529-531

LANYAR, FRANZ [1896-1976]
KURSCHNER 12,1829; POGG 7a(3),25
Anon., Münch. med. Wchschr. 118 (1976), 26

LAPCHINSKI, MIKHAIL DEMYANOVICH [1841-1889]
VENGEROV 3,393; ZMEEV 4,186-187; RSC 10,515; SG [2]9,212
Anon., Leopoldina 26 (1890), 109
A.S. Tauber, Meditsinski Sbornik Varshavskovo...Gospitala 3(1) (1890), 1-6

LAPICQUE, LOUIS EDOUARD [1866-1952]
DSB 8,28-29; CHARLE(2), 175-178; IB, 168; WWWS, 998
R. Leriche, C.R. Soc. Biol. 147 (1953), 4-8
A.V.L. Giroud, Bull. Acad. Med. 137 (1953), 159-161
J.F. Fulton, J. Neurophysiol. 16 (1953), 95-100
O.A.M. Wyss, Experientia 9 (1953), 198-200
J. Lescure, Lapicque. Paris 1956
P. Chauchard, Rev. Hist. Sci. 13 (1960), 246-264

LA PROVOSTAYE, FERDINAND HERVÉ de [1812-1863]
HOEFER 29,568; LAROUSSE 10,191; FERCHL, 232; POGG 1,1093-1094; 3,624

LAPWORTH, ARTHUR [1872-1941]
DSB 8,31-32; WW 1939, p.1828; FINDLAY, 353-368; SRS, 15; ABBOTT-C, 80-81; POTSCH, 259; W, 310; POGG 5,708-709; 6,1465; 7b,2728-2729
R. Robinson, Obit. Not. Fell. Roy. Soc. 5 (1947), 555-572; J. Chem. Soc. 1947, pp.989-996
M. Saltzman, J. Chem. Ed. 49 (1972), 750-752

LAQUER, BENNO HERMANN [1862-1925]
BERLIN, 620; SG [2]9,215; [3]7,453
Anon., Deutsche med. Wchschr. 51 (1925), 752

LAQUER, FRITZ [1888-1954]
FISCHER 2,866; KALLMORGEN, 335; LDGS, 60; IB, 168; TLK 3,381; KOREN, 207; POGG 6,1465; 7a(3),26

LAQUEUR, AUGUST [1875-1954]
KURSCHNER 5,779; FISCHER 2,867; WIDMANN, 155,274; LDGS, 66; KOREN, 206
Anon., Münch. med. Wchschr. 96 (1954), 1470

LAQUEUR, ERNST [1880-1947]
NDB 13,633-634; LINDEBOOM, 1146-1148; HES, 91-92; FISCHER 2,867; IB, 168; STC 2,150-151; BK, 254-255; WWWS, 999; POTSCH, 259; POGG 6,1465; 7b,2729-2730
E.C. Dodds, Nature 160 (1947), 459
O. Sackur, Bull. Soc. Chim. Biol. 29 (1947), 1115-1116
O. Riesser, Arch. exp. Path. Pharm. 206 (1949), 117-123

LARDY, HENRY ARNOLD [1917-]
AMS 15(4),583; WWA 1980-1981, p.1950; IWW 1982-1983, p.746; WWWS, 999; MH 2,312-313; MSE 2,207
H.A. Lardy, in Selected Topics in the History of Biochemistry (G. Semenza, Ed.), pp.297-325. Amsterdam 1986

LA RIVE, AUGUSTE ARTHUR de [1801-1873]
DSB 8,35-37; BRIQUET, 405-407; STELLING-MICHAUD 3,48-49; FERCHL, 296,330; POTSCH, 423; WWWS, 1000; POGG 2,657-659; 3,1126
Anon., Proc. Roy. Soc. 24 (1876), xxxvii-xl
J.L. Soret, Auguste de la Rive. Geneva 1877
E.H. Huntress, Proc. Am. Acad. Arts Sci. 79 (1951), 13-14

LA RIVE, CHARLES GASPARD de [1770-1834]
DSB 8,37-39; HOEFER 28,603-606; FERCHL, 296; WWWS, 1000; POGG 2,657
M. Cramer and G. de Morsier, Gesnerus 28 (1971), 234-245
P.A. Tunbridge, Notes Roy. Soc. 27 (1973), 263-298

LARK-HOROVITZ, KARL [1892-1958]
DSB 17,527-528; AMS 9(I),1112; WWAH, 346; WWWS, 1000; POGG 6,1466; 7b,2731-2733
H.M. James, Science 127 (1958), 1487-1488
V.A. Johnson, Karl Lark-Horovitz. Oxford 1969

LARMOR, JOSEPH [1857-1942]
DSB 8,39-41; DNB [1941-1950], 480-483; WWWS, 1000
A.S. Eddington, Obit. Not. Fell. Roy. Soc. 4 (1953), 197-207
E. Appleton, Proc. Roy. Irish Acad. 51A (1961), 55-66

LARNER, JOSEPH [1921-]
AMS 15(4),585; WWA 1980-1981, p.1851

LARSON, BRUCE LINDER [1927-]
AMS 15(4),589; WWWS, 1001

LARSON, ERIK LORENS [1899-1985]
VAD 1981, p.592; POGG 6,1468-1469; 7b,2736-2738

LARUE, GEORGE ROGER [1882-1967]
AMS 11,3015; NCAB 53,606
H.W. Stunkard, J. Parasit. 54 (1968), 544

LA RIVE, AUGUSTE ARTHUR de [1801-1873]
DSB 8,35-37; BRIQUET, 405-407; STELLING-MICHAUD 3,48-49; FERCHL, 296,330; POTSCH, 423; WWWS, 1000; POGG 2,657-659; 3,1126
Anon., Proc. Roy. Soc. 24 (1876), xxxvii-xl
J.L. Soret, Auguste de la Rive. Geneva 1877
E.H. Huntress, Proc. Am. Acad. Arts Sci. 79 (1951), 13-14

LA RIVE, CHARLES GASPARD de [1770-1834]
DSB 8,37-39; HOEFER 28,603-606; FERCHL, 296; WWWS, 1000; POGG 2,657
M. Cramer and G. de Morsier, Gesnerus 28 (1971), 234-245
P.A. Tunbridge, Notes Roy. Soc. 27 (1973), 263-298

LARK-HOROVITZ, KARL [1892-1958]
DSB 17,527-528; AMS 9(I),1112; WWAH, 346; WWWS, 1000; POGG 6,1466; 7b,2731-2733
H.M. James, Science 127 (1958), 1487-1488
V.A. Johnson, Karl Lark-Horovitz. Oxford 1969

LARMOR, JOSEPH [1857-1942]
DSB 8,39-41; DNB [1941-1950], 480-483; WWWS, 1000
A.S. Eddington, Obit. Not. Fell. Roy. Soc. 4 (1953), 197-207
E. Appleton, Proc. Roy. Irish Acad. 51A (1961), 55-66

LARNER, JOSEPH [1921-]
AMS 15(4),585; WWA 1980-1981, p.1851

LARSON, BRUCE LINDER [1927-]
AMS 15(4),589; WWWS, 1001

LARSON, ERIK LORENS [1899-1985]
VAD 1981, p.592; POGG 6,1468-1469; 7b,2736-2738

LARUE, GEORGE ROGER [1882-1967]
AMS 11,3015; NCAB 53,606
H.W. Stunkard, J. Parasit. 54 (1968), 544

LASER, HANS [1899-1980]
BHDE 2,695; LDGS, 75; NUC 317,146
Anon., Lancet 1980(I), p.270

LASHAS, VLADAS LAURINOVICH [1892-1966]
WWR, 338; IB, 169; WWWS, 1002

LASKOWSKI, MICHAEL [1905-1981]
AMS 14,2864
A. Kosloske, Marquette Medical Review 26 (1961), 113-116
M. Laskowski, in Proteinase Inhibitors (G. Fritz et al., Eds.), pp.3-10. Berlin 1974
L.F. Kress, Toxicon 20 (1982), 531-532
D. Shugar and Z. Zielinska, Acta Biochem. Pol. 30 (1983), 113-114

LASKOWSKI, MICHAEL, Jr. [1930-]
AMS 18(4),618

LASKOWSKI (LYASKOVSKI), NIKOLAI ERASTOVICH [1816-1871]
HEIN-E, 266-267; HIRSCH 3,646-647; POGG 1,1380
A.K. Batalin, Biokhimia 20 (1955), 507-510

LASNITZKI, ARTHUR [1896-]
LDGS, 75; NUC 317,230
Who's Who in British Science 1953, p.162

LASSAIGNE, JEAN LOUIS [1800-1859]
GE 21,988; LAROUSSE 10,223; BLOKH 1,419; HIRSCH 3,685; FERCHL, 297; POTSCH, 259; PHILIPPE, 792-793; DECHAMBRE [3]2,6-7; COLE, 300-301; WWWS, 1003; CALLISEN 11,94-103; 29,457-458; POGG 1,1380-1382
E.H. Huntress, Proc. Acad. Arts Sci. 78 (1950), 30

LASSAR, OSCAR [1849-1907]
NDB 13,669-670; FISCHER 2,868; PAGEL, 962-964; KOREN, 207; WWWS, 1003
J. Heiler, Deutsche med. Wchschr. 34 (1908), 70-71
J. Pagel, Virchows Jahresber. 1907, pp.452-453
O. Rosenthal, Dermatologische Zeitschrift 15 (1908), 113-120

LASZT, LADISLAS [1908-1981]
POGG 7a(3),29-30
Anon., Naturw. Rund. 35 (1982), 45

LATARJET, RAYMOND [1911-]
QQ 1989-1990, p.963

LATHAM, PETER WALLWORK [1832-1923]
WW 1923, pp.1505-1506; HIRSCH 3,687; WWWS, 1003
Anon., Nature 112 (1923), 733; Brit. Med. J. 1923(II), p.902; Lancet 1923(II), p.1059

LATIMER, WENDELL MITCHELL [1893-1955]
DSB 8,46-48; DAB [Suppl.5],413-414; AMS 9(I),1117; WWAH, 348; WWWS, 1004; POTSCH, 259-260; POGG 6,1471-1472; 7b,2740-2742
W.F. Giauque, Science 122 (1955), 406-407
J.H. Hildebrand, Biog. Mem. Nat. Acad. Sci. 32 (1958), 221-237
D. Quane, Bull. Hist. Chem. 7 (1990), 3-13

LATNER, ALBERT LOUIS [1912-]
WW 1981, p.1500; WWWS, 1004

LATSCHENBERGER, JOHANN [1847-1905]
BJN 10,250*; FISCHER 2,869
A. Kreidl, Z. Physiol. 19 (1905), 197-198

LATTES, LEONE [1887-1954]
FISCHER 2,870
Chi É 1936, p.199
G. di Guglielmo, Haematologica 38 (1954), iii-vi
P. Introzzi, Sci. Med. Ital. 6 (1958), 133-138

LAUBENHEIMER, AUGUST [1848-1904]
NDB 13,692-693; NB, 229; BLOKH 1,425; POGG 5,712
E. Buchka, Ber. chem. Ges. 37 (1904), 2885
H.W. Flemming (Ed.), Dokumente aus Hoechster Archiven, No.35. Frankfurt a.M. 1968

LAUBENHEIMER, KURT [1877-1955]
KURSCHNER 7,1172; KALLMORGEN, 335-336; FISCHER 2,870; DRULL, 156-157; TLK 3,382; IB, 169

LAUBINGER, CARL AUGUST LUDWIG [1838-1926]
HEIN-E, 269; WAGENITZ, 107

LAUBMANN, HEINRICH [1865-1951]
JV 4,284
E.O. Teutscher, Geologica Bavarica 14 (1952), 175-178

LAUCH, RICHARD [1860-1925]
JV 2,193; BLOKH 1,421-422
W. Traube, Ber. chem. Ges. 58A (1925), 49-51

LAUDENBACH, YULI PETROVICH [1863-1910]
VORONTSOV, 112; RSC 16,623

LAUE, MAX von [1879-1960]
DSB 8,50-53; NDB 13,702-705; SCHWERTE 1,56-62; STC 2,151-152; POTSCH, 260; W, 311-312; WWWS, 1005; POGG 5,712-713; 6,1473-1474; 7a(3),32-34
P.P. Ewald, Acta Cryst. 13 (1960), 513-515; Biog. Mem. Fell. Roy. Soc. 6 (1960), 135-156; Phys. Bl. 35 (1979), 337-349; Acta Hist. Leop. 14 (1980), 23-29
J. Franck, Am. Phil. Soc. Year Book 1960, pp.155-159
W. Meissner, Sitz. Bayer. Akad. 1960, pp.101-121
G. Menzer, Z. Krist. 114 (1960), 163-169
P. Herneck, Max von Laue. Leipzig 1979
K.H. Wiederkehr, Gesnerus 38 (1981), 351-369

LAUFBERGER, VILÉM [1890-1986]
IWW 1982-1983, p.749; IB, 169
V. Kruta and J. Krecek, Cesk. Fysiol. 19 (1970), 205-212
Who's Who in the Socialist Countries 1978, pp.345-346
L. Dostalek, Cesk. Fysiol. 30 (1981), 97-99
E. Travnickova and S. Trojan, Cesk. Fysiol. 39 (1990), 1083-1088

LAUFFER, MAX AUGUSTUS [1914-]
AMS 15(4),604; WWA 1980-1981, p.1961; WWWS, 1005
M.A. Lauffer, TIBS 9 (1984), 369-371

LAUGIER, HENRI [1888-1973]
CHARLE, 178-181; WWWS, 1005
Anon., Ann. Univ. Paris 22 (1952), 222

LAULANIÉ, FERDINAND [1850-1906]
J. Andrieu, Bibliographie Generale de l'Agenais 2,64-66. Paris 1889
A. Gueniot, Bull. Acad. Med. 58 (1906), 8-9
C. Cadeac and S. Arloing, J. Med. Vet. [4]10 (1906), 321-330,514-550

LAUNOIS, PIERRE ÉMILE [1856-1914]
FISCHER 2,872; 872; WWWS, 1005
P. Pagniez, Presse Med. 32 (1914), 381

LAUNOY, LEON LOUIS [1876-1971]
GORIS, 465-468; IB, 169; WWWS, 1005
J. Tréfouël, C.R. Soc. Biol. 166 (1972), 8-9

LAURENT, AUGUSTE [1807-1853]
DSB 8,54-61; BUGGE 2,98-114; GE 21,1037; BLOKH 1,422-425; TSCHIRCH, 1078; POTSCH, 260-261; FERCHL, 298-299; W, 312-313; WWWS, 1006; POGG 1,1386-1390
P. Yorke, J. Chem. Soc. 7 (1855), 149-157
C. de Milt, J. Chem. Ed. 28 (1951), 198-204; Chymia 4 (1953), 85-114
O. Potter, Ann. Sci. 9 (1953), 271-280
R. Stumper, La Vie et l'Oeuvre d'un Grand Chimiste. Luxembourg 1953
J. Jacques, Bull. Soc. Chim. 1954, pp.D31-D39; Institut Grand Ducal de Luxembourg, Section des Sciences etc. 32 (1955), 11-35
P. Viard et al., Bull. Soc. Langres 14 (1968), 277-292
S.C. Kapoor, Isis 60 (1969), 477-527
T.H. Levere, Ambix 17 (1970), 111-126
J.H. Brooke, Hist. Stud. Phys. Sci. 6 (1975), 405-429

LAURENT, ÉMILE [1861-1904]
L. Errera, Bull. Soc. Bot. Belg. 42 (1905), 7-20
A. Gravis, Ann. Acad. Roy. Belg. 75 (1909), 47-119

LAURENT, JOSEPH [1881-]
JV 27,428; NUC 318,436

LAURITZEN, MARIUS [1864-1949]
KBB 1949, p.842; VEIBEL 2,265

LAUTENSCHLÄGER, CARL LUDWIG [1888-1962]
NDB 13,731-732; HEIN-E, 270-271; KALLMORGEN, 336; POGG 6,1474; 7a(3),40
Anon., Chem. Z. 87 (1963), 29; Nachr. Chem. Techn. 11 (1963), 49
G. Ehrhart, Arzneimitt. 13 (1963), 159-160

LAUTH, CHARLES [1836-1913]
BLOKH 1,425; POGG 4,845; 5,714
E. Noelting, Chem. Z. 38 (1914), 17-18
A. Haller, Bull. Soc. Chim. [4]21 (1917), i-xviii

LAUTH, ERNEST ALEXANDRE [1803-1837]
HOEFER 29,951-952; HIRSCH 3,695; BERGER-LEVRAULT, 140; WWWS, 1006

LAUTSCH, WILLY [1912-1963]
KURSCHNER 9,1160; POGG 7a(3),41-42
Anon., Chem. Z. 87 (1963), 857; Nachr. Chem. Techn. 11 (1963), 402

LAUX, JULIUS [1886-]
JV 27,396

LA VALETTE SAINT-GEORGE, ADOLF von [1831-1910]
BJN 15,54*-55*; FISCHER 2,872; PAGEL, 966-967; DZ, 849-850
M. Nussbaum, Münch. med. Wchschr. 52 (1905), 708-710
O. Hertwig and W. Waldeyer, Arch. mikr. Anat. 76 (1910-1911), i-ii

LAVERAN, ALPHONSE [1845-1922]
DSB 8,65-66; GENTY 4,321-336; FISCHER 2,873; PAGEL, 967; OLPP, 225-227; HD 2,446-449; ABBOTT, 81-82; W, 313; WWWS, 1007
E. Roux, Ann. Inst. Pasteur 29 (1915), 405-414
E. Brumpt, Bull. Acad. Med. 87 (1922), 553-556
E. Roubaud, Rev. Gen. Sci. 33 (1922), 353-355
R.R., Proc. Roy. Soc. B94 (1923), xlix-liii
M. Phisalix, Alphonse Laveran. Paris 1923
C. Achard, Bull. Acad. Med. 102 (1929), 596-609
F. Mesnil, Not. Acad. Sci. 1 (1937), 300-310
J. Théodoridès, Hist. Sci. Med. 7 (1973), 225-231
M. Camelin, Hist. Sci. Med. 14 (1980), 377-382
S. Jarcho, Bull. Hist. Med. 58 (1984), 215-224

LAVES, ERNST [1863-1927]
HEIN, 360-361; HANNOVER, 43; TLK 2,413; POGG 4,845; 6,1475
Anon., Chem. Z. 51 (1927), 808; Pharm. Z. 72 (1927), 1274

LAVES, FRITZ [1906-1978]
DSB 17,529-532; NDB 14,3; KURSCHNER 12,1840; POGG 6,1475; 7a(3),42-43
H. Jagozinski, Bayer. Akad. Wiss. Jahrbuch 1979, pp.279-282

LAVIN, GEORGE ISRAEL [1903-1978]
AMS 13,2526; POGG 6,1475; 7b,2744-2745
Anon., Chem. Eng. News 56(48) (1978), 46

LAVRENTIEV, BORIS INNOKENTIEVICH [1892-1944]
DSB 8,92; BSE 24,201-202; BME 15,169-171; KUZNETSOV(1963), 448-456
A.N. Mislavski, Zhur. Ob. Biol. 5 (1944), 199-204
N.G. Kolosov, Arkh. Anat. Gist. Emb. 56(6) (1967), 113-121
E.K. Plechkova and E.M. Krokhina, Arkh. Anat. Gist. Emb. 66(6) (1974), 111-119

LAVROV, DAVID MELITONOVICH [1867-1929]
IB, 170; RSC 16,637-638
C.V. Isiganov, Odesski Meditsinski Zhurnal 3 (1928), 665-667

LAWACZECK, HEINZ [1891-1963]
Ludwigs Universität Justus Liebig Hochschule 1607-1957, pp.471-472. Giessen 1957

LAWES, JOHN BENNET [1814-1900]
DSB 8,92-93; DNB 22,951-954; DESMOND, 373-374; W, 315; WWWS, 1008
F.W., J. Chem. Soc. 79 (1901), 890-897
R.W., Proc. Roy. Soc. 75 (1905), 228-236, 242-245
F.C. Bawden, J. Nutrition 90 (1966), 3-12
N.W. Pirie, TIBS 5(7) (1980), iii-vi

LAWRENCE, ARTHUR STUART CLARKE [1902-1971]
Who's Who in British Science 1953, p.163
Anon., Chem. Brit. 7 (1971), 307

LAWRENCE, ERNEST ORLANDO [1901-1958]
DSB 8,93-96; DAB [Suppl.6],369-372; AMS 9(I),1122; MH 1,286-288; MH 1,286-288; STC 2,152-154; MSE 2,210-211; CB 1952, pp.329-332; NCAB G,473-474; 48,137-138; W, 315-316; WWAH, 349-350; WWWS,1008; POTSCH, 262; POGG 6,1476; 7b,2749-2751
A.H. Compton, Am. Phil. Soc. Year Book 1958, pp.127-131
H. Childs, An American Genius. New York 1968
L.W. Alvarez, Biog. Mem. Nat. Acad. Sci. 41 (1970), 251-294

LAWRENCE, JOHN HUNDALE [1904-]
AMS 14,2880; WWWS, 1008
W.G. Myers, J. Nucl. Med. 20 (1979), 560-564

LAWRENCE, WILLIAM [1783-1867]
DSB 8,96-98; DNB 11,727-728; BETTANY 1,303-311; WWWS, 1009
W.S.S., Proc. Roy. Soc. 16 (1868), xxv-xxx
J. Goodfield, J. Hist. Biol. 2 (1969), 283-320
K.D. Wells, J. Hist. Biol. 4 (1971), 319-362

LAWRENCE, WILLIAM TREVOR [1870-1934]
WW 1934, pp.1922-1923; DESMOND, 375; JV 11,20; RSC 16,640

LAWRIE, NORMAN RISBOROUGH [1905-1959]
Anon., Brit. Med. J. 1960(I), p.65

LAWSON, ALEXANDER [1906-]
Who's Who in British Science 1953, p.163

LAWSON, THOMAS ATKINSON [1862-1903]
JV 1,151; RSC 16,643; NUC 319,619
Anon., Ber. chem. Ges. 36 (1903), 4395

LAYCOCK, WILLIAM FREDERICK [1866-1912]
JV 5,293; RSC 16,643; NUC 320,10
Anon., Chem. Z. 36 (1912), 1514

LAZARENKO, FEDOR MIKHAILOVICH [1888-1953]
BME 15,181-182; WWR, 339-340
Anon., Arkh. Anat. Gist. Emb. 31(1) (1954), 94-95
P.G. Svetlov et al., Vest. Akad. Med. Nauk 1954(2), pp.77-78

LAZAREV, PETR PETROVICH [1878-1942]
DSB 8,101-102; BSE 24,225-226; BME 15,179-180; WWR, 340; STC 2,154
Anon., Biofizika 23 (1978), 1122-1123

LAZAROW, ARNOLD [1916-1975]
AMS 11,2043; WWWS, 1010
A.M. Carpenter, J. Histochem. 23 (1975), 791-792
B. Scharrer, Biol. Bull. 151 (1976), 16-17

LAZARUS, ADOLF [1867-1925]
FISCHER 2,874; KOREN, 208; WININGER 3,611
H. Hirschfeld, Deutsche med. Wchschr. 51 (1925), 1375

LAZARUS, MAURICE JULIUS [1863-]
RSC 13,234

LAZARUS, PAUL [1873-1957]
KURSCHNER 4,1681; FISCHER 2,875; LDGS, 67; KOREN, 208; WININGER 7,223

LEA, ARTHUR SHERIDAN [1853-1915]
WW 1915, p.1258; O'CONNOR, 180-181
J.N. L[angley], Proc. Roy. Soc. B89 (1917), xxv-xxvii

LEADER, ALFRED JOHN [1875-1947]
J. Needham and E. Baldwin, Hopkins and Biochemistry, p.344. Cambridge 1949

LEAKE, CHAUNCEY DEPEW [1896-1978]
AMS 13,2538; CHEN,81-82; WWWS, 1012
C.D. Leake, Anesthesiology 25 (1964), 428-435; Ann. Rev. Pharm. 16 (1976), 1-14
G.H. Brieger, Bull. Hist. Med. 51 (1977), 121-123
J.C. Krantz, Pharmacologist 20 (1978), 14-17
T.E. Keys, J. Hist. Med. 33 (1978), 428-431

LEARED, ARTHUR [1822-1879]
DNB 11,787-788; DESMOND, 377; RSC 3,903; 8,180

LEATHES, JOHN BERESFORD [1864-1956]
WW 1949, p.1618; FISCHER 2,876; YOUNG, 44-45; O'CONNOR(2), 247-249; IB, 170; WWWS, 1012
C.L. Evans, Nature 178 (1956), 833-834
R. Peters, Biog. Mem. Fell. Roy. Soc. 4 (1958), 185-191
D. Graham, Proc. Roy. Soc. Canada [3]52 (1958), 89-93

LEAVENWORTH, CHARLES STANLEY [1879-1948]
AMS 7,1040
Obituary Record of Yale Graduates, No.106, p.106. New Haven 1950

LEBACH, HANS [1881-]
JV 20,261

LEBEAU, PAUL [1868-1959]
IWW 1958, p.537; POTSCH, 262; POGG 4,848-849; 5,716-717; 6,1478; 7b,2758-2759
C. Chaudron, C.R. Acad. Sci. 249 (1959), 2431-2434
A. Morette, Ann. Univ. Paris 30 (1960), 79-83
W.H. Linnell, Nature 185 (1960), 429
L. Hackspill, Bull. Soc. Chim. 1961, pp.1-2
C. Champetier, Not. Acad. Sci. 4 (1964), 509-532
G. Dillemain, Produits et Problèmes Pharmaceutiques 28 (1973), 379-380

LEBEDEV, ALEKSANDR NIKOLAEVICH [1854-]
ZMEEV 4,190; RSC 10,535; 12,434; 16,648

LEBEDEV, ALEKSANDR NIKOLAEVICH [1881-1938]
DSB 18,533-534; BSE 24,378; BME 15,281; WWR, 340-341; IB, 170; POGG 6,1478-1479
N.D. Zelinski and A.A. Dikanova, Usp. Khim. 7 (1938), 1907-1910
A. Bezkorovainy, J. Hist. Med. 28 (1973), 388-392

LEBEDEV, SERGEI VASILIEVICH [1874-1934]
DSB 8,108; BSE 24,381-382; WWR, 341; KUZNETSOV(1961), 582-587; POTSCH, 262-263; ARBUZOV, 138-142,220-223; POGG 6,1479; 7b,2759-2761
A.I. Yakubchik, Zhur. Ob. Khim. 5 (1935), 1-17
A.E. Arbuzov, Usp. Khim. 13 (1944), 253-264
K.B. Piotrovsky, Priroda 37(6) (1948), 74-77

LEBEDEV, SERGEI VASILIEVICH [1876-1938]
POGG 6,1478; 7b,1761-1763

LEBEDINSKI, ANDREI VLADIMIROVICH [1902-1965]
BSE(3rd Ed.) 14,230; BME 15,283-284; WWR, 342-343
Anon., Fiziol. Zhur. 51 (1965), 634-635

LEBEDINTSEV, ALEKSANDR NIKANDROVICH [1878-1941]
KUZNETSOV(1963), 884-890
A.N. Lebedintsev, Izbrannie Trudy. Moscow 1960

LE BEL, JOSEPH ACHILLE [1847-1930]
DSB 8,109-110; SCHAEDLER, 68; W, 316-317; WWWS, 1013; POTSCH, 263; POGG 3,783-784; 4,849; 5,717; 6,1479
W.J. Pope, J. Chem. Soc. 1930, pp.2789-2791
M. Delépine, Cull. Soc. Chim. [4]47 (1930), 1344-1346; Vie et Oeuvres de J.A. Le Bel. Paris 1949
E. Wedekind, Z. angew. Chem. 43 (1930), 985-986
P. Federlin and J.P. Vigneron, Saisons d'Alsace NS52 (1974), 84-122
H.A.M. Snelders, in van't Hoff-Le Bel Centennial (O.B. Ramsey, Ed.), pp.66-73. Washington, D.C. 1975
R.B. Grossman, J. Chem. Ed. 66 (1989), 30-33

LEBER, ALFRED THEODOR [1881-]
KURSCHNER 4,1682; JV 21,264

LEBER, CHARLES THEODORE [1878-]
JV 21,274; NUC 321,575

LEBER, THEODOR [1840-1917]
NDB 14,19-20; DBJ 2,663; FISCHER 2,876; PAGEL, 969-970; DRULL, 157-158
A. Wagenmann, Arch. Ophthalm. 93 (1917), i-vi
W. Jaeger, Documenta Ophthalmologica 68 (1988), 71-77

LEBERT, HERMANN [1813-1878]
ADB 18,94-97; HIRSCH 3,705-706; Suppl.,392; PAGEL, 970-972; ZURICH-D, 31; KOHUT 2,281; WININGER 4,3-4; WWWS, 1013
E. Goldschmidt, Gesnerus 6 (1949), 17-33

LEBLANC, FÉLIX [1813-1886]
FERCHL, 302; WWWS, 1013; POGG 1,1389; 3,785
E.P. Bérard, Bull. Soc. Chim. [2]47 (1887), i-xi

LE BLANC, MAX [1865-1943]
DSB 8,112-113; NDB 14,21; TLK 3,384; POTSCH, 263; POGG 4,850; 5,718; 6,1480; 7a(3),44
M. Volmer, Z. Elektrochem. 41 (1935), 309-314
M. Born, Obit. Not. Fell. Roy. Soc. 6 (1948), 161-188
K.F. Bonhoeffer, Z. Elektrochem. 55 (1951), 74-75

LEBLOND, CHARLES PHILIPPE [1910-]
AMS 15(4),632; IWW 1982-1983, p.755; WWWS, 1013

LEBRUN, HECTOR [1866-1960]
GHENT 2,76-79

LECANU, LOUIS RENÉ [1800-1871]
VAPEREAU 1,1053; LAROUSSE 10,295; BOURQUELOT, 59-60; HIRSCH 3,708; TSCHIRCH, 1078; FERCHL, 303; CALLISEN 11,175-177; 29,485-486; POGG 1,1399-1400; 3,785
L. Figuier, Ann. Scient. 15 (1870-1871), 523-524
A. Marcel, Notice Biographique sur le Docteur le Canu. Paris 1872
G. Dillemann, Produits et Problèmes Pharmaceutiques 28 (1973), 36-37

LECCO, MARCO [1853-1932]
POGG 4,851-852; 5,719; 6,1481-1482
S.S. Miholic, Ber. chem. Ges. 65A (1932), 151

LECHARTIER, GEORGES VITAL [1837-1903]
VAPEREAU 6,945; POGG 3,785-786; 4,852

LE CHÂTELIER, HENRI LOUIS [1850-1936]
DSB 8,116-120; FREUND 3,89-98; FARBER, 909-920; W, 318; ABBOTT-C, 83; POTSCH, 264; WWWS, 1014; POGG 3,786; 4,852-853; 5,719-720; 6,1479; 7b,2765-2766
A. Labbé, Rev. Gen. Sci. 37 (1926), 38-43
R.E. Oesper, J. Chem. Ed. 8 (1931), 442-461
W.J. Pope, Nature 138 (1936), 711-712
P. Pascal, Bull. Soc. Chim. [5]4 (1937), 1557-1611
A. Silverman, J. Chem. Ed. 14 (1937), 555-560
C.H. Desch, J. Chem. Soc. 1938, pp.139-150
M. Bodenstein, Ber. chem. Ges. 72A (1939), 122-127
J. Perrin et al., Not. Acad. Sci. 2 (1949), 55-81
E.H. Huntress, Proc. Am. Acad. Arts Sci. 78 (1950), 30-32
P. Laffitte, Rev. Quest. Sci. 133 (1962), 457-480
R.S. Treplow, J. Chem. Ed. 57 (1980), 417-420

LECHER, HANS ZACHARIAS [1887-1970]
NDB 14,24; KURSCHNER 11,1705; AMS 11,3052; JV 29,726; TLK 3,384-385; POGG 5,720; 6,1482; 7a(3),45
Anon., Chem. Eng. News 48(53) (1970), 46

LECLAINCHE, EMMANUEL [1861-1953]
WWWS, 1014
G. Ramon, C.R. Acad. Sci. 237 (1953), 1457-1463
P. Bressou, C.R. Acad. Agr. 39 (1953), 748-750
R. Dujarric de la Rivière, Not. Acad. Sci. 3 (1957), 554-571

LECOMTE, HENRI [1856-1934]
IB, 171; WWWS, 1014
F. Gagnepain, Bull. Soc. Bot. 81 (1934), 472-474

LECOMTE du NOÜY, PIERRE [1883-1947]
IB, 171; WWWS, 171; POGG 6,1483-1484; 7b,2773-2775
A. Boutaric, Rev. Sci. 86 (1948), 3-15

LECOQ, RAOUL [1892-]
POGG 6,1484-1485; 7b,2775-2789

LECOQ DE BOISBAUDRAN, PAUL ÉMILE (dit FRANÇOIS) [1838-1912]
DSB 2,254-255; POTSCH, 53; WWWS, 426; POGG 3,788; 4,854-855; 5,721
G. Urbain, Chem. Z. 36 (1912), 929-933
N.D.L.R., Rev. Gen. Chim. 15 (1912), 369-372
J.H. Gardiner, Nature 138 (1912), 255-256
W. Ramsay, J. Chem. Soc. 103 (1913), 742-746
M.A. Gramont, Rev. Sci. 51 (1913), 97-109
M. Marche, Bulletin de l'Institut d'Histoire et d'Archéologie de Cognac 1(3) (1959), 233-262
C. Odouard, Lecoq de Boisbaudran. Paris 1985

LeCOUNT, EDWIN RAYMOND [1868-1935]
AMS 5,660; WWWS, 1014
L. Hektoen, Proc. Inst. Med. Chicago 10 (1935), 350-353; Arch. Path. 20 (1935), 816-819
P.S. Rhoads, Proc. Inst. Med. Chicago 29 (1972), 16-23

LE DANTEC, FÉLIX [1869-1917]
DSB 8,124-125; WWWS, 1015
E. R[abaut], Bull. Biol. 51 (1917), i-xi
Anon., Nature 99 (1917), 488-489
A. Diara and K. Wellman, Revue de Synthèse [3]100 (1979), 407-442; 102 (1981), 73-86

LEDDERHOSE, GEORG [1855-1925]
FISCHER 2,878-879; PAGEL, 972
Anon., Schw. med. Wchschr. 55 (1925), 268

LEDEBUR, JOACHIM von [1902-1944]
NUC 322,469
Genealogisches Handbuch des Adels 59 (1975), 288

LEDER, PHILIP [1934-]
AMS 15(4),635
P. Leder, BioEssays 1 (1984), 27-29,52-54

LEDERBERG, JOSHUA [1925-]
AMS 15(4),635; WWA 1980-1981, p.1978; IWW 1982-1983, p.756; COHEN, 139; MH 1,290-291; STC 2,159-160; MSE 2,213-214; CB 1959, pp.251-252; NCAB I,601-604; WWWS, 1015
J. Lederberg, Ann. Rev. Gen. 21 (1987), 23-46

LEDERER, EDGAR [1908-1988]
QQ 1975, p.1025; IWW 1982-1983, p.756; ETTRE, 237-245; STC 2,160-161; BHDE 2,619; LDGS, 22; WWWS, 1015; POTSCH, 264-265
Anon., Leopoldina [3]6/7 (1960-1961), 51-52; Chem. Z. 113 (1989), 192
E. Lederer, in Science and Scientists (M. Kageyama et al., Eds.), pp.315-322. Tokyo 1981; in Selected Topics in the History of Biochemistry (G. Semenza, Ed.), pp.437-489. Amsterdam 1986
M. Lederer, J. Chromat. 462 (1989), 1-2

LEDERER, LEONHARD [1860-1919]
RSC 16,668; NUC 322,489

LEDINGHAM, JOHN CHARLES GRANT [1875-1944]
WW 1944, p.1606; FISCHER 2,879; O'CONNOR(2), 458; IB, 171; WWWS, 1015
G.F. Petrie, J. Path. Bact. 58 (1946), 115-133
S.P. Bedson, Obit. Not. Fell. Roy. Soc. 5 (1947), 325-340

LEDUC, STÉPHANE [1853-1939]
IB, 171; SCHALI, 55-57
L. Desclaux, Annales d'Hygiène 17 (1939), 273-278
A.M. Baudouin, Bull. Acad. Med. 121 (1939), 576-577
R. Gauducheau, Presse Med. 47 (1939), 1040-1041

LEE, FERDINAND CHRISTIAN [1894-1974]
AMS 7,1042

LEE, FREDERIC SCHILLER [1859-1939]
DAB [Suppl.2],373-374; AMS 6,836; NCAB B,262-263; 29,24-25; IB, 171; BROBECK, 137-138; FISCHER 2,880; FLUELER, 78-79; WWAH, 352
H.B. Williams, Science 91 (1940), 133

LEE, JOHANNES van der [1901-1983]
POGG 6,1486; 7b,2792-2793
J. Bos, Chem. Wkbl. 79 (1983), 208

LEERS, LUDWIG [1812-1860]
DGB 17 (1910), 312

LEFÈVRE, AMÉDÉE [1798-1869]
HIRSCH 3,721-722; CALLISEN 29,495
C. Maisonneuve, Arch. Med. Navale 17 (1872), 128-146

LE FEVRE, RAYMOND JAMES WOOD [1905-1986]
WW 1983, pp.1320-1321; WWWS, 1018; POGG 6,1487; 7b,2794-2802
M.J. Aroney and A.D. Buckingham, Biog. Mem. Fell. Roy. Soc. 34 (1988), 375-403

LEFORT, JULES [1819-1896]
GORIS, 475-478; VAPEREAU 6,956; BOURQUELOT, 79-80; TSCHIRCH, 1079; FERCHL, 304; POGG 3,791
G. Planchon and H. Marty, J. Pharm. Chim. [6]3 (1896), 385-390

LEGAL, EMMO [1859-1922]
FISCHER 2,882

LEGALLOIS, CÉSAR [1770-1814]
DSB 8,132-134; HOEFER 30,370-372; HIRSCH 3,722-723; FERCHL, 304; WWWS, 1018
P. Huard, Hist. Med. 4 (1954), 23-25
G. Légee, Histoire et Nature 7 (1976), 59-73

LEGÉR, JEAN EUGÈNE [1849-1939]
GORIS, 478-480; SCHALI, 58; POGG 5,723; 6,1488
G. Dupont, Bull. Soc. Chim. [5]6 (1939), 595
A. Goris, Bull. Acad. Med. 121 (1939), 194-196
Anon., Bull. Soc. Pharm. 46 (1939), 122-125

LEGG, JOHN WICKHAM [1843-1921]
DNB [1912-1921], 330-331; MR 4,242-243; FISCHER 2,882; FRANKEN, 188-189
Anon., Lancet 1921(II), p.1027; Brit. Med. J. 1921(II), p.773
A.E. Garrod, St. Barts. Hosp. Rep. 55 (1922), 1-6

LEGROUX, RENÉ [1877-1951]
P. Lépine, Ann. Inst. Pasteur 80 (1951), 301-302,332-336
P. Vallery-Radot, Presse Med. 59 (1951), 341

LEHMAN, ISRAEL ROBERT [1924-]
AMS 15(4),661; WWA 1980-1981, p.1988

LEHMANN, CARL GOTTHELF [1812-1863]
ADB 18,147; HIRSCH 3,727; PAGEL, 975-976; POGG 1,1411-1412
A.W. Hofmann, J. Chem. Soc. 16 (1863), 433-434

LEHMANN, FRANZ [1860-1942]
NDB 14,78-79; KURSCHNER 4,1689

LEHMANN, FRANZ [1881-1961]
HEIN-E, 275-276; POGG 7a(3),49; (4),148*
H. Wollmann and C. Friedrich, Pharmazie 36 (1981), 139-146

LEHMANN, FRITZ ERICH [1902-1970]
KURSCHNER 11,1711; SBA 5,79
R. Weber, Verhandl. Schw. Nat. Ges. 150 (1970), 311-314
A. Bairati, Journal of Submicroscopic Cytology 13 (1981), 117-125

LEHMANN, FRITZ JULIUS [1874-]
JV 13,21; RSC 15,650

LEHMANN, GUNTHER [1897-1974]
KURSCHNER 11,1712-1713; IB, 171; POGG 7a(3),50-52

LEHMANN, HERMANN [1910-1985]
WW 1981, pp.1526-1527; BHDE 2,702-703; WWWS, 1019
Anon., Lancet 1985(II), pp.284,341; Brit. Med. J. 291 (1985), 288-289
R.W. Carrell, TIBS 10 (1985), 468-469
J. Dacie, Biog. Mem. Fell. Roy. Soc. 34 (1988), 407-449

LEHMANN, JØRGEN [1898-1989]
VAD 1981, p.603; SMK 4,510

LEHMANN, JOHANNES [1823-1899]
HEIN-E, 274-275
H. Schelenz, Pharm. Z. 44 (1899), 300

LEHMANN, JULIUS ALEXANDER [1825-1894]
SCHAEDLER, 68; RSC 3,937; 8,194; POGG 1,1412; 3,792
Anon., Chem. Z. 18 (1894), 89,2060

LEHMANN, KARL BERNHARD [1858-1940]
NDB 14,71-72; KURSCHNER 5,790; FISCHER 2,883; BOCK, 2-27; IB, 171; RAZINGER, 112-115; ZURICH-D, 64; WININGER 4,16-17
K. Kisskalt, Münch. med. Wchschr. 75 (1928), 1638-1641; Arch. Hyg. 123 (1940), 345-347
S. Pritze, Das Wirken von Prof. K.B. Lehmann in Würzburg. Würzburg 1984

LEHMANN, LUDWIG [1858-1939]
NDB 14,77; TLK 3,386
Anon., Z. angew. Chem. 53 (1940), 48

LEHMANN, OTTO [1855-1922]
DSB 8,148-149; DBJ 4,361; FREUND 3,261-271; BLOKH 1,432; TLK 2,416; DZ, 855; WWWS, 1019; POGG 3,792; 4,859-860; 5,724-726; 6,1490
A. Schleiermacher and R. Schachemacher, Phys. Z. 24 (1923), 289-291

LEHMANN, WILLY [1878-]
JV 17,192; NUC 324,417

LEHMSTEDT, KURT [1891-1969]
POGG 6,1490; 7a(3),55-56
Anon., Nachr. Chem. Techn. 17 (1969), 401

LEHN, JEAN MARIE [1939-]
QQ 1991-1992, pp.1035-1036; POTSCH, 265-266

LEHNARTZ, EMIL [1898-1979]
NDB 14,104-105; KURSCHNER 12,1852; KALLMORGEN, 337; WWWS, 1019; POGG 7a(3),56-57

LEHNE, ADOLF [1856-1930]
NDB 14,107-108; NB, 231; TLK 3,386-387; RSC 10,556; 16,689; POGG 6,1490
P. Krais, Z. angew. Chem. 39 (1926), 568-569; 43 (1930), 185

LEHNER, JOSEPH [1882-1938]
HEINDEL, 76-79; SKVARC, 137-139; IB, 172
J. Schaffer, Anat. Anz. 88 (1939), 358-365

LEHNERT, HERMANN [-1901]
RSC 16,75
Anon., Pharm. Z. 46 (1901), 93

LEHNINGER, ALBERT LESTER [1917-1986]
AMS 15(4),662; WWA 1980-1981,p.1989; IWW 1982-1983, p.760; MH 2,316-318; STC 2,166-167; MSE 2,218-219; WWWS, 1019; POTSCH, 266
A.M. Harvey, Johns Hopkins Med. J. 139 (1976), 257-263
E.C. Slater, Nature 321 (1986), 562
J.S. Fruton, Am. Phil. Soc. Year Book 1986, pp.141-144

LEHR, GUSTAV [-1892]
RSC 16,689
Anon., Pharm. Z. 37 (1892), 200

LEICESTER, HENRY MARSHALL [1906-1991]
AMS 11,3077; WWWS, 1020
G.B. Kauffman, Bull. Hist. Chem. 10 (1991), 15-21

LEICHER, HANS [1898-1989]
KURSCHNER 13,2270; KALLMORGEN, 337

LEIDIÉ, EMILÉ JULES [1855-1903]
GORIS, 481-482; TSCHIRCH, 1079; POGG 4,860-861;

5,726
Anon., J. Pharm. Chim. [6]18 (1903), 592

LEIDY, JOSEPH [1823-1891]
DSB 8,169-170; DAB 11,150-152; ELLIOTT, 155; SACKMANN, 201-203; WWAH, 354
W.S.W. Ruschenberger, Proc. Am. Phil. Soc. 30 (1892), 135-184
H.F. Osborn, Biog. Mem. Nat. Acad. Sci. 7 (1913), 339-396

LEIDY, JOSEPH, II [1866-1932]
AMS 4,576; WWWS, 1020
Anon., J. Parasit. 19 (1932), 148; J. Am. Med. Assn. 99 (1932), 239

LEIMBACH, ROBERT [1876-1914]
JV 15,132; NUC 325,45
Anon., Ber. chem. Ges. 47 (1914), 2666

LEINER, CARL [1871-1930]
OBL 5,108; FISCHER 2,884-885
F. Basch, Wiener klin. Wchschr. 43 (1930), 632

LEINER, LUDWIG [1830-1901]
HEIN, 367
Anon., Pharm. Z. 46 (1901), 335

LEININGEN-WESTERBURG, WILHELM (Graf zu) [1875-1956]
KURSCHNER 8,1373; POGG 7a(3),61-62

LEIPERT, THEODOR [1902-]
KURSCHNER 13,2273; POGG 7a(3),62

LEISHMAN, WILLIAM BOGG [1865-1926]
DNB [1922-1930], 502-503; FISCHER 2,885-886; OLPP, 235-236; W, 320-321; ABBOTT, 83-84; WWWS, 1022
Anon., Lancet 1926(I), pp.1171-1173
S.L.C., J. Path. Bact. 29 (1926), 515-528
C.J. M[artin], Proc. Roy. Soc. B102 (1927-1928), ix-xvii

LEITGELB, HUBERT [1835-1888]
ADB 51,627-629; OBL 5,114-115; FREUND 1,245-250
G. Haberlandt, Ber. bot. Ges. 6 (1888), (xxxix)-(xliv)
E. Heinricher, Mitt. Nat. Ver. Steier. 25 (1888), 159-181

LEITHE, WOLFGANG [1903-]
KURSCHNER 13,2276; POGG 6,1492; 7a(3),64

LEJWA, ARTHUR [1895-1972]
NCAB 57,721-722

LELLMANN, EUGEN [1856-1893]
NDB 14,179-180; BLOKH 1,432; SCHAEDLER, 69; POGG 3,793; 4,862
L. Meyer, Ber. chem. Ges. 26(4) (1893), 1033-1040
Anon., Pharm. Z. 38 (1893), 776-777

LELOIR, LUIS FEDERICO [1906-1987]
IWW 1982-1983, p.762; MH 2,318-320; STC 2,167-168; MSE 2,219-220
E. Cabib, Science 170 (1970), 608-609
L.F. Leloir, Ann. Rev. Biochem. 52 (1983), 1-15
L.F. Leloir and A.C. Paladini, in Selected Topics in the History of Biochemistry (G. Semenza, Ed.), pp.25-41. Amsterdam 1983
A.C. Paladini, FASEB Journal 2 (1988), 2751-2752
W.J. Whelan, Am. Phil. Soc. Year Book 1988, pp.191-194
S. Ochoa, Biog. Mem. Fell. Roy. Soc. 35 (1990), 203-208

LEMAIRE, FRANÇOIS JULES [1814-1886]
GORIS, 482; WWWS, 1023
H.A. Kelly, J. Am. Med. Assn. 36 (1901), 1083-1086

LEMBERG, RUDOLF [1896-1975]
KURSCHNER 10,1400; WW 1966, pp.1799-1800; BHDE 2,706-707; LDGS, 60; WWWS, 1023; POGG 6,1494-1495; 7a(3),66-67
R. Lemberg, Ann. Rev. Biochem. 34 (1965), 1-20
Anon., Nature 255 (1975), 347
C. Rimington and C.H. Gray, Biog. Mem. Fell. Roy. Soc. 22 (1976), 257-294
J. Barratt and R.N. Robinson, Rec. Austral. Acad. Sci. 4(1) (1979), 133-156

LEMBERT, MAX [1891-1925]
BLOKH 1,432-433
K. Freudenberg, Ber. chem. Ges. 58A (1925), 35-36

LEMIEUX, RAYMOND URGEL [1920-]
AMS 15(4),673; IWW 1982-1983, p.762; STC 2,169-170
Anon., Chem. Eng. News 44(43) (1966), 100
R.U. Lemieux, Explorations with Sugars. Washington, D.C. 1990

LEMME, GEORG [1865-1925]
JV 4,23
Anon., Chem. Z. 49 (1925), 1033

LEMMERMANN, OTTO [1869-1953]
TLK 3,388-389; IB, 172; POGG 7a(3),67-69
H. Karst, Forsch. Fort. 25 (1949), 220-221
Anon., Z. Pflanzenern. 62 (1953), 1-4

LEMOIGNE, MAURICE [1883-1967]
SACKMANN, 204-205; IB, 172; WWWS, 1023-1024; POGG 6,1495; 7b,2815-2818
R. Dujarric de la Rivière, C.R. Acad. Sci. 264 (1967), 136-140
M. Bustarret, C.R. Acad. Agr. 59 (1973), 791-793

LEMOINE, GEORGES [1841-1922]
WWWS, 1024; POGG 3,793-794; 4,864-865; 5,728; 6,1495
A. Haller, C.R. Acad. Sci. 175 (1922), 921-922
E. Blaise, Bull. Soc. Chim. [4]33 (1923), 30

LENDLE, LUDWIG [1899-1969]
NDB 14,201-202; NB, 233-234; KURSCHNER 10,1402; POGG 7a(3),69-71
Anon., Chem. Z. 93 (1969), 744
M. Vogt, Arzneimitt. 19 (1969), 243-246
F. Gross, Arzneimitt. 19 (1969), 2029-2030
H.J. Haike, Deutsche med. Wchschr. 94 (1969), 2455
R. Emmrich, Jahrb. Sachs. Akad. 1969-1970, pp.264-275

LENDLMAYER von LENDENFELD, ROBERT [1858-1913]
OBL 5,129; STURM 2,423-424; BJN 18,106*; LUDY, 195-210; RSC 10,563; 16,705-707
Anon., Leopoldina 49 (1913), 96; Nature 91 (1913), 535-536

LENGERKE, ERNST AUGUST KARL von [1823-1870]
Genealogisches Handbuch des Adels 73 (1980), 228

LENGFELD, FELIX [1863-1938]
AMS 5,666; WWAH, 355; WWWS, 1024; POGG 4,866; 6,1496; 7b,2819
Anon., Ind. Eng. Chem. News Ed. 16 (1938), 274

LENHARD, FRIEDRICH WOLFGANG [1877-]
JV 20,261; NUC 326,205

LENHARDT, SIGISMUND [1889-]
JV 28,284; NUC 326,206

LENK, ROBERT [1885-1966]
STURM 2,425; FISCHER 2,889; HASENEDER, 120-132
K. Weiss, Radiologica Austriaca 15 (1964), 1-3; 17 (1967), 89-92

LENNARD-JONES, JOHN EDWARD [1894-1954]
DSB 8,185-187; DNB [1951-1960], 621-622; WWWS, 1025; POTSCH, 267-268; POGG 6,1497; 7b,2820-2821
C.A. Coulson, Nature 174 (1954), 994-995
E.K. Rideal, J. Chem. Soc. 1955, pp.1047-1048
N.F. Mott, Biog. Mem. Fell. Roy. Soc. 1 (1955), 175-184

LENOBLE, ROBERT [1900-1959]
P. Costabel, Rev. Hist. Sci. 12 (1959), 167-169; Physis 2 (1960), 92-94
R.H. Popkin, Isis 51 (1960), 200-202

LENOIR, GEORG ANDREAS [1824-1909]
RSC 3,951
Anon., Chem. Z. 33 (1909), 1188

LENZ, WILHELM [1852-1916]
HEIN, 367-368; BLOKH 1,433-434; TSCHIRCH, 1079; POGG 4,867-868; 5,730
H. Thoms, Ber. chem. Ges. 49 (1916), 2841-2843

LEO, HANS [1854-1927]
FISCHER 2,891; PAGEL, 987-988; BONN, 323,350; IB, 173
B. Laquer, Münch. med. Wchschr. 71 (1924), 338
H. Fühner, Deutsche med. Wchschr. 53 (1927), 1956

LEONARD, ALFRED GODFREY GORDON [1885-1966]
SRS,40; NUC 327,83
Who's Who in British Science 1953, p.165
H.D. Thornton, Chem. Brit. 3 (1967), 224

LEONARD, NELSON JORDAN [1916-]
AMS 15(4),682; WWA 1980-1981, p.1999; IWW 1982-1983, pp.764-765; MH 2,320-321; MSE 2,220-221; WWWS, 1025-1026

LEONHARDI, GOTTFRIED [1915-]
KURSCHNER 13,2290; POGG 7a(3),75-76

LEONTIEV, IVAN FEDOROVICH [1892-1950]
G.I. Roskin and G.G. Leongardt, Priroda 40(4) (1951), 83-85

LEONTOVICH, ALEKSANDR VASILIEVICH [1869-1943]
BSE 24,582; WWR, 346; VORONTSOV, 118-122; UKRAINE 2,325-326; IB, 173
N.V. Bodrova et al., Fiziol. Zhur. 5 (1959), 689-696; 15 (1969), 579-590

LEOPOLD, GERHARD [1846-1911]
BJN 16,46*-47*; FISCHER 2,892; PAGEL, 988
A. Richter, Mon. Geburt. 34 (1911), vii-xiv
Anon., Leopoldina 47 (1911), 102

LEPEHNE, GEORG [1887-1967]
KURSCHNER 4,1717; FISCHER 2,892-893; TETZLAFF, 201

LEPESHINSKAYA, OLGA BORISOVNA [1871-1963]
BSE (3rd Ed.)14,344-345; WWR, 346; WWWS, 1027
R.C. Cook, J. Heredity 42 (1951), 121-123

LEPESHKIN, VLADIMIR VASILIEVICH [1876-1956]
IB, 173; POGG 6,1501; 7b,2827-2828

LÉPINE, PIERRE RAPHAEL [1901-1989]
QQ 1981-1982, p.925; IWW 1982-1983, p.766; WWWS, 1027
J.A. Thomas, Bull. Acad. Med. 174 (1990), 605-612

LÉPINE, RAPHAEL [1840-1919]
GE 22,57; FISCHER 2,893; PAGEL, 988-990; GUIART, 120-121; WWWS, 1027
A. Laveran, C.R. Acad. Sci. 169 (1919), 1065-1067
F.J. Collet, Rev. Med. 37 (1920), 1-6
C. Achard, Prog. Med. 40 (1925), 1927-1936

LEPKOVSKY, SAMUEL [1899-1984]
AMS 12,3657; KAGAN(JA), 351,742
S. Lepkovsky, Fed. Proc. 38 (1979), 2699-2700
T.H. Jukes, J. Nutrition 116 (1986), 329-340

LEPORSKI, NIKOLAI IVANOVICH [1877-1952]
BSE 24,590; BME 15,878; WWR, 347
A.I. Nestorov, Voprosi Nervatizma 12(6) (1972), 90

LEPOW, IRWIN HOWARD [1923-1984]
AMS 14,2938

LEPPERT, WLADYSLAW [1848-1920]
PSB 17,82-83; WE 6,460; KONOPKA 5,301-303

LEPPMANN, ARTHUR [1854-1921]
FISCHER 2,893-894; KOREN, 209
Anon., Aerztliche Sachverständige Zeitung 27 (1921), 125

LEPSIUS, BERNHARD [1854-1934]
NDB 14,309; KALLMORGEN, 338; TLK 3,391; POGG 4,869-870; 5,731; 6,1502-1503; 7a(3),77
H. Specketer, Z. angew. Chem. 37 (1924), 57
K.A. Hofmann, Ber. chem. Ges. 67A (1934), 167-169
Anon., Z. angew. Chem. 47 (1934), 740

LEPSIUS, RICHARD [1885-1969]
NDB 14,309; KURSCHNER 10,1410; POGG 6,1503; 7a(3),77
E. Baum, Chem. Z. 93 (1969), 744

LERCH, JOSEPH UDO [1816-1892]
OBL 5,151; STURM 2,425; HEIN-E, 277; WERSTLER 1,73-75; BLOKH 1,434; FERCHL, 310; POGG 3,798; 4,879
Anon., Prag. med. Wchschr. 17 (1892), 108

LERCHE, EBERHARD [1909-1980]
KURSCHNER 13,2293

LERMAN, LEONARD SALOMON [1925-]
AMS 15(4),686; WWWS, 1028

LERMAN, SIDNEY [1927-]
AMS 15(4),687; WWWS, 1028

LERNER, AARON BUNSEN [1920-]
AMS 15(4),687; WWA 1980-1981, p.2001; WWWS, 1028

LERNER, I. MICHAEL [1910-1977]
AMS 13,2580; WWA 1976-1977, p.1875; WWWS, 1028
E.R. Dempster et al., Genetics 88 (1978), s139-s140
B. Glass, Am. Phil. Soc. Year Book 1984, pp.130-135

LE ROYER, AUGUSTIN [1793-1863]
BRIQUET, 295-296; RSC 3,965; CALLISEN 11,273-274
F. Marcet, Mem. Soc. Phys. Hist. Nat. Gen. 17 (1863),

263-264
J.B.G. Galiffe, Notices Généalogiques sur les Familles Génèvoises 6 (1892), 380
F. Ducommun, Rev. Hist. Pharm. 48 (1960), 431-432

LESCHKE, ERICH [1887-]
KURSCHNER 4,1719; FISCHER 2,895-896; IB, 173

LESCOEUR, JEAN JOSEPH HENRI [1848-1935]
TSCHIRCH, 1080; POGG 3,799; 4,872; 6,1505

LEŚNIK, MAXIMILIAN [1858-1893]
KONOPKA 5,308; RSC 16,734; SG [2]9,466; NUC 328,284

LESPIEAU, ROBERT [1864-1947]
CHARLE(2), 182-184; WWWS, 1030; POGG 5,733; 6,1505; 7b,2831-2832
M. Bourguel, Bull. Soc. Chim. [4]53 (1933), 1145-1154
L. Blaringem, C.R. Acad. Sci. 224 (1947), 1193-1196
G. Dupont, Bull. Soc. Chim. [5]16 (1949), 1-9

LESSER, EDMUND [1852-1918]
NDB 14,336; PAGEL, 994; WININGER 4,36
A. Blaschko, Deutsche Wchschr. 44 (1918), 751-752
A. Buschke, Derm. Wchschr. 67 (1918), 520-523
J.J. Herzberg, Hautarzt 39 (1988), 598-601
G. Stuettgen, International Journal of Dermatology 27 (1988), 269-273

LESSER, ERNST JOSEPH [1879-1928]
NDB 14,336-337; DBJ 10,330; FISCHER 2,897; IB, 173; KOREN, 209-210; WININGER 7,234; TETZLAFF, 202; POGG 6,1505-1506
J.K. Parnas, Biochem. Z. 196 (1928), 1-2
R. Ammon, Z. med. Chem. 6 (1928), 44-48; Mannheimer Hefte 1 (1968), 29-37
K.O. Watzinger, Geschichte der Juden in Mannheim 1650-1842, pp.121-122. Stuttgart 1984
R. Ammon and H.D. Soling, in Naturwissenschaft und Medizin (R. Kattermann, Ed.), pp.59-84. Mannheim 1985

LESSER, FRITZ [1873-]
NDB 14,336*; FISCHER 2,897-898

LESSING, RUDOLF [1878-1964]
WW 1959, p.1794; JV 17,286; RSC 19,643
D.T.A. Townsend, Chem. Brit. 1 (1965), 381
A.H. Raine, Proc. Soc. Anal. Chem. 2 (1965), 26-27

LETHEBY, HENRY [1816-1876]
DNB 11,1010; FERCHL, 311
Anon., Chem. News 33 (1876), 146
J.M. Cameron and A.J. Hardy, Practitioner 208 (1972), 401-405

LETSCHE, EUGEN ALBERT [1878-1958]
JV 18,388

LETTERER, ERICH [1895-1982]
KURSCHNER 13,2298; DANIS, 49-65; IB, 173
Anon., Deutsche med. Wchschr. 107 (1982), 1120

LETTRÉ, HANS [1908-1971]
NDB 14,350; KURSCHNER 11,1742; POGG 7a(3),81-83
Anon., Nachr. Chem. Techn. 16 (1968), 412
D. Schmähl, Arzneimitt. 21 (1971), 1427

LETTS, EDMUND ALBERT [1852-1918]
BELFAST, 583; WWWS, 1031; POGG 3,802; 4,873; 5,734
A.W.S., J. Chem. Soc. 113 (1918), 314-316
T. Biurns, Anal. Proc. 14 (1977), 174-175

LEUBE, WILHELM von [1842-1922]
DBJ 4,361-362; LF 5,150-155; FISCHER 2,899; PAGEL, 995-996; DZ, 866-867; GIESE, 550-551; SCHWARTZ, 84-94; PITTROFF, 124-125; WWWS, 1031
H. Lüdke, Münch. med. Wchschr. 59 (1912), 2011-2013
F. Penzoldt, Münch. med. Wchschr. 69 (1922), 936-937
H. Strauss, Klin. Wchschr. 1 (1922), 1863-1864

LEUCHS, ERHARD FRIEDRICH [1800-1837]
FERCHL, 311; CALLISEN 30,28; POGG 1,1438

LEUCHS, HERMANN [1879-1945]
NDB 14,366*; TLK 3,392; POTSCH, 268; POGG 5,734; 6,1507-1508; 7a(3),83-84
Anon., Z. angew. Chem. 59 (1947), 36
F. Kröhnke, Chem. Ber. 85 (1952), lx-lxxxix

LEUCHS, KURT [1881-1949]
POGG 7a(3),84

LEUCHS, ROBERT FRIEDRICH [1881-]
JV 24,576; NUC 329,337

LEUCKART, RUDOLF [1822-1898]
DSB 8,269-271; NDB 14,372-373; ADB 51,672-675; BJN 5,39*; BLOKH 1,436; POTSCH, 268; HIRSCH 3,757; Suppl.,396; PAGEL, 996-998; OLPP, 236-240; WWWS, 1031
R. Leuckart, Arch. Parasit. 1 (1898), 185-190
J.V. Carus, Sitz. Akad. Wiss. Leipzig 1898, pp.49-62
R. von Hanstein, Naturw. Rund. 13 (1898), 242-246
A. Jacobi, Z. Bakt. 23 (1898), 1073-1081
O. Taschenburg, Leopoldina 35 (1899), 62-66,82-94,102-112
E.R. L[ankester], Proc. Roy. Soc. 75 (1905), 19-22
G. Grimpe, Sächsische Lebensbilder 1 (1930), 212-222
K. Wunderlich, Rudolph Leuckart. Jena 1978

LEUCKART, RUDOLF [1854-1889]
NDB 14,372; POGG 3,802; 4,873-874
K. Buchka, Ber. chem. Ges. 22(3) (1889), 855-860
Anon., Leopoldina 25 (1889), 169-170

LEUPOLD, ERNST [1884-1961]
KURSCHNER 9,1187; DANIS, 43-48
H. Heinlein, Deutsche med. Wchschr. 74 (1949), 1026

LEURET, FRANÇOIS [1797-1851]
DSB 8,272-273; GE 22,139; LAROUSSE 10,436; HIRSCH 3,759-760; FERCHL, 311; DECHAMBRE 54,403-404; WWWS, 1031; CALLISEN 11,302-305; 30,34-35
Anon., Arch. Gen. Med. 25 (1851), 492-494
M. Gourevitch, Inf. Psychiat. 44 (1968), 843-854
J.F. Noël, Mem. Acad. Stanislas [7]6 (1977-1978), 339-354

LEUTHARDT, FRANZ [1903-1985]
KURSCHNER 13,2300; PICCO, 31-38; POGG 7a(3),85-86
B. Glasson et al., Schw. med. Wchschr. 93 (1963), 1241-1245
G. Semenza, Chimia 27 (1973), 511-512; 37 (1983), 359-360; 39 (1985),368
J. Kägi, Viert. Nat. Ges. Zürich 131 (1986), 221-222

LEVADITI, CONSTANTIN [1874-1953]
DSB 8,273-274; FISCHER 2,900; IB, 174; WWWS, 1031-1032
P.L., Ann. Inst. Pasteur 85 (1953), 535-539

R. Leriche, C.R. Soc. Biol. 147 (1953), 1544-1545
R. Iftimovici, Hist. Sci. Med. 9 (1975-1976), 74-78; Virologie 30 (1979), 237-242

LEVASHOV, SERGEI VASILIEVICH [1857-]
KAZAN 2,241-246; ZMEEV 4,193-194; RSC 10,581; 12,444; 16,746; SG [2]9,493

LEVENE, PHOEBUS AARON [1869-1940]
DSB 8,275-276; DAB [Suppl.2],378-379; MILES, 295-296; FARBER, 1315-1324; AMS 6,845; NCAB E,83-84; CHITTENDEN, 116-124,134-137,220-224; W,235; POTSCH, 268-269; FISCHER 2,901; HUBER, 67-68; KAGAN(JA),332-33; WININGER 7,235-236; KOREN, 210; WWAH, 357; WWWS, 1032; POGG 6,1509-1512; 7b,2835-2838
L.W. Bass, Ind. Eng. Chem. News Ed. 12 (1934), 105; Science 92 (1940), 392-395
D.D. Van Slyke and W.A. Jacobs, Biog. Mem. Nat. Acad. Sci. 23 (1944), 75-126
S. Goldschmidt, Bayer. Akad. Wiss. Jahrbuch 1949, pp.139-140
R.S. Tipson, Adv. Carbohydrate Chem. 12 (1957), 1-12
J.S. Fruton, TIBS 4 (1979), 49-50

LEVERKUS, CARL FRIEDRICH WILHELM [1804-1889]
NDB 14,389-391; HEIN, 369-370; DGB 35 (1922), 202; FERCHL, 312
Anon., Leopoldina 25 (1889), 54

LEVERKUS, KARL OTTO [1883-1957]
JV 25,365; DGB 128 (1962), 92
Wer ist Wer 9 (1928), 940; 12 (1955), 709

LEVEY, STANLEY [1915-1967]
AMS 11,3107; NCAB 54,358; WWWS, 1032

LEVI, GUISEPPE [1872-1965]
DSB 8,282-283; FISCHER 2,901; IB, 174; KOREN, 210; WININGER 7,237
R. Amprino, Arch. Ital. Biol. 104 (1966), 134-138; Acta Anat. 66 (1967), 1-44
O.M. Olivo, Rend. Accad. Lincei [8]40 (1966), 954-972

LEVI, LEOPOLD [1868-1933]
FISCHER 2,901; STALDER, 52-53; KOREN, 210; WININGER 7,237-238
F. Jayle, Presse Med. 42 (1934), 244
R. Kohn, Rev. Hist. Med. Heb. 1952, p.142

LEVI-MONTALCINI, RITA [1909-]
AMS 14,2950; STC 2,176-178
R. Levi-Montalcini, Bioscience Reports 7 (1987), 681-699; In Praise of Imperfection. New York 1988

LEVIN, ISAAC [1866-1945]
AMS 7,1056; KOREN, 210; KAGAN(JA), 212,715-716; WWWS, 1033
I.I. Kaplan, Am. J. Roengtenol. 55 (1946), 773-775

LEVINE, MAX [1889-1967]
AMS 11,3112; FISCHER 2,901-902; WININGER 7,241; WWAH, 357-358

LEVINE, PHILIP [1900-1987]
AMS 14,2955; MH 1,293-294; STC 2,178-179; MSE 2,224-225; BIBEL, 224-228; IWW 1982-1983, p.771; IB,174; COHEN, 145; KAGAN(JA), 723; WWWS, 1034
Anon., BioScience 16 (1966), 726; Advances in Pathobiology 7 (1980), vii-xii New York Times, 20 October 1987

LEVINE, VICTOR EMANUEL [1891-1963]
AMS 10,2400; NCAB 54,93-94; IB, 174; WWAH, 357-358; WWWS, 1034; POGG 6,1516; 7b,2845-2846

LEVINSTEIN, IVAN [1845-1916]
W, 326; WININGER 4,62; POTSCH, 269
Anon., J. Soc. Chem. Ind. 35 (1916), 458

LEVINTHAL, CYRUS [1922-1990]
AMS 15(4),711; IWW 1982-1983, p.771
B. Honig, Proteins 11 (1991), 239-241

LEVITES, SIMON YAKOVLEVICH [1868-1911]
SG [3]7,624
P.P. von Weimarn, Zhur. Fiz. Khim. 43 (1911), 942-945

LEVITSKI, GRIGORI ANDREEVICH [1878-1942]
DSB 18,549-553

LEVY, ANTHONY LEWIS [1924-1954]
C.H. Li, Nature 174 (1954), 860-861
A.H. Cook, J. Chem. Soc. 1954, pp.4711-4712

LEVY, ERNST [1864-1919]
PAGEL, 998-999; WININGER 4,68-60

LEVY, FRITZ [1887-1957]
AMS 8,1484; LDGS, 57

LEVY, JEANNE [1895-]
POGG 6,1517; 7b,2855-2862

LEVY, LEO [1881-]
JV 19,182; NUC 330,201

LEVY, LUDWIG [1864-]
JV 4,243; RSC 16,755; NUC 330,205

LEVY, MAURICE [-1934?]
JV 10,235; RSC 16,755; SG [2]9,502; NUC 330,213

LEVY, MILTON [1903-1976]
AMS 13,2600; WWWS, 1035
Anon., Science 196 (1977), 745

LEVY, PAUL ERNST [1875-1956]
LDGS, 22; POGG 6,1517-1518; 7a(3),87-88

LEVY, ROBERT [1886-1956]
CHARLE, 184-186

LEVY, ROBERT LOUIS [1888-1974]
AMS 11,3120; NCAB I,279; 58,310-311; WWWS, 1035
I.S. Wright, Trans. Assoc. Am. Phys. 88 (1975), 27-28

LEVY, SIEGMUND [1857-1892]
Anon., Ber. chem. Ges. 25 (1892), 3674

LEWANDOWSKI, ALFRED [1864-1931]
KURSCHNER 4,1727; KOREN, 212; WININGER 7,247

LEWANDOWSKI, RUDOLF [1847-1902]
OBL 5,169; BJN 7,70*; FISCHER 2,904; KIRCHENBERGER, 109-110
L. Eisenberg, Das Geistige Wien 2,296-299. Vienna 1893

LEWANDOWSKY, MAX [1876-1918]
DBJ 2,697; FISCHER 2,904-905; DIEPGEN, 110-112; KOREN, 212
O. Kalischer, Z. ges. Neurol. 51 (1919), 1-44

LEWIN, CARL [1876-1930]
KURSCHNER 3,1389-1390; FISCHER 2,905; KOREN, 212; WININGER 7,247

LEWIN, GEORG RICHARD [1820-1896]
ADB 51,680-681; HIRSCH 3,769-770; WININGER 4,83-84
O. Lassar, Dermatologische Zeitschrift 3 (1896), 678-686
E. Lesser, Arch. Dermatol. 37 (1896), 318-320
J.H. Rille, Wiener klin. Wchschr. 9 (1896), 1068

LEWIN, LOUIS [1850-1929]
NDB 14,415-416; FISCHER 2,905-906; IB, 174; PAGEL, 1001-1002; TLK 3,394; TSCHIRCH, 1080; WININGER 4,85-86,175; 7,247; TETZLAFF, 205
S. Loewe, Deutsche med. Wchschr. 56 (1930), 151-152
W. Heubner, Münch. med. Wchschr. 77 (1930), 405-406
D.I. Macht, Ann. Med. Hist. NS3 (1931), 179-184
J. Kleeberg, Koroth 7 (1976), xviii-xxii
E.H. Ackerknecht, Gesnerus 36 (1979), 300-303
R.K. Müller et al., Der Toxikologe Louis Lewin. Leipzig 1985

LEWINSKI, JOHANN [1878-]
JV 19,51; SG [3]7,628; NUC 330,335

LEWIS, EDWARD B. [1918-]
AMS 15(4),722; WWWS, 1036-1037

LEWIS, FREDRICK THOMAS [1875-1951]
AMS 7,1060; IB, 174; WWAH, 359
E.A. Boyden, Anat. Rec. 112 (1952), 139-152

LEWIS, GILBERT NEWTON [1875-1946]
DSB 8,289-294; DAB [Suppl.4],487-489; AMS 7,1060; MILES, 296-297; NCAB 36,166-167; STC 2,180-181; ABBOTT-C, 83-84; WWAH, 359; WWWS, 1037; POTSCH, 269-270; POGG 4,879; 5,738-739; 6,1519-1520; 7b,2869-2870
W.F. Giauque, Am. Phil. Soc. Year Book 1946, pp.317-322
H. H. Hildebrand, Obit. Not. Fell. Roy Soc. 5 (1947), 491-506; Biog. Mem. Nat. Acad. Sci. 31 (1948), 210-235
R.E. Kohler, Hist. Stud. Phys. Sci. 3 (1971), 343-386; 4 (1974), 3-38; 6 (1975), 431-468; Brit. J. Hist. Sci. 8 (1975), 233-239
M. Calvin, Welch Foundation Conferences on Chemical Research 20 (1977), 116-149
D.A. Davenport et al., J. Chem. Ed. 61 (1984), 2-21,93-116

LEWIS, HERMAN WILLIAM [1923-]
AMS 15(4),723; WWA 1980-1981, p.2020; WWWS, 1037

LEWIS, HOWARD BISHOP [1887-1954]
AMS 8,1487-1488; NCAB 41,492; CHITTENDEN, 192-197; IB, 174; WWAH, 359
A.A. Christman, Science 120 (1954), 363-364; J. Nutrition 67 (1959), 7-18
W.C. Rose and M.J. Coon, Biog. Mem. Nat. Acad. Sci. 44 (1974), 139-173

LEWIS, ISAAC McKINNEY [1878-1943]
AMS 6,850

LEWIS, MARGARET REED [1881-1970]
AMS 10,2411; NCAB 58,142-143; IB, 174; NUC 330,565

LEWIS, PAUL ADIN [1879-1929]
AMS 4,583; FISCHER 2,906; IB, 174; WWWS, 1037
S. Flexner, Science 70 (1929), 133-134
Anon., Am. Rev. Tub. 21 (1930), 587-592

LEWIS, THOMAS [1881-1945]
DSB 8,294-296; DNB [1941-1950], 501-502; O'CONNOR(2), 170-172
A.N. Drury and R.T. Grant, Obit. Not. Fell. Roy. Soc. 5 (1945), 179-202
A. Hollman, J. Physiol. 263 (1976), 72P-74P

LEWIS, TIMOTHY RICHARDS [1841-1886]
DSB 8,296-297; HIRSCH 3,770; OLPP, 240-241; COLE, 104-106; W, 327; WWWS, 1038
Anon., Nature 34 (1886), 76-77; Brit. Med. J. 1886(I), pp.1242-1423; Lancet 1886(I), p.993
C. Dobell, Parasitology 14 (1922), 413-416

LEWIS, WARREN HARMON [1870-1964]
AMS 10,2413; FISCHER 2,906; IB, 174; WWAH, 359-360; WWWS, 1038
G.W. Corner, Am. Phil. Soc. Year Book 1965, pp.179-184; Biog. Mem. Nat. Acad. Sci. 39 (1967), 323-358
A.M. Harvey, Johns Hopkins Med. J. 136 (1975), 142-149

LEWIS, WILLIAM CUDMORE McCULLAGH [1885-1956]
WW 1949,1653; WWWS, 1038; POGG 5,739-740; 6,1521; 7b,2873-2874
E.A. Moelwyn-Hughes, Nature 177 (1956), 603-604
C.E.H. Bawn, Biog. Mem. Fell. Roy. Soc. 4 (1958), 193-203

LEWIS, WINFORD LEE [1878-1943]
AMS 6,851; NCAB A,369-370; MILES, 297-298; WWAH, 360; WWWS, 1038; POGG 6,1521; 7b,2874
O. Eisenschiml, Chem. Eng. News 21 (1943), 174

LEWKOWITSCH, JULIUS ISIDOR [1857-1913]
WININGER 7,250; POGG 4,880; 5,740
D. Holde, Chem. Z. 37 (1913), 1437-1438
C.A. Keane, Analyst 38 (1913), 493-495

LEWY, BERGNART CARL [1817-1863]
DBL 9,21; VEIBEL 2,167-169; POGG 1,1444; 3,805-806
Anon., Archiv for Pharmaci 20 (1863), 381-391

LEWY (LEWEY), FRITZ (FREDERIC HENRY) [1885-1950]
KURSCHNER 4,1728-1729; FISCHER 2,907; ASEN, 117; BHDE 2,720; LDGS, 71; IB, 175

LEY, HEINRICH [1872-1938]
KURSCHNER 4,1729; TLK 3,394; POGG 4,880-881; 5,740; 6,1522; 7a(3),89
M. Trautz, Ber. chem. Ges. 72A (1939), 44-45

LEYDEN, ERNST von [1832-1910]
NDB 14,428-429; NB, 236; BJN 15,66*; FISCHER 2,908-909; PAGEL, 1002-1003; FRANKEN, 189; HABERLING, 99-100; WREDE, 114-115; KROLLMANN, 394-395; DZ, 872; WWWS, 1039
E. von Leyden, Lebenserinnerungen. Stuttgart 1910
F. Blumenthal, Med. Klin. 6 (1910), 1764-1766
P. Lazarus, Münch. med. Wchschr. 57 (1910), 2588-2593
M. Lewandowsky, Z. ges. Neurol. 4 (1910), 1-11
F. Kraus, Deutsche med. Wchschr. 36 (1910), 2055-2061
R. von Jaksch, Wiener klin. Wchschr. 23 (1910), 1488-1490
H. Berndt, Charité Annalen NF5 (1985), 372-376

LEYDIG, FRANZ [1821-1908]
DSB 8,301-303; NDB 14,429-430; BJN 13,56*; HIRSCH 3,772; PAGEL,1003-1004; FREUND 2,215-220; WENIG, 177; ROEDER, 39-41; DZ, 873; WWWS, 1737
M. Nussbaum, Anat. Anz. 32 (1908), 503-506

R. von Hanstein, Naturw. Rund. 23 (1908), 347-352
O. Schultze, Münch. med. Wchschr. 55 (1908), 972-973
O. Taschenburg, Leopoldina 45 (1909), 37-44,47-52,57-64,70-76,82-88
G. Sticker, Fort. Med. 39 (1921), 802-805

LEYTON, ALBERT SIDNEY [see GRÜNBAUM]

L'HÉRITIER, PHILIPPE [1906-]
QQ 1971-1972, p.1027; BUICAN, 334-335

LHOTÁK, KAMIL [1876-1926]
FISCHER 2,909; NAVRATIL, 174-175; IB, 175
E. Babak, Biol. Listy 12 (1926), 2-15

LI, CHOH HAO [1913-1987]
AMS 15(4),713; WWA 1980-1981, p.2023; STC 2,182-183; MSE 2,228-229; CB 1963, pp.242-244; ABBOTT, 84; WWWS, 1039
C.H. Li, in Topics in the History of Biochemistry (G. Semenza, Ed.), pp.335-352. Amsterdam 1983
V.J. Hruby, Int. J. Pep. Res. 31 (1988), 253-254

LIBBY, WILLARD FRANK [1908-1980]
AMS 14,2978; WWA 1976-1977, p.1896; NCAB I,268-269; CB 1954, pp.406-407; MH 1,294-295; STC 2,183-184; MSE 2,229-230; AO 1980, pp.524-527; ABBOTT-C, 84-85; WWWS, 1040; POTSCH, 271
W.F. Libby, Ann. Rev. Phys. Chem. 15 (1964), 1-12
J.R. Arnold, Am. Phil. Soc. Year Book 1980, pp.608-612
H.E. Suess, Naturwiss. 68 (1981), 435-436

LICHTENSTEIN, LEON [1878-1933]
RSB 17,298-300; WE 6,498; POGG 5,741-742; 6,1523-1524; 7a(3),91

LICHTHEIM, LUDWIG [1845-1928]
FISCHER 2,911; PAGEL, 1004-1005; KROLLMANN, 1003; GIESE, 567; KOREN, 213; WININGER 4,101-102; WWWS, 1041
E. Neisser, Münch. med. Wchschr. 72 (1925), 2103
M. Matthes, Deutsches Arch. klin. Med. 159 (1928), i-iv
H. Sahli, Schw. med. Wchschr. 58 (1928), 226-228
C. Wegelin, Schw. Med. Wchschr. 86 (1956), 366-371

LICHTWITZ, LEOPOLD [1876-1943]
FISCHER 2,912; BHDE 2,726; LDGS, 67; KOREN, 213; WININGER 7,585-586; TETZLAFF, 208; KAGAN(JA), 82-83; WWWS, 1041
A. Jores, Klin. Wchschr. 24-25 (1946), 192

LICHTY, DAVID MARTIN [1862-1942]
AMS 6,852; JV 22,300
New York Times, 26 December 1942

LIDDELL, EDWARD GEORGE TANDY [1895-1981]
WW 1981, p.1549; IB, 175
C.G. Phillips, Biog. Mem. Fell. Roy. Soc. 29 (1983), 333-359

LIDDLE, LEONARD MERRITT [1885-1920]
W.A. Hamor, J. Ind. Eng. Chem. 12 (1920), 400-401

LIEB, CHARLES CHRISTIAN [1880-1956]
AMS 9(II),678; IB, 175; WWAH, 360; WWWS, 1042

LIEB, HANS von [1887-1979]
KURSCHNER 13,2307; WWWS, 1042; POGG 6,1524; 7a(3),91-92
A.A. Benedetti-Pichler, Mikrochemie 33 (1947), 107-121
Osterreicher der Gegenwart 1951, p.178
M.K. Zacher, Wiener klin. Wchschr. 69 (1957), 438
Anon., Nachr. Chem. Techn. 28 (1980), 247-248

LIEBEN, ADOLF [1836-1914]
NDB 14,473-474; NOB 15,119-125; OBL 5,192; STURM 2,448-449; DBJ 1,297; KNOLL, 46-48; BLOKH 1,441-442; SCHAEDLER, 70; WININGER 4,103; POTSCH, 271; POGG 3,809; 4,883-884; 5,742; 7aSuppl.,375
S. Zeisel, Chem. Z. 38 (1914), 829-831; Ber. chem. Ges. 49 (1916), 835-892
G. Goldschmidt, Alm. Akad. Wiss. Wien 65 (1915), 332-339

LIEBEN, FRITZ [1890-1965]
WIEN 1967, pp.48-49; TLK 3,396; IB, 175; WWWS, 1042; POGG 6,1524-1525; 7a(3),92-93
F. Brücke, Allgem. prakt. Chem. 1966, p.151
Anon., Chem. Z. 90 (1966), 224
E. Heischkel-Artelt, in F. Lieben, Geschichte der Physiologischen Chemie, pp.v-viii. Reprint of 1935 ed.; Hildesheim 1970

LIEBERKÜHN, NATHANIEL [1821-1887]
LKW 6,203-209; HIRSCH 3,779; PAGEL, 1006; GUNDLACH, 206
K. Bardeleben, Deutsche med. Wchschr. 13 (1887), 489-490
Anon., Leopoldina 23 (1887), 111

LIEBERMAN, SEYMOUR [1916-]
AMS 15(4),743; WWA 1980-1981, p.027; WWWS, 1042

LIEBERMANN, CARL [1842-1914]
DSB 8,328-329; NDB 14,481-482; DBJ 1,297; BLOKH 1,437-438; DZ, 875-876; SCHAEDLER, 70; KOHUT 2,231; WININGER 4,105-106; WWWS, 1042; POTSCH, 271-272; POGG 3,810-811; 4,885-886; 5,743
A. Bistrzycki, Chem. Z. 39 (1915), 165-167
O. Wallach and P. Jacobson, Ber. chem. Ges. 51 (1918), 1135-1204

LIEBERMANN, HANS [1876-1939?]
TLK 3,396; LDGS, 22; POGG 5,743; 7a(3),93

LIEBERMANN, LEO [1852-1926]
GROTE 6,205-251; FISCHER 2,913; KOREN,213; WININGER 4,106-107; 7,257-258; POGG 3,811; 4,886
Anon., Biochem. Z. 179 (1926), 1-2

LIEBERMEISTER, GUSTAV [1879-1943]
IB, 175
R. Bochalli, Deutsches Tuberculosis Blatt 17 (1943), 148

LIEBERMEISTER, KARL von [1833-1901]
NDB 14,486; BJN 6,361-363,64*; SL 5,424-439; HIRSCH 3,779-780; PAGEL, 1006-1007; FRANKEN, 189-190
F. Müller, Corr. Schw. Aerzte 32 (1902), 42-48
H. Abegg, Münch. med. Wchschr. 49 (1902), 194-196; Carl Liebermeister. Tübingen 1919
E. Reinert, Berl. klin. Wchschr. 39 (1902), 226-228,249-251,272-275, 294-296
H. Hentrich, Das Leben und das Wirken Carl Liebermeisters. Düsseldorf 1939
H.M. Koelbing, Gesnerus 26 (1969), 233-248
H.R. Baumberger, Carl Liebermeister. Zürich 1980

LIEBIG, GEORG [1827-1903]
BJN 8,103-104; HIRSCH 3,781-782; PAGEL, 1007-1008; POGG 3,811-812; 4,886
A. Schmid, Münch. med. Wchschr. 51 (1904), 218-219

LIEBIG, HANS von [1874-1931]
NDB 14,497; POGG 5,743; 6,1525
Anon., Chem. Z. 55 (1931), 155

LIEBIG, HERMANN [1831-1894]
HB 3,377-380
Anon., Chem. Z. 18 (1894), 2063

LIEBIG, JUSTUS von [1803-1873]
DSB 8,329-350; NDB 14,497-501; ADB 18,589-605; BUGGE 2,1-30; W, 328-329; PRANDTL, 79-134; BLOKH 1,442-449; FERCHL, 315-317; TSCHIRCH, 1080; SCHELENZ, 664-666; PHILIPPE, 763-766; SCHAEDLER, 70-72; POHL, 90-93; POTSCH, 272-273; MATSCHOSS, 157-158; ABBOTT-C, 85-86; CALLISEN 11, 352-356; 30,58-60; POGG 1,1455-1460; 3,811; 6,1525; 7aSuppl.,375-389
A.W. Hofmann, J. Chem. Soc. 28 (1875), 1065-1140; Ber. chem. Ges. 23(3) (1890), 792-816
W. Roth, Samml. chem. Vortr. 3 (1898), 165-200
A. Kohut, Justus von Liebig. Giessen 1904
J. Volhard, Justus von Liebig. Leipzig 1909
F. Haber, Z. angew. Chem. 41 (1928), 891-897
F.R. Moulton (Ed.), Liebig and after Liebig. Washington, D.C. 1942
H.S. van Klooster, J. Chem. Ed. 33 (1956), 493-497; 34 (1957), 27-30
H. von Dechend, Justus von Liebig in eigenen Zeugnissen und solchen seiner Zeitgenossen. Weinheim 1963
J.L. Dumas, Rev. Hist. Sci. 18 (1965), 73-108
C. Paolini, Justus von Liebig. Heidelberg 1968
J.B. Morrell, Ambix 19 (1972), 1-46
W.H. Brock, Ambix 19 (1972), 47-58; Hist. Sci. 19 (1981), 201-218; Ambix 37 (1990), 134-147
F. Kröhnke, Ann. Chem. 1973, pp.547-552
O. Sonntag, Ambix 24 (1977), 159-169
R.S. Turner, Hist. Stud. Phys. Sci. 13 (1982), 129-162
W.H. Brock (Ed.), Justus von Liebig und August Wilhelm Hofmann in ihren Briefen 1841-1873. Weinheim 1984
W. Conrad, Justus von Liebig und sein Einfluss auf die Entwicklung des Chemiestudiums etc. Darmstadt 1985
J.S. Fruton, Proc. Am. Phil. Soc. 132 (1988), 1-66
F.L. Holmes, Osiris [2] 5 (1989), 121-164
P. Munday, Ambix 37 (1990), 1-19; 38 (1991), 135-154
U. Schling-Brodersen, Ambix 39 (1992), 21-31

LIEBIG, (GEORG) KARL [1818-1870]
DGB 52,317

LIEBISCH, THEODOR [1852-1922]
DBJ 4,168-172,362; POGG 4,886-887; 5,743-744
K. Schulz, Zentralblatt für Mineralogie 22 (1922), 417-434
F. Becke, Alm. Akad. Wiss. Wien 72 (1922), 173-175
O. Mügge, Jahrb. Akad. Wiss. Gott. 1922, pp.79-85
A. Johnsen, Z. Kristall. 57 (1923), 443-448

LIEBLEIN, VICTOR [1869-1939]
STURM 2,453; KURSCHNER 4,1736; FISCHER 2,914; PUTZ,30-35
Anon., Deutsche med. Wchschr. 65 (1939), 936

LIEBMANN, LOUIS [1863-1929]
NUC 332,315

LIEBRECHT, ARTHUR [1862-]
JV 1,121; RSC 16,778; NUC 332,320

LIEBREICH, OSKAR [1839-1908]
NDB 14,511-512; BJN 13,56*; HIRSCH 3,782-783; PAGEL, 1009-1010; KROLLMANN, 1003-1004; BLOKH 1,449-450; REBER, 177-178; KOREN, 213; KOHUT 2,231-232; WININGER 4,113; DZ, 878-879; TSCHIRCH, 1081; WREDE, 116-117; POGG 3,812; 4,887; 5,745; 7aSuppl.,389-390
A. Langgaard, Ber. chem. Ges. 41 (1908), 4801-4804
L. Spiegel, Chem. Z. 32 (1908), 653; Med. Klin. 4 (1908), 1089-1091
C. Posner, Berl. klin. Wchschr. 45 (1908), 1387-1388
H. Thoms, Deutsche med. Wchschr. 34 (1908), 1588-1589
E. Saalfeld, Münch. med. Wchschr. 55 (1908), 1647-1648
K.F. Hoffmann, Med. Mon. 12 (1958), 475-477

LIECHTI, PAUL ROBERT [1866-1927]
DHB 4,519; STRAHLMANN, 473,481; RSC 16,779
Schweizerisches Zeitgenossen-Lexikon 1,410-411

LIEFMANN, ELSE [1881-1970]
KURSCHNER 11,1756; JV 24,159; SG [3]7,643; NUC 332,373

LIEPMANN, WILHELM [1878-1939]
KURSCHNER 5,810-811; FISCHER 2,915-916; GAUSS, 239; WININGER 7,260; KOREN, 213; BHDE 2,730; WIDMANN, 88-89,274; LDGS, 63

LIESCHE, OTTO [1878-1931]
POGG 6,1527-1528
P.W. Danckwortt, Chem. Z. 55 (1931), 941; Chemische Fabrik 4 (1931),461
G. Lockemann, Ber. chem. Ges. 65A (1932), 21

LIESCHING, FRANZ (FRANCIS) [1818-1903]
RSC 4,23
A. Wankmüller, Beitr. Wurtt. Apothekgesch. 7 (1966), 84

LIESEGANG, RAPHAEL [1869-1947]
NDB 14,538; NB, 238; PUDOR, 144-146; TLK 2,428-429; POTSCH, 273; POGG 5,745; 6,1529; 7a(3),97
Anon., Science 107 (1948), 267
E. Küster et al., Koll. Z. 117 (1950), 2-10
R. Jäger, Z. wiss. Mikr. 60 (1951), 45-48

LIESER, THEODOR [1900-]
POGG 6,1529-1530; 7a(3),98
Anon., Chem. Z. 84 (1960), 542

LIFSCHITZ, ISRAEL [1888-1953]
TLK 3,398; POGG 5,746; 6,1530; 7b,2876-2877
H.J. Backer, Chem. Wkbl. 49 (1953), 941

LIGNAC, GEORG OTTO EMILE [1891-1954]
LINDEBOOM, 1194-1195; IB, 175
H.Y. Doelman, Ned. Tijd. Gen. 98 (1954), 2650-2651

LIKHACHEV, ALEKSEI ALEKSEEVICH [1866-1942]
BSE 25,293; BME 16,110-112; WWR, 350-351; IB, 175
V.M. Karasik, Fiziol. Zhur. 31(1/2) (1945), 5

LIKIERNIK, ARTUR [1867-1937]
ZURICH-D, 162
K. Lemanczyk, Przemysl Chemiczny 21 (1937), 190-192

LILIENFELD, LEON [1869-1938]
RSC 16,783
Anon., Nature 142 (1938), 282; Ind. Eng. Chem. News Ed. 16 (1938), 386

LILJESTRAND, GOERAN [1886-1968]
SBL 23,46-49; FISCHER 2,916; IB, 175; WWWS, 1044
U.S. von Euler, Acta Physiol. Scand. 72 (1968), 1-8

LILLIE, FRANK RATTRAY [1870-1947]
DSB 8,354-360; DAB [Suppl.4],497-499; NCAB 36,32-33; WWAH,361; WWWS, 1045

R.G. Harrison, Am. Phil. Soc. Year Book 1947, pp.264-270
B.H. Willier et al., Biol. Bull. 95 (1948), 151-162
C.R. Moore, Science 107 (1948), 33-35
B.H. Willier, Biog. Mem. Nat. Acad. Sci. 30 (1957), 179-236
G. Allen, Mendel Newsletter 9 (1973), 1-6
R.L. Watterson, American Zoologist 19 (1979), 1275-1287

LILLIE, RALPH DOUGALL [1896-1979]
AMS 14,2991; WWWS,1045
G.G. Glenner, J. Histochem. 16 (1968), 3-16
F.H. Kasten, Stain Technology 55 (1980), 201-215
J.D. Longley, J. Histochem. 28 (1980), 291-296
G. Clark and F.H. Kasten, History of Staining, 3rd Ed., pp.1-34. Baltimore 1983

LILLIE, RALPH STAYNER [1875-1952]
AMS 8,1496; FISCHER 2,916; IB, 175; WWAH, 361; WWWS, 1045
R. Gerard, Science 116 (1952), 496-497
E.G. Conklin, Biol. Bull. 105 (1953), 14-16

LIM, ROBERT KHO-SENG [1897-1969]
AMS 11,3150; MH 2,321-322; MSE 2,231-232; IB, 175; WWWS, 1045
Anon., BioScience 16 (1966), 682
S.C. Wang, Pharmacologist 12 (1970), 24-25
H.W. Davenport, Biog. Mem. Nat. Acad. Sci. 51 (1980), 281-306

LIMBECK, RUDOLF von [1861-1900]
STURM 2,458; FISCHER 2,917; SCHIEBER, 65-69
W. Pauli, Wiener klin. Wchschr. 13 (1900), 439

LIMBOURG, PHILIPP MARIA [1860-]
JV 3,194; RSC 16,785; SG [2]9,558; NUC 333,447

LIMPACH, LEONHARD [1852-1933]
HEIN-E, 281-282; POTSCH, 273; POGG 4,889; 6,1531; 7a(3),101
M. Busch, Z. angew. Chem. 46 (1933), 99-100

LIMPRICHT, HEINRICH [1827-1909]
BJN 14,54*; BLOKH 1,450-451; DZ, 884; POTSCH, 273; POGG 1,1463; 4,889; 5,747
K. Auwers, Ber. chem. Ges. 42 (1909), 5001-5036
M. Scholtz, Chem. Z. 33 (1909)
G. Schneider, Heinrich Limpricht und sein Schülerkreis. Greifswald 1970

LIND, OTTO [1900-1972]
JV 53,248
Anon., Chem. Z. 94 (1970), 218; 96 (1972), 697

LIND, SAMUEL COLVILLE [1879-1965]
AMS 10,2419; NCAB 51,447-448; MILES, 299-300; WWAH, 362; WWWS, 1046; POGG 5,747-748; 6,1532-1533; 7b,2877-2879
A.D. McFadyen, Chem. Eng. News 25 (1947), 1664
G. Glockler, J. Chem. Ed. 36 (1959), 262-266
E.H. Taylor, J. Phys. Chem. 63 (1959), 773-776
P.S. Rudolph, Am. Phil. Soc. Year Book 1965, pp.184-191
P. Harteck, Ber. Bunsen Ges. 69 (1965), 561-562
S.C. Lind, Journal of the Tennessee Academy of Science 67 (1972), 1-40

LINDAHL, PER ERIC [1906-1991]
VAD 1981, p.623

LINDEGREN, CARL CLARENCE [1896-1986]
AMS 12,3730

LINDEMANN, HANS [1890-1932]
TLK 3,399; POGG 6,1534-1535
W.A. Roth, Ber. chem. Ges. 65A (1932), 68

LINDEMANN, LUDWIG [1868-1917]
FISCHER 2,348; BACHMANN, 99-101
Anon., Jahrbuch der Ludwigs-Maximilian-Universität München 1914-1919, pp.60-61

LINDEN, CHARLES FLORENT van der [1876-]
JV 19,182; NUC 334,175

LINDENBERG, EUGEN [1872-]
JV 13,225; RSC 16,791
Anon., Chem. Z. 61 (1937), 625

LINDENMANN, JEAN [1924-]
BIBEL, 208-212
Who's Who in Switzerland 1988-1989, p.334

LINDENMEYER, OSKAR [1839-1889]
Anon., Pharm. Z. 34 (1889), 657
A. Wankmüller, Beitr. Wurtt. Apothekgesch. 8 (1969), 112

LINDER, STEPHEN ERNEST [1868-1942]
Anon., Proc. Geol. Assn. 54 (1943), 44

LINDERSTRØM-LANG, KAJ ULRIC [1896-1959]
DSB 18,555-561; DBL (3rd Ed.) 9,68-70; VEIBEL 2,270-274; 3,115-151; POGG 6,1535; 7b,2882-2887
J.T. Edsall, Adv. Protein Chem. 14 (1959), xiii-xxiii
M. Otteson, Am. Phil. Soc. Year Book 1959, pp.133-138
H.M. Kalckar, Science 131 (1960), 1420-1425
A. Tiselius, Biog. Mem. Fell. Roy. Soc. 6 (1960), 157-168
H.K. Alber, Microchem. J. 4 (1960), 141-143
H. Neurath, Arch. Biochem. Biophys. 86 (1960), i-iv
F. Duspiva, Erg. Physiol. 51 (1961), 1-20
H. Holter, in The Carlsberg Laboratory 1876-1976 (H.Holter and K. Møller, Eds.), pp.88-117. Copenhagen 1976; TIBS 4 (1979), 239-240

LINDET, LÉON [1857-1929]
WWWS, 1047; POGG 4,891-892; 5,749; 6,1535-1536
P. Nottin, Bull. Soc. Chim. Biol. 11 (1929), 1254-1258
L. Mangin, C.R. Acad. Sci. 188 (1929), 1580-1582
P.N., Ann. Falsif. 22 (1929), 388-391
A.T. Schloesing, Not. Acad. Sci. 1 (1937), 264-266

LINDHAGEN, LEOPOLD [1884-1936]
SMK 4,663
S. Berg, Hygeia 98 (1936), 353-359

LINDHARD, JOHANNES [1870-1947]
DBL 14,396-398; KBB 1944-1946, p.797; FISCHER 2,919; VEIBEL 2,274-275
E. Hansen, Nord. Med. 36 (1947), 2423-2424
G. Liljestrand, Acta Physiol. Scand. 14 (1947), 291-295

LINDNER, FRITZ [1901-1977]
NDB 14,611-612; KURSCHNER 12,1902
Anon., Nachr. Chem. Techn. 8 (1960), 184-185; Chem. Z. 101 (1977), 517

LINDNER, JOSEF [1880-1951]
KERNBAUER, 390-395; MACHEK,203-205; TLK 3,400; POGG 6,1537; 7a(3),103-104
A. Luszczak, Ost. Chem. Z. 52 (1951), 101
E. Phillippi, Microchimica Acta 38 (1951), 189-193

LINDNER, PAUL [1861-1945]
KURSCHNER 4,1752; BK, 257; TLK 3,400; IB, 176
Anon., Chem. Z. 55 (1931), 309; 65 (1941), 177
H. Melchior, Ber. bot. Ges. 68a (1955), 298-302

LINEWEAVER, HANS [1907-]
AMS 12,3740

LING, ARTHUR ROBERT [1861-1937]
WW 1937, p.2006; WWWS, 1049; POGG 6,1538-1539; 7b,2893-2894
Anon., Chem. Ind. 1937, pp.744-745
J.L. Baker, J. Chem. Soc. 1937, pp.1748-1749
R.H. Hopkins, Nature 139 (1937), 954-955; Biochem. J. 31 (1937), 1439-1440

LINK, HEINRICH FRIEDRICH [1767-1851]
ADB 18,714-720; HIRSCH 3,796; TSCHIRCH, 1051; WWWS, 1049

LINK, KARL PAUL [1901-1978]
AMS 13,2638; WWWS, 1049; POGG 6,1539; 7b,2899-2901
K.P. Link, Circulation 19 (1959), 97-107
C.E. Ballou, Adv. Carbohydrate Chem. 39 (1981), 1-12
R.F. Schilling, J. Lab. Clin. Med. 109 (1987), 617-618

LINNEMANN, EDUARD [1841-1886]
OBL 5,228; STURM 2,465; BLOKH 1,452-453; SCHAEDLER, 72; POGG 3,817; 4,894
A.W. Hofmann, Ber. chem. Ges. 19 (1886), 1149-1151
E. Suess, Alm. Akad. Wiss. Wien 36 (1886), 177-180

LINNETT, JOHN WILFRID [1913-1975]
DNB [1971-1980], 508-509; WW 1975, p.1893; CAMPBELL, 139-140; WWWS, 1050; ABBOTT-C, 86-87
A.D. Buckingham, Biog. Mem. Fell. Roy. Soc. 23 (1977), 311-343

LINNEWEH, WILHELM [1903-]
POGG 6,1540; 7a(3),107-108

LINOSSIER, GEORGES [1857-1923]
FISCHER 2,920; GUIART, 160
L. Hallion, Bull. Acad. Med. 90 (1923), 414-420

LINSER, HANNS [1907-]
KURSCHNER 13,2330-2331; WWWS, 1050; POGG 7a(3),108-109

LINSTEAD, REGINALD PATRICK [1902-1966]
DNB [1961-1970], 658-659; W, 332-333; POTSCH, 274; POGG 6,1540; 7b,2903-2908
H.N. Rydon, Chem. Brit. 3 (1967), 126-127
D.H.R. Barton et al., Biog. Mem. Fell. Roy. Soc. 14 (1968), 309-347

LINTNER, CARL [1828-1900]
BJN 5,106*
Anon., Chem. Z. 24 (1900), 51

LINTNER, CARL JOSEPH [1855-1926]
BLOKH 1,453; TLK 2,433; POGG 6,1540-1541
K. Heim, Z. angew. Chem. 39 (1926), 1113-1115

LINTON, LAURA ALBERTA [1853-1915]
M.R.S. and T.M. Creese, Bull. Hist. Chem. 8 (1990), 15-18

LIPKIN, DAVID [1913-]
AMS 15(4),777; COHEN, 152; WWWS, 1051

LIPMAN, CHARLES BERNARD [1883-1944]
AMS 7,1072; HUMPHREY, 146-148; IB, 177; WWAH, 363; WWWS, 1051
Anon., Chem. Eng. News 22 (1944), 1954; Soil Science 59 (1945), 111-113
H.S. Reed, Science 100 (1944), 464-465

LIPMAN, JACOB GOODALE [1874-1939]
DAB [Suppl.2],387-388; AMS 6,859; NCAB E,154-155; SACKMANN, 209-212; IB, 177; WWAH, 363; WININGER 4,125-126
A.G. McCall, Science 89 (1939), 378-379
G. Pincus, Proc. Am. Acad. Arts Sci. 74(6) (1940), 142-143
S.A. Waksman, Chronica Botanica 6 (1941), 459-460
W. Sackmann, Soil Science 129 (1980), 135-137

LIPMANN, FRITZ ALBERT [1899-1986]
AMS 14,3012-3013; WWA 1978-1979, p.1974; IWW 1982-1983, pp.784-785; MH 1,296-298; STC 2,188-190; MSE 2,235-237; VEIBEL 2,275-276; TETZLAFF, 212; BHDE 2,734-735; CB 1954, pp.413-414; WWWS, 1051; POTSCH, 274-275
F. Lipmann, Wanderings of a Biochemist. New York 1971; Ann. Rev. Biochem. 53 (1984), 1-33
D. Richter and H. Hilz, TIBS 4 (1979), N123-N124
G.D. Novelli, in Chemical Recognition in Biology (F. Chapeville and A.L. Haenni, Eds.), pp.415-430. Berlin 1980
C. de Duve, Am. Phil. Soc. Year Book 1988, pp.197-202
H. Kleinkauf et al. (Eds.), The Roots of Modern Biochemistry. Berlin 1988

LIPP, ANDREAS [1855-1916]
BLOKH 1,453-454; DZ, 891-892; POGG 3,818-819; 4,894-895; 5,751; 6,1541
G. Rohde, Chem. Z. 40 (1916), 1081-1082
G. Rohde and P. Lipp, J. prakt. Chem. NF106 (1923), 77-107

LIPP (BREDT-SAVELSBERG), MARIA [1892-1966]
KURSCHNER 10,1440; BOEDEKER, 45; POGG 6,1541; 7a(3),110

LIPP, PETER [1885-1947]
TLK 3,402; POGG 5,751; 6,1541-1542; 7a(3),110

LIPPICH, FRIEDRICH [1875-1956]
KOERTING, 123

LIPPMAA, THEODOR [1892-1944]
Eesti Biograafiline Leksikon 1929, p.282; 1940, pp.182-183
L.S. Ettre, Chromatographia 20 (1985), 399-402

LIPPMANN, EDMUND OSKAR von [1857-1940]
NDB 14,666; TSCHIRCH, 1082; TLK 3,403; WWWS, 1051; WININGER 4,128; POTSCH, 275; TETZLAFF, 212; POGG 3,820; 4,896-897; 5,752; 6,1542; 7a(3),111
K. Sudhoff, Chem. Z. 51 (1927), 13-14
E. Farber, Naturwiss. 20 (1932), 25-28
J.R. Partington, Osiris 3 (1937), 5-21
R. Zaunick, Z. Zuckerind. 7 (1957), 29-32

LIPPMANN, EDUARD [1838-1919]
OBL 5,238-239; STURM 2,468; WININGER 7,264; POGG 4,895; 5,752
K.A. Hofmann, Ber. chem. Ges. 52A (1919), 165-166

LIPPMANN, GABRIEL [1845-1921]
DSB 8,387-388; VAPEREAU 6,1005; CHARLE(2), 186-187; W, 333; WWWS, 1051; POGG 3,819-820; 4,895-896; 5,752; 6,1542

A. Leduc, Rev. Gen. Sci. 32 (1921), 565-570
A.S., Proc. Roy. Soc. A101 (1922), i-iii
E. Picard, Rev. Sci. 70 (1932), 129-141

LIPSCHITZ (LINDLEY), WERNER [1892-1948]
KURSCHNER 5,819; ARNSBERG, 274; FISCHER 2,922; KALLMORGEN, 340; IB, 177; TLK 3,404; KOREN, 214; WININGER 7,264-265; BHDE 2,736; LDGS, 78; TETZLAFF, 213; WIDMANN, 274; POGG 6,1542-1543; 7a(3),111-112
Anon., Proc. Virchow Med. Soc. 7 (1948), 146-147; Science 107 (1948),219
W. Laubender, Arch. exp. Path. Pharm. 207 (1949), 243-255

LIPSCHÜTZ, ALEXANDER (ALEJANDRO) [1883-1980]
FISCHER 2,922; KOREN, 214; WININGER 4,132; 7,265; WWWS, 1052
J.J. Izquierdo, Gaceta Medica Mexicana 93 (1963), 935-941
A. Lipschütz, Persp. Biol. Med. 8 (1964), 3-14
R. Iglesias, Revista Medica de Chile 109 (1981), 1219-1229

LIPSCHÜTZ, BENJAMIN [1878-1931]
OBL 5,240; FISCHER 2,922-923; KOREN, 214; WININGER 7,265-266; WWWS, 1052
G. Nobl, Wiener med. Wchschr. 82 (1932), 38-39
L. Hess, Wiener klin. Wchschr. 45 (1932), 121-122

LIPSCHÜTZ, OSCAR [1864-]
JV 1,218; NUC 335,334

LIPSCOMB, WILLIAM NUNN [1919-]
AMS 15(4),780-781; WWA 1980-1981, p.2041; IWW 1982-1983, p.785; MH 2,326-327; MSE 2,237-238; McLACHLAN, 97-102; ABBOTT-C, 87; POTSCH, 275

LIPSON, HENRY SOLOMON [1910-1991]
WW 1991, p.1109
M.M. Wolfson, Acta Cryst. A47 (1991), 635-636

LIPSTEIN, ALFRED [1876-1942]
KALLMORGEN, 341; BHDE 2,736; SG [2]9,578; NUC 335,377

LISBONNE, MARCEL [1883-1946]
FISCHER 2,923; IB, 177
A. Boivin, Bull. Acad. Med. 130 (1946), 302-305
L. Nègre, Presse Med. 54 (1946), 460

LISTER, JOSEPH [1827-1912]
DSB 8,399-413; DNB [1912-1921], 339-343; HIRSCH 3,803-804; DESMOND, 388; PAGEL, 1018-1022; SACKMANN, 213-216; BETTANY 2,134-137; W, 334-335; ABBOTT, 85-86; WWWS, 1053
W.W.C., Proc. Roy. Soc. B86 (1913), i-xxi
C. Dukes, Lord Lister. London 1924
R. Truax, Joseph Lister. Indianapolis 1944
R.B. Fisher, Joseph Lister. London 1977
N.J. Fox, Hist. Sci. 26 (1988), 367-397

LITTEN, MORITZ [1845-1907]
BJN 12,53*; FISCHER 2,924; KROLLMANN, 1129; ASEN, 119; KOREN, 214

LITTLE, CLARENCE COOK [1888-1971]
DSB 18,562-564; AMS 11,3178; NCAB B,205-206; WWAH, 364; WWWS, 1054
Anon., Genetics 70 (1972), s90-s91
G.D. Snell, Biog. Mem. Nat. Acad. Sci. 46 (1975), 241-263

LITTLE, GEORGE [1838-1924]
AMS 3,415-416; RSC 4,54
Who Was Who in America 1,735

LIVEING, GEORGE DOWNING [1827-1924]
DNB [1922-1930], 510-522; POGG 3,822; 4,898; 5,753-754; 6,1545
W.J. Pope, Nature 115 (1925), 127-129
C.T. Heycock, J. Chem. Soc. 127 (1925), 2982-2984; Proc. Roy. Soc. A109 (1925), xxviii-xxix
B. Stephen, Girton College 1869-1932, p.184. Cambridge 1933

LIVERSIDGE, ARCHIBALD [1847-1927]
WWWS, 1055; POGG 3,822; 4,898; 6,1545
T.W.E. David, Proc. Roy. Soc. A126 (1930), xii-xiv; J. Chem. Soc. 1931, pp.1039-1042

LIVINGSTON, BURTON EDWARD [1875-1948]
DSB 8,425; DAB [Suppl.4],500-501; NCAB 36,334; HUMPHREY, 148-151; IB, 177-178; WWWS, 1056
D.T. MacDougal, Am. Phil. Soc. Year Book 1948, pp.278-280
W.B. Mack, Sci. Mon. 67 (1948), 34-38
C.A. Shull, Science 107 (1948), 558-560; Plant Physiology 23 (1948), iii-vii

LIVINGSTON, ROBERT STANLEY [1898-]
AMS 11,3186; WWWS, 1056; POGG 6,1545; 7b,2915-2917

LIVON, CHARLES MARIE [1850-1917]
FISCHER 2,926; SG [3]7,747
Anon., Marseilles Médical 53 (1917), 753-784

LLOYD, DAVID PIERCE CARADOC [1911-1985]
AMS 15(4),799; IWW 1982-1983, p.788; WWWS, 1057

LLOYD, DOROTHY JORDAN [1889-1946]
DNB [1941-1950], 511-512; WW 1945, p.1477
Anon., J. Int. Leather Chem. 30 (1946), 340-342
E.C. Bate-Smith, Biochem. J. 41 (1947), 481-482
E.R. Theis, J. Am. Leather Assn. 42 (1947), 40-41
R. Pickard, Chem. Ind. 1947, p.47

LLOYD, FRANCIS ERNEST [1868-1947]
AMS 7,1077; DESMOND, 390; HUMPHREY, 153-155; IB, 178; WWAH, 365; WWWS, 1057; POGG 6,1547; 7b,1923
G.W. Scarth, Plant Physiology 13 (1938), 878-880
Anon., Plant Physiology 23 (1948), 1-4

LLOYD, JOHN ALEXANDER [1878-1960]
SRS, 29; NUC 337,225
Anon., J. Roy. Inst. Chem. 1960, pp.344-345

LLOYD, JOHN URI [1849-1936]
DSB 8,427-428; DAB [Suppl.2],389-390; DAMB, 451-452; MILES, 3012-302; TSCHIRCH, 1082; OEHRI, 65-66; WWAH, 365-366; WWWS, 1057
G. Beal, Am. J. Pharm. Ed. 23 (1959), 202-206
C.M. Simons, John Uri Lloyd. Cincinnati 1972
V.E. Tyler and V.M. Tyler, Journal of Natural Products 50 (1987), 1-8

LLOYD, RACHEL [1839-1900]
ZURICH-D, 150
A.T. Tarbell and D.S. Tarbell, J. Chem. Ed. 59 (1982), 743-744

LOBINGER, KARL [1893-1938]
JV 43,415
Anon., Z. angew. Chem. 51 (1938), 818

LOBRY de BRUYN, CORNELIUS ADRIANN [1857-1904]
POTSCH, 71; WWWS, 1057; POGG 4,900-901; 5,755
E. Cohen and J.J. Blanksma, Ber. chem. Ges. 37 (1904), 4827-4860; Chem. Wkbl. 1 (1904), 971-1007
J.C.A. Simon Thomas, Rec. Trav. Chim. Pays-Bas 24 (1905), 223-255
W.A. Tilden, J. Chem. Soc. 87 (1905), 570-573

LOBSTEIN, ERNEST [1878-1936]
HEIN, 380; HUMBERT, 259-267
Anon., J. Pharm. Chim. [8]23 (1936), 636-638
C. Lapp, Bull. Soc. Pharm. 43 (1936), 436-437
A. Goris, Bull. Acad. Med. 115 (1936), 549-550

LOBSTEIN, JEAN FRÉDÉRIC [1777-1835]
NDB 14,738-740; ADB 19,54-56; HIRSCH 3,813-814; DECHAMBRE [2]2,755; WWWS, 1057; CALLISEN 11,418-422; 30,92-94
A. Brunschwig, Ann. Med. Hist. 5 (1933), 82-84

LOCK, GUNTHER [1900-]
KURSCHNER 12,1918; POGG 6,1548; 7a(3),115-115

LOCK, ROBERT HEATH [1879-1915]
DESMOND, 391; WWWS, 1058
Anon., Nature 95 (1915), 515
B.D.J., Proc. Linn. Soc. 128 (1916), 66-67

LOCKE, FRANK SPILLER [1866-1949]
O'CONNOR(2), 181-182
Anon., Nature 163 (1949), 757
J.A. Venn, Alumni Cantabrigienses II 4,195. Cambridge 1951

LOCKEMANN, GEORG [1871-1959]
NDB 15,6-7; KURSCHNER 8,1415; TSCHIRCH, 1082; TLK 3,405; IB, 178; POTSCH, 276; POGG 5,755-756; 6,1548-1549; 7a(3),116-118; (4),149*
Anon., Chem. Z. 55 (1931), 797-798; 80 (1956), 729; Z. angew. Chem. 54 (1941), 467-468
E. Pietsch, Ber. chem. Ges. 74A (1941), 229-231
R.E. Oesper, J. Chem. Ed. 27 (1950), 236
W. Ulrich, Arzneimitt. 10 (1960), 211-213

LOCQUIN, RENÉ [1876-1965]
WWWS, 1059; POGG 5,757; 6,1549-1550; 7b,2926
C. Dufraisse, C.R. Acad. Sci. 261 (1965), 5289-5291

LOCY, WILLIAM ALBERT [1857-1924]
DAB 11,345; NCAB 18,192-193; WWAH, 367; WWWS, 1059
H. Crew and F.R. Lillie, Science 60 (1924), 491-493

LODIBERT, JEAN ANTOINE [1772-1840]
LAROUSSE 10,615; BALLAND, 66-67; CALLISEN 11,427-428; 30,104-105
N. Nauroy, Hist. Sci. Med. 8 (1974), 789-793; Rev. Hist. Pharm. 22 (1975), 311-319

LODTER, WILHELM [1864-1895]
JV 2,193; RSC 13,276; NUC 338,115
Anon., Ber. chem. Ges. 28 (1895), 3303

LOEB, ADAM [1875-1931]
KALLMORGEN, 341

LOEB, JACQUES [1859-1924]
DSB 8,445-447; DAB 11,349-352; NDB 15,17-18; FISCHER 2,928; DAMB, 452; NCAB 11,72-73; CHITTENDEN, 105-109; BLOKH 1,413-414; KOREN, 214; KAGAN(JA),328-331; WININGER 4,140-141; WWAH, 367; WWWS, 1059; POGG 6,1550-1551
P.A. Levene and W.J.V. Osterhout, Science 59 (1924), 427-430
P.A. Levene et al., Proc. Soc. Exp. Biol. Med. 19 (1924), i-xiv
J.H. Northrop, Ind. Eng. Chem. 16 (1924), 318
C. Herbst, Naturwiss. 12 (1924), 397-406
H. Freundlich, Naturwiss. 12 (1924), 602-603
T.B. Robertson, Science Progress 43 (1926), 114-129
S. Flexner, Science 66 (1927), 333-337
H.E. Armstrong, J. Gen. Physiol. 8 (1927), 653-670
D.E. Palmer, J. Gen. Psychol. 2 (1928), 97-114
W.J.V. Osterhout, J. Gen. Physiol. 8 (1928), ix-xcii; Biog. Mem. Nat. Acad. Sci. 13 (1930), 318-401
D. Fleming, in J. Loeb, The Mechanistic Conception of Life (re-issue), pp.vii-xii. Cambridge, Mass. 1964
M. Rothberg, The Physiologist Jacques Loeb and his Research Activities. Zurich 1965
P.J. Pauly, Jacques Loeb and the Control of Life. Baltimore 1981; Journal of the History of the Behavioral Sciences 17 (1981),504-515; Controlling Life. New York 1987

LOEB, LEO [1869-1959]
DSB 8,447-448; DAB [Suppl.6],385-387; FISCHER 2,928; DAMB, 452-453; KAGAN(JA), 298-299,716; WININGER 7,270; IB, 178; WWAH,367; WWWS,1059
P.A. Shaffer, Arch. Path. 56 (1950), 661-675
L. Loeb, Persp. Biol. Med. 2 (1958), 1-23
W.S. Harcroft, Arch. Path. 70 (1960), 269-274
H.T. Blumenthal, Science 131 (1960), 907-908
C.F. Cori, Am. Phil. Soc. Year Book 1960, pp.159-162
E.W. Goodpasture, Biog. Mem. Nat. Acad. Sci. 35 (1961), 205-219
L.R. Rubin, Clio Medica 12 (1977), 33-56
J.A. Witkowski, Med. Hist. 27 (1983), 269-288

LOEB, ROBERT FREDERICK [1895-1973]
AMS 12,3777-3778; WWA 1974-1975, p.1904; MH 2,327-328; STC 2,192; MSE 2,240-241; DAMB, 453-454; WWWS, 1059-1060
C.A. Ragan, Trans. Assoc. Am. Phys. 87 (1974), 29-30
W.B. Castle, Am. Phil. Soc. Year Book 1974, pp.194-197
A.G. Bearn, Biog. Mem. Nat. Acad. Sci. 49 (1978), 149-183

LOEB, WALTHER [1872-1916]
DBJ 1,362; BLOKH 1,458-459; WININGER 4,142; POGG 4,903; 5,758
Anon., Chem. Z. 40 (1916), 145

LOEBEL, ROBERT OZIAS [1898-1960]
KAGAN(JA), 132
Anon., J. Am. Med. Assn. 172 (1960), 1957

LOEBEN, WOLF von [1869-1913]
RSC 14,1010
E. Beckmann, Ber. chem. Ges. 46 (1913), 1891

LÖBISCH, WILHELM FRANZ [1839-1912]
OBL 5,269; BJN 18,40*; HIRSCH 3,821-822; PAGEL, 1028-1029; HUTER, 238; SCHAEDLER, 72-73; WININGER 4,143
E. Ludwig, Wiener klin. Wchschr. 25 (1912), 254-255
Anon., Wiener med. Wchschr. 62 (1912), 222-223

LÖFFLER, FRIEDRICH [1852-1915]
DSB 8,448-451; NDB 15,33; DBJ 1,334; FISCHER 2,929-930; PAGEL, 1032-1035; BULLOCH, 381; DZ, 897; WWWS, 1060-1061
F. Goldschmidt, Münch. med. Wchschr. 42 (1895), 335-336
G.H.F. Nuttall, Parasitology 16 (1924), 214-238
P. Uhlenhuth, Z. Bakt. 125 (1932), i-xx
D.H. Howard, J. Hist. Med. 18 (1963), 272-281

H. Mochmann and W. Kohler, Z. ärzt. Fortbild. 84 (1990), 400-406

LÖFFLER, KARL WILHELM [1887-1972]
KURSCHNER 10,1450; FISCHER 2,930
M.L. Portmann, Gesnerus 36 (1979), 63-73

LÖHNER, LEOPOLD [1884-1958]
FISCHER 2,931; IB, 178; POGG 7a(3),120-122; (4),149*
R. Rigler, Wiener klin. Wchschr. 70 (1958), 565-566

LÖHNIS, FELIX [1874-1930]
AMS 4,595; IB, 178; NUC 338,241-242
Anon., Nature 127 (1931), 99-100
O. Amsterdamska, J. Hist. Biol. 24 (1991), 197-202

LÖHR, RICHARD [1863-]
JV 5,234; RSC 13,234

LOESCHCKE, HANS [1912-]
KURSCHNER 12,1924; WWWS, 1854; POGG 7a(3),122-123

LOESCHCKE, HERMANN [1882-1958]
KURSCHNER 8,1422; FISCHER 2,932; DRULL, 166; IB, 178
A. Terbrüggen, Verhandl. path. Ges. 43 (1959), 383-386

LOESER, ARNOLD [1902-1986]
KURSCHNER 13,2350; POGG 7a(3),124-125
S. Janssen, Arzneimitt. 12 (1962), 209-210
G. Bornemann, Arzneimitt. 17 (1967), 252-253
F. Kemper, Arzneimitt. 22 (1972), 475

LOEVENHART, ARTHUR SALOMON [1878-1929]
AMS 4,594; FISCHER 2,933; CHITTENDEN, 294-297; PARASCANDOLA, 40-42; CHEN, 26-27; KAGAN(JA),358; KOREN, 214; POGG 6,1553
C.D. Leake, J. Pharm. Exp. Ther. 36 (1929), 495-505
H.S. Gasser, Science 70 (1929), 317-321
C.W. Muehlberger, Ind. Eng. Chem. News Ed. 7(11) (1929), 10
C.R. Bardeen et al., Phi Beta Kappa Quarterly 26 (1929), 1-19

LØVENSKIOLD, HERMAN [1897-1982]
Hver er Hvem? 1979, p.412

LOEW, OSCAR [1844-1941]
NDB 15,72-74; TLK 3,411; TSCHIRCH, 1082; DZ, 902-903; WININGER 4,151; POTSCH, 276-277; BK, 259; POGG 3,826; 4,904; 5,759; 6,1554; 7a(3),126
A. Jacob, Z. angew. Chem. 42 (1929), 369-370
M. Klinkowski, Ber. chem. Ges. 74A (1941), 115-136
Anon., Chem. Z. 65 (1941), 69; Z. angew. Chem. 54 (1941), 246
S. Suzuki, Kagakushi Kenyu 11 (1972), 75-79

LÖW, WILHELM [1862-1940]
RSC 16,74; NUC 338,386
Anon., Z. angew. Chem. 53 (1940), 152

LOEW, WILHELM CHRISTIAN [1818-1908]
NDB 15,72*

LOEWE, HANS [1881-1962]
KURSCHNER 11,1790; POGG 7a(3),127
H. Stadlinger, Chem. Z. 80 (1956), 650
Anon., Chem. Z. 86 (1962), 828; Nachr. Chem. Techn. 10 (1962), 352

LÖWE, JULIUS [1823-1909]
KALLMORGEN, 341; RSC 4,97; POGG 1,1488-1489; 3,826-827

LOEWE, LOTTE [1900-1982]
KURSCHNER 14,2544; BOEDEKER, 47; LDGS, 22; POGG 7a(3),127-128

LOEWE, WALTER SIEGFRIED [1884-1963]
NDB 15,85-86; AMS 10,2461; KURSCHNER 9,1220; BHDE 2,741; LDGS, 79; DRULL, 166-167; KOREN, 214; WININGER 4,161; 7,272; WWWS, 1060
H.E. Voss and H.F. Zipf, Arzneimitt. 9 (1959), 533-535
H.E. Voss, Deutsche med. Wchschr. 89 (1964), 93-94
S.C. Harvey, Pharmacologist 6 (1964), 5-6
R. Kattermann, J. Clin. Chem. 22 (1984), 505-514
R. Kattermann and A. Butenandt, in Naturwissenschaft und Medizin (R. Kattermann, Ed.), pp.85-106. Mannheim 1985

LOEWEL, HENRI [1795-1856]
RSC 4,72; POGG 1,1489
R. Schmitt, Comptes Rendus du Congrès National des Sociétés Savantes 1968, vol.2, pp.175-188

LÖWENHERZ, RICHARD [1867-1929]
TLK 3,412; POGG 4,905; 6,1555

LÖWENSTEIN, ERNST [1878-1950]
NDB 15,101-102; OBL 5,291-292; STURM 2,487; KURSCHNER 4,1797; BHDE 2,743; FISCHER 2,934-935; WININGER 7,273

LOEWENTHAL, HANS [1899-]
IB, 179; LDGS, 57
Who's Who in British Science 1953, p.169

LOEWENTHAL, KARL [1892-1948]
IB, 179; KOREN, 215; WIDMANN, 275; LDGS, 54

LOEWENTHAL, WALDEMAR [1875-1928]
FISCHER 2,935; IB, 179; NUC 338,461
G. Sobernheim, Schw. med. Wchschr. 58 (1928), 562-563

LOEWI, OTTO [1873-1961]
DSB 8,451-457; NDB 15,108-109; DAB [Suppl.7],475-477; AMS 10,2462; ARNSBERG, 287-290; DECKER, 54-70; HAYMAKER, 293-296; W, 337-338; MEDVEI, 768-770; BHDE 2,774; IB, 179; KOREN, 215; POTSCH, 277; WININGER 7,275; WWAH, 367; WWWS, 1060; POGG 7a(3), 128-129
W.B. Cannon, Am. J. Med. Sci. 188 (1934), 145-159
O. Loewi, Ann. Rev. Physiol. 16 (1954), 1-10; J. Mt. Sinai Hosp. 24 (1957), 1014-1016; Persp. Biol. Med. 4 (1960), 3-25
H.H. Dale, Biog. Mem. Fell. Roy. Soc. 8 (1962), 67-90; Erg. Physiol. 52 (1963), 1-19
A.W. Forst, Bayer. Akad. Wiss. Jahrbuch 1962, pp.207-212
O. Krayer et al., Pharmacologist 4 (1962), 47-49
S.W. Kuffler, J. Neurophysiol. 25 (1962), 451-453
L. Lendle, Deutsche med. Wchschr. 87 (1962), 525-526
J.H. Gaddum, Nature 193 (1962), 525-526
F. Brücke, Alm. Akad. Wiss. Wien 113 (1963), 514-529
F. Lembeck and W. Giere, Otto Loewi, ein Lebensbild in Dokumenten. Berlin 1968
J. Cheymol, Rev. Hist. Med. Heb. 25 (1972), 37-42; Therapie 27 (1972), 57-65
U.S. von Euler, Wiener klin. Wchschr. 85 (1973), 721-724
U. Weiss and R.A. Brown, J. Chem. Ed. 64 (1987), 770-771

LÖWIG, CARL JACOB [1803-1890]
NDB 15,109-110; ADB 52,105-106; HEIN, 382; BLOKH 1,464-466; WWWS. 1073; ZURICH, 11-13; FERCHL, 321; TSCHIRCH, 1082; SCHAEDLER, 73; POTSCH, 281; POGG 1,1489-1490; 3,827
H. Landolt, Ber. chem. Ges. 23(2) (1890), 905-909
Anon., Leopoldina 26 (1890), 111; Pharm. Z. 35 (1890), 203-204

LÖWIT, MORITZ [1851-1918]
NDB 15,111-112; OBL 5,295; STURM 2,489; DBJ 2,697; FISCHER 2,936; HUTER 2,257-259; LOMMATZSCH, 54-58; KOREN, 215
Anon., Ber. Nat. Med. Innsb. 37 (1920), 16-31

LOEWY, ADOLF [1862-1937]
FISCHER 2,936; PAGEL, 1040; ZWEIFEL, 110-111; BERLIN, 616; IB, 179; KOREN, 215; WININGER 4,177; 7,276; TETZLAFF, 216
W. Hausmann, Schw. med. Wchschr. 68 (1938), 407

LOEWY, ARIEL GIDEON [1925-]
AMS 15(4),811; WWA 1980-1981, p.2056; WWWS, 1060

LÖWY, JULIUS [1885-1944]
KURSCHNER 4,1800; FISCHER 2,936-937; PELZNER, 76-83; KOERTING, 174
Anon., Lancet 1944(II), p.772

LOFLAND, HUGH [1921-1975]
AMS 12,3782; NCAB 62,93-94; WWWS, 1061
T.B. Clarkson and R.W. Pritchard, Experimental Molecular Pathology 24 (1976), 261-263

LOFTFIELD, ROBERT BERNER [1919-]
AMS 15(4),812; WWA 1980-1981, p.2056; WWWS, 1061

LOGAN, MILAN ALEXANDER [1897-1970]
AMS 11,3200; NCAB 62,104-105

LOHMANN, KARL [1898-1978]
NDB 15,128-129; KURSCHNER 8,1427; STC 2,192-194; POTSCH, 277; POGG 6,1556; 7a(3),130
Anon., Acta Biol. Med. Germ. 1 (1958), 365-367
K. Täufel and M. Ulmann, Ernährungforschung 8 (1963), 185
S. Rapaport, TIBS 3 (1978), 163
P. Langen, TIBS 3 (1978), N184

LOHMANN, WILHELM [1886-1926]
JV 28,209
F. König, Z. angew. Chem. 39 (1926), 1331

LOIR, JOSEPH JEAN ADRIEN [1816-1899]
HUMBERT, 210-212; BERGER-LEVRAULT, 148-149
Anon., Bull. Acad. Med. 41 (1899), 239

LOISEAU, DÉSIRÉ [-1910]
RSC 10,625; 16,853
Anon., Chem. Z. 34 (1910), 1335

LOISEAU, GEORGES [1872-1950]
Anon., Ann. Inst. Pasteur 81 (1950), 365-39

LOISELEUR, JEAN [1899-]
POGG 6,1557; 7b,2931-2935

LOMBARD, WARREN PLIMPTON [1855-1939]
DAB [Suppl.2],390-391; BROBECK, 138-139; FLUELER, 82-83; IB, 179
R. Gesell, Science 90 (1939), 345-346
J.F. Fulton, Nature 144 (1939), 1084-1085
Anon., Physiologist 8(2) (1965), 1-2
H.W. Davenport, Physiologist 25(1)Suppl. (1982), 50-76

LOMBROSO, UGO [1877-1952]
IB, 179; POGG 6,1557; 7b,2939-2941
C. Zumme, Arch. Fisiol. 52 (1952), 95-99
E. Meneghetti, Atti Accad. Lincei [8]5 (1953), 474-481
G. Orestano, Archivio Sci. Biol. 37 (1953), 514-522

LOMMEL, EUGEN [1837-1899]
NDB 15,144-145; BJN 4,94-96; LF 2,264-269; KLEIN, 91; POTSCH, 278; POGG 3,829-830; 4,908-909
Anon., Chem. Z. 23 (1899), 538
C. von Voit, Sitz. Bayer. Akad. 30 (1900), 324-329

LOMMEL, FELIX [1875-1968]
KURSCHNER 11,1797; FISCHER 2,939; GIESE, 574-575

LONDON, EFIM SEMENOVICH [1868-1939]
BSE 25,278; BME 16,310-311; FISCHER 2,939; IB, 179
F. Verzar, Schw. med. Wchschr. 69 (1939), 1314-1315
S.H. Mussaelyan et al., Arkh. Biol. Nauk 54(3) (1939), 3-9
N.N. Zaiko, Pat. Fiz. Eksp. Ter. 3(3) (1959), 3-7
N.N. Romanova, Istoria Biologii Isslevodania 9 (1983), 166-175

LONDON, FRITZ [1900-1954]
DSB 8,473-479; NDB 15,145-146; BHDE 2.747; LDGS, 99; POTSCH, 279; POGG 6,1559; 7a(3),132
H. Frohlich, Nature 174 (1954), 63
K. Mendelssohn, Naturwiss. 42 (1955), 617-619

LONDON, IRVING MYER [1918-]
AMS 15(4),818; WWA 1980-1981, pp.2059-2060; IWW 1982-1983, p.792; ICC, 420-422; WWWS, 1062

LONG, CYRIL NORMAN HUGH [1901-1970]
DSB 18,566-571; AMS 11,3207; DAMB, 456; ICC, 333-336; WWWS, 1063
P. Bondy, Yale J. Biol. Med. 41 (1968), 95-106; Endocrinology 88 (1971), 537-539
J.S. Fruton, Am. Phil. Soc. Year Book 1970, pp.143-145
O.L.K. Smith and J.D. Hardy, Biog. Mem. Nat. Acad. Sci. 46 (1975), 265-309

LONG, ESMOND RAY [1890-1979]
AMS 14,3047; WWA 1976-1977, p.1934; IB, 179; WWWS, 1063
K.R. Boucot, Arch. Env. Hth. 3 (1951), 543-544
R.E. Stowell, Am. J. Path. 100 (1980), 321-325
G.W. Corner, Am. Phil. Soc. Year Book 1980, pp.613-617
P.C. Nowell and L.B. Delpino, Biog. Mem. Nat. Acad. Sci.56 (1987), 285-310; Trans. Coll. Phys. Phila [5]11 (1989), 177-182

LONG, FRANKLIN ASBURY [1910-]
AMS 14,3047; WWA 1978-1979, p.1995; IWW 1982-1983, p.792; MH 2,329-331

LONG, JOHN HARPER [1856-1918]
DAB 11,378-379; NCAB 19,31; MILES, 303-304; WWWS, 1063; POGG 3,830; 4,909; 5,762
J. Stieglitz, Science 49 (1919), 31-38
F.B. Dains, J. Am. Chem. Soc. 41 (1919), 69-82

LONG, JOSEPH ABRAHAM [1879-1953]
AMS 8,1522; FISCHER 2,939-940; IB, 179

LONG, PERRIN HAMILTON [1899-1965]
DAB [Suppl.7],477-478; AMS 10,2470; DAMB, 458; WWAH, 368; WWWS, 1063

R. Austrian, Trans. Assoc. Am. Phys. 79 (1966), 59-61
A.M. Harvey, Johns Hopkins Med. J. 138 (1976), 54-60

LONGCOPE, WARFIELD THEOBALD [1877-1953]
DAB [Suppl.5],438-439; DAMB, 458-459; ICC, 42-44; IB, 179; WWWS, 1064
J. Bardley, Trans. Assoc. Am. Phys. 66 (1953), 9-11
W.S. Tillett, Biog. Mem. Nat. Acad. Sci. 33 (1959), 205-225
A.M. Harvey, Trans. Coll. Phys. Phila. [5]3 (1981), 161-173,314

LONGET, FRANÇOIS ACHILLE [1811-1871]
HIRSCH 3,834; WWWS, 1064
H. Larrey, Bull. Acad. Med. 36 (1871), 1063-1077

LONGINESCU, GHEORGHE [1869-1939]
POGG 5,762; 6,1559-1560; 7b,2944-2945
C. Belcot, Ber. chem. Ges. 73A (1940), 159-160

LONGLEY, ALBERT EDWARD [1893-]
AMS 11,3212; IB, 179; WWWS, 1064

LONGLEY, WILLIAM HARDING [1881-1937]
AMS 5,687; WWWS, 1064
R.E. Cleland, Science 85 (1937), 400-401

LONGMUIR, IAN STUART [1922-]
AMS 15(4),826

LONGSWORTH, LEWIS GIBSON [1904-1981]
AMS 14,3052; WWA 1976-1977, p.1937; WWWS, 1064-1065
Anon., Chem. Eng. News 46(4) (1968), 64; Rock. Univ. Notes 13(1) (1981), 3

LONSDALE, KATHLEEN [1903-1971]
DSB 8,484-486; DNB [1971-1980],517-518; WW 1965, p.1864; ABBOTT-C, 88-89; MH 1,301-302; MSE 2,245-246; CAMPBELL, 145-146; WWWS, 1065
D. Hodgkin, Chem. Brit. 7 (1971), 477-478; Biog. Mem. Fell. Roy. Soc. 21 (1975), 447-484
M.M. Julian, Physics Teacher 19 (1981), 159-165
J. Mason, Notes Roy. Soc. 46 (1992), 279-300

LOO, HENRI van [1859-]
JV 1,169; RSC 14,1017

LOOFBOUROW, JOHN ROBERT [1902-1951]
AMS 8,1524-1525; WWAH, 369; WWWS, 1065
G.R. Harrison, Proc. Phys. Soc. 64 (1951), 1144
Anon., Science 113 (1951), 196

LOOMIS, WILLIAM FARNSWORTH [1914-1973]
AMS 11,3215; WWWS, 1066
Anon., Harvard Med. Alumni Bull. 48(3) (1974), 39

LOONEY, JOSEPH MICHAEL [1896-1975]
AMS 11,3215; IB, 179
Anon., Harvard Med. Alumni Bull. 49(6) (1975), 59

LOOSE, ANTON [1867-1933]
JV 8,216; RSC 16,486
Anon., Chem. Z. 57 (1933), 657; Z. angew. Chem. 46 (1933), 555

LORAND, LASZLO [1923-]
AMS 15(4),832; WWA 1980-1981, p.2065; WWWS, 1066

LORBER, JOHN [1915-]
WWWS, 1066

LORBER, VICTOR [1912-]
AMS 15(4),832; WWA 1980-1981, p.2065; WWWS, 1067

LORD, FREDERICK TAYLOR [1875-1941]
AMS 6,871; NCAB 34,153-154; ICC, 19-20; WWAH, 369
Anon., Ann. Int. Med. 16 (1942), 381-383
J.H. Pratt, New Eng. J. Med. 226 (1942), 869-870

LORD, RICHARD COLLINS [1910-1989]
AMS 11,3218; WWWS, 1067
New York Times, 6 May 1989

LORENTZ-ANDREAE, GUIDO [1874-1935]
JV 18,192
Anon., Z. angew. Chem. 48 (1935), 448

LORENZ, HENRY WILLIAM FREDERICK [1871-]
JV 13,21; WWAH, 369; NUC 341,362

LORENZ, RICHARD [1863-1929]
DSB 8,502-503; NDB 15,172-174; DBJ 11,197-200,360; TLK 3,408; POTSCH, 279-280; POGG 4,912-913; 5,764-765; 6,1563-1564
A. Magnus, Ber. chem. Ges. 62A (1929), 88-90; Z. Elektrochem. 35 (1929), 815-822
J.J. van Laar, Chem. Wkbl. 26 (1929), 406-407
G. Hevesy, Helv. Chim. Acta 13 (1930), 13-17

LORENZEN, FERDINAND [1871-1913]
JV 11,167; NUC 341,394

LORENZEN, JULIUS [1864-1945]
JV 6,239; RSC 13,276

LORING, HUBERT SCOTT [1908-1974]
AMS 12,3807; WWWS, 1068
Anon., Chem. Eng. News 53(9) (1975), 34

LOSCHMIDT, JOSEPH [1821-1895]
DSB 8,507-511; NDB 15,195-196; ADB 52,82-84; OBL 5,326-327; STURM 2,501; NOB 3,63-71; KNOLL, 44-46; BLOKH 1,462-464; FERCHL, 322; POTSCH, 280; POGG 3,835; 4,916; 7aSuppl.,400
J. Hann, Alm. Akad. Wiss. Wien 46 (1896), 258-262
F. Exner, Naturwiss. 9 (1921), 177-180
R. Wegscheider, Chem. Z. 45 (1821), 321-322
H. Casselbaum, NTM 15 (1978), 23-26

LOSSEN, WILHELM [1838-1906]
NDB 15,202; BJN 11,42*; KROLLMANN, 407-408; BLOKH 1,464; SCHAEDLER, 73; POTSCH, 280; DRULL, 168; DZ, 899-900; POGG 3,835; 4,916-917; 5,768
Lassar-Cohn, Ber. chem. Ges. 40 (1907), 5079-5086
G. Zimmer, Chem. Z. 114 (1990), 197-204

LOSSOW, EMIL [1875-]
JV 15,215; RSC 16,377; NUC 342,98

LOTMAR, FRITZ [1878-1964]
JV 20,542
R. Bing et al., Schw. Arch. Neurol. 63 (1949), 1-4
M. Mumenthaler, Mitt. Nat. Ges. Bern NF22 (1965), 327-328
M. Minkowski, Schw. Arch. Neurol. 95 (1965), 320-327

LOTTERMOSER, ALFRED [1870-1945]
NDB 15,247-248; TLK 3,410; POTSCH, 280-281; POGG 5,769; 6,1567; 7a(3),140-141
A. Lottermoser, Koll. Z. 100 (1942), 58-64
E. Buchholz, Koll. Z. 115 (1949), 9-12

LOUIS, DAVID ALEXANDER [1856-1915]
G.T.H., J. Chem. Soc. 109 (1916), 385-386

LOURENÇO, AGOSTINHO VICENTO [1822-1893]
GEPB 15,498-499; BLOKH 1,466; DIERGART, 472
E. Fischer, Ber. chem. Ges. 26 (1893), 395

LOVEJOY, ARTHUR ONCKEN [1873-1962]
DSB 8,517-518; DAB [Suppl.7],480-483; NCAB F,130; WWWS, 1071
G. Boas et al., J. Hist. Ideas 9 (1948), 403-446
L.S. Feuer, American Scholar 46 (1977), 358-366

LOVÉN, JOHANN MARTIN [1856-1920]
SMK 5,76; POGG 4,918; 5,770; 6,1568
H. Johanssen, Sven. Kem. Tid. 32 (1920), 208-210

LOVÉN, OTTO CHRISTIAN [1835-1904]
SMK 5,79-80; HIRSCH 3,850-851
C.G.S., Hygeia [2]4 (1904), 775-778
R. Tigerstedt, Sven. Vet. Akad. Arbok 1905, pp.241-255

LOVIBOND, JOSEPH [1833-1918]
VERSO, 102; NUC 343,222
Who Was Who 1916-1928, p.646
M.L. Verso, Med. Hist. 15 (1971), 64

LOW, BARBARA WHARTON [1920-]
AMS 15(4),845; WWA 1980-1981, p.2073; WWWS, 1071

LOWMAN, OSCAR [1861-1939]
JV 2,193; RSC 14,241; NUC 343,484

LOWRY, CHARLES DOAK, Jr. [1896-1981]
AMS 13,2695; NUC 343,527
Anon., Chem. Eng. News 59(43) (1961), 53

LOWRY, OLIVER HOWE [1910-]
AMS 14,3073; WWA 1978-1979, p.3012; IWW 1982-1983, p.800; WWWS, 1074
Anon., Chem. Eng. News 40(48) (1962), 79
O.H. Lowry, Ann. Rev. Biochem. 59 (1990), 1-27

LOWRY, THOMAS MARTIN [1874-1936]
DNB [1931-1940], 547-548; FINDLAY, 402-418; W, 340; WWWS, 1074; POTSCH, 282; POGG 4,919-920; 5,771-772; 6,1569-1571; 7b,2959-2960
W.J. Pope, J. Chem. Soc. 1937, pp.701-705; Obit. Not. Fell. Roy. Soc. 2 (1938), 287-293

LOWY, ALEXANDER [1889-1941]
AMS 5,692; NCAB 37,158-159; WWAH, 372; WWWS, 1074; POGG 6,1571; 7b,2960-2961
A. Silverman, Science 95 (1942), 62

LU, GWEI-DJEN [1904-1991]
WW 1983, p.1383

LUBARSCH, OTTO [1860-1933]
NDB 15,261-262; FISCHER 2,945-946; FREUND 2,221-228; FRANKENTHAL, 298; HABERLING, 101; KIEL, 83; KOREN, 215; WININGER 4,192; 7,279; IB, 180; WWWS, 1074
O. Lubarsch, Ein Bewegtes Gelehrtenleben. Berlin 1931
O. Rössle, Verhandl. path. Ges. 27 (1934), 341-349

LUBAVIN, NIKOLAI NIKOLAEVICH [1845-1918]
BSE 25,525; BLOKH 1,467-468; POGG 3,823; 4,899-900
M.A. Blokh, Zhur. Fiz. Khim. 60Suppl. (1928), 139-140

LUBLIN, ALFRED [1875-]
JV 16,144; NUC 344,54

LUBOSCH, WILHELM [1875-1938]
KURSCHNER 4,1802; FISCHER 2,946; GIESE, 479-482; EBERT, 61-71; IB, 180
H. von Eggeling, Anat. Nachr. 1 (1949), 27-46

LUBOWSKI, ROBERT [1874-]
KONOPKA 5,374; SG [2]9,765; RSC 16,890; NUC 344,69

LUBS, HERBERT AUGUST [1891-1970]
AMS 10,2494; GOULD, 33

LUCA, SEBASTIANO [1820-1880]
BLOKH 1,469; FERCHL, 323-324; TSCHIRCH, 1082; RSC 4,107-110; 8,268-270; 10,645; 16,890; POGG 3,838

LUCAS, COLIN CAMERON [1903-1981]
AMS 14,3078; YOUNG, 84; WWWS, 1074
Anon., Chem. Eng. News 59(30) (1981), 96

LUCAS, HOWARD JOHNSON [1885-1963]
AMS 10,2495; MILES, 304-305; WWWS, 1075; POGG 6,1572; 7b,2963-2964
W.G. Young and S. Winstein, Biog. Mem. Nat. Acad. Sci. 43 (1973), 163-176

LUCAS, KEITH [1879-1916]
DSB 8,532-535; DNB [1912-1921], 347; WW 1915, p.1330; O'CONNOR(2), 50-52; WWWS, 1075
A. Forbes, Science 44 (1916), 808-810
H. Darwin and W.M. Bayliss, Proc. Roy. Soc. B90 (1919), xxxi-xlii
W. Fletcher et al., Keith Lucas. Cambridge 1934

LUCHSINGER, BALTHASAR [1849-1886]
HIRSCH 3,856; PAGEL, 1053; BOSSART, 120-121; ZURICH-D, 54
L. Hermann, Pflügers Arch. 38 (1886), 417-427
M. Flesch, Verhandl. Schw. Nat. Ges. 69 (1886), 138-151
H.R. Thier, Johann Balthasar Luchsinger. Basle 1953

LUCIANI, LUIGI [1840-1919]
DSB 8,535-536; FISCHER 2,947-948; HAYMAKER, 233-237; WWWS, 1075
S. Baglioni, Erg. Physiol. 18 (1920), xiv-xxvii; Arch. Ital. Biol. 70 (1921), 228-244

LUCIEN, MAURICE [1880-1947]
IB, 180; NUC 344,386

LUCIUS, EUGEN [1834-1903]
NDB 15,277-278; BJN 8,71*-72*; NL 6,248-251; NB, 247
Anon., Z. angew. Chem. 16 (1903), 687; Pharm. Z. 48 (1903), 402

LUCK, EDUARD [1819-1889]
RAUSCH, 381-382; RSC 4,132; 8,276; 10,653
Anon., Ber. chem. Ges. 22 (1889), 3373; Chem. Z. 13 (1889), 1727

LUCK, JAMES MURRAY [1899-1993]
AMS 12,3834; SRS, 60; POGG 6,1573; 7b,2966-2968
J.M. Luck, Ann. Rev. Biochem. 50 (1981), 1-22

LUCKHARDT, ARNO BENEDICT [1885-1957]
AMS 9(II),699-700; NCAB 42,626-627; FISCHER 2,948; BROBECK, 145-146 ;IB, 181; WWAH, 373; WWWS, 1076
F.C. McLean, Science 127 (1958), 509
F.C. Bing, Fed. Proc. 36 (1977), 2506-2509

LUCKING, HUBERT LESLIE [1887-]
JV 25,365

LUCKNER, HERBERT [1907-1985]
KURSCHNER 13,2377

LUDEWIG, STEPHAN [1901-1969]
AMS 11,3247
Anon., Chem. Eng. News 47(24) (1969), 95

LUDWIG, ALFRED [1879-1964]
IB, 181
A. Schumacher, Decheniana 118 (1967), 119-124

LUDWIG, CARL [1816-1895]
DSB 8,540-542; NDB 15,429-430; ADB 52,123-131; 55,895-901; LKW 3,279-288; LESKY, 268-272; HIRSCH 3,860-862; Suppl.,399-400; PAGEL, 1055-1058; LEIPZIG 2,73-86; BRAZIER, 100-104,110-111; JOHN,139-145; POTSCH,283; ABBOTT,87-88; W,341; WWWS,1076; POGG 1,1514; 3,840; 7aSuppl.,401-403
A. Fick and R. Tigerstedt, Biographische Blätter 1 (1895), 265-279
J. Burdon-Sanderson, Proc. Roy. Soc. B59 (1895), i-viii
F.S. Lee, Science 1 (1895), 630-632
W. His, Sitz. Akad. Wiss. Leipzig 47 (1895), 627-638
C. von Voit, Sitz. Bayer. Akad. 26 (1896), 326-338
W.P. Lombard, Science 44 (1916), 363-375
J. von Kries, Naturwiss. 11 (1923), 1-4
K.E. Rothschuh, Z. Kreis. 49 (1960), 2-19
H. Schröer, Carl Ludwig. Stuttgart 1967
W.B. Fye, Circulation 74 (1986), 920-928
T. Lenoir, in The Investigative Enterprise (W. Coleman and F.L. Holmes, Eds.), pp.139-178. Berkeley 1988
P.F. Cranefield, Gesnerus 45 (1988), 271-282
W. Gerabeck, Sudhoffs Arch. 75 (1991), 171-179

LUDWIG, ERNST [1842-1915]
NDB 15,427-428; OBL 5,347-348; HEIN, 388-389; PAGEL, 1058-1059; BLOKH 1,470-471; LESKY, 522-523; POTSCH, 283; POGG 3,840-841; 4,921; 5,773
J. Mauthner et al., Wiener med. Wchschr. 62 (1912), 353-358
T. Panzer, Wiener med. Wchschr. 65 (1915), 1825-1828; Chem. Z. 39 (1915), 857-858
F. Becke, Alm. Akad. Wiss. Wien 66 (1916), 323-327
R. Zeynek, Wiener klin. Wchschr. 46 (1933), 22-24

LUDWIG, HERMANN [1819-1873]
HEIN, 389-390; FERCHL, 325-326; TSCHIRCH, 1083; SCHELENZ, 689; SCHAEDLER, 73; POGG 1,1514-1516; 3,840
A. Geuther, Ber. chem. Ges. 6 (1873), 1578-1581
E. Reichardt, Arch. Pharm. 203 (1873), 97-102

LUDWIG, WILHELM [1901-1959]
KURSCHNER 8,1439; POGG 6,1576; 7a(3),146-147; (4),149*
R. Wette, Biometrische Zeitschrift 1 (1959), 147-149
Anon., Zool. Anz. 23Suppl. (1960), 535-536

LÜCKE, ALBERT [1829-1894]
HIRSCH 3,862-863; PAGEL, 1059-1060
G. Ledderhose, Berl. klin. Wchschr. 31 (1894), 251-255
Anon., Brit. Med. J. 1894(I), p.554; Leopoldina 30 (1894), 104-105

LUECKE, RICHARD WILLIAM [1917-]
AMS 15(4),866; WWWS, 1076

LÜDECKE, KARL [1880-1955]
Anon., Chem. Z. 46 (1922), 451; 64 (1940), 113; 79 (1955), 714; Nachr. Chem. Techn. 3 (1955), 179

LUEDEKING, ROBERT [1853-1908]
NCAB 15,124; SG [1]8,382; [3]7,781
New York Times, 1 March 1908, p.9

LÜDERSDORFF, FRIEDRICH WILHELM [1801-1886]
FERCHL, 326; SCHAEDLER, 74; POTSCH, 282-283; POGG 1,1516
Anon., Leopoldina 22 (1886), 213
L. Eck, Gummi-Zeitung 43 (1929), 2868-2869

LÜDY, ERNST [1862-1901]
RSC 16,902; NUC 345,104
Anon., Centralblatt des Zofingervereins 41 (1901), 100

LÜERS, HEINRICH [1890-1967]
KURSCHNER 11,1817; WWWS,1077; POGG 6,1578; 7a(3),152-153
Anon., Chem. Z. 91 (1967), 909

LÜERS, HERBERT [1910-1978]
KURSCHNER 12,1956
Anon., Naturw. Rund. 31 (1978), 130

LUEKEN, BERND [1908-1978]
NDB 15,467-468; KURSCHNER 12,1956; POGG 7a(3),154-155
Anon., Deutsche med. Wchschr. 103 (1978), 1628

LÜPPO-CRAMER, HINRICUS [1871-1943]
NDB 15,471-472; TLK 3,415-416; POGG 5,774; 6,1579-1580; 7a(3),155-156
K. Kieser, Chemie 54 (1941), 117-119
E. Stenger, Z. wiss. Phot. 48 (1941), 1-4

LÜSCHER, ERHARD [1894-1979]
KURSCHNER 13,2390; NUC 345,136
Who's Who in Europe 1972, pp.1939-1940

LÜSCHER, ERNST FRIEDRICH [1916-]
KURSCHNER 13,2390
A. von Muralt, Chimia 30 (1976), 436-437
Anon., Chimia 33 (1979), 225-226

LÜSCHER, FRIEDRICH (FRITZ) [1862-1934]
FISCHER 2,950; GRUTTER, 71
A. Denker, Arch. Ohrenheilkunde 139 (1935), vi

LÜTHJE, HUGO [1870-1915]
DBJ 1,335; FISCHER 2,950; KALLMORGEN, 345; HAGEL, 143-148; KIEL, 82; PITTROFF, 156-157
L. Krehl, Münch. med. Wchschr. 62 (1915), 1033
F. Müller, Deutsches Arch. klin. Med. 118 (1915), i-viii
L. Michaud, Erg. Physiol. 15 (1916), vi-x

LÜTTRINGHAUS, ARTHUR [1873-1945]
NDB 15,485-486; POTSCH, 286; POGG 7a(3),158
H. Neresheimer, Chem. Ber. 89 (1956), xi-xvii

LÜTTRINGHAUS, ARTHUR [1906-]
KURSCHNER 13,2394; POTSCH, 286-287; POGG 7a(3),158-159

LUFT, ULRICH CAMERON [1910-]
AMS 14,3085; KURSCHNER 13,2395; WWWS, 1077

LUGG, JOSEPH WILLIAM HENRY [1907-]
Who's Who in Australia 1977, pp.683-684

LUGININ, VLADIMIR FEDOROVICH [1834-1911]
DSB 8,545; BSE 25,448; KUZNETSOV(1961), 472-479; BLOKH 1,471-472
Y.I. Soloviev and P.I. Staroselski, Vladimir Fedorovich Luginin. Moscow 1963

LUKENS, FRANCIS DRING WETHERILL [1899-1978]
AMS 14,3087; ICC, 246-247; WWWS, 1077-1078
A.I. Winegrad, Trans. Assoc. Am. Phys. 92 (1979), 35-37; Endocrinology 105 (1979), 574

LUKES, RUDOLF [1897-1960]
POGG 6,1582; 7b,2972-2977
F. Šorm and F. Santary, Coll. Czech. Chem. Comm. 22 (1957), i-viii
K. Blaha and M. Ferles, Chem. Listy 55 (1961), 1-15

LUKYANOV, SERGEI MIKHAILOVICH [1855-]
FISCHER 2,952; ZMEEV 4,204; VENGEROV 4,24; NUC 345,376-377

LULLIES, HANS [1898-1982]
KURSCHNER 13,2398; KIEL, 91; POGG 7a(3),161-162
W. Rudolph, Christiana Albertina NF19 (1984), 235-239

LUMIÈRE, AUGUSTE [1862-1954]
IB, 181; WWWS, 1078; POGG 5,775; 6,1582-1586; 7b,2978-2982

LUMSDEN, JOHN SCOTT [1867-1950]
JV 10,213; RSC 14,804; 16,911; NUC 345,514
A. McKenzie, J. Chem. Soc. 1950, p.3697

LUND, HAKON [1898-1979]
DBL (3rd Ed.) 9,161-162; KBB 1979, pp.691-692; VEIBEL 2,280-282; POGG 6,1587; 7b,2982-2983

LUNDBERG, OSKAR YULIEVICH [1866-]
SG [2]9,780

LUNDE, GULBRAND [1901-1942]
Hver er Hvem? 1938, p.330

LUNDEGÅRDH, HENRIK GUNNAR [1888-1969]
SBL 24,256-260; MH 2,333-335; STC 2,200-202; MSE 2,249-251; IB, 181; POGG 6,1588; 7b,2986-2988

LUNDIN, ERIK HARRY [1896-1973]
SBL 24,325-328; SMK 5,127; VAD 1973, P.645; POGG 6,1589; 7b,2989-2991
Anon., Naturw. Rund. 26 (1973), 274

LUNDQUIST, FRANK [1916-]
KBB 1981, p.715; WWWS, 1079

LUNDSGAARD, CHRISTEN [1883-1930]
DBL 14,583-584; FISCHER 2,953; VEIBEL 2,283-288
Anon., Hosp. Tid. 73 (1930), 693-696

LUNDSGAARD, EINAR [1899-1968]
KBB 1968, p.848; VEIBEL 2,288-290
P. Kruhøffer and C. Crone, Erg. Physiol. 65 (1972), 1-4

LUNGE, GEORG [1839-1923]
DSB 8,553; DBJ 5,434; BUGGE 2,351-359; BLOKH 1,473-474; SCHAEDLER, 74; KOHUT 2,232; WININGER 4,201-202; TETZLAFF, 220; DZ, 910; WWWS, 1079; POTSCH, 284-285; POGG 3,843-844; 4,926-927; 6,1591; 7aSuppl.,403
P.P. B[edson], J. Chem. Soc. 123 (1923), 948-950
E. Bosshard, Verhandl. Schw. Nat. Ges. 104 (1923), 25-43
E. Berl, J. Chem. Ed. 16 (1939), 453-460

LUNIAK, ANDREI IVANOVICH [1881-1957]
V.S. Abramov et al., Zhur. Ob. Khim. 28 (1958), 1118-1119

LUNIN, NIKOLAI IVANOVICH [1853-1937]
BSE 25,475; BME 16,356-357; WWR, 357; POTSCH, 285
L.M. Krasnyanski, Biokhimia 14 (1949), 382-387
H.E. Voss, J. Am. Diet. Assn. 32 (1956), 317-320
V.V. Efremov, Vop. Ist. Est. Tekh. 1981(4), pp.92-96

LUPTON, NATHANIEL THOMAS [1830-1893]
NCAB 12,294-295

LURIA, SALVADOR EDWARD [1912-1991]
AMS 15(4),877; WWA 1980-1981, p.2088; IWW 1982-1983, p.804; COHEN, 157; MH 1,305; STC 2,202-204; MSE 2,251-252; CB 1970, pp.258-260; WWWS, 1079-1080
H.F. Judson, Lancet 337 (1991), 606
J.D. Watson, Nature 350 (1991), 113
G. Bertani, Genetics 131 (1992), 1-4

LUSK, GRAHAM [1866-1932]
DSB 8,555-556; DAB 21,517-518; FISCHER 2,954; DAMB, 464-465; OEHRI,67-68; CHITTENDEN, 65-70; W, 342-343; WWWS, 1080; POGG 6,1598; 7b,2999
F. Müller, Münch. med. Wchschr. 79 (1932), 1572-1573
E.F. DuBois, Science 76 (1932), 113-115; Erg. Physiol. 35 (1933), 10-12; Biog. Mem. Nat. Acad. Sci. 21 (1940), 95-142
L.B. Mendel, Sci. Mon. 35 (1932), 290-283
E.P. Cathcart, Obit. Not. Fell. Roy. Soc. 1 (1933), 143-148
A.E. Light, Yale J. Biol. Med. 6 (1934), 487-506
H.J. Deuel, J. Nutrition 41 (1950), 1-12

LUSSANA, FILIPPO [1820-1897]
EI 21,678
Anon., Chem. Z. 22 (1898), 1095

LUSTGARTEN, SIGMUND [1857-1911]
OBL 5,375; FISCHER 2,954-955; SIEGL, 28-30; KOREN, 216; WININGER 7,285; WWWS, 1080
Anon., Journal of Cutaneous Diseases 29 (1911), 254-256

LUSTIG, ALESSANDRO [1857-1937]
DSB 8,556-557; EI 21,689-690; FISCHER 2,955; OLPP, 249-251; IB, 182; SIGRIST, 62-63; KOREN, 216; WININGER 7,285-286; WWWS, 1080
G. Favilli and E. Morelli, Sperimentale 86 (1932), 651-700
G. Guerrini, Bioch. Ter. Sper. 24 (1937), 473

LUSTIG, BERNARD [1902-]
AMS 14,3095; BHDE 2,758; POGG 6,1593; 7b,2999-3001

LUSTIG, FRITZ [1879-1951]
JV 22,520; NUC 346,212
Anon., Aufbau 17(8) (1951), 27

LUTHER, ROBERT [1868-1945]
NDB 15,541-542; WELDING, 481; TLK 3,416; POTSCH, 285; POGG 4,928-929; 5,778-779; 6,1593-1594; 7a(3),357-359
J. Eggert, Z. Naturforsch. 1 (1946), 357-359
J. Brockhausen, Album Fratrum Academicorum, p.53. Weidenau 1960

LUTWAK-MANN, CECILIA [1909-]
Who's Who of British Scientists 1971-1972, p.537

LUTWAK-MANN, CECILIA [1909-1987]
Who's Who in British Science 1953, p.171

LUTZ, LOUIS CHARLES [1871-1952]
S. Lambin, Bull. Soc. Bot. (Mem.) 101 (1954), 122-135

LUTZ, OSCAR [1871-1950]
WELDING, 481; RIGA, 710; POGG 4,929; 5,779-780; 6,1594; 7a(3),165
B. Jirgensons, Z. angew. Chem. 64 (1952), 64

LUTZ, ROBERT ELIOT [1900-]
AMS 12,3857; POGG 6,1594; 7b,3002-3005

LUYNES, VICTOR HIPPOLYTE de [1828-1904]
POGG 3,845

LUZZATTO, ALBERTO MICHELANGELO [1874-1924]
FISCHER 2,956; WININGER 7,286
F. Rietti, Sperimentale 78 (1924), 843
L. Limentani, Bioch. Ter. Sper. 11 (1924), 296

LUZZATTO, RICCARDO [1876-1922]
A. Ascoli, Bioch. Ter. Sper. 9 (1922), 69
E. Cavazzani, Atti Accad. Ferrara 96 (1922), 9-12
G. Coronedi, Arch. Inter. Pharmacodyn. 26 (1922), 441
P. Colombini and G. Coronedi, Bioch. Ter. Sper. 10 (1923), 35-58

LVOV, SERGEI DMITRIEVICH [1879-1959]
BSE 25,505; WWR, 359; IB, 182

LWOFF, ANDRÉ MICHEL [1902-]
QQ 1978-1979, p.1020; IWW 1982-1983, p.806; MH 1,306-307; STC 2,204-205; MSE 2,254-255; ABBOTT, 88; IB, 182
A. Lwoff, Ann. Rev. Microbiol. 25 (1971), 1-26

LWOFF, MARGUERITE [1905-]
IB, 182

LYMAN, HENRY [1866-1934]
AMS 3,427
Anon., J. Am. Med. Assn. 103 (1934), 505

LYMAN, RUFUS ASHLEY [1875-1957]
AMS 9(II),704; NCAB H,141-142; WWAH, 375
J.B. Burt et al., Am. J. Pharm. Ed. 20 (1956), 1-4
J. Tom, Pharm. Hist. 14 (1972), 90-94,111

LYNEN, FEODOR [1911-1979]
NDB 15,588-590; KURSCHNER 12,1972; MH 1,307-308; STC 2,205-206; MSE 2,255-256; CB 1967, pp.263-265; BOHM, 149-150; ABBOTT-C, 88-89; POTSCH, 286; WWWS, 1083; POGG 7a(3),167-168
P. Karlson, Deutsche med. Wchschr. 90 (1965), 180-182
F. Lynen, Persp. Biol. Med. 12 (1969), 204-218; in Reflections on Biochemistry (A. Kornberg et al., Eds.), pp.151-160. Oxford 1975
H.G. Wood, TIBS 4 (1979), N300,N302
W. Wilmanns, Fort. Med. 97 (1979), 2191-2194
J.R. Grant, Nature 285 (1980), 177
K. Bloch and K. Decker, Max-Planck Ges. Ber. 1980(2), pp.21-40
K. Bloch et al., Naturw. Rund. 33 (1980), 213-232
F. Lüst et al., Feodor Lynen. Stuttgart 1980
T. Bücher, Bayer. Akad. Wiss. Jahrbuch 1980, pp.241-245
K. Decker, Erg. Physiol. 90 (1981), 1-11; Ann. Chem. 1984(9), pp.i-xl
K. Bloch, Am. Phil. Soc. Year Book 1981, pp.463-467
F. Lynen, Nova Acta Leop. NF55 (1982), 9-20
H. Krebs and K. Decker, Biog. Mem. Fell. Roy. Soc. 28 (1982), 261-317

LYONS, WILLIAM REGINALD [1901-]
AMS 11,3277, WWWS, 1084

LYSENKO, TROFIM DENISOVICH [1898-1976]
DSB 18,574-578; W (3rd Ed.),598-599; UKRAINE 2,328-329; ABBOTT, 89-90; WWWS, 1085
F.D. Skaskin and R.I. Lerman, T.D. Lysenko. Berlin 1952
T.D. Darlington, Nature 266 (1977), 287-288
V.P. Efroimson, Vop. Ist. Est. Tekhn. 44 (1989), 79-93
V.N. Soyfer, Nature 339 (1989), 415-420

LYTHGOE, RICHARD JAMES [1896-1940]
HUBER, 67-68
Anon., Lancet 1940(I), p.258; Brit. Med. J. 1940(I), p.640
Who Was Who 1929-1940, p.838

LYTTLETON, JOHN WESCOTE [1919-1975]
G.W. Butler, Proc. Roy. Soc. N.Z. 103 (1975), 111-114

LYUBIMENKO, VLADIMIR NIKOLAEVICH [1873-1937]
BSE 25,527; WWR, 361; UKRAINE 2,336-337; IB, 180
A.N. Danilov, Sov. Bot. 1937(6), pp.139-155
Anon., Chronica Botanica 4 (1938), 177
E.M. Senchenkova, in Life Phenomena: A Historical Survey, pp.116-171. Jerusalem 1966 [translation of Fizicheskie i Khimicheskie Osnovy Zhizennykh Yavlenii. Moscow 1963]

LYUBIMOVA, MILITSA NIKOLAEVNA [1898-1976]
see ENGELHARDT, VLADIMIR ALEKSANDROVICH

M

MAALØE, OLE [1914-1988]
KBB 1980, pp.783-784; IWW 1982-1983, p.809
N.O. Kjelgaard, TIBS 14 (1989), 51

MAAS, JOHANNA [1885-]
JV 24,576; NUC 348,141

MAAS, OTTO [1867-1916]
W. Roux, Arch. Entwickl. 42 (1917), 508-512

MAAS, OTTO [1871-1942?]
GEDENKBUCH 2,951; JV 14,249; RSC 16,934; NUC 348,142

MAAS, WERNER KARL [1921-]
AMS 15(5),2; BHDE 2,760

MAASS, THEODOR ALFRED [1877-]
LDGS, 22; NUC 348,166

MACADAM, STEVENSON [1829-1901]
RSC 4,145; 8,287; 16,937; NUC 348,260
Anon., J. Chem. Soc. 79 (1901), 897-898; J. Soc. Chem. Ind.20 (1901), 105; Proc. Roy. Soc. Edin. 24 (1901), 4

MACAIRE, ISAAC FRANÇOIS [1796-1869]
DHB 4,623; BRIQUET, 301-303; STELLING-MICHAUD 4,387; FERCHL, 331;POGG 2,1-2; 3,849; CALLISEN 12,29-30; 30,158-159

MACALLUM, ARCHIBALD BRUCE [1885-1976]
AMS 11,3282; YOUNG, 26; IB, 182
M.L. Barr, Proc. Roy. Soc. Canada [4]15 (1977), 97-100

MACALLUM, ARCHIBALD BYRON [1858-1934]
DSB 8,583-584; YOUNG, 5-7; ULLMANN, 97-98;

O'CONNOR(2), 511-512;WWWS, 1085; POTSCH, 286
J.B. L[eathes], Obit. Not. Fell. Roy. Soc. 1 (1934), 287-291
J.J.R.M., Nature 133 (1934), 711-712
Anon., Trans. Roy. Soc. Canada [3]28 (1934), xix-xxi
J.M. Neelin, Can. J. Biochem. 62(6) (1984), viii-xi

MacARTHUR, CHARLES GEORGE [1883-1959]
AMS 7,1107

McBAIN, JAMES WILLIAM [1882-1953]
JV 25,365; ABBOTT-C, 90; WWWS, 1137; POTSCH, 294-295; POGG 5,781-782; 6,1599-1600; 7b,3020-3024
E. Rideal, Obit. Not. Fell. Roy. Soc. 8 (1953), 529-547
H. Taylor, J. Chem. Soc. 1956, pp.1918-1920

MACBETH, ALEXANDER KILLEN [1889-1957]
POGG 5,782; 6,1600-1601; 7b,3024-3025
G.M. Badger and H.J. Rodda, Proc. Chem. Soc. 1958, pp.121-122

MacCALLUM, WILLIAM GEORGE [1874-1944]
DAB [Suppl.3],482-483; AMS 7,1110; NCAB 32,499; FISCHER 2,959; OLPP, 251;DAMB, 470; WWAH, 377; WWWS, 1086
S. Flexner, Science 99 (1944), 290-291
W.T. Longcope, Biog. Mem. Nat. Acad. Sci. 23 (1945), 339-364; Bull. Hist. Med. 18 (1945), 207-212
A.M. Harvey, Adventures in Medical Research, pp.24-31. Baltimore 1976

McCANCE, ROBERT ALEXANDER [1898-1993]
WW 1981, p.1617; IWW 1982-1983, p.811

McCANN, WILLIAM SHARP [1889-1971]
AMS 11,3291; NCAB 57,166-167; IB, 183; WWAH, 397; WWWS, 1138
L.E. Young, Trans. Assoc. Am. Phys. 85 (1972), 35-37

McCARTY, MACLYN [1911-]
AMS 15(5),28; WWA 1980-1981, p.2206; MH 2,355-356; STC 2,227; MSE 2,259-260; ICC, 412-413
M. McCarty, Ann. Rev. Gen. 14 (1980), 1-15; The Transforming Principle. New York 1985

McCAY, CLIVE MAINE [1898-1967]
AMS 11,3297
F.C. Bing, Persp. Biol. Med. 13 (1970), 563-581
L.A. Maynard, in C.M. McCay, Notes on the History of Nutrition Research, pp.9-12. Berne 1973
J.K. Loosli, J. Nutrition 103 (1973), 3-10

McCLEAN, DOUGLAS [1896-1967]
Who's Who in British Science 1953, p.175
Anon., Lancet 1967(II), pp.424-425

McCLENDON, JESSE FRANCIS [1880-1976]
AMS 11,3300; FISCHER 2,1737; IB, 183; WWWS, 1139; POGG 6,1601-1602; 7b,3026-3028
Anon., Chem. Eng. News 55(9) (1977), 31

McCLINTOCK, BARBARA [1902-]
AMS 14,3283; IWW 1982-1983, p.812; SCHULZ-SCHAEFFER, 212
E.F. Keller, A Feeling for the Organism. San Francisco 1983
B. Burr and F.A. Burr, TIBS 8 (1983), 3-10

McCLUNG, CLARENCE ERWIN [1870-1946]
DSB 8,586-590; DAB [Suppl.4],515-516; AMS 7,1114; FREUND 1,149-160; WWAH, 399; WWWS, 1139
D.H. Wenrich, J. Morphol. 66 (1940),635-688; Science 103 (1946), 551-552; Am. Nat. 80 (1946), 295-296; Stain Technology 21 (1946), 43-48;Am. Phil. Soc. Year Book 1946, pp.322-325
W.R. Green, Bios 17 (1946), 74-83

McCOLLUM, ELMER VERNER [1879-1967]
DSB 8,590-591; AMS 11,3303-3304; NCAB C,477-478; MILES, 323-324; IB, 183; CHITTENDEN, 344-350,365-368; STC 2,228-229; GOULD, 31-32; DAMB, 473; ABBOTT, 90-91; WWAH, 399; WWWS, 1140; POTSCH, 295
E.V. McCollum, Ann. Rev. Biochem. 22 (1953), 1-16; From Kansas Farm Boy to Scientist. Lawrence, Kan. 1964
G. Adams, Am. Phil. Soc. Year Book 1968, pp.152-156
H. Chick and R. Peters, Biog. Mem. Fell. Roy. Soc. 15 (1969), 159-171
A.A. Rider, J. Nutrition 100 (1970), 1-10
H.G. Day, Biog. Mem. Nat. Acad. Sci. 45 (1974), 263-335; Nutr. Revs. 37(3) (1979), 65-71
F.C. Bing, Chemistry 52(3) (1979), 5-7
D.R. Davis, Trans. Kansas Acad. Sci. 82 (1979), 133-145
R.M. Herriott, Fed. Proc. 39 (1980), 2713-2715
F.C. Bing and H. Perbluda, Agricultural History 54 (1980), 157-166
A.E. Harper, J. Agr. Food Chem. 29 (1981), 429-435
H. Perbluda, CHEM TECH 14 (1984), 266-268
G.W. Rafter, Persp. Biol. Med. 30 (1987), 527-534

McCOMBIE, HAMILTON [1880-1962]
POGG 5,783; 6,1602; 7b,3029
F.G. Mann, Proc. Chem. Soc. 1962, pp.122-123

McCOMBIE, JOHN TALBOT [1920-]
Who's Who of British Scientists 1971-1972, p.541

McCONKEY, ALFRED THEODORE [1861-1931]
WWWS, 1086
Anon., Lancet 1931(I), p.1213
G.F.P., Nature 127 (1931), 980-981
J.A.A., J. Path. Bact. 34 (1931), 697-699

McCONNAN, JAMES [1881-1916]
JV 19,202; NUC 349,358
A.W. Titherley, J. Chem. Soc. 111 (1917), 316-318

McCONNELL, HARDEN MARSDEN [1927-]
AMS 15(5),28; IWW 1982-1983, p.813; MH 2,356-358

McCORMICK, DONALD BRUCE [1932-]
AMS 15(5),31; WWWS, 1141

McCREA, EDWARD d'ARCY [1895-1940]
HUBER, 72-73; NUC 349,539
Anon., Lancet 1941(I), p.166; Brit. Med. J. 1941(I), pp.102-103

McCREA, THOMAS [1870-1935]
DAB [Suppl.1], 525; AMS 5,709; SRS, 51; WWWS, 1141
Anon., Lancet 1935(II), p.107; Brit. Med. J. 1935(II), p.91;
Trans. Coll. Phys. Phila. 3 (1935-1936), xv-xviii

McCRUDDEN, FRANCIS HENRY [1879-1962]
AMS 10,2548; IB, 183

McCULLAGH, DOUGLAS ROY [1903-1949]
AMS 8,1572; SRS, 63
Anon., J. Roy. Inst. Chem. 1949, pp.557-558

McCULLAGH, ERNEST PERRY [1901-]
WWWS, 1142

McDONALD, IAN WILBUR [1909-]
NUC 350,130
Who's Who in Australia 1977, p.701

MACDONALD, JAMES LESLIE AULD [1887-1952]
E.W. Shann, J. Chem. Soc. 1953, p.2192

MACDONALD, JOHN [1862-1933]
SRS, 13; JV 9,134; NUC 350,152
W. Johnson, Roll of the Graduates of the University of Aberdeen 1860-1900, p.307. Aberdeen 1906
Anon., Aberdeen University Review 20 (1933), 284

MACDONALD, JOHN SMYTH [1867-1941]
O'CONNOR(2), 335-336
H.S. Raper, Obit. Not. Fell. Roy. Soc. 3 (1941), 853-856

McDONALD (PRYTZ), MARGARET RITCHIE [1910-1988]
AMS 14,3304; WWWS, 1143

MacDONNELL, ROBERT [1828-1889]
DNB 12,503; HIRSCH 4,7
Anon., Lancet 1889(I), p.965; Brit. Med. J. 1889(I), p.1092

MacDOWELL, EDWIN CARLETON [1887-1973]
AMS 10,2560; IB, 184

MacDOUGAL, DANIEL TREMBLY [1865-1958]
NCAB 13,125-126; IB, 184; WWWS, 1087
D.T. MacDougal, Annals of Missouri Botanical Garden 19 (1932), 31-43
G.T. Moore et al., Plant Physiology 14 (1939), 191-202
Who Was Who in America 3, 541
S.E. Kingsland, Isis 82 (1991), 479-509

MACÉ, EUGÈNE [1856-1938]
DENOTH, 38
F. Jayle, Presse Med. 47 (1939), 75-76

McEACHERN, DONALD SNELL [1904-1951]
AMS 8,1578
W. Penfield, Arch. Neurol. 67 (1952), 267-268
D. Denny-Brown, Trans. Assoc. Am. Phys. 65 (1952), 33-34

McELLROY, WILLIAM SWINDLER [1893-1981]
AMS 11,3332

McELROY, WILLIAM DAVID [1917-]
AMS 15(5),55; WWA 1980-1981, p.2228; IWW 1982-1983, pp.817-818; WWWS,1144

McELVAIN, SAMUEL MARION [1897-1973]
AMS 12,3933; NCAB 58,264-265; POGG 6,1604; 7b,3035-3038
G. Stork, Biog. Mem. Nat. Acad. Sci. 54 (1983), 221-248

MACFADYEN, ALLAN [1860-1907]
DNB [1901-1911]2, 519-520; FISCHER 2,962; O'CONNOR(2), 452; BULLOCH, 382; DESMOND, 405; WWWS, 1088
H.R.T., J. Hygiene 7 (1907), 319-322; Nature 75 (1907), 443
Anon., Lancet 1907(I), pp.696-697; Brit. Med. J. 1907(I), p.601; Science 25 (1907), 635-636

McFARLAND, JOSEPH [1868-1945]
AMS 7,1127; WWAH, 402; WWWS, 1145
I.M. McMillan, Ann. Int. Med. 23 (1945), 913-914
I. Davidson, Arch. Path. 41 (1946), 338-340
S.P. Reimann, Trans. Coll. Phys. Phila. [4]14 (1946), 133-135

McFARLANE, ARTHUR SPROUL [1905-1978]
J.H. Humphrey, Nature 277 (1979), 335

MACFARLANE, ROBERT GWYN [1907-1987]
WW 1981, p.1639; WWWS, 1088
A.L. Copley, Biorheology 26 (1988), 11-14
G.V.R. Born and D.J. Weatherall, Biog. Mem. Fell. Roy. Soc. 35 (1990), 211-245

MACFARLANE, WALTER VICTOR [1913-1982]
A.K. McIntyre, Historical Records of Australian Science 6 (1985), 247-265

McFARLANE, WILLIAM DOUGLAS [1900-1975]
AMS 11,3337; YOUNG, 48-49

McGEE, LEMUEL CLYDE [1904-1975]
AMS 11,3340; DAMB, 479-480

McGREGOR, JAMES HOWARD [1872-1954]
AMS 8,1585; IB, 184; WWAH, 403; WWWS, 1146
L.C. Dunn, Am. Phil. Soc. Year Book 1955, pp.480-482
B. Schaeffer, News Bulletin of the Society of Vertebrate Paleontologists 43 (1955), 33-34

McGREGORY, JOSEPH FRANK [1855-1934]
AMS 5,717; RSC 16,75; NUC 351,168

MACH, FELIX [1868-1940]
KURSCHNER 4,1827; IB, 184; POGG 6,1605-1606

MACH, WILHELM von [1862-1923]
JV 3,166; RSC 16,961; SG [2]10,20; NUC 351,220
Genealogisches Handbuch des Adels 64 (1977), 301

MACHEBOEUF, MICHEL [1900-1953]
DSB 8,607-608; IB, 184; POTSCH, 287
Anon., Ann. Univ. Paris 20 (1950), 213-215
J. Polonovski, Bull. Soc. Chim. Biol. 35 (1953), 1279-1286; Bull. Acad. Med. 137 (1953), 494-499; Biochemie 56 (1974), 195-197
M. Javillier, Ann. Inst. Pasteur 85 (1953), 683-689
A. Chatelet, Ann. Univ. Paris 23 (1953), 707-709

McHENRY, EARLE WILLARD [1899-1961]
AMS 10,2576; YOUNG, 24-25
D.A. Scott, Proc. Roy. Soc. Canada [3]56 (1962), 219-224

MACHT, DAVID ISRAEL [1882-1961]
AMS 10,2576-2577; IB, 184; KOREN, 216
J.C. Krantz, Pharmacologist 4 (1962), 49
D. Wilk, Koroth 8 (1983), 305-317

McILWAIN, HENRY [1912-1992]
WW 1981, p.1647; IWW 1982-1983, p.822; WWWS, 1147
H.S. Bachelard, Neurochemistry International 14 (1989), 233-234

MacINNES, DUNCAN ARTHUR [1885-1965]
AMS 10,2578; MILES, 309-310; WWAH, 379; POGG 5,785; 6,1607-1608;7b,3040-3042
L.G. Longsworth and T. Shedlovsky, Am. Phil. Soc. Year Book 1967, pp.127-133; Biog. Mem. Nat. Acad. Sci. 41 (1970), 295-317

McINTOSH, JAMES [1882-1948]
Anon., Brit. Med. J. 1948(I), pp.759-760; Chem. Ind. 1948, p.255
E.C. Dodds, Nature 161 (1948), 713-714
P. Fildes, J. Path. Bact. 61 (1949), 285-299

MACINTYRE, ALFRED EDGAR [1863-1936]
AMS 5,719; JV 16,159; NUC 351,522

A. Tingle, J. Chem. Soc. 1936, pp.1573-1575
Anon., Ind. Eng. Chem. News Ed. 14 (1936), 136

MacINTYRE, WILLIAM [1792-1857]
MR 4,63
J.R. Clamp, Lancet 1967(II), pp.1354-1356

MacKAY, EATON MacLEOD [1900-1973]
AMS 9(II),723

McKENDRICK, JOHN GRAY [1841-1926]
O'CONNOR, 202-205; WWWS, 1193; POGG 4,939; 5,786; 6,1609
D.N.P., Proc. Roy. Soc. B100 (1926), xiv-xvii
R. Bayliss, Med. Hist. 17 (1973), 288-303

McKENZIE, ALEXANDER [1869-1951]
DNB [1951-1960], 671-672; POGG 5,786; 6,1609; 7b,3044-3045
D.H. Everett, Nature 168 (1951), 143-144
H. Wren, J. Chem. Soc. 1952, pp.270-271
J. Reed and R. Roger, Obit. Not. Fell. Roy. Soc. 8 (1952-1953), 207-228

MACKENZIE, COSMO GLENN [1907-]
AMS 14,3123; WWA 1976-1977, p.1981; WWWS, 1090

MACKENZIE, HECTOR [1856-1929]
WW 1929, p.952
Anon., Lancet 1929(I), p.524; Brit. Med. J. 1929(I), p.482

MACKENZIE, JAMES [1853-1925]
DNB [1922-1930], 543-544; FISCHER 2,965; O'CONNOR(2), 290-292; WWWS, 1090
A.K., Proc. Roy. Soc. B101 (1927), xxii-xxiv
A. Mair, Sir James Mackenzie. Edinburgh 1973

MacKENZIE, JOHN EDWIN ' [1868-1955]
POGG 4,939; 5,786; 6,1610; 7b,3045
J. Kendall, Roy. Soc. Edin. Year Book 1954-1955, pp.41-43
T.R. Bolan, J. Chem. Soc. 1955, p.3565

MACKENZIE, KENNETH SMITH [1832-1900]
The Times, 12 February 1900
F. Boase, Modern English Biography, Suppl.3, p.122. Truro 1921

McKIE, DOUGLAS [1896-1967]
DNB [1961-1970], 695-697; WW 1965, p.1950
W.A. Smeaton, Nature 216 (1967), 1053-1054
V.A. Eyles, Arch. Inter. Hist. Sci. 21 (1968), 303-305
E.H. Robinson, Isis 59 (1968), 319-327
H. Hartley, Notes Roy. Soc. 23 (1968), 101-103
Anon., Ann. Sci. 24 (1968), 1-5

McKINLEY, EARL BALDWIN [1894-1938]
AMS 5,723; WWAH, 404-405; WWWS, 1148
V. du Vigneaud, Science 88 (1938), 344-345
E.R. Long, Arch. Path. 26 (1938), 1085-1089
P. Lépine, Presse Med. 46 (1938), 1497-1498
M.H. Soule, Am. J. Trop. Med. 19 (1939), 97-101

MACKINNEY, GORDON [1905-1976]
AMS 13,2740; WWWS, 1090

McLEAN, FRANKLIN CHAMBERS [1888-1968]
AMS 11,3377; WWAH, 405; WWWS, 1149
F.C. McLean, Ann. Rev. Physiol. 22 (1960), 1-16
A.B. Hastings, Trans. Assoc. Am. Phys. 82 (1969), 40-42
A.M. Budy, Pharmacologist 11 (1969), 38-40
M.R. Urist, Persp. Biol. Med. 19 (1975), 23-58
W.L. Palmer, Persp. Biol. Med. 22 (1979), S2-S32

MacLEAN, HUGH [1879-1957]
O'CONNOR(2), 419-420; FISCHER 2,966; IB, 185
Anon., Lancet 1957(I), p.649

MacLEAN, IDA SMEDLEY [1877-1944]
POGG 6,1612; 7b,3051
Anon., Chemical Age 50 (1944), 339; J. Roy. Inst. Chem. 1944, p.74
M.A. Whiteley, J. Chem. Soc. 1946, pp.65-67
M.R.S. Creese, Brit. J. Hist. Sci. 24 (1991), 282-284

McLEAN, JAY [1890-1957]
AMS 9(II),727
J.H. Talbott, A Biographical History of Medicine, pp.928-929. New York 1970
C.R. Lam, Henry Ford Hospital Medical Journal 33 (1985), 18-23
J. Marcum, Persp. Biol. Med. 33 (1990), 214-230

MacLEOD, COLIN MUNRO [1909-1972]
DSB 18,587-589; AMS 11,3380; MH 2,340-341; STC 2,210; MSE 2,268; DAMB, 485-486; ICC, 326-328; WWAH, 380
M. McCarty, Am. Phil. Soc. Year Book 1972, pp.222-230; Trans. Assoc. Am. Phys. 85 (1972), 31-35; The Transforming Principle. New York 1985
R. Austrian, Journal of Infectious Diseases 127 (1973), 211-214
W. McDermott, Biog. Mem. Nat. Acad. Sci. 54 (1983), 183-219

McLEOD, JAMES WALTER [1887-1978]
DNB [1971-1980], 532-533; WW 1978, p.1576; IB, 185
Anon., Brit. Med. J. 1978(I), p.926; Lancet 1978(I), p.886
K. Zinnemann, Nature 274 (1978), 195-196
K.I. Johnstone, J. Gen. Microbiol. 109 (1978), 1-4
W. Graham and K. Zinnemann, Biog. Mem. Fell. Roy. Soc. 25 (1979),421-444
J. Howie, Brit. Med. J. 294 (1987), 1532-1533

MacLEOD, JOHN JAMES RICKARD [1876-1935]
DSB 8,614-615; DNB [1931-1940],585-586; FISCHER 2,967; W, 347; WWAH, 380; BROBECK, 140-141; O'CONNOR(2), 417-419; POTSCH,287; CHITTENDEN, 310-314; WWWS, 1091; POGG 6,1614-1615; 7b,3053-3054
Anon., Brit. Med. J. 1935(I), p.624; Chem. Ind. 1935, p.292
E.P. Cathcart, Obit. Not. Fell. Roy. Soc. 1 (1935), 585-589
J.B.C., Biochem. J. 29 (1935), 1253-1256
V.E. Henderson, Science 81 (1935), 355
J.J. Macleod, Bull. Hist. Med. 52 (1978), 295-312
L.G. Stevenson, TIBS 4 (1979), N158-N160
M. Bliss, Quart. J. Exp. Biol. 74 (1989), 87-96

McLESTER, JAMES SOMERVILLE [1877-1954]
AMS 8,1603; WWWS, 1150
R.M. Wilder, Science 120 (1954), 242-243
E.B. Carmichael, Bull. Hist. Med. 36 (1962), 141-147

McMANUS, IVY ROSABELLE [1923-1982]
AMS 15(5),106
Anon., Chem. Eng. News 61(5) (1983), 38

McMASTER, PHILIP DURYEE [1891-1973]
AMS 12,3991; DAMB, 486; WWAH, 406; WWWS, 1150
W.F. Goebel, Biog. Mem. Nat. Acad. Sci. 50 (1979), 287-308

McMEEKIN, THOMAS LEROY [1900-1979]
AMS 12,3992
J.T. Edsall, Nature 285 (1980), 58
Anon., Chem. Eng. News 58(5) (1980), 34

MacMUNN, CHARLES ALEXANDER [1852-1911]
O'CONNOR(2), 526-527
Anon., Lancet 1911(I), pp.551-552; Brit. Med. J. 1911(I), p.531
D. Keilin, History of Cell Respiration and Cytochrome, pp.355-357.Cambridge 1966

McMURTRIE, WILLIAM [1851-1913]
DAB 12,146-147; NCAB 12,206; MILES, 328; WWWS, 1150-1151
H.W. Wiley, J. Ind. Eng. Chem. 5 (1913), 616-618

MACNAIR, DUNCAN SCOTT [1861-1937]
JV 5,293; RSC 16,983-984
Anon., Chemical Age 37 (1937), 450
A.S. Macnair, J. Chem. Soc. 1938, p.160

McNEE, JOHN WILLIAM [1887-1984]
WW 1981, p.1681; IWW 1982-1983, p.829; AO 1984, pp.54-55
Anon., Lancet 1984(I), p.408; Brit. Med. J. 288 (1984), 294

MacNIDER, WILLIAM de BERNIERE [1881-1951]
AMS 7,1149; NCAB 40,153-154; DAMB, 487; PARASCANDOLA, 42-43; WWAH, 381; WWWS, 1092
W.C. George, Science 115 (1952), 489-490
A.N. Richards, Biog. Mem. Nat. Acad. Sci. 32 (1958), 238-272

McPHAIL, MURCHIE KILBURN [1907-]
AMS 14,3351; WWWS, 1151-1152

MacPHERSON, HERBERT TAYLOR [1909-]
Who's Who of British Scientists 1971-1972, p.557

MACRAE, THOMAS FOTHERINGHAM [1906-]
Who's Who in British Science 1953, p.179
Directory of British Scientists 1966-1967, vol.2, p.44

MacWILLIAM, JOHN ALEXANDER [1857-1937]
FISCHER 2,969; O'CONNOR(2), 415-417; WWWS, 1093; NUC 354,92
Who Was Who 1929-1940, p.888
A. Keith, Obit. Not. Fell. Roy. Soc. 2 (1938), 335-338

MADDEN, SIDNEY CLARENCE [1907-]
AMS 14,3133; WWA 1978-1979, p.2049; NCAB I,316-317; WWWS, 1093-1094

MADDOCK, STEVEN JAMES [1897-1967]
Anon., J. Am. Med. Assn. 199 (1967), 1021

MADELUNG, WALTHER [1879-1963]
KURSCHNER 10,1498; TLK 3,418-419; POGG 6,1617; 7a(3),177
R. Wizinger, Nachr. Chem. Techn. 2 (1954), 134
Anon., Chem. Z. 87 (1963), 298; Nachr. Chem. Techn. 11 (1963), 107

MADINAVEITIA-TABUYO, ANTONIO [1888-]
NUC 354,298; POGG 6,1618; 7b,3062

MADSEN, THORVALD [1870-1957]
DBL (3rd Ed.) 9,331-335; FISCHER 2,970; VEIBEL 2,295-296; WWWS, 1094; POGG 4,941; 5,790; 6,1618-1619; 7b,3062-3063
Anon., Acta Path. Scand., Suppl.3 (1950), 555-561
J. Parnas, Dan. Med. Bull. 28 (1981), 82-86

MÄDER, HORST [1885-1970]
JV 28,236; NUC 354,536
Anon., Nachr. Chem. Techn. 13 (1965), 182; 18 (1970), 179,346

MAEDA, MINORU [1891-]
HAKUSHI 4,855-856

MÄHLY-EGLINGER, JACOB [1850-1920]
RSC 15,136; NUC 354,549
F. Baur, Basler Bibliographie 1921, p.259

MAENGWYN-DAVIES, GERTRUDE DIANE [1910-1984]
AMS 15(5),130; BHDE 2,762

MAERCKER, HEINRICH MAXIMILIAN [1842-1901]
NDB 15,639-640; BJN 6,303-309,68*-69*; BLOKH 1,479-480; POGG 3,854; 4,941-942; 7aSuppl.,409
P. Behrend, Naturw. Rund. 16 (1901),658-659; Land. Jahrb. 31 (1902),1-54
M. Delbrück, Ber. chem. Ges. 34 (1902), 4457-4465
E. Schulze, Land. Vers. 56 (1902), 265-275

MAGASANIK, BORIS [1918-]
AMS 15(5),131; WWA 1980-1981, p.2116

MAGAT, MICHEL [1908-1978]
QQ 1977-1978, p.1096; LDGS, 22; WWWS, 1094-1095

MAGENDIE, FRANÇOIS [1783-1855]
DSB 9,6-11; GENTY 4,113-144; DECHAMBRE 55,681-686; LAROUSSE 10,913; HIRSCH 4,28-29; PAGEL, 1075-1076; HAYMAKER, 237-240; FERCHL, 333; BRAZIER, 42-47,60-61; W, 348-349; CALLISEN 12,104-122; 30,184-191
J.M.D. Olmsted, Francois Magendie. New York 1944
L. Deloyers, Francois Magendie. Brussels 1970
W.R. Albury, Stud. Hist. Biol. 1 (1977), 47-131
C. Lichtenthaler, Z. ärzt. Fortbild. 77 (1983), 785-788
J. Théodoridès et al., Hist. Sci. Med. 17 (1983), 321-380

MAGIDSON, ONISIM YULIEVICH [1890-1971]
WWWS, 1095; POGG 6,1620; 7b,3064-3067

MAGNUS, ALFRED [1880-1960]
TLK 3,419; POGG 5,792; 6,1621; 7a(3),179-180; (4),149*
Anon., Chem. Z. 84 (1960), 751

MAGNUS, GUSTAV [1802-1870]
DSB 9,18-19; NDB 15,673-674; ADB 20,77-90; PRANDTL, 303-314; FERCHL, 333; BLOKH 1,474-476; SCHAEDLER, 75; MATSCHOSS, 166-167; KOHUT 2,232-233;WININGER 4,222-223; POGG 2,14-15; 3,856; 6,1621; 7aSuppl.,410-411
A.W. Hofmann, Ber. chem. Ges. 3 (1870), 993-1101
A.W. Williamson, J. Chem. Soc. 24 (1871), 610-615
F. von Kobell, Sitz. Bayer. Akad. 1 (1871), 148-151
E.H. Huntress, Proc. Am. Acad. Arts Sci. 81 (1952), 70-71

MAGNUS, HUGO [1841-1907]
FISCHER 2,972-973; WININGER 4,223-224; 7,290,587

MAGNUS, PREBEN von [1912-1973]
DBL (3rd Ed.)9,358-359; KBB 1973, pp.882-883
Anon., Lancet 1974(I), p.35

MAGNUS, RUDOLF [1873-1927]
DSB 9,19-21; LINDEBOOM, 1261-1263; HES, 100-101; FISCHER 2,973; IB, 186; DRULL, 169-170; HAYMAKER, 240-243; KOREN, 216; WININGER 7,291
W. Heubner, Klin. Wchschr. 6 (1927), 2022-2024; Arch. exp. Path. Pharm.128 (1928), 16-23
W. Straub, Münch. med. Wchschr. 74 (1927), 1676-1677
G. Liljestrand, Erg. Physiol. 29 (1929), 646-654
U.G. Bijlsma, Acta Physiol. Neerl. 15 (1969), 111-116

MAGNUS, VILHELM [1871-1929]
NBL 9,11-12
H.H. Simmer, Sudhoffs Arch. 56 (1972), 76-99

MAGNUS, WERNER [1876-1942]
KURSCHNER 4,1832; ASEN, 123; LDGS, 12; IB, 186

MAGNUS-ALSLEBEN, ERNST [1879-1936]
KURSCHNER 4,1832; FISCHER 2,973; ZWEIFEL, 171; LDGS, 67; WIDMANN, 275
Anon., Deutsche med. Wchschr. 62 (1936), 2104
M. Meyer, Anadoglu Klinigi (Istanbul) 5 (1937), 101

MAGNUS-LEVY, ADOLF [1865-1955]
DSB 18,589-592; KURSCHNER 4,1832; FISCHER 2,973-974; LDGS,67; KOREN, 216; WININGER 4,225; WWWS, 1096
G. Haller, Deutsche med. Wchschr. 80 (1955), 729
L. Zadek, Münch. med. Wchschr. 97 (1955), 834-835
M.G. Goldner, Proc. Virchow Med. Soc. 14 (1955), 29-35; Diabetes 4 (1955), 422-424

MAGNUSSON, STAFFAN [1933-1990]
VAD 1990, p.760
H. Jornvall et al., TIBS 16 (1991), 3-4
L. Sottrup-Jennsen, Thrombosis and Haemostasis 65 (1991), 2

MAGOUN, HORACE WINCHELL [1907-1991]
AMS 17(5),140; WWWS, 1096

MAGROU, JOSEPH [1883-1951]
WWWS, 1096
Anon., Ann. Inst. Pasteur 80 (1951), 327-331,545-546
L. Plantefol, Presse Med. 59 (1951), 514
R. Leriche, C.R. Soc. Biol. 145 (1951), 213
F. Mariat and G. Secretain, Bull. Soc. Bot. (Mem.) 99 (1952), 177-181

MAHDIHASSAN, SYED [1892-]
JV 54,240; NUC 355,598

MAHLER, HENRY RALPH [1921-1983]
AMS 15(5),139; BHDE 2,763; WWWS, 1096
C.W. Cotman, J. Neurochem. 42 (1984), 1200-1202

MAI, KARL [1865-]
JV 8,216; RSC 16,377
Anon., Chem. Z. 44 (1920), 608; 46 (1922), 771; 67 (1943), 188

MAI, LUDWIG [1874-]
JV 11,287; TLK 3,419; NUC 356,24

MAIER, RUDOLF [1824-1888]
HIRSCH 4,37; WWWS, 1097
E. Ziegler, Beitr. path. Anat. 4 (1889), 473

MAIER-LEIBNITZ, HEINZ [1911-]
KURSCHNER 13,2426; POGG 7a(3),185
Anon., Leopoldina [3]10 (1964), 35-37

MAILHE, ALPHONSE [1872-1932]
CHARLE(2), 190-191; POGG 5,793; 6,1624-1626; 7b,3073
F. Caujolle, Bull. Soc. Chim. [4]53 (1933), 341-362

MAILLARD, LOUIS CAMILLE [1878-1936]
POGG 5,793-794; 6,1626; 7b,3073
C. Achard, C.R. Soc. Biol. 122 (1936), 347-348
C. Sannié, Bull. Soc. Chim. [5]4 (1937), 21-22

MAIMERI, CARLO [1886-]
NUC 356,175

MAIR, LEOPOLD [1885-]
JV 27,651; NUC 356,408

MAISCH, JOHN (JOHANN) MICHAEL [1831-1893]
DAB 12,213-214; HEIN, 394-395; TSCHIRCH, 1084; SILLIMAN, 126-128; WWAH, 382; POGG 3,859
J.F. Remington, Am. J. Pharm. 66 (1894), 1-9
C.S. Dodley, Proc. Am. Phil. Soc. 33 (1894), 345-352
M.I. Wilbert, Am. J. Pharm. 75 (1903), 351-377

MAITLAND, HUGH BETHUNE [1895-1972]
WW 1972, p.2083
A.W. Downie, J. Med. Microbiol. 6 (1973), 253-258

MAITLAND, PETER [1903-]
Who's Who of British Scientists 1971-1972, p.561

MAIXNER, EMERICH [1847-1920]
FISCHER 2,975; NAVRATIL, 178-179; WERSTLER, 149-158

MAIZELS, MONTAGUE [1899-1976]
WW 1976, p.1554
R.G. Macfarlane, Biog. Mem. Fell. Roy. Soc. 23 (1977), 345-366

MAJIMA, RIKO [1874-1962]
IB, 187; POGG 5,796; 6,1627; 7b,3073-3074
M. Kotake et al., Kagaku no Ryoiki 16 (1962), 790-792

MAJOR, RANDOLPH HERMON [1884-1970]
AMS 11,3432; FISCHER 2,976; DAMB, 490-491; IB, 187
W.S. Middleton, Persp. Biol. Med. 14 (1971), 651-658
C.S. Keefer, Trans. Assoc. Am. Phys. 84 (1971), 28-29

MAKINO, SAJIRO [1906-]
JBE, 780; WWWS, 1099

MAKSIMOV, NIKOLAI ALEKSANDROVICH [1880-1952]
DSB 9,43-45; BSE 26,124-125; WWR, 363-364; IB, 193
S.D. Lvov. Bot. Zhur. 35 (1950), 544-555; 37 (1952), 733-737

MALACHOWSKI, ROMAN [1887-1944]
POGG 6,1628; 7b,3079-3080
Z. Jerzmanowska, Roczniki Chemii 24 (1950), 24-42
R.D. Lillie, Stain Technology 53 (1978), 23-28

MALAGUTI, FAUSTINO [1802-1878]
DFI, 82-83; PROVENZAL, 153-160; BLOKH 1,476-477; FERCHL, 334-335; TSCHIRCH, 1084; SCHAEDLER, 75-76; WWWS, 1099; POGG 3,859
Anon., J. Chem. Soc. 35 (1879), 266-267

MALAPRADE, LÉON [1903-]
WWWS, 1099; POGG 6,1628; 7b,3080

MALASSEZ, LOUIS CHARLES [1842-1909]
FISCHER 2,976; VERSO, 96; WWWS, 1099
L. Labbé, Bull. Acad. Med. 62 (1909), 604-605
J. Jolly, C.R. Soc. Biol. 68 (1910),1-18; Fol. Haemat. 9 (1910),101-104
M.L. Verso, Med. Hist. 15 (1971), 58

MALCOLM, JOHN [1873-1954]
O'CONNOR(2), 509-510; IB, 187
Who's Who in British Science 1953, p.181
M. Bell, New Zealand Medical Journal 53 (1954), 436-439

MALENGREAU, FERNAND [1880-1958]
FISCHER 2,977; IB, 187
Anon., Bibl. Acad. Louvain 6 (1937), 355-356; Bull.

Acad. Med. Belg.23 (1958), 89-94

MALENYUK, VASILI DIOMEDOVICH [1867-]
VENGEROV 4,131

MALFATTI, HANS [1864-1945]
KURSCHNER 6(II),117; HUTER, 239; TLK 3,421

MALKIN, THOMAS [1897-1961]
POGG 6,1630; 7b,3082-3083
W. Baker and T.P. Hilditch, Proc. Chem. Soc. 1962, pp.161-163

MALL, FRANKLIN PAINE [1862-1917]
DSB 9,55-58; DAB 12,220-221; NCAB 14,309; FISCHER 2,978; WWWS, 1100
F.R. Sabin, Biog. Mem. Nat. Acad. Sci. 16 (1934), 65-122; Franklin Paine Mall, the Story of a Mind. Baltimore 1934

MALLARD, ERNEST [1833-1894]
DSB 9,58-60; RSC 16,1028; WWWS, 1100; POGG 3,860-861; 4,948
M. Loewy, C.R. Acad. Sci. 119 (1894), 1042-1044
P. Termier, Bulletin de la Société Géologique de France [3]23 (1895), 179-191

MALLET, JOHN WILLIAM [1832-1912]
DAB 12,223-224; MILES, 311-312; HIRSCH 4,46; ELLIOTT, 167; WWAH, 383; SILLIMAN, 108-110; POGG 3,862; 4,948-949; 5,797
F.P. Dunnington, J. Am. Chem. Soc. 35 (1913), 40-44; J. Chem. Ed.5 (1928), 183-188
D. Reilly, J. Chem. Ed. 25 (1948), 634-636; Endeavour 12 (1953), 48-51
O. Eisenschiml, Chem. Eng. News 29 (1951), 110-111

MALLINCKRODT, GUSTAV [1829-1904]
Genealogisches Handbuch des Adels 34 (1965), 188

MALLISON, HEINRICH [1886-1959]
NDB 15,936-737; TLK 3,421; POGG 6,1631; 7a(3),188-189
Anon., Chem. Z. 75 (1951), 607; 80 (1956), 757; 83 (1959), 521;
Nachr. Chem. Techn. 7 (1959), 267,294

MALLORY, FRANK BURR [1862-1941]
DAB [Suppl.3],502-503; FISCHER 2, 978; FRANKEN, 190; DAMB, 493; IB, 187;
SG [4]10,204; WWWS, 1101
F. Parker, Science 94 (1941), 430-431
S.R. Haythorn, J. Path. Bact. 54 (1942), 263-267
T. Leary, New Eng. J. Med. 226 (1942), 279-283
W. Freeman, New Eng. J. Med. 231 (1944), 824-828

MALMROS, HAQVIN [1895-]
VAD 1987, p.754; SMK 5,213; NUC 357,608

MALMSTROM, BO G. [1911-]
VAD 1981, p.711

MALORNY, GÜNTHER [1912-1978]
KURSCHNER 12,1990; KIEL, 106; WWWS, 1101; POGG 7a(3),189-190
Anon., Chem. Z. 102 (1978), 240; Deutsche med. Wchschr. 103 (1978),798 Münch. med. Wchschr. 120 (1978), 876

MALOWAN, LAWRENCE SIEGFRIED [1892-]
AMS 11,3433; BHDE 2,765; POGG 6,1632; 7a(3),190

MALTESOS, CHRISTOS [1908-]
NUC 358,67
Ellenikon Who's Who 1962, pp.310-311

MALUS, ÉTIENNE LOUIS [1775-1812]
DSB 9,72-74; HOEFER 33,117-121; BLOKH 1,478; WWWS, 1102; POGG 2,30
E. Fränkel, Centaurus 18 (1974), 223-245

MALY, RICHARD [1839-1891]
NDB 16,1; OBL 6,43-44; STURM 2,557; HIRSCH 4,49; PAGEL, 1973-1974; HUTER, 236-237; KERNBAUER, 42-50; BLOKH 1,478; SCHAEDLER, 76; POGG 3,863-964; 4,949
F. Emich, Ber. chem. Ges. 24 (1891), 1079-1088
E. Suess, Alm. Akad. Wiss. Wien 41 (1891), 185-187

MAMPEL, JULIUS [1883-1915]
JV 22,300
Anon., Z. angew. Chem. 28(III) (1915), 112; 29(III) (1916), 284

MANASSE, OTTO [1861-1942]
GEDENKBUCH 2,961; RSC 16,1034; NUC 358,249; POGG 4,949-950

MANASSE, PAUL [1866-1927]
FISCHER 2,979-980; GRAML, 140-149; IB, 188; RSC 16,1034-1035; KOREN, 217; WININGER 4,249; 7,292
W. Kümmel, Deutsche med. Wchschr. 53 (1927), 2041
O. Mayer, Mon. Ohren. 61 (1927), 1149-1152

MANASSE, WILHELM [1881-]
JV 22,29; NUC 358,249

MANASSEIN, MARIE MIKHAILOVNA [1843-1903]
RSC 16,1035
C.I. Vinokurov and R.V. Chagovets, Biokhimia 15 (1950), 558-562

MANASSEIN, VYACHESLAV AVKSENTIEVICH [1841-1901]
BSE 26,196; BME 16,815-816; KUZNETSOV(1963), 534-541; HIRSCH 4,51-52; PAGEL, 1084; KOREN, 217
H. Loeventhal, Berl. klin. Wchschr. 38 (1901), 328; Gaz. Med. Paris [12]1 (1901), 101
A. Kallmeyer, Deutsche med. Wchschr. 27 (1901), 302-304
L.N. Karlik, Klin. Med. 40 (1962), 140-146

MANCHÉ, EDUARD [1860-]
JV 3,194; RSC 16,1035; SG [2]10,99; MUC 358,274

MANCHOT, WILHELM [1869-1945]
NDB 16,7-8; TLK 3,421; POGG 4,950; 5.800; 6,1635; 7a(3),192
Anon., Chem. Z. 63 (1939), 529
W. Hieber, Bayer. Akad. Wiss. Jahrbuch 1944-1948, pp.214-216

MANCK, PHILIPP [1872-1929]
JV 10,213; RSC 17,757

MANDEL, ARTHUR RUDOLF [1876-1937]
AMS 5,735; KOREN, 217
Anon., Science 85 (1937), 281

MANDEL, JOHN ALFRED [1865-1929]
AMS 4,638; CHITTENDEN, 174-175; IB, 188; WWAH, 383; POGG 6,1636
C. Neuberg, Ber. chem. Ges. 62A (1929), 80-82; Biochem. Z. 216 (1929),242
W.C. McTavish, Science 70 (1929), 29-30
Anon., Ind. Eng. Chem. News Ed. 7(11) (1929), 10

MANDELSTAM, JOEL [1919-]
WW 1981, p.1707

MANDL, INES [1917-]
AMS 15(5),165; BHDE 2,767

MANDL, LOUIS LAZAR [1812-1881]
ADB 20,178; WURZBACH 16,364-367; HIRSCH 4,52-53; KOREN, 217
I. Csillag, Rev. Hist. Med. Heb. 31(4) (1978), 79-84

MANEGOLD, ERICH [1895-1972]
KURSCHNER 11,1853; POGG 6,1637; 7a(3),193
Anon., Chem. Z. 96 (1972), 293

MANGELSDORF, PAUL CHRISTOPH [1899-1989]
AMS 14,3168; MH 1,3140315; MSE 2,2730274; IB, 188; WWWS, 1103
New York Times, 28 July 1989
G. Wilkes, Journal of Ethnobiology 9 (1989), 257-263
K.V. Thimann and W.C. Galinat, Proc. Am. Phil. Soc. 135 (1991), 469-472

MANGIN, LOUIS ALEXANDRE [1852-1937]
DSB 9,78-79; WWWS, 1104
M.H. Jumelle, Rev. Gen. Bot. 36 (1924), 289-307
J.R., Nature 139 (1937), 828-829
R. Heim, Bull. Soc. Mycol. 54 (1938), 11-22
H. Colin, Not. Acad. Sci. 2 (1949), 256-281

MANGOLD, ERNST [1879-1961]
NDB 16,29-30; FISCHER 2,981; GIESE, 500; IB, 188; KOREN, 217; POGG 7a(3),193-195
J. Weitz, Voeding 25 (1964), 406-407

MANGOLD, FRIEDRICH WILHELM [1827-1898]
DGB 69,331

MANGOLD, HILDE PRÖSCHOLDT [1898-1924]
V. Hamburger, J. Hist. Biol. 17 (1984), 1-11

MANGOLD, OTTO [1891-1962]
KURSCHNER 9,1263; IB, 188

MANN, FRANK CHARLES [1887-1962]
AMS 10,2465; FISCHER 2,982; CHITTENDEN, 184-185; DAMB, 494; WWAH, 384
F.C. Mann, Ann. Rev. Physiol. 17 (1955), 1-16
J.L. Bollman, Gastroenterology 44 (1963), 700-702
Anon., Physiologist 6 (1963), 66-69
R.G. Sprague, Trans. Assoc. Am. Phys. 77 (1964), 22-26
M.B. Visscher, Biog. Mem. Nat. Acad. Sci. 38 (1965), 161-204
S. Sterioff and N. Rucker-Johnson, Mayo Clin. Proc. 62 (1987), 1051-1055

MANN, FREDERICK GEORGE [1897-1982]
WW 1981, p.1709; AO 1982, pp.150-151; WWWS, 1104; POGG 6,1104; 7b,3099-3104
I.T. Miller, Biog. Mem. Fell. Roy. Soc. 30 (1984), 409-441

MANN, GUSTAV [1864-1921]
O'CONNOR(2), 86; FISCHER 2,982; WWAH, 384
Anon., Brit. Med. J. 1921(II), pp.465-466
L. Arnold, Stain Technology 2 (1927), 4-7

MANN, THADDEUS [1908-1993]
WW 1981, p.1710; IWW 1982-1983, p.846; WWWS, 1105

MANNERS, DAVID JOHN [1928-]
WWWS, 1105
Who's Who of British Scientists 1980-1981, p.323

MANNHEIM, EMIL [1870-1924]
HEIN, 397; TSCHIRCH, 1084; TLK 3,422
G. Frerichs, Apoth. Z. 39 (1924), 998-999

MANNICH, CARL [1877-1947]
NDB 16,71-73; HEIN, 397-398; AUERBACH, 862; TSCHIRCH, 1084; TLK 3,422; DRUM, 284-286; POTSCH, 288-289; WWWS, 1105; POGG 5,801-802; 6,1640; 7a(3),196
K.W. Merz, Naturw. Rund. 1 (1948), 184-185
H. Böhme, Chem. Ber. 88 (1955), i-xxvi; Pharm. Z. 137 (1992), 43-44
W. Schneider, Pharm. Z. 118 (1972), 295-297
C. Friedrich, Beiträge zur Geschichte der Pharmazie 40 (1988), 33-36
B. Göber, Wiss. Z. Berlin 39 (1990), 238-246

MANROSS, NEWTON SPAULDING [1828-1862]
RSC 4,250
Anon., Am. J. Sci. 84 (1862), 452

MANSFELD, GÉZA [1882-1950]
MEL 2,135; FISCHER 2,984
B. Berde, Z. Vit. Forsch. 3 (1950), 426-428
Anon., Leopoldina [3]1 (1955), 15

MANSFIELD, CHARLES BLACHFORD [1819-1855]
DNB 12,975-976; POTSCH, 289; POGG 2,36
Anon., J. Chem. Soc. 8 (1855), 110-112
E.R. Ward, Chem. Brit. 15 (1979), 297-304; Ambix 31 (1984), 68-69

MANSKE, RICHARD HELMUTH FREDERICK [1901-1977]
AMS 13,2785; SRS, 61; POGG 6,1640; 7b,3106-3109
Anon., Chem. Ind. 1961, p.1421; Chemistry in Canada 15 (1963), 17
M. Kulka, Proc. Roy. Soc. Canada [4]16 (1978), 91-94
D.B. MacLean, The Alkaloids 17 (1979), xi-xiii

MANSON, PATRICK [1844-1922]
DSB 9,81-83; DNB [1922-1930], 560-562; FISCHER 2,984-985; OLPP, 262-269; W, 351-352; WWWS, 1106
A.A., Nature 109 (1922), 587-588
J. Cantlie et al., J. Trop. Med. 25 (1922), 101-102,115-117,127-129,155-208
E.J. Wood, Am. J. Trop. Med. 2 (1922), 361-368
A.W.A., Proc. Roy. Soc. B94 (1923), xliii-xlviii
P. Manson-Bahr and A. Alcock, The Life and Work of Sir Patrick Manson.London 1927
R. Ross, Memories of Sir Patrick Manson. London 1930
R.L. Pittfield, Ann. Med. Hist. 2 (1940), 22-29
P.H. Manson-Bahr, Proc. Roy. Soc. Med. 54 (1961), 91-100;
Patrick Manson: The Father of Tropical Medicine. London 1962
E. Chernin, Med. Hist. 36 (1992), 320-331

MANSON-BAHR, PHILIP HENRY [1881-1966]
WW 1965, p.2019; FISCHER 2,985; OLPP, 269-271; IB, 189
Anon., Lancet 1966(II), pp.1198-1199; Brit. Med. J. 1966(II), p.1332

MANTEGAZZA, PAOLO [1831-1910]
DSB 9,85-86; EI 22,159-160; HIRSCH 4,62; PAGEL, 1087-1088; WWWS, 1106
F. Starr, Pop. Sci. Mon. 41 (1892), 64-65
E. Ehrenfreund, Arch. Antr. Etnol. 56 (1926), 11-176
G. Bizzarini, Riv. Storia Sci. Med. 29 (1938), 148-152

MANTEUFEL, PAUL [1879-1941]
KURSCHNER 4,1851-1852; FISCHER 2,985; OLPP, 272;

IB, 189
H. Reiter, Reichsgesundheitsblatt 16 (1941), 87

MANTON, IRENE [1904-1988]
WW 1987, p.1159
R.D. Preston, Biog. Mem. Fell. Roy. Soc. 35 (1990), 249-261

MANWARING, WILFRID HAMILTON [1871-1960]
AMS 9(II),740; FISCHER 2,985-986; IB, 189
Anon., J. Am. Med. Assn. 174 (1960), 428

MANZ, HERMANN [1888-]
JV 28,318; NUC 360,208
Anon., Nachr. Chem. Techn. 11 (1963), 48

MAPES, CHARLES VICTOR [1836-1916]
NCAB 3,178; ELLIOTT, 167-168; WWAH, 385

MAPES, JAMES JAY [1806-1866]
DAB 12,264; MILES, 314-315; NCAB 3,178; ELLIOTT, 168; WWAH. 385

MAPSON, LESLIE WILLIAM [1907-1970]
DNB [1961-1970], 721-722; POTSCH, 289
R. Hill, Biog. Mem. Fell. Roy. Soc. 18 (1972), 427-444

MAQUENNE, LÉON [1853-1925]
WWWS, 1107; POGG 4,954-955; 5,803; 6,1641
C. Moureu, Bull. Soc. Chim. [4]37 (1925), 358-360
M. Bridel, Bull. Soc. Chim. Biol. 9 (1927), 624-628
G. André, Not. Acad. Sci. 1 (1937), 67-90

MARASSE, SIEGFRIED [1844-1896]
BLOKH 1,478-479; BERLIN, 683; RSC 8,323; NUC 360,445
H. Landolt, Ber. chem. Ges. 29 (1896), 2669

MARBURG, EDUARD CARL [1874-1925]
JV 14,226; RSC 15,897
Anon., Chem. Z. 49 (1925), 778; Z. angew. Chem. 38 (1925), 976

MARBURG, OTTO [1874-1948]
NDB 16,105-106; OBL 6,68; STURM 2,569; FISCHER 2,987; POINTNER, 147-148 BHDE 2,776; KOREN, 217; KAGAN, 69; WININGER 7,294-295; TETZLAFF, 224-225; WWWS, 1107-1108
O. Kauders, Wiener klin. Wchschr. 60 (1948), 461-462
E. Stransky, Wiener med. Wchschr. 98 (1948), 375

MARCELIN, RENÉ [1885-1914]
K.J. Laidler, J. Chem. Ed. 62 (1985), 1012-1014
K.J. Mysels, J. Chem. Ed. 63 (1986), 740

MARCET, ALEXANDER [1770-1822]
DNB 12,1007; HIRSCH 4,69-70; BLOKH 1,479; FERCHL, 337; WWWS, 1100; BOSSART, 111-112; WILKS, 207-208; CALLISEN 30,224; POGG 2,40
Anon., Am. J. Sci. 17 (1829-1830), 363-364
N.G. Coley, Med. Hist. 12 (1968), 394-402

MARCET, FRANÇOIS [1803-1883]
BRIQUET, 304-305; STELLING-MICHAUD 4,428; GENEVA 3,8-9; 4,22-26; FERCHL, 338; CALLISEN 12,195-196; 30,224; POGG 4,41; 3,868
C. Cellérier, Mem. Soc. Phys. Hist. Nat. Gen. 28 (1883), ii-vi

MARCET, JANE HALDIMAND [1785-1858]
DNB 12,1007-1008; OGILVIE, 125-127; WWWS, 1108; NUC 360,608-616
E.V. Armstrong, J. Chem. Ed. 15 (1938), 53-57
J.K. Crellin, J. Chem. Ed. 56 (1979), 459-460
M.S. Lindee, Isis 82 (1991), 8-23

MARCET, WILLIAM [1828-1900]
PICOT, 97-99; BOSSART, 114-115; STELLING-MICHAUD 4,429; WWWS, 1108;O'CONNOR, 157-159
Anon., J. Chem. Soc. 77 (1900), 594-595
C. Picot, Verhandl. Schw. Nat. Ges. 84 (1901), viii-xiii
E.A.S., Proc. Roy. Soc. 75 (1905), 165-169

MARCHAL, ÉMILE [1871-1954]
IB, 189; WWWS, 1108
Anon., Bulletin de l'Institut Agronomique de Gembloux 23 (1955), 3-12

MARCHAL, PAUL [1862-1942]
IB, 189; WWWS, 1108
P. Vayssière, Ann. Soc. Entomol. 111 (1942), 149-165; Ann. Inst. Agron. 33 (1942), 5-33
L. Fage, Not. Acad. Sci. 2 (1949), 446-466
M. Caullery, Proc. Linn. Soc. 162 (1949-1950), 106-108
C. Perez, C.R. Acad. Sci. 214 (1952), 449-452

MARCHAND, FELIX [1846-1928]
DBJ 10,167-172,331; GROTE 1,59-104; FISCHER 2,987; PAGEL, 1091-1092; FRANKEN, 190; IB, 189; WWWS, 1108
E. Korschelt, Naturwiss. 14 (1926), 953-956
W. Hueck, Verhandl. path. Ges. 23 (1928), 533-545; Sitz. Akad. Wiss. Leipzig 80 (1928), 337-356
A. Hecht, Z. allgem. Path. 122 (1978), 553-559

MARCHAND, JEAN [1816-1895]
TSCHIRCH, 1084; FERCHL, 338; POGG 3,868

MARCHAND, RICHARD FELIX [1813-1850]
DSB 9,91-92; ADB 20,296; VOIGT 28.452-453; HIRSCH 4,71; BLOKH 1,479;
PHILIPPE, 771-772; TSCHIRCH, 1085; FERCHL, 338; SCHAEDLER, 77;
POTSCH, 289; CALLISEN 30,225; POGG 2,41-43

MARCHI, VITTORIO [1851-1908]
DSB 9,93-94; EI 22,242; HAYMAKER, 131-133
L. Luciani, Arch. Ital. Biol. 49 (1908), 149-152

MARCHIAFAVA, ETTORE [1847-1935]
DSB 9,94-95; EI 22,242-243; FISCHER 2,988; PAGEL, 1092; OLPP, 272-273
P. Verga, Riforma Medica 51 (1935), 1736-1737
G. Bomponi, Pathologica 28 (1936), 93-99
G. Marchiafava, Sci. Med. Ital. 7 (1958), 157-198

MARCHIONINI, ALFRED [1899-1965]
NDB 16,114-115; KURSCHNER 10,1514-1515; KROLLMANN, 1013; BHDE 2,776; WIDMANN, 154,275
A. Goetz, Münch. med. Wchschr. 118 (1976), 341-342

MARCHLEWSKI, LEON [1869-1946]
DSB 9,95-96; PSB 19,542-545; OBL 6,71-72; KONOPKA 5,576-580; POTSCH, 289-290; URICH-D,166; POGG 4,956-957; 5,804; 6,1645-1646; 7b,3118-3119
B. Starzynski, Nature 157 (1946), 650-651; Roczniki Chemii 22 (1948),1-18; Szeczsetlecie Medycyny Krakowskie, pp.315-333. Cracow 1963
W. Ostrowski, Fol. Biol. 14 (1966), 341-356; Acta Physiol. Polon.38 (1987), 109-113

MARCHOUX, ÉMILE GABRIEL [1862-1943]
FISCHER 2,988-989; IB, 189
G. Ramon, Bull. Acad. Med. 127 (1943), 586-592
C. Mathis, Presse Med. 51 (1943),563-564; Paris Médical

33 (1943),217-218
Anon., Ann. Inst. Pasteur 70 (1944), 1-6

MARCKWALD, LEO [1866-1926]
Anon., Chem. Z. 51 (1927), 17

MARCKWALD, WILLY [1864-1942]
KURSCHNER 4,1854; BHDE 1,473; 2,777; ASEN, 124; WREDE,123; TSCHIRCH,1060; TLK 3,423; WININGER 7,295; TETZLAFF, 225; POGG 4,957; 5,804; 6,1646; 7a(3),197-198
Anon., Chem. Z. 58 (1934), 1003
E. Pilgrim, Entdeckung der Elemente, p.366. Stuttgart 1950

MARCUS, ERNST [1893-1968]
KURSCHNER 4,1854-1855; BHDE 2,777; IB, 189

MARCUS, HARRY [1880-1976]
KURSCHNER 11,1862; EGERER, 71-79; LDGS, 55; IB, 189
T. von Lanz, Münch. med. Wchschr. 92 (1950), 459-462

MARCUSE, WILHELM [1859-1900]
BJN 5,106*; RSC 17,13
Anon., Leopoldina 36 (1900), 133-134

MARCUSSON, JULIUS [1871-1934]
TLK 3,423-424; WININGER 4,267; POGG 6,1647; 7a(3),198

MARDASHEV, SERGEI RUFOVICH [1906-1974]
BSE (3rd Ed.) 15,360; WWWS, 1109-1110
Anon., Vop. Med. Khim. 20 (1974), 341-343

MAREN, THOMAS HARTLEY [1918-]
AMS 17(5),197
T.H. Maren, Ann. Rev. Pharmacol. 22 (1982), 1-18

MARENZI, AGUSTIN DOMINGO [1900-]
Quien es Quien en la Argentina, p.449. Buenos Aires 1968

MAREŠ, FRANTISEK [1857-1942]
OBL 6,79-80; STURM 2,572-573; FISCHER 2,990; NAVRATIL, 183-185
V. Laufberger, Cas. Lek. Cesk. 76 (1937), 1713-1714; Biol. Listy 22 (1937), 149-160
W. Ziegenfuss, Philosophen-Lexikon 2 (1950), 171

MARESCH, RUDOLF [1868-1936]
NDB 16,150; STURM 2,574; FISCHER 2,990; IB, 189
H. Chiari, Wiener med. Wchschr. 86 (1936), 149-150; Verhandl. path. Ges 29 (1936), 393-400; Wiener klin. Wchschr. 65 (1953), 703-704

MAREY, ÉTIENNE JULES [1830-1904]
DSB 9,101-103; VAPEREAU 6,1055; HIRSCH 4.79; PAGEL, 1093-1094; W, 352-353
G. Paris et al., Rev. Sci. [4]17 (1902), 129-138
O. Frank, Münch. med. Wchschr. 51 (1904), 2011-2013
P. Schultz, Berl. klin. Wchschr. 41 (1904), 658-659
H. Boruttau, Deutsche med. Wchschr. 30 (1904), 959-960
P. Tillaux, Bull. Acad. Med. 51 (1904), 426-430
Anon., Arch. Ital. Biol. 41 (1904), 489-498; Nature 70 (1904), 57-58;
Lancet 1904(I), pp.1530-1533
C.E.F. Franck, C.R. Soc. Biol. 59 (1905), 1-22; L'Oeuvre de E.J. Marey.Paris 1906
C. Richet, Bull. Acad. Med. 103 (1930), 705-714
A.R. Michaelis, Med. Hist. 10 (1966), 201-203
H.A. Snellen, E.J. Marey and Cardiology. Rotterdam 1980
F. Dagognet, Étienne Jules Marey: La Passion de la Trace. Paris 1987

MARFORI, PIO [1861-1942]
FISCHER 2,991
E. Trabucchi, Minerva Med. 1957, pp.3942-3943

MARGARIA, RODOLFO [1901-1983]
EI (App.2)2,263; VACCARO, 945; IB, 189; WWWS, 1110
Accademia Nazionale dei Lincei, Biografie e Bibliografie degli Academici Lincei, pp.415-417. Rome 1976

MARGOLIASH, EMANUEL [1920-]
AMS 15(5),190; WWA 1980-1981, p.2146

MARGOSHES, MAX [1876-1928]
STURM 2,575; WININGER 4,275-276; POGG 5,806-807; 6, 1648
H. Ditz, Z. angew. Chem. 41 (1928), 1213-1214
D. Holde, Ber. chem. Ges. 61A (1928), 143-145

MARGULIES, MAX [1883-]
JV 25,428; SG [3]7,966; NUC 361,509

MARIAM, THEODOR [1884-1977]
ARNSBERG, 301-302; NUC 361,557

MARIE, AUGUSTE CHARLES [1864-1935]
Anon., Ann. Inst. Pasteur 54 (1935), 513-517; Presse Med. 43 (1935),675

MARIE, PIERRE [1853-1940]
DSB 9,108-109; FISCHER 2,992; PAGEL, 1095-1096; HAYMAKER, 476-479
G. Guillain, Bull. Acad. Med. 123 (1940), 524-535

MARIGNAC, JEAN CHARLES GALISSARD de [1817-1894]
DSB 9,109-111; DHB 3,308-309; STELLING-MICHAUD 3,386-387; FERCHL, 341;BLOKH 1,480-481; SCHAEDLER, 77; GENEVA 3,1517; 4,102-109; W,353-354;POTSCH, 291; WWWS, 441; POGG 3,872; 4,960
E. Ador, Bull. Soc. Chi,. [3]11 (1894), i-xvi; Ber. chem. Ges. 27(4) (1894), 979-1021; Arch. Sci. Phys. Nat. 32 (1894), 183-215
P.T. Cleve, J. Chem. Soc. 67 (1895), 468-489

MARINE, DAVID [1880-1976]
AMS 10,2659; NCAB F,484; MEDVEI, 511-514; DAMB, 495-496; KOREN, 217-218
A.M. Harvey, Adventures in Medical Research, pp.316-317. Baltimore 1976;American Journal of Medicine 70 (1981), 483-485
R.W. Rawson, Trans. Assoc. Am. Phys. 91 (1978), 38-39
J. Matovinovic, Persp. Biol. Med. 21 (1978), 565-589

MARINESCU, GHEORGE [1863-1938]
FISCHER 2,993; PAGEL, 1096; HAYMAKER, 345-347; WWWS, 1111
L. Ribadeau-Dumas, Bull. Acad. Med. 120 (1938), 91-93
L.S. Copelman, Rev. Hist. Med. Heb. 57 (1962), 101-105
A. Petresco, Ann. Histochim. 9 (1964), 145-153
M. Marinescu, Leopoldina [3]11 (1965), 226-240
M. Marinescu et al., Hist. Sci. Med. 18 (1984), 13-18

MARINETTI, GUIDO V. [1918-]
AMS 15(5),193; WWWS, 1111

MARION, LÉO EDMOND
[1899-1979] WW 1979, p.1658; WWWS, 1111-1112
O.E. Edwards, Proc. Roy. Soc. Canada [4]18 (1980), 107-110
R.U. Lemieux and O.E. Edwards, Biog. Mem. Fell. Roy. Soc. 26 (1980), 357-380

MARK, EDWARD LAURENS [1847-1946]
AMS 7,1167; NCAB 9,271-272; IB, 189; WWAH,386
H.B. Bigelow and J.H. Welsh, Am. Phil. Soc. Year Book 1947, pp.275-278

MARK, HERMANN FRANCIS [1895-1992]
AMS 14,3192-3193; KURSCHNER 11,1865; WWA 1978-1979, p.2083; WWWS, 1112;IWW 1982-1983, p.852; CB 1961, pp.297-300; MH 1,315-316; MSE 2,277; POTSCH, 291; BHDE 2,780-781; POGG 6,1650-1651; 7a(3),201-204
G. Oster, J. Chem. Ed. 29 (1952), 544
Anon., Nachr. Chem. Techn. 3 (1955), 94-95
R.L. Rawls, Chem. Eng. News 53(9) (1975), 33
H. Mark, From Small Organic Molecules to Large: A Century of Progress.Washington, D.C. 1991

MARK, ROBERT [1898-1981]
KURSCHNER 13,2444; LDGS, 80

MARKER, RUSSELL EARL [1902-]
P.A. Lehmann-Fettler et al., J. Chem. Ed. 50 (1973), 195-199

MARKERT, CLEMENT LAWRENCE [1917-]
AMS 14,3193; WWA 1980-1981, p.2150; IWW 1982-1983, p.853; WWWS, 1112

MARKHAM, ROY [1916-1979]
WW 1979, p.1659
R.W. Horne, Nature 285 (1980), 57
S.R. Elsden, Biog. Mem. Fell. Roy. Soc. 28 (1982), 319-345

MARKLEY, KLARE STEPHEN [1895-1973]
AMS 11,3466-3467; IB, 190; WWWS, 1112-1113
Anon., Chem. Eng. News 51(33) (1973), 20

MARKLIN, GUSTAV FRIEDRICH [1803-1871]
HEIN, 394
A. Wankmüller, Beitr. Wurtt. Apothekges. 8 (1969), 100-107

MARKOVNIKOV, VLADIMIR VASILIEVICH [1838-1904]
DSB 9,130-132; BSE 26,296-297; KUZNETSOV(1961), 480-488; KAZAN 1,424-428; BLOKH 1,482-485; ARBUZOV, 143-150; POTSCH, 291-292; POGG 3,873; 4,961-962; 5,808
Anon., Chem. Z. 28 (1904), 199-200
H. Decker, Ber. chem. Ges. 38 (1905), 4249-4262
E.J. Mills, J. Chem. Soc. 87 (1905), 597-600
I.A. Kablukov, Zhur. Fiz. Khim. 37 (1905), 247-303
N.A. Kablulov, Usp. Khim. 8 (1939), 300-320
H.M. Leicester, J. Chem. Ed. 18 (1941), 53-57
G. Jones, J. Chem. Ed. 38 (1961), 297-300
J. Tierney, J. Chem. Ed. 65 (1988), 1053-1054

MARKS, LEWIS HART [1883-1958]
AMS 9(I),1270
Anon., J. Am. Med. Assn. 167 (1958), 240

MARKS, PAUL ALAN [1926-]
AMS 15(5),199-200; WWA 1980-1981, p.2151; WWWS, 1113

MARMIER, LOUIS [1865-1944]
IB, 190
C. Gernez-Rieux, Ann. Inst. Pasteur 72 (1946), 849-851

MARMOREK, ALEXANDER [1865-1923]
FISCHER 2,994; KOREN, 218; KOHUT 2,282-283; WININGER 4,283-284; KAGAN, 162; WWWS, 1114
Anon., Chem. Z. 47 (1923), 687; Z. angew. Chem. 36 (1923), 443

MARQUARDT, PETER [1910-]
KURSCHNER 13,2448-2449; WWWS, 1114; POGG 7a(3),206-208

MARQUART, LUDWIG CLAMOR [1804-1881]
HEIN, 403-404; POHL, 136-141; FERCHL, 340-341; TSCHIRCH, 1033,1085; POGG 3,874
C.J. Andrae, Leopoldina 19 (1883), 114-117,131-135
G. Bayer, Dr. Ludwig Clamor Marquart. Bonn 1962

MARRACK, JOHN RICHARDSON [1886-1976]
DSB 18, 592-595; WW 1965, pp.2028-2029
R. Augustin, Immunology 6 (1963), 1-2
R.G.W., Lancet 1976(II), p.378
J.H. Humphrey, Nature 263 (1976), 535

MARRIAN, GUY FREDERICK [1904-1981]
WW 1981, p.1722; POTSCH, 292; POGG 6,1654; 7b,3137-3139
G.F. Marrian, J. Endocrin. 35 (1966), vi-xvi
Anon., Lancet 1981(II), pp.375-376
J.K. Grant, Biog. Mem. Fell. Roy. Soc. 28 (1982), 347-378

MARRIOTT, WILLIAMS McKIM [1885-1936]
DAB [Suppl.2],432-433; AMS 5,741; NCAB 36,140-141; OEHRI, 70-71
B. Veeder, J. Pediat. 13 (1938), 619-626; 47 (1955), 791-801

MARSH, JAMES [1794-1846]
DNB 12,1100; HIRSCH 4,87; W, 354; FERCHL, 342; SCHAEDLER, 78; WWWS, 1115; POTSCH, 292
G. Lockemann, Chem. Z. 36 (1912), 1465-1466
A. Cardoso-Pereira, Chem. Z. 37 (1913), 41
W.A. Campbell, Chem. Brit. 1 (1965), 198-202

MARSH, JAMES ERNEST [1860-1938]
POGG 4,964; 5,810; 6,1655; 7b,3142
J.A. Gardner, J. Chem. Soc. 1938, pp.1130-1135
F. Soddy, Nature 141 (1938), 903-904; Obit. Not. Fell. Roy. Soc. 2 (1939), 549-556

MARSH, JULIAN BUNSICK [1926-]
AMS 15(5),208; WWA 1980-1981, p.2155; WWWS, 1115

MARSHALL, ELI KENNERLY, Jr. [1889-1966]
AMS 10,2671; DAMB, 497-498; CHEN, 52-53; PARASCANDOLA, 43-45; IB, 190; WWAH, 387; WWWS, 1116; POGG 5,810; 6,1656; 7b,3144-3146
T.H. Maren, Pharmacologist 8 (1966), 90-94; Bull. Johns Hopkins Hosp.119 (1966), 247-254; Biog. Mem. Nat. Acad. Sci. 56 (1987), 313-352
A.M. Harvey, Johns Hopkins Med. J. 138 (1976), 54-60

MARSHALL, FRANCIS HUGH ADAM [1878-1949]
WW 1949, p.1848; MEDVEI, 776-777; O'CONNOR(2), 42-44; IB, 190
A.S. Parkes, Obit. Not. Fell. Roy. Soc. 7 (1950-1951), 239-251

MARSHALL, HARRY TAYLOR [1875-1929]
AMS 4,644; FISCHER 2,995
Anon., J. Am. Med. Assn. 93 (1929), 1907
H.E. Jordan, Arch. Path. 9 (1930), 98

MARSHALL, HUGH [1868-1913]
WWWS, 1116; POGG 4,964; 5,810
L. Dobbin, Nature 92 (1913), 138-139
A.C.B., Proc. Roy. Soc. A90 (1914), xiv-xvi

MARSHALL, JOHN [1855-1925]
AMS 3,454
New York Times, 6 January 1925
E.F. Smith, Catalyst 10(2) (1925), 6

MARSHALL, LAWRENCE MARCELLUS [1910-1978]
AMS 13,2810; WWA 1978-1979, p.2092; WWWS, 1117

MARSLAND, DOUGLAS ALFRED [1899-]
AMS 13,2811-2812; WWWS, 1117

MARSSON, MAXIMILIAN [1845-1909]
HEIN, 405-406
R. Kolkowitz, Ber. bot. Ges. 27 (1909), (91)-(96)
Anon., Chem. Z. 33 (1909), 1333

MARSSON, THEODOR [1816-1892]
ADB 52,218-219; HEIN, 406-407; TSCHIRCH, 1085; FERCHL, 342; POGG 2,60
B. Ascherson, Ber. bot. Ges. 10 (1892), (30)-(33)
Anon., Mitt. Nat. Ver. Neu-Vorpomm. 24 (1892), 1-14 (1892); Chem. Z.16 (1892), 1974; Leopoldina 28 (1892), 57

MARSTON, HEDLEY RALPH [1900-1965]
WW 1965, p.2068
E.G. Holmes, Nature 208 (1965), 524-525
R.L.M. Synge, Biog. Mem. Fell. Roy. Soc. 13 (1967), 267-293
E.J. Underwood, Rec. Austral. Acad. Sci. 1(2) (1967). 73-86

MARTENS, ADOLF [1850-1914]
DSB 9,138-140; DBJ 1,69-71,300; FREUND 3,273-291
E. Heyn, Stahl und Eisen 34 (1914), 1393-1395

MARTENS, MARTIN [1797-1863]
BNB 13,876-879; FERCHL, 342; POGG 2,61,1427
L.P., Bull. Soc. Bot. Belg. 2 (1863), 67-71
P.J. van Beneden, Ann. Acad. Roy. Belg. 30 (1864), 115-141

MARTIN, ALOIS [1818-1891]
HIRSCH 4,95; KUKULA, 553; DUELUND, 56-61
Anon., Leopoldina 27 (1891), 157

MARTIN, ARCHER JOHN PORTER [1910-]
WW 1981, p.1732; IWW 1982-1983, p.857; MH 1,316-317; STC 2,220-221;MSE 2,280-281; CB 1953, pp.417-419; CAMPBELL, 149-150; WWWS, 1118; ETTRE, 285-296; ABBOTT-C, 91; POTSCH, 292
G.A. Stahl, J. Chem. Ed. 54 (1977), 80-83

MARTIN, CHARLES JAMES [1866-1955]
DSB 18,597-599; DNB [1951-1960]. 702-703; WW 1949, p.1854; FISCHER 2,997; O'CONNOR(2), 452-458; WWWS, 1118; BULLOCH, 382-383
H. Chick et al., Nature 175 (1955), 577-578
S.P. Bedson, J. Gen. Microbiol. 14 (1956), 487-493
M. Robertson, J. Path. Bact. 71 (1956), 519-534
H. Chick, Biog. Mem. Fell. Roy. Soc. 2 (1956), 173-208

MARTIN, HENRY NEWELL [1848-1896]
DSB 9,142-143; DAB 12,337-338; KELLY, 814-815; DAMB, 500-501; WWAH, 389;ELLIOTT, 170-171; O'CONNOR, 171-174
M. Foster, Proc. Roy. Soc. 60 (1897), xx-xxiii
H. Sewall, Bull. Johns Hopkins Hosp. 32 (1911), 327-333
C.S. Breathnach, Med. Hist. 13 (1969), 271-279
A.M. Harvey, Johns Hopkins Med. J. 136 (1975), 38-46
W.B. Fye, J. Hist. Med. 40 (1985), 133-166; The Development of American Physiology, pp.129-167. Baltimore 1987

MARTIN, KARL [1886-1935]
JV 29,549; NUC 364,566
Anon., Chem. Z. 59 (1935), 520

MARTIN, LOUIS [1864-1946]
FISCHER 2,998; WWWS, 1120
J. Tréfouël, Bull. Acad. Med. 130 (1946), 471-473
Anon., Ann. Inst. Pasteur 72 (1946), 705-707

MARTIN, RICHARD [1908-]
KURSCHNER 8,1482-1483; KALLMORGEN, 348

MARTIN, SIDNEY HARRIS COX [1860-1924]
FISCHER 2,998; O'CONNOR(2), 161; BULLOCH, 383; RSC 17,59
Who Was Who 1916-1928, p.708
Anon., Brit. Med. J. 1924(II), pp.647-648
Anon., J. Path. Bact. 28 (1925), 698-699
W.D.H., Proc. Roy. Soc. B99 (1926), xliv-xlv

MARTINDALE, WILLIAM [1840-1902]
TSCHIRCH, 1085
Anon., Science 15 (1902), 396
F. Hoffmann, Pharm. Z. 47 (1902), 154

MARTINO, GAETANO [1900-1967]
EI (App.2),43
Anon., Brit. Med. J. 1967(II), p.622
S. Cerquiglini, Arch. Fisiol. 65 (1967), 141-153

MARTINOTTI, GIOVANNI [1857-1928]
FISCHER 2,999; IB, 191
D. Giordano, Riv. Storia Sci. Med. 19 (1928), 305
G. Montalenti, Archeion 11 (1928), 234-235

MARTIUS, CARL [1906-]
KURSCHNER 13,2456; MEISTER, 37-47; POGG 7a(3),211-212
F.L., Chimia 20 (1966), 54-55
C. Martius, in Of Oxygen, Fuels and Living Matter, Part 2 (G. Semenza, Ed.), pp.1-57. Chichester 1982
G. Semenza, Chimia 40 (1986), 67-68

MARTIUS, CARL ALEXANDER von [1838-1920]
NDB 16,312-313; DBJ 2,754-755; BLOKH 1,486; POTSCH, 292-293; POGG 5,813
H. Wichelhaus, Ber. chem. Ges. 53A (1920), 72-75

MARTIUS, ERNST WILHELM [1756-1849]
HEIN, 409-410; TSCHIRCH, 1088; POGG 2,67; 7a Suppl., 415-416
E.W. Martius, Erinnerungen aus meinem neunzigjährigen Leben.Lepzig 1847 (reprint Mittenwald 1932)

MARTIUS, KARL FRIEDRICH [1794-1868]
DSB 9,148-149; NDB 16,310-312; ADB 20,517-527; GLOSAUER, 57-71; FERCHL, 343-344
A.W. Eichler, Flora 52 (1869), 3-13,17-24
W.H.B., Am. J. Sci. [2]47 (1869), 288-291
Anon., Proc. Roy. Soc. 18 (1870), vi-xi
H. Merxmüller, Sitz. Bayer. Akad. 1968, pp.79-96

MARTIUS, THEODOR [1796-1863]
HEIN, 410-411; HIRSCH 4,103-104; TSCHIRCH, 1086; FERCHL, 344; POGG 2,67; 3,879; 7sSuppl.,417; CALLISEN 12,285-288; 30,262

MARTON, LADISLAUS LASZLO [1901-1979]
AMS 12,4122; WWWS, 1121
L.L. Marton, Early History of the Electron Microscope. San Francisco 1968

MARTY, RUDOLPH [1829-1909]
NUC 365,636
A. Wankmüller, Der Schweizer Familienforscher 33 (1966), 46-50

MARTZ, ERWIN [1901-]
JV 44,450; NUC 365,672

MARTZ, FREDERIC [1871-]
RSC 17,65; SG [2]10,174; NUC 365,672

MARVEL, CARL SHIPP [1894-1988]
DSB 18,601-603; AMS 14,3223-3224; WWA 1976-1977, p.2040; IWW 1982-1983, pp.859-860; MH 1,317-318; MSE 2,282-283; POGG 6,1662-1663; 7b,3154-3162
J.E. Mulvaney, J. Chem. Ed. 53 (1976), 609-613
H. Mark, Journal of Macromolecular Chemistry A21 (1984), 1567-1606
B.C. Anderson and R.D. Lipscomb, Macromolecules 17 (1984), 1641-1643
N.J. Leonard, Am. Phil. Soc. Year Book 1988, pp.205-213
D.S. Tarbell and A.T. Tarbell, J. Chem. Ed. 68 (1991), 539-542

MARX, ALFRED VALENTIN [1880-1942]
KALLMORGEN, 348; SG [3]7,985; NUC 366,162

MARX, ERNST [1870-1951]
KALLMORGEN, 348-349
Anon., Natur und Volk 80 (1951), 171-172

MARX, MAX [1868-]
JV 6,298; RSC 17,67; NUC 366,162

MARX, WALTER [1907-1984]
AMS 14,3225; BHDE 2,785; LDGS, 23; WWWS, 1122

MASAMUNE, HAJIME [1896-1959]
JBE, 2113
Anon., Science 131 (1960), 1725
G. Kikuchi and Z. Yosizawa, Tohoku J. Exp. Med. 72 (1960), 1-4

MASCARELLI, LUIGI [1877-1941]
TSCHIRCH, 1086; POGG 5,813-814; 6,1663-1664; 7b,3163-3164
G. Isoglio, Annali della Reale Accademia Agricoltura di Torino 85 (1941-1942), 511-520

MASCART, ÉLEUTHÈRE ÉLIE NICOLAS [1837-1908]
DSB 9,154-156; VAPEREAU 6, 1069; WWWS, 1122-1123; POGG 3,879-880; 4,968-969; 5,814
Anon., Nature 78 (1908), 446-448
P. Janet, Rev. Gen. Sci. 20 (1909), 574-593

MASCHKA, JOSEF [1820-1899]
STURM 2,575; BJN 4,163*; HIRSCH 4,109; MOTSCH, 1-16; NAVRATIL, 186
P. Dittrich, Prag. med. Wchschr. 24 (1899), 73-75
M. Richter, Wiener klin. Wchschr. 7 (1899), 164

MASCHKE, OTTO [1824-1900]
HEIN, 412-413; TSCHIRCH, 1086; SCHELENZ, 691
Anon., Jahresb. Schles. Ges. 78 (1901), 9-11

MASCHMANN, ERNST [1894-1943]
NDB 16,352-353; POGG 6,1664; 7a(3),215
T. Wagner-Jauregg, Ber. chem. Ges. 76A (1943), 27-28
Anon., Z. angew. Chem. 56 (1943), 82

MASHEVSKI, NIKOLAI MARTINIANOVICH [1855-]
RSC 17,70; SG [2]10,179

MASING, EMIL [1839-1898]
HEIN, 413; LEVITSKI 2,255-256
Anon., St. Petersburg medizinische Wochenschrift 23 (1898), 94

MASING, ERNST [1879-1956]
FISCHER 2,1001; WELDING, 490-491; IB, 191
J. Brennsohn, Die Aerzte Estlands, p.479. Riga 1922
Anon., Baltische Briefe 9(6) (1956), 8-12

MASIUS, JEAN BAPTISTE VOLTAIRE [1836-1912]
HIRSCH 4.110; BLUM 2,26-27
L. Beco, Scalpel 65 (1912-1913), 443
C. Vanlair, Ann. Acad. Roy. Belg. 80 (1914), 79-116

MASLOVSKI, ALEKSEI FRANTSOVICH [1831-1888]
VORONTSOV, 7; RSC 4,276

MASON, FREDERICK ALFRED [1888-1947]
JV 28,318; POGG 6,1666; 7b,3165
P.C.L. Thorne, J. Chem. Soc. 1949, pp.260-261

MASON, HAROLD LAWRENCE [1901-1992]
AMS 11,3505; WWA 1968-1969, p.1425; WWWS, 1123-1124

MASON, THOMAS GODFREY [1890-1959]
WW 1959, p.2036; DESMOND, 426; WWWS, 1124
T.A. Bennet-Clark, Biog. Mem. Fell. Roy. Soc. 6 (1960), 183-189

MASON, VERNE RHEEM [1889-1965]
AMS 9(II),751; NCAB 51,477
T.H. Brem, Trans. Assoc. Am. Phys. 79 (1966), 62-63

MASSART, EDWARD MARIE LUCIEN [1908-]
IWW 1982-1983, p.682; GHENT 4,217-224
M. Mammerickx, Mem. Acad. Med. Belg. 5 (1967), 615-616

MASSART, JEAN [1865-1925]
BNB 38,561-569
E. Marchal, Bull. Soc. Bot. Belg. 59 (1926), 7-10; Ann. Acad. Roy. Belg.93 (1927), 69-158

MASSEY, VINCENT [1926-]
AMS 15(5),243; WWA 1980-1981, p.2174; WWWS, 1125

MASSINI, RUDOLF [1880-1955]
KURSCHNER 4,1868; SBA 3,92
A. Gigon, Basler Jahrbuch 1955, pp.201-205

MASSOL, GUSTAVE [1857-1951]
TSCHIRCH, 1086; DORVEAUX, 56; POGG 4,970; 5,815; 7b,3175
E. Canals, Ann. Pharm. Fran. 9 (1951), 89-90

MASSOL, LÉON [1838-1909]
FISCHER 2,1002-1003; GENEVA 5,91-92; 6,100
H. Maillart, Rev. Med. Suisse Rom. 29 (1909), 896-900
Anon., Ann. Inst. Pasteur 7 (1909), 1049-1051
A. Gottschalk, Hippocrate 3 (1935), 151-156
M. Cramer, Gesnerus 34 (1977), 203-206

MASSON, ANTOINE [1806-1860]
POGG 2,75-76; 3,881

MASSON, FRANÇOIS [1848-1915]
RSC 10,743; SG [4]10,357
H.M., Rev. Med. Suisse Rom. 35 (1915), 292-294

MASSON, JAMES IRVINE ORME [1887-1962]
SRS, 45; NUC 368,466; POGG 5,816; 6,1668; 7b,3176

Anon., Nature 141 (1938),1089-1090; 169 (1952),396; 196 (1962),1151-1152
R.D. Haworth and A.H. Lamberton, Proc. Chem. Soc. 1963, pp.120-121; Biog. Mem. Fell. Roy. Soc. 9 (1963), 205-221

MAST, SAMUEL OTTMAR [1871-1947]
DSB 9,167; AMS 7,1183; NCAB 33,492-493; IB, 191; WWAH, 390; WWWS, 1125
C.G. Wilbur, Trans. Am. Micr. Soc. 67 (1948), 82-83

MATHEWS, ALBERT PRESCOTT [1871-1957]
AMS 8,1659; CHITTENDEN, 225-231; IB, 191; WWAH, 391; WWWS, 1127; POGG 5,816; 6,1668; 7b,3182
E.N. Harvey, Science 127 (1958), 743-744
H.M. Sheaff, Proc. Inst. Med. Chicago 23 (1960), 65-67

MATIGNON, CAMILLE [1967-1934]
CHARLE, 165-166; WWWS, 1127-1128
F. Bourion, Bull. Soc. Chim. [5]2 (1935), 377-436
W.J. Pope, J. Chem. Soc. 1937, pp.705-708

MATRUCHOT, LOUIS [1863-1921]
DSB 9,175-176; SACKMANN, 217-218; CHARLE, 193-194
P. Portier, C.R. Soc. Biol. 85 (1921), 322-323
M.J. Constantin, Bull. Soc. Mycol. 38 (1922), 127-139

MATSON, GUSTAVE ALBIN [1899-]
AMS 12,4146; WWWS, 1128

MATSUBARA, KOICHI [1872-1955]
JBE,795
Y. Shibata, Kogaku no Ryoiki 10 (1956), 13-15

MATSUO, IWAO [1881-]
HAKUSHI 2,234-235

MATTENHEIMER, HERMANN [1921-]
AMS 14,3244-3245; KURSCHNER 13,2465-2466; POGG 7a(3),220-221

MATTEUCCI, CARLO [1811-1868]
DSB 9,176-177; EI 22,596; DFI, 86-97; HIRSCH 4,116; BRAZIER, 28-34; FERCHL, 245-246; WWWS, 1129; POGG 2,79-81; 3,883
G. Maruzzi, Physis 6 (1964), 101-140
R. Bernebeo, Carlo Matteucci, Profilo della Vita e dell'Opera. Ferrara 1972

MATTHAEI, RUPPRECHT [1895-1976]
NDB 16,390-392; KURSCHNER 11,1884; BERWIND, 163-170; IB, 192; POGG 7a(3),221-222

MATTHES, HERMANN [1869-1931]
NDB 16,399-400; HEIN, 413-414; TSCHIRCH,1086-1087; TLK 3,427; POGG 4,972-973; 5,819; 6,1672
A. Petrenz, Apoth. Z. 46 (1931), 336
W. Meyer, Chem. Z. 55 (1931), 237
F. Paneth, Ber. chem. Ges. 64A (1931), 107-108

MATTHES, KARL [1905-1962]
NDB 16,400-401; KURSCHNER 9,1283; HAGEL, 44-51
M. Bürger, Münch. med. Wchschr. 105 (1963), 260
K. Mechelke, Deutsche med. Wchschr. 88 (1963), 1251-1254
H. Schäfer, Z. Kreis. 52 (1963), 105-107

MATTHES, MAX [1865-1930]
FISCHER 2,1005; PAGEL, 1106; KROLLMANN, 1015-1016; GIESE, 571-572;
KALIN, 129-131; RSC 17,93
R. Stintzing, Münch. med. Wchschr. 77 (1930), 769-770

F. Moritz, Arch. klin. Med. 167 (1930), i-ii
P. Morawitz, Klin. Wchschr. 9 (1930), 1007
F. Kiewitz, Deutsche med. Wchschr. 56 (1931), 886-887

MATTHIESSEN, AUGUSTUS [1831-1870]
DSB 9,178; DNB 13,71-72; WWWS, 1129; POTSCH, 293; POGG 2,82; 3,885-886
Anon., Nature 2 (1870), 517-518; J. Chem. Soc. 24 (1871), 615-617
E. Sabine, Proc. Roy. Soc. 18 (1870), 111-112

MATTILL, HENRY ALBRIGHT [1883-1953]
AMS 7,1187; BERG, 55-58; IB, 192; WWAH, 393
C.P. Berg, Proc. Iowa Acad. Sci. 60 (1953), 57-59; Science 119 (1954),
274-275; J. Nutrition 66 (1958), 3-14

MATULA, JOHANN [1890-1952]
TLK 3,428; POGG 7a(3),226
E. Seelich, Ost. Chem. Z. 53 (1952), 148

MAUGUIN, CHARLES VICTOR [1878-1958]
DSB 9,182-183; CHARLE(2), 194-196; WWWS, 1130; POGG 5,820; 6,1675 7b,3193-3194
C. Jacob, C.R. Acad. Sci. 247 (1958), 5-9

MAULL, CARL [1870-]
JV 9,210; RSC 17,100; NUC 370,354

MAUMENÉ, EDMÉ JULES [1818-1898]
BLOKH 1,486-487; FERCHL, 346-347; POGG 2,84; 3,886-887; 4,974
M. Riban, Bull. Soc. Chim. [3]19 (1898), 257-258

MAUPAS, FRANCOIS ÉMILE [1842-1916]
DSB 9,185-186; WWWS, 1131
M. Caullery, Bull. Soc. Hist. Nat. Afr. 23 (1932), 339-347; Rev. Sci.
70 (1932), 405-407
E. Sergent, Arch. Inst. Pasteur Alg. 33 (1955), 69-70

MAURER, KURT [1900-1945]
POGG 6,1677; 7a(3),231
G. Reinäcker and G. Drefahl, Chem. Ber. 86 (1953), i-xiv

MAURITZ, ALFRED [1867-1938]
JV 6,298; NUC 370,650
Anon., Mauritz-Zeitung 1 (1924), 14-15; Chem. Z. 56 (1932), 805

MAURIZIO, ADAMO [1862-1941]
DHB 4,694; STRAHLMANN, 468,480
A. Volkhart, Verhandl. Schw. Nat. Ges. 121 (1941), 389-394

MAURO, ALEXANDER [1921-1989]
AMS 17(5),278

MAUSS, HANS [1901-1953]
NDB 16,449; POGG 7a(3),233
Anon., Chemische Technik 1 (1956), 113

MAUTHNER, JULIUS [1852-1917]
OBL 6,162; DBJ 2,665; FISCHER 2,1007; LESKY, 525-526; BLOKH 1,487; KOREN, 218; WININGER 7,302; POGG 3,888; 4,976-977; 5,823
W. Suida, Ber. chem. Ges. 51 (1918), 1025-1029
R. von Zeynek, Ost. Chem. Z. 21 (1918), 43-45

MAUTHNER, LUDWIG [1840-1894]
FISCHER 2,1007; KOREN, 218; WININGER 4,303; WWWS, 1132

MAUTHNER, LUDWIG WILHELM [1806-1858]
ADB 20,713; OBL 6,163; HIRSCH 4,127-128;
ATUYANA, 1-2; KOREN, 218
K. von Hassinger, Zeitschrift der Gesellschaft der Aerzte zu Wien 14 (1858), 366-368
Anon., Medical Directory 1859, p.979

MAUTNER, HANS [1886-1963]
KURSCHNER 4,1880; FISCHER 2,1007
Anon., Pirquet Bulletin of Clinical Medicine 10(7) (1963), 134-135

MAX, JULES [1882-]
JV 21,29; NUC 371,153

MAXIMOW, ALEXANDER [1874-1928]
FISCHER 2,1007-1008; IB, 193
F. Weidenreich, Anat. Anz. 67 (1928), 360-368; Arch. exp. Zellforsch. 9 (1930), 1-5
W. Bloom, Z. Zellforsch. 8 (1929), 801-805
N.G. Chlopin, Arch. exp. Zellforsch. 8 (1929), 183-188

MAXWELL, JAMES CLERK [1831-1879]
DSB 9,198-230; DNB 13,118-121; BLOKH 1,490-491;
MATSCHOSS, 171-172; POTSCH, 293-294; W, 357-358;
WWWS, 1133; POGG 3,889-890; 6,1679
L. Campbell and W. Garnett, The Life of James Clerk Maxwell, 2nd Ed. London 1882
Anon., Proc. Roy. Soc. 33 (1882), i-xvi
J.J. Thomson et al., J.C. Maxwell 1831-1931. Cambridge 1931
R.L. Smith-Rose, James Clerk Maxwell. London 1948
J. Randall, Nature 195 (1962), 427-434
C. Domb (Ed.), Clerk Maxwell and Modern Science. London 1963
R.V. Jones, Notes Roy. Soc. 28 (1973), 57-81
I. Tolstoy, James Clerk Maxwell. Edinburgh 1981
M. Goldman, The Demon in the Ether. Bristol 1983

MAXWELL, SAMUEL STEEN [1860-1939]
AMS 6,953
J.M.D. Olmsted, Science 89 (1939), 259-260

MAXWELL, WALTER [1854-1931]
RSC 17,110; NUC 371,309-310
C.A. Browne, Facts about Sugar 27 (1932), 24
J.G. Traynham, Essays on the History of Organic Chemistry, pp.86-88. Baton Rouge, La. 1987

MAY, FRIEDRICH JULIUS [1898-1969]
GERNETH, 163-168; POGG 7a(3),234

MAY, RICHARD [1863-1936]
KURSCHNER 4,1881; FISCHER 2,1008; BACHMANN, 94-98
H. Kürten, Münch. med. Wchschr. 84 (1937), 430-431

MAY, WILLIAM PAGE [1863-1910]
FISCHER, 1009
E.H. Starling, Erg. Physiol. 10 (1910), v-vii
Anon., Lancet 1910(I), p.339; Brit. Med. J. 1910(I), p.297

MAYDL, KAREL [1853-1903]
OBL 6,170-171; STURM 2,615; FISCHER 2,1009;
NAVRATIL, 191-192; WWWS, 1134
A. Fraenkel, Wiener klin. Wchschr. 16 (1903), 979
Anon., Cas. Lek. Cesk. 42 (1913), 1383-1387

MAYEDA, MATSUNAYE [1877-]
HAKUSHI 2,131-132

MAYER (MAYER-GMELIN), ADOLF [1843-1942]
NDB 16,533-534; KURSCHNER 4,1881-1882; DRULL, 174; TLK 3,429; SCHAEDLER, 79; POGG 3,892; 6,1679; 7a(3),234
A. Mayer, Naturwiss. 12 (1924), 905-911
Anon., Z. angew. Chem. 55 (1942),282; 56 (1943),20; Chem. Z. 67 (1943),37

MAYER, ANDRÉ [1875-1956]
CHARLE, 168-171; IB, 193; WWWS, 1134; POGG 6,1679-1682; 7b,3199-3202
E.F. Terroine, Bull. Soc. Chim. Biol. 38 (1956), 1389-1395
J. Roche, Bull. Acad. Med. 140 (1956), 408-410
R. Courrier, Not. Acad. Sci. 3 (1957), 660-663
J. Mayer, J. Nutrition 99 (1969), 308

MAYER, ERWIN WILHELM [1885-]
NUC 371,542

MAYER, EUGEN ARTHUR [1882-1930]
JV 20,470; NUC 371,543
Anon., Chem. Z. 54 (1930), 766; Chemische Fabrik 3 (1930), 408

MAYER, FERDINAND [-1869]
RSC 4,310; 8,365
Anon., Proceedings of the American Pharmaceutical Association
32 (1884), 612

MAYER, FRANZ XAVER [1904-1955]
WIEN 1956, pp.64-71; LIMLEY, 74-83; POGG 7a(3),235
H. Jansch, Ost. Chem. Z. 56 (1955), 333-336
H. Svejda, Microchemica Acta 1955, pp.1090-1092

MAYER, FRIEDRICH (FRITZ) MICHAEL [1880-]
JB 18,177MAYER, FRITZ [1876-1940]
LDGS, 23; TLK 3,430; POGG 5,823-824; 6,1682; 7a(3),235

MAYER, HEINRICH [1863-]
JV 1,27; RSC 17,114; SG [1]8,733; NUC 371,560

MAYER, JULIUS ROBERT [1814-1878]
DSB 9,235-240; NDB 16,546-548; ADB 21,126-128; SL 4,101-133; POTSCH,294;BLOKH 1,487-490; W, 359;
MATSCHOSS, 172; PAGEL, 1110-1112;SCHAEDLER, 79;
POGG 2,94; 3,890; 4,977; 6,1683; 7aSuppl.,420-429
S. Friedländer, Julius Robert Mayer. Leipzig 1905
B. Hell, Robert Mayer und das Gesetz der Erhaltung der Energie.Stuttgart 1925
A. Mittasch, Z. angew. Chem. 53 (1940), 113-118
W. Gerlach, Z. angew. Chem. 55 (1942), 369-375
H. Schmolz and H. Weckbach, Robert Mayer. Weissenhorn 1964
W. Schütz, Robert Mayer. Leipzig 1969
G. Eistert, Robert-Mayer-Bibliographie. Heilbronn 1978
R.B. Lindsay, Julius Robert Mayer. Oxford 1973
F. Kober, Chem. Z. 106 (1982), 397-405

MAYER, MANFRED MARTIN [1916-1984]
AMS 15(5),275; WWA 1980-1981, p.2192; BHDE 2,791;
WWWS, 1135
E.A. Kabat, J. Immunol. 134 (1985), 654-656
K.F. Austen, Biog. Mem. Nat. Acad. Sci. 59 (1990), 257-280

MAYER, MARTIN [1875-1951]
FISCHER 2,1010; LDGS, 57; IB, 193
New York Times, 18 February 1951
E.G. Nauck, Z. Tropenmed. 3 (1951), 1-3

MAYER, MAX [1881-]
JV 18,177; NUC 371,611

MAYER, OTTO von [1872-]
JV 23,309

MAYER, PAUL [1848-1923]
NDB 16,552-553; HEIN, 417-418; FISCHER 2,1011; FREUND 1,251-261
H. von Eggeling, Anat. Anz. 58 (1924), 88-93; 59 (1925), 139-140
T. Peterfi, Z. wiss. Mikr. 41 (1924), 145-154

MAYER, SIEGMUND [1842-1910]
STURM 2,620; BJN 15,60*; FISCHER 2,1011; PAGEL, 1112-1113; KOREN, 218;WININGER 4,306-308; LOMMATZSCH, 18-26; KOERTING, 108-109; RSC 8,366;
10,756; 12,494; 17,114-115
A. Kohn, Anat. Anz. 38 (1911), 87-93

MAYER, WILHELM [1827-1891]
POGG 3,891
Anon., Chem. Z. 15 (1891), 1901; Pharm. Z. 36 (1891), 576

MAYER, WILHELM [1833-1906]
BJN 11,43*

MAYERSON, HYMEN SAMUEL [1900-1085]
AMS 17(5),289; BROBECK, 174-175; WWWS, 1135

MAYNARD, LEONARD AMBY [1887-1972]
AMS 11,3540; MILES, 321-323; MH 2,351-352; STC 2,222-223; MSE 2,291
Anon., Chem. Eng. News 50(35) (1972), 22
J.K. Loosli, Journal of Animal Science 68 (1990), 1-4

MAYR, CARL [1881-1951]
WIEN 1952, pp.54-55

MAYR, ERNST [1872-1930]
JV 14,226; RSC 19,74; NUC 372,142
Anon., Chem. Z. 54 (1930), 506; Z. angew. Chem. 43 (1930), 220

MAYR, ERNST [1904-]
AMS 14,3264; WWA 1978-1979, p.2130; IWW 1982-1983, pp.872-873; IB, 194;MH 2,353-354; STC 2,223-225; MSE 2,293-294; BHDE 2,794; WWWS, 1136
S.J. Gould, Science 223 (1984), 255-257

MAYR, FRANZ [1814-1863]
WURZBACH 18,104-105; HIRSCH 4,140; ATUYANA, 3-4
H. Widerhofer, Jahrb. Kinderheil. 6(4) (1863), 59-66

MAYZEL, WACLAW [1847-1916]
PSB 20,285-286; KOSMINSKI, 311-313,613; KONOPKA 5,630-633
Z. Kramsztyk, Medycyna i Kronika Lekarska 1916, pp.254-257

MAZÉ, PIERRE [1868-1947]
IB, 194; RSC 17,119-120; NUC 372,225

MAZIA, DANIEL [1912-]
AMS 15(5),281; WWA 1980-1981, p.2195; IWW 1982-1983, p.873; COHEN, 164;
MH 2,354-355; STC 2,225-226; MSE 2,294-295; WWWS, 1136

MAZZA, FRANCESCO [1905-1943]
POGG 6,1684; 7b,3206-3208
G. Quagliarello, Chimica e Industria 25 (1943), 93-94; Enzymologia 12 (1946-1948), 138
L. Losana, Atti Accad. Torino 78 (1943), 326-369

MEAD, JAMES FRANKLYN [1916-1987]
AMS 15(5),282-283; WWWS, 1152

MEANS, JAMES HOWARD [1885-1967]
AMS 11,3547; NCAB 53,497-498; FISCHER 2,1014; WWAH, 408; WWWS, 1153
C.M. James, Harvard Med. Alumni Bull. 42(1) (1967), 33-34
W.B. Castle, Trans. Assoc. Am. Phys. 81 (1968), 16-19
A.M. Harvey, American Journal of Medicine 70 (1981), 1158-1160

MEARS, LEVERETT [1850-1917]
AMS 2,316; RSC 10,760; NUC 372,539
Anon., Science 45 (1917), 657

MECHEL, LUCAS von [1893-1956]
JV 33,239; NUC 372,587

MECHNIKOV, ILYA ILYICH [1845-1916]
DSB 9,331-335; BSE 27,396-397; BME 18,233-238; KUZNETSOV(1963), 192-200;FREUND 2,229-245; FISCHER 2,1029-1030; BULLOCH, 383; OLPP. 281-284; SACKMANN. 217-218; WININGER 7,315; W, 361-362; ABBOTT, 92; WWWS,1168
E.R. Lankester, Nature 97 (1916), 443-446; Proc. Roy. Soc. B89 (1917), li-lix
O. Metchnikoff, Vie d'Elia Metchnikoff. Paris 1920
G.F. Petrie, Nature 149 (1942), 547-548
S. Zalkind, Ilya Mechnikov, his Life and Work. Moscow 1959
J.G. Hirsch, Bact. Revs. 23 (1959), 48-60
R.B. Vaughan, Med. Hist. 9 (1965), 201-215
P. Lépine, Elie Metchnikoff et l'Immunologie. Vichy 1966
D. Wrotnowska, Arch. Inter. Hist. Sci. 21 (1968), 115-136
A.M. Silverstein, Cellular Immunology 48 (1979), 208-221
M. Baggiolini, Schw. med. Wchschr. 112 (1982), 1403-1411

MECKE, REINHARD [1895-1969]
KURSCHNER 10,1554; POGG 6,1687; 7a(3),239-240
Anon., Chem. Z. 94 (1970), 132

MECKEL, JOHANN FRIEDRICH [1781-1833]
DSB 9,252-253; ADB 21,160-162; HIRSCH 4,145-146; WWWS, 1153;CALLISEN 12,378-390; 30,306-308
Anon., Proc. Roy. Soc. 3 (1830-1837), 232-233
R. Benecke, Johann Friedrich Meckel der Jüngere. Halle 1934

MECKLENBURG, WERNER [1880-1968]
KURSCHNER 11,1911-1912; JV 23,334; POGG 5,826; 6,1687-1688; 7a(3),241-242
Anon., Chem. Z. 57 (1933), 206

MEDAWAR, PETER [1915-1987]
WW 1981, pp.1762-1763; IWW 1982-1983, p.875; MH 1,323-324; STC 2,232-233;MSE 2,296-297; ABBOTT, 92; WWWS, 92
P. Medawar, Memoir of a Thinking Radish: An Autobiography. Oxford 1986
N.A. Mitchison, Nature 330 (1987), 112; Biog. Mem. Fell. Roy. Soc.35 (1990), 283-301
J. Medawar, A Very Decided Preference. Oxford 1990

MEDER, OSKAR [1877-1944]
JV 16,159; NUC 373,1
J.F. Heydemann, Entomol. Z. 58 (1944), 33-34

MEDES, GRACE [1886-1967]
AMS 11,3550; MILES, 330-331

Anon., Chem. Eng. News 46(10) (1968), 68
A.J. Donnelly, Am. J. Clin. Path. 57 (1969), 401

MEDICUS, LUDWIG [1847-1915]
NDB 16,599; DBJ 1,335; BLOKH 1,491-492; TSCHIRCH, 1087; SCHAEDLER, 79;DZ, 944; POGG 3,893; 4,981; 5,826
F. Reitzenstein, Ber. chem. Ges. 48 (1915), 1744-1748; Chem. Z. 39 (1915), 837-838

MEDVEDEV, ANATOLI KONSTANTINOVICH [1863-]
RSC 17,128; SG [2]10,634; [3]7,1112

MEDVEDEV, SERGEI SERGEIEVICH [1891-1970]
BSE (3rd Ed.)15,555; WWWS, 1154
P.N. Demichev et al., Vest. Akad. Nauk 40(11) (1970), 127-128

MEEK, WALTER JOSEPH [1878-1963]
AMS 10,2225; BROBECK, 144-145
C.M. Brooks, Biog. Mem. Nat. Acad. Sci. 54 (1983), 251-268

MEER, EDMUND ter [1852-1931]
NDB 16,605-606
Anon., Chem. Z. 50 (1926), 448; 51 (1927), 581; 55 (1931), 866; Chemische Fabrik 4 (1931), 444
M. Bodenstein, Ber. chem. Ges. 64A (1931), 211

MEER, FRITZ ter [1884-1967]
NDB 16,606-608
Anon., Nachr. Chem. Techn. 7 (1959), 257; 15 (1967), 418

MEERWEIN, HANS [1879-1965]
NDB 16,608-610; SCHNACK, 323-331; MEINEL, 370-407,509; SCHMITZ, 311-320; POTSCH, 195-296; AUERBACH, 864; POGG 5,827; 6,1688-1689; 7a(3),243
G. Hesse, Z. angew. Chem. 61 (1949), 161-168
T. Bersin, Arzneimitt. 15 (1965), 1476-1477
R. Criegee and K. Dimroth, Angew. Chem. (Int. Ed.) 5 (1966), 331-341
S. Honig, Chem. Z. 90 (1966), 301-303
K. Dimroth, Chem. Ber. 100(1) (1967), lv-xciv

MEES, CHARLES EDWARD KENNETH [1882-1960]
DAB [Suppl.6],441-443; AMS 9(I),1302; NCAB 52,338-339; WWAH,409; WWWS, 1155; POGG 6,1688-1690; 7b,3211-3213
H. Baines, Proc. Chem. Soc. 1961, pp.269-271
W. Clark, Biog. Mem. Fell. Roy. Soc. 7 (1961), 173-197
H.T. Clarke, Am. Phil. Soc. Year Book 1961, pp.140-145; Biog. Mem. Nat. Acad. Sci. 42 (1971), 175-199

MEHL, JOHN WILBUR [1910-1987]
AMS 13,1951MEHLER, ALAN HASKELL [1922-]
AMS 15(5),293; WWWS, 1155

MEHLITZ, ALFRED [1899-1966]
KURSCHNER 10,1558; POGG 6,1690-1691; 7a(3),243-244
Anon., Nachr. Chem. Techn. 14 (1966), 270

MEHLTRETTER, CHARLES LOUIS [1905-]
AMS 12,4187

MÉHU, CAMILLE [1835-1887]
BOURQUELOT, 101-102; HIRSCH 4,150-151; GORIS, 503-505
G.H. Marty, Bull. Acad. Med. 18 (1887), 697-701; J. Pharm. Chim. [5]16 (1887), 574-577
E.V. McCollum, J. Chem. Ed. 33 (1956), 507

MEIDINGER, JOHANN HEINRICH [1831-1905]
POGG 2,102; 3,895; 4,983; 5,828

MEIER, ROLF [1897-1966]
IB, 194; POGG 7a(3),247
Anon., Chem. Z. 81 (1957), 302; Int. Arch. Allergy 10 (1957), 3-4
H.J. Bein, Pharmacologist 9 (1967), 109

MEIGEN, WILHELM [1873-1934]
TLK 3,433; POGG 4,983-984; 5,828; 6,1692
F. Weitz, Ber. chem. Ges. 68A (1935), 29

MEIGS, ARTHUR VINCENT [1850-1912]
DAB 12,502-503; NCAB 25,181; FISCHER 2,1016; DAMB, 510; WWWS, 1156
E.B. Meigs, Trans. Coll. Phys. Phila. 36 (1914), lxxxiii-xciii
A. Levinson, Ann. Med. Hist. 10 (1928), 138-148

MEIGS, EDWARD BROWNING [1879-1940]
AMS 6,959; NCAB 30,358-359; IB, 194
R.S. Lillie, Biol. Bull. 83 (1942), 15-18
P.E. Howe, J. Biol. Chem. 142 (1942), 1-2

MEIMBERG, FRANZ [1866-]
JV 10,214; RSC 13,276; 16,377

MEINCKE, KARL PETER [1893-]
JV 34,217; SG [3]7,1115; NUC 374,23

MEINWALD, JERROLD [1927-]
AMS 15(5),298; IWW 1982-1983, pp.876-877; WWWS, 1157

MEIROWSKY, EMIL [1876-1960]
KURSCHNER 4,1903-1904; FISCHER 2,1015; LDGS, 62; KOREN, 219
Deutscher Dermatologen Kalender 1929, pp.151-153
A. Hollander, Arch. Dermatol. 82 (1960), 644
H.H. Biberstein, Proc. Virchow Med. Soc. 19 (1960), 212-213

MEISEL, (HIRSCHA)N(ISEL) MORDEKHELEVICH [1868-]
SG [2]10,640

MEISENHEIMER, JAKOB [1876-1934]
NDB 16,685-686; NB, 258; TLK 3,433; POGG 5,829-830; 6,1692-1693
H. Wieland, Bayer. Akad. Wiss. Jahrbuch 1934-1935, pp.66-68
W. Merz, Ber. chem. Ges. 68A (1935), 32-33
W.H. Mills, J. Chem. Soc. 1935, pp.1355-1359
Anon., Z. angew. Chem. 48 (1935), 55-56

MEISER, WERNER [1906-]
JV 46,483; NUC 374,116

MEISSL, EMERICH [1855-1905]
OBL 6,200; BJN 10,215*
Anon., Chem. Z. 29 (1905), 213
W. Bersch, Z. Land. Ver. Ost. 8 (1905), 141-152

MEISSNER, GEORG [1829-1905]
DSB 9,258-260; NDB 16,698-699; BJN 10,215*; HIRSCH 4,155; BULLOCH, 383; WAGENITZ, 119-120; PAGEL, 1114-1115; POGG 3,896; 7a(3),251
O. Weiss, Naturw. Rund. 20 (1905), 349-350; Münch. med. Wchschr. 52 (1905), 1206-1207
H. Boruttau, Pflügers Arch. 110 (1905), 351-399
M. Verworn, Jahrb. Akad. Wiss. Gott. 1905, pp.45-54
C. von Voit, Sitz. Bayer. Akad. 36 (1906), 456-468
G. Müller, Georg Meissner. Düsseldorf 1935

MEISSNER, KARL FRIEDRICH [1800-1874]
ADB 21,248-249; DHB 4,708; BOSSART, 22-23; BRIQUET, 313-315; TSCHIRCH, 1087; WAGENITZ, 119
A. de Candolle, Bull. Soc. Bot. 21 (1874), 279-283
W.T. Stearn, Journal of the Society for the Bibliography of Natural History 4 (1967), 291-297

MEISSNER, PAUL TRAUGOTT [1778-1864]
ADB 21,248-251; NDB 16,703-705; OBL 6,202-203; WURZBACH 17,309-312; HEIN, 421-422; FERCHL, 250-251; TSCHIRCH, 1087; COLE, 381-382; CALLISEN 12,420-421; 30,321-322; POGG 2,106; 3,896

MEISSNER, WILHELM [1792-1853]
NDB 16,707; HEIN, 420-421; FERCHL, 251; POGG 2,106-107; 7aSuppl.,430
F.C. Bucholz, Arch. Pharm. [2]76 (1853), 209-211
H. Gittner, Pharm. Z. 89 (1953), 285-286

MEISTER, ALTON [1922-]
AMS 15(5),300; WWA 1980-1981, pp.2277-2278; IWW 1982-1983, p.877; MH 2,362-364; MSE 2,298-300

MEISTER, HERBERT von [1866-1919]
NB, 258; NL 6,253-254; BLOKH 1,493; POGG 6,1696
Anon., Chem. Z. 43 (1919), 23
W. von Rath, Ber. chem. Ges. 52A (1919), 52-53

MEISTER, WILHELM [1827-1895]
NDB 16,728-730; NB, 258; NL 6,251-253

MEITES, JOSEPH [1913-]
AMS 15(5),300; WWWS, 1157

MEIXNER, KARL [1879-1955]
FISCHER 2,1018; LIMLEY, 8-22
F.J. Holzer, Deutsche Z. ger. Med. 44 (1955), 341-342

MELCHERS, GEORG [1906-]
KURSCHNER 13,2516; WAGENITZ, 120-121; WWWS, 1158; POGG 7a(3),260-261

MELLANBY, EDWARD [1884-1955]
DSB 15,417-420; DNB [1951-1960], 731-732; WW 1949, pp.1897-1898; W, 362-363; WWWS, 1159
H. Dale, Brit. Med. J. 1955(I), pp.355-358; Biog. Mem. Fell. Roy. Soc. 1 (1955), 193-222
H.P. Himsworth, Trans. Assoc. Am. Phys. 68 (1955), 13-14
B.S. Platt, Ann. Rev. Biochem. 25 (1956), 1-28
C. Harington, Brit. Med. Bull. 12 (1956), 3-4
J. Parascandola and A.J. Ihde, Bull. Hist. Med. 51 (1977), 507-515
D.F. Smith and M. Nicolson, Proc. Roy. Coll. Med. Edin. 19 (1989), 51-60

MELLANBY, JOHN [1878-1939]
DNB [1931-1940], 610-611; WW 1939, p.2176; O'CONNOR(2), 250-252; FLUELER, 94-96; WWWS, 1159
V.J. Woolley, Nature 144 (1939), 358
J.B. Leathes, Obit. Not. Fell. Roy. Soc. 3 (1940), 173-195

MELLISS, DAVID ERNEST [1848-1913]
NUC 374,567
Who Was Who in America 1,828

MELLONI, MACEDONIO [1798-1854]
DSB 9,264-265; EI 22,814; WWWS, 1159; POGG 2,112-114
J. Nicklès, Am. J. Sci. [2]19 (1885), 414-415; 20 (1885), 143
E.S. Cornell, Ann. Sci. 3 (1938), 402-406
P. Codastefano and E. Schettino, Physis 26 (1984), 271-301
E. Schettino, Nuncius 2 (1987), 111-124

MELLOR, JOSEPH WILLIAM [1869-1938]
SRS, 24; WWWS, 1159; POGG 5,832; 6,1698; 7b,3224
W. Bragg, Proc. Roy. Soc. A169 (1938-1939), 7-8
A.T. Green, Nature 142 (1938), 281-282; Obit. Not. Fell. Roy. Soc. 2 (1939), 573-576; J. Chem. Soc. 1943, pp.341-343
F. Habashi, Bull. Hist. Chem. 7 (1990), 13-16

MELSBACH, HEINRICH [1877-1947]
JV 16,144; NUC 374,668

MELSENS, LOUIS HENRI [1814-1886]
BNB 30,570-572; HIRSCH 4,160; BLOKH 1,492-493; FERCHL, 351; POGG 2,114-115; 3,898; 4,987
P. de Heen, Ann. Acad. Roy. Belg. 59 (1893), 483-506

MELTZER, SAMUEL JAMES [1851-1920]
DSB 9,265-266; DAB 12,519-520; FISCHER 2,1020; DAMB, 511-512; CHEN, 7-8; PARASCANDOLA, 45-46; KELLY, 830-832; KOREN, 218; WININGER 7,310; KAGAN(JA), 323-327,746-747; WWAH, 411; WWWS, 1159
W.H. Howell, Science 53 (1921), 99-106; Mem. Nat. Acad. Sci. 21 No.9 (1927), 1-23
H.C. Jackson et al., Proc. Soc. Exp. Biol. Med. 18 Suppl. (1921), 7-42
A.M. Harvey, Persp. Biol. Med. 21 (1978), 431-440
A. Meltzer, Proc. Soc. Exp. Biol. Med. 184 (1987), 370-374

MELVILLE, HARRY WORK [1907-]
WW 1981, p.1769; IWW 1982-1983, p.878; SRS, 89; MH 1,325-326; STC 2,238-239; MSE 2,302-303; WWWS, 1159

MELVILLE, JAMES [1908-1985]
Who's Who in Australia 1983, pp.602-603
J.P. Quirk, Chemistry in Australia 52(4) (1985), 143

MEMMEN, FRIEDRICH [1898-]
JV 40,738

MENDEL, BRUNO [1897-1959]
LINDEBOOM, 1305-1306; POTSCH, 296
Anon., Chem. Wkbl. 55 (1959), B190
W.S. Feldberg, Biog. Mem. Fell. Roy. Soc. 6 (1960), 191-199

MENDEL, GREGOR [1822-1884]
DSB 9,277-283; OBL 6,218-219; STURM 2,640; HINK 1,239-271; ABBOTT, 93-94; W, 363-364; WWWS, 1160
H. Iltis, Life of Mendel (trans. by E. and C. Paul). London 1932
E. Tschermak-Seysenegg, Verhandl. zool. bot. Ges. Wien 92 (1951), 25-35
J. Krizenecky, Johann Gregor Mendel. Leipzig 1965
L.C. Dunn et al., Proc. Am. Phil. Soc. 109 (1965), 189-226
J. Sagner, Med. Hist. 2 (1967), 78-91
A. Gustafsson, Hereditas 62 (1969), 239-258
U. Witte and B. Hoppe, Folia Mendeliana 6 (1971), 117-138
V. Orel, J. Heredity 64 (1973), 314-318; Mendel. Oxford 1984
M. Campbell, Centaurus 20 (1976), 159-174; 26 (1982), 38-69
A. Baxter and J. Farley, J. Hist. Biol. 12 (1979), 137-173
A. Brannigan, Soc. Stud. Sci. 9 (1979), 423-454
H. Kalmus, Hist. Sci. 21 (1983), 61-83
R.S. Root-Bernstein, Hist. Sci. 21 (1983), 275-295

A.W.F. Edwards, Biol. Revs. 61 (1986), 295-312
L.A. Callender, Hist. Sci. 26 (1988), 41-75
J. Sapp, in Experimental Inquiries (H.E. LeGrand, ed.), pp.137-166. Dordrecht 1990
F. di Trocchio, J. Hist. Biol. 24 (1991), 485-519

MENDEL, LAFAYETTE BENEDICT [1872-1935]
DSB 9,284-285; DAB 21,549-550; AMS 5,761; FISCHER 2,1022; DAMB, 512-513; CHITTENDEN, 330-330; IB, 195; OEHRI, 75-76; WWAH, 411; KOREN, 218; KAGAN(JA), 334-336,735-736; WININGER 7,312; WWWS, 1160; POTSCH, 296-297
G. Lusk, Science 65 (1927), 555-558
Anon., Ind. Eng. Chem. News Ed. 13 (1935), 485
W.C. Rose, J. Nutrition 11 (1936), 606-613
H.C. Sherman, Science 83 (1936), 45-47
A.H. Smith, Yale J. Biol. Med. 8 (1936), 387-388; J. Nutrition 60 (1956), 3-12
R.H. Chittenden, Biog. Mem. Nat. Acad. Sci. 18 (1937), 123-156

MENDELEEV, DMITRI IVANOVICH [1834-1907]
DSB 9,286-295; BSE 27,137-141; KUZNETSOV(1961), 456-471; BUGGE 2,241-250;ARBUZOV, 101-102; FARBER, 719-732; BLOKH 1,494-497; ABBOTT-C, 91-93; POTSCH, 197-198; W, 364-265; WWWS, 1160; POGG 3,899-900; 4,987; 5,832; 6,1699
P. Walden, Chem. Z. 31 (1907), 167-172; Ber, chem. Ges. 41 (1908), 4719-4800
W.A. Tilden, J. Chem. Soc. 95 (1909), 2077-2105; Proc. Roy. Soc. A84 (1910-1911), xvii-xx
B. Brauner, Coll. Czech. Chem. Comm. 2 ((1930), 219-243
W. Winicov, J. Chem. Ed. 14 (1937), 372-375
H.M. Leicester, Chymia 1 (1948), 67-74
D.Q. Posin, Mendeleyev. New York 1948
P. Kolodkine, Mendeleev et la Loi Periodique. Paris 1963
R. Kargon, J. Chem. Ed. 42 (1965), 388-389
J.W. Spronson, J. Chem. Ed. 46 (1969), 139-141
Y.I. Soloviev, J, Chem. Ed. 61 (1984), 1069-1071
F. Kober, Chem. Z. 109 (1985), 419-427

MENDELSOHN, MARTIN [1860-1930]
FISCHER 2,1022; KOREN, 219; WININGER 4,336

MENDELSSOHN (LEYDEN), ALEXANDER [1886-1968]
SG [3]17,1139; NUC 375,534

MENDELSSOHN, HEINRICH [1910-]
BHDE 2,802
Who's Who in Israel 1980-1981, p.222

MENDIUS, OTTO [1831-1885]
RSC 4,339
A.W. Hofmann, Ber. chem. Ges. 18 (1885), 1607

MENEGHETTI, EGIDIO [1892-1961]
DSB 9,295-296; EI (App.2),283
A. Cestari, Ric. Sci. 1 (1961), 123-128
R. Santi, Arch. Ital. Sci, Farm. 11 (1961), 552-560
M. Aloisi, Atti Accad. Lincei 1962, pp.78-80

MENGARINI, MARGARETE [1856-1912]
FISCHER 2,1023; NUC 376,223
K. Sudhoff, Virchows Jahresber. 1913, pp.395-396

MENGEL, ALFRED [1873-]
JV 20,470; NUC 376,233

MENK, WALTHER [1892-1980]
KURSCHNER 13,2524; WWWS, 1161
Anon., Naturw. Rund. 34 (1981), 82MENKIN, VALY [1901-1960]
AMS 9(II),763; COHEN, 165
M.B. Lurie, Anat. Rec. 140 (1961), 234-236
A. Delaunay, Vie Médicale 42 (1961), 560-564

MENNE, FRITZ [1910-1979]
KURSCHNER 13,2525; WWWS, 1161; POGG 7a(3),264
Anon., Deutsche med. Wchschr. 105 (1980), 388

MENSHUTKIN, BORIS NIKOLAEVICH [1874-1938]
BSE 27,159; ARBUZOV, 104-105; WWWS, 1162; POGG 5,833-834; 6,1700-1701; 7b,3229-3231
S. Pogodin, Usp. Khim. 7 (1938), 1896-1906
T.L. Davis, J. Chem. Ed. 15 (1938), 203-209
M.A. Blokh, Zhur. Ob. Khim. 9 (1939), 2104-2112

MENSHUTKIN, NIKOLAI ALEKSANDROVICH [1842-1907]
DSB 9,304-305; BSE 27,159-160; ARBUZOV, 102-103; BLOKH 1,498-499; KOZLOV, 161-167; POTSCH, 298; POGG 3,901; 4,988; 5,834
B. Menshutkin, Ber. chem. Ges. 40 (1907), 5087-5098
O. Lütz, Z. angew. Chem. 20 (1907), 609-610
W.A. Tilden, J. Chem. Soc. 99 (1911), 1660-1666
P.I. Staroselski and Y.I. Soloviev, N.A. Menshutkin. Moscow 1969

MENTEN, MAUD LEONORA [1879-1960]
AMS 9(II),763; IB, 195
A.H. Stock and A.M. Carpenter, Nature 189 (1961), 965
D.B. Smith, TIBS 4 (1979), N150
G.H. Fetterman, Perspectives in Pediatric Pathology 1 (1984), 5-7
A. Gjedde, Journal of Cerebral Flow Metabolism 9 (1989), 243-246

MENZBIR, MIKHAIL ALEKSANDROVICH [1855-1935]
BSE 27,148; BME 17,996-997; KUZNETSOV(1963), 268-273; IB, 195
S.I. Ognev, Bull. Soc. Nat. Moscou NS51(1) (1946), 5-15
L.S. Tsetlin, Bull. Soc. Nat. Moscou NS65(6) (1960), 129-140

MENZIES, JAMES ACWORTH [1869-1921]
O'CONNOR(2), 339-340; RSC 17,164
Anon., Lancet 1921(II), p.155; Brit. Med. J. 1921(II), p.133

MERCER, JOHN [1791-1866]
DNB 13,265-267; BLOKH 1,499-500; FERCHL, 353; W, 365-366; WWWS, 1163; POTSCH,298
E.A. Parnell, The Life and Labours of John Mercer. London 1886
T.E. Thorpe, Nature 35 (1886), 145-147
A.W. Baldwin, Endeavour 3 (1944), 138-143

MERCK, EMANUEL AUGUST [1855-1923]
DBJ 5,435; BLOKH 1,501-502; POTSCH, 298; POGG 7aSuppl.,435
F. Haber, Ber. chem. Ges. 56A (1923), 57-58

MERCK, GEORG FRANZ [1825-1873]
HEIN, 424-425; BLOKH 1,501; POGG 7aSuppl.,435
A.W.H[ofmann], Ber. chem. Ges. 6 (1873), 1582-1583

MERCK, (HEINRICH) EMANUEL [1794-1855]
HB 2,369-371; HEIN, 425-426; HUHLE-KREUTZER, 121-150; BLOKH 1,500-501; POTSCH, 298-299; FERCHL,353; POGG 2,123; 7aSuppl.,435-436
C. Löw, Heinrich Emanuel Merck. Darmstadt 1951

MERCK, KARL [1886-1968]
POGG 7aSuppl.,438

MERCK, LOUIS [1854-1913]
BJN 18,109*; MATSCHOSS, 174; POGG 7aSuppl.,438
W. Will, Ber. Chem. Ges. 46 (1913), 2981-2982

MERCK, WILHELM [1833-1899]
BJN 4,164*; POGG 7aSuppl.,438

MERCKER, HERMANN [1912-]
KURSCHNER 13,2530; POGG 7a(3),267-268

MERCKLE, ELSA [1874-]
JV 25,576; NUC 377,10

MERING, JOSEPH von [1849-1908]
BJN 13,62*; FISCHER 2,1025-1026; PAGEL, 1119-1120; FRANKEN, 191; DZ, 952; BLOKH 1,501; WWWS, 1738; POGG 7aSuppl.,440-441
H. Winternitz und N. Zuntz, Münch. med. Wchschr. 55 (1908), 400-402
O. Frese, Berl. klin. Wchschr. 45 (1908), 178-179
E. Harnack, Med. Klin. 4 (1908), 171-172
H.M. Dittrich and H. Hahn von Drosche, Anat. Anz. 143 (1978), 509-517

MERLING, GEORG [1856-1939]
POGG 7a(3),268-269
A. Skita, Ber. chem. Ges. 72A (1939), 77-88

MERRIFIELD, ROBERT BRUCE [1921-]
AMS 15(5),323; WWA 1980-1981, p.2289; IWW 1982-1983, p.883-884; POTSCH, 299
J.F. Henehan, Chem. Eng. News 49(31) (1971), 22-26
R.B. Merrifield, The Concept and Development of Solid Phase Peptide Synthesis. Washington, D.C. 1992

MERRILL, ELMER DREW [1876-1956]
DSB 15,421-422; DAB [Suppl.6],449-450; NCAB 45,220-221; HUMPHREY,169-174;IB, 196; WWAH, 413; WWWS, 1165
R.C. Rollins, Science 123 (1956), 831-832
J. Ewan, J. Wash. Acad. Sci. 46 (1956), 267-268
W.J. Robbins, Am. Phil. Soc. Year Book 1956, pp.117-119
W.J. Robbins and L. Schwarten, Biog. Mem. Nat. Acad. Sci. 32 (1958), 273-333

MERRITT, ERNEST GEORGE [1865-1948]
AMS 7,1206; NCAB 15,195-196; WWAH, 414; POGG 4,991; 6,1707; 7b,3244
Anon., Science 107 (1948), 645
C.S. Barr, Am. J. Phys. 33 (1965), 87-88

MERRY, ERNEST WYNDHAM [1889-1972]
SRS, 44; NUC 377,607
Who's Who of British Science 1953, p.188
Anon., Chem. Brit. 8 (1972), 268

MERTZ, EDWIN THEODORE [1909-]
AMS 14,3395; WWA 1978-1979, p.2230; WWWS, 1166

MERTZDORFF, CHARLES [1818-1883]
L.D. Froissart, Charles Metzdorff, un Industriel Alsacien. Paris 1983

MERUNOWICZ, JÓZEF [1849-1912]
PSB 20,454-455; KOSMINSKI, 313-314; KONOPKA 6,43-49
S. Ciechanowski, Przeglad Lekarski 51 (1912), 265-267
K. Krzymanowski, Amtsarzt 4 (1912), 121-125

MERZ, JOHN THOEDORE [1840-1922]
Anon., Nature 109 (1922), 451-452

MERZ, KURT WALTER [1900-1967]
KURSCHNER 10,1584; HEIN-E, 308-309; DRUM, 297-299; POGG 7a(3),269-270
Anon., Chem. Z. 91 (1967), 615
G. Schenck, Mitt. Pharm. Ges. 37 (1967), 3-5; Deutsche Apoth. Z. 108 (1968), 109-113

MERZ, OTTO [1900-1967]
POGG 6,1708-1709; 7a(3),271

MERZ, VICTOR [1839-1904]
BJN 10,74*; BLOKH 1,502-503; ZURICH, 28-39; ZURICH-D, 113; SCHAEDLER, 80; POGG 3,904-905; 4,992; 7aSuppl.,441
E. Buchner, Ber. chem. Ges. 37(3) (1904), 2529-2530
A. Werner and O. Meister, Verhandl. Schw. Nat. Ges. 87 (1905), lx-cii

MERZBACHER, SIEGFRIED [1883-]
JV 24,576; NUC 378,56

MESELSON, MATTHEW STANLEY [1930-]
AMS 15(5),330; WWA 1980-1981, P.2293; IWW 1982-1983, p.885; WWWS, 1167; MH 2,367-368; STC 2,241-242; MSE 2,305-306

MESNIL, FÉLIX [1868-1938]
DSB 9,328-329; FISCHER 2,1028; HD 2,528-531; OLPP, 281; IB,281; WWWS,1167; G. Ramon, Bull. Acad. Med. 119 (1938), 241-247
M. Caullery, Presse Med. 46 (1938), 401-402
Anon., Ann. Inst. Pasteur 60 (1938), 223-226; Bull. Inst. Pasteur 36 (1938), 177-179
C. Rabaud, Bull. Soc. Path. Exot. 31 (1938), 173-177

MESROBEANU, LYDIA [1908-1978]
SG [5]11,339; NUC 378,138
A. Marx, Rivista de Igiene (Bact.) 23(2) (1978), 123-124
D. Draghici, Archives Roumaines de Pathologie Expérimentale et de Microbiologie 37(2) (1978), 83

MESSERSCHMIDT, THEODOR [1886-1971]
KURSCHNER 11,1948; FISCHER 2,1029; TLK 3,438-439; IB, 196

MESTER, BRUNO [1863-1895]
PAGEL, 1124
Anon., Leopoldina 31 (1895), 102

MESTREZAT, WILLIAM [1883-1928]
IB, 196; POGG 6,1710
H. Roger, Presse Med. 36 (1928), 1388
Anon., Bull. Inst. Pasteur 26 (1928), 985
E. Derrien, Bull. Soc. Chim. Biol. 11 (1929), 1067-1083

METALNIKOW, SERGE [1870-1946]
BIBEL, 312-316; IB, 196
L. Nègre, Ann. Inst. Pasteur 72 (1946), 860-861

METCHNIKOFF [see MECHNIKOV]

METRIONE, ROBERT M. [1933-]
AMS 18(5),349

METT, SIMON GRIGORIEVICH [1861-1929]
RSC 17,188
P.P. Evdokimov, Sov. Zdrav. 1982(3), pp.70-71

METTENHEIMER, KARL FRIEDRICH CHRISTIAN [1824-1898]
ADB 52,330-331; BJN 5,42*; HIRSCH 4,184; PAGEL, 1124-1125; KALLMORGEN,353
Anon., Leopoldina 34 (1898), 142-143
F. Dornblüth, Jahrb. Kinderheil. 49 (1899), 135-136

METTENHEIMER, WILHELM [1802-1864]
HEIN, 427; TSCHIRCH, 1088; FERCHL, 354; POGG 2,130
Anon., Neues Jahrbuch für Pharmazie 21 (1864), 369-371

METTLER, KARL [1877-]
JV 17,286

METZ, CHARLES WILLIAM [1889-1975]
AMS 11,3593; MH 2,368-370; FISCHER 2,1030; IB, 197; WWWS, 1169
J.R. Preer, Am. Phil. Soc. Year Book 1977, pp.75-79

METZ, EMIL [1906-]
JV 47,432; SG [4]10,974

METZELER, KARL [1865-1919]
JV 4,225; RSC 17,189

METZENER, WALTHER [1880-]
JV 22,521
Anon., Nachr. Chem. Techn. 13 (1965), 7

METZGER, HEINRICH [1897-1984]
Anon., Nachr. Chem. Techn. 33 (1985), 741

METZGER, HÉLÈNE [1889-1944]
DSB 9,340-342
C. Singer, Nature 157 (1946), 472
P. Brunet, Rev. Hist. Sci. 1 (1947), 68-70
M. Boas, Arch. Inter. Hist. Sci. 8 (1955), 432-444
V.P. Zoubov, Scientia 97 (1962), 233-238
J.R.R. Christie et al., Hist. Sci. 25 (1987), 71-109
G. Freudenthal et al., Etudes sur Hélène Metzger. Leiden 1990

METZLER, AUGUST [1883-1916]
JV 26,641; NUC 379,91

METZNER, PAUL [1893-1968]
U. Ruge and H. Sagromsky, Kulturpflanze 17 (1969), 9-20

METZNER, RUDOLF [1858-1935]
KURSCHNER 4,1937; FISCHER 2,1031; WEBER, 73; IB, 197
Anon., Deutsche med. Wchschr. 61 (1935), 112

MEUNIER, JEAN [1856-1937]
POGG 4,995-996; 5,839

MEUNIER, PAUL [1908-1954]
C. Mentzer, Bull. Soc. Chim. 1954, pp.869-875
Y. Raoul, Bull. Soc. Chim. Biol. 36 (1954), 1187-1190
Y. Raoul and L.J. Harris, in Paul Meunier 1908-1954, pp.7-23. Lons-le-Saunier 1954

MEVES, FRIEDRICH [1868-1923]
KIEL, 95
I. Broman, Anat. Anz. 59 (1925), 341-349
F.W. Pehlemann, Anat. Anz. 123 (1968), 405-422

MEYEN, FRANZ JULIUS FERDINAND [1804-1840]
DSB 9,344-345; ADB 21,549-533; TSCHIRCH, 1088; POGG 2,133; 3,907
J.T.C. Ratzeburg, Nova Acta Academiae Caesareae Leopoldina etc. 19 (1843),xiii-xxxii
K.M., Bot. Z. 2 (1844), 793-797

MEYENBERG, ALEXANDER [1871-]
JV 10,214; RSC 14,804; 17,198

MEYER, ARTHUR [1850-1922]
DBJ 4,363; HEIN, 432-434; LKW 6,222-226; GUNDLACH, 489-490; BK, 267-268; AUERBACH, 866-867; REBER, 47-48,365; WAGENITZ, 123; TSCHIRCH, 1088
K. Kroemer, Pharm. Z. 55 (1910), 219-220
F.J. Meyer, Ber. bot. Ges. 40 (1922), (100)-(111)

MEYER, ARTHUR [1874-1942?]
GEDENKBUCH 2,1023; JV 12,149; NUC 380,440

MEYER, CARL [1872-1922]
JV 10,214; NUC 380,456

MEYER, ERICH [1874-1927]
FISCHER 2,1033; KOREN, 220; WININGER 4,364; 7,316
P. Jungmann, Z. klin. Med. 106 (1927), i-ix

MEYER, ERNST von [1847-1916]
DBJ 1,364; BLOKH 1,503-504; SCHAEDLER, 80; DZ, 965; POGG 3,910; 4,997; 5,840; 7aSuppl.,444
R. von Walther, Chem. Z. 40 (1916), 477
F. Heinrich, Mitt. Gesch. Med. 15 (1916), 277-281
E. Mohr, J. prakt. Chem. [2]95 (1917), 1-36

MEYER, FERDINAND [1882-]
JV 23,309

MEYER (MEYER-ERLACH), GEORG [1877-1961]
JV 20,282; NUC 380,552
A. Angerer, Mainfränkisches Jahrbuch für Geschichte und Kunst 14 (1962), 344-359

MEYER, GUSTAVE MORRIS [1875-1945]
AMS 7,1210; FISCHER 2,1033-1034; KOREN, 220; KAGAN(JA), 337-338
Anon., Chem. Eng. News 23 (1945), 924; Science 101 (1945), 502

MEYER, HANS [1877-1964]
FISCHER 2,1034; AUERBACH, 318-319; KIEL, 117-118; WWWS, 1170
A. Proppe, Hautarzt 8 (1957), 333-335
R. Birkner, Münch. med. Wchschr. 99 (1957), 1081-1082
H. Holthusen, Fort. Geb. Röngtenstr. 101 (1964), 102-103

MEYER, HANS HORST [1853-1939]
FISCHER 2,1034; PAGEL, 1130; GROTE 2,139-168; DECKER, 43-53; DZ, 958; KNOLL, 105-106; KROLLMANN, 434; LEVITSKI 2,193-194; GUNDLACH, 251; WWWS, 1170; POGG 3,910; 4,1000-1001; 5,842-843; 6,1715; 7a(3),280
A. Jarisch, Erg. Physiol. 43 (1940), 1-8
H. Molitor, Arch. Inter. Pharmacodyn. 64 (1940), 257-264
E.P. Pick, J. Pharm. Exp. Ther. 71 (1940), 301-304
H. Chiari, Alm. Akad. Wiss. Wien 95 (1945), 313-319
R.L. Lipnick, Trends in Pharmacological Sciences 10 (1989), 265-269

MEYER, HANS LEOPOLD [1871-1944]
OBL 5,427; STURM 2,652; TETZLAFF, 235; TLK 3,443; POGG 4,1003-1004;5,842; 6,1715; 7a(3),279
F. Bock, Öst. Chem. Z. 48 (1947), 212

MEYER, JACOB [1863-]
JV 5,294; RSC 14,1010; 17,205; NUC 380,626

MEYER, JEAN [1830-1886]
Histoire Documentaire de l'Industrie de Mulhouse et ses Environs aux XIXe Siècle, vol.1, p.480. Mulhouse 1902

MEYER, JEAN de [1878-1934]
BNB 41,523-533; FISCHER 2,1035; IB, 197
M. Beckus, Bruxelles Médical 14 (1934), 1199

MEYER, JOHANN LUDWIG [1819-1894]
A. Wankmüller, Der Schweizer Familienforscher 33 (1966), 46-50

MEYER, JOHANN PETER FRIEDRICH [1886-1933]
POGG 6,1714
E. Tiede, Ber. chem. Ges. 67A (1934), 15-16

MEYER, JULIUS [1876-1960]
TLK 3,443; POGG 5,843; 6,1716-1717; 7a(3),281-282; (4),150*
Anon., Nachr. Chem. Techn. 4 (1956), 22; 8 (1960), 322

MEYER, KARL [1899-1990]
AMS 14,3407; WWA 1978-1979, p.2236; IWW 1982-1983, p.888; COHEN, 166;MH 2,370-371; STC 2,244-246; MSE 2,307-308; WWWS, 1170
Anon., Conn. Tiss. Res. 3 (1975), 3
New York Times, 22 May 1990

MEYER, KARL FRIEDRICH [1884-1974]
AMS 12,4237; DAMB, 519-520; IB, 197; WWWS, 1170-1171
A.B. Sabin, Biog. Mem. Nat. Acad. Sci. 52 (1980), 269-332

MEYER, KURT [1882-1942?]
GEDENKBUCH 2,1033; LDGS,57; IB, 197; NUC 380,673
Anon., Gior. Batt. Immun. 14 (1937), 685-686

MEYER, KURT HEINRICH [1883-1952]
DSB 9,354; STC 2,246-247; POGG 5,843-844; 6,1717; 7a(3),283-285
R.A. Boissonas, Chimia 6 (1952), 285-286

H. Mark, Z. angew. Chem. 64 (1952), 521-523
A.J. van Wyck, Helv. Chim. Acta 35 (1952), 1418-1422
R. Pummerer, Bayer. Akad. Wiss. Jahrbuch 1952, pp.208-211
E. Cherbuliez, Archives des Sciences (Geneva) 6 (1953), 35-39
R.W. Jeanloz, Adv. Carbohydrate Chem. 11 (1956), xiii-xviii
H. Hopff, Chem. Ber. 92 (1959), cxxi-cxxxvi

MEYER, LEOPOLD [1852-1918]
DBL (3rd Ed.) 9,550-552; FISCHER 2,1035; KOREN, 220; WININGER 7,316

MEYER, LOTHAR [1830-1895]
DSB 9,347-353; ADB 55,830-832; BUGGE 2,229-241; BLOKH 1,504-506; W, 367; SCHAEDLER, 80-81; ABBOTT-C, 93-94; WWWS, 1171; POTSCH, 299-300; POGG 2,141; 3,907; 4,996; 7aSuppl.,444-445
K. Seubert, Ber. chem. Ges. 28 (1895), 1109-1146
J.H. Long, J. Am. Chem. Soc. 17 (1895), 664-666
P.P. Bedson, J. Chem. Soc. 69 (1896), 1403-1439
F. Kober, Chem. Z. 109 (1985), 419-427

MEYER, LOTHAR [1906-1971]
AMS 12,4237; BHDE 2,811

MEYER, LUDWIG FERDINAND [1879-1954]
NB, 262-263; BHDE 2,811-812; LDGS, 77; FISCHER 2,1035-1036; ASEN, 130; KOREN, 220; COHEN, 166-167
Anon., Münch. med. Wchschr. 96 (1954), 1470

MEYER, MAX [1890-1954]
KURSCHNER 4,1949; FISCHER 2,1036; BHDE 2,812; LDGS, 74; WIDMANN, 155,276
H. Leicher, Z. Laryngol. 33 (1954), 717-721

MEYER, MORITZ [1821-1893]
HIRSCH 4,190; KOREN, 220; WININGER 4,367

MEYER, OSKAR EMIL [1834-1909]
BJN 14,157-160,61*; BLOKH 1,506; POGG 3,907-908; 4,996-997; 5,844
O. Lummer, Jahresb. Schles. Ges. 1909, pp.24-28

MEYER, OTTO [1860-]
SG [1]9,238; NUC 381,12

MEYER, OTTO MAX EMIL [1866-]
Schweizerisches Geschlechterbuch 3 (1910), 264

MEYER, PAUL [-1914]
JV 8,274; NUC 381,16; RSC 17,208
Anon., Z. angew. Chem. 27(III) (1914), 680

MEYER, PAUL [1902-1975]
KALLMORGEN, 354-355; LDGS, 67; IB, 198
Anon., Proc. Virchow-Pirquet Med. Soc. 31 (1977), 60

MEYER, RICHARD EMIL [1846-1927]
DZ, 960-961; TLK 2,476; WININGER 7,316; POGG 3,908; 4,997-998; 5,844-845; 6,1718
O. Spengler. Z. angew. Chem. 39 (1926), 841-842; 40 (1927), 421-422
H. Freundlich, Ber. chem. Ges. 60A (1927), 65-70

MEYER, RICHARD JOSEF [1865-]
ASEN, 130; WININGER 4,369; POGG 4,1003; 5,845; 6,1718; 7a(3),286

MEYER, RUDOLF [1885-1961]
KALLMORGEN, 355; SG [3]7,1253; NUC 381,39

MEYER, SELMA [1881-1958]
KURSCHNER 4,1952; FISCHER 2,1036; LDGS, 77; BOEDEKER, 94; KOREN, 220
New York Times, 14 November 1958

MEYER, VICTOR [1848-1897]
DSB 9,354-358; ADB 55,832-841; BJN 3,386-387; 4,57*; BAD 3,872-874; BUGGE 2,374-390; BLOKH 1,507-511; DRULL, 179-180; SCHAEDLER, 81; KOHUT 2,233; WININGER 4,371-372; W, 367-368; ABBOTT-C, 94-95; POTSCH, 300-301; POGG 3,908-909; 4,998-1000; 7aSuppl.,446-447
C. Liebermann, Ber. chem. Ges. 30 (1897), 2157-2168
P. Jacobson, Naturw. Rund. 12 (1897), 553-556,564-567
C. von Voit, Sitz. Bayer. Akad. 28 (1898), 455-460
H. Biltz, Z. anorg. Chem. 16 (1898), 1-14
T.E. Thorpe, J. Chem. Soc. 77 (1900), 169-206
R. Meyer, Ber. chem. Ges. 41 (1908), 4505-4718; Victor Meyer. Leipzig 1917
J. Sudborough, Proc. Chem. Soc. 1959, pp.137-141

MEYER, WILHELM [1827-1891] (see MAYER, WILHELM [1827-1891])

MEYER-ERLACH (see MEYER, GEORG)

MEYERHOF, OTTO [1884-1951]
DSB 9,359; DAB [Suppl.5],488-490; MILES, 332-333; DRULL, 180; W, 368-369; STC 2,247-248; FISCHER 2,1038; SCHWERTE 2,246-255; KOREN, 220-221; WININGER 4,377-378; TETZLAFF, 235; BHDE 2,814; ABBOTT-C, 95-96; POTSCH, 301; WWAH, 416; IB, 198; WWWS,1171; POGG 6,1720-1721; 7a(3),296-297
D. Needham, Nature 168 (1951), 895-896
H.H. Weber, Naturwiss. 39 (1952), 217-218
A. von Muralt, Erg. Physiol. 47 (1952), i-xx
D. Nachmansohn et al., Biochem. Biophys. Acta 4 (1950), 1-3; Science 115 (1952), 365-368; Biog. Mem. Nat. Acad. Sci. 34 (1960), 153-182; TIBS 5 (1980), 170-172
R. Peters, Obit. Not. Fell. Roy. Soc. 9 (1954), 175-200
C.L. Gemmill, Medical College of Virginia Quarterly 2 (1966), 141-142
H.H. Weber and H.A. Krebs, Molecular Energetics and Macromolecular Biochemistry (H.H. Weber, Ed.), pp.3-15. Berlin 1972
D. Nachmansohn, German-Jewish Pioneers in Science, pp. 268-311. Berlin 1979
H.G. Schweiger, Semper Apertus (W. Doerr, Ed.), vol.3, pp.359-375. Berlin 1985

MEYERHOFFER, WILHELM [1864-1906]
BJN 11,44*; BLOKH 1,511; POGG 4,1004-1005; 5,847
J.H. van't Hoff, Ber. chem. Ges. 39 (1906), 4471-4478
G. Bruni, Z. Elektrochem. 12 (1906), 385-386

MEYEROWITZ, LOUIS [1861-]
JV 4,225; RSC 17,189

MEYERSON, ÉMILE [1859-1933]
DSB 15,422-425
A. Reymond, Archeion 16 (1934), 107-108

MEYTHALER, FRIEDRICH [1898-1967]
KURSCHNER 10,1609; WWWS, 1172
G. Opitz, Personalbibliographien der Professoren der Inneren Medizin an der Universität Erlangen-Nürnberg, pp.3-38. Erlangen-Nürnberg 1969

MIALHE, LOUIS [1807-1886]
VAPEREAU 5,1283; LAROUSSE 11,211; BOURQUELOT, 87-88; GORIS, 512-515; HIRSCH 4,196-197; PAGEL, 1133; TSCHIRCH, 1089; FERCHL, 356; WWWS,1172; CALLISEN 13,43; 30,371-372; POGG 2,142-143; 3,911
L. Figuier, Ann. Scient. 30 (1886), 575-576
P. Chabert, Bull. Soc. Sci. Tarn 32 (1973), 752-755

MICHAEL, ARTHUR [1853-1942]
DSB 9,360; DAB [Suppl.3],520-521; AMS 6,972; NCAB 15,172; MILES, 334-335; POTSCH, 301-302; WWAH, 416; POGG 3,911; 4,1005-1006; 5,847-848;7b,3254-3255
A.B. Costa, J. Chem. Soc. 48 (1971), 243-246
L.F. Fieser, Biog. Mem. Nat. Acad. Sci. 46 (1975), 331-366

MICHAEL, GERHARD [1911-]
KURSCHNER 13,2569; WWWS, 1172; POGG 7a(3),298-299

MICHAELIS, AUGUST [1847-1916]
DBJ 1,364; BLOKH 1,511-512; TSCHIRCH, 1089; DZ, 965; TLK 3,445; POGG 3,912; 4,1006-1007; 5,848-849
H. Wichelhaus, Ber. chem. Ges. 49 (1916), 468

MICHAELIS, LEONOR [1875-1949]
DSB 18,620-625; DAB [Suppl.4],572-574; AMS 8,1698; MILES, 335-336; SANDRITTER, 103-105; FISCHER 2,1039-1040; BK, 269; IB, 198; TLK 3,445; KOREN, 221; KAGAN(JA), 736; WININGER 7,318-319; POTSCH, 302; WWAH, 416; WWWS, 1172; POGG 5,849-850; 6,1722; 7b,3255-3256
S. Granick, Nature 165 (1950), 299-300
W.M. Clark, Science 111 (1950), 55
L. Michaelis, Biog. Mem. Nat. Acad. Sci. 31 (1958), 282-321
E.S.G. Barron, Biol. Bull. 101 (1951), 13-16; Modern Trends in Physiology and Biochemistry, pp.xvii-xxii. New York 1952
D.W. Scheuch, Zeitschrift für medizinische Laboratoriumdiagnostik 20 (1979), 252-257

MICHAELIS, LUDWIG [1869-1942]
GEDENKBUCH 2,1043; JV 8,26; RSC 17,223; NUC 381,457

MICHAELIS, MAX [1869-1933]
KURSCHNER 4,1958; FISCHER 2,1040; KOREN, 221; WININGER 7,319
Anon., Münch. med. Wchschr. 80 (1933), 754

MICHAELIS, MORITZ [1906-1975]
AMS 12,4246; BHDE 2,816

MICHAELIS, PETER [1900-1975]
DSB 18,625-626; IB, 198; WWWS, 1173

MICHAILOVSKI, BENJAMIN [1880-1971]
New York Times, 28 July 1971

MICHAUD, LOUIS [1880-1956]
KIEL, 117
E. Jéquier, Verhandl. Schw. Nat. Ges. 136 (1950), 377-382

MICHEEL, FRITZ [1900-1982]
KURSCHNER 13,2571; WWWS, 1173; POGG 6,1725; 7a(3),299-301
Anon., Nachr. Chem. Techn. 13 (1965), 263; Chem. Z. 106 (1982), 350

MICHEL, ANDREAS [1861-1921]
DBJ 3,309; FISCHER 2,1040
K. Lehmann, Correspondenzblatt für Zahnärzte 48 (1922), 63-67

MICHEL, JULIUS [1843-1911]
ASEN, 131; FISCHER 2,1040-1041; SCHWARZ, 158-167

MICHELOTTI, VITTORIO [1774-1842]
FERCHL, 357; POGG 2,146-147

MICHLER, WILHELM [1846-1889]
BLOKH 1,512; ZURICH-D, 123; POTSCH, 302
E.A. Goldi, Ber. chem. Ges. 22(3) (1889), 867-873
Anon., Chem. Z. 14 (1889), 1780-1781

MICHURIN, IVAN VLADIMIROVICH [1855-1935]
DSB 15,425-427; BSE 27,620-629; WWR, 385; KUZNETSOV(1963), 233-267
I.T. Vasilchenko, I.V. Michurin, 2nd Ed. Moscow 1963

MICKEY, GEORGE HENRY [1910-]
AMS 14,3418; WWWS, 1174

MICKLETHWAIT, FRANCES MARY GORE [1868-1950]
Who Was Who 1941-1950, p.791
F.H. Bustall, J. Chem. Soc. 1952, pp.2946-2947

MIDDLETON, WILLIAM SHAINLINE [1890-1975]
AMS 12,4253-4254; WWA 1974-1975, p.2144; NCAB 58,261-262; DAMB, 521-522; WWWS, 1175
C.B. Leake et al., Bull. Hist. Med. 50 (1976), 122-137

MIEG, WALTER [1878-1958]
JV 22,521; NUC 383,3
Anon., Nachr. Chem. Techn. 7 (1959), 27

MIEKELEY, ARTHUR [1897-]
Jahrbuch der Dissertationen der Philosophischen Fakultät...Berlin 1921-1922, p.332

MIELCK, WILHELM HILDEMAR [1840-1896]
R. Schmitz, Geschichte der Hamburger Apotheken 1818-1965, pp.257-258. Frankfurt a.M. 1966

MIES, HEINZ [1902-1976]
KURSCHNER 12,2111; POGG 7a(3),305
Anon., Deutsche med. Wchschr. 101 (1976), 392

MIESCHER, FRIEDRICH [1844-1895]
DSB 9,380-381; PAGEL, 1139; HIS, 226-228; SANDRITTER, 107-109; BOSSART, 118-119; SCHULZ-SCHAEFFER, 198-199; WWWS, 1176; POTSCH, 303
A. Jaquet, Verhandl. Nat. Ges. Basel 11 (1897), 399-417
F. Suter et al., Helv. Physiol. Acta Suppl.II (1944), pp.5-43
H. Buess, Yale J. Biol. Med. 25 (1953), 250-261
M. Staehelin, Basler Stadtbuch 1962, pp.134-162
J. Reiner, Der Beitrag von Friedrich Miescher...zur Geschichte der Zellbiologie. Basle 1963
M. de Meuron-Landolt, Bull. Schw. Akad. Med. Wiss. 25 (1970), 9-24
F. Merke, Gesnerus 30 (1973), 47-52

MIESCHER, KARL [1892-1974]
KURSCHNER 11,1981; SBA 5,87
Anon., Chimia 26 (1972), 99; 28 (1974), 269; Chem. Z. 98 (1974), 373

MIETZSCH, FRITZ [1896-1958]
DUNSCHELE, 144-145; WWWS, 1176; POTSCH, 303; POGG 7a(3),306
Anon., Nachr. Chem. Techn. 4 (1956), 138-139; 6 (1958), 356
H. Breitschneider, Öst. Chem. Z. 57 (1956), 125-126; 60 (1959), 22-23
G. Hecht, Arzneimitt. 9 (1959), 79-80

MIHOLIČ, STANKO [1891-1960]
IB, 199
Enciklopedia Jugoslavije 6,102
Anon., Chem. Eng. News 39(3) (1961), 104

MIKESKA, LOUIS ALOIS [1889-1976]
AMS 11,3617; POGG 6,1732; 7b,3278-3279
Anon., Chem. Eng. News 55(14) (1977), 85

MIKHLIN, DAVID MIKHAILOVICH [1887-1960]
POGG 6,1727; 7b,3267-3269
P.A. Kolesnikov, Biokhimia 25 (1960), 383

MIKULASZEK, EDMUND JULIUS [1895-1978]
WE 7,323; WWWS, 1177

MILAS, NICHOLAS ATHANASIUS [1897-1971]
AMS 11,3618; WWWS, 1177; POGG 6,1733; 7b,3280-3282

MILES, ARNOLD ASHLEY [1904-1988]
WW 1981, p.1785; WWWS, 1177
A. Neuberger, Biog. Mem. Fell. Roy. Soc. 35 (1990), 305-326

MILLAR, WILLIAM SOMERVILLE [1883-1931]
SRS, 37; JV 26,395

MILLER, BENJAMIN FRANK [1907-1971]
AMS 11,3625-3626; COHEN, 168
Anon., J. Am. Med. Assn. 218 (1971), 751; Harvard Med. Alumni Bull. 46(1) (1971), 31

MILLER, CHARLES PHILIPP [1894-1985]
AMS 14,3432; MH 2,373-374; MSE 2,310-311; WWWS, 1179
L.T. Coggeshall, Trans. Assoc. Am. Phys. 99 (1986), xcix-ci

MILLER, EDGAR GRIM [1893-1955]
AMS 8,1707; WWAH, 418
Anon., Science 122 (1955), 280

MILLER, ELIZABETH CAVERT [1920-1987]
AMS 15(5),375; WWA 1980-1981, p.2312; WWWS, 1179

MILLER, MAX [1910-1978]
AMS 13,3022; WWWS, 1181
New York Times, 28 March 1978

MILLER, STANLEY LLOYD [1930-]
AMS 15(5),392; WWA 1980-1981, p.2322; WWWS, 1183

MILLER, WILHELM von [1848-1899]
BJN 4,115-116; BLOKH 1,512-513; POTSCH, 303-304; POGG 3,915-916; 4,1012-1013
O. Doebner, Ber. chem. Ges. 32 (1899), 3756-3776
C. von Voit, Sitz. Bayer. Akad. 30 (1900), 316-324

MILLER, WILLIAM ALLEN [1817-1870]
DSB 9,391-392; DNB 13,429-430; FERCHL, 357; COLE, 387; WWWS, 1183; POTSCH, 304; POGG 2,152; 3,915
C.T., Proc. Roy. Soc. 19 (1871), xix-xxvi
A. Williamson, J. Chem. Soc. 24 (1871), 617-620
C.W. Adams, Isis 34 (1943), 337-339
F. Trusell, J. Chem. Ed. 40 (1963), 612-613

MILLER, WILLIAM LASH [1866-1940]
DSB 9,393-395; AMS 6,982; WWWS, 1183; POTSCH, 304; POGG 4,1013-1014; 5,857-858; 6,1738; 7b,3292-3293
F.B. Kenrick, Proc. Roy. Soc. Canada [3]35 (1941), 131-134; J. Chem. Soc. 1942, pp.334-336
W.D. Bancroft, J. Am. Chem. Soc. 63 (1941), 1-2

MILLIKAN, GLENN ALLAN [1906-1947]
AMS 7,1228

MILLON, EUGÈNE [1812-1867]
DSB 9,401-402; GE 23,997; BALLAND, 69-75; BALLAND(2), 178-179; WWWS,1185; POTSCH, 304-305; BLOKH 1,513-514; FERCHL, 358; POGG 2,152-153; 3,917
L. Figuier, Ann. Scient. 12 (1867), 529-530
J. Reiset et al., E. Millon, sa Vie et ses Travaux de Chimie. Paris 1870
P. Delga and P. Malangeau, Rev. Hist. Pharm. 19 (1968), 69-82

MILLS, WILLIAM HOBSON [1873-1959]
DSB 9,402-404; DNB [1951-1960],739-740; WW 1958, pp.2093-2094; WWWS,1186; POTSCH, 305;DESMOND, 440; ABBOTT-C, 97-98; POGG 5,859; 6,1741; 7b,3298-3299
A.G. Sharpe, Nature 183 (1959), 929-930
F.G. Mann, Biog. Mem. Fell. Roy. Soc. 6 (1960), 201-225

MILLS, WILLIAM SLOAN [1876-1929]
SRS, 27

MILNE-EDWARDS, HENRI [1800-1885]
DSB 9,407-409; GE 15,585-586; LAROUSSE 7,216; HIRSCH 2, 379; WWWS, 1186; CALLISEN 5,525-529; 27,419-422
C. von Voit, Sitz. Bayer. Akad. 16 (1886), 38-44
M. Berthelot, Mem. Acad. Sci. 47 (1904), i-xxxviii

MILROY, JOHN ALEXANDER [1871-1934]
O'CONNOR(2), 436-437; BELFAST, 581; ULLMANN, 107
A. Hunter, Biochem. J. 29 (1935), 983-984

MILROY, THOMAS HUGH [1869-1950]
O'CONNOR(2), 435-436; BELFAST, 580; IB, 200; NUC 385,226
P. Eggleton, Biochem. J. 47 (1950), 385

MILSTEIN, CESAR [1927-]
WW 1983, p.1558; BIBEL, 304-308; ABBOTT, 95-96

MILSTONE, JACOB HASKELL [1912-1976]
AMS 13,3036; WWWS, 1187
Anon., J. Am. Med. Assn. 238 (1977), 169

MINAMI, SEIGO [1893-1975]
T. Yasuda, Japanese Journal of Dermatology 86 (1976), 811-814

MINCHIN, EDWARD ALFRED [1866-1915]
FISCHER 2,1046-1047
Anon., Nature 96 (1915), 148-150
J.H. Ashworth, J. Path. Bact. 20 (1915-1916), 361-365
S.J.H., Proc. Roy. Soc. B89 (1917), xxxviii-xlii
H.M. Woodcock, Parasitology 25 (1925), 157-162

MINES, GEORGE RALPH [1886-1914]
O'CONNOR(2), 45-46; WWWS, 1188
Who Was Who 1897-1915, pp.494-495
Anon., Brit. Med. J. 1915(I), p.142

MINKH (MÜNCH), GRIGORI NIKOLAEVICH [1836-1896]
BSE 27,554; BME 18,636-637; KUZNETSOV(1963), 149-156; FISCHER 2,1088; SKOROKHODOV, 90-94
P. Podiapolsky, Sudhoffs Arch. 18 (1926), 361-368; Janus 43 (1939), 203-213

MINKOWSKI, OSCAR [1858-1931]
DSB 18,626-633; FISCHER 2,1048; PAGEL, 1144; GREIFSWALD 2,378-381; FRANKEN, 191; KOREN, 221; WININGER 4,392-393; 7,321; TETZLAFF, 236-237; WWWS, 1188
W. Stepp, Deutsches Arch. klin. Med. 171 (1931), i-viii
L. Krehl, Arch. exp. Path. Pharm. 163 (1932), 621-634
B.A. Houssay, Diabetes 1 (1952), 112-116
H.M. Dittrich and H. Hahn von Drosche, Anat. Anz. 143 (1978), 509-517
W. Kaiser and A. Voelker, Z. ges. inn. Med. 36 (1981), 973-979

MINOT, CHARLES SEDGWICK [1852-1914]
DSB 9,416; DAB 13,30-31; DAMB, 527; FREUND 1,263-272; WWAH, 421-422; WWWS, 1188
C.W. Eliot, Science 41 (1915), 701-704
F.T. Lewis, Anat. Rec. 10 (1915-1916), 133-164
E.S. Morse, Biog. Mem. Nat. Acad. Sci. 9 (1920), 263-285

MINOT, GEORGE RICHARDS [1885-1950]
DSB 9,416-417; DAB [Suppl.4],580-583; AMS 8,1724; NCAB 38,548-549; FISCHER 2,1049; FORBES, 333-342; DAMB, 527-528; ICC, 122-124; W, 370-371; WWAH, 422; WWWS, 1188; POGG 6,1747; 7b,3317-3318
E.J. Cohn, Am. Phil. Soc. Year Book 1950, pp.313-319
W.B. Castle, Trans. Assoc. Am. Phys. 63 (1950), 11-13; New Eng. J. Med. 247 (1952), 585-592; Biog. Mem. Nat. Acad. Sci. 45 (1974), 337-383
F.M. Rackemann, The Inquisitive Physician. Cambridge, Mass. 1956

MINOVICI, STEFAN [1867-1935]
ILEA, 148-149,530; POGG 5,862; 6,1747; 7b,3318-3319
Anon., Chem. Z. 56 (1932), 385
R. Delaby, Bull. Soc. Chim. [5]3 (1936), 341-344; J. Chem. Soc. 1938, pp.161-163

MINTZ, NAUM [1867-]
RIGA, 711; JV 6,200; NUC 386,500

MINZ, BRUNO [1905-1965]
LDGS, 80
L. Goldstein and E.J. Walaszak, Pharmacologist 8 (1966), 19-20

MIOLATI, ARTURO [1869-1956]
ZURICH-D, 162; POGG 4,1016-1017; 5,862; 6,1748; 7b,3319
C. Sandonnini, Atti Ist. Veneto 115 (1960), 33-41
A. Nasini, Atti Accad. Torino 94 (1960), 121-126
G.B. Kauffman, Isis 61 (1970), 241-253

MIQUEL, FRIEDRICH ANTON WILHELM [1811-1871]
DSB 9,417; NNBW 4,986-987; LINDEBOOM, 1342-1344; UTRECHT, 79-80
F. von Kobell, Sitz. Bayer. Akad. 1 (1871), 151-153
C.J. Matthes, Jaarb. Akad. Wet. 1872, pp.29-49
F.A. Stafleu, Wentia 16 (1966), 1-98; Essays in Biochemistry (P. Smit and R.J.C.V. ter Laage, Eds.), pp.295-341. Utrecht 1970

MIQUEL, PIERRE [1850-1922]
BULLOCH, 384; RSC 17,272-273; WWWS, 1188-1189
A. Girauld, Paris Médical 44 (1922), 259-260

MIRBEL, CHARLES de [1776-1854]
DSB 9,418-419; GE 23,1100; LAROUSSE 11,320; HOEFER 35,657-660; WWWS, 441; TSCHIRCH, 1090; CALLISEN 3,177-180; 26,449-450
W. Margolie, Vie et Travaux de M. de Mirbel. St. Germain 1863

MIROMANOFF, ANDRÉ [1902-]
POGG 7a(3),311-312

MIRSKY, ALFRED EZRA [1900-1974]
DSB 18,633-636; AMS 12,4315; COHEN, 169; WWWS, 1189
P.F. Cranefield, J. Gen. Physiol. 64 (1974), 131-133
B.S. McEwen, Am. Phil. Soc. Year Book 1976, pp.100-103
C. Kopp, Mendel Newsletter 23 (1983), 1-5

MISLAVSKI, NIKOLAI ALEKSANDROVICH [1854-1928]
BSE 27,589; BME 18,763-764; WWR, 391; ZMEEV 5,14-15; IB, 201
A.V. Kibyakov and K.V. Lebedev, N.A. Mislavski. Moscow 1951

MISLOW, KURT MARTIN [1923-]
AMS 15(5),415; BHDE 2,822; WWWS, 1189

MISLOWITZER, ERNST [see MYLON]

MITCHELL, ALEXANDER MONCRIEFF [1822-1874]
W.I. Addison, Matriculation Albums of the University of Glasgow from 1728 to 1858, p.419. Glasgow 1913

MITCHELL, FREDERICK LEWIS [1921-]
WWWS, 1190
Who's Who of British Scientists 1980-1981, p.340

MITCHELL, HAROLD HANSON [1886-1966]
AMS 10,2815; IB, 201
Anon., Chem. Eng. News 44(19) (1966), 86
M. Edman et al., J. Nutrition 96 (1969), 1-9
R.M. Forbes, Journal of Animal Science 69 (1991), 1-4

MITCHELL, HELEN SWIFT [1905-1984]
AMS 11,3672
New York Times, 13 December 1984

MITCHELL, HERSCHEL KENWORTHY [1913-]
AMS 15(5),418; POTSCH, 305

MITCHELL, JOSEPH STANLEY [1909-1987]
WW 1985, p.1342; WWWS, 1190
D.H. Marrian, Biog. Mem. Fell. Roy. Soc. 34 (1988), 583-607

MITCHELL, MARK LEDINGHAM [1902-1977]
WW 1978, p.1708
Anon., Austral. J. Exp. Biol. 56 (1978), 383-384

MITCHELL, PETER CHALMERS [1864-1945]
DNB [1941-1950], 599-600; WW 1945, pp.1907-1908; WWWS, 1190
P.C. Mitchell, My Fill of Days. London 1937
E. Hindle, Obit. Not. Fell. Roy. Soc. 5 (1947), 367-372

MITCHELL, PETER DENNIS [1920-1992]
WW 1981, p.1810; IWW 1982-1983, pp.900-901; MSE 2,317-318; ABBOTT, 96
P. Mitchell, in Of Oxygen, Fuels and Living Matter, Part 1 (G. Semenza, Ed.), pp.1-160. Chichester 1981
J.D. Robinson, Persp. Biol. Med. 27 (1984), 367-383
P. Garland, Nature 356 (1992), 747

MITCHELL, PHILIP HENRY [1883-1955]
AMS 9(I),1343; IB, 201; WWAH, 423; WWWS, 1190

MITCHELL, SILAS WEIR [1829-1914]
DSB 9,422-423; DAB 13,62-65; HIRSCH 4,221-222; DAMB, 530-531; WWWS, 1191
A.R. Burr, Weir Mitchell: His Life and Letters. New York 1929
J.M. Taylor, Ann. Med. Hist. NS1 (1929), 583-598
A.J. Carlson, Science 87 (1938), 474-478
P. Bailey, Biog. Mem. Nat. Acad. Sci. 32 (1958), 334-353
R.D. Walter, S. Weir Mitchell M.D. Springfield, Ill. 1970
W.B. Fye, Bull. Hist. Med. 57 (1983), 188-202; The Development of American Physiology, pp.54-91. Baltimore 1987

MITCHISON, (NICHOLAS) AVRION [1928-]
WW 1987, p.1224; IWW 1987-1988, p.1018

MITLACHER, WILHELM [1872-1913]
FISCHER 2,1050; DECKER, 36-42; TSCHIRCH, 1090

MITSCHERLICH, ALEXANDER [1836-1918]
DBJ 2,698; BLOKH 2,517-518; MATSCHOSS, 177; WWWS, 1191; POTSCH, 305-306; POGG 3,921; 4,1017; 5,863-864
E. Wiedemann, Z. angew. Chem. 29 (1916), 229
B. Lepsius, Ber. chem. Ges. 51 (1918), 1030-1033
R. Lorenz, Jahresb. Phys. Ver. Frank. 1919, pp.5-12

MITSCHERLICH, EILHARD [1794-1863]
DSB 9,423-426; ADB 22,15-22; BUGGE 1,450-457; PRANDTL, 242-285; W, 371; FARBER, 483-493; BLOKH 2,514-517; FERCHL, 359-360; TSCHIRCH, 1090; SCHAEDLER, 81-82; MATSCHOSS, 177-178; COLE, 388-389; ABBOTT-C, 98; POTSCH, 306-307; WWWS, 1191; CALLISEN 13,113; 30,396-397; POGG 2,160-162
D.F.P. Martius, Sitz. Bayer. Akad. 1863(II), pp.388-391
Anon., Proc. Roy. Soc. 13 (1864), ix-xvi; Arch. Pharm. 172 (1865), 1-25
G. Rose, Z. geol. Ges. 16 (1864), 21-72
A. Williamson, J. Chem. Soc. 17 (1864), 440-442
E. Mitscherlich, Gesammelte Schriften. Berlin 1896
E.O. von Lippmann, Abhandlungen und Vorträge, pp.306-322. Leipzig 1900
W. Ostwald, Abhandlungen und Vorträge, pp.384-388. Leipzig 1904
P. Henrich, Chem. Z. 37 (1913), 1369-1370,1398-1399
R. Winderlich, J. Chem. Ed. 26 (1949), 358-361
H.W. Schuett, Physis 16 (1974), 5-22
E.M. Melhado, Hist. Stud. Phys. Sci. 11 (1980), 87-123

MITSCHERLICH, EILHARD ALFRED [1874-1956]
KROLLMANN, 1024-1025; KIEL, 214-215; IB, 201
E. Boguslawski, Z. Ackerbau 91(3) (1949), i-vii
R. Hoffmann, Z. Pflanzenern. 49 (1950), 1-6
Anon., Chem. Z. 80 (1956), 145; Leopoldina [3]2 (1956), 23-24

MITSCHERLICH, KARL GUSTAV [1805-1871]
ADB 22,22; HIRSCH 4,223-224; PAGEL, 1146; FERCHL, 360; WWWS, 1191; CALLISEN 13,111-112; 30,395-396; POGG 2,162; 3,921

MITTASCH, PAUL ALWIN [1869-1953]
DSB 9,427-428; POTSCH, 307; POGG 4,1018; 5,864; 6,1754; 7a(3),317-318
E. Pietsch, Z. angew. Chem. 52 (1939), 719-720
M. Enger, Naturw. Rund. 3 (1950), 284-285
W. Frankenberg, Adv. Catalysis 6 (1954), vii-x
K. Holdermann, Chem. Ber. 90 (1957), xli-lvi
E. Farber, Chymia 11 (1966), 157-178

MIURA, KINNOSUKE [1864-1950]
FISCHER 2,1050; RSC 17,280
S. Katsanuma, Folia Psychiat. Jap. 4 (1951), 371-373

MIXTER, WILLIAM GILBERT [1846-1936]
AMS 5,784; WWAH, 423
Anon., Ind. Eng. Chem. News Ed. 14 (1936), 115

MIYAKE, SUGURU [1894-1967]
JBE, 879,2260; NUC 388,360

MIZUSHIMA, SAN-ICHIRO [1899-1983]
JBE, 895-896; IWW 1982-1983, p.903; STC 2,257-258; WWWS, 1193; POGG 6,1755; 7b,3331-3336
Anon., Science 223 (1984), 616

MOCHIZUKI, JUN-ICHI [1859-]
HAKUSHI 2,126-127

MOCHNACKA, IRENA [1905-1979]
T. Szymczyk and Z. Porembska, Acta Biochem. Polon. 27 (1980), 177-180

MODRZEJEWSKI, EDMUND [1849-1893]
PSB 21,534-535; KOSMINSKI, 331-332,616; KONOPKA 6,190-191
Anon., Gazeta Lekarska [2]3 (1893), 413

MÖBIUS, KARL AUGUST [1825-1908]
DSB 9,431-432; BJN 13,64*; WREDE, 131-132; KIEL, 114; DZ, 974; WWWS, 1193
F. Dahl, Zool. Jahrb. Suppl.8 (1905), 1-22
Anon., Nature 78 (1908), 82-83
R. von Hansteen, Naturw. Rund. 23 (1908), 361-363,373-375

MÖBIUS, PAUL JULIUS [1853-1907]
BJN 12,58*; FISCHER 2,1053-1054; PAGEL, 1146; WWWS, 1193
H. Kron, Deutsche med. Wchschr. 33 (1907), 351-352
A. Weygandt, Münch. med. Wchschr. 54 (1907), 477-480
F. Jentsch, Zum Andenken an Paul Julius Möbius. Halle 1907
F. Windscheid, Jahrbuch der gesamten Medizin 293 (1907), 225-231
E.K. Waldeck-Semadini, Paul Julius Möbius. Berne 1980
W. Theobald, Med. Hist. J. 18 (1983), 100-117

MÖHLAU, RICHARD [1857-1940]
DZ, 976; TLK 3,450; POTSCH, 307; POGG 3,923; 4,1019; 5,865; 6,1756; 7a(3),319
Anon., Z. angew. Chem. 50 (1937), 830; 53 (1940), 595
W. Konig, Ber. chem. Ges. 73A (1940), 121-122

MÖLLENDORFF, WILHELM von [1887-1944]
KURSCHNER 4,1988; FISCHER 2,1054; KIEL, 86; IB, 202
B.M. Zurgligen, Der Anatom Wilhelm von Möllendorff. Zurich 1991

MÖLLER, ERNST FRIEDRICH [1906-]
POGG 7a(3),320-321

MØLLGAARD, HOLGER [1885-1973]
DBL (3rd Ed.) 10,283-284; FISCHER 2,1056
Anon., Z. Tierphysiol. 33 (1974), 177-179

MOELWYN-HUGHES, EMYR ALUN [1905-1978]
SRS, 89
J.M. Thomas, Nature 277 (1979), 334
M. Davies, Chem. Brit. 15 (1979), 397

MÖRICKE, EMIL [1822-1897]
A. Wankmüller and K.D. Möricke, Beitr. Württ. Apothekgesch. 11 (1976),65-90

MÖRICKE, MARTIN [1824-1881]
A. Wankmüller and K.D. Möricke, Beitr. Württ. Apothekgesch. 11 (1976), 65-90

MÖRING, WALTHER [1877-]
JV 21,275

MÖRNER, CARL THORE [1864-1940]
SMK 5,392; FISCHER 2,1057; IB, 202; POGG 4,1020-1021; 7b,3342-3343
M. Lindqvist, Sven. Farm. Tid. 44 (1940), 482
H. Theorell, Lev. Sven Vet. 10 (1968), 203-211

MÖRNER, KARL AXEL HAMPUS [1854-1917]
SMK 5,392
Anon., Chem. Z. 41 (1917), 301; Sven. Kem. Tid. 29 (1917), 230-231

MOEWUS, FRANZ [1908-1959]
KURSCHNER 8,1588
Anon., Science 130 (1959), 381; Nachr. Chem. Techn. 7 (1959), 224
J. Sapp, Hist. Phil. Life Sci. 9 (1987), 277-308; Where the Truth Lies. Cambridge 1990

MOFFITT, HERBERT CHARLES [1867-1951]
AMS 8,1732
W.J. Kerr, Trans. Assoc. Am. Phys. 64 (1951), 12-14

MOHL, HUGO von [1805-1872]
DSB 9,441-442; ADB 22,55-57; SL 10,375-387; FREUND 1,273-280; WWWS, 1194; VAPEREAU 4,1285-1286; LEHMANN, 180; TSCHIRCH, 1090; POGG 3,926
A. de Bary, Bot. Z. 30 (1872), 561-580
A. Schrötter, Alm. Akad. Wiss. Wien 23 (1873), 156-169
W. Ahles, Jahreshefte Württ. 29 (1873), 41-65
Anon., Proc. Roy. Soc. 23 (1875), i-vi
F.K. Studnicka, Protoplasma 27 (1937), 619-629
K. Ulshofer, Med. Welt 1964, pp.981-985
W. Pelz, Zellenlehre: Der Einfluss Hugo von Mohls auf die Entwicklung der Zellenlehre. Frankfurt a.M. 1987

MOHR, EDUARD CARL JULIUS [1873-1969]
RSC 17,300; NUC 389,401-402
Wie is Dat? 1956, p.421
F.A. Baren, Jaarboek der Rijksuniversiteit te Utrecht 1969-1970, pp.14-15

MOHR, ERNST [1873-1926]
DRULL, 183; BLOKH 2,519; POGG 4,1022; 5,867-868; 6,1760-1761
R. Stolle, Ber. chem. Ges. 59A (1926), 39-41

MOHR, KARL FRIEDRICH [1806-1879]
DSB 9,445-446; ADB 22,67-69; HEIN, 441-443; BLOKH 2,519-520; WWWS, 1195; FERCHL, 362-363; SCHAEDLER, 82; SCHELENZ, 679; POTSCH, 307-308; POGG 2,171-172; 3,927-928; 6,1761
C. von Voit, Sitz. Bayer. Akad. 10 (1880), 266-268
R. Hasenclever, Ber. chem. Ges. 33 (1900), 3827-3838
R.E. Oesper, J. Chem. Ed. 4 (1927), 1357-1363
J.M. Scott, Chymia 3 (1950), 191-203
R. Lotze, Der Einfluss von C.F. Mohr auf die Entwicklung der Massanalyse. Frankfurt a.M. 1968
N.A. Figurovski and V.J. Zacharias, NTM 14(2) (1977), 37-54

MOHR, PHILIPP [-1885]
Anon., Pharm. Z. 30 (1885), 469

MOHS, RUDOLF [1839-]
RSC 8,421

MOILIN, JULES ANTOINE (TONY) [1831-1871]
SG [1]9,361
P. Ganière, Presse Med. 74 (1966), 2775-2778

MOISSAN, HENRI [1852-1907]
DSB 9,450-452; CHARLE, 202-204; WWWS, 1195; POTSCH, 306-307; POGG 3,928;4,1022-1024; 5,868-869
A. Stock, Ber. chem. Ges. 40 (1907), 5099-5130
P. Lebeau, Bull. Soc. Chim. [4]3 (1908), i-xxxviii
W. Ramsay, J. Chem. Soc. 101 (1912), 477-488

MOKRUSHIN, SERGEI GRIGORIEVICH [1896-]
POGG 6,1762-1763; 7b,3355-3359

MOLDAVE, KIVIE [1923-]
AMS 15(5),439; WWWS, 1196

MOLDENHAUER, FERDINAND [1829-1913]
HEIN-E, 319

MOLDENHAUER, KARL AUGUST FRIEDRICH [1797-1866]
HB 3,249-251; POGG 2,176

MOLDENHAUER, WILHELM [1874-1933]
TLK 3,451; POGG 6,1763
E. Berl, Ber. chem. Ges. 66A (1933), 41

MOLESCHOTT, JACOB [1822-1893]
DSB 9,456-457; ADB 52,435-438; NNBW 3,874; LINDEBOOM, 1350-1353; HIRSCH 4,232; PAGEL, 1148-1149; WWWS, 1196; POGG 2,176; 3,939
H. Vierordt, Münch. med. Wchschr. 40 (1893), 483-484
Anon., Arch. Ital. Biol. 20 (1894), 1-14
E.P. Evans, Pop. Sci. Mon. 49 (1896), 399-406
J. Moleschott, Für meine Freunde, 2nd Ed. Giessen 1901
M.A. van Herwerden, Janus 20 (1915), 174-201,409-436
P. Daglio, Pag. Stor. Med. 5(4) (1961), 3-11
W. Moser, Der Physiologe Jacob Moleschott und seine Philosophie. Zurich 1967
J. O'Hara-May, J. Hist. Biol. 4 (1971), 249-273
R.C.V.J. ter Laage, Janus 63 (1976), 95-109
F. Gregory, Scientific Materialism in Nineteenth Century Germany. Dordrecht 1977
U. Hagelgans, Jacob Moleschott als Physiologe. Frankfurt a.M. 1985

MOLISCH, HANS [1856-1937]
OBL 6,251; STURM 2,688; FREUND 1,281-290; SACKMANN, 225-229; WWWS, 1196; PLANER, 230; BK, 271; IB, 203; TLK 3,451; POGG 6,1765; 7a(3),341
H. Molisch, Erinnerungen und Welteindrücke einer Naturforschers. Vienna 1934
K. Höfler, Ber. bot. Ges. 56 (1938), (161)-(199); Phyton 7 (1957), 199-205
O. Richter, Alm. Akad. Wiss. Wien 88 (1938), 221-234
O. Gertz, Bot. Not. 1938, pp.318-326
H. Gest, Photosynthesis Research 30 (1991), 49-59

MOLITOR, HANS [1895-1970]
STURM 2,688; DECKER, 106-115; WWAH, 425
A. Lindner, Wiener klin. Wchschr. 82 (1970), 871
E.B. Cook, Pharmacologist 13 (1971), 33

MOLL, KARL FRIEDRICH AUGUST [1897-]
JV 41,429; NUC 390,186

MOLL, LEOPOLD [1877-1933]
KURSCHNER 3,1603; FISCHER 2,1059; PELZNER, 192-202; ATUYANA, 143-152; KOERTING, 183
J. Zappert, Wiener med. Wchschr. 83 (1933), 294-295
E. Stransky, Jahrb. Kinderheil. 139 (1933), 271-275

MOLL, WILLEM JAN HENRI [1876-1947]
BWN 1,400-401; POGG 6,1765; 7b,3365

MOLLIARD, MARIN [1866-1944]
DSB 9,460-461; CHARLE(2), 204-205; WWWS, 1197
C. Maurin, C.R. Acad. Sci. 219 (1944), 144-147
R. Combes, Rev. Gen. Bot. 53 (1946), 145-157
J. Magrou, Not. Acad. Sci. 2 (1949), 489-512

MOLLIER, SIEGFRIED [1866-1954]
IB, 202; EGERER, 130-134; JV, 208
B. Romeis, Bayer Akad. Wiss. Jahrbuch 1955, pp.172-175
T. von Lanz, Anat. Anz. 106 (1959), 130-143

MOLONEY, PETER JOSEPH [1891-]
AMS 11,3691; WWWS, 1197

MOND, ALFRED MORITZ [1868-1930]
DNB [1922-1930], 602-605; WININGER 4,412-413; 7,325
Anon., Proc. Roy. Soc. A131 (1931), ii-v

MOND, LUDWIG [1839-1909]
DSB 9,466-467; DNB [1901-1911] 2,641-644; LKW 6,226-232; BLOKH 2,525-527; MATSCHOSS, 179; WININGER 4,413; ABBOTT-C, 99; W, 373-374; POTSCH, 309; WWWS,1198; POGG 4, 1026; 5,871
Anon., Nature 82 (1909), 221-223
E. Hatschek, Chem. Z. 33 (1909), 1326
C. Langer, Ber. chem. Ges. 43 (1910), 3665-3682
J.I. Watts, J. Chem. Soc. 113 (1918), 318-334
F.G. Donnan, Proc. Roy. Inst. 30 (1939), 709-736
J.M. Cohen, The Life of Ludwig Mond. London 1956
P.J.T. Morris, Endeavour NS13 (1989), 34-40

MOND, RUDOLF [1894-1960]
FISCHER 2,1061; KIEL, 101; IB, 203; POGG 6,1768; 7a(3),342-343; (4),150*
Anon., Nachr. Chem. Techn. 9 (1961), 8

MONJÉ, MANFRED [1901-1981]
KURSCHNER 13,2613; KIEL, 105; POGG 7a(3),343-344
Anon., Naturw. Rund. 35 (1982), 45
C. Weiss, Christiana Albertina NF16 (1982), 211-212

MONNIER, ALFRED [1874-1917]
BRIQUET, 335-336; GENEVA 5,295-296; 6,45-47
A. Pictet, Mem. Soc. Phys. Hist. Nat. Gen. 39 (1919), 55-59

MONNIER, DENYS [1834-1898]
DHB 4,780; GENEVA 3,34-35; 4,111-113; 5,40
Anon., Chem. Z. 22 (1898)
A. Rilliet, Mem. Soc. Phys. Nat. Gen. 33 (1898-1901), ix-xii

MONOD, JACQUES [1910-1976]
DSB 18,636-649; CB 1971, pp.276-279; MH 1,331-332; STC 2,258-262; MSE 2,320; ABBOTT, 96-97; WWWS, 1199; POTSCH, 309-310
B. F[antini], Scientia 110 (1975), 907-912
Anon., Lancet 1976(I), p.1421
M.R. Pollock, TIBS 1 (1976), N208-N209
A. Lwoff, Nouv. Pr. Med. 5 (1976), 2002-2004; Biog. Mem. Fell. Roy. Soc. 23 (1977), 385-412
R.Y. Stanier, J. Gen. Microbiol. 10 (1977), 1-12
M. Cohn, in Operon (J.H. Miller and W.S. Reznikoff, Eds.), pp.1-9. Cold Spring Harbor 1978
A. Lwoff and A. Ullmann (Eds.), Origins of Molecular Biology. New York 1979
M.D. Grmek and B. Fantini, Rev. Hist. Sci. 35 (1982), 193-215
M. Morange, Fundamenta Scientiae 3 (1982), 396-404

MONTAGNE, PIETER JOHANNES [1867-1925]
POGG 5,871; 6,1769
W.P. Jorissen, Chem. Wkbl. 21 (1924), 585-591; 22 (1925), 453-454

MONTEVERDE, NIKOLAI AUGUSTINOVICH [1856-1929]
IB, 2-3
G.A. Nadson, Priroda 18 (1929), 1076-1077

MONTGOMERY, EDMUND DUNCAN [1835-1911]
DSB 9,494-495; DAB 13,95-96
M.T. Kenton, J. Hist. Ideas 8 (1947), 309-341; The Philosophy of Edmund Montgomery. Dallas 1950

MONTGOMERY, THOMAS HARRISON [1873-1912]
DSB 9,495-497; DAB 13,99-100; NCAB 15,216-217
E.G. Conklin, Science 38 (1913), 207-214

MONTMOLLIN, GUILLAUME de [1884-]
JV 26,641; NUC 392,490
Schweizerisches Geschlechterbuch 6 (1936), 429

MOORE, ARTHUR RUSSELL [1882-1962]
AMS 10,2838
B.T. Scheer, Science 137 (1962), 411-412

MOORE, BENJAMIN [1867-1922]
O'CONNOR(2), 372-374; SRS, 13; FISCHER 2,1063-1064
F.G. H[opkins], Proc. Roy. Soc. B101 (1927), xvii-xix; Biochem. J. 22 (1928), 1-3

MOORE, CARL RICHARD [1892-1955]
AMS 9(II),788; IB, 203; WWAH, 426; WWWS, 1202
L.D. Dragstedt, Science 123 (1956), 496-497
D. Price, Endocrinology 58 (1956), 529-530; Biog. Mem. Nat. Acad. Sci. 45 (1974), 385-412

MOORE, CARL VERNON [1908-1972]
AMS 12,4358; DAMB, 534; WWAH, 426; WWWS, 1202
W.J. Harrington et al., Blood 30 (1972), 771-775

MOORE, CHARLES WATSON [1879-1956]
JV 22,521; SG [5]1,348; NUC 393,136
Anon., J. Roy. Inst. Chem. 1956, p.488

MOORE, FORRIS JEWETT [1867-1926]
MILES, 343-344; WWAH, 427; WWWS, 1203; POGG 5,874; 6,1772
A.H. Gill, Ber. chem. Ges. 60A (1927), 53-54
T.L. Davis, Ind. Eng. Chem. 19 (1927), 1066

MOORE, NORMAN SLAWSON [1901-]
AMS 11,3712; WWA 1978-1979, p.2298; WWWS, 1204

MOORE, STANFORD [1913-1982]
AMS 15(5),468; WWA 1980-1981, p.2360; IWW 1982-1983, pp.912-913; AO 1982, pp.400-402; STC 2,262-264; MSE 2,322-323; ETTRE, 297-308; POTSCH, 310
D.G. Smyth, TIBS 8 (1983), 44-45
C.H.W. Hirs, Anal. Biochem. 136 (1984), 3-6
E.L. Smith and C.H.W. Hirs, Biog. Mem. Nat. Acad. Sci. 56 (1987),355-385

MORA, PETER TIBOR [1924-]
AMS 12,4374; WWWS, 1206

MORACZEWSKI, WACLAW DAMIAN [1867-1950]
PSB 21,691-692; ZURICH-D, 93; IB, 204; KONOPKA 6,200-202
Anon., Tow. Nauk. Lwow. 1 (1921), 288-289; 13 (1933), 211-212

MORAHT, HERMANN [1864-1901]
JV 5,234; RSC 16,486; 17,342
Anon., Ber; chem. Ges. 34 (1901), 4385

MORALES, FRANK MANUEL [1919-]
AMS 15(5),472

MORAT, JEAN PIERRE [1846-1920]
DSB 9,505-507; FISCHER 2,1065; GENTY 6,257-272; GUIART, 128
A. Laveran, Bull. Acad. Med. 84 (1920), 77-78

MORAVEK, VLADIMIR [1896-]
POGG 6,1776; 7b,3388-3390

MORAWETZ, HERBERT [1915-]
AMS 15(5),474
M.A. Winnek, Macromolecules 18 (1985), 1797-1798

MORAWITZ, HUGO [1882-]
JV 25,530; NUC 394,137

MORAWITZ, PAUL [1879-1936]
KURSCHNER 4,1995; FISCHER 2,1066; GEISSLER, 54-70; GREIFSWALD 2,381-385
M. Hochrein, Deutsche med. Wchschr. 62 (1936), 1233-1234
L. Krehl, Münch. med. Wchschr. 83 (1936), 1397-1398
R. Schoen, Deutsches Arch. klin. Med. 179 (1936), i-x
G. Schoene, Deutsche med. Wchschr. 84 (1959), 692-696
A.M. Mingers, Würzburger Medizinischhtorische Mitteilungen 8 (1990), 53-72

MORAX, VICTOR [1866-1935]
FISCHER 2,1066; SACKMANN, 229-231; WWWS, 1206
F. de Lapersonne, Bull. Acad. Med. 113 (1935), 686-691; Presse Med. 43 (1935), 889-890
Anon., Ann. Inst. Pasteur 54 (1935), 649-652
A. Magitot, Ann. Ocul. 172 (1935), 529-563
P. Baillart, Ann. Ocul. 199 (1966), 225-237
P. Bregeat, Historia Ophthalmologica Internationals 3 (1984), 201-211

MOREAU, FRANÇOIS ARMAND [1823-1881]
RSC 4,462-463; 7,435; 10,845
C. Sappey, Bull. Acad. Med. 10 (1881), 936-940

MOREL, ALBERT [1875-1953]
IB, 204; POGG 6,1778-1779; 7b,3397-3398
M. Chambon, Ann. Pharm. Fran. 11 (1953), 238-241

MOREL, CHARLES BASILE [1822-1884]
BERGER-LEVRAULT, 167-168

MORENZ, PAUL [1870-]
JV 13,78; NUC 394,529

MORESCHI, ANNIBALE [1886-1931]
Anon., Chimica e Industria 13 (1931), 388

MORESCHI, CARLO [1876-1921]
EI 23,818; FISCHER 2,1067
A. Ascoli, Bioch. Ter. Sper. 8 (1921), 195-198; Policlinica 30 (1923), 1135-1140
D. Cesa-Bianchi, Atti Soc. Lomb. Sci. Med. 10 (1921), 312-316
G. Giunchi, Sci. Med. Ital. 7 (1958), 213-218

MORGAN, AGNES FAY [1884-1968]
AMS 11,3725-3726; MILES, 348-349; NAW 4,495-497; IB, 204; WWWS, 1207
G. Emerson, Fed. Proc. 36 (1977), 1911-1914

MORGAN, EDWARD JOSEPH [1885-]
AMS 6,1004; NUC 394,678

MORGAN, GILBERT THOMAS [1872-1940]
DNB [1931-1940], 629-630; FINDLAY, 316-352; WWWS, 1208; POTSCH, 310; POGG 5,876; 6,1779-1780; 7b,3402-3403
W. Wardlaw, J. Chem. Soc. 1941, pp.689-697
J.C. Irvine, Obit. Not. Fell. Roy. Soc. 3 (1941), 355-362

MORGAN, JOHN [1797-1847]
HIRSCH 4,266; DECHAMBRE [2]9,468-469; CALLISEN 13,228; 30,440

MORGAN, JOHN LIVINGSTON RUTGERS [1872-1935]
AMS 5,794; WWAH, 430; POGG 4,1030-1031; 5,876; 6,1781; 7b,3404
Anon., Ind. Eng. Chem. News Ed. 13 (1935), 204; Science 81 (1935), 396

MORGAN, LILLIAN VAUGHAN [1870-1952]
AMS 8,1753
M. Keenan, American Zoologist 23 (1983), 867-876

MORGAN, THOMAS HUNT [1866-1945]
DSB 9,515-526; DAB [Suppl.3],538-541; AMS 7,1254; NCAB 35,184; WWAH, 431; SCHULZ-SCHAEFFER, 205-206; DAMB, 537-538; ABBOTT, 97; IB, 204; W, 377-378; WWWS, 1209
K. von Frisch, Bayer. Akad. Wiss. Jahrbuch 1944-1948, pp.267-268
G.R. deBeer, Obit. Not. Fell. Roy. Soc. 5 (1947), 451-466
O. Storch, Alm. Akad. Wiss. Wien 100 (1950), 358-367
A.H. Sturtevant, Biog. Mem. Nat. Acad. Sci. 33 (1959), 283-325
G.E. Allen, Proc. Am. Phil. Soc. 110 (1966), 48-57; J. Hist. Biol. 1 (1968), 113-140; Quart. Rev. Biol. 44 (1969), 168-188; homas Hunt Morgan. Princeton 1978; Social Research 51 (1984), 709-738; Trends in Genetics 1 (1985), 151-154,186-190
I. Shine and S. Wrobel, Thomas Hunt Morgan. Lexington, Ky. 1976
J.A. Ramage, Filson Club Historical Quarterly 53 (1979), 5-25
I.M. Mountain et al., American Zoologist 23 (1983), 825-865
M. Lederman, J. Hist. Biol. 22 (1989), 163-176

MORGAN, WALTER THOMAS JAMES [1900-]
WW 1981, p.1839; IWW 1982-1983, pp.914-915
R.D. Marshall, Bioch. Soc. Trans. 9 (1981), 185-186
W.T.J. Morgan, TIBS 8 (1983), 220-222

MORGENROTH, JULIUS [1871-1924]
FISCHER 2,1068; KALLMORGEN, 356-357; BULLOCH, 384; WININGER 4,427-428;KOREN, 222; WWWS, 1209
F. Neufeld, Deutsche med. Wchschr. 51 (1925), 159
H. Sachs, Klin. Wchschr. 4 (1925), 239
Anon., Chem. Z. 49 (1925), 9; Z. angew. Chem. 38 (1925), 20

MORGULIS, SERGIUS [1885-1971]
AMS 10,2860; FISCHER 2,1068-1069; CHITTENDEN, 307; IB, 204; KOREN, 222;KAGAN(JA), 738; WININGER 7,330; POGG 6,1781; 7b,3405

MORI, TAKAJIRO [1897-1978]
JBE, 910

MORIN, ANTOINE [1800-1879]
STELLING-MICHAUD 4,594; FERCHL, 369; TSCHIRCH, 1092; POGG 2,208; 3,935; CALLISEN 13,236; 30,442

MORIN, LOUIS CÉSAR [1833-1867]
BALLAND, 76; GORIS, 517

MORIN, PYRAME LOUIS [1815-1864]
BRIQUET, 336; FERCHL. 369; POGG 2,208-209; 3,935
E. Plantamour, Mem. Soc. Phys. Hist. Nat. Gen. 18 (1865), 161-164

MORITZ, FRIEDRICH [1861-1938]
KURSCHNER 4,1997; FISCHER 2,1069-1070; PAGEL, 1159-1160; BACHMANN, 36-50; DZ, 985-986; SKVARC, 153-155; RSC 17,356-357
E. Schott, Deutsche med. Wchschr. 57 (1931), 2118-2119
H. Deilen, Arch. klin. Med. 181 (1938), 531-538
E. Weichmann, Deutsche med. Wchschr. 64 (1938), 279
G. Wüllenweber, Münch. med. Wchschr. 85 (1938), 257

MORKOTUN, KONSTANTIN STEPANOVICH [1861-1910]
SG [3]7,1354
M. Gorodenski, Zhurnal Russkovo Obshestvo Okhraniena Narodnovo Zdravia 20(3/4) (1910), 78-82
A.A. Sharov, Sov. Zdrav. 1986(6), pp.67-69

MORKOVIN, NIKOLAI VASILIEVICH [1873-1904]
VENGEROV 2,130

MORLEY, EDWARD WILLIAMS [1838-1923]
DSB 9,530-531; DAB 13,192-193; NCAB 4,520; MILES, 349-350; WWAH, 431; POTSCH, 310-311; W, 378; WWWS, 1210; POGG 3,936; 4,1031-1032; 5,877; 6,1782
O.F. Tower, Science 57 (1923), 431-434
C.F. Thwing, Science 73 (1931), 276-277
H.R. Williams, Edward Williams Morley. Easton, Pa. 1957

MORLEY, HENRY FORSTER [1855-1943]
POGG 3,936; 4,1032; 6,1782; 7b,3406
F.H. Carr, Nature 151 (1943), 608-609
J.A. Gardner, J. Chem. Soc. 1944, pp.43-46

MORO, ERNST [1874-1951]
KURSCHNER 7,1391; FISCHER 2,1070; WWWS, 1210
H. Kleinschmidt, Mon. Kind. 99 (1951), 311-312

MOROKHOVETS, LEV ZAKHAROVICH [1848-1918]
BME 19,38-39; ZMEEV 5,21-22; RSC 17,359; SG [1]9,461; [2]11,92; [3]7,1355; NUC 395,561

MORRELL, CLARENCE ALLISON [1899-1983]
AMS 12,4392; WWWS, 1211

MORRELL, ROBERT SELBY [1867-1946]
RSC 17,362
L.A. Jordan, J. Chem. Soc. 1947, pp.432-433
Anon., J. Roy. Inst. Chem. 1947, p.32

MORREN, CHARLES FRANÇOIS ANTOINE [1807-1858]
BNB 15,275-280; HOEFER 36,650-651; FERCHL, 370; POGG 2,211-212
E. Morren, Ann. Acad. Roy. Belg. 26 (1860), 167-251
J.R. Baker, Isis 42 (1951), 285-287

MORREN, JEAN FRANÇOIS AUGUSTE [1804-1870]
FERCHL, 370; POGG 3,937

MORRIS, COLIN JOHN OWEN RHONABWY [1910-1981]
Who's Who of British Scientists 1971-1972, p.603
R. Bonnett, Chem. Brit. 18 (1982), 729

MORRIS, GEORGE HARRIS [1858-1902]
Anon., Analyst 27 (1902), 76

MORRIS, HAROLD PAUL [1900-1982]
AMS 14,3532; WWWS, 1211

MORRIS, JAMES LUCIEN [1885-1926]
AMS 3,490; NCAB 20,438-439

MORRISON, MARTIN [1921-1987]
AMS 17(5),518
H.S. Mason, in Oxidases and Related Redox Sysytems (T.E. King et al., Eds.), pp.xxv-xxvi. New York 1988

MORSE, EDWARD SYLVESTER [1838-1925]
DSB 9,536-537; DAB 13,242-243; NCAB 24,407-408; WWWS, 1213
J.S. Kingsley, Proc. Am. Acad. Arts Sci. 61 (1926), 549-555
L.O. Howard, Biog. Mem. Nat. Acad. Sci. 17 (1935), 3-29

MORSE, HARMON NORTHROP [1848-1920]
DAB 13,243-244; NCAB 16,69; WWAH, 433; WWWS, 1213; POGG 4,1033; 5,878; 6,1783
I. Remsen, Science 52 (1920), 497-500; Proc. Am. Acad. Arts Sci. 58 (1923), 607-613; Mem. Nat. Acad. Sci. 21, No.11 (1927), 1-14

MORSE, MAX WITHROW [1880-1959]
AMS 7,1260; IB, 205

MORTENSEN, JAMES LEO [1924-1963]
AMS 10,2874
F.L. Himes, Soil Science 96 (1963), 367

MORTON, RICHARD ALAN [1899-1977]
DNB [1971-1980], 599-600; WW 1976, p.1702; WWWS, 1215; POGG

6,1784-1785; 7b,3441-3445
R.A. Morton, Med. Hist. 16 (1972), 321-353; TIBS 2 (1977), N75-N76
T.W. Goodwin, Nature 266 (1977), 394; Bioch. Soc. Symp. 6 (1978), 1-3
J. Glover et al., Biog. Mem. Fell. Roy. Soc. 24 (1978), 409-442

MORTON, ROBERT KERFORD [1920-1963]
M. Dixon, Nature 201 (1964), 133-134

MORTON, SAMUEL GEORGE [1799-1851]
DSB 9,540-541; DAB 13,265-266; ELLIOTT, 186-187; HIRSCH 4,270-271; KELLY, 874-877; DAMB, 539-540; FERCHL, 370; WWAH, 434; WWWS, 1215
Anon., Am. J. Sci. [2]12 (1851), 144-146
C.D. Meigs, Am. J. Sci. [2]13 (1852), 153-178
G.B. Wood, Trans. Coll. Phys. Phila. NS1 (1853), 372-388
R. Gerstner, in Beyond History of Science (E. Garber, ed), pp.126-136. Bethlehem, Pa. 1990

MORTON, WILLIAM THOMAS GREEN [1819-1868]
DAB 13,268-271; HIRSCH 4,271; DAMB, 540; ABBOTT, 98; W, 379; WWAH, 434; WWWS, 1215
W.D. Sharpe, Trans. Coll. Phys. Phila. [4]29 (1961), 137-152
G.S. Woodward, The Man who Conquered Pain: A Biography of William Thomas Green Morton. Boston, Mass. 1962

MORUZZI, GIOVANNI [1904-]
Who's Who in Italy 1980, p.352

MORUZZI, GIUSEPPE [1910-1986]
IWW 1982-1983, p.919; WWWS, 1215
O. Pompeiano, Arch. Ital. Biol. 124 (1986), 197-220
L.H. Marshall, Exp. Neurology 97 (1987), 225-242
A. Brodel, Arch. Ital. Biol. 126 (1988), 359-362
R.S. Dow, Am. Phil. Soc. Year Book 1989, pp.249-251

MOSANDER, CARL GUSTAV [1797-1858]
DSB 9,541; SMK 5,338-339; SCHAEDLER, 83; WWWS, 1215; POTSCH, 311; POGG 2,214
A. Erdmann, Lev. Sven. Vet. 1 (1869-1873), 163-183
E. Jorpes, Acta Chem. Scand. 14 (1960), 1681-1683

MOSCA, LUIGI [1822-1909]
DFI, 92

MOSCHELES, ROBERT [1863-]
JV 4,67; RSC 17,370; NUC 396,692

MOSCHKOWITZ, ELI [1879-1964]
AMS 10,2876; KOREN, 222
Anon., J. Am. Med. Assn. 188 (1964), 623

MOSER, LUDWIG [1879-1930]
OBL 6,389; TLK 3,455; POGG 5,879; 6,1785-1786
W.J. Müller, Ber. chem. Ges. 63A (1930), 165-168
F. Bock, Z. angew. Chem. 43 (1930), 1001-1002

MOSER, LUDWIG FERDINAND [1805-1880]
POGG 2,215; 3,939MOSETTIG, ERICH [1898-1962]
AMS 10,2879; POGG 6,1786; 7a(3),350-351
B. Witkop, Proc. Chem. Soc. 1964, pp.376-377

MOSS, WILLIAM LORENZO [1876-1957]
DSB 9,545-546
V.P. Sydenstryker, Trans. Assoc. Am. Phys. 71 (1958), 34

MOSSLER, GUSTAV [1876-1947]
TSCHIRCH, 1092; POGG 5,879-880; 6,1787; 7a(3),351

MOSSO, ANGELO [1846-1910]
DSB 9,546-547; EI 23,935; FISCHER 2,1074-1075; PAGEL, 1163-1164
A. Herlitzka, Arch. Ital. Biol. 54 (1910), i-xxiv; Arch. Fisiol. 9 (1911), 123-136
Anon., Nature 85 (1911), 174-175
F. Verzar, Naturw. Rund. 26 (1911), 117-119
V. Aducco, Rend. Accad. Lincei [5]20 (1911), 841-878

MOSTINSKI, VASILI ANDREEVICH [1870-1913]
SG [3]17,1368

MOSZKOWSKI, MAX [1873-1939]
FISCHER 2,1076; LDGS, 12; IB, 205; WININGER 7,334-335

MOTHES, KURT [1900-1983]
KURSCHNER 12,2155; HEIN-E, 323-325; IWW 1982-1983, p.921; IB, 205; POTSCH, 311-312; WWWS, 1218; POGG 7a(3),351-352
E. Lesky, Nova Acta Leop. NF36 (1970), 333-338
H.R. Schütte, Chem. Z. 94 (1970), 853; Naturw. Rund. 36 (1983), 491-494
V.E. Tyler, Journal of Natural Products 43 (1980), 541-545
A. Butenandt, Naturw. Rund. 33 (1980), 453-458
B. Parthier, Biochemie und Physiologie der Pflanzen 178 (1983), 695-768
K.V. Thimann, Biog. Mem. Fell. Roy. Soc. 30 (1984), 515-543
A. Pirson, Ber. bot. Ges. 98 (1985), 249-259

MOTHWURF, ARTHUR [1878-]
JV 19,342; NUC 397,660
New York Times, 24 August 1918, p.7

MOTT, OWEN EDWIN [1874-]
JV 17,157; NUC 398,51

MOTTRAM, JAMES CECIL [1879-1945]
DSB 9,549-551; WW 1945, pp.1961-1962
S.R., Lancet 1945(II), p.581
R.J. Ludford, Nature 157 (1946), 399-400

MOTTRAM, VERNON HENRY [1882-1976]
DNB [1971-1980], 604-605; WW 1975, pp.2260-2261; O'CONNOR(2), 39-41

MOTULSKY, ARNO GUNTHER [1923-]
AMS 15(5),515; BHDE 2,837

MOTZFELDT, KETIL [1883-1950]
J.H. Vogt, Nord. Med. Hist. Arsbok 1978, pp.143-155

MOUNEYRAT, ANTOINE [1870-1952]
POGG 6,1788
Anon., Presse Med. 60 (1952),1837; Languedoc Médical 26(2) (1953), 86-88

MOURANT, ARTHUR ERNEST [1904-]
WW 1981, p.1862; WWWS, 1219

MOUREU, CHARLES [1863-1929]
GORIS, 517-522; DORVEAUX, 95; POTSCH, 312; POGG 4,1036; 5,881; 6,1789-1791
M. Sommelet, Bull. Sci. Pharm. 36 (1929), 619-637
E. Fourneau, Ber. chem. Ges. 62A (1929), 93-99
C. Dufraisse, Bull. Soc. Chim. [4]49 (1931), 741-825
H. Le Chatelier et al., Not. Acad. Sci. 1 (1937), 267-287,341-373
G. Dillemain, Produits et Problèmes Pharmaceutiques 28 (1973), 377-378
H.D. Schwarz, Deutsche Apoth. Z. 128 (1988), 864

MOURIZ y RIESGO, JOSÉ [1884-1934]
NUC 398,350
Anon., Gaceta Medica Espanola 8 (1934), 223P
J. Alvarez-Sierra, Siglo Medico 93 (1934), 390-391

MOUSSU, GUSTAVE [1864-1945]
WWWS, 1219
A. Demolon, Not. Acad. Sci. 2 (1949), 526-548

MOUTON, HENRI [1869-1935]
CHARLE(2), 208-209

Anon., Ann. Inst. Pasteur 55 (1935), 5-7

MOYER, ANDREW JACKSON [1899-1959]
AMS 8,1771-1772; WWAH, 435; WWWS, 1120
Anon., Science 129 (1959), 712

MOYER, WENDELL WILLIAM [1903-]
AMS 11,3770

MOYLE, JENNIFER MARICE [1921-]
Who's Who of British Scientists 1980-1981, p.349

MOZOLOWSKI, WLODZIMIERZ [1895-1975]
IB, 206
M. Zydowo and L. Zelewski, Post. Bioch. 22 (1976), 3-10

MUCHA, VICTOR [1877-1933]
OBL 6,402; STURM 2,701; FISCHER 2,1079; CASARETTO, 90-94
G. Nobl, Wiener med. Wchschr. 83 (1933), 742-743

MUCKERMANN, ERNST [1874-1943]
DRULL, 186; JV 18,177; NUC 399,271

MUDD, STUART [1893-1975]
AMS 12,4436; NCAB E,53-54; IB, 206; WWWS, 1221
S. Mudd, Ann. Rev. Microbiol. 23 (1969), 1-28

MUDGE, GILBERT HORTON [1915-]
AMS 15(5),524; ICC, 413-414; WWWS, 1221

MÜHLHÄUSER, FRIEDRICH [-1887]
RSC 4,503
Anon., Pharm. Z. 32 (1887), 41

MÜHLHÄUSER, WILLI [1898-]
JV 41,429

MÜHLMANN, MOISSEI [1867-]
FISCHER 2,1080; WININGER 7,335; RSC 17,393

MÜLLER, ADOLF [1894-1955]
KURSCHNER 8,1612; POGG 6,1794; 7a(3),358
F. Wessely, Ost. Chem. Z. 56 (1955), 243-244

MÜLLER, ALBRECHT [1852-1888]
A. Beer, Ber. chem. Ges. 22(3) (1889), 861-864

MÜLLER, ALEXANDER [1828-1906]
BJN 11,45*; POGG 2,230; 3,944-945; 5,882
A. Morgen and S. Fingerling, Jahrbuch der Chemie 16 (1906), 294-295

MÜLLER, ALFRED [1876-]
JV 21,173; NUC 399,417

MÜLLER, BERNHARD WILHELM [1897-]
SG [5]1,358; NUC 399,443

MULLER, CARL [1855-1907]
BJN 12,60*; TSCHIRCH, 1093
L. Kny, Ber. bot. Ges. 25 (1907), (40)-(47)

MULLER, CARL [1889-]
JV 30,736; NUC 399,450

MÜLLER, CHRISTIAN [1816-1881]
STRAHLMANN, 481; TSCHIRCH, 1093; RSC 4,514; 8,450

MÜLLER, DETLEV [1899-]
KBB 1981, p.776; VEIBEL 2,301-302; IB, 206; WWWS, 1224

MÜLLER, ERICH [1870-1948]
TLK 3,459; POTSCH, 312-313; POGG 4,1043-1044; 5,883-884; 6,1796-1797; 7a(3),360-361

F. Foerster, Z. Elektrochem. 36 (1930), 113-114
G. Grube, Z. Elektrochem. 53 (1949), 337-338

MÜLLER, ERICH ALBERT [1898-1977]
KURSCHNER 12,2165; IB, 206; POGG 7a(3),361-362

MÜLLER, ERNST [1881-1945]
DRULL, 186-187; JV 19,183; NUC 399,501; POGG 5,884; 6,1797; 7a(3),362

MÜLLER, ERNST [1901-]
KURSCHNER 13,2641; MEISTER, 30-36; POGG 7a(3),362
Anon., Chem. Z. 100 (1976), 95

MÜLLER, EUGEN [1905-1976]
KURSCHNER 11,2036-2037; POTSCH, 313; POGG 7a(3),364-365
Anon., Chem. Z. 100 (1976), 397
T. Wieland, Heid. Akad. Jahrb. 1977, pp.59-60

MÜLLER, FERDINAND [1864-]
JV 2,193; RSC 15,136

MÜLLER, FRANZ [1871-]
KURSCHNER 4,2016; FISCHER 2,1082-1083; AD 3,70; ASEN, 135; TLK 3,459-460; IB, 206; POGG 6,1797-1798; 7a(3),365

MÜLLER, FRANZ [1874-]
JV 14,71; RSC 14,1010; NUC 399,525

MÜLLER, FRANZ KARL [1860-1913]
NUC 399,526-527

MÜLLER, FRIEDRICH von [1858-1941]
FISCHER 2,1083; PAGEL, 1171-1172; BACHMANN, 125-144
S.J. Thannhauser, Deutsche med. Wchschr. 54 (1928), 1566-1569
A.P. Martini, Deutsches Arch. klin. Med. 188 (1942), 453-472
F. von Müller, Lebenserinnerungen, 2nd Ed. Munich 1953
A. Pfarrweiler-Stieve, Friedrich von Müller und seine Stoffwechsel-untersuchungen. Zürich 1983
T.N. Bonner, J. Hist. Med. 45 (1990), 556-569

MÜLLER, FRIEDRICH [1912-]
KURSCHNER 12,2183; POGG 7a(3),366

MÜLLER, FRIEDRICH HORST [1907-1986]
MEINEL, 509; WWWS, 1225; POGG 7a(3),369-370
Anon., Chem. Z. 110 (1986), 394

MÜLLER, FRITZ (JOHANN FRIEDRICH THEODOR) [1822-1897]
DSB 9,559-561; BJN 4,58*; WWWS, 1224
E. Loew, Ber. bot. Ges. 15 (1897), (12)-(29)
W.F.H. B[lanford], Nature 56 (1897), 546-548
R. von Hanstein, Naturw. Rund. 12 (1897), 385-387
F. Ludwig, Bot. Z. 71 (1897), 291-302,347-363,401-408
E. Haeckel, Jena Z. Naturw. 31 (1898), 156-173
A. Möller, Fritz Müller: Werke, Briefe und Leben. Jena 1915-1921

MÜLLER, FRITZ [1877-]
JV 17,171; NUC 399,577

MÜLLER, FRITZ [1877-]
JV 21,297; NUC 399,577

MÜLLER, FRITZ [1880-1958]
JV 23,471; NUC 399,577

MÜLLER, HANS EDUARD [1881-]
JV 24,577; NUC 399,627
Anon., Chem. Z. 85 (1961), 721; Nachr. Chem. Techn. 14 (1966), 346

MÜLLER, HANS KARL [1899-1977]
KURSCHNER 12,2173
W. Best, Klin. Mon. Augenheilk. 171 (1977), 650-651

MÜLLER, HEINRICH [1820-1864]
ADB 22,557-558; PAGEL, 1169-1170; ROEDER, 23-38; WWWS, 1224
A. Kölliker, Würzb. Nat. Z. 5 (1864), xxix-xlvi

MÜLLER, HERMANN [1862-1938]
JV 5,77; RSC 17,404,757

MÜLLER, HERMANN [1867-]
JV 9,19; RSC 17,404; NUC 399,657

MÜLLER, HERMANN FRANZ [1866-1898]
BJN 4,331-341; 5,44*; FISCHER 2,1084; STUMPF, 18-22
H. Nothnagel, Wiener klin. Wchschr. 11 (1898), 1004-1006

MÜLLER, HUGO [1833-1915]
DESMOND, 453; BLOKH 2,529-531; POGG 3,945
Anon., Nature 95 (1915), 376-377; Chem. Z. 39 (1915), 434
J. Müller, Ber. chem. Ges. 48 (1915), 1023-1026
H.E.A., Proc. Roy. Soc. A95 (1919), xii-xxv

MÜLLER, JENS [1867-]
KALLMORGEN, 358; JV 8,216; RSC 13,276

MÜLLER, JOHANN [1806-]
FERCHL, 372

MÜLLER, JOHANNES [1801-1858]
DSB 9,567-574; ADB 22,625-628; HIRSCH 4,285-290; PAGEL, 1168-1169; FREUND 2,247-269; HAYMAKER, 243-247; BRAZIER, 56-59,63-64; FRANKEN,192; W, 381-382; POGG 2,227; CALLISEN 13,310-314; 30,472-476
Anon., Proc. Roy. Soc. 9 (1858), 556-563
A. Schrötter, Alm. Akad. Wiss. Wien 9 (1859), 119-134
E. du Bois-Reymond, Reden, vol.2, pp.143-334. Leipzig 1887
M. Nussbaum, Naturw. Rund. 12 (1897), 267-271
M. Lühe, Arch. Parasit. 5 (1902), 95-115
W. Haberling, Johannes Müller. Leipzig 1924
M. Müller, Sudhoffs Arch. 18 (1926), 130-150,209-234,328-350
U. Ebbecke, J. Müller, der Grosse Rheinischer Physiologe. Hannover 1951
G.C. Hirsch, Sudhoffs Arch. 26 (1953), 166-190
R. Bochalli, Naturw. Rund. 5 (1952), 345-347
G. Koller, Johannes Müller. Stuttgart 1958
J. Steudel, Rheinische Lebensbilder 1 (1961), 152-167
W. Riese and G.E. Arrington, Bull. Hist. Med. 17 (1963), 179-183,281

MUELLER, JOHN HOWARD [1891-1954]
AMS 8,1774; WWAH, 436
J.F. Enders, Harvard Med. Alumni Bull. 28(3) (1954), 36-39
A.M. Pappenheimer, Jr., Biog. Mem. Nat. Acad. Sci. 57 (1987), 307-321

MÜLLER, LUDWIG ROBERT [1870-1962]
KURSCHNER 8,1624; FISCHER 2,1085; HAGEL, 24-37; PITTROFF, 129-131; WWWS, 1225; NUC 400,103-104
L.R. Müller, Münch. med. Wchschr. 87 (1940), 494-496; Lebenserinnerungen. Munich 1957
V. Behr, Wiener klin. Wchschr. 90 (1940), 1940
F. Hoff, Deutsche med. Wchschr. 85 (1960), 823-824

MÜLLER, MAX [1863-1927]
TLK 3,461; JV 8,150; NUC 400,110
Anon., Ber. chem. Ges. 61A (1928), 78

MÜLLER, MAX [1873-1915]
JV 26,419; NUC 400,112
Anon., Z. angew. Chem. 28(III) (1915), 276

MÜLLER, NICOLAUS JACOB CARL [1842-1901]
BJN 6,75*; 7,365-366; TSCHIRCH, 1093
Anon., Leopoldina 37 (1901), 36
W. Wiese, Mündener Forstliche Hefte 17 (1901), 179

MÜLLER, OSKAR [1909-]
NUC 400,126

MÜLLER, PAUL HERMANN [1899-1965]
DSB 9,576-577; MH 1,341-342; MSE 2,336-337; ABBOTT-C, 99-100; WWWS, 1225; POTSCH, 313; POGG 7a(3), 372-373
Anon., Nature 208 (1965), 1043-1044
M. Spindler, Verhandl. Schw. Nat. Ges. 145 (1965), 277-280

MÜLLER, PAUL THEODOR [1849-1918]
RSC 17,408; SG [2]11,180; [3]7,1400; [5]1,358; NUC 400,149-150
Anon., Chem. Z. 33 (1909), 193

MÜLLER, RICHARD [1912-]
KURSCHNER 13,2661; POGG 7a(3),375

MÜLLER, ROBERT [1897-1951]
KERNBAUER, 397-401; TLK 3,462; POGG 6,1800; 7a(3),376
A. Pongratz, Öst. Chem. Z. 52 (1951), 219

MÜLLER, RUDOLF [1860-1915]
JV 4,67; RSC 13,276; NUC 400,183

MÜLLER, RUDOLF [1877-1934]
OBL 6,427; STURM 2,713; FISCHER 2,1086-1087; CASARETTO, 95-101

MÜLLER, SEBALD [1887-]
JV 27,652; NUC 400,189

MÜLLER, WILHELM [1832-1909]
BJN 14,65*; FISCHER 2,1087; GIESE, 508-510; KIEL, 79
P. Fürbringer, Münch. med. Wchschr. 56 (1909), 1850

MÜLLER, WILHELM [1866-]
JV 14,226; RSC 19,643

MÜLLER, WOLF [1874-1941]
OBL 6,429; TLK 3,463; POGG 4,1044; 5,885; 6,1801-1802; 7a(3),378
F. Ebert, Alm. Akad. Wiss. Wien 92 (1942), 208-211

MÜLLER-SCHÖLLHORN, FRITZ [1885-1950]
A. Stieger, Chimia 5 (1951), 61

MÜLLER-THURGAU, HERMANN [1850-1927]
STRAHLMANN, 468,480
A. Osterwalder, Verhandl. Schw. Nat. Ges. 108 (1927), 14-31
H. Schinz and A. Wolfer, Viert. Nat. Ges. Zürich 72 (1927), 433-451

MÜNTZ, ACHILLE [1846-1917]
POGG 3,947; 4,1045; 5,886
L. François, Rev. Gen. Bot. 32 (1920), 5-14

MÜNZEL, HEINRICH [1889-]
JV 29,630; NUC 400,420

MÜNZER, EGMONT [1865-1924]
STURM 2,717; FISCHER 2,1089; PAGEL, 1173-1174; SCHIEBER, 116-126; KOREN, 223; WININGER 4,476; 7,338
C. N[euberg], Biochem. Z. 153 (1924), 1

MUHLFELD, MARIE [1898-]
SG [5]1,363; NUC 400,539-540

MUIR, MATTHEW MONCRIEFF PATTISON [1848-1931]
DSB 9,557; WWWS, 1223; POGG 3,948-949; 4,1046-1047; 5,886; 6,1803
R.S. Morrell, J. Chem. Soc. 1932, pp.1330-1334
R.E. Oesper, J. Chem. Ed. 22 (1945), 458-459

MUIR, ROBERT [1864-1959]
DNB [1951-1960], 754-755; FISCHER 2,1089; O'CONNOR(2), 412-413; WWWS,1223

R. Cameron, Biog. Mem. Fell. Roy. Soc. 5 (1960), 149-173

MUIRHEAD, ARCHIBALD LAWRENCE [1863-1921]
SG [3]7,1404; NUC 400,597
A. West, Medical Review of Omaha 26 (1921), 269

MULDER, GERRIT JAN [1802-1880]
DSB 9,557-559; BWN 2,396-398; LINDEBOOM, 1373-1375; UTRECHT, 53-61; HIRSCH 4,296-297; BLOKH 2,528-529; PHILIPPE, 775-776; WWWS, 1223; POTSCH, 312; CALLISEN 13,358-359; 30,489-490; POGG 2,233-236; 3,949
T.H.N., Nature 22 (1880), 108-109
Anon., J. Chem. Soc. 39 (1881), 181-182
J.M. van Bemmelen, Verhand. Akad. Weten. Amst. 7, No.7 (1901), 1-33
H. Schlossberger, Mitt. Gesch. Med. 6 (1907), 385-391
E. Cohen, Rec. Trav. Chim. Pays-Bas 57 (1938), 729-736; Verhand. Akad. Weten. Amst. 19, No.2 (1948), 1-33
W. Labruyère, G.J. Mulder 1802-1880. Leiden 1938
E. Glas, Janus 62 (1975), 289-308; 63 (1976), 24-26,275-288
A. DeKnecht-van Lekelen, Voeding 42 (1981), 292-295
H.A.M. Snelders, Janus 69 (1982), 199-221; Tijd. Gesch. Geneesk. 7 (1984), 129-156; 13 (1990), 253-264; The Letters from Gerrit Jan Mulder to Justus Liebig 1838-1846. Amsterdam 1986

MULLER, HERMANN JOSEPH [1890-1967]
DSB 9,564-565; AMS 11,3782-3783; NCAB H,328-329; W,380-381; MH 1,340-341; STC 2,278-279; MSE 2,335-336; CB 1947, pp.458-460; DAMB, 544-545; SCHULZ-SCHAEFFER, 207-207; ABBOTT, 98-99; WWAH, 437; WWWS, 1225
L.C. Dunn, Nature 215 (1967), 108-109
E.A. Carlson, Am. Phil. Soc, Year Book 1967, pp.137-142; J. Hist. Biol. 4 (1971), 149-170; Genetics 70 (1972), 1-30; Genes, Radiation and Society. Ithaca, N.Y. 1981; Social Research 51 (1984), 763-782
T.M. Sonneborn, Science 162 (1968), 772-776
G. Pontecorvo, Biog. Mem. Fell. Roy. Soc. 14 (1968), 349-389; Ann. Rev. Gen. 2 (1968), 1-10
C. Auerbach, Mutation Research 5 (1968), 201-207

MULLIKEN, ROBERT SANDERSON [1896-1986]
AMS 14,3573; WWA 1978-1979, p.2337; IWW 1982-1983, p.929; MH 1,342-343; MSE 2,337-338; WWWS, 1225-1226; POTSCH, 314; POGG 6,1805-1806; 7b,3441-3445
J.H. Van Vleck, J. Phys. Chem. 84 (1980), 2091-2095
R.S. Berry, Am. Phil. Soc. Year Book 1987, pp.187-194
H.C. Longuet-Higgins, Biog. Mem. Fell. Roy. Soc. 35 (1990), 329-354

MULLIKEN, SAMUEL PARSONS [1864-1934]
DAB [Suppl.1],570-571; NCAB 25,73; MILES, 353-354; WWAH, 437; WWWS, 1226; POGG 6,1806; 7b,3445
T.L. Davis, Chem. Eng. News 12 (1934), 197-198; Proc. Am. Acad. Arts Sci. 70 (1935-1936), 560-565

MULVANIA, MAURICE [1878-]
AMS 7,1273; NUC 401,159

MUMM, OTTO [1877-1972]
KURSCHNER 10,1691; KIEL,163; TLK 3,464; POGG 5,889; 6,1808; 7a(3),383-384
Anon., Chem. Z. 97 (1973), 40; Christiana Albertina 15 (1973), 93-94

MUNK, HERMANN [1839-1912]
BJN 18,47*; FISCHER 2,1091; KOREN, 233; WININGER 4,471

MUNK, IMMANUEL [1852-1903]
BJN 8,79*; FISCHER 2,1091-1092; PAGEL, 1178-1179; BLOKH 2,531; KOREN, 223; WININGER 4,471
R. duBois-Reymond, Berl. klin. Wchschr. 40 (1903), 770-772
P. Schultz, Z. Physiol. 17 (1903), 251-257

MUNRO, HAMISH NISBET [1915-]
AMS 15(5),545; WWA 1980-1981, p.2403

MUNTWYLER, EDWARD [1903-1971]
AMS 11,3792

MURACHI, TAKASHI [1926-1990]
H. Neurath, TIBS 15 (1990), 368-369

MURALT, ALEXANDER von [1903-1990]
KURSCHNER 13,2685; SBA 5,91-92; POGG 7a(3),385-386
H. Aebi et al., Schw. med. Wchschr. 93 (1963), 1107-1110
Anon., Chimia 17 (1963), 276-277
A. von Muralt, Ann. Rev. Physiol. 46 (1984), 1-13
J.T. Edsall, Proc. Am. Phil. Soc. 135 (1991), 615-621

MURLIN, JOHN RAYMOND [1874-1960]
AMS 9(II),806; NCAB 47,406-407; IB,208
E.S. Nasset, J. Nutrition 31 (1946), 5-12; 71 (1960), 338
J.R. Murlin and B. Kramer, J. Hist. Med. 11 (1956), 288-298
W.O. Fenn, Am. Phil. Soc. Year Book 1961, pp.145-152

MURPHY, JAMES BUMGARDNER [1884-1950]
DSB 9,586-587; DAB [Suppl.4],615-617; NCAB 38,69; FISCHER 2,1093; DAMB, 547; IB, 206; WWAH, 438; WWWS, 1229
W.T. Longcope, Trans. Assoc. Am. Phys. 44 (1951), 15-18
C.C. Little, Biog. Mem. Nat. Acad. Sci. 34 (1960), 183-203
I. Lowy, Bull. Hist. Med. 63 (1989), 356-377

MURPHY, WILLIAM PARRY [1892-]
AMS 11,3801; WWA 1978-1979, p.2348; IWW 1982-1983, pp.933-934; WWWS,1230; STC 2,283-284; POGG 6,1811; 7b,3457-3458

MURRAY, GEORGE REDMAYNE [1865-1939]
DNB [1931-1940], 638-639; WW 1939, p.2309; FLUELER, 101-102; WWWS, 1230
Anon., Lancet 1939(II), p.767; Brit. Med. J. 1939(II), pp.707-708

MURRAY, HENRY ALEXANDER [1893-1988]
AMS 9(III),492; WWA 1978-1979, p.2349; WWWS, 1230
New York Times, 24 June 1988

MURRAY, JOHN [1778-1820]
DNB 13,1285-1286; HIRSCH 4,307; TSCHIRCH, 1093; FERCHL, 375; POGG 2,243-244,1430

MURRAY, MARGARET MARY ALBERTA [1899-1974]
WW 1965, p.1708; NUC 402,541
Who Was Who 1971-1980, p.572

MURRAY, MARGARET RANSONE [1901-]
AMS 14,3591; WWWS, 1231

MURSCHHAUSER, HANS [1878-]
JV 23,548; NUC 402,625

MUSCULUS, FRÉDÉRIC ALPHONSE [1829-1888]
SITZMANN 2,356-357; HEIN, 456-457; GORIS, 522-523; HUMBERT, 212-214; BALLAND, 76-77; TSCHIRCH, 1093; SCHELENZ, 694; POGG 3,952
A. Sartory, J. Pharm. Alsace 56 (1929), 168-171

MUSKAT, IRVING ELKIN [1905-]
AMS 12,4475-4476; WWWS, 1231-1232
A.D. McFadyen, Chemical Industries 49 (1941), 314-315,349

MUSPRATT, EDMUND KNOWLES [1833-1923]
M.D. Stephens and G.W. Roderick, Ann. Sci. 29 (1972), 300-309

MUSPRATT, FREDERICK [1825-1872]
Anon., J. Chem. Soc. 26 (1873), 780
M.D. Stephens and G.W. Roderick, Ann. Sci. 29 (1972), 309

MUSPRATT, JAMES SHERIDAN [1821-1871]
DNB 13,1331; BLOKH 2,512; COLE, 397-398; SCHAEDLER, 83; WWWS, 1232; POTSCH, 315; POGG 2,246; 3,952

Anon., J. Chem. Soc. 24 (1871), 620-621
D.W.F. Hardie, Endeavour 14 (1955), 29-33
M.D. Stephens and G.W. Roderick, Ann. Sci. 29 (1972), 287-311

MUSPRATT, RICHARD [1822-1885]
M.D. Stephens and G.W. Roderick, Ann. Sci. 29 (1972), 309-310

MUSSO, HANS [1925-]
KURSCHNER 13,2668; WWWS, 1232; POGG 7a(3),387

MUTHMANN, FRIEDRICH WILHELM [1861-1913]
BJN 18,111*; BLOKH 2,532-533; JV 1,170; POTSCH, 316; POGG 4,1050-1051; 5,890-891
E. Baur, Chem. Z. 37 (1913), 1253
H. Will, Ber. chem. Ges. 46 (1913), 2982
P. Groth, Alm. Bayer. Akad. Wiss. 1914, pp.80-82

MUTZENBECHER, PAUL von [1904-1940]
Anon., Chem. Z. 64 (1940), 320; Z. angew. Chem. 53 (1940), 336

MYERS, VICTOR CARYL [1883-1948]
BERG, 53-55; NCAB C,238-239; CHITTENDEN, 175; IB, 208; WWAH, 440; WWWS, 1234
Anon., Chem. Eng. News 26 (1948), 3357

MYERS, WALTER [1872-1901]
DIEPGEN, 90-92; RSC 17,438
J. Lea et al., J. Path. Bact. 7 (1901), 481-487

MYLIUS, FRANZ BENNO [1854-1931]
HEIN, 458-459; BERLIN, 710; POGG 4,1051; 5,891; 6,1812
F. Foerster, Ber. chem. Ges. 64A (1931), 167-194

MYLO, BRUNO [1884-1915]
JV 21,30; NUC 403,670
Anon., Z. angew. Chem. 28(III) (1915), 140

MYLON (MISLOWITZER), ERNST [1895-1985]
AMS 11,3817; IB, 208; TLK 3,449; LDGS, 67; WININGER 7,323; WWWS, 1235; POGG 6,1749; 7a(3),314

MYRBÄCK, KARL [1900-1985]
KURSCHNER 13,2691; VAD 1981, p.737; WWWS, 1235; POGG 6,1812-1813; 7a(3),3465-3470
Anon., Leopoldina [3]6/7 (1960-1961), 64

MYSTKOWSKI, EDMUND MARCELI [1905-1943]
J. Rostowski, History of the Polish School of Medicine at the University of Edinburgh, p.48. Edinburgh 1955

N

NABARRO, DAVID NUNES [1874-1958]
WW 1958, p.2192; FISCHER 2,1097; O'CONNOR(2), 162-163; WININGER 7,340; RSC 17,440; WWWS, 1235
A.G. Signey, J. Path. Bact. 79 (1960), 429-435

NACHMANSOHN, DAVID [1899-1983]
AMS 14,3606; IWW 1982-1983, p.938; BHDE 2,842; LDGS, 60; MH 2,386-387; STC 2,285-288; MSE 2,345-346; KOREN, 223; COHEN, 173; WWWS, 1236
D. Nachmansohn, Ann. Rev. Biochem. 41 (1972), 1-28
F. Hucho et al., Nachr. Chem. Techn. 32 (1984), 34-36
E. Neumann, Z. physiol. Chem. 365 (1984), (15)-(16)
S. Ochoa, Biog. Mem. Nat. Acad. Sci. 58 (1989), 357-404

NADENHEIM, FANNY [1894-1935]
JV 34,272; NUC 404,239

NADLER, SAMUEL BERNARD [1905-]
AMS 11,3820

NAEGELI, CARL von [1817-1891]
DSB 9,600-602; ADB 52,573-582; FREUND 1,291-297; BRIQUET, 352-364; TSCHIRCH, 1094; FERCHL, 377; W, 385; WWWS, 1237; POGG 3, 953-954
D.H. Scott, Nature 44 (1891), 580-583
S. Schwendener, Ber. bot. Ges. 9 (1891), (26)-(42)
S.H. V[ines], Proc. Roy. Soc. 51 (1892), xxvii-xxxvi
C. Cramer, Leben und Wirken von C.W. von Naegeli. Zurich 1896
J.W. Wilkie, Ann. Sci. 16 (1960),11-41,171-207,209-329; 17 (1961),27-62; Nature 190 (1961), 1145-1159
U. Witte and B. Hoppe, Folia Mendeliana 6 (1971), 117-138

NAEGELI, ERNST [1867-1916]
ZURICH-D, 143; RSC 17,443; 19,124; NUC 404,283
Anon., Ber. chem. Ges. 49 (1916), 1222

NAEGELI, KARL WILHELM [1895-1942]
POGG 6,1815; 7a(3),391
P. Karrer, Viert. Nat. Ges. Zürich 87 (1942), 525-527; Helv. Chim. Acta 26 (1943), 730-733

NAEGELI, OTTO [1871-1938]
FISCHER 2,1098; ZURICH-D, 104; SKVARC, 159-164; WWWS, 1237
H. Schinz and K. Ulrich, Viert. Nat. Ges. Zurich 83 (1938), 366-382
K. Rohr, Wiener med. Wchschr. 88 (1938), 1232-1234
F. von Müller, Münch. med. Wchschr. 85 (1938), 675-678
E. Hanhart, Deutsche med. Wchschr. 64 (1938), 511-512
A. von Domarus, Verhandl. path. Ges. 31 (1938), 543-553
H. Schmid and G. Kummer, Mitt. Nat. Ges. Schaffhausen 14 (1938), 185-210
A.U. Däniker, Verhandl. Schw. Nat. Ges. 1939, pp.297-302
W.M. Dufek, Der Internist Otto Naegeli. Zurich 1983

NAGAI, KAZUO [1873-1923]
HAKUSHI 1(Pharm.), 9
Y. Tsuzuki, J. Chem. Ed. 47 (1970), 695-696

NAGAI, NAGAYOSHI [1845-1929]
JBE, 952; IB, 209; BERLIN, 704; WWWS, 1237; POTSCH, 316; POGG 6,1817
B. Lepsius, Chem. Z. 52 (1928), 82
H. Thoms, Ber. chem. Ges. 62A (1929), 53-54
Y. Tsuzuki, J. Chem. Ed. 47 (1970), 695-696

NAGAYAMA, TAKEYOSHI [1885-1977]
JBE, 966; HAKUSHI 3,592-593

NAGEL, ERNEST [1901-1985]
IWW 1985-1986, p.1069
C.G. Hempel, Am. Phil. Soc. Year Book 1989, pp.267-270

NAGEL, WILHELM [1872-1916]
JV 14,226; RSC 17,436
Anon., Chem. Z. 40 (1916), 950

NAGEL, WILLIBALD [1870-1911]
BJN 16,56*; FISCHER 2,1099
W. Trendelenburg, Deutsche med. Wchschr. 37 (1911), 461-462

NAGEOTTE, JEAN [1866-1948]
FISCHER 2,1099; HAYMAKER, 133-136; CHARLE, 191-193; POGG 6,1819; 7b,3484
F. Thiébaut, Presse Med. 56 (1948), 837-838

NAGLER, FRIEDRICH PAUL [1908-1975]
AMS 12,4494; BHDE 2,842

NAGY, STEPHEN MEARS [1911-1972]
AMS 11,3823
Anon., Chem. Eng. News 50(31) (1972), 24

NAJJAR, VICTOR ASSAD [1914-]
AMS 15(5),586; WWWS, 1238

NAKAGAWA, TOMOICHI [1891-]
HAKUSHI 4,684-685

NAKANISHI, KOJI [1925-]
AMS 17(5),613
K. Nakanishi, A Wandering Natural Products Chemist. Washington, D,C. 1991

NAKASHIMA, MINORU [1892-]
HAKUSHI 4,1021-1022

NAMETKIN, SERGEI SEMENOVICH [1876-1950]
DSB 9,608-609; BSE 29,94; WWR, 403; KUZNETSOV(1961), 588-595; ARBUZOV, 166-168; POTSCH, 316-317; POGG 6,1822-1823; 7b,3492-3496
V.M. Rodionov et al., Zhur. Ob. Khim. 21 (1951), 2101-2146

NAMUR, FRANÇOIS PIERRE JOSEPH [1823-1892]
BLUM 2,155-156NAPOLI, RAFFAELE [1817-1866]
DFI, 92-93; RSC 4,568

NAQUET, ALFRED JOSEPH [1834-1916]
GE 24,804-805; LAROUSSE 11,835; WININGER 4,492-493; POGG 3,955
F. Lieben, Archeion 18 (1936), 190-191

NARBUTT, JOHANNES von [1879-1937]
WELDING, 541

NARDIN, LÉON AUGUSTE LOUIS [1857-1930]
GORIS, 524
Anon., Rev. Hist. Pharm. 1 (1930), 212-213

NARODNY, ALEKSANDR VASILIEVICH [1887-1953]
VORONTSOV, 58-69; UKRAINE 2,591-592

NASH, THOMAS PALMER [1890-1970]
AMS 11,3832; IB, 210

NASINI, RAFFAELO [1854-1931]
WWWS, 1240; POGG 4,1054-1055; 5,894; 6,1826
M. Delépine, Bull. Soc. Chim. [4]49 (1931), 828-830
M.G. Levi, Gazz. Chim. Ital. 62 (1932), 727-745
G. Provenzal, Rass. Clin. Terap. 37 (1938), 54-57

NASON, ALVIN [1919-1978]
AMS 13,3180; WWWS, 1240
S.P. Colowick et al., Anal. Biochem. 95 (1979), 1

NASON, HENRY BRADFORD [1831-1895]
DAB 13,390; MILES, 360-361; ELLIOTT, 188; WWAH, 442
W.P. Mason, J. Am. Chem. Soc. 17 (1895), 339-341
T.C. Chamberlin, Bulletin of the American Geological Society 7 (1896), 479-481

NASONOV, DMITRI NIKOLAEVICH [1895-1957]
BSE 29,202; WWR, 404; IB, 210
W.P. Michailow and A.W. Schirmanski, Anat. Anz. 106 (1959), 257-264
I.I. Polyanski, Tsitologia 7 (1965), 441-447

NASONOV, NIKOLAI VIKTOROVICH [1855-1939]
BSE(3rd Ed.) 17,304; RSC 12,532; 17,452-453
I.I. Shmalhausen and D.M. Fedotov, Vest. Akad. Nauk 1939, pp.81-85

NASSE, CHRISTIAN FRIEDRICH [1778-1851]
ADB 23,265-270; WL 2,274-288; HIRSCH 4,325; SCHULTE, 217-218; WWWS, 1241; CALLISEN 13,415-425; 31,9-11
W. von Noorden, Der Kliniker Christian Friedrich Nasse. Jena 1929; Christian Friedrich Nasse. Berlin 1936
W. Kaiser and U. Trenckmann, Z. ges. inn. Med. 33 (1978), 317-325; Wiss. Z. Halle M29 (1980), 65-89
H. Schipperges, Gesnerus 38 (1981), 105-118

NASSE, HERMANN [1807-1892]
GUNDLACH, 244-245; HIRSCH 4,326; PAGEL, 1187-1188
Anon., Berl. klin. Wchschr. 29 (1892), 712

NASSE, OTTO [1839-1903]
BJN 8,80*; HIRSCH 4,326; PAGEL, 1188-1189; POGG 3,957; 4,1055
O. Langendorff, Pflügers Arch. 101 (1904), 1-22

NASSET, EDMUND SIGURD [1900-1985]
AMS 14,3622; WWWS, 1241
New York Times, 23 March 1985

NASTIUKOV, ALEKSANDR MIKHAILOVICH [1868-1941]
BSE 29,223; WWR, 405; POGG 6,1827; 7b,3518
B.N. Rutovski, Zhurnal Khimicheskoi Promyshlenosti 18(11) (1941), 36

NASTVOGEL, OSCAR [1863-1910]
RIGA, 712; JV 3,275; RSC 17,454; NUC 405,534
Anon., Chem. Z. 34 (1910), 411

NATANSON, JACOB [1832-1884]
PSB 22,601-603; WE 7,636; WININGER 4,504; KONOPKA 6,319; POTSCH, 317; POGG 3,957

NATHAN, ERNST [1889-]
LDGS, 62; KOREN, 224; SG [3]8,25; NUC 405,593-594
Deutscher Dermatologen Kalender 1929, pp.167-169

NATHANS, DANIEL [1928-]
AMS 15(5),598; WWA 1980-1981, p.2429; IWW 1982-1983, pp.944-945; MSE 2,348-349; POTSCH, 317

NATIVELLE, CLAUDE ADOLPHE [1812-1889]
TSCHIRCH, 1094
A. Cahuet, Claude Adolphe Nativelle. Paris 1937
R. Paris, Figures Pharmaceutiques Françaises, pp.101-106. Paris 1953

NATTA, GIULIO [1903-1979]
DSB 18,661-663; MH 1,348-349; STC 2,292-293; MSE 2,349; WWWS, 1241; CB 1964, pp.312-314; ABBOTT-C, 100; POTSCH, 317; POGG 6,1829; 7b,3520-3530
G. Dal'Astia, Chem. Z. 87 (1963), 131
Anon., Nachr. Chem. Techn. 11 (1963), 160
C.E.H. Bawn, Nature 280 (1979), 707
C.H. Bamford, Chem. Brit. 17 (1981), 298-300
S. Carra et al. (Eds.), Giulio Natta. Milan 1982

NAUCK, ERNST [1819-1875]
WELDING, 541-542; RIGA, 712; POGG 3,958

NAUDIN, CHARLES [1815-1899]
DSB 9,618-619; VAPEREAU 5,1354; GE 24,849; W, 387; TSCHIRCH, 1094
E. Bornet, Rev. Gen. Bot. 11 (1897), 161-167; C.R. Acad. Sci. 128 (1899), 127-128
I.H.B., Nature 60 (1899), 58-59
L. Blaringhem, Prog. Rei Bot. 4 (1913), 27-108

NAUEN, OTTO [1857-1942]
GEDENKBUCH 2,1082; RSC 17,490-491; NUC 408,357

NAUMANN, ALEXANDER NICOLAUS [1837-1922]
DSB 9,619-620; DBJ 4,364; GUNDEL, 680-687; SCHAEDLER, 84; POGG 3,958-959; 4,1059; 5,895

NAUMANN, HANS NORBERT [1901-]
AMS 14,3625; LDGS, 67; POGG 6,1829-1830; 7a(3),398

NAUNYN, BERNHARD [1839-1925]
SCHOLZ, 47-49; PAGEL, 1190-1191; LEVITSKI 2,135-138; KROLLMANN, 456; FRANKEN, 192; DZ, 1008; WWWS, 1242
B. Naunyn, Erinnerungen, Gedanken und Meinungen. Munich 1925
E. Meyer, Deutsche med. Wchschr. 51 (1925), 1493-1495
F. Müller, Deutsches Arch. klin. Med. 150 (1926), 1-12
J. Paul, Bernhard Naunyn, eine Biographische Studie. Göttingen 1949
R. Abderhalden, Ciba Z. 5 (1951), 1616-1640

NAVASHIN, SERGEI GAVRILOVICH [1857-1930]
DSB 10,1-2; BSE 29,16-17; WWR, 406; KUZNETSOV(1963), 302-312; UKRAINE 2,350-351
G.A. Lewitsky, Ber. bot. Ges. 49 (1931), (149)-(163)
V.V. Finn, Bull. Acad. Sci. URSS 1931(7), pp.881-901
R.R. Gates, Proc. Linn. Soc. 143 (1931), 188-191
D.A. Trankovsky, S.G. Navashin. Moscow 1947

NAWROCKI, FELIX [1837-1902]
PSB 22,627-629; HIRSCH 4,331; KOSMINSKI, 343-344; KONOPKA 6,345-349; RSC 4,581; 8,485; 12,534; 15,462
S. Lagowski, Medycyna 30 (1902), 495-496
K. Kosmala, Arch. Hist. Fil. Med. 15 (1935), 1-39
A. Trzebski, Acta Physiol. Polon. 38 (1987), 66-73

NAZAROV, IVAN NIKOLAEVICH [1906-1957]
BSE 29,54-55; WWR, 406; POTSCH, 318
I.V. Torgov, Zhur. Ob. Khim. 29 (1959), 702-723

NEBELTHAU, EBERHARD [1864-1914]
DBJ 1,302; FISCHER 2,1103; PAGEL, 1192; RSC 17,463-464
Anon., Münch. med. Wchschr. 61 (1914), 1544

NEBER, PETER [1883-1960]
POGG 6,1830; 7a(3),399; (4),151*
Anon., Z. angew. Chem. 56 (1943), 280; Chem. Z. 82 (1958), 628

NEEDHAM, DOROTHY MOYLE [1896-1987]
WW 1981, p.1891; IB, 210; WWWS, 1243; POGG 6,1831; 7b,3536-3538
The Times, 26 December 1987

NEEDHAM, JAMES GEORGE [1868-1957]
AMS 9(II),816; NCAB B,289; IB, 210; WWWS, 1243
H.H. Schwardt, Ann. Ent. Soc. Am. 52 (1959), 338-339

NEEDHAM, JOSEPH [1900-]
WW 1981, p.1891; IWW 1982-1983, p.947; STC 2,293-295; ABBOTT, 100; IB, 210; WWWS, 1243; POGG 6,1831-1832; 7b,3538-3544
M. Teich and R. Young, Changing Perspectives in the History of Science, pp.1-20,472-478. London 1973
M. Elvin et al., Past and Present 87 (1980), 17-53
J. Kovel, Science as Culture 5 (1989), 50-70

NEEFF, CARL THEODOR [1898-1940]
RAZINGER, 129-131; POGG 6,1832; 7a(3),399-400
C.J. Gauss, Strahlentherapie 70 (1941), 1-10

NEEL, JAMES van GUNDIA [1915-]
AMS 15(5),608; WWA 1980-1981, p.2433; IWW 1982-1983, p.947; WWWS, 1243; MH 2,387-388; MSE 2,349-351
Anon., J. Heredity 74 (1983), 3

NEES von ESENBECK, CHRISTIAN GOTTFRIED [1776-1858]
DSB 10,11-14; ADB 23,368-376; WWWS, 1243; CALLISEN 13,450-452; 31,19-20; TSCHIRCH, 1094; POGG 7aSuppl.,452-453
H. Winkler, Schlesische Lebensbilder 2 (1926), 203-208
B. Kalbers-Sauer, Personalbibliographien der Professoren an der Medizinischen Fakultät der Universität Erlangen...1792-1850; pp.142-155. Erlangen-Nürnberg 1969

NEES von ESENBECK, THEODOR [1787-1837]
ABD 23,376-380; HEIN,460-461; TSCHIRCH, 1045,1094; FERCHL,239; WWWS,1244; CALLISEN 13,452-457;31,20-22; POGG 2,264; 7bSuppl.,453-454
L.F. Bley, Arch. Pharm. 91 (1845), 81-112

NEF, JOHN ULRIC [1862-1915]
DSB 10,14-15; MILES, 361-362; FARBER, 1131-1143; BLOKH 2,534; WWAH, 442; POTSCH, 318; WWWS, 1244; POGG 4,1060-1061; 5,896
Anon., Chem. Z. 39 (1915), 719
J.W.E. Glattfield and H.A. Spoehr, University Record (Chicago) NS16 (1930), 110-114,173-179
M.L. Wolfrom, Biog. Mem. Nat. Acad. Sci. 34 (1960), 204-227

NEGELEIN, ERWIN [1897-1979]
KURSCHNER 12,2229; POGG 6,1832; 7a(3),401

NEGER, FRANZ WILHELM [1868-1923]
DBJ 5,282-285,436; JV 8,216; NUC 209,514-515
O. Drude, Ber. bot. Ges. 41 (1923), (84)-(92)
L. Vanino, Chem. Z. 47 (1923), 423

NEGRI, ADELCHI [1876-1912]
DSB 10,15-16; FISCHER 2,1104; DIEPGEN, 87-89
G.F.H. Nuttall, Parasitology 5 (1912), 151-154
C. Golgi, Boll. Sci. Med. Pavia 26 (1912), 87-124
C. Joyeux, Arch. Parasit. 16 (1913), 161-167
E. Veratti, Riv. Biol. 16 (1934), 577-601

NEGUS, SIDNEY STEVENS [1892-1963]
AMS 10,2939; WWAH, 442-443

NEILL, JAMES MAFFETT [1894-1964]
AMS 10,2941; WWAH, 443

NEILSON, CHARLES HUGH [1872-1958]
AMS 3,502; WWAH, 443

NEISH, ARTHUR CHARLES [1916-1973]
WW 1973, p.2359; YOUNG, 58
J.K.N. Jones, Biog. Mem. Fell. Roy. Soc. 20 (1974), 295-315

NEISSER, ALBERT [1855-1916]
DSB 10,17-19; DBJ 1,364-365; LINDEBOOM, 1417-1418; SACKMANN, 243-245; PAGEL, 1193-1196; OLPP, 290-294; KOREN, 224; WININGER 4,510-512; KOHUT 2,183-184; WWWS, 1244
C. Bruck, Naturwiss. 4 (1916), 609-613
T. Veiel, Arch. Dermatol. 123 (1916), ix-lxiv
J. Jadassohn, Jahresber. Schles. Ges. 94 (1916), 15-36; Schlesische Lebensbilder 1 (1922), 111-115
S. Schmitz, Albert Neisser. Düsseldorf 1968
F. Wasik et al., Hautarzt 31 (1980), 328-333
A. Scholz and G. Sebastian, Hautarzt 36 (1985), 586-590
A. Scholz and F. Wasik, International Journal of Dermatology 24 (1985), 373-377

NEISSER, ERNST [1863-1942]
FISCHER 2,1105; KOREN, 224
G. Schoene, Deutsche med. Wchschr. 84 (1959), 692-696

NEISSER, MAX [1869-1938]
KURSCHNER 4,2063; ARNSBERG, 315-316; NB, 287; FISCHER 2,1105; IB, 211; KALLMORGEN, 361; SKVARC, 169-171; WININGER 7,344; TETZLAFF, 246
E. Klieneberger-Nobel, Z. Bakt. 215 (1970), 279-285

NELSON, ELMER MARTIN [1892-1958]
AMS 9(I),1403; WWAH, 443
C.D. Tolle, Science 129 (1959), 1344-1345
F.C. Bing, J. Nutrition 73 (1961), 5-13

NELSON, ERWIN ELLIS [1891-1966]
AMS 10,2945; PARASCANDOLA, 46-47; CHEN, 53-54; IB, 211

R.G. Smith, Pharmacologist 9 (1967), 109-110

NELSON, JOHN MAURICE [1876-1965]
AMS 10,2947; MILES, 362-363; IB, 211; POGG 6,1835; 7b,3549-3550
M.T. Bogert and J.H. Northrop, Columbia University Quarterly 33 (1941), 209-221
R.M. Herriott, J. Chem. Ed. 32 (1955), 513-517

NELSON, LOUIS [1878-1912]
SG [3]8,44
C.J. Blake, Transactions of the American Therapeutic Society 13 (1913), 9

NELSON, WARREN OTTO [1906-1964]
AMS 9(II),819; WWAH, 444
C.W. Lloyd, Endocrinology 77 (1965), 1-4

NEMEC, ANTONIN [1894-1958]
IB, 211; POGG 6,1836-1837; 7b,3551-3556
S.A. Wilde, Forestry Chronicle 34 (1958), 438

NEMEC, BOHUMIL [1873-1966]
STURM 3,21-22; IB, 211
Anon., Prestlia 2 (1922), 5-12
S.P. Sestavil, Prestlia 11 (1932), 3-10
S. Prat, Prestlia 38 (1966), 341-346
Z. Cernahorsky, Plant Science Bulletin 12(3) (1966), 9-10
O. Matousek, Arch. Inter. Hist. Sci. 20 (1967), 91-97
I. Klatersky, Rev. Hist. Sci. 20 (1967), 69-71

NENCKI, LEON [1848-1904]
PSB 22,669-671; FISCHER 2,1106; KOSMINSKI, 346-347,616; KONOPKA 6,351-355; RSC 17,482
J. Peszke, Kryt. Lek. 8 (1904), 165-168
A. Szwejcerowa, Arch. Hist. Med. 39 (1976), 81-91

NENCKI, MARCELI (MIKHAIL VILGELMOVICH) [1847-1901]
DSB 10,22-23; PSB 22,671-674; WE 7,677-678; BJN 6,76*; FISCHER 2,1106; PAGEL, 1198-1199; BLOKH 2,534-535; REBER, 291-294; TSCHIRCH, 1094; SCHAEDLER, 84; BULLOCH,385; KOSMINSKI, 345-346,616; KONOPKA 355-369; POTSCH, 318; POGG 3,961; 4,1062-1063
M. Hahn, Münch. med. Wchschr. 48 (1901), 1971-1973; Ber. chem. Ges. 35 (1902), 4503-4521
J. Pruszynski et al., Gazeta Lekarska 21 (1901), 1150-1199
F. Rohmann, Naturw. Rund. 17 (1902), 49-50
J.F. Heymans, Arch. Inter. Pharmacodyn. 10 (1902), 1-24
O. Loew, Jahresb. Fort. Tierchem. 31 (1902), ix-xxxvii
N.O. Sieber, Arch. Sci. Biol. 11 (1905), 167-195
S.B. Schryver, Science Progress 1 (1907), 512-530
M. Mastyncka, Arch. Hist. Fil. Med. 19 (1948), 152-200
V.A. Engelhardt, Biokhimia 16 (1951), 486-494
G.A. Vladimirov and E.E. Martinson, Fiziol. Zhur. 37 (1951), 672-687
W. Lampe, Wiad. Chem. 6 (1952), 141-150
A. Szwejcerowa and J. Groszynska, Marceli Nencki. Warsaw 1956
M.H. Bickel, Marceli Nencki 1847-1901. Berne 1972
A. Szwejcerowa, Arch. Hist. Med. 38 (1975), 207-220; 39 (1976), 81-91; Marceli Nencki. Warsaw 1977
S. Niemerko, Acta Physiol. Polon. 38 (1987), 149-157

NENITZESCU, COSTIN [1902-1970]
WWWS, 1246; POTSCH, 318-319; POGG 6,1837; 7b,3557-3564
F. Runge, Jahrb. Sachs. Akad. 1969-1970, pp.281-298
R. Huisgen, Bayer. Akad. Wiss. Jahrbuch 1970-1971, pp.221-227
M. Avram, Chem. Ber. 104 (1971), xvii-lxv
P.F.G. Praill, Chem. Brit. 7 (1971), 252-253

NERESHEIMER, JULIUS [1880-1918]
JV 24,577; NUC 410,375

NERKING, JOSEPH [1871-]
JV 11,140; 20,52; RSC 17,484; SG [3]8,48; NUC 410,393

NERNST, WALTHER [1864-1941]
DSB 15,432-453; SCHWERTE 1,129-138; KROLLMANN, 458; DZ, 1012; TLK 3,468; ABBOTT-C, 100-102; W, 388; WWWS, 1246; POTSCH, 319; POGG 4,1063-1064; 5,896-897; 6,1838; 7a(3),405
E.H. Riesenfeld, Z. angew. Chem. 37 (1924), 437-439
A. Einstein, Sci. Mon. 54 (1942), 195-196
Cherwell and F. Simon, Obit. Not. Fell. Roy. Soc. 4 (1942), 101-112
M. Bodenstein, Ber. chem. Ges. 75A (1942), 79-104
F. Hoffmann, Phys. Z. 43 (1942), 109-116
J. Eggert, Naturwiss. 31 (1943), 412-415
W. Meissner, Bayer. Akad. Wiss. Jahrbuch 1944-1948, pp.244-248
J.R. Partington, J. Chem. Soc. 1953, pp.2853-2872
K. Mendelssohn, Cryogenics 4 (1964), 129-135; The World of Walther Nernst. London 1973
E.S. Barr, Am. J. Phys. 32 (1964), 302-305
L. Suhling, Rete 1 (1972), 331-346
H.G. Bartel, Z. Chem. 23 (1983), 277-287

NES, WILLIAM ROBERT [1926-1988]
AMS 15(5),629; WWWS, 1246
E. Caspi, Steroids 53 (1989), v-x

NESMEYANOV, ALEKSANDR NIKOLAEVICH [1899-1980]
MH 2,390-392; STC 2,300-301; MSE 2,353-355; WWWS, 1246; POTSCH, 319-320;POGG 6,1839; 7b,3564-3592
J. Chatt and M.I. Rybinskaya, Biog. Mem. Fell. Roy. Soc. 29 (1983), 399-480

NESSLER, JULIUS [1827-1905]
BJN 10,220*; BAD 6,551-554; HEIN, 462-463; BLOKH 2,536; WWWS, 1247; POTSCH, 320
J. Behrens, Land. Vers. 62 (1905), 241-250
F. Mach, Chem. Z. 51 (1927), 434

NESTLER, ANTON [1854-1932]
STURM 3,26-27

NESTLER, CHRÉTIEN GEOFFROY [1778-1832]
HEIN, 463; BALLAND, 77-78; BERGER-LEVRAULT, 172

NETTER, ARNOLD [1855-1936]
FISCHER 2,1107; KOREN, 224-225; WININGER 4,514; 7,345
A.R. Debré, Bull. Acad. Med. 115 (1936), 419-422
R. Kohn, Rev. Hist. Med. Heb. 1952, pp.94-96

NETTER, HANS [1899-1977]
KURSCHNER 12,2233; KIEL, 89; IB, 212; POGG 7a(3),407
Anon., Nachr. Chem. Techn. 7 (1959), 241; Deutsche med. Wchschr. 102 (1977), 1792
H.D. Ohlenbusch, Christiana Albertina NF9 (1978), 187-189

NETZ, ERICH [1882-1951]
JV 22,324; NUC 411,253

NEUBAUER, CARL [1830-1879]
NB, 288; HEIN, 464; SCHAEDLER, 84; POGG 2,271-272; 3,962
R. Fresenius, Z. anal. Chem. 18Suppl. (1879), 1-8
C. Liebermann, Ber. chem. Ges. 12 (1879), 1931

NEUBAUER, OTTO [1874-1957]
DSB 18,663-664; FISCHER 2,1107-1108; LDGS, 68; BACHMANN, 145-147; KOREN, 225; WININGER 7,345
F. Valentin, Münch. med. Wchschr. 100 (1958), 354-355
B. Schepartz, Trans. Coll. Phys. Phila. [5]6 (1984), 139-154

NEUBECK, CARL [1867-1925]
JV 13,155; NUC 411,281

NEUBERG, CARL [1877-1956]
AMS 9(I),1407; FISCHER 2,1108; BHDE 2,853; LDGS, 60; TLK 3,469; IB, 212; WWAH, 444; KOREN, 225; WININGER 4,516; 7,345; TETZLAFF, 247; POTSCH, 320; POGG 4,1065-1066; 5,898; 6,1839-1841; 7a(3),409-411

A. Gottschalk, Nature 178 (1956), 722-723
R. Ammon and W. Dirscherl, Arzneimitt. 6 (1956), 411-412
D. Nachmansohn, Proc. Virchow Soc. 75 (1956), 75-82
F. Lipmann et al., Science 124 (1956), 1244-1245
Anon., Leopoldina [3]2 (1956), 26-27
E. Auhagen, Biochem. Z. 328 (1956), 323-324)
F.F. Nord, Adv. Carbohydrate Chem. 13 (1958), 1-7; Chem. Ber. 94 (1961), i-v
A. Butenandt, Bayer. Akad. Wiss. Jahrbuch 1958, pp.180-183
A. Nordwig, TIBS 9 (1984), 498-499

NEUBERGER, ALBERT [1908-]
WW 1981, p.1897; IWW 1982-1983, pp.951-952; BHDE 1,525; 2,853; LDGS, 60
A.K. Allen and T.K. Palmer, Biochem. Soc. Trans. 7 (1979), 781-782
A. Neuberger, in Selected Topics in the History of Biochemistry (G. Semenza and R. Jaenicke, eds.), pp.21-65. Amsterdam 1990

NEUBURGER, MAX [1868-1955]
FISCHER 2,1109; BHDE 2,854; WIEN 1956, pp.44-45; KOREN, 225; WININGER 4,516-517; 7,346; TETZLAFF, 247
S.R. Kagan, Bull. Hist. Med. 14 (1943), 423-448
E. Berghoff, Max Neuburger. Vienna 1948
J.F. Fulton, J. Hist. Med. 10 (1955), 228-229
E.A. Underwood, Centaurus 4 (1955), 67-79
E. Rossner, Med. Hist. J. 3 (1968), 328-332
E. Lesky, Clio Medica 4 (1969), 54-56

NEUFELD, ELIZABETH FONDAL [1928-]
AMS 15(5),634

NEUFELD, FRED [1869-1945]
FISCHER 2,1109; IB, 212; BULLOCH, 386; KOREN, 225
Anon., Med. Klin. 25 (1929), 287-289
F.K. Kleine, Z. Hyg. 127 (1947), 185-186

NEUFELD, KARL ALBERT [1866-1914]
DBJ 1,302; RSC 17,490
F. Wirthle, Chem. Z. 38 (1914), 121

NEUFVILLE, RUDOLF de [1866-]
KALLMORGEN, 362; JV 5,234; RSC 17,490-491

NEUKOMM, JAKOB [1833-1862]
ZURICH-D, 44; SG [1]9,823

NEUMANN, ALBERT [1866-1907]
BLOKH 2,536; RSC 17,492; NUC 411,501
W. Pagel, Virchows Jahresber. 1907(I), p.456
C. Liebermann, Ber. chem. Ges. 40 (1907), 609

NEUMANN, ERNST [1834-1918]
KROLLMANN, 461-462; HIRSCH 4,352
P. Baumgarten, Berl. klin. Wchschr. 55 (1918), 364-368
M. Askanazy, Z. allgem. Path. 29 (1918), 409-421; Verhandl. path. Ges. 28 (1935), 362-371
J. Kühbuck, Clio Medica 4 (1969), 121-125
E. Neumann-Redling von Meding, Pathologe 5 (1984), 53-56

NEUMANN, KAREL CYRIL [1856-1919]
STURM 3,38-39

NEUMANN, KARL AUGUST [1771-1866]
OBL 7,93; STURM 3,39; WRANY, 183-185; VOGEL, 14-19; FERCHL, 380-381; CALLISEN 13,478; POGG 2,274

NEUMANN, RUDOLF OTTO [1868-1952]
HEIN-E, 335-336; FISCHER 2,1111; DRULL, 192; OLPP, 295; KIEL, 114; TLK 3,472; IB, 212; POGG 7a(3),400
K. Süpfle, Deutsche med. Wchschr. 64 (1938), 941-942
J. Bürgers, Z. Bakt. 150 (1952), 1-2

NEUMANN, WENZEL [1816-1885]
STURM 3,41

NEUMANN, WILHELM [1877-1944]
STURM 3,41

NEUMANN, WILHELM [1898-1965]
KURSCHNER 9.1439; BOCK, 110-114; POGG 7a(3),420-421
Anon., Nachr. Chem. Techn. 13 (1965), 222

NEUMAYER, LUDWIG [1866-1934]
EGERER, 35-44; IB, 212
H. Marcus, Anat. Anz. 80 (1935), 295-299

NEUMEISTER, RICHARD [1854-1905]
FISCHER 2,1112; GIESE, 504-505
E. Weinland, Z. Biol. 48 (1906), 141-143
M. Matthes, Münch. med. Wchschr. 53 (1906), 367-368

NEURATH, HANS [1909-]
AMS 14,3655; IWW 1982-1983, pp.952-953; BHDE 2,858; MH 2,392-394; STC 2,305-306; MSE 2,357-358; WWWS, 1248

NEUSSER, EDMUND von [1852-1912]
OBL 7,104-105; BJN 18,48*; FISCHER 2,1113; PAGEL, 1204; STUMPF, 116-121; LESKY, 327-330; WWWS, 1738
A. Kronfeld, Wiener med. Wchschr. 62 (1912), 2129-2139
R. Fleckseder, Wiener med. Wchschr. 62 (1912), 2274-2278
N. Ortner, Wiener med. Wchschr. 62 (1912), 2945-2951

NEVEN, OTTO [1875-1963]
KALLMORGEN, 362; SG [3]8,244; NUC 412,63

NEVILLE, HENRY ALLEN DUGDALE [1880-1952]
WW 1949, p.2040
Anon., Chem. Ind. 1952, p.625NEVINNY, JOSEPH [1853-1923]
STURM 3,46-47; PAGEL, 1204-1205; DECKER, 30-35; NAVRATIL, 214
C. Ipsen, Ber. Nat. Med. Innsb. 39 (1924), xxi-xxx

NEVOLE, MILAN [1846-1907]
OBL 7,109; STURM 3,47

NEWBURGH, LOUIS HARRY [1883-1956]
AMS 9(II),822
J.H. Means, Trans. Assoc. Am. Phys. 70 (1957), 19-22
A.M. Beewkes and M.W. Johnston, J. Nutrition 85 (1965), 3-7
A.M. Harvey, American Journal of Medicine 70 (1981), 759-761

NEWBURGH, ROBERT WARREN [1922-]
AMS 15(5),641; WWWS, 1249

NEWELL, LYMAN CHURCHILL [1867-1933]
AMS 5,821-822; NCAB 26,432; POGG 6,1847; 7b,3599
T.L. Davis, Ind. Eng. Chem. 24 (1933), 1082-1083
Anon., Bostonia 7 (1934), 3-9
C.A. Browne, J. Chem. Ed. 11 (1934), 66-67

NEWLANDS, JOHN ALEXANDER REINA [1837-1898]
DSB 10,37-39; W, 390; BLOKH 2,536-537; ABBOTT-C, 102-103; WWWS, 1250; POTSCH, 321
W.A. Tilden, Nature 58 (1898), 395-396
J.A. Cameron, Chemical Age 59 (1948), 354-356
W.H. Taylor, J. Chem. Ed. 26 (1949), 491-496
W.A. Smeaton, J. Roy. Inst. Chem. 88 (1964), 271-274
J.W. van Spronson, Chymia 11 (1966), 125-137

NEWMAN, HORATIO HACKETT [1875-1957]
DSB 18,665-668; IB,213; WWWS, 1250
H.H. Strandskov, Science 127 (1958), 74

NEWMAN, LLOYD HENRY [1892-]
AMS 12,4560

NEWMAN, MELVIN SPENCER [1908-]
AMS 14,3664; WWA 1978-1979, p.2392; IWW 1982-1983, p.954; COHEN, 176; WWWS, 1250-1251

NEWPORT, GEORGE [1803-1854]
DSB 10,39-40; DNB 14,357-358; WWWS, 1251
Anon., Proc. Roy. Soc. 7 (1854-1855), 278-285

NEWTON, GUY GEOFFREY FREDERICK [1919-1969]
E.P. Abraham, Chem. Brit. 5 (1969), 368

NEYMANN, HANS von SPLAWA [1885-1916]
JV 25,421
Anon., Chem. Z. 40 (1916), 669; Z. angew. Chem. 29(III) (1916), 456

NICHOLAS, JOHN SPANGLER [1895-1963]
DSB 10,104-105; AMS 10,2971; NCAB 48,605-606; IB, 213; WWWS, 1253
D. Hooker, Am. Phil. Soc. Year Book 1963, pp.185-190
J.M. Oppenheimer, Biog. Mem. Nat. Acad. Sci. 40 (1969), 239-289

NICHOLLS, DORIS MARGARET McEWEN [1927-]
AMS 15(5),653; WWWS, 1254

NICHOLS, EDWARD LEAMINGTON [1854-1937]
DAB 22,487-488; AMS 5,824; WWAH, 446; WWWS, 1254; POGG 3,969; 4,1072; 5,902-903; 6,1849-1850; 7b,3602
E.G. Merritt, Science 86 (1937), 483-485; Biog. Mem. Nat. Acad. Sci. 21 (1940), 343-360
F.K. Richtmeyer, Am. Phil. Soc. Year Book 1938, pp.382-386

NICHOLS, JAMES BURTON [1902-]
AMS 11,3887; POGG 6,1849; 7b,3601-3602

NICHOLSON, EDWARD CHAMBERS [1827-1890]
W.J. Russell, J. Chem. Soc. 59 (1891), 464-465

NICHOLSON, WILLIAM [1753-1815]
DSB 10,107-108; DNB 14,473-475; COLE, 401-404; FERCHL, 381-382; WWWS, 1255; CALLISEN 31,40; POGG 2,280-281
A. Lilley, Ann. Sci. 6 (1948), 78-101

NICKLÈS, JERÔME [1820-1869]
LAROUSSE 11,984-985; BOURQUELOT, 94; DORVEAUX, 77-78; TSCHIRCH, 1095; POTSCH, 322; FERCHL, 382; POGG 2,281-282; 3,969
L. Figuier, Ann. Scient. 14 (1869), 581-583
J. Aubry, Mem. Acad. Stanislas [7]7 (1978-1979), 87-98

NICLOUX, MAURICE [1873-1945]
POGG 5,904; 6,1852-1853; 7b,3604-3606
J. Roche, Bull. Soc. Chim. Biol. 23 (1945), 136-139
M. Tiffeneau, Bull. Soc. Chim. [5]12 (1945), 6-7
A. Boivin, Bull. Acad. Med. 129 (1945), 141-145
M. Javillier, C.R. Acad. Sci. 220 (1945), 153-156

NICOLAIER, ARTHUR [1862-1942]
KURSCHNER 4,2082; FISCHER 2,1116; PAGEL, 1206-1207; WWWS, 1255-1256
C. Bley, Arthur Nicolaier. Göttingen 1948

NICOLET, BEN HARRY [1890-1959]
AMS 7,1304-1305; POGG 6,1953-1854; 7b,3609-3610
Anon., Chem. Eng. News 37(22) (1959), 122-123

NICOLLE, CHARLES [1866-1936]
DSB 15,453-455; FISCHER 2,1117; CHARLE, 193-195; BULLOCH, 386; W, 392-393; WWWS, 1256
F. Mesnil, Bull. Acad. Med. 115 (1936),541-549; Presse Med. 44 (1936), 595-598
Anon., Ann. Inst. Pasteur 56 (1936), 353-358
P. Giroud, Bull. Acad. Med. 145 (1961), 714-722
G. Lot, Charles Nicolle. Paris 1961
H. Diacono, Biol. Med. 55 (1966), 437-443
L.A. Robin, Revues Médicales Normandes 1967, pp.7-21
F. Nicolle et al., Précis Anal. Acad. Rouen 1967, pp.21-67
A. Chadli et al., Bull. Acad. Med. 170 (1986), 273-304

NICOLLE, MAURICE [1862-1932]
HD 4,537-544; POGG 6,1854; 7b,3610
A. Calmette, Ann. Inst. Pasteur 49 (1932), 275-279
J. Nicolle, Maurice Nicolle. Paris 1957

NIEBEL, WILHELM [-1901]
RSC 17,524
Anon., Ber. chem. Ges. 34 (1901), 4385

NIEDERHOFF, PAUL [1890-1954]
KURSCHNER 8,1677; POGG 7a(3),426-427

NIELSEN, NIELS [1900-1960]
KBB 1943, P.915; VEIBEL 2,314-315; IB, 214
P. Larson, in The Carlsberg Laboratory (H. Holter and K.M. Moller, Eds.), pp.231-238. Copenhagen 1976

NIEMANN, ALBERT [1834-1861]
HEIN, 468-469; DIEPGEN, 96-99; POTSCH, 322
R. Zaunick, Pharmazie 4 (1949), 475-478

NIEMANN, ALBERT [1880-1921]
DBJ 3,310; FISCHER 2,1118-1119; WWWS, 1257
E. Müller, Arch. Kind. 69 (1921), 398

NIEMANN, CARL [1908-1964]
AMS 10,2981; NCAB 50,366-367; WWAH, 448
J.D. Roberts and G.S. Hammond, Proc. Chem. Soc. 1964, p.433
J.D. Roberts, Biog. Mem. Nat. Acad. Sci. 40 (1969), 291-319

NIEME, ALEXANDER [1863-1930]
JV 3,62; RSC 17,527; NUC 419,109
Anon., Chem. Z. 53 (1929), 91; 54 (1930), 311; Z. angew. Chem. 43 (1930), 324

NIEMENTOWSKI, STEFAN [1866-1925]
PSB 22,789-790; POTSCH, 322; POGG 4,1074; 5,906; 6,1855
E. Sucharda, Roczniki Chemii 5 (1925), 405

NIEMER, HELMUT [1900-1988]
KURSCHNER 13,2747; POGG 7a(3),428-429
Anon., Chem. Z. 99 (1975), 249; 112 (1988), 116

NIEMEYER, FELIX [1820-1871]
ADB 23,680-682; HIRSCH 4,367-368; PAGEL, 1207-1209
H. von Ziemssen, Arch. klin. Med. 8 (1871), 427-444
Anon., Berl. klin. Wchschr. 8 (1871), 189-191
R. Thren, Felix von Niemeyer. Tübingen 1946

NIEMIERKO, WLODIMIERZ [1897-1985]
Who's Who in Socialist Countries 1978, p.435
S. Niemierko, Acta Biochim. Polon. 34 (1987), 239-252
Z. Zielinska, Acta Physiol. Polon. 38 (1987), 100-108

NIEMILOWICZ, WLADYSLAW [1863-1904]
PSB 22,801-802; KONOPKA 6,462; RSC 17,528; SG [2]11,687
G, Gittelmacher, Nowiny Lekarskie (Poznan) 16 (1904), 397
J. Pagel, Virchows Jahresber. 1904, p.475

NIENHAUS, KASIMIR [1838-1910]
TSCHIRCH, 1095
E. Beuttner, Verhandl. Schw. Nat. Ges. 94 (1911), 51-52

NIER, ALFRED OTTO CARL [1911-]
AMS 15(5),667; WWA 1980-1981, p.2465; IWW 1981-1982, p.932; WWWS, 1257; MH 2,396-397; MSE 2,361-363
A.O. Nier, Annual Review of Earth and Planetary Science 9 (1981), 1-17

NIERENSTEIN, MAXIMILIAN [1878-1946]
IB, 214; POGG 5,906; 6,1855-1856; 7b,3616-3617
Anon., Chem. Ind. 1946, p.67; Isis 38 (1947), 108

NIETZKI, RUDOLF [1847-1917]
HEIN-E, 338; BLOKH 2,528-530; SCHAEDLER, 85; WWWS, 1258; POTSCH, 322-323; POGG 3,972; 4,1075-1076; 5,906
H. Ost, Z. angew. Chem. 30 (1917), 285-287
E. Noelting, Helv. Chim. Acta 1 (1918), 343-432
H. Rupe, Chem. Z. 42 (1918), 101-104; Ber. chem. Ges. 52A (1919), 1-28

NIEUWLAND, JULIUS ARTHUR [1878-1936]
DSB 10,121-123; DAB [Suppl.2],488-489; NCAB 26,357-358; W, 393-394; POTSCH, 323; WWWS, 1258; POGG 6,1859-1860; 7b,3620-3621
M.W. Lyon, Science 84 (1936), 7-8
Anon., Ind. Eng. Chem. News Ed. 14 (1936), 248; American Midland Naturalist 17 (1936), iii-xv
W.S. Calcott, J. Chem. Soc. 1937, pp.708-709

NIGG, CLARA IDA [1897-]
AMS 11,3899

NIGGLI, PAUL [1888-1953]
DSB 10,124-127; AARGAU, 580; STC 2,307-308; TLK 3,474; POGG 5,906-907; 6,1860-1862; 7a(3),431-432
F. Laves, Experientia 9 (1953), 197-202
F. Machatski, Alm. Akad. Wiss. Wien 103 (1953), 466-471
E. Widmer, Argovia 65 (1953), 469-472
K.H. Schaumann, Geologie 2 (1953), 124-130
H. O'Daniel et al., Neues Jahrbuch für Mineralogie 1953, pp.51-67
H. Steinmetz, Bayer. Akad. Wiss. Jahrbuch 1953, pp.185-188
A. Streckeisen, Mitt. Nat. Ges. Bern 11 (1954), 109-113
C. Burri, Verhandl. Schw. Nat. Ges. 133 (1954), 345-351

NIKANOROV, PAVEL IVANOVICH [1863-]
RSC 17,534; SG [2]11,691

NIKITIN, BORIS ALEKSANDROVICH [1906-1952]
BSE 29,620; WWR, 412
V.M. Vdovenko, Usp. Khim. 22 (1953), 1030-1031

NIKITIN, NIKOLAI IGNATIEVICH [1890-1975]
BSE (3rd Ed.)17,613; POGG 6,1862-1863; 7b,3626-3629
F.P. Komarov and S.D. Antonovski, Zhur. Ob. Khim. 20 (1950), 557-562
S.I. Volkovich et al., Zhur. Prikl. Khim. 48 (1975), 1421-1423

NIKLEWSKI, BRONISLAW [1879-1961]
PSB 23,118-120; IB, 214

NIKOLSKI, ALEKSANDR MIKHAILOVICH [1850-1902]
UKRAINE 2,352-353
Anon., Travaux de la Société des Naturalistes Charkow 50 (1927), 1-15

NIKOLSKI, VLADIMIR IVANOVICH [1850-]
KAZAN 2,277-279; ZMEEV 5,34-35; VENGEROV 2,177; SG [1]11,693; [3]8,258

NILSON, LARS FREDRIK [1840-1899]
SMK 5,434-435; BLOKH 2,541; WWWS, 1259; POTSCH, 323; POGG 3,973; 4,1077
P. Klason, Ber. chem. Ges. 32 (1899), 1643-1646
O. Petterson, J. Chem. Soc. 77 (1900), 1277-1294
A.G. Ekstrand, Lev. Sven. Vet. 6 (1921), 1-102

NILSSON, RAGNAR [1903-1981]
VAD 1981, p.760; POGG 6,1863-1864; 7b,3632-3633

NILSSON-EHLE, HERMAN [1873-1949]
DSB 10,129-130; SMK 5,459-460
A. Müntzing, Z. Pflanzensucht. 29 (1950), 110-114
E. Tschermak-Seysenegg, Alm. Akad. Wiss. Wien 100 (1950), 367-377
G.D.H. Bell, Nature 165 (1950), 709-710

NIRENBERG, MARSHALL WARREN [1927-]
AMS 15(5),671; WWA 1980-1981, pp.2466-2467; IWW 1982-1983, p.962; MH 2,398-399; STC 2,310-311; MSE 2,363-365; WWWS, 1259; POTSCH, 323

NISHI, MORINOSHIN [1879-]
HAKUSHI 2,162

NISHIKAWA, SHOJI [1884-1952]
POGG 6,1865; 7b,3635
I. Nitta, in Fifty Years of X-Ray Diffraction (P.P. Ewald, ed.), pp.328-334. Dordrecht 1962

NISONOFF, ALFRED [1923-]
AMS 16(3),692

NISSL, FRANZ [1860-1919]
DSB 10,130-132; DBJ 2,729; FISCHER 2,1122; KALLMORGEN, 363; DRULL, 193; HAYMAKER, 351-355; WWWS, 1260
E. Kraepelin, Münch. med. Wchschr. 66 (1919), 1058-1060
H. Spatz, Berl. klin. Wchschr. 56 (1919), 1006-1007
F. Jahnel, Arch. Psychiat. 61 (1919-1820), 751-759
U. Spatz, in 50 Jahre Neuropathologie in Deutschland (W. Scholz, Ed.), pp.43-66. Stuttgart 1961

NITSCHKE, ALFRED [1898-1960]
KURSCHNER 8,1687
E. Rominger, Z. Kinderheil. 85 (1961), 1-6
A. Windorfer, Deutsche med. Wchschr. 86 (1961), 360-361

NITSCHMANN, HANS [1907-]
KURSCHNER 13,2757; POGG 7a(3),435-436

NITTI, FRÉDÉRIC [1905-1947]
J. Tréfouël, Ann. Inst. Pasteur 73 (1947), 711-712

NOACK, KURT [1888-1963]
TLK 3,475-476; IB, 215; POGG 6,1867; 7a(3),437
A. Pirson, Ber. Bot. Ges. 78 (1965), (182)-(190)

NOAD, HENRY MINCHIN [1815-1877]
DNB 14,522-523; COLE, 405; RSC 4,628; POGG 3,974

NOBILI, LEOPOLDO [1784-1835]
DSB 10,134-136; EI 24,869; FERCHL, 384; W, 394; WWWS,1261; POGG 2,291-292

NOCARD, EDMOND [1850-1903]
FISCHER 2,1124; SACKMANN, 248-250; BULLOCH, 386; WWWS, 1261
A.M. Bloch, C.R. Soc. Biol. 55 (1903), 1141-1143
G.H.F. Nuttall, J. Hygiene 3 (1903), 517-519
T. Ménard, Bull. Acad. Med. 50 (1903), 115-119
E. Metschnikoff, Deutsche med. Wchschr. 29 (1903), 712
S. Jaccoud, Mem. Acad. Med. 41 (1906), 1-23
S. Simonet, Edmond Nocard. Paris 1947
D. Wrotnowska, Bull. Soc. Vet. 48 (1973), 21-28

NOCHT, BERNHARD [1857-1945]
KURSCHNER 4,2097; FISCHER 2,1124-1125; FREUND 2,271-275; OLPP, 298-299
E. Martini, Münch. med. Wchschr. 99 (1957), 1985

NODDACK, WALTER [1893-1960]
DSB 10,136; KURSCHNER 9,1457; WWWS, 1261; POTSCH, 324-325; POGG 6,1868; 7a(3),437-438
O. Bayer et al., Chem. Ber. 96 (1963), xxvii-li

NOEGGERATH, CARL [1876-1952]
FISCHER 2,1125; KURSCHNER 7,1466-1467

H. Kleinschmidt, Mon. Kind. 100 (1952), 383-384
H. Hungerland, Klinische Paediatrie 189 (1977), 105-110

NOELL, WERNER K. [1913-]
AMS 15(5),678
W.K. Noell, Am. J. Ophthalm. [3]48 (1959), 345-347

NOELLNER, CARL [1808-1877]
HEIN, 469-470; FERCHL, 384; RSC 4,638; 8,513; 12,539; POGG 3,976
Anon., Jahreshefte Lüneburg 7 (1878), 42-49

NOELTING, EMILIO [1851-1922]
BLOKH 2,542-543; ZURICH-D, 125; POTSCH, 325; POGG 3,976; 4,1080; 5,909; 6,1869
P. Friedländer, Ber. chem. Ges. 55A (1922), 137-140
A. Haller, Bull. Soc. Chim. [4]33 (1923), 1-5
H. Rupe, Verhandl. Nat. Ges. Basel 34 (1923), 69-77
F. Reverdin and A. Pictet, Helv. Chim. Acta 6 (1923), 110-128
E.H. Huntress, Proc. Am. Acad. Arts Sci. 79 (1951), 30-31

NOGUCHI, HIDEYO [1876-1928]
DSB 10,141-145; DAB 13,542-543; FISCHER 2,1125-1126; OLPP, 299-301; STC 2,314-316; DAMB, 553-554; IB, 215; WWAH, 449; WWWS, 1262
S. Flexner, Science 69 (1929), 653-660
T. Smith, Bull. N.Y. Acad. Med. [2]5 (1929), 877-884
P.F. Clark, Bull. Hist. Med. 33 (1959), 1-20
C.E. Dolman, Clio Medica 12 (1977), 131-145
I.R. Plesset, Noguchi and his Patrons. Rutherford, N.J. 1980

NOGUCHI, TAKAMI [1894-1951]

NOLAN, LAURENCE S. [1890-1984]

NOLAN, OWEN L. [1888-1958]

NOLAN, THOMAS JOSEPH [1888-1945]
POGG 6,1871; 7b,3643-3644
J. Algar, Nature 155 (1945), 663-664
D. Reilly, J. Chem. Soc. 1946, pp.315-316

NOLF, PIERRE [1873-1953]
FISCHER 2,1127-1128; IB, 215
H. Fredericq, Ann. Acad. Roy. Belg. 121 (1955), 3-63; Liber Memorialis Universite de Liège 1936-1966, vol.2, pp.518-530. Liège 1967

NOLL, ALFRED [1870-1956]
GIESE, 499-500; IB, 215; POGG 7a(3),440-441

NOLL, FRIEDRICH [1858-1908]
BJN 13,185-186,67*; DZ, 1029
M. Koernicke, Ber. Bot. Ges. 26A (1908), (77)-(94)
H. Ullrich, Decheniana 111 (1959), 81-87

NOLLER, CARL ROBERT [1900-1980]
AMS 11,3911; WWWS, 1263; POGG 6,1871-1872; 7b,3644-3645
Anon., Chem. Eng. News 59(5) (1981), 35
W.S. Johnson, Organic Syntheses 61 (1983), vii-ix

NOMURA, MASAYASU [1927-]
AMS 15(5),681-682; WWA 1980-1981, p.2473; MSE 2,365-366
M. Nomura, in Ribosome (H.W. Hill, ed.), pp.3-55. Washington, D.C. 1990

NONIDEZ, JOSÉ FERNANDEZ [1892-1947]
AMS 7,1310; IB, 216

NONNENBRUCH, WILHELM [1887-1955]
KURSCHNER 7,1470; STURM 3,61; FISCHER 2,1127; PELZNER, 84-101; IB, 216; FRANKEN, 192; KOERTING, 170; POGG 7a(3),442
R. Zaunick, Leopoldina [3]1 (1955), 14

F. Hoff, Deutsche med. Wchschr. 80 (1955), 580-581
H. Lampert, Münch. med. Wchschr. 97 (1955), 305-306

NOON, LEONARD [1878-1913]
O'CONNOR(2), 44-45
Anon., Brit. Med. J. 1913(I), p.368; St. Barts Hosp. J. 20 (1913), 99

NOORDEN, CARL HARKO von [1858-1944]
OBL 7,147; FISCHER 2,1127-1128; PAGEL, 1212-1213; KALLMORGEN, 363-364; STUMPF, 38-71; WWWS, 1738
W. Huepke, Münch. med. Wchschr. 100 (1958), 1394-1395
A. Durig et al., Wiener med. Wchschr. 108 (1958), 719-723,754-755
H. Diebold, Wiener klin. Wchschr. 77 (1965), 854-856
J. Hauk, Carl Harko von Noorden. Mainz 1980

NOORDEN, KARL von [1888-]
JV 27,447; SG [3]8,277; NUC 421,134

NORD, FRIEDRICH FRANZ [1889-1973]
AMS 11,3913; KURSCHNER 10,1748; BHDE 2,866-867; TLK 2,511; POGG 6,1873-1874; 7a(3),442-446
F. Lynen, Bayer. Akad. Wiss. Jahrbuch 1974, pp.197-199

NORDENSKIÖLD, NILS ERIK [1872-1933]
DSB 10,149; SMK 5,500
A. Mieli, Archeion 17 (1935), 77
R. Zaunick, Mitt. Gesch. Med. 34 (1935), 303
N. von Hofsten, Isis 38 (1947), 103-106

NORDLIE, ROBERT CONRAD [1930-]
AMS 15(5),684; WWWS, 1264

NORMAN, ARTHUR GEOFFREY [1905-1982]
AMS 13,3248; WWWS, 1264; POGG 6,1875; 7b,3651-3654

NORMANN, WILHELM [1870-1939]
POGG 6,1876; 7a(3),447
W. Pflücke, Z. angew. Chem. 52 (1939), 433-434
J. Hefele, Chem. Z. 76 (1952), 769-771

NORPOTH, LEO [1901-1973]
KURSCHNER 13,2770; POGG 7a(3),447-448

NORRIS, EARL RALPH [1895-1952]
AMS 8,1833; POGG 6,1876; 7b,3654-3655
Anon., Chem. Eng. News 30 (1952), 1907

NORRIS, JAMES FLACK [1871-1940]
MILES, 365-366; NCAB 30,161-162; WWAH, 450; WWWS, 1265; POGG 4,1082; 5,912; 6,1876-1877; 7b,3655-3656
T.L. Davis, Ind. Eng. Chem. 14 (1936), 325-326
J.D. Roberts, Biog. Mem. Nat. Acad. Sci. 45 (1974), 413-426

NORRISH, RONALD GEORGE WREYFORD [1897-1978]
DNB [1971-1980],635-637; WW 1978, pp.1820-1821; CAMPBELL, 171-172; MH 1,352-353; STC 2,316-317; MSE 2,367-368; W(3rd Ed.), 601-602; POTSCH, 325-326; ABBOTT-C, 104-105; POGG 6,1877; 7b,3656-3660
R. Wayne, Journal of Photochemistry 9 (1978), 75-76
J.H. Purnell, Chem. Brit. 15 (1979), 85-86
F. Dainton and B.A. Thrush, Biog. Mem. Fell. Roy. Soc. 27 (1981), 379-424

NORTHROP, JOHN HOWARD [1891-1987]
AMS 14,3704; WWA 1978-1979, p.2416; IWW 1982-1983, p.968; NCAB G,272-273; CB 1947, pp.472-474; MH 1,353-354; STC 2,317-319; MSE 2,368-369; POTSCH, 326; WWWS, 1266; POGG 6,1877-1878; 7b,3660-3663
J.H. Northrop, Sci. Mon. 35 (1932), 333-340; Ann. Rev. Biochem. 30 (1961), 1-10
R.M. Herriott, J. Gen. Physiol. 45Suppl. (1962), 1-16,255-265; 77 (1981), 597-599; TIBS 8 (1983), 296-297
J.T. Edsall, Nature 329 (1987), 326

F.C. Robbins, Proc. Am. Phil. Soc. 135 (1991), 315-320

NORTON, JOHN PITKIN [1822-1852]
DSB 10,150-151; DAB 13,574-575; MILES, 366-368; ELLIOTT, 191; SILLIMAN, 58-59; WWAH, 451; WWWS, 1266; POGG 2.301
Anon., Am. J. Sci. [2]14 (1852), 448-449
L.I. Kuslan, J. Hist. Med. 24 (1969), 430-451
M.W. Rossiter, Liebig and the Americans: A Study in the Transit of Science 1840-1880, pp.180-259. New Haven 1971

NORTON, LEWIS MILLS [1855-1893]
MILES, 368; NCAB 4,301
T.M. Drown, Proc. Acad. Arts Sci. 28 (1893), 348-351

NOSSAL, GUSTAV JOSEPH VICTOR [1931-]
WW 1987, p.1305; IWW 1987-1988, p.1089

NOSSAL, PETER MARIA JOSEPH [1925-1958]
SRS, 75; BHDE 2,867

NOTH, HARTMUT [1892-]
JV 33,240; NUC 423,470

NOTHMANN, MARTIN [1894-1978]
AMS 11,3925; LDGS, 68; KOREN, 226; WININGER 7,592

NOTHNAGEL, HERMANN [1841-1905]
OBL 7,158; BJN 10,222*; 11,296-308; FISCHER 2,1129-1130; PAGEL,1213-1214; GIESE, 551-559; PETRY, 47-68; WWWS, 1267
E. von Leyden, Z. klin. Med. 57 (1905), i-vii
J. Mannaberg, Münch. med. Wchschr. 52 (1905), 1687-1689
R. Jaksch, Wiener med. Wchschr. 60 (1910), 2681-2688
H. Adler, Wiener med. Wchschr. 55 (1905), 1465-1470
M. Neuburger, Hermann Nothnagel. Vienna 1922
H. Wyklicky, Wiener klin. Wchschr. 74 (1962), 120-122; Med. Mon. 22 (1968), 168-172; Oesterreichische Aerztezeitung 31 (1977),487-490

NOTKIN, IGNATI ADOLFOVICH [1860-]
VENGEROV 2,189; RSC 17,566; SG [2]11,815-816

NOURI, OSMAN [1885-]
JV 34,238; NUC 423,648

NOVELLI, GUERINO DAVID [1918-1983]
AMS 15(5),697; WWWS, 1267

NOVICK, AARON [1919-]
AMS 15(5),697; WWA 1980-1981, p.2481

NOVICK, RICHARD [1932-]
AMS 15(5),697

NOVIKOFF, ALEX BENJAMIN [1913-1987]
AMS 15(5),697; WWA 1980-1981, p.2481; WWWS, 1268
C. deDuve et al., J. Histochem. 35 (1987), 931-938
D.R. Holmes, Stalking the Academic Communist. Hannover, N.H. 1989

NOVOTNY, FRANZ [1839-1879]
NAVRATIL, 216-217; WERSTLER 1,63-65

NOVY, FREDERICK GEORGE [1864-1957]
DSB 10,154-155; DAB [Suppl.6],481-482; NCAB 16,93; 54,427-428; OLPP, 301; AMS 9(II),832; FISCHER 2,1131; SACKMANN, 250-253; DAMB, 557-558; IB, 216; WWAH, 452; WWWS, 1268
W. Good, Univ. Mich. Med. Bull. 16 (1950), 257-268
W.J. Nungester, J. Bact. 74 (1957), 545-547; Science 127 (1958), 274
T. Francis, Trans. Assoc. Am. Phys. 71 (1958), 35-37
E.R. Long, Trans. Coll. Phys. Phila. 26 (1958); Biog. Mem. Nat. Acad. Sci. 33 (1959), 326-350
S.E. Gould, Am. J. Clin. Path. 29 (1958), 297-309

NOWINSKI, WIKTOR WACLAW [1903-1972]
AMS 11,3928; WWWS, 1268-1269

NOYES, ARTHUR AMOS [1866-1936]
DSB 10,156-157; DAB [Suppl.2],493-494; AMS 5,834; NCAB 13,284; WWAH, 452; MILES, 371-372; WWWS,1269; POGG 4,1083-1084; 5,912-913; 6,1883-1884; 7b,3670-3671
M.S. Sherrill, Ind. Eng. Chem. 25 (1931), 443-445; Science 84 (1936), 217-220; Proc. Am. Acad. Arts Sci. 74 (1940), 150-155
L. Pauling, Biog. Mem. Nat. Acad. Sci. 31 (1958), 322-346; Welch Foundation Conferences on Chemical Research 20 (1977), 88-101
J.W. Servos, Ambix 23 (1976), 175-186

NOYES, WILLIAM ALBERT [1857-1941]
DSB 10,157-158; DAB [Suppl.3],565-566; AMS 6,1054; NCAB B,314;44,258-259; MILES, 372-373; WWAH, 452; WWWS, 1269; POGG 4,1083; 5,913-914; 6,1884; 7b,3671-3672
A.M. Patterson, Science 94 (1941), 477-479
B.S. Hopkins, J. Am. Chem. Soc. 66 (1944), 1045-1056
R. Adams, Biog. Mem. Nat. Acad. Sci. 27 (1952), 179-208

NOYES, WILLIAM ALBERT, Jr. [1898-1980]
AMS 14,3711-3712; WWA 1976-1977, pp.2345-2346; WWWS, 1269; POGG 6,1884-1885; 7b,3672-3675
Anon., Chem. Eng. News 58(49) (1980), 8
N. Hackerman, Am. Phil. Soc. Year Book 1981, pp.479-485

NOYONS, ADRIAAN KAREL [1878-1941]
LINDEBOOM, 1441; IB, 216
Bibl. Acad. Louvain 6 (1937), 358-359
C. Heymans, Arch. Inter. Pharmacodyn. 66 (1941), 243-244
A. Gemelli, Acta Pont. Acad. Sci. 6 (1942), 193-207

NOYONS, EDUARD CHRISTIAAN [1900-1960]
LINDEBOOM, 1441-1442
Wie is Dat? 1956, p.445
R.J.F. Nivard, Chem. Wkbl. 56 (1960), 245-246

NOZOE, TETSURO [1902-]
JBE, 1094; WWWS, 1269
T. Mukai, Heterocycles 11 (1978), 11-58
T. Nozoe, Seventy Years in Organic Chemistry. Washington, D.C. 1991

NUHN, ANTON [1814-1889]
HIRSCH 4,391; PAGEL, 1214-1215
Anon., Leopoldina 25 (1889), 170

NUNGESTER, WALTER JAMES [1901-1985]
AMS 12,4620; WWWS, 1270

NUSSBAUM, JOHANN NEPOMUK [1829-1890]
ADB 52,667-668; HIRSCH 4,394-395; PAGEL, 1215-1216
J. Lindpainter, Münch. med. Wchschr. 37 (1890), 816-818
J. Fessler, Z. Chir. 56 (1929), 2178-2183

NUSSBAUM, MORITZ [1850-1915]
DBJ 1,336; FISCHER 2,1133; PAGEL, 1216; WININGER 4,550; 7,353
R. Bonnet, Anat. Anz. 48 (1915), 489-495

NUTTALL, GEORGE HENRY FALKINER [1862-1937]
DNB [1931-1940], 653-655; OLPP, 301-303; DAMB, 558-559; WWWS, 1270
G.S. Graham-Smith and D. Keilin, Parasitology 30 (1938), 403-418
E. Brumpt, Bull. Acad. Med. 119 (1938), 112-116
A. Lwoff, Presse Med. 46 (1938), 481-482
H.B. Ward, Science 87 (1938), 337-338
R. Matheson, J. Parasit. 24 (1938), 180-183
D.K[eilin], Nature 141 (1938), 318-319
G.S. Graham-Smith, J. Path. Bact. 46 (1938), 389-394; J. Hygiene 38 (1938), 129-139; Obit. Not. Fell. Roy. Soc. 2 (1939), 493-499

NYE, ROBERT NASON [1892-1947]
AMS 7,1317; IB, 217

Anon., J. Am. Med. Assn. 135 (1947), 300; New Eng. J. Med. 237 (1947),751

NYGAARD, AGNAR [1919-1979]
B.D. Hall et al., TIBS 5 (1980), 254-256

NYHOLM, RONALD SYDNEY [1917-1971]
DNB [1971-1980], 637-638; WW 1970, p.2318; CAMPBELL, 173-174; ABBOTT-C, 105; WWWS, 1271
D.P. Craig, Biog. Mem. Fell. Roy. Soc. 18 (1972), 445-475

NYLÉN, PAUL [1892-1976]
VAD 1977, p.768; POGG 6,1887-1888; 7b,3679-3680
Anon., Uppsala Universitets Matrikel 1937-1950, pp.428-430

NYSTEN, PIERRE HUBERT [1771-1818]
HOEFER 38,378-380; HIRSCH 4,397-398; WWWS, 1271; POGG 2,305
M. Florkin, Rev. Med. Liège 6 (1951), 300-310,373-383

O

OAKLEY, CYRIL LESLIE [1907-1975]
Who Was Who 1971-1980, p.589
G.G. Gray, Biog. Mem. Fell. Roy. Soc. 22 (1976), 295-305

OBERDISSE, KARL [1903-]
KURSCHNER 13,2781; GEISSLER, 134-139

OBERLIN, IGNACE LÉON [1810-1884]
HEIN, 473; HUMBERT, 216-217

OBERMAYER, FRIEDRICH [1861-1925]
OBL 7,190; FISCHER 2,1135; STUMPF, 24-29
E.P. Pick, Wiener klin. Wchschr. 38 (1925), 542-543

OBERMEIER, OTTO [1843-1873]
HIRSCH 4,400-401; OLPP, 303-305; WWWS, 1272
H. Zeiss, Sudhoffs Arch. 15 (1923), 161-164
G.A. Rost, Berl. Med. 14 (1963), 175-176

OBERMILLER, JULIUS [1873-1930]
KURSCHNER 3,1708; HEIN-E, 339; TLK 3,478; POGG 5,914-915; 6,1891
H. Freundlich, Ber. chem. Ges. 63A (1930), 137
E. Büttner, Z. angew. Chem. 43 (1930), 340

OBERMÜLLER, KUNO [1861-]
JV 8,26; RSC 17,581; SG [2]12,3; NUC 425,516

OBERNDORFER, SIEGFRIED [1876-1944]
KURSCHNER 5,984; FISCHER 2,1135; ECKERT, 90-99; IB, 217; BHDE 2,870; KOREN, 226; WIDMANN, 80,278-279; LDGS, 76; WININGER 7,354

OBOLENSKY, IVAN NIKOLAEVICH [1840-]
KHARKOV, 148-151; ZMEEV 5,39; RSC 8,524; 12,544

OBPACHER, HEINZ [1890-]
JV 32,329; NUC 425,593

OBRAZTSOV, VASILI PARMENOVICH [1851-1920]
BSE 30,376; BME 21,343-345; WWR, 419-420
A.I. Gubergrits, Ter. Arkh. 32 (1960), 80-82
V.P. Brevnov et al., Ter. Arkh. 49 (1977), 150-153

OCHOA, SEVERO [1905-1993]
AMS 14,3724; WWA 1978-1979, p.2430; IWW 1982-1983, p.976; ABBOTT-C, 106; MH 1,355-357; STC 2,324-326; MSE 2,374-376; CB 1962, pp.327-329; POTSCH, 326-327
Anon., Leopoldina [3]4/5 (1958-1959), 22-25
F. Grande and C. Asensio, in Reflections on Biochemistry (A. Kornberg et al., Eds.), pp.1-14. Oxford 1976
S. Ochoa, Ann. Rev. Biochem. 49 (1980), 1-30; Comp. Physiol. Biochem. 67B (1980), 359-365

ODDI, RUGGERO [1864-1913]
DSB 10,175-176
L. Belloni, Med. Hist. J. 1 (1966), 96-109
F. Sorcetti, Riv. Biol. 78 (1985), 133-140

ODDO, BERNARDO [1882-1941]
TSCHIRCH, 1096; POGG 5,915-916; 6,1893; 7b,3691-3692
G. Oddo, Ber. chem. Ges. 74A (1941), 183-183
Q. Mingoia, Gazz. Chim. Ital. 71 (1941), 737-752

ODDO, GIUSEPPE [1865-1954]
POGG 4,1088-1089; 5,916-917; 6,1893-1894; 7b,3692-3693
E. Oliveri-Mandola, Gazz. Chim. Ital. 85 (1955), 469-489

ODÉN, SVEN [1887-1934]
SMK 5,601-602; POGG 5,918-919; 6,1894-1895; 7b,3693
E.J. Russell, Nature 133 (1934), 438-439
C. Kjellin, Sven. Kem. Tid. 46 (1934), 297-299
T. Svedberg, J. Chem. Soc. 1935, pp.862-863

ODENIUS, MAXIMILIAN [1828-1913]
SMK 5,603; HIRSCH 4,406
M. Forssman, Hygeia 76 (1914), 85-92

ODERMATT-MARIOTTI, WILHELM [1854-1936]
RSC 10,947; SG [1]10,77; NUC 426,532
Anon., Schw. med. Wchschr. 66 (1936), 816

ODIER, AUGUSTE [1802-1870]
RSC 4,657
J. Théodoridès, Atti del Symposium Internazianole di Storia delle Scienze 1960, pp.1-7

ODLING, WILLIAM [1829-1921]
DSB 10,177-179; BLOKH 2,543-544; W, 396; WWWS, 1275; POTSCH, 327; POGG 3,982-983
J.E. Marsh, J. Chem. Soc. 119 (1921), 553-564
H.B. D[ixon], Proc. Roy. Soc. A100 (1922), i-vii
J.L. Thornton and A. Wiles, Ann. Sci. 12 (1956), 288-295
S. Rosen, J. Chem. Ed. 34 (1957), 517-519
H. Cassebaum, Chem. Z. 93 (1969), 929-935

OECONOMIDES, SPIRIDON [-1894]
RSC 10,948
Anon., Ber. chem. Ges. 27 (1894), 3550

OEFFINGER, HEINRICH CARL [1840-]
A. Wankmüller, Beitr. Wurtt. Apothekgesch. 7 (1966), 60

OEHLER, CARL GOTTLIEB REINHARD [1797-1874]
HB 1,195-196

OEHLER, EUGEN [1871-1950]
JV 11,234; RSC 17,593; NUC 427,16

OEHLER, RUDOLF [1859-1926]
KALLMORGEN, 364

OEHME, CURT [1883-1963]
KURSCHNER 9,1475; FISCHER 2,1139; DRULL, 195-196; IB, 217

ÖHRVALL, HJALMAR [1851-1929]
SMK 8,534-535; FISCHER 2,1139; PAGEL, 1221-1222
T. Thunberg, Skand. Arch. Physiol. 61 (1931), 1-19

OELKERS, HANS ADOLF [1901-]
POGG 7a(3),464-465

OERTEL, MAX JOSEPH [1835-1897]
ADB 52,713; BJN 2,97-98,81*; HIRSCH 4,423-424; PAGEL, 1222-1223
J. Bauer, Münch. med. Wchschr. 44 (1897), 919-921

OERTMANN, ERNST [1853-]
BONN, 300; RSC 10,950; SG [2]12,91; NUC 427,138

OESER, RICHARD [1891-1974]
JV 33,166; NUC 427,152

OESPER, PETER [1917-]
AMS 15(5),723

OESPER, RALPH EDWARD [1886-1977]
AMS 10,3018
G.B. Kauffman, Isis 70 (1979), 141-143

OESTERLE, OTTO [1866-1932]
HEIN, 476; HUMBERT, 245-246; TSCHIRCH, 1096; POGG 6,1899
A. Tschirch, Schw. Apoth. Z. 70 (1932), 342-343

OETKER, RUDOLF [1889-1916]
JV 30,517; NUC 427,283
Anon., Ber. chem. Ges. 49 (1916),1222; Z. angew. Chem. 29(III) (1916),203

OETTINGEN, KARL JOHANN von [1891-1953]
DRULL, 196; GAUSS, 202-203

OETTINGER, BENNO [1870-]
JV 10,139; RSC 17,597; NUC 427,138O'FARRELLY, ALFONS [1878-1968]
Anon., Chem. Brit. 4 (1968), 316

OGSTON, ALEXANDER [1844-1929]
FISCHER 2,1142; PAGEL, 1226-1227; BULLOCH, 387; WWWS, 1277
Anon. and W. Bulloch, Lancet 1929(I), pp.309-310
G. Smith, Brit. J. Surg. 52 (1965), 917-920
A. Lyell, Scottish Medical Journal 22 (1977), 277-278
L.G. Wilson, Med. Hist. 31 (1987), 403-414

OGSTON, ALEXANDER GEORGE [1911-]
WW 1981, p.1946; IWW 1982-1983, p.979

OHLE, HEINZ [1894-1959]
KURSCHNER 6(II),304; TLK 3,479; POGG 6,1902-1903; 7a(3),469-470
Anon., Nachr. Chem. Techn. 7 (1959), 294

OHLGART, CHRISTIAN [1877-]
JV 19,183; NUC 428,416

OHLMEYER, PAUL [1908-1977]
KURSCHNER 11,2168; POGG 7a(3),470-471
Anon., Chem. Z. 101 (1977), 221

OHLSSON, CARL ERIK [1891-1975]
SMK 5,615; VAD 1973, p.771; IB, 218

OHNMAIS, KARL [1865-1916]
JV 6,73; RSC 16,486; NUC 428,486
Anon., Chem. Z. 40 (1916), 250; Z. angew. Chem. 29(III) (1916), 203

OHTA, KOHSHI [1879-1953]
HAKUSHI 2,253

OKAZAKI, REIJI [1930-1975]
S. Suzuki, TIBS 1 (1976), N39-N40

OKEN, LORENZ [1779-1851]
DSB 10,194-196; ADB 24,216-226; HIRSCH 4,420-421; GLOSAUER, 105-111; TSCHIRCH, 1096; WWWS, 1278-1279; CALLISEN 14,113-116; 31,81-83; POGG 2,319; 7aSuppl.,468-471
R. Zaunick, Sudhoffs Arch. 33 (1941), 113-173

OKERBLOM, IVAN IVANOVICH [1870-]
RSC 17,605; SG [2]12,141

OKEY, RUTH ELIZA [1893-1987]
AMS 11,3961
T.H. Jukes, J. Nutrition 118 (1988), 1425-1431

OKKELBERG, PETER CLAUS [1880-1960]
AMS 9(II),838; NCAB 48, 577; IB, 218
H. von der Schalie, Nautilus 74(4) (1961), 163-164

OLAH, GEORGE ANDREW [1927-]
AMS 15(5),736; WWA 1980-1981, p.2504; POTSCH, 328

OLBRYCHT, JAN STANISLAW [1886-1968]
PSB 23,721-723; WE 8,201
J.S. Kobiela, Przeglad Lekarski 22 (1966), 570-573
J. Bogusz, Przeglad Lekarski 36 (1979), 148-152

OLEWINSKI, WLADYSLAW [1832-1862]
RSC 4,672
J. Roziewicz et al., Kwart. Hist. Nauki 34 (1989), 549-603

OLINGER, JOSEF [1881-]
NUC 429,400

OLITSKY, PETER KOSCIUSKO [1886-1964]
AMS 10,3028; IB, 219; KOREN, 226; KAGAN(JA), 309,733
Anon., Lancet 1964, p.423
W.F. Goebel, Rock. Inst. Rev. 2(4) (1964), 9-11

OLITZKI, ARIEH LEO [1898-1983]
KOREN, 226
Who's Who in Israel 1972, p.251

OLIVER, FRANCIS WALL [1864-1951]
DNB [1951-1960], 779-780; DESMOND, 472; IB, 219; WWWS, 1281
E. Salisbury, Obit. Not. Fell. Roy. Soc. 8 (1952), 229-240

OLIVER, GEORGE [1841-1915]
DSB 10,204-206; FISCHER 2,1144; O'CONNOR(2), 127-129; VERSO, 102-103
Anon., Lancet 1916(I), p.105; Brit. Med. J. 1916(I), p.73
M.L. Verso, Med. Hist. 15 (1971), 64-65

OLIVER, JEAN REDMAN [1889-1976]
AMS 11,3967; FISCHER 2,1144-1145; IB, 219
S.E. Bradley, Kidney International 5 (1974), 77-95
M.D. Stanford, Lancet 1976(II), p.1365

OLIVER, WADE WRIGHT [1890-1964]
AMS 10,3030; NCAB 53,84-85; FISCHER 2,1145; IB, 210

OLIVIER, SIMON [1879-1961]
POGG 5,922; 6,1905-1906; 7b,3722-3723
J.J.C. Tendeloo and G.B.R. de Graaf, Chem. Wkbl. 45 (1949), 385-387

OLSEN, CARSTEN ERIK [1891-1974]
KBB 1974, p.808; VEIBEL 2,321-322; IB, 210
P. Larson, in The Carlsberg Laboratory (H. Holter and K.M. Møller, Eds.) pp.130-138. Copenhagen 1976

OLSHAUSEN, OTTO [1840-1922]
DBJ 4,209-211,365; SHBL 7,153-154
F. Rathgen, Ber. chem. Ges. 55A (1922), 22

OLSON, AXEL RAGNAR [1889-1954]
NCAB 43,347-348; POGG 6,1907-1908; 7b,3724-3725
Anon., Chem. Eng. News 33 (1955), 534

OLSON, ROBERT EUGENE [1919-]
AMS 15(5),751-752; WWA 1980-1981, p.2512; WWWS, 1283

OLSZEWSKI, KAROL STANISLAUS [1846-1915]
DSB 10,206; PSB 24,27-30; WE 8,225; BLOKH 2,547-548; KONOPKA 7,181-183; POTSCH, 328; POGG 4,1095; 5,922-923
T. Estreicher, Ber. chem. Ges. 48 (1915), 739-741
H. Kammerling-Onnes, Chem. Z. 39 (1915), 517-519
H. von Smoluchowski, Naturwiss. 5 (1917), 738-740
Z. Wojtaszek, Studia z Dziejow Katedr Wydzial Matematyki, Fizyki i Chemii Universitet Jagellonski (S. Gotaba, Ed.), pp.166-178. Cracow 1964

OMELIANSKI, VASILI LEONIDOVICH [1867-1928]
BSE (3rd Ed.)18,393; BME 21,343-345; KUZNETSOV(1963), 337-344; IB, 219
S.A. Waksman, Soil Science 26 (1928), 255-256
B.L. Isachenko, Priroda 17 (1928), 771-776
L.D. Shturm, Mikrobiologia 22 (1953), 363-375
S.I. Kuznetsov, Mikrobiologia 36 (1967), 717-723

OMOROKOV, LEONID IVANOVICH [1881-1971]
Anon., Zhur. Nevropat. Psikh. 71 (1971), 1750-1751

ONAKA, MORIZO [1877-1920]
HAKUSHI 2,187

ONCLEY, JOHN LAWRENCE [1910-]
AMS 14,3760; WWA 1976-1977, p.2376; WWWS, 1285

ONO, SUMINOSUKE [1886-]
NUC 430,672

ONODERA, NAOSUKE [1883-1968]
JBE, 1204,2274; HAKUSHI 3,346

ONSAGER, LARS [1903-1976]
DSB 18,690-695; AMS 13,3304; MH 1,357-359; STC 2,326-327; MSE 2,278-280; WWWS, 1285; CB 1958, pp.321-322; W (3rd Ed.), 602; POTSCH, 328-329; ABBOTT-C, 106-107; POGG 6,1911; 7b,3731-3733
J.B. Kirkwood, Proc. Am. Acad. Arts Sci. 82 (1953), 298-300
C. Domb, Nature 264 (1976), 819-820
R. Schlögl, Ber. Bunsen Ges. 81 (1977), 253-254
H.C. Longuet-Higgins and M.E. Fischer, Biog. Mem. Fell. Roy. Soc. 24 (1978), 443-471; Biog. Mem. Nat. Acad. Sci. 60 (1991), 183-232
P. Lyons, Am. Phil. Soc. Year Book 1981, pp.485-495

ONSLOW, HUIA [1890-1922]
T.D.A. Cockerell, Science 56 (1922), 185-186
F.G. H[opkins], Biochem. J. 17 (1923), 1-4
M. Onslow, Huia Onslow. London 1924

ONSLOW, MURIEL WHELDALE [see WHELDALE]

ONSUM, IVAR [1834-1881]
KOBRO 2,218

ONUKI, MOTOI [1895-1960]
NUC 431,150

OPARIN, ALEKSANDR IVANOVICH [1894-1980]
DSB 18,695-700; BSE (3rd Ed.)18,409-410; STC 2,329-330; WWWS, 1286; AO 1981, pp.781-783; ABBOTT, 100-101; POTSCH, 329; POGG 6,1912; 7b,3737-3752
Anon., Leopoldina [3]2 (1956), 5-6
E. Broda, TIBS 5(11) (1980), iv-v

OPIE, EUGENE LINDSAY [1873-1971]
AMS 11,3983; NCAB 57,249-250; MH 1,360-361; MSE 2,381-382; DAMB, 563-564; FISCHER 2,1148; IB, 219; WWAH, 457; WWWS, 1286
P. Rous, Arch. Path. 34 (1942), 1-12
J.H. Talbott, Biographical History of Medicine, pp.940-943. New York 1970
J.G. Kidd, Am. J. Path. 65 (1971), 483-492
A.M. Harvey, Johns Hopkins Med. J. 134 (1974), 330-345
E.R. Long, Biog. Mem. Nat. Acad. Sci. 47 (1975), 293-320
M.O. Blackstone, Pancreas 3 (1988), 340-342

OPITZ, ERICH [1871-1926]
FISCHER 2,1148
R.T. von Jaschke, Z. Gyn. 50 (1927), 2679-2687

OPITZ, ERICH [1909-1953]
POGG 7a(3),473
R. Thauer, Pflügers Arch. 259 (1954), 21-29

OPPÉ, ALFRED [1873-1931]
JV 24,364; NUC 431,355
Anon., Chem. Z. 40 (1916), 503; 55 (1931), 511

OPPENHEIM, ALPHONS [1833-1877]
ADB 24,394-396; BLOKH 2,548-550; WININGER 7,357; POGG 3,987-988
A.W. Hofmann, Ber. chem. Ges. 10 (1877), 2262-2291

OPPENHEIM, FRANZ [1852-1929]
DBJ 11,363; POGG 6,1913
R. Willstätter, Ber. chem. Ges. 64A (1931), 133-149

OPPENHEIM, HERMANN [1858-1919]
DBJ 2,729; FISCHER 2,1149-1150; PAGEL, 1232-1233; TETZLAFF, 253; KOREN, 227; WWWS, 1286
Anon., Brit. Med. J. 1919(II), p.156
H. Liepmann, Z. ges. Neurol. 52 (1919), 1-6
R. Cassirer, Berl. klin. Wchschr. 56 (1919), 669-671
H.S. Decker, Bull. Hist. Med. 45 (1971), 461-481

OPPENHEIM, MORITZ [1876-1949]
KURSCHNER 4,2131-2132; FISCHER 2,1150; BHDE 2,876

OPPENHEIM, PAUL [1885-1977]
BHDE 2,876
New York Times, 24 June 1977

OPPENHEIMER, CARL [1874-1941]
KURSCHNER 5,991; FISCHER 2,1150-1151; BHDE 2,876-877; LDGS, 60; IB, 120; BK, 278-279; TLK 3,481; KOREN, 227; WININGER 4,575-577; 7,358; TETZLAFF, 254; WWWS, 1286; POGG 6,1913-1914; 7a(3),476
E. Simon, Palestine Journal of Botany R4 (1941), 47-50
W. Roman, Nature 150 (1942), 569-570
Anon., Enzymologia 12 (1946), 1-2

OPPENHEIMER, ERNST ADOLPH [1888-1962]
AMS 10,3041; IB, 220
F.F. Yonkman, Pharmacologist 4 (1962), 128-129

OPPENHEIMER, FRANZ [1864-1943]
FISCHER 2,1151; BHDE 2,877; BERLIN, 616; KOREN, 227; WWWS, 1286-1287; NUC 431,423-428
F. Oppenheimer, Mein Wissenschaftlicher Weg. Berlin 1929

OPPENHEIMER, GERTRUD [1893-1948]
LDGS, 23
Anon., Science 109 (1949), 129; Z. angew. Chem. 61 (1949), 48

OPPENHEIMER, HUGO [1866-1887]
JV 2,191; RSC 17,624
Anon., Ber. chem. Ges. 20 (1887), 3426

OPPENHEIMER, JANE MARION [1911-]
AMS 15(5),763; WWA 1980-1981, p.2519; WWWS, 1287

OPPENHEIMER, MAX [1875-1918]
JV 15,215; RSC 14,804; NUC 431,438

OPPENHEIMER, MAX [1886-]
JV 30,407; SG [3]8,445; NUC 431,438

OPPENHEIMER, SIEGFRIED [1882-1959]
KALLMORGEN, 366; BHDE 2,345
Dokumente zur Geschichte der Frankfurter Juden 1933-1945, pp.31,548. Frankfurt a.M. 1963

OPPENHEIMER, ZACHARIAS [1830-1904]
BJN 9,328; 10,81*; DRULL, 198; WININGER 7,359

OPPERMANN, CHARLES FRÉDÉRIC [1805-1872]
BALLAND(2), 206; HUMBERT, 217; BERGER-LEVRAULT, 178; TSCHIRCH, 1096-1097; FERCHL, 389; RSC 4,686-687; POGG 2,329-330

OPPLER, BERTHOLD [1871-1943]
GEDENKBUCH 2,1126

OPPOLZER, JOHANNES [1808-1871]
ADB 24,405-407; WURZBACH 21,76-79; STURM 3,105-106; HIRSCH 4,435-436; PAGEL, 1233-1235; LESKY, 149-152; FROHNE, 56-82; FRANKEN, 192-193
J. Seegen, Wiener med. Wchschr. 21 (1871), 379-381
M. Benedikt, Wiener klin. Wchschr. 21 (1908), 1109-1115
W. Schütte, Johann Ritter von Oppolzer, Leben und Werk. Düsseldorf 1937

ORBELI, LEON ABGAROVICH [1882-1958]
DSB 10,220-221; BSE 31,134-135; BME 21,1117-1120; STC 2,333-334
L.G. Leibson, Leon Abgarovich Orbeli. Leningrad 1973
N.A. Grigorian, Bull. Soc. Nat. Moscou NS80(6) (1975), 140-146

ORD, WILLIAM MILLER [1834-1902]
DNB [1901-1911] 3,52-53; HIRSCH 4,436-437; WWWS, 1287
Anon., Lancet 1902(I), pp.1494-1497; Brit. Med. J. 1902(I), pp.1315-1317

OREKHOV, ALEKSANDR PAVLOVICH [1881-1939]
WWR, 426; POGG 6,1916; 7b,3763-3765
Anon., Nature 144 (1939), 891
M.S. Rabinovich, Zhur. Ob. Khim. 10 (1940), 855-862

OREKHOVICH, VASILI NIKOLAEVICH [1905-]
BME 22,29-30; WWWS, 1288

ORFILA, MATTHIEU JOSEPH BONAVENTURE [1787-1853]
GENTY 1,245-256; GE 25,533; LAROUSSE 11,1445; HIRSCH 4,438-440; TSCHIRCH, 1097; FERCHL, 389; SCHAEDLER, 86; COLE, 407-408; CALLISEN 14,158-176; 31,95-100
A. Fayol, La Vie et l'Oeuvre d'Orfila. Paris 1930
C.E. Prelat and A.G. Velarde, Chymia 3 (1950), 77-93
A. Delmas et al., Bull. Acad. Med. 171 (1987), 447-483

ORGLER, ARNOLD [1874-]
FISCHER 2,1151; KOREN, 227; LDGS, 78

ORIENT, JULIUS (IULIU) [1869-1940]
TSCHIRCH, 1097
S. Iszak, Revista Medicale (Turgu-Mures) 16 (1970), 446-448

ORLA-JENSEN, SIGURD [1870-1949]
DBL 17,478-480; KBB 1949, pp.1056-1057; VEIBEL 1,234; 2,323-328; SACKMANN, 260-263
P.S. Arup, Chem. Ind. 1949, p.746
E. Olsen, J. Gen. Microbiol. 4 (1950), 107-109

ORMSBEE, RICHARD ARMSTRONG [1915-]
AMS 15(5),769; WWWS, 1289

ORNDORFF, WILLIAM RIDGELY [1862-1927]
AMS 4,736; WWAH, 458; POGG 4,1099; 5,924; 6,1918

ORTEN, JAMES M. [1905-1991]
AMS 14,3774; WWWS, 1289

ORTH, JOHANNES [1847-1923]
DBJ 5,285-290,437; NB, 296; FISCHER 2,1152-1153; PAGEL, 1236
W. Heubner, Jahrb. Akad. Wiss. Gott. 1922-1923, pp.49-56
R. Fick, Sitz. Preuss. Akad. Wiss. 1924, pp.xcii-xcvii
G.B. Gruber, Verhandl. path. Ges. 29 (1937), 400-406

ORTHNER, HERMANN LUDWIG [1897-1971]
KURSCHNER 10,1781; TLK 3,482; POGG 6,1920; 7a(3),478
Anon., Chem. Z. 95 (1971), 531

ORTIGOSA, JOSÉ VICENTE [1817-1877]
RSC 4,701
Enciclopedia de Mexico 10,18-19
V.S. Albis-Gonzalez, Quipu 1 (1984), 401-405

ORTON, KENNEDY JOSEPH PREVITÉ [1872-1930]
WW 1927, pp.2243-2244; POGG 5,925-926; 6,1921
F.D.C., Proc. Roy. Soc. A129 (1930), xi-xiv
A. Ferguson, Nature 125 (1930), 898-899
H.K., J. Chem. Soc. 1931, pp.1042-1048

ORUDSHIEV, DSHEVAD [1885-]
SG [3]8,466; NUC 433,570

OSA, ADOLPHE de [1873-]
JV 19,21; NUC 433,608OSANN, GOTTFRIED WILHELM [1797-1866] ADB 24,461; LANGHANS, 208-220; BLOKH 2,550-551; LEVITSKI 1,242-244; POTSCH, 329; FERCHL, 390; SCHAEDLER, 86; POGG 2,335-336; 3,990-991
F. Rinecker, Würzb. Nat. Ges. 6 (1867), xlv-liii

OSBORN, MARY JANE [1927-]
AMS 15(5),777; WWA 1980-1981, p.2527

OSBORNE, THOMAS BURR [1859-1929]
DSB 10,241-244; DAB 14,74-75; CHITTENDEN, 95-95; DAMB, 567-568; IB, 221; WWAH, 460; WWWS, 1291; POGG 5,926; 6,1923-1924
H.B. Vickery and L.B. Mendel, Science 69 (1929), 385-388
H.B. Vickery, Biog. Mem. Nat. Acad. Sci. 14 (1931), 261-304; J. Nutrition 59 (1956), 3-26

OSBORNE, WILHELM [1871-]
JV 13,226; RSC 19,74
Anon., Chem. Z. 42 (1918), 541

OSBORNE, WILLIAM ALEXANDER [1873-1967]
WW 1965, p.2319; FISCHER 2,1154; O'CONNOR(2), 501-502; SRS, 22
Who Was Who 1961-1970, p.860
Anon., Med. J. Austral. 2 (1967), 1143-1145

OSEKI, SAKAYE [1881-]
HAKUSHI 3,600

OSGOOD, EDWIN EUGENE [1899-1969]
AMS 11,3999; WWAH, 460; WWWS, 1291
B. Pirofsky, Blood 33 (1969), 265-267
R.D. Koler, Blood 35 (1970), 149-150

OSIPOV, IVAN PAVLOVICH [1855-1918]
BSE 31,283; ARBUZOV, 177-178; KHARKOV-F, 120-125
M.A. Blokh, Zhur. Fiz. Khim. 60Suppl. (1928), 141

SIPOV, VIKTOR PETROVICH [1871-1947]
BSE 31,283; BME 22,186-187; WWR, 429
G.I. Malis, Zhur. Nevropat. Psikh. 57 (1957), 1169-1170
N.N. Timofeev, Zhur. Nevropat. Psikh. 71 (1971), 1394-1397

OSLER, WILLIAM [1849-1919]
DNB [1912-1921], 417-419; DAMB, 569; ICC, 54-56; WWWS, 1291-1292
H. Cushing, The Life of William Osler. Oxford 1925
R.L. Golden and C.G. Roland, Sir William Osler: An Annotated Bibliography. San Francisco 1988

OSSERMAN, ELLIOTT FREDERICK [1925-1989]
AMS 17(5),817; WWWS, 1292
New York Times, 14 April 1989

OST, HERMANN [1852-1931]
KURSCHNER 4,2139; HANNOVER, 29-30; DZ, 1053; TLK 3,482; POTSCH, 330; POGG 3,991; 4,1101; 5,928; 6,1924; 7a(3),481
F. Quincke, Z. angew. Chem. 44 (1931), 557
G. Keppeler, Chem. Z. 55 (1931), 517

OSTACHOWSKI, EMILIAN FRANCICZEK [1890-1962]
B. Rausch, Kwart. Hist. Nauki 7 (1962), 531-535

OSTER, GERALD [1918-1993]
AMS 15(5),782

OSTERHOUT, WINTHROP JOHN VAN LEUVEN [1871-1964]
DSB 18,702-704; DAB [Suppl.7],592-593; AMS 10,3055; FISCHER 2,1156-1157; WWAH, 460; WWWS, 1292; POGG 6,1925-1927; 7b,3785-3787
W.J.V. Osterhout, Ann. Rev. Physiol. 20 (1958), 1-12
Anon., Nature 202 (1964), 952-953
T. Shedlovsky, Am. Phil. Soc. Year Book 1964, pp.143-147; Biol. Bull. 129 (1965), 9-11
L.R. Blinks, Biog. Mem. Nat. Acad. Sci. 44 (1974), 213-249

OSTERMEYER, EUGEN [1849-1903]
BJN 8,84*; HEIN, 478; BLOKH 2,552
C. Liebermann, Ber. chem. Ges. 36 (1903), 2371
P. Braun, Beitr. Württ. Apothekgesch. 11 (1976), 101-102

OSTERN, PAWEL [1902-1943?]
J.K. Parnas, Nature 154 (1944), 695-696
Anon., Roczniki Chemii 20 (1946), xxii-xxiii

OSTMANN, PAUL [1893-1965]
Jahrbuch der Dissertationen der Philosophischen Fakultät...Berlin 1919-1920, p.279

OSTROMISLENSKI, IVAN IVANOVICH [1880-1939]
DAB 22,505-506; AMS 7,847; MILES, 377-379; POTSCH, 330

OSTROUMOV, ALEKSEI ALEKSANDROVICH [1844-1908]
BSE 31,351; BME 22,536-539; KAZAN 1,447-450; KUZNETSOV(1963), 550-555; FISCHER 2,1157-1158; PAGEL, 1238
V. Vorobyev, Meditsinskoye Obozrenie 70 (1908), 497-501

OSTWALD, WILHELM [1853-1932]
DSB 15,455-469; LEIPZIG 2,51-62; FARBER, 1021-1030; WELDING, 568-569; LEVITSKI 1,255-257; TLK 3,483-484; SCHAEDLER, 86; W, 398-399; RIGA, 713; ABBOTT-C, 107-108; WWWS, 1293; POTSCH, 330-331; POGG 3,991; 4,1101-1103; 5,929-930; 6,1928-1929; 7aSuppl.,476-482
F. Haber, Z. angew. Chem. 36 (1923), 460
W. Ostwald, Lebenslinien. Leipzig 1926-1927
P. Walden, Ber. chem. Ges. 65A (1932), 101-141; Wilhelm Ostwald. Leipzig 1933
A. Findlay, Nature 129 (1932), 750-751
W. Nernst, Z. Elektrochem. 38 (1932), 337-341
R. Wegscheider, Alm. Akad. Wiss. Wien 82 (1932), 315-325
P. Gunther, Z. angew. Chem. 45 (1932), 489-496; Z. Elektrochem. 57 (1953), 868-874
W.D. Bancroft, Science 75 (1932), 454-455; J. Chem. Ed. 10 (1933), 539-542,609-613
F.G. Donnan, J. Chem. Soc. 136 (1933), 316-332
F.E. Wall, J. Chem. Ed. 25 (1948), 2-10
E.P. Hillpern, Chymia 2 (1949), 57-64
J.R. Partington, Nature 172 (1953), 380-381
G. Ostwald, Wilhelm Ostwald, mein Vater. Stuttgart 1953
Y.U. Soloviev, Arch. Inter. Hist. Sci. 27 (1977), 222-230
M.I. Rodnyj and J.I. Solowjew, Wilhelm Ostwald. Leipzig 1977
L. Dunsch, Chemie Unserer Zeit 16(6) (1982), 186-196
H. Remane, Kwart. Hist. Nauki 30 (1985), 289-296
A. Leegwater, Centaurus 29 (1986), 314-337

OSTWALD, WOLFGANG [1883-1943]
DSB 10,251-252; TLK 3,484; WWWS, 1293; POTSCH, 331; POGG 4,1103; 5,930-931; 6,1929-1931; 7a(3),484-486
A. Lottermoser, Koll. Z. 103 (1943), 89-94
R.E. Oesper, J. Chem. Ed. 22 (1945), 263-264
H. Erbring, Koll. Z. 115 (1949), 3-5
E.A. Hauser, J. Chem. Ed. 32 (1955), 2-9
M.H. Fischer, Koll. Z. 145 (1956), 1-2
E. Wolfram, Colloid and Polymer Science 260 (1982), 353-355
K. Sühnel, NTM 26 (1989), 31-45

O'SULLIVAN, CORNELIUS [1847-1907]
DNB [1901-1911] 3,59; REILLY, 47-48
Anon., Nature 75 (1907), 277; J. Soc. Chem. Ind. 26 (1907), 139-140
H.D. O'Sullivan, The Life and Work of Cornelius O'Sullivan FRS. Guernsey 1934
H.E. Armstrong, J. Soc. Chem. Ind. 53 (1934), 342-344
D. Reilly, Wallerstein Laboratory Communications 14 (1951), 101-110
W.J. Davis, Proc. Roy. Irish Acad. 77B (1977), 309-315

O'SULLIVAN, DESMOND GERARD [1925-]
WWWS, 1293
Who's Who of British Scientists 1980-1981, p.366

OSWALD, ADOLF [1870-1956]
ZURICH-D, 170; IB, 221; TLK 3,484; POGG 6,1931; 7a(3),486
A. Grumbach, Viert. Nat. Ges. Zürich 101 (1956), 226-231

OTT, EMIL [1902-1963]
AMS 10,3058; MILES, 378-379; WWAH, 461

OTT, ERWIN [1886-1977]
KURSCHNER 13,2827; TLK 3,484; POGG 5,931; 6,1932; 7a(3),487-488
Anon., Chem. Z. 80 (1956), 457; Naturw. Rund. 30 (1977), 114

OTT, ISAAC [1847-1916]
DSB 10,252-253; NCAB 20,188; FISCHER 2,1159; WWWS, 1293

OTT, KARL [1885-]
JV 24,578; NUC 434,688

OTTE, RUDOLF [1858-1977]
DGB 143 (1967), 260; JV 5,234; RSC 17,652,757

OTTENSTEIN, BERTA [1891-1956]
KURSCHNER 8,1724; BHDE 2,881; LDGS, 62; WIDMANN, 279

OTTESON, MARTIN [1920-]
KBB 1981, p.865
B. Foltmann, in The Carlsberg Laboratory 1876-1976 (H. Holter and K.M. Møller, Eds.), pp.151-161. Copenhagen 1976

OTTO, FRIEDRICH JULIUS [1809-1870]
ADB 24,747-751; HEIN, 479-480; ALBRECHT, 62; BLOKH 2,553; FERCHL, 391; TSCHIRCH, 1097; SCHAEDLER, 86-87; POGG 2,339-340; 3,992
W. Schneider, Pharm. Z. 105 (1960), 1489-1492

OTTO, JACOB GOTTFRIED [1859-1888]
NBL 10,575-576; NUC 435,118

OTTO, RICHARD [1872-1952]
KURSCHNER 4,2150; FISCHER 2,1160; KALLMORGEN, 367;

OLPP, 306; IB, 221; POGG 7a(3),490
G. Blumenthal, Z. Bakt. 159 (1952), 5-10
R. Siegert, Arzneimitt. 2 (1952), 596-597

OTTO, ROBERT [1837-1907]
BJN 12,63*; ALBRECHT, 62-63; BLOKH 2,553; TSCHIRCH, 1097; SCHAEDLER, 87; POTSCH, 331; POGG 3,992-993; 4,1103; 5,931
C. Graebe, Ber. chem. Ges. 40 (1907), 1103
Anon., Chem. Z. 31 (1907), 191

OU, CHING-KO [1885-]
JV 27,50; NUC 435,202

OUCHTERLONY, ÖRJAN [1914-]
VAD 1981, p.818; BIBEL, 287-290

OUDEMANS, ANTONIE CORNEILLE, Jr. [1831-1895]
NNBW 1,1394-1395; RSC 17,656-657; NUC 435,236; POGG 3,994; 4,1105
S. Hoogewerff, Rec. Trav. Chim. Pays-Bas 15 (1896), 288-321; Jaarb. Akad. Wet. 1896, pp.69-94

OUDEMANS, CORNEILLE [1825-1906]
DSB 10,253-254; NNBW 1,1396-1307; LINDEBOOM, 1481-1483; HIRSCH 4,461; REBER, 165-169; TSCHIRCH, 19097-1098
J.M. Moll, Ber. bot. Ges. 26A (1908), (12)-(33)

OUDIN, JACQUES [1908-1985]
QQ 1984-1985, p.1102
G. Bordenave, Nature 319 (1986), 174

OURISSON, GUY [1926-]
QQ 1981-1982, p.1134; IWW 1982-1983, p.995

d'OUTREPONT, JOSEPH SERVAZ [1776-1845]
ADB 24,780-781; HIRSCH 4,462-463; DECHAMBRE 70, 647-648; CALLISEN 14,240-244; 31,118-120
G. Burckhard, Joseph Servatius von d'Outrepont. Jena 1913
H. Hauser, Joseph Servatius von d'Outrepont. Würzburg 1927

OVARY, ZOLTAN [1907-]
AMS 15(5),792; WWWS, 1295

OVCHINNIKOV, YURI ANATOLIEVICH [1934-1988]
IWW 1982-1983, p.996; POTSCH, 331
G. Semenza, FEBS Letters 230(1) (1988), v
V. Lipkin and E. Grishin, TIBS 13 (1988), 470-471
E. Blout, Am. Phil. Soc. Year Book 1988, pp.239-244

OVERBECK, OTTO [1842-1919]
RSC 8,545
G. Luntowski, Die kommunale Selbstverwaltung. p.150. Dortmund 1977

OVERBECK, OTTO [-1937]
RSC 17,661; NUC 435,453
Anon., Proc. Chem. Soc. 1937, p.63

OVERBERGER, CHARLES [1920-]
AMS 15(5),793; WWA 1980-1981, p.2535; WWWS, 1295

OVERTON, ERNEST [1865-1933]
DSB 10,256-257; FISCHER 2,1162; ZURICH-D, 158
R.R. Collander, Protoplasma 20 (1933), 228-231; Leopoldina [3]8/9 (1962-1963), 242-254
T. Thunberg, Skand. Arch. Physiol. 70 (1934), 1-9
W. Url, Verhandl. zool. bot. Ges. Wien 115 (1976), 24-33
R.L. Lipnick (ed.), Studies of Narcosis: Charles Ernest Overton. London 1991

OVSYANNIKOV, FILIPP VASILIEVICH [1827-1906]
BSE 30,479-480; BME 21,419-420; HIRSCH 4,466; GOCZOL, 36-40; ZMEEV 2,39; 3,43

M.E. Kuzmin, Zhur. Nevropat. Psikh. 54 (1954), 948-950

OWEN, BENTON BROOKS [1900-1989]
AMS 11,4013-4014
New York Times, 10 May 1989

OWEN, RAY DAVID [1915-]
AMS 15(5),797; WWA 1980-1981, p.2537; IWW 1982-1983, p.996; WWWS, 1296

OZANAM, JEAN ANTOINE FRANÇOIS [1773-1837]
HOEFER 38,1018; HIRSCH 4,466-467; DECHAMBRE [2]19,548-549; CALLISEN 14,248-251; 31,122-123

P

PAAL, ARPAD [1889-1943]
MEL 2,335; IB, 222

PAAL, CARL [1860-1935]
OBL 7,274-275; KURSCHNER 5,1001; LADIS, 61-86; STUPP-KUGA, 189-213; POTSCH, 332; TLK 3,486; TSCHIRCH, 1098; POGG 4,1106-1107; 5,932-933; 6,1934
M. Busch, Z. angew. Chem. 43 (1930), 631-632
Anon., Chem. Z. 59 (1935), 169; Z. angew. Chem. 48 (1935), 206
K.H. Bauer, Ber. chem. Ges. 68A (1935), 43-45
B. Helferich, Sitz. Akad. Wiss. Leipzig 87 (1935), 207-212

PAAL, HERMANN [1899-1965]
KURSCHNER 10,1792; NUC 436,499

PABST, ALBERT [1852-1901]
BLOKH 2,557; POGG 3,996; 4,1107-1108
Anon., Chem. Z. 25 (1901), 1168; Ber. chem. Ges. 34 (1901), 1083; Leopoldina 37 (1901), 55

PACHON, VICTOR [1867-1939]
FISCHER 2,1163; SCHALI, 84-86
M. Tiffeneau, Paris Médical 114 (1939), 145-148
A. Baudouin, Bull. Acad. Med. 121 (1939), 653-654

PACINI, FILIPPO [1812-1883]
DSB 10,266-268; EI 25,880; HIRSCH 4,469-471; PAGEL, 1243-1244; WWWS,1297; DECHAMBRE [2]19,644-645; CALLISEN 31,124
A. Tafani, Arch. Ital. Biol. 4 (1883), 123-126
L. Castaldi, Riv. Storia Sci. Med. 14 (1923), 182-212
P. Franceschini, Physis 13 (1971), 324-332; 18 (1976), 349-365

PACSU, EUGENE (JENŐ) [1891-1972]
AMS 11,4024; MEL 3,583-584; NCAB 56,310; POGG 6,1935; 7b,3799-3801

PADOA, MAURIZIO [1881-1945]
MODENA 1,281; POGG 5,933-934; 6,1935-1936; 7b,3801

PAECH, KARL [1908-1955]
KURSCHNER 8,1728; POGG 7a(3),494
Anon., Science 122 (1955), 757PAGANO, GIUSEPPE [1872-1959]
DSB 15,469-470
J. Pick, Am. Pract. 10 (1959), 451-460,464

PAGE, HAROLD JAMES [1890-1972]
SRS, 48; IB, 222-223; NUC 437,355
Who's Who in British Science 1953, p.206
Anon., Chem. Brit. 8 (1972), 183

PAGE, IRVINE HEINLY [1901-1991]
AMS 14,3806; WWA 1978-1979, pp.2481-2482; MH 1,361-362; MSE 2,387-388; APPEL, 213-218; WWWS, 1298; POGG 6,1937; 7b,3806-3822
I.H. Page, Persp. Biol. Med. 20 (1976), 1-8
Anon., J. Am. Med. Assn. 244 (1980), 1765-1766,1771-1772;

New York Times, 12 June 1991; Cornell University Medical College Quarterly 51(4) (1991), 53

PAGEL, JULIUS LEOPOLD [1851-1912]
BJN 18,49*-50*; FISCHER 2,1164; WREDE, 142; KOREN, 228; WININGER 4,497
P. Richter, Arch. Gesch. Med. 6 (1912), 71-79
A. Pagel, Janus 30 (1926), 179-191
W. Pagel, Bull. Hist. Med. 25 (1951), 207-225; Deutsches Med. J. 22 (1971), 567-570
I. Winter, NTM 14 (1977), 99-102
J. Gromer, Julius Leopold Pagel. Cologne 1985

PAGEL, WALTER [1898-1983]
KURSCHNER 13,2841-2842; BHDE 2,883-884; LDGS, 64; KOREN, 228; WWWS, 1299
H. Buess, Deutsche med. Wchschr. 93 (1968), 2186-2188; Clio Medica 18 (1983), 233-239
A. Debus, Ed., Science, Medicine and Society in the Renaissance, pp.289-326. New York 1972
F.A. Yates, Hist. Sci. 11 (1973), 286-298
M. Winder and R. Burgess, Med. Hist. 27 (1983), 310-311
A.B. Davis, Isis 75 (1984), 363-365

PAGENSTECHER, HEINRICH ALEXANDER [1825-1889]
ADB 53,789-790; DRULL, 200
Anon., Lancet 1889(I), p.152; Nature 39 (1889), 280-281; Leopoldina 25 (1889), 52; Zool. Anz. 12 (1889), 80

PAGENSTECHER, JOHANN SAMUEL FRIEDRICH [1783-1856]
FERCHL, 391-392; TSCHIRCH, 1098; CALLISEN 14,259-261; POGG 2,346

PAGÈS, CALIXTE [1857-]
RSC 17,674; SG [2]12,401; [3]8,552; NUC 437,456-457

PAGET, JAMES [1814-1899]
DNB 22,1112-1114; HIRSCH 4,475; BETTANY 2,167-177; O'CONNOR, 49-52; ABBOTT, 101-102; WWWS, 1299
Anon., Lancet 1900(I), pp.52-56; Brit. Med. J. 1900(I), pp.49-54
S. Paget, Memoirs and Letters of Sir James Paget. London 1901
J.H., Proc. Roy. Soc. 75 (1905), 136-140

PAILER, MATTHIAS [1911-]
KURSCHNER 13,2843; WWWS, 1299; POGG 7a(3),498

PAINTER, THEOPHILUS SHICKEL [1889-1969]
DSB 10,276-277; AMS 11,4031; NCAB 55,554-555; IB, 223; WWWS, 1300
R.P. Wagner, Genetics 64 (1970), s87-s88
B. Glass, Am. Phil. Soc. Year Book 1970, pp.151-155; Biog. Mem. Nat. Acad. Sci. 59 (1990), 309-337

PAL, JAKOB [1863-1936]
FISCHER 2,1165; STUMPF, 72-89; IB, 223
S. Bondi, Wiener klin. Wchschr. 49 (1936), 946-947
L. Popper, Wiener klin. Wchschr. 75 (1963), 606-608

PALADE, GEORGE EMIL [1912-]
AMS 15(5),815; WWA 1980-1981, p.2546; IWW 1982-1983, p.1002; WWWS, 1300; MH 1,362-364; STC 2,335-336; MSE 2,388-389; CB 1967, pp.324-326
K.R. Porter, J. Cell. Biol. 97(1) (1983), D1-D7

PALADINI, ALEJANDRO CONSTANTINO [1919-]
Quien es Quien en la Argentina 1968, p.548

PALLADIN, ALEKSANDR VLADIMIROVICH [1885-1972]
DSB 18,704-705; BSE (3rd Ed.)19,121; UKRAINE 2,367-368; IB, 223; WWWS, 1301; POGG 6,1942; 7b,3833-3843
Anon., Ukrainski Biokhimichni Zhurnal 45 (1973), 3-6
E. Kreps and D.B. Tower, J. Neurochem. 21 (1973), 723-724
A.M. Utevski, A.V. Palladin. Kiev 1975

PALLADIN, VLADIMIR IVANOVICH [1859-1922]
DSB 10,281-283; BSE 31,610-611; KHARKOV-F, 225-227; BLOKH 2,555; POGG 6,1942-1943
C. N[euberg], Biochem. Z. 130 (1922), 321-322
S. Kostychev, Priroda 11 (1922), 78-85
S.D. Lvov, Sov. Bot. 1937(6), pp.155-164
N.A. Maksimov and S.M. Manskaya, Biokhimia 17 (1952), 249-254
B.A. Rubin, Russian Review of Biology 48 (1959), 112-116
D.M. Mikhlin, Biokhimia 24 (1959), 766-767

PALMAER, KNUT WILHELM [1868-1942]
SMK 6,13; POGG 4,1114; 5,938; 6,1943; 7b,3843-3844
G. Hägg, Sven. Vet. Akad. Arbok 1945, pp.291-314

PALME, FRANZ [1907-1960]
KURSCHNER 8,1731
H. Julich, Z. ges. inn. Med. 16 (1961), 349

PALMER, ARTHUR WILLIAM [1861-1904]
ELLIOTT, 198
Anon., Science 19 (1904), 276

PALMER, LEROY SHELDON [1887-1944]
AMS 7,1347; CHITTENDEN, 362-363; ETTRE, 490-491; IB, 223; WWAH, 464; POGG 6,1944; 7b,3847-3849
C. Kennedy, Science 99 (1944), 442-444
R. Jenness and R. Luecke, J. Nutrition 63 (1957), 3-18

PALMER, WALTER WALKER [1882-1950]
DAB [Suppl.4],642-643; AMS 8,1887; NCAB B,447; IB, 223; WWWS, 1303
D.W. Atchley, Trans. Assoc. Am. Phys. 64 (1951), 19-22

PALMER, WILLIAM GEORGE [1892-1969]
POGG 6,1944-1945; 7b,3850
H.J. Emeleus, Chem. Brit. 6 (1970), 394

PALMSTEDT, CARL [1785-1870]
SMK 6,28; BLOKH 2,555-556; FERCHL, 392; POGG 2,349-350; 3,1002
C.W. Blomstrand, Ber. chem. Ges. 3 (1870), 579-581
C. Rydqvist, Lev. Sven. Vet. 1 (1873), 555-570

PALOMAA, MATTI HERMAN [1871-1947]
ENKVIST, 105-107; POGG 6,1945; 7b,3850-3851
S. Kilpi, Sitz. Finn. Akad. Wiss. 1948, pp.45-55

PALTAUF, ARNOLD [1860-1893]
STURM 3,130; FISCHER 2,1167; MOTSCH, 17-18; BINDSEIL, 62-66; KOERTING,160
M. Richter, Prag. med. Wchschr. 18 (1893), 273-275
Anon., Wiener med. Wchschr. 43 (1893), 1030; Leopoldina 29 (1893), 159

PALTAUF, RICHARD [1858-1924]
OBL 7,307-308; FISCHER 2,1167-1168; PAGEL, 1354; LESKY, 577-582
R. Kraus, Wiener med. Wchschr. 74 (1924), 930-931
H. Meyer, Alm. Akad. Wiss. Wien 74 (1924), 166-172
C. Sternberg, Verhandl. path. Ges. 20 (1925), 430-434

PALYUTA, GEORGI ANDREEVICH [1820-1897]
VORONTSOV, 56

PANCERI, PAOLO [1833-1877]
EI 26,173; HIRSCH 4,487-488; DECHAMBRE [2]20,100
P. Pavesi, Atti. Soc. Ital. Sci. Nat. 20 (1877), 28-48
E.N. Harvey, A History of Luminescence, pp.241-242. Philadephia 1957

PANDER, CHRISTIAN HEINRICH [1794-1865]
DSB 10,286-288; ADB 25,117-119; BSE 31,642; WELDING, 576; WWWS, 1303; HIRSCH

4,489-490; CALLISEN 14,286; 31,138; POGG 2,351-352; 3,1002; 7aSuppl.,487
E. Loesch, Biol. Z. 40 (1920), 481-502
B.E. Raikov, Christian Pander. Moscow 1964 (German transl., Frankfurt a.M. 1984)

PANETH, FRIEDRICH ADOLF [1887-1958]
DSB 10,288-289; NOB 19,113-119; BHDE 2,885-886; LDGS, 23; W, 400-401; ABBOTT-C, 108-109; WNINIGER 4,603; TETZLAFF, 257; POTSCH, 332-333; POGG 5,938-939; 6,1946-1947; 7a(3),500-502
G.R. Martin, Nature 182 (1958), 1274-1275
E. Hayek, Alm. Akad. Wiss. Wien 108 (1958), 480-484
E. Glueckauf, Chem. Ind. 1958, p.1358; Proc. Chem. Soc. 1959, 103-105
H.J. Emeleus, Biog. Mem. Fell. Roy. Soc. 6 (1960), 227-246

PANETH, JOSEF [1857-1890]
OBL 7,311; FISCHER 2,1169; KOREN, 228; WININGER 7,365

PANGBORN, MARY CANDACE [1907-]
AMS 12,4756

PANORMOV, ALEKSEI ALEKSANDROVICH [1859-1927]
KAZAN 2,286-287; IB, 224; RSC 17,695-696

PANTIN, CARL FREDERICK ABEL [1899-1967]
DNB [1961-1970]. 820-821; WW 1965, p.2340; MH 1,367-368; MSE 2,391-392
F. Russell, Biog. Mem. Fell. Roy. Soc. 14 (1968), 417-434

PANUM, PETER LUDVIG [1820-1885]
DBL(3rd Ed.), 11,153-156; MEISEN, 134-137; KIEL, 79; HIRSCH 4,492-494; PAGEL, 1256-1258; VEIBEL 2,332-333; BULLOCH, 388; FERCHL, 393; WWWS, 1305; POGG 2,353; 3,1005
C. Bohr, Biol. Z. 5 (1886), 257-259
L. Gafafer, Bull. Inst. Hist. Med. 2 (1934), 259-280
L. Delhoume, Concours Médical 84 (1962), 255-258
J. Carstensen, Peter Ludvig Panum, Professor der Physiologie in Kiel. Neumünster 1967
A. Gjedde, Peter Ludvig Panums Videnskabelige Indsatz. Copenhagen 1971
J. Parnas, Dan. Med. Bull. 23 (1976), 143-146
B. Havsteen, Christiana Albertina NF15 (1981), 27-34

PAOLINI, VINCENZO [1876-1964]
TSCHIRCH, 1098; POGG 5,940; 6,940; 7b,3862
Anon., Chimica e Industria 46 (1964), 843

PAPANICOLAOU, GEORGE NICHOLAS [1883-1962]
DSB 10,291-292; DAB [Suppl.7],598-599; AMS 10,3083; NCAB 50,52-53; FREUND 2,277-288; DAMB, 576-577; IB, 224; WWWS, 1305
R.C. Swan, Anat. Rec. 143 (1962), 276-278
J.G. Kidd, Am. J. Clin. Path. 39 (1963), 400-405
D.E. Carmichael, The Pap Smear: The Life of George N. Papanicolaou. Springfield, Ill. 1973
H.H. Simmer, in Medizinische Diagnostik in Geschichte und Gegenwart (C. Habrich et al., Eds.), pp.341-356. Munich 1978

PAPE, CARL [1857-1926]
RSC 10,989; 17,699; NUC 440,385
Anon., Chem. Z. 50 (1926), 448; Z. angew. Chem. 39 (1926), 736

PAPENDIECK, AUGUST [1867-]
JV 8,216; RSC 13,890

PAPON, JAKOB [1827-1860]
DHB 5,224; STELLING-MICHAUD 5,82; RSC 4,753; NUC 440,497
Anon., Jahresb. Nat. Ges. Graub. 6 (1861), 274-278

PAPPENHEIM, ARTUR [1870-1916]
FISCHER 2,1172; KOREN, 228; WININGER 7,366-367; WWWS, 1306
T. Brugsch, Fol. Haemat. 21 (1917), 79-90

PAPPENHEIM, PAUL [1878-1945]
IB, 224

PAPPENHEIM, SAMUEL MORITZ [1811-1882]
ADB 25,152-163; HIRSCH 4,496-497; GRAETZER, 188-189; KOREN, 228; WININGER 4,607-608
Anon., Leopoldina 18 (1882), 48,122

PAPPENHEIMER, ALWYN MAX, Jr. [1908-]
AMS 16(5),861

PAPPENHEIMER, JOHN RICHARD [1915-]
AMS 17(5),873; BROBECK, 181-184
J.R. Pappenheimer, Ann. Rev. Physiol. 49 (1987), 1-15

PARAT, MAURICE [1899-1936]
IB, 224
P. Macquot, Presse Med. 45 (1937), 241-242

PARCUS, EUGEN [1857-]
RSC 10,989; 17,702; SG [1]10,490; NUC 441,40

PARDEE, ARTHUR BECK [1921-]
AMS 15(5),836; WWA 1980-1981, p.2555; IWW 1982-1983, pp.1008-1009; MH 2,406; MSE 2,393-394; WWWS, 1306

PARISI, ERNESTO [1891-1944]
POGG 6,1950; 7b,3863-3864
U. Pratalongo, Chimica e Industria 27 (1945), 25

PARK, CHARLES RAWLINSON [1916-]
AMS 15(5),839; WWWS, 1307

PARK, EDWARDS ALBERT [1877-1969]
AMS 11,4049; ICC, 172-174; WWAH, 466; WWWS, 1307
H.B. Taussig, J. Pediat. 77 (1970), 722-731
J.E. Howard, Trans. Assoc. Am. Phys. 83 (1970), 28-29

PARK, JAMES THEODORE [1922-]
AMS 15(5),840; WWA 1980-1981, p.1557; WWWS, 1307

PARK, WILLIAM HALLOCK [1863-1939]
DAB [Suppl.2],513-514; AMS 6,1082; NCAB C,314; FISCHER 2,1174; IB, 224; DAMB, 578; FLUELER, 107-110; WWAH, 466; WWWS, 1308
G. Ramon, J. Immunol. 37 (1939), 179-183
H. Zinsser, J. Bact. 38 (1939), 1-3
W.W. Oliver, The Man Who Lived For Tomorrow. New York 1941
M. Schaeffer, Am. J. Pub. Hlth. 75 (1985), 1296-1302

PARKE, JOHN LATIMER [1825-1907]
RSC 8,562
Anon., Brit. Med. J. 1907(I), p.845

PARKER, FREDERIC [1890-1969]
AMS 11,4053
S.L. Robbins, Am. J. Path. 59 (1970), 197-201

PARKER, GEORGE HOWARD [1864-1955]
DAB [Suppl.5].535-536; AMS 8,1892; IB, 225; WWAH, 466; WWWS, 1308
G.H. Parker, The World Expands. Cambridge, Mass. 1946
A.S. Romer, Biog. Mem. Nat. Acad. Sci. 39 (1967), 359-390

PARKER, RAYMOND CRANDALL [1903-1974]
AMS 12,4776; WWWS, 1309
A.M. Fisher and R.J. Wilson, In Vitro 9 (1974), 2-3; Anat. Rec. 183 (1975), 142-143

PARKES, ALAN STERLING [1900-1990]
WW 1981, p.1989; IWW 1982-1983, p.1010; IB, 225; WWWS, 1309
A.S. Parkes, The Off-beat Biologist. Cambridge 1985

PARKES, EDMUND ALEXANDER [1818-1876]
DNB 15,294-296; HIRSCH 4,508-509; BETTANY 2,296-302; WWWS, 1309
W. Jenner, Brit. Med. J. 1876(II), pp.33-34

PARKINSON, ROBERT [1831-1913]
RSC 4,760
Anon., Pharm. J. 91 (1913), 236

PARKMAN, THEODORE [1837-1862]
RSC 4,760
Anon., Am. J. Sci. 85 (1863), 155-156

PARKS, GEORGE SUTTON [1894-1966]
AMS 11,4060; POGG 6,1951; 7b,3864-3866
Anon., Chem. Eng. News 45(7) (1967), 85

PARKS, THOMAS BRANSON [1894-1972]
AMS 10,3096

PARMENTIER, JEAN ANTOINE AUGUSTIN [1737-1813]
DSB 10,325-326; GE 25,1177-1178; LAROUSSE 12,308; BOURQUELOT, 29-30; BALLAND, 79-87; BALLAND(2), 40-46; HIRSCH 4,510; FERCHL, 394; DECHAMBRE 73, 350-353; WWWS, 1310; CALLISEN 31,155; POGG 2,362-364
A. Vitalis, Précis Anal. Acad. Rouen 16 (1814), 36-40
A. Balland, Notices Biographiques sur les Anciens Pharmaciens Inspecteurs de l'Armée, pp.11-15. Paris 1892; La Chimie Alimentaire dans l'Oeuvre de Parmentier. Paris 1913
F.F.C. Seince, Parmentier et l'Hygiène Alimentaire. Bordeaux 1935
A. Guillon, Rev. Hist. Pharm. 25 (1978), 121-125
E. Kahane, Parmentier ou la Dignité de la Pomme de Terre. Paris 1978
H. Bonnemain and P. Julien, Rev. Hist. Pharm. 34 (1987), 299-318
M. Bertrand, Antoine Auguste Parmentier. Toulouse 1987

PARNAS, JACOB KAROL [1884-1949]
DSB 10,326-327; PSB 25,218-221; WE 8,483; WWR, 434-435; WARSAW, 148; FISCHER 2,1174-1175; KOREN, 228; WININGER 7,367-368; POTSCH, 334
Anon., Nachr. Chem. Techn. 3 (1955), 101
I. Mochnacka, Acta Biochim. Polon. 3 (1956), 3-39
J. Heller and W. Mozolowski, Post. Bioch. 4 (1958), 5-65
C. Lutwak-Mann and T. Mann, TIBS 6 (1981), 309-310
W.S. Ostrowski, Post. Bioch. 32 (1986), 247-260
Z. Zielinska, Acta Physiol. Polon. 38 (1987), 91-99

PARPART, ARTHUR KEMBLE [1903-1965]
AMS 10,3097; NCAB 51,2-3; NUC 443,79-80
Anon., Journal of Cellular and Comparative Physiology 66 (1965), 261-265

PARR, SAMUEL WILSON [1857-1931]
DAB 14,252-253; MILES, 381-382; NCAB C,153; WWWS,1311; POGG 5,942; 6,1952
R. Adams, Science 74 (1931), 8-9
W.A. Noyes, J. Am. Chem. Soc. 54 (1932), 1-7

PARRAVANO, NICOLA [1883-1938]
POGG 5,942; 6,1952-1953; 7b,3868-3869
G. Morselli, Chimica e Industria 20 (1938), 585-586
Anon., Ber. chem. Ges. 71A (1938), 191; Chem. Wkbl. 35 (1938), 612
G. Dupont, Bull. Soc. Chim. [5]6 (1939), 30-31

PARROT, GEORG FRIEDRICH (EGOR IVANOVICH) [1767-1852]
ADB 25,184-186; BSE 32,156; WELDING, 579; FERCHL, 395; POGG 2,365-367

PARROT, JOSEPH MARIE JULES [1829-1883]
HIRSCH 4,513-514; PAGEL, 1260; DECHAMBRE [2]21,488-490; WWWS, 1311
L. Gavarret, Bull. Acad. Med. 12 (1883), 1458-1460
S. Thieffry, Bull. Acad. Med. 164 (1980), 725-729

PARRY, CALEB HILLIARD [1775-1822]
DNB 15,371-372; HIRSCH 4,514; WWWS, 1311
B. Livesey, Med. Hist. 19 (1975), 158-171

PARSONS, THOMAS RICHARD [1892-]
Who's Who in British Science 1953, p.207

PARTHEIL, ALFRED [1861-1909]
BJN 14,69*; HEIN, 480-481; BLOKH 2,557; GUNDLACH, 484; TSCHIRCH, 1099; POGG 4,1118-1119; 5,943
A. Stutzer, Chem. Z. 33 (1909), 469
H. Nothnagel, Ber. pharm. Ges. 19 (1909), 249-254

PARTINGTON, JAMES RIDDICK [1886-1965]
DSB 10,329-330; DNB [1961-1970],822-823; WW 1965, pp.2536-2537; SRS, 46; POTSCH, 334-335; W, 404-405; POGG 5,943; 6,1953-1954; 7b,3870-3873
F.H.C. Butler, Brit. J. Hist. Sci. 3 (1966), 70-72
F.W. Gibbs, Chem. Ind. 1966, p.151
Z.I. Sheptunova, Vop. Ist. Est. Tekh. 1981(3), pp.107-110

PARTRIDGE, STANLEY MILES [1913-1992]
WW 1991, p.1413
Anon., Chem. Brit. 28 (1992), 660

PASCAL, PAUL [1880-1968]
CHARLE(2), 214-216; POTSCH, 335; WWWS, 1313; POGG 5,945; 6,1956; 7b,3875-3877
C. Duval and R.E. Oesper, J. Chem. Ed. 29 (1952), 40-41
A. Couder and A. Pascault, C.R. Acad. Sci. 267 (1968), 104-105; 268 (1968), 68-75,146

PASCHKIS, HEINRICH [1849-1923]
OBL 7,331; STURM 3,139; FISCHER 2,1177; DECKER, 22-29; KOREN, 228; WININGER 4,610
Anon., Chem. Z. 47 (1923), 599

PASCHKIS, KARL ERNST [1896-1961]
AMS 9(II),860; BHDE 2,889
Anon., Medical Circle Bulletin 8(6) (1961), 82

PASCHKIS, RUDOLF [1879-1964]
KURSCHNER 4,2166; FISCHER 2,1177
S. Peller, Pirquet Bulletin of Clinical Medicine 11(4) (1964), 228-229

PASCUAL-VILA, JOSÉ [1897-1977]
POGG 6,1957; 7b,3877-3879

PASHUTIN, VIKTOR VASILIEVICH [1845-1901]
BSE 32,256; BME 23,594-595; KAZAN 2,287-289; FISCHER 2,1177; ZMEEV 5,54-55
B. Kallmeyer, Deutsche med. Wchschr. 27 (1901), 127-128
H. Loeventhal, Berl. klin. Wchschr. 38 (1901), 228
P.N. Veselkin, V.V. Pashutin. Moscow 1950
A.D. Ado and A.M. Khomyakov, Arkh. Pat. 13(2) (1951), 83

PASSERINI, MARIO [1891-]
POGG 6,1958; 7b,3880-3881

PASSMORE, FRANCIS WILLIAM [1867-1921]
JV 6,298; RSC 17,726
C.R.H., Z. angew. Chem. 34 (1921), 595
E.T. Brewis, J. Soc. Chem. Ind. 40 (1921), 441R-442R
Anon., Chemical Age 5 (1921), 552; J. Roy. Inst. Chem. 1921, pp.356-357
H. Ballantyne, Analyst 47 (1922), 194-195

PASSMORE, REGINALD [1910-]
NUC 444,244
Who's Who in British Science 1953, p.208

PASTAN, IRA HARRY [1931-]
AMS 15(5),862

PASTEUR, LOUIS [1822-1895]
DSB 10,350-416; BUGGE 2,154-172; HIRSCH 4,522-524; PAGEL, 1262-1264; BULLOCH, 388-390; OLPP, 313-315; BLOKH 2,558-562; ABBOTT, 103-104; FERCHL, 396; SCHAEDLER, 89; TSCHIRCH, 1099; BERGER-LEVRAULT,180-181; POTSCH, 335-336; W, 406-408; WWWS, 1313; POGG 2,372-373; 3,1007; 6,1959
R.J. Dubos, Louis Pasteur, Free Lance of Science. Boston 1950
J. Nicolle, Rev. Hist. Sci. 6 (1953), 135-149
D. Wrotnowska, Hist. Med. 5(7) (1955), 65-81; Louis Pasteur, Professeur et Doyen de la Faculté des Sciences de Lille. Paris 1975; Clio Medica 15 (1981), 191-199
F. Dagognet, Methodes et Doctrines dans l'Oeuvre de Pasteur. Paris 1967
M. Flanzy, C.R. Acad. Agr. 59 (1973), 1397-1428
J. Farley, Ann. Rev. Microbiol. 32 (1978), 143-154
G.L. Geison, Isis 72 (1981), 425-445; J. Hist. Med. 45 (1990), 341-365
G.L. Geison and J.A. Secord, Isis 79 (1988), 6-36
K.C. Carter, Bull. Hist. Med. 62 (1988), 42-59

PASTIA, CONSTANTIN [1883-1926]
N. Marcu, Revista de Igiena etc. (Bacteriol.) 28 (1983), 89-95

PATAT, FRANZ [1906-1982]
KURSCHNER 13,2856; HANNOVER, 35; POGG 7a(3),511-513
Anon., Chem. Z. 105 (1981), 154; 107 (1983), 44
E. Hayek, Alm. Akad. Wiss. Wien 133 (1984), 335-338

PATEIN, GUSTAVE CONSTANT [1857-1928]
GORIS, 527-528; TSCHIRCH, 1095; POGG 4,1122; 5,946; 6,1959
E. Perrot, Bull. Acad. Med. 99 (1928), 72-74

PATERNÓ, EMANUELE [1847-1935]
PROVENZAL, 223-246; WWWS, 1314; POTSCH, 336; POGG 3,1007-1008; 4,1122-1123; 5,946-947; 6,1959-1960; 7b,3882-3883
G. Bargellini, Ber. chem. Ges. 68A (1935), 47-50
W.J. Pope, J. Chem. Soc. 1937, pp.181-183
G. Provenzal, Rass. Clin. Terap. 37 (1938), 48-53
F.C. Pallozzo, Gazz. Chim. Ital. 81 (1951), 3-22

PATON, DIARMID NOËL [1859-1928]
DNB [1922-1930], 657; FISCHER 2,1179; O'CONNOR(2), 406-408;IB, 226; WWWS, 1315
Anon., Nature 122 (1928), 656-657; Glasgow Med. J. 110 (1928), 270-278; Proc. Roy. Soc. Edin. 48 (1928), 220-224
E.P.C., Biochem. J. 23 (1929), 3; Proc. Roy. Soc. B104 (1929), ix-xii
D. Smith and M. Nicholson, Soc. Stud. Sci. 19 (1989), 195-238

PATON, WILLIAM DRUMMOND MACDONALD [1917-]
WW 1989, pp.1383-1384
W.D.M. Paton, Ann. Rev. Pharmacol. 26 (1986), 1-22

PATTEN, ANDREW JARVIS [1874-]
AMS 10,3107; NUC 445,43

PATTERSON, ARTHUR LINDO [1902-1966]
AMS 10,3108; McLACHLAN, 103-107; WWAH, 470; WWWS, 1315; POGG 6,1960-1961 7b,3888-3890
A.J.C. Wilson, Nature 212 (1966), 1414
J.R. Clark, American Mineralogist 53 (1968), 576-586
J.P. Glusker, TIBS 9 (1984), 328-330

PATTERSON, THOMAS STEWART [1872-1949]
WW 1949, p.2157; POGG 5,947; 6,1961-1962; 7b,3891-3892
J.D. Loudon, J. Chem. Soc. 1949, pp.1667-1668
A. McKenzie, Nature 163 (1949), 433
A. Kent, Chem. Ind. 1949, p.209

PATZELT, VICTOR [1887-1956]
STURM 3,145; HEINDEL, 80-88; PLANER, 253; WIEN 1957, pp.59-63; IB, 226
C. Zawisch, Anat. Anz. 104 (1957), 298-303
H. Hayek, Alm. Akad. Wiss. Wien 106 (1957), 397-403

PAUL, BENJAMIN HORATIO [1827-1917]
RSC 4,780; 17,736-737; NUC 445,283
Anon., J. Chem. Soc. 113 (1918), 334-336

PAUL, JOHN RODMAN [1893-1971]
AMS 11,4085; NCAB 55,215-216; MH 2,407-408; MSE 2,398; DAMB, 585-586; IB, 226; WWAH, 471; WWWS, 1316
C.P. Miller, Trans. Assoc. Am. Phys. 85 (1972), 40-43
D.M. Horstmann and P.B. Beeson, Biog. Mem. Nat. Acad. Sci. 47 (1975), 323-368
A.J. Viseltear, Yale J. Biol. Med. 55 (1982), 167-172

PAUL, KARL GUSTAV [1919-1990]
VAD 1983, p.812

PAUL, THEODOR [1862-1928]
DBJ 10,333; KURSCHNER 3,1757; HEIN, 483-484; FISCHER 2,1181; TLK 2,526; TSCHIRCH, 1099; SCHELENZ, 706; POGG 4,1123; 5,947-948; 6,1962
K. Täufel, Z. angew. Chem. 47 (1928), 1253-1254
E. Heiduschka, Chem. Z. 52 (1928), 841
O. Hönigschmid, Bayer. Akad. Wiss. Jahrbuch 1928-1929, pp.82-84
R. Dietzel, Ber. chem. Ges. 62A (1929), 7-11
C. N[euberg], Biochem. Z. 209 (1929), 249-250
A. Wankmüller, Beitr. Wurtt. Apothekgesch. 10 (1974), 65-68

PAULESCO, NICOLAS CONSTANTIN [1869-1931]
ILEA, 193-194
V. Trifu, Presse Med. 39 (1931), 1704-1705
I. Murray, J. Hist. Med. 26 (1971), 150-157
I. Pavel et al., Israel J. Med. Sci. 8 (1972), 488-494; Münch. med. Wchschr. 115 (1973), 729-730
C. Bart, Arch. Inter. Cl. Bern. 9 (1976), 33-55
I. Pavel, The Priority of N.C. Paulesco in the Discovery of Insulin. Bucharest 1976
S.L. Teichmann and P.A. Alden, Physiologie 22 (1985), 121-134

PAULI, EDUARD [1882-1950]
KURSCHNER 4,2173; DRULL, 201; POGG 5,948

PAULI, HERMANN [1874-1913]
JV 17,158; NUC 445,425
Anon., Chem. Z. 37 (1913), 689

PAULI (PASCHELES), WOLFGANG [1869-1955]
STURM 3,149; FISCHER 2,1182; HASENEDER,75-89; WIEN 1957, p.44; KOREN,228; BHDE 2,891; RSC 17,721; POGG 5,948; 6,1962-1963; 7a(3),517
Anon., Chem. Z. 78 (1954),527; 79 (1955),861; Leopoldina [3]2 (1956), 20
F. Lieben, Ost. Chem. Z. 56 (1955); Wiener med. Wchschr. 57 (1955), 970
G.M. Schwab, Bayer. Akad. Wiss. Jahrbuch 1956, pp.235-236
E.I. Valko, J. Coll. Sci. 12 (1957), 241-242
W. Schwarzacher, Alm. Akad. Wiss. Wien 108 (1958), 458-460
F. Smutny, Gesnerus 46 (1989), 185-186

PAULING, LINUS CARL [1901-1994]
AMS 14,3865; WWA 1978-1979, p.2518; IWW 1982-1983, p.1018; WWWS, 1317; CB 1949,pp.473-475; 1964,pp.339-342; MH 1,370-372; STC 2,342-343; MSE 2,400-401; McLACHLAN, 27-30; ABBOTT-C, 109-110; POTSCH, 336-337; POGG 6,1693-1694; 7b,3892-3902
Anon., Chem. Eng. News 27 (1949), 28; 32 (1954), 4486
M.L. Huggins, Chem. Eng. News 33 (1955), 242-244
L. Pauling, Ann. Rev. Phys. Chem. 16 (1965), 1-14; Proc. Am. Acad. Arts Sci. 99 (1970), 988-1014; Ann. Rev. Biophys. 15 (1986), 1-9
R.J. Paradowski, The Structural Chemistry of Linus Pauling. Madison 1972
L. Pauling and D. Ridgway, J. Chem. Ed. 53 (1976), 471-476
F. Husain et al., New Scientist 75 (1977), 216-220; 76 (1978), 66,98-101
T.G. Goertzel et al., Antioch Review 38 (1980), 371-382
Y. Abe, Historia Scientarum 20 (1981), 107-124
R. Baum, Chem. Eng. News 61(28) (1983), 18-20

J.R. Goldstein, Social Research 51 (1984), 691-708
A. Serafini, Linus Pauling: A Man and his Science. New York 1989
L.E. Kay, Hist. Phil. Life Sci. 11 (1989), 211-219

PAULMIER, FREDERICK CLARK [1873-1906]
E.B.W., Science 23 (1906), 556-557

PAULY, AUGUST [1850-1914]
DSB 10,427-428; DBJ 1,303
K. Escherich, Z. angew. Ent. 1 (1914), 370-373
A. Röhre, Entomol. Bl. 10 (1914), 129-135

PAULY, HERMANN [1870-1950]
TLK 3,490; POGG 4,1124; 5,949; 6,1694; 7a(3),520
L. Anschütz, Chem. Ber. 86 (1953), xv-xxi

PAVLINOV, KONSTANTIN MIKHAILOVICH [1845-1933]
BSE 31,513-514; WWR, 436; ZMEEV 5,47-48; SG [2]12,703
V.V. Snegirova, Sov. Med. 17(6) (1953), 44-45

PAVLOV, DMITRI PETROVICH [1851-1903]
BSE 31,515; BLOKH 2,554-555
D.P. Konovalov, Zhur. Fiz. Khim. 35 (1903), 78-80

PAVLOV, IVAN PETROVICH [1849-1936]
DSB 10,431-436; BSE 31,516-519; BME 22,981-1000; WWR, 437; WWWS, 1318; KUZNETSOV(1963), 201-211; FISCHER 2,1183-1184; HAYMAKER, 250-254; KOLLE 1,200-215; SCHWERTE 2,77-85; GOCZOL, 43-61; ABBOTT, 104-105; W, 409; POGG 6,1966
G.V. Anrep, Obit. Not. Fell. Roy. Soc. 2 (1936), 1-18
W.N. Boldyreff, Erg. Physiol. 39 (1937), 1-9
B.P. Babkin, Pavlov. Chicago 1949
E.H. Huntress, Proc. Am. Acad. Arts Sci. 77 (1949), 49

PAVLOVSKI, ALEKSANDR DMITRIEVICH [1857-1944]
BME 22,1003; WWR, 438; ZMEEV 5,49-50
T.G. Lichtenstein, Zhur. Mikr. Ep. Imm. 39(2) (1959), 152-154

PAVY, FREDERICK WILLIAM [1829-1911]
DNB [1901-1911] 3,84-85; HIRSCH 4,536-537; PAGEL, 1265; WWWS, 1318; O'CONNOR, 209-211; POGG 3,1009
Anon., Lancet 1911(II), pp.976-980
F. Taylor, Nature 87 (1911), 421-422
H.W. Bywaters, Biochem. J. 10 (1916), 1-4
D. Adlersberg, Diabetes 5 (1956), 491-492
S.P.W. Chave, Brit. Med. J. 1957(II), pp.300-301

PAWLEWSKI, BRONISLAW [1852-1917]
PSB 25,419-421; POGG 6,1966

PAYEN, ANSELME [1795-1871]
DSB 10,436; GE 26,155-156; LAROUSSE 12,452; VAPEREAU 4,411; WWWS, 1319; DECHAMBRE 73,734; BLOKH 2,562; COLE, 420-422; POTSCH, 337; POGG 2,380-382; 3,1009-1010; CALLISEN 14,363-367; 31,169-170
J.D. Reid and E.C. Dryden, Textile Colorist 62 (1940), 43-45
M. Phillips, J. Wash. Acad. Sci. 30 (1940), 65-71

PAYNE, FERNANDUS [1881-1977]
DSB 18,709-710
F. Payne, Memories and Reflections. Bloomington, Ind. 1975

PAYR, ERWIN [1871-1946]
OBL 7,377; FISCHER 2,1184-1185; KILLIAN, 141; WWWS, 1319
U. Paul, Z. Chir. 99 (1974), 1172-1174
T. Becker, Z. Chir. 106 (1981), 1563-1568

PEABODY, FRANCIS WELD [1881-1927]
AMS 4,758; FISCHER 2,1185
Anon., Journal of Clinical Investigation 5 (1927), 1-6
A.S. Warthin, Trans. Assoc. Am. Phys. 48 (1928), 12-13

T.F. Williams, Bull. Hist. Med. 24 (1950), 462-481
A.M. Harvey, Pharos 44(3) (1981), 6-8

PÉAN de SAINT-GILLES, LÉON [1832-1862]
DSB 10,440-441; POGG 3,1010-1011
M. Berthelot, Bull. Soc. Chim. 5 (1863), 226-227

PEARCE, LOUISE [1885-1959]
AMS 9(II),865; DAMB, 586-587; CHEN, 145-146; WWAH, 472; WWWS, 1320
Anon., Science 130 (1959), 495
M. Fay, J. Path. Bact. 82 (1961), 542-545

PEARCE, RICHARD MILLS [1874-1930]
DAB 14,354-355; AMS 4,759; NCAB 15,204; OEHRI, 85-86; WWWS, 1320
Anon., Arch. Path. 9 (1930), 714-716
S. Flexner, Science 71 (1930), 331-332

PEARL, RAYMOND [1879-1940]
DSB 10,444-445; DAB [Suppl.2],521-522; NCAB 15,382; HUBER, 95-97; DAMB, 537-538; IB, 227; WWAH, 472-473; WWWS, 1320
L.J. Reed, Science 92 (1940), 595-597
L.J. Henderson, Am. Phil. Soc. Year Book 1940, pp.431-433
H.S. Jennings, Biog. Mem. Nat. Acad. Sci. 22 (1943), 295-347
M.D. Smith, Mendel Newsletter 10 (1974), 5-7

PEARSE, ANTHONY GUY EVERSON [1916-]
WW 1981, p.2011; WWWS, 1320

PEARSON, PAUL BROWN [1905-]
AMS 14,3879; WWA 1980-1981, p.2587; WWWS, 1321

PEARSON, RICHARD [1765-1836]
DNB 15,620-621; HIRSCH 4,540; CALLISEN 14,376-379; 31,175

PEART, WILLIAM STANLEY [1922-]
WW 1981, pp.2013-2014; IWW 1982-1983, p.1021; WWWS, 1322

PEAT, STANLEY [1902-1969]
DNB [1961-1970], 832; WW 1965, p.2381; POTSCH, 337
J.R. Turvey, Chem. Ind. 1969, p.575; Adv. Carbohydrate Chem. 25 (1970), 1-12
E.L. Hirst and J.L. Turvey, Biog. Mem. Fell. Roy. Soc. 16 (1970),441-462

PEBAL, LEOPOLD von [1826-1887]
KERNBAUER, 24-26,58-106; BLOKH 2,563; SCHAEDLER,89-90; POGG 2,385; 3,1011
A.W. Hofmann, Ber. chem. Ges. 20 (1887), 467-473

PECHMANN, HANS von [1850-1902]
BJN 7,87'; BLOKH 2,564; SCHAEDLER, 90; WWWS, 1738; POTSCH, 337-338; POGG 3,1011; 4,1126-1128; 5,951-952
W. Koenigs, Ber. chem. Ges. 36 (1903), 4417-4511
E.H. Huntress, Proc. Am. Acad. Arts Sci. 78 (1950), 14-15

PECHSTEIN, HEINRICH [1886-]
JV 26,513; SG [3]8,747; NUC 447,36

PECK, ERNEST LAWRENCE [1874-1935]
JV 13,155; NUC 447,52
Anon., J. Roy. Inst. Chem. 1935, p.275

PECK, EUGENE CURTIS [1895-1946]
NUC 447,52
J. Am. Med. Assn. 132 (1946), 1024

PECKOLT, THEODOR [1822-1912]
BJN 18,50*; HEIN, 485-486; SCHELENZ,712-713; REBER,125-129; TSCHIRCH,1100
Anon., Leopoldina 48 (1912), 102-103; Chem. Z. 36 (1912), 1163
P. Siedler, Ber. pharm. Ges. 24 (1914), 81-97

PEDERSEN, CHARLES JOHN [1904-1989]
AMS 17(5),941; WWA 1988-1989, p.2412; POTSCH, 338
Anon., Chem. Eng. News 68(8) (1990), 38
M.R. Truter, Chem. Brit. 26 (1990), 363

PEDERSEN, KAI JULIUS [1899-1984]
KBB 1981, p.883; VEIBEL 2,334-335

PEDERSEN, KAI OLUF [1901-]
VAD 1981, p.828; VEIBEL 2,335-336

PEDERSEN, RASMUS [1840-1905]
W. Johanssen, Bot. Tid. 27 (1906), li-liv
D. Müller, in The Carlsberg Laboratory 1876-1976 (H. Holter and K. Møller, Eds.), pp.162-167. Copenhagen 1976

PEKELHARING, CORNELIS ADRIANUS [1848-1922]
DSB 10,492-493; LINDEBOOM, 1508-1510; FISCHER 2,1187; UTRECHT, 120-121; OLPP, 318; POGG 6,1874-1975
H. Zwaardemaker, Arch. Neerl. Physiol. 2 (1918), 451-464
A.M. Erdman, J. Nutrition 83 (1964), 3-9

PELIGOT, EUGÈNE MELCHIOR [1811-1890]
GE 26,155-156; LAROUSSE 12,522; VAPEREAU 5,1420; BLOKH 2,564-566; W, 410; FERCHL, 399-400; SCHAEDLER, 90; TSCHIRCH, 1100; WWWS, 1325; POTSCH, 338; POGG 2,389-390; 3,1013-1014
L. Figuier, Ann. Scient. 34 (1890), 592-595
E. Jungfleisch, Bull. Soc. Chim. [3]5 (1891), xxi-xlvii

PELIKAN, EVGENI VENTSESLAVOVICH [1824-1884]
BME 23,631-632; HIRSCH 4,545
Anon., Arkhiv Veterinarnikh Nauk 14 (1884), i-x

PELLETIER, JOSEPH [1788-1842]
DSB 10,497-499; GE 26,275; LAROUSSE 12,528; HOEFER 39,503; BLOKH 2,566; DECHAMBRE 74,396-397; PHILIPPE, 795-796; FERCHL, 400-401; W, 410; POTSCH, 339; TSCHIRCH, 1100; POURQUELOT, 41-42; ABBOTT-C, 110-111; WWWS, 1325; POGG 2,392-393; CALLISEN 14,394-405; 31,182-183
A. Bussy, J. Pharm. Chim. [3]3 (1843), 48-59
M. Delépine, J. Chem. Ed. 28 (1951), 454-461
M. Javillier et al., Not. Acad. Sci. 3 (1957), 185-201
G. Dillemann et al., Rev. Hist. Pharm. 36 (1989), 128-214

PELLINI, GIOVANNI [1874-1926]
TSCHIRCH, 1100; POGG 5,954; 6,1975

PELLIZZARI, GUIDO [1858-1938]
POTSCH, 339; TSCHIRCH, 1100; POGG 4,1131-1132; 5,954-955; 6,1975-1976; 7b,3925
M. Passerini, Chemica e Industria 20 (1938), 498-499
Anon., Ber. chem. Ges. 71A (1938), 167-168
M. Betti, Atti Accad. Lincei [6]29 (1939), 353-361

PELOUZE, THÉOPHILE JULES [1807-1867]
DSB 10,499; GE 26,281; LAROUSSE 12,534; VAPEREAU 4,1357-1358; WWWS, 1326; BLOKH 2,566-567; HIRSCH 4, 548-549; FERCHL, 401; PHILIPPE, 797-798; SCHAEDLER, 91; TSCHIRCH, 1100; POTSCH, 339; GORIS, 531-532; COLE, 425-426; POGG 2,394-395; 3,1015; CALLISEN 14,406-407; 31,184
L. Figuier, Ann. Scient. 12 (1867), 519-521
Anon., J. Chem. Soc. 21 (1868), xxv-xxix
K.F.P. Martius, Sitz. Bayer. Akad. 1868(I), pp.434-439
K.R. Webb, Chem. Ind. 1945, pp.163-165

PEMBREY, MARCUS SEYMOUR [1868-1934]
DNB [1931-1940], 686-687; O'CONNOR(2), 223-225; FISCHER 2,1188
Who Was Who 1929-1940, pp.1062-1063
C.G. Douglas, Obit. Not. Fell. Roy. Soc. 1 (1935), 563-567

PEMSEL, HERMANN [1872-1957]
JV 13,155
Anon., Nachr. Chem. Techn. 4 (1956),57; 5 (1957),185; Chem. Z. (1957),406

PEMSEL, WILHELM [1873-]
JV 13,143; RSC 18,882; NUC 448,201

PÉNAU, HENRY [1884-1970]
POGG 6,1976; 7b,3925-3927

PENNELL, ROBERT BROWN [1909-1993]
AMS 14,3892; WWWS, 1327

PENNY, FREDERICK [1816-1869]
DSB 10,510-511; FERCHL, 402
J. Adams, Glasgow Med. J. 2 (1870), 258-270
A.W. Williamson, J. Chem. Soc. 23 (1870), 301-306
A.J. Berry, Chem. Ind. 51 (1932), 453-454

PENROSE, LIONEL SHARPLES [1898-1972]
DNB [1971-1980],664-665; WW 1965, p.2392
H. Harris, Biog. Mem. Fell. Roy. Soc. 19 (1973), 521-561;
J. Med. Gen. 11 (1974), 1-24
C. Clarke, J. Roy. Coll. Phys. 8 (1974), 237-250

PENTMANN, NOAH ELIE [1887-1967]
NUC 449,414

PENZOLDT, FRANZ [1849-1927]
FISCHER 2,1180; PAGEL, 1270-1271; GROTE 2,169-186; HAGEL, 11-23; PITTROFF, 127-128; DZ,1080
F. Jamin, Münch. med. Wchschr. 74 (1927), 1649,1883
A. Schittenhelm, Deutsche med. Wchschr. 53 (1927), 1831

PEPYS, WILLIAM HASLEDINE [1775-1856]
DNB 15,811-812; FERCHL, 402; WWWS, 1328; CALLISEN 14,413-414; POGG 2,400-401
D. Gooding, Notes Roy. Soc. 39 (1985), 229-244

PERATONER, ALBERTO ANTONIO [1862-1925]
TSCHIRCH, 1101; POGG 4,1135; 5,956; 6,1980
F.C. Pallazzo, Gazz. Chim. Ital. 56 (1926), 227-246

PERCIVAL, EDUMND GEORGE VINCENT [1907-1951]
E.L. Hirst, Chem. Ind. 1951, pp.909-910
E.L. Hirst and A.J. Ross, Adv. Carbohydrate Chem. 10 (1955), xiii-xx

PERCIVAL, JOHN [1863-1949]
WW 1949, p.2184; DESMOND, 480; IB, 228
W.B. Brierley, Nature 163 (1949),275; Proc. Linn. Soc. 161 (1949),248-251

PERCY, JOHN [1817-1889]
DSB 10,511-512; DNB 15,870-872; BLOKH 2,567-571; FERCHL, 403; SCHAEDLER, 91-92; WWWS, 1329; POGG 2,402; 3,1018
W.C. Roberts-Austen, Proc. Roy. Soc. 46 (1889), xxxv-xl
Anon., J. Soc. Chem. Ind. 8 (1889), 448-449

PEREIRA, JONATHAN [1804-1853]
DNB 15,887-888; TSCHIRCH, 1101; FERCHL, 403; POGG 2,403
Anon., Proc. Roy. Soc. 6 (1854), 357-359

PEREMESHKO, PETR IVANOVICH [1833-1893]
BME 23,884-885; HIRSCH 4,556; PAGEL, 1271-1272; KAZAN 2,290-291; IKONNIKOV, 540-542; ZMEEV 5,57
Y.N. Kvitnitski-Ryzhov, Usp. Sov. Biol. 40 (1955), 364-371

M.K. Kuzmin, Arkh. Pat. 1957, pp.66-69

PÉREZ, CHARLES [1873-1952]
CHARLE, 219-221; WWWS, 1329
A. Chatelet, Ann. Univ. Paris 23 (1953), 566-570

PERGER, HUGO von [1844-1901]
OBL 7,417; BJN 6,79*; POGG 3,1019; 4,1136
Anon., Chem. Z. 26 (1902), 63

PERKIN, ARTHUR GEORGE [1861-1937]
DNB [1931-1940], 687; FINDLAY, 219-246; POTSCH, 340; POGG 4,1138-1139; 6,1981; 7b,3934
E.J. Cross and F.M. Rowe, Chem. Ind. 1937, pp.744-745
F.M. Rowe and R. Robinson, J. Chem. Soc. 1938, pp.1738-1754
R. Robinson, Obit. Not. Fell. Roy. Soc. 2 (1939), 445-450
H.S. van Klooster, J. Chem. Ed. 28 (1951), 359-363

PERKIN, WILLIAM HENRY [1838-1907]
DSB 10,515-517; DNB [1901-1911] 3,104-106; FARBER, 759-772; W, 411-412; POTSCH, 340-341; ABBOTT-C, 111-113; WWWS, 1330; POGG 3,1019; 4,1137; 5,957
R. Meldola, J. Chem. Soc. 93 (1908), 2214-2257; Ber. chem. Ges. 44 (1908), 911-956; Proc. Roy. Soc. A80 (1908), xxxviii-lix
D.H. Leaback, Chem. Brit. 24 (1988), 787-790

PERKIN, WILLIAM HENRY, Jr. [1860-1929]
DSB 10,517; DNB [1922-1930], 665-667; FINDLAY, 176-218; W, 412-413; POTSCH, 341; TSCHIRCH, 1101; POGG 4,1137-1138; 5,957; 6,1981-1982 J.F.T., Proc. Roy. Soc. A130 (1930-1931), i-xii
A.J. Greenaway et al., Life and Work of William Henry Perkin. London 1932

PERKINS, MAURICE [1836-1901]
NUC 451,10PERKOWSKI, ZYGMUNT [1884-]
JB 28,319; NUC 451,59

PERLMAN, DAVID [1920-1980]
AMS 14,3901; WWWS, 1330
Anon., Chem. Eng. News 58(16) (1980), 80
H.J. Peppler, Ann. Rep. Ferm. Proc. 4 (1980), xiii-xiv

PERLMANN, GERTRUDE [1912-1974]
AMS 12,4858-4859; MSE 2,409-410; BHDE 2,897
Anon., Chem. Eng. News 52(38) (1974), 67; Science 186 (1974), 1058

PERLS, MAX [1843-1881]
HIRSCH 4,560

PERLZWEIG, WILLIAM ALEXANDRE [1891-1949]
AMS 8,1927; IB, 229; KOREN, 229; WININGER 5,10; WWAH, 478
Anon., Science 110 (1949), 700; Chem. Eng. News 28 (1950), 119

PERRENOUD, PAUL [1846-1889]
TSCHIRCH, 1101
A. Tschirch, Ber. Chem. Ges. 23 (1890), 2111-2112

PERRET, ADRIEN [1901-1962]
POGG 7a(3),529
R. Perrot, Chimia 17 (1963), 17-18

PERRIN, JEAN BAPTISTE [1870-1942]
DSB 10,524-526; W, 413-414; CHARLE(2), 223-225; WWWS, 1331-1332; POTSCH, 341; POGG 6,1986; 7b,3939-3941
E. Esclangon, C.R. Acad. Sci. 214 (1942), 725-729
R. Audubert, J. Chim. Phys. 39 (1942), 45-47
J.S. Townsend, Obit. Not. Fell. Roy. Soc. 4 (1943), 301-305
L. de Broglie, Mem. Acad. Sci. [2]67 (1949), 1-29; Rev. Hist. Sci. 24 (1971), 99-105
F. Lot, Jean Perrin. Paris 1963
M.J. Nye, Molecular Reality. London 1972

PERROT, ADOLPHE [1833-1887]
DHB 5,253; STELLING-MICHAUD 5,132; POGG 3,1022
A. Wartmann-Perrot, Verhandl. Schw. Nat. Ges. 70 (1887), 167-171

PERRY, SAMUEL VICTOR [1918-]
WW 1981, p.2030

PERSIEL, HANS [1896-1981]
JV 41,668
Anon., Nachr. Chem. Techn. 24 (1976), 477; 29 (1981), 505

PERSON, PHILIP [1919-]
AMS 14,3909; WWWS, 1333

PERSONNE, JACQUES [1816-1880]
DSB 10,530; GORIS, 536-538; FERCHL, 404; TSCHIRCH, 1101; POGG 3,1024
C.J.M. Méhu, Bull. Acad. Med. 9 (1880), 1320-1322
E. Jungfleisch, J. Pharm. Chim. [5]3 (1881), 108-112
J.E. Courtois, Rev. Hist. Pharm. 27 (1980), 263-271

PERSOON, CHRISTIAAN HENDRIK [1762-1836]
DSB 10,530-532; TSCHIRCH, 1101; WWWS, 1333; WAGENITZ, 136; CALLISEN 15,433-434; 31,194
G.C. Ainsworth, Nature 193 (1962), 22-23
J. MacLean, Janus 63 (1976), 155-166

PERSOZ, JEAN FRANÇOIS [1805-1868]
DSB 10,532; GE 26,493-494; LAROUSSE 12,680; VAPEREAU 1,1368; WWWS, 1333; BALLAND(2),206; BLOKH 2,571; BERGER-LEVRAULT, 181; FERCHL, 404-405; HUMBERT, 217-219; DECHAMBRE 75,604; SCHAEDLER, 92; TSCHIRCH, 1101; POGG 2,408-410; 3,1024; CALLISEN 14,434; 31,194
R. Sartory, Figures Pharmaceutiques Françaises, pp.95-100. Paris 1953

PERTY, MAXIMILIAN [1804-1884]
DHB 5,255-256; BULLOCH, 389
C. von Voit, Sitz. Bayer. Akad. 15 (1885), 170-174
J. Kurmann, Gesnerus 16 (1959), 139-143
R.H. Laeng, Mitt. Nat. Ges. Bern NF30 (1973), 8-10

PERUTZ, MAX FERDINAND [1914-]
WW 1981, p.2031; IWW 1982-1983, pp.1030-1031; MH 1,373-374; STC 2,348-350; MSE 2,410-411; ABBOTT-C, 113-114; BHDE 2,898; WWWS, 1333; POTSCH, 341
M.F. Perutz, Ann. Rev. Physiol. 52 (1990), 1-25

PERUTZ, OTTO [1847-1922]
STURM 3,175

PESCH, KARL LUDWIG [1889-1941]
STURM 3,176; KURSCHNER 4,2185; PELZNER, 211-213; IB, 229
F. Zeil, Med. Klin. 37 (1941), 647-648

PESCHIER, JACQUES [1769-1832]
BRIQUET, 366-369; TSCHIRCH, 1101; FERCHL, 405; POGG 2,411; CALLISEN 31,196-198
Anon., Mem. Soc. Phys. Hist. Nat. Gen. 5 (1832), ix-xi
A. de Montet, Dictionnaire Biographique des Genevois et des Vaudois, vol.2, p.278. Lausanne 1878

PESCI, LEONE [1852-1917]
TSCHIRCH, 1102; POGG 3,1025; 4,1144; 5,962; 6,1989
G. Ciamician, Rend. Accad. Bologna NS21 (1917), 63-68

PETER, KARL [1870-1955]
FISCHER 2,1195-1196; BUSCHHUTER, 147-167; IB, 229
W. Pfuhl, Anat. Anz. 102 (1955), 224-244

PÉTERFI, TIBOR [1883-1953]
MEL 2,400; BHDE 2,898; WIDMANN, 84-85,280; LDGS, 15; WININGER 7,374
R. Chambers and U. Masker, Acta Anatomica 19 (1953), 1-7

S. Baginski, Folia Morphologica 5 (1954), 229-230
F. Kiss, Anat. Anz. 103 (1956), 225-232

PETERMANN, MARY LOCKE [1908-1975]
AMS 12,4869-4870; WWWS, 1333
Anon., Chem. Eng. News 44(12) (1966), 87; 54(4) (1976), 30

PETERS, JOHN PUNNETT [1887-1955]
AMS 9(II),874; NCAB 42,95; CHITTENDEN, 266-272; ICC, 138-140
J.F. Fulton, Science 123 (1956), 1023
D.D. Van Slyke, Trans. Assoc. Am. Phys. 69 (1956), 22-23; Clin. Chem. 3 (1957), 287-293
P.H. Lavietes, Yale J. Biol. Med. 29 (1956), 175-190
M.B. Strauss, New Eng. J. Med. 254 (1956), 344
J.R. Paul and C.N.H. Long, Biog. Mem. Nat. Acad. Sci. 31 (1958), 347- 375

PETERS, RUDOLF [1869-]
RSC 17,817; NUC 452,597

PETERS, RUDOLPH ALBERT [1889-1982]
WW 1981, p.2033; MH 1,374-376; MSE 2,411-412
R.A. Peters, Ann. Rev. Biochem. 26 (1957), 1-16; TIBS 6(2) (1981), 62-63
R.H. Thompson, Biochem. Pharm. 20 (1971), 513-517; TIBS 8 (1983), 460- 461
Anon., Brit. Med. J. 284 (1982), 589
P. Ward, Chem. Brit. 18 (1982), 874
A.M. Copping, Brit. J. Nutrition 49 (1983), 1-2
R.H.S. Thompson and A.G. Ogston, Biog. Mem. Fell. Roy. Soc. 29 (1983), 495-523

PETERSEN, EMIL [1856-1907]
DBL 18,199-200; VEIBEL 2,337-340; POGG 4,1146; 5,963
Anon., Chem. Z. 31 (1907), 752

PETERSEN, HANS [1885-1946]
EBERT, 82-88; IB, 229
B. Romeis, Bayer. Akad. Wiss. Jahrbuch 1944-1948, pp.216-219
C. Elze, Z. Anat. Ent. 122 (1961), 445-458

PETERSEN, THEODOR [1836-1918]
POGG 3,1028-1029; 4,1145-1146; 5,963
Anon., Chem. Z. 42 (1918), 633
W. Boller, Ber. Senck. Nat. Ges. 50 (1920), 187-188

PETERSON, WILLIAM HAROLD [1880-1960]
AMS 9(I),1506; IB, 230; WWAH, 480; POGG 6,1991; 7b,3948-3953
Anon., Chem. Eng. News 38(32) (1960), 153

PETIT, ALEXIS THÉRÈSE [1791-1820]
DSB 10,545-546; HOEFER 39,718; FELLER 6,488; GE 26,526; LAROUSSE 12,715; BLOKH 2,571-572; FERCHL, 405; SCHAEDLER, 92; ABBOTT-C, 114; POTSCH, 341-342; W, 414-415; POGG 2,415-416; CALLISEN 31,202
J.B. Biot, Ann. Chim. [2]16 (1821), 327-335
J.C. Jamin, Revue des Deux Mondes 11 (1855), 375-412
R. Fox, Brit. J. Hist. Sci. 4 (1968), 1-22
W.J. Hughes, Chem. Brit. 6 (1970), 490-491

PETIT, ARTHUR [1837-1912]
GORIS, 538-539
E. Bourquelot, J. Pharm. Chim. [7]5 (1912), 328
L. Maquenne, Bull. Soc. Chim. [4]11 (1912), 474

PETIT, PAUL ÉMILE [1862-1936]
POGG 4,1147; 5,963; 6,1992

PETRÉN, KARL [1868-1927]
SMK 6,94-95; VAD 1927, p.634; FISCHER 2,1198-1199; GROTE 3,165-169
G. Bergmark, Acta med. Scand. 67 (1927), ix-xvii
F. Umber, Deutsche med. Wchschr. 53 (1927), 2133-2134

R.W. Wilder, Diabetes 4 (1955), 159-160
E. Ask-Upmark, Sydsvenska Medicinhistoriska Sallskapets Årsskrift 14 (1977), 13-18

PETRENKO-KRICHENKO, PAVEL IVANOVICH [1866-1944]
POTSCH, 342; POGG 4,1148; 5,964; 6,1992; 7b,3954
O.S. Stepanova and A.V. Mozharovskaya, Ukr. Khim. Zhur. 23 (1957),122-127

PÉTREQUIN, JOSEPH [1810-1876]
VAPEREAU 5,1438; GENTY 7,81-96; HIRSCH 4,571-572; DECHAMBRE [2]23,763-764
C. Gayet, Ann. Ocul. 76 (1876), 104-110

PETRI, JULIUS RICHARD [1852-1921]
FISCHER 2,1200; PAGEL, 1281-1282; BULLOCH, 389; WWWS, 1336
R.C. Benedict, Torreya 29 (1929), 9-12
V.T. Babes, Lancet 1974(II), p.666

PETRIE, ARTHUR HILL KELVIN [1903-1942]
SRS,63
Who's Who in Australia 1941, p.541

PETRIE, GEORGE FORD [1874-1955]
D. McClean, J. Path. Bact. 70 (1955), 559-562

PETRIEV, VASILI MOISEEVICH [1845-1908]
BLOKH 2,572-573; ARBUZOV, 192
P. Melikov, Ber. chem. Ges. 41 (1908), 4873-4874; Zhur. Fiz. Khim. 41 (1909), 119-130

PETROV, ALEKSANDR DMITRIEVICH [1895-1964]
WWWS, 1337; POGG 6,1994; 7b,3960-3972

PETROV, GRIGORI SEMENOVICH [1886-1957]
POGG 6,1994-1995; 7b,3972-3976
V.V. Lebedinski, Usp. Khim. 12 (1943), 252
Anon., Khimicheskaya Promyshlenost 1956(7), p.53

PETROZ, CLAUDE HENRI [1788-1867]
GORIS, 540; RSC 4,860; CALLISEN 14,458-459; 31,205-206
Anon., J. Pharm. Chim. [4]5 (1867), 135

PETRUNKEVITCH, ALEXANDER [1875-1964]
AMS 10,3162; NCAB A,4490450; IB, 230; WWWS, 1337-1338
G.E. Hutchinson, Trans. Conn. Acad. Arts Sci. 36 (1945), 9-15; Biog. Mem. Nat. Acad. Sci. 60 (1991), 235-248

PETRY, EUGEN [1873-1945]
RSC 17,830

PETRY, PHILIP [-1896]
Anon., Pharm. Z. 41 (1896), 457

PETT, LIONEL BRADLEY [1909-]
AMS 11,4152; YOUNG, 67; SRS, 67

PETTENKOFER, FRANZ XAVER [1783-1850]
HEIN, 490; TSCHIRCH, 1102; FERCHL, 406; POGG 2,418; 7aSuppl.,502
A. Lauer, Deutsche Apoth. Z. 93 (1953), 822-824

PETTENKOFER, MAX von [1818-1901]
DSB 10,556-563; BJN 6,3-8; HIRSCH 4,576-577; PAGEL, 1282-1285; WWWS,1738; BLOKH 2,573-574; HEIN, 491; SCHELENZ, 687-688; FERCHL, 406-407; TSCHIRCH, 1102; DUELUND, 29-54; W, 536-537; POTSCH, 342; POGG 2,418-419; 3,1030; 4,1159; 6,1995; 7aSuppl., 502-506
E. Sell, Chem. Z. 17 (1893), 947-949
K.B. Lehmann, Münch. med. Wchschr. 48 (1901), 464-473
L. Pfeiffer, Hygienische Rundschau 11 (1901), 717-732
F. Erismann, Deutsche med. Wchschr. 27 (1901), 209-211,253-255,285-

287, 299-302,323-327
M. Rubner, Berl. klin. Wchschr. 38 (1901), 268-270,301-303,321-326
M. Gruber, Ber. chem. Ges. 36 (1903), 4512-4572
O. Neustetter, Dr. Max Pettenkofer. Vienna 1925
E.E. Hume, Max Josef von Pettenkofer. New York 1927
A. Fischer, Münch. med. Wchschr. 80 (1933), 1665-1671
W. Zimmermann, Süddeutsche Apoth. Z. 82 (1942), 91-92
K. Kisskalt, Max von Pettenkofer. Stuttgart 1948
A.S. Evans, Yale J. Biol. Med. 46 (1973), 161-176
D.L. Trout, J. Nutrition 107 (1977), 1567-1574
H. Breyer, Max von Pettenkofer. Leipzig 1980

PETTERS, WILHELM (VILEM) [1826-1875]
OBL 8,13; STURM 3,194; NAVRATIL, 233-234; ROTH, 78-86; KOERTING, 245

PETTIBONE, CHAUNCEY J. VALLETTE [1884-1929]
AMS 4,769

PETTIT, AUGUSTE [1869-1939]
SCHAL I, 92-94; IB, 220
J. Jolly, Bull. Acad. Med. 122 (1939), 421-426
B. Kolochina-Erber, Ann. Inst. Pasteur 63 (1939), 525-530
L. Lapicque, C.R. Soc. Biol. 132 (1939), 184-186
E. Roubaud, Bull. Soc. Path. Exot. 32 (1939), 845-848

PETZHOLD, ALEXANDER [1810-1889]
FERCHL, 407; POGG 2,420-421; 3,1031; 7aSuppl.,506

PEYRONE, MICHELE [1814-1885]
RSC 4,865; NUC 454,108
V. Fino, Annali della R. Accademia d'Agricoltora (Torino) 27 (1884),23-32

PÉZARD, ALBERT [1875-1927]
DSB 10,569-571
P. Pagniez, C.R. Soc. Biol. 97 (1927), 1442-1444

PFÄHLER, ERNST [1890-1981]
JV 29,488; NUC 454,161
Anon., Nachr. Chem. Techn. 13 (1965), 53; 29 (1982), 314

PFAFF, AUGUST [1881-]
JV 20; NUC 454,166

PFAFF, CHRISTOPH (CHRISTIAN) HEINRICH [1773-1852]
ADB 25.582-587; SHBL 5,206-210; HIRSCH 4,581-582; KIEL, 76; WWWS, 1339; FERCHL, 407-408; SCHAEDLER, 93; TSCHIRCH, 1102; CALLISEN 14,468-483; 31,210-211; POGG 2,425-428; 7aSuppl.,510-511
L.F. Bley et al., Arch. Pharm. 155 (1861), 90-364
W. Ipsen, Deutsche Apoth. Z. 102 (1962), 901-904
H.R. Wiedemann, Christiana Albertina NF23 (1986), 25-34

PFAFF, FRIEDRICH [1825-1886]
RSC 4,871; 8,609; POGG 1,430; 3,1032; 4,1151

PFAFF, SIEGFRIED [1851-1928]
POGG 6,1998
Anon., Chem. Z. 52 (1928), 422; Öst. Chem. Z. 31 (1928), 106
H. Thoms, Ber. chem. Ges. 61A (1928), 131-132

PFALTZ (FEJOS), MIMOSA HORTENSE [1896-]
NUC 454,189

PFANDER, FRIEDRICH [1908-]
KURSCHNER 13,2904

PFANNENSTIEL, ADOLF [1880-1957]
JV 21,486; NUC 454,200

PFANNENSTIEL, JOHANNES [1862-1909]
FISCHER 2,1203; GAUSS, 169; DZ, 1089; KIEL, 82; BERLIN, 621
A. Martin and A. von Rosthorn, Mon. Geburt. 30 (1909), 137-145
C. Frisch, Verhandl. path. Ges. 14 (1910), 366-368
G. Rudolph, Fortschritte der Fertilität, pp.23-29. Berlin 1980
L. Easton, British Journal of Obstetrics and Gynecology 91 (1984),538-541
K. Semm, Z. Gyn. 110 (1988), 1593-1597

PFANNENSTIEL, WILHELM [1890-]
KURSCHNER 13,2905; FISCHER 2,1203-1204; AUERBACH, 338; IB, 231; POGG 7a(3),547-548

PFAU, ALEXANDRE STANISLAS [1889-1938]
Y.R. Naves, Helv. Chim. Acta 21 (1938), 1562-1570

PFAUNDLER, LEOPOLD von [1839-1920]
OBL 8,26-27; DBJ 2,756; BLOKH 2,574-575; SCHAEDLER, 93; TLK 2,532; WWWS, 1738; POGG 3,1033; 4,1151; 5,966
Anon., Alm. Akad. Wiss. Wien 70 (1920), 117-120
E.W. Lund, J. Chem. Ed. 45 (1968), 125-128

PFAUNDLER, MEINHARD von [1872-1947]
OBL 8,27; KURSCHNER 4,2205-2206; FISCHER 2,1204
R. Priesel, Wiener klin. Wchschr. 59 (1947), 673-674
R. Wagner, Bayer. Akad. Wiss. Jahrbuch 1944-1948, pp.222-225
A. Wiskott, Z. Kinderheil. 96 (1966), 97-105

PFEFFER, OTTO [1885-1918]
JV 26,549
B. Lepsius, Ber. chem. Ges. 52A (1919), 103
Anon., Chem. Z. 43 (1919), 375

PFEFFER, WILHELM [1814-]
RSC 8,610

PFEFFER, WILHELM [1845-1920]
DSB 10,574-578; DBJ 2,578-582,750-751; LKW 5,227-238; HEIN, 495-496; BLOKH 2,575-576; LEHMANN, 103-104; TSCHIRCH, 1102; W,415; WWWS,1340; POTSCH, 342-343; ABBOTT, 105-106; POGG 3,1033-1034; 4,1152; 6,1998; 7aSuppl.,512-513
G. Haberlandt et al., Naturwiss. 3 (1915), 115-131
F. Czapek, Alm. Akad. Wiss. Wien 70 (1920), 167-173
H. and E.G. Pringsheim, Ber. chem. Ges. 53A (1920), 36-39
W. Ostwald, Chem. Z. 44 (1920), 145
H. Fitting, Ber. bot. Ges. 38 (1920), (30)-(63)
W. Ruhland, Sitz. Akad. Wiss. Leipzig 75 (1923), 107-124
F.M. Andrews, Plant Physiology 4 (1929), 285-288
E. Bünning, Wilhelm Pfeffer. Stuttgart 1975 (English translation, Ottawa 1989); Ann. Rev. Plant Physiol. 28 (1977), 1-22
U. Sucker, NTM 25(2) (1988), 43-57

PFEIFFER, EMIL [1846-1921]
NB, 304-305; FISCHER 2,1205; RSC 17,839-840; WWWS, 1340
H. Lehndorff, Archives of Pediatrics 63 (1946), 218-223

PFEIFFER, HERMANN [1874-1915]
JV 14,227; RSC 14,804
Anon., Chem. Z. 39 (1915), 474

PFEIFFER, HERMANN [1877-1929]
OBL 8,31; KURSCHNER 4,2207-2208; FISCHER 2,1205-1206

PFEIFFER, PAUL [1875-1951]
DSB 10,578; TLK 3,494; ZURICH, 87-95; POTSCH, 343; POGG 5,966-967; 6,2000-2001; 7a(3),552-553
R. Wizinger, Z. angew. Chem. 62 (1950), 201-205; Helv. Chim. Acta 36 (1953), 2032-2037
G.B. Kauffman, J. Chem. Ed. 50 (1973), 277-278

PFEIFFER, RICHARD [1858-1945]
FISCHER 2,1206-1207; PAGEL, 1288-1289; BULLOCH, 389-390; W, 415-416; DZ, 1091; TLK 3,494; IB, 231; WWWS, 1340
P. Fildes, Biog. Mem. Fell. Roy. Soc. 2 (1956), 237-247
D.H. Groeschel and R.B. Hornick, Review of Infectious Diseases 3 (1981),1251-1254

PFEIFFER, THEODOR [1856-1923]
RSC 17,841-842; NUC 454,296-299
E. Blanck, Z. Pflanzenern. 28 (1923), 337-341
P. Ehrenberg, Land. Vers. 102 (1924), 36-42

PFEIFFER, THEODOR [1867-1916]
W. Scholz, Mitt. Ver. Aerzte Steier. 53 (1916), 117-125

PFEIFFER, WILHELM (WILLY) [1879-1937]
KALLMORGEN, 372; WAGENITZ, 138; ZWEIFEL, 127; JV 22,347; SG [2]13,157; NUC 454,301
A. Güttich, Archiv für Ohren...Heilkunde 144 (1938), 1

PFEILSTICKER, KARL [1897-1988]
POGG 7a(3),553
Anon., Chem. Z. 101 (1977), 219; 106 (1982), 147; 122 (1988), 180

PFENNINGER, URS [1883-1946]
A. Stettbacher, Schw. Chem. Z. 1946, pp.405-406

PFIFFNER, JOSEPH JOHN [1903-1975]
AMS 11,4157
Anon., Science 190 (1975), 966

PFISTER, KARL [1881-]
AMS 9(I),1509; JV 24,578; NUC 454,343

PFITZNER, WILHELM [1853-1903]
BJN 8,82-83,87*; RSC 17,842-843
G. Schwalbe, Anat. Anz. 22 (1903), 481-487

PFLEIDERER, GERHARD [1921-]
KURSCHNER 16,2760; POTSCH, 343

PFLÜGER, EDUARD WILHELM [1829-1910]
DSB 10,578-581; BJN 15,66*; LKW 4,253-263; WWWS, 1341; HIRSCH 4,586-587; PAGEL, 1290-1291; BRAZIER, 97-99.108-109; DZ, 1095-1096; POTSCH, 343-344; POGG 2,433; 3,1034; 6,2002
M. Nussbaum, E.F.W. Pflüger als Naturforscher. Bonn 1909
E. von Cyon, Pflügers Arch. 132 (1910), 1-19
M. Arthus, Rev. Gen. Sci. 21 (1910), 366-367
L. Bleibtreu, Med. Klin. 6 (1910), 1158-1160
H. Boruttau, Deutsche med. Wchschr. 36 (1910), 851-852
H. Leo, Münch. med. Wchschr. 57 (1910), 1128-1130
A. Walker, Nature 83 (1910), 314
E. du Bois-Reymond, Berl. klin. Wchschr. 47 (1910), 658-659
W.D.H., Proc. Roy. Soc. B82 (1910), i-iii
M. Dastre, C.R. Soc. Biol. 68 (1910), 648-650
L. Asher, Naturwiss. 17 (1929), 555-557
R. Rosemann et al., Pflügers Arch. 222 (1929), 548-574
U. Ebbecke, Deutsche med. Wchschr. 55 (1929), 967-969
C.A. Culotta, Bull. Hist. Med. 44 (1970), 109-140
H.H. Simmer, Clio Medica 12 (1977), 57-90
H. Moschner, Wissenschaftliche Arbeit und Lehre des Bonner Physiologen Eduard Wilhelm Pflüger auf dem Gebiet des Stoffwechsels. Bonn 1984

PFLUG, LUDWIG [1861-1940]
JV 6,73; NUC 454,412

PFORDTEN, OTTO von der [1861-1918]
DBJ 2,700; NUC 454,428-429; POGG 4,1154; 6,2002
Anon., Chem. Z. 31 (1907), 247; 42 (1918), 630; Z,. angew. Chem. 31(III) (1918), 132

PFÜLF, AUGUST [1862-1921]
JV 3,275; RSC 17,845; NUC 454,437
Anon., Chem. Z. 45 (1921), 243; Z. angew. Chem. 34 (1921), 98,100

PFYL, BALTHASAR [1873-1933]
JV 14,227; RSC 14,804
W. Preiss, Chem. Z. 57 (1933), 216

PHELPS, EDWARD PARKHURST [1889-1932]
AMS 4,770

PHILIP, JAMES CHARLES [1873-1941]
POGG 5,968; 6,2002; 7b,3991-3992
E.F. Armstrong, Nature 148 (1941), 249-250
A.E. Egerton, Obit. Not. Fell. Roy. Soc. 4 (1942), 51-62

PHILIP, MAX [1861-1918]
RSC 13,276; 17,847; NUC 455,357
Anon., Chem. Z. 42 (1918), 498-499; Z. angew. Chem. 31(III) (1918), 460

PHILIP, WILSON [1770-1851]
DSB 10,581-583; DNB 15,1041-1042; HIRSCH 4,588-589; CALLISEN 15,9-17; 31,216-218
W.H. McMenemy, J. Hist. Med. 13 (1958), 289-328

PHILIPPI, ERNST [1888-1969]
KERNBAUER, 283-296; MACHEK, 195-197; TLK 3,496; POGG 5,968-969; 6,2003; 7a(3),562-563

PHILIPPSON, PAULA [1874-1949]
KALLMORGEN, 372-373; JV 19,51
Basler Bibliographie 1949-1951, p.101
K. Schulte, Bonner Juden und ihre Nachkommen bis um 1930, p.425. Bonn 1976

PHILLIPS, DAVID CHILTON [1924-]
WW 1981, p.2039; IWW 1982-1983, p>1037; WWWS, 1342

PHILLIPS, MAX [1894-1957]
AMS 9(I),1512; POGG 6,2004-2005; 7b,3992-3993

PHILLIPS, PAUL HORRELL [1898-1977]
AMS 11,4166; WWA 1966-1967, p.1674; NCAB 60,23

PHILLIPS, RICHARD [1778-1851]
DNB 15,1097-1098; FERCHL, 410; WWWS, 1343; POGG 2,434-436
Anon., Phil. Mag. [4]1 (1851), 514-515
C. Daubney, J. Chem. Soc. 5 (1852), 155-156

PHILPOT, JOHN ST. LEGER [1907-]
NUC 456,412

PHIPSON, THOMAS LAMB [1833-1908]
TSCHIRCH, 1103; POGG 2,538-439; 3,1036; 4,1155
Anon., Chem. Z. 34 (1908), 247

PHISALIX, CÉSAIRE AUGUSTE [1852-1906]
FISCHER 2,1210; RSC 17,855-857
J. Noir, Prog. Med. [3]22 (1906), 218-220
A. Desgrez, Arch. Parasit. 14 (1910-1911), 54-153
J. des Cilleuls, Hist. Sci. Med. 6 (1972), 237-241

PHOEBUS, PHILIPP [1804-1880]
ADB 26,89-91; STURM 3,205; HIRSCH 4,593-595; TSCHIRCH, 1103; FERCHL,410; POGG 2,439; 3,1036
F. Siebert, Arch. Pharm. 17 (1880), 241-252
E. Heischkel-Artelt, Nachr. Giessener Ges. 32 (1963), 213-224

PIANESE, GIUSEPPE [1864-1933]
DSB 15,479-480; EI 27,103; FISCHER 2,1211
P. Rondoni, Ann. Accad. Ital. 7-9 (1938), 382-393

PICARD, MARTIN [1879-1945]
NUC 457,56
Anon., Chem. Z. 51 (1927), 104

PICARD, MAX [1888-1965]
TETZLAFF, 260; JV 28,86; NUC 457,57-59
G. von Wilpert, Deutsches Dichterlexikon, vol.2, p.546. Stuttgart 1976

PICCARD, JEAN FELIX [1884-1963]
DAB [Suppl.7],618-620; AMS 10,3176; NUC 457,90; WWAH, 482; POGG 5,971; 6,2007-2008; 7b,4001-4002
C.E. Chang, Nature 199 (1963), 229-230

PICCARD, JULES [1840-1933]
DHB 5,286; BLOKH 2,576; SCHAEDLER, 93; TSCHIRCH, 1103; POGG 3,1038;4,1158; 6,2008
F. Fichter, Verhandl. Schw. Nat. Ges. 114 (1933), 497-499; Verhandl. Nat. Ges. Basel 45 (1934), 35-42

PICCINI, AUGUSTO [1854-1905]
PROVENZAL, 249-254; EI 27,153; TSCHIRCH, 1103; POGG 4,1158-1159; 5,971
L. Balbiani, Rend. Accad. Lincei [5]14 (1905), 730-732
G. Provenzal, Riv. Storia Sci. Med. 12 (1930), 189-208

PICCININI, GALEAZZO [1879-1910]
BLOKH 2,576-577; POGG 5,972
I. Guareschi, Galeazzo Piccinini. Turin 1911

PICK, ERNST PETER [1872-1960]
STURM 3,209; KURSCHNER 8,1778; BHDE 2,903-904; DECKER, 77-94; IB, 232; KOREN, 229; COHEN, 185; WININGER 5,30; TLK 3,497; PLANER, 262; TETZLAFF, 261; WWWS, 1345; POGG 6,2009-2010; 7a(3),569; (4),152*
F. von Brücke, Arzneimitt. 2 (1952), 251-252; 7 (1957), 332-333; 10 (1960), 414-415; Alm. Akad. Wiss. Wien 110 (1960), 446-459
D. Lehr, Proc. Virchow Med. Soc. 19 (1960), 42-49
F.T. Brücke et al., Wiener klin. Wchschr. 72 (1960), 109-110
H. Molitor, Arch. Inter. Pharmacodyn. 132 (1961), 205-221

PICK, FRIEDRICH (GOTTFRIED) [1867-1926]
OBL 8,60-61; STURM 3,209-210; KURSCHNER 2,1458; FISCHER 2,1213; PELZNER, 7-19; KOERTING, 242-243; KOREN, 230; WININGER 5,30
J. von Löschner, Med. Klin. 21 (1925), 1784-1785

PICK, LUDWIG [1868-1944]
KURSCHNER 4,2221; FISCHER 2,1214; PAGEL, 1293; TETZLAFF, 261

PICK, PHILIPP JOSEPH [1834-1910]
STURM 3,210; FISCHER 2,1214; RUSTLER, 44-52

PICKARD, ROBERT HOWSON [1874-1949]
DNB [1941-1950],672-673; WW 1949, p.2206; SRS, 18; JV 13,226; NUC 457,229; POGG 5,972; 6,2011; 7b,4005-4006
F.C. Toy, Nature 164 (1949), 946
L.A. Jordan, Chem. Ind. 1949, pp.907-908
J. Kenyon, Obit. Not. Fell. Roy. Soc. 7 (1950), 253-263; J. Chem. Soc. 1950, pp.2253-2259

PICKARDT, MAX [1868-1917]
JV 7,13; RSC 17,874; SG [2]13,358; NUC 457,233
Anon., Z. angew. Chem. 30(III), 599

PICKEL, JOHANN GEORG [1751-1838]
FERCHL, 411; POGG 2,443-444

PICKEL, MAX [1861-1933]
RSC 17,875; NUC 457,236
Anon., Z. angew. Chem. 46 (1933), 123; Chemische Fabrik 6 (1933), 100

PICKELS, EDWARD GREYDON [1911-]
AMS 11,4172

PICKERING, GEORGE WHITE [1904-1980]
DNB [1971-1980], 668-669; WW 1979, p.1978; AO 1980, pp.501-503
J. McMichael and W.S. Peart, Biog. Mem. Fell. Roy. Soc. 28 (1982),431-449

PICKERING, PERCIVAL SPENCER UMFREVILLE [1858-1920]
BLOKH 2,577; POGG 4,1162-1163; 5,973
A.D.H., Nature 106 (1920), 509-510
E.J. Russell, Biochem. J. 15 (1921),1-3; J. Chem. Soc. 119 (1921),564-569
A. Harden, Proc. Roy. Soc. A111 (1926), viii-xii
T.M. Lowry et al., P.S.U. Pickering. London 1927

PICOT, CONSTANT [1844-1931]
DHB 5,287; FISCHER 2,1215
R. Mayer, Rev. Med. Suisse Rom. 100 (1980), 1020-1025

PICTET, AMÉ [1857-1937]
DHB 5,290; STELLING-MICHAUD 5,171; GENEVA 4,120-122; 5,43-45; 6,23-25; 7,106-110; 8,103-104; W, 418; TSCHIRCH, 1103; POTSCH, 344; POGG 4,1163-1164; 5,974-975; 6,2013-2014; 7a(3),570
E. Cherbuliez, Ber. chem. Ges. 70A (1937), 79-82
G. Barger, J. Chem. Soc. 141 (1938), 1113-1125
A. Pictet, Souvenirs et Travaux d'un Chimiste. Neuchatel 1941

PICTET, RAOUL [1846-1929]
DSB 10,603-604; DHB 5,289; TSCHIRCH, 1103; SCHAEDLER, 94; WWWS, 1346; POGG 3,1040; 4,1163; 5,975; 6,2014
Anon., Verhandl. Schw. Nat. Ges. 111 (1930), 411-414
C.E. Guye, C.R. Soc. Phys. Hist. Nat. Gen. 47 (1930), 18-20

PICTON, HAROLD [1867-1956]
Anon., Chem. Ind. 1956, p.287
G.M. Bennett, Proc. Chem. Soc. 1957, pp.102-103

PICTON, NORMAN [1884-1960]
SRS,35; NUC 457,387
Anon., J. Roy. Inst. Chem. 1960, p.452

PIERCE, JOHN GRISSIM [1920-]
AMS 15(5),978; WWA 1980-1981, p.2635

PIERONI, ANTONIO [1885-1973]
WWWS, 1347; POGG 5,976; 6,2015; 7b,4019

PIERRE, JOACHIM ISIDORE [1812-1881]
FERCHL, 412; POGG 2,448-449; 3, 1040-1041
A. de St. Germain, Mémoires de l'Académie des Sciences, Arts et Belles-Lettres de Caen 1891, pp.102-104
P. Gossart, Bulletin de la Societe Linnéenne de Normandie [4]7 (1893), 52-56

PIETSCH, ERICH [1902-1979]
KURSCHNER 12,2411; TLK 3,497; WWWS, 1347; POGG 6,2017; 7a(3),573-574

PIETSCH, PAUL ANDREW [1929-]
AMS 15(5),982; WWWS, 1347

PIETTRE, MAURICE [1878-1954]
POGG 6,2017; 7b,4024
M. Bressou, C.R. Acad. Agr. 40 (1954), 89-90

PIEZ, KARL ANTON [1924-]
AMS 15(5),982; MSE 2,419-420; WWWS, 1347

PIGMAN, WILLIAM WARD [1910-1977]
AMS 13,3470; NCAB 60,105; WWWS, 1347-1348
A. Herp, Adv. Carbohydrate Chem. 37 (1980), 1-5

PIGULEVSKI, GEORGI VASILIEVICH [1888-1964]
POGG 6,2018; 7b,4028-4032
Anon., Zhur. Prikl. Khim. 31 (1958), 1780-1782

PIHL, ALEXANDER [1920-]
WWWS, 1348
Hver er Hvem? 1979, p.504

PILLAI, RAMAN KOCHIKRISHNA [1906-1946]
J. Needham and E. Baldwin, Hopkins and Biochemistry, p.348. Cambridge 1949

PILLEMER, LOUIS [1908-1957]
AMS 9(I),1520; WWAH, 483
B. Cinader, Nature 181 (1958), 234
E.E. Ecker, J. Immunol. 80 (1958), 415-416; Science 127 (1958),328-329; Z. Immunitätsforsch. 118 (1959), 225-227
T.H. Lebow, J. Immunol. 125 (1980), 471-478
W.D. Ratnoff, Persp. Biol. Med. 23 (1980), 638-657

PILOTY, OSKAR [1866-1915]
BLOKH 2,577-579; POGG 4,1166; 5,976-977
L. Vanino, Chem. Z. 40 (1916), 517
C. Harries, Ber. chem. Ges. 53A (1920), 152-168

PIMENTEL, GEORGE CLAUDE [1922-1989]
AMS 16(5),1020; IWW 1987-1988, p.1171; WWWS, 1349
F. Basolo, Chem. Brit. 25 (1989), 1129
G.T. Seaborg, Am. Phil. Soc. Year Book 1989, pp.281-286
W. Andrews et al., J. Phys. Chem. 95 (1991), 2607-2608

PINCOFFS, MAURICE CHARLES [1886-1960]
NUC 458,618
G.M. Piersol, Arch. Int. Med. 53 (1960), 633-635
T.E. Woodward, Trans. Assoc. Am. Phys. 74 (1961), 33-37

PINCUS, GREGORY [1903-1967]
DSB 10,610-611; AMS 11,4186-4187; DAMB, 597-598; MH 2,416-417; STC 2,360-362; MSE 2,420-421; COHEN, 185; KAGAN(JA), 749; WWAH, 484; WWWS, 1349
H. Hoagland, Nature 215 (1967), 1316
D.J. Ingle et al., Persp. Biol. Med. 11 (1968), 337-338,358-370,422-426
A. White, Endocrinology 82 (1968), 651-654
D.J. Ingle, Biog. Mem. Nat. Acad. Sci. 42 (1971), 229-270

PINCUSSEN, LUDWIG [1873-1942]
FISCHER 2,1220; IB, 232; NUC 458,622-623; POGG 6,2019-2020; 7a(3),577
Anon., Chem. Eng. News 20 (1942), 348

PINELES, FRIEDRICH [1868-1936]
OBL 8,81; FISCHER 2,1220; KOREN, 230
O. Marburg, Jahrb. Psychiat. 53 (1936), 10-12
L. Hess, Wiener med. Wchschr. 86 (1936), 338

PINHEY, KATHLEEN FRANCES [1901-]
IB, 233

PINKUS, FELIX [1868-1947]
KURSCHNER 4,2227; FISCHER 2,1220; IB, 233; BHDE 2,905; WININGER 5,37
H.E. Michelson, Arch. Dermatol. 58 (1948), 92-94
A. Binger, Derm. Wchschr. 120 (1949), 129-131
A.H. Mehregan, J. Am. Acad. Dermatol. 18 (1988), 1158-1168

PINKUS, GEORG [1870-1943]
JV 8,27; RSC 17,901; NUC 459,125
New York Times, 22 May 1943

PINKUS, STANISLAW [1872-1929]
Anon., Chem. Z. 53 (1929). 518; Z. angew. Chem. 42 (1929), 727

PINNER, ADOLF [1842-1909]
BJN 14,72*; BLOKH 2,579-580; DZ,1042-1043; WREDE, 144-145; TSCHIRCH,1103; POTSCH, 345; WININGER 5,38; POGG 3,1042-1043; 4,1167; 5,977-978
G. von Kramer, Ber. chem. Ges. 42 (1909), 4989-5000
C. Oppenheimer, Chem. Z. 33 (1909), 661

PINOY, PIERRE ERNEST [1873-1948]
IB, 233

J. Magrou, Ann. Inst. Pasteur 76 (1949), 63-66

PIORRY, PIERRE ADOLPHE [1794-1879]
VAPEREAU 2,1456; GE 26,957; LAROUSSE 12,1044; HIRSCH 4,611-612; WWWS,1350
D. Richet and P. Tillaux, Bull. Acad. Med. 8 (1879), 591-595
P. Legendre, Bull. Soc. Hist. Med. 21 (1927), 436-459; 22 (1928), 57-82
Anon., Prog. Med. 1928, pp.1437-1445
A. Sakula, Thorax 34 (1979), 575-581
F. Simon, Med. Hist. J. 24 (1989), 138-146

PIOTROVSKI, ANTON OSIPOVICH [1825-1905]
ZMEEV 5,62
T.Y. Shchesno, Biokhimia 19 (1954), 111-115

PIOTROWSKI, GEORGES [1893-1974]
SBA 4,98; POGG 7a(3),579-580

PIOTROWSKI, GUSTAW [1833-1884]
OBL 8,87; PSB 26,469-470; HIRSCH 4,612; KONOPKA 7,509-514; KOSMINSKI, 389,619; POGG 3,1034; 4,1168
E. Ostachowski, Kwart. Hist. Nauki 2 (1957), 515-528

PIOTROWSKI, GUSTAW [1863-1905]
OBL 8,87; PSB 26,470-471

PIPER, HANS [1877-1915]
DBJ 1,337; FISCHER 2,1221; KIEL, 116; BK, 185
A. Kreidl, Z. Physiol. 30 (1915), 501
H. Boruttau, Berl. klin. Wchschr. 53 (1916), 175

PIRIA, RAFFAELE [1815-1865]
PROVENZAL, 171-174; BLOKH 2,580-581; FERCHL, 413-414; POTSCH, 346; POGG 3,1043
J.B. Dumas and C. Matteucci, Bull. Soc. Chim. [2]4 (1865), 182-186
Anon., J. Chem. Soc. 19 (1866), 512-513S. Cannizzaro, Discorso sulla Vita e sulle Opere di Raffaele Piria.Turin 1883
D. Marotta, Raffaele Piria: Labori Scientifici e Scritti Vari. Rome 1932
D. Cavanna and S. Richietta, Minerva Med. 52 (1961), 2233-2236

PIRIE, ANTOINETTE [1905-1991]
NUC 459,489
The Times, 21 October 1991

PIRIE, NORMAN WINGATE [1907-]
WW 1981, p.2056; MSE 2,422-423
N.W. Pirie, in Selected Topics in the History of Biochemistry (G. Semenza, Ed.), pp.491-522. Amsterdam 1986

PIROGOV, NIKOLAI IVANOVICH [1810-1881]
DSB 10,619-621; BSE 33,75-77; BME 24,424-437; KUZNETSOV(1963), 495-505; HIRSCH 4,614-616; PAGEL, 1298-1300; LEVITSKI 2,261-268; KILLIAN,164; ZMEEV 2,55-60; 3,47-49; WWWS, 1351
A. Fraenkel, Wiener klin. Wchschr. 23 (1910), 1684-1685
E. Kontorowicz. Deutsche med. Wchschr. 58 (1932), 541-542
F. Muravina, Nikolai Ivanovich Pirogov. Moscow 1951
G. Halperin, Bull. Hist. Med. 30 (1956), 347-355
Anon., J. Am. Med. Assn. 202 (1967), 648-649

PIRQUET, CLEMENS von [1874-1929]
OBL 8,95-96; DBJ 11,364; FISCHER 2,1221-1222; ATUYANA, 104-116; PLANER, 264; IB, 233; WWWS, 1739
M. Pfaundler, Münch. med. Wchschr. 76 (1929), 581-583
R. Wagner, Clemens von Pirquet. Baltimore 1968
H.G. Rapaport, Annals of Allergy 31 (1973), 467-475
H. Wyklicky, Wiener med. Wchschr. 130 (1980), 123-125

PIRSON, ANDRÉ [1910-]
KURSCHNER 13,2940; AUERBACH, 879; POGG 7a(3),582

PISARZHEVSKI, LEV VLADIMIROVICH [1874-1938]
BSE 33,90-91; WWR, 448; KUZNETSOV(1963), 574-581; POGG 4,1169; 5,980; 6,2023; 7b,4036-4037
A.I. Brodsky, Zhur. Ob. Khim. 9 (1939), 86-96; Usp. Khim. 17 (1948), 501-515

PISCHINGER, ALFRED [1899-1983]
KURSCHNER 13,2941; HEINDEL, 132-140
H.G. Schwarzacher, Z. mikr. anat. Forsch. 98 (1984), 801-804

PISENTI, GUSTAVO [1861-1945]
RSC 17,909; NUC 459,596

PISTOR, HERMANN [1822-1883]
DGB 54 (1927), 414

PITT-RIVERS, ROSALIND VENETIA [1907-1990]
WW 1981, p.2058
J.R. Tata, TIBS 15 (1990), 282-284

PITTS, ROBERT FRANKLIN [1908-1977]
AMS 13,3483; MH 2,419-420; STC 2,362-363; MSE 2,424-425; WWWS, 1353
E.E. Selkurt et al., Physiologist 20(5) (1977), 9-11
R.S. Alexander, Physiologist 26 (1983), 364-366
R.W. Berliner and G.H. Giebisch, Biog. Mem. Nat. Acad. Sci. 57 (1987), 323-344

PIUTTI, ARNALDO [1857-1928]
PROVENZAL, 273-278; EI 27,464; TSCHIRCH, 1104; POGG 4,1170-1171; 5,980-981; 6,2025-2026
G. Pellizzari, Gazz. Chim. Ital. 59 (1929), 225-231
G. Provenzal, Rass. Clin. Terap. 37 (1938), 114-118

PLANCHE, LOUIS ANTOINE [1776-1840]
LAROUSSE 12,1127; BOURQUELOT, 33-34; TSCHIRCH, 1104; FERCHL, 415; POGG 2,463-464; CALLISEN 15,99-104; 31,252
J. Girardin, Précis Anal. Acad. Rouen 42 (1840), 41-43
F. Boudet, Éloge de Louis Antoine Planche. Paris 1841

PLANCHER, GIUSEPPE [1870-1929]
EI 27,475; POGG 5,981-982; 6,2026-2027
M. Betti, Rend. Accad. Bologna NS33 (1929), 75-81
G. Bruni, Chimica e Industria 11 (1929), 182

PLANCHON, GUSTAVE [1833-1900]
VAPEREAU 6,1258; HIRSCH 4,622; PAGEL,1303; REBER, 273-279; TSCHIRCH,1104; BOURQUELOT, 81-82; DORVEAUX, 41-42
H. Moissan et al., J. Pharm. Chim. [6]11 (1900), 405-427
G. Dillemann, Produits et Problèmes Pharmaceutiques 26 (1971), 560

PLANCHON, JULES ÉMILE [1823-1888]
REBER, 387-388; DESMOND, 497; TSCHIRCH, 1104; WWWS, 1353
Anon., Proc. Linn. Soc. 102 (1888), 95-96; Annals of Botany 2 (1889), 423-428
J. Susplugas, Figures Pharmaceutiques Françaises, pp.131-136. Paris 1953

PLANCHON, LOUIS [1858-1915]
REBER, 388-389; TSCHIRCH, 1104; SCHELENZ, 719; DORVEAUX, 70,74,95
F. Jadin, Rev. Gen. Bot. 28 (1916), 1-10

PLANCK, MAX [1858-1947]
DSB 11,7-17; SCHWERTE 1,38-46; DZ, 1108-1109; TLK 3,499; KIEL, 181; W, 418-420; WWWS, 1353; POGG 3,1046-1047; 4,1172; 5, 982-983; 6,2027-2028; 7a(3),587-588
W. Meissner, Z. Naturforsch. 2a (1947), 587-595; Sitz. Bayer. Akad. 1948, pp.1-20; Science 113 (1951), 75-81
J. Franck, Am. Phil. Soc. Year Book 1947, pp.284-292
Max Planck in seinen Akademie-Ansprachen. Berlin 1948
M. Planck, Vorträge und Erinnerungen. Stuttgart 1949
M. von Laue et al., Z. angew. Chem. 61 (1949), 114-150
H. Hartmann, Max Planck als Mensch und Denker. Thun 1953
H. Kretzschmar, Max Planck als Philosoph. Munich 1967
E. Garber, Hist. Stud. Phys. Sci. 7 (1976), 89-126

PLANT, SYDNEY GLENN PRESTON [1896-1955]
POGG 6,2029; 7b,4042-4043
J.C. Smith and M. Tomlinson, Nature 176 (1955), 628
M. Tomlinson, J. Chem. Soc. 1956, pp.1920-1922

PLANTA, ADOLF von [1820-1895]
DHB 5,305; RSC 4,933; POGG 5,983-984
P. Lorenz, Jahresb. Nat. Ges. Graub. NF38 (1895), 88-102
E. Bosshard, Verhandl. Schw. Nat. Ges. 78 (1895), 256-271

PLANTAMOUR, PHILIPPE [1816-1898]
DHB 5,306; STELLING-MICHAUD 5,203-204; FERCHL, 415; RSC 4,935; POGG 2,466; 3,1047; 4,1172
Anon., Chem. Z. 22 (1898), 1096

PLATEAU, FÉLIX [1841-1911]
BNB 39,719-725
Anon., Entomol. News 22 (1911), 239-240
V. Villem, Liber Memorialis Gand 2,221-232; Ann. Acad. Roy. Belg. 107 (1941), 21-82

PLATEAU, JOSEPH [1801-1883]
DSB 11,20-21; BNB 17,786-788; POGG 2,466-468; 3,1048; 4,1173
J. Delsaulx, Rev. Quest. Sci. 15 (1884), 114-158,518-577; 16,383-437
Anon., Ann. Acad. Roy. Belg. 51 (1885), 389-486
S.B. Herrick, Pop. Sci. Mon. 36 (1890), 693-698
G. van Mensbrugge, Liber Memorialis Gand 2,54-71. Ghent 1913

PLATTNER, FRIEDRICH [1896-]
KURSCHNER 13,2947; IB, 233; POGG 7a(3),590-591

PLATTNER, PLACIDUS ANDREAS [1904-1975]
SBA 1,112-113; POGG 7a(3),591-593
A. Fürst, Helv. Chim. Acta 60 (1977), 2109-2121

PLATZ, LUDWIG WILHELM [1878-1914]
JV 21,486; NUC 461,315
Anon., Z. angew. Chem. 27(III) (1914), 684

PLAUT, GERHARD [1921-]
AMS 15(5),1006; BHDE 2,910-911; WWWS, 1355

PLAUT, HUGO [1858-1928]
FISCHER 2,1224-1225; TETZLAFF, 263; WININGER 5,50; 7,380
Anon., Z. Bakt. 107 (1928), i-ii
W.R. Bett, Medical Press 240 (1958), 993-994

PLAUT, MAX [1880-1938]
KALLMORGEN, 373; JV 20,149; NUC 461,330

PLAYFAIR, LYON [1818-1898]
DSB 11,36-37; DNB 22,1142-1144; PAGEL, 1304; FERCHL, 415-416; W, 420-421; POTSCH, 346-347; WWWS, 1355; POGG 2,470-471; 3,1049; 4,1173; 5,985
H.E. Roscoe, Nature 58 (1898), 128-129
A.C.B., Proc. Roy. Soc. 64 (1899), ix-xi
T.W. Reid, Memoirs and Correspondence of Lyon Playfair. London 1899
A. Scott, J. Chem. Soc. 87 (1905), 600-605
W.H.G. Armytage, Nature 161 (1948), 752-753
R.G.W. Norrish, J. Roy. Soc. Arts 99 (1951), 537-548
R.V. Jones, Nature 200 (1963), 105-111

PLAZEK, EDWIN [1898-1964]
POGG 6,2030; 7b,4047-4048
Z. Skrowaczewska, Wiad. Chem. 19 (1965), 647-672

PLEISCHL, ADOLPH MARTIN [1787-1867]
OBL 8,121-122; STURM 3,239; WURZBACH 22,415; VOGEL, 24-28; WRANY,178-179; FERCHL, 416; POGG 2,471; 3,1049; CALLISEN 15,111-114; 31,256
K. Ganzinger, Pharm. Z. 116 (1971), 1308-1311

PLESCH, JOHANN (JANOS) [1878-1957]
KURSCHNER 4,2240; FISCHER 2,1226; LDGS,68; KOREN, 230
J. Plesch, Janos, the Story of a Doctor. London 1947
R. Prigge, Deutsche med. Wchschr. 82 (1957), 1019

PLESS, FRANZ [1819-1905]
OBL 8,124; STURM 3,241

PLIENINGER, HANS [1914-1984]
KURSCHNER 13,2951; WWWS, 1355; POGG 7a(3),594-595
Anon., Chem. Z. 102 (1978), 457; 109 (1985), 18

PLIMMER, HENRY GEORGE [1856-1918]
O'CONNOR(2), 463-464
J.B.F., Nature 101 (1918), 328
Anon., Lancet 1918(II), p.128; Brit. Med. J. 1918(I), p.738; J. Roy. Micr. Soc. 1918, pp.349-357

PLIMMER, ROBERT HENRY ADERS [1877-1955]
DNB [1951-1960],818-819; WW 1949, p.2221-2222; O'CONNOR(2), 155- 157; POGG 6,2031; 7b,4050
J. Lowndes, Nature 176 (1955), 283-284; Biochem. J. 62 (1956), 353-357
P. Haas, J. Chem. Soc. 1956, p.798

PLIMPTON, RICHARD TAYLER [1856-1899]
H.F.M., J. Chem. Soc. 77 (1900), 595-596

PLISSON, AUGUSTE ARTHUR [-1832]
LAROUSSE 12,1187; GORIS, 542; TSCHIRCH, 1104; FERCHL, 416; POGG 2,473; CALLISEN 15,117-119; 31,257-258
O.H., J. Pharm. Chim. [2]8 (1832), 359

PLÖCHL, WILHELM [1853-1923]
M. Rubenhauer, Chem. Z. 47 (1923), 803

PLÖSSL, SIMON [1794-1868]
ADB 26,311; OBL 8,128; WURZBACH 22,441-443; POGG 2,473-474; 3,1050
E. Bancher and J. Holzl, Simon Plössl. Vienna 1968
F. Kotlan, Schriften des Vereins zur Verbreitung Naturwissenschaftlichen Kenntnisse in Wien 109 (1969), 1-24
J. Holzl et al., Blätter für Technikgeschichte 31 (1969), 45-89; Microscopy 32 (1972), 173-188

PLOETZ, THEODOR [1912-1975]
KURSCHNER 12,2428; POGG 7a(3),595-596

PLÓSZ, PAL [1844-1902]
OBL 8,129; BJN 7,90*; MEL 2,420-421; FISCHER 2,1227
Anon., Chem. Z. 26 (1902), 1236; Ung. med. Presse 7 (1902), 513-514
O. von Krücken and J. Parlagi, Das Geistige Ungarn 2,380. Vienna 1918

PLOTNIKOW, IVAN [1878-1955]
TLK 3,501; POGG 5,986; 6,2031-2033; 7b,4050-4051
Anon., Chem. Z. 54 (1930), 455; 79 (1955), 561
V. Njegovan and K. Weber, Croat. Chim. Acta 28 (1956), 131-140

PLOTZ, HARRY [1890-1947]
DAB [Suppl.4],667-669; AMS 7,1401; IB, 234; WININGER 5,51-52
Anon., Ann. Inst. Pasteur 73 (1947), 584-586

PLOUGH, HAROLD HENRY [1892-1985]
AMS 11,4205; IB, 234
New York Times, 17 November 1985

PLÜCKER, JULIUS [1801-1868]
DSB 11,44-47; ADB 26,321-323; W, 421-422; WWWS, 1356; POGG 2,475-477; 3,1050; 7aSuppl.,513-514
F. von Kobell, Sitz. Bayer. Akad. 1869(I), pp.393-395
A. Clebsch. Zum Gedächtnis an Julius Plücker. Gottingen 1872
W. Ernst, Julius Plücker. Bonn 1933
E.H. Huntress, Proc. Am. Acad. Arts Sci. 79 (1951), 33-35

PLUGGE, PIETER CORNELIS [1847-1897]
LINDEBOOM, 1549-1550; FISCHER 2,1227; REBER, 235-241,369-370; BLOKH 2,581-582; TSCHIRCH, 1104; POGG 3,1050; 4,1174
J.F. Heymans, Arch. Inter. Pharmacodyn. 4 (1897-1898), 185-193

PLUMBRIDGE, DOUGLAS VICTOR [1887-]
JV 27,653; NUC 462,134

PLUMMER, HENRY STANLEY [1874-1936]
AMS 5,888; FISCHER 2,1227; DAMB, 599-600; OEHRI, 90-91; WWWS, 1356
S.W. Harrington, J. Thor. Surg. 7 (1937), 110-111
A.S. Jackson and J.H. Means, J. Clin. Endocrin. 9 (1949), 967-973
F.A. Willius, Henry Stanley Plummer. Springfield, Ill. 1960
W.M. NcConahey and D.S. Pady, Endocrinology 129 (1991), 2271-2273

POCKELS, AGNES [1862-1935]
W. Ostwald, Koll. Z. 58 (1932), 1-8
E. Pockels, Berichte der Oberhessischen Gesellschaft für Natur- und Heilkunde 24 (1949), 303-307
C.H. Giles and S.D. Forester, Chem. Ind. 1971, pp.43-53
G. Beisswanger, Chemie in unserer Zeit 25 (1991), 96-101

PODCZASKI, TEODOR [1873-1930]
M. Demel, Arch. Hist. Med. 29 (1966), 450

PODOLINSKI, SERGEI ANDREEVICH [1850-1891]
RSC 8,637; 11,37; SG [1]11,462
Ukrainska Radyanska Entsiklopedia 8,449
A. Kokhan, Sov. Zdrav. 32 (1973), 75-76; Fiziol. Zhur. 59 (1973), 1296-1298; Vrachebnoe Delo 1977, pp.152-154

PODVYSSOTSKI, VALERIAN OSIPOVICH [1822-1892]
KAZAN 2,297-300; LEVITSKI 2,194-195; HIRSCH 4,638; TSCHIRCH, 1105; ZMEEV 5,65-66; POGG 3,1051
Anon., Chem. Z. 16 (1892), 1977

PODVYSSOTSKI, VLADIMIR VALERIANOVICH [1857-1913]
FISCHER 2,1228; PAGEL, 1306-1307; ZMEEV 5,66; RSC 17,935
L. Aschoff and F. Marchand, Beitr. path. Anat. 57 (1913), 1
V.N. Klimenko, Arch. Sci. Biol. 18 (1914), 1-14

POEHL, ALEKSANDR VASILIEVICH [1850-1908]
HEIN, 501-502; FISCHER 2,1228; TSCHIRCH, 1105; POGG 3,1051; 4,1176; 5,988
W. Nernst, Ber. chem. Ges. 41 (1908), 3278
Anon., Chem. Z. 32 (1908), 893
A.A. Souchoff, Rev. Endocrin. 5 (1927), 339-342

PÖTZL, OTTO [1877-1962]
FISCHER 2,1229
H. Hoff, Wiener klin. Wchschr. 69 (1957), 905-907
E. Pichler, Wiener med. Wchschr. 112 (1962), 579-580

POGGENDORFF, JOHANN CHRISTIAN [1796-1877]
DSB 11,49-51; ADB 26,364-366; HEIN, 502-503; BLOKH 2,582-583; W, 422; SCHAEDLER, 95-96; FERCHL, 417; WWWS, 1357; POGG 2,480-482; 3,1052-1053; 7aSuppl.,514-516
A.W. Hofmann, Ber. chem. Ges. 10 (1877), 103-104
Anon., Nature 15 (1877), 314-315
W. Barentin, Ann. Phys. 160 (1877), v-xxiv
H. Salié, Isis 57 (1966), 389-392

POGGI, RAOUL [1899-]
POGG 6,2036-2037; 7b,4061-4062

POGGIALE, ANTOINE [1808-1879]
DSB 11,51; BALLAND, 90-95; BALLAND(2), 66-70; FERCHL, 417-418; BLOKH 2,583; BOURQUELOT, 75-76; SCHAEDLER, 96; TSCHIRCH, 1105; POGG 2,482-483; 3,1053
P. Blondeau, J. Pharm. Chim. [4]30 (1879), 383-385
E.A. Bourgoin, Bull. Acad. Med. 8 (1879), 921-924

POHL, GEORG FRIEDRICH [1788-1849]
ADB 26,368-369; POGG 2,484-485

POHL, JULIUS [1861-1942]
OBL 8,155; STURM 3,258; FISCHER 2,1229; LOMMATZSCH, 95-100; TLK 3,502; KOERTING, 134; POGG 6,2037; 7a(3),600
O. Reisser, Deutsche med. Wchschr. 57 (1931), 1869

POHL, OTTO [1874-]
JV 20,262

POHL, RICHARD [1910-]
KURSCHNER 13,2963; POGG 7a(3),600-601

POHLE, KONRAD [1899-]
KURSCHNER 12,2437; POGG 7a(3),603
G. Urban, Arzneimitt. 14 (1964), 1069-1070

POHLMANN, JACOB [1875-1940]
NUC 463,185
Anon., Chem. Wkbl. 37 (1940), 663

POISEUILLE, JEAN [1797-1869]
DSB 11,62-64; HIRSCH 4,642; WWWS, 1357; POGG 2,487; 3,1054; CALLISEN 31,265
M. Brillouin, Journal of Rheology 1 (1930), 345-348; Annales de Physique [10]15 (1931), 411-417

POITEVIN, ALPHONSE LOUIS [1819-1882]
BLOKH 2,583; POTSCH, 347; POGG 3,1054

POKROVSKI, VASILI TIMOFEEVICH [1838-1877]
IKONNIKOV, 554-556; HIRSCH 4,643; ZMEEV 2,65; 3,50; RSC 4,971; 8,641; 12,581-582; NUC 463,336
L.I. Zhukovski, Ter. Arkh. 25 (1953), 71-76

POLANYI, MICHAEL [1891-1976]
DSB 18,718-719; DNB [1971-1980],677-678; WW 1976, p.1898; MEL 3,622-623; BHDE 2,914-915; LDGS, 23; KOREN, 231; WININGER 7,381; TLK 3,502;TETZLAFF, 265;WWWS, 1358; ABBOTT-C, 114-115; POGG 5,991; 6,2041-2042; 7b,4071-4076
M. Polanyi, Science 141 (1963), 1010-1013
E.P. Wigner et al., Nature 261 (1976), 83-84
M. Davies, Chem. Brit. 12 (1976), 323-324
[E. Shils], Minerva 14 (1976), 1-5
D.D. Eley, Adv. Catalysis 26 (1977), xvii-xix
E.P. Wigner and R.A. Hodgkin, Biog. Mem. Fell. Roy. Soc. 23(1977),413-448

POLECK, THEODOR [1821-1906]
BJN 11,50*; HEIN, 503-504; REBER, 89-92,370-374; BLOKH 2,583-584; DZ, 1119-1120; SCHAEDLER, 96; POGG 3,1055-1056; 4,1180-1181; 5,992
G. Kassner, Ber. pharm. Ges. 16 (1906), 195-200

POLIS, BERYL DAVID [1914-1977]
AMS 13,3498
F.W. Cope, Physiological Chemistry and Physics 12 (1980), 541-544

POLITZER, ADAM [1835-1920]
DBJ 2,757; JOHN, 204-224; KUKULA, 712; WININGER 5,57-58
F.L. Lederer, Arch. Otolaryng. 74 (1961), 130-133

POLITZER, GEORG [1898-1956]
HEINDEL, 183-195; IB, 235
H. Hayek, Wiener klin. Wchschr. 69 (1957), 86-87

POLL, HEINRICH [1877-1939]
KURSCHNER 4,2251; FISCHER 2,1231-1232; BHDE 2,916; LDGS, 55; IB, 235
E. Mayer, Bio-Morphosis 1 (1939), 586-600

POLLACCI, EGIDIO [1833-1913]
TSCHIRCH, 1105; POGG 3,1056; 3,1182-1183; 5,993
T. Guglielmini, Riv. Stor. Sci. 30 (1939), 99-102

POLLACCI, GINO LUIGI [1872-1963]
Chi É? 1957, p.439
R. Ciferri, Atti Ist. Bot. Pavia 21 (1964), 25-38

POLLAK, HERMANN [1884-1914]
Anon., Z. angew. Chem. 28(III) (1915), 667

POLLAK, JAKOB (JACQUES) [1872-1942]
OBL 8,170; KURSCHNER 4,2251; TLK 3,503; WININGER 7,382; POGG 5,993; 6,2043; 7a(3),605
E. Reisz, Ost. Chem. Z. 53 (1952), 25-31POLLAK, LEO [1878-1946]
OBL 8,171; STURM 3,273; KURSCHNER 4,2252; FISCHER 2,1232; KOREN, 231; WININGER 7,382
J.S.H., Brit. Med. J. 1946(II), p.519
H. Schur, Wiener klin. Wchschr. 58 (1946), 739

POLLENDER, ALOYS [1800-1879]
DSB 11,68-71; HIRSCH 4,647; BULLOCH, 390-391; OLPP, 324-327; WWWS, 1359
R. Müller, Münch. med. Wchschr. 67 (1920), 114-117
H. Hiddemann, Naturw. Rund. 3 (1950), 483

POLLINGER, ADOLF [1894-]
JV 37,1358

POLLISTER, ARTHUR WAGG [1903-]
AMS 12,4967

POLLITZER, FRANZ [1885-1942]
STURM 3,276; TLK 3,504; POGG 6,2045; 7a(3),607

POLLITZER, SIGMUND [1859-1937]
FISCHER 2,1232; OEHRI, 93-94; KOREN, 231; WININGER 5,65; RSC 17,954
Anon., Arch. Dermatol. 37 (1938), 499-503

POLLOCK, MARTIN RIVERS [1914-]
WW 1981, p.2069; WWWS, 1359

POLLOK, JAMES HOLMS [1868-1915]
POGG 5,993
Anon., Nature 96 (1915), 376
G.T. Morgan, J. Chem. Soc. 109 (1916), 389-394

POLONOVSKI, MAX [1861-1939]
POGG 5,993-994; 6,2045-2046; 7b,4077
R. Delaby, Bull. Soc. Chim. [5]6 (1939), 1269-1270

POLONOVSKI, MICHEL [1889-1954]
WWWS, 1369; WWWS, 6,2046; 7b,4077-4084
R. Hazard, Ann. Univ. Paris 24 (1954), 387-394
R. Lépine, Bull. Acad. Med. 138 (1954), 347-349
M. Javillier, Presse Med. 62 (1954), 1283-1284
M.F. Jayle, Rev. Quest. Sci. 125 (1954), 419-424; Bull. Soc. Chim. Biol. 36 (1954), 1379-1389
P. Boulanger, Enzymologia 17 (1954), 113-115
R. Leriche, C.R. Soc. Biol. 148 (1954), 1021-1023

POLUNIN, ALEKSEI IVANOVICH [1820-1888]
BSE 33,642; BME 25,973; HIRSCH 4,648; ZMEEV 2,67-68; 3,51

POMERANZ, CESAR [1860-1926]
HEIN, 505-506; POGG 4,1183; 5,995; 6,2048

POMMER, GUSTAV [1851-1935]
FISCHER 2,1233; PAGEL, 1311-1312
G.B. Gruber, Münch. med. Wchschr. 83 (1936), 612-613
F. Lang, Wiener klin. Wchschr. 49 (1936), 186-188; Verhandl. path. Ges. 29 (1937), 406-417; Wiener med. Wchschr. 86 (1946), 173-174

POMORSKI, JOZEF [1868-]
RIGA, 180

POND, GEORGE GILBERT [1861-1920]
AMS 2,374; MILES, 392; RSC 17,960; NUC 465,1
W.E. Walker, Ind. Eng. Chem. 12 (1920), 718; J. Chem. Ed. 4 (1927), 150-157

PONDER, ERIC HALDANE [1898-1970]
AMS 10,3213; WWA 1954-1955, p.2143
R.I. Weed, in Formation and Destruction of Blood Cells (T. Greenwalt and G.A. Jamieson, Eds.), pp.15-33. Philadelphia 1970

PONFICK, EMIL [1844-1913]
BJN 18,115*; FISCHER 2,1234; PAGEL, 1312; FRANKEN, 193-194; WWWS, 1361; DZ, 1120-1121; RSC 8,643; 12,582; 17,961
E. Kaufmann, Münch. med. Wchschr. 60 (1913); Deutsche med. Wchschr. 40 (1914), 86
R. Stumpf, Z. allgem. Path. 25 (1914)
Anon., Verhandl. path. Ges. 17 (1914), 598-600

PONNDORF, GEORG [1881-1921]
JV 21,297; NUC 465,32
Anon., Chem. Z. 45 (1921), 760; Z. angew. Chem. 34 (1921), 412

PONNDORF, WILHELM [1864-1948]
FISCHER 2,1234
K. Engelbrecht, Med. Klin. 44 (1949), 1260-1261

PONOMAREV, ALEKSEI PETROVICH [1886-1939]
A.M. Alekseev et al., Bot. Zhur. 1940, pp.178-180

PONOMAREV, IVAN MIKHAILOVICH [1848-1905]
BLOKH 2,584-585; RSC 8,643; 12,582; 17,961
N.A. Chernay, Zhur. Fiz. Khim. 44 (1912), 485-492

PONS, CHARLES [1876-1952]
J. Rodhaim, Ann. Soc. Belges Med. Trop. 33 (1953), 1-2

PONTECORVO, GUIDO [1907-]
WW 1981, p.2071; IWW 1982-1983, p.1056; MH 2,422-423; STC 2,369-370; MSE 2,428-430; WWWS, 1361
J.A. Roper and D.A. Hopwood, Cancer Survey 7 (1988), 229-237

POPE, WILLIAM JACKSON [1870-1939]
DSB 11,84-92; DNB [1931-1940],716-717; FINDLAY, 285-316; W, 424- 425; POTSCH, 348; WWWS, 1362; POGG 4,1186; 5,996-997; 6,2053-2054; 7b,4100
Anon., J. Roy. Inst. Chem. 1939, pp.436-437
W.H. Mills, J. Chem. Soc. 1941, pp.697-715; J. Roy. Soc. Arts 94 (1946), 668-674
C.S. Gibson, Obit. Not. Fell. Roy. Soc. 3 (1941), 291-324

POPIELSKI, LEON [1866-1920]
OBL 8,197; PSB 27,585-586; KONOPKA 8,139-140
Z. Steusing, Tow. Nauk Lwow 1 (1921), 62-64
W. Koskowski, Kosmos 56 (1931), vii-xvi
J. Zakrzewska, Acta Physiol. Pol. 12 (1961), 331-345
A. Kaskiewicz, Pol. Med. Sci. 10 (1967), 147-148

POPJAK, GEORGE JOSEPH [1914-]
AMS 15(5),1034; WWA 1980-1981, p.2662; IWW 1982-1983, p.1057; WWWS, 1362-1363
G. Popjak, J. Am. Oil Chem. Soc. 54 (1977), 647A-655A

POPOFF, METHODI [1881-1954]
IB, 235; WWWS, 1363

POPOV, ALEKSANDR NIKOFOROVICH [1840?-1881]
DSB 11,92-93; BSE 34,157; ARBUZOV, 184-185; POGG 3,1060
G.V. Bykov, Trudy Inst. Ist. Est. 12 (1956), 200-245

POPOV, LEV VASILIEVICH [1845-1906]
BSE 34,161; FISCHER 2,1234; ZMEEV 5,97
N.Y. Khristovich, Russki Vrach 5 (1906), 1429-1432

POPOV, VASILI NIKOLAEVICH [1862-1895]
LEVITSKI 2,310-311
J. Brennsohn, Die Arzte Estlands, p.496. Riga 1922

POPPENBERG, OTTO [1876-1956]
KURSCHNER 8,1806; POGG 6,2055; 7a(3),610

POPPER, HANS [1903-1988]
AMS 15(5),1036; BHDE 2,919; WWWS, 1363-1364
R. Schmid and S. Schenker, Hepatology 9 (1989), 669-674
R.D. Berk, Hans Popper. Tunbridge Wells 1991

POPPER, KARL RAIMUND [1902-]
WW 1981, p.2074; KURSCHNER 13,2973-2974; IWW 1982-1983, p.1058; BHDE 2,919-920; TETZLAFF, 266-267
B. Magee, Karl Popper. London 1973
P.A. Schilpp, Ed., The Philosophy of Karl Popper. La Salle, Ill. 1974

PORCHER, CHARLES [1872-1933]
GUIART, 171-172; STALDER, 62-63; IB, 236; POGG 6,2055-2060
E. Nicholas, Bull. Soc. Chim. Biol. 16 (1934), 160-165
C. Achard, C.R. Soc. Biol. 115 (1934), 3-5
A. Morel, Bull. Soc. Chim. [5]1 (1934), 356-357

PORGES, OTTO [1879-1968]
STURM 3,285; KURSCHNER 4,2259-2260; AMS 10,3217; BHDE 2,920; WWWS, 1364; PLANER, 268; KOREN, 231-232; WININGER 7,383
J. Novak, Wiener klin. Wchschr. 80 (1968), 559-560
O. Sternberg, Pirquet Bull. Clin. Med. 15(2) (1968), 12

PORRETT, ROBERT [1783-1868]
DNB 16,153-154; FERCHL, 420; WWWS, 1364; POTSCH, 348; POGG 2,503; 3,1060; CALLISEN 15,163-164
W. De La Rue, J. Chem. Soc. 22 (1869), vii-x
Anon., Proc. Roy. Soc. 18 (1870), iv-v

PORTER, CHARLES WALTER [1880-1971]
POGG 6,2061; 7b,4107-4108

PORTER, JOHN ADDISON [1822-1866]
DAB 15,96-97; ELLIOTT, 207; SILLIMAN, 134; WWWS, 1364

PORTER, JOHN ROGER [1909-1979]
AMS 14,4002-4003; WWWS, 1364-1365

PORTER, JOHN WILLARD [1915-]
AMS 15(5),1039; WWA 1980-1981, p.2665; WWWS, 1365

PORTER, KEITH ROBERTS [1912-]
AMS 15(5),1039; WWA 1980-1981, p.2665; MH 1,377-378; STC 2,373- 374; MSE 2,432-433; WWWS, 1365
G. Palade, J. Cell. Biol. 75 (1977), D4-D19
M.A. Bonneville, Ultrastructural Pathology 4 (1983), 401-408

PORTER, RODNEY ROBERT [1917-1985]
WW 1981, p.1079; IWW 1982-1983, p.1059; STC 2,374-375; MSE 2,433-434; ABBOTT, 106-107; POTSCH, 348-349
L.A. Steiner, Nature 317 (1985), 383
J. Humphrey, TIBS 10 (1985), 470-471
C.A. Pasternak, Bioscience Reports 5 (1985), 809-813
S.V. Perry, Biog. Mem. Fell. Roy. Soc. 33 (1987), 445-489

PORTER, WILLIAM TOWNSEND [1862-1949]
DAB [Suppl.4],675-677; NCAB 15,288; FISCHER 2,1238; DAMB,

603; IB, 236
E.M. Landis, Am. J. Physiol. 158 (1949), v-vii
A.J. Carlson, Science 110 (1949), 111-112
A.C. Barger, Physiologist 14 (1971), 277-285; 22 (1979), 20; 25 (1982), 407-413

PORTIER, PAUL JULES [1866-1962]
DSB 11,101-102; FISCHER 2,1238; CHARLE(2), 233-235; IB,236; WWWS, 1364
L. Binet, Ann. Univ. Paris 32 (1962), 152-160; Biol. Med. 51 (1962), 113-123
R. Courrier, Vies Laborieuses, Vies Fécondes 1,191-213. Paris 1968; Not. Acad. Sci. 5 (1972), 348-388
B. Masse, Paul Portier. Paris 1969
H. Benard, Bull. Acad. Med. 154 (1970), 842-847
C.D. May, Journal of Allergy 75 (1985), 485-495

PORTNER, EDUARD [1874-]
JV 11,168; NUC 466,514

POSENER, KARL [1897-1945]
BHDE 2,921; LDGS, 81; NUC 467,225

POSER, GOTTLIEB von [1896-]
JV 46,486; NUC 467,228

POSNER, CARL [1854-1928]
DBJ 10,219-221; KURSCHNER 3,1829; GROTE 7,151-184; FISCHER 2,1239; PAGEL, 1315-1316; BHDE 2,922; KOREN, 232; WININGER 5,84
M.W. Schwalbe, Deutsche med. Wchschr. 44 (1929), 30-31
H. Kupsch-Petzel, Louis und Carl Posner. Berlin 1969
H. Hausmann, Z. Urol. 80 (1987), 601-604

POSNER, THEODOR [1871-1929]
WREDE, 148; TLK 3,506; POGG 4,1188; 5,999; 6,2063
H. Thoms, Ber. chem. Ges. 62A (1929), 58-59

POSSELT, CARL [1837-1916]
FISCHER 2,1240; KAMP, 5-8; DUELUND, 113
A. Jesionek, Derm. Wchschr. 64 (1917), 63-65
L. von Zumbusch, Münch. med. Wchschr. 64 (1917), 11

POSSELT, CHRISTIAN WILHELM [1806-1877]
DRULL, 208; TSCHIRCH, 1106; FERCHL, 420; POGG 2,508
P.F. Koenig, Die Entdeckung des reinen Nicotins. Bremen 1940

POSSELT, FRIEDRICH LUDWIG (LOUIS) [1817-1880]
HEIN, 507-508; DRULL, 208; FERCHL, 420-421; POGG 2,508

POSTERNAK, JEAN MARC [1913-]
SBA 4,100; WWWS, 1366; POGG 7a(3),613

POSTERNAK, SWIGEL [1871-1932]
POGG 6,2064-2065
R. Fabre, Bull. Soc. Chim. Biol. 14 (1932), 1104-1106
E. Cherbuliez, Verhandl. Schw. Nat. Ges. 114 (1933), 453-455

POSTERNAK, THÉODORE [1903-1982]
WWWS, 1366; POGG 7a(3),613-614
Anon., Chimia 17 (1963), 299-300
J.M.J. Tronchet, Chimia 27 (1973), 512
J. Deshusses, Chimia 36 (1982), 218
R. Monnier and E. Charollais, Archives des Sciences (Geneva), 36 (1983), 181-185

POTILITSIN, ALEKSEI LAVRENTIEVICH [1845-1905]
BLOKH 2,585-587
Anon., Chem. Z. 29 (1905), 303
F.F. Selivanov, Zhur. Fiz. Khim. 40 (1908), 1149-1182

POTT, EMIL [1851-1913]
BJN 18,115*

A. Morgen and C. Begar, Jahrbuch der Chemie 23 (1913), 328

POTTER, LEY FRANCIS [1882-]
JV 22,301; NUC 467,611

POTTER, VAN RENSSELAER [1911-]
AMS 15(5),1050; WWA 1980-1981, p.2670; MH 2,425-426; MSE 2,434-
435; WWWS, 1367

POTTEVIN, HENRI [-1928]
Anon., Presse Med. 36 (1928), 1388

POTTHAST, JOHANN [1854-]
RSC 11,53; NUC 467,661

POUCHET, FÉLIX ARCHIMÈDE [1800-1872]
DSB 11,109-110; GENTY 5,77-92; VAPEREAU 4,1473; LAROUSSE 12, 1527; GE 27,463; DECHAMBRE 78,786; HIRSCH 4,665; BULLOCH, 39; WWWS, 1368; CALLISEN 31,291
J. Roger, Les Médecins Normands 1,221-229. Paris 1890
J. Farley and G.L. Geison, Bull. Hist. Med. 48 (1974), 161-198
N. Roll-Hansen, J. Hist. Med. 34 (1979), 273-292

POUCHET, GABRIEL [1851-1938]
FISCHER 2,1241; RSC 17,985-986
A. Tiffeneau, Bull. Acad. Med. 120 (1938), 9-15

POULSSON, POUL EDVARD [1858-1935]
NBL 11,150-152; FISCHER 2,1241-1242; IB, 237; POGG 6,2066
G. Liljestrand, Arch. Inter. Pharmacodyn. 52 (1936), 123-128

POULTON, EDWARD BAGNALL [1856-1943]
DSB 18,721-727; DNB [1941-1950], 687-689; O'CONNOR(2), 94-95; WWWS, 1368
G.D.H. Carpenter, Nature 153 (1944), 15-17; Proc. Linn. Soc. 156 (1944), 219-223; Obit. Not. Fell. Roy. Soc. 4 (1944), 655-680

POULTON, EDWARD PALMER [1883-1939]
WW 1939, p.2573; O'CONNOR(2), 230-231; WWWS, 1368
Anon., Lancet 1939(II), pp.960,1052; Brit. Med. J. 1939(II), p.886
M. Campbell, Nature 144 (1939), 969; Guy's Hosp. Rep. 89 (1939), 251-261

POWELL, GARFIELD [1893-1943]
AMS 6,1133
Anon., Science 97 (1943), 347

POWELL, HERBERT MARCUS [1906-]
WW 1981, p.2086; WWWS, 1369

POWELL, RICHARD [1767-1834]
DNB 16,246-247; HIRSCH 4,666-667; MR 2,456-457; WWWS, 1369
Anon., J. Am. Med. Assn. 208 (1969), 353-354

POWER, FREDERICK BELDING [1853-1927]
DSB 11,120-121; DAB 15,154-155; MILES, 393; REBER, 119-124; WWWS, 1369; TSCHIRCH, 1106; WWAH, 489; POGG 5,1001; 6,2067-2068
I. Griffith, Am. J. Pharm. 96 (1924), 601-614
Anon., Ind. Eng. Chem. News Ed. 5(7) (1927), 5
M. Phillips, J. Chem. Ed. 31 (1954), 258-261

POWERS, GROVER FRANCIS [1887-1968]
AMS 11,4249; WWWS, 1370

POWERS, HARRY HENRY [1900-]
AMS 10,3233

POZERSKI, ÉDOUARD ALEXANDRE [1875-1964]
IB, 237
Anon., Ann. Inst. Pasteur 106 (1964), 103-108
NUC 468,576-577

J.C.L., Ann. Inst. Pasteur 106 (1964), 813-818
J. Brossollet, Hist. Sci. Med. 23 (1989), 45-50

POZZI-ESCOT, MARIUS EMMANUEL [1880-1963]
POGG 5,1001; 6,2068; 7b,4119-4121
Anon., Chem. Eng. News 41(13) (1963), 77

PRABHAKAR, MORESHWAR [1885-1922]
JV 28,194; NUC 468,621
Anon., Ber. chem. Ges. 55A (1923), 70

PRAGER, BERNHARD [1867-1934]
POGG 6,2068-2069
F. Richter, Ber. chem. Ges. 67A (1934), 166-167

PRANDTL, WILHELM [1878-1956]
TLK 3,507; JV 16,253; NUC 469,221-222; POTSCH, 349; POGG 5,1002-1003; 6,2070; 7a(3),620-621
R.E. Oesper, J. Chem. Ed. 26 (1949), 398-399
Anon., Chem. Z. 77 (1953), 220; 80 (1956), 788; Arch. Inter. Hist. Sci. 10 (1957), 91-92

PRAŠEK, EMIL [1884-1934]
FISCHER 2,1243; IB, 237
M. Prica, Lijecnički Vjestnik 1934, pp.55-57

PRATT, FREDERICK HAVEN [1873-1958]
DSB 11,125-126; AMS 9(II),897; NCAB 46,556

PRATT, HENRY SHERRING [1859-1946]
AMS 7,1419; IB, 237; WWAH, 490; WWWS, 1372
Anon., Science 104 (1946), 422

PRAUSNITZ, CARL [1876-1963]
KURSCHNER 4,2271; BHDE 2,924; LDGS, 58; WININGER 7,384; NUC 469,431
A.W. Downie, J. Path. Bact. 92 (1966), 241-252

PRAUSNITZ, WILHELM [1861-1933]
OBL 8,248; FISCHER 2,1243-1244; PAGEL, 1321-1322; TLK 3,507-508;DZ, 1127; KOREN, 232
E. Glaser, Wiener med. Wchschr. 83 (1933), 1190-1191

PRAZMOWSKI, ADAM [1853-1920]
OBL 8,250-251; PSB 28,376-378; KONOPKA 8,280; RSC 17,1001
B. Hryniewski, Acta Soc. Bot. Pol. 23 (1954), 217-227
W. Kunicki-Goldfinger, Acta Microbiol. Pol. 7 (1958), 251-258
J. Marszewska-Ziemecka and J. Golebiewska, Rev. Pol. Acad. Sci. 6(3) (1961), 25-35

PRECHT, HEINRICH [1852-1924]
BLOKH 2,588; POTSCH, 349-350; POGG 3,1065; 4,1190-1191
C. Przibylla, Z. angew. Chem. 35 (1922), 401-404

PRECHTL, JOHANN JOSEPH [1778-1854]
ADB 26,539; BLOKH 2,588-589; SCHAEDLER, 97; POGG 2,519-520
A. Schrötter, Alm. Akad. Wiss. Wien 6 (1856), 77-118

PREGL, FRITZ [1869-1930]
DSB 11,128-129; OBL 8,254; NOB 8,117-124; KALIN, 161-164; POTSCH, 350; IB, 238; W, 425-426; WWWS, 1372; POGG 5,1004; 6,2075; 7a(3),622
R. Wegscheider, Alm. Akad. Wiss. Wien 81 (1931), 317-320
R. Strebinger, Chem. Z. 55 (1931), 29-30; Ost. Chem. Z. 34 (1931), 10-11
H. Lieb, Ber. chem. Ges. 64A (1931); Mikrochemie 3 (1931), 105-116; Microchimica Acta 35 (1950), 123-129
S. Edlbacher, Z. angew. Chem. 44 (1931), 29-30
E. Philippi, Microchem. J. 6 (1962), 5-16

PREIS, KAREL [1846-1916]
OBL 8,256; STURM 3,304; POGG 3,1066
Anon., Chem. Z. 40 (1916), 401PREISLER, PAUL [1902-1971]

AMS 11,4258

PRELOG, VLADIMIR [1906-]
KURSCHNER 13,2989-2990; IWW 1982-1983, p.1064; WWWS, 1373; MH 2,428-429; STC 2,379-380; MSE 2,437-438; ABBOTT-C, 116-117; POTSCH, 350; POGG 6,2075-2076; 7a(3),624-626
V. Prelog, My 128 Semesters of Studies of Chemistry. Washington,D.C. 1991

PRENANT, AUGUSTE [1861-1927]
FISCHER 2,1245; IB, 238; WWWS, 1373
L. Lapicque, Bull. Acad. Med. 98 (1927), 342-344
P. Bouin, Arch. Anat. Micr. 26 (1930), 1-42

PRENANT, MARCEL [1893-1983]
CHARLE, 235-238; NUC 469,657-658

PRENTICE, BERTRAM [1867-1938]
JV 10,214
E. Clark, J. Chem. Soc. 1939, pp.219-220

PRESCOTT, ALBERT BENJAMIN [1832-1905]
DAB 15,192-193; MILES, 393-394; NCAB 13,53; ELLIOTT, 209; KELLY, 988-989; TSCHIRCH, 1106; WWAH, 490
J.H. Long, Proc. Am. Chem. Soc. 27 (1905), 76-78
O. Oldberg, Am. J. Pharm. 77 (1905), 251-255
E.D. Campbell, History of the Chemical Laboratory of the University of Michigan, pp.101-124. Ann Arbor 1916
H.R. Manasse, Pharm. Hist. 15 (1973), 22-28

PRESCOTT, JOHN MACK [1921-]
AMS 15(5),1071; WWA 1980-1981, p.2680; WWWS, 1373

PRESSMAN, DAVID [1916-1980]
AMS 14,4030; WWA 1978-1979, p.2617; WWWS, 1373-1374
Anon., Chem. Eng. News 58(38) (1980), 44

PRETTNER, AUGUST [1875-1931]
JV 19,342; NUC 470,500
Anon., Chem. Z. 55 (1931), 755

PREUSSER, FERDINAND [1878-]
JV 24,365; NUC 470,542

PREVOST, CHARLES PAUL [1899-]
WWWS, 1374; POGG 6,2077; 7b,4145-4148
PRÉVOST, JEAN LOUIS [1790-1850]
DSB 11,132-133; LAROUSSE 13,134; PICOT, 82-83; BOSSART, 33-35; WWWS,1374; HIRSCH 4,673; FERCHL, 423; POGG 2,527; CALLISEN 15,210-212; 31,302
H. Lebert, C.R. Soc. Biol. 2 (1850), 60-66
V. Brunner, Der Genfer Arzt Jean Louis Prévost. Zurich 1966
G. de Morsier, Gesnerus 23 (1966), 117-121
J. Théodoridès, Gesnerus 34 (1977), 82-89

PRÉVOST, JEAN LOUIS [1838-1927]
DHB 5,345; FISCHER 2,1246; STELLING-MICHAUD 5,239; POGG 6,2077-2078
M. Roch, Verhandl. Schw. Nat. Ges. 109 (1928), 39-47
G. de Morsier, Gesnerus 31 (1974), 19-38
R. Mayer, Rev. Med. Suisse Rom. 100 (1980), 1017-1020

PREYER, WILHELM [1841-1897]
DSB 11,135-136; ADB 53,116-119; NB, 311; FISCHER 2,1246; PAGEL,1323-1325; GIESE, 493-494; VERSO, 97-98; WWWS, 1375; POGG 3,1069-1070; 4,1192-1193
Anon., Science 6 (1897), 252-253
S. Fuchs, Wiener klin. Wchschr. 10 (1897), 703-704
A. Eulenburg, Jahrb. Nass. Ver. Nat. 51 (1898), xi-xii
M.L. Verso, Med. Hist. 15 (1971), 59-60
J. Carmichael, Persp. Biol. Med. 16 (1973), 411-417

PREYSZ, MÓRICA [1829-1877]
OBL 8,273-274; DIERGART, 268

PRIANISHNIKOV, DMITRI NIKOLAEVICH [1865-1948]
DSB 11,179-180; BSE 35,221-222; WWR, 465-466; KUZNETSOV(1963), 795-814; IB, 238; WWWS,1375
O.K. Kedrov-Zikhman, Usp. Khim. 8 (1939), 1-10
J.S. Joffe, Soil Science 66 (1948), 165-169
K. Mothes, Ber. bot. Ges. 68a (1955), 311-313

PRIBRAM, ALFRED [1841-1912]
OBL 8,274-275; STURM 3,312; FISCHER 2,1247; SCHIEBER,44-54; WININGER 5,95
A. Burdach, Deutsche med. Wchschr. 38 (1912), 1198-1199

PRIBRAM, BRUNO OSKAR [1887-]
KURSCHNER 4,2279; FISCHER 2,1247; ASEN, 152; IB, 238; LDGS, 83; WININGER 7,385

PRIBRAM, ERNST [1879-1940]
OBL 8,275; STURM 3,312; KURSCHNER 4,2280; FISCHER 2,1247-1248; WWWS, 1375; AMS 6,1138; RAZINGER, 147-148; KOREN, 232; WININGER 5,95-96; 7,385
Anon., J. Am. Med. Assn. 115 (1940), 1470

PRIBRAM, HUGO [1881-1943]
OBL 8,275-276; STURM 3,312; KURSCHNER 5,1060; FISCHER 2,1248; PELZNER, 56-64; WININGER 7,385

PRIBRAM, RICHARD [1847-1928]
OBL 8,276; STURM 3,313; DBJ 10,334; WININGER 5,96; POGG 3,1070; 4,1193; 5,1005; 6,2078
Anon., Chem. Z. 52 (1928), 47; Ost. Chem. Z. 31 (1928), 17-18
W. Schlenk, Ber. chem. Ges. 61A (1928), 47

PRICE, CHARLES COALE [1913-]
AMS 15(5),1077; WWA 1980-1981, p.2683; WWWS, 1375

PRICE, DAVID SIMPSON [1823-1888]
RSC 5,18; 8,662
Anon., J. Chem. Soc. 55 (1889), 294

PRICE, THOMAS SLATER [1875-1948]
SRS,19; POGG 4,1193; 5,1005; 6,2078
J. Kendall, Obit. Not. Fell. Roy. Soc. 7 (1952), 469-474

PRICE, WINSTON HARVEY [1923-1981]
AMS 13,3543-3544; WWWS, 1376

PRICE-JONES, CECIL [1863-1943]
CULE, 146
G.W. Goodhart, J. Path. Bact. 58 (1946), 301-309

PRICHARD, JAMES COWLES [1786-1848]
DSB 11,136-138; DNB 16,344-346; CULE, 146-149,221; WWWS, 1376; CALLISEN 15,218-219; 31,304-305
Earl of Rosse, Proc. Roy. Soc. 5 (1849), 886-888
E.B. Poulton, Science Progress NS1 (1897), 278-296

PRIESTLEY, JOHN GILLIES [1879-1941]
DSB 11,138-139; O'CONNOR(2), 103-105
C.G. Douglas, Nature 147 (1941), 319-320
G.E. Allen, J. Hist. Med. 22 (1967), 392-412

PRIESTLEY, JOSEPH HUBERT [1883-1944]
DESMOND, 505; SRS,31; WWWS, 1376; NUC 471,286
W.H. Pearsall et al., Nature 154 (1944), 694-695; Proc. Linn. Soc. 1943-1944, pp.231-232

PRIGGE, RICHARD [1896-1967]
KURSCHNER 10,1895-1896; FISCHER 2,1248-1249; KALLMORGEN, 375; POGG 7a(3),629-631
F. Klose, Arb. Ehrlich Inst. 52 (1956), 5-20
E. Eissner, Arzneimitt. 6 (1956), 231-232
W.H. Wagner, Arzneimitt. 16 (1966), 579-580
Anon., Chem. Z. 91 (1967), 163; Nachr. Chem. Techn. 15 (1967), 64
G. Heyman, Arzneimitt. 17 (1967), 350-351

PRIGOGINE, ILYA [1917-]
IWW 1982-1983, p.1068; POTSCH, 351

PRILEZHAYEV, NIKOLAI ALEKSANDROVICH [1872-1944]
BSE 34,510-511; WWR, 462; ARBUZOV, 189
A.A. Akhrem et al., Zhur. Ob. Khim. 21 (1951), 1925-1932

PRINGLE, HAROLD [1876-1935]
WW 1934, p.2698; O'CONNOR(2), 440
Anon., Lancet 1935(II), pp.1261,1381; Brit. Med. J. 1935(II), pp.1235-1236,1286

PRINGLE, JOHN WILLIAM SUTTON [1912-1982]
WW 1981,p.2104; WWWS, 1377
V. Wigglesworth, Biog. Mem. Fell. Roy. Soc. 29 (1983), 525-551

PRINGSHEIM, ERNST GEORG [1881-1970]
STURM 3,318; KURSCHNER 10,1897; BHDE 2,929; IB, 238; WININGER 5,99; WWWS, 1377; POGG 6,2081; 7a(3),632-633
E.G. Pringsheim, Ann. Rev. Microbiol. 24 (1970), 1-16; Med. Hist. J. 5 (1970), 125-137
A. Pirson, Ber. bot. Ges. 85 (1972), 651-659

PRINGSHEIM, HANS [1876-1940]
KURSCHNER 4,2283; LDGS, 60; TLK 3,509; KOREN, 232; TETZLAFF, 269; BHDE 2,929; POGG 5,1006; 6,2081-2082; 7a(3),633

PRINGSHEIM, NATHANIEL [1823-1894]
DSB 11,151-155; ADB 53,120-124; TSCHIRCH, 1107; WWWS, 1377; KOHUT 2,229-230; WININGER 5,99-100
F. Cohn, Ber. Bot. Ges. 13 (1895), (10)-(33)
D.H. Scott, Nature 51 (1895), 399-402
C. von Voit, Sitz. Bayer. Akad. 25 (1895), 180-183

PRINGSHEIM, PETER [1881-1963]
AMS 10,3251; BHDE 2,929-930; LDGS, 100; WININGER 7,386; TETZLAFF, 270; POGG 5,1006-1007; 6,2082-2083; 7a(3),633
W. Hanle, Phys. Bl. 12 (1956), 126-127

PRINS, HENDRIK JACOBUS [1889-1958]
POGG 6,2083; 7b,4150-4151
H. Gerding, Chem. Wkbl. 54 (1958), 401-403

PRINZ, OTTO [-1899]
Anon, Ber. chem. Ges. 32 (1899), 3706

PRITZKOW, ARTHUR [1866-1945]
JV 7,149; POGG 6,2084-2085; 7a(3),634
Anon., Chem. Z. 60 (1936), 798

PRITZKOW, WILHELM [1874-]
JV 14,155; NUC 472,90

PROBST, JOHANN MAX [1812-1842]
DRULL, 210; TSCHIRCH, 1107; POGG 2,533

PROCHOWNIK, LUDWIG [1851-1923]
FISCHER 2,1250-1251

PROCTER, HENRY RICHARDSON [1848-1927]
WW 1927, p.2417; POTSCH, 351-352; POGG 6,2086-2087
A.S., J. Chem. Soc. 1928, pp.3300-3307; Proc. Roy. Soc. A122 (1929), I-vi

PROCTER, WILLIAM [1817-1874]
DAB 15,242; DAMB, 610-611; TSCHIRCH, 1107
J.P. Remington, Am. J. Pharm. 72 (1900), 255

PRÖSCHER, FRIEDRICH (FREDERICK) [1875-1967]
AMS 10,3254; RSC 17,1028
Anon., Chem. Eng. News 46(10) (1968), 71

PROFFT, ELMAR [1905-1978]
KURSCHNER 12,2468; POGG 7a(3),636-637
Anon., Chem. Z. 102 (1978), 116

PROOST, WILLEM FREDERIK [1867-1939]
RSC 17,1030

PROPFE, ALEXANDER [1877-1914]
JV 15,133; NUC 472,64
Anon., Z. angew. Chem. 27(III) (1914), 621

PROSKAUER, BERNHARD [1851-1915]
DBJ 1,338; FISCHER 2,1252; BLOKH 2,591-592; BULLOCH, 391; KOREN, 232; WININGER 5,101-102
G. Sobernheim, Deutsche med. Wchschr. 41 (1915), 1224
H. Schelenz, Mitt. Gesch. Med. 14 (1915), 296

PROSSER, CLIFFORD LADD [1907-]
AMS 14,4047; WWA 1976-1977, p.2544; BROBECK, 201-205; WWWS, 1379
C.L. Prosser, Ann. Rev. Physiol. 48 (1986), 1-6

PROUST, JOSEPH LOUIS [1754-1826]
DSB 11,166-172; GE 27,830; LAROUSSE 13,320; BUGGE 1,350-355; WWWS, 1380; POTSCH, 352; BLOKH 2,592-593; SCHAEDLER, 98-99; FERCHL, 427-428; W, 428; ABBOTT-C, 118-119; POGG 2,536-538; CALLISEN 31,312
E. Färber, Z. angew. Chem. 34 (1921), 245-246
S.C. Kapoor, Chymia 10 (1965), 53-110

PROUT, WILLIAM [1785-1850]
DSB 11,172-174; DNB 16,426-427; HIRSCH 4,680-681; BLOKH 2,593; WWWS,1380; PHILIPPE, 709-710; FERCHL, 427; ABBOTT-C, 119; POTSCH, 352-353; COLE, 452-453; POGG 2,539; CALLISEN 15,237-243; 31,312-313
A.M. Kasich, Bull. Hist. Med. 20 (1946), 340-358
S. Glasstone, J. Chem. Ed. 24 (1947), 478-481
O.T. Benfey, J. Chem. Ed. 29 (1952), 78-81
W.H. Brock, J. Chem. Ed. 40 (1963), 652-653; Med. Hist. 9 (1965),101-126; Ann. Sci. 25 (1969), 127-137; Notes Roy. Soc. 24 (1970), 281-294; From Protyle to Proton. Bristol 1985
W.S. Copeman, Notes Roy. Soc. 24 (1970), 273-280
D.F. Larder, Centaurus 15 (1970), 44-50
R. Ahrens, J. Nutrition 107 (1977), 17-23

PROVENZAL, GIULIO [1872-1954]
DFI, 107-108

PROWAZEK, STANISLAUS [1875-1915]
DSB 11,174-175; DBJ 1,338; FISCHER 2,1252-1253; STC 2,383-384; OLPP, 330; SACKMANN, 274-276
M. Hartmann, Arch. Protist. 36 (1916), i-xix
V. Dyk and W. Eichler, Angewandte Parasitologie 19 (1978), 230-232

PRUDDEN, THEOPHIL MITCHELL [1849-1924]
DSB 11,175-177; DAB 15,252-253; NCAB 9,347-348; DAMB, 611-612; WWAH, 492; WWWS, 1380
L. Hektoen, Biog. Mem. Nat. Acad. Sci. 12 (1929), 73-98

PRUNIER, LOUIS LÉON [1841-1906]
VAPEREAU 6,1285; BOURQUELOT, 89-90; FISCHER 2,1253; PAGEL, 1327; GORIS, 545-546; TSCHIRCH, 1107; POGG 3,1074-1075; 4,1198
P. Yvon et al., J. Pharm. Chim. [6]24 (1906), 232-240
G. Dillemann, Produits et Problèmes Pharmaceutiques 28 (1973), 296-297

PRUNTY, FRANCIS THOMAS GARNET [1910-1979]
WW 1979, p.2039; WWWS, 1380

PRUSZYNSKI, JAN [1861-1918]
PSB 28,616-617; KONOPKA 8,315-318; RSC 17,1037
H. Nusbaum, Gazeta Lekarska [3]3 (1918), 437-442

PRYDE, JOHN [1897-1973]
SRS, 19; POGG 6,2088; 7b,4160
Anon., Biochem. Soc. Trans. 2 (1974), 589-590

PRYM, OSKAR [1873-1964]
KURSCHNER 5,1064; FISCHER 2,1254; WENIG, 232; JV 13,32; NUC 474,163

PRZIBRAM, HANS LEO [1874-1944]
OBL 8,314-315; NOB 13,184-191; FISCHER 2,1254-1255; BHDE 2,930; IB, 239; WININGER 5,103; TETZLAFF, 270; POGG 6,2089; 7a(3),638
W. Thompson, Nature 155 (1945), 782
Anon., Science 102 (1945), 217

PRZIBRAM, KARL [1878-1973]
TLK 3,511; POGG 5,1008-1009
B. Karlik, Alm. Akad. Wiss. Wien 124 (1975), 379-387

PRZYLECKI, STANISLAW [1891-1944]
PSB 29,201-204; WE 9,568; WARSAW, 149; IB, 239; POGG 6,2090; 7b,4161-4163
A. Sródka, Acta Physiol. Polon. 38 (1987), 158-167

PSCHORR, ROBERT [1868-1930]
TLK 3,511; POTSCH, 353; POGG 4,1198-1199; 5,1009; 6,2090
K.A. Hofmann, Ber. chem. Ges. 63A (1930), 108-110
P. Duden, Z. angew. Chem. 43 (1930), 245-245
O. Gerngross, Chem. Z. 54 (1930), 181

PTASHNE, MARK STEPHAN [1940-]
AMS 15(5),1097

PUCHER, GEORGE WALTER [1898-1947]
AMS 7,1427; POGG 6,2090; 7b,4163-4165
Anon., Science 106 (1947), 615-616; Chem. Eng. News 26 (1948), 178; Plant Physiology 23 (1948), 258-259

PUCK, THEODORE THOMAS [1916-]
AMS 15(5),1098; WWA 1980-1981, p.2695; IWW 1982-1983, pp.1071-1072; MH 2,432-433; MSE 2,443-445; NCAB 63,79-80

PÜTTER, AUGUST [1879-1929]
KURSCHNER 3,1854-1855; DRULL, 210-211; FISCHER 2,1255; WAGENITZ, 140; KIEL, 86; IB, 239
F.W. Frölich, Deutsche med. Wchschr. 55 (1929), 587-588
E. Wöhlisch, Erg. Physiol. 28 (1929), 690-692

PUGH, EVAN [1828-1864]
DAB 15,257-258; MILES, 395-396; NCAB 11,320; ELLIOTT, 210-211; SILLIMAN, 59-63; WWAH, 492
Anon., Am. J. Sci. [2]38 (1864), 301-302
C.A. Browne, J. Chem. Ed. 7 (1930), 499-517
P.W. Wilson, Bact. Revs. 22 (1958), 143-144

PUGH, WILLIAM [1897-1955]
POGG 6,2091; 7b,4165
A.H. Spong, J. Chem. Soc. 1955, pp.3566-3567

PUGIN, (JOHANN) MICHAEL HEINRICH [1874-]
JV 17,158; NUC 475,132

PUKALL, WILHELM [1860-1937]
POGG 6,2091-2092
K. Pukall, Ber. chem. Ges. 70A (1937), 147-148

PULEWKA, PAUL [1896-1989]
KURSCHNER 13,3009; BHDE 2,931; WIDMANN, 156-157,281-282; LDGS, 78; IB, 239; POGG 7a(3),640-641

Anon., Chem. Z. 80 (1956), 144
W. Schmid, Arzneimitt. 11 (1961), 134
P. Pulewka, Therapie der Gegenwart 119 (1980), 199-211

PULFRICH, CARL [1858-1927]
DSB 11,207-209; WWWS, 1382; POGG 4,1199-1200; 5,1010-1011; 6,2092; 7aSuppl.,517
F. Lowe, Z. Instrum. 47 (1927), 561-567

PULVERMACHER, GEORG [1866-]
JV 2,24; NUC 475,308

PUMMERER, RUDOLF [1882-1973]
KURSCHNER 11,2326; TLK 3,511; POGG 5,1011; 6,2093; 7a(3),641-642
Anon., Nachr. Chem. Techn. 5 (1957), 216
G. Hesse, Bayer. Akad. Wiss. Jahrbuch 1974, pp.217-220

PUNNETT, REGINALD CRUNDALL [1875-1967]
DSB 11,211-212; WW 1965, p.2493; WWWS, 1382
F.A.E. Crew, Biog. Mem. Fell. Roy. Soc. 13 (1967), 309-326; Genetics 58 (1968), 1-7

PUPPE, GEORG [1867-1925]
FISCHER 2,1256; AD 3,297
F. Strassmann, Deutsche med. Wchschr. 51 (1925), 2129

PURDIE, THOMAS [1843-1916]
DSB 18,729; BLOKH 2,593-594; WWWS, 1383; POGG 4,1201-1202; 5,1011
J.C. Irvine, J. Chem. Soc. 111 (1917), 359-369
P.F.F., Proc. Roy. Soc. A101 (1922), iv-x

PURGOTTI, SEBASTIANO [1799-1879]
PROVENZAL, 147-150; FERCHL, 428; POGG 3,1077-1078

PURKYNĚ, JAN EVANGELISTA [1787-1869]
DSB 11,213-217; ADB 26,717-731; 28,808; OBL 8,339; HIRSCH 4,688-689; PAGEL, 1328-1329; NAVRATIL, 251-255; FREUND 2,299-309; W, 419-430; KOERTING, 114-116; WERSTLER, 34-59; SKULINA,115-119; ABBOTT,107-108; HAYMAKER, 254-258; STURM 3,358-359; POGG 2,544; 3,1078; CALLISEN 15,264-266; 31,321
Anon., Proc. Roy. Soc. 19 (1871), ix-xii
H. Winterstein, Schlesische Lebensbilder 4 (1931), 240-251
H.J. John, Jan Evangelista Purkyně. Phiakdelphia 1959
V. Kruta, Physiol. Bohem. 7 (1958), 1-8; J.E, Purkyně, Physiologist. Prague 1969; Clio Medica 6 (1971), 109-120
M. Teich, Zeitschrift für Slawistik 5 (1960), 87-100
B. Tichacek et al., Physiol. Bohem. 36 (1987), 129-202
W. Coleman, in The Investigative Enterprise (W. Coleman and F.L. Holmes,Eds.), pp.15-64. Berkeley, Cal. 1988
J. Purs (ed.), Jan Evangelista Purkyně in Science and Culture. Prague 1988

PURRMANN, ROBERT [1914-]
POGG 7a(3),643PURVES, CLIFFORD BURROUGH [1902-1965]
AMS 10,3264; WWAH, 493
C.A. Winkler, Proc. Roy. Soc. Canada [4]4 (1966), 111-115
A.S. Berlin, Adv. Carbohydrate Chem. 23 (1968), 1-10

PURVIS, JOHN EDWARD [1862-1930]
POGG 5,1012; 6,2095
W.H.M., J. Chem. Soc. 1931, pp.3380-3382

PUTNAM, FRANK WILLIAM [1917-]
AMS 15(5),1106; WWA 1980-1981, p.2699

PUTZEYS, FÉLIX [1847-1932]
NUC 476,127
F. Schoofs, Liège Medical 25 (1932), 612-613; Bull. Acad. Med. Belg. [5]12 (1932), 379-385; Liber Memorialis Université de Liège 1867-1935, vol.3, pp.66-71. Liège 1936

PYL, GOTTFRIED [1897-1956]
POGG 7a(3),644-645

PYMAN, FRANK LEE [1882-1944]
WW 1944,p.2252; ABBOTT-C, 119-120; POGG 5,1013; 6,2097-2098; 7b,4177-4178
H. King, Obit. Not. Fell. Roy. Soc. 4 (1944), 681-697
F.H. Carr and T.A. Henry, J. Chem. Soc. 1944, pp.563-570
Anon., Chemical Age 50 (1944), 86-87; Chem. Ind. 1944, p.54
L. Anderson, Biochem. J. 38 (1944), 283-284

Q

QUADRAT, BERNHARD [1821-1895]
OBL 8,350-351; STURM 3,361-362; POGG 2,547; 3,1079

QUADRAT, OTTOKAR [1886-1963]
STURM 3,362; POGG 6,2098; 7b,4178-4179

QUAGLIARIELLO, GAETANO [1883-1957]
IB, 240; POGG 6,2098-2099; 7b,4179-4181
F. Cedrangolo, Enzymologia 19 (1958), 201-210
A. Rossi-Fanelli, Atti Accad. Lincei [8]26 (1959), 298-306
A. Orru, Atti Accad. Pontiniana NS9 (1959), 363-370

QUAGLIO, JULIUS [1833-1899]
E. Renatus, Chemie in Unserer Zeit 17(3) (1983), 96-102

QUAST, PAUL [1894-]
KURSCHNER 4,2293; WENIG, 233; LDGS, 55

QUASTEL, JUDA
HIRSCH [1899-1987] WW 1981, p.2117; IWW 1982-1983, p.1076; SRS, 84; MSE 2,447-448; YOUNG, 76-77; COHEN, 189; WWWS, 1385; POGG 6,2099; 7b,4183-4189
J.H. Quastel, Can. J. Biochem. 52 (1974), 71-82; TIBS 5 (1980),199-200; 8 (1983), 103-104; 9 (1984), 117-118; Bull. Can. Biochem. Soc. 18 (1981), 13-34; 25 (1987), 5-6; Selected Topics in the History of Biochemistry (G. Semenza, Ed.), pp.135-187. Amsterdam 1983
S.C. Sung, TIBS 4 (1979), N101-N102
H.F. Bradford, The Biochemist 10(2) (1988), 4-7
F.C. MacIntosh and T.L. Sourkes, Biog. Mem. Fell. Roy. Soc. 36 (1990), 381-418

QUATREFAGES DE BRÉAU, ARMAND de [1810-1892]
DSB 11,233-235; VAPEREAU 6,1287; HIRSCH 4,697-698; WWWS, 445-446
A.E. Malard, Nouv. Arch. Hist. Nat. [3]4 (1892), i-xlix
G. Malloizel, Armand de Quatrefages de Bréau. Autun 1893
G. Herve and L. de Quatrefages, Bull. Soc. Hist. Med. 20 (1926),209-330; 21 (1927), 17-35,200-231

QUEDENFELDT, EDWIN THEODOR [1869-]
JV 11,168; NUC 476,629

QUELET, RAYMOND PIERRE [1897-1967]
POGG 6,2100; 7b,4191-4194

QUENTEL, EDUARD [1823-1865]
W. Fugman, Die Deutschen in Parana, p.269. Curityba 1929

QUESNEVILLE, GUSTAVE AUGUSTIN [1810-1889]
BLOKH 2,594; TSCHIRCH, 1107; FERCHL, 429; SCHAEDLER, 100; POGG 2,551; 3,1079; 4,1203
Anon., Mon. Sci. [4]3 (1889), 1401-1404

QUESNEVILLE, GUSTAVE GEORGES [1846-1927]
TSCHIRCH, 1107; RSC 18, 6; POGG 3,1079-1080; 4,1203; 6,2101
G.M., J. Pharm. Chim. [8]5 (1927), 140

QUEVENNE, THÉODORE AUGUSTE [1805-1855]
HIRSCH 4,699-700; GORIS, 547; DECHAMBRE 80,182; CALLISEN

31,330-331
A. Bouchardat, J. Pharm. Chim. [3]31 (1857), 53-63

QUICK, ARMAND JAMES [1894-1978]
AMS 12,5064; NCAB 61,38-39; WWWS, 1385
H.F. Hardman and C.O. Haavik, Pharmacologist 21 (1979), 386
K.M. Brinkhous, Thromb. Diath. Haemorrh. 40 (1979), 267-269
J.H. Dirckx, Annals of Internal Medicine 92 (1980), 553-558
A.V. Pisciotta, J. Lab. Clin. Sci. 115 (1990), 388-389

QUINCKE, GEORG HERMANN [1834-1924]
DSB 11,241-242; DRULL, 211; BLOKH 2,594-595; DZ, 1136-1137; TLK 2,551; SCHAEDLER, 100; WWWS, 1386; POGG 2,553-554; 3,1080-1081; 4,1203-1204; 5,1015; 6,2101-2102
A.S., Proc. Roy. Soc. A105 (1924), xiii-xv
W. Koenig, Naturwiss. 12 (1924), 621-627

QUINCKE, HEINRICH [1842-1922]
DBJ 4,215-218,366; FISCHER 2,1260-1261; PAGEL, 1337-1338; KALLMORGEN,376; HAYMAKER, 499-503; WWWS, 1386
B. Naunyn, Arch. exp. Path. Pharm. 93 (1922), i-iii
G. von Bergmann, Z. klin. Med. 96 (1923), 1-22
G. Goldschmid, Schw. med. Wchschr. 75 (1945), 973-978
H. Bethe, Heinrich Quincke. Neumünster 1968

QUINQUAUD, CHARLES EUGÈNE [1841-1894]
FISCHER 2,1261; WWWS, 1386
A.Robin, Bull. Acad. Med. 31 (1894), 73-75

QUINTON, RENÉ [1866-1925]
J. de Gaultier, Mercure de France 1925, pp.695-702
C.L. Julliot, René Quinton. Paris 1926
J. Deguilly, Vie en Champagne 23 (1975), 9-14; 24 (1976), 9-13

QUITMANN, EUGEN [1884-]
JV 26,643; NUC 477,664

R

RAAB, WILHELM [1895-1970]
AMS 11,4302; KURSCHNER 10,1913; DAMB, 621-622; BHDE 2,932-933; IB, 240
M.J. Halhuber, Med. Klin. 66 (1971), 1318-1320

RAABE, FELIX [1877-1955]
JV 25,392; NUC 478,85

RAADTS, JOSEPH [1888-1945]
JV 30,665; NUC 478,109

RABATÉ, JACQUES [1907-1941]
POGG 6,2103; 7b,4201-4202
J.E. Courtois, Actualités Chimiques 1978(4), pp.47-48

RABAUD, JULES ETIENNE [1868-1956]
CHARLE(2), 245-247; BUICAN, 139-169; IB, 240

RABE, ARTUR [1889-]
JV 36,625; NUC 478,168

RABE, FRITZ [1884-1977]
KURSCHNER 12,2482; FISCHER 2,1261RABE, PAUL [1869-1952]
KURSCHNER 7,1607; STURM 3,367; POGG 4,1204; 5, 1015; 6,2103; 7a(3),656
H. Albers and W. Hochstätter, Chem. Ber. 99 (1966), xci-cxi

RABE, WILHELM OTTO [1873-]
JV 14,227; RSC 15,897; NUC 478,175

RABINERSON, ALEKSANDR IGNATIEVICH [1896-1942]
IB, 240; POGG 6,2103; 7b,4202-4203
G.I. Fuks, Kolloidnoi Zhurnal 10 (1948), 69-71

RABINOVICH, ADOLF IOSIFOVICH [1893-1942]
BSE 35,418; WWR, 468-469; POGG 6,2103-2104; 7b,4203-4205
Anon., Acta Physicochimica URSS 17 (1942), 125-126
V.A. Kargin, Bull. Acad. Sci. URSS 1943(2), pp.81-86

RABINOWITCH, EUGENE [1901-1973]
AMS 11,4305; BHDE 2,935; WWWS, 1387
Anon., Physics Today 26(8) (1973), 78-79

RABINOWITCH, ISRAEL MORDECAI [1890-1983]
L.H. Bensley, Clin. Chem. 33 (1987), 196-197

RABINOWITSCH-KEMPNER, LYDIA [1871-1935]
FISCHER 2,1262; BOEDEKER, 93; BHDE 1,360; WWWS, 1387; KOREN, 233; WININGER 5,120; 7,387; TETZLAFF, 271
Anon., Tubercle 16 (1935), 567; Deutsche med. Wchschr. 61 (1935), 1336

RABINOWITZ, JESSE CHARLES [1925-]
AMS 15(6),14

RABINOWITZ, JOSEPH LOSHAK [1923-]
AMS 15(6),14; WWWS, 1387

RABINOWITZ, MURRAY [1927-1983]
AMS 15(6),14
A.H. Rubenstein, Trans. Assoc. Am. Phys. 97 (1984), cxxvi-cxxviii

RABL, CARL [1853-1917]
DSB 11,254-256; OBL 8,361; DBJ 2,668; FISCHER 2,1262; KOERTING, 102-103; STURM 3,368; STOBER, 76-81; LOMMATZSCH, 12-17
A. Fischel, Anat. Anz. 51 (1918), 54-79
H. Held, Sitz. Akad. Wiss. Leipzig 70 (1918), 363-378
F. Hochstetter, Alm. Akad. Wiss. Wien 68 (1918), 260-274

RABL, HANS [1868-1936]
OBL 8,360; HEINDEL, 31-38
A. Pischinger, Anat. Anz. 84 (1937), 272-285
J. Schaffer, Alm. Akad. Wiss. Wien 87 (1937), 238-244

RABOW, SIEGFRIED [1848-1931]
FISCHER 2,1263

RACE, ROBERT RUSSELL [1907-1984]
WW 1983, p.1835; NUC 478,329-330
C. Clarke, Biog. Mem. Fell. Roy. Soc. 31 (1985), 455-492

RACHFORD, BENJAMIN KNOX [1857-1929]
NUC 478,344
M. Dreyfoos, Cincinnati Journal of Medicine 10 (1929), 228-233
Anon., American Journal of Diseases of Children 38 (1929), 141-142

RACIBORSKI, MARIAN [1863-1917]
OBL 8,363-364; PSB 29,598-601; WE 9,645-646; KONOPKA 8,462
K. Goebel, Ber. bot. Ges. 35 (1917), (97)-(107)

RACKE, FRITZ [1897-1945]
JV 36,704; NUC 478,462

RACKER, EFRAIM [1913-1991]
AMS 15(6),16; WWA 1980-1981, p.2708; IWW 1982-1983, p.1081; BHDE 2,935; MH 2,434-435; MSE 3,2-3
E. Racker, Reflections on Biochemistry (A. Kornberg et al., Eds.), pp.39-44. Oxford 1976; Of Oxygen, Fuels and Living Matter, Part I (G. Semenza, Ed.), pp.265-273. Chichester 1981
G. Hause, Z. physiol. Chem. 373 (1992), (3)-(5)
N. Nelson, Photosynthesis Research 31 (1992), 165-166

RACKY, GEORG [1889-]
JV 29,730; NUC 478,474

RACZYNSKI, JAN [1865-1918]
OBL 8,366; PSB 29,639-640; KONOPKA 8,463-465; WWWS, 1387
P. Groer, Pol. Med. Sci. 5 (1962), 152-154

RADEMACHER, FERDINAND [1871-1932]
STURM 3,370-371

RADEMACHER, JOHANN GOTTFRIED [1772-1850]
ADB 27,116-118; WL 3,405-421; HIRSCH 4,708-709; PAGEL, 1341-1342; FERCHL, 430; DECHAMBRE [3]1,717; CALLISEN 15,295-298; 31,335-336
F. Oehmann, Johann Gottfried Rademacher. Bonn 1900
H. Paal, Johann Gottfried Rademacher. Jena 1932
N. Krack, Dr. Johann Gottfried Rademacher. Heidelberg 1984

RADEMAKER, GYSBERTUS GODEFRIEDUS JOHANNES [1887-1957]
LINDEBOOM, 1584; FISCHER 2,1263
H. Verbiest, Ned. Tijd, Gen. 101 (1957), 849-851

RADENHAUSEN, RUDOLF [1871-1943]
JV 9,146; NUC 478,578

RÁDL, EMMANUEL [1873-1942]
OBL 8,377-378; STURM 3,373-374; IB, 241

RADLKOFER, LUDWIG [1829-1927]
KURSCHNER 2,1512; BK, 290; TSCHIRCH, 1108; IB, 241; RSC 18,22
Anon., Alm. Bayer. Akad. Wiss. 1909, pp.314-322
T. Herzog, Ber. bot. Ges. 45 (1927), (79)-(88)
K. von Goebel, Bayer. Akad. Wiss. Jahrbuch 1927, pp.26-29

RADULESCU, DAN [1884-1969]
POGG 6,2106-2107; 7b,4208-4209

RADZIEJEWSKI, ZYGMUNT [1841-1874]
PSB 30,44-45; HIRSCH 4,710; KOSMINSKI, 410-411; BLOKH 2,595; KONOPKA 8,477
C. Salkowski, Ber. chem. Ges. 7 (1874), 1801-1802

RADZISZEWSKI, BRONISLAW [1838-1914]
OBL 8,384; PSB 30,114-118; WE 9,677; KOSMINSKI, 411; BLOKH 2,595-596; KONOPKA 8,478-483; TSCHIRCH, 1108; POGG 3,1083; 4,1206; 5,1017
S. Opolski, Kosmos 33 (1910), 437-447
W. Will, Ber. chem. Ges. 47 (1914), 1521
M. Wroblewska, Wiad. Chem. 30 (1976), 215-222

RAEHLMANN, EDUARD [1848-1917]
DBJ 2,668; FISCHER 2,1264; LEVITSKI 2,90-91
J. Hirschberg, Cent. prakt. Augenheil. 41 (1917), 134-141
E. Ischreyt, Klin. Mon. Augenheilk. 60 (1918), 272-274

RAEVSKI, ARKADI ALEKSANDROVICH [1848-1916]
A.A. Zvorykin, Biograficheski Slovar Deyateli Esteestvoznania i Tekhniki 2,161-162. Moscow 1959

RAEVSKI, IGOR SVYATOSLAVOVICH [1854-1879]
A.D. Nekrassov, Trudy Inst. Ist. Est. 5 (1953), 275-280

RAFFEL, SIDNEY [1911-]
AMS 17(6),23
S. Raffel, Ann. Rev. Microbiol. 36 (1982), 1-26

RAGOZIN, VIKTOR IVANOVICH [1833-1901]
BSE(3rd Ed.) 21,329
Anon., Pharm. Z. 46 (1901), 692

RAHM, FRITZ [1878-]
JV 20,283; NUC 479,344

RAHN, HERMANN [1912-1990]
AMS 15(6),25; IWW 1982-1983, p.1083; WWWS, 1389
Anon., Physiologist 33 (1990), 181

RAHN, OTTO [1881-1957]
AMS 9(II),909; NCAB 49,586-587; KIEL, 219; IB, 241; WWAH, 495; POGG 6,2108; 7b,4210-4211
Anon., Science 126 (1957), 744,968

RAHNENFÜHRER, CARL [1859-1921]
RSC 18,31
Anon., Chem. Z. 45 (1921), 784; Z. angew. Chem. 34 (1921), 424

RAIFORD, LEMUEL CHARLES [1872-1944]
AMS 7,1436; POGG 6,2108; 7b,4211-4212
G. Glockler, Science 99 (1944), 72-73
S.E. Hazlet, Journal of Organic Chemistry 10 (1945), 87-100

RAIKOV, PENTCHO NIKOLOV [1864-1940]
POGG 6,2108-2109; 7b,4214-4215
J. Colakov et al., Chimija i Industrija 19 (1941), 329-395
S. Tchorbadjev, Chymia 12 (1967), 171-181

RAINEY, GEORGE [1801-1884]
DNB 16,621-622; SG [1]11,998; [2]14,282
W. Wagstaffe, St. Thomas's Hospital Reports NS22 (1894), xxiii-xl
P. Wildy, St. Thomas's Hospital Gazette 55 (1957), 91-94

RAISTRICK, HAROLD [1890-1971]
DNB [1971-1980], 698-699; WW 1965, pp.2511-2512; POTSCH, 353
J.H. Birkinshaw, Biog. Mem. Fell. Roy. Soc. 18 (1972), 489-509
J.C. Dacre, TIBS 8 (1983), 70-71

RAIZISS, GEORGE [1884-1945]
AMS 7,1437; IB, 241; POGG 6,2109-2110; 7b,4213-4214

RAJEWSKI, BORIS [1893-1974]
KURSCHNER 11,2345; POGG 6,2110; 7a(3),661-662
R. Bauer, Deutsche med. Wchschr. 88 (1963), 1493-1494
Anon., Deutsche med. Wchschr. 99 (1974), 2594

RAKE, GEOFFREY WILLIAM [1904-1958]
AMS 9(II),910; WWAH, 495; WWWS, 1390
H. Blank, Arch. Dermatol. 78 (1958), 511

RAKESTRAW, NORRIS WATSON [1895-1982]
AMS 11,4317; IB, 241; WWWS, 1390
W.F. Kieffer, J. Chem. Ed. 60 (1983), 108

RAKOVSKI, ADAM VLADISLAVOVICH [1879-1941]
BSE 36,1; POTSCH, 353-354; POGG 6,2110; 7b,4215-4216
Y.I. Gerasimov and D.N. Tarasenkov, Usp. Khim. 11 (1942), 75-77

RAKUSIN, MOISEI ABRAMOVICH [1869-1932]
NUC 479,621; POGG 6,2111-2112; 7b,4216
J. Tausz and A. Rabl, Chem. Z. 56 (1932), 341
M. Bodenstein, Ber. chem. Ges. 65A (1932), 77

RALL, JOSEPH EDWARD [1920-]
AMS 15(6),31; WWA 1980-1981, p.2713

RAMACHANDRAN, GOPALASAMUDRAM NARAYANA [1922-]
WW 1981, p.2129; IWW 1982-1983, p.1085

RAMAGE, HUGH [1865-1938]
W. Lincolne-Sutton, J. Chem. Soc. 1938, pp.1135-1136
Anon., Chemical Age 38 (1938), 330; J. Roy. Inst. Chem. 1938, p.285

RAMAN, CHANDRASEKHAR VENKATA [1888-1970]
DSB 11,264-267; DNB [1961-1970], 864-866; DNBI 3,470-475; W, 588-589; POTSCH, 354; POGG 5,1018; 6,2112-2113; 7b,4217-4222
S. Bhagavantam, Biog. Mem. Fell. Roy. Soc. 17 (1971), 565-592
G. Venkataraman, Journey into Light, Life and Science. Bangalore 1988

RAMART-LUCAS, PAULINE [1880-1953]
CHARLE(2), 245-247; POGG 5,1018; 6,2114-2115; 7b,4224-4226
Anon., Ann. Univ. Paris 23 (1953), 225-230
P. Denis and M. Martynoff, Bull. Soc. Chim. 1954, pp.269-280

RAMBERG, LUDWIG [1874-1940]
SMK 6,205; POGG 5,1019-1020; 6,2115; 7b,4226-4227
A. Fredga, Sven. Kem. Tid. 52 (1940), 331-334; Lev. Sven. Vet. 7 (1944-1948), 535-560
K. Myrbäck, Ber. chem. Ges. 74A (1941), 109

RAMMELSBERG, KARL FRIEDRICH [1813-1899]
DSB 11,270-271; BJN 4,172; HEIN, 511-512; BLOKH 2,597-599; WWWS, 1391; FERCHL, 430-432; TSCHIRCH, 1108; SCHAEDLER, 100; WREDE, 152; POTSCH, 354; POGG 2,562-563; 3,1085-1086; 4,1209; 7aSuppl.,520-521
H.A. Miers, J. Chem. Soc. 79 (1901), 1-43
H. Landolt, Ber. chem. Ges. 42 (1910), 4942-4970

RAMON, GASTON [1886-1963]
DSB 11,271-272; STC 2,397-398; IB, 242; WWWS, 1392
C. Gernez-Rieux, Bull. Acad. Med. 147 (1963), 610-620
Anon., Ann. Inst. Pasteur 105 (1963), 809-812
T. Monod, Not. Acad. Sci. 5 (1972), 252-297
E. Gilbrin, Hist. Sci. Med. 18 (1984), 53-60
J. Théodoridès, Hist. Sci. Med. 20 (1986), 475-485

RAMON y CAJAL, SANTIAGO [1852-1934]
DSB 11,273-276; GROTE 5,131-175; FREUND 2,311-325; FISCHER 2,1265-1266; PAGEL, 1343-1344; HAYMAKER, 147-151; BRAZIER, 143-444,149; IB, 242; GRUETTER, 86-89; W, 431-432; ABBOTT, 108-109; WWWS, 1392
C.S. Sherrington, Obit. Not. Fell. Roy. Soc. 1 (1935), 425-441
U. D'Ancona, Riv. Biol. 19 (1935), 208-220
J.F. Tello, Anat. Anz. 80 (1935), 46-75
S. Ramon y Cajal, Recollections of my Life. Philadelphia 1937
E.H. Huntress, Proc. Am. Acad. Arts Sci. 81 (1952), 81-84
A.D. Loewy, Persp. Biol. Med. 15 (1971), 7-36
V. Hamburger, Persp. Biol. Med. 23 (1980), 600-616
H.H. Simmer, Med. Hist. J. 16 (1981), 414-423
F. Reinoso-Suarez et al., in Ramon y Cajal's Contribution to the Neurosciences (S. Grisolia et al., Eds.), pp.3-31. Amsterdam 1983

RAMSAY, HENRIK [1886-1951]
NUC 480,406
Kuka Kukin On 1950, p.597

RAMSAY, WILLIAM [1852-1916]
DSB 11,277-284; DNB [1912-1921],444-446; FINDLAY, 146-175; W, 432- 433; BUGGE 2,250-263; FARBER, 999-1012; BLOKH 2,599-601; MATSCHOSS, 214; POTSCH, 354-355; ABBOTT-C,120-121; POGG 3,1087; 4,1209-1210; 5,1020-1021; 6,2117
F.G. Donnan, J. Chem. Soc. 111 (1917), 369-376
J.N.C., Proc. Roy. Soc. A93 (1917), xliii-liv
W.A. Tilden, Sir William Ramsay. London 1918
R.B. Moore, J. Franklin Inst. 186 (1918), 29-55
C. Moureau, Bull. Soc. Chim. [4]25 (1919), 401-426
E.H. Huntress, Proc. Am. Acad. Arts Sci. 81 (1952), 84-88
M.W. Travers, A Life of Sir William Ramsay. London 1956

RAMSDEN, WALTER [1868-1947]
WW 1945, p.2256; O'CONNOR(2), 85; IB, 242; POGG 6,2117; 7b,4229
Anon., Chem. Ind. 1947, p.183
R. Coope, Nature 159 (1947), 801
R.A. Peters, Biochem. J. 42 (1948), 321-322

RAMSDEN, WILLIAM BATES [1876-1906]
Anon., Brit. Med. J. 1906(II), p.55; Lancet 1906(II), p.56
J. Pagel, Virchows Jahresber. 1907, p.485

RANDALL, HARRISON McALLISTER [1870-1969]
AMS 10,3291; WWWS, 1393; POGG 5,1021; 6,2118; 7b,4230-2432
O. Laporte and D.M. Dennison, J. Opt. Soc. Am. 60 (1970), 429-431

RANDALL, JOHN HERMAN [1899-1980]
WWA 1978-1979, p.2653; AO 1980, pp.749-750
J.A. Anton (Ed.), Naturalism and Historical Understanding. Albany, N.Y. 1967
P.O. Kristeller, Renaissance Quarterly 34 (1981), 298-299
P.O. Kristeller et al., J. Hist. Ideas 42 (1981), 489-501

RANDALL, JOHN TURTON [1905-1984]
WW 1981, pp.2133-2134; IWW 1982-1983, p.1088; AO 1984, pp.343-344
M.H.F. Wilkins, Biog. Mem. Fell. Roy. Soc. 33 (1987), 493-535

RANDALL, LOWELL ORLANDO [1910-]
AMS 14,4099; WWWS, 1393

RANDALL, MERLE [1888-1950]
AMS 8,2016; NCAB 39,200; WWAH, 496; POGG 6,2118; 7b,4232-4233
Anon., Chem. Eng. News 28 (1950), 1136; Science 111 (1950), 395

RANDLE, PHILIP JOHN [1926-]
WW 1981, p.2134; WWWS, 1394

RANDOIN, LUCIE [1888-1960]
POGG 6,2119; 7b,4233-4238
J. Raoul, Bull. Soc. Chim. Biol. 43 (1961), 15-17
J. Fabianek and P. Fournier, J. Nutrition 91 (1967), 3-8

RANEY, MURRAY [1885-1966]
DAVIS, 491-503; POTSCH, 355
D.S. Tarbell and A.T. Tarbell, J. Chem. Ed. 54 (1977), 26-28

RANKE, HEINRICH von [1830-1909]
BJN 14,74*; FISCHER 2,1267-1268; DUELUND, 104-112; AD 3,132

RANKE, JOHANNES [1836-1916]
DBJ 1,366; LF 4,309-311; PAGEL, 1345; DZ, 1144
Anon., Alm. Bayer. Akad. Wiss. 1909, pp.322-328
J. Rückert, Bayer. Akad. Wiss. Jahrbuch 1917, pp.67-77

RANKE, OTTO FRIEDRICH [1899-1959]
KURSCHNER 8,1848; BERWIND, 171-177; POGG 7a(3),672-673
W.D. Keidel, Deutsche med. Wchschr. 85 (1960), 1020-1022; Erg. Physiol. 51 (1961), 21-37

RANKINE, WILLIAM JOHN MACQUORN [1820-1872]
DSB 11,291-295; DNB 16,733-735; BLOKH 2,601; MATSCHOSS, 284-285; WWWS, 1394; POGG 2,568-569; 3,1097-1088
Anon., Proc. Roy. Soc. 21 (1873), i-iv
E.E. Daub, Isis 58 (1967), 293-303; 61 (1970), 105-106
K. Hutchinson, Ann. Sci. 30 (1973), 341-364; Brit. J. Hist. Sci. 14 (1981), 1-26
V.V. Raman, J. Chem. Ed. 50 (1973), 274-276
D.F. Channel, in Beyond History of Science (E. Garber, ed.), pp.194-203.
Bethlehem, Pa. 1990

RANSON, STEPHEN WALTER [1880-1942]
DAB [Suppl.3],619-620; WWAH,498; WWWS, 1394
L.B. Arey, Anat. Rec. 86 (1943), 611-618
F.R. Sabin, Biog. Mem. Nat. Acad. Sci. 23 (1945), 365-397

RANVIER, LOUIS ANTOINE [1835-1922]
DSB 11,295-297; HIRSCH 4,722, GUIART, 222
J. Nageotte, C.R. Soc. Biol. 36 (1922), 1144-1153
F. Sireday, Bull. Acad. Med. 87 (1922), 344-348
J. Jolly, Arch. Anat. Micr. 19 (1923), i-lxxii

RANWEZ, FERNAND [1866-1925]
TSCHIRCH, 1108; SCHELENZ, 822

F. Schoofs, Bull. Acad. Med. Belg. [5]7 (1927), 463-476

RAOULT, FRANÇOIS MARIE [1830-1901]
DSB 11,297-300; GE 28,139; BLOKH 2,601-602; W, 434; WWWS, 1395; POTSCH, 355; POGG 3,1089; 4,1212
H.C. Jones, Science 13 (1901), 881-883
W.R., Nature 64 (1901), 17-18
J.H. van't Hoff, J. Chem. Soc. 81 (1902), 969-981
F.H. Getman, J. Chem. Ed. 13 (1936), 153-156

RAPAPORT, MAX [1889-1915]
JV 30,519; NUC 481,359
Anon., Chem. Z. 39 (1915), 434; Z. angew. Chem. 28(III) (1915), 332

RAPAPORT, SAMUEL [1912-]
KURSCHNER 12,2495; BHDE 2,940; POTSCH, 355; POGG 7a(3),674-676

RAPER, HENRY STANLEY [1882-1951]
WW 1949, p.2299; SRS, 32; FISCHER 2,1269-1270; O'CONNOR(2), 348- 350; POGG 6,2122; 7b,4242-4243
Anon., Brit. Med. J. 1951(I),pp.1527-1528; Lancet 1951(I),pp.1229-1230
W. Schlapp, Nature 169 (1952), 177-178
L.P. Kendal, Biochem. J. 52 (1952), 353-356
P. Hartley, Obit. Not. Fell. Roy. Soc. 8 (1953), 567-582
S.J. Folley, J. Chem. Soc. 1955, pp.2987-2988

RAPER, JOHN ROBERT [1911-1974]
AMS 11,4333; WWA 1974-1975, p.2527; MH 2,438-440; STC 2,399-402; MSE 3,10-11; WWWS, 1395
K.B. Raper, Biog. Mem. Nat. Acad. Sci. 57 (1987), 347-370

RAPER, KENNETH BRYAN [1908-1987]
AMS 14,4107; WWA 1976-1977, p.2572; IWW 1982-1983, p.1090; WWWS, 1395
R.M. Burris, Am. Phil. Soc. Year Book 1987, pp.196-201
R.M. Burris and E.H. Newcomb, Biog. Mem. Nat. Acad. Sci. 60 (1991), 251-270

RAPKINE, LOUIS [1904-1948]
R. Wurmser, Bull. Soc. Chim. Biol. 30 (1948), 716-720
A. Lwoff, Ann. Inst. Pasteur 76 (1949), 271-275
V. and E. Karp (eds.), Louis Rapkine. North Bennington, Vt. 1988
D.T. Zallen, French Historical Studies 17 (1991), 6-37

RAPP, RUDOLF [1866-1941]
HEIN, 513; TSCHIRCH, 1108-1109; TLK 2,552-553
R. Dietzel, Pharm. Z. 76 (1931), 89-90
H. Kaiser, Deutsche Apoth. Z. 106 (1966), 834-835

RAPP, WILHELM von [1794-1868]
ADB 27,299-300; HIRSCH 4,723-724

RAPPORT, MAURICE [1919-]
AMS 15(6),54; WWA 1980-1981, p.2722
M.M. Rapport, Journal of Lipid Research 25 (1984), 1522-1527

RAPS, AUGUST [1865-1920]
DBJ 2,582-587,757; MATSCHOSS, 215; RSC 18,54; NUC 481,500

RASCHEN, JULIUS [1865-]
JV 2,245; RSC 18,54; NUC 481,537
Who's Who of Science International 1914, p.482

RASCHIG, FRIEDRICH (FRITZ) [1863-1928]
DBJ 10,334; TLK 2,553; BERLIN, 720; WWWS, 1396; POTSCH, 356; POGG 4,1213; 5,1022; 6,2123-2124; 7aSuppl.,521
A. Rosenheim, Ber. chem. Ges. 62A (1929), 109-126

RASCHIG, KURT [1898-1969]
JV 40,466
Anon., Chem. Z. 93 (1969), 690

RASHEVSKY, NICOLAS [1899-1972]
AMS 11,4337; WWWS, 1396; POGG 6,2123; 7b,4246-4250

RASKE, KARL AUGUST HEINRICH [1863-]
JV 1,5; 20,37; NUC 481,579

RASMUSSEN, FRITS VALDEMAR [1837-1877]
HIRSCH 4,725
Anon., Hosp. Tid. [2]4 (1877), 129-131

RASPAIL, FRANÇOIS VINCENT [1794-1878]
DSB 11,300-302; VAPEREAU 5,1502-1503; GE 28,160-161; LAROUSSE 13,713-714; HIRSCH 4,727; POGG 2,571; 3,1089; CALLISEN 15,351-352; 31,356-359
R. Blanchard, Arch. Parsit. 8 (1903), 5-87
L. Cayeux, Rev. Sci. 52 (1914), 776-778
X. Raspail, Vie et l'Oeuvre Scientifique de F.V. Raspail. Paris 1926
H. Harms, Apoth. Z. 46 (1931), 1454-1458; 47 (1932), 1274-1275, 1293-1294,1307-1310,1324-1326,1433-1436
J. Schonfeld, Raspail et la Médecine. Paris 1933
A. Delaunay, Hist. Med. 10(12) (1960), 20-31
D.B. Weiner, Raspail, Scientist and Reformer. New York 1968
A. Colard, Rev. Med. Brux. 32 (1976), 131-152
S. Raspail and L. Dubief, François Vincent Raspail. Paris 1978
L. Velluz, Raspail, un Contestaire du 19e Siècle. Perigueux 1974
R. Hildebrandt, Anat. Anz. 154 (1983), 353-358

RASSOW, BERTHOLD [1866-1954]
TLK 3,516; NUC 481,657; POGG 4,1213-1214; 5,1022-1023; 6,2125; 7a(3),676
A. Binz and F. Scherf, Z. angew. Chem. 49 (1936), 707-709
R.E. Oesper, J. Chem. Ed. 29 (1952), 76-77
W.K. Schwarze, Chem. Z. 79 (1955), 52
R. Zaunick, Chemische Technologie 7 (1955), 699-703

RASSOW, HERMANN [1858-]
BONN, 346

RATHKE, HEINRICH BERTHOLD [1840-1923]
AUERBACH, 879; GUNDLACH, 470; MEINEL, 217-218,510; SCHMITZ, 282; DZ,1147; POTSCH, 356; BLOKH 2,602-603; POGG 3,1092; 4,1214; 5,1023; 7aSuppl.,522
K. Schaum, Ber. chem. Ges. 57A (1924), 83-92

RATHKE, MARTIN HEINRICH [1793-1860]
DSB 11,307-308; ADB 27,352-355; HIRSCH 4,728; LEVITSKI 2,300-301
Anon., Proc. Roy. Soc. 11 (1862), xxxvii-xl

RATHMANN, FRANZ [1904-]
AMS 14,4114; POGG 7a(3),681

RATNER, SARAH [1903-]
AMS 14,4114; WWWS, 1397
S. Ratner, Ann. Rev. Biochem. 46 (1977), 1-24
M.B. Bentley, Trans. N.Y. Acad. Sci. 41 (1983), 1-4

RATNOFF, OSCAR DAVIS [1916-]
AMS 15(6),61; WWA 1980-1981, p.2725

RAU, ALBRECHT [1843-1918]
HEIN-E, 353-354; RSC 11,112; NUC 482,178-179
Kürschner's Deutscher Literatur Kalender 39 (1917), 1335
Anon., Chem. Z. 42 (1918), 141; Z. angew. Chem. 31(III) (1918), 114
Y.U. Solovyev, Arch. Inter. Hist. Sci. 27 (1977), 222-230

RAUBER, AUGUST [1841-1917]
FISCHER 2,1271-1272; EGERER, 15-16; WELDING, 608; WWWS, 1397
W. Lubosch, Anat. Anz. 58 (1924), 129-172
V. Pawlow, Z. ärzt. Fortbild. 85 (1991), 591-593

RAUBITSCHEK, HUGO [1889-1918]
BIRK, 37
Anon., Chem. Z. 33 (1909), 247

RAUCH, HANS [1817-1890]
PFISTER, 320

RAUCH, HANS [1896-]
JV 37,560; NUC 482,221

RAUCH, KONRAD [1905-]
NUC 482,226

RAUCHENBICHLER, RUDOLF von [1881-]
F. Martin, Mitteilungen der Gesellschaft für Salzburger Landeskunde 76 (1936), 133

RAUDENBUSCH, WILHELM [1901-1970]
JV 42,338

RAUDNITZ, ROBERT WOLF [1856-1921]
FISCHER 2,1272; MOTSCH, 177-185; KOERTING, 182; KOREN, 234
R. Fischl, Lotos 69 (1921), 289-294; Jahrb. Kinderheil. 47 (1922), 110

RAUEN, HERMANN [1913-]
KURSCHNER 13,3048; POGG 7a(3),686-687

RAULIN, JULES [1836-1896]
DSB 11,310-311; GUIART, 174
L. Grandeau, Ann. Sci. Agron. [2]2 (1896), 387-389
P. Cazeneuve, Ann. Soc. Agr. Lyon [7]4 (1897), 351-355
J. Javillier and W.H. Schopfer, Ann. Univ. Lyon 1948-1949, pp.61-158

RAUTERBERG, FERDINAND [1862-1944]
JV 7,159; NUC 482,348

RAVENEL, ST. JULIEN [1819-1882]
DAB 15,397; NCAB 10,272-273; ELLIOTT, 215; KELLY, 1013; WWWS, 1398
Anon., Proc. Am. Acad. Arts Sci. NS9 (1882), 437
C.A. Browne, J. Am. Chem. Soc. 48(8a) (1926), 177-201

RAVIN, ARNOLD WARREN [1921-1981]
AMS 14,4119; WWA 1980-1981, p.2728; WWWS, 1398

RAWITSCHER, FELIX [1890-1957]
IB, 243; LDGS, 12
C.B. Aust, Staden Jahrbuch 11/12 (1963-1964), 197-211

RAWITZ, BERNHARD [1857-1932]
FISCHER 2,1274
A. Friedel, Anat. Anz. 76 (1933), 230-237

RAWSON, RULON WELLS [1908-1989]
AMS 14,4121; WWWS, 1399

RAY, JNANENDRA NATH [1898-1970]
POGG 6,2130; 7b,4256-4257
Anon., Chem. Brit. 6 (1970), 272

RAY, PRAFULLA CHANDRA [1861-1944]
DSB 11.318-319; POGG 5,1024-1025; 6,2130-2131; 7b,4257-4258
P.C. Ray, Life and Experiences of a Bengali Chemist. Calcutta 1932
J.L. Simonsen, Nature 154 (1944), 76
G.D. Kellogg, J. Ind. Chem. Soc. 21 (1944), 253-260
J.N. Mukerjee, J. Chem. Soc. 1946, pp.216-218
S.S. Bhatnagar, J. Ind. Chem. Soc. 29 (1952), 714-720

RAY, SURENDRA NATH [1908-1981]
Anon., Fellows of the Indian National Science Academy 1935-1984, p.417. New Delhi 1984

RAY, THOMAS WILLIAM [1883-1958]
AMS 9(I),1576; NCAB 48,293

RAYER, PIERRE [1793-1867]
GE 28,187; LAROUSSE 13,742; GENTY 3,33-48; HIRSCH 4,736-737; WWWS, 1399; DECHAMBRE 81,576-579; BULLOCH, 391; CALLISEN 15,384-389; 31,373-377
J. Roger, Les Médecins Normands 2,121-126. Paris 1895
G. Richet et al., Hist. Sci. Med. 25 (1991), 261-307

RAYMAN, BOHUSLAV [1852-1910]
OBL 8,445-446; STURM 3,392; POGG 4,1219
Anon., Chem. Z. 34 (1910), 1041; Cas. Lek. Cesk. 29 (1910), 480-481
S. Strbanova, Dejiny ved a Techniky 11 (1978), 82-96; 12 (1979), 129-144

RAYMOND, ALBERT L. [1901-1989]
AMS 11,4350

RAZENKOV, IVAN PETROVICH [1888-1954]
BSE 35,618; BME 27,864-866; WWR, 474
Anon., Fiziol. Zhur. 41 (1955), 157-159; Voprosi Pitania 14(2) (1955),61
G.N. Zilov, Arkh. Pat. 17(2) (1955), 93-95

RAZUVAEV, GRIGORI ALEKSEEVICH [1895-]
BSE(3rd Ed.), 21,433; WWWS, 1400; POGG 6,2125-2126; 7b,4263-4378
Y.A. Oldekop and N.A. Maier, Zhur. Ob. Khim. 36 (1966), 173-177

REACH, FELIX [1872-1943?]
KURSCHNER 4,2324-2325; STURM 3,393; FISCHER 2,1275; IB, 243; NUC 483,231

READ, JOHN [1884-1963]
DNB [1961-1970],872-874; WW 1963, p.2526; POGG 6,2134-2135; 7b,4278-4280
E.L. Hirst, Nature 198 (1963), 336-337; Proc. Chem. Soc. 1963, 353-355;
Biog. Mem. Fell. Roy. Soc. 9 (1963), 237-260
I.G.M. Campbell and K.R. Webb, Chem. Ind. 1963, pp.829-830

REAGH, ARTHUR LINCOLN [1871-1949]
Anon., Auk 67 (1950), 429

REAY, GEORGE ADAM [1901-1971]
WW 1971, p.2611
Anon., Chem. Brit. 7 (1971), 217

REBER, BURKHARD [1848-1926]
DHB 5,400; HEIN, 518
P. Röthlisberger, Gesnerus 34 (1977), 213-231
P. Jaroschinsky, Burkhard Reber, ein Vorläufer der Schweizerischen Pharmaziegeschichte. Stuttgart 1988

REBER, RUFUS KING [1903-]
AMS 12,5140RECHNITZ, HEINRICH [1876-1910]
JV 16,145; NUC 484,15
Anon., Ber. chem. Ges. 43 (1910), 3620

RECKLEBEN, HANS [1864-1920]
BLOKH 2,604-605; POGG 5,1029
J. Scheiber, Ber. chem. Ges. 53A (1920), 116-117

RECKLINGHAUSEN, FRIEDRICH von [1833-1910]
BJN 15,71*; HIRSCH 4,742; PAGEL, 1351-1352; ROEDER, 94-108; WWWS, 1739; FRANKEN, 194-195; SCHULTE, 257-258; DZ, 1152
W. Waldeyer, Anat. Anz. 37 (1910), 509-511
G. Hauser, Sitz. Phys. Med. Erlangen 42 (1910), 1-10
H. Chiari, Verhandl. path. Ges. 15 (1912), 478-488
W.T. Councilman, Proc. Am. Acad. Arts Sci. 53 (1918), 872-875

REDDELIEN, GUSTAV [1882-1938]
TLK 3,518; POGG 5,1030; 6,2137; 7a(3),693

B. Helferich, Ber. chem. Ges. 71A (1938), 86-88

REDDICK, DONALD [1883-1955]
AMS 7,1450; IB, 243-244; WWAH, 550
L.C. Peterson and D.S. Welch, Phytopathology 46 (1956), 299

REDENZ, ERNST [1898-1940]
KURSCHNER 4,2327; GAUSS, 169-170; EBERT, 92-95; IB, 244

REDFIELD, ALFRED CLARENCE [1890-1983]
AMS 14,4133; WWA 1978-1979, p.2672; IWW 1982-1983, p.1099; WWWS, 1402
B.H. Ketchum and M. Sears, Limnology and Oceanography 19Suppl. (1965), R1-R8
F.M. Carpenter et al., Harvard Gazette 83(18) (1988), 6

REDLICH, OTTO [1896-1978]
BHDE 2,946-947; POGG 6,2137-2138; 7a(3), 694-695

REDTENBACHER, JOSEPH [1810-1870]
ADB 27,542-543; OBL 9,13-14; WURZBACH 25,116; BLOKH 2,605-606; STURM 3,397; VOGEL, 49-54; SKULINA, 122-124; TSCHIRCH, 1109; FERCHL, 435; SCHAEDLER, 101; POGG 2,585; 3,1098; 7aSuppl.,529
A. Williamson, J. Chem. Soc. 23 (1870), 311
F. von Kobell, Sitz. Bayer. Akad. 1870(I), pp.418-420
A. Schrötter, Alm. Akad. Wiss. Wien 20 (1870), 230-247
M. Kohn, J. Chem. Ed. 24 (1947), 366-368

REDWOOD, THEOPHILUS [1806-1892]
TSCHIRCH, 1110; FERCHL, 435; POGG 2,585
Anon., Analyst 17 (1892), 61; J. Soc. Chem. Ind. 11 (1892), 228-229; Pharm. J. 22 (1892), 763-766; Am. J. Pharm. 64 (1892), 223-224
P.H. Thomas, Pharmaceutical Historian 13(2) (1983), 9-12

RÉE, ALFRED [1863-1933]
A.L., J. Chem. Soc. 1933, pp.471-472
Anon., J. Roy. Inst. Chem. 1933, pp.144-145; Chem. Ind. 1933, p.185

REED, HOWARD SPRAGUE [1876-1950]
AMS 8,2032; NCAB 40,359; IB, 244; WWAH, 501
F.C. Steward and J. Dufrenoy, Nature 166 (1950), 586-587

REED, LESTER JAMES [1925-]
AMS 15(6),89; WWA 1980-1981, p.2741; WWWS, 1403

REED, WALTER [1851-1902]
DSB 11,345-347; DAB 15,459-461; NCAB 33,143-144; OLPP, 336-341; WWAH,501; DAMB, 628-629; W, 436-437, WWWS, 1403

REES, GEORGE OWEN [1813-1889]
DNB 16,842-843; WILKS, 251-261; CULE, 159-160
S. Wilks, Guy's Hosp. Rep. 31 (1889), xxiii-xxxiii
L. Rosenfeld, Clin. Chem. 31 (1984), 1068-1070
N.G. Coley, Med. Hist. 30 (1986), 173-190

REES, MAURICE WILLIAM [1915-1978]
R.L.M. Synge, John Innes Annual Report 69 (1978), 17-20

REESE, CHARLES LEE [1862-1940]
DAB [Suppl.2],550-551; MILES, 401-402; NCAB 30,238-239; WWWS,1404
R.E. Curtin, Am. Phil. Soc. Year Book 1940, pp.433-436; J. Am. Chem. Soc. 62 (1940), 1889-1891
Anon., Ind. Eng. Chem. News Ed. 18 (1940), 364-365

REESE, HEINRICH [1879-1951]
NUC 485,189
Anon., Schw. med. Wchschr. 81 (1951), 716
C. Haffter, Schw. Arch. Neurol. 70 (1952), 189-190

REESE, LUDWIG [1885-1916]
RSC 18,93

Anon., Chem. Z. 40 (1916), 920; Z. angew. Chem. 29(III) (1916), 596

REESS, MAX FERDINAND FRIEDRICH [1845-1901]
BJN 6,84*; 7,435-437; TSCHIRCH, 1110
Anon., Leopoldina 37 (1901), 95-96

REFORMATSKI, ALEKSANDR NIKOLAEVICH [1864-1937]
BSE 36,429; POGG 4,1223; 5,1031; 7b,4308
I. Kablukov, Usp. Khim. 7 (1938), 321-322

REFORMATSKI, SERGEI NIKOLAEVICH [1860-1934]
BSE 36,429; WWR, 474-475; ARBUZOV, 182-184; KAZAN 1,467-468; POTSCH, 357; POGG 4,1222-1223; 6,2138; 7b,4308
A. Semenzow, Ber. chem. Ges. 68A (1935), 61; Zhur. Ob. Khim. [2]5 (1935), 583-601; J. Chem. Ed. 34 (1957), 530-532
A.I. Kiprianov, Ukr. Khim. Zhur. 26 (1960), 471-475

REGAUD, CLAUDE [1870-1940]
FISCHER 2,1278; GUIART, 162,226; IB, 244
E. Sergeant, Presse Med. 49 (1941), 252-254
A. Lacassagne, Ann. Inst. Pasteur 66 (1941), 181-186
J. Jolly, Bull. Acad. Med. 124 (1941), 44-48
J. Regaud, Claudius Regaud. Paris 1982

REGNAULD, JULES ANTOINE [1820-1895]
GORIS, 549-550; BOURQUELOT, 77-78; TSCHIRCH, 1110; FERCHL, 436; POGG 3,1099; 4,1223
A.R. et al., J. Pharm. Chim. [6]1 (1895), 169,271-273
J. Cheymol and A. Soubiran, Presse Med. 76 (1968), 2366-2368

REGNAULT, HENRI VICTOR [1810-1878]
DSB 11,352-354; VAPEREAU 4,1515-1516; GE 28,292; LAROUSSE 13,862; W, 438;MATSCHOSS, 219-220; PHILIPPE, 798-799; TSCHIRCH, 1110; FERCHL, 436; POTSCH, 357-358; ABBOTT-C, 121-122; WWWS, 1405; POGG 2,588-590; 3,1099
H.J. Debray, C.R. Acad. Sci. 86 (1878), 131-143
Anon., J. Chem. Soc. 33 (1878), 235-239
T.H.N., Nature 17 (1878), 263-264

REH, ALFRED [1878-1959]
JV 19,391; SG [2]14,415; NUC 486,83
Anon., Toute l'Histoire de Napoleon 6 (1951), 89-91

REHBERG, POUL BRANDT [1895-1989]
KBB 1981, p.943; IWW 1982-1983, p.1101; VEIBEL 2,356-357; IB, 244

REHE, JOHANN AUGUST [-1892]
Anon., Pharm. Z. 37 (1892), 539

REHLÄNDER, PAUL [1869-]
TLK 3,520; RSC 18,101; NUC 486,115

REHN, LUDWIG [1849-1930]
FISCHER 2,1279; GROTE 3,201-244; KALLMORGEN, 379; KILLIAN, 406-407; KALIN, 176-178; WWWS, 1405
A. Bier, Münch. med. Wchschr. 76 (1929), 583
V. Schmieden, Deutsche med. Wchschr. 56 (1930), 1185-1187
F. Koenig, Münch. med. Wchschr. 77 (1930), 1330-1332

REHNS, JULES [1871-]
RSC 18,102; SG [2]14,416; NUC 486,138

REHORST, CURT [1888-]
POGG 6,2140; 7a(3),699
Anon., Chem. Z. 92 (1968), 366REHSTEINER, HUGO [1864-1947]
NUC 486,143
K. Rehsteiner, Verhandl. Schw. Nat. Ges. 127 (1947), 272-274

REICHARD, PETER [1909-]
VAD 1981, p.861

REICHARD, PETER ADOLF [1928-]
VAD 1983, p.847

REICHARDT, EDUARD [1827-1891]
BLOKH 2,606-607; TSCHIRCH, 1110; POGG 2,591-592; 3,1100-1101
T. von Goltz, Leopoldina 27 (1891), 196-199
A.W. Hofmann, Ber. chem. Ges. 24 (1891), 3167-3168

REICHEL, HANS [1911-]
KURSCHNER 13,3071; POGG 7a(3),703-704

REICHEL, HERMANN [1876-1943]
OBL 9,29; KURSCHNER 4,2337; FISCHER 2,1280
R. Grassburger, Wiener klin. Wchschr. 56 (1943), 335-336

REICHEL, LUDWIG [1900-]
KURSCHNER 12,2524; POGG 6,2142; 7a(3),704

REICHENBACH, CARL von [1788-1869]
DSB 11,359-360; ADB 27,670-671; BLOKH 2,607-608; POTSCH, 358-359; SCHAEDLER, 103; FERCHL, 436-437; TSCHIRCH, 1110; WWWS, 1739; CALLISEN 31,394-395; POGG 2,593-594; 3,1101; 7aSuppl.,534-535
Anon., Alm. Akad. Wiss. Wien 19 (1869), 326-369
M. Kohn, J. Chem. Ed. 32 (1955), 188-189

REICHER, LODEWIJK THEODOR [1857-1943]
POGG 4,1224; 6,2143; 7b,4309-4310
J.H. van't Hoff et al., Chem. Wkbl. 5 (1908), 517-526
W.P. Jorissen, Chem. Wkbl. 32 (1955), 188-189

REICHERT, BENNO [1906-1970]
KURSCHNER 10,1950; HEIN-E, 356-357; WWWS, 1406; POGG 7a(3),708-709
Anon., Chem. Z. 94 (1970), 132; Deutsche Apoth. Z. 110 (1970), 7-11
W. Saenger, Arzneimitt. 20 (1970), 419-420

REICHERT, CARL BOGUSLAUS [1811-1883]
DSB 11,360-361; ADB 27,679-681; HIRSCH 4,752-753; PAGEL, 1357-1359; LEVITSKI 2,15-19
G. B[roesike], Berl. klin. Wchschr. 21 (1884), 45-46

REICHERT, EDWARD TYSON [1855-1931]
AMS 3,566; NCAB 23,206; CHITTENDEN, 150-153

REICHINSTEIN, DAVID [1882-1955]
POGG 5,1032; 6,2143; 7a(3),709-710

REICHOLD, ALBERT [1864-]
JV 5,78; NUC 486,325

REICHSTEIN, TADEUS [1897-]
KURSCHNER 13,3076; SBA 2,99; IWW 1982-1983, p.1102; CB 1951, pp.512-514; MH 1,389-390; STC 2,410-411; MSE 3,17-18; BOHM, 161- 162; COHEN, 193; POTSCH, 359; TETZLAFF, 274; WWWS, 1406; POGG 6,2143; 7a(3),710-714
Anon., Leopoldina [3]12 (1966), 61-64
C. Tamm, Chimia 21 (1967), 432

REID, ALBERT [1911-1942]
Anon., Chemische Technik 15 (1942), 204
W. Foerst, Z. angew. Chem. 56 (1943), 66

REID, EBENEZER EMMET [1872-1973]
AMS 11,4377; GOULD, 30-31; WWWS, 1406; POGG 5,1032-1033; 6,2143-2144; 7b,4310-4311
Anon., Chem. Eng. News 48(5) (1970), 42-43
E.E. Reid, My First One Hundred Years. New York 1972

REID, EDWARD WAYMOUTH [1862-1948]
WW 1945, p.2281; O'CONNOR(2), 424-425
E.P. Cathcart and R.C. Garcy, Obit. Not. Fell. Roy. Soc. 6 (1948),213-218
P.T. Herring, Nature 161 (1948), 591-592
G.H. Bell and D.S. Parsons, J. Physiol. 263 (1976), 75P-78P

REID, JOHN HERBERT [1881-]
JV 20,283; NUC 486,467

REID, ROGER DELBERT [1905-]
AMS 13,3641; WWWS, 1407; NUC 486,518

REIF, JOHANN GEORG [1880-1964]
POGG 6,2144-2145; 7a(3),714-715
Anon., Z. Leb. Unt. 92 (1951), 106

REIGHARD, JACOB ELLSWORTH [1861-1942]
AMS 6,1166; NCAB 16,432-433; IB, 245; WWAH, 503
A.F. Shull, Science 95 (1942), 344-346

REIHLEN, HANS [1892-1950]
KURSCHNER 7,1640; TLK 3,522; POGG 6,2145-2146; 7a(3),716
W. Rüdorff, Z. angew. Chem. 62 (1950), 545

REIL, JOHANN CHRISTIAN [1759-1813]
DSB 11,363-365; ADB 27,700-701; HIRSCH 4,755-756; FERCHL, 437; CALLISEN 31,400; POGG 2,595; 7aSuppl.,535-536
H. Steffens, Johann Christian Reil. Halle 1815
M. Neuburger, Johann Christian Reil. Stuttgart 1913
R. Zaunick, Johann Christian Reil. Leipzig 1960
R. Zaunick et al., Nova Acta Leop. 22 (1960), 5-169

REILLY, JOSEPH [1889-1965]
SRS, 53; POGG 6,2146; 7b,4314-4315
Who Was Who 1961-1970, p.946

REIMANN, FRIEDRICH [1897-]
KURSCHNER 14,3327; BHDE 2,954-955; WIDMANN, 283; PELZNER, 173-181
Anon., New Istanbul Contributions to Medical Science 9 (1967), 141-154

REIMANN, HOBART ANSTETH [1897-1986]
AMS 14,4155; WWWS, 1405
J.H. Hodges, Trans. Assoc. Am. Phys. 99 (1986), cvi-cviii

REIMANN, KARL LUDWIG [1804-1872]
HEIN, 521; TSCHIRCH, 1110; FERCHL, 437; POGG 2,595; 3,1101-1102
P.F. Koenig, Die Entdeckung des Reinen Nicotins. Bremen 1940
K. Oberdorffer, Ludwigshafener Chemiker, pp.14-31. Ludwigshafen 1960

REIMANN, STANLEY PHILIP [1891-1968]
AMS 11,4382; NCAB 54,281-282
T.R. Talbot, Verhandl. path. Ges. 55 (1971), 729-735

REIMER, CARL LUDWIG [1845-1883]
BLOKH 2,608-609; BERLIN, 686
A.W. Hofmann, Ber. chem. Ges. 16 (1883), 99-102
A.J. Rocke and A.J. Ihde, J. Chem. Ed. 63 (1986), 309-310

REIMER, CARL LUDWIG [1856-1921]
BLOKH 2,609; POTSCH, 359; POGG 3,1102; 4,1225; 6,2146
B. Lepsius, Ber. chem. Ges. 54A (1921), 159-160
A.J. Rocke and A.J. Ihde, J. Chem. Ed. 63 (1986), 309-310

REIMER, MARIE [1875-1962]
AMS 9(I),1589; WWAH, 503; POGG 6,2146-2147; 7b,4315

REIN, HERMANN [1898-1953]
FISCHER 2,1282; POGG 7a(3),717-719
D.H. Smyth, Nature 172 (1953), 381
M. Schneider, Pflügers Arch. 259 (1954), 14-20
A. von Muralt, Erg. Physiol. 48 (1955), 1-12

REINBOLD, BÉLA [1875-1927]
MEL 2,497; IB, 245
Anon., Chem. Z. 30 (1906), 1277
E. Veress, Orv. Het. 71 (1927), 1457

REINDEL, FRIEDRICH [1891-1973]
KURSCHNER 11,2381-2382; TLK 3,522; POGG 7a(3),720-721
Anon., Chem. Z. 97 (1973), 210

REINDERS, WILLEM [1874-1951]
POGG 4,1226; 5,1034-1035; 6,2147; 7b,4315-4316
M.C.F. Beukers, Chem. Wkbl. 36 (1939), 451-456
J.G. Hoogland, Chem. Wkbl. 47 (1951), 865-866

REINEKE, EZRA PAUL [1909-1985]
AMS 13,3645; WWWS, 1408

REINER, LASZLO [1894-1955]
AMS 9(II),925; POGG 6,2147-2148; 7b,4316-4317
M. Green, Science 124 (1956), 434

REINFURTH, ELSA [1889-]
JV 35,474; NUC 487,136

REINHOLD, JOHN GUNTHER [1900-]
AMS 12,5176; IB, 245; WWWS, 1408
J. Reinhold, Clin. Chem. 28 (1984), 2314-2323

REINITZER, BENJAMIN [1855-1928]
STURM 3,413; POGG 6,2149
G. Jantsch, Ost. Chem. Z. 31 (1928), 24-25

REINITZER, FRIEDRICH [1857-1927]
STURM 3,413; OBL 9,51; PRAG, 348-349; TLK 2,562; POGG 4,1227; 5,1035; 6,2149
H. Pabisch, Ost. Chem. Z. 30 (1927), 66

REINKE, JOHANNES [1849-1931]
SHBL 1,227-229; TSCHIRCH, 1111; KIEL, 150; WAGENITZ, 144-145; DZ, 1164-1165; WWWS, 1408; POGG 4,1227; 6,2149
J. Reinke, Mein Tagewerk. Freiburg i.B. 1925
W. Benecke, Ber. bot. Ges. 50 (1932), (171)-(202)

REINSCH, HUGO [1808-1884]
HEIN, 523-524; TSCHIRCH, 1111; FERCHL, 438; POGG 3,1102-1103
J. Vogel, Erlanger Heimatsblätter 17 (1934), 89-102; 18 (1935), 2-3

REINSCH, SIGMUND [1874-]
JV 13,226; RSC 15,897

REINWEIN, HELMUTH [1895-1966]
KURSCHNER 10,1960; SHBL 5,231-233; GEISSLER, 114-121; KIEL, 90; POGG 6,2150; 7a(3),724-725
A.W. Fischer, Christiana Albertina 3 (1967), 85-87
A. Bernsmier, Münch. med. Wchschr. 109 (1967), 477-478
W. Siede, Med. Klin. 62 (1967), 835-855

REIS, ALFRED [1882-1951]
TLK 3,524; LDGS, 25; POGG 5,1035; 6,2150-2151; 7a(3),725
Anon., Ost. Chem. Z. 53 (1952), 18

REISENEGGER, CURT [1888-]
JV 29,730; NUC 487,330
Wer Ist's 10,1284

REISENEGGER, HERMANN [1861-1930]
TLK 3,524; POGG 6,2151
K.A. Hofmann, Ber. chem. Ges. 63A (1930), 177-179
H. Laubmann, Chem. Z. 55 (1931), 181; Z. angew. Chem. 44 (1931), 213- 215

REISET, JULES [1818-1896]
VAPEREAU 6,1515-1516; GE 28,329; FERCHL, 438; SCHAEDLER, 103; WWWS, 1409; POTSCH, 360; POGG 2,600-601; 3.1103; 4,1227
E. Gautier, Ann. Scient. 40 (1896), 492-493

REISS, EMIL [1878-1923]
FISCHER 2,1283; KALLMORGEN, 381

REISS, FRANZ [1808-1861]
STURM 3,418; WURZBACH 25,253; HIRSCH 4,761; ROTH, 98-105

REISS, PAUL [1901-1944]
HEIM, 119-121; IB, 245; NUC 487,374
Anon., Bulletin d'Histologie Appliquée 22 (1945), 16

REISSERT, ARNOLD [1860-1945]
AUERBACH, 882; GUNDLACH, 471; MEINEL, 222-224,511; TLK 3,525; BERLIN,717; POGG 4,1228; 5,1035-1036; 6,2151; 7a(3),726

REISSIG, WILHELM [1829-1901]
RSC 5,156; 8,727
Anon., Pharm. Z. 46 (1901), 959

REITSTÖTTER, JOSEF [1894-1982]
KURSCHNER 13,3097; TLK 3,525; POGG 6,2152; 7a(3),728
Anon., Naturw. Rund. 35 (1982), 507; Chem. Z. 107 (1983), 44

REITTER, HANS [1865-1912]
POGG 5,1036
Anon., Ber. chem. Ges. 45 (1912), 2839

REITZENSTEIN, FRIEDRICH [1868-1940]
JV 8,274; POGG 4,1228; 5,1036-1037; 6,2152; 7a(3),729

REMAK, ROBERT [1815-1865]
DSB 11,367-370; ADB 28,191-192; HIRSCH 4,764-765; PAGEL, 1361-1362; BRAZIER, 136-138,148; KOHUT 2,286; KOREN, 234; WININGER 5,183; W, 438-439; KOSMINSKI, 414-415; CALLISEN 31,406-407
W. His, Berl. klin. Wchschr. 2 (1865), 372
B. Kisch, Trans. Am. Phil. Soc. 44 (1954), 227-296
C.T. Anderson, Bull. Hist. Med. 60 (1986), 523-543
H.P. Schmiedebach, Berichte zur Wissenschaftsgeschichte 13 (1990), 37-41

REMBOLD, OTTO [1834-1904]
BJN 9,327; 10,31*; FROHNE, 83-85; KUKULA, 692
F. Knappitsch, Mitt. Ver. Aerzte Steier. 41 (1904), 313-316

REMICK, ARTHUR EDWARD [1899-1980]
AMS 11,4391; WWWS, 1410
Anon., Chem. Eng. News 58(44) (1980), 62

REMSEN, IRA [1846-1927]
DSB 11,370-371; DAB 15,500-502; MILES, 402-403; FARBER, 819-822; W, 439; BLOKH 2,610; SILLIMAN, 169-170; GOULD, 26-28; SCHAEDLER, 103; WWAH, 504; WWWS, 1410-1411; POTSCH, 360; POGG 3,1105-1106; 4,1229-1230; 5,1038; 6,2154
W.A. Noyes, J. Chem. Soc. 1927, pp.3182-3189
J.F. Norris, J. Am. Chem. Soc. 50 (1928), 67-86
W.A. Noyes and J.F. Norris, Biog. Mem. Nat. Acad. Sci. 14 (1931),207-257
F.H. Getman, J. Chem. Ed. 16 (1939); The Life of Ira Remsen. Easton 1940
O. Hannaway, Ambix 23 (1976), 145-164
A.H. Corwin, Welch Foundation Conferences on Chemical Research 20 (1977), 46-49
G.B. Kauffman and P.M. Priebe, Ambix 25 (1978), 191-207
D.S. Tarbell et al., Isis 71 (1980), 620-626

RENARD, GUILLAUME ADOLPHE [1846-1919]
TSCHIRCH, 1111; POGG 3,1106-1107; 4,1231

RENAUT, JOSEPH LOUIS [1844-1917]
DSB 11,373-374; FISCHER 2,1284-1285; WWWS, 1411
J. Mollard. Lyon Médical 127 (1918), 49-59
G. Hayem, Bull. Acad. Med. 78 (1918), 795-798

RENNER, OTTO [1883-1960]
DSB 18,731-732; KURSCHNER 8,1888; IB, 246; WWAH, 504
L. Brauner, Bayer. Akad. Wiss. Jahrbuch 1960, pp.181-185
K. Buder, Jahrb. Sachs. Akad. 1960-1962, pp.375-385
F. Oehlkers, Ber. bot. Ges. 74 (1961), (82)-(94)
K. Mägdefrau, Ber. Bayer. bot. Ges. 34 (1961), 101-113
F. Knoll, Alm. Akad. Wiss. Wien 112 (1962), 429-435
R.E. Cleland, Genetics 53 (1966), 1-6

RENNING, JULIUS [1886-1964]
JV 28,320; NUC 488,563

RENOUF, EDWARD [1846-1934]
AMS 5,923; GOULD, 29; WWAH, 504
Anon., Chem. Z. 58 (1934), 1033; Science 80 (1934), 471

RENQVIST (REENPÄA), YRJÖ [1894-1976]
FISCHER 2,1285; IB, 246
Who's Who in Europe 1972, p.2502
Vem Och Vad 1975, pp.519-520

RENSHAW, RAEMER REX [1880-1938]
AMS 7,1169; POGG 6,2156; 7b,4323-4324
H.G. Lindwall, Science 88 (1938), 394

RENTZ, EDUARD [1898-1962]
KURSCHNER 9,1641; 7a(3),734-736
Anon., Chem. Z. 87 (1963), 331; Nachr. Chem. Techn. 11 (1963), 245

RENZ, JANY [1907-]
A. Huber-Morath, Bauhinia 6 (1978), 259-261

REPOND, PAUL [1856-1919]
DHB 5,444

REPPE, WALTER [1892-1969]
KURSCHNER 10,1970; WWWS, 1412; POTSCH, 360; POGG 7a(3),736- 737
Anon., Nachr. Chem. Techn. 5 (1957), 231-232
E. Baum, Chem. Z. 93 (1969), 639
P.J.T. Morris, Brit. J. Hist. Sci. 25 (1992), 145-167

RESENSCHECK, FRIEDRICH [1879-]
JV 19,77; NUC 489,380

RETTGER, LEO FREDERICK [1874-1954]
AMS 8,2048; NCAB E,125; 42,618-619; FISCHER 2,1874-1875; WWAH, 504-505; SACKMANN, 275-280
G. Valley, J. Bact. 69 (1955), 1-2

RETZIUS, ANDERS ADOLF [1796-1860]
DSB 11,379-381; SMK 6,244-245; HIRSCH 4,772-773; WWWS, 1413
S. Lovén, Lev. Sven. Vet. 2 (1878-1885), 1-36
O. Larsell, Ann. Med. Hist. 6 (1924), 16-24
V. Kruta, Lychnos 1956, pp.96-131

RETZIUS, MAGNUS GUSTAV [1842-1919]
DSB 11,381-383; SMK 6,246-247; FISCHER 2,1287; PAGEL,1368-1369; WWWS,1413
W. von Waldeyer-Hartz, Anat. Anz. 52 (1919-1920), 261-268; Deutsche med. Wchschr. 45 (1919), 942-943
O. Larsell, Sci. Mon. 10 (1920), 559-569
E.A.S.S., Proc. Roy. Soc. B91 (1920), xxxvi-xxxviii
M.G. Retzius, Biol. Unter. NF19 (1921), 1-100

REULING, LUDWIG [1811-1879]
DGB 69 (1930), 557; RSC 5,172

REULING, ROBERT [1808-1852]
DGB 69 (1930), 555

REUSS, AUGUST EMANUEL von [1811-1873]
DSB 11,385-387; ADB 28,303-305; OBL 9,97; WURZBACH 25,350-354; HIENSTOFER, 144-161; POGG 2,615; 3,1113; 7aSuppl.,541-542
H.B. Geinitz, Leopoldina 9 (1874), 67-72
A. Schrötter, Alm. Akad. Wiss. Wien 24 (1874), 129-152

REUSS, FERDINAND FRIEDRICH [1778-1852]
BSE 36,307; FERCHL, 441; POGG 2,614-615
E.A. Steinberg, Uspekhi Fizicheskikh Nauk 45 (1951), 439-444
N.A. Figurovsky et al., Vest. Mosk. Univ. 20(1) (1979), 87-90

REUTER, BAPTIST [1863-]
JV 9,134; RSC 16,344; NUC 490,57
Anon., Chem. Z. 44 (1920), 232

REUTER, FERDINAND [1879-1942]
JV 19,21; NUC 490,65

REVERDIN, FRÉDÉRIC [1849-1931]
STELLING-MICHAUD 5,309; WWWS, 1413; POGG 3,1113-1114; 4,1236-1237;
5,1040-1041; 6,2157
A. Pictet, Verhandl. Schw. Nat. Ges. 112 (1931), 428-437;
Helv. Chim. Acta 14 (1931), 1046-1067
F. Ullmann, Ber. chem. Ges. 64A (1931), 106-107
E.H. Huntress, Proc. Am. Acad. Arts Sci. 77 (1949), 45-46

REVERDIN, JACQUES LOUIS [1842-1929]
STELLING-MICHAUD 5,309; BOSSART, 63-65; FISCHER 2,1288-1289; IB, 246; WWWS, 1413
A. Gosset, Bull. Acad. Med. 101 (1929), 332-334
M. Michler and J. Benedum, Gesnerus 27 (1970), 169-184
E. Martin, Rev. Med. Suisse Rom. 91 (1971), 923-928
H. Reverdin, Jacques Louis Reverdin. Aarau 1971
R. Mayer, Rev. Med. Suisse Rom. 100 (1980), 1020-1025

REWALD, BRUNO [1883-1947]
BHDE 2,964; TLK 3,527-528; POGG 6,2157-2158; 7a(3),742
Anon., Chem. Ind. 1947, p.710

REY-PAILHADE, JOSEPH de [1850-1936]
POGG 4,1237; 5,1419; 6,2159; 7b,4327
V. Brustier, Bull. Soc. Chim. [5]3 (1936), 943-944

REYCHLER, ALBERT [1854-1938]
POGG 4,1237; 5,1041-1042; 6,2160; 7b.4331
H. Wuyts, Bull. Soc. Chim. Belg. 48 (1939), v-xv

REYE, WILHELM [1871-1916]
JV 13,247; SG [2]14,519; NUC 490,682

REYERSON, LLOYD HILTON [1893-1969]
AMS 11,4402; NCAB 55,61-62; MILES, 404-405; WWAH, 505; WWWS, 1414; POGG 6,2160-2161; 7b,4331-4333
Anon., Chem. Eng. News 33 (1955), 3292; 47(44) (1969), 74

REYHER, CARL PETER [1846-1890]
WELDING, 626; LEVITSKI 2,283-287
J. Brennsohn, Die Aerzte Livlands, p.333. Mitau 1905
O.H. Wangensteen, Surgery 74 (1973), 641-649

REYNIER, HENRI FRÉDÉRIC [1824-1902]
Schweizerisches Geschlechterbuch 2,431

REYNOLDS, JAMES EMERSON [1844-1920]
DNB [1912-1921], pp.455-456; WWWS, 1414; POGG 3,1116; 4,1239; 5,1042
T.E.T., Proc. Roy. Soc. A97 (1920), iii-vi; J. Chem. Soc. 117 (1920), 1633-1637

REYNOSO, ALVARO [1829-1889]
SG [1]12,111; POGG 3,1116

REZNIKOFF, PAUL [1896-1984]
AMS 11,4408; IB, 246; KAGAN(JA), 128-129

RHEINBOLDT, HEINRICH [1891-1955]
BHDE 2,965; LDGS, 24; POTSCH, 361; POGG 6,2162; 7a(3),744
Q. Mingoia et al., Selecta Chimica 15 (1956), 7-37; 16 (1957), 5-26
L. Giesbrecht, Chem. Ber. 93(3) (1960), i-xii

RHOADES, MARCUS MORTON [1903-1991]
AMS 14,4182
M.M. Rhoades, Ann. Rev. Gen. 18 (1984), 1-29

RHOADS, CORNELIUS PACKARD [1898-1959]
DAB [Suppl.6],537-538; AMS 9(II),930; ICC, 238-240; WWAH, 506; WWWS, 1415
Anon., Brit. Med. J. 1959(II), pp.309-310
C.C. Stock, Cancer Research 20 (1960), 409-411
W.B. Castle, Trans. Assoc. Am. Phys. 73 (1960), 25-28

RHODIUS, GUSTAV [1830-1901]
Anon., Ber. chem. Ges. 34 (1901), 4386; Pharm. Z. 46 (1901), 1024

RHORER, LASZLO (LADISLAUS) von [1874-1937]
MEL 2,518-519
G. Orban, Magyar Radiologia 7 (1955), 1-6

RHUMBLER, LUDWIG [1864-1939]
KURSCHNER 4,2363; BK, 293-294; WAGENITZ, 145-146; IB, 247; RSC 18,164
J. Spek, Protoplasma 33 (1939), i-iv
E. Küster, Z. wiss. Mikr. 56 (1939), 257-258
F. Schwerdtfeger and H. Eidman, Forstarchiv 15 (1939), 260-268

RIBAN, ALEXANDRE JOSEPH [1838-1917]
POGG 3,1116-1117; 4,1239; 5,1042
Anon., Nature 99 (1917), 247
C. Poulenc, Bull. Soc. Chim. [4]21 (1917), 106

RIBBERT, HUGO [1855-1920]
DBJ 2,758; FISCHER 2,1290-1291; PAGEL, 1372-1373; BONN, 321; WWWS, 1416; SG [2]14,574; [3]9,211-212
P. Prym, Deutsche med. Wchschr. 47 (1921), 22-24
B. Fischer-Wasels, Verhandl. path. Ges. 28 (1935), 372-376
S. Jarcho, Am. J. Card. 30 (1972), 865-867
M. Mettler, Der Pathologe Hugo Ribbert. Zurich 1991

RICCI, CARLO [1923-1989]
B.L. Horecker, Ital. J. Biochem. 40 (1991), Suppl.1, 1-14

RICE, FRANCIS OWEN [1890-1989]
AMS 14,4197; POGG 6,2165; 7b,4345-4346
G.N. Kowkabany, Capitol Chemist 7 (1957), 74-75
Anon., Chem. Eng. News 67(22) (1989), 62

RICE, OSCAR KNEFLER [1903-1978]
AMS 13,3672; BURSEY, 150-153; POGG 6,2166; 7b,4347-4350
B. Widom and R.A. Marcus, Biog. Mem. Nat. Acad. Sci. 58 (1989), 425- 456

RICH, ALEXANDER [1924-]
AMS 15(6),150; WWA 1980-1981, p.1767; IWW 1982-1983, p.1109; WWWS, 1417

RICH, ARNOLD RICE [1893-1968]
AMS 11,4418; KOREN, 234; WWAH, 508; WWWS, 1417
I.L. Bennett, Trans. Assoc. Am. Phys. 81 (1968), 20-21
A.M. Harvey, Johns Hopkins Med. J. 136 (1975), 212-219
E.H. Oppenheimer, Biog. Mem. Nat. Acad. Sci. 50 (1979), 331-350

RICHARDS, ALFRED NEWTON [1876-1966]
DSB 18,734-736; AMS 10,3361; CB 1950, pp.488-490; DAMB, 632-633; PARASCANDOLA, 47-50; ICC, 234-236; WWAH, 508; WWWS, 1418
J.T. Wearn, Trans. Assoc. Am. Phys. 79 (1966), 68-73
C.F. Schmidt, Pharmacologist 8 (1967), 95-97; Biog. Mem. Fell. Roy. Soc. 13 (1967), 327-342; Biog. Mem. Nat. Acad. Sci. 42 (1971), 271-318
I. Starr et al., Ann. Int. Med. 71 Suppl.8 (1969), 1-89
D.W. Bronk, Am. Phil. Soc. Year Book 1971, pp.143-153; Persp. Biol. Med. 19 (1976), 413-422
H.W. Davenport, Physiologist 21(6) (1978), 25-30
D.Y. Cooper, Trans. Coll. Phys. Phila. [5]6 (1984), 63-73

RICHARDS, DICKINSON WOODRUFF [1895-1973]
AMS 12,5218; NCAB I,336-337; MH 1,391-392; MSE 3,24; DAMB, 633-634; ICC, 233-234; WWAH, 508; WWWS, 1418
A.F. Cournand, American Journal of Medicine 57 (1974), 312-313; Biog. Mem. Nat. Acad. Sci. 58 (1989), 459-487

RICHARDS, FRANCIS JOHN [1901-1965]
DSB 11,416; DNB [1961-1970],877-878; WW 1965, p.2573; DESMOND, 518
W.W. Schwabe, Nature 205 (1965), 853-854
H.K. Porter, Biog. Mem. Fell. Roy. Soc. 12 (1966), 423-436

RICHARDS, FRANK FREDERICK [1928-]
AMS 15(6),153

RICHARDS, FREDERIC MIDDLEBROOK [1925-]
AMS 15(6),153; IWW 1982-1983, p.1100

RICHARDS, THEODORE WILLIAM [1868-1928]
DSB 11,416-418; DAB 15,556-559; MILES, 407-408; FORBES, 167-170; W, 443; POTSCH, 361-362; ABBOTT-C,122; WWWS,1419; POGG 4,1242-1244; 5,1043-1044; 6,2166-2167
H.B.D., Proc. Roy. Soc. A121 (1928), xxix-xxxiv
G.P. Baxter, Science 68 (1928), 333-339
H. Hartley, J. Chem. Soc. 133 (1930), 1937-1969
G.S. Forbes, J. Chem. Ed. 9 (1932), 453-458
A.J. Ihde, Science 164 (1969), 647-651
J.B. Conant, Biog. Mem. Nat. Acad. Sci. 44 (1974), 251-286
S.J. Kopperl, Ambix 23 (1976), 165-174; J. Chem. Ed. 60 (1983), 738-739
E.B. Wilson, Welch Foundation Conferences on Chemical Research 20 (1977), 106-112

RICHARDSON, BENJAMIN WARD [1828-1896]
DSB 11,418-419; DNB 22,1169-1170; HIRSCH 4,793; PAGEL,1374-1375; O'CONNOR, 65-67; WWWS,1419
B.W. Richardson, Vita Medica. London 1897
T.C.A., Nature 57 (1898), 265-266; Proc. Roy. Soc. 75 (1905), 51-52
A.S. MacNalty, A Biography of Sir Benjamin Ward Richardson. London 1950
W.F. Bynum, Bull. Hist. Med. 44 (1970), 518-538
J. Parascandola, Pharm. Hist. 13 (1971), 3-10
E.H. Ackerknecht, Gesnerus 45 (1988), 317-321

RICHARDSON, CHARLES CLIFTON [1935-]
AMS 15(6),156; WWWS, 1419

RICHARDSON, HENRY BARBER [1889-1963]
AMS 9(II),934
P. Reznikoff, Trans. Assoc. Am. Phys. 77 (1964), 27-28

RICHARDSON, THOMAS [1816-1867]
DNB 16,1135-1136; RSC 5,190
Anon., Proc. Roy. Soc. Edin. 6 (1867), 198-199

RICHAUD, ALBERT [1867-1925]
POGG 6,2170
G.M., J. Pharm. Chim. [8]2 (1925), 311-314
H. Busquet, Paris Médical 58 (1925), 416

RICHE, ALFRED [1829-1908]
VAPEREAU 6,1355; BOURQUELOT, 83-84; POGG 2,630-631; 3,1119; 4,1245
M. Hanriot, Bull. Soc. Chim. [4]3 (1908), i-xiv
G. Dillemann, Produits et Problèmes Pharmaceutiques 27 (1972), 416-417

RICHERAND, ANTHELME BALTHASAR [1779-1840]
HOEFER 42,253-255; HIRSCH 4,796-797; DECHAMBRE [3]5,21-24; CALLISEN 16,66-76; 31,437-439
F. Dubois, Bull. Acad. Med. 16 (1852), lxxix-civ

RICHERT, DAN ARNOLD [1915-1971]
AMS 11,4429; WWWS, 1420

RICHET, CHARLES [1850-1935]
DSB 11,425-432; GROTE 7,185-220; GENTY 5,157-188; FISCHER 2,1293; PAGEL, 1377-1378; W, 443; WWWS, 1420
C.R. Richet, Souvenirs d'un Physiologiste. Paris 1933
H. Roger, Presse Med. 43 (1935), 2043-2045
A. Mayer, Bull. Acad. Med. 115 (1936), 51-64
G. Roussy, Bull. Acad. Med. 129 (1945), 725-731
E.H. Huntress, Proc. Acad. Arts Sci. 78 (1950), 27-28
M. Juri, Charles Richet Physiologiste. Zurich 1965

RICHTER, DEREK [1907-]
WWWS, 1421
Who's Who of British Scientists 1971-1972, p.712

RICHTER, ERNST [1861-1917]
JV 2,26; NUC 493,623
Anon., Chem. Z. 41 (1917), 918

RICHTER, MAX [1867-1932]
FISCHER 2,1294

RICHTER, VICTOR von [1841-1891]
WELDING, 628-629; BLOKH 2,611-612; POTSCH, 362-363; POGG 3,1120-1121; 4,1246
G. Prausnitz, Ber. chem. Ges. 24(3) (1891), 1123-1130

RICKENBERG, HOWARD [1922-]
AMS 15(6),165; BHDE 2,068

RICKER, (ALBIN) HERMANN [1811-1852]
NB,323-324; HB 3,79

RICKER, GUSTAV [1870-1948]
FREUND 2,327-331
M. Nordmann, Verhandl. path. Ges. 32 (1950), 449-452
R. Hofmann, NTM 1(4) (1963), 122-123
H. Eggers, Med. Klin. 63 (1968), 693-695
A. Hecht and W. Kühne, Pathologe 11 (1990), 313-315

RICKES, EDWARD LAWRENCE [1912-]
AMS 14,4203

RICKETTS, HOWARD TAYLOR [1871-1910]
DSB 11,442-443; NCAB 34,543; DAMB, 637-638; SACKMANN, 280-282; ABBOTT, 110; W, 444; WWWS, 1422
Anon., Chicago Medicine 64 (1962), 21-22
V.A. Harden, Rocky Mountain Spotted Fever, pp.47-71. Baltimore 1990

RICKMANN, WILHELM [1870-1916]
NUC 494,155
Anon., Chem. Z. 40 (1916), 250; Z. angew. Chem. 29(III) (1916), 203

RICORD, PHILIPPE [1800-1889]
VAPEREAU 5,1542; HIRSCH 4,806-807; PAGEL, 1380-1382; WWWS, 1422; CALLISEN 16,101; 31,454-456
J.E. Péan, Bull. Acad. Med. 22 (1889), 377-382
P. Huard and M.J. Imbert-Huard, Gaz. Med. Fran. 82 (1975), 1687-1692

RIDDLE, OSCAR [1877-1968]
DSB 18,736-738; AMS 11,4435; IB, 248; WWAH, 511; WWWS, 1422
R.W. Bates, Endocrinology 85 (1969), 185-193
G.W. Corner, Am. Phil. Soc. Year Book 1971, pp.154-158; Biog. Mem. Nat. Acad. Sci. 45 (1974), 427-465

RIDEAL, ERIC KEIGHTLEY [1890-1974]
DSB 18,738-743; DNB [1971-1980],723; WW 1973, p.2729; CAMPBELL, 191-192; MH 1,393-395; WWWS, 1422; POTSCH, 363; POGG 5,1048; 6,2172-2173; 7b,4362-4367
J.F. Padday, J. Coll. Sci. 53 (1975), 1-5
D.D. Eley, Chem. Ind. 1975, pp.800-806; Biog. Mem. Fell. Roy. Soc. 22 (1976), 381-413; Adv. Catalysis 26 (1977), xiii-xv

RIEBELING, CARL [1900-1961]
KURSCHNER 9,1654-1655; POGG 7a(3),756
Anon., Deutsche med. Wchschr. 86 (1961), 1367

RIECKE, EDUARD [1845-1915]
DSB 11,445-447; DBJ 1,339; WWWS, 1423; POGG 3,1122; 4,1247-1248; 5,1049; 6,2174; 7aSuppl.,546-547
W. Voigt, Phys. Z. 16 (1915), 219-221
A. Sommerfeld, Bayer. Akad. Wiss. Jahrbuch 1916, pp.115-118

RIECKE, WOLDEMAR [1872-1937]
JV 13,155; NUC 494,405
Anon., Z. angew. Chem. 51 (1938), 448

RIECKHER, THEODOR [1818-1888]
HEIN, 526-527; TSCHIRCH, 1112; SCHELENZ, 688; FERCHL, 444; POGG 2,639-640
A. Wankmüller, Beitr. Wurtt. Apothekgesch. 7 (1966), 83

RIED, WALTER [1920-]
KURSCHNER 13,3131; WWWS, 1423; POGG 7a(3),760-761
G. Oromek and B. Heinz, Chem. Z. 104 (1980), 75; 109 (1985), 86

RIEDEL, ADOLF [1877-]
JV 17,158; NUC 494,412

RIEDEL, CARL [1856-1943]
RSC 11,178
Anon., Chem. Z. 67 (1943), 354

RIEDEL, (JOHANN DANIEL) GUSTAV [1816-1886]
HUHLE-KREUTZER, 176-183
Anon., Chem. Z. 10 (1886), 1619; Pharm. Z. 26 (1886), 206

RIEDER, HERMANN [1858-1932]
KURSCHNER 4,2374; FISCHER 2,1296-1297
G. Hammer, Münch. med. Wchschr. 101 (1959), 441-449

RIEGEL, BYRON [1906-1975]
AMS 12,5238; WWWS, 1423
Anon., Chem. Eng. News 53(21) (1975), 5

RIEGEL, EMIL [1817-1873]
HEIN, 529-530; FERCHL, 445
A. Wankmüller, Deutsche Apoth. Z. 113 (1973), 1127-1131

RIEGEL, FRANZ [1843-1904]
BJN 9,330-334; 10,92*; FISCHER 1,1298; PAGEL, 1298

RIEHL, GUSTAV [1855-1943]
OBL 9,1299; FISCHER 2,1299; SIEGL, 44-52
O. Kren, Wiener med. Wchschr. 75 (1925), 326-327
L. Arzt, Wiener klin. Wchschr. 67 (1955), 101-103

RIEHMANN, KURT [-1919]
Anon., Chem. Z. 43 (1919), 89

RIEMERSCHMID, CARL [1860-1915]
RSC 9,869; NUC 494,587

Anon., Chem. Z. 39 (1915), 993

RIEMSCHNEIDER, RANDOLPH [1920-]
KURSCHNER 13,3139; POGG 7a(3),769-771

RIESELL, ALBERT [1844-1889]
RSC 8,749; SG [1]12,220; NUC 494,652
Anon., Pharm. Z. 34 (1889), 455

RIESSER, OTTO [1882-1949]
NB, 325; KALLMORGEN, 383; AUERBACH, 347; BHDE 2,971; LDGS, 78; IB, 248; TLK 3,532; KOREN, 235; POGG 6,2177-2178; 7a(3),774-775
G. Taubman, Arch. exp. Path. Pharm. 209 (1950), i-viii

RIGLER, RUDOLF [1898-]
KURSCHNER 13,3143; KALLMORGEN, 383; IB, 248; POGG 7a(3),776-777

RIIBER, CLAUS NISSEN [1867-1936]
NBL 11,451-453; POGG 5,1053; 6,2180; 7b,4372
E. Berner, Norske Vid. Akad. Arbok 1936, pp.69-72
S. Schmidt-Nielsen, Kongelige Norske Videnskabernes Selskabs Foreningen 8 (1936), 99-110

RILLIET, ALBERT [1848-1904]
STELLING-MICHAUD 5,340

RILLIET, AUGUSTE ROBERT [1880-]
NUC 495,379
A. Choisy, Généalogies Genevoises, p.363. Geneva 1947

RILLIET, FRÉDÉRIC [1814-1861]
STELLING-MICHAUD 5,342-343; HIRSCH 4,816; PAGEL, 1390; DECHAMBRE [3]5,52-53; CALLISEN 31,466-467
P. Fischer, F. Rilliet und E. Barthez und ihre Traité. Zurich 1966

RILLIEUX, NORBERT [1806-1894]
MATSCHOSS, 226-227; FERCHL, 446
Anon., Chem. Z. 18 (1894), 2063

RIMBACH, EBERHARD [1852-1933]
HEIN, 531-532; TLK 3,532; POGG 4,1253-1254; 5,1053; 7a(3),777-778
R. Anschütz, Ber. chem. Ges. 67A (1934), 73-79

RIMELE, EUGEN [1883-1966]
JV 24,339; NUC 495,414

RIMINGTON, CLAUDE [1902-1993]
WW 1981, pp.2193-2194; IWW 982-1983, p.1116; WWWS, 1425
A.M. Battle, Int. J. Biochem. 9(12) (1978), i-v

RIMPAU, WILHELM [1842-1903]
BJN 8,259-261,93*; HINK 1,505-507
K. von Rümker, Mitteldeutsche Lebensbilder 1 (1926), 376-389
K. Meyer, Zeitschrift für Pflanzenzüchtung 32 (1953), 225-232

RINDFLEISCH, GEORG EDUARD [1836-1908]
BJN 13,76*; HIRSCH 4,817-818; PAGEL, 1390-1392; ROEDER, 112-125
M. Borst, Münch. med. Wchschr. 53 (1906),2448-2450; 55 (1908),2682-2683; Verhandl. Phys. Med. Würz. NF41 (1909), 103-130; Verhandl. path.Ges. 28 (1935), 546-548
W. von Brunn, Münch. med. Wchschr. 83 (1936), 2058-2061

RINECKER, FRANZ von [1811-1883]
ADB 28,628-629; LF 5,279-288; HIRSCH 4,818-819; CALLISEN 31,467
C. Gerhardt, Sitz. Phys. Med. Würzburg 1883, pp.120-131
R. Pfeffer, Professor Franz von Rinecker. Würzburg 1981

RINGER, ADOLPH IRVING [1882-]
AMS 6,1186; KAGAN(JA), 97.571-572

RINGER, SYDNEY [1835-1910]
DSB 11,462-463; DNB [1901-1911] 3,200-201; HIRSCH 4,819; WWWS, 1436; O'CONNOR, 153-155
Anon., Lancet 1910(II),pp.1386-1387; Brit. Med. J. 1910(II),pp.1384-1386
B. Moore, Biochem. J. 5 (1911), i-xix
E.A.S., Proc. Roy. Soc. B84 (1912), i-iii
J.A. Lee, Anaesthesia 36 (1981), 1115-1121
W.G. Naylor, Journal of Molecular and Cellular Cardiology 16 (1984), 113-116
W.B. Fye, Circulation 69 (1984), 849-853

RINGER, WILHELM EDUARD [1874-1953]
POGG 6,2181; 7b,4375
W.P. Jorissen, Chem. Wkbl. 50 (1954), 265-271

RINNE, FRIEDRICH [1863-1933]
FREUND 3,333-341; KIEL, 157; WWWS, 1426; POGG 4,1254; 5,1054-1055; 6,2182-2183; 7a(3),778-779

RINTOUL, WILLIAM [1870-1936]
Anon., Chem. Ind. 1936, pp.709-710

RIPPEL-BALDES, AUGUST [1888-1970]
KURSCHNER 13,3149; WAGENITZ, 146-147; WWWS, 1426; POGG 7a(3),779-781
H. Bortels, Ber. bot. Ges. 84 (1971), 289-298

RIPPER, MAXIMILIAN [1864-1928]
OBL 9,173-174
Anon., Ost. Chem. Z. 31 (1928), 132; Chem. Z. 52 (1928), 566-567

RIS, HANS [1914-]
AMS 15(6),188; WWA 1980-1981, p.2782; WWWS, 1426

RISSOM, JOHANNES [1868-1954]
DRULL, 220; JV 13,37; NUC 496,246
Wer Ist Wer 9,1275

RIST, ÉDOUARD [1871-1956]
J. Rolland, Presse Med. 64 (1956), 1787-1790
L. Bezanquen et al., Maroc Médical 36 (1957), 57-59

RISTENPART, EUGEN [1873-1953]
TLK 3,533; POGG 6,2186; 7a(3),781-782

RITCHIE, ARTHUR DAVID [1891-1967]
IB, 248
Who Was Who 1961-1970, p.960

RITTENBERG, DAVID [1906-1970]
AMS 11,4456; WWWS, 1427
Anon., Chem. Eng. News 48(24) (1970), 87
R.M. Caprioli, in Biochemical Applications of Mass Spectrometry (G.R. Waller, Ed.), pp.i-x. New York 1972

RITTER, BERNHARD [1804-1893]
HIRSCH 4,827; PAGEL, 1394; CALLISEN 31,474-476
Anon., Leopoldina 29 (1893), 162

RITTER, CHARLES ÉMILE EUGÈNE [1837-1884]
BERGER-LEVRAULT, 198

RITTER, ERNST [1875-1920]
RSC 18,261; NUC 496,403

RITTER, JOHANN WILHELM [1776-1810]
DSB 11,473-475; ADB 28,675-678; HEIN, 534; BRAZIER, 21-22,25; W, 445; POTSCH, 365; WWWS, 1428; POGG 2,652-653; 7aSuppl.,553-555
W. Ostwald, Abhandlungen und Vorträge, pp.359-383. Leipzig 1904
C. Klinckenstroem, Arch. Gesch. Naturwiss. 9 (1922), 68-85
P. Hahn, Schlesische Lebensbilder 3 (1928), 202-210

D. Hufmeier, Sudhoffs Arch. 45 (1961), 225-234
W.D. Wetzels, Johann Wilhelm Ritter. Berlin 1973
S. Dietzsch, Berichte zur Wissenschaftgeschichte 9 (1986), 191-197

RITTER, KARL [1880-]
JV 21,298; NUC 496,442
Anon., Nachr. Chem. Techn. 9 (1961), 212; 14 (1966), 249

RITTER, WILLIAM EMERSON [1856-1944]
DSB 18,745-746; DAB [Suppl.3],635-636; AMS 7,1486; IB, 249
F.B. Sumner, Science 99 (1944), 335-338
W.E. Ritter, Charles Darwin and the Golden Rule, pp.vii-xx,381-392. Washington, D.C. 1954

RITTER, WOLFGANG [1898-]
JV 38,440

RITTERSHAUSEN, FRIEDRICH [-1875]
Anon., Ber. chem. Ges. 8 (1875), 1607,1693

RITTHAUSEN, HEINRICH [1826-1912]
DSB 18,747-748; BJN 18,54*; BLOKH 2,612; SCHAEDLER, 105; DZ, 1192; POGG 2,655; 3,1125; 4,1256
T.B. Osborne, Bioch. Bull. 2 (1913), 335-346
A. Stutzer, Ber. chem. Ges. 47 (1914), 591-593

RITZ, HANS [1884-]
JV 27,377; SG [3]9,244; NUC 496,500

RIVA, ALBERTO [1844-1916]
EI 29,488; FISCHER 2,1304-1305
Anon., Brit. Med. J. 1916(I), p.869

RIVA-ROCCI, SCIPIONE [1863-1937]
DSB 11,481-482; EI 29,489; FISCHER 2,1305; SIGRIST, 83-84
E. Benassi, Minerva Med. 54 (1963), 3766-3371

RIVERS, THOMAS MILTON [1888-1962]
DAB [Suppl.7],648-649; AMS 10,3391; FISCHER 2,1305; DAMB, 639- 640; ICC, 193-195; WWAH, 514
R.E. Shope, J. Bact. 84 (1962), 385-388
F.L. Horsfall, Am. Phil. Soc. Year Book 1963, pp.190-196; Trans. Assoc.
Am. Phys. 76 (1063), 16-19; Biog. Mem. Nat. Acad. Sci. 38 (1965), 263-294
S. Benison, Tom Rivers: Reflections on a Life in Medicine and Science. Cambridge, Mass. 1967

RIVETT, DAVID [1885-1961]
WW 1959, p.2580; POGG 6,2188; 7b,4388
N.S. Bayliss, Proc. Chem. Soc. 1962, pp.91-93
H.R. Marston, Biog. Mem. Fell. Roy. Soc. 12 (1966), 437-455

ROAF, HERBERT ELDON [1881-1952]
WW 1952, p.2434; O'CONNOR(2), 365-366; IB, 249; WWWS, 1429
Anon., Nature 170 (1952), 910-911; Lancet 1952(II), p.784;
Brit. Med. J. 1952(II), p.782

ROBBINS, FREDERICK CHAPMAN [1916-]
AMS 15(6),199; WWA 1908-1981, p.2782; IWW 1982-1983, p.1119; WWWS, 1429; MH 1,396; STC 2,421-422; MSE 3,32-33

ROBBINS, PHILLIPS WESLEY [1930-]
AMS 15(6),200

ROBBINS, WILLIAM JACOB [1890-1978]
AMS 13,3711; NCAB I,178-179; IB, 249; WWWS, 1429
D.R. Goddard, Am. Phil. Soc. Year Book 1979, pp.100-102
F. Kavanaugh and A. Harvey, Biog. Mem. Nat. Acad. Sci. 60 (1991), pp.293-328

ROBERT, JULIUS [1826-1888]
MATSCHOSS, 229-230; POTSCH, 365-366

Anon., Chem. Z. 12 (1888), 1743

ROBERTS, JOHN D. [1918-]
AMS 15(6),206; WWA 1980-1981, p.2791; IWW 1982-1983, p.1121; WWWS, 1430; MH 2,452-454; MSE 3,33-34
J.D. Roberts, The Right Place at the Right Time. Washington, D.C. 1990

ROBERTS, RICHARD BROOKE [1910-1980]
AMS 14,4238; WWA 1976-1977, p.2647; AO 1980, pp.219-220; WWWS, 1431

ROBERTS, WILLIAM [1830-1899]
DNB 22,1170-1172; FISCHER 2,1306; PAGEL, 1396; O'CONNOR, 230-
231; WWWS, 1431; RSC 11,193; 18,236
Anon., Lancet 1899(I), pp.1184-1188; Brit. Med. J. 1899(I), pp.1063-1066
D.L. Leech, Manchester Medical Chronicle [3]1 (1899), 157-189
J.R.B., Proc. Roy. Soc. 75 (1905), 68-71

ROBERTSON, ALEXANDER [1896-1970]
DNB [1961-1970],883-885; WW 1965, p.2609; POGG 6,2189-2190; 7b,4392-4395
Anon., Nature 226 (1970), 193-194
R.D. Haworth and W.H. Whalley, Biog. Mem. Fell. Roy. Soc. 17 (1971), 617-642

ROBERTSON, GEORGE JAMES [1898-1939]
Anon., Chem. Ind. 1939, p.863
J. Read, J. Chem. Soc. 1940, pp.104-105

ROBERTSON, JOHN MONTEATH [1900-1989]
WW 1981, p.2215; IWW 1982-1983, p.1122; CAMPBELL, 195-196; WWWS, 1431; MH 1,396-397; MSE 3,34-35
G.A. Sim, Chem. Brit. 26 (1990), 871

ROBERTSON, MURIEL [1883-1973]
WW 1970, p.2654; IB, 249
A. Bishop and A.A. Miles, Biog. Mem. Fell. Roy. Soc. 20 (1974), 317-347
A.A. Miles, J. Gen. Microbiol. 95 (1976), 1-8

ROBERTSON, OSWALD HOPE [1886-1966]
AMS 10,3406; WWWS, 1432
M. Hamburger, Trans. Assoc. Am. Phys. 79 (1966), 76-78
L.T. Coggeshall, Biog. Mem. Nat. Acad. Sci. 42 (1971), 319-338
J.C. Allen, Lab World 1981(Feb.), pp.20-23

ROBERTSON, RUTHERFORD NESS [1913-]
WW 1982, p.1888; IWW 1982-1983, p.1122; SRS, 69; MH 2,454-455; WWWS, 1432

ROBERTSON, THORBURN BRAILSFORD [1884-1930]
WW 1929, p.2607; CHITTENDEN, 314-316; IB, 249; POGG 5,1057-1058; 6,2190
W. Ostwald, Koll. Z. 53 (1930), 384
H.W., Biochem. J. 24 (1930), 577-578
H.R. Marston and M.D. Dawbarn, Austral. J. Exp. Biol. 9 (1932), 1-21
A.V. Everitt, Gerontologia 4 (1960), 60-75
A.W. Burgess, TIBS 14 (1989), 117-120

ROBIN, ALBERT [1847-1928]
FISCHER 2,1307; WININGER 7,399; WWWS, 1432
G. Pouchet, Bull. Acad. Med. 100 (1928), 943-947

ROBIN, CHARLES [1821-1885]
DSB 11,491-492; VAPEREAU 5,1554-1555; GE 28,754; LAROUSSE 13,1264-1265; GENTY 3,1-16; HIRSCH 4,838-839; PAGEL, 1397-1398; BULLOCH, 392-393; GUIART, 225; WWWS, 1432
A. Laboulbène, Ann. Soc. Entomol. [6]5 (1885), 467-472
G. Pouchet, J. Anat. Physiol. 22 (1886), i-clxxxiv
A. Cartaz, Les Médecins Bressans, pp.229-238. Paris 1902
V. Genty, Un Grand Biologiste: Charles Robin 1821-1885. Lyon 1931

ROBIN, ÉDOUARD [1808-]
TSCHIRCH, 1112; FERCHL, 448; RSC 5,235; 8,762; 11,195; NUC 498,291; CALLISEN 31,487; POGG 2,665
A. Lacoste, Essai Biographique sur les Travaux en Chemie de M. Édouard Robin. Paris 1853

ROBINET, STÉPHANE [1796-1869]
LAROUSSE 13,1266; FERCHL, 448-449; COLE, 466; POGG 2,665-666; 3,1128
L. Figuier, Ann. Scient. 14 (1869), 583

ROBINOW, CARL [1909-]
AMS 15(6),216; BHDE 2,973-974

ROBINSON, GERTRUDE MAUD [1886-1954]
POGG 6,2191-2192; 7b,4399-4400
W. Baker, Nature 173 (1954), 566-567
J.L. Simonsen, J. Chem. Soc. 1954, pp.2667-2668

ROBINSON, ROBERT [1886-1975]
DNB [1971-1980],729-731; WW 1975, p.2703; MH 1,397-399; STC 2,425-426; MSE 3,36-37; CAMPBELL, 197-109; SCHWERTE 1,184-190; WWWS, 1434; POTSCH, 366; SRS, 38; W(3rd Ed.), 607-608; POGG 6,2192-2193; 7b,4403-4412
R. Robinson, Chem. Brit. 10 (1974), 54-57; Memoirs of a Minor Prophet.70 Years of Organic Chemistry. New York 1976
A.R. Todd, Nature 253 (1975), 761
R.N. Chakravarti, Science and Culture 42 (1976), 208-210
A.R. Todd and J.W. Cornforth, Biog. Mem. Fell. Roy. Soc. 22 (1976), 415-527
A.J. Birch, Journal and Proceedings of the Royal Society of New South Wales 109 (1976), 151-160
M.D. Salzmann, Chem. Brit. 22 (1986), 543-544
A.R. Todd et al., Natural Product Reports 4 (1987), 3-87
T.I. Williams, Robert Robinson, Chemist Extraordinary. Oxford 1990

ROBIQUET, HENRI EDMOND [1822-1860]
BLOKH 2,613-614; TSCHIRCH, 1112-1113; DECHAMBRE [3]5,94; POGG 2,667

ROBIQUET, JEAN PIERRE [1780-1840]
DSB 11,494-495; GE 28,759; HOEFER 42, 444; BOURQUELOT, 51-52; WWWS, 1434; POTSCH, 366; BLOKH 2,614; PHILIPPE, 799-800; BALLAND,98-99; BALLAND(2), 181-182; CALLISEN 16,205-210; 31,488-489; POGG 2,666-667
J. Girardin, Précis Anal. Acad. Rouen 42 (1840), 36-40
B. Bussy, J. Pharm. Chim. 27 (1841), 220-242
G. Dillemann, Produits et Problèmes Pharmaceutiques 26 (1971), 427-428

ROBISON, ROBERT [1883-1941]
DNB [1941-1950],730-731; WW 1941, p.2696; STC 2,425-426; POGG 6,2193; 7b,4412
W.T.J. M[organ], Biochem. J. 35 (1941), 1081-1087; J. Chem. Soc. 1942, pp.67-69
C.R. Harington, Obit. Not. Fell. Roy. Soc. 3 (1941), 929-939
J.C.G. Ledingham, Nature 148 (1941), 77-78

ROBLIN, RICHARD OWEN [1907-1985]
AMS 12,5299; WWWS, 1435

ROBSCHEIT-ROBBINS, FRIEDA [1895-1973]
AMS 10,3415

ROBSON, JOHN MICHAEL [1900-1982]
WW 1981, p.2225

ROBSON, WILLIAM [1893-1975]
WW 1975, p.2707

ROCHA e SILVA, MAURICIO [1910-1983]
IWW 1982-1983, p.1125

ROCHE, ANDRÉE [1900-1936]
M. Nicloux, Bull. Soc. Chim. Biol. 18 (1936), 1888-1894
L. Cornil and G. Jayle, Presse Med. 44 (1936), 1662

ROCHE, JEAN [1901-1991]
QQ 1978-1979, pp.1364-1365; IWW 1982-1983, p.1125; STC 2,426-427; POGG 6,2195; 7b,4416-4435
J. Roche, Comp. Bioch. Physiol. 47B (1974), 521-529

ROCHLEDER, FRIEDRICH [1819-1874]
ADB 28,726-727; OBL 9,192; WURZBACH 26,216; BLOKH 2,614-615; KNOLL,43-45; FERCHL, 449-450; TSCHIRCH, 1113; SCHAEDLER, 106; VOGEL, 61-78; POTSCH, 366-367; POGG 2,669-670; 3,1130; 7aSuppl.,555
W.F. Gintl, J. prakt. Chem. 118 (1874), 457-473
H. Hlasiwetz, Ber. chem. Ges. 8 (1875), 1702-1712
J. Stefan, Alm. Akad. Wiss. Wien 25 (1875), 195-212
U. Egert, Personalbibliographien ...der Philosophischen Fakultät zu Prag von 1800 bis 1860, pp.119-133. Erlangen-Nürnberg 1970
H. Cassebaum, NTM 15 (1978), 23-36

ROCHOW, EUGENE GEORGE [1909-]
AMS 17(6),242; POTSCH, 367

ROCHUSSEN, FRANK HERMANN [1873-1955]
NUC 449,410-411; POGG 7a(3),785-786

ROCK, HENRY JOSEPH [1864-1946]
Anon., J. Am. Med. Assn. 132 (1946), 876

ROCK, JOHN [1890-1984]
T. Ziporyn, J. Am. Med. Assn. 253 (1985), 18

ROCKWOOD, ELBERT WILLIAM [1860-1935]
AMS 5,945; BERG, 38-48; IB, 250; WWAH, 517
J.N. Pearce, Science 82 (1935), 294-295; Ind. Eng. Chem. News Ed. 13 (1935), 356

RODEBUSH, WORTH HUFF [1887-1959]
AMS 9(I),1628; NCAB 50,460-461; POGG 6,2195; 7b,4435-4437
W.H. Rodebush, Frontiers in Chemistry 3 (1945), 137-161
C.S. Marvel and F.T. Wall, Biog. Mem. Nat. Acad. Sci. 36 (1962), 277-288
D. Quane, Bull. Hist. Chem. 7 (1990), 3-13

RODER, ANTON [1864-1931]
JV 2,245; RSC 18,250; NUC 499,658
Anon., Chem. Z. 55 (1931), 684

RODIONOV, VLADIMIR MIKHAILOVICH [1878-1954]
BSE 36,600; WWR, 478; WWWS, 1436; POGG 6,2195-2196; 7b,4437-4442
M.M. Shemyakin, Zhur. Ob. Khim. 23 (1953), 1785-1794

ROE, JOSEPH HYRAM [1892-1967]
AMS 11,4500; NCAB D,175; WWAH, 518; WWWS, 1436; POGG 6,2196; 7b,4442-4443
E.W. Rice, Clin. Chem. 30 (1984), 1575-1578

ROEDER, GEORG [1874-]
JV 15,14; RSC 15,650; NUC 500,434
Anon., Chem. Z. 38 (1914), 65

ROEHL, WILHELM [1881-1929]
KALLMORGEN, 385; OLPP, 344-346; DUNSCHELE, 146-150; POGG 7a(3),788
H. Horlein, Therapeutische Berichte 6 (1929), 259-263

RÖHMANN, FRANZ [1856-1919]
DBJ 2,731; FISCHER 2,1310-1311; BLOKH 2,615-616; BK, 295; WININGER 7,400; POGG 4,1261; 6,2196
Anon., Chem. Z. 43 (1919), 441; Z. angew. Chem. 32(III) (1919), 394
K.A. Hofmann, Ber. chem. Ges. 52A (1919), 114

ROELIG, HERMANN [1871-]
JV 14,227; RSC 17,436; NUC 500,516

RÖMER, ADOLF [1859-1924]
JV 1,197; RSC 18,258
F. Hundeshagen, Chem. Z. 48 (1924), 606

RÖMHELD, JULIUS [1827-1901]
DGB 98,393-394
Anon., Pharm. Z. 46 (1901), 357

RÖNTGEN, WILHELM CONRAD [1845-1923]
DSB 11,529-531; DBJ 5,317-326; LF 4,319-340; BLOKH 2,616-618; KLEIN, 111; MATSCHOSS, 232; DZ, 1201; ZURICH-D, 116; W, 448- 449; WWWS, 1442; POGG 3,1132-1133; 4,1261; 5,1058-1059; 6,2196-2197; 7aSuppl.,556-561
L. Zehnder, Helv. Chim. Acta 6 (1933), 608-629
O. Glasser, Wilhelm Conrad Röntgen. Springfield, Ill. 1934
W.O. Nitske, The Life of Wilhelm Conrad Röntgen. Tucson, Ariz. 1971

RÖSCH, HANS [1894-1945]
GAUSS, 308

ROESNER, HANS [1896-]
JV 26,50; NUC 500, 664

RÖSSEL, WOLFGANG [1911-1967]
KURSCHNER 10,2015

RÖSSLE, ROBERT [1876-1956]
FISCHER 2,1312; FREUND 2,333-339; ECKERT, 100-119; KIEL, 116; WWWS, 1448
W. Fischer, Deutsche med. Wchschr. 72 (1947), 40-41
W. Hueck, Münch. med. Wchschr. 98 (1956), 1098-1100
W. Doerr, Deutsches med. J. 7 (1956), 524-532; Virchows Arch. 371 (1976), 1-4
A. Werthemann, Schw. med. Wchschr. 87 (1957), 115-118
A. Linzbach, Verhandl. path. Ges. 42 (1959), 492-493
B. Krempien, Münch. med. Wchschr. 119 (1977), 325-328

RÖSSLER, FRITZ [1870-1937]
KALLMORGEN, 385; POGG 6,2198
A. Stock, Ber. chem. Ges. 71A (1938), 68-69

RÖSSLER, PAUL [1873-1936]
JV 17,172; NUC 500,689
Anon., Chemische Fabrik 9 (1936), 339

RÖSSLER, RICHARD [1897-1945]
OBL 9,209-210; DECKER, 116-119
F. Brücke, Wiener klin. Wchschr. 67 (1955), 305-306

RÖSSNER, HEINRICH [1876-1962]
JV 14,227; RSC 19,74

RÖTHIG, PAUL [1874-1940]
KURSCHNER 4,2439-2440; KALLMORGEN, 387; ASEN, 161; LDGS, 55; IB, 250

ROGER, GEORGES HENRI [1860-1946]
FISCHER 2,1313; RSC 18,265-267; WWWS, 1437
F. Trémolières, Bull. Acad. Med. 130 (1946), 316-321
G.H. Roger, Entre Deux Siècles. Paris 1947

ROGER, JACQUES [1920-1990]
J.J. Dreifuss, Gesnerus 47 (1990), 345-346

ROGER, MURIEL [1922-1981]
AO 1981, pp.768-769
Anon., Rock. Univ. Notes 13(3) (1982), 3

ROGERS, LEONARD [1868-1962]
DNB [1961-1970], 891-893; WW 1959, p.2617; WWWS, 1438
C. Wilcocks, Nature 196 (1962), 517-518
Anon., Lancet 1962(II), pp.666-667
J.S.K. Boyd, Biog. Mem. Fell. Roy. Soc. 9 (1963), 261-285
P.C.C. Garnham, Trans. Soc. Trop. Med. 62 (1968), 161-165

ROGERS, ROBERT EMPIE [1813-1884]
DAB 16,109-110; MILES, 415-416; ELLIOTT, 219-220; FERCHL, 451; WWAH, 520; WWWS, 1438-1439
W.S.W. Ruschenberger, Proc. Am. Phil. Soc. 23 (1886), 104-146
E.F. Smith, Biog. Mem. Nat. Acad. Sci. 5 (1905), 291-306

ROGET, PETER MARK [1779-1869]
DNB 17,149-151; HIRSCH 4,853-854; DESMOND, 528; FERCHL, 451; WWWS, 1439
Anon., Proc. Roy. Soc. 18 (1870), xxviii-xl
D.L. Emblen, Peter Mark Roget. London 1970
W.E. Swinton, Can. Med. Assn. J. 123 (1980), 916-921

ROGINSKI, SIMON ZALMANOVICH [1900-1970]
BSE(3rd Ed.) 22,154; WWWS, 1439; POGG 6,2204; 7b,4453-4463
O.V. Krylov, Kinetika i Kataliz 21 (1980), 821-831

ROGOFF, JULIUS MOSES [1884-1966]
AMS 10,3429-3430; IB, 251; KOREN, 235; COHEN, 197; WWWS, 1439
B.M. Fisher, Pharmacologist 9 (1967), 23-24

ROGOWICZ, JAKOB [1839-1896]
PSB 31,436-437; HIRSCH 4,855; KOSMINSKI, 419-420; KONOPKA 9,81-87
H. Dobrzycki, Medycyna 24 (1896), 1238-1242
M.M. Zweigenbaum, Gazeta Lekarska 31 (1896), 1325-1326

ROGOZINSKI, FELIX [1879-1940]
PSB 31,464-466; IB, 251
Anon., Science 91 (1940), 444; Roczniki Chemii 20 (1946), xv
W. Lampe, Zarys Historii Chemii w Polsce, pp.14-15. Cracow 1948

ROHDE (DAVIS), ALICE [1882-1933]
AMS 5,263ROHDE, ERWIN [1881-1915]
STUBLER, 330
Anon., Chem. Z. 39 (1915), 501; Z. angew. Chem. 30(III) (1917), 474
R. Gottlieb, Verhandl. Nat. Med. Heidel. NF13 (1914-1917), 482-484

ROHDE, GEORG [1853-1942]
KURSCHNER 4,2402; TLK 3,535; POGG 5,1060; 6,2205; 7a(3),794

ROHDEWALD, MARGARETE [1900-]
KURSCHNER 13,3183; WENIG, 250; BOEDEKER, 46-47; POGG 7a(3),794-795

ROHMANN, CARL [1897-1966]
KURSCHNER 10,2020; HEIN-E, 367; POGG 7a(3),795-796
Anon., Chem. Z. 90 (1966), 780; Nachr. Chem. Techn. 14 (1966), 447

ROHMER, MARTIN [1878-1941]
JV 15,14; NUC 501,465
Anon., Z. angew. Chem. 54 (1941), 272; Ber. chem. Ges. 74A (1941), 185

ROHRER, FRITZ [1888-1926]
FISCHER 2,1314-1315; DIEPGEN, 38-40
R. Metzner, Schw. med. Wchschr. 56 (1926), 505-507

ROJAHN, CARL [1889-1938]
HEIN, 536-537; TLK 3,537; POGG 6,2207-2208; 7a(3),801
Anon., Chem. Z. 62 (1938), 229-230
K. Ziegler, Ber. chem. Ges. 71A (1938), 124-125

ROKITANSKY, KARL von [1804-1878]
ADB 29,69-72; OBL 9,221-222; WURZBACH 26,288-295; LESKY, 129-141; W, 447; HIRSCH 4,855-859; Suppl.,414; PAGEL, 1403-1408; NAVRATIL, 261-262; FRANKEN, 195-196; WWWS, 1739;

CALLISEN 31,510-511
J. Stefan, Alm. Akad. Wiss. Wien 29 (1879), 157-172
F.R. Menne, Ann. Med. Hist. 7 (1925), 379-386
M. Neuburger, Wiener klin. Wchschr. 47 (1934), 348-360; Medical Life 41 (1934), 542-546
Z. Hornof, Clio Medica 3 (1968), 373-378
L.J. Rather, Clio Medica 4 (1969), 127-140; 7 (1972), 215-227

ROLF, IDA PAULINE [1896-1979]
NUC 501,693
New York Times, 21 March 1979

ROLLESTON, HUMPHRY DAVY [1862-1944]
DNB [1941-1950],733-734; WW 1944, pp.1368-2369; O'CONNOR(2), 22-23; FISCHER 2,1315; WWWS,1440
W.L.-B., Lancet 1944(II), pp.487-488
J.H. Pratt, Trans. Assoc. Am. Phys. 59 (1946), 29-31

ROLLETT, ALEXANDER [1834-1903]
OBL 9,227-228; BJN 8,249-255,93*; HIRSCH 4,865; PAGEL, 1410-1412; POGG 3,1138; 4,1266
V. Ebner, Wiener klin. Wchschr. 16 (1903), 1332-1335
O. Zoth, Pflügers Arch. 101 (1904), 103-153
C. von Voit, Sitz. Bayer. Akad. 34 (1904), 260-271

ROMAN, BENJAMIN [1876-1929]
FISCHER 2,1316

ROMAN, HERSCHEL LEWIS [1914-1989]
AMS 17(6),271
H. Roman, Ann. Rev. Gen. 20 (1986), 1-12
New York Times, 6 July 1989
S. Gartler and D. Stadler, Genetics 126 (1990), 1-3

ROMANES, GEORGE JOHN [1848-1894]
DSB 11,516-520; DNB 17,177-180; O'CONNOR, 249-251; WWWS, 1440-1441
E.R. Lankester, Nature 50 (1894), 108-109
J. Burdon-Sanderson, Proc. Roy. Soc. 57 (1895), vii-xiv
E. Romanes, Life and Letters of George Romanes. London 1896
R.D. French, J. Hist. Biol. 3 (1970), 253-274
J.E. Lesch, Isis 66 (1975), 483-503

ROMANOFF, ALEXIS LAWRENCE [1892-1980]
AMS 11,4516; CB 1953, pp.542-543; WWWS, 1441
Anon., Poultry Science 59 (1980), 943-944

ROMANOVSKI, DMITRI LEONIDOVICH [1861-1921]
BSE 36,650; BME 28,1101-1102; WWR, 479; OLPP, 346-347
C.A. Hoare, Trans. Soc. Trop. Med. 54 (1960), 292-293
J. Hatcher, Medical Technologist 4(3) (1974), 30-31

ROMBERG, ERNST von [1865-1933]
FISCHER 2,1317; BACHMANN, 83-93
F. Moritz, Deutsches Arch. klin. Med. 176 (1934), 1-4
L. Krehl, Deutsche med. Wchschr. 60 (1934), 113-114
F. Lange and W.H. Veil, Münch. med. Wchschr. 81 (1934), 79-82
K. Lydtin, Z. Tub. 69 (1934), 321-325

ROMEIS, BENNO [1888-1971]
KURSCHNER 11,2461-2462; FISCHER 2,1317; EGERER, 115-125; IB, 251
W. Bargmann, Münch. med. Wchschr. 100 (1958), 1498-1499

ROMENY, JOANNES [1851-1940]
RSC 11,213; NUC 502,690
Anon., Chem. Wkbl. 37 (1940), 167

ROMER, ALFRED SHERWOOD [1894-1973]
DSB 18,752-753; WWWS, 1441
T.S. Westall and F.R. Parrington, Biog. Mem. Fell. Roy. Soc. 21 (1975), 497-516
G.E. Erickson, Anat. Rec. 189 (1977), 314-324
E.H. Colbert, Biog. Mem. Nat. Acad. Sci. 53 (1982), 264-294

RONA, PETER [1871-1945]
KURSCHNER 4,2413; FISCHER 2,1318; BHDE 2,979; LDGS, 24; TLK 3,538; IB, 251; FRANKENTHAL, 303-304; KOREN, 236; WININGER 5,222; 7,401;TETZLAFF, 279; POGG 6,2211-2212; 7a(3),807
R. Ammon, Arzneimitt. 10 (1960), 321-327
H.J. Trumit, Arzneimitt. 10 (1960), 400-404

RONALDS, EDMUND [1819-1889]
DNB 17,201; COLE, 467-468; RSC 5,268; 8,775
Anon., J. Chem. Soc. 57 (1890), 436
J.Y. Buchanan, Proc. Roy. Soc. Edin. 17 (1890), xxviii-xxix

RONDONI, PIETRO [1882-1956]
STC 2,429-430; FISCHER 2,1319; IB, 252; POGG 6,2213; 7b,4480-4484
L. Califano, Atti Accad. Lincei [8]22 (1957), 777-800
F. Perussia, Sci. Med. Ital. 6 (1957), 201-211
A. Giordano, Verhandl. path. Ges. 41 (1958), 417-426
Anon., Leopoldina [3]4/5 (1958-1959), 100-101

RONGGER, NIKOLAUS [1874-1957]
RSC 18,326; NUC 505,367

RONTALER, STEFAN AUGUST KLEMENTOVICH [1867-]
KONOPKA 9,125; RSC 18,287; SG [2]14,726

RONZONI, ETHEL [1890-1975]
AMS 11,4520

ROOS, ERNST [1866-1926]
KURSCHNER 2,1588

ROOS, ISRAEL [1864-1913]
ARNSBERG, 375-376
Anon., Chem. Z. 37 (1913), 1606

ROOZEBOOM, HENDRIK WILLEM BAKHUIS [1854-1907]
DSB 11,534-535; BWN 1,22-23; BLOKH 2,618-619; WWWS,1442; POTSCH, 368; POGG 4,1267-1268
J. van Bemmelen et al., Ber. chem. Ges. 40 (1907), 5141-5174
J.H. van't Hoff, Chem. Z. 31 (1907), 199
F.D. Chattaway, Nature 75 (1907), 464-465
W. Stortenbeker, Rec. Trav. Chim. Pays-Bas 27 (1908), 360-410
H.R. Kruyt et al., Chem. Wkbl. 50 (1954), 749-763

ROSA, EDWARD BENNETT [1861-1921]
DAB 17,154-155; NCAB 26,312-313; WWWS, 1442; POGG 4,1268; 5,1063-1064
W.W. Coblentz, Biog. Mem. Nat. Acad. Sci. 16 (1935), 355-368

ROSANOFF, MARTIN ANDRÉ [1874-1951]
AMS 8,2108; NCAB C,285; MILES, 416-417; WININGER 5,225-226; WWAH, 522; POGG 5,1064; 6,2214; 7b,4485-4486
Anon., Science 114 (1951), 270; School and Society 74 (1951), 94

ROSCOE, HENRY ENFIELD [1833-1915]
DSB 11,536-539; DNB [1912-1921],478-479; BLOKH 2,619-620; WWWS, 1442; W, 449; SCHAEDLER, 106-107; MATSCHOSS, 233; POTSCH, 368-369; POGG 2,686; 3,1140-1141; 4,1269; 5,1065; 6,2215
H.E. Roscoe, Life and Experiences. London 1906
T.E. Thorpe, The Rt. Hon. Sir Henry Roscoe. London 1916; Proc. Roy. Soc. A93 (1917), i-xxi
A.W. Walters, Mem. Lit. Phil. Soc. 60 (1916), lii-lxii
C. Thesing, Chem. Z. 40 (1916), 189-192

ROSE, EMBREE RECTOR [1893-]
AMS 11, 4526

ROSE, FERDINAND [1809-1861]
FERCHL, 453; POGG 2,694

ROSE, FRANCIS LESLIE [1909-1988]
WW 1987, p.1513; WWWS, 1443
C.W. Suckling and B.W. Langley, Biog. Mem. Fell. Roy. Soc. 36 (1990), 491-524

ROSE, FREDERICK [1867-1932]
JV 9,20; NUC 504,96

ROSE, FRIEDRICH [1839-1925]
BLOKH 2,620; SCHAEDLER, 107
B. Lepsius, Ber. chem. Ges. 58A (1925), 19-20

ROSE, GUSTAV [1798-1873]
DSB 11,539-540; ADB 29,175-177; BLOKH 2,620-621; SCHAEDLER, 107; POTSCH, 269; POGG 2,692-694; 3,1141-1142; 7aSuppl.,561-562
C. Rammelsberg, Ber. chem. Ges. 6 (1873), 1573-1577; Z. geol. Ges. 25 (1873), i-xx
Anon., Nature 8 (1873), 277-279; Am. J. Sci. [3]6 (1873), 238-240; Proc. Roy. Soc. 24 (1876), iii-v

ROSE, HEINRICH [1795-1864]
DSB 11,540-542; ADB 29,177-181; PRANDTL, 287-301; HEIN, 538-540; BLOKH 2,621-622; FERCHL, 453-454; SCHAEDLER, 107-108; TSCHIRCH,1113; PHILIPPE, 761; WWWS, 1443; POTSCH, 369; POGG 2,687-692; 3,1141; 7aSuppl.,562; CALLISEN 16,304-306; 32,3-4
D., Am. J. Sci. [2]38 (1864), 305-330
A. Remelé, Mon. Sci. [2]6 (1864), 385-389
C.F.P. Martius, Sitz. Bayer. Akad. 1864(I), pp.193-196

ROSE, IRWIN ALLAN [1926-]
AMS 15(6),268

ROSE, MARY SWARTZ [1874-1941]
DAB [Suppl.3],670-672; IB, 252
H.C. Sherman, J. Biol. Chem. 140 (1941), 687-688
J.A. Eagle et al., Mary Swartz Rose. New York 1979

ROSE, WILLIAM CUMMING [1887-1985]
AMS 14,4290; WWA 1976-1977, p.2681; IWW 1982-1983, pp.1134-1135; IB, 252; MH 2,456-457; STC 2,432-433; MSE 3,39-41; CB 1953, pp.545-547; CHITTENDEN, 96-97; WWWS, 1443; POTSCH, 370
W.C. Rose, Ann. N.Y. Acad. Sci. 325 (1979), 229-234
D.A. Roe, J. Nutrition 111 (1981), 1313-1320
New York Times, 27 September 1985

ROSEEU, ALEXANDER [1886-]
JV 29,730; NUC 504,214

ROSELL, MAX [1876-1938]
JV 17,324; SG [2]14,733; NUC 504,244

ROSEMAN, SAUL [1921-]
AMS 15(6),271; WWA 1980-1981, p.2825; IWW 1982-1983, p.1135; WWWS, 1444

ROSEMANN, HANS ULRICH [1904-]
KURSCHNER 13,3199; AUERBACH, 352; POGG 7a(3),810-811

ROSEMANN, RUDOLF [1870-1943]
FISCHER 2,1320-3121; POGG 6,2216-2217; 7a(3),811
E. Schütz, Med. Klin. 39 (1943), 389-390

ROSEN, FELIX [1863-1925]
TSCHIRCH, 1113
H. Winkler, Ber. Bot. Ges. 43 (1925), (65)-(73)

ROSEN, ORA MENDELSOHN [1935-1990]
AMS 17(6),289

ROSENAU, MILTON JOSEPH [1869-1946]
DAB [Suppl.4],700-702; AMS 7,1510; NCAB 42,690-692; DAMB, 649-650; FISCHER 2,1321; BARRY, 66-70; IB, 252; KOREN, 236; WININGER 5,228; KAGAN(JA), 387-389,764-765; WWAH, 523
Anon., Am. J. Pub. Hlth. 36 (1946), 530-531
S.B. Wolbach, Trans. Assoc. Am. Phys. 59 (1946), 32-33
J.D. Felton, J. Bact. 53 (1947), 1-3

ROSENBACH, JULIUS FRIEDRICH [1842-1923]
DBJ 5,326-329,439; GROTE 2,187-192; FISCHER 2,1321; PAGEL, 1417-1418; BULLOCH, 393; GRUTTER, 94; KOREN, 236; WWWS, 1444
R. Stich, Deutsche med. Wchschr. 50 (1924), 184

ROSENBAUM, SIEGFRIED [1890-1969]
BHDE 2,982; LDGS, 78; KOREN, 237

ROSENBERG, EMIL [1842-1925]
FISCHER 2.1323; KOREN, 237; WININGER 7,402-403

ROSENBERG, HANS [1890-]
KURSCHNER 4,2420; LDGS, 80; IB, 252
Who's Who in British Science 1953, p.230

ROSENBERG, MAX [1887-1943]
KURSCHNER 4,2421; FISCHER 2,1323; LDGS, 68; KOREN, 237

ROSENBLUETH, ARTURO STEARNS [1900-1970]
DSB 11,545-547; AMS 11,4538; WWAH, 523; WWWS, 1445
J.C. White, Harvard Med. Alumni Bull. 45(4) (1971), 43

ROSENFELD, BRUNO [1903-]
LDGS, 24

ROSENFELD, GEORG [1861-1933]
KURSCHNER 4,2421; FISCHER 2,1324; PAGEL, 1423-1424; WININGER 7,403-404
F. Ehrlich, Jahrb. Schles. Ges. 106 (1933), 2-4

ROSENGARTEN, GEORGE DAVID [1869-1936]
MILES, 418-419; JV 8,151; WWA 1934-1935, p.2048; NCAB 35,272-273

ROSENGARTEN, SAMUEL GEORGE [1827-1908]
RSC 5,293
Anon., Chem. Z. 32 (1908), 610

ROSENHEIM, ARTHUR [1865-1942]
DSB 11,550-552; KURSCHNER 4,2422; TLK 3,539; POGG 4,1271; 5,1066-1067; 6,2220-2221; 7a(3),813

ROSENHEIM, MAX LEONARD [1908-1972]
DNB [1971-1980], 735-736; WW 1972, p.2760; KOREN, 237; COHEN, 200
Anon., Brit. Med. J. 1972(4), pp.675-676
G. Pickering, Biog. Mem. Fell. Roy. Soc. 20 (1974), 349-358

ROSENHEIM, SIGMUND OTTO [1871-1955]
DNB [1951-1960], 848-849; WW 1949, p.2400; O'CONNOR(2), 187-188; WWWS, 1445; POGG 6,2220; 7a(3),813
H. King, Biog. Mem. Fell. Roy. Soc. 2 (1956), 257-267; J. Chem. Soc. 1956, pp.799-801

ROSENKRANTZ, HARRIS [1922-1991]
AMS 15(6),283; WWWS, 1445

ROSENMUND, KARL [1884-1965]
KURSCHNER 9,1693; HEIN-E, 369-370; TLK 3,539; KIEL, 166-167; POTSCH, 370; POGG 5,1067; 6,2221; 7a(3),814
Anon., Nachr. Chem. Techn. 3 (1955), 74
H. Vogt, Arzneimitt. 9 (1959), 794

ROSENOW, EDWARD CARL [1875-1966]
AMS 11,4541; NCAB A,332-333; FISCHER 2,1325; IB, 252-253; WWWS, 1445
J. Eckman, Am. J. Clin. Path. 46 (1966), 123-124

ROSENSTIEHL, DANIEL AUGUSTE [1839-1916]
POGG 3,1143-1144; 4,1271-1272; 5,1067
E. Wild, Chem. Z. 40 (1916), 1009-1011
A. Haller, Bull. Soc. Chim. [4]21 (1917), i-xxiv

ROSENTHAL, FELIX [1885-1952]
KURSCHNER 4,2425; FISCHER 2,1325; KOREN, 237

ROSENTHAL, ISIDOR [1836-1915]
DBJ 1,339; HIRSCH 4,882-883; PAGEL, 1426-1427; SCHWARTZ,27-40; WWWS,1446; KOREN, 237; KOHUT 2,236-237; POGG 3,1144; 4,1272; 5,1067; 6,2221
R. Höber, Münch. med. Wchschr. 62 (1915), 293

ROSENTHAL, JOSEF [1867-1938]
KURSCHNER 4,2425; TLK 3,540; POGG 6,2221-2222

ROSENTHAL, MORITZ [1833-1889]
ADB 53,496; HIRSCH 4,882; FROHNE, 86-98; WININGER 5,250
Anon., Wiener med. Wchschr. 40 (1890), 29-30

ROSENTHAL, OTTO [1881-]
JV 22,523; NUC 504,573

ROSENTHAL, OTTO [1898-1981]
AMS 11,4543; BHDE 2,991; LDGS, 60; NUC 504,573

ROSENTHAL, WERNER [1870-1934]
KURSCHNER 4,2425-2426; EBEL, 84; DZ, 1207; LDGS, 58; KOREN, 238; WININGER 5,252; NUC 504,581

ROSENTHALER, LEOPOLD [1875-1962]
HEIN-E, 370-371; LEHMANN, 147; TSCHIRCH, 1114; IB, 253; TLK 3,540; POGG 5,1068; 6,2222; 7a(3),815-817
M. Zafir, Folia Pharmaceutica 4 (1963), 579-580
A. Wankmüller, Beitr. Wurtt. Apothekgesch. 13 (1981), 46-49

ROSER, GUSTAV [1823-1860]
RSC 5,295
A. Wankmüller, Beitr. Wurtt. Apothekgesch. 7 (1966), 84

ROSER, WILHELM [1858-1923]
MEINEL, 219-221,511; SCHMITZ, 282-283; POGG 4,1272; 6,2222-2223
E. Bryk, Z. angew. Chem. 36 (1923), 557-558
E. Baum, Chem. Z. 47 (1923), 714

ROSIN, HEINRICH [1863-1934]
KURSCHNER 4,2426; FISCHER 2,1326; PAGEL, 1429-1430; TETZLAFF, 284; KOREN,238; RSC 18,301

ROSING, ANTON [1827-1867]
NBL 11,573-577; POGG 2,697; 3,1144

ROSS, RONALD [1857-1932]
DSB 11,555-557; DNB [1931-1940],752-754; FISCHER 2,1327; OLPP, 347-355; FREUND 2,341-349; W, 449-450; ABBOTT, 111-112; WWWS, 1447; POGG 6,2225; 7b,4501-4502
R. Ross, Memoirs. London 1923
R.L. Megroz, Sir Ronald Ross. London 1931
G.H.F. Nuttall, Obit. Not. Fell. Roy. Soc. 1 (1933), 108-115
M. Yoeli, Bull. N.Y. Acad. Med. 49 (1973), 722-735
E. Chernin, Med. Hist. 32 (1988), 119-141

ROSSBACH, MICHAEL JOSEPH [1842-1894]
ADB 53,514; FISCHER 2,1327-1328; PAGEL, 1430-1431; ROEDER, 200-209; GIESE, 559-560
Anon., Leopoldina 30 (1894), 210-211; Münch. med. Wchschr. 41 (1894), 872

ROSSENBECK, HEINRICH [1895-1945]
GAUSS, 270

ROSSI, HEINRICH [1872-1938]
JV 18,326; NUC 505,367
Anon., Z. angew. Chem. 51 (1938), 72

ROSSI FANELLI, ALLESANDRO [1906-1991]
N. Siliprandi and W.E. Blumberg, in Structure and Function Relationship in Biochemical Analysis (F. Brossa, Ed.), pp.1-19. New York 1982
W.E. Blumberg, Advances in Experimental Medicine and Biology 148 (1982), 7-19

ROSSINI, FREDERICK DOMINIC [1899-]
AMS 14,4315; WWA 1978-1979, p.2784; IWW 1982-1983, p.1138; WWWS, 1448; MH 2,459-461; MSE 3,45-46; POGG 6,2227; 7b,4510-4515
Anon., Chem. Eng. News 48(30) (1970), 71
F.D. Rossini, J. Chem. Thermodyn. 8 (1976), 805-834

ROSSITER, ROGER JAMES [1913-1976]
DSB 18,753-755; AMS 13,3785; YOUNG, 27; MH 2,461-462; MSE 2,46-47; WWWS, 1448
R.G.E. Murray, Proc. Roy. Soc. Canada [4]14 (1976), 109-113
W.C. Murray, J. Neurochem. 27 (1976), 827-828

ROSSMANN, MICHAEL [1930-]
AMS 15(6),300; WWA 1980-1981, p.2839; BHDE 2,994-995

ROST, EUGEN [1870-1953]
TSCHIRCH, 1114; IB, 253; TLK 3,541; POGG 6,2227; 7a(3),823
Anon., Chem. Z. 77 (1953), 485
A. Fleckenstein, Arzneimitt. 3 (1953), 595

ROST, GEORG ALEXANDER [1877-1970]
KURSCHNER 11,2477; FISCHER 2,1330; LDGS, 62
G.A. Rost, Hautarzt 15 (1964), 441-448,512-513,559-560
K. Halter, Z. Haut Geschl. 46 (1971), 427-429

ROSTAND, JEAN [1894-1977]
DSB 18,755-759; QQ 1977-1978, p.1446; CB 1954, pp.545-547; WWWS, 1445; BUICAN, 209-226
A. Delaunay, Jean Rostand. Paris 1956
J. Théodoridès, Hist. Sci. Med. 13 (1979), 87-90

ROSTENBERG, ADOLPH, Jr. [1905-]
AMS 14,4316
L. Rostenberg, J. Am. Acad. Dermatol. 23 (1990), 1163-1165

ROSTOSKI, OTTO [1872-1962]
GEISSLER, 22-24; JV 15,562; RSC 18,311; SG [2]14,729
R. Oehme, Deutsche med. Wchschr. 77 (1952), 1097-1098
M. Gülzow, Jahrb. Akad. Wiss. Berlin 1964, pp.217-218

ROTCH, THOMAS MORGAN [1849-1914]
AMS 2,402; FISCHER 2,1331; NCAB 19,350-351; WWWS, 1449
H. Bloch, Pediatrics 50 (1972), 112-117

ROTH, OTTO [1853-1927]
STRAHLMANN, 464,479
H. Schinz and A. Wolfer, Viert. Nat. Ges. Zürich 72 (1927), 460-465

ROTH, PAUL BERNHARD [1881-1970]
JV 21,298; NUC 506,123

ROTH, RUDOLF [1887-]
JV 26,644; NUC 506,127

ROTH, WALTER [1874-1954]
POGG 4,1274
H. Stadliger, Allgemeine Öl- und Fett-Zeitung 28 (1931), 97

Anon., Chem. Z. 78 (1954), 273; Chem. Eng. News 32 (1954), 1605

ROTH, WALTHER ADOLF [1873-1950]
TLK 3,542; POGG 5,1070; 6,2228-2229; 7a(3),825
W.A. Roth, Naturwiss. 36 (1949), 225-229
R. Suhrmann, Abh. Braunschw. Wiss. Ges. 2 (1950), 141-148

ROTHE, HEINRICH AUGUST [1773-1842]
ADB 29,349-350; POGG 2,702
M. Nierenstein, Isis 21 (1934), 123-130

ROTHE, OTTO [1887-]
JV 28,210; NUC 506,171

ROTHE, WALTER [1888-1954]
HEIN-E, 372; DRUM, 324-325; POGG 7a(3),827

ROTHEN, ALEXANDRE [1900-1987]
AMS 13,3790
Anon., Rock. Univ. Notes 19(3) (1988), 3

ROTHENBURG, RUDOLF von [1870-1949]
POGG 4,1275; 6,2230; 7a(3),827

ROTHENFUSSER, SIMON [1873-]
JV 19,342; NUC 506,188

ROTHERA, ARTHUR CECIL HAMEL [1880-1915]
O'CONNOR(2), 504
F.G. Hopkins, Biochem. J. 10 (1915-1916), 11-13

ROTHLAUF, LEO [1877-1942]
JV 26,644; NUC 506,224

ROTHLIN, ERNST [1888-1972]
KURSCHNER 11,2483; SBA 3,114; IB, 254; POGG 7a(3),826-831
Anon., Chimia 13 (1959), 31; Naturw. Rund. 25 (1972), 491
E. Rothlin, Ann. Rev. Pharm. 4 (1964), ix-xxxii
M. Strasser and M. Taeschler, Pharmacologist 15 (1973), 29

ROTHMANN, ALBERT [1879-]
JV 20,262

ROTHMUND, LUDWIG VIKTOR [1870-1927]
OBL 9,287-288; JV 10,215; POGG 4,1275; 5,1072; 6,2230
H. Meyer, Ber. chem. Ges. 60A (1927), 153-154

ROTHSCHILD, PAUL [1901-1965]
LDGS, 68; NUC 506,259
A.V. Hill, Nature 209 (1966), 348-349

ROTHSCHUH, KARL EDUARD [1908-1984]
KURSCHNER 13,3219; POGG 7a(3),632-833
G. Polaczek, Münster Beiträge zur Geschichte und Theorie der Medizin 8 (1973), 1-37

ROTHSTEIN, MATHILDE [1893-]
JV 35,280; NUC 506,268ROTKY, HANS [1876-1965]
KURSCHNER 10,2043; PELZNER, 44-48

ROTSCHY, ARNOLD [1874-1966]
GENEVA 5,51,302; RSC 17,483; SG [2]14,758; NUC 506,298
P. Boymond, Schw. Apoth. Z. 104 (1966), 740

ROTTA, WALTER [1886-1970]
JV 26,421; NUC 506,305
Anon., Chem. Z. 80 (1956), 210; Nachr. Chem. Techn. 18 (1970), 53

ROUGET, CHARLES [1824-1904]
DSB 11,565-567
N. Gréhant, Nouv. Arch. Hist. Nat. [4]6 (1904), iii-xii

ROUGHTON, FRANCIS JOHN WORSLEY [1899-1972]
DSB 18,759-762; DNB [1971-1980], 737-738; WW 1965, p.2656; IB, 254
Anon., Nature 238 (1972), 297
Q.H. Gibson, Biog. Mem. Fell. Roy. Soc. 19 (1973), 563-582

ROUILLER, CHARLES AUGUST [1883-1968]
AMS 10,3470
Anon., Chem. Eng. News 46(35) (1968), 58

ROUILLIER, KARL FRANTSOVICH [1814-1858]
DSB 11,567-568; BSE 37,319-320; HIRSCH 4,896-897; KUZNETSOV(1963), 89-104
S.R. Mikulinski, K.F. Rouillier. Moscow 1957
G.A. Novikov, Bull. Soc. Nat. Moscou 65(2) (1960), 135-143

ROULIN, FRANÇOIS DESIRÉ [1796-1874]
LAROUSSE 13,1451; HIRSCH 4,896; DECHAMBRE [3]5,476; RSC 5,307; 8,789; POGG 2,705; CALLISEN 16,352; 32,18
M. Combes, Pauvre et Avantureuse Bourgeoisie: Roulin et ses Amis. Paris 1928

ROUS, PEYTON [1879-1970]
AMS 11,4566; NCAB 55,156-157; CB 1967, pp.354-357; FISCHER 2,1133; MH 1,405-407; STC 2,440-441; MSE 3,48-49; DAMB, 651-652; IB, 254; W, 590; WWAH, 524; WWWS, 1450
C. Huggins, Persp. Biol. Med. 13 (1970), 465-468
J.S. Henderson, Am. Phil. Soc. Year Book 1971, pp.168-179; A Notable Career in Finding Out: Peyton Rous 1879-1970. New York 1971
C.H. Andrewes, Biog. Mem. Fell. Roy. Soc. 17 (1971), 643-662
R. Dulbecco, Biog. Mem. Nat. Acad. Sci. 48 (1976), 275-306
C.B. Huggins and R. Dubos, J. Exp. Med. 150 (1979), 733-737

ROUSSEL, THÉOPHILE [1816-1903]
VAPEREAU 5,1584; HIRSCH 4,898-899; PAGEL, 1438
E. Lancereaux, Bull. Acad. Med. 50 (1903), 119-123
A.A. Motet, Bull. Acad. Med. 50 (1903), 538-540
G. Picot, Notices Historiques 2,219-258,358-361. Paris 1907
F.S. Jaccoud, Mem. Acad. Med. 41 (1910), 29-48

ROUSSIN, FRANÇOIS ZACHARIE [1827-1894]
BALLAND, 102-105; BALLAND(2), 193-194; BLOKH 2,623; GORIS, 559-562; TSCHIRCH, 1114; WWWS, 1451; POTSCH, 371
A. Balland, J. Pharm. Chim. [5]29 (1894), 486-487; Bull. Soc. Hist. Pharm. 4 (1927), 373-376
A. Balland and D. Luiset, Le Chimiste Z. Roussin. Paris 1908
C. Matignon, Chimie et Industrie 3 (1920), 821-827

ROUX, ÉMILE [1853-1933]
DSB 11,569; GENTY 6,369-384; FISCHER 2,1334-1335; PAGEL, 1441; OLPP, 357; BULLOCH, 393; STALDER, 77-80; IB, 254; W, 450-451; WWWS, 1451
C. Regaud, Paris Médical 90 (1933), 452-455
F. Mesnil, Bull. Acad. Med. 110 (1933), 708-721
F. Bezancon, Presse Med. 41 (1933), 1893-1895; L'Oeuvre de M. Émile Roux. Paris 1933
Anon., Obit. Not. Fell. Roy. Soc. 1 (1933), 197-204
H. Sachs, Münch. med. Wchschr. 81 (1934), 27-29
L. Launoy, Bull. Soc. Chim. Biol. 16 (1934), 467-470
W. Bulloch, J. Path. Bact. 38 (1934), 99-105
A. Delaunay, Prog. Med. 81 (1953), 531-534
E. Lagrange, Monsieur Roux. Brussels 1955
A. Delaunay and E. Dahl, Hist. Med. 7 (1957), 53-73
G.L. Geison, J. Hist. Med. 45 (1990), 341-365

ROUX, GABRIEL [1853-1914]
RSC 18,333
C. Lesieur, Lyon Médical 123 (1914), 380-382
P. Hassenforder, Hist. Med. 7(4) (1957), 35-40
J. Archimbaud, Bull. Hist. Sci. Auvergne 88 (1976), 111-131

ROUX, WILHELM [1850-1924]
DSB 11,570-575; FISCHER 2,1335; PAGEL, 1438-1441; BK,297-300;

WWWS,1451; DZ, 1215-1216; GROTE 1,141-206; ABBOTT, 112; SCHULZ-SCHAEFFER, 200
D. Barfurth, Arch. Entwickl. 30 (1910), i-xxxvii; Naturwiss. 8 (1920), 431-549; Anat. Anz. 59 (1925), 153-176; Arch. mikr. Anat. 104 (1925), i-xxii
B. Dürken, Biol. Gen. 1 (1925), 507-521
G. Pommer, Wiener klin. Wchschr. 36 (1925), 963-967,993-995
L. Castaldi, Riv. Biol. 7 (1925), 97-104
H. Stieve, Mitteldeutsche Lebensbilder 2 (1927), 452-461
R. Koch, Sudhoffs Arch. 22 (1929), 114-150
R. Mocek, Wilhelm Roux - Hans Driesch. Zur Geschichte der Entwicklungsgeschichte der Tiere. Jena 1975

ROUXEAU, ALFRED [1854-1925]
RSC 19,335; SG [1]12,350; NUC 507,208-209

ROWE, ALLAN WINTER [1879-1934]
DSB 11,576-577; AMS 5,956; IB, 254
A.A. Ashdown, Ind. Eng. Chem. News Ed. 12 (1934), 454-455
W. Goodwin, J. Chem. Soc. 1935, pp.863-864
J.F. Norris, Proc. Am. Acad. Arts Sci. 72 (1938), 385-387

ROWE, FREDERICK MAURICE [1891-1946]
WW 1945, p.2372; POGG 6,2232; 7b,4521-4523
E.J. Cross and J.B. Speakman, Nature 159 (1947), 53
E.J. Cross, Chem. Ind. 1947, pp.46-47
E.H. Rodd, J. Chem. Soc. 1948, pp.2323-2330; Obit. Not. Fell. Roy. Soc. 6 (1948), 231-250

ROWE, WALLACE PRESCOTT [1926-1983]
AMS 15(6),318

ROWLAND, SYDNEY DONVILLE [1872-1917]
BULLOCH, 393-394
C.J. M[artin], Nature 99 (1917), 67-68

ROWNEY, THOMAS HENRY [1817-1894]
FERCHL, 458; RSC 5,314; 8,793; POGG 3,1149-1150

ROWNTREE, LEONARD GEORGE [1883-1959]
AMS 10,3477; NCAB 44,546-547; FISCHER 2,1336; WWWS, 1453
N.M. Keith and P.S. Hench, Trans. Assoc. Am. Phys. 73 (1960), 29-31

ROY, CHARLES SMART [1854-1897]
O'CONNOR, 183-185; PAGEL, 1441-1442
Anon., Lancet 1897(II), p.954; Brit. Med. J. 1897(II), pp.1031,1124
J.J.G. Brown, J. Path. Bact. 5 (1898), 143-146
S.C.S., Proc. Roy. Soc. B75 (1905), 131-136

RÓZYCKI, LEON [1872-]
KONOPKA 9,265; RSC 18,342

RUBEN, SAMUEL [1913-1943]
AMS 7,1521
Anon., Chem. Eng. News 21 (1943), 1766; Science 98 (1943), 337
M.D. Kamen, J. Chem. Ed. 40 (1963), 234-242

RUBENHAUER, HANS [1902-]
JV 44,453

RUBINSTEIN, SALOMON [1869-1915]
J. Brennsohn, Die Aerzte Livlands, p.341. Mitau 1905

RUBNER, MAX [1854-1932]
DSB 11,585-586; FISCHER 2,1337-1338; PAGEL, 1442-1444; GUNDLACH, 270; WREDE, 162-163; DZ, 1216; TLK 3,544; IB, 254; WININGER 5,291; POGG 6,2234-2235; 7a(3),836
O. Frank, Bayer. Akad. Wiss. Jahrbuch 1931-1932, pp.71-74
M. Cremer, Ber. chem. Ges. 65A (1932), 95-98
A. Durig, Alm. Akad. Wiss. Wien 82 (1932), 325-330
K. Thomas, Klin. Wchschr. 50 (1932), 926
K.B. Lehmann, Münch. med. Wchschr. 79 (1932), 1038-1042
G. Lusk, Science 76 (1932), 129-135
O. Kestner, Deutsche med. Wchschr. 58 (1932), 786-788
W.H. Chambers, J. Nutrition 48 (1952), 3-12
H.C. Knowles, Diabetes 6 (1957), 369-371

RUBOW, VICTOR [1871-1929]
VEIBEL 2,361-362; SG [2]14,789; NUC 508,550
C. Sonne, Hosp. Tid. 72 (1929), 1029-1031

RÜBSAMEN, KARL [1826-1902]
RAUSCH, 289
R. Schrotzenberger, Francofurtensia, 2nd Ed., p.7. Frankfurt a.M. 1884
Anon., Pharm. Z. 38 (1902), 651

RÜCKER, HANS [1874-]
JV 17,325; SG [2]14,793; NUC 509,90

RÜCKERT, JOHANNES [1854-1923]
DBJ 5,439-440; FISCHER 2,1339; EGERER, 59-65; DZ, 1217
S. Mollier, Bayer. Akad. Wiss. Jahrbuch 1922-1923, pp.89-90
H. Stieve, Anat. Anz. 57 (1924), 305-352
F. Wassermann, Ergebnisse der Anatomie 25 (1924), i-x

RUDINGER, JOSEF [1924-1975]
R. Schwyzer, Chimia 29 (1975), 376-377
G.T. Young, Chem. Brit. 12 (1976), 226-227
M. Brenner and V. Pliska, in Peptides 1976 (E. Loffet, Ed.), pp.5-51. Brussels 1976
I.L. Schwarz, Ann. N.Y. Acad. Sci. 394 (1982), xiii-xvii

RÜDINGER, NICOLAUS [1832-1896]
ADB 53,580-582; BJN 1,353-354; 3,127*; HIRSCH 4,913-914; PAGEL, 1444-1445; EGERER, 45-58
J. Rückert, Münch. med. Wchschr. 43 (1896), 1017-1019
C. von Kupffer, Anat. Anz. 13 (1897), 219-232
C. von Voit, Sitz. Bayer. Akad. 27 (1898), 390-401

RUDNEV, MIKHAIL MATVEEVICH [1837-1878]
BSE 37,295-296; BME 28,1232; ZMEEV 2,87
A.K. Ageev, Arkh. Pat. 41 (1979), 79-81; 49 (1987), 75-77
V.P. Mikhailov, Arkh. Anat. Gist. Emb. 78(2) (1980), 104-115

RUDNEV, VLADIMIR MATVEEVICH [1850-1898]
BSE 37,294; KAZAN 1,474-475

RUDOLPH, CHRISTIAN [-1922]
RSC 9,869; 18,346
Anon., Ber. chem. Ges. 55A (1922), 70

RUDOLPH, OTTO [1864-]
JV 4,284; RSC 18,347; NUC 509,33

RUDOLPHI, KARL ASMUND [1771-1832]
DSB 11,592-593; ADB 29,577-579; SMK 6,400; HIRSCH 4,911-913; WWWS, 1456; CALLISEN 32,28-33
M. Lühe, Arch. Parasit. 3 (1900), 549-577

RÜDORFF, FRIEDRICH [1832-1902]
BJN 7,96*-97*; BLOKH 2,623; SCHAEDLER, 109; POGG 3,1151-1152; 4,1282
A. Stavenhagen, Ber. chem. Ges. 35 (1902), 4536-4541

RUDY, HERMANN [1904-1966]
KURSCHNER 10,2049; POGG 7a(3),839
Anon., Chem. Z. 90 (1966), 192; Nachr. Chem. Techn. 14 (1966), 79

RUEGAMER, WILLIAM RAYMOND [1922-]
AMS 15(6),338; WWA 1980-1981, p.2858; WWWS, 1456

RÜGHEIMER, LEOPOLD [1850-1917]
DBJ 2,669; BLOKH 2,624; KIEL, 181-182; POTSCH, 372; POGG 3,1152; 4,1282-1283; 5,1076
L.B. Berend, Ber. chem. Ges. 50 (1917), 849-851

RUFF, OTTO [1871-1939]
HEIN-E, 373; TLK 3,546; POTSCH, 371; POGG 4,1283; 5,1077-1078; 6,2239-2240; 7a(3),846
W. Hückel, Ber. chem. Ges. 73A (1940), 125-156

RUFFER, MARC ARMAND [1859-1917]
DSB 11,595-596; FISCHER 2,1340-1341; O'CONNOR(2), 451-452
A.T. Sandison, Med. Hist. 11 (1967), 150-156

RUFFI, HANS [1863-1025]
STRAHLMANN, 473,481; RSC 18,354

RUFFINI, ANGELO [1864-1929]
DSB 11,596-598; EI 30,219; FISCHER 2,1341
E. Giacomini, Mon. Zool. Ital. 40 (1929), 277-292
G. Cotronei, Riv. Biol. 12 (1930), 198-202
J. Eccles, Notes Roy. Soc. 30 (1975), 69-88

RUGGLI, PAUL [1884-1945]
TLK 3,546; POTSCH, 371-372; POGG 5,1078; 6,2240-2241; 7a(3),847-848
H. Rupe, Helv. Chim. Acta 29 (1946), 796-811

RUHEMANN, SIEGFRIED [1859-1943]
BERLIN, 704; POGG 4,1284-1285; 5,1078; 6,2241; 7a(3),848
F. Frank, Chem. Z. 53 (1929), 12
R.S. Morrell, J. Chem. Soc. 1944, pp.46-48
R. West, J. Chem. Ed. 42 (1965), 386-387

RUHLAND, REINHOLD LUDWIG [1786-1827]
ADB 29,608; FERCHL, 459; POGG 2,717

RUHLAND, WILHELM [1878-1960]
SACKMANN, 289-290; IB, 255; POGG 6,2241-2242; 7a(3),849
O. Renner, Z. Naturforsch. 3b (1948), 141-144
H. Ulrich, Handbuch der Pflanzenphysiologie 12 (1960), v-xxxii
L. Brauner, Bayer. Akad. Wiss. Jahrbuch 1961, pp.173-178

RULE, ALEXANDER [1880-1960]
JV 20,283; NUC 509,540
Anon., J. Roy. Inst. Chem. 1960, pp.198-199

RULE, HAROLD GORDON [1887-1943]
JV 29,731; POGG 6,2242; 7b,4532-4533
J.E. Mackenzie and N. Campbell, J. Chem. Soc. 1943, pp.509-510
N. Campbell, Nature 151 (1943), 665-666

RÜLING, EDUARD [1811-1875]
DGB 130 (1962), 211; RSC 5,326

RUMFORD, COUNT (BENJAMIN THOMPSON) [1753-1814]
DSB 13,350-352; DAB 18,449-452; BLOKH 2,624-625; ELLIOTT, 248- 249; MATSCHOSS, 235; FERCHL, 460-481; SCHAEDLER, 109-110; W, 452-453; POTSCH, 421-422; WWAH, 601; WWWS, 1666; POGG 2,718-720; 3,1153; 4,1286; 6,2243
G.E. Ellis, Memoir of Sir Benjamin Thompson, Count Rumford. Boston 1871
S.C. Brown, Benjamin Thompson, Count Rumford. Cambridge, Mass. 1979

RUMPF, THEODOR [1851-1934]
KURSCHNER 4,2458
P. Horn, Deutsche med. Wchschr. 47 (1921), 1565
T. Rumpf, Lebenserinnerungen. Bonn 1925

RÜMPLER, ALWIN [1844-1907]
BJN 12,73*; BLOKH 2,625-626
R. Woy, Jahresb. Schles. Ges. 1907, pp.36-38

RUND (SCHMIDT-RUND), CHARLOTTE [1890-1970]
BHDE 2,1038; JV 32,246; NUC 509,659

RUNDLE, ROBERT EUGENE [1915-1963]
AMS 10,3491; McLACHLAN, 120
Anon., Chem. Eng. News 41(44) (1963), 169; Proc. Iowa Acad. Sci. 71 (1964), 44-45

RUNEBERG, JOHANN WILHELM [1843-1918]
FISCHER 2,1344; PAGEL, 1450-1451; SG [1]12,394; [2]14,803
S.E. Henschen, Hygeia 80 (1918), 577-629

RUNGE, FRIEDLIEB FERDINAND [1794-1867]
DSB 11,615-616; ADB 29,684-687; HEIN, 548-549; FERCHL 2,460-461; POTSCH, 372-373; W,453-454; WWWS,1457; CALLISEN 16,424-427; 32,47-48; POGG 2,721-722; 3,1153; 7aSuppl.,569-569
H. Schelenz, Pharm. Zent. 48 (1907), 301-312,324-333
B. Anft, Friedlieb Ferdinand Runge. Berlin 1937; J. Chem. Ed. 32 (1955), 566-574
T.I. Williams and H. Weil, Arkiv for Kemi 5 (1953), 283-299

RUNGE, PAUL [1869-1953]
HEIN-E, 374-375; DRUM, 291-293; JV 10,215; RSC 17,757; NUC 510,9
Anon., Nachr. Chem. Techn. 1 (1953), 75
R. Schmitz, Geschichte der Hamburger Apotheken 1818-1965, pp.258-259. Frankfurt a.M. 1966

RUNNSTRÖM, JOHN AXEL MAURITZ [1888-1971]
SMK 6,411-412; VAD 1971, p.830; IB, 255
P. Pasquini, Acta Embryol. Exp. 1971, pp.120-136
G. Reverberi, Acta Embryol. Exp. 1972, pp.5-6

RUPE, HANS [1866-1951]
TLK 3,548; WWWS, 1457-1458; POGG 4,1287-1288; 5,1079-1080; 6,2245-2246; 7a(3),856-857
A. Ebert, Chimia 1 (1947), 12-14
H. Dahn and T. Reichstein, Helv. Chim. Acta 35 (1952), 1-28

RUPP, ERWIN [1872-1956]
HEIN-E, 375-376; TLK 3,548; POGG 5,1080; 6,2246-2247; 7a(3),857
A. Wankmüller, Beitr. Wurtt. Apothekgesch. 12 (1979), 79-90
C. Friedrich et al., Pharmazie 42 (1987), 36-43

RUPPERT, EDUARD [1877-]
JV 17,288; NUC 510,95

RUSCHIG, HEINRICH [1906-]
KURSCHNER 13,3252
Anon., Nachr. Chem. Techn. 8 (1960), 185

RUSHTON, WILLIAM ALBERT HUGH [1901-1980]
DNB [1971-1980],741-742; WW 1980, pp.2223-2224; MH 2,462-463; MSE 3,51-52
Anon. and A.L. Hodgkin, Vision Research 22 (1982), 611-625
H.B. Barlow, Biog. Mem. Fell. Roy. Soc. 32 (1986), 423-459

RUSKA, ERNST [1906-1988]
KURSCHNER 13,3252; STC 2,441-443; SCHULZ-SCHAEFFER, 209-210; WWWS, 1459; POGG 7a(3),857-858
Anon., Leopoldina [3]12 (1966), 65-67; Naturw. Rund. 41 (1988), 299
D. Gabor, Optik 24 (1966-1967), 370-374
E. Ruska, BioScience Reports 7 (1987), 607-629

RUSKA, HELMUTH [1908-1973]
KURSCHNER 11,2510; WWWS, 1459; POGG 7a(3),858-859

RUSS, VIKTOR KARL [1879-1956]
FISCHER 2,1345

RUSSELL, EDWARD JOHN [1872-1965]
DSB 15,492-493; DNB [1961-1970],908-909; WW 1965, p.2674; DESMOND, 534; W, 454-455; POGG 6,2250; 7b,4555-4558
E.J. Russell, The Land Called Me. London 1956
F.C. Bawden, Nature 207 (1965), 1031-1032
A.R. Prevot, C.R. Acad. Sci. 261 (1965), 2999-3001

W.K. Slater, Chem. Brit. 2 (1966), 97-98
H.G. Thornton, Biog. Mem. Fell. Roy. Soc. 12 (1966), 457-477

RUSSELL, FREDERICK FULLER [1870-1960]
AMS 9(II),969; DAMB, 657-658; WWAH, 527-528; WWWS, 1459
J.M. Kinsman, Trans. Assoc. Am. Phys. 74 (1961), 44-46
T.F. Whayne, Trans. Coll. Phys. Phila. [4]30 (1962), 43-44

RUSSELL, HARRY LUMAN [1866-1954]
AMS 9(II),969; WWAH, 528; WWWS, 1460
E.B. Fred, J. Bact. 68 (1954), 133-134
E.H. Beardsley, Harry L. Russell and Agricultural Science in Wisconsin. Madison, Wis. 1969

RUSSELL, JANE ANNE [1911-1967]
AMS 11,4600; NAW 4,610-611
C.N.H. Long, Endocrinology 81 (1967), 689-692

RUSSELL, WALTER CHARLES [1892-1954]
AMS 8,2141

RUSSELL, WILLIAM [1852-1940]
WW 1940, p.2785; FISCHER 2,1346
D.F. King, American Journal of Dermopathology 3 (1981), 55-58

RUSSELL, WILLIAM JAMES [1830-1909]
DNB [1901-1911]3,243-244; POGG 3,1154; 4,1288; 5,1082
G.C. Foster, Nature 82 (1909), 101-102; Proc. Roy. Soc. A84 (1910), xxx-xxxi
J.R. Brown and J.L. Thornton, Ann. Sci. 11 (1955), 331-336
W. Templeton, Chem. Brit. 18 (1982), 266

RUSSOW, EDMUND [1841-1897]
BJN 4,59*; WELDING, 657; LEVITSKI 1,360-363; TSCHIRCH, 1115
N.J. Kuznetsov, Bot. Zh. 71 (1897), 265-269
C. Winkler, Ber. bot. Ges. 15 (1897), (46)-(55)

RUSZNYAK, ISTVAN [1889-1974]
MEL 3,668
M. Foldi, Lymphology 7(4) (1974), 185-186
A. Szent-Gyorgyi, Acta Medica Academiae Scientarum Hungaricae 32 (1975), 3-4

RÜTGERS, JULIUS [1830-1903]
BJN 8,96*; BLOKH 2,627-628
G. Kramer, Ber. chem. Ges. 36 (1903), 4582-4584

RUTHERFORD, ERNEST [1871-1937]
DSB 12,25-36; DNB [1931-1940],765-774; SCHWERTE 1,63-70; W, 455- 456; POTSCH, 373-374; SRS, 19; POGG 4,1290; 5,1082-1083; 6,2251-2252; 7b,4561-4566
H. Geiger, Jahrb. Akad. Wiss. Gott. 1937-1938, pp.19-25
A.S. Eve and J. Chadwick, Obit. Not. Fell. Roy. Soc. 2 (1938), 395-423
A.S. Eve, Rutherford. Cambridge 1939
N. Feather, Lord Rutherford. Glasgow 1940
H. Tizard, Notes Roy. Soc. 4 (1946), 103-108
J.B. Birks, Ed., Rutherford at Manchester. London 1962
M. Oliphant, Recollections of the Cambridge Days. Amsterdam 1972
T.J. Trenn, The Self-Splitting Atom. London 1977; Ambix 26 (1979),134-136
D. Wilson, Rutherford, Simple Genius. Cambridge, Mass. 1983

RUTHERFORD, WILLIAM [1839-1899]
DNB 22,1205-1206; HIRSCH 4,933; O'CONNOR, 198-200; WWWS, 1461
Anon., Nature 59 (1899), 590-591; Science 9 (1899), 381-382; Leopoldina 35 (1899), 77; Lancet 1899(I), pp.538-541; Brit. Med. J. 1899(I), pp.564-567
S. Richards, Notes Roy. Soc. 40 (1986), 193-217

RUTTAN, ROBERT FULFORD [1856-1930]
YOUNG, 9-10
A.B. Macallum, Trans. Roy. Soc. Canada [3]24 (1930), vii-xi

RUTTER, WILLIAM [1928-]
AMS 15(6),360; WWA 1980-1981, p.2870

RUZICKA, LEOPOLD [1887-1976]
DSB 18,764-765; SBA 1,119; MH 2,468-470; STC 2,445-446; MSE 3,56-57; POTSCH, 374; TLK 3,550; WWWS, 1462; POGG 5,1084; 6,2253-2254; 7a(3),863-868
L. Ruzicka, Ann. Rev. Biochem. 42 (1973), 1-20
G.W. Kenner, Nature 266 (1977), 393-394
F. Lynen, Bayer. Akad. Wiss. Jahrbuch 1978, pp.222-227
V. Prelog and O. Jeger, Biog. Mem. Fell. Roy. Soc. 26 (1980), 411-501; Helv. Chim. Acta 66 (1983), 1307-1342
A. Eschenmoser, Chimia 44 (1990), 1-21

RUZICKA, STANISLAW [1872-1946]
IB, 256
V. Mucha, Wiss. Z. Halle 11 (1962), 345-353

RUZICKA, VLADISLAV [1870-1934]
OBL 9,344; FISCHER 2,1346-1347; GRUETTER, 98; IB, 256
M. Rydl, Sbornik pro Dejiny Prirodnich ved a Techniky 4 (1958), 33-79

RYAN, FRANCIS JOSEPH [1916-1963]
AMS 10,3503; WWAH, 529
J.A. Moore, Genetics 50 (1964), s15-s17
A.W. Ravin, Genetics 84 (1976), 1-25

RYAN, HUGH [1873-1931]
WW 1931, p.2780; REILLY, 54; SRS, 24; POGG 4,1291; 5,1084; 6,2254
J. Algar, Nature 127 (1931), 635-636
Anon., Proc. Roy. Irish Acad. 1931-1932, pp.4-5

RYAZANTSEV, MIKHAIL VLADIMIROVICH [1856-1930]
VORONTSOV, 57

RYDBERG, JOHANNES ROBERT [1854-1919]
DSB 12,42-45; SMK 6,424; W, 456-457; WWWS, 1462; POTSCH, 374-375; POGG 4,1291-1292; 5,1084; 6,2255
St. John Nepomucene, Chymia 6 (1960), 127-145

RYDER, JOHN ADAM [1852-1895]
H. Allen, Science 2 (1895), 334-336
H. Allen and H.F. Moore, Proc. Acad. Nat. Sci. Phila. 1896, pp.222-256

RYDON, HENRY NORMAN [1912-1991]
WW 1981, pp.2272-2273; SRS, 91
Anon., Chem. Brit. 27 (1991), 1155

RYSER, WALTER [1888-1984]
JV 28,194; NUC 512,505

RYZHKOV, VITALI LEONIDOVICH [1896-]
BSE (3rd Ed.)22,448; WWWS, 1462

S

SAAKE, WILHELM [1865-]
JV 8,132

SABALITSCHKA, THEODOR [1889-1971]
KURSCHNER 11,2514-2515; HEIN-E, 377-378; TSCHIRCH, 1115-1116; TLK 3,550 IB, 256; POGG 6,2257-2259; 7a(4),1-3
W. Erdmann, Arzneimitt. 9 (1959), 335-337

SABANEEV, ALEKSANDR PAVLOVICH [1842-1923]
BSE 37,557; POGG 4,1293-1294

SABATIER, PAUL [1854-1941]
DSB 12,46-47; W, 457; ABBOTT-C, 124-125; WWWS, 1463; POTSCH, 375; POGG 3,1157; 4,1294; 5,1085-1086; 6,2259; 7b,4574-4575

H. Vincent, C.R. Acad. Sci. 213 (1941), 281-283
E.K. Rideal, Obit. Not. Fell. Roy. Soc. 4 (1942), 63-66
J.R. Partington, Nature 174 (1954), 859-860
C. Camichel et al., Bull. Soc. Chim. 1955, pp.465-475
G. Bertrand et al., Not. Acad. Sci. 3 (1957), 572-598
M.J. Nye, Isis 68 (1977), 375-391SABBATANI, LUIGI [1863-1928]
POGG 6,2259; 7b,4575
I. Simon, Arch. Fisiol. 26(3) (1928), i-iv; Annuario della Universita Studi di Padova 1928-1929, pp.308-315

SABETAY, SÉBASTIEN [1897-1978]
QQ 1977-1978, p.1467; POGG 6,2259-2260; 7b,4575-4579

SABIN, ALBERT BRUCE [1906-1993]
AMS 14,4372; IWW 1982-1983, p.1155; WW 1980, p.2233; ABBOTT, 113-114
S. Benison, Bull. Hist. Med. 56 (1982), 460-483

SABIN, FLORENCE RENA [1871-1953]
DSB 12,48-49; DAB [Suppl.5],600-601; AMS 8,2146; CB 1945, pp.527-529; FISCHER 2,1348; NAW 4,614-617; DAMB, 659; OGILVIE, 153-156; IB, 257; WWAH, 529; WWWS, 1463
V.T. Andriole, J. Hist. Med. 14 (1959), 320-350
E. Bluemel, Florence Sabin. Boulder, Colo. 1959
P.D. McMaster and M. Heidelberger, Biog. Mem. Nat. Acad. Sci. 34 (1960), 271-319
L.S. Kubie, Persp. Biol. Med. 4 (1961), 306-315
J. Kronstadt, Florence Sabin, Medical Researcher. New York 1990

SACC, FRÉDÉRIC [1819-1890]
DHB 5,629; FERCHL, 462; RSC 5,355-356; 8,806; 18,391; NUC 513,215-216; POGG 2,730; 3,1158; 4,1295

SACHDEV, GOVERDHAN PAL [1941-]
AMS 17(6),391

SACHS, BERNARD [1858-1944]
DAB [Suppl.3],682-683; AMS 7,1524; NCAB 34,127; FISCHER 2,1349; DAMB, 659-660

SACHS, FRANZ [1875-1919]
BLOKH 2,628-629; POGG 4,1296; 5,1086
P. Jacobson, Ber. chem. Ges. 52A (1919), 92-96
J. Meyer, Chem. Z. 43 (1919), 321

SACHS, FRITZ [1881-]
JV 22,285; SG [5]1,359; NUC 513,304

SACHS, HANS [1877-1945]
FISCHER 2,1349-1350; KALLMORGEN, 392; BHDE 2,1007; LDGS, 58; DRULL, 229; BK, 300-302; BULLOCH, 394; IB, 257; WININGER 5,308; TETZLAFF, 287; POGG 6,2262-2265; 7a(4),6
C.H. Browning, Nature 155 (1945), 600

SACHS, JULIUS von [1832-1897]
DSB 12,58-60; BJN 2,262-275; LF 2,372-385; HINK 1,589-594; KOERTING, 116; TSCHIRCH, 1116; KOHUT 2,230; WININGER 5,310; WWWS, 1739
K. Goebel, Flora 84 (1897), 101-130; Science 7 (1898), 662-668,695-702
F. Noll, Naturw. Rund. 12 (1897), 460-464,472-475
C. von Voit, Sitz. Bayer. Akad. 28 (1898), 478-487
S.H. Vines, Proc. Roy. Soc. 62 (1898), xxiv-xxix
E.G. Pringsheim, Julius Sachs, der Begründer der neuen Pflanzenphysiologie. Jena 1932
W.O. James, Endeavour 28 (1969), 61-64
H. Gimmler, Julius Sachs, Würzburger Botaniker und Pflanzenphysiologe. Würzburg 1982
F. Weiling, Bonner Geschichtsblätter 35 (1984), 137-177

SACHS, MORITZ [1865-1940]
OBL 9,369; FISCHER 2,1350; WININGER 7,416

SACHS, PAULA [1886-1970]
JV 29,731; NUC 513,350

SACHSE, HERMANN [1862-1893]
JV 4,26; POGG 4,1296
Anon., Chem. Z. 17 (1893), 1929

SACHSE, ULRICH [1854-1911]
C. Liebermann, Ber. chem. Ges. 44 (1911), 2810

SACHSSE, GEORG ROBERT [1840-1895]
POGG 3,1159; 4,1296
Anon., Leopoldina 31 (1895), 107

SACKS, JACOB [1901-1978]
AMS 11,4618; KAGAN(JA), 755; WWWS, 1464
Anon., Chem. Eng. News 41(52) (1963), 51; Pharmacologist 21 (1979), 106

SACKUR, OTTO [1880-1914]
DBJ 1,309; BLOKH 2,629-631; WININGER 5,314; POGG 4,1296-1297;
5,1086-1087; 6,2266
E. Beckmann, Ber. chem. Ges. 48 (1915), 1-4
H. Pick, Chem. Z. 39 (1915), 13

SADIKOV, VLADIMIR SERGEEVICH [1874-1942]
IB, 283-284; POGG 6,2266-2267; 7b,4582-4585
Y.M. Gefter, Vop. Med. Khim. 9 (1963), 441-442

SADRON, CHARLES LOUIS [1902-]
QQ 1979-1980, p.1401; IWW 982-1983, p.1156; MH 2,471-473; MSE 3,62-64; WWWS, 1465

SADTLER, SAMUEL PHILIP [1847-1923]
AMS 3,594-595; NCAB 5,350; RSC 18,398
Who Was Who in America 1,1072
C.H. LaWall, Science 59 (1924), 183-184; Am. J. Pharm. 96 (1924), 134-138

SÄNGER, MAX [1853-1903]
BJN 8,104-105,97*; FISCHER 2,1351; KOREN 240; WNINIGER 7,416-417

SAFFRAN, MURRAY [1924-]
AMS 15(6),378; WWWS, 1465
M. Saffran, Steroids 56 (1991), 298-310

SAGER, RUTH [1918-]
AMS 15(6),378; WWWS, 1465
J. Sapp, Beyond the Gene, pp.204-208. New York 1987

SAH, PETER PEN TIEH [1900-]
AMS 10,3518; WWWS, 1466

SAHLI, HERMANN [1856-1933]
KURSCHNER 4,2477; FISCHER 2,1352-1353; PAGEL, 1465; GROTE 5,177-235
L. Michaud, Verhandl. Schw. Nat. Ges. 114 (1933), 506-510
M. Frey, Schw. med. Wchschr. 63 (1933), 698-700,886-889; Gesnerus 14 (1957), 51-64
W. Hadorn, Schw. med. Wchschr. 86 (1956), 694-698

SAHYUN, MELVILLE [1895-1977]
AMS 13,3845-3846
Anon., Chem. Eng. News 55(44) (1977), 33; Science 198 (1977), 1131

SAINTE-CLAIRE DEVILLE [see DEVILLE]

SAINT-ÈVRE, ÉDOUARD [1817-1890?]
RSC 5,365
A. Maire, Catalogue des Thèses Soutenues en France de 1810 à 1890 Inclusivement, p.29. Paris 1892

SAINT-GILLES [see PÉAN de SAINT-GILLES]

SAINT-PIERRE, CAMILLE [1834-1881]
G. Pécholier, Montpell. Med. 47 (1881), 565-568

SAITO, KENDO [1878-1960]
JBE, 1275,2355; IB, 257; WWWS, 1467

SAJOUS, CHARLES [1852-1929]
DAB 16,306-307; NCAB 9,351-352; FISCHER 2,1353; WWWS, 1468
V. Robinson, Medical Life 32 (1925), 3-21
J.M. Anders, Trans. Coll. Phys. Phlia. [3]52 (1930), lxv-lxx

SAKAGUCHI, KIN-ICHIRO [1897-]
JBE, 1284
H. Yukawa, Profiles of Japanese Science and Scientists, p.89. Tokyo 1970

SAKAMI, WARWICK [1916-1986]
AMS 15(6),385; WWA 1980-1981, p.2881; WWWS, 1468

SAKELLARIOS, EUKLIDES [1894-]
POGG 6,2270; 7b,4597

SAKHARIN (ZAKHARIN), GRIGORI ANTONOVICH [1829-1897]
BSE 16,512; HIRSCH 4,942; VENGEROV 2,427; RSC 19,753
Anon., Lancet 1898(I), p.193; Leopoldina 34 (1898), 57
A.G. Gukasyan, G.A. Zakharin. Moscow 1948

SAKHAROV, GAVRIL PETROVICH [1873-1953]
BSE(3rd Ed.) 23,13-14; BME 29,446; IB, 251
S.M. Pavlenko, Arkh. Pat. 16(2) (1954), 90-92

SAKHAROV, VLADIMIR VLADIMIROVICH [1902-1969]
DSB 12,76-77; BSE(3rd Ed.) 23,13
B.L. Astaurov et al., Genetika 5 (1969), 177-182
Y.I. Polanski, Tsitologia 2 (1969), 398-400

SAKURAI, JOJI [1858-1939]
JBE, 1302; WWWS, 1468; POGG 4,1300; 5,1088-1089; 6,2271; 7b,4612
F.G. Donnan, Nature 144 (1939), 234-235
K. Matsubara, J. Am. Chem. Soc. 61 (1939), 2255-2257
R.J.H. Clark and M. Tasumi, Chem. Brit. 20 (1984), 1000-1001

SALAMAN, REDCLIFFE NATHAN [1874-1955]
DNB [1951-1960],860-861; WW 1949, p.2442; DESMOND, 537; WWWS, 1469; O'CONNOR(2), 296-297; WININGER 7,417-418
P.S. Hudson, Nature 176 (1955), 97
Anon., Lancet 1955(I), pp.1383-1384
K.M. Smith, Biog. Mem. Fell. Roy. Soc. 1 (1955), 239-245

SALANT, WILLIAM [1870-1943]
AMS 7,1537; PARASCANDOLA, 51-52; CHEN, 9; WININGER 7,419; WWAH, 530
Anon., Science 98 (1943), 531; J. Am. Med. Assn. 124 (1944), 455

SALAZKIN, SERGEI SERGEEVICH [1862-1932]
BSE 37,616; BME 29,73-74; IB, 258
Anon., Arkh. Biol. Nauk 32 (1932), 347-351
C. N[euberg], Biochem. Z. 256 (1932), 1
L.T. Sobolev, Vop. Med. Khim. 1 (1949), 3-17
V.A. Bazanov, Sov. Zdrav. 1988(5), pp.67-70

SALIMBENI, ALESSANDRO [1867-1942]
L.M., Ann. Inst. Pasteur 68 (1942), 369-372

SALK, JONAS EDWARD [1914-]
AMS 15(6),390; WWA 1980-1981, p.2883; IWW 1982-1983, p.1162; WWWS, 1470 MH 1,413-414; MSE 3,64-66; ABBOTT, 114; KOREN, 240

SALKOWSKI, ERNST LEOPOLD [1844-1923]
DBJ 5,440; FISCHER 2,1355-1356; PAGEL, 1466-1468; BLOKH 2,634-635; DZ, 1226; TLK 2,592; SCHAEDLER, 110; KOREN, 240; WININGER 5,322-323; POGG 5,1089-1090; 6,2273-2274
C. Neuberg, Ber. chem. Ges. 56A (1923), 58-60; Biochem. Z. 138 (1923),1-4

SALKOWSKI, HEINRICH [1846-1929]
DBJ 11,366; DZ, 1226-1227; TLK 3,553; SCHAEDLER, 110; WININGER 5,323; POGG 3,1166; 4,1301; 5,1090; 6,2274
H. Grossmann, Ber. chem. Ges. 62A (1929), 127-128

SALMON, DANIEL ELMER [1850-1914]
DAB 16,311-312; DAMB, 660-661; SACKMANN, 298-300; WWWS, 1470
Anon., Science 40 (1914), 848; Entomol. News 26 (1915), 96

SALMON, WILLIAM DAVIS [1895-1966]
AMS 10,3523; IB,258; WWWS, 1470
Anon., J. Nutrition 108 (1978), 17-21

SALOMON, GEORG [1849-1916]
FISCHER 2,1356; PAGEL, 1469-1470
J.S., Deutsche med. Wchschr. 42 (1916), 1172
Anon., Chem. Z. 40 (1916), 793; Z. angew. Chem. 29(III) (1916), 539

SALOMON, HUGO [1872-1954]
KALLMORGEN, 393; FISCHER 2,1357; STANGL, 104-120; WININGER 5,328
Anon., Deutsche med. Wchschr. 79 (1954), 1772
A. da S. Mello, Revista Brasiliera de Medicina 12 (1955), 54-56

SALOMON, KURT [1898-1974]
AMS 12,5472; BHDE 2,1012
Anon., Science 186 (1974), 813

SALOMON, MAX [1837-1912]
FISCHER 2,1357; PAGEL, 1468-1469; WININGER 5,336

SALOMONSEN, CARL JULIUS [1847-1924]
DSB 12,87-89; DBL 20,516-519; FISCHER 2,1357; BULLOCH, 394-395; WININGER 5,339; 7,420
T. Madsen, J. Path. Bact. 28 (1925), 702-708
J. Parnas, Dan. Med. Bull. 25 (1978), 220-222

SALOMONSEN, KNUD EJNAR [1883-1950]
J. Krag, Militarlaegen 56 (1950), 33-34

SALT, GEORGE [1903-]
WW 1981, p.2283; IWW 1982-1983, p.1163; WWWS, 1471

SALTER, WILLIAM THOMAS [1901-1952]
AMS 8,2153; WWAH, 531; WWWS, 1471
Anon., Yale J. Biol. Med. 25 (1952), 149-150
J.H. Means, J. Clin. Endocrin. 12 (1952), 1507-1508

SALVETAT, ALPHONSE [1820-1882]
POGG 2,744-745; 3,1168
F. Leblanc, Bulletin de la Société d'Encouragement pour l'Industrie Nationale 81 (1882), 444-450

SALVIOLI, GAETANO [1852-1888]
G. Bizzozero, Arch. Ital. Biol. 11 (1889), 205-211

SALWAY, ARTHUR HENRY [1881-1966]
SRS, 32; SG [5]1,530; POGG 5,1091-1092; 6,2275

SALZER, THEODOR [1833-1900]
HEIN, 555; HB 1,346-347
Anon., Chem. Z. 24 (1900), 1146; Pharm. Z. 45 (1900), 157

SALZMANN, LEO [1904-]
AMS 9(I),1675; LDGS, 24

SAMEC, MAX (MAKS) [1881-1964]
IB, 258; POGG 6,2275-2276; 7b,4623-4626
Enciklopedija Jugoslavije 7,127
K. Ulmann, Koll. Z. 176 (1961),97-98; Ernährungforschung 6 (1961),204-212
M. Rebek, Ost. Chem. Z. 62 (1961), 216-217; 65 (1964), 357-359

SAMES, KARL [1812-]
DGB 138 (1964), 88

SAMOILOV, ALEKSANDR FILIPPOVICH [1867-1930]
DSB 12,95; BSE 37,663; BME 29,137-138; WWR, 492-493; IB, 258; KUZNETSOV(1963), 345-353; KAZAN 1,479-481
S. Genos, Vrachebnoe Delo 1930, pp.1537-1548
I.P. Pavlov, Kazan. Med. Zhur. 31 (1931), 331-332
K.S. Koshtoyants, Essays on the History of Physiology in Russia, pp.265-274. Washington, D.C. 1964
D.M. Krikler, Brit. Med. J. 295 (1987), 1624-1627

SAMOILOV, YAKOV VLADIMIROVICH [1870-1925]
BSE 37,665; BLOKH 2,636-638; POGG 6,2277

SAMTER, MAX [1908-]
AMS 16(6),416-417; BHDE 2,1017; TETZLAFF, 291; WWWS, 1472

SAMUEL, ERNST [1871-1934]
JV 14,227; RSC 16,1034
Anon., Z. angew. Chem. 47 (1934), 389

SAMUELS, LEO TOLSTOY [1899-1978]
AMS 13,3859; WWWS, 1472-1473
L.L. Engel, Endocrinology 103 (1978), 997-998

SAMUELSON, OLOF [1914-]
VAD 1981, p.907

SAMUELSSON, BENGT [1912-]
VAD 1983, p.891; POTSCH, 376

SAMUELY, FRANZ [1879-1913]
BJN 18,120*
Anon., Leopoldina 49 (1913), 71; Münch. med. Wchschr. 60 (1913), 1584

SANARELLI, GIUSEPPE [1864-1940]
DSB 12,96-97; EI 30,618; FISCHER 2,1360; AGRIFOGLIO, 92-99; WWWS, 1473; ARCIERI, 299-312
J.D. Rolleston, Nature 146 (1940), 54-55
A. Boquet, Ann. Inst. Pasteur 64 (1940), 357-358
P. Ambrosioni, Bioch. Ter. Sper. 27 (1940), 101-102

SAND, JULIUS [1878-1917]
RSC 18,430; POGG 5,1093
Anon., Chem. Z. 32 (1908), 658; 41 (1917), 750

SANDBERGER, FRIDOLIN [1826-1898]
ADB 53,701-702; BJN 3,121; 5,54*; NB, 337; RSC 5,389-390; 8,824-826; POGG 2,747; 3,1168-1169; 4,1303-1304; 7aSuppl.,570

SANDER, ALBERT [1887-1963]
JV 29,731; NUC 518,691
Anon., Nachr. Chem. Techn. 10 (1962), 119; 11 (1963), 363

SANDER, FRITZ [1898-1959]
KURSCHNER 8,1987-1988; POGG 7a(4),21-22

SANDER, WILHELM [1812-1881]
TSCHIRCH, 1116

SANDHAAS, WILHELM [1893-]
JV 40,466

SANDMANN, FRIEDRICH [1818-1876]
RSC 5,393
C. Pagenstert, Lohner Familien, p.248. Vechta/Oldenburg 1927

SANDMEYER, TRAUGOTT [1854-1922]
DHB 5,709; AARGAU,651-652; BLOKH 2,638-640; POTSCH, 376-377; POGG 6,2281; 7aSuppl.,570-571
H. Hagenbach, Helv. Chim. Acta 6 (1923), 134-186

SANDMEYER, WILHELM [1863-]
FISCHER 2,1361; PAGEL, 1475; GUNDLACH, 248; POGG 7aSuppl.,571

SANDOW, ERNST [1846-1904]
HEIN, 557-558
Anon., Pharm. Z. 49 (1904), 985

SANDQVIST, HÅKON [1882-1930]
SMK 6,512; POGG 5,1093-1094; 6,2281
Anon., Sven. Farm. Tid. 34 (1931), 633-634
C. Mannich, Ber. chem. Ges. 64A (1931), 1

SANDRAS, CLAUDE MARIE [1802-1856]
LAROUSSE 14,172; HIRSCH 5,13; DECHAMBRE 85,441-442; CALLISEN 17,13-14; 32,91

SANDRITTER, WALTER [1920-1980]
KURSCHNER 13,3278
Anon., Verhandl. path. Ges. 65 (1981), 533-548

SANDSTRÖM, IVAR VICTOR [1852-1889]
SMK 6,517; FISCHER 2,1361; MEDVEI, 793-794; WWWS, 1474
L. Breimer and P. Sourander, Bull. Hist. Med. 55 (1981), 558-563

SANFELICE, FRANCESCO [1861-1945]
EI 30,640; FISCHER 2,1361; AGRIFOGLIO, 69-74; IB, 259
Anon., Rivista Italiana d'Igiene 4-5(3/4) (1945), 3-15

SANGER, CHARLES ROBERT [1860-1912]
DAB 16,350; NCAB 28,150-151; MILES, 423; WWAH, 532; POGG 4,1305; 5,1095
T.W. Richards et al., Science 35 (1912), 532
C.L. Jackson, Proc. Am. Acad. Arts Sci. 48 (1913), 813-822

SANGER, FREDERICK [1918-]
WW 1981, p.2291; IWW 1982-1983, p.1167; MH 1,415-416; STC 2,456-457; MSE 3,67-68; WWWS, 1474; POTSCH, 377
Anon., Nachr. Chem. Techn. 6 (1958), 331
F. Sanger, Ann. Rev. Biochem. 57 (1988), 1-28

SANGER, RUTH ANN [1918-]
WW 1982, p.1953; IWW 1982-1983, p.1167

SANNIÉ, CHARLES [1896-1957]
POGG 6,2284; 7b,4642-4646
R. Truhaut, Bull. Soc. Chim. 1958, pp.1041-1047

SANSON, ANDRÉ [1826-1902]
RSC 18,440-441; WWWS, 1475
V.L., Tribune Medicale 35 (1902), 880-881

SANTESSON, CARL GUSTAF [1862-1939]
SMK 6,524-525; FISCHER 2,1361-1362; IB, 259
G. Liljestrand, Skand. Arch. Physiol. 83 (1936), 1-6

SAPOZHNIKOV, ALEKSEI VASILIEVICH [1868-1935]
BSE 38,88-89; POGG 6,2285; 7b,4653
A. Okatov, Zhur. Ob. Khim. 6 (1936), 785-790

SAPOZHNIKOV, VASILI VASILIEVICH [1861-1924]
BSE 38,89
E. Nikitina, Journal de la Societe Botanique de Russie 10 (1925), 205-208

SARCIRON, RENÉ [1906-1946]
A. Boivin, Ann. Inst. Pasteur 72 (1946), 859-860

SARETT, LEWIS HASTINGS [1917-]
AMS 15(6),421-422; WWA 1980-1981, p.2897; MH 2,473-475; STC 2,457-459; MSE 3,68-70

SARMA, PADUBIDRI SUBBARAYA [1917-1970]
S.C. Pillai and H.R. Cama, Biographical Memoirs of Fellows of the Indian National Science Academy 3 (1973), 89-95

SARTON, GEORGE [1884-1956]
DSB 12,107-114; DAB [Suppl.6].564-566; BNB 38,713-733; NCAB 45,430-431;
CB 1942, pp.734-735; WWWS, 1477; POGG 6,2286; 7b,4663-4667
I.B. Cohen, Am. Phil. Soc. Year Book 1956, pp.124-128
E.J. Dijksterhuis, Centaurus 4 (1956), 369-378
D. Stimson et al., Isis 48 (1957), 283-350
C.D. Hellman, Science 128 (1958), 641-644
A. Thackray and R.K. Merton, Isis 63 (1972), 473-495
A. Thackray et al., Isis 75 (1984), 6-62
H. Elkhaden, Sartoniana 1 (1988), 95-103

SARTORY, AUGUSTE THÉODORE [1881-1950]
GORIS, 563-568; HUMBERT, 268-272; POGG 6,2286-2287; 7b,4667-4671
L. Launoy, Bull. Acad. Med. 135 (1951), 100-104
S. Lambin, Ann. Univ. Paris 21 (1951), 97-99; Ann. Pharm. Fran. 11 (1953), 310-313
J. Meyer, Mycopathologia 8 (1957), 239-242
G. Dillemann, Produits et Problèmes Pharmaceutiques 26 (1971), 352-353

SARX, HANS FRIEDRICH [1912-]
POGG 7a(4),26

SASAKI, TAKAOKI [1878-1966]
JBE, 1323-1324; FISCHER 2,1363-1364; IB, 260

SATAVA, JAN [1878-1938]
Anon., Chem. Ind. 1938, p.1235
O. Miskovsky and P. Rach, Chem. Listy 32 (1938), 233-237
J. Korinek, Coll. Czech. Chem. Comm. 11 (1939), 107-111

SATTLER, WILHELM [1859-1917]
JV 6,298; NUC 521,499
Anon., Z. angew. Chem. 30(III) (1917), 203

SAUERBERG, HANS [1895-]
JV 41,429

SAUNDERS, ARTHUR PERCY [1869-1953]
AMS 8,2166; RSC 18,456
C.K. Chase, Hamilton Alumni Review 4 (1937), 185-186
Anon., Hamilton Alumni Review 19 (1953), 32,46

SAUNDERS, BERNARD CHARLES [1903-1983]
Who's Who of British Scientists 1971-1972, p.746
N.B. Chapman, Chem. Brit. 20 (1984), 917

SAUNDERS, EDITH REBECCA [1865-1945]
DESMOND, 543
G.L. Elles et al., Nature 156 (1945), 198-199,385
H.H. Thomas, Proc. Linn. Soc. 158 (1947), 75-76

SAUSSURE, NICOLAS THÉODORE de [1767-1845]
DSB 12,123-124; DHB 5,730; BRIQUET, 425-428; BLOKH 2,640-641; COLE, 480; TSCHIRCH, 1117; FERCHL, 469-470; SCHAEDLER, 111; POTSCH, 378; POGG 2,756-758; CALLISEN 17,45-49; 32,105
I.F. Macaire, Notice sur la Vie...de Théodore de Saussure. Geneva 1845
A. de Candolle, Mem. Soc. Phys. Hist. Nat. Gen. 11 (1846), viii-xi
Anon., Proc. Roy. Soc. 5 (1843-1850), 583-586

SAUVIN, FRÉDÉRIC (AUGUSTE FRANÇOIS) [1872-]
JV 19,183; NUC 522,161

SAVARD, FRANCIS GERALD KENNETH [1918-1990]
AMS 17(6),462; WWWS, 1481

SAVICH, VLADIMIR VASILIEVICH [1874-1936]
BME 29,59-60; WWR, 495; IB, 260
A.I. Kuznetsov, Farm. Toks. 10 (1947), 3-17

SAVORY, THEODORE HORACE [1896-]
Directory of British Scientists 1966-1967, vol.2, p.454

SAVYALOV [see ZAVYALOV]

SAWYER, WILBUR AUGUSTUS [1879-1951]
DAB [Suppl.5],604-605; AMS 8,2170; NCAB 52,625-626; IB, 160; WWWS, 1483
Anon., Lancet 1951(II), p.992
G.M. Findlay, Brit. Med. J. 1951(II), pp.1347-1348; Trans. Soc. Trop. Med. 46 (1952), 109-111
W.G. Smillie, Trans. Assoc. Am. Phys. 65 (1952), 40-41

SAX, KARL [1892-1973]
DSB 18,775-776; AMS 12,5524; MH 1,416-417; MSE 3,71-72; WWWS, 1483
C.P. Swanson, Genetics 83 (1976), 1-4

C.P. Swanson and N.H. Giles, Biog. Mem. Nat. Acad. Sci. 57 (1987),373-397

SAXER, FRANZ [1864-1903]
BJN 8,98*
F. Marchand, Deutsche med. Wchschr. 29 (1903); Z. allgem. Path. 14 (1903), 417-420

SAXL, PAUL [1880-1932]
OBL 10,8; KURSCHNER 4,2499-2500; FISCHER 2,1368; KOREN, 241;WININGER 7,425; WWWS, 1483
C. Noorden, Wiener med. Wchschr. 82 (1932), 458-459
A. Müller-Deham, Wiener klin. Wchschr. 45 (1932), 443

SAYERS, GEORGE [1914-]
AMS 17(6),469

SAYRE, DAVID [1924-]
AMS 17(6),470

SAYRE, LEWIS ALBERT [1820-1900]
DAB 16,403-404; HIRSCH 5,41-42; KELLY, 1079-1080; WWWS, 1484
Anon., Lancet 1900(II), p.1246; Brit. Med. J. 1900(II), p.1753
D.P. Hall, Am. J. Surg. 103 (1962), 406-407
A.R. Shands, Current Practice in Orthopedic Surgery 4 (1969), 22-42

SAYTZEV, ALEKSANDR MIKHAILOVICH [1841-1910]
BSE 16,336-337; KUZNETSOV(1961), 489-494; ARBUZOV, 124-127; KOZLOV, 160-161; BLOKH 2,631-634; KAZAN 1,323-332; POTSCH, 454; POGG 3,1176; 4,1310; 4,1310; 5,1098; 6,2292
L.A. Chugaev, Zhur. Fiz. Khim. 42 (1910), 1318-1323

SAYTZEV, MIKHAIL MIKHAILOVICH [1845-1904]
BSE 16,337; BLOKH 2,634; KAZAN 1,333-334; POGG 3,1176
Anon., Zhur. Fiz. Khim. 36 (1904), 459-462

SCAFFIDI, VITTORIO [1877-1936]
IB, 261
E. Moracci, Riv. Biol. 23 (1937), 341-350

SCARBOROUGH, HAROLD ARCHIBALD [1891-1969]
SRS, 52; POGG 6,2293
W.A. Waters, Chem. Brit. 6 (1970), 124

SCARTH, GEORGE WILLIAM [1881-1951]
AMS 8,2173; HUMPHREY, 218-219; IB, 261; POGG 6,2294; 7b,4684
Anon., Plant Physiology 26 (1951), 855
J. Levitt, Science 115 (1952), 509
R.D. Gibbs, Trans. Roy. Soc. Canada [3]46 (1952), 99-101

SCATCHARD, GEORGE [1892-1973]
DSB 18,776-779; AMS 11,4681; MH 1,417-418; MSE 3,72-73; WWWS, 1486; POGG 6,2294-2295; 7b,4684-4687
[C. Tanford], Nature 248 (1974), 367
G. Scatchard, Equilibrium in Solutions etc., pp.xix-xxxiv,283-293. Cambridge, Mass. 1976
J.T. Edsall and W.H. Stockmayer, Biog. Mem. Nat. Acad. Sci. 52 (1980), 335-377

SCHAAL, EUGEN [1842-1928]
HEIN-E, 380-381; RSC 8,843
Anon., Chem. Z. 46 (1922), 387; Z. angew. Chem. 41 (1928), 752
F. Katz, Chem. Z. 52 (1928), 531

SCHAARSCHMIDT, ALFRED [1883-1932]
KURSCHNER 4,2501; TLK 3,555; POGG 5,1100; 6,2296
U. Hofmann, Ber. chem. Ges. 66A (1933), 33

SCHACHERL, GUSTAV [1853-1937]
OBL 10,16; HEIN-E, 381-382; KERNBAUER, 110-115; POGG 4,1311

SCHACHMAN, HOWARD KAPNEK [1918-]
AMS 15(6),451; WWA 1980-1981, p.2910; IWW 1982-1983, p.1174; WWWS, 1485

SCHACHT, HERMANN [1814-1864]
ADB 30,482-486; HEIN, 561-562; TSCHIRCH, 1117
Anon., Bull. Soc. Bot. Belg. 3 (1864), 421-423
J. Groenland, Bull. Soc. Bot. 11 (1864), 235-240
E. Höxtermann, NTM 27 (1990), 45-56

SCHACHT, KARL [1836-1905]
BJN 10,240*; HEIN, 563-564; REBER, 339-341; TSCHIRCH, 1117
Anon., Chem. Z. 29 (1905), 1183-1184
H. Böttcher, Pharm. Z. 50 (1905), 945-946

SCHADE, ARTHUR LINCOLN [1912-]
AMS 13,3895

SCHADE, HEINRICH [1876-1935]
FISCHER 2,1369-1370; WEBER, 92-93; POGG 5,1101-1102; 6,2297; 7a(3),3
F. Häbler, Deutsche med. Wchschr. 61 (1935), 2027-2028
F. Hoff, Münch. med. Wchschr. 83 (1936), 110-111
F. Claussen, Koll. Z. 75 (1936), 257-262
J. Hadjamu, Professor Dr.med. Heinrich Schade, der Begründer der Molekularpathologie. Düsseldorf 1974; Deutsche med. Wchschr. 101 (1976), 464-466

SCHADOW, GOTTFRIED [-1885]
NUC 523,480; SG [1]12,626
Anon., Pharm. Z. 31 (1886), 24

SCHAEFER, HANS [1906-]
KURSCHNER 13,3303; POGG 7a(4),40-42

SCHAEFER, KONRAD [1874-1922]
DBJ 4,368; POGG 5,1102; 6,2298
F. Hein, Z. angew. Chem. 35 (1922), 444

SCHAEFFER, ALFRED [1879-]
JV 22,523; NUC 523,597
Anon., Chem. Z. 34 (1910), 232; 35 (1911), 409

SCHAEFFER, FRANZ LOUIS [1840-1906]
E. Fischer, Ber. chem. Ges. 39 (1906), 3790

SCHAEFFER, GEORGES [1882-1953]
R. Leriche, C.R. Soc. Biol. 147 (1953), 1541-1542
A. Mayer, Presse Med. 61 (1953), 1565-1566

SCHAER, EDUARD [1842-1913]
BJN 18,121*; HEIN, 565-566; REBER, 39-41,363-364; BLOKH 2,641-642; HUMBERT, 241-244; BOSSART, 116-117; DZ, 1243-1244; TSCHIRCH, 1117; SCHELENZ, 699; SCHAEDLER, 111; POGG 3,1177-1178; 4,1312-1313; 5,1102; 7aSuppl.,572-573
L. Rosenthaler, Chem. Z. 37 (1913), 1269
C. Hartwich, Verhandl. Schw. Nat. Ges. 1914, pp.106-125

SCHÄFER, ARTHUR [1886-]
JV 40,264; NUC 523,504

SCHÄFER, JOSEF [1867-]
JV 8,217; RSC 17,436

SCHÄFER, WILHELM [1912-1981]
DSB 18,779-780

SCHÄFFNER, ANTON [1900-1945]
KURSCHNER 6(II),558
Anon., Z. angew. Chem. 59 (1947), 64

SCHÄTZLEIN, CHRISTIAN [1882-1956]
JV 19,183; NUC 523,678
J. Keller, Deutscher Weinbau 11 (1956), 498

SCHAFFER, FRIEDRICH [1855-1932]
DHB 5,756; TLK 3,557
B. Strahlmann, Mitt. Lebensmitt. 53 (1962), 480

SCHAFFER, JOSEF [1861-1939]
OBL 10,27-28; HEINDEL, 39-55
V. von Exner, Anat. Anz. 64 (1927), 1-46
S. Schumacher, Alm. Akad. Wiss. Wien 89 (1940), 194-199
V. Patzelt, Anat. Anz. 98 (1951), 185-192

SCHAFFER, NORWOOD KORTER [1905-1980]
AMS 12,5539; WWWS, 1486SCHAIBLE, PHILIP JOHN [1899-1971] AMS 11,4688; NCAB 57,653

SCHALES, OTTO [1910-1988]
AMS 15(6),458; BHDE 2,1022

SCHALL, KARL [1856-1939]
ZURICH, 50-56; POGG 3,1178; 4,1313-1314; 5,1103; 6,2299; 7a(4),52
Anon., Z. angew. Chem. 53 (1940), 24

SCHALLY, ANDREW VICTOR [1925-]
AMS 15(6),459; WWA 1980-1981, p.2914; IWW 1982-1983, p.1176; MSE 3,75-76; POTSCH, 378
N. Wade, The Nobel Duel. Garden City, N.Y. 1981

SCHAMBERG, EDUARD [1884-]
JV 33,304; NUC 524,104

SCHAPER, ALFRED [1863-1905]
DIEPGEN, 28-34; FISCHER 2, 1374
G. Wetzel, Anat. Anz. 29 (1906), 529-538

SCHARDINGER, FRANZ [1853-1920]
DBJ 2,670; RSC 18,490
Anon., Chem. Z. 44 (1920), 774,991; Z. angew. Chem. 33(II) (1920), 396

SCHARLING, EDVARD AUGUST [1807-1866]
DBL 21,61-62; VEIBEL 1,188-193; 2,369-375; 3,29-45; FERCHL, 472; POGG 2,773-774; 3,1179

SCHARRER, BERTA VOGEL [1906-]
AMS 14,4447-4448; WWA 1978-1979, p.2857; IWW 1982-1983,

p.1176;BHDE 2,1024; MEITES, 256-265; WWWS,1487
B. Scharrer, Annual Review of Neuroscience 10 (1987), 1-17

SCHARRER, ERNST ALBERT [1905-1965]
AMS 10,3567; BHDE 2,1024; KALLMORGEN, 396; MEITES, 256-265
W. Bargmann, Anat. Anz. 119 (1966), 119-127

SCHARRER, KARL [1892-1959]
KURSCHNER 8,2016; GUNDEL, 800-808; TLK 3,559; POGG 6,2300-2301; 7a(4),56
Anon., Chem. Z. 81 (1957), 467

SCHATZ, ALBERT ISRAEL [1920-]
AMS 11,4692
M. Wainwright, Hist. Phil. Life Sci. 13 (1991), 97-124

SCHAUDINN, FRITZ [1871-1906]
DSB 12,141-143; BJN 11,56*-57*; FISCHER 2,1376; OLPP, 360-364; W,460-461; FREUND 2,351-360; WREDE, 166-167; WWWS, 1487
S. von Prowazek, Wiener klin. Wchschr. 19 (1906), 880-882
F.W. Winter, Zool. Anz. 30 (1906), 825-846
M. Langeron, Arch. Parasit. 11 (1906-1907), 388-408
F. Neufeld, Deutsche med. Wchschr. 56 (1930), 710-712
M. Hartmann, Naturwiss. 18 (1930), 573-576
J.H. Stokes, Science 74 (1931), 502-506
C. Kuhn, Aus dem Leben Fritz Richard Schaudinns. Stuttgart 1949

SCHAUENSTEIN, ERWIN [1918-]
KURSCHNER 13,3327-3328; POGG 7a(4),61-63

SCHAUM, KARL [1870-1947]
GUNDEL, 820-828; FREUND 3,349-358; TLK 3,559; POGG 4,1315-1316; 5,1104; 6,2301-2302; 7(4),63
L. Hock and M. Volmer, Z. Elektrochem. 46 (1940), 377-378
L. Hock, Nachr. Giessener Ges. 16 (1948), 170-181

SCHAUMANN, OTTO [1891-1977]
KURSCHNER 12,2730; POGG 7a(3),63-64
J. Bock and F. Brücke, Wiener klin. Wchschr. 73 (1961), 253
F. Brücke, Deutsche med. Wchschr. 91 (1966), 724
H. Konzett, Alm. Akad. Wiss. Wien 127 (1978), 500-509

SCHAXEL, JULIUS [1887-1943]
KURSCHNER 4,2521-2522; BHDE 2,1026; LDGS, 15

SCHEDEL, HENRY EDWARD [1804-1856]
NUC 524,371

SCHEFF, GEORGE JULIUS [1897-]
AMS 15(6),564; POGG 6,2303; 7b,4700-4701

SCHEFFER, EMIL [1821-1902]
HEIN, 576
C.L. Diehl, Am. J. Pharm. 74 (1902), 209-217

SCHEFFER, FRANZ EPPO CORNELIS [1883-1954]
BWN 2,495-496; POGG 5,1105; 6.2303-2304; 7b,4701
G. Meyer, Chem. Wkbl. 50 (1954), 625-629

SCHEIBE, ANTON [1856-1948]
JV 6,74; RSC 11,301; NUC 524,503
Anon., Chem. Z. 60 (1936), 79

SCHEIBE, GÜNTER [1893-1980]
KURSCHNER 13,3337; TLK 3,562; POTSCH, 379-380; POGG 6,2304-2305; 7a(4),71-72
Anon., Chem. Z. 77 (1953), 829; 92 (1968), 791; 104 (1980), 247; Naturw. Rund. 33 (1980), 350
A. van Dormael, Chemie Magazine (Ghent) 9(4) (1983), 29-30

SCHEIBER, JOHANNES [1879-1961]
KURSCHNER 6(II),573; TLK 3,562; TSCHIRCH, 1118; POGG 5,1106; 6,2305; 7a(4),72-73
Anon., Chem. Z. 86 (1962), 196

SCHEIBLER, CARL [1827-1899]
BJN 4,177*; BLOKH 2,645-647; WREDE, 227-229; TSCHIRCH, 1118; WWWS, 1488; SCHAEDLER, 112; POGG 3,1180-1181; 4,1318; 7aSuppl.,578
Anon., Chem. Z. 23 (1899), 309-310
P. Degener, Ber. chem. Ges. 33 (1900), 3839-3846

SCHEIBLER, HELMUTH [1882-1966]
KURSCHNER 10,2113; TLK 3,562; POGG 5,1106-1107; 6,2305-2306; 7a(4),73
Anon., Nachr. Chem. Techn. 14 (1966), 308; Chem. Z. 90 (1966), 330
H.W. Wanzlick, Chem. Ber. 102 (1969), xxvii-xxxix

SCHEIDEGGER, JAKOB [1900-]
NUC 524,536

SCHEIDT, MAX [-1894]
JV 6,299; RSC 16,344; 18,499; NUC 524,552
Anon., Ber. chem. Ges. 27 (1894), 3550

SCHEINBERG, ISRAEL HERBERT [1919-]
AMS 15(6),467; ICC, 524-525

SCHEINER, HERMANN [1895-]
LDGS, 79

SCHEINFINKEL, NATHAN [1893-]
KURSCHNER 4,2529; NUC 524,580

SCHELENZ, HERMANN [1848-1922]
DBJ 4,368; HEIN, 576-577; BLOKH 2,647; TSCHIRCH, 1118; TLK 2,603; POGG 6,2306-2307; 7aSuppl.,580
J.W.S. Johnsson, Janus 23 (1918), 1-4
P. Diergart, Z. angew. Chem. 31(III) (1918), 168
H. Thoms, Pharm. Z. 67 (1922), 841-842
G. Urdang, Ber. pharm. Ges. 32 (1922), 225-228
G.E. Dann, Pharm. Z. 84 (1948), 149-151

SCHELLING, FRIEDRICH WILHELM JOSEPH von [1775-1854]
DSB 12,153-159; ADB 31,427-429; LANGHANS, 89-99; WWWS, 1488
G. Schneeberger, F.W.J. Schelling. Berne 1954
O. Temkin, Gesnerus 23 (1966), 188-195
H.J. Sandkühler, F.W.J. Schelling. Stuttgart 1970
G.B. Risse, J. Hist. Med. 27 (1972), 145-158; Bull. Hist. Med. 50 (1976), 321-324

SCHELLMAN, JOHN ANTHONY [1924-]
AMS 15(6),469

SCHELZ, HERBERT [1902-]
JV 43,418; NUC 525,7

SCHEMINZKY, FERDINAND [1899-1973]
KURSCHNER 10,2117; HUTER, 224-227; IB, 262; POGG 6,2307-2308; 7a(4),78-79
H. Schröcksnadel, Ber. Nat. Med. Innsb. 61 (1974), 133-148
W. Auerswald, Alm. Akad. Wiss. Wien 124 (1975), 358-379

SCHENCK, ERNST GÜNTHER [1904-]
KURSCHNER 14,3616; NUC 525,34

SCHENCK, FRIEDRICH [1862-1916]
DBJ 1,368; FISCHER 2,1380; ROEDER, 172-182; AUERBACH, 359; DZ, 1254
A. Gürber, Erg. Physiol. 19 (1921), vii-xix

SCHENCK, GERHARD [1904-]
KURSCHNER 13,3343; DRUM, 302-303; POGG 7a(4),79-80
Anon., Chem. Z. 103 (1979), 158

SCHENCK, MARTIN [1876-1960]
AUERBACH, 359-360; TLK 3,562; POGG 6,2308-2309; 7a(4),82-83,153*
Anon., Tierische Umschau 15 (1960), 261-262

SCHENCK, OTTO von [1878-1965]
JV 20,263

SCHENCK, RUDOLF [1870-1965]
POTSCH, 380; POGG 4,1321; 5,1108; 6,2309-2310; 7a(4),83-84
Z. Elektrochem. 46 (1940), 101-105

SCHENK, FELIX [1850-1900]
DHB 6,1; FISCHER 2,1380-1381

SCHENK, SAMUEL LEOPOLD [1842-1902]
BJN 7,99*; FISCHER 2,1381; JOHN, 38-47
L. Schenk, Aus meinem Universitätsleben. Halle 1900
Anon., Lancet 1902(II), p.562

SCHENKEL, JULIUS [1840-1917]
Anon., Z, angew. Chem. 30(III) (1917), 600

SCHENKEL, JULIUS WILHELM [1878-1967]
JV 21,34; NUC 525,115

SCHERAGA, HAROLD ABRAHAM [1921-]
AMS 15(6),472; WWA 1980-1981, p.2919; IWW 1982-1983, p.1178

SCHERER, ALEXANDER NIKOLAUS von [1771-1824]
ADB 31,99-102; BLOKH 2,547-648; LEVITSKI 1,236-238; FERCHL, 476-477; TSCHIRCH, 1118; SCHAEDLER, 113; COLE, 486; WWWS, 1739; POGG 2,788-789; 7aSuppl.,581-582
E. Kremers, J. Am. Pharm. Assn. 19 (1930), 1245-1247
R. Moller, NTM 2(6) (1965), 37-55

SCHERER, JOHANN BAPTIST ANDREAS von [1775-1844]
COLE, 486-488; POGG 2,787-788

SCHERER, JOHANN JOSEPH [1814-1869]
ADB 31,115-116; HIRSCH 5,67-68; BLOKH 2,648; ROEDER, 185-194; FERCHL,477; TSCHIRCH, 1118; POGG 2,790; 3,1182-1183; 7aSuppl.,582
R. Wagner, Ber. chem. Ges. 2 (1869), 108-110
J. Büttner, Z. klin. Chem. 16 (1978), 478-483

SCHERING, ERNST FRIEDRICH CHRISTIAN [1824-1889]
HEIN, 579-580; HUHLE-KREUTZER, 184-206; REBER, 376-377; BLOKH 2,648-649; POTSCH, 380; TSCHIRCH, 1118; POGG 3,1183; 4,1321-1322
Anon., Pharm. Z. 35 (1890), 3-4
J.F. Holtz, Ber. chem. Ges. 23(3) (1890), 900-904

SCHERRER, PAUL HERMANN [1890-1969]
DSB 18,784-785; KURSCHNER 10,2122; WWWS, 1490; POTSCH, 380- 381; POGG 5,1109; 6,2313; 7a(4),88-89
J. Weigle, Helv. Phys. Acta 23 (1950), 4-6
W. Känzig, Viert, Nat. Ges. Zürich 114 (1969), 507-509
R. Huber, Helv. Phys. Acta 43 (1970), 5-8
B. Glaus, Gesnerus 43 (1986), 133-134

SCHERTZ, FRANK MILTON [1889-1946]
AMS 7,1557; IB, 262

SCHEUING, GEORG [1895-1949]
POGG 7a(4),91-92
Anon., Deutsche Apoth. Z. 94 (1954), 281

SCHEUNERT, ARTHUR [1879-1957]
FISCHER 2,1382-1383; IB, 262; POGG 6,2314-2316; 7a(4),93-96
K. Täufel and W, Stepp, Pharmazie 9 (1954), 443-446
H.K. Grafe, C.A. Scheunert. Berlin 1954
W. Stepp, Münch. med. Wchschr. 99 (1957), 678-689
H. Haenel, Int. Z. Vitaminforsch. 27 (1957), 423-425
K. Täufel, Jahrb. Sachs. Akad. 1957-1959, pp.329-385

SCHEURER, WILHELM [1887-]
JV 27,655; NUC 525,316

SCHEURER-KESTNER, AUGUSTE [1833-1899]
BLOKH 2,649; POGG 2,794; 3,1184; 4,1323
C. Lauth, Bull. Soc. Chim. [3]25 (1901), i-xxxi

SCHICK, BELA [1877-1967]
NCAB E,444-445; 53,425; FISCHER 2,1383-1384; ATUYANA, 153-168; DAMB, 665; WININGER 7,431; WWAH, 537; WWWS, 1490
A. Gronowicz, Bela Schick and the World of Children. New York 1954
H.L. Hodes, Pediatrics 41 (1968), 379-381
S. Karelitz, J. Mt. Sinai Hosp. 35 (1968), 211-213

SCHICKELÉ, GUSTAVE [1875-1927]
FISCHER 2,1384
E. Reeb, Gynécologie 26 (1927), 257-264
A. Couvelaire, Presse Med. 35(I) (1927), 604-605

SCHIEDT, ULRICH [1925-1957]
POGG 7a(4),98-99
Anon., Naturw. Rund. 11 (1958), 76

SCHIEFFELIN, WILLIAM JAY [1866-1955]
AMS 8,2183; NCAB 44,52-53; MILES, 426-427; JV 4,226; WWAH, 537
Anon., Chem. Eng. News 33 (1955), 2536

SCHIEFFERDECKER, PAUL [1849-1931]
KURSCHNER 4,2544-2545; FISCHER 2,1384; FREUND 2,361-366; IB, 263
J. Sobotta, Anat. Anz. 73 (1932), 487-494

SCHIEL, JACOB (JAMES) HEINRICH WILHELM [1813-1889]
SILLIMAN, 121,130; FERCHL, 478; RSC 5,461-462; 11,306; POGG 3,1187

SCHIEMANN, ELISABETH [1881-1972]
KURSCHNER 11,2590; BOEDEKER, 22-23; IB, 263
H. Kuckuck, Ber. bot. Ges. 93 (1980), 517-537

SCHIEMANN, GÜNTHER [1899-1967]
KURSCHNER 10,2129; HANNOVER, 36; WWWS, 1490; POTSCH, 381; POGG 6,2319; 7a(4),99-100
E. Baum, Chem. Z. 91 (1967), 719

SCHIERGE, MANFRED [1893-]
POGG 6,2319; 7a(4),100

SCHIFF, ERWIN [1891-1971]
FISCHER 2,1385; KOREN, 243; LDGS, 78
K. Stehr, Klinische Paediatrie 184 (1972), 81-82

SCHIFF, FRITZ [1889-1940]
KURSCHNER 4,2548; BHDE 2,1031; LDGS, 58; IB, 263; NUC 525,475

SCHIFF, HUGO [1834-1915]
DSB 12,163-164; BLOKH 2,650; WININGER 5,416; POTSCH, 381; POGG 2,796-797; 3,1187-1188; 4,1324-1325; 5,1110; 6,2319-2320; 7aSuppl.,586
H. Wichelhaus, Ber. chem. Ges. 48 (1915), 1566-1567
M. Betti, J. Chem. Soc. 109 (1916), 424-428; Chem. Z. 40 (1916), 37-38
W. McPherson, Science 43 (1916), 921-922

SCHIFF, MORITZ [1823-1896]
DSB 12,164-165; ADB 54,8-11; BJN 1,159; 3,94*,127*; HIRSCH 5,72-73; PAGEL, 1497-1498; ARNSBERG, 470-471; KALLMORGEN, 399; BRIQUET, 429; HAYMAKER, 258-264; FRANKEN, 196;

GENEVA 3,81-87; 4,334-344; W, 462; KOREN, 243; KOHUT 2,237; WININGER 5,418; WWWS, 1491; POGG 3,1188; 4,1324; 7aSuppl.,586-587
A. Biedl, Wiener klin. Wchschr. 9 (1896), 1007-1010
L. Langlois, Rev. Sci. 6 (1896), 548-552
J.L. Prévost, Rev. Med. Suisse Rom. 16 (1896), 585-589
H. Friedenwald, Bull. Inst. Hist. Med. 5 (1937), 589-602
P. Riedo, Der Physiologe Moritz Schiff. Zurich 1970
J. Starobinski, Gesnerus 34 (1977), 2-20
H. Heintzel, Med. Hist. J. 15 (1980), 378-384
J.J. Dreifuss, Gesnerus 42 (1985), 289-303; Rev. Med. Suisse Rom. 104 (1985), 957-968

SCHIFF, ROBERT [1854-1940]
MODENA, 305-306; POGG 3,1188-1189; 4,1325-1326; 5,1110; 6,2320

SCHILD, HEINZ OTTO [1906-1984]
WW 1981, p.2306; WWWS, 1941
H.O. Schild, Trends in Pharmacological Sciences 1 (1979), 1-19
Anon., Brit. Med. J. 289 (1984), 118
D.H. Jenkinson, Receptor Biochemistry and Methodology 6 (1987), 1-10

SCHILD, JOSEF [1824-1866]
H. Brugger, Die Schweizerische Landschaft 1850 bis 1914, pp.396-397. Zurich 1978

SCHILLER, JOSEPH [1906-1977]
J.D. Burchfield et al., Isis 69 (1978), 75-76

SCHILLER, LUDWIG [1882-1961]
TLK 3,566-567; NUC 525,687; POGG 5,1110-1111; 6,2321-2322; 7a(4),104,153*
A. Naumann, Phys. Bl. 13 (1957), 571-572

SCHILLING, ROBERT FREDERICK [1919-]
AMS 17(6),504; WWA 1988-1989, p.2741

SCHILLING, VICTOR [1883-1960]
KURSCHNER 8,2044; FISCHER 2,1386; WWWS, 1501
H. Hirscher, Blut 6 (1960), 364-366
I. Günther et al., Z. ärzt. Frtbild. 65 (1971), 395-396; Klin. Med. 50(6) (1972), 141-142
G. Bast and H. Stobbe, Fol. Haemat. 110 (1983), 617-633

SCHILLINGER, ALBIN [-1916]
RSC 11,310; NUC 526,42
Anon., Pharm. Z. 61 (1916), 601; Chem. Z. 43 (1919), 924

SCHIMPER, GUILLAUME PHILIPPE [1808-1880]
BERGER-LEVRAULT, 213-2314

SCHIMPER, KARL FRIEDRICH [1803-1867]
DSB 12,167-168; ADB 31,274-277; BAD 2,257; WWWS, 1492; POGG 7aSuppl.,587
W. Hofmeister, Bot. Zt. 26 (1868), 33-40
L. Eyrich, Jahresber. Ver. Nat. Mann. 1883-1884, pp.37-64
R. Lauterborn, Ber. Nat. Ges. Freiburg 33 (1934), 269-324
L. Jost, Ber. bot. Ges. 58 (1940), 306-327

SCHIMPER, WILHELM [1856-1901]
DSB 12,165-167; BJN 6,309-311; TSCHIRCH, 1110; W, 462-463; WWWS, 1492
H. Schenck, Ber. bot. Ges. 19 (1901),(54)-(70); Naturw. Rund. 17 (1902),36-39
P. Groom, Nature 64 (1901), 551-552

SCHINNER, ANDREAS [1885-]
JV 30,738; NUC 526,135

SCHIRMACHER, KARL [1868-1931]
POGG 6,2324
E. Bryk, Z. angew. Chem. 44 (1931), 179-180

SCHITTENHELM, ALFRED [1874-1954]
FISCHER 2,1389; SCHOLZ, 52-53; HAGEL, 149-164; PITTROFF, 157-
158; KIEL, 84; IB, 263
M. Bürger, Deutsche med. Wchschr. 79 (1954), 1573-1574
G. Bodechtel, Münch. med. Wchschr. 97 (1955), 145-146
E. Schütz, Klin. Wchschr. 33 (1955), 343-344

SCHJERNING, HENRIK [1862-1914]
DBL 21,175-176; VEIBEL 2,378-380; POGG 4,1329; 6,2325
R. Koefoed, in The Carlsberg Laboratory (H.Holter and K.M. Møller, Eds.), pp.82-87. Copenhagen 1976

SCHLACK, PAUL [1897-1987]
KURSCHNER 13,3385-3386; POTSCH, 381-382; POGG 7a(4),114-115
Anon., Nachr. Chem. Techn. 6 (1958), 4-5; Chem. Z. 81 (1957), 799; 92 (1968), 13; 111 (1987), 284

SCHLAEPFER, JOHANN JAKOB [1867-1919]
JV 9,135; NUC 526,242
Anon., Chem. Z. 43 (1919), 924

SCHLAGDENHAUFFEN, FRÉDÉRIC [1830-1907]
HEIN, 582-583; REBER, 352-354; HUMBERT, 220-221; DORVEAUX, 78-79; BERGER-LEVRAULT, 215; TSCHIRCH, 1119; POGG 2,801;
3,1192; 4,1329-1330; 5,1112
J. Godfrin, Bull. Aacad. Sci. Nancy [3]9 (1908), 1-87

SCHLAGENHAUFER, FRIEDRICH [1866-1930]
OBL 10,167; FISCHER 2,1390; LESKY, 573; KALIN, 184-185
R. Maresch, Wiener klin. Wchschr. 43 (1930), 889-890

SCHLAPP, MAX GUSTAV [1869-1928]
AMS 4,861; WWWS, 1492; SG [2]15,208; NUC 526,271

SCHLEICH, HANS [1904-]
AMS 11,4712

SCHLEIDEN, MATTHIAS JAKOB [1804-1881]
DSB 12,173-176; ADB 31,417-421; FREUND 1,299-302; GIESE, 488-490; LEVITSKI 1,373-375; KALLMORGEN, 400; TSCHIRCH, 1119; FERCHL, 479; W, 463; ABBOTT, 114-115; WWWS, 1493; CALLISEN 32,153
W. Behrens, Bot. Zb. 7 (1881), 150-156,183-190
L. Errera, Rev. Sci. [3]2 (1881), 289-298
M. Moebius, Matthias Jakob Schleiden. Leipzig 1904
F.K. Studnicka, Anat. Anz. 76 (1933), 80-95
A.N. Khazanov, Ist. Est. Tekhn. Pribalt. 2 (1970), 193-204
W.W. Franke, European Journal of Cell Biology 47 (1988), 145-156

SCHLEMM, FRIEDRICH [1795-1858]
ADB 31,462-464; HIRSCH 5,86

SCHLENK, FRITZ [1909-]
AMS 14,462-463; BHDE 2,1033-1034

SCHLENK, WILHELM [1879-1943]
KURSCHNER 4,2564; TLK 3,569; POTSCH, 382; POGG 5,1113; 6,2327; 7a(4),122
E. Späth, Alm. Akad. Wiss. Wien 93 (1943), 208-212
H. Wieland, Bayer. Akad. Wiss. Jahrbuch 1948-1949, pp.249-251

SCHLENK, WILHELM [1907-1974]
KURSCHNER 13,3392; BHDE 2,1033; POGG 7a(4),122

SCHLENTHER, EMIL [-1889]
Anon., Pharm. Z. 34 (1889), 511

SCHLESINGER, BERNARD [1896-1984]
WW 1983, p.1995; WWWS, 1493

SCHLESINGER, EUGEN [1869-]
KALLMORGEN, 400; SG [2]15,210

SCHLESINGER, HERMANN [1866-1934]
OBL 10,191; KURSCHNER 4,2565; FISCHER 2,1392; STUMPF, 140-167; PLANER, 297-298; GRUETTER, 102-103; WININGER 5,429-430; 7,434; WWWS, 1493
A. Maller, Münch. med. Wchschr. 84 (1934), 450-451

SCHLESINGER, HERMANN IRVING [1882-1960]
MILES, 427-428; POGG 6,2327-2328; 7b,4710-4711
H.I. Schlesinger, Chemist 28 (1951), 463-472
E. Wiberg, Bayer. Akad. Wiss. Jahrbuch 1961, pp.191-194

SCHLESINGER, MAX [1905-1937]
LDGS, 25
Anon., Lancet 1937(II), p.413

SCHLESINGER, ROBERT WALTER [1913-]
AMS 14,4468; BHDE 2,1035-1036

SCHLEUSSNER, KARL [1868-1928]
KALLMORGEN, 400; JV 8,188; NUC 526,487
Anon., Chem. Z. 52 (1928), 303

SCHLIEP, LEOPOLD [1876-]
JV 17,340; SG [2]15,212; NUC 526,527

SCHLIEPER, ADOLF [1825-1887]
ADB 31,785-786; BLOKH 2,653
A.W. Hofmann, Ber. chem. Ges. 20 (1887), 3167-3169
Anon., Leopoldina 23 (1887), 217; Chem. Z. 11 (1887), 1637-1638

SCHLIEPER, ADOLF, Jr. [1865-1945]
JV 3,276; RSC 15,530; NUC 526,528
Anon., Chemische Fabrik 10 (1937), 120
G. Grote, Wuppertaler Biographien 12 (1974), 49-56

SCHLITTLER, EMIL [1906-1979]
KURSCHNER 12,2785; POGG 7a(4),126-127
Anon., Chem. Z. 103 (1979), 274

SCHLÖGL, KARL [1924-]
KURSCHNER 13,3400; POGG 7a(4),127

SCHLOESING, ALPHONSE THÉOPHILE [1856-1930]
WWWS, 1493; POGG 4,1333-1334; 5,1114; 6,2329-2330
L. Lecornu, C.R. Acad. Sci. 191 (1930), 85-87
M. Javillier, Bull. Sci. Pharm. 38 (1931), 643-662

SCHLOESING, JEAN JACQUES THÉOPHILE [1824-1919]
DOETSCH, 103-105; WWWS, 1493; POGG 3,1196; 4,1333; 5,1114
L. Guignard, C.R. Acad. Sci. 168 (1919), 293-296
P. Ehrenberg, Chem. Z. 43 (1919), 305

SCHLÖSSER, HANS RUDOLF [1888-]
JV 30,738; NUC 526,570

SCHLOSS, ERNST [1882-1918]
FISCHER 2,1393

SCHLOSS, OSCAR MENDERSON [1882-1952]
AMS 8,2187; FISCHER 2,1394; DAMB, 665-666; ICC, 202-203; KOREN, 243
L.E. Holt, Jr., Arch. Ped. 70 (1953), 157-159

SCHLOSSBERGER, HANS [1887-1960]
KURSCHNER 8,2058; FISCHER 2,1394; KALLMORGEN, 401; POGG 7a(4),130
H. Brandis, Z. Immunitätsforsch. 114 (1957), 421-422; 120 (1960), 2-5
W. Kikuth, Z. Hyg. 146 (1960), 285-286
B. Schmidt, Z. Bakt. 178 (1960), 409-412

SCHLOSSBERGER, JULIUS EUGEN [1819-1860]
ADB 31,531-533; HIRSCH 5,88; BLOKH 2,654; FERCHL, 479-480; SCHAEDLER, 113-114; POGG 2,809-810; 7aSuppl.,590
F. Reusch, Jahreshefte Wurtt. 19 (1863), 26-30
H. Simmer, Sudhoffs Arch. 39 (1955), 216-236
F. Hesse, Julius Eugen Schlossberger. Düsseldorf 1976; Med. Welt 29 (1978), 242-245

SCHLOSSER, THEODOR [1822-1907]
HEIN-E,391; RSC 5,494,950; NUC 526,609
Anon., Chem. Z. 31 (1907), 669

SCHLOSSMANN, ARTHUR [1867-1932]
FISCHER 2,1394-1395; HABERLING, 120-121; PUSCHEL, 135-136; TETZLAFF, 295-296; POGG 6,2330-2331
W. Haberling, Arthur Schlossmann. Düsseldorf 1927
A. Kaess, Z. angew. Chem. 45 (1932), 455
M. Pfaundler, Klin. Wchschr. 11 (1932), 1246-1247
P. Wunderlich, Kinderärztliche Praxis 50 (1982), 589-595

SCHLOSSMANN, HANS [1894-1956]
LDGS, 79; IB, 264; POGG 7a(4),131

SCHLOTTERBECK, FRITZ [1876-1940]
JV 24,693
Anon., Chem. Z. 39 (1915), 879; 53 (1940), 440

SCHLOTTERBECK, JULIUS OTTO [1865-1917]
WWAH, 538; TSCHIRCH, 1119
Anon., J. Am. Med. Assn. 68 (1917), 1862

SCHLUBACH, HANS HEINRICH [1889-1975]
KURSCHNER 12,2790; POGG 6,2331; 7a(4),131-132
Anon., Nachr. Chem. Techn. 3 (1955), 187; Chem. Z. 83 (1959), 601-602

SCHMAEDEL, WOLFGANG von [1877-1950]
JV 20,474; NUC 526,663
Genealogisches Handbuch des Adels 46 (1970), 333

SCHMALFUSS, HANS THEODOR [1894-1955]
POGG 6,2331-2332; 7a(4),134-136
Anon., Chem. Z. 66 (1942), 207
H.P. Kaufmann, Fette und Seife 57 (1955), 322-323

SCHMID, ALFRED [1863-1926]
STRAHLMANN, 473,481SCHMID, ALFRED [1886-]
JV 24,44; NUC 527,59

SCHMID, ERNST ERHARD [1815-1885]
ADB 31,659-661; FERCHL, 480; POGG 2,812-813; 3,1107
Anon., Leopoldina 21 (1885), 59

SCHMID, HANS EDUARD [1917-1976]
KURSCHNER 11,2630; POGG 7a(4),147-148
H.J. Hansen et al., Helv. Chim. Acta 61 (1978), 1-29
C.H. Eugster, Chimia 37 (1983), 226-229

SCHMID, JAKOB [1862-1918]
AARGAU, 673-674; ZURICH-D, 149; POGG 6,2335
E. Noelting, Helv. Chim. Acta 2 (1919), 39-59
F. Fichter, Verhandl. Schw. Nat. Ges. 121 (1941), 17-19

SCHMID, JOHANN FRIEDRICH [1850-1916]
STRAHLMANN, 474-475,482

SCHMID, LEOPOLD [1898-1975]
KURSCHNER 12,2802; POGG 6,2335; 7a(4),150-151
Anon., Chem. Z. 99 (1975), 471

SCHMID, RUDI RUDOLF [1922-]
AMS 15(6),491-492; WWA 1980-1981, p.2917; WWWS, 1494

SCHMID, WALTER [1909-]
KURSCHNER 14,3695; POGG 7a(4),151-152
W. Kunz and M. Süss, Arzneimitt. 24 (1974), 358-359

SCHMID, WILHELM [1858-1939]
RSC 10,909; NUC 527,173

SCHMIDLIN, JULIUS [1880-1962]
TLK 2,615; POGG 5,1116
Eidgenossische Technische Hochschule 1855-1955, p.251. Zurich 1955
Anon., Chem. Z. 86 (1962), 581; Naturw. Rund. 15 (1962), 454

SCHMIDMER, EDUARD [1861-1933]
JV 7,264; RSC 14,1010; NUC 527,203

SCHMIDT, ADOLF [1865-1918]
DBJ 2,703; FISCHER 2,1395-1396
A. Strasburger, Münch. med. Wchschr. 65 (1918), 1334,1412-1413
W. Kaiser, Allergie und Asthma 15 (1969), 106-110

SCHMIDT, ALBRECHT [1874-1945]
NB, 350; POGG 6,2336-2337; 7a(4),155

SCHMIDT, ALEXANDER [1831-1894]
BSE 48,123; BME 34,1014; HIRSCH 5,99-100; PAGEL, 1508-1509; WELDING, 688; BLOKH 2,654; LEVITSKI 2,303-305
E., Berl. klin. Wchschr. 31 (1894), 461-462
F. Krüger, Münch. med. Wchschr. 41 (1894), 826-827
I. Munk, Deutsche med. Wchschr. 20 (1894), 411-412
J. Brennsohn, Die Aerzte Estlands, p.513. Riga 1922
E. Jorpes, J. Chem. Ed. 28 (1951), 578-579

SCHMIDT, AUGUST [1844-1907]
RSC 12,659
A. Wankmüller, Beitr. Wurtt. Apothekgesch. 7 (1966), 63; 8 (1969), 114

SCHMIDT, CARL [1822-1894]
HEIN, 588-589; HIRSCH 5,98; PAGEL, 1509; WELDING, 685; BLOKH 2,654-655; ARBUZOV, 136; LEVITSKI 1,246-255; FRANKEN, 196; TSCHIRCH, 1120; POTSCH, 382; POGG 2,818-819; 3,1198; 4,1335
S.S. Zaleski, Ber. chem. Ges. 27(4) (1894), 963-978
Anon., Chem. Z. 18 (1894), 467; Leopoldina 30 (1894), 105-106
U.V. Palm, Ist. Est. Tekhn. Pribalt. 2 (1970), 169-178

SCHMIDT, CARL FREDERICK [1893-1988]
AMS 14,4474; WWA 1976-1977, p.2776; IWW 1982-1983, p.1181; WWWS, 1494
New York Times, 21 April 1988
G.B. Koelle, Trans. Coll. Phys. Phila. [5]11 (1989), 261-265

SCHMIDT, CARL LOUIS AUGUST [1885-1946]
DAB [Suppl.4],719-720; AMS 7,1561; NCAB 34,471-472; WWAH, 538; IB, 264; POGG 6,2337; 7b,4711-4713
D.M. Greenberg, Science 104 (1946), 387; Ann. Rev. Biochem. 15 (1946), vii-x

SCHMIDT, ERICH [1890-1975]
KURSCHNER 11,2637; TLK 3,573; POGG 6,2337-2338; 7a(4),157-158
Anon., Chem. Z. 99 (1975), 295

SCHMIDT, ERNST [1845-1921]
DBJ 3,315; HEIN, 586-588; LKW 5,340-352; GUNDLACH, 484; SCHMITZ, 371-374; AUERBACH, 368-369; BLOKH 2,655-656; SCHAEDLER, 114; REBER, 159-163; TSCHIRCH, 1120; SCHELENZ, 700; POGG 3,1200-1201; 4,1336; 5,1118
H. Thoms, Ber. chem. Ges. 54A (1921), 190-192
J. Gadamer, Chem. Z. 45 (1921), 729-730; Arch. Pharm. 260 (1922), 1-8
C. Friedrich and G. Meltzer, Pharmazie 43 (1988), 642-647

SCHMIDT, FRIEDRICH HAUBOLD [1877-1954]
JV 20,456; SG [2]15,222; NUC 527,332

SCHMIDT, FRIEDRICH WILHELM [1866-]
KALLMORGEN, 401; JV 3,221; RSC 16,486

SCHMIDT, GERHARD [1901-1981]
AMS 14,4475; WWA 1976-1977, p.2777; BHDE 2,1038; LDGS, 60
C.F. Cori, Mol. Cell. Biochem. 6 (1975), 167-169
H. Kalckar, Biog. Mem. Nat. Acad. Sci. 57 (1987), 399-429

SCHMIDT, GERHARD CARL [1865-1949]
DSB 12,191-192; POGG 4,1339; 5,1118; 6,2339-2340; 7a(4),160
A. Kratzer, Phys. Bl. 6 (1950), 30
K. Kuhn, Naturw. Rund. 4 (1951), 41
L. Badash, J. Chem. Ed. 43 (1966), 219-220

SCHMIDT, GERHARD MARTIN JULIUS [1919-1971]
BHDE 2,1039-1040
D. Ginsburg, Israel J. Chem. 10 (1972), 59-72

SCHMIDT, HANS [1882-1975]
FISCHER 2,1396-1397; AUERBACH, 369-370; IB, 265; POGG 7a(4),161
G. Poetschke, Z. Immunitätsforsch. 114 (1957), 303-307
Anon., Nachr. Chem. Techn. 5 (1957), 291-292; Deutsche med. Wchschr. 100 (1975), 714

SCHMIDT, HANS [1886-1959]
DUNSCHELE, 150-151; TLK 3,575; POGG 6,2340; 7a(4),161-162
Anon., Nachr. Chem. Techn. 7 (1959), 157

SCHMIDT, INGEBORG [1899-]
AMS 14,4475; WWWS, 1495

SCHMIDT (SCHMIDT-THOMÉ), JOSEF [1909-]
KURSCHNER 13,3449; POGG 7a(4),178-179
Anon., Chem. Z. 103 (1979), 304

SCHMIDT, JULIUS [1872-1933]
TLK 3,575-576; POGG 4,1339-1340; 5,1119; 6,2341
R. Glauner and G. Glauner, Ber. chem. Ges. 66A (1933), 51-53
Anon., Chem. Z. 57 (1933), 266

SCHMIDT, KARL FRIEDRICH [1887-1971]
BAD NF2,241-243; DRULL, 238; POTSCH, 382-383; POGG 7a(4),165-166
Anon., Chem. Z. 81 (1957), 573; 95 (1971), 966

SCHMIDT, LUDWIG [1879-]
JV 20,263

SCHMIDT, MARTIN BENNO [1863-1949]
KURSCHNER 6(II),625-626; AUERBACH, 370-371; HABERLING, 121; DANIS, 28-42
E. Kirch, Sitz. Phys. Med. Würzburg 65 (1951), 90-109

SCHMIDT, OTTMAR [1835-1903]
BJN 8,101*-102*; HEIN, 590; BLOKH 2,656-657; REBER, 379-380; TSCHIRCH, 1120
O. Hesse, Ber. chem. Ges. 36 (1903), 4585-4590

SCHMIDT, OTTO [1874-1943]
WOLF, 183; POTSCH, 383; POGG 7a(4),167-168
O. Hecht, Ber. chem. Ges. 76A (1943), 121-125

SCHMIDT, OTTO THEODOR [1894-1972]
KURSCHNER 11,2649; POGG 6,2342; 7a(4),168-169
W. Mayer, Ann. Chem. 1973, pp.1758-1776
K. Freudenberg and W. Mayer, Heid. Akad. Jahrb. 1973, pp.79-83

SCHMIDT, ROBERT EMANUEL [1864-1938]
POTSCH, 383; POGG 6,2342-2343; 7a(4),171
G. Holste, Z. angew. Chem. 40 (1927), 242-244
A. Stock, Ber. chem. Ges. 71A (1938), 121-122
R. Schmitt, Comptes Rendus du 94e Congrès National des Sociétés

Savantes, Pau 1969, Section des Sciences, 1 (1970), 163-172

SCHMIDT, RUDOLF [1873-1947]
FISCHER 2,1398; STANGL, 128-144; PELZNER, 20-21; KOERTING, 171
G. Holler, Wiener klin. Wchschr. 60 (1948), 53

SCHMIDT, ULRICH [1924-]
KURSCHNER 16,3264; POTSCH, 383-384; POGG 7a(4),172

SCHMIDT, WERNER [1874-]
JV 13,155; RSC 16,344; NUC 527,578

SCHMIDT, WILHELM AUGUST [1870-1927]
POGG 6,2343
G. Bredig, Z. angew. Chem. 40 (1927), 1071-1072

SCHMIDT, WILHELM JOSEPH [1884-1974]
KURSCHNER 11,2652; GUNDEL, 847-855; POGG 6,2343-2344; 7a(4),174-176
R.E. Liesegang, Koll. Z. 106 (1944), 135-137

SCHMIDT-MÜLHEIM, ADOLF [1841-1890]
RSC 11,327; 18,550; NUC 527,615
Anon., Archiv für animalische Nahrungsmittelkunde 6 (1890), 47-48; Chem. Z. 14 (1890), 1783; Pharm. Z. 35 (1890), 471

SCHMIDT-NIELSEN, KNUT [1915-]
AMS 15(6),497; WWA 1980-1981, p.2929; IWW 1982-1983, p.1182; MH 2,475-476; STC 2,469-470

SCHMIDT-NIELSEN, SIGVAL [1877-1956]
NBL 12,467-469; IB, 265; POGG 6,2345-2346; 7b,4713-4716
Hver er Hvem? 1950, p.533
A.W. Owe, Norske Vid. Akad. Arbok 1957, pp.43-48

SCHMIDT-OTT, ALBRECHT [1899-]
KALLMORGEN, 403

SCHMIDTMANN, HERMANN [1868-1919]
JV 11,23; NUC 527,640
Anon., Chem. Z. 43 (1919), 816

SCHMIEDEBERG, OSWALD [1838-1921]
DSB 18,789-791; DBJ 3,224-228; WELDING, 691; HIRSCH 5,101-102; PAGEL, 1513-1514; BK, 306-307; DZ, 1287-1288; LEVITSKI 2,189-191; WWWS, 1496; POGG 3,1201; 4,1340; 6,2346; 7aSuppl.,592
R. Luzzatto, Bioch. Ter. Sper. 8 (1921), 257-260
Anon., Naturwiss. 10,105-107 (1922)
H.H. Meyer, Arch. exp. Path. Pharm. 92 (1922), i-xvii
J. Koch-Weser and P.J. Schachter, Life Sciences 22 (1978), 1361-1371

SCHMIEDEL, ROLAND [1888-1967]
HEIN-E,393-394; LEHMANN, 148-149; TSCHIRCH, 1120; POGG 7a(4),180
Anon., Arzneimitt. 8 (1958), 306; Deutsche Apoth. Z. 103 (1963), 527
F. Schlemm, Deutsche Apoth. Z. 98 (1958), 379-380

SCHMITT, FRANCIS OTTO [1903-]
AMS 14,4479; WWA 1978-1979, p.2873; IWW 1982-1983, pp.1182-1183; MH 2,476-477; STC 2,470-471; MSE 3,80-81; WWWS, 1496
F.O. Schmitt, Ann. Rev. Biophys. 14 (1985), 1-22; The Never-ending Search. Philadelphia 1990

SCHMITT, FRIDA [1897-1971]
KURSCHNER 11,2660; BOEDEKER, 99-100

SCHMITT, RUDOLF [1830-1898]
BJN 5,56*; GUNDLACH, 469; MEINEL, 116-117,513; SCHMITZ, 165- 166; POTSCH, 384; BLOKH 2,657-658; POGG 3,1201; 4,1340-1341
Anon., Chem. Z. 22 (1898), 179-180; Leopoldina 34 (1898), 57-58
W. Hempel, Ber. chem. Ges. 31 (1898), 3359-3367
E. von Meyer, Sitz. Akad. Wiss. Leipzig 50 (1898), 39-48; J. prakt. Chem. NF57 (1898), 397-408

SCHMITT, THEODOR FRIEDRICH [1859-]
JV 4,285; RSC 18,552; NUC 528,44

SCHMITTMANN, JOSEF [1875-1958]
JV 19,184; NUC 528,60

SCHMITZ, ALOYS JOSEPH [1876-]
JV 18,178; NUC 528,62

SCHMITZ, ERNST [1882-1960]
FISCHER 2,1400; IB, 265; POTSCH, 384; POGG 6,2347; 7a(4),187-188
J. Kühnau, Deutsche med. Wchschr. 77 (1952), 787
Anon., Nachr. Chem. Techn. 8 (1960), 773

SCHMITZ, FRIEDRICH [1850-1895]
ADB 54,126-128; TSCHIRCH, 1120; WWWS, 1497
P. Falkenberg, Ber. bot. Ges. 13 (1895), (47)-(53)
P. Hauptfleisch, Hedwigia 34 (1895), 132-138

SCHMITZ, HANNS [1922-1967]
AUERBACH, 371-372; POGG 7a(4),188-189
Anon., Chem. Z. 91 (1967), 455; Nachr. Chem. Techn. 15 (1967), 229

SCHMITZ, PETER [1855-]
JV 1,203; RSC 13,355; NUC 528,112
Anon., Chem. Z. 56 (1932), 276

SCHMITZ, WILHELM [1879-1939]
JV 20,183
Anon., Z. angew. Chem. 52 (1939), 614

SCHMORL, GEORG [1861-1932]
FISCHER 2,1400-1401; PAGEL, 1515; IB, 265; SG [3]9,445
P. Geipel, Verhandl. path. Ges. 27 (1934), 326-339

SCHNEDERMANN, GEORG HEINRICH [1818-1881]
FERCHL, 482; RSC 5,511-512; 8,874; POGG 2,825; 3,1202

SCHNEIDER, ALBERT JOHN [1863-1928]
AMS 4,863-864; WWAH, 538; WWWS, 1497

SCHNEIDER, ANTON [1831-1890]
DSB 12,192-194; FREUND 1,303-312
Anon., Anat. Anz. 5 (1890), 322-323
G. Limpricht, Jahresb. Schles. Ges. 68 (1891), 9-13

SCHNEIDER, ERICH GERHARD [1903-]
IB, 265; LDGS, 12

SCHNEIDER, ERNST ROBERT [1825-1900]
BJN 5,116*-117*; TSCHIRCH, 1120; POGG 2,827-828; 3,1203; 4,1341
Anon., Chem. Z. 24 (1900), 307; Z. angew. Chem. 13 (1900), 379-380; Leopoldina 36 (1900), 154-155

SCHNEIDER, FERDINAND [1911-1984]
KURSCHNER 14,3754; POGG 7a(4),194-195

SCHNEIDER, FRANZ CÖLESTIN [1813-1897]
ADB 54, 135-136; BJN 4,22*; WURZBACH 31,20-21; HIRSCH 5,110-111; PAGEL, 1515; REBER, 115-117; TSCHIRCH, 1120; SCHAEDLER, 114; FERCHL, 482; POGG 2,826-827; 3,1203; 4,1341
M. Gruber, Wiener klin. Wchschr. 10 (1897), 1081-1083
F. Kratschmer, Wiener med. Wchschr. 47 (1897), 2263-2266

SCHNEIDER, HEINRICH [1886-1971]
JV 25,614; NUC 528,342
Anon., Chem. Z. 96 (1972), 116

SCHNEIDER, HUGO [1883-]
JV 25,506; NUC 528,358

SCHNEIDER, KARL [1895-1958]
JV 40,743
Anon., Nachr. Chem. Techn. 6 (1958), 303

SCHNEIDER, MAX [1871-1910]
RSC 16,430,559; NUC 528,403
Anon., Chem. Z. 34 (1910), 980

SCHNEIDER, MAX [1904-1979]
KURSCHNER 13,3483; POGG 7a(4),196
H. Hirsch, Deutsche med. Wchschr. 104 (1979), 1613

SCHNEIDER, WALTER [1909-]
JV 55,617; NUC 528,445

SCHNEIDER, WALTER CARL [1919-]
AMS 15(6),506

SCHNEIDER, WILHELM [1882-1939]
TLK 3,580; POGG 5,1121-1122; 6,2349; 7a(4),197
Anon., Chem. Z. 63 (1939), 518

SCHNEIDER, WOLFGANG [1912-]
KURSCHNER 13,3487-3488; POGG 7a(4),198-200
W. Hein, Pharm. Z. 122 (1977), 1302-1303
E. Hickel, Deutsche Apoth. Z. 117 (1977),1210-1211; 122 (1982),1533-1534

SCHNEIDERMAN, HOWARD [1927-1990]
AMS 17(6),532; WWWS, 1498
P.J. Bryant, Dev. Biol. 146 (1991), 1-3

SCHNEIDERS, FRANZ [1879-1933]
JV 19,184; NUC 528,467
Anon., Chem. Z. 57 (1933), 206

SCHNELL, FERDINAND [1820-]
NUC 528,487

SCHNITZER, ROBERT JULIUS [1894-]
AMS 10,3594; BHDE 2,1044

SCHNITZLEIN, ADALBERT [1814-1888]
ADB 32,177-179; HEIN, 593-594; TSCHIRCH, 1121
B. Beyerlein, Pharmazie als Hochschuldisziplin, pp.155-159. Stuttgart 1991

SCHOBIG, EUGEN [1856-1941]
RSC 11,334
J. Walzberg, Ber. chem. Ges. 74A (1941), 221-222

SCHOCKEN, VICTOR [1921-]
AMS 8,2193

SCHODER, ROBERT [1866-]
JV 6,239; RSC 18,563

SCHÖBERL, ALFONS [1903-]
KURSCHNER 13,3494-3495; POGG 7a(4),207-209
A. Schoberl, Stationen auf dem Wege eines Hochschullehrers 1933 bis 1971. Hannover 1971

SCHOEDEL, WOLFGANG [1905-1973]
KURSCHNER 11,2692; POGG 7a(4),209-210

SCHÖDLER, FRIEDRICH [1813-1884]
ADB 32,213; HEIN, 595-596; BLOKH 2,639; FERCHL, 483; TSCHIRCH, 1121; SCHAEDLER, 114-115; POGG 2,828; 3,1204; 4,1342-1343
Anon., Chem. Z. 8 (1884), 633; Leopoldina 20 (1884), 115

E. Hickel, Jahrbuch der Freunde der Universität Mainz 25/26 (1976-1977), 21-37

SCHÖFER, GEORG [1866-]
JV 7,159; NUC 528,641

SCHOELLER, WALTER [1880-1965]
TLK 3,582; POGG 5,1122; 6,2352; 7a(4),210
A. Butenandt, Z. Naturforsch. 5b (1950), 449-450
Anon., Chem. Z. 79 (1955), 784; Nachr. Chem. Techn. 8 (1960), 368; 13 (1965), 337
K. Junkmann, Arzneimitt. 5 (1955), 672-674

SCHOEN, MOISE [1884-1938]
POGG 6,2353; 7b,4718-4719
Anon., Ann. Inst. Pasteur 62 (1939), 129-132; Ann. Ferment. 5 (1939),1-12

SCHÖNBEIN, CHRISTIAN FRIEDRICH [1799-1868]
DSB 12,196-199; ADB 32,256-259; SL 2,415-430; BUGGE 1,456-468; HIS,89-94; BLOKH 2,483-484; PRANDTL, 193-241; TSCHIRCH, 1121; SCHAEDLER, 115; POTSCH, 384-385; FERCHL, 483-484; COLE,489; W,464-465; WWWS,1499; CALLISEN 32,190; 2,829-832; 3,1205; 6,2353; 7aSuppl.,592-594
E. Hagenbach, Verhandl. Schw. Nat. Ges. 52 (1868), 207-220
F. von Kobell, Sitz. Bayer. Akad. 1869(I), pp.389-393
G.W.A. Kahlbaum and E. Schaar, Schönbein. Leipzig 1899
G.W.A. Kahlbaum et al., Verhandl. Nat. Ges. Basel 12Suppl. (1900), 11-58
M. Philip, Z. angew. Chem. 23 (1910), 913-916
R.E. Oesper, J. Chem. Ed. 6 (1929), 432-440,677-685
E.H. Huntress, Proc. Am. Acad. Arts Sci. 77 (1959), 50

SCHÖNBERG, ALEXANDER [1892-1985]
KURSCHNER 12,2865; TLK 3,583; LDGS, 25; POGG 6,2353-2354; 7a(4),213-216
Anon., Nachr. Chem. Techn. 15 (1967), 397; Chem. Z. 109 (1985), 134
K. Praefke and M. Sidky, Egyptian Journal of Chemistry 20 (1977),423-425
A.A. Nada, J. Chem. Ed. 60 (1983), 451-452
E. Singer, EPA Newsletter 26 (1986), 1-11

SCHÖNDORFF, BERNHARD [1865-1934]
IB, 266; RSC 18,564; NUC 529,60

SCHÖNE, EMIL [1838-1896]
BLOKH 2,662-663; POGG 3,1205; 4,1343
H. Landolt, Ber. chem. Ges. 29 (1896), 1537-1539
I. Miklashevski, Zhur. Fiz. Khim. 28 (1896), 835-841

SCHÖNE, GEORG [1875-1960]
AUERBACH, 375-376
E. Gohrbandt, Z. Chir. 85 (!960), 609-610
R. Pichlmayr, Münch. med. Wchschr. 120 (1978), 480-482

SCHOENEBECK, OTTO von [1902-]
JV 46,488; GGTA 14 (1920), 813

SCHÖNEWALD, HANS [1874-]
JV 18,326; NUC 529,100

SCHOENHEIMER, RUDOLF [1898-1941]
DSB 18,791-795; DAB [Suppl.3],693-694; MILES, 428; BHDE 2,1048; LDGS, 60; ABBOTT-C, 127; IB, 266; KOREN, 244; KAGAN(JA), 741-742; POTSCH, 384; TETZLAFF, 298; POGG 6,2355-2356; 7a(4),225-226
H.T. Clarke, Science 94 (1941), 553-554
J.H. Quastel, Nature 149 (1942), 15-16
U. Peyer, Rudolf Schoenheimer. Zurich 1972
R.E. Kohler, Hist. Stud. Phys. Sci. 8 (1977), 257-298
D. Stetten, Persp. Biol. Med. 25 (1982), 354-368

SCHÖNHEYDER, FRITZ [1905-1979]
KBB 1979, pp.991-992; VEIBEL 2,394

SCHOENLEIN, JOHANN LUKAS [1793-1864]
DSB 12,202-203; ADB 32,315-319; LF 5,332-339; DECHAMBRE [3]7,512-514; HIRSCH 5,123-127; Suppl.,417; PAGEL, 1522-1524; WWWS, 1500; CALLISEN 17,300; 32,195-195
W. Griesinger, Berl. klin. Wchschr. 1 (1864), 276-279
E. von Leyden, Deutsche med. Wchschr. 19 (1893), 1249-1253
E.H. Ackerknecht, J. Hist. Med. 19 (1964), 131-138

SCHOEPF, CLEMENS [1899-1970]
KURSCHNER 10,2216-2217; BOHM, 163-164; WOLF, 186; TLK 3,585; POTSCH, 385-386; POGG 6,2357; 7a(4),229-231
Anon., Nachr. Chem. Techn. 7 (1959), 256
G. Hesse, Bayer. Akad. Wiss. Jahrbuch 1971, pp.236-238
J. Thesing, Chem. Ber. 112 (1979), i-xix

SCHØYEN, ARNULF BERNHARD [1837-1902]
NBL 13,105-106; BLOKH 2,665

SCHOLANDER, PER FREDRIK [1905-1980]
AMS 14,4491; WWWS, 1499
H.T. Hammel and T.H. Bullock, Am. Phil. Soc. Year Book 1981, pp.505-509
K. Schmidt-Nielsen, Biog. Mem. Nat. Acad. Sci. 56 (1987), 387-412
P.F. Scholander, Enjoying a Life in Science. Fairbanks, Alaska 1990

SCHOLL, FRANZ [1867-1942]
KALLMORGEN, 405-406; JV 8,151
Anon., Chem. Z. 66 (1942), 375

SCHOLL, ROLAND [1865-1945]
KERNBAUER, 176-204; TLK 3,582; POGG 4,1344-1345; 5,1124; 6,2359; 7a(4),232-233
A. Zinke and O. Dischendorfer, Z. angew. Chem. 38 (1925), 901-903
A. Skrabal, Alm. Akad. Wiss. Wien 99 (1949), 306-312

SCHOLTZ, MAX [1861-1919]
DBJ 2,733; HEIN, 598; BLOKH 2,660; POGG 4,1345-1346; 5,1124-1125
J. Meisenheimer, Ber. chem. Ges. 52A (1919), 89-91
W. Herz, Chem. Z. 43 (1919), 209
T. Posner, Ber. Pharm. Ges. 29 (1919), 373-376

SCHOLZ, ERICH [1903-]
JV 44,536; NUC 529,328

SCHOLZ, WILHELM [1864-1934]
KURSCHNER 4,2645-2646; IB, 266; GRUETTER, 104

SCHOMBURGK, RICHARD [1811-1891]
TSCHIRCH, 1121; WWWS, 1499
Anon., Timehri NS3 (1889), 1-29; Leopoldina 27 (1891), 107

SCHOOFS, FRANÇOIS [1875-1959]
POGG 6,2360; 7b,4724-4725
Anon., Liber Memorialis Université de Liège 1867-1935, vol.3, pp.268-273. Liège 1936
C. Heusghem, Liber Memorialis Université de Liège 1936-1966, vol.2, pp.543-546. Liège 1967

SCHOORL, NICOLAAS [1872-1942]
RSC 15,572; NUC 529,476; POGG 5,1125-1126; 6,2360-2361; 7b,4725-4726
D. van Os, Chem. Wkbl. 39 (1942), 414-415
J.J.L. Zwicker, Pharm. Weekbl. 79 (1942), 490-495

SCHOPFER, WILLIAM HENRI [1900-1962]
DSB 12,207-208; IB, 266
K.H. Erismann, Verhandl. Schw. Nat. Ges. 142 (1962), 252-258; Mitt. Nat. Ges. Bern 20 (1963), 86-102
F. Chodat, Archives des Sciences (Geneva) 16 (1963), 157-159

SCHORLEMMER, CARL [1834-1892]
DSB 12,208-209; DNB 17,928-929; HB 1,111-113; HEIN, 599-600; WWWS, 1500;BLOKH 2,657-658; POTSCH, 386; POGG 3,1208; 4,1346; 7aSuppl.,599-600
H.E. Roscoe, Nature 46 (1892), 394-395; Proc. Roy. Soc. 52 (1893), vii-ix
A. Spiegel, Ber. chem. Ges. 25 (1892), 1107-1123
A.H., J. Chem. Soc. 63 (1893), 756-763
H.B. Dixon, Mem. Lit. Phil. Soc. [4]7 (1893), 191-198
K. Heinig, NTM 1 (1960), 62-71; Carl Schorlemmer. Leipzig 1974
C. Duschek, Z. Chem. 24 (1984), 313-325
H.J. Bittrich et al., Carl Schorlemmer. Leipzig 1984

SCHORMÜLLER, ANTON [1905-]
AMS 12,5602

SCHORMÜLLER, JOSEF [1903-1974]
POGG 7a(4),237-239
F. Lynen, Nachr. Chem. Techn. 21 (1973), 307-308

SCHOTT, EHRHART [1879-1968]
JV 23,309

SCHOTT, HERMANN F. [1904-]
AMS 12,5602

SCHOTTE, HERBERT [1897-1950]
JV 36,625; IB, 266
Anon., Z. angew. Chem. 63 (1951), 132

SCHOTTELIUS, MAX [1849-1919]
FISCHER 2,1408; PAGEL, 1526-1527; BK, 308; DZ, 1305-1306

SCHOTTEN, CARL [1853-1910]
BJN 15,77; BLOKH 2,664; WWWS, 1500; POTSCH, 386-387; POGG 3,1209; 4,1347; 5,1126
W. Will, Ber. chem. Ges. 43 (1910), 3703-3714

SCHOTTMÜLLER, ARNOLD [1879-]
JV 23,42; NUC 529,624

SCHOTTMÜLLER, HUGO [1867-1936]
KURSCHNER 4,2661; FISCHER 2,1409; IB, 266; WWWS, 1500
H. Lehnartz, Deutsche med. Wchschr. 62 (1936), 1144-1145
H. Schulten, Münch. med. Wchschr. 83 (1936), 1097-1099
G. Budelmann, Internist (Berlin) 10 (1969), 92-101

SCHRADER, FRANZ [1891-1962]
DSB 18,795-799; AMS 10,3601-3602; IB, 266; WWAH, 540
D.E. Lancefield, Biol. Bull. 125 (1963), 9-10
K.W. Cooper, J. Cell. Biol. 16(3) (1963), ix-xix

SCHRADER, FRIEDRICH FRANZ [1868-]
JV 9,146; NUC 529,687

SCHRADER, HANS [1887-1982]
POGG 6,2366; 7a(4),245
Anon., Nachr. Chem. Techn. 15 (1967), 30

SCHRADER, JOHANN [1762-1826]
PHILIPPE, 737-738; TSCHIRCH, 1121-1122; FERCHL, 485; CALLISEN 32,205; POGG 2,839-840

SCHRADER, MAX [1860-1892]
PAGEL, 1527; JV 1,185; RSC 18,579; NUC 530,1
Anon., Leopoldina 28 (1892), 105; Pharm. Z. 37 (1892), 261

SCHRAMM, GERHARD [1910-1969]
KURSCHNER 10,2226; BOHM, 181-182; WWWS, 1501; POTSCH, 387; POGG 7a(4),247-248
Anon., Nachr. Chem. Techn. 6 (1958), 315; Chem. Z. 93 (1969), 152

SCHRAUBE, CONRAD [1849-1923]
HEIN, 602-603; POGG 3,1210; 4,1351; 6,2367

SCHRAUTH, WALTHER [1881-1939]
KURSCHNER 5,1249; TLK 3,586; POTSCH, 387; POGG 5,1129; 6,2367; 7a(4),251
M. Pflücke, Z. angew. Chem. 52 (1939), 433-436

SCHREIBER, HANS [1902-1968]
KURSCHNER 10,2228; WWWS, 1501; POGG 7a(4),252-253
H. Langendorff and H. Monig, Strahlentherapie 133 (1967), 312-313
H. Pfleiderer et al., Strahlentherapie 136 (1968), 512-514

SCHREIBER, HERMANN [1908-]
JV 55,562; NUC 530,116

SCHREIBER, KLAUS [1927-]
KURSCHNER 16,3348; POTSCH, 387-388; WWWS, 1501

SCHREIDER, MIKHAIL (MICHEL) NIKOLAEVICH [1854-]
RSC 18,583; SG [2]15,276; NUC 530,146

SCHREINER, OSWALD [1875-1965]
AMS 10,3603; NCAB E,380-381; MILES, 428-429; IB, 266

SCHRIEVER, HANS [1898-1979]
KURSCHNER 12,2893; EBERT, 137-142; POGG 7a(4),260-261
Anon., Münch. med. Wchschr. 120 (1978), 710; Naturw. Rund. 32 (1979), 431

SCHRÖCKSNADEL, HANS [1912-1985]
KURSCHNER 13,3534; HUTER, 226
Anon., Naturw. Rund. 38 (1985), 255

SCHRÖDER, ERNST [1872-1914]
JV 14,228; RSC 17,436

SCHROEDER, HEINRICH [1810-1885]
BLOKH 2,666; BULLOCH, 396; FERCHL, 487; POGG 3,1212
K. Birnbaum, Ber. chem. Ges. 18(3) (1885), 843-846
C. von Voit, Sitz. Bayer. Akad. 16 (1885), 57-62

SCHROEDER, HENRY (HEINRICH HARRY STANISLAUS) [1873-1945]
KURSCHNER 4,2672; KIEL, 162; IB, 267; TLK 3,588; POGG 6,2370; 7a(4),263

SCHROEDER, HERMANN [1902-]
KURSCHNER 8,2139; POGG 7a(4),263-264

SCHRÖDER, KARL [1838-1887]
ADB 32,523-524; HIRSCH 5,143; ASEN, 178

SCHROEDER, RICHARD [1867-]
JV 12,171; NUC 530,364

SCHROEDER, WALTER [1930-1961]
New York Times, 21 October 1961

SCHRÖDER, WILHELM [1911-]
KURSCHNER 13,3542; AUERBACH, 378; POGG 7a(4),267-268

SCHROEDER, WOLDEMAR von [1850-1898]
BJN 5,57*; BAD 5,920; DRULL, 246; STUBLER, 329; FISCHER 2,1412; BLOKH 2,667; POGG 3,1213; 4,1354
R. Gottlieb, Ber. chem. Ges. 31 (1898), 227-231

SCHROEDER van der KOLK, HENDRIK WILLEM [1836-1867]
POGG 2,845; 3,1213
H.A.M. Snelders, Scientarum Historia 13 (1971), 184-197

SCHROEDER van der KOLK, LODEWIJK CONRADUS [1865-1905]
NNBW 2,698-700; POGG 4,1355; 5,1132
J.A. Gutterink, Chem. Wkbl. 2 (1905), 601-612

SCHRÖDINGER, ERWIN [1887-1961]
DSB 12,217-223; NOB 18,63-68; STC 2,473-475; BHDE 2,1053; TLK 3,588;W, 465-466; WWWS, 1502; POGG 5,1132-1133; 6,2372-2373; 7a(4),268-269
W. Heitler, Biog. Mem. Fell. Roy. Soc. 7 (1961), 221-228
L. Flamm, Alm. Akad. Wiss. Wien 111 (1961), 402-411
R. Olby, J. Hist. Biol. 4 (1971), 119-148
E.J. Yoxen, Hist. Sci. 17 (1979), 17-52
E.P. Fischer, Social Research 51 (1984), 809-835
N. Symonds, Quart. Rev. Biol. 61 (1986), 221-226
J.A. Witkowski, TIBS 11 (1986), 266-268
M.F. Perutz, Nature 226 (1987), 555-558; in Selected Topics in the History of Biochemistry (G. Semenza and R. Jaenicke, eds.), pp.1-20. Amsterdam 1990
W. Moore, Schrödinger, Life and Thought. Cambridge 1989

SCHRÖTER, GEORG [1869-1943]
TLK 3,589; POGG 4,1356; 5,1133; 6,2373; 7a(4),271
P. Brigl, Ber. chem. Ges. 72A (1939), 131
Anon., Z. angew. Chem. 56 (1943), 304; Chemie 57 (1944), 81-82

SCHRÖTER, JOSEPH [1835-1894]
ADB 54,218-219; HIRSCH 5,146
H. Kionka, Jahresb. Schles. Ges. 72 (1894), 9-16
P. Magnus, Ber. bot. Ges. 13 (1895), (34)-(42)

SCHRÖTTER, ANTON [1802-1875]
DSB 12,227; ADB 32,575-577; WURZBACH 32,1-7; BLOKH 2,667-668; FERCHL,488; TSCHIRCH, 1122; SCHAEDLER, 115; KNOLL, 55-57; WWWS, 1503; POTSCH, 388; POGG 2,848-849; 3,1214; 7aSuppl.,604
J. Loschmidt, Alm. Akad. Wiss. Wien 25 (1875), 216-234
R. Schneider, J. prakt. Chem. 120 (1875), 449-455
A. Lieben, Ber. chem. Ges. 9 (1876), 90-108
F.A. Abel, J. Chem. Soc. 29 (1876), 622-625
A. Bauer, Ost. Chem. Z. 10 (1907), 17-22; Anton Schrotter. Vienna 1917
H. Lagler, Blätter für Technikgeschichte 29 (1967), 1-140

SCHRÖTTER, HERMANN [1870-1928]
FISCHER 2,1412; STUMPF, 190-201
A. Loewy, Deutsche med. Wchschr. 54 (1928), 365
M. Weinberger, Wiener klin. Wchschr. 41 (1928), 136

SCHRÖTTER, HUGO [1856-1911]
BJN 16,136-138,70*; HEIN, 605; KERNBAUER, 115-117,161-166; TSCHIRCH,1122; SCHAEDLLER, 115; POGG 3,1214; 4,1356; 5,1133
C. Liebermann, Ber. chem. Ges. 44 (1911), 2270
R. Kremann, Ost. Chem. Z. 16 (1911), 186-187

SCHRÖTTER, LEOPOLD [1837-1908]
HIRSCH 5,146-147; FROHNE, 118-131
O. Chiari, Wiener klin. Wchschr. 21 (1908), 641-642

SCHROFF, CARL von [1844-1892]
ADB 54,216-217; PAGEL, 1534; STAHL, 77-78; TSCHIRCH, 1122
Anon., Leopoldina 28 (1892), 104

SCHROFF, KARL DAMIAN von [1802-1887]
ADB 54,216; WURZBACH 32,12-15; HIRSCH 5,147; PAGEL, 1534-1535; STAHL, 44-76; BOEGERSHAUSEN, 96-97; TSCHIRCH, 1122; FERCHL, 488
Anon., Brit. Med. J. 1887(II), p.158; Leopoldina 23 (1887), 114-115

SCHROHE, ADAM [1860-1933]
TLK 3,588
Anon., Chem. Z. 57 (1933), 145; Z. angew. Chem. 46 (1933), 134
M. Speter, Zeitschrift für Spiritusindustrie 59 (1936), 442-443

SCHRUMPF (SCHRUMPF-PIERRON), PIERRE [1882-]
JV 20,544; SG [2]15,282; [3]9,472; NUC 530,481-482

SCHRYVER, SAMUEL BARNETT [1869-1929]
WW 1929, p.2720; O'CONNOR(2), 155; POGG 6,2374
A.R. Ling, Nature 124 (1929), 490-491
A. H[arden], Biochem. J. 24 (1930), 229-232; J. Chem. Soc. 133 (1930), 901-905
V.H.B., Proc. Roy. Soc. B110 (1932), xxii-xxiv

SCHUBERT, GUSTAV [1897-1976]
KURSCHNER 12,2907; MAASS, 129-141; KOERTING, 121; POGG 7a(4),278-279

SCHUBERT, HANS [1906-1951]
POGG 7a(4),279
T. Bürgers, Z. Bakt. 157 (1952), 467

SCHUBERT, HERMANN [1884-1915]
JV 27,433
Anon., Chem. Z. 39 (1915), 862; Z. angew. Chem. 28(III) (1915), 608

SCHUBERT, HERMANN [1921-]
KURSCHNER 11,2733; POGG 7a(4),280

SCHUBERT, MAXWELL [1902-]
AMS 11,4749

SCHUBERT, PAUL BERNHARD [1873-]
JV 19,107; SG [2]15,283; NUC 530,512
Anon., Deutsche Tierärztliche Wochenschrift 61 (1954), 23

SCHUBERT, RENÉ [1910-1976]
POGG 7a(4),282-283

SCHUDEL, GUSTAV [1891-1918]
NUC 531,50

SCHÜBEL, KONRAD [1885-1978]
KURSCHNER 13,3552; BERWIND, 145-151; PITTROFF, 194-195; TLK 3,589; IB, 267; POGG 7a(4),283-284
Anon., Naturw. Rund. 8 (1955), 371

SCHÜBLER, GUSTAV [1787-1834]
ADB 32,639-640; HIRSCH 5,151; BLOKH 2,668-669; FERCHL,489; SCHAEDLER,116; CALLISEN 17,355-360; 32,220; POGG 2,853-855; 7aSuppl.,609

SCHÜFFNER, WILHELM [1867-1949]
KURSCHNER 7,1882; FISCHER 2,1414; OLPP,368-371

SCHÜLER, HERBERT [1909-]
JV 49,216; NUC 531,97

SCHÜLLER, JOSEPH [1888-1968]
KURSCHNER 10,2248; FISCHER 2,1415; TLK 3,591; IB, 267; POGG 7a(4),287

SCHÜMMELFEDER, NORBERT [1916-1965]
KURSCHNER 10,2249; WWWS, 1505; POGG 7a(4),287-288

SCHÜRMANN, PAUL [1895-1941]
KURSCHNER 4,2722; FISCHER 2,1415
R. Rössle, Verhandl. path. Ges. 32 (1950), 457
A.H. Murken, Gütersloher Beiträge zur Heimat- und Landeskunde 1977, pp.927-933

SCHÜRMANN, WALTER [1880-1974]
KURSCHNER 11,2740-2741; FISCHER 2,1415-1416; POGG 7a(4),289- 290

SCHÜTT, FRANZ [1859-1921]
DBJ 3,316; KIEL, 212

SCHÜTTE, ERNST [1908-1985]
KURSCHNER 13,3559; POGG 7a(4),290
Anon., Chem. Z. 102 (1978), 271

SCHÜTZ, EMIL [1853-1941]
FISCHER 2,1416; SCHIEBER, 127-136; PLANER, 307; KOREN, 244; RSC 18,597

SCHÜTZ, FRANZ [1887-1951]
Anon., Papier 6 (1952), 105-106

SCHÜTZ, JAKOB [1816-1898]
BJN 5,56*; RAGEL, 1545; WERSTLER, 136-140

SCHÜTZ, JULIUS [1876-1923]
DECKER, 95-100
Anon., Wiener med. Wchschr. 73 (1923), 2353-2354

SCHÜTZ, LUDWIG AUGUST [1887-]
JV 30,667; NUC 531,209

SCHÜTZE, ALBERT [1872-1912]
BJN 18,58*; FISCHER 2,1417
Anon., Deutsche med. Wchschr. 38 (1912), 969-970; Münch. med. Wchschr. 59 (1912), 624,736

SCHÜTZENBERGER, CHARLES [1809-1881]
HIRSCH 5,154
R. Mourgue, Revue de Médecine 35 (1916), 691-696
M. Klein et al., Comptes Rendus du Congres National des Sociétés Savantes, Section des Sciences 1,111-134. Paris 1969

SCHÜTZENBERGER, PAUL [1829-1897]
VAPEREAU 6,1425; GE 29,785; BLOKH 2,672-673; BERGER-LEVRAULT, 221-222; POTSCH, 389; WWWS, 1506; POGG 3,1216-1217; 4,1360; 6,2378
C. Friedel, Bull. Soc. Chim. [3]19 (1898), i-xliii
T.L. Davis, J. Chem. Ed. 6 (1929), 1403-1414
G. Urbain et al., Rev. Sci. 67 (1929), 705-713; Centenaire Paul Schützenberger. Macon 1931

SCHULEK, ELEMÉR [1893-1964]
MEL 2,601-602; POGG 6,2379; 7b,4737-4747
J. Laszlovsky
, Talanta 10 (1963), 429-431
E. Pungor and A. Vegh, Acta Chim. Acad. Sci. Hung. 41 (1964), 1-36

SCHULEMANN, WERNER [1888-1975]
KURSCHNER 11,2749; DUNSCHELE, 152-154; POGG 6,2379-2380; 7a(4),295
O.R. Klimmer, Arzneimitt. 8 (1958), 305-306
W. Mohr, Z. Tropenmed. 24 (1973), 129-130
H. Mückter, Arzneimitt. 25 (1975), 1466-1467

SCHULER, FRIDOLIN [1832-1903]
E. Auer and H. Buess, Gesnerus 16 (1959), 66-75

SCHULER, JOSEF [1883-1963]
JV 25,41; NUC 531,278
Anon., Ciba-Blätter 64 (1949), 21-22

SCHULER, LEONHARD [1899-]
KURSCHNER 13,3566; POGG 7a(4),295-296

SCHULER, WERNER [1900-1966]
KURSCHNER 10,2257; HAGEL, 104-108; POGG 7a(4),296-297
R.M., Schw. med. Wchschr. 90 (1960), 833

SCHULMAN, JACK HENRY [1904-1967]
E.K. Rideal, Chem. Ind. 1967, p.1484

SCHULTE, KARL ERNST [1911-]
KURSCHNER 13,3569-3570; WWWS, 1504; POGG 7a(4),297-299

SCHULTHESS, EDMUND [1826-1906]
AARGAU, 696
Schweizerisches Geschlechterbuch 1,528

SCHULTZ, EDWIN WILLIAM [1887-1971]
AMS 11,4755; NCAB E,115-116; IB, 267

SCHULTZ, GUSTAV [1851-1928]
DZ, 1321; SCHAEDLER, 116; WWWS, 1604; POGG 3,1219; 4,1362; 5,1137; 6,2380
Anon., Z. angew. Chem. 34 (1921), 613
H. Bucherer, Ber. chem. Ges. 61A (1928), 82-83
E.H. Huntress, Proc. Am. Acad. Arts Sci. 79 (1951), 35-37

SCHULTZ, IGNAZ [1877-1960]
JV 20,263

SCHULTZ, JACK [1904-1971]
AMS 10,3612
G.T. Rudkin, Genetics 68Suppl. (1971), 67-68
T.F. Anderson, Genetics 81 (1975), 1-7; Biog. Mem. Nat. Acad. Sci. 47 (1975), 393-422

SCHULTZ, JOHANNES HEINRICH [1884-1970]
KURSCHNER 11,2755; FISCHER 2,1418; WWWS, 1504
G. Crosa, Minerva Med. 63 (1972), 2970

SCHULTZ, JULIUS [1914-1985]
AMS 15(6),535; WWWS, 1504

SCHULTZ, OTTO ERICH [1908-1985]
KURSCHNER 13,3574; HEIN-E,399-400; LEHMANN, 191-192; POGG 7a(4),299
Anon., Chem. Z. 109 (1985), 214

SCHULTZ, PAUL [1864-1905]
BJN 10,234-237,248*-249*; FISCHER 2,1419
G.F. Nicolai, Archiv für Anatomie und Physiologie 1906, pp.376-384

SCHULTZ, WILLIAM HENRY [1873-1955]
NCAB 42,354-355
Anon., Science 122 (1955), 280

SCHULTZ-SCHULTZENSTEIN, CARL HEINRICH [1798-1871]
ADB 32,723-725; HIRSCH 5,159-160; FERCHL, 490; TSCHIRCH, 1122; RSC 5,569-571; 8,893; 12,666; CALLISEN 17,376-378; 32,227-229; POGG 2,861; 3,1219
Anon., Bot. Zt. 29 (1871), 270

SCHULTZE, CARL AUGUST SIGISMUND [1795-1877]
HIRSCH 5,162; PAGEL, 1549-1550; CALLISEN 17,380-381; 32,230-231
B.S., Leopoldina 13 (1877), 145-146
K.E. Rothschuh, Sudhoffs Arch. 47 (1963), 334-359
M. Dittrich, Med. Hist. J. 4 (1969), 271-276

SCHULTZE, FRIEDRICH [1848-1934]
DRULL, 248-249; FISCHER 2,1419-1420; GROTE 2,193-215; LEVITSKI 2,160-162
J. Brennsohn, Die Aerzte Livlands, p.365. Mitau 1905

SCHULTZE, HERMANN [1899-1985]
KURSCHNER 13,3576; AUERBACH, 381; POGG 7a(4),302-304
Anon., Chem. Z. 93 (1969), 479; 103 (1979), 235; 109 (1985), 214

SCHULTZE, MAX [1825-1874]
DSB 12,230-233; ADB 54,256-257; HIRSCH 5,162-163; PAGEL, 1550-1551; FREUND 2,367-375; W, 466; WWWS, 1505; POGG 3,1220
G. Schwalbe, Arch. mikr. Anat. 10 (1874), iii-xxiii
T.H. Bast, Ann. Med. Hist. 3 (1931), 166-178
M. Dittrich, NTM Beiheft 1964, pp.164-179
R.R. Lücker, Max J.S. Schultze und die Zellenlehre des 19. Jahrhunderts. Bonn 1977

SCHULTZE, OSKAR [1859-1920]
DBJ 2,760; EBERT, 27-36
W. Lubosch, Verhandl. Phys. Med. Wurz. NF46 (1921), 19-45; Anat. Anz. 54 (1921), 411-428

SCHULTZE, WALTER HANS [1880-1964]
KURSCHNER 9,1888; FISCHER 2,1420
R. Poche, Verhandl. path. Ges. 49 (1965), 386-389

SCHULTZEN, OTTO [1837-1875]
HIRSCH 5,164; PAGEL, 1554; LEVITSKI 2,158-160; FRANKEN, 196
Anon., Berl. klin. Wchschr. 12 (1875), 684

SCHULZ, FRIEDRICH NIKOLAUS [1871-1956]
KURSCHNER 4,2704; FISCHER 2,1420; GIESE, 505-506; BK, 310; IB, 267; TLK 3,593; POGG 6,2382; 7a(4),307

SCHULZ, GÜNTER VIKTOR [1905-]
KURSCHNER 13,3580; POGG 7a(4),307-310

SCHULZ, HEINRICH [1861-]
JV 6,74; NUC 531,571

SCHULZ, HUGO [1853-1932]
GROTE 2,217-250; FISCHER 2,1420-1421; PAGEL, 1554-1555; TSCHIRCH, 1123; DZ, 1327; TLK 3,594
P. Weis, Arch. exp. Path. Pharm. 170 (1933), 744-757

SCHULZ, MAX [1905-1982]
KURSCHNER 13,3581-3582; POGG 7a(4),311-313

SCHULZ, OSKAR [1858-1944]
KURSCHNER 6(II),1268; FISCHER 2,1421; BERWIND, 178-189; PITTROFF,207-208

SCHULZE (SCHULZE-FORSTER), ARNOLD [1882-1946]
JV 22,36; NUC 531,622; POGG 6,2384-2385; 7a(4),323
A. Heller, Gas- und Wasserfach 92 (1951), 72

SCHULZE, BERNHARD [1857-]
NUC 531,623-624

SCHULZE, ERNST [1840-1912]
BJN 18,58*; BLOKH 2,670-671; POGG 3,1220-1221; 4,1363; 5,1139
E. Winterstein, Z. physiol. Chem. 79 (1912), 353-358; Biol. Bull. 2 (1912), 1-20; Verhandl. Schw. Nat. Ges. 95 (1912), 54-71; Viert. Nat. Ges. Zurich 57 (1912), 604-612; Ber. chem. Ges. 47 (1914), 429-449

SCHULZE, ERNST [1908-1970]
KURSCHNER 11,2765; POGG 7a(4),319-320

SCHULZE, FRANZ EILHARD [1840-1921]
DBJ 3,232-236,316; PAGEL, 1556-1557; WREDE, 176-178; DZ, 1327
K. Heider, Sitz. Preuss. Akad. Wiss. 1922, pp.lxxxvii-xcv
K. Grabben, Alm. Akad. Wiss. Wien 72 (1922), 164-168
R. Hesse, Pommersche Lebensbilder 2 (1936), 281-287

SCHULZE, FRANZ FERDINAND [1815-1873]
DSB 12,233-234; ADB 34,749-751; BLOKH 2,671; FERCHL, 490; SCHAEDLER, 117; POGG 2,865; 3,1220

SCHULZE, FRIEDRICH GOTTLOB [1795-1860]
FERCHL, 490-491; POGG 2,865

SCHULZE, HANS OSKAR [1853-1892]
Anon., Leopoldina 29 (1893), 52; Anales Universidad Chile 82 (1892), 714-716

SCHULZE, HEINRICH [1874-1926]
HEIN, 608-609; POGG 6,2384-2385
H. Thoms, Ber. chem. Ges. 60A (1927), 37

SCHUMACHER, HERTHA [1902-]
JV 45,136; SG [3]9,479; NUC 532,30

SCHUMACHER, JOSEPH [1885-]
KURSCHNER 8,2176; JV 29,264; IB, 268; NUC 532,33

SCHUMACHER-KOPP, EMIL [1850-1927]
STRAHLMANN, 469,481; RSC 18,615
H. Bachmann, Verhandl. Schw. Nat. Ges. 108 (1927), 42-44

SCHUMAKER, VERNE NORMAN [1929-]
AMS 17(6),567

SCHUMANN, GOTTHILF DAVID [1788-1865]
HEIN, 610; POGG 2,868
A. Wankmüller, Beitr. Württ. Apothekgesch. 11 (1977), 97-100

SCHUMM, OTTO [1874-1958]
KURSCHNER 8,2179; HEIN-E,401-402; FISCHER 2,1422; TLK 3,597; IB, 268; POGG 6,2387-2389; 7a(4),330-331
Anon., Chem. Z. 78 (1954), 540

SCHUNCK, EDWARD [1820-1903]
DSB 12,236-237; DNB [1901-1911] 3,274-275; BLOKH 2,672; DESMOND, 546; POTSCH, 388; FERCHL, 491-492; POGG 2,869; 3,1221; 4,1366; 6,2389
Anon., Chem. News 87 (1903), 34-35
C. Liebermann, Ber. chem. Ges. 36 (1903), 305-306
C.L.B., Mem. Lit. Phil. Soc. 47 (1903), xlix-liii
H.B.D., Proc. Roy. Soc. 75 (1905), 261-265
J.R. Partington, J. Soc. Chem. Ind. 52 (1933), 478-482
W.V. Farrar, Notes Roy. Soc. 31 (1977), 273-296

SCHUPP, GUSTAV [1880-]
JV 21,488; NUC 532,215

SCHUPPLI, OTTO [1888-1948]
NUC 532,224

SCHUSTER, PHILIPP [1904-]
JV 48,96

SCHWAB, GEORG MARIA [1899-1984]
KURSCHNER 13,3600-3601; BHDE 2,1057-1058; ETTRE, 375-380; TLK 3,600; WWWS, 1506; POGG 6,2390-2391; 7a(4),335-337
Anon., Nachr. Chem. Techn. 7 (1959), 23; Leopoldina [3]12 (1966), 72

SCHWAB, LEONHARDT CONRAD [1855-]
RSC 11,361
Anon., Wissenschaftliche Zeitschrift der Technischen Universität Dresden 14 (1965), 248-255

SCHWABE, KURT [1905-1983]
POTSCH, 389; POGG 7a(4),337-339
Anon., Leopoldina [3]2 (1956), 7-8; Naturw. Rund. 37 (1984), 209
J. Koryta, Journal of Electroanalytical Chemistry 180 (1984), 124

SCHWABE, OTTO [1878-1951]
JV 19,203; NUC 532,422

SCHWALBE, (GUSTAV) ALBERT [1844-1916]
FISCHER 2,1424; GIESE, 463-464; WININGER 5,473; 7,440
F. Keibel, Anat. Anz. 49 (1916), 210-211

SCHWALBE, CARL GUSTAV [1871-1938]
KURSCHNER 4,2732; WOLF, 190; TLK 3,601; POGG 4,1370; 5,1142; 6,2391-2392; 7a(4),339-340
Anon., Chem. Z. 55 (1931), 817; Z. angew. Chem. 51 (1938), 778; Zellstoff und Papier 18 (1938), 389-390

SCHWALBE, ERNST [1871-1920]
DBJ 2,760; FISCHER 2,1424; DRULL, 250
P. Ernst, Beitr. path. Anat. 67 (1920), i-iv

R. Hanser, Verhandl. path. Ges. 30 (1937), 549-555

SCHWAN, NICOLAUS [1868-1903]
JV 7,159; NUC 532,470
Anon., Pharm. Z. 48 (1903), 583

SCHWANERT, HUGO [1828-1902]
BJN 7,106*; BLOKH 2,673; TSCHIRCH, 1123; SCHELENZ, 694; SCHAEDLER, 117; POTSCH, 389; POGG 3,1223; 4,1370
H. Limpricht, Ber. chem. Ges. 35 (1902), 4522-4527
Anon., Chem. Z. 26 (1902), 1053; Pharm. Z. 47 (1902), 835

SCHWANN, THEODOR [1810-1882]
DSB 12,240-245; ADB 33,188-190; BNB 22,77-98; FREUND 1,313-321; HIRSCH 5,173-175; PAGEL, 1560-1562; BLOKH 2,673-674; FERCHL, 492; TSCHIRCH, 1123; SCHAEDLER, 117-118; W, 466-467; ABBOTT, 115; WWWS,1507; CALLISEN 32,241-242; POGG 2,872; 3,1223;
7aSuppl.,613-616
A. Kossel, Z. physiol. Chem. 6 (1882), 280-285
J. Henle, Arch. mikr. Anat. 21 (1882), i-xlix
R. Virchow, Virchows Arch. 87 (1882), 389-392
L. Frédericq, Theodore Schwann, sa Vie et ses Travaux. Liège 1884
O. Hertwig and W. Waldeyer, Arch. mikr. Anat. 74 (1909), 469-473
M. Florkin, Naissance et Deviation dans l'Oeuvre de Theodore Schwann. Paris 1960
R. Watermann, Theodor Schwann, Leben und Werk. Düsseldorf 1960; Z. mikr. anat. Forsch. 96 (1982), 1032-1043
R.C. Maulitz, J. Hist. Med. 26 (1971), 422-436

SCHWANTKE, ARTHUR [1872-1939]
KURSCHNER 5,1280; POGG 6,2392; 7a(4),342

SCHWARTZ, MORTON K. [1925-]
AMS 15(6),550; WWWS, 1507

SCHWARTZ, PHILIPP [1894-1977]
KURSCHNER 12,2959; FISCHER 2,1425-1426; IB, 268; WIDMANN, 79-80,268

SCHWARZ, (CARL LEONHARDT) HEINRICH [1824-1890]
KERNBAUER, 90-91; KUKULA, 799; SCHAEDLER, 118; POGG 2,872; 3,1224; 4,1371
Anon., Chem. Z. 14 (1890), 1784; Pharm. Z. 35 (1890), 602

SCHWARZ, KARL RUDOLF [1887-]
JV 29,732; NUC 532,646

SCHWARZ, LEO [1872-1903]
RSC 18,631; NUC 532,647
J. Pagel, Virchows Jahresber. 1903, p.423

SCHWARZ, LEOPOLD [1877-1962]
KURSCHNER 4,2740; JV 19,9; IB, 268

SCHWARZ, OSWALD [1883-1945]
KURSCHNER 4,2741; FISCHER 2,1427; KOREN, 245

SCHWARZ, ROBERT [1887-1963]
TLK 3,603; POTSCH, 390; POGG 6,2394-2395; 7a(4),346-348
B. Helferich, Z. angew. Chem. 69 (1959), 362-363
E. Wiberg, Bayer. Akad. Wiss. Jahrbuch 1963, pp.229-234

SCHWARZ-WENDL, CARL [1876-1953]
FISCHER 2,1426; WIEN 1954, pp.41-42; PLANER, 307-308; POGG 7a(4),349-350
W. Auerswald, Wiener med. Wchschr. 103 (1953), 180

SCHWARZE, (MAX ALEXIS) GUSTAV [1857-
BERLIN, 579; SG [1]12,760; [5]1,571; NUC 532,685

SCHWARZENBACH, GEROLD KARL [1904-1978]
DSB 18,799-801; KURSCHNER 12,2967; MH 2,477-479; STC 2,475-

476; MSE 3,86-87; WWWS, 1507-1508; POGG 6,2395; 7a(4),352-354
Anon., Chimia 18 (1964), 111-112; 28 (1974), 95
G. Giovanni, Helv. Chim. Acta 61 (1978), 1949-1961

SCHWARZENBACH, VALENTIN [1830-1890]
ROEDER, 195-199; BLOKH 2,574; TSCHIRCH, 1123;
SCHAEDLER, 118; POTSCH, 390; POGG 2,873; 3,1225; 4,1371
Anon., Chem. Z. 14 (1890), 1782

SCHWARZSCHILD, MORITZ [1881-1933]
JV 19,392; SG [2]115,307; NUC 533,12

SCHWEET, RICHARD S. [1918-1967]
AMS 11,4773; NCAB 53,568-569; WWWS, 1508
G.W. Schwert and R.L. Lester, Arch. Biochem. 125 (1968), v-xii

SCHWEIGGER, JOHANN [1779-1857]
DSB 12,253-255; ADB 33,335-339; SCHAEDLER, 118-119; WWWS, 1508; POTSCH, 390; POGG 2,873-875; 7aSuppl.,616-617
C.F.P. Martius, Akademische Denkreden, pp.345-364. Leipzig 1866
H. Degen, Naturw. Rund. 8 (1955), 421-427,472-480
H.A.M. Snelders, Isis 62 (1971), 328-338

SCHWEIGGER-SEIDEL, FRANZ [1834-1871]
ADB 33,342; HIRSCH 5,181

SCHWEIGGER-SEIDEL, FRANZ WILHELM [1795-1838]
ADB 33,340-342; HEIN, 616-617; POHL, 109-120; FERCHL, 493; POGG 2,875
L.F. Bley, Arch. Pharm. 74 (1840), 121-125

SCHWEINITZ, EMIL ALEXANDER de [1866-1904]
DAB 16,483; ELLIOTT, 231; WWAH, 541
Anon., Brit. Med. J. 1904(I), P.761; Washington Medical Annals 3 (1904-1905), 137-143

SCHWEIZER, MATTHIAS EDUARD [1818-1860]
ZURICH, 14-16; FERCHL, 493; POTSCH, 390-391; POGG 2,877
G.B. Kauffman, J. Chem. Ed. 61 (1984), 1095-1097

SCHWENDENER, SIMON [1829-1919]
DSB 12,255-256; DBJ 2,733; DHB 6,103-104; HINK 1,429-439; TSCHIRCH, 1124; STELLING-MICHAUD 5,537-538; LEHMANN, 183; WREDE, 179-180; WWWS,1508; DZ, 1339-1340; ZURICH-D, 111; POGG 3,1227; 4,1373; 5,1145; 6,2396
K. von Goebel, Bayer Akad. Wiss. Jahrbuch 1919, pp.57-61
G. Haberlandt, Alm. Akad. Wiss. Wien 70 (1920), 149-155; Ber. bot. Ges. 47 (1929), (1)-(20)
A. Zimmermann, Ber. bot. Ges. 40 (1922), (53)-(76)
B. Hoppe, Ber. bot. Ges. 100 (1987), 302-326

SCHWENK, ERWIN [1887-1976]
AMS 11,4775; BHDE 2,1064; POGG 6,2397; 7a(4),355-357

SCHWERIN, BOTHO (Graf) [1865-1917]
BLOKH 2,674-676
M. Moest, Ber. chem. Ges. 50 (1917), 477-478
Anon., Chem. Z. 41 (1917), 62; Z. angew. Chem. 30(III) (1917), 117,120

SCHWERIN, ERNST ALFRED [1846-]
NUC 533,272

SCHWERT, GEORGE WILLIAM [1919-]
AMS 17(6),586

SCHWIENING, HEINRICH [1870-1920]
FISCHER 2,1429-1430; RSC 18,639
J. Schwalbe, Deutsche med. Wchschr. 46 (1930), 247

SCHWIMMER, MAX [1863-1941]
JV 7,149; NUC 533,318
Anon., Chem. Z. 65 (1941), 249; Z. angew Chem. 54 (1941), 329

SCHWYZER, ROBERT [1920-]
KURSCHNER 13,3638-3639; IWW 1982-1983, p.1190; POTSCH, 391; POGG 7a(4),363-364
H. Zuber, Chimia 34 (1980), 473-474; 39 (1985), 402-403

SCOFFONE, ERNESTO [1923-1973]
C. Toniolo, Int. J. Pep. Res. 6 (1974), 361-370

SCOTT, ALEXANDER [1853-1947]
WW 1945, p.2438; POGG 4,1375; 5,1147; 6,2400
R. Robertson and H.J. Plenderleith, Obit. Not. Fell. Roy. Soc. 6 (1948), 251-262
R. Robertson, J. Chem. Soc. 1950, pp.762-767

SCOTT, ALISTAIR IAN [1928-]
WW 1981, p.2313; AMS 14,4523; WWA 1980-1981, p.2955

SCOTT, DAVID AYLMER [1892-1971]
POTSCH, 391
Who Was Who 1971-1980, p.709
C.H. Best and A.M. Fisher, Biog. Mem. Fell. Roy. Soc. 18 (1972), 511-524

SCOTT, ERNEST LYMAN [1877-1966]
AMS 9(II), 1003; IB, 269
D.W. Richards, Persp. Biol. Med. 10 (1966), 84-95
A. Scott, Great Scott. Bogota, N.J. 1972
L.N. Magner, Pharmacy in History 19 (1977), 103-108
W.A. Sawyer, Persp. Biol. Med. 29 (1986), 611-618

SCOTT, FREDERICK HUGHES [1876-1951]
AMS 8,2214; O'CONNOR(2), 518-519
H. Blaschko, Notes Roy. Soc. 37 (1983), 235-247

SCOTT, JESSE FRIEND [1917-1990]
AMS 14,4535; WWWS, 1511

SCOTT-MONCRIEFF, ROSE [1903-1991]
R. Scott-Moncrieff, Notes Roy. Soc. 36 (1981), 125-154

SCOVELL, MELVILLE AMASA [1855-1912]
DAB 16,512-513; NCAB 16,32-33; MILES, 431-432; WWAH, 543; WWWS, 1512

SCRIBA, CARL [1854-1929]
HEIN-E, 404-405

SCRIBA, EMIL [1814-1886]
RAUSCH, 248; RSC 5,607
Anon., Chem. Z. 10 (1886), 1619; Pharm. Z. 31 (1886), 113

SCRIBA, THEODOR [-1886]
Anon., Pharm. Z. 31 (1886), 487

SCUDAMORE, CHARLES [1799-1849]
DNB 17,1090-1091; HIRSCH 5,192-193; WWWS, 1512; CALLISEN 17,459-462; 32,254-255

SEBELIEN, JOHN [1858-1932]
NBL 13,135-138; VEIBEL 2,394-398; POGG 5,1149; 6,2403; 7b,4760
Anon., Science 76 (1932), 186
S. Schmidt-Nielsen, Norske Vid. Akad. Arbok 1933, pp.61-64

SEBENING, WALTER [1893-1942]
KURSCHNER 4,2754-2755; KALLMORGEN, 410; ASEN, 184
V. Schmieden, Z. Chir. 69 (1942), 2021-2022

SEBRELL, WILLIAM HENRY [1901-1992]
AMS 14,4547; WWA 1976-1977, p.2813; WWWS, 1515
W.H. Sebrell, J. Nutrition 115 (1985), 23-38

SECCHI, ANGELO [1818-1878]
DSB 12,266-270; BLOKH 2,676; SCHAEDLER, 119; WWWS, 1515;

POGG 2,844-845; 3,1229-1230
R.P. van Tricht, Rev. Quest. Sci. 4 (1878), 353-402
J. Boccardi, Rev. Quest. Sci. [4]39 (1928), 393-395

SECHENOV, IVAN MIKHAILOVICH [1829-1905]
DSB 12,270-271; BSE 38,623-625; BME 30,43-49; KUZNETSOV(1963), 132-148; HIRSCH 5,381-382; BRAZIER, 212-218,242-243; HAYMAKER, 264-267; BLOKH 2,681-682; VORONTSOV, 156-170; ZMEEV, 2,99-100; WWWS, 1515
K.S. Koshtoyants, I.M. Sechenov. Moscow 1950
I.M. Sechenov, Autobiographical Notes (publ. 1907, trans. by K. Hamer). Washington, D.C. 1965
M.G. Yaroslavski, I.M. Sechenov. Leningrad 1958

SÉCRETAN, ALFRED [1851-1920]
RSC 11,379; NUC 536,81

SEDGWICK, ADAM [1854-1913]
DNB [1912-1921], 487-488; O'CONNOR, 248-249
Anon., Nature 91 (1913), 14-15
E.W.M., Proc. Roy. Soc. B86 (1913), xxiv-xxix

SEDGWICK, WULLIAM THOMPSON [1855-1921]
DAB 16,552-553; NCAB 13,290-291; DAMB, 669-670; WWAH, 544; WWWS, 1515
G.C. Whipple, Science 53 (1921), 171-178

SÉDILLOT, CHARLES EMMANUEL [1804-1883]
HOEFER 43,680681; GE 29,870; GENTY 3,277-292; HIRSCH 5,198-200;PAGEL, 1567-1568; DECHAMBRE [3]8,469-479; BERGER-LEVRAULT, 225-226; WWWS,1115; CALLISEN 17,473-474; 32,260-261
M. Larrey, Bull. Acad. Med. 12 (1883), 169-173
P. Harteloup, Bull. Soc. Chir. NS11 (1885), 45-62; Gaz. Hop. Paris 58 (1885), 89-94

SEEBECK, THOMAS [1770-1831]
DSB 12,281-282; ADB 33,564-565; WELDING, 717; BLOKH 2,676-677; WWWS,1516; POTSCH, 392; SCHAEDLER, 119; POGG 2,889-890; 7aSuppl.,618-619

SEEBERGER, LUDWIG [1867-]
JV 8,217; RSC 13,276; 17,757

SEEGEN, JOSEF [1822-1904]
BJN 9,57-61; 10,107*; HIRSCH 5,201-202; PAGEL, 1570-1571; FROHNE,137-149
R. Kolisch, Wiener klin. Wchschr. 17 (1904), 113-114
E. von Leyden, Deutsche med. Wchschr. 30 (1904), 329
S. Exner, Alm. Akad. Wiss. Wien 54 (1904), 335-336
E. Ludwig, Wiener med. Wchschr, 60 (1910), 561-568

SEEGERS, WALTER HENRY [1910-]
AMS 14,4550; WWA 1976-1977, p.2815; WWWS, 1516
R.L. Henry, Seminar in Thrombosis and Hemostasis 6 (1980), 325-327

SEEL, HANS [1898-1961]
KURSCHNER 9,1923; POGG 6,2404; 7a(4),368-371
W. Poethke, Pharmazie 16 (1961), 337-338
Anon., Nachr. Chem. Techn. 9 (1961), 163

SEELICH, FRANZ [1902-1985]
KURSCHNER 13,3646; KIEL, 124; POGG 7a(4),371-372
Anon., Chem. Z. 109 (1985), 393

SEEMANN, JOHN [1874-1913]
BJN 18,127*; FISCHER 2,1433
M. Cremer, Erg. Physiol. 14 (1914), 657-664

SEEMANN, LORENZ [1877-1957]
JV 24,143; RSC 19,292; NUC 536,383

SEGAL, HARRY LOUIS [1900-]
AMS 14,4553; WWWS, 1517

SÉGALAS, PIERRE SALOMON [1792-1875]
VAPEREAU 5,1552; HIRSCH 5,204-205; DECHAMBRE [3]8,474; CALLISEN 17,488-492; 32,266-267

SÉGUIN, ARMAND [1767-1835]
DSB 12,286-287; GE 29,882; LAROUSSE 14,485; HIRSCH 5,208-209; WWWS,1518; FERCHL,496; CALLISEN 17, 494-497; POGG 2,895-896

SÉGUIN, MARC [1786-1875]
DSB 12,287-288; VAPEREAU 5,1659; GE 29,882; LAROUSSE 14,485; WWWS, 1518; MATSCHOSS, 245-246; W, 468; POGG 3,1233-1234
E. Picard, Rev. Gen. Sci. 34 (1923), 453-456
C. Bechetoille, Marc Seguin, Grand Savant Meconnu. Bourges 1975

SEIBERT, FLORENCE BARBARA [1897-1991]
AMS 14,4556; WWA 1976-1977, p.2818; NCAB G,481-482; IB, 270; WWWS, 1518
F.B. Seibert, Pebbles on a Hill of a Scientist. St. Petersburg, Fla. 1986
E.S. Leake, Trans. Coll. Phys. Phila. [5]3 (1981), 256-263
B. Lambert, New York Times, 31 August 1991

SEIDELL, ATHERTON [1878-1961]
AMS 9(I),1733; MILES, 432-433; BARRY, 73-75; POGG 5,1152; 6,2408; 7b,4773
Anon., Chem. Eng. News 30 (1952), 1285; 39(32) (1961), 91

SEIFERT, BRUNO RICHARD [1861-1919]
BLOKH 2,678; WWWS, 1519; POGG 6,2410
R. Möhlau, Ber. chem. Ges. 53A (1920), 1-7

SEIFERT, OTTO [1853-1933]
KURSCHNER 4,2767; FISCHER 2,1435-1436; PAGEL, 1575; TLK 3,608
Anon., Z. Laryngol. 24 (1933), 82-83

SEIFFERT, GUSTAV [1884-1964]
JV 25,166
B. Freytag, Münch. med. Wchschr. 106 (1964), 1415-1416

SEIFRIZ, WILLIAM [1888-1955]
AMS 8,2227; IB, 270; WWAH, 545; WWWS, 1519
N. Kamiya, Protoplasma 45 (1956), 514-524
K. Höfler, Verhandl. zool. bot. Ges. Wien 96 (1956), 12-13

SEILER, FREDERIC [1863-1913]
STRAHLMANN, 468,480

SEILER, KARL [1882-]
JV 22,524

SEITTER, EDUARD [1870-1953]
JV 12,247; RSC 17,436; NUC 537,227
Anon., Nachr. Chem. Techn. 2 (1954), 53

SEITZ, EUGEN [1817-1899]
NB, 370; HB 3,294-299; PAGEL, 1576-1578

SEITZ, FRANZ [1811-1892]
HIRSCH 5,215-216; DUELUND, 83-89
M. von Pettenkoffer, Münch. med. Wchschr. 38 (1891), 848
K.F. Hoffman, Medizinische Monatsschrift 16 (1962), 181-183

SEITZ, FRANZ [1895-]
JV 40,744; SG [5]1,577; NUC 537,234

SEITZ, WALTER [1905-]
KURSCHNER 13,3667; POGG 7a(4),388-389

SEKA, REINHARD [1898-1946]
POGG 6,2411; 7a(4),389-390
G. Jantsch, Ost. Chem. Z. 48 (1947), 176

SELA, MICHAEL [1924-]
IWW 1982-1983, p.1190
M. Sela, Annual Review of Immunology 5 (1987), 1-20

SELENKA, EMIL [1842-1902]
BJN 7,296,107*-108*; FISCHER 2,1438; KOHUT 2,241-242; WAGENITZ, 164
C. Voit, Sitz. Bayer. Akad. 32 (1902), 241-248

SELIGMAN, ARNOLD MAX [1912-1976]
AMS 13,4002; WWWS, 1520
R.J. Barrnett, J. Histochem. 23 (1975), 871-872; 24 (1976), 1045

SELIGMANN, ERICH [1880-1954]
FISCHER 2,1438; IB, 270

SELIVANOV, FEDOR FEDOROVICH [1859-1938]
POGG 4,1382; 5,1153
Anon., Zhur. prikl. Khim. 11 (1938),1195-1196

SELL, ERNST [1808-1854]
Dokumente aus Hoechster Archiven No.26. Frankfurt a.M. 1967
F.W. Weitersheim, Mitteilungen des Oberhessischen Geschichtsverein NF62 (1977), 209

SELL, EUGEN [1842-1896]
BJN 1,209-210; 3,94*; BLOKH 2,678-679; BONN, 208; TSCHIRCH, 1124-1125; SCHAEDLER, 120; POGG 3,1235
K. Windisch, Ber. chem. Ges. 29(4) (1896), 1199-1208
Anon., Leopoldina 32 (1896), 184

SELLA, QUINTINO [1827-1884]
EI 31,328-329; BLOKH 2,679; RSC 18,684; POGG 2,899-900; 3,1235
Anon., Nature 29 (1884), 551
A.W. Hofmann, Ber. chem. Ges. 17(3) (1885), 731-823

SELLHEIM, HUGO [1871-1936]
KURSCHNER 4,2773-2774; FISCHER 2,1439; HABERLING, 124; LDGS, 64; WWWS, 1520; SG [2]15,396; [3]9,531
A. Mayer, Z. Gyn. 60 (1936), 1506-1531

SELMI, FRANCESCO [1817-1881]
DFI, 119-120; PROVENZAL, 177-191; HIRSCH 5,220-221; PAGEL, 1580-1581; BLOKH 2,679-680; FERCHL, 497; SCHAEDLER, 120-121; BULLOCH, 397; POTSCH, 393; WWWS, 1521; POGG 2,900-901; 3,1235-1236
I. Guareschi, Chem. Z. 34 (1910), 1189-1190; Francesco Selmi e la sua Opera Scientifica. Turin 1911
P. DiPietro, Riv. Storia Med. 5 (1961), 158-166
R. Bernebeo, Riv. Storia Med. 14 (1970), 43-50

SELTER, GEORG EMIL [1901-1976]
KALLMORGEN, 411-412

SELYE, HANS [1907-1982]
AMS 14,4565-4566; WWA 1976-1977, p.2824; IWW 1982-1983, pp.1199-1200;
AO 1982, pp.498-502; YOUNG, 30-31; WWWS, 1521
H. Selye, The Stress of my Life. Toronto 1977
H.G. Classen and P. Marquardt, Arzneimitt. 22 (1972), 475-476; 27 (1977), 892-893
J. Kangilaski, J. Am. Med. Assn. 248 (1982), 3084-3085
R. Guillemin et al., Experientia 41 (1985), 560-568

SEMASHKO, NIKOLAI ALEKSANDROVICH [1874-1949]
BSE 38,458-459; BME 29,699-703; WWR, 497; KUZNETSOV(1963), 660-668; FISCHER 2,1440
L.M. Geizer, N.A. Semashko. Moscow 1950
M.A. Shagov, Vest. Oftalm. 33(6) (1954), 40-44
D.V. Gorfin, N.A. Semashko. Moscow 1967
K.H. Karbe, NTM 5(11) (1968), 79-82
L.G. Lekarev, Santé Publique 18 (1975), 3-17,489-500

SEMBRITZKI, KURT [1878-1943]
JV 12,23; RSC 18,686; NUC 538,47
Anon., Z. angew. Chem. 56Suppl. (1943), 105

SEMENOV, NIKOLAI NIKOLAEVICH [1896-1986]
BSE (3rd Ed.) 23,229-230; WWWS,1521; POTSCH, 383; POGG 6,2414; 7b,4781-4789
F.S. Dainton, Biog. Mem. Fell. Roy. Soc. 36 (1990), 527-546

SEMENZA, GIORGIO [1928-]
Who's Who in Switzerland 1978-1979, p.583

SEMMELWEIS, IGNAC FÜLOP [1818-1865]
DSB 12,294-297; ADB 33,704-706; HIRSCH 5,222-223; WININGER 5,501;W, 468-469; WWWS, 1522
W.J. Sinclair, Semmelweis, his Life and Doctrine. Manchester 1909
F.P. Murphy, Bull. Hist. Med. 20 (1946), 653-707
G. Gortvay and I. Zoltan, Semmelweis, his Life and Work. Budapest 1968
E. Lesky, Deutsche med. Wchschr. 97 (1972), 627-632
S.B. Nuland, J. Hist. Med. 34 (1979), 255-272
I. Benedek, Ignaz Fülop Semmelweis. Vienna 1983

SEMMLER, FRIEDRICH WILHELM [1860-1931]
TSCHIRCH, 1125; POTSCH, 393; POGG 4,1383-1384; 5,1153; 6,2414
Anon., Chem. Z. 54 (1930), 367
H. Becker-Rose, Z. angew. Chem. 44 (1931), 301-302

SEMMOLA, GIOVANNI [1793-1865]
EI 31,356; HIRSCH 5,223-224; DECHAMBRE [3]8,600-601; CALLISEN 18,3; 32,277

SEMMOLA, MARIANO [1831-1896]
HIRSCH 5,224; PAGEL, 1583-1584
Anon., Lancet 1896(I), p.1179; Brit. Med. J. 1896(I), p.1123
G. Gautier and J. Larat, Rev. Int. Electr. 6 (1895-1896), 33-44
A. Capparoni, Pag. Stor. Med. 8 (1964), 36-39

SEMON, FELIX [1849-1921]
FISCHER 2,1440; PAGEL, 1584-1585; WININGER 5,501-502; WWWS, 1522
F. Semon, Autobiography (H.C. Semon and T.A. McIntyre, Eds.). London 1927
E. Huizinga, Arch. Otolaryng. 84 (1966), 473-478

SEMON, RICHARD WOLFGANG [1859-1918]
DSB 12,297-298; DBJ 2,704-705; FISCHER 2,1441; GIESE, 466-468; WININGER 5,502; 7,443; TETZLAFF, 305
J. Schatzmann, Richard Semon (1859-1918) und seine Mnemetheorie. Zurich 1968

SEMPER, AUGUST [-1915]
RSC 13,485
Anon., Pharm. Z. 60 (1915), 790

SEMPER, CARL GOTTFRIED [1832-1893]
DSB 12,299; ADB 54,315-316
J. Beard, Nature 48 (1893), 271-272
Anon., Sitz. Phys. Med. Würzburg 1893, pp.109-134; Pop. Sci. Mon. 52 (1897-1898), 837-842; Leopoldina 29 (1893), 111

SEMPER, LEOPOLD [1882-1915]
JV 23,550
Anon., Chem. Z. 39 (1915), 485

SEN, KSHITISH CHANDRA [1899-1971]
NUC 538,174; POGG 6,2415-2416; 7b,4794-4795

SENATOR, HERMANN [1834-1911]
BJN 16,72*; HIRSCH 5,226; PAGEL, 1585-1586; FRANKEN, 196-197; DZ, 1358; WREDE, 181; KOHUT 2,286
H. Strauss, Berl. klin. Wchschr. 48 (1911), 1406-1407
A. Goldscheider, Berl. klin. Wchschr. 48 (1911), 1961-1968; Deutsche med. Wchschr. 37 (1911), 1444-1447
A. Wolff-Eisner, Münch. med. Wchschr. 58 (1911), 1733-1735

SENDERENS, JEAN BAPTISTE [1856-1937]
WWWS, 1522; POTSCH, 393-394; POGG 4,1384; 5,1154; 6,2416; 7b,4795-4796
J.B. Maxted, J. Chem. Soc. 1938, pp.168-170
H. Palfray, Bull. Soc. Chim. [5]6 (1939), 1-29
M. Mousseron, Rev. Gen. Sci. 63 (1956), 245-247

SENDROY, JULIUS [1900-1982]
AMS 14,4567; WWWS, 1523

SENDTNER, RUDOLF [1853-1933]
RSC 11,392; NUC 538,235
Anon., Chem. Z. 57 (1933), 726; Z. angew. Chem. 46 (1933), 614

SENEBIER, JEAN [1742-1809]
DSB 12,308-309; DHB 6,160; HIRSCH 5,227-228; BRIQUET, 433-436; WWWS,1523; BLOKH 2,680; STELLING-MICHAUD 5,553; FERCHL, 498-499; COLE, 495-496; SCHAEDLER, 121; POGG 2,904-905
J.C. Bay, Plant Physiology 6 (1931), 189-193
P.F. Pilet, Arch. Inter. Hist. Sci. 15 (1962), 303-313
D. Kottler, Jean Senebier and the Emergence of Plant Physiology 1775-1802. From Natural History to Chemical Science. Ph.D. Dissertation, Johns Hopkins Univ. 1973
J. Marx, Janus 61 (1974), 201-220
G. Legée, Gesnerus 49 (1991), 307-322

SENHOFER, KARL [1841-1904]
HEIN, 621-622; MACHEK, 184-190; SCHAEDLER, 121; POGG 3,1236; 5,1154

SENIOR, JAMES KUHN [1889-1976]
AMS 9(I),1737
Anon., Chem. Eng. News 54(34) (1976), 31

SENN, NICHOLAS [1844-1908]
DAB 16,584-585; FISCHER 2,1442; DAMB, 672-673; WWWS, 1523; SG [1]12,876; [2]15,415-418
H. Buess, Gesnerus 36 (1979), 238-245
H.M. Brown, Milwaukee History 4(3/4) (1981), 87-94

SENTER, GEORGE [1874-1942]
SRS, 28; POGG 4,1385; 5,1154; 6,2417; 7b,4796
Anon., Chemical Age 46 (1942), 158; J. Roy. Inst. Chem. 1942, pp.92-93
W. Wardlaw, Nature 149 (1942), 405-406
S. Sugden, J. Chem. Soc. 1944, pp.359-360

SEQUEIRA, JAMES HARRY [1865-1948]
DNB [1941-1950], 771; WW 1945, p.2461; FISCHER 2,1442
Anon., Lancet 1948(II), pp.911-912; Brit. Med. J. 1948(II), p.1040; Brit. J. Dermatol. 61 (1949), 117-118

SEREBROVSKI, ALEKSANDR SERGEEVICH [1892-1948]
DSB 18,803-811

SERENI, ENRICO [1900-1931]
IB, 271; WININGER 7,443
A.V. Hill, Nature 127 (1931), 562-563
P.S. Israel, Riv. Biol. 13 (1931), 541-549
G. Colosi, Boll. Soc. Nat. Napoli 44 (1932), 81-91

SERONO, CESARE [1871-1952]
EI 31,445; POGG 6,2419; 7b,4800-4801
R. Mattei et al., Rass. Clin. Terap. 52(1) (1953), vi-xxii
D. Marotta, Chimica e Industria 35 (1953), 103-104

SERTOLI, ENRICO [1842-1910]
DSB 12,319-320; WWWS, 1524
A. Pugliese, Arch. Ital. Biol. 53 (1910), 161-164

SERTÜRNER, FRIEDRICH WILHELM [1783-1841]
DSB 12,320-321; WL 2,128-141; HEIN, 623-625; HIRSCH 5,236; W, 469-470; BLOKH 2,680-681; FERCHL, 500; TSCHIRCH, 1125; SCHAEDLER, 121-122; POTSCH, 394-395; COLE, 496-497; CALLISEN 18,26-31; 32,285; POGG 2,911-912; 6,2420; 7aSuppl.,626-628
G. Lockemann, Z. angew. Chem. 37 (1924), 525-532; J. Chem. Ed. 28 (1951), 277-279
F. Kromecke, Friedrich Wilhelm Sertürner. Jena 1925
R. Schmitz, Pharm. Hist. 27 (1985), 61-74

SÉRULLAS, GEORGES SIMON [1774-1832]
DSB 12,321-322; GE 29,1085; LAROUSSE 14,618; HOEFER 43,802-804; BLOKH 2,681-682; FERCHL, 500; TSCHIRCH, 1125; SCHAEDLER, 122; BALLAND, 107-120; BALLAND(2), 183-184; POTSCH, 395; WWWS, 1524; DECHAMBRE [3]9,431-432; CALLISEN 32,285-287; POGG 2,912-913
J.J. Virey, J. Pharm. Chim. 18 (1832), 318-321

SESHADRI, TIRUVENKATA RAJENDRA [1900-1975]
IWW 1975-1976, p.1587; WWWS, 1525; POTSCH, 395
S. Rangaswami, Current Science 44 (1975), 801-802
R.N. Chakravarti, Science and Culture 42 (1976), 208-210
W. Baker and S. Rangaswami, Biog. Mem. Fell. Roy. Soc. 25 (1979), 505-533

SESTINI, FAUSTO [1839-1904]
POGG 3,1239; 4,1387
I. Giglioli, Rend. Soc. Chim. 2 (1904), 165-176
Annuario Università Pisa 1904-1905, pp.312-314

SETLOW, RICHARD BURTON [1921-]
AMS 15(6),610; WWA 1980-1981, p.2981; WWWS, 1525

SEUBERT, KARL [1851-1942]
HEIN, 625-626; HANNOVER, 30; TSCHIRCH, 1125; SCHAEDLER, 122; TLK 3,611; POGG 3,1239; 4,1387; 5,1155; 6,2420-2421; 7a(4),394
W. Biltz, Z. angew. Chem. 44 (1931), 269
A. Wankmüller, Beitr. Württ. Apothekgesch. 10 (1973), 17-26

SEUBERT, WERNER [1928-1975]
KURSCHNER 11,2835

SEUFFERT, OTTO [1875-1952]
JV 16,253

SEUFFERT, RUDOLF WILHELM [1884-]
KURSCHNER 4,2783; JV 25,615; IB, 271-272; NUC 539,548

SEVAG, MANASSEH GIRAGOS [1897-1967]
AMS 11,4827
Anon., Chem. Eng. News 46(7) (1968), 166

SEVERIN, JOSEF [1866-]
JV 6,38; SG [2]15,499; NUC 539,654

SEVERIN, JOSEPH BERNHARD [1881-1948]
JV 24,186; SG [2]15,500; NUC 539,654

SEVERIN, SERGEI EVGENIEVICH [1901-]
BSE (3rd Ed.)23,110; BME 29,589-590; POTSCH, 395
Who's Who in Socialist Countries 1978, p.549
S.E. Severin, in Selected Topics in the History of Biochemistry (G. Semenza, Ed.), pp.365-390. Amsterdam 1983

SEVERINGHAUS, ELMER LOUIS [1894-1980]
AMS 11,4827; WWA 1976-1977, p.2829; IB, 272

K.W. Thompson, J. Clin. Endocrin. 52 (1981), 829-830

SEVERTSOV, ALEKSEI NIKOLAEVICH [1866-1936]
DSB 12,336-339; BSE 38,357-358; BME 29,590; KUZNETSOV(1963), 330-336; UKRAINE 2,411-412; LEVITSKI 1,289-290
H. Marcus, Bio-Morphosis 1 (1939), 404-416
K.M. Zavadski, Vestnik Leningradskovo Universiteta 1953(7), pp.3-23
A.A. Makhotin, Zhur. Ob. Biol. 27 (1966), 513-521
G.V. Nikolski, Zhur. Ob. Biol. 28 (1967), 413-422

SEVERTSOV, NIKOLAI ALEKSEEVICH [1827-1885]
BSE 38,358; KUZNETSOV(1963), 126-131
H. von Paucker, Leopoldina 21 (1885), 158-159
N.A. Dementyev, N.A. Severtsov, 2nd Ed. Moscow 1948
P.L. Zolotnitskaya, N.A. Severtsov. Moscow 1953

SEWALL, HENRY [1855-1936]
AMS 5,1000-1001; NCAB A,199; 26,323; DAMB, 673; WWAH, 546-547
G.B. Webb and D. Powell, Henry Sewall. Baltimore 1946
A.M. Harvey, Johns Hopkins Med. J. 142 (1978), 47-51
H.W. Davenport, Physiologist 25Suppl. (1982), 25-44

SEYBOLD, AUGUST [1901-1965]
KURSCHNER 10,2325; IB, 272

SEYDERHELM, RICHARD [1888-1940]
KALLMORGEN, 413; FISCHER 2,1445; RAZINGER, 167-169
W.H. Veil, Deutsche med. Wchschr. 66 (1940), 1053-1055

SEYDOUX, FRANÇOIS [1942-1979]
J. Yon, Biochimie 63 (1981), 87-88

SEYFFERTH, ERICH [1881-1937]
JV 23,337; NUC 540,218
Anon., Chem. Z. 61 (1937), 686; Z. angew. Chem. 50 (1937), 702

SCHACKELL, LEON FRANCIS [1887-1967]
AMS 9(II),1015

SHAFFER, PHILIP ANDERSON [1881-1960]
AMS 9(II),1016; FISCHER 2,1447; CHITTENDEN, 187-191; DAMB, 674-675; WWAH, 547; WWWS, 1527
E.A. Doisy, Am. Phil. Soc. Year Book 1961, pp.184-189; Biog. Mem. Nat. Acad. Sci. 40 (1969), 321-336

SHALFEYEV, MIKHAIL IVANOVICH [1845-1910]
BLOKH 2,642-643; ZMEEV 5,196
V.N. Platyev, Zhur. Fiz. Khim. 42 (1910), 367

SHAPIRO, DAVID [1903-]
BHDE 2,1076

SHAPOSHNIKOV, VLADIMIR NIKOLAEVICH [1884-1968]
SACKMANN, 307-309
N.S. Egorov, Mikrobiologia 33 (1964), 555-557

SHARP, THOMAS MARVEL [1897-1959]
W. Solomon, Proc. Chem. Soc. 1960, pp.35-36

SHARPEY, WILLIAM [1802-1880]
DSB 12,354; DNB 17,1365-1367; HIRSCH 5,247-248; O'CONNOR, 78-87
Anon., Nature 21 (1880), 567-568; Proc. Roy. Soc. 31 (1881), x-xix
J.C. Brougher, Ann. Med. Hist. 9 (1927), 124-128
D.W. Taylor, Med. Hist. 15 (1971), 126-153,241-259

SHARPEY-SCHÄFER, EDWARD ALBERT [1850-1935]
DSB 12,355-357; DNB [1931-1940],788-790; FISCHER 2,1447; W, 470-471; O'CONNOR, 147-150; WWWS, 1531
J.A.C., Nature 135 (1935), 608-610
L. Hill, Obit. Not. Fell. Roy. Soc. 1 (1935), 401-407
C.S. Sherrington, Edinburgh Med. J. 42 (1935), 393-406; Quart. J. Exp. Biol. 25 (1935), 99-104
A.V. Hill, J. Physiol. 263 (1976), 54P-56P
M. Borell, Med. Hist. 22 (1978), 282-290

SHATTUCK, FREDERICK CHEEVER [1847-1929]
DAB 17,30-31; AMS 4,881-882; NCAB 12, 272; WWWS, 1531
W.B. Cannon., Science 69 (1929), 207-209
G.G. Sears et al., New. Eng. J. Med. 201 (1929), 48-50

SHAW, LOUIS AGASSIZ [1886-1940]
AMS 6,1279; NCAB 33,233-234; IB, 272; WWWS, 1532

SHAW, TREVOR IAN [1928-1972]
Who Was Who 1971-1980, p.720
E.J. Denton, Biog. Mem. Fell. Roy. Soc. 20 (1974), 359-380

SHCHELKOV (SCZELKOW), IVAN PETROVICH [1833-1909]
KHARKOV, 35-41; VORONTSOV, 14-23; RSC 5,609; 8,916; 12,270; 18,655
E.A. Finkelstein, Trudy Inst. Ist. Est. 4 (1955), 133-148

SHCHERBAKOV, ALEKSEI YAKOVLEVICH [1842-1901]
KAZAN 1,208; RSC 12,677; 18,474
W. Pagel, Virchows Jahresb. 1901(I), p.343

SHEAR, MURRAY JACOB [1899-1983]
AMS 11,4856; WWWS, 1533
New York Times, 29 September 1983

SHEARER, CRESSWELL [1874-1941]
IB, 272
J. Gray, Obit. Not. Fell. Roy. Soc. 4 (1942), 15-19

SHEDLOVSKY, THEODORE [1898-1976]
AMS 11,4857
Anon., Rock. Univ. Notes 8(3) (1976), 2
R.M. Fuoss, Biog. Mem. Nat. Acad. Sci. 52 (1980), 379-408

SHEEHAN, JOHN CLARK [1915-1992]
AMS 15(6),647; WWA 1980-1981, p.3002; IWW 1982-1983, p.1210; WWWS, 1533; MH 2,490-491; MSE 3,105-106
J.C. Sheehan, Biopolymers 25 (1986), S1-S10

SHEFFER (SCHAFFER). ALEKSANDR ALEKSANDROVICH [1831-]
IKONNOKOV, 740-741

SHELESNYAK, MOSES CHAIM [1909-]
AMS 15(6),651; WWWS, 1534
M.C. Shelesnyak, Ann. N.Y. Acad. Sci. 476 (1986), 5-24

SHELFORD, VICTOR ERNEST [1877-1968]
AMS 10,3694; IB, 273; WWWS, 1535
S.C. Kendeigh, Bull. Ecol. Soc. Am. 49 (1968), 97-100
S.T. Miller, Mendel Newsletter 13 (1977), 4-6
R.A. Crocker, A Pioneer Ecologist. Washington, D.C. 1991

SHEMIN, DAVID [1911-1991]
AMS 15(6),655; WWA 1980-1981, p.3006; IWW 1982-1983, p.1211; WWWS, 1535; MH 2,491-492; MSE 3,106-107
D. Shemin, BioEssays 10 (1898), 30-35
New York Times, 28 November 1991
Anon., Chem. Eng. News 70(5) (1992), 55

SHEMYAKIN, MIKHAIL MIKHAILOVICH [1908-1970]
DSB 18,813-814; MH 2,492-494; MSE 3,107-109; WWWS, 1535
Anon., Leopoldina [3]4 (1968), 33-36; Nature 228 (1970), 98; in Peptides 1969 (E. Scoffone, Ed.), pp.xv-xvi. Amsterdam 1971
Y.U. Ovchinnikov, Pure and Applied Chemistry 25 (1971), 1-3

SHENSTONE, WILLIAM ASHWELL [1850-1908]
DNB [1901-1911] 3,305-306; POGG 4,1391-1392; 5,1159
W.A. Tilden, Nature 77 (1908), 348-349; J. Chem. Soc. 95 (1909), 2206-2209; Proc. Roy. Soc. A82 (1909), xxii-xxiv

SHEPARD, CHARLES UPHAM [1804-1886]
DAB 17,71-72; FERCHL, 502; WWWS, 1535; COLE, 499-500; POGG 2,919-920
Anon., Pop. Sci. Mon. 47 (1895), 548-553

SHEPARD, JAMES HENRY [1850-1918]
DAB 17,73-74; WWA 1916-1917, p.2226; NCAB 17,218-220; ELLIOTT, 235
L.F. Kebler and E.R. Series, J. Am. Pharm. Assn. 18 (1929), 1197-1204

SHEPOVALNIKOV, NIKOLAI PETROVICH [1872-1945]
D.G. Kvasov and A. Federova-Grot, Fiziol. Zhur. 46 (1960), 126-132

SHEPPARD, PHILIP MACDONALD [1921-1976]
DNB [1971-1980], 769-770; WW 1976, pp.2162-2163
C. Clarke, Biog. Mem. Fell. Roy. Soc. 23 (1977), 465-500

SHEPPARD, SAMUEL EDWARD [1882-1948]
DAB [Suppl.4],743-744; AMS 7,1611; NCAB 52,340; MILES, 433-434; SRS, 36; WWWS, 1536; NUC 543,202-204; POGG 5,1159; 6,2430-2432; 7b,4830-4832
J. Eggert, Camera 27 (1948), 380-381
C.E.K. Mees, J. Chem. Soc. 1949, pp.261-263

SHERESHEVSKI, NIKOLAI ADOLFOVICH [1885-1961]
BME 34,917-918; WWR, 506
Anon., Klin. Med. 1961(2), p.12

SHERMAN, HENRY CLAPP [1875-1955]
DAB [Suppl.5],622-623; AMS 9(I),1758; MILES, 434-436; NCAB 45,170-171; CB 1949, pp.565-567; IB, 273; WWAH, 551-552; WWWS, 1536-1537; POGG 5,1159-1160; 6,2431-2433; 7b,4832-4835
H.C. Sherman, Selected Works. New York 1948
E.C. Kendall, J. Chem. Ed. 32 (1955), 510-513
P.L. Day, J. Nutrition 61 (1957), 1-11
C.G. King, Biog. Mem. Nat. Acad. Sci. 46 (1975), 396-429

SHERMAN, PENOYER LEVI, Jr. [1867-]
JV 10,215; BARNHART 3,270; RSC 14,805; 15,114; NUC 539,378

SHERNDAL, ALFRED EINAR [1884-1958]
AMS 9(I),1758; JV 28,321; NUC 543,400

SHERRINGTON, CHARLES SCOTT [1857-1952]
DSB 12,395-403; DNB [1951-1960],881-883; FISCHER 2,1449; KOLLE 1,245-253; O'CONNOR(2), 353-362; HAYMAKER, 267-272; ABBOTT, 116-117; W, 471-472; WWWS, 1537
J.F. Fulton, J. Neurophysiol. 15 (1952), 167-190
E.G.T. Liddell, Obit. Not. Fell. Roy. Soc. 8 (1952), 241-270
W. Penfield, Brain 80 (1957), 402-410; Notes Roy. Soc. 17 (1962), 163-168
R. Granit, Charles Scott Sherrington, an Appraisal. London 1966
J.P. Swazey, J. Hist. Biol. 1 (1968), 57-89
R.D. French, Med. Hist. 14 (1970), 154-165
L. Volicer, Persp. Biol. Med. 16 (1973), 381-393
C.E.R. Sherrington et al., Notes Roy. Soc. 30 (1975), 45-88
J.C. Eccles and W.C. Gibson, Sherrington, his Life and Thought. Berlin 1979
E. Jokl, Trans. Coll. Phys. Phila. [5]2 (1980), 223-235
J.C. Eccles. Trends in Neurosciences 5 (1982), 108-110
H. McIlwain, Journal of the Royal Society of Medicine 77 (1984), 417-425

SHERVINSKI, VASILI DMITRIEVICH [1850-1941]
BSE 48,1; BME 34,916-917; WWR, 506; FISCHER 2,1382; ZMEEV 5,198
E.N. Artemiev, Ter. Arkh. 33 (11) (1961), 107-113
A.M. Tsikdik, V.D. Shervinski. Moscow 1972

SHERWIN, CARL PAXSON [1885-1974]
AMS 9(II),1025; NCAB C,220-221; IB, 273

SHIBATA, KEITA [1877-1949]
IB, 273; POGG 6,2434; 7b,4836-4837
S. Hattori, Botanical Magazine (Tokyo) 62 (1949), 155-158; Science (Japan) 20 (1950), 50-51
H. Tamiya, Ber. bot. Ges. 68a (1955), 13-16

SHIBATA, YUJI [1882-1979]
JBE, 1377,2358; POTSCH, 396; POGG 6,2434-2435; 7b,4837-4839
Anon., Gendai Kagaku 108 (1980), 10-13

SHIGA, KIYOSHI [1870-1957]
JBE, 1383; FISCHER 2,1449; OLPP, 377-378; SACKMANN, 311-312
O. Felsenfeld, Science 126 (1957), 113

SHILOV, NIKOLAI ALEKSANDROVICH [1872-1930]
DSB 12,404-405; BSE 48,40-41; KUZNETSOV(1961), 564-573; POTSCH, 397; POGG 4,1328; 5,1111; 6,2322-2333
S. Voskressenski, Koll. Z. 53 (1930), 383; Zhur. Fiz. Khim. 62 (1930), 2101-2112
Anon., Usp. Khim. 15 (1946), 233-264
N.N. Ushakova, N.A. Shilov. Moscow 1966

SHIMIZU, TOMIHIDE [1889-1958]
JBE, 1406,2359; HAKUSHI 3,511
Anon., Asian Medical Journal 1 (1958), 8SHINODA, JUNZO [1889-1975]
JBE, 1420,2284; POGG 6,2435-2436; 7b,4841-4842

SHISHKIN, BORIS KONSTANTINOVICH [1886-1963]
BSE 48,71-72; WWR, 508; IB, 263; WWWS, 1539
E.G. Bobrov, Bot. Zhur. 41 (1956), 925-930
E.G. Pobedimova, Bull. Soc. Nat. Moscou NS70 (1965), 107-109

SHISHKOV, LEON NIKOLAEVICH [1830-1908]
BSE 48,74; BLOKH 2,651-653; ARBUZOV, 107-109; KOZLOV, 184-186
V. Ipatiev, Zhur. Fiz. Khim. 41 (1910), 1335-1349

SHOHL, ALFRED THEODORE [1889-1946]
AMS 7,1619
F.C. Bing, J. Nutrition 110 (1980), 1083-1088

SHOJI, KINNOSUKE [1886-1962]
JBE, 1443; IB, 274

SHONLE, HORACE ABBOT [1892-1947]
AMS 7,1619; NCAB 36,84
Anon., Science 105 (1947), 251,281

SHOOTER, ERIC MANNERS [1924-]
AMS 15(6),687; WWWS, 1541

SHOPE, RICHARD EDWIN [1901-1966]
AMS 10,3715; WWAH, 553; DAMB, 679
C.H. Andrewes, Am. Phil. Soc. Year Book 1971, pp.179-184; Biog. Mem. Nat. Acad. Sci. 50 (1979), 353-375

SHOPPEE, CHARLES WILLIAM [1904-]
WW 1981, pp.2365-2366; IWW 1982-1983, p.1216; SRS, 86; WWWS, 1541; POGG 6,2437-2438; 7b,4844-4848
Anon., Nature 161 (1948), 1004; 178 (1956), 777

SHORB, MARY SHAW [1907-1990]
AMS 14,4644
New York Times, 21 August 1990

SHORE, LEWIS ERLE [1863-1944]
O'CONNOR(2), 14-15; IB, 274; RSC 18,732
Anon., Lancet 1944(II), p.360; Brit. Med. J. 1944(II), p.226

SHORR, EPHRAIM [1897-1956]
AMS 9(II),1029; KOREN, 247
D.P. Barr, Trans. Assoc. Am. Phys. 69 (1956), 28-29

SHORT, WALLACE FRANK [1898-1955]
POGG 6,2438; 7b,4848-4850
J.C. Smith, J. Chem. Soc. 1956, pp.2569-2572

SHORYGIN, PAVEL POLIEKTOVICH [1881-1939]
BSE 48,147; WWR, 511; ARBUZOV, 173-174; POTSCH, 402; POGG 6,2361
B.U. Belov et al., Usp. Khim. 8 (1939), 768-777
S.N. Danilov, Zhur. Ob. Khim. 10 (1940), 176-192
Z.E. Gelman, Vop. Ist. Est. Tekh. 1983(1), pp.48-58

SHREEVE, WALTON WALLACE [1921-]
AMS 15(6),693; WWWS, 1542

SHRINER, RALPH LLOYD [1899-]
AMS 14,4649; IB, 274; WWWS, 1542; POGG 6,2438; 7b,4850-4853
Anon., Chem. Eng. News 28 (1950), 104

SHRYOCK, RICHARD HARRISON [1893-1972]
NCAB 56,374-375; DAMB, 651
L.G. Stevenson, J. Hist. Med. 23 (1968), 1-7
W.J. Bell, Bull. Hist. Med. 46 (1972), 499-503; J. Hist. Med. 29 (1974), 15-31
N. Reingold, Isis 64 (1973), 96-100
E.R. Long, Am. Phil. Soc. Year Book 1973, pp.150-156

SHUBENKO, GRIGORI STEPANOVICH [1857-]
SG [2]15,608

SHULL, AARON FRANKLIN [1881-1961]
DSB 12,416-418; AMS 10,3721; IB, 274; WWAH, 553; WWWS, 1542-1543

SHULL, CHARLES ALBERT [1879-1962]
AMS 9(II),1030; IB, 274

SHULL, GEORGE HARRISON [1874-1954]
DAB [Suppl.5],628-629; AMS 8,2271; IB, 274; WWAH, 553-554; WWWS, 1543
G.H. Shull, Science 103 (1946), 547-550
E.N. Harvey, Am. Phil. Soc. Year Book 1954, pp.446-449
H.P. Riley, J. Heredity 46 (1955), 65-66
H.K. Hayes, A Professor's Story of Hybrid Corn. Minneapolis 1963

SHULMAN, ROBERT GERSON [1924-]
AMS 15(6),698; WWA 1980-1981, p.3026

SHUMOVA-SIMANOVSKAYA, EKATERINA OLIMPIEVNA [1853-1905]
ZMEEV 5,205-205; KONOPKA 10,436
V. Sirotinin, Russki Vrach 4 (1905), 927

SHWARTZMAN, GREGORY [1896-1965]
AMS 9(II),1031; IB, 274; KOREN, 247

SICHERER, WALTHER von [1876-]
JV 16,295; RSC 19,643; NUC 545,177

SICKEL, HANS [1896-1929]
E. Abderhalden, Chem. Z. 53 (1929), 918,938

SIDGWICK, NEVIL VINCENT [1873-1952]
DSB 12,418-420; DNB [1951-1952],885-886; WW 1949, pp.2535-2536; W, 473; POTSCH, 396; ABBOTT-C, 129; POGG 5,1161; 6,2440; 7b,4856-4857
H. Tizard, Obit. Not. Fell. Roy. Soc. 9 (1954), 237-258
J.E. Sutton, Proc. Chem. Soc. 1958, pp.310-319

SIEBECK, RICHARD [1883-1965]
KURSCHNER 10,2327; DRULL, 254-255; FISCHER 2,1452; IB, 274
C. Korth, Med. Klin. 60 (1965), 1418-1419
O.H. Arnold, Deutsche med. Wchschr. 90 (1965), 1443-1446
F. Curtius, Med. Welt 34 (1983), 347-353

SIEBER, NADINA (ZIBER-SHUMOVA, NADEZHDA OLIMPEIVNA) [1856-1916]
KONOPKA 9,481-484; RSC 18,471
S. Dziergowski, Rocznik Towarzystwo Naukowe Warszawske 1918, pp.163-166

SIEBER, WILHELM [1871-1948]
JV 12,84; NUC 545,342

SIEBERT, GÜNTHER [1920-]
KURSCHNER 13,3689; KLEIN, 121; POGG 7a(4),401-403

SIEBOLD, CARL THEODOR ERNST von [1804-1885]
DSB 12,420-422; ADB 34,186-188; HIRSCH 5,262-263; PAGEL, 1590-1591; WAGENITZ, 166-167; OLPP, 378-379; W, 473-474; WWWS, 1739-1740
E. Ehlers, Z. wiss. Zool. 42 (1885), i-xxxiv
H. Korner, Die Würzburger Siebold. Leipzig 1967

SIEBOLD, GEORG von [1812-1873]
GGTA 8 (1914), 898

SIEBURG, ERNST [1885-1937]
IB, 274; POGG 6,2441-2442

SIEDEL, WALTER [1906-1968]
KURSCHNER 10,2330; POGG 7a(4),403
Anon., Arzneimitt. 16 (1966), 451-452; Chem. Z. 93 (1969), 82; Nachr. Chem. Techn. 17 (1969), 9

SIEDENTOPF, HENRY [1872-1940]
DSB 12,422-423; TLK 3,612; WWWS, 1545; POTSCH, 396; POGG 4,1394; 5,1162; 6,2442; 7a(4),405

SIEDLECKI, MICHAL [1873-1940]
DSB 12,423-424; WE 10,496; IB, 274
C. Dobell, Parasitology 33 (1941), 1-7
Z. Federowicz, Memorabilia Zoologica 17 (1966), 7-162

SIEGERT, FERDINAND [1865-1946]
KURSCHNER 4,2792; FISCHER 2,1454; PUSCHEL, 136-137

SIEGFELD, MORITZ [1870-1913]
A. Burr, Chem. Z. 37 (1913), 633

SIEGFRIED, KURT [1873-1945]
AARGAU, 716-717; HEIN-E, 409-410; TSCHIRCH, 1126; POGG 4,1394; 6,2444; 7a(4),408
Anon., Schw. Apoth. Z. 83 (1945), 175

SIEGFRIED, MAX AUGUST [1864-1920]
DBJ 2,761; FISCHER 2,1454; BLOKH 2,683-684; POGG 4,1394; 5,1163; 6,2444
C. Neuberg, Ber. chem. Ges. 53A (1920), 77-78
S. Garten, Chem. Z. 44 (1920), 221; Sitz. Akad. Wiss. Leipzig 74 (1922), 145-156; Erg. Physiol. 21 (1923), v-viii

SIEGMUND, AUGUST GUSTAV [1820-1903]
RSC 5,689; SG [2]15,623
J. Pagel, Virchows Jahresber. 1903, p.425

SIEGMUND, PETER [1920-1982]
KURSCHNER 13,3697

SIGALAS, CLÉMENT [1866-1944]
IB, 275
R. Fabre, Bull. Acad. Med. 129 (1945), 90-92

SIGERIST, HENRY ERNEST [1891-1957]
DAB [Suppl.6],580-581; NCAB 46,436-437; FISCHER 2,1457; TSCHIRCH, 1126; DAMB, 682; POGG 7a(4),412-413
E.H. Ackerknecht, Gesnerus 14 (1957), 65-68
W. Pagel, Med. Hist. 1 (1957), 285-289

R.H. Shryock, Am. Phil. Soc. Year Book 1957, pp.159-162
O. Temkin, Bull. Hist. Med. 32 (1958), 485-499
H. Fischer, Schaffhauser Biographien 3 (1969), 47-51
H.E. Sigerist, Autobiographical Writings. Montreal 1968
B. Gebhart, Med. Hist. J. 4 (1969), 89-98
E. Berg-Schorn, Henry E. Sigerist. Cologne 1978
F.G. Vescia, Med. Hist. J. 14 (1979), 218-232

SIGMUND, WILHELM [1859-]
KURSCHNER 4,2783; PRAG, 392; IB, 275

SIGNER, RUDOLF [1903-]
KURSCHNER 13,3702; SBA 3,118-119; WWWS, 1547; POGG 6,2448; 7a(4),413-414
H. Nitschmann, Chem. Z. 87 (1963), 170
Anon., Chimia 17 (1963), 91-92; 37 (1983), 101

SIGURDSON, BJØRN [1913-1959]
A.P. Waterson and L. Wilkinson, An Introduction to the History of Virology, pp.205-206. Cambridge 1978

SIGWART, LUDWIG [1784-1864]
ADB 34,304-305; HIRSCH 5,271; FERCHL, 503; POGG 2,928; 3, 1246; CALLISEN 18,99-100; 32, 317
E.E. Reusch, Jahreshefte Wurtt. 22 (1866), 22-24
H. Simmer, Sudhoffs Arch. 39 (1955), 216-236

SIGWART, WALTER [1876-1948]
KALLMORGEN, 414; GAUSS, 243
W. Stoeckel, Z. Gyn. 71 (1949), 1047

SILBER, GUSTAV [1826-1904]
A. Wankmüller, Beitr. Wurtt. Apothekgesch. 7 (1966), 84

SILBER, PAUL [1851-1932]
M. Betti, Chimica e Industria 15 (1933), 33
N.D. Heindel and M.A. Pfau, J. Chem. Ed. 42 (1965), 383-386

SILBERMANN, JEAN THIÉBAUT [1806-1865]
LAROUSSE 14,714; WWWS, 1547; POGG 2,928-929; 3,1246
J. Nicklès, Am. J. Sci. [2]41 (1866), 103-105

SILBERSCHMIDT, KARL [1903-]
LDGS, 12; NUC 546,110

SILBERSCHMIDT, ROBERT [1901-1979]

SILBERSCHMIDT, WILLIAM [1869-1947]
KURSCHNER 4,2801; FISCHER 2,1458; KOREN, 248
M. Loretan, William Silberschmidt. Zurich 1988

SILBERSTEIN, FRIEDRICH [1888-1975]
KURSCHNER 4,2801; FISCHER 2,1458; BHDE 2,1082; KOREN,248; WININGER 7,445; PLANER, 315; NUC 546,113

SILLÉN, LARS GUNNAR [1916-1970]
VAD 1969, p.857
F.J.C. Rossotti, Chem. Brit. 7 (1971), 253

SILLIMAN, BENJAMIN [1779-1864]
DSB 12,432-434; DAB 17,160-163; ELLIOTT, 236-237; FARBER, 405-417; MILES, 437-438; HIRSCH 5,272; KELLY, 1111-1112; DAMB, 682-683; FERCHL, 503-504; SCHAEDLER, 123; COLE, 503-504; WWAH, 555; POTSCH, 397; WWWS, 1547; POGG 2,931-932; CALLISEN 18,103; 32,319
Anon., Am. J. Sci. [2]39 (1865), 1-9
G.P. Fisher, Life of Benjamin Silliman. Philadelphia 1866
A. Caswell, Biog. Mem. Nat. Acad. Sci. 1 (1877), 101-112
J.F. Fulton and E.H. Thomson, Benjamin Silliman. New York 1947
C.M. Brown, Benjamin Silliman. Princeton 1989

SILLIMAN, BENJAMIN, Jr. [1816-1885]
DSB 12,434-436; DAB 17,160; MILES, 438-440; ELLIOTT, 237; WWWS, 1547; POGG 2,932-933; 3,1247
J.D. Dana, Am. J. Sci. [3]29 (1885), 85-92
A.W. Wright, Biog. Mem. Nat. Acad. Sci. 7 (1913), 115-141

SIMANOVSKI, NIKOLAI PETROVICH [1854-1922]
BSE 39,52; BME 30,129
I.B. Soldatov, Nikolai Petrovich Simanovski. Moscow 1951

SIMHA, ROBERT [1912-]
AMS 15(6),730; WWWS, 1548; BHDE 2,1083

SIMMICH, PAUL [1869-]
JV 11,153; NUC 546,508

SIMMONDS, MORRIS [1855-1925]
FISCHER 2,1459-1460
W.E. Griesbach, J. Clin. Endocrin. 25 (1965), 1671-1673

SIMMONDS, NINA [1893-]
AMS 10,3741

SIMMONDS, PETER LUND [1814-1897]
Anon., Am. J. Pharm. 69 (1897), 616

SIMMONDS, SOFIA [1917-]
AMS 15(6),731

SIMMS, HENRY SWAIN [1896-1978]
AMS 11,4930; IB, 275

SIMOLA, PAAVO EEVERTI [1902-1961]
Kuka Kukin On 1950, p.705
J. Jaernefelt, Duodecim (Helsinki) 97 (1981), 928-930

SIMON, CHARLES EDMUND [1866-1927]
AMS 3,627
W.H. Howell, American Journal of Hygiene 8 (1928), i-vi
A.M. Harvey, Johns Hopkins Med. J. 142 (1978), 161-186

SIMON, ERIC JACOB [1924-]
AMS 15(6),735; WWA 1980-1981, p.3044; BHDE 2,1084; WWWS, 1549

SIMON, ERNST EYTAN [1902-1973]
BHDE 2, 1084-1085; LDGS, 58

SIMON, FRANZ (FRANCIS) EUGEN [1893-1956]
DSB 12,437-439; WW 1955, p.2711; TETZLAFF, 309; LDGS, 101; POGG 6,2453-2454; 7b,4887-4889
N. Kurti, Biog. Mem. Fell. Roy. Soc. 4 (1958), 225-256

SIMON, GUSTAV [1824-1876]
ADB 34,369-371; DRULL, 255; HIRSCH 5,279-281; PAGEL, 1595-1596; WWWS,1549
M. Jantsch, Wiener med. Wchschr. 119 (1969), 663-664
J. Keller, Med. Welt 37 (1969), 2043-2046
L. Lauridsen, Acta Chir. Scand. Suppl.433 (1973), 31-41

SIMON, JOHANN EDUARD [1789-1856]
HEIN, 635-636; TSCHIRCH, 1126; FERCHL, 504; SCHELENZ, 770; POGG 2,935-936,1441

SIMON, JOHANN FRANZ [1807-1843]
ADB 34,377; HIRSCH 5,277; HEIN, 636; FERCHL, 504; CALLISEN 18,118-119; 32,322-327; POGG 2,936; 7aSuppl.,643
J. M[inding], Beitr. physiol. path. Chem. 1 (1844), 547-552
A. Nowak, Wiss. Fort. 11 (1961), 60-61

SIMON, JOHN [1816-1904]
DNB [1901-1911] 3,316-318; HIRSCH 5,278-279; BETTANY 2,304-306; O'CONNOR, 40-42; WWWS,1549
Anon., Lancet 1904(II), pp.320-325; Brit. Med. J. 1904(II), pp.265-267
R.T. Hewlett, Nature 70 (1904), 326-327

J.B.S., Proc. Roy. Soc. 75 (1905), 336-346
R. Lambert, Sir John Simon. London 1963
J.L. Brand, Bull. Hist. Med. 37 (1963), 184-194
T.N. Stokes, Med. Hist. 33 (1989), 343-359

SIMON, LOUIS JACQUES [1867-1925]
POGG 4,1399; 5,1167; 6,2454-2455
C. Mauguin, Bull. Soc. Chim. [4]39 (1926), 1653-1674
A. Mayer, Bull. Soc. Chim. Biol. 8 (1926), 211-213

SIMON, MAX [1863-1942]
FISCHER 2,1461; KALLMORGEN, 415; GEDENKBUCH 2,1397; NUC 547,63

SIMON, MAX [1868-1930]
KALLMORGEN, 415

SIMON, SIEGFRIED VEIT [1877-1934]
KURSCHNER 4,2804-2805; WAGENITZ, 167; TETZLAFF, 310; IB, 276
L. Jost, Ber. bot. Ges. 53 (1935), (71)-(84)

SIMON, WILLIAM [1844-1916]
AMS 2,429; MILES, 442; HEIN-E, 412; GOULD, 21-22

SIMONART, ANDRÉ [1903-]
Anon., Bibl. Acad. Louvain 6 (1937), 412-413; 8 (1956), 65-66; 9 (1957), 52; 10 (1963), 210-211; 12 (1972), 645-648

SIMONART, PAUL [1907-]
BAF, 106; WWWS, 1549
Anon., Bibl. Acad. Louvain 8 (1956), 166; 9 (1957), 154-155; 11 (1969), 217-221; 12 (1972), 782,976-977
Qui est Qui en Belgique Francophone 1981-1985, p.793

SIMONIS, HUGO [1874-1949]
TLK 3,615; POGG 4,1399-1400; 5,1168; 6,2455-2456; 7a(4),418
H. Scheibler, Z. angew. Chem. 63 (1951), 36

SIMONNET, HENRI [1891-1965]
IB, 276
R. Lépine, Bull. Acad. Med. 150 (1966), 370-374

SIMONSEN, JOHN LIONEL [1884-1957]
DNB [1951-1960],895-897; WW 1949, p.2544; WWWS, 1550; POGG 5,1168; 6,2456; 7b,4893-4896
R.S. Cahn, Nature 179 (1957), 697-698
E.R.H. Jones, Proc. Chem. Soc. 1958, pp.86-89
R. Robinson, Biog. Mem. Fell. Roy. Soc. 5 (1959), 237-252

SIMPSON, GEORGE GAYLORD [1902-1984]
AMS 15(6),741; WWWS, 1550
H.B. Whittington, Biog. Mem. Fell. Roy. Soc. 32 (1986), 525-539
E.O. Olson, Biog. Mem. Nat. Acad. Sci. 60 (1991), 331-353

SIMPSON, JAMES CRAWFORD [1876-1944]
AMS 7,1631; SRS,39

SIMPSON, MAXWELL [1815-1902]
DNB [1901-1911] 3,319-320; FERCHL, 504; WWWS, 1551; POTSCH, 397-398; POGG 2,938; 3,1252; 5,1168
A.E. Dixon, Nature 65 (1902), 515-516
H.D., Proc. Roy. Soc. 75 (1905), 175-181
D. Reilly, Chymia 4 (1953), 159-170

SIMPSON, MELVIN VERNON [1921-]
AMS 18(6),779

SIMPSON, SAMUEL LEONARD (LEVY) [1900-1983]
WW 1981, p.2383; STC 2,504-506; NUC 547,369
Anon., Lancet 1983(II), p.526

SIMPSON, SUTHERLAND [1863-1926]
AMS 3,628; O'CONNOR(2), 392-393; NUC 547,372-373
E. Sharpey-Schafer, Quarterly Journal of Experimental Physiology 17 (1927), 1-14

SIMSON, CLARA von [1897-1983]
KURSCHNER 13,3712; BOEDEKER, 46; POGG 7a(4),418
W.H. Westphal, Phys. Bl. 13 (1957), 470

SINCLAIR, HUGH MACDONALD [1910-1990]
WW 1990, pp.1666-1667; WWWS, 1552
T.H. Jukes, J. Nutrition 121 (1991), 1297-1304

SINCLAIR, ROBERT GORDON [1903-1949]
AMS 8,2286; YOUNG, 32
Anon., Science 110 (1949), 385; Chem. Eng. News 27 (1949), 2948
G.H. Ellinger, Trans. Roy. Soc. Canada [3]44 (1950), 105-106

SINCLAIR, WILLIAM [1865-]
JV 8,217

SINELNIKOV, EVGENI IVANOVICH [1885-1951]
VORONTSOV, 187-189

SINGER, CHARLES [1876-1960]
DNB [1951-1960],897-899; MR 5,379-381; WININGER 5,544; 7,447-448
Z. Cope, Bull. Hist. Med. 34 (1960), 471-473
A.R. Hall, Isis 51 (1960), 558-560
E.A. Underwood, Med. Hist. 4 (1960), 353-358; Proc. Roy. Soc. Med. 55 (1962), 853-859
T.I. Williams, Science 132 (1960), 1296-1297
R.H. Shryock, Am. Phil. Soc. Year Book 1961, pp.189-191

SINGER, FRITZ [1879-]
JV 19,343; NUC 547,668

SINGER, JAKOB [1853-1926]
FISCHER 2,1463; SCHIEBER, 70-75; KOREN, 249

SINGER, MAXINE FRANK [1931-]
AMS 15(6),749; WWA 1980-1981, p.3053

SINGER, SEYMOUR JONATHAN [1924-]
AMS 15(6),749; WWA 1980-1981, p.3053

SINNOTT, EDMUND WARE [1888-1968]
AMS 11,4947; NCAB H,191-192; MH 1,434-435; MSE 3,121-122; IB, 276; WWAH, 557; WWWS, 1553-1554
G.S. Avery, Bull. Torrey Bot. Club 95 (1968), 647-652
L.C. Dunn, Am. Phil. Soc. Year Book 1968, pp.157-161
K.S. Wilson, Plant Science Bulletin 14(1) (1968), 6-7
W.G. Whaley, Biog. Mem. Nat. Acad. Sci. 54 (1983), 351-372

SINSHEIMER, ROBERT LOUIS [1920-]
AMS 15(6),757; WWA 1980-1981, p.3054; IWW 1982-1983, p.1228; WWWS, 1554; CB 1968, pp.368-370

SISAKYAN, NORAYR MARTISOVICH [1907-1966]
BSE 39,155-156; WWR, 515; WWWS, 1554
Anon., Izvestia Akademii Nauk (Biol.) 3 (1966), 457-460; Biochimia 31 (1966), 235-236

SIVÉN, VALTER OSVALD [1868-1918]
NUC 548,422
H. Fabritius, Finks Läkaresätskapets Handlingar 61 (1919), 264-270

SIZER, IRWIN WHITING [1910-]
AMS 14,4706; WWA 1976-1977, p.2901; WWWS, 1555

SJÖBERG, KNUT [1896-1975]
VAD 1975, p.921; POGG 6,2643; 7b,4916-4917

SJÖQVIST, JOHN AUGUST [1863-1934]
SMK 7.70; FISCHER 2,1465
J.E. Johanssen, Hygeia 97 (1935), 1-8

SJÖSTRAND, FRITIOF STIG [1912-]
VAD 1981, p.948; WWWS, 1555

SJOLLEMA, BOUWE [1868-1962]
IB, 276; POGG 4,1401-1402; 6,2463; 7b,4917-4720
J. Smit et al., Chem. Wkbl. 35 (1938), 448-474
J. Grashuis, Chem. Wkbl. 58 (1962), 217-219
L. Karsemeyer, Deutsche Tierärztliche Wochenschrift 69 (1962), 533-539

SKARZYNSKI, BOLESLAW [1901-1963]
WE 10,549-550
S. Konopka, Arch. Hist. Med. 26 (1963), 161-162
N. Ostrowski, Przglad Lekarski 19 (1963), 363-368; Postepy Biochemii 10 (1964), 171-193; Folia Biologica 12 (1964), 1-15

SKEGGS, LEONARD TUCKER [1918-]
AMS 15(6),766; IWW 1982-1983, p.1231
L.A. Lewis, Clin. Chem. 27 (1981), 1465-1468

SKITA, ALADAR [1876-1953]
HANNOVER, 33; KIEL, 193-194; TLK 3,616-617; POTSCH, 398; POGG 5,1172; 6,2464; 7a(4),420
Anon., Chem. Z. 75 (1951), 135; 77 (1953), 830

SKODA, JOSEF [1805-1881]
DSB 12,450-451; ADB 34,446-447; WURZBACH 35,66-72; HIRSCH 5,300-301; PAGEL, 1605-1607; LESKY, 142-152; FROHNE, 155-175; WWWS, 1556; CALLISEN 32,332-333
M. Heitler, Wiener Klinik 7 (1881), 279-294
J. Stefan, Alm. Akad. Wiss. Wien 32 (1882), 265-270
L. Schroetter, Wiener klin. Wchschr. 18 (1905), 1315-1323; 36 (1923), 278-280
M. Sternberg, Josef Skoda. Vienna 1924
E. Lesky, Wiener klin. Wchschr. 68 (1956), 726-728
Z. Hornof, Clio Medica 2 (1967), 55-62

SKOOG, FOLKE KARL [1908-]
AMS 14,4713-4714; WWA 1976-1977, p.2905; IWW 1982-1983, p.1232; WWWS,1557SKRABAL, ANTON [1877-1955]
KERNBAUER, 214-252,265-266; POGG 5,1172-1173; 6,2465-2466; 7a(4),421
H. Schmid, Ost. Chem. Z. 58 (1955), 285-289

SKRAMLIK, EMIL von [1886-1970]
KURSCHNER 10,2345; FISCHER 2,1465-1466; IB, 277; POGG 6,2466-2467; 7a(4),422-423

SKRAUP, SIEGFRIED [1890-1972]
KURSCHNER 11,2863; POGG 6,2467; 7a(4),423
Anon., Chem. Z. 96 (1972), 697

SKRAUP, ZDENKO HANS [1850-1910]
DSB 12,452; BJN 15,80*-81*; KERNBAUER, 125-150; KNOLL, 49-51; W, 478; BLOKH 2, 688-691; SCHAEDLER, 123; WWWS, 1557; POTSCH, 398; POGG 3,1254; 4,1402; 5,1173; 7aSuppl.,643
H. Schroetter, Ber. chem. Ges. 43 (1910), 3683-3702
R. Wegscheider, Chem. Z. 34 (1910), 1013-1014
M. Kohn, J. Chem. Ed. 20 (1943), 471-473
E.H. Huntress, Proc. Am. Acad. Arts Sci. 78 (1950), 13-14

SKVIRSKY, PETR [1883-]
IB, 277; NUC 549,258

SKVORTSOV, MIKHAIL ALEKSANDROVICH [1876-1963]
BSE 39,219; BME 30,472-473; WWR, 517; IB, 277
A.P. Avtsin et al., Arkh. Pat. 38(10) (1976), 86-80

SLATER, EDWARD CHARLES [1917-]
WW 1981, p.2395; IWW 1982-1983, p.1231
P. Borst and G.S. van den Bergh, Structure and Function of Energy-Transducing Systems, pp.xvii-xxvii. Amsterdam 1977

SLATER, WILLIAM KERSHAW [1893-1970]
DNB [1961-1970], 955-956; IB, 277; WWAH, 558
H.D. Kay, Biog. Mem. Fell. Roy. Soc. 17 (1971), 663-680

SLATOR, ARTHUR [1879-1953]
SRS, 27; POGG 4,1403; 5,1174; 6,2467; 7b,4933
J.H.S. Johnston, Chem. Ind. 1953, pp.943-944
Anon., J. Roy. Inst. Chem. 1953, pp.460-461

SLIMMER, MAX DARWIN [1877-]
JV 18,27; NUC 529,508

SLONIMSKI, PIOTR [1893-1944]
IB, 278
R. Michailowski, Folia Morphologica 28 (1969), 405-413

SLOSSE, AUGUSTE [1863-1930]
FISCHER 2,1467
A.P. Dustin, Bull. Acad. Med. Belg. [6]3 (1938), 247-260

SLOTIN, LOUIS [1911-1946]
H.L. Anderson et al., Science 104 (1946), 182-183
Anon., Chem. Eng. News 24 (1946), 1565

SLOTTA, KARL HEINRICH [1895-1987]
KURSCHNER 13,3723; AMS 12,5886; BHDE 2,1090-1091; LDGS, 25; WWWS, 1560; POGG 6,2469-2470; 7a(4),425-426
R. Ruschig and R. Tschesche, Arzneimitt. 25 (1975), 838
K.H. Slotta, TIBS 8 (1983), 417-419
New York Times, 21 July 1987

SLOVTSOV, BORIS IVANOVICH [1874-1924]
BME 30,682
V. Veselkin, Arkh. Klin. Eksp. Med. 4 (1924), 1-6
Y.M. Gefter, Vop. Med. Khim. 2 (1950), 3-11

SLYE, MAUD [1879-1954]
AMS 8,2300; NCAB F,440; NAW 4,651-652; WWAH, 559; WWWS, 1560
Anon., Proc. Inst. Med. Chicago 20 (1955), 330-331
J.J. McCoy, The Cancer Lady. Nashville, Tenn. 1977

SMADEL, JOSEPH EDWIN [1907-1963]
AMS 10,3775; WWAH, 559-560
T.E. Woodward, Trans. Assoc. Am. Phys. 77 (1964), 29-32

SMAKULA, ALEXANDER [1900-1983]
AMS 14,4727; WWWS, 1560
Anon., Naturw. Rund. 36 (1983), 467

SMALL, LYNDON FREDERICK [1897-1957]
AMS 8,2301; NCAB 44,562-563; MILES, 442-444
E. Mosettig, Biog. Mem. Nat. Acad. Sci. 33 (1959), 397-412

SMELLIE, ROBERT MARTIN STUART [1927-1988]
WW 1981, p.2402; WWWS, 1561
A. Campbell and R. Thomson, The Biochemist 10(2) (1988), 15-16

SMIDT, HANS [1886-1955]
KURSCHNER 4,2818; FISCHER 2,1468
W. Rieder, Bremische Biographien 1912-1962, pp.486-487. Bremen 1969

SMILES, SAMUEL [1877-1953]
SRS, 26; WWWS, 1561-1562; POGG 5,1176-1177; 6,2474; 7b,4951
G.M. Bennett, J. Chem. Soc. 1953, pp.4192-4200; Obit. Not. Fell. Roy. Soc. 8 (1953), 583-600
R. Child, Nature 172 (1953), 13-14

SMIRNOV, ALEKSEI EFIMOVICH [1859-1910]
SG [1]13,205
K.A. Kitmanov, Protokoly Obshchestva Estestvoispitalei i Vrachi pri Tomskom Universiteta 1908-1910, pp.127-128
J. Pagel, Virchows Jahresb. 1910, p.439SMIT, JAN [1885-1957]
POGG 6,2474; 7b,4957-4958
K.G. Mulder, Chem. Wkbl. 52 (1956), 730-735; 53 (1957), 405-406

SMITH, ALEXANDER [1865-1922]
DAB 17,235; NCAB 20,421-422; MILES, 444-445; BLOKH 2,693; WWAH, 559; WWWS, 1563; POGG 4,1406-1407; 5,1177-1178; 6,2474
J. Kendall, J. Am. Chem. Soc. 44 (1922), 113-117
W.A. Noyes, Mem. Nat. Acad. Sci. 21(No.12) (1927), 1-7
R.H. McKee and J. Kendall, J. Chem. Ed. 9 (1932), 246-260

SMITH, ARTHUR HENRY [1893-1976]
AMS 11,4980; WWWS, 1563
J.M. Orten, J. Nutrition 108 (1978), 733-738

SMITH, EDGAR FAHS [1854-1928]
DSB 12,465; DAB 17,255-256; MILES, 445-446; FARBER, 826-828; WWAH, 561; POTSCH, 398-399; WWWS, 1564; POGG 3,1259-1260; 4,1405-1406; 5,1178-1179; 6,2476-2477
C.A. Browne, Isis 11 (1928), 375-384; J. Chem. Ed. 5 (1928), 656-663
M.T. Bogert, Science 69 (1929), 557-565
W.T. Taggart et al., J. Chem. Ed. 9 (1932), 607-665
G.H. Meeker, Biog. Mem. Nat. Acad. Sci. 17 (1936), 103-149
H.S. Klickstein, Chymia 5 (1959), 11-30

SMITH, EDWARD [1819-1874]
DSB 12,465-467; DNB 18,439-440; HIRSCH 5,317; WWWS, 1554; RSC 5,719-720
C.B. Chapman, J. Hist. Med. 22 (1967), 1-26
K.J. Carpenter, J. Nutrition 121 (1991), 1515-1521

SMITH, EMIL L. [1911-]
AMS 15(6),807; IWW 1982-1983, p.1237; WWWS, 1565
E.L. Smith, in Of Oxygen, Fuels and Living Matter, Part 2 (G. Semenza, Ed.), pp.361-445. Chichester 1982

SMITH, ERASTUS GILBERT [1855-1937]
AMS 5,1033-1034; RSC 18,800-801; NUC 550,692
New York Times, 20 June 1937

SMITH, ERNEST LESTER [1904-1992]
WW 1981, p.2409; CAMPBELL, 207-208; WWWS, 1565

SMITH, ERWIN FRINK [1854-1927]
DSB 12,467-468; DAB 17,262-263; DOETSCH, 206-208; SACKMANN, 321-323; HUMPHREY, 234-236; IB, 278; WWWS, 1565
R.H. True, Phytopathology 17 (1927), 675-688
L.R. Jones, Biog. Mem. Nat. Acad. Sci. 21 (1939), 1-71
A.D. Rodgers, Edwin Frink Smith. Philadelphia 1952
C.L. Campbell, Annual Review of Phytopathology 21 (1983), 21-27

SMITH, FRED [1911-1965]
AMS 10,3791
R. Montgomery, Adv. Carbohydrate Chem. 22 (1967), 1-10

SMITH, GEORGE FREDERICK [1891-1976]
AMS 11,4994; WWWS, 1566; POGG 6,2477; 7b,4965-4969
H. Diehl, Talanta 13 (1966), 867-894
R. Belcher, Proc. Soc. Anal. Chem. 14 (1977), 41-42

SMITH, HAMILTON OTHANIEL [1931-]
AMS 15(6),812; WWA 1980-1981, p.3080; IWW 1982-1983, p.1238; POTSCH, 399

SMITH, HERBERT EUGENE [1857-1933]
AMS 2,436; NCAB 24,429-430
Obituary Record of Graduates of Yale University 1934, pp.147-148. New Haven 1934

SMITH, HOMER WILLIAM [1895-1962]
DSB 12,470-471; DAB [Suppl.7],699-700; AMS 10,3795; DAMB, 692-693; WWAH, 563; POGG 6,2478-2479; 7b,4970-4973
J.A. Shannon, Trans. Assoc. Am. Phys. 75 (1962), 40-44
H. Chasis and W. Goldring (Eds.), Homer William Smith. New York 1965
R.F. Pitts, Biog. Mem. Nat. Acad. Sci. 39 (1967), 445-470

SMITH, JAMES HOLLINGWORTH CLEMMER [1895-1970]
AMS 11,5000; POGG 6,2479; 7b,4973-4974

SMITH, JAMES LORRAIN [1862-1931]
BELFAST, 617
J.S.H., Biochem. J. 25 (1931), 1849-1850
R.M., J. Path. Bact. 34 (1931), 683-696
A. Haddow, Persp. Biol. Med. 18 (1975), 433-455

SMITH, JOHN LAWRENCE [1818-1883]
DAB 17,304-305; MILES, 447-448; SILLIMAN, 92-94; FERCHL, 506; WWAH, 564; POTSCH, 399; WWWS, 1568; POGG 3,1257-1258
B. Silliman, Biog. Mem. Nat. Acad. Sci. 2 (1886), 217-248
J.P. Phillips, J. Chem. Ed. 42 (1965), 390-391
C.L. Gemmill and M.J. Jones, Pharmacology at the University of Virginia School of Medicine, pp.51-58. Charlottesville 1966

SMITH, KENNETH MANLEY [1892-1981]
WW 1981, p.2414
B. Kassanis, Biog. Mem. Fell. Roy. Soc. 28 (1982), 451-477
B. Hull, J. Gen. Microbiol. 128 (1982), 431-432
M.A. Lauffer and K. Maramorosch, Advances in Virus Research 27 (1982), ix-xi

SMITH, LEE IRVIN [1891-1973]
AMS 12,5925; POGG 6,2479; 7b,4976-4977
Anon., Chem. Eng. News 36(47) (1958), 90; 51(26) (1973), 35

SMITH, LONGFIELD [1875-1959]
SRS, 21; NUC 551,599

SMITH, MAURICE ISADORE [1887-1951]
AMS 8,2320
G.W. McCoy, J. Wash. Acad. Sci. 42 (1952), 136

SMITH, NORMAN [1877-1963]
WW 1959, p.2828; SRS, 26

SMITH, OLIVE WATKINS [1901-]
AMS 14,4176

SMITH, PHILIP EDWARD [1884-1970]
DSB 12,472-477; AMS 11,5011; FISCHER 2,1469; MEDVEI, 518-519; WWAH, 565; DAMB, 698-699; IB, 279
F.J. Agate, Anat. Rec. 171 (1971), 134-136
A.E. Severinghaus, Am. J. Anat. 135 (1972), 161-163
N.P. Christy, Endocrinology 90 (1972), 1415-1416
J.H. Leathem, Hormonal Proteins and Peptides 4 (1977), 175-192

SMITH, ROBERT ANGUS [1817-1884]
DSB 12,478-479; DNB 18,520-522; BLOKH 2,692-693; FERCHL, 506-507; WWWS, 1570
T.E. Thorpe, Nature 30 (1884), 104-105
C. von Voit, Sitz. Bayer. Akad. 15 (1885), 167-170
H.E. Schunck, Mem. Lit. Phil. Soc. [3]10 (1887), 90-102
A. Gibson and W.V. Farrar, Notes Roy. Soc. 28 (1974), 241-262
J.M. Eyler, Bull. Hist. Med. 54 (1980), 216-234
E. Gorham, Notes Roy. Soc. 36 (1982), 267-272

SMITH, SYDNEY [1886-1962]
POGG 6,2481; 7b,4983
W.M. Duffin, Proc. Chem. Soc. 1963, pp.29-30

SMITH, THEOBALD [1859-1934]
DSB 12,480-486; DAB 21,665-667; NCAB D,133-134; 35,5-6; FISCHER 2,1470; DAMB, 700-701; OLPP, 380; ULLMANN, 128-130; WWAH, 565; WWWS, 1570
W. Bulloch, J. Path. Bact. 40 (1935), 621-635
E.B. McKinley, Science 82 (1935), 575-586
M.C. Hall, J. Parasit. 21 (1935), 231-243
S.H. Gage, Science 84 (1936), 365-371
H. Zinsser, Biog. Mem. Nat. Acad. Sci. 17 (1936), 261-303
P.F. Clark, J. Hist. Med. 14 (1959), 496-514
C.E. Dolman, Clio Medica 4 (1969), 1-31; N.Y. J. Med. 69 (1969), 2801-2816; Persp. Biol. Med. 25 (1982), 417-427
E. Chernin and H. Zinsser, Review of Infectious Diseases 9 (1987), 625-654

SMITH, THOMAS SOUTHWOOD [1788-1861]
DNB 18,543-545; HIRSCH 5,310; DECHAMBRE [3]10,84; CALLISEN 18,173-174; 32,340
F.N.L. Poynter, Proc. Roy. Soc. Med. 55 (1962), 381-392

SMITH, WILLIAM STANLEY [1863-]
JV 6,299; NUC 552,397

SMITH, WILSON [1897-1965]
DSB 12,492-493; WW 1965, p.2851; WWWS, 1571
C.H. Andrewes, Nature 207 (1965), 1130-1131
D.G. Evans, Biog. Mem. Fell. Roy. Soc. 12 (1966), 479-487
G.R. Cameron, J. Path. Bact. 95 (1968), 326-336

SMITHBURN, KENNETH C. [1904-1973]
AMS 11,5023; WWWS, 1571
Anon., J. Am. Med. Assn. 228 (1974), 642

SMITHELLS, ARTHUR [1860-1939]
DSB 12,493-494; DNB [1931-1940],820-821; POGG 4,1407; 5,1181; 6,2482; 7b,4987
J.W. Cobb, J. Chem. Soc. 1939, pp.1234-1236
H.S. Raper, Obit. Not. Fell. Roy. Soc. 3 (1940), 97-107
R.K. Dekosky, Ambix 27 (1980), 103-123

SMITS, ANDREAS [1870-1948]
DSB 12,495-496; POGG 4,1407-1408; 5,1181-1182; 6,2482-2483; 7b,4988
J.M. Bijvoet et al., Chem. Wkbl. 28 (1931), 555-556; 37 (1940), 430-436; 45 (1949), 149-151

SMOLUCHOWSKI, MARIAN von [1872-1917]
DSB 12,496-498; WE 10,618-619; BLOKH 2,693-695; POTSCH, 399; POGG 4,1408; 5,1183-1184; 6,2483
A. Sommerfeld, Phys. Z. 18 (1917), 533-539
A. Teske, Sudhoffs Arch. 53 (1969), 292-305

SMORODINTSEV, IVAN ANDREEVICH [1881-1946]
BME 30,753; POGG 6,2485; 7b,4992-4995
N.S. Drozdov, Zhur. Ob. Khim. 19 (1949), 200-203

SMYTH, DAVID HENRY [1908-1979]
WW 1978, p.2287; WWWS, 1572
H. Barcroft and D.M. Matthews, Biog. Mem. Fell. Roy. Soc. 27 (1981), 525-561

SMYTH, MORLAND [1878-1949]
JV 17,172; NUC 552,620
The Green Can, July 1949, p.123

SMYTHE, CARL VINCENT [1903-1989]
AMS 11,5027

SNELL, ESMOND EMERSON [1914-]
AMS 15(6),846; WWA 1980-1981, p.3099; IWW 1982-1983, p.1241; WWWS, 1573; MH 2,507-508; STC 2,514
E.E. Snell, Fed. Proc. 38 (1979), 2690-2693; Annual Review of Nutrition 9 (1989), 1-19SNELL, GEORGE DAVIS [1903-]
AMS 15(6),846; IWW 1982-1983, pp.1241-1242; BIBEL, 282-287

ŚNIADECKI, JĘDRZEJ (ANDREAS) [1768-1838]
HIRSCH 5,325; BLOKH 2,695-696; KOSMINSKI, 466-467; KONOPKA 10,499-505; POTSCH, 399; FERCHL, 507; POGG 2,950
K. Slawinski and W. Lampe, Roczniki Chemii 19 (1939), 1-16
W. Ostrowski, Arch. Hist. Med. 32 (1969), 19-28
W. Hubicki, Organon 7 (1970), 231-245
W. Zacharewicz, Jędrzej Śniadecki. Warsaw 1975
E. Mietkiewski, Acta Physiol. Polon. 38 (1987), 52-65
R. Soloniewicz, Kwart. Hist. Nauki 36 (1991), 133-143

SNOW, JOHN [1813-1858]
DSB 12,502-503; DNB 18,615-616; HIRSCH 5,325-326; WWWS, 1574
P.E. Brown, Bull. Hist. Med. 35 (1961), 519-528
H. Cohen, Proc. Roy. Soc. Med. 62 (1969), 99-106

SNYDER, LAURENCE HASBROUCK [1901-1986]
AMS 11,5036; IB, 280
Anon., American Journal of Medical Genetics 8 (1981), 449-468
E.L. Green, American Journal of Medical Genetics 41 (1987), 276-285

SNYDER, SOLOMON HALBERT [1938-]
AMS 15(6),855; WWA 1980-1981, p.3103
E.L. Green, American Journal of Human Genetics 41 (1987), 276-285

SOBER, HERBERT ALEXANDER [1918-1974]
AMS 12,5964; WWWS, 1675
Anon., Nature 253 (1975), 486; Chem. Eng. News 53(3) (1975), 55
A. Meister, Anal. Biochem. 66 (1975), 298-301

SOBERNHEIM, GEORG [1865-1963]
KURSCHNER 9,1963-1964; FISCHER 2,1472; IB, 280; WININGER 7,451-452

SOBOLEV, LEONID VASILIEVICH [1876-1919]
BSE 39,458; BME 30,772-773
C. van Beek, Diabetes 7 (1958), 245-248

SOBOTKA, HARRY HERMAN [1899-1965]
AMS 10,3826; KAGAN(JA), 742; POGG 6,2487-2488; 7b,5009-5013
H.D. Appleton and M.M. Friedman, Clin. Chem. 12 (1966), 115-119

SOBOTTA, JOHANNES [1869-1945]
KURSCHNER 6(II),797; PAGEL, 1613-1614; EBERT, 14-26; IB, 280
F. Wagenseil, Anat. Anz. 101 (1954-1955), 265-280

SOBRERO, ASCANIO [1812-1888]
EI 31,986; PROVENZAL, 163-167; BLOKH 2,697; FERCHL, 207; WWWS, 1575; POTSCH, 399-400; SCHAEDLER, 123; POGG 2,952; 3.1263
Anon., Leopoldina 24 (1888), 223; Chem. Z. 12 (1888), 1745
A. Cossa, Atti Accad. Torino 24 (1889), 158-163
I. Guareschi, Isis 1 (1913), 351-358
G. Provenzal, Rass. Clin. Terp. 34 (1935), 50-54

SOCIN, CARL ANDREAS [1866-1933]
JV 9,235; RSC 18,831; SG [2]16,132
M. Siegfried, L'Alsace Française 13 (1933), 702-703
J.K., Vierteljahrschrift für Schweizerische Sanitätsoffiziere 10 (1933), 37-39

SODDY, FREDERICK [1877-1956]
DSB 12,504-509; DNB [1951-1960],904-905; STC 2,515-517; ABBOTT-C,129-130; W, 482; WWWS, 1575; POTSCH, 400; POGG 5,1185; 6,2488-2489; 7b,5013-5015
A.S. Russell, Science 124 (1956), 1069-1070
F.A, Paneth, Nature 180 (1957), 1085-1087
A. Fleck, Biog. Mem. Fell. Roy. Soc. 3 (1957), 203-216
M. Howorth, Pioneer Research on the Atom. London 1958
A. Kent, Proc. Chem. Soc. 1963, pp.327-330
T.J. Trenn, Rete 1 (1971),51-70
L. Badash et al., Brit. J. Hist. Sci. 12 (1980), 245-288

G.B. Kauffman, Frederick Soddy. Dordrecht 1986

SODEN, HUGO von [1858-1954]
POGG 4,1411; 5,1185-1186; 6,2489
Y.R. Naves, Perfumery and Essential Oil Record 45 (1954), 131

SÖHNGEN, NICOLAS LOUIS [1878-1934]
SACKMANN, 324-325

SOEHRING, KLAUS [1911-1975]
KURSCHNER 12,3055; POGG 7a(4),431-433
Anon., Naturw. Rund. 29 (1976), 34

SÖLL, DIETER GERHARD [1935-]
AMS 14,4792

SÖLLNER, KARL [1903-]
AMS 12,5973; BHDE 2,1092; LDGS, 25; WWWS, 1577

SOEMMERRING, SAMUEL THOMAS von [1755-1830]
DSB 12,509-511; ADB 34,610-615; HIRSCH 5,329-331;
KALLMORGEN, 416-417; WAGENITZ, 168-169; FRANKEN, 197;
MATSCHOSS, 256-257; WWWS, 1740;
CALLISEN 32,348-349
T.H. Bast, Ann. Med. Hist. 6 (1924), 369-387
E. Görge, S.T. von Soemmerring. Düsseldorf 1938
J. Karcher, Gesnerus 10 (1953), 26-36
H. Schierhorn, Z. mikr. anat. Forsch. 94 (1980), 1051-1076
R. Bachmann, Verhandlungen der anatomischen Gesellschaft 75 (1981),33-46

SØRENSEN, MARGRETHE [1884-1954]
VEIBEL 2,410-411

SØRENSEN, SØREN PETER LAURITZ [1868-1939]
DSB 12,546-547; VEIBEL 1,229-230; 2,411-421; 3,115-121; W,484;
WWWS,1580; POTSCH, 401; POGG 4,1412; 5,1186-1187; 6,2496;
7b,5015-5016
K. Linderstrøm-Lang, C.R. Lab. Carlsberg 23 (1939), i-xxi; Koll. Z.
88 (1939), 129-139
H. Jørgensen, Ber. chem. Ges. 72A (1939), 67-70
W.M. Clark, Science 89 (1939), 282-283
E.J. Cohn, J. Am. Chem. Soc. 61 (1939), 2573-2574
E.K. Rideal, J. Chem. Soc. 143 (1940), 554-561
P. Wery, Scalpel 122 (1969), 171-175

SOHNCKE, LEONHARD [1842-1897]
DSB 12,511-512; ADB 54,377-379; WWWS, 1576; POGG 3,1263-1264; 4,1412
C. von Voit, Sitz. Bayer. Akad. 28 (1898), 440-449

SOKHEY, SAHIB SINGH [1887-1971]
IB, 280; NUC 555,211
M.L. Ahuja, Indian Journal of Medical Science 59(11) (1971), vii-viii

SOKOLOV, NIKOLAI NIKOLAEVICH [1826-1877]
BSE 40,7; ARBUZOV, 190-191; BLOKH 2,697-699; POGG 2,955;
3,1264
Russki Biograficheski Slovar 19,62-63
P. Lachinov, Zhur. Fiz. Khim. 10 (1878), 8-15
Y.S. Musabekov, Zhur. Prikl. Khim. 22 (1949), 1133-1142
N.S. Kozlov, Usp. Khim. 22 (1953), 119-128

SOKOLOV, NIL IVANOVICH [1844-1894]
PAGEL, 1615; ZMEEV 5,135-136; RSC 11,447; 12,692; 17,836

SOLDAN, FRIEDRICH [1817-1881]
R. Sommer, Familienforschung und Vererbungskunde, Appendix.
Leipzig 1907

SOLEREDER, HANS [1860-1920]
DBJ 2,761; TSCHIRCH, 1127; RSC 16,486; 18,839; NUC 555,404
L. Radlkofer, Ber. bot. Ges. 38 (1920), (92)-(102)

SOLLMANN, TORALD HERMANN [1874-1965]
AMS 10,3830; PARASCANDOLA, 50-51; CHEN, 23; DAMB, 705;
WWAH, 568-569
J.M. Hayman, Trans. Assoc. Am. Phys. 78 (1965), 37-38
J. Seifter, Pharmacologist 7 (1965), 68-69

SOLLNER, KARL [1903-1986]
AMS 14,4792-4793; BHDE 2,1092; LDGS, 25; WWWS, 1577

SOLLY, EDWARD [1819-1886]
DNB 18,622; DESMOND, 573; FERCHL, 508; POGG 3,1265
Anon., Nature 33 (1886), 536; Leopoldina 22 (1886), 113

SOLMSSEN, ULRICH VOLCKMAR [1909-]
AMS 15(6),865; BHDE 1,710

SOLOMON, ARTHUR KASKEL [1912-]
AMS 15(6),866; IWW 1982-1983, p.1246

SOLS, ALBERTO [1917-1987]
Who's Who in Science in Europe, 4th ed., p.2050
A. Sols, in Selected Topics in the History of Biochemistry (G. Semenza
and R. Jaenicke, eds.), pp.177-199. Amsterdam 1990
M. Sapag-Hagar, Archivos de Biologia y Medicina Experimentrales
22 (1989), 335-337

SOMERS, GEORGE FREDERICK [1914-]
AMS 15(6),871; WWA 1980-1981, p.3110; WWWS, 1578

SOMMARUGA, ERWIN von [1844-1897]
WURZBACH 35,285; SCHAEDLER, 123; POGG 4,1266

SOMMELET, MARCEL [1877-1952]
GORIS, 571; TSCHIRCH, 1127; POGG 5,1187-1188; 6,2492; 7b,5021
J.A. Gautier, Ann. Univ. Paris 22 (1952), 556-561; Bull. Soc. Chim.
1953, pp.231-232; Ann. Pharm. Fran. 11 (1953), 473-478
G. Dillemann, Produits et Problèmes Pharmaceutiques 27 (1972), 921-922

SOMOGYI, MICHAEL [1883-1971]
AMS 7,2340
H. Walker, Ital. J. Biochem. 20 (1971), 124-127; Metabolism 21 (1972),
589-590

SONDEREGGER, LAURENZ [1825-1896]
DHB 6,271; STRAHLMANN, 465,479
E. Haffter, Dr. L. Sonderegger. Frauenfeld 1898

SONDERHOFF, ROBERT [1908-1937]
JV 49,763

SONDHEIMER, ALBERT [1876-1942]
BHDE 1,711

SONDHEIMER, FRANZ [1926-1981]
WW 1980, pp.2385-2386; AO 1981, pp.106-107; MSE 3,135-136;
BHDE 2,1094; CAMPBELL, 209-210; POTSCH, 400-401
Anon., Naturw. Rund. 34 (1981), 177; Chem. Z. 105 (1981), 155
R.A. Raphael, Chem. Brit. 18 (1982), 274
E. Jones and P. Garrett, Biog. Mem. Fell. Roy. Soc. 28 (1982), 505-536

SONN, ADOLF [1882-1957]
TLK 3,620; POGG 5,1190; 6,2494; 7a(4),441
Anon., Chem. Z. 81 (1957), 219

SONNE, CARL OLAF [1882-1948]
DBL(3rd Ed.) 13,551-553; FISCHER 2,1476; IB, 281

SONNE, WILHELM [1857-1941]
TLK 3,620; POGG 6,2494-2495; 7a(4),441

SONNEBORN, TRACY MORTON [1905-1981]
DSB 18,840-841; AMS 14,4801; WWA 1978-1979, pp.3057-3058; MH

1,435-437; STC 2,517-519; MSE 3,136-137; COHEN, 229; WWWS, 1579
B. Glass, Am. Phil. Soc. Year Book 1981, pp.509-518
D.L. Nanney, Ann. Rev. Gen. 15 (1981), 1-9; Genetics 102 (1982), 1-7
G.H. Beale, Biog. Mem. Fell. Roy. Soc. 28 (1982), 537-574
J. Sapp, Beyond the Gene, pp.87-122. New York 1987

SONNENFELD, EUGEN [1890-]
NUC 556,333

SONNENSCHEIN, CURT [1894-1986]
KURSCHNER 14,4035
Anon., Naturw. Rund. 40 (1987), 81

SONNENSCHEIN, FRANZ LEOPOLD [1817-1879]
HEIN, 639-640; BLOKH 2,701; TSCHIRCH, 1127; SCHAEDLER, 123-124; POTSCH, 401; FERCHL, 509; POGG 2,959; 3,1267; 7aSuppl.,648
H.D. Schwarz, Deutsche Apoth. Z. 107 (1967), 1004-1005

SORBY, HENRY CLIFTON [1826-1908]
DSB 12,542-546; DNB [1901-1911] 3,355-356; FREUND 3,409-416; WWWS, 1579; W, 483-484; POGG 3,1268-1269; 4,1415-1416; 6,2495-2496
J.W. Judd, Geological Magazine [5]5 (1908), 193-204
A. Geikie, Proc. Roy. Soc. B80 (1908), lvi-lxvi
J.O.A., Nature 77 (1908), 465-467
N. Higham, A Very Scientific Gentleman. Oxford 1963
R.H. Nuttall, Tech. Cult. 22 (1981), 275-280

SORET, JACQUES LOUIS [1827-1890]
DHB 6,278; STELLING-MICHAUD 5,595; GENEVA 3,31-34; 4,82-89; POTSCH, 401; POGG 3,1269-1270; 4,1416
A. Rilliet, Arch. Sci. Phys. Nat. 24 (1890), 305-346
L. de la Rive, Verhandl. Schw. Nat. Ges. 73 (1890), 251-255; Mem. Soc. Phys. Hist. Nat. Gen. 31 (1890-1891), xxviii-xxxv

SORG, FRANZ LOTHAR AUGUST [1773-1827]
HIRSCH 5,345; FERCHL, 509; POGG 2,961

SORG, KURT [1885-]
JV 40,610; SG [3]9,755; NUC 556,680

SORGE, HERMANN [1877-1958]
JV 31,300; NUC 556,685
J. Mader, Dr. Sorge-Report, p.528. Berlin 1984

ŠORM, FRANTISEK [1913-1980]
IWW 1976-1977, p.1625; STC 2,520-521; WWWS, 1580; POTSCH, 401-402
Anon., Leopoldina [3]8/9 (1962-1963), 80-81
V. Herout, Chem. Brit. 18 (1982), 116
K. Sebesta, Int. J. Pep. Res. 19 (1982), 436
D. Barton et al., Tetrahedron 38 (1982), 2223-2224

SOROKIN, VASILI IVANOVICH [1848-1919]
KAZAN 1,499-501; SKOROKHODOV, 127-133; RSC 18,857-858; POGG 4,1417
L.M. Litvichenko, Ukr. Khim. Zhur. 19 (1953), 337-340

SOUBEIRAN, EUGÈNE [1797-1858]
LAROUSSE 14,906; HIRSCH 5,347; GORIS, 572-575; BOURQUELOT, 73-74; BLOKH 2,701; FERCHL, 509-510; TSCHIRCH, 1127; POTSCH, 402; WWWS, 1581; POGG 2,962-963; CALLISEN 18,218-223; 32,370-372
A. Wurtz, J. Pharm. Chim. [3]36 (1859), 426-440
E. Robiquet, J. Pharm. Chim. [3]37 (1860), 39-55
M. Speter, Chem. Z. 55 (1931), 781-782
J. Cheymol and A. Soubeiran, Presse Med. 73 (1965), 505-508

SOUÈGES, RENÉ [1876-1967]
IB, 281
R. Heim, C.R. Acad. Sci. 265 (1967), 96-101
A. Lebegue, Bull. Soc. Bot. 114 (1967), 361-363
G. Deysson and J.L. Guignard, Ann. Pharm. Fran. 26 (1968), 391-404
A. Aubreville, Not. Acad. Sci. 5 (1972), 605-616

SOURKES, THEODORE LIONEL [1919-]
AMS 15(6),884; WWA 1980-1981, p.3117; WWWS, 1581

SOUTHWORTH, THOMAS SHEPARD [1861-1940]
RSC 18,864; SG [2]16,198; NUC 558,594
New York Times, 15 November 1940

SOWDEN, JOHN CLINTON [1910-1963]
AMS 10,3842
S.M. Cantor, Adv. Carbohydrate Chem. 20 (1965), 1-10

SOXHLET, FRANZ von [1848-1926]
FISCHER 2,1479; BLOKH 2,701; SCHAEDLER, 124; DZ, 1387; TLK 2,667; POGG 3,1271; 4,1417; 7aSuppl.,648
O. Rommel, Münch. med. Wchschr. 73 (1926), 994-995

SOYKA, ISIDOR [1850-1889]
DIEPGEN, 81-85; FISCHER 2,1479; MOTSCH, 116-124; KOERTING, 124-125
H. Sattler, Prag. med. Wchschr. 14 (1889), 103-104

SPÄTH, ERNST [1886-1946]
FARBER, 1553-1562; KNOLL, 55-57; PLANER, 320; TLK 3,622; POGG 5,1191; 6,2499-2500; 7a(4),445-447
F. Wessely, Alm. Akad. Wiss. Wien 97 (1947), 304-317; Ost. Chem. Z. 48 (1947), 58-65S

SPALTEHOLZ, WERNER [1861-1940]
BERLIN, 714; POGG 6,2501; 7a(4),449
H. Voss, Ber. Chem. Ges. 73A (1940), 72
M. Clara, Anat. Anz. 90 (1940), 102-111
C. Elze, Z. wiss. Mikr. 57 (1940), 1-4

SPANNAGEL, HANS OSKAR [1886-]
JV 27,398

SPEAKMAN, HORACE BRADBURY [1893-1974]
AMS 10,3847
A.E.R. Westman, Proc. Roy. Soc. Canada [4]13 (1975), 83-87

SPEAKMAN, JOHN BAMBER [1896-1969]
WW 1965, p.2876; POGG 6,2503-2504; 7b,5038-5042
Anon., Nature 144 (1939), 15-16; 223 (1969), 545
C.S. Whewell, Chem. Ind. 1970, p.336

SPECHT, HERMANN [1911-1966]
KURSCHNER 10,2360; POGG 7a(4),452-453
Anon., Chem. Z. 90 (1966), 675; Nachr. Chem. Techn. 14 (1966), 365

SPECK, JOHN FREDERICK [1920-1949]
AMS 8,2348

SPECK, KARL [1828-1916]
HIRSCH 5,356-357; PAGEL, 1627
P. Schenck, Sitz. Ges. Nat. Marburg 63 (1928), 193-212

SPECKETER, HEINRICH [1873-1933]
POTSCH, 402-403; POGG 6,2504
H. Suchy, Ber. chem. Ges. 66A (1933), 41
P. Siedler, Z. angew. Chem. 46 (1933), 239-240

SPEDDING, FRANK HAROLD [1902-1984]
AMS 14,4819; IWW 1982-1983, p.1253; MH 2,509-511; MSE 3,137-139; WWWS,1584
Anon., Chem. Eng. News 63(15) (1985), 78

SPEK, JOSEF [1895-1964]
KURSCHNER 9,1975; DRULL, 257; IB, 282
E.A. Arndt, Protoplasma 53 (1961), 291-293

E.A. Arndt and H. Penzilin, Zool. Anz. 1967, pp.610-612

SPEMANN, HANS [1869-1941]
DSB 12,567-569; FISCHER 2,1481; SCHWERTE 2,228-236; ABBOTT, 120; IB, 282; W, 486-487; WWWS, 1584
O. Mangold, Naturwiss. 17 (1929), 453-478; Arch. Entwickl. 141 (1942), 385-423; Hans Spemann. Stuttgart 1953
H. Bautzmann, Jahrb. Morph. 87 (1942), 1-26
F. Baltzer, Naturwiss. 30 (1942), 229-239
H. Spemann, Forschung und Leben (F.W. Spemann, Ed.). Stuttgart 1943
K. von Frisch, Bayer. Akad. Wiss. Jahrbuch 1944-1948, pp.242-244
V. Hamburger, Experientia 25 (1969), 1121-1125; The Heritage of Experimental Embryology: Hans Spemann and the Organizer. Oxford 1988
J.F. Holtfreter, Dev. Biol. 5 (1988), 127-150
R.G. Rinard, J. Hist. Biol. 21 (1988), 95-116; Synthese 91 (1992), 73-91

SPENCE, DAVID [1881-1957]
AMS 8,2350; MILES, 449-450; JV 20,284; NUC 561,253-254; POGG 5,1192
Anon., Chem. Ind. 1957, p.1423; Science 126 (1957), 744

SPENCER, HERBERT [1820-1903]
DSB 12,569-572; DNB [1901-1911] 3,360-359; W, 487; WWWS, 1583
H. Spencer, An Autobiography. London 1904
D. Duncan, Life and Letters of Herbert Spencer. New York 1908
H.S.R. Elliott, Herbert Spencer. London 1917
G. Sarton, Isis 3 (1921), 375-390

SPENCER, JAMES FREDERICK [1881-1950]
WW 1949, p.2616; SRS, 31; POGG 6,2505; 7b,5044
V.C.G. Trew and E.E. Turner, J. Chem. Soc. 1955, pp.3311-3312

SPENCER, THOMAS [1793-1857]
KELLY, 1146-1147; FERCHL, 510; NUC 561,409; POGG 2,970

SPENGLER, ALEKSANDR EDUARDOVICH [1865-]
FISCHER 2,1482; SG [2]16,238

SPERANSKI, ALEKSEI DMITRIEVICH [1888-1961]
BSE 40,282; BME 30,1178-1179; WWR, 523
D.F. Pletsiki, A.D. Speranski. Moscow 1967

SPERANSKI, GEORGI NESTOROVICH [1873-1969]
BSE 40,282; BME 30,1179-1180; WWR, 523; WWWS, 1585
M.I. Studenkin et al., Pediatria 51 (1972), 11-13,32-36
M.P. Matveev, Klin. Med. 51(8) (1973), 143-146

SPERANSKI, VLADIMIR NIKOLAEVICH [1884-]
NUC 561,503
M.M. Klavdeeva, Vop. Ist. Est. Tekhn. 1989(1), pp.57-62

SPERLING, RUDOLF [1888-1914]
JV 30,523; NUC 561,529
Anon., Chem. Z. 39 (1915), 145; Ber. chem. Ges. 48 (1915), 731

SPERRY, WARREN MYRON [1900-1990]
AMS 12,6015
New York Times, 21 July 1990
M.M. Rapport, J. Neurochem. 56 (1991), 1820-1821

SPETER, MAX [1883-1942]
TLK 2,669; WWWS, 1586; POGG 6,2506-2507; 7a(4),459-460
M.E. Weeks, Isis 34 (1943), 340-344

SPEYER, EDMUND [1878-1942?]
GEDENKBUCH 2,1422; LDGS, 25; POGG 6,2507-2508; 7a(4),460

SPICA, PIETRO [1854-1929]
EI 32,369; FISCHER 2,1483; TSCHIRCH, 1128; REBER, 175-176; POGG 4,1420; 5,1193; 6,2508
R.L. Vanzetti, Chimica e Industria 11 (1929), 324-325
A. Miolati, Ber. chem. Ges. 62A (1929), 87

E. Mameli, Gazz. Chim. Ital. 69 (1939), 479-498

SPIEGEL, ADOLF [1856-1938]
POTSCH, 403; POGG 4,1421; 6,2508
L. Anschütz, Ber. chem. Ges. 73A (1940), 31-38

SPIEGEL, ERNST ADOLF [1895-]
KURSCHNER 4,2845; IB, 283; WININGER 7,456

SPIEGEL, LEOPOLD [1865-1927]
TSCHIRCH, 1128; TLK 2,670; POGG 4,1421; 5,1193; 6,2508
A. Rosenheim, Ber. chem. Ges. 60A (1927), 41-44

SPIEGEL-ADOLF, MONA (ANNA SIMONA) [1893-1983]
AMS 12,6017; FISCHER 2,1484; KOREN, 251; KAGAN(JA), 740; COHEN, 231; WWWS, 1586; POGG 6,2508-2509; 7b,5068-5071
Anon., J. Am. Med. Assn. 253 (1985), 97

SPIEGELMAN, SOLOMON [1914-1983]
AMS 15(6),906-907; WWA 1980-1981, p.3130; IWW 1982-1983, p.1255; SCHULZ-SCHAEFFER, 214-215; CB 1983, p.474; COHEN, 231-232
New York Times, 22 January 1983

SPIEGLER, EDUARD [1860-1908]
BJN 13,90*
I. Pollak, Chem. Z. 32 (1908), 729

SPIELMANN, CHARLES FRÉDÉRIC [1789-1854]
BERGER-LEVRAULT, 230

SPIELMANN, PERCY EDWARD [1881-1964]
WININGER 2,587; POGG 6,2510; 7b,5071-5072
Who Was Who 1961-1970, p.1064

SPIERS, HENRY MICHAEL [1893-]
NUC 561,694-695
Directory of British Scientists 1966-1967, vol.1, p.561

SPIES, JOSEPH REUBEN [1904-]
AMS 14,4829; WWWS, 1587

SPIES, TOM DOUGLAS [1902-1960]
AMS 9(II),1067; NCAB 43,320-321; WWAH, 570; WWWS, 1587
Anon., Science 131 (1960), 819; J. Am. Med. Assn. 172 (1960), 2101-2102
W.B. Bean, Trans. Assoc. Am. Phys. 73 (1960), 36-38
T.H. Jukes, J. Nutrition 102 (1972), 1395-1399

SPIETHOFF, HEINRICH [1854-]
BERLIN, 612; SG [1]13,392; NUC 562,34

SPILKER, ADOLF [1863-1954]
HEIN-E, 416; TLK 3,623; POGG 6,2510-2511; 7a(4),460
H. Ihlder, Z. angew. Chem. 46 (1933), 457-459
Anon., Oel Kohle Erdoel Teer 14 (1938), 517-518; Chem. Z. 77 (1953), 484; 78 (1954), 302; Nachr. Chem. Techn. 2 (1954), 94

SPILLING, EMIL [1857-]
BERLIN, 579; SG [1]13,393; NUC 562,64

SPILLMAN, WILLIAM JASPER [1863-1931]
DAB 17,458-459; AMS 4,925
L.P.V. Johnson and R. Spillman, J. Heredity 39 (1948), 247-254

SPINA, ARNOLD [1850-1918]
NUC 562,82

SPIRGATIS, (JOHANN JULIUS) HERMANN [1822-1899]
HEIN, 645; TSCHIRCH, 1128; RSC 5,775; 8,989-990; POGG 3,1272; 4,1421

SPIRO, KARL [1867-1932]
FISCHER 2,1486; IB, 283; KOREN, 251; WININGER 5,596; 7,459; TETZLAFF, 316; WWWS, 1588; POGG 6,2511
L. Asher, Erg. Physiol. 34 (1932), 1-17
F. Leuthardt, Koll. Z. 59 (1932), 257-263
A. Roos and W.F. Boron, Respiration Physiology 40 (1980), 1-32

SPIRO, PETR ANTONOVICH [1844-1893]
VORONTSOV, 170-173; RSC 12,697
V.R. Faitelberg-Blank, Fiziol. Zhur. 1 (1955), 140-143

SPIRO, ROBERT GUNTER [1929-]
AMS 15(6),911; BHDE 2,1103; WWWS, 1588

SPITZER, WILHELM [1865-1901]
BJN 6,102*
W. Pagel, Virchows Jahresber. 1901(I), p.394
F. Rohmann, Jahresb. Schles. Ges. 79 (1901), 12-13

SPOEHR, HERMANN AUGUSTUS [1885-1954]
AMS 8, NCAB 43,171-172; IB, 283; WWAH, 571; WWWS, 1589; POGG 6,2514; 7b,5072-5073
W. Stiles, Nature 174 (1954), 534-535
G.W. Beadle, Am. Phil. Soc. Year Book 1954, pp.450-454
J.H.C. Smith and C.S. French, Ann. Rev. Biochem. 24 (1955), xi-xvi
C.S. French, Ber. bot. Ges. 68a (1955), 21-25

SPONSLER, OLENUS LEE [1879-1953]
AMS 8,2357; POGG 6,2515; 7b,5073
C. Epling and K.C. Hamner, Plant Physiology 28 (1953), 553-554

SPOONER, EDWARD TENNEY CASSWELL [1904-]
WW 1981, p.2445

SPRENGEL, HERMANN JOHANN PHILIPP [1834-1906]
W, 488-489; WWWS, 1590; POGG 3,2175; 5,1194; 7aSuppl.,650
R. Messel, J. Chem. Soc. 91 (1907), 661-663
K.R. Webb, Chem. Brit. 1965, pp.569-571

SPRENGEL, KARL [1787-1859]
ADB 35,293; ALBRECHT, 87; TSCHIRCH, 1128; FERCHL, 512; POGG 2,976-977

SPRENGER, GUSTAV [1878-1959]
JV 16,146; NUC 562,626

SPRING, FRANK STUART [1907-]
WW 1990, p.1712

SPRING, WALTHÈRE VICTOR [1848-1911]
DSB 12,592-594; BNB 32,675-678; BLOKH 2,702; POGG 3,1275-1276; 4,1423-1424; 5,1194
F. Swarts, Chem. Z. 35 (1911), 949-950
Anon., Nature 87 (1911), 252-253
F. Schwers, J. Chem. Soc. 101 (1912), 692-696
L. Crismer, Bull. Soc. Chim. Belg. 26 (1912), 157-185
F. Lionetti and M. Mager, J. Chem. Ed. 28 (1951), 604-605

SPRINGALL, HAROLD DOUGLAS [1910-1982]
Who's Who of British Scientists 1971-1972, p.797
I.T. Miller, Chem. Brit. 20 (1984), 55

SPRINGER, GEORGE FERDINAND [1924-]
AMS 15(6),921; WWA 1980-1981, p.3136; WWWS, 1590

SRB, ADRIAN MORRIS [1917-]
AMS 15(6),926; WWA 1980-1981, p.3138; IWW 1982-1983, p.1258

SREENIVASAYA, MONTAHALLI [1895-1969]
IB, 283; NUC 266,321

STAAB, HEINZ [1926-]
KURSCHNER 12,3085; POTSCH, 403-404; POGG 7a(4),473

STACEY, MAURICE [1907-]
WW 1981, p.2540; IWW 1982-1983, p.1258

STACEY, REGINALD STEPHEN [1905-1974]
WW 1973, pp.3051-3052
Anon., Lancet 1974(I), p.371

STADELMANN, ERNST [1853-1941]
FISCHER 2,1490-1491; PAGEL, 1636; LEVITSKI 2,146-148; WREDE, 185; RSC 18,895-896
J. Brennsohn, Die Aerzte Livlands, pp.379-380. Mitau 1905

STADIE, WILLIAM CHRISTOPHER [1886-1959]
AMS 9(II),1071; POGG 6,2518; 7b,5081-5082
F.W. Lukens, Diabetes 8 (1959), 476-478; Trans. Assoc. Am. Phys. 73 (1960), 39-41
I. Starr, Trans. Coll. Phys. Phila. [3]28 (1960), 50-51; Biog. Mem. Nat. Acad. Sci. 58 (1989), 513-528
J. Stokes, Am. Phil. Soc. Year Book 1962, 176-178

STADLER, LEWIS JOHN [1896-1954]
AMS 8,2363; NCAB 45,63-64; IB, 284; WWAH, 573
M.M. Rhoades, Science 120 (1954), 553-554; Genetics 41 (1956), 1-3; Biog. Mem. Nat. Acad. Sci. 30 (1957), 329-347
H. Roman, Am. Phil. Soc. Year Book 1957, pp.162-164; Genetics 119 (1988), 739-741
G.P. Redei, Stadler Symposium, pp.5-20. Columbia, Mo. 1971

STADLER, PAUL [1901-1984]
NUC 563,464
Anon., Nachr. Chem. Techn. 33 (1985), 247

STADLMAYR, FRANZ [1872-1932]
HEIN, 649-650; POGG 6,2518-2519
A. Binz, Ber. chem. Ges. 66A (1933), 31

STADNIKOV, GEORGI LEONTOVICH [1880-1974]
POTSCH, 404; POGG 6,2519; 7b,5083-5084
A.I. Kamneva, Khimia Tverdogo Topliva 1980(1), pp.151-152

STADT, HENDRIK JUSTUS van de [1868-1954]
RSC 18,897
Kamper Almanak 1934, p.132
Anon., Chem. Wkbl. 50 (1954), 293

STADTHAGEN, MAX [1850-]
RSC 18,897; NUC 563,475

STADTMAN, EARL REECE [1919-]
AMS 15(6),931; WWA 1980-1981, p.3140; IWW 1982-1983, p.1258; WWWS, 1592

STADTMAN, THRESSA CAMPBELL [1920-]
AMS 15(6),931STAEDEL, WILHELM [1843-1919]
DBJ 2,734; BLOKH 2,703; DZ, 1398; SCHAEDLER, 124-125; POTSCH, 404; POGG 3,1278-1279; 4,1428; 5,1195
L. Wöhler, Ber. chem. Ges. 52A (1919), 109-114; Chem. Z. 43 (1919), 393-394

STAEDELER, GEORG [1821-1871]
ADB 35,778-780; BLOKH 2,703-704; SCHAEDLER, 125; ZURICH, 17-20; POGG 2,978-979; 3,1279
K. Kraut, Ber. chem. Ges. 4 (1871), 425-428

STÄHELIN, CHRISTOPH [1804-1870]
STELLING-MICHAUD 5,609-610; RSC 5,789; POGG 2,979; 3,1279

STÄHLER, ARTHUR [1877-1950]
TLK 3,626; POGG 5,1195; 6,2520; 7a(4),480
Anon., Z. angew. Chem. 62 (1950), 470

STAHEL, RUDOLF [1866-]
JV 7,264; RSC 14,1010; NUC 563,616

STAHL, EGON [1924-1986]
KURSCHNER 14,4079-4080; POGG 7a(4),481-482
K. Macek, J. Chromat. 391 (1987), 1-2

STAHL, FRANKLIN WILLIAM [1929-]
AMS 15(6),933

STAHL, OTTO [1887-]
KURSCHNER 4,2863; FISCHER 2,1492

STAHLSCHMIDT, ALEX [1882-1966]
JV 26,53; NUC 563,672

STAHLSCHMIDT, FRIEDRICH [1831-1902]
BLOKH 2,704; POGG 2,981; 3,1279; 4,1429
Anon., Chem. Z. 26 (1902), 861; Z. angew. Chem. 15 (1902), 977-978; Ber. chem. Ges. 35 (1902), 3364; Leopoldina 38 (1902), 108-109

STAKMAN, ELVIN CHARLES [1885-1979]
AMS 13,4261; IB, 284; WWWS, 1593
J.G. Harrar, Am. Phil. Soc. Year Book 1979, pp.107-112
L.C. Cochran and E.P. Imle, Phytopathology 69 (1979), 195
C.M. Christensen, E.C. Stakman, Statesman of Science. St. Paul, Minn. 1984; Biog. Mem. Nat. Acad. Sci. 61 (1992), 331-349

STALKER, HARRISON DAILEY [1915-1982]
AMS 15(6),936; WWWS, 1593
H.L. Carson and J.V. Neel, Genetics 107 (1984), s139-s140

STAMM, JOHANNES [1881-1969]
KURSCHNER 10,2382-2383; HEIN-E, 420-421; WELDING, 758-759; IB, 284; TSCHIRCH, 1129; POGG 7a(4),484
Anon., Chem. Z. 80 (1956), 680; 93 (1969), 907

STAMM, WILHELM [1813-1902]
Anon., Pharm. Z. 47 (1902), 691

STAMMER, KARL [1828-1893]
POGG 2,982-983; 3,1280-1281
Anon., Leopoldina 29 (1893), 160

STANDFAST, ARTHUR FRANCIS BULMER [1905-]
Who's Who in British Science 1953, p.249

STANEK, JAN [1826-1868]
B. Hayek et al., Acta Hist. Rerum Not., Special Issue 9 (1977), 118-119; Chem. Listy 72 (1978), 1249-1256

STANEK, VLADIMIR [1879-1940]
POGG 6,2521-2523; 7b,5091-5093
J. Vondrák et al., Chem. Listy 33 (1939), 121-128

STANGASSINGER, RICHARD [1878-1965]
JV 21,276; NUC 564,428

STANGE, OTTO [1870-1941]
JV 10,216; RSC 19,74
Anon., Z. angew. Chem. 53 (1940), 535; 54 (1941), 452

STANIER, ROGER YATE [1916-1982]
AMS 12,6058; MH 2,513-514; STC 2,525-526; MSE 3,145-146
R.Y. Stanier, Ann. Rev. Microbiol. 34 (1980), 1-48
J.G. Morris, J. Gen. Microbiol. 129 (1983), 255-261
P.H. Clarke, Biog. Mem. Fell. Roy. Soc. 32 (1986), 543-568

STANKOVICH, RADENKO [1880-1956]
SG [3]9,895
New York Times, 7 December 1956

STANLEY, WENDELL MEREDITH [1904-1971]
DSB 18,841-848; AMS 11,5116; NCAB G,410-411; MILES, 451-452; WWWS, 1595; MH 1,440-441; STC 2,526-527; MSE 3,146-147; CB 1947, pp.604-607; DAMB, 712-713; POTSCH, 405; WWAH, 574; POGG 6,2523; 7b,5093-5097
Anon., Nature 233 (1971), 149-150; Chem. Eng. News 49(26) (1971), 35
J.T. Edsall, Am. Phil. Soc. Year Book 1971, pp.184-190
H. Fraenkel-Conrat, Welch Foundation Conferences on Chemical Research 20 (1977), 254-260
L.E. Kay, Isis 77 (1986), 450-472

STANNIUS, HERMANN FRIEDRICH [1808-1883]
DSB 12,611-612; ADB 35,446-448; HIRSCH 5,390; PAGEL, 1638; DECHAMBRE [3]11,433-434; CALLISEN 18,302-304; 32,412-414
W. Stieda, Jahrb. Ver. Meckl. Gesch. 93 (1929), 1-36

STARE, FREDERICK JOHN [1910-]
AMS 14,4862; WWA 1978-1978, p.3093; WWWS, 1595
F.J. Stare, Annual Review of Nutrition 11 (1991), 1-20

STARK, KARL WILHELM [1787-1845]
ADB 35,491-492; HIRSCH 5,394; DECHAMBRE [3]11,518; CALLISEN 18,311-313; 32,415-416

STARKENSTEIN, EMIL [1884-1942]
KURSCHNER 4,2873-2874; FISCHER 2,1494; HARTMANN, 103-118; BHDE 2,1108; TSCHIRCH, 1129; IB, 285; KOERTING, 135; KOREN, 251; WININGER 5,603; TETZLAFF, 317; TLK 3,627; POGG 7a(4),488-490
C. Neuberg, Exp. Med. Surg. 4 (1946), 184-185
R. Junkmann, Arch. exp. Path. Pharm. 204 (1947), 13-19
M. Matowski and J. Kok, Arzneimitt. 14 (1964), 1367-1368
K.E. Senius, Arch. exp. Path. Pharm. 328 (1984), 95-102
I.M. Hais, J. Chromat. 376 (1986), 5-9

STARKEY, ROBERT LYMAN [1899-]
AMS 11,5122; IB, 285

STARLING, ERNEST HENRY [1866-1927]
DSB 12,617-619; DNB [1922-1930],807-809; FISCHER 2,1494-1495; W, 490-491; O'CONNOR(2), 139-149; MEDVEI, 799-801; ABBOTT, 120-121; WWWS, 1596; POTSCH, 405-406
C.J. M[artin], Proc. Roy. Soc. B102 (1928), xvii-xxvii
M. von Frey, Münch. med. Wchschr. 74 (1927), 878-879
C.L.E., Biochem. J. 22 (1928), 618-620
R. Colp, J. Hist. Med. 7 (1952), 280-294
E.B. Verney, Ann. Sci. 12 (1956), 30-47
C.B. Chapman, Ann. Int. Med. 57 Suppl.2 (1962), 1-43
L.G. Wilson, Episteme 2 (1968), 3-25
A.V. Hill, J. Physiol. 204 (1969), 1-13
H.H. Simmer, Med. Welt 29 (1978), 1991-1996

STARR, ISAAC [1895-1989]
AMS 12,6069; WWA 1966-1967, p.2031; IWW 1982-1983, p.1262; ICC, 303-304
E.A. Stead, New Eng. J. Med. 300 (1979), 930-931
New York Times, 28 June 1989

STARY, ZDENKO [1899-1968]
KURSCHNER 10,2388; IB, 285; KOERTING, 123; POGG 6,2524-2525; 7b,5098-5100
Anon., Nature 218 (1968), 1283-1284; Chem. Z. 92 (1968), 556

STAS, JEAN SERVAIS [1813-1891]
DSB 12,619-620; BNB 23,654-684; BLOKH 2,705-706; W, 491; WWWS, 1597; ABBOTT-C,133-134; FERCHL, 515; TSCHIRCH, 1129; SCHAEDLER, 127; POTSCH, 406; POGG 2,986-987; 3,1282; 4,1431-1432
A.W. Hofmann, Ber. chem. Ges. 25 (1892), 1-14
Anon., Nature 46 (1892), 81-83,130-131
W. Spring, Ann. Acad. Roy. Belg. 59 (1893), 217-376
A. Scott, J. Chem. Soc. 111 (1917), 288-312
A. Timmermans, J. Chem. Ed. 15 (1938), 353-356

STATHER, FRIEDRICH [1901-1974]
KURSCHNER 12,3104; POGG 6,2525; 7a(4),491-494

U. Freimuth et al., Jahrb. Sachs. Akad. 1973-1974, pp.252-274

STAUB, HANS [1890-1967]
KURSCHNER 10,2389; STC 3,14-15; FISCHER 2,1496
R. Nissen, Schw. med. Wchschr. 97 (1967), 1157-1158

STAUB, MAXIMILIAN [1899-]
SBA 3,131; POGG 7a(4),495

STAUBACH, FRANZ [1885-1958]
JV 24,368; NUC 565,472

STAUCH, HANS [1871-]
JV 13,156; NUC 565,475

STAUDENMAIER, LUDWIG [1865-1933]
JV 11,235; POGG 4,1432; 6,2525
Anon., Ber. chem. Ges. 66A (1933), 73

STAUDINGER, HANSJÜRGEN [1914-1990]
KURSCHNER 13,3789; POGG 7a(4),498-499
Anon., Chem. Z. 103 (1979), 375; J. Clin. Chem. 18 (1980), 929-936
H. Kersten, in Naturwissenschaft und Medizin (R. Kattermann, Ed.), pp.125-145. Mannheim 1985
Anon., Chem. Z. 114 (1990), 114-115

STAUDINGER, HERMANN [1881-1965]
DSB 13,1-4; BAD NF2,265-267; SCHWERTE 1,199-206; STC 3,15-16; MSE 3,150; CB 1954, pp.588-589; TLK 3,628; W,491-492; ABBOTT-C, 132-133; POTSCH, 406; WWWS, 1597; POGG 5,1199-1200; 6,2525-2527; 7a(4),499-503
H. Batzer et al., Chem. Z. 75 (1951), 159-163
W. Kern, Z. angew. Chem. 63 (1951), 229-231
W. Quarles, J. Chem. Ed. 28 (1951), 120-122
G.S. Whitby, Chem. Z. 91 (1967), 696-700
H. Staudinger, Arbeitserinnerungen. Heidelberg 1961 (English translation, New York 1970)
V.E. Yarsley, Chem. Ind. 1967, pp.250-271
D.H. Napper, Chemistry in Australia 48 (1981), 384-387
Y. Furukawa, Historia Scientarum 22 (1982), 1-18
I. Strube, NTM 24 (1987), 87-92STAUDT, WALTER [1895-1928]
JV 38,419
Anon., Ber. chem. Ges. 61A (1928), 78

STAVENHAGEN, ALFRED [1859-1931]
TLK 3,628-629; POGG 4,1432-1433; 5,1200; 6,2527
H. Wolbling, Ber. chem. Ges. 64A (1931), 10-12; Chem. Z. 55 (1931),145; Z. angew. Chem. 44 (1931), 178

STEACIE, EDGAR WILLIAM RICHARD [1900-1962]
DSB 13,6-7; STC 3,16-17; WWAH, 575; WWWS, 1598; POTSCH, 407; POGG 6,2527; 7b,5100-5105
Anon., Nature 196 (1962), 110-111
H.E. Gunning, Proc. Chem. Soc. 1964, pp.73-79
L. Marion, Biog. Mem. Fell. Roy. Soc. 10 (1964), 257-281
M.C. King, E.W.R. Steacie and Science in Canada. Toronto 1989

STEARN, ALLEN EDWIN [1894-]
AMS 12,6075; IB, 285; WWWS, 1598; POGG 6,2527-2528; 7b,5105-5106

STEARNS, GENEVIEVE [1892-]
AMS 11,5130; BERG, 66-67; WWWS, 1598

STEBBINS, GEORGE LEONARD [1906-]
AMS 14,4871; IWW 1982-1983,p.1264; MH 2,516-517; MSE 3,150-152; WWWS,1598

STECHE, ALBERT [1862-]
JV 2,246; RSC 14,1010; NUC 566,3

STEDMAN, EDGAR [1890-1975]
WW 1974, p.3106; POGG 6,2528; 7b,5108-5109

H.J. Cruft, Biog. Mem. Fell. Roy. Soc. 22 (1976), 529-553

STEELE, JOHN MURRAY [1900-1969]
AMS 11,5133-5134; WWAH, 576; WWWS, 1599
C.M. MacLeod, Trans. Assoc. Am. Phys. 84 (1971), 41-42

STEENBOCK, HARRY [1886-1967]
DSB 18,849-851; AMS 11,5135; DAMB, 715; WWAH, 576; WWWS, 1599; POGG 6,2529; 7b,5110-5112
H.A. Schneider, J. Nutrition 103 (1973), 1234-1247
R.D. Apple, Isis 80 (1989), 375-394

STEENSTRUP, JOHANN JAPETUS [1813-1897]
DSB 13,9-10; HIRSCH 5,400-401; PAGEL, 1639-1640; W, 482; WWWS, 1599
Anon., Leopoldina 33 (1897), 114; Proc. Linn. Soc. 110 (1898), 50-51;

Sitz. Bayer. Akad. 28 (1899), 476-478
R. Spärch, Vid. Med. Nat. For. 95 (1933), 56-90

STEFFENS, HENRIK [1773-1845]
ADB 35,555-558; HIRSCH 5,402; VEIBEL 2,404; KIEL, 203; WWWS, 1600; CALLISEN 18,332-335; 32,421; POGG 2,988-989; 7aSuppl.,655-657
E.R. Hussy, Schlesische Lebensbilder 3 (1931), 264-280

STEGER, ALPHONSUS MARIA ANTONIUS ALOYSIUS [1874-1953]
RSC 18,927; POGG 5,1202; 6,2530; 7b,5113-5114
J. van Loon, Chem. Wkbl., 34 (1937), 429-431
Wie is Dat? 1948, p.479
Anon., Chem. Wkbl. 49 (1953), 699

STEGGERDA, FREDERIC RUSSELL [1903-1971]
AMS 11,5138; WWWS, 1600

STEIBELT, WERNER [1891-1932]
JV 31,335; NUC 566,424
Anon., Chem. Z. 57 (1933), 726

STEIGER, EDUARD [1859-1922]
DHB 6,341-342; ZURICH-D, 150; RSC 18,927-928
H. Rehsteiner, Jahrbuch der St. Gallischen Naturwissenschaftlichen Gesellschaft 58(I) (1922), 74-76

STEIGER, ROBERT [1900-]
NUC 566,451

STEIN, HEINRICH WILHELM [1811-1889]
TSCHIRCH, 1129; RSC 5,810-811; 8,1004-1006; POGG 2,992-993; 3,1286
Anon., Leopoldina 25 (1889), 219

STEIN, MAX [1860-1942]
GEDENKBUCH 2,1437; JV 3,276

STEIN, RICHARD [1877-1922]
JV 20,475; NUC 566,556
Anon., Chem. Z. 46 (1922), 412

STEIN, SIEGMUND THEODOR [1840-1891]
KALLMORGEN, 422
Anon., Pharm. Z. 36 (1891), 621

STEIN, WERNER [1913-]
KURSCHNER 13,3803; POGG 7a(4),512

STEIN, WILLIAM HOWARD [1911-1980]
DSB 18,851-855; AMS 14,4882; WWA 1978-1979, p.3106; AO 1980, pp.75-77; ETTRE, 297-308; STC 3,18-19; MSE 3,153-154; WWWS, 1601; POTSCH, 407
S. Moore, J. Biol. Chem. 255 (1980), 9517-9518; Biog. Mem. Nat. Acad. Sci. 56 (1987), 415-440

STEINACH, EUGEN [1861-1944]
BHDE 2,1113; FISCHER 2,1499; LOMMATZSCH, 116-122; IB, 286; WWWS, 1601
J. Novak, Wiener med. Wchschr. 81a (1931), 173-174
H. Benjamin, Sci. Mon. 61 (1945), 427-442
E. Harms, Bull. N.Y. Acad. Med. 45 (1969), 761-766

STEINBACH, HENRY BURR [1905-1981]
AMS 14,4882; WWA 1980-1981, p.3163
D.W. Bishop, Society of General Physiologists Series 36 (1981), xi-xiv

STEINBOCK, HERMANN [1873-]
JV 15,15; NUC 566,646

STEINDORFF, KURT [1875-1942]
KURSCHNER 5,1351; NUC 566,673

STEINER, ANTAL [1840-1905]
RSC 11,486; 17,932
Anon., Ber. chem. Ges. 38 (1905), 3212

STEINER, DONALD FREDERICK [1930-]
AMS 15(6),974; WWA 1980-1981, p.3162

STEINER, ISIDOR [1849-1914]
DBJ 1,313; FISCHER 2,1500; PAGEL, 1644-1655; KOVACSICS, 96-98; KOREN, 262

STEINER, MAXIMILIAN [1904-]
KURSCHNER 13,3811; WAGENITZ, 173; POGG 7a(4),515-516

STEINER, ROBERT FRANK [1926-]
AMS 15(6),975; WWA 1980-1981, p.3163; WWWS, 1602

STEINGROEVER, JOSEPH [1884-]
JV 23,44; NUC 567,99
Anon., Z. angew. Chem. 30(III) (1917), 518

STEINHARDT, JACINTO [1906-1985]
AMS 14,4886

STEINHEIL, KARL AUGUST [1801-1870]
DSB 13,22-23; ADB 35,720-724; GLOSAUER, 140-154; FERCHL, 516; POGG 2,996-998; 3,1288; 7aSuppl.,659-661
Anon., Alm. Bayer. Akad. Wiss. 1867, pp.259-268
F. Seidel, Alm. Akad. Wiss. Wien 21 (1871), 205-222
H. Pieper, Technikgeschichte 31 (1970), 323-352

STEINKOPF, WILHELM [1879-1949]
TLK 3,631; POGG 5,1203-1204; 6,2534; 7a(4),521

STEINLE, RUDOLF [1877-]
JV 25,366STEINMANN, JOSEPH JOHANN [1779-1833]
WURZBACH 38,143-145; HEIN, 654; WRANY, 185-187; TSCHIRCH, 1129; FERCHL, 516POGG 2,999; CALLISEN 18,367-368; 32,429

STEINMETZ, HERMANN [1879-1964]
POGG 5,1204; 6,2535; 7a(4),522
G. Steinmetz, Bayer. Akad. Wiss. Jahrbuch 1966, pp.203-206

STEKOL, JACOB A. [1905-1969]
AMS 11,5149
Anon., Chem. Eng. News 47(14) (1969), 51

STELLER, WILHELM [1870-1929]
JV 12,183; NUC 567,269
Anon., Chem. Z. 53 (1929), 751

STELZNER, ROBERT [1869-1943]
TLK 3,633; POGG 6,2536; 7a(4),524
M. Pflücke and W. Merz, Ber. chem. Ges. 76A (1943), 18-20

STENBERG, STEN [1824-1884]
SMK 7,196-197; HIRSCH 5,414-415; SG [1]13,661-662
Anon., Hygeia 46 (1884), 535-546

STENDER, HEDWIG [1886-]
JV 35,438; NUC 567,336

STENHAGEN, EINAR AUGUST [1911-1973]
VAD 1973, p.923
B.A. Andersson, Progress in Chemistry of Fats and Lipids 1978(16), pp.1-7

STENHOUSE, JOHN [1809-1880]
DNB 18,1036; BLOKH 2,705; DESMOND, 582; TSCHIRCH, 1129; SCHAEDLER, 128; FERCHL, 517; POGG 2,1001-1002; 3,1290-1291
H.E. Roscoe, J. Chem. Soc. 39 (1881), 185-188
Anon., Nature 23 (1881), 244-245; Proc. Roy. Soc. 31 (1881), xix-xxi

STENT, GUNTHER SIEGMUND [1924-]
AMS 15(6),982; WWA 1980-1981, p.3166; BHDE 2,1120

STENZL, HANS KARL MARIA [1880-1980]
JV 24,583; NUC 567,423

STEPHEN, HENRY [1889-1965]
SRS, 52; POGG 6,2539-2540; 7b,5137-5138
Anon., Tetrahedron 22 Suppl.7 (1966), v-vi

STEPHENSON, MARJORY [1885-1948]
DSB 18,857-860; DNB [1941-1950], 835-836
S.R. Elsden and N.W. Pirie, J. Gen. Microbiol. 3 (1949), 329-339
M. Robertson, Obit. Not. Fell. Roy. Soc. 6 (1949), 563-575
D.D. Woods, Biochem. J. 46 (1950), 377-383
R.E. Kohler, Isis 76 (1985), 162-181
J. Mason, Notes Roy. Soc. 46 (1992), 279-300

STEPP, WILHELM [1882-1963]
KURSCHNER 6(II),856-857; WWWS, 1606; POGG 6,2540; 7a(4),525-526
J. Kühnau, Deutsche med. Wchschr. 89 (1964), 1911-1912

STERN, ADOLPH [1900-]
AMS 14,4897; WWWS, 1606

STERN, CURT [1902-1981]
DSB 18,860-867; AMS 14,4897; WWA 1976-1977, p.3013; MH 1,445-446; MSE 3,155-156; IB, 287; AO 1981, pp.663-664; BHDE 2,1122; LDGS,15; SCHULZ-SCHAEFFER,208-209; WWWS, 1606
C. Stern, in Chromosomes and Cancer (J. German, Ed.), pp.xiii-xxv. New York 1974
B. Glass, Am. Phil. Soc. Year Book 1982, pp.514-520
J.V. Neel, Ann. Rev. Gen. 17 (1983), 1-10; Biog. Mem. Nat. Acad. Sci. 56 (1987), 443-473
J.C. Lucchesi, Genetics 103 (1983), 1-4

STERN, FRITZ [1902-]
LDGS, 62; NUC 568,80

STERN, JOSEPH RICHARD [1919-1974]
AMS 12,6111
Anon., Science 185 (1974), 340; Chem. Eng. News 52(10) (1974), 35

STERN, KURT GUENTER [1904-1956]
DAB [Suppl.6],595-596; AMS 9(I),1858; BHDE 2,1127; LDGS,60; COHEN, 235; KAGAN(JA), 743; TETZLAFF, 322-323; WWAH, 578; WWWS, 1606;POGG 6,2541; 7a(4),526-528
H. Mark, Nature 177 (1956), 556
Anon., Chem. Z. 80 (1956), 255; Chem. Eng. News 34 (1956), 1360

STERN, LINA SALOMONOVNA [1878-1968]
BSE 48,196; BME 34,1063-1064; GENEVA 5,392-394; 6,373-377; 7,309-311
D. Kvasov, Fiziol. Zhur. 45 (1959), 199-203

STERN, RUDOLF [1895-1962]
KURSCHNER 4,1907; BHDE 2,1123; LDGS, 69; KOREN, 253
H. Biberstein, Proc. Virchow Med. Soc. 22 (1963), 46-49

STERNBACH, LEO HENRYK [1908-]
AMS 14,4899
P.J.T. Morris, CHOC News 4(1) (1986), 6-8

STERNBERG, CARL [1872-1935]
KURSCHNER 4,2908; FISCHER 2,1505-1506; TETZLAFF, 324
R. Paltauf, Verhandl. path. Ges. 29 (1937), 417-425

STERNBERG, CHARLES HAZELIUS [1850-1943]
AMS 6,1357; WWAH, 578; WWWS, 1606-1607
Anon., Science 98 (1943), 101

STERNBERG, GEORGE MILLER [1838-1915]
DAB 17,590-592; FISCHER 2,1506; DAMB, 716-717; WWWS, 1607
M.L. Sternberg, George Miller Sternberg. Chicago 1920
E.E. Hume, Ann. Med. Hist. 10 (1938), 266-272
J.M. Gibson, Soldier in White. Durham, N.C. 1958

STERNBERG, MAXIMILIAN [1863-1934]
KURSCHNER 4,2909; FISCHER 2,1506; HASENEDER, 10-12
Z. Forschner, Wiener med. Wchschr. 83 (1933), 1221-1222
Anon., Wiener med. Wchschr. 84 (1934), 1063

STERNITZKI, HERMANN [1869-]
JV 8,217; RSC 13,276

STETTEN, DeWITT, Jr. [1909-1990]
AMS 14,4901; WWA 1976-1977, p.3016; WWWS, 1607
D. Stetten, Persp. Biol. Med. 25 (1982), 354-368
New York Times, 31 August 1990

STETTEN, MARJORIE ROLOFF [1915-1983]
AMS 15(6),995; WWWS, 1607

STETTENHEIMER, LUDWIG [1866-1932]
JV 5,235; RSC 13,276; NUC 568,283
Anon., Chem. Z. 56 (1932), 787

STETTER, HERMANN [1917-]
KURSCHNER 13,3829; WWWS, 1607; POTSCH, 407; POGG 7a(4),529-530

STEUDEL, HERMANN [1871-1967]
KURSCHNER 10,2415-2416; DRULL, 262; FISCHER 2,1508; TLK 3,635; IB, 287; POGG 6,2542-2544; 7a(4),531-532
Anon., Chem. Z. 91 (1967), 276

STEUDEL, JOHANNES [1901-1973]
KURSCHNER 11,2944-2945; POGG 7a(4),532-533
N. Mani, Clio Medica 8 (1973), 324-326
H. Schipperges, Sudhoffs Arch. 57 (1973), 225-227

STEUER, EUGEN [1880-]
JV 19,204; NUC 568,330

STEVENS, CARL MANTLE [1915-1987]
AMS 13,4304

STEVENS, HENRY POTTER [1875-]
JV 15,134; NUC 568,442-443

STEVENS, NETTIE MARIA [1861-1912]
DSB 18,867-869; AMS 2,450; NAW 3,372-373; OGILVIE, 167-169
T.H. Morgan, Science 36 (1912), 468-470
S.G. Brush, Isis 69 (1978), 163-172
M.B. Ogilvie and C.J. Choquette, Proc. Am. Phil. Soc. 125 (1981),292-311

STEVENS, PHILIP GREELEY [1902-]
AMS 11,5168

STEVENS, THOMAS STEVENS [1900-]
WW 1981, p.2476; CAMPBELL, 221-222; WWWS, 1608; POGG 6,2544; 7b,5142-5143

STEVENS, WILLIAM [1786-1868]
HIRSCH 5,422; DECHAMBRE [3]12,83; CALLISEN 18,399-401; 32,436

STEVENSON, LLOYD GRENFELL [1918-1988]
F.L. Holmes, J. Hist. Med. 45 (1990), 277-284

STEWARD, FREDERICK CAMPION [1904-1993]
AMS 14,4908; WWA 1976-1977, p.3020; MH 2,517-519; MSE 3,159-160

STEWART, ALFRED WALTER [1880-1947]
BELFAST, 583; WWWS, 1610; POGG 5,1208-1209; 6,2545; 7b,5143-5144
R.E. Oesper, J. Chem. Ed. 18 (1941), 492
S. Smiles, J. Chem. Soc. 1948, pp.396-398
G.B. Kauffman, J. Chem. Ed. 60 (1983), 38-40

STEWART, CORBET PAGE [1897-1972]
POGG 6,2545; 7b,5144-5146
S.C. Fraser and T. Strengers, Clinica Chimica Acta 38 (1972), i-iii
A.L. Latner, Advances in Clinical Chemistry 15 (1972), ix-xii

STEWART, GEORGE NEIL [1860-1930]
DSB 13,53-54; FISCHER 2,1509; O'CONNOR(2), 517-518; OEHRI, 111-112; WWWS, 1610
J.J.R. MacLeod, Nature 125 (1930), 980-981
T. Sollmann, Science 72 (1930), 157-162
C.J. Wiggers, Bulletin of the Cleveland Medical Library 10 (1963), 5-18

STEWART, HAROLD JULIAN [1896-1975]
AMS 11,5176
J.A. Barondess, Trans. Assoc. Am. Phys. 89 (1976), 32-33

STEWART, MATTHEW JOHN [1885-1957]
WW 1949, p.2659; FISCHER 2,1510
J.H. Dible et al., J. Path. Bact. 76 (1958), 295-313

STHAMER, JOHANN GEORG BERNHARD [1817-1903]
HEIN, 655-656

STHAMER, KARL [1828-1893]
Anon., Pharm. Z. 38 (1893), 466

STIASNY, EDMUND [1872-1965]
KURSCHNER 10,2416; WOLF, 200; BHDE 2,1131; LDGS, 25; TLK 3,635; POGG 6,2546; 7a(4),533-534
Anon., Chem. Z. 89 (1965), 871

STICKER, ANTON [1861-]
FISCHER 2,1510-1511; NUC 569,416

STICKER, GEORG [1860-1960]
KURSCHNER 8,2312-2313; FISCHER 2,1511; PAGEL, 1650-1651; OLPP, 385-386; TLK 3,636; WWWS, 1612; POGG 7a(4),535-537
J. Steudel, Sudhoffs Arch. 44 (1960), 81-82
E. Wickersheimer, Arch. Inter. Hist. Sci. 13 (1960), 277-278
M. Quick, Med. Hist. J. 22 (1987), 382-386

STICKLAND, LEONARD HUBERT [1905-]
Who's Who of British Scientists 1971-1972, p.807

STIEDA, CHRISTIAN HERMANN LUDWIG [1837-1918]
DBJ 2,706; HIRSCH 5,427; LEVITSKI 2,25-36; WELDING, 771
P. Eisler, Anat. Anz. 55 (1919), 131-144; Leopoldina 55 (1919), 21-24

STIEGLITZ, JULIUS OSCAR [1867-1937]
DAB [Suppl.2],630-631; MILES, 456-457; KAGAN(JA), 331-332; WININGER 6,36; WWWS, 1612; POGG 4,1440-1441; 5,1209-120; 6,2547; 7b,5149-5150
Anon., Ind. Eng. Chem. News Ed. 15 (1937), 39
M. Gomberg, Am. Phil. Soc. Year Book 1937, pp.409-411
H.N. McCoy, J. Am. Chem. Soc. 60 (1938), 3-21
W.A. Noyes, Biog. Mem. Nat. Acad. Sci. 21 (1939), 275-314

STIERLIN, EMANUEL FRIEDRICH ROBERT [1844-1913]
STRAHLMANN, 469,481

STIEVE, HERMANN [1886-1952]
IB, 287-288; EGERER, 91-114
B. Romeis, Anat. Anz. 99 (1953), 401-440; Bayer. Akad. Wiss. Jahrbuch 1953, pp.172-178

STIGLER, ROBERT [1878-1975]
FISCHER 2,1513-1514; PLANER, 328-329; IB, 288; POGG 7a(4),539-540
P. Deetjen, Wiener med. Wchschr. 124 (1974), 325

STILES, CHARLES WARDELL [1867-1941]
DSB 13,62-63; DAB [Suppl.3],737-739; NCAB D,62-63; OLPP, 386-387; IB,288; DAMB, 719; WWAH, 581; WWWS, 1612
W.H. Wright, J. Parsit. 27 (1941), 195-201
F.G. Brooks, Systematic Zoology 13 (1964), 220-226

STILES, WALTER [1886-1966]
DNB [1961-1970],983-984; WW 1965, pp.2932-2933; DESMOND, 585; IB, 288; POGG 6,2548; 7b,5150-5151
W.O. James, Biog. Mem. Fell. Roy. Soc. 13 (1967), 343-357

STILES, WALTER STANLEY [1901-1985]
WW 1983, p.2152
M. Alpern, Biog. Mem. Fell. Roy. Soc. 34 (1988), 817-885

STILL, GEORGE FREDERICK [1868-1941]
DNB [1941-1950], pp.839-840; FISCHER 2,1515; WWWS, 1612-1613
W. Sheldon, J. Pediatrics 49 (1956), 229-233

STILL, JACK LESLIE [1911-]
Who's Who in Australia 1977, p.998

STILLER, ERIC THOMAS [1907-]
AMS 12,6138; WWWS, 1613

STILLING, BENEDICT [1810-1879]
ADB 36,247-249; HIRSCH 5,431-432; PAGEL, 1652-1655; DECHAMBRE [3]12,97; WININGER 6,37-38; CALLISEN 32,443-445
A. Kussmaul, Rev. Sci. [2]17 (1879), 585-587
Anon., Leopoldina 15 (1879), 114-117
L. Straus, Münch. med. Wchschr. 57 (1910), 699
E.H. Ackerknecht, Gesnerus 30 (1973), 143-149
G. Aumüller, Med. Hist. J. 19 (1984), 53-67
B. Ottermann, Würzburg medizinisch-historische Mitteilungen 4 (1986), 253-287

STILLING, HEINRICH [1853-1911]
FISCHER 2,1515-1516; KOREN, 253; WININGER 7,467
Anon., Leopoldina 47 (1911), 88; J. Path. Bact. 16 (1911-1912), 380-383

STILLING, JAKOB [1842-1915]
DBJ 1,341; FISCHER 2,1516; KOREN, 253-254; WININGER 6,38; 7,467

STILLMAN, JOHN MAXSON [1852-1923]
NCAB 20,145-146; MILES, 457-458; WWAH,581; WWWS,1613; POGG 5,1210; 6,2548
H.M. Elsey, J. Chem. Ed. 6 (1929), 466-472
F.O. Koenig, Isis 34 (1942), 142-146
E.H. Huntress, Proc. Am. Acad. Arts Sci. 81 (1952), 88-90

STILLMARK, HERMANN [1860-]
RSC 18,964; SG [1]13,695; NUC 569,671
J. Brennsohn, Die Aerzte Livlands, pp.383-384. Mitau 1905; Die Aerzte Estlands, p.341. Riga 1922

STIMMEL, BENJAMIN FRANKLIN [1904-1966]
AMS 10,3933; WWWS, 1613

STINTZING, HUGO [1888-1970]
KURSCHNER 11,2950; WOLF, 202; POGG 6,2548-2549; 7a(4),546
E. Brüche, Phys. Bl. 14 (1958), 370

STINTZING, RODERICH [1854-1933]
KURSCHNER 4,1924; FISCHER 2,1516-1517; PAGEL, 1656-1657; BACHMANN, 25-34; GIESE, 561-564; BONN, 323; DZ, 1417; RSC 11,501; 18,965
M. Matthes, Münch. med. Wchschr. 71 (1924), 153

STIRLING, JOHN DEMPSTER [1907-1933]
N.C.W., Biochem. J. 27 (1933), 611

STIRLING, WILLIAM [1851-1932]
HIRSCH 5,432-433; WWWS, 1614
Anon., Lancet 1932(II), pp.815,871; Brit. Med. J. 1932(II), pp.695-696

STOBBE, JOHANN (HANS) [1860-1938]
TLK 3,637; POTSCH, 407-408; POGG 4,1441; 5,1211-1212; 6,2549-2550; 7a(4),547
C. Weygand, Ber. chem. Ges. 71A (1938), 188-190; Koll. Z. 87 (1939),1-3

STOCK, ALFRED [1876-1946]
DSB 13,70-71; KROLLMANN, 1158-1159; TLK 3,637; WWWS, 1614; POTSCH, 408; POGG 4,1442; 5,1212-1213; 6,2550; 7a(4),547-548
E. Wiberg, Chem. Ber. 83 (1950), xix-lxxvi

STOCK, JOSEF [1888-1931]
JV 31,381; NUC 570,175
Anon., Ber. chem. Ges. 65A (1932), 89

STOCK, ROBERT [1858-]
JV 5,80; RSC 14,421; NUC 570,181

STOCKARD, CHARLES RUPERT [1879-1939]
DAB [Suppl.2],631-633; AMS 6,1365; NCAB 30,130-131; FISCHER 2,1517; FLUELER, 134-136; IB, 288; WWAH, 582; WWWS, 1614
F. Gudernatsch, Anat. Anz. 89 (1939), 131-142
D.J. Edwards, Am. Phil. Soc. Year Book 1939, pp.464-467
J.M. Oppenheimer, Bull. Hist. Med. 58 (1984), 236-240

STOCKBRIDGE, HORACE EDWARD [1857-1930]
DAB 18,37-38; AMS 4,943; RSC 18,968; NUC 570,194-195

STOCKER, BRUCE ARNOLD DUNBAR [1917-]
WW 1983, p.2154; AMS 15(6),1020-1021

STOCKHAUSEN, FERDINAND [1875-1949]
JV 18,327; NUC 570,225

STOCKINGER, LEOPOLD [1919-]
KURSCHNER 13,3840; HEINDEL, 114-127

STOCKMAN, RALPH [1861-1946]
WW 1945, p.2621; FISCHER 2,1517; O'CONNOR(2), 411-412; RSC 18,969-970; NUC 570,331
Who Was Who 1941-1950, p.1111

STOCKMAYER, WALTER HUGO [1914-]
AMS 15(6),1022; WWA 1980-1981, p.3185; WWWS, 1614
W.H. Stockmayer and B. Zimm, Ann. Rev. Phys. Chem. 35 (1984), 1-21

STODDARD, JOHN TAPPAN [1852-1919]
DAB 18,56; AMS 2,454; MILES, 460-461; NUC 570,432-433; WWWS, 1614-1615; POGG 3,1296; 4,1443; 5,1213; 6,2551

STOECKENIUS, WALTHER [1921-]
AMS 15(6), 1023

STÖCKHARDT, JULIUS ADOLF [1809-1886]
ADB 36,288-290; HEIN, 657-658; BLOKH 2,708; FERCHL, 518; TSCHIRCH, 1130; POTSCH, 408; WWWS, 1615; POGG 2,1014; 3,1296; 4,1443; 7aSuppl.,667
W.O. Atwater, Pop. Sci. Mon. 19 (1881), 261-264
F. Nobbe, Land. Vers. 33 (1887), 424-433
A. von Langsdorff, Ill. Land. Z. 27 (1907), 561-562

STÖHR, PHILIPP [1849-1911]
BJN 16,75*-76*; FISCHER 2,1518; PAGEL, 1658; FREUND 2,384-388; EBERT, 7-13; DZ, 1419-1420
J. Sobotta, Münch. med. Wchschr. 58 (1911), 2747-2749
W. Felix, Verhandl. Schw. Nat. Ges. 95 (1912), 32-39
O. Schultze, Verhandl. Phys. Med. Wurz. NF42 (1912), 1-12; Anat. Anz. 40 (1912), 551-556

STÖHR, PHILIPP [1891-1979]
KURSCHNER 13,3845; FISCHER 2,1518; IB, 288; NUC 570,523-524
P. Stohr, Hippocrates 33 (1962), 168-171
K. Fleischhauer, Anat. Anz. 150 (1981), 239-247

STÖHR, RICHARD [1902-]
KURSCHNER 13,3845; POGG 7a(4),552-553
Anon., Chem. Z. 101 (1977), 411

STOELZEL, KARL [1826-1896]
BJN 1,415; 3,81*; RSC 5,842; NUC 579,539
Anon., Leopoldina 32 (1896), 58-59; Chem. Z. 20 (1896), 1036; Ber. chem. Ges. 29 (1896), 3044

STOERK, OSKAR [1870-1926]
FISCHER 2,1519
R. Wiesner, Verhandl. path. Ges. 29 (1936), 425-428

STOERMER, RICHARD [1870-1940]
KURSCHNER 4,2936; TLK 3,639; POGG 4,1444; 5,1215-1216; 6,2552; 7a(4),553
Anon., Z. angew. Chem. 53 (1940), 264

STOETZER, WALTER [1890-1980]
JV 37,766
Anon., Chem. Z. 95 (1971), 41STOFFEL, WILHELM [1928-]
KURSCHNER 16,3653-3654; POTSCH, 408-409

STOHMANN, FRIEDRICH [1832-1897]
ADB 54,543-546; BJN 4,42*; BLOKH 2,709-710; SCHAEDLER, 129; POGG 2,1015; 3,1297; 4,1445
W. Ostwald, Ber. chem. Ges. 30 (1897), 3214-3226
Anon., Chem. Z. 21 (1897), 931,1081; Leopoldina 33 (1897), 161

STOKES, ADRIAN [1887-1927]
DNB [1922-1930], 812-813; FISCHER 2,1519; VERSO, 148-149
A.F. Hurst and J.A.R. Hurst, Guy's Hosp. Rep. 78 (1928), 1-17
N.P. Hudson, Trans. Soc. Trop. Med. 60 (1966), 170-174

STOKES, ALEXANDER RAWSON [1919-]
Who's Who of British Scientists 1980-1981, p.466

STOKES, GEORGE GABRIEL [1819-1903]
DSB 13,74-79; DNB [1901-1911] 3,421-424; FERCHL, 519; W, 496; WWWS, 1616; POGG 2,1016-1017; 3,1297-1298; 4,1445-1446; 5,1216; 6,2533
C. von Voit, Sitz. Bayer. Akad. 33 (1903), 550-556
R., Proc. Roy. Soc. 75 (1905), 199-216
C.S. Breathnach, Irish J. Med. Sci. 6 (1966), 121-125

STOKES, HENRY NEWLIN [1859-1942]
AMS 7,1717; RSC 17,757; 18,978; NUC 570,652-653; POGG 4,1446; 5,1216

STOKES, JOSEPH [1896-1972]
AMS 11,5195; DAMB, 721; WWAH, 582
J.E. Rhoads, Am. Phil. Soc. Year Book 1972, pp.239-244
A.M. Bongiovanni, Trans. Assoc. Am. Phys. 85 (1972), 44-45

STOKES, WILLIAM [1804-1878]
DNB 18,1288-1290; HIRSCH 5,439-440; Suppl.,420; PAGEL, 1661-1662; LYONS, 84-87; BETTANY 2,188-193; FRANKEN, 197; VERSO, 84-85; DECHAMBRE [3]12,140-142; CALLISEN 18,442-445; 32,449-451
Anon., Lancet 1878(I), p.108
W. Stokes, Jr., William Stokes. New York 1898

STOKLASA, JULIUS [1857-1936]
BK, 320; IB, 288; POGG 4,1446; 5,1216-1217; 6,2553-2554; 7b,5156-5157
F. Bornemann, Chem. Z. 56 (1932), 709
A. Ernest, Chem. Listy 30 (1936), 89-91; Archeion 23 (1941), 206-210
O. Schmatolla, Pharm. Z. 81 (1936), 418-419

STOKSTAD, EVAN LUDVIG ROBERT [1913-]
AMS 15(6),1026; WWA 1980-1981, p.3188; WWWS, 1616

STOKVIS, BAREND JOSEPH [1834-1902]
LINDEBOOM, 1890-1892; HES, 157-158; HIRSCH 5,440-441; PAGEL, 1663-1664; REBER, 406-407; WININGER 6,39-40; WWWS, 1616
Anon., Arch. Inter. Pharmacodyn. 11 (1902), 1-8; Lancet 1902(II), pp.1022-1023
A. Riche, Bull. Acad. Med. 48 (1902), 519-521
C.A. Pekelharing, Münch. med. Wchschr. 49 (1902), 1920-1922
K.P. Pel, Deutsche med. Wchschr. 28 (1902), 749-750

STOLBA, FRANTISEK [1839-1910]
POGG 3,1298; 4,1446-1447; 5,1217

STOLL, ARTHUR [1887-1971]
SBA 4,131-132; STC 3,22-23; WWWS, 1616; POTSCH, 409; POGG 7a(4),553-557
C. Schöpf, Z. angew. Chem. 69 (1957), 1-5; Festschrift Prof. Dr. Arthur Stoll, pp.1-34. Basle 1957
C.M. Jacottet, Chimia 11 (1957), 85-87; Arzneimitt. 17 (1967), 109-110
C. Henze, Pharmacologist 13 (1971), 111-113
L. Ruzicka, Biog. Mem. Fell. Roy. Soc. 18 (1972), 567-593

STOLL, NORMAN RUDOLPH [1892-1977]
AMS 12,6151; WWWS, 1616
M.S. Ferguson, Experimental Parasitology 41 (1977), 253-271

STOLL, WILHELM [1897-1969]
JV 41,430; NUC 571,37

STOLLÉ, ROBERT [1869-1938]
DRULL, 262-263; POGG 4,1447-1448; 5,1271-1218; 6,2554-2555; 7a(4),558
E. Müller, Ber. chem. Ges. 71A (1938), 195-196

STOLNIKOV, YAKOV YAKOVLEVICH [1850-]
ZMEEV 5,147-148; RSC 18,931; SG [2]16,619
Russki Biograficheski Slovar 19,440-441

STOLTE, KARL [1881-1951]
FISCHER 2,1520
H.W. Ocklitz, Arch. Kind. 143 (1951), 161-162
J. Külz et al., Kinderärztliche Praxis 49 (1981), 583-595,617-627

STOLTZE, GEORG HEINRICH [1784-1826]
ADB 36,419; HEIN, 658; FERCHL, 519; TSCHIRCH, 1130; POGG 2,1017-1018; CALLISEN 32,452

STOLZ, FRIEDRICH [1860-1936]
HEIN, 659; KALLMORGEN, 425; FISCHER 2,1520; POTSCH, 409; POGG 6,2555
M. Bockmühl, Z. angew. Chem. 43 (1930), 285-286
Dokumente aus Hoechster Archiven 12 (1965), 1-64; 13 (1966), 1-80

STONE, WILSON STUART [1907-1968]
AMS 11,5203; WWWS, 1618
P.R. Wagner and C.P. Oliver, Genetics 60 (1968), s29-s30
J.F. Crow, Biog. Mem. Nat. Acad. Sci. 52 (1980), 451-468

STONE, WINTHROP ELLSWORTH [1862-1921]
AMS 3,662; JV 3,103; RSC 18,986-987; NUC 571,266-267; POGG 4,1449; 5,1218

STONEY, GEORGE JOHNSTONE [1826-1911]
DSB 13,82; DNB [1901-1911] 3,419-431; W, 496-497; WWWS, 1618; POGG 4,1449-1450; 5,1218-1219
F.T. Trouton, Nature 87 (1911), 50-51
J.J., Proc. Roy. Soc. A86 (1912), xx-xxxv
J.G. O'Hara, Notes Roy. Soc. 29 (1975), 265-279
N. Robotti, Physis 21 (1979), 103-143

STOOKEY, LYMAN BRUMBAUGH [1878-1940]
AMS 6,1370; WWAH, 584

STOPCZANSKI, ALEKSANDER [1835-1912]
WE 11,28
M. Sarnecka-Keller, Stud. Nauk. Pol. C25 (1981), 87-105

STOPPANI, ANDRES OSCAR [1915-]
WWWS, 1619

STORCH, LUDWIG [1859-1938]
PRAG, 385-386; TLK 3,639; POGG 4,1450; 6,2557; 7a(4),560
G.F. Hüttig, Chem. Z. 53 (1929), 197

STORER, FRANCIS HUMPHREYS [1832-1914]
DAB 18,94-95; NCAB 11,337-338; MILES, 461-462; SILLIMAN, 141-143; WWWS, 1619; POGG 3,1302-1303; 4,1450; 5,1219
L.W. Fetzer, Biochem. Bull. 4 (1915), 1-17
C.W. Eliot, Proc. Am. Acad. Arts Sci. 54 (1918-1919), 415-418

STOREY, HAROLD HAYDON [1894-1969]
WW 1965, p.2944; DESMOND, 588; IB, 289; WWWS, 1619
K.M. Smith, Biog. Mem. Fell. Roy. Soc. 15 (1969), 239-246

STORK, GILBERT JOSSE [1921-]
AMS 17(6),1090; WWA 1986-1987, p.2700; IWW 1987-1988, p.1427; WWWS, 1619; POTSCH, 409-410
J. Tsuji, Heterocycles 25 (1987), 1-6

STORM van LEEUWEN, WILLEM [1882-1933]
LINDEBOOM, 1160-1162; FISCHER 2,1521
Anon., Brit. Med. J. 1933(II), pp.318-319
F. Hansen, Deutsche med. Wchschr. 59 (1933), 1516
H. Varenkamp, Ned. Tijd. Gen. 110 (1966), 2014-2016
H.M. Beumer, Willem Storm van Leeuwen und seine Bedeutung für die Asthmaforschung. Düsseldorf 1968; Med. Welt 39 (1971), 1530-1536

STOTZ, ELMER HENRY [1911-1987]
AMS 15(6),1040; WWA 1980-1981, p.3195; WWWS, 1619

STRACK, ERICH [1897-1988]
KURSCHNER 12,3158; WWWS, 1620; POGG 7a(4),562-563

STRADA, FERDINANDO [1872-1969]
IB, 289; NUC 122,567
F. Garzon Maceda, Historia de la Facultad de Ciencias Medicas vol.3, pp.221-224. Cordoba 1928

STRAHL, HANS [1857-1920]
DBJ 2,762; FISCHER 2,1522; PAGEL, 1665; GUNDEL, 939-954
B. Henneberg, Anat. Anz. 55 (1922), 211-220

STRAIN, HAROLD HENRY [1904-]
AMS 12,6170; ETTRÉ, 437-442; WWWS, 1621; POGG 6,2558-2559; 7b,5167-5170

STRANGEWAYS, THOMAS STRANGEWAYS PIGG [1866-1926]
FISCHER 2,1522
E.D. Strangeways et al., History of the Strangeways Research Laboratory. Cambridge 1962

STRASBURGER, EDUARD ADOLF [1844-1912]
DSB 13,87-90; BJN 17,25-39; SCHULZ-SCHAEFFER, 200-201; WAGENITZ, 174; TSCHIRCH, 1130; BK, 366; W, 497; WININGER 6,42-43; 7,468;
WWWS, 1621-1622
J.B.F., Nature 89 (1912), 379-380
G. Karsten, Ber. bot. Ges. 30 (1912), (61)-(86)
J. Beauverie, Rev. Gen. Bot. 24 (1912), 417-452,479-493
E. Küster, Sitz. Nat. Ver. Rhein.-Westf. 1912, pp.5-18
G. Tischler, Arch. Zellforsch. 9 (1913), 1-40
B.M. Davis, Genetics 36 (1951), 1-3

STRASBURGER, JULIUS [1871-1934]
KURSCHNER 4,2939-2940; FISCHER 2,1523; WININGER 6,43; 7,468

STRASSBURG, GUSTAV ADOLF [1848-]
BONN, 278; RSC 8,1031; 11,515; SG [1]13,812; NUC 572,406

STRASSER, ALOIS [1867-1945]
FISCHER 2,1523; PAGEL, 1665; HASENEDER, 57-74; PLANER,333; WININGER 7,468
W. Lowenstein, Wiener klin Wchschr. 50 (1937), 1625-1627

STRASSER, LUDWIG [1865-1933]
JV 5,235; RSC 13,276-277; 18,998
Anon., Chem. Z. 57 (1933), 676; Z. angew. Chem. 46 (1933), 530

STRASSMANN, FRITZ [1858-1935]
KURSCHNER 4,2940-2941; FISCHER 2,1524-1525; ASEN, 195; WININGER 6,44
Anon., Deutsche Z. ger. Med. 12 (1928), i-viii
P. Fraenckel, Deutsche med. Wchschr. 54 (1928), 1427-1428

STRASSMANN, PAUL FERDINAND [1866-1938]
KURSCHNER 4,2941; FISCHER 2,1525; KOREN, 254; WININGER 6,44-45; TETZLAFF, 326

STRASSNER, HORST [1882-1957]
KALLMORGEN, 411-412; SG [2]16,832; NUC 572,450

STRATINGH, SIBRANDUS [1785-1841]
LINDEBOOM, 1899-1900; HIRSCH 5,451; TSCHIRCH, 1130; POGG 2,1021-1022
Anon., Arch. Pharm. [2]36 (1843), 340-344

STRAUB, FERENC BRUNO [1914-]
IWW 1982-1983, p.1274
Anon., Leopoldina [3]8/9 (1962-1963), 82-83
F.B. Straub, in Of Oxygen, Fuels and Living Matter, Part 1 (G. Semenza, Ed.), pp.325-344. Chichester 1981

STRAUB, HERMANN [1882-1938]
KURSCHNER 4,2942; FISCHER 2,1525-1526; SKVARC, 204-206; IB, 289
W.H. Veil, Münch. med. Wchschr. 85 (1938), 1118-1119
K. Beckmann, Deutsches Arch. klin. Med. 183(3) (1938), i-iv

STRAUB, JAN [1888-1975]
POGG 6,2560; 7b,5175-5176
Anon., Chem. Wkbl. 46 (1950), 554-555
T. Hekker, Chem. Wkbl. 71(6) (1975), 8

STRAUB, WALTHER [1874-1944]
FISCHER 2,1526; BOCK, 72-86; TLK 3,640; POGG 6,2560-2561; 7a(4),569-570
B. Romeis, Bayer. Akad. Wiss. Jahrbuch 1944-1948, pp.199-202
H. Gremels, Arch. exp. Path. Pharm. 204 (1947), 1-12

STRAUS, FRITZ LUDWIG [1877-1942]
KURSCHNER 4,2944; BHDE 2,1135-1136; LDGS,25; TLK 3,640; POGG 5,1220-1221; 6,2561; 7a(4),573
H. Hauptmann, Selecta Chimica 2 (1945), 7-42; Chem. Ber. 83 (1950), I-v

STRAUS, ISIDORE [1845-1896]
FISCHER 2,1527; PAGEL, 1668-1669; WININGER 6,46; RSC 18,1000-1001
E. Hervieux, Bull. Acad. Med. 36 (1896), 774
A. Wurtz, C.R. Soc. Biol. 49 (1897), v-xiv
R. Kohn, Rev. Hist. Med. Heb. 1952, pp.43-44

STRAUSS, EDUARD [1876-1952]
KALLMORGEN, 426; ARNSBERG, 501-502; BHDE 2,1137; LDGS, 60; WININGER 6,51; TETZLAFF, 326-327; POGG 6,2562; 7a(4),573
Anon., Science 116 (1952), 389; Aufbau (New York) 1952(10), p.29
Dokumente zur Geschichte der Frankfurter Juden 1933-1945, pp.400,551. Frankfurt a.M. 1963

STRAUSS, HERMANN [1868-1942]
FISCHER 2,1527-1528; WININGER 6,52; TETZLAFF, 327; NUC 572,619-620
E.G. Lowenthal, Bewährung im Untergang: Ein Gedenkbuch, 2nd Ed., pp.171-172
H. Keller, Deutsche med. Wchschr. 93 (1968), 2237-2238
H. Schmolz, Jahrbuch für Schwäbisch-Fränkische Geschichte 26 (1969), 211-212

STRAUSS, HERMANN [1884-1942]
GEDENKBUCH 2,1478; JV 24,633; SG [2]116,835; NUC 572,621

STRAUSS, JOSEF [1872-1944]
KALLMORGEN, 426; SG [2]116,835; NUC 572,622
W. Strauss, Lebenszeichen: Juden in Württemburg nach 1933, pp.311-312. Gerlingen 1982

STRAUSS, JOSEF [1873-1935]
JV 12,34; SG [2]116,835; NUC 572,622

STRAUSS, LEOPOLD [1873-1944]
GEDENKBUCH 2,1480; JV 9,234; SG [2]116,835; NUC 572,668

STRAUSS, MAURICE BENJAMIN [1904-1974]
AMS 11,2550; NCAB 58,182
W.B. Castle, Trans. Assoc. Am. Phys. 87 (1974), 36-39

STRECKER, ADOLF [1822-1871]
ADB 36,555-560; BLOKH 2,711-712; TSCHIRCH, 1131; SCHAEDLER, 129-130; POTSCH, 410-411; POGG 2,1024-1025; 3,1304; 7aSuppl.,668
R. Wagner, Ber. chem. Ges. 5 (1872), 125-131
F. von Kobell, Sitz. Bayer. Akad. 2 (1872), 99-100
H. Cassebaum, NTM 8(1) (1971), 46-57

STREETER, GEORGE LINIUS [1873-1948]
DSB 13,96-98; DAB [Suppl.4],800-802; NCAB 37,356-357; DAMB, 723-724; IB, 289-290; WWAH, 586; WWWS, 1623
G.W. Corner, Biog. Mem. Nat. Acad. Sci. 28 (1954), 261-287

STREHLER, BERNARD LOUIS [1925-]
AMS 15(6),1055; WWWS, 1623

STREISINGER, GEORGE [1927-1984]
AMS 15(6),1056; WWA 1980-1981, p.3203

STRICKER, SALOMON [1834-1898]
ADB 56,622-623; BJN 3,58-58; 5,81*; HIRSCH 5,456-457; PAGEL, 1671-1673; HEINDEL, 162-167; KOREN, 255; WWWS, 1624
A. Humber, Alm. Akad. Wiss. Wien 48 (1898), 328-330
G. Kapsamer and A. Biedl, Wiener klin. Wchschr. 11 (1898), 461-466,911-919

STROGONOV, NIKOLAI ALEKSEEVICH [1842-]
ZMEEV 5,148-149; RSC 11,520; 12,709; SG [1]13,822; [2]16,855

STROHL, ANDRÉ [1887-1977]
POGG 6,2568; 7b,5182-5184
H. Desgrez, Bull. Acad. Med. 161 (1977), 563-571

STROHL, GEORGES ÉMILE [1827-1882]
HEIN, 662-663; BALLAND, 112-113; HUMBERT, 221-222; DORVEAUX, 79; BERGER-LEVRAULT, 236-237

STROMEYER, FRIEDRICH [1776-1835]
HIRSCH 5,458-459; BLOKH 2,713; WAGENITZ, 175-176; FERCHL, 520-521; TSCHIRCH, 1131; SCHAEDLER, 130; PHILIPPE, 718; COLE, 518; WWWS, 1624; POTSCH, 411; POGG 2,1031-1032; 7aSuppl.,669
R. Zaunick, Sudhoffs Arch. 35 (1942), 243-254
G. Lockemann and R. Oseper, J. Chem. Ed. 30 (1953), 202-204

STROMINGER, JACK LEONARD [1925-]
MSE 3,169-170

STRONG, FRANK MORGAN [1908-]
AMS 13,4355

STRONG, LEONELL CLARENCE [1894-1982]
AMS 11,5231; IB, 290; WWWS, 1624-1625
L.C. Strong, Cancer Research 36 (1976), 3545-3553

STRONG, RICHARD PEARSON [1872-1948]
DAB [Suppl.4],802-803; NCAB A,93; FORBES, 215-226; IB, 290; WWWS, 1625
Anon., Science 108 (1948), 36; Brit. Med. J. 1948(II), p.880; J. Trop. Med. Hyg. 51 (1948), 241-242
J.S. Simmonds, Trans. Assoc. Am. Phys. 62 (1949), 20-21
E. Chernin, J. Hist. Med. 44 (1989), 296-319

STROOF, IGNAZ [1838-1920]
DBJ 2,762; POTSCH, 411; POGG 6,2568-2569
B. Lepsius, Ber. chem. Ges. 54A (1921), 101-107

STROSCHEIN, FRITZ [1885-]
JV 25,366; HUHLE-KREUTZER, 356

STRUGGER, SIEGFRIED [1906-1961]
SANDRITTER, 111-114; FREUND 1,331-346
E. Perner, Z. wiss. Mikr. 55 (1963), 129-131
K. Hofler, Alm. Akad. Wiss. Wien 115 (1965), 401-408

STRUGHOLD, HUBERTUS [1898-1986]
AMS 11,5234; EBERT, 131-136; WWWS, 1625

STRUNZ, FRANZ [1875-1953]
TSCHIRCH, 1131; TLK 3,643; POGG 5,1224; 6,2569; 7a(4),586
G. Wagner, Ost. Chem. Z. 54 (1953), 188-189
W. Huth, Nova Acta Paracelsica 7 (1954), 103-120

STRUTTON, WILLIAM RUSSEL [1896-1967]
Anon., Bull. Rockland County Med. Soc. 6(1) (1962), 10; N.Y. J. Med. 67 (1967), 3030

STRUVE, GUSTAV ADOLF [1866-1917]
JV 6,178; NUC 574,149
Anon., Z. angew. Chem. 30(III) (1917), 250

STRUVE, HEINRICH WILHELM [1822-1908]
BSE 41,148; POGG 2,1038-1039; 3,1308; 4,1457
G. Kraemer, Ber. Chem. Ges. 41 (1908), 1103

STRYER, LUBERT [1938-]
AMS 15(6),1067; WWA 1980-1981, p.3209; MSE 3,171-172

STUART, AMBROSE PASCAL SEVILON [1820-1899]
Anon., Science 10 (1899), 423

STUBBE, HANS [1902-1989]
KURSCHNER 12,3180; IWW 1982-1983, p.1277

STUCKY, CHARLES JOSEPH [1896-1938]
AMS 5,1084; NCAB 29,252-253

STUDINICKA, FRANTISEK KAREL [1870-1955]
FISCHER 2,1531; NAVRATIL, 303-305; IB, 290; POGG 6,2573; 7b,5196-5198

STÜBEL, HANS (BRNO JOHANNES) [1885-1961]
KURSCHNER 9,2056; FISCHER 2,1531; IB, 290

STÜTZEL, LUDWIG [1873-]
JV 14,228; RSC 17,436

STUMPF, PAUL KARL [1919-]
AMS 15(6),1073; WWA 1980-1981, p.3211

STURGEON, WILLIAM [1783-1850]
DSB 13,126; DNB 19,131-135; FERCHL, 522; W, 498; WWWS, 1628; POGG 2,1042-1043
Anon., Am. J. Sci. [2]11 (1851), 444-446
J.P. Joule, Mem. Lit. Phil. Soc. 14 (1857), 53-83
S.P. Thompson, Science [1]16 (1890), 199-202
I.R. Morus, Hist. Sci. 30 (1992), 1-28

STURLI, ADRIANO [1873-1964]
L. Premuda, in Medizinische Diagnostik in Geschichte und Gegenwart (C. Habrich et al., Eds.), pp.327-339. Munich 1978

STURTEVANT, ALFRED HENRY [1891-1970]
DSB 13,133-138; AMS 11,5244; MH 1,453-454; STC 3,28-29; MSE 3,172-173; B, 291; WWAH, 588; WWWS, 1628
G.W. Beadle, Am. Phil. Soc. Year Book 1970, pp.166-171
D.E. Lancefield, Biol. Bull. 141 (1971), 17-18
S. Emerson, Ann. Rev. Gen. 5 (1971), 1-4

STURTEVANT, JULIAN MUNSON [1908-]
AMS 14,4969; WWWS, 1628

STURTON, STEPHEN DOUGLAS [1896-]
NUC 575,85

STUTZER, ALBERT [1849-1923]
TLK 2,691-692; DZ, 1448; RSC 18,1028-1030
Anon., Chem. Z. 47 (1923), 760

STYLOS, NIKOLAOS [1865-]
JV 3,221; RSC 14,241

SUBBAROW, YELLAPRAGADA [1895-1948]
AMS 7,1735
S.P.K. Gupta, Bull. Indian Inst. Hist. Med. 6 (1976), 128-143; In Search of Panacea. New Delhi 1987

SUBBOTIN, VIKTOR ANDREEVICH [1844-1898]
PAGEL, 1680; IKONNIKOV, 634-636; ZMEEV 5,152
R.Y. Beniumov and I.M. Makarenko, Gigiena i Sanitaria 21(5) (1956), 38-43
A. Mager, Med. Hist. J. 9 (1974), 63-76

SUBRAHMANYAN, VAIDEJANATHA [1902-1979]
R.N. Chakravarti, Journal of the Institute of Chemistry 51 (1979), 96-97

SUCK, OSKAR [1864-1928]
J. Brennsohn, Die Aerzte Kurlands, p.388. Riga 1929

SUCKOW, GUSTAV [1803-1867]
ADB 37,106; FERCHL, 523; POGG 2,1047; 3,1313

SUDBOROUGH, JOHN JOSEPH [1869-1963]
WW 1959, p.2944; SRS, 12; NUC 575,392-393; POGG 4,1463; 5,1229; 6,2578; 7b,5214-2515
T.C. James and C.W. Davies, J. Roy. Inst. Chem. 80 (1956), 569-571
T.C. James, Nature 199 (1963), 1136; Proc. Chem. Soc. 1964, pp.68-70

SUDHOFF, KARL [1853-1938]
DSB 13,141-143; KALLMORGEN, 430; FISCHER 2,1533-1534; TSCHIRCH, 1131; SKVARC, 209-214; DZ, 1447; TLK 3,645-646; WWWS, 1629; POGG 5,1229; 6,2578; 7a(4),602
K. Sudhoff, Sudhoffs Arch. 21 (1929), 333-387
G. Harbrand-Hochmuth, Sudhoffs Arch. 27 (1934), 131-186
H.E. Sigerist et al, Bull. Inst. Hist. Med. 2 (1934), 1-25
W. Artelt, Janus 43 (1939), 84-91
H.E. Sigerist, Bull. Hist. Med. 7 (1939), 800-804; Sudhoffs Arch. 37 (1953), 97-103
P. Diepgen, Arch. Inter. Hist. Sci. 6 (1953), 260-265

SÜS, OSKAR [1903-1978]
KURSCHNER 12,3171; POGG 7a(4),604
Anon., Chem. Z. 102 (1978), 158

SÜSSENGUTH, OTTO [-1893]
RSC 8,1046
Anon., Pharm. Z. 38 (1893), 443

SÜSSER, ARTUR [1879-]
JV 28,321; NUC 575,545

SUGASAWA, SHIGEHIKO [1898-1991]
JBE, 1481
T. Shioiri and Y. Ban, Tetrahedron 48(10) (1992), v-viii

SUGDEN, SAMUEL [1892-1950]
WW 1949, p.2694; W, 498; WWWS, 1630; POTSCH, 411; POGG 6,2580; 7b,5215-5216
W. Wardlow, Nature 166 (1950), 933
L.E. Sutton, Obit. Not. Fell. Roy. Soc. 7 (1951), 493-503; J. Chem. Soc. 1952, pp.1987-1992

SUGINOME, HARUSADA [1892-1973]
JBE, 1488; WWWS, 1630
Anon., Chem. Brit. 9 (1973), 190

SUGIURA, JUGO [1855-1924]
JBE, 1490
T. Doke, Jap. Stud. Hist. Sci. 8 (1969), 148,153

SUGIURA, KANEMATSU [1890-1979]
D.J. Hutchison, Cancer Research 40 (1980), 2625-2626

SUIDA, WILHELM [1853-1922]
POGG 6,2582-2583
C. Oettinger, Ost. Chem. Z. 25 (1922), 111-115

ŠULC, OTAKAR [1869-1901]
POGG 4,1464
Anon., Leopoldina 37 (1901), 71; Chem. Z. 25 (1901), 1168

SULLIVAN, MICHAEL XAVIER [1875-1963]
AMS 10,3981; WWAH, 588-589
Anon., Chem. Eng. News 41(22) (1963), 104

SULLIVAN, WILLIAM DANIEL [1918-]
AMS 15(6),1090; WWA 1980-1981, p.3219; WWWS, 1631

SULLIVAN, WILLIAM KIRBY [1821-1893]
RSC 5,885; 8,1045; NUC 576,199-200; POGG 3,1314
Anon., Chem. Z. 14 (1893), 1783; Journal of the Cork Historical and Archaeological Society [2]46 (1941), 43

SULZBERGER, MARION BALDUR [1895-1983]
AMS 11,5256; KAGA(JA), 203,652; COHEN,238; WWWS, 1631-1632
R.L. Baer et al., International Journal of Dermatology 16 (1977), 306-309,535-387; 422-432
Anon., American Journal of Dermopathology 6 (1984), 344-370
L. Forman, Brit. J. Dermatol. 111 (1984), 367-369

SULZBERGER, NATHAN [1874-1954]
JV 16,254
Anon., Chem. Eng. News 32 (1954), 2632; New York Times, 7 June 1954

SULZE, WALTER [1879-]
KURSCHNER 10,2461; IB,291; NUC 576,250

SULZER, ROBERT [1894-]
SG [3]9,1103; NUC 576,258

SUMIKI, YUSUKE [1901-]
JBE, 1497,2286,2361; WWWS, 1632

SUMMER, THOMAS JEFFERSON [-1850?]
RSC 5,886; NUC 576,307

SUMNER, FRANCIS BERTODY [1874-1945]
DSB 13,150-151; DAB [Suppl.3],752-753; NCAB 34,333-334; IB, 291; WWAH, 589; WWWS, 1632
F.B. Sumner, The Life History of an American Naturalist. Lancaster, Pa. 1945
W.R. Coe, Science 102 (1945), 344-346
D.L. Fox, Am. Phil. Soc. Year Book 1945, pp.416-420
C.M. Child, Biog. Mem. Nat. Acad. Sci. 25 (1948), 147-153
W.B. Provine, Stud. Hist. Biol. 3 (1979), 211-240

SUMNER, JAMES BATCHELLER [1887-1955]
DSB 13,152-153; DAB [Suppl.5],669-671; AMS 9(I),1891; MH 1,456-457; STC 3,32-34; MSE 3,174-175; MILES, 464-465; CB 1947, pp.620-622; W, 498-499; WWAH, 589; WWWS, 1632; POTSCH, 412; POGG 6,2583-2584; 7b,5217-5219
J.B. Sumner, J. Chem. Ed. 14 (1937), 255-259
A.L. Dounce, Nature 176 (1955), 859
L.A. Maynard, Biog. Mem. Nat. Acad. Sci. 31 (1958), 376-396
C.F. Cori, TIBS 6 (1981), 194-196
A.B. Costa, Chem. Brit. 25 (1989), 788-790

SUNDBERG, CARL [1859-1931]
SMK 7,319; FISCHER 2,1535-1536; IB, 291

SUNDBERG, CARL GUSTAF [1892-1963]
VAD 1963, p.1018; IB, 291

SUNDVIK, ERNST [1850-1918]
ENKVIST, 110-111; RSC 18,1038

SUNDWALL, JOHN [1880-1950]
AMS 8,2436; NCAB 39,77-78; IB, 291
W.J. Murray, Univ. Mich. Med. Bull. 17 (1951), 126

SUPNIEWSKI, JANUSZ WIKTOR [1899-1964]
POGG 6,2586; 7b,5224-5229
S. Misztal, Wiad. Chem. 19 (1965), 421-426

SURÁNYI, GYULA [1899-1958]
MEL 2,667
K. Gergely, Orv. Het. 99 (1958), 361-362

SURE, BARNETT [1891-1960]
AMS 9(I),1892; WWAH, 589; NUC 577,87-88

SUREDA y BLANES, JOSÉ [1890-]
EUI Suppl. 1936-1939, p.558

SURGENOR, DOUGLAS MacNEVIN [1918-]
AMS 15(6),1099; WWA 1980-1981, p.3223; WWWS, 1634

SUSHKIN, PETR PETROVICH [1868-1928]
KUZNETSOV(1963), 354-359; IB, 292
N.A. Dementyev, P.P. Shishkin. Moscow 1940; Bull. Soc. Nat. Moscou NS59(5) (1954), 99-105
N.N. Danilov, Bull. Soc. Nat. Moscou NS73(6) (1968), 149-153

SUTER, FRIEDRICH [1870-1961]
FISCHER 2,1536; RSC 18,1041
E. Hagenbach et al., Schw. med. Wchschr. 71 (1940), 225-226,243
H. Heuser, Schw. med. Wchschr. 91 (1961), 1240-1241
R. Howald, Schw. med. Jahrb. 1962, pp.xxxix-xlii
A.L. Vischer, Basler Stadtbuch 1963, pp.98-104

SUTHERLAND, EARL WILBUR [1915-1974]
AMS 12,6230; STC 3,34-35; MSE 3,175-177; DAMB, 727-728; ABBOTT, 123; POTSCH, 412
C.R. Park, Pharmacologist 17 (1975), 34-36
R.W. Butcher and G.A. Robinson, Pharmacologist 17 (1975), 110-112
C.F. Cori, Biog. Mem. Nat. Acad. Sci. 49 (1978), 319-350

SUTHERLAND, GORDON BRIMS BLACK McIVOR [1907-1980]
DNB [1971-1980],823-824; WW 1980, p.2474; WWWS, 1634
N. Sheppard, Biog. Mem. Fell. Roy. Soc. 28 (1982), 589-626
S. Krimm, Am. Phil. Soc. Year Book 1985, pp.193-197

SUTHERLAND, WILLIAM [1859-1911]
DSB 13,155-156; POGG 4,1467; 5,1231-1232
W.A. Osborne, William Sutherland. Melbourne 1920

SUTTER, HERMANN [1905-1982]
KURSCHNER 13,3906; POGG 7a(4),612
Anon., Naturw. Rund. 35 (1982), 507

SUTTON (ZORTMANN), HERBERT JACKSON [1883-1969]
SRS, 34; NUC 685,178

SUTTON, LESLIE ERNEST [1906-1992]
WW 1990, p.1762

SUTTON, WALTER STANBOROUGH [1877-1916]
DSB 13,156-158; SCHULZ-SCHAEFFER, 204
Anon., Walter Stanborough Sutton. Kansas City, Kan. 1917
V.A. McKusick, Bull. Hist. Med. 34 (1960), 487-497

SUZUKI, BUNSUKE [1888-1949]
POGG 6,2588; 7b,5232

SUZUKI, UMETARO [1874-1943]
JBE, 1518; IB, 292; WWWS, 1636
Anon., Science 98 (1943), 295
B. Suzuki, Science (Japan) 14 (1944), 37-38

SVANBERG, LARS FREDRIK [1805-1878]
FERCHL, 524; PHILIPPE, 778-779; POGG 2,1052-1053; 3,1315
H.G. Söderbaum, Lev. Sven. Vet 5 (1915-1920), 1-110

SVANBERG, OLOF [1896-1975]
POGG 6,2588; 7b,5233-5237

SVEDBERG, THEODOR [1884-1971]
DSB 13,158-164; SCHWERTE 1,191-198; STC 3,36-38; ABBOTT-C, 133-134; POTSCH, 412-413; WWWS, 1636; POGG 5,1232; 6,2589; 7b,5237-5240
A. Tiselius and S. Claesson, An. Rev. Phys. Chem. 18 (1967), 1-8

S. Claesson and K.O. Pedersen, Biog. Mem. Fell. Roy. Soc. 18 (1972), 595-627
K.O. Pedersen, Fractions 1974, No.1, pp.1-8; Biophysical Chemistry 5 (1976), 3-18; Horm. Prot. Pep. 9 (1980),239-249; in Selected Topics in the History of Biochemistry (G. Semenza, Ed.), pp.233-281. Amsterdam 1983
J.W. Williams, Fractions 1874, No.2, pp.1-10
M. Kerker, Isis 67 (1976), 190-216; 77 (1986), 278-282
J.T. Edsall, TIBS 3 (1978), 114-115

SVEHLA, KAREL [1866-1929]
FISCHER 2,1537; NAVRATIL, 327-328; SG [2]17,309
F. Luska, Cas. Lek. Cesk. 68 (1929), 1601

SVIRSKI, GEORGI PETROVICH [1853-1910]
LEVITSKI 2,203-204
J. Brennsohn, Die Aerzte Livlands, pp.390-391. Mitau 1905

SWAIN, ROBERT ECKLES [1875-1961]
AMS 8,2440

SWAN, WILLIAM [1818-1894]
DNB 19,196; FERCHL, 524

SWANSON, PEARL PAULINE [1895-1980]
AMS 11,5777; WWWS, 1638

SWARTS, FRÉDÉRIC [1866-1940]
DSB 13,177; GHENT 4,35-38; POGG 5,1234; 6,2591-2592; 7b,5250
Anon., Bull. Soc. Chim. Belg. 49 (1940), 33-35
M. Delépine, C.R. Acad. Sci. 212 (1941), 1067-1049
J. Timmermans, J. Chem. Soc. 1946, pp.559-560
G.B. Kauffman, J. Chem. Ed. 32 (1955), 301-303

SWARTS, THEODORE [1839-1911]
NBW 9,718-722; POGG 4,1467
J. MacLeod, Liber Memorialis Gand 2,181-186. Ghent 1913

SWEET, JOSHUA EDWIN [1876-1957]
AMS 8,2446
Anon., Science 125 (1957), 880

SWIETOSLAWSKI, WOJCIECH [1881-1968]
POTSCH, 413-414; POGG 5,1234; 6,2592-2593; 7b,5251-5257
W. Malesinski, Przemysl Chemiczni 47 (1968), 385-386
A. Dorabialska, Wojciech Swietoslawski. Warsaw 1974
J. Swietoslawska-Zolkiewska, Kwart. Hist. Nauki 26 (1981), 279-301; 30 (1985), 247-271
W. Swietoslawski, Kwart. Hist. Nauki 30 (1985), 489-549

SWIFT, ERNEST HAYWOOD [1897-1987]
AMS 13,4399; WWA 1976-1977, p.3078; WWWS,1640; POGG 6,2593; 7b,5257-5259

SWIFT, HOMER FORDYCE [1881-1953]
AMS 8,2448; DAMB, 729-730; WWAH, 591; WWWS, 1640
A.R. Dochez, Trans. Assoc. Am. Phys. 67 (1954), 25-27

SWINGLE, WALTER TENNYSON [1871-1952]
AMS 8,2450; NCAB 54,13; HUMPHREY, 241-244; IB, 293; WWAH, 591; WWWS, 1640
W. Seifriz, Science 118 (1953), 288-289

SWINGLE, WILBUR WILLIS [1891-1975]
AMS 11,5290; WWWS, 1640
Anon., Science 189 (1975), 275

SWORN, SYDNEY AUGUSTUS [1866-1899]
Anon., J. Chem. Soc. 77 (1900), 598-599

SYDENSTRICKER, VIRGIL PRESTON [1889-1964]
AMS 10,4010; WWAH, 592
P. Gyorgy, Trans. Assoc. Am. Phys. 79 (1966), 79-80

K.J. Meador et al., Southern Medical Journal 81 (1988), 1042-1046

SYM, ERNEST ALEXANDER [1893-1950]
POGG 6,2594-2595; 7b,5266-5267
B. Skarzynski, Acta Physiol. Polon. 2 (1951), 3-8

SYNGE, RICHARD LAURENCE MILLINGTON [1914-]
WW 1981, p.2536; IWW 1982-1983, p.1292; MH 1,458-459; STC 3,40-41; MSE 3, 180-181; ETTRE, 447-451; CB 1953, pp.611-612; WWWS, 1642; POTSCH, 414

SYNIEWSKI, WIKTOR [1865-1927]
POGG 4,1669-1470; 5,1235; 6,2595
A. Joszt, Roczniki Chemii 7 (1927), 381-396

SZABÓ, DENES [1856-1918]
MEL 2,674; FISCHER 2,1540

SZECSI, STEPHAN [1887-1914]
NUC 580,462
K.G. Boroviczeny et al., Einführung in die Geschichte der Hämatologie, p. 146. Stuttgart 1944

SZENT-GYÖRGYI, ALBERT [1893-1986]
AMS 14,5018; WWA 1978-1979, p.3186; IWW 1982-1983, p.1294; MH 2,531-532; STC 3,42-44; MSE 3,181; SCHWERTE 2,135-142; CB 1955, pp.596-599; POTSCH, 414-415; ABBOTT, 135; WWWS, 1643; POGG 6,2598; 7b,5291-5296
R. Wurmser, in Horizons in Biochemistry (M. Kasha and B. Pullman, Eds.), pp.1-7. New York 1962
A. Szent-Györgyi, Ann. Rev. Biochem. 32 (1963), 1-14; Persp. Biol. Med. 15 (1971), 1-5
I. Banga, Naturw. Rund. 36 (1983), 381-394
M. Florkin, TIBS 10 (1985), 35-38
H.F. Bradford, TIBS 12 (1987), 344-347
F.B. Straub and S.S. Cohen, Acta Biochem. Biophys. Hung. 22(1987),135-148
R.W. Moss, Free Radical. New York 1988

SZENT-GYÖRGYI, ANDREW GABRIEL [1924-]
AMS 15(6),1135; WWWS, 1643

SZILARD, LEO [1898-1964]
DSB 13,225-228; DAB [Suppl.7],732-733; AMS 10,4014; MH 1,459; STC 3,44; MSE 3,181-182; BHDE 2,1149; LDGS, 101; WWAH, 592; WWWS, 1643
E.P. Wigner, Biog. Mem. Nat. Acad. Sci. 40 (1969), 337-347
L. Szilard, His Version of the Facts (S.R. Weart and G.W. Szilard, Eds.). Cambridge, Mass. 1980
B. Feld, Social Research 51 (1984), 675-689

SZILY, ADOLF von [1848-1920]
MEL 2,781; FISCHER 2,1541; KOREN, 255; WININGER 6,77

SZILY, AUREL von [1880-1945]
BHDE 2,1149; LDGS, 74; FISCHER 2,1541; KOREN, 255
E. Engelking, Klin. Mon. Augenheil. 111 (1946), 65-68

SZILY, PÁL [1878-1945]
DSB 13,228-229; MEL 2,782
F. Szabadváry, Orv. Kozl. 38 (1966), 121-130

SZOKALSKI, HIPOLIT (1868-1936]
K. Milak, Kwart. Hist. Nauki 2 (1957), 304

SZPERL, LUDWIK [1879-1944]
POTSCH, 415
W. Polaczkowa and J. Böhm, Roczniki Chemii 22 (1948), 97-104

SZPILMAN, JÓZEF [1855-1920]
KONOPKA 10,384
Anon., Przegl. Lek. Lwow 8 (1909), 298-300

SZYBALSKI, WACLAW [1921-]
AMS 17(6), 1197; WWWS, 1643

SZYMONOWICZ, WLADYSLAW (LADISLAUS) [1869-1939]
WE 11,292; FISCHER 2,1542; KONOPKA 10,453-454; IB, 294; KOREN, 255
H.H. Loetzke, Arch. Hist. Med. 42 (1979), 359-364

T

TABOR, HERBERT [1918-]
AMS 15(7),3; WWA 1980-1981, p.3241

TABORSKY, GEORGE [1928-]
AMS 18(7),3

TACHAU, HERMANN [1884-]
JV 24,222; SG [2]17,583; NUC 581,134

TACHAU, PAUL [1887-1967]
BHDE 2,1150; SG [3]10,152; NUC 581, 134
Deutscher Dermatologen Kalender 1929, pp.236-237
Anon., J. Am. Med. Assn. 203 (1968), 316

TADDEI, GIOACCHINO [1792-1860]
PROVENZAL, 134-144; HIRSCH 5,505-506; POGG 2,1065; CALLISEN 19,70-72; 32,495
A. Corsini, Chim. Ind. Agr. Biol. 17 (1941), 493-498
A. Esposito Vitolo, L'Opera di Gioacchino Taddei. Pisa 1946
P. Antoniotti, Nuncius 3 (1988), 71-100

TÄUBER, ERNST [1861-1944]
POGG 4,1471-1472; 6,2600; 7a(4),617

TÄUFEL, KURT [1892-1970]
KURSCHNER 11,3006-3007; TLK 3,649; POGG 6,2600-2801; 7a(4),617-620
U. Freimuth and K. Rauscher, Nahrung 2 (1958), 1-5
U. Freimuth, Jahrb. Sachs. Akad. 1969-1970, pp.299-334

TAFEL, JULIUS [1862-1918]
DBJ 2,706; BLOKH 2,714; DZ, 1451-1452; POTSCH, 416; POGG 4,1472; 5,1238; 6,2601
B. Emmert, Chem. Z. 42 (1918), 481-482; Ber. chem. Ges. 51 (1918), 1686-1687

TAHARA, YOSHISUMI [1855-1935]
HAKUSHI 1(Pharm.), 2
Y. Tsuzuki, J. Chem. Ed. 47 (1970), 695-696
M. Onuma and T. Doke, Jap. Stud. Hist. Sci. 12 (1973), 5-14

TAINTER, MAURICE LANE [1899-1991]
AMS 11,5303; WWWS, 1645

TAKAGI (TAKAKI), KENJI [1881-1919]
HAKUSHI 2,128-129

TAKAHASHI, DENGO [1866-1917]
HAKUSHI 2,207

TAKAHASHI, TEIZO [1875-1952]
JBE, 1561,2363; POGG 6,2603; 7b,5302
K. Sakaguchi, J. Agr. Chem. Soc. Japan 26(7) (1952), 1-2

TAKAHASHI, TEIZO [1897-]
JBE, 1561,2290

TAKAKI, KANEHIRO [1849-1920]
Y. Itokawa, J. Nutrition 106 (1976), 581-588

TAKAMINE, JOKICHI [1854-1922]
DAB 18,275-276; MILES, 468-469; FISCHER 2,1542; DAMB, 731-732; WWAH, 593; POTSCH, 416
K.K. Kawakami, Jokichi Takamine. New York 1928
F.G. Creech, Chemistry 50(4) (1977), 5-6
A. DeMille, Where the Wings Grow, pp.175-186. Garden City, N.Y. 1978

TAKAYAMA, MASAO [1871-]
FISCHER 2,1544; HAKUSHI 2,88

TAKEDA, KEN-ICHI [1907-1991]
W. Nagata, Tetrahedron 47(32) (1991), v-xi

TAKEI, SANKICHI [1896-1982]
JBE, 1599,2291; IB, 294; POGG 6,2605; 7b,5306-5308
Anon., Leopoldina [3]22 (1980), 34-35

TAKEMURA, MASARO [1876-]
HAKUSHI 2,181

TALALAY, PAUL [1923-]
AMS 15(7),9; BHDE 2,1152; WWWS, 1646

TALAMON, CHARLES [1850-1929]
FISCHER 2,1544; RSC 19,11; WWWS, 1646
E. Rist, Bull. Soc. Med. Paris 53 (1929), 1551-1554
Anon., Nourisson 17 (1929), 126-128

TALBOT, FRITZ BRADLEY [1878-1964]
AMS 10,4019; ICC, 125-126
L.E. Holt, American Journal of Diseases of Children 110 (1965), 333-334

TALBOT, WILLIAM HENRY FOX [1800-1877]
DSB 13,237-239; DNB 19,339-341; BLOKH 2,714-715; MATSCHOSS, 270; POTSCH, 416; FERCHL, 527-528; SCHAEDLER, 132; W, 503; WWWS, 1646
E.H. Huntress, Proc. Am. Acad. Arts Sci. 78 (1950), 8-9
H. Gernsheim, Endeavour NS1 (1977), 18-22

TALIAFERRO, WILLIAM HAY [1895-1973]
AMS 12,6277; NCAB F,83-84; 58,456-457; IB, 295; WWWS, 1646-1647
W.H. Taliaferro, Ann. Rev. Microbiol. 22 (1968), 1-14
B.N. Jarolow, Am. Phil. Soc. Year Book 1974, pp.203-207
D.W. Talmage, Biog. Mem. Nat. Acad. Sci. 54 (1983), 375-407

TALLAN, HARRIS HERMAN [1924-]
AMS 18(7),11

TALLQVIST, THEODOR WALDEMAR [1871-1927]
FISCHER 2,1545; OLPP, 394
J. Hagelstam, Acta Med. Scand. 67 (1927), i-vii

TALMA, SAPE [1847-1918]
LINDEBOOM, 1946-1948; FISCHER 2,1545; UTRECHT, 137-140
Anon., Brit. Med. J. 1918(II), p.591

TALMAGE, DAVID WILSON [1919-]
AMS 17(7),13; WWA 1986-1987, p.2744; WWWS, 1647
D.W. Talmage, Ann. Rev. Immunol. 4 (1986), 1-11

TALMUD, DAVID LVOVICH [1900-1973]
BSE 41,561; WWWS, 1647; POGG 6,2607; 7b,5313-5315

TAMBACH, RUDOLF [1862-1925]
NUC 582,339
Anon., Chem. Z. 49 (1925), 315

TAMBOR, JOSEPH [1867-1934]
HEIN, 670-671; TLK 3,648; POGG 5,1239-1240; 6,2608-2609
V. Lampe, Helv. Chim. Acta 18 (1935), 1243-1247

TAMBURINI, AUGUSTO [1848-1919]
EI 33,213; FISCHER 2,1545-1546; PAGEL, 1688-1689; SG [1]14,200-

201; [3]10,156
S.E. Jelliffe, J. Nerv. Ment. Dis. 51 (1920), 205-207

TAMIYA, HIROSHI [1903-1984]
JBE, 1626; IWW 1982-1983, p.1299; WWWS, 1647; MH 2,535-535; STC 3,46-48; MSE 3,185-186; POGG 6,2609; 7b,5319-5322
Anon., Leopoldina [3]4/5 (1961), 19
A. Takamiya, in Studies on Microalgae and Photosynthetic Bacteria, pp.15-19. Tokyo 1963
S. Miyachi, Ber. bot. Ges. 99 (1986), 133-135

TAMM, CHRISTOPH [1923-]
KURSCHNER 13,3915; POGG 7a(4),621-622
U. Seguin, Chimia 37 (1983), 60-61

TAMM, IGOR [1922-]
AMS 15(7),14

TAMMANN, GUSTAV [1861-1938]
DSB 13,242-248; WELDING, 783; LEVITSKI 1,257-259; TLK 3,648; WWWS, 1647; POTSCH, 416-417; POGG 4,1474; 5,1240; 6,2610-2612; 7a(4),623-625
U. Dehlinger, Z. angew. Chem. 52 (1939), 229-231
W. Biltz, Ber. chem. Ges. 72A (1939), 43-44
G. Masing, Z. Elektrochem. 45 (1939), 121-124; Ber. chem. Ges. 73A (1940), 25-28
W.E. Garner, J. Chem. Soc. 1952, pp.1961-1973
A.N. Shamin, Ist. Est. Tekhn. Pribalt. 2 (1970), 185-191
U.V. Palm, Ist. Est. Tekhn. Pribalt. 5 (1976), 134-141

TAMURA, SAKAE [1884-]
HAKUSHI 2,254-255

TANAKA, KANICHI [1882-]
JBE, 1640

TANAKA, MASAHIKO [1876-]
HAKUSHI 3,274

TANAKA, YOSHIMARO [1884-1972]
JBE, 1650; IB, 295
S. Nakazawa, Folia Mendeliana 11 (1976), 47-52

TANANAEV, NIKOLAI ALEKSANDROVICH [1878-1959]
BSE 41,586; WWR, 533; POGG 6,2614; 7b,5329-5333
Anon., Zhurnal Analiticheskoi Khimii 3 (1948),267-270; 14 (1959),749-750

TĂNĂSESCU, ION [1892-1959]
POTSCH, 417; POGG 6,2614; 7b,5333-5336

TANATAR, SEBASTIAN MOISSEEVICH [1849-1917]
BSE 41,587; ARBUZOV, 192-193; BLOKH 2,715-716; POGG 3,1323; 4,1475; 5,1241
L.A. Chugaev, Zhur. Fiz. Khim. 49 (1917), 637-644
M.A. Blokh, Zhur. Fiz. Khim. 60Suppl. (1928), 146

TANDLER, JULIUS [1869-1936]
FISCHER 2,1546-1547; STOBER, 85-94; PLANER, 338-339; BHDE 1,754-755; IB, 295; WININGER 6,81; 7,473; TETZLAFF, 333
A. Hafferl, Wiener klin. Wchschr. 49 (1936), 1265-1267
A. Schick, Pirquet Bull. Clin. Med. 16(5) (1969), 8-10
K. Sablik, Julius Tandler. Vienna 1983

TANFORD, CHARLES [1921-]
AMS 15(7),17; WWA 1980-1981, p.3247; IWW 1982-1983, p.1300; WWWS, 1648

TANGL, FRANZ [1866-1917]
MEL 2,819; FISCHER 2,1547
Anon., Chem. Z. 42 (1918), 45; Z. angew. Chem. 31(III) (1918), 8
J. Sós, Orv. Kozl. 51-53 (1969), 143-154

TANNENBERG, JOSEPH [1895-]
KURSCHNER 14,4230; FISCHER 2,1547-1548; IB, 295; LDGS, 76; KOREN, 256

TANNER, FRED WILBUR [1888-1957]
AMS 9(II),1117; IB, 295; WWAH, 594

TANNERY, PAUL [1843-1904]
DSB 13,251-256; CHARLE(2), 250-252; POGG 3,1323-1324; 4,1426; 5,1242; 6,2615
H. Bosmans, Rev. Quest. Sci. [3]8 (1905), 544-574
J. Tannery, Mem. Soc. Sci. Bordeaux [6]4 (1908), 269-382
M. Tannery et al., Osiris 4 (1938), 633-705
G. Sarton, Isis 38 (1947), 33-51; Arch. Inter. Hist. Sci. 4 (1951), 324-356

TANRET, CHARLES [1847-1917]
GORIS, 576-578; TSCHIRCH, 1132; REBER, 385; WWWS, 1649; POGG 3,1324;
4,1476-1477; 5,1242
G. André, Bull. Soc. Chim. [4]23 (1917), i-xxxvi
R. Delvincourt, Rev. Hist. Pharm. 19 (1968), 3-15,121-130

TANRET, GEORGES [1878-1937]
POGG 5,1242; 6,2615-2616; 7b,5336
R. Fabre, Bull. Soc. Chim. Biol. 19 (1937), 1585-1587

TAPPEINER, HERMANN [1847-1927]
FISCHER 2,1548-1549; PAGEL, 1690; TSCHIRCH, 1132; DZ, 1452-1453; TLK 2,696; POGG 3,1324; 4,1477; 6,2616
A. Jodlbauer, Deutsche med. Wchschr. 53 (1927), 377

TAPPEN, HANS [1879-1969]
JV 19,25; TLK 3,649; NUC 583,63
Anon., Nachr. Chem. Techn. 12 (1964), 458; 17 (1969), 401

TARASEVICH, LEV ALEKSANDROVICH [1868-1927]
BSE 41,615-616; BME 31,1141-1142; WWR, 534
V. Lubavski, Russkaya Klinika 8 (1928), 157-162
H. Zeiss, Münch. med. Wchschr. 75 (1928), 227
D.D. Pletner, Sudhoffs Arch. 23 (1930), 91-96

TARBELL, DEAN STANLEY [1913-]
AMS 15(7),24; WWA 1980-1981, p.3249; IWW 1981-1982, p.1266; WWWS, 1649; MH 2,535-537; MSE 3,187-1888

TARKHANOV, IVAN ROMANOVICH [1846-1908]
BSE 41,635; BME 31,1145-1146; FISCHER 2,1549-1550
N. Cybulski, Gazeta Lekarska 28 (1908), 987-992,1013-1017
G.I. Mchedshvili, Fiziol. Zhur. 40 (1954), 368-371

TARR, HUGH LEWIS AUBREY [1905-]
AMS 13,4426; SRS, 66; YOUNG, 63; WWWS, 1649

TARVER, HAROLD [1908-]
AMS 14,5042; WWWS, 1649

TASCHNER, EMIL [1900-1982]
B. Liberek, in Peptides 1982 (K. Blaha and P. Malon, Eds.), pp.xlviii-xlxi. Berlin 1983

TASHIRO, SHIRO [1883-1963]
DSB 13,262-263; AMS 10,4029; IB, 296; WWWS, 1649

TATTERSALL, GEORGE [1881-1943]
SRS, 31
Anon., J. Roy. Inst. Chem. 1943, pp.128-129

TATUM, ARTHUR LAWRIE [1884-1955]
AMS 9(II),1118; PARASCANDOLA, 52-53; CHEN, 42; IB, 296
F.E. Shideman, Science 123 (1956), 449
J.P. Swann, J. Hist. Med. 40 (1985), 167-187

TATUM, EDWARD LAWRIE [1909-1975]
AMS 12,6296; NCAB I,475-476; MH 1,464-465; STC 3,51-52; MSE 3,188-189; CB 1959, pp.437-439; DAMB, 732-733; WWWS, 1650; POTSCH, 417
J. Lederberg, Am. Phil. Soc. Year Book 1977, pp.97-101; Ann. Rev. Gen. 13 (1979), 1-5; Biog. Mem. Nat. Acad. Sci. 59 (1990), 357-386
C. Kopp, Mendel Newsletter 20 (1981), 1-2

TAUB, LUDWIG [1877-]
JV 21,579; NUC 583,675

TAUB, WILLIAM [1914-]
BHDE 2,1154

TAUBE, HENRY [1915-]
AMS 15(7),31; WWA 1980-1981, p.3252; IWW 1982-1983, p.1304; WWWS, 1650; MSE 3,190-191; POTSCH, 417-418

TAUBENHAUS, JACOB JOSEPH [1884-1937]
AMS 5,1099; WWA 1934-1935, p.2319; IB, 296; WININGER 6,84

TAUFKIRCH, HEINRICH [1866-]
JV 8,152; RSC 16,344; NUC 584,25

TAYLOR, ALFRED SWAINE [1806-1880]
DNB 19,402-403; BETTANY 2,291-294; HIRSCH 5,525
N.G. Coley, Med. Hist. 35 (1991), 409-427

TAYLOR, ALONZO ENGELBERT [1870-1949]
AMS 7,1758; CHITTENDEN, 171-172; IB, 296
J.P. Baumberger and B.O. Raulston, Trans. Assoc. Am. Phys. 63 (1950), 17-18

TAYLOR, CHARLES VINCENT [1885-1946]
DSB 13,268-269; AMS 7,1759; WWWS, 1652
W.C. Twitty, Anat. Rec. 98 (1947), 242-243
C.H. Danforth, Biog. Mem. Nat. Acad. Sci. 25 (1948), 205-225

TAYLOR, CLARA [1885-1940]
I.D. Rae, Chem. Brit. 27 (1991), 145-148

TAYLOR, FRANK SHERWOOD [1897-1956]
DNB [1951-1960], 957-958; POGG 6,2620; 7b,5363-5365
E.J. Holmyard, Ambix 5 (1956), 57-58
D. McKie, Nature 177 (1956), 774
E.F. Caldin, Proc. Chem. Soc. 1957, pp.151-152
A.V. Simcock, Ambix 34 (1987), 129-139

TAYLOR, FRED ANDERSON [1893-1986]
AMS 11,5329

TAYLOR, HUGH STOTT [1890-1974]
DSB 18,898-899; AMS 11,5330; WW 1966, p.3044; NCAB 58,49-50; SRS, 48; DAVIS, 33-44; MH 1,469-470; STC 3,54-55; MSE 3,193-194; WWWS, 1653; POGG 5,1244; 6,2621-2622; 7b,5350-5355
H.S. Taylor, Ann. Rev. Phys. Chem. 13 (1962), 1-18
C. Kemball, Biog. Mem. Fell. Roy. Soc. 21 (1975), 517-547
J. Turkevich, Am. Phil. Soc. Year Book 1977, pp.101-104

TAYLOR, JOHN FULLER [1912-]
AMS 15(7),41; WWWS, 1653

TAYLOR, THOMAS WESTON JOHNS [1895-1953]
POGG 6,2624; 7b,5365-5366
Who Was Who 1951-1960, pp.1071-1072
N. Millott, Nature 172 (1953), 652-653
D.L. Hammick, J. Chem. Soc. 1954, pp.767-768

TCHEN, TCHE TSING [1924-]
AMS 15(7),48; WWWS, 1654-1655

TEBB, MARY CHRISTINE [1868-1953]
O'CONNOR(2), 188; RSC 19,43
H. King, Biog. Mem. Fell. Roy. Soc. 2 (1956), p.257
M.R.S. Creese, Brit. J. Hist. Sci. 24 (1991), 281-282

TECHOW, WALTER [1870-]
JV 9,21; RSC 19,46; NUC 585,378

TEICHMANN, LUDWIG [1823-1895]
DSB 13,273-274; HIRSCH 5,529; PAGEL, 1696-1697; KOSMINSKI, 512
K. von Kostanecki, Anat. Anz. 11 (1895), 422-424
F. Lejars, Rev. Sci. [4]5 (1896), 481-487
L. Wachholz, Arch. Hist. Fil. Med. 10 (1930), 34-62
T. Rogalski, Arch. Hist. Med. 20 (1957), 45-126
S. Kohmann, Szescetlecie Medycyny Krakowskiej, pp.141-156. Cracow 1963
J. Stahnke, Med. Hist. J. 16 (1981), 391-413; Würzburg medizinisch-historische Mitteilungen 2 (1984), 205-267; 5 (1987), 381

TEISSIER, GEORGES [1900-1972]
DSB 18,901-904; QQ 1971-1972, p.1472; BUICAN, 308-334; IB, 297
J. Bergerard, Cahiers de la Biologie Marine 13 (1972), 689-715

TELETOV, IVAN SERGEEVICH [1878-1947]
RIGA, 722; JV 22,302; POGG 6,2627; 7b,5366-5368

TEMIN, HOWARD MARTIN [1934-1994]
AMS 15(7),55-56; WWA 1980-1981, p.3264; IWW 1982-1983, p.1309; MSE 3,198-199

TEMKIN, OWSEI [1902-]
AMS 15(7),56
L.G. Stevenson and R. Multhauf, Eds., Medicine, Science and Culture, pp.303-312. Baltimore 1968

TENBROEK, CARL [1885-1966]
AMS 10,4051; IB, 36; WWAH, 597
Anon., Rock. Univ. Rev. 1966(Sept.-Oct.), p.20

TENDELOO, HENRICUS JACOBUS CHARLES [1896-1984]
BWN 3,585-586

TENDRON, EDMOND [1867-1942]
G. Loiseau, Ann. Inst. Pasteur 68 (1942), 367-368

TENNENT, DAVID HILT [1873-1941]
DSB 13,281-282; AMS 6,1407; NCAB 36,354; IB, 297; WWWS, 1657
M.H. Jacobs, Am. Phil. Soc. Year Book 1941, pp.409-411
W.S. Gardiner, Biog. Mem. Nat. Acad. Sci. 26 (1951), 99-119

TENNER, ALFONS [1829-1898]
RAUSCH, 277-278

THEORELL, ERIC TORSTEN [1905-]
VAD 1983, p.1016; POGG 6,2679; 7b,5379-5382

TEREG, JOSEPH [1850-1915]
DZ, 1455-1456
H. Arnold, Deutsche Tierärztliche Wochenschrift 23 (1915), 49

TERENIN, ALEKSANDR NIKOLAEVICH [1896-1967]
BSE(3rd Ed.) 25,470; WWR, 538; WWWS, 1658; POGG 6,2630; 7b,5383-5892
R.I. Goryacheva and O. Rumyantseva, A.N. Terenin. Moscow 1971

TERRAY, PAUL von [1861-1926]
MEL 2,845-846; FISCHER 2,1556; RSC 19,61
Z. Ritock, Orv. Het. 70 (1926), 1313

TERROINE, ÉMILE FLORENT [1882-1974]
QQ 1974-1975, p.1525; IB, 297; POGG 6,2633-2634; 7b,5412-5415
E.F. Terroine, Ann. Rev. Biochem. 28 (1959), 1-14
J. Roche, C.R. Soc. Biol. 168 (1974), 930-932
R. Jacquet, Annales Nutritives et Alimenteuses 28(4) (1974), 3-12

M. Florkin, Bull. Acad. Roy. Belg. 42 (1975), 821-824
J. Bustarret, C.R. Acad. Agr. 61 (1975), 51-53

TERRY, BENJAMIN TAYLOR [1876-1955]
AMS 8,2477
Anon., J. Am. Med. Assn. 158 (1955), 1293

TERRY, ROBERT JAMES [1934-]
AMS 15(7),66

TERU-UCHI, YUKATA [1873-1936]
JBE, 1683; HAKUSHI 2,120-121; IB, 297

TESKE, ARMIN [1910-1967]
W. Hubicki, Kwart. Hist. Nauki 12 (1967), 801-808; Arch. Inter. Hist. Sci. 21 (1968), 143-145

TEUBER, HANS JOACHIM [1918-]
KURSCHNER 13,2931; WWWS, 1660; POGG 7a(4),639-640

THACKRAH, CHARLES TURNER [1795-1833]
HIRSCH 5,540; CALLISEN 19,138-139; 33,1
A. Meiklejohn, The Life, Works and Times of Charles Turner Thackrah. Baltimore 1957
R.I. McCallum, Pharmaceutical Historian 15(4) (1985), 2-4

THAER, ALBRECHT DANIEL [1752-1828]
ADB 37,636-641; HIRSCH 5,541; FERCHL, 531; WWWS, 1660; POGG 2,1087
W. Korte, Albrecht Thaer. Leipzig 1839
M. Jumpertz, Med. Welt 2 (1928), 1589-1590
V. Klemm et al., Wiss. Z. Berlin 27(1) (1978), 5-18

THAL, ALEXANDER [1889-]
RIGA, 390; JV 30,739; NUC 588,583

THAN, KAROLY (CARL) [1834-1908]
DSB 13,298-299; MEL 2,852-854; POTSCH, 418-419; POGG 3,1333; 4,1487; 5,1248
F. von Konek-Norwall, Chem. Z. 32 (1908), 673
G.B. Kauffman, J. Chem. Ed. 66 (1988), 213-216

THANNHAUSER, SIEGFRIED JOSEPH [1885-1962]
KURSCHNER 9,2086; FISCHER 2,1559; BHDE 2,1161; LDGS, 69; DRULL, 267; FRANKEN, 198; KOREN, 256-257; WININGER 7,474-475; COHEN, 241; POGG 6,2638-2639; 7a(4),644-645
L. Heilmeyer, Münch. med. Wchschr. 97 (1955), 1183-1184
G.A. Martini and G. Schmidt, Deutsche med. Wchschr. 80 (1955), 987-989
J. Siegfried, New Eng. J. Med. 268 (1963), 680-681
M.G. Goldner, Proc. Virchow Med. Soc. 22 (1963), 138-143
G. Hoffmann, Münch. med. Wchschr. 105 (1963), 357-359
N. Zöllner, Deutsche med. Wchschr. 88 (1963), 337-340; Metabolism 12 (1963), 261-263; Internist (Berlin) 10 (1969), 106-109
T. Lechner, Med. Welt 1965, pp.226-227

THAUER, RUDOLF [1906-1986]
KURSCHNER 13,3934; KALLMORGEN, 432; WWWS, 1661; POGG 7a(4),647-648
Anon., Naturw. Rund. 39 (1986), 233

THAULOW, HARALD CONRAD [1815-1881]
NBL 16,177-178; HEIN, 674; FERCHL, 531; POGG 3,1333-1334
J.B. Halvorsen, Norsk Forfatter-Lexikon 1814-1880 5,676-677. Christiania 1901

THAULOW, (MORITZ CHRISTIAN) JULIUS [1812-1850]
FERCHL, 531; RSC 5,944; POGG 2,1088

THAXTER, ROLAND [1858-1932]
DSB 13,299-300; DAB 18,398-399; NCAB 30,482-483; HUMPHREY, 244-246; IB, 298; WWAH, 599; WWWS, 1661
C.W. Dodge, Annales de Cryptogamie Exotique 6 (1933), 1-12
W.H. Weston, Mycologia 25 (1933), 69-89
W.A. Setchell, Proc. Am. Acad. Arts Sci. 68 (1933), 678-682
G.P. Clinton, Biog. Mem. Nat. Acad. Sci. 17 (1936), 55-68
J.G. Horsfall, Annual Review of Phytopathology 17 (1979), 29-35

THAYER, SIDNEY ALLEN [1902-1969]
AMS 11,5363; NCAB 54,524-525; WWWS, 1661
Anon., Chem. Eng. News 47(14), 51

THAYER, WILLIAM SYDNEY [1864-1932]
DSB 13,300-301; DAB 18,414-415; NCAB 24,409-410; OEHRI, 116-118; DAMB, 737; ICC, 44-46; WWAH, 599; WWWS, 1661
L.P. Barker, Science 76 (1932), 617-619
E.I. Reid, The Life and Convictions of Sydney William Thayer. London 1936
E.S. Hunley, Bull. Inst. Hist. Med. 4 (1936), 751-781

THEILACKER, WALTER [1903-1968]
KURSCHNER 12,3223; HANNOVER, 35; POGG 7a(4),648-650
G. Wittig, Chem. Ber. 103(8) (1970), xli-lii

THEILER, MAX [1899-1972]
AMS 11,5363; MH 1,473-474; STC 3,60-61; MSE 3,200-201; DAMB, 738-739; ABBOTT, 126-127; WWAH, 599; WWWS, 1661
Anon., J. South Afr. Vet. Assn. 44 (1973), 460-462

THENARD, LOUIS JACQUES [1777-1857]
DSB 13,309-314; GE 29,1180; LAROUSSE 15,74; BUGGE 1,386-404; W, 507-508; FRANKEN, 198; FARBER, 345-347; BLOKH 2,718-720; TSCHIRCH, 1132-1133; FERCHL, 532; SCHAEDLER, 133; COLE, 522-528; WWWS, 1661; POTSCH, 419; POGG 2,1088-1090; CALLISEN 19,152-160; 33,5-6
Anon., Proc. Roy. Soc. 9 (1857), 60-64
P. Kelland, Proc. Roy. Soc. Edin. 4 (1857-1862), 30-33
P. Flourens, Mem. Acad. Sci. 32 (1864), i-xxxv
K.R. Webb, Chem. Ind. 1945, pp.2-4
P. Thenard, Un Grand Français: Le Chimiste Thenard. Dijon 1950

THENARD, PAUL [1819-1884]
FERCHL, 531-532; POGG 3,1334
P.P. Dehérain, Rev. Sci. [3]8 (1884), 613-518
F. Bouley and E. Fremy, C.R. Acad. Sci. 99 (1884), 293-302

THEOBALD, ERNST [1889-1958]
JV 27,395

THEORELL, AXEL HUGO THEODOR [1903-1982]
VAD 1981, P.1037; IWW 1982-1983, pp.1313-1314; AO 1982, pp.385-387; MH 1,474-475; STC 3,61-62; MSE 3,201-202; CB 1956, pp.622-624; POTSCH, 419-420
Anon., Nachr. Chem. Techn. 3 (1955), 204-205
E.C. Slater, Vitamins and Hormones 28 (1970), 147-150
H. Theorell, in Proteolysis and Physiological Regulation (D.W. Robbins and K. Brew, eds.), pp.1-27. New York 1976
B. Chance and B.L. Vallee, TIBS 8 (1983), 45
K. Dalziel, Biog. Mem. Fell. Roy. Soc. 29 (1983), 585-621
G. Braunitzer, Bayer. Akad. Wiss. Jahrbuch 1984, pp.244-246
M.D. Kamen, Am. Phil. Soc. Year Book 1985, pp.198-206

THIEL, ALFRED [1879-1942]
KURSCHNER 4,3007; MEINEL, 321-324,514; AUERBACH, 916; TLK 3,653; POGG 4,1488; 5,1248-1249; 6,2642; 7a(4),654-655
Anon., Z. Elektrochem. 48 (1942), 451

THIEL, KARL EUGEN [1830-1915]
WOLF, 205; RSC 5,952; POGG 3,1335

THIELE, EDMUND [1867-1927]
JV 9,211; NUC 589,642; POGG 6,2643
W.A. Dyes, Chem. Z. 51 (1927), 481-482

THIELE, JOHANNES [1865-1918]
DSB 13,337-338; DBJ 2,706; BLOKH 2,720-722; DZ, 1460; W, 508-509; POTSCH, 420; WWWS, 1662; POGG 4,1489-1490; 5,1249-1260; 6,2643
H. Wieland, Bayer. Akad. Wiss. Jahrbuch 1918, pp.65-69
E. Wedekind, Chem. Z. 42 (1918), 217
F. Straus, Ber. chem. Ges. 60A (1927), 75-132

THIELE, JOHANNES [1903-]
JV 42,664; NUC 589,682

THIEMANN, HERMANN [1885-1953]
JV 27,398; NUC 589,679

THIERFELDER, HANS [1858-1930]
DSB 18,904-906; FISCHER 2,1562; BK, 325; IB, 298; POGG 6,2644-2645
P. Brigl, Ber. chem. Ges. 63A (1930), 176-177
E. Klenk, Z. physiol. Chem. 203 (1931), 1-9

THIERSCH, KARL [1822-1895]
ADB 55,254-263; HIRSCH 5,556; PAGEL, 1704-1705; SCHWARTZ, 113-119; PITTROFF, 36-38; KILLIAN, 139-140; WWWS, 1662
A. von Bardeleben, Deutsche med. Wchschr. 21 (1895), 311
A. Landerer, Münch. med. Wchschr. 42 (1895), 472-475
H. Helfereich, Deutsche Z. Chir. 41 (1895), 617-633
W. His, Pop. Sci. Mon. 52 (1897-1898), 338-353
J. Thiersch, Carl Thiersch. Leipzig 1922
K. Sudhoff, Sächsische Lebensbilder 1 (1930), 377-386

THIMANN, KENNETH VIVIAN [1904-]
AMS 14,5087; WWA 1978-1979, p.3320; IWW 1982-1983, p.1315; MH 1,475-476; STC 3,62-63; MSE 3,202-203; WWWS, 1662-1663

THIRY, LUDWIG [1817-1897]
RSC 5,956; 8,1074; 12,728

THÖRNER, WALTER [1886-1969]
KURSCHNER 10,2490; FISCHER 2,1562; IB, 298

THOM, CHARLES [1872-1956]
AMS 9(II),1127; NCAB 44,330-331; IB, 298; WWAH, 599
K.B. Raper, J. Bact. 72 (1956), 725-727; Mycologia 49 (1957), 134-150; Biog. Mem. Nat. Acad. Sci. 38 (1965), 309-344

THOMA, RICHARD [1847-1923]
FISCHER 2,1562-1563; PAGEL, 1706; FREUND 2,389-396; LEVITSKI 2,102-107; VERSO, 100
P. Ernst, Deutsches Arch. klin. Med. 145 (1924), i-iv
M.L. Verso, Med. Hist. 15 (1971), 62

THOMAS, KARL [1883-1969]
DSB 18,906-908; FISCHER 2,1563; BERWIND, 199-204; IB, 298-299; TLK 3,656; POGG 6,2649; 7a(4),670-671
K. Thomas, Ann. Rev. Biochem. 23 (1954), 1-16
E. Schütte, Naturwiss. 50 (1963), 701-702; Z.physiol. Chem. 364(11) (1983), (1)-(3)
E. Strack, Jahrb. Sachs. Akad. 1969-1970, pp.275-281
G. Weitzel, Z. physiol. Chem. 351 (1970), 1-14; Deutsche med. Wchschr. 95 (1970), 291-296
H. Fisher, J. Nutrition 101 (1971), 1109-1115
M. Liefländer, Göttinger Jahrbuch 1984, pp.257-259

THOMAS, LYELL JAY [1892-1977]
AMS 11,5377; IB, 299
B.B. Babero, J. Parasit. 64 (1978), 122

THOMAS, MEIRION [1894-1977]
DNB [1971-1980], 839-840; WW 1977, p.2394
H.K. Porter and S.L. Ranson, Biog. Mem. Fell. Roy. Soc. 24 (1978), 547-568

THOMAS, PIERRE [1876-1964]
POGG 6,2650; 7b,5433

THOMPSON, ALBERTO FREDERIC, Jr. [1907-]
AMS 9(I),1934

THOMPSON, BENJAMIN [see RUMFORD]

THOMPSON, D'ARCY WENTWORTH [1860-1948]
DSB 13,352-353; DNB [1941-1950],877-879; O'CONNOR(2), 429-430
W.T. Calman and J.L. Myers, Nature 162 (1948), 93-94
R. Chambers, Science 109 (1949), 138-139,151
C. Dobell, Obit. Not. Fell. Roy. Soc. 6 (1949), 599-617
G. Sarton, Isis 41 (1950), 3-8
R.D. Thompson, D'Arcy Wentworth Thompson. London 1958

THOMPSON, EDWARD OWEN PAUL [1925-]
Who's Who in Australia 1991, pp.1147-1148

THOMPSON, JAMES [1876-1956]
JV 22,303; NUC 591,539
Anon., J. Roy. Inst. Chem. 80 (1956), 604

THOMPSON, KENWORTHY JAMES [1881-1933]
SRS, 30; NCAB 29,37; NUC 591,595

THOMPSON, LEWIS [1810?-1889]
RSC 5,959; 11,584; 19,96
W.V. Farrar, Ambix 18 (1971), 123-128

THOMPSON, ROBERT HENRY STEWART [1912-]
WW 1981, p.2581; IWW 1982-1983, p.1318

THOMPSON, WALTER PALMER [1889-1970]
AMS 11,5394; SRS, 44
J.G. Rempel, Proc. Roy. Soc. Canada [4]9 (1971), 111-114

THOMPSON, WILLIAM HENRY [1860-1918]
WW 1918, p.2361; O'CONNOR(2), 434-435; NUC 592,56
Anon., Nature 102 (1918), 170

THOMS, GEORGE [1843-1902]
WELDING, 790-791; BJN 7,116*-117*; RIGA, 723; POGG 4,1494

THOMS, HERMANN [1859-1931]
KURSCHNER 4,3017; HEIN, 675-677; FISCHER 2,1564; REBER, 309-314; IB, 299; DRUM, 273-282; WREDE, 194-195; TSCHIRCH, 1133; SCHELENZ, 704; TLK 3,656-657; POGG 4,1494-1495; 5,1252-1253; 6,2653; 7a(4),671
C. Mannich, Ber. chem. Ges. 62A (1929), 65-66
P. Siedler, Arch. Pharm. 270 (1932), 1-14
M. Bodenstein, Ber. chem. Ges. 65A (1932), 22-23
E. Urban, Pharm. Z. 85 (1949), 123-126
K.W. Merz, Pharm. Z. 104 (1959), 1061-1053
B. Reichert, Arzneimitt. 9 (1959), 279-280
C. Friedrich, Beiträge zur Geschichte der Pharmazie 40 (1988), 33-36

THOMSEN, JULIUS [1826-1909]
DSB 13,358-359; DBL 23,568-575; MEISEN, 143-147; VEIBEL 1,202-210; 2,426-444; 3,46-68; BLOKH 2,723-726; SCHAEDLER, 134; WWWS, 1668; POTSCH, 422; POGG 2,1097; 3,1338-1340; 4,1495-1496; 5,1253
N. Bjerrum, Ber. chem. Ges. 42 (1909), 4971-4988
I. Traube, Chem. Z. 33 (1909), 189
E. Thorpe, Nature 82 (1910), 501-505; J. Chem. Soc. 97 (1910), 161-172
H. Kragh, Ann. Sci. 39 (1982), 37-60

THOMSON, DAVID LANDSBOROUGH [1901-1964]
WW 1965, p.3043; AMS 10,4086; YOUNG, 11; SRS, 85
J.S.L. Browne and O.F. Denstedt, Nature 206 (1965), 136
K.A.C. Elliott, Proc. Roy. Soc. Canada [4]3 (1965), 177-181

THOMSON, JOSEPH JOHN [1856-1940]
DSB 13,362-372; DNB [1931-1940],857-863; W, 510-511; WWWS, 1669; POTSCH, 422; POGG 3,1344-1345; 4,1498-1499; 5,1253-1254; 6,2654; 7b,5450-5451
J.J. Thomson, Recollections and Reflections. New York 1936
Rayleigh, Obit. Not. Fell. Roy. Soc. 3 (1941), 587-609; J.J. Thomson. Cambridge 1942
G. Thomson and J. Thomson, Notes Roy. Soc. 12 (1957), 201-210
G.P. Thomson, J.J. Thomson and the Cavendish Laboratory in his Day. New York 1965

THOMSON, ROBERT DUNDAS [1810-1864]
DNB 19,748; FERCHL, 534; WWWS, 1669-1670; POGG 3,1344
Anon., Proc. Roy. Soc. Edin. 5 (1862-1866), 307-309
A.W. Williamson, J. Chem. Soc. 18 (1865), 344-345

THOMSON, THOMAS [1773-1852]
DSB 13,372-374; DNB 19,751-752; BLOKH 2,726-727; FERCHL, 534-535; TSCHIRCH, 1133; SCHAEDLER, 134-135; COLE, 530-536; WWWS, 1670; POTSCH, 422-423; POGG 2,1097-1100; CALLISEN 19,213-220; 33,25-26
R.D. Thomson, Glasgow Med. J. 5 (1857), 69-80,121-153,379-380
H.S. Klickstein, Chymia 1 (1948), 37-53
J.R. Partington, Ann. Sci. 6 (1949), 115-126
E.H. Huntress, Proc. Am. Acad. Arts Sci. 81 (1952), 90-92
A. Kent, Brit. J. Hist. Sci. 2 (1964), 59-63
J.B. Morrell, Brit. J. Hist. Sci. 4 (1969), 245-265; Ambix 19 (1972), 1-46
D.F. Larder, Notes Roy. Soc. 24 (1969), 295-304
S.H. Mauskopf, Ann. Sci. 25 (1969), 229-242

THOMSON, THOMAS [1817-1878]
DNB 19,752-753; DESMOND, 609-610; NUC 592,327
M.J. Berkeley, Nature 18 (1878), 15-16
Anon., Annals of Botany 1878, p.160; 1899, pp.461-462

THOMSON, WILLIAM [see KELVIN]

THORELL, BO [1921-1982]
VAD 1979, p.1029
E. Kohen, Nouv. Rev. Hem. 26 (1984), 53-55

THORN, GEORGE WIDMER [1906-]
AMS 12,6385; MH 2,544-546; STC 3,65-66; MSE 3,211-212; ICC, 248-250; WWWS, 1670

THORNDIKE, LYNN [1882-1965]
DAB [Suppl.7],741-742; NCAB 51,214-215
D.B. Durand, Isis 33 (1942), 691-712
P. Kibre, Osiris 11 (1954), 5-22; Arch. Inter. Hist. Sci. 20 (1967), 285-288
M. Clagett, Isis 57 (1966), 85-89
K.M. Setton, Am. Phil. Soc. Year Book 1967, pp.150-155

THORNE, LEONARD TEMPLE [1855-1941]
RSC 11,595-596; 19,108
Anon., J. Roy. Inst. Chem. 1941, pp.240-241; Chem. Ind. 1941, p.709
J.L. Baker, J. Chem. Soc. 1942, p.336

THORNE, PERCY CYRIL LESLIE [1890-1957]
A.M. Ward, Proc. Chem. Soc. 1958, p.174

THORNTON, HENRY GERARD [1892-1977]
WW 1977, p.2408; IB, 299; WWWS, 1671
P.S. Nutman, Biog. Mem. Fell. Roy. Soc. 23 (1977), 557-574

THORPE, JOCELYN FIELD [1872-1940]
DSB 13,388-389; FINDLAY, 369-401; WWWS, 1671; POTSCH, 423; POGG 4,1500; 5,1255; 6,1657-2658; 7b,5458-5459
Anon., J. Roy. Inst. Chem. 1940, pp.273-275; Chem. Ind. 1940, p.424
G.A.R. Kon and R.P. Linstead, J. Chem. Soc. 1941, pp.444-464

THORPE, THOMAS EDWARD [1845-1925]
DSB 13,389-390; DNB [1922-1930],842-843; BLOKH 2,727-730; WWWS, 1671; W, 513; POGG 3,1345; 4,1499-1500; 5,1255; 6,2658
A.E.H. Tutton, Proc. Roy. Soc. A109 (1925), xviii-xxiv
P.P.B., J. Chem. Soc. 129 (1926), 1031-1050

THUDICHUM, LUDWIG [1829-1901]
BJN 6,107; HIRSCH 5,578-579; PAGEL, 1709-1710; HAYMAKER, 297-302; FRANKEN, 199; POGG 3,1346
Anon., Nature 64 (1901), 527-528; Brit. Med. J. 1901, p.726
K. Sudhoff, Nachr. Giessener Ges. 9 (1932), 33-45
D.L. Drabkin, Thudichum. Philadelphia 1958
H. McIlwain, Proc. Roy. Soc. Med. 51 (1958), 127-132
C. Chatagnon and P. Chatagnon, Ann. Med. Psych. 116(I) (1958), 267-282
R.S. Sparkman, Brit. Med. J. 1959(II), pp.753-754
H. Debuch and R.M. Dawson, Nature 207 (1965), 814
J.D. Spillane, Brit. Med. J. 1974(IV), pp.701-706,757-759
O.J. Rafaelson, in Historical Aspects of the Neurosciences (F.C. Rose and W.F. Bynum, Eds.), pp.293-305. New York 1982
B. Schulz, in De Novo Inventis (A.H.M. Kerkhoff et al., eds.), pp.389-400. Amsterdam 1984

THUN, KARL [1866-1915]
JV 5,81; NUC 593,340

THUNBERG, THORSTEN LUDVIG [1873-1952]
DSB 13,393-394; FISCHER 2,1567-1568; IB, 299; POGG 6,2550
S. Lindroth, Swedish Men of Science, pp.151-159. Stockholm 1952
H. Wieland, Bayer. Akad. Wiss. Jahrbuch 1953, pp.180-181
G. Kahlson, Acta Physiol. Scand. 30 Suppl.111 (1953), 1-24
F.G. Young, Nature 172 (1953), 1079
Anon., Leopoldina [3]2 (1956), 17-18

THURET, GUSTAVE ADOLPHE [1817-1875]
DSB 13,394-396; TSCHIRCH, 1134; WWWS, 1672
E. Bornet, Ann. Sci. Nat. [6]2 (1875), 308-360
W.G. Farlow, Journal of Botany 14 (1876), 4-9

THURN, GEORG WILHELM [1813-]
DGB 66 (1929), 494

THYSSEN, HEINRICH [1875-1947]
JV 15,134; NUC 593,539

TIDY, CHARLES MEYMOTT [1843-1892]
DNB 19,864-865; HIRSCH 5,585-586; WWWS, 1673; NUC 593,682-683
Anon., Lancet 1892(I), p.650; Brit. Med. J. 1892(I), p.619; Chem. News 65 (1892), 143; J. Chem. Soc. 63 (1893), 766-769

TIEDE, ERICH [1884-1951]
JV 23,66; TLK 3,657; POGG 5,1257; 6,2661-2662; 7a(4),679-680
A. Schleede, Z. angew. Chem. 64 (1952), 577-579

TIEDEMANN, FRIEDRICH [1781-1861]
DSB 13,402-404; ADB 38,277-278; BAD 2,352-358; STUBLER, 247-253; DRULL, 269-270; HIRSCH 5,586-597; FERCHL, 537; WWWS, 1673; CALLISEN 19,240-245; 33,33-35; POGG 2,1106; 7aSuppl.,688-670
Anon., Proc. Roy. Soc. 12 (1863), xxvii-xxxii
K. Licksteig, Die Verdienste des Heidelberger Professoren Friedrich Tiedemann für die Heilkunde. Düsseldorf 1939
N. Mani, Gesnerus 13 (1956), 190-214

TIEDEMANN, HEINZ [1923-]
KURSCHNER 13,3967; POGG 7a(4),680-681

TIEGEL, ERNST [1849-1889]
RSC 8,1087; 11,603; 19,120
Anon., Japanese Physiology Past and Present, pp.41-42. Tokyo 1965
O. Keller, Schaffhauser Beiträge zur Vaterlandischen Geschichte 62 (1985), 71

TIEGHEM, PHILIPPE van [1839-1914]
DSB 13,405-406; VAPEREAU 6,1548; TSCHIRCH, 1140
G. Bonnier, Rev. Gen. Bot. 26 (1914), 353-440
J. Constantin, Ann. Sci. Nat. [9]19 (1914), i-viii; Nouv. Arch. Sci. Nat. [6]2 (1927), 1-19
R. Chodat, Ber. Bot. Ges. 33 (1915), (5)-(24)

TIEGS, OSCAR WERNER [1897-1956]
WW 1956, p.2969
C.F.A. Pantin, Biog. Mem. Fell. Roy. Soc. 3 (1957), 247-255

TIEMANN, FERDINAND [1848-1899]
DSB 13,406-407; BJN 4,185*; HEIN, 678-679; BLOKH 2,730-732; WWWS, 1673; POTSCH, 424; POGG 3,1348-1349; 4,1503-1504; 7aSuppl.,690
O.N. Witt, Ber. chem. Ges. 34 (1901), 4403-4455
T. Kunzmann, Naturw. Rund. 2 (1949), 564-565

TIETZE, HERMANN [1864-]
JV 9,211; RSC 19,126

TIFFENEAU, MARC [1873-1945]
CHARLE(2), 252-254; GORIS, 581-584; WWWS, 1673; POGG 5,1258-1259; 6,2663-2664; 7b,5471-5477
M. Delépine, Bull. Acad. Med. 129 (1945), 407-413
G. Roussy, Presse Med. 53 (1945), 438-439
A. McKenzie, Nature 156 (1945), 656-657; J. Chem. Soc. 1949, pp.1668-1669
J. Levy, Bull. Soc. Chim. Biol. 28 (1946), 196-200
E. Fourneau, Bull. Soc. Chim. [5]15 (1948), 905-932

TIGERSTEDT, ARTUR [1865-1908]
RIGA, 209

TIGERSTEDT, CARL [1882-1930]
FISCHER 2,1569; IB, 300; POGG 6,2664
C.G. Santesson, Skand. Arch. Physiol. 61 (1931), 20-22

TIGERSTEDT, ROBERT [1853-1923]
FISCHER 2,1569-1570; PAGEL, 1710-1711; MEDVEI, 809-810; POGG 6,2554
C.G. Santesson, Skand. Arch. Physiol. 45 (1924), 1-6
L.S. Marks and M.H. Maxwell, Hypertension 1 (1979), 384-388

TIKHOMANDRITSKI, ALEKSEI NIKITICH [1814-1853]
IKONNIKOV, 647-651

TIKHOMIROV, ALEKSANDR ANDREEVICH [1850-1931]
BSE 42,498-499; WWR, 540
B.M. Shitkov, Anat. Anz. 73 (1932), 417-424

TILANUS, CHRISTIAN BERNARD [1796-1883]
NNBW 1,1498; HIRSCH 5,588-589
Anon., Leopoldina 19 (1883), 167

TILDEN, WILLIAM AUGUSTUS [1842-1926]
DSB 13,410-411; W, 514-515; WWWS, 1673; POTSCH, 424; POGG 3,1350; 4,1504-1505; 5,1259; 6,2664-2665
M.O. Forster, J. Chem. Soc. 129 (1927), 3190-3202
J.C.P., Proc. Roy. Soc. A117 (1928), i-v

TILLETT, WILLIAM SMITH [1892-1974]
AMS 11,5414; WWA 1974-1975, p.3081; MH 2,546-547; MSE 3,213-215; DAMB, 743-744; ICC, 201-203; WWWS, 1674
S. Sherry, Trans. Assoc. Am. Phys. 88 (1975), 32-34

TILLEY, THOMAS GEORGE [-1849]
FERCHL, 537; RSC 5,996; NUC 594,319; POGG 2,1108-1109

TILLMANN, SAMUEL DYER [1815-1875]
FERCHL, 537-538; POGG 3,1350

TILLMANNS, HEINRICH [1831-1907]
DGB 96 (1937), 155

TILLMANS, JOSEF [1876-1935]
KALLMORGEN, 433; TLK 3,658; TSCHIRCH, 1134; POGG 6,2665-2666
E. Merres, Z. angew. Chem. 48 (1935), 157-160

TIMIRYAZEV, KLIMENT ARKADYEVICH [1843-1920]
DSB 13,416-417; BSE 42,430-434; WWR, 542; KUZENTSOV(1963), 173-191; BLOKH 2,732-733; WWWS, 1674; RSC 8,1092; 11,607; 12,733; 19,132
V.L. Komarov et al., Kliment Arkadyevich Timiryazev. Moscow 1945
G.H. Beale, Nature 159 (1947), 51-53
G. Platonov, Kliment Arkadyevich Timiryazev. Moscow 1955
A.E. Gaissinovich, Folia Mendeliana 6 (1971), 305-320;
Hist. Phil. Life Sci. 7 (1985), 257-286
S.P. Landau-Tylkina, Kliment Timiryazwv. Moscow 1988
R.L. Berg, Quart. Rev. Biol. 65 (1990), 457-479

TIMMERMANS, JEAN [1882-1971]
POGG 5,1259; 6,2667; 7b,5475-5477

TIMOFÉEFF-RESSOVSKY, NIKOLAI VLADIMIROVICH [1900-1981]
DSB 18,919-926; SG [5]1,646
N.N. Vorontsov, Bull. Soc. Nat. Moscou NS75(5) (1970), 144-158
L.A. Blumenfeld et al., Molekularnaya Biologia 15 (1981), 1194-1199
Z.A. Medvedev, Genetics 100 (1982), 1-5
D.R. Paul and C.B. Krimbes, Scientific American 266(2) (1992), 64-70

TIMOFEEV, VLADIMIR FEDOROVICH [1858-1923]
BSE 42,438; KHARKOV-F, 190-191
M.A. Blokh, Zhur. Fiz. Khim. 60Suppl. (1928), 144

TINDALL, WALTER JOSEPH [1907-1974]
Anon., Brit. Med. J. 1974(II), p.337

TINGLE, JOHN BISHOP [1866-1918]
AMS 2,471; JV 5,236; NUC 594,645-646; WWAH, 606; POGG 4,1506-1507; 5,1250; 6,2668
W.R.L., J. Chem. Soc. 115 (1919), 453-454

TIPSON, ROBERT STUART [1906-1991]
AMS 12,6410; WWWS, 1675
H.S. El Khadem and D. Horton, Carbohydrate Res. 169 (1987), xi-xiii

TISELIUS, ARNE [1902-1971]
DSB 13,418-421; MH 1,481-482; STC 3,70-71; MSE 3,219-220; ETTRE, 494-496; CB 1949, pp.603-604; ABBOTT-C, 136; W, 591; WWWS, 1675; POTSCH, 425; POGG 6,2669; 7b,5481-5486
A. Tiselius, Ann. Rev. Biochem. 37 (1968), 1-24
S. Hjerten, J. Chromat. 65 (1972), 345-348
R.A. Kekwick and K.O. Pedersen, Biog. Mem. Fell. Roy. Soc. 20 (1974), 401-428
J. Porath, Horm. Prot. Pep. 5 (1978), 159-185
K.O. Pedersen, in Selected Topics in the History of Biochemistry (G. Semenza, Ed.), pp.233-281. Amsterdam 1983
L.E. Kay, Hist. Phil. Life Sci. 10 (1988), 51-72

TISHCHENKO, VYACHESLAV EVGENIEVICH [1861-1941]
BSE 42,508-509; WWR, 543; ARBUZOV, 134-135; POTSCH, 425
A.P. Okatov, Zhur. Ob. Khim. 18 (1948), 3-13
N.I. Nikitin, Zhur. Ob. Khim. 31 (1961), 3849-3851

TISHLER, MAX [1906-1989]
AMS 14,5135; WWA 1978-1979, p.3248; IWW 1982-1983, pp.1324-1325; MH 2,550-551; MSE 3,220-221
G.B. Kauffman, CHEM TECH 20 (1990), 268-274
J.C. Sheehan, Organic Syntheses 69 (1990), xiii-xv

TISSIER, HENRI [1866-1926]
SG [2]18,284; [3]10,268
A. Besredka, Bulletin of Battle Creek Sanatorium 24 (1929), 73-82

TITHERLEY, ARTHUR WALSH [1874-1966]
SRS, 15; NUC 595,210; POGG 5,1260; 6,2669
E. Mather, Chem. Brit. 3 (1967), 171

TIZZONI, GUIDO [1853-1932]
FISCHER 2,1738-1739; IB, 301; RSC 19,140-141
E. Centanni, Riv. Biol. 14 (1932), 572-577; Bioch. Ter. Sper. 19 (1932), 353-361

TOBLER, FRIEDRICH [1879-1957]
KURSCHNER 8,2402-2403; IB, 301
H. Ulbricht, Ber. bot. Ges. 70 (1957), (43)-(50)

TOBLER, LUDWIG [1877-1915]
DBJ 1,342; FISCHER 2,1573; DRULL, 270
L. Bessau, Jahresb. Schles. Ges. 93 (1915), 37-41
E. Moro, Münch. med. Wchschr. 62 (1915), 880

TODA, SHOZO [1885-1961]
JBE, 688,2367; IB, 301

TODD, ALEXANDER ROBERTUS [1907-]
WW 1981, pp.2598-2599; IWW 1982-1983, p.1326; MH 1,482-483; STC 3,72-73; MSE 3,223; CAMPBELL, 243-244; CB 1958, pp.437-439; WWWS, 1677; SRS, 88; ABBOTT-C, 136-138; POTSCH, 425
Anon., Chem. Eng. News 58(40) (1980), 28-33
A. Todd, A Time to Remember. Cambridge 1983

TODD, CHARLES [1869-1957]
WW 1949, p.2780
C.H. Andrewes, Biog. Mem. Fell. Roy. Soc. 4 (1958), 281-290

TODD, EDGAR WILLIAM [1889-1950]
L.F. Hewitt, J. Path. Bact. 75 (1958), 229-234

TODD, ROBERT BENTLEY [1809-1860]
DNB 19,912-814; HIRSCH 5,599; MR 4,15-16; O'CONNOR, 36-37; FRANKEN, 199; WWWS, 1678
Anon., Lancet 1860(I),p.151; Brit. Med. J. 1860(I), p.111; Proc. Roy. Soc. 11 (1862), xxxii-xxxvi
L.S. Beale, Brit. Med. J. 1870(I), p.21
F.F. Cartwright, Proc. Roy. Soc. Med. 67 (1974), 893-897

TÖDT, FRITZ [1897-1984]
KURSCHNER 13,3980; WWWS, 1678; POGG 6,2671; 7a(4),690-692
Anon., Chem. Z. 108 (1984), 334

TOENNIES, GERRIT [1898-1978]
AMS 11,5429
Anon., Chem. Eng. News 57(4) (1979), 39

TOENNIESSEN, ERICH [1883-1958]
FISCHER 2,1574; HAGEL, 74-79; PITTROFF, 137-138; IB, 301
Anon., Naturw. Rund. 12 (1959), 117

TÖPLER, AUGUST [1836-1912]
BJN 17,159-168; 18,66*; RIGA, 723; POGG 3,1355-1356; 4,1510; 5,1261; 7aSuppl.,692
F. Becke, Alm. Akad. Wiss. Wien 62 (1912), 343-344

TOIVONEN, NIILO JOHANNES [1888-1961]
POGG 6,2672; 7b,5496-5498
W. Hückel, Chem. Ber. 99 (1966), i-xxxiii

TOKIN, BORIS PETROVICH [1900-]
NUC 596,233-234
Anon., Arkh. Anat. Gistol. Embriol. 59 (1970), 121-125

TOLLENS, BERNHARD [1841-1918]
DBJ 2,706; HEIN, 682-683; BLOKH 2,734-736; POTSCH, 426; WWWS, 1678; POGG 3,1356-1357; 4,1512-1513; 5,1262; 6,2673; 7aSuppl.,692
O. Wallach, Ber. chem. Ges. 51 (1918), 1539-1555

P. Ehrenberg, Chem. Z. 42 (1918), 109
C.A. Browne, J. Chem. Ed. 19 (1942), 253-259
Z.E. Gelman, Ambix 25 (1978), 56-62

TOLMACHEV, NIKOLAI ALEKSANDROVICH [1823-1901]
KAZAN 2,346-348; HIRSCH 5,604; ZMEEV 5,163
L., Kazan. Med. Zhur. 1 (1901), 111

TOLMAN, RICHARD CHACE [1881-1948]
DSB 13,429-430; DAB [Suppl.4],837-840; MILES, 474-475; WWWS, 1679; POGG 5,1262-1263; 6,2674; 7b,5505-5506
H.P. Robertson, Am. Phil. Soc. Year Book 1948, pp.295-299
J.G. Kirkwood et al., Biog. Mem. Nat. Acad. Sci. 27 (1952), 139-153

TOMASI, THOMAS B., Jr. [1927-]
AMS 17(7),156; BIBEL, 107-111

TOMITA, MASAJI [1889-1967]
JBE, 1716-1717,2368; POGG 6,2675; 7b,5509
K. Ichihara, Seikagaku 40 (1968), 1-3

TOMKINS, GORDON MAYER [1926-1975]
AMS 12,6426; WWWS, 1679
B.N. Ames, Biochemical Action of Hormones 4 (1977), xvii-xxv
E.B. Thompson, Molecular Endocrinology 1 (1987), 1-3

TOMMASI, SALVATORE [1813-1888]
EI 33,1011-1012; HIRSCH 5,606-607; PAGEL, 1717-1718
A. Mosso, Arch. Ital. Biol. 10 (1888), 139-140
Anon., Leopoldina 24 (1888), 167-168; Sperimentale 62 (1888), 104-105
M. di Giandomenico, Episteme 3 (1969), 17-30
A. Pazzini, Pag. Stor. Med. 13 (1969), 38-50
G. Baldi, Pag. Stor. Med. 15 (1971), 43-49

TOMPSON, FREDERICK WILLIAM [1859-1930]
Anon., J. Chem. Soc. 134 (1931), 1031

TOMSA, VLADIMIR BOGUMILOVICH [1831-1895]
VORONTSOV, 102-105
E.I. Slivko, Fiziol. Zhur. (Ukraine) 1(4) (1955), 136-138

TONKOV, VLADIMIR NIKOLAEVICH [1872-1954]
BSE 42,623; BME 32,411; WWR, 546
E.N. Melman, Arkh. Anat. Gist. Emb. 39(7) (1960), 101-108
V.V. Kuprianov, Arkh. Anat. Gist. Emb. 43(10) (1962), 109-114
G.Z. Roginski, Arkh. Anat. Gist. Emb. 47(10) (1964), 103-105

TOPCHIEV, ALEKSANDR VASILIEVICH [1907-1962]
BSE (3rd Ed.)26,94; WWR, 546
R.I. Goryacheva and V.A. Zaytseva, A.V. Topchiev. Moscow 1964

TOPLEY, WILLIAM WHITEMAN CARLETON [1887-1944]
WW 1944, p.1757
Anon., Brit. Med. J. 1944(I), pp.201-202
M. Greenwood, Obit. Not. Fell. Roy. Soc. 4 (1944), 699-712

TORLAIS, JEAN [1897-1964]
R. Hahn, Isis 57 (1966), 260-261
P. Huard, Clio Medica 2 (1967), 63

TORREY, HENRY AUGUSTUS [1871-1910]
RSC 19,169; POGG 5,1264
T.W. Richards et al., Science 32 (1910), 50-51
Anon., Chem. Z. 34 (1910), 400

TORREY, JOHN [1796-1873]
DSB 13,432-433; COLE,536-537
C.C. Robbins, Bull. Torrey Bot. Club 95 (1968), 515-645

TOSTESON, DANIEL CHARLES [1925-]
AMS 17(7),169-170; BROBECK, 217-221; WWWS, 1682

TOTANI, GINZABURO [1883-]
JBE, 1689,2296; HAKUSHI 3,323

TOTH, GÉZA [1907-]
M. Fekete, Prominent Hungarians at Home and Abroad, 3rd Ed., p.489. London 1979

TOURDES, GABRIEL [1810-1900]
HIRSCH 5,615-616; RSC 6,14; CALLISEN 33,53
P. P[arisot] et al., Rev. Med. Est 32 (1900), 65-86
E.J. Marey, Bull. Acad. Med. 43 (1900), 113-114
Anon., Leopoldina 36 (1900), 54-55

TRACEY, MICHAEL VINCENT [1918-]
Who's Who in British Science 1953, p.262

TRACHMANN, OTTO [1860-1933]
JB 8,164; NUC 599,327
Anon., Chem. Z. 57 (1933), 317

TRAETTA-MOSCA, FILIPPO [1875-1940]
POGG 5,1265-1266; 6,1682; 7b,5529
A. De Dominicis, Chimica e Industria 22 (1940), 304

TRANTOM, WILLIAM [1877-1939]
JV 13,156; NUC 600,22
Anon., Chem. Ind. 1939, p.517

TRAPMANN, HEINZ [1926-]
KURSCHNER 13,3991; WWWS, 1684; POGG 7a(4),701-702TRAPP, RUDOLF [1877-1965]
JV 25,366

TRAPPE, PAUL [1872-1956]
JV 18,194; NUC 600,52
Anon., Nachr. Chem. Techn. 4 (1956), 116

TRAUBE, ISIDOR [1860-1943]
BERLIN,707; LDGS, 25; TLK 3,662; WININGER 6,125; WWWS, 1685; LDGS, 25; POTSCH, 426; POGG 4,1520-1521; 5,1266-1267; 7a(4),702
R.E. Liesegang, Chem. Z. 54 (1930), 249
L.J. Weber, Z. angew. Chem. 43 (1930), 272-274
H. Freundlich, Koll. Z. 50 (1930), 194-196
D.H. Bangham, Nature 152 (1943), 743-744
J.T. Edsall, Proc. Am. Phil. Soc. 129 (1985), 371-406

TRAUBE, LUDWIG [1818-1876]
ADB 38,504-507; HIRSCH 5,625-626; PAGEL, 1721-1724; GRAETZER, 125-133; KOHUT 2,289-290; KOREN, 257; WININGER 6,125-126; WWWS, 1685; POGG 7aSuppl.,693-694
R. Virchow, Berl. klin. Wchschr. 13 (1876), 209
H. Morrison, Boston Medical and Surgical Journal 196 (1927), 1097-1101

TRAUBE, MORITZ [1826-1894]
DSB 13,451-453; BLOKH 2,737-739; BERLIN, 646; WININGER 6,126; WWWS, 1685; W, 518; POTSCH, 426-427; POGG 2,1126; 3,1363; 4,1519; 7aSuppl.,694
G. Bodländer, Ber. chem. Ges. 28(4) (1895), 1085-1108
F. Müller (ed.), Der Geistvolle Weinhändler von Ratibor: Briefwechsel. Munich 1933
T.L. Sourkes, J. Hist. Med. 10 (1955), 379-391
K. Müller, Moritz Traube und seine Theorie der Fermente. Zurich 1970

TRAUBE, WILHELM [1866-1942]
TETZLAFF, 338; TLK 3,662; POTSCH, 427; POGG 4,1520; 5,1266; 6,2683; 7a(4),702,154*
H. Pringsheim et al., Z. angew. Chem. 39 (1926), 61-67
F. Kröhnke, Nachr. Chem. Techn. 9 (1961), 257

TRAUMANN, KARL [1882-1941]
GEDENKBUCH 2,1511; JV 26,397; NUC 600,129

TRAUTSCHOLD, HERMANN [1817-1902]
HEIN, 687-688; POGG 2,1126; 3,1363-1364; 4,1521; 7aSuppl.,694
Anon., Jahresb. Schles. Ges. 80 (1902), 16-19

TRAUTWEIN, JAKOB BERNHARD [1793-1855]
HEIN, 688; TSCHIRCH, 1135; FERCHL,541; POGG 2,1126-1127; 7aSuppl.,694-695

TRAUTWEIN, KURT [1881-1958]
KURSCHNER 8,2411; LDGS, 26

TRAUTZ, MAX [1880-1960]
DRULL, 270-271; POGG 4,1521; 5,1267-1268; 6,2683-2684; 7a(4),705-706
G.M. Schwab, Z. Elektrochem. 59 (1955), 139-140; Rete 1 (1971), 125-134
I.N. Stranski, Z. Elektrochem. 65 (1961), 401-402

TRAVERS, MORRIS WILLIAM [1872-1961]
DSB 13,453-455; DNB [1961-1970],1014-1015; POGG 4,1521-1522; 5,1268; 6,2685; 7b,5531-5532
D.H. Everett, Nature 192 (1961), 1127-1128
C.E. Bawn, Biog. Mem. Fell. Roy. Soc. 9 (1963), 301-313
D. McKie, Proc. Chem. Soc. 1964, pp.377-378

TREADWELL, CARLETON RAYMOND [1911-1989]
AMS 15(7),179; WWWS, 1685

TREADWELL, FREDERICK PEARSON [1857-1918]
BLOKH 2,739; POTSCH, 427; POGG 5,1268-1268; 6,2685
E. Bosshard, Viert. Nat. Ges. Zurich 63 (1918), 576-579; Chem. Z. 42 (1918), 341-342
K.A. Hofmann, Ber. chem. Ges. 51 (1918), 1206-1207

TRÉCUL, AUGUSTE [1818-1896]
VAPEREAU 6,1521; GE 31,336; GORIS, 584-588; TSCHIRCH, 1135; WWWS, 1685
E. Gautier, Ann. Scient. 40 (1896), 505-508
Anon., Bull. Soc. Bot. 43 (1896), 431-432
G. Dillemann, Les Informations Pharmaceutiques 1982, pp.759-767

TREFF, WALTER [1874-]
JV 16,159; NUC 600,436

TRÉFOUËL, JACQUES [1897-1977]
DSB 18,932-933; QQ 1975-1976, p.1587; MH 2,554-555; STC 3,78-79; MSE 3,228-229; WWWS,1685
J. Roche, C.R. Soc. Biol. 171 (1977), 989-990
E. Chain, Nature 270 (1977), 647-648
L. Aublant, Bull. Acad. Med. 161 (1977), 517
J.A. Gautier, Bull. Soc. Chim. 1978, pp.7-11; Eur. J. Med. Chem. 13 (1978), 11-15

TREIBS, ALFRED [1899-1983]
KURSCHNER 13,3995; POTSCH, 428; POGG 6,2687; 7a(4),710-711
Anon., Chem. Z. 98 (1974), 373; 107 (1983), 248

TREIBS, WILHELM [1890-1978]
POTSCH, 428; POGG 6,2687; 7a(4),711-714
Anon., Nachr. Chem. Techn. 8 (1960), 336; Chem. Z. 94 (1970), 853; 99 (1975), 509; 102 (1978), 202-203
W. Ziegenbein, Chem. Ber. 115(9) (1982), xxvii-l

TRENDELENBURG, PAUL [1884-1931]
FISCHER 2,1581; IB, 303; TLK 3,663
G. Stroomann, Erg. Physiol. 32 (1931), v-ix

TRENDELENBURG, WILHELM [1877-1946]
FISCHER 2,1581; IB, 303; POGG 7a(4),715
R. Wagner, Bayer. Akad. Wiss. Jahrbuch 1948, pp.269-271
A. Schittenhelm, Z. ges. inn. Med. 115 (1949), 1-6
E. Schütz, Erg. Physiol. 46 (1950), 6-21

TRESKIN, FEDOR VASILIEVICH [1836-]
ZMEEV 5,166-167; SG [1]14,749

TRETYAKOV, DMITRI KONSTANTINOVICH [1878-1950]
BSE 43,206; WWR, 548; IB, 303

TREUB, MELCHIOR [1851-1910]
DSB 13,458-460; SACKMANN, 343-345; WWWS, 1686
K. Goebel, Ber. bot. Ges. 28 (1910), (21)-(31); Sitz. Bayer. Akad. 1911, pp.42-53
Anon., Nature 84 (1910), 539-540
F.A.F.C. Went, Bull. Soc. Bot. Belg. 48 (1911), 285-325
C. Schröter, Verhandl. Schw. Nat. Ges. 94 (1911), 154-164
H.H. Zeijlstra, Melchior Treub. Amsterdam 1959

TREUPEL, ERNST WILHELM [1828-1871]
DGB 49,401

TREVAN, JOHN WILLIAM [1887-1956]
WW 1954, p.2958; WWWS, 1686
J.H. Gaddum, Biog. Mem. Fell. Roy. Soc. 3 (1957), 273-288

TREVIRANUS, GOTTFRIED REINHOLD [1776-1837]
DSB 13,460-462; ADB 38,588; HIRSCH 5,633; WAGENITZ, 180-181; WWWS, 1687; CALLISEN 19,376-383; 33,67-69; POGG 2,1132-1133; 7aSuppl.,697-698
W.O. Focke, Abh. Nat. Ver. Bremen 6 (1880), 11-48
I. Schunke, Sudhoffs Arch. 30 (1937), 115-132

TREVIRANUS, LUDOLPH CHRISTIAN [1779-1864]
DSB 13,462-464; ADB 38,588-591; WWWS, 1687; CALLISEN 19,383-386; 33,69-70
C.F.P. von Martius, Sitz. Bayer. Akad. 1865(I), pp.264-287

TREY, HEINRICH [1851-1916]
WELDING, 812; RIGA, 723; POGG 4,1524; 5,1269

TRIBE, ALFRED [1840-1885]
POGG 3,1366; 4,1524
J.H. Gladstone, Nature 33 (1885), 180
H. Müller, J. Chem. Soc. 49 (1886), 352-353

TRIBE, MARGARET [1890-]
IB, 303
Who's Who in British Science 1953, p.263

TRIBOLET, GEORGES de [1830-1873]
RSC 6,39; 8,1115; POGG 3,1366TRIER, GEORG [1884-1944]
TLK 3,664; POGG 6,2689; 7a(4),717

TRIFANOVSKI, DMITRI SEMENOVICH [1845-]
ZMEEV 5,168; RSC 11,644; SG [1]14,769

TRIKOJUS, VICTOR MARTIN [1902-1985]
SRS, 61
Who's Who in Australia 1977, pp.1039-1040
J.W. Legge and F. Gibson, Historical Records of Australian Science 6 (1987), 519-531

TRILLAT, AUGUSTE [1861-1944]
POGG 5,1269-1270
J. Tréfouël, Bull. Acad. Med. 128 (1944), 477-478
Anon., Ann. Chim. Anal. 27 (1945), 39-40
A. Kling, Bull. Soc. Chim. 1947, pp.1-7
M. Macheboeuf, Ann. Inst. Pasteur 73 (1947), 622-623

TRIM, ARTHUR REGINALD HENRY [1915-]
Who's Who in British Science 1953, p.263

TRIMBLE, HARRY CLYDE [1889-1962]
AMS 10,4136-4137
E.G. Ball, Harvard Med. Alumni Bull. 37 (1962), 2-3

TRIMBLE, HENRY [1853-1898]
NCAB 5,350-351
S.P. Sadtler et al., J. Franklin Inst. 146 (1898), 307-312
Anon., Am. J. Pharm. 70 (1898), 537-543; Leopoldina 34 (1898), 144; J. Soc. Chem. Ind. 17 (1898), 904

TRINCHESE, SALVATORE [1836-1897]
RSC 19,204
C.E., Arch. Ital. Biol. 26 (1896), 497-500

TRISTRAM, GEORGE ROLAND [1912-]
Who's Who in British Science 1953, p.263

TROBRIDGE, FREDERICK GEORGE [1887-1954]
SRS, 39; NUC 602,7

TRÖGER, JULIUS [1862-1942]
TSCHIRCH, 1136; POGG 4,1524-1525; 5,1270-1271; 6,2691-2692; 7a(4),721
W. Kern, Pharm. Zent. 83 (1942), 431-432

TRØNSEGAARD, NIELS [1873-1961]
DBL 24,289; KBB 1960, p.1563; VEIBEL 2,455-456; POGG 6,2692; 7b,5552

TROISIER, CHARLES ÉMILE [1844-1919]
FRANKEN, 199; WWWS, 1688
A. Laveran, Bull. Acad. Med. 82 (1919), 490-491
A. Gilbert, Presse Med. 36 (1920), 91

TROISIER, JEAN [1881-1945]
P. Ameuille, Bull. Acad. Med. 130 (1946), 24-27
R. Deschiens, Ann. Inst. Pasteur 72 (1946), 851-852
F. Besançon, Presse Med. 54 (1946), 43-44

TROLAND, LEONARD THOMPSON [1889-1932]
DAB 18,646-647; AMS 4,994; OEHRI, 121; WWAH, 612; POGG 6,2692
C. Murchison, Psychological Register 3 (1932), 503-505
A.A. Roback, Science 76 (1932), 26-27
J.P.C.S., J. Opt. Soc. Am. 22 (1932), 509-511

TROLL, WALTER [1922-]
AMS 15(7),191; BHDE 2,1175

TROMMER, CARL [1806-1879]
HIRSCH 5,643
E. Ebstein, Z. Urol. 23 (1929), 913-915

TROMMSDORFF, (CHRISTIAN WILHELM) HERMANN [1811-1884]
HEIN, 690-691; HUHLE-KREUTZER, 151-171; TSCHIRCH, 1136; FERCHL, 542-543;
POGG 2,1138; 7aSuppl.,699
A. Biltz, Arch. Pharm. 222 (1884), 593-605

TROMMSDORFF, JOHANN BARTHOLOMÄUS [1770-1837]
DSB 13,465-466; ADB 38,641-644; HEIN, 692-695; POHL, 38-71; WWWS, 1688; BLOKH 2,740-741; BLOKH 2,740-741; PHILIPPE, 739-748; POTSCH, 428; FERCHL, 543-544; SCHAEDLER, 137-138; COLE, 539-540; WWWS, 1688; CALLISEN 19,403-438; 33,74-77; POGG 2,1136-1138; 7aSuppl.,699-701
L.F. Bley, Arch. Pharm. 68 (1839), 1-20,113-145,225-237
O. Rosenheimer and H. Trommsdorff, Das Lebensbild einer der grössten Pharmaceuten und Chemiker etc. Jena 1913
H. Trommsdorff, Mitteldeutsche Lebensbilder 3 (1928), 270-285
H. Patzer et al., Johann Bartholomäus Trommsdorff und die Begründung der modernen Pharmazie. Leipzig 1972
H.R. Abe, Leopoldina [3]19 (1975), 178-200
W. Götz, Zum Leben und Werk von J.B. Trommsdorff. Würzburg 1977; Bibliographie der Schriften von Johann Bartholomäus Trommsdorff. Stuttgart 1985

TROMMSDORFF, RICHARD [1874-1944]
FISCHER 2,1584; BK, 327-328; IB. 303; NUC 602, 205

TROOST, LOUIS JOSEPH [1825-1911]
DSB 13,467-468; BLOKH 2,741; SCHAEDLER, 138; WWWS, 1688; POGG 2,1139; 3,1367-1368; 4,1525; 5,1271
A. Gautier, C.R. Acad. Sci. 153 (1911), 611-615
Anon., Nature 87 (1911), 491-492; Chem. Z. 35 (1911), 1145

TROPP, CASPAR [1899-]
GEISSLER, 140-146; NUC 602,264

TROSCHEL, FRANZ HERMANN [1810-1882]
RSC 6,54-56; 8,1119; 11,650
Anon., Leopoldina 19 (1882), 209-210; Zool. Anz. 5 (1882), 644
H. von Dechen, Corr. Nat. Ver. Bonn 40 (1883), 35-44

TROSCHKE, HERMANN OSWALD [1851-]
RSC 11,650; NUC 602,271

TROUSSEAU, ARMAND [1801-1867]
HOEFER 45,673-674; HIRSCH 5,646-647; PAGEL, 1730-1732; FRANKEN, 199-200; DECHAMBRE [3]18,299; WWWS, 1689; CALLISEN 19,445-450; 33,79-81
C. Lasegue, France Médicale 16 (1869), 495-499
J. Beclard, Notices et Portraits, pp.197-225. Paris 1878
C. Paul, Bull. Acad. Med. 18 (1887), 621-625
J. Mayer, Nutr. Revs. 15 (1957), 321-323
M. Bariety, Bull. Acad. Med. 151 (1967), 627-635

TROUTON, FREDERICK THOMAS [1863-1922]
DSB 13,471-472; POGG 4,1525; 5,1271; 6,2694
E.N. daC. A., Nature 110 (1922), 490-491
A.W.P., Proc. Roy. Aoc. A110 (1926), iv-ix

TRUCHE, CHARLES [1871-1951]
Anon., Ann. Inst. Pasteur 82 (1952), 629-630

TRUE, ALFRED CHARLES [1853-1929]
DAB 19,4; AMS 4,995; WWWS, 1689
Anon., Science 69 (1929), 491

TRUE, RODNEY HOWARD [1866-1940]
AMS 6,1440; HUMPHREY, 253-255; IB, 304; WWAH, 613; WWWS, 1689
H.L. Schantz, Science 92 (1940), 546-547
R.B. Harvey, Chronica Botanica 6 (1941), 424-425

TRURNIT, HANS JOACHIM [1907-]
AMS 10,4144; KIEL, 123; POGG 7a(4),729

TRZCZINSKI, WAWRZYNIEC [-1891]
KONOPKA 11,293; RSC 19,219
Anon., Chem. Z. 15 (1891), 1901

TSCHERMAK von SEYSENEGG, ARMIN [1870-1952]
KURSCHNER 7,2126-2127; FISCHER 2,1587; MAASS, 60-81; KOERTING, 120-121; WWWS, 1690; POGG 6,2696; 7a(4),730-731
A. Düring, Alm. Akad. Wiss. Wien 102 (1952), 374-385
M.H. Fischer, Pflügers Arch. 258 (1953), 87-89

TSCHERMAK von SEYSENEGG, ERICH [1871-1962]
DSB 13,477-479; FISCHER 2,1587-1588; PLANER, 349-350; TLK 3,665; WWWS, 1690; POGG 6,2696-2697; 7a(4),731
E. Tschermak von Seysenegg, J. Heredity 42 (1951), 163-171; Leben und Wirken eines Oesterreichischen Pflanzenzüchters. Berlin 1958
G.H. Shull, Genetics 37 (1952), 1-7
F. Knoll, Alm. Akad. Wiss. Wien 112 (1962), 412-419

TSCHESCHE, RUDOLF [1905-1981]
KURSCHNER 13,4011-4012; WWWS, 1690; POGG 7a(4),731-733
K. Soehring, Arzneimitt. 15 (1965), 584
G. Grimmer, Arzneimitt. 20 (1970), 730
Anon., Chem. Z. 94 (1970), 335; 105 (1981), 197

TSCHIRCH, ALEXANDER [1856-1939]
HEIN, 696-697; FISCHER 2,1588; PAGEL, 1732; REBER, 43-46,355-360; IB,304; TSCHIRCH, 1136-1137; DZ, 1480-1481; TLK 3,665-666; WWWS, 1690; POTSCH, 429; POGG 4,1528; 5,1273; 6,2698-2699; 7a(4),733
A. Tschirch, Erlebtes und Erstrebtes: Lebenserinnerungen. Bonn 1921
P.N. Schurhoff, Apoth. Z. 41 (1926), 1146-1148
H. Flück, Mitt. Nat. Ges. Bern 1939, pp.102-111; Verhandl. Schw. Nat. Ges. 120 (1940), 503-523
T. Sabalitschka, Ber. bot. Ges. 59 (1941), (67)-(108)
P. Casparis, Arch. Inter. Hist. Sci. 2 (1948), 210-211
G. Schramm, Pharmaceutica Acta Helvetiae 52 (1977), 143-147; Schw. Apoth. Z. 119(3) (1981), 51-55; Gesnerus 44 (1987), 134-134

TSCHUDI, JOACHIM [1822-1893]
DHB 6,698
Schweizerisches Geschlechterbuch 5,655

TSENKOVSKI, LEV SEMENOVICH [1822-1887]
BSE 46,519; BME 34,346-347; KUZENTSOV(1963), 105-115; SKOROKHODOV, 78-83
B.E. Raikov, Mikrobiologia 21 (1952), 360-366
A.I. Metelkin, Zhur. Mikr. Ep. Imm. 46 (1969), 141-144
N.D. Revenok, Zhur. Mikr. Ep. Imm. 50 (1973), 131-136

TSITOVICH, IVAN SERGEEVICH [1876-1955]
BME 34,507-508; WWR, 552; IB, 335
Anon., Farm Toks. 19(1) (1956), 63-64

TSOU, CHEN-LU (ZOU, CHENGLU) [1923-]
W. Bartke, Who's Who in the People's Republic of China, p.717. Munich 1987
C.L. Tsou, in Selected Topics in the History of Biochemistry (G. Semenza and R. Jaenicke, eds.), pp.349-386. Amsterdam 1990

TSUJIMOTO, MITSUMARU [1877-1940]
JBE, 1750; POTSCH, 430; POGG 6,2701-2702; 7b,5568-5569
Y. Tsuzuki, J. Chem. Ed. 47 (1970), 695-696

TSWETT, MIKHAIL SEMENOVICH [1872-1919]
DSB 13,486-488; KUZNETSOV(1963), 374-380; BRIQUET, 463-466; W, 520-521; ETTRE, 483-490; ABBOTT-C, 139; WWWS, 1691; POTSCH, 104
C. Dhéré, Candollea 10 (1943), 23-63
T. Robinson, Chymia 6 (1960), 146-161
R.L.M. Synge, Arch. Biochem. Biophys. Suppl.1 (1962), 1-6
H.H. Strain and J. Sherma, J. Chem. Ed. 44 (1967), 235-242
K. Sakodynski and K.V. Chmutov, Chromatographia 5 (1972), 471-476
K. Sakodynski, J. Chromat. 73 (1972), 303-360
E.M. Sencenkova, NTM 12 (1975), 54-69
V.G. Berezkin, Chem. Revs. 89 (1989), 279-285

TUBANDT, CARL [1878-1942]
TLK 3,666; POGG 5,1275; 6,2702; 7a(4),734

TÜLLNER, HERMANN [1879-1848]
JV 18,28; NUC 604,54

TÜRCK, LUDWIG [1810-1868]
DSB 13,492-493; ADB 39,2; WURZBACH 48,79-82; LESKY,186-194; JOHN,179-185; HIRSCH 5,653-654; PAGEL, 1732-1733; BOEGERSHAUSEN,145-152; KOREN,285
M. Neuburger, Jahrb. Psychiat. 31 (1910), 1-21,25-194
H. von Schrötter, Wiener med. Wchschr. 70 (1928), 103-105

TÜRK, WILHELM [1871-1916]
FISCHER 2,1590; STUMPF, 122-127
R. Fleckseder, Wiener klin. Wchschr. 29 (1916), 720-721
H. Lehndorf, Blood 9 (1954), 642-647

TÜXEN, REINHOLD [1899-1980]
KURSCHNER 13,4015; JV 42,338; NUC 604,104
J. Braun-Blanquet, Vegetatio 8 (1959), 271-279
J. Backman, Vegetatio 48 (1981), 87-91
H. Ellenberg, Ber. bot. Ges. 95 (1982), 387-391

TUNNICLIFFE, HUBERT ERLIN [1899-1964]
Anon., Lancet 1964(II), p.1187

TUPPER, CHARLES JOHN [1920-]
AMS 15(7),218; WWA 1980-1981, pp.3336-3337; WWWS, 1693

TUPPY, HANS [1924-]
KURSCHNER 13,4016-4017; WWWS, 1693-1694; POGG 7a(4),734-735
G. Krel, Ost. Chem. Z. 87 (1986), 238-239

TURBA, FRITZ [1917-1965]
KURSCHNER 10,2528; MEISTER, 48-61; POGG 7a(4),735-736
Anon., Nachr. Chem. Techn. 13 (1965), 431

TURNER, BENJAMIN BERNARD [1871-1945]
AMS 6,1445TURNER, EDNA MAY [1903-]
NUC 605,146

TURNER, EDWARD [1796-1837]
DSB 13,499-500; DNB 19,1262-1263; FERCHL, 545-546; COLE, 541-542; POGG 2,1146-1147
H. Terrey, Ann. Sci. 2 (1937), 137-152
W.A. Campbell, Anal. Proc. 18 (1981), 381-383
W.H. Brock, Ambix 33 (1986), 31-42

TURNER, EUSTACE EBENEZER [1893-1966]
DNB [1961-1970],1020-1021; WW 1965, p.3104; POGG 6,2704; 7b,5573-5575
M.M. Harris, Chem. Ind. 1966, pp.1953-1955
D.M. Hall, Chem. Brit. 3 (1967), 74-75
C.K. Ingold, Biog. Mem. Fell. Roy. Soc. 14 (1968), 449-467

TURNER, RICHARD BALDWIN [1916-1971]
AMS 11,5500
M. Gates, Biog. Mem. Nat. Acad. Sci. 53 (1982), 351-365

TURPIN, PIERRE JEAN FRANÇOIS [1775-1840]
DSB 13,506-507; FELLER 8,265-266; HOEFER 45,742-743; LAROUSSE 15,598; WWWS, 1696; CALLISEN 19,487-489; 33,94
E. Lagrange, Scalpel 123 (1970), 39-44

TURPIN, RAYMOND ALEXANDRE [1895-1988]
QQ 1979-1980, p.1527; IWW 1982-1983, pp.1343-1344L WWWS, 1696
R. Laplane, Bull. Acad. Med. 173 (1989), 535-543

TUSHNOV, MIKHAIL PAVLOVICH [1879-1935]
BSE 43,521; BME 33,37; WWR, 556; IB, 306
N.A. Krylova, Veterinaria 49 (1973), 56-58

TUSON, RICHARD VINE [1832-1888]
Anon., Chem. News 58 (1888), 230

TUTEIN, FRIEDRICH [1862-1930]
JV 5,81; RSC 13,234; NUC 605,463
Anon., Chem. Z. 54 (1930), 690

TUTTLE, DAVID KITCHELL [1835-1915]
AMS 2,478; RSC 6,73; NUC 605,501
Who Was Who in America 1,1260

TUTTON, ALFRED [1864-1938]
DSB 13,517-518; DNB [1931-1940],875-876; WW 1938, p.3428; POGG 4,1532-1533; 5,1280; 6,2709-2710; 7b,5583
J.R. Partington, Nature 142 (1938), 321-323
Anon., Obit. Not. Fell. Roy. Soc. 2 (1939), 621-626

TWITTY, VICTOR CHANDLER [1901-1967]
AMS 10,4164; MH 2,557-558; MSE 3,235-236; WWAH, 617; WWWS, 1697

TWORT, FEREDERICK WILLIAM [1877-1950]
DSB 13,519-521; WW 1950, p.2839; STC 3,88; BULLOCH, 400; ABBOTT, 129; W, 521-522
Anon., Lancet 1950(I), p.1839; Brit. Med. J. 1950, pp.788-789
P. Fildes, Obit. Not. Fell. Roy. Soc. 7 (1951), 505-517
P. Nicolle, Presse Med. 59 (1951), 120
N.W. Pirie, Nature 343 (1990), 504

TYLER, ALBERT [1906-1968]
AMS 11,5507; WWWS, 1697
C.B. Metz, Biol. Bull. 139 (1970), 9-11

TYLER, JOHN MASON [1851-1929]
AMS 3,697; WAGENITZ, 182; NUC 606,156-157

TYNDALL, JOHN [1820-1893]
DSB 13,521-524; DNB 19,1358-1363; BLOKH 2,747; SCHAEDLER, 139; W,522-523; POTSCH, 430; POGG 2,1149-1150; 3,1375-1376; 4,1533
E. Frankland, Proc. Roy. Soc. 55 (1894), xviii-xxxiv
L.A. Weed, Ann. Med. Hist. 4 (1942), 55-62
A.S. Eve and C.H. Creasey, Life and Work of John Tyndall. London 1945
W.H. Brock et al., Eds., John Tyndall. Dublin 1981

TYRODE, MAURICE VEJUX [1878-1930]
AMS 4,1001-1002; FISCHER 2,1592; CHEN, 10

U

UBBELOHDE, LEO [1876-1964]
KURSCHNER 9,2130; TLK 3,666-667; POGG 6,2714-2715; 7a(4),735-737
Anon., Chem. Z. 85 (1961), 18

UBISCH, GERTA von [1876-1965]
KURSCHNER 4,3059; BOEDEKER, 23; LDGS, 13; IB, 306

UBISCH, LEOPOLD von [1886-1965]
LDGS, 15; IB, 306
B. Føyn, Norske Vid. Akad Årbok 1986, pp.100-110

UDENFRIEND, SIDNEY [1918-]
AMS 15(7),237; WWA 1980-1981, p.3348; WWWS, 1698

UDRÁNSZKY, LASZLÓ von [1862-1914]
MEL 2,922; FISCHER 2,1593; PAGEL, 1737
A. Strasser, Wiener klin. Wchschr. 27 (1914), 445

UEXKÜLL, JAKOB [1864-1944]
WELDING, 817; IB, 306

UFFELMANN, JULIUS [1837-1894]
ADB 39,131-132; HIRSCH 5,672; PAGEL, 1737-1738
Anon., Leopoldina 30 (1894), 103-104; Chem. Z. 18 (1894), 277

UHLENHUTH, EDUARD [1885-1961]
KURSCHNER 10,2533; AMS 9(II), 1156
Anon., Bulletin School of Medicine University of Maryland 43(3) (1958), 5-7
H.J.F. Figge, Anat. Rec. 143 (1962), 278-280

UHLENHUTH, PAUL [1870-1957]
KURSCHNER 8,2431; FISCHER 2,1594-1595; AUERBACH, 405-406; OLPP, 398-401; FREUND 2,397-412; IB, 306; POGG 7a(4),741-743
W. Fromme and J. Kathe, Z. Immunitätsforsch. 115 (1958), 229-276
K.E. Schonherr, Arzneimitt. 8 (1958), 249-250

UHLFEDER, EMIL [1871-1935]
JV 11,235; RSC 19,74,292; NUC 607,65
Anon., Chem. Z. 59 (1935), 717; Z. angew. Chem. 48 (1935), 592

UHLIRZ, RUDOLF [1880-]
NUC 607,82
Oesterreicher der Gegenwart 1951, p.317

UKHTOMSKI, ALEKSEI ALEKSEEVICH [1875-1942]
DSB 13,529-530; BSE 44,445-446; BME 33,399-401; WWR, 560; KUZNETSOV(1963), 400-414
Anon., Nature 150 (1942), 541-542
M.I. Vinogradov, Usp. Sov. Biol. 16(2) (1943), 201-211
P.G. Terekhov, Fiziol. Zhur. 41 (1955), 709-711
S.A. Kosilov, Gigiena i Sanitaria 10 (1955), 7-12
D.G. Kvasov, Fiziol. Zhur. 51 (1965), 637-645

ULBRICH, ERNST [1888-]
NUC 607,192

ULEX, GEORG LUDWIG [1811-1885]
HEIN, 701; TSCHIRCH, 1137; FERCHL, 547; POGG 2,1153; 3,1376

ULLMANN, FRITZ [1875-1939]
TLK 3,668; POTSCH, 431; POGG 4,1535-1536; 5,1282-1283; 6,2719; 7a(4),745
K.H. Meyer, Helv. Chim. Acta 23 (1940), 93-100

ULMER, WERNER [1895-]
JV 39,518

ULPTS, REINHOLD [1896-]
JV 39,63; SG [5]11,660; NUC 607,331

ULRICH, GUSTAV [1963-1943]
POGG 7a(4),748
L. Anschütz, Ber. chem. Ges. 76A (1943), 129-143

ULRICH, KARL [1876-1960]
TLK 2,716
Anon., Chem. Z. 63 (1939), 453; Z. Zuckerind. NF10 (1960), 208-209

ULTZMANN, ROBERT [1842-1889]
FISCHER 2,1597; PAGEL, 1740; SG [2]19,29
L. Eisenberg, Das Geistige Wien 2,608-609. Vienna 1893

UMBACH, CARL [1856-1926]
JV 2,117

UMBARGER, HAROLD EDWIN [1921-]
Directory of Graduate Research 1979, p.856

UMBER, FRIEDRICH [1871-1946]
KURSCHNER 4,3070; FISCHER 2,1597-1598
F.K. Störring, Deutsche med. Wchschr. 67 (1941), 1185-1186
E. Frommelt, Deutsche med. Wchschr. 71 (1946), 272

UMBREIT, WAYNE WILLIAM [1913-]
AMS 15(7),245; WWA 1980-1981, p.3351; WWWS, 1701

UMEZAWA, HAMAO [1914-1986]
JBE, 1801
E.P. Abraham et al., Heterocycles 13 (1979), 1-139
T. Tsuchiya et al., Adv. Carbohydrate Chem. 48 (1990), 1-20

UMLAUFT, WENZEL [-1878]
ZURICH-D, 127; RSC 11,667; NUC 607,450

UNDERHILL, FRANK PELL [1877-1932]
AMS 4,1003; NCAB C,304; 25,23-24; CHITTENDEN, 333-335,387-389; IB, 307; PARASCANDOLA, 54; OEHRI, 122-123; POGG 6,1721
R.H. Chittenden, Science 76 (1932), 457
Anon., Ind. Eng. Chem. News Ed. 10 (1932), 174; Obituary Records of Yale Graduates, No.92, pp.241-242. New Haven 1933
H.G. Barbour, Yale J. Biol. Med. 5 (1933), 301-321

UNDERKOFLER, LELAND ALFRED [1906-]
AMS 14,5224; WWWS, 1701

UNGAR, EMIL [1849-1934]
FISCHER 2,1598; PAGEL, 1740-1741

UNGAR, ENDRE [1890-]
NUC 607,649

UNGER, ERNST [1875-1938]
E.A. Winkler, Ernst Unger. Berlin 1975; J. Hist. Med. 37 (1982), 269-286
H. Hausmann, Z. ärzt. Fortbild. 82 (1988), 843-845

UNGER, FRANZ [1800-1870]
DSB 13,542-543; ADB 39,286-289; WURZBACH 49,44-59; TSCHIRCH, 1137; FERCHL, 548; WWWS, 1701; POGG 3,1379
H. Leitgelb, Bot. Z. 28 (1870), 241-264
A. Schrötter, Alm. Akad. Wiss. Wien 30 (1870), 201-229
F. von Kobell, Sitz. Bayer. Akad. 1870(I), pp.420-423
A. Reyer, Leben und Wirken des Naturhistoriker Franz Unger. Graz 1871
A. Rollett and G. Haberlandt, Mitt. Nat. Ver. Steier. 37 (1901), xlvi-lxviii

UNGER, JULIUS BODO [1819-1885]
FERCHL, 548; POGG 2,1157-1158
Anon., Pharm. Z. 30 (1885), 441

UNGER, LUDWIG [1848-1923]
DBJ 5,444; FISCHER 2,1598; ATUYANA, 29-38

UNGER, OSKAR [1870-1937]
JV 10,216; NUC 608,8; POGG 4,1537; 7a(4),753
Anon., Z. angew. Chem. 50 (1937), 90,112

UNNA, EUGEN [1885-1958]
HEIN-E,430-431; DRUM, 355-357; JV 26,55; NUC 625,330
Anon., Chem. Z. 78 (1954), 301,694; 82 (1958), 116

UNNA, PAUL GERSON [1850-1929]
DBJ 11,371; KURSCHNER 3,2499; GROTE 8,175-219; SANDRITTER, 115-118;
FISCHER 2,1598-1599; PAGEL, 1741-1743;
OLPP, 401-404; KOREN, 259; WININGER 6,163-165; TETZLAFF, 341; IB, 307; WWWS, 1701
L. Culmann, Medical Life 35 (1928), 74-84
E. Unna, Stain Technology 4 (1929), 101-103
Deutscher Dermatologen Kalender 1929, pp.245-251
C. Schirren, Münch. med. Wchschr. 118Suppl. (1976), 35-38
A. Hollander, American Journal of Dermopathology 2 (1980), 137-142;
Z. Haut. Geschl. 59 (1984), 680-687

UNTERZAUCHER, JOSEF [1901-1973]
POGG 7a(4),754-755
G. Kainz, Proc. Soc. Anal. Chem. 12 (1975), 102

UNVERDORBEN, OTTO [1806-1873]
ADB 54,735-735; BLOKH 2,748; TSCHIRCH, 1137-1138; FERCHL, 548; W, 523; POTSCH, 431; WWWS, 1702; POGG 2,1159; 6,2722; 7aSuppl.,713-714
H. Schelenz, Z. angew. Chem. 34 (1921), 31-32
O. Schlenk, Z. angew. Chem. 39 (1926), 757-759

URANO, FUMIHIKO [1870-]
JV 23,650; SG [2]19,165; NUC 625,669

URBAIN, GEORGES [1872-1938]
DSB 13,546-547; W, 523; CHARLE(2), 255-257; POTSCH, 431-432; WWWS, 1702; POGG 4,1539; 5,1283-1284; 6,2724; 7b,5612-5613

P. Job, Bull. Soc. Chim. [5]6 (1939), 745-766
G. Champetier and C.H. Boatner, J. Chem. Ed. 17 (1940), 103-109
R. Courrier, Not. Acad. Sci. 6 (1974), 1-28

URBANSZKI, TADEUSZ [1901-1985]
POTSCH, 432
Anon., Leopoldina [3]27 (1983), 30-31; 31 (1985), 67
T. Urbanszki, Kwart. Hist. Nauki 29 (1984), 3-34

URDANG, GEORG [1882-1960]
AMS 8,2554; NCAB 54,158-159; HEIN-E, 433-435; BHDE 2,1186; TLK 3,670
G. Sonnedecker, Isis 51 (1960), 562-564
H.G. Wolfe, Pharm. Hist. 5 (1960), 33-42; Acta Pharmaceutica Historicae 1961(2), pp.39-84
I. Greenberg, Pharm. Hist. 11 (1969), 95-99

URE, ANDREW [1778-1857]
DSB 13,547-548; DNB 20,40-41; BLOKH 2,748-749; FERCHL, 548-549; W, 523; POTSCH, 432; WWWS, 1702; COLE, 543-546; POGG 2,1159-1161
W.S.C. Copeman, Proc. Roy. Soc. Med. 44 (1951), 655-662
W.R. Farrar, Notes Roy. Soc. 27 (1973), 299-324

UREY, HAROLD CLAYTON [1893-1981]
DSB 18,943-948; AMS 14,5229-5230; WWA 1978-1979, pp.3301-3302; NCAB E,475; WWWS, 1702; AO 1981, pp.13-16; MH 1,490-492; STC 3,94-96; MSE 3,239-241; CB 1941, pp.877-878; 1960, pp.441-442; W(3rd Ed.), 611-612; ABBOTT-C, 139-140; POTSCH, 432-433; POGG 6,2724; 7b,5613-5621
B. Garrett, J. Chem. Ed. 39 (1962), 583-584
C.C. Addison, Chem. Brit. 17 (1981), 383
J.R. Arnold, Am. Phil. Soc. Year Book 1981, pp.518-522
F.G. Brickwedde, Physics Today 35(9) (1982), 34-39
K.P. Cohen et al., Biog. Mem. Fell. Roy. Soc. 29 (1983), 623-659

URSPRUNG, ALFRED [1876-1952]
IB, 307
G. Blum, Bull. Soc. Frib. Sci. Nat. 41 (1952), 195-207; Verhandl. Schw. Nat. Ges. 132 (1952), 381-385

USHAKOV, SERGEI NIKOLAEVICH [1893-1964]
BSE(3rd Ed.) 27,164; WWR, 563; WWWS, 1703

USHER, FRANCIS LAWRY [1885-1969]
SRS, 38
Who's Who in British Science 1953, p.266
Anon., Chem. Brit. 5 (1969), 376

USLAR, LUDWIG von [1828-1894]
HEIN, 703-704; TSCHIRCH, 1138; SCHAEDLER, 140; RSC 6,91; POGG 2,1162
Anon., Chem. Z. 18 (1894), 585

USSING, HANS [1911-]
KBB 1981, p.1145
H. Ussing, Ann. Rev. Physiol. 42 (1980), 1-16

USUI, RYUTA [1877-]
HAKUSHI 3,317-318

UTHEMANN, WALTHER [1863-1944]
SG [1]15,443; NUC 626,615
A. Caanitz, Deutsche med. Wchschr. 69 (1943), 704-705
Anon., Deutsche med. Wchschr. 70 (1944), 224

UTTER, MERTON FRANKLIN [1917-1980]
AMS 14,5232; WWA 1978-1979, p.3303
H.G. Wood, TIBS 6(4) (1981), v-vi
H.G. Wood and R.W. Hanson, Biog. Mem. Nat. Acad. Sci. 56 (1987), 475-499

UTZINGER, MAX [1886-1954]
NUC 627,39

UTZINO, SENJI [1894-1957]

UVAROV, BORIS PETROVICH [1889-1070]
DNB [1961-1970], 1032-1034; WW 1965, p.3126
V.B. Wigglesworth, Biog. Mem. Fell. Roy. Soc. 17 (1971), 713-740

UVNÄS, BORJE [1913-]
VAD 1991, p.1113
B. Uvnäs, Ann. Rev. Pharmacol. 24 (1984), 1-18

V

VAGELOS, PINDAROS ROY [1929-]
AMS 15(7),258; WWA 1981-1982, p.3357; WWWS, 1704

VAGNER (WAGNER), EGOR EGOROVICH (GEORG) [1849-1903]
DSB 13,549-550; BSE 6,499-500; BJN 8,118*; KUZNETSOV(1961), 495-503; ARBUZOV, 185-189; BLOKH 2,760-764; VENGEROV 4(2),17-18; POTSCH, 433; TSCHIRCH, 1142; POGG 4,1586-1587
J. Bewad, Ber. chem. Ges. 36 (1903), 4591-4613
V. Lavrov, Zhur. Fiz. Khim. 36 (1904), 1337-1486
E.H. Huntress, Proc. Am. Acad. Arts Sci. 77 (1949), 53-54
L.I. Bazhenova, Usp. Khim. 19 (1950), 619-631
A.E. Arbuzov, Trudy Inst. Ist. Est. 1952(4), pp.46-61
A. Sementsov, Chymia 11 (1966), 151-155
P.I. Staroselski and E.P. Nikulina, Egor Egorovich Vagner. Moscow 1977

VAGNER (WAGNER), VLADIMIR ALEKSANDROVICH [1849-1934]
BSE 6,499; WWR, 565; VENGEROV 4(2),14-18; IB, 314
G.V. Roginski, Priroda 38(11) (1949), 72-73
B.E. Raikov, Vop. Ist. Est. Tekh. 1961(11), pp.148-149

VAHLEN, ERNST [1865-1941]
KURSCHNER 4,3081; FISCHER 2,1602-1603; POGG 4,1541; 6,2728; 7a(4),757
R. Vahlen, Ber. chem. Ges. 74A (1941), 183

VAILLARD, LOUIS [1850-1935]
FISCHER 2,1603
E. Sacquepée, Bull. Acad. Med. 113 (1935), 426-434
Anon., Ann. Inst. Pasteur 54 (1935), 269-272
R. Debré, Presse Med. 43 (1935), 441-442

VALENCIENNES, ACHILLE [1794-1865]
DSB 13,554-555; GE 31,664-665; LAROUSSE 15,735-736; DECHAMBRE 99,549; WWWS, 1705; CALLISEN 20,22; 33,115
L. Figuier, Ann. Scient. 10 (1865), 474-477
A. Milne-Edwards, J. Pharm. Chim. [4]5 (1867), 5-17
J. Théodoridès, Biol. Med. 54Suppl.1 (1965), i-cxxix
T. Monod et al., Nouv. Arch. Hist. Nat. [7]9 (1965-1966), 9-109
G. Dillemann, Produits et Problèmes Pharmaceutiques 26 (1971), 711-712

VALENTIN, GABRIEL GUSTAV [1810-1883]
DSB 13,555-558; ADB 39,463-464; HIRSCH 5,692-693; PAGEL, 1749-1750; FREUND 2,413-422; GRAETZER, 166-167; BRAZIER, 131-134,147; FERCHL, 549; KOHUT 2,237-238; KOREN, 259; WININGER 6,171-172; WWWW, 1705; POGG 2,1165-1166; 3,1381-1382; 7aSuppl.,715-716
E. Hintzsche, Gabriel Gustav Valentin. Berne 1953
B. Kisch, Trans. Am. Phil. Soc. 44 (1954), 142-192
R.H. Laeng, Mitt. Nat. Ges. Bern NF30 (1973), 12-14
G. Rudolph, Hist. Sci. Med. 19 (1985), 367-375
P. Müller, Gesnerus 45 (1988), 191-200

VALENTIN, WILLIAM GEORGE [1829-1879]
BLOKH 2,751; POGG 3,1382
A.W. Hofmann, Ber. chem. Ges. 12 (1879), 1041-1042

VALENTINER, WILHELM [1830-1893]
HIRSCH 5,693
Anon., Leopoldina 29 (1893), 58

VALEUR, AMAND [1870-1927]
GORIS, 590-591; POGG 4,1544; 5,1287; 6,2731
M. Javillier, Bull. Soc. Pharm. 34 (1927), 221-233
C. Moureu, Bull. Soc. Chim. [4]43 (1928), 492-504

VALLEE, BERT LESTER [1919-]
AMS 15(7),262; WWA 1980-1981, p.3360

VALTER (WALTER), ALEKSANDR PETROVICH [1817-1889]
BME 4,919-920; VENGEROV 4(2),71-73; IKONNIKOV, 84-89; HIRSCH 5,836-837
Y.V. Bukin, Meditsinski Zhurnal 24 (1934), 107

VALTIS, JEAN [1888-1950]
FISCHER 2,1605
Anon., Ann. Inst. Pasteur 80 (1951), 1-3

VAN BRUGGEN, JOHN TIMOTHY [1913-1989]
AMS 15(7),266; WWWS, 1707

VANDEL, ALBERT [1894-1981]
BUICAN, 181-187; IB, 308
P.P. Grasse, C.R. Acad. Sci. 294 (1982), 23

VAN DER HOEVEN, BERNARD JACOB CORNELIS [1899-1972]
AMS 10,4190
Anon., Chem. Eng. News 51(14) (1973), 17

VAN DER SCHEER, JAMES [1888-]
AMS 9(II), 1162

VAN DYKE, HARRY BENJAMIN [1895-1971]
AMS 11,5547; PARASCANDOLA, 54-55; CHEN, 87-89; IB,72; WWAH,622; WWWS,1709
B.F. Hoffman, Pharmacologist 13 (1971), 113-114
R.O. Greep, Horm. Prot. Pep. 8 (1980), 199-224

VAN DYKE, KARL SKILLMAN [1892-1966]
AMS 10,4194; WWAH, 622

VANE, JOHN ROBERT [1927-]
WW 1991, p.1874; POTSCH, 433-434

VAN HEYNINGEN, WILLIAM EDWARD [1911-]
WW 1981, p.2654; SRS, 91

VANINO, LUDWIG [1861-1944]
KURSCHNER 4,3084-3085; HEIN, 704-705; TLK 3,672; POTSCH, 434; POGG 4,1551-1553; 5,1294-1295; 6,2735; 7a(4),760
Anon., Chem. Z. 55 (1931), 765; 65 (1941), 402

VANLAIR, CONSTANT [1839-1914]
BNB 26,442-445; HIRSCH 5,704-705; PAGEL, 1752-1753
P. Nolf, Ann. Acad. Roy. Belg. 89 (1923), 125-150

VAN NIEL, CORNELIS BERNARDUS [1897-1985]
AMS 14,5257; WWA 1976-1977, p.3218; MH 2,563-564; STC 3,104-105; MSE 3,248-249; SACKMANN, 245-247; IB, 214; WWWS, 1711
C.B. Van Niel, Ann. Rev. Microbiol. 21 (1967), 1-30
H. Kamminga, TIBS 6 (1981), 164-165
W.D. McElroy, Am. Phil. Soc. Year Book 1985, pp.168-170
H.A. Barker and R.E. Hungate, Biog. Mem. Nat. Acad. Sci. 59 (1990), 389-423

VAN PILSUM, JOHN FRANKLIN [1922-]
AMS 15(7),286; WWWS, 1711

VAN SLYKE, DONALD DEXTER [1883-1971]
DSB 13,574-575; AMS 11,5558; NCAB 56,68-69; MILES, 483; DAMB, 760-761; MH 1,495-496; STC 3,105-106; MSE 3,249-250; CB 1943, pp.781-782; FISCHER 2,1467-1468; CHITTENDEN, 99-101,159,162; WWAH, 623; WWWS, 1712; POGG 6,2470-2471; 7b,4944-4948
Anon., Clin. Chem. 9 (1963), 645-663
A.B. Hastings, Fed. Proc. 23 (1964), 586-591; J. Biol. Chem. 247 (1972), 1635-1640; Biog. Mem. Nat. Acad. Sci. 48 (1976), 309-360
J. Sendroy, Clin. Chem. 17 (1971), 670-682
R.M. Archibald, Am. Phil. Soc. Year Book 1971, pp.194-200
J.T. Edsall, Hist. Phil. Life Sci. 7 (1985), 105-120

VAN SLYKE, LUCIUS LINCOLN [1859-1931]
AMS 4,1009; WWAH, 623; WWWS, 1712; POGG 5,1298-1299
R.W. Thatcher, Ind. Eng. Chem. 17 (1925), 1203
Anon., Ind. Eng. Chem. News Ed. 9 (1931), 903

VAN SYCKEL, BENJAMIN MILLER [1857-1903]
Anon., Boston Medical and Surgical Journal 148 (1903), 356

VAN TAMELEN, EUGENE EARLE [1925-]
AMS 15(7),289; WWA 1980-1981, p.3371; IWW 1982-1983, p.1355

VAN WAGENEN, GERTRUDE [1893-1978]
AMS 12,6571; IB, 313
Anon., Science 201 (1978), 428; Yale Medicine 13(2) (1978), 21
van't HOFF [see HOFF]

VAQUEZ, HENRI [1860-1936]
FISCHER 2,1606; RSC 19,294; WWWS, 1713
C. Laubry, Bull. Acad. Med. 142 (1958), 846-854

VARGHA, LASZLÓ [1903-1971]
MEL 3,820-821
V. Bruckner, Acta Chim. Acad. Sci. Hung. 74 (1972), 53-77
J. Kuszmann, Adv. Carbohydrate Chem. 28 (1973), 1-10

VARNEK [see WARNECK]

VARRENTRAPP, FRANZ [1818-1877]
HEIN, 706; BLOKH 2,752; FERCHL, 551; POGG 2,1178; 7aSuppl.,716
F. Knapp, Ber. chem. Ges. 10 (1877), 2291-2297
W.C. von Arnswald, Aus der Geschichte der Familie Varrentrapp, pp.108-123. Frankfurt a.M. 1908

VARS, HARRY MORTON [1903-1983]
AMS 15(7),297
F.C. Bing, J. Nutrition 116 (1986), 711-713

VARVOGLIS, GEORG [1907-]
International Chemical Directory 1969-1970, p.531

VASILEVSKI, VIKTOR MIKHAILOVICH [1907-1954]
VORONTSOV, 81-83

VASILIU, HARALAMB [1880-1953]
JV 21,90; NUC 630,455
L.M. Buruiana, Revue Roumaine de Biochimie 18 (1981), 77-84

VASSALE, GIULIO [1862-1913]
DSB 13,589-590; FISCHER 2,1607
E. Centanni, Bioch. Ter. Sper. 4 (1913), 193-198

VASSILIEV, NIKOLAI PETROVICH [1852-1891]
VENGEROV 4,165-167; RSC 11,68; 12,751; 19,301
Anon., Leopoldina 27 (1891), 107

VASSILIEV, STEPAN MIKHAILOVICH [1854-1903]
LEVITSKI 2,167-173
Anon., Leopoldina 39 (1903), 88
J. Brennsohn, Die Aerzte Livlands, p.420. Mitau 1905

VAUBEL, JOHANN WILHELM [1864-1957]
KURSCHNER 4,3087; WOLF, 210; TLK 3,672; POGG 4,1556-1557; 5,1301-1302; 6,2738; 7a(4),760
Anon., Chem. Z. 82 (1958), 50

VAUCHER, JEAN PIERRE ÉTIENNE [1763-1841]
DSB 13,595-596; DHB 7,51; BRIQUET, 470-471; TSCHIRCH, 1140; WWWS, 1714
Anon., Verhandl. Schw. Nat. Ges. 26 (1841), 308-313; Mem. Soc. Phys. Hist. Nat. Gen. 10 (1843), xxiv-xxvi
A. de Candolle, Ann. Mag. Nat. Hist. 10 (1842), 161-168,241-248
G. de Morsier, Physis 7 (1965), 497-500

VAUGHAN, VICTOR CLARENCE [1851-1929]
DAB 19,236-237; AMS 4,1011; NCAB 29,434-435; MILES, 484-485; WWAH, 624; DAMB, 762-763; FISCHER 2,1608; WWWS, 1715; POGG 6,2738
V.C. Vaughan, A Doctor's Memories. Indianapolis 1926
F.G. Novy, Science 70 (1929), 624-626
W.T. Vaughan, J. Lab. Clin. Med. 15 (1930), 817-821
C.H. McIntyre, J. Mich. Med. Soc. 31 (1932), 410-418
E.H. Huntress, Proc. Am. Acad. Arts Sci. 79 (1951), 40-42

VAUGHAN, WARREN TAYLOR [1893-1944]
AMS 7,1838; NCAB 33,120-121; WWAH, 624-625
W.R. Graham, South. Med. J. 38 (1945), 30-33

VAUQUELIN, NICOLAS LOUIS [1763-1829]
DSB 13,596-598; GE 31,757; LAROUSSE 15,814-815; BUGGE 1,356-368; W, 528; HIRSCH 5,714; BOURQUELOT, 47-48; BLOKH 2,752-753; FERCHL, 552-554; PHILIPPE, 710-711; SCHAEDLER, 140-141; TSCHIRCH, 1140; WWWS, 1715; COLE, 548-549; HAYMAKER, 302-306; ABBOTT-C, 143; POTSCH, 434-435; POGG 2,1182-1190; CALLISEN 33,128-129
G. Kersaint, Bull. Soc. Chim. 1958, pp.1603-1619
J. Cheymol and A. Soubiran, Presse Med. 70 (1962), 1960-1962
M. Schofield, Pharm. J. 190 (1963), 453-454
M. Bouvet and G. Kersaint, Rev. Hist. Pharm. 51 (1963), 17-25
H.G. Williams-Ashman, Inv. Urol. 2 (1965), 605-613
J.P. Poirier, Hist. Sci. Med. 24 (1990), 121-126
T.L. Sourkes, Ambix, 39 (1992), 11-16

VAVILOV, NIKOLAI IVANOVICH [1887-1943]
DSB 15,505-513; WWR, 572; KUZNETSOV(1963), 434-447; UKRAINE 2,207-208; W, 528-529; IB, 317
T. Dobzhansky, J. Heredity 38 (1947), 227-232
P.C. Mangelsdorf, Genetics 38 (1953),1-4
P.M. Khudovski, Bot. Zhur. 43 (1958), 905-911
Z.A. Medvedev, Bull. Soc. Nat. Moscou NS68 (1963), 138-142
E.T. Vasina-Popova, Genetika 23 (1987), 2002-2006

VAVON, GUSTAVE [1884-1953]
POGG 5,1302; 6,2738; 7b,5673-5674
G. Dupont, Bull. Soc. Chim. 1953, pp.657-664
A. Chatelet, Ann. Univ. Paris 23 (1953), 231-234

VEDDER, EDWARD BRIGHT [1878-1952]
DAB [Suppl.5],710-711; NCAB 41,229-230; FISCHER 2,1609; WWWS, 1716
R.R. Williams, J. Nutrition 77 (1962), 3-6

VEEN, ANDRÉ GERARD van [1903-1986]
Anon., Chem. Wkbl. 44 (1948), 367-368
R. Luyken, Voeding 87 (1987), 100-101

VEHSE, ADALBERT [1865-]
JV 7,150; NUC 631,532

VEIBEL, STIG ERIK [1898-1976]
KBB 1976, p.1129; IWW 1976-1977, p.1779; VEIBEL 2,461-463; WWWS, 1716; POGG 6,2741; 7b,5681-5684
A. Kjaer, Chem. Brit. 14 (1978), 89

VEIL, SUZANNE [1886-]
POGG 6,2741; 7b,5684-5685

VEJDOVSKY, FRANTISEK [1849-1939]
DSB 15,513-514
C. Dobell, Nature 156 (1945), 530

VELDE, ALBERT JACQUES JOSEPH van de [1871-1956]
BNB 38,766-792; GHENT 2,205-225; POGG 6,2743; 7b,5688-5693
B.V.J. Cuvelier, Chem. Wkbl. 48 (1952), 337-338
S. Delorme, Rev. Hist. Sci. 9 (1956), 171
P. van Oye, Jaarb. Vlaam. Acad. 1956, pp.225-228

VELDE, JEAN JACQUES ALBERT van de [1897-]
GHENT 2,240-244

VELDE, JOSEPH van de [1889-1974]
BNB 43,685-698

VELDEN, REINHARD von den [1851-1903]
FISCHER 2,1611; KALLMORGEN, 437

VELDEN, REINHARD von den [1880-1944]
KURSCHNER 4,3088-3089; FISCHER 2,1611-1612; AUERBACH, 406-407; IB, 309; GUNDLACH, 224-225; TLK 3,672
Anon., Deutsche med. Wchschr. 68 (1942), 123
P. Beckmann, Deutsche med. Wchschr. 75 (1950), 1699

VELEY, VICTOR HERBERT [1856-1933]
WW 1933, p.3326; POGG 3,1386; 4,1558; 5,1303; 6,1743; 7b,5693
J.A. Gardner, Obit. Not. Fell. Roy. Soc. 1 (1933), 229-235; J. Chem. Soc. 1934, pp.570-573

VELIAMINOV, NIKOLAI ALEKSANDROVICH [1855-1920]
BSE 7,354; BMS 5,93-94; WWR, 575-576; ZMEEV 4,55
M.I. Tikotin and V.I. Lapin, Vest. Khir. 75(4) (1955), 137-139
G.L. Magazanik, Voprosy Kurortologii 27 (1962), 163-167
K.K. Silvai, Problemy Tuberkuleza 49 (1971), 87-89

VELICK, SIDNEY FREDERICK [1913-]
AMS 15(7),306; WWA 1980-1981, p.3378

VELLA, LUIGI [1825-1886]
EI 35,28; HIRSCH 5,719-720; RSC 6,131; 11,685
Anon., Rend. Accad. Bologna 1886-1887, pp.9-11

VELLUZ, LÉON [1904-1981]
QQ 1978-1979, p.1548
M. Vigneron, Ann. Pharm. Fran. 39 (1981), 301-303
J. Roche, C.R. Soc. Biol. 176 (1982), 5-7; Bull. Acad. Med. 166 (1982), 165-169
J. Mathieu, Tetrahedron 38 (1982), 565-566

VELTEN, WILHELM [1848-1876]
A. Burgerstein, Ost. Bot. Z. 26 (1876), 373-375

VENABLE, FRANCIS PRESTON [1856-1934]
DAB 19,246-247; AMS 5,1150; MILES, 487; RSC 11,686-687; 19,316; WWWS,1717; POTSCH, 435
J.M. Bell, J. Chem. Ed. 7 (1930), 1300-1304

VENABLES, ROBERT [1788-]
HIRSCH 5,724; SG [1]15,634; RSC 6,132; NUC 632,185

VENDRELY, ROGER [1910-]
QQ 1981-1982, p.1462

VENNESLAND, BIRGIT [1913-]
AMS 14,5276-5277; MSE 3,253-254; WWWS, 1717
B. Vennesland, Ann. Rev. Plant Physiol. 32 (1981), 1-20

VENTZKE, CARL AUGUST EDUARD [1797-1865]
W. Wöhlert, Z. Zuckerind. 17 (1967), 581-585

VERAGUTH, HANS [1879-1960]
JV 21,489; NUC 632,649
Anon., CIBA-Chronik No.167 (1960)

VERATTI, EMILIO [1872-]
EI (1938-1948) 2,1104; IB, 309

VERDA, ANTONIO [1876-1949]
TSCHIRCH, 1140; POGG 6,2748; 7b,5706-5707
A. Buchner, Neue Schweizer Biographie, p.547. Basle 1938
A.R., Schw. Apoth. Z. 87 (1949), 721-722

VERDEIL, FRANÇOIS [1826-]
RSC 6,137; NUC 633,110

VERDON, ÉMILE [1884-1950]
GORIS, 524
R. Weitz, Ann. Pharm. Fran. 8 (1950), 65-66

VERGUIN, EMMANUEL [1814-1864]
E. Sack, Chimie et Industrie 80 (1958), 720-725

VERIGO, ALEKSANDR ANDREEVICH [1837-1905]
BSE 7,494-495; ARBUZOV, 192; BLOKH 2,773-774
P. Melikov, Zhur. Fiz. Khim. 37 (1905), 469-475

VERIGO, BRONISLAV FORTUNATOVICH [1860-1925]
BSE 7,495; BME 5,239-240; VORONTSOV, 173-179; BRAZIER, 225-227,245-246; WWWS, 1718
B.I. Khodorov, Zhur. Ob. Biol. 12 (1951), 55-69
K. Gamboroglu, Fiziol. Zhur. 46 (1960), 1422-1423
N.V. Chernusenko, Zhur. Mikr. Ep. Imm. 40(2) (1963), 117-119

VERKADE, PIETER EDUARD [1891-1979]
BWN 2,574-575; POGG 5,1305; 6,2748-2749; 7b,5707-5712
Anon., Chem. Wkbl. 35 (1938), 613-616
P.E. Verkade et al., Chem. Wkbl. 57 (1961), 353-376
Who's Who in the Netherlands 1962-1963, p.733
B.M. Wepster, Jaarb. Akad. Wet. 1979, pp.172-177

VERNADSKY, VLADIMIR IVANOVICH [1863-1945]
BSE 13,616-620; BSE(3rd Ed.) 4,536-537; KUZNETSOV(1962), 135-157, UKRAINE 2,213-214; WWWS, 1719; POTSCH, 436; POGG 4,1619; 5,1305-1306; 6,2749-2750; 7b,5712-5719
V.C. Asmous, Science 102 (1945), 439-440
V.V. Tikhomirov, Physis 4 (1962), 371-378
I.I. Machalov, Vladimir I. Vernadsky. Moscow 1970
K.E. Bailes, Science and Russian Culture in Age of Revolutions. Bloomington, Ind. 1990

VERNE, CLAUDE MARIE JEAN [1890-1982]
QQ 1975-1976, p.1622; POGG 6,2750; 7b,5719-5725
J.L. Parrot, Bull. Acad. Med. 167 (1983), 387-392

VERNEY, ERNEST BASIL [1894-1967]
DSB 18,960-962; DNB [1961-1970],1035; WW 1965, pp.3140-3141; STC 3,114-115; IB, 309
I. deB. Daly and L.M. Pickford, Biog. Mem. Fell. Roy. Soc. 16 (1970), 523-542
J.T. Fitzsimmons and W.J. O'Connor, J. Physiol. 263 (1976), 92P-93P

VERNOIS, MAXIME [1809-1877]
VAPEREAU 2,1807; HIRSCH 5,734-735; PAGEL, 1762-1763
L. Hémar and A. Delpech, Ann. Hyg. Pub. [2]47 (1877), 533-574
A. Delpech, Bull. Acad. Med. 6 (1877), 250-259
L. Figuier, Ann. Scient. 21 (1877), 537-538

VERNON, HORACE MIDDLETON [1870-1951]
WW 1949, p.2848; FISCHER 2,1615; O'CONNOR(2), 83-85; IB, 309-310; WWWS, 1719; POGG 4,1560
M. Smith, Nature 167 (1951), 383-384
Anon., Lancet 1951(I), p.477; Brit. Med. J. 1951(I), p.419
T. Bedford, Brit. J. Ind. Med. 8 (1951), 96-97

VERSCHAFFELT, JULES ÉMILE [1870-1955]
NBW 2,903-906; POGG 4,1561; 5,1306-1307; 6,2751-2752; 7b,5728-5730
E. Henriot, Ann. Acad. Roy. Belg. 123 (1957), 12-114

VERSÉ, MAX [1877-1947]
KURSCHNER 4,3091; IB, 310
E. Rix, Verhandl. path. Ges. 33 (1950), 420-425

VERWORN, MAX [1863-1921]
DSB 14,2-3; DBJ 3,265-267,320-321; FISCHER 2,1616-1617; PAGEL, 1764-1765; WAGENITZ, 184; GIESE, 496-499; DZ, 1498-1499; BK,330; WWWS, 1720; POGG 7aSuppl.,719
P. Jensen, Jahrb. Akad. Wiss. Gott. 1922, pp.61-78
R. Matthaei, Deutsche med. Wchschr. 48 (1922), 102-103
S. Baglioni, Riv. Biol. 4 (1922), 126-133
F.W. Frohlich, Z. allgem. Physiol. 20 (1923), 185-192
R. Wüllenberger, Der Physiologe Max Verworn. Bonn 1968

VERZÁR, FREDERIC [1886-1979]
KURSCHNER 12,3324; FISCHER 2,1617; IB, 310; WWWS, 1720
B. Rex-Kiss and S. Szabo, Orv. Kozl. 62-63 (1971), 159-173
H. M[islin] et al., Experientia 37 (1981), 1039-1059

VESQUE, JULIEN [1848-1895]
E. Gilg, Ber. bot. Ges. 13 (1895), (59)-(66)
C.E. Bertrand, Bull. Soc. Bot. 42 (1895), 472-478

VESTERBERG, KARL ALBERT [1863-1927]
POGG 5,1309; 6,2753-2754
H. von Euler, Ber. chem. Ges. 60A (1927), 152-153

VIALA, JULES [1871-1940]
L. Cruveilhier, Ann. Inst. Pasteur 64 (1940), 173-174

VIALE, GAETANO [1889-1934]
POGG 6,2755; 7b,5735-5737

VICKERY, HUBERT BRADFORD [1893-1978]
DSB 18,964-965; AMS 13,4637-4638; SRS, 56; MH 2,567-5??; MSE 3,255-256; IB, 310; POGG 6,2756-2757; 7b,5737-5740
H.B. Vickery, Ann. Rev. Plant Physiol. 23 (1972), 1-28
I. Zelitch, Biog. Mem. Nat. Acad. Sci. 55 (1985), 473-504

VIERORDT, CARL [1818-1884]
ADB 39,678-689; BAD 4,474-479; HIRSCH 5,752-753; PAGEL, 1768-1770; BLOKH 2,754; FERCHL, 556; SCHAEDLER, 141; VERSO, 98-100; POGG 2,1204; 3,1389-1390; 7aSuppl.,719-720
C. von Voit, Sitz. Bayer. Akad. 15 (1885), 180-185
N. Zuntz, Fort. Med., Suppl.3 (1885), 3-5
R.H. Major, Ann. Med. Hist. 10 (1938), 463-473
I. Krahn, Karl von Vierordt. Tübingen 1948
M.L. Vesco, Med. Hist. 15 (1971), 60-62

VIERORDT, HERMANN [1853-1944]
KURSCHNER 6(II),977; FISCHER 2,1619; PAGEL, 1770-1771; DZ, 1500

VIFANSKI, NIKOLAI MIKHAILOVICH [1862-]
SG [2]20,247

VIGIER, PIERRE VICTOR [1833-1905]
GORIS, 594-595
Anon., J. Pharm. Chim. [6]21 (1905), 576
J.A. Gautier, Figures Pharmaceutiques Françaises, pp.161-166. Paris 1953

VIGNAL, WILLIAM [1852-1893]
L. Malassez et al., C.R. Soc. Biol. 46 (1894), 845-855

VIGNON, LÉO [1850-1923]
GUIART, 175; POGG 4,1565-1566; 5,1310; 6,2758

VILLE, GEORGES [1824-1897]
GORIS, 597-598
L. Maquenne, Nouv. Arch. Hist. Nat. [3]9 (1897), iii-xiv
E. Gautier, Ann. Scient. 41 (1897), 394-398
F.W.J. McCosh, Ann. Sci. 32 (1975), 475-490

VILLEMIN, JEAN ANTOINE [1827-1892]
GENTY 6,289-320; HIRSCH 5,761-762; PAGEL, 1772-1773; BULLOCH, 402; W, 532-533; WWWS, 1723
Anon., Brit. Med. J. 1892(II), p.1091
F.S. Jaccoud, Mem. Acad. Med. 40 (1906), 1-18
L. Lereboullet, Paris Médical 2 (1912), 251-257
S.L. Cummins, in Science, Medicine and History (E.A. Underwood, Ed.), vol.2, pp.331-340. London 1953
M. Bariéty, Bull. Acad. Med. 149 (1965), 761-767

VILLIGER, VICTOR [1868-1934]
POGG 6,1761
M.A. Kunz, Ber. chem. Ges. 67A (1934), 111-113

VILMORIN, LOUIS de [1816-1850]
DSB 14,33-34

VINCENT, CAMILLE PHILIPPE [1839-1910]
POGG 3,1392-1393; 4,1569; 5,1312
B. Delachanal, Bull. Soc. Chim. [4]9 (1911), i-vi

VINCENT, JEAN HYACINTHE [1862-1950]
GENTY 5,189-224; FISCHER 2,1621; CHARLE, 236-238; SACKMANN, 347-349; WWWS, 1723
Anon., Brit. Med. J. 1950(II), p.1395
G. Julia, C.R. Acad. Sci. 231 (1950), 1181-1184
L. Jame, Presse Med. 59 (1951), 477-478
L. Tanon, Bull. Acad. Med. 135 (1951), 119-127
H.M. Pickard, Proc. Roy. Soc. Med. 66 (1973), 695-698

VINCENT, THOMAS SWALE [1868-1933]
O'CONNOR(2), 513-515; IB, 311
W.C., Nature 133 (1934), 128-129VINCKE, ERICH [1905-1968]
KURSCHNER 10,2564; POGG 7a(4),771-772
Anon., Chem. Z. 92 (1968), 650

VINES, SYDNEY HOWARD [1849-1934]
DNB [1931-1940],881-882; WW 1929, p.3137; O'CONNOR, 255; DESMOND, 631; BK, 331; WWWS, 1723
A.B. Rendle, Obit. Not. Fell. Roy. Soc. 1 (1934), 185-188; Journal of Botany 72 (1934), 139-141
F.E.W., Nature 133 (1934), 675-677

VINOGRAD, JEROME RUBEN [1913-1976]
DSB 18,965-966; AMS 13,4643; MSE 3,256-257
Anon., Chem. Eng. News 48(4) (1970), 123; 54(34) (1976), 31
N. Davidson, TIBS 1 (1976), N282

VINOGRADOV, ALEKSANDR PAVLOVICH [1895-1975]
BSE(3rd Ed.) 5,81; WWWS, 1723-1724
M.Y. Marov, Icarus 30 (1977), 239-241

VINOGRADOV, NIKOLAI ANDREEVICH [1831-1885]
BME 5,411-412; HIRSCH 5,960; KAZAN 2,148-151; ZMEEV 1,49-50; 3,8-9; 4,58
V.Y. Albitski, Klin. Med. 60(9) (1982), 121-123

VINSON, CARL GEORGE [1891-1964]
AMS 7,1844; IB, 311

VINTILESCU, JEAN (ION) [1881-1954]
ILEA, 477

VINTSCHGAU, MAXIMILIAN von [1832-1913]
HIRSCH 5,567; HUTER, 217-219; WERSTLER 1,66-72; KOERTING, 117-118; POGG 3,1398; 4,1569
W. Trendelenburg, Ber. Nat. Med. Innsb. 35 (1914), x-xvi

L. Bizzotto and G. Rialdi, Acta Med. Hist. Pat. 22 (1975-1976), 9-20

VIRCHOW, HANS [1852-]
IB, 311; WININGER 6,187; 7,483

VIRCHOW, RUDOLF [1821-1902]
DSB 14,39-44; BJN 7,352-361; LF 2,465-475; HIRSCH 5,768-772; W, 533-534; PAGEL, 1774-1777; FREUND 2,423-433; HAYMAKER, 380-384; FRANKEN, 200; SANDRITTER, 119-122; WREDE, 201-204; WININGER 6,187-189; 7,483; WWS, 1724; POGG 7aSuppl.,720-731
J. Schwalbe, Virchow Bibliographie 1843-1901. Berlin 1901
H. Ribbert, Deutsche med. Wchschr. 27 (1901), 702-704; 28 (1902),657-658
J. Pagel, Virchows Jahresber. 1902, pp.426-427
W. Waldeyer, Berl. klin. Wchschr. 39 (1902), 861-864
J. Orth, Berl. klin. Wchschr. 39 (1902), 1021-1027
O. Bollinger, Münch. med. Wchschr. 49 (1902), 1621-1624
A. von Kölliker, Anat. Anz. 22 (1902), 59-62
G.S. Woodhead, J. Path. Bact. 8 (1903), 374-378
C. von Voit, Sitz. Bayer. Akad. 33 (1903), 515-535
M. Verworn, Z. allgem. Physiol. 2 (1903), i-vii
R. Beneke, Naturw. Rund. 18 (1903), 25-27,35-39,49-50; Pommersche Lebensbilder 2 (1936), 198-216
P.H.P.S., Proc. Roy. Soc. 75 (1905), 297-300
R. Virchow, Briefe an seine Eltern 1839-1864. Leipzig 1907
C. Posner, Rudolf Virchow. Vienna 1921
R. Bochalli, Naturw. Rund. 5 (1952), 352-354
I. Hesche-Klünder, Centaurus 2 (1952), 205-250
P. Diepgen, Arch. path. Anat. 322 (1952), 221-232
E.H. Ackerknecht, R. Virchow, Doctor, Statesman, Anthropologist. Madison, Wis. 1953 (German revision, Stuttgart 1957)
H.M. Koelbing, Münch. med. Wchschr. 110 (1968), 349-354
K. Winter, Rudolph Virchow. Leipzig 1976
R.C. Maulitz, Bull. Hist. Med. 52 (1978), 162-173
A. Bauer, Rudolph Virchow, der politische Arzt. Berlin 1982
M. Vasold, Rudolf Virchow, der grosse Arzt und Politiker. Stuttgart 1988

VIREY, JULIEN JOSEPH [1775-1846]
DSB 14,44-45; LAROUSSE 15,1100-1101; BALLAND, 118-125; BALLAND(2), 185; HIRSCH 5,772-773; BOURQUELOT, 39-40; FERCHL, 557; WWWS, 1724; POGG 2,1210-1211; CALLISEN 20,157-177; 33,159-162
J. Soubeiran, J. Pharm. Chim. [3]9 (1846), 277-282
A. Berman, Bull. Hist. Med. 39 (1965), 134-142

VIRTANEN, ARTTURI ILMARI [1895-1973]
DSB 14,45-46; MH 1,497-499; STC 3,117-118; MSE 3,257-258; WWWS, 1724; POTSCH, 438; POGG 6,2762-2764; 7b,5762-5776
A. Butenandt, Bayer. Akad. Wiss. Jahrbuch 1974, pp.221-229

VISCONTINI, MAX [1913-]
KURSCHNER 13,4067; WWWS, 1724; POGG 7a(4),772-773
C.H. Eugster, Chimia 27 (1973), 46-47
W. von Philipsborn, Chimia 37 (1983), 25

VISHNEGRADSKI, ALEKSEI NIKOLAEVICH [1851-1880]
BLOKH 2,789-791; ARBUZOV, 132; RSC 11,710; 12,762; POGG 3,1472
D.P. Pavlov, Zhur. Fiz. Khim. 13 (1881), 370-375,380
A.N. Kost and Y.G. Yashunski, Usp. Khim. 21 (1952), 260-264
Y.S. Musabekov, Zhur. Prikl. Khim. 25 (1952), 681-686

VISHNIAC, WOLF VLADIMIR [1922-1973]
AMS 12,6609; BHDE 2,1193; WWWS, 1725
S.L. Thomas, Men of Space 6,260-279. New York 1963
Anon., Science 183 (1974), 1181

VISSCHER, MAURICE BOLKS [1901-1983]
AMS 14,5294; WWA 1976-1977, p.3235; IWW 1982-1093, p.1386; MH 2,569-570; BROBECK, 155; WWWS, 1725
M.B. Visscher, Ann. Rev. Physiol. 31 (1969), 1-18
I.H. Page, Am. Phil. Soc. Year Book 1985, pp.215-217

VISSER, LOUIS EDUARD OTTO de [1865-1904]
NNBW 2,1500; POGG 4,1571; 5,286
W.P. Jorissen and W.E. Ringer, Ber. chem. Ges. 37 (1904), 4947-4950

VIVENOT, RUDOLF von [1834-1870]
ADB 40,85; WURZBACH 51,90-96; HIRSCH 5,777-778; SG [1]15,789
Wahrmann, Wiener medizinische Presse 12 (1871), 131-133

VLADESCU, RADU [1886-1964]
POGG 6,2766; 7b,5785-5787

VLADIMIROV, ALEKSANDR ALEKSANDROVICH [1862-1942]
BME 5,650-651
K.I. Skryabin et al., Mikrobiologia 16 (1947), 545

VLADIMIROV, IVAN PROKLOVICH [1857-]
RSC 19,382; SG [2]20,331

VLÈS, FRED [1885-1944]
HEIM, 125-129; IB, 311; POGG 6,2767; 7b,5790-5792
E. Fauré-Fremiet, Bull. Soc. Chim. Biol. 27 (1945), 459-460
A. Strohl, Bull. Acad. Med. 129 (1945),677-678; Presse Med. 53 (1945),718
G. Achard, Bull. Soc. Zool. 72 (1948), 125-127

VÖCHTING, HERMANN von [1847-1917]
DBJ 1,675; LEHMANN, 184-185; WAGENITZ, 184; TSCHIRCH, 1141
Anon., Württemberger Nekrolog 1917, pp.151-160
H. Fitting, Ber. bot. Ges. 37 (1919), (41)-(77)
G. Senn, Verhandl. Nat. Ges. Basel 30 (1919), 1-9

VOEGTLIN, CARL [1879-1960]
AMS 10,4226; FISCHER 2M1623; DAMB. 766-767; PARASCANDOLA, 56-57; CHEN, 29-30; BARRY, 90-92; WWWS, 1726; POGG 6,1768; 7b,5793-5795
Anon., J. Nat. Cancer Inst. 25 (1960), v-xv
H. Ritter, Schw. med. Wchschr. 91 (1961), 216-217
J. Parascandola, J. Hist. Med. 32 (1977), 151-171

VÖLCKEL, FRIEDRICH CARL [1819-1880]
POGG 2,1215; 3,1394-1395

VOELCKER, JOHN CHRISTOPHER AUGUSTUS [1822-1884]
DNB 20,386-387; BLOKH 2,754; WWWS, 1726; POGG 2,1215; 3,1394
J.H.G., Proc. Roy. Soc. 38 (1885), xviii-xxiii

VÖLKER, OTTO [1907-]
KURSCHNER 13,4070; POGG 7a(4),775-776

VÖLTZ, WILHELM [1872-1930]
BK, 339; IB, 311; NUC 640,669-671

VOERMANN, GERARDUS LEONARDUS [1879-1950]
POGG 6,2768; 7b,5795
W.P. Jorissen, Chem. Wkbl. 47 (1951), 533-537

VOGEL, AUGUST [1817-1889]
ADB 40,95-96; GLOSAUER, 235-252; BLOKH 2,764-755; FERCHL, 558-559; RSC 6,179-183; POGG 2,1220-1221; 3,1395; 7aSuppl.,731
Anon., Chem. Z. 13 (1889), 1125-1126
C. von Voit, Sitz. Bayer. Akad. 20 (1891), 391-396

VOGEL, HANS (JOHANNES) [1900-1980]
KURSCHNER 13,4074; KALLMORGEN, 438; POGG 6,2769; 7a(4),776-777
Anon., Zeitschrift für Parasitenkunde 46 (1975), 1
Wer ist Wer 1976-1977, p.1039

VOGEL, HEINRICH AUGUST [1778-1867]
HEIN, 710-712; GLOSAUER, 92-99; BOURQUELOT, 43-44; TSCHIRCH, 1141; FERCHL, 559; WWWS, 1740; CALLISEN 20,190-200; 33,167; POGG 2,1217-1220; 7aSuppl.,732
Anon., Alm. Bayer. Akad. Wiss. 1867, pp.268-270
A. Vogel, Denkrede auf Heinrich August Vogel. Munich 1868

VOGEL, HERMANN WILHELM [1834-1898]
BJN 3,157-158; 5,64*; BLOKH 2,755-756; SCHAEDLER, 142-143; W, 534; POTSCH, 439; WWWS, 1726; POGG 2,1221-1222; 3,1395-1396 4,1573 7aSuppl.,732-733
Anon., Nature 19 (1898), 204-205; Leopoldina 34 (1898), 194-195
H. Landolt, Ber. chem. Ges. 32 (1899), 1-4
B. Schwalbe, Verhandl. phys. Ges. 1 (1899), 60-64
E. Stenger, Naturwiss. 22 (1934), 177-181; Z. wiss. Phot. 34 (1935), 1-8
F. Herneck, Hermann Wilhelm Vogel. Leipzig 1984

VOGEL, JULIUS [1814-1880]
ADB 40,114-115; HIRSCH 5,785-786; PAGEL, 1780-1781; RSC 5,189
W. Kaiser and W. Piechoski, Z. ges. inn. Med. 24 (1969), 142-151

VOGEL, SAMUEL GOTTLIEB [1750-1837]
ADB 40,124-125; HIRSCH 5,782-783; CALLISEN 20,205-206; 33,169
J. Landmann, Gesnerus 28 (1971), 168-195

VOGEL, WILHELM [1878-1943]
JV 18,328; TLK 3,676; POGG 6,2770; 7a(4),779

VOGL, AUGUST EMIL [1833-1909]
BJN 14,97*; WURZBACH 51,165; HIRSCH 5,786-787; PAGEL, 1782; DECKER, 9-21; REBER, 9-11,352; PRAG, 347-348; KIRCHENBERGER, 213-215
E. Ludwig, Wiener klin. Wchschr. 17 (1904), 774
K. Ganzinger, Ost. Apoth. Z. 19 (1965), 316-318

VOGT, CARL [1817-1895]
DSB 14,57-58; ADB 40,181-189; HIRSCH 5,789-790; PAGEL, 1782-1783; BLOKH 2,756; BOSSART, 40-42; SCHAEDLER, 143; GENEVA 3,24-28; 4,161-169; POGG 2,1223; 3,1398; 7aSuppl., 735-736
Anon., Nature 52 (1895), 108-110; Leopoldina 31 (1895), 108-109
E. Yung, Rev. Sci. [4]3 (1895), 769-779
H. de Varigny, Nature 54 (1896), 386-388
A.S. Packard, Science 4 (1896), 947-954
C. Vogt, Aus meinem Leben. Stuttgart 1896
W. May, Naturwiss. 5 (1917), 449-452
O. Taschenburg, Leopoldina 56 (1920), 10-12,18-24,51-54,57-62,73-74; 57 (1921), 24
F. Gregory, Scientific Materialism in Nineteenth Century Germany, pp.51-79. Dordrecht 1977
G. Bernbeck, Mitteilungen des Oberhessischen Geschichtsvereins NF62 (1977), 221-236
M.D. Grmek, Rev. Med. Suisse Rom. 109 (1989), 1013-1021

VOGT, HANS [1874-1963]
KURSCHNER 4,3108; FISCHER 2,1624
U. Köttgen, Mon. Kind. 112 (1964), 106-107

VOGT, HANS [1913-]
KURSCHNER 13,4082; KIEL, 201; POGG 7a(4),780-781
Anon., Chem. Z. 102 (1978), 365

VOGT, MARGUERITE [1919-]
AMS 13,4650

VOGT, MARTHE LOUISE [1903-]
WW 1981, p.2671; MEITES, 314-321; BHDE 2,1195; WWWS, 1726

VOGT, OSKAR [1870-1959]
FREUND 2,435-443; BHDE 2,1195; IB, 312; WWWS, 1726
W. Haymaker, Neurology 1 (1951), 179-204
H. Meesen, Deutsche med. Wchschr. 84 (1959), 1796-1797

VOGT, WALTHER [1888-1941]
FISCHER 2,1625; EGERER, 150-157; IB, 312
H. Spemann, Arch. Entwickl. 141 (1941-1942), 1-14

A. Dabelow, Münch. med. Wchschr. 88 (1941), 1309-1311
K. Goerttler, Anat. Anz. 92 (1941-1942), 184-196

VOHL, EDUARD HERMANN LUDWIG [1823-1878]
SCHAEDLER, 143; RSC 6,192; 8,1167-1168; POGG 2,1223-1224; 3,1398

VOIGES, PAUL [1878-1939]
JV 24,368; NUC 641,265

VOIT, CARL [1831-1908]
DSB 14,63-67; BJN 13,3-15,96*; HIRSCH 5,795-796; PAGEL, 1787-1788; DZ, 1510; WWWS, 1740; POGG 7aSuppl.,739-740
M. Cremer, Münch. med. Wchschr. 55 (1908), 1437-1442
A. Lesser, Naturw. Rund. 23 (1908), 180-183
A. Durig, Wiener klin. Wchschr. 21 (1908), 261-262
F. Müller, Med. Klin. 4 (1908), 337-339
O. Frank, Z. Biol. 51 (1908), i-xxiv
H. Boruttau, Deutsche med. Wchschr. 34 (1908), 340-341
G. Lusk, Ann. Med. Hist. NS3 (1931), 583-594
H.H. Mitchell, J. Nutrition 13 (1937), 3-13
F.L. Holmes, in Transformation and Tradition in the Sciences (E. Mendelsohn, Ed.), pp.455-470. Cambridge, 1984

VOIT, ERWIN [1852-1932]
KURSCHNER 4,3112; FISCHER 2,1625; DZ, 1509; BK, 332; IB, 312
O. Frank, Z. Biol. 93 (1932), 11-13

VOIT, FRITZ [1863-1944]
KURSCHNER 4,3112; FISCHER 2,1626; BACHMANN, 67-70; PITTROFF, 148-149

VOIT, MAX [1876-1949]
KURSCHNER 6(II),991; FISCHER 2,1626; WAGENITZ, 185
F. Stadtmüller, Anat. Nachr. 1 (1951), 318-335

VOITINOVICI, ARTHUR [1881-]
JV 22,39; NUC 641,351

VOLCK, CONRAD [1867-]
JV 8,217; RSC 16,486
Anon., Chem. Z. 57 (1933), 286

VOLHARD, FRANZ [1872-1950]
KURSCHNER 7,2172-2173; FISCHER 2,1626; KALLMORGEN, 439; IB, 312
W. Kaiser, Med. Welt. 23 (1972),694-701; Z. ärzt. Fortbild. 66 (1972), 580-585
O. Hoevel et al., Med. Welt 24 (1973), 9-42
H.E. Bock et al.(Eds.), Franz Volhard: Erinnerungen. Stuttgart 1982

VOLHARD, JACOB [1834-1910]
BJN 15,87*; BLOKH 2,756-758; STUPP-KUGA, 4-16; SCHAEDLER, 143-144; ABBOTT-C, 143-144; DZ, 1510-1511; WWWS, 1727; POTSCH, 439; POGG 3,1399-1400; 4,1577-1578; 5, 1317; 7aSuppl.,740-741
E. Renouf, Am. Chem. J. 43 (1910), 281-283
C. Tubandt, Chem. Z. 34 (1910), 73
J. Thiele, Ann. Chem. 372Suppl. (1910), 1-17
A. von Baeyer, Sitz. Bayer. Akad. 1910, pp.28-32
D. Vorländer, Z. angew. Chem. 23 (1910), 337-340; Leopoldina 46 (1910), 45-48; Ber. chem. Ges. 45 (1912), 1855-1902

VOLKENSTEIN, MIKHAIL VLADIMIROVICH [1912-]
IWW 1982-1983, p.1371

VOLKMANN, ALFRED WILHELM [1800-1877]
ADB 40,236-237; HIRSCH 5,797; PAGEL, 1788-1789; LEVITSKI 2,301-303; DECHAMBRE [5]3,79; WWWS, 1727; CALLISEN 20,244; 33,180; POGG 2,1229-1230; 3,1400; 7aSuppl.,741-742
Anon., Leopoldina 14 (1878), 98-100
F. von Kobell, Sitz. Bayer. Akad. 8 (1878), 103-104

VOLKONSKY, MICHEL [1907-1942]
A. Lwoff, Ann. Inst. Pasteur 68 (1942), 499-501

VOLKOV, ALEKSANDR NIKOLAEVICH [1849-1928]
LIPSHITS 2,148-149

VOLKOV, ALEKSEI ALEKSEEVICH [1863-1903]
BLOKH 2,799-800
B.N. Menshutkin, Zhur. Fiz. Khim. 35 (1904), 1261-1263

VOLKOV, MIKHAIL MATVEEVICH [1861-1913]
ZMEEV 4,65; 5Suppl.,2; RSC 19,400
A.E. Greiser, Sov. Med. 26 (1962), 155-158

VOLTA, ALESSANDRO [1745-1827]
DSB 14,69-82; EI 35,572-576; BLOKH 2,758-759; BRAZIER, 15-19,23-24; MATSCHOSS, 284-285; FERCHL, 561-562; COLE, 551-552; W, 535; POTSCH, 440; WWWS, 1727; POGG 2,1230-1233; CALLISEN 33,182
F. Arago, Ann. Chim. [2]54 (1833), 396-444
F. Scolari, A. Volta: Guide Bibliographique. Rome 1927
A. Mieli, Alessandro Volta. Rome 1927
G. Bilancioni, Archeion 8 (1927), 351-363
B. Duschnitz, Z. Elektrochem. 35 (1929), 822-825
A. Mauro, J. Hist. Med. 24 (1969), 140-150
G. Pancaldi, Hist. Stud. Phys. Sci. 21 (1990), 123-160

VONGERICHTEN, EDUARD [1852-1930]
LADIS, 87-97; TLK 3,680; RSC 9,989; 15,268; POGG 4,1581; 5,1321; 6,2776
H.P. Kauffman, Ber. chem. Ges. 64A (1931), 201-210

VONK, HUBERTUS JOHANNES [1897-1982]
Who's Who in the Netherlands 1962-1963, pp.747-748
Anon., Vakblad voor Biologen 82 (1982), 336-337

VON OETTINGEN, WOLFGANG FELIX [1888-1976]
AMS 10,4235; WWWS, 1738; POGG 6,1900-1901; 7b,3709-3711
H.W. Hays, Toxicology and Applied Pharmacology 38 (1976), 651-653

VONWILLER, PAUL [1885-1962]
KURSCHNER 9,2176; SBA 2,129; FREUND 2,445-463; IB, 312
F. Bruman, Acta Anat. 52 (1963), 163-174

VOORHEES, CLARK GREENWOOD [1871-1933]
Obituary Record of Yale Graduates, No.93, pp.158-159. New Haven 1934

VORBACH, KARL [1900-1935]
JV 42,338; NUC 642,407
Anon., Chem. Z. 59 (1935), 858; Z. angew. Chem. 48 (1935), 656

VORLÄNDER, DANIEL [1867-1941]
KURSCHNER 4,3126; POTSCH, 440-441; POGG 4,1581-1582; 5,1321; 6,2776-2777; 7a(4),792
C. Weygand, Ber. chem. Ges. 76A (1943), 41-58

VORMS, VLADIMIR VILGEMOVICH [1868-]
KAZAN 2,155-156

VOROBYOV, ANATOLI MARKOVICH [1900-1955]
WWR, 592; VORONTSOV, 83-84

VOROBYOV, VLADIMIR PETROVICH [1876-1937]
BSE 9,97-98; BME 5,1034-1035; WWR, 593; UKRAINE 2,221-222; WWWS, 1741
N.I. Grashchenkov, Acta Medica URSS 1 (1938), 263-270
R.D. Sinelnikov, Zhizn v Nauke. Moscow 1969; Arkh. Anat. Gist. Emb. 7(12) (1976), 5-10

VORONIN, MIKHAIL STEPANOVICH [1838-1903]
DSB 14,93-94; BSE 9,121; LIPSHITS 2,163-158; SKOROKHODOV, 83

S. Navaschin, Ber. bot. Ges. 21 (1903), (35)-(47)
E.F. Smith, Phytopathology 2 (1912), 1-4
V.A. Parnes, Mikhail Stepanovich Voronin. Moscow 1976

VORONTSOV, DANIEL SEMENOVICH [1886-1965]
BSE(3rd Ed.) 5,369; BME 5,1038-1039; WWR,593; VORONTSOV, 133-136,188-189; UKRAINE 2,223-224; IB, 329; WWWS, 1741
P.O. Makarov, Fiziol. Zhur. (Kiev) 12 (1966), 703-711

VOROSHILOV, KONSTANTIN VASILIEVICH [1842-1899]
KAZAN 1,295-296; VENGEROV 1,648; ZMEEV 4,67

VOROZHTSOV, NIKOLAI NIKOLAEVICH [1881-1941]
BSE 9,104; WWR, 594; POGG 6,2930; 7b,5813-5815
V.A. Izmailski, Zhur. Ob. Khim. 13 (1943), 525-539
V.N. Ufimtsev, Usp. Khim. 21 (1952), 110-115

VORTMANN, GEORG [1854-1932]
TLK 3,680; POGG 3,1402; 4,1582; 5,1321-1322; 6,2778
F.Feigl, Ber. chem. Ges. 65A (1932), 149

VOS, PETRUS ARNOLDUS [1858-1934]
NUC 642,514
P. van der Weilen, Pharm. Weekbl. 71 (1934), 707

VOSKRESSENSKI, ALEKSANDR ABRAMOVICH [1809-1880]
BSE 9,158-159; ARBUZOV, 100-101; KUZENTSOV(1961), 434-440; FERCHL, 589; BLOKH 2,801-802; POTSCH, 441; POGG 2,1368; 3,1466
G. Steinberg, J. Chem. Ed. 42 (1965), 675-681

VOSS, ARTHUR [1882-1940]
JV 23,46
Anon., Chem. Z. 64 (1940), 320; Z. angew. Chem. 53 (1940), 336
H.W. Flemming, Ed., Arthur Voss. Dokumente aus Hoechster Archiven No.24. Frankfurt a.M. 1967

VOSS, HERMANN [1888-1979]
POGG 7a(4),793-796
M. Hardebeck, Der Boehringer Kreis 1968(1), pp.16-17
R. Kattermann, in Naturwissenschaft und Medizin (R. Kattermann, Ed.), pp.107-123. Mannheim 1985

VOSS, OTTO [1869-1959]
KURSCHNER 8,2480; KALLMORGEN, 440
L. Bablik, Mon. Ohren. 94 (1960), 65-66

VOSS, WALTER [1899-]
KURSCHNER 12,3361; POGG 6,2779; 7a(4),796
Anon., Chem. Z. 98 (1974), 515

VOTCHAL, EVGENI FILIPOVICH [1864-1937]
BSE 9,226; WWR, 595; LIPSHITS 2,186-189
A.M. Levshin, Zhurnal Instituta Botaniki Akademii Nauk SSSR 16(24) (1938), 197-212

VOTOČEK, EMIL [1872-1950]
POGG 5,1322; 6,2779; 7b,5817-5818
B. Strehlik, Coll. Czech. Chem. Comm. 4 (1932), 377-387
V. Vesely, Chem. Listy 26 (1932), 435-442; 36 (1942), 257-260
S. Malachta, Chem. Ind. 1951, p.113
G.B. Kauffman and F. Jursik, Chem. Brit. 25 (1989), 495-498

VREVSKI, MIKHAIL STEPANOVICH [1871-1929]
BSE 9,249; WWR, 596-597; IB, 313; POGG 6,2933
K.P. Mishchenko, Zhur. Prikl. Khim. 2(6) (1929), i-vi
S.A. Shukarev, Zhur. Ob. Khim. 1 (1931), 1145-1157

VRIENS, JOHANNES GERARDUS CORNELIS [1866-1933]
POGG 4,1583; 5,1323
Anon., Chem. Z. 37 (1913), 227
P.G. Groenen, Koloniaal Missie Tijdschrift 17 (1934), 193-195

VRIES, HUGO de [1848-1935]
DSB 14,95-105; BWN 1,631-635; GERRITS, 317-333; SCHULZ-SCHAEFFER, 203; TSCHIRCH, 1142; IB, 313; ABBOTT, 36-37; W, 537-538; WWWS, 453; POGG 4,324; 6,2781; 7b,5825-5827
K. Höfler, Protoplasma 3 (1928), 605-610
T. Stomps et al., Hugo de Vries. Stuttgart 1929 (Tübinger Naturwissenschaftliche Abhandlungen 12)
T.J. Stomps, Ber. bot. Ges. 53 (1935), (85)-(96)
A.F. Blakeslee, Science 81 (1935), 581-582
A.D. Hall, Obit. Not. Fell. Roy. Soc. 1 (1935), 371-373
H. Molisch, Alm. Akad. Wiss. Wien 85 (1935), 242-247
R.E. Cleland, J. Heredity 26 (1935), 289-297; Sci. Mon. 68 (1949),35-41
J. Heimans, Am. Nat. 96 (1962), 93-102
G.E. Allen, J. Hist. Biol. 2 (1969), 55-87
P. de Veer, Hugo de Vries. Groningen 1969
P.W. van der Pas, Folia Mendeliana 11 (1978), 3-18
M. Campbell, Ann. Sci. 37 (1980), 639-655
O.G. Meijer, Ann. Sci. 42 (1985), 189-232

VRIJ, JOHAN ELIZA de [1813-1898]
LINDEBOOM, 2105-2106; REBER, 195-212; FERCHL, 562; TSCHIRCH, 1142
Anon., Pharm. Z. 43 (1898), 567-568
H. van Gelder, Ber. pharm. Ges. 9 (1899), 1-25

VROBLEVSKI, EDUARD ANTONOVICH [1848-1892]
BSE 9,279; RSC 8,1279; 11,858; 12,761-762; 19,414
I.Z. Siemion, Wiad. Chem. 38 (1984), 289-309

VULF (WULFF), GEORGI VIKTOROVICH [1863-1925]
DSB 14,525-526; BSE 9,403-404; WWR, 597; WWWS, 1742; POGG 5,1396; 6,2938
E.E. Flint, Trudy Instituta Prikladnoi Mineralogii 34 (1925), 1-49
A.V. Shubnikov, Priroda 15 (1926), 6-7
J.J. Spencer, Min. Mag. 21 (1927), 255-256
M. von Laue, Z. Kristall. 105 (1943), 124-133

VULPIAN, ALFRED [1826-1887]
HIRSCH 5,807; PAGEL, 1794; LACROIX 2,325-329; BOURQUELOT, 99-100; HAYMAKER, 272-275
J. Camus, Paris Médical 1913, pp.733-747
E. Gley, Bull. Acad. Med. 97 (1927), 733-748; Prog. Med. 42 (1927), 881-891
M. Laignol-Lavastine, Bull. Soc. Hist. Med. 21 (1927), 287-303
A. Ebner, Edme Félix Alfred Vulpian. Zurich 1967
H. Duclohier, Hist. Phil. Life Sci. 8 (1986), 27-40

VVEDENSKI, NIKOLAI EVGENIEVICH [1852-1922]
DSB 14,105; BSE 7,74-77; BME 4,1032-1034; WWR, 598; BRAZIER, 239-242; KUZENTSOV(1963), 212-222
A.A. Ukhtomski, Fiziol. Zhur. 23 (1937), 183-186
I.A. Arshavski, N.E. Vvedenski. Moscow 1950; Priroda 1964(12), pp.83-86
E.K. Zhukov, Fiziol. Zhur. 48 (1962), 505-509

VYROBOV, GRIGORI NIKOLAEVICH [1843-1913]
BLOKH 2,789; POGG 4,1675; 5,1397
H. Copaux, Bull. Soc. Chim. [4]16 (1914), i-xxi

VYSOKOVICH, VLADIMIR KONSTANTINOVICH [1854-1912]
BSE 9,464; BME 6,123-124; KHARKOV, 65-67; SKOROKHODOV, 162-166
A.I. Smirnova-Zamkova, Arkh. Pat. 16(3) (1954), 68-74
S. Jarcho, American Journal of Cardiology 24 (1969), 876-879
A.F. Kiseleva, Arkh. Pat. 41(11) (1979), 81-84

VYSOTSKI, GEORGI NIKOLAEVICH [1865-1940]
DSB 14,106-108; BSE 9,490-491; WWR, 599; KUZNETSOV(1963), 815-824; LIPSHITS 2,200-207
A.S. Skorodymov, Bot. Zhur. 36 (1951), 106-110

VYSOTSKI, NIKOLAI FEDOROVICH [1843-1922]
KAZAN 2,158-161; ZMEEV 4,66-67; 5Suppl.,2-3

W

WAAGE, PETER [1833-1900]
DSB 14,108-109; NBL 18,255-268; W, 538; WWWS, 1742; POTSCH, 441; POGG 3,1403-1404
W. Ramsay, J. Chem. Soc. 77 (1900), 591-592
H. Haraldsen, The Law of Mass Action: Centenary Volume, Oslo 1964
P. Øhstrøm, Centaurus 28 (1985), 277-287

WAAL, HERMANUS LAMBERTUS van de [1907-]
JV 51,658; NUC 643,517

WAALS, JOHANNES DIDERIK van der [1837-1923]
DSB 14,109-111; BWN 1,638-640; GERRITS, 275-315; W, 524-525L
WWWS, 1708; ABBOTT-C, 140-141; POTSCH, 433; POGG 3,1404; 4,1547-1548; 5,1291-1292; 6,2785
P. Zeeman, Naturwiss. 5 (1917), 701-703
J.H. Jeans, J. Chem. Soc. 123 (1923), 3398-3414
H. Kamerlingh-Innes, Nature 111 (1923), 609-610
S.G. Brush, Physics Teacher 11 (1973), 261-270
J.R. Rowlinson, Chem. Brit. 16 (1980), 32-35

WACHHOLDER, KURT [1893-1961]
FISCHER 2,1630; IB, 313; POGG 7a(4),800-802
H. Klensch, Z. Biol. 114 (1964), 321-323

WACHHOLZ, LEON [1867-1942]
WE 12,70; FISCHER 2,1630; KONOPKA 11,426-432
J. Olbrycht, Pol. Tyg. Lek. 2 (1947), 1377-1382
J.A. Mezyk, Pol. Med. Sci. Hist. Bull. 8 (1965), 51-53

WACHS, WERNER [1907-1969]
KURSCHNER 12,3364; POGG 7a(4),802-803
Anon., Chem. Z. 93 (1969), 690

WACHTER, VINCENZ [1865-]
JV 5,294; NUC 643,595

WACKENRODER, HEINRICH [1798-1854]
DSB 14,111-112; ADB 40,443-444; HEIN, 717-719; POHL, 100-108; BLOKH 2,759-760; FERCHL, 562-563; TSCHIRCH, 1142; PHILIPPE, 762-763; SCHELENZ, 619; SCHAEDLER, 144; POTSCH, 441-442; CALLISEN 20,274-276; 33,189-190; POGG 2,1237; 7aSuppl.,742-743
H. Ludwig and E. Reichardt, Arch. Pharm. 135 (1856), 101-110
H. Helmuth, Pharmazie 34 (1980), 321-323

WACKER, ADOLF [1919-1984]
KURSCHNER 13,4107; POGG 7a(4),806
Anon., Chem. Z. 108 (1984), 261

WACKER, LEONHARD [1864-1936]
POGG 5,1324; 6,2785; 7a(4),806

WADA, MITSUNORI [1896-1987]

WADDINGTON, CONRAD HAL [1905-1975]
DNB [1971-1980],877-878; WW 1975, p.3521; MH 2,572-573; STC 3,121-122; MSE 3,263-264; CB 1962, pp.440-442; SRS, 90; WWWS, 1743
C.H. Waddington, The Evolution of an Evolutionist. Edinburgh 1975
D.R. Newth, Nature 258 (1975), 371-372
A. Robertson, Biog. Mem. Fell. Roy. Soc. 23 (1977), 575-622

WADE, JOHN [1864-1912]
F.G. Hopkins, J. Chem. Soc. 103 (1913), 767-774; Biochem. J. 10 (1916), 5-7

WAELCHLI, GUSTAV [1855-1922]
RSC 11,732; SG [1]16,6; NUC 644,120

WAELSCH, HEINRICH [1905-1966]
AMS 10,4244; BHDE 2,1198-1199; WWWS, 1744
Anon., Int. J. Neuropharm. 5(3) (1966), i-ii
A. Lajtha, Brain Research 2 (1966), 1-2
J. Elkes, Pharmacologist 9 (1967), 25-26; Psychopharmacologia 10 (1967), 285-288

WAELSCH, LUDWIG [1867-1924]
FISCHER 2,1630; RUSTLER, 56-64
P. Sobotka, Med. Klin. 20 (1924), 659

WAENTIG, RUDOLF [1883-1914]
JV 24,369; NUC 644,131
Anon., Chem. Z. 38 (1914), 1165; Z.angew. Chem. 27(III) (1914), 677

WAGER, HAROLD [1862-1929]
WW 1930, p.3200; DESMOND, 632
Anon., Nature 124 (1929), 953-954
A.C.S., Proc. Roy. Soc. B106 (1930), xix-xxii
F.A. Mason, Journal of Botany 68 (1930), 18-20

WAGNER, GEORG [see VAGNER]

WAGNER, HANS [1887-1948]
KURSCHNER 6(II),1011-1012; NUC 644,327

WAGNER, HEINRICH [1880-]
JV 25,617; NUC 644,333

WAGNER, HERMANN [1876-1932]
HEIN-E, 441
A. Wankmüller, Beitr. Wurtt. Apothekgesch. 12 (1980), 140-142

WAGNER, JULIUS EUGEN [1857-1924]
BLOKH 2,764-765; POGG 2,1407; 4,1587; 5,1326; 6,2790
W. Böttger, Z. angew. Chem. 38 (1925), 309-310

WAGNER, PAUL [1843-1930]
HEIN, 722-723; SCHAEDLER, 145; TLK 3,683; POGG 6,2791; 7aSuppl.,745
O. Eckstein, Z. angew. Chem. 43 (1930), 839
H. Rössler, Chem. Z. 54 (1930), 697
L. Schmitt, Forschungsdienst 15 (1943), 138-151

WAGNER, PHILIPP [1863-]
JV 3,276; RSC 19,433; NUC 644,415
Anon., Chem. Z. 57 (1933), 963

WAGNER, RICHARD [1887-1974]
AMS 11,5622; KURSCHNER 4,3146; FISCHER 2,1632; BHDE 2,1200; WWWS, 1745
Anon., Science 186 (1974), 129; J. Am. Med. Assn. 229 (1974), 858

WAGNER, RICHARD [1893-1970]
KURSCHNER 10,2599; FISCHER 2,1632; PITTROFF, 202; IB, 314; WWWS, 1745; POGG 7a(4),824-826
E. Bauereisen, Deutsche med. Wchschr. 96 (1971), 398-399
H. Reichel, Z. Biol. 116 (1971), 518-520
W. Gerlach, Bayer. Akad. Wiss. Jahrbuch 1972, pp.275-280
H. Bornschein, Alm. Akad. Wiss. Wien 121 (1972), 316-327

WAGNER, RUDOLPH [1805-1864]
DSB 14,113-114; ADB 40,573-574; HIRSCH 5,816; WAGENITZ, 188-189; WWWS, 1745-1746
A. Wagner, Jahrb. Akad. Wiss. Gott. 1864, pp.375-399
C.F.P. Martius, Sitz. Bayer. Akad. 1865(I), pp.287-294
B. Kaulbers-Sauer, Personalbibliographien der Professoren der Medizinischen Fakultät der Universität Erlangen von 1792-1850, pp.116-123. Erlangen-Nürnberg 1969

WAGNER, THEODORE BRENTANO [1867-1936]
AMS 5,1158; NCAB 27,231-232; NUC 644,629
Anon., Chem. Z. 60 (1936), 1028

WAGNER-JAUREGG, JULIUS [1857-1940]
DSB 14,114-116; FISCHER 2,1632-1633; KOLLE 1,254-266; HAYMAKER, 528-531; PLANER, 357; RAZINGER, 193-197; WWWS, 1746
A. Pilcz, Wiener med. Wchschr. 78 (1928), 892-894; 87 (1937), 253-255
H. Hoff, Wiener med. Wchschr. 107 (1957), 618-621

WAGNER-JAUREGG, THEODOR [1903-]
KURSCHNER 13,4124; POTSCH, 442; POGG 6,2792; 7a(4),827-829
C.E. Barrelet, Arzneimitt. 23 (1973), 741
T. Wagner-Jauregg, J. Chem. Ed. 62 (1985), 592-600; Mein Lebensweg als Bio-organischer Chemiker. Stuttgart 1985

WAGSTAFFE, ERNEST ARTHUR [1870-1928]
JV 8,217; RSC 16,377
Anon., J. Roy. Inst. Chem. 1929, p.214

WAHL, ANDRÉ [1872-1944]
HEIM, 103-106; POGG 5,1327-1328; 6,1792; 7b,5847
R. Locquin, Bull. Soc. Chim. 1946, pp.441-458

WAHLENBERG, GORAN (GEORG) [1780-1851]
DSB 14,116-117; TSCHIRCH, 1142-1143; POGG 2,1242-1243

WAITE, FREDERICK CLAYTON [1870-1956]
AMS 9(II),1177; NCAB 44,97-98; IB, 314
Anon., Ohio Journal of Science 56 (1956), 255
S.W. Chase, Anat. Rec. 125 (1956), 302

WAKEMAN, ALFRED JOHN [1865-1955]
AMS 7,1856

WAKIL, SALIH JAWAD [1927-]
AMS 15(7),371; WWWS, 1747

WAKSMAN, SELMAN ABRAHAM [1888-1973]
DSB 18,970-974; AMS 11,5629-5630; NCAB I,312-313; MH 1,505-506; STC 3,124-125; KOREN,260; MSE 3,265-266; CB 1946, pp.615-617; SACKMANN, 350-354; DAMB,769-770;IB, 314; WWWS, 1747; POGG 6,2794-2796; 7b,5849-5858
S.A. Waksman, My Life with Microbes. New York 1954; Persp. Biol. Med. 7 (1964), 377-398
H.B. Woodruff, Ed., Scientific Contributions of Selman A. Waksman. New Brunswick, N.J. 1968
T.M. Daniel, J. Lab. Clin. Med. 111 (1988), 133-134

WALBAUM, HERMANN [1864-1946]
POGG 4,1588; 5,1328; 6,2796; 7a(4),830-831
Anon., Z. angew. Chem. 47 (1934), 219-220

WALBUM, LUDVIG EMIL [1879-1943]
DBL 25,22-23; KBB 1942, p.1230; VEIBEL 2,466-469; IB, 314

WALCHNER, FRIEDRICH AUGUST [1799-1865]
ADB 40,656-657; BAD 2,421; FERCHL, 564; CALLISEN 20,322; 33,204; POGG 2,1244-1245WALD, FRANTISEK [1861-1930]
DSB 14,123-124; POTSCH, 442-443; POGG 4,1589; 5,1329; 6,2796
B. Baborovsky et al., Chem. Z. 54 (1930), 905-906; Coll. Czech. Chem. Comm. 3 (1931), 3-52
G. Druce, Two Czech Chemists. London 1944
J. Thiele, Ann. Sci. 30 (1973), 417-433
F. Cuta, Chem. Listy 70 (1976), 950-960
J. Pinkava, Acta Hist. Rerum Nat., Special Issue 9 (1977), pp.133-148

WALD, GEORGE [1906-]
AMS 14,5329; WWA 1978-1979, p.3355; MH 1,506-507; STC 3,126-127; MSE 3,266-267; CB 1968, pp.412-414; ABBOTT-C, 144; WWWS, 1748; POTSCH, 443

WALDEN, PAUL [1863-1957]
DSB 14,124-125; KURSCHNER 8,2497; WELDING, 845-846; TLK 3,538-539; RIGA, 724; W, 538-539; WWWS, 1748; POTSCH, 443; POGG 4,1589-1590; 5,1329-1330; 6,2796-2798; 7a(4),833-834
P. Günther, Z. angew. Chem. 46 (1933), 497-498
H. Stadlinger, Chem. Z. 67 (1943), 222-223
P. Walden, Naturwiss. 37 (1950), 73-81; J. Chem. Ed. 28 (1951), 160-163; Wege und Herbergen: Mein Leben (G. Kerstein, Ed.). Wiesbaden 1974
W. Hückel, Chem. Ber. 91 (1958), xix-lxv
C.W. Davies, Proc. Chem. Soc. 1960, pp.186-189
D.S. Tarbell, J. Chem. Ed. 51 (1974), 7-9
Y.I. Soloviev and Y.P. Stradin, Ist. Est. Tekhn. Pribalt. 5(1976),111-133
U. Wirth, NTM 20(2) (1983), 39-49; Ist. Est. Tekhn. Pribalt. 7 (1984), 63-72
I. Andersone, Akademik P.I. Valden (Bibliography). Riga 1983

WALDENBURG, LOUIS [1837-1881]
ADB 40,688-689; HIRSCH 5,824-825; PAGEL, 1804-1805; KOREN, 260
F. Salzmann, Berl. klin. Wchschr. 18 (1881), 245
E. Waldenburg, Med. Welt 5 (1931), 1087

WALDENSTRÖM, JAN GOSTA [1906-]
VAD 1981, pp.1081-1082; IWW 1982-1983, p.1377; STC 3,127-128; WWWS, 1748
Anon., Leopoldina [3]10 (1964), 52-53
S.E. Bjorkmann, Acta Med. Scand. 179 Suppl.445 (1966), 5-8

WALDEYER (WALDEYER-HARTZ), WILHELM von [1836-1921]
DSB 14,125-127; DBJ 3,267-271; WL 6,166-175; W, 539; FISCHER 2,1635; PAGEL, 1805-1807; FREUND 2,455-461; WREDE, 208-209; DZ, 1527-1528; WAGENITZ, 189-190; BK, 336-338; WWWS, 1748
H.W.G. Waldeyer, Lebenserinnerungen. Bonn 1920
G.E. Smith, Nature 107 (1921), 368-369
J. Sobotta, Anat. Anz. 56 (1923), 1-53
G. Herxheimer, Verhandl. path. Ges. 28 (1935), 381-385

WALDMANN, AUGUST [1883-1959]
JV 25,396; NUC 645,389

WALDMANN, EDMUND [1889-1960]
KURSCHNER 8,2497-2498; POGG 7a(4),834-835
V. Prey, Ost. Chem. Z. 61 (1960), 333-334

WALDSCHMIDT-LEITZ, ERNST [1894-1972]
KURSCHNER 10,2604; BIRK, 121; TLK 3,685; POGG 6,2798; 7a(4),840-841
Anon., Chem. Z. 97 (1973), 160

WALKER, ALEXANDER [1779-1852]
DSB 14,128-131; CALLISEN 20,205-206; 33,328-329

WALKER, BURNHAM SARLE [1901-1980]
AMS 12,6660; WWWS, 1749

WALKER, ERNEST [1900-1942]
R.A. Peters, Biochem. J. 37 (1943), 449-450
Anon., J. Roy. Inst. Chem. 1943, p.54

WALKER, ISAAC CHANDLER [1883-1950]
AMS 7,1860; NCAB 43,533-534
F.R. Rackemann, Trans. Assoc. Am. Phys. 64 (1951), 23-24

WALKER, JAMES [1863-1935]
WW 1935, p.3439; W, 539; WWWS, 1749-1750; POGG 6,2800; 7b,5861
J. Kendall, Obit. Not. Fell. Roy. Soc. 1 (1935), 537-549; J. Chem. Soc. 1935, pp.1347-1354

WALKER, JAMES [1909-]
SRS, 90
Who's Who of British Scientists 1969-1970, p.836

WALKO, KARL [1872-1954]
KURSCHNER 5,1483; FISCHER 2,1637; PELZNER, 22-27; KOERTING, 172-173

WALLACE, ALFRED RUSSEL [1823-1913]
DSB 14,133-140; DNB [1912-1921],546-549; W, 539-540; ABBOTT, 132-133; WWWS, 1751; POGG 3,1409-1410
E.B. Poulton, Nature 92 (1913), 347-349; Proc. Roy. Soc. B95 (1923-1924), i-xxxv
J. Marchant, Alfred Russel Wallace. London 1916
G. Wichler, Sudhoffs Arch. 30 (1938), 364-400
C.F.A. Pantin, Notes Roy. Soc. 14 (1959), 67-84
H.L. McKinney, Wallace and Natural Selection. New Haven 1972
R. Smith, Brit. J. Hist. Sci. 6 (1972), 178-199
M.J. Kottler, Isis 65 (1974), 145-192
H. Clements, Alfred Russel Wallace. London 1981

WALLACE, GEORGE BARCLAY [1875-1948]
AMS 7,1862; PARASCANDOLA, 57-58; CHEN, 35; IB, 314
Anon., Science 107 (1948), 138; J. Am. Med. Assn. 136 (1948), 996
W. deB. MacNider, J. Pharm. Exp. Ther. 93 (1948), 127-128

WALLACE, WILLIAM [1832-1888]
NUC 646,406; POGG 3,1410
D. Maclagan, Proc. Roy. Soc. Edin. 16 (1888), 6-7
Anon., J. Soc. Chem. Ind. 7 (1888), 737; J. Chem. Soc. 55 (1889), 296-298
J. Mayer, Proceedings of the Glasgow Philosophocal Society 20 (1889), 314-321

WALLACH, OTTO [1847-1931]
DSB 14,141-142; TSCHIRCH, 1143; DZ, 1528; TLK 3,685-686; TETZLAFF,345; W,540-541; WWWS,1752; POTSCH, 443; POGG 3,1410-1411; 4,1592-1593; 5,1331; 6,2800
E. Gildemeister, Z. angew. Chem. 40 (1927), 365-366
A. Windaus, Jahrb. Akad. Wiss. Gott. 1930-1931, pp.58-65
H.E. Armstrong, Nature 127 (1931), 601-602
A. Ellmer, Z. angew. Chem. 44 (1931), 929-932
L. Ruzicka, J. Chem. Soc. 1932, pp.1582-1597
W. Hückel, Chem. Ber. 94 (1961), vii-cviii
E. Blumann, Proc. Chem. Soc. 1964, pp.387-389

WALLENFELS, KURT [1910-]
KURSCHNER 13,4133; POTSCH, 443-444; POGG 7a(4),841-843
Anon., Nachr. Chem. Techn. 18 (1970), 271-272; Chem. Z. 104 (1980), 210

WALLER, AUGUSTUS DÉSIRÉ [1856-1922]
FISCHER 2,1638; WWWS, 1752
W.D.H., Proc. Roy. Soc. B93 (1922), xxvii-xxx
Z. Cope, Med. Hist. 17 (1973), 380-385
E. Besterman and R. Creese, British Heart Journal 42 (1979), 61-64
A.H. Sykes, Med. Hist. 33 (1989), 217-234

WALLER, AUGUSTUS VOLNEY [1816-1870]
DSB 14,142-144; DNB 20,579-580; HIRSCH 5,831-832; O'CONNOR, 63-64; FERCHL, 565; WWWS, 1752
Anon., Proc. Roy. Soc. 20 (1872), xi-xiii
E. Wallach, Sudhoffs Arch. 22 (1929), 105-113,344-351
R. Gertler-Samuel, Augustus Volney Waller als Experimentalforscher. Zurich 1965

WALLERSTEIN, SALY [1878-]
JV 17,326; SG [2]20,404; NUC 646,531

WALLICH, GEORGE CHARLES [1815-1899]
DSB 14,145-146; DESMOND, 637
Anon., Nature 60 (1899), 13; Lancet 1899(I), p.997; J. Roy. Micr. Soc. 1899, pp.263-264

WALLIN, IVAN EMANUEL [1883-1969]
AMS 8,2610; IB, 315
T.S. Eliot, Anat. Rec. 171 (1971), 137-139

WALLING, CHEVES THOMSON [1916-]
AMS 17(7),407; WWA 1988-1989, p.3210; MH 2,576-577; MSE 3,270-271; WWWS, 1752-1753
C. Walling, Fifty Years of Free Radicals. Washington, D.C. 1991

WALLIS, EVERETT STANLEY [1899-1965]
AMS 10,4270; WWAH, 634; POGG 6,2802; 7b,5868-5869

WALLRAFF, JOSEF [1904-]
KURSCHNER 13,4135; BUSCHHUTER, 23-30

WALSHE, FRANCIS MARTIN ROUSE [1886-1973]
DNB [1971-1980], 882-883; WW 1973, pp.3365-3366
C.G. Phillips, Biog. Mem. Fell. Roy. Soc. 20 (1974), 457-481

WALTER, FRIEDRICH [1850-1905]
SG [1]16,34; NUC 647,200
J. Brennsohn, Die Aerzte Livlands, p.418. Mitau 1905

WALTER, PHILIPPE [1810-1847]
DSB 14,156-157; WE 12,96; BLOKH 2,766-767; FERCHL, 566; W, 542

WALTER, RODERICH [1937-1979]
AMS 14,5352
J. Meienhofer, Int. J. Pep. Res. 16 (1980), 355-358

WALTER, WILHELM [1877-1960]
JV 16,228; NUC 647,261
Anon., Nachr. Chem. Techn. 8 (1960), 371

WALTER, WOLFGANG [1919-]
KURSCHNER 13,4135; POGG 7a(4),846

WALTHER, ALFRED [1888-]
JV 32,331; NUC 647,324

WALTHER, ANTON ANTONOVICH [1870-1902]
DIEPGEN, 41-44; RSC 19,462
R.A. and K.S., Jahresb. Fort. Tierchem. 31 (1902), xxxix

WALTI, ALPHONSE [1897-]
AMS 11,5661

WALZ, GEORG FRIEDRICH [1813-1862]
DRULL, 285-286; HEIN, 725-726; POHL, 142-151; POGG 3,1413-1414
A. Wankmüller, Deutsche Apoth. Z. 110 (1970), 1521-1524
W.U. Eckart, G.F. Walz. Stuttgart 1990

WANG, JUI HSIN [1921-]
AMS 15(7),410; WWWS, 1755WANG, SHIH CHUN [1910-]
AMS 14,5361-5362; WWA 1976-1977, p.3279; WWWS, 1755-1756

WANGERIN, (CARL) ALBERT [1873-1903]
BJN 8,119*; HEIN, 727; TSCHIRCH, 1143; POGG 4,1597
Anon., Leopoldina 39 (1903), 132; Pharm. Z. 48 (1903), 863

WANKLYN, JAMES ALFRED [1834-1906]
DSB 14,168-170; DNB [1901-1911] 3,587-588; SCHAEDLER, 146; POGG 2,1259-1260; 3,1414-1415; 4,1597
Anon., Brit. Med. J. 1906(II), pp.278-279

WARBURG, EMIL [1846-1931]
DSB 14,170-172; WREDE, 209; DZ, 1535; TLK 3,689; WININGER 6,208-209; TETZLAFF, 346; POGG 3,1415-1416; 4,1598; 5,1334-1335; 6,2806; 7aSuppl.,747-748
A. Einstein, Naturwiss. 10 (1922), 823-828
J. Franck, Naturwiss. 19 (1931), 993-997

WARBURG, ERIK JOHAN [1892-1963]
KBB 1963, pp.1592-1593; VEIBEL 2,469-470; IB, 315
E. Gotfreden, Acta Med. Scand., Suppl.266 (1952), 13-17

WARBURG, OTTO [1859-1938]
WININGER 6,210; TLK 3,689; IB, 315; WWWS, 1756
Anon., Nature 141 (1938), 191
M. Plaut, Ber. bot. Ges. 72 (1959), 43-47

WARBURG, OTTO [1883-1970]
DSB 14,172-177; MH 2,577-579; STC 3,133-134; MSE 3,274-275; WWWS, 1756; SCHWERTE 2,127-134; FISCHER 2,1642; KOREN, 260; WININGER 6,209; POTSCH, 444-445; TETZLAFF, 347; ABBOTT, 133-134; POGG 6,2806-2807; 7a(4),857-859
O. Warburg, Ann. Rev. Biochem. 33 (1964), 1-14
R.E. Kohler, J. Hist. Biol. 6 (1973), 171-192
H.A. Krebs, Naturw. Rund. 24 (1971), 1-4; 31 (1978), 349-356; Biog. Mem. Fell. Roy. Soc. 18 (1972), 629-699; Otto Warburg. Stuttgart 1979 (English version, Fairlawn 1981)
E. Höxtermann, NTM 20(2) (1983), 1-14; Otto Heinrich Warburg. Leipzig 1988
E. Jokl, Trans. Coll. Phys. Phila. [5]5 (1983), 67-75
T. Bücher, 34. Colloquium Mosbach, pp.1-29. Berlin 1983
P. Werner, Z. ärzt. Fortbild. 77 (1983), 947-949; Otto Warburg. Berlin 1988; Ein Genie Irrt Seltener. Berlin 1991
M. von Ardenne, NTM 23 (1986), 61-77

WARD, FRED WILBERT [1891-1938]
AMS 5,1167

WARD, HARRY MARSHALL [1854-1906]
DNB [1901-1911] 3,589-591; DESMOND, 639; WWWS, 1757
S.H. Vines, Nature 74 (1906),493-495; Journal of Botany 21 (1907),ix-xiii
B. Balfour, Trans. Bot. Soc. Edin. 23 (1907), 218-232
W.T.T.I., Proc. Roy. Soc. B83 (1911), i-xiv

WARD, HENRY BALDWIN [1865-1945]
DAB [Suppl3],802-803; NCAB 35,174-175; WAGENITZ, 190; IB, 315; WWAH, 637
W.W. Cort, Science 102 (1945), 658-660
J.R. Christie, J. Parasit. 32 (1946), 323-324

WARINGTON, ROBERT [1807-1867]
DNB 20,844; FERCHL, 567; WWWS, 1758; POGG 2,1263-1264; 3,1416
Anon., Proc. Roy. Soc. 16 (1868), xlix-l; J. Chem. Soc. 21 (1868), xxii-xxiv
J.H.S. Green, Proc. Chem. Soc. 1957, pp.241-246

WARINGTON, ROBERT [1838-1907]
DNB [1901-1911] 3,593-594; DOETSCH, 155-156; WWWS, 1758
Anon., J. Soc. Chem. Ind. 26 (1907), 394-395
A.D.H., Nature 75 (1907), 511-512
S.U. Pickering, J. Chem. Soc. 93 (1908), 2258-2260; Proc. Roy. Soc. B80 (1908), xv-xxiv

WARNECK (VARNEK), NIKOLAI ALEKSANDROVICH [1821-1876]
BSE 7,9
T.P. Platova, Trudy Inst. Ist. Est. 5 (1953), 317-362

WARNECKE, THEODOR SOPHUS [1820-1890]
DBL 25,146-147; VEIBEL 2,470-471

WARREN, JOHN COLLINS [1842-1927]
DAB 19,481-482; DAMB, 779-780; WWWS, 1759
R. Truax, The Doctors Warren of Boston. Boston 1968

WARTHIN, ALDRED SCOTT [1866-1931]
AMS 4.1030; FISCHER 2,1643-1644; OEHRI, 128-129; IB, 316; WWWS, 1760
W.M. Simpson, Am. J. Surg. 14 (1931), 502-504
C.V. Weller, Arch. Path. 12 (1931), 276-279
M.H. Soule, J. Lab. Clin. Med. 16 (1931), 1043-1046
Anon., J. Path. Bact. 35 (1932), 133-135

WASER, ERNST [1887-1941]
IB, 316; TLK 3,690; POGG 6,2810; 7a(4),860
P. Karrer, Viert. Nat. Ges. Zürich 86 (1941), 367-371; Helv. Chim. Acta 24 (1941), 852-861

WASER, PETER [1918-]
KURSCHNER 13,4153; IWW 1982-1983, pp.1353-1354; POGG 7a(4),860-861

WASHBOURN, JOHN WICKENFORD [1863-1902]
NUC 649,455
Anon., Brit. Med. J. 1902(II), 85-86

WASHBURN, EDWARD WIGHT [1881-1934]
DSB 14,182-183; DAB 19,498-499; AMS 5,1171; MILES, 493-494; WWWS, 1761; POTSCH, 445-446; POGG 5,1335-1336; 6,2810-2811; 7b,5882
T.M. Lowry, Nature 133 (1934), 712-713
L.J. Briggs, Science 79 (1934), 221-222
W.A. Noyes, Biog. Mem. Nat. Acad. Sci. 17 (1936), 69-81

WASHBURN, JOHN HOSEA [1859-1932]
AMS 4,1031; JV 4,105; RSC 19,481-482; NUC 649,480

WASICKY, RICHARD [1884-1970]
KURSCHNER 11,3185; HEIN-E, 443-444; BHDE 2,1209; TSCHIRCH, 1143-1144;
IB, 316; POGG 6,2811-2812; 7a(4),861-863
L. Fuchs, Scientia Pharmaceutica 21 (1953), 225-236

WASMANN, ADOLPH [1807-1853]
RSC 6,273; CALLISEN 23,224
H. Schröder, Lexikon der Hamburgischen Schriftsteller 7 (1879), 581

WASSERMANN, ALBERT [1901-1971]
BHDE 2,1210; LDGS, 26; POGG 6,2812; 7a(4),865-868

WASSERMANN, AUGUST von [1866-1925]
DSB 15,521-524; FISCHER 2,1644-1645; KOREN, 260; WININGER 6,214-215; TETZLAFF, 347; W, 543; WWWS, 1740; POGG 6,1812
E. Friedberger, Z. Immunitätsforsch. 43 (1925), i-xii; Biochim. Ter. Sper. 12 (1925), 225-231
H. Sachs, Klin. Wchschr. 4 (1925), 902-903
R. Müller, Wiener klin. Wchschr. 38 (1925), 365-366
H. Schadewaldt, Deutsche med. Wchschr. 100 (1975), 2506-2508

WASSERMANN, ERNST [1880-1925]
BLOKH 2,768; SG [2]20,430
Anon., Z. angew. Chem. 38 (1925), 1139-1140

WASSERMANN, FRIEDRICH [1884-1969]
AMS 11,5692; KURSCHNER 10,2615; BHDE 2,1210; LDGS, 55; FISCHER 2,1645; EGERER, 82-90; IB, 316; KOREN, 260; WWWS, 1761
W. Bargmann, Münch. med. Wchschr. 106 (1964), 1536-1537
F. Wassermann, Persp. Biol. Med. 13 (1970), 537-562
G.B. Gruber, Leopoldina [3]17 (1971), 287-302

WASSERMANN, WALTER [1898-]
JV 40,746

WASSERMEYER, HANS [1899-1976]
NUC 650,343

WASTENEYS, HARDOLPH [1881-1965]
AMS 10,4298; YOUNG, 21-22; FISCHER 2,1646; POGG 6,2812-2813; 7b,5883
A.M. Wynne, Proc. Roy. Soc. Canada [4]5 (1967), 125-129

WASTL, HELENE [1896-1948]
KURSCHNER 4,3176-3177

WATCHORN, ELSIE [1893-]
Who's Who in British Science 1953, p.274

WATERMAN, NATHANIEL [1883-1965]
FISCHER 2,1646; HES, 169-170

WATERS, WILLIAM ALEXANDER [1903-1985]
WW 1984, p.2386; CAMPBELL, 259-260
R.O.C. Norman and J.H. Jones, Biog. Mem. Fell. Roy. Soc. 32 (1986), 599-627

WATERS, WILLIAM HORSCROFT [1855-1887]
O'CONNOR, 235; RSC 19,488; NUC 650,532

WATERSTON, JOHN JAMES [1811-1883]
DSB 14,184-186; W, 543
J.J. Waterston, Collected Scientific Papers. Edinburgh 1928
S. Brush, Ann. Sci. 13 (1957), 273-282; Am. Scient. 49 (1961), 202-214

WATKINS, WINIFRED MAY [1924-]
WW 1983, p.2352

WATSON, CECIL JAMES [1901-1983]
AMS 14,5389; WWA 1978-1979, p.3393; IWW 1982-1983, p.1389; WWWS, 1763;
AO 1983, pp.178-179
A.B. Lerner, Trans. Assoc. Am. Phys. 97 (1984), cxxix-cxli

WATSON, DAVID MEREDITH SEARES [1886-1973]
WW 1973, pp.3394-3395; IB, 317; WWWS, 1763
F.R. Parrington and T.S. Westall, Biog. Mem. Fell. Roy. Soc. 20 (1974), 483-504

WATSON, HERBERT BEN [1894-1975]
POGG 6,2817; 7b,5903-5904
F.J.J. Dippy, Chem. Brit. 12 (1976), 227-228

WATSON, JAMES DEWEY [1928-]
AMS 15(7),446; WWA 1980-1981, p.3445; IWW 1982-1983, p.1389; WWWS, 1764; MH 1,513; STC 3,136-137; MSE 3,281-282; CB 1963, pp.458-459; POTSCH, 446

WATSON, THOMAS [1792-1882]
WWWS, 1764
G.J., Proc. Roy. Soc. 38 (1885), v-ix

WATTIEZ, NESTOR [1886-1972]
POGG 6,2818; 7b,5907
Anon., Ann. Pharm. Fran. 36 (1973), 17-18
L. Maricq, Bull. Acad. Med. Belg. 1973, pp.235-241

WATZKA, MAXIMILIAN [1905-1981]
KURSCHNER 13,4156; MAASS, 142-152; KOERTING, 113-114
Anon., Naturw. Rund. 34 (1981), 439

WEAVER, WARREN [1894-1978]
AMS 13,4742; WWA 1978-1979, p.3401; WWWS, 1767; POGG 6,2821; 7b,5913-5915
W. Weaver, Scene of Change. New York 1970
R.E. Kohler, Minerva 14 (1976), 279-306
J.G. Harrar, Am. Phil. Soc. Year Book 1979, pp.113-117
M. Rees, Biog. Mem. Nat. Acad. Sci. 57 (1987), 493-530

WEBB, EDWIN CLIFFORD [1921-]
WW 1981, p.2730

WEBER, ADOLF [1847-1888]
STRAHLMANN, 471,481

WEBER, CARL OTTO [1860-1905]
POGG 6,1822
A.D. Little, J. Am. Chem. Soc. [Proc.] 27 (1905), 39-43

WEBER, EDUARD FRIEDRICH [1806-1871]
ADB 41,287; HIRSCH 5,868; PAGEL, 1814-1815; WWWS, 1768; POGG 2,1274; CALLISEN 33,230
H.E. Hoff, Ann. Med. Hist. 8 (1936), 138-144

WEBER, ERNST [1875-1925]
O. Kalischer, Klin. Wchschr. 4 (1925), 479

WEBER, ERNST HEINRICH [1795-1878]
DSB 14,199-202; ADB 41,290; HIRSCH 5,866-867; PAGEL, 1813-1814; WWWS, 1768; CALLISEN 20,450-454; 33,230-232; POGG 2,1274; 3,1421; 7aSuppl.,750-751
Anon., Leopoldina 14 (1878), 34-37; Proc. Roy. Soc. 29 (1879), xxviii-xxxii
P.M. Dawson, Phi Beta Kappa Quarterly 25 (1928), 86-116
U. Bück-Rich, Ernst Heinrich Weber und der Anfang der Physiologie der Hautsinne. Zurich 1970

WEBER, FRIEDL [1886-1960]
IB, 317
I. Thaler, Protoplasma 46 (1956), 835-846
F. Widder, Phyton 9 (1960), 1-14
K. Höfler, Protoplasma 55 (1962), 1-9

WEBER, GREGORIO [1916-]
AMS 15(7),467WEBER, HANS [1895-]
JV 41,671

WEBER, HANS HERMANN [1896-1974]
DSB 18,978-983; KURSCHNER 11,3191; MH 2,583-584; STC 3,139-140; MSE 3,284-285; BOHM, 65; POGG 6,2822; 7a(4),875-876
R. Ammon, Arzneimitt. 24 (1974), 1358
W. Hasselbach, Erg. Physiol. 73 (1975), 1-7
W. Hasselbach and H. Schaffer, Heid. Akad. Jahrb. 1975, pp.81-84

WEBER, IONE ETHEL [1898-1965]
AMS 10,4320

WEBER, JOHANNES [1898-1965]
JV 42,442; SG [3]10,1208; NUC 652,250

WEBER, JULIUS [1864-1924]
STRAHLMANN, 472,481; RSC 19,500
H. Schardt, Verhandl. Schw. Nat. Ges. 105 (1924), 53-57

WEBER, RUDOLF [1829-1894]
RSC 6,292-293; 8,1209; 19,503
Anon., Chem. Z. 18 (1894), 1111; Leopoldina 30 (1894), 157-158; Pharm. Z. 39 (1894), 500

WEBSTER, JOHN WHITE [1793-1850]
DAB 19,592-593; MILES, 499-500; ELLIOTT, 269; FERCHL, 568; COLE, 565-567

WEBSTER, LESLIE TILLOTSON [1894-1943]
AMS 5,1179; NCAB 32,313-314; IB, 318; WWAH, 646
J. Casals and T.M. Rivers, Science 98 (1943), 167
P.K. Olitsky, Arch. Path. 36 (1943), 536-537

WECHSBERG, FRIEDRICH [1873-1929]
FISCHER 2,1649-1650; STANGL, 154-157; KOREN, 260
A. Eiselsberg, Wiener klin. Wchschr. 42 (1929), 1391

WECKER, ERNST ROBERT [1884-1961]
JV 26,647; NUC 652,686
Anon., Heilbronner Stimme, 30 December 1954, p.3

WEDDIGE, ANTON [1843-1930]
SCHAEDLER, 147; POGG 3,1424; 4,1605; 6,2824
B. Rassow, Z. angew. Chem. 45 (1932), 750

WEDEKIND, EDGAR [1870-1938]
KURSCHNER 4,3189; KALLMORGEN, 443; RIGA, 724; TLK 3,694-695; POGG 4,1605-1606; 5,1341-1342; 6,2824-2825; 7a(4),878
Anon., Chem. Z. 54 (1930), 90; 62 (1938), 814
H. Wienhaus, Ber. chem. Ges. 71A (1938), 196-198

WEDL, KARL [1815-1891]
ADB 41,417; WURZBACH 53,228-231; HIRSCH 5,877; HEINDEL, 6-13
E. Suess, Alm. Akad. Wiss. Wien 42 (1892), 189-192
V. Patzelt, Anat. Anz. 100 (1954), 147-156

WEECH, ALEXANDER ASHLEY [1895-1977]
AMS 11,5229
S. Krugman, Texas Association of American Physicians 91 (1978), 48-50

WEED, LEWIS HILL [1886-1952]
DAB [Suppl.5],733-734; DAMB, 784-785; IB, 318; WWAH, 646; WWWS, 1770
G.W. Corner, Am. Phil. Soc. Year Book 1953, pp.375-381
A.M. Harvey, Johns Hopkins Med. J. 139 (1978), 77-83

WEERMANN, RUDOLF ADRIAAN [1880-1931]
I.J. Rinkes, Chem. Wkbl. 28 (1931), 326-328

WEESE, HELMUT [1897-1954]
IB, 318; POGG 7a(4),879-880
F. Brücke, Ost. Chem. Z. 55 (1954), 88-89
G. Hecht and W. Schulemann, Arzneimitt. 4 (1954), 218-220

WEGLER, RICHARD [1906-]
KURSCHNER 13,4178; POGG 7a(4),884

WEGSCHEIDER, RUDOLF [1859-1935]
PLANER, 362; TLK 3,696; POGG 4,1607; 5,1343-1344; 6,2828; 7a(4),887
A. Skrabal, Ber. chem. Ges. 68A (1935), 45-47
E. Späth, Alm. Akad. Wiss. Wien 85 (1935), 231-237

WEHMER, CARL [1858-1935]
KURSCHNER 4,3196; TLK 3,696; WAGENITZ, 192; IB, 318; POGG 4,1608; 5,1344-1345; 6,2828; 7aSuppl.,754
Anon., Chem. Z. 52 (1928), 769; Z. angew. Chem. 48 (1935), 132; Ber. chem. Ges. 68A (1935), 45
E. Jahn, Ber. bot. Ges. 52 (1934), (223)-(234)
R. Koch, Z. Bakt. 93 (1936), 417-420

WEHSARG, KARL [1862-1887]
RSC 17,757; 19,515
Anon., Ber. chem. Ges. 20 (1887), 3426

WEICHARDT, WOLFGANG [1875-1945]
KURSCHNER 6(II),1501; FISCHER 2,1652-1653; BERWIND, 72-86; TLK 3,696-697; PITTROFF, 111-112; AD 3,70; IB, 318
R. Doerr and H. Schlossberger, Erg. Hyg. 26 (1949), 1-4

WEICHHOLD, OSKAR [1883-1965]
JV 24,297; NUC 653,442

WEICHSELBAUM, ANTON [1845-1920]
DSB 14,218-219; DBJ 2,764; FISCHER 2,1653-1654; PAGEL, 1824-1825; LESKY, 568-574; SCHWARZ, 1-19; BULLOCH, 402-403; WWWS, 1771; WININGER 6,221; 7,487
J. Wiesner, Wiener klin. Wchschr. 33 (1920), 979-981
Anon., Lancet 1920(II), p.921; Nature 106 (1920), 317
S. Exner, Alm. Akad. Wiss. Wien 71 (1921), 152-155
R. Wiesner, Verhandl. path. Ges. 29 (1937), 429-434

WEICKEL, TOBIAS FRIEDRICH [1883-]
JV 26,648; NUC 653,450

WEICKER, BRUNO [1899-1988]
KURSCHNER 13,4184

WEIDEL, ARNO [1883-1960]
JV 25,396; NUC 653,460
Anon., Nachr. Chem. Techn. 6 (1958), 303; 8 (1960), 267

WEIDEL, HUGO [1849-1899]
BJN 4,191*; BLOKH 2,770-771; POGG 4,1609-1610; 7aSuppl.,754
J. Herzig, Ber. chem. Ges. 32 (1899), 3745-3755
V. von Lang, Alm. Akad. Wiss. Wien 50 (1900), 290-293
M. Kohn, J. Chem. Ed. 21 (1944), 374-376,379

WEIDEL, WOLFHARD [1916-1964]
KURSCHNER 9,2221; POGG 7a(4),891
Anon., Nachr. Chem. Techn. 12 (1964), 331; Attempto 14 (1964), 62-63

WEIDENHAGEN, RUDOLF [1900-1979]
KURSCHNER 12,3428; POGG 6,2830; 7a(4),892-893

WEIDENREICH, FRANZ [1873-1948]
KURSCHNER 5,1503; DRULL, 290; ARNSBERG, 514-515; FISCHER 2,1654; IB, 318; BHDE 2,1213-1214; LDGS, 15; WWWS,1771; WININGER 7,487; TETZLAFF, 348
W.E.L. Clark, Nature 162 (1948), 805
H. von Eggeling, Anat. Nachr. 1 (1950), 149-167

WEIGERT, CARL [1845-1904]
DSB 14,227-230; BJN 9,313-314; ARNSBERG, 515-516; KALLMORGEN, 444-445; FREUND 2,463-473; FISCHER 2,1655; PAGEL, 1825-1826; BULLOCH, 403; SANDRITTER, 123-127; HAYMAKER, 388-391; WWWS, 1772; KOREN, 261; WININGER 6,221-222
C. Helbing, Münch. med. Wchschr. 51 (1904), 1747-1748
O. Lubarsch, Deutsche med. Wchschr. 30 (1904), 1318-1319
R. Rieder, Carl Weigert. Berlin 1906
H. Morrison, Ann. Med. Hist. 6 (1924), 163-177
W. Krücke, in 50 Jahre Neuropathologie in Deutschland (W. Scholz, Ed.), pp.5-19. Stuttgart 1961
J.T. Lie, Mayo Clin. Proc. 55 (1980), 716-720

WEIGERT, FRITZ [1876-1947]
BHDE 2,1216; LDGS,26; TLK 3,698; POGG 5,1345-1346; 6,2832-2833; 7a(4),897
I. Berenblum and H. Halban, Nature 159 (1947), 733

WEIGLE, JEAN JACQUES [1901-1958]
POGG 6,2833; 7a(4),897-898

WEIL, ADOLF [1848-1916]
DBJ 1,371; DRULL, 290; FISCHER 2,1656; OLPP, 410-411; LEVITSKI 2,140-141; FRANKEN, 200; KOREN, 261; WININGER 6,222; 7,487; WWWS, 1772
F. Schultze, Münch. med. Wchschr. 63 (1916), 1293

WEIL, ALBERT [1867-]
JV 4,84; NUC 653,622
W. Zorn, Tradition 4 (1959), 197-204

WEIL, ARTHUR [1887-1982]
AMS 11,5737; IB, 219; KOREN, 261; KAGAN(JA), 629-630

WEIL, EDMUND [1879-1922]
DBJ 4,374; FISCHER 2,1656-1657; DIEPGEN, 85-87; KOERTING, 128-129; KOREN, 261
H. Braun, Klin. Wchschr. 1 (1922), 1583
D. Bail, Z. Immunitätsforsch. 35 (1923), 2-24

WEIL, FRIEDRICH JOSEF [1888-1917]
BLOKH 2,771; JV 30,740; NUC 653,633
H. Wichelhaus, Ber. chem. Ges. 50 (1917), 851
Anon., Chem. Z. 41 (1917), 484; Z. angew. Chem. 30(III) (1917), 311

WEIL, HERMANN [1875-1934]
TLK 3,699; RSC 19,519; NUC 653,639
Anon., Chem. Z. 58 (1934), 48; Z. angew. Chem. 48 (1935), 447

WEIL, HUGO [1863-1942]
GEDENKBUCH 2,1556; NUC 653,639; POGG 5,1346

WEIL, LEOPOLD [1906-1964]
AMS 10,4333
Anon., Science 147 (1965), 279

WEIL, RICHARD [1876-1917]
AMS 2,499; FISCHER 2,1657; KAGAN(JA), 78-79; WININGER 7,488; WWWS, 1773
Anon., Science 46 (1917), 557-558; J. Cancer Res. 3 (1918), i-v
E. Beer and C. Eggleston, J. Immunol. 2 (1918), i-vii

WEIL-MALHERBE, HANS [1905-]
AMS 13,4761; BHDE 2,1219-1220; LDGS, 61; WWWS, 1773
T.L. Sourkes, Neuropharmacology 15 (1976), 443-448

WEILAND, WALTER [1881-1937]
KURSCHNER 4,3207

WEILER, JULIUS [1850-1904]
E. Buchner, Ber. chem. Ges. 37 (1904), 3530
Anon., Pharm. Z. 49 (1904), 681

WEILINGER, KARL [1870-]
JV 18,194; NUC 653,678
Anon., Nachr. Chem. Techn. 3 (1955), 166

WEIMARN, PETR PETROVICH [1879-1935]
POTSCH, 447-448; POGG 5,1346-1347; 6,2833-2836; 7b,5923
W. Ostwald, Koll. Z. 74 (1936), 1-10

WEINBERG, ARTHUR von [1860-1943]
ARNSBERG, 519-520; KALLMORGEN, 445; BHDE 1,804; POTSCH, 448; POGG 6,2836; 7a(4),901
L. Gans and P. Walden, Z. angew. Chem. 43 (1930), 703-708
R. Richter, Natur und Volk 80 (1950), 209-216
H. Ritter and H. Zerweck, Chem. Ber. 89 (1956), xix-xli

WEINBERG, MICHEL [1868-1940]
IB, 319
G. Ramon, Presse Med. 48 (1940), 581-582; Bull. Acad. Med. 123 (1940), 387-390; Ann. Inst. Pasteur 64 (1940), 461-465

WEINBERG, WILHELM [1862-1937]
DSB 14,230-231; KURSCHNER 4,3210; FISCHER 2,1658; ZWEIFEL, 158-159; IB, 319
C. Stern, Genetics 47 (1962), 1-5
E. Hübler, Jahreshefte Wurtt. 118-119 (1964), 57-67

WEINDEL, ANTON ERWIN [1882-1950]
JV 20,264

WEINDLING, RICHARD [1899-]
AMS 11,5743

WEINHOUSE, SIDNEY [1909-]
AMS 14,5428; WWA 1978-1979, p.3415; COHEN, 250; WWWS, 1773-1774

WEINLAND, ERNST [1869-1932]
BERWIND, 154-159; PITTROFF, 200-201; IB, 319
K. Gross, Sitz. Phys. Med. Erlangen 63-64 (1933), 357-370

WEINLAND, RUDOLF [1865-1936]
KURSCHNER 5,1508; HEIN, 730-731; TSCHIRCH, 1144; TLK 3,699; POGG 4,1613-1614; 5,1349-1350; 6,2837
W. Hieber, Ber. chem. Ges. 69A (1936), 210-211
A. Wankmüller, Beitr. Wurtt. Apothekgesch. 11 (1975), 4-11

WEINSCHENCK, ERNST [1865-1921]
POGG 4,1614-1615; 5,1350WEINTRAUD, WILHELM [1866-1920]
DBJ 2,764; FISCHER 2,1658-1659; PAGEL, 1827-1828
K. Brandenburg, Med. Klin. 16 (1920), 968
F. Blumenfeld, Berl. klin. Wchschr. 57 (1920), 1183

WEIR, JOHN [1885-1973]
SRS, 37; NUC 654,212
Anon., Chem. Brit. 10 (1974), 71

WEIS, FREDERIK ANTON [1871-1933]
VEIBEL 2,472-474; IB, 319

WEIS, JULIUS [1867-1954]
FISCHER 2,1660; PAGEL, 1828; HASENEDER, 45-56

WEISBURGER, JOHN HANS [1921-]
AMS 15(7),499; WWA 1980-1981, p.3468; WWWS, 1774

WEISKE, HUGO [1843-1907]
BJN 12,91*; RSC 8,1213; 11,776; 19,530-531

WEISMANN, AUGUST [1834-1914]
DSB 14,232-239; DBJ 1,97-103,318; FISCHER 2,1659-1660; KALLMORGEN, 446; WAGENITZ, 192-193; SCHULZ-SCHAEFFER, 199-200; DZ, 1549-1550; BK, 342; W, 547-548; ABBOTT, 135-136; WWWS, 1775; POGG 6,2839
R. Hertwig, Bayer. Akad. Wiss. Jahrbuch 1915, pp.118-127
E.G. Conklin, Proc. Am. Phil. Soc. 54 (1915), iii-xii
E.R.P., Proc. Roy. Soc. B89 (1916), xxvii-xxxiv
E. Gaupp, August Weismann, sein Leben und Werk. Jena 1917
H. Spemann, Ber. Nat. Ges. Freiburg 34 (1934), 81-94
A. Petrunkevich, J. Hist. Med. 18 (1963), 20-35
F.B. Churchill, J. Hist. Biol. 1 (1968), 91-112; Isis 61 (1970), 429-457
E. Mayr, J. Hist. Biol. 18 (1985), 295-329
K. Sander et al., Freiburger Universitätsblätter 87-88 (1985), 21-203

WEISS, CHARLES [1894-]
AMS 13,4771; IB, 319; KOREN, 262; KAGAN(JA), 317-318

WEISS, CHRISTIAN SAMUEL [1780-1856]
DSB 14,239-242; ADB 41,599-600; FERCHL, 572-573; POTSCH, 448; POGG 2,1287-1289; 7aSuppl.,760-761
E. Fischer, Wiss. Z. Berlin 11 (1962), 249-255

WEISS, FRANZ [1868-1946]
JV 10,230; RSC 19,534; NUC 564,393

WEISS, GIOVANNI [1844-1917]
FISCHER 2,1660
E. Padovani, Atti Accad. Ferrara 94 (1920), viii-xi

WEISS, GUSTAV ADOLF [1837-1894]
ADB 41,556-558; WURZBACH 54,82-87; LUDY, 150-156; TSCHIRCH, 1144; POGG 2,1290; 3,1429
B. Molisch, Ber. bot. Ges. 12 (1894), (28)-(34)

WEISS, HERMANN [1894-1977]
JV 37,948; NUC 654,412

WEISS, JOSEPH JOSHUA [1905-1972]
WW 1973, p.3416; CAMPBELL, 263-264; BHDE 2,1228; LDGS, 26
Anon., Chem. Brit. 8 (1972), 268

WEISS, MORIZ [1877-1945]
FISCHER 2,1661-1662

WEISS, NATHAN [1851-1883]
FISCHER 2,1662
Anon., Leopoldina 19 (1883), 168

WEISS, OTTO [1871-1943]
FISCHER 2,1662; IB, 319; POGG 6,2840-2841; 7a(4),912-913
H. Lullies, Erg. Physiol. 45 (1944), 464-481; Pflügers Arch. 247 (1944), 611-517

WEISS, PAUL ALFRED [1898-1989]
AMS 14,5439; WWA 1978-1979, p.3420; IWW 1982-1983, p.1394; WWWS, 1775; MH 2,585-587; STC 3,146-148; MSE 3,288-289; CB

1970, pp.438-441
P. Weiss, Dev. Biol. 7 (1963), vii-xix; Leopoldina [3]12 (1966), 80-82
J. Overton, Am. Phil. Soc. Year Book 1989, pp.343-346

WEISS, RICHARD [1884-]
WIDMANN, 292; POGG 6,2841; 7a(4),913

WEISS, SAMUEL BERNARD [1926-]
AMS 15(7),507; WWWS, 1776

WEISS, SOMA [1899-1942]
DAB [Suppl.3],805-806; AMS 6,1511; DAMB, 787-788; KOREN, 262
R. Fitz, Science 95 (1942), 215-216
W.B. Castle, Trans. Assoc. Am. Phys. 57 (1942), 36-38
J.V. Warren, New Eng. J. Med. 286 (1972), 658-659

WEISSBACH, HERBERT [1932-]
AMS 15(7),508

WEISSBERGER, ARNOLD [1898-1984]
AMS 11,5757; BHDE 2,1260; LDGS, 26; POGG 6,2841-2842; 7a(4),913-914
A. Weissberger, CHEM TECH 8 (1978), 583-587
Anon., Chem. Eng. News 61(47) (1984), 67
R.A. Jeffreys, Chem. Brit. 21 (1985), 1098

WEISSENBERG, KARL [1893-1976]
TLK 3,701; LDGS, 102; POGG 6,2842; 7a(4),914-915
Anon., Naturw. Rund. 29 (1976), 445

WEISSENBERG, (JULIUS) RICHARD [1882-1974]
AMS 10,4347; KURSCHNER 4,3223; IB, 319; LDGS, 15
J.R. Weissenberg, Ann. N.Y. Acad. Sci. 126 (1965), 362-374
V. Sprague, Anat. Rec. 183 (1975), 148-149

WEISSGERBER, RUDOLF [1869-1928]
POGG 6,2844
A. Spilker, Ber. chem. Ges. 61A (1928), 141-142

WEISSMAN, SHERMAN MORTON [1930-]
AMS 15(7),510; WWWS, 1776

WEISSMANN, CHARLES [1931-]
KURSCHNER 13,4213

WEITH, WILHELM [1846-1881]
ADB 41,624; NB, 423; BLOKH 2,771-773; ZURICH, 40-46; SCHAEDLER, 148; POTSCH, 448-449; POGG 3,1430; 7aSuppl.,761
V. Meyer, Ber. chem. Ges. 15 (1882), 3291-3309

WEITZ, BERNARD GEORGE FELIX [1919-]
Who's Who of British Scientists 1969-1970, p.854

WEITZ, ERNST [1883-1954]
GUNDEL, 1027-1040; TLK 3,701; POGG 5,1352; 6,2844; 7a(4),917
R. Kuhn, Z. angew. Chem. 66 (1954), 657
F. Schmidt, Chem. Ber. 97 (1964), i-xx

WEITZEL, GÜNTER [1915-1984]
KURSCHNER 13,4214-4215; POGG 7a(4),917-918
Anon., Nachr. Chem. Techn. 23 (1975), 170; Z. physiol. Chem. 356 (1975), 624a; Naturw. Rund. 37 (1984), 342
F. Schneider, Z. physiol. Chem. 366 (1985), 609-616

WEIZMANN, CHAIM [1874-1952]
DSB 14,247-248; WNININGER 6,251-253; WWAH, 648; WWWS, 1777; POTSCH, 449
C. Weizmann, Trial and Error. New York 1949
E.D. Bergmann, J. Chem. Soc. 1953, pp.2840-2844
M.W. Weisgal and J. Carmichael, Chaim Weizmann: A Biography by Several Hands. London 1962

WEIZSÄCKER, VIKTOR von [1886-1957]
KURSCHNER 8,2549; FISCHER 2,1663; IB, 319; WWWS, 1740
R. Siebeck, Deutsche med. Wchschr. 82 (1957), 924-928

WELCH, ARNOLD DeMERRITT [1908-]
AMS 14,5444; WWA 1976-1977, p.3324; WWWS, 1777
A.D. Welch, Ann. Rev. Pharm. 25 (1985), 1-26

WELCH, WILLIAM HENRY [1850-1934]
DSB 14,248-250; DAB 19,621-624; NCAB 26,6-8; FISCHER 2,1664; IB, 320; DAMB, 788-789; FREUND 2,475-482; ULLMANN, 189-192; W, 548-549; WWAH, 648-649; WWWS, 1777-1778
S. Flexner, Science 79 (1934), 529-533; Biog. Mem. Nat. Acad. Sci. 22 (1943), 215-231
S. Bayne-Jones, J. Bact. 28 (1934), 433-446
S. Flexner and J.T. Flexner, William Henry Welch and the Heroic Age of American Medicine. New York 1941
G. Rosen, J. Hist. Med. 5 (1950), 233-235
E.H. Huntress, Proc. Am. Acad. Arts Sci. 78 (1950), 15-17
D. Fleming, William H. Welch and the Rise of Modern Medicine. Boston 1954
A.M. Harvey, Johns Hopkins Med. J. 135 (1974), 178-190

WELCKER, HERMANN [1822-1897]
ADB 55,38-41; BJN 2,115-116; 4,61*-62*; HIRSCH 5,888-890; PAGEL, 1830-1831; FREUND 2,483-493; WWWS, 1778
M. Heidenhain, Münch. med. Wchschr. 44 (1897), 1353
B. Solger, Anat. Anz. 14 (1898), 102-112
A.E. Best, Proc. Roy. Micr. Soc. 3 (1968), 210-219
P. Tautz, Anat. Anz. 131 (1972), 204-224

WELDE, ERNST [1883-1915]
JV 24,340; NUC 655,6
Anon., Z. angew. Chem. 28(III) (1915), 568

WELDON, WALTER FRANK RAPHAEL [1860-1906]
DSB 14,251-252; DNB [1901-1911] 3,629-631; DESMOND, 650; WWWS, 1778
K. Pearson, Biometrica 5 (1906), 1-50; Proc. Roy. Soc. B80 (1908), xxv-xl
G.C.B., Proc. Linn. Soc. 118 (1906), 109-114

WELLER, HEINRICH [1853-1923]
HEIN, 735
Anon., Chem. Z. 47 (1923), 275
G. Dann, Apoth. Z. 41 (1926), 1173-1174

WELLINGTON, CHARLES [1853-1926]
AMS 3,726-727; JV 1,74; RSC 19,541; NUC 655,175

WELLS, HARRY GIDEON [1875-1943]
DSB 14,252-253; DAB [Suppl.3],806-808; AMS 6,1514; NCAB 37,110-111; FISCHER 2,1165; CHITTENDEN, 233-241; DAMB, 789; IB, 320; WWAH, 650
A.B. Luckhardt, Am. J. Path. 17 (1941), 643-644
P.R. Cannon., Arch. Path. 36 (1943), 331-334; Trans. Assoc. Am. Phys. 58 (1944), 40-42
E.R. Long, Biog. Mem. Nat. Acad. Sci. 26 (1949), 233-261
L.W. Mayron, Annals of Allergy 42 (1979), 177-182

WELLS, HORACE [1815-1848]
DAB 19,640-641; NCAB 6,438-439; HIRSCH 5,892; DAMB, 789-790; W, 549-550;
WWWS, 1779

WELLS, HORACE LEMUEL [1855-1924]
NUC 655,387-390
Obituary Record of Yale Graduates 1924-1925, pp.1474-1475. New Haven 1925
R.H. Chittenden, Biog. Mem. Nat. Acad. Sci. 12 (1929), 273-285

WELS, PAUL [1890-1963]
KURSCHNER 9,2242; KIEL, 120; TLK 3, 702; POGG 6,2846-2847;

7a(4),926
G. Urban, Strahlentherapie 111 (1960), 321-324; Arzneimitt. 13 (1963), 1119-1120
P. Holtz, Arzneimitt. 10 (1960),142; Deutsche med. Wchschr. 89 (1964),390
F. Hauschild, Jahrb. Akad. Wiss. Berlin 1964, pp.235-236
K. Repke, Forsch. Fort. 38 (1964), 156

WELSH, DAVID ARTHUR [1865-1948]
WW 1945, p.2883; O'CONNOR(2), 507-508
Anon., Med. J. Australia 1948(II), pp.24-28; Brit. Med. J. 1948(II),p.273

WELTER, JEAN JOSEPH [1763-1852]
SCHAEDLER, 148; POGG 2,1294

WELTZIEN, CARL [1813-1870]
ADB 41,698; BLOKH 2,773; FERCHL, 573-574; POGG 2,1294-1295; 3,1431
K. Birnbaum, Ber. chem. Ges. 8 (1875), 1698-1702
C. deMilt, Chymia 1 (1948), 153-169

WENCKEBACH, KAREL FREDERIK [1864-1940]
LINDEBOOM, 2141-2143; FISCHER 2,1665-1666; STUMPF, 90-105; PLANER, 366; RAZINGER, 199-201
W. Falta, Wiener med. Wchschr. 53 (1940), 1067-1073
N. von Jagic, Münch. med. Wchschr. 87 (1940), 1421-1422
J.D. Rolleston, Nature 147 (1941), 260

WENDER, NEUMANN [1865-1934]
HEIN-E, 448-449; WININGER 6,255-257

WENDT, GEORG WILLEHAD von [1876-1954]
FISCHER 2,1666; IB, 320
Kuka Kukin On 1950, p.841

WENGLEIN, LUDWIG [1867-1901]
JV 11,153; NUC 656,52
Anon., Pharm. Z. 46 (1901), 199

WENKERT, ERNEST [1925-]
AMS 15(7),528; BHDE 2,1235; WWWS, 1780

WENRICH, DAVID HENRY [1885-1968]
AMS 11,5776; IB, 320; WWWS, 1780-1781
W.F. Diller, J. Protozool. 15 (1968), 400-401
R.M. Stabler, J. Parasit. 55 (1969), 429-430

WENSE, THEODOR von der [1904-1977]
KURSCHNER 12,3464; POGG 7a(4),936-937

WENT, FRIEDRICH AUGUST FERDINAND CHRISTIAN [1863-1935]
DSB 14,255; BWN 1,650-651; UTRECHT, 82-83; IB, 320
E. Küster, Ber. bot. Ges. 53 (1935), (97)-(120)
G.E. Briggs, Obit. Not. Fell. Roy. Soc. 2 (1936), 161-164

WENT, FRITS WARMOLT [1903-]
AMS 14,5458; IWW 1982-1983, p.1397; MH 1,517-519; MSE 3,294-295; WWWS, 1781

WENZ, JOSEPH [1862-]
RSC 19,548

WENZEL, ERWIN [1900-]
JV 40,747

WERDER, FRITZ von [1901-]
JV 44,457; NUC 656,246

WERDER, JOHANN ULRICH [1870-1943]
STRAHLMANN, 474,482

WERDT, FELIX von [1880-1923]
DIEPGEN, 68-69; FISCHER 2,1667; HUTER, 250

G. Pommer, Verhandl. path. Ges. 27 (1934), 367-369

WERKMAN, CHESTER HAMLIN [1893-1962]
AMS 10,4362; IB, 320; WWAH, 651; POGG 6,2850-2851; 7b,5933-5936
R.W. Brown, Biog. Mem. Nat. Acad. Sci. 44 (1974), 329-370

WERLE, EUGEN [1902-1975]
KURSCHNER 11,3251; WWWS, 1751; POGG 7a(4),941-944
W. Appel, Arzneimitt. 17 (1967), 1468
R. Huber, Advances in Experimental Medicine and Biology 156 (1983),29-36

WERNER, ALFRED [1866-1919]
DSB 14,264-272; DBJ 2,484-489,737; ZURICH, 60-84; BLOKH 2,774-776; MATSCHOSS, 290-291; ZURICH-D, 162; ABBOTT-C, 144-146; W, 551-552; POTSCH, 449-450; WWWS, 1781; POGG 4,1620; 5,1355-1356; 6,2851-2852
P. Karrer, Helv. Chim. Acta 3 (1920), 196-232; Gesnerus 23 (1966),275-300
P. Pfeiffer, Z. angew. Chem. 33 (1920), 37-39; J. Chem. Ed. 5 (1928), 1090-1098
G.T. Morgan, J. Chem. Soc. 117 (1920), 1639-1648
J. Lifschitz, Z. Elektrochem. 26 (1920), 514-529
G.B. Kauffman, Alfred Werner, Founder of Coordination Chemistry. Berlin 1966; Naturwiss. 63 (1976), 324-337
G. Schwarzenbach, Experientia 22 (1966), 633-646
C.H. Eugster, Chimia 37 (1983), 203-210

WERNER, EMIL ALPHONSE [1864?-1951]
POGG 5,1356; 6,2852; 7b,5937
W.R. Fearon, J. Chem. Soc. 1952, pp.2947-2952

WERNHER, CARL CHRISTIAN [1830-1889]
DGB 15 (1909), 464
Anon., Pharm. Z. 34 (1889), 209

WERTHEIM, THEODOR [1820-1864]
ADB 42,111; KERNBAUER, 32-41; BLOKH 2,776; FERCHL, 575; TSCHIRCH, 1145; WININGER 6,263-264; POGG 2,1303; 3,1432

WERTHEIMER, ÉMILE [1852-1924]
FISCHER 2,1671; KOREN, 262-263; WININGER 7,493-494; RSC 19,554-555; WWWS, 1782
J. Camus, Bull. Acad. Med. 92 (1924), 1143-1145
C. Richet, C.R. Soc. Biol. 91 (1924), 1060

WERTHEIMER, ERNST [1893-1978]
KURSCHNER 4,3243; BHDE 2,1238-1239; LDGS, 61; IB, 321; KOREN, 253

WERTHER, GUSTAV [1815-1869]
BLOKH 2,776; FERCHL, 575-576; SCHAEDLER, 149; POGG 2,1303-1304; 3,1432; 7aSuppl.,768-769
O. Liebreich, Ber. chem. Ges. 3 (1870), 372-374

WERTHESSEN, NICHOLAS THEODORE [1911-1981]
AMS 14,5463

WESSELY, FRIEDRICH [1897-1967]
KURSCHNER 10,2681-2682; POGG 6,2854; 7a(4),949-951
F. Galinovsky, J. Chem. Ed. 31 (1954), 663-664
Anon., Chem. Z. 81 (1957), 505-506; 92 (1968), 53; Leopoldina [3]8/9 (1962-1963), 100-101; Nachr. Chem. Techn. 15 (1967), 374-375
U. Schmidt, Alm. Akad. Wiss. Wien 127 (1978), 449-463

WEST, CHARLES [1816-1893]
DNB 20,1238-1239; HIRSCH 5,912

WEST, CLARENCE JAY [1886-1953]
AMS 8,2671; MILES, 501
Anon., Chemical Abstracts 47 (1953), 1381

WEST, EDWARD STAUNTON [1896-1984]
AMS 14,5466; IB, 321; WWWS, 1783; POGG 6,2854; 7b,5940-5942

WEST, RANDOLPH [1890-1949]
AMS 8,2671; ICC, 198-199
F.M. Hanger, Trans. Assoc. Am. Phys. 63 (1950), 19-20

WESTENBRINK, HENDRIK GERRIT KOLB [1901-1964]
LINDEBOOM, 2148-2149; POGG 6,2854; 7b,5942-5944
M.Gruber et al., Bioch. Biophys. Acta 97 (1965), i-xviii
E.P. Steyn-Perle, Jaarb. Akad. Wet. 1965-1966, pp.357-369

WESTERFELD, WILFRED WIEDEY [1913-]
AMS 14,5469; WWA 1980-1981, p.3487; WWWS, 1783

WESTERGAARD, MOGENS [1912-1975]
KBB 1975, pp.1122-1123; WWWS, 1783
D. von Wettstein, in The Carlsberg Laboratory 1876-1976 (H. Holter and K.M. Møller, Eds.), pp.139-150. Copenhagen 1976

WESTHEIMER, FRANK HENRY [1912-]
AMS 15(7),543; WWA 1980-1981, p.3488; MH 2,588-590; MSE 3,295-296

WESTPHAL, (ERNST FRIEDRICH) EUGEN [1856-]
BERLIN, 572; SG [1]16,404; NUC 658,266

WESTPHAL, OTTO [1913-]
KURSCHNER 13,4250; WWWS, 1784; POGG 7a(4),952-953
Anon., Nachr. Chem. Techn. 10 (1962), 279; Chem. Z. 102 (1978), 115

WESTPHAL, ULRICH [1910-1993]
AMS 14,5472; WWWS, 1784; POGG 7a(4),954-955
U. Westphal, Ligand Review 2 Suppl.1 (1980), 37-45

WETHERILL, CHARLES MAYER [1825-1871]
DAB 20,22-23; MILES, 502; ELLIOTT, 271-272; SILLIMAN, 63-65; WWAH, 653; POGG 3,1433
Anon., Am. J. Sci. [3]1 (1871), 478-479
E.F. Smith, J. Chem. Ed. 6 (1929), 1076-1089,1215-1225,1461-1477, 1668-1680,1916-1927,2160-2177

WETTERER, ERIK [1909-]
KURSCHNER 13,4251-4252; WWWS, 1785; POGG 7a(4),956
B. Rollmann, Personalbibliographien von Professoren der Physiologie... an der Medizinischen Fakultät der Universität Erlangen-Nürnberg... 1919-1967, pp.39-57. Erlangen-Nürnberg 1969

WETTSTEIN, ALBERT [1907-1974]
KURSCHNER 12,3482; SBA 4,445-446; WWWS, 1785
Anon., Chimia 21 (1967), 99-100; 28 (1974), 269

WETTSTEIN, FRITZ von {1895-1945]
KURSCHNER 6(II),1080-1081; IB, 321; WAGENITZ, 195
O. Renner, Bayer. Akad. Wiss. Jahrbuch 1944-1948, pp.261-265
A. Kuhn, Z. Naturforsch. 1 (1946), 48-50
H. Stubbe, Jahrb. Akad. Wiss. Berlin 1950-1951, pp.168-179
G. Melchers, Ber. bot. Ges. 100 (1987), 373-405

WETZEL, GEORG [1871-1951]
KURSCHNER 6(II),1081; FISCHER 2,1673-1674; AUERBACH, 416; IB, 321
H. Voos, Anat. Anz. 99 (1952), 21-28

WETZEL, ROBERT [1898-1962]
EBERT, 89-91
O. Paret, Jahreshefte Württ. 117 (1962), 67-73
Anon., Naturw. Rund. 15 (1962), 292; Attempto 10 (1962), 46

WEURINGH, PIERRE GUILLAUME [1886-]
JV 28,322; NUC 658,431

WEWER, HERMANN [1874-1925]
JV 19,185; NUC 658,437
Anon., Chem. Z. 49 (1925), 463

WEYGAND, CONRAD [1890-1945]
TLK 3,706; POGG 6,2859; 7a(4),962-963
Anon., Z. angew. Chem. 59 (1947), 124

WEYGAND, FRIEDRICH [1911-1969]
KURSCHNER 10,2689; POGG 7a(4),963-965
E. Baum, Chem. Z. 93 (1969), 788
H. Simon and W. Steglich, Z. Naturforsch. 25B (1970), 127-133
H.N. Rydon, Chem. Brit. 6 (1970), 439
S. Goldschmidt, Bayer. Akad. Wiss. Jahrbuch 1970, pp.229-232
Anon., Peptides 1969 (E. Scoffone, Ed.), pp.xiii-xiv. Amsterdam 1971

WEYL, THEODOR [1851-1913]
BJN 18,136*; FISCHER 2,1674-1675; PAGEL, 1845-1846; KOVACSICS, 99-103; BLOKH 2,777; WREDE, 214; BK, 345; WININGER 7,494-495; POGG 6,2860
E. Börnstein, Ber. chem. Ges. 47 (1914), 2395-2404
E.H. Huntress, Proc. Am. Acad. Arts Sci. 79 (1951), 42-43

WEYRICH, ADAM [1875-1933]
JV 24,340

WEYRICH, KARL RUFUS VICTOR [1819-1876]
ADB 42,285; HIRSCH 5,917-918; WELDING, 863; LEVITSKI 2,128
J. Brennsohn, Die Aerzte Livlands, pp.425-426. Mitau 1905

WEZLER, KARL [1900-1987]
KURSCHNER 13,4257-4258; POGG 7a(4),970-971
K. Greven, Arzneimitt. 15 (1965), 583-584

WHATLEY, FREDERICK ROBERT [1924-]
WW 1981, p.2754

WHEELER, ALVIN SAWYER [1866-1940]
AMS 6,1523; WWA 1940-1941, p.2720; BURSEY, 87-89; POGG 6,2860-2861; 7b,5951
R.W. Bost, Ind. Eng. Chem. News Ed. 12 (1934), 385

WHEELER, HENRY LORD [1867-1914]
POGG 4,1624-1625; 5,1359-1360
Obituary Record of Yale Graduates 1914-1915, pp.887-888. New Haven 1915

WHEELER, THOMAS SHERLOCK [1899-1962]
POGG 6,2863; 7b,5954-5957
E.M. Philbin, Proc. Chem. Soc. 1963, pp.154-156
W. Baker, Nature 197 (1963), 1152

WHEELER, WILLIAM MORTON [1865-1937]
DSB 14,291-292; DAB [Suppl.2],707-708; NCAB 27,395-396; IB,321; WWWS,1787
L.J. Henderson et al., Science 85 (1937), 533-535
A.D. Imms, Nature 139 (1937), 827-828
G.H. Parker, Biog. Mem. Nat. Acad. Sci. 19 (1938), 203-241
M.A. Evans and H.E. Evans, William Morton Wheeler, Biologist. Cambridge, Mass. 1970

WHEELWRIGHT, EDWIN WHITFIELD [1868-1916]
JV 8,218
Anon., J. Roy. Inst. Chem. 1916(4), p.35
R. Threlfall, J. Chem. Soc. 111 (1917), 377-378

WHELAND, GEORGE WILLARD [1907-1972]
AMS 11,5805; WWWS, 1787
Anon., Chem. Eng. News 51(14) (1973), 17

WHELDALE (ONSLOW), MURIEL [1880-1932]
DESMOND, 472; POGG 6,1911; 7b,3733
M. S[tephenson], Biochem. J. 26 (1932), 915-916

Anon. [F.G. Hopkins], Nature 129 (1932), 859
R. Scott-Moncrieff, Notes Roy. Soc. 36 (1981), 125-154

WHETHAM, MARGARET DAMPIER (Mrs. A.B. ANDERSON) [1900-]
NUC 132,131

WHEWELL, WILLIAM [1794-1866]
DSB 14,291-295; DNB 20,1365-1374; FERCHL, 577-578; WWWS, 1787; POGG 2,1309-1311; 3,1437
J.W. Herschel, Proc. Roy. Soc. 16 (1868), li-lxi
J.C. Maxwell, Nature 14 (1876), 206-208
T.G. Bonney, Nature 24 (1881), 137-139
R. Robson and W.F. Cannon, Notes Roy. Soc. 19 (1964), 168-191
M. Ruse, Centaurus 20 (1976), 227-257
Y. Elkana, Rivista di Storia delle Scienze 1 (1984), 149-199
M. Fisch and S. Schaeffer (eds.), William Whewell. Oxford 1991
J. Morrell, Hist. Sci. 30 (1992), 97-114

WHIPPLE, GEORGE HOYT [1878-1976]
AMS 12,6854; FISCHER 2,1675; MH 2,595-597; STC 3,163-164; MSE 3,305-306; W(3rd Ed.), 614; DAMB, 795-796; IB, 322; WWWS, 1787
G.H. Whipple, Persp. Biol. Med. 2 (1959), 253-289
G.W. Corner, George Hoyt Whipple and his Friends. Philadelphia 1963; Am. Phil. Soc. Year Book 1976, pp.135-140
L.E. Young, Trans. Assoc. Am. Phys. 89 (1976), 34-37
L.W. Diggs, Johns Hopkins Med. J. 139 (1976), 196-200
H.W. Davenport, Physiologist 24(2) (1981), 1-5

WHITAKER, DOUGLAS MERRITT [1904-1973]
AMS 10,4385; NCAB 58,489-490

WHITBY, LIONEL ERNEST HOWARD [1895-1956]
DNB [1951-1960], 1044-1046
Anon., Lancet 1956(II), pp.1165-1167; Brit. Med. J. 1956(II),pp.1306-1309
H.H. Thomas, Nature 179 (1957), 16-17

WHITE, ABRAHAM [1908-1980]
AMS 14,5485; WWA 1978-1979, p.3449; WWWS, 1788
M.H. Makman, Biochemical Action of Hormones 7 (1980), xiii-xv
E.L. Smith, Biog. Mem. Nat. Acad. Sci. 55 (1985), 507-536

WHITE, ADAM CAIRNS [1901-1962]
IB, 322

WHITE, ALAN GEORGE CASTLE [1916-1987]
AMS 15(7),561; WWA 1980-1981, p.3498; WWWS, 1788

WHITE, BENJAMIN [1879-1938]
AMS 6,1527; IB, 322
Yale University Obituary Record for 1937-1938, pp.184-185. New Haven 1939
A.H. Eggerth, The History of the Hoagland Laboratory. Brooklyn 1960

WHITE, JAMES CLARKE [1833-1916]
DAB 19,108-109; NCAB 19,358; FISCHER 2,1676; DAMB, 796-797
J.C. White, Sketches from my Life. Cambridge, Mass. 1914
F.C. Shattuck, Proc. Am. Acad. Arts Sci. 52 (1917), 873-876
B.B. Bechet, Ann. Med. Hist. NS7 (1935), 503-508
R.G. Ojeman, Clinical Neurosurgery 13 (1965), xii-xxii

WHITE, JULIUS [1904-1985]
AMS 14,5492; WWA 1976-1977, p.3355; WWWS, 1789

WHITE, MICHAEL JAMES DENHAM [1910-1983]
WW 1981, p.2765; IWW 1982-1983, p.1402; MH 2,597-598; MSE 3,307-308;WWWS, 1790
W.R. Atchley, Evolution and Speciation, pp.3-20. Cambridge 1981
E. Mayr, Am. Phil. Soc. Year Book 1984, pp.156-159

WHITE, PHILIP BRUCE [1891-1949]
P. Hartley, J. Path. Bact. 62 (1950), 468-481
W. Smith, Obit. Not. Fell. Roy. Soc. 7 (1950), 279-292

WHITE, PHILIP RODNEY [1901-1968]
AMS 11,5618; IB, 322; WWAH, 657; WWWS, 1790
D. de Torok, Plant Science Bulletin 14(2) (1968), 6-7

WHITE, THOMAS PHILIP [1855-1901]
SG [2]21,79
W.H.W., Cincinnati Lancet-Clinic NS47 (1901), 71

WHITE, WILLIAM HALE [1857-1949]
M. Campbell, Guy's Hosp. Rep. 98 (1949), 1-17
Anon., Lancet 1949(I), p.421; Brit. Med. J. 1949(I), pp.414-415

WHITELEY, ANNIE MARTHA [1866-1956]
Who's Who in British Science 1953, p.279
Who Was Who 1951-1960, p.1162
M.R.S. Creese, Brit. J. Hist. Sci. 24 (1991), 289

WHITFIELD, ARTHUR [1868-1947]
O'CONNOR(2), 180-181; RSC 19,586
Anon., Brit. J. Dermatol. 59 (1947), 173-176

WHITING, PHINEAS WESCOTT [1887-1978]
AMS 12,6875; IB, 322; WWWS, 1792
D.S. Grosch and C.H. Bastian, Genetics 89 (1978), 1-4

WHITLEY, EDWARD [1879-1945]
O'CONNOR(2), 374-375
R.A. Morton, Med. Hist. 16 (1972), 325

WHITMAN, CHARLES OTIS [1842-1910]
DSB 14,313-315; DAB 20,139-140; NCAB 11,73; WWAH, 658; WWWS, 1792
F.R. Lillie, Science 33 (1911), 54-56; J. Morphol. 22 (1911), xv-lxxvii
E.S. Morse, Biog. Mem. Nat. Acad. Sci. 7 (1912), 269-288
C.B. Davenport, Am. Nat. 51 (1917), 5-30
J. Maienschein, Mendel Newsletter 13 (1977), 1-4
R.W. Dexter, American Zoologist 19 (1979), 1251-1253
R.E. Brown, Mendel Newsletter 20 (1981), 2-4

WHITMORE, FRANK CLIFFORD [1886-1947]
DAB [Suppl.4],888-889; NCAB 39,359; MILES, 506-507; WWAH, 658; POGG 6,2868-2869; 7b,5971-5976
M.R. Fenske, J. Chem. Soc. 1948, pp.1090-1091
C.S. Marvel, Biog. Mem. Nat. Acad. Sci. 28 (1954), 289-311

WHITNEY, JOSIAH DWIGHT [1819-1896]
DAB 20,161-163; MILES, 507-508; ELLIOTT, 273-274; WWAH, 659; WWWS, 1792; POGG 2,1312-1313; 3,1438-1439
Anon., Am. J. Sci. [4]2 (1896), 312-313
E.T. Brewster, The Life and Letters of Josiah Dwight Whitney. Boston 1909

WHYTLAW-GRAY, ROBERT [1877-1958]
DSB 14,318-319; WW 1958, p.3237; POGG 5,447; 6,2870-2871; 7b,5983-5984
F. Challenger, Nature 181 (1958), 527
E.G. Cox and J. Hume, Biog. Mem. Fell. Roy. Soc. 4 (1958), 327-339
R.S. Bradley, Proc. Chem. Soc. 1959, pp.18-20

WIBAUT, JOHAN PIETER [1886-1967]
POGG 5,1362; 6,2871; 7b,5984-5991
Anon., Chem. Wkbl. 22 (1925), 170-171; Chem. Brit. 4 (1968), 34
A.F. Holleman et al., Chem. Wkbl. 34 (1937), 739-748
H. Gerding et al., Chem. Wkbl. 52 (1956), 693-705
H.J. den Hartog, Chem. Wkbl. 58 (1962), 561-566
S. Goldschmidt, Bayer. Akad. Wiss. Jahrbuch 1968, pp.220-223

WIBERG, KENNETH BERLE [1927-]
AMS 15(7),588; WWA 1980-1981, p.3513; IWW 1982-1983, p.1404; WWWS, 1794

WIBMER, KARL AUGUST [1805-1885]
ADB 42,303-304; HIRSCH 5,922-923; CALLISEN 21,119-121; 33,286-287
Anon., Leopoldina 21 (1885), 163
F. Seitz, Aerzt. Int. Bl. 32 (1885), 561-563

WICHELHAUS, HERMANN [1842-1927]
DZ, 1564; SCHAEDLER, 150; WWWS, 1794; POGG 3,1439-1440; 4,1628; 5,1362-1363; 6,2872
W. Schlenk, Ber. chem. Ges. 60A (1927), 59-61

WICHMANN, ARTHUR [1851-1927]
POGG 6,1872
L. Rutten, Zeitschrift für Vulkanologie 11 (1928), 153-162

WICKE, WILHELM [1822-1871]
RSC 11,800; POGG 2,1313-1314
T. Mithoff and G. Drechsler, Land. Vers. 15 (1872), 306-318

WIDAL, FERNAND [1862-1929]
FISCHER 2,1680; HD 5,544; WININGER 6,275; 7,495; WWWS, 1795
J.A. Sicard, Presse Med. 37 (1928), 105-107
F.J. Besançon, Bull. Acad. Med. 101 (1929), 93-103
A.A. Lemierre, Bull. Acad. Med. 135 (1951), 647-655
R. Kohn, Rev. Hist. Med. Heb. 1954, pp.97-100
P.R. Hunter, Med. Hist. 7 (1963), 56-61
L.J. Pasteur Vallery Radot, Bull. Acad. Med. 154 (1970), 834-841
H. Baruk, Rev. Hist. Med. Heb. 24 (1971), 69-72

WIDMAN, OSCAR [1852-1930]
SMK 8,347-348; POGG 3,1441; 4,1629-1630; 5,1363; 6,2873
L. Ramberg, Ber. chem. Ges. 63A (1930), 161-162

WIDMARK, ERIK MATTEO [1889-1945]
SMK 8,350; FISCHER 2,1681; POGG 6,2873; 7b, 5994-5996
G. Liljestrand, Acta Physiol. Scand. 10 (1945), 193-194

WIECHOWSKI, WILHELM [1873-1928]
FISCHER 2,1681; HARTMANN, 95-102; KOERTING, 134-135; TSCHIRCH, 1145; IB, 323; TLK 3,707; POGG 6,2873
H. Langecker, Lotos 77 (1929), 65-86
E. Starkenstein, Deutsche med. Wchschr. 55 (1929), 199-200
K. Spiro, Med. Welt 3 (1929), 220-222
C. Neuberg, Biochem. Z. 216 (1929), 241

WIEDE, OTTO FRITZ [1871-1905]
JV 12,248; RSC 15,818
Anon., Ber. chem. Ges. 38 (1905), 4202

WIEDEMANN, EILHARD [1852-1928]
BLOKH 2,778; WOLF, 228; WWWS, 1795; POGG 3,1441-1442; 4,1631-1632; 5,1364-1365; 7aSuppl.,769-770
H.J. Seemann, Isis 14 (1930), 166-186
E.H. Huntress, Proc. Am. Acad. Arts Sci. 81 (1952), 96-97

WIEDEMANN, GUSTAV HEINRICH [1826-1899]
DSB 14,329-331; ADB 55,67-70; BJN 4,188*; BAD 5,812-813; BLOKH 2,778; BERLIN, 647; ALBRECHT, 98; SCHAEDLER, 150; WWWS, 1795; POGG 2,1319; 3,1441; 4,1631; 7aSuppl.,770
F. Kohlrausch, Verhandl. phys. Ges. 1 (1899), 155-167
P. van Tieghem, C.R. Acad. Sci. 129 (1899), 1062-1063
C. von Voit, Sitz. Bayer. Akad. 30 (1901), 353-359
W. Ostwald, Abhandlungen und Vorträge, pp.395-403. Leipzig 1904
C.C.F., Proc. Roy. Soc. 75 (1905), 41-43

WIEDERHOLD, JACOB ERNST EDUARD [1822-1898]
BJN 5,66*; SCHAEDLER, 151; POGG 4,1632-1633
Anon., Leopoldina 34 (1898), 60

WIEDERSHEIM, (HANS) VOLKER [1903-]
JV 44,457; NUC 662,191

WIEDERSHEIM, ROBERT [1848-1923]
DSB 14,331; DBJ 5,382-386,445; GROTE 1,207-227; FISCHER 2,1681-1682
R. Wiedersheim, Lebenserinnerungen. Tübingen 1919

WIELAND, HEINRICH [1877-1957]
DSB 14,334-335; STC 3,169-170; FARBER, 1443-1451; ABBOTT-C, 146-147; TLK 3,708; W, 557-558; WWWS, 1796; POTSCH, 452-453; POGG 5,1366-1367; 6,2876-2877; 7a(4),985-987
G. Lunde, J. Chem. Ed. 7 (1930), 1763-1777
E. Dane et al., Naturwiss. 30 (1942), 333-373
R. Huisgen, Proc. Chem. Soc. 1958, pp.210-219
P. Karrer, Biog. Mem. Fell. Roy. Soc. 4 (1958), 341-352
C. Schöpf and R. Huisgen, Z. angew. Chem. 71 (1959), 1-6
F.G. Fischer, Bayer. Akad. Wiss. Jahrbuch 1959, pp.160-170
G. Hesse, Ann. Chem. 1977, pp.1058-1063
B. Witkop, Angew. Chem. (Int. Ed.) 16 (1977), 559-572; Medicinal Research Reviews 12 (1992), 193-265; Ann. Chem. 1992, i-xxxii
K. Bloch, TIBS 7 (1982), 334-336

WIELAND, HERMANN [1885-1929]
DBJ 11,328-332,372; DRULL, 296-297; DIEPGEN, 99-101; FISCHER 2,1682; TSCHIRCH, 1146; IB, 323; TLK 3,708; POGG 6,2877
B. Behrens, Der Schmerz 2 (1929), 83-85
E. Oppenheimer, Klin. Wchschr. 8 (1929), 1286-1287
H. Freundlich, Ber. chem. Ges. 62A (1929), 82

WIELAND, THEODOR [1846-1928]
HEIN-E, 450-451
Anon., Chem. Z. 52 (1928), 779
A. Wankmüller, Beitr. Wurtt. Apothekgesch. 10 (1973), 42-46

WIELAND, THEODOR [1913-]
KURSCHNER 13,4271; WWWS, 1796; POTSCH, 453; POGG 7a(4),988-990
Anon., Nachr. Chem. Techn. 16 (1968), 243-244
B. Witkop, Naturw. Rund. 36 (1983), 261-275

WIELAND, ULRICH [1911-]
JV 54,555; NUC 662,277

WIENER, ALEXANDER SALOMON [1907-1976]
AMS 13,5892; NCAB G,469-470; MH 1,527-528; MSE 3,314-315; KOREN, 263; KAGAN(JA), 726; COHEN, 253; W(3rd Ed.), 614-615; WWWS, 1796
Anon., Leopoldina [3]16 (1970), 64-66; Haematologia 6 (1972), 11-45

WIENER, HUGO [1868-1930]
FISCHER 2,1682-1683; KOERTING, 153; IB, 323

WIENER, LUDWIG CHRISTIAN [1826-1896]
DSB 14,343-344; ADB 42,790-792; BJN 1,207-208; 3,88*; HB 2,321-324; BAD 5,814-817; WWWS, 1796; POGG 2,1322; 3,1442; 4,1634; 5,1369; 6,2879; 7aSuppl.,771
Anon., Leopoldina 32 (1896), 155-159,166-169
O. Wiener, Naturwiss. 15 (1927), 81-84

WIENER, NORBERT [1894-1964]
DSB 14,344-347; AMS 10,4410-4411; MH 1,528-529; WWWS, 1796; POGG 5,1369; 6,2879-2880; 7b,6000-6004
I.E. Segal, Biog. Mem. Nat. Acad. Sci. 61 (1992), 389-436

WIENGREEN, FRIEDRICH [1875-]
JV 19,186; NUC 662,367

WIENHAUS, HEINRICH [1882-1959]
KURSCHNER 8,2587; TLK 3,709; POGG 6,2880; 7a(4),992
Anon., Chem. Z. 83 (1959), 659

WIERZUCHOWSKI, MIECZYSLAW [1895-1967]
IB, 323
T. Toczynski, Acta Physiol. Pol. 21 (1970), 155-165
J. Kiersz, Acta Physiol. Pol. 38 (1987), 264-268

WIESEL, RUDOLF [1887-]
JV 26,377; SG [2]21,113; NUC 662,455

WIESNER, JULIUS [1838-1916]
DSB 14,349-350; NOB 5,141-161; DBJ 1,371-372; FREUND 1,347-356; BK, 345; WININGER 6,282-283; REBER, 93-98,366-367; SCHAEDLER, 151-152; TSCHIRCH, 1146; WWWS, 1740-1741; POGG 2,1322-1323; 3,1443; 4,1635; 5,1370; 6,2882
K. Linsbauer, Wiesner und seine Schule. Vienna 1903,1910
E. Hanausek, Chem. Z. 32 (1908), 61-62
H. Molisch, Ber. bot. Ges. 34 (1917), (71)-(99); Alm. Akad. Wiss. Wien 67 (1917), 362-368

WIESNER, KAREL [1919-1986]
Z. Valenta, Tetrahedron 44 (1988), iii-v

WIGAND, ALBERT [1821-1886]
DSB 14,350; ADB 42,445-449; LKW 2,408-415; SCHMITZ, 116-132,381-390; TSCHIRCH, 1146; WWWS, 1797
Anon., Chem. Z. 10 (1886), 1357; Pharm. Z. 31 (1886), 679-682
E. Dennert, Flora 69 (1886), 531-539
F.G. Kohl, Bot. Zt. 28 (1886), 350-352,381-384
A. Tschirch, Ber. bot. Ges. 5 (1887), (xli)-(li); Arch. Pharm. 225 (1887), 1-13
B. Lehmann, Julius Wilhelm Albert Wigand. Marburg 1973

WIGAND, FRIEDRICH [1788-1855]
HEIN, 746-747
A. Wigand, Arch. Pharm. 135 (1856), 209-215

WIGGERS, CARL JOHN [1883-1963]
AMS 10,4412; NCAB 50,615-616; FISCHER 2,1685; DAMB, 803-804; WWAH, 661
C.J. Wiggers, Reminiscences and Adventures in Circulation. New York 1958
P. Rijlant, Bull. Acad. Med. Belg. [7]3 (1963), 383-391
A.A. Luisada, Am. J. Card. 12 (1963), 135-136
J.W. McCubbin, Pharmacologist 6 (1964), 7-8
E.M. Landis, Biog. Mem. Nat. Acad. Sci. 48 (1976), 363-397
W.C. Randall, Physiologist 21(3) (1978), 1-5
B. Silverman, Clinical Cardiology 12 (1989), 731-734

WIGGERS, HEINRICH AUGUST LUDWIG [1803-1880]
ADB 42,465; HEIN, 747-748; HIRSCH 5,931-932; SCHELENZ, 708-709; WAGENITZ, 196; FERCHL, 582; SCHAEDLER, 152; CALLISEN 21,148; 33,293; POGG 2,1323; 3,1444; 7aSuppl.,771-772
T. Husemann, Arch. Pharm. 216 (1880), 401-415
N.P. Hamburg, Hygiea 17 (1880), 490-492

WIGGLESWORTH, VINCENT BRIAN [1899-1994]
WW 1981, p.2779; STC 3,172-174; MSE 3,316-317; ABBOTT, 137; WWWS, 1797

WIGNER, EUGENE PAUL [1902-]
AMS 14,5519; WWA 1978-1979, p.3471; IWW 1982-1983, pp.1406-1407; KURSCHNER 13,4280-4281; MH 1,531-532; STC 3,174-175; MSE 3,317-318; CB 1953, pp.657-659; LDGS, 102; POGG 6,2884; 7a(4),995-997
V. Bargmann et al., Rev. Mod. Phys. 34 (1962), 587-591

WIIS, JACOB JAN ALEXANDER [1864-1942]
RSC 19,610
J.F. Carrière, Chem. Wkbl. 40 (1943), 140-141
Anon., Ann. Chim. Anal. 25 (1943), 87; Chem. Ind. 1944, p.34

WIISMAN, HENDRIK PAULUS [1862-1916]
RSC 19,610; NUC 662,693
Anon., Chem. Wkbl. 11 (1914), 670-671
N. Schoorl, Chem. Wkbl. 13 (1916), 810-816

WILBRAND, JOHANN BERNHARD [1779-1846]
DSB 14,351-352; ADB 44,520-521; CALLISEN 21,151-158
H. Karlheim, Joh. B. Wilbrand, sein Leben uns sein Werk. Münster 1943
C. Probst, Sudhoffs Arch. 50 (1966), 157-178
A. Murken, Med. Mon. 24 (1970), 165-170

WILBRANDT, WALTER [1907-1979]
KURSCHNER 13,4281; BHDE 2,1245; POGG 7a(4),998-1000
H. Reuter, Journal of Membrane Biology 57(2) (1980), 85

WILDER, BURT GREEN [1841-1925]
AMS 3,740; NCAB 4,481; WWAH, 662; WWWS, 1798
E.H. Beardsley, Isis 64 (1973), 50-66
C.G. Roland, Physiologist 26 (1983), 361-363

WILDER, RUSSELL MORSE [1885-1959]
AMS 8,2706; NCAB 45,544-545; IB, 324; WWAH, 662; WWWS, 1798
H.T. Ricketts, Proc. Inst. Med. Chicago 23 (1960), 61-63
R.G. Sprague, Diabetes 9 (1960), 419-420
G.A. Goldsmith, J. Nutrition 74 (1961), 3-8

WILEY, HARVEY WASHINGTON [1844-1930]
DSB 14,357-358; DAB 20,215-216; NCAB 21,72-73; MILES, 510-512; IB, 324; DAMB, 805; OEHRI, 132-133; WWAH, 663; POGG 3,1446; 4,1639-1640; 5,1372-1373; 6,2885-2886
W.D. Bigelow, Science 72 (1930), 311-312; J. Am. Chem. Soc. 54 (1932), 33-43
H.E. Armstrong, Nature 126 (1930), 444-445
C.A. Browne et al., J. Assoc. Agr. Chem. 15 (1931), iii-xxii
J.H. Young, J. Hist. Med. 23 (1968), 86-104

WILHELMI, ALFRED ELLIS [1910-]
AMS 13,4852; WWA 1976-1977, p.3379; WWWS, 1799

WILHELMI, DELIA [1906-]
JV 49,587

WILHELMY, LUDWIG FERDINAND [1812-1864]
DSB 14,359-360; HEIN, 751; BLOKH 2,779-780; FERCHL, 581; POGG 2,1327-1328; 3,1446; 4,1640
E. Farber, Chymia 7 (1961), 135-148

WILKE, ERNST [1882-1934]
DRULL, 297-298; POGG 6,2886-2887

WILKE, KARL [1887-]
JV 31,382; NUC 663,647

WILKENS, HERMANN [1816-1886]
Anon., Pharm. Z. 31 (1886), 144

WILKINS, MAURICE HUGH FREDERICK [1916-]
WW 1981, pp.2784-2785; IWW 1982-1983, p.1408; CB 1963, pp.465-466; MH 1,533-534; STC 3,175-177; MSE 3,320-321; WWWS, 1799; POTSCH, 453
H.R. Wilson, TIBS 13 (1988), 275-288

WILKINSON, GEOFFREY [1921-]
WW 1991, p.1967; CAMPBELL, 275-276; POTSCH, 453-454

WILKINSON, JOHN FREDERICK [1897-]
WW 1983, p.2411; NUC 664,93
M. Hampton, World Medicine 10(5) (1974), 27-31

WILKS, SAMUEL [1824-1911]
DNB [1901-1911] 3,668-669; HIRSCH 5,938
Anon., Lancet 1911(II), pp.1441-1445; Brit. Med. J. 1911(II), pp.1384-1390; Guy's Hosp. Gaz. 25 (1911), 508-520
W.H. White, Guy's Hosp. Rep. [3]67 (1913), 1-39
S.D. Clippingdale, Brit. Med. J. 1920(I), p.104

WILL, HANNS [1891-]
NUC 664,164

WILL, HEINRICH [1812-1890]
BLOKH 2,780-781; FERCHL, 581; PHILIPPE, 773-774; SCHAEDLER, 152; POTSCH, 454; TSCHIRCH, 1146; POGG 2,1329; 3,1447; 7aSuppl.,773
Anon., Chem. Z. 14 (1890), 1784
A.W. Hofmann, Ber. chem. Ges. 23(3) (1890), 852-899
C. von Voit, Sitz. Bayer. Akad. 21 (1891), 154-160

WILL, WILHELM [1854-1919]
BLOKH 2,781-782; TLK 2,763-764; POGG 4,1640-1641; 5,1373
B. Lepsius, Chem. Z. 44 (1920), 113-115; Ber. chem. Ges. 54A (1921), 205-268

WILLCOCK, EDITH GERTRUDE (Mrs. JOHN S. GARDINER) [1879-1953]
O'CONNOR(2), 37-38
Register, Newnham College, p.33

WILLE, FRANZ [1909-1986]
KURSCHNER 13,4291; POGG 7a(4),1011
Anon., Chem. Z. 103 (1979), 121; Naturw. Rund. 39 (1986), 233

WILLGERODT, CONRAD [1841-1930]
TLK 3,712; SCHAEDLER, 153; POTSCH, 454; POGG 3,1447; 4,1641-1642; 5,1373-1374; 6,2889
E. Riesenfeld, Ber. chem. Ges. 64A (1931), 5-6
O. Hinsberg, Chem. Z. 55 (1931), 85

WILLIAMS, ALAN FREDERICK [1945-1992]
WW 1992, p.1999

WILLIAMS, CARROLL MILTON [1916-1991]
AMS 15(7),622; WWA 1980-1981, p.3528; IWW 1982-1983, p.1411; WWWS, 1801; MH 2,603-605; STC 3,178-179; MSE 3,326-328

WILLIAMS, CHARLES GREVILLE [1829-1910]
DNB [1901-1911] 3,672-673; WWWS, 1801; POTSCH, 454-455;POGG 3,1447-1448; 5,1374
A.H.C., Proc. Roy. Soc. A85 (1911), xvii-xx; J. Chem. Soc. 99 (1911), 606-609
H.J. Stern, Chem. Brit. 15 (1979), 457-458

WILLIAMS, CHARLES JAMES [1805-1889]
DNB 21,383-384; HIRSCH 5,942-943; PAGEL, 1855-1856; WWWS, 1801; CALLISEN 21,200-201; 33,302-304
C.J.B. Williams, Memoirs of Life and Work. London 1884
Anon., Leopoldina 25 (1889), 115-116
A.B.G., Proc. Roy. Soc. 46 (1890), xxvi-xxxiii

WILLIAMS, CURTIS ALVIN, Jr. [1927-]
AMS 17(7),644; BIBEL, 290-293

WILLIAMS, HAROLD HENDERSON [1907-1991]
AMS 14,5542; WWWS,1802

WILLIAMS, HORATIO BURT [1877-1955]
AMS 8,2716; NCAB 43,242-243; WWAH, 666; POGG 6,2891; 7b,6016
W.S. Root et al., Science 124 (1956), 527

WILLIAMS, JOHN WARREN [1898-1988]
AMS 14,5544; WWA 1976-1977, p.3390; MH 2,606-608; MSE 3,328-330; WWWS, 1802; POGG 6,2891; 7b,6016-6018
J.D. Ferry, Langmuir 5 (1989), 2-3

WILLIAMS, OWEN THOMAS [1877-1913]
O'CONNOR(2), 375
Who Was Who 1897-1915, p.545
Anon., Brit. Med. J. 1913(I), p.201

WILLIAMS, PERCY [1873-1942]
RSC 19,636
R. Seligman, J. Chem. Soc. 1944, p.48

WILLIAMS, RICHARD TECWYN [1909-1979]
DNB [1971-1980], 907-908; WW 1979, p.2705; WWWS, 1803
D.V. Parke, Chem. Brit. 17 (1981), 240
A. Neuberger and R.L. Smith, Biog. Mem. Fell. Roy. Soc. 28 (1982), 685-717; Drug Metabolism Review 14 (1983), 559-607

WILLIAMS, ROBERT JOSEPH PATON [1926-]
WW 1985, p.2079
R.J.P. Williams, Journal of Inorganic Biochemistry 28 (1986), 81-84
A.G. Sykes, Advances in Inorganic Chemistry 36 (1991), xi-xiii

WILLIAMS, ROBERT RUNNELS (RAMPATHNAM) [1886-1965]
DSB 14,392-394; AMS 10,4438; NCAB F,204-205; 58,411-413; MH 1,537; STC 3,180-181; MSE 3,330; CHITTENDEN, 353-355; CB 1951, pp.659-661; POTSCH, 455; WWAH, 667; WWWS, 1803; POGG 6,2892; 7b,6019-6022
J.W. Barker, Am. Phil. Soc. Year Book 1966, pp.206-209
R.S. Baldwin, J. Nutrition 105 (1975), 3-14

WILLIAMS, ROBLEY COOK [1908-]
AMS 14,5549; WWA 1978-1979, p.3491; IWW 1982-1983, p.1412; MH 2,608-609; MSE 3,330-332; WWWS, 1804
R.C. Williams, Ann. Rev. Microbiol. 32 (1978), 1-18

WILLIAMS, ROGER JOHN [1893-1988]
AMS 14,5549; WWA 1978-1979, p.3491; IWW 1982-1983, pp.1412-1413; MH 2,609-610; MSE 3,332-333; CB 1957, pp.594-596; WWWS, 1804; POTSCH, 455; POGG 6, 2892-2893; 7b,6022-6026

WILLIAMSON, ALEXANDER WILLIAM [1824-1904]
DSB 14,394-396; DNB [1901-1911] 3,678-680; BLOKH 2,783-786; W, 561-562; POTSCH, 455-456; ABBOTT-C, 147-148; WWWS, 1804; POGG 2,1331; 3,1448; 4,1643
G.C. Foster, J. Chem. Soc. 87 (1905), 605-618; Ber. chem. Ges. 44 (1911), 2253-2269
E. Divers, Proc. Roy. Soc. A78 (1907), xxiv-xliv
J. Harris and W.H. Brock, Ann. Sci. 31 (1974), 95-130
E.R. Paul, Ann. Sci. 35 (1978), 17-31

WILLIAMSON, SIDNEY [1867-1939]
JV 7,215; RSC 13,277; 19,638
R.A. Cooper, J. Chem. Soc. 1939, pp.1236-1237
Anon., J. Roy. Inst. Chem. 1939, pp.333-334

WILLIER, BENJAMIN HARRISON [1890-1972]
AMS 11,5886; IB,325; WWWS, 1805
J. Oppenheimer, Am. Phil. Soc. Year Book 1973, pp.174-179
R.L. Watterson, Dev. Biol. 34 (1973), 1-19; Biog. Mem. Nat. Acad. Sci. 55 (1985), 539-628

WILLIGK, ERWIN JULIUS [1826-1887]
BLOKH 2,786; SCHAEDLER, 153-154; PRAG, 345

WILLIMOTT, STANLEY GORDON [1899-]
POGG 6,2894; 7b,6028-6029
Who's Who in British Science 1953, p.282

WILLKOMM, MORITZ [1821-1895]
ADB 43,298-300; LUDY, 157-175; LEVITSKI 1,356-360; TSCHIRCH, 1147
E. Roth, Leopoldina 32 (1896), 94-96
R. von Wettstein, Ber. bot. Ges. 14 (1896), (19)-(25)

WILLMER, EDWARD NEVILL [1902-]
WW 1981, p.2808; IB, 235; WWWS, 1805

WILLSTÄDT, HARRY [1904-1947]
LDGS, 26
K. Myrbäck, Sven. Kem. Tid. 60 (1948), 20-21

WILLSTÄTTER, RICHARD [1872-1942]
DSB 14,411-412; SCHWERTE 1,167-174; FARBER, 1367-1374; TLK 3,712-713; TSCHIRCH, 1147; W, 563; WININGER 6,286-287;

TETZLAFF, 357; BHDE 2,1248; ABBOTT-C, 148; POTSCH, 456; POGG 4,1644-1645; 5,1375-1376; 6,2894-2896; 7a(4),1014
H.E. Armstrong, Nature 120 (1927), 1-5
R. Willstätter, Science 78 (1933), 271-274; Aus meinem Leben. Weinheim 1949 (English translation, New York 1965)
T.R. Seshadri, Proc. Indian Acad. Sci. 17A (1943), 143-157
H. Wieland, Bayer. Akad. Wiss. Jahrbuch 1944-1948, pp.194-198
F. Wessely, Alm. Akad. Wiss. Wien 99 (1949), 296-306
R. Kuhn, Naturwiss. 36 (1949), 1-5
Anon., Z. angew. Chem. 61 (1949), 349-351
R. Robinson, Obit. Not. Fell. Roy. Soc. 8 (1953), 609-634; J. Chem. Soc. 1953, pp.999-1026
R. Huisgen, J. Chem. Ed. 38 (1961), 10-15
A. Dees de Sterio, Münch. med. Wchschr. 109 (1967), 2018-2019
A. Schellenberger, Leopoldina [3]18 (1972), 121-128
J. Renz, Helv. Chim. Acta 56 (1973), 1-14
I. Strube, NTM 10(2) (1973), 87-99
J.S. Fruton, TIBS 2 (1977), 210-211

WILMANNS, GUSTAV FRIEDRICH [1881-1965]
JV 20,264; DGB 82 (1934), 577
Anon., Chem. Z. 80 (1956), 757; Nachr. Chem. Techn. 9 (1961), 277

WILSON, DAVID WRIGHT [1889-1965]
AMS 10,4448; FISCHER 2,1690; IB, 325; WWWS, 1807
E.G. Ball and J.M. Buchanan, Biog. Mem. Nat. Acad. Sci. 43 (1973),261-284

WILSON, EDMUND BEECHER [1856-1939]
DSB 14,423-426; DAB [Suppl.2],724-725; AMS 6,1553-1554; FREUND 1,357-362; FLEULER, 159-160; SCHULZ-SCHAEFFER, 202-203; WWAH, 669; WWWS, 1807
C. Herbst, Naturwiss. 15 (1927), 257-259
T.H. Morgan, Science 89 (1939), 258-259; 96 (1942), 239-242; Biog. Mem. Nat. Acad. Sci. 21 (1940), 315-342
H.J. Muller, Am. Nat. 77 (1943), 5-37,142-172; Genetics 34 (1949), 1-9
A.L. Baxter, E.B. Wilson and the Problem of Development. New Haven 1974; J. Hist. Biol. 9 (1976), 29-57; Isis 68 (1977), 363-374

WILSON, FORSYTH JAMES [1880-1944]
POGG 6,2897; 7b,6033-6034
Anon., J. Roy. Inst. Chem. 1944, p.188
M.M.J. Sutherland, J. Chem. Soc. 1945, pp.723-724

WILSON, GEORGE [1818-1859]
BLOKH 2,787; FERCHL, 582; COLE, 577; WWWS, 1807; POGG 2,1334
Lord Neaves, Proc. Roy. Soc. Edin. 4 (1857-1862), 227-229

WILSON, HAROLD ALBERT [1874-1964]
POGG 5,1378-1379; 6,2897; 7b,6034-6035
G. Thomson, Biog. Mem. Fell. Roy. Soc. 11 (1965), 187-201

WILSON, IRWIN B. [1921-]
AMS 15(7),654; WWWS, 1807

WILSON, JAMES THOMAS [1861-1945]
WW 1945, p.2951
J.P. Hill, Obit. Not. Fell. Roy. Soc. 6 (1948), 643-660

WILSON, PERRY WILLIAM [1902-1981]
AMS 14,5568-5569; WWA 1976-1977, p.3408; IWW 1982-1983, p.1417; WWWS,1808
P.W. Wilson, Ann. Rev. Microbiol. 26 (1972), 1-22
R.H. Burris, Biog. Mem. Nat. Acad. Sci. 61 (1992), 439-467

WILSON, WILLIAM POWELL [1844-1927]
AMS 3,751; WWA 1924-1925, p.3465; RSC 11,822; 19,653; NUC 667,308

WIMMER, JOHANNES [-1890]
Anon., Pharm. Z. 35 (1890), 147

WINCKLER, FERDINAND LUDWIG [1801-1868]
HB 1,165-167; HEIN, 755; TSCHIRCH, 1147; FERCHL, 582; PHILIPPE, 810; SCHELENZ, 675; POGG 2,1335

WINDAUS, ADOLF [1876-1959]
DSB 14,443-446; SCHWERTE 1,175-183; STC 3,183-184; TLK 3,713; WWWS, 1810; W, 564-565; POTSCH, 457; POGG 5,1380; 6,2901-2902; 7a(4),1016-1018
G. Lunde, J. Chem. Ed. 7 (1930), 1763-1777
L. Josephson, J. Am. Pharm. Assn. 22 (1933), 309-312
O. Dalmer, Naturwiss. 30 (1942), 1-4
A. Butenandt and H. Brockmann, Jahrb. Akad. Wiss. Gott. 1944-1960, pp.173-186
W. Hückel, Z. angew. Chem. 59 (1947), 185-188; Naturw. Rund. 1 (1948), 127-129
H. Hauptmann and A. Lüttringhaus, Chem. Z. 83 (1959), 812-814
A. Butenandt, Bayer. Akad. Wiss. Jahrbuch 1960, pp.157-164; Münch. med. Wchschr. 102 (1960), 1755-1757; Proc. Chem. Soc. 1961, pp.131-138
K. Dimroth, Chem. Ber. 119(11) (1986), xxxi-lviii
H.W. Schütt, Sudhoffs Arch. 72 (1988), 98-105

WINDISCH, FRIEDRICH [1898-1961]
KURSCHNER 9,2283; POGG 7a(4),1019-1021
H. Haehn, Brauwelt 95 (1955), 1693
Anon., Brauwissenschaft 9 (1956), 20-23

WINGE, ØJVIND [1886-1964]
KBB 1963, pp.1629-1630; VEIBEL 2,476-477; IB, 325; WWWS, 1810
E.A. Bevan, Proc. Linn. Soc. 176 (1965), 228-229
M. Westergaard, Genetics 82 (1976), 1-7; in The Carlsberg Laboratory 1876-1976 (H. Holter and K.M. Møller, Eds.), pp.210-230. Copenhagen 1976

WINGLER, AUGUST [1898-1960]
POGG 7a(4),1022
Anon., Nachr. Chem. Techn. 8 (1960), 72,95

WINKELBLECH, KARL GEORG [1810-1865]
LKW 2,422-428; HEIN, 755-756; MEINEL, 23-25,514-515; GUNDLACH, 465-466; SCHMITZ, 230-231; FERCHL, 582; POGG 2,1337; 3,1452; 7aSuppl.,774
W.E. Biermann, K.G. Winkelblech. Leipzig 1909

WINKLER, CLEMENS ALEXANDER [1838-1904]
DSB 14,447; BJN 10,129*; BUGGE 2,336-350; BLOKH 2,787-789; WWWS, 1810; W, 565; POTSCH, 457-458; POGG 3,1452-1453; 4,1650-1651; 5,1381; 7aSuppl.,774-775
T. Döring, Z. angew. Chem. 18(I) (1905), 1-7
O. Brunck, Ber. chem. Ges. 39 (1906), 4491-4548
B. Sorms, History and Technology 7 (1990), 51-61

WINKLER, FERDINAND [1870-1936]
FISCHER 2,1673

WINKLER, HANS [1877-1945]
KURSCHNER 4,1931; BK, 348; IB, 326
F. Brabec, Ber. bot. Ges. 68a (1955), 27-32

WINKLER, LAJOS WILHELM [1863-1939]
DSB 14,447-448; MEL 2,1047-1048; POGG 4,1651; 5,1381; 6,2903-2804; 7b,6043-6044
E. Schulek, Talanta 10 (1963), 423-428

WINKLER von MOHRENFELS, KARL WOLF [1820-1888]
GGTA 6 (1912), 1038

WINOGRADSKY, SERGEI NIKOLAEVICH [1856-1953]
DSB 14,36-38; BSE 8,123-124; BME 5,412-413; KUZENTSOV(1963), 274-287; SKOROKHODOV, 252-256; SACKMANN, 364-365; DOETSCH, 132-134; WWWS,1724
V.L. Omelianski, Arkh. Biol. Nauk 27 (1927), 11-36
S.A. Waksman, Soil Science 62 (1946), 197-226; Science 118 (1953), 36-

37;
Sergei N. Winogradsky, his Life and Work. New Brunswick, N.J. 1953
H.G. Thornton, Obit. Not. Fell. Roy. Soc. 8 (1953), 635-644
R. Courrier, Not. Acad. Sci. 3 (1957), 677-713
V.N. Gutina, Vop. Ist. Est. Tekh. 1982(1), pp.117-123
G.A. Zaparzin, Priroda 1985(2), pp.71-85

WINSSINGER, CAMILLE [-1923]
RSC 13,665; NUC 668,343

WINSTEIN, SAUL [1912-1969]
DSB 18,994-996; AMS 11,5919; MILES, 514-515; MSE 3,338-339; WWWS, 1811
Anon., Chem. Eng. News 47(51) (1969),73; Nachr. Chem. Techn. 18 (1970),7
P.D. Bartlett, J. Am. Chem. Soc. 94 (1972), 2161-2170
A. Streitweiser, Progress in Physical Organic Chemistry 9 (1972), 1-24
W.G. Young and D.J. Cram, Biog. Mem. Nat. Acad. Sci. 43 (1973), 321-353

WINTER, ERNST [1874-]
JV 15,217; RSC 19,74; NUC 668,406

WINTERBERG, HEINRICH [1867-1929]
FISCHER 2,1695; STUMPF, 106-112
T. Scherrer, Wiener klin. Wchschr. 42 (1929), 154-155

WINTERFELD, KARL ANTON [1891-1971]
KURSCHNER 10,2720; HEIN-E, 455-456; TSCHIRCH, 1147; WWWS, 1812; POTSCH, 458; POGG 6,2905; 7a(4),1030-1031
M. Grünthal, Deutsche Apoth. Z. 101 (1961), 1599-1600
Anon., Chem. Z. 96 (1972), 42

WINTERNITZ, HUGO [1868-1934]
FISCHER 2,1695; GRUETTER, 130; RSC 19,667
W. Zinn, Deutsche med. Wchschr. 60 (1934), 1731

WINTERNITZ, MILTON CHARLES [1885-1959]
AMS 10,4468; DAMB, 812-813; KOREN, 263; WWAH, 673
A.A. Liebow and L.L. Waters, Yale J. Biol. Med. 32 (1959), 143-172
L.L. Waters, Science 111 (1960), 1029-1030
J.R. Paul, Yale J. Biol. Med. 43 (1970), 110-119

WINTERNITZ, RUDOLF [1859-1933?]
KURSCHNER 4,3297; FISCHER 2,1696; RUSTLER, 65-71; KOERTING, 248; KOREN, 263; WININGER 7,496-497

WINTERNITZ, WILHELM [1835-1917]
DBJ 2,678; HIRSCH 5,965; WININGER 6,291

WINTERSTEIN, ALFRED [1899-1960]
KURSCHNER 8,2610; POGG 6,2905; 7a(4),1032-1034
P.M. Kunz, Thromb. Diath. Haemorrh. 5 (1960), 145-147
Anon., Nachr. Chem. Techn. 8 (1960), 301

WINTERSTEIN, ERNST [1865-1949]
TLK 3,715; IB, 326; POGG 4,1652; 6,2905; 7a(4),1034
E. Crasemann, Viert. Nat. Ges. Zürich 94 (1949), 268-269

WINTERSTEIN, HANS [1879-1963]
KURSCHNER 9,2291-2292; FISCHER 2,1696; BOHM, 61-62; HAYMAKER, 307-311; BHDE 2,1251-1252; WIDMANN, 80-81,292; LDGS, 80; BK, 348; IB, 326; KOREN, 263; WININGER 6,292; 7,497; TETZLAFF, 358
H.H. Weber and H.H. Loeschke, Erg. Physiol. 55 (1964), 1-27

WINTERSTEINER, OSCAR [1898-1971]
AMS 11,5923; MH 2,612-614; MSE 3,39-341; WWAH, 673-674; WWWS, 1812; POGG 6,2905; 7b,6044-6046
Anon., BioScience 16 (1966), 695

WINTON, FRANK ROBERT [1894-1985]
WW 1982, p.2412; IB, 326

WINTREBERT, PAUL MARIE JOSEPH [1867-1966]
CHARLE(2), 265-266; BUICAN, 171-180; IB, 326
P.R. Grasse, C.R. Acad. Sci. 263 (1966), 152-157

WINTROBE, MAXWELL MYER [1901-1986]
AMS 14,5586; KOREN, 264
M.M. Wintrobe, Hematology, the Blossoming of a Science. Philadelphia 1985
W.N. Valentine, Biog. Mem. Nat. Acad. Sci. 59 (1990), 447-472

WINZLER, RICHARD JOHN [1914-1972]
AMS 11,5924; WWWS, 1812-1813
E. Frieden, Cancer Research 33 (1973), 436-437

WIRBATZ, WILHELM [1895-]
JV 39,518

WIRSING, (ANDREAS) FRIEDRICH [1869-1934]
JV 8,164; NUC 668,662
Anon., Chem. Z. 58 (1934), 169

WIRTH, ERNST [1859-1927]
RSC 13,234
Anon., Chem. Z. 51 (1927), 541; Z. angew. Chem. 40 (1927), 730

WIRTH, JOSEPH [1878-1937]
KALLMORGEN, 452

WIRTH, THEODOR [1883-1960]
NUC 669,14

WIRTH, WOLFGANG [1898-]
KURSCHNER 13,4317; BOCK, 106-109; POGG 7a(4),1035-1036

WIRTHLE, FERDINAND [1873-1936]
JV 5,82; RSC 19,669-670; NUC 669,17
Anon., Chem. Z. 35 (1911), 52 (1928), 827

WISCHIN, CARL [1871-]
JV 8,73; RSC 17,342; NUC 669,64

WISE, LOUIS ELSBERG [1888-1980]
AMS 11,5928; POGG 6,2910; 7b,6054-6056
E.C. John, J. Polymer Sci. C36 (1971), iii-v

WISHART, GEORGE MACFEAT [1895-1958]
WW 1958, p.3299
R.C. Garry, Roy. Soc. Edin. Year Book 1959-1960, pp.77-78

WISLICENUS, JOHANNES [1835-1902]
DSB 14,454-455; BJN 7,126; LF 2,500-512; LEIPZIG 2,37-49; ZURICH-D, 112; BLOKH 2,791-792; SCHAEDLER, 155-156; TSCHIRCH, 1148; W, 565-566; POTSCH, 458-459; WWWS, 1813; POGG 2,1342; 3,1455; 4,1653-1654; 7aSuppl.,776-777
C. von Voit, Sitz. Bayer. Akad. 33 (1903), 539-550
J. Biehringer, Naturw. Rund. 18 (1903), 192-194,204-207
E. Beckmann, Ber. chem. Ges. 37 (1904), 4861-4946
W. Ostwald, Abhandlungen und Vorträge, pp.444-445. Leipzig 1904
W.H. Perkin, J. Chem. Soc. 87 (1905), 501-534
P.F.F., Proc. Roy. Soc. A78 (1907), iii-xii
N.W. Fisher, in van't Hoff-LeBel Centennial (O.B. Ramsey, Ed.), pp.33-54. Washington, D.C. 1975

WISLICENUS, WILHELM [1861-1922]
DBJ 4,374-375; BLOKH 2,792-793; DZ, 1585; TLK 2,770; POGG 4,1654; 5,1383-1384; 6,2911
R. Weinland, Ber. chem. Ges. 55A (1922), 120-123

WISS, OSWALD [1916-]
KURSCHNER 13,4319; POGG 7a(4),1038-1040

WISTINGHAUSEN, KARL von [1826-1883]
WELDING, 872

J. Brennsohn, Die Aerzte Estlands, p.368. Riga 1922

WITEBSKY, ERNEST [1901-1969]
AMS 11,5930; WWA 1964-1965, p.2195; BHDE 2,1252; LDGS, 58; WWWS, 1814
N.R. Rose, Immunology Today 12 (1991), 167-168

WITKOP, BERNHARD [1917-]
AMS 15(7),688; KURSCHNER 13,4322; WWA 1980-1981, p.3565; IWW 1982-1983, pp.1421-1422; POTSCH, 459; POGG 7a(4), 1041-1042
B. Witkop, Heterocycles 20 (1983), 2059-2075; Medicinal Research Reviews 12 (1992), 275-276

WITSCHI, EMIL [1890-1971]
AMS 11,5932; WWAH, 675; WWWS, 1814
A. Gorbman, American Zoologist 19 (1979), 1261-1270

WITT, DAN HITER [1890-1942]
Anon., J. Am. Med. Assn. 119 (1942), 97

WITT, OTTO NIKOLAUS [1853-1915]
DBJ 1,345; BLOKH 2,793-794; ZURICH-D, 125; MATSCHOSS, 297-298; DZ, 1587; WREDE, 215-216; WWWS, 1814-1815; POTSCH, 459-460; POGG 3,1456 4,1656-1657; 5,1384; 7aSuppl.,777-778
E. Noelting, Chem. Z. 39 (1915), 441-449; J. Chem. Soc. 109 (1916), 428-431; Ber. chem. Ges. 49 (1916), 1751-1832
J. d'Ans, Chem. Z. 77 (1953), 279-281

WITTE, FRIEDRICH [1829-1893]
HEIN, 759-760; POGG 7aSuppl.,778
J. Ohage, Mecklenburgische Monatshefte 7 (1931), 479-483
W. Schneider, Deutsche Apoth. Z. 96 (1956), 944-945
M. Hamann, Pharmazie 1959(7), pp.5-25

WITTER, HUGO [1862-1908]
JV 7,228; RSC 13,277,890
Anon., Chem. Z. 32 (1908), 272

WITTHAUS, RUDOLF AUGUST [1846-1915]
DAB 20,439; NCAB 11,60-61; MILES, 517-518; FISCHER 2,1698; DAMB, 815-816; KELLY, 1320-1321; WWAH, 675; WWWS, 1815
W. Coleman et al., Bioch. Bull. 5 (1916), 216-217
Anon., Science 43 (1916), 527-528
F.L. Kozelka, Ciba Symposia 11 (1950-1951), 1305-1312

WITTICH, WILHELM HEINRICH von [1821-1884]
ADB 43,638; HIRSCH 5,974-975; PAGEL, 1869; POGG 3,1456
L. Hermann, Berl. klin. Wchschr. 22 (1885), 207-208,222-224

WITTIG, GEORG [1897-1987]
KURSCHNER 13,4328; IWW 1982-1983, p.1422; AUERBACH, 932-933; TLK 3,718; MEINEL, 311-315,515; SCHMITZ, 303-304; STC 3,185-187; MSE 3,341-342; POTSCH, 460; ABBOTT-C, 148-149; POGG 6,2912; 7a(4),1047-1048
R.E. Oesper, J. Chem. Ed. 31 (1954), 357-358
Anon., Chem. Z. 81 (1957), 405; 111 (1987), 316
U. Schöllkopf, Chemie in Unserer Zeit 1 (1961), 1047-1049
R.A. Shaw, Nature 282 (1979), 231-232; Chem. Brit. 24 (1988), 63
E. Vedejs, Science 207 (1980), 42-44
J.J. Eisch, Journal of Organometallic Chemistry 356 (1988), 271-283

WITTMACK, LUDWIG [1839-1929]
WAGENITZ, 198; WREDE, 216-217; TSCHIRCH, 1148; IB,317; TLK 3,718; RSC 11,834; 19,678-679
H. Harms, Ber. bot. Ges. 49 (1931), (200)-(219)

WITTMANN, HEINZ GÜNTHER [1927-1990]
POTSCH, 460
Wer ist Wer 1988-1989, p.1496
I.G. Wool, TIBS 15 (1990), 332

WITTSTEIN, GEORG CHRISTIAN [1810-1887]
HEIN, 761; TSCHIRCH, 1148; FERCHL, 584-585; RSC 6,406-407; 8,1257-1258; 11,834-835; 12,789-790; POGG 2,1345-1346; 3,1456-1457; 7aSuppl.,779
Anon., Leopoldina 23 (1887), 114; Pharm. Z. 32 (1887), 319-320; Chem. Z. 11 (1887), 711; Annals of Botany 1 (1887-1888), 415
A. Tschirch, Arch. Pharm. 226 (1888), 1-29

WITTSTOCK, CHRISTIAN [1791-1867]
HEIN, 762; TSCHIRCH, 1148; FERCHL, 585; CALLISEN 21,302-303; 33,325; POGG 3,1457
V. Weber, Arch. Pharm. 180 (1867), 193-200
J. Berendes, Das Apothekerwesen, p.95. Stuttgart 1907

WITZEL, HERBERT [1924-]
KURSCHNER 13,4332; AUERBACH, 933

WITZEMANN, EDGAR JOHN [1884-1947]
AMS 7,1966; IB, 327; WWAH, 675; POGG 6,2914; 7b,6061-6062

WIZINGER (WIZINGER-AUST), ROBERT [1896-1973]
KURSCHNER 11,3325; BHDE 2,1255; WWWS, 1815; POGG 6,2914; 7a(4),1054
M. Coenen, Chem. Z. 80 (1956), 287-288
Anon., Nachr. Chem. Techn. 9 (1961), 177; Chem. Z. 97 (1973), 393; Chimia 27 (1973), 302-303
W. Jenny, Chimia 20 (1966), 265-266

WLEÜGEL, SEVERIN SEGELCKE [1852-]
RSC 11,836; 19,681
Hver er Hvem? 1912, p.287
Who's Who of Science International 1914, p.609

WÖHLER, FRIEDRICH [1800-1882]
DSB 14,474-479; ADB 43,711-717; LKW 3,410-420; BUGGE 2,31-52; W, 566-567; KALLMORGEN, 453; BLOKH 2,795-798; PRANDTL, 135-192; FARBER, 507-520; MATSCHOSS, 298-299; FERCHL, 585-587; PHILIPPE, 766-768; WWWS, 1816; TSCHIRCH, 1148; SCHAEDLER, 156-157; SCHELENZ, 666-667; ABBOTT-C,149; POTSCH, 460-461; POGG 2,1348-1352; 3,1458; 7aSuppl.,779-783
A.W. Hofmann, Ber. chem. Ges. 15 (1882), 3127-3190; 23(3) (1890),833-848
C. von Voit, Sitz. Bayer. Akad. 13 (1883), 231-242
Anon., Proc. Roy. Soc. 35 (1883), xii-xx
W.H. Warren and E.F. Smith, J. Chem. Ed. 5 (1928), 1539-1557
F.G. Hopkins, Biochem. J. 22 (1928), 1341-1348
B. Lepsius, Ber. chem. Ges. 65A (1932), 89-94
H.S. van Klooster, J. Chem. Ed. 21 (1944), 158-170
J. Valentin, Friedrich Wöhler. Stuttgart 1949
E.H. Huntress, Proc. Am. Acad. Arts Sci. 78 (1950), 25-26
H. Wolter, Chem. Z. 82 (1958), 419-422
R. Keen, The Life and Work of Friedrich Wöhler. London 1976
G. Schwedt, Naturw. Rund. 35 (1982), 406-413
H. Teichmann, Z. Chem. 23 (1983), 125-136
R. Schmitz, Sudhoffs Arch. Beihefte 24 (1984), 105-111

WÖHLER, LOTHAR [1870-1952]
KURSCHNER 7,2299; WOLF, 232; TLK 3,718; POGG 5,1385-1386; 6,2914-2915; 7a(4),1056-1057
Anon., Chem. Z. 74 (1950), 696

WÖHLISCH, EDGAR [1890-1960]
KURSCHNER 9,2304; DRULL, 303; FISCHER 2,1700; EBERT, 143-156; KIEL, 120; POGG 6,2915; 7a(4),1057-1058
R. Marx, Blut 7 (1961), 115-116

WÖLFL, VALENTIN [1879-1932]
JV 22,525; NUC 670,438
Anon., Z. angew. Chem. 45 (1932), 230

WOHL, ALFRED [1863-1939]
KURSCHNER 4,3315-3316; BHDE 2,1266; TLK 3,718; POGG

4,1660; 5,1386; 6,2916; 7a(4),1060
C. Neuberg, Ber. chem. Ges. 66A (1933), 78-79
Anon., Nature 145 (1940), 290

WOHLGEMUTH, JULIUS [1874-1948]
FISCHER 2,1701; BHDE 2,1256; LDGS, 75; KOREN, 264; IB, 327; POGG 6,2916; 7a(4),1061
Anon., Z. angew. Chem. 60 (1948), 288; Deutsche med. Wchschr. 73 (1948), 420

WOHLWILL, FRIEDRICH [1881-1958]
KURSCHNER 8,2626; BHDE 2,1256; LDGS, 76; IB, 327
A. Schuback, Verhandl. path. Ges. 44 (1960), 360-363

WOKER, GERTRUD [1878-1968]
SBA 6,123-124; TLK 3,719; IB, 327; POGG 5,1387; 6,2916-2017; 7a(4),1062-1063
G. Woker, in Führende Frauen in Europa (H. Kern, Ed.), pp.138-169. Munich 1928
R.E. Oesper, J. Chem. Ed. 30 (1953), 435-437
H. Schaltegger, Verhandl. Schw. Ges. Aarau 149 (1969), 300-302

WOKES, FRANK [1892-1974]
Who's Who in British Science 1953, p.285
Anon., Chem. Brit. 11 (1975), 239

WOLBACH, SIMEON BURT [1880-1954]
AMS 8,2749; FISCHER 2,1702; DAMB, 816; SACKMANN, 366-367; KOREN, 264; KAGAN(JA), 301-302; WWAH, 675-676
C.A. Janeway, Trans. Assoc. Am. Phys. 67 (1954), 30-35
M.C. Sosman, J. Path. Bact. 68 (1954), 656-657

WOLD, FINN [1928-]
AMS 15(7),698; WWA 1980-1981, p.3568; WWWS, 1816

WOLF, ABRAHAM [1876-1948]
Who Was Who 1941-1950, p.1255

WOLF, FRITZ [1890-]
JV 29,632; NUC 670,683

WOLF, GEORGE [1922-]
AMS 15(7),700; BHDE 2,1257; WWWS, 1817

WOLF, KARL LOTHAR [1901-1969]
KURSCHNER 10,2742; KIEL, 169; POGG 6,2917-2918; 7a(4),1068-1069
Anon., Chem. Z. 93 (1969), 225

WOLFE, RALPH STONER [1921-]
AMS 17(7),728
R.S. Wolfe, Ann. Rev. Microbiol. 45 (1991), 1-35

WOLFES, OTTO [1877-1942]
POGG 7aSuppl.,785
Anon., Chem. Z. 66 (1942), 482; Chemische Technik 15 (1942), 234

WOLFF, ALFRED [1850-1916]
DBJ 1,372; FISCHER 2,1702; NUC 671,174

WOLFF, ALFRED [1877-1942?]
JV 16,293; NUC 671,174

WOLFF, EMIL THEODOR von [1818-1896]
ADB 55,115-117; BJN 1,100-101; 3,68*; BLOKH 2,799; SCHAEDLER, 157-158; FERCHL, 587; KLEIN, 137; POGG 2,1360-1361; 3,1462
Anon., Leopoldina 32 (1896), 190; Chem. Z. 20 (1896), 1038

WOLFF, ÉTIENNE [1904-]
QQ 1979-1980, p.1601; IWW 1982-1983, p.1423; STC 3,187-188; WWWS, 1818
E. Wolff, Trois Pattes pour un Canard. Souvenirs d'un Biologiste. Paris 1990

WOLFF, FRITZ [1866-1931]
JV 12,248; RSC 15,809; 16,377
Anon., Chem. Z. 55 (1931), 704

WOLFF, HANS [1853-1891]
STRAHLMANN, 473,481

WOLFF, JULES [1878-1940]
NUC 671,269; POGG 5,1389

WOLFF, JULIUS AUGUST [1830-1898]
DGB 83 (1935), 625; RSC 6,427; 8,1264

WOLFF, LUDWIG [1857-1919]
BLOKH 2,799; POTSCH, 461; POGG 4,1664-1665; 5,1389-1390; 6,2923
L. Knorr, Ber. chem. Ges. 52A (1919), 67-68
R. Krüche, Chem. Z. 43 (1919), 121
W. Meckdenburg and W. Schneider, Ber. chem. Ges. 62A (1929), 145-159

WOLFF, LUDWIG KARL [1879-1938]
LINDEBOOM, 2202-2203; POGG 6,2923; 7b,6065-6066
A. Emmerie, Voeding 14 (1953), 397-398

WOLFF, PAUL [1894-]
KURSCHNER 4,3330-3331; FISCHER 2,1703; ASEN, 221; LDGS, 79; IB, 328; NUC 671,309-310

WOLFF-EISNER, ALFRED [1877-1948]
KURSCHNER 5,1563; FISCHER 2,1703; LDGS, 58; WININGER 6,317; TETZLAFF,362
Anon., Lancet 1948(II), pp.793-794
P. Voswinckel, Allergologie 11(2) (1986), 41-46

WOLFFBERG, SIEGFRIED [1853-]
FISCHER 2,1704; PAGEL, 1874-1875; KUKULA, 970-971; WENIG, 671; RSC 8,1265; 11,841; NUC 671,342

WOLFFENSTEIN, RICHARD [1864-1929]
TLK 3,722; POTSCH, 462; POGG 4,1665; 5,1390; 6,2924
H. Scheibler, Z. angew. Chem. 42 (1929), 1149-1151

WOLFFHÜGEL, GUSTAV [1845-1899]
BJN 4,189*; BLOKH 2,799
K. von Buchka, Ber. chem. Ges. 32 (1899), 299-303

WOLFROM, MELVILLE LAWRENCE [1900-1969]
DSB 18,997-999; AMS 11,5950; MILES, 519-520; MH 2,614-616; MSE 3,343-344; WWAH, 676; WWWS, 1819; POGG 6,2924-2925; 7b,6067-6072
D. Horton and W.Z. Hassid, Biog. Mem. Nat. Acad. Sci. 47 (1975), 487-549
D. Horton, Adv. Carbohydrate Chem. 26 (1977), 1-47

WOLLASTON, WILLIAM HYDE [1766-1828]
DSB 14,486-494; DNB 21,782-787; HIRSCH 5,987; BLOKH 2,800-801; W, 568; SCHAEDLER, 158-159; FERCHL, 588; TSCHIRCH, 1149; ABBOTT-C, 149-150; POTSCH, 461-462; COLE, 580; WWWS, 1819; CALLISEN 33,341; POGG 2,1362-1363
H.G. Wayling, Science Progress 22 (1927), 81-95
E.G. Ferguson, J. Chem. Ed. 18 (1941), 3-7
P.T. Hinde, J. Chem. Ed. 43 (1966), 673-676
R.H. Cragg, Chem. Brit. 2 (1966), 525-527
D.C. Goodman, Hist. Stud. Phys. Sci. 1 (1969), 37-59

WOLLMAN, ÉLIE [1917-]
QQ 1988-1989, p.1580

WOLLMAN, ELISABETH [1888-1943]
A.P. Waterson and L. Wilkinson, An Introduction to the History of Virology, p.209. Cambridge 1978

WOLLMAN, EUGÈNE [1883-1943]
HEIM, 133-145; IB,328
P. Nicolle, Ann. Inst. Pasteur 72 (1946), 855-858

WOLLMAN, SEYMOUR HORACE [1915-]
AMS 15(7),712

WOLVEKAMP, HENDRIK PIETER [1904-1980]
IB, 328
Wie is Dat? 1956, p.688
M. Jeuken, Vakblad voor Biologen 80 (1980), 128

WOOD, ALEXANDER [1817-1884]
DNB 21,818-819; HIRSCH 5,990-991; WWWS, 1820
N. Howard-Jones, J. Hist. Med. 2 (1947), 201-249

WOOD, EDWARD STICKNEY [1846-1905]
DAB 20,455-456; MILES, 520-521; ELLIOTT, 280
J.C. Warren, Proc. Am. Acad. Arts Sci. 51 (1916), 929-930

WOOD, GEORGE EDWARD CARTWRIGHT [1864-1941]
O'CONNOR(2), 447-448; RSC 19,700
Anon., Brit. Med. J. 1942(I), 203-204

[illegible], HARLAND GOFF [1907-1991]
[illegible]S 14,5621; WWA 1978-1979, p.3524; IWW 1982-1983, p.1426; MH 2,616-618; STC 3,188-190; MSE 3,345-346; WWWS, 1820
H.G. Wood, in The Molecular Basis of Biological Transport (J.F. Woessner, Ed.), pp.1-54. New York 1972; Reflections on Biochemistry (A. Kornberg et al., Eds.), pp.105-115. Oxford 1976; Of Oxygen, Fuels and Living Matter, Part 1 (G. Semenza, Ed.), pp.173-250. Chichester 1982; Ann. Rev. Biochem. 54 (1985), 1-41
D.A. Goldthwait, The FASEB Journal 1 (1987), 259-261; TIBS 17 (1992), 52-53
R.W. Hanson, The FASEB Journal 5 (1991), 3015-3017
L. Jaenicke, Z. physiol. Chem. 373 (1992), (5)-(7)

WOOD, HORATIO C [1841-1920]
DSB 14,495-497; DAB 20,459-460; NCAB 13,569-570; FISCHER 2,1706; DAMB, 820-821; PARASCANDOLA, 58-59
G.E. de Schweinitz, Trans. Coll. Phys. Phila. [3]42 (1920), 155-186,235-241
H. Skinner, Entomol. News 31 (1920), 115-117
H. Gunthorp, Canadian Entomologist 52 (1920), 112-114
G.B. Roth, Isis 30 (1939), 37-45; Biog. Mem. Nat. Acad. Sci. 33 (1959), 462-484

WOOD, HORATIO CHARLES [1874-1958]
AMS 9(II),1248; CHEN, 9-10; IB, 328; WWAH, 677-678

WOOD, JOSEPH GARNETT [1900-1959]
SRS, 61; NUC 672,313

WOOD, THOMAS BARLOW [1869-1929]
E.J. Russell, Biochem. J. 24 (1930), 1-3

WOOD, WILLIAM BARRY, Jr. [1910-1971]
AMS 11,5964; MH 2,618; ICC, 398-401; DAMB, 822-823; WWWS, 1821
R.C. Tilghman, Johns Hopkins Med. J. 129 (1971), 111-120
J.G. Hirsch, Biog. Mem. Nat. Acad. Sci. 51 (1980), 387-418

WOODALL, CHARLES SIMPSON [1893-1939]
AMS 6,1571

WOODGER, JOSEPH HENRY [1894-1981]
WW 1981, pp.2843-2844; IB, 327; ABBOTT, 138-139
M. Woodger, in Form and Strategy in Science (J.R. Gregg, Ed.), pp.473-476. Dordrecht 1961
K. Popper, Brit. J. Phil. Sci. 32 (1981), 328-330
N. Roll-Hansen, J. Hist. Biol. 17 (1984), 399-428

WOODMAN, HERBERT ERNEST [1889-1956]
SRS, 47; NUC 672,622-623
R.E. Evans, Nature 177 (1956), 648-649

WOODRUFF, HAROLD BOYD [1917-]
AMS 15(7),733
H.B. Woodruff, Ann. Rev. Microbiol. 35 (1981), 1-28

WOODS, ALBERT FREDERICK [1866-1948]
AMS 7,1977; NCAB B,467-468; 46,457-458
F.V. Rand, J. Wash. Acad. Sci. 39 (1949), 313-315

WOODS, DONALD DEVEREUX [1912-1964]
DNB [1961-1970], 1102-1103; WW 1959, p.3325
F.E. Dixon, Nature 205 (1965), 447-448
E.F. Gale and P. Fildes, Biog. Mem. Fell. Roy. Soc. 11 (1965), 203-219

WOODWARD, ROBERT BURNS [1917-1979]
AMS 13,4946; WWA 1978-1979, p.3543; MH 1,538-540; STC 3,190-191; MSE 3,347-348; CB 1952, pp.647-649; ABBOTT-C, 150; WWWS, 1823; POTSCH, 462
H. Wasserman and D. Dolphin, Heterocycles 7 (1977), 1-35
J. Gostelli, Chimia 33 (1979), 348-349
G. Stork, Nature 284 (1980), 383-384
W.D. Ollis, Chem. Brit. 16 (1980), 210-216
A.R. Todd and J. Cornforth, Biog. Mem. Fell. Roy. Soc. 27 (1981), 629-695
D.M.S. Wheeler, Chemie in Unserer Zeit 18 (1984), 109-119

WOODYATT, ROLLIN TURNER [1878-1953]
AMS 6,1576; FISCHER 2,1707; WWAH, 680
H.F. Root, Trans. Am. Clin. Assn. 66 (1954), liv-lv
R.W. Keeton, Trans. Assoc. Am. Phys. 67 (1954), 36-38
E.A. Graham, Quarterly Journal Northwestern Medical School 30 (1956), 286-289

WOOLDRIDGE, LEONARD CHARLES [1857-1889]
WILKS, 313-316; O'CONNOR, 212-214
Anon., Guy's Hosp. Rep. 46 (1889), xxxv-xlii; Lancet 1889(I), 1281-1282
V. Horsley, in L.C. Wooldridge, On the Chemistry of the Blood etc., pp.1-41. London 1893

WOOLDRIDGE, WALTER REGINALD [1900-1966]
WW 1965, p.3364
J. Clabby, Chem. Brit. 3 (1967), 223

WOOLF, BARNET [1902-]
SRS, 87

WOOLLARD, HERBERT HENRY [1889-1939]
WW 1939, p.3490
Anon., St. Barts. Hosp. Rep. 72 (1939), 11-15
D. de Lange, Bio-Morphosis 1 (1939), 417-419
W.E. LeGros Clark, Obit. Not. Fell. Roy. Soc. 3 (1940), 89-95

WOOLLEY, DILWORTH WAYNE [1914-1966]
AMS 10,4508; WWAH, 680; WWWS, 1824
E.N. Todhunter, J. Am. Diet. Assn. 51 (1967), 472
J.M. Stewart, Pharmacologist 9 (1967), 28-30
T.H. Jukes, J. Nutrition 104 (1974), 509-511
D.A. Roe, J. Nutrition 117 (1987), 1324

WORDEN, ALASTAIR NORMAN [1916-1987]
WW 1981, p.2852

WORK, ELIZABETH EDGAR [1912-]
Who's Who of British Scientists 1971-1972, p.929

WORK, THOMAS [1912-]
Who's Who of British Scientists 1971-1972, p.929

WORM-MÜLLER, JACOB [1834-1889]
HIRSCH 5,996-997; PAGEL, 1879-1880; POGG 3,1465; 4,1669
F. Kaier, Norges Laeger, p.297. Christiania 1887
Anon., Chem. Z. 13 (1889), 1723
W. Stricker, Virchows Arch. 119 (1890), 372

WORMALL, ARTHUR [1900-1964]
DNB [1961-1970], 1110-1111; WW 1959, p.3335
W.T. Morgan and G.E. Francis, Biog. Mem. Fell. Roy. Soc. 12 (1966), 543-564

WORMLEY, THEODORE GEORGE [1826-1897]
DAB 20,535-536; NCAB 13,104; MILES, 525-526; HIRSCH 5,997; TSCHIRCH,1149; KELLY, 1335-1336; SILLIMAN, 128-129; WWAH, 681; POGG 3,1465; 4,1659
E.F. Smith, J. Am. Chem. Soc. 19 (1897), 275-279
J. Ashhurst, Trans. Coll. Phys. Phila. [3]19 (1897), lxxix-lxxxviii
W.M. Neven, J. Chem. Ed. 25 (1948), 182-186
B. Rosenfeld, Microchem. J. 3 (1959), 135-136

WORTMANN, JULIUS [1856-1924]
NB, 435; BK, 352-354; TSCHIRCH, 1149
E. Muth, Ber. bot. Ges. 43 (1925), (112)-(142)

WOTIZ, HERBERT HENRY [1922-]
AMS 15(7),748; BHDE 2,1268; WWWS, 1826

WRANGELL, MARGARETHE (DAISY) von [1876-1932]
KURSCHNER 4,3341-3342; WELDING, 889; BOEDEKER, 22; IB, 329; JV, 25,713; POGG 6,2931-2932; 7a(4),1077
A. Mayer, Naturwiss. 20 (1932), 322-324
E. Wedekind, Chem. Z. 56 (1932), 401-402
M. Bodenstein, Ber. chem. Ges. 65A (1932), 95
W. Andronikow, Margarethe von Wrangell, das Leben einer Frau. Munich 1935

WRANÝ, ADALBERT [1836-1902]
ROTH, 152-157; NAVRATIL, 361-362
O. Matousek, Cas. Lek. Cesk. 91 (1952), 1079
J. Kops, Zdrav. Prac. 25 (1975), 740-741

WREDE, FRANZ [1877-1946]
DGB 185 (1981), 393; JV 18,30; NUC 675,24

WREDE, FRITZ [1891-]
KURSCHNER 5,1568; TLK 3,724; IB, 329; POGG 6,2932-2933; 7a(4),1077-1078
F.A. Hoppe-Seyler, in 100 Jahre Medizinische Forschung in Greifswald, pp.69-71. Greifswald 1938

WREDEN, FELIKS ROMANOVICH [1841-1878]
BSE 9,250; WELDING, 891; BLOKH 2,802; RSC 8,1275; 11,851; POGG 2,1467
N. Menshutkin, Zhur. Fiz. Khim. 11 (1879), xii-xiv

WREDEN, ROBERT ROBERTOVICH (ROMAN ROMANOVICH) [1867-1934]
BSE 9,249-250; WELDING, 892; FISCHER 2,1708; VENGEROV 1,153; RSC 19,720
M.I. Kuslik, Vest. Khir. 75(5) (1955), 130-134

WREN, HENRY [1881-1955]
POGG 5,1393; 6,2933; 7b,6089-6090
F. Bell, J. Chem. Soc. 1955, pp.3567-3568
Anon., Chem. Ind. 1955, p.570

WRIGHT, ALMROTH [1861-1947]
DSB 14,511-513; DNB [1941-1950],976-977; O'CONNOR(2), 215-217; BULLOCH, 404; ABBOTT, 139; IB, 329; WWWS, 1826
L. Colebrook, Obit. Not. Fell. Roy. Soc. 17 (1948), 297-313; Almroth Wright. London 1954
Z. Cope, Almroth Wright, Founder of Modern Vaccine Therapy. London 1966

WRIGHT, BARBARA EVELYN [1926-]
AMS 15(7),750; WWWS, 1826

WRIGHT, CHARLES ROMLEY ALDER [1844-1894]
TSCHIRCH, 1149; NUC 675,146; POGG 3,1467-1468; 4,1671-1672
Anon., Analyst 19 (1894), 193-194; Ber. chem. Ges. 27 (1894),2654-2655; J. Chem. Soc. 67 (1895), 1113-1115
T.E.T., Proc. Roy. Soc. 57 (1895), v-vii

WRIGHT, GEORGE F. [1904-1976]
AMS 13,4958; WWWS, 1827

WRIGHT, JAMES HOMER [1869-1928]
AMS 4,1089; FISCHER 2,1709; WWAH, 682
A.S. Warthin, Trans. Assoc. Am. Phys. 43 (1928), 9-10

WRIGHT, SEWALL [1889-1988]
AMS 14,5649; WWA 1976-1977, p.3456; IWW 1982-1983, p.1430; WWWS, 1828; MH 1,540-541; STC 3,191-192; MSE 3,348-349; IB, 329
J.F. Crow, Persp. Biol. Med. 25 (1982), 279-294; J. Hist. Biol. 23 (1990), 57-89
W.B. Provine, Sewall Wright and Evolutionary Biology. Chicago 1986
W.L. Russell, Am. Phil. Soc. Year Book 1989, pp.349-353
E.S. Russell, Ann. Rev. Gen. 23 (1989), 1-18
W.G. Hill, Biog. Mem. Fell. Roy. Soc. 36 (1990), 569-579
T. Park and W.L. Russell, Persp. Biol. Med. 34 (1991), 497-515

WRINCH, DOROTHY [1894-1976]
AMS 11,5995; CB 1947, pp.693-695; WWWS, 1828-1829; POGG 6,2935; 7b,6093-6095
D.C. Hodgkin and H. Jeffreys, Nature 260 (1976), 564
M.M. Julian, J. Chem. Ed. 61 (1984), 890-892

WRÓBLEWSKI, AUGUSTYN [1866-after 1913]
KONOPKA 12,362-364; RIGA, 724; SG [2]21,350; RSC 19,725-726
Z. Wojtaszek, in Studia z Dziejow Katedr Wydzialu Matematiki, Fiziki i Chemii Universitetu Jagiellonskiego (S. Gotaba, Ed.), p.175. Cracow 1964
A. Skrabek-Solowieska, Czas Udreki i Czas Radosci, pp.247-264. Wroclaw 1977
E. Ruzevich and I.Y. Grosvald, Ist. Est. Tekhn. Pribalt. 7 (1984), 87-88

WRÓBLEWSKI, EDUARD [see VROBLEVSKI]

WRÓBLEWSKI, ZYGMUNT FLORENTY [1845-1888]
DSB 14,522-523; WE 12,512; BLOKH 2,802-803; SCHAEDLER, 153; WWWS, 1741; W, 570; POGG 3,1468; 4,1672
E. Suess, Alm. Akad. Wiss. Wien 38 (1888), 190-192

WU, HSIEN [1893-1959]
DSB 14,523-524; AMS 9(II),1257; IB, 330; POGG 6,2935-2936; 7b,6095-6097
D.Y. Wu, Hsien Wu. Boston 1959; Chinese Medical Journal 6 (1959), 207-219
E.B. Carmichael, Nature 185 (1960), 809-810
C. Bishop, Clin. Chem. 28 (1982), 378-380
J.A. Anderson, The Study of Change: Chemistry in China 1840-1949, pp.140-148. Cambridge 1991

WÜNSCH, EDWIN [1881-1973]
JV 22,303

WÜNSCH, ERICH [1925-]
KURSCHNER 16,4165

WÜRTZ, ADOLF [1873-]
SG [2]21,358; NUC 676,139

WÜSTENFELD, RICHARD [1879-1943]
JV 18,180; NUC 676,178
Anon., Chem. Z. 67 (1943), 332

WULFF, OTTO [1885-]
JV 25,367

WULLING, FREDERICK JOHN [1866-1947]
AMS 7,1988; NCAB E,131; WWAH, 683
C.H. Rogers, Ind. Eng. Chem. News Ed. 13 (1935), 13
Anon., J. Am. Pharm. Assn. 8 (1947), 613

WULZ, PAUL [1867-1933]
JV 6,76; RSC 13,277; NUC 676,236

WUNDERLICH, KARL REINHOLD AUGUST [1815-1877]
ADB 44,313-314; SL 2,481-487; HIRSCH 5,1002-1004; PAGEL, 1881-1883; LEIPZIG 2,63-71; WWWS, 1829
O. Heubner and W. Roser, Archiv für Heilkunde 4 (1878), 289-339
A. Strumpell, Med. Klin. 11 (1915), 901-903
K. Sudhoff, Mitt. Gesch. Med. 14 (1915), 243-245
G. Sticker, Sudhoffs Arch. 32 (1939), 217-174; 33 (1940), 1-54
O. Temkin, Gesnerus 23 (1966), 188-195
H.M. Koelbing, Carl August Wunderlich, Wien und Paris. Berne 1974
I. Kastner, Z. ärzt. Fortbild. 84 (1990), 783-786

WURMSER, RENÉ [1890-]
QQ 1978-1979, pp.1602-1603; IWW 1982-1983, p.1431; IB, 330; POGG 6,2938-2939; 7b,6098-6100

WURSTER, CASIMIR [1854-1913]
BLOKH 2,803-804; ZURICH-D, 120; POTSCH, 462-463; POGG 4,1674; 5,1397
W. Will, Ber. chem. Ges. 47 (1914), 1
C.G. Schwalbe, Chem. Z. 38 (1914), 257

WURTZ, CHARLES ADOLPHE [1817-1884]
DSB 14,529-532; GE 31,1241-1242; LAROUSSE 15,1385-1386; BUGGE 2,115-123; BLOKH 2,804-808; HIRSCH 5,1006-1007; BOURQUELOT, 93-94; W, 570-571; FERCHL,590-591; TSCHIRCH,1149; SCHAEDLER,159-160; ABBOTT-C,150-151; POTSCH, 463; WWWS, 1829-1830; POGG 2,1374-1375; 3,1470-1471
M. Friedel, Bull. Soc. Chim. [2]43 (1885), i-lxxx
A.W.W., Proc. Roy. Soc. 38 (1885), xxiii-xxxiv
C. von Voit, Sitz. Bayer. Akad. 15 (1885), 153-160
W. Perkin, J. Chem. Soc. 47 (1885), 328-329
A.W. Hofmann, Ber. chem. Ges. 20 (1887), 815-996
A. Hanriot, Bull. Acad. Med. 86 (1921), 21-26
G. Urbain, Bull. Soc. Chim. [5]1 (1934), 1425-1447
E.C. Achard, Bull. Acad. Med. 120 (1938), 469-478
M. Durand, Rev. Gen. Sci. 75 (1968), 33-45
N. Hjelm-Hansen, Rev. Hist. Sci. 28 (1975), 259-265

WURZER, FERDINAND [1765-1844]
ADB 44,367; GUNDLACH, 462-463; MEINEL, 2-18,515-516; SCHMITZ, 221-230; FERCHL, 591; POGG 2,1377-1379; 7aSuppl.,785-786
P. Diergart, Sudhoffs Arch. 29 (1936), 113-122

WYATT, GERARD ROBERT [1925-]
AMS 15(7),766; WWA 1980-1981, p.3600; WWWS, 1830

WYCKOFF, RALPH WALTER GRAYSTONE [1897-]
AMS 14,5658; WWA 1976-1977, p.3460; IWW 1982-1983, p.1431; WWWS, 1830; MH 2,618-620; STC 3,192; MSE 3,351-352; McLACHLAN, 35-36; POGG 5,1397; 6,1940-2941; 7b,6105-6111
R.W.G. Wyckoff, TIBS 3 (1978), N104-N106

WYDLER, FERDINAND [1792-1854]
DHB 7,589; NUC 676,466

WYDLER, FERDINAND [1821-1873]
STELLING-MICHAUD 6,266

WYDLER, FRANZ WILHELM [1818-1877]
A. Wankmüller, Der Schweizer Familienforscher 33 (1966), 46-50

WYMAN, JEFFRIES [1814-1874]
DSB 14,532-534; DAB 20,583-584; NCAB 2,254-255; ELLIOTT, 283-284; DAMB, 828-829
Anon., Proc. Am. Acad. Arts Sci. NS10 (1875), 496-505
A.S. Packard, Biog. Mem. Nat. Acad. Sci. 2 (1886), 75-126
R.N. Doesch, J. Hist. Med. 17 (1962), 326-332
G.E. Gifford, J. Hist. Med. 22 (1967), 274-285; 23 (1968), 387-389
T.A. Appel, J. Hist. Biol. 21 (1988), 69-94

WYMAN, JEFFRIES [1901-]
AMS 14,5660; IWW 1982-1983, p.1432; WWWS, 1830
J.T. Edsall, Journal of Molecular Biology 108 (1976), 269-270; in Selected Topics in the History of Biochemistry (G. Semenza, Ed.). pp.99-195. Amsterdam 1986; Biophysical Chemistry 37 (1990), 7-14

WYNNE, ARTHUR MARSHALL [1892-1972]
AMS 11,6005; YOUNG, 22

WYRZYKOWSKI, JULIAN [1838-]
KOSMINSKI, 564; KONOPKA 12,431-436

WYZNIKIEWICZ (WYZNIKIEWICZ-TURCZYNOWICZ), WLADYSLAW [1865-1904]
RSC 17,483
Anon., Przeglad Weterynarski 19 (1904), 187-190
K. Milak, Kwart. Hist. Nauki 2 (1957), 304

Y

YABUTA, TEIJIRO [1888-1977]
JBE, 1842,2301; IB, 330; WWWS, 1831; POGG 6,2941; 7b,6114-6115
Anon., Japanese Journal of Antibiotics 30 (1977), 875-876

YAKIMOV, VASILI LARIONOVICH [1870-1940]
BSE 49,521; BME 35,1172; OLPP, 418-421; IB, 330
V.K. Pelevin, Med. Parazit. 9(3) (1940), 318-320

YALOW, ROSALIND SUSSMAN [1921-]
AMS 15(7),776; IWW 1982-1983, p.1436; POTSCH, 463

YAMAGAWA, MAKOTO [1883-]
NUC 677,598

YAMAGIWA, KATSUSABURO [1863-1930]
JBE, 1859; FISCHER 2,1713; OLPP, 421
K. Sugiura, J. Cancer Research 14 (1930), 568-569
F. Henschen, Gann 59 (1968), 447-451

YANG, PETER SUHSUN [1900-]
NUC 678,34

YANOFSKY, CHARLES [1925-]
AMS 15(7),784; WWA 1980-1981, p.3607; IWW 1982-1983, p.1438; WWWS, 1833; MH 2,621-622; STC 3,194-196; MSE 3,355-356

YARMOLINSKY, MICHAEL BEZALEL [1929-]
AMS 15(7),787; WWWS, 1833

YASUNOBU, KERRY TSUYOSHI [1925-]
AMS 15(7),788; WWA 1980-1981, p.3609; WWWS, 1834

YEMM, EDMUND WILLIAM [1909-]
WW 1981, p.2871

YERSIN, ALEXANDRE [1863-1943]
DSB 14,551-552; FISCHER 2,1713; PAGEL, 1887-1888; SACKMANN, 368-371; BULLOCH, 405; OLPP, 422; BOSSART, 79-

81; WWWS, 1835
A. Lacroix, C.R. Acad. Sci. 216 (1943), 361-364
G. Ramon, Bull. Acad. Med. 127 (1943), 382-388
N. Bernard, Ann. Inst. Pasteur 69 (1943), 129-134
Y. Mafart, Médecine Tropicale 25 (1965), 427-438
N. Howard-Jones, Persp. Biol. Med. 16 (1973), 292-307
H.H. Mollaret and J. Brosselat, Alexandre Yersin. Zurich 1987

YOFFEY, JOSEPH MENDEL [1902-]
WW 1991, p.2030; WWWS, 1836

YOSHIDA, HIKOROKURO [1859-1928]
POGG 4,1676

YOSHIDA, TOMIZO [1903-1973]
JBE, 1931
H. Druckrey, Arzneimitt. 23 (1973), 1008

YOSHIMOTO, MISAO [1882-]
FISCHER 2,1714; HAKUSHI 2,611

YOSHIMURA, KIYOHISA [1871-1958]
POGG 6,2944; 7b,6124-6125

YOUMANS, JOHN BARLOW [1893-1979]
AMS 11,6030
R.H. Kampmeier, Trans. Assoc. Am. Phys. 93 (1980), 25-27; J. Nutrition 116 (1986), 19-35

YOUNG, ELRID GORDON [1897-1976]
AMS 12,7114; SRS, 54; IB, 331; WWWS, 1837
C.C. Lucas, Proc. Roy. Soc. Canada [4]14 (1977), 125-128

YOUNG, FRANK GEORGE [1908-1988]
WW 1981, p.2876; IWW 1982-1983, p.1445
P.J. Randle, Chem. Brit. 25 (1989), 403; Biog. Mem. Fell. Roy. Soc. 36 (1990), 583-599

YOUNG, GEOFFREY TYNDALE [1915-]
Who's Who of British Scientists 1980-1981, p.546

YOUNG, JOHN RICHARDSON [1782-1804]
DAB 20,630; WWWS, 1838
L.F. Kebler, J. Chem. Ed. 17 (1940), 573-575
W.C. Rose, Introduction to J.R. Young, An Experimental Inquiry into the Principles of Nutrition etc., pp.v-xxvi. Urbana, Ill. 1958
D.G. Bates, Bull. Hist. Med. 36 (1962), 341-361

YOUNG, SYDNEY [1857-1937]
DSB 14,560-562; DNB [1931-1940],932-933; WWWS, 1838; POTSCH, 463-464; POGG 4,1676-1677; 5,1397-1400; 6,2946
F. Francis, J. Chem. Soc. 1937, pp.1332-1336
W.R.G. Atkins, Obit. Not. Fell. Roy. Soc. 2 (1938), 371-379
J. Timmermans, Endeavour 6 (1947), 11-14

YOUNG, THOMAS [1773-1829]
DSB 14,562-572; DNB 21,1308-1314; HIRSCH 5,1017; DESMOND, 682; W, 572; BRAZIER, 7-9,11; FERCHL, 592; SCHAEDLER, 161; WWWS, 1838-1839; POGG 2,1383-1385
H. Gurney, Memoir of the Life of Thomas Young. London 1831
H.S. Robinson, Medical Life 36 (1929), 527-540
A. Wood and F. Oldham, Thomas Young. Cambridge 1954
G.N. Cantor, Notes Roy. Soc. 25 (1970), 87-112
Y. Oldham, Brit. Med. J. 1974(IV), pp.150-152
S. Behrman, Clio Medica 10 (1975), 277-284

YOUNG, WILLIAM CALDWELL [1899-1965]
AMS 10,4552; WWWS, 1839
R.W. Goy, Anat. Rec. 157 (1967), 3-11

YOUNG, WILLIAM GOULD [1902-1980]
AMS 14,5700; WWWS, 1839
Anon., Chem. Eng. News 58(38) (1980), 43

YOUNG, WILLIAM JOHN [1878-1942]
O'CONNOR(2), 502-503; POTSCH, 464
I. Smedley-MacLean, Nature 149 (1942), 725; Biochem. J. 37 (1943),165-166
D. Maxwell, J. Chem. Soc. 1943, pp.44-45

YOUNGBURG, GUY EDGAR [1884-1952]
AMS 8,2794YUDKIN, JOHN [1910-]
WW 1981, p.2882; COHEN, 260

YVON, PAUL [1848-1913]
GORIS, 600-602; FISCHER 2,1715
L. Martin, C.R. Soc. Biol. 74 (1913), 905-907
J. Courtois, in Figures Pharmaceutiques Françaises, pp.179-184. Paris 1953

Z

ZABIN, IRVING [1919-]
AMS 15(7),831

ZACH, KARL [1888-1968]
JV 27,60; NUC 680,630
Anon., Chem. Z. 46 (1922), 918

ZACHARIAS, EDUARD [1852-1911]
BJN 16,87*; TSCHIRCH, 1149
C. Brick, Ber. bot. Ges. 29 (1911), (26)-(48)
Anon., Chem. Z. 35 (1911), 385

ZACHARIAS, GOTTHARD [1886-1918]
H. Vogt, Jahresb. Fort. Tierchem. 47 (1919), i-ii

ZAHN, FRIEDRICH WILHELM [1845-1905]
BJN 9,336; 10,132*; FISCHER 2,1716-1717; PAGEL, 1888; GENEVA 3,74-75; 4,316-320; 5,221; RSC 8,129; 11,875; 19,756-757
J.L. Prévost and C. Picot, Verhandl. Schw. Nat. Ges. 87 (1904), cxlii-cl
A. Esternod, Anat. Anz. 25 (1904), 574-576
M. Askenazy, Verhandl. path. Ges. 9 (1905), 331-341

ZAHN, HELMUT GUSTAV [1916-]
KURSCHNER 13, 4382-4383; WWWS, 1840; POGG 7a(4),1101-1103
Anon., Nachr. Chem. Techn. 24 (1976), 260-261

ZAHN, KARL [1885-]
JV 24,533; NUC 681,115

ZAHN, VICTOR [1886-]
JV 25,397; NUC 681,125

ZAIKOVSKI, DMITRI DMITRIEVICH [1838-1867]
ZMEEV 1,113; 3,19

ZAK, EMIL [1877-1949]
KURSCHNER 4,3364; BHDE 2,1274; FISCHER 2,1717
E.P. Pick, Wiener klin. Wchschr. 61 (1949), 410

ZAKHARIN, GRIGORI ANTONOVICH [1829-1897]
BSE 16,512; BME 10,706-709; HIRSCH 5,942; KUZENTSOV(1963), 511-517; ZMEEV 1,114-115; 3,20
D.N. Zhbankov, Vrach 18 (1897), 97-108
Anon., Lancet 1898(I), p.193
N. Volkov, Klin. Med. 6(2) (1928), 74-81
E.M. Tareev, Sov. Med. 19 (1955), 78-84
G.P. Shultsev, Klin. Med. 57(8) (1979), 103-107

ZALESKI, JAN [1868-1932]
WE 12,622; IB, 332; KONOPKA 13,47-48; POGG 6,2950-2951
W. Olszewski, Bull. Soc. Chim. [4]53 (1933), 138
A. Koss, Roczniki Chemii 13 (1933), 625-628

ZALESKI, NIKOLAI LAVRENTIEVICH [1835-]
KHARKOV, 101-106; VENGEROV 1,291; ZMEEV 4,119

ZALESKI, STANISLAW [1858-1918]
KONOPKA 13,51-53; RSC 19,760

ZALESKI, VYACHESLAV KONSTANTINOVICH [1871-1936]
WWR, 622; UKRAINE 2,534-535; KHARKOV-F, 233; LIPSHITS 3,328-331; IB, 332
A.V. Reingard and V.P. Israelski, Priroda 6 (1937), 133-134

ZALKIND, YULI SIGISMUNDOVICH [1875-1948]
BSE 16,392; WWR, 622; POGG 6,2973; 7b, 6144-6147
K.B. Balyan, Zhur. Ob. Khim. 23 (1953), 1965-1981

ZAMECNIK, PAUL CHARLES [1912-]
AMS 15(7),838; WWA 1980-1981, pp.3626-3627; IWW 1982-1983, p.1449; ICC, 396-398; WWWS, 1841
J.T. Edsall, Fed. Proc. 29 (1970), 1309-1311

ZAMENHOF, STEPHEN [1911-]
AMS 15(7),838; WWA 1980-1981, p.3627

ZANDER, EMIL [1867-1929]
SMK 8,464; FISCHER 2,1717
G. Liljestrand, Hygeia 91 (1929), 673-678

ZANGEMEISTER, WILHELM [1871-1930]
FISCHER 2,1718; AUERBACH, 421-422
G. Winter, Münch. med. Wchschr. 77 (1930), 452-453

ZANGGER, HEINRICH [1874-1957]
POGG 7a(4),1104-1105
F. Schwarz, Viert. Nat. Ges. Zürich 102 (1957), 376-378

ZAPOLSKI, NIKOLAI VASILIEVICH [1835-1883]
ZMEEV 4,120; RSC 8,1294
Russki Biograficheski Slovar 7,228

ZARETSKI, SERGEI GRIGORIEVICH [1880-]
SG [2]21,404

ZAUNICK, RUDOLF [1893-1967]
KURSCHNER 10,2770-2771; TLK 3,727; POGG 7a(4),1108-1112
F. Klemm, Sudhoffs Arch. 47 (1963), 193-198
G. Uschmann, Clio Medica 3 (1968), 282-286
K. Mothes and G. Uschmann, Leopoldina [3]14 (1968), 143-153

ZAVADOVSKI, MIKHAIL MIKHAILOVICH [1891-1957]
DSB 14,595-596; BSE(3rd Ed.) 9,265; IB, 333
L.Y. Blacher, Bull. Soc. Nat. Moscou NS62(4) (1957), 105-109

ZAVARZIN, ALEKSEI ALEKSEEVICH [1886-1945]
DSB 14,596-597; BSE 16,282; WWR, 626; IB, 333
D.N. Nasonov and G.S. Strelin, Vest. Akad. Nauk 1946(7), pp.41-46
G.A. Nevmyvaka, Arkh. Anat. Gist. Emb. 28 (1948), 5-13; 50 (1966), 3-7,25-40,58-66; A.A. Zavarzin. Leningrad 1971

ZAVYALOV, VASILI VASILIEVICH [1873-]
LEVITSKI 2,313-314; IB, 332
J. Brennsohn, Die Aerzte Estlands. p.511. Riga 1922

ZAWADSKI, JÓZEF [1865-1935]
RSC 19,766
G. Kurztelak, Arch. Hist. Med. 29 (1966), 149-154

ZAWADSKI, JÓZEF [1886-1951]
POTSCH, 464-465; POGG 6,2955-2956;7b, 6157-6158
W.P. Jorissen, Chem. Wkbl. 48 (1952), 1011
A. Bretsznajder ans S. Weychart, Przemysl Chemiczni 37 (1958), 298-305

ZAWIDSKI, JAN [1866-1928]
WE 12,661; RIGA, 723; JV 16,215; POGG 4,1681-1682; 5,1404; 6,2956

ZAYTSEV [see SAYTSEV]

ZBARSKY, BORIS ILYICH [1885-1954]
BME 10,723-724; WWR, 627; IB, 261
S.R. Mardashev, Vest. Akad. Med. Nauk 1955(1), pp.62-63
A.E. Braunstein, Vop. Med. Khim. 1(2) (1955), 152-154

ZECHMEISTER, LASZLO [1889-1972]
AMS 11,5059; NCAB 57,583; MEL 3,858-859; ETTRE, 491-494; IB, 333; WWAH, 687; WWWS, 1843; POTSCH, 465; POGG 6,2956-2957; 7b, 6159-6163
Anon., Chem. Eng. News 50(13) (1972), 54

ZEDEL, WILHELM [1863-]
JV 4,226; RSC 19,768

ZEDTWITZ, ARMIN von [1886-]
JV 26,648; NUC 682,249ZEIDLER, OTHMAR [1859-1911]
HEIN, 770-771; RSC 19,878
Anon., Chem. Z. 35 (1991), 718; Pharmazeutische Post 44 (1911), 519-520

ZEILE, KARL [1905-1980]
KURSCHNER 13,4394; POGG 6,2958; 7a(4),1118-1119
Anon., Arzneimitt. 15 (1965), 191-192; Chem. Z. 104 (1980), 75,281

ZEINE, RUDOLF [1882-1963]
JV 22,329; NUC 682,366

ZEISE, WILLIAM CHRISTOPHER [1789-1847]
DSB 14,600-601; MEISEN, 104-106; BLOKH 2,808-809; VEIBEL 1,180-188; 2,488-494; 3,29-45; FERCHL, 594; TSCHIRCH, 1149; POTSCH, 465; POGG 2,1401-1402
K.A. Jensen, Centaurus 32 (1989), 324-335

ZEISEL, SIMON [1854-1933]
POTSCH, 465; POGG 3,1479; 4,1683; 5,1406-1407; 6,2959
A. Franke, Ost. Chem. Z. 36 (1933), 75-78

ZEITSCHEL, OTTO [1893-1954]
POGG 6,2959; 7a(4),1119-1120
Anon., Pharmazie 10 (1955), 139

ZEKERT, OTTO [1893-1968]
KURSCHNER 10,2775; HEIN-E, 461-462; TSCHIRCH, 1149-1150; POGG 6,2959; 7a(4),1120-1121
K. Ganzinger, Deutsche Apoth. Z. 103 (1963), 598-599; Materia Therapeutica (Wien) 15 (1969), 70-80
Anon., Chem. Z. 92 (1968), 650

ZELENY, CHARLES [1878-1939]
AMS 6,1594-1595; NCAB 42,247-248; IB, 333; WWAH, 687
F. Payne, Science 91 (1940), 160-161

ZELINSKI, NIKOLAI DMITRIEVICH [1861-1963]
DSB 14,601-603; BSE 16,624-625; WWR, 628-629; KUZNETSOV(1961), 530-545; ARBUZOV, 160-163; WWWS, 1843; POTSCH, 465-466; POGG 4,1684-1685; 5,1407; 6,2959-2960; 7b,6167-6176
Y. Yuriev and R. Levina, Life and Work of Academician Nikolai Zelinski. Moscow 1958

ZELLER, ALFRED [1908-1976]
KURSCHNER 12,3601; WWWS, 1843; POGG 7a(4),1121

ZEMPLÉN, GÉZA [1883-1956]
DSB 14,603; MEL 2,1068; IB, 333; POGG 5,1408; 6,2961; 7b,6177-6179
Anon., Leopoldina [3]4/5 (1958-1959), 102-103

O.T. Schmidt, Chem. Ber. 92 (1959), i-xix
L. Mester, Adv. Carbohydrate Chem. 14 (1959), 1-8
R. Bognar, Acta Chim. Acad. Sci. Hung. 19 (1959), 121-142; 113 (1983), 341-353

ZENGIREV, APPOLONARI ALEKSEEVICH [1852-1881]
ZMEEV 4,124
T.Y. Shchesno, Biokhimia 19 (1954), 111-115

ZENKER, FRIEDRICH ALBERT von [1825-1898]
ADB 45,62; BJN 5,68*-69*; HIRSCH 5,1034-1035; SCHWARTZ, 42-47; PITTROFF, 173-174; WWWS, 1741
G. Hauser, Münch. med. Wchschr. 42 (1895), 854
H. Schroder, Münch. med. Wchschr. 72 (1925), 436

ZENNECK, LUDWIG HEINRICH [1779-1859]
KLEIN, 139; FERCHL, 594-595; POGG 2,1404-1405
F. Dückert, Jahreshefte Wurtt. 16 (1860), 26-29

ZERBAN, FRITZ (FREDERICK WILLIAM) [1880-1956]
AMS 9(I),2168; MILES, 531-532; JV 18,328; WWAH, 687
Anon., Chem. Z. 37 (1913), 301; Chem. Eng. News 34 (1956), 4643;
J. Assn. Agr. Chem. 40 (1957), 101-103
L. Sattler, International Sugar Journal 59 (1957), 212

ZEREVITINOV (TSEREVITINOV), FEDOR VASILIEVICH [1874-1947]
BSE 46,570; WWR, 551; POTSCH, 82

ZERFAS, LEON GROTIUS [1897-1978]
AMS 10,4567; WWWS, 1844

ZERNER, ERNST [1884-1966]
AMS 10,4567; POGG 5,1409; 6,2962; 7a(4),1126
Anon., Chem. Eng. News 45(6) (1967), 112

ZERNIKE, FRITS [1888-1966]
DSB 14,616-617; BWN 1,674-677; FREUND 3,509-513; W, 574; MH 1,548-549; STC 3,203-205; WWAH, 687; WWWS, 1844; POGG 5,1409-1410; 6,2963; 7b,6181-6183
F. Zernike, Science 121 (1955), 345-348
J.A. Prins, Jaarb. Akad. Wet. 1965-1966, pp.370-377
S. Tolansky, Biog. Mem. Fell. Roy. Soc. 13 (1967), 393-402

ZERVAS, LEONIDAS [1902-1980]
DSB 18,1007-1008; POTSCH, 466
Ellenikon (Who's Who) 1962, pp.162-163
P.G. Katsoyannis, The Chemistry of Polypeptides, pp.1-20. New York 1973
I. Photaki, Peptides 1980 (E. Brunfeldt, Ed.), pp.23-25. Copenhagen 1981

ZETZSCHE, FRITZ [1892-1945]
TLK 3,730; POGG 6,2963; 7a(4),1128ZEUTHEN, ERIK [1914-1980]
KBB 1979, p.1177; WWWS, 1844
H. Holter, J. Protozool. 27 (1980), 145-146

ZEYNEK, RICHARD von [1869-1945]
KURSCHNER 4,3379; FISCHER 2,1721; KOERTING, 122-123; IB, 334; POGG 6,2964; 7a(4),1130

ZHIVAGO, PETR IVANOVICH [1883-1948]
BSE 16,87; WWR, 634; IB, 274
A.F. Ivanitskaya et al., P.I. Zhivago. Moscow 1975

ZIEGENBRUCH, LUDWIG [1861-]
JV 7,150; NUC 683,451

ZIEGLER, ERNST [1849-1905]
BJN 10,272-276,277*; FISCHER 2,1722; PAGEL, 1898-1899; ROEDER, 126-129
C. Nauwerk, Beitr. path. Anat. 38 (1905), iii-xxvii
E. Gierke, Münch. med. Wchschr. 52 (1905), 2532-2534
H. Ribbert, Deutsche med. Wchschr. 31 (1905), 2069-2070
L. Aschoff, Verhandl. path. Ges. 10 (1906), 284-289
J. Miller, J. Path. Bact. 11 (1906), 378-382
C. Hodel and H. Buess, Clio Medica 1 (1956), 303-318

ZIEGLER, HEINRICH ERNST [1858-1925]
KLEIN, 139
F. Keibel, Anat. Anz. 60 (1925), 235-238
M. Rauther, Jahreshefte Wurtt. 81 (1925), xl-lii

ZIEGLER, JOHANN HEINRICH [1857-1936]
DHB 7,435; STELLING-MICHAUD 6,291-292
A. Saager, Der Winterthurer Naturphilosoph Johann Heinrich Ziegler. Winterthur 1930

ZIEGLER, JOSEPH [1853-]
JV 3,65; RSC 19,885; NUC 683,885

ZIEGLER, KARL [1898-1973]
DSB 18,1008-1010; KURSCHNER 11,3384-3385; AUERBACH, 935-936; DRULL, 310; MEINEL, 304-311,516-517; SCHMITZ, 302-303; MH 1,549-551; STC 3,205-206; MSE 3,366-367; W(3rd Ed.),615-616; ABBOTT-C, 151-152; POTSCH, 466-467; TLK 3,730-731; WWWS, 1844-1845; POGG 6,2964-2965; 7a(4),1133-1136
C.E.H. Bawn, Biog. Mem. Fell. Roy. Soc. 21 (1975), 569-584
G. Wilke, Ann. Chem. 1975, pp.805-883
J.J. Eisch, J. Chem. Ed. 60 (1983), 1009-1014

ZIEMKE, ERNST [1867-1935]
KURSCHNER 4,3388; FISCHER 2,1724; KIEL, 84-85
H. Merkel, Zeitschrift für gerichtliche Medizin 25 (1936), I-iv

ZIEMSSEN, HUGO [1829-1902]
BJN 7,43-48,130*; HIRSCH 5,1040; PAGEL, 1899-1901; GREIFSWALD 2,374-376 BACHMANN, 1-24; SCHWARTZ, 74-83; PITTROFF, 122-124; KOSMINSKI, 577; WWWS, 1741
A. Schmid, Deutsches Arch. klin. Med. 66 (1899), 1-20
W. Leube, Deutsche med. Wchschr. 28 (1902), 105-107
H. Rieder, Berl. klin. Wchschr. 39 (1902), 176-178
F. Moritz, Deutsches Arch. klin. Med. 72 (1902), v-viii; Münch. med. Wchschr. 49 (1902), 238-242; 77 (1930), 1-5

ZILLESEN, HERMANN [1871-]
JV 14,152; RSC 19,786; SG [2]21,425

ZILLIKEN, FRIEDRICH [1920-]
KURSCHNER 13,4416; AUERBACH, 423

ZILVA, SOLOMON SILVESTER [1885-1956]
E.D. Todhunter, J. Am. Diet. Assn. 45 (1964), 529

ZILVERSMIT, DONALD BERTHOLD [1919-]
AMS 15(7),858; WWA 1980-1981, p.3634; WWWS, 1845

ZIMM, BRUNO HASBROUCK [1920-]
AMS 15(7),858; WWA 1980-1981, p.3634; IWW 1982-1983, p.1456; WWWS, 1845
W.H. Stockmayer and B.H. Zimm, Ann. Rev. Phys. Chem. 35 (1984), 1-21

ZIMMER, KARL GUENTER [1911-1988]
KURSCHNER 15,2564; WWWS, 1846
P. Herrlich, Radiation Research 116 (1988), 178-180

ZIMMERLI, FRITZ [1874-1928]
T. Gränacher and G. Gross, Die Ortsbürger von Zofingen, p.291. Zofingen 1931

ZIMMERLI, HEDWIG [1864-1924]
SG [1]16,778; NUC 684,66

ZIMMERLI, WILLIAM FREDERICK [1888-]
AMS 9(I),2171; JV 28,398; NUC 684,67
Chemical Who's Who 1956, p.1197

ZIMMERMAN, HOWARD ELLIOTT [1926-]
AMS 15(7),861; WWA 1980-1981, p.3636; WWWS, 1846

ZIMMERMANN, ALBRECHT [1860-1931]
TSCHIRCH, 1150; IB, 334
C. Correns and H. Morstatt, Ber. bot. Ges. 49 (1931), (220)-(243)

ZIMMERMANN, CLEMENS [1856-1885]
BLOKH 2,809; POGG 4,1690
A. von Baeyer, Ber. chem. Ges. 18(3) (1885), 825-833
Anon., Leopoldina 22 (1886), 110

ZIMMERMANN, HEINRICH [1880-1955]
JV 22,329; NUC 684,137

ZIMMERMANN, KARL WILHELM [1861-1935]
FISCHER 2,1725; PAGEL, 1903; WEBER, 123-124
E. Hintzsche, Anat. Anz. 82 (1936), 300-313

ZIMMERMANN, MARGARETHE (Mrs. LEHNARTZ) [1899-1948]
NDB 14,104

ZIMMERMANN, THEODOR FRIEDRICH [1881-]
JV 21,276

ZIMMERMANN, WILHELM [1910-1982]
KURSCHNER 13,4424; POGG 7a(4),1139-1140

ZINCKE, THEODOR [1843-1928]
HEIN, 779-780; GUNDLACH, 465; AUERBACH, 936; MEINEL, 151-213,229-232,517; SCHMITZ, 272-279; SCHAEDLER, 161; TLK 2,786; POTSCH, 467; POGG 3,1484-1485; 4,1691-1692; 5,1411-1412; 6,2966; 7aSuppl.,793
K. Schaum, Z. angew. Chem. 36 (1923), 261-262
F. Krollpfeiffer, Z. angew. Chem. 41 (1928), 367-368
K. Fries, Ber. chem. Ges. 62A (1929), 17-45

ZINDER, NORTON DAVID [1928-]
AMS 15(7),864; WWA 1980-1981, p.3637; IWW 1982-1983, p.1456; WWWS, 1846; MH 2,626-627; STC 3,206-207; MSE 3,367-368

ZININ, NIKOLAI NIKOLAEVICH [1812-1880]
DSB 14,622; BSE 17,92-93; KUZNETSOV(1961), 441-447; ARBUZOV,110-115; BLOKH 2,685-687; KOZLOV, 120-122; KAZAN 1,338-341; FERCHL, 597; POTSCH, 467; SCHAEDLER, 161-162; POGG 2,1415; 3,1486
T.H.N., Nature 21 (1880), 572-573
A.M. Butlerov and A.P. Borodin, Zhur. Fiz. Khim. 12 (1880), 215-254; Ber. chem. Ges. 14 (1881), 2887-2908
H.M. Leicester, J. Chem. Ed. 17 (1940), 303-306
A.E. Arbuzov, Usp. Khim. 12 (1943), 81-93

ZINKE, ALOIS [1892-1963]
KURSCHNER 9,2351; KERNBAUER, 299-325; TLK 3,732; POGG 6,2966-2967; 7a(4),1140-1142
Anon., Chem. Z. 87 (1963), 332

ZINKEISEN, EDUARD [1873-1951]
JV 11,169; NUC 684,287

ZINNEMANN, KURT SALLY [1907-]
BHDE 2,1280; NUC 684,300
Who's Who in British Science 1953, p.291

ZINNER, GUSTAV [1921-]
KURSCHNER 12,3625; POGG 7a(4),1145

ZINOFFSKY, OSCAR [1848-1889]
Anon., Leopoldina 25 (1889), 113; Pharm. Z. 34 (1889), 333
J. Brennsohn, Die Aerzte Estlands, p.373. Riga 1922
J. Brockhausen, Album Fratrum Academicorum, p.28. Weidenau 1960

ZINSSER, GUSTAV [1878-1929]
JV 17,160; NUC 684,326

ZINSSER, HANS [1878-1940]
DSB 14,622-625; DAB [Suppl.2],744-745; AMS 6,1597-1598; NCAB 36,35-36; FISCHER 2,1726; HUBER, 140-142; FORBES, 373-390; DAMB, 835-836; IB, 334; WWAH, 688; WWWS, 1847
H. Zinsser, As I Remember Him. Boston 1940
J.H. Mueller, J. Bact. 40 (1940), 747-753
R.P. Strong, Science 92 (1940), 276-279
L.J. Henderson, Am. Phil. Soc. Year Book 1940, pp.451-454
S. Bayne-Jones, Arch. Path. 31 (1941), 269-280
S.B. Wolbach, Trans. Assoc. Am. Phys. 56 (1941), 28-30; Biog. Mem. Nat. Acad. Sci. 24 (1947), 323-360

ZINTL, EDUARD [1898-1941]
WOLF, 237; TLK 3,732; POTSCH, 468; POGG 6,2967-2968; 7a(4),1146-1147
W. Klemm, Z. anorg. Chem. 247 (1941), 1-21
F. Laves, Naturwiss. 29 (1941), 244-245
C. Wagner, Z. angew. Chem. 54 (1941), 525-527
O. Hönigschmid et al., Ber. chem. Ges. 75A (1942), 40-74

ZIPF, HANS [1911-1969]
KURSCHNER 12,3626; POGG 7a(4),1148
Anon., Münch. med. Wchschr. 111 (1969), 1448
W. Bruns, Arzneimitt. 19 (1969), 1156

ZIPF, KARL [1895-1990]
KURSCHNER 14,4793; IB, 334; POGG 7a(4),1148-1149
K.W. Merz, Arzneimitt. 15 (1965), 95
Anon., Chem. Z. 114 (1990), 182

ZIPPERER, PAUL [1857-1903]
NUC 684,387
Anon., Chem. Z. 27 (1903), 1066; Pharm. Z. 48 (1903), 864

ZIRKLE, CONWAY [1895-1972]
AMS 11,6077; IB, 335; WWAH, 688; WWWS, 1847
R.O. Erikson, Stain Technology 48 (1972), 3-4

ZIRKLE, RAYMOND ELLIOTT [1902-1988]
AMS 14,5736; WWA 1976-1977, p.3497; WWWS, 1847

ZIUREK, OTTO [1821-1886]
HEIN, 780-781; BLOKH 2,809-810; TSCHIRCH, 1150; SCHAEDLER, 162
Anon., Leopoldina 22 (1886), 114; Pharm. Z. 31 (1886), 317-319
A. Tschirch, Ber. chem. Ges. 19(3) (1886), 893-901
W. Blum, Der Apotheker und Chemiker Otto Ziurek. Braunschweig 1980

ZLATAROV, ASEN KHRISTOV [1885-1936]
Kratka Bulgarska Entsiklopediya 2,419
V.Y. Gaila and S.S. Krivobokova, Sov. Zdrav. 1986(2), pp.68-71

ZOCHER, HANS [1893-1973]
KURSCHNER 11,3999; TLK 3,733; POGG 6,2969; 7a(4),1150-1161

ZÖLLER, PHILIPP [1831-1885]
WURZBACH 60,230-231; HEIN, 782-783; BLOKH 2,810-811; SCHAEDLER, 162; POGG 3,1488
J. Seissl, Ber. chem. Ges. 18(3) (1885), 839-842

ZÖLLNER, CLEMENS [1884-]
JV 25,46; NUC 684,563

ZOEPPRITZ, JOHANN FRIEDRICH [1814-1861]
STELLING-MICHAUD 6,294; DGB 69,35

ZÖRNIG, HEINRICH [1866-1942]
HEIN, 783-784; TLK 3,733; TSCHIRCH, 1150; POGG 7a(4),1152
A. Leupin, Schw. Apoth. Z. 80 (1942), 721-724

ZOJA, LUIGI [1866-1959]
EI 35,973; FISCHER 2,1727-1723
A. Omodei-Zorini, Riforma Medica 74 (1959), 867-869

ZOLLINGER, ERNEST [1891-1971]
NUC 685,39

ZONDEK, BERNHARD [1891-1966]
FISCHER 2,1728; MEDVEI, 818-819; BHDE 2,1282-1283; LDGS, 64; KOREN, 267; WININGER 6,369; 7,505; COHEN, 262; FRANKENTHAL, 316; TETZLAFF, 368
M. Finkelstein et al., Israel J. Med. Sci. 2 (1966), 805-806; Rhode Isl. Med. J. 50 (1967), 406-408
M. Finkelstein, J. Repr. Fert. 12 (1966), 3-19
G. Goretzlehner and K. Rudolph, Z. Gyn. 100 (1978), 638-641

ZONDEK, HERMANN [1887-1979]
FISCHER 2,1728; BHDE 2,1283; LDGS, 69; WININGER 6,369; 7,505; WWWS, 1848
H.E. Leszynsky, Koroth 6 (1972), 131-132
Anon., Lancet 1979(II), p.213

ZOPF, WILHELM FRIEDRICH [1846-1909]
BJN 14,104*; FISCHER 2,1729; TSCHIRCH, 1150; DZ, 1621; POGG 4,1694; 5,1413; 6,2969-2970
Anon., Leopoldina 45 (1909), 120; Chem. Z. 33 (1909), 705
F. Tobler, Ber. bot. Ges. 27 (1909), (58)-(72)

ZORN, BERNHARD [1891-1958]
POGG 7a(4),1154-1155
Anon., Nachr. Chem. Techn. 7 (1959), 28

ZOTH, OSKAR [1864-1933]
FISCHER 2,1729; TLK 3,733; IB, 335; POGG 7a(4),1156-1157
L. Lohner, Erg. Physiol. 36 (1934), 1-10; Pflügers Arch. 234 (1934), 273-275

ZOTTERMAN, YNGVE [1898-1982]
VAD 1981, p.1163
Y. Zotterman, Touch, Tickle and Pain. Oxford, 1969,1971
R. Granit, Trends in Neurosciences 5 (1982), 265

ZSCHEILE, FREDERICK PAUL, Jr. [1907-]
AMS 14,5740; WWWS, 1848

ZSIGMONDY, RICHARD [1865-1929]
DSB 14,632-634; DBJ 11,335-338; FISCHER 2,1729; FREUND 3,525-535; TLK 3,734; W, 574-575; ABBOTT-C, 152; WWWS, 1848; POTSCH, 469; POGG 4,1695-1696; 5,1414; 6,2971; 7aSuppl.,796-797
A. Lottermoser, Z. angew. Chem. 42 (1929), 1069-1070
A. Thiessen, Chem. Z. 53 (1929), 849-850
A. Chwala, Ost. Chem. Z. 32 (1929), 298-299
H. Freundlich, Ber. chem. Ges. 63A (1930), 171-175
R. Wegscheider, Alm. Akad. Wiss. Wien 80 (1930), 262-268
L.B. Hunt, Endeavour NS5 (1981), 61-67

ZUCKER, ALFRED [1871-]
KURSCHNER 4,3409; TLK 3,734; WININGER 6,371; NUC 685,333

ZUCKER, EUGEN [1883-]
JV 21,490; NUC 685,334

ZUCKERKANDL, EMIL [1849-1910]
BJN 15,156-158,94*; FISCHER 2,1729-1730; PAGEL, 1907-1908; STOBER, 41-54; BHDE 2,1284; WININGER 6,371-372
J. Tandler, Anat. Anz. 37 (1910), 86-96; Wiener klin. Wchschr. 23 (1910), 798-800

ZUCKERKANDL, OTTO [1861-1921]
DBJ 3,323; FISCHER 2,1730; WININGER 6,372-373

ZUELZER, GEORG [1870-1949]
FISCHER 2,1730-1731; BHDE 2,1285; WININGER 7,506
O.W. Gross, Deutsche med. Wchschr. 75 (1950), 153-154
K.H. Mellinghoff, Med. Welt 23 (1972), 622-626
P. Stein, Gesnerus 31 (1974), 107-112

ZUELZER, MARGARETE [1877-]
LDGS, 16; NUC 685,371
Directory of British Scientists 1966-1967 (2), p.891

ZUELZER, WILHELM [1834-1893]
HIRSCH 5,1052-1053; WININGER 6,376

ZULKOWSKI, KARL [1833-1907]
POGG 3,1490; 4,1697; 5,1414
W. Nernst, Ber. chem. Ges. 41 (1908), 133

ZUMBUSCH, EMILIE [1880-1923]
JV 26,649; NUC 685,528
Anon., Chem. Z. 47 (1923), 211

ZUMBUSCH, LEO von [1874-1940]
FISCHER 2,1731; KAMP, 52-69
C. Moncorps, Münch. med. Wchschr. 87 (1940), 516-517

ZUNTZ, NATHAN [1847-1920]
DBJ 2,636-637; FISCHER 2,1731-1732; PAGEL, 1910; BLOKH 2,811-812; DZ,1623; TLK 2,790; WININGER 6,378; TETZLAFF,369; POGG 7aSuppl.,797
C. Neuberg, Ber. chem. Ges. 53A (1920), 83-88; Chem. Z. 44 (1920),281-282
A. Durig, Wiener klin. Wchschr. 33 (1920), 344-345
A. Loewy, Pflügers Arch. 194 (1922), 1-19
R.M. Forbes, J. Nutrition 57 (1955), 3-15

ZUNZ, EDGARD VICTOR [1874-1939]
BNB 44,777-780; FISCHER 2,1732; SCHALI, 110-112; IB, 335; KOREN, 268; WININGER 7,506; POGG 6,2973; 7b,6191-6194
Anon., Arch. Inter. Pharmacodyn. 63 (1939), 1-9; Bull. Acad. Med. Belg. [6]4 (1939), 252-257; Bull. Soc. Chim. Biol. 21 (1942), 1040-1042

ZWAARDEMAKER, HENDRIK [1857-1930]
LINDEBOOM, 2240-2241; UTRECHT, 118-119; FISCHER 2,1733; KALIN, 215-217; IB, 335; POGG 6,2974
A.K.M. Noyons, Arch. Neer. Physiol. 7 (1922), 1-30; Erg. Physiol. 33 (1931), v-xii

ZWEIFEL, PAUL [1848-1927]
FISCHER 2,1733-1734; GAUSS, 214
B. Schweizer, Z. Gyn. 51 (1927), 2586-2600
A. Hoop, Z. Gyn. 109 (1987), 979-980

ZWENGER, CONSTANTIN [1814-1884]
ADB 45,526; GUNDLACH, 483; MEINEL, 45; SCHMITZ, 365-370; TSCHIRCH, 1151; FERCHL, 598; POGG 2,1422-1423; 3,1490
Anon., Chem. Z. 8 (1884), 432; Leopoldina 20 (1884), 60

www.ingramcontent.com/pod-product-compliance
Lightning Source LLC
Chambersburg PA
CBHW081339190326
41458CB00018B/6053

* 9 7 8 0 8 7 1 6 9 9 8 3 1 *